S0-ASW-672

TABLE OF ION ENERGIES FOR METASTABLE TRANSITIONS IN MASS SPECTROMETRY

TABLE OF ION ENERGIES FOR METASTABLE TRANSITIONS IN MASS SPECTROMETRY

BY J. H. BEYNON, R. M. CAPRIOLI, A. W. KUNDERD AND R. B. SPENCER

DEPARTMENT OF CHEMISTRY, PURDUE UNIVERSITY, W. LAFAYETTE, IND. 47907

ELSEVIER PUBLISHING COMPANY/AMSTERDAM/LONDON/NEW YORK/1970

ELSEVIER PUBLISHING COMPANY
335 JAN VAN GALENSTRAAT, P.O. BOX 211,
AMSTERDAM, THE NETHERLANDS

ELSEVIER PUBLISHING CO. LTD.
BARKING, ESSEX, ENGLAND

AMERICAN ELSEVIER PUBLISHING COMPANY, INC.
52 VANDERBILT AVENUE, NEW YORK, NEW YORK 10017

LIBRARY OF CONGRESS CARD NUMBER 75-118250
STANDARD BOOK NUMBER 444-40874-6

PRINTED IN THE NETHERLANDS

Preface

The table presented in this book was arranged and produced at Purdue University, West Lafayette, Indiana. It was computed by a Sigma 5 computer using a program written in XDS Fortran-IV-H language. The printer output was produced by an XDS model 7440 line printer.

We wish to thank the National Institutes of Health, Grant No. 5-P07-FR00354-02, for their support in the production of this book.

West Lafayette, Indiana, 1970 JHB/RMC/AWK/RBS

Introduction

Ions formed in the ionization chamber of a mass spectrometer which decompose during or after acceleration are known as metastable ions. These ions will decompose throughout the length of the analyzer tube. In a double-focusing mass spectrometer, only those decomposing between the electric sector and the magnetic sector (the second field-free region) are normally seen in the mass spectrum and give rise to the well known "metastable peaks". These appear as small diffuse peaks which are usually at non-integral masses. "Metastable peaks" are extremely useful in the study of the fragmentation of molecules since they show that certain ions are linked together in a unimolecular decomposition.

Let us consider the metastable ion decomposition,

$$m_1^{+} \rightarrow m_2^{+} + (m_1 - m_2)$$

where m_1 is the metastable ion, m_2 the daughter ion, and $(m_1 - m_2)$ a neutral fragment. The kinetic energy of the metastable ion will be shared by the fragments in proportion to their masses. The daughter ion, therefore, will contain a fraction of the energy m_2/m_1 of the original ion. In the mass spectrum, the daughter ion from a decomposition taking place in the second field-free region will give rise to a "metastable peak" at an apparent mass,

$$m^* \cong \frac{m_2^2}{m_1}$$

Decompositions taking place before this region will not appear in the mass spectrum since the product ions will not have sufficient energy to pass through the electric sector. It would be advantageous, however, to record those decompositions occurring between the accelerating region and the electric sector (*i.e.*, the first field-free region) because in passing through the electric sector, the product ions would be better focused and give sharper peaks. This can be accomplished by two techniques. In the first, the voltage on the electric sector is altered while keeping the accelerating voltage constant. In the second, the voltage on the electric sector is kept constant and the accelerating voltage changed. This allows decompositions taking place in the first field-free region to be recorded.

Using these techniques, a beam of ions which contains a fraction of the kinetic energy of the main ion beam will be focused at the energy-resolving slit (β-slit). If a detector is placed at this position, then sweeping the voltage on the electric sector will produce an ion kinetic energy (IKE) spectrum.

For example, let us again consider the transition mentioned above. An ion of mass m_1 will enter the field-free region with a certain kinetic energy; the product ion, m_2, will carry a fraction m_2/m_1 of this energy. If an electric sector voltage E is required to pass ions that have received the full energy of acceleration, then reducing this voltage to m_2E/m_1 will allow an ion beam to pass which is due only to ions that have a fraction m_2/m_1 of the normal ion energy. Alternatively, the accelerating voltage can be varied with the magnet tuned to detect ions of mass m_2. This has the advantage of not altering the mass scale (as in the case of varying the electric sector voltage), but the disadvantage of changing the tuning conditions in the ionization chamber. For the transition under consideration, a peak will appear at m_1V/m_2, where V is the accelerating voltage necessary to pass ions in the main beam. This peak will only be due to ions of mass m_2 which have the fraction m_2/m_1 of the normal ion energy. Since all the energy peaks in an IKE spectrum except the main ion beam arise from metastable transitions, it is of great value to relate their position with the possible transitions through which they arose.

The table presented in this book has been designed to aid in the interpretation of ion kinetic energy spectra. The data is presented in terms of the per cent energy ($\% E$) which a product ion would have for a given metastable transition when the electric sector voltage is scanned. In the case where the accelerating voltage is scanned, $\% E$ is the value $100\ V/V_1$. The mass of m_1 varies from 1–500. For each 0.1% energy change, all the transitions which could possibly give rise to a peak having this fraction of the energy are given in the form of values for m_1, the original ion; m_2, the product ion; DM, the neutral fragment; and m^* the position of the corresponding "metastable peak" in the mass spectrum.

This table is intended to supplement the information given in the *Table of Metastable Transitions for Use in Mass Spectrometry* by J. H. Beynon, R. A. Saunders and A. E. Williams, Elsevier Publishing Company, Amsterdam, 1965.

%E	M1	M2	DM	M*
99.8	500	499	1	498.0
99.8	499	498	1	497.0
99.8	498	497	1	496.0
99.8	497	496	1	495.0
99.8	496	495	1	494.0
99.8	495	494	1	493.0
99.8	494	493	1	492.0
99.8	493	492	1	491.0
99.8	492	491	1	490.0
99.8	491	490	1	489.0
99.8	490	489	1	488.0
99.8	489	488	1	487.0
99.8	488	487	1	486.0
99.8	487	486	1	485.0
99.8	486	485	1	484.0
99.8	485	484	1	483.0
99.8	484	483	1	482.0
99.8	483	482	1	481.0
99.8	482	481	1	480.0
99.8	481	480	1	479.0
99.8	480	479	1	478.0
99.8	479	478	1	477.0
99.8	478	477	1	476.0
99.8	477	476	1	475.0
99.8	476	475	1	474.0
99.8	475	474	1	473.0
99.8	474	473	1	472.0
99.8	473	472	1	471.0
99.8	472	471	1	470.0
99.8	471	470	1	469.0
99.8	470	469	1	468.0
99.8	469	468	1	467.0
99.8	468	467	1	466.0
99.8	467	466	1	465.0
99.8	466	465	1	464.0
99.8	465	464	1	463.0
99.8	464	463	1	462.0
99.8	463	462	1	461.0
99.8	462	461	1	460.0
99.8	461	460	1	459.0
99.8	460	459	1	458.0
99.8	459	458	1	457.0
99.8	458	457	1	456.0
99.8	457	456	1	455.0
99.8	456	455	1	454.0
99.8	455	454	1	453.0
99.8	454	453	1	452.0
99.8	453	452	1	451.0
99.8	452	451	1	450.0
99.8	451	450	1	449.0

%E	M1	M2	DM	M*
99.8	450	449	1	448.0
99.8	449	448	1	447.0
99.8	448	447	1	446.0
99.8	447	446	1	445.0
99.8	446	445	1	444.0
99.8	445	444	1	443.0
99.8	444	443	1	442.0
99.8	443	442	1	441.0
99.8	442	441	1	440.0
99.8	441	440	1	439.0
99.8	440	439	1	438.0
99.8	439	438	1	437.0
99.8	438	437	1	436.0
99.8	437	436	1	435.0
99.8	436	435	1	434.0
99.8	435	434	1	433.0
99.8	434	433	1	432.0
99.8	433	432	1	431.0
99.8	432	431	1	430.0
99.8	431	430	1	429.0
99.8	430	429	1	428.0
99.8	429	428	1	427.0
99.8	428	427	1	426.0
99.8	427	426	1	425.0
99.8	426	425	1	424.0
99.8	425	424	1	423.0
99.8	424	423	1	422.0
99.8	423	422	1	421.0
99.8	422	421	1	420.0
99.8	421	420	1	419.0
99.8	420	419	1	418.0
99.8	419	418	1	417.0
99.8	418	417	1	416.0
99.8	417	416	1	415.0
99.8	416	415	1	414.0
99.8	415	414	1	413.0
99.8	414	413	1	412.0
99.8	413	412	1	411.0
99.8	412	411	1	410.0
99.8	411	410	1	409.0
99.8	410	409	1	408.0
99.8	409	408	1	407.0
99.8	408	407	1	406.0
99.8	407	406	1	405.0
99.8	406	405	1	404.0
99.8	405	404	1	403.0
99.8	404	403	1	402.0
99.8	403	402	1	401.0
99.8	402	401	1	400.0
99.8	401	400	1	399.0

%E	M1	M2	DM	M*
99.8	400	399	1	398.0
99.7	399	398	1	397.0
99.7	398	397	1	396.0
99.7	397	396	1	395.0
99.7	396	395	1	394.0
99.7	395	394	1	393.0
99.7	394	393	1	392.0
99.7	393	392	1	391.0
99.7	392	391	1	390.0
99.7	391	390	1	389.0
99.7	390	389	1	388.0
99.7	389	388	1	387.0
99.7	388	387	1	386.0
99.7	387	386	1	385.0
99.7	386	385	1	384.0
99.7	385	384	1	383.0
99.7	384	383	1	382.0
99.7	383	382	1	381.0
99.7	382	381	1	380.0
99.7	381	380	1	379.0
99.7	380	379	1	378.0
99.7	379	378	1	377.0
99.7	378	377	1	376.0
99.7	377	376	1	375.0
99.7	376	375	1	374.0
99.7	375	374	1	373.0
99.7	374	373	1	372.0
99.7	373	372	1	371.0
99.7	372	371	1	370.0
99.7	371	370	1	369.0
99.7	370	369	1	368.0
99.7	369	368	1	367.0
99.7	368	367	1	366.0
99.7	367	366	1	365.0
99.7	366	365	1	364.0
99.7	365	364	1	363.0
99.7	364	363	1	362.0
99.7	363	362	1	361.0
99.7	362	361	1	360.0
99.7	361	360	1	359.0
99.7	360	359	1	358.0
99.7	359	358	1	357.0
99.7	358	357	1	356.0
99.7	357	356	1	355.0
99.7	356	355	1	354.0
99.7	355	354	1	353.0
99.7	354	353	1	352.0
99.7	353	352	1	351.0
99.7	352	351	1	350.0
99.7	351	350	1	349.0

%E	M1	M2	DM	M*
99.7	350	349	1	348.0
99.7	349	348	1	347.0
99.7	348	347	1	346.0
99.7	347	346	1	345.0
99.7	346	345	1	344.0
99.7	345	344	1	343.0
99.7	344	343	1	342.0
99.7	343	342	1	341.0
99.7	342	341	1	340.0
99.7	341	340	1	339.0
99.7	340	339	1	338.0
99.7	339	338	1	337.0
99.7	338	337	1	336.0
99.7	337	336	1	335.0
99.7	336	335	1	334.0
99.7	335	334	1	333.0
99.7	334	333	1	332.0
99.7	333	332	1	331.0
99.7	332	331	1	330.0
99.7	331	330	1	329.0
99.7	330	329	1	328.0
99.7	329	328	1	327.0
99.7	328	327	1	326.0
99.7	327	326	1	325.0
99.7	326	325	1	324.0
99.7	325	324	1	323.0
99.7	324	323	1	322.0
99.7	323	322	1	321.0
99.7	322	321	1	320.0
99.7	321	320	1	319.0
99.7	320	319	1	318.0
99.7	319	318	1	317.0
99.7	318	317	1	316.0
99.7	317	316	1	315.0
99.7	316	315	1	314.0
99.7	315	314	1	313.0
99.7	314	313	1	312.0
99.7	313	312	1	311.0
99.7	312	311	1	310.0
99.7	311	310	1	309.0
99.7	310	309	1	308.0
99.7	309	308	1	307.0
99.7	308	307	1	306.0
99.7	307	306	1	305.0
99.7	306	305	1	304.0
99.7	305	304	1	303.0
99.7	304	303	1	302.0
99.7	303	302	1	301.0
99.7	302	301	1	300.0
99.7	301	300	1	299.0

%E	M1	M2	DM	M*
99.7	300	299	1	298.0
99.7	299	298	1	297.0
99.7	298	297	1	296.0
99.7	297	296	1	295.0
99.7	296	295	1	294.0
99.7	295	294	1	293.0
99.7	294	293	1	292.0
99.7	293	292	1	291.0
99.7	292	291	1	290.0
99.7	291	290	1	289.0
99.7	290	289	1	288.0
99.7	289	288	1	287.0
99.7	288	287	1	286.0
99.7	287	286	1	285.0
99.7	286	285	1	284.0
99.6	500	498	2	496.0
99.6	499	497	2	495.0
99.6	498	496	2	494.0
99.6	497	495	2	493.0
99.6	496	494	2	492.0
99.6	495	493	2	491.0
99.6	494	492	2	490.0
99.6	493	491	2	489.0
99.6	492	490	2	488.0
99.6	491	489	2	487.0
99.6	490	488	2	486.0
99.6	489	487	2	485.0
99.6	488	486	2	484.0
99.6	487	485	2	483.0
99.6	486	484	2	482.0
99.6	485	483	2	481.0
99.6	484	482	2	480.0
99.6	483	481	2	479.0
99.6	482	480	2	478.0
99.6	481	479	2	477.0
99.6	480	478	2	476.0
99.6	479	477	2	475.0
99.6	478	476	2	474.0
99.6	477	475	2	473.0
99.6	476	474	2	472.0
99.6	475	473	2	471.0
99.6	474	472	2	470.0
99.6	473	471	2	469.0
99.6	472	470	2	468.0
99.6	471	469	2	467.0
99.6	470	468	2	466.0
99.6	469	467	2	465.0
99.6	468	466	2	464.0
99.6	467	465	2	463.0
99.6	466	464	2	462.0

%E	M1	M2	DM	M*
99.6	465	463	2	461.0
99.6	464	462	2	460.0
99.6	463	461	2	459.0
99.6	462	460	2	458.0
99.6	461	459	2	457.0
99.6	460	458	2	456.0
99.6	459	457	2	455.0
99.6	458	456	2	454.0
99.6	457	455	2	453.0
99.6	456	454	2	452.0
99.6	455	453	2	451.0
99.6	454	452	2	450.0
99.6	453	451	2	449.0
99.6	452	450	2	448.0
99.6	451	449	2	447.0
99.6	450	448	2	446.0
99.6	449	447	2	445.0
99.6	448	446	2	444.0
99.6	447	445	2	443.0
99.6	446	444	2	442.0
99.6	445	443	2	441.0
99.6	444	442	2	440.0
99.6	285	284	1	283.0
99.6	284	283	1	282.0
99.6	283	282	1	281.0
99.6	282	281	1	280.0
99.6	281	280	1	279.0
99.6	280	279	1	278.0
99.6	279	278	1	277.0
99.6	278	277	1	276.0
99.6	277	276	1	275.0
99.6	276	275	1	274.0
99.6	275	274	1	273.0
99.6	274	273	1	272.0
99.6	273	272	1	271.0
99.6	272	271	1	270.0
99.6	271	270	1	269.0
99.6	270	269	1	268.0
99.6	269	268	1	267.0
99.6	268	267	1	266.0
99.6	267	266	1	265.0
99.6	266	265	1	264.0
99.6	265	264	1	263.0
99.6	264	263	1	262.0
99.6	263	262	1	261.0
99.6	262	261	1	260.0
99.6	261	260	1	259.0
99.6	260	259	1	258.0
99.6	259	258	1	257.0
99.6	258	257	1	256.0

%E	M1	M2	DM	M*
99.6	257	256	1	255.0
99.6	256	255	1	254.0
99.6	255	254	1	253.0
99.6	254	253	1	252.0
99.6	253	252	1	251.0
99.6	252	251	1	250.0
99.6	251	250	1	249.0
99.6	250	249	1	248.0
99.6	249	248	1	247.0
99.6	248	247	1	246.0
99.6	247	246	1	245.0
99.6	246	245	1	244.0
99.6	245	244	1	243.0
99.6	244	243	1	242.0
99.6	243	242	1	241.0
99.6	242	241	1	240.0
99.6	241	240	1	239.0
99.6	240	239	1	238.0
99.6	239	238	1	237.0
99.6	238	237	1	236.0
99.6	237	236	1	235.0
99.6	236	235	1	234.0
99.6	235	234	1	233.0
99.6	234	233	1	232.0
99.6	233	232	1	231.0
99.6	232	231	1	230.0
99.6	231	230	1	229.0
99.6	230	229	1	228.0
99.6	229	228	1	227.0
99.6	228	227	1	226.0
99.6	227	226	1	225.0
99.6	226	225	1	224.0
99.6	225	224	1	223.0
99.6	224	223	1	222.0
99.6	223	222	1	221.0
99.6	222	221	1	220.0
99.5	443	441	2	439.0
99.5	442	440	2	438.0
99.5	441	439	2	437.0
99.5	440	438	2	436.0
99.5	439	437	2	435.0
99.5	438	436	2	434.0
99.5	437	435	2	433.0
99.5	436	434	2	432.0
99.5	435	433	2	431.0
99.5	434	432	2	430.0
99.5	433	431	2	429.0
99.5	432	430	2	428.0
99.5	431	429	2	427.0
99.5	430	428	2	426.0

%E	M1	M2	DM	M*
99.5	429	427	2	425.0
99.5	428	426	2	424.0
99.5	427	425	2	423.0
99.5	426	424	2	422.0
99.5	425	423	2	421.0
99.5	424	422	2	420.0
99.5	423	421	2	419.0
99.5	422	420	2	418.0
99.5	421	419	2	417.0
99.5	420	418	2	416.0
99.5	419	417	2	415.0
99.5	418	416	2	414.0
99.5	417	415	2	413.0
99.5	416	414	2	412.0
99.5	415	413	2	411.0
99.5	414	412	2	410.0
99.5	413	411	2	409.0
99.5	412	410	2	408.0
99.5	411	409	2	407.0
99.5	410	408	2	406.0
99.5	409	407	2	405.0
99.5	408	406	2	404.0
99.5	407	405	2	403.0
99.5	406	404	2	402.0
99.5	405	403	2	401.0
99.5	404	402	2	400.0
99.5	403	401	2	399.0
99.5	402	400	2	398.0
99.5	401	399	2	397.0
99.5	400	398	2	396.0
99.5	399	397	2	395.0
99.5	398	396	2	394.0
99.5	397	395	2	393.0
99.5	396	394	2	392.0
99.5	395	393	2	391.0
99.5	394	392	2	390.0
99.5	393	391	2	389.0
99.5	392	390	2	388.0
99.5	391	389	2	387.0
99.5	390	388	2	386.0
99.5	389	387	2	385.0
99.5	388	386	2	384.0
99.5	387	385	2	383.0
99.5	386	384	2	382.0
99.5	385	383	2	381.0
99.5	384	382	2	380.0
99.5	383	381	2	379.0
99.5	382	380	2	378.0
99.5	381	379	2	377.0
99.5	380	378	2	376.0

%E	M1	M2	DM	M*
99.5	379	377	2	375.0
99.5	378	376	2	374.0
99.5	377	375	2	373.0
99.5	376	374	2	372.0
99.5	375	373	2	371.0
99.5	374	372	2	370.0
99.5	373	371	2	369.0
99.5	372	370	2	368.0
99.5	371	369	2	367.0
99.5	370	368	2	366.0
99.5	369	367	2	365.0
99.5	368	366	2	364.0
99.5	367	365	2	363.0
99.5	366	364	2	362.0
99.5	365	363	2	361.0
99.5	364	362	2	360.0
99.5	221	220	1	219.0
99.5	220	219	1	218.0
99.5	219	218	1	217.0
99.5	218	217	1	216.0
99.5	217	216	1	215.0
99.5	216	215	1	214.0
99.5	215	214	1	213.0
99.5	214	213	1	212.0
99.5	213	212	1	211.0
99.5	212	211	1	210.0
99.5	211	210	1	209.0
99.5	210	209	1	208.0
99.5	209	208	1	207.0
99.5	208	207	1	206.0
99.5	207	206	1	205.0
99.5	206	205	1	204.0
99.5	205	204	1	203.0
99.5	204	203	1	202.0
99.5	203	202	1	201.0
99.5	202	201	1	200.0
99.5	201	200	1	199.0
99.5	200	199	1	198.0
99.5	199	198	1	197.0
99.5	198	197	1	196.0
99.5	197	196	1	195.0
99.5	196	195	1	194.0
99.5	195	194	1	193.0
99.5	194	193	1	192.0
99.5	193	192	1	191.0
99.5	192	191	1	190.0
99.5	191	190	1	189.0
99.5	190	189	1	188.0
99.5	189	188	1	187.0
99.5	188	187	1	186.0

%E	M1	M2	DM	M*
99.5	187	186	1	185.0
99.5	186	185	1	184.0
99.5	185	184	1	183.0
99.5	184	183	1	182.0
99.5	183	182	1	181.0
99.5	182	181	1	180.0
99.4	500	497	3	494.0
99.4	499	496	3	493.0
99.4	498	495	3	492.0
99.4	497	494	3	491.0
99.4	496	493	3	490.0
99.4	495	492	3	489.0
99.4	494	491	3	488.0
99.4	493	490	3	487.0
99.4	492	489	3	486.0
99.4	491	488	3	485.0
99.4	490	487	3	484.0
99.4	489	486	3	483.0
99.4	488	485	3	482.0
99.4	487	484	3	481.0
99.4	486	483	3	480.0
99.4	485	482	3	479.0
99.4	484	481	3	478.0
99.4	483	480	3	477.0
99.4	482	479	3	476.0
99.4	481	478	3	475.0
99.4	480	477	3	474.0
99.4	479	476	3	473.0
99.4	478	475	3	472.0
99.4	477	474	3	471.0
99.4	476	473	3	470.0
99.4	475	472	3	469.0
99.4	474	471	3	468.0
99.4	473	470	3	467.0
99.4	472	469	3	466.0
99.4	471	468	3	465.0
99.4	470	467	3	464.0
99.4	469	466	3	463.0
99.4	468	465	3	462.0
99.4	467	464	3	461.0
99.4	466	463	3	460.0
99.4	465	462	3	459.0
99.4	464	461	3	458.0
99.4	463	460	3	457.0
99.4	462	459	3	456.0
99.4	363	361	2	359.0
99.4	362	360	2	358.0
99.4	361	359	2	357.0
99.4	360	358	2	356.0
99.4	359	357	2	355.0

%E	M1	M2	DM	M*
99.4	358	356	2	354.0
99.4	357	355	2	353.0
99.4	356	354	2	352.0
99.4	355	353	2	351.0
99.4	354	352	2	350.0
99.4	353	351	2	349.0
99.4	352	350	2	348.0
99.4	351	349	2	347.0
99.4	350	348	2	346.0
99.4	349	347	2	345.0
99.4	348	346	2	344.0
99.4	347	345	2	343.0
99.4	346	344	2	342.0
99.4	345	343	2	341.0
99.4	344	342	2	340.0
99.4	343	341	2	339.0
99.4	342	340	2	338.0
99.4	341	339	2	337.0
99.4	340	338	2	336.0
99.4	339	337	2	335.0
99.4	338	336	2	334.0
99.4	337	335	2	333.0
99.4	336	334	2	332.0
99.4	335	333	2	331.0
99.4	334	332	2	330.0
99.4	333	331	2	329.0
99.4	332	330	2	328.0
99.4	331	329	2	327.0
99.4	330	328	2	326.0
99.4	329	327	2	325.0
99.4	328	326	2	324.0
99.4	327	325	2	323.0
99.4	326	324	2	322.0
99.4	325	323	2	321.0
99.4	324	322	2	320.0
99.4	323	321	2	319.0
99.4	322	320	2	318.0
99.4	321	319	2	317.0
99.4	320	318	2	316.0
99.4	319	317	2	315.0
99.4	318	316	2	314.0
99.4	317	315	2	313.0
99.4	316	314	2	312.0
99.4	315	313	2	311.0
99.4	314	312	2	310.0
99.4	313	311	2	309.0
99.4	312	310	2	308.0
99.4	311	309	2	307.0
99.4	310	308	2	306.0
99.4	309	307	2	305.0
99.4	308	306	2	304.0
99.4	181	180	1	179.0
99.4	180	179	1	178.0
99.4	179	178	1	177.0
99.4	178	177	1	176.0
99.4	177	176	1	175.0
99.4	176	175	1	174.0
99.4	175	174	1	173.0
99.4	174	173	1	172.0
99.4	173	172	1	171.0
99.4	172	171	1	170.0
99.4	171	170	1	169.0
99.4	170	169	1	168.0
99.4	169	168	1	167.0
99.4	168	167	1	166.0
99.4	167	166	1	165.0
99.4	166	165	1	164.0
99.4	165	164	1	163.0
99.4	164	163	1	162.0
99.4	163	162	1	161.0
99.4	162	161	1	160.0
99.4	161	160	1	159.0
99.4	160	159	1	158.0
99.4	159	158	1	157.0
99.4	158	157	1	156.0
99.4	157	156	1	155.0
99.4	156	155	1	154.0
99.4	155	154	1	153.0
99.4	154	153	1	152.0
99.3	461	458	3	455.0
99.3	460	457	3	454.0
99.3	459	456	3	453.0
99.3	458	455	3	452.0
99.3	457	454	3	451.0
99.3	456	453	3	450.0
99.3	455	452	3	449.0
99.3	454	451	3	448.0
99.3	453	450	3	447.0
99.3	452	449	3	446.0
99.3	451	448	3	445.0
99.3	450	447	3	444.0
99.3	449	446	3	443.0
99.3	448	445	3	442.0
99.3	447	444	3	441.0
99.3	446	443	3	440.0
99.3	445	442	3	439.0
99.3	444	441	3	438.0
99.3	443	440	3	437.0
99.3	442	439	3	436.0
99.3	441	438	3	435.0
99.3	440	437	3	434.0
99.3	439	436	3	433.0
99.3	438	435	3	432.0
99.3	437	434	3	431.0
99.3	436	433	3	430.0
99.3	435	432	3	429.0
99.3	434	431	3	428.0
99.3	433	430	3	427.0
99.3	432	429	3	426.0
99.3	431	428	3	425.0
99.3	430	427	3	424.0
99.3	429	426	3	423.0
99.3	428	425	3	422.0
99.3	427	424	3	421.0
99.3	426	423	3	420.0
99.3	425	422	3	419.0
99.3	424	421	3	418.0
99.3	423	420	3	417.0
99.3	422	419	3	416.0
99.3	421	418	3	415.0
99.3	420	417	3	414.0
99.3	419	416	3	413.0
99.3	418	415	3	412.0
99.3	417	414	3	411.0
99.3	416	413	3	410.0
99.3	415	412	3	409.0
99.3	414	411	3	408.0
99.3	413	410	3	407.0
99.3	412	409	3	406.0
99.3	411	408	3	405.0
99.3	410	407	3	404.0
99.3	409	406	3	403.0
99.3	408	405	3	402.0
99.3	407	404	3	401.0
99.3	406	403	3	400.0
99.3	405	402	3	399.0
99.3	404	401	3	398.0
99.3	403	400	3	397.0
99.3	402	399	3	396.0
99.3	401	398	3	395.0
99.3	400	397	3	394.0
99.3	307	305	2	303.0
99.3	306	304	2	302.0
99.3	305	303	2	301.0
99.3	304	302	2	300.0
99.3	303	301	2	299.0
99.3	302	300	2	298.0
99.3	301	299	2	297.0
99.3	300	298	2	296.0
99.3	299	297	2	295.0
99.3	298	296	2	294.0
99.3	297	295	2	293.0
99.3	296	294	2	292.0
99.3	295	293	2	291.0
99.3	294	292	2	290.0
99.3	293	291	2	289.0
99.3	292	290	2	288.0
99.3	291	289	2	287.0
99.3	290	288	2	286.0
99.3	289	287	2	285.0
99.3	288	286	2	284.0
99.3	287	285	2	283.0
99.3	286	284	2	282.0
99.3	285	283	2	281.0
99.3	284	282	2	280.0
99.3	283	281	2	279.0
99.3	282	280	2	278.0
99.3	281	279	2	277.0
99.3	280	278	2	276.0
99.3	279	277	2	275.0
99.3	278	276	2	274.0
99.3	277	275	2	273.0
99.3	276	274	2	272.0
99.3	275	273	2	271.0
99.3	274	272	2	270.0
99.3	273	271	2	269.0
99.3	272	270	2	268.0
99.3	271	269	2	267.0
99.3	270	268	2	266.0
99.3	269	267	2	265.0
99.3	268	266	2	264.0
99.3	267	265	2	263.0
99.3	153	152	1	151.0
99.3	152	151	1	150.0
99.3	151	150	1	149.0
99.3	150	149	1	148.0
99.3	149	148	1	147.0
99.3	148	147	1	146.0
99.3	147	146	1	145.0
99.3	146	145	1	144.0
99.3	145	144	1	143.0
99.3	144	143	1	142.0
99.3	143	142	1	141.0
99.3	142	141	1	140.0
99.3	141	140	1	139.0
99.3	140	139	1	138.0
99.3	139	138	1	137.0
99.3	138	137	1	136.0
99.3	137	136	1	135.0
99.3	136	135	1	134.0
99.3	135	134	1	133.0
99.3	134	133	1	132.0
99.2	500	496	4	492.0
99.2	499	495	4	491.0
99.2	498	494	4	490.0
99.2	497	493	4	489.0
99.2	496	492	4	488.0
99.2	495	491	4	487.0
99.2	494	490	4	486.0
99.2	493	489	4	485.0
99.2	492	488	4	484.0
99.2	491	487	4	483.0
99.2	490	486	4	482.0
99.2	489	485	4	481.0
99.2	488	484	4	480.0
99.2	487	483	4	479.0
99.2	486	482	4	478.0
99.2	485	481	4	477.0
99.2	484	480	4	476.0
99.2	483	479	4	475.0
99.2	482	478	4	474.0
99.2	481	477	4	473.0
99.2	480	476	4	472.0
99.2	479	475	4	471.0
99.2	478	474	4	470.0
99.2	477	473	4	469.0
99.2	476	472	4	468.0
99.2	475	471	4	467.0
99.2	474	470	4	466.0
99.2	473	469	4	465.0
99.2	472	468	4	464.0
99.2	471	467	4	463.0
99.2	399	396	3	393.0
99.2	398	395	3	392.0
99.2	397	394	3	391.0
99.2	396	393	3	390.0
99.2	395	392	3	389.0
99.2	394	391	3	388.0
99.2	393	390	3	387.0
99.2	392	389	3	386.0
99.2	391	388	3	385.0
99.2	390	387	3	384.0
99.2	389	386	3	383.0
99.2	388	385	3	382.0
99.2	387	384	3	381.0
99.2	386	383	3	380.0
99.2	385	382	3	379.0
99.2	384	381	3	378.0
99.2	383	380	3	377.0
99.2	382	379	3	376.0

%E	M1	M2	DM	M*
99.2	381	378	3	375.0
99.2	380	377	3	374.0
99.2	379	376	3	373.0
99.2	378	375	3	372.0
99.2	377	374	3	371.0
99.2	376	373	3	370.0
99.2	375	372	3	369.0
99.2	374	371	3	368.0
99.2	373	370	3	367.0
99.2	372	369	3	366.0
99.2	371	368	3	365.0
99.2	370	367	3	364.0
99.2	369	366	3	363.0
99.2	368	365	3	362.0
99.2	367	364	3	361.0
99.2	366	363	3	360.0
99.2	365	362	3	359.0
99.2	364	361	3	358.0
99.2	363	360	3	357.0
99.2	362	359	3	356.0
99.2	361	358	3	355.0
99.2	360	357	3	354.0
99.2	359	356	3	353.0
99.2	358	355	3	352.0
99.2	357	354	3	351.0
99.2	356	353	3	350.0
99.2	355	352	3	349.0
99.2	354	351	3	348.0
99.2	353	350	3	347.0
99.2	266	264	2	262.0
99.2	265	263	2	261.0
99.2	264	262	2	260.0
99.2	263	261	2	259.0
99.2	262	260	2	258.0
99.2	261	259	2	257.0
99.2	260	258	2	256.0
99.2	259	257	2	255.0
99.2	258	256	2	254.0
99.2	257	255	2	253.0
99.2	256	254	2	252.0
99.2	255	253	2	251.0
99.2	254	252	2	250.0
99.2	253	251	2	249.0
99.2	252	250	2	248.0
99.2	251	249	2	247.0
99.2	250	248	2	246.0
99.2	249	247	2	245.0
99.2	248	246	2	244.0
99.2	247	245	2	243.0
99.2	246	244	2	242.0

%E	M1	M2	DM	M*
99.2	245	243	2	241.0
99.2	244	242	2	240.0
99.2	243	241	2	239.0
99.2	242	240	2	238.0
99.2	241	239	2	237.0
99.2	240	238	2	236.0
99.2	239	237	2	235.0
99.2	238	236	2	234.0
99.2	237	235	2	233.0
99.2	236	234	2	232.0
99.2	133	132	1	131.0
99.2	132	131	1	130.0
99.2	131	130	1	129.0
99.2	130	129	1	128.0
99.2	129	128	1	127.0
99.2	128	127	1	126.0
99.2	127	126	1	125.0
99.2	126	125	1	124.0
99.2	125	124	1	123.0
99.2	124	123	1	122.0
99.2	123	122	1	121.0
99.2	122	121	1	120.0
99.2	121	120	1	119.0
99.2	120	119	1	118.0
99.2	119	118	1	117.0
99.2	118	117	1	116.0
99.1	470	466	4	462.0
99.1	469	465	4	461.0
99.1	468	464	4	460.0
99.1	467	463	4	459.0
99.1	466	462	4	458.0
99.1	465	461	4	457.0
99.1	464	460	4	456.0
99.1	463	459	4	455.0
99.1	462	458	4	454.0
99.1	461	457	4	453.0
99.1	460	456	4	452.0
99.1	459	455	4	451.0
99.1	458	454	4	450.0
99.1	457	453	4	449.0
99.1	456	452	4	448.0
99.1	455	451	4	447.0
99.1	454	450	4	446.0
99.1	453	449	4	445.0
99.1	452	448	4	444.0
99.1	451	447	4	443.0
99.1	450	446	4	442.0
99.1	449	445	4	441.0
99.1	448	444	4	440.0
99.1	447	443	4	439.0

%E	M1	M2	DM	M*
99.1	446	442	4	438.0
99.1	445	441	4	437.0
99.1	444	440	4	436.0
99.1	443	439	4	435.0
99.1	442	438	4	434.0
99.1	441	437	4	433.0
99.1	440	436	4	432.0
99.1	439	435	4	431.0
99.1	438	434	4	430.0
99.1	437	433	4	429.0
99.1	436	432	4	428.0
99.1	435	431	4	427.0
99.1	434	430	4	426.0
99.1	433	429	4	425.0
99.1	432	428	4	424.0
99.1	431	427	4	423.0
99.1	430	426	4	422.0
99.1	429	425	4	421.0
99.1	428	424	4	420.0
99.1	427	423	4	419.0
99.1	426	422	4	418.0
99.1	425	421	4	417.0
99.1	424	420	4	416.0
99.1	423	419	4	415.0
99.1	422	418	4	414.0
99.1	421	417	4	413.0
99.1	352	349	3	346.0
99.1	351	348	3	345.0
99.1	350	347	3	344.0
99.1	349	346	3	343.0
99.1	348	345	3	342.0
99.1	347	344	3	341.0
99.1	346	343	3	340.0
99.1	345	342	3	339.0
99.1	344	341	3	338.0
99.1	343	340	3	337.0
99.1	342	339	3	336.0
99.1	341	338	3	335.0
99.1	340	337	3	334.0
99.1	339	336	3	333.0
99.1	338	335	3	332.0
99.1	337	334	3	331.0
99.1	336	333	3	330.0
99.1	335	332	3	329.0
99.1	334	331	3	328.0
99.1	333	330	3	327.0
99.1	332	329	3	326.0
99.1	331	328	3	325.0
99.1	330	327	3	324.0
99.1	329	326	3	323.0

%E	M1	M2	DM	M*
99.1	328	325	3	322.0
99.1	327	324	3	321.0
99.1	326	323	3	320.0
99.1	325	322	3	319.0
99.1	324	321	3	318.0
99.1	323	320	3	317.0
99.1	322	319	3	316.0
99.1	321	318	3	315.0
99.1	320	317	3	314.0
99.1	319	316	3	313.0
99.1	318	315	3	312.0
99.1	317	314	3	311.0
99.1	316	313	3	310.0
99.1	235	233	2	231.0
99.1	234	232	2	230.0
99.1	233	231	2	229.0
99.1	232	230	2	228.0
99.1	231	229	2	227.0
99.1	230	228	2	226.0
99.1	229	227	2	225.0
99.1	228	226	2	224.0
99.1	227	225	2	223.0
99.1	226	224	2	222.0
99.1	225	223	2	221.0
99.1	224	222	2	220.0
99.1	223	221	2	219.0
99.1	222	220	2	218.0
99.1	221	219	2	217.0
99.1	220	218	2	216.0
99.1	219	217	2	215.0
99.1	218	216	2	214.0
99.1	217	215	2	213.0
99.1	216	214	2	212.0
99.1	215	213	2	211.0
99.1	214	212	2	210.0
99.1	213	211	2	209.0
99.1	212	210	2	208.0
99.1	211	209	2	207.0
99.1	117	116	1	115.0
99.1	116	115	1	114.0
99.1	115	114	1	113.0
99.1	114	113	1	112.0
99.1	113	112	1	111.0
99.1	112	111	1	110.0
99.1	111	110	1	109.0
99.1	110	109	1	108.0
99.1	109	108	1	107.0
99.1	108	107	1	106.0
99.1	107	106	1	105.0
99.1	106	105	1	104.0

%E	M1	M2	DM	M*
99.0	500	495	5	490.0
99.0	499	494	5	489.1
99.0	498	493	5	488.1
99.0	497	492	5	487.1
99.0	496	491	5	486.1
99.0	495	490	5	485.1
99.0	494	489	5	484.1
99.0	493	488	5	483.1
99.0	492	487	5	482.1
99.0	491	486	5	481.1
99.0	490	485	5	480.1
99.0	489	484	5	479.1
99.0	488	483	5	478.1
99.0	487	482	5	477.1
99.0	486	481	5	476.1
99.0	485	480	5	475.1
99.0	484	479	5	474.1
99.0	483	478	5	473.1
99.0	482	477	5	472.1
99.0	481	476	5	471.1
99.0	480	475	5	470.1
99.0	479	474	5	469.1
99.0	478	473	5	468.1
99.0	477	472	5	467.1
99.0	476	471	5	466.1
99.0	420	416	4	412.0
99.0	419	415	4	411.0
99.0	418	414	4	410.0
99.0	417	413	4	409.0
99.0	416	412	4	408.0
99.0	415	411	4	407.0
99.0	414	410	4	406.0
99.0	413	409	4	405.0
99.0	412	408	4	404.0
99.0	411	407	4	403.0
99.0	410	406	4	402.0
99.0	409	405	4	401.0
99.0	408	404	4	400.0
99.0	407	403	4	399.0
99.0	406	402	4	398.0
99.0	405	401	4	397.0
99.0	404	400	4	396.0
99.0	403	399	4	395.0
99.0	402	398	4	394.0
99.0	401	397	4	393.0
99.0	400	396	4	392.0
99.0	399	395	4	391.0
99.0	398	394	4	390.0
99.0	397	393	4	389.0
99.0	396	392	4	388.0

%E	M1	M2	DM	M*	%E	M1	M2	DM	M*	%E	M1	M2	DM	M*	%E	M1	M2	DM	M*	%E	M1	M2	DM	M*
99.0	395	391	4	387.0	99.0	205	203	2	201.0	98.9	450	445	5	440.1	98.9	284	281	3	278.0	98.8	500	494	6	488.1
99.0	394	390	4	386.0	99.0	204	202	2	200.0	98.9	449	444	5	439.1	98.9	283	280	3	277.0	98.8	499	493	6	487.1
99.0	393	389	4	385.0	99.0	203	201	2	199.0	98.9	448	443	5	438.1	98.9	282	279	3	276.0	98.8	498	492	6	486.1
99.0	392	388	4	384.0	99.0	202	200	2	198.0	98.9	447	442	5	437.1	98.9	281	278	3	275.0	98.8	497	491	6	485.1
99.0	391	387	4	383.0	99.0	201	199	2	197.0	98.9	446	441	5	436.1	98.9	280	277	3	274.0	98.8	496	490	6	484.1
99.0	390	386	4	382.0	99.0	200	198	2	196.0	98.9	445	440	5	435.1	98.9	279	276	3	273.0	98.8	495	489	6	483.1
99.0	389	385	4	381.0	99.0	199	197	2	195.0	98.9	444	439	5	434.1	98.9	278	275	3	272.0	98.8	494	488	6	482.1
99.0	388	384	4	380.0	99.0	198	196	2	194.0	98.9	443	438	5	433.1	98.9	277	274	3	271.0	98.8	493	487	6	481.1
99.0	387	383	4	379.0	99.0	197	195	2	193.0	98.9	442	437	5	432.1	98.9	276	273	3	270.0	98.8	492	486	6	480.1
99.0	386	382	4	378.0	99.0	196	194	2	192.0	98.9	441	436	5	431.1	98.9	275	272	3	269.0	98.8	491	485	6	479.1
99.0	385	381	4	377.0	99.0	195	193	2	191.0	98.9	440	435	5	430.1	98.9	274	271	3	268.0	98.8	490	484	6	478.1
99.0	384	380	4	376.0	99.0	194	192	2	190.0	98.9	439	434	5	429.1	98.9	273	270	3	267.0	98.8	489	483	6	477.1
99.0	383	379	4	375.0	99.0	193	191	2	189.0	98.9	438	433	5	428.1	98.9	272	269	3	266.0	98.8	488	482	6	476.1
99.0	382	378	4	374.0	99.0	192	190	2	188.0	98.9	437	432	5	427.1	98.9	271	268	3	265.0	98.8	487	481	6	475.1
99.0	381	377	4	373.0	99.0	191	189	2	187.0	98.9	436	431	5	426.1	98.9	270	267	3	264.0	98.8	486	480	6	474.1
99.0	315	312	3	309.0	99.0	105	104	1	103.0	98.9	435	430	5	425.1	98.9	269	266	3	263.0	98.8	485	479	6	473.1
99.0	314	311	3	308.0	99.0	104	103	1	102.0	98.9	380	376	4	372.0	98.9	268	265	3	262.0	98.8	484	478	6	472.1
99.0	313	310	3	307.0	99.0	103	102	1	101.0	98.9	379	375	4	371.0	98.9	267	264	3	261.0	98.8	483	477	6	471.1
99.0	312	309	3	306.0	99.0	102	101	1	100.0	98.9	378	374	4	370.0	98.9	266	263	3	260.0	98.8	482	476	6	470.1
99.0	311	308	3	305.0	99.0	101	100	1	99.0	98.9	377	373	4	369.0	98.9	265	262	3	259.0	98.8	481	475	6	469.1
99.0	310	307	3	304.0	99.0	100	99	1	98.0	98.9	376	372	4	368.0	98.9	264	261	3	258.0	98.8	480	474	6	468.1
99.0	309	306	3	303.0	99.0	99	98	1	97.0	98.9	375	371	4	367.0	98.9	263	260	3	257.0	98.8	434	429	5	424.1
99.0	308	305	3	302.0	99.0	98	97	1	96.0	98.9	374	370	4	366.0	98.9	262	259	3	256.0	98.8	433	428	5	423.1
99.0	307	304	3	301.0	99.0	97	96	1	95.0	98.9	373	369	4	365.0	98.9	261	258	3	255.0	98.8	432	427	5	422.1
99.0	306	303	3	300.0	99.0	96	95	1	94.0	98.9	372	368	4	364.0	98.9	190	188	2	186.0	98.8	431	426	5	421.1
99.0	305	302	3	299.0	98.9	475	470	5	465.1	98.9	371	367	4	363.0	98.9	189	187	2	185.0	98.8	430	425	5	420.1
99.0	304	301	3	298.0	98.9	474	469	5	464.1	98.9	370	366	4	362.0	98.9	188	186	2	184.0	98.8	429	424	5	419.1
99.0	303	300	3	297.0	98.9	473	468	5	463.1	98.9	369	365	4	361.0	98.9	187	185	2	183.0	98.8	428	423	5	418.1
99.0	302	299	3	296.0	98.9	472	467	5	462.1	98.9	368	364	4	360.0	98.9	186	184	2	182.0	98.8	427	422	5	417.1
99.0	301	298	3	295.0	98.9	471	466	5	461.1	98.9	367	363	4	359.0	98.9	185	183	2	181.0	98.8	426	421	5	416.1
99.0	300	297	3	294.0	98.9	470	465	5	460.1	98.9	366	362	4	358.0	98.9	184	182	2	180.0	98.8	425	420	5	415.1
99.0	299	296	3	293.0	98.9	469	464	5	459.1	98.9	365	361	4	357.0	98.9	183	181	2	179.0	98.8	424	419	5	414.1
99.0	298	295	3	292.0	98.9	468	463	5	458.1	98.9	364	360	4	356.0	98.9	182	180	2	178.0	98.8	423	418	5	413.1
99.0	297	294	3	291.0	98.9	467	462	5	457.1	98.9	363	359	4	355.0	98.9	181	179	2	177.0	98.8	422	417	5	412.1
99.0	296	293	3	290.0	98.9	466	461	5	456.1	98.9	362	358	4	354.0	98.9	180	178	2	176.0	98.8	421	416	5	411.1
99.0	295	292	3	289.0	98.9	465	460	5	455.1	98.9	361	357	4	353.0	98.9	179	177	2	175.0	98.8	420	415	5	410.1
99.0	294	291	3	288.0	98.9	464	459	5	454.1	98.9	360	356	4	352.0	98.9	178	176	2	174.0	98.8	419	414	5	409.1
99.0	293	290	3	287.0	98.9	463	458	5	453.1	98.9	359	355	4	351.0	98.9	177	175	2	173.0	98.8	418	413	5	408.1
99.0	292	289	3	286.0	98.9	462	457	5	452.1	98.9	358	354	4	350.0	98.9	176	174	2	172.0	98.8	417	412	5	407.1
99.0	291	288	3	285.0	98.9	461	456	5	451.1	98.9	357	353	4	349.0	98.9	175	173	2	171.0	98.8	416	411	5	406.1
99.0	290	287	3	284.0	98.9	460	455	5	450.1	98.9	356	352	4	348.0	98.9	174	172	2	170.0	98.8	415	410	5	405.1
99.0	289	286	3	283.0	98.9	459	454	5	449.1	98.9	355	351	4	347.0	98.9	95	94	1	93.0	98.8	414	409	5	404.1
99.0	288	285	3	282.0	98.9	458	453	5	448.1	98.9	354	350	4	346.0	98.9	94	93	1	92.0	98.8	413	408	5	403.1
99.0	287	284	3	281.0	98.9	457	452	5	447.1	98.9	353	349	4	345.0	98.9	93	92	1	91.0	98.8	412	407	5	402.1
99.0	286	283	3	280.0	98.9	456	451	5	446.1	98.9	352	348	4	344.0	98.9	92	91	1	90.0	98.8	411	406	5	401.1
99.0	210	208	2	206.0	98.9	455	450	5	445.1	98.9	351	347	4	343.0	98.9	91	90	1	89.0	98.8	410	405	5	400.1
99.0	209	207	2	205.0	98.9	454	449	5	444.1	98.9	350	346	4	342.0	98.9	90	89	1	88.0	98.8	409	404	5	399.1
99.0	208	206	2	204.0	98.9	453	448	5	443.1	98.9	349	345	4	341.0	98.9	89	88	1	87.0	98.8	408	403	5	398.1
99.0	207	205	2	203.0	98.9	452	447	5	442.1	98.9	348	344	4	340.0	98.9	88	87	1	86.0	98.8	407	402	5	397.1
99.0	206	204	2	202.0	98.9	451	446	5	441.1	98.9	285	282	3	279.0	98.9	87	86	1	85.0	98.8	406	401	5	396.1

%E	M1	M2	DM	M*
98.8	405	400	5	395.1
98.8	404	399	5	394.1
98.8	403	398	5	393.1
98.8	402	397	5	392.1
98.8	401	396	5	391.1
98.8	400	395	5	390.1
98.8	347	343	4	339.0
98.8	346	342	4	338.0
98.8	345	341	4	337.0
98.8	344	340	4	336.0
98.8	343	339	4	335.0
98.8	342	338	4	334.0
98.8	341	337	4	333.0
98.8	340	336	4	332.0
98.8	339	335	4	331.0
98.8	338	334	4	330.0
98.8	337	333	4	329.0
98.8	336	332	4	328.0
98.8	335	331	4	327.0
98.8	334	330	4	326.0
98.8	333	329	4	325.0
98.8	332	328	4	324.0
98.8	331	327	4	323.0
98.8	330	326	4	322.0
98.8	329	325	4	321.0
98.8	328	324	4	320.0
98.8	327	323	4	319.0
98.8	326	322	4	318.0
98.8	325	321	4	317.0
98.8	324	320	4	316.0
98.8	323	319	4	315.0
98.8	322	318	4	314.0
98.8	321	317	4	313.0
98.8	320	316	4	312.0
98.8	260	257	3	254.0
98.8	259	256	3	253.0
98.8	258	255	3	252.0
98.8	257	254	3	251.0
98.8	256	253	3	250.0
98.8	255	252	3	249.0
98.8	254	251	3	248.0
98.8	253	250	3	247.0
98.8	252	249	3	246.0
98.8	251	248	3	245.0
98.8	250	247	3	244.0
98.8	249	246	3	243.0
98.8	248	245	3	242.0
98.8	247	244	3	241.0
98.8	246	243	3	240.0
98.8	245	242	3	239.0

%E	M1	M2	DM	M*
98.8	244	241	3	238.0
98.8	243	240	3	237.0
98.8	242	239	3	236.0
98.8	241	238	3	235.0
98.8	240	237	3	234.0
98.8	173	171	2	169.0
98.8	172	170	2	168.0
98.8	171	169	2	167.0
98.8	170	168	2	166.0
98.8	169	167	2	165.0
98.8	168	166	2	164.0
98.8	167	165	2	163.0
98.8	166	164	2	162.0
98.8	165	163	2	161.0
98.8	164	162	2	160.0
98.8	163	161	2	159.0
98.8	162	160	2	158.0
98.8	161	159	2	157.0
98.8	160	158	2	156.0
98.8	86	85	1	84.0
98.8	85	84	1	83.0
98.8	84	83	1	82.0
98.8	83	82	1	81.0
98.8	82	81	1	80.0
98.8	81	80	1	79.0
98.8	80	79	1	78.0
98.7	479	473	6	467.1
98.7	478	472	6	466.1
98.7	477	471	6	465.1
98.7	476	470	6	464.1
98.7	475	469	6	463.1
98.7	474	468	6	462.1
98.7	473	467	6	461.1
98.7	472	466	6	460.1
98.7	471	465	6	459.1
98.7	470	464	6	458.1
98.7	469	463	6	457.1
98.7	468	462	6	456.1
98.7	467	461	6	455.1
98.7	466	460	6	454.1
98.7	465	459	6	453.1
98.7	464	458	6	452.1
98.7	463	457	6	451.1
98.7	462	456	6	450.1
98.7	461	455	6	449.1
98.7	460	454	6	448.1
98.7	459	453	6	447.1
98.7	458	452	6	446.1
98.7	457	451	6	445.1
98.7	456	450	6	444.1

%E	M1	M2	DM	M*
98.7	455	449	6	443.1
98.7	454	448	6	442.1
98.7	453	447	6	441.1
98.7	452	446	6	440.1
98.7	451	445	6	439.1
98.7	450	444	6	438.1
98.7	449	443	6	437.1
98.7	448	442	6	436.1
98.7	447	441	6	435.1
98.7	446	440	6	434.1
98.7	445	439	6	433.1
98.7	399	394	5	389.1
98.7	398	393	5	388.1
98.7	397	392	5	387.1
98.7	396	391	5	386.1
98.7	395	390	5	385.1
98.7	394	389	5	384.1
98.7	393	388	5	383.1
98.7	392	387	5	382.1
98.7	391	386	5	381.1
98.7	390	385	5	380.1
98.7	389	384	5	379.1
98.7	388	383	5	378.1
98.7	387	382	5	377.1
98.7	386	381	5	376.1
98.7	385	380	5	375.1
98.7	384	379	5	374.1
98.7	383	378	5	373.1
98.7	382	377	5	372.1
98.7	381	376	5	371.1
98.7	380	375	5	370.1
98.7	379	374	5	369.1
98.7	378	373	5	368.1
98.7	377	372	5	367.1
98.7	376	371	5	366.1
98.7	375	370	5	365.1
98.7	374	369	5	364.1
98.7	373	368	5	363.1
98.7	372	367	5	362.1
98.7	371	366	5	361.1
98.7	319	315	4	311.1
98.7	318	314	4	310.1
98.7	317	313	4	309.1
98.7	316	312	4	308.1
98.7	315	311	4	307.1
98.7	314	310	4	306.1
98.7	313	309	4	305.1
98.7	312	308	4	304.1
98.7	311	307	4	303.1
98.7	310	306	4	302.1

%E	M1	M2	DM	M*
98.7	309	305	4	301.1
98.7	308	304	4	300.1
98.7	307	303	4	299.1
98.7	306	302	4	298.1
98.7	305	301	4	297.1
98.7	304	300	4	296.1
98.7	303	299	4	295.1
98.7	302	298	4	294.1
98.7	301	297	4	293.1
98.7	300	296	4	292.1
98.7	299	295	4	291.1
98.7	298	294	4	290.1
98.7	297	293	4	289.1
98.7	239	236	3	233.0
98.7	238	235	3	232.0
98.7	237	234	3	231.0
98.7	236	233	3	230.0
98.7	235	232	3	229.0
98.7	234	231	3	228.0
98.7	233	230	3	227.0
98.7	232	229	3	226.0
98.7	231	228	3	225.0
98.7	230	227	3	224.0
98.7	229	226	3	223.0
98.7	228	225	3	222.0
98.7	227	224	3	221.0
98.7	226	223	3	220.0
98.7	225	222	3	219.0
98.7	224	221	3	218.0
98.7	223	220	3	217.0
98.7	159	157	2	155.0
98.7	158	156	2	154.0
98.7	157	155	2	153.0
98.7	156	154	2	152.0
98.7	155	153	2	151.0
98.7	154	152	2	150.0
98.7	153	151	2	149.0
98.7	152	150	2	148.0
98.7	151	149	2	147.0
98.7	150	148	2	146.0
98.7	149	147	2	145.0
98.7	79	78	1	77.0
98.7	78	77	1	76.0
98.7	77	76	1	75.0
98.7	76	75	1	74.0
98.7	75	74	1	73.0
98.6	500	493	7	486.1
98.6	499	492	7	485.1
98.6	498	491	7	484.1
98.6	497	490	7	483.1

%E	M1	M2	DM	M*
98.6	496	489	7	482.1
98.6	495	488	7	481.1
98.6	494	487	7	480.1
98.6	493	486	7	479.1
98.6	492	485	7	478.1
98.6	491	484	7	477.1
98.6	490	483	7	476.1
98.6	489	482	7	475.1
98.6	488	481	7	474.1
98.6	487	480	7	473.1
98.6	486	479	7	472.1
98.6	485	478	7	471.1
98.6	484	477	7	470.1
98.6	483	476	7	469.1
98.6	444	438	6	432.1
98.6	443	437	6	431.1
98.6	442	436	6	430.1
98.6	441	435	6	429.1
98.6	440	434	6	428.1
98.6	439	433	6	427.1
98.6	438	432	6	426.1
98.6	437	431	6	425.1
98.6	436	430	6	424.1
98.6	435	429	6	423.1
98.6	434	428	6	422.1
98.6	433	427	6	421.1
98.6	432	426	6	420.1
98.6	431	425	6	419.1
98.6	430	424	6	418.1
98.6	429	423	6	417.1
98.6	428	422	6	416.1
98.6	427	421	6	415.1
98.6	426	420	6	414.1
98.6	425	419	6	413.1
98.6	424	418	6	412.1
98.6	423	417	6	411.1
98.6	422	416	6	410.1
98.6	421	415	6	409.1
98.6	420	414	6	408.1
98.6	419	413	6	407.1
98.6	418	412	6	406.1
98.6	417	411	6	405.1
98.6	416	410	6	404.1
98.6	415	409	6	403.1
98.6	414	408	6	402.1
98.6	370	365	5	360.1
98.6	369	364	5	359.1
98.6	368	363	5	358.1
98.6	367	362	5	357.1
98.6	366	361	5	356.1

%E	M1	M2	DM	M*
98.6	365	360	5	355.1
98.6	364	359	5	354.1
98.6	363	358	5	353.1
98.6	362	357	5	352.1
98.6	361	356	5	351.1
98.6	360	355	5	350.1
98.6	359	354	5	349.1
98.6	358	353	5	348.1
98.6	357	352	5	347.1
98.6	356	351	5	346.1
98.6	355	350	5	345.1
98.6	354	349	5	344.1
98.6	353	348	5	343.1
98.6	352	347	5	342.1
98.6	351	346	5	341.1
98.6	350	345	5	340.1
98.6	349	344	5	339.1
98.6	348	343	5	338.1
98.6	347	342	5	337.1
98.6	346	341	5	336.1
98.6	345	340	5	335.1
98.6	296	292	4	288.1
98.6	295	291	4	287.1
98.6	294	290	4	286.1
98.6	293	289	4	285.1
98.6	292	288	4	284.1
98.6	291	287	4	283.1
98.6	290	286	4	282.1
98.6	289	285	4	281.1
98.6	288	284	4	280.1
98.6	287	283	4	279.1
98.6	286	282	4	278.1
98.6	285	281	4	277.1
98.6	284	280	4	276.1
98.6	283	279	4	275.1
98.6	282	278	4	274.1
98.6	281	277	4	273.1
98.6	280	276	4	272.1
98.6	279	275	4	271.1
98.6	278	274	4	270.1
98.6	277	273	4	269.1
98.6	276	272	4	268.1
98.6	222	219	3	216.0
98.6	221	218	3	215.0
98.6	220	217	3	214.0
98.6	219	216	3	213.0
98.6	218	215	3	212.0
98.6	217	214	3	211.0
98.6	216	213	3	210.0
98.6	215	212	3	209.0

%E	M1	M2	DM	M*
98.6	214	211	3	208.0
98.6	213	210	3	207.0
98.6	212	209	3	206.0
98.6	211	208	3	205.0
98.6	210	207	3	204.0
98.6	209	206	3	203.0
98.6	208	205	3	202.0
98.6	207	204	3	201.0
98.6	148	146	2	144.0
98.6	147	145	2	143.0
98.6	146	144	2	142.0
98.6	145	143	2	141.0
98.6	144	142	2	140.0
98.6	143	141	2	139.0
98.6	142	140	2	138.0
98.6	141	139	2	137.0
98.6	140	138	2	136.0
98.6	139	137	2	135.0
98.6	138	136	2	134.0
98.6	74	73	1	72.0
98.6	73	72	1	71.0
98.6	72	71	1	70.0
98.6	71	70	1	69.0
98.6	70	69	1	68.0
98.6	69	68	1	67.0
98.5	482	475	7	468.1
98.5	481	474	7	467.1
98.5	480	473	7	466.1
98.5	479	472	7	465.1
98.5	478	471	7	464.1
98.5	477	470	7	463.1
98.5	476	469	7	462.1
98.5	475	468	7	461.1
98.5	474	467	7	460.1
98.5	473	466	7	459.1
98.5	472	465	7	458.1
98.5	471	464	7	457.1
98.5	470	463	7	456.1
98.5	469	462	7	455.1
98.5	468	461	7	454.1
98.5	467	460	7	453.1
98.5	466	459	7	452.1
98.5	465	458	7	451.1
98.5	464	457	7	450.1
98.5	463	456	7	449.1
98.5	462	455	7	448.1
98.5	461	454	7	447.1
98.5	460	453	7	446.1
98.5	459	452	7	445.1
98.5	458	451	7	444.1

%E	M1	M2	DM	M*
98.5	457	450	7	443.1
98.5	456	449	7	442.1
98.5	455	448	7	441.1
98.5	454	447	7	440.1
98.5	453	446	7	439.1
98.5	452	445	7	438.1
98.5	413	407	6	401.1
98.5	412	406	6	400.1
98.5	411	405	6	399.1
98.5	410	404	6	398.1
98.5	409	403	6	397.1
98.5	408	402	6	396.1
98.5	407	401	6	395.1
98.5	406	400	6	394.1
98.5	405	399	6	393.1
98.5	404	398	6	392.1
98.5	403	397	6	391.1
98.5	402	396	6	390.1
98.5	401	395	6	389.1
98.5	400	394	6	388.1
98.5	399	393	6	387.1
98.5	398	392	6	386.1
98.5	397	391	6	385.1
98.5	396	390	6	384.1
98.5	395	389	6	383.1
98.5	394	388	6	382.1
98.5	393	387	6	381.1
98.5	392	386	6	380.1
98.5	391	385	6	379.1
98.5	390	384	6	378.1
98.5	389	383	6	377.1
98.5	388	382	6	376.1
98.5	387	381	6	375.1
98.5	344	339	5	334.1
98.5	343	338	5	333.1
98.5	342	337	5	332.1
98.5	341	336	5	331.1
98.5	340	335	5	330.1
98.5	339	334	5	329.1
98.5	338	333	5	328.1
98.5	337	332	5	327.1
98.5	336	331	5	326.1
98.5	335	330	5	325.1
98.5	334	329	5	324.1
98.5	333	328	5	323.1
98.5	332	327	5	322.1
98.5	331	326	5	321.1
98.5	330	325	5	320.1
98.5	329	324	5	319.1
98.5	328	323	5	318.1

%E	M1	M2	DM	M*
98.5	327	322	5	317.1
98.5	326	321	5	316.1
98.5	325	320	5	315.1
98.5	324	319	5	314.1
98.5	323	318	5	313.1
98.5	275	271	4	267.1
98.5	274	270	4	266.1
98.5	273	269	4	265.1
98.5	272	268	4	264.1
98.5	271	267	4	263.1
98.5	270	266	4	262.1
98.5	269	265	4	261.1
98.5	268	264	4	260.1
98.5	267	263	4	259.1
98.5	266	262	4	258.1
98.5	265	261	4	257.1
98.5	264	260	4	256.1
98.5	263	259	4	255.1
98.5	262	258	4	254.1
98.5	261	257	4	253.1
98.5	260	256	4	252.1
98.5	259	255	4	251.1
98.5	258	254	4	250.1
98.5	206	203	3	200.0
98.5	205	202	3	199.0
98.5	204	201	3	198.0
98.5	203	200	3	197.0
98.5	202	199	3	196.0
98.5	201	198	3	195.0
98.5	200	197	3	194.0
98.5	199	196	3	193.0
98.5	198	195	3	192.0
98.5	197	194	3	191.0
98.5	196	193	3	190.0
98.5	195	192	3	189.0
98.5	194	191	3	188.0
98.5	137	135	2	133.0
98.5	136	134	2	132.0
98.5	135	133	2	131.0
98.5	134	132	2	130.0
98.5	133	131	2	129.0
98.5	132	130	2	128.0
98.5	131	129	2	127.0
98.5	130	128	2	126.0
98.5	129	127	2	125.0
98.5	68	67	1	66.0
98.5	67	66	1	65.0
98.5	66	65	1	64.0
98.5	65	64	1	63.0
98.4	500	492	8	484.1

%E	M1	M2	DM	M*
98.4	499	491	8	483.1
98.4	498	490	8	482.1
98.4	497	489	8	481.1
98.4	496	488	8	480.1
98.4	495	487	8	479.1
98.4	494	486	8	478.1
98.4	493	485	8	477.1
98.4	492	484	8	476.1
98.4	491	483	8	475.1
98.4	490	482	8	474.1
98.4	489	481	8	473.1
98.4	488	480	8	472.1
98.4	487	479	8	471.1
98.4	486	478	8	470.1
98.4	485	477	8	469.1
98.4	451	444	7	437.1
98.4	450	443	7	436.1
98.4	449	442	7	435.1
98.4	448	441	7	434.1
98.4	447	440	7	433.1
98.4	446	439	7	432.1
98.4	445	438	7	431.1
98.4	444	437	7	430.1
98.4	443	436	7	429.1
98.4	442	435	7	428.1
98.4	441	434	7	427.1
98.4	440	433	7	426.1
98.4	439	432	7	425.1
98.4	438	431	7	424.1
98.4	437	430	7	423.1
98.4	436	429	7	422.1
98.4	435	428	7	421.1
98.4	434	427	7	420.1
98.4	433	426	7	419.1
98.4	432	425	7	418.1
98.4	431	424	7	417.1
98.4	430	423	7	416.1
98.4	429	422	7	415.1
98.4	428	421	7	414.1
98.4	427	420	7	413.1
98.4	426	419	7	412.1
98.4	425	418	7	411.1
98.4	386	380	6	374.1
98.4	385	379	6	373.1
98.4	384	378	6	372.1
98.4	383	377	6	371.1
98.4	382	376	6	370.1
98.4	381	375	6	369.1
98.4	380	374	6	368.1
98.4	379	373	6	367.1

%E	M1	M2	DM	M*
98.4	378	372	6	366.1
98.4	377	371	6	365.1
98.4	376	370	6	364.1
98.4	375	369	6	363.1
98.4	374	368	6	362.1
98.4	373	367	6	361.1
98.4	372	366	6	360.1
98.4	371	365	6	359.1
98.4	370	364	6	358.1
98.4	369	363	6	357.1
98.4	368	362	6	356.1
98.4	367	361	6	355.1
98.4	366	360	6	354.1
98.4	365	359	6	353.1
98.4	364	358	6	352.1
98.4	322	317	5	312.1
98.4	321	316	5	311.1
98.4	320	315	5	310.1
98.4	319	314	5	309.1
98.4	318	313	5	308.1
98.4	317	312	5	307.1
98.4	316	311	5	306.1
98.4	315	310	5	305.1
98.4	314	309	5	304.1
98.4	313	308	5	303.1
98.4	312	307	5	302.1
98.4	311	306	5	301.1
98.4	310	305	5	300.1
98.4	309	304	5	299.1
98.4	308	303	5	298.1
98.4	307	302	5	297.1
98.4	306	301	5	296.1
98.4	305	300	5	295.1
98.4	304	299	5	294.1
98.4	303	298	5	293.1
98.4	257	253	4	249.1
98.4	256	252	4	248.1
98.4	255	251	4	247.1
98.4	254	250	4	246.1
98.4	253	249	4	245.1
98.4	252	248	4	244.1
98.4	251	247	4	243.1
98.4	250	246	4	242.1
98.4	249	245	4	241.1
98.4	248	244	4	240.1
98.4	247	243	4	239.1
98.4	246	242	4	238.1
98.4	245	241	4	237.1
98.4	244	240	4	236.1
98.4	243	239	4	235.1

%E	M1	M2	DM	M*
98.4	193	190	3	187.0
98.4	192	189	3	186.0
98.4	191	188	3	185.0
98.4	190	187	3	184.0
98.4	189	186	3	183.0
98.4	188	185	3	182.0
98.4	187	184	3	181.0
98.4	186	183	3	180.0
98.4	185	182	3	179.0
98.4	184	181	3	178.0
98.4	183	180	3	177.0
98.4	182	179	3	176.0
98.4	128	126	2	124.0
98.4	127	125	2	123.0
98.4	126	124	2	122.0
98.4	125	123	2	121.0
98.4	124	122	2	120.0
98.4	123	121	2	119.0
98.4	122	120	2	118.0
98.4	64	63	1	62.0
98.4	63	62	1	61.0
98.4	62	61	1	60.0
98.4	61	60	1	59.0
98.3	484	476	8	468.1
98.3	483	475	8	467.1
98.3	482	474	8	466.1
98.3	481	473	8	465.1
98.3	480	472	8	464.1
98.3	479	471	8	463.1
98.3	478	470	8	462.1
98.3	477	469	8	461.1
98.3	476	468	8	460.1
98.3	475	467	8	459.1
98.3	474	466	8	458.1
98.3	473	465	8	457.1
98.3	472	464	8	456.1
98.3	471	463	8	455.1
98.3	470	462	8	454.1
98.3	469	461	8	453.1
98.3	468	460	8	452.1
98.3	467	459	8	451.1
98.3	466	458	8	450.1
98.3	465	457	8	449.1
98.3	464	456	8	448.1
98.3	463	455	8	447.1
98.3	462	454	8	446.1
98.3	461	453	8	445.1
98.3	460	452	8	444.1
98.3	459	451	8	443.1
98.3	458	450	8	442.1

%E	M1	M2	DM	M*
98.3	424	417	7	410.1
98.3	423	416	7	409.1
98.3	422	415	7	408.1
98.3	421	414	7	407.1
98.3	420	413	7	406.1
98.3	419	412	7	405.1
98.3	418	411	7	404.1
98.3	417	410	7	403.1
98.3	416	409	7	402.1
98.3	415	408	7	401.1
98.3	414	407	7	400.1
98.3	413	406	7	399.1
98.3	412	405	7	398.1
98.3	411	404	7	397.1
98.3	410	403	7	396.1
98.3	409	402	7	395.1
98.3	408	401	7	394.1
98.3	407	400	7	393.1
98.3	406	399	7	392.1
98.3	405	398	7	391.1
98.3	404	397	7	390.1
98.3	403	396	7	389.1
98.3	402	395	7	388.1
98.3	401	394	7	387.1
98.3	400	393	7	386.1
98.3	363	357	6	351.1
98.3	362	356	6	350.1
98.3	361	355	6	349.1
98.3	360	354	6	348.1
98.3	359	353	6	347.1
98.3	358	352	6	346.1
98.3	357	351	6	345.1
98.3	356	350	6	344.1
98.3	355	349	6	343.1
98.3	354	348	6	342.1
98.3	353	347	6	341.1
98.3	352	346	6	340.1
98.3	351	345	6	339.1
98.3	350	344	6	338.1
98.3	349	343	6	337.1
98.3	348	342	6	336.1
98.3	347	341	6	335.1
98.3	346	340	6	334.1
98.3	345	339	6	333.1
98.3	344	338	6	332.1
98.3	343	337	6	331.1
98.3	302	297	5	292.1
98.3	301	296	5	291.1
98.3	300	295	5	290.1
98.3	299	294	5	289.1

%E	M1	M2	DM	M*
98.3	298	293	5	288.1
98.3	297	292	5	287.1
98.3	296	291	5	286.1
98.3	295	290	5	285.1
98.3	294	289	5	284.1
98.3	293	288	5	283.1
98.3	292	287	5	282.1
98.3	291	286	5	281.1
98.3	290	285	5	280.1
98.3	289	284	5	279.1
98.3	288	283	5	278.1
98.3	287	282	5	277.1
98.3	286	281	5	276.1
98.3	242	238	4	234.1
98.3	241	237	4	233.1
98.3	240	236	4	232.1
98.3	239	235	4	231.1
98.3	238	234	4	230.1
98.3	237	233	4	229.1
98.3	236	232	4	228.1
98.3	235	231	4	227.1
98.3	234	230	4	226.1
98.3	233	229	4	225.1
98.3	232	228	4	224.1
98.3	231	227	4	223.1
98.3	230	226	4	222.1
98.3	229	225	4	221.1
98.3	181	178	3	175.0
98.3	180	177	3	174.0
98.3	179	176	3	173.1
98.3	178	175	3	172.1
98.3	177	174	3	171.1
98.3	176	173	3	170.1
98.3	175	172	3	169.1
98.3	174	171	3	168.1
98.3	173	170	3	167.1
98.3	172	169	3	166.1
98.3	121	119	2	117.0
98.3	120	118	2	116.0
98.3	119	117	2	115.0
98.3	118	116	2	114.0
98.3	117	115	2	113.0
98.3	116	114	2	112.0
98.3	115	113	2	111.0
98.3	60	59	1	58.0
98.3	59	58	1	57.0
98.3	58	57	1	56.0
98.2	500	491	9	482.2
98.2	499	490	9	481.2
98.2	498	489	9	480.2

%E	M1	M2	DM	M*
98.2	497	488	9	479.2
98.2	496	487	9	478.2
98.2	495	486	9	477.2
98.2	494	485	9	476.2
98.2	493	484	9	475.2
98.2	492	483	9	474.2
98.2	491	482	9	473.2
98.2	490	481	9	472.2
98.2	489	480	9	471.2
98.2	488	479	9	470.2
98.2	487	478	9	469.2
98.2	457	449	8	441.1
98.2	456	448	8	440.1
98.2	455	447	8	439.1
98.2	454	446	8	438.1
98.2	453	445	8	437.1
98.2	452	444	8	436.1
98.2	451	443	8	435.1
98.2	450	442	8	434.1
98.2	449	441	8	433.1
98.2	448	440	8	432.1
98.2	447	439	8	431.1
98.2	446	438	8	430.1
98.2	445	437	8	429.1
98.2	444	436	8	428.1
98.2	443	435	8	427.1
98.2	442	434	8	426.1
98.2	441	433	8	425.1
98.2	440	432	8	424.1
98.2	439	431	8	423.1
98.2	438	430	8	422.1
98.2	437	429	8	421.1
98.2	436	428	8	420.1
98.2	435	427	8	419.1
98.2	434	426	8	418.1
98.2	433	425	8	417.1
98.2	399	392	7	385.1
98.2	398	391	7	384.1
98.2	397	390	7	383.1
98.2	396	389	7	382.1
98.2	395	388	7	381.1
98.2	394	387	7	380.1
98.2	393	386	7	379.1
98.2	392	385	7	378.1
98.2	391	384	7	377.1
98.2	390	383	7	376.1
98.2	389	382	7	375.1
98.2	388	381	7	374.1
98.2	387	380	7	373.1
98.2	386	379	7	372.1

%E	M1	M2	DM	M*
98.2	385	378	7	371.1
98.2	384	377	7	370.1
98.2	383	376	7	369.1
98.2	382	375	7	368.1
98.2	381	374	7	367.1
98.2	380	373	7	366.1
98.2	379	372	7	365.1
98.2	342	336	6	330.1
98.2	341	335	6	329.1
98.2	340	334	6	328.1
98.2	339	333	6	327.1
98.2	338	332	6	326.1
98.2	337	331	6	325.1
98.2	336	330	6	324.1
98.2	335	329	6	323.1
98.2	334	328	6	322.1
98.2	333	327	6	321.1
98.2	332	326	6	320.1
98.2	331	325	6	319.1
98.2	330	324	6	318.1
98.2	329	323	6	317.1
98.2	328	322	6	316.1
98.2	327	321	6	315.1
98.2	326	320	6	314.1
98.2	325	319	6	313.1
98.2	285	280	5	275.1
98.2	284	279	5	274.1
98.2	283	278	5	273.1
98.2	282	277	5	272.1
98.2	281	276	5	271.1
98.2	280	275	5	270.1
98.2	279	274	5	269.1
98.2	278	273	5	268.1
98.2	277	272	5	267.1
98.2	276	271	5	266.1
98.2	275	270	5	265.1
98.2	274	269	5	264.1
98.2	273	268	5	263.1
98.2	272	267	5	262.1
98.2	271	266	5	261.1
98.2	228	224	4	220.1
98.2	227	223	4	219.1
98.2	226	222	4	218.1
98.2	225	221	4	217.1
98.2	224	220	4	216.1
98.2	223	219	4	215.1
98.2	222	218	4	214.1
98.2	221	217	4	213.1
98.2	220	216	4	212.1
98.2	219	215	4	211.1

%E	M1	M2	DM	M*
98.2	218	214	4	210.1
98.2	217	213	4	209.1
98.2	171	168	3	165.1
98.2	170	167	3	164.1
98.2	169	166	3	163.1
98.2	168	165	3	162.1
98.2	167	164	3	161.1
98.2	166	163	3	160.1
98.2	165	162	3	159.1
98.2	164	161	3	158.1
98.2	163	160	3	157.1
98.2	114	112	2	110.0
98.2	113	111	2	109.0
98.2	112	110	2	108.0
98.2	111	109	2	107.0
98.2	110	108	2	106.0
98.2	109	107	2	105.0
98.2	57	56	1	55.0
98.2	56	55	1	54.0
98.2	55	54	1	53.0
98.1	486	477	9	468.2
98.1	485	476	9	467.2
98.1	484	475	9	466.2
98.1	483	474	9	465.2
98.1	482	473	9	464.2
98.1	481	472	9	463.2
98.1	480	471	9	462.2
98.1	479	470	9	461.2
98.1	478	469	9	460.2
98.1	477	468	9	459.2
98.1	476	467	9	458.2
98.1	475	466	9	457.2
98.1	474	465	9	456.2
98.1	473	464	9	455.2
98.1	472	463	9	454.2
98.1	471	462	9	453.2
98.1	470	461	9	452.2
98.1	469	460	9	451.2
98.1	468	459	9	450.2
98.1	467	458	9	449.2
98.1	466	457	9	448.2
98.1	465	456	9	447.2
98.1	464	455	9	446.2
98.1	463	454	9	445.2
98.1	462	453	9	444.2
98.1	432	424	8	416.1
98.1	431	423	8	415.1
98.1	430	422	8	414.1
98.1	429	421	8	413.1
98.1	428	420	8	412.1

%E	M1	M2	DM	M*
98.1	427	419	8	411.1
98.1	426	418	8	410.2
98.1	425	417	8	409.2
98.1	424	416	8	408.2
98.1	423	415	8	407.2
98.1	422	414	8	406.2
98.1	421	413	8	405.2
98.1	420	412	8	404.2
98.1	419	411	8	403.2
98.1	418	410	8	402.2
98.1	417	409	8	401.2
98.1	416	408	8	400.2
98.1	415	407	8	399.2
98.1	414	406	8	398.2
98.1	413	405	8	397.2
98.1	412	404	8	396.2
98.1	411	403	8	395.2
98.1	378	371	7	364.1
98.1	377	370	7	363.1
98.1	376	369	7	362.1
98.1	375	368	7	361.1
98.1	374	367	7	360.1
98.1	373	366	7	359.1
98.1	372	365	7	358.1
98.1	371	364	7	357.1
98.1	370	363	7	356.1
98.1	369	362	7	355.1
98.1	368	361	7	354.1
98.1	367	360	7	353.1
98.1	366	359	7	352.1
98.1	365	358	7	351.1
98.1	364	357	7	350.1
98.1	363	356	7	349.1
98.1	362	355	7	348.1
98.1	361	354	7	347.1
98.1	360	353	7	346.1
98.1	359	352	7	345.1
98.1	324	318	6	312.1
98.1	323	317	6	311.1
98.1	322	316	6	310.1
98.1	321	315	6	309.1
98.1	320	314	6	308.1
98.1	319	313	6	307.1
98.1	318	312	6	306.1
98.1	317	311	6	305.1
98.1	316	310	6	304.1
98.1	315	309	6	303.1
98.1	314	308	6	302.1
98.1	313	307	6	301.1
98.1	312	306	6	300.1

%E	M1	M2	DM	M*
98.1	311	305	6	299.1
98.1	310	304	6	298.1
98.1	309	303	6	297.1
98.1	308	302	6	296.1
98.1	270	265	5	260.1
98.1	269	264	5	259.1
98.1	268	263	5	258.1
98.1	267	262	5	257.1
98.1	266	261	5	256.1
98.1	265	260	5	255.1
98.1	264	259	5	254.1
98.1	263	258	5	253.1
98.1	262	257	5	252.1
98.1	261	256	5	251.1
98.1	260	255	5	250.1
98.1	259	254	5	249.1
98.1	258	253	5	248.1
98.1	257	252	5	247.1
98.1	216	212	4	208.1
98.1	215	211	4	207.1
98.1	214	210	4	206.1
98.1	213	209	4	205.1
98.1	212	208	4	204.1
98.1	211	207	4	203.1
98.1	210	206	4	202.1
98.1	209	205	4	201.1
98.1	208	204	4	200.1
98.1	207	203	4	199.1
98.1	206	202	4	198.1
98.1	162	159	3	156.1
98.1	161	158	3	155.1
98.1	160	157	3	154.1
98.1	159	156	3	153.1
98.1	158	155	3	152.1
98.1	157	154	3	151.1
98.1	156	153	3	150.1
98.1	155	152	3	149.1
98.1	154	151	3	148.1
98.1	108	106	2	104.0
98.1	107	105	2	103.0
98.1	106	104	2	102.0
98.1	105	103	2	101.0
98.1	104	102	2	100.0
98.1	103	101	2	99.0
98.1	54	53	1	52.0
98.1	53	52	1	51.0
98.1	52	51	1	50.0
98.0	500	490	10	480.2
98.0	499	489	10	479.2
98.0	498	488	10	478.2

%E	M1	M2	DM	M*
98.0	497	487	10	477.2
98.0	496	486	10	476.2
98.0	495	485	10	475.2
98.0	494	484	10	474.2
98.0	493	483	10	473.2
98.0	492	482	10	472.2
98.0	491	481	10	471.2
98.0	490	480	10	470.2
98.0	489	479	10	469.2
98.0	488	478	10	468.2
98.0	461	452	9	443.2
98.0	460	451	9	442.2
98.0	459	450	9	441.2
98.0	458	449	9	440.2
98.0	457	448	9	439.2
98.0	456	447	9	438.2
98.0	455	446	9	437.2
98.0	454	445	9	436.2
98.0	453	444	9	435.2
98.0	452	443	9	434.2
98.0	451	442	9	433.2
98.0	450	441	9	432.2
98.0	449	440	9	431.2
98.0	448	439	9	430.2
98.0	447	438	9	429.2
98.0	446	437	9	428.2
98.0	445	436	9	427.2
98.0	444	435	9	426.2
98.0	443	434	9	425.2
98.0	442	433	9	424.2
98.0	441	432	9	423.2
98.0	440	431	9	422.2
98.0	439	430	9	421.2
98.0	410	402	8	394.2
98.0	409	401	8	393.2
98.0	408	400	8	392.2
98.0	407	399	8	391.2
98.0	406	398	8	390.2
98.0	405	397	8	389.2
98.0	404	396	8	388.2
98.0	403	395	8	387.2
98.0	402	394	8	386.2
98.0	401	393	8	385.2
98.0	400	392	8	384.2
98.0	399	391	8	383.2
98.0	398	390	8	382.2
98.0	397	389	8	381.2
98.0	396	388	8	380.2
98.0	395	387	8	379.2
98.0	394	386	8	378.2

%E	M1	M2	DM	M*
98.0	393	385	8	377.2
98.0	392	384	8	376.2
98.0	391	383	8	375.2
98.0	358	351	7	344.1
98.0	357	350	7	343.1
98.0	356	349	7	342.1
98.0	355	348	7	341.1
98.0	354	347	7	340.1
98.0	353	346	7	339.1
98.0	352	345	7	338.1
98.0	351	344	7	337.1
98.0	350	343	7	336.1
98.0	349	342	7	335.1
98.0	348	341	7	334.1
98.0	347	340	7	333.1
98.0	346	339	7	332.1
98.0	345	338	7	331.1
98.0	344	337	7	330.1
98.0	343	336	7	329.1
98.0	342	335	7	328.1
98.0	307	301	6	295.1
98.0	306	300	6	294.1
98.0	305	299	6	293.1
98.0	304	298	6	292.1
98.0	303	297	6	291.1
98.0	302	296	6	290.1
98.0	301	295	6	289.1
98.0	300	294	6	288.1
98.0	299	293	6	287.1
98.0	298	292	6	286.1
98.0	297	291	6	285.1
98.0	296	290	6	284.1
98.0	295	289	6	283.1
98.0	294	288	6	282.1
98.0	293	287	6	281.1
98.0	256	251	5	246.1
98.0	255	250	5	245.1
98.0	254	249	5	244.1
98.0	253	248	5	243.1
98.0	252	247	5	242.1
98.0	251	246	5	241.1
98.0	250	245	5	240.1
98.0	249	244	5	239.1
98.0	248	243	5	238.1
98.0	247	242	5	237.1
98.0	246	241	5	236.1
98.0	245	240	5	235.1
98.0	244	239	5	234.1
98.0	205	201	4	197.1
98.0	204	200	4	196.1

%E	M1	M2	DM	M*
98.0	203	199	4	195.1
98.0	202	198	4	194.1
98.0	201	197	4	193.1
98.0	200	196	4	192.1
98.0	199	195	4	191.1
98.0	198	194	4	190.1
98.0	197	193	4	189.1
98.0	196	192	4	188.1
98.0	153	150	3	147.1
98.0	152	149	3	146.1
98.0	151	148	3	145.1
98.0	150	147	3	144.1
98.0	149	146	3	143.1
98.0	148	145	3	142.1
98.0	147	144	3	141.1
98.0	102	100	2	98.0
98.0	101	99	2	97.0
98.0	100	98	2	96.0
98.0	99	97	2	95.0
98.0	98	96	2	94.0
98.0	51	50	1	49.0
98.0	50	49	1	48.0
98.0	49	48	1	47.0
97.9	487	477	10	467.2
97.9	486	476	10	466.2
97.9	485	475	10	465.2
97.9	484	474	10	464.2
97.9	483	473	10	463.2
97.9	482	472	10	462.2
97.9	481	471	10	461.2
97.9	480	470	10	460.2
97.9	479	469	10	459.2
97.9	478	468	10	458.2
97.9	477	467	10	457.2
97.9	476	466	10	456.2
97.9	475	465	10	455.2
97.9	474	464	10	454.2
97.9	473	463	10	453.2
97.9	472	462	10	452.2
97.9	471	461	10	451.2
97.9	470	460	10	450.2
97.9	469	459	10	449.2
97.9	468	458	10	448.2
97.9	467	457	10	447.2
97.9	466	456	10	446.2
97.9	438	429	9	420.2
97.9	437	428	9	419.2
97.9	436	427	9	418.2
97.9	435	426	9	417.2
97.9	434	425	9	416.2

%E	M1	M2	DM	M*
97.9	433	424	9	415.2
97.9	432	423	9	414.2
97.9	431	422	9	413.2
97.9	430	421	9	412.2
97.9	429	420	9	411.2
97.9	428	419	9	410.2
97.9	427	418	9	409.2
97.9	426	417	9	408.2
97.9	425	416	9	407.2
97.9	424	415	9	406.2
97.9	423	414	9	405.2
97.9	422	413	9	404.2
97.9	421	412	9	403.2
97.9	420	411	9	402.2
97.9	419	410	9	401.2
97.9	390	382	8	374.2
97.9	389	381	8	373.2
97.9	388	380	8	372.2
97.9	387	379	8	371.2
97.9	386	378	8	370.2
97.9	385	377	8	369.2
97.9	384	376	8	368.2
97.9	383	375	8	367.2
97.9	382	374	8	366.2
97.9	381	373	8	365.2
97.9	380	372	8	364.2
97.9	379	371	8	363.2
97.9	378	370	8	362.2
97.9	377	369	8	361.2
97.9	376	368	8	360.2
97.9	375	367	8	359.2
97.9	374	366	8	358.2
97.9	373	365	8	357.2
97.9	341	334	7	327.1
97.9	340	333	7	326.1
97.9	339	332	7	325.1
97.9	338	331	7	324.1
97.9	337	330	7	323.1
97.9	336	329	7	322.1
97.9	335	328	7	321.1
97.9	334	327	7	320.1
97.9	333	326	7	319.1
97.9	332	325	7	318.1
97.9	331	324	7	317.1
97.9	330	323	7	316.1
97.9	329	322	7	315.1
97.9	328	321	7	314.1
97.9	327	320	7	313.1
97.9	326	319	7	312.2
97.9	292	286	6	280.1

%E	M1	M2	DM	M*
97.9	291	285	6	279.1
97.9	290	284	6	278.1
97.9	289	283	6	277.1
97.9	288	282	6	276.1
97.9	287	281	6	275.1
97.9	286	280	6	274.1
97.9	285	279	6	273.1
97.9	284	278	6	272.1
97.9	283	277	6	271.1
97.9	282	276	6	270.1
97.9	281	275	6	269.1
97.9	280	274	6	268.1
97.9	243	238	5	233.1
97.9	242	237	5	232.1
97.9	241	236	5	231.1
97.9	240	235	5	230.1
97.9	239	234	5	229.1
97.9	238	233	5	228.1
97.9	237	232	5	227.1
97.9	236	231	5	226.1
97.9	235	230	5	225.1
97.9	234	229	5	224.1
97.9	233	228	5	223.1
97.9	195	191	4	187.1
97.9	194	190	4	186.1
97.9	193	189	4	185.1
97.9	192	188	4	184.1
97.9	191	187	4	183.1
97.9	190	186	4	182.1
97.9	189	185	4	181.1
97.9	188	184	4	180.1
97.9	187	183	4	179.1
97.9	146	143	3	140.1
97.9	145	142	3	139.1
97.9	144	141	3	138.1
97.9	143	140	3	137.1
97.9	142	139	3	136.1
97.9	141	138	3	135.1
97.9	140	137	3	134.1
97.9	97	95	2	93.0
97.9	96	94	2	92.0
97.9	95	93	2	91.0
97.9	94	92	2	90.0
97.9	48	47	1	46.0
97.9	47	46	1	45.0
97.8	500	489	11	478.2
97.8	499	488	11	477.2
97.8	498	487	11	476.2
97.8	497	486	11	475.2
97.8	496	485	11	474.2

%E	M1	M2	DM	M*
97.8	495	484	11	473.2
97.8	494	483	11	472.2
97.8	493	482	11	471.2
97.8	492	481	11	470.2
97.8	491	480	11	469.2
97.8	490	479	11	468.2
97.8	489	478	11	467.2
97.8	465	455	10	445.2
97.8	464	454	10	444.2
97.8	463	453	10	443.2
97.8	462	452	10	442.2
97.8	461	451	10	441.2
97.8	460	450	10	440.2
97.8	459	449	10	439.2
97.8	458	448	10	438.2
97.8	457	447	10	437.2
97.8	456	446	10	436.2
97.8	455	445	10	435.2
97.8	454	444	10	434.2
97.8	453	443	10	433.2
97.8	452	442	10	432.2
97.8	451	441	10	431.2
97.8	450	440	10	430.2
97.8	449	439	10	429.2
97.8	448	438	10	428.2
97.8	447	437	10	427.2
97.8	446	436	10	426.2
97.8	445	435	10	425.2
97.8	418	409	9	400.2
97.8	417	408	9	399.2
97.8	416	407	9	398.2
97.8	415	406	9	397.2
97.8	414	405	9	396.2
97.8	413	404	9	395.2
97.8	412	403	9	394.2
97.8	411	402	9	393.2
97.8	410	401	9	392.2
97.8	409	400	9	391.2
97.8	408	399	9	390.2
97.8	407	398	9	389.2
97.8	406	397	9	388.2
97.8	405	396	9	387.2
97.8	404	395	9	386.2
97.8	403	394	9	385.2
97.8	402	393	9	384.2
97.8	401	392	9	383.2
97.8	400	391	9	382.2
97.8	372	364	8	356.2
97.8	371	363	8	355.2
97.8	370	362	8	354.2

%E	M1	M2	DM	M*
97.8	369	361	8	353.2
97.8	368	360	8	352.2
97.8	367	359	8	351.2
97.8	366	358	8	350.2
97.8	365	357	8	349.2
97.8	364	356	8	348.2
97.8	363	355	8	347.2
97.8	362	354	8	346.2
97.8	361	353	8	345.2
97.8	360	352	8	344.2
97.8	359	351	8	343.2
97.8	358	350	8	342.2
97.8	357	349	8	341.2
97.8	356	348	8	340.2
97.8	325	318	7	311.2
97.8	324	317	7	310.2
97.8	323	316	7	309.2
97.8	322	315	7	308.2
97.8	321	314	7	307.2
97.8	320	313	7	306.2
97.8	319	312	7	305.2
97.8	318	311	7	304.2
97.8	317	310	7	303.2
97.8	316	309	7	302.2
97.8	315	308	7	301.2
97.8	314	307	7	300.2
97.8	313	306	7	299.2
97.8	312	305	7	298.2
97.8	279	273	6	267.1
97.8	278	272	6	266.1
97.8	277	271	6	265.1
97.8	276	270	6	264.1
97.8	275	269	6	263.1
97.8	274	268	6	262.1
97.8	273	267	6	261.1
97.8	272	266	6	260.1
97.8	271	265	6	259.1
97.8	270	264	6	258.1
97.8	269	263	6	257.1
97.8	268	262	6	256.1
97.8	267	261	6	255.1
97.8	232	227	5	222.1
97.8	231	226	5	221.1
97.8	230	225	5	220.1
97.8	229	224	5	219.1
97.8	228	223	5	218.1
97.8	227	222	5	217.1
97.8	226	221	5	216.1
97.8	225	220	5	215.1
97.8	224	219	5	214.1

%E	M1	M2	DM	M*
97.8	223	218	5	213.1
97.8	186	182	4	178.1
97.8	185	181	4	177.1
97.8	184	180	4	176.1
97.8	183	179	4	175.1
97.8	182	178	4	174.1
97.8	181	177	4	173.1
97.8	180	176	4	172.1
97.8	179	175	4	171.1
97.8	178	174	4	170.1
97.8	139	136	3	133.1
97.8	138	135	3	132.1
97.8	137	134	3	131.1
97.8	136	133	3	130.1
97.8	135	132	3	129.1
97.8	134	131	3	128.1
97.8	93	91	2	89.0
97.8	92	90	2	88.0
97.8	91	89	2	87.0
97.8	90	88	2	86.0
97.8	89	87	2	85.0
97.8	46	45	1	44.0
97.8	45	44	1	43.0
97.7	488	477	11	466.2
97.7	487	476	11	465.2
97.7	486	475	11	464.2
97.7	485	474	11	463.2
97.7	484	473	11	462.3
97.7	483	472	11	461.3
97.7	482	471	11	460.3
97.7	481	470	11	459.3
97.7	480	469	11	458.3
97.7	479	468	11	457.3
97.7	478	467	11	456.3
97.7	477	466	11	455.3
97.7	476	465	11	454.3
97.7	475	464	11	453.3
97.7	474	463	11	452.3
97.7	473	462	11	451.3
97.7	472	461	11	450.3
97.7	471	460	11	449.3
97.7	470	459	11	448.3
97.7	469	458	11	447.3
97.7	468	457	11	446.3
97.7	444	434	10	424.2
97.7	443	433	10	423.2
97.7	442	432	10	422.2
97.7	441	431	10	421.2
97.7	440	430	10	420.2
97.7	439	429	10	419.2

%E	M1	M2	DM	M*
97.7	438	428	10	418.2
97.7	437	427	10	417.2
97.7	436	426	10	416.2
97.7	435	425	10	415.2
97.7	434	424	10	414.2
97.7	433	423	10	413.2
97.7	432	422	10	412.2
97.7	431	421	10	411.2
97.7	430	420	10	410.2
97.7	429	419	10	409.2
97.7	428	418	10	408.2
97.7	427	417	10	407.2
97.7	426	416	10	406.2
97.7	399	390	9	381.2
97.7	398	389	9	380.2
97.7	397	388	9	379.2
97.7	396	387	9	378.2
97.7	395	386	9	377.2
97.7	394	385	9	376.2
97.7	393	384	9	375.2
97.7	392	383	9	374.2
97.7	391	382	9	373.2
97.7	390	381	9	372.2
97.7	389	380	9	371.2
97.7	388	379	9	370.2
97.7	387	378	9	369.2
97.7	386	377	9	368.2
97.7	385	376	9	367.2
97.7	384	375	9	366.2
97.7	383	374	9	365.2
97.7	355	347	8	339.2
97.7	354	346	8	338.2
97.7	353	345	8	337.2
97.7	352	344	8	336.2
97.7	351	343	8	335.2
97.7	350	342	8	334.2
97.7	349	341	8	333.2
97.7	348	340	8	332.2
97.7	347	339	8	331.2
97.7	346	338	8	330.2
97.7	345	337	8	329.2
97.7	344	336	8	328.2
97.7	343	335	8	327.2
97.7	342	334	8	326.2
97.7	341	333	8	325.2
97.7	311	304	7	297.2
97.7	310	303	7	296.2
97.7	309	302	7	295.2
97.7	308	301	7	294.2
97.7	307	300	7	293.2

%E	M1	M2	DM	M*
97.7	306	299	7	292.2
97.7	305	298	7	291.2
97.7	304	297	7	290.2
97.7	303	296	7	289.2
97.7	302	295	7	288.2
97.7	301	294	7	287.2
97.7	300	293	7	286.2
97.7	299	292	7	285.2
97.7	298	291	7	284.2
97.7	266	260	6	254.1
97.7	265	259	6	253.1
97.7	264	258	6	252.1
97.7	263	257	6	251.1
97.7	262	256	6	250.1
97.7	261	255	6	249.1
97.7	260	254	6	248.1
97.7	259	253	6	247.1
97.7	258	252	6	246.1
97.7	257	251	6	245.1
97.7	256	250	6	244.1
97.7	222	217	5	212.1
97.7	221	216	5	211.1
97.7	220	215	5	210.1
97.7	219	214	5	209.1
97.7	218	213	5	208.1
97.7	217	212	5	207.1
97.7	216	211	5	206.1
97.7	215	210	5	205.1
97.7	214	209	5	204.1
97.7	213	208	5	203.1
97.7	177	173	4	169.1
97.7	176	172	4	168.1
97.7	175	171	4	167.1
97.7	174	170	4	166.1
97.7	173	169	4	165.1
97.7	172	168	4	164.1
97.7	171	167	4	163.1
97.7	133	130	3	127.1
97.7	132	129	3	126.1
97.7	131	128	3	125.1
97.7	130	127	3	124.1
97.7	129	126	3	123.1
97.7	128	125	3	122.1
97.7	88	86	2	84.0
97.7	87	85	2	83.0
97.7	86	84	2	82.0
97.7	44	43	1	42.0
97.7	43	42	1	41.0
97.6	500	488	12	476.3
97.6	499	487	12	475.3

%E	M1	M2	DM	M*
97.6	498	486	12	474.3
97.6	497	485	12	473.3
97.6	496	484	12	472.3
97.6	495	483	12	471.3
97.6	494	482	12	470.3
97.6	493	481	12	469.3
97.6	492	480	12	468.3
97.6	491	479	12	467.3
97.6	490	478	12	466.3
97.6	467	456	11	445.3
97.6	466	455	11	444.3
97.6	465	454	11	443.3
97.6	464	453	11	442.3
97.6	463	452	11	441.3
97.6	462	451	11	440.3
97.6	461	450	11	439.3
97.6	460	449	11	438.3
97.6	459	448	11	437.3
97.6	458	447	11	436.3
97.6	457	446	11	435.3
97.6	456	445	11	434.3
97.6	455	444	11	433.3
97.6	454	443	11	432.3
97.6	453	442	11	431.3
97.6	452	441	11	430.3
97.6	451	440	11	429.3
97.6	450	439	11	428.3
97.6	449	438	11	427.3
97.6	425	415	10	405.2
97.6	424	414	10	404.2
97.6	423	413	10	403.2
97.6	422	412	10	402.2
97.6	421	411	10	401.2
97.6	420	410	10	400.2
97.6	419	409	10	399.2
97.6	418	408	10	398.2
97.6	417	407	10	397.2
97.6	416	406	10	396.2
97.6	415	405	10	395.2
97.6	414	404	10	394.2
97.6	413	403	10	393.2
97.6	412	402	10	392.2
97.6	411	401	10	391.2
97.6	410	400	10	390.2
97.6	409	399	10	389.2
97.6	382	373	9	364.2
97.6	381	372	9	363.2
97.6	380	371	9	362.2
97.6	379	370	9	361.2
97.6	378	369	9	360.2

%E	M1	M2	DM	M*
97.6	377	368	9	359.2
97.6	376	367	9	358.2
97.6	375	366	9	357.2
97.6	374	365	9	356.2
97.6	373	364	9	355.2
97.6	372	363	9	354.2
97.6	371	362	9	353.2
97.6	370	361	9	352.2
97.6	369	360	9	351.2
97.6	368	359	9	350.2
97.6	340	332	8	324.2
97.6	339	331	8	323.2
97.6	338	330	8	322.2
97.6	337	329	8	321.2
97.6	336	328	8	320.2
97.6	335	327	8	319.2
97.6	334	326	8	318.2
97.6	333	325	8	317.2
97.6	332	324	8	316.2
97.6	331	323	8	315.2
97.6	330	322	8	314.2
97.6	329	321	8	313.2
97.6	328	320	8	312.2
97.6	327	319	8	311.2
97.6	297	290	7	283.2
97.6	296	289	7	282.2
97.6	295	288	7	281.2
97.6	294	287	7	280.2
97.6	293	286	7	279.2
97.6	292	285	7	278.2
97.6	291	284	7	277.2
97.6	290	283	7	276.2
97.6	289	282	7	275.2
97.6	288	281	7	274.2
97.6	287	280	7	273.2
97.6	286	279	7	272.2
97.6	255	249	6	243.1
97.6	254	248	6	242.1
97.6	253	247	6	241.1
97.6	252	246	6	240.1
97.6	251	245	6	239.1
97.6	250	244	6	238.1
97.6	249	243	6	237.1
97.6	248	242	6	236.1
97.6	247	241	6	235.1
97.6	246	240	6	234.1
97.6	245	239	6	233.1
97.6	212	207	5	202.1
97.6	211	206	5	201.1
97.6	210	205	5	200.1

%E	M1	M2	DM	M*
97.6	209	204	5	199.1
97.6	208	203	5	198.1
97.6	207	202	5	197.1
97.6	206	201	5	196.1
97.6	205	200	5	195.1
97.6	170	166	4	162.1
97.6	169	165	4	161.1
97.6	168	164	4	160.1
97.6	167	163	4	159.1
97.6	166	162	4	158.1
97.6	165	161	4	157.1
97.6	164	160	4	156.1
97.6	127	124	3	121.1
97.6	126	123	3	120.1
97.6	125	122	3	119.1
97.6	124	121	3	118.1
97.6	123	120	3	117.1
97.6	85	83	2	81.0
97.6	84	82	2	80.0
97.6	83	81	2	79.0
97.6	82	80	2	78.0
97.6	42	41	1	40.0
97.6	41	40	1	39.0
97.5	489	477	12	465.3
97.5	488	476	12	464.3
97.5	487	475	12	463.3
97.5	486	474	12	462.3
97.5	485	473	12	461.3
97.5	484	472	12	460.3
97.5	483	471	12	459.3
97.5	482	470	12	458.3
97.5	481	469	12	457.3
97.5	480	468	12	456.3
97.5	479	467	12	455.3
97.5	478	466	12	454.3
97.5	477	465	12	453.3
97.5	476	464	12	452.3
97.5	475	463	12	451.3
97.5	474	462	12	450.3
97.5	473	461	12	449.3
97.5	472	460	12	448.3
97.5	471	459	12	447.3
97.5	448	437	11	426.3
97.5	447	436	11	425.3
97.5	446	435	11	424.3
97.5	445	434	11	423.3
97.5	444	433	11	422.3
97.5	443	432	11	421.3
97.5	442	431	11	420.3
97.5	441	430	11	419.3

%E	M1	M2	DM	M*
97.5	440	429	11	418.3
97.5	439	428	11	417.3
97.5	438	427	11	416.3
97.5	437	426	11	415.3
97.5	436	425	11	414.3
97.5	435	424	11	413.3
97.5	434	423	11	412.3
97.5	433	422	11	411.3
97.5	432	421	11	410.3
97.5	408	398	10	388.2
97.5	407	397	10	387.2
97.5	406	396	10	386.2
97.5	405	395	10	385.2
97.5	404	394	10	384.2
97.5	403	393	10	383.2
97.5	402	392	10	382.2
97.5	401	391	10	381.2
97.5	400	390	10	380.3
97.5	399	389	10	379.3
97.5	398	388	10	378.3
97.5	397	387	10	377.3
97.5	396	386	10	376.3
97.5	395	385	10	375.3
97.5	394	384	10	374.3
97.5	393	383	10	373.3
97.5	367	358	9	349.2
97.5	366	357	9	348.2
97.5	365	356	9	347.2
97.5	364	355	9	346.2
97.5	363	354	9	345.2
97.5	362	353	9	344.2
97.5	361	352	9	343.2
97.5	360	351	9	342.2
97.5	359	350	9	341.2
97.5	358	349	9	340.2
97.5	357	348	9	339.2
97.5	356	347	9	338.2
97.5	355	346	9	337.2
97.5	354	345	9	336.2
97.5	353	344	9	335.2
97.5	326	318	8	310.2
97.5	325	317	8	309.2
97.5	324	316	8	308.2
97.5	323	315	8	307.2
97.5	322	314	8	306.2
97.5	321	313	8	305.2
97.5	320	312	8	304.2
97.5	319	311	8	303.2
97.5	318	310	8	302.2
97.5	317	309	8	301.2

%E	M1	M2	DM	M*
97.5	316	308	8	300.2
97.5	315	307	8	299.2
97.5	314	306	8	298.2
97.5	285	278	7	271.2
97.5	284	277	7	270.2
97.5	283	276	7	269.2
97.5	282	275	7	268.2
97.5	281	274	7	267.2
97.5	280	273	7	266.2
97.5	279	272	7	265.2
97.5	278	271	7	264.2
97.5	277	270	7	263.2
97.5	276	269	7	262.2
97.5	275	268	7	261.2
97.5	244	238	6	232.1
97.5	243	237	6	231.1
97.5	242	236	6	230.1
97.5	241	235	6	229.1
97.5	240	234	6	228.1
97.5	239	233	6	227.2
97.5	238	232	6	226.2
97.5	237	231	6	225.2
97.5	236	230	6	224.2
97.5	204	199	5	194.1
97.5	203	198	5	193.1
97.5	202	197	5	192.1
97.5	201	196	5	191.1
97.5	200	195	5	190.1
97.5	199	194	5	189.1
97.5	198	193	5	188.1
97.5	197	192	5	187.1
97.5	163	159	4	155.1
97.5	162	158	4	154.1
97.5	161	157	4	153.1
97.5	160	156	4	152.1
97.5	159	155	4	151.1
97.5	158	154	4	150.1
97.5	157	153	4	149.1
97.5	122	119	3	116.1
97.5	121	118	3	115.1
97.5	120	117	3	114.1
97.5	119	116	3	113.1
97.5	118	115	3	112.1
97.5	81	79	2	77.0
97.5	80	78	2	76.0
97.5	79	77	2	75.1
97.5	40	39	1	38.0
97.4	500	487	13	474.3
97.4	499	486	13	473.3
97.4	498	485	13	472.3

%E	M1	M2	DM	M*
97.4	497	484	13	471.3
97.4	496	483	13	470.3
97.4	495	482	13	469.3
97.4	494	481	13	468.3
97.4	493	480	13	467.3
97.4	492	479	13	466.3
97.4	491	478	13	465.3
97.4	470	458	12	446.3
97.4	469	457	12	445.3
97.4	468	456	12	444.3
97.4	467	455	12	443.3
97.4	466	454	12	442.3
97.4	465	453	12	441.3
97.4	464	452	12	440.3
97.4	463	451	12	439.3
97.4	462	450	12	438.3
97.4	461	449	12	437.3
97.4	460	448	12	436.3
97.4	459	447	12	435.3
97.4	458	446	12	434.3
97.4	457	445	12	433.3
97.4	456	444	12	432.3
97.4	455	443	12	431.3
97.4	454	442	12	430.3
97.4	453	441	12	429.3
97.4	431	420	11	409.3
97.4	430	419	11	408.3
97.4	429	418	11	407.3
97.4	428	417	11	406.3
97.4	427	416	11	405.3
97.4	426	415	11	404.3
97.4	425	414	11	403.3
97.4	424	413	11	402.3
97.4	423	412	11	401.3
97.4	422	411	11	400.3
97.4	421	410	11	399.3
97.4	420	409	11	398.3
97.4	419	408	11	397.3
97.4	418	407	11	396.3
97.4	417	406	11	395.3
97.4	416	405	11	394.3
97.4	392	382	10	372.3
97.4	391	381	10	371.3
97.4	390	380	10	370.3
97.4	389	379	10	369.3
97.4	388	378	10	368.3
97.4	387	377	10	367.3
97.4	386	376	10	366.3
97.4	385	375	10	365.3
97.4	384	374	10	364.3

%E	M1	M2	DM	M*
97.4	383	373	10	363.3
97.4	382	372	10	362.3
97.4	381	371	10	361.3
97.4	380	370	10	360.3
97.4	379	369	10	359.3
97.4	378	368	10	358.3
97.4	352	343	9	334.2
97.4	351	342	9	333.2
97.4	350	341	9	332.2
97.4	349	340	9	331.2
97.4	348	339	9	330.2
97.4	347	338	9	329.2
97.4	346	337	9	328.2
97.4	345	336	9	327.2
97.4	344	335	9	326.2
97.4	343	334	9	325.2
97.4	342	333	9	324.2
97.4	341	332	9	323.2
97.4	340	331	9	322.2
97.4	313	305	8	297.2
97.4	312	304	8	296.2
97.4	311	303	8	295.2
97.4	310	302	8	294.2
97.4	309	301	8	293.2
97.4	308	300	8	292.2
97.4	307	299	8	291.2
97.4	306	298	8	290.2
97.4	305	297	8	289.2
97.4	304	296	8	288.2
97.4	303	295	8	287.2
97.4	302	294	8	286.2
97.4	274	267	7	260.2
97.4	273	266	7	259.2
97.4	272	265	7	258.2
97.4	271	264	7	257.2
97.4	270	263	7	256.2
97.4	269	262	7	255.2
97.4	268	261	7	254.2
97.4	267	260	7	253.2
97.4	266	259	7	252.2
97.4	265	258	7	251.2
97.4	235	229	6	223.2
97.4	234	228	6	222.2
97.4	233	227	6	221.2
97.4	232	226	6	220.2
97.4	231	225	6	219.2
97.4	230	224	6	218.2
97.4	229	223	6	217.2
97.4	228	222	6	216.2
97.4	227	221	6	215.2
97.4	196	191	5	186.1
97.4	195	190	5	185.1
97.4	194	189	5	184.1
97.4	193	188	5	183.1
97.4	192	187	5	182.1
97.4	191	186	5	181.1
97.4	190	185	5	180.1
97.4	189	184	5	179.1
97.4	156	152	4	148.1
97.4	155	151	4	147.1
97.4	154	150	4	146.1
97.4	153	149	4	145.1
97.4	152	148	4	144.1
97.4	151	147	4	143.1
97.4	117	114	3	111.1
97.4	116	113	3	110.1
97.4	115	112	3	109.1
97.4	114	111	3	108.1
97.4	78	76	2	74.1
97.4	77	75	2	73.1
97.4	76	74	2	72.1
97.4	39	38	1	37.0
97.4	38	37	1	36.0
97.3	490	477	13	464.3
97.3	489	476	13	463.3
97.3	488	475	13	462.3
97.3	487	474	13	461.3
97.3	486	473	13	460.3
97.3	485	472	13	459.3
97.3	484	471	13	458.3
97.3	483	470	13	457.3
97.3	482	469	13	456.4
97.3	481	468	13	455.4
97.3	480	467	13	454.4
97.3	479	466	13	453.4
97.3	478	465	13	452.4
97.3	477	464	13	451.4
97.3	476	463	13	450.4
97.3	475	462	13	449.4
97.3	474	461	13	448.4
97.3	473	460	13	447.4
97.3	452	440	12	428.3
97.3	451	439	12	427.3
97.3	450	438	12	426.3
97.3	449	437	12	425.3
97.3	448	436	12	424.3
97.3	447	435	12	423.3
97.3	446	434	12	422.3
97.3	445	433	12	421.3
97.3	444	432	12	420.3
97.3	443	431	12	419.3
97.3	442	430	12	418.3
97.3	441	429	12	417.3
97.3	440	428	12	416.3
97.3	439	427	12	415.3
97.3	438	426	12	414.3
97.3	437	425	12	413.3
97.3	415	404	11	393.3
97.3	414	403	11	392.3
97.3	413	402	11	391.3
97.3	412	401	11	390.3
97.3	411	400	11	389.3
97.3	410	399	11	388.3
97.3	409	398	11	387.3
97.3	408	397	11	386.3
97.3	407	396	11	385.3
97.3	406	395	11	384.3
97.3	405	394	11	383.3
97.3	404	393	11	382.3
97.3	403	392	11	381.3
97.3	402	391	11	380.3
97.3	401	390	11	379.3
97.3	400	389	11	378.3
97.3	377	367	10	357.3
97.3	376	366	10	356.3
97.3	375	365	10	355.3
97.3	374	364	10	354.3
97.3	373	363	10	353.3
97.3	372	362	10	352.3
97.3	371	361	10	351.3
97.3	370	360	10	350.3
97.3	369	359	10	349.3
97.3	368	358	10	348.3
97.3	367	357	10	347.3
97.3	366	356	10	346.3
97.3	365	355	10	345.3
97.3	364	354	10	344.3
97.3	339	330	9	321.2
97.3	338	329	9	320.2
97.3	337	328	9	319.2
97.3	336	327	9	318.2
97.3	335	326	9	317.2
97.3	334	325	9	316.2
97.3	333	324	9	315.2
97.3	332	323	9	314.2
97.3	331	322	9	313.2
97.3	330	321	9	312.2
97.3	329	320	9	311.2
97.3	328	319	9	310.2
97.3	301	293	8	285.2
97.3	300	292	8	284.2
97.3	299	291	8	283.2
97.3	298	290	8	282.2
97.3	297	289	8	281.2
97.3	296	288	8	280.2
97.3	295	287	8	279.2
97.3	294	286	8	278.2
97.3	293	285	8	277.2
97.3	292	284	8	276.2
97.3	291	283	8	275.2
97.3	264	257	7	250.2
97.3	263	256	7	249.2
97.3	262	255	7	248.2
97.3	261	254	7	247.2
97.3	260	253	7	246.2
97.3	259	252	7	245.2
97.3	258	251	7	244.2
97.3	257	250	7	243.2
97.3	256	249	7	242.2
97.3	255	248	7	241.2
97.3	226	220	6	214.2
97.3	225	219	6	213.2
97.3	224	218	6	212.2
97.3	223	217	6	211.2
97.3	222	216	6	210.2
97.3	221	215	6	209.2
97.3	220	214	6	208.2
97.3	219	213	6	207.2
97.3	188	183	5	178.1
97.3	187	182	5	177.1
97.3	186	181	5	176.1
97.3	185	180	5	175.1
97.3	184	179	5	174.1
97.3	183	178	5	173.1
97.3	182	177	5	172.1
97.3	150	146	4	142.1
97.3	149	145	4	141.1
97.3	148	144	4	140.1
97.3	147	143	4	139.1
97.3	146	142	4	138.1
97.3	113	110	3	107.1
97.3	112	109	3	106.1
97.3	111	108	3	105.1
97.3	110	107	3	104.1
97.3	75	73	2	71.1
97.3	74	72	2	70.1
97.3	73	71	2	69.1
97.3	37	36	1	35.0
97.2	500	486	14	472.4
97.2	499	485	14	471.4
97.2	498	484	14	470.4
97.2	497	483	14	469.4
97.2	496	482	14	468.4
97.2	495	481	14	467.4
97.2	494	480	14	466.4
97.2	493	479	14	465.4
97.2	492	478	14	464.4
97.2	472	459	13	446.4
97.2	471	458	13	445.4
97.2	470	457	13	444.4
97.2	469	456	13	443.4
97.2	468	455	13	442.4
97.2	467	454	13	441.4
97.2	466	453	13	440.4
97.2	465	452	13	439.4
97.2	464	451	13	438.4
97.2	463	450	13	437.4
97.2	462	449	13	436.4
97.2	461	448	13	435.4
97.2	460	447	13	434.4
97.2	459	446	13	433.4
97.2	458	445	13	432.4
97.2	457	444	13	431.4
97.2	436	424	12	412.3
97.2	435	423	12	411.3
97.2	434	422	12	410.3
97.2	433	421	12	409.3
97.2	432	420	12	408.3
97.2	431	419	12	407.3
97.2	430	418	12	406.3
97.2	429	417	12	405.3
97.2	428	416	12	404.3
97.2	427	415	12	403.3
97.2	426	414	12	402.3
97.2	425	413	12	401.3
97.2	424	412	12	400.3
97.2	423	411	12	399.3
97.2	422	410	12	398.3
97.2	421	409	12	397.3
97.2	399	388	11	377.3
97.2	398	387	11	376.3
97.2	397	386	11	375.3
97.2	396	385	11	374.3
97.2	395	384	11	373.3
97.2	394	383	11	372.3
97.2	393	382	11	371.3
97.2	392	381	11	370.3
97.2	391	380	11	369.3
97.2	390	379	11	368.3
97.2	389	378	11	367.3

%E	M1	M2	DM	M*
97.2	388	377	11	366.3
97.2	387	376	11	365.3
97.2	386	375	11	364.3
97.2	363	353	10	343.3
97.2	362	352	10	342.3
97.2	361	351	10	341.3
97.2	360	350	10	340.3
97.2	359	349	10	339.3
97.2	358	348	10	338.3
97.2	357	347	10	337.3
97.2	356	346	10	336.3
97.2	355	345	10	335.3
97.2	354	344	10	334.3
97.2	353	343	10	333.3
97.2	352	342	10	332.3
97.2	351	341	10	331.3
97.2	327	318	9	309.2
97.2	326	317	9	308.2
97.2	325	316	9	307.2
97.2	324	315	9	306.3
97.2	323	314	9	305.3
97.2	322	313	9	304.3
97.2	321	312	9	303.3
97.2	320	311	9	302.3
97.2	319	310	9	301.3
97.2	318	309	9	300.3
97.2	317	308	9	299.3
97.2	316	307	9	298.3
97.2	290	282	8	274.2
97.2	289	281	8	273.2
97.2	288	280	8	272.2
97.2	287	279	8	271.2
97.2	286	278	8	270.2
97.2	285	277	8	269.2
97.2	284	276	8	268.2
97.2	283	275	8	267.2
97.2	282	274	8	266.2
97.2	281	273	8	265.2
97.2	254	247	7	240.2
97.2	253	246	7	239.2
97.2	252	245	7	238.2
97.2	251	244	7	237.2
97.2	250	243	7	236.2
97.2	249	242	7	235.2
97.2	248	241	7	234.2
97.2	247	240	7	233.2
97.2	246	239	7	232.2
97.2	218	212	6	206.2
97.2	217	211	6	205.2
97.2	216	210	6	204.2

%E	M1	M2	DM	M*
97.2	215	209	6	203.2
97.2	214	208	6	202.2
97.2	213	207	6	201.2
97.2	212	206	6	200.2
97.2	211	205	6	199.2
97.2	181	176	5	171.1
97.2	180	175	5	170.1
97.2	179	174	5	169.1
97.2	178	173	5	168.1
97.2	177	172	5	167.1
97.2	176	171	5	166.1
97.2	145	141	4	137.1
97.2	144	140	4	136.1
97.2	143	139	4	135.1
97.2	142	138	4	134.1
97.2	141	137	4	133.1
97.2	109	106	3	103.1
97.2	108	105	3	102.1
97.2	107	104	3	101.1
97.2	106	103	3	100.1
97.2	72	70	2	68.1
97.2	71	69	2	67.1
97.2	36	35	1	34.0
97.1	491	477	14	463.4
97.1	490	476	14	462.4
97.1	489	475	14	461.4
97.1	488	474	14	460.4
97.1	487	473	14	459.4
97.1	486	472	14	458.4
97.1	485	471	14	457.4
97.1	484	470	14	456.4
97.1	483	469	14	455.4
97.1	482	468	14	454.4
97.1	481	467	14	453.4
97.1	480	466	14	452.4
97.1	479	465	14	451.4
97.1	478	464	14	450.4
97.1	477	463	14	449.4
97.1	476	462	14	448.4
97.1	475	461	14	447.4
97.1	456	443	13	430.4
97.1	455	442	13	429.4
97.1	454	441	13	428.4
97.1	453	440	13	427.4
97.1	452	439	13	426.4
97.1	451	438	13	425.4
97.1	450	437	13	424.4
97.1	449	436	13	423.4
97.1	448	435	13	422.4
97.1	447	434	13	421.4

%E	M1	M2	DM	M*
97.1	446	433	13	420.4
97.1	445	432	13	419.4
97.1	444	431	13	418.4
97.1	443	430	13	417.4
97.1	442	429	13	416.4
97.1	441	428	13	415.4
97.1	420	408	12	396.3
97.1	419	407	12	395.3
97.1	418	406	12	394.3
97.1	417	405	12	393.3
97.1	416	404	12	392.3
97.1	415	403	12	391.3
97.1	414	402	12	390.3
97.1	413	401	12	389.3
97.1	412	400	12	388.3
97.1	411	399	12	387.4
97.1	410	398	12	386.4
97.1	409	397	12	385.4
97.1	408	396	12	384.4
97.1	407	395	12	383.4
97.1	385	374	11	363.3
97.1	384	373	11	362.3
97.1	383	372	11	361.3
97.1	382	371	11	360.3
97.1	381	370	11	359.3
97.1	380	369	11	358.3
97.1	379	368	11	357.3
97.1	378	367	11	356.3
97.1	377	366	11	355.3
97.1	376	365	11	354.3
97.1	375	364	11	353.3
97.1	374	363	11	352.3
97.1	373	362	11	351.3
97.1	350	340	10	330.3
97.1	349	339	10	329.3
97.1	348	338	10	328.3
97.1	347	337	10	327.3
97.1	346	336	10	326.3
97.1	345	335	10	325.3
97.1	344	334	10	324.3
97.1	343	333	10	323.3
97.1	342	332	10	322.3
97.1	341	331	10	321.3
97.1	340	330	10	320.3
97.1	339	329	10	319.3
97.1	315	306	9	297.3
97.1	314	305	9	296.3
97.1	313	304	9	295.3
97.1	312	303	9	294.3
97.1	311	302	9	293.3

%E	M1	M2	DM	M*
97.1	310	301	9	292.3
97.1	309	300	9	291.3
97.1	308	299	9	290.3
97.1	307	298	9	289.3
97.1	306	297	9	288.3
97.1	280	272	8	264.2
97.1	279	271	8	263.2
97.1	278	270	8	262.2
97.1	277	269	8	261.2
97.1	276	268	8	260.2
97.1	275	267	8	259.2
97.1	274	266	8	258.2
97.1	273	265	8	257.2
97.1	272	264	8	256.2
97.1	245	238	7	231.2
97.1	244	237	7	230.2
97.1	243	236	7	229.2
97.1	242	235	7	228.2
97.1	241	234	7	227.2
97.1	240	233	7	226.2
97.1	239	232	7	225.2
97.1	238	231	7	224.2
97.1	210	204	6	198.2
97.1	209	203	6	197.2
97.1	208	202	6	196.2
97.1	207	201	6	195.2
97.1	206	200	6	194.2
97.1	205	199	6	193.2
97.1	204	198	6	192.2
97.1	175	170	5	165.1
97.1	174	169	5	164.1
97.1	173	168	5	163.1
97.1	172	167	5	162.1
97.1	171	166	5	161.1
97.1	170	165	5	160.1
97.1	140	136	4	132.1
97.1	139	135	4	131.1
97.1	138	134	4	130.1
97.1	137	133	4	129.1
97.1	136	132	4	128.1
97.1	105	102	3	99.1
97.1	104	101	3	98.1
97.1	103	100	3	97.1
97.1	102	99	3	96.1
97.1	70	68	2	66.1
97.1	69	67	2	65.1
97.1	68	66	2	64.1
97.1	35	34	1	33.0
97.1	34	33	1	32.0
97.0	500	485	15	470.4

%E	M1	M2	DM	M*
97.0	499	484	15	469.5
97.0	498	483	15	468.5
97.0	497	482	15	467.5
97.0	496	481	15	466.5
97.0	495	480	15	465.5
97.0	494	479	15	464.5
97.0	493	478	15	463.5
97.0	492	477	15	462.5
97.0	474	460	14	446.4
97.0	473	459	14	445.4
97.0	472	458	14	444.4
97.0	471	457	14	443.4
97.0	470	456	14	442.4
97.0	469	455	14	441.4
97.0	468	454	14	440.4
97.0	467	453	14	439.4
97.0	466	452	14	438.4
97.0	465	451	14	437.4
97.0	464	450	14	436.4
97.0	463	449	14	435.4
97.0	462	448	14	434.4
97.0	461	447	14	433.4
97.0	460	446	14	432.4
97.0	459	445	14	431.4
97.0	440	427	13	414.4
97.0	439	426	13	413.4
97.0	438	425	13	412.4
97.0	437	424	13	411.4
97.0	436	423	13	410.4
97.0	435	422	13	409.4
97.0	434	421	13	408.4
97.0	433	420	13	407.4
97.0	432	419	13	406.4
97.0	431	418	13	405.4
97.0	430	417	13	404.4
97.0	429	416	13	403.4
97.0	428	415	13	402.4
97.0	427	414	13	401.4
97.0	406	394	12	382.4
97.0	405	393	12	381.4
97.0	404	392	12	380.4
97.0	403	391	12	379.4
97.0	402	390	12	378.4
97.0	401	389	12	377.4
97.0	400	388	12	376.4
97.0	399	387	12	375.4
97.0	398	386	12	374.4
97.0	397	385	12	373.4
97.0	396	384	12	372.4
97.0	395	383	12	371.4

%E	M1	M2	DM	M*
97.0	394	382	12	370.4
97.0	372	361	11	350.3
97.0	371	360	11	349.3
97.0	370	359	11	348.3
97.0	369	358	11	347.3
97.0	368	357	11	346.3
97.0	367	356	11	345.3
97.0	366	355	11	344.3
97.0	365	354	11	343.3
97.0	364	353	11	342.3
97.0	363	352	11	341.3
97.0	362	351	11	340.3
97.0	361	350	11	339.3
97.0	338	328	10	318.3
97.0	337	327	10	317.3
97.0	336	326	10	316.3
97.0	335	325	10	315.3
97.0	334	324	10	314.3
97.0	333	323	10	313.3
97.0	332	322	10	312.3
97.0	331	321	10	311.3
97.0	330	320	10	310.3
97.0	329	319	10	309.3
97.0	328	318	10	308.3
97.0	305	296	9	287.3
97.0	304	295	9	286.3
97.0	303	294	9	285.3
97.0	302	293	9	284.3
97.0	301	292	9	283.3
97.0	300	291	9	282.3
97.0	299	290	9	281.3
97.0	298	289	9	280.3
97.0	297	288	9	279.3
97.0	296	287	9	278.3
97.0	271	263	8	255.2
97.0	270	262	8	254.2
97.0	269	261	8	253.2
97.0	268	260	8	252.2
97.0	267	259	8	251.2
97.0	266	258	8	250.2
97.0	265	257	8	249.2
97.0	264	256	8	248.2
97.0	263	255	8	247.2
97.0	237	230	7	223.2
97.0	236	229	7	222.2
97.0	235	228	7	221.2
97.0	234	227	7	220.2
97.0	233	226	7	219.2
97.0	232	225	7	218.2
97.0	231	224	7	217.2
97.0	230	223	7	216.2
97.0	203	197	6	191.2
97.0	202	196	6	190.2
97.0	201	195	6	189.2
97.0	200	194	6	188.2
97.0	199	193	6	187.2
97.0	198	192	6	186.2
97.0	197	191	6	185.2
97.0	169	164	5	159.1
97.0	168	163	5	158.1
97.0	167	162	5	157.1
97.0	166	161	5	156.2
97.0	165	160	5	155.2
97.0	164	159	5	154.2
97.0	135	131	4	127.1
97.0	134	130	4	126.1
97.0	133	129	4	125.1
97.0	132	128	4	124.1
97.0	101	98	3	95.1
97.0	100	97	3	94.1
97.0	99	96	3	93.1
97.0	67	65	2	63.1
97.0	66	64	2	62.1
97.0	33	32	1	31.0
96.9	491	476	15	461.5
96.9	490	475	15	460.5
96.9	489	474	15	459.5
96.9	488	473	15	458.5
96.9	487	472	15	457.5
96.9	486	471	15	456.5
96.9	485	470	15	455.5
96.9	484	469	15	454.5
96.9	483	468	15	453.5
96.9	482	467	15	452.5
96.9	481	466	15	451.5
96.9	480	465	15	450.5
96.9	479	464	15	449.5
96.9	478	463	15	448.5
96.9	477	462	15	447.5
96.9	458	444	14	430.4
96.9	457	443	14	429.4
96.9	456	442	14	428.4
96.9	455	441	14	427.4
96.9	454	440	14	426.4
96.9	453	439	14	425.4
96.9	452	438	14	424.4
96.9	451	437	14	423.4
96.9	450	436	14	422.4
96.9	449	435	14	421.4
96.9	448	434	14	420.4
96.9	447	433	14	419.4
96.9	446	432	14	418.4
96.9	445	431	14	417.4
96.9	426	413	13	400.4
96.9	425	412	13	399.4
96.9	424	411	13	398.4
96.9	423	410	13	397.4
96.9	422	409	13	396.4
96.9	421	408	13	395.4
96.9	420	407	13	394.4
96.9	419	406	13	393.4
96.9	418	405	13	392.4
96.9	417	404	13	391.4
96.9	416	403	13	390.4
96.9	415	402	13	389.4
96.9	414	401	13	388.4
96.9	413	400	13	387.4
96.9	393	381	12	369.4
96.9	392	380	12	368.4
96.9	391	379	12	367.4
96.9	390	378	12	366.4
96.9	389	377	12	365.4
96.9	388	376	12	364.4
96.9	387	375	12	363.4
96.9	386	374	12	362.4
96.9	385	373	12	361.4
96.9	384	372	12	360.4
96.9	383	371	12	359.4
96.9	382	370	12	358.4
96.9	381	369	12	357.4
96.9	360	349	11	338.3
96.9	359	348	11	337.3
96.9	358	347	11	336.3
96.9	357	346	11	335.3
96.9	356	345	11	334.3
96.9	355	344	11	333.3
96.9	354	343	11	332.3
96.9	353	342	11	331.3
96.9	352	341	11	330.3
96.9	351	340	11	329.3
96.9	350	339	11	328.3
96.9	327	317	10	307.3
96.9	326	316	10	306.3
96.9	325	315	10	305.3
96.9	324	314	10	304.3
96.9	323	313	10	303.3
96.9	322	312	10	302.3
96.9	321	311	10	301.3
96.9	320	310	10	300.3
96.9	319	309	10	299.3
96.9	318	308	10	298.3
96.9	295	286	9	277.3
96.9	294	285	9	276.3
96.9	293	284	9	275.3
96.9	292	283	9	274.3
96.9	291	282	9	273.3
96.9	290	281	9	272.3
96.9	289	280	9	271.3
96.9	288	279	9	270.3
96.9	287	278	9	269.3
96.9	286	277	9	268.3
96.9	262	254	8	246.2
96.9	261	253	8	245.2
96.9	260	252	8	244.2
96.9	259	251	8	243.2
96.9	258	250	8	242.2
96.9	257	249	8	241.2
96.9	256	248	8	240.3
96.9	255	247	8	239.3
96.9	254	246	8	238.3
96.9	229	222	7	215.2
96.9	228	221	7	214.2
96.9	227	220	7	213.2
96.9	226	219	7	212.2
96.9	225	218	7	211.2
96.9	224	217	7	210.2
96.9	223	216	7	209.2
96.9	196	190	6	184.2
96.9	195	189	6	183.2
96.9	194	188	6	182.2
96.9	193	187	6	181.2
96.9	192	186	6	180.2
96.9	191	185	6	179.2
96.9	163	158	5	153.2
96.9	162	157	5	152.2
96.9	161	156	5	151.2
96.9	160	155	5	150.2
96.9	159	154	5	149.2
96.9	131	127	4	123.1
96.9	130	126	4	122.1
96.9	129	125	4	121.1
96.9	128	124	4	120.1
96.9	127	123	4	119.1
96.9	98	95	3	92.1
96.9	97	94	3	91.1
96.9	96	93	3	90.1
96.9	65	63	2	61.1
96.9	64	62	2	60.1
96.9	32	31	1	30.0
96.8	500	484	16	468.5
96.8	499	483	16	467.5
96.8	498	482	16	466.5
96.8	497	481	16	465.5
96.8	496	480	16	464.5
96.8	495	479	16	463.5
96.8	494	478	16	462.5
96.8	493	477	16	461.5
96.8	476	461	15	446.5
96.8	475	460	15	445.5
96.8	474	459	15	444.5
96.8	473	458	15	443.5
96.8	472	457	15	442.5
96.8	471	456	15	441.5
96.8	470	455	15	440.5
96.8	469	454	15	439.5
96.8	468	453	15	438.5
96.8	467	452	15	437.5
96.8	466	451	15	436.5
96.8	465	450	15	435.5
96.8	464	449	15	434.5
96.8	463	448	15	433.5
96.8	462	447	15	432.5
96.8	444	430	14	416.4
96.8	443	429	14	415.4
96.8	442	428	14	414.4
96.8	441	427	14	413.4
96.8	440	426	14	412.4
96.8	439	425	14	411.4
96.8	438	424	14	410.4
96.8	437	423	14	409.4
96.8	436	422	14	408.4
96.8	435	421	14	407.5
96.8	434	420	14	406.5
96.8	433	419	14	405.5
96.8	432	418	14	404.5
96.8	431	417	14	403.5
96.8	412	399	13	386.4
96.8	411	398	13	385.4
96.8	410	397	13	384.4
96.8	409	396	13	383.4
96.8	408	395	13	382.4
96.8	407	394	13	381.4
96.8	406	393	13	380.4
96.8	405	392	13	379.4
96.8	404	391	13	378.4
96.8	403	390	13	377.4
96.8	402	389	13	376.4
96.8	401	388	13	375.4
96.8	400	387	13	374.4
96.8	380	368	12	356.4

%E	M1	M2	DM	M*
96.8	379	367	12	355.4
96.8	378	366	12	354.4
96.8	377	365	12	353.4
96.8	376	364	12	352.4
96.8	375	363	12	351.4
96.8	374	362	12	350.4
96.8	373	361	12	349.4
96.8	372	360	12	348.4
96.8	371	359	12	347.4
96.8	370	358	12	346.4
96.8	349	338	11	327.3
96.8	348	337	11	326.3
96.8	347	336	11	325.3
96.8	346	335	11	324.3
96.8	345	334	11	323.4
96.8	344	333	11	322.4
96.8	343	332	11	321.4
96.8	342	331	11	320.4
96.8	341	330	11	319.4
96.8	340	329	11	318.4
96.8	339	328	11	317.4
96.8	317	307	10	297.3
96.8	316	306	10	296.3
96.8	315	305	10	295.3
96.8	314	304	10	294.3
96.8	313	303	10	293.3
96.8	312	302	10	292.3
96.8	311	301	10	291.3
96.8	310	300	10	290.3
96.8	309	299	10	289.3
96.8	308	298	10	288.3
96.8	285	276	9	267.3
96.8	284	275	9	266.3
96.8	283	274	9	265.3
96.8	282	273	9	264.3
96.8	281	272	9	263.3
96.8	280	271	9	262.3
96.8	279	270	9	261.3
96.8	278	269	9	260.3
96.8	277	268	9	259.3
96.8	253	245	8	237.3
96.8	252	244	8	236.3
96.8	251	243	8	235.3
96.8	250	242	8	234.3
96.8	249	241	8	233.3
96.8	248	240	8	232.3
96.8	247	239	8	231.3
96.8	222	215	7	208.2
96.8	221	214	7	207.2
96.8	220	213	7	206.2

%E	M1	M2	DM	M*
96.8	219	212	7	205.2
96.8	218	211	7	204.2
96.8	217	210	7	203.2
96.8	216	209	7	202.2
96.8	190	184	6	178.2
96.8	189	183	6	177.2
96.8	188	182	6	176.2
96.8	187	181	6	175.2
96.8	186	180	6	174.2
96.8	185	179	6	173.2
96.8	158	153	5	148.2
96.8	157	152	5	147.2
96.8	156	151	5	146.2
96.8	155	150	5	145.2
96.8	154	149	5	144.2
96.8	126	122	4	118.1
96.8	125	121	4	117.1
96.8	124	120	4	116.1
96.8	95	92	3	89.1
96.8	94	91	3	88.1
96.8	93	90	3	87.1
96.8	63	61	2	59.1
96.8	62	60	2	58.1
96.8	31	30	1	29.0
96.7	492	476	16	460.5
96.7	491	475	16	459.5
96.7	490	474	16	458.5
96.7	489	473	16	457.5
96.7	488	472	16	456.5
96.7	487	471	16	455.5
96.7	486	470	16	454.5
96.7	485	469	16	453.5
96.7	484	468	16	452.5
96.7	483	467	16	451.5
96.7	482	466	16	450.5
96.7	481	465	16	449.5
96.7	480	464	16	448.5
96.7	479	463	16	447.5
96.7	478	462	16	446.5
96.7	461	446	15	431.5
96.7	460	445	15	430.5
96.7	459	444	15	429.5
96.7	458	443	15	428.5
96.7	457	442	15	427.5
96.7	456	441	15	426.5
96.7	455	440	15	425.5
96.7	454	439	15	424.5
96.7	453	438	15	423.5
96.7	452	437	15	422.5
96.7	451	436	15	421.5

%E	M1	M2	DM	M*
96.7	450	435	15	420.5
96.7	449	434	15	419.5
96.7	448	433	15	418.5
96.7	430	416	14	402.5
96.7	429	415	14	401.5
96.7	428	414	14	400.5
96.7	427	413	14	399.5
96.7	426	412	14	398.5
96.7	425	411	14	397.5
96.7	424	410	14	396.5
96.7	423	409	14	395.5
96.7	422	408	14	394.5
96.7	421	407	14	393.5
96.7	420	406	14	392.5
96.7	419	405	14	391.5
96.7	418	404	14	390.5
96.7	399	386	13	373.4
96.7	398	385	13	372.4
96.7	397	384	13	371.4
96.7	396	383	13	370.4
96.7	395	382	13	369.4
96.7	394	381	13	368.4
96.7	393	380	13	367.4
96.7	392	379	13	366.4
96.7	391	378	13	365.4
96.7	390	377	13	364.4
96.7	389	376	13	363.4
96.7	369	357	12	345.4
96.7	368	356	12	344.4
96.7	367	355	12	343.4
96.7	366	354	12	342.4
96.7	365	353	12	341.4
96.7	364	352	12	340.4
96.7	363	351	12	339.4
96.7	362	350	12	338.4
96.7	361	349	12	337.4
96.7	360	348	12	336.4
96.7	359	347	12	335.4
96.7	338	327	11	316.4
96.7	337	326	11	315.4
96.7	336	325	11	314.4
96.7	335	324	11	313.4
96.7	334	323	11	312.4
96.7	333	322	11	311.4
96.7	332	321	11	310.4
96.7	331	320	11	309.4
96.7	330	319	11	308.4
96.7	329	318	11	307.4
96.7	307	297	10	287.3
96.7	306	296	10	286.3

%E	M1	M2	DM	M*
96.7	305	295	10	285.3
96.7	304	294	10	284.3
96.7	303	293	10	283.3
96.7	302	292	10	282.3
96.7	301	291	10	281.3
96.7	300	290	10	280.3
96.7	299	289	10	279.3
96.7	276	267	9	258.3
96.7	275	266	9	257.3
96.7	274	265	9	256.3
96.7	273	264	9	255.3
96.7	272	263	9	254.3
96.7	271	262	9	253.3
96.7	270	261	9	252.3
96.7	269	260	9	251.3
96.7	246	238	8	230.3
96.7	245	237	8	229.3
96.7	244	236	8	228.3
96.7	243	235	8	227.3
96.7	242	234	8	226.3
96.7	241	233	8	225.3
96.7	240	232	8	224.3
96.7	239	231	8	223.3
96.7	215	208	7	201.2
96.7	214	207	7	200.2
96.7	213	206	7	199.2
96.7	212	205	7	198.2
96.7	211	204	7	197.2
96.7	210	203	7	196.2
96.7	209	202	7	195.2
96.7	184	178	6	172.2
96.7	183	177	6	171.2
96.7	182	176	6	170.2
96.7	181	175	6	169.2
96.7	180	174	6	168.2
96.7	153	148	5	143.2
96.7	152	147	5	142.2
96.7	151	146	5	141.2
96.7	150	145	5	140.2
96.7	123	119	4	115.1
96.7	122	118	4	114.1
96.7	121	117	4	113.1
96.7	120	116	4	112.1
96.7	92	89	3	86.1
96.7	91	88	3	85.1
96.7	90	87	3	84.1
96.7	61	59	2	57.1
96.7	60	58	2	56.1
96.7	30	29	1	28.0
96.6	500	483	17	466.6

%E	M1	M2	DM	M*
96.6	499	482	17	465.6
96.6	498	481	17	464.6
96.6	497	480	17	463.6
96.6	496	479	17	462.6
96.6	495	478	17	461.6
96.6	494	477	17	460.6
96.6	493	476	17	459.6
96.6	477	461	16	445.5
96.6	476	460	16	444.5
96.6	475	459	16	443.5
96.6	474	458	16	442.5
96.6	473	457	16	441.5
96.6	472	456	16	440.5
96.6	471	455	16	439.5
96.6	470	454	16	438.5
96.6	469	453	16	437.5
96.6	468	452	16	436.5
96.6	467	451	16	435.5
96.6	466	450	16	434.5
96.6	465	449	16	433.6
96.6	464	448	16	432.6
96.6	447	432	15	417.5
96.6	446	431	15	416.5
96.6	445	430	15	415.5
96.6	444	429	15	414.5
96.6	443	428	15	413.5
96.6	442	427	15	412.5
96.6	441	426	15	411.5
96.6	440	425	15	410.5
96.6	439	424	15	409.5
96.6	438	423	15	408.5
96.6	437	422	15	407.5
96.6	436	421	15	406.5
96.6	435	420	15	405.5
96.6	417	403	14	389.5
96.6	416	402	14	388.5
96.6	415	401	14	387.5
96.6	414	400	14	386.5
96.6	413	399	14	385.5
96.6	412	398	14	384.5
96.6	411	397	14	383.5
96.6	410	396	14	382.5
96.6	409	395	14	381.5
96.6	408	394	14	380.5
96.6	407	393	14	379.5
96.6	406	392	14	378.5
96.6	388	375	13	362.4
96.6	387	374	13	361.4
96.6	386	373	13	360.4
96.6	385	372	13	359.4

%E	M1	M2	DM	M*
96.6	384	371	13	358.4
96.6	383	370	13	357.4
96.6	382	369	13	356.4
96.6	381	368	13	355.4
96.6	380	367	13	354.4
96.6	379	366	13	353.4
96.6	378	365	13	352.4
96.6	377	364	13	351.4
96.6	358	346	12	334.4
96.6	357	345	12	333.4
96.6	356	344	12	332.4
96.6	355	343	12	331.4
96.6	354	342	12	330.4
96.6	353	341	12	329.4
96.6	352	340	12	328.4
96.6	351	339	12	327.4
96.6	350	338	12	326.4
96.6	349	337	12	325.4
96.6	348	336	12	324.4
96.6	328	317	11	306.4
96.6	327	316	11	305.4
96.6	326	315	11	304.4
96.6	325	314	11	303.4
96.6	324	313	11	302.4
96.6	323	312	11	301.4
96.6	322	311	11	300.4
96.6	321	310	11	299.4
96.6	320	309	11	298.4
96.6	319	308	11	297.4
96.6	298	288	10	278.3
96.6	297	287	10	277.3
96.6	296	286	10	276.3
96.6	295	285	10	275.3
96.6	294	284	10	274.3
96.6	293	283	10	273.3
96.6	292	282	10	272.3
96.6	291	281	10	271.3
96.6	290	280	10	270.3
96.6	268	259	9	250.3
96.6	267	258	9	249.3
96.6	266	257	9	248.3
96.6	265	256	9	247.3
96.6	264	255	9	246.3
96.6	263	254	9	245.3
96.6	262	253	9	244.3
96.6	261	252	9	243.3
96.6	238	230	8	222.3
96.6	237	229	8	221.3
96.6	236	228	8	220.3
96.6	235	227	8	219.3

%E	M1	M2	DM	M*
96.6	234	226	8	218.3
96.6	233	225	8	217.3
96.6	232	224	8	216.3
96.6	208	201	7	194.2
96.6	207	200	7	193.2
96.6	206	199	7	192.2
96.6	205	198	7	191.2
96.6	204	197	7	190.2
96.6	203	196	7	189.2
96.6	179	173	6	167.2
96.6	178	172	6	166.2
96.6	177	171	6	165.2
96.6	176	170	6	164.2
96.6	175	169	6	163.2
96.6	174	168	6	162.2
96.6	149	144	5	139.2
96.6	148	143	5	138.2
96.6	147	142	5	137.2
96.6	146	141	5	136.2
96.6	145	140	5	135.2
96.6	119	115	4	111.1
96.6	118	114	4	110.1
96.6	117	113	4	109.1
96.6	116	112	4	108.1
96.6	89	86	3	83.1
96.6	88	85	3	82.1
96.6	87	84	3	81.1
96.6	59	57	2	55.1
96.6	58	56	2	54.1
96.6	29	28	1	27.0
96.5	492	475	17	458.6
96.5	491	474	17	457.6
96.5	490	473	17	456.6
96.5	489	472	17	455.6
96.5	488	471	17	454.6
96.5	487	470	17	453.6
96.5	486	469	17	452.6
96.5	485	468	17	451.6
96.5	484	467	17	450.6
96.5	483	466	17	449.6
96.5	482	465	17	448.6
96.5	481	464	17	447.6
96.5	480	463	17	446.6
96.5	479	462	17	445.6
96.5	463	447	16	431.6
96.5	462	446	16	430.6
96.5	461	445	16	429.6
96.5	460	444	16	428.6
96.5	459	443	16	427.6
96.5	458	442	16	426.6

%E	M1	M2	DM	M*
96.5	457	441	16	425.6
96.5	456	440	16	424.6
96.5	455	439	16	423.6
96.5	454	438	16	422.6
96.5	453	437	16	421.6
96.5	452	436	16	420.6
96.5	451	435	16	419.6
96.5	434	419	15	404.5
96.5	433	418	15	403.5
96.5	432	417	15	402.5
96.5	431	416	15	401.5
96.5	430	415	15	400.5
96.5	429	414	15	399.5
96.5	428	413	15	398.5
96.5	427	412	15	397.5
96.5	426	411	15	396.5
96.5	425	410	15	395.5
96.5	424	409	15	394.5
96.5	423	408	15	393.5
96.5	405	391	14	377.5
96.5	404	390	14	376.5
96.5	403	389	14	375.5
96.5	402	388	14	374.5
96.5	401	387	14	373.5
96.5	400	386	14	372.5
96.5	399	385	14	371.5
96.5	398	384	14	370.5
96.5	397	383	14	369.5
96.5	396	382	14	368.5
96.5	395	381	14	367.5
96.5	376	363	13	350.4
96.5	375	362	13	349.5
96.5	374	361	13	348.5
96.5	373	360	13	347.5
96.5	372	359	13	346.5
96.5	371	358	13	345.5
96.5	370	357	13	344.5
96.5	369	356	13	343.5
96.5	368	355	13	342.5
96.5	367	354	13	341.5
96.5	347	335	12	323.4
96.5	346	334	12	322.4
96.5	345	333	12	321.4
96.5	344	332	12	320.4
96.5	343	331	12	319.4
96.5	342	330	12	318.4
96.5	341	329	12	317.4
96.5	340	328	12	316.4
96.5	339	327	12	315.4
96.5	338	326	12	314.4

%E	M1	M2	DM	M*
96.5	318	307	11	296.4
96.5	317	306	11	295.4
96.5	316	305	11	294.4
96.5	315	304	11	293.4
96.5	314	303	11	292.4
96.5	313	302	11	291.4
96.5	312	301	11	290.4
96.5	311	300	11	289.4
96.5	310	299	11	288.4
96.5	289	279	10	269.3
96.5	288	278	10	268.3
96.5	287	277	10	267.3
96.5	286	276	10	266.3
96.5	285	275	10	265.4
96.5	284	274	10	264.4
96.5	283	273	10	263.4
96.5	282	272	10	262.4
96.5	260	251	9	242.3
96.5	259	250	9	241.3
96.5	258	249	9	240.3
96.5	257	248	9	239.3
96.5	256	247	9	238.3
96.5	255	246	9	237.3
96.5	254	245	9	236.3
96.5	231	223	8	215.3
96.5	230	222	8	214.3
96.5	229	221	8	213.3
96.5	228	220	8	212.3
96.5	227	219	8	211.3
96.5	226	218	8	210.3
96.5	202	195	7	188.2
96.5	201	194	7	187.2
96.5	200	193	7	186.2
96.5	199	192	7	185.2
96.5	198	191	7	184.2
96.5	173	167	6	161.2
96.5	172	166	6	160.2
96.5	171	165	6	159.2
96.5	170	164	6	158.2
96.5	169	163	6	157.2
96.5	144	139	5	134.2
96.5	143	138	5	133.2
96.5	142	137	5	132.2
96.5	141	136	5	131.2
96.5	115	111	4	107.1
96.5	114	110	4	106.1
96.5	113	109	4	105.1
96.5	86	83	3	80.1
96.5	85	82	3	79.1
96.5	57	55	2	53.1

%E	M1	M2	DM	M*
96.4	500	482	18	464.6
96.4	499	481	18	463.6
96.4	498	480	18	462.7
96.4	497	479	18	461.7
96.4	496	478	18	460.7
96.4	495	477	18	459.7
96.4	494	476	18	458.7
96.4	478	461	17	444.6
96.4	477	460	17	443.6
96.4	476	459	17	442.6
96.4	475	458	17	441.6
96.4	474	457	17	440.6
96.4	473	456	17	439.6
96.4	472	455	17	438.6
96.4	471	454	17	437.6
96.4	470	453	17	436.6
96.4	469	452	17	435.6
96.4	468	451	17	434.6
96.4	467	450	17	433.6
96.4	466	449	17	432.6
96.4	450	434	16	418.6
96.4	449	433	16	417.6
96.4	448	432	16	416.6
96.4	447	431	16	415.6
96.4	446	430	16	414.6
96.4	445	429	16	413.6
96.4	444	428	16	412.6
96.4	443	427	16	411.6
96.4	442	426	16	410.6
96.4	441	425	16	409.6
96.4	440	424	16	408.6
96.4	439	423	16	407.6
96.4	422	407	15	392.5
96.4	421	406	15	391.5
96.4	420	405	15	390.5
96.4	419	404	15	389.5
96.4	418	403	15	388.5
96.4	417	402	15	387.5
96.4	416	401	15	386.5
96.4	415	400	15	385.5
96.4	414	399	15	384.5
96.4	413	398	15	383.5
96.4	412	397	15	382.5
96.4	411	396	15	381.5
96.4	394	380	14	366.5
96.4	393	379	14	365.5
96.4	392	378	14	364.5
96.4	391	377	14	363.5
96.4	390	376	14	362.5
96.4	389	375	14	361.5

%E	M1	M2	DM	M*
96.4	388	374	14	360.5
96.4	387	373	14	359.5
96.4	386	372	14	358.5
96.4	385	371	14	357.5
96.4	384	370	14	356.5
96.4	366	353	13	340.5
96.4	365	352	13	339.5
96.4	364	351	13	338.5
96.4	363	350	13	337.5
96.4	362	349	13	336.5
96.4	361	348	13	335.5
96.4	360	347	13	334.5
96.4	359	346	13	333.5
96.4	358	345	13	332.5
96.4	357	344	13	331.5
96.4	337	325	12	313.4
96.4	336	324	12	312.4
96.4	335	323	12	311.4
96.4	334	322	12	310.4
96.4	333	321	12	309.4
96.4	332	320	12	308.4
96.4	331	319	12	307.4
96.4	330	318	12	306.4
96.4	329	317	12	305.4
96.4	309	298	11	287.4
96.4	308	297	11	286.4
96.4	307	296	11	285.4
96.4	306	295	11	284.4
96.4	305	294	11	283.4
96.4	304	293	11	282.4
96.4	303	292	11	281.4
96.4	302	291	11	280.4
96.4	281	271	10	261.4
96.4	280	270	10	260.4
96.4	279	269	10	259.4
96.4	278	268	10	258.4
96.4	277	267	10	257.4
96.4	276	266	10	256.4
96.4	275	265	10	255.4
96.4	274	264	10	254.4
96.4	253	244	9	235.3
96.4	252	243	9	234.3
96.4	251	242	9	233.3
96.4	250	241	9	232.3
96.4	249	240	9	231.3
96.4	248	239	9	230.3
96.4	247	238	9	229.3
96.4	225	217	8	209.3
96.4	224	216	8	208.3
96.4	223	215	8	207.3

%E	M1	M2	DM	M*
96.4	222	214	8	206.3
96.4	221	213	8	205.3
96.4	220	212	8	204.3
96.4	197	190	7	183.2
96.4	196	189	7	182.3
96.4	195	188	7	181.3
96.4	194	187	7	180.3
96.4	193	186	7	179.3
96.4	192	185	7	178.3
96.4	168	162	6	156.2
96.4	167	161	6	155.2
96.4	166	160	6	154.2
96.4	165	159	6	153.2
96.4	140	135	5	130.2
96.4	139	134	5	129.2
96.4	138	133	5	128.2
96.4	137	132	5	127.2
96.4	112	108	4	104.1
96.4	111	107	4	103.1
96.4	110	106	4	102.1
96.4	84	81	3	78.1
96.4	83	80	3	77.1
96.4	56	54	2	52.1
96.4	55	53	2	51.1
96.4	28	27	1	26.0
96.3	493	475	18	457.7
96.3	492	474	18	456.7
96.3	491	473	18	455.7
96.3	490	472	18	454.7
96.3	489	471	18	453.7
96.3	488	470	18	452.7
96.3	487	469	18	451.7
96.3	486	468	18	450.7
96.3	485	467	18	449.7
96.3	484	466	18	448.7
96.3	483	465	18	447.7
96.3	482	464	18	446.7
96.3	481	463	18	445.7
96.3	480	462	18	444.7
96.3	465	448	17	431.6
96.3	464	447	17	430.6
96.3	463	446	17	429.6
96.3	462	445	17	428.6
96.3	461	444	17	427.6
96.3	460	443	17	426.6
96.3	459	442	17	425.6
96.3	458	441	17	424.6
96.3	457	440	17	423.6
96.3	456	439	17	422.6
96.3	455	438	17	421.6

%E	M1	M2	DM	M*
96.3	454	437	17	420.6
96.3	438	422	16	406.6
96.3	437	421	16	405.6
96.3	436	420	16	404.6
96.3	435	419	16	403.6
96.3	434	418	16	402.6
96.3	433	417	16	401.6
96.3	432	416	16	400.6
96.3	431	415	16	399.6
96.3	430	414	16	398.6
96.3	429	413	16	397.6
96.3	428	412	16	396.6
96.3	427	411	16	395.6
96.3	410	395	15	380.5
96.3	409	394	15	379.6
96.3	408	393	15	378.6
96.3	407	392	15	377.6
96.3	406	391	15	376.6
96.3	405	390	15	375.6
96.3	404	389	15	374.6
96.3	403	388	15	373.6
96.3	402	387	15	372.6
96.3	401	386	15	371.6
96.3	400	385	15	370.6
96.3	383	369	14	355.5
96.3	382	368	14	354.5
96.3	381	367	14	353.5
96.3	380	366	14	352.5
96.3	379	365	14	351.5
96.3	378	364	14	350.5
96.3	377	363	14	349.5
96.3	376	362	14	348.5
96.3	375	361	14	347.5
96.3	374	360	14	346.5
96.3	356	343	13	330.5
96.3	355	342	13	329.5
96.3	354	341	13	328.5
96.3	353	340	13	327.5
96.3	352	339	13	326.5
96.3	351	338	13	325.5
96.3	350	337	13	324.5
96.3	349	336	13	323.5
96.3	348	335	13	322.5
96.3	347	334	13	321.5
96.3	328	316	12	304.4
96.3	327	315	12	303.4
96.3	326	314	12	302.4
96.3	325	313	12	301.4
96.3	324	312	12	300.4
96.3	323	311	12	299.4

%E	M1	M2	DM	M*
96.3	322	310	12	298.4
96.3	321	309	12	297.4
96.3	320	308	12	296.4
96.3	301	290	11	279.4
96.3	300	289	11	278.4
96.3	299	288	11	277.4
96.3	298	287	11	276.4
96.3	297	286	11	275.4
96.3	296	285	11	274.4
96.3	295	284	11	273.4
96.3	294	283	11	272.4
96.3	273	263	10	253.4
96.3	272	262	10	252.4
96.3	271	261	10	251.4
96.3	270	260	10	250.4
96.3	269	259	10	249.4
96.3	268	258	10	248.4
96.3	267	257	10	247.4
96.3	246	237	9	228.3
96.3	245	236	9	227.3
96.3	244	235	9	226.3
96.3	243	234	9	225.3
96.3	242	233	9	224.3
96.3	241	232	9	223.3
96.3	240	231	9	222.3
96.3	219	211	8	203.3
96.3	218	210	8	202.3
96.3	217	209	8	201.3
96.3	216	208	8	200.3
96.3	215	207	8	199.3
96.3	214	206	8	198.3
96.3	191	184	7	177.3
96.3	190	183	7	176.3
96.3	189	182	7	175.3
96.3	188	181	7	174.3
96.3	187	180	7	173.3
96.3	164	158	6	152.2
96.3	163	157	6	151.2
96.3	162	156	6	150.2
96.3	161	155	6	149.2
96.3	160	154	6	148.2
96.3	136	131	5	126.2
96.3	135	130	5	125.2
96.3	134	129	5	124.2
96.3	109	105	4	101.1
96.3	108	104	4	100.1
96.3	107	103	4	99.1
96.3	82	79	3	76.1
96.3	81	78	3	75.1
96.3	80	77	3	74.1

%E	M1	M2	DM	M*
96.3	54	52	2	50.1
96.3	27	26	1	25.0
96.2	500	481	19	462.7
96.2	499	480	19	461.7
96.2	498	479	19	460.7
96.2	497	478	19	459.7
96.2	496	477	19	458.7
96.2	495	476	19	457.7
96.2	494	475	19	456.7
96.2	479	461	18	443.7
96.2	478	460	18	442.7
96.2	477	459	18	441.7
96.2	476	458	18	440.7
96.2	475	457	18	439.7
96.2	474	456	18	438.7
96.2	473	455	18	437.7
96.2	472	454	18	436.7
96.2	471	453	18	435.7
96.2	470	452	18	434.7
96.2	469	451	18	433.7
96.2	468	450	18	432.7
96.2	453	436	17	419.6
96.2	452	435	17	418.6
96.2	451	434	17	417.6
96.2	450	433	17	416.6
96.2	449	432	17	415.6
96.2	448	431	17	414.6
96.2	447	430	17	413.6
96.2	446	429	17	412.6
96.2	445	428	17	411.6
96.2	444	427	17	410.7
96.2	443	426	17	409.7
96.2	442	425	17	408.7
96.2	426	410	16	394.6
96.2	425	409	16	393.6
96.2	424	408	16	392.6
96.2	423	407	16	391.6
96.2	422	406	16	390.6
96.2	421	405	16	389.6
96.2	420	404	16	388.6
96.2	419	403	16	387.6
96.2	418	402	16	386.6
96.2	417	401	16	385.6
96.2	416	400	16	384.6
96.2	399	384	15	369.6
96.2	398	383	15	368.6
96.2	397	382	15	367.6
96.2	396	381	15	366.6
96.2	395	380	15	365.6
96.2	394	379	15	364.6

%E	M1	M2	DM	M*
96.2	393	378	15	363.6
96.2	392	377	15	362.6
96.2	391	376	15	361.6
96.2	390	375	15	360.6
96.2	373	359	14	345.5
96.2	372	358	14	344.5
96.2	371	357	14	343.5
96.2	370	356	14	342.5
96.2	369	355	14	341.5
96.2	368	354	14	340.5
96.2	367	353	14	339.5
96.2	366	352	14	338.5
96.2	365	351	14	337.5
96.2	364	350	14	336.5
96.2	346	333	13	320.5
96.2	345	332	13	319.5
96.2	344	331	13	318.5
96.2	343	330	13	317.5
96.2	342	329	13	316.5
96.2	341	328	13	315.5
96.2	340	327	13	314.5
96.2	339	326	13	313.5
96.2	338	325	13	312.5
96.2	319	307	12	295.5
96.2	318	306	12	294.5
96.2	317	305	12	293.5
96.2	316	304	12	292.5
96.2	315	303	12	291.5
96.2	314	302	12	290.5
96.2	313	301	12	289.5
96.2	312	300	12	288.5
96.2	293	282	11	271.4
96.2	292	281	11	270.4
96.2	291	280	11	269.4
96.2	290	279	11	268.4
96.2	289	278	11	267.4
96.2	288	277	11	266.4
96.2	287	276	11	265.4
96.2	286	275	11	264.4
96.2	266	256	10	246.4
96.2	265	255	10	245.4
96.2	264	254	10	244.4
96.2	263	253	10	243.4
96.2	262	252	10	242.4
96.2	261	251	10	241.4
96.2	260	250	10	240.4
96.2	239	230	9	221.3
96.2	238	229	9	220.3
96.2	237	228	9	219.3
96.2	236	227	9	218.3
96.2	235	226	9	217.3
96.2	234	225	9	216.3
96.2	213	205	8	197.3
96.2	212	204	8	196.3
96.2	211	203	8	195.3
96.2	210	202	8	194.3
96.2	209	201	8	193.3
96.2	208	200	8	192.3
96.2	186	179	7	172.3
96.2	185	178	7	171.3
96.2	184	177	7	170.3
96.2	183	176	7	169.3
96.2	182	175	7	168.3
96.2	159	153	6	147.2
96.2	158	152	6	146.2
96.2	157	151	6	145.2
96.2	156	150	6	144.2
96.2	133	128	5	123.2
96.2	132	127	5	122.2
96.2	131	126	5	121.2
96.2	130	125	5	120.2
96.2	106	102	4	98.2
96.2	105	101	4	97.2
96.2	104	100	4	96.2
96.2	79	76	3	73.1
96.2	78	75	3	72.1
96.2	53	51	2	49.1
96.2	52	50	2	48.1
96.2	26	25	1	24.0
96.1	493	474	19	455.7
96.1	492	473	19	454.7
96.1	491	472	19	453.7
96.1	490	471	19	452.7
96.1	489	470	19	451.7
96.1	488	469	19	450.7
96.1	487	468	19	449.7
96.1	486	467	19	448.7
96.1	485	466	19	447.7
96.1	484	465	19	446.7
96.1	483	464	19	445.7
96.1	482	463	19	444.7
96.1	481	462	19	443.8
96.1	467	449	18	431.7
96.1	466	448	18	430.7
96.1	465	447	18	429.7
96.1	464	446	18	428.7
96.1	463	445	18	427.7
96.1	462	444	18	426.7
96.1	461	443	18	425.7
96.1	460	442	18	424.7
96.1	459	441	18	423.7
96.1	458	440	18	422.7
96.1	457	439	18	421.7
96.1	456	438	18	420.7
96.1	441	424	17	407.7
96.1	440	423	17	406.7
96.1	439	422	17	405.7
96.1	438	421	17	404.7
96.1	437	420	17	403.7
96.1	436	419	17	402.7
96.1	435	418	17	401.7
96.1	434	417	17	400.7
96.1	433	416	17	399.7
96.1	432	415	17	398.7
96.1	431	414	17	397.7
96.1	415	399	16	383.6
96.1	414	398	16	382.6
96.1	413	397	16	381.6
96.1	412	396	16	380.6
96.1	411	395	16	379.6
96.1	410	394	16	378.6
96.1	409	393	16	377.6
96.1	408	392	16	376.6
96.1	407	391	16	375.6
96.1	406	390	16	374.6
96.1	389	374	15	359.6
96.1	388	373	15	358.6
96.1	387	372	15	357.6
96.1	386	371	15	356.6
96.1	385	370	15	355.6
96.1	384	369	15	354.6
96.1	383	368	15	353.6
96.1	382	367	15	352.6
96.1	381	366	15	351.6
96.1	380	365	15	350.6
96.1	363	349	14	335.5
96.1	362	348	14	334.5
96.1	361	347	14	333.5
96.1	360	346	14	332.5
96.1	359	345	14	331.5
96.1	358	344	14	330.5
96.1	357	343	14	329.5
96.1	356	342	14	328.6
96.1	355	341	14	327.6
96.1	337	324	13	311.5
96.1	336	323	13	310.5
96.1	335	322	13	309.5
96.1	334	321	13	308.5
96.1	333	320	13	307.5
96.1	332	319	13	306.5
96.1	331	318	13	305.5
96.1	330	317	13	304.5
96.1	311	299	12	287.5
96.1	310	298	12	286.5
96.1	309	297	12	285.5
96.1	308	296	12	284.5
96.1	307	295	12	283.5
96.1	306	294	12	282.5
96.1	305	293	12	281.5
96.1	304	292	12	280.5
96.1	285	274	11	263.4
96.1	284	273	11	262.4
96.1	283	272	11	261.4
96.1	282	271	11	260.4
96.1	281	270	11	259.4
96.1	280	269	11	258.4
96.1	279	268	11	257.4
96.1	259	249	10	239.4
96.1	258	248	10	238.4
96.1	257	247	10	237.4
96.1	256	246	10	236.4
96.1	255	245	10	235.4
96.1	254	244	10	234.4
96.1	233	224	9	215.3
96.1	232	223	9	214.3
96.1	231	222	9	213.4
96.1	230	221	9	212.4
96.1	229	220	9	211.4
96.1	228	219	9	210.4
96.1	207	199	8	191.3
96.1	206	198	8	190.3
96.1	205	197	8	189.3
96.1	204	196	8	188.3
96.1	203	195	8	187.3
96.1	181	174	7	167.3
96.1	180	173	7	166.3
96.1	179	172	7	165.3
96.1	178	171	7	164.3
96.1	155	149	6	143.2
96.1	154	148	6	142.2
96.1	153	147	6	141.2
96.1	152	146	6	140.2
96.1	129	124	5	119.2
96.1	128	123	5	118.2
96.1	127	122	5	117.2
96.1	103	99	4	95.2
96.1	102	98	4	94.2
96.1	77	74	3	71.1
96.1	76	73	3	70.1
96.1	51	49	2	47.1
96.0	500	480	20	460.8
96.0	499	479	20	459.8
96.0	498	478	20	458.8
96.0	497	477	20	457.8
96.0	496	476	20	456.8
96.0	495	475	20	455.8
96.0	494	474	20	454.8
96.0	480	461	19	442.8
96.0	479	460	19	441.8
96.0	478	459	19	440.8
96.0	477	458	19	439.8
96.0	476	457	19	438.8
96.0	475	456	19	437.8
96.0	474	455	19	436.8
96.0	473	454	19	435.8
96.0	472	453	19	434.8
96.0	471	452	19	433.8
96.0	470	451	19	432.8
96.0	455	437	18	419.7
96.0	454	436	18	418.7
96.0	453	435	18	417.7
96.0	452	434	18	416.7
96.0	451	433	18	415.7
96.0	450	432	18	414.7
96.0	449	431	18	413.7
96.0	448	430	18	412.7
96.0	447	429	18	411.7
96.0	446	428	18	410.7
96.0	445	427	18	409.7
96.0	430	413	17	396.7
96.0	429	412	17	395.7
96.0	428	411	17	394.7
96.0	427	410	17	393.7
96.0	426	409	17	392.7
96.0	425	408	17	391.7
96.0	424	407	17	390.7
96.0	423	406	17	389.7
96.0	422	405	17	388.7
96.0	421	404	17	387.7
96.0	420	403	17	386.7
96.0	405	389	16	373.6
96.0	404	388	16	372.6
96.0	403	387	16	371.6
96.0	402	386	16	370.6
96.0	401	385	16	369.6
96.0	400	384	16	368.6
96.0	399	383	16	367.6
96.0	398	382	16	366.6
96.0	397	381	16	365.6
96.0	396	380	16	364.6

%E	M1	M2	DM	M*
96.0	379	364	15	349.6
96.0	378	363	15	348.6
96.0	377	362	15	347.6
96.0	376	361	15	346.6
96.0	375	360	15	345.6
96.0	374	359	15	344.6
96.0	373	358	15	343.6
96.0	372	357	15	342.6
96.0	371	356	15	341.6
96.0	354	340	14	326.6
96.0	353	339	14	325.6
96.0	352	338	14	324.6
96.0	351	337	14	323.6
96.0	350	336	14	322.6
96.0	349	335	14	321.6
96.0	348	334	14	320.6
96.0	347	333	14	319.6
96.0	346	332	14	318.6
96.0	329	316	13	303.5
96.0	328	315	13	302.5
96.0	327	314	13	301.5
96.0	326	313	13	300.5
96.0	325	312	13	299.5
96.0	324	311	13	298.5
96.0	323	310	13	297.5
96.0	322	309	13	296.5
96.0	321	308	13	295.5
96.0	303	291	12	279.5
96.0	302	290	12	278.5
96.0	301	289	12	277.5
96.0	300	288	12	276.5
96.0	299	287	12	275.5
96.0	298	286	12	274.5
96.0	297	285	12	273.5
96.0	278	267	11	256.4
96.0	277	266	11	255.4
96.0	276	265	11	254.4
96.0	275	264	11	253.4
96.0	274	263	11	252.4
96.0	273	262	11	251.4
96.0	272	261	11	250.4
96.0	253	243	10	233.4
96.0	252	242	10	232.4
96.0	251	241	10	231.4
96.0	250	240	10	230.4
96.0	249	239	10	229.4
96.0	248	238	10	228.4
96.0	247	237	10	227.4
96.0	227	218	9	209.4
96.0	226	217	9	208.4

%E	M1	M2	DM	M*
96.0	225	216	9	207.4
96.0	224	215	9	206.4
96.0	223	214	9	205.4
96.0	202	194	8	186.3
96.0	201	193	8	185.3
96.0	200	192	8	184.3
96.0	199	191	8	183.3
96.0	198	190	8	182.3
96.0	177	170	7	163.3
96.0	176	169	7	162.3
96.0	175	168	7	161.3
96.0	174	167	7	160.3
96.0	173	166	7	159.3
96.0	151	145	6	139.2
96.0	150	144	6	138.2
96.0	149	143	6	137.2
96.0	126	121	5	116.2
96.0	125	120	5	115.2
96.0	124	119	5	114.2
96.0	101	97	4	93.2
96.0	100	96	4	92.2
96.0	99	95	4	91.2
96.0	75	72	3	69.1
96.0	50	48	2	46.1
96.0	25	24	1	23.0
95.9	493	473	20	453.8
95.9	492	472	20	452.8
95.9	491	471	20	451.8
95.9	490	470	20	450.8
95.9	489	469	20	449.8
95.9	488	468	20	448.8
95.9	487	467	20	447.8
95.9	486	466	20	446.8
95.9	485	465	20	445.8
95.9	484	464	20	444.8
95.9	483	463	20	443.8
95.9	482	462	20	442.8
95.9	469	450	19	431.8
95.9	468	449	19	430.8
95.9	467	448	19	429.8
95.9	466	447	19	428.8
95.9	465	446	19	427.8
95.9	464	445	19	426.8
95.9	463	444	19	425.8
95.9	462	443	19	424.8
95.9	461	442	19	423.8
95.9	460	441	19	422.8
95.9	459	440	19	421.8
95.9	458	439	19	420.8
95.9	444	426	18	408.7

%E	M1	M2	DM	M*
95.9	443	425	18	407.7
95.9	442	424	18	406.7
95.9	441	423	18	405.7
95.9	440	422	18	404.7
95.9	439	421	18	403.7
95.9	438	420	18	402.7
95.9	437	419	18	401.7
95.9	436	418	18	400.7
95.9	435	417	18	399.7
95.9	434	416	18	398.7
95.9	419	402	17	385.7
95.9	418	401	17	384.7
95.9	417	400	17	383.7
95.9	416	399	17	382.7
95.9	415	398	17	381.7
95.9	414	397	17	380.7
95.9	413	396	17	379.7
95.9	412	395	17	378.7
95.9	411	394	17	377.7
95.9	410	393	17	376.7
95.9	395	379	16	363.6
95.9	394	378	16	362.6
95.9	393	377	16	361.7
95.9	392	376	16	360.7
95.9	391	375	16	359.7
95.9	390	374	16	358.7
95.9	389	373	16	357.7
95.9	388	372	16	356.7
95.9	387	371	16	355.7
95.9	386	370	16	354.7
95.9	370	355	15	340.6
95.9	369	354	15	339.6
95.9	368	353	15	338.6
95.9	367	352	15	337.6
95.9	366	351	15	336.6
95.9	365	350	15	335.6
95.9	364	349	15	334.6
95.9	363	348	15	333.6
95.9	362	347	15	332.6
95.9	345	331	14	317.6
95.9	344	330	14	316.6
95.9	343	329	14	315.6
95.9	342	328	14	314.6
95.9	341	327	14	313.6
95.9	340	326	14	312.6
95.9	339	325	14	311.6
95.9	338	324	14	310.6
95.9	320	307	13	294.5
95.9	319	306	13	293.5
95.9	318	305	13	292.5

%E	M1	M2	DM	M*
95.9	317	304	13	291.5
95.9	316	303	13	290.5
95.9	315	302	13	289.5
95.9	314	301	13	288.5
95.9	296	284	12	272.5
95.9	295	283	12	271.5
95.9	294	282	12	270.5
95.9	293	281	12	269.5
95.9	292	280	12	268.5
95.9	291	279	12	267.5
95.9	290	278	12	266.5
95.9	271	260	11	249.4
95.9	270	259	11	248.4
95.9	269	258	11	247.4
95.9	268	257	11	246.5
95.9	267	256	11	245.5
95.9	266	255	11	244.5
95.9	246	236	10	226.4
95.9	245	235	10	225.4
95.9	244	234	10	224.4
95.9	243	233	10	223.4
95.9	242	232	10	222.4
95.9	241	231	10	221.4
95.9	222	213	9	204.4
95.9	221	212	9	203.4
95.9	220	211	9	202.4
95.9	219	210	9	201.4
95.9	218	209	9	200.4
95.9	217	208	9	199.4
95.9	197	189	8	181.3
95.9	196	188	8	180.3
95.9	195	187	8	179.3
95.9	194	186	8	178.3
95.9	193	185	8	177.3
95.9	172	165	7	158.3
95.9	171	164	7	157.3
95.9	170	163	7	156.3
95.9	169	162	7	155.3
95.9	148	142	6	136.2
95.9	147	141	6	135.2
95.9	146	140	6	134.2
95.9	145	139	6	133.2
95.9	123	118	5	113.2
95.9	122	117	5	112.2
95.9	121	116	5	111.2
95.9	98	94	4	90.2
95.9	97	93	4	89.2
95.9	74	71	3	68.1
95.9	73	70	3	67.1
95.9	49	47	2	45.1

%E	M1	M2	DM	M*
95.8	500	479	21	458.9
95.8	499	478	21	457.9
95.8	498	477	21	456.9
95.8	497	476	21	455.9
95.8	496	475	21	454.9
95.8	495	474	21	453.9
95.8	481	461	20	441.8
95.8	480	460	20	440.8
95.8	479	459	20	439.8
95.8	478	458	20	438.8
95.8	477	457	20	437.8
95.8	476	456	20	436.8
95.8	475	455	20	435.8
95.8	474	454	20	434.8
95.8	473	453	20	433.8
95.8	472	452	20	432.8
95.8	471	451	20	431.8
95.8	457	438	19	419.8
95.8	456	437	19	418.8
95.8	455	436	19	417.8
95.8	454	435	19	416.8
95.8	453	434	19	415.8
95.8	452	433	19	414.8
95.8	451	432	19	413.8
95.8	450	431	19	412.8
95.8	449	430	19	411.8
95.8	448	429	19	410.8
95.8	433	415	18	397.7
95.8	432	414	18	396.8
95.8	431	413	18	395.8
95.8	430	412	18	394.8
95.8	429	411	18	393.8
95.8	428	410	18	392.8
95.8	427	409	18	391.8
95.8	426	408	18	390.8
95.8	425	407	18	389.8
95.8	424	406	18	388.8
95.8	409	392	17	375.7
95.8	408	391	17	374.7
95.8	407	390	17	373.7
95.8	406	389	17	372.7
95.8	405	388	17	371.7
95.8	404	387	17	370.7
95.8	403	386	17	369.7
95.8	402	385	17	368.7
95.8	401	384	17	367.7
95.8	400	383	17	366.7
95.8	385	369	16	353.7
95.8	384	368	16	352.7
95.8	383	367	16	351.7

%E	M1	M2	DM	M*
95.8	382	366	16	350.7
95.8	381	365	16	349.7
95.8	380	364	16	348.7
95.8	379	363	16	347.7
95.8	378	362	16	346.7
95.8	377	361	16	345.7
95.8	361	346	15	331.6
95.8	360	345	15	330.6
95.8	359	344	15	329.6
95.8	358	343	15	328.6
95.8	357	342	15	327.6
95.8	356	341	15	326.6
95.8	355	340	15	325.6
95.8	354	339	15	324.6
95.8	353	338	15	323.6
95.8	337	323	14	309.6
95.8	336	322	14	308.6
95.8	335	321	14	307.6
95.8	334	320	14	306.6
95.8	333	319	14	305.6
95.8	332	318	14	304.6
95.8	331	317	14	303.6
95.8	330	316	14	302.6
95.8	313	300	13	287.5
95.8	312	299	13	286.5
95.8	311	298	13	285.5
95.8	310	297	13	284.5
95.8	309	296	13	283.5
95.8	308	295	13	282.5
95.8	307	294	13	281.6
95.8	306	293	13	280.6
95.8	289	277	12	265.5
95.8	288	276	12	264.5
95.8	287	275	12	263.5
95.8	286	274	12	262.5
95.8	285	273	12	261.5
95.8	284	272	12	260.5
95.8	283	271	12	259.5
95.8	265	254	11	243.5
95.8	264	253	11	242.5
95.8	263	252	11	241.5
95.8	262	251	11	240.5
95.8	261	250	11	239.5
95.8	260	249	11	238.5
95.8	259	248	11	237.5
95.8	240	230	10	220.4
95.8	239	229	10	219.4
95.8	238	228	10	218.4
95.8	237	227	10	217.4
95.8	236	226	10	216.4

%E	M1	M2	DM	M*
95.8	216	207	9	198.4
95.8	215	206	9	197.4
95.8	214	205	9	196.4
95.8	213	204	9	195.4
95.8	212	203	9	194.4
95.8	192	184	8	176.3
95.8	191	183	8	175.3
95.8	190	182	8	174.3
95.8	189	181	8	173.3
95.8	168	161	7	154.3
95.8	167	160	7	153.3
95.8	166	159	7	152.3
95.8	165	158	7	151.3
95.8	144	138	6	132.3
95.8	143	137	6	131.3
95.8	142	136	6	130.3
95.8	120	115	5	110.2
95.8	119	114	5	109.2
95.8	118	113	5	108.2
95.8	96	92	4	88.2
95.8	95	91	4	87.2
95.8	72	69	3	66.1
95.8	71	68	3	65.1
95.8	48	46	2	44.1
95.8	24	23	1	22.0
95.7	494	473	21	452.9
95.7	493	472	21	451.9
95.7	492	471	21	450.9
95.7	491	470	21	449.9
95.7	490	469	21	448.9
95.7	489	468	21	447.9
95.7	488	467	21	446.9
95.7	487	466	21	445.9
95.7	486	465	21	444.9
95.7	485	464	21	443.9
95.7	484	463	21	442.9
95.7	483	462	21	441.9
95.7	470	450	20	430.9
95.7	469	449	20	429.9
95.7	468	448	20	428.9
95.7	467	447	20	427.9
95.7	466	446	20	426.9
95.7	465	445	20	425.9
95.7	464	444	20	424.9
95.7	463	443	20	423.9
95.7	462	442	20	422.9
95.7	461	441	20	421.9
95.7	460	440	20	420.9
95.7	447	428	19	409.8
95.7	446	427	19	408.8

%E	M1	M2	DM	M*
95.7	445	426	19	407.8
95.7	444	425	19	406.8
95.7	443	424	19	405.8
95.7	442	423	19	404.8
95.7	441	422	19	403.8
95.7	440	421	19	402.8
95.7	439	420	19	401.8
95.7	438	419	19	400.8
95.7	437	418	19	399.8
95.7	423	405	18	387.8
95.7	422	404	18	386.8
95.7	421	403	18	385.8
95.7	420	402	18	384.8
95.7	419	401	18	383.8
95.7	418	400	18	382.8
95.7	417	399	18	381.8
95.7	416	398	18	380.8
95.7	415	397	18	379.8
95.7	414	396	18	378.8
95.7	399	382	17	365.7
95.7	398	381	17	364.7
95.7	397	380	17	363.7
95.7	396	379	17	362.7
95.7	395	378	17	361.7
95.7	394	377	17	360.7
95.7	393	376	17	359.7
95.7	392	375	17	358.7
95.7	391	374	17	357.7
95.7	376	360	16	344.7
95.7	375	359	16	343.7
95.7	374	358	16	342.7
95.7	373	357	16	341.7
95.7	372	356	16	340.7
95.7	371	355	16	339.7
95.7	370	354	16	338.7
95.7	369	353	16	337.7
95.7	368	352	16	336.7
95.7	352	337	15	322.6
95.7	351	336	15	321.6
95.7	350	335	15	320.6
95.7	349	334	15	319.6
95.7	348	333	15	318.6
95.7	347	332	15	317.6
95.7	346	331	15	316.7
95.7	345	330	15	315.7
95.7	329	315	14	301.6
95.7	328	314	14	300.6
95.7	327	313	14	299.6
95.7	326	312	14	298.6
95.7	325	311	14	297.6

%E	M1	M2	DM	M*
95.7	324	310	14	296.6
95.7	323	309	14	295.6
95.7	322	308	14	294.6
95.7	305	292	13	279.6
95.7	304	291	13	278.6
95.7	303	290	13	277.6
95.7	302	289	13	276.6
95.7	301	288	13	275.6
95.7	300	287	13	274.6
95.7	299	286	13	273.6
95.7	282	270	12	258.5
95.7	281	269	12	257.5
95.7	280	268	12	256.5
95.7	279	267	12	255.5
95.7	278	266	12	254.5
95.7	277	265	12	253.5
95.7	276	264	12	252.5
95.7	258	247	11	236.5
95.7	257	246	11	235.5
95.7	256	245	11	234.5
95.7	255	244	11	233.5
95.7	254	243	11	232.5
95.7	253	242	11	231.5
95.7	235	225	10	215.4
95.7	234	224	10	214.4
95.7	233	223	10	213.4
95.7	232	222	10	212.4
95.7	231	221	10	211.4
95.7	230	220	10	210.4
95.7	211	202	9	193.4
95.7	210	201	9	192.4
95.7	209	200	9	191.4
95.7	208	199	9	190.4
95.7	207	198	9	189.4
95.7	188	180	8	172.3
95.7	187	179	8	171.3
95.7	186	178	8	170.3
95.7	185	177	8	169.3
95.7	184	176	8	168.3
95.7	164	157	7	150.3
95.7	163	156	7	149.3
95.7	162	155	7	148.3
95.7	161	154	7	147.3
95.7	141	135	6	129.3
95.7	140	134	6	128.3
95.7	139	133	6	127.3
95.7	138	132	6	126.3
95.7	117	112	5	107.2
95.7	116	111	5	106.2
95.7	115	110	5	105.2

%E	M1	M2	DM	M*
95.7	94	90	4	86.2
95.7	93	89	4	85.2
95.7	92	88	4	84.2
95.7	70	67	3	64.1
95.7	69	66	3	63.1
95.7	47	45	2	43.1
95.7	46	44	2	42.1
95.7	23	22	1	21.0
95.6	500	478	22	457.0
95.6	499	477	22	456.0
95.6	498	476	22	455.0
95.6	497	475	22	454.0
95.6	496	474	22	453.0
95.6	495	473	22	452.0
95.6	482	461	21	440.9
95.6	481	460	21	439.9
95.6	480	459	21	438.9
95.6	479	458	21	437.9
95.6	478	457	21	436.9
95.6	477	456	21	435.9
95.6	476	455	21	434.9
95.6	475	454	21	433.9
95.6	474	453	21	432.9
95.6	473	452	21	431.9
95.6	472	451	21	430.9
95.6	459	439	20	419.9
95.6	458	438	20	418.9
95.6	457	437	20	417.9
95.6	456	436	20	416.9
95.6	455	435	20	415.9
95.6	454	434	20	414.9
95.6	453	433	20	413.9
95.6	452	432	20	412.9
95.6	451	431	20	411.9
95.6	450	430	20	410.9
95.6	436	417	19	398.8
95.6	435	416	19	397.8
95.6	434	415	19	396.8
95.6	433	414	19	395.8
95.6	432	413	19	394.8
95.6	431	412	19	393.8
95.6	430	411	19	392.8
95.6	429	410	19	391.8
95.6	428	409	19	390.8
95.6	427	408	19	389.8
95.6	413	395	18	377.8
95.6	412	394	18	376.8
95.6	411	393	18	375.8
95.6	410	392	18	374.8
95.6	409	391	18	373.8

%E	M1	M2	DM	M*
95.6	408	390	18	372.8
95.6	407	389	18	371.8
95.6	406	388	18	370.8
95.6	405	387	18	369.8
95.6	390	373	17	356.7
95.6	389	372	17	355.7
95.6	388	371	17	354.7
95.6	387	370	17	353.7
95.6	386	369	17	352.7
95.6	385	368	17	351.8
95.6	384	367	17	350.8
95.6	383	366	17	349.8
95.6	382	365	17	348.8
95.6	367	351	16	335.7
95.6	366	350	16	334.7
95.6	365	349	16	333.7
95.6	364	348	16	332.7
95.6	363	347	16	331.7
95.6	362	346	16	330.7
95.6	361	345	16	329.7
95.6	360	344	16	328.7
95.6	344	329	15	314.7
95.6	343	328	15	313.7
95.6	342	327	15	312.7
95.6	341	326	15	311.7
95.6	340	325	15	310.7
95.6	339	324	15	309.7
95.6	338	323	15	308.7
95.6	321	307	14	293.6
95.6	320	306	14	292.6
95.6	319	305	14	291.6
95.6	318	304	14	290.6
95.6	317	303	14	289.6
95.6	316	302	14	288.6
95.6	315	301	14	287.6
95.6	298	285	13	272.6
95.6	297	284	13	271.6
95.6	296	283	13	270.6
95.6	295	282	13	269.6
95.6	294	281	13	268.6
95.6	293	280	13	267.6
95.6	275	263	12	251.5
95.6	274	262	12	250.5
95.6	273	261	12	249.5
95.6	272	260	12	248.5
95.6	271	259	12	247.5
95.6	270	258	12	246.5
95.6	252	241	11	230.5
95.6	251	240	11	229.5
95.6	250	239	11	228.5

%E	M1	M2	DM	M*
95.6	249	238	11	227.5
95.6	248	237	11	226.5
95.6	229	219	10	209.4
95.6	228	218	10	208.4
95.6	227	217	10	207.4
95.6	226	216	10	206.4
95.6	225	215	10	205.4
95.6	206	197	9	188.4
95.6	205	196	9	187.4
95.6	204	195	9	186.4
95.6	203	194	9	185.4
95.6	183	175	8	167.3
95.6	182	174	8	166.4
95.6	181	173	8	165.4
95.6	180	172	8	164.4
95.6	160	153	7	146.3
95.6	159	152	7	145.3
95.6	158	151	7	144.3
95.6	137	131	6	125.3
95.6	136	130	6	124.3
95.6	135	129	6	123.3
95.6	114	109	5	104.2
95.6	113	108	5	103.2
95.6	91	87	4	83.2
95.6	90	86	4	82.2
95.6	68	65	3	62.1
95.6	45	43	2	41.1
95.5	494	472	22	451.0
95.5	493	471	22	450.0
95.5	492	470	22	449.0
95.5	491	469	22	448.0
95.5	490	468	22	447.0
95.5	489	467	22	446.0
95.5	488	466	22	445.0
95.5	487	465	22	444.0
95.5	486	464	22	443.0
95.5	485	463	22	442.0
95.5	484	462	22	441.0
95.5	471	450	21	429.9
95.5	470	449	21	428.9
95.5	469	448	21	427.9
95.5	468	447	21	426.9
95.5	467	446	21	425.9
95.5	466	445	21	424.9
95.5	465	444	21	423.9
95.5	464	443	21	423.0
95.5	463	442	21	422.0
95.5	462	441	21	421.0
95.5	449	429	20	409.9
95.5	448	428	20	408.9

%E	M1	M2	DM	M*
95.5	447	427	20	407.9
95.5	446	426	20	406.9
95.5	445	425	20	405.9
95.5	444	424	20	404.9
95.5	443	423	20	403.9
95.5	442	422	20	402.9
95.5	441	421	20	401.9
95.5	440	420	20	400.9
95.5	426	407	19	388.8
95.5	425	406	19	387.8
95.5	424	405	19	386.9
95.5	423	404	19	385.9
95.5	422	403	19	384.9
95.5	421	402	19	383.9
95.5	420	401	19	382.9
95.5	419	400	19	381.9
95.5	418	399	19	380.9
95.5	404	386	18	368.8
95.5	403	385	18	367.8
95.5	402	384	18	366.8
95.5	401	383	18	365.8
95.5	400	382	18	364.8
95.5	399	381	18	363.8
95.5	398	380	18	362.8
95.5	397	379	18	361.8
95.5	396	378	18	360.8
95.5	381	364	17	347.8
95.5	380	363	17	346.8
95.5	379	362	17	345.8
95.5	378	361	17	344.8
95.5	377	360	17	343.8
95.5	376	359	17	342.8
95.5	375	358	17	341.8
95.5	374	357	17	340.8
95.5	359	343	16	327.7
95.5	358	342	16	326.7
95.5	357	341	16	325.7
95.5	356	340	16	324.7
95.5	355	339	16	323.7
95.5	354	338	16	322.7
95.5	353	337	16	321.7
95.5	352	336	16	320.7
95.5	337	322	15	307.7
95.5	336	321	15	306.7
95.5	335	320	15	305.7
95.5	334	319	15	304.7
95.5	333	318	15	303.7
95.5	332	317	15	302.7
95.5	331	316	15	301.7
95.5	330	315	15	300.7

%E	M1	M2	DM	M*
95.5	314	300	14	286.6
95.5	313	299	14	285.6
95.5	312	298	14	284.6
95.5	311	297	14	283.6
95.5	310	296	14	282.6
95.5	309	295	14	281.6
95.5	308	294	14	280.6
95.5	292	279	13	266.6
95.5	291	278	13	265.6
95.5	290	277	13	264.6
95.5	289	276	13	263.6
95.5	288	275	13	262.6
95.5	287	274	13	261.6
95.5	286	273	13	260.6
95.5	269	257	12	245.5
95.5	268	256	12	244.5
95.5	267	255	12	243.5
95.5	266	254	12	242.5
95.5	265	253	12	241.5
95.5	264	252	12	240.5
95.5	247	236	11	225.5
95.5	246	235	11	224.5
95.5	245	234	11	223.5
95.5	244	233	11	222.5
95.5	243	232	11	221.5
95.5	242	231	11	220.5
95.5	224	214	10	204.4
95.5	223	213	10	203.4
95.5	222	212	10	202.5
95.5	221	211	10	201.5
95.5	220	210	10	200.5
95.5	202	193	9	184.4
95.5	201	192	9	183.4
95.5	200	191	9	182.4
95.5	199	190	9	181.4
95.5	198	189	9	180.4
95.5	179	171	8	163.4
95.5	178	170	8	162.4
95.5	177	169	8	161.4
95.5	176	168	8	160.4
95.5	157	150	7	143.3
95.5	156	149	7	142.3
95.5	155	148	7	141.3
95.5	154	147	7	140.3
95.5	134	128	6	122.3
95.5	133	127	6	121.3
95.5	132	126	6	120.3
95.5	112	107	5	102.2
95.5	111	106	5	101.2
95.5	110	105	5	100.2

%E	M1	M2	DM	M*
95.5	89	85	4	81.2
95.5	88	84	4	80.2
95.5	67	64	3	61.1
95.5	66	63	3	60.1
95.5	44	42	2	40.1
95.5	22	21	1	20.0
95.4	500	477	23	455.1
95.4	499	476	23	454.1
95.4	498	475	23	453.1
95.4	497	474	23	452.1
95.4	496	473	23	451.1
95.4	495	472	23	450.1
95.4	483	461	22	440.0
95.4	482	460	22	439.0
95.4	481	459	22	438.0
95.4	480	458	22	437.0
95.4	479	457	22	436.0
95.4	478	456	22	435.0
95.4	477	455	22	434.0
95.4	476	454	22	433.0
95.4	475	453	22	432.0
95.4	474	452	22	431.0
95.4	461	440	21	420.0
95.4	460	439	21	419.0
95.4	459	438	21	418.0
95.4	458	437	21	417.0
95.4	457	436	21	416.0
95.4	456	435	21	415.0
95.4	455	434	21	414.0
95.4	454	433	21	413.0
95.4	453	432	21	412.0
95.4	452	431	21	411.0
95.4	439	419	20	399.9
95.4	438	418	20	398.9
95.4	437	417	20	397.9
95.4	436	416	20	396.9
95.4	435	415	20	395.9
95.4	434	414	20	394.9
95.4	433	413	20	393.9
95.4	432	412	20	392.9
95.4	431	411	20	391.9
95.4	417	398	19	379.9
95.4	416	397	19	378.9
95.4	415	396	19	377.9
95.4	414	395	19	376.9
95.4	413	394	19	375.9
95.4	412	393	19	374.9
95.4	411	392	19	373.9
95.4	410	391	19	372.9
95.4	409	390	19	371.9

%E	M1	M2	DM	M*
95.4	395	377	18	359.8
95.4	394	376	18	358.8
95.4	393	375	18	357.8
95.4	392	374	18	356.8
95.4	391	373	18	355.8
95.4	390	372	18	354.8
95.4	389	371	18	353.8
95.4	388	370	18	352.8
95.4	373	356	17	339.8
95.4	372	355	17	338.8
95.4	371	354	17	337.8
95.4	370	353	17	336.8
95.4	369	352	17	335.8
95.4	368	351	17	334.8
95.4	367	350	17	333.8
95.4	366	349	17	332.8
95.4	351	335	16	319.7
95.4	350	334	16	318.7
95.4	349	333	16	317.7
95.4	348	332	16	316.7
95.4	347	331	16	315.7
95.4	346	330	16	314.7
95.4	345	329	16	313.7
95.4	329	314	15	299.7
95.4	328	313	15	298.7
95.4	327	312	15	297.7
95.4	326	311	15	296.7
95.4	325	310	15	295.7
95.4	324	309	15	294.7
95.4	323	308	15	293.7
95.4	307	293	14	279.6
95.4	306	292	14	278.6
95.4	305	291	14	277.6
95.4	304	290	14	276.6
95.4	303	289	14	275.6
95.4	302	288	14	274.6
95.4	285	272	13	259.6
95.4	284	271	13	258.6
95.4	283	270	13	257.6
95.4	282	269	13	256.6
95.4	281	268	13	255.6
95.4	280	267	13	254.6
95.4	263	251	12	239.5
95.4	262	250	12	238.5
95.4	261	249	12	237.6
95.4	260	248	12	236.6
95.4	259	247	12	235.6
95.4	241	230	11	219.5
95.4	240	229	11	218.5
95.4	239	228	11	217.5

%E	M1	M2	DM	M*
95.4	238	227	11	216.5
95.4	237	226	11	215.5
95.4	219	209	10	199.5
95.4	218	208	10	198.5
95.4	217	207	10	197.5
95.4	216	206	10	196.5
95.4	197	188	9	179.4
95.4	196	187	9	178.4
95.4	195	186	9	177.4
95.4	194	185	9	176.4
95.4	175	167	8	159.4
95.4	174	166	8	158.4
95.4	173	165	8	157.4
95.4	153	146	7	139.3
95.4	152	145	7	138.3
95.4	151	144	7	137.3
95.4	131	125	6	119.3
95.4	130	124	6	118.3
95.4	109	104	5	99.2
95.4	108	103	5	98.2
95.4	87	83	4	79.2
95.4	65	62	3	59.1
95.3	494	471	23	449.1
95.3	493	470	23	448.1
95.3	492	469	23	447.1
95.3	491	468	23	446.1
95.3	490	467	23	445.1
95.3	489	466	23	444.1
95.3	488	465	23	443.1
95.3	487	464	23	442.1
95.3	486	463	23	441.1
95.3	485	462	23	440.1
95.3	473	451	22	430.0
95.3	472	450	22	429.0
95.3	471	449	22	428.0
95.3	470	448	22	427.0
95.3	469	447	22	426.0
95.3	468	446	22	425.0
95.3	467	445	22	424.0
95.3	466	444	22	423.0
95.3	465	443	22	422.0
95.3	464	442	22	421.0
95.3	451	430	21	410.0
95.3	450	429	21	409.0
95.3	449	428	21	408.0
95.3	448	427	21	407.0
95.3	447	426	21	406.0
95.3	446	425	21	405.0
95.3	445	424	21	404.0
95.3	444	423	21	403.0

%E	M1	M2	DM	M*
95.3	443	422	21	402.0
95.3	430	410	20	390.9
95.3	429	409	20	389.9
95.3	428	408	20	388.9
95.3	427	407	20	387.9
95.3	426	406	20	386.9
95.3	425	405	20	385.9
95.3	424	404	20	384.9
95.3	423	403	20	383.9
95.3	422	402	20	382.9
95.3	408	389	19	370.9
95.3	407	388	19	369.9
95.3	406	387	19	368.9
95.3	405	386	19	367.9
95.3	404	385	19	366.9
95.3	403	384	19	365.9
95.3	402	383	19	364.9
95.3	401	382	19	363.9
95.3	400	381	19	362.9
95.3	387	369	18	351.8
95.3	386	368	18	350.8
95.3	385	367	18	349.8
95.3	384	366	18	348.8
95.3	383	365	18	347.8
95.3	382	364	18	346.8
95.3	381	363	18	345.9
95.3	380	362	18	344.9
95.3	379	361	18	343.9
95.3	365	348	17	331.8
95.3	364	347	17	330.8
95.3	363	346	17	329.8
95.3	362	345	17	328.8
95.3	361	344	17	327.8
95.3	360	343	17	326.8
95.3	359	342	17	325.8
95.3	358	341	17	324.8
95.3	344	328	16	312.7
95.3	343	327	16	311.7
95.3	342	326	16	310.7
95.3	341	325	16	309.8
95.3	340	324	16	308.8
95.3	339	323	16	307.8
95.3	338	322	16	306.8
95.3	337	321	16	305.8
95.3	322	307	15	292.7
95.3	321	306	15	291.7
95.3	320	305	15	290.7
95.3	319	304	15	289.7
95.3	318	303	15	288.7
95.3	317	302	15	287.7

%E	M1	M2	DM	M*
95.3	316	301	15	286.7
95.3	301	287	14	273.7
95.3	300	286	14	272.7
95.3	299	285	14	271.7
95.3	298	284	14	270.7
95.3	297	283	14	269.7
95.3	296	282	14	268.7
95.3	295	281	14	267.7
95.3	279	266	13	253.6
95.3	278	265	13	252.6
95.3	277	264	13	251.6
95.3	276	263	13	250.6
95.3	275	262	13	249.6
95.3	274	261	13	248.6
95.3	258	246	12	234.6
95.3	257	245	12	233.6
95.3	256	244	12	232.6
95.3	255	243	12	231.6
95.3	254	242	12	230.6
95.3	253	241	12	229.6
95.3	236	225	11	214.5
95.3	235	224	11	213.5
95.3	234	223	11	212.5
95.3	233	222	11	211.5
95.3	232	221	11	210.5
95.3	215	205	10	195.5
95.3	214	204	10	194.5
95.3	213	203	10	193.5
95.3	212	202	10	192.5
95.3	211	201	10	191.5
95.3	193	184	9	175.4
95.3	192	183	9	174.4
95.3	191	182	9	173.4
95.3	190	181	9	172.4
95.3	172	164	8	156.4
95.3	171	163	8	155.4
95.3	170	162	8	154.4
95.3	169	161	8	153.4
95.3	150	143	7	136.3
95.3	149	142	7	135.3
95.3	148	141	7	134.3
95.3	129	123	6	117.3
95.3	128	122	6	116.3
95.3	127	121	6	115.3
95.3	107	102	5	97.2
95.3	106	101	5	96.2
95.3	86	82	4	78.2
95.3	85	81	4	77.2
95.3	64	61	3	58.1
95.3	43	41	2	39.1

%E	M1	M2	DM	M*
95.2	500	476	24	453.2
95.2	499	475	24	452.2
95.2	498	474	24	451.2
95.2	497	473	24	450.2
95.2	496	472	24	449.2
95.2	495	471	24	448.2
95.2	484	461	23	439.1
95.2	483	460	23	438.1
95.2	482	459	23	437.1
95.2	481	458	23	436.1
95.2	480	457	23	435.1
95.2	479	456	23	434.1
95.2	478	455	23	433.1
95.2	477	454	23	432.1
95.2	476	453	23	431.1
95.2	475	452	23	430.1
95.2	463	441	22	420.0
95.2	462	440	22	419.0
95.2	461	439	22	418.0
95.2	460	438	22	417.1
95.2	459	437	22	416.1
95.2	458	436	22	415.1
95.2	457	435	22	414.1
95.2	456	434	22	413.1
95.2	455	433	22	412.1
95.2	454	432	22	411.1
95.2	442	421	21	401.0
95.2	441	420	21	400.0
95.2	440	419	21	399.0
95.2	439	418	21	398.0
95.2	438	417	21	397.0
95.2	437	416	21	396.0
95.2	436	415	21	395.0
95.2	435	414	21	394.0
95.2	434	413	21	393.0
95.2	433	412	21	392.0
95.2	421	401	20	381.9
95.2	420	400	20	381.0
95.2	419	399	20	380.0
95.2	418	398	20	379.0
95.2	417	397	20	378.0
95.2	416	396	20	377.0
95.2	415	395	20	376.0
95.2	414	394	20	375.0
95.2	413	393	20	374.0
95.2	399	380	19	361.9
95.2	398	379	19	360.9
95.2	397	378	19	359.9
95.2	396	377	19	358.9
95.2	395	376	19	357.9

%E	M1	M2	DM	M*
95.2	394	375	19	356.9
95.2	393	374	19	355.9
95.2	392	373	19	354.9
95.2	378	360	18	342.9
95.2	377	359	18	341.9
95.2	376	358	18	340.9
95.2	375	357	18	339.9
95.2	374	356	18	338.9
95.2	373	355	18	337.9
95.2	372	354	18	336.9
95.2	357	340	17	323.8
95.2	356	339	17	322.8
95.2	355	338	17	321.8
95.2	354	337	17	320.8
95.2	353	336	17	319.8
95.2	352	335	17	318.8
95.2	351	334	17	317.8
95.2	336	320	16	304.8
95.2	335	319	16	303.8
95.2	334	318	16	302.8
95.2	333	317	16	301.8
95.2	332	316	16	300.8
95.2	331	315	16	299.8
95.2	330	314	16	298.8
95.2	315	300	15	285.7
95.2	314	299	15	284.7
95.2	313	298	15	283.7
95.2	312	297	15	282.7
95.2	311	296	15	281.7
95.2	310	295	15	280.7
95.2	294	280	14	266.7
95.2	293	279	14	265.7
95.2	292	278	14	264.7
95.2	291	277	14	263.7
95.2	290	276	14	262.7
95.2	289	275	14	261.7
95.2	273	260	13	247.6
95.2	272	259	13	246.6
95.2	271	258	13	245.6
95.2	270	257	13	244.6
95.2	269	256	13	243.6
95.2	252	240	12	228.6
95.2	251	239	12	227.6
95.2	250	238	12	226.6
95.2	249	237	12	225.6
95.2	248	236	12	224.6
95.2	231	220	11	209.5
95.2	230	219	11	208.5
95.2	229	218	11	207.5
95.2	228	217	11	206.5

%E	M1	M2	DM	M*
95.2	227	216	11	205.5
95.2	210	200	10	190.5
95.2	209	199	10	189.5
95.2	208	198	10	188.5
95.2	207	197	10	187.5
95.2	189	180	9	171.4
95.2	188	179	9	170.4
95.2	187	178	9	169.4
95.2	186	177	9	168.4
95.2	168	160	8	152.4
95.2	167	159	8	151.4
95.2	166	158	8	150.4
95.2	165	157	8	149.4
95.2	147	140	7	133.3
95.2	146	139	7	132.3
95.2	145	138	7	131.3
95.2	126	120	6	114.3
95.2	125	119	6	113.3
95.2	124	118	6	112.3
95.2	105	100	5	95.2
95.2	104	99	5	94.2
95.2	84	80	4	76.2
95.2	83	79	4	75.2
95.2	63	60	3	57.1
95.2	62	59	3	56.1
95.2	42	40	2	38.1
95.2	21	20	1	19.0
95.1	494	470	24	447.2
95.1	493	469	24	446.2
95.1	492	468	24	445.2
95.1	491	467	24	444.2
95.1	490	466	24	443.2
95.1	489	465	24	442.2
95.1	488	464	24	441.2
95.1	487	463	24	440.2
95.1	486	462	24	439.2
95.1	485	461	24	438.2
95.1	474	451	23	429.1
95.1	473	450	23	428.1
95.1	472	449	23	427.1
95.1	471	448	23	426.1
95.1	470	447	23	425.1
95.1	469	446	23	424.1
95.1	468	445	23	423.1
95.1	467	444	23	422.1
95.1	466	443	23	421.1
95.1	465	442	23	420.1
95.1	453	431	22	410.1
95.1	452	430	22	409.1
95.1	451	429	22	408.1

%E	M1	M2	DM	M*
95.1	450	428	22	407.1
95.1	449	427	22	406.1
95.1	448	426	22	405.1
95.1	447	425	22	404.1
95.1	446	424	22	403.1
95.1	445	423	22	402.1
95.1	432	411	21	391.0
95.1	431	410	21	390.0
95.1	430	409	21	389.0
95.1	429	408	21	388.0
95.1	428	407	21	387.0
95.1	427	406	21	386.0
95.1	426	405	21	385.0
95.1	425	404	21	384.0
95.1	412	392	20	373.0
95.1	411	391	20	372.0
95.1	410	390	20	371.0
95.1	409	389	20	370.0
95.1	408	388	20	369.0
95.1	407	387	20	368.0
95.1	406	386	20	367.0
95.1	405	385	20	366.0
95.1	404	384	20	365.0
95.1	391	372	19	353.9
95.1	390	371	19	352.9
95.1	389	370	19	351.9
95.1	388	369	19	350.9
95.1	387	368	19	349.9
95.1	386	367	19	348.9
95.1	385	366	19	347.9
95.1	384	365	19	346.9
95.1	371	353	18	335.9
95.1	370	352	18	334.9
95.1	369	351	18	333.9
95.1	368	350	18	332.9
95.1	367	349	18	331.9
95.1	366	348	18	330.9
95.1	365	347	18	329.9
95.1	364	346	18	328.9
95.1	350	333	17	316.8
95.1	349	332	17	315.8
95.1	348	331	17	314.8
95.1	347	330	17	313.8
95.1	346	329	17	312.8
95.1	345	328	17	311.8
95.1	344	327	17	310.8
95.1	329	313	16	297.8
95.1	328	312	16	296.8
95.1	327	311	16	295.8
95.1	326	310	16	294.8

%E	M1	M2	DM	M*
95.1	325	309	16	293.8
95.1	324	308	16	292.8
95.1	309	294	15	279.7
95.1	308	293	15	278.7
95.1	307	292	15	277.7
95.1	306	291	15	276.7
95.1	305	290	15	275.7
95.1	304	289	15	274.7
95.1	303	288	15	273.7
95.1	288	274	14	260.7
95.1	287	273	14	259.7
95.1	286	272	14	258.7
95.1	285	271	14	257.7
95.1	284	270	14	256.7
95.1	283	269	14	255.7
95.1	268	255	13	242.6
95.1	267	254	13	241.6
95.1	266	253	13	240.6
95.1	265	252	13	239.6
95.1	264	251	13	238.6
95.1	263	250	13	237.6
95.1	247	235	12	223.6
95.1	246	234	12	222.6
95.1	245	233	12	221.6
95.1	244	232	12	220.6
95.1	243	231	12	219.6
95.1	226	215	11	204.5
95.1	225	214	11	203.5
95.1	224	213	11	202.5
95.1	223	212	11	201.5
95.1	206	196	10	186.5
95.1	205	195	10	185.5
95.1	204	194	10	184.5
95.1	203	193	10	183.5
95.1	202	192	10	182.5
95.1	185	176	9	167.4
95.1	184	175	9	166.4
95.1	183	174	9	165.4
95.1	182	173	9	164.4
95.1	164	156	8	148.4
95.1	163	155	8	147.4
95.1	162	154	8	146.4
95.1	144	137	7	130.3
95.1	143	136	7	129.3
95.1	142	135	7	128.3
95.1	123	117	6	111.3
95.1	122	116	6	110.3
95.1	103	98	5	93.2
95.1	102	97	5	92.2
95.1	101	96	5	91.2

%E	M1	M2	DM	M*
95.1	82	78	4	74.2
95.1	81	77	4	73.2
95.1	61	58	3	55.1
95.1	41	39	2	37.1
95.0	500	475	25	451.3
95.0	499	474	25	450.3
95.0	498	473	25	449.3
95.0	497	472	25	448.3
95.0	496	471	25	447.3
95.0	484	460	24	437.2
95.0	483	459	24	436.2
95.0	482	458	24	435.2
95.0	481	457	24	434.2
95.0	480	456	24	433.2
95.0	479	455	24	432.2
95.0	478	454	24	431.2
95.0	477	453	24	430.2
95.0	476	452	24	429.2
95.0	464	441	23	419.1
95.0	463	440	23	418.1
95.0	462	439	23	417.1
95.0	461	438	23	416.1
95.0	460	437	23	415.1
95.0	459	436	23	414.2
95.0	458	435	23	413.2
95.0	457	434	23	412.2
95.0	456	433	23	411.2
95.0	444	422	22	401.1
95.0	443	421	22	400.1
95.0	442	420	22	399.1
95.0	441	419	22	398.1
95.0	440	418	22	397.1
95.0	439	417	22	396.1
95.0	438	416	22	395.1
95.0	437	415	22	394.1
95.0	436	414	22	393.1
95.0	424	403	21	383.0
95.0	423	402	21	382.0
95.0	422	401	21	381.0
95.0	421	400	21	380.0
95.0	420	399	21	379.0
95.0	419	398	21	378.1
95.0	418	397	21	377.1
95.0	417	396	21	376.1
95.0	416	395	21	375.1
95.0	403	383	20	364.0
95.0	402	382	20	363.0
95.0	401	381	20	362.0
95.0	400	380	20	361.0
95.0	399	379	20	360.0

%E	M1	M2	DM	M*
95.0	398	378	20	359.0
95.0	397	377	20	358.0
95.0	383	364	19	345.9
95.0	382	363	19	344.9
95.0	381	362	19	343.9
95.0	380	361	19	342.9
95.0	379	360	19	342.0
95.0	378	359	19	341.0
95.0	377	358	19	340.0
95.0	363	345	18	327.9
95.0	362	344	18	326.9
95.0	361	343	18	325.9
95.0	360	342	18	324.9
95.0	359	341	18	323.9
95.0	358	340	18	322.9
95.0	357	339	18	321.9
95.0	343	326	17	309.8
95.0	342	325	17	308.8
95.0	341	324	17	307.8
95.0	340	323	17	306.8
95.0	339	322	17	305.9
95.0	338	321	17	304.9
95.0	337	320	17	303.9
95.0	323	307	16	291.8
95.0	322	306	16	290.8
95.0	321	305	16	289.8
95.0	320	304	16	288.8
95.0	319	303	16	287.8
95.0	318	302	16	286.8
95.0	317	301	16	285.8
95.0	302	287	15	272.7
95.0	301	286	15	271.7
95.0	300	285	15	270.8
95.0	299	284	15	269.8
95.0	298	283	15	268.8
95.0	282	268	14	254.7
95.0	281	267	14	253.7
95.0	280	266	14	252.7
95.0	279	265	14	251.7
95.0	278	264	14	250.7
95.0	262	249	13	236.6
95.0	261	248	13	235.6
95.0	260	247	13	234.6
95.0	259	246	13	233.7
95.0	258	245	13	232.7
95.0	242	230	12	218.6
95.0	241	229	12	217.6
95.0	240	228	12	216.6
95.0	239	227	12	215.6
95.0	238	226	12	214.6

%E	M1	M2	DM	M*
95.0	222	211	11	200.5
95.0	221	210	11	199.5
95.0	220	209	11	198.5
95.0	219	208	11	197.6
95.0	218	207	11	196.6
95.0	201	191	10	181.5
95.0	200	190	10	180.5
95.0	199	189	10	179.5
95.0	181	172	9	163.4
95.0	180	171	9	162.4
95.0	179	170	9	161.5
95.0	161	153	8	145.4
95.0	160	152	8	144.4
95.0	159	151	8	143.4
95.0	141	134	7	127.3
95.0	140	133	7	126.3
95.0	139	132	7	125.4
95.0	121	115	6	109.3
95.0	120	114	6	108.3
95.0	119	113	6	107.3
95.0	100	95	5	90.3
95.0	80	76	4	72.2
95.0	60	57	3	54.1
95.0	40	38	2	36.1
95.0	20	19	1	18.0
94.9	495	470	25	446.3
94.9	494	469	25	445.3
94.9	493	468	25	444.3
94.9	492	467	25	443.3
94.9	491	466	25	442.3
94.9	490	465	25	441.3
94.9	489	464	25	440.3
94.9	488	463	25	439.3
94.9	487	462	25	438.3
94.9	486	461	25	437.3
94.9	475	451	24	428.2
94.9	474	450	24	427.2
94.9	473	449	24	426.2
94.9	472	448	24	425.2
94.9	471	447	24	424.2
94.9	470	446	24	423.2
94.9	469	445	24	422.2
94.9	468	444	24	421.2
94.9	467	443	24	420.2
94.9	466	442	24	419.2
94.9	455	432	23	410.2
94.9	454	431	23	409.2
94.9	453	430	23	408.2
94.9	452	429	23	407.2
94.9	451	428	23	406.2

%E	M1	M2	DM	M*
94.9	450	427	23	405.2
94.9	449	426	23	404.2
94.9	448	425	23	403.2
94.9	447	424	23	402.2
94.9	435	413	22	392.1
94.9	434	412	22	391.1
94.9	433	411	22	390.1
94.9	432	410	22	389.1
94.9	431	409	22	388.1
94.9	430	408	22	387.1
94.9	429	407	22	386.1
94.9	428	406	22	385.1
94.9	415	394	21	374.1
94.9	414	393	21	373.1
94.9	413	392	21	372.1
94.9	412	391	21	371.1
94.9	411	390	21	370.1
94.9	410	389	21	369.1
94.9	409	388	21	368.1
94.9	408	387	21	367.1
94.9	396	376	20	357.0
94.9	395	375	20	356.0
94.9	394	374	20	355.0
94.9	393	373	20	354.0
94.9	392	372	20	353.0
94.9	391	371	20	352.0
94.9	390	370	20	351.0
94.9	389	369	20	350.0
94.9	376	357	19	339.0
94.9	375	356	19	338.0
94.9	374	355	19	337.0
94.9	373	354	19	336.0
94.9	372	353	19	335.0
94.9	371	352	19	334.0
94.9	370	351	19	333.0
94.9	369	350	19	332.0
94.9	356	338	18	320.9
94.9	355	337	18	319.9
94.9	354	336	18	318.9
94.9	353	335	18	317.9
94.9	352	334	18	316.9
94.9	351	333	18	315.9
94.9	350	332	18	314.9
94.9	336	319	17	302.9
94.9	335	318	17	301.9
94.9	334	317	17	300.9
94.9	333	316	17	299.9
94.9	332	315	17	298.9
94.9	331	314	17	297.9
94.9	316	300	16	284.8

%E	M1	M2	DM	M*
94.9	315	299	16	283.8
94.9	314	298	16	282.8
94.9	313	297	16	281.8
94.9	312	296	16	280.8
94.9	311	295	16	279.8
94.9	297	282	15	267.8
94.9	296	281	15	266.8
94.9	295	280	15	265.8
94.9	294	279	15	264.8
94.9	293	278	15	263.8
94.9	292	277	15	262.8
94.9	277	263	14	249.7
94.9	276	262	14	248.7
94.9	275	261	14	247.7
94.9	274	260	14	246.7
94.9	273	259	14	245.7
94.9	272	258	14	244.7
94.9	257	244	13	231.7
94.9	256	243	13	230.7
94.9	255	242	13	229.7
94.9	254	241	13	228.7
94.9	253	240	13	227.7
94.9	237	225	12	213.6
94.9	236	224	12	212.6
94.9	235	223	12	211.6
94.9	234	222	12	210.6
94.9	233	221	12	209.6
94.9	217	206	11	195.6
94.9	216	205	11	194.6
94.9	215	204	11	193.6
94.9	214	203	11	192.6
94.9	198	188	10	178.5
94.9	197	187	10	177.5
94.9	196	186	10	176.5
94.9	195	185	10	175.5
94.9	178	169	9	160.5
94.9	177	168	9	159.5
94.9	176	167	9	158.5
94.9	175	166	9	157.5
94.9	158	150	8	142.4
94.9	157	149	8	141.4
94.9	156	148	8	140.4
94.9	138	131	7	124.4
94.9	137	130	7	123.4
94.9	136	129	7	122.4
94.9	118	112	6	106.3
94.9	117	111	6	105.3
94.9	99	94	5	89.3
94.9	98	93	5	88.3
94.9	79	75	4	71.2

%E	M1	M2	DM	M*
94.9	78	74	4	70.2
94.9	59	56	3	53.2
94.9	39	37	2	35.1
94.8	500	474	26	449.4
94.8	499	473	26	448.4
94.8	498	472	26	447.4
94.8	497	471	26	446.4
94.8	496	470	26	445.4
94.8	485	460	25	436.3
94.8	484	459	25	435.3
94.8	483	458	25	434.3
94.8	482	457	25	433.3
94.8	481	456	25	432.3
94.8	480	455	25	431.3
94.8	479	454	25	430.3
94.8	478	453	25	429.3
94.8	477	452	25	428.3
94.8	465	441	24	418.2
94.8	464	440	24	417.2
94.8	463	439	24	416.2
94.8	462	438	24	415.2
94.8	461	437	24	414.2
94.8	460	436	24	413.3
94.8	459	435	24	412.3
94.8	458	434	24	411.3
94.8	446	423	23	401.2
94.8	445	422	23	400.2
94.8	444	421	23	399.2
94.8	443	420	23	398.2
94.8	442	419	23	397.2
94.8	441	418	23	396.2
94.8	440	417	23	395.2
94.8	439	416	23	394.2
94.8	427	405	22	384.1
94.8	426	404	22	383.1
94.8	425	403	22	382.1
94.8	424	402	22	381.1
94.8	423	401	22	380.1
94.8	422	400	22	379.1
94.8	421	399	22	378.1
94.8	420	398	22	377.2
94.8	407	386	21	366.1
94.8	406	385	21	365.1
94.8	405	384	21	364.1
94.8	404	383	21	363.1
94.8	403	382	21	362.1
94.8	402	381	21	361.1
94.8	401	380	21	360.1
94.8	400	379	21	359.1
94.8	388	368	20	349.0

%E	M1	M2	DM	M*
94.8	387	367	20	348.0
94.8	386	366	20	347.0
94.8	385	365	20	346.0
94.8	384	364	20	345.0
94.8	383	363	20	344.0
94.8	382	362	20	343.0
94.8	381	361	20	342.0
94.8	368	349	19	331.0
94.8	367	348	19	330.0
94.8	366	347	19	329.0
94.8	365	346	19	328.0
94.8	364	345	19	327.0
94.8	363	344	19	326.0
94.8	362	343	19	325.0
94.8	349	331	18	313.9
94.8	348	330	18	312.9
94.8	347	329	18	311.9
94.8	346	328	18	310.9
94.8	345	327	18	309.9
94.8	344	326	18	308.9
94.8	343	325	18	307.9
94.8	330	313	17	296.9
94.8	329	312	17	295.9
94.8	328	311	17	294.9
94.8	327	310	17	293.9
94.8	326	309	17	292.9
94.8	325	308	17	291.9
94.8	324	307	17	290.9
94.8	310	294	16	278.8
94.8	309	293	16	277.8
94.8	308	292	16	276.8
94.8	307	291	16	275.8
94.8	306	290	16	274.8
94.8	305	289	16	273.8
94.8	291	276	15	261.8
94.8	290	275	15	260.8
94.8	289	274	15	259.8
94.8	288	273	15	258.8
94.8	287	272	15	257.8
94.8	286	271	15	256.8
94.8	271	257	14	243.7
94.8	270	256	14	242.7
94.8	269	255	14	241.7
94.8	268	254	14	240.7
94.8	267	253	14	239.7
94.8	252	239	13	226.7
94.8	251	238	13	225.7
94.8	250	237	13	224.7
94.8	249	236	13	223.7
94.8	248	235	13	222.7

%E	M1	M2	DM	M*
94.8	232	220	12	208.6
94.8	231	219	12	207.6
94.8	230	218	12	206.6
94.8	229	217	12	205.6
94.8	213	202	11	191.6
94.8	212	201	11	190.6
94.8	211	200	11	189.6
94.8	210	199	11	188.6
94.8	194	184	10	174.5
94.8	193	183	10	173.5
94.8	192	182	10	172.5
94.8	191	181	10	171.5
94.8	174	165	9	156.5
94.8	173	164	9	155.5
94.8	172	163	9	154.5
94.8	155	147	8	139.4
94.8	154	146	8	138.4
94.8	153	145	8	137.4
94.8	135	128	7	121.4
94.8	134	127	7	120.4
94.8	116	110	6	104.3
94.8	115	109	6	103.3
94.8	97	92	5	87.3
94.8	96	91	5	86.3
94.8	77	73	4	69.2
94.8	58	55	3	52.2
94.7	495	469	26	444.4
94.7	494	468	26	443.4
94.7	493	467	26	442.4
94.7	492	466	26	441.4
94.7	491	465	26	440.4
94.7	490	464	26	439.4
94.7	489	463	26	438.4
94.7	488	462	26	437.4
94.7	487	461	26	436.4
94.7	486	460	26	435.4
94.7	476	451	25	427.3
94.7	475	450	25	426.3
94.7	474	449	25	425.3
94.7	473	448	25	424.3
94.7	472	447	25	423.3
94.7	471	446	25	422.3
94.7	470	445	25	421.3
94.7	469	444	25	420.3
94.7	468	443	25	419.3
94.7	457	433	24	410.3
94.7	456	432	24	409.3
94.7	455	431	24	408.3
94.7	454	430	24	407.3
94.7	453	429	24	406.3

%E	M1	M2	DM	M*
94.7	452	428	24	405.3
94.7	451	427	24	404.3
94.7	450	426	24	403.3
94.7	449	425	24	402.3
94.7	438	415	23	393.2
94.7	437	414	23	392.2
94.7	436	413	23	391.2
94.7	435	412	23	390.2
94.7	434	411	23	389.2
94.7	433	410	23	388.2
94.7	432	409	23	387.2
94.7	431	408	23	386.2
94.7	430	407	23	385.2
94.7	419	397	22	376.2
94.7	418	396	22	375.2
94.7	417	395	22	374.2
94.7	416	394	22	373.2
94.7	415	393	22	372.2
94.7	414	392	22	371.2
94.7	413	391	22	370.2
94.7	412	390	22	369.2
94.7	399	378	21	358.1
94.7	398	377	21	357.1
94.7	397	376	21	356.1
94.7	396	375	21	355.1
94.7	395	374	21	354.1
94.7	394	373	21	353.1
94.7	393	372	21	352.1
94.7	380	360	20	341.1
94.7	379	359	20	340.1
94.7	378	358	20	339.1
94.7	377	357	20	338.1
94.7	376	356	20	337.1
94.7	375	355	20	336.1
94.7	374	354	20	335.1
94.7	361	342	19	324.0
94.7	360	341	19	323.0
94.7	359	340	19	322.0
94.7	358	339	19	321.0
94.7	357	338	19	320.0
94.7	356	337	19	319.0
94.7	342	324	18	306.9
94.7	341	323	18	305.9
94.7	340	322	18	305.0
94.7	339	321	18	304.0
94.7	338	320	18	303.0
94.7	337	319	18	302.0
94.7	323	306	17	289.9
94.7	322	305	17	288.9
94.7	321	304	17	287.9

%E	M1	M2	DM	M*
94.7	320	303	17	286.9
94.7	319	302	17	285.9
94.7	318	301	17	284.9
94.7	304	288	16	272.8
94.7	303	287	16	271.8
94.7	302	286	16	270.8
94.7	301	285	16	269.9
94.7	300	284	16	268.9
94.7	285	270	15	255.8
94.7	284	269	15	254.8
94.7	283	268	15	253.8
94.7	282	267	15	252.8
94.7	281	266	15	251.8
94.7	266	252	14	238.7
94.7	265	251	14	237.7
94.7	264	250	14	236.7
94.7	263	249	14	235.7
94.7	262	248	14	234.7
94.7	247	234	13	221.7
94.7	246	233	13	220.7
94.7	245	232	13	219.7
94.7	244	231	13	218.7
94.7	243	230	13	217.7
94.7	228	216	12	204.6
94.7	227	215	12	203.6
94.7	226	214	12	202.6
94.7	225	213	12	201.6
94.7	209	198	11	187.6
94.7	208	197	11	186.6
94.7	207	196	11	185.6
94.7	206	195	11	184.6
94.7	190	180	10	170.5
94.7	189	179	10	169.5
94.7	188	178	10	168.5
94.7	187	177	10	167.5
94.7	171	162	9	153.5
94.7	170	161	9	152.5
94.7	169	160	9	151.5
94.7	152	144	8	136.4
94.7	151	143	8	135.4
94.7	150	142	8	134.4
94.7	133	126	7	119.4
94.7	132	125	7	118.4
94.7	131	124	7	117.4
94.7	114	108	6	102.3
94.7	113	107	6	101.3
94.7	95	90	5	85.3
94.7	94	89	5	84.3
94.7	76	72	4	68.2
94.7	75	71	4	67.2

%E	M1	M2	DM	M*
94.7	57	54	3	51.2
94.7	38	36	2	34.1
94.7	19	18	1	17.1
94.6	500	473	27	447.5
94.6	499	472	27	446.5
94.6	498	471	27	445.5
94.6	497	470	27	444.5
94.6	496	469	27	443.5
94.6	485	459	26	434.4
94.6	484	458	26	433.4
94.6	483	457	26	432.4
94.6	482	456	26	431.4
94.6	481	455	26	430.4
94.6	480	454	26	429.4
94.6	479	453	26	428.4
94.6	478	452	26	427.4
94.6	467	442	25	418.3
94.6	466	441	25	417.3
94.6	465	440	25	416.3
94.6	464	439	25	415.3
94.6	463	438	25	414.3
94.6	462	437	25	413.4
94.6	461	436	25	412.4
94.6	460	435	25	411.4
94.6	459	434	25	410.4
94.6	448	424	24	401.3
94.6	447	423	24	400.3
94.6	446	422	24	399.3
94.6	445	421	24	398.3
94.6	444	420	24	397.3
94.6	443	419	24	396.3
94.6	442	418	24	395.3
94.6	441	417	24	394.3
94.6	429	406	23	384.2
94.6	428	405	23	383.2
94.6	427	404	23	382.2
94.6	426	403	23	381.2
94.6	425	402	23	380.2
94.6	424	401	23	379.2
94.6	423	400	23	378.3
94.6	422	399	23	377.3
94.6	411	389	22	368.2
94.6	410	388	22	367.2
94.6	409	387	22	366.2
94.6	408	386	22	365.2
94.6	407	385	22	364.2
94.6	406	384	22	363.2
94.6	405	383	22	362.2
94.6	404	382	22	361.2
94.6	392	371	21	351.1

%E	M1	M2	DM	M*
94.6	391	370	21	350.1
94.6	390	369	21	349.1
94.6	389	368	21	348.1
94.6	388	367	21	347.1
94.6	387	366	21	346.1
94.6	386	365	21	345.1
94.6	373	353	20	334.1
94.6	372	352	20	333.1
94.6	371	351	20	332.1
94.6	370	350	20	331.1
94.6	369	349	20	330.1
94.6	368	348	20	329.1
94.6	367	347	20	328.1
94.6	355	336	19	318.0
94.6	354	335	19	317.0
94.6	353	334	19	316.0
94.6	352	333	19	315.0
94.6	351	332	19	314.0
94.6	350	331	19	313.0
94.6	349	330	19	312.0
94.6	336	318	18	301.0
94.6	335	317	18	300.0
94.6	334	316	18	299.0
94.6	333	315	18	298.0
94.6	332	314	18	297.0
94.6	331	313	18	296.0
94.6	317	300	17	283.9
94.6	316	299	17	282.9
94.6	315	298	17	281.9
94.6	314	297	17	230.9
94.6	313	296	17	279.9
94.6	312	295	17	278.9
94.6	299	283	16	267.9
94.6	298	282	16	266.9
94.6	297	281	16	265.9
94.6	296	280	16	264.9
94.6	295	279	16	263.9
94.6	294	278	16	262.9
94.6	280	265	15	250.8
94.6	279	264	15	249.8
94.6	278	263	15	248.8
94.6	277	262	15	247.8
94.6	276	261	15	246.8
94.6	261	247	14	233.8
94.6	260	246	14	232.8
94.6	259	245	14	231.8
94.6	258	244	14	230.8
94.6	257	243	14	229.8
94.6	242	229	13	216.7
94.6	241	228	13	215.7
94.6	240	227	13	214.7
94.6	239	226	13	213.7
94.6	224	212	12	200.6
94.6	223	211	12	199.6
94.6	222	210	12	198.6
94.6	221	209	12	197.7
94.6	205	194	11	183.6
94.6	204	193	11	182.6
94.6	203	192	11	181.6
94.6	202	191	11	180.6
94.6	186	176	10	166.5
94.6	185	175	10	165.5
94.6	184	174	10	164.5
94.6	168	159	9	150.5
94.6	167	158	9	149.5
94.6	166	157	9	148.5
94.6	149	141	8	133.4
94.6	148	140	8	132.4
94.6	147	139	8	131.4
94.6	130	123	7	116.4
94.6	129	122	7	115.4
94.6	112	106	6	100.3
94.6	111	105	6	99.3
94.6	93	88	5	83.3
94.6	92	87	5	82.3
94.6	74	70	4	66.2
94.6	56	53	3	50.2
94.6	37	35	2	33.1
94.5	495	468	27	442.5
94.5	494	467	27	441.5
94.5	493	466	27	440.5
94.5	492	465	27	439.5
94.5	491	464	27	438.5
94.5	490	463	27	437.5
94.5	489	462	27	436.5
94.5	488	461	27	435.5
94.5	487	460	27	434.5
94.5	477	451	26	426.4
94.5	476	450	26	425.4
94.5	475	449	26	424.4
94.5	474	448	26	423.4
94.5	473	447	26	422.4
94.5	472	446	26	421.4
94.5	471	445	26	420.4
94.5	470	444	26	419.4
94.5	469	443	26	418.4
94.5	458	433	25	409.4
94.5	457	432	25	408.4
94.5	456	431	25	407.4
94.5	455	430	25	406.4
94.5	454	429	25	405.4
94.5	453	428	25	404.4
94.5	452	427	25	403.4
94.5	451	426	25	402.4
94.5	440	416	24	393.3
94.5	439	415	24	392.3
94.5	438	414	24	391.3
94.5	437	413	24	390.3
94.5	436	412	24	389.3
94.5	435	411	24	388.3
94.5	434	410	24	387.3
94.5	433	409	24	386.3
94.5	421	398	23	376.3
94.5	420	397	23	375.3
94.5	419	396	23	374.3
94.5	418	395	23	373.3
94.5	417	394	23	372.3
94.5	416	393	23	371.3
94.5	415	392	23	370.3
94.5	403	381	22	360.2
94.5	402	380	22	359.2
94.5	401	379	22	358.2
94.5	400	378	22	357.2
94.5	399	377	22	356.2
94.5	398	376	22	355.2
94.5	397	375	22	354.2
94.5	385	364	21	344.1
94.5	384	363	21	343.1
94.5	383	362	21	342.2
94.5	382	361	21	341.2
94.5	381	360	21	340.2
94.5	380	359	21	339.2
94.5	379	358	21	338.2
94.5	366	346	20	327.1
94.5	365	345	20	326.1
94.5	364	344	20	325.1
94.5	363	343	20	324.1
94.5	362	342	20	323.1
94.5	361	341	20	322.1
94.5	348	329	19	311.0
94.5	347	328	19	310.0
94.5	346	327	19	309.0
94.5	345	326	19	308.0
94.5	344	325	19	307.0
94.5	343	324	19	306.1
94.5	330	312	18	295.0
94.5	329	311	18	294.0
94.5	328	310	18	293.0
94.5	327	309	18	292.0
94.5	326	308	18	291.0
94.5	325	307	18	290.0
94.5	311	294	17	277.9
94.5	310	293	17	276.9
94.5	309	292	17	275.9
94.5	308	291	17	274.9
94.5	307	290	17	273.9
94.5	293	277	16	261.9
94.5	292	276	16	260.9
94.5	291	275	16	259.9
94.5	290	274	16	258.9
94.5	289	273	16	257.9
94.5	275	260	15	245.8
94.5	274	259	15	244.8
94.5	273	258	15	243.8
94.5	272	257	15	242.8
94.5	271	256	15	241.8
94.5	256	242	14	228.8
94.5	255	241	14	227.8
94.5	254	240	14	226.8
94.5	253	239	14	225.8
94.5	238	225	13	212.7
94.5	237	224	13	211.7
94.5	236	223	13	210.7
94.5	235	222	13	209.7
94.5	220	208	12	196.7
94.5	219	207	12	195.7
94.5	218	206	12	194.7
94.5	217	205	12	193.7
94.5	201	190	11	179.6
94.5	200	189	11	178.6
94.5	199	188	11	177.6
94.5	183	173	10	163.5
94.5	182	172	10	162.5
94.5	181	171	10	161.6
94.5	165	156	9	147.5
94.5	164	155	9	146.5
94.5	163	154	9	145.5
94.5	146	138	8	130.4
94.5	145	137	8	129.4
94.5	128	121	7	114.4
94.5	127	120	7	113.4
94.5	110	104	6	98.3
94.5	109	103	6	97.3
94.5	91	86	5	81.3
94.5	73	69	4	65.2
94.5	55	52	3	49.2
94.4	500	472	28	445.6
94.4	499	471	28	444.6
94.4	498	470	28	443.6
94.4	497	469	28	442.6
94.4	496	468	28	441.6
94.4	486	459	27	433.5
94.4	485	458	27	432.5
94.4	484	457	27	431.5
94.4	483	456	27	430.5
94.4	482	455	27	429.5
94.4	481	454	27	428.5
94.4	480	453	27	427.5
94.4	479	452	27	426.5
94.4	478	451	27	425.5
94.4	468	442	26	417.4
94.4	467	441	26	416.4
94.4	466	440	26	415.5
94.4	465	439	26	414.5
94.4	464	438	26	413.5
94.4	463	437	26	412.5
94.4	462	436	26	411.5
94.4	461	435	26	410.5
94.4	450	425	25	401.4
94.4	449	424	25	400.4
94.4	448	423	25	399.4
94.4	447	422	25	398.4
94.4	446	421	25	397.4
94.4	445	420	25	396.4
94.4	444	419	25	395.4
94.4	443	418	25	394.4
94.4	432	408	24	385.3
94.4	431	407	24	384.3
94.4	430	406	24	383.3
94.4	429	405	24	382.3
94.4	428	404	24	381.3
94.4	427	403	24	380.3
94.4	426	402	24	379.4
94.4	425	401	24	378.4
94.4	414	391	23	369.3
94.4	413	390	23	368.3
94.4	412	389	23	367.3
94.4	411	388	23	366.3
94.4	410	387	23	365.3
94.4	409	386	23	364.3
94.4	408	385	23	363.3
94.4	396	374	22	353.2
94.4	395	373	22	352.2
94.4	394	372	22	351.2
94.4	393	371	22	350.2
94.4	392	370	22	349.2
94.4	391	369	22	348.2
94.4	390	368	22	347.2
94.4	378	357	21	337.2
94.4	377	356	21	336.2

%E	M1	M2	DM	M*
94.4	376	355	21	335.2
94.4	375	354	21	334.2
94.4	374	353	21	333.2
94.4	373	352	21	332.2
94.4	372	351	21	331.2
94.4	360	340	20	321.1
94.4	359	339	20	320.1
94.4	358	338	20	319.1
94.4	357	337	20	318.1
94.4	356	336	20	317.1
94.4	355	335	20	316.1
94.4	354	334	20	315.1
94.4	342	323	19	305.1
94.4	341	322	19	304.1
94.4	340	321	19	303.1
94.4	339	320	19	302.1
94.4	338	319	19	301.1
94.4	337	318	19	300.1
94.4	324	306	18	289.0
94.4	323	305	18	288.0
94.4	322	304	18	287.0
94.4	321	303	18	286.0
94.4	320	302	18	285.0
94.4	319	301	18	284.0
94.4	306	289	17	272.9
94.4	305	288	17	271.9
94.4	304	287	17	271.0
94.4	303	286	17	270.0
94.4	302	285	17	269.0
94.4	301	284	17	268.0
94.4	288	272	16	256.9
94.4	287	271	16	255.9
94.4	286	270	16	254.9
94.4	285	269	16	253.9
94.4	284	268	16	252.9
94.4	270	255	15	240.8
94.4	269	254	15	239.8
94.4	268	253	15	238.8
94.4	267	252	15	237.8
94.4	266	251	15	236.8
94.4	252	238	14	224.8
94.4	251	237	14	223.8
94.4	250	236	14	222.8
94.4	249	235	14	221.8
94.4	248	234	14	220.8
94.4	234	221	13	208.7
94.4	233	220	13	207.7
94.4	232	219	13	206.7
94.4	231	218	13	205.7
94.4	216	204	12	192.7

%E	M1	M2	DM	M*
94.4	215	203	12	191.7
94.4	214	202	12	190.7
94.4	213	201	12	189.7
94.4	198	187	11	176.6
94.4	197	186	11	175.6
94.4	196	185	11	174.6
94.4	195	184	11	173.6
94.4	180	170	10	160.6
94.4	179	169	10	159.6
94.4	178	168	10	158.6
94.4	177	167	10	157.6
94.4	162	153	9	144.5
94.4	161	152	9	143.5
94.4	160	151	9	142.5
94.4	144	136	8	128.4
94.4	143	135	8	127.4
94.4	142	134	8	126.5
94.4	126	119	7	112.4
94.4	125	118	7	111.4
94.4	124	117	7	110.4
94.4	108	102	6	96.3
94.4	107	101	6	95.3
94.4	90	85	5	80.3
94.4	89	84	5	79.3
94.4	72	68	4	64.2
94.4	71	67	4	63.2
94.4	54	51	3	48.2
94.4	36	34	2	32.1
94.4	18	17	1	16.1
94.3	495	467	28	440.6
94.3	494	466	28	439.6
94.3	493	465	28	438.6
94.3	492	464	28	437.6
94.3	491	463	28	436.6
94.3	490	462	28	435.6
94.3	489	461	28	434.6
94.3	488	460	28	433.6
94.3	487	459	28	432.6
94.3	477	450	27	424.5
94.3	476	449	27	423.5
94.3	475	448	27	422.5
94.3	474	447	27	421.5
94.3	473	446	27	420.5
94.3	472	445	27	419.5
94.3	471	444	27	418.5
94.3	470	443	27	417.6
94.3	460	434	26	409.5
94.3	459	433	26	408.5
94.3	458	432	26	407.5
94.3	457	431	26	406.5

%E	M1	M2	DM	M*
94.3	456	430	26	405.5
94.3	455	429	26	404.5
94.3	454	428	26	403.5
94.3	453	427	26	402.5
94.3	442	417	25	393.4
94.3	441	416	25	392.4
94.3	440	415	25	391.4
94.3	439	414	25	390.4
94.3	438	413	25	389.4
94.3	437	412	25	388.4
94.3	436	411	25	387.4
94.3	435	410	25	386.4
94.3	424	400	24	377.4
94.3	423	399	24	376.4
94.3	422	398	24	375.4
94.3	421	397	24	374.4
94.3	420	396	24	373.4
94.3	419	395	24	372.4
94.3	418	394	24	371.4
94.3	407	384	23	362.3
94.3	406	383	23	361.3
94.3	405	382	23	360.3
94.3	404	381	23	359.3
94.3	403	380	23	358.3
94.3	402	379	23	357.3
94.3	401	378	23	356.3
94.3	400	377	23	355.3
94.3	389	367	22	346.2
94.3	388	366	22	345.2
94.3	387	365	22	344.3
94.3	386	364	22	343.3
94.3	385	363	22	342.3
94.3	384	362	22	341.3
94.3	383	361	22	340.3
94.3	371	350	21	330.2
94.3	370	349	21	329.2
94.3	369	348	21	328.2
94.3	368	347	21	327.2
94.3	367	346	21	326.2
94.3	366	345	21	325.2
94.3	353	333	20	314.1
94.3	352	332	20	313.1
94.3	351	331	20	312.1
94.3	350	330	20	311.1
94.3	349	329	20	310.1
94.3	348	328	20	309.1
94.3	336	317	19	299.1
94.3	335	316	19	298.1
94.3	334	315	19	297.1
94.3	333	314	19	296.1

%E	M1	M2	DM	M*
94.3	332	313	19	295.1
94.3	331	312	19	294.1
94.3	318	300	18	283.0
94.3	317	299	18	282.0
94.3	316	298	18	281.0
94.3	315	297	18	280.0
94.3	314	296	18	279.0
94.3	300	283	17	267.0
94.3	299	282	17	266.0
94.3	298	281	17	265.0
94.3	297	280	17	264.0
94.3	296	279	17	263.0
94.3	283	267	16	251.9
94.3	282	266	16	250.9
94.3	281	265	16	249.9
94.3	280	264	16	248.9
94.3	279	263	16	247.9
94.3	265	250	15	235.8
94.3	264	249	15	234.9
94.3	263	248	15	233.9
94.3	262	247	15	232.9
94.3	261	246	15	231.9
94.3	247	233	14	219.8
94.3	246	232	14	218.8
94.3	245	231	14	217.8
94.3	244	230	14	216.8
94.3	230	217	13	204.7
94.3	229	216	13	203.7
94.3	228	215	13	202.7
94.3	227	214	13	201.7
94.3	212	200	12	188.7
94.3	211	199	12	187.7
94.3	210	198	12	186.7
94.3	209	197	12	185.7
94.3	194	183	11	172.6
94.3	193	182	11	171.6
94.3	192	181	11	170.6
94.3	176	166	10	156.6
94.3	175	165	10	155.6
94.3	174	164	10	154.6
94.3	159	150	9	141.5
94.3	158	149	9	140.5
94.3	157	148	9	139.5
94.3	141	133	8	125.5
94.3	140	132	8	124.5
94.3	123	116	7	109.4
94.3	122	115	7	108.4
94.3	106	100	6	94.3
94.3	105	99	6	93.3
94.3	88	83	5	78.3

%E	M1	M2	DM	M*
94.3	87	82	5	77.3
94.3	70	66	4	62.2
94.3	53	50	3	47.2
94.3	35	33	2	31.1
94.2	500	471	29	443.7
94.2	499	470	29	442.7
94.2	498	469	29	441.7
94.2	497	468	29	440.7
94.2	496	467	29	439.7
94.2	486	458	28	431.6
94.2	485	457	28	430.6
94.2	484	456	28	429.6
94.2	483	455	28	428.6
94.2	482	454	28	427.6
94.2	481	453	28	426.6
94.2	480	452	28	425.6
94.2	479	451	28	424.6
94.2	469	442	27	416.6
94.2	468	441	27	415.6
94.2	467	440	27	414.6
94.2	466	439	27	413.6
94.2	465	438	27	412.6
94.2	464	437	27	411.6
94.2	463	436	27	410.6
94.2	462	435	27	409.6
94.2	452	426	26	401.5
94.2	451	425	26	400.5
94.2	450	424	26	399.5
94.2	449	423	26	398.5
94.2	448	422	26	397.5
94.2	447	421	26	396.5
94.2	446	420	26	395.5
94.2	445	419	26	394.5
94.2	434	409	25	385.4
94.2	433	408	25	384.4
94.2	432	407	25	383.4
94.2	431	406	25	382.4
94.2	430	405	25	381.5
94.2	429	404	25	380.5
94.2	428	403	25	379.5
94.2	417	393	24	370.4
94.2	416	392	24	369.4
94.2	415	391	24	368.4
94.2	414	390	24	367.4
94.2	413	389	24	366.4
94.2	412	388	24	365.4
94.2	411	387	24	364.4
94.2	399	376	23	354.3
94.2	398	375	23	353.3
94.2	397	374	23	352.3

%E	M1	M2	DM	M*
94.2	396	373	23	351.3
94.2	395	372	23	350.3
94.2	394	371	23	349.3
94.2	382	360	22	339.3
94.2	381	359	22	338.3
94.2	380	358	22	337.3
94.2	379	357	22	336.3
94.2	378	356	22	335.3
94.2	377	355	22	334.3
94.2	365	344	21	324.2
94.2	364	343	21	323.2
94.2	363	342	21	322.2
94.2	362	341	21	321.2
94.2	361	340	21	320.2
94.2	360	339	21	319.2
94.2	359	338	21	318.2
94.2	347	327	20	308.2
94.2	346	326	20	307.2
94.2	345	325	20	306.2
94.2	344	324	20	305.2
94.2	343	323	20	304.2
94.2	342	322	20	303.2
94.2	330	311	19	293.1
94.2	329	310	19	292.1
94.2	328	309	19	291.1
94.2	327	308	19	290.1
94.2	326	307	19	289.1
94.2	325	306	19	288.1
94.2	313	295	18	278.0
94.2	312	294	18	277.0
94.2	311	293	18	276.0
94.2	310	292	18	275.0
94.2	309	291	18	274.0
94.2	308	290	18	273.1
94.2	295	278	17	262.0
94.2	294	277	17	261.0
94.2	293	276	17	260.0
94.2	292	275	17	259.0
94.2	291	274	17	258.0
94.2	278	262	16	246.9
94.2	277	261	16	245.9
94.2	276	260	16	244.9
94.2	275	259	16	243.9
94.2	274	258	16	242.9
94.2	260	245	15	230.9
94.2	259	244	15	229.9
94.2	258	243	15	228.9
94.2	257	242	15	227.9
94.2	243	229	14	215.8
94.2	242	228	14	214.8

%E	M1	M2	DM	M*
94.2	241	227	14	213.8
94.2	240	226	14	212.8
94.2	226	213	13	200.7
94.2	225	212	13	199.8
94.2	224	211	13	198.8
94.2	223	210	13	197.8
94.2	208	196	12	184.7
94.2	207	195	12	183.7
94.2	206	194	12	182.7
94.2	191	180	11	169.6
94.2	190	179	11	168.6
94.2	189	178	11	167.6
94.2	173	163	10	153.6
94.2	172	162	10	152.6
94.2	171	161	10	151.6
94.2	156	147	9	138.5
94.2	155	146	9	137.5
94.2	154	145	9	136.5
94.2	139	131	8	123.5
94.2	138	130	8	122.5
94.2	137	129	8	121.5
94.2	121	114	7	107.4
94.2	120	113	7	106.4
94.2	104	98	6	92.3
94.2	103	97	6	91.3
94.2	86	81	5	76.3
94.2	69	65	4	61.2
94.2	52	49	3	46.2
94.1	495	466	29	438.7
94.1	494	465	29	437.7
94.1	493	464	29	436.7
94.1	492	463	29	435.7
94.1	491	462	29	434.7
94.1	490	461	29	433.7
94.1	489	460	29	432.7
94.1	488	459	29	431.7
94.1	478	450	28	423.6
94.1	477	449	28	422.6
94.1	476	448	28	421.6
94.1	475	447	28	420.7
94.1	474	446	28	419.7
94.1	473	445	28	418.7
94.1	472	444	28	417.7
94.1	471	443	28	416.7
94.1	461	434	27	408.6
94.1	460	433	27	407.6
94.1	459	432	27	406.6
94.1	458	431	27	405.6
94.1	457	430	27	404.6
94.1	456	429	27	403.6

%E	M1	M2	DM	M*
94.1	455	428	27	402.6
94.1	454	427	27	401.6
94.1	444	418	26	393.5
94.1	443	417	26	392.5
94.1	442	416	26	391.5
94.1	441	415	26	390.5
94.1	440	414	26	389.5
94.1	439	413	26	388.5
94.1	438	412	26	387.5
94.1	437	411	26	386.5
94.1	427	402	25	378.5
94.1	426	401	25	377.5
94.1	425	400	25	376.5
94.1	424	399	25	375.5
94.1	423	398	25	374.5
94.1	422	397	25	373.5
94.1	421	396	25	372.5
94.1	410	386	24	363.4
94.1	409	385	24	362.4
94.1	408	384	24	361.4
94.1	407	383	24	360.4
94.1	406	382	24	359.4
94.1	405	381	24	358.4
94.1	404	380	24	357.4
94.1	393	370	23	348.3
94.1	392	369	23	347.3
94.1	391	368	23	346.4
94.1	390	367	23	345.4
94.1	389	366	23	344.4
94.1	388	365	23	343.4
94.1	387	364	23	342.4
94.1	376	354	22	333.3
94.1	375	353	22	332.3
94.1	374	352	22	331.3
94.1	373	351	22	330.3
94.1	372	350	22	329.3
94.1	371	349	22	328.3
94.1	370	348	22	327.3
94.1	358	337	21	317.2
94.1	357	336	21	316.2
94.1	356	335	21	315.2
94.1	355	334	21	314.2
94.1	354	333	21	313.2
94.1	353	332	21	312.2
94.1	341	321	20	302.2
94.1	340	320	20	301.2
94.1	339	319	20	300.2
94.1	338	318	20	299.2
94.1	337	317	20	298.2
94.1	324	305	19	287.1

%E	M1	M2	DM	M*
94.1	323	304	19	286.1
94.1	322	303	19	285.1
94.1	321	302	19	284.1
94.1	320	301	19	283.1
94.1	307	289	18	272.1
94.1	306	288	18	271.1
94.1	305	287	18	270.1
94.1	304	286	18	269.1
94.1	303	285	18	268.1
94.1	290	273	17	257.0
94.1	289	272	17	256.0
94.1	288	271	17	255.0
94.1	287	270	17	254.0
94.1	286	269	17	253.0
94.1	273	257	16	241.9
94.1	272	256	16	240.9
94.1	271	255	16	239.9
94.1	270	254	16	238.9
94.1	269	253	16	238.0
94.1	256	241	15	226.9
94.1	255	240	15	225.9
94.1	254	239	15	224.9
94.1	253	238	15	223.9
94.1	239	225	14	211.8
94.1	238	224	14	210.8
94.1	237	223	14	209.8
94.1	236	222	14	208.8
94.1	222	209	13	196.8
94.1	221	208	13	195.8
94.1	220	207	13	194.8
94.1	219	206	13	193.8
94.1	205	193	12	181.7
94.1	204	192	12	180.7
94.1	203	191	12	179.7
94.1	202	190	12	178.7
94.1	188	177	11	166.6
94.1	187	176	11	165.6
94.1	186	175	11	164.7
94.1	185	174	11	163.7
94.1	170	160	10	150.6
94.1	169	159	10	149.6
94.1	153	144	9	135.5
94.1	152	143	9	134.5
94.1	136	128	8	120.5
94.1	135	127	8	119.5
94.1	119	112	7	105.4
94.1	118	111	7	104.4
94.1	102	96	6	90.4
94.1	101	95	6	89.4
94.1	85	80	5	75.3

%E	M1	M2	DM	M*
94.1	68	64	4	60.2
94.1	51	48	3	45.2
94.1	34	32	2	30.1
94.1	17	16	1	15.1
94.0	500	470	30	441.8
94.0	499	469	30	440.8
94.0	498	468	30	439.8
94.0	497	467	30	438.8
94.0	496	466	30	437.8
94.0	487	458	29	430.7
94.0	486	457	29	429.7
94.0	485	456	29	428.7
94.0	484	455	29	427.7
94.0	483	454	29	426.7
94.0	482	453	29	425.7
94.0	481	452	29	424.7
94.0	480	451	29	423.8
94.0	470	442	28	415.7
94.0	469	441	28	414.7
94.0	468	440	28	413.7
94.0	467	439	28	412.7
94.0	466	438	28	411.7
94.0	465	437	28	410.7
94.0	464	436	28	409.7
94.0	463	435	28	408.7
94.0	453	426	27	400.6
94.0	452	425	27	399.6
94.0	451	424	27	398.6
94.0	450	423	27	397.6
94.0	449	422	27	396.6
94.0	448	421	27	395.6
94.0	447	420	27	394.6
94.0	436	410	26	385.6
94.0	435	409	26	384.6
94.0	434	408	26	383.6
94.0	433	407	26	382.6
94.0	432	406	26	381.6
94.0	431	405	26	380.6
94.0	430	404	26	379.6
94.0	420	395	25	371.5
94.0	419	394	25	370.5
94.0	418	393	25	369.5
94.0	417	392	25	368.5
94.0	416	391	25	367.5
94.0	415	390	25	366.5
94.0	414	389	25	365.5
94.0	403	379	24	356.4
94.0	402	378	24	355.4
94.0	401	377	24	354.4
94.0	400	376	24	353.4

%E	M1	M2	DM	M*
94.0	399	375	24	352.4
94.0	398	374	24	351.4
94.0	397	373	24	350.5
94.0	386	363	23	341.4
94.0	385	362	23	340.4
94.0	384	361	23	339.4
94.0	383	360	23	338.4
94.0	382	359	23	337.4
94.0	381	358	23	336.4
94.0	369	347	22	326.3
94.0	368	346	22	325.3
94.0	367	345	22	324.3
94.0	366	344	22	323.3
94.0	365	343	22	322.3
94.0	364	342	22	321.3
94.0	352	331	21	311.3
94.0	351	330	21	310.3
94.0	350	329	21	309.3
94.0	349	328	21	308.3
94.0	348	327	21	307.3
94.0	336	316	20	297.2
94.0	335	315	20	296.2
94.0	334	314	20	295.2
94.0	333	313	20	294.2
94.0	332	312	20	293.2
94.0	331	311	20	292.2
94.0	319	300	19	282.1
94.0	318	299	19	281.1
94.0	317	298	19	280.1
94.0	316	297	19	279.1
94.0	315	296	19	278.1
94.0	302	284	18	267.1
94.0	301	283	18	266.1
94.0	300	282	18	265.1
94.0	299	281	18	264.1
94.0	298	280	18	263.1
94.0	285	268	17	252.0
94.0	284	267	17	251.0
94.0	283	266	17	250.0
94.0	282	265	17	249.0
94.0	281	264	17	248.0
94.0	268	252	16	237.0
94.0	267	251	16	236.0
94.0	266	250	16	235.0
94.0	265	249	16	234.0
94.0	252	237	15	222.9
94.0	251	236	15	221.9
94.0	250	235	15	220.9
94.0	249	234	15	219.9
94.0	248	233	15	218.9

%E	M1	M2	DM	M*
94.0	235	221	14	207.8
94.0	234	220	14	206.8
94.0	233	219	14	205.8
94.0	232	218	14	204.8
94.0	218	205	13	192.8
94.0	217	204	13	191.8
94.0	216	203	13	190.8
94.0	215	202	13	189.8
94.0	201	189	12	177.7
94.0	200	188	12	176.7
94.0	199	187	12	175.7
94.0	184	173	11	162.7
94.0	183	172	11	161.7
94.0	182	171	11	160.7
94.0	168	158	10	148.6
94.0	167	157	10	147.6
94.0	166	156	10	146.6
94.0	151	142	9	133.5
94.0	150	141	9	132.5
94.0	149	140	9	131.5
94.0	134	126	8	118.5
94.0	133	125	8	117.5
94.0	117	110	7	103.4
94.0	116	109	7	102.4
94.0	100	94	6	88.4
94.0	84	79	5	74.3
94.0	83	78	5	73.3
94.0	67	63	4	59.2
94.0	50	47	3	44.2
93.9	495	465	30	436.8
93.9	494	464	30	435.8
93.9	493	463	30	434.8
93.9	492	462	30	433.8
93.9	491	461	30	432.8
93.9	490	460	30	431.8
93.9	489	459	30	430.8
93.9	488	458	30	429.8
93.9	479	450	29	422.8
93.9	478	449	29	421.8
93.9	477	448	29	420.8
93.9	476	447	29	419.8
93.9	475	446	29	418.8
93.9	474	445	29	417.8
93.9	473	444	29	416.8
93.9	472	443	29	415.8
93.9	462	434	28	407.7
93.9	461	433	28	406.7
93.9	460	432	28	405.7
93.9	459	431	28	404.7
93.9	458	430	28	403.7

%E	M1	M2	DM	M*
93.9	457	429	28	402.7
93.9	456	428	28	401.7
93.9	446	419	27	393.6
93.9	445	418	27	392.6
93.9	444	417	27	391.6
93.9	443	416	27	390.6
93.9	442	415	27	389.6
93.9	441	414	27	388.7
93.9	440	413	27	387.7
93.9	439	412	27	386.7
93.9	429	403	26	378.6
93.9	428	402	26	377.6
93.9	427	401	26	376.6
93.9	426	400	26	375.6
93.9	425	399	26	374.6
93.9	424	398	26	373.6
93.9	423	397	26	372.6
93.9	413	388	25	364.5
93.9	412	387	25	363.5
93.9	411	386	25	362.5
93.9	410	385	25	361.5
93.9	409	384	25	360.5
93.9	408	383	25	359.5
93.9	407	382	25	358.5
93.9	396	372	24	349.5
93.9	395	371	24	348.5
93.9	394	370	24	347.5
93.9	393	369	24	346.5
93.9	392	368	24	345.5
93.9	391	367	24	344.5
93.9	380	357	23	335.4
93.9	379	356	23	334.4
93.9	378	355	23	333.4
93.9	377	354	23	332.4
93.9	376	353	23	331.4
93.9	375	352	23	330.4
93.9	374	351	23	329.4
93.9	363	341	22	320.3
93.9	362	340	22	319.3
93.9	361	339	22	318.3
93.9	360	338	22	317.3
93.9	359	337	22	316.3
93.9	358	336	22	315.4
93.9	347	326	21	306.3
93.9	346	325	21	305.3
93.9	345	324	21	304.3
93.9	344	323	21	303.3
93.9	343	322	21	302.3
93.9	342	321	21	301.3
93.9	330	310	20	291.2

%E	M1	M2	DM	M*
93.9	329	309	20	290.2
93.9	328	308	20	289.2
93.9	327	307	20	288.2
93.9	326	306	20	287.2
93.9	314	295	19	277.1
93.9	313	294	19	276.2
93.9	312	293	19	275.2
93.9	311	292	19	274.2
93.9	310	291	19	273.2
93.9	309	290	19	272.2
93.9	297	279	18	262.1
93.9	296	278	18	261.1
93.9	295	277	18	260.1
93.9	294	276	18	259.1
93.9	293	275	18	258.1
93.9	280	263	17	247.0
93.9	279	262	17	246.0
93.9	278	261	17	245.0
93.9	277	260	17	244.0
93.9	264	248	16	233.0
93.9	263	247	16	232.0
93.9	262	246	16	231.0
93.9	261	245	16	230.0
93.9	247	232	15	217.9
93.9	246	231	15	216.9
93.9	245	230	15	215.9
93.9	244	229	15	214.9
93.9	231	217	14	203.8
93.9	230	216	14	202.9
93.9	229	215	14	201.9
93.9	228	214	14	200.9
93.9	214	201	13	188.8
93.9	213	200	13	187.8
93.9	212	199	13	186.8
93.9	198	186	12	174.7
93.9	197	185	12	173.7
93.9	196	184	12	172.7
93.9	181	170	11	159.7
93.9	180	169	11	158.7
93.9	179	168	11	157.7
93.9	165	155	10	145.6
93.9	164	154	10	144.6
93.9	163	153	10	143.6
93.9	148	139	9	130.5
93.9	147	138	9	129.6
93.9	132	124	8	116.5
93.9	131	123	8	115.5
93.9	115	108	7	101.4
93.9	114	107	7	100.4
93.9	99	93	6	87.4

%E	M1	M2	DM	M*
93.9	98	92	6	86.4
93.9	82	77	5	72.3
93.9	66	62	4	58.2
93.9	49	46	3	43.2
93.9	33	31	2	29.1
93.8	500	469	31	439.9
93.8	499	468	31	438.9
93.8	498	467	31	437.9
93.8	497	466	31	436.9
93.8	496	465	31	435.9
93.8	487	457	30	428.8
93.8	486	456	30	427.9
93.8	485	455	30	426.9
93.8	484	454	30	425.9
93.8	483	453	30	424.9
93.8	482	452	30	423.9
93.8	481	451	30	422.9
93.8	480	450	30	421.9
93.8	471	442	29	414.8
93.8	470	441	29	413.8
93.8	469	440	29	412.8
93.8	468	439	29	411.8
93.8	467	438	29	410.8
93.8	466	437	29	409.8
93.8	465	436	29	408.8
93.8	464	435	29	407.8
93.8	455	427	28	400.7
93.8	454	426	28	399.7
93.8	453	425	28	398.7
93.8	452	424	28	397.7
93.8	451	423	28	396.7
93.8	450	422	28	395.7
93.8	449	421	28	394.7
93.8	448	420	28	393.8
93.8	438	411	27	385.7
93.8	437	410	27	384.7
93.8	436	409	27	383.7
93.8	435	408	27	382.7
93.8	434	407	27	381.7
93.8	433	406	27	380.7
93.8	432	405	27	379.7
93.8	422	396	26	371.6
93.8	421	395	26	370.6
93.8	420	394	26	369.6
93.8	419	393	26	368.6
93.8	418	392	26	367.6
93.8	417	391	26	366.6
93.8	416	390	26	365.6
93.8	406	381	25	357.5
93.8	405	380	25	356.5

%E	M1	M2	DM	M*
93.8	404	379	25	355.5
93.8	403	378	25	354.6
93.8	402	377	25	353.6
93.8	401	376	25	352.6
93.8	400	375	25	351.6
93.8	390	366	24	343.5
93.8	389	365	24	342.5
93.8	388	364	24	341.5
93.8	387	363	24	340.5
93.8	386	362	24	339.5
93.8	385	361	24	338.5
93.8	384	360	24	337.5
93.8	373	350	23	328.4
93.8	372	349	23	327.4
93.8	371	348	23	326.4
93.8	370	347	23	325.4
93.8	369	346	23	324.4
93.8	368	345	23	323.4
93.8	357	335	22	314.4
93.8	356	334	22	313.4
93.8	355	333	22	312.4
93.8	354	332	22	311.4
93.8	353	331	22	310.4
93.8	352	330	22	309.4
93.8	341	320	21	300.3
93.8	340	319	21	299.3
93.8	339	318	21	298.3
93.8	338	317	21	297.3
93.8	337	316	21	296.3
93.8	336	315	21	295.3
93.8	325	305	20	286.2
93.8	324	304	20	285.2
93.8	323	303	20	284.2
93.8	322	302	20	283.2
93.8	321	301	20	282.2
93.8	320	300	20	281.3
93.8	308	289	19	271.2
93.8	307	288	19	270.2
93.8	306	287	19	269.2
93.8	305	286	19	268.2
93.8	304	285	19	267.2
93.8	292	274	18	257.1
93.8	291	273	18	256.1
93.8	290	272	18	255.1
93.8	289	271	18	254.1
93.8	288	270	18	253.1
93.8	276	259	17	243.0
93.8	275	258	17	242.1
93.8	274	257	17	241.1
93.8	273	256	17	240.1

%E	M1	M2	DM	M*
93.8	272	255	17	239.1
93.8	260	244	16	229.0
93.8	259	243	16	228.0
93.8	258	242	16	227.0
93.8	257	241	16	226.0
93.8	256	240	16	225.0
93.8	243	228	15	213.9
93.8	242	227	15	212.9
93.8	241	226	15	211.9
93.8	240	225	15	210.9
93.8	227	213	14	199.9
93.8	226	212	14	198.9
93.8	225	211	14	197.9
93.8	224	210	14	196.9
93.8	211	198	13	185.8
93.8	210	197	13	184.8
93.8	209	196	13	183.8
93.8	208	195	13	182.8
93.8	195	183	12	171.7
93.8	194	182	12	170.7
93.8	193	181	12	169.7
93.8	192	180	12	168.8
93.8	178	167	11	156.7
93.8	177	166	11	155.7
93.8	176	165	11	154.7
93.8	162	152	10	142.6
93.8	161	151	10	141.6
93.8	160	150	10	140.6
93.8	146	137	9	128.6
93.8	145	136	9	127.6
93.8	144	135	9	126.6
93.8	130	122	8	114.5
93.8	129	121	8	113.5
93.8	128	120	8	112.5
93.8	113	106	7	99.4
93.8	112	105	7	98.4
93.8	97	91	6	85.4
93.8	96	90	6	84.4
93.8	81	76	5	71.3
93.8	80	75	5	70.3
93.8	65	61	4	57.2
93.8	64	60	4	56.3
93.8	48	45	3	42.2
93.8	32	30	2	28.1
93.8	16	15	1	14.1
93.7	495	464	31	434.9
93.7	494	463	31	433.9
93.7	493	462	31	432.9
93.7	492	461	31	432.0
93.7	491	460	31	431.0

%E	M1	M2	DM	M*
93.7	490	459	31	430.0
93.7	489	458	31	429.0
93.7	479	449	30	420.9
93.7	478	448	30	419.9
93.7	477	447	30	418.9
93.7	476	446	30	417.9
93.7	475	445	30	416.9
93.7	474	444	30	415.9
93.7	473	443	30	414.9
93.7	463	434	29	406.8
93.7	462	433	29	405.8
93.7	461	432	29	404.8
93.7	460	431	29	403.8
93.7	459	430	29	402.8
93.7	458	429	29	401.8
93.7	457	428	29	400.8
93.7	447	419	28	392.8
93.7	446	418	28	391.8
93.7	445	417	28	390.8
93.7	444	416	28	389.8
93.7	443	415	28	388.8
93.7	442	414	28	387.8
93.7	441	413	28	386.8
93.7	431	404	27	378.7
93.7	430	403	27	377.7
93.7	429	402	27	376.7
93.7	428	401	27	375.7
93.7	427	400	27	374.7
93.7	426	399	27	373.7
93.7	415	389	26	364.6
93.7	414	388	26	363.6
93.7	413	387	26	362.6
93.7	412	386	26	361.6
93.7	411	385	26	360.6
93.7	410	384	26	359.6
93.7	399	374	25	350.6
93.7	398	373	25	349.6
93.7	397	372	25	348.6
93.7	396	371	25	347.6
93.7	395	370	25	346.6
93.7	394	369	25	345.6
93.7	383	359	24	336.5
93.7	382	358	24	335.5
93.7	381	357	24	334.5
93.7	380	356	24	333.5
93.7	379	355	24	332.5
93.7	378	354	24	331.5
93.7	367	344	23	322.4
93.7	366	343	23	321.4
93.7	365	342	23	320.4

%E	M1	M2	DM	M*
93.7	364	341	23	319.5
93.7	363	340	23	318.5
93.7	351	329	22	308.4
93.7	350	328	22	307.4
93.7	349	327	22	306.4
93.7	348	326	22	305.4
93.7	347	325	22	304.4
93.7	335	314	21	294.3
93.7	334	313	21	293.3
93.7	333	312	21	292.3
93.7	332	311	21	291.3
93.7	331	310	21	290.3
93.7	319	299	20	280.3
93.7	318	298	20	279.3
93.7	317	297	20	278.3
93.7	316	296	20	277.3
93.7	315	295	20	276.3
93.7	303	284	19	266.2
93.7	302	283	19	265.2
93.7	301	282	19	264.2
93.7	300	281	19	263.2
93.7	287	269	18	252.1
93.7	286	268	18	251.1
93.7	285	267	18	250.1
93.7	284	266	18	249.1
93.7	271	254	17	238.1
93.7	270	253	17	237.1
93.7	269	252	17	236.1
93.7	268	251	17	235.1
93.7	255	239	16	224.0
93.7	254	238	16	223.0
93.7	253	237	16	222.0
93.7	252	236	16	221.0
93.7	239	224	15	209.9
93.7	238	223	15	208.9
93.7	237	222	15	207.9
93.7	223	209	14	195.9
93.7	222	208	14	194.9
93.7	221	207	14	193.9
93.7	207	194	13	181.8
93.7	206	193	13	180.8
93.7	205	192	13	179.8
93.7	191	179	12	167.8
93.7	190	178	12	166.8
93.7	189	177	12	165.8
93.7	175	164	11	153.7
93.7	174	163	11	152.7
93.7	159	149	10	139.6
93.7	158	148	10	138.6
93.7	143	134	9	125.6

%E	M1	M2	DM	M*
93.7	142	133	9	124.6
93.7	127	119	8	111.5
93.7	126	118	8	110.5
93.7	111	104	7	97.4
93.7	95	89	6	83.4
93.7	79	74	5	69.3
93.7	63	59	4	55.3
93.6	500	468	32	438.0
93.6	499	467	32	437.1
93.6	498	466	32	436.1
93.6	497	465	32	435.1
93.6	488	457	31	428.0
93.6	487	456	31	427.0
93.6	486	455	31	426.0
93.6	485	454	31	425.0
93.6	484	453	31	424.0
93.6	483	452	31	423.0
93.6	482	451	31	422.0
93.6	481	450	31	421.0
93.6	472	442	30	413.9
93.6	471	441	30	412.9
93.6	470	440	30	411.9
93.6	469	439	30	410.9
93.6	468	438	30	409.9
93.6	467	437	30	408.9
93.6	466	436	30	407.9
93.6	456	427	29	399.8
93.6	455	426	29	398.8
93.6	454	425	29	397.9
93.6	453	424	29	396.9
93.6	452	423	29	395.9
93.6	451	422	29	394.9
93.6	450	421	29	393.9
93.6	440	412	28	385.8
93.6	439	411	28	384.8
93.6	438	410	28	383.8
93.6	437	409	28	382.8
93.6	436	408	28	381.8
93.6	435	407	28	380.8
93.6	425	398	27	372.7
93.6	424	397	27	371.7
93.6	423	396	27	370.7
93.6	422	395	27	369.7
93.6	421	394	27	368.7
93.6	420	393	27	367.7
93.6	419	392	27	366.7
93.6	409	383	26	358.7
93.6	408	382	26	357.7
93.6	407	381	26	356.7
93.6	406	380	26	355.7

%E	M1	M2	DM	M*
93.6	405	379	26	354.7
93.6	404	378	26	353.7
93.6	393	368	25	344.6
93.6	392	367	25	343.6
93.6	391	366	25	342.6
93.6	390	365	25	341.6
93.6	389	364	25	340.6
93.6	388	363	25	339.6
93.6	377	353	24	330.5
93.6	376	352	24	329.5
93.6	375	351	24	328.5
93.6	374	350	24	327.5
93.6	373	349	24	326.5
93.6	362	339	23	317.5
93.6	361	338	23	316.5
93.6	360	337	23	315.5
93.6	359	336	23	314.5
93.6	358	335	23	313.5
93.6	357	334	23	312.5
93.6	346	324	22	303.4
93.6	345	323	22	302.4
93.6	344	322	22	301.4
93.6	343	321	22	300.4
93.6	342	320	22	299.4
93.6	330	309	21	289.3
93.6	329	308	21	288.3
93.6	328	307	21	287.3
93.6	327	306	21	286.3
93.6	326	305	21	285.4
93.6	314	294	20	275.3
93.6	313	293	20	274.3
93.6	312	292	20	273.3
93.6	311	291	20	272.3
93.6	299	280	19	262.2
93.6	298	279	19	261.2
93.6	297	278	19	260.2
93.6	296	277	19	259.2
93.6	295	276	19	258.2
93.6	283	265	18	248.1
93.6	282	264	18	247.1
93.6	281	263	18	246.2
93.6	280	262	18	245.2
93.6	267	250	17	234.1
93.6	266	249	17	233.1
93.6	265	248	17	232.1
93.6	264	247	17	231.1
93.6	251	235	16	220.0
93.6	250	234	16	219.0
93.6	249	233	16	218.0
93.6	236	221	15	207.0

%E	M1	M2	DM	M*
93.6	235	220	15	206.0
93.6	234	219	15	205.0
93.6	233	218	15	204.0
93.6	220	206	14	192.9
93.6	219	205	14	191.9
93.6	218	204	14	190.9
93.6	204	191	13	178.8
93.6	203	190	13	177.8
93.6	202	189	13	176.8
93.6	188	176	12	164.8
93.6	187	175	12	163.8
93.6	173	162	11	151.7
93.6	172	161	11	150.7
93.6	171	160	11	149.7
93.6	157	147	10	137.6
93.6	156	146	10	136.6
93.6	141	132	9	123.6
93.6	140	131	9	122.6
93.6	125	117	8	109.5
93.6	110	103	7	96.4
93.6	109	102	7	95.4
93.6	94	88	6	82.4
93.6	78	73	5	68.3
93.6	47	44	3	41.2
93.5	496	464	32	434.1
93.5	495	463	32	433.1
93.5	494	462	32	432.1
93.5	493	461	32	431.1
93.5	492	460	32	430.1
93.5	491	459	32	429.1
93.5	490	458	32	428.1
93.5	489	457	32	427.1
93.5	480	449	31	420.0
93.5	479	448	31	419.0
93.5	478	447	31	418.0
93.5	477	446	31	417.0
93.5	476	445	31	416.0
93.5	475	444	31	415.0
93.5	474	443	31	414.0
93.5	465	435	30	406.9
93.5	464	434	30	405.9
93.5	463	433	30	404.9
93.5	462	432	30	403.9
93.5	461	431	30	403.0
93.5	460	430	30	402.0
93.5	459	429	30	401.0
93.5	458	428	30	400.0
93.5	449	420	29	392.9
93.5	448	419	29	391.9
93.5	447	418	29	390.9

%E	M1	M2	DM	M*
93.5	446	417	29	389.9
93.5	445	416	29	388.9
93.5	444	415	29	387.9
93.5	443	414	29	386.9
93.5	434	406	28	379.8
93.5	433	405	28	378.8
93.5	432	404	28	377.8
93.5	431	403	28	376.8
93.5	430	402	28	375.8
93.5	429	401	28	374.8
93.5	428	400	28	373.8
93.5	418	391	27	365.7
93.5	417	390	27	364.7
93.5	416	389	27	363.8
93.5	415	388	27	362.8
93.5	414	387	27	361.8
93.5	413	386	27	360.8
93.5	403	377	26	352.7
93.5	402	376	26	351.7
93.5	401	375	26	350.7
93.5	400	374	26	349.7
93.5	399	373	26	348.7
93.5	398	372	26	347.7
93.5	397	371	26	346.7
93.5	387	362	25	338.6
93.5	386	361	25	337.6
93.5	385	360	25	336.6
93.5	384	359	25	335.6
93.5	383	358	25	334.6
93.5	382	357	25	333.6
93.5	372	348	24	325.5
93.5	371	347	24	324.6
93.5	370	346	24	323.6
93.5	369	345	24	322.6
93.5	368	344	24	321.6
93.5	367	343	24	320.6
93.5	356	333	23	311.5
93.5	355	332	23	310.5
93.5	354	331	23	309.5
93.5	353	330	23	308.5
93.5	352	329	23	307.5
93.5	341	319	22	298.4
93.5	340	318	22	297.4
93.5	339	317	22	296.4
93.5	338	316	22	295.4
93.5	337	315	22	294.4
93.5	336	314	22	293.4
93.5	325	304	21	284.4
93.5	324	303	21	283.4
93.5	323	302	21	282.4

%E	M1	M2	DM	M*
93.5	322	301	21	281.4
93.5	321	300	21	280.4
93.5	310	290	20	271.3
93.5	309	289	20	270.3
93.5	308	288	20	269.3
93.5	307	287	20	268.3
93.5	306	286	20	267.3
93.5	294	275	19	257.2
93.5	293	274	19	256.2
93.5	292	273	19	255.2
93.5	291	272	19	254.2
93.5	279	261	18	244.2
93.5	278	260	18	243.2
93.5	277	259	18	242.2
93.5	276	258	18	241.2
93.5	275	257	18	240.2
93.5	263	246	17	230.1
93.5	262	245	17	229.1
93.5	261	244	17	228.1
93.5	260	243	17	227.1
93.5	248	232	16	217.0
93.5	247	231	16	216.0
93.5	246	230	16	215.0
93.5	245	229	16	214.0
93.5	232	217	15	203.0
93.5	231	216	15	202.0
93.5	230	215	15	201.0
93.5	229	214	15	200.0
93.5	217	203	14	189.9
93.5	216	202	14	188.9
93.5	215	201	14	187.9
93.5	214	200	14	186.9
93.5	201	188	13	175.8
93.5	200	187	13	174.8
93.5	199	186	13	173.8
93.5	186	174	12	162.8
93.5	185	173	12	161.8
93.5	184	172	12	160.8
93.5	170	159	11	148.7
93.5	169	158	11	147.7
93.5	168	157	11	146.7
93.5	155	145	10	135.6
93.5	154	144	10	134.6
93.5	153	143	10	133.7
93.5	139	130	9	121.6
93.5	138	129	9	120.6
93.5	124	116	8	108.5
93.5	123	115	8	107.5
93.5	108	101	7	94.5
93.5	107	100	7	93.5

%E	M1	M2	DM	M*
93.5	93	87	6	81.4
93.5	92	86	6	80.4
93.5	77	72	5	67.3
93.5	62	58	4	54.3
93.5	46	43	3	40.2
93.5	31	29	2	27.1
93.4	500	467	33	436.2
93.4	499	466	33	435.2
93.4	498	465	33	434.2
93.4	497	464	33	433.2
93.4	488	456	32	426.1
93.4	487	455	32	425.1
93.4	486	454	32	424.1
93.4	485	453	32	423.1
93.4	484	452	32	422.1
93.4	483	451	32	421.1
93.4	482	450	32	420.1
93.4	473	442	31	413.0
93.4	472	441	31	412.0
93.4	471	440	31	411.0
93.4	470	439	31	410.0
93.4	469	438	31	409.0
93.4	468	437	31	408.1
93.4	467	436	31	407.1
93.4	457	427	30	399.0
93.4	456	426	30	398.0
93.4	455	425	30	397.0
93.4	454	424	30	396.0
93.4	453	423	30	395.0
93.4	452	422	30	394.0
93.4	442	413	29	385.9
93.4	441	412	29	384.9
93.4	440	411	29	383.9
93.4	439	410	29	382.9
93.4	438	409	29	381.9
93.4	437	408	29	380.9
93.4	427	399	28	372.8
93.4	426	398	28	371.8
93.4	425	397	28	370.8
93.4	424	396	28	369.8
93.4	423	395	28	368.9
93.4	422	394	28	367.9
93.4	412	385	27	359.8
93.4	411	384	27	358.8
93.4	410	383	27	357.8
93.4	409	382	27	356.8
93.4	408	381	27	355.8
93.4	407	380	27	354.8
93.4	406	379	27	353.8
93.4	396	370	26	345.7

%E	M1	M2	DM	M*
93.4	395	369	26	344.7
93.4	394	368	26	343.7
93.4	393	367	26	342.7
93.4	392	366	26	341.7
93.4	391	365	26	340.7
93.4	381	356	25	332.6
93.4	380	355	25	331.6
93.4	379	354	25	330.6
93.4	378	353	25	329.7
93.4	377	352	25	328.7
93.4	376	351	25	327.7
93.4	366	342	24	319.6
93.4	365	341	24	318.6
93.4	364	340	24	317.6
93.4	363	339	24	316.6
93.4	362	338	24	315.6
93.4	361	337	24	314.6
93.4	351	328	23	306.5
93.4	350	327	23	305.5
93.4	349	326	23	304.5
93.4	348	325	23	303.5
93.4	347	324	23	302.5
93.4	346	323	23	301.5
93.4	335	313	22	292.4
93.4	334	312	22	291.4
93.4	333	311	22	290.5
93.4	332	310	22	289.5
93.4	331	309	22	288.5
93.4	320	299	21	279.4
93.4	319	298	21	278.4
93.4	318	297	21	277.4
93.4	317	296	21	276.4
93.4	316	295	21	275.4
93.4	305	285	20	266.3
93.4	304	284	20	265.3
93.4	303	283	20	264.3
93.4	302	282	20	263.3
93.4	301	281	20	262.3
93.4	290	271	19	253.2
93.4	289	270	19	252.2
93.4	288	269	19	251.3
93.4	287	268	19	250.3
93.4	286	267	19	249.3
93.4	274	256	18	239.2
93.4	273	255	18	238.2
93.4	272	254	18	237.2
93.4	271	253	18	236.2
93.4	259	242	17	226.1
93.4	258	241	17	225.1
93.4	257	240	17	224.1

%E	M1	M2	DM	M*
93.4	256	239	17	223.1
93.4	244	228	16	213.0
93.4	243	227	16	212.1
93.4	242	226	16	211.1
93.4	241	225	16	210.1
93.4	228	213	15	199.0
93.4	227	212	15	198.0
93.4	226	211	15	197.0
93.4	213	199	14	185.9
93.4	212	198	14	184.9
93.4	211	197	14	183.9
93.4	198	185	13	172.9
93.4	197	184	13	171.9
93.4	196	183	13	170.9
93.4	183	171	12	159.8
93.4	182	170	12	158.8
93.4	181	169	12	157.8
93.4	167	156	11	145.7
93.4	166	155	11	144.7
93.4	152	142	10	132.7
93.4	151	141	10	131.7
93.4	137	128	9	119.6
93.4	136	127	9	118.6
93.4	122	114	8	106.5
93.4	121	113	8	105.5
93.4	106	99	7	92.5
93.4	91	85	6	79.4
93.4	76	71	5	66.3
93.4	61	57	4	53.3
93.3	496	463	33	432.2
93.3	495	462	33	431.2
93.3	494	461	33	430.2
93.3	493	460	33	429.2
93.3	492	459	33	428.2
93.3	491	458	33	427.2
93.3	490	457	33	426.2
93.3	489	456	33	425.2
93.3	481	449	32	419.1
93.3	480	448	32	418.1
93.3	479	447	32	417.1
93.3	478	446	32	416.1
93.3	477	445	32	415.1
93.3	476	444	32	414.2
93.3	475	443	32	413.2
93.3	466	435	31	406.1
93.3	465	434	31	405.1
93.3	464	433	31	404.1
93.3	463	432	31	403.1
93.3	462	431	31	402.1
93.3	461	430	31	401.1

%E	M1	M2	DM	M*
93.3	460	429	31	400.1
93.3	451	421	30	393.0
93.3	450	420	30	392.0
93.3	449	419	30	391.0
93.3	448	418	30	390.0
93.3	447	417	30	389.0
93.3	446	416	30	388.0
93.3	445	415	30	387.0
93.3	436	407	29	379.9
93.3	435	406	29	378.9
93.3	434	405	29	377.9
93.3	433	404	29	376.9
93.3	432	403	29	375.9
93.3	431	402	29	375.0
93.3	430	401	29	374.0
93.3	421	393	28	366.9
93.3	420	392	28	365.9
93.3	419	391	28	364.9
93.3	418	390	28	363.9
93.3	417	389	28	362.9
93.3	416	388	28	361.9
93.3	415	387	28	360.9
93.3	405	378	27	352.8
93.3	404	377	27	351.8
93.3	403	376	27	350.8
93.3	402	375	27	349.8
93.3	401	374	27	348.8
93.3	400	373	27	347.8
93.3	390	364	26	339.7
93.3	389	363	26	338.7
93.3	388	362	26	337.7
93.3	387	361	26	336.7
93.3	386	360	26	335.8
93.3	375	350	25	326.7
93.3	374	349	25	325.7
93.3	373	348	25	324.7
93.3	372	347	25	323.7
93.3	371	346	25	322.7
93.3	360	336	24	313.6
93.3	359	335	24	312.6
93.3	358	334	24	311.6
93.3	357	333	24	310.6
93.3	356	332	24	309.6
93.3	345	322	23	300.5
93.3	344	321	23	299.5
93.3	343	320	23	298.5
93.3	342	319	23	297.5
93.3	341	318	23	296.6
93.3	330	308	22	287.5
93.3	329	307	22	286.5

%E	M1	M2	DM	M*
93.3	328	306	22	285.5
93.3	327	305	22	284.5
93.3	326	304	22	283.5
93.3	315	294	21	274.4
93.3	314	293	21	273.4
93.3	313	292	21	272.4
93.3	312	291	21	271.4
93.3	300	280	20	261.3
93.3	299	279	20	260.3
93.3	298	278	20	259.3
93.3	297	277	20	258.3
93.3	285	266	19	248.3
93.3	284	265	19	247.3
93.3	283	264	19	246.3
93.3	282	263	19	245.3
93.3	270	252	18	235.2
93.3	269	251	18	234.2
93.3	268	250	18	233.2
93.3	267	249	18	232.2
93.3	255	238	17	222.1
93.3	254	237	17	221.1
93.3	253	236	17	220.1
93.3	252	235	17	219.1
93.3	240	224	16	209.1
93.3	239	223	16	208.1
93.3	238	222	16	207.1
93.3	225	210	15	196.0
93.3	224	209	15	195.0
93.3	223	208	15	194.0
93.3	210	196	14	182.9
93.3	209	195	14	181.9
93.3	208	194	14	180.9
93.3	195	182	13	169.9
93.3	194	181	13	168.9
93.3	193	180	13	167.9
93.3	180	168	12	156.8
93.3	179	167	12	155.8
93.3	178	166	12	154.8
93.3	165	154	11	143.7
93.3	164	153	11	142.7
93.3	163	152	11	141.7
93.3	150	140	10	130.7
93.3	149	139	10	129.7
93.3	135	126	9	117.6
93.3	134	125	9	116.6
93.3	120	112	8	104.5
93.3	119	111	8	103.5
93.3	105	98	7	91.5
93.3	104	97	7	90.5
93.3	90	84	6	78.4

%E	M1	M2	DM	M*
93.3	89	83	6	77.4
93.3	75	70	5	65.3
93.3	60	56	4	52.3
93.3	45	42	3	39.2
93.3	30	28	2	26.1
93.3	15	14	1	13.1
93.2	500	466	34	434.3
93.2	499	465	34	433.3
93.2	498	464	34	432.3
93.2	497	463	34	431.3
93.2	488	455	33	424.2
93.2	487	454	33	423.2
93.2	486	453	33	422.2
93.2	485	452	33	421.2
93.2	484	451	33	420.3
93.2	483	450	33	419.3
93.2	482	449	33	418.3
93.2	474	442	32	412.2
93.2	473	441	32	411.2
93.2	472	440	32	410.2
93.2	471	439	32	409.2
93.2	470	438	32	408.2
93.2	469	437	32	407.2
93.2	468	436	32	406.2
93.2	459	428	31	399.1
93.2	458	427	31	398.1
93.2	457	426	31	397.1
93.2	456	425	31	396.1
93.2	455	424	31	395.1
93.2	454	423	31	394.1
93.2	453	422	31	393.1
93.2	444	414	30	386.0
93.2	443	413	30	385.0
93.2	442	412	30	384.0
93.2	441	411	30	383.0
93.2	440	410	30	382.0
93.2	439	409	30	381.1
93.2	438	408	30	380.1
93.2	429	400	29	373.0
93.2	428	399	29	372.0
93.2	427	398	29	371.0
93.2	426	397	29	370.0
93.2	425	396	29	369.0
93.2	424	395	29	368.0
93.2	414	386	28	359.9
93.2	413	385	28	358.9
93.2	412	384	28	357.9
93.2	411	383	28	356.9
93.2	410	382	28	355.9
93.2	409	381	28	354.9

%E	M1	M2	DM	M*
93.2	399	372	27	346.8
93.2	398	371	27	345.8
93.2	397	370	27	344.8
93.2	396	369	27	343.8
93.2	395	368	27	342.8
93.2	385	359	26	334.8
93.2	384	358	26	333.8
93.2	383	357	26	332.8
93.2	382	356	26	331.8
93.2	381	355	26	330.8
93.2	380	354	26	329.8
93.2	370	345	25	321.7
93.2	369	344	25	320.7
93.2	368	343	25	319.7
93.2	367	342	25	318.7
93.2	366	341	25	317.7
93.2	365	340	25	316.7
93.2	355	331	24	308.6
93.2	354	330	24	307.6
93.2	353	329	24	306.6
93.2	352	328	24	305.6
93.2	351	327	24	304.6
93.2	340	317	23	295.6
93.2	339	316	23	294.6
93.2	338	315	23	293.6
93.2	337	314	23	292.6
93.2	336	313	23	291.6
93.2	325	303	22	282.5
93.2	324	302	22	281.5
93.2	323	301	22	280.5
93.2	322	300	22	279.5
93.2	311	290	21	270.4
93.2	310	289	21	269.4
93.2	309	288	21	268.4
93.2	308	287	21	267.4
93.2	307	286	21	266.4
93.2	296	276	20	257.4
93.2	295	275	20	256.4
93.2	294	274	20	255.4
93.2	293	273	20	254.4
93.2	292	272	20	253.4
93.2	281	262	19	244.3
93.2	280	261	19	243.3
93.2	279	260	19	242.3
93.2	278	259	19	241.3
93.2	266	248	18	231.2
93.2	265	247	18	230.2
93.2	264	246	18	229.2
93.2	263	245	18	228.2
93.2	251	234	17	218.2
93.2	250	233	17	217.2
93.2	249	232	17	216.2
93.2	237	221	16	206.1
93.2	236	220	16	205.1
93.2	235	219	16	204.1
93.2	234	218	16	203.1
93.2	222	207	15	193.0
93.2	221	206	15	192.0
93.2	220	205	15	191.0
93.2	219	204	15	190.0
93.2	207	193	14	179.9
93.2	206	192	14	179.0
93.2	205	191	14	178.0
93.2	192	179	13	166.9
93.2	191	178	13	165.9
93.2	190	177	13	164.9
93.2	177	165	12	153.8
93.2	176	164	12	152.8
93.2	162	151	11	140.7
93.2	161	150	11	139.8
93.2	148	138	10	128.7
93.2	147	137	10	127.7
93.2	146	136	10	126.7
93.2	133	124	9	115.6
93.2	132	123	9	114.6
93.2	118	110	8	102.5
93.2	117	109	8	101.5
93.2	103	96	7	89.5
93.2	88	82	6	76.4
93.2	74	69	5	64.3
93.2	73	68	5	63.3
93.2	59	55	4	51.3
93.2	44	41	3	38.2
93.1	496	462	34	430.3
93.1	495	461	34	429.3
93.1	494	460	34	428.3
93.1	493	459	34	427.3
93.1	492	458	34	426.3
93.1	491	457	34	425.4
93.1	490	456	34	424.4
93.1	481	448	33	417.3
93.1	480	447	33	416.3
93.1	479	446	33	415.3
93.1	478	445	33	414.3
93.1	477	444	33	413.3
93.1	476	443	33	412.3
93.1	475	442	33	411.3
93.1	467	435	32	405.2
93.1	466	434	32	404.2
93.1	465	433	32	403.2
93.1	464	432	32	402.2
93.1	463	431	32	401.2
93.1	462	430	32	400.2
93.1	461	429	32	399.2
93.1	452	421	31	392.1
93.1	451	420	31	391.1
93.1	450	419	31	390.1
93.1	449	418	31	389.1
93.1	448	417	31	388.1
93.1	447	416	31	387.1
93.1	437	407	30	379.1
93.1	436	406	30	378.1
93.1	435	405	30	377.1
93.1	434	404	30	376.1
93.1	433	403	30	375.1
93.1	432	402	30	374.1
93.1	423	394	29	367.0
93.1	422	393	29	366.0
93.1	421	392	29	365.0
93.1	420	391	29	364.0
93.1	419	390	29	363.0
93.1	418	389	29	362.0
93.1	408	380	28	353.9
93.1	407	379	28	352.9
93.1	406	378	28	351.9
93.1	405	377	28	350.9
93.1	404	376	28	349.9
93.1	403	375	28	348.9
93.1	394	367	27	341.9
93.1	393	366	27	340.9
93.1	392	365	27	339.9
93.1	391	364	27	338.9
93.1	390	363	27	337.9
93.1	389	362	27	336.9
93.1	379	353	26	328.8
93.1	378	352	26	327.8
93.1	377	351	26	326.8
93.1	376	350	26	325.8
93.1	375	349	26	324.8
93.1	364	339	25	315.7
93.1	363	338	25	314.7
93.1	362	337	25	313.7
93.1	361	336	25	312.7
93.1	360	335	25	311.7
93.1	350	326	24	303.6
93.1	349	325	24	302.7
93.1	348	324	24	301.7
93.1	347	323	24	300.7
93.1	346	322	24	299.7
93.1	335	312	23	290.6
93.1	334	311	23	289.6
93.1	333	310	23	288.6
93.1	332	309	23	287.6
93.1	331	308	23	286.6
93.1	321	299	22	278.5
93.1	320	298	22	277.5
93.1	319	297	22	276.5
93.1	318	296	22	275.5
93.1	317	295	22	274.5
93.1	306	285	21	265.4
93.1	305	284	21	264.4
93.1	304	283	21	263.5
93.1	303	282	21	262.5
93.1	291	271	20	252.4
93.1	290	270	20	251.4
93.1	289	269	20	250.4
93.1	288	268	20	249.4
93.1	277	258	19	240.3
93.1	276	257	19	239.3
93.1	275	256	19	238.3
93.1	274	255	19	237.3
93.1	262	244	18	227.2
93.1	261	243	18	226.2
93.1	260	242	18	225.2
93.1	259	241	18	224.3
93.1	248	231	17	215.2
93.1	247	230	17	214.2
93.1	246	229	17	213.2
93.1	245	228	17	212.2
93.1	233	217	16	202.1
93.1	232	216	16	201.1
93.1	231	215	16	200.1
93.1	218	203	15	189.0
93.1	217	202	15	188.0
93.1	216	201	15	187.0
93.1	204	190	14	177.0
93.1	203	189	14	176.0
93.1	202	188	14	175.0
93.1	189	176	13	163.9
93.1	188	175	13	162.9
93.1	175	163	12	151.8
93.1	174	162	12	150.8
93.1	173	161	12	149.8
93.1	160	149	11	138.8
93.1	159	148	11	137.8
93.1	145	135	10	125.7
93.1	144	134	10	124.7
93.1	131	122	9	113.6
93.1	130	121	9	112.6
93.1	116	108	8	100.6
93.1	102	95	7	88.5
93.1	101	94	7	87.5
93.1	87	81	6	75.4
93.1	72	67	5	62.3
93.1	58	54	4	50.3
93.1	29	27	2	25.1
93.0	500	465	35	432.4
93.0	499	464	35	431.5
93.0	498	463	35	430.5
93.0	497	462	35	429.5
93.0	489	455	34	423.4
93.0	488	454	34	422.4
93.0	487	453	34	421.4
93.0	486	452	34	420.4
93.0	485	451	34	419.4
93.0	484	450	34	418.4
93.0	483	449	34	417.4
93.0	474	441	33	410.3
93.0	473	440	33	409.3
93.0	472	439	33	408.3
93.0	471	438	33	407.3
93.0	470	437	33	406.3
93.0	469	436	33	405.3
93.0	460	428	32	398.2
93.0	459	427	32	397.2
93.0	458	426	32	396.2
93.0	457	425	32	395.2
93.0	456	424	32	394.2
93.0	455	423	32	393.3
93.0	454	422	32	392.3
93.0	446	415	31	386.2
93.0	445	414	31	385.2
93.0	444	413	31	384.2
93.0	443	412	31	383.2
93.0	442	411	31	382.2
93.0	441	410	31	381.2
93.0	440	409	31	380.2
93.0	431	401	30	373.1
93.0	430	400	30	372.1
93.0	429	399	30	371.1
93.0	428	398	30	370.1
93.0	427	397	30	369.1
93.0	426	396	30	368.1
93.0	417	388	29	361.0
93.0	416	387	29	360.0
93.0	415	386	29	359.0
93.0	414	385	29	358.0
93.0	413	384	29	357.0
93.0	412	383	29	356.0
93.0	402	374	28	348.0

%E	M1	M2	DM	M*
93.0	401	373	28	347.0
93.0	400	372	28	346.0
93.0	399	371	28	345.0
93.0	398	370	28	344.0
93.0	388	361	27	335.9
93.0	387	360	27	334.9
93.0	386	359	27	333.9
93.0	385	358	27	332.9
93.0	384	357	27	331.9
93.0	383	356	27	330.9
93.0	374	348	26	323.8
93.0	373	347	26	322.8
93.0	372	346	26	321.8
93.0	371	345	26	320.8
93.0	370	344	26	319.8
93.0	369	343	26	318.8
93.0	359	334	25	310.7
93.0	358	333	25	309.7
93.0	357	332	25	308.8
93.0	356	331	25	307.8
93.0	355	330	25	306.8
93.0	345	321	24	298.7
93.0	344	320	24	297.7
93.0	343	319	24	296.7
93.0	342	318	24	295.7
93.0	341	317	24	294.7
93.0	330	307	23	285.6
93.0	329	306	23	284.6
93.0	328	305	23	283.6
93.0	327	304	23	282.6
93.0	316	294	22	273.5
93.0	315	293	22	272.5
93.0	314	292	22	271.5
93.0	313	291	22	270.5
93.0	302	281	21	261.5
93.0	301	280	21	260.5
93.0	300	279	21	259.5
93.0	299	278	21	258.5
93.0	298	277	21	257.5
93.0	287	267	20	248.4
93.0	286	266	20	247.4
93.0	285	265	20	246.4
93.0	284	264	20	245.4
93.0	273	254	19	236.3
93.0	272	253	19	235.3
93.0	271	252	19	234.3
93.0	270	251	19	233.3
93.0	258	240	18	223.3
93.0	257	239	18	222.3
93.0	256	238	18	221.3

%E	M1	M2	DM	M*
93.0	244	227	17	211.2
93.0	243	226	17	210.2
93.0	242	225	17	209.2
93.0	230	214	16	199.1
93.0	229	213	16	198.1
93.0	228	212	16	197.1
93.0	227	211	16	196.1
93.0	215	200	15	186.0
93.0	214	199	15	185.1
93.0	213	198	15	184.1
93.0	201	187	14	174.0
93.0	200	186	14	173.0
93.0	199	185	14	172.0
93.0	187	174	13	161.9
93.0	186	173	13	160.9
93.0	185	172	13	159.9
93.0	172	160	12	148.8
93.0	171	159	12	147.8
93.0	158	147	11	136.8
93.0	157	146	11	135.8
93.0	143	133	10	123.7
93.0	142	132	10	122.7
93.0	129	120	9	111.6
93.0	128	119	9	110.6
93.0	115	107	8	99.6
93.0	114	106	8	98.6
93.0	100	93	7	86.5
93.0	86	80	6	74.4
93.0	71	66	5	61.4
93.0	57	53	4	49.3
93.0	43	40	3	37.2
92.9	496	461	35	428.5
92.9	495	460	35	427.5
92.9	494	459	35	426.5
92.9	493	458	35	425.5
92.9	492	457	35	424.5
92.9	491	456	35	423.5
92.9	490	455	35	422.5
92.9	482	448	34	416.4
92.9	481	447	34	415.4
92.9	480	446	34	414.4
92.9	479	445	34	413.4
92.9	478	444	34	412.4
92.9	477	443	34	411.4
92.9	476	442	34	410.4
92.9	468	435	33	404.3
92.9	467	434	33	403.3
92.9	466	433	33	402.3
92.9	465	432	33	401.3
92.9	464	431	33	400.3

%E	M1	M2	DM	M*
92.9	463	430	33	399.4
92.9	462	429	33	398.4
92.9	453	421	32	391.3
92.9	452	420	32	390.3
92.9	451	419	32	389.3
92.9	450	418	32	388.3
92.9	449	417	32	387.3
92.9	448	416	32	386.3
92.9	439	408	31	379.2
92.9	438	407	31	378.2
92.9	437	406	31	377.2
92.9	436	405	31	376.2
92.9	435	404	31	375.2
92.9	434	403	31	374.2
92.9	425	395	30	367.1
92.9	424	394	30	366.1
92.9	423	393	30	365.1
92.9	422	392	30	364.1
92.9	421	391	30	363.1
92.9	420	390	30	362.1
92.9	411	382	29	355.0
92.9	410	381	29	354.1
92.9	409	380	29	353.1
92.9	408	379	29	352.1
92.9	407	378	29	351.1
92.9	406	377	29	350.1
92.9	397	369	28	343.0
92.9	396	368	28	342.0
92.9	395	367	28	341.0
92.9	394	366	28	340.0
92.9	393	365	28	339.0
92.9	392	364	28	338.0
92.9	382	355	27	329.9
92.9	381	354	27	328.9
92.9	380	353	27	327.9
92.9	379	352	27	326.9
92.9	378	351	27	325.9
92.9	368	342	26	317.8
92.9	367	341	26	316.8
92.9	366	340	26	315.8
92.9	365	339	26	314.9
92.9	364	338	26	313.9
92.9	354	329	25	305.8
92.9	353	328	25	304.8
92.9	352	327	25	303.8
92.9	351	326	25	302.8
92.9	350	325	25	301.8
92.9	340	316	24	293.7
92.9	339	315	24	292.7
92.9	338	314	24	291.7

%E	M1	M2	DM	M*
92.9	337	313	24	290.7
92.9	336	312	24	289.7
92.9	326	303	23	281.6
92.9	325	302	23	280.6
92.9	324	301	23	279.6
92.9	323	300	23	278.6
92.9	322	299	23	277.6
92.9	312	290	22	269.6
92.9	311	289	22	268.6
92.9	310	288	22	267.6
92.9	309	287	22	266.6
92.9	308	286	22	265.6
92.9	297	276	21	256.5
92.9	296	275	21	255.5
92.9	295	274	21	254.5
92.9	294	273	21	253.5
92.9	283	263	20	244.4
92.9	282	262	20	243.4
92.9	281	261	20	242.4
92.9	280	260	20	241.4
92.9	269	250	19	232.3
92.9	268	249	19	231.3
92.9	267	248	19	230.4
92.9	266	247	19	229.4
92.9	255	237	18	220.3
92.9	254	236	18	219.3
92.9	253	235	18	218.3
92.9	252	234	18	217.3
92.9	241	224	17	208.2
92.9	240	223	17	207.2
92.9	239	222	17	206.2
92.9	238	221	17	205.2
92.9	226	210	16	195.1
92.9	225	209	16	194.1
92.9	224	208	16	193.1
92.9	212	197	15	183.1
92.9	211	196	15	182.1
92.9	210	195	15	181.1
92.9	198	184	14	171.0
92.9	197	183	14	170.0
92.9	196	182	14	169.0
92.9	184	171	13	158.9
92.9	183	170	13	157.9
92.9	182	169	13	156.9
92.9	170	158	12	146.8
92.9	169	157	12	145.9
92.9	168	156	12	144.9
92.9	156	145	11	134.8
92.9	155	144	11	133.8
92.9	154	143	11	132.8

%E	M1	M2	DM	M*
92.9	141	131	10	121.7
92.9	140	130	10	120.7
92.9	127	118	9	109.6
92.9	126	117	9	108.6
92.9	113	105	8	97.6
92.9	112	104	8	96.6
92.9	99	92	7	85.5
92.9	98	91	7	84.5
92.9	85	79	6	73.4
92.9	84	78	6	72.4
92.9	70	65	5	60.4
92.9	56	52	4	48.3
92.9	42	39	3	36.2
92.9	28	26	2	24.1
92.9	14	13	1	12.1
92.8	500	464	36	430.6
92.8	499	463	36	429.6
92.8	498	462	36	428.6
92.8	497	461	36	427.6
92.8	489	454	35	421.5
92.8	488	453	35	420.5
92.8	487	452	35	419.5
92.8	486	451	35	418.5
92.8	485	450	35	417.5
92.8	484	449	35	416.5
92.8	483	448	35	415.5
92.8	475	441	34	409.4
92.8	474	440	34	408.4
92.8	473	439	34	407.4
92.8	472	438	34	406.4
92.8	471	437	34	405.5
92.8	470	436	34	404.5
92.8	469	435	34	403.5
92.8	461	428	33	397.4
92.8	460	427	33	396.4
92.8	459	426	33	395.4
92.8	458	425	33	394.4
92.8	457	424	33	393.4
92.8	456	423	33	392.4
92.8	447	415	32	385.3
92.8	446	414	32	384.3
92.8	445	413	32	383.3
92.8	444	412	32	382.3
92.8	443	411	32	381.3
92.8	442	410	32	380.3
92.8	433	402	31	373.2
92.8	432	401	31	372.2
92.8	431	400	31	371.2
92.8	430	399	31	370.2
92.8	429	398	31	369.2

%E	M1	M2	DM	M*
92.8	428	397	31	368.2
92.8	419	389	30	361.1
92.8	418	388	30	360.2
92.8	417	387	30	359.2
92.8	416	386	30	358.2
92.8	415	385	30	357.2
92.8	414	384	30	356.2
92.8	405	376	29	349.1
92.8	404	375	29	348.1
92.8	403	374	29	347.1
92.8	402	373	29	346.1
92.8	401	372	29	345.1
92.8	400	371	29	344.1
92.8	391	363	28	337.0
92.8	390	362	28	336.0
92.8	389	361	28	335.0
92.8	388	360	28	334.0
92.8	387	359	28	333.0
92.8	377	350	27	324.9
92.8	376	349	27	323.9
92.8	375	348	27	322.9
92.8	374	347	27	321.9
92.8	373	346	27	321.0
92.8	363	337	26	312.9
92.8	362	336	26	311.9
92.8	361	335	26	310.9
92.8	360	334	26	309.9
92.8	359	333	26	308.9
92.8	349	324	25	300.8
92.8	348	323	25	299.8
92.8	347	322	25	298.8
92.8	346	321	25	297.8
92.8	345	320	25	296.8
92.8	335	311	24	288.7
92.8	334	310	24	287.7
92.8	333	309	24	286.7
92.8	332	308	24	285.7
92.8	321	298	23	276.6
92.8	320	297	23	275.7
92.8	319	296	23	274.7
92.8	318	295	23	273.7
92.8	307	285	22	264.6
92.8	306	284	22	263.6
92.8	305	283	22	262.6
92.8	304	282	22	261.6
92.8	293	272	21	252.5
92.8	292	271	21	251.5
92.8	291	270	21	250.5
92.8	290	269	21	249.5
92.8	279	259	20	240.4

%E	M1	M2	DM	M*
92.8	278	258	20	239.4
92.8	277	257	20	238.4
92.8	276	256	20	237.4
92.8	265	246	19	228.4
92.8	264	245	19	227.4
92.8	263	244	19	226.4
92.8	251	233	18	216.3
92.8	250	232	18	215.3
92.8	249	231	18	214.3
92.8	237	220	17	204.2
92.8	236	219	17	203.2
92.8	235	218	17	202.2
92.8	223	207	16	192.1
92.8	222	206	16	191.2
92.8	221	205	16	190.2
92.8	209	194	15	180.1
92.8	208	193	15	179.1
92.8	207	192	15	178.1
92.8	195	181	14	168.0
92.8	194	180	14	167.0
92.8	181	168	13	155.9
92.8	180	167	13	154.9
92.8	167	155	12	143.9
92.8	166	154	12	142.9
92.8	153	142	11	131.8
92.8	152	141	11	130.8
92.8	139	129	10	119.7
92.8	138	128	10	118.7
92.8	125	116	9	107.6
92.8	111	103	8	95.6
92.8	97	90	7	83.5
92.8	83	77	6	71.4
92.8	69	64	5	59.4
92.7	496	460	36	426.6
92.7	495	459	36	425.6
92.7	494	458	36	424.6
92.7	493	457	36	423.6
92.7	492	456	36	422.6
92.7	491	455	36	421.6
92.7	490	454	36	420.6
92.7	482	447	35	414.5
92.7	481	446	35	413.5
92.7	480	445	35	412.6
92.7	479	444	35	411.6
92.7	478	443	35	410.6
92.7	477	442	35	409.6
92.7	468	434	34	402.5
92.7	467	433	34	401.5
92.7	466	432	34	400.5
92.7	465	431	34	399.5

%E	M1	M2	DM	M*
92.7	464	430	34	398.5
92.7	463	429	34	397.5
92.7	455	422	33	391.4
92.7	454	421	33	390.4
92.7	453	420	33	389.4
92.7	452	419	33	388.4
92.7	451	418	33	387.4
92.7	450	417	33	386.4
92.7	449	416	33	385.4
92.7	441	409	32	379.3
92.7	440	408	32	378.3
92.7	439	407	32	377.3
92.7	438	406	32	376.3
92.7	437	405	32	375.3
92.7	436	404	32	374.3
92.7	427	396	31	367.3
92.7	426	395	31	366.3
92.7	425	394	31	365.3
92.7	424	393	31	364.3
92.7	423	392	31	363.3
92.7	422	391	31	362.3
92.7	413	383	30	355.2
92.7	412	382	30	354.2
92.7	411	381	30	353.2
92.7	410	380	30	352.2
92.7	409	379	30	351.2
92.7	399	370	29	343.1
92.7	398	369	29	342.1
92.7	397	368	29	341.1
92.7	396	367	29	340.1
92.7	395	366	29	339.1
92.7	386	358	28	332.0
92.7	385	357	28	331.0
92.7	384	356	28	330.0
92.7	383	355	28	329.0
92.7	382	354	28	328.1
92.7	381	353	28	327.1
92.7	372	345	27	320.0
92.7	371	344	27	319.0
92.7	370	343	27	318.0
92.7	369	342	27	317.0
92.7	368	341	27	316.0
92.7	358	332	26	307.9
92.7	357	331	26	306.9
92.7	356	330	26	305.9
92.7	355	329	26	304.9
92.7	354	328	26	303.9
92.7	344	319	25	295.8
92.7	343	318	25	294.8
92.7	342	317	25	293.8

%E	M1	M2	DM	M*
92.7	341	316	25	292.8
92.7	331	307	24	284.7
92.7	330	306	24	283.7
92.7	329	305	24	282.8
92.7	328	304	24	281.8
92.7	327	303	24	280.8
92.7	317	294	23	272.7
92.7	316	293	23	271.7
92.7	315	292	23	270.7
92.7	314	291	23	269.7
92.7	313	290	23	268.7
92.7	303	281	22	260.6
92.7	302	280	22	259.6
92.7	301	279	22	258.6
92.7	300	278	22	257.6
92.7	289	268	21	248.5
92.7	288	267	21	247.5
92.7	287	266	21	246.5
92.7	286	265	21	245.5
92.7	275	255	20	236.5
92.7	274	254	20	235.5
92.7	273	253	20	234.5
92.7	262	243	19	225.4
92.7	261	242	19	224.4
92.7	260	241	19	223.4
92.7	259	240	19	222.4
92.7	248	230	18	213.3
92.7	247	229	18	212.3
92.7	246	228	18	211.3
92.7	245	227	18	210.3
92.7	234	217	17	201.2
92.7	233	216	17	200.2
92.7	232	215	17	199.2
92.7	220	204	16	189.2
92.7	219	203	16	188.2
92.7	218	202	16	187.2
92.7	206	191	15	177.1
92.7	205	190	15	176.1
92.7	193	179	14	166.0
92.7	192	178	14	165.0
92.7	191	177	14	164.0
92.7	179	166	13	153.9
92.7	178	165	13	152.9
92.7	177	164	13	152.0
92.7	165	153	12	141.9
92.7	164	152	12	140.9
92.7	151	140	11	129.8
92.7	150	139	11	128.8
92.7	137	127	10	117.7
92.7	124	115	9	106.7

%E	M1	M2	DM	M*
92.7	123	114	9	105.7
92.7	110	102	8	94.6
92.7	109	101	8	93.6
92.7	96	89	7	82.5
92.7	82	76	6	70.4
92.7	55	51	4	47.3
92.7	41	38	3	35.2
92.6	500	463	37	428.7
92.6	499	462	37	427.7
92.6	498	461	37	426.7
92.6	497	460	37	425.8
92.6	489	453	36	419.7
92.6	488	452	36	418.7
92.6	487	451	36	417.7
92.6	486	450	36	416.7
92.6	485	449	36	415.7
92.6	484	448	36	414.7
92.6	476	441	35	408.6
92.6	475	440	35	407.6
92.6	474	439	35	406.6
92.6	473	438	35	405.6
92.6	472	437	35	404.6
92.6	471	436	35	403.6
92.6	470	435	35	402.6
92.6	462	428	34	396.5
92.6	461	427	34	395.5
92.6	460	426	34	394.5
92.6	459	425	34	393.5
92.6	458	424	34	392.5
92.6	457	423	34	391.5
92.6	448	415	33	384.4
92.6	447	414	33	383.4
92.6	446	413	33	382.4
92.6	445	412	33	381.4
92.6	444	411	33	380.5
92.6	443	410	33	379.5
92.6	435	403	32	373.4
92.6	434	402	32	372.4
92.6	433	401	32	371.4
92.6	432	400	32	370.4
92.6	431	399	32	369.4
92.6	430	398	32	368.4
92.6	421	390	31	361.3
92.6	420	389	31	360.3
92.6	419	388	31	359.3
92.6	418	387	31	358.3
92.6	417	386	31	357.3
92.6	408	378	30	350.2
92.6	407	377	30	349.2
92.6	406	376	30	348.2

%E	M1	M2	DM	M*
92.6	405	375	30	347.2
92.6	404	374	30	346.2
92.6	403	373	30	345.2
92.6	394	365	29	338.1
92.6	393	364	29	337.1
92.6	392	363	29	336.1
92.6	391	362	29	335.2
92.6	390	361	29	334.2
92.6	380	352	28	326.1
92.6	379	351	28	325.1
92.6	378	350	28	324.1
92.6	377	349	28	323.1
92.6	376	348	28	322.1
92.6	367	340	27	315.0
92.6	366	339	27	314.0
92.6	365	338	27	313.0
92.6	364	337	27	312.0
92.6	363	336	27	311.0
92.6	353	327	26	302.9
92.6	352	326	26	301.9
92.6	351	325	26	300.9
92.6	350	324	26	299.9
92.6	349	323	26	298.9
92.6	340	315	25	291.8
92.6	339	314	25	290.8
92.6	338	313	25	289.8
92.6	337	312	25	288.9
92.6	336	311	25	287.9
92.6	326	302	24	279.8
92.6	325	301	24	278.8
92.6	324	300	24	277.8
92.6	323	299	24	276.8
92.6	312	289	23	267.7
92.6	311	288	23	266.7
92.6	310	287	23	265.7
92.6	309	286	23	264.7
92.6	299	277	22	256.6
92.6	298	276	22	255.6
92.6	297	275	22	254.6
92.6	296	274	22	253.6
92.6	285	264	21	244.5
92.6	284	263	21	243.6
92.6	283	262	21	242.6
92.6	282	261	21	241.6
92.6	272	252	20	233.5
92.6	271	251	20	232.5
92.6	270	250	20	231.5
92.6	269	249	20	230.5
92.6	258	239	19	221.4
92.6	257	238	19	220.4

%E	M1	M2	DM	M*
92.6	256	237	19	219.4
92.6	244	226	18	209.3
92.6	243	225	18	208.3
92.6	242	224	18	207.3
92.6	231	214	17	198.3
92.6	230	213	17	197.3
92.6	229	212	17	196.3
92.6	217	201	16	186.2
92.6	216	200	16	185.2
92.6	215	199	16	184.2
92.6	204	189	15	175.1
92.6	203	188	15	174.1
92.6	202	187	15	173.1
92.6	190	176	14	163.0
92.6	189	175	14	162.0
92.6	188	174	14	161.0
92.6	176	163	13	151.0
92.6	175	162	13	150.0
92.6	163	151	12	139.9
92.6	162	150	12	138.9
92.6	149	138	11	127.8
92.6	148	137	11	126.8
92.6	136	126	10	116.7
92.6	135	125	10	115.7
92.6	122	113	9	104.7
92.6	121	112	9	103.7
92.6	108	100	8	92.6
92.6	95	88	7	81.5
92.6	94	87	7	80.5
92.6	81	75	6	69.4
92.6	68	63	5	58.4
92.6	54	50	4	46.3
92.6	27	25	2	23.1
92.5	496	459	37	424.8
92.5	495	458	37	423.8
92.5	494	457	37	422.8
92.5	493	456	37	421.8
92.5	492	455	37	420.8
92.5	491	454	37	419.8
92.5	483	447	36	413.7
92.5	482	446	36	412.7
92.5	481	445	36	411.7
92.5	480	444	36	410.7
92.5	479	443	36	409.7
92.5	478	442	36	408.7
92.5	477	441	36	407.7
92.5	469	434	35	401.6
92.5	468	433	35	400.6
92.5	467	432	35	399.6
92.5	466	431	35	398.6

%E	M1	M2	DM	M*
92.5	465	430	35	397.6
92.5	464	429	35	396.6
92.5	456	422	34	390.5
92.5	455	421	34	389.5
92.5	454	420	34	388.5
92.5	453	419	34	387.6
92.5	452	418	34	386.6
92.5	451	417	34	385.6
92.5	442	409	33	378.5
92.5	441	408	33	377.5
92.5	440	407	33	376.5
92.5	439	406	33	375.5
92.5	438	405	33	374.5
92.5	429	397	32	367.4
92.5	428	396	32	366.4
92.5	427	395	32	365.4
92.5	426	394	32	364.4
92.5	425	393	32	363.4
92.5	424	392	32	362.4
92.5	416	385	31	356.3
92.5	415	384	31	355.3
92.5	414	383	31	354.3
92.5	413	382	31	353.3
92.5	412	381	31	352.3
92.5	411	380	31	351.3
92.5	402	372	30	344.2
92.5	401	371	30	343.2
92.5	400	370	30	342.3
92.5	399	369	30	341.3
92.5	398	368	30	340.3
92.5	389	360	29	333.2
92.5	388	359	29	332.2
92.5	387	358	29	331.2
92.5	386	357	29	330.2
92.5	385	356	29	329.2
92.5	375	347	28	321.1
92.5	374	346	28	320.1
92.5	373	345	28	319.1
92.5	372	344	28	318.1
92.5	371	343	28	317.1
92.5	362	335	27	310.0
92.5	361	334	27	309.0
92.5	360	333	27	308.0
92.5	359	332	27	307.0
92.5	358	331	27	306.0
92.5	348	322	26	297.9
92.5	347	321	26	296.9
92.5	346	320	26	296.0
92.5	345	319	26	295.0
92.5	335	310	25	286.9

%E	M1	M2	DM	M*
92.5	334	309	25	285.9
92.5	333	308	25	284.9
92.5	332	307	25	283.9
92.5	322	298	24	275.8
92.5	321	297	24	274.8
92.5	320	296	24	273.8
92.5	319	295	24	272.8
92.5	318	294	24	271.8
92.5	308	285	23	263.7
92.5	307	284	23	262.7
92.5	306	283	23	261.7
92.5	305	282	23	260.7
92.5	295	273	22	252.6
92.5	294	272	22	251.6
92.5	293	271	22	250.7
92.5	292	270	22	249.7
92.5	281	260	21	240.6
92.5	280	259	21	239.6
92.5	279	258	21	238.6
92.5	268	248	20	229.5
92.5	267	247	20	228.5
92.5	266	246	20	227.5
92.5	265	245	20	226.5
92.5	255	236	19	218.4
92.5	254	235	19	217.4
92.5	253	234	19	216.4
92.5	252	233	19	215.4
92.5	241	223	18	206.3
92.5	240	222	18	205.3
92.5	239	221	18	204.4
92.5	228	211	17	195.3
92.5	227	210	17	194.3
92.5	226	209	17	193.3
92.5	214	198	16	183.2
92.5	213	197	16	182.2
92.5	212	196	16	181.2
92.5	201	186	15	172.1
92.5	200	185	15	171.1
92.5	199	184	15	170.1
92.5	187	173	14	160.0
92.5	186	172	14	159.1
92.5	174	161	13	149.0
92.5	173	160	13	148.0
92.5	161	149	12	137.9
92.5	160	148	12	136.9
92.5	159	147	12	135.9
92.5	147	136	11	125.8
92.5	146	135	11	124.8
92.5	134	124	10	114.7
92.5	133	123	10	113.8

%E	M1	M2	DM	M*
92.5	120	111	9	102.7
92.5	107	99	8	91.6
92.5	106	98	8	90.6
92.5	93	86	7	79.5
92.5	80	74	6	68.4
92.5	67	62	5	57.4
92.5	53	49	4	45.3
92.5	40	37	3	34.2
92.4	500	462	38	426.9
92.4	499	461	38	425.9
92.4	498	460	38	424.9
92.4	497	459	38	423.9
92.4	490	453	37	418.8
92.4	489	452	37	417.8
92.4	488	451	37	416.8
92.4	487	450	37	415.8
92.4	486	449	37	414.8
92.4	485	448	37	413.8
92.4	484	447	37	412.8
92.4	476	440	36	406.7
92.4	475	439	36	405.7
92.4	474	438	36	404.7
92.4	473	437	36	403.7
92.4	472	436	36	402.7
92.4	471	435	36	401.8
92.4	463	428	35	395.6
92.4	462	427	35	394.7
92.4	461	426	35	393.7
92.4	460	425	35	392.7
92.4	459	424	35	391.7
92.4	458	423	35	390.7
92.4	450	416	34	384.6
92.4	449	415	34	383.6
92.4	448	414	34	382.6
92.4	447	413	34	381.6
92.4	446	412	34	380.6
92.4	445	411	34	379.6
92.4	437	404	33	373.5
92.4	436	403	33	372.5
92.4	435	402	33	371.5
92.4	434	401	33	370.5
92.4	433	400	33	369.5
92.4	432	399	33	368.5
92.4	423	391	32	361.4
92.4	422	390	32	360.4
92.4	421	389	32	359.4
92.4	420	388	32	358.4
92.4	419	387	32	357.4
92.4	410	379	31	350.3
92.4	409	378	31	349.3

%E	M1	M2	DM	M*
92.4	408	377	31	348.4
92.4	407	376	31	347.4
92.4	406	375	31	346.4
92.4	397	367	30	339.3
92.4	396	366	30	338.3
92.4	395	365	30	337.3
92.4	394	364	30	336.3
92.4	393	363	30	335.3
92.4	384	355	29	328.2
92.4	383	354	29	327.2
92.4	382	353	29	326.2
92.4	381	352	29	325.2
92.4	380	351	29	324.2
92.4	370	342	28	316.1
92.4	369	341	28	315.1
92.4	368	340	28	314.1
92.4	367	339	28	313.1
92.4	366	338	28	312.1
92.4	357	330	27	305.0
92.4	356	329	27	304.0
92.4	355	328	27	303.1
92.4	354	327	27	302.1
92.4	353	326	27	301.1
92.4	344	318	26	294.0
92.4	343	317	26	293.0
92.4	342	316	26	292.0
92.4	341	315	26	291.0
92.4	340	314	26	290.0
92.4	331	306	25	282.9
92.4	330	305	25	281.9
92.4	329	304	25	280.9
92.4	328	303	25	279.9
92.4	327	302	25	278.9
92.4	317	293	24	270.8
92.4	316	292	24	269.8
92.4	315	291	24	268.8
92.4	314	290	24	267.8
92.4	304	281	23	259.7
92.4	303	280	23	258.7
92.4	302	279	23	257.8
92.4	301	278	23	256.8
92.4	291	269	22	248.7
92.4	290	268	22	247.7
92.4	289	267	22	246.7
92.4	288	266	22	245.7
92.4	278	257	21	237.6
92.4	277	256	21	236.6
92.4	276	255	21	235.6
92.4	275	254	21	234.6
92.4	264	244	20	225.5

%E	M1	M2	DM	M*
92.4	263	243	20	224.5
92.4	262	242	20	223.5
92.4	251	232	19	214.4
92.4	250	231	19	213.4
92.4	249	230	19	212.4
92.4	238	220	18	203.4
92.4	237	219	18	202.4
92.4	236	218	18	201.4
92.4	225	208	17	192.3
92.4	224	207	17	191.3
92.4	223	206	17	190.3
92.4	211	195	16	180.2
92.4	210	194	16	179.2
92.4	198	183	15	169.1
92.4	197	182	15	168.1
92.4	185	171	14	158.1
92.4	184	170	14	157.1
92.4	183	169	14	156.1
92.4	172	159	13	147.0
92.4	171	158	13	146.0
92.4	170	157	13	145.0
92.4	158	146	12	134.9
92.4	157	145	12	133.9
92.4	145	134	11	123.8
92.4	144	133	11	122.8
92.4	132	122	10	112.8
92.4	131	121	10	111.8
92.4	119	110	9	101.7
92.4	118	109	9	100.7
92.4	105	97	8	89.6
92.4	92	85	7	78.5
92.4	79	73	6	67.5
92.4	66	61	5	56.4
92.3	496	458	38	422.9
92.3	495	457	38	421.9
92.3	494	456	38	420.9
92.3	493	455	38	419.9
92.3	492	454	38	418.9
92.3	491	453	38	417.9
92.3	483	446	37	411.8
92.3	482	445	37	410.8
92.3	481	444	37	409.8
92.3	480	443	37	408.9
92.3	479	442	37	407.9
92.3	478	441	37	406.9
92.3	470	434	36	400.8
92.3	469	433	36	399.8
92.3	468	432	36	398.8
92.3	467	431	36	397.8
92.3	466	430	36	396.8

%E	M1	M2	DM	M*
92.3	465	429	36	395.8
92.3	457	422	35	389.7
92.3	456	421	35	388.7
92.3	455	420	35	387.7
92.3	454	419	35	386.7
92.3	453	418	35	385.7
92.3	452	417	35	384.7
92.3	444	410	34	378.6
92.3	443	409	34	377.6
92.3	442	408	34	376.6
92.3	441	407	34	375.6
92.3	440	406	34	374.6
92.3	439	405	34	373.6
92.3	431	398	33	367.5
92.3	430	397	33	366.5
92.3	429	396	33	365.5
92.3	428	395	33	364.5
92.3	427	394	33	363.6
92.3	426	393	33	362.6
92.3	418	386	32	356.4
92.3	417	385	32	355.5
92.3	416	384	32	354.5
92.3	415	383	32	353.5
92.3	414	382	32	352.5
92.3	413	381	32	351.5
92.3	405	374	31	345.4
92.3	404	373	31	344.4
92.3	403	372	31	343.4
92.3	402	371	31	342.4
92.3	401	370	31	341.4
92.3	400	369	31	340.4
92.3	392	362	30	334.3
92.3	391	361	30	333.3
92.3	390	360	30	332.3
92.3	389	359	30	331.3
92.3	388	358	30	330.3
92.3	379	350	29	323.2
92.3	378	349	29	322.2
92.3	377	348	29	321.2
92.3	376	347	29	320.2
92.3	375	346	29	319.2
92.3	365	337	28	311.1
92.3	364	336	28	310.2
92.3	363	335	28	309.2
92.3	362	334	28	308.2
92.3	352	325	27	300.1
92.3	351	324	27	299.1
92.3	350	323	27	298.1
92.3	349	322	27	297.1
92.3	339	313	26	289.0

%E	M1	M2	DM	M*
92.3	338	312	26	288.0
92.3	337	311	26	287.0
92.3	336	310	26	286.0
92.3	326	301	25	277.9
92.3	325	300	25	276.9
92.3	324	299	25	275.9
92.3	323	298	25	274.9
92.3	313	289	24	266.8
92.3	312	288	24	265.8
92.3	311	287	24	264.9
92.3	310	286	24	263.9
92.3	300	277	23	255.8
92.3	299	276	23	254.8
92.3	298	275	23	253.8
92.3	297	274	23	252.8
92.3	287	265	22	244.7
92.3	286	264	22	243.7
92.3	285	263	22	242.7
92.3	284	262	22	241.7
92.3	274	253	21	233.6
92.3	273	252	21	232.6
92.3	272	251	21	231.6
92.3	271	250	21	230.6
92.3	261	241	20	222.5
92.3	260	240	20	221.5
92.3	259	239	20	220.5
92.3	248	229	19	211.5
92.3	247	228	19	210.5
92.3	246	227	19	209.5
92.3	235	217	18	200.4
92.3	234	216	18	199.4
92.3	233	215	18	198.4
92.3	222	205	17	189.3
92.3	221	204	17	188.3
92.3	220	203	17	187.3
92.3	209	193	16	178.2
92.3	208	192	16	177.2
92.3	207	191	16	176.2
92.3	196	181	15	167.1
92.3	195	180	15	166.2
92.3	194	179	15	165.2
92.3	182	168	14	155.1
92.3	181	167	14	154.1
92.3	169	156	13	144.0
92.3	168	155	13	143.0
92.3	156	144	12	132.9
92.3	155	143	12	131.9
92.3	143	132	11	121.8
92.3	142	131	11	120.9
92.3	130	120	10	110.8

%E	M1	M2	DM	M*
92.3	117	108	9	99.7
92.3	104	96	8	88.6
92.3	91	84	7	77.5
92.3	78	72	6	66.5
92.3	65	60	5	55.4
92.3	52	48	4	44.3
92.3	39	36	3	33.2
92.3	26	24	2	22.2
92.3	13	12	1	11.1
92.2	500	461	39	425.0
92.2	499	460	39	424.0
92.2	498	459	39	423.1
92.2	497	458	39	422.1
92.2	490	452	38	416.9
92.2	489	451	38	416.0
92.2	488	450	38	415.0
92.2	487	449	38	414.0
92.2	486	448	38	413.0
92.2	485	447	38	412.0
92.2	477	440	37	405.9
92.2	476	439	37	404.9
92.2	475	438	37	403.9
92.2	474	437	37	402.9
92.2	473	436	37	401.9
92.2	472	435	37	400.9
92.2	464	428	36	394.8
92.2	463	427	36	393.8
92.2	462	426	36	392.8
92.2	461	425	36	391.8
92.2	460	424	36	390.8
92.2	459	423	36	389.8
92.2	451	416	35	383.7
92.2	450	415	35	382.7
92.2	449	414	35	381.7
92.2	448	413	35	380.7
92.2	447	412	35	379.7
92.2	446	411	35	378.7
92.2	438	404	34	372.6
92.2	437	403	34	371.6
92.2	436	402	34	370.7
92.2	435	401	34	369.7
92.2	434	400	34	368.7
92.2	425	392	33	361.6
92.2	424	391	33	360.6
92.2	423	390	33	359.6
92.2	422	389	33	358.6
92.2	421	388	33	357.6
92.2	412	380	32	350.5
92.2	411	379	32	349.5
92.2	410	378	32	348.5

%E	M1	M2	DM	M*
92.2	409	377	32	347.5
92.2	408	376	32	346.5
92.2	399	368	31	339.4
92.2	398	367	31	338.4
92.2	397	366	31	337.4
92.2	396	365	31	336.4
92.2	395	364	31	335.4
92.2	387	357	30	329.3
92.2	386	356	30	328.3
92.2	385	355	30	327.3
92.2	384	354	30	326.3
92.2	383	353	30	325.3
92.2	374	345	29	318.2
92.2	373	344	29	317.3
92.2	372	343	29	316.3
92.2	371	342	29	315.3
92.2	370	341	29	314.3
92.2	361	333	28	307.2
92.2	360	332	28	306.2
92.2	359	331	28	305.2
92.2	358	330	28	304.2
92.2	357	329	28	303.2
92.2	348	321	27	296.1
92.2	347	320	27	295.1
92.2	346	319	27	294.1
92.2	345	318	27	293.1
92.2	344	317	27	292.1
92.2	335	309	26	285.0
92.2	334	308	26	284.0
92.2	333	307	26	283.0
92.2	332	306	26	282.0
92.2	322	297	25	273.9
92.2	321	296	25	272.9
92.2	320	295	25	272.0
92.2	319	294	25	271.0
92.2	309	285	24	262.9
92.2	308	284	24	261.9
92.2	307	283	24	260.9
92.2	306	282	24	259.9
92.2	296	273	23	251.8
92.2	295	272	23	250.8
92.2	294	271	23	249.8
92.2	293	270	23	248.8
92.2	283	261	22	240.7
92.2	282	260	22	239.7
92.2	281	259	22	238.7
92.2	270	249	21	229.6
92.2	269	248	21	228.6
92.2	268	247	21	227.6
92.2	258	238	20	219.6
92.2	257	237	20	218.6
92.2	256	236	20	217.6
92.2	255	235	20	216.6
92.2	245	226	19	208.5
92.2	244	225	19	207.5
92.2	243	224	19	206.5
92.2	232	214	18	197.4
92.2	231	213	18	196.4
92.2	230	212	18	195.4
92.2	219	202	17	186.3
92.2	218	201	17	185.3
92.2	217	200	17	184.3
92.2	206	190	16	175.2
92.2	205	189	16	174.2
92.2	204	188	16	173.3
92.2	193	178	15	164.2
92.2	192	177	15	163.2
92.2	180	166	14	153.1
92.2	179	165	14	152.1
92.2	167	154	13	142.0
92.2	166	153	13	141.0
92.2	154	142	12	130.9
92.2	153	141	12	129.9
92.2	141	130	11	119.9
92.2	129	119	10	109.8
92.2	128	118	10	108.8
92.2	116	107	9	98.7
92.2	115	106	9	97.7
92.2	103	95	8	87.6
92.2	102	94	8	86.6
92.2	90	83	7	76.5
92.2	77	71	6	65.5
92.2	64	59	5	54.4
92.2	51	47	4	43.3
92.1	496	457	39	421.1
92.1	495	456	39	420.1
92.1	494	455	39	419.1
92.1	493	454	39	418.1
92.1	492	453	39	417.1
92.1	491	452	39	416.1
92.1	484	446	38	411.0
92.1	483	445	38	410.0
92.1	482	444	38	409.0
92.1	481	443	38	408.0
92.1	480	442	38	407.0
92.1	479	441	38	406.0
92.1	478	440	38	405.0
92.1	471	434	37	399.9
92.1	470	433	37	398.9
92.1	469	432	37	397.9
92.1	468	431	37	396.9
92.1	467	430	37	395.9
92.1	466	429	37	394.9
92.1	458	422	36	388.8
92.1	457	421	36	387.8
92.1	456	420	36	386.8
92.1	455	419	36	385.8
92.1	454	418	36	384.9
92.1	453	417	36	383.9
92.1	445	410	35	377.8
92.1	444	409	35	376.8
92.1	443	408	35	375.8
92.1	442	407	35	374.8
92.1	441	406	35	373.8
92.1	433	399	34	367.7
92.1	432	398	34	366.7
92.1	431	397	34	365.7
92.1	430	396	34	364.7
92.1	429	395	34	363.7
92.1	428	394	34	362.7
92.1	420	387	33	356.6
92.1	419	386	33	355.6
92.1	418	385	33	354.6
92.1	417	384	33	353.6
92.1	416	383	33	352.6
92.1	407	375	32	345.5
92.1	406	374	32	344.5
92.1	405	373	32	343.5
92.1	404	372	32	342.5
92.1	403	371	32	341.5
92.1	394	363	31	334.4
92.1	393	362	31	333.4
92.1	392	361	31	332.5
92.1	391	360	31	331.5
92.1	390	359	31	330.5
92.1	382	352	30	324.4
92.1	381	351	30	323.4
92.1	380	350	30	322.4
92.1	379	349	30	321.4
92.1	378	348	30	320.4
92.1	369	340	29	313.3
92.1	368	339	29	312.3
92.1	367	338	29	311.3
92.1	366	337	29	310.3
92.1	365	336	29	309.3
92.1	356	328	28	302.2
92.1	355	327	28	301.2
92.1	354	326	28	300.2
92.1	353	325	28	299.2
92.1	343	316	27	291.1
92.1	342	315	27	290.1
92.1	341	314	27	289.1
92.1	340	313	27	288.1
92.1	331	305	26	281.0
92.1	330	304	26	280.0
92.1	329	303	26	279.1
92.1	328	302	26	278.1
92.1	318	293	25	270.0
92.1	317	292	25	269.0
92.1	316	291	25	268.0
92.1	315	290	25	267.0
92.1	305	281	24	258.9
92.1	304	280	24	257.9
92.1	303	279	24	256.9
92.1	302	278	24	255.9
92.1	292	269	23	247.8
92.1	291	268	23	246.8
92.1	290	267	23	245.8
92.1	280	258	22	237.7
92.1	279	257	22	236.7
92.1	278	256	22	235.7
92.1	277	255	22	234.7
92.1	267	246	21	226.7
92.1	266	245	21	225.7
92.1	265	244	21	224.7
92.1	254	234	20	215.6
92.1	253	233	20	214.6
92.1	252	232	20	213.6
92.1	242	223	19	205.5
92.1	241	222	19	204.5
92.1	240	221	19	203.5
92.1	239	220	19	202.5
92.1	229	211	18	194.4
92.1	228	210	18	193.4
92.1	227	209	18	192.4
92.1	216	199	17	183.3
92.1	215	198	17	182.3
92.1	214	197	17	181.4
92.1	203	187	16	172.3
92.1	202	186	16	171.3
92.1	191	176	15	162.2
92.1	190	175	15	161.2
92.1	189	174	15	160.2
92.1	178	164	14	151.1
92.1	177	163	14	150.1
92.1	165	152	13	140.0
92.1	164	151	13	139.0
92.1	152	140	12	128.9
92.1	151	139	12	128.0
92.1	140	129	11	118.9
92.1	139	128	11	117.9
92.1	127	117	10	107.8
92.1	126	116	10	106.8
92.1	114	105	9	96.7
92.1	101	93	8	85.6
92.1	89	82	7	75.6
92.1	76	70	6	64.5
92.1	63	58	5	53.4
92.1	38	35	3	32.2
92.0	500	460	40	423.2
92.0	499	459	40	422.2
92.0	498	458	40	421.2
92.0	497	457	40	420.2
92.0	490	451	39	415.1
92.0	489	450	39	414.1
92.0	488	449	39	413.1
92.0	487	448	39	412.1
92.0	486	447	39	411.1
92.0	485	446	39	410.1
92.0	477	439	38	404.0
92.0	476	438	38	403.0
92.0	475	437	38	402.0
92.0	474	436	38	401.0
92.0	473	435	38	400.1
92.0	465	428	37	393.9
92.0	464	427	37	393.0
92.0	463	426	37	392.0
92.0	462	425	37	391.0
92.0	461	424	37	390.0
92.0	460	423	37	389.0
92.0	452	416	36	382.9
92.0	451	415	36	381.9
92.0	450	414	36	380.9
92.0	449	413	36	379.9
92.0	448	412	36	378.9
92.0	440	405	35	372.8
92.0	439	404	35	371.8
92.0	438	403	35	370.8
92.0	437	402	35	369.8
92.0	436	401	35	368.8
92.0	435	400	35	367.8
92.0	427	393	34	361.7
92.0	426	392	34	360.7
92.0	425	391	34	359.7
92.0	424	390	34	358.7
92.0	423	389	34	357.7
92.0	415	382	33	351.6
92.0	414	381	33	350.6
92.0	413	380	33	349.6
92.0	412	379	33	348.6

%E	M1	M2	DM	M*
92.0	411	378	33	347.6
92.0	410	377	33	346.7
92.0	402	370	32	340.5
92.0	401	369	32	339.6
92.0	400	368	32	338.6
92.0	399	367	32	337.6
92.0	398	366	32	336.6
92.0	389	358	31	329.5
92.0	388	357	31	328.5
92.0	387	356	31	327.5
92.0	386	355	31	326.5
92.0	377	347	30	319.4
92.0	376	346	30	318.4
92.0	375	345	30	317.4
92.0	374	344	30	316.4
92.0	373	343	30	315.4
92.0	364	335	29	308.3
92.0	363	334	29	307.3
92.0	362	333	29	306.3
92.0	361	332	29	305.3
92.0	352	324	28	298.2
92.0	351	323	28	297.2
92.0	350	322	28	296.2
92.0	349	321	28	295.2
92.0	348	320	28	294.3
92.0	339	312	27	287.2
92.0	338	311	27	286.2
92.0	337	310	27	285.2
92.0	336	309	27	284.2
92.0	327	301	26	277.1
92.0	326	300	26	276.1
92.0	325	299	26	275.1
92.0	324	298	26	274.1
92.0	323	297	26	273.1
92.0	314	289	25	266.0
92.0	313	288	25	265.0
92.0	312	287	25	264.0
92.0	311	286	25	263.0
92.0	301	277	24	254.9
92.0	300	276	24	253.9
92.0	299	275	24	252.9
92.0	289	266	23	244.8
92.0	288	265	23	243.8
92.0	287	264	23	242.8
92.0	286	263	23	241.8
92.0	276	254	22	233.8
92.0	275	253	22	232.8
92.0	274	252	22	231.8
92.0	264	243	21	223.7
92.0	263	242	21	222.7

%E	M1	M2	DM	M*
92.0	262	241	21	221.7
92.0	261	240	21	220.7
92.0	251	231	20	212.6
92.0	250	230	20	211.6
92.0	249	229	20	210.6
92.0	238	219	19	201.5
92.0	237	218	19	200.5
92.0	226	208	18	191.4
92.0	225	207	18	190.4
92.0	224	206	18	189.4
92.0	213	196	17	180.4
92.0	212	195	17	179.4
92.0	201	185	16	170.3
92.0	200	184	16	169.3
92.0	199	183	16	168.3
92.0	188	173	15	159.2
92.0	187	172	15	158.2
92.0	176	162	14	149.1
92.0	175	161	14	148.1
92.0	174	160	14	147.1
92.0	163	150	13	138.0
92.0	162	149	13	137.0
92.0	150	138	12	127.0
92.0	138	127	11	116.9
92.0	137	126	11	115.9
92.0	125	115	10	105.8
92.0	113	104	9	95.7
92.0	112	103	9	94.7
92.0	100	92	8	84.6
92.0	88	81	7	74.6
92.0	87	80	7	73.6
92.0	75	69	6	63.5
92.0	50	46	4	42.3
92.0	25	23	2	21.2
91.9	496	456	40	419.2
91.9	495	455	40	418.2
91.9	494	454	40	417.2
91.9	493	453	40	416.2
91.9	492	452	40	415.3
91.9	491	451	40	414.3
91.9	484	445	39	409.1
91.9	483	444	39	408.1
91.9	482	443	39	407.2
91.9	481	442	39	406.2
91.9	480	441	39	405.2
91.9	479	440	39	404.2
91.9	472	434	38	399.1
91.9	471	433	38	398.1
91.9	470	432	38	397.1
91.9	469	431	38	396.1

%E	M1	M2	DM	M*
91.9	468	430	38	395.1
91.9	467	429	38	394.1
91.9	459	422	37	388.0
91.9	458	421	37	387.0
91.9	457	420	37	386.0
91.9	456	419	37	385.0
91.9	455	418	37	384.0
91.9	454	417	37	383.0
91.9	447	411	36	377.9
91.9	446	410	36	376.9
91.9	445	409	36	375.9
91.9	444	408	36	374.9
91.9	443	407	36	373.9
91.9	442	406	36	372.9
91.9	434	399	35	366.8
91.9	433	398	35	365.8
91.9	432	397	35	364.8
91.9	431	396	35	363.8
91.9	430	395	35	362.8
91.9	422	388	34	356.7
91.9	421	387	34	355.7
91.9	420	386	34	354.8
91.9	419	385	34	353.8
91.9	418	384	34	352.8
91.9	409	376	33	345.7
91.9	408	375	33	344.7
91.9	407	374	33	343.7
91.9	406	373	33	342.7
91.9	405	372	33	341.7
91.9	397	365	32	335.6
91.9	396	364	32	334.6
91.9	395	363	32	333.6
91.9	394	362	32	332.6
91.9	393	361	32	331.6
91.9	385	354	31	325.5
91.9	384	353	31	324.5
91.9	383	352	31	323.5
91.9	382	351	31	322.5
91.9	381	350	31	321.5
91.9	372	342	30	314.4
91.9	371	341	30	313.4
91.9	370	340	30	312.4
91.9	369	339	30	311.4
91.9	360	331	29	304.3
91.9	359	330	29	303.3
91.9	358	329	29	302.3
91.9	357	328	29	301.4
91.9	356	327	29	300.4
91.9	347	319	28	293.3
91.9	346	318	28	292.3

%E	M1	M2	DM	M*
91.9	345	317	28	291.3
91.9	344	316	28	290.3
91.9	335	308	27	283.2
91.9	334	307	27	282.2
91.9	333	306	27	281.2
91.9	332	305	27	280.2
91.9	322	296	26	272.1
91.9	321	295	26	271.1
91.9	320	294	26	270.1
91.9	319	293	26	269.1
91.9	310	285	25	262.0
91.9	309	284	25	261.0
91.9	308	283	25	260.0
91.9	307	282	25	259.0
91.9	298	274	24	251.9
91.9	297	273	24	250.9
91.9	296	272	24	249.9
91.9	295	271	24	249.0
91.9	285	262	23	240.9
91.9	284	261	23	239.9
91.9	283	260	23	238.9
91.9	273	251	22	230.8
91.9	272	250	22	229.8
91.9	271	249	22	228.8
91.9	270	248	22	227.8
91.9	260	239	21	219.7
91.9	259	238	21	218.7
91.9	258	237	21	217.7
91.9	248	228	20	209.6
91.9	247	227	20	208.6
91.9	246	226	20	207.6
91.9	236	217	19	199.5
91.9	235	216	19	198.5
91.9	234	215	19	197.5
91.9	223	205	18	188.5
91.9	222	204	18	187.5
91.9	221	203	18	186.5
91.9	211	194	17	178.4
91.9	210	193	17	177.4
91.9	209	192	17	176.4
91.9	198	182	16	167.3
91.9	197	181	16	166.3
91.9	186	171	15	157.2
91.9	185	170	15	156.2
91.9	173	159	14	146.1
91.9	172	158	14	145.1
91.9	161	148	13	136.0
91.9	160	147	13	135.1
91.9	149	137	12	126.0
91.9	148	136	12	125.0

%E	M1	M2	DM	M*
91.9	136	125	11	114.9
91.9	135	124	11	113.9
91.9	124	114	10	104.8
91.9	123	113	10	103.8
91.9	111	102	9	93.7
91.9	99	91	8	83.6
91.9	86	79	7	72.6
91.9	74	68	6	62.5
91.9	62	57	5	52.4
91.9	37	34	3	31.2
91.8	500	459	41	421.4
91.8	499	458	41	420.4
91.8	498	457	41	419.4
91.8	497	456	41	418.4
91.8	490	450	40	413.3
91.8	489	449	40	412.3
91.8	488	448	40	411.3
91.8	487	447	40	410.3
91.8	486	446	40	409.3
91.8	485	445	40	408.3
91.8	478	439	39	403.2
91.8	477	438	39	402.2
91.8	476	437	39	401.2
91.8	475	436	39	400.2
91.8	474	435	39	399.2
91.8	473	434	39	398.2
91.8	466	428	38	393.1
91.8	465	427	38	392.1
91.8	464	426	38	391.1
91.8	463	425	38	390.1
91.8	462	424	38	389.1
91.8	461	423	38	388.1
91.8	453	416	37	382.0
91.8	452	415	37	381.0
91.8	451	414	37	380.0
91.8	450	413	37	379.0
91.8	449	412	37	378.0
91.8	441	405	36	371.9
91.8	440	404	36	370.9
91.8	439	403	36	370.0
91.8	438	402	36	369.0
91.8	437	401	36	368.0
91.8	429	394	35	361.9
91.8	428	393	35	360.9
91.8	427	392	35	359.9
91.8	426	391	35	358.9
91.8	425	390	35	357.9
91.8	417	383	34	351.8
91.8	416	382	34	350.8
91.8	415	381	34	349.8

%E	M1	M2	DM	M*
91.8	414	380	34	348.8
91.8	413	379	34	347.8
91.8	404	371	33	340.7
91.8	403	370	33	339.7
91.8	402	369	33	338.7
91.8	401	368	33	337.7
91.8	400	367	33	336.7
91.8	392	360	32	330.6
91.8	391	359	32	329.6
91.8	390	358	32	328.6
91.8	389	357	32	327.6
91.8	388	356	32	326.6
91.8	380	349	31	320.5
91.8	379	348	31	319.5
91.8	378	347	31	318.5
91.8	377	346	31	317.5
91.8	376	345	31	316.6
91.8	368	338	30	310.4
91.8	367	337	30	309.5
91.8	366	336	30	308.5
91.8	365	335	30	307.5
91.8	364	334	30	306.5
91.8	355	326	29	299.4
91.8	354	325	29	298.4
91.8	353	324	29	297.4
91.8	352	323	29	296.4
91.8	343	315	28	289.3
91.8	342	314	28	288.3
91.8	341	313	28	287.3
91.8	340	312	28	286.3
91.8	331	304	27	279.2
91.8	330	303	27	278.2
91.8	329	302	27	277.2
91.8	328	301	27	276.2
91.8	318	292	26	268.1
91.8	317	291	26	267.1
91.8	316	290	26	266.1
91.8	306	281	25	258.0
91.8	305	280	25	257.0
91.8	304	279	25	256.1
91.8	294	270	24	248.0
91.8	293	269	24	247.0
91.8	292	268	24	246.0
91.8	291	267	24	245.0
91.8	282	259	23	237.9
91.8	281	258	23	236.9
91.8	280	257	23	235.9
91.8	279	256	23	234.9
91.8	269	247	22	226.8
91.8	268	246	22	225.8

%E	M1	M2	DM	M*
91.8	267	245	22	224.8
91.8	257	236	21	216.7
91.8	256	235	21	215.7
91.8	255	234	21	214.7
91.8	245	225	20	206.6
91.8	244	224	20	205.6
91.8	243	223	20	204.6
91.8	233	214	19	196.5
91.8	232	213	19	195.6
91.8	231	212	19	194.6
91.8	220	202	18	185.5
91.8	219	201	18	184.5
91.8	208	191	17	175.4
91.8	207	190	17	174.4
91.8	196	180	16	165.3
91.8	195	179	16	164.3
91.8	194	178	16	163.3
91.8	184	169	15	155.2
91.8	183	168	15	154.2
91.8	182	167	15	153.2
91.8	171	157	14	144.1
91.8	170	156	14	143.2
91.8	159	146	13	134.1
91.8	158	145	13	133.1
91.8	147	135	12	124.0
91.8	146	134	12	123.0
91.8	134	123	11	112.9
91.8	122	112	10	102.8
91.8	110	101	9	92.7
91.8	98	90	8	82.7
91.8	97	89	8	81.7
91.8	85	78	7	71.6
91.8	73	67	6	61.5
91.8	61	56	5	51.4
91.8	49	45	4	41.3
91.7	496	455	41	417.4
91.7	495	454	41	416.4
91.7	494	453	41	415.4
91.7	493	452	41	414.4
91.7	492	451	41	413.4
91.7	491	450	41	412.4
91.7	484	444	40	407.3
91.7	483	443	40	406.3
91.7	482	442	40	405.3
91.7	481	441	40	404.3
91.7	480	440	40	403.3
91.7	472	433	39	397.2
91.7	471	432	39	396.2
91.7	470	431	39	395.2
91.7	469	430	39	394.2

%E	M1	M2	DM	M*
91.7	468	429	39	393.3
91.7	460	422	38	387.1
91.7	459	421	38	386.1
91.7	458	420	38	385.2
91.7	457	419	38	384.2
91.7	456	418	38	383.2
91.7	448	411	37	377.1
91.7	447	410	37	376.1
91.7	446	409	37	375.1
91.7	445	408	37	374.1
91.7	444	407	37	373.1
91.7	436	400	36	367.0
91.7	435	399	36	366.0
91.7	434	398	36	365.0
91.7	433	397	36	364.0
91.7	432	396	36	363.0
91.7	424	389	35	356.9
91.7	423	388	35	355.9
91.7	422	387	35	354.9
91.7	421	386	35	353.9
91.7	420	385	35	352.9
91.7	412	378	34	346.8
91.7	411	377	34	345.8
91.7	410	376	34	344.8
91.7	409	375	34	343.8
91.7	408	374	34	342.8
91.7	399	366	33	335.7
91.7	398	365	33	334.7
91.7	397	364	33	333.7
91.7	396	363	33	332.8
91.7	387	355	32	325.6
91.7	386	354	32	324.7
91.7	385	353	32	323.7
91.7	384	352	32	322.7
91.7	375	344	31	315.6
91.7	374	343	31	314.6
91.7	373	342	31	313.6
91.7	372	341	31	312.6
91.7	363	333	30	305.5
91.7	362	332	30	304.5
91.7	361	331	30	303.5
91.7	360	330	30	302.5
91.7	351	322	29	295.4
91.7	350	321	29	294.4
91.7	349	320	29	293.4
91.7	348	319	29	292.4
91.7	339	311	28	285.3
91.7	338	310	28	284.3
91.7	337	309	28	283.3
91.7	336	308	28	282.3

%E	M1	M2	DM	M*
91.7	327	300	27	275.2
91.7	326	299	27	274.2
91.7	325	298	27	273.2
91.7	324	297	27	272.3
91.7	315	289	26	265.1
91.7	314	288	26	264.2
91.7	313	287	26	263.2
91.7	312	286	26	262.2
91.7	303	278	25	255.1
91.7	302	277	25	254.1
91.7	301	276	25	253.1
91.7	300	275	25	252.1
91.7	290	266	24	244.0
91.7	289	265	24	243.0
91.7	288	264	24	242.0
91.7	278	255	23	233.9
91.7	277	254	23	232.9
91.7	276	253	23	231.9
91.7	266	244	22	223.8
91.7	265	243	22	222.8
91.7	264	242	22	221.8
91.7	254	233	21	213.7
91.7	253	232	21	212.7
91.7	252	231	21	211.8
91.7	242	222	20	203.7
91.7	241	221	20	202.7
91.7	240	220	20	201.7
91.7	230	211	19	193.6
91.7	229	210	19	192.6
91.7	228	209	19	191.6
91.7	218	200	18	183.5
91.7	217	199	18	182.5
91.7	216	198	18	181.5
91.7	206	189	17	173.4
91.7	205	188	17	172.4
91.7	204	187	17	171.4
91.7	193	177	16	162.3
91.7	192	176	16	161.3
91.7	181	166	15	152.2
91.7	180	165	15	151.3
91.7	169	155	14	142.2
91.7	168	154	14	141.2
91.7	157	144	13	132.1
91.7	156	143	13	131.1
91.7	145	133	12	122.0
91.7	144	132	12	121.0
91.7	133	122	11	111.9
91.7	132	121	11	110.9
91.7	121	111	10	101.8
91.7	120	110	10	100.8

%E	M1	M2	DM	M*
91.7	109	100	9	91.7
91.7	108	99	9	90.8
91.7	96	88	8	80.7
91.7	84	77	7	70.6
91.7	72	66	6	60.5
91.7	60	55	5	50.4
91.7	48	44	4	40.3
91.7	36	33	3	30.3
91.7	24	22	2	20.2
91.7	12	11	1	10.1
91.6	500	458	42	419.5
91.6	499	457	42	418.5
91.6	498	456	42	417.5
91.6	490	449	41	411.4
91.6	489	448	41	410.4
91.6	488	447	41	409.4
91.6	487	446	41	408.5
91.6	486	445	41	407.5
91.6	479	439	40	402.3
91.6	478	438	40	401.3
91.6	477	437	40	400.4
91.6	476	436	40	399.4
91.6	475	435	40	398.4
91.6	474	434	40	397.4
91.6	467	428	39	392.3
91.6	466	427	39	391.3
91.6	465	426	39	390.3
91.6	464	425	39	389.3
91.6	463	424	39	388.3
91.6	462	423	39	387.3
91.6	455	417	38	382.2
91.6	454	416	38	381.2
91.6	453	415	38	380.2
91.6	452	414	38	379.2
91.6	451	413	38	378.2
91.6	450	412	38	377.2
91.6	443	406	37	372.1
91.6	442	405	37	371.1
91.6	441	404	37	370.1
91.6	440	403	37	369.1
91.6	439	402	37	368.1
91.6	438	401	37	367.1
91.6	431	395	36	362.0
91.6	430	394	36	361.0
91.6	429	393	36	360.0
91.6	428	392	36	359.0
91.6	427	391	36	358.0
91.6	419	384	35	351.9
91.6	418	383	35	350.9
91.6	417	382	35	349.9

%E	M1	M2	DM	M*
91.6	416	381	35	348.9
91.6	415	380	35	348.0
91.6	407	373	34	341.8
91.6	406	372	34	340.8
91.6	405	371	34	339.9
91.6	404	370	34	338.9
91.6	403	369	34	337.9
91.6	395	362	33	331.8
91.6	394	361	33	330.8
91.6	393	360	33	329.8
91.6	392	359	33	328.8
91.6	391	358	33	327.8
91.6	383	351	32	321.7
91.6	382	350	32	320.7
91.6	381	349	32	319.7
91.6	380	348	32	318.7
91.6	379	347	32	317.7
91.6	371	340	31	311.6
91.6	370	339	31	310.6
91.6	369	338	31	309.6
91.6	368	337	31	308.6
91.6	367	336	31	307.6
91.6	359	329	30	301.5
91.6	358	328	30	300.5
91.6	357	327	30	299.5
91.6	356	326	30	298.5
91.6	347	318	29	291.4
91.6	346	317	29	290.4
91.6	345	316	29	289.4
91.6	344	315	29	288.4
91.6	335	307	28	281.3
91.6	334	306	28	280.3
91.6	333	305	28	279.4
91.6	332	304	28	278.4
91.6	323	296	27	271.3
91.6	322	295	27	270.3
91.6	321	294	27	269.3
91.6	320	293	27	268.3
91.6	311	285	26	261.2
91.6	310	284	26	260.2
91.6	309	283	26	259.2
91.6	308	282	26	258.2
91.6	299	274	25	251.1
91.6	298	273	25	250.1
91.6	297	272	25	249.1
91.6	296	271	25	248.1
91.6	287	263	24	241.0
91.6	286	262	24	240.0
91.6	285	261	24	239.0
91.6	275	252	23	230.9
91.6	274	251	23	229.9
91.6	273	250	23	228.9
91.6	263	241	22	220.8
91.6	262	240	22	219.8
91.6	261	239	22	218.9
91.6	251	230	21	210.8
91.6	250	229	21	209.8
91.6	249	228	21	208.8
91.6	239	219	20	200.7
91.6	238	218	20	199.7
91.6	237	217	20	198.7
91.6	227	208	19	190.6
91.6	226	207	19	189.6
91.6	225	206	19	188.6
91.6	215	197	18	180.5
91.6	214	196	18	179.5
91.6	203	186	17	170.4
91.6	202	185	17	169.4
91.6	191	175	16	160.3
91.6	190	174	16	159.3
91.6	179	164	15	150.3
91.6	178	163	15	149.3
91.6	167	153	14	140.2
91.6	166	152	14	139.2
91.6	155	142	13	130.1
91.6	154	141	13	129.1
91.6	143	131	12	120.0
91.6	131	120	11	109.9
91.6	119	109	10	99.8
91.6	107	98	9	89.8
91.6	95	87	8	79.7
91.6	83	76	7	69.6
91.5	497	455	42	416.5
91.5	496	454	42	415.6
91.5	495	453	42	414.6
91.5	494	452	42	413.6
91.5	493	451	42	412.6
91.5	492	450	42	411.6
91.5	485	444	41	406.5
91.5	484	443	41	405.5
91.5	483	442	41	404.5
91.5	482	441	41	403.5
91.5	481	440	41	402.5
91.5	480	439	41	401.5
91.5	473	433	40	396.4
91.5	472	432	40	395.4
91.5	471	431	40	394.4
91.5	470	430	40	393.4
91.5	469	429	40	392.4
91.5	468	428	40	391.4
91.5	461	422	39	386.3
91.5	460	421	39	385.3
91.5	459	420	39	384.3
91.5	458	419	39	383.3
91.5	457	418	39	382.3
91.5	449	411	38	376.2
91.5	448	410	38	375.2
91.5	447	409	38	374.2
91.5	446	408	38	373.2
91.5	445	407	38	372.2
91.5	437	400	37	366.1
91.5	436	399	37	365.1
91.5	435	398	37	364.1
91.5	434	397	37	363.2
91.5	433	396	37	362.2
91.5	426	390	36	357.0
91.5	425	389	36	356.0
91.5	424	388	36	355.1
91.5	423	387	36	354.1
91.5	422	386	36	353.1
91.5	414	379	35	347.0
91.5	413	378	35	346.0
91.5	412	377	35	345.0
91.5	411	376	35	344.0
91.5	410	375	35	343.0
91.5	402	368	34	336.9
91.5	401	367	34	335.9
91.5	400	366	34	334.9
91.5	399	365	34	333.9
91.5	398	364	34	332.9
91.5	390	357	33	326.8
91.5	389	356	33	325.8
91.5	388	355	33	324.8
91.5	387	354	33	323.8
91.5	386	353	33	322.8
91.5	378	346	32	316.7
91.5	377	345	32	315.7
91.5	376	344	32	314.7
91.5	375	343	32	313.7
91.5	366	335	31	306.6
91.5	365	334	31	305.6
91.5	364	333	31	304.6
91.5	363	332	31	303.6
91.5	355	325	30	297.5
91.5	354	324	30	296.5
91.5	353	323	30	295.5
91.5	352	322	30	294.6
91.5	351	321	30	293.6
91.5	343	314	29	287.5
91.5	342	313	29	286.5
91.5	341	312	29	285.5
91.5	340	311	29	284.5
91.5	331	303	28	277.4
91.5	330	302	28	276.4
91.5	329	301	28	275.4
91.5	328	300	28	274.4
91.5	319	292	27	267.3
91.5	318	291	27	266.3
91.5	317	290	27	265.3
91.5	316	289	27	264.3
91.5	307	281	26	257.2
91.5	306	280	26	256.2
91.5	305	279	26	255.2
91.5	295	270	25	247.1
91.5	294	269	25	246.1
91.5	293	268	25	245.1
91.5	284	260	24	238.0
91.5	283	259	24	237.0
91.5	282	258	24	236.0
91.5	281	257	24	235.0
91.5	272	249	23	227.9
91.5	271	248	23	227.0
91.5	270	247	23	226.0
91.5	269	246	23	225.0
91.5	260	238	22	217.9
91.5	259	237	22	216.9
91.5	258	236	22	215.9
91.5	248	227	21	207.8
91.5	247	226	21	206.8
91.5	246	225	21	205.8
91.5	236	216	20	197.7
91.5	235	215	20	196.7
91.5	234	214	20	195.7
91.5	224	205	19	187.6
91.5	223	204	19	186.6
91.5	213	195	18	178.5
91.5	212	194	18	177.5
91.5	211	193	18	176.5
91.5	201	184	17	168.4
91.5	200	183	17	167.4
91.5	199	182	17	166.5
91.5	189	173	16	158.4
91.5	188	172	16	157.4
91.5	177	162	15	148.3
91.5	176	161	15	147.3
91.5	165	151	14	138.2
91.5	164	150	14	137.2
91.5	153	140	13	128.1
91.5	142	130	12	119.0
91.5	141	129	12	118.0
91.5	130	119	11	108.9
91.5	129	118	11	107.9
91.5	118	108	10	98.8
91.5	117	107	10	97.9
91.5	106	97	9	88.8
91.5	94	86	8	78.7
91.5	82	75	7	68.6
91.5	71	65	6	59.5
91.5	59	54	5	49.4
91.5	47	43	4	39.3
91.4	500	457	43	417.7
91.4	499	456	43	416.7
91.4	498	455	43	415.7
91.4	491	449	42	410.6
91.4	490	448	42	409.6
91.4	489	447	42	408.6
91.4	488	446	42	407.6
91.4	487	445	42	406.6
91.4	486	444	42	405.6
91.4	479	438	41	400.5
91.4	478	437	41	399.5
91.4	477	436	41	398.5
91.4	476	435	41	397.5
91.4	475	434	41	396.5
91.4	474	433	41	395.5
91.4	467	427	40	390.4
91.4	466	426	40	389.4
91.4	465	425	40	388.4
91.4	464	424	40	387.4
91.4	463	423	40	386.5
91.4	456	417	39	381.3
91.4	455	416	39	380.3
91.4	454	415	39	379.4
91.4	453	414	39	378.4
91.4	452	413	39	377.4
91.4	451	412	39	376.4
91.4	444	406	38	371.3
91.4	443	405	38	370.3
91.4	442	404	38	369.3
91.4	441	403	38	368.3
91.4	440	402	38	367.3
91.4	432	395	37	361.2
91.4	431	394	37	360.2
91.4	430	393	37	359.2
91.4	429	392	37	358.2
91.4	428	391	37	357.2
91.4	421	385	36	352.1
91.4	420	384	36	351.1
91.4	419	383	36	350.1
91.4	418	382	36	349.1

%E	M1	M2	DM	M*
91.4	417	381	36	348.1
91.4	409	374	35	342.0
91.4	408	373	35	341.0
91.4	407	372	35	340.0
91.4	406	371	35	339.0
91.4	405	370	35	338.0
91.4	397	363	34	331.9
91.4	396	362	34	330.9
91.4	395	361	34	329.9
91.4	394	360	34	328.9
91.4	385	352	33	321.8
91.4	384	351	33	320.8
91.4	383	350	33	319.8
91.4	382	349	33	318.9
91.4	374	342	32	312.7
91.4	373	341	32	311.7
91.4	372	340	32	310.8
91.4	371	339	32	309.8
91.4	370	338	32	308.8
91.4	362	331	31	302.7
91.4	361	330	31	301.7
91.4	360	329	31	300.7
91.4	359	328	31	299.7
91.4	350	320	30	292.6
91.4	349	319	30	291.6
91.4	348	318	30	290.6
91.4	347	317	30	289.6
91.4	339	310	29	283.5
91.4	338	309	29	282.5
91.4	337	308	29	281.5
91.4	336	307	29	280.5
91.4	327	299	28	273.4
91.4	326	298	28	272.4
91.4	325	297	28	271.4
91.4	324	296	28	270.4
91.4	315	288	27	263.3
91.4	314	287	27	262.3
91.4	313	286	27	261.3
91.4	304	278	26	254.2
91.4	303	277	26	253.2
91.4	302	276	26	252.2
91.4	301	275	26	251.2
91.4	292	267	25	244.1
91.4	291	266	25	243.1
91.4	290	265	25	242.2
91.4	280	256	24	234.1
91.4	279	255	24	233.1
91.4	278	254	24	232.1
91.4	268	245	23	224.0
91.4	267	244	23	223.0

%E	M1	M2	DM	M*
91.4	266	243	23	222.0
91.4	257	235	22	214.9
91.4	256	234	22	213.9
91.4	255	233	22	212.9
91.4	245	224	21	204.8
91.4	244	223	21	203.8
91.4	243	222	21	202.8
91.4	233	213	20	194.7
91.4	232	212	20	193.7
91.4	222	203	19	185.6
91.4	221	202	19	184.6
91.4	220	201	19	183.6
91.4	210	192	18	175.5
91.4	209	191	18	174.6
91.4	198	181	17	165.5
91.4	197	180	17	164.5
91.4	187	171	16	156.4
91.4	186	170	16	155.4
91.4	185	169	16	154.4
91.4	175	160	15	146.3
91.4	174	159	15	145.3
91.4	163	149	14	136.2
91.4	162	148	14	135.2
91.4	152	139	13	127.1
91.4	151	138	13	126.1
91.4	140	128	12	117.0
91.4	139	127	12	116.0
91.4	128	117	11	106.9
91.4	116	106	10	96.9
91.4	105	96	9	87.8
91.4	93	85	8	77.7
91.4	81	74	7	67.6
91.4	70	64	6	58.5
91.4	58	53	5	48.4
91.4	35	32	3	29.3
91.3	497	454	43	414.7
91.3	496	453	43	413.7
91.3	495	452	43	412.7
91.3	494	451	43	411.7
91.3	493	450	43	410.8
91.3	492	449	43	409.8
91.3	485	443	42	404.6
91.3	484	442	42	403.6
91.3	483	441	42	402.7
91.3	482	440	42	401.7
91.3	481	439	42	400.7
91.3	480	438	42	399.7
91.3	473	432	41	394.6
91.3	472	431	41	393.6
91.3	471	430	41	392.6

%E	M1	M2	DM	M*
91.3	470	429	41	391.6
91.3	469	428	41	390.6
91.3	462	422	40	385.5
91.3	461	421	40	384.5
91.3	460	420	40	383.5
91.3	459	419	40	382.5
91.3	458	418	40	381.5
91.3	450	411	39	375.4
91.3	449	410	39	374.4
91.3	448	409	39	373.4
91.3	447	408	39	372.4
91.3	446	407	39	371.4
91.3	439	401	38	366.3
91.3	438	400	38	365.3
91.3	437	399	38	364.3
91.3	436	398	38	363.3
91.3	435	397	38	362.3
91.3	427	390	37	356.2
91.3	426	389	37	355.2
91.3	425	388	37	354.2
91.3	424	387	37	353.2
91.3	423	386	37	352.2
91.3	416	380	36	347.1
91.3	415	379	36	346.1
91.3	414	378	36	345.1
91.3	413	377	36	344.1
91.3	412	376	36	343.1
91.3	404	369	35	337.0
91.3	403	368	35	336.0
91.3	402	367	35	335.0
91.3	401	366	35	334.1
91.3	400	365	35	333.1
91.3	393	359	34	327.9
91.3	392	358	34	326.9
91.3	391	357	34	326.0
91.3	390	356	34	325.0
91.3	389	355	34	324.0
91.3	381	348	33	317.9
91.3	380	347	33	316.9
91.3	379	346	33	315.9
91.3	378	345	33	314.9
91.3	369	337	32	307.8
91.3	368	336	32	306.8
91.3	367	335	32	305.8
91.3	366	334	32	304.8
91.3	358	327	31	298.7
91.3	357	326	31	297.7
91.3	356	325	31	296.7
91.3	355	324	31	295.7
91.3	346	316	30	288.6

%E	M1	M2	DM	M*
91.3	345	315	30	287.6
91.3	344	314	30	286.6
91.3	343	313	30	285.6
91.3	335	306	29	279.5
91.3	334	305	29	278.5
91.3	333	304	29	277.5
91.3	332	303	29	276.5
91.3	323	295	28	269.4
91.3	322	294	28	268.4
91.3	321	293	28	267.4
91.3	320	292	28	266.4
91.3	312	285	27	260.3
91.3	311	284	27	259.3
91.3	310	283	27	258.4
91.3	309	282	27	257.4
91.3	300	274	26	250.3
91.3	299	273	26	249.3
91.3	298	272	26	248.3
91.3	289	264	25	241.2
91.3	288	263	25	240.2
91.3	287	262	25	239.2
91.3	286	261	25	238.2
91.3	277	253	24	231.1
91.3	276	252	24	230.1
91.3	275	251	24	229.1
91.3	265	242	23	221.0
91.3	264	241	23	220.0
91.3	263	240	23	219.0
91.3	254	232	22	211.9
91.3	253	231	22	210.9
91.3	252	230	22	209.9
91.3	242	221	21	201.8
91.3	241	220	21	200.8
91.3	240	219	21	199.8
91.3	231	211	20	192.7
91.3	230	210	20	191.7
91.3	229	209	20	190.7
91.3	219	200	19	182.6
91.3	218	199	19	181.7
91.3	208	190	18	173.6
91.3	207	189	18	172.6
91.3	206	188	18	171.6
91.3	196	179	17	163.5
91.3	195	178	17	162.5
91.3	184	168	16	153.4
91.3	183	167	16	152.4
91.3	173	158	15	144.3
91.3	172	157	15	143.3
91.3	161	147	14	134.2
91.3	160	146	14	133.2

%E	M1	M2	DM	M*
91.3	150	137	13	125.1
91.3	149	136	13	124.1
91.3	138	126	12	115.0
91.3	127	116	11	106.0
91.3	126	115	11	105.0
91.3	115	105	10	95.9
91.3	104	95	9	86.8
91.3	103	94	9	85.8
91.3	92	84	8	76.7
91.3	80	73	7	66.6
91.3	69	63	6	57.5
91.3	46	42	4	38.3
91.3	23	21	2	19.2
91.2	500	456	44	415.9
91.2	499	455	44	414.9
91.2	498	454	44	413.9
91.2	491	448	43	408.8
91.2	490	447	43	407.8
91.2	489	446	43	406.8
91.2	488	445	43	405.8
91.2	487	444	43	404.8
91.2	486	443	43	403.8
91.2	479	437	42	398.7
91.2	478	436	42	397.7
91.2	477	435	42	396.7
91.2	476	434	42	395.7
91.2	475	433	42	394.7
91.2	468	427	41	389.6
91.2	467	426	41	388.6
91.2	466	425	41	387.6
91.2	465	424	41	386.6
91.2	464	423	41	385.6
91.2	457	417	40	380.5
91.2	456	416	40	379.5
91.2	455	415	40	378.5
91.2	454	414	40	377.5
91.2	453	413	40	376.5
91.2	452	412	40	375.5
91.2	445	406	39	370.4
91.2	444	405	39	369.4
91.2	443	404	39	368.4
91.2	442	403	39	367.4
91.2	441	402	39	366.4
91.2	434	396	38	361.3
91.2	433	395	38	360.3
91.2	432	394	38	359.3
91.2	431	393	38	358.4
91.2	430	392	38	357.4
91.2	422	385	37	351.2
91.2	421	384	37	350.3

%E	M1	M2	DM	M*
91.2	420	383	37	349.3
91.2	419	382	37	348.3
91.2	411	375	36	342.2
91.2	410	374	36	341.2
91.2	409	373	36	340.2
91.2	408	372	36	339.2
91.2	407	371	36	338.2
91.2	399	364	35	332.1
91.2	398	363	35	331.1
91.2	397	362	35	330.1
91.2	396	361	35	329.1
91.2	388	354	34	323.0
91.2	387	353	34	322.0
91.2	386	352	34	321.0
91.2	385	351	34	320.0
91.2	377	344	33	313.9
91.2	376	343	33	312.9
91.2	375	342	33	311.9
91.2	374	341	33	310.9
91.2	373	340	33	309.9
91.2	365	333	32	303.8
91.2	364	332	32	302.8
91.2	363	331	32	301.8
91.2	362	330	32	300.8
91.2	354	323	31	294.7
91.2	353	322	31	293.7
91.2	352	321	31	292.7
91.2	351	320	31	291.7
91.2	342	312	30	284.6
91.2	341	311	30	283.6
91.2	340	310	30	282.6
91.2	339	309	30	281.7
91.2	331	302	29	275.5
91.2	330	301	29	274.5
91.2	329	300	29	273.6
91.2	328	299	29	272.6
91.2	319	291	28	265.5
91.2	318	290	28	264.5
91.2	317	289	28	263.5
91.2	308	281	27	256.4
91.2	307	280	27	255.4
91.2	306	279	27	254.4
91.2	297	271	26	247.3
91.2	296	270	26	246.3
91.2	295	269	26	245.3
91.2	294	268	26	244.3
91.2	285	260	25	237.2
91.2	284	259	25	236.2
91.2	283	258	25	235.2
91.2	274	250	24	228.1

%E	M1	M2	DM	M*
91.2	273	249	24	227.1
91.2	272	248	24	226.1
91.2	262	239	23	218.0
91.2	261	238	23	217.0
91.2	260	237	23	216.0
91.2	251	229	22	208.9
91.2	250	228	22	207.9
91.2	249	227	22	206.9
91.2	239	218	21	198.8
91.2	238	217	21	197.9
91.2	228	208	20	189.8
91.2	227	207	20	188.8
91.2	226	206	20	187.8
91.2	217	198	19	180.7
91.2	216	197	19	179.7
91.2	215	196	19	178.7
91.2	205	187	18	170.6
91.2	204	186	18	169.6
91.2	194	177	17	161.5
91.2	193	176	17	160.5
91.2	182	166	16	151.4
91.2	181	165	16	150.4
91.2	171	156	15	142.3
91.2	170	155	15	141.3
91.2	159	145	14	132.2
91.2	148	135	13	123.1
91.2	147	134	13	122.1
91.2	137	125	12	114.1
91.2	136	124	12	113.1
91.2	125	114	11	104.0
91.2	114	104	10	94.9
91.2	113	103	10	93.9
91.2	102	93	9	84.8
91.2	91	83	8	75.7
91.2	68	62	6	56.5
91.2	57	52	5	47.4
91.2	34	31	3	28.3
91.1	497	453	44	412.9
91.1	496	452	44	411.9
91.1	495	451	44	410.9
91.1	494	450	44	409.9
91.1	493	449	44	408.9
91.1	492	448	44	407.9
91.1	485	442	43	402.8
91.1	484	441	43	401.8
91.1	483	440	43	400.8
91.1	482	439	43	399.8
91.1	481	438	43	398.8
91.1	474	432	42	393.7
91.1	473	431	42	392.7

%E	M1	M2	DM	M*
91.1	472	430	42	391.7
91.1	471	429	42	390.7
91.1	470	428	42	389.8
91.1	463	422	41	384.6
91.1	462	421	41	383.6
91.1	461	420	41	382.6
91.1	460	419	41	381.7
91.1	459	418	41	380.7
91.1	451	411	40	374.5
91.1	450	410	40	373.6
91.1	449	409	40	372.6
91.1	448	408	40	371.6
91.1	447	407	40	370.6
91.1	440	401	39	365.5
91.1	439	400	39	364.5
91.1	438	399	39	363.5
91.1	437	398	39	362.5
91.1	436	397	39	361.5
91.1	429	391	38	356.4
91.1	428	390	38	355.4
91.1	427	389	38	354.4
91.1	426	388	38	353.4
91.1	425	387	38	352.4
91.1	418	381	37	347.3
91.1	417	380	37	346.3
91.1	416	379	37	345.3
91.1	415	378	37	344.3
91.1	414	377	37	343.3
91.1	406	370	36	337.2
91.1	405	369	36	336.2
91.1	404	368	36	335.2
91.1	403	367	36	334.2
91.1	395	360	35	328.1
91.1	394	359	35	327.1
91.1	393	358	35	326.1
91.1	392	357	35	325.1
91.1	384	350	34	319.0
91.1	383	349	34	318.0
91.1	382	348	34	317.0
91.1	381	347	34	316.0
91.1	380	346	34	315.0
91.1	372	339	33	308.9
91.1	371	338	33	307.9
91.1	370	337	33	306.9
91.1	369	336	33	306.0
91.1	361	329	32	299.8
91.1	360	328	32	298.8
91.1	359	327	32	297.9
91.1	358	326	32	296.9
91.1	350	319	31	290.7

%E	M1	M2	DM	M*
91.1	349	318	31	289.8
91.1	348	317	31	288.8
91.1	347	316	31	287.8
91.1	338	308	30	280.7
91.1	337	307	30	279.7
91.1	336	306	30	278.7
91.1	327	298	29	271.6
91.1	326	297	29	270.6
91.1	325	296	29	269.6
91.1	316	288	28	262.5
91.1	315	287	28	261.5
91.1	314	286	28	260.5
91.1	313	285	28	259.5
91.1	305	278	27	253.4
91.1	304	277	27	252.4
91.1	303	276	27	251.4
91.1	302	275	27	250.4
91.1	293	267	26	243.3
91.1	292	266	26	242.3
91.1	291	265	26	241.3
91.1	282	257	25	234.2
91.1	281	256	25	233.2
91.1	280	255	25	232.2
91.1	271	247	24	225.1
91.1	270	246	24	224.1
91.1	269	245	24	223.1
91.1	259	236	23	215.0
91.1	258	235	23	214.1
91.1	257	234	23	213.1
91.1	248	226	22	206.0
91.1	247	225	22	205.0
91.1	246	224	22	204.0
91.1	237	216	21	196.9
91.1	236	215	21	195.9
91.1	235	214	21	194.9
91.1	225	205	20	186.8
91.1	224	204	20	185.8
91.1	214	195	19	177.7
91.1	213	194	19	176.7
91.1	203	185	18	168.6
91.1	202	184	18	167.6
91.1	192	175	17	159.5
91.1	191	174	17	158.5
91.1	190	173	17	157.5
91.1	180	164	16	149.4
91.1	179	163	16	148.4
91.1	169	154	15	140.3
91.1	168	153	15	139.3
91.1	158	144	14	131.2
91.1	157	143	14	130.2

%E	M1	M2	DM	M*
91.1	146	133	13	121.2
91.1	135	123	12	112.1
91.1	124	113	11	103.0
91.1	123	112	11	102.0
91.1	112	102	10	92.9
91.1	101	92	9	83.8
91.1	90	82	8	74.7
91.1	79	72	7	65.6
91.1	56	51	5	46.4
91.1	45	41	4	37.4
91.0	500	455	45	414.0
91.0	499	454	45	413.1
91.0	498	453	45	412.1
91.0	491	447	44	406.9
91.0	490	446	44	406.0
91.0	489	445	44	405.0
91.0	488	444	44	404.0
91.0	487	443	44	403.0
91.0	480	437	43	397.9
91.0	479	436	43	396.9
91.0	478	435	43	395.9
91.0	477	434	43	394.9
91.0	476	433	43	393.9
91.0	469	427	42	388.8
91.0	468	426	42	387.8
91.0	467	425	42	386.8
91.0	466	424	42	385.8
91.0	465	423	42	384.8
91.0	458	417	41	379.7
91.0	457	416	41	378.7
91.0	456	415	41	377.7
91.0	455	414	41	376.7
91.0	454	413	41	375.7
91.0	446	406	40	369.6
91.0	445	405	40	368.6
91.0	444	404	40	367.6
91.0	443	403	40	366.6
91.0	442	402	40	365.6
91.0	435	396	39	360.5
91.0	434	395	39	359.5
91.0	433	394	39	358.5
91.0	432	393	39	357.5
91.0	431	392	39	356.5
91.0	424	386	38	351.4
91.0	423	385	38	350.4
91.0	422	384	38	349.4
91.0	421	383	38	348.4
91.0	420	382	38	347.4
91.0	413	376	37	342.3
91.0	412	375	37	341.3

%E	M1	M2	DM	M*
91.0	411	374	37	340.3
91.0	410	373	37	339.3
91.0	409	372	37	338.3
91.0	402	366	36	333.2
91.0	401	365	36	332.2
91.0	400	364	36	331.2
91.0	399	363	36	330.2
91.0	398	362	36	329.3
91.0	391	356	35	324.1
91.0	390	355	35	323.1
91.0	389	354	35	322.1
91.0	388	353	35	321.2
91.0	387	352	35	320.2
91.0	379	345	34	314.1
91.0	378	344	34	313.1
91.0	377	343	34	312.1
91.0	376	342	34	311.1
91.0	368	335	33	305.0
91.0	367	334	33	304.0
91.0	366	333	33	303.0
91.0	365	332	33	302.0
91.0	357	325	32	295.9
91.0	356	324	32	294.9
91.0	355	323	32	293.9
91.0	354	322	32	292.9
91.0	346	315	31	286.8
91.0	345	314	31	285.8
91.0	344	313	31	284.8
91.0	343	312	31	283.8
91.0	335	305	30	277.7
91.0	334	304	30	276.7
91.0	333	303	30	275.7
91.0	332	302	30	274.7
91.0	324	295	29	268.6
91.0	323	294	29	267.6
91.0	322	293	29	266.6
91.0	321	292	29	265.6
91.0	312	284	28	258.5
91.0	311	283	28	257.5
91.0	310	282	28	256.5
91.0	301	274	27	249.4
91.0	300	273	27	248.4
91.0	299	272	27	247.4
91.0	290	264	26	240.3
91.0	289	263	26	239.3
91.0	288	262	26	238.3
91.0	279	254	25	231.2
91.0	278	253	25	230.2
91.0	277	252	25	229.3
91.0	268	244	24	222.1

%E	M1	M2	DM	M*
91.0	267	243	24	221.2
91.0	266	242	24	220.2
91.0	256	233	23	212.1
91.0	255	232	23	211.1
91.0	245	223	22	203.0
91.0	244	222	22	202.0
91.0	234	213	21	193.9
91.0	233	212	21	192.9
91.0	223	203	20	184.8
91.0	222	202	20	183.8
91.0	221	201	20	182.8
91.0	212	193	19	175.7
91.0	211	192	19	174.7
91.0	210	191	19	173.7
91.0	201	183	18	166.6
91.0	200	182	18	165.6
91.0	199	181	18	164.6
91.0	189	172	17	156.5
91.0	188	171	17	155.5
91.0	178	162	16	147.4
91.0	177	161	16	146.4
91.0	167	152	15	138.3
91.0	166	151	15	137.4
91.0	156	142	14	129.3
91.0	155	141	14	128.3
91.0	145	132	13	120.2
91.0	144	131	13	119.2
91.0	134	122	12	111.1
91.0	133	121	12	110.1
91.0	122	111	11	101.0
91.0	111	101	10	91.9
91.0	100	91	9	82.8
91.0	89	81	8	73.7
91.0	78	71	7	64.6
91.0	67	61	6	55.5
90.9	497	452	45	411.1
90.9	496	451	45	410.1
90.9	495	450	45	409.1
90.9	494	449	45	408.1
90.9	493	448	45	407.1
90.9	492	447	45	406.1
90.9	486	442	44	402.0
90.9	485	441	44	401.0
90.9	484	440	44	400.0
90.9	483	439	44	399.0
90.9	482	438	44	398.0
90.9	481	437	44	397.0
90.9	475	432	43	392.9
90.9	474	431	43	391.9
90.9	473	430	43	390.9

%E	M1	M2	DM	M*
90.9	472	429	43	389.9
90.9	471	428	43	388.9
90.9	470	427	43	387.9
90.9	464	422	42	383.8
90.9	463	421	42	382.8
90.9	462	420	42	381.8
90.9	461	419	42	380.8
90.9	460	418	42	379.8
90.9	459	417	42	378.8
90.9	453	412	41	374.7
90.9	452	411	41	373.7
90.9	451	410	41	372.7
90.9	450	409	41	371.7
90.9	449	408	41	370.7
90.9	441	401	40	364.6
90.9	440	400	40	363.6
90.9	439	399	40	362.6
90.9	438	398	40	361.7
90.9	430	391	39	355.5
90.9	429	390	39	354.5
90.9	428	389	39	353.6
90.9	427	388	39	352.6
90.9	419	381	38	346.4
90.9	418	380	38	345.5
90.9	417	379	38	344.5
90.9	416	378	38	343.5
90.9	408	371	37	337.4
90.9	407	370	37	336.4
90.9	406	369	37	335.4
90.9	405	368	37	334.4
90.9	397	361	36	328.3
90.9	396	360	36	327.3
90.9	395	359	36	326.3
90.9	394	358	36	325.3
90.9	386	351	35	319.2
90.9	385	350	35	318.2
90.9	384	349	35	317.2
90.9	383	348	35	316.2
90.9	375	341	34	310.1
90.9	374	340	34	309.1
90.9	373	339	34	308.1
90.9	372	338	34	307.1
90.9	364	331	33	301.0
90.9	363	330	33	300.0
90.9	362	329	33	299.0
90.9	361	328	33	298.0
90.9	353	321	32	291.9
90.9	352	320	32	290.9
90.9	351	319	32	289.9
90.9	350	318	32	288.9

%E	M1	M2	DM	M*
90.9	342	311	31	282.8
90.9	341	310	31	281.8
90.9	340	309	31	280.8
90.9	339	308	31	279.8
90.9	331	301	30	273.7
90.9	330	300	30	272.7
90.9	329	299	30	271.7
90.9	328	298	30	270.7
90.9	320	291	29	264.6
90.9	319	290	29	263.6
90.9	318	289	29	262.6
90.9	317	288	29	261.7
90.9	309	281	28	255.5
90.9	308	280	28	254.5
90.9	307	279	28	253.6
90.9	306	278	28	252.6
90.9	298	271	27	246.4
90.9	297	270	27	245.5
90.9	296	269	27	244.5
90.9	287	261	26	237.4
90.9	286	260	26	236.4
90.9	285	259	26	235.4
90.9	276	251	25	228.3
90.9	275	250	25	227.3
90.9	274	249	25	226.3
90.9	265	241	24	219.2
90.9	264	240	24	218.2
90.9	263	239	24	217.2
90.9	254	231	23	210.1
90.9	253	230	23	209.1
90.9	252	229	23	208.1
90.9	243	221	22	201.0
90.9	242	220	22	200.0
90.9	241	219	22	199.0
90.9	232	211	21	191.9
90.9	231	210	21	190.9
90.9	230	209	21	189.9
90.9	220	200	20	181.8
90.9	219	199	20	180.8
90.9	209	190	19	172.7
90.9	208	189	19	171.7
90.9	198	180	18	163.6
90.9	197	179	18	162.6
90.9	187	170	17	154.5
90.9	186	169	17	153.6
90.9	176	160	16	145.5
90.9	175	159	16	144.5
90.9	165	150	15	136.4
90.9	164	149	15	135.4
90.9	154	140	14	127.3

%E	M1	M2	DM	M*
90.9	153	139	14	126.3
90.9	143	130	13	118.2
90.9	132	120	12	109.1
90.9	121	110	11	100.0
90.9	110	100	10	90.9
90.9	99	90	9	81.8
90.9	88	80	8	72.7
90.9	77	70	7	63.6
90.9	66	60	6	54.5
90.9	55	50	5	45.5
90.9	44	40	4	36.4
90.9	33	30	3	27.3
90.9	22	20	2	18.2
90.9	11	10	1	9.1
90.8	500	454	46	412.2
90.8	499	453	46	411.2
90.8	498	452	46	410.2
90.8	491	446	45	405.1
90.8	490	445	45	404.1
90.8	489	444	45	403.1
90.8	488	443	45	402.1
90.8	487	442	45	401.2
90.8	480	436	44	396.0
90.8	479	435	44	395.0
90.8	478	434	44	394.1
90.8	477	433	44	393.1
90.8	476	432	44	392.1
90.8	469	426	43	386.9
90.8	468	425	43	386.0
90.8	467	424	43	385.0
90.8	466	423	43	384.0
90.8	465	422	43	383.0
90.8	458	416	42	377.9
90.8	457	415	42	376.9
90.8	456	414	42	375.9
90.8	455	413	42	374.9
90.8	448	407	41	369.8
90.8	447	406	41	368.8
90.8	446	405	41	367.8
90.8	445	404	41	366.8
90.8	444	403	41	365.8
90.8	437	397	40	360.7
90.8	436	396	40	359.7
90.8	435	395	40	358.7
90.8	434	394	40	357.7
90.8	433	393	40	356.7
90.8	426	387	39	351.6
90.8	425	386	39	350.6
90.8	424	385	39	349.6
90.8	423	384	39	348.6

%E	M1	M2	DM	M*
90.8	422	383	39	347.6
90.8	415	377	38	342.5
90.8	414	376	38	341.5
90.8	413	375	38	340.5
90.8	412	374	38	339.5
90.8	411	373	38	338.5
90.8	404	367	37	333.4
90.8	403	366	37	332.4
90.8	402	365	37	331.4
90.8	401	364	37	330.4
90.8	400	363	37	329.4
90.8	393	357	36	324.3
90.8	392	356	36	323.3
90.8	391	355	36	322.3
90.8	390	354	36	321.3
90.8	382	347	35	315.2
90.8	381	346	35	314.2
90.8	380	345	35	313.2
90.8	379	344	35	312.2
90.8	371	337	34	306.1
90.8	370	336	34	305.1
90.8	369	335	34	304.1
90.8	368	334	34	303.1
90.8	360	327	33	297.0
90.8	359	326	33	296.0
90.8	358	325	33	295.0
90.8	357	324	33	294.1
90.8	349	317	32	287.9
90.8	348	316	32	286.9
90.8	347	315	32	286.0
90.8	346	314	32	285.0
90.8	338	307	31	278.8
90.8	337	306	31	277.9
90.8	336	305	31	276.9
90.8	327	297	30	269.8
90.8	326	296	30	268.8
90.8	325	295	30	267.8
90.8	316	287	29	260.7
90.8	315	286	29	259.7
90.8	314	285	29	258.7
90.8	305	277	28	251.6
90.8	304	276	28	250.6
90.8	303	275	28	249.6
90.8	295	268	27	243.5
90.8	294	267	27	242.5
90.8	293	266	27	241.5
90.8	292	265	27	240.5
90.8	284	258	26	234.4
90.8	283	257	26	233.4
90.8	282	256	26	232.4

%E	M1	M2	DM	M*
90.8	273	248	25	225.3
90.8	272	247	25	224.3
90.8	271	246	25	223.3
90.8	262	238	24	216.2
90.8	261	237	24	215.2
90.8	260	236	24	214.2
90.8	251	228	23	207.1
90.8	250	227	23	206.1
90.8	249	226	23	205.1
90.8	240	218	22	198.0
90.8	239	217	22	197.0
90.8	238	216	22	196.0
90.8	229	208	21	188.9
90.8	228	207	21	187.9
90.8	218	198	20	179.8
90.8	217	197	20	178.8
90.8	207	188	19	170.7
90.8	206	187	19	169.8
90.8	196	178	18	161.7
90.8	195	177	18	160.7
90.8	185	168	17	152.6
90.8	184	167	17	151.6
90.8	174	158	16	143.5
90.8	173	157	16	142.5
90.8	163	148	15	134.4
90.8	152	138	14	125.3
90.8	142	129	13	117.2
90.8	141	128	13	116.2
90.8	131	119	12	108.1
90.8	130	118	12	107.1
90.8	120	109	11	99.0
90.8	119	108	11	98.0
90.8	109	99	10	89.9
90.8	98	89	9	80.8
90.8	87	79	8	71.7
90.8	76	69	7	62.6
90.8	65	59	6	53.6
90.7	497	451	46	409.3
90.7	496	450	46	408.3
90.7	495	449	46	407.3
90.7	494	448	46	406.3
90.7	493	447	46	405.3
90.7	492	446	46	404.3
90.7	486	441	45	400.2
90.7	485	440	45	399.2
90.7	484	439	45	398.2
90.7	483	438	45	397.2
90.7	482	437	45	396.2
90.7	475	431	44	391.1
90.7	474	430	44	390.1

%E	M1	M2	DM	M*
90.7	473	429	44	389.1
90.7	472	428	44	388.1
90.7	471	427	44	387.1
90.7	464	421	43	382.0
90.7	463	420	43	381.0
90.7	462	419	43	380.0
90.7	461	418	43	379.0
90.7	460	417	43	378.0
90.7	454	412	42	373.9
90.7	453	411	42	372.9
90.7	452	410	42	371.9
90.7	451	409	42	370.9
90.7	450	408	42	369.9
90.7	443	402	41	364.8
90.7	442	401	41	363.8
90.7	441	400	41	362.8
90.7	440	399	41	361.8
90.7	439	398	41	360.8
90.7	432	392	40	355.7
90.7	431	391	40	354.7
90.7	430	390	40	353.7
90.7	429	389	40	352.7
90.7	428	388	40	351.7
90.7	421	382	39	346.6
90.7	420	381	39	345.6
90.7	419	380	39	344.6
90.7	418	379	39	343.6
90.7	410	372	38	337.5
90.7	409	371	38	336.5
90.7	408	370	38	335.5
90.7	407	369	38	334.5
90.7	399	362	37	328.4
90.7	398	361	37	327.4
90.7	397	360	37	326.4
90.7	396	359	37	325.5
90.7	389	353	36	320.3
90.7	388	352	36	319.3
90.7	387	351	36	318.3
90.7	386	350	36	317.4
90.7	378	343	35	311.2
90.7	377	342	35	310.2
90.7	376	341	35	309.3
90.7	375	340	35	308.3
90.7	367	333	34	302.1
90.7	366	332	34	301.2
90.7	365	331	34	300.2
90.7	364	330	34	299.2
90.7	356	323	33	293.1
90.7	355	322	33	292.1
90.7	354	321	33	291.1

%E	M1	M2	DM	M*
90.7	353	320	33	290.1
90.7	345	313	32	284.0
90.7	344	312	32	283.0
90.7	343	311	32	282.0
90.7	335	304	31	275.9
90.7	334	303	31	274.9
90.7	333	302	31	273.9
90.7	332	301	31	272.9
90.7	324	294	30	266.8
90.7	323	293	30	265.8
90.7	322	292	30	264.8
90.7	321	291	30	263.8
90.7	313	284	29	257.7
90.7	312	283	29	256.7
90.7	311	282	29	255.7
90.7	302	274	28	248.6
90.7	301	273	28	247.6
90.7	300	272	28	246.6
90.7	291	264	27	239.5
90.7	290	263	27	238.5
90.7	289	262	27	237.5
90.7	281	255	26	231.4
90.7	280	254	26	230.4
90.7	279	253	26	229.4
90.7	270	245	25	222.3
90.7	269	244	25	221.3
90.7	268	243	25	220.3
90.7	259	235	24	213.2
90.7	258	234	24	212.2
90.7	257	233	24	211.2
90.7	248	225	23	204.1
90.7	247	224	23	203.1
90.7	246	223	23	202.2
90.7	237	215	22	195.0
90.7	236	214	22	194.1
90.7	227	206	21	186.9
90.7	226	205	21	186.0
90.7	225	204	21	185.0
90.7	216	196	20	177.9
90.7	215	195	20	176.9
90.7	214	194	20	175.9
90.7	205	186	19	168.8
90.7	204	185	19	167.8
90.7	194	176	18	159.7
90.7	193	175	18	158.7
90.7	183	166	17	150.6
90.7	182	165	17	149.6
90.7	172	156	16	141.5
90.7	162	147	15	133.4
90.7	161	146	15	132.4

%E	M1	M2	DM	M*
90.7	151	137	14	124.3
90.7	150	136	14	123.3
90.7	140	127	13	115.2
90.7	129	117	12	106.1
90.7	118	107	11	97.0
90.7	108	98	10	88.9
90.7	107	97	10	87.9
90.7	97	88	9	79.8
90.7	86	78	8	70.7
90.7	75	68	7	61.7
90.7	54	49	5	44.5
90.7	43	39	4	35.4
90.6	500	453	47	410.4
90.6	499	452	47	409.4
90.6	498	451	47	408.4
90.6	491	445	46	403.3
90.6	490	444	46	402.3
90.6	489	443	46	401.3
90.6	488	442	46	400.3
90.6	487	441	46	399.3
90.6	481	436	45	395.2
90.6	480	435	45	394.2
90.6	479	434	45	393.2
90.6	478	433	45	392.2
90.6	477	432	45	391.2
90.6	470	426	44	386.1
90.6	469	425	44	385.1
90.6	468	424	44	384.1
90.6	467	423	44	383.1
90.6	466	422	44	382.2
90.6	459	416	43	377.0
90.6	458	415	43	376.0
90.6	457	414	43	375.0
90.6	456	413	43	374.1
90.6	449	407	42	368.9
90.6	448	406	42	367.9
90.6	447	405	42	366.9
90.6	446	404	42	366.0
90.6	445	403	42	365.0
90.6	438	397	41	359.8
90.6	437	396	41	358.8
90.6	436	395	41	357.9
90.6	435	394	41	356.9
90.6	434	393	41	355.9
90.6	427	387	40	350.7
90.6	426	386	40	349.8
90.6	425	385	40	348.8
90.6	424	384	40	347.8
90.6	417	378	39	342.6
90.6	416	377	39	341.7

%E	M1	M2	DM	M*
90.6	415	376	39	340.7
90.6	414	375	39	339.7
90.6	413	374	39	338.7
90.6	406	368	38	333.6
90.6	405	367	38	332.6
90.6	404	366	38	331.6
90.6	403	365	38	330.6
90.6	395	358	37	324.5
90.6	394	357	37	323.5
90.6	393	356	37	322.5
90.6	392	355	37	321.5
90.6	385	349	36	316.4
90.6	384	348	36	315.4
90.6	383	347	36	314.4
90.6	382	346	36	313.4
90.6	381	345	36	312.4
90.6	374	339	35	307.3
90.6	373	338	35	306.3
90.6	372	337	35	305.3
90.6	371	336	35	304.3
90.6	363	329	34	298.2
90.6	362	328	34	297.2
90.6	361	327	34	296.2
90.6	360	326	34	295.2
90.6	352	319	33	289.1
90.6	351	318	33	288.1
90.6	350	317	33	287.1
90.6	342	310	32	281.0
90.6	341	309	32	280.0
90.6	340	308	32	279.0
90.6	339	307	32	278.0
90.6	331	300	31	271.9
90.6	330	299	31	270.9
90.6	329	298	31	269.9
90.6	320	290	30	262.8
90.6	319	289	30	261.8
90.6	318	288	30	260.8
90.6	310	281	29	254.7
90.6	309	280	29	253.7
90.6	308	279	29	252.7
90.6	307	278	29	251.7
90.6	299	271	28	245.6
90.6	298	270	28	244.6
90.6	297	269	28	243.6
90.6	288	261	27	236.5
90.6	287	260	27	235.5
90.6	286	259	27	234.5
90.6	278	252	26	228.4
90.6	277	251	26	227.4
90.6	276	250	26	226.4

%E	M1	M2	DM	M*
90.6	267	242	25	219.3
90.6	266	241	25	218.3
90.6	265	240	25	217.4
90.6	256	232	24	210.3
90.6	255	231	24	209.3
90.6	254	230	24	208.3
90.6	245	222	23	201.2
90.6	244	221	23	200.2
90.6	235	213	22	193.1
90.6	234	212	22	192.1
90.6	233	211	22	191.1
90.6	224	203	21	184.0
90.6	223	202	21	183.0
90.6	213	193	20	174.9
90.6	212	192	20	173.9
90.6	203	184	19	166.8
90.6	202	183	19	165.8
90.6	192	174	18	157.7
90.6	191	173	18	156.7
90.6	181	164	17	148.6
90.6	180	163	17	147.6
90.6	171	155	16	140.5
90.6	170	154	16	139.5
90.6	160	145	15	131.4
90.6	159	144	15	130.4
90.6	149	135	14	122.3
90.6	139	126	13	114.2
90.6	138	125	13	113.2
90.6	128	116	12	105.1
90.6	127	115	12	104.1
90.6	117	106	11	96.0
90.6	106	96	10	86.9
90.6	96	87	9	78.8
90.6	85	77	8	69.8
90.6	64	58	6	52.6
90.6	53	48	5	43.5
90.6	32	29	3	26.3
90.5	497	450	47	407.4
90.5	496	449	47	406.5
90.5	495	448	47	405.5
90.5	494	447	47	404.5
90.5	493	446	47	403.5
90.5	486	440	46	398.4
90.5	485	439	46	397.4
90.5	484	438	46	396.4
90.5	483	437	46	395.4
90.5	482	436	46	394.4
90.5	476	431	45	390.3
90.5	475	430	45	389.3
90.5	474	429	45	388.3

%E	M1	M2	DM	M*
90.5	473	428	45	387.3
90.5	472	427	45	386.3
90.5	465	421	44	381.2
90.5	464	420	44	380.2
90.5	463	419	44	379.2
90.5	462	418	44	378.2
90.5	461	417	44	377.2
90.5	455	412	43	373.1
90.5	454	411	43	372.1
90.5	453	410	43	371.1
90.5	452	409	43	370.1
90.5	451	408	43	369.1
90.5	444	402	42	364.0
90.5	443	401	42	363.0
90.5	442	400	42	362.0
90.5	441	399	42	361.0
90.5	440	398	42	360.0
90.5	433	392	41	354.9
90.5	432	391	41	353.9
90.5	431	390	41	352.9
90.5	430	389	41	351.9
90.5	423	383	40	346.8
90.5	422	382	40	345.8
90.5	421	381	40	344.8
90.5	420	380	40	343.8
90.5	419	379	40	342.8
90.5	412	373	39	337.7
90.5	411	372	39	336.7
90.5	410	371	39	335.7
90.5	409	370	39	334.7
90.5	402	364	38	329.6
90.5	401	363	38	328.6
90.5	400	362	38	327.6
90.5	399	361	38	326.6
90.5	398	360	38	325.6
90.5	391	354	37	320.5
90.5	390	353	37	319.5
90.5	389	352	37	318.5
90.5	388	351	37	317.5
90.5	380	344	36	311.4
90.5	379	343	36	310.4
90.5	378	342	36	309.4
90.5	377	341	36	308.4
90.5	370	335	35	303.3
90.5	369	334	35	302.3
90.5	368	333	35	301.3
90.5	367	332	35	300.3
90.5	359	325	34	294.2
90.5	358	324	34	293.2
90.5	357	323	34	292.2

%E	M1	M2	DM	M*
90.5	349	316	33	286.1
90.5	348	315	33	285.1
90.5	347	314	33	284.1
90.5	346	313	33	283.1
90.5	338	306	32	277.0
90.5	337	305	32	276.0
90.5	336	304	32	275.0
90.5	328	297	31	268.9
90.5	327	296	31	267.9
90.5	326	295	31	266.9
90.5	325	294	31	266.0
90.5	317	287	30	259.8
90.5	316	286	30	258.8
90.5	315	285	30	257.9
90.5	306	277	29	250.7
90.5	305	276	29	249.8
90.5	304	275	29	248.8
90.5	296	268	28	242.6
90.5	295	267	28	241.7
90.5	294	266	28	240.7
90.5	285	258	27	233.6
90.5	284	257	27	232.6
90.5	283	256	27	231.6
90.5	275	249	26	225.5
90.5	274	248	26	224.5
90.5	273	247	26	223.5
90.5	264	239	25	216.4
90.5	263	238	25	215.4
90.5	262	237	25	214.4
90.5	253	229	24	207.3
90.5	252	228	24	206.3
90.5	243	220	23	199.2
90.5	242	219	23	198.2
90.5	241	218	23	197.2
90.5	232	210	22	190.1
90.5	231	209	22	189.1
90.5	222	201	21	182.0
90.5	221	200	21	181.0
90.5	220	199	21	180.0
90.5	211	191	20	172.9
90.5	210	190	20	171.9
90.5	201	182	19	164.8
90.5	200	181	19	163.8
90.5	199	180	19	162.8
90.5	190	172	18	155.7
90.5	189	171	18	154.7
90.5	179	162	17	146.6
90.5	169	153	16	138.5
90.5	168	152	16	137.5
90.5	158	143	15	129.4

%E	M1	M2	DM	M*
90.5	148	134	14	121.3
90.5	147	133	14	120.3
90.5	137	124	13	112.2
90.5	126	114	12	103.1
90.5	116	105	11	95.0
90.5	105	95	10	86.0
90.5	95	86	9	77.9
90.5	84	76	8	68.8
90.5	74	67	7	60.7
90.5	63	57	6	51.6
90.5	42	38	4	34.4
90.5	21	19	2	17.2
90.4	500	452	48	408.6
90.4	499	451	48	407.6
90.4	498	450	48	406.6
90.4	492	445	47	402.5
90.4	491	444	47	401.5
90.4	490	443	47	400.5
90.4	489	442	47	399.5
90.4	488	441	47	398.5
90.4	481	435	46	393.4
90.4	480	434	46	392.4
90.4	479	433	46	391.4
90.4	478	432	46	390.4
90.4	477	431	46	389.4
90.4	471	426	45	385.3
90.4	470	425	45	384.3
90.4	469	424	45	383.3
90.4	468	423	45	382.3
90.4	467	422	45	381.3
90.4	460	416	44	376.2
90.4	459	415	44	375.2
90.4	458	414	44	374.2
90.4	457	413	44	373.2
90.4	456	412	44	372.2
90.4	450	407	43	368.1
90.4	449	406	43	367.1
90.4	448	405	43	366.1
90.4	447	404	43	365.1
90.4	446	403	43	364.1
90.4	439	397	42	359.0
90.4	438	396	42	358.0
90.4	437	395	42	357.0
90.4	436	394	42	356.0
90.4	429	388	41	350.9
90.4	428	387	41	349.9
90.4	427	386	41	348.9
90.4	426	385	41	347.9
90.4	425	384	41	347.0
90.4	418	378	40	341.8

%E	M1	M2	DM	M*
90.4	417	377	40	340.8
90.4	416	376	40	339.8
90.4	415	375	40	338.9
90.4	408	369	39	333.7
90.4	407	368	39	332.7
90.4	406	367	39	331.7
90.4	405	366	39	330.8
90.4	397	359	38	324.6
90.4	396	358	38	323.6
90.4	395	357	38	322.7
90.4	394	356	38	321.7
90.4	387	350	37	316.5
90.4	386	349	37	315.5
90.4	385	348	37	314.6
90.4	384	347	37	313.6
90.4	376	340	36	307.4
90.4	375	339	36	306.5
90.4	374	338	36	305.5
90.4	366	331	35	299.3
90.4	365	330	35	298.4
90.4	364	329	35	297.4
90.4	363	328	35	296.4
90.4	356	322	34	291.2
90.4	355	321	34	290.3
90.4	354	320	34	289.3
90.4	353	319	34	288.3
90.4	345	312	33	282.2
90.4	344	311	33	281.2
90.4	343	310	33	280.2
90.4	342	309	33	279.2
90.4	335	303	32	274.1
90.4	334	302	32	273.1
90.4	333	301	32	272.1
90.4	332	300	32	271.1
90.4	324	293	31	265.0
90.4	323	292	31	264.0
90.4	322	291	31	263.0
90.4	314	284	30	256.9
90.4	313	283	30	255.9
90.4	312	282	30	254.9
90.4	311	281	30	253.9
90.4	303	274	29	247.8
90.4	302	273	29	246.8
90.4	301	272	29	245.8
90.4	293	265	28	239.7
90.4	292	264	28	238.7
90.4	291	263	28	237.7
90.4	282	255	27	230.6
90.4	281	254	27	229.6
90.4	280	253	27	228.6

%E	M1	M2	DM	M*
90.4	272	246	26	222.5
90.4	271	245	26	221.5
90.4	270	244	26	220.5
90.4	261	236	25	213.4
90.4	260	235	25	212.4
90.4	251	227	24	205.3
90.4	250	226	24	204.3
90.4	249	225	24	203.3
90.4	240	217	23	196.2
90.4	239	216	23	195.2
90.4	230	208	22	188.1
90.4	229	207	22	187.1
90.4	228	206	22	186.1
90.4	219	198	21	179.0
90.4	218	197	21	178.0
90.4	209	189	20	170.9
90.4	208	188	20	169.9
90.4	198	179	19	161.8
90.4	197	178	19	160.8
90.4	188	170	18	153.7
90.4	187	169	18	152.7
90.4	178	161	17	145.6
90.4	177	160	17	144.6
90.4	167	151	16	136.5
90.4	166	150	16	135.5
90.4	157	142	15	128.4
90.4	156	141	15	127.4
90.4	146	132	14	119.3
90.4	136	123	13	111.2
90.4	135	122	13	110.3
90.4	125	113	12	102.2
90.4	115	104	11	94.1
90.4	114	103	11	93.1
90.4	104	94	10	85.0
90.4	94	85	9	76.9
90.4	83	75	8	67.8
90.4	73	66	7	59.7
90.4	52	47	5	42.5
90.3	497	449	48	405.6
90.3	496	448	48	404.6
90.3	495	447	48	403.7
90.3	494	446	48	402.7
90.3	493	445	48	401.7
90.3	487	440	47	397.5
90.3	486	439	47	396.5
90.3	485	438	47	395.6
90.3	484	437	47	394.6
90.3	483	436	47	393.6
90.3	476	430	46	388.4
90.3	475	429	46	387.5

%E	M1	M2	DM	M*
90.3	474	428	46	386.5
90.3	473	427	46	385.5
90.3	472	426	46	384.5
90.3	466	421	45	380.3
90.3	465	420	45	379.4
90.3	464	419	45	378.4
90.3	463	418	45	377.4
90.3	462	417	45	376.4
90.3	455	411	44	371.3
90.3	454	410	44	370.3
90.3	453	409	44	369.3
90.3	452	408	44	368.3
90.3	445	402	43	363.2
90.3	444	401	43	362.2
90.3	443	400	43	361.2
90.3	442	399	43	360.2
90.3	435	393	42	355.1
90.3	434	392	42	354.1
90.3	433	391	42	353.1
90.3	432	390	42	352.1
90.3	431	389	42	351.1
90.3	424	383	41	346.0
90.3	423	382	41	345.0
90.3	422	381	41	344.0
90.3	421	380	41	343.0
90.3	414	374	40	337.9
90.3	413	373	40	336.9
90.3	412	372	40	335.9
90.3	411	371	40	334.9
90.3	404	365	39	329.8
90.3	403	364	39	328.8
90.3	402	363	39	327.8
90.3	401	362	39	326.8
90.3	400	361	39	325.8
90.3	393	355	38	320.7
90.3	392	354	38	319.7
90.3	391	353	38	318.7
90.3	390	352	38	317.7
90.3	383	346	37	312.6
90.3	382	345	37	311.6
90.3	381	344	37	310.6
90.3	380	343	37	309.6
90.3	373	337	36	304.5
90.3	372	336	36	303.5
90.3	371	335	36	302.5
90.3	370	334	36	301.5
90.3	362	327	35	295.4
90.3	361	326	35	294.4
90.3	360	325	35	293.4
90.3	359	324	35	292.4

%E	M1	M2	DM	M*
90.3	352	318	34	287.3
90.3	351	317	34	286.3
90.3	350	316	34	285.3
90.3	349	315	34	284.3
90.3	341	308	33	278.2
90.3	340	307	33	277.2
90.3	339	306	33	276.2
90.3	331	299	32	270.1
90.3	330	298	32	269.1
90.3	329	297	32	268.1
90.3	321	290	31	262.0
90.3	320	289	31	261.0
90.3	319	288	31	260.0
90.3	318	287	31	259.0
90.3	310	280	30	252.9
90.3	309	279	30	251.9
90.3	308	278	30	250.9
90.3	300	271	29	244.8
90.3	299	270	29	243.8
90.3	298	269	29	242.8
90.3	290	262	28	236.7
90.3	289	261	28	235.7
90.3	288	260	28	234.7
90.3	279	252	27	227.6
90.3	278	251	27	226.6
90.3	277	250	27	225.6
90.3	269	243	26	219.5
90.3	268	242	26	218.5
90.3	267	241	26	217.5
90.3	259	234	25	211.4
90.3	258	233	25	210.4
90.3	257	232	25	209.4
90.3	248	224	24	202.3
90.3	247	223	24	201.3
90.3	238	215	23	194.2
90.3	237	214	23	193.2
90.3	236	213	23	192.2
90.3	227	205	22	185.1
90.3	226	204	22	184.1
90.3	217	196	21	177.0
90.3	216	195	21	176.0
90.3	207	187	20	168.9
90.3	206	186	20	167.9
90.3	196	177	19	159.8
90.3	195	176	19	158.9
90.3	186	168	18	151.7
90.3	185	167	18	150.8
90.3	176	159	17	143.6
90.3	175	158	17	142.7
90.3	165	149	16	134.6

%E	M1	M2	DM	M*
90.3	155	140	15	126.5
90.3	154	139	15	125.5
90.3	145	131	14	118.4
90.3	144	130	14	117.4
90.3	134	121	13	109.3
90.3	124	112	12	101.2
90.3	113	102	11	92.1
90.3	103	93	10	84.0
90.3	93	84	9	75.9
90.3	72	65	7	58.7
90.3	62	56	6	50.6
90.3	31	28	3	25.3
90.2	500	451	49	406.8
90.2	499	450	49	405.8
90.2	498	449	49	404.8
90.2	492	444	48	400.7
90.2	491	443	48	399.7
90.2	490	442	48	398.7
90.2	489	441	48	397.7
90.2	488	440	48	396.7
90.2	482	435	47	392.6
90.2	481	434	47	391.6
90.2	480	433	47	390.6
90.2	479	432	47	389.6
90.2	478	431	47	388.6
90.2	471	425	46	383.5
90.2	470	424	46	382.5
90.2	469	423	46	381.5
90.2	468	422	46	380.5
90.2	467	421	46	379.5
90.2	461	416	45	375.4
90.2	460	415	45	374.4
90.2	459	414	45	373.4
90.2	458	413	45	372.4
90.2	457	412	45	371.4
90.2	451	407	44	367.3
90.2	450	406	44	366.3
90.2	449	405	44	365.3
90.2	448	404	44	364.3
90.2	447	403	44	363.3
90.2	441	398	43	359.2
90.2	440	397	43	358.2
90.2	439	396	43	357.2
90.2	438	395	43	356.2
90.2	437	394	43	355.2
90.2	430	388	42	350.1
90.2	429	387	42	349.1
90.2	428	386	42	348.1
90.2	427	385	42	347.1
90.2	420	379	41	342.0

%E	M1	M2	DM	M*
90.2	419	378	41	341.0
90.2	418	377	41	340.0
90.2	417	376	41	339.0
90.2	410	370	40	333.9
90.2	409	369	40	332.9
90.2	408	368	40	331.9
90.2	407	367	40	330.9
90.2	399	360	39	324.8
90.2	398	359	39	323.8
90.2	397	358	39	322.8
90.2	396	357	39	321.8
90.2	389	351	38	316.7
90.2	388	350	38	315.7
90.2	387	349	38	314.7
90.2	386	348	38	313.7
90.2	379	342	37	308.6
90.2	378	341	37	307.6
90.2	377	340	37	306.6
90.2	376	339	37	305.6
90.2	369	333	36	300.5
90.2	368	332	36	299.5
90.2	367	331	36	298.5
90.2	366	330	36	297.5
90.2	358	323	35	291.4
90.2	357	322	35	290.4
90.2	356	321	35	289.4
90.2	348	314	34	283.3
90.2	347	313	34	282.3
90.2	346	312	34	281.3
90.2	338	305	33	275.2
90.2	337	304	33	274.2
90.2	336	303	33	273.2
90.2	328	296	32	267.1
90.2	327	295	32	266.1
90.2	326	294	32	265.1
90.2	325	293	32	264.2
90.2	317	286	31	258.0
90.2	316	285	31	257.0
90.2	315	284	31	256.1
90.2	307	277	30	249.9
90.2	306	276	30	248.9
90.2	305	275	30	248.0
90.2	297	268	29	241.8
90.2	296	267	29	240.8
90.2	295	266	29	239.9
90.2	287	259	28	233.7
90.2	286	258	28	232.7
90.2	285	257	28	231.8
90.2	276	249	27	224.6
90.2	275	248	27	223.7

%E	M1	M2	DM	M*
90.2	266	240	26	216.5
90.2	265	239	26	215.6
90.2	264	238	26	214.6
90.2	256	231	25	208.4
90.2	255	230	25	207.5
90.2	254	229	25	206.5
90.2	246	222	24	200.3
90.2	245	221	24	199.4
90.2	244	220	24	198.4
90.2	235	212	23	191.3
90.2	234	211	23	190.3
90.2	225	203	22	183.2
90.2	224	202	22	182.2
90.2	215	194	21	175.1
90.2	214	193	21	174.1
90.2	205	185	20	167.0
90.2	204	184	20	166.0
90.2	194	175	19	157.9
90.2	193	174	19	156.9
90.2	184	166	18	149.8
90.2	183	165	18	148.8
90.2	174	157	17	141.7
90.2	173	156	17	140.7
90.2	164	148	16	133.6
90.2	163	147	16	132.6
90.2	153	138	15	124.5
90.2	143	129	14	116.4
90.2	133	120	13	108.3
90.2	132	119	13	107.3
90.2	123	111	12	100.2
90.2	122	110	12	99.2
90.2	112	101	11	91.1
90.2	102	92	10	83.0
90.2	92	83	9	74.9
90.2	82	74	8	66.8
90.2	61	55	6	49.6
90.2	51	46	5	41.5
90.2	41	37	4	33.4
90.1	497	448	49	403.8
90.1	496	447	49	402.8
90.1	495	446	49	401.9
90.1	494	445	49	400.9
90.1	493	444	49	399.9
90.1	487	439	48	395.7
90.1	486	438	48	394.7
90.1	485	437	48	393.8
90.1	484	436	48	392.8
90.1	483	435	48	391.8
90.1	477	430	47	387.6
90.1	476	429	47	386.6

%E	M1	M2	DM	M*
90.1	475	428	47	385.7
90.1	474	427	47	384.7
90.1	473	426	47	383.7
90.1	466	420	46	378.5
90.1	465	419	46	377.6
90.1	464	418	46	376.6
90.1	463	417	46	375.6
90.1	456	411	45	370.4
90.1	455	410	45	369.5
90.1	454	409	45	368.5
90.1	453	408	45	367.5
90.1	446	402	44	362.3
90.1	445	401	44	361.4
90.1	444	400	44	360.4
90.1	443	399	44	359.4
90.1	436	393	43	354.2
90.1	435	392	43	353.3
90.1	434	391	43	352.3
90.1	433	390	43	351.3
90.1	426	384	42	346.1
90.1	425	383	42	345.2
90.1	424	382	42	344.2
90.1	423	381	42	343.2
90.1	416	375	41	338.0
90.1	415	374	41	337.1
90.1	414	373	41	336.1
90.1	413	372	41	335.1
90.1	406	366	40	329.9
90.1	405	365	40	329.0
90.1	404	364	40	328.0
90.1	403	363	40	327.0
90.1	402	362	40	326.0
90.1	395	356	39	320.9
90.1	394	355	39	319.9
90.1	393	354	39	318.9
90.1	392	353	39	317.9
90.1	385	347	38	312.8
90.1	384	346	38	311.8
90.1	383	345	38	310.8
90.1	382	344	38	309.8
90.1	375	338	37	304.7
90.1	374	337	37	303.7
90.1	373	336	37	302.7
90.1	372	335	37	301.7
90.1	365	329	36	296.6
90.1	364	328	36	295.6
90.1	363	327	36	294.6
90.1	362	326	36	293.6
90.1	355	320	35	288.5
90.1	354	319	35	287.5

%E	M1	M2	DM	M*
90.1	353	318	35	286.5
90.1	352	317	35	285.5
90.1	345	311	34	280.4
90.1	344	310	34	279.4
90.1	343	309	34	278.4
90.1	342	308	34	277.4
90.1	335	302	33	272.3
90.1	334	301	33	271.3
90.1	333	300	33	270.3
90.1	332	299	33	269.3
90.1	324	292	32	263.2
90.1	323	291	32	262.2
90.1	322	290	32	261.2
90.1	314	283	31	255.1
90.1	313	282	31	254.1
90.1	312	281	31	253.1
90.1	304	274	30	247.0
90.1	303	273	30	246.0
90.1	302	272	30	245.0
90.1	294	265	29	238.9
90.1	293	264	29	237.9
90.1	292	263	29	236.9
90.1	284	256	28	230.8
90.1	283	255	28	229.8
90.1	282	254	28	228.8
90.1	274	247	27	222.7
90.1	273	246	27	221.7
90.1	272	245	27	220.7
90.1	263	237	26	213.6
90.1	262	236	26	212.6
90.1	253	228	25	205.5
90.1	252	227	25	204.5
90.1	243	219	24	197.4
90.1	242	218	24	196.4
90.1	233	210	23	189.3
90.1	232	209	23	188.3
90.1	223	201	22	181.2
90.1	222	200	22	180.2
90.1	213	192	21	173.1
90.1	212	191	21	172.1
90.1	203	183	20	165.0
90.1	202	182	20	164.0
90.1	201	181	20	163.0
90.1	192	173	19	155.9
90.1	191	172	19	154.9
90.1	182	164	18	147.8
90.1	181	163	18	146.8
90.1	172	155	17	139.7
90.1	171	154	17	138.7
90.1	162	146	16	131.6

%E	M1	M2	DM	M*
90.1	161	145	16	130.6
90.1	152	137	15	123.5
90.1	151	136	15	122.5
90.1	142	128	14	115.4
90.1	141	127	14	114.4
90.1	131	118	13	106.3
90.1	121	109	12	98.2
90.1	111	100	11	90.1
90.1	101	91	10	82.0
90.1	91	82	9	73.9
90.1	81	73	8	65.8
90.1	71	64	7	57.7
90.0	500	450	50	405.0
90.0	499	449	50	404.0
90.0	498	448	50	403.0
90.0	492	443	49	398.9
90.0	491	442	49	397.9
90.0	490	441	49	396.9
90.0	489	440	49	395.9
90.0	488	439	49	394.9
90.0	482	434	48	390.8
90.0	481	433	48	389.8
90.0	480	432	48	388.8
90.0	479	431	48	387.8
90.0	478	430	48	386.8
90.0	472	425	47	382.7
90.0	471	424	47	381.7
90.0	470	423	47	380.7
90.0	469	422	47	379.7
90.0	468	421	47	378.7
90.0	462	416	46	374.6
90.0	461	415	46	373.6
90.0	460	414	46	372.6
90.0	459	413	46	371.6
90.0	458	412	46	370.6
90.0	452	407	45	366.5
90.0	451	406	45	365.5
90.0	450	405	45	364.5
90.0	449	404	45	363.5
90.0	448	403	45	362.5
90.0	442	398	44	358.4
90.0	441	397	44	357.4
90.0	440	396	44	356.4
90.0	439	395	44	355.4
90.0	438	394	44	354.4
90.0	432	389	43	350.3
90.0	431	388	43	349.3
90.0	430	387	43	348.3
90.0	429	386	43	347.3
90.0	428	385	43	346.3

%E	M1	M2	DM	M*
90.0	422	380	42	342.2
90.0	421	379	42	341.2
90.0	420	378	42	340.2
90.0	419	377	42	339.2
90.0	418	376	42	338.2
90.0	412	371	41	334.1
90.0	411	370	41	333.1
90.0	410	369	41	332.1
90.0	409	368	41	331.1
90.0	408	367	41	330.1
90.0	401	361	40	325.0
90.0	400	360	40	324.0
90.0	399	359	40	323.0
90.0	398	358	40	322.0
90.0	391	352	39	316.9
90.0	390	351	39	315.9
90.0	389	350	39	314.9
90.0	381	343	38	308.8
90.0	380	342	38	307.8
90.0	379	341	38	306.8
90.0	371	334	37	300.7
90.0	370	333	37	299.7
90.0	369	332	37	298.7
90.0	361	325	36	292.6
90.0	360	324	36	291.6
90.0	359	323	36	290.6
90.0	351	316	35	284.5
90.0	350	315	35	283.5
90.0	349	314	35	282.5
90.0	341	307	34	276.4
90.0	340	306	34	275.4
90.0	339	305	34	274.4
90.0	331	298	33	268.3
90.0	330	297	33	267.3
90.0	329	296	33	266.3
90.0	321	289	32	260.2
90.0	320	288	32	259.2
90.0	319	287	32	258.2
90.0	311	280	31	252.1
90.0	310	279	31	251.1
90.0	309	278	31	250.1
90.0	301	271	30	244.0
90.0	300	270	30	243.0
90.0	299	269	30	242.0
90.0	291	262	29	235.9
90.0	290	261	29	234.9
90.0	289	260	29	233.9
90.0	281	253	28	227.8
90.0	280	252	28	226.8
90.0	279	251	28	225.8

%E	M1	M2	DM	M*
90.0	271	244	27	219.7
90.0	270	243	27	218.7
90.0	269	242	27	217.7
90.0	261	235	26	211.6
90.0	260	234	26	210.6
90.0	259	233	26	209.6
90.0	251	226	25	203.5
90.0	250	225	25	202.5
90.0	249	224	25	201.5
90.0	241	217	24	195.4
90.0	240	216	24	194.4
90.0	239	215	24	193.4
90.0	231	208	23	187.3
90.0	230	207	23	186.3
90.0	229	206	23	185.3
90.0	221	199	22	179.2
90.0	220	198	22	178.2
90.0	219	197	22	177.2
90.0	211	190	21	171.1
90.0	210	189	21	170.1
90.0	209	188	21	169.1
90.0	200	180	20	162.0
90.0	199	179	20	161.0
90.0	190	171	19	153.9
90.0	180	162	18	145.8
90.0	170	153	17	137.7
90.0	160	144	16	129.6
90.0	150	135	15	121.5
90.0	140	126	14	113.4
90.0	130	117	13	105.3
90.0	120	108	12	97.2
90.0	110	99	11	89.1
90.0	100	90	10	81.0
90.0	90	81	9	72.9
90.0	80	72	8	64.8
90.0	70	63	7	56.7
90.0	60	54	6	48.6
90.0	50	45	5	40.5
90.0	40	36	4	32.4
90.0	30	27	3	24.3
90.0	20	18	2	16.2
90.0	10	9	1	8.1
89.9	497	447	50	402.0
89.9	496	446	50	401.0
89.9	495	445	50	400.1
89.9	494	444	50	399.1
89.9	493	443	50	398.1
89.9	487	438	49	393.9
89.9	486	437	49	392.9
89.9	485	436	49	392.0

%E	M1	M2	DM	M*
89.9	484	435	49	391.0
89.9	483	434	49	390.0
89.9	477	429	48	385.8
89.9	476	428	48	384.8
89.9	475	427	48	383.9
89.9	474	426	48	382.9
89.9	473	425	48	381.9
89.9	467	420	47	377.7
89.9	466	419	47	376.7
89.9	465	418	47	375.8
89.9	464	417	47	374.8
89.9	457	411	46	369.6
89.9	456	410	46	368.6
89.9	455	409	46	367.7
89.9	454	408	46	366.7
89.9	447	402	45	361.5
89.9	446	401	45	360.5
89.9	445	400	45	359.6
89.9	444	399	45	358.6
89.9	437	393	44	353.4
89.9	436	392	44	352.4
89.9	435	391	44	351.5
89.9	434	390	44	350.5
89.9	427	384	43	345.3
89.9	426	383	43	344.3
89.9	425	382	43	343.4
89.9	424	381	43	342.4
89.9	417	375	42	337.2
89.9	416	374	42	336.2
89.9	415	373	42	335.3
89.9	414	372	42	334.3
89.9	407	366	41	329.1
89.9	406	365	41	328.1
89.9	405	364	41	327.2
89.9	404	363	41	326.2
89.9	397	357	40	321.0
89.9	396	356	40	320.0
89.9	395	355	40	319.1
89.9	388	349	39	313.9
89.9	387	348	39	312.9
89.9	386	347	39	311.9
89.9	385	346	39	311.0
89.9	378	340	38	305.8
89.9	377	339	38	304.8
89.9	376	338	38	303.8
89.9	375	337	38	302.9
89.9	368	331	37	297.7
89.9	367	330	37	296.7
89.9	366	329	37	295.7
89.9	365	328	37	294.8

%E	M1	M2	DM	M*
89.9	358	322	36	289.6
89.9	357	321	36	288.6
89.9	356	320	36	287.6
89.9	355	319	36	286.7
89.9	348	313	35	281.5
89.9	347	312	35	280.5
89.9	346	311	35	279.5
89.9	345	310	35	278.6
89.9	338	304	34	273.4
89.9	337	303	34	272.4
89.9	336	302	34	271.4
89.9	335	301	34	270.5
89.9	328	295	33	265.3
89.9	327	294	33	264.3
89.9	326	293	33	263.3
89.9	318	286	32	257.2
89.9	317	285	32	256.2
89.9	316	284	32	255.2
89.9	308	277	31	249.1
89.9	307	276	31	248.1
89.9	306	275	31	247.1
89.9	298	268	30	241.0
89.9	297	267	30	240.0
89.9	296	266	30	239.0
89.9	288	259	29	232.9
89.9	287	258	29	231.9
89.9	286	257	29	230.9
89.9	278	250	28	224.8
89.9	277	249	28	223.8
89.9	276	248	28	222.8
89.9	268	241	27	216.7
89.9	267	240	27	215.7
89.9	266	239	27	214.7
89.9	258	232	26	208.6
89.9	257	231	26	207.6
89.9	248	223	25	200.5
89.9	247	222	25	199.5
89.9	238	214	24	192.4
89.9	237	213	24	191.4
89.9	228	205	23	184.3
89.9	227	204	23	183.3
89.9	218	196	22	176.2
89.9	217	195	22	175.2
89.9	208	187	21	168.1
89.9	207	186	21	167.1
89.9	198	178	20	160.0
89.9	189	170	19	152.9
89.9	188	169	19	151.9
89.9	179	161	18	144.8
89.9	178	160	18	143.8

%E	M1	M2	DM	M*
89.9	169	152	17	136.7
89.9	168	151	17	135.7
89.9	159	143	16	128.6
89.9	158	142	16	127.6
89.9	149	134	15	120.5
89.9	148	133	15	119.5
89.9	139	125	14	112.4
89.9	138	124	14	111.4
89.9	129	116	13	104.3
89.9	119	107	12	96.2
89.9	109	98	11	88.1
89.9	99	89	10	80.0
89.9	89	80	9	71.9
89.9	79	71	8	63.8
89.9	69	62	7	55.7
89.8	500	449	51	403.2
89.8	499	448	51	402.2
89.8	498	447	51	401.2
89.8	492	442	50	397.1
89.8	491	441	50	396.1
89.8	490	440	50	395.1
89.8	489	439	50	394.1
89.8	488	438	50	393.1
89.8	482	433	49	389.0
89.8	481	432	49	388.0
89.8	480	431	49	387.0
89.8	479	430	49	386.0
89.8	472	424	48	380.9
89.8	471	423	48	379.9
89.8	470	422	48	378.9
89.8	469	421	48	377.9
89.8	463	416	47	373.8
89.8	462	415	47	372.8
89.8	461	414	47	371.8
89.8	460	413	47	370.8
89.8	459	412	47	369.8
89.8	453	407	46	365.7
89.8	452	406	46	364.7
89.8	451	405	46	363.7
89.8	450	404	46	362.7
89.8	449	403	46	361.7
89.8	443	398	45	357.6
89.8	442	397	45	356.6
89.8	441	396	45	355.6
89.8	440	395	45	354.6
89.8	433	389	44	349.5
89.8	432	388	44	348.5
89.8	431	387	44	347.5
89.8	430	386	44	346.5
89.8	423	380	43	341.4

%E	M1	M2	DM	M*
89.8	422	379	43	340.4
89.8	421	378	43	339.4
89.8	420	377	43	338.4
89.8	413	371	42	333.3
89.8	412	370	42	332.3
89.8	411	369	42	331.3
89.8	410	368	42	330.3
89.8	403	362	41	325.2
89.8	402	361	41	324.2
89.8	401	360	41	323.2
89.8	400	359	41	322.2
89.8	394	354	40	318.1
89.8	393	353	40	317.1
89.8	392	352	40	316.1
89.8	391	351	40	315.1
89.8	384	345	39	310.0
89.8	383	344	39	309.0
89.8	382	343	39	308.0
89.8	381	342	39	307.0
89.8	374	336	38	301.9
89.8	373	335	38	300.9
89.8	372	334	38	299.9
89.8	371	333	38	298.9
89.8	364	327	37	293.8
89.8	363	326	37	292.8
89.8	362	325	37	291.8
89.8	361	324	37	290.8
89.8	354	318	36	285.7
89.8	353	317	36	284.7
89.8	352	316	36	283.7
89.8	344	309	35	277.6
89.8	343	308	35	276.6
89.8	342	307	35	275.6
89.8	334	300	34	269.5
89.8	333	299	34	268.5
89.8	332	298	34	267.5
89.8	325	292	33	262.4
89.8	324	291	33	261.4
89.8	323	290	33	260.4
89.8	322	289	33	259.4
89.8	315	283	32	254.3
89.8	314	282	32	253.3
89.8	313	281	32	252.3
89.8	305	274	31	246.2
89.8	304	273	31	245.2
89.8	303	272	31	244.2
89.8	295	265	30	238.1
89.8	294	264	30	237.1
89.8	293	263	30	236.1
89.8	285	256	29	230.0

%E	M1	M2	DM	M*
89.8	284	255	29	229.0
89.8	283	254	29	228.0
89.8	275	247	28	221.9
89.8	274	246	28	220.9
89.8	265	238	27	213.8
89.8	264	237	27	212.8
89.8	256	230	26	206.6
89.8	255	229	26	205.7
89.8	254	228	26	204.7
89.8	246	221	25	198.5
89.8	245	220	25	197.6
89.8	244	219	25	196.6
89.8	236	212	24	190.4
89.8	235	211	24	189.5
89.8	226	203	23	182.3
89.8	225	202	23	181.4
89.8	216	194	22	174.2
89.8	215	193	22	173.3
89.8	206	185	21	166.1
89.8	205	184	21	165.2
89.8	197	177	20	159.0
89.8	196	176	20	158.0
89.8	187	168	19	150.9
89.8	186	167	19	149.9
89.8	177	159	18	142.8
89.8	176	158	18	141.8
89.8	167	150	17	134.7
89.8	166	149	17	133.7
89.8	157	141	16	126.6
89.8	147	132	15	118.5
89.8	137	123	14	110.4
89.8	128	115	13	103.3
89.8	127	114	13	102.3
89.8	118	106	12	95.2
89.8	108	97	11	87.1
89.8	98	88	10	79.0
89.8	88	79	9	70.9
89.8	59	53	6	47.6
89.8	49	44	5	39.5
89.7	497	446	51	400.2
89.7	496	445	51	399.2
89.7	495	444	51	398.3
89.7	494	443	51	397.3
89.7	493	442	51	396.3
89.7	487	437	50	392.1
89.7	486	436	50	391.1
89.7	485	435	50	390.2
89.7	484	434	50	389.2
89.7	478	429	49	385.0
89.7	477	428	49	384.0

%E	M1	M2	DM	M*
89.7	476	427	49	383.0
89.7	475	426	49	382.1
89.7	474	425	49	381.1
89.7	468	420	48	376.9
89.7	467	419	48	375.9
89.7	466	418	48	374.9
89.7	465	417	48	374.0
89.7	464	416	48	373.0
89.7	458	411	47	368.8
89.7	457	410	47	367.8
89.7	456	409	47	366.8
89.7	455	408	47	365.9
89.7	448	402	46	360.7
89.7	447	401	46	359.7
89.7	446	400	46	358.7
89.7	445	399	46	357.8
89.7	439	394	45	353.6
89.7	438	393	45	352.6
89.7	437	392	45	351.6
89.7	436	391	45	350.6
89.7	435	390	45	349.7
89.7	429	385	44	345.5
89.7	428	384	44	344.5
89.7	427	383	44	343.5
89.7	426	382	44	342.5
89.7	419	376	43	337.4
89.7	418	375	43	336.4
89.7	417	374	43	335.4
89.7	416	373	43	334.4
89.7	409	367	42	329.3
89.7	408	366	42	328.3
89.7	407	365	42	327.3
89.7	406	364	42	326.3
89.7	399	358	41	321.2
89.7	398	357	41	320.2
89.7	397	356	41	319.2
89.7	390	350	40	314.1
89.7	389	349	40	313.1
89.7	388	348	40	312.1
89.7	387	347	40	311.1
89.7	380	341	39	306.0
89.7	379	340	39	305.0
89.7	378	339	39	304.0
89.7	377	338	39	303.0
89.7	370	332	38	297.9
89.7	369	331	38	296.9
89.7	368	330	38	295.9
89.7	360	323	37	289.8
89.7	359	322	37	288.8
89.7	358	321	37	287.8

%E	M1	M2	DM	M*
89.7	351	315	36	282.7
89.7	350	314	36	281.7
89.7	349	313	36	280.7
89.7	348	312	36	279.7
89.7	341	306	35	274.6
89.7	340	305	35	273.6
89.7	339	304	35	272.6
89.7	331	297	34	266.5
89.7	330	296	34	265.5
89.7	329	295	34	264.5
89.7	321	288	33	258.4
89.7	320	287	33	257.4
89.7	319	286	33	256.4
89.7	312	280	32	251.3
89.7	311	279	32	250.3
89.7	310	278	32	249.3
89.7	302	271	31	243.2
89.7	301	270	31	242.2
89.7	300	269	31	241.2
89.7	292	262	30	235.1
89.7	291	261	30	234.1
89.7	290	260	30	233.1
89.7	282	253	29	227.0
89.7	281	252	29	226.0
89.7	273	245	28	219.9
89.7	272	244	28	218.9
89.7	271	243	28	217.9
89.7	263	236	27	211.8
89.7	262	235	27	210.8
89.7	261	234	27	209.8
89.7	253	227	26	203.7
89.7	252	226	26	202.7
89.7	243	218	25	195.6
89.7	242	217	25	194.6
89.7	234	210	24	188.5
89.7	233	209	24	187.5
89.7	232	208	24	186.5
89.7	224	201	23	180.4
89.7	223	200	23	179.4
89.7	214	192	22	172.3
89.7	213	191	22	171.3
89.7	204	183	21	164.2
89.7	203	182	21	163.2
89.7	195	175	20	157.1
89.7	194	174	20	156.1
89.7	185	166	19	149.0
89.7	184	165	19	148.0
89.7	175	157	18	140.9
89.7	174	156	18	139.9
89.7	165	148	17	132.8

%E	M1	M2	DM	M*
89.7	156	140	16	125.6
89.7	155	139	16	124.7
89.7	146	131	15	117.5
89.7	145	130	15	116.6
89.7	136	122	14	109.4
89.7	126	113	13	101.3
89.7	117	105	12	94.2
89.7	116	104	12	93.2
89.7	107	96	11	86.1
89.7	97	87	10	78.0
89.7	87	78	9	69.9
89.7	78	70	8	62.8
89.7	68	61	7	54.7
89.7	58	52	6	46.6
89.7	39	35	4	31.4
89.7	29	26	3	23.3
89.6	500	448	52	401.4
89.6	499	447	52	400.4
89.6	498	446	52	399.4
89.6	492	441	51	395.3
89.6	491	440	51	394.3
89.6	490	439	51	393.3
89.6	489	438	51	392.3
89.6	483	433	50	388.2
89.6	482	432	50	387.2
89.6	481	431	50	386.2
89.6	480	430	50	385.2
89.6	479	429	50	384.2
89.6	473	424	49	380.1
89.6	472	423	49	379.1
89.6	471	422	49	378.1
89.6	470	421	49	377.1
89.6	469	420	49	376.1
89.6	463	415	48	372.0
89.6	462	414	48	371.0
89.6	461	413	48	370.0
89.6	460	412	48	369.0
89.6	454	407	47	364.9
89.6	453	406	47	363.9
89.6	452	405	47	362.9
89.6	451	404	47	361.9
89.6	450	403	47	360.9
89.6	444	398	46	356.8
89.6	443	397	46	355.8
89.6	442	396	46	354.8
89.6	441	395	46	353.8
89.6	434	389	45	348.7
89.6	433	388	45	347.7
89.6	432	387	45	346.7
89.6	431	386	45	345.7

%E	M1	M2	DM	M*
89.6	425	381	44	341.6
89.6	424	380	44	340.6
89.6	423	379	44	339.6
89.6	422	378	44	338.6
89.6	415	372	43	333.5
89.6	414	371	43	332.5
89.6	413	370	43	331.5
89.6	412	369	43	330.5
89.6	405	363	42	325.4
89.6	404	362	42	324.4
89.6	403	361	42	323.4
89.6	402	360	42	322.4
89.6	396	355	41	318.2
89.6	395	354	41	317.3
89.6	394	353	41	316.3
89.6	393	352	41	315.3
89.6	386	346	40	310.1
89.6	385	345	40	309.2
89.6	384	344	40	308.2
89.6	383	343	40	307.2
89.6	376	337	39	302.0
89.6	375	336	39	301.1
89.6	374	335	39	300.1
89.6	367	329	38	294.9
89.6	366	328	38	293.9
89.6	365	327	38	293.0
89.6	364	326	38	292.0
89.6	357	320	37	286.8
89.6	356	319	37	285.8
89.6	355	318	37	284.9
89.6	347	311	36	278.7
89.6	346	310	36	277.7
89.6	345	309	36	276.8
89.6	338	303	35	271.6
89.6	337	302	35	270.6
89.6	336	301	35	269.6
89.6	335	300	35	268.7
89.6	328	294	34	263.5
89.6	327	293	34	262.5
89.6	326	292	34	261.5
89.6	318	285	33	255.4
89.6	317	284	33	254.4
89.6	316	283	33	253.4
89.6	309	277	32	248.3
89.6	308	276	32	247.3
89.6	307	275	32	246.3
89.6	299	268	31	240.2
89.6	298	267	31	239.2
89.6	297	266	31	238.2
89.6	289	259	30	232.1

%E	M1	M2	DM	M*
89.6	288	258	30	231.1
89.6	280	251	29	225.0
89.6	279	250	29	224.0
89.6	278	249	29	223.0
89.6	270	242	28	216.9
89.6	269	241	28	215.9
89.6	268	240	28	214.9
89.6	260	233	27	208.8
89.6	259	232	27	207.8
89.6	251	225	26	201.7
89.6	250	224	26	200.7
89.6	249	223	26	199.7
89.6	241	216	25	193.6
89.6	240	215	25	192.6
89.6	231	207	24	185.5
89.6	230	206	24	184.5
89.6	222	199	23	178.4
89.6	221	198	23	177.4
89.6	212	190	22	170.3
89.6	211	189	22	169.3
89.6	202	181	21	162.2
89.6	201	180	21	161.2
89.6	193	173	20	155.1
89.6	192	172	20	154.1
89.6	183	164	19	147.0
89.6	182	163	19	146.0
89.6	173	155	18	138.9
89.6	164	147	17	131.8
89.6	163	146	17	130.8
89.6	154	138	16	123.7
89.6	144	129	15	115.6
89.6	135	121	14	108.5
89.6	134	120	14	107.5
89.6	125	112	13	100.4
89.6	115	103	12	92.3
89.6	106	95	11	85.1
89.6	96	86	10	77.0
89.6	77	69	8	61.8
89.6	67	60	7	53.7
89.6	48	43	5	38.5
89.5	497	445	52	398.4
89.5	496	444	52	397.5
89.5	495	443	52	396.5
89.5	494	442	52	395.5
89.5	493	441	52	394.5
89.5	488	437	51	391.3
89.5	487	436	51	390.3
89.5	486	435	51	389.4
89.5	485	434	51	388.4
89.5	484	433	51	387.4

%E	M1	M2	DM	M*
89.5	478	428	50	383.2
89.5	477	427	50	382.2
89.5	476	426	50	381.3
89.5	475	425	50	380.3
89.5	474	424	50	379.3
89.5	468	419	49	375.1
89.5	467	418	49	374.1
89.5	466	417	49	373.2
89.5	465	416	49	372.2
89.5	459	411	48	368.0
89.5	458	410	48	367.0
89.5	457	409	48	366.0
89.5	456	408	48	365.1
89.5	455	407	48	364.1
89.5	449	402	47	359.9
89.5	448	401	47	358.9
89.5	447	400	47	357.9
89.5	446	399	47	357.0
89.5	440	394	46	352.8
89.5	439	393	46	351.8
89.5	438	392	46	350.8
89.5	437	391	46	349.8
89.5	436	390	46	348.9
89.5	430	385	45	344.7
89.5	429	384	45	343.7
89.5	428	383	45	342.7
89.5	427	382	45	341.7
89.5	421	377	44	337.6
89.5	420	376	44	336.6
89.5	419	375	44	335.6
89.5	418	374	44	334.6
89.5	411	368	43	329.5
89.5	410	367	43	328.5
89.5	409	366	43	327.5
89.5	408	365	43	326.5
89.5	401	359	42	321.4
89.5	400	358	42	320.4
89.5	399	357	42	319.4
89.5	392	351	41	314.3
89.5	391	350	41	313.3
89.5	390	349	41	312.3
89.5	389	348	41	311.3
89.5	382	342	40	306.2
89.5	381	341	40	305.2
89.5	380	340	40	304.2
89.5	373	334	39	299.1
89.5	372	333	39	298.1
89.5	371	332	39	297.1
89.5	370	331	39	296.1
89.5	363	325	38	291.0

%E	M1	M2	DM	M*
89.5	362	324	38	290.0
89.5	361	323	38	289.0
89.5	354	317	37	283.9
89.5	353	316	37	282.9
89.5	352	315	37	281.9
89.5	351	314	37	280.9
89.5	344	308	36	275.8
89.5	343	307	36	274.8
89.5	342	306	36	273.8
89.5	334	299	35	267.7
89.5	333	298	35	266.7
89.5	332	297	35	265.7
89.5	325	291	34	260.6
89.5	324	290	34	259.6
89.5	323	289	34	258.6
89.5	315	282	33	252.5
89.5	314	281	33	251.5
89.5	313	280	33	250.5
89.5	306	274	32	245.3
89.5	305	273	32	244.4
89.5	304	272	32	243.4
89.5	296	265	31	237.2
89.5	295	264	31	236.3
89.5	294	263	31	235.3
89.5	287	257	30	230.1
89.5	286	256	30	229.1
89.5	285	255	30	228.2
89.5	277	248	29	222.0
89.5	276	247	29	221.0
89.5	275	246	29	220.1
89.5	267	239	28	213.9
89.5	266	238	28	212.9
89.5	258	231	27	206.8
89.5	257	230	27	205.8
89.5	256	229	27	204.8
89.5	248	222	26	198.7
89.5	247	221	26	197.7
89.5	239	214	25	191.6
89.5	238	213	25	190.6
89.5	237	212	25	189.6
89.5	229	205	24	183.5
89.5	228	204	24	182.5
89.5	220	197	23	176.4
89.5	219	196	23	175.4
89.5	218	195	23	174.4
89.5	210	188	22	168.3
89.5	209	187	22	167.3
89.5	200	179	21	160.2
89.5	191	171	20	153.1
89.5	190	170	20	152.1

%E	M1	M2	DM	M*
89.5	181	162	19	145.0
89.5	172	154	18	137.9
89.5	171	153	18	136.9
89.5	162	145	17	129.8
89.5	153	137	16	122.7
89.5	152	136	16	121.7
89.5	143	128	15	114.6
89.5	133	119	14	106.5
89.5	124	111	13	99.4
89.5	114	102	12	91.3
89.5	105	94	11	84.2
89.5	95	85	10	76.1
89.5	86	77	9	68.9
89.5	76	68	8	60.8
89.5	57	51	6	45.6
89.5	38	34	4	30.4
89.5	19	17	2	15.2
89.4	500	447	53	399.6
89.4	499	446	53	398.6
89.4	498	445	53	397.6
89.4	492	440	52	393.5
89.4	491	439	52	392.5
89.4	490	438	52	391.5
89.4	489	437	52	390.5
89.4	483	432	51	386.4
89.4	482	431	51	385.4
89.4	481	430	51	384.4
89.4	480	429	51	383.4
89.4	479	428	51	382.4
89.4	473	423	50	378.3
89.4	472	422	50	377.3
89.4	471	421	50	376.3
89.4	470	420	50	375.3
89.4	464	415	49	371.2
89.4	463	414	49	370.2
89.4	462	413	49	369.2
89.4	461	412	49	368.2
89.4	454	406	48	363.1
89.4	453	405	48	362.1
89.4	452	404	48	361.1
89.4	451	403	48	360.1
89.4	445	398	47	356.0
89.4	444	397	47	355.0
89.4	443	396	47	354.0
89.4	442	395	47	353.0
89.4	435	389	46	347.9
89.4	434	388	46	346.9
89.4	433	387	46	345.9
89.4	432	386	46	344.9
89.4	426	381	45	340.8

%E	M1	M2	DM	M*
89.4	425	380	45	339.8
89.4	424	379	45	338.8
89.4	423	378	45	337.8
89.4	417	373	44	333.6
89.4	416	372	44	332.7
89.4	415	371	44	331.7
89.4	414	370	44	330.7
89.4	407	364	43	325.5
89.4	406	363	43	324.6
89.4	405	362	43	323.6
89.4	404	361	43	322.6
89.4	398	356	42	318.4
89.4	397	355	42	317.4
89.4	396	354	42	316.5
89.4	395	353	42	315.5
89.4	388	347	41	310.3
89.4	387	346	41	309.3
89.4	386	345	41	308.4
89.4	385	344	41	307.4
89.4	379	339	40	303.2
89.4	378	338	40	302.2
89.4	377	337	40	301.2
89.4	376	336	40	300.3
89.4	369	330	39	295.1
89.4	368	329	39	294.1
89.4	367	328	39	293.1
89.4	360	322	38	288.0
89.4	359	321	38	287.0
89.4	358	320	38	286.0
89.4	357	319	38	285.0
89.4	350	313	37	279.9
89.4	349	312	37	278.9
89.4	348	311	37	277.9
89.4	341	305	36	272.8
89.4	340	304	36	271.8
89.4	339	303	36	270.8
89.4	331	296	35	264.7
89.4	330	295	35	263.7
89.4	329	294	35	262.7
89.4	322	288	34	257.6
89.4	321	287	34	256.6
89.4	320	286	34	255.6
89.4	312	279	33	249.5
89.4	311	278	33	248.5
89.4	310	277	33	247.5
89.4	303	271	32	242.4
89.4	302	270	32	241.4
89.4	301	269	32	240.4
89.4	293	262	31	234.3
89.4	292	261	31	233.3
89.4	284	254	30	227.2
89.4	283	253	30	226.2
89.4	282	252	30	225.2
89.4	274	245	29	219.1
89.4	273	244	29	218.1
89.4	265	237	28	212.0
89.4	264	236	28	211.0
89.4	263	235	28	210.0
89.4	255	228	27	203.9
89.4	254	227	27	202.9
89.4	246	220	26	196.7
89.4	245	219	26	195.8
89.4	236	211	25	188.6
89.4	235	210	25	187.7
89.4	227	203	24	181.5
89.4	226	202	24	180.5
89.4	217	194	23	173.4
89.4	216	193	23	172.4
89.4	208	186	22	166.3
89.4	207	185	22	165.3
89.4	199	178	21	159.2
89.4	198	177	21	158.2
89.4	189	169	20	151.1
89.4	188	168	20	150.1
89.4	180	161	19	144.0
89.4	179	160	19	143.0
89.4	170	152	18	135.9
89.4	161	144	17	128.8
89.4	160	143	17	127.8
89.4	151	135	16	120.7
89.4	142	127	15	113.6
89.4	141	126	15	112.6
89.4	132	118	14	105.5
89.4	123	110	13	98.4
89.4	113	101	12	90.3
89.4	104	93	11	83.2
89.4	94	84	10	75.1
89.4	85	76	9	68.0
89.4	66	59	7	52.7
89.4	47	42	5	37.5
89.3	497	444	53	396.7
89.3	496	443	53	395.7
89.3	495	442	53	394.7
89.3	494	441	53	393.7
89.3	488	436	52	389.5
89.3	487	435	52	388.6
89.3	486	434	52	387.6
89.3	485	433	52	386.6
89.3	484	432	52	385.6
89.3	478	427	51	381.4
89.3	477	426	51	380.5
89.3	476	425	51	379.5
89.3	475	424	51	378.5
89.3	469	419	50	374.3
89.3	468	418	50	373.3
89.3	467	417	50	372.4
89.3	466	416	50	371.4
89.3	460	411	49	367.2
89.3	459	410	49	366.2
89.3	458	409	49	365.2
89.3	457	408	49	364.3
89.3	456	407	49	363.3
89.3	450	402	48	359.1
89.3	449	401	48	358.1
89.3	448	400	48	357.1
89.3	447	399	48	356.2
89.3	441	394	47	352.0
89.3	440	393	47	351.0
89.3	439	392	47	350.0
89.3	438	391	47	349.0
89.3	431	385	46	343.9
89.3	430	384	46	342.9
89.3	429	383	46	341.9
89.3	428	382	46	340.9
89.3	422	377	45	336.8
89.3	421	376	45	335.8
89.3	420	375	45	334.8
89.3	419	374	45	333.8
89.3	413	369	44	329.7
89.3	412	368	44	328.7
89.3	411	367	44	327.7
89.3	410	366	44	326.7
89.3	403	360	43	321.6
89.3	402	359	43	320.6
89.3	401	358	43	319.6
89.3	400	357	43	318.6
89.3	394	352	42	314.5
89.3	393	351	42	313.5
89.3	392	350	42	312.5
89.3	391	349	42	311.5
89.3	384	343	41	306.4
89.3	383	342	41	305.4
89.3	382	341	41	304.4
89.3	375	335	40	299.3
89.3	374	334	40	298.3
89.3	373	333	40	297.3
89.3	366	327	39	292.2
89.3	365	326	39	291.2
89.3	364	325	39	290.2
89.3	363	324	39	289.2
89.3	356	318	38	284.1
89.3	355	317	38	283.1
89.3	354	316	38	282.1
89.3	347	310	37	276.9
89.3	346	309	37	276.0
89.3	345	308	37	275.0
89.3	338	302	36	269.8
89.3	337	301	36	268.8
89.3	336	300	36	267.9
89.3	335	299	36	266.9
89.3	328	293	35	261.7
89.3	327	292	35	260.7
89.3	326	291	35	259.8
89.3	319	285	34	254.6
89.3	318	284	34	253.6
89.3	317	283	34	252.6
89.3	309	276	33	246.5
89.3	308	275	33	245.5
89.3	307	274	33	244.5
89.3	300	268	32	239.4
89.3	299	267	32	238.4
89.3	298	266	32	237.4
89.3	291	260	31	232.3
89.3	290	259	31	231.3
89.3	289	258	31	230.3
89.3	281	251	30	224.2
89.3	280	250	30	223.2
89.3	272	243	29	217.1
89.3	271	242	29	216.1
89.3	270	241	29	215.1
89.3	262	234	28	209.0
89.3	261	233	28	208.0
89.3	253	226	27	201.9
89.3	252	225	27	200.9
89.3	244	218	26	194.8
89.3	243	217	26	193.8
89.3	242	216	26	192.8
89.3	234	209	25	186.7
89.3	233	208	25	185.7
89.3	225	201	24	179.6
89.3	224	200	24	178.6
89.3	215	192	23	171.5
89.3	214	191	23	170.5
89.3	206	184	22	164.3
89.3	205	183	22	163.4
89.3	197	176	21	157.2
89.3	196	175	21	156.3
89.3	187	167	20	149.1
89.3	178	159	19	142.0
89.3	177	158	19	141.0
89.3	169	151	18	134.9
89.3	168	150	18	133.9
89.3	159	142	17	126.8
89.3	150	134	16	119.7
89.3	149	133	16	118.7
89.3	140	125	15	111.6
89.3	131	117	14	104.5
89.3	122	109	13	97.4
89.3	121	108	13	96.4
89.3	112	100	12	89.3
89.3	103	92	11	82.2
89.3	84	75	9	67.0
89.3	75	67	8	59.9
89.3	56	50	6	44.6
89.3	28	25	3	22.3
89.2	500	446	54	397.8
89.2	499	445	54	396.8
89.2	498	444	54	395.9
89.2	493	440	53	392.7
89.2	492	439	53	391.7
89.2	491	438	53	390.7
89.2	490	437	53	389.7
89.2	489	436	53	388.7
89.2	483	431	52	384.6
89.2	482	430	52	383.6
89.2	481	429	52	382.6
89.2	480	428	52	381.6
89.2	474	423	51	377.5
89.2	473	422	51	376.5
89.2	472	421	51	375.5
89.2	471	420	51	374.5
89.2	465	415	50	370.4
89.2	464	414	50	369.4
89.2	463	413	50	368.4
89.2	462	412	50	367.4
89.2	461	411	50	366.4
89.2	455	406	49	362.3
89.2	454	405	49	361.3
89.2	453	404	49	360.3
89.2	452	403	49	359.3
89.2	446	398	48	355.2
89.2	445	397	48	354.2
89.2	444	396	48	353.2
89.2	443	395	48	352.2
89.2	437	390	47	348.1
89.2	436	389	47	347.1
89.2	435	388	47	346.1
89.2	434	387	47	345.1
89.2	427	381	46	340.0
89.2	426	380	46	339.0

%E	M1	M2	DM	M*
89.2	425	379	46	338.0
89.2	424	378	46	337.0
89.2	418	373	45	332.8
89.2	417	372	45	331.9
89.2	416	371	45	330.9
89.2	415	370	45	329.9
89.2	409	365	44	325.7
89.2	408	364	44	324.7
89.2	407	363	44	323.8
89.2	406	362	44	322.8
89.2	399	356	43	317.6
89.2	398	355	43	316.6
89.2	397	354	43	315.7
89.2	390	348	42	310.5
89.2	389	347	42	309.5
89.2	388	346	42	308.5
89.2	381	340	41	303.4
89.2	380	339	41	302.4
89.2	379	338	41	301.4
89.2	378	337	41	300.4
89.2	372	332	40	296.3
89.2	371	331	40	295.3
89.2	370	330	40	294.3
89.2	369	329	40	293.3
89.2	362	323	39	288.2
89.2	361	322	39	287.2
89.2	360	321	39	286.2
89.2	353	315	38	281.1
89.2	352	314	38	280.1
89.2	351	313	38	279.1
89.2	344	307	37	274.0
89.2	343	306	37	273.0
89.2	342	305	37	272.0
89.2	341	304	37	271.0
89.2	334	298	36	265.9
89.2	333	297	36	264.9
89.2	332	296	36	263.9
89.2	325	290	35	258.8
89.2	324	289	35	257.8
89.2	323	288	35	256.8
89.2	316	282	34	251.7
89.2	315	281	34	250.7
89.2	314	280	34	249.7
89.2	306	273	33	243.6
89.2	305	272	33	242.6
89.2	297	265	32	236.4
89.2	296	264	32	235.5
89.2	295	263	32	234.5
89.2	288	257	31	229.3
89.2	287	256	31	228.3

%E	M1	M2	DM	M*
89.2	286	255	31	227.4
89.2	279	249	30	222.2
89.2	278	248	30	221.2
89.2	277	247	30	220.2
89.2	269	240	29	214.1
89.2	268	239	29	213.1
89.2	260	232	28	207.0
89.2	259	231	28	206.0
89.2	251	224	27	199.9
89.2	250	223	27	198.9
89.2	249	222	27	197.9
89.2	241	215	26	191.8
89.2	240	214	26	190.8
89.2	232	207	25	184.7
89.2	231	206	25	183.7
89.2	223	199	24	177.6
89.2	222	198	24	176.6
89.2	213	190	23	169.5
89.2	212	189	23	168.5
89.2	204	182	22	162.4
89.2	203	181	22	161.4
89.2	195	174	21	155.3
89.2	194	173	21	154.3
89.2	186	166	20	148.2
89.2	185	165	20	147.2
89.2	176	157	19	140.1
89.2	167	149	18	132.9
89.2	166	148	18	132.0
89.2	158	141	17	125.8
89.2	157	140	17	124.8
89.2	148	132	16	117.7
89.2	139	124	15	110.6
89.2	130	116	14	103.5
89.2	120	107	13	95.4
89.2	111	99	12	88.3
89.2	102	91	11	81.2
89.2	93	83	10	74.1
89.2	83	74	9	66.0
89.2	74	66	8	58.9
89.2	65	58	7	51.8
89.2	37	33	4	29.4
89.1	497	443	54	394.9
89.1	496	442	54	393.9
89.1	495	441	54	392.9
89.1	494	440	54	391.9
89.1	488	435	53	387.8
89.1	487	434	53	386.8
89.1	486	433	53	385.8
89.1	485	432	53	384.8
89.1	484	431	53	383.8

%E	M1	M2	DM	M*
89.1	479	427	52	380.6
89.1	478	426	52	379.7
89.1	477	425	52	378.7
89.1	476	424	52	377.7
89.1	475	423	52	376.7
89.1	470	419	51	373.5
89.1	469	418	51	372.5
89.1	468	417	51	371.6
89.1	467	416	51	370.6
89.1	466	415	51	369.6
89.1	460	410	50	365.4
89.1	459	409	50	364.4
89.1	458	408	50	363.5
89.1	457	407	50	362.5
89.1	451	402	49	358.3
89.1	450	401	49	357.3
89.1	449	400	49	356.3
89.1	448	399	49	355.4
89.1	442	394	48	351.2
89.1	441	393	48	350.2
89.1	440	392	48	349.2
89.1	439	391	48	348.2
89.1	433	386	47	344.1
89.1	432	385	47	343.1
89.1	431	384	47	342.1
89.1	430	383	47	341.1
89.1	423	377	46	336.0
89.1	422	376	46	335.0
89.1	421	375	46	334.0
89.1	414	369	45	328.9
89.1	413	368	45	327.9
89.1	412	367	45	326.9
89.1	411	366	45	325.9
89.1	405	361	44	321.8
89.1	404	360	44	320.8
89.1	403	359	44	319.8
89.1	402	358	44	318.8
89.1	396	353	43	314.7
89.1	395	352	43	313.7
89.1	394	351	43	312.7
89.1	393	350	43	311.7
89.1	387	345	42	307.6
89.1	386	344	42	306.6
89.1	385	343	42	305.6
89.1	384	342	42	304.6
89.1	377	336	41	299.5
89.1	376	335	41	298.5
89.1	375	334	41	297.5
89.1	368	328	40	292.3
89.1	367	327	40	291.4

%E	M1	M2	DM	M*
89.1	366	326	40	290.4
89.1	359	320	39	285.2
89.1	358	319	39	284.2
89.1	357	318	39	283.3
89.1	350	312	38	278.1
89.1	349	311	38	277.1
89.1	348	310	38	276.1
89.1	340	303	37	270.0
89.1	339	302	37	269.0
89.1	338	301	37	268.1
89.1	331	295	36	262.9
89.1	330	294	36	261.9
89.1	329	293	36	260.9
89.1	322	287	35	255.8
89.1	321	286	35	254.8
89.1	320	285	35	253.8
89.1	313	279	34	248.7
89.1	312	278	34	247.7
89.1	311	277	34	246.7
89.1	304	271	33	241.6
89.1	303	270	33	240.6
89.1	302	269	33	239.6
89.1	294	262	32	233.5
89.1	293	261	32	232.5
89.1	285	254	31	226.4
89.1	284	253	31	225.4
89.1	276	246	30	219.3
89.1	275	245	30	218.3
89.1	274	244	30	217.3
89.1	267	238	29	212.1
89.1	266	237	29	211.2
89.1	265	236	29	210.2
89.1	258	230	28	205.0
89.1	257	229	28	204.1
89.1	256	228	28	203.1
89.1	248	221	27	196.9
89.1	247	220	27	196.0
89.1	239	213	26	189.8
89.1	238	212	26	188.8
89.1	230	205	25	182.7
89.1	229	204	25	181.7
89.1	221	197	24	175.6
89.1	220	196	24	174.6
89.1	211	188	23	167.5
89.1	202	180	22	160.4
89.1	201	179	22	159.4
89.1	193	172	21	153.3
89.1	192	171	21	152.3
89.1	184	164	20	146.2
89.1	183	163	20	145.2

%E	M1	M2	DM	M*
89.1	175	156	19	139.1
89.1	174	155	19	138.1
89.1	165	147	18	131.0
89.1	156	139	17	123.9
89.1	147	131	16	116.7
89.1	138	123	15	109.6
89.1	137	122	15	108.6
89.1	129	115	14	102.5
89.1	128	114	14	101.5
89.1	119	106	13	94.4
89.1	110	98	12	87.3
89.1	101	90	11	80.2
89.1	92	82	10	73.1
89.1	64	57	7	50.8
89.1	55	49	6	43.7
89.1	46	41	5	36.5
89.0	500	445	55	396.0
89.0	499	444	55	395.1
89.0	498	443	55	394.1
89.0	493	439	54	390.9
89.0	492	438	54	389.9
89.0	491	437	54	388.9
89.0	490	436	54	388.0
89.0	489	435	54	387.0
89.0	483	430	53	382.8
89.0	482	429	53	381.8
89.0	481	428	53	380.8
89.0	480	427	53	379.9
89.0	474	422	52	375.7
89.0	473	421	52	374.7
89.0	472	420	52	373.7
89.0	471	419	52	372.7
89.0	465	414	51	368.6
89.0	464	413	51	367.6
89.0	463	412	51	366.6
89.0	462	411	51	365.6
89.0	456	406	50	361.5
89.0	455	405	50	360.5
89.0	454	404	50	359.5
89.0	453	403	50	358.5
89.0	447	398	49	354.4
89.0	446	397	49	353.4
89.0	445	396	49	352.4
89.0	444	395	49	351.4
89.0	438	390	48	347.3
89.0	437	389	48	346.3
89.0	436	388	48	345.3
89.0	435	387	48	344.3
89.0	429	382	47	340.1
89.0	428	381	47	339.2

%E	M1	M2	DM	M*
89.0	427	380	47	338.2
89.0	426	379	47	337.2
89.0	420	374	46	333.0
89.0	419	373	46	332.1
89.0	418	372	46	331.1
89.0	417	371	46	330.1
89.0	410	365	45	324.9
89.0	409	364	45	324.0
89.0	408	363	45	323.0
89.0	401	357	44	317.8
89.0	400	356	44	316.8
89.0	399	355	44	315.9
89.0	392	349	43	310.7
89.0	391	348	43	309.7
89.0	390	347	43	308.7
89.0	383	341	42	303.6
89.0	382	340	42	302.6
89.0	381	339	42	301.6
89.0	374	333	41	296.5
89.0	373	332	41	295.5
89.0	372	331	41	294.5
89.0	365	325	40	289.4
89.0	364	324	40	288.4
89.0	363	323	40	287.4
89.0	362	322	40	286.4
89.0	356	317	39	282.3
89.0	355	316	39	281.3
89.0	354	315	39	280.3
89.0	353	314	39	279.3
89.0	347	309	38	275.2
89.0	346	308	38	274.2
89.0	345	307	38	273.2
89.0	344	306	38	272.2
89.0	337	300	37	267.1
89.0	336	299	37	266.1
89.0	335	298	37	265.1
89.0	328	292	36	260.0
89.0	327	291	36	259.0
89.0	326	290	36	258.0
89.0	319	284	35	252.8
89.0	318	283	35	251.9
89.0	317	282	35	250.9
89.0	310	276	34	245.7
89.0	309	275	34	244.7
89.0	308	274	34	243.8
89.0	301	268	33	238.6
89.0	300	267	33	237.6
89.0	299	266	33	236.6
89.0	292	260	32	231.5
89.0	291	259	32	230.5

%E	M1	M2	DM	M*
89.0	290	258	32	229.5
89.0	283	252	31	224.4
89.0	282	251	31	223.4
89.0	281	250	31	222.4
89.0	273	243	30	216.3
89.0	272	242	30	215.3
89.0	264	235	29	209.2
89.0	263	234	29	208.2
89.0	255	227	28	202.1
89.0	254	226	28	201.1
89.0	246	219	27	195.0
89.0	245	218	27	194.0
89.0	237	211	26	187.9
89.0	236	210	26	186.9
89.0	228	203	25	180.7
89.0	227	202	25	179.8
89.0	219	195	24	173.6
89.0	218	194	24	172.6
89.0	210	187	23	166.5
89.0	209	186	23	165.5
89.0	200	178	22	158.4
89.0	191	170	21	151.3
89.0	182	162	20	144.2
89.0	181	161	20	143.2
89.0	173	154	19	137.1
89.0	172	153	19	136.1
89.0	164	146	18	130.0
89.0	163	145	18	129.0
89.0	155	138	17	122.9
89.0	154	137	17	121.9
89.0	146	130	16	115.8
89.0	145	129	16	114.8
89.0	136	121	15	107.7
89.0	127	113	14	100.5
89.0	118	105	13	93.4
89.0	109	97	12	86.3
89.0	100	89	11	79.2
89.0	91	81	10	72.1
89.0	82	73	9	65.0
89.0	73	65	8	57.9
88.9	497	442	55	393.1
88.9	496	441	55	392.1
88.9	495	440	55	391.1
88.9	494	439	55	390.1
88.9	488	434	54	386.0
88.9	487	433	54	385.0
88.9	486	432	54	384.0
88.9	485	431	54	383.0
88.9	479	426	53	378.9
88.9	478	425	53	377.9

%E	M1	M2	DM	M*
88.9	477	424	53	376.9
88.9	476	423	53	375.9
88.9	470	418	52	371.8
88.9	469	417	52	370.8
88.9	468	416	52	369.8
88.9	467	415	52	368.8
88.9	461	410	51	364.6
88.9	460	409	51	363.7
88.9	459	408	51	362.7
88.9	458	407	51	361.7
88.9	452	402	50	357.5
88.9	451	401	50	356.5
88.9	450	400	50	355.6
88.9	449	399	50	354.6
88.9	443	394	49	350.4
88.9	442	393	49	349.4
88.9	441	392	49	348.4
88.9	440	391	49	347.5
88.9	434	386	48	343.3
88.9	433	385	48	342.3
88.9	432	384	48	341.3
88.9	431	383	48	340.3
88.9	425	378	47	336.2
88.9	424	377	47	335.2
88.9	423	376	47	334.2
88.9	422	375	47	333.2
88.9	416	370	46	329.1
88.9	415	369	46	328.1
88.9	414	368	46	327.1
88.9	413	367	46	326.1
88.9	407	362	45	322.0
88.9	406	361	45	321.0
88.9	405	360	45	320.0
88.9	404	359	45	319.0
88.9	398	354	44	314.9
88.9	397	353	44	313.9
88.9	396	352	44	312.9
88.9	395	351	44	311.9
88.9	389	346	43	307.8
88.9	388	345	43	306.8
88.9	387	344	43	305.8
88.9	386	343	43	304.8
88.9	380	338	42	300.6
88.9	379	337	42	299.7
88.9	378	336	42	298.7
88.9	377	335	42	297.7
88.9	371	330	41	293.5
88.9	370	329	41	292.5
88.9	369	328	41	291.6
88.9	368	327	41	290.6

%E	M1	M2	DM	M*
88.9	361	321	40	285.4
88.9	360	320	40	284.4
88.9	359	319	40	283.5
88.9	352	313	39	278.3
88.9	351	312	39	277.3
88.9	350	311	39	276.3
88.9	343	305	38	271.2
88.9	342	304	38	270.2
88.9	341	303	38	269.2
88.9	334	297	37	264.1
88.9	333	296	37	263.1
88.9	332	295	37	262.1
88.9	325	289	36	257.0
88.9	324	288	36	256.0
88.9	323	287	36	255.0
88.9	316	281	35	249.9
88.9	315	280	35	248.9
88.9	314	279	35	247.9
88.9	307	273	34	242.8
88.9	306	272	34	241.8
88.9	305	271	34	240.8
88.9	298	265	33	235.7
88.9	297	264	33	234.7
88.9	296	263	33	233.7
88.9	289	257	32	228.5
88.9	288	256	32	227.6
88.9	287	255	32	226.6
88.9	280	249	31	221.4
88.9	279	248	31	220.4
88.9	271	241	30	214.3
88.9	270	240	30	213.3
88.9	262	233	29	207.2
88.9	261	232	29	206.2
88.9	253	225	28	200.1
88.9	252	224	28	199.1
88.9	244	217	27	193.0
88.9	243	216	27	192.0
88.9	235	209	26	185.9
88.9	234	208	26	184.9
88.9	226	201	25	178.8
88.9	225	200	25	177.8
88.9	217	193	24	171.7
88.9	216	192	24	170.7
88.9	208	185	23	164.5
88.9	207	184	23	163.6
88.9	199	177	22	157.4
88.9	198	176	22	156.4
88.9	190	169	21	150.3
88.9	189	168	21	149.3
88.9	180	160	20	142.2

%E	M1	M2	DM	M*
88.9	171	152	19	135.1
88.9	162	144	18	128.0
88.9	153	136	17	120.9
88.9	144	128	16	113.8
88.9	135	120	15	106.7
88.9	126	112	14	99.6
88.9	117	104	13	92.4
88.9	108	96	12	85.3
88.9	99	88	11	78.2
88.9	90	80	10	71.1
88.9	81	72	9	64.0
88.9	72	64	8	56.9
88.9	63	56	7	49.8
88.9	54	48	6	42.7
88.9	45	40	5	35.6
88.9	36	32	4	28.4
88.9	27	24	3	21.3
88.9	18	16	2	14.2
88.9	9	8	1	7.1
88.8	500	444	56	394.3
88.8	499	443	56	393.3
88.8	498	442	56	392.3
88.8	493	438	55	389.1
88.8	492	437	55	388.1
88.8	491	436	55	387.2
88.8	490	435	55	386.2
88.8	489	434	55	385.2
88.8	484	430	54	382.0
88.8	483	429	54	381.0
88.8	482	428	54	380.0
88.8	481	427	54	379.1
88.8	480	426	54	378.1
88.8	475	422	53	374.9
88.8	474	421	53	373.9
88.8	473	420	53	372.9
88.8	472	419	53	372.0
88.8	466	414	52	367.8
88.8	465	413	52	366.8
88.8	464	412	52	365.8
88.8	463	411	52	364.8
88.8	457	406	51	360.7
88.8	456	405	51	359.7
88.8	455	404	51	358.7
88.8	454	403	51	357.7
88.8	448	398	50	353.6
88.8	447	397	50	352.6
88.8	446	396	50	351.6
88.8	445	395	50	350.6
88.8	439	390	49	346.5
88.8	438	389	49	345.5

%E	M1	M2	DM	M*
88.8	437	388	49	344.5
88.8	436	387	49	343.5
88.8	430	382	48	339.4
88.8	429	381	48	338.4
88.8	428	380	48	337.4
88.8	427	379	48	336.4
88.8	421	374	47	332.2
88.8	420	373	47	331.3
88.8	419	372	47	330.3
88.8	418	371	47	329.3
88.8	412	366	46	325.1
88.8	411	365	46	324.1
88.8	410	364	46	323.2
88.8	409	363	46	322.2
88.8	403	358	45	318.0
88.8	402	357	45	317.0
88.8	401	356	45	316.0
88.8	400	355	45	315.1
88.8	394	350	44	310.9
88.8	393	349	44	309.9
88.8	392	348	44	308.9
88.8	385	342	43	303.8
88.8	384	341	43	302.8
88.8	383	340	43	301.8
88.8	376	334	42	296.7
88.8	375	333	42	295.7
88.8	374	332	42	294.7
88.8	367	326	41	289.6
88.8	366	325	41	288.6
88.8	365	324	41	287.6
88.8	358	318	40	282.5
88.8	357	317	40	281.5
88.8	356	316	40	280.5
88.8	349	310	39	275.4
88.8	348	309	39	274.4
88.8	347	308	39	273.4
88.8	340	302	38	268.2
88.8	339	301	38	267.3
88.8	338	300	38	266.3
88.8	331	294	37	261.1
88.8	330	293	37	260.1
88.8	329	292	37	259.2
88.8	322	286	36	254.0
88.8	321	285	36	253.0
88.8	320	284	36	252.0
88.8	313	278	35	246.9
88.8	312	277	35	245.9
88.8	304	270	34	239.8
88.8	303	269	34	238.8
88.8	295	262	33	232.7

%E	M1	M2	DM	M*
88.8	294	261	33	231.7
88.8	286	254	32	225.6
88.8	285	253	32	224.6
88.8	278	247	31	219.5
88.8	277	246	31	218.5
88.8	276	245	31	217.5
88.8	269	239	30	212.3
88.8	268	238	30	211.4
88.8	267	237	30	210.4
88.8	260	231	29	205.2
88.8	259	230	29	204.2
88.8	258	229	29	203.3
88.8	251	223	28	198.1
88.8	250	222	28	197.1
88.8	249	221	28	196.1
88.8	242	215	27	191.0
88.8	241	214	27	190.0
88.8	240	213	27	189.0
88.8	233	207	26	183.9
88.8	232	206	26	182.9
88.8	224	199	25	176.8
88.8	223	198	25	175.8
88.8	215	191	24	169.7
88.8	214	190	24	168.7
88.8	206	183	23	162.6
88.8	205	182	23	161.6
88.8	197	175	22	155.5
88.8	196	174	22	154.5
88.8	188	167	21	148.3
88.8	187	166	21	147.4
88.8	179	159	20	141.2
88.8	178	158	20	140.2
88.8	170	151	19	134.1
88.8	169	150	19	133.1
88.8	161	143	18	127.0
88.8	160	142	18	126.0
88.8	152	135	17	119.9
88.8	143	127	16	112.8
88.8	134	119	15	105.7
88.8	125	111	14	98.6
88.8	116	103	13	91.5
88.8	107	95	12	84.3
88.8	98	87	11	77.2
88.8	89	79	10	70.1
88.8	80	71	9	63.0
88.7	497	441	56	391.3
88.7	496	440	56	390.3
88.7	495	439	56	389.3
88.7	494	438	56	388.3
88.7	488	433	55	384.2

%E	M1	M2	DM	M*
88.7	487	432	55	383.2
88.7	486	431	55	382.2
88.7	485	430	55	381.2
88.7	479	425	54	377.1
88.7	478	424	54	376.1
88.7	477	423	54	375.1
88.7	476	422	54	374.1
88.7	471	418	53	371.0
88.7	470	417	53	370.0
88.7	469	416	53	369.0
88.7	468	415	53	368.0
88.7	467	414	53	367.0
88.7	462	410	52	363.9
88.7	461	409	52	362.9
88.7	460	408	52	361.9
88.7	459	407	52	360.9
88.7	453	402	51	356.7
88.7	452	401	51	355.8
88.7	451	400	51	354.8
88.7	450	399	51	353.8
88.7	444	394	50	349.6
88.7	443	393	50	348.6
88.7	442	392	50	347.7
88.7	441	391	50	346.7
88.7	435	386	49	342.5
88.7	434	385	49	341.5
88.7	433	384	49	340.5
88.7	432	383	49	339.6
88.7	426	378	48	335.4
88.7	425	377	48	334.4
88.7	424	376	48	333.4
88.7	423	375	48	332.4
88.7	417	370	47	328.3
88.7	416	369	47	327.3
88.7	415	368	47	326.3
88.7	408	362	46	321.2
88.7	407	361	46	320.2
88.7	406	360	46	319.2
88.7	399	354	45	314.1
88.7	398	353	45	313.1
88.7	397	352	45	312.1
88.7	391	347	44	308.0
88.7	390	346	44	307.0
88.7	389	345	44	306.0
88.7	388	344	44	305.0
88.7	382	339	43	300.8
88.7	381	338	43	299.9
88.7	380	337	43	298.9
88.7	379	336	43	297.9
88.7	373	331	42	293.7

%E	M1	M2	DM	M*
88.7	372	330	42	292.7
88.7	371	329	42	291.8
88.7	364	323	41	286.6
88.7	363	322	41	285.6
88.7	362	321	41	284.6
88.7	355	315	40	279.5
88.7	354	314	40	278.5
88.7	353	313	40	277.5
88.7	346	307	39	272.4
88.7	345	306	39	271.4
88.7	344	305	39	270.4
88.7	337	299	38	265.3
88.7	336	298	38	264.3
88.7	335	297	38	263.3
88.7	328	291	37	258.2
88.7	327	290	37	257.2
88.7	326	289	37	256.2
88.7	319	283	36	251.1
88.7	318	282	36	250.1
88.7	311	276	35	244.9
88.7	310	275	35	244.0
88.7	309	274	35	243.0
88.7	302	268	34	237.8
88.7	301	267	34	236.8
88.7	300	266	34	235.9
88.7	293	260	33	230.7
88.7	292	259	33	229.7
88.7	291	258	33	228.7
88.7	284	252	32	223.6
88.7	283	251	32	222.6
88.7	282	250	32	221.6
88.7	275	244	31	216.5
88.7	274	243	31	215.5
88.7	266	236	30	209.4
88.7	265	235	30	208.4
88.7	257	228	29	202.3
88.7	256	227	29	201.3
88.7	248	220	28	195.2
88.7	247	219	28	194.2
88.7	239	212	27	188.1
88.7	238	211	27	187.1
88.7	231	205	26	181.9
88.7	230	204	26	180.9
88.7	222	197	25	174.8
88.7	221	196	25	173.8
88.7	213	189	24	167.7
88.7	212	188	24	166.7
88.7	204	181	23	160.6
88.7	203	180	23	159.6
88.7	195	173	22	153.5

%E	M1	M2	DM	M*
88.7	194	172	22	152.5
88.7	186	165	21	146.4
88.7	177	157	20	139.3
88.7	168	149	19	132.1
88.7	159	141	18	125.0
88.7	151	134	17	118.9
88.7	150	133	17	117.9
88.7	142	126	16	111.8
88.7	141	125	16	110.8
88.7	133	118	15	104.7
88.7	124	110	14	97.6
88.7	115	102	13	90.5
88.7	106	94	12	83.4
88.7	97	86	11	76.2
88.7	71	63	8	55.9
88.7	62	55	7	48.8
88.7	53	47	6	41.7
88.6	500	443	57	392.5
88.6	499	442	57	391.5
88.6	498	441	57	390.5
88.6	493	437	56	387.4
88.6	492	436	56	386.4
88.6	491	435	56	385.4
88.6	490	434	56	384.4
88.6	484	429	55	380.3
88.6	483	428	55	379.3
88.6	482	427	55	378.3
88.6	481	426	55	377.3
88.6	475	421	54	373.1
88.6	474	420	54	372.2
88.6	473	419	54	371.2
88.6	472	418	54	370.2
88.6	466	413	53	366.0
88.6	465	412	53	365.0
88.6	464	411	53	364.1
88.6	463	410	53	363.1
88.6	458	406	52	359.9
88.6	457	405	52	358.9
88.6	456	404	52	357.9
88.6	455	403	52	356.9
88.6	449	398	51	352.8
88.6	448	397	51	351.8
88.6	447	396	51	350.8
88.6	446	395	51	349.8
88.6	440	390	50	345.7
88.6	439	389	50	344.7
88.6	438	388	50	343.7
88.6	437	387	50	342.7
88.6	431	382	49	338.6
88.6	430	381	49	337.6

%E	M1	M2	DM	M*
88.6	429	380	49	336.6
88.6	428	379	49	335.6
88.6	422	374	48	331.5
88.6	421	373	48	330.5
88.6	420	372	48	329.5
88.6	414	367	47	325.3
88.6	413	366	47	324.3
88.6	412	365	47	323.4
88.6	411	364	47	322.4
88.6	405	359	46	318.2
88.6	404	358	46	317.2
88.6	403	357	46	316.3
88.6	402	356	46	315.3
88.6	396	351	45	311.1
88.6	395	350	45	310.1
88.6	394	349	45	309.1
88.6	393	348	45	308.2
88.6	387	343	44	304.0
88.6	386	342	44	303.0
88.6	385	341	44	302.0
88.6	378	335	43	296.9
88.6	377	334	43	295.9
88.6	376	333	43	294.9
88.6	370	328	42	290.8
88.6	369	327	42	289.8
88.6	368	326	42	288.8
88.6	367	325	42	287.8
88.6	361	320	41	283.7
88.6	360	319	41	282.7
88.6	359	318	41	281.7
88.6	352	312	40	276.5
88.6	351	311	40	275.6
88.6	350	310	40	274.6
88.6	343	304	39	269.4
88.6	342	303	39	268.4
88.6	341	302	39	267.5
88.6	334	296	38	262.3
88.6	333	295	38	261.3
88.6	332	294	38	260.3
88.6	325	288	37	255.2
88.6	324	287	37	254.2
88.6	317	281	36	249.1
88.6	316	280	36	248.1
88.6	315	279	36	247.1
88.6	308	273	35	242.0
88.6	307	272	35	241.0
88.6	306	271	35	240.0
88.6	299	265	34	234.9
88.6	298	264	34	233.9
88.6	297	263	34	232.9

%E	M1	M2	DM	M*
88.6	290	257	33	227.8
88.6	289	256	33	226.8
88.6	281	249	32	220.6
88.6	280	248	32	219.7
88.6	273	242	31	214.5
88.6	272	241	31	213.5
88.6	271	240	31	212.5
88.6	264	234	30	207.4
88.6	263	233	30	206.4
88.6	262	232	30	205.4
88.6	255	226	29	200.3
88.6	254	225	29	199.3
88.6	246	218	28	193.2
88.6	245	217	28	192.2
88.6	237	210	27	186.1
88.6	236	209	27	185.1
88.6	229	203	26	180.0
88.6	228	202	26	179.0
88.6	220	195	25	172.8
88.6	219	194	25	171.9
88.6	211	187	24	165.7
88.6	210	186	24	164.7
88.6	202	179	23	158.6
88.6	201	178	23	157.6
88.6	193	171	22	151.5
88.6	185	164	21	145.4
88.6	184	163	21	144.4
88.6	176	156	20	138.3
88.6	175	155	20	137.3
88.6	167	148	19	131.2
88.6	166	147	19	130.2
88.6	158	140	18	124.1
88.6	149	132	17	116.9
88.6	140	124	16	109.8
88.6	132	117	15	103.7
88.6	131	116	15	102.7
88.6	123	109	14	96.6
88.6	114	101	13	89.5
88.6	105	93	12	82.4
88.6	88	78	10	69.1
88.6	79	70	9	62.0
88.6	70	62	8	54.9
88.6	44	39	5	34.6
88.6	35	31	4	27.5
88.5	497	440	57	389.5
88.5	496	439	57	388.6
88.5	495	438	57	387.6
88.5	494	437	57	386.6
88.5	489	433	56	383.4
88.5	488	432	56	382.4

%E	M1	M2	DM	M*
88.5	487	431	56	381.4
88.5	486	430	56	380.5
88.5	485	429	56	379.5
88.5	480	425	55	376.3
88.5	479	424	55	375.3
88.5	478	423	55	374.3
88.5	477	422	55	373.3
88.5	471	417	54	369.2
88.5	470	416	54	368.2
88.5	469	415	54	367.2
88.5	468	414	54	366.2
88.5	462	409	53	362.1
88.5	461	408	53	361.1
88.5	460	407	53	360.1
88.5	459	406	53	359.1
88.5	454	402	52	356.0
88.5	453	401	52	355.0
88.5	452	400	52	354.0
88.5	451	399	52	353.0
88.5	445	394	51	348.8
88.5	444	393	51	347.9
88.5	443	392	51	346.9
88.5	442	391	51	345.9
88.5	436	386	50	341.7
88.5	435	385	50	340.7
88.5	434	384	50	339.8
88.5	433	383	50	338.8
88.5	427	378	49	334.6
88.5	426	377	49	333.6
88.5	425	376	49	332.6
88.5	419	371	48	328.5
88.5	418	370	48	327.5
88.5	417	369	48	326.5
88.5	416	368	48	325.5
88.5	410	363	47	321.4
88.5	409	362	47	320.4
88.5	408	361	47	319.4
88.5	407	360	47	318.4
88.5	401	355	46	314.3
88.5	400	354	46	313.3
88.5	399	353	46	312.3
88.5	392	347	45	307.2
88.5	391	346	45	306.2
88.5	390	345	45	305.2
88.5	384	340	44	301.0
88.5	383	339	44	300.1
88.5	382	338	44	299.1
88.5	381	337	44	298.1
88.5	375	332	43	293.9
88.5	374	331	43	292.9

%E	M1	M2	DM	M*
88.5	373	330	43	292.0
88.5	366	324	42	286.8
88.5	365	323	42	285.8
88.5	364	322	42	284.8
88.5	358	317	41	280.7
88.5	357	316	41	279.7
88.5	356	315	41	278.7
88.5	355	314	41	277.7
88.5	349	309	40	273.6
88.5	348	308	40	272.6
88.5	347	307	40	271.6
88.5	340	301	39	266.5
88.5	339	300	39	265.5
88.5	338	299	39	264.5
88.5	331	293	38	259.4
88.5	330	292	38	258.4
88.5	329	291	38	257.4
88.5	323	286	37	253.2
88.5	322	285	37	252.3
88.5	321	284	37	251.3
88.5	314	278	36	246.1
88.5	313	277	36	245.1
88.5	312	276	36	244.2
88.5	305	270	35	239.0
88.5	304	269	35	238.0
88.5	296	262	34	231.9
88.5	295	261	34	230.9
88.5	288	255	33	225.8
88.5	287	254	33	224.8
88.5	286	253	33	223.8
88.5	279	247	32	218.7
88.5	278	246	32	217.7
88.5	270	239	31	211.6
88.5	269	238	31	210.6
88.5	261	231	30	204.4
88.5	260	230	30	203.5
88.5	253	224	29	198.3
88.5	252	223	29	197.3
88.5	244	216	28	191.2
88.5	243	215	28	190.2
88.5	235	208	27	184.1
88.5	234	207	27	183.1
88.5	227	201	26	178.0
88.5	226	200	26	177.0
88.5	218	193	25	170.9
88.5	217	192	25	169.9
88.5	209	185	24	163.8
88.5	208	184	24	162.8
88.5	200	177	23	156.6
88.5	192	170	22	150.5

%E	M1	M2	DM	M*
88.5	191	169	22	149.5
88.5	183	162	21	143.4
88.5	182	161	21	142.4
88.5	174	154	20	136.3
88.5	165	146	19	129.2
88.5	157	139	18	123.1
88.5	156	138	18	122.1
88.5	148	131	17	116.0
88.5	139	123	16	108.8
88.5	130	115	15	101.7
88.5	122	108	14	95.6
88.5	113	100	13	88.5
88.5	104	92	12	81.4
88.5	96	85	11	75.3
88.5	87	77	10	68.1
88.5	78	69	9	61.0
88.5	61	54	7	47.8
88.5	52	46	6	40.7
88.5	26	23	3	20.3
88.4	500	442	58	390.7
88.4	499	441	58	389.7
88.4	498	440	58	388.8
88.4	493	436	57	385.6
88.4	492	435	57	384.6
88.4	491	434	57	383.6
88.4	490	433	57	382.6
88.4	484	428	56	378.5
88.4	483	427	56	377.5
88.4	482	426	56	376.5
88.4	481	425	56	375.5
88.4	476	421	55	372.4
88.4	475	420	55	371.4
88.4	474	419	55	370.4
88.4	473	418	55	369.4
88.4	467	413	54	365.2
88.4	466	412	54	364.3
88.4	465	411	54	363.3
88.4	464	410	54	362.3
88.4	458	405	53	358.1
88.4	457	404	53	357.1
88.4	456	403	53	356.2
88.4	455	402	53	355.2
88.4	450	398	52	352.0
88.4	449	397	52	351.0
88.4	448	396	52	350.0
88.4	447	395	52	349.0
88.4	441	390	51	344.9
88.4	440	389	51	343.9
88.4	439	388	51	342.9
88.4	438	387	51	341.9

%E	M1	M2	DM	M*
88.4	432	382	50	337.8
88.4	431	381	50	336.8
88.4	430	380	50	335.8
88.4	424	375	49	331.7
88.4	423	374	49	330.7
88.4	422	373	49	329.7
88.4	421	372	49	328.7
88.4	415	367	48	324.6
88.4	414	366	48	323.6
88.4	413	365	48	322.6
88.4	412	364	48	321.6
88.4	406	359	47	317.4
88.4	405	358	47	316.5
88.4	404	357	47	315.5
88.4	398	352	46	311.3
88.4	397	351	46	310.3
88.4	396	350	46	309.3
88.4	395	349	46	308.4
88.4	389	344	45	304.2
88.4	388	343	45	303.2
88.4	387	342	45	302.2
88.4	380	336	44	297.1
88.4	379	335	44	296.1
88.4	378	334	44	295.1
88.4	372	329	43	291.0
88.4	371	328	43	290.0
88.4	370	327	43	289.0
88.4	363	321	42	283.9
88.4	362	320	42	282.9
88.4	361	319	42	281.9
88.4	354	313	41	276.7
88.4	353	312	41	275.8
88.4	352	311	41	274.8
88.4	346	306	40	270.6
88.4	345	305	40	269.6
88.4	344	304	40	268.7
88.4	337	298	39	263.5
88.4	336	297	39	262.5
88.4	335	296	39	261.5
88.4	328	290	38	256.4
88.4	327	289	38	255.4
88.4	320	283	37	250.3
88.4	319	282	37	249.3
88.4	318	281	37	248.3
88.4	311	275	36	243.2
88.4	310	274	36	242.2
88.4	309	273	36	241.2
88.4	303	268	35	237.0
88.4	302	267	35	236.1
88.4	301	266	35	235.1

%E	M1	M2	DM	M*
88.4	294	260	34	229.9
88.4	293	259	34	228.9
88.4	292	258	34	228.0
88.4	285	252	33	222.8
88.4	284	251	33	221.8
88.4	277	245	32	216.7
88.4	276	244	32	215.7
88.4	275	243	32	214.7
88.4	268	237	31	209.6
88.4	267	236	31	208.6
88.4	259	229	30	202.5
88.4	258	228	30	201.5
88.4	251	222	29	196.4
88.4	250	221	29	195.4
88.4	249	220	29	194.4
88.4	242	214	28	189.2
88.4	241	213	28	188.3
88.4	233	206	27	182.1
88.4	232	205	27	181.1
88.4	225	199	26	176.0
88.4	224	198	26	175.0
88.4	216	191	25	168.9
88.4	215	190	25	167.9
88.4	207	183	24	161.8
88.4	206	182	24	160.8
88.4	199	176	23	155.7
88.4	198	175	23	154.7
88.4	190	168	22	148.5
88.4	189	167	22	147.6
88.4	181	160	21	141.4
88.4	173	153	20	135.3
88.4	172	152	20	134.3
88.4	164	145	19	128.2
88.4	155	137	18	121.1
88.4	147	130	17	115.0
88.4	146	129	17	114.0
88.4	138	122	16	107.9
88.4	129	114	15	100.7
88.4	121	107	14	94.6
88.4	112	99	13	87.5
88.4	103	91	12	80.4
88.4	95	84	11	74.3
88.4	86	76	10	67.2
88.4	69	61	8	53.9
88.4	43	38	5	33.6
88.3	497	439	58	387.8
88.3	496	438	58	386.8
88.3	495	437	58	385.8
88.3	494	436	58	384.8
88.3	489	432	57	381.6

%E	M1	M2	DM	M*
88.3	488	431	57	380.7
88.3	487	430	57	379.7
88.3	486	429	57	378.7
88.3	480	424	56	374.5
88.3	479	423	56	373.5
88.3	478	422	56	372.6
88.3	477	421	56	371.6
88.3	472	417	55	368.4
88.3	471	416	55	367.4
88.3	470	415	55	366.4
88.3	469	414	55	365.4
88.3	463	409	54	361.3
88.3	462	408	54	360.3
88.3	461	407	54	359.3
88.3	460	406	54	358.3
88.3	454	401	53	354.2
88.3	453	400	53	353.2
88.3	452	399	53	352.2
88.3	446	394	52	348.1
88.3	445	393	52	347.1
88.3	444	392	52	346.1
88.3	443	391	52	345.1
88.3	437	386	51	341.0
88.3	436	385	51	340.0
88.3	435	384	51	339.0
88.3	429	379	50	334.8
88.3	428	378	50	333.8
88.3	427	377	50	332.9
88.3	426	376	50	331.9
88.3	420	371	49	327.7
88.3	419	370	49	326.7
88.3	418	369	49	325.7
88.3	411	363	48	320.6
88.3	410	362	48	319.6
88.3	409	361	48	318.6
88.3	403	356	47	314.5
88.3	402	355	47	313.5
88.3	401	354	47	312.5
88.3	400	353	47	311.5
88.3	394	348	46	307.4
88.3	393	347	46	306.4
88.3	392	346	46	305.4
88.3	386	341	45	301.2
88.3	385	340	45	300.3
88.3	384	339	45	299.3
88.3	383	338	45	298.3
88.3	377	333	44	294.1
88.3	376	332	44	293.1
88.3	375	331	44	292.2
88.3	369	326	43	288.0

%E	M1	M2	DM	M*
88.3	368	325	43	287.0
88.3	367	324	43	286.0
88.3	366	323	43	285.1
88.3	360	318	42	280.9
88.3	359	317	42	279.9
88.3	358	316	42	278.9
88.3	351	310	41	273.8
88.3	350	309	41	272.8
88.3	349	308	41	271.8
88.3	343	303	40	267.7
88.3	342	302	40	266.7
88.3	341	301	40	265.7
88.3	334	295	39	260.6
88.3	333	294	39	259.6
88.3	332	293	39	258.6
88.3	326	288	38	254.4
88.3	325	287	38	253.4
88.3	324	286	38	252.5
88.3	317	280	37	247.3
88.3	316	279	37	246.3
88.3	315	278	37	245.3
88.3	308	272	36	240.2
88.3	307	271	36	239.2
88.3	300	265	35	234.1
88.3	299	264	35	233.1
88.3	298	263	35	232.1
88.3	291	257	34	227.0
88.3	290	256	34	226.0
88.3	283	250	33	220.8
88.3	282	249	33	219.9
88.3	281	248	33	218.9
88.3	274	242	32	213.7
88.3	273	241	32	212.8
88.3	266	235	31	207.6
88.3	265	234	31	206.6
88.3	264	233	31	205.6
88.3	257	227	30	200.5
88.3	256	226	30	199.5
88.3	248	219	29	193.4
88.3	247	218	29	192.4
88.3	240	212	28	187.3
88.3	239	211	28	186.3
88.3	231	204	27	180.2
88.3	230	203	27	179.2
88.3	223	197	26	174.0
88.3	222	196	26	173.0
88.3	214	189	25	166.9
88.3	213	188	25	165.9
88.3	205	181	24	159.8
88.3	197	174	23	153.7

%E	M1	M2	DM	M*
88.3	196	173	23	152.7
88.3	188	166	22	146.6
88.3	180	159	21	140.4
88.3	179	158	21	139.5
88.3	171	151	20	133.3
88.3	163	144	19	127.2
88.3	162	143	19	126.2
88.3	154	136	18	120.1
88.3	145	128	17	113.0
88.3	137	121	16	106.9
88.3	128	113	15	99.8
88.3	120	106	14	93.6
88.3	111	98	13	86.5
88.3	94	83	11	73.3
88.3	77	68	9	60.1
88.3	60	53	7	46.8
88.2	500	441	59	389.0
88.2	499	440	59	388.0
88.2	498	439	59	387.0
88.2	493	435	58	383.8
88.2	492	434	58	382.8
88.2	491	433	58	381.9
88.2	490	432	58	380.9
88.2	485	428	57	377.7
88.2	484	427	57	376.7
88.2	483	426	57	375.7
88.2	482	425	57	374.7
88.2	481	424	57	373.8
88.2	476	420	56	370.6
88.2	475	419	56	369.6
88.2	474	418	56	368.6
88.2	473	417	56	367.6
88.2	468	413	55	364.5
88.2	467	412	55	363.5
88.2	466	411	55	362.5
88.2	465	410	55	361.5
88.2	459	405	54	357.4
88.2	458	404	54	356.4
88.2	457	403	54	355.4
88.2	456	402	54	354.4
88.2	451	398	53	351.2
88.2	450	397	53	350.2
88.2	449	396	53	349.3
88.2	448	395	53	348.3
88.2	442	390	52	344.1
88.2	441	389	52	343.1
88.2	440	388	52	342.1
88.2	439	387	52	341.2
88.2	434	383	51	338.0
88.2	433	382	51	337.0

%E	M1	M2	DM	M*
88.2	432	381	51	336.0
88.2	431	380	51	335.0
88.2	425	375	50	330.9
88.2	424	374	50	329.9
88.2	423	373	50	328.9
88.2	422	372	50	327.9
88.2	417	368	49	324.8
88.2	416	367	49	323.8
88.2	415	366	49	322.8
88.2	414	365	49	321.8
88.2	408	360	48	317.6
88.2	407	359	48	316.7
88.2	406	358	48	315.7
88.2	399	352	47	310.5
88.2	398	351	47	309.6
88.2	397	350	47	308.6
88.2	391	345	46	304.4
88.2	390	344	46	303.4
88.2	389	343	46	302.4
88.2	382	337	45	297.3
88.2	381	336	45	296.3
88.2	380	335	45	295.3
88.2	374	330	44	291.2
88.2	373	329	44	290.2
88.2	372	328	44	289.2
88.2	365	322	43	284.1
88.2	364	321	43	283.1
88.2	363	320	43	282.1
88.2	357	315	42	277.9
88.2	356	314	42	277.0
88.2	355	313	42	276.0
88.2	348	307	41	270.8
88.2	347	306	41	269.8
88.2	346	305	41	268.9
88.2	340	300	40	264.7
88.2	339	299	40	263.7
88.2	338	298	40	262.7
88.2	331	292	39	257.6
88.2	330	291	39	256.6
88.2	323	285	38	251.5
88.2	322	284	38	250.5
88.2	321	283	38	249.5
88.2	314	277	37	244.4
88.2	313	276	37	243.4
88.2	306	270	36	238.2
88.2	305	269	36	237.2
88.2	304	268	36	236.3
88.2	297	262	35	231.1
88.2	296	261	35	230.1
88.2	289	255	34	225.0

%E	M1	M2	DM	M*
88.2	288	254	34	224.0
88.2	287	253	34	223.0
88.2	280	247	33	217.9
88.2	279	246	33	216.9
88.2	272	240	32	211.8
88.2	271	239	32	210.8
88.2	263	232	31	204.7
88.2	262	231	31	203.7
88.2	255	225	30	198.5
88.2	254	224	30	197.5
88.2	246	217	29	191.4
88.2	245	216	29	190.4
88.2	238	210	28	185.3
88.2	237	209	28	184.3
88.2	229	202	27	178.2
88.2	228	201	27	177.2
88.2	221	195	26	172.1
88.2	220	194	26	171.1
88.2	212	187	25	164.9
88.2	211	186	25	164.0
88.2	204	180	24	158.8
88.2	203	179	24	157.8
88.2	195	172	23	151.7
88.2	187	165	22	145.6
88.2	186	164	22	144.6
88.2	178	157	21	138.5
88.2	170	150	20	132.4
88.2	169	149	20	131.4
88.2	161	142	19	125.2
88.2	153	135	18	119.1
88.2	152	134	18	118.1
88.2	144	127	17	112.0
88.2	136	120	16	105.9
88.2	127	112	15	98.8
88.2	119	105	14	92.6
88.2	110	97	13	85.5
88.2	102	90	12	79.4
88.2	93	82	11	72.3
88.2	85	75	10	66.2
88.2	76	67	9	59.1
88.2	68	60	8	52.9
88.2	51	45	6	39.7
88.2	34	30	4	26.5
88.2	17	15	2	13.2
88.1	497	438	59	386.0
88.1	496	437	59	385.0
88.1	495	436	59	384.0
88.1	494	435	59	383.0
88.1	489	431	58	379.9
88.1	488	430	58	378.9

%E	M1	M2	DM	M*
88.1	487	429	58	377.9
88.1	486	428	58	376.9
88.1	480	423	57	372.8
88.1	479	422	57	371.8
88.1	478	421	57	370.8
88.1	477	420	57	369.8
88.1	472	416	56	366.6
88.1	471	415	56	365.7
88.1	470	414	56	364.7
88.1	469	413	56	363.7
88.1	464	409	55	360.5
88.1	463	408	55	359.5
88.1	462	407	55	358.5
88.1	461	406	55	357.6
88.1	455	401	54	353.4
88.1	454	400	54	352.4
88.1	453	399	54	351.4
88.1	452	398	54	350.5
88.1	447	394	53	347.3
88.1	446	393	53	346.3
88.1	445	392	53	345.3
88.1	444	391	53	344.3
88.1	438	386	52	340.2
88.1	437	385	52	339.2
88.1	436	384	52	338.2
88.1	430	379	51	334.0
88.1	429	378	51	333.1
88.1	428	377	51	332.1
88.1	427	376	51	331.1
88.1	421	371	50	326.9
88.1	420	370	50	326.0
88.1	419	369	50	325.0
88.1	413	364	49	320.8
88.1	412	363	49	319.8
88.1	411	362	49	318.8
88.1	405	357	48	314.7
88.1	404	356	48	313.7
88.1	403	355	48	312.7
88.1	402	354	48	311.7
88.1	396	349	47	307.6
88.1	395	348	47	306.6
88.1	394	347	47	305.6
88.1	388	342	46	301.5
88.1	387	341	46	300.5
88.1	386	340	46	299.5
88.1	385	339	46	298.5
88.1	379	334	45	294.3
88.1	378	333	45	293.4
88.1	377	332	45	292.4
88.1	371	327	44	288.2

%E	M1	M2	DM	M*
88.1	370	326	44	287.2
88.1	369	325	44	286.2
88.1	362	319	43	281.1
88.1	361	318	43	280.1
88.1	360	317	43	279.1
88.1	354	312	42	275.0
88.1	353	311	42	274.0
88.1	352	310	42	273.0
88.1	345	304	41	267.9
88.1	344	303	41	266.9
88.1	337	297	40	261.7
88.1	336	296	40	260.8
88.1	335	295	40	259.8
88.1	329	290	39	255.6
88.1	328	289	39	254.6
88.1	327	288	39	253.7
88.1	320	282	38	248.5
88.1	319	281	38	247.5
88.1	318	280	38	246.5
88.1	312	275	37	242.4
88.1	311	274	37	241.4
88.1	310	273	37	240.4
88.1	303	267	36	235.3
88.1	302	266	36	234.3
88.1	295	260	35	229.2
88.1	294	259	35	228.2
88.1	293	258	35	227.2
88.1	286	252	34	222.0
88.1	285	251	34	221.1
88.1	278	245	33	215.9
88.1	277	244	33	214.9
88.1	270	238	32	209.8
88.1	269	237	32	208.8
88.1	268	236	32	207.8
88.1	261	230	31	202.7
88.1	260	229	31	201.7
88.1	253	223	30	196.6
88.1	252	222	30	195.6
88.1	244	215	29	189.4
88.1	243	214	29	188.5
88.1	236	208	28	183.3
88.1	235	207	28	182.3
88.1	227	200	27	176.2
88.1	226	199	27	175.2
88.1	219	193	26	170.1
88.1	218	192	26	169.1
88.1	210	185	25	163.0
88.1	202	178	24	156.9
88.1	201	177	24	155.9
88.1	194	171	23	150.7

%E	M1	M2	DM	M*
88.1	193	170	23	149.7
88.1	185	163	22	143.6
88.1	177	156	21	137.5
88.1	176	155	21	136.5
88.1	168	148	20	130.4
88.1	160	141	19	124.3
88.1	159	140	19	123.3
88.1	151	133	18	117.1
88.1	143	126	17	111.0
88.1	135	119	16	104.9
88.1	134	118	16	103.9
88.1	126	111	15	97.8
88.1	118	104	14	91.7
88.1	109	96	13	84.6
88.1	101	89	12	78.4
88.1	84	74	10	65.2
88.1	67	59	8	52.0
88.1	59	52	7	45.8
88.1	42	37	5	32.6
88.0	500	440	60	387.2
88.0	499	439	60	386.2
88.0	498	438	60	385.2
88.0	493	434	59	382.1
88.0	492	433	59	381.1
88.0	491	432	59	380.1
88.0	490	431	59	379.1
88.0	485	427	58	375.9
88.0	484	426	58	375.0
88.0	483	425	58	374.0
88.0	482	424	58	373.0
88.0	476	419	57	368.8
88.0	475	418	57	367.8
88.0	474	417	57	366.9
88.0	468	412	56	362.7
88.0	467	411	56	361.7
88.0	466	410	56	360.7
88.0	465	409	56	359.7
88.0	460	405	55	356.6
88.0	459	404	55	355.6
88.0	458	403	55	354.6
88.0	457	402	55	353.6
88.0	451	397	54	349.5
88.0	450	396	54	348.5
88.0	449	395	54	347.5
88.0	443	390	53	343.3
88.0	442	389	53	342.4
88.0	441	388	53	341.4
88.0	440	387	53	340.4
88.0	435	383	52	337.2
88.0	434	382	52	336.2

%E	M1	M2	DM	M*
88.0	433	381	52	335.2
88.0	432	380	52	334.3
88.0	426	375	51	330.1
88.0	425	374	51	329.1
88.0	424	373	51	328.1
88.0	418	368	50	324.0
88.0	417	367	50	323.0
88.0	416	366	50	322.0
88.0	415	365	50	321.0
88.0	410	361	49	317.9
88.0	409	360	49	316.9
88.0	408	359	49	315.9
88.0	407	358	49	314.9
88.0	401	353	48	310.7
88.0	400	352	48	309.8
88.0	399	351	48	308.8
88.0	393	346	47	304.6
88.0	392	345	47	303.6
88.0	391	344	47	302.6
88.0	384	338	46	297.5
88.0	383	337	46	296.5
88.0	382	336	46	295.5
88.0	376	331	45	291.4
88.0	375	330	45	290.4
88.0	374	329	45	289.4
88.0	368	324	44	285.3
88.0	367	323	44	284.3
88.0	366	322	44	283.3
88.0	359	316	43	278.2
88.0	358	315	43	277.2
88.0	357	314	43	276.2
88.0	351	309	42	272.0
88.0	350	308	42	271.0
88.0	349	307	42	270.1
88.0	343	302	41	265.9
88.0	342	301	41	264.9
88.0	341	300	41	263.9
88.0	334	294	40	258.8
88.0	333	293	40	257.8
88.0	332	292	40	256.8
88.0	326	287	39	252.7
88.0	325	286	39	251.7
88.0	324	285	39	250.7
88.0	317	279	38	245.6
88.0	316	278	38	244.6
88.0	309	272	37	239.4
88.0	308	271	37	238.4
88.0	301	265	36	233.3
88.0	300	264	36	232.3
88.0	299	263	36	231.3

%E	M1	M2	DM	M*
88.0	292	257	35	226.2
88.0	291	256	35	225.2
88.0	284	250	34	220.1
88.0	283	249	34	219.1
88.0	276	243	33	213.9
88.0	275	242	33	213.0
88.0	274	241	33	212.0
88.0	267	235	32	206.8
88.0	266	234	32	205.8
88.0	259	228	31	200.7
88.0	258	227	31	199.7
88.0	251	221	30	194.6
88.0	250	220	30	193.6
88.0	249	219	30	192.6
88.0	242	213	29	187.5
88.0	241	212	29	186.5
88.0	234	206	28	181.4
88.0	233	205	28	180.4
88.0	225	198	27	174.2
88.0	217	191	26	168.1
88.0	216	190	26	167.1
88.0	209	184	25	162.0
88.0	208	183	25	161.0
88.0	200	176	24	154.9
88.0	192	169	23	148.8
88.0	191	168	23	147.8
88.0	184	162	22	142.6
88.0	183	161	22	141.6
88.0	175	154	21	135.5
88.0	167	147	20	129.4
88.0	166	146	20	128.4
88.0	158	139	19	122.3
88.0	150	132	18	116.2
88.0	142	125	17	110.0
88.0	133	117	16	102.9
88.0	125	110	15	96.8
88.0	117	103	14	90.7
88.0	108	95	13	83.6
88.0	100	88	12	77.4
88.0	92	81	11	71.3
88.0	83	73	10	64.2
88.0	75	66	9	58.1
88.0	50	44	6	38.7
88.0	25	22	3	19.4
87.9	497	437	60	384.2
87.9	496	436	60	383.3
87.9	495	435	60	382.3
87.9	494	434	60	381.3
87.9	489	430	59	378.1
87.9	488	429	59	377.1

%E	M1	M2	DM	M*
87.9	487	428	59	376.1
87.9	486	427	59	375.2
87.9	481	423	58	372.0
87.9	480	422	58	371.0
87.9	479	421	58	370.0
87.9	478	420	58	369.0
87.9	473	416	57	365.9
87.9	472	415	57	364.9
87.9	471	414	57	363.9
87.9	470	413	57	362.9
87.9	464	408	56	358.8
87.9	463	407	56	357.8
87.9	462	406	56	356.8
87.9	461	405	56	355.8
87.9	456	401	55	352.6
87.9	455	400	55	351.6
87.9	454	399	55	350.7
87.9	453	398	55	349.7
87.9	448	394	54	346.5
87.9	447	393	54	345.5
87.9	446	392	54	344.5
87.9	445	391	54	343.6
87.9	439	386	53	339.4
87.9	438	385	53	338.4
87.9	437	384	53	337.4
87.9	431	379	52	333.3
87.9	430	378	52	332.3
87.9	429	377	52	331.3
87.9	428	376	52	330.3
87.9	423	372	51	327.1
87.9	422	371	51	326.2
87.9	421	370	51	325.2
87.9	420	369	51	324.2
87.9	414	364	50	320.0
87.9	413	363	50	319.1
87.9	412	362	50	318.1
87.9	406	357	49	313.9
87.9	405	356	49	312.9
87.9	404	355	49	311.9
87.9	398	350	48	307.8
87.9	397	349	48	306.8
87.9	396	348	48	305.8
87.9	390	343	47	301.7
87.9	389	342	47	300.7
87.9	388	341	47	299.7
87.9	387	340	47	298.7
87.9	381	335	46	294.6
87.9	380	334	46	293.6
87.9	379	333	46	292.6
87.9	373	328	45	288.4

%E	M1	M2	DM	M*
87.9	372	327	45	287.4
87.9	371	326	45	286.5
87.9	365	321	44	282.3
87.9	364	320	44	281.3
87.9	363	319	44	280.3
87.9	356	313	43	275.2
87.9	355	312	43	274.2
87.9	354	311	43	273.2
87.9	348	306	42	269.1
87.9	347	305	42	268.1
87.9	346	304	42	267.1
87.9	340	299	41	262.9
87.9	339	298	41	262.0
87.9	338	297	41	261.0
87.9	331	291	40	255.8
87.9	330	290	40	254.8
87.9	323	284	39	249.7
87.9	322	283	39	248.7
87.9	321	282	39	247.7
87.9	315	277	38	243.6
87.9	314	276	38	242.6
87.9	313	275	38	241.6
87.9	307	270	37	237.5
87.9	306	269	37	236.5
87.9	305	268	37	235.5
87.9	298	262	36	230.3
87.9	297	261	36	229.4
87.9	290	255	35	224.2
87.9	289	254	35	223.2
87.9	282	248	34	218.1
87.9	281	247	34	217.1
87.9	280	246	34	216.1
87.9	273	240	33	211.0
87.9	272	239	33	210.0
87.9	265	233	32	204.9
87.9	264	232	32	203.9
87.9	257	226	31	198.7
87.9	256	225	31	197.8
87.9	248	218	30	191.6
87.9	247	217	30	190.6
87.9	240	211	29	185.5
87.9	239	210	29	184.5
87.9	232	204	28	179.4
87.9	231	203	28	178.4
87.9	224	197	27	173.3
87.9	223	196	27	172.3
87.9	215	189	26	166.1
87.9	214	188	26	165.2
87.9	207	182	25	160.0
87.9	206	181	25	159.0

%E	M1	M2	DM	M*
87.9	199	175	24	153.9
87.9	198	174	24	152.9
87.9	190	167	23	146.8
87.9	182	160	22	140.7
87.9	174	153	21	134.5
87.9	173	152	21	133.5
87.9	165	145	20	127.4
87.9	157	138	19	121.3
87.9	149	131	18	115.2
87.9	141	124	17	109.0
87.9	140	123	17	108.1
87.9	132	116	16	101.9
87.9	124	109	15	95.8
87.9	116	102	14	89.7
87.9	107	94	13	82.6
87.9	99	87	12	76.5
87.9	91	80	11	70.3
87.9	66	58	8	51.0
87.9	58	51	7	44.8
87.9	33	29	4	25.5
87.8	500	439	61	385.4
87.8	499	438	61	384.5
87.8	498	437	61	383.5
87.8	493	433	60	380.3
87.8	492	432	60	379.3
87.8	491	431	60	378.3
87.8	490	430	60	377.3
87.8	485	426	59	374.2
87.8	484	425	59	373.2
87.8	483	424	59	372.2
87.8	482	423	59	371.2
87.8	477	419	58	368.1
87.8	476	418	58	367.1
87.8	475	417	58	366.1
87.8	474	416	58	365.1
87.8	469	412	57	361.9
87.8	468	411	57	360.9
87.8	467	410	57	360.0
87.8	466	409	57	359.0
87.8	460	404	56	354.8
87.8	459	403	56	353.8
87.8	458	402	56	352.8
87.8	452	397	55	348.7
87.8	451	396	55	347.7
87.8	450	395	55	346.7
87.8	449	394	55	345.7
87.8	444	390	54	342.6
87.8	443	389	54	341.6
87.8	442	388	54	340.6
87.8	441	387	54	339.6

%E	M1	M2	DM	M*
87.8	436	383	53	336.4
87.8	435	382	53	335.5
87.8	434	381	53	334.5
87.8	433	380	53	333.5
87.8	427	375	52	329.3
87.8	426	374	52	328.3
87.8	425	373	52	327.4
87.8	419	368	51	323.2
87.8	418	367	51	322.2
87.8	417	366	51	321.2
87.8	411	361	50	317.1
87.8	410	360	50	316.1
87.8	409	359	50	315.1
87.8	403	354	49	311.0
87.8	402	353	49	310.0
87.8	401	352	49	309.0
87.8	400	351	49	308.0
87.8	395	347	48	304.8
87.8	394	346	48	303.8
87.8	393	345	48	302.9
87.8	392	344	48	301.9
87.8	386	339	47	297.7
87.8	385	338	47	296.7
87.8	384	337	47	295.8
87.8	378	332	46	291.6
87.8	377	331	46	290.6
87.8	376	330	46	289.6
87.8	370	325	45	285.5
87.8	369	324	45	284.5
87.8	368	323	45	283.5
87.8	362	318	44	279.3
87.8	361	317	44	278.4
87.8	360	316	44	277.4
87.8	353	310	43	272.2
87.8	352	309	43	271.3
87.8	345	303	42	266.1
87.8	344	302	42	265.1
87.8	343	301	42	264.1
87.8	337	296	41	260.0
87.8	336	295	41	259.0
87.8	335	294	41	258.0
87.8	329	289	40	253.9
87.8	328	288	40	252.9
87.8	327	287	40	251.9
87.8	320	281	39	246.8
87.8	319	280	39	245.8
87.8	312	274	38	240.6
87.8	311	273	38	239.6
87.8	304	267	37	234.5
87.8	303	266	37	233.5

%E	M1	M2	DM	M*
87.8	296	260	36	228.4
87.8	295	259	36	227.4
87.8	294	258	36	226.4
87.8	288	253	35	222.3
87.8	287	252	35	221.3
87.8	286	251	35	220.3
87.8	279	245	34	215.1
87.8	278	244	34	214.2
87.8	271	238	33	209.0
87.8	270	237	33	208.0
87.8	263	231	32	202.9
87.8	262	230	32	201.9
87.8	255	224	31	196.8
87.8	254	223	31	195.8
87.8	246	216	30	189.7
87.8	245	215	30	188.7
87.8	238	209	29	183.5
87.8	237	208	29	182.5
87.8	230	202	28	177.4
87.8	229	201	28	176.4
87.8	222	195	27	171.3
87.8	221	194	27	170.3
87.8	213	187	26	164.2
87.8	205	180	25	158.0
87.8	197	173	24	151.9
87.8	196	172	24	150.9
87.8	189	166	23	145.8
87.8	188	165	23	144.8
87.8	181	159	22	139.7
87.8	180	158	22	138.7
87.8	172	151	21	132.6
87.8	164	144	20	126.4
87.8	156	137	19	120.3
87.8	148	130	18	114.2
87.8	147	129	18	113.2
87.8	139	122	17	107.1
87.8	131	115	16	101.0
87.8	123	108	15	94.8
87.8	115	101	14	88.7
87.8	98	86	12	75.5
87.8	90	79	11	69.3
87.8	82	72	10	63.2
87.8	74	65	9	57.1
87.8	49	43	6	37.7
87.8	41	36	5	31.6
87.7	497	436	61	382.5
87.7	496	435	61	381.5
87.7	495	434	61	380.5
87.7	494	433	61	379.5
87.7	489	429	60	376.4

%E	M1	M2	DM	M*
87.7	488	428	60	375.4
87.7	487	427	60	374.4
87.7	486	426	60	373.4
87.7	481	422	59	370.2
87.7	480	421	59	369.3
87.7	479	420	59	368.3
87.7	478	419	59	367.3
87.7	473	415	58	364.1
87.7	472	414	58	363.1
87.7	471	413	58	362.1
87.7	470	412	58	361.2
87.7	465	408	57	358.0
87.7	464	407	57	357.0
87.7	463	406	57	356.0
87.7	462	405	57	355.0
87.7	457	401	56	351.9
87.7	456	400	56	350.9
87.7	455	399	56	349.9
87.7	454	398	56	348.9
87.7	448	393	55	344.8
87.7	447	392	55	343.8
87.7	446	391	55	342.8
87.7	440	386	54	338.6
87.7	439	385	54	337.6
87.7	438	384	54	336.7
87.7	432	379	53	332.5
87.7	431	378	53	331.5
87.7	430	377	53	330.5
87.7	424	372	52	326.4
87.7	423	371	52	325.4
87.7	422	370	52	324.4
87.7	416	365	51	320.3
87.7	415	364	51	319.3
87.7	414	363	51	318.3
87.7	413	362	51	317.3
87.7	408	358	50	314.1
87.7	407	357	50	313.1
87.7	406	356	50	312.2
87.7	405	355	50	311.2
87.7	399	350	49	307.0
87.7	398	349	49	306.0
87.7	397	348	49	305.0
87.7	391	343	48	300.9
87.7	390	342	48	299.9
87.7	389	341	48	298.9
87.7	383	336	47	294.8
87.7	382	335	47	293.8
87.7	381	334	47	292.8
87.7	375	329	46	288.6
87.7	374	328	46	287.7

%E	M1	M2	DM	M*
87.7	373	327	46	286.7
87.7	367	322	45	282.5
87.7	366	321	45	281.5
87.7	365	320	45	280.5
87.7	359	315	44	276.4
87.7	358	314	44	275.4
87.7	357	313	44	274.4
87.7	351	308	43	270.3
87.7	350	307	43	269.3
87.7	349	306	43	268.3
87.7	342	300	42	263.2
87.7	341	299	42	262.2
87.7	334	293	41	257.0
87.7	333	292	41	256.0
87.7	332	291	41	255.1
87.7	326	286	40	250.9
87.7	325	285	40	249.9
87.7	324	284	40	248.9
87.7	318	279	39	244.8
87.7	317	278	39	243.8
87.7	316	277	39	242.8
87.7	310	272	38	238.7
87.7	309	271	38	237.7
87.7	308	270	38	236.7
87.7	302	265	37	232.5
87.7	301	264	37	231.5
87.7	300	263	37	230.6
87.7	293	257	36	225.4
87.7	292	256	36	224.4
87.7	285	250	35	219.3
87.7	284	249	35	218.3
87.7	277	243	34	213.2
87.7	276	242	34	212.2
87.7	269	236	33	207.0
87.7	268	235	33	206.1
87.7	261	229	32	200.9
87.7	260	228	32	199.9
87.7	253	222	31	194.8
87.7	252	221	31	193.8
87.7	244	214	30	187.7
87.7	243	213	30	186.7
87.7	236	207	29	181.6
87.7	235	206	29	180.6
87.7	228	200	28	175.4
87.7	227	199	28	174.5
87.7	220	193	27	169.3
87.7	219	192	27	168.3
87.7	212	186	26	163.2
87.7	211	185	26	162.2
87.7	204	179	25	157.1

%E	M1	M2	DM	M*
87.7	203	178	25	156.1
87.7	195	171	24	150.0
87.7	187	164	23	143.8
87.7	179	157	22	137.7
87.7	171	150	21	131.6
87.7	163	143	20	125.5
87.7	162	142	20	124.5
87.7	155	136	19	119.3
87.7	154	135	19	118.3
87.7	146	128	18	112.2
87.7	138	121	17	106.1
87.7	130	114	16	100.0
87.7	122	107	15	93.8
87.7	114	100	14	87.7
87.7	106	93	13	81.6
87.7	81	71	10	62.2
87.7	73	64	9	56.1
87.7	65	57	8	50.0
87.7	57	50	7	43.9
87.6	500	438	62	383.7
87.6	499	437	62	382.7
87.6	498	436	62	381.7
87.6	493	432	61	378.5
87.6	492	431	61	377.6
87.6	491	430	61	376.6
87.6	490	429	61	375.6
87.6	485	425	60	372.4
87.6	484	424	60	371.4
87.6	483	423	60	370.5
87.6	482	422	60	369.5
87.6	477	418	59	366.3
87.6	476	417	59	365.3
87.6	475	416	59	364.3
87.6	474	415	59	363.3
87.6	469	411	58	360.2
87.6	468	410	58	359.2
87.6	467	409	58	358.2
87.6	466	408	58	357.2
87.6	461	404	57	354.0
87.6	460	403	57	353.1
87.6	459	402	57	352.1
87.6	458	401	57	351.1
87.6	453	397	56	347.9
87.6	452	396	56	346.9
87.6	451	395	56	346.0
87.6	450	394	56	345.0
87.6	445	390	55	341.8
87.6	444	389	55	340.8
87.6	443	388	55	339.8
87.6	442	387	55	338.8

%E	M1	M2	DM	M*
87.6	437	383	54	335.7
87.6	436	382	54	334.7
87.6	435	381	54	333.7
87.6	434	380	54	332.7
87.6	429	376	53	329.5
87.6	428	375	53	328.6
87.6	427	374	53	327.6
87.6	426	373	53	326.6
87.6	421	369	52	323.4
87.6	420	368	52	322.4
87.6	419	367	52	321.5
87.6	418	366	52	320.5
87.6	412	361	51	316.3
87.6	411	360	51	315.3
87.6	410	359	51	314.3
87.6	404	354	50	310.2
87.6	403	353	50	309.2
87.6	402	352	50	308.2
87.6	396	347	49	304.1
87.6	395	346	49	303.1
87.6	394	345	49	302.1
87.6	388	340	48	297.9
87.6	387	339	48	297.0
87.6	386	338	48	296.0
87.6	380	333	47	291.8
87.6	379	332	47	290.8
87.6	378	331	47	289.8
87.6	372	326	46	285.7
87.6	371	325	46	284.7
87.6	370	324	46	283.7
87.6	364	319	45	279.6
87.6	363	318	45	278.6
87.6	362	317	45	277.6
87.6	356	312	44	273.4
87.6	355	311	44	272.5
87.6	354	310	44	271.5
87.6	348	305	43	267.3
87.6	347	304	43	266.3
87.6	346	303	43	265.3
87.6	340	298	42	261.2
87.6	339	297	42	260.2
87.6	338	296	42	259.2
87.6	331	290	41	254.1
87.6	330	289	41	253.1
87.6	323	283	40	248.0
87.6	322	282	40	247.0
87.6	315	276	39	241.8
87.6	314	275	39	240.8
87.6	307	269	38	235.7
87.6	306	268	38	234.7

%E	M1	M2	DM	M*
87.6	299	262	37	229.6
87.6	298	261	37	228.6
87.6	291	255	36	223.5
87.6	290	254	36	222.5
87.6	283	248	35	217.3
87.6	282	247	35	216.3
87.6	275	241	34	211.2
87.6	274	240	34	210.2
87.6	267	234	33	205.1
87.6	266	233	33	204.1
87.6	259	227	32	199.0
87.6	258	226	32	198.0
87.6	251	220	31	192.8
87.6	250	219	31	191.8
87.6	249	218	31	190.9
87.6	242	212	30	185.7
87.6	241	211	30	184.7
87.6	234	205	29	179.6
87.6	233	204	29	178.6
87.6	226	198	28	173.5
87.6	225	197	28	172.5
87.6	218	191	27	167.3
87.6	217	190	27	166.4
87.6	210	184	26	161.2
87.6	209	183	26	160.2
87.6	202	177	25	155.1
87.6	201	176	25	154.1
87.6	194	170	24	149.0
87.6	193	169	24	148.0
87.6	186	163	23	142.8
87.6	185	162	23	141.9
87.6	178	156	22	136.7
87.6	177	155	22	135.7
87.6	170	149	21	130.6
87.6	169	148	21	129.6
87.6	161	141	20	123.5
87.6	153	134	19	117.4
87.6	145	127	18	111.2
87.6	137	120	17	105.1
87.6	129	113	16	99.0
87.6	121	106	15	92.9
87.6	113	99	14	86.7
87.6	105	92	13	80.6
87.6	97	85	12	74.5
87.6	89	78	11	68.4
87.5	497	435	62	380.7
87.5	496	434	62	379.8
87.5	495	433	62	378.8
87.5	489	428	61	374.6
87.5	488	427	61	373.6

%E	M1	M2	DM	M*
87.5	487	426	61	372.6
87.5	481	421	60	368.5
87.5	480	420	60	367.5
87.5	479	419	60	366.5
87.5	473	414	59	362.4
87.5	472	413	59	361.4
87.5	471	412	59	360.4
87.5	465	407	58	356.2
87.5	464	406	58	355.3
87.5	463	405	58	354.3
87.5	457	400	57	350.1
87.5	456	399	57	349.1
87.5	455	398	57	348.1
87.5	449	393	56	344.0
87.5	448	392	56	343.0
87.5	447	391	56	342.0
87.5	441	386	55	337.9
87.5	440	385	55	336.9
87.5	439	384	55	335.9
87.5	433	379	54	331.7
87.5	432	378	54	330.8
87.5	431	377	54	329.8
87.5	425	372	53	325.6
87.5	424	371	53	324.6
87.5	423	370	53	323.6
87.5	417	365	52	319.5
87.5	416	364	52	318.5
87.5	415	363	52	317.5
87.5	409	358	51	313.4
87.5	408	357	51	312.4
87.5	407	356	51	311.4
87.5	401	351	50	307.2
87.5	400	350	50	306.3
87.5	399	349	50	305.3
87.5	393	344	49	301.1
87.5	392	343	49	300.1
87.5	391	342	49	299.1
87.5	385	337	48	295.0
87.5	384	336	48	294.0
87.5	383	335	48	293.0
87.5	377	330	47	288.9
87.5	376	329	47	287.9
87.5	375	328	47	286.9
87.5	369	323	46	282.7
87.5	368	322	46	281.8
87.5	367	321	46	280.8
87.5	361	316	45	276.6
87.5	360	315	45	275.6
87.5	359	314	45	274.6
87.5	353	309	44	270.5

%E	M1	M2	DM	M*
87.5	352	308	44	269.5
87.5	351	307	44	268.5
87.5	345	302	43	264.4
87.5	344	301	43	263.4
87.5	343	300	43	262.4
87.5	337	295	42	258.2
87.5	336	294	42	257.3
87.5	335	293	42	256.3
87.5	329	288	41	252.1
87.5	328	287	41	251.1
87.5	327	286	41	250.1
87.5	321	281	40	246.0
87.5	320	280	40	245.0
87.5	319	279	40	244.0
87.5	313	274	39	239.9
87.5	312	273	39	238.9
87.5	311	272	39	237.9
87.5	305	267	38	233.7
87.5	304	266	38	232.8
87.5	303	265	38	231.8
87.5	297	260	37	227.6
87.5	296	259	37	226.6
87.5	295	258	37	225.6
87.5	289	253	36	221.5
87.5	288	252	36	220.5
87.5	287	251	36	219.5
87.5	281	246	35	215.4
87.5	280	245	35	214.4
87.5	279	244	35	213.4
87.5	273	239	34	209.2
87.5	272	238	34	208.3
87.5	271	237	34	207.3
87.5	265	232	33	203.1
87.5	264	231	33	202.1
87.5	263	230	33	201.1
87.5	257	225	32	197.0
87.5	256	224	32	196.0
87.5	255	223	32	195.0
87.5	248	217	31	189.9
87.5	240	210	30	183.8
87.5	232	203	29	177.6
87.5	224	196	28	171.5
87.5	216	189	27	165.4
87.5	208	182	26	159.3
87.5	200	175	25	153.1
87.5	192	168	24	147.0
87.5	184	161	23	140.9
87.5	176	154	22	134.8
87.5	168	147	21	128.6
87.5	160	140	20	122.5

%E	M1	M2	DM	M*
87.5	152	133	19	116.4
87.5	144	126	18	110.3
87.5	136	119	17	104.1
87.5	128	112	16	98.0
87.5	120	105	15	91.9
87.5	112	98	14	85.8
87.5	104	91	13	79.6
87.5	96	84	12	73.5
87.5	88	77	11	67.4
87.5	80	70	10	61.3
87.5	72	63	9	55.1
87.5	64	56	8	49.0
87.5	56	49	7	42.9
87.5	48	42	6	36.8
87.5	40	35	5	30.6
87.5	32	28	4	24.5
87.5	24	21	3	18.4
87.5	16	14	2	12.3
87.5	8	7	1	6.1
87.4	500	437	63	381.9
87.4	499	436	63	381.0
87.4	494	432	62	377.8
87.4	493	431	62	376.8
87.4	492	430	62	375.8
87.4	491	429	62	374.8
87.4	486	425	61	371.7
87.4	485	424	61	370.7
87.4	484	423	61	369.7
87.4	483	422	61	368.7
87.4	478	418	60	365.5
87.4	477	417	60	364.5
87.4	476	416	60	363.6
87.4	475	415	60	362.6
87.4	470	411	59	359.4
87.4	469	410	59	358.4
87.4	468	409	59	357.4
87.4	467	408	59	356.5
87.4	462	404	58	353.3
87.4	461	403	58	352.3
87.4	460	402	58	351.3
87.4	459	401	58	350.3
87.4	454	397	57	347.2
87.4	453	396	57	346.2
87.4	452	395	57	345.2
87.4	451	394	57	344.2
87.4	446	390	56	341.0
87.4	445	389	56	340.0
87.4	444	388	56	339.1
87.4	443	387	56	338.1
87.4	438	383	55	334.9

%E	M1	M2	DM	M*
87.4	437	382	55	333.9
87.4	436	381	55	332.9
87.4	435	380	55	332.0
87.4	430	376	54	328.8
87.4	429	375	54	327.8
87.4	428	374	54	326.8
87.4	427	373	54	325.8
87.4	422	369	53	322.7
87.4	421	368	53	321.7
87.4	420	367	53	320.7
87.4	419	366	53	319.7
87.4	414	362	52	316.5
87.4	413	361	52	315.5
87.4	412	360	52	314.6
87.4	406	355	51	310.4
87.4	405	354	51	309.4
87.4	404	353	51	308.4
87.4	398	348	50	304.3
87.4	397	347	50	303.3
87.4	396	346	50	302.3
87.4	390	341	49	298.2
87.4	389	340	49	297.2
87.4	388	339	49	296.2
87.4	382	334	48	292.0
87.4	381	333	48	291.0
87.4	380	332	48	290.1
87.4	374	327	47	285.9
87.4	373	326	47	284.9
87.4	372	325	47	283.9
87.4	366	320	46	279.8
87.4	365	319	46	278.8
87.4	364	318	46	277.8
87.4	358	313	45	273.7
87.4	357	312	45	272.7
87.4	356	311	45	271.7
87.4	350	306	44	267.5
87.4	349	305	44	266.5
87.4	348	304	44	265.6
87.4	342	299	43	261.4
87.4	341	298	43	260.4
87.4	340	297	43	259.4
87.4	334	292	42	255.3
87.4	333	291	42	254.3
87.4	326	285	41	249.2
87.4	325	284	41	248.2
87.4	318	278	40	243.0
87.4	317	277	40	242.0
87.4	310	271	39	236.9
87.4	309	270	39	235.9
87.4	302	264	38	230.8
87.4	301	263	38	229.8
87.4	294	257	37	224.7
87.4	293	256	37	223.7
87.4	286	250	36	218.5
87.4	285	249	36	217.5
87.4	278	243	35	212.4
87.4	277	242	35	211.4
87.4	270	236	34	206.3
87.4	269	235	34	205.3
87.4	262	229	33	200.2
87.4	261	228	33	199.2
87.4	254	222	32	194.0
87.4	253	221	32	193.0
87.4	247	216	31	188.9
87.4	246	215	31	187.9
87.4	239	209	30	182.8
87.4	238	208	30	181.8
87.4	231	202	29	176.6
87.4	230	201	29	175.7
87.4	223	195	28	170.5
87.4	222	194	28	169.5
87.4	215	188	27	164.4
87.4	214	187	27	163.4
87.4	207	181	26	158.3
87.4	206	180	26	157.3
87.4	199	174	25	152.1
87.4	198	173	25	151.2
87.4	191	167	24	146.0
87.4	190	166	24	145.0
87.4	183	160	23	139.9
87.4	182	159	23	138.9
87.4	175	153	22	133.8
87.4	174	152	22	132.8
87.4	167	146	21	127.6
87.4	159	139	20	121.5
87.4	151	132	19	115.4
87.4	143	125	18	109.3
87.4	135	118	17	103.1
87.4	127	111	16	97.0
87.4	119	104	15	90.9
87.4	111	97	14	84.8
87.4	103	90	13	78.6
87.4	95	83	12	72.5
87.4	87	76	11	66.4
87.3	498	435	63	380.0
87.3	497	434	63	379.0
87.3	496	433	63	378.0
87.3	495	432	63	377.0
87.3	490	428	62	373.8
87.3	489	427	62	372.9
87.3	488	426	62	371.9
87.3	487	425	62	370.9
87.3	482	421	61	367.7
87.3	481	420	61	366.7
87.3	480	419	61	365.8
87.3	479	418	61	364.8
87.3	474	414	60	361.6
87.3	473	413	60	360.6
87.3	472	412	60	359.6
87.3	471	411	60	358.6
87.3	466	407	59	355.5
87.3	465	406	59	354.5
87.3	464	405	59	353.5
87.3	463	404	59	352.5
87.3	458	400	58	349.3
87.3	457	399	58	348.4
87.3	456	398	58	347.4
87.3	455	397	58	346.4
87.3	450	393	57	343.2
87.3	449	392	57	342.2
87.3	448	391	57	341.3
87.3	442	386	56	337.1
87.3	441	385	56	336.1
87.3	440	384	56	335.1
87.3	434	379	55	331.0
87.3	433	378	55	330.0
87.3	432	377	55	329.0
87.3	426	372	54	324.8
87.3	425	371	54	323.9
87.3	424	370	54	322.9
87.3	418	365	53	318.7
87.3	417	364	53	317.7
87.3	416	363	53	316.8
87.3	411	359	52	313.6
87.3	410	358	52	312.6
87.3	409	357	52	311.6
87.3	408	356	52	310.6
87.3	403	352	51	307.5
87.3	402	351	51	306.5
87.3	401	350	51	305.5
87.3	400	349	51	304.5
87.3	395	345	50	301.3
87.3	394	344	50	300.3
87.3	393	343	50	299.4
87.3	387	338	49	295.2
87.3	386	337	49	294.2
87.3	385	336	49	293.2
87.3	379	331	48	289.1
87.3	378	330	48	288.1
87.3	377	329	48	287.1
87.3	371	324	47	283.0
87.3	370	323	47	282.0
87.3	369	322	47	281.0
87.3	363	317	46	276.8
87.3	362	316	46	275.8
87.3	361	315	46	274.9
87.3	355	310	45	270.7
87.3	354	309	45	269.7
87.3	353	308	45	268.7
87.3	347	303	44	264.6
87.3	346	302	44	263.6
87.3	339	296	43	258.5
87.3	338	295	43	257.5
87.3	332	290	42	253.3
87.3	331	289	42	252.3
87.3	330	288	42	251.3
87.3	324	283	41	247.2
87.3	323	282	41	246.2
87.3	322	281	41	245.2
87.3	316	276	40	241.1
87.3	315	275	40	240.1
87.3	314	274	40	239.1
87.3	308	269	39	234.9
87.3	307	268	39	234.0
87.3	306	267	39	233.0
87.3	300	262	38	228.8
87.3	299	261	38	227.8
87.3	292	255	37	222.7
87.3	291	254	37	221.7
87.3	284	248	36	216.6
87.3	283	247	36	215.6
87.3	276	241	35	210.4
87.3	275	240	35	209.5
87.3	268	234	34	204.3
87.3	267	233	34	203.3
87.3	260	227	33	198.2
87.3	259	226	33	197.2
87.3	252	220	32	192.1
87.3	251	219	32	191.1
87.3	245	214	31	186.9
87.3	244	213	31	185.9
87.3	237	207	30	180.8
87.3	236	206	30	179.8
87.3	229	200	29	174.7
87.3	228	199	29	173.7
87.3	221	193	28	168.5
87.3	220	192	28	167.6
87.3	213	186	27	162.4
87.3	212	185	27	161.4
87.3	205	179	26	156.3
87.3	204	178	26	155.3
87.3	197	172	25	150.2
87.3	189	165	24	144.0
87.3	181	158	23	137.9
87.3	173	151	22	131.8
87.3	166	145	21	126.7
87.3	165	144	21	125.7
87.3	158	138	20	120.5
87.3	157	137	20	119.5
87.3	150	131	19	114.4
87.3	142	124	18	108.3
87.3	134	117	17	102.2
87.3	126	110	16	96.0
87.3	118	103	15	89.9
87.3	110	96	14	83.8
87.3	102	89	13	77.7
87.3	79	69	10	60.3
87.3	71	62	9	54.1
87.3	63	55	8	48.0
87.3	55	48	7	41.9
87.2	500	436	64	380.2
87.2	499	435	64	379.2
87.2	494	431	63	376.0
87.2	493	430	63	375.1
87.2	492	429	63	374.1
87.2	491	428	63	373.1
87.2	486	424	62	369.9
87.2	485	423	62	368.9
87.2	484	422	62	367.9
87.2	483	421	62	367.0
87.2	478	417	61	363.8
87.2	477	416	61	362.8
87.2	476	415	61	361.8
87.2	475	414	61	360.8
87.2	470	410	60	357.7
87.2	469	409	60	356.7
87.2	468	408	60	355.7
87.2	467	407	60	354.7
87.2	462	403	59	351.5
87.2	461	402	59	350.6
87.2	460	401	59	349.6
87.2	454	396	58	345.4
87.2	453	395	58	344.4
87.2	452	394	58	343.4
87.2	447	390	57	340.3
87.2	446	389	57	339.3
87.2	445	388	57	338.3
87.2	444	387	57	337.3
87.2	439	383	56	334.1
87.2	438	382	56	333.2

%E	M1	M2	DM	M*
87.2	437	381	56	332.2
87.2	436	380	56	331.2
87.2	431	376	55	328.0
87.2	430	375	55	327.0
87.2	429	374	55	326.1
87.2	428	373	55	325.1
87.2	423	369	54	321.9
87.2	422	368	54	320.9
87.2	421	367	54	319.9
87.2	415	362	53	315.8
87.2	414	361	53	314.8
87.2	413	360	53	313.8
87.2	407	355	52	309.6
87.2	406	354	52	308.7
87.2	405	353	52	307.7
87.2	399	348	51	303.5
87.2	398	347	51	302.5
87.2	397	346	51	301.6
87.2	392	342	50	298.4
87.2	391	341	50	297.4
87.2	390	340	50	296.4
87.2	384	335	49	292.3
87.2	383	334	49	291.3
87.2	382	333	49	290.3
87.2	376	328	48	286.1
87.2	375	327	48	285.1
87.2	374	326	48	284.2
87.2	368	321	47	280.0
87.2	367	320	47	279.0
87.2	366	319	47	278.0
87.2	360	314	46	273.9
87.2	359	313	46	272.9
87.2	358	312	46	271.9
87.2	352	307	45	267.8
87.2	351	306	45	266.8
87.2	345	301	44	262.6
87.2	344	300	44	261.6
87.2	343	299	44	260.6
87.2	337	294	43	256.5
87.2	336	293	43	255.5
87.2	335	292	43	254.5
87.2	329	287	42	250.4
87.2	328	286	42	249.4
87.2	327	285	42	248.4
87.2	321	280	41	244.2
87.2	320	279	41	243.3
87.2	313	273	40	238.1
87.2	312	272	40	237.1
87.2	305	266	39	232.0
87.2	304	265	39	231.0

%E	M1	M2	DM	M*
87.2	298	260	38	226.8
87.2	297	259	38	225.9
87.2	296	258	38	224.9
87.2	290	253	37	220.7
87.2	289	252	37	219.7
87.2	288	251	37	218.8
87.2	282	246	36	214.6
87.2	281	245	36	213.6
87.2	274	239	35	208.5
87.2	273	238	35	207.5
87.2	266	232	34	202.3
87.2	265	231	34	201.4
87.2	258	225	33	196.2
87.2	257	224	33	195.2
87.2	250	218	32	190.1
87.2	243	212	31	185.0
87.2	242	211	31	184.0
87.2	235	205	30	178.8
87.2	234	204	30	177.8
87.2	227	198	29	172.7
87.2	226	197	29	171.7
87.2	219	191	28	166.6
87.2	218	190	28	165.6
87.2	211	184	27	160.5
87.2	203	177	26	154.3
87.2	196	171	25	149.2
87.2	195	170	25	148.2
87.2	188	164	24	143.1
87.2	187	163	24	142.1
87.2	180	157	23	136.9
87.2	179	156	23	136.0
87.2	172	150	22	130.8
87.2	164	143	21	124.7
87.2	156	136	20	118.6
87.2	149	130	19	113.4
87.2	148	129	19	112.4
87.2	141	123	18	107.3
87.2	133	116	17	101.2
87.2	125	109	16	95.0
87.2	117	102	15	88.9
87.2	109	95	14	82.8
87.2	94	82	12	71.5
87.2	86	75	11	65.4
87.2	78	68	10	59.3
87.2	47	41	6	35.8
87.2	39	34	5	29.6
87.1	498	434	64	378.2
87.1	497	433	64	377.2
87.1	496	432	64	376.3
87.1	495	431	64	375.3

%E	M1	M2	DM	M*
87.1	490	427	63	372.1
87.1	489	426	63	371.1
87.1	488	425	63	370.1
87.1	487	424	63	369.1
87.1	482	420	62	366.0
87.1	481	419	62	365.0
87.1	480	418	62	364.0
87.1	479	417	62	363.0
87.1	474	413	61	359.9
87.1	473	412	61	358.9
87.1	472	411	61	357.9
87.1	466	406	60	353.7
87.1	465	405	60	352.7
87.1	464	404	60	351.8
87.1	459	400	59	348.6
87.1	458	399	59	347.6
87.1	457	398	59	346.6
87.1	456	397	59	345.6
87.1	451	393	58	342.5
87.1	450	392	58	341.5
87.1	449	391	58	340.5
87.1	448	390	58	339.5
87.1	443	386	57	336.3
87.1	442	385	57	335.4
87.1	441	384	57	334.4
87.1	435	379	56	330.2
87.1	434	378	56	329.2
87.1	433	377	56	328.2
87.1	427	372	55	324.1
87.1	426	371	55	323.1
87.1	425	370	55	322.1
87.1	420	366	54	318.9
87.1	419	365	54	318.0
87.1	418	364	54	317.0
87.1	417	363	54	316.0
87.1	412	359	53	312.8
87.1	411	358	53	311.8
87.1	410	357	53	310.9
87.1	404	352	52	306.7
87.1	403	351	52	305.7
87.1	402	350	52	304.7
87.1	396	345	51	300.6
87.1	395	344	51	299.6
87.1	394	343	51	298.6
87.1	389	339	50	295.4
87.1	388	338	50	294.4
87.1	387	337	50	293.5
87.1	381	332	49	289.3
87.1	380	331	49	288.3
87.1	379	330	49	287.3

%E	M1	M2	DM	M*
87.1	373	325	48	283.2
87.1	372	324	48	282.2
87.1	371	323	48	281.2
87.1	365	318	47	277.1
87.1	364	317	47	276.1
87.1	363	316	47	275.1
87.1	357	311	46	270.9
87.1	356	310	46	269.9
87.1	350	305	45	265.8
87.1	349	304	45	264.8
87.1	348	303	45	263.8
87.1	342	298	44	259.7
87.1	341	297	44	258.7
87.1	340	296	44	257.7
87.1	334	291	43	253.5
87.1	333	290	43	252.6
87.1	326	284	42	247.4
87.1	325	283	42	246.4
87.1	319	278	41	242.3
87.1	318	277	41	241.3
87.1	317	276	41	240.3
87.1	311	271	40	236.1
87.1	310	270	40	235.2
87.1	309	269	40	234.2
87.1	303	264	39	230.0
87.1	302	263	39	229.0
87.1	295	257	38	223.9
87.1	294	256	38	222.9
87.1	287	250	37	217.8
87.1	286	249	37	216.8
87.1	280	244	36	212.6
87.1	279	243	36	211.6
87.1	278	242	36	210.7
87.1	272	237	35	206.5
87.1	271	236	35	205.5
87.1	264	230	34	200.4
87.1	263	229	34	199.4
87.1	256	223	33	194.3
87.1	255	222	33	193.3
87.1	249	217	32	189.1
87.1	248	216	32	188.1
87.1	241	210	31	183.0
87.1	240	209	31	182.0
87.1	233	203	30	176.9
87.1	232	202	30	175.9
87.1	225	196	29	170.7
87.1	224	195	29	169.8
87.1	217	189	28	164.6
87.1	210	183	27	159.5
87.1	209	182	27	158.5

%E	M1	M2	DM	M*
87.1	202	176	26	153.3
87.1	201	175	26	152.4
87.1	194	169	25	147.2
87.1	186	162	24	141.1
87.1	178	155	23	135.0
87.1	171	149	22	129.8
87.1	170	148	22	128.8
87.1	163	142	21	123.7
87.1	155	135	20	117.6
87.1	147	128	19	111.5
87.1	140	122	18	106.3
87.1	139	121	18	105.3
87.1	132	115	17	100.2
87.1	124	108	16	94.1
87.1	116	101	15	87.9
87.1	101	88	13	76.7
87.1	93	81	12	70.5
87.1	85	74	11	64.4
87.1	70	61	9	53.2
87.1	62	54	8	47.0
87.1	31	27	4	23.5
87.0	500	435	65	378.4
87.0	499	434	65	377.5
87.0	494	430	64	374.3
87.0	493	429	64	373.3
87.0	492	428	64	372.3
87.0	491	427	64	371.3
87.0	486	423	63	368.2
87.0	485	422	63	367.2
87.0	484	421	63	366.2
87.0	483	420	63	365.2
87.0	478	416	62	362.0
87.0	477	415	62	361.1
87.0	476	414	62	360.1
87.0	471	410	61	356.9
87.0	470	409	61	355.9
87.0	469	408	61	354.9
87.0	468	407	61	354.0
87.0	463	403	60	350.8
87.0	462	402	60	349.8
87.0	461	401	60	348.8
87.0	460	400	60	347.8
87.0	455	396	59	344.7
87.0	454	395	59	343.7
87.0	453	394	59	342.7
87.0	447	389	58	338.5
87.0	446	388	58	337.5
87.0	445	387	58	336.6
87.0	440	383	57	333.4
87.0	439	382	57	332.4

%E	M1	M2	DM	M*
87.0	438	381	57	331.4
87.0	437	380	57	330.4
87.0	432	376	56	327.3
87.0	431	375	56	326.3
87.0	430	374	56	325.3
87.0	424	369	55	321.1
87.0	423	368	55	320.2
87.0	422	367	55	319.2
87.0	416	362	54	315.0
87.0	415	361	54	314.0
87.0	414	360	54	313.0
87.0	409	356	53	309.9
87.0	408	355	53	308.9
87.0	407	354	53	307.9
87.0	401	349	52	303.7
87.0	400	348	52	302.8
87.0	399	347	52	301.8
87.0	393	342	51	297.6
87.0	392	341	51	296.6
87.0	391	340	51	295.7
87.0	386	336	50	292.5
87.0	385	335	50	291.5
87.0	384	334	50	290.5
87.0	378	329	49	286.4
87.0	377	328	49	285.4
87.0	376	327	49	284.4
87.0	370	322	48	280.2
87.0	369	321	48	279.2
87.0	368	320	48	278.3
87.0	362	315	47	274.1
87.0	361	314	47	273.1
87.0	355	309	46	269.0
87.0	354	308	46	268.0
87.0	353	307	46	267.0
87.0	347	302	45	262.8
87.0	346	301	45	261.9
87.0	345	300	45	260.9
87.0	339	295	44	256.7
87.0	338	294	44	255.7
87.0	332	289	43	251.6
87.0	331	288	43	250.6
87.0	330	287	43	249.6
87.0	324	282	42	245.4
87.0	323	281	42	244.5
87.0	322	280	42	243.5
87.0	316	275	41	239.3
87.0	315	274	41	238.3
87.0	308	268	40	233.2
87.0	307	267	40	232.2
87.0	301	262	39	228.1

%E	M1	M2	DM	M*
87.0	300	261	39	227.1
87.0	299	260	39	226.1
87.0	293	255	38	221.9
87.0	292	254	38	220.9
87.0	285	248	37	215.8
87.0	284	247	37	214.8
87.0	277	241	36	209.7
87.0	276	240	36	208.7
87.0	270	235	35	204.5
87.0	269	234	35	203.6
87.0	262	228	34	198.4
87.0	261	227	34	197.4
87.0	254	221	33	192.3
87.0	253	220	33	191.3
87.0	247	215	32	187.1
87.0	246	214	32	186.2
87.0	239	208	31	181.0
87.0	238	207	31	180.0
87.0	231	201	30	174.9
87.0	230	200	30	173.9
87.0	223	194	29	168.8
87.0	216	188	28	163.6
87.0	215	187	28	162.6
87.0	208	181	27	157.5
87.0	207	180	27	156.5
87.0	200	174	26	151.4
87.0	193	168	25	146.2
87.0	192	167	25	145.3
87.0	185	161	24	140.1
87.0	184	160	24	139.1
87.0	177	154	23	134.0
87.0	169	147	22	127.9
87.0	162	141	21	122.7
87.0	161	140	21	121.7
87.0	154	134	20	116.6
87.0	146	127	19	110.5
87.0	138	120	18	104.3
87.0	131	114	17	99.2
87.0	123	107	16	93.1
87.0	115	100	15	87.0
87.0	108	94	14	81.8
87.0	100	87	13	75.7
87.0	92	80	12	69.6
87.0	77	67	10	58.3
87.0	69	60	9	52.2
87.0	54	47	7	40.9
87.0	46	40	6	34.8
87.0	23	20	3	17.4
86.9	498	433	65	376.5
86.9	497	432	65	375.5

%E	M1	M2	DM	M*
86.9	496	431	65	374.5
86.9	495	430	65	373.5
86.9	490	426	64	370.4
86.9	489	425	64	369.4
86.9	488	424	64	368.4
86.9	487	423	64	367.4
86.9	482	419	63	364.2
86.9	481	418	63	363.3
86.9	480	417	63	362.3
86.9	475	413	62	359.1
86.9	474	412	62	358.1
86.9	473	411	62	357.1
86.9	472	410	62	356.1
86.9	467	406	61	353.0
86.9	466	405	61	352.0
86.9	465	404	61	351.0
86.9	464	403	61	350.0
86.9	459	399	60	346.8
86.9	458	398	60	345.9
86.9	457	397	60	344.9
86.9	452	393	59	341.7
86.9	451	392	59	340.7
86.9	450	391	59	339.7
86.9	449	390	59	338.8
86.9	444	386	58	335.6
86.9	443	385	58	334.6
86.9	442	384	58	333.6
86.9	436	379	57	329.5
86.9	435	378	57	328.5
86.9	434	377	57	327.5
86.9	429	373	56	324.3
86.9	428	372	56	323.3
86.9	427	371	56	322.3
86.9	426	370	56	321.4
86.9	421	366	55	318.2
86.9	420	365	55	317.2
86.9	419	364	55	316.2
86.9	413	359	54	312.1
86.9	412	358	54	311.1
86.9	411	357	54	310.1
86.9	406	353	53	306.9
86.9	405	352	53	305.9
86.9	404	351	53	305.0
86.9	398	346	52	300.8
86.9	397	345	52	299.8
86.9	396	344	52	298.8
86.9	390	339	51	294.7
86.9	389	338	51	293.7
86.9	388	337	51	292.7
86.9	383	333	50	289.5

%E	M1	M2	DM	M*
86.9	382	332	50	288.5
86.9	381	331	50	287.6
86.9	375	326	49	283.4
86.9	374	325	49	282.4
86.9	373	324	49	281.4
86.9	367	319	48	277.3
86.9	366	318	48	276.3
86.9	360	313	47	272.1
86.9	359	312	47	271.2
86.9	358	311	47	270.2
86.9	352	306	46	266.0
86.9	351	305	46	265.0
86.9	350	304	46	264.0
86.9	344	299	45	259.9
86.9	343	298	45	258.9
86.9	337	293	44	254.7
86.9	336	292	44	253.8
86.9	335	291	44	252.8
86.9	329	286	43	248.6
86.9	328	285	43	247.6
86.9	327	284	43	246.7
86.9	321	279	42	242.5
86.9	320	278	42	241.5
86.9	314	273	41	237.4
86.9	313	272	41	236.4
86.9	312	271	41	235.4
86.9	306	266	40	231.2
86.9	305	265	40	230.2
86.9	298	259	39	225.1
86.9	297	258	39	224.1
86.9	291	253	38	220.0
86.9	290	252	38	219.0
86.9	289	251	38	218.0
86.9	283	246	37	213.8
86.9	282	245	37	212.9
86.9	275	239	36	207.7
86.9	274	238	36	206.7
86.9	268	233	35	202.6
86.9	267	232	35	201.6
86.9	260	226	34	196.4
86.9	259	225	34	195.5
86.9	252	219	33	190.3
86.9	251	218	33	189.3
86.9	245	213	32	185.2
86.9	244	212	32	184.2
86.9	237	206	31	179.1
86.9	236	205	31	178.1
86.9	229	199	30	172.9
86.9	222	193	29	167.8
86.9	221	192	29	166.8

%E	M1	M2	DM	M*
86.9	214	186	28	161.7
86.9	213	185	28	160.7
86.9	206	179	27	155.5
86.9	199	173	26	150.4
86.9	198	172	26	149.4
86.9	191	166	25	144.3
86.9	183	159	24	138.1
86.9	176	153	23	133.0
86.9	175	152	23	132.0
86.9	168	146	22	126.9
86.9	160	139	21	120.8
86.9	153	133	20	115.6
86.9	145	126	19	109.5
86.9	137	119	18	103.4
86.9	130	113	17	98.2
86.9	122	106	16	92.1
86.9	107	93	14	80.8
86.9	99	86	13	74.7
86.9	84	73	11	63.4
86.9	61	53	8	46.0
86.8	500	434	66	376.7
86.8	499	433	66	375.7
86.8	494	429	65	372.6
86.8	493	428	65	371.6
86.8	492	427	65	370.6
86.8	491	426	65	369.6
86.8	486	422	64	366.4
86.8	485	421	64	365.4
86.8	484	420	64	364.5
86.8	479	416	63	361.3
86.8	478	415	63	360.3
86.8	477	414	63	359.3
86.8	476	413	63	358.3
86.8	471	409	62	355.2
86.8	470	408	62	354.2
86.8	469	407	62	353.2
86.8	468	406	62	352.2
86.8	463	402	61	349.0
86.8	462	401	61	348.1
86.8	461	400	61	347.1
86.8	456	396	60	343.9
86.8	455	395	60	342.9
86.8	454	394	60	341.9
86.8	453	393	60	340.9
86.8	448	389	59	337.8
86.8	447	388	59	336.8
86.8	446	387	59	335.8
86.8	441	383	58	332.6
86.8	440	382	58	331.6
86.8	439	381	58	330.7

%E	M_1	M_2	DM	M*
86.8	438	380	58	329.7
86.8	433	376	57	326.5
86.8	432	375	57	325.5
86.8	431	374	57	324.5
86.8	425	369	56	320.4
86.8	424	368	56	319.4
86.8	423	367	56	318.4
86.8	418	363	55	315.2
86.8	417	362	55	314.3
86.8	416	361	55	313.3
86.8	410	356	54	309.1
86.8	409	355	54	308.1
86.8	408	354	54	307.1
86.8	403	350	53	304.0
86.8	402	349	53	303.0
86.8	401	348	53	302.0
86.8	400	347	53	301.0
86.8	395	343	52	297.8
86.8	394	342	52	296.9
86.8	393	341	52	295.9
86.8	387	336	51	291.7
86.8	386	335	51	290.7
86.8	385	334	51	289.8
86.8	380	330	50	286.6
86.8	379	329	50	285.6
86.8	378	328	50	284.6
86.8	372	323	49	280.5
86.8	371	322	49	279.5
86.8	370	321	49	278.5
86.8	365	317	48	275.3
86.8	364	316	48	274.3
86.8	363	315	48	273.3
86.8	357	310	47	269.2
86.8	356	309	47	268.2
86.8	355	308	47	267.2
86.8	349	303	46	263.1
86.8	348	302	46	262.1
86.8	342	297	45	257.9
86.8	341	296	45	256.9
86.8	340	295	45	256.0
86.8	334	290	44	251.8
86.8	333	289	44	250.8
86.8	326	283	43	245.7
86.8	325	282	43	244.7
86.8	319	277	42	240.5
86.8	318	276	42	239.5
86.8	317	275	42	238.6
86.8	311	270	41	234.4
86.8	310	269	41	233.4
86.8	304	264	40	229.3

%E	M_1	M_2	DM	M*
86.8	303	263	40	228.3
86.8	302	262	40	227.3
86.8	296	257	39	223.1
86.8	295	256	39	222.2
86.8	288	250	38	217.0
86.8	287	249	38	216.0
86.8	281	244	37	211.9
86.8	280	243	37	210.9
86.8	273	237	36	205.7
86.8	272	236	36	204.8
86.8	266	231	35	200.6
86.8	265	230	35	199.6
86.8	258	224	34	194.5
86.8	257	223	34	193.5
86.8	250	217	33	188.4
86.8	243	211	32	183.2
86.8	242	210	32	182.2
86.8	235	204	31	177.1
86.8	234	203	31	176.1
86.8	228	198	30	171.9
86.8	227	197	30	171.0
86.8	220	191	29	165.8
86.8	219	190	29	164.8
86.8	212	184	28	159.7
86.8	205	178	27	154.6
86.8	204	177	27	153.6
86.8	197	171	26	148.4
86.8	190	165	25	143.3
86.8	189	164	25	142.3
86.8	182	158	24	137.2
86.8	174	151	23	131.0
86.8	167	145	22	125.9
86.8	159	138	21	119.8
86.8	152	132	20	114.6
86.8	151	131	20	113.6
86.8	144	125	19	108.5
86.8	136	118	18	102.4
86.8	129	112	17	97.2
86.8	121	105	16	91.1
86.8	114	99	15	86.0
86.8	106	92	14	79.8
86.8	91	79	12	68.6
86.8	76	66	10	57.3
86.8	68	59	9	51.2
86.8	53	46	7	39.9
86.8	38	33	5	28.7
86.7	498	432	66	374.7
86.7	497	431	66	373.8
86.7	496	430	66	372.8
86.7	495	429	66	371.8

%E	M_1	M_2	DM	M*
86.7	490	425	65	368.6
86.7	489	424	65	367.6
86.7	488	423	65	366.7
86.7	487	422	65	365.7
86.7	483	419	64	363.5
86.7	482	418	64	362.5
86.7	481	417	64	361.5
86.7	480	416	64	360.5
86.7	475	412	63	357.4
86.7	474	411	63	356.4
86.7	473	410	63	355.4
86.7	472	409	63	354.4
86.7	467	405	62	351.2
86.7	466	404	62	350.2
86.7	465	403	62	349.3
86.7	460	399	61	346.1
86.7	459	398	61	345.1
86.7	458	397	61	344.1
86.7	457	396	61	343.1
86.7	452	392	60	340.0
86.7	451	391	60	339.0
86.7	450	390	60	338.0
86.7	445	386	59	334.8
86.7	444	385	59	333.8
86.7	443	384	59	332.9
86.7	442	383	59	331.9
86.7	437	379	58	328.7
86.7	436	378	58	327.7
86.7	435	377	58	326.7
86.7	430	373	57	323.6
86.7	429	372	57	322.6
86.7	428	371	57	321.6
86.7	427	370	57	320.6
86.7	422	366	56	317.4
86.7	421	365	56	316.4
86.7	420	364	56	315.5
86.7	415	360	55	312.3
86.7	414	359	55	311.3
86.7	413	358	55	310.3
86.7	412	357	55	309.3
86.7	407	353	54	306.2
86.7	406	352	54	305.2
86.7	405	351	54	304.2
86.7	399	346	53	300.0
86.7	398	345	53	299.1
86.7	397	344	53	298.1
86.7	392	340	52	294.9
86.7	391	339	52	293.9
86.7	390	338	52	292.9
86.7	384	333	51	288.8

%E	M_1	M_2	DM	M*
86.7	383	332	51	287.8
86.7	377	327	50	283.6
86.7	376	326	50	282.6
86.7	375	325	50	281.7
86.7	369	320	49	277.5
86.7	368	319	49	276.5
86.7	362	314	48	272.4
86.7	361	313	48	271.4
86.7	360	312	48	270.4
86.7	354	307	47	266.2
86.7	353	306	47	265.3
86.7	347	301	46	261.1
86.7	346	300	46	260.1
86.7	345	299	46	259.1
86.7	339	294	45	255.0
86.7	338	293	45	254.0
86.7	332	288	44	249.8
86.7	331	287	44	248.8
86.7	330	286	44	247.9
86.7	324	281	43	243.7
86.7	323	280	43	242.7
86.7	316	274	42	237.6
86.7	315	273	42	236.6
86.7	309	268	41	232.4
86.7	308	267	41	231.5
86.7	301	261	40	226.3
86.7	300	260	40	225.3
86.7	294	255	39	221.2
86.7	293	254	39	220.2
86.7	286	248	38	215.0
86.7	285	247	38	214.1
86.7	279	242	37	209.9
86.7	278	241	37	208.9
86.7	271	235	36	203.8
86.7	270	234	36	202.8
86.7	264	229	35	198.6
86.7	263	228	35	197.7
86.7	256	222	34	192.5
86.7	255	221	34	191.5
86.7	249	216	33	187.4
86.7	248	215	33	186.4
86.7	241	209	32	181.2
86.7	240	208	32	180.3
86.7	233	202	31	175.1
86.7	226	196	30	170.0
86.7	225	195	30	169.0
86.7	218	189	29	163.9
86.7	211	183	28	158.7
86.7	210	182	28	157.7
86.7	203	176	27	152.6

%E	M_1	M_2	DM	M*
86.7	196	170	26	147.4
86.7	195	169	26	146.5
86.7	188	163	25	141.3
86.7	181	157	24	136.2
86.7	180	156	24	135.2
86.7	173	150	23	130.1
86.7	166	144	22	124.9
86.7	165	143	22	123.9
86.7	158	137	21	118.8
86.7	150	130	20	112.7
86.7	143	124	19	107.5
86.7	135	117	18	101.4
86.7	128	111	17	96.3
86.7	120	104	16	90.1
86.7	113	98	15	85.0
86.7	105	91	14	78.9
86.7	98	85	13	73.7
86.7	90	78	12	67.6
86.7	83	72	11	62.5
86.7	75	65	10	56.3
86.7	60	52	8	45.1
86.7	45	39	6	33.8
86.7	30	26	4	22.5
86.7	15	13	2	11.3
86.6	500	433	67	375.0
86.6	499	432	67	374.0
86.6	494	428	66	370.8
86.6	493	427	66	369.8
86.6	492	426	66	368.9
86.6	491	425	66	367.9
86.6	486	421	65	364.7
86.6	485	420	65	363.7
86.6	484	419	65	362.7
86.6	479	415	64	359.6
86.6	478	414	64	358.6
86.6	477	413	64	357.6
86.6	476	412	64	356.6
86.6	471	408	63	353.4
86.6	470	407	63	352.4
86.6	469	406	63	351.5
86.6	464	402	62	348.3
86.6	463	401	62	347.3
86.6	462	400	62	346.3
86.6	461	399	62	345.3
86.6	456	395	61	342.2
86.6	455	394	61	341.2
86.6	454	393	61	340.2
86.6	449	389	60	337.0
86.6	448	388	60	336.0
86.6	447	387	60	335.1

%E	M1	M2	DM	M*
86.6	441	382	59	330.9
86.6	440	381	59	329.9
86.6	439	380	59	328.9
86.6	434	376	58	325.8
86.6	433	375	58	324.8
86.6	432	374	58	323.8
86.6	426	369	57	319.6
86.6	425	368	57	318.6
86.6	424	367	57	317.7
86.6	419	363	56	314.5
86.6	418	362	56	313.5
86.6	417	361	56	312.5
86.6	411	356	55	308.4
86.6	410	355	55	307.4
86.6	409	354	55	306.4
86.6	404	350	54	303.2
86.6	403	349	54	302.2
86.6	402	348	54	301.3
86.6	396	343	53	297.1
86.6	395	342	53	296.1
86.6	389	337	52	292.0
86.6	388	336	52	291.0
86.6	387	335	52	290.0
86.6	382	331	51	286.8
86.6	381	330	51	285.8
86.6	380	329	51	284.8
86.6	374	324	50	280.7
86.6	373	323	50	279.7
86.6	372	322	50	278.7
86.6	367	318	49	275.5
86.6	366	317	49	274.6
86.6	365	316	49	273.6
86.6	359	311	48	269.4
86.6	358	310	48	268.4
86.6	357	309	48	267.5
86.6	352	305	47	264.3
86.6	351	304	47	263.3
86.6	350	303	47	262.3
86.6	344	298	46	258.2
86.6	343	297	46	257.2
86.6	342	296	46	256.2
86.6	337	292	45	253.0
86.6	336	291	45	252.0
86.6	335	290	45	251.0
86.6	329	285	44	246.9
86.6	328	284	44	245.9
86.6	322	279	43	241.7
86.6	321	278	43	240.8
86.6	320	277	43	239.8
86.6	314	272	42	235.6

%E	M1	M2	DM	M*
86.6	313	271	42	234.6
86.6	307	266	41	230.5
86.6	306	265	41	229.5
86.6	305	264	41	228.5
86.6	299	259	40	224.4
86.6	298	258	40	223.4
86.6	292	253	39	219.2
86.6	291	252	39	218.2
86.6	290	251	39	217.2
86.6	284	246	38	213.1
86.6	283	245	38	212.1
86.6	277	240	37	207.9
86.6	276	239	37	207.0
86.6	269	233	36	201.8
86.6	268	232	36	200.8
86.6	262	227	35	196.7
86.6	261	226	35	195.7
86.6	254	220	34	190.6
86.6	253	219	34	189.6
86.6	247	214	33	185.4
86.6	246	213	33	184.4
86.6	239	207	32	179.3
86.6	238	206	32	178.3
86.6	232	201	31	174.1
86.6	231	200	31	173.2
86.6	224	194	30	168.0
86.6	217	188	29	162.9
86.6	216	187	29	161.9
86.6	209	181	28	156.8
86.6	202	175	27	151.6
86.6	201	174	27	150.6
86.6	194	168	26	145.5
86.6	187	162	25	140.3
86.6	186	161	25	139.4
86.6	179	155	24	134.2
86.6	172	149	23	129.1
86.6	171	148	23	128.1
86.6	164	142	22	123.0
86.6	157	136	21	117.8
86.6	149	129	20	111.7
86.6	142	123	19	106.5
86.6	134	116	18	100.4
86.6	127	110	17	95.3
86.6	119	103	16	89.2
86.6	112	97	15	84.0
86.6	97	84	13	72.7
86.6	82	71	11	61.5
86.6	67	58	9	50.2
86.5	498	431	67	373.0
86.5	497	430	67	372.0

%E	M1	M2	DM	M*
86.5	496	429	67	371.1
86.5	495	428	67	370.1
86.5	490	424	66	366.9
86.5	489	423	66	365.9
86.5	488	422	66	364.9
86.5	483	418	65	361.7
86.5	482	417	65	360.8
86.5	481	416	65	359.8
86.5	480	415	65	358.8
86.5	475	411	64	355.6
86.5	474	410	64	354.6
86.5	473	409	64	353.7
86.5	468	405	63	350.5
86.5	467	404	63	349.5
86.5	466	403	63	348.5
86.5	465	402	63	347.5
86.5	460	398	62	344.4
86.5	459	397	62	343.4
86.5	458	396	62	342.4
86.5	453	392	61	339.2
86.5	452	391	61	338.2
86.5	451	390	61	337.3
86.5	446	386	60	334.1
86.5	445	385	60	333.1
86.5	444	384	60	332.1
86.5	443	383	60	331.1
86.5	438	379	59	327.9
86.5	437	378	59	327.0
86.5	436	377	59	326.0
86.5	431	373	58	322.8
86.5	430	372	58	321.8
86.5	429	371	58	320.8
86.5	423	366	57	316.7
86.5	422	365	57	315.7
86.5	421	364	57	314.7
86.5	416	360	56	311.5
86.5	415	359	56	310.6
86.5	414	358	56	309.6
86.5	408	353	55	305.4
86.5	407	352	55	304.4
86.5	406	351	55	303.5
86.5	401	347	54	300.3
86.5	400	346	54	299.3
86.5	399	345	54	298.3
86.5	394	341	53	295.1
86.5	393	340	53	294.1
86.5	392	339	53	293.2
86.5	386	334	52	289.0
86.5	385	333	52	288.0
86.5	384	332	52	287.0

%E	M1	M2	DM	M*
86.5	379	328	51	283.9
86.5	378	327	51	282.9
86.5	377	326	51	281.9
86.5	371	321	50	277.7
86.5	370	320	50	276.8
86.5	369	319	50	275.8
86.5	364	315	49	272.6
86.5	363	314	49	271.6
86.5	362	313	49	270.6
86.5	356	308	48	266.5
86.5	355	307	48	265.5
86.5	349	302	47	261.3
86.5	348	301	47	260.3
86.5	347	300	47	259.4
86.5	341	295	46	255.2
86.5	340	294	46	254.2
86.5	334	289	45	250.1
86.5	333	288	45	249.1
86.5	327	283	44	244.9
86.5	326	282	44	243.9
86.5	325	281	44	243.0
86.5	319	276	43	238.8
86.5	318	275	43	237.8
86.5	312	270	42	233.7
86.5	311	269	42	232.7
86.5	310	268	42	231.7
86.5	304	263	41	227.5
86.5	303	262	41	226.5
86.5	297	257	40	222.4
86.5	296	256	40	221.4
86.5	289	250	39	216.3
86.5	288	249	39	215.3
86.5	282	244	38	211.1
86.5	281	243	38	210.1
86.5	275	238	37	206.0
86.5	274	237	37	205.0
86.5	267	231	36	199.9
86.5	266	230	36	198.9
86.5	260	225	35	194.7
86.5	259	224	35	193.7
86.5	252	218	34	188.6
86.5	251	217	34	187.6
86.5	245	212	33	183.4
86.5	244	211	33	182.5
86.5	237	205	32	177.3
86.5	230	199	31	172.2
86.5	229	198	31	171.2
86.5	223	193	30	167.0
86.5	222	192	30	166.1
86.5	215	186	29	160.9

%E	M1	M2	DM	M*
86.5	208	180	28	155.8
86.5	207	179	28	154.8
86.5	200	173	27	149.6
86.5	193	167	26	144.5
86.5	192	166	26	143.5
86.5	185	160	25	138.4
86.5	178	154	24	133.2
86.5	170	147	23	127.1
86.5	163	141	22	122.0
86.5	156	135	21	116.8
86.5	155	134	21	115.8
86.5	148	128	20	110.7
86.5	141	122	19	105.6
86.5	133	115	18	99.4
86.5	126	109	17	94.3
86.5	111	96	15	83.0
86.5	104	90	14	77.9
86.5	96	83	13	71.8
86.5	89	77	12	66.6
86.5	74	64	10	55.4
86.5	52	45	7	38.9
86.5	37	32	5	27.7
86.4	500	432	68	373.2
86.4	499	431	68	372.3
86.4	494	427	67	369.1
86.4	493	426	67	368.1
86.4	492	425	67	367.1
86.4	491	424	67	366.1
86.4	487	421	66	363.9
86.4	486	420	66	363.0
86.4	485	419	66	362.0
86.4	484	418	66	361.0
86.4	479	414	65	357.8
86.4	478	413	65	356.8
86.4	477	412	65	355.9
86.4	472	408	64	352.7
86.4	471	407	64	351.7
86.4	470	406	64	350.7
86.4	469	405	64	349.7
86.4	464	401	63	346.6
86.4	463	400	63	345.6
86.4	462	399	63	344.6
86.4	457	395	62	341.4
86.4	456	394	62	340.4
86.4	455	393	62	339.4
86.4	450	389	61	336.3
86.4	449	388	61	335.3
86.4	448	387	61	334.3
86.4	447	386	61	333.3
86.4	442	382	60	330.1

%E	M1	M2	DM	M*
86.4	441	381	60	329.2
86.4	440	380	60	328.2
86.4	435	376	59	325.0
86.4	434	375	59	324.0
86.4	433	374	59	323.0
86.4	428	370	58	319.9
86.4	427	369	58	318.9
86.4	426	368	58	317.9
86.4	425	367	58	316.9
86.4	420	363	57	313.7
86.4	419	362	57	312.8
86.4	418	361	57	311.8
86.4	413	357	56	308.6
86.4	412	356	56	307.6
86.4	411	355	56	306.6
86.4	405	350	55	302.5
86.4	404	349	55	301.5
86.4	403	348	55	300.5
86.4	398	344	54	297.3
86.4	397	343	54	296.3
86.4	396	342	54	295.4
86.4	391	338	53	292.2
86.4	390	337	53	291.2
86.4	389	336	53	290.2
86.4	383	331	52	286.1
86.4	382	330	52	285.1
86.4	381	329	52	284.1
86.4	376	325	51	280.9
86.4	375	324	51	279.9
86.4	374	323	51	279.0
86.4	368	318	50	274.8
86.4	367	317	50	273.8
86.4	361	312	49	269.7
86.4	360	311	49	268.7
86.4	359	310	49	267.7
86.4	354	306	48	264.5
86.4	353	305	48	263.5
86.4	352	304	48	262.5
86.4	346	299	47	258.4
86.4	345	298	47	257.4
86.4	339	293	46	253.2
86.4	338	292	46	252.3
86.4	337	291	46	251.3
86.4	332	287	45	248.1
86.4	331	286	45	247.1
86.4	330	285	45	246.1
86.4	324	280	44	242.0
86.4	323	279	44	241.0
86.4	317	274	43	236.8
86.4	316	273	43	235.9

%E	M1	M2	DM	M*
86.4	309	267	42	230.7
86.4	308	266	42	229.7
86.4	302	261	41	225.6
86.4	301	260	41	224.6
86.4	295	255	40	220.4
86.4	294	254	40	219.4
86.4	287	248	39	214.3
86.4	286	247	39	213.3
86.4	280	242	38	209.2
86.4	279	241	38	208.2
86.4	273	236	37	204.0
86.4	272	235	37	203.0
86.4	265	229	36	197.9
86.4	264	228	36	196.9
86.4	258	223	35	192.7
86.4	257	222	35	191.8
86.4	250	216	34	186.6
86.4	243	210	33	181.5
86.4	242	209	33	180.5
86.4	236	204	32	176.3
86.4	235	203	32	175.4
86.4	228	197	31	170.2
86.4	221	191	30	165.1
86.4	220	190	30	164.1
86.4	214	185	29	159.9
86.4	213	184	29	158.9
86.4	206	178	28	153.8
86.4	199	172	27	148.7
86.4	198	171	27	147.7
86.4	191	165	26	142.5
86.4	184	159	25	137.4
86.4	177	153	24	132.3
86.4	176	152	24	131.3
86.4	169	146	23	126.1
86.4	162	140	22	121.0
86.4	154	133	21	114.9
86.4	147	127	20	109.7
86.4	140	121	19	104.6
86.4	132	114	18	98.5
86.4	125	108	17	93.3
86.4	118	102	16	88.2
86.4	110	95	15	82.0
86.4	103	89	14	76.9
86.4	88	76	12	65.6
86.4	81	70	11	60.5
86.4	66	57	9	49.2
86.4	59	51	8	44.1
86.4	44	38	6	32.8
86.4	22	19	3	16.4
86.3	498	430	68	371.3

%E	M1	M2	DM	M*
86.3	497	429	68	370.3
86.3	496	428	68	369.3
86.3	495	427	68	368.3
86.3	490	423	67	365.2
86.3	489	422	67	364.2
86.3	488	421	67	363.2
86.3	483	417	66	360.0
86.3	482	416	66	359.0
86.3	481	415	66	358.1
86.3	480	414	66	357.1
86.3	476	411	65	354.9
86.3	475	410	65	353.9
86.3	474	409	65	352.9
86.3	473	408	65	351.9
86.3	468	404	64	348.8
86.3	467	403	64	347.8
86.3	466	402	64	346.8
86.3	461	398	63	343.6
86.3	460	397	63	342.6
86.3	459	396	63	341.6
86.3	454	392	62	338.5
86.3	453	391	62	337.5
86.3	452	390	62	336.5
86.3	451	389	62	335.5
86.3	446	385	61	332.3
86.3	445	384	61	331.4
86.3	444	383	61	330.4
86.3	439	379	60	327.2
86.3	438	378	60	326.2
86.3	437	377	60	325.2
86.3	432	373	59	322.1
86.3	431	372	59	321.1
86.3	430	371	59	320.1
86.3	424	366	58	315.9
86.3	423	365	58	315.0
86.3	422	364	58	314.0
86.3	417	360	57	310.8
86.3	416	359	57	309.8
86.3	415	358	57	308.8
86.3	410	354	56	305.6
86.3	409	353	56	304.7
86.3	408	352	56	303.7
86.3	402	347	55	299.5
86.3	401	346	55	298.5
86.3	400	345	55	297.6
86.3	395	341	54	294.4
86.3	394	340	54	293.4
86.3	393	339	54	292.4
86.3	388	335	53	289.2
86.3	387	334	53	288.3

%E	M1	M2	DM	M*
86.3	386	333	53	287.3
86.3	380	328	52	283.1
86.3	379	327	52	282.1
86.3	373	322	51	278.0
86.3	372	321	51	277.0
86.3	371	320	51	276.0
86.3	366	316	50	272.8
86.3	365	315	50	271.8
86.3	364	314	50	270.9
86.3	358	309	49	266.7
86.3	357	308	49	265.7
86.3	351	303	48	261.6
86.3	350	302	48	260.6
86.3	344	297	47	256.4
86.3	343	296	47	255.4
86.3	342	295	47	254.5
86.3	336	290	46	250.3
86.3	335	289	46	249.3
86.3	329	284	45	245.2
86.3	328	283	45	244.2
86.3	322	278	44	240.0
86.3	321	277	44	239.0
86.3	320	276	44	238.0
86.3	315	272	43	234.9
86.3	314	271	43	233.9
86.3	313	270	43	232.9
86.3	307	265	42	228.7
86.3	306	264	42	227.8
86.3	300	259	41	223.6
86.3	299	258	41	222.6
86.3	293	253	40	218.5
86.3	292	252	40	217.5
86.3	291	251	40	216.5
86.3	285	246	39	212.3
86.3	284	245	39	211.4
86.3	278	240	38	207.2
86.3	277	239	38	206.2
86.3	271	234	37	202.1
86.3	270	233	37	201.1
86.3	263	227	36	195.9
86.3	262	226	36	194.9
86.3	256	221	35	190.8
86.3	255	220	35	189.8
86.3	249	215	34	185.6
86.3	248	214	34	184.7
86.3	241	208	33	179.5
86.3	240	207	33	178.5
86.3	234	202	32	174.4
86.3	233	201	32	173.4
86.3	227	196	31	169.2

%E	M1	M2	DM	M*
86.3	226	195	31	168.3
86.3	219	189	30	163.1
86.3	212	183	29	158.0
86.3	211	182	29	157.0
86.3	205	177	28	152.8
86.3	204	176	28	151.8
86.3	197	170	27	146.7
86.3	190	164	26	141.6
86.3	183	158	25	136.4
86.3	182	157	25	135.4
86.3	175	151	24	130.3
86.3	168	145	23	125.1
86.3	161	139	22	120.0
86.3	160	138	22	119.0
86.3	153	132	21	113.9
86.3	146	126	20	108.7
86.3	139	120	19	103.6
86.3	131	113	18	97.5
86.3	124	107	17	92.3
86.3	117	101	16	87.2
86.3	102	88	14	75.9
86.3	95	82	13	70.8
86.3	80	69	11	59.5
86.3	73	63	10	54.4
86.3	51	44	7	38.0
86.2	500	431	69	371.5
86.2	499	430	69	370.5
86.2	494	426	68	367.4
86.2	493	425	68	366.4
86.2	492	424	68	365.4
86.2	491	423	68	364.4
86.2	487	420	67	362.2
86.2	486	419	67	361.2
86.2	485	418	67	360.3
86.2	484	417	67	359.3
86.2	479	413	66	356.1
86.2	478	412	66	355.1
86.2	477	411	66	354.1
86.2	472	407	65	351.0
86.2	471	406	65	350.0
86.2	470	405	65	349.0
86.2	465	401	64	345.8
86.2	464	400	64	344.8
86.2	463	399	64	343.8
86.2	458	395	63	340.7
86.2	457	394	63	339.7
86.2	456	393	63	338.7
86.2	455	392	63	337.7
86.2	450	388	62	334.5
86.2	449	387	62	333.6

%E	M1	M2	DM	M*
86.2	448	386	62	332.6
86.2	443	382	61	329.4
86.2	442	381	61	328.4
86.2	441	380	61	327.4
86.2	436	376	60	324.3
86.2	435	375	60	323.3
86.2	434	374	60	322.3
86.2	429	370	59	319.1
86.2	428	369	59	318.1
86.2	427	368	59	317.2
86.2	426	367	59	316.2
86.2	421	363	58	313.0
86.2	420	362	58	312.0
86.2	419	361	58	311.0
86.2	414	357	57	307.8
86.2	413	356	57	306.9
86.2	412	355	57	305.9
86.2	407	351	56	302.7
86.2	406	350	56	301.7
86.2	405	349	56	300.7
86.2	399	344	55	296.6
86.2	398	343	55	295.6
86.2	392	338	54	291.4
86.2	391	337	54	290.5
86.2	390	336	54	289.5
86.2	385	332	53	286.3
86.2	384	331	53	285.3
86.2	383	330	53	284.3
86.2	378	326	52	281.2
86.2	377	325	52	280.2
86.2	376	324	52	279.2
86.2	370	319	51	275.0
86.2	369	318	51	274.0
86.2	363	313	50	269.9
86.2	362	312	50	268.9
86.2	361	311	50	267.9
86.2	356	307	49	264.7
86.2	355	306	49	263.8
86.2	354	305	49	262.8
86.2	349	301	48	259.6
86.2	348	300	48	258.6
86.2	347	299	48	257.6
86.2	341	294	47	253.5
86.2	340	293	47	252.5
86.2	334	288	46	248.3
86.2	333	287	46	247.4
86.2	327	282	45	243.2
86.2	326	281	45	242.2
86.2	325	280	45	241.2
86.2	319	275	44	237.1

%E	M1	M2	DM	M*
86.2	318	274	44	236.1
86.2	312	269	43	231.9
86.2	311	268	43	230.9
86.2	305	263	42	226.8
86.2	304	262	42	225.8
86.2	298	257	41	221.6
86.2	297	256	41	220.7
86.2	290	250	40	215.5
86.2	289	249	40	214.5
86.2	283	244	39	210.4
86.2	282	243	39	209.4
86.2	276	238	38	205.2
86.2	275	237	38	204.3
86.2	269	232	37	200.1
86.2	268	231	37	199.1
86.2	261	225	36	194.0
86.2	260	224	36	193.0
86.2	254	219	35	188.8
86.2	253	218	35	187.8
86.2	247	213	34	183.7
86.2	246	212	34	182.7
86.2	239	206	33	177.6
86.2	232	200	32	172.4
86.2	225	194	31	167.3
86.2	224	193	31	166.3
86.2	218	188	30	162.1
86.2	217	187	30	161.1
86.2	210	181	29	156.0
86.2	203	175	28	150.9
86.2	196	169	27	145.7
86.2	195	168	27	144.7
86.2	189	163	26	140.6
86.2	188	162	26	139.6
86.2	181	156	25	134.5
86.2	174	150	24	129.3
86.2	167	144	23	124.2
86.2	159	137	22	118.0
86.2	152	131	21	112.9
86.2	145	125	20	107.8
86.2	138	119	19	102.6
86.2	130	112	18	96.5
86.2	123	106	17	91.3
86.2	116	100	16	86.2
86.2	109	94	15	81.1
86.2	94	81	13	69.8
86.2	87	75	12	64.7
86.2	65	56	9	48.2
86.2	58	50	8	43.1
86.2	29	25	4	21.6
86.1	498	429	69	369.6

%E	M1	M2	DM	M*
86.1	497	428	69	368.6
86.1	496	427	69	367.6
86.1	495	426	69	366.6
86.1	490	422	68	363.4
86.1	489	421	68	362.5
86.1	488	420	68	361.5
86.1	483	416	67	358.3
86.1	482	415	67	357.3
86.1	481	414	67	356.3
86.1	476	410	66	353.2
86.1	475	409	66	352.2
86.1	474	408	66	351.2
86.1	469	404	65	348.0
86.1	468	403	65	347.0
86.1	467	402	65	346.0
86.1	466	401	65	345.1
86.1	462	398	64	342.9
86.1	461	397	64	341.9
86.1	460	396	64	340.9
86.1	459	395	64	339.9
86.1	454	391	63	336.7
86.1	453	390	63	335.8
86.1	452	389	63	334.8
86.1	447	385	62	331.6
86.1	446	384	62	330.6
86.1	445	383	62	329.6
86.1	440	379	61	326.5
86.1	439	378	61	325.5
86.1	438	377	61	324.5
86.1	433	373	60	321.3
86.1	432	372	60	320.3
86.1	431	371	60	319.4
86.1	425	366	59	315.2
86.1	424	365	59	314.2
86.1	423	364	59	313.2
86.1	418	360	58	310.0
86.1	417	359	58	309.1
86.1	416	358	58	308.1
86.1	411	354	57	304.9
86.1	410	353	57	303.9
86.1	409	352	57	302.9
86.1	404	348	56	299.8
86.1	403	347	56	298.8
86.1	402	346	56	297.8
86.1	397	342	55	294.6
86.1	396	341	55	293.6
86.1	395	340	55	292.7
86.1	389	335	54	288.5
86.1	388	334	54	287.5
86.1	382	329	53	283.4

%E	M1	M2	DM	M*
86.1	381	328	53	282.4
86.1	380	327	53	281.4
86.1	375	323	52	278.2
86.1	374	322	52	277.2
86.1	373	321	52	276.2
86.1	368	317	51	273.1
86.1	367	316	51	272.1
86.1	366	315	51	271.1
86.1	360	310	50	266.9
86.1	359	309	50	266.0
86.1	353	304	49	261.8
86.1	352	303	49	260.8
86.1	346	298	48	256.7
86.1	345	297	48	255.7
86.1	339	292	47	251.5
86.1	338	291	47	250.5
86.1	337	290	47	249.6
86.1	332	286	46	246.4
86.1	331	285	46	245.4
86.1	330	284	46	244.4
86.1	324	279	45	240.3
86.1	323	278	45	239.3
86.1	317	273	44	235.1
86.1	316	272	44	234.1
86.1	310	267	43	230.0
86.1	309	266	43	229.0
86.1	303	261	42	224.8
86.1	302	260	42	223.8
86.1	296	255	41	219.7
86.1	295	254	41	218.7
86.1	294	253	41	217.7
86.1	288	248	40	213.6
86.1	287	247	40	212.6
86.1	281	242	39	208.4
86.1	280	241	39	207.4
86.1	274	236	38	203.3
86.1	273	235	38	202.3
86.1	267	230	37	198.1
86.1	266	229	37	197.1
86.1	259	223	36	192.0
86.1	252	217	35	186.9
86.1	251	216	35	185.9
86.1	245	211	34	181.7
86.1	244	210	34	180.7
86.1	238	205	33	176.6
86.1	237	204	33	175.6
86.1	231	199	32	171.4
86.1	230	198	32	170.5
86.1	223	192	31	165.3
86.1	216	186	30	160.2

%E	M1	M2	DM	M*
86.1	209	180	29	155.0
86.1	208	179	29	154.0
86.1	202	174	28	149.9
86.1	201	173	28	148.9
86.1	194	167	27	143.8
86.1	187	161	26	138.6
86.1	180	155	25	133.5
86.1	173	149	24	128.3
86.1	166	143	23	123.2
86.1	165	142	23	122.2
86.1	158	136	22	117.1
86.1	151	130	21	111.9
86.1	144	124	20	106.8
86.1	137	118	19	101.6
86.1	122	105	17	90.4
86.1	115	99	16	85.2
86.1	108	93	15	80.1
86.1	101	87	14	74.9
86.1	79	68	11	58.5
86.1	72	62	10	53.4
86.1	36	31	5	26.7
86.0	500	430	70	369.8
86.0	499	429	70	368.8
86.0	494	425	69	365.6
86.0	493	424	69	364.7
86.0	492	423	69	363.7
86.0	487	419	68	360.5
86.0	486	418	68	359.5
86.0	485	417	68	358.5
86.0	484	416	68	357.6
86.0	480	413	67	355.4
86.0	479	412	67	354.4
86.0	478	411	67	353.4
86.0	477	410	67	352.4
86.0	473	407	66	350.2
86.0	472	406	66	349.2
86.0	471	405	66	348.2
86.0	470	404	66	347.3
86.0	465	400	65	344.1
86.0	464	399	65	343.1
86.0	463	398	65	342.1
86.0	458	394	64	338.9
86.0	457	393	64	338.0
86.0	456	392	64	337.0
86.0	451	388	63	333.8
86.0	450	387	63	332.8
86.0	449	386	63	331.8
86.0	444	382	62	328.7
86.0	443	381	62	327.7
86.0	442	380	62	326.7

%E	M1	M2	DM	M*
86.0	437	376	61	323.5
86.0	436	375	61	322.5
86.0	435	374	61	321.6
86.0	430	370	60	318.4
86.0	429	369	60	317.4
86.0	428	368	60	316.4
86.0	422	363	59	312.2
86.0	421	362	59	311.3
86.0	420	361	59	310.3
86.0	415	357	58	307.1
86.0	414	356	58	306.1
86.0	413	355	58	305.1
86.0	408	351	57	302.0
86.0	407	350	57	301.0
86.0	406	349	57	300.0
86.0	401	345	56	296.8
86.0	400	344	56	295.8
86.0	399	343	56	294.9
86.0	394	339	55	291.7
86.0	393	338	55	290.7
86.0	392	337	55	289.7
86.0	387	333	54	286.5
86.0	386	332	54	285.6
86.0	385	331	54	284.6
86.0	379	326	53	280.4
86.0	378	325	53	279.4
86.0	372	320	52	275.3
86.0	371	319	52	274.3
86.0	365	314	51	270.1
86.0	364	313	51	269.1
86.0	363	312	51	268.2
86.0	358	308	50	265.0
86.0	357	307	50	264.0
86.0	356	306	50	263.0
86.0	351	302	49	259.8
86.0	350	301	49	258.9
86.0	349	300	49	257.9
86.0	344	296	48	254.7
86.0	343	295	48	253.7
86.0	342	294	48	252.7
86.0	336	289	47	248.6
86.0	335	288	47	247.6
86.0	329	283	46	243.4
86.0	328	282	46	242.5
86.0	322	277	45	238.3
86.0	321	276	45	237.3
86.0	315	271	44	233.1
86.0	314	270	44	232.2
86.0	308	265	43	228.0
86.0	307	264	43	227.0

%E	M1	M2	DM	M*
86.0	301	259	42	222.9
86.0	300	258	42	221.9
86.0	299	257	42	220.9
86.0	293	252	41	216.7
86.0	292	251	41	215.8
86.0	286	246	40	211.6
86.0	285	245	40	210.6
86.0	279	240	39	206.5
86.0	278	239	39	205.5
86.0	272	234	38	201.3
86.0	271	233	38	200.3
86.0	265	228	37	196.2
86.0	264	227	37	195.2
86.0	258	222	36	191.0
86.0	257	221	36	190.0
86.0	250	215	35	184.9
86.0	243	209	34	179.8
86.0	242	208	34	178.8
86.0	236	203	33	174.6
86.0	235	202	33	173.6
86.0	229	197	32	169.5
86.0	228	196	32	168.5
86.0	222	191	31	164.3
86.0	221	190	31	163.3
86.0	215	185	30	159.2
86.0	214	184	30	158.2
86.0	207	178	29	153.1
86.0	200	172	28	147.9
86.0	193	166	27	142.8
86.0	186	160	26	137.6
86.0	179	154	25	132.5
86.0	178	153	25	131.5
86.0	172	148	24	127.3
86.0	171	147	24	126.4
86.0	164	141	23	121.2
86.0	157	135	22	116.1
86.0	150	129	21	110.9
86.0	143	123	20	105.8
86.0	136	117	19	100.7
86.0	129	111	18	95.5
86.0	121	104	17	89.4
86.0	114	98	16	84.2
86.0	107	92	15	79.1
86.0	100	86	14	74.0
86.0	93	80	13	68.8
86.0	86	74	12	63.7
86.0	57	49	8	42.1
86.0	50	43	7	37.0
86.0	43	37	6	31.8
85.9	498	428	70	367.8

%E	M1	M2	DM	M*
85.9	497	427	70	366.9
85.9	496	426	70	365.9
85.9	495	425	70	364.9
85.9	491	422	69	362.7
85.9	490	421	69	361.7
85.9	489	420	69	360.7
85.9	488	419	69	359.8
85.9	483	415	68	356.6
85.9	482	414	68	355.6
85.9	481	413	68	354.6
85.9	476	409	67	351.4
85.9	475	408	67	350.5
85.9	474	407	67	349.5
85.9	469	403	66	346.3
85.9	468	402	66	345.3
85.9	467	401	66	344.3
85.9	462	397	65	341.1
85.9	461	396	65	340.2
85.9	460	395	65	339.2
85.9	455	391	64	336.0
85.9	454	390	64	335.0
85.9	453	389	64	334.0
85.9	448	385	63	330.9
85.9	447	384	63	329.9
85.9	446	383	63	328.9
85.9	441	379	62	325.7
85.9	440	378	62	324.7
85.9	439	377	62	323.8
85.9	434	373	61	320.6
85.9	433	372	61	319.6
85.9	432	371	61	318.6
85.9	427	367	60	315.4
85.9	426	366	60	314.5
85.9	425	365	60	313.5
85.9	419	360	59	309.3
85.9	418	359	59	308.3
85.9	417	358	59	307.3
85.9	412	354	58	304.2
85.9	411	353	58	303.2
85.9	410	352	58	302.2
85.9	405	348	57	299.0
85.9	404	347	57	298.0
85.9	403	346	57	297.1
85.9	398	342	56	293.9
85.9	397	341	56	292.9
85.9	396	340	56	291.9
85.9	391	336	55	288.7
85.9	390	335	55	287.8
85.9	389	334	55	286.8
85.9	384	330	54	283.6

%E	M1	M2	DM	M*
85.9	383	329	54	282.6
85.9	382	328	54	281.6
85.9	377	324	53	278.5
85.9	376	323	53	277.5
85.9	375	322	53	276.5
85.9	370	318	52	273.3
85.9	369	317	52	272.3
85.9	368	316	52	271.3
85.9	362	311	51	267.2
85.9	361	310	51	266.2
85.9	355	305	50	262.0
85.9	354	304	50	261.1
85.9	348	299	49	256.9
85.9	347	298	49	255.9
85.9	341	293	48	251.8
85.9	340	292	48	250.8
85.9	334	287	47	246.6
85.9	333	286	47	245.6
85.9	327	281	46	241.5
85.9	326	280	46	240.5
85.9	320	275	45	236.3
85.9	319	274	45	235.3
85.9	313	269	44	231.2
85.9	312	268	44	230.2
85.9	311	267	44	229.2
85.9	306	263	43	226.0
85.9	305	262	43	225.1
85.9	304	261	43	224.1
85.9	298	256	42	219.9
85.9	297	255	42	218.9
85.9	291	250	41	214.8
85.9	290	249	41	213.8
85.9	284	244	40	209.6
85.9	283	243	40	208.7
85.9	277	238	39	204.5
85.9	276	237	39	203.5
85.9	270	232	38	199.3
85.9	269	231	38	198.4
85.9	263	226	37	194.2
85.9	262	225	37	193.2
85.9	256	220	36	189.1
85.9	255	219	36	188.1
85.9	249	214	35	183.9
85.9	248	213	35	182.9
85.9	241	207	34	177.8
85.9	234	201	33	172.7
85.9	227	195	32	167.5
85.9	220	189	31	162.4
85.9	213	183	30	157.2
85.9	206	177	29	152.1

%E	M1	M2	DM	M*
85.9	205	176	29	151.1
85.9	199	171	28	146.9
85.9	198	170	28	146.0
85.9	192	165	27	141.8
85.9	191	164	27	140.8
85.9	185	159	26	136.7
85.9	184	158	26	135.7
85.9	177	152	25	130.5
85.9	170	146	24	125.4
85.9	163	140	23	120.2
85.9	156	134	22	115.1
85.9	149	128	21	110.0
85.9	142	122	20	104.8
85.9	135	116	19	99.7
85.9	128	110	18	94.5
85.9	99	85	14	73.0
85.9	92	79	13	67.8
85.9	85	73	12	62.7
85.9	78	67	11	57.6
85.9	71	61	10	52.4
85.9	64	55	9	47.3
85.8	500	429	71	368.1
85.8	499	428	71	367.1
85.8	494	424	70	363.9
85.8	493	423	70	362.9
85.8	492	422	70	362.0
85.8	487	418	69	358.8
85.8	486	417	69	357.8
85.8	485	416	69	356.8
85.8	480	412	68	353.6
85.8	479	411	68	352.7
85.8	478	410	68	351.7
85.8	473	406	67	348.5
85.8	472	405	67	347.5
85.8	471	404	67	346.5
85.8	466	400	66	343.3
85.8	465	399	66	342.4
85.8	464	398	66	341.4
85.8	459	394	65	338.2
85.8	458	393	65	337.2
85.8	457	392	65	336.2
85.8	452	388	64	333.1
85.8	451	387	64	332.1
85.8	450	386	64	331.1
85.8	445	382	63	327.9
85.8	444	381	63	326.9
85.8	443	380	63	326.0
85.8	438	376	62	322.8
85.8	437	375	62	321.8
85.8	436	374	62	320.8

%E	M1	M2	DM	M*
85.8	431	370	61	317.6
85.8	430	369	61	316.7
85.8	429	368	61	315.7
85.8	424	364	60	312.5
85.8	423	363	60	311.5
85.8	422	362	60	310.5
85.8	416	357	59	306.4
85.8	415	356	59	305.4
85.8	409	351	58	301.2
85.8	408	350	58	300.2
85.8	402	345	57	296.1
85.8	401	344	57	295.1
85.8	400	343	57	294.1
85.8	395	339	56	290.9
85.8	394	338	56	290.0
85.8	393	337	56	289.0
85.8	388	333	55	285.8
85.8	387	332	55	284.8
85.8	386	331	55	283.8
85.8	381	327	54	280.7
85.8	380	326	54	279.7
85.8	379	325	54	278.7
85.8	374	321	53	275.5
85.8	373	320	53	274.5
85.8	372	319	53	273.6
85.8	367	315	52	270.4
85.8	366	314	52	269.4
85.8	365	313	52	268.4
85.8	360	309	51	265.2
85.8	359	308	51	264.2
85.8	358	307	51	263.3
85.8	353	303	50	260.1
85.8	352	302	50	259.1
85.8	351	301	50	258.1
85.8	346	297	49	254.9
85.8	345	296	49	254.0
85.8	344	295	49	253.0
85.8	339	291	48	249.8
85.8	338	290	48	248.8
85.8	337	289	48	247.8
85.8	332	285	47	244.7
85.8	331	284	47	243.7
85.8	330	283	47	242.7
85.8	325	279	46	239.5
85.8	324	278	46	238.5
85.8	323	277	46	237.6
85.8	318	273	45	234.4
85.8	317	272	45	233.4
85.8	316	271	45	232.4
85.8	310	266	44	228.2

%E	M1	M2	DM	M*
85.8	309	265	44	227.3
85.8	303	260	43	223.1
85.8	302	259	43	222.1
85.8	296	254	42	218.0
85.8	295	253	42	217.0
85.8	289	248	41	212.8
85.8	288	247	41	211.8
85.8	282	242	40	207.7
85.8	281	241	40	206.7
85.8	275	236	39	202.5
85.8	274	235	39	201.6
85.8	268	230	38	197.4
85.8	267	229	38	196.4
85.8	261	224	37	192.2
85.8	260	223	37	191.3
85.8	254	218	36	187.1
85.8	253	217	36	186.1
85.8	247	212	35	182.0
85.8	246	211	35	181.0
85.8	240	206	34	176.8
85.8	239	205	34	175.8
85.8	233	200	33	171.7
85.8	232	199	33	170.7
85.8	226	194	32	166.5
85.8	225	193	32	165.6
85.8	219	188	31	161.4
85.8	218	187	31	160.4
85.8	212	182	30	156.2
85.8	211	181	30	155.3
85.8	204	175	29	150.1
85.8	197	169	28	145.0
85.8	190	163	27	139.8
85.8	183	157	26	134.7
85.8	176	151	25	129.6
85.8	169	145	24	124.4
85.8	162	139	23	119.3
85.8	155	133	22	114.1
85.8	148	127	21	109.0
85.8	141	121	20	103.8
85.8	134	115	19	98.7
85.8	127	109	18	93.6
85.8	120	103	17	88.4
85.8	113	97	16	83.3
85.8	106	91	15	78.1
85.7	498	427	71	366.1
85.7	497	426	71	365.1
85.7	496	425	71	364.2
85.7	495	424	71	363.2
85.7	491	421	70	361.0
85.7	490	420	70	360.0

%E	M1	M2	DM	M*
85.7	489	419	70	359.0
85.7	488	418	70	358.0
85.7	484	415	69	355.8
85.7	483	414	69	354.9
85.7	482	413	69	353.9
85.7	481	412	69	352.9
85.7	477	409	68	350.7
85.7	476	408	68	349.7
85.7	475	407	68	348.7
85.7	474	406	68	347.8
85.7	470	403	67	345.6
85.7	469	402	67	344.6
85.7	468	401	67	343.6
85.7	467	400	67	342.6
85.7	463	397	66	340.4
85.7	462	396	66	339.4
85.7	461	395	66	338.4
85.7	460	394	66	337.5
85.7	456	391	65	335.3
85.7	455	390	65	334.3
85.7	454	389	65	333.3
85.7	453	388	65	332.3
85.7	449	385	64	330.1
85.7	448	384	64	329.1
85.7	447	383	64	328.2
85.7	446	382	64	327.2
85.7	442	379	63	325.0
85.7	441	378	63	324.0
85.7	440	377	63	323.0
85.7	435	373	62	319.8
85.7	434	372	62	318.9
85.7	433	371	62	317.9
85.7	428	367	61	314.7
85.7	427	366	61	313.7
85.7	426	365	61	312.7
85.7	421	361	60	309.6
85.7	420	360	60	308.6
85.7	419	359	60	307.6
85.7	414	355	59	304.4
85.7	413	354	59	303.4
85.7	412	353	59	302.4
85.7	407	349	58	299.3
85.7	406	348	58	298.3
85.7	405	347	58	297.3
85.7	399	342	57	293.1
85.7	398	341	57	292.2
85.7	392	336	56	288.0
85.7	391	335	56	287.0
85.7	385	330	55	282.9
85.7	384	329	55	281.9

%E	M1	M2	DM	M*
85.7	378	324	54	277.7
85.7	377	323	54	276.7
85.7	371	318	53	272.6
85.7	370	317	53	271.6
85.7	364	312	52	267.4
85.7	363	311	52	266.4
85.7	357	306	51	262.3
85.7	356	305	51	261.3
85.7	350	300	50	257.1
85.7	349	299	50	256.2
85.7	343	294	49	252.0
85.7	342	293	49	251.0
85.7	336	288	48	246.9
85.7	335	287	48	245.9
85.7	329	282	47	241.7
85.7	328	281	47	240.7
85.7	322	276	46	236.6
85.7	321	275	46	235.6
85.7	315	270	45	231.4
85.7	314	269	45	230.4
85.7	308	264	44	226.3
85.7	307	263	44	225.3
85.7	301	258	43	221.1
85.7	300	257	43	220.2
85.7	294	252	42	216.0
85.7	293	251	42	215.0
85.7	287	246	41	210.9
85.7	286	245	41	209.9
85.7	280	240	40	205.7
85.7	279	239	40	204.7
85.7	273	234	39	200.6
85.7	272	233	39	199.6
85.7	266	228	38	195.4
85.7	265	227	38	194.4
85.7	259	222	37	190.3
85.7	258	221	37	189.3
85.7	252	216	36	185.1
85.7	251	215	36	184.2
85.7	245	210	35	180.0
85.7	244	209	35	179.0
85.7	238	204	34	174.9
85.7	237	203	34	173.9
85.7	231	198	33	169.7
85.7	230	197	33	168.7
85.7	224	192	32	164.6
85.7	223	191	32	163.6
85.7	217	186	31	159.4
85.7	210	180	30	154.3
85.7	203	174	29	149.1
85.7	196	168	28	144.0

%E	M1	M2	DM	M*
85.7	189	162	27	138.9
85.7	182	156	26	133.7
85.7	175	150	25	128.6
85.7	168	144	24	123.4
85.7	161	138	23	118.3
85.7	154	132	22	113.1
85.7	147	126	21	108.0
85.7	140	120	20	102.9
85.7	133	114	19	97.7
85.7	126	108	18	92.6
85.7	119	102	17	87.4
85.7	112	96	16	82.3
85.7	105	90	15	77.1
85.7	98	84	14	72.0
85.7	91	78	13	66.9
85.7	84	72	12	61.7
85.7	77	66	11	56.6
85.7	70	60	10	51.4
85.7	63	54	9	46.3
85.7	56	48	8	41.1
85.7	49	42	7	36.0
85.7	42	36	6	30.9
85.7	35	30	5	25.7
85.7	28	24	4	20.6
85.7	21	18	3	15.4
85.7	14	12	2	10.3
85.7	7	6	1	5.1
85.6	500	428	72	366.4
85.6	499	427	72	365.4
85.6	494	423	71	362.2
85.6	493	422	71	361.2
85.6	492	421	71	360.2
85.6	487	417	70	357.1
85.6	486	416	70	356.1
85.6	485	415	70	355.1
85.6	480	411	69	351.9
85.6	479	410	69	350.9
85.6	478	409	69	350.0
85.6	473	405	68	346.8
85.6	472	404	68	345.8
85.6	471	403	68	344.8
85.6	466	399	67	341.6
85.6	465	398	67	340.7
85.6	464	397	67	339.7
85.6	459	393	66	336.5
85.6	458	392	66	335.5
85.6	457	391	66	334.5
85.6	452	387	65	331.3
85.6	451	386	65	330.4
85.6	450	385	65	329.4

%E	M1	M2	DM	M*
85.6	445	381	64	326.2
85.6	444	380	64	325.2
85.6	443	379	64	324.2
85.6	439	376	63	322.0
85.6	438	375	63	321.1
85.6	437	374	63	320.1
85.6	436	373	63	319.1
85.6	432	370	62	316.9
85.6	431	369	62	315.9
85.6	430	368	62	314.9
85.6	425	364	61	311.8
85.6	424	363	61	310.8
85.6	423	362	61	309.8
85.6	418	358	60	306.6
85.6	417	357	60	305.6
85.6	416	356	60	304.7
85.6	411	352	59	301.5
85.6	410	351	59	300.5
85.6	409	350	59	299.5
85.6	404	346	58	296.3
85.6	403	345	58	295.3
85.6	402	344	58	294.4
85.6	397	340	57	291.2
85.6	396	339	57	290.2
85.6	395	338	57	289.2
85.6	390	334	56	286.0
85.6	389	333	56	285.1
85.6	388	332	56	284.1
85.6	383	328	55	280.9
85.6	382	327	55	279.9
85.6	381	326	55	278.9
85.6	376	322	54	275.8
85.6	375	321	54	274.8
85.6	374	320	54	273.8
85.6	369	316	53	270.6
85.6	368	315	53	269.6
85.6	367	314	53	268.7
85.6	362	310	52	265.5
85.6	361	309	52	264.5
85.6	360	308	52	263.5
85.6	355	304	51	260.3
85.6	354	303	51	259.3
85.6	353	302	51	258.4
85.6	348	298	50	255.2
85.6	347	297	50	254.2
85.6	341	292	49	250.0
85.6	340	291	49	249.1
85.6	334	286	48	244.9
85.6	333	285	48	243.9
85.6	327	280	47	239.8

%E	M1	M2	DM	M*
85.6	326	279	47	238.8
85.6	320	274	46	234.6
85.6	319	273	46	233.6
85.6	313	268	45	229.5
85.6	312	267	45	228.5
85.6	306	262	44	224.3
85.6	305	261	44	223.3
85.6	299	256	43	219.2
85.6	298	255	43	218.2
85.6	292	250	42	214.0
85.6	291	249	42	213.1
85.6	285	244	41	208.9
85.6	284	243	41	207.9
85.6	278	238	40	203.8
85.6	277	237	40	202.8
85.6	271	232	39	198.6
85.6	270	231	39	197.6
85.6	264	226	38	193.5
85.6	263	225	38	192.5
85.6	257	220	37	188.3
85.6	250	214	36	183.2
85.6	243	208	35	178.0
85.6	236	202	34	172.9
85.6	229	196	33	167.8
85.6	222	190	32	162.6
85.6	216	185	31	158.4
85.6	215	184	31	157.5
85.6	209	179	30	153.3
85.6	208	178	30	152.3
85.6	202	173	29	148.2
85.6	201	172	29	147.2
85.6	195	167	28	143.0
85.6	194	166	28	142.0
85.6	188	161	27	137.9
85.6	187	160	27	136.9
85.6	181	155	26	132.7
85.6	180	154	26	131.8
85.6	174	149	25	127.6
85.6	167	143	24	122.4
85.6	160	137	23	117.3
85.6	153	131	22	112.2
85.6	146	125	21	107.0
85.6	139	119	20	101.9
85.6	132	113	19	96.7
85.6	125	107	18	91.6
85.6	118	101	17	86.4
85.6	111	95	16	81.3
85.6	104	89	15	76.2
85.6	97	83	14	71.0
85.6	90	77	13	65.9

%E	M1	M2	DM	M*
85.5	498	426	72	364.4
85.5	497	425	72	363.4
85.5	496	424	72	362.5
85.5	495	423	72	361.5
85.5	491	420	71	359.3
85.5	490	419	71	358.3
85.5	489	418	71	357.3
85.5	488	417	71	356.3
85.5	484	414	70	354.1
85.5	483	413	70	353.1
85.5	482	412	70	352.2
85.5	477	408	69	349.0
85.5	476	407	69	348.0
85.5	475	406	69	347.0
85.5	470	402	68	343.8
85.5	469	401	68	342.9
85.5	468	400	68	341.9
85.5	463	396	67	338.7
85.5	462	395	67	337.7
85.5	461	394	67	336.7
85.5	456	390	66	333.6
85.5	455	389	66	332.6
85.5	454	388	66	331.6
85.5	449	384	65	328.4
85.5	448	383	65	327.4
85.5	447	382	65	326.5
85.5	442	378	64	323.3
85.5	441	377	64	322.3
85.5	440	376	64	321.3
85.5	435	372	63	318.1
85.5	434	371	63	317.1
85.5	433	370	63	316.2
85.5	429	367	62	314.0
85.5	428	366	62	313.0
85.5	427	365	62	312.0
85.5	422	361	61	308.8
85.5	421	360	61	307.8
85.5	420	359	61	306.9
85.5	415	355	60	303.7
85.5	414	354	60	302.7
85.5	413	353	60	301.7
85.5	408	349	59	298.5
85.5	407	348	59	297.6
85.5	406	347	59	296.6
85.5	401	343	58	293.4
85.5	400	342	58	292.4
85.5	399	341	58	291.4
85.5	394	337	57	288.2
85.5	393	336	57	287.3
85.5	392	335	57	286.3

%E	M1	M2	DM	M*
85.5	387	331	56	283.1
85.5	386	330	56	282.1
85.5	385	329	56	281.1
85.5	380	325	55	278.0
85.5	379	324	55	277.0
85.5	378	323	55	276.0
85.5	373	319	54	272.8
85.5	372	318	54	271.8
85.5	366	313	53	267.7
85.5	365	312	53	266.7
85.5	359	307	52	262.5
85.5	358	306	52	261.6
85.5	352	301	51	257.4
85.5	351	300	51	256.4
85.5	346	296	50	253.2
85.5	345	295	50	252.2
85.5	344	294	50	251.3
85.5	339	290	49	248.1
85.5	338	289	49	247.1
85.5	337	288	49	246.1
85.5	332	284	48	242.9
85.5	331	283	48	242.0
85.5	330	282	48	241.0
85.5	325	278	47	237.8
85.5	324	277	47	236.8
85.5	318	272	46	232.7
85.5	317	271	46	231.7
85.5	311	266	45	227.5
85.5	310	265	45	226.5
85.5	304	260	44	222.4
85.5	303	259	44	221.4
85.5	297	254	43	217.2
85.5	296	253	43	216.2
85.5	290	248	42	212.1
85.5	289	247	42	211.1
85.5	283	242	41	206.9
85.5	282	241	41	206.0
85.5	276	236	40	201.8
85.5	275	235	40	200.8
85.5	269	230	39	196.7
85.5	262	224	38	191.5
85.5	256	219	37	187.3
85.5	255	218	37	186.4
85.5	249	213	36	182.2
85.5	248	212	36	181.2
85.5	242	207	35	177.1
85.5	241	206	35	176.1
85.5	235	201	34	171.9
85.5	234	200	34	170.9
85.5	228	195	33	166.8

%E	M1	M2	DM	M*
85.5	227	194	33	165.8
85.5	221	189	32	161.6
85.5	220	188	32	160.7
85.5	214	183	31	156.5
85.5	207	177	30	151.3
85.5	200	171	29	146.2
85.5	193	165	28	141.1
85.5	186	159	27	135.9
85.5	179	153	26	130.8
85.5	173	148	25	126.6
85.5	172	147	25	125.6
85.5	166	142	24	121.5
85.5	165	141	24	120.5
85.5	159	136	23	116.3
85.5	152	130	22	111.2
85.5	145	124	21	106.0
85.5	138	118	20	100.9
85.5	131	112	19	95.8
85.5	124	106	18	90.6
85.5	117	100	17	85.5
85.5	110	94	16	80.3
85.5	83	71	12	60.7
85.5	76	65	11	55.6
85.5	69	59	10	50.4
85.5	62	53	9	45.3
85.5	55	47	8	40.2
85.4	500	427	73	364.7
85.4	499	426	73	363.7
85.4	494	422	72	360.5
85.4	493	421	72	359.5
85.4	492	420	72	358.5
85.4	487	416	71	355.4
85.4	486	415	71	354.4
85.4	485	414	71	353.4
85.4	481	411	70	351.2
85.4	480	410	70	350.2
85.4	479	409	70	349.2
85.4	478	408	70	348.3
85.4	474	405	69	346.0
85.4	473	404	69	345.1
85.4	472	403	69	344.1
85.4	471	402	69	343.1
85.4	467	399	68	340.9
85.4	466	398	68	339.9
85.4	465	397	68	338.9
85.4	460	393	67	335.8
85.4	459	392	67	334.8
85.4	458	391	67	333.8
85.4	453	387	66	330.6
85.4	452	386	66	329.6

%E	M1	M2	DM	M*
85.4	451	385	66	328.7
85.4	446	381	65	325.5
85.4	445	380	65	324.5
85.4	444	379	65	323.5
85.4	439	375	64	320.3
85.4	438	374	64	319.4
85.4	437	373	64	318.4
85.4	432	369	63	315.2
85.4	431	368	63	314.2
85.4	426	364	62	311.0
85.4	425	363	62	310.0
85.4	424	362	62	309.1
85.4	419	358	61	305.9
85.4	418	357	61	304.9
85.4	417	356	61	303.9
85.4	412	352	60	300.7
85.4	411	351	60	299.8
85.4	410	350	60	298.8
85.4	405	346	59	295.6
85.4	404	345	59	294.6
85.4	403	344	59	293.6
85.4	398	340	58	290.5
85.4	397	339	58	289.5
85.4	396	338	58	288.5
85.4	391	334	57	285.3
85.4	390	333	57	284.3
85.4	384	328	56	280.2
85.4	383	327	56	279.2
85.4	377	322	55	275.0
85.4	376	321	55	274.0
85.4	371	317	54	270.9
85.4	370	316	54	269.9
85.4	369	315	54	268.9
85.4	364	311	53	265.7
85.4	363	310	53	264.7
85.4	362	309	53	263.8
85.4	357	305	52	260.6
85.4	356	304	52	259.6
85.4	355	303	52	258.6
85.4	350	299	51	255.4
85.4	349	298	51	254.5
85.4	343	293	50	250.3
85.4	342	292	50	249.3
85.4	336	287	49	245.1
85.4	335	286	49	244.2
85.4	329	281	48	240.0
85.4	328	280	48	239.0
85.4	323	276	47	235.8
85.4	322	275	47	234.9
85.4	321	274	47	233.9
85.4	316	270	46	230.7
85.4	315	269	46	229.7
85.4	314	268	46	228.7
85.4	309	264	45	225.6
85.4	308	263	45	224.6
85.4	302	258	44	220.4
85.4	301	257	44	219.4
85.4	295	252	43	215.3
85.4	294	251	43	214.3
85.4	288	246	42	210.1
85.4	287	245	42	209.1
85.4	281	240	41	205.0
85.4	280	239	41	204.0
85.4	274	234	40	199.8
85.4	268	229	39	195.7
85.4	267	228	39	194.7
85.4	261	223	38	190.5
85.4	260	222	38	189.6
85.4	254	217	37	185.4
85.4	253	216	37	184.4
85.4	247	211	36	180.2
85.4	246	210	36	179.3
85.4	240	205	35	175.1
85.4	239	204	35	174.1
85.4	233	199	34	170.0
85.4	226	193	33	164.8
85.4	219	187	32	159.7
85.4	213	182	31	155.5
85.4	212	181	31	154.5
85.4	206	176	30	150.4
85.4	205	175	30	149.4
85.4	199	170	29	145.2
85.4	198	169	29	144.2
85.4	192	164	28	140.1
85.4	185	158	27	134.9
85.4	178	152	26	129.8
85.4	171	146	25	124.7
85.4	164	140	24	119.5
85.4	158	135	23	115.3
85.4	157	134	23	114.4
85.4	151	129	22	110.2
85.4	144	123	21	105.1
85.4	137	117	20	99.9
85.4	130	111	19	94.8
85.4	123	105	18	89.6
85.4	103	88	15	75.2
85.4	96	82	14	70.0
85.4	89	76	13	64.9
85.4	82	70	12	59.8
85.4	48	41	7	35.0
85.4	41	35	6	29.9
85.3	498	425	73	362.7
85.3	497	424	73	361.7
85.3	496	423	73	360.7
85.3	495	422	73	359.8
85.3	491	419	72	357.6
85.3	490	418	72	356.6
85.3	489	417	72	355.6
85.3	484	413	71	352.4
85.3	483	412	71	351.4
85.3	482	411	71	350.5
85.3	477	407	70	347.3
85.3	476	406	70	346.3
85.3	475	405	70	345.3
85.3	470	401	69	342.1
85.3	469	400	69	341.2
85.3	468	399	69	340.2
85.3	464	396	68	338.0
85.3	463	395	68	337.0
85.3	462	394	68	336.0
85.3	457	390	67	332.8
85.3	456	389	67	331.8
85.3	455	388	67	330.9
85.3	450	384	66	327.7
85.3	449	383	66	326.7
85.3	448	382	66	325.7
85.3	443	378	65	322.5
85.3	442	377	65	321.6
85.3	441	376	65	320.6
85.3	436	372	64	317.4
85.3	435	371	64	316.4
85.3	434	370	64	315.4
85.3	430	367	63	313.2
85.3	429	366	63	312.3
85.3	428	365	63	311.3
85.3	423	361	62	308.1
85.3	422	360	62	307.1
85.3	421	359	62	306.1
85.3	416	355	61	302.9
85.3	415	354	61	302.0
85.3	414	353	61	301.0
85.3	409	349	60	297.8
85.3	408	348	60	296.8
85.3	407	347	60	295.8
85.3	402	343	59	292.7
85.3	401	342	59	291.7
85.3	400	341	59	290.7
85.3	395	337	58	287.5
85.3	394	336	58	286.5
85.3	389	332	57	283.4
85.3	388	331	57	282.4
85.3	387	330	57	281.4
85.3	382	326	56	278.2
85.3	381	325	56	277.2
85.3	380	324	56	276.3
85.3	375	320	55	273.1
85.3	374	319	55	272.1
85.3	373	318	55	271.1
85.3	368	314	54	267.9
85.3	367	313	54	266.9
85.3	361	308	53	262.8
85.3	360	307	53	261.8
85.3	354	302	52	257.6
85.3	353	301	52	256.7
85.3	348	297	51	253.5
85.3	347	296	51	252.5
85.3	346	295	51	251.5
85.3	341	291	50	248.3
85.3	340	290	50	247.4
85.3	339	289	50	246.4
85.3	334	285	49	243.2
85.3	333	284	49	242.2
85.3	327	279	48	238.0
85.3	326	278	48	237.1
85.3	320	273	47	232.9
85.3	319	272	47	231.9
85.3	313	267	46	227.8
85.3	312	266	46	226.8
85.3	307	262	45	223.6
85.3	306	261	45	222.6
85.3	300	256	44	218.5
85.3	299	255	44	217.5
85.3	293	250	43	213.3
85.3	292	249	43	212.3
85.3	286	244	42	208.2
85.3	285	243	42	207.2
85.3	279	238	41	203.0
85.3	278	237	41	202.0
85.3	273	233	40	198.9
85.3	272	232	40	197.9
85.3	266	227	39	193.7
85.3	265	226	39	192.7
85.3	259	221	38	188.6
85.3	258	220	38	187.6
85.3	252	215	37	183.4
85.3	251	214	37	182.5
85.3	245	209	36	178.3
85.3	238	203	35	173.1
85.3	232	198	34	169.0
85.3	231	197	34	168.0
85.3	225	192	33	163.8
85.3	224	191	33	162.9
85.3	218	186	32	158.7
85.3	217	185	32	157.7
85.3	211	180	31	153.6
85.3	204	174	30	148.4
85.3	197	168	29	143.3
85.3	191	163	28	139.1
85.3	190	162	28	138.1
85.3	184	157	27	134.0
85.3	177	151	26	128.8
85.3	170	145	25	123.7
85.3	163	139	24	118.5
85.3	156	133	23	113.4
85.3	150	128	22	109.2
85.3	143	122	21	104.1
85.3	136	116	20	98.9
85.3	129	110	19	93.8
85.3	116	99	17	84.5
85.3	109	93	16	79.3
85.3	102	87	15	74.2
85.3	95	81	14	69.1
85.3	75	64	11	54.6
85.3	68	58	10	49.5
85.3	34	29	5	24.7
85.2	500	426	74	363.0
85.2	499	425	74	362.0
85.2	494	421	73	358.8
85.2	493	420	73	357.8
85.2	492	419	73	356.8
85.2	488	416	72	354.6
85.2	487	415	72	353.6
85.2	486	414	72	352.7
85.2	485	413	72	351.7
85.2	481	410	71	349.5
85.2	480	409	71	348.5
85.2	479	408	71	347.5
85.2	474	404	70	344.3
85.2	473	403	70	343.4
85.2	472	402	70	342.4
85.2	467	398	69	339.2
85.2	466	397	69	338.2
85.2	465	396	69	337.2
85.2	461	393	68	335.0
85.2	460	392	68	334.1
85.2	459	391	68	333.1
85.2	458	390	68	332.1
85.2	454	387	67	329.9
85.2	453	386	67	328.9
85.2	452	385	67	327.9

%E	M1	M2	DM	M*
85.2	447	381	66	324.7
85.2	446	380	66	323.8
85.2	445	379	66	322.8
85.2	440	375	65	319.6
85.2	439	374	65	318.6
85.2	438	373	65	317.6
85.2	433	369	64	314.5
85.2	432	368	64	313.5
85.2	431	367	64	312.5
85.2	427	364	63	310.3
85.2	426	363	63	309.3
85.2	425	362	63	308.3
85.2	420	358	62	305.2
85.2	419	357	62	304.2
85.2	418	356	62	303.2
85.2	413	352	61	300.0
85.2	412	351	61	299.0
85.2	411	350	61	298.1
85.2	406	346	60	294.9
85.2	405	345	60	293.9
85.2	399	340	59	289.7
85.2	398	339	59	288.7
85.2	393	335	58	285.6
85.2	392	334	58	284.6
85.2	391	333	58	283.6
85.2	386	329	57	280.4
85.2	385	328	57	279.4
85.2	384	327	57	278.5
85.2	379	323	56	275.3
85.2	378	322	56	274.3
85.2	372	317	55	270.1
85.2	371	316	55	269.2
85.2	366	312	54	266.0
85.2	365	311	54	265.0
85.2	364	310	54	264.0
85.2	359	306	53	260.8
85.2	358	305	53	259.8
85.2	357	304	53	258.9
85.2	352	300	52	255.7
85.2	351	299	52	254.7
85.2	345	294	51	250.5
85.2	344	293	51	249.6
85.2	338	288	50	245.4
85.2	337	287	50	244.4
85.2	332	283	49	241.2
85.2	331	282	49	240.3
85.2	330	281	49	239.3
85.2	325	277	48	236.1
85.2	324	276	48	235.1
85.2	318	271	47	230.9

%E	M1	M2	DM	M*
85.2	317	270	47	230.0
85.2	311	265	46	225.8
85.2	310	264	46	224.8
85.2	305	260	45	221.6
85.2	304	259	45	220.7
85.2	298	254	44	216.5
85.2	297	253	44	215.5
85.2	291	248	43	211.4
85.2	290	247	43	210.4
85.2	284	242	42	206.2
85.2	283	241	42	205.2
85.2	277	236	41	201.1
85.2	271	231	40	196.9
85.2	270	230	40	195.9
85.2	264	225	39	191.8
85.2	263	224	39	190.8
85.2	257	219	38	186.6
85.2	256	218	38	185.6
85.2	250	213	37	181.5
85.2	244	208	36	177.3
85.2	243	207	36	176.3
85.2	237	202	35	172.2
85.2	236	201	35	171.2
85.2	230	196	34	167.0
85.2	229	195	34	166.0
85.2	223	190	33	161.9
85.2	216	184	32	156.7
85.2	210	179	31	152.6
85.2	209	178	31	151.6
85.2	203	173	30	147.4
85.2	196	167	29	142.3
85.2	189	161	28	137.1
85.2	183	156	27	133.0
85.2	182	155	27	132.0
85.2	176	150	26	127.8
85.2	169	144	25	122.7
85.2	162	138	24	117.6
85.2	155	132	23	112.4
85.2	149	127	22	108.2
85.2	142	121	21	103.1
85.2	135	115	20	98.0
85.2	128	109	19	92.8
85.2	122	104	18	88.7
85.2	115	98	17	83.5
85.2	108	92	16	78.4
85.2	88	75	13	63.9
85.2	81	69	12	58.8
85.2	61	52	9	44.3
85.2	54	46	8	39.2
85.2	27	23	4	19.6

%E	M1	M2	DM	M*
85.1	498	424	74	361.0
85.1	497	423	74	360.0
85.1	496	422	74	359.0
85.1	495	421	74	358.1
85.1	491	418	73	355.9
85.1	490	417	73	354.9
85.1	489	416	73	353.9
85.1	484	412	72	350.7
85.1	483	411	72	349.7
85.1	482	410	72	348.8
85.1	478	407	71	346.5
85.1	477	406	71	345.6
85.1	476	405	71	344.6
85.1	475	404	71	343.6
85.1	471	401	70	341.4
85.1	470	400	70	340.4
85.1	469	399	70	339.4
85.1	464	395	69	336.3
85.1	463	394	69	335.3
85.1	462	393	69	334.3
85.1	457	389	68	331.1
85.1	456	388	68	330.1
85.1	455	387	68	329.2
85.1	451	384	67	327.0
85.1	450	383	67	326.0
85.1	449	382	67	325.0
85.1	444	378	66	321.8
85.1	443	377	66	320.8
85.1	442	376	66	319.9
85.1	437	372	65	316.7
85.1	436	371	65	315.7
85.1	435	370	65	314.7
85.1	430	366	64	311.5
85.1	429	365	64	310.5
85.1	424	361	63	307.4
85.1	423	360	63	306.4
85.1	422	359	63	305.4
85.1	417	355	62	302.2
85.1	416	354	62	301.2
85.1	415	353	62	300.3
85.1	410	349	61	297.1
85.1	409	348	61	296.1
85.1	404	344	60	292.9
85.1	403	343	60	291.9
85.1	402	342	60	291.0
85.1	397	338	59	287.8
85.1	396	337	59	286.8
85.1	395	336	59	285.8
85.1	390	332	58	282.6
85.1	389	331	58	281.6

%E	M1	M2	DM	M*
85.1	388	330	58	280.7
85.1	383	326	57	277.5
85.1	382	325	57	276.5
85.1	377	321	56	273.3
85.1	376	320	56	272.3
85.1	375	319	56	271.4
85.1	370	315	55	268.2
85.1	369	314	55	267.2
85.1	368	313	55	266.2
85.1	363	309	54	263.0
85.1	362	308	54	262.1
85.1	356	303	53	257.9
85.1	355	302	53	256.9
85.1	350	298	52	253.7
85.1	349	297	52	252.7
85.1	348	296	52	251.8
85.1	343	292	51	248.6
85.1	342	291	51	247.6
85.1	336	286	50	243.4
85.1	335	285	50	242.5
85.1	329	280	49	238.3
85.1	328	279	49	237.3
85.1	323	275	48	234.1
85.1	322	274	48	233.2
85.1	316	269	47	229.0
85.1	315	268	47	228.0
85.1	309	263	46	223.8
85.1	308	262	46	222.9
85.1	303	258	45	219.7
85.1	302	257	45	218.7
85.1	301	256	45	217.7
85.1	296	252	44	214.5
85.1	295	251	44	213.6
85.1	289	246	43	209.4
85.1	288	245	43	208.4
85.1	282	240	42	204.3
85.1	281	239	42	203.3
85.1	276	235	41	200.1
85.1	275	234	41	199.1
85.1	269	229	40	194.9
85.1	268	228	40	194.0
85.1	262	223	39	189.8
85.1	261	222	39	188.8
85.1	255	217	38	184.7
85.1	249	212	37	180.5
85.1	248	211	37	179.5
85.1	242	206	36	175.4
85.1	241	205	36	174.4
85.1	235	200	35	170.2
85.1	228	194	34	165.1

%E	M1	M2	DM	M*
85.1	222	189	33	160.9
85.1	221	188	33	159.9
85.1	215	183	32	155.8
85.1	208	177	31	150.6
85.1	202	172	30	146.5
85.1	201	171	30	145.5
85.1	195	166	29	141.3
85.1	194	165	29	140.3
85.1	188	160	28	136.2
85.1	181	154	27	131.0
85.1	175	149	26	126.9
85.1	174	148	26	125.9
85.1	168	143	25	121.7
85.1	161	137	24	116.6
85.1	154	131	23	111.4
85.1	148	126	22	107.3
85.1	141	120	21	102.1
85.1	134	114	20	97.0
85.1	121	103	18	87.7
85.1	114	97	17	82.5
85.1	101	86	15	73.2
85.1	94	80	14	68.1
85.1	87	74	13	62.9
85.1	74	63	11	53.6
85.1	67	57	10	48.5
85.1	47	40	7	34.0
85.0	500	425	75	361.3
85.0	499	424	75	360.3
85.0	494	420	74	357.1
85.0	493	419	74	356.1
85.0	492	418	74	355.1
85.0	488	415	73	352.9
85.0	487	414	73	351.9
85.0	486	413	73	351.0
85.0	481	409	72	347.8
85.0	480	408	72	346.8
85.0	479	407	72	345.8
85.0	474	403	71	342.6
85.0	473	402	71	341.7
85.0	472	401	71	340.7
85.0	468	398	70	338.5
85.0	467	397	70	337.5
85.0	466	396	70	336.5
85.0	461	392	69	333.3
85.0	460	391	69	332.3
85.0	459	390	69	331.4
85.0	454	386	68	328.2
85.0	453	385	68	327.2
85.0	452	384	68	326.2
85.0	448	381	67	324.0

%E	M1	M2	DM	M*
85.0	447	380	67	323.0
85.0	446	379	67	322.1
85.0	441	375	66	318.9
85.0	440	374	66	317.9
85.0	439	373	66	316.9
85.0	434	369	65	313.7
85.0	433	368	65	312.8
85.0	432	367	65	311.8
85.0	428	364	64	309.6
85.0	427	363	64	308.6
85.0	426	362	64	307.6
85.0	421	358	63	304.4
85.0	420	357	63	303.4
85.0	419	356	63	302.5
85.0	414	352	62	299.3
85.0	413	351	62	298.3
85.0	412	350	62	297.3
85.0	408	347	61	295.1
85.0	407	346	61	294.1
85.0	406	345	61	293.2
85.0	401	341	60	290.0
85.0	400	340	60	289.0
85.0	399	339	60	288.0
85.0	394	335	59	284.8
85.0	393	334	59	283.9
85.0	387	329	58	279.7
85.0	386	328	58	278.7
85.0	381	324	57	275.5
85.0	380	323	57	274.5
85.0	379	322	57	273.6
85.0	374	318	56	270.4
85.0	373	317	56	269.4
85.0	367	312	55	265.2
85.0	366	311	55	264.3
85.0	361	307	54	261.1
85.0	360	306	54	260.1
85.0	359	305	54	259.1
85.0	354	301	53	255.9
85.0	353	300	53	255.0
85.0	347	295	52	250.8
85.0	346	294	52	249.8
85.0	341	290	51	246.6
85.0	340	289	51	245.6
85.0	339	288	51	244.7
85.0	334	284	50	241.5
85.0	333	283	50	240.5
85.0	327	278	49	236.3
85.0	326	277	49	235.4
85.0	321	273	48	232.2
85.0	320	272	48	231.2

%E	M1	M2	DM	M*
85.0	319	271	48	230.2
85.0	314	267	47	227.0
85.0	313	266	47	226.1
85.0	307	261	46	221.9
85.0	306	260	46	220.9
85.0	300	255	45	216.8
85.0	299	254	45	215.8
85.0	294	250	44	212.6
85.0	293	249	44	211.6
85.0	287	244	43	207.4
85.0	286	243	43	206.5
85.0	280	238	42	202.3
85.0	274	233	41	198.1
85.0	273	232	41	197.2
85.0	267	227	40	193.0
85.0	266	226	40	192.0
85.0	260	221	39	187.8
85.0	254	216	38	183.7
85.0	253	215	38	182.7
85.0	247	210	37	178.5
85.0	246	209	37	177.6
85.0	240	204	36	173.4
85.0	234	199	35	169.2
85.0	233	198	35	168.3
85.0	227	193	34	164.1
85.0	226	192	34	163.1
85.0	220	187	33	158.9
85.0	214	182	32	154.8
85.0	213	181	32	153.8
85.0	207	176	31	149.6
85.0	206	175	31	148.7
85.0	200	170	30	144.5
85.0	193	164	29	139.4
85.0	187	159	28	135.2
85.0	180	153	27	130.0
85.0	173	147	26	124.9
85.0	167	142	25	120.7
85.0	160	136	24	115.6
85.0	153	130	23	110.5
85.0	147	125	22	106.3
85.0	140	119	21	101.1
85.0	133	113	20	96.0
85.0	127	108	19	91.8
85.0	120	102	18	86.7
85.0	113	96	17	81.6
85.0	107	91	16	77.4
85.0	100	85	15	72.3
85.0	80	68	12	57.8
85.0	60	51	9	43.3
85.0	40	34	6	28.9

%E	M1	M2	DM	M*
85.0	20	17	3	14.4
84.9	498	423	75	359.3
84.9	497	422	75	358.3
84.9	496	421	75	357.3
84.9	491	417	74	354.2
84.9	490	416	74	353.2
84.9	489	415	74	352.2
84.9	485	412	73	350.0
84.9	484	411	73	349.0
84.9	483	410	73	348.0
84.9	482	409	73	347.1
84.9	478	406	72	344.8
84.9	477	405	72	343.9
84.9	476	404	72	342.9
84.9	471	400	71	339.7
84.9	470	399	71	338.7
84.9	469	398	71	337.7
84.9	465	395	70	335.5
84.9	464	394	70	334.6
84.9	463	393	70	333.6
84.9	458	389	69	330.4
84.9	457	388	69	329.4
84.9	456	387	69	328.4
84.9	451	383	68	325.3
84.9	450	382	68	324.3
84.9	449	381	68	323.3
84.9	445	378	67	321.1
84.9	444	377	67	320.1
84.9	443	376	67	319.1
84.9	438	372	66	315.9
84.9	437	371	66	315.0
84.9	436	370	66	314.0
84.9	431	366	65	310.8
84.9	430	365	65	309.8
84.9	425	361	64	306.6
84.9	424	360	64	305.7
84.9	423	359	64	304.7
84.9	418	355	63	301.5
84.9	417	354	63	300.5
84.9	416	353	63	299.5
84.9	411	349	62	296.4
84.9	410	348	62	295.4
84.9	405	344	61	292.2
84.9	404	343	61	291.2
84.9	403	342	61	290.2
84.9	398	338	60	287.0
84.9	397	337	60	286.1
84.9	392	333	59	282.9
84.9	391	332	59	281.9
84.9	390	331	59	280.9

%E	M1	M2	DM	M*
84.9	385	327	58	277.7
84.9	384	326	58	276.8
84.9	383	325	58	275.8
84.9	378	321	57	272.6
84.9	377	320	57	271.6
84.9	372	316	56	268.4
84.9	371	315	56	267.5
84.9	370	314	56	266.5
84.9	365	310	55	263.3
84.9	364	309	55	262.3
84.9	358	304	54	258.1
84.9	357	303	54	257.2
84.9	352	299	53	254.0
84.9	351	298	53	253.0
84.9	350	297	53	252.0
84.9	345	293	52	248.8
84.9	344	292	52	247.9
84.9	338	287	51	243.7
84.9	337	286	51	242.7
84.9	332	282	50	239.5
84.9	331	281	50	238.6
84.9	325	276	49	234.4
84.9	324	275	49	233.4
84.9	318	270	48	229.2
84.9	317	269	48	228.3
84.9	312	265	47	225.1
84.9	311	264	47	224.1
84.9	305	259	46	219.9
84.9	304	258	46	219.0
84.9	298	253	45	214.8
84.9	292	248	44	210.6
84.9	291	247	44	209.7
84.9	285	242	43	205.5
84.9	284	241	43	204.5
84.9	279	237	42	201.3
84.9	278	236	42	200.3
84.9	272	231	41	196.2
84.9	271	230	41	195.2
84.9	265	225	40	191.0
84.9	259	220	39	186.9
84.9	258	219	39	185.9
84.9	252	214	38	181.7
84.9	251	213	38	180.8
84.9	245	208	37	176.6
84.9	239	203	36	172.4
84.9	238	202	36	171.4
84.9	232	197	35	167.3
84.9	225	191	34	162.1
84.9	219	186	33	158.0
84.9	218	185	33	157.0

%E	M1	M2	DM	M*
84.9	212	180	32	152.8
84.9	205	174	31	147.7
84.9	199	169	30	143.5
84.9	192	163	29	138.4
84.9	186	158	28	134.2
84.9	185	157	28	133.2
84.9	179	152	27	129.1
84.9	172	146	26	123.9
84.9	166	141	25	119.8
84.9	159	135	24	114.6
84.9	152	129	23	109.5
84.9	146	124	22	105.3
84.9	139	118	21	100.2
84.9	126	107	19	90.9
84.9	119	101	18	85.7
84.9	106	90	16	76.4
84.9	93	79	14	67.1
84.9	86	73	13	62.0
84.9	73	62	11	52.7
84.9	53	45	8	38.2
84.8	500	424	76	359.6
84.8	499	423	76	358.6
84.8	495	420	75	356.4
84.8	494	419	75	355.4
84.8	493	418	75	354.4
84.8	492	417	75	353.4
84.8	488	414	74	351.2
84.8	487	413	74	350.2
84.8	486	412	74	349.3
84.8	481	408	73	346.1
84.8	480	407	73	345.1
84.8	479	406	73	344.1
84.8	475	403	72	341.9
84.8	474	402	72	340.9
84.8	473	401	72	340.0
84.8	468	397	71	336.8
84.8	467	396	71	335.8
84.8	466	395	71	334.8
84.8	462	392	70	332.6
84.8	461	391	70	331.6
84.8	460	390	70	330.7
84.8	455	386	69	327.5
84.8	454	385	69	326.5
84.8	453	384	69	325.5
84.8	448	380	68	322.3
84.8	447	379	68	321.3
84.8	446	378	68	320.4
84.8	442	375	67	318.2
84.8	441	374	67	317.2
84.8	440	373	67	316.2

%E	M1	M2	DM	M*
84.8	435	369	66	313.0
84.8	434	368	66	312.0
84.8	433	367	66	311.1
84.8	429	364	65	308.8
84.8	428	363	65	307.9
84.8	427	362	65	306.9
84.8	422	358	64	303.7
84.8	421	357	64	302.7
84.8	420	356	64	301.8
84.8	415	352	63	298.6
84.8	414	351	63	297.6
84.8	409	347	62	294.4
84.8	408	346	62	293.4
84.8	407	345	62	292.4
84.8	402	341	61	289.3
84.8	401	340	61	288.3
84.8	400	339	61	287.3
84.8	396	336	60	285.1
84.8	395	335	60	284.1
84.8	394	334	60	283.1
84.8	389	330	59	279.9
84.8	388	329	59	279.0
84.8	387	328	59	278.0
84.8	382	324	58	274.8
84.8	381	323	58	273.8
84.8	376	319	57	270.6
84.8	375	318	57	269.7
84.8	374	317	57	268.7
84.8	369	313	56	265.5
84.8	368	312	56	264.5
84.8	363	308	55	261.3
84.8	362	307	55	260.4
84.8	361	306	55	259.4
84.8	356	302	54	256.2
84.8	355	301	54	255.2
84.8	349	296	53	251.0
84.8	348	295	53	250.1
84.8	343	291	52	246.9
84.8	342	290	52	245.9
84.8	341	289	52	244.9
84.8	336	285	51	241.7
84.8	335	284	51	240.8
84.8	330	280	50	237.6
84.8	329	279	50	236.6
84.8	328	278	50	235.6
84.8	323	274	49	232.4
84.8	322	273	49	231.5
84.8	316	268	48	227.3
84.8	315	267	48	226.3
84.8	310	263	47	223.1

%E	M1	M2	DM	M*
84.8	309	262	47	222.1
84.8	303	257	46	218.0
84.8	302	256	46	217.0
84.8	297	252	45	213.8
84.8	296	251	45	212.8
84.8	290	246	44	208.7
84.8	289	245	44	207.7
84.8	283	240	43	203.5
84.8	282	239	43	202.6
84.8	277	235	42	199.4
84.8	276	234	42	198.4
84.8	270	229	41	194.2
84.8	269	228	41	193.2
84.8	264	224	40	190.1
84.8	263	223	40	189.1
84.8	257	218	39	184.9
84.8	256	217	39	183.9
84.8	250	212	38	179.8
84.8	244	207	37	175.6
84.8	243	206	37	174.6
84.8	237	201	36	170.5
84.8	231	196	35	166.3
84.8	230	195	35	165.3
84.8	224	190	34	161.2
84.8	223	189	34	160.2
84.8	217	184	33	156.0
84.8	211	179	32	151.9
84.8	210	178	32	150.9
84.8	204	173	31	146.7
84.8	198	168	30	142.5
84.8	197	167	30	141.6
84.8	191	162	29	137.4
84.8	184	156	28	132.3
84.8	178	151	27	128.1
84.8	171	145	26	123.0
84.8	165	140	25	118.8
84.8	164	139	25	117.8
84.8	158	134	24	113.6
84.8	151	128	23	108.5
84.8	145	123	22	104.3
84.8	138	117	21	99.2
84.8	132	112	20	95.0
84.8	125	106	19	89.9
84.8	112	95	17	80.6
84.8	105	89	16	75.4
84.8	99	84	15	71.3
84.8	92	78	14	66.1
84.8	79	67	12	56.8
84.8	66	56	10	47.5
84.8	46	39	7	33.1

%E	M1	M2	DM	M*
84.8	33	28	5	23.8
84.7	498	422	76	357.6
84.7	497	421	76	356.6
84.7	496	420	76	355.6
84.7	491	416	75	352.5
84.7	490	415	75	351.5
84.7	489	414	75	350.5
84.7	485	411	74	348.3
84.7	484	410	74	347.3
84.7	483	409	74	346.3
84.7	478	405	73	343.1
84.7	477	404	73	342.2
84.7	476	403	73	341.2
84.7	472	400	72	339.0
84.7	471	399	72	338.0
84.7	470	398	72	337.0
84.7	465	394	71	333.8
84.7	464	393	71	332.9
84.7	463	392	71	331.9
84.7	459	389	70	329.7
84.7	458	388	70	328.7
84.7	457	387	70	327.7
84.7	452	383	69	324.5
84.7	451	382	69	323.6
84.7	450	381	69	322.6
84.7	445	377	68	319.4
84.7	444	376	68	318.4
84.7	443	375	68	317.4
84.7	439	372	67	315.2
84.7	438	371	67	314.2
84.7	437	370	67	313.3
84.7	432	366	66	310.1
84.7	431	365	66	309.1
84.7	430	364	66	308.1
84.7	426	361	65	305.9
84.7	425	360	65	304.9
84.7	424	359	65	304.0
84.7	419	355	64	300.8
84.7	418	354	64	299.8
84.7	417	353	64	298.8
84.7	413	350	63	296.6
84.7	412	349	63	295.6
84.7	411	348	63	294.7
84.7	406	344	62	291.5
84.7	405	343	62	290.5
84.7	404	342	62	289.5
84.7	399	338	61	286.3
84.7	398	337	61	285.3
84.7	393	333	60	282.2
84.7	392	332	60	281.2

%E	M1	M2	DM	M*
84.7	391	331	60	280.2
84.7	386	327	59	277.0
84.7	385	326	59	276.0
84.7	380	322	58	272.9
84.7	379	321	58	271.9
84.7	378	320	58	270.9
84.7	373	316	57	267.7
84.7	372	315	57	266.7
84.7	367	311	56	263.5
84.7	366	310	56	262.6
84.7	365	309	56	261.6
84.7	360	305	55	258.4
84.7	359	304	55	257.4
84.7	354	300	54	254.2
84.7	353	299	54	253.3
84.7	352	298	54	252.3
84.7	347	294	53	249.1
84.7	346	293	53	248.1
84.7	340	288	52	244.0
84.7	339	287	52	243.0
84.7	334	283	51	239.8
84.7	333	282	51	238.8
84.7	327	277	50	234.6
84.7	326	276	50	233.7
84.7	321	272	49	230.5
84.7	320	271	49	229.5
84.7	314	266	48	225.3
84.7	313	265	48	224.4
84.7	308	261	47	221.2
84.7	307	260	47	220.2
84.7	301	255	46	216.0
84.7	300	254	46	215.1
84.7	295	250	45	211.9
84.7	294	249	45	210.9
84.7	288	244	44	206.7
84.7	287	243	44	205.7
84.7	281	238	43	201.6
84.7	275	233	42	197.4
84.7	274	232	42	196.4
84.7	268	227	41	192.3
84.7	262	222	40	188.1
84.7	261	221	40	187.1
84.7	255	216	39	183.0
84.7	249	211	38	178.8
84.7	248	210	38	177.8
84.7	242	205	37	173.7
84.7	236	200	36	169.5
84.7	235	199	36	168.5
84.7	229	194	35	164.3
84.7	222	188	34	159.2

%E	M1	M2	DM	M*
84.7	216	183	33	155.0
84.7	215	182	33	154.1
84.7	209	177	32	149.9
84.7	203	172	31	145.7
84.7	202	171	31	144.8
84.7	196	166	30	140.6
84.7	190	161	29	136.4
84.7	189	160	29	135.4
84.7	183	155	28	131.3
84.7	177	150	27	127.1
84.7	176	149	27	126.1
84.7	170	144	26	122.0
84.7	163	138	25	116.8
84.7	157	133	24	112.7
84.7	150	127	23	107.5
84.7	144	122	22	103.4
84.7	137	116	21	98.2
84.7	131	111	20	94.1
84.7	124	105	19	88.9
84.7	118	100	18	84.7
84.7	111	94	17	79.6
84.7	98	83	15	70.3
84.7	85	72	13	61.0
84.7	72	61	11	51.7
84.7	59	50	9	42.4
84.6	500	423	77	357.9
84.6	499	422	77	356.9
84.6	495	419	76	354.7
84.6	494	418	76	353.7
84.6	493	417	76	352.7
84.6	492	416	76	351.7
84.6	488	413	75	349.5
84.6	487	412	75	348.6
84.6	486	411	75	347.6
84.6	482	408	74	345.4
84.6	481	407	74	344.4
84.6	480	406	74	343.4
84.6	479	405	74	342.4
84.6	475	402	73	340.2
84.6	474	401	73	339.2
84.6	473	400	73	338.3
84.6	469	397	72	336.1
84.6	468	396	72	335.1
84.6	467	395	72	334.1
84.6	462	391	71	330.9
84.6	461	390	71	329.9
84.6	460	389	71	329.0
84.6	456	386	70	326.7
84.6	455	385	70	325.8
84.6	454	384	70	324.8

%E	M1	M2	DM	M*
84.6	449	380	69	321.6
84.6	448	379	69	320.6
84.6	447	378	69	319.7
84.6	442	374	68	316.5
84.6	441	373	68	315.5
84.6	436	369	67	312.3
84.6	435	368	67	311.3
84.6	434	367	67	310.3
84.6	429	363	66	307.2
84.6	428	362	66	306.2
84.6	423	358	65	303.0
84.6	422	357	65	302.0
84.6	421	356	65	301.0
84.6	416	352	64	297.8
84.6	415	351	64	296.9
84.6	410	347	63	293.7
84.6	409	346	63	292.7
84.6	408	345	63	291.7
84.6	403	341	62	288.5
84.6	402	340	62	287.6
84.6	397	336	61	284.4
84.6	396	335	61	283.4
84.6	395	334	61	282.4
84.6	390	330	60	279.2
84.6	389	329	60	278.3
84.6	384	325	59	275.1
84.6	383	324	59	274.1
84.6	382	323	59	273.1
84.6	377	319	58	269.9
84.6	376	318	58	268.9
84.6	371	314	57	265.8
84.6	370	313	57	264.8
84.6	369	312	57	263.8
84.6	364	308	56	260.6
84.6	363	307	56	259.6
84.6	358	303	55	256.4
84.6	357	302	55	255.5
84.6	356	301	55	254.5
84.6	351	297	54	251.3
84.6	350	296	54	250.3
84.6	345	292	53	247.1
84.6	344	291	53	246.2
84.6	338	286	52	242.0
84.6	337	285	52	241.0
84.6	332	281	51	237.8
84.6	331	280	51	236.9
84.6	325	275	50	232.7
84.6	324	274	50	231.7
84.6	319	270	49	228.5
84.6	318	269	49	227.6

%E	M1	M2	DM	M*
84.6	312	264	48	223.4
84.6	311	263	48	222.4
84.6	306	259	47	219.2
84.6	305	258	47	218.2
84.6	299	253	46	214.1
84.6	298	252	46	213.1
84.6	293	248	45	209.9
84.6	292	247	45	208.9
84.6	286	242	44	204.8
84.6	285	241	44	203.8
84.6	280	237	43	200.6
84.6	279	236	43	199.6
84.6	273	231	42	195.5
84.6	272	230	42	194.5
84.6	267	226	41	191.3
84.6	266	225	41	190.3
84.6	260	220	40	186.2
84.6	259	219	40	185.2
84.6	254	215	39	182.0
84.6	253	214	39	181.0
84.6	247	209	38	176.8
84.6	246	208	38	175.9
84.6	241	204	37	172.7
84.6	240	203	37	171.7
84.6	234	198	36	167.5
84.6	228	193	35	163.4
84.6	227	192	35	162.4
84.6	221	187	34	158.2
84.6	214	181	33	153.1
84.6	208	176	32	148.9
84.6	201	170	31	143.8
84.6	195	165	30	139.6
84.6	188	159	29	134.5
84.6	182	154	28	130.3
84.6	175	148	27	125.2
84.6	169	143	26	121.0
84.6	162	137	25	115.9
84.6	156	132	24	111.7
84.6	149	126	23	106.6
84.6	143	121	22	102.4
84.6	136	115	21	97.2
84.6	130	110	20	93.1
84.6	123	104	19	87.9
84.6	117	99	18	83.8
84.6	104	88	16	74.5
84.6	91	77	14	65.2
84.6	78	66	12	55.8
84.6	65	55	10	46.5
84.6	52	44	8	37.2
84.6	39	33	6	27.9

%E	M1	M2	DM	M*
84.6	26	22	4	18.6
84.6	13	11	2	9.3
84.5	498	421	77	355.9
84.5	497	420	77	354.9
84.5	496	419	77	354.0
84.5	491	415	76	350.8
84.5	490	414	76	349.8
84.5	489	413	76	348.8
84.5	485	410	75	346.6
84.5	484	409	75	345.6
84.5	483	408	75	344.6
84.5	478	404	74	341.5
84.5	477	403	74	340.5
84.5	476	402	74	339.5
84.5	472	399	73	337.3
84.5	471	398	73	336.3
84.5	470	397	73	335.3
84.5	466	394	72	333.1
84.5	465	393	72	332.1
84.5	464	392	72	331.2
84.5	459	388	71	328.0
84.5	458	387	71	327.0
84.5	457	386	71	326.0
84.5	453	383	70	323.8
84.5	452	382	70	322.8
84.5	451	381	70	321.9
84.5	446	377	69	318.7
84.5	445	376	69	317.7
84.5	444	375	69	316.7
84.5	440	372	68	314.5
84.5	439	371	68	313.5
84.5	438	370	68	312.6
84.5	433	366	67	309.4
84.5	432	365	67	308.4
84.5	431	364	67	307.4
84.5	427	361	66	305.2
84.5	426	360	66	304.2
84.5	425	359	66	303.2
84.5	420	355	65	300.1
84.5	419	354	65	299.1
84.5	418	353	65	298.1
84.5	414	350	64	295.9
84.5	413	349	64	294.9
84.5	412	348	64	293.9
84.5	407	344	63	290.8
84.5	406	343	63	289.8
84.5	401	339	62	286.6
84.5	400	338	62	285.6
84.5	399	337	62	284.6
84.5	394	333	61	281.4

%E	M1	M2	DM	M*
84.5	393	332	61	280.5
84.5	388	328	60	277.3
84.5	387	327	60	276.3
84.5	386	326	60	275.3
84.5	381	322	59	272.1
84.5	380	321	59	271.2
84.5	375	317	58	268.0
84.5	374	316	58	267.0
84.5	373	315	58	266.0
84.5	368	311	57	262.8
84.5	367	310	57	261.9
84.5	362	306	56	258.7
84.5	361	305	56	257.7
84.5	355	300	55	253.5
84.5	354	299	55	252.5
84.5	349	295	54	249.4
84.5	348	294	54	248.4
84.5	343	290	53	245.2
84.5	342	289	53	244.2
84.5	341	288	53	243.2
84.5	336	284	52	240.0
84.5	335	283	52	239.1
84.5	330	279	51	235.9
84.5	329	278	51	234.9
84.5	328	277	51	233.9
84.5	323	273	50	230.7
84.5	322	272	50	229.8
84.5	317	268	49	226.6
84.5	316	267	49	225.6
84.5	310	262	48	221.4
84.5	309	261	48	220.5
84.5	304	257	47	217.3
84.5	303	256	47	216.3
84.5	297	251	46	212.1
84.5	296	250	46	211.1
84.5	291	246	45	208.0
84.5	290	245	45	207.0
84.5	284	240	44	202.8
84.5	283	239	44	201.8
84.5	278	235	43	198.7
84.5	277	234	43	197.7
84.5	271	229	42	193.5
84.5	265	224	41	189.3
84.5	264	223	41	188.4
84.5	258	218	40	184.2
84.5	252	213	39	180.0
84.5	251	212	39	179.1
84.5	245	207	38	174.9
84.5	239	202	37	170.7
84.5	238	201	37	169.8

%E	M1	M2	DM	M*
84.5	233	197	36	166.6
84.5	232	196	36	165.6
84.5	226	191	35	161.4
84.5	220	186	34	157.3
84.5	219	185	34	156.3
84.5	213	180	33	152.1
84.5	207	175	32	147.9
84.5	206	174	32	147.0
84.5	200	169	31	142.8
84.5	194	164	30	138.6
84.5	193	163	30	137.7
84.5	187	158	29	133.5
84.5	181	153	28	129.3
84.5	174	147	27	124.2
84.5	168	142	26	120.0
84.5	161	136	25	114.9
84.5	155	131	24	110.7
84.5	148	125	23	105.6
84.5	142	120	22	101.4
84.5	129	109	20	92.1
84.5	116	98	18	82.8
84.5	110	93	17	78.6
84.5	103	87	16	73.5
84.5	97	82	15	69.3
84.5	84	71	13	60.0
84.5	71	60	11	50.7
84.5	58	49	9	41.4
84.4	500	422	78	356.2
84.4	499	421	78	355.2
84.4	495	418	77	353.0
84.4	494	417	77	352.0
84.4	493	416	77	351.0
84.4	492	415	77	350.1
84.4	488	412	76	347.8
84.4	487	411	76	346.9
84.4	486	410	76	345.9
84.4	482	407	75	343.7
84.4	481	406	75	342.7
84.4	480	405	75	341.7
84.4	475	401	74	338.5
84.4	474	400	74	337.6
84.4	473	399	74	336.6
84.4	469	396	73	334.4
84.4	468	395	73	333.4
84.4	467	394	73	332.4
84.4	463	391	72	330.2
84.4	462	390	72	329.2
84.4	461	389	72	328.2
84.4	456	385	71	325.1
84.4	455	384	71	324.1

%E	M1	M2	DM	M*
84.4	454	383	71	323.1
84.4	450	380	70	320.9
84.4	449	379	70	319.9
84.4	448	378	70	318.9
84.4	443	374	69	315.7
84.4	442	373	69	314.8
84.4	441	372	69	313.8
84.4	437	369	68	311.6
84.4	436	368	68	310.6
84.4	435	367	68	309.6
84.4	430	363	67	306.4
84.4	429	362	67	305.5
84.4	424	358	66	302.3
84.4	423	357	66	301.3
84.4	422	356	66	300.3
84.4	417	352	65	297.1
84.4	416	351	65	296.2
84.4	411	347	64	293.0
84.4	410	346	64	292.0
84.4	409	345	64	291.0
84.4	405	342	63	288.8
84.4	404	341	63	287.8
84.4	403	340	63	286.8
84.4	398	336	62	283.7
84.4	397	335	62	282.7
84.4	392	331	61	279.5
84.4	391	330	61	278.5
84.4	390	329	61	277.5
84.4	385	325	60	274.4
84.4	384	324	60	273.4
84.4	379	320	59	270.2
84.4	378	319	59	269.2
84.4	377	318	59	268.2
84.4	372	314	58	265.0
84.4	371	313	58	264.1
84.4	366	309	57	260.9
84.4	365	308	57	259.9
84.4	360	304	56	256.7
84.4	359	303	56	255.7
84.4	358	302	56	254.8
84.4	353	298	55	251.6
84.4	352	297	55	250.6
84.4	347	293	54	247.4
84.4	346	292	54	246.4
84.4	340	287	53	242.3
84.4	339	286	53	241.3
84.4	334	282	52	238.1
84.4	333	281	52	237.1
84.4	327	276	51	233.0
84.4	326	275	51	232.0

%E	M1	M2	DM	M*
84.4	321	271	50	228.8
84.4	320	270	50	227.8
84.4	315	266	49	224.6
84.4	314	265	49	223.6
84.4	308	260	48	219.5
84.4	307	259	48	218.5
84.4	302	255	47	215.3
84.4	301	254	47	214.3
84.4	295	249	46	210.2
84.4	294	248	46	209.2
84.4	289	244	45	206.0
84.4	288	243	45	205.0
84.4	282	238	44	200.9
84.4	276	233	43	196.7
84.4	275	232	43	195.7
84.4	270	228	42	192.5
84.4	269	227	42	191.6
84.4	263	222	41	187.4
84.4	262	221	41	186.4
84.4	257	217	40	183.2
84.4	256	216	40	182.3
84.4	250	211	39	178.1
84.4	244	206	38	173.9
84.4	243	205	38	172.9
84.4	237	200	37	168.8
84.4	231	195	36	164.6
84.4	225	190	35	160.4
84.4	224	189	35	159.5
84.4	218	184	34	155.3
84.4	212	179	33	151.1
84.4	211	178	33	150.2
84.4	205	173	32	146.0
84.4	199	168	31	141.8
84.4	192	162	30	136.7
84.4	186	157	29	132.5
84.4	180	152	28	128.4
84.4	179	151	28	127.4
84.4	173	146	27	123.2
84.4	167	141	26	119.0
84.4	160	135	25	113.9
84.4	154	130	24	109.7
84.4	147	124	23	104.6
84.4	141	119	22	100.4
84.4	135	114	21	96.3
84.4	128	108	20	91.1
84.4	122	103	19	87.0
84.4	109	92	17	77.7
84.4	96	81	15	68.3
84.4	90	76	14	64.2
84.4	77	65	12	54.9

%E	M1	M2	DM	M*
84.4	64	54	10	45.6
84.4	45	38	7	32.1
84.4	32	27	5	22.8
84.3	498	420	78	354.2
84.3	497	419	78	353.2
84.3	496	418	78	352.3
84.3	491	414	77	349.1
84.3	490	413	77	348.1
84.3	489	412	77	347.1
84.3	485	409	76	344.9
84.3	484	408	76	343.9
84.3	483	407	76	343.0
84.3	479	404	75	340.7
84.3	478	403	75	339.8
84.3	477	402	75	338.8
84.3	472	398	74	335.6
84.3	471	397	74	334.6
84.3	470	396	74	333.7
84.3	466	393	73	331.4
84.3	465	392	73	330.5
84.3	464	391	73	329.5
84.3	460	388	72	327.3
84.3	459	387	72	326.3
84.3	458	386	72	325.3
84.3	453	382	71	322.1
84.3	452	381	71	321.2
84.3	451	380	71	320.2
84.3	447	377	70	318.0
84.3	446	376	70	317.0
84.3	445	375	70	316.0
84.3	440	371	69	312.8
84.3	439	370	69	311.8
84.3	434	366	68	308.7
84.3	433	365	68	307.7
84.3	432	364	68	306.7
84.3	428	361	67	304.5
84.3	427	360	67	303.5
84.3	426	359	67	302.5
84.3	421	355	66	299.3
84.3	420	354	66	298.4
84.3	415	350	65	295.2
84.3	414	349	65	294.2
84.3	413	348	65	293.2
84.3	408	344	64	290.0
84.3	407	343	64	289.1
84.3	402	339	63	285.9
84.3	401	338	63	284.9
84.3	400	337	63	283.9
84.3	396	334	62	281.7
84.3	395	333	62	280.7

%E	M1	M2	DM	M*
84.3	394	332	62	279.8
84.3	389	328	61	276.6
84.3	388	327	61	275.6
84.3	383	323	60	272.4
84.3	382	322	60	271.4
84.3	381	321	60	270.4
84.3	376	317	59	267.3
84.3	375	316	59	266.3
84.3	370	312	58	263.1
84.3	369	311	58	262.1
84.3	364	307	57	258.9
84.3	363	306	57	258.0
84.3	362	305	57	257.0
84.3	357	301	56	253.8
84.3	356	300	56	252.8
84.3	351	296	55	249.6
84.3	350	295	55	248.6
84.3	345	291	54	245.5
84.3	344	290	54	244.5
84.3	343	289	54	243.5
84.3	338	285	53	240.3
84.3	337	284	53	239.3
84.3	332	280	52	236.1
84.3	331	279	52	235.2
84.3	325	274	51	231.0
84.3	324	273	51	230.0
84.3	319	269	50	226.8
84.3	318	268	50	225.9
84.3	313	264	49	222.7
84.3	312	263	49	221.7
84.3	306	258	48	217.5
84.3	305	257	48	216.6
84.3	300	253	47	213.4
84.3	299	252	47	212.4
84.3	293	247	46	208.2
84.3	287	242	45	204.1
84.3	286	241	45	203.1
84.3	281	237	44	199.9
84.3	280	236	44	198.9
84.3	274	231	43	194.7
84.3	268	226	42	190.6
84.3	267	225	42	189.6
84.3	261	220	41	185.4
84.3	255	215	40	181.3
84.3	254	214	40	180.3
84.3	249	210	39	177.1
84.3	248	209	39	176.1
84.3	242	204	38	172.0
84.3	236	199	37	167.8
84.3	235	198	37	166.8

%E	M1	M2	DM	M*
84.3	230	194	36	163.6
84.3	229	193	36	162.7
84.3	223	188	35	158.5
84.3	217	183	34	154.3
84.3	216	182	34	153.4
84.3	210	177	33	149.2
84.3	204	172	32	145.0
84.3	198	167	31	140.9
84.3	197	166	31	139.9
84.3	191	161	30	135.7
84.3	185	156	29	131.5
84.3	178	150	28	126.4
84.3	172	145	27	122.2
84.3	166	140	26	118.1
84.3	159	134	25	112.9
84.3	153	129	24	108.8
84.3	140	118	22	99.5
84.3	134	113	21	95.3
84.3	127	107	20	90.1
84.3	121	102	19	86.0
84.3	115	97	18	81.8
84.3	108	91	17	76.7
84.3	102	86	16	72.5
84.3	89	75	14	63.2
84.3	83	70	13	59.0
84.3	70	59	11	49.7
84.3	51	43	8	36.3
84.2	500	421	79	354.5
84.2	499	420	79	353.5
84.2	495	417	78	351.3
84.2	494	416	78	350.3
84.2	493	415	78	349.3
84.2	488	411	77	346.1
84.2	487	410	77	345.2
84.2	486	409	77	344.2
84.2	482	406	76	342.0
84.2	481	405	76	341.0
84.2	480	404	76	340.0
84.2	476	401	75	337.8
84.2	475	400	75	336.8
84.2	474	399	75	335.9
84.2	469	395	74	332.7
84.2	468	394	74	331.7
84.2	467	393	74	330.7
84.2	463	390	73	328.5
84.2	462	389	73	327.5
84.2	461	388	73	326.6
84.2	457	385	72	324.3
84.2	456	384	72	323.4
84.2	455	383	72	322.4

%E	M1	M2	DM	M*
84.2	450	379	71	319.2
84.2	449	378	71	318.2
84.2	448	377	71	317.3
84.2	444	374	70	315.0
84.2	443	373	70	314.1
84.2	442	372	70	313.1
84.2	438	369	69	310.9
84.2	437	368	69	309.9
84.2	436	367	69	308.9
84.2	431	363	68	305.7
84.2	430	362	68	304.8
84.2	425	358	67	301.6
84.2	424	357	67	300.6
84.2	423	356	67	299.6
84.2	419	353	66	297.4
84.2	418	352	66	296.4
84.2	417	351	66	295.4
84.2	412	347	65	292.3
84.2	411	346	65	291.3
84.2	406	342	64	288.1
84.2	405	341	64	287.1
84.2	404	340	64	286.1
84.2	399	336	63	282.9
84.2	398	335	63	282.0
84.2	393	331	62	278.8
84.2	392	330	62	277.8
84.2	387	326	61	274.6
84.2	386	325	61	273.6
84.2	385	324	61	272.7
84.2	380	320	60	269.5
84.2	379	319	60	268.5
84.2	374	315	59	265.3
84.2	373	314	59	264.3
84.2	368	310	58	261.1
84.2	367	309	58	260.2
84.2	366	308	58	259.2
84.2	361	304	57	256.0
84.2	360	303	57	255.0
84.2	355	299	56	251.8
84.2	354	298	56	250.9
84.2	349	294	55	247.7
84.2	348	293	55	246.7
84.2	347	292	55	245.7
84.2	342	288	54	242.5
84.2	341	287	54	241.6
84.2	336	283	53	238.4
84.2	335	282	53	237.4
84.2	330	278	52	234.2
84.2	329	277	52	233.2
84.2	323	272	51	229.1

%E	M1	M2	DM	M*
84.2	322	271	51	228.1
84.2	317	267	50	224.9
84.2	316	266	50	223.9
84.2	311	262	49	220.7
84.2	310	261	49	219.7
84.2	304	256	48	215.6
84.2	303	255	48	214.6
84.2	298	251	47	211.4
84.2	297	250	47	210.4
84.2	292	246	46	207.2
84.2	291	245	46	206.3
84.2	285	240	45	202.1
84.2	284	239	45	201.1
84.2	279	235	44	197.9
84.2	278	234	44	197.0
84.2	273	230	43	193.8
84.2	272	229	43	192.8
84.2	266	224	42	188.6
84.2	265	223	42	187.7
84.2	260	219	41	184.5
84.2	259	218	41	183.5
84.2	253	213	40	179.3
84.2	247	208	39	175.2
84.2	241	203	38	171.0
84.2	240	202	38	170.0
84.2	234	197	37	165.9
84.2	228	192	36	161.7
84.2	222	187	35	157.5
84.2	221	186	35	156.5
84.2	215	181	34	152.4
84.2	209	176	33	148.2
84.2	203	171	32	144.0
84.2	202	170	32	143.1
84.2	196	165	31	138.9
84.2	190	160	30	134.7
84.2	184	155	29	130.6
84.2	183	154	29	129.6
84.2	177	149	28	125.4
84.2	171	144	27	121.3
84.2	165	139	26	117.1
84.2	158	133	25	112.0
84.2	152	128	24	107.8
84.2	146	123	23	103.6
84.2	139	117	22	98.5
84.2	133	112	21	94.3
84.2	120	101	19	85.0
84.2	114	96	18	80.8
84.2	101	85	16	71.5
84.2	95	80	15	67.4
84.2	76	64	12	53.9

%E	M1	M2	DM	M*
84.2	57	48	9	40.4
84.2	38	32	6	26.9
84.2	19	16	3	13.5
84.1	498	419	79	352.5
84.1	497	418	79	351.6
84.1	496	417	79	350.6
84.1	492	414	78	348.4
84.1	491	413	78	347.4
84.1	490	412	78	346.4
84.1	485	408	77	343.2
84.1	484	407	77	342.3
84.1	483	406	77	341.3
84.1	479	403	76	339.1
84.1	478	402	76	338.1
84.1	477	401	76	337.1
84.1	473	398	75	334.9
84.1	472	397	75	333.9
84.1	471	396	75	332.9
84.1	466	392	74	329.8
84.1	465	391	74	328.8
84.1	464	390	74	327.8
84.1	460	387	73	325.6
84.1	459	386	73	324.6
84.1	458	385	73	323.6
84.1	454	382	72	321.4
84.1	453	381	72	320.4
84.1	452	380	72	319.5
84.1	447	376	71	316.3
84.1	446	375	71	315.3
84.1	441	371	70	312.1
84.1	440	370	70	311.1
84.1	439	369	70	310.2
84.1	435	366	69	307.9
84.1	434	365	69	307.0
84.1	433	364	69	306.0
84.1	429	361	68	303.8
84.1	428	360	68	302.8
84.1	427	359	68	301.8
84.1	422	355	67	298.6
84.1	421	354	67	297.7
84.1	416	350	66	294.5
84.1	415	349	66	293.5
84.1	414	348	66	292.5
84.1	410	345	65	290.3
84.1	409	344	65	289.3
84.1	408	343	65	288.4
84.1	403	339	64	285.2
84.1	402	338	64	284.2
84.1	397	334	63	281.0
84.1	396	333	63	280.0

%E	M1	M2	DM	M*
84.1	395	332	63	279.0
84.1	391	329	62	276.8
84.1	390	328	62	275.9
84.1	389	327	62	274.9
84.1	384	323	61	271.7
84.1	383	322	61	270.7
84.1	378	318	60	267.5
84.1	377	317	60	266.5
84.1	372	313	59	263.4
84.1	371	312	59	262.4
84.1	370	311	59	261.4
84.1	365	307	58	258.2
84.1	364	306	58	257.2
84.1	359	302	57	254.1
84.1	358	301	57	253.1
84.1	353	297	56	249.9
84.1	352	296	56	248.9
84.1	346	291	55	244.7
84.1	345	290	55	243.8
84.1	340	286	54	240.6
84.1	339	285	54	239.6
84.1	334	281	53	236.4
84.1	333	280	53	235.4
84.1	328	276	52	232.2
84.1	327	275	52	231.3
84.1	321	270	51	227.1
84.1	320	269	51	226.1
84.1	315	265	50	222.9
84.1	314	264	50	222.0
84.1	309	260	49	218.8
84.1	308	259	49	217.8
84.1	302	254	48	213.6
84.1	301	253	48	212.7
84.1	296	249	47	209.5
84.1	295	248	47	208.5
84.1	290	244	46	205.3
84.1	289	243	46	204.3
84.1	283	238	45	200.2
84.1	277	233	44	196.0
84.1	276	232	44	195.0
84.1	271	228	43	191.8
84.1	270	227	43	190.8
84.1	264	222	42	186.7
84.1	258	217	41	182.5
84.1	252	212	40	178.3
84.1	251	211	40	177.4
84.1	246	207	39	174.2
84.1	245	206	39	173.2
84.1	239	201	38	169.0
84.1	233	196	37	164.9

%E	M1	M2	DM	M*
84.1	232	195	37	163.9
84.1	227	191	36	160.7
84.1	226	190	36	159.7
84.1	220	185	35	155.6
84.1	214	180	34	151.4
84.1	208	175	33	147.2
84.1	207	174	33	146.3
84.1	201	169	32	142.1
84.1	195	164	31	137.9
84.1	189	159	30	133.8
84.1	182	153	29	128.6
84.1	176	148	28	124.5
84.1	170	143	27	120.3
84.1	164	138	26	116.1
84.1	157	132	25	111.0
84.1	151	127	24	106.8
84.1	145	122	23	102.6
84.1	138	116	22	97.5
84.1	132	111	21	93.3
84.1	126	106	20	89.2
84.1	113	95	18	79.9
84.1	107	90	17	75.7
84.1	88	74	14	62.2
84.1	82	69	13	58.1
84.1	69	58	11	48.8
84.1	63	53	10	44.6
84.1	44	37	7	31.1
84.0	500	420	80	352.8
84.0	499	419	80	351.8
84.0	495	416	79	349.6
84.0	494	415	79	348.6
84.0	493	414	79	347.7
84.0	489	411	78	345.4
84.0	488	410	78	344.5
84.0	487	409	78	343.5
84.0	486	408	78	342.5
84.0	482	405	77	340.3
84.0	481	404	77	339.3
84.0	480	403	77	338.4
84.0	476	400	76	336.1
84.0	475	399	76	335.2
84.0	474	398	76	334.2
84.0	470	395	75	332.0
84.0	469	394	75	331.0
84.0	468	393	75	330.0
84.0	463	389	74	326.8
84.0	462	388	74	325.9
84.0	457	384	73	322.7
84.0	456	383	73	321.7
84.0	455	382	73	320.7

%E	M1	M2	DM	M*
84.0	451	379	72	318.5
84.0	450	378	72	317.5
84.0	449	377	72	316.5
84.0	445	374	71	314.3
84.0	444	373	71	313.4
84.0	443	372	71	312.4
84.0	438	368	70	309.2
84.0	437	367	70	308.2
84.0	432	363	69	305.0
84.0	431	362	69	304.0
84.0	430	361	69	303.1
84.0	426	358	68	300.9
84.0	425	357	68	299.9
84.0	424	356	68	298.9
84.0	420	353	67	296.7
84.0	419	352	67	295.7
84.0	418	351	67	294.7
84.0	413	347	66	291.5
84.0	412	346	66	290.6
84.0	407	342	65	287.4
84.0	406	341	65	286.4
84.0	405	340	65	285.4
84.0	401	337	64	283.2
84.0	400	336	64	282.2
84.0	399	335	64	281.3
84.0	394	331	63	278.1
84.0	393	330	63	277.1
84.0	388	326	62	273.9
84.0	387	325	62	272.9
84.0	382	321	61	269.7
84.0	381	320	61	268.8
84.0	376	316	60	265.6
84.0	375	315	60	264.6
84.0	374	314	60	263.6
84.0	369	310	59	260.4
84.0	368	309	59	259.5
84.0	363	305	58	256.3
84.0	362	304	58	255.3
84.0	357	300	57	252.1
84.0	356	299	57	251.1
84.0	351	295	56	247.9
84.0	350	294	56	247.0
84.0	349	293	56	246.0
84.0	344	289	55	242.8
84.0	343	288	55	241.8
84.0	338	284	54	238.6
84.0	337	283	54	237.7
84.0	332	279	53	234.5
84.0	331	278	53	233.5
84.0	326	274	52	230.3
84.0	325	273	52	229.3
84.0	324	272	52	228.3
84.0	319	268	51	225.2
84.0	318	267	51	224.2
84.0	313	263	50	221.0
84.0	312	262	50	220.0
84.0	307	258	49	216.8
84.0	306	257	49	215.8
84.0	300	252	48	211.7
84.0	294	247	47	207.5
84.0	293	246	47	206.5
84.0	288	242	46	203.3
84.0	287	241	46	202.4
84.0	282	237	45	199.2
84.0	281	236	45	198.2
84.0	275	231	44	194.0
84.0	269	226	43	189.9
84.0	268	225	43	188.9
84.0	263	221	42	185.7
84.0	262	220	42	184.7
84.0	257	216	41	181.5
84.0	256	215	41	180.6
84.0	250	210	40	176.4
84.0	244	205	39	172.2
84.0	243	204	39	171.3
84.0	238	200	38	168.1
84.0	237	199	38	167.1
84.0	231	194	37	162.9
84.0	225	189	36	158.8
84.0	219	184	35	154.6
84.0	213	179	34	150.4
84.0	212	178	34	149.5
84.0	206	173	33	145.3
84.0	200	168	32	141.1
84.0	194	163	31	137.0
84.0	188	158	30	132.8
84.0	187	157	30	131.8
84.0	181	152	29	127.6
84.0	175	147	28	123.5
84.0	169	142	27	119.3
84.0	163	137	26	115.1
84.0	162	136	26	114.2
84.0	156	131	25	110.0
84.0	150	126	24	105.8
84.0	144	121	23	101.7
84.0	131	110	21	92.4
84.0	125	105	20	88.2
84.0	119	100	19	84.0
84.0	106	89	17	74.7
84.0	100	84	16	70.6
84.0	94	79	15	66.4
84.0	81	68	13	57.1
84.0	75	63	12	52.9
84.0	50	42	8	35.3
84.0	25	21	4	17.6
83.9	498	418	80	350.9
83.9	497	417	80	349.9
83.9	496	416	80	348.9
83.9	492	413	79	346.7
83.9	491	412	79	345.7
83.9	490	411	79	344.7
83.9	485	407	78	341.5
83.9	484	406	78	340.6
83.9	483	405	78	339.6
83.9	479	402	77	337.4
83.9	478	401	77	336.4
83.9	477	400	77	335.4
83.9	473	397	76	333.2
83.9	472	396	76	332.2
83.9	471	395	76	331.3
83.9	467	392	75	329.0
83.9	466	391	75	328.1
83.9	465	390	75	327.1
83.9	461	387	74	324.9
83.9	460	386	74	323.9
83.9	459	385	74	322.9
83.9	454	381	73	319.7
83.9	453	380	73	318.8
83.9	452	379	73	317.8
83.9	448	376	72	315.6
83.9	447	375	72	314.6
83.9	446	374	72	313.6
83.9	442	371	71	311.4
83.9	441	370	71	310.4
83.9	440	369	71	309.5
83.9	436	366	70	307.2
83.9	435	365	70	306.3
83.9	434	364	70	305.3
83.9	429	360	69	302.1
83.9	428	359	69	301.1
83.9	423	355	68	297.9
83.9	422	354	68	297.0
83.9	417	350	67	293.8
83.9	416	349	67	292.8
83.9	415	348	67	291.8
83.9	411	345	66	289.6
83.9	410	344	66	288.6
83.9	409	343	66	287.7
83.9	404	339	65	284.5
83.9	403	338	65	283.5
83.9	398	334	64	280.3
83.9	397	333	64	279.3
83.9	392	329	63	276.1
83.9	391	328	63	275.2
83.9	386	324	62	272.0
83.9	385	323	62	271.0
83.9	384	322	62	270.0
83.9	380	319	61	267.8
83.9	379	318	61	266.8
83.9	378	317	61	265.8
83.9	373	313	60	262.7
83.9	372	312	60	261.7
83.9	367	308	59	258.5
83.9	366	307	59	257.5
83.9	361	303	58	254.3
83.9	360	302	58	253.3
83.9	355	298	57	250.2
83.9	354	297	57	249.2
83.9	353	296	57	248.2
83.9	348	292	56	245.0
83.9	347	291	56	244.0
83.9	342	287	55	240.8
83.9	341	286	55	239.9
83.9	336	282	54	236.7
83.9	335	281	54	235.7
83.9	330	277	53	232.5
83.9	329	276	53	231.5
83.9	323	271	52	227.4
83.9	322	270	52	226.4
83.9	317	266	51	223.2
83.9	316	265	51	222.2
83.9	311	261	50	219.0
83.9	310	260	50	218.1
83.9	305	256	49	214.9
83.9	304	255	49	213.9
83.9	299	251	48	210.7
83.9	298	250	48	209.7
83.9	292	245	47	205.6
83.9	286	240	46	201.4
83.9	285	239	46	200.4
83.9	280	235	45	197.2
83.9	279	234	45	196.3
83.9	274	230	44	193.1
83.9	273	229	44	192.1
83.9	267	224	43	187.9
83.9	261	219	42	183.8
83.9	255	214	41	179.6
83.9	254	213	41	178.6
83.9	249	209	40	175.4
83.9	248	208	40	174.5
83.9	242	203	39	170.3
83.9	236	198	38	166.1
83.9	230	193	37	162.0
83.9	224	188	36	157.8
83.9	223	187	36	156.8
83.9	218	183	35	153.6
83.9	217	182	35	152.6
83.9	211	177	34	148.5
83.9	205	172	33	144.3
83.9	199	167	32	140.1
83.9	193	162	31	136.0
83.9	192	161	31	135.0
83.9	186	156	30	130.8
83.9	180	151	29	126.7
83.9	174	146	28	122.5
83.9	168	141	27	118.3
83.9	161	135	26	113.2
83.9	155	130	25	109.0
83.9	149	125	24	104.9
83.9	143	120	23	100.7
83.9	137	115	22	96.5
83.9	124	104	20	87.2
83.9	118	99	19	83.1
83.9	112	94	18	78.9
83.9	93	78	15	65.4
83.9	87	73	14	61.3
83.9	62	52	10	43.6
83.9	56	47	9	39.4
83.9	31	26	5	21.8
83.8	500	419	81	351.1
83.8	499	418	81	350.1
83.8	495	415	80	347.9
83.8	494	414	80	347.0
83.8	493	413	80	346.0
83.8	489	410	79	343.8
83.8	488	409	79	342.8
83.8	487	408	79	341.8
83.8	482	404	78	338.6
83.8	481	403	78	337.6
83.8	480	402	78	336.7
83.8	476	399	77	334.5
83.8	475	398	77	333.5
83.8	474	397	77	332.5
83.8	470	394	76	330.3
83.8	469	393	76	329.3
83.8	468	392	76	328.3
83.8	464	389	75	326.1
83.8	463	388	75	325.1
83.8	462	387	75	324.2
83.8	458	384	74	322.0

%E	M1	M2	DM	M*
83.8	457	383	74	321.0
83.8	456	382	74	320.0
83.8	451	378	73	316.8
83.8	450	377	73	315.8
83.8	445	373	72	312.6
83.8	444	372	72	311.7
83.8	439	368	71	308.5
83.8	438	367	71	307.5
83.8	437	366	71	306.5
83.8	433	363	70	304.3
83.8	432	362	70	303.3
83.8	431	361	70	302.4
83.8	427	358	69	300.1
83.8	426	357	69	299.2
83.8	425	356	69	298.2
83.8	421	353	68	296.0
83.8	420	352	68	295.0
83.8	419	351	68	294.0
83.8	414	347	67	290.8
83.8	413	346	67	289.9
83.8	408	342	66	286.7
83.8	407	341	66	285.7
83.8	402	337	65	282.5
83.8	401	336	65	281.5
83.8	400	335	65	280.6
83.8	396	332	64	278.3
83.8	395	331	64	277.4
83.8	394	330	64	276.4
83.8	390	327	63	274.2
83.8	389	326	63	273.2
83.8	388	325	63	272.2
83.8	383	321	62	269.0
83.8	382	320	62	268.1
83.8	377	316	61	264.9
83.8	376	315	61	263.9
83.8	371	311	60	260.7
83.8	370	310	60	259.7
83.8	365	306	59	256.5
83.8	364	305	59	255.6
83.8	359	301	58	252.4
83.8	358	300	58	251.4
83.8	357	299	58	250.4
83.8	352	295	57	247.2
83.8	351	294	57	246.3
83.8	346	290	56	243.1
83.8	345	289	56	242.1
83.8	340	285	55	238.9
83.8	339	284	55	237.9
83.8	334	280	54	234.7
83.8	333	279	54	233.8

%E	M1	M2	DM	M*
83.8	328	275	53	230.6
83.8	327	274	53	229.6
83.8	321	269	52	225.4
83.8	320	268	52	224.4
83.8	315	264	51	221.3
83.8	314	263	51	220.3
83.8	309	259	50	217.1
83.8	308	258	50	216.1
83.8	303	254	49	212.9
83.8	302	253	49	212.0
83.8	297	249	48	208.8
83.8	296	248	48	207.8
83.8	291	244	47	204.6
83.8	290	243	47	203.6
83.8	284	238	46	199.5
83.8	278	233	45	195.3
83.8	277	232	45	194.3
83.8	272	228	44	191.1
83.8	271	227	44	190.1
83.8	266	223	43	187.0
83.8	265	222	43	186.0
83.8	260	218	42	182.8
83.8	259	217	42	181.8
83.8	253	212	41	177.6
83.8	247	207	40	173.5
83.8	241	202	39	169.3
83.8	240	201	39	168.3
83.8	235	197	38	165.1
83.8	234	196	38	164.2
83.8	229	192	37	161.0
83.8	228	191	37	160.0
83.8	222	186	36	155.8
83.8	216	181	35	151.7
83.8	210	176	34	147.5
83.8	204	171	33	143.3
83.8	198	166	32	139.2
83.8	197	165	32	138.2
83.8	191	160	31	134.0
83.8	185	155	30	129.9
83.8	179	150	29	125.7
83.8	173	145	28	121.5
83.8	167	140	27	117.4
83.8	160	134	26	112.2
83.8	154	129	25	108.1
83.8	148	124	24	103.9
83.8	142	119	23	99.7
83.8	136	114	22	95.6
83.8	130	109	21	91.4
83.8	117	98	19	82.1
83.8	111	93	18	77.9

%E	M1	M2	DM	M*
83.8	105	88	17	73.8
83.8	99	83	16	69.6
83.8	80	67	13	56.1
83.8	74	62	12	51.9
83.8	68	57	11	47.8
83.8	37	31	6	26.0
83.7	498	417	81	349.2
83.7	497	416	81	348.2
83.7	496	415	81	347.2
83.7	492	412	80	345.0
83.7	491	411	80	344.0
83.7	490	410	80	343.1
83.7	486	407	79	340.8
83.7	485	406	79	339.9
83.7	484	405	79	338.9
83.7	479	401	78	335.7
83.7	478	400	78	334.7
83.7	473	396	77	331.5
83.7	472	395	77	330.6
83.7	471	394	77	329.6
83.7	467	391	76	327.4
83.7	466	390	76	326.4
83.7	465	389	76	325.4
83.7	461	386	75	323.2
83.7	460	385	75	322.2
83.7	459	384	75	321.3
83.7	455	381	74	319.0
83.7	454	380	74	318.1
83.7	453	379	74	317.1
83.7	449	376	73	314.9
83.7	448	375	73	313.9
83.7	447	374	73	312.9
83.7	443	371	72	310.7
83.7	442	370	72	309.7
83.7	441	369	72	308.8
83.7	436	365	71	305.6
83.7	435	364	71	304.6
83.7	430	360	70	301.4
83.7	429	359	70	300.4
83.7	424	355	69	297.2
83.7	423	354	69	296.3
83.7	418	350	68	293.1
83.7	417	349	68	292.1
83.7	416	348	68	291.1
83.7	412	345	67	288.9
83.7	411	344	67	287.9
83.7	410	343	67	286.9
83.7	406	340	66	284.7
83.7	405	339	66	283.8
83.7	404	338	66	282.8

%E	M1	M2	DM	M*
83.7	399	334	65	279.6
83.7	398	333	65	278.6
83.7	393	329	64	275.4
83.7	392	328	64	274.4
83.7	387	324	63	271.3
83.7	386	323	63	270.3
83.7	381	319	62	267.1
83.7	380	318	62	266.1
83.7	375	314	61	262.9
83.7	374	313	61	261.9
83.7	369	309	60	258.8
83.7	368	308	60	257.8
83.7	367	307	60	256.8
83.7	363	304	59	254.6
83.7	362	303	59	253.6
83.7	361	302	59	252.6
83.7	356	298	58	249.4
83.7	355	297	58	248.5
83.7	350	293	57	245.3
83.7	349	292	57	244.3
83.7	344	288	56	241.1
83.7	343	287	56	240.1
83.7	338	283	55	236.9
83.7	337	282	55	236.0
83.7	332	278	54	232.8
83.7	331	277	54	231.8
83.7	326	273	53	228.6
83.7	325	272	53	227.6
83.7	319	267	52	223.5
83.7	313	262	51	219.3
83.7	312	261	51	218.3
83.7	307	257	50	215.1
83.7	306	256	50	214.2
83.7	301	252	49	211.0
83.7	300	251	49	210.0
83.7	295	247	48	206.8
83.7	294	246	48	205.8
83.7	289	242	47	202.6
83.7	288	241	47	201.7
83.7	283	237	46	198.5
83.7	282	236	46	197.5
83.7	276	231	45	193.3
83.7	270	226	44	189.2
83.7	264	221	43	185.0
83.7	263	220	43	184.0
83.7	258	216	42	180.8
83.7	257	215	42	179.9
83.7	252	211	41	176.7
83.7	251	210	41	175.7
83.7	246	206	40	172.5

%E	M1	M2	DM	M*
83.7	245	205	40	171.5
83.7	239	200	39	167.4
83.7	233	195	38	163.2
83.7	227	190	37	159.0
83.7	221	185	36	154.9
83.7	215	180	35	150.7
83.7	209	175	34	146.5
83.7	208	174	34	145.6
83.7	203	170	33	142.4
83.7	202	169	33	141.4
83.7	196	164	32	137.2
83.7	190	159	31	133.1
83.7	184	154	30	128.9
83.7	178	149	29	124.7
83.7	172	144	28	120.6
83.7	166	139	27	116.4
83.7	153	128	25	107.1
83.7	147	123	24	102.9
83.7	141	118	23	98.8
83.7	135	113	22	94.6
83.7	129	108	21	90.4
83.7	123	103	20	86.3
83.7	104	87	17	72.8
83.7	98	82	16	68.6
83.7	92	77	15	64.4
83.7	86	72	14	60.3
83.7	49	41	8	34.3
83.7	43	36	7	30.1
83.6	500	418	82	349.4
83.6	499	417	82	348.5
83.6	495	414	81	346.3
83.6	494	413	81	345.3
83.6	493	412	81	344.3
83.6	489	409	80	342.1
83.6	488	408	80	341.1
83.6	487	407	80	340.1
83.6	483	404	79	337.9
83.6	482	403	79	336.9
83.6	481	402	79	336.0
83.6	477	399	78	333.8
83.6	476	398	78	332.8
83.6	475	397	78	331.8
83.6	470	393	77	328.6
83.6	469	392	77	327.6
83.6	464	388	76	324.4
83.6	463	387	76	323.5
83.6	462	386	76	322.5
83.6	458	383	75	320.3
83.6	457	382	75	319.3
83.6	456	381	75	318.3

%E	M1	M2	DM	M*
83.6	452	378	74	316.1
83.6	451	377	74	315.1
83.6	450	376	74	314.2
83.6	446	373	73	311.9
83.6	445	372	73	311.0
83.6	444	371	73	310.0
83.6	440	368	72	307.8
83.6	439	367	72	306.8
83.6	438	366	72	305.8
83.6	434	363	71	303.6
83.6	433	362	71	302.6
83.6	432	361	71	301.7
83.6	428	358	70	299.4
83.6	427	357	70	298.5
83.6	426	356	70	297.5
83.6	422	353	69	295.3
83.6	421	352	69	294.3
83.6	420	351	69	293.3
83.6	415	347	68	290.1
83.6	414	346	68	289.2
83.6	409	342	67	286.0
83.6	408	341	67	285.0
83.6	403	337	66	281.8
83.6	402	336	66	280.8
83.6	397	332	65	277.6
83.6	396	331	65	276.7
83.6	391	327	64	273.5
83.6	390	326	64	272.5
83.6	385	322	63	269.3
83.6	384	321	63	268.3
83.6	383	320	63	267.4
83.6	379	317	62	265.1
83.6	378	316	62	264.2
83.6	377	315	62	263.2
83.6	373	312	61	261.0
83.6	372	311	61	260.0
83.6	371	310	61	259.0
83.6	366	306	60	255.8
83.6	365	305	60	254.9
83.6	360	301	59	251.7
83.6	359	300	59	250.7
83.6	354	296	58	247.5
83.6	353	295	58	246.5
83.6	348	291	57	243.3
83.6	347	290	57	242.4
83.6	342	286	56	239.2
83.6	341	285	56	238.2
83.6	336	281	55	235.0
83.6	335	280	55	234.0
83.6	330	276	54	230.8

%E	M1	M2	DM	M*
83.6	329	275	54	229.9
83.6	324	271	53	226.7
83.6	323	270	53	225.7
83.6	318	266	52	222.5
83.6	317	265	52	221.5
83.6	311	260	51	217.4
83.6	305	255	50	213.2
83.6	304	254	50	212.2
83.6	299	250	49	209.0
83.6	298	249	49	208.1
83.6	293	245	48	204.9
83.6	292	244	48	203.9
83.6	287	240	47	200.7
83.6	286	239	47	199.7
83.6	281	235	46	196.5
83.6	280	234	46	195.6
83.6	275	230	45	192.4
83.6	274	229	45	191.4
83.6	269	225	44	188.2
83.6	268	224	44	187.2
83.6	262	219	43	183.1
83.6	256	214	42	178.9
83.6	250	209	41	174.7
83.6	244	204	40	170.6
83.6	238	199	39	166.4
83.6	232	194	38	162.2
83.6	231	193	38	161.3
83.6	226	189	37	158.1
83.6	225	188	37	157.1
83.6	220	184	36	153.9
83.6	219	183	36	152.9
83.6	214	179	35	149.7
83.6	213	178	35	148.8
83.6	207	173	34	144.6
83.6	201	168	33	140.4
83.6	195	163	32	136.3
83.6	189	158	31	132.1
83.6	183	153	30	127.9
83.6	177	148	29	123.8
83.6	171	143	28	119.6
83.6	165	138	27	115.4
83.6	159	133	26	111.3
83.6	152	127	25	106.1
83.6	146	122	24	101.9
83.6	140	117	23	97.8
83.6	134	112	22	93.6
83.6	128	107	21	89.4
83.6	122	102	20	85.3
83.6	116	97	19	81.1
83.6	110	92	18	76.9

%E	M1	M2	DM	M*
83.6	73	61	12	51.0
83.6	67	56	11	46.8
83.6	61	51	10	42.6
83.6	55	46	9	38.5
83.5	498	416	82	347.5
83.5	497	415	82	346.5
83.5	496	414	82	345.6
83.5	492	411	81	343.3
83.5	491	410	81	342.4
83.5	490	409	81	341.4
83.5	486	406	80	339.2
83.5	485	405	80	338.2
83.5	484	404	80	337.2
83.5	480	401	79	335.0
83.5	479	400	79	334.0
83.5	478	399	79	333.1
83.5	474	396	78	330.8
83.5	473	395	78	329.9
83.5	472	394	78	328.9
83.5	468	391	77	326.7
83.5	467	390	77	325.7
83.5	466	389	77	324.7
83.5	461	385	76	321.5
83.5	460	384	76	320.6
83.5	455	380	75	317.4
83.5	454	379	75	316.4
83.5	449	375	74	313.2
83.5	448	374	74	312.2
83.5	443	370	73	309.0
83.5	442	369	73	308.1
83.5	437	365	72	304.9
83.5	436	364	72	303.9
83.5	431	360	71	300.7
83.5	430	359	71	299.7
83.5	429	358	71	298.8
83.5	425	355	70	296.5
83.5	424	354	70	295.6
83.5	423	353	70	294.6
83.5	419	350	69	292.4
83.5	418	349	69	291.4
83.5	417	348	69	290.4
83.5	413	345	68	288.2
83.5	412	344	68	287.2
83.5	411	343	68	286.3
83.5	407	340	67	284.0
83.5	406	339	67	283.1
83.5	405	338	67	282.1
83.5	401	335	66	279.9
83.5	400	334	66	278.9
83.5	399	333	66	277.9

%E	M1	M2	DM	M*
83.5	395	330	65	275.7
83.5	394	329	65	274.7
83.5	393	328	65	273.8
83.5	389	325	64	271.5
83.5	388	324	64	270.6
83.5	387	323	64	269.6
83.5	382	319	63	266.4
83.5	381	318	63	265.4
83.5	376	314	62	262.2
83.5	375	313	62	261.3
83.5	370	309	61	258.1
83.5	369	308	61	257.1
83.5	364	304	60	253.9
83.5	363	303	60	252.9
83.5	358	299	59	249.7
83.5	357	298	59	248.8
83.5	352	294	58	245.6
83.5	351	293	58	244.6
83.5	346	289	57	241.4
83.5	345	288	57	240.4
83.5	340	284	56	237.2
83.5	339	283	56	236.3
83.5	334	279	55	233.1
83.5	333	278	55	232.1
83.5	328	274	54	228.9
83.5	327	273	54	227.9
83.5	322	269	53	224.7
83.5	321	268	53	223.8
83.5	316	264	52	220.6
83.5	315	263	52	219.6
83.5	310	259	51	216.4
83.5	309	258	51	215.4
83.5	303	253	50	211.3
83.5	297	248	49	207.1
83.5	291	243	48	202.9
83.5	285	238	47	198.8
83.5	284	237	47	197.8
83.5	279	233	46	194.6
83.5	278	232	46	193.6
83.5	273	228	45	190.4
83.5	272	227	45	189.4
83.5	267	223	44	186.3
83.5	266	222	44	185.3
83.5	261	218	43	182.1
83.5	260	217	43	181.1
83.5	255	213	42	177.9
83.5	254	212	42	176.9
83.5	249	208	41	173.8
83.5	248	207	41	172.8
83.5	243	203	40	169.6

%E	M1	M2	DM	M*
83.5	242	202	40	168.6
83.5	237	198	39	165.4
83.5	236	197	39	164.4
83.5	230	192	38	160.3
83.5	224	187	37	156.1
83.5	218	182	36	151.9
83.5	212	177	35	147.8
83.5	206	172	34	143.6
83.5	200	167	33	139.4
83.5	194	162	32	135.3
83.5	188	157	31	131.1
83.5	182	152	30	126.9
83.5	176	147	29	122.8
83.5	170	142	28	118.6
83.5	164	137	27	114.4
83.5	158	132	26	110.3
83.5	139	116	23	96.8
83.5	133	111	22	92.6
83.5	127	106	21	88.5
83.5	121	101	20	84.3
83.5	115	96	19	80.1
83.5	109	91	18	76.0
83.5	103	86	17	71.8
83.5	97	81	16	67.6
83.5	91	76	15	63.5
83.5	85	71	14	59.3
83.5	79	66	13	55.1
83.4	500	417	83	347.8
83.4	499	416	83	346.8
83.4	495	413	82	344.6
83.4	494	412	82	343.6
83.4	493	411	82	342.6
83.4	489	408	81	340.4
83.4	488	407	81	339.4
83.4	487	406	81	338.5
83.4	483	403	80	336.3
83.4	482	402	80	335.3
83.4	481	401	80	334.3
83.4	477	398	79	332.1
83.4	476	397	79	331.1
83.4	475	396	79	330.1
83.4	471	393	78	327.9
83.4	470	392	78	326.9
83.4	469	391	78	326.0
83.4	465	388	77	323.8
83.4	464	387	77	322.8
83.4	463	386	77	321.8
83.4	459	383	76	319.6
83.4	458	382	76	318.6
83.4	457	381	76	317.6

%E	M1	M2	DM	M*
83.4	453	378	75	315.4
83.4	452	377	75	314.4
83.4	451	376	75	313.5
83.4	447	373	74	311.3
83.4	446	372	74	310.3
83.4	445	371	74	309.3
83.4	441	368	73	307.1
83.4	440	367	73	306.1
83.4	439	366	73	305.1
83.4	435	363	72	302.9
83.4	434	362	72	301.9
83.4	433	361	72	301.0
83.4	428	357	71	297.8
83.4	427	356	71	296.8
83.4	422	352	70	293.6
83.4	421	351	70	292.6
83.4	416	347	69	289.4
83.4	415	346	69	288.5
83.4	410	342	68	285.3
83.4	409	341	68	284.3
83.4	404	337	67	281.1
83.4	403	336	67	280.1
83.4	398	332	66	276.9
83.4	397	331	66	276.0
83.4	392	327	65	272.8
83.4	391	326	65	271.8
83.4	386	322	64	268.6
83.4	385	321	64	267.6
83.4	380	317	63	264.4
83.4	379	316	63	263.5
83.4	374	312	62	260.3
83.4	373	311	62	259.3
83.4	368	307	61	256.1
83.4	367	306	61	255.1
83.4	362	302	60	251.9
83.4	361	301	60	251.0
83.4	356	297	59	247.8
83.4	355	296	59	246.8
83.4	350	292	58	243.6
83.4	349	291	58	242.6
83.4	344	287	57	239.4
83.4	343	286	57	238.5
83.4	338	282	56	235.3
83.4	337	281	56	234.3
83.4	332	277	55	231.1
83.4	331	276	55	230.1
83.4	326	272	54	226.9
83.4	325	271	54	226.0
83.4	320	267	53	222.8
83.4	319	266	53	221.8

%E	M1	M2	DM	M*
83.4	314	262	52	218.6
83.4	313	261	52	217.6
83.4	308	257	51	214.4
83.4	307	256	51	213.5
83.4	302	252	50	210.3
83.4	301	251	50	209.3
83.4	296	247	49	206.1
83.4	295	246	49	205.1
83.4	290	242	48	201.9
83.4	289	241	48	201.0
83.4	283	236	47	196.8
83.4	277	231	46	192.6
83.4	271	226	45	188.5
83.4	265	221	44	184.3
83.4	259	216	43	180.1
83.4	253	211	42	176.0
83.4	247	206	41	171.8
83.4	241	201	40	167.6
83.4	235	196	39	163.5
83.4	229	191	38	159.3
83.4	223	186	37	155.1
83.4	217	181	36	151.0
83.4	211	176	35	146.8
83.4	205	171	34	142.6
83.4	199	166	33	138.5
83.4	193	161	32	134.3
83.4	187	156	31	130.1
83.4	181	151	30	126.0
83.4	175	146	29	121.8
83.4	169	141	28	117.6
83.4	163	136	27	113.5
83.4	157	131	26	109.3
83.4	151	126	25	105.1
83.4	145	121	24	101.0
83.3	498	415	83	345.8
83.3	497	414	83	344.9
83.3	496	413	83	343.9
83.3	492	410	82	341.7
83.3	491	409	82	340.7
83.3	490	408	82	339.7
83.3	486	405	81	337.5
83.3	485	404	81	336.5
83.3	484	403	81	335.6
83.3	480	400	80	333.3
83.3	479	399	80	332.4
83.3	478	398	80	331.4
83.3	474	395	79	329.2
83.3	473	394	79	328.2
83.3	472	393	79	327.2
83.3	468	390	78	325.0

%E	M1	M2	DM	M*
83.3	467	389	78	324.0
83.3	466	388	78	323.1
83.3	462	385	77	320.8
83.3	461	384	77	319.9
83.3	460	383	77	318.9
83.3	456	380	76	316.7
83.3	455	379	76	315.7
83.3	454	378	76	314.7
83.3	450	375	75	312.5
83.3	449	374	75	311.5
83.3	448	373	75	310.6
83.3	444	370	74	308.3
83.3	443	369	74	307.4
83.3	442	368	74	306.4
83.3	438	365	73	304.2
83.3	437	364	73	303.2
83.3	436	363	73	302.2
83.3	432	360	72	300.0
83.3	431	359	72	299.0
83.3	430	358	72	298.1
83.3	426	355	71	295.8
83.3	425	354	71	294.9
83.3	424	353	71	293.9
83.3	420	350	70	291.7
83.3	419	349	70	290.7
83.3	418	348	70	289.7
83.3	414	345	69	287.5
83.3	413	344	69	286.5
83.3	412	343	69	285.6
83.3	408	340	68	283.3
83.3	407	339	68	282.4
83.3	406	338	68	281.4
83.3	402	335	67	279.2
83.3	401	334	67	278.2
83.3	400	333	67	277.2
83.3	396	330	66	275.0
83.3	395	329	66	274.0
83.3	390	325	65	270.8
83.3	389	324	65	269.9
83.3	384	320	64	266.7
83.3	383	319	64	265.7
83.3	378	315	63	262.5
83.3	377	314	63	261.5
83.3	372	310	62	258.3
83.3	371	309	62	257.4
83.3	366	305	61	254.2
83.3	365	304	61	253.2
83.3	360	300	60	250.0
83.3	359	299	60	249.0
83.3	354	295	59	245.8

%E	M1	M2	DM	M*
83.3	353	294	59	244.9
83.3	348	290	58	241.7
83.3	347	289	58	240.7
83.3	342	285	57	237.5
83.3	341	284	57	236.5
83.3	336	280	56	233.3
83.3	335	279	56	232.4
83.3	330	275	55	229.2
83.3	329	274	55	228.2
83.3	324	270	54	225.0
83.3	323	269	54	224.0
83.3	318	265	53	220.8
83.3	317	264	53	219.9
83.3	312	260	52	216.7
83.3	311	259	52	215.7
83.3	306	255	51	212.5
83.3	305	254	51	211.5
83.3	300	250	50	208.3
83.3	299	249	50	207.4
83.3	294	245	49	204.2
83.3	293	244	49	203.2
83.3	288	240	48	200.0
83.3	287	239	48	199.0
83.3	282	235	47	195.8
83.3	281	234	47	194.9
83.3	276	230	46	191.7
83.3	275	229	46	190.7
83.3	270	225	45	187.5
83.3	269	224	45	186.5
83.3	264	220	44	183.3
83.3	263	219	44	182.4
83.3	258	215	43	179.2
83.3	257	214	43	178.2
83.3	252	210	42	175.0
83.3	251	209	42	174.0
83.3	246	205	41	170.8
83.3	245	204	41	169.9
83.3	240	200	40	166.7
83.3	239	199	40	165.7
83.3	234	195	39	162.5
83.3	233	194	39	161.5
83.3	228	190	38	158.3
83.3	227	189	38	157.4
83.3	222	185	37	154.2
83.3	221	184	37	153.2
83.3	216	180	36	150.0
83.3	215	179	36	149.0
83.3	210	175	35	145.8
83.3	209	174	35	144.9
83.3	204	170	34	141.7

%E	M1	M2	DM	M*
83.3	203	169	34	140.7
83.3	198	165	33	137.5
83.3	192	160	32	133.3
83.3	186	155	31	129.2
83.3	180	150	30	125.0
83.3	174	145	29	120.8
83.3	168	140	28	116.7
83.3	162	135	27	112.5
83.3	156	130	26	108.3
83.3	150	125	25	104.2
83.3	144	120	24	100.0
83.3	138	115	23	95.8
83.3	132	110	22	91.7
83.3	126	105	21	87.5
83.3	120	100	20	83.3
83.3	114	95	19	79.2
83.3	108	90	18	75.0
83.3	102	85	17	70.8
83.3	96	80	16	66.7
83.3	90	75	15	62.5
83.3	84	70	14	58.3
83.3	78	65	13	54.2
83.3	72	60	12	50.0
83.3	66	55	11	45.8
83.3	60	50	10	41.7
83.3	54	45	9	37.5
83.3	48	40	8	33.3
83.3	42	35	7	29.2
83.3	36	30	6	25.0
83.3	30	25	5	20.8
83.3	24	20	4	16.7
83.3	18	15	3	12.5
83.3	12	10	2	8.3
83.3	6	5	1	4.2
83.2	500	416	84	346.1
83.2	499	415	84	345.1
83.2	495	412	83	342.9
83.2	494	411	83	341.9
83.2	493	410	83	341.0
83.2	489	407	82	338.8
83.2	488	406	82	337.8
83.2	487	405	82	336.8
83.2	483	402	81	334.6
83.2	482	401	81	333.6
83.2	481	400	81	332.6
83.2	477	397	80	330.4
83.2	476	396	80	329.4
83.2	475	395	80	328.5
83.2	471	392	79	326.3
83.2	470	391	79	325.3

%E	M1	M2	DM	M*
83.2	469	390	79	324.3
83.2	465	387	78	322.1
83.2	464	386	78	321.1
83.2	463	385	78	320.1
83.2	459	382	77	317.9
83.2	458	381	77	316.9
83.2	457	380	77	316.0
83.2	453	377	76	313.8
83.2	452	376	76	312.8
83.2	447	372	75	309.6
83.2	446	371	75	308.6
83.2	441	367	74	305.4
83.2	440	366	74	304.4
83.2	435	362	73	301.3
83.2	434	361	73	300.3
83.2	429	357	72	297.1
83.2	428	356	72	296.1
83.2	423	352	71	292.9
83.2	422	351	71	291.9
83.2	417	347	70	288.8
83.2	416	346	70	287.8
83.2	411	342	69	284.6
83.2	410	341	69	283.6
83.2	405	337	68	280.4
83.2	404	336	68	279.4
83.2	399	332	67	276.3
83.2	398	331	67	275.3
83.2	394	328	66	273.1
83.2	393	327	66	272.1
83.2	392	326	66	271.1
83.2	388	323	65	268.9
83.2	387	322	65	267.9
83.2	386	321	65	266.9
83.2	382	318	64	264.7
83.2	381	317	64	263.8
83.2	380	316	64	262.8
83.2	376	313	63	260.6
83.2	375	312	63	259.6
83.2	374	311	63	258.6
83.2	370	308	62	256.4
83.2	369	307	62	255.4
83.2	368	306	62	254.4
83.2	364	303	61	252.2
83.2	363	302	61	251.3
83.2	358	298	60	248.1
83.2	357	297	60	247.1
83.2	352	293	59	243.9
83.2	351	292	59	242.9
83.2	346	288	58	239.7
83.2	345	287	58	238.8

%E	M1	M2	DM	M*
83.2	340	283	57	235.6
83.2	339	282	57	234.6
83.2	334	278	56	231.4
83.2	333	277	56	230.4
83.2	328	273	55	227.2
83.2	327	272	55	226.3
83.2	322	268	54	223.1
83.2	321	267	54	222.1
83.2	316	263	53	218.9
83.2	315	262	53	217.9
83.2	310	258	52	214.7
83.2	309	257	52	213.8
83.2	304	253	51	210.6
83.2	303	252	51	209.6
83.2	298	248	50	206.4
83.2	297	247	50	205.4
83.2	292	243	49	202.2
83.2	291	242	49	201.3
83.2	286	238	48	198.1
83.2	285	237	48	197.1
83.2	280	233	47	193.9
83.2	279	232	47	192.9
83.2	274	228	46	189.7
83.2	273	227	46	188.8
83.2	268	223	45	185.6
83.2	262	218	44	181.4
83.2	256	213	43	177.2
83.2	250	208	42	173.1
83.2	244	203	41	168.9
83.2	238	198	40	164.7
83.2	232	193	39	160.6
83.2	226	188	38	156.4
83.2	220	183	37	152.2
83.2	214	178	36	148.1
83.2	208	173	35	143.9
83.2	202	168	34	139.7
83.2	197	164	33	136.5
83.2	196	163	33	135.6
83.2	191	159	32	132.4
83.2	190	158	32	131.4
83.2	185	154	31	128.2
83.2	184	153	31	127.2
83.2	179	149	30	124.0
83.2	173	144	29	119.9
83.2	167	139	28	115.7
83.2	161	134	27	111.5
83.2	155	129	26	107.4
83.2	149	124	25	103.2
83.2	143	119	24	99.0
83.2	137	114	23	94.9

%E	M1	M2	DM	M*
83.2	131	109	22	90.7
83.2	125	104	21	86.5
83.2	119	99	20	82.4
83.2	113	94	19	78.2
83.2	107	89	18	74.0
83.2	101	84	17	69.9
83.2	95	79	16	65.7
83.1	498	414	84	344.2
83.1	497	413	84	343.2
83.1	496	412	84	342.2
83.1	492	409	83	340.0
83.1	491	408	83	339.0
83.1	490	407	83	338.1
83.1	486	404	82	335.8
83.1	485	403	82	334.9
83.1	484	402	82	333.9
83.1	480	399	81	331.7
83.1	479	398	81	330.7
83.1	478	397	81	329.7
83.1	474	394	80	327.5
83.1	473	393	80	326.5
83.1	472	392	80	325.6
83.1	468	389	79	323.3
83.1	467	388	79	322.4
83.1	462	384	78	319.2
83.1	461	383	78	318.2
83.1	456	379	77	315.0
83.1	455	378	77	314.0
83.1	451	375	76	311.8
83.1	450	374	76	310.8
83.1	449	373	76	309.9
83.1	445	370	75	307.6
83.1	444	369	75	306.7
83.1	443	368	75	305.7
83.1	439	365	74	303.5
83.1	438	364	74	302.5
83.1	437	363	74	301.5
83.1	433	360	73	299.3
83.1	432	359	73	298.3
83.1	431	358	73	297.4
83.1	427	355	72	295.1
83.1	426	354	72	294.2
83.1	425	353	72	293.2
83.1	421	350	71	291.0
83.1	420	349	71	290.0
83.1	419	348	71	289.0
83.1	415	345	70	286.8
83.1	414	344	70	285.8
83.1	413	343	70	284.9
83.1	409	340	69	282.6

%E	M1	M2	DM	M*
83.1	408	339	69	281.7
83.1	403	335	68	278.5
83.1	402	334	68	277.5
83.1	397	330	67	274.3
83.1	396	329	67	273.3
83.1	391	325	66	270.1
83.1	390	324	66	269.2
83.1	385	320	65	266.0
83.1	384	319	65	265.0
83.1	379	315	64	261.8
83.1	378	314	64	260.8
83.1	373	310	63	257.6
83.1	372	309	63	256.7
83.1	367	305	62	253.5
83.1	366	304	62	252.5
83.1	362	301	61	250.3
83.1	361	300	61	249.3
83.1	360	299	61	248.3
83.1	356	296	60	246.1
83.1	355	295	60	245.1
83.1	354	294	60	244.2
83.1	350	291	59	241.9
83.1	349	290	59	241.0
83.1	344	286	58	237.8
83.1	343	285	58	236.8
83.1	338	281	57	233.6
83.1	337	280	57	232.6
83.1	332	276	56	229.4
83.1	331	275	56	228.5
83.1	326	271	55	225.3
83.1	325	270	55	224.3
83.1	320	266	54	221.1
83.1	319	265	54	220.1
83.1	314	261	53	216.9
83.1	313	260	53	216.0
83.1	308	256	52	212.8
83.1	307	255	52	211.8
83.1	302	251	51	208.6
83.1	301	250	51	207.6
83.1	296	246	50	204.4
83.1	295	245	50	203.5
83.1	290	241	49	200.3
83.1	284	236	48	196.1
83.1	278	231	47	191.9
83.1	272	226	46	187.8
83.1	267	222	45	184.6
83.1	266	221	45	183.6
83.1	261	217	44	180.4
83.1	260	216	44	179.4
83.1	255	212	43	176.3

%E	M1	M2	DM	M*
83.1	254	211	43	175.3
83.1	249	207	42	172.1
83.1	248	206	42	171.1
83.1	243	202	41	167.9
83.1	242	201	41	166.9
83.1	237	197	40	163.8
83.1	236	196	40	162.8
83.1	231	192	39	159.6
83.1	225	187	38	155.4
83.1	219	182	37	151.3
83.1	213	177	36	147.1
83.1	207	172	35	142.9
83.1	201	167	34	138.8
83.1	195	162	33	134.6
83.1	189	157	32	130.4
83.1	183	152	31	126.3
83.1	178	148	30	123.1
83.1	177	147	30	122.1
83.1	172	143	29	118.9
83.1	166	138	28	114.7
83.1	160	133	27	110.6
83.1	154	128	26	106.4
83.1	148	123	25	102.2
83.1	142	118	24	98.1
83.1	136	113	23	93.9
83.1	130	108	22	89.7
83.1	124	103	21	85.6
83.1	118	98	20	81.4
83.1	89	74	15	61.5
83.1	83	69	14	57.4
83.1	77	64	13	53.2
83.1	71	59	12	49.0
83.1	65	54	11	44.9
83.1	59	49	10	40.7
83.0	500	415	85	344.4
83.0	499	414	85	343.5
83.0	495	411	84	341.3
83.0	494	410	84	340.3
83.0	493	409	84	339.3
83.0	489	406	83	337.1
83.0	488	405	83	336.1
83.0	487	404	83	335.1
83.0	483	401	82	332.9
83.0	482	400	82	332.0
83.0	481	399	82	331.0
83.0	477	396	81	328.8
83.0	476	395	81	327.8
83.0	471	391	80	324.6
83.0	470	390	80	323.6
83.0	466	387	79	321.4

%E	M1	M2	DM	M*
83.0	465	386	79	320.4
83.0	464	385	79	319.5
83.0	460	382	78	317.2
83.0	459	381	78	316.3
83.0	458	380	78	315.3
83.0	454	377	77	313.1
83.0	453	376	77	312.1
83.0	452	375	77	311.1
83.0	448	372	76	308.9
83.0	447	371	76	307.9
83.0	446	370	76	307.0
83.0	442	367	75	304.7
83.0	441	366	75	303.8
83.0	440	365	75	302.8
83.0	436	362	74	300.6
83.0	435	361	74	299.6
83.0	430	357	73	296.4
83.0	429	356	73	295.4
83.0	424	352	72	292.2
83.0	423	351	72	291.3
83.0	418	347	71	288.1
83.0	417	346	71	287.1
83.0	412	342	70	283.9
83.0	411	341	70	282.9
83.0	407	338	69	280.7
83.0	406	337	69	279.7
83.0	405	336	69	278.8
83.0	401	333	68	276.5
83.0	400	332	68	275.6
83.0	399	331	68	274.6
83.0	395	328	67	272.4
83.0	394	327	67	271.4
83.0	393	326	67	270.4
83.0	389	323	66	268.2
83.0	388	322	66	267.2
83.0	383	318	65	264.0
83.0	382	317	65	263.1
83.0	377	313	64	259.9
83.0	376	312	64	258.9
83.0	371	308	63	255.7
83.0	370	307	63	254.7
83.0	365	303	62	251.5
83.0	364	302	62	250.6
83.0	359	298	61	247.4
83.0	358	297	61	246.4
83.0	353	293	60	243.2
83.0	352	292	60	242.2
83.0	348	289	59	240.0
83.0	347	288	59	239.0
83.0	342	284	58	235.8

%E	M1	M2	DM	M*
83.0	341	283	58	234.9
83.0	336	279	57	231.7
83.0	335	278	57	230.7
83.0	330	274	56	227.5
83.0	329	273	56	226.5
83.0	324	269	55	223.3
83.0	323	268	55	222.4
83.0	318	264	54	219.2
83.0	317	263	54	218.2
83.0	312	259	53	215.0
83.0	311	258	53	214.0
83.0	306	254	52	210.8
83.0	305	253	52	209.9
83.0	300	249	51	206.7
83.0	294	244	50	202.5
83.0	289	240	49	199.3
83.0	288	239	49	198.3
83.0	283	235	48	195.1
83.0	282	234	48	194.2
83.0	277	230	47	191.0
83.0	276	229	47	190.0
83.0	271	225	46	186.8
83.0	270	224	46	185.8
83.0	265	220	45	182.6
83.0	264	219	45	181.7
83.0	259	215	44	178.5
83.0	253	210	43	174.3
83.0	247	205	42	170.1
83.0	241	200	41	166.0
83.0	235	195	40	161.8
83.0	230	191	39	158.6
83.0	229	190	39	157.6
83.0	224	186	38	154.4
83.0	223	185	38	153.5
83.0	218	181	37	150.3
83.0	212	176	36	146.1
83.0	206	171	35	141.9
83.0	200	166	34	137.8
83.0	194	161	33	133.6
83.0	188	156	32	129.4
83.0	182	151	31	125.3
83.0	176	146	30	121.1
83.0	171	142	29	117.9
83.0	165	137	28	113.8
83.0	159	132	27	109.6
83.0	153	127	26	105.4
83.0	147	122	25	101.3
83.0	141	117	24	97.1
83.0	135	112	23	92.9
83.0	112	93	19	77.2

%E	M1	M2	DM	M*
83.0	106	88	18	73.1
83.0	100	83	17	68.9
83.0	94	78	16	64.7
83.0	88	73	15	60.6
83.0	53	44	9	36.5
83.0	47	39	8	32.4
82.9	498	413	85	342.5
82.9	497	412	85	341.5
82.9	496	411	85	340.6
82.9	492	408	84	338.3
82.9	491	407	84	337.4
82.9	490	406	84	336.4
82.9	486	403	83	334.2
82.9	485	402	83	333.2
82.9	484	401	83	332.2
82.9	480	398	82	330.0
82.9	479	397	82	329.0
82.9	475	394	81	326.8
82.9	474	393	81	325.8
82.9	473	392	81	324.9
82.9	469	389	80	322.6
82.9	468	388	80	321.7
82.9	467	387	80	320.7
82.9	463	384	79	318.5
82.9	462	383	79	317.5
82.9	461	382	79	316.5
82.9	457	379	78	314.3
82.9	456	378	78	313.3
82.9	455	377	78	312.4
82.9	451	374	77	310.1
82.9	450	373	77	309.2
82.9	449	372	77	308.2
82.9	445	369	76	306.0
82.9	444	368	76	305.0
82.9	439	364	75	301.8
82.9	438	363	75	300.8
82.9	434	360	74	298.6
82.9	433	359	74	297.6
82.9	432	358	74	296.7
82.9	428	355	73	294.5
82.9	427	354	73	293.5
82.9	426	353	73	292.5
82.9	422	350	72	290.3
82.9	421	349	72	289.3
82.9	420	348	72	288.3
82.9	416	345	71	286.1
82.9	415	344	71	285.1
82.9	414	343	71	284.2
82.9	410	340	70	282.0
82.9	409	339	70	281.0

%E	M1	M2	DM	M*
82.9	404	335	69	277.8
82.9	403	334	69	276.8
82.9	398	330	68	273.6
82.9	397	329	68	272.6
82.9	392	325	67	269.5
82.9	391	324	67	268.5
82.9	387	321	66	266.3
82.9	386	320	66	265.3
82.9	385	319	66	264.3
82.9	381	316	65	262.1
82.9	380	315	65	261.1
82.9	379	314	65	260.1
82.9	375	311	64	257.9
82.9	374	310	64	257.0
82.9	369	306	63	253.8
82.9	368	305	63	252.8
82.9	363	301	62	249.6
82.9	362	300	62	248.6
82.9	357	296	61	245.4
82.9	356	295	61	244.5
82.9	351	291	60	241.3
82.9	350	290	60	240.3
82.9	346	287	59	238.1
82.9	345	286	59	237.1
82.9	340	282	58	233.9
82.9	339	281	58	232.9
82.9	334	277	57	229.7
82.9	333	276	57	228.8
82.9	328	272	56	225.6
82.9	327	271	56	224.6
82.9	322	267	55	221.4
82.9	321	266	55	220.4
82.9	316	262	54	217.2
82.9	315	261	54	216.3
82.9	310	257	53	213.1
82.9	304	252	52	208.9
82.9	299	248	51	205.7
82.9	298	247	51	204.7
82.9	293	243	50	201.5
82.9	292	242	50	200.6
82.9	287	238	49	197.4
82.9	286	237	49	196.4
82.9	281	233	48	193.2
82.9	280	232	48	192.2
82.9	275	228	47	189.0
82.9	269	223	46	184.9
82.9	263	218	45	180.7
82.9	258	214	44	177.5
82.9	257	213	44	176.5
82.9	252	209	43	173.3

%E	M1	M2	DM	M*
82.9	251	208	43	172.4
82.9	246	204	42	169.2
82.9	245	203	42	168.2
82.9	240	199	41	165.0
82.9	234	194	40	160.8
82.9	228	189	39	156.7
82.9	222	184	38	152.5
82.9	217	180	37	149.3
82.9	216	179	37	148.3
82.9	211	175	36	145.1
82.9	210	174	36	144.2
82.9	205	170	35	141.0
82.9	199	165	34	136.8
82.9	193	160	33	132.6
82.9	187	155	32	128.5
82.9	181	150	31	124.3
82.9	175	145	30	120.1
82.9	170	141	29	116.9
82.9	164	136	28	112.8
82.9	158	131	27	108.6
82.9	152	126	26	104.4
82.9	146	121	25	100.3
82.9	140	116	24	96.1
82.9	129	107	22	88.8
82.9	123	102	21	84.6
82.9	117	97	20	80.4
82.9	111	92	19	76.3
82.9	105	87	18	72.1
82.9	82	68	14	56.4
82.9	76	63	13	52.2
82.9	70	58	12	48.1
82.9	41	34	7	28.2
82.9	35	29	6	24.0
82.8	500	414	86	342.8
82.8	499	413	86	341.8
82.8	495	410	85	339.6
82.8	494	409	85	338.6
82.8	493	408	85	337.7
82.8	489	405	84	335.4
82.8	488	404	84	334.5
82.8	487	403	84	333.5
82.8	483	400	83	331.3
82.8	482	399	83	330.3
82.8	478	396	82	328.1
82.8	477	395	82	327.1
82.8	476	394	82	326.1
82.8	472	391	81	323.9
82.8	471	390	81	322.9
82.8	470	389	81	322.0
82.8	466	386	80	319.7

%E	M1	M2	DM	M*
82.8	465	385	80	318.8
82.8	464	384	80	317.8
82.8	460	381	79	315.6
82.8	459	380	79	314.6
82.8	458	379	79	313.6
82.8	454	376	78	311.4
82.8	453	375	78	310.4
82.8	448	371	77	307.2
82.8	447	370	77	306.3
82.8	443	367	76	304.0
82.8	442	366	76	303.1
82.8	441	365	76	302.1
82.8	437	362	75	299.9
82.8	436	361	75	298.9
82.8	435	360	75	297.9
82.8	431	357	74	295.7
82.8	430	356	74	294.7
82.8	429	355	74	293.8
82.8	425	352	73	291.5
82.8	424	351	73	290.6
82.8	419	347	72	287.4
82.8	418	346	72	286.4
82.8	413	342	71	283.2
82.8	412	341	71	282.2
82.8	408	338	70	280.0
82.8	407	337	70	279.0
82.8	406	336	70	278.1
82.8	402	333	69	275.8
82.8	401	332	69	274.9
82.8	400	331	69	273.9
82.8	396	328	68	271.7
82.8	395	327	68	270.7
82.8	390	323	67	267.5
82.8	389	322	67	266.5
82.8	384	318	66	263.3
82.8	383	317	66	262.4
82.8	378	313	65	259.2
82.8	377	312	65	258.2
82.8	373	309	64	256.0
82.8	372	308	64	255.0
82.8	367	304	63	251.8
82.8	366	303	63	250.8
82.8	361	299	62	247.6
82.8	360	298	62	246.7
82.8	355	294	61	243.5
82.8	354	293	61	242.5
82.8	349	289	60	239.3
82.8	348	288	60	238.3
82.8	344	285	59	236.1
82.8	343	284	59	235.1

%E	M1	M2	DM	M*
82.8	338	280	58	232.0
82.8	337	279	58	231.0
82.8	332	275	57	227.8
82.8	331	274	57	226.8
82.8	326	270	56	223.6
82.8	325	269	56	222.6
82.8	320	265	55	219.5
82.8	319	264	55	218.5
82.8	314	260	54	215.3
82.8	309	256	53	212.1
82.8	308	255	53	211.1
82.8	303	251	52	207.9
82.8	302	250	52	207.0
82.8	297	246	51	203.8
82.8	296	245	51	202.8
82.8	291	241	50	199.6
82.8	290	240	50	198.6
82.8	285	236	49	195.4
82.8	279	231	48	191.3
82.8	274	227	47	188.1
82.8	273	226	47	187.1
82.8	268	222	46	183.9
82.8	267	221	46	182.9
82.8	262	217	45	179.7
82.8	261	216	45	178.8
82.8	256	212	44	175.6
82.8	250	207	43	171.4
82.8	244	202	42	167.2
82.8	239	198	41	164.0
82.8	238	197	41	163.1
82.8	233	193	40	159.9
82.8	232	192	40	158.9
82.8	227	188	39	155.7
82.8	221	183	38	151.5
82.8	215	178	37	147.4
82.8	209	173	36	143.2
82.8	204	169	35	140.0
82.8	203	168	35	139.0
82.8	198	164	34	135.8
82.8	192	159	33	131.7
82.8	186	154	32	127.5
82.8	180	149	31	123.3
82.8	174	144	30	119.2
82.8	169	140	29	116.0
82.8	163	135	28	111.8
82.8	157	130	27	107.6
82.8	151	125	26	103.5
82.8	145	120	25	99.3
82.8	134	111	23	91.9
82.8	128	106	22	87.8

%E	M1	M2	DM	M*
82.8	122	101	21	83.6
82.8	116	96	20	79.4
82.8	99	82	17	67.9
82.8	93	77	16	63.8
82.8	87	72	15	59.6
82.8	64	53	11	43.9
82.8	58	48	10	39.7
82.8	29	24	5	19.9
82.7	498	412	86	340.9
82.7	497	411	86	339.9
82.7	496	410	86	338.9
82.7	492	407	85	336.7
82.7	491	406	85	335.7
82.7	490	405	85	334.7
82.7	486	402	84	332.5
82.7	485	401	84	331.5
82.7	481	398	83	329.3
82.7	480	397	83	328.4
82.7	479	396	83	327.4
82.7	475	393	82	325.2
82.7	474	392	82	324.2
82.7	473	391	82	323.2
82.7	469	388	81	321.0
82.7	468	387	81	320.0
82.7	467	386	81	319.0
82.7	463	383	80	316.8
82.7	462	382	80	315.9
82.7	457	378	79	312.7
82.7	456	377	79	311.7
82.7	452	374	78	309.5
82.7	451	373	78	308.5
82.7	450	372	78	307.5
82.7	446	369	77	305.3
82.7	445	368	77	304.3
82.7	444	367	77	303.4
82.7	440	364	76	301.1
82.7	439	363	76	300.2
82.7	434	359	75	297.0
82.7	433	358	75	296.0
82.7	428	354	74	292.8
82.7	427	353	74	291.8
82.7	423	350	73	289.6
82.7	422	349	73	288.6
82.7	421	348	73	287.7
82.7	417	345	72	285.4
82.7	416	344	72	284.5
82.7	415	343	72	283.5
82.7	411	340	71	281.3
82.7	410	339	71	280.3
82.7	405	335	70	277.1

%E	M1	M2	DM	M*
82.7	404	334	70	276.1
82.7	399	330	69	272.9
82.7	398	329	69	272.0
82.7	394	326	68	269.7
82.7	393	325	68	268.8
82.7	392	324	68	267.8
82.7	388	321	67	265.6
82.7	387	320	67	264.6
82.7	382	316	66	261.4
82.7	381	315	66	260.4
82.7	376	311	65	257.2
82.7	375	310	65	256.3
82.7	371	307	64	254.0
82.7	370	306	64	253.1
82.7	369	305	64	252.1
82.7	365	302	63	249.9
82.7	364	301	63	248.9
82.7	359	297	62	245.7
82.7	358	296	62	244.7
82.7	353	292	61	241.5
82.7	352	291	61	240.6
82.7	347	287	60	237.4
82.7	346	286	60	236.4
82.7	342	283	59	234.2
82.7	341	282	59	233.2
82.7	336	278	58	230.0
82.7	335	277	58	229.0
82.7	330	273	57	225.8
82.7	329	272	57	224.9
82.7	324	268	56	221.7
82.7	323	267	56	220.7
82.7	318	263	55	217.5
82.7	317	262	55	216.5
82.7	313	259	54	214.3
82.7	312	258	54	213.3
82.7	307	254	53	210.1
82.7	306	253	53	209.2
82.7	301	249	52	206.0
82.7	300	248	52	205.0
82.7	295	244	51	201.8
82.7	294	243	51	200.8
82.7	289	239	50	197.7
82.7	284	235	49	194.5
82.7	283	234	49	193.5
82.7	278	230	48	190.3
82.7	277	229	48	189.3
82.7	272	225	47	186.1
82.7	271	224	47	185.2
82.7	266	220	46	182.0
82.7	260	215	45	177.8

%E	M1	M2	DM	M*
82.7	255	211	44	174.6
82.7	254	210	44	173.6
82.7	249	206	43	170.4
82.7	248	205	43	169.5
82.7	243	201	42	166.3
82.7	237	196	41	162.1
82.7	231	191	40	157.9
82.7	226	187	39	154.7
82.7	225	186	39	153.8
82.7	220	182	38	150.6
82.7	214	177	37	146.4
82.7	208	172	36	142.2
82.7	202	167	35	138.1
82.7	197	163	34	134.9
82.7	196	162	34	133.9
82.7	191	158	33	130.7
82.7	185	153	32	126.5
82.7	179	148	31	122.4
82.7	173	143	30	118.2
82.7	168	139	29	115.0
82.7	162	134	28	110.8
82.7	156	129	27	106.7
82.7	150	124	26	102.5
82.7	139	115	24	95.1
82.7	133	110	23	91.0
82.7	127	105	22	86.8
82.7	110	91	19	75.3
82.7	104	86	18	71.1
82.7	98	81	17	66.9
82.7	81	67	14	55.4
82.7	75	62	13	51.3
82.7	52	43	9	35.6
82.6	500	413	87	341.1
82.6	499	412	87	340.2
82.6	495	409	86	337.9
82.6	494	408	86	337.0
82.6	493	407	86	336.0
82.6	489	404	85	333.8
82.6	488	403	85	332.8
82.6	484	400	84	330.6
82.6	483	399	84	329.6
82.6	482	398	84	328.6
82.6	478	395	83	326.4
82.6	477	394	83	325.4
82.6	476	393	83	324.5
82.6	472	390	82	322.2
82.6	471	389	82	321.3
82.6	470	388	82	320.3
82.6	466	385	81	318.1
82.6	465	384	81	317.1

%E	M1	M2	DM	M*
82.6	461	381	80	314.9
82.6	460	380	80	313.9
82.6	459	379	80	312.9
82.6	455	376	79	310.7
82.6	454	375	79	309.7
82.6	453	374	79	308.8
82.6	449	371	78	306.6
82.6	448	370	78	305.6
82.6	447	369	78	304.6
82.6	443	366	77	302.4
82.6	442	365	77	301.4
82.6	438	362	76	299.2
82.6	437	361	76	298.2
82.6	436	360	76	297.2
82.6	432	357	75	295.0
82.6	431	356	75	294.1
82.6	430	355	75	293.1
82.6	426	352	74	290.9
82.6	425	351	74	289.9
82.6	420	347	73	286.7
82.6	419	346	73	285.7
82.6	414	342	72	282.5
82.6	413	341	72	281.6
82.6	409	338	71	279.3
82.6	408	337	71	278.4
82.6	407	336	71	277.4
82.6	403	333	70	275.2
82.6	402	332	70	274.2
82.6	397	328	69	271.0
82.6	396	327	69	270.0
82.6	391	323	68	266.8
82.6	390	322	68	265.9
82.6	386	319	67	263.6
82.6	385	318	67	262.7
82.6	384	317	67	261.7
82.6	380	314	66	259.5
82.6	379	313	66	258.5
82.6	374	309	65	255.3
82.6	373	308	65	254.3
82.6	368	304	64	251.1
82.6	367	303	64	250.2
82.6	363	300	63	247.9
82.6	362	299	63	247.0
82.6	357	295	62	243.8
82.6	356	294	62	242.8
82.6	351	290	61	239.6
82.6	350	289	61	238.6
82.6	345	285	60	235.4
82.6	344	284	60	234.5
82.6	340	281	59	232.2

%E	M1	M2	DM	M*
82.6	339	280	59	231.3
82.6	334	276	58	228.1
82.6	333	275	58	227.1
82.6	328	271	57	223.9
82.6	327	270	57	222.9
82.6	322	266	56	219.7
82.6	321	265	56	218.8
82.6	316	261	55	215.6
82.6	311	257	54	212.4
82.6	310	256	54	211.4
82.6	305	252	53	208.2
82.6	304	251	53	207.2
82.6	299	247	52	204.0
82.6	298	246	52	203.1
82.6	293	242	51	199.9
82.6	288	238	50	196.7
82.6	287	237	50	195.7
82.6	282	233	49	192.5
82.6	281	232	49	191.5
82.6	276	228	48	188.3
82.6	270	223	47	184.2
82.6	265	219	46	181.0
82.6	264	218	46	180.0
82.6	259	214	45	176.8
82.6	258	213	45	175.8
82.6	253	209	44	172.7
82.6	247	204	43	168.5
82.6	242	200	42	165.3
82.6	241	199	42	164.3
82.6	236	195	41	161.1
82.6	235	194	41	160.2
82.6	230	190	40	157.0
82.6	224	185	39	152.8
82.6	219	181	38	149.6
82.6	218	180	38	148.6
82.6	213	176	37	145.4
82.6	207	171	36	141.3
82.6	201	166	35	137.1
82.6	195	161	34	132.9
82.6	190	157	33	129.7
82.6	184	152	32	125.6
82.6	178	147	31	121.4
82.6	172	142	30	117.2
82.6	167	138	29	114.0
82.6	161	133	28	109.9
82.6	155	128	27	105.7
82.6	149	123	26	101.5
82.6	144	119	25	98.3
82.6	138	114	24	94.2
82.6	132	109	23	90.0

%E	M1	M2	DM	M*
82.6	121	100	21	82.6
82.6	115	95	20	78.5
82.6	109	90	19	74.3
82.6	92	76	16	62.8
82.6	86	71	15	58.6
82.6	69	57	12	47.1
82.6	46	38	8	31.4
82.6	23	19	4	15.7
82.5	498	411	87	339.2
82.5	497	410	87	338.2
82.5	496	409	87	337.3
82.5	492	406	86	335.0
82.5	491	405	86	334.1
82.5	487	402	85	331.8
82.5	486	401	85	330.9
82.5	485	400	85	329.9
82.5	481	397	84	327.7
82.5	480	396	84	326.7
82.5	479	395	84	325.7
82.5	475	392	83	323.5
82.5	474	391	83	322.5
82.5	473	390	83	321.6
82.5	469	387	82	319.3
82.5	468	386	82	318.4
82.5	464	383	81	316.1
82.5	463	382	81	315.2
82.5	462	381	81	314.2
82.5	458	378	80	312.0
82.5	457	377	80	311.0
82.5	456	376	80	310.0
82.5	452	373	79	307.8
82.5	451	372	79	306.8
82.5	446	368	78	303.6
82.5	445	367	78	302.7
82.5	441	364	77	300.4
82.5	440	363	77	299.5
82.5	439	362	77	298.5
82.5	435	359	76	296.3
82.5	434	358	76	295.3
82.5	429	354	75	292.1
82.5	428	353	75	291.1
82.5	424	350	74	288.9
82.5	423	349	74	287.9
82.5	422	348	74	287.0
82.5	418	345	73	284.7
82.5	417	344	73	283.8
82.5	416	343	73	282.8
82.5	412	340	72	280.6
82.5	411	339	72	279.6
82.5	406	335	71	276.4

%E	M1	M2	DM	M*
82.5	405	334	71	275.4
82.5	401	331	70	273.2
82.5	400	330	70	272.3
82.5	399	329	70	271.3
82.5	395	326	69	269.1
82.5	394	325	69	268.1
82.5	389	321	68	264.9
82.5	388	320	68	263.9
82.5	383	316	67	260.7
82.5	382	315	67	259.8
82.5	378	312	66	257.5
82.5	377	311	66	256.6
82.5	372	307	65	253.4
82.5	371	306	65	252.4
82.5	366	302	64	249.2
82.5	365	301	64	248.2
82.5	361	298	63	246.0
82.5	360	297	63	245.0
82.5	359	296	63	244.1
82.5	355	293	62	241.8
82.5	354	292	62	240.9
82.5	349	288	61	237.7
82.5	348	287	61	236.7
82.5	343	283	60	233.5
82.5	342	282	60	232.5
82.5	338	279	59	230.3
82.5	337	278	59	229.3
82.5	332	274	58	226.1
82.5	331	273	58	225.2
82.5	326	269	57	222.0
82.5	325	268	57	221.0
82.5	320	264	56	217.8
82.5	315	260	55	214.6
82.5	314	259	55	213.6
82.5	309	255	54	210.4
82.5	308	254	54	209.5
82.5	303	250	53	206.3
82.5	302	249	53	205.3
82.5	297	245	52	202.1
82.5	292	241	51	198.9
82.5	291	240	51	197.9
82.5	286	236	50	194.7
82.5	285	235	50	193.8
82.5	280	231	49	190.6
82.5	275	227	48	187.4
82.5	274	226	48	186.4
82.5	269	222	47	183.2
82.5	268	221	47	182.2
82.5	263	217	46	179.0
82.5	257	212	45	174.9

%E	M1	M2	DM	M*
82.5	252	208	44	171.7
82.5	251	207	44	170.7
82.5	246	203	43	167.5
82.5	240	198	42	163.3
82.5	234	193	41	159.2
82.5	229	189	40	156.0
82.5	228	188	40	155.0
82.5	223	184	39	151.8
82.5	217	179	38	147.7
82.5	212	175	37	144.5
82.5	211	174	37	143.5
82.5	206	170	36	140.3
82.5	200	165	35	136.1
82.5	194	160	34	132.0
82.5	189	156	33	128.8
82.5	183	151	32	124.6
82.5	177	146	31	120.4
82.5	171	141	30	116.3
82.5	166	137	29	113.1
82.5	160	132	28	108.9
82.5	154	127	27	104.7
82.5	143	118	25	97.4
82.5	137	113	24	93.2
82.5	126	104	22	85.8
82.5	120	99	21	81.7
82.5	114	94	20	77.5
82.5	103	85	18	70.1
82.5	97	80	17	66.0
82.5	80	66	14	54.4
82.5	63	52	11	42.9
82.5	57	47	10	38.8
82.5	40	33	7	27.2
82.4	500	412	88	339.5
82.4	499	411	88	338.5
82.4	495	408	87	336.3
82.4	494	407	87	335.3
82.4	493	406	87	334.4
82.4	490	404	86	333.1
82.4	489	403	86	332.1
82.4	488	402	86	331.2
82.4	484	399	85	328.9
82.4	483	398	85	328.0
82.4	482	397	85	327.0
82.4	478	394	84	324.8
82.4	477	393	84	323.8
82.4	476	392	84	322.8
82.4	472	389	83	320.6
82.4	471	388	83	319.6
82.4	467	385	82	317.4
82.4	466	384	82	316.4

%E	M1	M2	DM	M*
82.4	465	383	82	315.5
82.4	461	380	81	313.2
82.4	460	379	81	312.3
82.4	459	378	81	311.3
82.4	455	375	80	309.1
82.4	454	374	80	308.1
82.4	450	371	79	305.9
82.4	449	370	79	304.9
82.4	448	369	79	303.9
82.4	444	366	78	301.7
82.4	443	365	78	300.7
82.4	442	364	78	299.8
82.4	438	361	77	297.5
82.4	437	360	77	296.6
82.4	433	357	76	294.3
82.4	432	356	76	293.4
82.4	431	355	76	292.4
82.4	427	352	75	290.2
82.4	426	351	75	289.2
82.4	425	350	75	288.2
82.4	421	347	74	286.0
82.4	420	346	74	285.0
82.4	415	342	73	281.8
82.4	414	341	73	280.9
82.4	410	338	72	278.6
82.4	409	337	72	277.7
82.4	408	336	72	276.7
82.4	404	333	71	274.5
82.4	403	332	71	273.5
82.4	398	328	70	270.3
82.4	397	327	70	269.3
82.4	393	324	69	267.1
82.4	392	323	69	266.1
82.4	391	322	69	265.2
82.4	387	319	68	262.9
82.4	386	318	68	262.0
82.4	381	314	67	258.8
82.4	380	313	67	257.8
82.4	376	310	66	255.6
82.4	375	309	66	254.6
82.4	374	308	66	253.6
82.4	370	305	65	251.4
82.4	369	304	65	250.4
82.4	364	300	64	247.3
82.4	363	299	64	246.3
82.4	358	295	63	243.1
82.4	357	294	63	242.1
82.4	353	291	62	239.9
82.4	352	290	62	238.9
82.4	347	286	61	235.7

%E	M1	M2	DM	M*
82.4	346	285	61	234.8
82.4	341	281	60	231.6
82.4	340	280	60	230.6
82.4	336	277	59	228.4
82.4	335	276	59	227.4
82.4	330	272	58	224.2
82.4	329	271	58	223.2
82.4	324	267	57	220.0
82.4	323	266	57	219.1
82.4	319	263	56	216.8
82.4	318	262	56	215.9
82.4	313	258	55	212.7
82.4	312	257	55	211.7
82.4	307	253	54	208.5
82.4	306	252	54	207.5
82.4	301	248	53	204.3
82.4	296	244	52	201.1
82.4	295	243	52	200.2
82.4	290	239	51	197.0
82.4	289	238	51	196.0
82.4	284	234	50	192.8
82.4	279	230	49	189.6
82.4	278	229	49	188.6
82.4	273	225	48	185.4
82.4	272	224	48	184.5
82.4	267	220	47	181.3
82.4	262	216	46	178.1
82.4	261	215	46	177.1
82.4	256	211	45	173.9
82.4	255	210	45	172.9
82.4	250	206	44	169.7
82.4	245	202	43	166.5
82.4	244	201	43	165.6
82.4	239	197	42	162.4
82.4	238	196	42	161.4
82.4	233	192	41	158.2
82.4	227	187	40	154.0
82.4	222	183	39	150.9
82.4	221	182	39	149.9
82.4	216	178	38	146.7
82.4	210	173	37	142.5
82.4	205	169	36	139.3
82.4	204	168	36	138.4
82.4	199	164	35	135.2
82.4	193	159	34	131.0
82.4	188	155	33	127.8
82.4	187	154	33	126.8
82.4	182	150	32	123.6
82.4	176	145	31	119.5
82.4	170	140	30	115.3

%E	M1	M2	DM	M*
82.4	165	136	29	112.1
82.4	159	131	28	107.9
82.4	153	126	27	103.8
82.4	148	122	26	100.6
82.4	142	117	25	96.4
82.4	136	112	24	92.2
82.4	131	108	23	89.0
82.4	125	103	22	84.9
82.4	119	98	21	80.7
82.4	108	89	19	73.3
82.4	102	84	18	69.2
82.4	91	75	16	61.8
82.4	85	70	15	57.6
82.4	74	61	13	50.3
82.4	68	56	12	46.1
82.4	51	42	9	34.6
82.4	34	28	6	23.1
82.4	17	14	3	11.5
82.3	498	410	88	337.6
82.3	497	409	88	336.6
82.3	496	408	88	335.6
82.3	492	405	87	333.4
82.3	491	404	87	332.4
82.3	487	401	86	330.2
82.3	486	400	86	329.2
82.3	485	399	86	328.2
82.3	481	396	85	326.0
82.3	480	395	85	325.1
82.3	479	394	85	324.1
82.3	475	391	84	321.9
82.3	474	390	84	320.9
82.3	470	387	83	318.7
82.3	469	386	83	317.7
82.3	468	385	83	316.7
82.3	464	382	82	314.5
82.3	463	381	82	313.5
82.3	462	380	82	312.6
82.3	458	377	81	310.3
82.3	457	376	81	309.4
82.3	453	373	80	307.1
82.3	452	372	80	306.2
82.3	451	371	80	305.2
82.3	447	368	79	303.0
82.3	446	367	79	302.0
82.3	441	363	78	298.8
82.3	440	362	78	297.8
82.3	436	359	77	295.6
82.3	435	358	77	294.6
82.3	434	357	77	293.7
82.3	430	354	76	291.4

%E	M1	M2	DM	M*
82.3	429	353	76	290.5
82.3	424	349	75	287.3
82.3	423	348	75	286.3
82.3	419	345	74	284.1
82.3	418	344	74	283.1
82.3	417	343	74	282.1
82.3	413	340	73	279.9
82.3	412	339	73	278.9
82.3	407	335	72	275.7
82.3	406	334	72	274.8
82.3	402	331	71	272.5
82.3	401	330	71	271.6
82.3	400	329	71	270.6
82.3	396	326	70	268.4
82.3	395	325	70	267.4
82.3	390	321	69	264.2
82.3	389	320	69	263.2
82.3	385	317	68	261.0
82.3	384	316	68	260.0
82.3	379	312	67	256.8
82.3	378	311	67	255.9
82.3	373	307	66	252.7
82.3	372	306	66	251.7
82.3	368	303	65	249.5
82.3	367	302	65	248.5
82.3	362	298	64	245.3
82.3	361	297	64	244.3
82.3	356	293	63	241.1
82.3	355	292	63	240.2
82.3	351	289	62	238.0
82.3	350	288	62	237.0
82.3	345	284	61	233.8
82.3	344	283	61	232.8
82.3	339	279	60	229.6
82.3	334	275	59	226.4
82.3	333	274	59	225.5
82.3	328	270	58	222.3
82.3	327	269	58	221.3
82.3	322	265	57	218.1
82.3	317	261	56	214.9
82.3	316	260	56	213.9
82.3	311	256	55	210.7
82.3	310	255	55	209.8
82.3	305	251	54	206.6
82.3	300	247	53	203.4
82.3	299	246	53	202.4
82.3	294	242	52	199.2
82.3	293	241	52	198.2
82.3	288	237	51	195.0
82.3	283	233	50	191.8

%E	M1	M2	DM	M*
82.3	282	232	50	190.9
82.3	277	228	49	187.7
82.3	271	223	48	183.5
82.3	266	219	47	180.3
82.3	265	218	47	179.3
82.3	260	214	46	176.1
82.3	254	209	45	172.0
82.3	249	205	44	168.8
82.3	248	204	44	167.8
82.3	243	200	43	164.6
82.3	237	195	42	160.4
82.3	232	191	41	157.2
82.3	231	190	41	156.3
82.3	226	186	40	153.1
82.3	220	181	39	148.9
82.3	215	177	38	145.7
82.3	209	172	37	141.6
82.3	203	167	36	137.4
82.3	198	163	35	134.2
82.3	192	158	34	130.0
82.3	186	153	33	125.9
82.3	181	149	32	122.7
82.3	175	144	31	118.5
82.3	164	135	29	111.1
82.3	158	130	28	107.0
82.3	147	121	26	99.6
82.3	141	116	25	95.4
82.3	130	107	23	88.1
82.3	124	102	22	83.9
82.3	113	93	20	76.5
82.3	96	79	17	65.0
82.3	79	65	14	53.5
82.3	62	51	11	42.0
82.2	500	411	89	337.8
82.2	499	410	89	336.9
82.2	495	407	88	334.6
82.2	494	406	88	333.7
82.2	493	405	88	332.7
82.2	490	403	87	331.4
82.2	489	402	87	330.5
82.2	488	401	87	329.5
82.2	484	398	86	327.3
82.2	483	397	86	326.3
82.2	482	396	86	325.3
82.2	478	393	85	323.1
82.2	477	392	85	322.1
82.2	473	389	84	319.9
82.2	472	388	84	318.9
82.2	471	387	84	318.0
82.2	467	384	83	315.8

%E	M1	M2	DM	M*
82.2	466	383	83	314.8
82.2	465	382	83	313.8
82.2	461	379	82	311.6
82.2	460	378	82	310.6
82.2	456	375	81	308.4
82.2	455	374	81	307.4
82.2	454	373	81	306.5
82.2	450	370	80	304.2
82.2	449	369	80	303.3
82.2	445	366	79	301.0
82.2	444	365	79	300.1
82.2	443	364	79	299.1
82.2	439	361	78	296.9
82.2	438	360	78	295.9
82.2	437	359	78	294.9
82.2	433	356	77	292.7
82.2	432	355	77	291.7
82.2	428	352	76	289.5
82.2	427	351	76	288.5
82.2	426	350	76	287.6
82.2	422	347	75	285.3
82.2	421	346	75	284.4
82.2	416	342	74	281.2
82.2	415	341	74	280.2
82.2	411	338	73	278.0
82.2	410	337	73	277.0
82.2	409	336	73	276.0
82.2	405	333	72	273.8
82.2	404	332	72	272.8
82.2	399	328	71	269.6
82.2	398	327	71	268.7
82.2	394	324	70	266.4
82.2	393	323	70	265.5
82.2	388	319	69	262.3
82.2	387	318	69	261.3
82.2	383	315	68	259.1
82.2	382	314	68	258.1
82.2	381	313	68	257.1
82.2	377	310	67	254.9
82.2	376	309	67	253.9
82.2	371	305	66	250.7
82.2	370	304	66	249.8
82.2	366	301	65	247.5
82.2	365	300	65	246.6
82.2	360	296	64	243.4
82.2	359	295	64	242.4
82.2	354	291	63	239.2
82.2	353	290	63	238.2
82.2	349	287	62	236.0
82.2	348	286	62	235.0

%E	M1	M2	DM	M*
82.2	343	282	61	231.8
82.2	342	281	61	230.9
82.2	338	278	60	228.7
82.2	337	277	60	227.7
82.2	332	273	59	224.5
82.2	331	272	59	223.5
82.2	326	268	58	220.3
82.2	325	267	58	219.4
82.2	321	264	57	217.1
82.2	320	263	57	216.2
82.2	315	259	56	213.0
82.2	314	258	56	212.0
82.2	309	254	55	208.8
82.2	304	250	54	205.6
82.2	303	249	54	204.6
82.2	298	245	53	201.4
82.2	297	244	53	200.5
82.2	292	240	52	197.3
82.2	287	236	51	194.1
82.2	286	235	51	193.1
82.2	281	231	50	189.9
82.2	276	227	49	186.7
82.2	275	226	49	185.7
82.2	270	222	48	182.5
82.2	269	221	48	181.6
82.2	264	217	47	178.4
82.2	259	213	46	175.2
82.2	258	212	46	174.2
82.2	253	208	45	171.0
82.2	247	203	44	166.8
82.2	242	199	43	163.6
82.2	241	198	43	162.7
82.2	236	194	42	159.5
82.2	230	189	41	155.3
82.2	225	185	40	152.1
82.2	219	180	39	147.9
82.2	214	176	38	144.7
82.2	213	175	38	143.8
82.2	208	171	37	140.6
82.2	202	166	36	136.4
82.2	197	162	35	133.2
82.2	191	157	34	129.1
82.2	185	152	33	124.9
82.2	180	148	32	121.7
82.2	174	143	31	117.5
82.2	169	139	30	114.3
82.2	163	134	29	110.2
82.2	157	129	28	106.0
82.2	152	125	27	102.8
82.2	146	120	26	98.6

%E	M1	M2	DM	M*
82.2	135	111	24	91.3
82.2	129	106	23	87.1
82.2	118	97	21	79.7
82.2	107	88	19	72.4
82.2	101	83	18	68.2
82.2	90	74	16	60.8
82.2	73	60	13	49.3
82.2	45	37	8	30.4
82.1	498	409	89	335.9
82.1	497	408	89	334.9
82.1	496	407	89	334.0
82.1	492	404	88	331.7
82.1	491	403	88	330.8
82.1	487	400	87	328.5
82.1	486	399	87	327.6
82.1	485	398	87	326.6
82.1	481	395	86	324.4
82.1	480	394	86	323.4
82.1	476	391	85	321.2
82.1	475	390	85	320.2
82.1	474	389	85	319.2
82.1	470	386	84	317.0
82.1	469	385	84	316.0
82.1	468	384	84	315.1
82.1	464	381	83	312.8
82.1	463	380	83	311.9
82.1	459	377	82	309.6
82.1	458	376	82	308.7
82.1	457	375	82	307.7
82.1	453	372	81	305.5
82.1	452	371	81	304.5
82.1	448	368	80	302.3
82.1	447	367	80	301.3
82.1	446	366	80	300.3
82.1	442	363	79	298.1
82.1	441	362	79	297.2
82.1	436	358	78	294.0
82.1	435	357	78	293.0
82.1	431	354	77	290.8
82.1	430	353	77	289.8
82.1	429	352	77	288.8
82.1	425	349	76	286.6
82.1	424	348	76	285.6
82.1	420	345	75	283.4
82.1	419	344	75	282.4
82.1	418	343	75	281.5
82.1	414	340	74	279.2
82.1	413	339	74	278.3
82.1	408	335	73	275.1
82.1	407	334	73	274.1

%E	M1	M2	DM	M*
82.1	403	331	72	271.9
82.1	402	330	72	270.9
82.1	397	326	71	267.7
82.1	396	325	71	266.7
82.1	392	322	70	264.5
82.1	391	321	70	263.5
82.1	390	320	70	262.6
82.1	386	317	69	260.3
82.1	385	316	69	259.4
82.1	380	312	68	256.2
82.1	379	311	68	255.2
82.1	375	308	67	253.0
82.1	374	307	67	252.0
82.1	369	303	66	248.8
82.1	368	302	66	247.8
82.1	364	299	65	245.6
82.1	363	298	65	244.6
82.1	358	294	64	241.4
82.1	357	293	64	240.5
82.1	352	289	63	237.3
82.1	351	288	63	236.3
82.1	347	285	62	234.1
82.1	346	284	62	233.1
82.1	341	280	61	229.9
82.1	340	279	61	228.9
82.1	336	276	60	226.7
82.1	335	275	60	225.7
82.1	330	271	59	222.5
82.1	329	270	59	221.6
82.1	324	266	58	218.4
82.1	319	262	57	215.2
82.1	318	261	57	214.2
82.1	313	257	56	211.0
82.1	312	256	56	210.1
82.1	308	253	55	207.8
82.1	307	252	55	206.9
82.1	302	248	54	203.7
82.1	301	247	54	202.7
82.1	296	243	53	199.5
82.1	291	239	52	196.3
82.1	290	238	52	195.3
82.1	285	234	51	192.1
82.1	280	230	50	188.9
82.1	279	229	50	188.0
82.1	274	225	49	184.8
82.1	273	224	49	183.8
82.1	268	220	48	180.6
82.1	263	216	47	177.4
82.1	262	215	47	176.4
82.1	257	211	46	173.2

%E	M1	M2	DM	M*
82.1	252	207	45	170.0
82.1	251	206	45	169.1
82.1	246	202	44	165.9
82.1	240	197	43	161.7
82.1	235	193	42	158.5
82.1	234	192	42	157.5
82.1	229	188	41	154.3
82.1	224	184	40	151.1
82.1	223	183	40	150.2
82.1	218	179	39	147.0
82.1	212	174	38	142.8
82.1	207	170	37	139.6
82.1	201	165	36	135.4
82.1	196	161	35	132.3
82.1	195	160	35	131.3
82.1	190	156	34	128.1
82.1	184	151	33	123.9
82.1	179	147	32	120.7
82.1	173	142	31	116.6
82.1	168	138	30	113.4
82.1	162	133	29	109.2
82.1	156	128	28	105.0
82.1	151	124	27	101.8
82.1	145	119	26	97.7
82.1	140	115	25	94.5
82.1	134	110	24	90.3
82.1	123	101	22	82.9
82.1	117	96	21	78.8
82.1	112	92	20	75.6
82.1	106	87	19	71.4
82.1	95	78	17	64.0
82.1	84	69	15	56.7
82.1	78	64	14	52.5
82.1	67	55	12	45.1
82.1	56	46	10	37.8
82.1	39	32	7	26.3
82.1	28	23	5	18.9
82.0	500	410	90	336.2
82.0	499	409	90	335.2
82.0	495	406	89	333.0
82.0	494	405	89	332.0
82.0	490	402	88	329.8
82.0	489	401	88	328.8
82.0	488	400	88	327.9
82.0	484	397	87	325.6
82.0	483	396	87	324.7
82.0	482	395	87	323.7
82.0	479	393	86	322.4
82.0	478	392	86	321.5
82.0	477	391	86	320.5

%E	M1	M2	DM	M*
82.0	473	388	85	318.3
82.0	472	387	85	317.3
82.0	471	386	85	316.3
82.0	467	383	84	314.1
82.0	466	382	84	313.1
82.0	462	379	83	310.9
82.0	461	378	83	309.9
82.0	460	377	83	309.0
82.0	456	374	82	306.7
82.0	455	373	82	305.8
82.0	451	370	81	303.5
82.0	450	369	81	302.6
82.0	449	368	81	301.6
82.0	445	365	80	299.4
82.0	444	364	80	298.4
82.0	440	361	79	296.2
82.0	439	360	79	295.2
82.0	438	359	79	294.2
82.0	434	356	78	292.0
82.0	433	355	78	291.1
82.0	428	351	77	287.9
82.0	427	350	77	286.9
82.0	423	347	76	284.7
82.0	422	346	76	283.7
82.0	417	342	75	280.5
82.0	416	341	75	279.5
82.0	412	338	74	277.3
82.0	411	337	74	276.3
82.0	410	336	74	275.4
82.0	406	333	73	273.1
82.0	405	332	73	272.2
82.0	401	329	72	269.9
82.0	400	328	72	269.0
82.0	399	327	72	268.0
82.0	395	324	71	265.8
82.0	394	323	71	264.8
82.0	389	319	70	261.6
82.0	388	318	70	260.6
82.0	384	315	69	258.4
82.0	383	314	69	257.4
82.0	378	310	68	254.2
82.0	377	309	68	253.3
82.0	373	306	67	251.0
82.0	372	305	67	250.1
82.0	367	301	66	246.9
82.0	366	300	66	245.9
82.0	362	297	65	243.7
82.0	361	296	65	242.7
82.0	356	292	64	239.5
82.0	355	291	64	238.5

%E	M1	M2	DM	M*
82.0	350	287	63	235.3
82.0	345	283	62	232.1
82.0	344	282	62	231.2
82.0	339	278	61	228.0
82.0	338	277	61	227.0
82.0	334	274	60	224.8
82.0	333	273	60	223.8
82.0	328	269	59	220.6
82.0	327	268	59	219.6
82.0	323	265	58	217.4
82.0	322	264	58	216.4
82.0	317	260	57	213.2
82.0	316	259	57	212.3
82.0	311	255	56	209.1
82.0	306	251	55	205.9
82.0	305	250	55	204.9
82.0	300	246	54	201.7
82.0	295	242	53	198.5
82.0	294	241	53	197.6
82.0	289	237	52	194.4
82.0	284	233	51	191.2
82.0	283	232	51	190.2
82.0	278	228	50	187.0
82.0	272	223	49	182.8
82.0	267	219	48	179.6
82.0	266	218	48	178.7
82.0	261	214	47	175.5
82.0	256	210	46	172.3
82.0	255	209	46	171.3
82.0	250	205	45	168.1
82.0	245	201	44	164.9
82.0	244	200	44	163.9
82.0	239	196	43	160.7
82.0	233	191	42	156.6
82.0	228	187	41	153.4
82.0	222	182	40	149.2
82.0	217	178	39	146.0
82.0	211	173	38	141.8
82.0	206	169	37	138.6
82.0	205	168	37	137.7
82.0	200	164	36	134.5
82.0	194	159	35	130.3
82.0	189	155	34	127.1
82.0	183	150	33	123.0
82.0	178	146	32	119.8
82.0	172	141	31	115.6
82.0	167	137	30	112.4
82.0	161	132	29	108.2
82.0	150	123	27	100.9
82.0	139	114	25	93.5

%E	M1	M2	DM	M*
82.0	133	109	24	89.3
82.0	128	105	23	86.1
82.0	122	100	22	82.0
82.0	111	91	20	74.6
82.0	100	82	18	67.2
82.0	89	73	16	59.9
82.0	61	50	11	41.0
82.0	50	41	9	33.6
81.9	498	408	90	334.3
81.9	497	407	90	333.3
81.9	496	406	90	332.3
81.9	493	404	89	331.1
81.9	492	403	89	330.1
81.9	491	402	89	329.1
81.9	487	399	88	326.9
81.9	486	398	88	325.9
81.9	485	397	88	325.0
81.9	481	394	87	322.7
81.9	480	393	87	321.8
81.9	476	390	86	319.5
81.9	475	389	86	318.6
81.9	474	388	86	317.6
81.9	470	385	85	315.4
81.9	469	384	85	314.4
81.9	465	381	84	312.2
81.9	464	380	84	311.2
81.9	463	379	84	310.2
81.9	459	376	83	308.0
81.9	458	375	83	307.0
81.9	454	372	82	304.8
81.9	453	371	82	303.8
81.9	452	370	82	302.9
81.9	448	367	81	300.6
81.9	447	366	81	299.7
81.9	443	363	80	297.4
81.9	442	362	80	296.5
81.9	441	361	80	295.5
81.9	437	358	79	293.3
81.9	436	357	79	292.3
81.9	432	354	78	290.1
81.9	431	353	78	289.1
81.9	430	352	78	288.1
81.9	426	349	77	285.9
81.9	425	348	77	285.0
81.9	421	345	76	282.7
81.9	420	344	76	281.8
81.9	419	343	76	280.8
81.9	415	340	75	278.6
81.9	414	339	75	277.6
81.9	409	335	74	274.4

%E	M1	M2	DM	M*
81.9	408	334	74	273.4
81.9	404	331	73	271.2
81.9	403	330	73	270.2
81.9	398	326	72	267.0
81.9	397	325	72	266.1
81.9	393	322	71	263.8
81.9	392	321	71	262.9
81.9	387	317	70	259.7
81.9	386	316	70	258.7
81.9	382	313	69	256.5
81.9	381	312	69	255.5
81.9	376	308	68	252.3
81.9	375	307	68	251.3
81.9	371	304	67	249.1
81.9	370	303	67	248.1
81.9	365	299	66	244.9
81.9	364	298	66	244.0
81.9	360	295	65	241.7
81.9	359	294	65	240.8
81.9	354	290	64	237.6
81.9	353	289	64	236.6
81.9	349	286	63	234.4
81.9	348	285	63	233.4
81.9	343	281	62	230.2
81.9	342	280	62	229.2
81.9	337	276	61	226.0
81.9	332	272	60	222.8
81.9	331	271	60	221.9
81.9	326	267	59	218.7
81.9	321	263	58	215.5
81.9	320	262	58	214.5
81.9	315	258	57	211.3
81.9	310	254	56	208.1
81.9	309	253	56	207.1
81.9	304	249	55	204.0
81.9	299	245	54	200.8
81.9	298	244	54	199.8
81.9	293	240	53	196.6
81.9	288	236	52	193.4
81.9	287	235	52	192.4
81.9	282	231	51	189.2
81.9	281	230	51	188.3
81.9	277	227	50	186.0
81.9	276	226	50	185.1
81.9	271	222	49	181.9
81.9	270	221	49	180.9
81.9	265	217	48	177.7
81.9	260	213	47	174.5
81.9	259	212	47	173.5
81.9	254	208	46	170.3

%E	M1	M2	DM	M*
81.9	249	204	45	167.1
81.9	248	203	45	166.2
81.9	243	199	44	163.0
81.9	238	195	43	159.8
81.9	237	194	43	158.8
81.9	232	190	42	155.6
81.9	227	186	41	152.4
81.9	226	185	41	151.4
81.9	221	181	40	148.2
81.9	216	177	39	145.0
81.9	215	176	39	144.1
81.9	210	172	38	140.9
81.9	204	167	37	136.7
81.9	199	163	36	133.5
81.9	193	158	35	129.3
81.9	188	154	34	126.1
81.9	182	149	33	122.0
81.9	177	145	32	118.8
81.9	171	140	31	114.6
81.9	166	136	30	111.4
81.9	160	131	29	107.3
81.9	155	127	28	104.1
81.9	149	122	27	99.9
81.9	144	118	26	96.7
81.9	138	113	25	92.5
81.9	127	104	23	85.2
81.9	116	95	21	77.8
81.9	105	86	19	70.4
81.9	94	77	17	63.1
81.9	83	68	15	55.7
81.9	72	59	13	48.3
81.8	500	409	91	334.6
81.8	499	408	91	333.6
81.8	495	405	90	331.4
81.8	494	404	90	330.4
81.8	490	401	89	328.2
81.8	489	400	89	327.2
81.8	488	399	89	326.2
81.8	484	396	88	324.0
81.8	483	395	88	323.0
81.8	479	392	87	320.8
81.8	478	391	87	319.8
81.8	477	390	87	318.9
81.8	473	387	86	316.6
81.8	472	386	86	315.7
81.8	468	383	85	313.4
81.8	467	382	85	312.5
81.8	466	381	85	311.5
81.8	462	378	84	309.3
81.8	461	377	84	308.3

%E	M1	M2	DM	M*
81.8	457	374	83	306.1
81.8	456	373	83	305.1
81.8	455	372	83	304.1
81.8	451	369	82	301.9
81.8	450	368	82	300.9
81.8	446	365	81	298.7
81.8	445	364	81	297.7
81.8	444	363	81	296.8
81.8	440	360	80	294.5
81.8	439	359	80	293.6
81.8	435	356	79	291.3
81.8	434	355	79	290.4
81.8	433	354	79	289.4
81.8	429	351	78	287.2
81.8	428	350	78	286.2
81.8	424	347	77	284.0
81.8	423	346	77	283.0
81.8	422	345	77	282.0
81.8	418	342	76	279.8
81.8	417	341	76	278.9
81.8	413	338	75	276.6
81.8	412	337	75	275.7
81.8	411	336	75	274.7
81.8	407	333	74	272.5
81.8	406	332	74	271.5
81.8	402	329	73	269.3
81.8	401	328	73	268.3
81.8	400	327	73	267.3
81.8	396	324	72	265.1
81.8	395	323	72	264.1
81.8	391	320	71	261.9
81.8	390	319	71	260.9
81.8	385	315	70	257.7
81.8	384	314	70	256.8
81.8	380	311	69	254.5
81.8	379	310	69	253.6
81.8	374	306	68	250.4
81.8	373	305	68	249.4
81.8	369	302	67	247.2
81.8	368	301	67	246.2
81.8	363	297	66	243.0
81.8	362	296	66	242.0
81.8	358	293	65	239.8
81.8	357	292	65	238.8
81.8	352	288	64	235.6
81.8	351	287	64	234.7
81.8	347	284	63	232.4
81.8	346	283	63	231.5
81.8	341	279	62	228.3
81.8	340	278	62	227.3

%E	M1	M2	DM	M*
81.8	336	275	61	225.1
81.8	335	274	61	224.1
81.8	330	270	60	220.9
81.8	329	269	60	219.9
81.8	325	266	59	217.7
81.8	324	265	59	216.7
81.8	319	261	58	213.5
81.8	318	260	58	212.6
81.8	314	257	57	210.3
81.8	313	256	57	209.4
81.8	308	252	56	206.2
81.8	307	251	56	205.2
81.8	303	248	55	203.0
81.8	302	247	55	202.0
81.8	297	243	54	198.8
81.8	296	242	54	197.9
81.8	292	239	53	195.6
81.8	291	238	53	194.7
81.8	286	234	52	191.5
81.8	285	233	52	190.5
81.8	280	229	51	187.3
81.8	275	225	50	184.1
81.8	274	224	50	183.1
81.8	269	220	49	179.9
81.8	264	216	48	176.7
81.8	258	211	47	172.6
81.8	253	207	46	169.4
81.8	247	202	45	165.2
81.8	242	198	44	162.0
81.8	236	193	43	157.8
81.8	231	189	42	154.6
81.8	225	184	41	150.5
81.8	220	180	40	147.3
81.8	214	175	39	143.1
81.8	209	171	38	139.9
81.8	203	166	37	135.7
81.8	198	162	36	132.5
81.8	192	157	35	128.4
81.8	187	153	34	125.2
81.8	181	148	33	121.0
81.8	176	144	32	117.8
81.8	170	139	31	113.7
81.8	165	135	30	110.5
81.8	159	130	29	106.3
81.8	154	126	28	103.1
81.8	148	121	27	98.9
81.8	143	117	26	95.7
81.8	137	112	25	91.6
81.8	132	108	24	88.4
81.8	121	99	22	81.0

%E	M1	M2	DM	M*
81.8	110	90	20	73.6
81.8	99	81	18	66.3
81.8	88	72	16	58.9
81.8	77	63	14	51.5
81.8	66	54	12	44.2
81.8	55	45	10	36.8
81.8	44	36	8	29.5
81.8	33	27	6	22.1
81.8	22	18	4	14.7
81.8	11	9	2	7.4
81.7	498	407	91	332.6
81.7	497	406	91	331.7
81.7	496	405	91	330.7
81.7	493	403	90	329.4
81.7	492	402	90	328.5
81.7	491	401	90	327.5
81.7	487	398	89	325.3
81.7	486	397	89	324.3
81.7	482	394	88	322.1
81.7	481	393	88	321.1
81.7	480	392	88	320.1
81.7	476	389	87	317.9
81.7	475	388	87	316.9
81.7	471	385	86	314.7
81.7	470	384	86	313.7
81.7	469	383	86	312.8
81.7	465	380	85	310.5
81.7	464	379	85	309.6
81.7	460	376	84	307.3
81.7	459	375	84	306.4
81.7	458	374	84	305.4
81.7	454	371	83	303.2
81.7	453	370	83	302.2
81.7	449	367	82	300.0
81.7	448	366	82	299.0
81.7	447	365	82	298.0
81.7	443	362	81	295.8
81.7	442	361	81	294.8
81.7	438	358	80	292.6
81.7	437	357	80	291.6
81.7	436	356	80	290.7
81.7	432	353	79	288.4
81.7	431	352	79	287.5
81.7	427	349	78	285.2
81.7	426	348	78	284.3
81.7	421	344	77	281.1
81.7	420	343	77	280.1
81.7	416	340	76	277.9
81.7	415	339	76	276.9
81.7	410	335	75	273.7

%E	M1	M2	DM	M*
81.7	409	334	75	272.8
81.7	405	331	74	270.5
81.7	404	330	74	269.6
81.7	399	326	73	266.4
81.7	398	325	73	265.4
81.7	394	322	72	263.2
81.7	393	321	72	262.2
81.7	389	318	71	260.0
81.7	388	317	71	259.0
81.7	387	316	71	258.0
81.7	383	313	70	255.8
81.7	382	312	70	254.8
81.7	378	309	69	252.6
81.7	377	308	69	251.6
81.7	372	304	68	248.4
81.7	371	303	68	247.5
81.7	367	300	67	245.2
81.7	366	299	67	244.3
81.7	361	295	66	241.1
81.7	360	294	66	240.1
81.7	356	291	65	237.9
81.7	355	290	65	236.9
81.7	350	286	64	233.7
81.7	349	285	64	232.7
81.7	345	282	63	230.5
81.7	344	281	63	229.5
81.7	339	277	62	226.3
81.7	338	276	62	225.4
81.7	334	273	61	223.1
81.7	333	272	61	222.2
81.7	328	268	60	219.0
81.7	327	267	60	218.0
81.7	323	264	59	215.8
81.7	322	263	59	214.8
81.7	317	259	58	211.6
81.7	312	255	57	208.4
81.7	311	254	57	207.4
81.7	306	250	56	204.2
81.7	301	246	55	201.0
81.7	300	245	55	200.1
81.7	295	241	54	196.9
81.7	290	237	53	193.7
81.7	289	236	53	192.7
81.7	284	232	52	189.5
81.7	279	228	51	186.3
81.7	278	227	51	185.4
81.7	273	223	50	182.2
81.7	268	219	49	179.0
81.7	263	215	48	175.8
81.7	262	214	48	174.8

%E	M1	M2	DM	M*
81.7	257	210	47	171.6
81.7	252	206	46	168.4
81.7	251	205	46	167.4
81.7	246	201	45	164.2
81.7	241	197	44	161.0
81.7	240	196	44	160.1
81.7	235	192	43	156.9
81.7	230	188	42	153.7
81.7	229	187	42	152.7
81.7	224	183	41	149.5
81.7	219	179	40	146.3
81.7	218	178	40	145.3
81.7	213	174	39	142.1
81.7	208	170	38	138.9
81.7	202	165	37	134.8
81.7	197	161	36	131.6
81.7	191	156	35	127.4
81.7	186	152	34	124.2
81.7	180	147	33	120.0
81.7	175	143	32	116.9
81.7	169	138	31	112.7
81.7	164	134	30	109.5
81.7	153	125	28	102.1
81.7	142	116	26	94.8
81.7	131	107	24	87.4
81.7	126	103	23	84.2
81.7	120	98	22	80.0
81.7	115	94	21	76.8
81.7	109	89	20	72.7
81.7	104	85	19	69.5
81.7	93	76	17	62.1
81.7	82	67	15	54.7
81.7	71	58	13	47.4
81.7	60	49	11	40.0
81.6	500	408	92	332.9
81.6	499	407	92	332.0
81.6	495	404	91	329.7
81.6	494	403	91	328.8
81.6	490	400	90	326.5
81.6	489	399	90	325.6
81.6	488	398	90	324.6
81.6	485	396	89	323.3
81.6	484	395	89	322.4
81.6	483	394	89	321.4
81.6	479	391	88	319.2
81.6	478	390	88	318.2
81.6	477	389	88	317.2
81.6	474	387	87	316.0
81.6	473	386	87	315.0
81.6	472	385	87	314.0

%E	M1	M2	DM	M*
81.6	468	382	86	311.8
81.6	467	381	86	310.8
81.6	463	378	85	308.6
81.6	462	377	85	307.6
81.6	461	376	85	306.7
81.6	457	373	84	304.4
81.6	456	372	84	303.5
81.6	452	369	83	301.2
81.6	451	368	83	300.3
81.6	450	367	83	299.3
81.6	446	364	82	297.1
81.6	445	363	82	296.1
81.6	441	360	81	293.9
81.6	440	359	81	292.9
81.6	435	355	80	289.7
81.6	434	354	80	288.7
81.6	430	351	79	286.5
81.6	429	350	79	285.5
81.6	425	347	78	283.3
81.6	424	346	78	282.3
81.6	423	345	78	281.4
81.6	419	342	77	279.2
81.6	418	341	77	278.2
81.6	414	338	76	276.0
81.6	413	337	76	275.0
81.6	412	336	76	274.0
81.6	408	333	75	271.8
81.6	407	332	75	270.8
81.6	403	329	74	268.6
81.6	402	328	74	267.6
81.6	397	324	73	264.4
81.6	396	323	73	263.5
81.6	392	320	72	261.2
81.6	391	319	72	260.3
81.6	386	315	71	257.1
81.6	385	314	71	256.1
81.6	381	311	70	253.9
81.6	380	310	70	252.9
81.6	376	307	69	250.7
81.6	375	306	69	249.7
81.6	374	305	69	248.7
81.6	370	302	68	246.5
81.6	369	301	68	245.5
81.6	365	298	67	243.3
81.6	364	297	67	242.3
81.6	359	293	66	239.1
81.6	358	292	66	238.2
81.6	354	289	65	235.9
81.6	353	288	65	235.0
81.6	348	284	64	231.8

%E	M1	M2	DM	M*
81.6	347	283	64	230.8
81.6	343	280	63	228.6
81.6	342	279	63	227.6
81.6	337	275	62	224.4
81.6	332	271	61	221.2
81.6	331	270	61	220.2
81.6	326	266	60	217.0
81.6	321	262	59	213.8
81.6	320	261	59	212.9
81.6	316	258	58	210.6
81.6	315	257	58	209.7
81.6	310	253	57	206.5
81.6	309	252	57	205.5
81.6	305	249	56	203.3
81.6	304	248	56	202.3
81.6	299	244	55	199.1
81.6	294	240	54	195.9
81.6	293	239	54	195.0
81.6	288	235	53	191.8
81.6	283	231	52	188.6
81.6	282	230	52	187.6
81.6	277	226	51	184.4
81.6	272	222	50	181.2
81.6	271	221	50	180.2
81.6	267	218	49	178.0
81.6	266	217	49	177.0
81.6	261	213	48	173.8
81.6	256	209	47	170.6
81.6	255	208	47	169.7
81.6	250	204	46	166.5
81.6	245	200	45	163.3
81.6	244	199	45	162.3
81.6	239	195	44	159.1
81.6	234	191	43	155.9
81.6	228	186	42	151.7
81.6	223	182	41	148.5
81.6	217	177	40	144.4
81.6	212	173	39	141.2
81.6	207	169	38	138.0
81.6	206	168	38	137.0
81.6	201	164	37	133.8
81.6	196	160	36	130.6
81.6	190	155	35	126.4
81.6	185	151	34	123.2
81.6	179	146	33	119.1
81.6	174	142	32	115.9
81.6	163	133	30	108.5
81.6	158	129	29	105.3
81.6	152	124	28	101.2
81.6	147	120	27	98.0

%E	M1	M2	DM	M*
81.6	141	115	26	93.8
81.6	136	111	25	90.6
81.6	125	102	23	83.2
81.6	114	93	21	75.9
81.6	103	84	19	68.5
81.6	98	80	18	65.3
81.6	87	71	16	57.9
81.6	76	62	14	50.6
81.6	49	40	9	32.7
81.6	38	31	7	25.3
81.5	498	406	92	331.0
81.5	497	405	92	330.0
81.5	496	404	92	329.1
81.5	493	402	91	327.8
81.5	492	401	91	326.8
81.5	491	400	91	325.9
81.5	487	397	90	323.6
81.5	486	396	90	322.7
81.5	482	393	89	320.4
81.5	481	392	89	319.5
81.5	480	391	89	318.5
81.5	476	388	88	316.3
81.5	475	387	88	315.3
81.5	471	384	87	313.1
81.5	470	383	87	312.1
81.5	469	382	87	311.1
81.5	466	380	86	309.9
81.5	465	379	86	308.9
81.5	464	378	86	307.9
81.5	460	375	85	305.7
81.5	459	374	85	304.7
81.5	455	371	84	302.5
81.5	454	370	84	301.5
81.5	453	369	84	300.6
81.5	449	366	83	298.3
81.5	448	365	83	297.4
81.5	444	362	82	295.1
81.5	443	361	82	294.2
81.5	439	358	81	291.9
81.5	438	357	81	291.0
81.5	437	356	81	290.0
81.5	433	353	80	287.8
81.5	432	352	80	286.8
81.5	428	349	79	284.6
81.5	427	348	79	283.6
81.5	426	347	79	282.7
81.5	422	344	78	280.4
81.5	421	343	78	279.5
81.5	417	340	77	277.2
81.5	416	339	77	276.3

%E	M1	M2	DM	M*
81.5	411	335	76	273.1
81.5	410	334	76	272.1
81.5	406	331	75	269.9
81.5	405	330	75	268.9
81.5	401	327	74	266.7
81.5	400	326	74	265.7
81.5	399	325	74	264.7
81.5	395	322	73	262.5
81.5	394	321	73	261.5
81.5	390	318	72	259.3
81.5	389	317	72	258.3
81.5	384	313	71	255.1
81.5	383	312	71	254.2
81.5	379	309	70	251.9
81.5	378	308	70	251.0
81.5	373	304	69	247.8
81.5	372	303	69	246.8
81.5	368	300	68	244.6
81.5	367	299	68	243.6
81.5	363	296	67	241.4
81.5	362	295	67	240.4
81.5	357	291	66	237.2
81.5	356	290	66	236.2
81.5	352	287	65	234.0
81.5	351	286	65	233.0
81.5	346	282	64	229.8
81.5	341	278	63	226.6
81.5	340	277	63	225.7
81.5	336	274	62	223.4
81.5	335	273	62	222.5
81.5	330	269	61	219.3
81.5	329	268	61	218.3
81.5	325	265	60	216.1
81.5	324	264	60	215.1
81.5	319	260	59	211.9
81.5	314	256	58	208.7
81.5	313	255	58	207.7
81.5	308	251	57	204.5
81.5	303	247	56	201.3
81.5	302	246	56	200.4
81.5	298	243	55	198.2
81.5	297	242	55	197.2
81.5	292	238	54	194.0
81.5	287	234	53	190.8
81.5	286	233	53	189.8
81.5	281	229	52	186.6
81.5	276	225	51	183.4
81.5	275	224	51	182.5
81.5	270	220	50	179.3
81.5	265	216	49	176.1

%E	M1	M2	DM	M*
81.5	260	212	48	172.9
81.5	259	211	48	171.9
81.5	254	207	47	168.7
81.5	249	203	46	165.5
81.5	248	202	46	164.5
81.5	243	198	45	161.3
81.5	238	194	44	158.1
81.5	233	190	43	154.9
81.5	232	189	43	154.0
81.5	227	185	42	150.8
81.5	222	181	41	147.6
81.5	216	176	40	143.4
81.5	211	172	39	140.2
81.5	205	167	38	136.0
81.5	200	163	37	132.8
81.5	195	159	36	129.6
81.5	189	154	35	125.5
81.5	184	150	34	122.3
81.5	178	145	33	118.1
81.5	173	141	32	114.9
81.5	168	137	31	111.7
81.5	162	132	30	107.6
81.5	157	128	29	104.4
81.5	151	123	28	100.2
81.5	146	119	27	97.0
81.5	135	110	25	89.6
81.5	130	106	24	86.4
81.5	124	101	23	82.3
81.5	119	97	22	79.1
81.5	108	88	20	71.7
81.5	92	75	17	61.1
81.5	81	66	15	53.8
81.5	65	53	12	43.2
81.5	54	44	10	35.9
81.5	27	22	5	17.9
81.4	500	407	93	331.3
81.4	499	406	93	330.3
81.4	495	403	92	328.1
81.4	494	402	92	327.1
81.4	490	399	91	324.9
81.4	489	398	91	323.9
81.4	488	397	91	323.0
81.4	485	395	90	321.7
81.4	484	394	90	320.7
81.4	483	393	90	319.8
81.4	479	390	89	317.5
81.4	478	389	89	316.6
81.4	474	386	88	314.3
81.4	473	385	88	313.4
81.4	472	384	88	312.4

%E	M1	M2	DM	M*
81.4	468	381	87	310.2
81.4	467	380	87	309.2
81.4	463	377	86	307.0
81.4	462	376	86	306.0
81.4	458	373	85	303.8
81.4	457	372	85	302.8
81.4	456	371	85	301.8
81.4	452	368	84	299.6
81.4	451	367	84	298.6
81.4	447	364	83	296.4
81.4	446	363	83	295.4
81.4	442	360	82	293.2
81.4	441	359	82	292.2
81.4	440	358	82	291.3
81.4	436	355	81	289.0
81.4	435	354	81	288.1
81.4	431	351	80	285.8
81.4	430	350	80	284.9
81.4	429	349	80	283.9
81.4	425	346	79	281.7
81.4	424	345	79	280.7
81.4	420	342	78	278.5
81.4	419	341	78	277.5
81.4	415	338	77	275.3
81.4	414	337	77	274.3
81.4	413	336	77	273.4
81.4	409	333	76	271.1
81.4	408	332	76	270.2
81.4	404	329	75	267.9
81.4	403	328	75	267.0
81.4	398	324	74	263.8
81.4	397	323	74	262.8
81.4	393	320	73	260.6
81.4	392	319	73	259.6
81.4	388	316	72	257.4
81.4	387	315	72	256.4
81.4	382	311	71	253.2
81.4	381	310	71	252.2
81.4	377	307	70	250.0
81.4	376	306	70	249.0
81.4	371	302	69	245.8
81.4	370	301	69	244.9
81.4	366	298	68	242.6
81.4	365	297	68	241.7
81.4	361	294	67	239.4
81.4	360	293	67	238.5
81.4	355	289	66	235.3
81.4	354	288	66	234.3
81.4	350	285	65	232.1
81.4	349	284	65	231.1

%E	M1	M2	DM	M*
81.4	345	281	64	228.9
81.4	344	280	64	227.9
81.4	339	276	63	224.7
81.4	338	275	63	223.7
81.4	334	272	62	221.5
81.4	333	271	62	220.5
81.4	328	267	61	217.3
81.4	323	263	60	214.1
81.4	322	262	60	213.2
81.4	318	259	59	210.9
81.4	317	258	59	210.0
81.4	312	254	58	206.8
81.4	311	253	58	205.8
81.4	307	250	57	203.6
81.4	306	249	57	202.6
81.4	301	245	56	199.4
81.4	296	241	55	196.2
81.4	295	240	55	195.3
81.4	291	237	54	193.0
81.4	290	236	54	192.1
81.4	285	232	53	188.9
81.4	280	228	52	185.7
81.4	279	227	52	184.7
81.4	274	223	51	181.5
81.4	269	219	50	178.3
81.4	264	215	49	175.1
81.4	263	214	49	174.1
81.4	258	210	48	170.9
81.4	253	206	47	167.7
81.4	247	201	46	163.6
81.4	242	197	45	160.4
81.4	237	193	44	157.2
81.4	236	192	44	156.2
81.4	231	188	43	153.0
81.4	226	184	42	149.8
81.4	221	180	41	146.6
81.4	220	179	41	145.6
81.4	215	175	40	142.4
81.4	210	171	39	139.2
81.4	204	166	38	135.1
81.4	199	162	37	131.9
81.4	194	158	36	128.7
81.4	188	153	35	124.5
81.4	183	149	34	121.3
81.4	177	144	33	117.2
81.4	172	140	32	114.0
81.4	167	136	31	110.8
81.4	161	131	30	106.6
81.4	156	127	29	103.4
81.4	145	118	27	96.0

%E	M1	M2	DM	M*
81.4	140	114	26	92.8
81.4	129	105	24	85.5
81.4	118	96	22	78.1
81.4	113	92	21	74.9
81.4	102	83	19	67.5
81.4	97	79	18	64.3
81.4	86	70	16	57.0
81.4	70	57	13	46.4
81.4	59	48	11	39.1
81.4	43	35	8	28.5
81.3	498	405	93	329.4
81.3	497	404	93	328.4
81.3	496	403	93	327.4
81.3	493	401	92	326.2
81.3	492	400	92	325.2
81.3	491	399	92	324.2
81.3	487	396	91	322.0
81.3	486	395	91	321.0
81.3	482	392	90	318.8
81.3	481	391	90	317.8
81.3	480	390	90	316.9
81.3	477	388	89	315.6
81.3	476	387	89	314.6
81.3	475	386	89	313.7
81.3	471	383	88	311.4
81.3	470	382	88	310.5
81.3	466	379	87	308.2
81.3	465	378	87	307.3
81.3	464	377	87	306.3
81.3	461	375	86	305.0
81.3	460	374	86	304.1
81.3	459	373	86	303.1
81.3	455	370	85	300.9
81.3	454	369	85	299.9
81.3	450	366	84	297.7
81.3	449	365	84	296.7
81.3	448	364	84	295.8
81.3	445	362	83	294.5
81.3	444	361	83	293.5
81.3	443	360	83	292.6
81.3	439	357	82	290.3
81.3	438	356	82	289.4
81.3	434	353	81	287.1
81.3	433	352	81	286.2
81.3	432	351	81	285.2
81.3	428	348	80	283.0
81.3	427	347	80	282.0
81.3	423	344	79	279.8
81.3	422	343	79	278.8
81.3	418	340	78	276.6

%E	M1	M2	DM	M*
81.3	417	339	78	275.6
81.3	416	338	78	274.6
81.3	412	335	77	272.4
81.3	411	334	77	271.4
81.3	407	331	76	269.2
81.3	406	330	76	268.2
81.3	402	327	75	266.0
81.3	401	326	75	265.0
81.3	400	325	75	264.1
81.3	396	322	74	261.8
81.3	395	321	74	260.9
81.3	391	318	73	258.6
81.3	390	317	73	257.7
81.3	386	314	72	255.4
81.3	385	313	72	254.5
81.3	384	312	72	253.5
81.3	380	309	71	251.3
81.3	379	308	71	250.3
81.3	375	305	70	248.1
81.3	374	304	70	247.1
81.3	369	300	69	243.9
81.3	368	299	69	242.9
81.3	364	296	68	240.7
81.3	363	295	68	239.7
81.3	359	292	67	237.5
81.3	358	291	67	236.5
81.3	353	287	66	233.3
81.3	352	286	66	232.4
81.3	348	283	65	230.1
81.3	347	282	65	229.2
81.3	343	279	64	226.9
81.3	342	278	64	226.0
81.3	337	274	63	222.8
81.3	336	273	63	221.8
81.3	332	270	62	219.6
81.3	331	269	62	218.6
81.3	327	266	61	216.4
81.3	326	265	61	215.4
81.3	321	261	60	212.2
81.3	320	260	60	211.3
81.3	316	257	59	209.0
81.3	315	256	59	208.1
81.3	310	252	58	204.9
81.3	305	248	57	201.7
81.3	304	247	57	200.7
81.3	300	244	56	198.5
81.3	299	243	56	197.5
81.3	294	239	55	194.3
81.3	289	235	54	191.1
81.3	288	234	54	190.1

%E	M1	M2	DM	M*
81.3	284	231	53	187.9
81.3	283	230	53	186.9
81.3	278	226	52	183.7
81.3	273	222	51	180.5
81.3	272	221	51	179.6
81.3	268	[illegible]8	50	177.3
81.3	267	217	50	176.4
81.3	262	213	49	173.2
81.3	257	209	48	170.0
81.3	256	208	48	169.0
81.3	252	205	47	166.8
81.3	251	204	47	165.8
81.3	246	200	46	162.6
81.3	241	196	45	159.4
81.3	240	195	45	158.4
81.3	235	191	44	155.2
81.3	230	187	43	152.0
81.3	225	183	42	148.8
81.3	224	182	42	147.9
81.3	219	178	41	144.7
81.3	214	174	40	141.5
81.3	209	170	39	138.3
81.3	208	169	39	137.3
81.3	203	165	38	134.1
81.3	198	161	37	130.9
81.3	193	157	36	127.7
81.3	192	156	36	126.8
81.3	187	152	35	123.6
81.3	182	148	34	120.4
81.3	176	143	33	116.2
81.3	171	139	32	113.0
81.3	166	135	31	109.8
81.3	160	130	30	105.6
81.3	155	126	29	102.4
81.3	150	122	28	99.2
81.3	144	117	27	95.1
81.3	139	113	26	91.9
81.3	134	109	25	88.7
81.3	128	104	24	84.5
81.3	123	100	23	81.3
81.3	112	91	21	73.9
81.3	107	87	20	70.7
81.3	96	78	18	63.4
81.3	91	74	17	60.2
81.3	80	65	15	52.8
81.3	75	61	14	49.6
81.3	64	52	12	42.3
81.3	48	39	9	31.7
81.3	32	26	6	21.1
81.3	16	13	3	10.6

%E	M1	M2	DM	M*
81.2	500	406	94	329.7
81.2	499	405	94	328.7
81.2	495	402	93	326.5
81.2	494	401	93	325.5
81.2	490	398	92	323.3
81.2	489	397	92	322.3
81.2	485	394	91	320.1
81.2	484	393	91	319.1
81.2	483	392	91	318.1
81.2	479	389	90	315.9
81.2	478	388	90	314.9
81.2	474	385	89	312.7
81.2	473	384	89	311.7
81.2	469	381	88	309.5
81.2	468	380	88	308.5
81.2	467	379	88	307.6
81.2	463	376	87	305.3
81.2	462	375	87	304.4
81.2	458	372	86	302.1
81.2	457	371	86	301.2
81.2	453	368	85	298.9
81.2	452	367	85	298.0
81.2	451	366	85	297.0
81.2	447	363	84	294.8
81.2	446	362	84	293.8
81.2	442	359	83	291.6
81.2	441	358	83	290.6
81.2	437	355	82	288.4
81.2	436	354	82	287.4
81.2	431	350	81	284.2
81.2	430	349	81	283.3
81.2	426	346	80	281.0
81.2	425	345	80	280.1
81.2	421	342	79	277.8
81.2	420	341	79	276.9
81.2	415	337	78	273.7
81.2	414	336	78	272.7
81.2	410	333	77	270.5
81.2	409	332	77	269.5
81.2	405	329	76	267.3
81.2	404	328	76	266.3
81.2	399	324	75	263.1
81.2	398	323	75	262.1
81.2	394	320	74	259.9
81.2	393	319	74	258.9
81.2	389	316	73	256.7
81.2	388	315	73	255.7
81.2	383	311	72	252.5
81.2	382	310	72	251.6
81.2	378	307	71	249.3

%E	M1	M2	DM	M*
81.2	377	306	71	248.4
81.2	373	303	70	246.1
81.2	372	302	70	245.2
81.2	367	298	69	242.0
81.2	362	294	68	238.8
81.2	361	293	68	237.8
81.2	357	290	67	235.6
81.2	356	289	67	234.6
81.2	351	285	66	231.4
81.2	346	281	65	228.2
81.2	345	280	65	227.2
81.2	341	277	64	225.0
81.2	340	276	64	224.0
81.2	335	272	63	220.8
81.2	330	268	62	217.6
81.2	329	267	62	216.7
81.2	325	264	61	214.4
81.2	324	263	61	213.5
81.2	319	259	60	210.3
81.2	314	255	59	207.1
81.2	313	254	59	206.1
81.2	309	251	58	203.9
81.2	308	250	58	202.9
81.2	303	246	57	199.7
81.2	298	242	56	196.5
81.2	293	238	55	193.3
81.2	292	237	55	192.4
81.2	287	233	54	189.2
81.2	282	229	53	186.0
81.2	277	225	52	182.8
81.2	276	224	52	181.8
81.2	271	220	51	178.6
81.2	266	216	50	175.4
81.2	261	212	49	172.2
81.2	260	211	49	171.2
81.2	255	207	48	168.0
81.2	250	203	47	164.8
81.2	245	199	46	161.6
81.2	239	194	45	157.5
81.2	234	190	44	154.3
81.2	229	186	43	151.1
81.2	223	181	42	146.9
81.2	218	177	41	143.7
81.2	213	173	40	140.5
81.2	207	168	39	136.3
81.2	202	164	38	133.1
81.2	197	160	37	129.9
81.2	191	155	36	125.8
81.2	186	151	35	122.6
81.2	181	147	34	119.4

%E	M1	M2	DM	M*
81.2	170	138	32	112.0
81.2	165	134	31	108.8
81.2	154	125	29	101.5
81.2	149	121	28	98.3
81.2	138	112	26	90.9
81.2	133	108	25	87.7
81.2	117	95	22	77.1
81.2	101	82	19	66.6
81.2	85	69	16	56.0
81.2	69	56	13	45.4
81.1	498	404	94	327.7
81.1	497	403	94	326.8
81.1	493	400	93	324.5
81.1	492	399	93	323.6
81.1	491	398	93	322.6
81.1	488	396	92	321.3
81.1	487	395	92	320.4
81.1	486	394	92	319.4
81.1	482	391	91	317.2
81.1	481	390	91	316.2
81.1	477	387	90	314.0
81.1	476	386	90	313.0
81.1	475	385	90	312.1
81.1	472	383	89	310.8
81.1	471	382	89	309.8
81.1	470	381	89	308.9
81.1	466	378	88	306.6
81.1	465	377	88	305.7
81.1	461	374	87	303.4
81.1	460	373	87	302.5
81.1	456	370	86	300.2
81.1	455	369	86	299.3
81.1	454	368	86	298.3
81.1	450	365	85	296.1
81.1	449	364	85	295.1
81.1	445	361	84	292.9
81.1	444	360	84	291.9
81.1	440	357	83	289.7
81.1	439	356	83	288.7
81.1	438	355	83	287.7
81.1	435	353	82	286.5
81.1	434	352	82	285.5
81.1	433	351	82	284.5
81.1	429	348	81	282.3
81.1	428	347	81	281.3
81.1	424	344	80	279.1
81.1	423	343	80	278.1
81.1	419	340	79	275.9
81.1	418	339	79	274.9
81.1	417	338	79	274.0

%E	M1	M2	DM	M*
81.1	413	335	78	271.7
81.1	412	334	78	270.8
81.1	408	331	77	268.5
81.1	407	330	77	267.6
81.1	403	327	76	265.3
81.1	402	326	76	264.4
81.1	397	322	75	261.2
81.1	396	321	75	260.2
81.1	392	318	74	258.0
81.1	391	317	74	257.0
81.1	387	314	73	254.8
81.1	386	313	73	253.8
81.1	381	309	72	250.6
81.1	380	308	72	249.6
81.1	376	305	71	247.4
81.1	375	304	71	246.4
81.1	371	301	70	244.2
81.1	370	300	70	243.2
81.1	366	297	69	241.0
81.1	365	296	69	240.0
81.1	360	292	68	236.8
81.1	359	291	68	235.9
81.1	355	288	67	233.6
81.1	354	287	67	232.7
81.1	350	284	66	230.4
81.1	349	283	66	229.5
81.1	344	279	65	226.3
81.1	343	278	65	225.3
81.1	339	275	64	223.1
81.1	338	274	64	222.1
81.1	334	271	63	219.9
81.1	333	270	63	218.9
81.1	328	266	62	215.7
81.1	323	262	61	212.5
81.1	322	261	61	211.6
81.1	318	258	60	209.3
81.1	317	257	60	208.4
81.1	312	253	59	205.2
81.1	307	249	58	202.0
81.1	302	245	57	198.8
81.1	301	244	57	197.8
81.1	297	241	56	195.6
81.1	296	240	56	194.6
81.1	291	236	55	191.4
81.1	286	232	54	188.2
81.1	285	231	54	187.2
81.1	281	228	53	185.0
81.1	280	227	53	184.0
81.1	275	223	52	180.8
81.1	270	219	51	177.6

%E	M1	M2	DM	M*
81.1	265	215	50	174.4
81.1	264	214	50	173.5
81.1	259	210	49	170.3
81.1	254	206	48	167.1
81.1	249	202	47	163.9
81.1	244	198	46	160.7
81.1	243	197	46	159.7
81.1	238	193	45	156.5
81.1	233	189	44	153.3
81.1	228	185	43	150.1
81.1	227	184	43	149.1
81.1	222	180	42	145.9
81.1	217	176	41	142.7
81.1	212	172	40	139.5
81.1	206	167	39	135.4
81.1	201	163	38	132.2
81.1	196	159	37	129.0
81.1	190	154	36	124.8
81.1	185	150	35	121.6
81.1	180	146	34	118.4
81.1	175	142	33	115.2
81.1	169	137	32	111.1
81.1	164	133	31	107.9
81.1	159	129	30	104.7
81.1	148	120	28	97.3
81.1	143	116	27	94.1
81.1	132	107	25	86.7
81.1	127	103	24	83.5
81.1	122	99	23	80.3
81.1	111	90	21	73.0
81.1	106	86	20	69.8
81.1	95	77	18	62.4
81.1	90	73	17	59.2
81.1	74	60	14	48.6
81.1	53	43	10	34.9
81.1	37	30	7	24.3
81.0	500	405	95	328.0
81.0	499	404	95	327.1
81.0	496	402	94	325.8
81.0	495	401	94	324.9
81.0	494	400	94	323.9
81.0	490	397	93	321.7
81.0	489	396	93	320.7
81.0	485	393	92	318.5
81.0	484	392	92	317.5
81.0	483	391	92	316.5
81.0	480	389	91	315.3
81.0	479	388	91	314.3
81.0	478	387	91	313.3
81.0	474	384	90	311.1

%E	M1	M2	DM	M*
81.0	473	383	90	310.1
81.0	469	380	89	307.9
81.0	468	379	89	306.9
81.0	464	376	88	304.7
81.0	463	375	88	303.7
81.0	462	374	88	302.8
81.0	459	372	87	301.5
81.0	458	371	87	300.5
81.0	457	370	87	299.6
81.0	453	367	86	297.3
81.0	452	366	86	296.4
81.0	448	363	85	294.1
81.0	447	362	85	293.2
81.0	443	359	84	290.9
81.0	442	358	84	290.0
81.0	441	357	84	289.0
81.0	437	354	83	286.8
81.0	436	353	83	285.8
81.0	432	350	82	283.6
81.0	431	349	82	282.6
81.0	427	346	81	280.4
81.0	426	345	81	279.4
81.0	422	342	80	277.2
81.0	421	341	80	276.2
81.0	420	340	80	275.2
81.0	416	337	79	273.0
81.0	415	336	79	272.0
81.0	411	333	78	269.8
81.0	410	332	78	268.8
81.0	406	329	77	266.6
81.0	405	328	77	265.6
81.0	401	325	76	263.4
81.0	400	324	76	262.4
81.0	399	323	76	261.5
81.0	395	320	75	259.2
81.0	394	319	75	258.3
81.0	390	316	74	256.0
81.0	389	315	74	255.1
81.0	385	312	73	252.8
81.0	384	311	73	251.9
81.0	379	307	72	248.7
81.0	378	306	72	247.7
81.0	374	303	71	245.5
81.0	373	302	71	244.5
81.0	369	299	70	242.3
81.0	368	298	70	241.3
81.0	364	295	69	239.1
81.0	363	294	69	238.1
81.0	358	290	68	234.9
81.0	357	289	68	234.0

%E	M1	M2	DM	M*
81.0	353	286	67	231.7
81.0	352	285	67	230.8
81.0	348	282	66	228.5
81.0	347	281	66	227.6
81.0	342	277	65	224.4
81.0	337	273	64	221.2
81.0	336	272	64	220.2
81.0	332	269	63	218.0
81.0	331	268	63	217.0
81.0	327	265	62	214.8
81.0	326	264	62	213.8
81.0	321	260	61	210.6
81.0	316	256	60	207.4
81.0	315	255	60	206.4
81.0	311	252	59	204.2
81.0	310	251	59	203.2
81.0	306	248	58	201.0
81.0	305	247	58	200.0
81.0	300	243	57	196.8
81.0	295	239	56	193.6
81.0	294	238	56	192.7
81.0	290	235	55	190.4
81.0	289	234	55	189.5
81.0	284	230	54	186.3
81.0	279	226	53	183.1
81.0	274	222	52	179.9
81.0	273	221	52	178.9
81.0	269	218	51	176.7
81.0	268	217	51	175.7
81.0	263	213	50	172.5
81.0	258	209	49	169.3
81.0	253	205	48	166.1
81.0	252	204	48	165.1
81.0	248	201	47	162.9
81.0	247	200	47	161.9
81.0	242	196	46	158.7
81.0	237	192	45	155.5
81.0	232	188	44	152.3
81.0	231	187	44	151.4
81.0	226	183	43	148.2
81.0	221	179	42	145.0
81.0	216	175	41	141.8
81.0	211	171	40	138.6
81.0	210	170	40	137.6
81.0	205	166	39	134.4
81.0	200	162	38	131.2
81.0	195	158	37	128.0
81.0	189	153	36	123.9
81.0	184	149	35	120.7
81.0	179	145	34	117.5

%E	M1	M2	DM	M*
81.0	174	141	33	114.3
81.0	168	136	32	110.1
81.0	163	132	31	106.9
81.0	158	128	30	103.7
81.0	153	124	29	100.5
81.0	147	119	28	96.3
81.0	142	115	27	93.1
81.0	137	111	26	89.9
81.0	126	102	24	82.6
81.0	121	98	23	79.4
81.0	116	94	22	76.2
81.0	105	85	20	68.8
81.0	100	81	19	65.6
81.0	84	68	16	55.0
81.0	79	64	15	51.8
81.0	63	51	12	41.3
81.0	58	47	11	38.1
81.0	42	34	8	27.5
81.0	21	17	4	13.8
80.9	498	403	95	326.1
80.9	497	402	95	325.2
80.9	493	399	94	322.9
80.9	492	398	94	322.0
80.9	491	397	94	321.0
80.9	488	395	93	319.7
80.9	487	394	93	318.8
80.9	486	393	93	317.8
80.9	482	390	92	315.6
80.9	481	389	92	314.6
80.9	477	386	91	312.4
80.9	476	385	91	311.4
80.9	472	382	90	309.2
80.9	471	381	90	308.2
80.9	470	380	90	307.2
80.9	467	378	89	306.0
80.9	466	377	89	305.0
80.9	465	376	89	304.0
80.9	461	373	88	301.8
80.9	460	372	88	300.8
80.9	456	369	87	298.6
80.9	455	368	87	297.6
80.9	451	365	86	295.4
80.9	450	364	86	294.4
80.9	446	361	85	292.2
80.9	445	360	85	291.2
80.9	444	359	85	290.3
80.9	440	356	84	288.0
80.9	439	355	84	287.1
80.9	435	352	83	284.8
80.9	434	351	83	283.9

%E	M1	M2	DM	M*
80.9	430	348	82	281.6
80.9	429	347	82	280.7
80.9	425	344	81	278.4
80.9	424	343	81	277.5
80.9	423	342	81	276.5
80.9	419	339	80	274.3
80.9	418	338	80	273.3
80.9	414	335	79	271.1
80.9	413	334	79	270.1
80.9	409	331	78	267.9
80.9	408	330	78	266.9
80.9	404	327	77	264.7
80.9	403	326	77	263.7
80.9	398	322	76	260.5
80.9	397	321	76	259.5
80.9	393	318	75	257.3
80.9	392	317	75	256.3
80.9	388	314	74	254.1
80.9	387	313	74	253.1
80.9	383	310	73	250.9
80.9	382	309	73	250.0
80.9	377	305	72	246.8
80.9	376	304	72	245.8
80.9	372	301	71	243.6
80.9	371	300	71	242.6
80.9	367	297	70	240.4
80.9	366	296	70	239.4
80.9	362	293	69	237.2
80.9	361	292	69	236.2
80.9	356	288	68	233.0
80.9	351	284	67	229.8
80.9	350	283	67	228.8
80.9	346	280	66	226.6
80.9	345	279	66	225.6
80.9	341	276	65	223.4
80.9	340	275	65	222.4
80.9	335	271	64	219.2
80.9	330	267	63	216.0
80.9	329	266	63	215.1
80.9	325	263	62	212.8
80.9	324	262	62	211.9
80.9	320	259	61	209.6
80.9	319	258	61	208.7
80.9	314	254	60	205.5
80.9	309	250	59	202.3
80.9	304	246	58	199.1
80.9	303	245	58	198.1
80.9	299	242	57	195.9
80.9	298	241	57	194.9
80.9	293	237	56	191.7

%E	M1	M2	DM	M*
80.9	288	233	55	188.5
80.9	283	229	54	185.3
80.9	282	228	54	184.3
80.9	278	225	53	182.1
80.9	277	224	53	181.1
80.9	272	220	52	177.9
80.9	267	216	51	174.7
80.9	262	212	50	171.5
80.9	257	208	49	168.3
80.9	256	207	49	167.4
80.9	251	203	48	164.2
80.9	246	199	47	161.0
80.9	241	195	46	157.8
80.9	236	191	45	154.6
80.9	235	190	45	153.6
80.9	230	186	44	150.4
80.9	225	182	43	147.2
80.9	220	178	42	144.0
80.9	215	174	41	140.8
80.9	209	169	40	136.7
80.9	204	165	39	133.5
80.9	199	161	38	130.3
80.9	194	157	37	127.1
80.9	188	152	36	122.9
80.9	183	148	35	119.7
80.9	178	144	34	116.5
80.9	173	140	33	113.3
80.9	162	131	31	105.9
80.9	157	127	30	102.7
80.9	152	123	29	99.5
80.9	141	114	27	92.2
80.9	136	110	26	89.0
80.9	131	106	25	85.8
80.9	115	93	22	75.2
80.9	110	89	21	72.0
80.9	94	76	18	61.4
80.9	89	72	17	58.2
80.9	68	55	13	44.5
80.9	47	38	9	30.7
80.8	500	404	96	326.4
80.8	499	403	96	325.5
80.8	496	401	95	324.2
80.8	495	400	95	323.2
80.8	494	399	95	322.3
80.8	490	396	94	320.0
80.8	489	395	94	319.1
80.8	485	392	93	316.8
80.8	484	391	93	315.9
80.8	480	388	92	313.6
80.8	479	387	92	312.7

%E	M1	M2	DM	M*
80.8	478	386	92	311.7
80.8	475	384	91	310.4
80.8	474	383	91	309.5
80.8	473	382	91	308.5
80.8	469	379	90	306.3
80.8	468	378	90	305.3
80.8	464	375	89	303.1
80.8	463	374	89	302.1
80.8	459	371	88	299.9
80.8	458	370	88	298.9
80.8	454	367	87	296.7
80.8	453	366	87	295.7
80.8	452	365	87	294.7
80.8	449	363	86	293.5
80.8	448	362	86	292.5
80.8	447	361	86	291.5
80.8	443	358	85	289.3
80.8	442	357	85	288.3
80.8	438	354	84	286.1
80.8	437	353	84	285.1
80.8	433	350	83	282.9
80.8	432	349	83	281.9
80.8	428	346	82	279.7
80.8	427	345	82	278.7
80.8	426	344	82	277.8
80.8	422	341	81	275.5
80.8	421	340	81	274.6
80.8	417	337	80	272.3
80.8	416	336	80	271.4
80.8	412	333	79	269.1
80.8	411	332	79	268.2
80.8	407	329	78	265.9
80.8	406	328	78	265.0
80.8	402	325	77	262.7
80.8	401	324	77	261.8
80.8	400	323	77	260.8
80.8	396	320	76	258.6
80.8	395	319	76	257.6
80.8	391	316	75	255.4
80.8	390	315	75	254.4
80.8	386	312	74	252.2
80.8	385	311	74	251.2
80.8	381	308	73	249.0
80.8	380	307	73	248.0
80.8	375	303	72	244.8
80.8	370	299	71	241.6
80.8	369	298	71	240.7
80.8	365	295	70	238.4
80.8	364	294	70	237.5
80.8	360	291	69	235.2

%E	M1	M2	DM	M*
80.8	359	290	69	234.3
80.8	355	287	68	232.0
80.8	354	286	68	231.1
80.8	349	282	67	227.9
80.8	344	278	66	224.7
80.8	343	277	66	223.7
80.8	339	274	65	221.5
80.8	338	273	65	220.5
80.8	334	270	64	218.3
80.8	333	269	64	217.3
80.8	328	265	63	214.1
80.8	323	261	62	210.9
80.8	318	257	61	207.7
80.8	317	256	61	206.7
80.8	313	253	60	204.5
80.8	312	252	60	203.5
80.8	308	249	59	201.3
80.8	307	248	59	200.3
80.8	302	244	58	197.1
80.8	297	240	57	193.9
80.8	292	236	56	190.7
80.8	291	235	56	189.8
80.8	287	232	55	187.5
80.8	286	231	55	186.6
80.8	281	227	54	183.4
80.8	276	223	53	180.2
80.8	271	219	52	177.0
80.8	266	215	51	173.8
80.8	265	214	51	172.8
80.8	261	211	50	170.6
80.8	260	210	50	169.6
80.8	255	206	49	166.4
80.8	250	202	48	163.2
80.8	245	198	47	160.0
80.8	240	194	46	156.8
80.8	239	193	46	155.9
80.8	234	189	45	152.7
80.8	229	185	44	149.5
80.8	224	181	43	146.3
80.8	219	177	42	143.1
80.8	214	173	41	139.9
80.8	213	172	41	138.9
80.8	208	168	40	135.7
80.8	203	164	39	132.5
80.8	198	160	38	129.3
80.8	193	156	37	126.1
80.8	182	147	35	118.7
80.8	177	143	34	115.5
80.8	172	139	33	112.3
80.8	167	135	32	109.1

%E	M1	M2	DM	M*
80.8	156	126	30	101.8
80.8	151	122	29	98.6
80.8	146	118	28	95.4
80.8	130	105	25	84.8
80.8	125	101	24	81.6
80.8	120	97	23	78.4
80.8	104	84	20	67.8
80.8	99	80	19	64.6
80.8	78	63	15	50.9
80.8	73	59	14	47.7
80.8	52	42	10	33.9
80.8	26	21	5	17.0
80.7	498	402	96	324.5
80.7	497	401	96	323.5
80.7	493	398	95	321.3
80.7	492	397	95	320.3
80.7	491	396	95	319.4
80.7	488	394	94	318.1
80.7	487	393	94	317.1
80.7	486	392	94	316.2
80.7	483	390	93	314.9
80.7	482	389	93	313.9
80.7	481	388	93	313.0
80.7	477	385	92	310.7
80.7	476	384	92	309.8
80.7	472	381	91	307.5
80.7	471	380	91	306.6
80.7	467	377	90	304.3
80.7	466	376	90	303.4
80.7	462	373	89	301.1
80.7	461	372	89	300.2
80.7	460	371	89	299.2
80.7	457	369	88	297.9
80.7	456	368	88	297.0
80.7	455	367	88	296.0
80.7	451	364	87	293.8
80.7	450	363	87	292.8
80.7	446	360	86	290.6
80.7	445	359	86	289.6
80.7	441	356	85	287.4
80.7	440	355	85	286.4
80.7	436	352	84	284.2
80.7	435	351	84	283.2
80.7	431	348	83	281.0
80.7	430	347	83	280.0
80.7	429	346	83	279.1
80.7	425	343	82	276.8
80.7	424	342	82	275.9
80.7	420	339	81	273.6
80.7	419	338	81	272.7

%E	M1	M2	DM	M*
80.7	415	335	80	270.4
80.7	414	334	80	269.5
80.7	410	331	79	267.2
80.7	409	330	79	266.3
80.7	405	327	78	264.0
80.7	404	326	78	263.1
80.7	399	322	77	259.9
80.7	398	321	77	258.9
80.7	394	318	76	256.7
80.7	393	317	76	255.7
80.7	389	314	75	253.5
80.7	388	313	75	252.5
80.7	384	310	74	250.3
80.7	383	309	74	249.3
80.7	379	306	73	247.1
80.7	378	305	73	246.1
80.7	374	302	72	243.9
80.7	373	301	72	242.9
80.7	368	297	71	239.7
80.7	367	296	71	238.7
80.7	363	293	70	236.5
80.7	362	292	70	235.5
80.7	358	289	69	233.3
80.7	357	288	69	232.3
80.7	353	285	68	230.1
80.7	352	284	68	229.1
80.7	348	281	67	226.9
80.7	347	280	67	225.9
80.7	342	276	66	222.7
80.7	337	272	65	219.5
80.7	336	271	65	218.6
80.7	332	268	64	216.3
80.7	331	267	64	215.4
80.7	327	264	63	213.1
80.7	326	263	63	212.2
80.7	322	260	62	209.9
80.7	321	259	62	209.0
80.7	316	255	61	205.8
80.7	311	251	60	202.6
80.7	306	247	59	199.4
80.7	305	246	59	198.4
80.7	301	243	58	196.2
80.7	300	242	58	195.2
80.7	296	239	57	193.0
80.7	295	238	57	192.0
80.7	290	234	56	188.8
80.7	285	230	55	185.6
80.7	280	226	54	182.4
80.7	275	222	53	179.2
80.7	274	221	53	178.3

%E	M1	M2	DM	M*
80.7	270	218	52	176.0
80.7	269	217	52	175.1
80.7	264	213	51	171.9
80.7	259	209	50	168.7
80.7	254	205	49	165.5
80.7	249	201	48	162.3
80.7	244	197	47	159.1
80.7	243	196	47	158.1
80.7	238	192	46	154.9
80.7	233	188	45	151.7
80.7	228	184	44	148.5
80.7	223	180	43	145.3
80.7	218	176	42	142.1
80.7	212	171	41	137.9
80.7	207	167	40	134.7
80.7	202	163	39	131.5
80.7	197	159	38	128.3
80.7	192	155	37	125.1
80.7	187	151	36	121.9
80.7	181	146	35	117.8
80.7	176	142	34	114.6
80.7	171	138	33	111.4
80.7	166	134	32	108.2
80.7	161	130	31	105.0
80.7	150	121	29	97.6
80.7	145	117	28	94.4
80.7	140	113	27	91.2
80.7	135	109	26	88.0
80.7	119	96	23	77.4
80.7	114	92	22	74.2
80.7	109	88	21	71.0
80.7	88	71	17	57.3
80.7	83	67	16	54.1
80.7	57	46	11	37.1
80.6	500	403	97	324.8
80.6	499	402	97	323.9
80.6	496	400	96	322.6
80.6	495	399	96	321.6
80.6	494	398	96	320.7
80.6	490	395	95	318.4
80.6	489	394	95	317.5
80.6	485	391	94	315.2
80.6	484	390	94	314.3
80.6	480	387	93	312.0
80.6	479	386	93	311.1
80.6	475	383	92	308.8
80.6	474	382	92	307.9
80.6	473	381	92	306.9
80.6	470	379	91	305.6
80.6	469	378	91	304.7

%E	M1	M2	DM	M*
80·6	468	377	91	303·7
80·6	465	375	90	302·4
80·6	464	374	90	301·5
80·6	463	373	90	300·5
80·6	459	370	89	298·3
80·6	458	369	89	297·3
80·6	454	366	88	295·1
80·6	453	365	88	294·1
80·6	449	362	87	291·9
80·6	448	361	87	290·9
80·6	444	358	86	288·7
80·6	443	357	86	287·7
80·6	439	354	85	285·5
80·6	438	353	85	284·5
80·6	434	350	84	282·3
80·6	433	349	84	281·3
80·6	432	348	84	280·3
80·6	428	345	83	278·1
80·6	427	344	83	277·1
80·6	423	341	82	274·9
80·6	422	340	82	273·9
80·6	418	337	81	271·7
80·6	417	336	81	270·7
80·6	413	333	80	268·5
80·6	412	332	80	267·5
80·6	408	329	79	265·3
80·6	407	328	79	264·3
80·6	403	325	78	262·1
80·6	402	324	78	261·1
80·6	397	320	77	257·9
80·6	396	319	77	257·0
80·6	392	316	76	254·7
80·6	391	315	76	253·8
80·6	387	312	75	251·5
80·6	386	311	75	250·6
80·6	382	308	74	248·3
80·6	381	307	74	247·4
80·6	377	304	73	245·1
80·6	376	303	73	244·2
80·6	372	300	72	241·9
80·6	371	299	72	241·0
80·6	366	295	71	237·8
80·6	361	291	70	234·6
80·6	360	290	70	233·6
80·6	356	287	69	231·4
80·6	355	286	69	230·4
80·6	351	283	68	228·2
80·6	350	282	68	227·2
80·6	346	279	67	225·0
80·6	345	278	67	224·0

%E	M1	M2	DM	M*
80·6	341	275	66	221·8
80·6	340	274	66	220·8
80·6	335	270	65	217·6
80·6	330	266	64	214·4
80·6	325	262	63	211·2
80·6	324	261	63	210·3
80·6	320	258	62	208·0
80·6	319	257	62	207·1
80·6	315	254	61	204·8
80·6	314	253	61	203·9
80·6	310	250	60	201·6
80·6	309	249	60	200·7
80·6	304	245	59	197·5
80·6	299	241	58	194·3
80·6	294	237	57	191·1
80·6	289	233	56	187·9
80·6	288	232	56	186·9
80·6	284	229	55	184·7
80·6	283	228	55	183·7
80·6	279	225	54	181·5
80·6	278	224	54	180·5
80·6	273	220	53	177·3
80·6	268	216	52	174·1
80·6	263	212	51	170·9
80·6	258	208	50	167·7
80·6	253	204	49	164·5
80·6	252	203	49	163·5
80·6	248	200	48	161·3
80·6	247	199	48	160·3
80·6	242	195	47	157·1
80·6	237	191	46	153·9
80·6	232	187	45	150·7
80·6	227	183	44	147·5
80·6	222	179	43	144·3
80·6	217	175	42	141·1
80·6	216	174	42	140·2
80·6	211	170	41	137·0
80·6	206	166	40	133·8
80·6	201	162	39	130·6
80·6	196	158	38	127·4
80·6	191	154	37	124·2
80·6	186	150	36	121·0
80·6	180	145	35	116·8
80·6	175	141	34	113·6
80·6	170	137	33	110·4
80·6	165	133	32	107·2
80·6	160	129	31	104·0
80·6	155	125	30	100·8
80·6	144	116	28	93·4
80·6	139	112	27	90·2

%E	M1	M2	DM	M*
80·6	134	108	26	87·0
80·6	129	104	25	83·8
80·6	124	100	24	80·6
80·6	108	87	21	70·1
80·6	103	83	20	66·9
80·6	98	79	19	63·7
80·6	93	75	18	60·5
80·6	72	58	14	46·7
80·6	67	54	13	43·5
80·6	62	50	12	40·3
80·6	36	29	7	23·4
80·6	31	25	6	20·2
80·5	498	401	97	322·9
80·5	497	400	97	321·9
80·5	493	397	96	319·7
80·5	492	396	96	318·7
80·5	488	393	95	316·5
80·5	487	392	95	315·5
80·5	486	391	95	314·6
80·5	483	389	94	313·3
80·5	482	388	94	312·3
80·5	481	387	94	311·4
80·5	478	385	93	310·1
80·5	477	384	93	309·1
80·5	476	383	93	308·2
80·5	472	380	92	305·9
80·5	471	379	92	305·0
80·5	467	376	91	302·7
80·5	466	375	91	301·8
80·5	462	372	90	299·5
80·5	461	371	90	298·6
80·5	457	368	89	296·3
80·5	456	367	89	295·4
80·5	452	364	88	293·1
80·5	451	363	88	292·2
80·5	447	360	87	289·9
80·5	446	359	87	289·0
80·5	442	356	86	286·7
80·5	441	355	86	285·8
80·5	440	354	86	284·8
80·5	437	352	85	283·5
80·5	436	351	85	282·6
80·5	435	350	85	281·6
80·5	431	347	84	279·4
80·5	430	346	84	278·4
80·5	426	343	83	276·2
80·5	425	342	83	275·2
80·5	421	339	82	273·0
80·5	420	338	82	272·0
80·5	416	335	81	269·8

%E	M1	M2	DM	M*
80·5	415	334	81	268·8
80·5	411	331	80	266·6
80·5	410	330	80	265·6
80·5	406	327	79	263·4
80·5	405	326	79	262·4
80·5	401	323	78	260·2
80·5	400	322	78	259·2
80·5	399	321	78	258·2
80·5	395	318	77	256·0
80·5	394	317	77	255·0
80·5	390	314	76	252·8
80·5	389	313	76	251·8
80·5	385	310	75	249·6
80·5	384	309	75	248·6
80·5	380	306	74	246·4
80·5	379	305	74	245·4
80·5	375	302	73	243·2
80·5	374	301	73	242·2
80·5	370	298	72	240·0
80·5	369	297	72	239·0
80·5	365	294	71	236·8
80·5	364	293	71	235·8
80·5	359	289	70	232·6
80·5	354	285	69	229·4
80·5	353	284	69	228·5
80·5	349	281	68	226·2
80·5	348	280	68	225·3
80·5	344	277	67	223·0
80·5	343	276	67	222·1
80·5	339	273	66	219·8
80·5	338	272	66	218·9
80·5	334	269	65	216·6
80·5	333	268	65	215·7
80·5	329	265	64	213·4
80·5	328	264	64	212·5
80·5	323	260	63	209·3
80·5	318	256	62	206·1
80·5	313	252	61	202·9
80·5	308	248	60	199·7
80·5	307	247	60	198·7
80·5	303	244	59	196·5
80·5	302	243	59	195·5
80·5	298	240	58	193·3
80·5	297	239	58	192·3
80·5	293	236	57	190·1
80·5	292	235	57	189·1
80·5	287	231	56	185·9
80·5	282	227	55	182·7
80·5	277	223	54	179·5
80·5	272	219	53	176·3

%E	M1	M2	DM	M*
80·5	267	215	52	173·1
80·5	266	214	52	172·2
80·5	262	211	51	169·9
80·5	261	210	51	169·0
80·5	257	207	50	166·7
80·5	256	206	50	165·8
80·5	251	202	49	162·6
80·5	246	198	48	159·4
80·5	241	194	47	156·2
80·5	236	190	46	153·0
80·5	231	186	45	149·8
80·5	226	182	44	146·6
80·5	221	178	43	143·4
80·5	220	177	43	142·4
80·5	215	173	42	139·2
80·5	210	169	41	136·0
80·5	205	165	40	132·8
80·5	200	161	39	129·6
80·5	195	157	38	126·4
80·5	190	153	37	123·2
80·5	185	149	36	120·0
80·5	174	140	34	112·6
80·5	169	136	33	109·4
80·5	164	132	32	106·2
80·5	159	128	31	103·0
80·5	154	124	30	99·8
80·5	149	120	29	96·6
80·5	133	107	26	86·1
80·5	128	103	25	82·9
80·5	123	99	24	79·7
80·5	118	95	23	76·5
80·5	113	91	22	73·3
80·5	87	70	17	56·3
80·5	82	66	16	53·1
80·5	77	62	15	49·9
80·5	41	33	8	26·6
80·4	500	402	98	323·2
80·4	499	401	98	322·2
80·4	496	399	97	321·0
80·4	495	398	97	320·0
80·4	494	397	97	319·0
80·4	491	395	96	317·8
80·4	490	394	96	316·8
80·4	489	393	96	315·8
80·4	485	390	95	313·6
80·4	484	389	95	312·6
80·4	480	386	94	310·4
80·4	479	385	94	309·4
80·4	475	382	93	307·2
80·4	474	381	93	306·2

%E	M1	M2	DM	M*
80.4	470	378	92	304.0
80.4	469	377	92	303.0
80.4	465	374	91	300.8
80.4	464	373	91	299.8
80.4	460	370	90	297.6
80.4	459	369	90	296.6
80.4	455	366	89	294.4
80.4	454	365	89	293.4
80.4	453	364	89	292.5
80.4	450	362	88	291.2
80.4	449	361	88	290.2
80.4	448	360	88	289.3
80.4	445	358	87	288.0
80.4	444	357	87	287.0
80.4	443	356	87	286.1
80.4	439	353	86	283.8
80.4	438	352	86	282.9
80.4	434	349	85	280.6
80.4	433	348	85	279.7
80.4	429	345	84	277.4
80.4	428	344	84	276.5
80.4	424	341	83	274.2
80.4	423	340	83	273.3
80.4	419	337	82	271.0
80.4	418	336	82	270.1
80.4	414	333	81	267.8
80.4	413	332	81	266.9
80.4	409	329	80	264.6
80.4	408	328	80	263.7
80.4	404	325	79	261.4
80.4	403	324	79	260.5
80.4	398	320	78	257.3
80.4	397	319	78	256.3
80.4	393	316	77	254.1
80.4	392	315	77	253.1
80.4	388	312	76	250.9
80.4	387	311	76	249.9
80.4	383	308	75	247.7
80.4	382	307	75	246.7
80.4	378	304	74	244.5
80.4	377	303	74	243.5
80.4	373	300	73	241.3
80.4	372	299	73	240.3
80.4	368	296	72	238.1
80.4	367	295	72	237.1
80.4	363	292	71	234.9
80.4	362	291	71	233.9
80.4	358	288	70	231.7
80.4	357	287	70	230.7
80.4	352	283	69	227.5

%E	M1	M2	DM	M*
80.4	347	279	68	224.3
80.4	342	275	67	221.1
80.4	341	274	67	220.2
80.4	337	271	66	217.9
80.4	336	270	66	217.0
80.4	332	267	65	214.7
80.4	331	266	65	213.8
80.4	327	263	64	211.5
80.4	326	262	64	210.6
80.4	322	259	63	208.3
80.4	321	258	63	207.4
80.4	317	255	62	205.1
80.4	316	254	62	204.2
80.4	312	251	61	201.9
80.4	311	250	61	201.0
80.4	306	246	60	197.8
80.4	301	242	59	194.6
80.4	296	238	58	191.4
80.4	291	234	57	188.2
80.4	286	230	56	185.0
80.4	285	229	56	184.0
80.4	281	226	55	181.8
80.4	280	225	55	180.8
80.4	276	222	54	178.6
80.4	275	221	54	177.6
80.4	271	218	53	175.4
80.4	270	217	53	174.4
80.4	265	213	52	171.2
80.4	260	209	51	168.0
80.4	255	205	50	164.8
80.4	250	201	49	161.6
80.4	245	197	48	158.4
80.4	240	193	47	155.2
80.4	235	189	46	152.0
80.4	230	185	45	148.8
80.4	225	181	44	145.6
80.4	224	180	44	144.6
80.4	219	176	43	141.4
80.4	214	172	42	138.2
80.4	209	168	41	135.0
80.4	204	164	40	131.8
80.4	199	160	39	128.6
80.4	194	156	38	125.4
80.4	189	152	37	122.2
80.4	184	148	36	119.0
80.4	179	144	35	115.8
80.4	168	135	33	108.5
80.4	163	131	32	105.3
80.4	158	127	31	102.1
80.4	153	123	30	98.9

%E	M1	M2	DM	M*
80.4	148	119	29	95.7
80.4	143	115	28	92.5
80.4	138	111	27	89.3
80.4	112	90	22	72.3
80.4	107	86	21	69.1
80.4	102	82	20	65.9
80.4	97	78	19	62.7
80.4	92	74	18	59.5
80.4	56	45	11	36.2
80.4	51	41	10	33.0
80.4	46	37	9	29.8
80.3	498	400	98	321.3
80.3	497	399	98	320.3
80.3	493	396	97	318.1
80.3	492	395	97	317.1
80.3	488	392	96	314.9
80.3	487	391	96	313.9
80.3	483	388	95	311.7
80.3	482	387	95	310.7
80.3	478	384	94	308.5
80.3	477	383	94	307.5
80.3	476	382	94	306.6
80.3	473	380	93	305.3
80.3	472	379	93	304.3
80.3	471	378	93	303.4
80.3	468	376	92	302.1
80.3	467	375	92	301.1
80.3	466	374	92	300.2
80.3	463	372	91	298.9
80.3	462	371	91	297.9
80.3	461	370	91	297.0
80.3	458	368	90	295.7
80.3	457	367	90	294.7
80.3	456	366	90	293.8
80.3	452	363	89	291.5
80.3	451	362	89	290.6
80.3	447	359	88	288.3
80.3	446	358	88	287.4
80.3	442	355	87	285.1
80.3	441	354	87	284.2
80.3	437	351	86	281.9
80.3	436	350	86	281.0
80.3	432	347	85	278.7
80.3	431	346	85	277.8
80.3	427	343	84	275.5
80.3	426	342	84	274.6
80.3	422	339	83	272.3
80.3	421	338	83	271.4
80.3	417	335	82	269.1
80.3	416	334	82	268.2

%E	M1	M2	DM	M*
80.3	412	331	81	265.9
80.3	411	330	81	265.0
80.3	407	327	80	262.7
80.3	406	326	80	261.8
80.3	402	323	79	259.5
80.3	401	322	79	258.6
80.3	400	321	79	257.6
80.3	396	318	78	255.4
80.3	395	317	78	254.4
80.3	391	314	77	252.2
80.3	390	313	77	251.2
80.3	386	310	76	249.0
80.3	385	309	76	248.0
80.3	381	306	75	245.8
80.3	380	305	75	244.8
80.3	376	302	74	242.6
80.3	375	301	74	241.6
80.3	371	298	73	239.4
80.3	370	297	73	238.4
80.3	366	294	72	236.2
80.3	365	293	72	235.2
80.3	361	290	71	233.0
80.3	360	289	71	232.0
80.3	356	286	70	229.8
80.3	355	285	70	228.8
80.3	351	282	69	226.6
80.3	350	281	69	225.6
80.3	346	278	68	223.4
80.3	345	277	68	222.4
80.3	340	273	67	219.2
80.3	335	269	66	216.0
80.3	330	265	65	212.8
80.3	325	261	64	209.6
80.3	320	257	63	206.4
80.3	319	256	63	205.4
80.3	315	253	62	203.2
80.3	314	252	62	202.2
80.3	310	249	61	200.0
80.3	309	248	61	199.0
80.3	305	245	60	196.8
80.3	304	244	60	195.8
80.3	300	241	59	193.6
80.3	299	240	59	192.6
80.3	295	237	58	190.4
80.3	294	236	58	189.4
80.3	290	233	57	187.2
80.3	289	232	57	186.2
80.3	284	228	56	183.0
80.3	279	224	55	179.8
80.3	274	220	54	176.6

%E	M1	M2	DM	M*
80.3	269	216	53	173.4
80.3	264	212	52	170.2
80.3	259	208	51	167.0
80.3	254	204	50	163.8
80.3	249	200	49	160.6
80.3	244	196	48	157.4
80.3	239	192	47	154.2
80.3	238	191	47	153.3
80.3	234	188	46	151.0
80.3	233	187	46	150.1
80.3	229	184	45	147.8
80.3	228	183	45	146.9
80.3	223	179	44	143.7
80.3	218	175	43	140.5
80.3	213	171	42	137.3
80.3	208	167	41	134.1
80.3	203	163	40	130.9
80.3	198	159	39	127.7
80.3	193	155	38	124.5
80.3	188	151	37	121.3
80.3	183	147	36	118.1
80.3	178	143	35	114.9
80.3	173	139	34	111.7
80.3	157	126	31	101.1
80.3	152	122	30	97.9
80.3	147	118	29	94.7
80.3	142	114	28	91.5
80.3	137	110	27	88.3
80.3	132	106	26	85.1
80.3	127	102	25	81.9
80.3	122	98	24	78.7
80.3	117	94	23	75.5
80.3	76	61	15	49.0
80.3	71	57	14	45.8
80.3	66	53	13	42.6
80.3	61	49	12	39.4
80.2	500	401	99	321.6
80.2	499	400	99	320.6
80.2	496	398	98	319.4
80.2	495	397	98	318.4
80.2	494	396	98	317.4
80.2	491	394	97	316.2
80.2	490	393	97	315.2
80.2	489	392	97	314.2
80.2	486	390	96	313.0
80.2	485	389	96	312.0
80.2	484	388	96	311.0
80.2	481	386	95	309.8
80.2	480	385	95	308.8
80.2	479	384	95	307.8

%E	M1	M2	DM	M*
80.2	475	381	94	305.6
80.2	474	380	94	304.6
80.2	470	377	93	302.4
80.2	469	376	93	301.4
80.2	465	373	92	299.2
80.2	464	372	92	298.2
80.2	460	369	91	296.0
80.2	459	368	91	295.0
80.2	455	365	90	292.8
80.2	454	364	90	291.8
80.2	450	361	89	289.6
80.2	449	360	89	288.6
80.2	445	357	88	286.4
80.2	444	356	88	285.4
80.2	440	353	87	283.2
80.2	439	352	87	282.2
80.2	435	349	86	280.0
80.2	434	348	86	279.0
80.2	430	345	85	276.8
80.2	429	344	85	275.8
80.2	425	341	84	273.6
80.2	424	340	84	272.6
80.2	420	337	83	270.4
80.2	419	336	83	269.4
80.2	415	333	82	267.2
80.2	414	332	82	266.2
80.2	410	329	81	264.0
80.2	409	328	81	263.0
80.2	405	325	80	260.8
80.2	404	324	80	259.8
80.2	399	320	79	256.6
80.2	398	319	79	255.7
80.2	394	316	78	253.4
80.2	393	315	78	252.5
80.2	389	312	77	250.2
80.2	388	311	77	249.3
80.2	384	308	76	247.0
80.2	383	307	76	246.1
80.2	379	304	75	243.8
80.2	378	303	75	242.9
80.2	374	300	74	240.6
80.2	373	299	74	239.7
80.2	369	296	73	237.4
80.2	368	295	73	236.5
80.2	364	292	72	234.2
80.2	363	291	72	233.3
80.2	359	288	71	231.0
80.2	358	287	71	230.1
80.2	354	284	70	227.8
80.2	353	283	70	226.9

%E	M1	M2	DM	M*
80.2	349	280	69	224.6
80.2	348	279	69	223.7
80.2	344	276	68	221.4
80.2	343	275	68	220.5
80.2	339	272	67	218.2
80.2	338	271	67	217.3
80.2	334	268	66	215.0
80.2	333	267	66	214.1
80.2	329	264	65	211.8
80.2	328	263	65	210.9
80.2	324	260	64	208.6
80.2	323	259	64	207.7
80.2	318	255	63	204.5
80.2	313	251	62	201.3
80.2	308	247	61	198.1
80.2	303	243	60	194.9
80.2	298	239	59	191.7
80.2	293	235	58	188.5
80.2	288	231	57	185.3
80.2	283	227	56	182.1
80.2	278	223	55	178.9
80.2	273	219	54	175.7
80.2	268	215	53	172.5
80.2	267	214	53	171.5
80.2	263	211	52	169.3
80.2	262	210	52	168.3
80.2	258	207	51	166.1
80.2	257	206	51	165.1
80.2	253	203	50	162.9
80.2	252	202	50	161.9
80.2	248	199	49	159.7
80.2	247	198	49	158.7
80.2	243	195	48	156.5
80.2	242	194	48	155.5
80.2	237	190	47	152.3
80.2	232	186	46	149.1
80.2	227	182	45	145.9
80.2	222	178	44	142.7
80.2	217	174	43	139.5
80.2	212	170	42	136.3
80.2	207	166	41	133.1
80.2	202	162	40	129.9
80.2	197	158	39	126.7
80.2	192	154	38	123.5
80.2	187	150	37	120.3
80.2	182	146	36	117.1
80.2	177	142	35	113.9
80.2	172	138	34	110.7
80.2	167	134	33	107.5
80.2	162	130	32	104.3

%E	M1	M2	DM	M*
80.2	131	105	26	84.2
80.2	126	101	25	81.0
80.2	121	97	24	77.8
80.2	116	93	23	74.6
80.2	111	89	22	71.4
80.2	106	85	21	68.2
80.2	101	81	20	65.0
80.2	96	77	19	61.8
80.2	91	73	18	58.6
80.2	86	69	17	55.4
80.2	81	65	16	52.2
80.1	498	399	99	319.7
80.1	497	398	99	318.7
80.1	493	395	98	316.5
80.1	492	394	98	315.5
80.1	488	391	97	313.3
80.1	487	390	97	312.3
80.1	483	387	96	310.1
80.1	482	386	96	309.1
80.1	478	383	95	306.9
80.1	477	382	95	305.9
80.1	473	379	94	303.7
80.1	472	378	94	302.7
80.1	468	375	93	300.5
80.1	467	374	93	299.5
80.1	463	371	92	297.3
80.1	462	370	92	296.3
80.1	458	367	91	294.1
80.1	457	366	91	293.1
80.1	453	363	90	290.9
80.1	452	362	90	289.9
80.1	448	359	89	287.7
80.1	447	358	89	286.7
80.1	443	355	88	284.5
80.1	442	354	88	283.5
80.1	438	351	87	281.3
80.1	437	350	87	280.3
80.1	433	347	86	278.1
80.1	432	346	86	277.1
80.1	428	343	85	274.9
80.1	427	342	85	273.9
80.1	423	339	84	271.7
80.1	422	338	84	270.7
80.1	418	335	83	268.5
80.1	417	334	83	267.5
80.1	413	331	82	265.3
80.1	412	330	82	264.3
80.1	408	327	81	262.1
80.1	407	326	81	261.1
80.1	403	323	80	258.9

%E	M1	M2	DM	M*
80.1	402	322	80	257.9
80.1	401	321	80	257.0
80.1	397	318	79	254.7
80.1	396	317	79	253.8
80.1	392	314	78	251.5
80.1	391	313	78	250.6
80.1	387	310	77	248.3
80.1	386	309	77	247.4
80.1	382	306	76	245.1
80.1	381	305	76	244.2
80.1	377	302	75	241.9
80.1	376	301	75	241.0
80.1	372	298	74	238.7
80.1	371	297	74	237.8
80.1	367	294	73	235.5
80.1	366	293	73	234.6
80.1	362	290	72	232.3
80.1	361	289	72	231.4
80.1	357	286	71	229.1
80.1	356	285	71	228.2
80.1	352	282	70	225.9
80.1	351	281	70	225.0
80.1	347	278	69	222.7
80.1	346	277	69	221.8
80.1	342	274	68	219.5
80.1	341	273	68	218.6
80.1	337	270	67	216.3
80.1	336	269	67	215.4
80.1	332	266	66	213.1
80.1	331	265	66	212.2
80.1	327	262	65	209.9
80.1	326	261	65	209.0
80.1	322	258	64	206.7
80.1	321	257	64	205.8
80.1	317	254	63	203.5
80.1	316	253	63	202.6
80.1	312	250	62	200.3
80.1	311	249	62	199.4
80.1	307	246	61	197.1
80.1	306	245	61	196.2
80.1	302	242	60	193.9
80.1	301	241	60	193.0
80.1	297	238	59	190.7
80.1	296	237	59	189.8
80.1	292	234	58	187.5
80.1	291	233	58	186.6
80.1	287	230	57	184.3
80.1	286	229	57	183.4
80.1	282	226	56	181.1
80.1	281	225	56	180.2

%E	M1	M2	DM	M*
80.1	277	222	55	177.9
80.1	276	221	55	177.0
80.1	272	218	54	174.7
80.1	271	217	54	173.8
80.1	266	213	53	170.6
80.1	261	209	52	167.4
80.1	256	205	51	164.2
80.1	251	201	50	161.0
80.1	246	197	49	157.8
80.1	241	193	48	154.6
80.1	236	189	47	151.4
80.1	231	185	46	148.2
80.1	226	181	45	145.0
80.1	221	177	44	141.8
80.1	216	173	43	138.6
80.1	211	169	42	135.4
80.1	206	165	41	132.2
80.1	201	161	40	129.0
80.1	196	157	39	125.8
80.1	191	153	38	122.6
80.1	186	149	37	119.4
80.1	181	145	36	116.2
80.1	176	141	35	113.0
80.1	171	137	34	109.8
80.1	166	133	33	106.6
80.1	161	129	32	103.4
80.1	156	125	31	100.2
80.1	151	121	30	97.0
80.1	146	117	29	93.8
80.1	141	113	28	90.6
80.1	136	109	27	87.4
80.0	500	400	100	320.0
80.0	499	399	100	319.0
80.0	496	397	99	317.8
80.0	495	396	99	316.8
80.0	494	395	99	315.8
80.0	491	393	98	314.6
80.0	490	392	98	313.6
80.0	489	391	98	312.6
80.0	486	389	97	311.4
80.0	485	388	97	310.4
80.0	484	387	97	309.4
80.0	481	385	96	308.2
80.0	480	384	96	307.2
80.0	479	383	96	306.2
80.0	476	381	95	305.0
80.0	475	380	95	304.0
80.0	474	379	95	303.0
80.0	471	377	94	301.8
80.0	470	376	94	300.8

%E	M1	M2	DM	M*
80.0	469	375	94	299.8
80.0	466	373	93	298.6
80.0	465	372	93	297.6
80.0	464	371	93	296.6
80.0	461	369	92	295.4
80.0	460	368	92	294.4
80.0	459	367	92	293.4
80.0	456	365	91	292.2
80.0	455	364	91	291.2
80.0	454	363	91	290.2
80.0	451	361	90	289.0
80.0	450	360	90	288.0
80.0	449	359	90	287.0
80.0	446	357	89	285.8
80.0	445	356	89	284.8
80.0	444	355	89	283.8
80.0	441	353	88	282.6
80.0	440	352	88	281.6
80.0	439	351	88	280.6
80.0	436	349	87	279.4
80.0	435	348	87	278.4
80.0	434	347	87	277.4
80.0	431	345	86	276.2
80.0	430	344	86	275.2
80.0	429	343	86	274.2
80.0	426	341	85	273.0
80.0	425	340	85	272.0
80.0	424	339	85	271.0
80.0	421	337	84	269.8
80.0	420	336	84	268.8
80.0	419	335	84	267.8
80.0	416	333	83	266.6
80.0	415	332	83	265.6
80.0	414	331	83	264.6
80.0	411	329	82	263.4
80.0	410	328	82	262.4
80.0	409	327	82	261.4
80.0	406	325	81	260.2
80.0	405	324	81	259.2
80.0	404	323	81	258.2
80.0	400	320	80	256.0
80.0	399	319	80	255.0
80.0	395	316	79	252.8
80.0	390	312	78	249.6
80.0	385	308	77	246.4
80.0	380	304	76	243.2
80.0	375	300	75	240.0
80.0	370	296	74	236.8
80.0	365	292	73	233.6
80.0	360	288	72	230.4

%E	M1	M2	DM	M*
80.0	355	284	71	227.2
80.0	350	280	70	224.0
80.0	345	276	69	220.8
80.0	340	272	68	217.6
80.0	335	268	67	214.4
80.0	330	264	66	211.2
80.0	325	260	65	208.0
80.0	320	256	64	204.8
80.0	315	252	63	201.6
80.0	310	248	62	198.4
80.0	305	244	61	195.2
80.0	300	240	60	192.0
80.0	295	236	59	188.8
80.0	290	232	58	185.6
80.0	285	228	57	182.4
80.0	280	224	56	179.2
80.0	275	220	55	176.0
80.0	270	216	54	172.8
80.0	265	212	53	169.6
80.0	260	208	52	166.4
80.0	255	204	51	163.2
80.0	250	200	50	160.0
80.0	245	196	49	156.8
80.0	240	192	48	153.6
80.0	235	188	47	150.4
80.0	230	184	46	147.2
80.0	225	180	45	144.0
80.0	220	176	44	140.8
80.0	215	172	43	137.6
80.0	210	168	42	134.4
80.0	205	164	41	131.2
80.0	200	160	40	128.0
80.0	195	156	39	124.8
80.0	190	152	38	121.6
80.0	185	148	37	118.4
80.0	180	144	36	115.2
80.0	175	140	35	112.0
80.0	170	136	34	108.8
80.0	165	132	33	105.6
80.0	160	128	32	102.4
80.0	155	124	31	99.2
80.0	150	120	30	96.0
80.0	145	116	29	92.8
80.0	140	112	28	89.6
80.0	135	108	27	86.4
80.0	130	104	26	83.2
80.0	125	100	25	80.0
80.0	120	96	24	76.8
80.0	115	92	23	73.6
80.0	110	88	22	70.4

%E	M1	M2	DM	M*
80.0	105	84	21	67.2
80.0	100	80	20	64.0
80.0	95	76	19	60.8
80.0	90	72	18	57.6
80.0	85	68	17	54.4
80.0	80	64	16	51.2
80.0	75	60	15	48.0
80.0	70	56	14	44.8
80.0	65	52	13	41.6
80.0	60	48	12	38.4
80.0	55	44	11	35.2
80.0	50	40	10	32.0
80.0	45	36	9	28.8
80.0	40	32	8	25.6
80.0	35	28	7	22.4
80.0	30	24	6	19.2
80.0	25	20	5	16.0
80.0	20	16	4	12.8
80.0	15	12	3	9.6
80.0	10	8	2	6.4
80.0	5	4	1	3.2
79.9	498	398	100	318.1
79.9	497	397	100	317.1
79.9	493	394	99	314.9
79.9	492	393	99	313.9
79.9	488	390	98	311.7
79.9	487	389	98	310.7
79.9	483	386	97	308.5
79.9	482	385	97	307.5
79.9	478	382	96	305.3
79.9	477	381	96	304.3
79.9	473	378	95	302.1
79.9	472	377	95	301.1
79.9	468	374	94	298.9
79.9	467	373	94	297.9
79.9	463	370	93	295.7
79.9	462	369	93	294.7
79.9	458	366	92	292.5
79.9	457	365	92	291.5
79.9	453	362	91	289.3
79.9	452	361	91	288.3
79.9	448	358	90	286.1
79.9	447	357	90	285.1
79.9	443	354	89	282.9
79.9	442	353	89	281.9
79.9	438	350	88	279.7
79.9	437	349	88	278.7
79.9	433	346	87	276.5
79.9	432	345	87	275.5
79.9	428	342	86	273.3

%E	M1	M2	DM	M*
79.9	427	341	86	272.3
79.9	423	338	85	270.1
79.9	422	337	85	269.1
79.9	418	334	84	266.9
79.9	417	333	84	265.9
79.9	413	330	83	263.7
79.9	412	329	83	262.7
79.9	408	326	82	260.5
79.9	407	325	82	259.5
79.9	403	322	81	257.3
79.9	402	321	81	256.3
79.9	398	318	80	254.1
79.9	394	315	79	251.8
79.9	393	314	79	250.9
79.9	389	311	78	248.6
79.9	388	310	78	247.7
79.9	384	307	77	245.4
79.9	383	306	77	244.5
79.9	379	303	76	242.2
79.9	378	302	76	241.3
79.9	374	299	75	239.0
79.9	373	298	75	238.1
79.9	369	295	74	235.8
79.9	368	294	74	234.9
79.9	364	291	73	232.6
79.9	363	290	73	231.7
79.9	359	287	72	229.4
79.9	358	286	72	228.5
79.9	354	283	71	226.2
79.9	353	282	71	225.3
79.9	349	279	70	223.0
79.9	348	278	70	222.1
79.9	344	275	69	219.8
79.9	343	274	69	218.9
79.9	339	271	68	216.6
79.9	338	270	68	215.7
79.9	334	267	67	213.4
79.9	333	266	67	212.5
79.9	329	263	66	210.2
79.9	328	262	66	209.3
79.9	324	259	65	207.0
79.9	323	258	65	206.1
79.9	319	255	64	203.8
79.9	318	254	64	202.9
79.9	314	251	63	200.6
79.9	313	250	63	199.7
79.9	309	247	62	197.4
79.9	308	246	62	196.5
79.9	304	243	61	194.2
79.9	303	242	61	193.3

%E	M1	M2	DM	M*
79.9	299	239	60	191.0
79.9	298	238	60	190.1
79.9	294	235	59	187.8
79.9	293	234	59	186.9
79.9	289	231	58	184.6
79.9	288	230	58	183.7
79.9	284	227	57	181.4
79.9	283	226	57	180.5
79.9	279	223	56	178.2
79.9	278	222	56	177.3
79.9	274	219	55	175.0
79.9	273	218	55	174.1
79.9	269	215	54	171.8
79.9	268	214	54	170.9
79.9	264	211	53	168.6
79.9	259	207	52	165.4
79.9	254	203	51	162.2
79.9	249	199	50	159.0
79.9	244	195	49	155.8
79.9	239	191	48	152.6
79.9	234	187	47	149.4
79.9	229	183	46	146.2
79.9	224	179	45	143.0
79.9	219	175	44	139.8
79.9	214	171	43	136.6
79.9	209	167	42	133.4
79.9	204	163	41	130.2
79.9	199	159	40	127.0
79.9	194	155	39	123.8
79.9	189	151	38	120.6
79.9	184	147	37	117.4
79.9	179	143	36	114.2
79.9	174	139	35	111.0
79.9	169	135	34	107.8
79.9	164	131	33	104.6
79.9	159	127	32	101.4
79.9	154	123	31	98.2
79.9	149	119	30	95.0
79.9	144	115	29	91.8
79.9	139	111	28	88.6
79.9	134	107	27	85.4
79.8	500	399	101	318.4
79.8	499	398	101	317.4
79.8	496	396	100	316.2
79.8	495	395	100	315.2
79.8	494	394	100	314.2
79.8	491	392	99	313.0
79.8	490	391	99	312.0
79.8	489	390	99	311.0
79.8	486	388	98	309.8

%E	M1	M2	DM	M*
79.8	485	387	98	308.8
79.8	484	386	98	307.8
79.8	481	384	97	306.6
79.8	480	383	97	305.6
79.8	476	380	96	303.4
79.8	475	379	96	302.4
79.8	471	376	95	300.2
79.8	470	375	95	299.2
79.8	466	372	94	297.0
79.8	465	371	94	296.0
79.8	461	368	93	293.8
79.8	460	367	93	292.8
79.8	456	364	92	290.6
79.8	455	363	92	289.6
79.8	451	360	91	287.4
79.8	450	359	91	286.4
79.8	446	356	90	284.2
79.8	445	355	90	283.2
79.8	441	352	89	281.0
79.8	440	351	89	280.0
79.8	436	348	88	277.8
79.8	435	347	88	276.8
79.8	431	344	87	274.6
79.8	430	343	87	273.6
79.8	426	340	86	271.4
79.8	425	339	86	270.4
79.8	421	336	85	268.2
79.8	420	335	85	267.2
79.8	416	332	84	265.0
79.8	415	331	84	264.0
79.8	411	328	83	261.8
79.8	410	327	83	260.8
79.8	406	324	82	258.6
79.8	405	323	82	257.6
79.8	401	320	81	255.4
79.8	400	319	81	254.4
79.8	397	317	80	253.1
79.8	396	316	80	252.2
79.8	392	313	79	249.9
79.8	391	312	79	249.0
79.8	387	309	78	246.7
79.8	386	308	78	245.8
79.8	382	305	77	243.5
79.8	381	304	77	242.6
79.8	377	301	76	240.3
79.8	376	300	76	239.4
79.8	372	297	75	237.1
79.8	371	296	75	236.2
79.8	367	293	74	233.9
79.8	366	292	74	233.0

%E	M1	M2	DM	M*
79.8	362	289	73	230.7
79.8	361	288	73	229.8
79.8	357	285	72	227.5
79.8	356	284	72	226.6
79.8	352	281	71	224.3
79.8	351	280	71	223.4
79.8	347	277	70	221.1
79.8	346	276	70	220.2
79.8	342	273	69	217.9
79.8	341	272	69	217.0
79.8	337	269	68	214.7
79.8	336	268	68	213.8
79.8	332	265	67	211.5
79.8	331	264	67	210.6
79.8	327	261	66	208.3
79.8	326	260	66	207.4
79.8	322	257	65	205.1
79.8	321	256	65	204.2
79.8	317	253	64	201.9
79.8	312	249	63	198.7
79.8	307	245	62	195.5
79.8	302	241	61	192.3
79.8	297	237	60	189.1
79.8	292	233	59	185.9
79.8	287	229	58	182.7
79.8	282	225	57	179.5
79.8	277	221	56	176.3
79.8	272	217	55	173.1
79.8	267	213	54	169.9
79.8	263	210	53	167.7
79.8	262	209	53	166.7
79.8	258	206	52	164.5
79.8	257	205	52	163.5
79.8	253	202	51	161.3
79.8	252	201	51	160.3
79.8	248	198	50	158.1
79.8	247	197	50	157.1
79.8	243	194	49	154.9
79.8	242	193	49	153.9
79.8	238	190	48	151.7
79.8	233	186	47	148.5
79.8	228	182	46	145.3
79.8	223	178	45	142.1
79.8	218	174	44	138.9
79.8	213	170	43	135.7
79.8	208	166	42	132.5
79.8	203	162	41	129.3
79.8	198	158	40	126.1
79.8	193	154	39	122.9
79.8	188	150	38	119.7

%E	M1	M2	DM	M*
79.8	183	146	37	116.5
79.8	178	142	36	113.3
79.8	173	138	35	110.1
79.8	168	134	34	106.9
79.8	163	130	33	103.7
79.8	129	103	26	82.2
79.8	124	99	25	79.0
79.8	119	95	24	75.8
79.8	114	91	23	72.6
79.8	109	87	22	69.4
79.8	104	83	21	66.2
79.8	99	79	20	63.0
79.8	94	75	19	59.8
79.8	89	71	18	56.6
79.8	84	67	17	53.4
79.7	498	397	101	316.5
79.7	497	396	101	315.5
79.7	493	393	100	313.3
79.7	492	392	100	312.3
79.7	488	389	99	310.1
79.7	487	388	99	309.1
79.7	483	385	98	306.9
79.7	482	384	98	305.9
79.7	479	382	97	304.6
79.7	478	381	97	303.7
79.7	477	380	97	302.7
79.7	474	378	96	301.4
79.7	473	377	96	300.5
79.7	472	376	96	299.5
79.7	469	374	95	298.2
79.7	468	373	95	297.3
79.7	467	372	95	296.3
79.7	464	370	94	295.0
79.7	463	369	94	294.1
79.7	462	368	94	293.1
79.7	459	366	93	291.8
79.7	458	365	93	290.9
79.7	457	364	93	289.9
79.7	454	362	92	288.6
79.7	453	361	92	287.7
79.7	449	358	91	285.4
79.7	448	357	91	284.5
79.7	444	354	90	282.2
79.7	443	353	90	281.3
79.7	439	350	89	279.0
79.7	438	349	89	278.1
79.7	434	346	88	275.8
79.7	433	345	88	274.9
79.7	429	342	87	272.6
79.7	428	341	87	271.7

%E	M1	M2	DM	M*
79.7	424	338	86	269.4
79.7	423	337	86	268.5
79.7	419	334	85	266.2
79.7	418	333	85	265.3
79.7	414	330	84	263.0
79.7	413	329	84	262.1
79.7	409	326	83	259.8
79.7	408	325	83	258.9
79.7	404	322	82	256.6
79.7	403	321	82	255.7
79.7	399	318	81	253.4
79.7	395	315	80	251.2
79.7	394	314	80	250.2
79.7	390	311	79	248.0
79.7	389	310	79	247.0
79.7	385	307	78	244.8
79.7	384	306	78	243.8
79.7	380	303	77	241.6
79.7	379	302	77	240.6
79.7	375	299	76	238.4
79.7	374	298	76	237.4
79.7	370	295	75	235.2
79.7	369	294	75	234.2
79.7	365	291	74	232.0
79.7	364	290	74	231.0
79.7	360	287	73	228.8
79.7	359	286	73	227.8
79.7	355	283	72	225.6
79.7	354	282	72	224.6
79.7	350	279	71	222.4
79.7	349	278	71	221.4
79.7	345	275	70	219.2
79.7	344	274	70	218.2
79.7	340	271	69	216.0
79.7	335	267	68	212.8
79.7	330	263	67	209.6
79.7	325	259	66	206.4
79.7	320	255	65	203.2
79.7	316	252	64	201.0
79.7	315	251	64	200.0
79.7	311	248	63	197.8
79.7	310	247	63	196.8
79.7	306	244	62	194.6
79.7	305	243	62	193.6
79.7	301	240	61	191.4
79.7	300	239	61	190.4
79.7	296	236	60	188.2
79.7	295	235	60	187.2
79.7	291	232	59	185.0
79.7	290	231	59	184.0

%E	M1	M2	DM	M*
79.7	286	228	58	181.8
79.7	281	224	57	178.6
79.7	276	220	56	175.4
79.7	271	216	55	172.2
79.7	266	212	54	169.0
79.7	261	208	53	165.8
79.7	256	204	52	162.6
79.7	251	200	51	159.4
79.7	246	196	50	156.2
79.7	241	192	49	153.0
79.7	237	189	48	150.7
79.7	236	188	48	149.8
79.7	232	185	47	147.5
79.7	231	184	47	146.6
79.7	227	181	46	144.3
79.7	222	177	45	141.1
79.7	217	173	44	137.9
79.7	212	169	43	134.7
79.7	207	165	42	131.5
79.7	202	161	41	128.3
79.7	197	157	40	125.1
79.7	192	153	39	121.9
79.7	187	149	38	118.7
79.7	182	145	37	115.5
79.7	177	141	36	112.3
79.7	172	137	35	109.1
79.7	158	126	32	100.5
79.7	153	122	31	97.3
79.7	148	118	30	94.1
79.7	143	114	29	90.9
79.7	138	110	28	87.7
79.7	133	106	27	84.5
79.7	128	102	26	81.3
79.7	123	98	25	78.1
79.7	118	94	24	74.9
79.7	79	63	16	50.2
79.7	74	59	15	47.0
79.7	69	55	14	43.8
79.7	64	51	13	40.6
79.7	59	47	12	37.4
79.6	500	398	102	316.8
79.6	499	397	102	315.8
79.6	496	395	101	314.6
79.6	495	394	101	313.6
79.6	494	393	101	312.6
79.6	491	391	100	311.4
79.6	490	390	100	310.4
79.6	489	389	100	309.4
79.6	486	387	99	308.2
79.6	485	386	99	307.2

%E	M1	M2	DM	M*
79.6	481	383	98	305.0
79.6	480	382	98	304.0
79.6	476	379	97	301.8
79.6	475	378	97	300.8
79.6	471	375	96	298.6
79.6	470	374	96	297.6
79.6	466	371	95	295.4
79.6	465	370	95	294.4
79.6	461	367	94	292.2
79.6	460	366	94	291.2
79.6	456	363	93	289.0
79.6	455	362	93	288.0
79.6	452	360	92	286.7
79.6	451	359	92	285.8
79.6	450	358	92	284.8
79.6	447	356	91	283.5
79.6	446	355	91	282.6
79.6	445	354	91	281.6
79.6	442	352	90	280.3
79.6	441	351	90	279.4
79.6	437	348	89	277.1
79.6	436	347	89	276.2
79.6	432	344	88	273.9
79.6	431	343	88	273.0
79.6	427	340	87	270.7
79.6	426	339	87	269.8
79.6	422	336	86	267.5
79.6	421	335	86	266.6
79.6	417	332	85	264.3
79.6	416	331	85	263.4
79.6	412	328	84	261.1
79.6	411	327	84	260.2
79.6	407	324	83	257.9
79.6	406	323	83	257.0
79.6	402	320	82	254.7
79.6	401	319	82	253.8
79.6	398	317	81	252.5
79.6	397	316	81	251.5
79.6	393	313	80	249.3
79.6	392	312	80	248.3
79.6	388	309	79	246.1
79.6	387	308	79	245.1
79.6	383	305	78	242.9
79.6	382	304	78	241.9
79.6	378	301	77	239.7
79.6	377	300	77	238.7
79.6	373	297	76	236.5
79.6	372	296	76	235.5
79.6	368	293	75	233.3
79.6	367	292	75	232.3

%E	M1	M2	DM	M*
79.6	363	289	74	230.1
79.6	362	288	74	229.1
79.6	358	285	73	226.9
79.6	357	284	73	225.9
79.6	353	281	72	223.7
79.6	348	277	71	220.5
79.6	343	273	70	217.3
79.6	339	270	69	215.0
79.6	338	269	69	214.1
79.6	334	266	68	211.8
79.6	333	265	68	210.9
79.6	329	262	67	208.6
79.6	328	261	67	207.7
79.6	324	258	66	205.4
79.6	323	257	66	204.5
79.6	319	254	65	202.2
79.6	318	253	65	201.3
79.6	314	250	64	199.0
79.6	313	249	64	198.1
79.6	309	246	63	195.8
79.6	304	242	62	192.6
79.6	299	238	61	189.4
79.6	294	234	60	186.2
79.6	289	230	59	183.0
79.6	285	227	58	180.8
79.6	284	226	58	179.8
79.6	280	223	57	177.6
79.6	279	222	57	176.6
79.6	275	219	56	174.4
79.6	274	218	56	173.4
79.6	270	215	55	171.2
79.6	269	214	55	170.2
79.6	265	211	54	168.0
79.6	260	207	53	164.8
79.6	255	203	52	161.6
79.6	250	199	51	158.4
79.6	245	195	50	155.2
79.6	240	191	49	152.0
79.6	235	187	48	148.8
79.6	230	183	47	145.6
79.6	226	180	46	143.4
79.6	225	179	46	142.4
79.6	221	176	45	140.2
79.6	216	172	44	137.0
79.6	211	168	43	133.8
79.6	206	164	42	130.6
79.6	201	160	41	127.4
79.6	196	156	40	124.2
79.6	191	152	39	121.0
79.6	186	148	38	117.8

%E	M1	M2	DM	M*
79.6	181	144	37	114.6
79.6	167	133	34	105.9
79.6	162	129	33	102.7
79.6	157	125	32	99.5
79.6	152	121	31	96.3
79.6	147	117	30	93.1
79.6	142	113	29	89.9
79.6	137	109	28	86.7
79.6	113	90	23	71.7
79.6	108	86	22	68.5
79.6	103	82	21	65.3
79.6	98	78	20	62.1
79.6	93	74	19	58.9
79.6	54	43	11	34.2
79.6	49	39	10	31.0
79.5	498	396	102	314.9
79.5	497	395	102	313.9
79.5	493	392	101	311.7
79.5	492	391	101	310.7
79.5	488	388	100	308.5
79.5	487	387	100	307.5
79.5	484	385	99	306.3
79.5	483	384	99	305.3
79.5	482	383	99	304.3
79.5	479	381	98	303.1
79.5	478	380	98	302.1
79.5	477	379	98	301.1
79.5	474	377	97	299.9
79.5	473	376	97	298.9
79.5	469	373	96	296.7
79.5	468	372	96	295.7
79.5	464	369	95	293.5
79.5	463	368	95	292.5
79.5	459	365	94	290.3
79.5	458	364	94	289.3
79.5	454	361	93	287.1
79.5	453	360	93	286.1
79.5	449	357	92	283.9
79.5	448	356	92	282.9
79.5	444	353	91	280.7
79.5	443	352	91	279.7
79.5	440	350	90	278.4
79.5	439	349	90	277.5
79.5	438	348	90	276.5
79.5	435	346	89	275.2
79.5	434	345	89	274.3
79.5	430	342	88	272.0
79.5	429	341	88	271.1
79.5	425	338	87	268.8
79.5	424	337	87	267.9

%E	M1	M2	DM	M*
79.5	420	334	86	265.6
79.5	419	333	86	264.7
79.5	415	330	85	262.4
79.5	414	329	85	261.5
79.5	410	326	84	259.2
79.5	409	325	84	258.3
79.5	405	322	83	256.0
79.5	404	321	83	255.1
79.5	400	318	82	252.8
79.5	396	315	81	250.6
79.5	395	314	81	249.6
79.5	391	311	80	247.4
79.5	390	310	80	246.4
79.5	386	307	79	244.2
79.5	385	306	79	243.2
79.5	381	303	78	241.0
79.5	380	302	78	240.0
79.5	376	299	77	237.8
79.5	375	298	77	236.8
79.5	371	295	76	234.6
79.5	370	294	76	233.6
79.5	366	291	75	231.4
79.5	365	290	75	230.4
79.5	361	287	74	228.2
79.5	356	283	73	225.0
79.5	352	280	72	222.7
79.5	351	279	72	221.8
79.5	347	276	71	219.5
79.5	346	275	71	218.6
79.5	342	272	70	216.3
79.5	341	271	70	215.4
79.5	337	268	69	213.1
79.5	336	267	69	212.2
79.5	332	264	68	209.9
79.5	331	263	68	209.0
79.5	327	260	67	206.7
79.5	322	256	66	203.5
79.5	317	252	65	200.3
79.5	312	248	64	197.1
79.5	308	245	63	194.9
79.5	307	244	63	193.9
79.5	303	241	62	191.7
79.5	302	240	62	190.7
79.5	298	237	61	188.5
79.5	297	236	61	187.5
79.5	293	233	60	185.3
79.5	292	232	60	184.3
79.5	288	229	59	182.1
79.5	283	225	58	178.9
79.5	278	221	57	175.7

%E	M1	M2	DM	M*
79.5	273	217	56	172.5
79.5	268	213	55	169.3
79.5	264	210	54	167.0
79.5	263	209	54	166.1
79.5	259	206	53	163.8
79.5	258	205	53	162.9
79.5	254	202	52	160.6
79.5	249	198	51	157.4
79.5	244	194	50	154.2
79.5	239	190	49	151.0
79.5	234	186	48	147.8
79.5	229	182	47	144.6
79.5	224	178	46	141.4
79.5	220	175	45	139.2
79.5	219	174	45	138.2
79.5	215	171	44	136.0
79.5	210	167	43	132.8
79.5	205	163	42	129.6
79.5	200	159	41	126.4
79.5	195	155	40	123.2
79.5	190	151	39	120.0
79.5	185	147	38	116.8
79.5	176	140	36	111.4
79.5	171	136	35	108.2
79.5	166	132	34	105.0
79.5	161	128	33	101.8
79.5	156	124	32	98.6
79.5	151	120	31	95.4
79.5	146	116	30	92.2
79.5	132	105	27	83.5
79.5	127	101	26	80.3
79.5	122	97	25	77.1
79.5	117	93	24	73.9
79.5	112	89	23	70.7
79.5	88	70	18	55.7
79.5	83	66	17	52.5
79.5	78	62	16	49.3
79.5	73	58	15	46.1
79.5	44	35	9	27.8
79.5	39	31	8	24.6
79.4	500	397	103	315.2
79.4	499	396	103	314.3
79.4	496	394	102	313.0
79.4	495	393	102	312.0
79.4	494	392	102	311.1
79.4	491	390	101	309.8
79.4	490	389	101	308.8
79.4	486	386	100	306.6
79.4	485	385	100	305.6
79.4	481	382	99	303.4

%E	M1	M2	DM	M*
79.4	480	381	99	302.4
79.4	476	378	98	300.2
79.4	475	377	98	299.2
79.4	472	375	97	297.9
79.4	471	374	97	297.0
79.4	470	373	97	296.0
79.4	467	371	96	294.7
79.4	466	370	96	293.8
79.4	465	369	96	292.8
79.4	462	367	95	291.5
79.4	461	366	95	290.6
79.4	457	363	94	288.3
79.4	456	362	94	287.4
79.4	452	359	93	285.1
79.4	451	358	93	284.2
79.4	447	355	92	281.9
79.4	446	354	92	281.0
79.4	442	351	91	278.7
79.4	441	350	91	277.8
79.4	437	347	90	275.5
79.4	436	346	90	274.6
79.4	433	344	89	273.3
79.4	432	343	89	272.3
79.4	431	342	89	271.4
79.4	428	340	88	270.1
79.4	427	339	88	269.1
79.4	423	336	87	266.9
79.4	422	335	87	265.9
79.4	418	332	86	263.7
79.4	417	331	86	262.7
79.4	413	328	85	260.5
79.4	412	327	85	259.5
79.4	408	324	84	257.3
79.4	407	323	84	256.3
79.4	403	320	83	254.1
79.4	402	319	83	253.1
79.4	399	317	82	251.9
79.4	398	316	82	250.9
79.4	394	313	81	248.7
79.4	393	312	81	247.7
79.4	389	309	80	245.5
79.4	388	308	80	244.5
79.4	384	305	79	242.3
79.4	383	304	79	241.3
79.4	379	301	78	239.1
79.4	378	300	78	238.1
79.4	374	297	77	235.9
79.4	373	296	77	234.9
79.4	369	293	76	232.7
79.4	364	289	75	229.5
79.4	360	286	74	227.2
79.4	359	285	74	226.3
79.4	355	282	73	224.0
79.4	354	281	73	223.1
79.4	350	278	72	220.8
79.4	349	277	72	219.9
79.4	345	274	71	217.6
79.4	344	273	71	216.7
79.4	340	270	70	214.4
79.4	339	269	70	213.5
79.4	335	266	69	211.2
79.4	330	262	68	208.0
79.4	326	259	67	205.8
79.4	325	258	67	204.8
79.4	321	255	66	202.6
79.4	320	254	66	201.6
79.4	316	251	65	199.4
79.4	315	250	65	198.4
79.4	311	247	64	196.2
79.4	310	246	64	195.2
79.4	306	243	63	193.0
79.4	301	239	62	189.8
79.4	296	235	61	186.6
79.4	291	231	60	183.4
79.4	287	228	59	181.1
79.4	286	227	59	180.2
79.4	282	224	58	177.9
79.4	281	223	58	177.0
79.4	277	220	57	174.7
79.4	272	216	56	171.5
79.4	267	212	55	168.3
79.4	262	208	54	165.1
79.4	257	204	53	161.9
79.4	253	201	52	159.7
79.4	252	200	52	158.7
79.4	248	197	51	156.5
79.4	247	196	51	155.5
79.4	243	193	50	153.3
79.4	238	189	49	150.1
79.4	233	185	48	146.9
79.4	228	181	47	143.7
79.4	223	177	46	140.5
79.4	218	173	45	137.3
79.4	214	170	44	135.0
79.4	209	166	43	131.8
79.4	204	162	42	128.6
79.4	199	158	41	125.4
79.4	194	154	40	122.2
79.4	189	150	39	119.0
79.4	180	143	37	113.6
79.4	175	139	36	110.4
79.4	170	135	35	107.2
79.4	165	131	34	104.0
79.4	160	127	33	100.8
79.4	155	123	32	97.6
79.4	141	112	29	89.0
79.4	136	108	28	85.8
79.4	131	104	27	82.6
79.4	126	100	26	79.4
79.4	107	85	22	67.5
79.4	102	81	21	64.3
79.4	97	77	20	61.1
79.4	68	54	14	42.9
79.4	63	50	13	39.7
79.4	34	27	7	21.4
79.3	498	395	103	313.3
79.3	497	394	103	312.3
79.3	493	391	102	310.1
79.3	492	390	102	309.1
79.3	489	388	101	307.9
79.3	488	387	101	306.9
79.3	487	386	101	305.9
79.3	484	384	100	304.7
79.3	483	383	100	303.7
79.3	482	382	100	302.7
79.3	479	380	99	301.5
79.3	478	379	99	300.5
79.3	474	376	98	298.3
79.3	473	375	98	297.3
79.3	469	372	97	295.1
79.3	468	371	97	294.1
79.3	464	368	96	291.9
79.3	463	367	96	290.9
79.3	460	365	95	289.6
79.3	459	364	95	288.7
79.3	458	363	95	287.7
79.3	455	361	94	286.4
79.3	454	360	94	285.5
79.3	450	357	93	283.2
79.3	449	356	93	282.3
79.3	445	353	92	280.0
79.3	444	352	92	279.1
79.3	440	349	91	276.8
79.3	439	348	91	275.9
79.3	435	345	90	273.6
79.3	434	344	90	272.7
79.3	430	341	89	270.4
79.3	429	340	89	269.5
79.3	426	338	88	268.2
79.3	425	337	88	267.2
79.3	421	334	87	265.0
79.3	420	333	87	264.0
79.3	416	330	86	261.8
79.3	415	329	86	260.8
79.3	411	326	85	258.6
79.3	410	325	85	257.6
79.3	406	322	84	255.4
79.3	405	321	84	254.4
79.3	401	318	83	252.2
79.3	400	317	83	251.2
79.3	397	315	82	249.9
79.3	396	314	82	249.0
79.3	392	311	81	246.7
79.3	391	310	81	245.8
79.3	387	307	80	243.5
79.3	386	306	80	242.6
79.3	382	303	79	240.3
79.3	381	302	79	239.4
79.3	377	299	78	237.1
79.3	376	298	78	236.2
79.3	372	295	77	233.9
79.3	368	292	76	231.7
79.3	367	291	76	230.7
79.3	363	288	75	228.5
79.3	362	287	75	227.5
79.3	358	284	74	225.3
79.3	357	283	74	224.3
79.3	353	280	73	222.1
79.3	352	279	73	221.1
79.3	348	276	72	218.9
79.3	347	275	72	217.9
79.3	343	272	71	215.7
79.3	338	268	70	212.5
79.3	334	265	69	210.3
79.3	333	264	69	209.3
79.3	329	261	68	207.1
79.3	328	260	68	206.1
79.3	324	257	67	203.9
79.3	323	256	67	202.9
79.3	319	253	66	200.7
79.3	314	249	65	197.5
79.3	309	245	64	194.3
79.3	305	242	63	192.0
79.3	304	241	63	191.1
79.3	300	238	62	188.8
79.3	299	237	62	187.9
79.3	295	234	61	185.6
79.3	294	233	61	184.7
79.3	290	230	60	182.4
79.3	285	226	59	179.2
79.3	280	222	58	176.0
79.3	276	219	57	173.8
79.3	275	218	57	172.8
79.3	271	215	56	170.6
79.3	270	214	56	169.6
79.3	266	211	55	167.4
79.3	261	207	54	164.2
79.3	256	203	53	161.0
79.3	251	199	52	157.8
79.3	246	195	51	154.6
79.3	242	192	50	152.3
79.3	241	191	50	151.4
79.3	237	188	49	149.1
79.3	232	184	48	145.9
79.3	227	180	47	142.7
79.3	222	176	46	139.5
79.3	217	172	45	136.3
79.3	213	169	44	134.1
79.3	208	165	43	130.9
79.3	203	161	42	127.7
79.3	198	157	41	124.5
79.3	193	153	40	121.3
79.3	188	149	39	118.1
79.3	184	146	38	115.8
79.3	179	142	37	112.6
79.3	174	138	36	109.4
79.3	169	134	35	106.2
79.3	164	130	34	103.0
79.3	150	119	31	94.4
79.3	145	115	30	91.2
79.3	140	111	29	88.0
79.3	135	107	28	84.8
79.3	121	96	25	76.2
79.3	116	92	24	73.0
79.3	111	88	23	69.8
79.3	92	73	19	57.9
79.3	87	69	18	54.7
79.3	82	65	17	51.5
79.3	58	46	12	36.5
79.3	29	23	6	18.2
79.2	500	396	104	313.6
79.2	499	395	104	312.7
79.2	496	393	103	311.4
79.2	495	392	103	310.4
79.2	494	391	103	309.5
79.2	491	389	102	308.2
79.2	490	388	102	307.2
79.2	486	385	101	305.0
79.2	485	384	101	304.0
79.2	481	381	100	301.8

%E	M1	M2	DM	M*
79.2	480	380	100	300.8
79.2	477	378	99	299.5
79.2	476	377	99	298.6
79.2	475	376	99	297.6
79.2	472	374	98	296.3
79.2	471	373	98	295.4
79.2	467	370	97	293.1
79.2	466	369	97	292.2
79.2	462	366	96	289.9
79.2	461	365	96	289.0
79.2	457	362	95	286.7
79.2	456	361	95	285.8
79.2	453	359	94	284.5
79.2	452	358	94	283.5
79.2	451	357	94	282.6
79.2	448	355	93	281.3
79.2	447	354	93	280.3
79.2	443	351	92	278.1
79.2	442	350	92	277.1
79.2	438	347	91	274.9
79.2	437	346	91	273.9
79.2	433	343	90	271.7
79.2	432	342	90	270.8
79.2	428	339	89	268.5
79.2	427	338	89	267.6
79.2	424	336	88	266.3
79.2	423	335	88	265.3
79.2	419	332	87	263.1
79.2	418	331	87	262.1
79.2	414	328	86	259.9
79.2	413	327	86	258.9
79.2	409	324	85	256.7
79.2	408	323	85	255.7
79.2	404	320	84	253.5
79.2	403	319	84	252.5
79.2	399	316	83	250.3
79.2	395	313	82	248.0
79.2	394	312	82	247.1
79.2	390	309	81	244.8
79.2	389	308	81	243.9
79.2	385	305	80	241.6
79.2	384	304	80	240.7
79.2	380	301	79	238.4
79.2	379	300	79	237.5
79.2	375	297	78	235.2
79.2	371	294	77	233.0
79.2	370	293	77	232.0
79.2	366	290	76	229.8
79.2	365	289	76	228.8
79.2	361	286	75	226.6

%E	M1	M2	DM	M*
79.2	360	285	75	225.6
79.2	356	282	74	223.4
79.2	355	281	74	222.4
79.2	351	278	73	220.2
79.2	346	274	72	217.0
79.2	342	271	71	214.7
79.2	341	270	71	213.8
79.2	337	267	70	211.5
79.2	336	266	70	210.6
79.2	332	263	69	208.3
79.2	331	262	69	207.4
79.2	327	259	68	205.1
79.2	322	255	67	201.9
79.2	318	252	66	199.7
79.2	317	251	66	198.7
79.2	313	248	65	196.5
79.2	312	247	65	195.5
79.2	308	244	64	193.3
79.2	307	243	64	192.3
79.2	303	240	63	190.1
79.2	298	236	62	186.9
79.2	293	232	61	183.7
79.2	289	229	60	181.5
79.2	288	228	60	180.5
79.2	284	225	59	178.3
79.2	283	224	59	177.3
79.2	279	221	58	175.1
79.2	274	217	57	171.9
79.2	269	213	56	168.7
79.2	265	210	55	166.4
79.2	264	209	55	165.5
79.2	260	206	54	163.2
79.2	259	205	54	162.3
79.2	255	202	53	160.0
79.2	250	198	52	156.8
79.2	245	194	51	153.6
79.2	240	190	50	150.4
79.2	236	187	49	148.2
79.2	231	183	48	145.0
79.2	226	179	47	141.8
79.2	221	175	46	138.6
79.2	216	171	45	135.4
79.2	212	168	44	133.1
79.2	207	164	43	129.9
79.2	202	160	42	126.7
79.2	197	156	41	123.5
79.2	192	152	40	120.3
79.2	183	145	38	114.9
79.2	178	141	37	111.7
79.2	173	137	36	108.5

%E	M1	M2	DM	M*
79.2	168	133	35	105.3
79.2	159	126	33	99.8
79.2	154	122	32	96.6
79.2	149	118	31	93.4
79.2	144	114	30	90.3
79.2	130	103	27	81.6
79.2	125	99	26	78.4
79.2	120	95	25	75.2
79.2	106	84	22	66.6
79.2	101	80	21	63.4
79.2	96	76	20	60.2
79.2	77	61	16	48.3
79.2	72	57	15	45.1
79.2	53	42	11	33.3
79.2	48	38	10	30.1
79.2	24	19	5	15.0
79.1	498	394	104	311.7
79.1	497	393	104	310.8
79.1	493	390	103	308.5
79.1	492	389	103	307.6
79.1	489	387	102	306.3
79.1	488	386	102	305.3
79.1	487	385	102	304.4
79.1	484	383	101	303.1
79.1	483	382	101	302.1
79.1	479	379	100	299.9
79.1	478	378	100	298.9
79.1	474	375	99	296.7
79.1	473	374	99	295.7
79.1	470	372	98	294.4
79.1	469	371	98	293.5
79.1	468	370	98	292.5
79.1	465	368	97	291.2
79.1	464	367	97	290.3
79.1	463	366	97	289.3
79.1	460	364	96	288.0
79.1	459	363	96	287.1
79.1	455	360	95	284.8
79.1	454	359	95	283.9
79.1	450	356	94	281.6
79.1	449	355	94	280.7
79.1	446	353	93	279.4
79.1	445	352	93	278.4
79.1	444	351	93	277.5
79.1	441	349	92	276.2
79.1	440	348	92	275.2
79.1	436	345	91	273.0
79.1	435	344	91	272.0
79.1	431	341	90	269.8
79.1	430	340	90	268.8

%E	M1	M2	DM	M*
79.1	426	337	89	266.6
79.1	425	336	89	265.6
79.1	422	334	88	264.4
79.1	421	333	88	263.4
79.1	417	330	87	261.2
79.1	416	329	87	260.2
79.1	412	326	86	258.0
79.1	411	325	86	257.0
79.1	407	322	85	254.8
79.1	406	321	85	253.8
79.1	402	318	84	251.6
79.1	401	317	84	250.6
79.1	398	315	83	249.3
79.1	397	314	83	248.4
79.1	393	311	82	246.1
79.1	392	310	82	245.2
79.1	388	307	81	242.9
79.1	387	306	81	242.0
79.1	383	303	80	239.7
79.1	382	302	80	238.8
79.1	378	299	79	236.5
79.1	374	296	78	234.3
79.1	373	295	78	233.3
79.1	369	292	77	231.1
79.1	368	291	77	230.1
79.1	364	288	76	227.9
79.1	363	287	76	226.9
79.1	359	284	75	224.7
79.1	358	283	75	223.7
79.1	354	280	74	221.5
79.1	350	277	73	219.2
79.1	349	276	73	218.3
79.1	345	273	72	216.0
79.1	344	272	72	215.1
79.1	340	269	71	212.8
79.1	339	268	71	211.9
79.1	335	265	70	209.6
79.1	330	261	69	206.4
79.1	326	258	68	204.2
79.1	325	257	68	203.2
79.1	321	254	67	201.0
79.1	320	253	67	200.0
79.1	316	250	66	197.8
79.1	311	246	65	194.6
79.1	306	242	64	191.4
79.1	302	239	63	189.1
79.1	301	238	63	188.2
79.1	297	235	62	185.9
79.1	296	234	62	185.0
79.1	292	231	61	182.7

%E	M1	M2	DM	M*
79.1	287	227	60	179.5
79.1	282	223	59	176.3
79.1	278	220	58	174.1
79.1	277	219	58	173.1
79.1	273	216	57	170.9
79.1	268	212	56	167.7
79.1	263	208	55	164.5
79.1	258	204	54	161.3
79.1	254	201	53	159.1
79.1	253	200	53	158.1
79.1	249	197	52	155.9
79.1	244	193	51	152.7
79.1	239	189	50	149.5
79.1	235	186	49	147.2
79.1	234	185	49	146.3
79.1	230	182	48	144.0
79.1	225	178	47	140.8
79.1	220	174	46	137.6
79.1	215	170	45	134.4
79.1	211	167	44	132.2
79.1	206	163	43	129.0
79.1	201	159	42	125.8
79.1	196	155	41	122.6
79.1	191	151	40	119.4
79.1	187	148	39	117.1
79.1	182	144	38	113.9
79.1	177	140	37	110.7
79.1	172	136	36	107.5
79.1	163	129	34	102.1
79.1	158	125	33	98.9
79.1	153	121	32	95.7
79.1	148	117	31	92.5
79.1	139	110	29	87.1
79.1	134	106	28	83.9
79.1	129	102	27	80.7
79.1	115	91	24	72.0
79.1	110	87	23	68.8
79.1	91	72	19	57.0
79.1	86	68	18	53.8
79.1	67	53	14	41.9
79.1	43	34	9	26.9
79.0	500	395	105	312.0
79.0	499	394	105	311.1
79.0	496	392	104	309.8
79.0	495	391	104	308.9
79.0	491	388	103	306.6
79.0	490	387	103	305.7
79.0	486	384	102	303.4
79.0	485	383	102	302.5
79.0	482	381	101	301.2

%E	M1	M2	DM	M*
79.0	481	380	101	300.2
79.0	480	379	101	299.3
79.0	477	377	100	298.0
79.0	476	376	100	297.0
79.0	472	373	99	294.8
79.0	471	372	99	293.8
79.0	467	369	98	291.6
79.0	466	368	98	290.6
79.0	462	365	97	288.4
79.0	461	364	97	287.4
79.0	458	362	96	286.1
79.0	457	361	96	285.2
79.0	453	358	95	282.9
79.0	452	357	95	282.0
79.0	448	354	94	279.7
79.0	447	353	94	278.8
79.0	443	350	93	276.5
79.0	442	349	93	275.6
79.0	439	347	92	274.3
79.0	438	346	92	273.3
79.0	434	343	91	271.1
79.0	433	342	91	270.1
79.0	429	339	90	267.9
79.0	428	338	90	266.9
79.0	424	335	89	264.7
79.0	423	334	89	263.7
79.0	420	332	88	262.4
79.0	419	331	88	261.5
79.0	415	328	87	259.2
79.0	414	327	87	258.3
79.0	410	324	86	256.0
79.0	409	323	86	255.1
79.0	405	320	85	252.8
79.0	404	319	85	251.9
79.0	400	316	84	249.6
79.0	396	313	83	247.4
79.0	395	312	83	246.4
79.0	391	309	82	244.2
79.0	390	308	82	243.2
79.0	386	305	81	241.0
79.0	385	304	81	240.0
79.0	381	301	80	237.8
79.0	377	298	79	235.6
79.0	376	297	79	234.6
79.0	372	294	78	232.4
79.0	371	293	78	231.4
79.0	367	290	77	229.2
79.0	366	289	77	228.2
79.0	362	286	76	226.0
79.0	357	282	75	222.8

%E	M1	M2	DM	M*
79.0	353	279	74	220.5
79.0	352	278	74	219.6
79.0	348	275	73	217.3
79.0	347	274	73	216.4
79.0	343	271	72	214.1
79.0	338	267	71	210.9
79.0	334	264	70	208.7
79.0	333	263	70	207.7
79.0	329	260	69	205.5
79.0	328	259	69	204.5
79.0	324	256	68	202.3
79.0	319	252	67	199.1
79.0	315	249	66	196.8
79.0	314	248	66	195.9
79.0	310	245	65	193.6
79.0	309	244	65	192.7
79.0	305	241	64	190.4
79.0	300	237	63	187.2
79.0	295	233	62	184.0
79.0	291	230	61	181.8
79.0	290	229	61	180.8
79.0	286	226	60	178.6
79.0	281	222	59	175.4
79.0	276	218	58	172.2
79.0	272	215	57	169.9
79.0	271	214	57	169.0
79.0	267	211	56	166.7
79.0	262	207	55	163.5
79.0	257	203	54	160.3
79.0	252	199	53	157.1
79.0	248	196	52	154.9
79.0	243	192	51	151.7
79.0	238	188	50	148.5
79.0	233	184	49	145.3
79.0	229	181	48	143.1
79.0	224	177	47	139.9
79.0	219	173	46	136.7
79.0	214	169	45	133.5
79.0	210	166	44	131.2
79.0	205	162	43	128.0
79.0	200	158	42	124.8
79.0	195	154	41	121.6
79.0	186	147	39	116.2
79.0	181	143	38	113.0
79.0	176	139	37	109.8
79.0	167	132	35	104.3
79.0	162	128	34	101.1
79.0	157	124	33	97.9
79.0	143	113	30	89.3
79.0	138	109	29	86.1

%E	M1	M2	DM	M*
79.0	124	98	26	77.5
79.0	119	94	25	74.3
79.0	105	83	22	65.6
79.0	100	79	21	62.4
79.0	81	64	17	50.6
79.0	62	49	13	38.7
78.9	498	393	105	310.1
78.9	497	392	105	309.2
78.9	494	390	104	307.9
78.9	493	389	104	306.9
78.9	492	388	104	306.0
78.9	489	386	103	304.7
78.9	488	385	103	303.7
78.9	487	384	103	302.8
78.9	484	382	102	301.5
78.9	483	381	102	300.5
78.9	479	378	101	298.3
78.9	478	377	101	297.3
78.9	475	375	100	296.1
78.9	474	374	100	295.1
78.9	473	373	100	294.1
78.9	470	371	99	292.9
78.9	469	370	99	291.9
78.9	465	367	98	289.7
78.9	464	366	98	288.7
78.9	460	363	97	286.5
78.9	459	362	97	285.5
78.9	456	360	96	284.2
78.9	455	359	96	283.3
78.9	454	358	96	282.3
78.9	451	356	95	281.0
78.9	450	355	95	280.1
78.9	446	352	94	277.8
78.9	445	351	94	276.9
78.9	441	348	93	274.6
78.9	440	347	93	273.7
78.9	437	345	92	272.4
78.9	436	344	92	271.4
78.9	435	343	92	270.5
78.9	432	341	91	269.2
78.9	431	340	91	268.2
78.9	427	337	90	266.0
78.9	426	336	90	265.0
78.9	422	333	89	262.8
78.9	421	332	89	261.8
78.9	418	330	88	260.5
78.9	417	329	88	259.6
78.9	413	326	87	257.3
78.9	412	325	87	256.4
78.9	408	322	86	254.1

%E	M1	M2	DM	M*
78.9	407	321	86	253.2
78.9	403	318	85	250.9
78.9	402	317	85	250.0
78.9	399	315	84	248.7
78.9	398	314	84	247.7
78.9	394	311	83	245.5
78.9	393	310	83	244.5
78.9	389	307	82	242.3
78.9	388	306	82	241.3
78.9	384	303	81	239.1
78.9	383	302	81	238.1
78.9	380	300	80	236.8
78.9	379	299	80	235.9
78.9	375	296	79	233.6
78.9	374	295	79	232.7
78.9	370	292	78	230.4
78.9	369	291	78	229.5
78.9	365	288	77	227.2
78.9	361	285	76	225.0
78.9	360	284	76	224.0
78.9	356	281	75	221.8
78.9	355	280	75	220.8
78.9	351	277	74	218.6
78.9	350	276	74	217.6
78.9	346	273	73	215.4
78.9	342	270	72	213.2
78.9	341	269	72	212.2
78.9	337	266	71	210.0
78.9	336	265	71	209.0
78.9	332	262	70	206.8
78.9	331	261	70	205.8
78.9	327	258	69	203.6
78.9	323	255	68	201.3
78.9	322	254	68	200.4
78.9	318	251	67	198.1
78.9	317	250	67	197.2
78.9	313	247	66	194.9
78.9	308	243	65	191.7
78.9	304	240	64	189.5
78.9	303	239	64	188.5
78.9	299	236	63	186.3
78.9	298	235	63	185.3
78.9	294	232	62	183.1
78.9	289	228	61	179.9
78.9	285	225	60	177.6
78.9	284	224	60	176.7
78.9	280	221	59	174.4
78.9	279	220	59	173.5
78.9	275	217	58	171.2
78.9	270	213	57	168.0

%E	M1	M2	DM	M*
78.9	266	210	56	165.8
78.9	265	209	56	164.8
78.9	261	206	55	162.6
78.9	256	202	54	159.4
78.9	251	198	53	156.2
78.9	247	195	52	153.9
78.9	246	194	52	153.0
78.9	242	191	51	150.7
78.9	237	187	50	147.5
78.9	232	183	49	144.3
78.9	228	180	48	142.1
78.9	227	179	48	141.1
78.9	223	176	47	138.9
78.9	218	172	46	135.7
78.9	213	168	45	132.5
78.9	209	165	44	130.3
78.9	204	161	43	127.1
78.9	199	157	42	123.9
78.9	194	153	41	120.7
78.9	190	150	40	118.4
78.9	185	146	39	115.2
78.9	180	142	38	112.0
78.9	175	138	37	108.8
78.9	171	135	36	106.6
78.9	166	131	35	103.4
78.9	161	127	34	100.2
78.9	152	120	32	94.7
78.9	147	116	31	91.5
78.9	142	112	30	88.3
78.9	133	105	28	82.9
78.9	128	101	27	79.7
78.9	123	97	26	76.5
78.9	114	90	24	71.1
78.9	109	86	23	67.9
78.9	95	75	20	59.2
78.9	90	71	19	56.0
78.9	76	60	16	47.4
78.9	71	56	15	44.2
78.9	57	45	12	35.5
78.9	38	30	8	23.7
78.9	19	15	4	11.8
78.8	500	394	106	310.5
78.8	499	393	106	309.5
78.8	496	391	105	308.2
78.8	495	390	105	307.3
78.8	491	387	104	305.0
78.8	490	386	104	304.1
78.8	486	383	103	301.8
78.8	485	382	103	300.9
78.8	482	380	102	299.6

%E	M1	M2	DM	M*
78.8	481	379	102	298.6
78.8	480	378	102	297.7
78.8	477	376	101	296.4
78.8	476	375	101	295.4
78.8	472	372	100	293.2
78.8	471	371	100	292.2
78.8	468	369	99	290.9
78.8	467	368	99	290.0
78.8	466	367	99	289.0
78.8	463	365	98	287.7
78.8	462	364	98	286.8
78.8	458	361	97	284.5
78.8	457	360	97	283.6
78.8	453	357	96	281.3
78.8	452	356	96	280.4
78.8	449	354	95	279.1
78.8	448	353	95	278.1
78.8	444	350	94	275.9
78.8	443	349	94	274.9
78.8	439	346	93	272.7
78.8	438	345	93	271.7
78.8	434	342	92	269.5
78.8	433	341	92	268.5
78.8	430	339	91	267.3
78.8	429	338	91	266.3
78.8	425	335	90	264.1
78.8	424	334	90	263.1
78.8	420	331	89	260.9
78.8	419	330	89	259.9
78.8	416	328	88	258.6
78.8	415	327	88	257.7
78.8	411	324	87	255.4
78.8	410	323	87	254.5
78.8	406	320	86	252.2
78.8	405	319	86	251.3
78.8	401	316	85	249.0
78.8	400	315	85	248.1
78.8	397	313	84	246.8
78.8	396	312	84	245.8
78.8	392	309	83	243.6
78.8	391	308	83	242.6
78.8	387	305	82	240.4
78.8	386	304	82	239.4
78.8	382	301	81	237.2
78.8	378	298	80	234.9
78.8	377	297	80	234.0
78.8	373	294	79	231.7
78.8	372	293	79	230.8
78.8	368	290	78	228.5
78.8	364	287	77	226.3

%E	M1	M2	DM	M*
78.8	363	286	77	225.3
78.8	359	283	76	223.1
78.8	358	282	76	222.1
78.8	354	279	75	219.9
78.8	353	278	75	218.9
78.8	349	275	74	216.7
78.8	345	272	73	214.4
78.8	344	271	73	213.5
78.8	340	268	72	211.2
78.8	339	267	72	210.3
78.8	335	264	71	208.0
78.8	330	260	70	204.8
78.8	326	257	69	202.6
78.8	325	256	69	201.6
78.8	321	253	68	199.4
78.8	320	252	68	198.4
78.8	316	249	67	196.2
78.8	312	246	66	194.0
78.8	311	245	66	193.0
78.8	307	242	65	190.8
78.8	306	241	65	189.8
78.8	302	238	64	187.6
78.8	297	234	63	184.4
78.8	293	231	62	182.1
78.8	292	230	62	181.2
78.8	288	227	61	178.9
78.8	283	223	60	175.7
78.8	278	219	59	172.5
78.8	274	216	58	170.3
78.8	273	215	58	169.3
78.8	269	212	57	167.1
78.8	264	208	56	163.9
78.8	260	205	55	161.6
78.8	259	204	55	160.7
78.8	255	201	54	158.4
78.8	250	197	53	155.2
78.8	245	193	52	152.0
78.8	241	190	51	149.8
78.8	240	189	51	148.8
78.8	236	186	50	146.6
78.8	231	182	49	143.4
78.8	226	178	48	140.2
78.8	222	175	47	138.0
78.8	217	171	46	134.8
78.8	212	167	45	131.6
78.8	208	164	44	129.3
78.8	203	160	43	126.1
78.8	198	156	42	122.9
78.8	193	152	41	119.7
78.8	189	149	40	117.5

%E	M1	M2	DM	M*
78.8	184	145	39	114.3
78.8	179	141	38	111.1
78.8	170	134	36	105.6
78.8	165	130	35	102.4
78.8	160	126	34	99.2
78.8	156	123	33	97.0
78.8	151	119	32	93.8
78.8	146	115	31	90.6
78.8	137	108	29	85.1
78.8	132	104	28	81.9
78.8	118	93	25	73.3
78.8	113	89	24	70.1
78.8	104	82	22	64.7
78.8	99	78	21	61.5
78.8	85	67	18	52.8
78.8	80	63	17	49.6
78.8	66	52	14	41.0
78.8	52	41	11	32.3
78.8	33	26	7	20.5
78.7	498	392	106	308.6
78.7	497	391	106	307.6
78.7	494	389	105	306.3
78.7	493	388	105	305.4
78.7	492	387	105	304.4
78.7	489	385	104	303.1
78.7	488	384	104	302.2
78.7	484	381	103	299.9
78.7	483	380	103	299.0
78.7	479	377	102	296.7
78.7	478	376	102	295.8
78.7	475	374	101	294.5
78.7	474	373	101	293.5
78.7	470	370	100	291.3
78.7	469	369	100	290.3
78.7	465	366	99	288.1
78.7	464	365	99	287.1
78.7	461	363	98	285.8
78.7	460	362	98	284.9
78.7	456	359	97	282.6
78.7	455	358	97	281.7
78.7	451	355	96	279.4
78.7	450	354	96	278.5
78.7	447	352	95	277.2
78.7	446	351	95	276.2
78.7	445	350	95	275.3
78.7	442	348	94	274.0
78.7	441	347	94	273.0
78.7	437	344	93	270.8
78.7	436	343	93	269.8
78.7	432	340	92	267.6

%E	M1	M2	DM	M*
78.7	431	339	92	266.6
78.7	428	337	91	265.3
78.7	427	336	91	264.4
78.7	423	333	90	262.1
78.7	422	332	90	261.2
78.7	418	329	89	258.9
78.7	417	328	89	258.0
78.7	414	326	88	256.7
78.7	413	325	88	255.8
78.7	409	322	87	253.5
78.7	408	321	87	252.6
78.7	404	318	86	250.3
78.7	403	317	86	249.4
78.7	399	314	85	247.1
78.7	395	311	84	244.9
78.7	394	310	84	243.9
78.7	390	307	83	241.7
78.7	389	306	83	240.7
78.7	385	303	82	238.5
78.7	381	300	81	236.2
78.7	380	299	81	235.3
78.7	376	296	80	233.0
78.7	375	295	80	232.1
78.7	371	292	79	229.8
78.7	367	289	78	227.6
78.7	366	288	78	226.6
78.7	362	285	77	224.4
78.7	361	284	77	223.4
78.7	357	281	76	221.2
78.7	356	280	76	220.2
78.7	352	277	75	218.0
78.7	348	274	74	215.7
78.7	347	273	74	214.8
78.7	343	270	73	212.5
78.7	342	269	73	211.6
78.7	338	266	72	209.3
78.7	334	263	71	207.1
78.7	333	262	71	206.1
78.7	329	259	70	203.9
78.7	328	258	70	202.9
78.7	324	255	69	200.7
78.7	319	251	68	197.5
78.7	315	248	67	195.3
78.7	314	247	67	194.3
78.7	310	244	66	192.1
78.7	305	240	65	188.9
78.7	301	237	64	186.6
78.7	300	236	64	185.7
78.7	296	233	63	183.4
78.7	291	229	62	180.2

%E	M1	M2	DM	M*
78.7	287	226	61	178.0
78.7	286	225	61	177.0
78.7	282	222	60	174.8
78.7	277	218	59	171.6
78.7	272	214	58	168.4
78.7	268	211	57	166.1
78.7	267	210	57	165.2
78.7	263	207	56	162.9
78.7	258	203	55	159.7
78.7	254	200	54	157.5
78.7	253	199	54	156.5
78.7	249	196	53	154.3
78.7	244	192	52	151.1
78.7	239	188	51	147.9
78.7	235	185	50	145.6
78.7	230	181	49	142.4
78.7	225	177	48	139.2
78.7	221	174	47	137.0
78.7	216	170	46	133.8
78.7	211	166	45	130.6
78.7	207	163	44	128.4
78.7	202	159	43	125.2
78.7	197	155	42	122.0
78.7	188	148	40	116.5
78.7	183	144	39	113.3
78.7	178	140	38	110.1
78.7	174	137	37	107.9
78.7	169	133	36	104.7
78.7	164	129	35	101.5
78.7	155	122	33	96.0
78.7	150	118	32	92.8
78.7	141	111	30	87.4
78.7	136	107	29	84.2
78.7	127	100	27	78.7
78.7	122	96	26	75.5
78.7	108	85	23	66.9
78.7	94	74	20	58.3
78.7	89	70	19	55.1
78.7	75	59	16	46.4
78.7	61	48	13	37.8
78.7	47	37	10	29.1
78.6	500	393	107	308.9
78.6	499	392	107	307.9
78.6	496	390	106	306.7
78.6	495	389	106	305.7
78.6	491	386	105	303.5
78.6	490	385	105	302.5
78.6	487	383	104	301.2
78.6	486	382	104	300.3
78.6	485	381	104	299.3

%E	M1	M2	DM	M*	%E	M1	M2	DM	M*	%E	M1	M2	DM	M*	%E	M1	M2	DM	M*	%E	M1	M2	DM	M*
78.6	482	379	103	298.0	78.6	370	291	79	228.9	78.6	187	147	40	115.6	78.5	433	340	93	267.0	78.5	293	230	63	180.5
78.6	481	378	103	297.1	78.6	369	290	79	227.9	78.6	182	143	39	112.4	78.5	432	339	93	266.0	78.5	289	227	62	178.3
78.6	477	375	102	294.8	78.6	365	287	78	225.7	78.6	173	136	37	106.9	78.5	428	336	92	263.8	78.5	288	226	62	177.3
78.6	476	374	102	293.9	78.6	364	286	78	224.7	78.6	168	132	36	103.7	78.5	427	335	92	262.8	78.5	284	223	61	175.1
78.6	473	372	101	292.6	78.6	360	283	77	222.5	78.6	159	125	34	98.3	78.5	424	333	91	261.5	78.5	279	219	60	171.9
78.6	472	371	101	291.6	78.6	359	282	77	221.5	78.6	154	121	33	95.1	78.5	423	332	91	260.6	78.5	275	216	59	169.7
78.6	471	370	101	290.7	78.6	355	279	76	219.3	78.6	145	114	31	89.6	78.5	419	329	90	258.3	78.5	274	215	59	168.7
78.6	468	368	100	289.4	78.6	351	276	75	217.0	78.6	140	110	30	86.4	78.5	418	328	90	257.4	78.5	270	212	58	166.5
78.6	467	367	100	288.4	78.6	350	275	75	216.1	78.6	131	103	28	81.0	78.5	414	325	89	255.1	78.5	265	208	57	163.3
78.6	463	364	99	286.2	78.6	346	272	74	213.8	78.6	126	99	27	77.8	78.5	413	324	89	254.2	78.5	261	205	56	161.0
78.6	462	363	99	285.2	78.6	345	271	74	212.9	78.6	117	92	25	72.3	78.5	410	322	88	252.9	78.5	260	204	56	160.1
78.6	459	361	98	283.9	78.6	341	268	73	210.6	78.6	112	88	24	69.1	78.5	409	321	88	251.9	78.5	256	201	55	157.8
78.6	458	360	98	283.0	78.6	337	265	72	208.4	78.6	103	81	22	63.7	78.5	405	318	87	249.7	78.5	251	197	54	154.6
78.6	457	359	98	282.0	78.6	336	264	72	207.4	78.6	98	77	21	60.5	78.5	404	317	87	248.7	78.5	247	194	53	152.4
78.6	454	357	97	280.7	78.6	332	261	71	205.2	78.6	84	66	18	51.9	78.5	400	314	86	246.5	78.5	246	193	53	151.4
78.6	453	356	97	279.8	78.6	331	260	71	204.2	78.6	70	55	15	43.2	78.5	396	311	85	244.2	78.5	242	190	52	149.2
78.6	449	353	96	277.5	78.6	327	257	70	202.0	78.6	56	44	12	34.6	78.5	395	310	85	243.3	78.5	237	186	51	146.0
78.6	448	352	96	276.6	78.6	323	254	69	199.7	78.6	42	33	9	25.9	78.5	391	307	84	241.0	78.5	233	183	50	143.7
78.6	444	349	95	274.3	78.6	322	253	69	198.8	78.6	28	22	6	17.3	78.5	390	306	84	240.1	78.5	228	179	49	140.5
78.6	443	348	95	273.4	78.6	318	250	68	196.5	78.6	14	11	3	8.6	78.5	386	303	83	237.8	78.5	223	175	48	137.3
78.6	440	346	94	272.1	78.6	313	246	67	193.3	78.5	498	391	107	307.0	78.5	382	300	82	235.6	78.5	219	172	47	135.1
78.6	439	345	94	271.1	78.6	309	243	66	191.1	78.5	497	390	107	306.0	78.5	381	299	82	234.6	78.5	214	168	46	131.9
78.6	435	342	93	268.9	78.6	308	242	66	190.1	78.5	494	388	106	304.7	78.5	377	296	81	232.4	78.5	209	164	45	128.7
78.6	434	341	93	267.9	78.6	304	239	65	187.9	78.5	493	387	106	303.8	78.5	376	295	81	231.4	78.5	205	161	44	126.4
78.6	430	338	92	265.7	78.6	299	235	64	184.7	78.5	492	386	106	302.8	78.5	372	292	80	229.2	78.5	200	157	43	123.2
78.6	429	337	92	264.7	78.6	295	232	63	182.5	78.5	489	384	105	301.5	78.5	368	289	79	227.0	78.5	195	153	42	120.0
78.6	426	335	91	263.4	78.6	294	231	63	181.5	78.5	488	383	105	300.6	78.5	367	288	79	226.0	78.5	191	150	41	117.8
78.6	425	334	91	262.5	78.6	290	228	62	179.3	78.5	484	380	104	298.3	78.5	363	285	78	223.8	78.5	186	146	40	114.6
78.6	421	331	90	260.2	78.6	285	224	61	176.1	78.5	483	379	104	297.4	78.5	362	284	78	222.8	78.5	181	142	39	111.4
78.6	420	330	90	259.3	78.6	281	221	60	173.8	78.5	480	377	103	296.1	78.5	358	281	77	220.6	78.5	177	139	38	109.2
78.6	416	327	89	257.0	78.6	280	220	60	172.9	78.5	479	376	103	295.1	78.5	354	278	76	218.3	78.5	172	135	37	106.0
78.6	415	326	89	256.1	78.6	276	217	59	170.6	78.5	478	375	103	294.2	78.5	353	277	76	217.4	78.5	163	128	35	100.5
78.6	412	324	88	254.8	78.6	271	213	58	167.4	78.5	475	373	102	292.9	78.5	349	274	75	215.1	78.5	158	124	34	97.3
78.6	411	323	88	253.8	78.6	266	209	57	164.2	78.5	474	372	102	291.9	78.5	344	270	74	211.9	78.5	149	117	32	91.9
78.6	407	320	87	251.6	78.6	262	206	56	162.0	78.5	470	369	101	289.7	78.5	340	267	73	209.7	78.5	144	113	31	88.7
78.6	406	319	87	250.6	78.6	257	202	55	158.8	78.5	469	368	101	288.8	78.5	339	266	73	208.7	78.5	135	106	29	83.2
78.6	402	316	86	248.4	78.6	252	198	54	155.6	78.5	466	366	100	287.5	78.5	335	263	72	206.5	78.5	130	102	28	80.0
78.6	401	315	86	247.4	78.6	248	195	53	153.3	78.5	465	365	100	286.5	78.5	330	259	71	203.3	78.5	121	95	26	74.6
78.6	398	313	85	246.2	78.6	243	191	52	150.1	78.5	461	362	99	284.3	78.5	326	256	70	201.0	78.5	107	84	23	65.9
78.6	397	312	85	245.2	78.6	238	187	51	146.9	78.5	460	361	99	283.3	78.5	325	255	70	200.1	78.5	93	73	20	57.3
78.6	393	309	84	243.0	78.6	234	184	50	144.7	78.5	456	358	98	281.1	78.5	321	252	69	197.8	78.5	79	62	17	48.7
78.6	392	308	84	242.0	78.6	229	180	49	141.5	78.5	455	357	98	280.1	78.5	317	249	68	195.6	78.5	65	51	14	40.0
78.6	388	305	83	239.8	78.6	224	176	48	138.3	78.5	452	355	97	278.8	78.5	316	248	68	194.6	78.4	500	392	108	307.3
78.6	387	304	83	238.8	78.6	220	173	47	136.0	78.5	451	354	97	277.9	78.5	312	245	67	192.4	78.4	499	391	108	306.4
78.6	384	302	82	237.5	78.6	215	169	46	132.8	78.5	447	351	96	275.6	78.5	311	244	67	191.4	78.4	496	389	107	305.1
78.6	383	301	82	236.6	78.6	210	165	45	129.6	78.5	446	350	96	274.7	78.5	307	241	66	189.2	78.4	495	388	107	304.1
78.6	379	298	81	234.3	78.6	206	162	44	127.4	78.5	442	347	95	272.4	78.5	303	238	65	186.9	78.4	491	385	106	301.9
78.6	378	297	81	233.4	78.6	201	158	43	124.2	78.5	441	346	95	271.5	78.5	302	237	65	186.0	78.4	490	384	106	300.9
78.6	374	294	80	231.1	78.6	196	154	42	121.0	78.5	438	344	94	270.2	78.5	298	234	64	183.7	78.4	487	382	105	299.6
78.6	373	293	80	230.2	78.6	192	151	41	118.8	78.5	437	343	94	269.2	78.5	297	233	64	182.8	78.4	486	381	105	298.7

%E	M1	M2	DM	M*
78.4	485	380	105	297.7
78.4	482	378	104	296.4
78.4	481	377	104	295.5
78.4	477	374	103	293.2
78.4	476	373	103	292.3
78.4	473	371	102	291.0
78.4	472	370	102	290.0
78.4	468	367	101	287.8
78.4	467	366	101	286.8
78.4	464	364	100	285.6
78.4	463	363	100	284.6
78.4	462	362	100	283.6
78.4	459	360	99	282.4
78.4	458	359	99	281.4
78.4	454	356	98	279.2
78.4	453	355	98	278.2
78.4	450	353	97	276.9
78.4	449	352	97	276.0
78.4	445	349	96	273.7
78.4	444	348	96	272.8
78.4	440	345	95	270.5
78.4	439	344	95	269.6
78.4	436	342	94	268.3
78.4	435	341	94	267.3
78.4	431	338	93	265.1
78.4	430	337	93	264.1
78.4	426	334	92	261.9
78.4	425	333	92	260.9
78.4	422	331	91	259.6
78.4	421	330	91	258.7
78.4	417	327	90	256.4
78.4	416	326	90	255.5
78.4	412	323	89	253.2
78.4	408	320	88	251.0
78.4	407	319	88	250.0
78.4	403	316	87	247.8
78.4	402	315	87	246.8
78.4	399	313	86	245.5
78.4	398	312	86	244.6
78.4	394	309	85	242.3
78.4	393	308	85	241.4
78.4	389	305	84	239.1
78.4	388	304	84	238.2
78.4	385	302	83	236.9
78.4	384	301	83	235.9
78.4	380	298	82	233.7
78.4	379	297	82	232.7
78.4	375	294	81	230.5
78.4	371	291	80	228.3
78.4	370	290	80	227.3
78.4	366	287	79	225.1
78.4	365	286	79	224.1
78.4	361	283	78	221.9
78.4	357	280	77	219.6
78.4	356	279	77	218.7
78.4	352	276	76	216.4
78.4	348	273	75	214.2
78.4	347	272	75	213.2
78.4	343	269	74	211.0
78.4	342	268	74	210.0
78.4	338	265	73	207.8
78.4	334	262	72	205.5
78.4	333	261	72	204.6
78.4	329	258	71	202.3
78.4	328	257	71	201.4
78.4	324	254	70	199.1
78.4	320	251	69	196.9
78.4	319	250	69	195.9
78.4	315	247	68	193.7
78.4	310	243	67	190.5
78.4	306	240	66	188.2
78.4	305	239	66	187.3
78.4	301	236	65	185.0
78.4	296	232	64	181.8
78.4	292	229	63	179.6
78.4	291	228	63	178.6
78.4	287	225	62	176.4
78.4	283	222	61	174.1
78.4	282	221	61	173.2
78.4	278	218	60	170.9
78.4	273	214	59	167.8
78.4	269	211	58	165.5
78.4	268	210	58	164.6
78.4	264	207	57	162.3
78.4	259	203	56	159.1
78.4	255	200	55	156.9
78.4	250	196	54	153.7
78.4	245	192	53	150.5
78.4	241	189	52	148.2
78.4	236	185	51	145.0
78.4	232	182	50	142.8
78.4	231	181	50	141.8
78.4	227	178	49	139.6
78.4	222	174	48	136.4
78.4	218	171	47	134.1
78.4	213	167	46	130.9
78.4	208	163	45	127.7
78.4	204	160	44	125.5
78.4	199	156	43	122.3
78.4	194	152	42	119.1
78.4	190	149	41	116.8
78.4	185	145	40	113.6
78.4	176	138	38	108.2
78.4	171	134	37	105.0
78.4	167	[illegible]31	36	102.8
78.4	162	127	35	99.6
78.4	153	120	33	94.1
78.4	148	116	32	90.9
78.4	139	109	30	85.5
78.4	134	105	29	82.3
78.4	125	98	27	76.8
78.4	116	91	25	71.4
78.4	111	87	24	68.2
78.4	102	80	22	62.7
78.4	97	76	21	59.5
78.4	88	69	19	54.1
78.4	74	58	16	45.5
78.4	51	40	11	31.4
78.4	37	29	8	22.7
78.3	498	390	108	305.4
78.3	497	389	108	304.5
78.3	494	387	107	303.2
78.3	493	386	107	302.2
78.3	492	385	107	301.3
78.3	489	383	106	300.0
78.3	488	382	106	299.0
78.3	484	379	105	296.8
78.3	483	378	105	295.8
78.3	480	376	104	294.5
78.3	479	375	104	293.6
78.3	475	372	103	291.3
78.3	474	371	103	290.4
78.3	471	369	102	289.1
78.3	470	368	102	288.1
78.3	469	367	102	287.2
78.3	466	365	101	285.9
78.3	465	364	101	284.9
78.3	461	361	100	282.7
78.3	460	360	100	281.7
78.3	457	358	99	280.4
78.3	456	357	99	279.5
78.3	452	354	98	277.2
78.3	451	353	98	276.3
78.3	448	351	97	275.0
78.3	447	350	97	274.0
78.3	446	349	97	273.1
78.3	443	347	96	271.8
78.3	442	346	96	270.9
78.3	438	343	95	268.6
78.3	437	342	95	267.7
78.3	434	340	94	266.4
78.3	433	339	94	265.4
78.3	429	336	93	263.2
78.3	428	335	93	262.2
78.3	424	332	92	260.0
78.3	423	331	92	259.0
78.3	420	329	91	257.7
78.3	419	328	91	256.8
78.3	415	325	90	254.5
78.3	414	324	90	253.6
78.3	411	322	89	252.3
78.3	410	321	89	251.3
78.3	406	318	88	249.1
78.3	405	317	88	248.1
78.3	401	314	87	245.9
78.3	400	313	87	244.9
78.3	397	311	86	243.6
78.3	396	310	86	242.7
78.3	392	307	85	240.4
78.3	391	306	85	239.5
78.3	387	303	84	237.2
78.3	383	300	83	235.0
78.3	382	299	83	234.0
78.3	378	296	82	231.8
78.3	374	293	81	229.5
78.3	373	292	81	228.6
78.3	369	289	80	226.3
78.3	368	288	80	225.4
78.3	364	285	79	223.1
78.3	360	282	78	220.9
78.3	359	281	78	219.9
78.3	355	278	77	217.7
78.3	351	275	76	215.5
78.3	350	274	76	214.5
78.3	346	271	75	212.3
78.3	345	270	75	211.3
78.3	341	267	74	209.1
78.3	337	264	73	206.8
78.3	336	263	73	205.9
78.3	332	260	72	203.6
78.3	327	256	71	200.4
78.3	323	253	70	198.2
78.3	322	252	70	197.2
78.3	318	249	69	195.0
78.3	314	246	68	192.7
78.3	313	245	68	191.8
78.3	309	242	67	189.5
78.3	304	238	66	186.3
78.3	300	235	65	184.1
78.3	299	234	65	183.1
78.3	295	231	64	180.9
78.3	290	227	63	177.7
78.3	286	224	62	175.4
78.3	281	220	61	172.2
78.3	277	217	60	170.0
78.3	276	216	60	169.0
78.3	272	213	59	166.8
78.3	267	209	58	163.6
78.3	263	206	57	161.4
78.3	258	202	56	158.2
78.3	254	199	55	155.9
78.3	253	198	55	155.0
78.3	249	195	54	152.7
78.3	244	191	53	149.5
78.3	240	188	52	147.3
78.3	235	184	51	144.1
78.3	230	180	50	140.9
78.3	226	177	49	138.6
78.3	221	173	48	135.4
78.3	217	170	47	133.2
78.3	212	166	46	130.0
78.3	207	162	45	126.8
78.3	203	159	44	124.5
78.3	198	155	43	121.3
78.3	189	148	41	115.9
78.3	184	144	40	112.7
78.3	180	141	39	110.4
78.3	175	137	38	107.3
78.3	166	130	36	101.8
78.3	161	126	35	98.6
78.3	157	123	34	96.4
78.3	152	119	33	93.2
78.3	143	112	31	87.7
78.3	138	108	30	84.5
78.3	129	101	28	79.1
78.3	120	94	26	73.6
78.3	115	90	25	70.4
78.3	106	83	23	65.0
78.3	92	72	20	56.3
78.3	83	65	18	50.9
78.3	69	54	15	42.3
78.3	60	47	13	36.8
78.3	46	36	10	28.2
78.3	23	18	5	14.1
78.2	500	391	109	305.8
78.2	499	390	109	304.8
78.2	496	388	108	303.5
78.2	495	387	108	302.6
78.2	491	384	107	300.3
78.2	490	383	107	299.4

%E	M1	M2	DM	M*
78.2	487	381	106	298.1
78.2	486	380	106	297.1
78.2	482	377	105	294.9
78.2	481	376	105	293.9
78.2	478	374	104	292.6
78.2	477	373	104	291.7
78.2	476	372	104	290.7
78.2	473	370	103	289.4
78.2	472	369	103	288.5
78.2	468	366	102	286.2
78.2	467	365	102	285.3
78.2	464	363	101	284.0
78.2	463	362	101	283.0
78.2	459	359	100	280.8
78.2	458	358	100	279.8
78.2	455	356	99	278.5
78.2	454	355	99	277.6
78.2	450	352	98	275.3
78.2	449	351	98	274.4
78.2	445	348	97	272.1
78.2	444	347	97	271.2
78.2	441	345	96	269.9
78.2	440	344	96	268.9
78.2	436	341	95	266.7
78.2	435	340	95	265.7
78.2	432	338	94	264.5
78.2	431	337	94	263.5
78.2	427	334	93	261.3
78.2	426	333	93	260.3
78.2	422	330	92	258.1
78.2	418	327	91	255.8
78.2	417	326	91	254.9
78.2	413	323	90	252.6
78.2	412	322	90	251.7
78.2	409	320	89	250.4
78.2	408	319	89	249.4
78.2	404	316	88	247.2
78.2	403	315	88	246.2
78.2	399	312	87	244.0
78.2	395	309	86	241.7
78.2	394	308	86	240.8
78.2	390	305	85	238.5
78.2	386	302	84	236.3
78.2	385	301	84	235.3
78.2	381	298	83	233.1
78.2	380	297	83	232.1
78.2	377	295	82	230.8
78.2	376	294	82	229.9
78.2	372	291	81	227.6
78.2	371	290	81	226.7
78.2	367	287	80	224.4
78.2	363	284	79	222.2
78.2	362	283	79	221.2
78.2	358	280	78	219.0
78.2	357	279	78	218.0
78.2	354	277	77	216.7
78.2	353	276	77	215.8
78.2	349	273	76	213.6
78.2	348	272	76	212.6
78.2	344	269	75	210.4
78.2	340	266	74	208.1
78.2	339	265	74	207.2
78.2	335	262	73	204.9
78.2	331	259	72	202.7
78.2	330	258	72	201.7
78.2	326	255	71	199.5
78.2	325	254	71	198.5
78.2	321	251	70	196.3
78.2	317	248	69	194.0
78.2	316	247	69	193.1
78.2	312	244	68	190.8
78.2	308	241	67	188.6
78.2	307	240	67	187.6
78.2	303	237	66	185.4
78.2	298	233	65	182.2
78.2	294	230	64	179.9
78.2	293	229	64	179.0
78.2	289	226	63	176.7
78.2	285	223	62	174.5
78.2	284	222	62	173.5
78.2	280	219	61	171.3
78.2	275	215	60	168.1
78.2	271	212	59	165.8
78.2	266	208	58	162.6
78.2	262	205	57	160.4
78.2	261	204	57	159.4
78.2	257	201	56	157.2
78.2	252	197	55	154.0
78.2	248	194	54	151.8
78.2	243	190	53	148.6
78.2	239	187	52	146.3
78.2	238	186	52	145.4
78.2	234	183	51	143.1
78.2	229	179	50	139.9
78.2	225	176	49	137.7
78.2	220	172	48	134.5
78.2	216	169	47	132.2
78.2	211	165	46	129.0
78.2	206	161	45	125.8
78.2	202	158	44	123.6
78.2	197	154	43	120.4
78.2	193	151	42	118.1
78.2	188	147	41	114.9
78.2	179	140	39	109.5
78.2	174	136	38	106.3
78.2	170	133	37	104.1
78.2	165	129	36	100.9
78.2	156	122	34	95.4
78.2	147	115	32	90.0
78.2	142	111	31	86.8
78.2	133	104	29	81.3
78.2	124	97	27	75.9
78.2	119	93	26	72.7
78.2	110	86	24	67.2
78.2	101	79	22	61.8
78.2	87	68	19	53.1
78.2	78	61	17	47.7
78.2	55	43	12	33.6
78.1	498	389	109	303.9
78.1	497	388	109	302.9
78.1	494	386	108	301.6
78.1	493	385	108	300.7
78.1	489	382	107	298.4
78.1	488	381	107	297.5
78.1	485	379	106	296.2
78.1	484	378	106	295.2
78.1	483	377	106	294.3
78.1	480	375	105	293.0
78.1	479	374	105	292.0
78.1	475	371	104	289.8
78.1	474	370	104	288.8
78.1	471	368	103	287.5
78.1	470	367	103	286.6
78.1	466	364	102	284.3
78.1	465	363	102	283.4
78.1	462	361	101	282.1
78.1	461	360	101	281.1
78.1	457	357	100	278.9
78.1	456	356	100	277.9
78.1	453	354	99	276.6
78.1	452	353	99	275.7
78.1	448	350	98	273.4
78.1	447	349	98	272.5
78.1	443	346	97	270.2
78.1	442	345	97	269.3
78.1	439	343	96	268.0
78.1	438	342	96	267.0
78.1	434	339	95	264.8
78.1	433	338	95	263.8
78.1	430	336	94	262.5
78.1	429	335	94	261.6
78.1	425	332	93	259.4
78.1	424	331	93	258.4
78.1	421	329	92	257.1
78.1	420	328	92	256.2
78.1	416	325	91	253.9
78.1	415	324	91	253.0
78.1	411	321	90	250.7
78.1	407	318	89	248.5
78.1	406	317	89	247.5
78.1	402	314	88	245.3
78.1	401	313	88	244.3
78.1	398	311	87	243.0
78.1	397	310	87	242.1
78.1	393	307	86	239.8
78.1	392	306	86	238.9
78.1	389	304	85	237.6
78.1	388	303	85	236.6
78.1	384	300	84	234.4
78.1	383	299	84	233.4
78.1	379	296	83	231.2
78.1	375	293	82	228.9
78.1	374	292	82	228.0
78.1	370	289	81	225.7
78.1	366	286	80	223.5
78.1	365	285	80	222.5
78.1	361	282	79	220.3
78.1	360	281	79	219.3
78.1	356	278	78	217.1
78.1	352	275	77	214.8
78.1	351	274	77	213.9
78.1	347	271	76	211.6
78.1	343	268	75	209.4
78.1	342	267	75	208.4
78.1	338	264	74	206.2
78.1	334	261	73	204.0
78.1	333	260	73	203.0
78.1	329	257	72	200.8
78.1	324	253	71	197.6
78.1	320	250	70	195.3
78.1	319	249	70	194.4
78.1	315	246	69	192.1
78.1	311	243	68	189.9
78.1	310	242	68	188.9
78.1	306	239	67	186.7
78.1	302	236	66	184.4
78.1	301	235	66	183.5
78.1	297	232	65	181.2
78.1	292	228	64	178.0
78.1	288	225	63	175.8
78.1	283	221	62	172.6
78.1	279	218	61	170.3
78.1	278	217	61	169.4
78.1	274	214	60	167.1
78.1	270	211	59	164.9
78.1	269	210	59	163.9
78.1	265	207	58	161.7
78.1	260	203	57	158.5
78.1	256	200	56	156.3
78.1	251	196	55	153.1
78.1	247	193	54	150.8
78.1	242	189	53	147.6
78.1	237	185	52	144.4
78.1	233	182	51	142.2
78.1	228	178	50	139.0
78.1	224	175	49	136.7
78.1	219	171	48	133.5
78.1	215	168	47	131.3
78.1	210	164	46	128.1
78.1	201	157	44	122.6
78.1	196	153	43	119.4
78.1	192	150	42	117.2
78.1	187	146	41	114.0
78.1	183	143	40	111.7
78.1	178	139	39	108.5
78.1	169	132	37	103.1
78.1	160	125	35	97.7
78.1	155	121	34	94.5
78.1	151	118	33	92.2
78.1	146	114	32	89.0
78.1	137	107	30	83.6
78.1	128	100	28	78.1
78.1	114	89	25	69.5
78.1	105	82	23	64.0
78.1	96	75	21	58.6
78.1	73	57	16	44.5
78.1	64	50	14	39.1
78.1	32	25	7	19.5
78.0	500	390	110	304.2
78.0	499	389	110	303.2
78.0	496	387	109	302.0
78.0	495	386	109	301.0
78.0	492	384	108	299.7
78.0	491	383	108	298.8
78.0	490	382	108	297.8
78.0	487	380	107	296.5
78.0	486	379	107	295.6
78.0	482	376	106	293.3
78.0	481	375	106	292.4
78.0	478	373	105	291.1

%E	M1	M2	DM	M*
78.0	477	372	105	290.1
78.0	473	369	104	287.9
78.0	472	368	104	286.9
78.0	469	366	103	285.6
78.0	468	365	103	284.7
78.0	464	362	102	282.4
78.0	463	361	102	281.5
78.0	460	359	101	280.2
78.0	459	358	101	279.2
78.0	455	355	100	277.0
78.0	454	354	100	276.0
78.0	451	352	99	274.7
78.0	450	351	99	273.8
78.0	449	350	99	272.8
78.0	446	348	98	271.5
78.0	445	347	98	270.6
78.0	441	344	97	268.3
78.0	440	343	97	267.4
78.0	437	341	96	266.1
78.0	436	340	96	265.1
78.0	432	337	95	262.9
78.0	431	336	95	261.9
78.0	428	334	94	260.6
78.0	427	333	94	259.7
78.0	423	330	93	257.4
78.0	422	329	93	256.5
78.0	419	327	92	255.2
78.0	418	326	92	254.2
78.0	414	323	91	252.0
78.0	413	322	91	251.1
78.0	410	320	90	249.8
78.0	409	319	90	248.8
78.0	405	316	89	246.6
78.0	404	315	89	245.6
78.0	400	312	88	243.4
78.0	396	309	87	241.1
78.0	395	308	87	240.2
78.0	391	305	86	237.9
78.0	387	302	85	235.7
78.0	386	301	85	234.7
78.0	382	298	84	232.5
78.0	381	297	84	231.5
78.0	378	295	83	230.2
78.0	377	294	83	229.3
78.0	373	291	82	227.0
78.0	372	290	82	226.1
78.0	369	288	81	224.8
78.0	368	287	81	223.8
78.0	364	284	80	221.6
78.0	363	283	80	220.6

%E	M1	M2	DM	M*
78.0	359	280	79	218.4
78.0	355	277	78	216.1
78.0	354	276	78	215.2
78.0	350	273	77	212.9
78.0	346	270	76	210.7
78.0	345	269	76	209.7
78.0	341	266	75	207.5
78.0	337	263	74	205.2
78.0	336	262	74	204.3
78.0	332	259	73	202.1
78.0	328	256	72	199.8
78.0	327	255	72	198.9
78.0	323	252	71	196.6
78.0	322	251	71	195.7
78.0	318	248	70	193.4
78.0	314	245	69	191.2
78.0	313	244	69	190.2
78.0	309	241	68	188.0
78.0	305	238	67	185.7
78.0	304	237	67	184.8
78.0	300	234	66	182.5
78.0	296	231	65	180.3
78.0	295	230	65	179.3
78.0	291	227	64	177.1
78.0	287	224	63	174.8
78.0	286	223	63	173.9
78.0	282	220	62	171.6
78.0	277	216	61	168.4
78.0	273	213	60	166.2
78.0	268	209	59	163.0
78.0	264	206	58	160.7
78.0	259	202	57	157.5
78.0	255	199	56	155.3
78.0	254	198	56	154.3
78.0	250	195	55	152.1
78.0	246	192	54	149.9
78.0	245	191	54	148.9
78.0	241	188	53	146.7
78.0	236	184	52	143.5
78.0	232	181	51	141.2
78.0	227	177	50	138.0
78.0	223	174	49	135.8
78.0	218	170	48	132.6
78.0	214	167	47	130.3
78.0	209	163	46	127.1
78.0	205	160	45	124.9
78.0	200	156	44	121.7
78.0	191	149	42	116.2
78.0	186	145	41	113.0
78.0	182	142	40	110.8

%E	M1	M2	DM	M*
78.0	177	138	39	107.6
78.0	173	135	38	105.3
78.0	168	131	37	102.1
78.0	164	128	36	99.9
78.0	159	124	35	96.7
78.0	150	117	33	91.3
78.0	141	110	31	85.8
78.0	132	103	29	80.4
78.0	127	99	28	77.2
78.0	123	96	27	74.9
78.0	118	92	26	71.7
78.0	109	85	24	66.3
78.0	100	78	22	60.8
78.0	91	71	20	55.4
78.0	82	64	18	50.0
78.0	59	46	13	35.9
78.0	50	39	11	30.4
78.0	41	32	9	25.0
77.9	498	388	110	302.3
77.9	497	387	110	301.3
77.9	494	385	109	300.1
77.9	493	384	109	299.1
77.9	489	381	108	296.9
77.9	488	380	108	295.9
77.9	485	378	107	294.6
77.9	484	377	107	293.7
77.9	480	374	106	291.4
77.9	479	373	106	290.5
77.9	476	371	105	289.2
77.9	475	370	105	288.2
77.9	471	367	104	286.0
77.9	470	366	104	285.0
77.9	467	364	103	283.7
77.9	466	363	103	282.8
77.9	462	360	102	280.5
77.9	461	359	102	279.6
77.9	458	357	101	278.3
77.9	457	356	101	277.3
77.9	456	355	101	276.4
77.9	453	353	100	275.1
77.9	452	352	100	274.1
77.9	448	349	99	271.9
77.9	447	348	99	270.9
77.9	444	346	98	269.6
77.9	443	345	98	268.7
77.9	439	342	97	266.4
77.9	438	341	97	265.5
77.9	435	339	96	264.2
77.9	434	338	96	263.2
77.9	430	335	95	261.0

%E	M1	M2	DM	M*
77.9	429	334	95	260.0
77.9	426	332	94	258.7
77.9	425	331	94	257.8
77.9	421	328	93	255.5
77.9	420	327	93	254.6
77.9	417	325	92	253.3
77.9	416	324	92	252.3
77.9	412	321	91	250.1
77.9	411	320	91	249.1
77.9	408	318	90	247.9
77.9	407	317	90	246.9
77.9	403	314	89	244.7
77.9	402	313	89	243.7
77.9	399	311	88	242.4
77.9	398	310	88	241.5
77.9	394	307	87	239.2
77.9	393	306	87	238.3
77.9	390	304	86	237.0
77.9	389	303	86	236.0
77.9	385	300	85	233.8
77.9	384	299	85	232.8
77.9	380	296	84	230.6
77.9	376	293	83	228.3
77.9	375	292	83	227.4
77.9	371	289	82	225.1
77.9	367	286	81	222.9
77.9	366	285	81	221.9
77.9	362	282	80	219.7
77.9	358	279	79	217.4
77.9	357	278	79	216.5
77.9	353	275	78	214.2
77.9	349	272	77	212.0
77.9	348	271	77	211.0
77.9	344	268	76	208.8
77.9	340	265	75	206.5
77.9	339	264	75	205.6
77.9	335	261	74	203.3
77.9	331	258	73	201.1
77.9	330	257	73	200.1
77.9	326	254	72	197.9
77.9	321	250	71	194.7
77.9	317	247	70	192.5
77.9	312	243	69	189.3
77.9	308	240	68	187.0
77.9	307	239	68	186.1
77.9	303	236	67	183.8
77.9	299	233	66	181.6
77.9	298	232	66	180.6
77.9	294	229	65	178.4
77.9	290	226	64	176.1

%E	M1	M2	DM	M*
77.9	289	225	64	175.2
77.9	285	222	63	172.9
77.9	281	219	62	170.7
77.9	280	218	62	169.7
77.9	276	215	61	167.5
77.9	272	212	60	165.2
77.9	271	211	60	164.3
77.9	267	208	59	162.0
77.9	263	205	58	159.8
77.9	262	204	58	158.8
77.9	258	201	57	156.6
77.9	253	197	56	153.4
77.9	249	194	55	151.1
77.9	244	190	54	148.0
77.9	240	187	53	145.7
77.9	235	183	52	142.5
77.9	231	180	51	140.3
77.9	226	176	50	137.1
77.9	222	173	49	134.8
77.9	217	169	48	131.6
77.9	213	166	47	129.4
77.9	208	162	46	126.2
77.9	204	159	45	123.9
77.9	199	155	44	120.7
77.9	195	152	43	118.5
77.9	190	148	42	115.3
77.9	181	141	40	109.8
77.9	172	134	38	104.4
77.9	163	127	36	99.0
77.9	154	120	34	93.5
77.9	149	116	33	90.3
77.9	145	113	32	88.1
77.9	140	109	31	84.9
77.9	136	106	30	82.6
77.9	131	102	29	79.4
77.9	122	95	27	74.0
77.9	113	88	25	68.5
77.9	104	81	23	63.1
77.9	95	74	21	57.6
77.9	86	67	19	52.2
77.9	77	60	17	46.8
77.9	68	53	15	41.3
77.8	500	389	111	302.6
77.8	499	388	111	301.7
77.8	496	386	110	300.4
77.8	495	385	110	299.4
77.8	492	383	109	298.1
77.8	491	382	109	297.2
77.8	490	381	109	296.2
77.8	487	379	108	295.0

%E	M1	M2	DM	M*
77.8	486	378	108	294.0
77.8	483	376	107	292.7
77.8	482	375	107	291.8
77.8	481	374	107	290.8
77.8	478	372	106	289.5
77.8	477	371	106	288.6
77.8	474	369	105	287.3
77.8	473	368	105	286.3
77.8	472	367	105	285.4
77.8	469	365	104	284.1
77.8	468	364	104	283.1
77.8	465	362	103	281.8
77.8	464	361	103	280.9
77.8	463	360	103	279.9
77.8	460	358	102	278.6
77.8	459	357	102	277.7
77.8	455	354	101	275.4
77.8	454	353	101	274.5
77.8	451	351	100	273.2
77.8	450	350	100	272.2
77.8	446	347	99	270.0
77.8	445	346	99	269.0
77.8	442	344	98	267.7
77.8	441	343	98	266.8
77.8	437	340	97	264.5
77.8	436	339	97	263.6
77.8	433	337	96	262.3
77.8	432	336	96	261.3
77.8	428	333	95	259.1
77.8	427	332	95	258.1
77.8	424	330	94	256.8
77.8	423	329	94	255.9
77.8	419	326	93	253.6
77.8	418	325	93	252.7
77.8	415	323	92	251.4
77.8	414	322	92	250.4
77.8	410	319	91	248.2
77.8	409	318	91	247.2
77.8	406	316	90	246.0
77.8	405	315	90	245.0
77.8	401	312	89	242.8
77.8	400	311	89	241.8
77.8	397	309	88	240.5
77.8	396	308	88	239.6
77.8	392	305	87	237.3
77.8	388	302	86	235.1
77.8	387	301	86	234.1
77.8	383	298	85	231.9
77.8	379	295	84	229.6
77.8	378	294	84	228.7
77.8	374	291	83	226.4
77.8	370	288	82	224.2
77.8	369	287	82	223.2
77.8	365	284	81	221.0
77.8	361	281	80	218.7
77.8	360	280	80	217.8
77.8	356	277	79	215.5
77.8	352	274	78	213.3
77.8	351	273	78	212.3
77.8	347	270	77	210.1
77.8	343	267	76	207.8
77.8	342	266	76	206.9
77.8	338	263	75	204.6
77.8	334	260	74	202.4
77.8	333	259	74	201.4
77.8	329	256	73	199.2
77.8	325	253	72	197.0
77.8	324	252	72	196.0
77.8	320	249	71	193.8
77.8	316	246	70	191.5
77.8	315	245	70	190.6
77.8	311	242	69	188.3
77.8	306	238	68	185.1
77.8	302	235	67	182.9
77.8	297	231	66	179.7
77.8	293	228	65	177.4
77.8	288	224	64	174.2
77.8	284	221	63	172.0
77.8	279	217	62	168.8
77.8	275	214	61	166.5
77.8	270	210	60	163.3
77.8	266	207	59	161.1
77.8	261	203	58	157.9
77.8	257	200	57	155.6
77.8	252	196	56	152.4
77.8	248	193	55	150.2
77.8	243	189	54	147.0
77.8	239	186	53	144.8
77.8	234	182	52	141.6
77.8	230	179	51	139.3
77.8	225	175	50	136.1
77.8	221	172	49	133.9
77.8	216	168	48	130.7
77.8	212	165	47	128.4
77.8	207	161	46	125.2
77.8	203	158	45	123.0
77.8	198	154	44	119.8
77.8	194	151	43	117.5
77.8	189	147	42	114.3
77.8	185	144	41	112.1
77.8	180	140	40	108.9
77.8	176	137	39	106.6
77.8	171	133	38	103.4
77.8	167	130	37	101.2
77.8	162	126	36	98.0
77.8	158	123	35	95.8
77.8	153	119	34	92.6
77.8	144	112	32	87.1
77.8	135	105	30	81.7
77.8	126	98	28	76.2
77.8	117	91	26	70.8
77.8	108	84	24	65.3
77.8	99	77	22	59.9
77.8	90	70	20	54.4
77.8	81	63	18	49.0
77.8	72	56	16	43.6
77.8	63	49	14	38.1
77.8	54	42	12	32.7
77.8	45	35	10	27.2
77.8	36	28	8	21.8
77.8	27	21	6	16.3
77.8	18	14	4	10.9
77.8	9	7	2	5.4
77.7	498	387	111	300.7
77.7	497	386	111	299.8
77.7	494	384	110	298.5
77.7	493	383	110	297.5
77.7	489	380	109	295.3
77.7	488	379	109	294.3
77.7	485	377	108	293.0
77.7	484	376	108	292.1
77.7	480	373	107	289.9
77.7	479	372	107	288.9
77.7	476	370	106	287.6
77.7	475	369	106	286.7
77.7	471	366	105	284.4
77.7	470	365	105	283.5
77.7	467	363	104	282.2
77.7	466	362	104	281.2
77.7	462	359	103	279.0
77.7	461	358	103	278.0
77.7	458	356	102	276.7
77.7	457	355	102	275.8
77.7	453	352	101	273.5
77.7	452	351	101	272.6
77.7	449	349	100	271.3
77.7	448	348	100	270.3
77.7	444	345	99	268.1
77.7	443	344	99	267.1
77.7	440	342	98	265.8
77.7	439	341	98	264.9
77.7	435	338	97	262.6
77.7	434	337	97	261.7
77.7	431	335	96	260.4
77.7	430	334	96	259.4
77.7	426	331	95	257.2
77.7	422	328	94	254.9
77.7	421	327	94	254.0
77.7	417	324	93	251.7
77.7	413	321	92	249.5
77.7	412	320	92	248.5
77.7	408	317	91	246.3
77.7	404	314	90	244.0
77.7	403	313	90	243.1
77.7	399	310	89	240.9
77.7	395	307	88	238.6
77.7	394	306	88	237.7
77.7	391	304	87	236.4
77.7	390	303	87	235.4
77.7	386	300	86	233.2
77.7	385	299	86	232.2
77.7	382	297	85	230.9
77.7	381	296	85	230.0
77.7	377	293	84	227.7
77.7	376	292	84	226.8
77.7	373	290	83	225.5
77.7	372	289	83	224.5
77.7	368	286	82	222.3
77.7	367	285	82	221.3
77.7	364	283	81	220.0
77.7	363	282	81	219.1
77.7	359	279	80	216.8
77.7	358	278	80	215.9
77.7	355	276	79	214.6
77.7	354	275	79	213.6
77.7	350	272	78	211.4
77.7	349	271	78	210.4
77.7	346	269	77	209.1
77.7	345	268	77	208.2
77.7	341	265	76	205.9
77.7	337	262	75	203.7
77.7	336	261	75	202.7
77.7	332	258	74	200.5
77.7	328	255	73	198.2
77.7	327	254	73	197.3
77.7	323	251	72	195.0
77.7	319	248	71	192.8
77.7	318	247	71	191.9
77.7	314	244	70	189.6
77.7	310	241	69	187.4
77.7	309	240	69	186.4
77.7	305	237	68	184.2
77.7	301	234	67	181.9
77.7	300	233	67	181.0
77.7	296	230	66	178.7
77.7	292	227	65	176.5
77.7	291	226	65	175.5
77.7	287	223	64	173.3
77.7	283	220	63	171.0
77.7	282	219	63	170.1
77.7	278	216	62	167.8
77.7	274	213	61	165.6
77.7	273	212	61	164.6
77.7	269	209	60	162.4
77.7	265	206	59	160.1
77.7	264	205	59	159.2
77.7	260	202	58	156.9
77.7	256	199	57	154.7
77.7	251	195	56	151.5
77.7	247	192	55	149.2
77.7	242	188	54	146.0
77.7	238	185	53	143.8
77.7	233	181	52	140.6
77.7	229	178	51	138.4
77.7	224	174	50	135.2
77.7	220	171	49	132.9
77.7	215	167	48	129.7
77.7	211	164	47	127.5
77.7	206	160	46	124.3
77.7	202	157	45	122.0
77.7	197	153	44	118.8
77.7	193	150	43	116.6
77.7	188	146	42	113.4
77.7	184	143	41	111.1
77.7	179	139	40	107.9
77.7	175	136	39	105.7
77.7	166	129	37	100.2
77.7	157	122	35	94.8
77.7	148	115	33	89.4
77.7	139	108	31	83.9
77.7	130	101	29	78.5
77.7	121	94	27	73.0
77.7	112	87	25	67.6
77.7	103	80	23	62.1
77.7	94	73	21	56.7
77.6	500	388	112	301.1
77.6	499	387	112	300.1
77.6	496	385	111	298.8
77.6	495	384	111	297.9
77.6	492	382	110	296.6

%E	M1	M2	DM	M*
77.6	491	381	110	295.6
77.6	490	380	110	294.7
77.6	487	378	109	293.4
77.6	486	377	109	292.4
77.6	483	375	108	291.1
77.6	482	374	108	290.2
77.6	478	371	107	288.0
77.6	477	370	107	287.0
77.6	474	368	106	285.7
77.6	473	367	106	284.8
77.6	469	364	105	282.5
77.6	468	363	105	281.6
77.6	465	361	104	280.3
77.6	464	360	104	279.3
77.6	460	357	103	277.1
77.6	459	356	103	276.1
77.6	456	354	102	274.8
77.6	455	353	102	273.9
77.6	451	350	101	271.6
77.6	450	349	101	270.7
77.6	447	347	100	269.4
77.6	446	346	100	268.4
77.6	442	343	99	266.2
77.6	441	342	99	265.2
77.6	438	340	98	263.9
77.6	437	339	98	263.0
77.6	433	336	97	260.7
77.6	429	333	96	258.5
77.6	428	332	96	257.5
77.6	425	330	95	256.2
77.6	424	329	95	255.3
77.6	420	326	94	253.0
77.6	419	325	94	252.1
77.6	416	323	93	250.8
77.6	415	322	93	249.8
77.6	411	319	92	247.6
77.6	410	318	92	246.6
77.6	407	316	91	245.3
77.6	406	315	91	244.4
77.6	402	312	90	242.1
77.6	401	311	90	241.2
77.6	398	309	89	239.9
77.6	397	308	89	239.0
77.6	393	305	88	236.7
77.6	392	304	88	235.8
77.6	389	302	87	234.5
77.6	388	301	87	233.5
77.6	384	298	86	231.3
77.6	380	295	85	229.0
77.6	379	294	85	228.1

%E	M1	M2	DM	M*
77.6	375	291	84	225.8
77.6	371	288	83	223.6
77.6	370	287	83	222.6
77.6	366	284	82	220.4
77.6	362	281	81	218.1
77.6	361	280	81	217.2
77.6	357	277	80	214.9
77.6	353	274	79	212.7
77.6	352	273	79	211.7
77.6	348	270	78	209.5
77.6	344	267	77	207.2
77.6	343	266	77	206.3
77.6	340	264	76	205.0
77.6	339	263	76	204.0
77.6	335	260	75	201.8
77.6	331	257	74	199.5
77.6	330	256	74	198.6
77.6	326	253	73	196.3
77.6	322	250	72	194.1
77.6	321	249	72	193.1
77.6	317	246	71	190.9
77.6	313	243	70	188.7
77.6	312	242	70	187.7
77.6	308	239	69	185.5
77.6	304	236	68	183.2
77.6	303	235	68	182.3
77.6	299	232	67	180.0
77.6	295	229	66	177.8
77.6	294	228	66	176.8
77.6	290	225	65	174.6
77.6	286	222	64	172.3
77.6	281	218	63	169.1
77.6	277	215	62	166.9
77.6	272	211	61	163.7
77.6	268	208	60	161.4
77.6	263	204	59	158.2
77.6	259	201	58	156.0
77.6	255	198	57	153.7
77.6	254	197	57	152.8
77.6	250	194	56	150.5
77.6	246	191	55	148.3
77.6	245	190	55	147.3
77.6	241	187	54	145.1
77.6	237	184	53	142.9
77.6	232	180	52	139.7
77.6	228	177	51	137.4
77.6	223	173	50	134.2
77.6	219	170	49	132.0
77.6	214	166	48	128.8
77.6	210	163	47	126.5

%E	M1	M2	DM	M*
77.6	205	159	46	123.3
77.6	201	156	45	121.1
77.6	196	152	44	117.9
77.6	192	149	43	115.6
77.6	183	142	41	110.2
77.6	174	135	39	104.7
77.6	170	132	38	102.5
77.6	165	128	37	99.3
77.6	161	125	36	97.0
77.6	156	121	35	93.9
77.6	152	118	34	91.6
77.6	147	114	33	88.4
77.6	143	111	32	86.2
77.6	134	104	30	80.7
77.6	125	97	28	75.3
77.6	116	90	26	69.8
77.6	107	83	24	64.4
77.6	98	76	22	58.9
77.6	85	66	19	51.2
77.6	76	59	17	45.8
77.6	67	52	15	40.4
77.6	58	45	13	34.9
77.6	49	38	11	29.5
77.5	498	386	112	299.2
77.5	497	385	112	298.2
77.5	494	383	111	296.9
77.5	493	382	111	296.0
77.5	489	379	110	293.7
77.5	488	378	110	292.8
77.5	485	376	109	291.5
77.5	484	375	109	290.5
77.5	481	373	108	289.2
77.5	480	372	108	288.3
77.5	479	371	108	287.4
77.5	476	369	107	286.1
77.5	475	368	107	285.1
77.5	472	366	106	283.8
77.5	471	365	106	282.9
77.5	467	362	105	280.6
77.5	466	361	105	279.7
77.5	463	359	104	278.4
77.5	462	358	104	277.4
77.5	458	355	103	275.2
77.5	457	354	103	274.2
77.5	454	352	102	272.9
77.5	453	351	102	272.0
77.5	449	348	101	269.7
77.5	448	347	101	268.8
77.5	445	345	100	267.5
77.5	444	344	100	266.5

%E	M1	M2	DM	M*
77.5	440	341	99	264.3
77.5	436	338	98	262.0
77.5	435	337	98	261.1
77.5	432	335	97	259.8
77.5	431	334	97	258.8
77.5	427	331	96	256.6
77.5	426	330	96	255.6
77.5	423	328	95	254.3
77.5	422	327	95	253.4
77.5	418	324	94	251.1
77.5	417	323	94	250.2
77.5	414	321	93	248.9
77.5	413	320	93	247.9
77.5	409	317	92	245.7
77.5	408	316	92	244.7
77.5	405	314	91	243.4
77.5	404	313	91	242.5
77.5	400	310	90	240.3
77.5	396	307	89	238.0
77.5	395	306	89	237.1
77.5	391	303	88	234.8
77.5	387	300	87	232.6
77.5	386	299	87	231.6
77.5	383	297	86	230.3
77.5	382	296	86	229.4
77.5	378	293	85	227.1
77.5	377	292	85	226.2
77.5	374	290	84	224.9
77.5	373	289	84	223.9
77.5	369	286	83	221.7
77.5	365	283	82	219.4
77.5	364	282	82	218.5
77.5	360	279	81	216.2
77.5	356	276	80	214.0
77.5	355	275	80	213.0
77.5	351	272	79	210.8
77.5	347	269	78	208.5
77.5	346	268	78	207.6
77.5	342	265	77	205.3
77.5	338	262	76	203.1
77.5	334	259	75	200.8
77.5	333	258	75	199.9
77.5	329	255	74	197.6
77.5	325	252	73	195.4
77.5	324	251	73	194.4
77.5	320	248	72	192.2
77.5	316	245	71	190.0
77.5	315	244	71	189.0
77.5	311	241	70	186.8
77.5	307	238	69	184.5

%E	M1	M2	DM	M*
77.5	306	237	69	183.6
77.5	302	234	68	181.3
77.5	298	231	67	179.1
77.5	293	227	66	175.9
77.5	289	224	65	173.6
77.5	285	221	64	171.4
77.5	284	220	64	170.4
77.5	280	217	63	168.2
77.5	276	214	62	165.9
77.5	275	213	62	165.0
77.5	271	210	61	162.7
77.5	267	207	60	160.5
77.5	262	203	59	157.3
77.5	258	200	58	155.0
77.5	253	196	57	151.8
77.5	249	193	56	149.6
77.5	244	189	55	146.4
77.5	240	186	54	144.1
77.5	236	183	53	141.9
77.5	231	179	52	138.7
77.5	227	176	51	136.5
77.5	222	172	50	133.3
77.5	218	169	49	131.0
77.5	213	165	48	127.8
77.5	209	162	47	125.6
77.5	204	158	46	122.4
77.5	200	155	45	120.1
77.5	191	148	43	114.7
77.5	187	145	42	112.4
77.5	182	141	41	109.2
77.5	178	138	40	107.0
77.5	173	134	39	103.8
77.5	169	131	38	101.5
77.5	160	124	36	96.1
77.5	151	117	34	90.7
77.5	142	110	32	85.2
77.5	138	107	31	83.0
77.5	129	100	29	77.5
77.5	120	93	27	72.1
77.5	111	86	25	66.6
77.5	102	79	23	61.2
77.5	89	69	20	53.5
77.5	80	62	18	48.0
77.5	71	55	16	42.6
77.5	40	31	9	24.0
77.4	500	387	113	299.5
77.4	499	386	113	298.6
77.4	496	384	112	297.3
77.4	495	383	112	296.3
77.4	492	381	111	295.0

%E	M1	M2	DM	M*
77.4	491	380	111	294.1
77.4	487	377	110	291.8
77.4	486	376	110	290.9
77.4	483	374	109	289.6
77.4	482	373	109	288.6
77.4	478	370	108	286.4
77.4	477	369	108	285.5
77.4	474	367	107	284.2
77.4	473	366	107	283.2
77.4	470	364	106	281.9
77.4	469	363	106	281.0
77.4	468	362	106	280.0
77.4	465	360	105	278.7
77.4	464	359	105	277.8
77.4	461	357	104	276.5
77.4	460	356	104	275.5
77.4	456	353	103	273.3
77.4	455	352	103	272.3
77.4	452	350	102	271.0
77.4	451	349	102	270.1
77.4	447	346	101	267.8
77.4	446	345	101	266.9
77.4	443	343	100	265.6
77.4	442	342	100	264.6
77.4	439	340	99	263.3
77.4	438	339	99	262.4
77.4	434	336	98	260.1
77.4	433	335	98	259.2
77.4	430	333	97	257.9
77.4	429	332	97	256.9
77.4	425	329	96	254.7
77.4	424	328	96	253.7
77.4	421	326	95	252.4
77.4	420	325	95	251.5
77.4	416	322	94	249.2
77.4	412	319	93	247.0
77.4	411	318	93	246.0
77.4	407	315	92	243.8
77.4	403	312	91	241.5
77.4	402	311	91	240.6
77.4	399	309	90	239.3
77.4	398	308	90	238.4
77.4	394	305	89	236.1
77.4	393	304	89	235.2
77.4	390	302	88	233.9
77.4	389	301	88	232.9
77.4	385	298	87	230.7
77.4	381	295	86	228.4
77.4	380	294	86	227.5
77.4	376	291	85	225.2

%E	M1	M2	DM	M*
77.4	372	288	84	223.0
77.4	371	287	84	222.0
77.4	368	285	83	220.7
77.4	367	284	83	219.8
77.4	363	281	82	217.5
77.4	359	278	81	215.3
77.4	358	277	81	214.3
77.4	354	274	80	212.1
77.4	350	271	79	209.8
77.4	349	270	79	208.9
77.4	345	267	78	206.6
77.4	341	264	77	204.4
77.4	340	263	77	203.4
77.4	337	261	76	202.1
77.4	336	260	76	201.2
77.4	332	257	75	198.9
77.4	328	254	74	196.7
77.4	327	253	74	195.7
77.4	323	250	73	193.5
77.4	319	247	72	191.3
77.4	318	246	72	190.3
77.4	314	243	71	188.1
77.4	310	240	70	185.8
77.4	305	236	69	182.6
77.4	301	233	68	180.4
77.4	297	230	67	178.1
77.4	296	229	67	177.2
77.4	292	226	66	174.9
77.4	288	223	65	172.7
77.4	287	222	65	171.7
77.4	283	219	64	169.5
77.4	279	216	63	167.2
77.4	274	212	62	164.0
77.4	270	209	61	161.8
77.4	266	206	60	159.5
77.4	265	205	60	158.6
77.4	261	202	59	156.3
77.4	257	199	58	154.1
77.4	252	195	57	150.9
77.4	248	192	56	148.6
77.4	243	188	55	145.4
77.4	239	185	54	143.2
77.4	235	182	53	141.0
77.4	234	181	53	140.0
77.4	230	178	52	137.8
77.4	226	175	51	135.5
77.4	221	171	50	132.3
77.4	217	168	49	130.1
77.4	212	164	48	126.9
77.4	208	161	47	124.6

%E	M1	M2	DM	M*
77.4	199	154	45	119.2
77.4	195	151	44	116.9
77.4	190	147	43	113.7
77.4	186	144	42	111.5
77.4	177	137	40	106.0
77.4	168	130	38	100.6
77.4	164	127	37	98.3
77.4	159	123	36	95.2
77.4	155	120	35	92.9
77.4	146	113	33	87.5
77.4	137	106	31	82.0
77.4	133	103	30	79.8
77.4	124	96	28	74.3
77.4	115	89	26	68.9
77.4	106	82	24	63.4
77.4	93	72	21	55.7
77.4	84	65	19	50.3
77.4	62	48	14	37.2
77.4	53	41	12	31.7
77.4	31	24	7	18.6
77.3	498	385	113	297.6
77.3	497	384	113	296.7
77.3	494	382	112	295.4
77.3	493	381	112	294.4
77.3	490	379	111	293.1
77.3	489	378	111	292.2
77.3	488	377	111	291.2
77.3	485	375	110	289.9
77.3	484	374	110	289.0
77.3	481	372	109	287.7
77.3	480	371	109	286.8
77.3	476	368	108	284.5
77.3	475	367	108	283.6
77.3	472	365	107	282.3
77.3	471	364	107	281.3
77.3	467	361	106	279.1
77.3	466	360	106	278.1
77.3	463	358	105	276.8
77.3	462	357	105	275.9
77.3	459	355	104	274.6
77.3	458	354	104	273.6
77.3	454	351	103	271.4
77.3	453	350	103	270.4
77.3	450	348	102	269.1
77.3	449	347	102	268.2
77.3	445	344	101	265.9
77.3	444	343	101	265.0
77.3	441	341	100	263.7
77.3	440	340	100	262.7
77.3	437	338	99	261.4

%E	M1	M2	DM	M*
77.3	436	337	99	260.5
77.3	432	334	98	258.2
77.3	431	333	98	257.3
77.3	428	331	97	256.0
77.3	427	330	97	255.0
77.3	423	327	96	252.8
77.3	422	326	96	251.8
77.3	419	324	95	250.5
77.3	418	323	95	249.6
77.3	415	321	94	248.3
77.3	414	320	94	247.3
77.3	410	317	93	245.1
77.3	409	316	93	244.1
77.3	406	314	92	242.8
77.3	405	313	92	241.9
77.3	401	310	91	239.7
77.3	400	309	91	238.7
77.3	397	307	90	237.4
77.3	396	306	90	236.5
77.3	392	303	89	234.2
77.3	388	300	88	232.0
77.3	387	299	88	231.0
77.3	384	297	87	229.7
77.3	383	296	87	228.8
77.3	379	293	86	226.5
77.3	375	290	85	224.3
77.3	374	289	85	223.3
77.3	370	286	84	221.1
77.3	366	283	83	218.8
77.3	365	282	83	217.9
77.3	362	280	82	216.6
77.3	361	279	82	215.6
77.3	357	276	81	213.4
77.3	353	273	80	211.1
77.3	352	272	80	210.2
77.3	348	269	79	207.9
77.3	344	266	78	205.7
77.3	343	265	78	204.7
77.3	339	262	77	202.5
77.3	335	259	76	200.2
77.3	331	256	75	198.0
77.3	330	255	75	197.0
77.3	326	252	74	194.8
77.3	322	249	73	192.5
77.3	321	248	73	191.6
77.3	317	245	72	189.4
77.3	313	242	71	187.1
77.3	309	239	70	184.9
77.3	308	238	70	183.9
77.3	304	235	69	181.7

%E	M1	M2	DM	M*
77.3	300	232	68	179.4
77.3	299	231	68	178.5
77.3	295	228	67	176.2
77.3	291	225	66	174.0
77.3	286	221	65	170.8
77.3	282	218	64	168.5
77.3	278	215	63	166.3
77.3	277	214	63	165.3
77.3	273	211	62	163.1
77.3	269	208	61	160.8
77.3	264	204	60	157.6
77.3	260	201	59	155.4
77.3	256	198	58	153.1
77.3	255	197	58	152.2
77.3	251	194	57	149.9
77.3	247	191	56	147.7
77.3	242	187	55	144.5
77.3	238	184	54	142.3
77.3	233	180	53	139.1
77.3	229	177	52	136.8
77.3	225	174	51	134.6
77.3	220	170	50	131.4
77.3	216	167	49	129.1
77.3	211	163	48	125.9
77.3	207	160	47	123.7
77.3	203	157	46	121.4
77.3	198	153	45	118.2
77.3	194	150	44	116.0
77.3	185	143	42	110.5
77.3	181	140	41	108.3
77.3	176	136	40	105.1
77.3	172	133	39	102.8
77.3	163	126	37	97.4
77.3	154	119	35	92.0
77.3	150	116	34	89.7
77.3	141	109	32	84.3
77.3	132	102	30	78.8
77.3	128	99	29	76.6
77.3	119	92	27	71.1
77.3	110	85	25	65.7
77.3	97	75	22	58.0
77.3	88	68	20	52.5
77.3	75	58	17	44.9
77.3	66	51	15	39.4
77.3	44	34	10	26.3
77.3	22	17	5	13.1
77.2	500	386	114	298.0
77.2	499	385	114	297.0
77.2	496	383	113	295.7
77.2	495	382	113	294.8

%E	M1	M2	DM	M*
77.2	492	380	112	293.5
77.2	491	379	112	292.5
77.2	487	376	111	290.3
77.2	486	375	111	289.4
77.2	483	373	110	288.1
77.2	482	372	110	287.1
77.2	479	370	109	285.8
77.2	478	369	109	284.9
77.2	474	366	108	282.6
77.2	473	365	108	281.7
77.2	470	363	107	280.4
77.2	469	362	107	279.4
77.2	465	359	106	277.2
77.2	464	358	106	276.2
77.2	461	356	105	274.9
77.2	460	355	105	274.0
77.2	457	353	104	272.7
77.2	456	352	104	271.7
77.2	452	349	103	269.5
77.2	451	348	103	268.5
77.2	448	346	102	267.2
77.2	447	345	102	266.3
77.2	443	342	101	264.0
77.2	439	339	100	261.8
77.2	438	338	100	260.8
77.2	435	336	99	259.5
77.2	434	335	99	258.6
77.2	430	332	98	256.3
77.2	429	331	98	255.4
77.2	426	329	97	254.1
77.2	425	328	97	253.1
77.2	421	325	96	250.9
77.2	417	322	95	248.6
77.2	416	321	95	247.7
77.2	413	319	94	246.4
77.2	412	318	94	245.4
77.2	408	315	93	243.2
77.2	407	314	93	242.3
77.2	404	312	92	241.0
77.2	403	311	92	240.0
77.2	399	308	91	237.8
77.2	395	305	90	235.5
77.2	394	304	90	234.6
77.2	391	302	89	233.3
77.2	390	301	89	232.3
77.2	386	298	88	230.1
77.2	382	295	87	227.8
77.2	381	294	87	226.9
77.2	378	292	86	225.6
77.2	377	291	86	224.6

%E	M1	M2	DM	M*
77.2	373	288	85	222.4
77.2	372	287	85	221.4
77.2	369	285	84	220.1
77.2	368	284	84	219.2
77.2	364	281	83	216.9
77.2	360	278	82	214.7
77.2	359	277	82	213.7
77.2	356	275	81	212.4
77.2	355	274	81	211.5
77.2	351	271	80	209.2
77.2	347	268	79	207.0
77.2	346	267	79	206.0
77.2	342	264	78	203.8
77.2	338	261	77	201.5
77.2	337	260	77	200.6
77.2	334	258	76	199.3
77.2	333	257	76	198.3
77.2	329	254	75	196.1
77.2	325	251	74	193.8
77.2	324	250	74	192.9
77.2	320	247	73	190.7
77.2	316	244	72	188.4
77.2	312	241	71	186.2
77.2	311	240	71	185.2
77.2	307	237	70	183.0
77.2	303	234	69	180.7
77.2	302	233	69	179.8
77.2	298	230	68	177.5
77.2	294	227	67	175.3
77.2	290	224	66	173.0
77.2	289	223	66	172.1
77.2	285	220	65	169.8
77.2	281	217	64	167.6
77.2	276	213	63	164.4
77.2	272	210	62	162.1
77.2	268	207	61	159.9
77.2	267	206	61	158.9
77.2	263	203	60	156.7
77.2	259	200	59	154.4
77.2	254	196	58	151.2
77.2	250	193	57	149.0
77.2	246	190	56	146.7
77.2	241	186	55	143.6
77.2	237	183	54	141.3
77.2	232	179	53	138.1
77.2	228	176	52	135.9
77.2	224	173	51	133.6
77.2	219	169	50	130.4
77.2	215	166	49	128.2
77.2	206	159	47	122.7

%E	M1	M2	DM	M*
77.2	202	156	46	120.5
77.2	197	152	45	117.3
77.2	193	149	44	115.0
77.2	189	146	43	112.8
77.2	184	142	42	109.6
77.2	180	139	41	107.3
77.2	171	132	39	101.9
77.2	167	129	38	99.6
77.2	162	125	37	96.5
77.2	158	122	36	94.2
77.2	149	115	34	88.8
77.2	145	112	33	86.5
77.2	136	105	31	81.1
77.2	127	98	29	75.6
77.2	123	95	28	73.4
77.2	114	88	26	67.9
77.2	101	78	23	60.2
77.2	92	71	21	54.8
77.2	79	61	18	47.1
77.2	57	44	13	34.0
77.1	498	384	114	296.1
77.1	497	383	114	295.1
77.1	494	381	113	293.8
77.1	493	380	113	292.9
77.1	490	378	112	291.6
77.1	489	377	112	290.7
77.1	485	374	111	288.4
77.1	484	373	111	287.5
77.1	481	371	110	286.2
77.1	480	370	110	285.2
77.1	477	368	109	283.9
77.1	476	367	109	283.0
77.1	475	366	109	282.0
77.1	472	364	108	280.7
77.1	471	363	108	279.8
77.1	468	361	107	278.5
77.1	467	360	107	277.5
77.1	463	357	106	275.3
77.1	462	356	106	274.3
77.1	459	354	105	273.0
77.1	458	353	105	272.1
77.1	455	351	104	270.8
77.1	454	350	104	269.8
77.1	450	347	103	267.6
77.1	449	346	103	266.6
77.1	446	344	102	265.3
77.1	445	343	102	264.4
77.1	442	341	101	263.1
77.1	441	340	101	262.1
77.1	437	337	100	259.9

%E	M1	M2	DM	M*
77.1	436	336	100	258.9
77.1	433	334	99	257.6
77.1	432	333	99	256.7
77.1	428	330	98	254.4
77.1	424	327	97	252.2
77.1	423	326	97	251.2
77.1	420	324	96	249.9
77.1	419	323	96	249.0
77.1	415	320	95	246.7
77.1	414	319	95	245.8
77.1	411	317	94	244.5
77.1	410	316	94	243.6
77.1	406	313	93	241.3
77.1	402	310	92	239.1
77.1	401	309	92	238.1
77.1	398	307	91	236.8
77.1	397	306	91	235.9
77.1	393	303	90	233.6
77.1	389	300	89	231.4
77.1	388	299	89	230.4
77.1	385	297	88	229.1
77.1	384	296	88	228.2
77.1	380	293	87	225.9
77.1	376	290	86	223.7
77.1	375	289	86	222.7
77.1	371	286	85	220.5
77.1	367	283	84	218.2
77.1	363	280	83	216.0
77.1	362	279	83	215.0
77.1	358	276	82	212.8
77.1	354	273	81	210.5
77.1	353	272	81	209.6
77.1	350	270	80	208.3
77.1	349	269	80	207.3
77.1	345	266	79	205.1
77.1	341	263	78	202.8
77.1	340	262	78	201.9
77.1	336	259	77	199.6
77.1	332	256	76	197.4
77.1	328	253	75	195.1
77.1	327	252	75	194.2
77.1	323	249	74	192.0
77.1	319	246	73	189.7
77.1	315	243	72	187.5
77.1	314	242	72	186.5
77.1	310	239	71	184.3
77.1	306	236	70	182.0
77.1	301	232	69	178.8
77.1	297	229	68	176.6
77.1	293	226	67	174.3

%E	M1	M2	DM	M*
77.1	292	225	67	173.4
77.1	288	222	66	171.1
77.1	284	219	65	168.9
77.1	280	216	64	166.6
77.1	279	215	64	165.7
77.1	275	212	63	163.4
77.1	271	209	62	161.2
77.1	266	205	61	158.0
77.1	262	202	60	155.7
77.1	258	199	59	153.5
77.1	253	195	58	150.3
77.1	249	192	57	148.0
77.1	245	189	56	145.8
77.1	240	185	55	142.6
77.1	236	182	54	140.4
77.1	231	178	53	137.2
77.1	227	175	52	134.9
77.1	223	172	51	132.7
77.1	218	168	50	129.5
77.1	214	165	49	127.2
77.1	210	162	48	125.0
77.1	205	158	47	121.8
77.1	201	155	46	119.5
77.1	192	148.	44	114.1
77.1	188	145	43	111.8
77.1	179	138	41	106.4
77.1	175	135	40	104.1
77.1	170	131	39	100.9
77.1	166	128	38	98.7
77.1	157	121	36	93.3
77.1	153	118	35	91.0
77.1	144	111	33	85.6
77.1	140	108	32	83.3
77.1	131	101	30	77.9
77.1	118	91	27	70.2
77.1	109	84	25	64.7
77.1	105	81	24	62.5
77.1	96	74	22	57.0
77.1	83	64	19	49.3
77.1	70	54	16	41.7
77.1	48	37	11	28.5
77.1	35	27	8	20.8
77.0	500	385	115	296.4
77.0	499	384	115	295.5
77.0	496	382	114	294.2
77.0	495	381	114	293.3
77.0	492	379	113	292.0
77.0	491	378	113	291.0
77.0	488	376	112	289.7
77.0	487	375	112	288.8

%E	M1	M2	DM	M*
77.0	486	374	112	287.8
77.0	483	372	111	286.5
77.0	482	371	111	285.6
77.0	479	369	110	284.3
77.0	478	368	110	283.3
77.0	474	365	109	281.1
77.0	473	364	109	280.1
77.0	470	362	108	278.8
77.0	469	361	108	277.9
77.0	466	359	107	276.6
77.0	465	358	107	275.6
77.0	461	355	106	273.4
77.0	460	354	106	272.4
77.0	457	352	105	271.1
77.0	456	351	105	270.2
77.0	453	349	104	268.9
77.0	452	348	104	267.9
77.0	448	345	103	265.7
77.0	447	344	103	264.7
77.0	444	342	102	263.4
77.0	443	341	102	262.5
77.0	440	339	101	261.2
77.0	439	338	101	260.2
77.0	435	335	100	258.0
77.0	434	334	100	257.0
77.0	431	332	99	255.7
77.0	430	331	99	254.8
77.0	427	329	98	253.5
77.0	426	328	98	252.5
77.0	422	325	97	250.3
77.0	421	324	97	249.3
77.0	418	322	96	248.0
77.0	417	321	96	247.1
77.0	413	318	95	244.9
77.0	409	315	94	242.6
77.0	408	314	94	241.7
77.0	405	312	93	240.4
77.0	404	311	93	239.4
77.0	400	308	92	237.2
77.0	396	305	91	234.9
77.0	395	304	91	234.0
77.0	392	302	90	232.7
77.0	391	301	90	231.7
77.0	387	298	89	229.5
77.0	383	295	88	227.2
77.0	382	294	88	226.3
77.0	379	292	87	225.0
77.0	378	291	87	224.0
77.0	374	288	86	221.8
77.0	370	285	85	219.5

%E	M1	M2	DM	M*
77.0	369	284	85	218.6
77.0	366	282	84	217.3
77.0	365	281	84	216.3
77.0	361	278	83	214.1
77.0	357	275	82	211.8
77.0	356	274	82	210.9
77.0	352	271	81	208.6
77.0	348	268	80	206.4
77.0	344	265	79	204.1
77.0	343	264	79	203.2
77.0	339	261	78	200.9
77.0	335	258	77	198.7
77.0	331	255	76	196.5
77.0	330	254	76	195.5
77.0	326	251	75	193.3
77.0	322	248	74	191.0
77.0	318	245	73	188.8
77.0	317	244	73	187.8
77.0	313	241	72	185.6
77.0	309	238	71	183.3
77.0	305	235	70	181.1
77.0	304	234	70	180.1
77.0	300	231	69	177.9
77.0	296	228	68	175.6
77.0	291	224	67	172.4
77.0	287	221	66	170.2
77.0	283	218	65	167.9
77.0	282	217	65	167.0
77.0	278	214	64	164.7
77.0	274	211	63	162.5
77.0	270	208	62	160.2
77.0	269	207	62	159.3
77.0	265	204	61	157.0
77.0	261	201	60	154.8
77.0	257	198	59	152.5
77.0	256	197	59	151.6
77.0	252	194	58	149.3
77.0	248	191	57	147.1
77.0	244	188	56	144.9
77.0	243	187	56	143.9
77.0	239	184	55	141.7
77.0	235	181	54	139.4
77.0	230	177	53	136.2
77.0	226	174	52	134.0
77.0	222	171	51	131.7
77.0	217	167	50	128.5
77.0	213	164	49	126.3
77.0	209	161	48	124.0
77.0	204	157	47	120.8
77.0	200	154	46	118.6

%E	M1	M2	DM	M*
77.0	196	151	45	116.3
77.0	191	147	44	113.1
77.0	187	144	43	110.9
77.0	183	141	42	108.6
77.0	178	137	41	105.4
77.0	174	134	40	103.2
77.0	165	127	38	97.8
77.0	161	124	37	95.5
77.0	152	117	35	90.1
77.0	148	114	34	87.8
77.0	139	107	32	82.4
77.0	135	104	31	80.1
77.0	126	97	29	74.7
77.0	122	94	28	72.4
77.0	113	87	26	67.0
77.0	100	77	23	59.3
77.0	87	67	20	51.6
77.0	74	57	17	43.9
77.0	61	47	14	36.2
76.9	498	383	115	294.6
76.9	497	382	115	293.6
76.9	494	380	114	292.3
76.9	493	379	114	291.4
76.9	490	377	113	290.1
76.9	489	376	113	289.1
76.9	485	373	112	286.9
76.9	484	372	112	285.9
76.9	481	370	111	284.6
76.9	480	369	111	283.7
76.9	477	367	110	282.4
76.9	476	366	110	281.4
76.9	472	363	109	279.2
76.9	471	362	109	278.2
76.9	468	360	108	276.9
76.9	467	359	108	276.0
76.9	464	357	107	274.7
76.9	463	356	107	273.7
76.9	459	353	106	271.5
76.9	458	352	106	270.5
76.9	455	350	105	269.2
76.9	454	349	105	268.3
76.9	451	347	104	267.0
76.9	450	346	104	266.0
76.9	446	343	103	263.8
76.9	445	342	103	262.8
76.9	442	340	102	261.5
76.9	441	339	102	260.6
76.9	438	337	101	259.3
76.9	437	336	101	258.3
76.9	433	333	100	256.1

%E	M1	M2	DM	M*
76.9	432	332	100	255.1
76.9	429	330	99	253.8
76.9	428	329	99	252.9
76.9	425	327	98	251.6
76.9	424	326	98	250.7
76.9	420	323	97	248.4
76.9	419	322	97	247.5
76.9	416	320	96	246.2
76.9	415	319	96	245.2
76.9	412	317	95	243.9
76.9	411	316	95	243.0
76.9	407	313	94	240.7
76.9	403	310	93	238.5
76.9	402	309	93	237.5
76.9	399	307	92	236.2
76.9	398	306	92	235.3
76.9	394	303	91	233.0
76.9	390	300	90	230.8
76.9	389	299	90	229.8
76.9	386	297	89	228.5
76.9	385	296	89	227.6
76.9	381	293	88	225.3
76.9	377	290	87	223.1
76.9	376	289	87	222.1
76.9	373	287	86	220.8
76.9	372	286	86	219.9
76.9	368	283	85	217.6
76.9	364	280	84	215.4
76.9	363	279	84	214.4
76.9	360	277	83	213.1
76.9	359	276	83	212.2
76.9	355	273	82	209.9
76.9	351	270	81	207.7
76.9	350	269	81	206.7
76.9	347	267	80	205.4
76.9	346	266	80	204.5
76.9	342	263	79	202.2
76.9	338	260	78	200.0
76.9	337	259	78	199.1
76.9	334	257	77	197.8
76.9	333	256	77	196.8
76.9	329	253	76	194.6
76.9	325	250	75	192.3
76.9	324	249	75	191.4
76.9	321	247	74	190.1
76.9	320	246	74	189.1
76.9	316	243	73	186.9
76.9	312	240	72	184.6
76.9	308	237	71	182.4
76.9	307	236	71	181.4

%E	M1	M2	DM	M*
76.9	303	233	70	179.2
76.9	299	230	69	176.9
76.9	295	227	68	174.7
76.9	294	226	68	173.7
76.9	290	223	67	171.5
76.9	286	220	66	169.2
76.9	281	216	65	166.0
76.9	277	213	64	163.8
76.9	273	210	63	161.5
76.9	268	206	62	158.3
76.9	264	203	61	156.1
76.9	260	200	60	153.8
76.9	255	196	59	150.7
76.9	251	193	58	148.4
76.9	247	190	57	146.2
76.9	242	186	56	143.0
76.9	238	183	55	140.7
76.9	234	180	54	138.5
76.9	229	176	53	135.3
76.9	225	173	52	133.0
76.9	221	170	51	130.8
76.9	216	166	50	127.6
76.9	212	163	49	125.3
76.9	208	160	48	123.1
76.9	199	153	46	117.6
76.9	195	150	45	115.4
76.9	186	143	43	109.9
76.9	182	140	42	107.7
76.9	173	133	40	102.2
76.9	169	130	39	100.0
76.9	160	123	37	94.6
76.9	156	120	36	92.3
76.9	147	113	34	86.9
76.9	143	110	33	84.6
76.9	134	103	31	79.2
76.9	130	100	30	76.9
76.9	121	93	28	71.5
76.9	117	90	27	69.2
76.9	108	83	25	63.8
76.9	104	80	24	61.5
76.9	91	70	21	53.8
76.9	78	60	18	46.2
76.9	65	50	15	38.5
76.9	52	40	12	30.8
76.9	39	30	9	23.1
76.9	26	20	6	15.4
76.9	13	10	3	7.7
76.8	500	384	116	294.9
76.8	499	383	116	294.0
76.8	496	381	115	292.7

%E	M1	M2	DM	M*
76.8	495	380	115	291.7
76.8	492	378	114	290.4
76.8	491	377	114	289.5
76.8	488	375	113	288.2
76.8	487	374	113	287.2
76.8	483	371	112	285.0
76.8	482	370	112	284.0
76.8	479	368	111	282.7
76.8	478	367	111	281.8
76.8	475	365	110	280.5
76.8	474	364	110	279.5
76.8	470	361	109	277.3
76.8	469	360	109	276.3
76.8	466	358	108	275.0
76.8	465	357	108	274.1
76.8	462	355	107	272.8
76.8	461	354	107	271.8
76.8	457	351	106	269.6
76.8	456	350	106	268.6
76.8	453	348	105	267.3
76.8	452	347	105	266.4
76.8	449	345	104	265.1
76.8	448	344	104	264.1
76.8	444	341	103	261.9
76.8	440	338	102	259.6
76.8	439	337	102	258.7
76.8	436	335	101	257.4
76.8	435	334	101	256.5
76.8	431	331	100	254.2
76.8	427	328	99	252.0
76.8	426	327	99	251.0
76.8	423	325	98	249.7
76.8	422	324	98	248.8
76.8	418	321	97	246.5
76.8	414	318	96	244.3
76.8	413	317	96	243.3
76.8	410	315	95	242.0
76.8	409	314	95	241.1
76.8	406	312	94	239.8
76.8	405	311	94	238.8
76.8	401	308	93	236.6
76.8	400	307	93	235.6
76.8	397	305	92	234.3
76.8	396	304	92	233.4
76.8	393	302	91	232.1
76.8	392	301	91	231.1
76.8	388	298	90	228.9
76.8	384	295	89	226.6
76.8	383	294	89	225.7
76.8	380	292	88	224.4

%E	M1	M2	DM	M*
76.8	379	291	88	223.4
76.8	375	288	87	221.2
76.8	371	285	86	218.9
76.8	370	284	86	218.0
76.8	367	282	85	216.7
76.8	366	281	85	215.7
76.8	362	278	84	213.5
76.8	358	275	83	211.2
76.8	357	274	83	210.3
76.8	354	272	82	209.0
76.8	353	271	82	208.0
76.8	349	268	81	205.8
76.8	345	265	80	203.6
76.8	341	262	79	201.3
76.8	340	261	79	200.4
76.8	336	258	78	198.1
76.8	332	255	77	195.9
76.8	328	252	76	193.6
76.8	327	251	76	192.7
76.8	323	248	75	190.4
76.8	319	245	74	188.2
76.8	315	242	73	185.9
76.8	314	241	73	185.0
76.8	311	239	72	183.7
76.8	310	238	72	182.7
76.8	306	235	71	180.5
76.8	302	232	70	178.2
76.8	298	229	69	176.0
76.8	297	228	69	175.0
76.8	293	225	68	172.8
76.8	289	222	67	170.5
76.8	285	219	66	168.3
76.8	284	218	66	167.3
76.8	280	215	65	165.1
76.8	276	212	64	162.8
76.8	272	209	63	160.6
76.8	271	208	63	159.6
76.8	267	205	62	157.4
76.8	263	202	61	155.1
76.8	259	199	60	152.9
76.8	254	195	59	149.7
76.8	250	192	58	147.5
76.8	246	189	57	145.2
76.8	241	185	56	142.0
76.8	237	182	55	139.8
76.8	233	179	54	137.5
76.8	228	175	53	134.3
76.8	224	172	52	132.1
76.8	220	169	51	129.8
76.8	211	162	49	124.4

%E	M1	M2	DM	M*
76.8	207	159	48	122.1
76.8	203	156	47	119.9
76.8	198	152	46	116.7
76.8	194	149	45	114.4
76.8	190	146	44	112.2
76.8	185	142	43	109.0
76.8	181	139	42	106.7
76.8	177	136	41	104.5
76.8	168	129	39	99.1
76.8	164	126	38	96.8
76.8	155	119	36	91.4
76.8	151	116	35	89.1
76.8	142	109	33	83.7
76.8	138	106	32	81.4
76.8	125	96	29	73.7
76.8	112	86	26	66.0
76.8	99	76	23	58.3
76.8	95	73	22	56.1
76.8	82	63	19	48.4
76.8	69	53	16	40.7
76.8	56	43	13	33.0
76.7	498	382	116	293.0
76.7	497	381	116	292.1
76.7	494	379	115	290.8
76.7	493	378	115	289.8
76.7	490	376	114	288.5
76.7	489	375	114	287.6
76.7	486	373	113	286.3
76.7	485	372	113	285.3
76.7	484	371	113	284.4
76.7	481	369	112	283.1
76.7	480	368	112	282.1
76.7	477	366	111	280.8
76.7	476	365	111	279.9
76.7	473	363	110	278.6
76.7	472	362	110	277.6
76.7	468	359	109	275.4
76.7	467	358	109	274.4
76.7	464	356	108	273.1
76.7	463	355	108	272.2
76.7	460	353	107	270.9
76.7	459	352	107	269.9
76.7	455	349	106	267.7
76.7	454	348	106	266.7
76.7	451	346	105	265.4
76.7	450	345	105	264.5
76.7	447	343	104	263.2
76.7	446	342	104	262.3
76.7	443	340	103	260.9
76.7	442	339	103	260.0

%E	M1	M2	DM	M*
76.7	438	336	102	257.8
76.7	437	335	102	256.8
76.7	434	333	101	255.5
76.7	433	332	101	254.6
76.7	430	330	100	253.3
76.7	429	329	100	252.3
76.7	425	326	99	250.1
76.7	424	325	99	249.1
76.7	421	323	98	247.8
76.7	420	322	98	246.9
76.7	417	320	97	245.6
76.7	416	319	97	244.6
76.7	412	316	96	242.4
76.7	408	313	95	240.1
76.7	407	312	95	239.2
76.7	404	310	94	237.9
76.7	403	309	94	236.9
76.7	399	306	93	234.7
76.7	395	303	92	232.4
76.7	394	302	92	231.5
76.7	391	300	91	230.2
76.7	390	299	91	229.2
76.7	387	297	90	227.9
76.7	386	296	90	227.0
76.7	382	293	89	224.7
76.7	378	290	88	222.5
76.7	377	289	88	221.5
76.7	374	287	87	220.2
76.7	373	286	87	219.3
76.7	369	283	86	217.0
76.7	365	280	85	214.8
76.7	361	277	84	212.5
76.7	360	276	84	211.6
76.7	356	273	83	209.4
76.7	352	270	82	207.1
76.7	348	267	81	204.9
76.7	347	266	81	203.9
76.7	344	264	80	202.6
76.7	343	263	80	201.7
76.7	339	260	79	199.4
76.7	335	257	78	197.2
76.7	331	254	77	194.9
76.7	330	253	77	194.0
76.7	326	250	76	191.7
76.7	322	247	75	189.5
76.7	318	244	74	187.2
76.7	317	243	74	186.3
76.7	313	240	73	184.0
76.7	309	237	72	181.8
76.7	305	234	71	179.5

%E	M1	M2	DM	M*
76.7	301	231	70	177.3
76.7	300	230	70	176.3
76.7	296	227	69	174.1
76.7	292	224	68	171.8
76.7	288	221	67	169.6
76.7	287	220	67	168.6
76.7	283	217	66	166.4
76.7	279	214	65	164.1
76.7	275	211	64	161.9
76.7	270	207	63	158.7
76.7	266	204	62	156.5
76.7	262	201	61	154.2
76.7	258	198	60	152.0
76.7	257	197	60	151.0
76.7	253	194	59	148.8
76.7	249	191	58	146.5
76.7	245	188	57	144.3
76.7	240	184	56	141.1
76.7	236	181	55	138.8
76.7	232	178	54	136.6
76.7	227	174	53	133.4
76.7	223	171	52	131.1
76.7	219	168	51	128.9
76.7	215	165	50	126.6
76.7	210	161	49	123.4
76.7	206	158	48	121.2
76.7	202	155	47	118.9
76.7	197	151	46	115.7
76.7	193	148	45	113.5
76.7	189	145	44	111.2
76.7	180	138	42	105.8
76.7	176	135	41	103.6
76.7	172	132	40	101.3
76.7	163	125	38	95.9
76.7	159	122	37	93.6
76.7	150	115	35	88.2
76.7	146	112	34	85.9
76.7	133	102	31	78.2
76.7	129	99	30	76.0
76.7	120	92	28	70.5
76.7	116	89	27	68.3
76.7	103	79	24	60.6
76.7	90	69	21	52.9
76.7	86	66	20	50.7
76.7	73	56	17	43.0
76.7	60	46	14	35.3
76.7	43	33	10	25.3
76.7	30	23	7	17.6
76.6	500	383	117	293.4
76.6	499	382	117	292.4

%E	M1	M2	DM	M*
76.6	496	380	116	291.1
76.6	495	379	116	290.2
76.6	492	377	115	288.9
76.6	491	376	115	287.9
76.6	488	374	114	286.6
76.6	487	373	114	285.7
76.6	483	370	113	283.4
76.6	482	369	113	282.5
76.6	479	367	112	281.2
76.6	478	366	112	280.2
76.6	475	364	111	278.9
76.6	474	363	111	278.0
76.6	471	361	110	276.7
76.6	470	360	110	275.7
76.6	466	357	109	273.5
76.6	465	356	109	272.6
76.6	462	354	108	271.2
76.6	461	353	108	270.3
76.6	458	351	107	269.0
76.6	457	350	107	268.1
76.6	453	347	106	265.8
76.6	449	344	105	263.6
76.6	448	343	105	262.6
76.6	445	341	104	261.3
76.6	444	340	104	260.4
76.6	441	338	103	259.1
76.6	440	337	103	258.1
76.6	436	334	102	255.9
76.6	435	333	102	254.9
76.6	432	331	101	253.6
76.6	431	330	101	252.7
76.6	428	328	100	251.4
76.6	427	327	100	250.4
76.6	423	324	99	248.2
76.6	419	321	98	245.9
76.6	418	320	98	245.0
76.6	415	318	97	243.7
76.6	414	317	97	242.7
76.6	411	315	96	241.4
76.6	410	314	96	240.5
76.6	406	311	95	238.2
76.6	402	308	94	236.0
76.6	401	307	94	235.0
76.6	398	305	93	233.7
76.6	397	304	93	232.8
76.6	393	301	92	230.5
76.6	389	298	91	228.3
76.6	385	295	90	226.0
76.6	384	294	90	225.1
76.6	381	292	89	223.8

%E	M1	M2	DM	M*
76.6	380	291	89	222.8
76.6	376	288	88	220.6
76.6	372	285	87	218.3
76.6	371	284	87	217.4
76.6	368	282	86	216.1
76.6	367	281	86	215.2
76.6	364	279	85	213.8
76.6	363	278	85	212.9
76.6	359	275	84	210.7
76.6	355	272	83	208.4
76.6	354	271	83	207.5
76.6	351	269	82	206.2
76.6	350	268	82	205.2
76.6	346	265	81	203.0
76.6	342	262	80	200.7
76.6	338	259	79	198.5
76.6	337	258	79	197.5
76.6	334	256	78	196.2
76.6	333	255	78	195.3
76.6	329	252	77	193.0
76.6	325	249	76	190.8
76.6	321	246	75	188.5
76.6	320	245	75	187.6
76.6	316	242	74	185.3
76.6	312	239	73	183.1
76.6	308	236	72	180.8
76.6	304	233	71	178.6
76.6	303	232	71	177.6
76.6	299	229	70	175.4
76.6	295	226	69	173.1
76.6	291	223	68	170.9
76.6	290	222	68	169.9
76.6	286	219	67	167.7
76.6	282	216	66	165.4
76.6	278	213	65	163.2
76.6	274	210	64	160.9
76.6	273	209	64	160.0
76.6	269	206	63	157.8
76.6	265	203	62	155.5
76.6	261	200	61	153.3
76.6	256	196	60	150.1
76.6	252	193	59	147.8
76.6	248	190	58	145.6
76.6	244	187	57	143.3
76.6	239	183	56	140.1
76.6	235	180	55	137.9
76.6	231	177	54	135.6
76.6	222	170	52	130.2
76.6	218	167	51	127.9
76.6	214	164	50	125.7

%E	M1	M2	DM	M*
76.6	209	160	49	122.5
76.6	205	157	48	120.2
76.6	201	154	47	118.0
76.6	192	147	45	112.5
76.6	188	144	44	110.3
76.6	184	141	43	108.0
76.6	175	134	41	102.6
76.6	171	131	40	100.4
76.6	167	128	39	98.1
76.6	158	121	37	92.7
76.6	154	118	36	90.4
76.6	145	111	34	85.0
76.6	141	108	33	82.7
76.6	137	105	32	80.5
76.6	128	98	30	75.0
76.6	124	95	29	72.8
76.6	111	85	26	65.1
76.6	107	82	25	62.8
76.6	94	72	22	55.1
76.6	77	59	18	45.2
76.6	64	49	15	37.5
76.6	47	36	11	27.6
76.5	498	381	117	291.5
76.5	497	380	117	290.5
76.5	494	378	116	289.2
76.5	493	377	116	288.3
76.5	490	375	115	287.0
76.5	489	374	115	286.0
76.5	486	372	114	284.7
76.5	485	371	114	283.8
76.5	481	368	113	281.5
76.5	480	367	113	280.6
76.5	477	365	112	279.3
76.5	476	364	112	278.4
76.5	473	362	111	277.0
76.5	472	361	111	276.1
76.5	469	359	110	274.8
76.5	468	358	110	273.9
76.5	464	355	109	271.6
76.5	463	354	109	270.7
76.5	460	352	108	269.4
76.5	459	351	108	268.4
76.5	456	349	107	267.1
76.5	455	348	107	266.2
76.5	452	346	106	264.9
76.5	451	345	106	263.9
76.5	447	342	105	261.7
76.5	446	341	105	260.7
76.5	443	339	104	259.4
76.5	442	338	104	258.5

%E	M1	M2	DM	M*
76.5	439	336	103	257.2
76.5	438	335	103	256.2
76.5	434	332	102	254.0
76.5	430	329	101	251.7
76.5	429	328	101	250.8
76.5	426	326	100	249.5
76.5	425	325	100	248.5
76.5	422	323	99	247.2
76.5	421	322	99	246.3
76.5	417	319	98	244.0
76.5	413	316	97	241.8
76.5	412	315	97	240.8
76.5	409	313	96	239.5
76.5	408	312	96	238.6
76.5	405	310	95	237.3
76.5	404	309	95	236.3
76.5	400	306	94	234.1
76.5	396	303	93	231.8
76.5	395	302	93	230.9
76.5	392	300	92	229.6
76.5	391	299	92	228.6
76.5	388	297	91	227.3
76.5	387	296	91	226.4
76.5	383	293	90	224.1
76.5	379	290	89	221.9
76.5	378	289	89	221.0
76.5	375	287	88	219.7
76.5	374	286	88	218.7
76.5	370	283	87	216.5
76.5	366	280	86	214.2
76.5	362	277	85	212.0
76.5	361	276	85	211.0
76.5	358	274	84	209.7
76.5	357	273	84	208.8
76.5	353	270	83	206.5
76.5	349	267	82	204.3
76.5	345	264	81	202.0
76.5	344	263	81	201.1
76.5	341	261	80	199.8
76.5	340	260	80	198.8
76.5	336	257	79	196.6
76.5	332	254	78	194.3
76.5	328	251	77	192.1
76.5	327	250	77	191.1
76.5	324	248	76	189.8
76.5	323	247	76	188.9
76.5	319	244	75	186.6
76.5	315	241	74	184.4
76.5	311	238	73	182.1
76.5	310	237	73	181.2

%E	M1	M2	DM	M*
76.5	307	235	72	179.9
76.5	306	234	72	178.9
76.5	302	231	71	176.7
76.5	298	228	70	174.4
76.5	294	225	69	172.2
76.5	293	224	69	171.2
76.5	289	221	68	169.0
76.5	285	218	67	166.8
76.5	281	215	66	164.5
76.5	277	212	65	162.3
76.5	272	208	64	159.1
76.5	268	205	63	156.8
76.5	264	202	62	154.6
76.5	260	199	61	152.3
76.5	255	195	60	149.1
76.5	251	192	59	146.9
76.5	247	189	58	144.6
76.5	243	186	57	142.4
76.5	238	182	56	139.2
76.5	234	179	55	136.9
76.5	230	176	54	134.7
76.5	226	173	53	132.4
76.5	221	169	52	129.2
76.5	217	166	51	127.0
76.5	213	163	50	124.7
76.5	204	156	48	119.3
76.5	200	153	47	117.0
76.5	196	150	46	114.8
76.5	187	143	44	109.4
76.5	183	140	43	107.1
76.5	179	137	42	104.9
76.5	170	130	40	99.4
76.5	166	127	39	97.2
76.5	162	124	38	94.9
76.5	153	117	36	89.5
76.5	149	114	35	87.2
76.5	136	104	32	79.5
76.5	132	101	31	77.3
76.5	119	91	28	69.6
76.5	115	88	27	67.3
76.5	102	78	24	59.6
76.5	98	75	23	57.4
76.5	85	65	20	49.7
76.5	81	62	19	47.5
76.5	68	52	16	39.8
76.5	51	39	12	29.8
76.5	34	26	8	19.9
76.5	17	13	4	9.9
76.4	500	382	118	291.8
76.4	499	381	118	290.9

%E	M1	M2	DM	M*
76.4	496	379	117	289.6
76.4	495	378	117	288.7
76.4	492	376	116	287.3
76.4	491	375	116	286.4
76.4	488	373	115	285.1
76.4	487	372	115	284.2
76.4	484	370	114	282.9
76.4	483	369	114	281.9
76.4	479	366	113	279.7
76.4	478	365	113	278.7
76.4	475	363	112	277.4
76.4	474	362	112	276.5
76.4	471	360	111	275.2
76.4	470	359	111	274.2
76.4	467	357	110	272.9
76.4	466	356	110	272.0
76.4	462	353	109	269.7
76.4	461	352	109	268.8
76.4	458	350	108	267.5
76.4	457	349	108	266.5
76.4	454	347	107	265.2
76.4	453	346	107	264.3
76.4	450	344	106	263.0
76.4	449	343	106	262.0
76.4	445	340	105	259.8
76.4	444	339	105	258.8
76.4	441	337	104	257.5
76.4	440	336	104	256.6
76.4	437	334	103	255.3
76.4	436	333	103	254.3
76.4	433	331	102	253.0
76.4	432	330	102	252.1
76.4	428	327	101	249.8
76.4	424	324	100	247.6
76.4	423	323	100	246.6
76.4	420	321	99	245.3
76.4	419	320	99	244.4
76.4	416	318	98	243.1
76.4	415	317	98	242.1
76.4	411	314	97	239.9
76.4	407	311	96	237.6
76.4	406	310	96	236.7
76.4	403	308	95	235.4
76.4	402	307	95	234.5
76.4	399	305	94	233.1
76.4	398	304	94	232.2
76.4	394	301	93	230.0
76.4	390	298	92	227.7
76.4	389	297	92	226.8
76.4	386	295	91	225.5
76.4	385	294	91	224.5
76.4	382	292	90	223.2
76.4	381	291	90	222.3
76.4	377	288	89	220.0
76.4	373	285	88	217.8
76.4	369	282	87	215.5
76.4	368	281	87	214.6
76.4	365	279	86	213.3
76.4	364	278	86	212.3
76.4	360	275	85	210.1
76.4	356	272	84	207.8
76.4	352	269	83	205.6
76.4	351	268	83	204.6
76.4	348	266	82	203.3
76.4	347	265	82	202.4
76.4	343	262	81	200.1
76.4	339	259	80	197.9
76.4	335	256	79	195.6
76.4	331	253	78	193.4
76.4	330	252	78	192.4
76.4	326	249	77	190.2
76.4	322	246	76	187.9
76.4	318	243	75	185.7
76.4	314	240	74	183.4
76.4	313	239	74	182.5
76.4	309	236	73	180.2
76.4	305	233	72	178.0
76.4	301	230	71	175.7
76.4	297	227	70	173.5
76.4	296	226	70	172.6
76.4	292	223	69	170.3
76.4	288	220	68	168.1
76.4	284	217	67	165.8
76.4	280	214	66	163.6
76.4	276	211	65	161.3
76.4	275	210	65	160.4
76.4	271	207	64	158.1
76.4	267	204	63	155.9
76.4	263	201	62	153.6
76.4	259	198	61	151.4
76.4	258	197	61	150.4
76.4	254	194	60	148.2
76.4	250	191	59	145.9
76.4	246	188	58	143.7
76.4	242	185	57	141.4
76.4	237	181	56	138.2
76.4	233	178	55	136.0
76.4	229	175	54	133.7
76.4	225	172	53	131.5
76.4	220	168	52	128.3
76.4	216	165	51	126.0
76.4	212	162	50	123.8
76.4	208	159	49	121.5
76.4	203	155	48	118.3
76.4	199	152	47	116.1
76.4	195	149	46	113.9
76.4	191	146	45	111.6
76.4	182	139	43	106.2
76.4	178	136	42	103.9
76.4	174	133	41	101.7
76.4	165	126	39	96.2
76.4	161	123	38	94.0
76.4	157	120	37	91.7
76.4	148	113	35	86.3
76.4	144	110	34	84.0
76.4	140	107	33	81.8
76.4	127	97	30	74.1
76.4	123	94	29	71.8
76.4	110	84	26	64.1
76.4	106	81	25	61.9
76.4	89	68	21	52.0
76.4	72	55	17	42.0
76.4	55	42	13	32.1
76.3	498	380	118	290.0
76.3	497	379	118	289.0
76.3	494	377	117	287.7
76.3	493	376	117	286.8
76.3	490	374	116	285.5
76.3	489	373	116	284.5
76.3	486	371	115	283.2
76.3	485	370	115	282.3
76.3	482	368	114	281.0
76.3	481	367	114	280.0
76.3	480	366	114	279.1
76.3	477	364	113	277.8
76.3	476	363	113	276.8
76.3	473	361	112	275.5
76.3	472	360	112	274.6
76.3	469	358	111	273.3
76.3	468	357	111	272.3
76.3	465	355	110	271.0
76.3	464	354	110	270.1
76.3	460	351	109	267.8
76.3	459	350	109	266.9
76.3	456	348	108	265.6
76.3	455	347	108	264.6
76.3	452	345	107	263.3
76.3	451	344	107	262.4
76.3	448	342	106	261.1
76.3	447	341	106	260.1
76.3	443	338	105	257.9
76.3	439	335	104	255.6
76.3	438	334	104	254.7
76.3	435	332	103	253.4
76.3	434	331	103	252.4
76.3	431	329	102	251.1
76.3	430	328	102	250.2
76.3	427	326	101	248.9
76.3	426	325	101	247.9
76.3	422	322	100	245.7
76.3	418	319	99	243.4
76.3	417	318	99	242.5
76.3	414	316	98	241.2
76.3	413	315	98	240.3
76.3	410	313	97	238.9
76.3	409	312	97	238.0
76.3	405	309	96	235.8
76.3	401	306	95	233.5
76.3	400	305	95	232.6
76.3	397	303	94	231.3
76.3	396	302	94	230.3
76.3	393	300	93	229.0
76.3	392	299	93	228.1
76.3	388	296	92	225.8
76.3	384	293	91	223.6
76.3	380	290	90	221.3
76.3	379	289	90	220.4
76.3	376	287	89	219.1
76.3	375	286	89	218.1
76.3	372	284	88	216.8
76.3	371	283	88	215.9
76.3	367	280	87	213.6
76.3	363	277	86	211.4
76.3	359	274	85	209.1
76.3	358	273	85	208.2
76.3	355	271	84	206.9
76.3	354	270	84	205.9
76.3	350	267	83	203.7
76.3	346	264	82	201.4
76.3	342	261	81	199.2
76.3	338	258	80	196.9
76.3	337	257	80	196.0
76.3	334	255	79	194.7
76.3	333	254	79	193.7
76.3	329	251	78	191.5
76.3	325	248	77	189.2
76.3	321	245	76	187.0
76.3	320	244	76	186.0
76.3	317	242	75	184.7
76.3	316	241	75	183.8
76.3	312	238	74	181.6
76.3	308	235	73	179.3
76.3	304	232	72	177.1
76.3	300	229	71	174.8
76.3	299	228	71	173.9
76.3	295	225	70	171.6
76.3	291	222	69	169.4
76.3	287	219	68	167.1
76.3	283	216	67	164.9
76.3	279	213	66	162.6
76.3	278	212	66	161.7
76.3	274	209	65	159.4
76.3	270	206	64	157.2
76.3	266	203	63	154.9
76.3	262	200	62	152.7
76.3	257	196	61	149.5
76.3	253	193	60	147.2
76.3	249	190	59	145.0
76.3	245	187	58	142.7
76.3	241	184	57	140.5
76.3	240	183	57	139.5
76.3	236	180	56	137.3
76.3	232	177	55	135.0
76.3	228	174	54	132.8
76.3	224	171	53	130.5
76.3	219	167	52	127.3
76.3	215	164	51	125.1
76.3	211	161	50	122.8
76.3	207	158	49	120.6
76.3	198	151	47	115.2
76.3	194	148	46	112.9
76.3	190	145	45	110.7
76.3	186	142	44	108.4
76.3	177	135	42	103.0
76.3	173	132	41	100.7
76.3	169	129	40	98.5
76.3	160	122	38	93.0
76.3	156	119	37	90.8
76.3	152	116	36	88.5
76.3	139	106	33	80.8
76.3	135	103	32	78.6
76.3	131	100	31	76.3
76.3	118	90	28	68.6
76.3	114	87	27	66.4
76.3	97	74	23	56.5
76.3	93	71	22	54.2
76.3	80	61	19	46.5
76.3	76	58	18	44.3
76.3	59	45	14	34.3
76.3	38	29	9	22.1

%E	M1	M2	DM	M*
76.2	500	381	119	290.3
76.2	499	380	119	289.4
76.2	496	378	118	288.1
76.2	495	377	118	287.1
76.2	492	375	117	285.8
76.2	491	374	117	284.9
76.2	488	372	116	283.6
76.2	487	371	116	282.6
76.2	484	369	115	281.3
76.2	483	368	115	280.4
76.2	479	365	114	278.1
76.2	478	364	114	277.2
76.2	475	362	113	275.9
76.2	474	361	113	274.9
76.2	471	359	112	273.6
76.2	470	358	112	272.7
76.2	467	356	111	271.4
76.2	466	355	111	270.4
76.2	463	353	110	269.1
76.2	462	352	110	268.2
76.2	458	349	109	265.9
76.2	454	346	108	263.7
76.2	453	345	108	262.7
76.2	450	343	107	261.4
76.2	449	342	107	260.5
76.2	446	340	106	259.2
76.2	445	339	106	258.2
76.2	442	337	105	256.9
76.2	441	336	105	256.0
76.2	437	333	104	253.8
76.2	433	330	103	251.5
76.2	432	329	103	250.6
76.2	429	327	102	249.3
76.2	428	326	102	248.3
76.2	425	324	101	247.0
76.2	424	323	101	246.1
76.2	421	321	100	244.8
76.2	420	320	100	243.8
76.2	416	317	99	241.6
76.2	412	314	98	239.3
76.2	411	313	98	238.4
76.2	408	311	97	237.1
76.2	407	310	97	236.1
76.2	404	308	96	234.8
76.2	403	307	96	233.9
76.2	399	304	95	231.6
76.2	395	301	94	229.4
76.2	391	298	93	227.1
76.2	390	297	93	226.2
76.2	387	295	92	224.9
76.2	386	294	92	223.9
76.2	383	292	91	222.6
76.2	382	291	91	221.7
76.2	378	288	90	219.4
76.2	374	285	89	217.2
76.2	370	282	88	214.9
76.2	369	281	88	214.0
76.2	366	279	87	212.7
76.2	365	278	87	211.7
76.2	362	276	86	210.4
76.2	361	275	86	209.5
76.2	357	272	85	207.2
76.2	353	269	84	205.0
76.2	349	266	83	202.7
76.2	345	263	82	200.5
76.2	344	262	82	199.5
76.2	341	260	81	198.2
76.2	340	259	81	197.3
76.2	336	256	80	195.0
76.2	332	253	79	192.8
76.2	328	250	78	190.5
76.2	324	247	77	188.3
76.2	323	246	77	187.4
76.2	319	243	76	185.1
76.2	315	240	75	182.9
76.2	311	237	74	180.6
76.2	307	234	73	178.4
76.2	303	231	72	176.1
76.2	302	230	72	175.2
76.2	298	227	71	172.9
76.2	294	224	70	170.7
76.2	290	221	69	168.4
76.2	286	218	68	166.2
76.2	282	215	67	163.9
76.2	281	214	67	163.0
76.2	277	211	66	160.7
76.2	273	208	65	158.5
76.2	269	205	64	156.2
76.2	265	202	63	154.0
76.2	261	199	62	151.7
76.2	260	198	62	150.8
76.2	256	195	61	148.5
76.2	252	192	60	146.3
76.2	248	189	59	144.0
76.2	244	186	58	141.8
76.2	239	182	57	138.6
76.2	235	179	56	136.3
76.2	231	176	55	134.1
76.2	227	173	54	131.8
76.2	223	170	53	129.6
76.2	214	163	51	124.2
76.2	210	160	50	121.9
76.2	206	157	49	119.7
76.2	202	154	48	117.4
76.2	193	147	46	112.0
76.2	189	144	45	109.7
76.2	185	141	44	107.5
76.2	181	138	43	105.2
76.2	172	131	41	99.8
76.2	168	128	40	97.5
76.2	164	125	39	95.3
76.2	151	115	36	87.6
76.2	147	112	35	85.3
76.2	143	109	34	83.1
76.2	130	99	31	75.4
76.2	126	96	30	73.1
76.2	122	93	29	70.9
76.2	105	80	25	61.0
76.2	101	77	24	58.7
76.2	84	64	20	48.8
76.2	63	48	15	36.6
76.2	42	32	10	24.4
76.2	21	16	5	12.2
76.1	498	379	119	288.4
76.1	497	378	119	287.5
76.1	494	376	118	286.2
76.1	493	375	118	285.2
76.1	490	373	117	283.9
76.1	489	372	117	283.0
76.1	486	370	116	281.7
76.1	485	369	116	280.7
76.1	482	367	115	279.4
76.1	481	366	115	278.5
76.1	477	363	114	276.2
76.1	476	362	114	275.3
76.1	473	360	113	274.0
76.1	472	359	113	273.1
76.1	469	357	112	271.7
76.1	468	356	112	270.8
76.1	465	354	111	269.5
76.1	464	353	111	268.6
76.1	461	351	110	267.2
76.1	460	350	110	266.3
76.1	457	348	109	265.0
76.1	456	347	109	264.1
76.1	452	344	108	261.8
76.1	451	343	108	260.9
76.1	448	341	107	259.6
76.1	447	340	107	258.6
76.1	444	338	106	257.3
76.1	443	337	106	256.4
76.1	440	335	105	255.1
76.1	439	334	105	254.1
76.1	436	332	104	252.8
76.1	435	331	104	251.9
76.1	431	328	103	249.6
76.1	427	325	102	247.4
76.1	426	324	102	246.4
76.1	423	322	101	245.1
76.1	422	321	101	244.2
76.1	419	319	100	242.9
76.1	418	318	100	241.9
76.1	415	316	99	240.6
76.1	414	315	99	239.7
76.1	410	312	98	237.4
76.1	406	309	97	235.2
76.1	402	306	96	232.9
76.1	401	305	96	232.0
76.1	398	303	95	230.7
76.1	397	302	95	229.7
76.1	394	300	94	228.4
76.1	393	299	94	227.5
76.1	389	296	93	225.2
76.1	385	293	92	223.0
76.1	381	290	91	220.7
76.1	380	289	91	219.8
76.1	377	287	90	218.5
76.1	376	286	90	217.5
76.1	373	284	89	216.2
76.1	372	283	89	215.3
76.1	368	280	88	213.0
76.1	364	277	87	210.8
76.1	360	274	86	208.5
76.1	356	271	85	206.3
76.1	355	270	85	205.4
76.1	352	268	84	204.0
76.1	351	267	84	203.1
76.1	348	265	83	201.8
76.1	347	264	83	200.9
76.1	343	261	82	198.6
76.1	339	258	81	196.4
76.1	335	255	80	194.1
76.1	331	252	79	191.9
76.1	330	251	79	190.9
76.1	327	249	78	189.6
76.1	326	248	78	188.7
76.1	322	245	77	186.4
76.1	318	242	76	184.2
76.1	314	239	75	181.9
76.1	310	236	74	179.7
76.1	309	235	74	178.7
76.1	306	233	73	177.4
76.1	305	232	73	176.5
76.1	301	229	72	174.2
76.1	297	226	71	172.0
76.1	293	223	70	169.7
76.1	289	220	69	167.5
76.1	285	217	68	165.2
76.1	284	216	68	164.3
76.1	280	213	67	162.0
76.1	276	210	66	159.8
76.1	272	207	65	157.5
76.1	268	204	64	155.3
76.1	264	201	63	153.0
76.1	259	197	62	149.8
76.1	255	194	61	147.6
76.1	251	191	60	145.3
76.1	247	188	59	143.1
76.1	243	185	58	140.8
76.1	238	181	57	137.7
76.1	234	178	56	135.4
76.1	230	175	55	133.2
76.1	226	172	54	130.9
76.1	222	169	53	128.7
76.1	218	166	52	126.4
76.1	213	162	51	123.2
76.1	209	159	50	121.0
76.1	205	156	49	118.7
76.1	201	153	48	116.5
76.1	197	150	47	114.2
76.1	188	143	45	108.8
76.1	184	140	44	106.5
76.1	180	137	43	104.3
76.1	176	134	42	102.0
76.1	163	124	39	94.3
76.1	159	121	38	92.1
76.1	155	118	37	89.8
76.1	142	108	34	82.1
76.1	138	105	33	79.9
76.1	134	102	32	77.6
76.1	117	89	28	67.7
76.1	113	86	27	65.5
76.1	109	83	26	63.2
76.1	92	70	22	53.3
76.1	88	67	21	51.0
76.1	71	54	17	41.1
76.1	67	51	16	38.8
76.1	46	35	11	26.6
76.0	500	380	120	288.8
76.0	499	379	120	287.9

%E	M1	M2	DM	M*
76.0	496	377	119	286.6
76.0	495	376	119	285.6
76.0	492	374	118	284.3
76.0	491	373	118	283.4
76.0	488	371	117	282.1
76.0	487	370	117	281.1
76.0	484	368	116	279.8
76.0	483	367	116	278.9
76.0	480	365	115	277.6
76.0	479	364	115	276.6
76.0	475	361	114	274.4
76.0	471	358	113	272.1
76.0	470	357	113	271.2
76.0	467	355	112	269.9
76.0	466	354	112	268.9
76.0	463	352	111	267.6
76.0	462	351	111	266.7
76.0	459	349	110	265.4
76.0	458	348	110	264.4
76.0	455	346	109	263.1
76.0	454	345	109	262.2
76.0	450	342	108	259.9
76.0	446	339	107	257.7
76.0	445	338	107	256.7
76.0	442	336	106	255.4
76.0	441	335	106	254.5
76.0	438	333	105	253.2
76.0	437	332	105	252.2
76.0	434	330	104	250.9
76.0	433	329	104	250.0
76.0	430	327	103	248.7
76.0	429	326	103	247.7
76.0	425	323	102	245.5
76.0	421	320	101	243.2
76.0	420	319	101	242.3
76.0	417	317	100	241.0
76.0	416	316	100	240.0
76.0	413	314	99	238.7
76.0	412	313	99	237.8
76.0	409	311	98	236.5
76.0	408	310	98	235.5
76.0	405	308	97	234.2
76.0	404	307	97	233.3
76.0	400	304	96	231.0
76.0	396	301	95	228.8
76.0	392	298	94	226.5
76.0	391	297	94	225.6
76.0	388	295	93	224.3
76.0	387	294	93	223.3
76.0	384	292	92	222.0

%E	M1	M2	DM	M*
76.0	383	291	92	221.1
76.0	379	288	91	218.8
76.0	375	285	90	216.6
76.0	371	282	89	214.4
76.0	367	279	88	212.1
76.0	366	278	88	211.2
76.0	363	276	87	209.9
76.0	362	275	87	208.9
76.0	359	273	86	207.6
76.0	358	272	86	206.7
76.0	354	269	85	204.4
76.0	350	266	84	202.2
76.0	346	263	83	199.9
76.0	342	260	82	197.7
76.0	341	259	82	196.7
76.0	338	257	81	195.4
76.0	337	256	81	194.5
76.0	334	254	80	193.2
76.0	333	253	80	192.2
76.0	329	250	79	190.0
76.0	325	247	78	187.7
76.0	321	244	77	185.5
76.0	317	241	76	183.2
76.0	313	238	75	181.0
76.0	312	237	75	180.0
76.0	308	234	74	177.8
76.0	304	231	73	175.5
76.0	300	228	72	173.3
76.0	296	225	71	171.0
76.0	292	222	70	168.8
76.0	288	219	69	166.5
76.0	287	218	69	165.6
76.0	283	215	68	163.3
76.0	279	212	67	161.1
76.0	275	209	66	158.8
76.0	271	206	65	156.6
76.0	267	203	64	154.3
76.0	263	200	63	152.1
76.0	262	199	63	151.1
76.0	258	196	62	148.9
76.0	254	193	61	146.6
76.0	250	190	60	144.4
76.0	246	187	59	142.2
76.0	242	184	58	139.9
76.0	233	177	56	134.5
76.0	229	174	55	132.2
76.0	225	171	54	130.0
76.0	221	168	53	127.7
76.0	217	165	52	125.5
76.0	208	158	50	120.0

%E	M1	M2	DM	M*
76.0	204	155	49	117.8
76.0	200	152	48	115.5
76.0	196	149	47	113.3
76.0	192	146	46	111.0
76.0	183	139	44	105.6
76.0	179	136	43	103.3
76.0	175	133	42	101.1
76.0	171	130	41	98.8
76.0	167	127	40	96.6
76.0	154	117	37	88.9
76.0	150	114	36	86.6
76.0	146	111	35	84.4
76.0	129	98	31	74.4
76.0	125	95	30	72.2
76.0	121	92	29	70.0
76.0	104	79	25	60.0
76.0	100	76	24	57.8
76.0	96	73	23	55.5
76.0	75	57	18	43.3
76.0	50	38	12	28.9
76.0	25	19	6	14.4
75.9	498	378	120	286.9
75.9	497	377	120	286.0
75.9	494	375	119	284.7
75.9	493	374	119	283.7
75.9	490	372	118	282.4
75.9	489	371	118	281.5
75.9	486	369	117	280.2
75.9	485	368	117	279.2
75.9	482	366	116	277.9
75.9	481	365	116	277.0
75.9	478	363	115	275.7
75.9	477	362	115	274.7
75.9	474	360	114	273.4
75.9	473	359	114	272.5
75.9	469	356	113	270.2
75.9	468	355	113	269.3
75.9	465	353	112	268.0
75.9	464	352	112	267.0
75.9	461	350	111	265.7
75.9	460	349	111	264.8
75.9	457	347	110	263.5
75.9	456	346	110	262.5
75.9	453	344	109	261.2
75.9	452	343	109	260.3
75.9	449	341	108	259.0
75.9	448	340	108	258.0
75.9	444	337	107	255.8
75.9	440	334	106	253.5
75.9	439	333	106	252.6

%E	M1	M2	DM	M*
75.9	436	331	105	251.3
75.9	435	330	105	250.3
75.9	432	328	104	249.0
75.9	431	327	104	248.1
75.9	428	325	103	246.8
75.9	427	324	103	245.8
75.9	424	322	102	244.5
75.9	423	321	102	243.6
75.9	419	318	101	241.3
75.9	415	315	100	239.1
75.9	411	312	99	236.8
75.9	410	311	99	235.9
75.9	407	309	98	234.6
75.9	406	308	98	233.7
75.9	403	306	97	232.3
75.9	402	305	97	231.4
75.9	399	303	96	230.1
75.9	398	302	96	229.2
75.9	395	300	95	227.8
75.9	394	299	95	226.9
75.9	390	296	94	224.7
75.9	386	293	93	222.4
75.9	382	290	92	220.2
75.9	381	289	92	219.2
75.9	378	287	91	217.9
75.9	377	286	91	217.0
75.9	374	284	90	215.7
75.9	373	283	90	214.7
75.9	370	281	89	213.4
75.9	369	280	89	212.5
75.9	365	277	88	210.2
75.9	361	274	87	208.0
75.9	357	271	86	205.7
75.9	353	268	85	203.5
75.9	352	267	85	202.5
75.9	349	265	84	201.2
75.9	348	264	84	200.3
75.9	345	262	83	199.0
75.9	344	261	83	198.0
75.9	340	258	82	195.8
75.9	336	255	81	193.5
75.9	332	252	80	191.3
75.9	328	249	79	189.0
75.9	324	246	78	186.8
75.9	323	245	78	185.8
75.9	320	243	77	184.5
75.9	319	242	77	183.6
75.9	316	240	76	182.3
75.9	315	239	76	181.3
75.9	311	236	75	179.1

%E	M1	M2	DM	M*
75.9	307	233	74	176.8
75.9	303	230	73	174.6
75.9	299	227	72	172.3
75.9	295	224	71	170.1
75.9	294	223	71	169.1
75.9	291	221	70	167.8
75.9	290	220	70	166.9
75.9	286	217	69	164.6
75.9	282	214	68	162.4
75.9	278	211	67	160.1
75.9	274	208	66	157.9
75.9	270	205	65	155.6
75.9	266	202	64	153.4
75.9	261	198	63	150.2
75.9	257	195	62	148.0
75.9	253	192	61	145.7
75.9	249	189	60	143.5
75.9	245	186	59	141.2
75.9	241	183	58	139.0
75.9	237	180	57	136.7
75.9	232	176	56	133.5
75.9	228	173	55	131.3
75.9	224	170	54	129.0
75.9	220	167	53	126.8
75.9	216	164	52	124.5
75.9	212	161	51	122.3
75.9	203	154	49	116.8
75.9	199	151	48	114.6
75.9	195	148	47	112.3
75.9	191	145	46	110.1
75.9	187	142	45	107.8
75.9	174	132	42	100.1
75.9	170	129	41	97.9
75.9	166	126	40	95.6
75.9	162	123	39	93.4
75.9	158	120	38	91.1
75.9	145	110	35	83.4
75.9	141	107	34	81.2
75.9	137	104	33	78.9
75.9	133	101	32	76.7
75.9	116	88	28	66.8
75.9	112	85	27	64.5
75.9	108	82	26	62.3
75.9	87	66	21	50.1
75.9	83	63	20	47.8
75.9	79	60	19	45.6
75.9	58	44	14	33.4
75.9	54	41	13	31.1
75.9	29	22	7	16.7
75.8	500	379	121	287.3

%E	M1	M2	DM	M*
75.8	499	378	121	286.3
75.8	496	376	120	285.0
75.8	495	375	120	284.1
75.8	492	373	119	282.8
75.8	491	372	119	281.8
75.8	488	370	118	280.5
75.8	487	369	118	279.6
75.8	484	367	117	278.3
75.8	483	366	117	277.3
75.8	480	364	116	276.0
75.8	479	363	116	275.1
75.8	476	361	115	273.8
75.8	475	360	115	272.8
75.8	472	358	114	271.5
75.8	471	357	114	270.6
75.8	467	354	113	268.3
75.8	466	353	113	267.4
75.8	463	351	112	266.1
75.8	462	350	112	265.2
75.8	459	348	111	263.8
75.8	458	347	111	262.9
75.8	455	345	110	261.6
75.8	454	344	110	260.7
75.8	451	342	109	259.3
75.8	450	341	109	258.4
75.8	447	339	108	257.1
75.8	446	338	108	256.2
75.8	443	336	107	254.8
75.8	442	335	107	253.9
75.8	438	332	106	251.7
75.8	434	329	105	249.4
75.8	433	328	105	248.5
75.8	430	326	104	247.2
75.8	429	325	104	246.2
75.8	426	323	103	244.9
75.8	425	322	103	244.0
75.8	422	320	102	242.7
75.8	421	319	102	241.7
75.8	418	317	101	240.4
75.8	417	316	101	239.5
75.8	414	314	100	238.2
75.8	413	313	100	237.2
75.8	409	310	99	235.0
75.8	405	307	98	232.7
75.8	401	304	97	230.5
75.8	400	303	97	229.5
75.8	397	301	96	228.2
75.8	396	300	96	227.3
75.8	393	298	95	226.0
75.8	392	297	95	225.0

%E	M1	M2	DM	M*
75.8	389	295	94	223.7
75.8	388	294	94	222.8
75.8	385	292	93	221.5
75.8	384	291	93	220.5
75.8	380	288	92	218.3
75.8	376	285	91	216.0
75.8	372	282	90	213.8
75.8	368	279	89	211.5
75.8	364	276	88	209.3
75.8	363	275	88	208.3
75.8	360	273	87	207.0
75.8	359	272	87	206.1
75.8	356	270	86	204.8
75.8	355	269	86	203.8
75.8	351	266	85	201.6
75.8	347	263	84	199.3
75.8	343	260	83	197.1
75.8	339	257	82	194.8
75.8	335	254	81	192.6
75.8	331	251	80	190.3
75.8	330	250	80	189.4
75.8	327	248	79	188.1
75.8	326	247	79	187.1
75.8	322	244	78	184.9
75.8	318	241	77	182.6
75.8	314	238	76	180.4
75.8	310	235	75	178.1
75.8	306	232	74	175.9
75.8	302	229	73	173.6
75.8	298	226	72	171.4
75.8	297	225	72	170.5
75.8	293	222	71	168.2
75.8	289	219	70	166.0
75.8	285	216	69	163.7
75.8	281	213	68	161.5
75.8	277	210	67	159.2
75.8	273	207	66	157.0
75.8	269	204	65	154.7
75.8	265	201	64	152.5
75.8	264	200	64	151.5
75.8	260	197	63	149.3
75.8	256	194	62	147.0
75.8	252	191	61	144.8
75.8	248	188	60	142.5
75.8	244	185	59	140.3
75.8	240	182	58	138.0
75.8	236	179	57	135.8
75.8	231	175	56	132.6
75.8	227	172	55	130.3
75.8	223	169	54	128.1

%E	M1	M2	DM	M*
75.8	219	166	53	125.8
75.8	215	163	52	123.6
75.8	211	160	51	121.3
75.8	207	157	50	119.1
75.8	198	150	48	113.6
75.8	194	147	47	111.4
75.8	190	144	46	109.1
75.8	186	141	45	106.9
75.8	182	138	44	104.6
75.8	178	135	43	102.4
75.8	165	125	40	94.7
75.8	161	122	39	92.4
75.8	157	119	38	90.2
75.8	153	116	37	87.9
75.8	149	113	36	85.7
75.8	132	100	32	75.8
75.8	128	97	31	73.5
75.8	124	94	30	71.3
75.8	120	91	29	69.0
75.8	99	75	24	56.8
75.8	95	72	23	54.6
75.8	91	69	22	52.3
75.8	66	50	16	37.9
75.8	62	47	15	35.6
75.8	33	25	8	18.9
75.7	498	377	121	285.4
75.7	497	376	121	284.5
75.7	494	374	120	283.1
75.7	493	373	120	282.2
75.7	490	371	119	280.9
75.7	489	370	119	280.0
75.7	486	368	118	278.7
75.7	485	367	118	277.7
75.7	482	365	117	276.4
75.7	481	364	117	275.5
75.7	478	362	116	274.2
75.7	477	361	116	273.2
75.7	474	359	115	271.9
75.7	473	358	115	271.0
75.7	470	356	114	269.7
75.7	469	355	114	268.7
75.7	465	352	113	266.5
75.7	461	349	112	264.2
75.7	460	348	112	263.3
75.7	457	346	111	262.0
75.7	456	345	111	261.0
75.7	453	343	110	259.7
75.7	452	342	110	258.8
75.7	449	340	109	257.5
75.7	448	339	109	256.5

%E	M1	M2	DM	M*
75.7	445	337	108	255.2
75.7	444	336	108	254.3
75.7	441	334	107	253.0
75.7	440	333	107	252.0
75.7	437	331	106	250.7
75.7	436	330	106	249.8
75.7	432	327	105	247.5
75.7	428	324	104	245.3
75.7	424	321	103	243.0
75.7	423	320	103	242.1
75.7	420	318	102	240.8
75.7	419	317	102	239.8
75.7	416	315	101	238.5
75.7	415	314	101	237.6
75.7	412	312	100	236.3
75.7	411	311	100	235.3
75.7	408	309	99	234.0
75.7	407	308	99	233.1
75.7	404	306	98	231.8
75.7	403	305	98	230.8
75.7	399	302	97	228.6
75.7	395	299	96	226.3
75.7	391	296	95	224.1
75.7	387	293	94	221.8
75.7	383	290	93	219.6
75.7	382	289	93	218.6
75.7	379	287	92	217.3
75.7	378	286	92	216.4
75.7	375	284	91	215.1
75.7	374	283	91	214.1
75.7	371	281	90	212.8
75.7	370	280	90	211.9
75.7	367	278	89	210.6
75.7	366	277	89	209.6
75.7	362	274	88	207.4
75.7	358	271	87	205.1
75.7	354	268	86	202.9
75.7	350	265	85	200.6
75.7	346	262	84	198.4
75.7	345	261	84	197.5
75.7	342	259	83	196.1
75.7	341	258	83	195.2
75.7	338	256	82	193.9
75.7	337	255	82	193.0
75.7	334	253	81	191.6
75.7	333	252	81	190.7
75.7	329	249	80	188.5
75.7	325	246	79	186.2
75.7	321	243	78	184.0
75.7	317	240	77	181.7

%E	M1	M2	DM	M*
75.7	313	237	76	179.5
75.7	309	234	75	177.2
75.7	305	231	74	175.0
75.7	304	230	74	174.0
75.7	301	228	73	172.7
75.7	300	227	73	171.8
75.7	296	224	72	169.5
75.7	292	221	71	167.3
75.7	288	218	70	165.0
75.7	284	215	69	162.8
75.7	280	212	68	160.5
75.7	276	209	67	158.3
75.7	272	206	66	156.0
75.7	268	203	65	153.8
75.7	267	202	65	152.8
75.7	263	199	64	150.6
75.7	259	196	63	148.3
75.7	255	193	62	146.1
75.7	251	190	61	143.8
75.7	247	187	60	141.6
75.7	243	184	59	139.3
75.7	239	181	58	137.1
75.7	235	178	57	134.8
75.7	230	174	56	131.6
75.7	226	171	55	129.4
75.7	222	168	54	127.1
75.7	218	165	53	124.9
75.7	214	162	52	122.6
75.7	210	159	51	120.4
75.7	206	156	50	118.1
75.7	202	153	49	115.9
75.7	189	143	46	108.2
75.7	185	140	45	105.9
75.7	181	137	44	103.7
75.7	177	134	43	101.4
75.7	173	131	42	99.2
75.7	169	128	41	96.9
75.7	152	115	37	87.0
75.7	148	112	36	84.8
75.7	144	109	35	82.5
75.7	140	106	34	80.3
75.7	136	103	33	78.0
75.7	115	87	28	65.8
75.7	111	84	27	63.6
75.7	107	81	26	61.3
75.7	103	78	25	59.1
75.7	74	56	18	42.4
75.7	70	53	17	40.1
75.7	37	28	9	21.2
75.6	500	378	122	285.8

%E	M1	M2	DM	M*	%E	M1	M2	DM	M*	%E	M1	M2	DM	M*	%E	M1	M2	DM	M*	%E	M1	M2	DM	M*
75.6	499	377	122	284.8	75.6	389	294	95	222.2	75.6	217	164	53	123.9	75.5	444	335	109	252.8	75.5	318	240	78	181.1
75.6	496	375	121	283.5	75.6	386	292	94	220.9	75.6	213	161	52	121.7	75.5	441	333	108	251.4	75.5	314	237	77	178.9
75.6	495	374	121	282.6	75.6	385	291	94	220.0	75.6	209	158	51	119.4	75.5	440	332	108	250.5	75.5	310	234	76	176.6
75.6	492	372	120	281.3	75.6	381	288	93	217.7	75.6	205	155	50	117.2	75.5	437	330	107	249.2	75.5	306	231	75	174.4
75.6	491	371	120	280.3	75.6	377	285	92	215.5	75.6	201	152	49	114.9	75.5	436	329	107	248.3	75.5	302	228	74	172.1
75.6	488	369	119	279.0	75.6	373	282	91	213.2	75.6	197	149	48	112.7	75.5	433	327	106	246.9	75.5	298	225	73	169.9
75.6	487	368	119	278.1	75.6	369	279	90	211.0	75.6	193	146	47	110.4	75.5	432	326	106	246.0	75.5	294	222	72	167.6
75.6	484	366	118	276.8	75.6	365	276	89	208.7	75.6	180	136	44	102.8	75.5	429	324	105	244.7	75.5	290	219	71	165.4
75.6	483	365	118	275.8	75.6	361	273	88	206.5	75.6	176	133	43	100.5	75.5	428	323	105	243.8	75.5	286	216	70	163.1
75.6	480	363	117	274.5	75.6	360	272	88	205.5	75.6	172	130	42	98.3	75.5	425	321	104	242.4	75.5	282	213	69	160.9
75.6	479	362	117	273.6	75.6	357	270	87	204.2	75.6	168	127	41	96.0	75.5	424	320	104	241.5	75.5	278	210	68	158.6
75.6	476	360	116	272.3	75.6	356	269	87	203.3	75.6	164	124	40	93.8	75.5	421	318	103	240.2	75.5	277	209	68	157.7
75.6	475	359	116	271.3	75.6	353	267	86	202.0	75.6	160	121	39	91.5	75.5	420	317	103	239.3	75.5	274	207	67	156.4
75.6	472	357	115	270.0	75.6	352	266	86	201.0	75.6	156	118	38	89.3	75.5	417	315	102	237.9	75.5	273	206	67	155.4
75.6	471	356	115	269.1	75.6	349	264	85	199.7	75.6	135	102	33	77.1	75.5	416	314	102	237.0	75.5	269	203	66	153.2
75.6	468	354	114	267.8	75.6	348	263	85	198.8	75.6	131	99	32	74.8	75.5	413	312	101	235.7	75.5	265	200	65	150.9
75.6	467	353	114	266.8	75.6	344	260	84	196.5	75.6	127	96	31	72.6	75.5	412	311	101	234.8	75.5	261	197	64	148.7
75.6	464	351	113	265.5	75.6	340	257	83	194.3	75.6	123	93	30	70.3	75.5	408	308	100	232.5	75.5	257	194	63	146.4
75.6	463	350	113	264.6	75.6	336	254	82	192.0	75.6	119	90	29	68.1	75.5	404	305	99	230.3	75.5	253	191	62	144.2
75.6	459	347	112	262.3	75.6	332	251	81	189.8	75.6	90	68	22	51.4	75.5	400	302	98	228.0	75.5	249	188	61	141.9
75.6	455	344	111	260.1	75.6	328	248	80	187.5	75.6	86	65	21	49.1	75.5	396	299	97	225.8	75.5	245	185	60	139.7
75.6	454	343	111	259.1	75.6	324	245	79	185.3	75.6	82	62	20	46.9	75.5	392	296	96	223.5	75.5	241	182	59	137.4
75.6	451	341	110	257.8	75.6	320	242	78	183.0	75.6	78	59	19	44.6	75.5	388	293	95	221.3	75.5	237	179	58	135.2
75.6	450	340	110	256.9	75.6	316	239	77	180.8	75.6	45	34	11	25.7	75.5	387	292	95	220.3	75.5	233	176	57	132.9
75.6	447	338	109	255.6	75.6	315	238	77	179.8	75.6	41	31	10	23.4	75.5	384	290	94	219.0	75.5	229	173	56	130.7
75.6	446	337	109	254.6	75.6	312	236	76	178.5	75.5	498	376	122	283.9	75.5	383	289	94	218.1	75.5	220	166	54	125.3
75.6	443	335	108	253.3	75.6	311	235	76	177.6	75.5	497	375	122	282.9	75.5	380	287	93	216.8	75.5	216	163	53	123.0
75.6	442	334	108	252.4	75.6	308	233	75	176.3	75.5	494	373	121	281.6	75.5	379	286	93	215.8	75.5	212	160	52	120.8
75.6	439	332	107	251.1	75.6	307	232	75	175.3	75.5	493	372	121	280.7	75.5	376	284	92	214.5	75.5	208	157	51	118.5
75.6	438	331	107	250.1	75.6	303	229	74	173.1	75.5	490	370	120	279.4	75.5	375	283	92	213.6	75.5	204	154	50	116.3
75.6	435	329	106	248.8	75.6	299	226	73	170.8	75.5	489	369	120	278.4	75.5	372	281	91	212.3	75.5	200	151	49	114.0
75.6	434	328	106	247.9	75.6	295	223	72	168.6	75.5	486	367	119	277.1	75.5	371	280	91	211.3	75.5	196	148	48	111.8
75.6	431	326	105	246.6	75.6	291	220	71	166.3	75.5	485	366	119	276.2	75.5	368	278	90	210.0	75.5	192	145	47	109.5
75.6	430	325	105	245.6	75.6	287	217	70	164.1	75.5	482	364	118	274.9	75.5	367	277	90	209.1	75.5	188	142	46	107.3
75.6	427	323	104	244.3	75.6	283	214	69	161.8	75.5	481	363	118	273.9	75.5	364	275	89	207.8	75.5	184	139	45	105.0
75.6	426	322	104	243.4	75.6	279	211	68	159.6	75.5	478	361	117	272.6	75.5	363	274	89	206.8	75.5	163	123	40	92.8
75.6	422	319	103	241.1	75.6	275	208	67	157.3	75.5	477	360	117	271.7	75.5	359	271	88	204.6	75.5	159	120	39	90.6
75.6	418	316	102	238.9	75.6	271	205	66	155.1	75.5	474	358	116	270.4	75.5	355	268	87	202.3	75.5	155	117	38	88.3
75.6	414	313	101	236.6	75.6	270	204	66	154.1	75.5	473	357	116	269.4	75.5	351	265	86	200.1	75.5	151	114	37	86.1
75.6	410	310	100	234.4	75.6	266	201	65	151.9	75.5	470	355	115	268.1	75.5	347	262	85	197.8	75.5	147	111	36	83.8
75.6	409	309	100	233.4	75.6	262	198	64	149.6	75.5	469	354	115	267.2	75.5	343	259	84	195.6	75.5	143	108	35	81.6
75.6	406	307	99	232.1	75.6	258	195	63	147.4	75.5	466	352	114	265.9	75.5	339	256	83	193.3	75.5	139	105	34	79.3
75.6	405	306	99	231.2	75.6	254	192	62	145.1	75.5	465	351	114	264.9	75.5	335	253	82	191.1	75.5	110	83	27	62.6
75.6	402	304	98	229.9	75.6	250	189	61	142.9	75.5	462	349	113	263.6	75.5	331	250	81	188.8	75.5	106	80	26	60.4
75.6	401	303	98	229.0	75.6	246	186	60	140.6	75.5	461	348	113	262.7	75.5	330	249	81	187.9	75.5	102	77	25	58.1
75.6	398	301	97	227.6	75.6	242	183	59	138.4	75.5	458	346	112	261.4	75.5	327	247	80	186.6	75.5	98	74	24	55.9
75.6	397	300	97	226.7	75.6	238	180	58	136.1	75.5	457	345	112	260.4	75.5	326	246	80	185.6	75.5	94	71	23	53.6
75.6	394	298	96	225.4	75.6	234	177	57	133.9	75.5	453	342	111	258.2	75.5	323	244	79	184.3	75.5	53	40	13	30.2
75.6	393	297	96	224.5	75.6	225	170	55	128.4	75.5	449	339	110	255.9	75.5	322	243	79	183.4	75.5	49	37	12	27.9
75.6	390	295	95	223.1	75.6	221	167	54	126.2	75.5	445	336	109	253.7	75.5	319	241	78	182.1	75.4	500	377	123	284.3

%E	M1	M2	DM	M*
75.4	499	376	123	283.3
75.4	496	374	122	282.0
75.4	495	373	122	281.1
75.4	492	371	121	279.8
75.4	491	370	121	278.8
75.4	488	368	120	277.5
75.4	487	367	120	276.6
75.4	484	365	119	275.3
75.4	483	364	119	274.3
75.4	480	362	118	273.0
75.4	479	361	118	272.1
75.4	476	359	117	270.8
75.4	475	358	117	269.8
75.4	472	356	116	268.5
75.4	471	355	116	267.6
75.4	468	353	115	266.3
75.4	467	352	115	265.3
75.4	464	350	114	264.0
75.4	463	349	114	263.1
75.4	460	347	113	261.8
75.4	459	346	113	260.8
75.4	456	344	112	259.5
75.4	455	343	112	258.6
75.4	452	341	111	257.3
75.4	451	340	111	256.3
75.4	448	338	110	255.0
75.4	447	337	110	254.1
75.4	443	334	109	251.8
75.4	439	331	108	249.6
75.4	435	328	107	247.3
75.4	431	325	106	245.1
75.4	427	322	105	242.8
75.4	426	321	105	241.9
75.4	423	319	104	240.6
75.4	422	318	104	239.6
75.4	419	316	103	238.3
75.4	418	315	103	237.4
75.4	415	313	102	236.1
75.4	414	312	102	235.1
75.4	411	310	101	233.8
75.4	410	309	101	232.9
75.4	407	307	100	231.6
75.4	406	306	100	230.6
75.4	403	304	99	229.3
75.4	402	303	99	228.4
75.4	399	301	98	227.1
75.4	398	300	98	226.1
75.4	395	298	97	224.8
75.4	394	297	97	223.9
75.4	391	295	96	222.6

%E	M1	M2	DM	M*
75.4	390	294	96	221.6
75.4	386	291	95	219.4
75.4	382	288	94	217.1
75.4	378	285	93	214.9
75.4	374	282	92	212.6
75.4	370	279	91	210.4
75.4	366	276	90	208.1
75.4	362	273	89	205.9
75.4	358	270	88	203.6
75.4	357	269	88	202.7
75.4	354	267	87	201.4
75.4	353	266	87	200.4
75.4	350	264	86	199.1
75.4	349	263	86	198.2
75.4	346	261	85	196.9
75.4	345	260	85	195.9
75.4	342	258	84	194.6
75.4	341	257	84	193.7
75.4	338	255	83	192.4
75.4	337	254	83	191.4
75.4	334	252	82	190.1
75.4	333	251	82	189.2
75.4	329	248	81	186.9
75.4	325	245	80	184.7
75.4	321	242	79	182.4
75.4	317	239	78	180.2
75.4	313	236	77	177.9
75.4	309	233	76	175.7
75.4	305	230	75	173.4
75.4	301	227	74	171.2
75.4	297	224	73	168.9
75.4	293	221	72	166.7
75.4	289	218	71	164.4
75.4	285	215	70	162.2
75.4	284	214	70	161.3
75.4	281	212	69	159.9
75.4	280	211	69	159.0
75.4	276	208	68	156.8
75.4	272	205	67	154.5
75.4	268	202	66	152.3
75.4	264	199	65	150.0
75.4	260	196	64	147.8
75.4	256	193	63	145.5
75.4	252	190	62	143.3
75.4	248	187	61	141.0
75.4	244	184	60	138.8
75.4	240	181	59	136.5
75.4	236	178	58	134.3
75.4	232	175	57	132.0
75.4	228	172	56	129.8

%E	M1	M2	DM	M*
75.4	224	169	55	127.5
75.4	211	159	52	119.8
75.4	207	156	51	117.6
75.4	203	153	50	115.3
75.4	199	150	49	113.1
75.4	195	147	48	110.8
75.4	191	144	47	108.6
75.4	187	141	46	106.3
75.4	183	138	45	104.1
75.4	179	135	44	101.8
75.4	175	132	43	99.6
75.4	171	129	42	97.3
75.4	167	126	41	95.1
75.4	142	107	35	80.6
75.4	138	104	34	78.4
75.4	134	101	33	76.1
75.4	130	98	32	73.9
75.4	126	95	31	71.6
75.4	122	92	30	69.4
75.4	118	89	29	67.1
75.4	114	86	28	64.9
75.4	69	52	17	39.2
75.4	65	49	16	36.9
75.4	61	46	15	34.7
75.4	57	43	14	32.4
75.3	498	375	123	282.4
75.3	497	374	123	281.4
75.3	494	372	122	280.1
75.3	493	371	122	279.2
75.3	490	369	121	277.9
75.3	489	368	121	276.9
75.3	486	366	120	275.6
75.3	485	365	120	274.7
75.3	482	363	119	273.4
75.3	481	362	119	272.4
75.3	478	360	118	271.1
75.3	477	359	118	270.2
75.3	474	357	117	268.9
75.3	473	356	117	267.9
75.3	470	354	116	266.6
75.3	469	353	116	265.7
75.3	466	351	115	264.4
75.3	465	350	115	263.4
75.3	462	348	114	262.1
75.3	461	347	114	261.2
75.3	458	345	113	259.9
75.3	457	344	113	258.9
75.3	454	342	112	257.6
75.3	453	341	112	256.7
75.3	450	339	111	255.4

%E	M1	M2	DM	M*
75.3	449	338	111	254.4
75.3	446	336	110	253.1
75.3	445	335	110	252.2
75.3	442	333	109	250.9
75.3	441	332	109	249.9
75.3	438	330	108	248.6
75.3	437	329	108	247.7
75.3	434	327	107	246.4
75.3	433	326	107	245.4
75.3	430	324	106	244.1
75.3	429	323	106	243.2
75.3	425	320	105	240.9
75.3	421	317	104	238.7
75.3	417	314	103	236.4
75.3	413	311	102	234.2
75.3	409	308	101	231.9
75.3	405	305	100	229.7
75.3	401	302	99	227.4
75.3	400	301	99	226.5
75.3	397	299	98	225.2
75.3	396	298	98	224.3
75.3	393	296	97	222.9
75.3	392	295	97	222.0
75.3	389	293	96	220.7
75.3	388	292	96	219.8
75.3	385	290	95	218.4
75.3	384	289	95	217.5
75.3	381	287	94	216.2
75.3	380	286	94	215.3
75.3	377	284	93	213.9
75.3	376	283	93	213.0
75.3	373	281	92	211.7
75.3	372	280	92	210.8
75.3	369	278	91	209.4
75.3	368	277	91	208.5
75.3	365	275	90	207.2
75.3	364	274	90	206.3
75.3	361	272	89	204.9
75.3	360	271	89	204.0
75.3	356	268	88	201.8
75.3	352	265	87	199.5
75.3	348	262	86	197.3
75.3	344	259	85	195.0
75.3	340	256	84	192.8
75.3	336	253	83	190.5
75.3	332	250	82	188.3
75.3	328	247	81	186.0
75.3	324	244	80	183.8
75.3	320	241	79	181.5
75.3	316	238	78	179.3

%E	M1	M2	DM	M*
75.3	312	235	77	177.0
75.3	308	232	76	174.8
75.3	304	229	75	172.5
75.3	300	226	74	170.3
75.3	299	225	74	169.3
75.3	296	223	73	168.0
75.3	295	222	73	167.1
75.3	292	220	72	165.8
75.3	291	219	72	164.8
75.3	288	217	71	163.5
75.3	287	216	71	162.6
75.3	283	213	70	160.3
75.3	279	210	69	158.1
75.3	275	207	68	155.8
75.3	271	204	67	153.6
75.3	267	201	66	151.3
75.3	263	198	65	149.1
75.3	259	195	64	146.8
75.3	255	192	63	144.6
75.3	251	189	62	142.3
75.3	247	186	61	140.1
75.3	243	183	60	137.8
75.3	239	180	59	135.6
75.3	235	177	58	133.3
75.3	231	174	57	131.1
75.3	227	171	56	128.8
75.3	223	168	55	126.6
75.3	219	165	54	124.3
75.3	215	162	53	122.1
75.3	198	149	49	112.1
75.3	194	146	48	109.9
75.3	190	143	47	107.6
75.3	186	140	46	105.4
75.3	182	137	45	103.1
75.3	178	134	44	100.9
75.3	174	131	43	98.6
75.3	170	128	42	96.4
75.3	166	125	41	94.1
75.3	162	122	40	91.9
75.3	158	119	39	89.6
75.3	154	116	38	87.4
75.3	150	113	37	85.1
75.3	146	110	36	82.9
75.3	97	73	24	54.9
75.3	93	70	23	52.7
75.3	89	67	22	50.4
75.3	85	64	21	48.2
75.3	81	61	20	45.9
75.3	77	58	19	43.7
75.3	73	55	18	41.4

%E	M1	M2	DM	M*
75.2	500	376	124	282.8
75.2	499	375	124	281.8
75.2	496	373	123	280.5
75.2	495	372	123	279.6
75.2	492	370	122	278.3
75.2	491	369	122	277.3
75.2	488	367	121	276.0
75.2	487	366	121	275.1
75.2	484	364	120	273.8
75.2	483	363	120	272.8
75.2	480	361	119	271.5
75.2	479	360	119	270.6
75.2	476	358	118	269.3
75.2	475	357	118	268.3
75.2	472	355	117	267.0
75.2	471	354	117	266.1
75.2	468	352	116	264.8
75.2	467	351	116	263.8
75.2	464	349	115	262.5
75.2	463	348	115	261.6
75.2	460	346	114	260.3
75.2	459	345	114	259.3
75.2	456	343	113	258.0
75.2	455	342	113	257.1
75.2	452	340	112	255.8
75.2	451	339	112	254.8
75.2	448	337	111	253.5
75.2	447	336	111	252.6
75.2	444	334	110	251.3
75.2	443	333	110	250.3
75.2	440	331	109	249.0
75.2	439	330	109	248.1
75.2	436	328	108	246.8
75.2	435	327	108	245.8
75.2	432	325	107	244.5
75.2	431	324	107	243.6
75.2	428	322	106	242.3
75.2	427	321	106	241.3
75.2	424	319	105	240.0
75.2	423	318	105	239.1
75.2	420	316	104	237.8
75.2	419	315	104	236.8
75.2	416	313	103	235.5
75.2	415	312	103	234.6
75.2	412	310	102	233.3
75.2	411	309	102	232.3
75.2	408	307	101	231.0
75.2	407	306	101	230.1
75.2	404	304	100	228.8
75.2	403	303	100	227.8

%E	M1	M2	DM	M*
75.2	399	300	99	225.6
75.2	395	297	98	223.3
75.2	391	294	97	221.1
75.2	387	291	96	218.8
75.2	383	288	95	216.6
75.2	379	285	94	214.3
75.2	375	282	93	212.1
75.2	371	279	92	209.8
75.2	367	276	91	207.6
75.2	363	273	90	205.3
75.2	359	270	89	203.1
75.2	355	267	88	200.8
75.2	351	264	87	198.6
75.2	347	261	86	196.3
75.2	343	258	85	194.1
75.2	339	255	84	191.8
75.2	335	252	83	189.6
75.2	334	251	83	188.6
75.2	331	249	82	187.3
75.2	330	248	82	186.4
75.2	327	246	81	185.1
75.2	326	245	81	184.1
75.2	323	243	80	182.8
75.2	322	242	80	181.9
75.2	319	240	79	180.6
75.2	318	239	79	179.6
75.2	315	237	78	178.3
75.2	314	236	78	177.4
75.2	311	234	77	176.1
75.2	310	233	77	175.1
75.2	307	231	76	173.8
75.2	306	230	76	172.9
75.2	303	228	75	171.6
75.2	302	227	75	170.6
75.2	298	224	74	168.4
75.2	294	221	73	166.1
75.2	290	218	72	163.9
75.2	286	215	71	161.6
75.2	282	212	70	159.4
75.2	278	209	69	157.1
75.2	274	206	68	154.9
75.2	270	203	67	152.6
75.2	266	200	66	150.4
75.2	262	197	65	148.1
75.2	258	194	64	145.9
75.2	254	191	63	143.6
75.2	250	188	62	141.4
75.2	246	185	61	139.1
75.2	242	182	60	136.9
75.2	238	179	59	134.6

%E	M1	M2	DM	M*
75.2	234	176	58	132.4
75.2	230	173	57	130.1
75.2	226	170	56	127.9
75.2	222	167	55	125.6
75.2	218	164	54	123.4
75.2	214	161	53	121.1
75.2	210	158	52	118.9
75.2	206	155	51	116.6
75.2	202	152	50	114.4
75.2	165	124	41	93.2
75.2	161	121	40	90.9
75.2	157	118	39	88.7
75.2	153	115	38	86.4
75.2	149	112	37	84.2
75.2	145	109	36	81.9
75.2	141	106	35	79.7
75.2	137	103	34	77.4
75.2	133	100	33	75.2
75.2	129	97	32	72.9
75.2	125	94	31	70.7
75.2	121	91	30	68.4
75.2	117	88	29	66.2
75.2	113	85	28	63.9
75.2	109	82	27	61.7
75.2	105	79	26	59.4
75.2	101	76	25	57.2
75.1	498	374	124	280.9
75.1	497	373	124	279.9
75.1	494	371	123	278.6
75.1	493	370	123	277.7
75.1	490	368	122	276.4
75.1	489	367	122	275.4
75.1	486	365	121	274.1
75.1	485	364	121	273.2
75.1	482	362	120	271.9
75.1	481	361	120	270.9
75.1	478	359	119	269.6
75.1	477	358	119	268.7
75.1	474	356	118	267.4
75.1	473	355	118	266.4
75.1	470	353	117	265.1
75.1	469	352	117	264.2
75.1	466	350	116	262.9
75.1	465	349	116	261.9
75.1	462	347	115	260.6
75.1	461	346	115	259.7
75.1	458	344	114	258.4
75.1	457	343	114	257.4
75.1	454	341	113	256.1
75.1	453	340	113	255.2

%E	M1	M2	DM	M*
75.1	450	338	112	253.9
75.1	449	337	112	252.9
75.1	446	335	111	251.6
75.1	445	334	111	250.7
75.1	442	332	110	249.4
75.1	441	331	110	248.4
75.1	438	329	109	247.1
75.1	437	328	109	246.2
75.1	434	326	108	244.9
75.1	433	325	108	243.9
75.1	430	323	107	242.6
75.1	429	322	107	241.7
75.1	426	320	106	240.4
75.1	425	319	106	239.4
75.1	422	317	105	238.1
75.1	421	316	105	237.2
75.1	418	314	104	235.9
75.1	417	313	104	234.9
75.1	414	311	103	233.6
75.1	413	310	103	232.7
75.1	410	308	102	231.4
75.1	409	307	102	230.4
75.1	406	305	101	229.1
75.1	405	304	101	228.2
75.1	402	302	100	226.9
75.1	401	301	100	225.9
75.1	398	299	99	224.6
75.1	397	298	99	223.7
75.1	394	296	98	222.4
75.1	393	295	98	221.4
75.1	390	293	97	220.1
75.1	389	292	97	219.2
75.1	386	290	96	217.9
75.1	385	289	96	216.9
75.1	382	287	95	215.6
75.1	381	286	95	214.7
75.1	378	284	94	213.4
75.1	377	283	94	212.4
75.1	374	281	93	211.1
75.1	373	280	93	210.2
75.1	370	278	92	208.9
75.1	369	277	92	207.9
75.1	366	275	91	206.6
75.1	365	274	91	205.7
75.1	362	272	90	204.4
75.1	361	271	90	203.4
75.1	358	269	89	202.1
75.1	357	268	89	201.2
75.1	354	266	88	199.9
75.1	353	265	88	198.9

%E	M1	M2	DM	M*
75.1	350	263	87	197.6
75.1	349	262	87	196.7
75.1	346	260	86	195.4
75.1	345	259	86	194.4
75.1	342	257	85	193.1
75.1	341	256	85	192.2
75.1	338	254	84	190.9
75.1	337	253	84	189.9
75.1	333	250	83	187.7
75.1	329	247	82	185.4
75.1	325	244	81	183.2
75.1	321	241	80	180.9
75.1	317	238	79	178.7
75.1	313	235	78	176.4
75.1	309	232	77	174.2
75.1	305	229	76	171.9
75.1	301	226	75	169.7
75.1	297	223	74	167.4
75.1	293	220	73	165.2
75.1	289	217	72	162.9
75.1	285	214	71	160.7
75.1	281	211	70	158.4
75.1	277	208	69	156.2
75.1	273	205	68	153.9
75.1	269	202	67	151.7
75.1	265	199	66	149.4
75.1	261	196	65	147.2
75.1	257	193	64	144.9
75.1	253	190	63	142.7
75.1	249	187	62	140.4
75.1	245	184	61	138.2
75.1	241	181	60	135.9
75.1	237	178	59	133.7
75.1	233	175	58	131.4
75.1	229	172	57	129.2
75.1	225	169	56	126.9
75.1	221	166	55	124.7
75.1	217	163	54	122.4
75.1	213	160	53	120.2
75.1	209	157	52	117.9
75.1	205	154	51	115.7
75.1	201	151	50	113.4
75.1	197	148	49	111.2
75.1	193	145	48	108.9
75.1	189	142	47	106.7
75.1	185	139	46	104.4
75.1	181	136	45	102.2
75.1	177	133	44	99.9
75.1	173	130	43	97.7
75.1	169	127	42	95.4

%E	M1	M2	DM	M*
75.0	500	375	125	281.3
75.0	499	374	125	280.3
75.0	496	372	124	279.0
75.0	492	369	123	276.8
75.0	488	366	122	274.5
75.0	484	363	121	272.3
75.0	480	360	120	270.0
75.0	476	357	119	267.8
75.0	472	354	118	265.5
75.0	468	351	117	263.3
75.0	464	348	116	261.0
75.0	460	345	115	258.8
75.0	456	342	114	256.5
75.0	452	339	113	254.3
75.0	448	336	112	252.0
75.0	444	333	111	249.8
75.0	440	330	110	247.5
75.0	436	327	109	245.3
75.0	432	324	108	243.0
75.0	428	321	107	240.8
75.0	424	318	106	238.5
75.0	420	315	105	236.3
75.0	416	312	104	234.0
75.0	412	309	103	231.8
75.0	408	306	102	229.5
75.0	404	303	101	227.3
75.0	400	300	100	225.0
75.0	396	297	99	222.8
75.0	392	294	98	220.5
75.0	388	291	97	218.3
75.0	384	288	96	216.0
75.0	380	285	95	213.8
75.0	376	282	94	211.5
75.0	372	279	93	209.3
75.0	368	276	92	207.0
75.0	364	273	91	204.8
75.0	360	270	90	202.5
75.0	356	267	89	200.3
75.0	352	264	88	198.0
75.0	348	261	87	195.8
75.0	344	258	86	193.5
75.0	340	255	85	191.3
75.0	336	252	84	189.0
75.0	332	249	83	186.8
75.0	328	246	82	184.5
75.0	324	243	81	182.3
75.0	320	240	80	180.0
75.0	316	237	79	177.8
75.0	312	234	78	175.5
75.0	308	231	77	173.3

%E	M1	M2	DM	M*
75.0	304	228	76	171.0
75.0	300	225	75	168.8
75.0	296	222	74	166.5
75.0	292	219	73	164.3
75.0	288	216	72	162.0
75.0	284	213	71	159.8
75.0	280	210	70	157.5
75.0	276	207	69	155.3
75.0	272	204	68	153.0
75.0	268	201	67	150.8
75.0	264	198	66	148.5
75.0	260	195	65	146.3
75.0	256	192	64	144.0
75.0	252	189	63	141.8
75.0	248	186	62	139.5
75.0	244	183	61	137.3
75.0	240	180	60	135.0
75.0	236	177	59	132.8
75.0	232	174	58	130.5
75.0	228	171	57	128.3
75.0	224	168	56	126.0
75.0	220	165	55	123.8
75.0	216	162	54	121.5
75.0	212	159	53	119.3
75.0	208	156	52	117.0
75.0	204	153	51	114.8
75.0	200	150	50	112.5
75.0	196	147	49	110.3
75.0	192	144	48	108.0
75.0	188	141	47	105.8
75.0	184	138	46	103.5
75.0	180	135	45	101.3
75.0	176	132	44	99.0
75.0	172	129	43	96.8
75.0	168	126	42	94.5
75.0	164	123	41	92.3
75.0	160	120	40	90.0
75.0	156	117	39	87.8
75.0	152	114	38	85.5
75.0	148	111	37	83.3
75.0	144	108	36	81.0
75.0	140	105	35	78.8
75.0	136	102	34	76.5
75.0	132	99	33	74.3
75.0	128	96	32	72.0
75.0	124	93	31	69.8
75.0	120	90	30	67.5
75.0	116	87	29	65.3
75.0	112	84	28	63.0
75.0	108	81	27	60.8

%E	M1	M2	DM	M*
75.0	104	78	26	58.5
75.0	100	75	25	56.3
75.0	96	72	24	54.0
75.0	92	69	23	51.8
75.0	88	66	22	49.5
75.0	84	63	21	47.3
75.0	80	60	20	45.0
75.0	76	57	19	42.8
75.0	72	54	18	40.5
75.0	68	51	17	38.3
75.0	64	48	16	36.0
75.0	60	45	15	33.8
75.0	56	42	14	31.5
75.0	52	39	13	29.3
75.0	48	36	12	27.0
75.0	44	33	11	24.8
75.0	40	30	10	22.5
75.0	36	27	9	20.3
75.0	32	24	8	18.0
75.0	28	21	7	15.8
75.0	24	18	6	13.5
75.0	20	15	5	11.3
75.0	16	12	4	9.0
75.0	12	9	3	6.8
75.0	8	6	2	4.5
75.0	4	3	1	2.3
74.9	498	373	125	279.4
74.9	495	371	124	278.1
74.9	494	370	124	277.1
74.9	491	368	123	275.8
74.9	490	367	123	274.9
74.9	487	365	122	273.6
74.9	486	364	122	272.6
74.9	483	362	121	271.3
74.9	482	361	121	270.4
74.9	479	359	120	269.1
74.9	478	358	120	268.1
74.9	475	356	119	266.8
74.9	474	355	119	265.9
74.9	471	353	118	264.6
74.9	470	352	118	263.6
74.9	467	350	117	262.3
74.9	466	349	117	261.4
74.9	463	347	116	260.1
74.9	462	346	116	259.1
74.9	459	344	115	257.8
74.9	458	343	115	256.9
74.9	455	341	114	255.6
74.9	454	340	114	254.6
74.9	451	338	113	253.3

%E	M1	M2	DM	M*
74.9	450	337	113	252.4
74.9	447	335	112	251.1
74.9	446	334	112	250.1
74.9	443	332	111	248.8
74.9	442	331	111	247.9
74.9	439	329	110	246.6
74.9	438	328	110	245.6
74.9	435	326	109	244.3
74.9	434	325	109	243.4
74.9	431	323	108	242.1
74.9	430	322	108	241.1
74.9	427	320	107	239.8
74.9	426	319	107	238.9
74.9	423	317	106	237.6
74.9	422	316	106	236.6
74.9	419	314	105	235.3
74.9	418	313	105	234.4
74.9	415	311	104	233.1
74.9	414	310	104	232.1
74.9	411	308	103	230.8
74.9	410	307	103	229.9
74.9	407	305	102	228.6
74.9	406	304	102	227.6
74.9	403	302	101	226.3
74.9	402	301	101	225.4
74.9	399	299	100	224.1
74.9	398	298	100	223.1
74.9	395	296	99	221.8
74.9	394	295	99	220.9
74.9	391	293	98	219.6
74.9	390	292	98	218.6
74.9	387	290	97	217.3
74.9	386	289	97	216.4
74.9	383	287	96	215.1
74.9	382	286	96	214.1
74.9	379	284	95	212.8
74.9	378	283	95	211.9
74.9	375	281	94	210.6
74.9	374	280	94	209.6
74.9	371	278	93	208.3
74.9	370	277	93	207.4
74.9	367	275	92	206.1
74.9	366	274	92	205.1
74.9	363	272	91	203.8
74.9	362	271	91	202.9
74.9	359	269	90	201.6
74.9	358	268	90	200.6
74.9	355	266	89	199.3
74.9	354	265	89	198.4
74.9	351	263	88	197.1

%E	M1	M2	DM	M*
74.9	350	262	88	196.1
74.9	347	260	87	194.8
74.9	346	259	87	193.9
74.9	343	257	86	192.6
74.9	342	256	86	191.6
74.9	339	254	85	190.3
74.9	338	253	85	189.4
74.9	335	251	84	188.1
74.9	334	250	84	187.1
74.9	331	248	83	185.8
74.9	327	245	82	183.6
74.9	323	242	81	181.3
74.9	319	239	80	179.1
74.9	315	236	79	176.8
74.9	311	233	78	174.6
74.9	307	230	77	172.3
74.9	303	227	76	170.1
74.9	299	224	75	167.8
74.9	295	221	74	165.6
74.9	291	218	73	163.3
74.9	287	215	72	161.1
74.9	283	212	71	158.8
74.9	279	209	70	156.6
74.9	275	206	69	154.3
74.9	271	203	68	152.1
74.9	267	200	67	149.8
74.9	263	197	66	147.6
74.9	259	194	65	145.3
74.9	255	191	64	143.1
74.9	251	188	63	140.8
74.9	247	185	62	138.6
74.9	243	182	61	136.3
74.9	239	179	60	134.1
74.9	235	176	59	131.8
74.9	231	173	58	129.6
74.9	227	170	57	127.3
74.9	223	167	56	125.1
74.9	219	164	55	122.8
74.9	215	161	54	120.6
74.9	211	158	53	118.3
74.9	207	155	52	116.1
74.9	203	152	51	113.8
74.9	199	149	50	111.6
74.9	195	146	49	109.3
74.9	191	143	48	107.1
74.9	187	140	47	104.8
74.9	183	137	46	102.6
74.9	179	134	45	100.3
74.9	175	131	44	98.1
74.9	171	128	43	95.8

%E	M1	M2	DM	M*
74.9	167	125	42	93.6
74.8	500	374	126	279.8
74.8	497	372	125	278.4
74.8	496	371	125	277.5
74.8	493	369	124	276.2
74.8	492	368	124	275.3
74.8	489	366	123	273.9
74.8	488	365	123	273.0
74.8	485	363	122	271.7
74.8	484	362	122	270.8
74.8	481	360	121	269.4
74.8	480	359	121	268.5
74.8	477	357	120	267.2
74.8	476	356	120	266.3
74.8	473	354	119	264.9
74.8	472	353	119	264.0
74.8	469	351	118	262.7
74.8	468	350	118	261.8
74.8	465	348	117	260.4
74.8	464	347	117	259.5
74.8	461	345	116	258.2
74.8	460	344	116	257.3
74.8	457	342	115	255.9
74.8	456	341	115	255.0
74.8	453	339	114	253.7
74.8	452	338	114	252.8
74.8	449	336	113	251.4
74.8	448	335	113	250.5
74.8	445	333	112	249.2
74.8	444	332	112	248.3
74.8	441	330	111	246.9
74.8	440	329	111	246.0
74.8	437	327	110	244.7
74.8	436	326	110	243.8
74.8	433	324	109	242.4
74.8	432	323	109	241.5
74.8	429	321	108	240.2
74.8	428	320	108	239.3
74.8	425	318	107	237.9
74.8	424	317	107	237.0
74.8	421	315	106	235.7
74.8	420	314	106	234.8
74.8	417	312	105	233.4
74.8	416	311	105	232.5
74.8	413	309	104	231.2
74.8	412	308	104	230.3
74.8	409	306	103	228.9
74.8	408	305	103	228.0
74.8	405	303	102	226.7
74.8	404	302	102	225.8

%E	M1	M2	DM	M*
74.8	401	300	101	224.4
74.8	400	299	101	223.5
74.8	397	297	100	222.2
74.8	393	294	99	219.9
74.8	389	291	98	217.7
74.8	385	288	97	215.4
74.8	381	285	96	213.2
74.8	377	282	95	210.9
74.8	373	279	94	208.7
74.8	369	276	93	206.4
74.8	365	273	92	204.2
74.8	361	270	91	201.9
74.8	357	267	90	199.7
74.8	353	264	89	197.4
74.8	349	261	88	195.2
74.8	345	258	87	192.9
74.8	341	255	86	190.7
74.8	337	252	85	188.4
74.8	333	249	84	186.2
74.8	330	247	83	184.9
74.8	329	246	83	183.9
74.8	326	244	82	182.6
74.8	325	243	82	181.7
74.8	322	241	81	180.4
74.8	321	240	81	179.4
74.8	318	238	80	178.1
74.8	317	237	80	177.2
74.8	314	235	79	175.9
74.8	313	234	79	174.9
74.8	310	232	78	173.6
74.8	309	231	78	172.7
74.8	306	229	77	171.4
74.8	305	228	77	170.4
74.8	302	226	76	169.1
74.8	301	225	76	168.2
74.8	298	223	75	166.9
74.8	294	220	74	164.6
74.8	290	217	73	162.4
74.8	286	214	72	160.1
74.8	282	211	71	157.9
74.8	278	208	70	155.6
74.8	274	205	69	153.4
74.8	270	202	68	151.1
74.8	266	199	67	148.9
74.8	262	196	66	146.6
74.8	258	193	65	144.4
74.8	254	190	64	142.1
74.8	250	187	63	139.9
74.8	246	184	62	137.6
74.8	242	181	61	135.4

%E	M1	M2	DM	M*
74.8	238	178	60	133.1
74.8	234	175	59	130.9
74.8	230	172	58	128.6
74.8	226	169	57	126.4
74.8	222	166	56	124.1
74.8	218	163	55	121.9
74.8	214	160	54	119.6
74.8	210	157	53	117.4
74.8	206	154	52	115.1
74.8	202	151	51	112.9
74.8	163	122	41	91.3
74.8	159	119	40	89.1
74.8	155	116	39	86.8
74.8	151	113	38	84.6
74.8	147	110	37	82.3
74.8	143	107	36	80.1
74.8	139	104	35	77.8
74.8	135	101	34	75.6
74.8	131	98	33	73.3
74.8	127	95	32	71.1
74.8	123	92	31	68.8
74.8	119	89	30	66.6
74.8	115	86	29	64.3
74.8	111	83	28	62.1
74.8	107	80	27	59.8
74.8	103	77	26	57.6
74.7	499	373	126	278.8
74.7	498	372	126	277.9
74.7	495	370	125	276.6
74.7	494	369	125	275.6
74.7	491	367	124	274.3
74.7	490	366	124	273.4
74.7	487	364	123	272.1
74.7	486	363	123	271.1
74.7	483	361	122	269.8
74.7	482	360	122	268.9
74.7	479	358	121	267.6
74.7	478	357	121	266.6
74.7	475	355	120	265.3
74.7	474	354	120	264.4
74.7	471	352	119	263.1
74.7	470	351	119	262.1
74.7	467	349	118	260.8
74.7	466	348	118	259.9
74.7	463	346	117	258.6
74.7	462	345	117	257.6
74.7	459	343	116	256.3
74.7	458	342	116	255.4
74.7	455	340	115	254.1
74.7	454	339	115	253.1

%E	M1	M2	DM	M*
74.7	451	337	114	251.8
74.7	450	336	114	250.9
74.7	447	334	113	249.6
74.7	446	333	113	248.6
74.7	443	331	112	247.3
74.7	442	330	112	246.4
74.7	439	328	111	245.1
74.7	438	327	111	244.1
74.7	435	325	110	242.8
74.7	434	324	110	241.9
74.7	431	322	109	240.6
74.7	430	321	109	239.6
74.7	427	319	108	238.3
74.7	423	316	107	236.1
74.7	419	313	106	233.8
74.7	415	310	105	231.6
74.7	411	307	104	229.3
74.7	407	304	103	227.1
74.7	403	301	102	224.8
74.7	399	298	101	222.6
74.7	396	296	100	221.3
74.7	395	295	100	220.3
74.7	392	293	99	219.0
74.7	391	292	99	218.1
74.7	388	290	98	216.8
74.7	387	289	98	215.8
74.7	384	287	97	214.5
74.7	383	286	97	213.6
74.7	380	284	96	212.3
74.7	379	283	96	211.3
74.7	376	281	95	210.0
74.7	375	280	95	209.1
74.7	372	278	94	207.8
74.7	371	277	94	206.8
74.7	368	275	93	205.5
74.7	367	274	93	204.6
74.7	364	272	92	203.3
74.7	363	271	92	202.3
74.7	360	269	91	201.0
74.7	359	268	91	200.1
74.7	356	266	90	198.8
74.7	352	263	89	196.5
74.7	348	260	88	194.3
74.7	344	257	87	192.0
74.7	340	254	86	189.8
74.7	336	251	85	187.5
74.7	332	248	84	185.3
74.7	328	245	83	183.0
74.7	324	242	82	180.8
74.7	320	239	81	178.5

%E	M1	M2	DM	M*
74.7	316	236	80	176.3
74.7	312	233	79	174.0
74.7	308	230	78	171.8
74.7	304	227	77	169.5
74.7	300	224	76	167.3
74.7	297	222	75	165.9
74.7	296	221	75	165.0
74.7	293	219	74	163.7
74.7	292	218	74	162.8
74.7	289	216	73	161.4
74.7	288	215	73	160.5
74.7	285	213	72	159.2
74.7	281	210	71	156.9
74.7	277	207	70	154.7
74.7	273	204	69	152.4
74.7	269	201	68	150.2
74.7	265	198	67	147.9
74.7	261	195	66	145.7
74.7	257	192	65	143.4
74.7	253	189	64	141.2
74.7	249	186	63	138.9
74.7	245	183	62	136.7
74.7	241	180	61	134.4
74.7	237	177	60	132.2
74.7	233	174	59	129.9
74.7	229	171	58	127.7
74.7	225	168	57	125.4
74.7	221	165	56	123.2
74.7	217	162	55	120.9
74.7	198	148	50	110.6
74.7	194	145	49	108.4
74.7	190	142	48	106.1
74.7	186	139	47	103.9
74.7	182	136	46	101.6
74.7	178	133	45	99.4
74.7	174	130	44	97.1
74.7	170	127	43	94.9
74.7	166	124	42	92.6
74.7	162	121	41	90.4
74.7	158	118	40	88.1
74.7	154	115	39	85.9
74.7	150	112	38	83.6
74.7	146	109	37	81.4
74.7	99	74	25	55.3
74.7	95	71	24	53.1
74.7	91	68	23	50.8
74.7	87	65	22	48.6
74.7	83	62	21	46.3
74.7	79	59	20	44.1
74.7	75	56	19	41.8

%E	M1	M2	DM	M*
74.6	500	373	127	278.3
74.6	497	371	126	276.9
74.6	496	370	126	276.0
74.6	493	368	125	274.7
74.6	492	367	125	273.8
74.6	489	365	124	272.4
74.6	488	364	124	271.5
74.6	485	362	123	270.2
74.6	484	361	123	269.3
74.6	481	359	122	267.9
74.6	480	358	122	267.0
74.6	477	356	121	265.7
74.6	476	355	121	264.8
74.6	473	353	120	263.4
74.6	472	352	120	262.5
74.6	469	350	119	261.2
74.6	468	349	119	260.3
74.6	465	347	118	258.9
74.6	464	346	118	258.0
74.6	461	344	117	256.7
74.6	460	343	117	255.8
74.6	457	341	116	254.4
74.6	456	340	116	253.5
74.6	453	338	115	252.2
74.6	452	337	115	251.3
74.6	449	335	114	249.9
74.6	448	334	114	249.0
74.6	445	332	113	247.7
74.6	444	331	113	246.8
74.6	441	329	112	245.4
74.6	437	326	111	243.2
74.6	433	323	110	240.9
74.6	429	320	109	238.7
74.6	426	318	108	237.4
74.6	425	317	108	236.4
74.6	422	315	107	235.1
74.6	421	314	107	234.2
74.6	418	312	106	232.9
74.6	417	311	106	231.9
74.6	414	309	105	230.6
74.6	413	308	105	229.7
74.6	410	306	104	228.4
74.6	409	305	104	227.4
74.6	406	303	103	226.1
74.6	405	302	103	225.2
74.6	402	300	102	223.9
74.6	401	299	102	222.9
74.6	398	297	101	221.6
74.6	397	296	101	220.7
74.6	394	294	100	219.4
74.6	393	293	100	218.4
74.6	390	291	99	217.1
74.6	389	290	99	216.2
74.6	386	288	98	214.9
74.6	382	285	97	212.6
74.6	378	282	96	210.4
74.6	374	279	95	208.1
74.6	370	276	94	205.9
74.6	366	273	93	203.6
74.6	362	270	92	201.4
74.6	358	267	91	199.1
74.6	355	265	90	197.8
74.6	354	264	90	196.9
74.6	351	262	89	195.6
74.6	350	261	89	194.6
74.6	347	259	88	193.3
74.6	346	258	88	192.4
74.6	343	256	87	191.1
74.6	342	255	87	190.1
74.6	339	253	86	188.8
74.6	338	252	86	187.9
74.6	335	250	85	186.6
74.6	334	249	85	185.6
74.6	331	247	84	184.3
74.6	327	244	83	182.1
74.6	323	241	82	179.8
74.6	319	238	81	177.6
74.6	315	235	80	175.3
74.6	311	232	79	173.1
74.6	307	229	78	170.8
74.6	303	226	77	168.6
74.6	299	223	76	166.3
74.6	295	220	75	164.1
74.6	291	217	74	161.8
74.6	287	214	73	159.6
74.6	284	212	72	158.3
74.6	283	211	72	157.3
74.6	280	209	71	156.0
74.6	279	208	71	155.1
74.6	276	206	70	153.8
74.6	272	203	69	151.5
74.6	268	200	68	149.3
74.6	264	197	67	147.0
74.6	260	194	66	144.8
74.6	256	191	65	142.5
74.6	252	188	64	140.3
74.6	248	185	63	138.0
74.6	244	182	62	135.8
74.6	240	179	61	133.5
74.6	236	176	60	131.3
74.6	232	173	59	129.0
74.6	228	170	58	126.8
74.6	224	167	57	124.5
74.6	213	159	54	118.7
74.6	209	156	53	116.4
74.6	205	153	52	114.2
74.6	201	150	51	111.9
74.6	197	147	50	109.7
74.6	193	144	49	107.4
74.6	189	141	48	105.2
74.6	185	138	47	102.9
74.6	181	135	46	100.7
74.6	177	132	45	98.4
74.6	173	129	44	96.2
74.6	169	126	43	93.9
74.6	142	106	36	79.1
74.6	138	103	35	76.9
74.6	134	100	34	74.6
74.6	130	97	33	72.4
74.6	126	94	32	70.1
74.6	122	91	31	67.9
74.6	118	88	30	65.6
74.6	114	85	29	63.4
74.6	71	53	18	39.6
74.6	67	50	17	37.3
74.6	63	47	16	35.1
74.6	59	44	15	32.8
74.5	499	372	127	277.3
74.5	498	371	127	276.4
74.5	495	369	126	275.1
74.5	494	368	126	274.1
74.5	491	366	125	272.8
74.5	490	365	125	271.9
74.5	487	363	124	270.6
74.5	486	362	124	269.6
74.5	483	360	123	268.3
74.5	482	359	123	267.4
74.5	479	357	122	266.1
74.5	478	356	122	265.1
74.5	475	354	121	263.8
74.5	474	353	121	262.9
74.5	471	351	120	261.6
74.5	470	350	120	260.6
74.5	467	348	119	259.3
74.5	466	347	119	258.4
74.5	463	345	118	257.1
74.5	462	344	118	256.1
74.5	459	342	117	254.8
74.5	458	341	117	253.9
74.5	455	339	116	252.6
74.5	451	336	115	250.3
74.5	447	333	114	248.1
74.5	443	330	113	245.8
74.5	440	328	112	244.5
74.5	439	327	112	243.6
74.5	436	325	111	242.3
74.5	435	324	111	241.3
74.5	432	322	110	240.0
74.5	431	321	110	239.1
74.5	428	319	109	237.8
74.5	427	318	109	236.8
74.5	424	316	108	235.5
74.5	423	315	108	234.6
74.5	420	313	107	233.3
74.5	419	312	107	232.3
74.5	416	310	106	231.0
74.5	415	309	106	230.1
74.5	412	307	105	228.8
74.5	411	306	105	227.8
74.5	408	304	104	226.5
74.5	404	301	103	224.3
74.5	400	298	102	222.0
74.5	396	295	101	219.8
74.5	392	292	100	217.5
74.5	388	289	99	215.3
74.5	385	287	98	213.9
74.5	384	286	98	213.0
74.5	381	284	97	211.7
74.5	380	283	97	210.8
74.5	377	281	96	209.4
74.5	376	280	96	208.5
74.5	373	278	95	207.2
74.5	372	277	95	206.3
74.5	369	275	94	204.9
74.5	368	274	94	204.0
74.5	365	272	93	202.7
74.5	364	271	93	201.8
74.5	361	269	92	200.4
74.5	357	266	91	198.2
74.5	353	263	90	195.9
74.5	349	260	89	193.7
74.5	345	257	88	191.4
74.5	341	254	87	189.2
74.5	337	251	86	186.9
74.5	333	248	85	184.7
74.5	330	246	84	183.4
74.5	329	245	84	182.4
74.5	326	243	83	181.1
74.5	325	242	83	180.2
74.5	322	240	82	178.9
74.5	321	239	82	177.9
74.5	318	237	81	176.6
74.5	314	234	80	174.4
74.5	310	231	79	172.1
74.5	306	228	78	169.9
74.5	302	225	77	167.6
74.5	298	222	76	165.4
74.5	294	219	75	163.1
74.5	290	216	74	160.9
74.5	286	213	73	158.6
74.5	282	210	72	156.4
74.5	278	207	71	154.1
74.5	275	205	70	152.8
74.5	274	204	70	151.9
74.5	271	202	69	150.6
74.5	267	199	68	148.3
74.5	263	196	67	146.1
74.5	259	193	66	143.8
74.5	255	190	65	141.6
74.5	251	187	64	139.3
74.5	247	184	63	137.1
74.5	243	181	62	134.8
74.5	239	178	61	132.6
74.5	235	175	60	130.3
74.5	231	172	59	128.1
74.5	220	164	56	122.3
74.5	216	161	55	120.0
74.5	212	158	54	117.8
74.5	208	155	53	115.5
74.5	204	152	52	113.3
74.5	200	149	51	111.0
74.5	196	146	50	108.8
74.5	192	143	49	106.5
74.5	188	140	48	104.3
74.5	184	137	47	102.0
74.5	165	123	42	91.7
74.5	161	120	41	89.4
74.5	157	117	40	87.2
74.5	153	114	39	84.9
74.5	149	111	38	82.7
74.5	145	108	37	80.4
74.5	141	105	36	78.2
74.5	137	102	35	75.9
74.5	110	82	28	61.1
74.5	106	79	27	58.9
74.5	102	76	26	56.6
74.5	98	73	25	54.4
74.5	94	70	24	52.1
74.5	55	41	14	30.6
74.5	51	38	13	28.3

%E	M1	M2	DM	M*
74.5	47	35	12	26.1
74.4	500	372	128	276.8
74.4	497	370	127	275.5
74.4	496	369	127	274.5
74.4	493	367	126	273.2
74.4	492	366	126	272.3
74.4	489	364	125	271.0
74.4	488	363	125	270.0
74.4	485	361	124	268.7
74.4	484	360	124	267.8
74.4	481	358	123	266.5
74.4	480	357	123	265.5
74.4	477	355	122	264.2
74.4	476	354	122	263.3
74.4	473	352	121	262.0
74.4	472	351	121	261.0
74.4	469	349	120	259.7
74.4	468	348	120	258.8
74.4	465	346	119	257.5
74.4	464	345	119	256.5
74.4	461	343	118	255.2
74.4	457	340	117	253.0
74.4	454	338	116	251.6
74.4	453	337	116	250.7
74.4	450	335	115	249.4
74.4	449	334	115	248.5
74.4	446	332	114	247.1
74.4	445	331	114	246.2
74.4	442	329	113	244.9
74.4	441	328	113	244.0
74.4	438	326	112	242.6
74.4	437	325	112	241.7
74.4	434	323	111	240.4
74.4	433	322	111	239.5
74.4	430	320	110	238.1
74.4	429	319	110	237.2
74.4	426	317	109	235.9
74.4	425	316	109	235.0
74.4	422	314	108	233.6
74.4	418	311	107	231.4
74.4	414	308	106	229.1
74.4	410	305	105	226.9
74.4	407	303	104	225.6
74.4	406	302	104	224.6
74.4	403	300	103	223.3
74.4	402	299	103	222.4
74.4	399	297	102	221.1
74.4	398	296	102	220.1
74.4	395	294	101	218.8
74.4	394	293	101	217.9
74.4	391	291	100	216.6
74.4	390	290	100	215.6
74.4	387	288	99	214.3
74.4	386	287	99	213.4
74.4	383	285	98	212.1
74.4	379	282	97	209.8
74.4	375	279	96	207.6
74.4	371	276	95	205.3
74.4	367	273	94	203.1
74.4	363	270	93	200.8
74.4	360	268	92	199.5
74.4	359	267	92	198.6
74.4	356	265	91	197.3
74.4	355	264	91	196.3
74.4	352	262	90	195.0
74.4	351	261	90	194.1
74.4	348	259	89	192.8
74.4	347	258	89	191.8
74.4	344	256	88	190.5
74.4	340	253	87	188.3
74.4	336	250	86	186.0
74.4	332	247	85	183.8
74.4	328	244	84	181.5
74.4	324	241	83	179.3
74.4	320	238	82	177.0
74.4	317	236	81	175.7
74.4	316	235	81	174.8
74.4	313	233	80	173.4
74.4	312	232	80	172.5
74.4	309	230	79	171.2
74.4	308	229	79	170.3
74.4	305	227	78	168.9
74.4	301	224	77	166.7
74.4	297	221	76	164.4
74.4	293	218	75	162.2
74.4	289	215	74	159.9
74.4	285	212	73	157.7
74.4	281	209	72	155.4
74.4	277	206	71	153.2
74.4	273	203	70	150.9
74.4	270	201	69	149.6
74.4	266	198	68	147.4
74.4	262	195	67	145.1
74.4	258	192	66	142.9
74.4	254	189	65	140.6
74.4	250	186	64	138.4
74.4	246	183	63	136.1
74.4	242	180	62	133.9
74.4	238	177	61	131.6
74.4	234	174	60	129.4
74.4	227	169	58	125.8
74.4	223	166	57	123.6
74.4	219	163	56	121.3
74.4	215	160	55	119.1
74.4	211	157	54	116.8
74.4	207	154	53	114.6
74.4	203	151	52	112.3
74.4	199	148	51	110.1
74.4	195	145	50	107.8
74.4	180	134	46	99.8
74.4	176	131	45	97.5
74.4	172	128	44	95.3
74.4	168	125	43	93.0
74.4	164	122	42	90.8
74.4	160	119	41	88.5
74.4	156	116	40	86.3
74.4	133	99	34	73.7
74.4	129	96	33	71.4
74.4	125	93	32	69.2
74.4	121	90	31	66.9
74.4	117	87	30	64.7
74.4	90	67	23	49.9
74.4	86	64	22	47.6
74.4	82	61	21	45.4
74.4	78	58	20	43.1
74.4	43	32	11	23.8
74.4	39	29	10	21.6
74.3	499	371	128	275.8
74.3	498	370	128	274.9
74.3	495	368	127	273.6
74.3	494	367	127	272.6
74.3	491	365	126	271.3
74.3	490	364	126	270.4
74.3	487	362	125	269.1
74.3	486	361	125	268.2
74.3	483	359	124	266.8
74.3	482	358	124	265.9
74.3	479	356	123	264.6
74.3	478	355	123	263.7
74.3	475	353	122	262.3
74.3	474	352	122	261.4
74.3	471	350	121	260.1
74.3	470	349	121	259.2
74.3	467	347	120	257.8
74.3	463	344	119	255.6
74.3	460	342	118	254.3
74.3	459	341	118	253.3
74.3	456	339	117	252.0
74.3	455	338	117	251.1
74.3	452	336	116	249.8
74.3	451	335	116	248.8
74.3	448	333	115	247.5
74.3	447	332	115	246.6
74.3	444	330	114	245.3
74.3	443	329	114	244.3
74.3	440	327	113	243.0
74.3	439	326	113	242.1
74.3	436	324	112	240.8
74.3	435	323	112	239.8
74.3	432	321	111	238.5
74.3	428	318	110	236.3
74.3	424	315	109	234.0
74.3	421	313	108	232.7
74.3	420	312	108	231.8
74.3	417	310	107	230.5
74.3	416	309	107	229.5
74.3	413	307	106	228.2
74.3	412	306	106	227.3
74.3	409	304	105	226.0
74.3	408	303	105	225.0
74.3	405	301	104	223.7
74.3	404	300	104	222.8
74.3	401	298	103	221.5
74.3	400	297	103	220.5
74.3	397	295	102	219.2
74.3	393	292	101	217.0
74.3	389	289	100	214.7
74.3	385	286	99	212.5
74.3	382	284	98	211.1
74.3	381	283	98	210.2
74.3	378	281	97	208.9
74.3	377	280	97	208.0
74.3	374	278	96	206.6
74.3	373	277	96	205.7
74.3	370	275	95	204.4
74.3	369	274	95	203.5
74.3	366	272	94	202.1
74.3	362	269	93	199.9
74.3	358	266	92	197.6
74.3	354	263	91	195.4
74.3	350	260	90	193.1
74.3	346	257	89	190.9
74.3	343	255	88	189.6
74.3	342	254	88	188.6
74.3	339	252	87	187.3
74.3	338	251	87	186.4
74.3	335	249	86	185.1
74.3	334	248	86	184.1
74.3	331	246	85	182.8
74.3	327	243	84	180.6
74.3	323	240	83	178.3
74.3	319	237	82	176.1
74.3	315	234	81	173.8
74.3	311	231	80	171.6
74.3	307	228	79	169.3
74.3	304	226	78	168.0
74.3	303	225	78	167.1
74.3	300	223	77	165.8
74.3	296	220	76	163.5
74.3	292	217	75	161.3
74.3	288	214	74	159.0
74.3	284	211	73	156.8
74.3	280	208	72	154.5
74.3	276	205	71	152.3
74.3	272	202	70	150.0
74.3	269	200	69	148.7
74.3	268	199	69	147.8
74.3	265	197	68	146.4
74.3	261	194	67	144.2
74.3	257	191	66	141.9
74.3	253	188	65	139.7
74.3	249	185	64	137.4
74.3	245	182	63	135.2
74.3	241	179	62	133.0
74.3	237	176	61	130.7
74.3	230	171	59	127.1
74.3	226	168	58	124.9
74.3	222	165	57	122.6
74.3	218	162	56	120.4
74.3	214	159	55	118.1
74.3	210	156	54	115.9
74.3	206	153	53	113.6
74.3	202	150	52	111.4
74.3	191	142	49	105.6
74.3	187	139	48	103.3
74.3	183	136	47	101.1
74.3	179	133	46	98.8
74.3	175	130	45	96.6
74.3	171	127	44	94.3
74.3	167	124	43	92.1
74.3	152	113	39	84.0
74.3	148	110	38	81.8
74.3	144	107	37	79.5
74.3	140	104	36	77.3
74.3	136	101	35	75.0
74.3	113	84	29	62.4
74.3	109	81	28	60.2
74.3	105	78	27	57.9
74.3	101	75	26	55.7
74.3	74	55	19	40.9

%E	M1	M2	DM	M*
74.3	70	52	18	38.6
74.3	35	26	9	19.3
74.2	500	371	129	275.3
74.2	497	369	128	274.0
74.2	496	368	128	273.0
74.2	493	366	127	271.7
74.2	492	365	127	270.8
74.2	489	363	126	269.5
74.2	488	362	126	268.5
74.2	485	360	125	267.2
74.2	484	359	125	266.3
74.2	481	357	124	265.0
74.2	480	356	124	264.0
74.2	477	354	123	262.7
74.2	476	353	123	261.8
74.2	473	351	122	260.5
74.2	472	350	122	259.5
74.2	469	348	121	258.2
74.2	466	346	120	256.9
74.2	465	345	120	256.0
74.2	462	343	119	254.7
74.2	461	342	119	253.7
74.2	458	340	118	252.4
74.2	457	339	118	251.5
74.2	454	337	117	250.2
74.2	453	336	117	249.2
74.2	450	334	116	247.9
74.2	449	333	116	247.0
74.2	446	331	115	245.7
74.2	445	330	115	244.7
74.2	442	328	114	243.4
74.2	441	327	114	242.5
74.2	438	325	113	241.2
74.2	434	322	112	238.9
74.2	431	320	111	237.6
74.2	430	319	111	236.7
74.2	427	317	110	235.3
74.2	426	316	110	234.4
74.2	423	314	109	233.1
74.2	422	313	109	232.2
74.2	419	311	108	230.8
74.2	418	310	108	229.9
74.2	415	308	107	228.6
74.2	414	307	107	227.7
74.2	411	305	106	226.3
74.2	407	302	105	224.1
74.2	403	299	104	221.8
74.2	399	296	103	219.6
74.2	396	294	102	218.3
74.2	395	293	102	217.3

%E	M1	M2	DM	M*
74.2	392	291	101	216.0
74.2	391	290	101	215.1
74.2	388	288	100	213.8
74.2	387	287	100	212.8
74.2	384	285	99	211.5
74.2	383	284	99	210.6
74.2	380	282	98	209.3
74.2	376	279	97	207.0
74.2	372	276	96	204.8
74.2	368	273	95	202.5
74.2	365	271	94	201.2
74.2	364	270	94	200.3
74.2	361	268	93	199.0
74.2	360	267	93	198.0
74.2	357	265	92	196.7
74.2	356	264	92	195.8
74.2	353	262	91	194.5
74.2	349	259	90	192.2
74.2	345	256	89	190.0
74.2	341	253	88	187.7
74.2	337	250	87	185.5
74.2	333	247	86	183.2
74.2	330	245	85	181.9
74.2	329	244	85	181.0
74.2	326	242	84	179.6
74.2	325	241	84	178.7
74.2	322	239	83	177.4
74.2	318	236	82	175.1
74.2	314	233	81	172.9
74.2	310	230	80	170.6
74.2	306	227	79	168.4
74.2	302	224	78	166.1
74.2	299	222	77	164.8
74.2	298	221	77	163.9
74.2	295	219	76	162.6
74.2	294	218	76	161.6
74.2	291	216	75	160.3
74.2	287	213	74	158.1
74.2	283	210	73	155.8
74.2	279	207	72	153.6
74.2	275	204	71	151.3
74.2	271	201	70	149.1
74.2	267	198	69	146.8
74.2	264	196	68	145.5
74.2	260	193	67	143.3
74.2	256	190	66	141.0
74.2	252	187	65	138.8
74.2	248	184	64	136.5
74.2	244	181	63	134.3
74.2	240	178	62	132.0

%E	M1	M2	DM	M*
74.2	236	175	61	129.8
74.2	233	173	60	128.5
74.2	229	170	59	126.2
74.2	225	167	58	124.0
74.2	221	164	57	121.7
74.2	217	161	56	119.5
74.2	213	158	55	117.2
74.2	209	155	54	115.0
74.2	198	147	51	109.1
74.2	194	144	50	106.9
74.2	190	141	49	104.6
74.2	186	138	48	102.4
74.2	182	135	47	100.1
74.2	178	132	46	97.9
74.2	163	121	42	89.8
74.2	159	118	41	87.6
74.2	155	115	40	85.3
74.2	151	112	39	83.1
74.2	147	109	38	80.8
74.2	132	98	34	72.8
74.2	128	95	33	70.5
74.2	124	92	32	68.3
74.2	120	89	31	66.0
74.2	97	72	25	53.4
74.2	93	69	24	51.2
74.2	89	66	23	48.9
74.2	66	49	17	36.4
74.2	62	46	16	34.1
74.2	31	23	8	17.1
74.1	499	370	129	274.3
74.1	498	369	129	273.4
74.1	495	367	128	272.1
74.1	494	366	128	271.2
74.1	491	364	127	269.8
74.1	490	363	127	268.9
74.1	487	361	126	267.6
74.1	486	360	126	266.7
74.1	483	358	125	265.3
74.1	482	357	125	264.4
74.1	479	355	124	263.1
74.1	478	354	124	262.2
74.1	475	352	123	260.9
74.1	474	351	123	259.9
74.1	471	349	122	258.6
74.1	468	347	121	257.3
74.1	467	346	121	256.4
74.1	464	344	120	255.0
74.1	463	343	120	254.1
74.1	460	341	119	252.8
74.1	459	340	119	251.9

%E	M1	M2	DM	M*
74.1	456	338	118	250.5
74.1	455	337	118	249.6
74.1	452	335	117	248.3
74.1	451	334	117	247.4
74.1	448	332	116	246.0
74.1	444	329	115	243.8
74.1	440	326	114	241.5
74.1	437	324	113	240.2
74.1	436	323	113	239.3
74.1	433	321	112	238.0
74.1	432	320	112	237.0
74.1	429	318	111	235.7
74.1	428	317	111	234.8
74.1	425	315	110	233.5
74.1	424	314	110	232.5
74.1	421	312	109	231.2
74.1	417	309	108	229.0
74.1	413	306	107	226.7
74.1	410	304	106	225.4
74.1	409	303	106	224.5
74.1	406	301	105	223.2
74.1	405	300	105	222.2
74.1	402	298	104	220.9
74.1	401	297	104	220.0
74.1	398	295	103	218.7
74.1	397	294	103	217.7
74.1	394	292	102	216.4
74.1	390	289	101	214.2
74.1	386	286	100	211.9
74.1	382	283	99	209.7
74.1	379	281	98	208.3
74.1	378	280	98	207.4
74.1	375	278	97	206.1
74.1	374	277	97	205.2
74.1	371	275	96	203.8
74.1	370	274	96	202.9
74.1	367	272	95	201.6
74.1	363	269	94	199.3
74.1	359	266	93	197.1
74.1	355	263	92	194.8
74.1	352	261	91	193.5
74.1	351	260	91	192.6
74.1	348	258	90	191.3
74.1	347	257	90	190.3
74.1	344	255	89	189.0
74.1	343	254	89	188.1
74.1	340	252	88	186.8
74.1	336	249	87	184.5
74.1	332	246	86	182.3
74.1	328	243	85	180.0

%E	M1	M2	DM	M*
74.1	324	240	84	177.8
74.1	321	238	83	176.5
74.1	320	237	83	175.5
74.1	317	235	82	174.2
74.1	316	234	82	173.3
74.1	313	232	81	172.0
74.1	309	229	80	169.7
74.1	305	226	79	167.5
74.1	301	223	78	165.2
74.1	297	220	77	163.0
74.1	293	217	76	160.7
74.1	290	215	75	159.4
74.1	286	212	74	157.1
74.1	282	209	73	154.9
74.1	278	206	72	152.6
74.1	274	203	71	150.4
74.1	270	200	70	148.1
74.1	266	197	69	145.9
74.1	263	195	68	144.6
74.1	259	192	67	142.3
74.1	255	189	66	140.1
74.1	251	186	65	137.8
74.1	247	183	64	135.6
74.1	243	180	63	133.3
74.1	239	177	62	131.1
74.1	232	172	60	127.5
74.1	228	169	59	125.3
74.1	224	166	58	123.0
74.1	220	163	57	120.8
74.1	216	160	56	118.5
74.1	212	157	55	116.3
74.1	205	152	53	112.7
74.1	201	149	52	110.5
74.1	197	146	51	108.2
74.1	193	143	50	106.0
74.1	189	140	49	103.7
74.1	185	137	48	101.5
74.1	174	129	45	95.6
74.1	170	126	44	93.4
74.1	166	123	43	91.1
74.1	162	120	42	88.9
74.1	158	117	41	86.6
74.1	143	106	37	78.6
74.1	139	103	36	76.3
74.1	135	100	35	74.1
74.1	116	86	30	63.8
74.1	112	83	29	61.5
74.1	108	80	28	59.3
74.1	85	63	22	46.7
74.1	81	60	21	44.4

%E	M1	M2	DM	M*
74.1	58	43	15	31.9
74.1	54	40	14	29.6
74.1	27	20	7	14.8
74.0	500	370	130	273.8
74.0	497	368	129	272.5
74.0	496	367	129	271.6
74.0	493	365	128	270.2
74.0	492	364	128	269.3
74.0	489	362	127	268.0
74.0	488	361	127	267.1
74.0	485	359	126	265.7
74.0	484	358	126	264.8
74.0	481	356	125	263.5
74.0	480	355	125	262.6
74.0	477	353	124	261.2
74.0	476	352	124	260.3
74.0	473	350	123	259.0
74.0	470	348	122	257.7
74.0	469	347	122	256.7
74.0	466	345	121	255.4
74.0	465	344	121	254.5
74.0	462	342	120	253.2
74.0	461	341	120	252.2
74.0	458	339	119	250.9
74.0	457	338	119	250.0
74.0	454	336	118	248.7
74.0	453	335	118	247.7
74.0	450	333	117	246.4
74.0	447	331	116	245.1
74.0	446	330	116	244.2
74.0	443	328	115	242.9
74.0	442	327	115	241.9
74.0	439	325	114	240.6
74.0	438	324	114	239.7
74.0	435	322	113	238.4
74.0	434	321	113	237.4
74.0	431	319	112	236.1
74.0	430	318	112	235.2
74.0	427	316	111	233.9
74.0	423	313	110	231.6
74.0	420	311	109	230.3
74.0	419	310	109	229.4
74.0	416	308	108	228.0
74.0	415	307	108	227.1
74.0	412	305	107	225.8
74.0	411	304	107	224.9
74.0	408	302	106	223.5
74.0	407	301	106	222.6
74.0	404	299	105	221.3
74.0	400	296	104	219.0

%E	M1	M2	DM	M*
74.0	396	293	103	216.8
74.0	393	291	102	215.5
74.0	392	290	102	214.5
74.0	389	288	101	213.2
74.0	388	287	101	212.3
74.0	385	285	100	211.0
74.0	384	284	100	210.0
74.0	381	282	99	208.7
74.0	377	279	98	206.5
74.0	373	276	97	204.2
74.0	369	273	96	202.0
74.0	366	271	95	200.7
74.0	365	270	95	199.7
74.0	362	268	94	198.4
74.0	361	267	94	197.5
74.0	358	265	93	196.2
74.0	357	264	93	195.2
74.0	354	262	92	193.9
74.0	350	259	91	191.7
74.0	346	256	90	189.4
74.0	342	253	89	187.2
74.0	339	251	88	185.8
74.0	338	250	88	184.9
74.0	335	248	87	183.6
74.0	334	247	87	182.7
74.0	331	245	86	181.3
74.0	327	242	85	179.1
74.0	323	239	84	176.8
74.0	319	236	83	174.6
74.0	315	233	82	172.3
74.0	312	231	81	171.0
74.0	311	230	81	170.1
74.0	308	228	80	168.8
74.0	304	225	79	166.5
74.0	300	222	78	164.3
74.0	296	219	77	162.0
74.0	292	216	76	159.8
74.0	289	214	75	158.5
74.0	288	213	75	157.5
74.0	285	211	74	156.2
74.0	281	208	73	154.0
74.0	277	205	72	151.7
74.0	273	202	71	149.5
74.0	269	199	70	147.2
74.0	265	196	69	145.0
74.0	262	194	68	143.6
74.0	258	191	67	141.4
74.0	254	188	66	139.1
74.0	250	185	65	136.9
74.0	246	182	64	134.7

%E	M1	M2	DM	M*
74.0	242	179	63	132.4
74.0	238	176	62	130.2
74.0	235	174	61	128.8
74.0	231	171	60	126.6
74.0	227	168	59	124.3
74.0	223	165	58	122.1
74.0	219	162	57	119.8
74.0	215	159	56	117.6
74.0	208	154	54	114.0
74.0	204	151	53	111.8
74.0	200	148	52	109.5
74.0	196	145	51	107.3
74.0	192	142	50	105.0
74.0	181	134	47	99.2
74.0	177	131	46	97.0
74.0	173	128	45	94.7
74.0	169	125	44	92.5
74.0	154	114	40	84.4
74.0	150	111	39	82.1
74.0	146	108	38	79.9
74.0	131	97	34	71.8
74.0	127	94	33	69.6
74.0	123	91	32	67.3
74.0	119	88	31	65.1
74.0	104	77	27	57.0
74.0	100	74	26	54.8
74.0	96	71	25	52.5
74.0	77	57	20	42.2
74.0	73	54	19	39.9
74.0	50	37	13	27.4
73.9	499	369	130	272.9
73.9	498	368	130	271.9
73.9	495	366	129	270.6
73.9	494	365	129	269.7
73.9	491	363	128	268.4
73.9	490	362	128	267.4
73.9	487	360	127	266.1
73.9	486	359	127	265.2
73.9	483	357	126	263.9
73.9	482	356	126	262.9
73.9	479	354	125	261.6
73.9	475	351	124	259.4
73.9	472	349	123	258.1
73.9	471	348	123	257.1
73.9	468	346	122	255.8
73.9	467	345	122	254.9
73.9	464	343	121	253.6
73.9	463	342	121	252.6
73.9	460	340	120	251.3
73.9	459	339	120	250.4

%E	M1	M2	DM	M*
73.9	456	337	119	249.1
73.9	452	334	118	246.8
73.9	449	332	117	245.5
73.9	448	331	117	244.6
73.9	445	329	116	243.2
73.9	444	328	116	242.3
73.9	441	326	115	241.0
73.9	440	325	115	240.1
73.9	437	323	114	238.7
73.9	436	322	114	237.8
73.9	433	320	113	236.5
73.9	429	317	112	234.2
73.9	426	315	111	232.9
73.9	425	314	111	232.0
73.9	422	312	110	230.7
73.9	421	311	110	229.7
73.9	418	309	109	228.4
73.9	417	308	109	227.5
73.9	414	306	108	226.2
73.9	413	305	108	225.2
73.9	410	303	107	223.9
73.9	406	300	106	221.7
73.9	403	298	105	220.4
73.9	402	297	105	219.4
73.9	399	295	104	218.1
73.9	398	294	104	217.2
73.9	395	292	103	215.9
73.9	394	291	103	214.9
73.9	391	289	102	213.6
73.9	387	286	101	211.4
73.9	383	283	100	209.1
73.9	380	281	99	207.8
73.9	379	280	99	206.9
73.9	376	278	98	205.5
73.9	375	277	98	204.6
73.9	372	275	97	203.3
73.9	371	274	97	202.4
73.9	368	272	96	201.0
73.9	364	269	95	198.8
73.9	360	266	94	196.5
73.9	356	263	93	194.3
73.9	353	261	92	193.0
73.9	352	260	92	192.0
73.9	349	258	91	190.7
73.9	348	257	91	189.8
73.9	345	255	90	188.5
73.9	341	252	89	186.2
73.9	337	249	88	184.0
73.9	333	246	87	181.7
73.9	330	244	86	180.4

%E	M1	M2	DM	M*
73.9	329	243	86	179.5
73.9	326	241	85	178.2
73.9	322	238	84	175.9
73.9	318	235	83	173.7
73.9	314	232	82	171.4
73.9	310	229	81	169.2
73.9	307	227	80	167.8
73.9	306	226	80	166.9
73.9	303	224	79	165.6
73.9	299	221	78	163.3
73.9	295	218	77	161.1
73.9	291	215	76	158.8
73.9	287	212	75	156.6
73.9	284	210	74	155.3
73.9	283	209	74	154.3
73.9	280	207	73	153.0
73.9	276	204	72	150.8
73.9	272	201	71	148.5
73.9	268	198	70	146.3
73.9	264	195	69	144.0
73.9	261	193	68	142.7
73.9	257	190	67	140.5
73.9	253	187	66	138.2
73.9	249	184	65	136.0
73.9	245	181	64	133.7
73.9	241	178	63	131.5
73.9	234	173	61	127.9
73.9	230	170	60	125.7
73.9	226	167	59	123.4
73.9	222	164	58	121.2
73.9	218	161	57	118.9
73.9	211	156	55	115.3
73.9	207	153	54	113.1
73.9	203	150	53	110.8
73.9	199	147	52	108.6
73.9	188	139	49	102.8
73.9	184	136	48	100.5
73.9	180	133	47	98.3
73.9	176	130	46	96.0
73.9	165	122	43	90.2
73.9	161	119	42	88.0
73.9	157	116	41	85.7
73.9	153	113	40	83.5
73.9	142	105	37	77.6
73.9	138	102	36	75.4
73.9	134	99	35	73.1
73.9	115	85	30	62.8
73.9	111	82	29	60.6
73.9	92	68	24	50.3
73.9	88	65	23	48.0

%E	M1	M2	DM	M*
73.9	69	51	18	37.7
73.9	46	34	12	25.1
73.9	23	17	6	12.6
73.8	500	369	131	272.3
73.8	497	367	130	271.0
73.8	496	366	130	270.1
73.8	493	364	129	268.8
73.8	492	363	129	267.8
73.8	489	361	128	266.5
73.8	488	360	128	265.6
73.8	485	358	127	264.3
73.8	484	357	127	263.3
73.8	481	355	126	262.0
73.8	480	354	126	261.1
73.8	478	353	125	260.7
73.8	477	352	125	259.8
73.8	474	350	124	258.4
73.8	473	349	124	257.5
73.8	470	347	123	256.2
73.8	469	346	123	255.3
73.8	466	344	122	253.9
73.8	465	343	122	253.0
73.8	462	341	121	251.7
73.8	461	340	121	250.8
73.8	458	338	120	249.4
73.8	455	336	119	248.1
73.8	454	335	119	247.2
73.8	451	333	118	245.9
73.8	450	332	118	244.9
73.8	447	330	117	243.6
73.8	446	329	117	242.7
73.8	443	327	116	241.4
73.8	442	326	116	240.4
73.8	439	324	115	239.1
73.8	435	321	114	236.9
73.8	432	319	113	235.6
73.8	431	318	113	234.6
73.8	428	316	112	233.3
73.8	427	315	112	232.4
73.8	424	313	111	231.1
73.8	423	312	111	230.1
73.8	420	310	110	228.8
73.8	416	307	109	226.6
73.8	412	304	108	224.3
73.8	409	302	107	223.0
73.8	408	301	107	222.1
73.8	405	299	106	220.7
73.8	404	298	106	219.8
73.8	401	296	105	218.5
73.8	400	295	105	217.6

%E	M1	M2	DM	M*
73.8	397	293	104	216.2
73.8	393	290	103	214.0
73.8	390	288	102	212.7
73.8	389	287	102	211.7
73.8	386	285	101	210.4
73.8	385	284	101	209.5
73.8	382	282	100	208.2
73.8	381	281	100	207.2
73.8	378	279	99	205.9
73.8	374	276	98	203.7
73.8	370	273	97	201.4
73.8	367	271	96	200.1
73.8	366	270	96	199.2
73.8	363	268	95	197.9
73.8	362	267	95	196.9
73.8	359	265	94	195.6
73.8	355	262	93	193.4
73.8	351	259	92	191.1
73.8	347	256	91	188.9
73.8	344	254	90	187.5
73.8	343	253	90	186.6
73.8	340	251	89	185.3
73.8	336	248	88	183.0
73.8	332	245	87	180.8
73.8	328	242	86	178.5
73.8	325	240	85	177.2
73.8	324	239	85	176.3
73.8	321	237	84	175.0
73.8	320	236	84	174.0
73.8	317	234	83	172.7
73.8	313	231	82	170.5
73.8	309	228	81	168.2
73.8	305	225	80	166.0
73.8	302	223	79	164.7
73.8	301	222	79	163.7
73.8	298	220	78	162.4
73.8	294	217	77	160.2
73.8	290	214	76	157.9
73.8	286	211	75	155.7
73.8	282	208	74	153.4
73.8	279	206	73	152.1
73.8	275	203	72	149.9
73.8	271	200	71	147.6
73.8	267	197	70	145.4
73.8	263	194	69	143.1
73.8	260	192	68	141.8
73.8	256	189	67	139.5
73.8	252	186	66	137.3
73.8	248	183	65	135.0
73.8	244	180	64	132.8

%E	M1	M2	DM	M*
73.8	240	177	63	130.5
73.8	237	175	62	129.2
73.8	233	172	61	127.0
73.8	229	169	60	124.7
73.8	225	166	59	122.5
73.8	221	163	58	120.2
73.8	214	158	56	116.7
73.8	210	155	55	114.4
73.8	206	152	54	112.2
73.8	202	149	53	109.9
73.8	195	144	51	106.3
73.8	191	141	50	104.1
73.8	187	138	49	101.8
73.8	183	135	48	99.6
73.8	172	127	45	93.8
73.8	168	124	44	91.5
73.8	164	121	43	89.3
73.8	160	118	42	87.0
73.8	149	110	39	81.2
73.8	145	107	38	79.0
73.8	141	104	37	76.7
73.8	130	96	34	70.9
73.8	126	93	33	68.6
73.8	122	90	32	66.4
73.8	107	79	28	58.3
73.8	103	76	27	56.1
73.8	84	62	22	45.8
73.8	80	59	21	43.5
73.8	65	48	17	35.4
73.8	61	45	16	33.2
73.8	42	31	11	22.9
73.7	499	368	131	271.4
73.7	498	367	131	270.5
73.7	495	365	130	269.1
73.7	494	364	130	268.2
73.7	491	362	129	266.9
73.7	490	361	129	266.0
73.7	487	359	128	264.6
73.7	486	358	128	263.7
73.7	483	356	127	262.4
73.7	482	355	127	261.5
73.7	479	353	126	260.1
73.7	476	351	125	258.8
73.7	475	350	125	257.9
73.7	472	348	124	256.6
73.7	471	347	124	255.6
73.7	468	345	123	254.3
73.7	467	344	123	253.4
73.7	464	342	122	252.1
73.7	463	341	122	251.1

%E	M1	M2	DM	M*
73.7	460	339	121	249.8
73.7	457	337	120	248.5
73.7	456	336	120	247.6
73.7	453	334	119	246.3
73.7	452	333	119	245.3
73.7	449	331	118	244.0
73.7	448	330	118	243.1
73.7	445	328	117	241.8
73.7	441	325	116	239.5
73.7	438	323	115	238.2
73.7	437	322	115	237.3
73.7	434	320	114	235.9
73.7	433	319	114	235.0
73.7	430	317	113	233.7
73.7	429	316	113	232.8
73.7	426	314	112	231.4
73.7	422	311	111	229.2
73.7	419	309	110	227.9
73.7	418	308	110	226.9
73.7	415	306	109	225.6
73.7	414	305	109	224.7
73.7	411	303	108	223.4
73.7	410	302	108	222.4
73.7	407	300	107	221.1
73.7	403	297	106	218.9
73.7	399	294	105	216.6
73.7	396	292	104	215.3
73.7	395	291	104	214.4
73.7	392	289	103	213.1
73.7	391	288	103	212.1
73.7	388	286	102	210.8
73.7	384	283	101	208.6
73.7	380	280	100	206.3
73.7	377	278	99	205.0
73.7	376	277	99	204.1
73.7	373	275	98	202.7
73.7	372	274	98	201.8
73.7	369	272	97	200.5
73.7	365	269	96	198.2
73.7	361	266	95	196.0
73.7	358	264	94	194.7
73.7	357	263	94	193.8
73.7	354	261	93	192.4
73.7	353	260	93	191.5
73.7	350	258	92	190.2
73.7	346	255	91	187.9
73.7	342	252	90	185.7
73.7	339	250	89	184.4
73.7	338	249	89	183.4
73.7	335	247	88	182.1

%E	M1	M2	DM	M*
73.7	334	246	88	181.2
73.7	331	244	87	179.9
73.7	327	241	86	177.6
73.7	323	238	85	175.4
73.7	319	235	84	173.1
73.7	316	233	83	171.8
73.7	315	232	83	170.9
73.7	312	230	82	169.6
73.7	308	227	81	167.3
73.7	304	224	80	165.1
73.7	300	221	79	162.8
73.7	297	219	78	161.5
73.7	293	216	77	159.2
73.7	289	213	76	157.0
73.7	285	210	75	154.7
73.7	281	207	74	152.5
73.7	278	205	73	151.2
73.7	274	202	72	148.9
73.7	270	199	71	146.7
73.7	266	196	70	144.4
73.7	262	193	69	142.2
73.7	259	191	68	140.9
73.7	255	188	67	138.6
73.7	251	185	66	136.4
73.7	247	182	65	134.1
73.7	243	179	64	131.9
73.7	236	174	62	128.3
73.7	232	171	61	126.0
73.7	228	168	60	123.8
73.7	224	165	59	121.5
73.7	217	160	57	118.0
73.7	213	157	56	115.7
73.7	209	154	55	113.5
73.7	205	151	54	111.2
73.7	198	146	52	107.7
73.7	194	143	51	105.4
73.7	190	140	50	103.2
73.7	186	137	49	100.9
73.7	179	132	47	97.3
73.7	175	129	46	95.1
73.7	171	126	45	92.8
73.7	167	123	44	90.6
73.7	156	115	41	84.8
73.7	152	112	40	82.5
73.7	137	101	36	74.5
73.7	133	98	35	72.2
73.7	118	87	31	64.1
73.7	114	84	30	61.9
73.7	99	73	26	53.8
73.7	95	70	25	51.6

%E	M1	M2	DM	M*
73.7	76	56	20	41.3
73.7	57	42	15	30.9
73.7	38	28	10	20.6
73.7	19	14	5	10.3
73.6	500	368	132	270.8
73.6	497	366	131	269.5
73.6	496	365	131	268.6
73.6	493	363	130	267.3
73.6	492	362	130	266.3
73.6	489	360	129	265.0
73.6	488	359	129	264.1
73.6	485	357	128	262.8
73.6	484	356	128	261.9
73.6	481	354	127	260.5
73.6	478	352	126	259.2
73.6	477	351	126	258.3
73.6	474	349	125	257.0
73.6	473	348	125	256.0
73.6	470	346	124	254.7
73.6	469	345	124	253.8
73.6	466	343	123	252.5
73.6	462	340	122	250.2
73.6	459	338	121	248.9
73.6	458	337	121	248.0
73.6	455	335	120	246.6
73.6	454	334	120	245.7
73.6	451	332	119	244.4
73.6	450	331	119	243.5
73.6	447	329	118	242.1
73.6	444	327	117	240.8
73.6	443	326	117	239.9
73.6	440	324	116	238.6
73.6	439	323	116	237.7
73.6	436	321	115	236.3
73.6	435	320	115	235.4
73.6	432	318	114	234.1
73.6	431	317	114	233.2
73.6	428	315	113	231.8
73.6	425	313	112	230.5
73.6	424	312	112	229.6
73.6	421	310	111	228.3
73.6	420	309	111	227.3
73.6	417	307	110	226.0
73.6	416	306	110	225.1
73.6	413	304	109	223.8
73.6	409	301	108	221.5
73.6	406	299	107	220.2
73.6	405	298	107	219.3
73.6	402	296	106	218.0
73.6	401	295	106	217.0

%E	M1	M2	DM	M*
73.6	398	293	105	215.7
73.6	397	292	105	214.8
73.6	394	290	104	213.5
73.6	390	287	103	211.2
73.6	387	285	102	209.9
73.6	386	284	102	209.0
73.6	383	282	101	207.6
73.6	382	281	101	206.7
73.6	379	279	100	205.4
73.6	375	276	99	203.1
73.6	371	273	98	200.9
73.6	368	271	97	199.6
73.6	367	270	97	198.6
73.6	364	268	96	197.3
73.6	363	267	96	196.4
73.6	360	265	95	195.1
73.6	356	262	94	192.8
73.6	352	259	93	190.6
73.6	349	257	92	189.3
73.6	348	256	92	188.3
73.6	345	254	91	187.0
73.6	341	251	90	184.8
73.6	337	248	89	182.5
73.6	333	245	88	180.3
73.6	330	243	87	178.9
73.6	329	242	87	178.0
73.6	326	240	86	176.7
73.6	322	237	85	174.4
73.6	318	234	84	172.2
73.6	314	231	83	169.9
73.6	311	229	82	168.6
73.6	307	226	81	166.4
73.6	303	223	80	164.1
73.6	299	220	79	161.9
73.6	296	218	78	160.6
73.6	295	217	78	159.6
73.6	292	215	77	158.3
73.6	288	212	76	156.1
73.6	284	209	75	153.8
73.6	280	206	74	151.6
73.6	277	204	73	150.2
73.6	276	203	73	149.3
73.6	273	201	72	148.0
73.6	269	198	71	145.7
73.6	265	195	70	143.5
73.6	261	192	69	141.2
73.6	258	190	68	139.9
73.6	254	187	67	137.7
73.6	250	184	66	135.4
73.6	246	181	65	133.2

%E	M1	M2	DM	M*
73.6	242	178	64	130.9
73.6	239	176	63	129.6
73.6	235	173	62	127.4
73.6	231	170	61	125.1
73.6	227	167	60	122.9
73.6	220	162	58	119.3
73.6	216	159	57	117.0
73.6	212	156	56	114.8
73.6	208	153	55	112.5
73.6	201	148	53	109.0
73.6	197	145	52	106.7
73.6	193	142	51	104.5
73.6	182	134	48	98.7
73.6	178	131	47	96.4
73.6	174	128	46	94.2
73.6	163	120	43	88.3
73.6	159	117	42	86.1
73.6	148	109	39	80.3
73.6	144	106	38	78.0
73.6	140	103	37	75.8
73.6	129	95	34	70.0
73.6	125	92	33	67.7
73.6	121	89	32	65.5
73.6	110	81	29	59.6
73.6	106	78	28	57.4
73.6	91	67	24	49.3
73.6	87	64	23	47.1
73.6	72	53	19	39.0
73.6	53	39	14	28.7
73.5	499	367	132	269.9
73.5	498	366	132	269.0
73.5	495	364	131	267.7
73.5	494	363	131	266.7
73.5	491	361	130	265.4
73.5	490	360	130	264.5
73.5	487	358	129	263.2
73.5	486	357	129	262.2
73.5	483	355	128	260.9
73.5	480	353	127	259.6
73.5	479	352	127	258.7
73.5	476	350	126	257.4
73.5	475	349	126	256.4
73.5	472	347	125	255.1
73.5	471	346	125	254.2
73.5	468	344	124	252.9
73.5	465	342	123	251.5
73.5	464	341	123	250.6
73.5	461	339	122	249.3
73.5	460	338	122	248.4
73.5	457	336	121	247.0

%E	M1	M2	DM	M*
73.5	456	335	121	246.1
73.5	453	333	120	244.8
73.5	452	332	120	243.9
73.5	449	330	119	242.5
73.5	446	328	118	241.2
73.5	445	327	118	240.3
73.5	442	325	117	239.0
73.5	441	324	117	238.0
73.5	438	322	116	236.7
73.5	437	321	116	235.8
73.5	434	319	115	234.5
73.5	430	316	114	232.2
73.5	427	314	113	230.9
73.5	426	313	113	230.0
73.5	423	311	112	228.7
73.5	422	310	112	227.7
73.5	419	308	111	226.4
73.5	415	305	110	224.2
73.5	412	303	109	222.8
73.5	411	302	109	221.9
73.5	408	300	108	220.6
73.5	407	299	108	219.7
73.5	404	297	107	218.3
73.5	400	294	106	216.1
73.5	396	291	105	213.8
73.5	393	289	104	212.5
73.5	392	288	104	211.6
73.5	389	286	103	210.3
73.5	388	285	103	209.3
73.5	385	283	102	208.0
73.5	381	280	101	205.8
73.5	378	278	100	204.5
73.5	377	277	100	203.5
73.5	374	275	99	202.2
73.5	373	274	99	201.3
73.5	370	272	98	200.0
73.5	366	269	97	197.7
73.5	362	266	96	195.5
73.5	359	264	95	194.1
73.5	358	263	95	193.2
73.5	355	261	94	191.9
73.5	351	258	93	189.6
73.5	347	255	92	187.4
73.5	344	253	91	186.1
73.5	343	252	91	185.1
73.5	340	250	90	183.8
73.5	339	249	90	182.9
73.5	336	247	89	181.6
73.5	332	244	88	179.3
73.5	328	241	87	177.1

%E	M1	M2	DM	M*
73.5	325	239	86	175.8
73.5	324	238	86	174.8
73.5	321	236	85	173.5
73.5	317	233	84	171.3
73.5	313	230	83	169.0
73.5	310	228	82	167.7
73.5	309	227	82	166.8
73.5	306	225	81	165.4
73.5	302	222	80	163.2
73.5	298	219	79	160.9
73.5	294	216	78	158.7
73.5	291	214	77	157.4
73.5	287	211	76	155.1
73.5	283	208	75	152.9
73.5	279	205	74	150.6
73.5	275	202	73	148.4
73.5	272	200	72	147.1
73.5	268	197	71	144.8
73.5	264	194	70	142.6
73.5	260	191	69	140.3
73.5	257	189	68	139.0
73.5	253	186	67	136.7
73.5	249	183	66	134.5
73.5	245	180	65	132.2
73.5	238	175	63	128.7
73.5	234	172	62	126.4
73.5	230	169	61	124.2
73.5	226	166	60	121.9
73.5	223	164	59	120.6
73.5	219	161	58	118.4
73.5	215	158	57	116.1
73.5	211	155	56	113.9
73.5	204	150	54	110.3
73.5	200	147	53	108.0
73.5	196	144	52	105.8
73.5	189	139	50	102.2
73.5	185	136	49	100.0
73.5	181	133	48	97.7
73.5	170	125	45	91.9
73.5	166	122	44	89.7
73.5	162	119	43	87.4
73.5	155	114	41	83.8
73.5	151	111	40	81.6
73.5	147	108	39	79.3
73.5	136	100	36	73.5
73.5	132	97	35	71.3
73.5	117	86	31	63.2
73.5	113	83	30	61.0
73.5	102	75	27	55.1
73.5	98	72	26	52.9

%E	M1	M2	DM	M*
73.5	83	61	22	44.8
73.5	68	50	18	36.8
73.5	49	36	13	26.4
73.5	34	25	9	18.4
73.4	500	367	133	269.4
73.4	497	365	132	268.1
73.4	496	364	132	267.1
73.4	493	362	131	265.8
73.4	492	361	131	264.9
73.4	489	359	130	263.6
73.4	488	358	130	262.6
73.4	485	356	129	261.3
73.4	482	354	128	260.0
73.4	481	353	128	259.1
73.4	478	351	127	257.7
73.4	477	350	127	256.8
73.4	474	348	126	255.5
73.4	473	347	126	254.6
73.4	470	345	125	253.2
73.4	467	343	124	251.9
73.4	466	342	124	251.0
73.4	463	340	123	249.7
73.4	462	339	123	248.7
73.4	459	337	122	247.4
73.4	458	336	122	246.5
73.4	455	334	121	245.2
73.4	451	331	120	242.9
73.4	448	329	119	241.6
73.4	447	328	119	240.7
73.4	444	326	118	239.4
73.4	443	325	118	238.4
73.4	440	323	117	237.1
73.4	436	320	116	234.9
73.4	433	318	115	233.5
73.4	432	317	115	232.6
73.4	429	315	114	231.3
73.4	428	314	114	230.4
73.4	425	312	113	229.0
73.4	421	309	112	226.8
73.4	418	307	111	225.5
73.4	417	306	111	224.5
73.4	414	304	110	223.2
73.4	413	303	110	222.3
73.4	410	301	109	221.0
73.4	409	300	109	220.0
73.4	406	298	108	218.7
73.4	403	296	107	217.4
73.4	402	295	107	216.5
73.4	399	293	106	215.2
73.4	398	292	106	214.2

%E	M1	M2	DM	M*
73.4	395	290	105	212.9
73.4	394	289	105	212.0
73.4	391	287	104	210.7
73.4	387	284	103	208.4
73.4	384	282	102	207.1
73.4	383	281	102	206.2
73.4	380	279	101	204.8
73.4	379	278	101	203.9
73.4	376	276	100	202.6
73.4	372	273	99	200.3
73.4	369	271	98	199.0
73.4	368	270	98	198.1
73.4	365	268	97	196.8
73.4	364	267	97	195.8
73.4	361	265	96	194.5
73.4	357	262	95	192.3
73.4	354	260	94	191.0
73.4	353	259	94	190.0
73.4	350	257	93	188.7
73.4	349	256	93	187.8
73.4	346	254	92	186.5
73.4	342	251	91	184.2
73.4	338	248	90	182.0
73.4	335	246	89	180.6
73.4	334	245	89	179.7
73.4	331	243	88	178.4
73.4	327	240	87	176.1
73.4	323	237	86	173.9
73.4	320	235	85	172.6
73.4	319	234	85	171.6
73.4	316	232	84	170.3
73.4	312	229	83	168.1
73.4	308	226	82	165.8
73.4	305	224	81	164.5
73.4	304	223	81	163.6
73.4	301	221	80	162.3
73.4	297	218	79	160.0
73.4	293	215	78	157.8
73.4	290	213	77	156.4
73.4	289	212	77	155.5
73.4	286	210	76	154.2
73.4	282	207	75	151.9
73.4	278	204	74	149.7
73.4	274	201	73	147.4
73.4	271	199	72	146.1
73.4	267	196	71	143.9
73.4	263	193	70	141.6
73.4	259	190	69	139.4
73.4	256	188	68	138.1
73.4	252	185	67	135.8

%E	M1	M2	DM	M*
73.4	248	182	66	133.6
73.4	244	179	65	131.3
73.4	241	177	64	130.0
73.4	237	174	63	127.7
73.4	233	171	62	125.5
73.4	229	168	61	123.2
73.4	222	163	59	119.7
73.4	218	160	58	117.4
73.4	214	157	57	115.2
73.4	207	152	55	111.6
73.4	203	149	54	109.4
73.4	199	146	53	107.1
73.4	192	141	51	103.5
73.4	188	138	50	101.3
73.4	184	135	49	99.0
73.4	177	130	47	95.5
73.4	173	127	46	93.2
73.4	169	124	45	91.0
73.4	158	116	42	85.2
73.4	154	113	41	82.9
73.4	143	105	38	77.1
73.4	139	102	37	74.8
73.4	128	94	34	69.0
73.4	124	91	33	66.8
73.4	109	80	29	58.7
73.4	94	69	25	50.6
73.4	79	58	21	42.6
73.4	64	47	17	34.5
73.3	499	366	133	268.4
73.3	498	365	133	267.5
73.3	495	363	132	266.2
73.3	494	362	132	265.3
73.3	491	360	131	264.0
73.3	490	359	131	263.0
73.3	487	357	130	261.7
73.3	486	356	130	260.8
73.3	484	355	129	260.4
73.3	483	354	129	259.5
73.3	480	352	128	258.1
73.3	479	351	128	257.2
73.3	476	349	127	255.9
73.3	475	348	127	255.0
73.3	472	346	126	253.6
73.3	469	344	125	252.3
73.3	468	343	125	251.4
73.3	465	341	124	250.1
73.3	464	340	124	249.1
73.3	461	338	123	247.8
73.3	460	337	123	246.9
73.3	457	335	122	245.6

%E	M1	M2	DM	M*
73.3	454	333	121	244.2
73.3	453	332	121	243.3
73.3	450	330	120	242.0
73.3	449	329	120	241.1
73.3	446	327	119	239.8
73.3	445	326	119	238.8
73.3	442.	324	118	237.5
73.3	439	322	117	236.2
73.3	438	321	117	235.3
73.3	435	319	116	233.9
73.3	434	318	116	233.0
73.3	431	316	115	231.7
73.3	430	315	115	230.8
73.3	427	313	114	229.4
73.3	424	311	113	228.1
73.3	423	310	113	227.2
73.3	420	308	112	225.9
73.3	419	307	112	224.9
73.3	416	305	111	223.6
73.3	415	304	111	222.7
73.3	412	302	110	221.4
73.3	408	299	109	219.1
73.3	405	297	108	217.8
73.3	404	296	108	216.9
73.3	401	294	107	215.6
73.3	400	293	107	214.6
73.3	397	291	106	213.3
73.3	393	288	105	211.1
73.3	390	286	104	209.7
73.3	389	285	104	208.8
73.3	386	283	103	207.5
73.3	382	280	102	205.2
73.3	378	277	101	203.0
73.3	375	275	100	201.7
73.3	374	274	100	200.7
73.3	371	272	99	199.4
73.3	367	269	98	197.2
73.3	363	266	97	194.9
73.3	360	264	96	193.6
73.3	359	263	96	192.7
73.3	356	261	95	191.4
73.3	352	258	94	189.1
73.3	348	255	93	186.9
73.3	345	253	92	185.5
73.3	344	252	92	184.6
73.3	341	250	91	183.3
73.3	337	247	90	181.0
73.3	333	244	89	178.8
73.3	330	242	88	177.5
73.3	329	241	88	176.5

%E	M1	M2	DM	M*
73.3	326	239	87	175.2
73.3	322	236	86	173.0
73.3	318	233	85	170.7
73.3	315	231	84	169.4
73.3	311	228	83	167.2
73.3	307	225	82	164.9
73.3	303	222	81	162.7
73.3	300	220	80	161.3
73.3	296	217	79	159.1
73.3	292	214	78	156.8
73.3	288	211	77	154.6
73.3	285	209	76	153.3
73.3	281	206	75	151.0
73.3	277	203	74	148.8
73.3	273	200	73	146.5
73.3	270	198	72	145.2
73.3	266	195	71	143.0
73.3	262	192	70	140.7
73.3	258	189	69	138.5
73.3	255	187	68	137.1
73.3	251	184	67	134.9
73.3	247	181	66	132.6
73.3	243	178	65	130.4
73.3	240	176	64	129.1
73.3	236	173	63	126.8
73.3	232	170	62	124.6
73.3	225	165	60	121.0
73.3	221	162	59	118.8
73.3	217	159	58	116.5
73.3	210	154	56	112.9
73.3	206	151	55	110.7
73.3	202	148	54	108.4
73.3	195	143	52	104.9
73.3	191	140	51	102.6
73.3	187	137	50	100.4
73.3	180	132	48	96.8
73.3	176	129	47	94.6
73.3	172	126	46	92.3
73.3	165	121	44	88.7
73.3	161	118	43	86.5
73.3	150	110	40	80.7
73.3	146	107	39	78.4
73.3	135	99	36	72.6
73.3	131	96	35	70.4
73.3	120	88	32	64.5
73.3	116	85	31	62.3
73.3	105	77	28	56.5
73.3	101	74	27	54.2
73.3	90	66	24	48.4
73.3	86	63	23	46.2

%E	M1	M2	DM	M*
73.3	75	55	20	40.3
73.3	60	44	16	32.3
73.3	45	33	12	24.2
73.3	30	22	8	16.1
73.3	15	11	4	8.1
73.2	500	366	134	267.9
73.2	497	364	133	266.6
73.2	496	363	133	265.7
73.2	493	361	132	264.3
73.2	492	360	132	263.4
73.2	489	358	131	262.1
73.2	488	357	131	261.2
73.2	485	355	130	259.8
73.2	482	353	129	258.5
73.2	481	352	129	257.6
73.2	478	350	128	256.3
73.2	477	349	128	255.3
73.2	474	347	127	254.0
73.2	473	346	127	253.1
73.2	471	345	126	252.7
73.2	470	344	126	251.8
73.2	467	342	125	250.5
73.2	466	341	125	249.5
73.2	463	339	124	248.2
73.2	462	338	124	247.3
73.2	459	336	123	246.0
73.2	456	334	122	244.6
73.2	455	333	122	243.7
73.2	452	331	121	242.4
73.2	451	330	121	241.5
73.2	448	328	120	240.1
73.2	447	327	120	239.2
73.2	444	325	119	237.9
73.2	441	323	118	236.6
73.2	440	322	118	235.6
73.2	437	320	117	234.3
73.2	436	319	117	233.4
73.2	433	317	116	232.1
73.2	429	314	115	229.8
73.2	426	312	114	228.5
73.2	425	311	114	227.6
73.2	422	309	113	226.3
73.2	421	308	113	225.3
73.2	418	306	112	224.0
73.2	414	303	111	221.8
73.2	411	301	110	220.4
73.2	410	300	110	219.5
73.2	407	298	109	218.2
73.2	406	297	109	217.3
73.2	403	295	108	215.9

%E	M1	M2	DM	M*
73.2	399	292	107	213.7
73.2	396	290	106	212.4
73.2	395	289	106	211.4
73.2	392	287	105	210.1
73.2	388	284	104	207.9
73.2	385	282	103	206.6
73.2	384	281	103	205.6
73.2	381	279	102	204.3
73.2	380	278	102	203.4
73.2	377	276	101	202.1
73.2	373	273	100	199.8
73.2	370	271	99	198.5
73.2	369	270	99	197.6
73.2	366	268	98	196.2
73.2	365	267	98	195.3
73.2	362	265	97	194.0
73.2	358	262	96	191.7
73.2	355	260	95	190.4
73.2	354	259	95	189.5
73.2	351	257	94	188.2
73.2	347	254	93	185.9
73.2	343	251	92	183.7
73.2	340	249	91	182.4
73.2	339	248	91	181.4
73.2	336	246	90	180.1
73.2	332	243	89	177.9
73.2	328	240	88	175.6
73.2	325	238	87	174.3
73.2	321	235	86	172.0
73.2	317	232	85	169.8
73.2	314	230	84	168.5
73.2	313	229	84	167.5
73.2	310	227	83	166.2
73.2	306	224	82	164.0
73.2	302	221	81	161.7
73.2	299	219	80	160.4
73.2	298	218	80	159.5
73.2	295	216	79	158.2
73.2	291	213	78	155.9
73.2	287	210	77	153.7
73.2	284	208	76	152.3
73.2	280	205	75	150.1
73.2	276	202	74	147.8
73.2	272	199	73	145.6
73.2	269	197	72	144.3
73.2	265	194	71	142.0
73.2	261	191	70	139.8
73.2	257	188	69	137.5
73.2	254	186	68	136.2
73.2	250	183	67	134.0

%E	M1	M2	DM	M*
73.2	246	180	66	131.7
73.2	239	175	64	128.1
73.2	235	172	63	125.9
73.2	231	169	62	123.6
73.2	228	167	61	122.3
73.2	224	164	60	120.1
73.2	220	161	59	117.8
73.2	213	156	57	114.3
73.2	209	153	56	112.0
73.2	205	150	55	109.8
73.2	198	145	53	106.2
73.2	194	142	52	103.9
73.2	190	139	51	101.7
73.2	183	134	49	98.1
73.2	179	131	48	95.9
73.2	168	123	45	90.1
73.2	164	120	44	87.8
73.2	157	115	42	84.2
73.2	153	112	41	82.0
73.2	149	109	40	79.7
73.2	142	104	38	76.2
73.2	138	101	37	73.9
73.2	127	93	34	68.1
73.2	123	90	33	65.9
73.2	112	82	30	60.0
73.2	97	71	26	52.0
73.2	82	60	22	43.9
73.2	71	52	19	38.1
73.2	56	41	15	30.0
73.2	41	30	11	22.0
73.1	499	365	134	267.0
73.1	498	364	134	266.1
73.1	495	362	133	264.7
73.1	494	361	133	263.8
73.1	491	359	132	262.5
73.1	490	358	132	261.6
73.1	487	356	131	260.2
73.1	484	354	130	258.9
73.1	483	353	130	258.0
73.1	480	351	129	256.7
73.1	479	350	129	255.7
73.1	476	348	128	254.4
73.1	475	347	128	253.5
73.1	472	345	127	252.2
73.1	469	343	126	250.9
73.1	468	342	126	249.9
73.1	465	340	125	248.6
73.1	464	339	125	247.7
73.1	461	337	124	246.4
73.1	458	335	123	245.0

%E	M1	M2	DM	M*
73.1	457	334	123	244.1
73.1	454	332	122	242.8
73.1	453	331	122	241.9
73.1	450	329	121	240.5
73.1	449	328	121	239.6
73.1	446	326	120	238.3
73.1	443	324	119	237.0
73.1	442	323	119	236.0
73.1	439	321	118	234.7
73.1	438	320	118	233.8
73.1	435	318	117	232.5
73.1	432	316	116	231.1
73.1	431	315	116	230.2
73.1	428	313	115	228.9
73.1	427	312	115	228.0
73.1	424	310	114	226.7
73.1	423	309	114	225.7
73.1	420	307	113	224.4
73.1	417	305	112	223.1
73.1	416	304	112	222.2
73.1	413	302	111	220.8
73.1	412	301	111	219.9
73.1	409	299	110	218.6
73.1	405	296	109	216.3
73.1	402	294	108	215.0
73.1	401	293	108	214.1
73.1	398	291	107	212.8
73.1	394	288	106	210.5
73.1	391	286	105	209.2
73.1	390	285	105	208.3
73.1	387	283	104	206.9
73.1	386	282	104	206.0
73.1	383	280	103	204.7
73.1	379	277	102	202.5
73.1	376	275	101	201.1
73.1	375	274	101	200.2
73.1	372	272	100	198.9
73.1	368	269	99	196.6
73.1	364	266	98	194.4
73.1	361	264	97	193.1
73.1	360	263	97	192.1
73.1	357	261	96	190.8
73.1	353	258	95	188.6
73.1	350	256	94	187.2
73.1	349	255	94	186.3
73.1	346	253	93	185.0
73.1	342	250	92	182.7
73.1	338	247	91	180.5
73.1	335	245	90	179.2
73.1	334	244	90	178.3

%E	M1	M2	DM	M*
73.1	331	242	89	176.9
73.1	327	239	88	174.7
73.1	324	237	87	173.4
73.1	323	236	87	172.4
73.1	320	234	86	171.1
73.1	316	231	85	168.9
73.1	312	228	84	166.6
73.1	309	226	83	165.3
73.1	308	225	83	164.4
73.1	305	223	82	163.0
73.1	301	220	81	160.8
73.1	297	217	80	158.5
73.1	294	215	79	157.2
73.1	290	212	78	155.0
73.1	286	209	77	152.7
73.1	283	207	76	151.4
73.1	282	206	76	150.5
73.1	279	204	75	149.2
73.1	275	201	74	146.9
73.1	271	198	73	144.7
73.1	268	196	72	143.3
73.1	264	193	71	141.1
73.1	260	190	70	138.8
73.1	253	185	68	135.3
73.1	249	182	67	133.0
73.1	245	179	66	130.8
73.1	242	177	65	129.5
73.1	238	174	64	127.2
73.1	234	171	63	125.0
73.1	227	166	61	121.4
73.1	223	163	60	119.1
73.1	219	160	59	116.9
73.1	216	158	58	115.6
73.1	212	155	57	113.3
73.1	208	152	56	111.1
73.1	201	147	54	107.5
73.1	197	144	53	105.3
73.1	193	141	52	103.0
73.1	186	136	50	99.4
73.1	182	133	49	97.2
73.1	175	128	47	93.6
73.1	171	125	46	91.4
73.1	167	122	45	89.1
73.1	160	117	43	85.6
73.1	156	114	42	83.3
73.1	145	106	39	77.5
73.1	141	103	38	75.2
73.1	134	98	36	71.7
73.1	130	95	35	69.4
73.1	119	87	32	63.6

%E	M1	M2	DM	M*
73.1	108	79	29	57.8
73.1	104	76	28	55.5
73.1	93	68	25	49.7
73.1	78	57	21	41.7
73.1	67	49	18	35.8
73.1	52	38	14	27.8
73.1	26	19	7	13.9
73.0	500	365	135	266.4
73.0	497	363	134	265.1
73.0	496	362	134	264.2
73.0	493	360	133	262.9
73.0	492	359	133	262.0
73.0	489	357	132	260.6
73.0	488	356	132	259.7
73.0	486	355	131	259.3
73.0	485	354	131	258.4
73.0	482	352	130	257.1
73.0	481	351	130	256.1
73.0	478	349	129	254.8
73.0	477	348	129	253.9
73.0	474	346	128	252.6
73.0	471	344	127	251.2
73.0	470	343	127	250.3
73.0	467	341	126	249.0
73.0	466	340	126	248.1
73.0	463	338	125	246.7
73.0	460	336	124	245.4
73.0	459	335	124	244.5
73.0	456	333	123	243.2
73.0	455	332	123	242.3
73.0	452	330	122	240.9
73.0	448	327	121	238.7
73.0	445	325	120	237.4
73.0	444	324	120	236.4
73.0	441	322	119	235.1
73.0	440	321	119	234.2
73.0	437	319	118	232.9
73.0	434	317	117	231.5
73.0	433	316	117	230.6
73.0	430	314	116	229.3
73.0	429	313	116	228.4
73.0	426	311	115	227.0
73.0	422	308	114	224.8
73.0	419	306	113	223.5
73.0	418	305	113	222.5
73.0	415	303	112	221.2
73.0	411	300	111	219.0
73.0	408	298	110	217.7
73.0	407	297	110	216.7
73.0	404	295	109	215.4
73.0	403	294	109	214.5
73.0	400	292	108	213.2
73.0	397	290	107	211.8
73.0	396	289	107	210.9
73.0	393	287	106	209.6
73.0	392	286	106	208.7
73.0	389	284	105	207.3
73.0	385	281	104	205.1
73.0	382	279	103	203.8
73.0	381	278	103	202.8
73.0	378	276	102	201.5
73.0	374	273	101	199.3
73.0	371	271	100	198.0
73.0	370	270	100	197.0
73.0	367	268	99	195.7
73.0	366	267	99	194.8
73.0	363	265	98	193.5
73.0	359	262	97	191.2
73.0	356	260	96	189.9
73.0	355	259	96	189.0
73.0	352	257	95	187.6
73.0	348	254	94	185.4
73.0	345	252	93	184.1
73.0	344	251	93	183.1
73.0	341	249	92	181.8
73.0	337	246	91	179.6
73.0	333	243	90	177.3
73.0	330	241	89	176.0
73.0	326	238	88	173.8
73.0	322	235	87	171.5
73.0	319	233	86	170.2
73.0	318	232	86	169.3
73.0	315	230	85	167.9
73.0	311	227	84	165.7
73.0	307	224	83	163.4
73.0	304	222	82	162.1
73.0	300	219	81	159.9
73.0	296	216	80	157.6
73.0	293	214	79	156.3
73.0	289	211	78	154.1
73.0	285	208	77	151.8
73.0	281	205	76	149.6
73.0	278	203	75	148.2
73.0	274	200	74	146.0
73.0	270	197	73	143.7
73.0	267	195	72	142.4
73.0	263	192	71	140.2
73.0	259	189	70	137.9
73.0	256	187	69	136.6
73.0	252	184	68	134.3
73.0	248	181	67	132.1
73.0	244	178	66	129.9
73.0	241	176	65	128.5
73.0	237	173	64	126.3
73.0	233	170	63	124.0
73.0	230	168	62	122.7
73.0	226	165	61	120.5
73.0	222	162	60	118.2
73.0	215	157	58	114.6
73.0	211	154	57	112.4
73.0	204	149	55	108.8
73.0	200	146	54	106.6
73.0	196	143	53	104.3
73.0	189	138	51	100.8
73.0	185	135	50	98.5
73.0	178	130	48	94.9
73.0	174	127	47	92.7
73.0	163	119	44	86.9
73.0	159	116	43	84.6
73.0	152	111	41	81.1
73.0	148	108	40	78.8
73.0	137	100	37	73.0
73.0	126	92	34	67.2
73.0	122	89	33	64.9
73.0	115	84	31	61.4
73.0	111	81	30	59.1
73.0	100	73	27	53.3
73.0	89	65	24	47.5
73.0	74	54	20	39.4
73.0	63	46	17	33.6
73.0	37	27	10	19.7
72.9	499	364	135	265.5
72.9	498	363	135	264.6
72.9	495	361	134	263.3
72.9	494	360	134	262.3
72.9	491	358	133	261.0
72.9	490	357	133	260.1
72.9	487	355	132	258.8
72.9	484	353	131	257.5
72.9	483	352	131	256.5
72.9	480	350	130	255.2
72.9	479	349	130	254.3
72.9	476	347	129	253.0
72.9	473	345	128	251.6
72.9	472	344	128	250.7
72.9	469	342	127	249.4
72.9	468	341	127	248.5
72.9	465	339	126	247.1
72.9	462	337	125	245.8
72.9	461	336	125	244.9
72.9	458	334	124	243.6
72.9	457	333	124	242.6
72.9	454	331	123	241.3
72.9	451	329	122	240.0
72.9	450	328	122	239.1
72.9	447	326	121	237.8
72.9	446	325	121	236.8
72.9	443	323	120	235.5
72.9	442	322	120	234.6
72.9	439	320	119	233.3
72.9	436	318	118	231.9
72.9	435	317	118	231.0
72.9	432	315	117	229.7
72.9	431	314	117	228.8
72.9	428	312	116	227.4
72.9	425	310	115	226.1
72.9	424	309	115	225.2
72.9	421	307	114	223.9
72.9	420	306	114	222.9
72.9	417	304	113	221.6
72.9	414	302	112	220.3
72.9	413	301	112	219.4
72.9	410	299	111	218.1
72.9	409	298	111	217.1
72.9	406	296	110	215.8
72.9	402	293	109	213.6
72.9	399	291	108	212.2
72.9	398	290	108	211.3
72.9	395	288	107	210.0
72.9	391	285	106	207.7
72.9	388	283	105	206.4
72.9	387	282	105	205.5
72.9	384	280	104	204.2
72.9	380	277	103	201.9
72.9	377	275	102	200.6
72.9	376	274	102	199.7
72.9	373	272	101	198.3
72.9	369	269	100	196.1
72.9	365	266	99	193.9
72.9	362	264	98	192.5
72.9	361	263	98	191.6
72.9	358	261	97	190.3
72.9	354	258	96	188.0
72.9	351	256	95	186.7
72.9	350	255	95	185.8
72.9	347	253	94	184.5
72.9	343	250	93	182.2
72.9	340	248	92	180.9
72.9	339	247	92	180.0
72.9	336	245	91	178.6
72.9	332	242	90	176.4
72.9	329	240	89	175.1
72.9	328	239	89	174.1
72.9	325	237	88	172.8
72.9	321	234	87	170.6
72.9	317	231	86	168.3
72.9	314	229	85	167.0
72.9	310	226	84	164.8
72.9	306	223	83	162.5
72.9	303	221	82	161.2
72.9	299	218	81	158.9
72.9	295	215	80	156.7
72.9	292	213	79	155.4
72.9	291	212	79	154.4
72.9	288	210	78	153.1
72.9	284	207	77	150.9
72.9	280	204	76	148.6
72.9	277	202	75	147.3
72.9	273	199	74	145.1
72.9	269	196	73	142.8
72.9	266	194	72	141.5
72.9	262	191	71	139.2
72.9	258	188	70	137.0
72.9	255	186	69	135.7
72.9	251	183	68	133.4
72.9	247	180	67	131.2
72.9	240	175	65	127.6
72.9	236	172	64	125.4
72.9	229	167	62	121.8
72.9	225	164	61	119.5
72.9	221	161	60	117.3
72.9	218	159	59	116.0
72.9	214	156	58	113.7
72.9	210	153	57	111.5
72.9	207	151	56	110.1
72.9	203	148	55	107.9
72.9	199	145	54	105.7
72.9	192	140	52	102.1
72.9	188	137	51	99.8
72.9	181	132	49	96.3
72.9	177	129	48	94.0
72.9	170	124	46	90.4
72.9	166	121	45	88.2
72.9	155	113	42	82.4
72.9	144	105	39	76.6
72.9	140	102	38	74.3
72.9	133	97	36	70.7
72.9	129	94	35	68.5
72.9	118	86	32	62.7
72.9	107	78	29	56.9

%E	M1	M2	DM	M*
72.9	96	70	26	51.0
72.9	85	62	23	45.2
72.9	70	51	19	37.2
72.9	59	43	16	31.3
72.9	48	35	13	25.5
72.8	500	364	136	265.0
72.8	497	362	135	263.7
72.8	496	361	135	262.7
72.8	493	359	134	261.4
72.8	492	358	134	260.5
72.8	489	356	133	259.2
72.8	486	354	132	257.9
72.8	485	353	132	256.9
72.8	482	351	131	255.6
72.8	481	350	131	254.7
72.8	478	348	130	253.4
72.8	475	346	129	252.0
72.8	474	345	129	251.1
72.8	471	343	128	249.8
72.8	470	342	128	248.9
72.8	467	340	127	247.5
72.8	464	338	126	246.2
72.8	463	337	126	245.3
72.8	460	335	125	244.0
72.8	459	334	125	243.0
72.8	456	332	124	241.7
72.8	453	330	123	240.4
72.8	452	329	123	239.5
72.8	449	327	122	238.1
72.8	448	326	122	237.2
72.8	445	324	121	235.9
72.8	441	321	120	233.7
72.8	438	319	119	232.3
72.8	437	318	119	231.4
72.8	434	316	118	230.1
72.8	430	313	117	227.8
72.8	427	311	116	226.5
72.8	426	310	116	225.6
72.8	423	308	115	224.3
72.8	419	305	114	222.0
72.8	416	303	113	220.7
72.8	415	302	113	219.8
72.8	412	300	112	218.4
72.8	408	297	111	216.2
72.8	405	295	110	214.9
72.8	404	294	110	214.0
72.8	401	292	109	212.6
72.8	400	291	109	211.7
72.8	397	289	108	210.4
72.8	394	287	107	209.1

%E	M1	M2	DM	M*
72.8	393	286	107	208.1
72.8	390	284	106	206.8
72.8	389	283	106	205.9
72.8	386	281	105	204.6
72.8	383	279	104	203.2
72.8	382	278	104	202.3
72.8	379	276	103	201.0
72.8	378	275	103	200.1
72.8	375	273	102	198.7
72.8	372	271	101	197.4
72.8	371	270	101	196.5
72.8	368	268	100	195.2
72.8	367	267	100	194.2
72.8	364	265	99	192.9
72.8	360	262	98	190.7
72.8	357	260	97	189.4
72.8	356	259	97	188.4
72.8	353	257	96	187.1
72.8	349	254	95	184.9
72.8	346	252	94	183.5
72.8	345	251	94	182.6
72.8	342	249	93	181.3
72.8	338	246	92	179.0
72.8	335	244	91	177.7
72.8	334	243	91	176.8
72.8	331	241	90	175.5
72.8	327	238	89	173.2
72.8	324	236	88	171.9
72.8	323	235	88	171.0
72.8	320	233	87	169.7
72.8	316	230	86	167.4
72.8	313	228	85	166.1
72.8	312	227	85	165.2
72.8	309	225	84	163.8
72.8	305	222	83	161.6
72.8	302	220	82	160.3
72.8	301	219	82	159.3
72.8	298	217	81	158.0
72.8	294	214	80	155.8
72.8	290	211	79	153.5
72.8	287	209	78	152.2
72.8	283	206	77	150.0
72.8	279	203	76	147.7
72.8	276	201	75	146.4
72.8	272	198	74	144.1
72.8	268	195	73	141.9
72.8	265	193	72	140.6
72.8	261	190	71	138.3
72.8	257	187	70	136.1
72.8	254	185	69	134.7

%E	M1	M2	DM	M*
72.8	250	182	68	132.5
72.8	246	179	67	130.2
72.8	243	177	66	128.9
72.8	239	174	65	126.7
72.8	235	171	64	124.4
72.8	232	169	63	123.1
72.8	228	166	62	120.9
72.8	224	163	61	118.6
72.8	217	158	59	115.0
72.8	213	155	58	112.8
72.8	206	150	56	109.2
72.8	202	147	55	107.0
72.8	195	142	53	103.4
72.8	191	139	52	101.2
72.8	184	134	50	97.6
72.8	180	131	49	95.3
72.8	173	126	47	91.8
72.8	169	123	46	89.5
72.8	162	118	44	86.0
72.8	158	115	43	83.7
72.8	151	110	41	80.1
72.8	147	107	40	77.9
72.8	136	99	37	72.1
72.8	125	91	34	66.2
72.8	114	83	31	60.4
72.8	103	75	28	54.6
72.8	92	67	25	48.8
72.8	81	59	22	43.0
72.7	499	363	136	264.1
72.7	498	362	136	263.1
72.7	495	360	135	261.8
72.7	494	359	135	260.9
72.7	491	357	134	259.6
72.7	490	356	134	258.6
72.7	488	355	133	258.2
72.7	487	354	133	257.3
72.7	484	352	132	256.0
72.7	483	351	132	255.1
72.7	480	349	131	253.8
72.7	479	348	131	252.8
72.7	477	347	130	252.4
72.7	476	346	130	251.5
72.7	473	344	129	250.2
72.7	472	343	129	249.3
72.7	469	341	128	247.9
72.7	468	340	128	247.0
72.7	466	339	127	246.6
72.7	465	338	127	245.7
72.7	462	336	126	244.4
72.7	461	335	126	243.4

%E	M1	M2	DM	M*
72.7	458	333	125	242.1
72.7	455	331	124	240.8
72.7	454	330	124	239.9
72.7	451	328	123	238.5
72.7	450	327	123	237.6
72.7	447	325	122	236.3
72.7	444	323	121	235.0
72.7	443	322	121	234.0
72.7	440	320	120	232.7
72.7	439	319	120	231.8
72.7	436	317	119	230.5
72.7	433	315	118	229.2
72.7	432	314	118	228.2
72.7	429	312	117	226.9
72.7	428	311	117	226.0
72.7	425	309	116	224.7
72.7	422	307	115	223.3
72.7	421	306	115	222.4
72.7	418	304	114	221.1
72.7	417	303	114	220.2
72.7	414	301	113	218.8
72.7	411	299	112	217.5
72.7	410	298	112	216.6
72.7	407	296	111	215.3
72.7	406	295	111	214.3
72.7	403	293	110	213.0
72.7	399	290	109	210.8
72.7	396	288	108	209.5
72.7	395	287	108	208.5
72.7	392	285	107	207.2
72.7	388	282	106	205.0
72.7	385	280	105	203.6
72.7	384	279	105	202.7
72.7	381	277	104	201.4
72.7	377	274	103	199.1
72.7	374	272	102	197.8
72.7	373	271	102	196.9
72.7	370	269	101	195.6
72.7	366	266	100	193.3
72.7	363	264	99	192.0
72.7	362	263	99	191.1
72.7	359	261	98	189.8
72.7	355	258	97	187.5
72.7	352	256	96	186.2
72.7	351	255	96	185.3
72.7	348	253	95	183.9
72.7	344	250	94	181.7
72.7	341	248	93	180.4
72.7	337	245	92	178.1
72.7	333	242	91	175.9

%E	M1	M2	DM	M*
72.7	330	240	90	174.5
72.7	326	237	89	172.3
72.7	322	234	88	170.0
72.7	319	232	87	168.7
72.7	315	229	86	166.5
72.7	311	226	85	164.2
72.7	308	224	84	162.9
72.7	304	221	83	160.7
72.7	300	218	82	158.4
72.7	297	216	81	157.1
72.7	293	213	80	154.8
72.7	289	210	79	152.6
72.7	286	208	78	151.3
72.7	282	205	77	149.0
72.7	278	202	76	146.8
72.7	275	200	75	145.5
72.7	271	197	74	143.2
72.7	267	194	73	141.0
72.7	264	192	72	139.6
72.7	260	189	71	137.4
72.7	256	186	70	135.1
72.7	253	184	69	133.8
72.7	249	181	68	131.6
72.7	245	178	67	129.3
72.7	242	176	66	128.0
72.7	238	173	65	125.8
72.7	234	170	64	123.5
72.7	231	168	63	122.2
72.7	227	165	62	119.9
72.7	220	160	60	116.4
72.7	216	157	59	114.1
72.7	209	152	57	110.5
72.7	205	149	56	108.3
72.7	198	144	54	104.7
72.7	194	141	53	102.5
72.7	187	136	51	98.9
72.7	183	133	50	96.7
72.7	176	128	48	93.1
72.7	172	125	47	90.8
72.7	165	120	45	87.3
72.7	161	117	44	85.0
72.7	154	112	42	81.5
72.7	150	109	41	79.2
72.7	143	104	39	75.6
72.7	139	101	38	73.4
72.7	132	96	36	69.8
72.7	128	93	35	67.6
72.7	121	88	33	64.0
72.7	117	85	32	61.8
72.7	110	80	30	58.2

%E	M1	M2	DM	M*	%E	M1	M2	DM	M*	%E	M1	M2	DM	M*	%E	M1	M2	DM	M*	%E	M1	M2	DM	M*
72.7	99	72	27	52.4	72.6	409	297	112	215.7	72.6	248	180	68	130.6	72.5	454	329	125	238.4	72.5	327	237	90	171.8
72.7	88	64	24	46.5	72.6	405	294	111	213.4	72.6	241	175	66	127.1	72.5	451	327	124	237.1	72.5	324	235	89	170.4
72.7	77	56	21	40.7	72.6	402	292	110	212.1	72.6	237	172	65	124.8	72.5	448	325	123	235.8	72.5	320	232	88	168.2
72.7	66	48	18	34.9	72.6	401	291	110	211.2	72.6	230	167	63	121.3	72.5	447	324	123	234.8	72.5	316	229	87	166.0
72.7	55	40	15	29.1	72.6	398	289	109	209.9	72.6	226	164	62	119.0	72.5	444	322	122	233.5	72.5	313	227	86	164.6
72.7	44	32	12	23.3	72.6	394	286	108	207.6	72.6	223	162	61	117.7	72.5	443	321	122	232.6	72.5	309	224	85	162.4
72.7	33	24	9	17.5	72.6	391	284	107	206.3	72.6	219	159	60	115.4	72.5	440	319	121	231.3	72.5	306	222	84	161.1
72.7	22	16	6	11.6	72.6	390	283	107	205.4	72.6	215	156	59	113.2	72.5	437	317	120	230.0	72.5	305	221	84	160.1
72.7	11	8	3	5.8	72.6	387	281	106	204.0	72.6	212	154	58	111.9	72.5	436	316	120	229.0	72.5	302	219	83	158.8
72.6	500	363	137	263.5	72.6	383	278	105	201.8	72.6	208	151	57	109.6	72.5	433	314	119	227.7	72.5	298	216	82	156.6
72.6	497	361	136	262.2	72.6	380	276	104	200.5	72.6	201	146	55	106.0	72.5	432	313	119	226.8	72.5	295	214	81	155.2
72.6	496	360	136	261.3	72.6	379	275	104	199.5	72.6	197	143	54	103.8	72.5	429	311	118	225.5	72.5	291	211	80	153.0
72.6	493	358	135	260.0	72.6	376	273	103	198.2	72.6	190	138	52	100.2	72.5	426	309	117	224.1	72.5	287	208	79	150.7
72.6	492	357	135	259.0	72.6	372	270	102	196.0	72.6	186	135	51	98.0	72.5	425	308	117	223.2	72.5	284	206	78	149.4
72.6	489	355	134	257.7	72.6	369	268	101	194.6	72.6	179	130	49	94.4	72.5	422	306	116	221.9	72.5	280	203	77	147.2
72.6	486	353	133	256.4	72.6	368	267	101	193.7	72.6	175	127	48	92.2	72.5	418	303	115	219.6	72.5	276	200	76	144.9
72.6	485	352	133	255.5	72.6	365	265	100	192.4	72.6	168	122	46	88.6	72.5	415	301	114	218.3	72.5	273	198	75	143.6
72.6	482	350	132	254.1	72.6	361	262	99	190.1	72.6	164	119	45	86.3	72.5	414	300	114	217.4	72.5	269	195	74	141.4
72.6	481	349	132	253.2	72.6	358	260	98	188.8	72.6	157	114	43	82.8	72.5	411	298	113	216.1	72.5	265	192	73	139.1
72.6	478	347	131	251.9	72.6	354	257	97	186.6	72.6	146	106	40	77.0	72.5	408	296	112	214.7	72.5	262	190	72	137.8
72.6	475	345	130	250.6	72.6	350	254	96	184.3	72.6	135	98	37	71.1	72.5	407	295	112	213.8	72.5	258	187	71	135.5
72.6	474	344	130	249.7	72.6	347	252	95	183.0	72.6	124	90	34	65.3	72.5	404	293	111	212.5	72.5	255	185	70	134.2
72.6	471	342	129	248.3	72.6	343	249	94	180.8	72.6	113	82	31	59.5	72.5	403	292	111	211.6	72.5	251	182	69	132.0
72.6	470	341	129	247.4	72.6	340	247	93	179.4	72.6	106	77	29	55.9	72.5	400	290	110	210.3	72.5	247	179	68	129.7
72.6	467	339	128	246.1	72.6	339	246	93	178.5	72.6	95	69	26	50.1	72.5	397	288	109	208.9	72.5	244	177	67	128.4
72.6	464	337	127	244.8	72.6	336	244	92	177.2	72.6	84	61	23	44.3	72.5	396	287	109	208.0	72.5	240	174	66	126.1
72.6	463	336	127	243.8	72.6	332	241	91	174.9	72.6	73	53	20	38.5	72.5	393	285	108	206.7	72.5	236	171	65	123.9
72.6	460	334	126	242.5	72.6	329	239	90	173.6	72.6	62	45	17	32.7	72.5	389	282	107	204.4	72.5	233	169	64	122.6
72.6	457	332	125	241.2	72.6	328	238	90	172.7	72.5	499	362	137	262.6	72.5	386	280	106	203.1	72.5	229	166	63	120.3
72.6	456	331	125	240.3	72.6	325	236	89	171.4	72.5	498	361	137	261.7	72.5	385	279	106	202.2	72.5	222	161	61	116.8
72.6	453	329	124	238.9	72.6	321	233	88	169.1	72.5	495	359	136	260.4	72.5	382	277	105	200.9	72.5	218	158	60	114.5
72.6	452	328	124	238.0	72.6	318	231	87	167.8	72.5	494	358	136	259.4	72.5	378	274	104	198.6	72.5	211	153	58	110.9
72.6	449	326	123	236.7	72.6	317	230	87	166.9	72.5	491	356	135	258.1	72.5	375	272	103	197.3	72.5	207	150	57	108.7
72.6	446	324	122	235.4	72.6	314	228	86	165.6	72.5	488	354	134	256.8	72.5	374	271	103	196.4	72.5	204	148	56	107.4
72.6	445	323	122	234.4	72.6	310	225	85	163.3	72.5	487	353	134	255.9	72.5	371	269	102	195.0	72.5	200	145	55	105.1
72.6	442	321	121	233.1	72.6	307	223	84	162.0	72.5	484	351	133	254.5	72.5	367	266	101	192.8	72.5	193	140	53	101.6
72.6	441	320	121	232.2	72.6	303	220	83	159.7	72.5	483	350	133	253.6	72.5	364	264	100	191.5	72.5	189	137	52	99.3
72.6	438	318	120	230.9	72.6	299	217	82	157.5	72.5	480	348	132	252.3	72.5	363	263	100	190.5	72.5	182	132	50	95.7
72.6	435	316	119	229.6	72.6	296	215	81	156.2	72.5	477	346	131	251.0	72.5	360	261	99	189.2	72.5	178	129	49	93.5
72.6	434	315	119	228.6	72.6	292	212	80	153.9	72.5	476	345	131	250.1	72.5	357	259	98	187.9	72.5	171	124	47	89.9
72.6	431	313	118	227.3	72.6	288	209	79	151.7	72.5	473	343	130	248.7	72.5	356	258	98	187.0	72.5	167	121	46	87.7
72.6	430	312	118	226.4	72.6	285	207	78	150.3	72.5	472	342	130	247.8	72.5	353	256	97	185.7	72.5	160	116	44	84.1
72.6	427	310	117	225.1	72.6	281	204	77	148.1	72.5	469	340	129	246.5	72.5	349	253	96	183.4	72.5	153	111	42	80.5
72.6	424	308	116	223.7	72.6	277	201	76	145.9	72.5	466	338	128	245.2	72.5	346	251	95	182.1	72.5	149	108	41	78.3
72.6	423	307	116	222.8	72.6	274	199	75	144.5	72.5	465	337	128	244.2	72.5	345	250	95	181.2	72.5	142	103	39	74.7
72.6	420	305	115	221.5	72.6	270	196	74	142.3	72.5	462	335	127	242.9	72.5	342	248	94	179.8	72.5	138	100	38	72.5
72.6	419	304	115	220.6	72.6	266	193	73	140.0	72.5	461	334	127	242.0	72.5	338	245	93	177.6	72.5	131	95	36	68.9
72.6	416	302	114	219.2	72.6	263	191	72	138.7	72.5	459	333	126	241.6	72.5	335	243	92	176.3	72.5	120	87	33	63.1
72.6	413	300	113	217.9	72.6	259	188	71	136.5	72.5	458	332	126	240.7	72.5	334	242	92	175.3	72.5	109	79	30	57.3
72.6	412	299	113	217.0	72.6	252	183	69	132.9	72.5	455	330	125	239.3	72.5	331	240	91	174.0	72.5	102	74	28	53.7

%E	M1	M2	DM	M*
72.5	91	66	25	47.9
72.5	80	58	22	42.0
72.5	69	50	19	36.2
72.5	51	37	14	26.8
72.5	40	29	11	21.0
72.4	500	362	138	262.1
72.4	497	360	137	260.8
72.4	496	359	137	259.8
72.4	493	357	136	258.5
72.4	492	356	136	257.6
72.4	490	355	135	257.2
72.4	489	354	135	256.3
72.4	486	352	134	254.9
72.4	485	351	134	254.0
72.4	482	349	133	252.7
72.4	479	347	132	251.4
72.4	478	346	132	250.5
72.4	475	344	131	249.1
72.4	474	343	131	248.2
72.4	471	341	130	246.9
72.4	468	339	129	245.6
72.4	467	338	129	244.6
72.4	464	336	128	243.3
72.4	463	335	128	242.4
72.4	460	333	127	241.1
72.4	457	331	126	239.7
72.4	456	330	126	238.8
72.4	453	328	125	237.5
72.4	450	326	124	236.2
72.4	449	325	124	235.2
72.4	446	323	123	233.9
72.4	445	322	123	233.0
72.4	442	320	122	231.7
72.4	439	318	121	230.4
72.4	438	317	121	229.4
72.4	435	315	120	228.1
72.4	434	314	120	227.2
72.4	431	312	119	225.9
72.4	428	310	118	224.5
72.4	427	309	118	223.6
72.4	424	307	117	222.3
72.4	421	305	116	221.0
72.4	420	304	116	220.0
72.4	417	302	115	218.7
72.4	416	301	115	217.8
72.4	413	299	114	216.5
72.4	410	297	113	215.1
72.4	409	296	113	214.2
72.4	406	294	112	212.9
72.4	402	291	111	210.6

%E	M1	M2	DM	M*
72.4	399	289	110	209.3
72.4	398	288	110	208.4
72.4	395	286	109	207.1
72.4	392	284	108	205.8
72.4	391	283	108	204.8
72.4	388	281	107	203.5
72.4	387	280	107	202.6
72.4	384	278	106	201.3
72.4	381	276	105	199.9
72.4	380	275	105	199.0
72.4	377	273	104	197.7
72.4	373	270	103	195.4
72.4	370	268	102	194.1
72.4	369	267	102	193.2
72.4	366	265	101	191.9
72.4	362	262	100	189.6
72.4	359	260	99	188.3
72.4	355	257	98	186.1
72.4	352	255	97	184.7
72.4	351	254	97	183.8
72.4	348	252	96	182.5
72.4	344	249	95	180.2
72.4	341	247	94	178.9
72.4	340	246	94	178.0
72.4	337	244	93	176.7
72.4	333	241	92	174.4
72.4	330	239	91	173.1
72.4	326	236	90	170.8
72.4	323	234	89	169.5
72.4	322	233	89	168.6
72.4	319	231	88	167.3
72.4	315	228	87	165.0
72.4	312	226	86	163.7
72.4	308	223	85	161.5
72.4	304	220	84	159.2
72.4	301	218	83	157.9
72.4	297	215	82	155.6
72.4	294	213	81	154.3
72.4	293	212	81	153.4
72.4	290	210	80	152.1
72.4	286	207	79	149.8
72.4	283	205	78	148.5
72.4	279	202	77	146.3
72.4	275	199	76	144.0
72.4	272	197	75	142.7
72.4	268	194	74	140.4
72.4	261	189	72	136.9
72.4	257	186	71	134.6
72.4	254	184	70	133.3
72.4	250	181	69	131.0

%E	M1	M2	DM	M*
72.4	246	178	68	128.8
72.4	243	176	67	127.5
72.4	239	173	66	125.2
72.4	232	168	64	121.7
72.4	228	165	63	119.4
72.4	225	163	62	118.1
72.4	221	160	61	115.8
72.4	217	157	60	113.6
72.4	214	155	59	112.3
72.4	210	152	58	110.0
72.4	203	147	56	106.4
72.4	199	144	55	104.2
72.4	196	142	54	102.9
72.4	192	139	53	100.6
72.4	185	134	51	97.1
72.4	181	131	50	94.8
72.4	174	126	48	91.2
72.4	170	123	47	89.0
72.4	163	118	45	85.4
72.4	156	113	43	81.9
72.4	152	110	42	79.6
72.4	145	105	40	76.0
72.4	134	97	37	70.2
72.4	127	92	35	66.6
72.4	123	89	34	64.4
72.4	116	84	32	60.8
72.4	105	76	29	55.0
72.4	98	71	27	51.4
72.4	87	63	24	45.6
72.4	76	55	21	39.8
72.4	58	42	16	30.4
72.4	29	21	8	15.2
72.3	499	361	138	261.2
72.3	498	360	138	260.2
72.3	495	358	137	258.9
72.3	494	357	137	258.0
72.3	491	355	136	256.7
72.3	488	353	135	255.3
72.3	487	352	135	254.4
72.3	484	350	134	253.1
72.3	483	349	134	252.2
72.3	481	348	133	251.8
72.3	480	347	133	250.9
72.3	477	345	132	249.5
72.3	476	344	132	248.6
72.3	473	342	131	247.3
72.3	470	340	130	246.0
72.3	469	339	130	245.0
72.3	466	337	129	243.7
72.3	465	336	129	242.8

%E	M1	M2	DM	M*
72.3	462	334	128	241.5
72.3	459	332	127	240.1
72.3	458	331	127	239.2
72.3	455	329	126	237.9
72.3	452	327	125	236.6
72.3	451	326	125	235.6
72.3	448	324	124	234.3
72.3	447	323	124	233.4
72.3	444	321	123	232.1
72.3	441	319	122	230.8
72.3	440	318	122	229.8
72.3	437	316	121	228.5
72.3	433	313	120	226.3
72.3	430	311	119	224.9
72.3	429	310	119	224.0
72.3	426	308	118	222.7
72.3	423	306	117	221.4
72.3	422	305	117	220.4
72.3	419	303	116	219.1
72.3	415	300	115	216.9
72.3	412	298	114	215.5
72.3	411	297	114	214.6
72.3	408	295	113	213.3
72.3	405	293	112	212.0
72.3	404	292	112	211.0
72.3	401	290	111	209.7
72.3	400	289	111	208.8
72.3	397	287	110	207.5
72.3	394	285	109	206.2
72.3	393	284	109	205.2
72.3	390	282	108	203.9
72.3	386	279	107	201.7
72.3	383	277	106	200.3
72.3	382	276	106	199.4
72.3	379	274	105	198.1
72.3	376	272	104	196.8
72.3	375	271	104	195.8
72.3	372	269	103	194.5
72.3	368	266	102	192.3
72.3	365	264	101	190.9
72.3	364	263	101	190.0
72.3	361	261	100	188.7
72.3	358	259	99	187.4
72.3	357	258	99	186.5
72.3	354	256	98	185.1
72.3	350	253	97	182.9
72.3	347	251	96	181.6
72.3	346	250	96	180.6
72.3	343	248	95	179.3
72.3	339	245	94	177.1

%E	M1	M2	DM	M*
72.3	336	243	93	175.7
72.3	332	240	92	173.5
72.3	329	238	91	172.2
72.3	328	237	91	171.2
72.3	325	235	90	169.9
72.3	321	232	89	167.7
72.3	318	230	88	166.4
72.3	314	227	87	164.1
72.3	311	225	86	162.8
72.3	310	224	86	161.9
72.3	307	222	85	160.5
72.3	303	219	84	158.3
72.3	300	217	83	157.0
72.3	296	214	82	154.7
72.3	292	211	81	152.5
72.3	289	209	80	151.1
72.3	285	206	79	148.9
72.3	282	204	78	147.6
72.3	278	201	77	145.3
72.3	274	198	76	143.1
72.3	271	196	75	141.8
72.3	267	193	74	139.5
72.3	264	191	73	138.2
72.3	260	188	72	135.9
72.3	256	185	71	133.7
72.3	253	183	70	132.4
72.3	249	180	69	130.1
72.3	242	175	67	126.5
72.3	238	172	66	124.3
72.3	235	170	65	123.0
72.3	231	167	64	120.7
72.3	224	162	62	117.2
72.3	220	159	61	114.9
72.3	213	154	59	111.3
72.3	206	149	57	107.8
72.3	202	146	56	105.5
72.3	195	141	54	102.0
72.3	191	138	53	99.7
72.3	188	136	52	98.4
72.3	184	133	51	96.1
72.3	177	128	49	92.6
72.3	173	125	48	90.3
72.3	166	120	46	86.7
72.3	159	115	44	83.2
72.3	155	112	43	80.9
72.3	148	107	41	77.4
72.3	141	102	39	73.8
72.3	137	99	38	71.5
72.3	130	94	36	68.0
72.3	119	86	33	62.2

%E	M1	M2	DM	M*
72.3	112	81	31	58.6
72.3	101	73	28	52.8
72.3	94	68	26	49.2
72.3	83	60	23	43.4
72.3	65	47	18	34.0
72.3	47	34	13	24.6
72.2	500	361	139	260.6
72.2	497	359	138	259.3
72.2	496	358	138	258.4
72.2	493	356	137	257.1
72.2	492	355	137	256.1
72.2	490	354	136	255.7
72.2	489	353	136	254.8
72.2	486	351	135	253.5
72.2	485	350	135	252.6
72.2	482	348	134	251.3
72.2	479	346	133	249.9
72.2	478	345	133	249.0
72.2	475	343	132	247.7
72.2	474	342	132	246.8
72.2	472	341	131	246.4
72.2	471	340	131	245.4
72.2	468	338	130	244.1
72.2	467	337	130	243.2
72.2	464	335	129	241.9
72.2	461	333	128	240.5
72.2	460	332	128	239.6
72.2	457	330	127	238.3
72.2	454	328	126	237.0
72.2	453	327	126	236.0
72.2	450	325	125	234.7
72.2	449	324	125	233.8
72.2	446	322	124	232.5
72.2	443	320	123	231.2
72.2	442	319	123	230.2
72.2	439	317	122	228.9
72.2	436	315	121	227.6
72.2	435	314	121	226.7
72.2	432	312	120	225.3
72.2	431	311	120	224.4
72.2	428	309	119	223.1
72.2	425	307	118	221.8
72.2	424	306	118	220.8
72.2	421	304	117	219.5
72.2	418	302	116	218.2
72.2	417	301	116	217.3
72.2	414	299	115	215.9
72.2	413	298	115	215.0
72.2	410	296	114	213.7
72.2	407	294	113	212.4

%E	M1	M2	DM	M*
72.2	406	293	113	211.5
72.2	403	291	112	210.1
72.2	399	288	111	207.9
72.2	396	286	110	206.6
72.2	395	285	110	205.6
72.2	392	283	109	204.3
72.2	389	281	108	203.0
72.2	388	280	108	202.1
72.2	385	278	107	200.7
72.2	381	275	106	198.5
72.2	378	273	105	197.2
72.2	374	270	104	194.9
72.2	371	268	103	193.6
72.2	370	267	103	192.7
72.2	367	265	102	191.3
72.2	363	262	101	189.1
72.2	360	260	100	187.8
72.2	356	257	99	185.5
72.2	353	255	98	184.2
72.2	352	254	98	183.3
72.2	349	252	97	182.0
72.2	345	249	96	179.7
72.2	342	247	95	178.4
72.2	338	244	94	176.1
72.2	335	242	93	174.8
72.2	334	241	93	173.9
72.2	331	239	92	172.6
72.2	327	236	91	170.3
72.2	324	234	90	169.0
72.2	320	231	89	166.8
72.2	317	229	88	165.4
72.2	316	228	88	164.5
72.2	313	226	87	163.2
72.2	309	223	86	160.9
72.2	306	221	85	159.6
72.2	302	218	84	157.4
72.2	299	216	83	156.0
72.2	295	213	82	153.8
72.2	291	210	81	151.5
72.2	288	208	80	150.2
72.2	284	205	79	148.0
72.2	281	203	78	146.7
72.2	277	200	77	144.4
72.2	273	197	76	142.2
72.2	270	195	75	140.8
72.2	266	192	74	138.6
72.2	263	190	73	137.3
72.2	259	187	72	135.0
72.2	255	184	71	132.8
72.2	252	182	70	131.4

%E	M1	M2	DM	M*
72.2	248	179	69	129.2
72.2	245	177	68	127.9
72.2	241	174	67	125.6
72.2	237	171	66	123.4
72.2	234	169	65	122.1
72.2	230	166	64	119.8
72.2	227	164	63	118.5
72.2	223	161	62	116.2
72.2	216	156	60	112.7
72.2	212	153	59	110.4
72.2	209	151	58	109.1
72.2	205	148	57	106.8
72.2	198	143	55	103.3
72.2	194	140	54	101.0
72.2	187	135	52	97.5
72.2	180	130	50	93.9
72.2	176	127	49	91.6
72.2	169	122	47	88.1
72.2	162	117	45	84.5
72.2	158	114	44	82.3
72.2	151	109	42	78.7
72.2	144	104	40	75.1
72.2	133	96	37	69.3
72.2	126	91	35	65.7
72.2	115	83	32	59.9
72.2	108	78	30	56.3
72.2	97	70	27	50.5
72.2	90	65	25	46.9
72.2	79	57	22	41.1
72.2	72	52	20	37.6
72.2	54	39	15	28.2
72.2	36	26	10	18.8
72.2	18	13	5	9.4
72.1	499	360	139	259.7
72.1	498	359	139	258.8
72.1	495	357	138	257.5
72.1	494	356	138	256.6
72.1	491	354	137	255.2
72.1	488	352	136	253.9
72.1	487	351	136	253.0
72.1	484	349	135	251.7
72.1	483	348	135	250.7
72.1	481	347	134	250.3
72.1	480	346	134	249.4
72.1	477	344	133	248.1
72.1	476	343	133	247.2
72.1	473	341	132	245.8
72.1	470	339	131	244.5
72.1	469	338	131	243.6
72.1	466	336	130	242.3

%E	M1	M2	DM	M*
72.1	463	334	129	240.9
72.1	462	333	129	240.0
72.1	459	331	128	238.7
72.1	458	330	128	237.8
72.1	456	329	127	237.4
72.1	455	328	127	236.4
72.1	452	326	126	235.1
72.1	451	325	126	234.2
72.1	448	323	125	232.9
72.1	445	321	124	231.6
72.1	444	320	124	230.6
72.1	441	318	123	229.3
72.1	438	316	122	228.0
72.1	437	315	122	227.1
72.1	434	313	121	225.7
72.1	433	312	121	224.8
72.1	430	310	120	223.5
72.1	427	308	119	222.2
72.1	426	307	119	221.2
72.1	423	305	118	219.9
72.1	420	303	117	218.6
72.1	419	302	117	217.7
72.1	416	300	116	216.3
72.1	412	297	115	214.1
72.1	409	295	114	212.8
72.1	408	294	114	211.9
72.1	405	292	113	210.5
72.1	402	290	112	209.2
72.1	401	289	112	208.3
72.1	398	287	111	207.0
72.1	394	284	110	204.7
72.1	391	282	109	203.4
72.1	390	281	109	202.5
72.1	387	279	108	201.1
72.1	384	277	107	199.8
72.1	383	276	107	198.9
72.1	380	274	106	197.6
72.1	377	272	105	196.2
72.1	376	271	105	195.3
72.1	373	269	104	194.0
72.1	369	266	103	191.8
72.1	366	264	102	190.4
72.1	365	263	102	189.5
72.1	362	261	101	188.2
72.1	359	259	100	186.9
72.1	358	258	100	185.9
72.1	355	256	99	184.6
72.1	351	253	98	182.4
72.1	348	251	97	181.0
72.1	344	248	96	178.8

%E	M1	M2	DM	M*
72.1	341	246	95	177.5
72.1	340	245	95	176.5
72.1	337	243	94	175.2
72.1	333	240	93	173.0
72.1	330	238	92	171.6
72.1	326	235	91	169.4
72.1	323	233	90	168.1
72.1	322	232	90	167.2
72.1	319	230	89	165.8
72.1	315	227	88	163.6
72.1	312	225	87	162.3
72.1	308	222	86	160.0
72.1	305	220	85	158.7
72.1	301	217	84	156.4
72.1	298	215	83	155.1
72.1	297	214	83	154.2
72.1	294	212	82	152.9
72.1	290	209	81	150.6
72.1	287	207	80	149.3
72.1	283	204	79	147.1
72.1	280	202	78	145.7
72.1	276	199	77	143.5
72.1	272	196	76	141.2
72.1	269	194	75	139.9
72.1	265	191	74	137.7
72.1	262	189	73	136.3
72.1	258	186	72	134.1
72.1	251	181	70	130.5
72.1	247	178	69	128.3
72.1	244	176	68	127.0
72.1	240	173	67	124.7
72.1	233	168	65	121.1
72.1	229	165	64	118.9
72.1	226	163	63	117.6
72.1	222	160	62	115.3
72.1	219	158	61	114.0
72.1	215	155	60	111.7
72.1	208	150	58	108.2
72.1	204	147	57	105.9
72.1	201	145	56	104.6
72.1	197	142	55	102.4
72.1	190	137	53	98.8
72.1	183	132	51	95.2
72.1	179	129	50	93.0
72.1	172	124	48	89.4
72.1	165	119	46	85.8
72.1	161	116	45	83.6
72.1	154	111	43	80.0
72.1	147	106	41	76.4
72.1	140	101	39	72.9

%E	M1	M2	DM	M*
72.1	136	98	38	70.6
72.1	129	93	36	67.0
72.1	122	88	34	63.5
72.1	111	80	31	57.7
72.1	104	75	29	54.1
72.1	86	62	24	44.7
72.1	68	49	19	35.3
72.1	61	44	17	31.7
72.1	43	31	12	22.3
72.0	500	360	140	259.2
72.0	497	358	139	257.9
72.0	496	357	139	257.0
72.0	493	355	138	255.6
72.0	492	354	138	254.7
72.0	490	353	137	254.3
72.0	489	352	137	253.4
72.0	486	350	136	252.1
72.0	485	349	136	251.1
72.0	482	347	135	249.8
72.0	479	345	134	248.5
72.0	478	344	134	247.6
72.0	475	342	133	246.2
72.0	472	340	132	244.9
72.0	471	339	132	244.0
72.0	468	337	131	242.7
72.0	465	335	130	241.3
72.0	464	334	130	240.4
72.0	461	332	129	239.1
72.0	460	331	129	238.2
72.0	457	329	128	236.9
72.0	454	327	127	235.5
72.0	453	326	127	234.6
72.0	450	324	126	233.3
72.0	447	322	125	232.0
72.0	446	321	125	231.0
72.0	443	319	124	229.7
72.0	440	317	123	228.4
72.0	439	316	123	227.5
72.0	436	314	122	226.1
72.0	435	313	122	225.2
72.0	432	311	121	223.9
72.0	429	309	120	222.6
72.0	428	308	120	221.6
72.0	425	306	119	220.3
72.0	422	304	118	219.0
72.0	421	303	118	218.1
72.0	418	301	117	216.7
72.0	415	299	116	215.4
72.0	414	298	116	214.5
72.0	411	296	115	213.2

%E	M1	M2	DM	M*
72.0	410	295	115	212.3
72.0	407	293	114	210.9
72.0	404	291	113	209.6
72.0	403	290	113	208.7
72.0	400	288	112	207.4
72.0	397	286	111	206.0
72.0	396	285	111	205.1
72.0	393	283	110	203.8
72.0	389	280	109	201.5
72.0	386	278	108	200.2
72.0	382	275	107	198.0
72.0	379	273	106	196.6
72.0	378	272	106	195.7
72.0	375	270	105	194.4
72.0	372	268	104	193.1
72.0	371	267	104	192.2
72.0	368	265	103	190.8
72.0	364	262	102	188.6
72.0	361	260	101	187.3
72.0	357	257	100	185.0
72.0	354	255	99	183.7
72.0	353	254	99	182.8
72.0	350	252	98	181.4
72.0	347	250	97	180.1
72.0	346	249	97	179.2
72.0	343	247	96	177.9
72.0	339	244	95	175.6
72.0	336	242	94	174.3
72.0	332	239	93	172.1
72.0	329	237	92	170.7
72.0	328	236	92	169.8
72.0	325	234	91	168.5
72.0	321	231	90	166.2
72.0	318	229	89	164.9
72.0	314	226	88	162.7
72.0	311	224	87	161.3
72.0	307	221	86	159.1
72.0	304	219	85	157.8
72.0	300	216	84	155.5
72.0	296	213	83	153.3
72.0	293	211	82	151.9
72.0	289	208	81	149.7
72.0	286	206	80	148.4
72.0	282	203	79	146.1
72.0	279	201	78	144.8
72.0	275	198	77	142.6
72.0	271	195	76	140.3
72.0	268	193	75	139.0
72.0	264	190	74	136.7
72.0	261	188	73	135.4

%E	M1	M2	DM	M*
72.0	257	185	72	133.2
72.0	254	183	71	131.8
72.0	250	180	70	129.6
72.0	246	177	69	127.4
72.0	243	175	68	126.0
72.0	239	172	67	123.8
72.0	236	170	66	122.5
72.0	232	167	65	120.2
72.0	225	162	63	116.6
72.0	218	157	61	113.1
72.0	214	154	60	110.8
72.0	211	152	59	109.5
72.0	207	149	58	107.3
72.0	200	144	56	103.7
72.0	193	139	54	100.1
72.0	189	136	53	97.9
72.0	186	134	52	96.5
72.0	182	131	51	94.3
72.0	175	126	49	90.7
72.0	168	121	47	87.1
72.0	164	118	46	84.9
72.0	157	113	44	81.3
72.0	150	108	42	77.8
72.0	143	103	40	74.2
72.0	132	95	37	68.4
72.0	125	90	35	64.8
72.0	118	85	33	61.2
72.0	107	77	30	55.4
72.0	100	72	28	51.8
72.0	93	67	26	48.3
72.0	82	59	23	42.5
72.0	75	54	21	38.9
72.0	50	36	14	25.9
72.0	25	18	7	13.0
71.9	499	359	140	258.3
71.9	498	358	140	257.4
71.9	495	356	139	256.0
71.9	494	355	139	255.1
71.9	491	353	138	253.8
71.9	488	351	137	252.5
71.9	487	350	137	251.5
71.9	484	348	136	250.2
71.9	481	346	135	248.9
71.9	480	345	135	248.0
71.9	477	343	134	246.6
71.9	474	341	133	245.3
71.9	473	340	133	244.4
71.9	470	338	132	243.1
71.9	469	337	132	242.2
71.9	467	336	131	241.7

%E	M1	M2	DM	M*
71.9	466	335	131	240.8
71.9	463	333	130	239.5
71.9	462	332	130	238.6
71.9	459	330	129	237.3
71.9	456	328	128	235.9
71.9	455	327	128	235.0
71.9	452	325	127	233.7
71.9	449	323	126	232.4
71.9	448	322	126	231.4
71.9	445	320	125	230.1
71.9	442	318	124	228.8
71.9	441	317	124	227.9
71.9	438	315	123	226.5
71.9	437	314	123	225.6
71.9	434	312	122	224.3
71.9	431	310	121	223.0
71.9	430	309	121	222.0
71.9	427	307	120	220.7
71.9	424	305	119	219.4
71.9	423	304	119	218.5
71.9	420	302	118	217.2
71.9	417	300	117	215.8
71.9	416	299	117	214.9
71.9	413	297	116	213.6
71.9	409	294	115	211.3
71.9	406	292	114	210.0
71.9	405	291	114	209.1
71.9	402	289	113	207.8
71.9	399	287	112	206.4
71.9	398	286	112	205.5
71.9	395	284	111	204.2
71.9	392	282	110	202.9
71.9	391	281	110	201.9
71.9	388	279	109	200.6
71.9	385	277	108	199.3
71.9	384	276	108	198.4
71.9	381	274	107	197.0
71.9	377	271	106	194.8
71.9	374	269	105	193.5
71.9	373	268	105	192.6
71.9	370	266	104	191.2
71.9	367	264	103	189.9
71.9	366	263	103	189.0
71.9	363	261	102	187.7
71.9	360	259	101	186.3
71.9	359	258	101	185.4
71.9	356	256	100	184.1
71.9	352	253	99	181.8
71.9	349	251	98	180.5
71.9	345	248	97	178.3

%E	M1	M2	DM	M*
71.9	342	246	96	176.9
71.9	338	243	95	174.7
71.9	335	241	94	173.4
71.9	334	240	94	172.5
71.9	331	238	93	171.1
71.9	327	235	92	168.9
71.9	324	233	91	167.6
71.9	320	230	90	165.3
71.9	317	228	89	164.0
71.9	313	225	88	161.7
71.9	310	223	87	160.4
71.9	306	220	86	158.2
71.9	303	218	85	156.8
71.9	302	217	85	155.9
71.9	299	215	84	154.6
71.9	295	212	83	152.4
71.9	292	210	82	151.0
71.9	288	207	81	148.8
71.9	285	205	80	147.5
71.9	281	202	79	145.2
71.9	278	200	78	143.9
71.9	274	197	77	141.6
71.9	270	194	76	139.4
71.9	267	192	75	138.1
71.9	263	189	74	135.8
71.9	260	187	73	134.5
71.9	256	184	72	132.3
71.9	253	182	71	130.9
71.9	249	179	70	128.7
71.9	242	174	68	125.1
71.9	235	169	66	121.5
71.9	231	166	65	119.3
71.9	228	164	64	118.0
71.9	224	161	63	115.7
71.9	221	159	62	114.4
71.9	217	156	61	112.1
71.9	210	151	59	108.6
71.9	203	146	57	105.0
71.9	199	143	56	102.8
71.9	196	141	55	101.4
71.9	192	138	54	99.2
71.9	185	133	52	95.6
71.9	178	128	50	92.0
71.9	171	123	48	88.5
71.9	167	120	47	86.2
71.9	160	115	45	82.7
71.9	153	110	43	79.1
71.9	146	105	41	75.5
71.9	139	100	39	71.9
71.9	135	97	38	69.7

%E	M1	M2	DM	M*
71.9	128	92	36	66.1
71.9	121	87	34	62.6
71.9	114	82	32	59.0
71.9	96	69	27	49.6
71.9	89	64	25	46.0
71.9	64	46	18	33.1
71.9	57	41	16	29.5
71.9	32	23	9	16.5
71.8	500	359	141	257.8
71.8	497	357	140	256.4
71.8	496	356	140	255.5
71.8	493	354	139	254.2
71.8	490	352	138	252.9
71.8	489	351	138	251.9
71.8	486	349	137	250.6
71.8	485	348	137	249.7
71.8	483	347	136	249.3
71.8	482	346	136	248.4
71.8	479	344	135	247.0
71.8	478	343	135	246.1
71.8	476	342	134	245.7
71.8	475	341	134	244.8
71.8	472	339	133	243.5
71.8	471	338	133	242.6
71.8	468	336	132	241.2
71.8	465	334	131	239.9
71.8	464	333	131	239.0
71.8	461	331	130	237.7
71.8	458	329	129	236.3
71.8	457	328	129	235.4
71.8	454	326	128	234.1
71.8	451	324	127	232.8
71.8	450	323	127	231.8
71.8	447	321	126	230.5
71.8	444	319	125	229.2
71.8	443	318	125	228.3
71.8	440	316	124	226.9
71.8	439	315	124	226.0
71.8	436	313	123	224.7
71.8	433	311	122	223.4
71.8	432	310	122	222.5
71.8	429	308	121	221.1
71.8	426	306	120	219.8
71.8	425	305	120	218.9
71.8	422	303	119	217.6
71.8	419	301	118	216.2
71.8	418	300	118	215.3
71.8	415	298	117	214.0
71.8	412	296	116	212.7
71.8	411	295	116	211.7

%E	M1	M2	DM	M*
71.8	408	293	115	210.4
71.8	404	290	114	208.2
71.8	401	288	113	206.8
71.8	400	287	113	205.9
71.8	397	285	112	204.6
71.8	394	283	111	203.3
71.8	393	282	111	202.4
71.8	390	280	110	201.0
71.8	387	278	109	199.7
71.8	386	277	109	198.8
71.8	383	275	108	197.5
71.8	380	273	107	196.1
71.8	379	272	107	195.2
71.8	376	270	106	193.9
71.8	372	267	105	191.6
71.8	369	265	104	190.3
71.8	365	262	103	188.1
71.8	362	260	102	186.7
71.8	358	257	101	184.5
71.8	355	255	100	183.2
71.8	354	254	100	182.2
71.8	351	252	99	180.9
71.8	348	250	98	179.6
71.8	347	249	98	178.7
71.8	344	247	97	177.4
71.8	341	245	96	176.0
71.8	340	244	96	175.1
71.8	337	242	95	173.8
71.8	333	239	94	171.5
71.8	330	237	93	170.2
71.8	326	234	92	168.0
71.8	323	232	91	166.6
71.8	319	229	90	164.4
71.8	316	227	89	163.1
71.8	312	224	88	160.8
71.8	309	222	87	159.5
71.8	308	221	87	158.6
71.8	305	219	86	157.2
71.8	301	216	85	155.0
71.8	298	214	84	153.7
71.8	294	211	83	151.4
71.8	291	209	82	150.1
71.8	287	206	81	147.9
71.8	284	204	80	146.5
71.8	280	201	79	144.3
71.8	277	199	78	143.0
71.8	273	196	77	140.7
71.8	266	191	75	137.1
71.8	262	188	74	134.9
71.8	259	186	73	133.6

%E	M1	M2	DM	M*
71.8	255	183	72	131.3
71.8	252	181	71	130.0
71.8	248	178	70	127.8
71.8	245	176	69	126.4
71.8	241	173	68	124.2
71.8	238	171	67	122.9
71.8	234	168	66	120.6
71.8	227	163	64	117.0
71.8	220	158	62	113.5
71.8	216	155	61	111.2
71.8	213	153	60	109.9
71.8	209	150	59	107.7
71.8	206	148	58	106.3
71.8	202	145	57	104.1
71.8	195	140	55	100.5
71.8	188	135	53	96.9
71.8	181	130	51	93.4
71.8	177	127	50	91.1
71.8	174	125	49	89.8
71.8	170	122	48	87.6
71.8	163	117	46	84.0
71.8	156	112	44	80.4
71.8	149	107	42	76.8
71.8	142	102	40	73.3
71.8	131	94	37	67.5
71.8	124	89	35	63.9
71.8	117	84	33	60.3
71.8	110	79	31	56.7
71.8	103	74	29	53.2
71.8	85	61	24	43.8
71.8	78	56	22	40.2
71.8	71	51	20	36.6
71.8	39	28	11	20.1
71.7	499	358	141	256.8
71.7	498	357	141	255.9
71.7	495	355	140	254.6
71.7	494	354	140	253.7
71.7	492	353	139	253.3
71.7	491	352	139	252.4
71.7	488	350	138	251.0
71.7	487	349	138	250.1
71.7	484	347	137	248.8
71.7	481	345	136	247.5
71.7	480	344	136	246.5
71.7	477	342	135	245.2
71.7	474	340	134	243.9
71.7	473	339	134	243.0
71.7	470	337	133	241.6
71.7	467	335	132	240.3
71.7	466	334	132	239.4

%E	M1	M2	DM	M*
71.7	463	332	131	238.1
71.7	460	330	130	236.7
71.7	459	329	130	235.8
71.7	456	327	129	234.5
71.7	453	325	128	233.2
71.7	452	324	128	232.2
71.7	449	322	127	230.9
71.7	448	321	127	230.0
71.7	446	320	126	229.6
71.7	445	319	126	228.7
71.7	442	317	125	227.4
71.7	441	316	125	226.4
71.7	438	314	124	225.1
71.7	435	312	123	223.8
71.7	434	311	123	222.9
71.7	431	309	122	221.5
71.7	428	307	121	220.2
71.7	427	306	121	219.3
71.7	424	304	120	218.0
71.7	421	302	119	216.6
71.7	420	301	119	215.7
71.7	417	299	118	214.4
71.7	414	297	117	213.1
71.7	413	296	117	212.1
71.7	410	294	116	210.8
71.7	407	292	115	209.5
71.7	406	291	115	208.6
71.7	403	289	114	207.2
71.7	399	286	113	205.0
71.7	396	284	112	203.7
71.7	392	281	111	201.4
71.7	389	279	110	200.1
71.7	385	276	109	197.9
71.7	382	274	108	196.5
71.7	381	273	108	195.6
71.7	378	271	107	194.3
71.7	375	269	106	193.0
71.7	374	268	106	192.0
71.7	371	266	105	190.7
71.7	368	264	104	189.4
71.7	367	263	104	188.5
71.7	364	261	103	187.1
71.7	361	259	102	185.8
71.7	360	258	102	184.9
71.7	357	256	101	183.6
71.7	353	253	100	181.3
71.7	350	251	99	180.0
71.7	346	248	98	177.8
71.7	343	246	97	176.4
71.7	339	243	96	174.2

%E	M1	M2	DM	M*
71.7	336	241	95	172.9
71.7	332	238	94	170.6
71.7	329	236	93	169.3
71.7	325	233	92	167.0
71.7	322	231	91	165.7
71.7	321	230	91	164.8
71.7	318	228	90	163.5
71.7	315	226	89	162.1
71.7	314	225	89	161.2
71.7	311	223	88	159.9
71.7	307	220	87	157.7
71.7	304	218	86	156.3
71.7	300	215	85	154.1
71.7	297	213	84	152.8
71.7	293	210	83	150.5
71.7	290	208	82	149.2
71.7	286	205	81	146.9
71.7	283	203	80	145.6
71.7	279	200	79	143.4
71.7	276	198	78	142.0
71.7	272	195	77	139.8
71.7	269	193	76	138.5
71.7	265	190	75	136.2
71.7	258	185	73	132.7
71.7	254	182	72	130.4
71.7	251	180	71	129.1
71.7	247	177	70	126.8
71.7	244	175	69	125.5
71.7	240	172	68	123.3
71.7	237	170	67	121.9
71.7	233	167	66	119.7
71.7	230	165	65	118.4
71.7	226	162	64	116.1
71.7	223	160	63	114.8
71.7	219	157	62	112.6
71.7	212	152	60	109.0
71.7	205	147	58	105.4
71.7	198	142	56	101.8
71.7	191	137	54	98.3
71.7	187	134	53	96.0
71.7	184	132	52	94.7
71.7	180	129	51	92.4
71.7	173	124	49	88.9
71.7	166	119	47	85.3
71.7	159	114	45	81.7
71.7	152	109	43	78.2
71.7	145	104	41	74.6
71.7	138	99	39	71.0
71.7	127	91	36	65.2
71.7	120	86	34	61.6

%E	M1	M2	DM	M*
71.7	113	81	32	58.1
71.7	106	76	30	54.5
71.7	99	71	28	50.9
71.7	92	66	26	47.3
71.7	60	43	17	30.8
71.7	53	38	15	27.2
71.7	46	33	13	23.7
71.6	500	358	142	256.3
71.6	497	356	141	255.0
71.6	496	355	141	254.1
71.6	493	353	140	252.8
71.6	490	351	139	251.4
71.6	489	350	139	250.5
71.6	486	348	138	249.2
71.6	483	346	137	247.9
71.6	482	345	137	246.9
71.6	479	343	136	245.6
71.6	476	341	135	244.3
71.6	475	340	135	243.4
71.6	472	338	134	242.0
71.6	471	337	134	241.1
71.6	469	336	133	240.7
71.6	468	335	133	239.8
71.6	465	333	132	238.5
71.6	464	332	132	237.6
71.6	462	331	131	237.1
71.6	461	330	131	236.2
71.6	458	328	130	234.9
71.6	457	327	130	234.0
71.6	455	326	129	233.6
71.6	454	325	129	232.7
71.6	451	323	128	231.3
71.6	450	322	128	230.4
71.6	447	320	127	229.1
71.6	444	318	126	227.8
71.6	443	317	126	226.8
71.6	440	315	125	225.5
71.6	437	313	124	224.2
71.6	436	312	124	223.3
71.6	433	310	123	221.9
71.6	430	308	122	220.6
71.6	429	307	122	219.7
71.6	426	305	121	218.4
71.6	423	303	120	217.0
71.6	422	302	120	216.1
71.6	419	300	119	214.8
71.6	416	298	118	213.5
71.6	415	297	118	212.6
71.6	412	295	117	211.2
71.6	409	293	116	209.9

%E	M1	M2	DM	M*
71.6	408	292	116	209.0
71.6	405	290	115	207.7
71.6	402	288	114	206.3
71.6	401	287	114	205.4
71.6	398	285	113	204.1
71.6	395	283	112	202.8
71.6	394	282	112	201.8
71.6	391	280	111	200.5
71.6	388	278	110	199.2
71.6	387	277	110	198.3
71.6	384	275	109	196.9
71.6	380	272	108	194.7
71.6	377	270	107	193.4
71.6	373	267	106	191.1
71.6	370	265	105	189.8
71.6	366	262	104	187.6
71.6	363	260	103	186.2
71.6	359	257	102	184.0
71.6	356	255	101	182.7
71.6	352	252	100	180.4
71.6	349	250	99	179.1
71.6	348	249	99	178.2
71.6	345	247	98	176.8
71.6	342	245	97	175.5
71.6	341	244	97	174.6
71.6	338	242	96	173.3
71.6	335	240	95	171.9
71.6	334	239	95	171.0
71.6	331	237	94	169.7
71.6	328	235	93	168.4
71.6	327	234	93	167.4
71.6	324	232	92	166.1
71.6	320	229	91	163.9
71.6	317	227	90	162.6
71.6	313	224	89	160.3
71.6	310	222	88	159.0
71.6	306	219	87	156.7
71.6	303	217	86	155.4
71.6	299	214	85	153.2
71.6	296	212	84	151.8
71.6	292	209	83	149.6
71.6	289	207	82	148.3
71.6	285	204	81	146.0
71.6	282	202	80	144.7
71.6	278	199	79	142.4
71.6	275	197	78	141.1
71.6	271	194	77	138.9
71.6	268	192	76	137.6
71.6	264	189	75	135.3
71.6	261	187	74	134.0

%E	M1	M2	DM	M*
71.6	257	184	73	131.7
71.6	250	179	71	128.2
71.6	243	174	69	124.6
71.6	236	169	67	121.0
71.6	232	166	66	118.8
71.6	229	164	65	117.4
71.6	225	161	64	115.2
71.6	222	159	63	113.9
71.6	218	156	62	111.6
71.6	215	154	61	110.3
71.6	211	151	60	108.1
71.6	208	149	59	106.7
71.6	204	146	58	104.5
71.6	201	144	57	103.2
71.6	197	141	56	100.9
71.6	194	139	55	99.6
71.6	190	136	54	97.3
71.6	183	131	52	93.8
71.6	176	126	50	90.2
71.6	169	121	48	86.6
71.6	162	116	46	83.1
71.6	155	111	44	79.5
71.6	148	106	42	75.9
71.6	141	101	40	72.3
71.6	134	96	38	68.8
71.6	116	83	33	59.4
71.6	109	78	31	55.8
71.6	102	73	29	52.2
71.6	95	68	27	48.7
71.6	88	63	25	45.1
71.6	81	58	23	41.5
71.6	74	53	21	38.0
71.6	67	48	19	34.4
71.5	499	357	142	255.4
71.5	498	356	142	254.5
71.5	495	354	141	253.2
71.5	494	353	141	252.2
71.5	492	352	140	251.8
71.5	491	351	140	250.9
71.5	488	349	139	249.6
71.5	487	348	139	248.7
71.5	485	347	138	248.3
71.5	484	346	138	247.3
71.5	481	344	137	246.0
71.5	480	343	137	245.1
71.5	478	342	136	244.7
71.5	477	341	136	243.8
71.5	474	339	135	242.4
71.5	473	338	135	241.5
71.5	[illegible]	336	134	240.2

%E	M1	M2	DM	M*
71.5	467	334	133	238.9
71.5	466	333	133	238.0
71.5	463	331	132	236.6
71.5	460	329	131	235.3
71.5	459	328	131	234.4
71.5	456	326	130	233.1
71.5	453	324	129	231.7
71.5	452	323	129	230.8
71.5	449	321	128	229.5
71.5	446	319	127	228.2
71.5	445	318	127	227.2
71.5	442	316	126	225.9
71.5	439	314	125	224.6
71.5	438	313	125	223.7
71.5	435	311	124	222.3
71.5	432	309	123	221.0
71.5	431	308	123	220.1
71.5	428	306	122	218.8
71.5	425	304	121	217.4
71.5	424	303	121	216.5
71.5	421	301	120	215.2
71.5	418	299	119	213.9
71.5	417	298	119	213.0
71.5	414	296	118	211.6
71.5	411	294	117	210.3
71.5	410	293	117	209.4
71.5	407	291	116	208.1
71.5	404	289	115	206.7
71.5	403	288	115	205.8
71.5	400	286	114	204.5
71.5	397	284	113	203.2
71.5	396	283	113	202.2
71.5	393	281	112	200.9
71.5	390	279	111	199.6
71.5	389	278	111	198.7
71.5	386	276	110	197.3
71.5	383	274	109	196.0
71.5	382	273	109	195.1
71.5	379	271	108	193.8
71.5	376	269	107	192.4
71.5	375	268	107	191.5
71.5	372	266	106	190.2
71.5	369	264	105	188.9
71.5	368	263	105	188.0
71.5	365	261	104	186.6
71.5	362	259	103	185.3
71.5	361	258	103	184.4
71.5	358	256	102	183.1
71.5	355	254	101	181.7
71.5	354	253	101	180.8

%E	M1	M2	DM	M*
71.5	351	251	100	179.5
71.5	347	248	99	177.2
71.5	344	246	98	175.9
71.5	340	243	97	173.7
71.5	337	241	96	172.3
71.5	333	238	95	170.1
71.5	330	236	94	168.8
71.5	326	233	93	166.5
71.5	323	231	92	165.2
71.5	319	228	91	163.0
71.5	316	226	90	161.6
71.5	312	223	89	159.4
71.5	309	221	88	158.1
71.5	305	218	87	155.8
71.5	302	216	86	154.5
71.5	298	213	85	152.2
71.5	295	211	84	150.9
71.5	291	208	83	148.7
71.5	288	206	82	147.3
71.5	284	203	81	145.1
71.5	281	201	80	143.8
71.5	277	198	79	141.5
71.5	274	196	78	140.2
71.5	270	193	77	138.0
71.5	267	191	76	136.6
71.5	263	188	75	134.4
71.5	260	186	74	133.1
71.5	256	183	73	130.8
71.5	253	181	72	129.5
71.5	249	178	71	127.2
71.5	246	176	70	125.9
71.5	242	173	69	123.7
71.5	239	171	68	122.3
71.5	235	168	67	120.1
71.5	228	163	65	116.5
71.5	221	158	63	113.0
71.5	214	153	61	109.4
71.5	207	148	59	105.8
71.5	200	143	57	102.2
71.5	193	138	55	98.7
71.5	186	133	53	95.1
71.5	179	128	51	91.5
71.5	172	123	49	88.0
71.5	165	118	47	84.4
71.5	158	113	45	80.8
71.5	151	108	43	77.2
71.5	144	103	41	73.7
71.5	137	98	39	70.1
71.5	130	93	37	66.5
71.5	123	88	35	63.0

%E	M1	M2	DM	M*
71.4	500	357	143	254.9
71.4	497	355	142	253.6
71.4	496	354	142	252.7
71.4	493	352	141	251.3
71.4	490	350	140	250.0
71.4	489	349	140	249.1
71.4	486	347	139	247.8
71.4	483	345	138	246.4
71.4	482	344	138	245.5
71.4	479	342	137	244.2
71.4	476	340	136	242.9
71.4	475	339	136	241.9
71.4	472	337	135	240.6
71.4	469	335	134	239.3
71.4	468	334	134	238.4
71.4	465	332	133	237.0
71.4	462	330	132	235.7
71.4	461	329	132	234.8
71.4	458	327	131	233.5
71.4	455	325	130	232.1
71.4	454	324	130	231.2
71.4	451	322	129	229.9
71.4	448	320	128	228.6
71.4	447	319	128	227.7
71.4	444	317	127	226.3
71.4	441	315	126	225.0
71.4	440	314	126	224.1
71.4	437	312	125	222.8
71.4	434	310	124	221.4
71.4	433	309	124	220.5
71.4	430	307	123	219.2
71.4	427	305	122	217.9
71.4	426	304	122	216.9
71.4	423	302	121	215.6
71.4	420	300	120	214.3
71.4	419	299	120	213.4
71.4	416	297	119	212.0
71.4	413	295	118	210.7
71.4	412	294	118	209.8
71.4	409	292	117	208.5
71.4	406	290	116	207.1
71.4	405	289	116	206.2
71.4	402	287	115	204.9
71.4	399	285	114	203.6
71.4	398	284	114	202.7
71.4	395	282	113	201.3
71.4	392	280	112	200.0
71.4	391	279	112	199.1
71.4	388	277	111	197.8
71.4	385	275	110	196.4

%E	M1	M2	DM	M*
71.4	384	274	110	195.5
71.4	381	272	109	194.2
71.4	378	270	108	192.9
71.4	377	269	108	191.9
71.4	374	267	107	190.6
71.4	371	265	106	189.3
71.4	370	264	106	188.4
71.4	367	262	105	187.0
71.4	364	260	104	185.7
71.4	363	259	104	184.8
71.4	360	257	103	183.5
71.4	357	255	102	182.1
71.4	353	252	101	179.9
71.4	350	250	100	178.6
71.4	346	247	99	176.3
71.4	343	245	98	175.0
71.4	339	242	97	172.8
71.4	336	240	96	171.4
71.4	332	237	95	169.2
71.4	329	235	94	167.9
71.4	325	232	93	165.6
71.4	322	230	92	164.3
71.4	318	227	91	162.0
71.4	315	225	90	160.7
71.4	311	222	89	158.5
71.4	308	220	88	157.1
71.4	304	217	87	154.9
71.4	301	215	86	153.6
71.4	297	212	85	151.3
71.4	294	210	84	150.0
71.4	290	207	83	147.8
71.4	287	205	82	146.4
71.4	283	202	81	144.2
71.4	280	200	80	142.9
71.4	276	197	79	140.6
71.4	273	195	78	139.3
71.4	269	192	77	137.0
71.4	266	190	76	135.7
71.4	262	187	75	133.5
71.4	259	185	74	132.1
71.4	255	182	73	129.9
71.4	252	180	72	128.6
71.4	248	177	71	126.3
71.4	245	175	70	125.0
71.4	241	172	69	122.8
71.4	238	170	68	121.4
71.4	234	167	67	119.2
71.4	231	165	66	117.9
71.4	227	162	65	115.6
71.4	224	160	64	114.3

%E	M1	M2	DM	M*
71.4	220	157	63	112.0
71.4	217	155	62	110.7
71.4	213	152	61	108.5
71.4	210	150	60	107.1
71.4	206	147	59	104.9
71.4	203	145	58	103.6
71.4	199	142	57	101.3
71.4	196	140	56	100.0
71.4	192	137	55	97.8
71.4	189	135	54	96.4
71.4	185	132	53	94.2
71.4	182	130	52	92.9
71.4	175	125	50	89.3
71.4	168	120	48	85.7
71.4	161	115	46	82.1
71.4	154	110	44	78.6
71.4	147	105	42	75.0
71.4	140	100	40	71.4
71.4	133	95	38	67.9
71.4	126	90	36	64.3
71.4	119	85	34	60.7
71.4	112	80	32	57.1
71.4	105	75	30	53.6
71.4	98	70	28	50.0
71.4	91	65	26	46.4
71.4	84	60	24	42.9
71.4	77	55	22	39.3
71.4	70	50	20	35.7
71.4	63	45	18	32.1
71.4	56	40	16	28.6
71.4	49	35	14	25.0
71.4	42	30	12	21.4
71.4	35	25	10	17.9
71.4	28	20	8	14.3
71.4	21	15	6	10.7
71.4	14	10	4	7.1
71.4	7	5	2	3.6
71.3	499	356	143	254.0
71.3	498	355	143	253.1
71.3	495	353	142	251.7
71.3	494	352	142	250.8
71.3	492	351	141	250.4
71.3	491	350	141	249.5
71.3	488	348	140	248.2
71.3	487	347	140	247.2
71.3	485	346	139	246.8
71.3	484	345	139	245.9
71.3	481	343	138	244.6
71.3	480	342	138	243.7
71.3	478	341	137	243.3

%E	M1	M2	DM	M*
71.3	477	340	137	242.3
71.3	474	338	136	241.0
71.3	471	336	135	239.7
71.3	470	335	135	238.8
71.3	467	333	134	237.4
71.3	464	331	133	236.1
71.3	463	330	133	235.2
71.3	460	328	132	233.9
71.3	457	326	131	232.6
71.3	456	325	131	231.6
71.3	453	323	130	230.3
71.3	450	321	129	229.0
71.3	449	320	129	228.1
71.3	446	318	128	226.7
71.3	443	316	127	225.4
71.3	442	315	127	224.5
71.3	439	313	126	223.2
71.3	436	311	125	221.8
71.3	435	310	125	220.9
71.3	432	308	124	219.6
71.3	429	306	123	218.3
71.3	428	305	123	217.3
71.3	425	303	122	216.0
71.3	422	301	121	214.7
71.3	421	300	121	213.8
71.3	418	298	120	212.4
71.3	415	296	119	211.1
71.3	414	295	119	210.2
71.3	411	293	118	208.9
71.3	408	291	117	207.6
71.3	407	290	117	206.6
71.3	404	288	116	205.3
71.3	401	286	115	204.0
71.3	400	285	115	203.1
71.3	397	283	114	201.7
71.3	394	281	113	200.4
71.3	390	278	112	198.2
71.3	387	276	111	196.8
71.3	383	273	110	194.6
71.3	380	271	109	193.3
71.3	376	268	108	191.0
71.3	373	266	107	189.7
71.3	369	263	106	187.4
71.3	366	261	105	186.1
71.3	362	258	104	183.9
71.3	359	256	103	182.6
71.3	356	254	102	181.2
71.3	355	253	102	180.3
71.3	352	251	101	179.0
71.3	349	249	100	177.7

%E	M1	M2	DM	M*
71.3	348	248	100	176.7
71.3	345	246	99	175.4
71.3	342	244	98	174.1
71.3	341	243	98	173.2
71.3	338	241	97	171.8
71.3	335	239	96	170.5
71.3	334	238	96	169.6
71.3	331	236	95	168.3
71.3	328	234	94	166.9
71.3	327	233	94	166.0
71.3	324	231	93	164.7
71.3	321	229	92	163.4
71.3	320	228	92	162.4
71.3	317	226	91	161.1
71.3	314	224	90	159.8
71.3	310	221	89	157.6
71.3	307	219	88	156.2
71.3	303	216	87	154.0
71.3	300	214	86	152.7
71.3	296	211	85	150.4
71.3	293	209	84	149.1
71.3	289	206	83	146.8
71.3	286	204	82	145.5
71.3	282	201	81	143.3
71.3	279	199	80	141.9
71.3	275	196	79	139.7
71.3	272	194	78	138.4
71.3	268	191	77	136.1
71.3	265	189	76	134.8
71.3	261	186	75	132.6
71.3	258	184	74	131.2
71.3	254	181	73	129.0
71.3	251	179	72	127.7
71.3	247	176	71	125.4
71.3	244	174	70	124.1
71.3	240	171	69	121.8
71.3	237	169	68	120.5
71.3	230	164	66	116.9
71.3	223	159	64	113.4
71.3	216	154	62	109.8
71.3	209	149	60	106.2
71.3	202	144	58	102.7
71.3	195	139	56	99.1
71.3	188	134	54	95.5
71.3	181	129	52	91.9
71.3	178	127	51	90.6
71.3	174	124	50	88.4
71.3	171	122	49	87.0
71.3	167	119	48	84.8
71.3	164	117	47	83.5

%E	M1	M2	DM	M*
71.3	160	114	46	81.2
71.3	157	112	45	79.9
71.3	150	107	43	76.3
71.3	143	102	41	72.8
71.3	136	97	39	69.2
71.3	129	92	37	65.6
71.3	122	87	35	62.0
71.3	115	82	33	58.5
71.3	108	77	31	54.9
71.3	101	72	29	51.3
71.3	94	67	27	47.8
71.3	87	62	25	44.2
71.3	80	57	23	40.6
71.2	500	356	144	253.5
71.2	497	354	143	252.1
71.2	496	353	143	251.2
71.2	493	351	142	249.9
71.2	490	349	141	248.6
71.2	489	348	141	247.7
71.2	486	346	140	246.3
71.2	483	344	139	245.0
71.2	482	343	139	244.1
71.2	479	341	138	242.8
71.2	476	339	137	241.4
71.2	475	338	137	240.5
71.2	473	337	136	240.1
71.2	472	336	136	239.2
71.2	469	334	135	237.9
71.2	468	333	135	236.9
71.2	466	332	134	236.5
71.2	465	331	134	235.6
71.2	462	329	133	234.3
71.2	461	328	133	233.4
71.2	459	327	132	233.0
71.2	458	326	132	232.0
71.2	455	324	131	230.7
71.2	452	322	130	229.4
71.2	451	321	130	228.5
71.2	448	319	129	227.1
71.2	445	317	128	225.8
71.2	444	316	128	224.9
71.2	441	314	127	223.6
71.2	438	312	126	222.2
71.2	437	311	126	221.3
71.2	434	309	125	220.0
71.2	431	307	124	218.7
71.2	430	306	124	217.8
71.2	427	304	123	216.4
71.2	424	302	122	215.1
71.2	423	301	122	214.2
71.2	420	299	121	212.9
71.2	417	297	120	211.5
71.2	416	296	120	210.6
71.2	413	294	119	209.3
71.2	410	292	118	208.0
71.2	406	289	117	205.7
71.2	403	287	116	204.4
71.2	399	284	115	202.1
71.2	396	282	114	200.8
71.2	393	280	113	199.5
71.2	392	279	113	198.6
71.2	389	277	112	197.2
71.2	386	275	111	195.9
71.2	385	274	111	195.0
71.2	382	272	110	193.7
71.2	379	270	109	192.3
71.2	378	269	109	191.4
71.2	375	267	108	190.1
71.2	372	265	107	188.8
71.2	371	264	107	187.9
71.2	368	262	106	186.5
71.2	365	260	105	185.2
71.2	364	259	105	184.3
71.2	361	257	104	183.0
71.2	358	255	103	181.6
71.2	354	252	102	179.4
71.2	351	250	101	178.1
71.2	347	247	100	175.8
71.2	344	245	99	174.5
71.2	340	242	98	172.2
71.2	337	240	97	170.9
71.2	333	237	96	168.7
71.2	330	235	95	167.3
71.2	326	232	94	165.1
71.2	323	230	93	163.8
71.2	319	227	92	161.5
71.2	316	225	91	160.2
71.2	313	223	90	158.9
71.2	312	222	90	158.0
71.2	309	220	89	156.6
71.2	306	218	88	155.3
71.2	302	215	87	153.1
71.2	299	213	86	151.7
71.2	295	210	85	149.5
71.2	292	208	84	148.2
71.2	288	205	83	145.9
71.2	285	203	82	144.6
71.2	281	200	81	142.3
71.2	278	198	80	141.0
71.2	274	195	79	138.8
71.2	271	193	78	137.5
71.2	267	190	77	135.2
71.2	264	188	76	133.9
71.2	260	185	75	131.6
71.2	257	183	74	130.3
71.2	250	178	72	126.7
71.2	243	173	70	123.2
71.2	236	168	68	119.6
71.2	233	166	67	118.3
71.2	229	163	66	116.0
71.2	226	161	65	114.7
71.2	222	158	64	112.5
71.2	219	156	63	111.1
71.2	215	153	62	108.9
71.2	212	151	61	107.6
71.2	208	148	60	105.3
71.2	205	146	59	104.0
71.2	198	141	57	100.4
71.2	191	136	55	96.8
71.2	184	131	53	93.3
71.2	177	126	51	89.7
71.2	170	121	49	86.1
71.2	163	116	47	82.6
71.2	156	111	45	79.0
71.2	153	109	44	77.7
71.2	146	104	42	74.1
71.2	139	99	40	70.5
71.2	132	94	38	66.9
71.2	125	89	36	63.4
71.2	118	84	34	59.8
71.2	111	79	32	56.2
71.2	104	74	30	52.7
71.2	73	52	21	37.0
71.2	66	47	19	33.5
71.2	59	42	17	29.9
71.2	52	37	15	26.3
71.1	499	355	144	252.6
71.1	498	354	144	251.6
71.1	495	352	143	250.3
71.1	494	351	143	249.4
71.1	492	350	142	249.0
71.1	491	349	142	248.1
71.1	488	347	141	246.7
71.1	485	345	140	245.4
71.1	484	344	140	244.5
71.1	481	342	139	243.2
71.1	478	340	138	241.8
71.1	477	339	138	240.9
71.1	474	337	137	239.6
71.1	471	335	136	238.3
71.1	470	334	136	237.4
71.1	467	332	135	236.0
71.1	464	330	134	234.7
71.1	463	329	134	233.8
71.1	460	327	133	232.5
71.1	457	325	132	231.1
71.1	456	324	132	230.2
71.1	454	323	131	229.8
71.1	453	322	131	228.9
71.1	450	320	130	227.6
71.1	447	318	129	226.2
71.1	446	317	129	225.3
71.1	443	315	128	224.0
71.1	440	313	127	222.7
71.1	439	312	127	221.7
71.1	436	310	126	220.4
71.1	433	308	125	219.1
71.1	432	307	125	218.2
71.1	429	305	124	216.8
71.1	426	303	123	215.5
71.1	425	302	123	214.6
71.1	422	300	122	213.3
71.1	419	298	121	211.9
71.1	418	297	121	211.0
71.1	415	295	120	209.7
71.1	412	293	119	208.4
71.1	409	291	118	207.0
71.1	408	290	118	206.1
71.1	405	288	117	204.8
71.1	402	286	116	203.5
71.1	401	285	116	202.6
71.1	398	283	115	201.2
71.1	395	281	114	199.9
71.1	394	280	114	199.0
71.1	391	278	113	197.7
71.1	388	276	112	196.3
71.1	387	275	112	195.4
71.1	384	273	111	194.1
71.1	381	271	110	192.8
71.1	380	270	110	191.8
71.1	377	268	109	190.5
71.1	374	266	108	189.2
71.1	370	263	107	186.9
71.1	367	261	106	185.6
71.1	363	258	105	183.4
71.1	360	256	104	182.0
71.1	357	254	103	180.7
71.1	356	253	103	179.8
71.1	353	251	102	178.5
71.1	350	249	101	177.1
71.1	349	248	101	176.2
71.1	346	246	100	174.9
71.1	343	244	99	173.6
71.1	342	243	99	172.7
71.1	339	241	98	171.3
71.1	336	239	97	170.0
71.1	332	236	96	167.8
71.1	329	234	95	166.4
71.1	325	231	94	164.2
71.1	322	229	93	162.9
71.1	318	226	92	160.6
71.1	315	224	91	159.3
71.1	311	221	90	157.0
71.1	308	219	89	155.7
71.1	305	217	88	154.4
71.1	304	216	88	153.5
71.1	301	214	87	152.1
71.1	298	212	86	150.8
71.1	294	209	85	148.6
71.1	291	207	84	147.2
71.1	287	204	83	145.0
71.1	284	202	82	143.7
71.1	280	199	81	141.4
71.1	277	197	80	140.1
71.1	273	194	79	137.9
71.1	270	192	78	136.5
71.1	266	189	77	134.3
71.1	263	187	76	133.0
71.1	256	182	74	129.4
71.1	253	180	73	128.1
71.1	249	177	72	125.8
71.1	246	175	71	124.5
71.1	242	172	70	122.2
71.1	239	170	69	120.9
71.1	235	167	68	118.7
71.1	232	165	67	117.3
71.1	228	162	66	115.1
71.1	225	160	65	113.8
71.1	218	155	63	110.2
71.1	211	150	61	106.6
71.1	204	145	59	103.1
71.1	201	143	58	101.7
71.1	197	140	57	99.5
71.1	194	138	56	98.2
71.1	190	135	55	95.9
71.1	187	133	54	94.6
71.1	180	128	52	91.0
71.1	173	123	50	87.5
71.1	166	118	48	83.9
71.1	159	113	46	80.3

%E	M1	M2	DM	M*
71.1	152	108	44	76.7
71.1	149	106	43	75.4
71.1	142	101	41	71.8
71.1	135	96	39	68.3
71.1	128	91	37	64.7
71.1	121	86	35	61.1
71.1	114	81	33	57.6
71.1	97	69	28	49.1
71.1	90	64	26	45.5
71.1	83	59	24	41.9
71.1	76	54	22	38.4
71.1	45	32	13	22.8
71.1	38	27	11	19.2
71.0	500	355	145	252.0
71.0	497	353	144	250.7
71.0	496	352	144	249.8
71.0	493	350	143	248.5
71.0	490	348	142	247.2
71.0	489	347	142	246.2
71.0	487	346	141	245.8
71.0	486	345	141	244.9
71.0	483	343	140	243.6
71.0	482	342	140	242.7
71.0	480	341	139	242.3
71.0	479	340	139	241.3
71.0	476	338	138	240.0
71.0	473	336	137	238.7
71.0	472	335	137	237.8
71.0	469	333	136	236.4
71.0	466	331	135	235.1
71.0	465	330	135	234.2
71.0	462	328	134	232.9
71.0	459	326	133	231.5
71.0	458	325	133	230.6
71.0	455	323	132	229.3
71.0	452	321	131	228.0
71.0	451	320	131	227.1
71.0	449	319	130	226.6
71.0	448	318	130	225.7
71.0	445	316	129	224.4
71.0	442	314	128	223.1
71.0	441	313	128	222.2
71.0	438	311	127	220.8
71.0	435	309	126	219.5
71.0	434	308	126	218.6
71.0	431	306	125	217.3
71.0	428	304	124	215.9
71.0	427	303	124	215.0
71.0	424	301	123	213.7
71.0	421	299	122	212.4

%E	M1	M2	DM	M*
71.0	420	298	122	211.4
71.0	417	296	121	210.1
71.0	414	294	120	208.8
71.0	411	292	119	207.5
71.0	410	291	119	206.5
71.0	407	289	118	205.2
71.0	404	287	117	203.9
71.0	403	286	117	203.0
71.0	400	284	116	201.6
71.0	397	282	115	200.3
71.0	396	281	115	199.4
71.0	393	279	114	198.1
71.0	390	277	113	196.7
71.0	389	276	113	195.8
71.0	386	274	112	194.5
71.0	383	272	111	193.2
71.0	379	269	110	190.9
71.0	376	267	109	189.6
71.0	373	265	108	188.3
71.0	372	264	108	187.4
71.0	369	262	107	186.0
71.0	366	260	106	184.7
71.0	365	259	106	183.8
71.0	362	257	105	182.5
71.0	359	255	104	181.1
71.0	358	254	104	180.2
71.0	355	252	103	178.9
71.0	352	250	102	177.6
71.0	348	247	101	175.3
71.0	345	245	100	174.0
71.0	341	242	99	171.7
71.0	338	240	98	170.4
71.0	335	238	97	169.1
71.0	334	237	97	168.2
71.0	331	235	96	166.8
71.0	328	233	95	165.5
71.0	324	230	94	163.3
71.0	321	228	93	161.9
71.0	317	225	92	159.7
71.0	314	223	91	158.4
71.0	310	220	90	156.1
71.0	307	218	89	154.8
71.0	303	215	88	152.6
71.0	300	213	87	151.2
71.0	297	211	86	149.9
71.0	293	208	85	147.7
71.0	290	206	84	146.3
71.0	286	203	83	144.1
71.0	283	201	82	142.8
71.0	279	198	81	140.5

%E	M1	M2	DM	M*
71.0	276	196	80	139.2
71.0	272	193	79	136.9
71.0	269	191	78	135.6
71.0	262	186	76	132.0
71.0	259	184	75	130.7
71.0	255	181	74	128.5
71.0	252	179	73	127.1
71.0	248	176	72	124.9
71.0	245	174	71	123.6
71.0	241	171	70	121.3
71.0	238	169	69	120.0
71.0	231	164	67	116.4
71.0	224	159	65	112.9
71.0	221	157	64	111.5
71.0	217	154	63	109.3
71.0	214	152	62	108.0
71.0	210	149	61	105.7
71.0	207	147	60	104.4
71.0	200	142	58	100.8
71.0	193	137	56	97.2
71.0	186	132	54	93.7
71.0	183	130	53	92.3
71.0	179	127	52	90.1
71.0	176	125	51	88.8
71.0	169	120	49	85.2
71.0	162	115	47	81.6
71.0	155	110	45	78.1
71.0	145	103	42	73.2
71.0	138	98	40	69.6
71.0	131	93	38	66.0
71.0	124	88	36	62.5
71.0	107	76	31	54.0
71.0	100	71	29	50.4
71.0	93	66	27	46.8
71.0	69	49	20	34.8
71.0	62	44	18	31.2
71.0	31	22	9	15.6
70.9	499	354	145	251.1
70.9	498	353	145	250.2
70.9	495	351	144	248.9
70.9	494	350	144	248.0
70.9	492	349	143	247.6
70.9	491	348	143	246.6
70.9	488	346	142	245.3
70.9	485	344	141	244.0
70.9	484	343	141	243.1
70.9	481	341	140	241.7
70.9	478	339	139	240.4
70.9	477	338	139	239.5
70.9	475	337	138	239.1

%E	M1	M2	DM	M*
70.9	474	336	138	238.2
70.9	471	334	137	236.8
70.9	470	333	137	235.9
70.9	468	332	136	235.5
70.9	467	331	136	234.6
70.9	464	329	135	233.3
70.9	461	327	134	232.0
70.9	460	326	134	231.0
70.9	457	324	133	229.7
70.9	454	322	132	228.4
70.9	453	321	132	227.5
70.9	450	319	131	226.1
70.9	447	317	130	224.8
70.9	446	316	130	223.9
70.9	444	315	129	223.5
70.9	443	314	129	222.6
70.9	440	312	128	221.2
70.9	437	310	127	219.9
70.9	436	309	127	219.0
70.9	433	307	126	217.7
70.9	430	305	125	216.3
70.9	429	304	125	215.4
70.9	426	302	124	214.1
70.9	423	300	123	212.8
70.9	422	299	123	211.9
70.9	419	297	122	210.5
70.9	416	295	121	209.2
70.9	413	293	120	207.9
70.9	412	292	120	207.0
70.9	409	290	119	205.6
70.9	406	288	118	204.3
70.9	405	287	118	203.4
70.9	402	285	117	202.1
70.9	399	283	116	200.7
70.9	398	282	116	199.8
70.9	395	280	115	198.5
70.9	392	278	114	197.2
70.9	388	275	113	194.9
70.9	385	273	112	193.6
70.9	382	271	111	192.3
70.9	381	270	111	191.3
70.9	378	268	110	190.0
70.9	375	266	109	188.7
70.9	374	265	109	187.8
70.9	371	263	108	186.4
70.9	368	261	107	185.1
70.9	364	258	106	182.9
70.9	361	256	105	181.5
70.9	357	253	104	179.3
70.9	354	251	103	178.0

%E	M1	M2	DM	M*
70.9	351	249	102	176.6
70.9	350	248	102	175.7
70.9	347	246	101	174.4
70.9	344	244	100	173.1
70.9	340	241	99	170.8
70.9	337	239	98	169.5
70.9	333	236	97	167.3
70.9	330	234	96	165.9
70.9	327	232	95	164.6
70.9	326	231	95	163.7
70.9	323	229	94	162.4
70.9	320	227	93	161.0
70.9	316	224	92	158.8
70.9	313	222	91	157.5
70.9	309	219	90	155.2
70.9	306	217	89	153.9
70.9	302	214	88	151.6
70.9	299	212	87	150.3
70.9	296	210	86	149.0
70.9	292	207	85	146.7
70.9	289	205	84	145.4
70.9	285	202	83	143.2
70.9	282	200	82	141.8
70.9	278	197	81	139.6
70.9	275	195	80	138.3
70.9	268	190	78	134.7
70.9	265	188	77	133.4
70.9	261	185	76	131.1
70.9	258	183	75	129.8
70.9	254	180	74	127.6
70.9	251	178	73	126.2
70.9	247	175	72	124.0
70.9	244	173	71	122.7
70.9	237	168	69	119.1
70.9	234	166	68	117.8
70.9	230	163	67	115.5
70.9	227	161	66	114.2
70.9	223	158	65	111.9
70.9	220	156	64	110.6
70.9	213	151	62	107.0
70.9	206	146	60	103.5
70.9	203	144	59	102.1
70.9	199	141	58	99.9
70.9	196	139	57	98.6
70.9	189	134	55	95.0
70.9	182	129	53	91.4
70.9	175	124	51	87.9
70.9	172	122	50	86.5
70.9	165	117	48	83.0
70.9	158	112	46	79.4

%E	M1	M2	DM	M*
70.9	151	107	44	75.8
70.9	148	105	43	74.5
70.9	141	100	41	70.9
70.9	134	95	39	67.4
70.9	127	90	37	63.8
70.9	117	83	34	58.9
70.9	110	78	32	55.3
70.9	103	73	30	51.7
70.9	86	61	25	43.3
70.9	79	56	23	39.7
70.9	55	39	16	27.7
70.8	500	354	146	250.6
70.8	497	352	145	249.3
70.8	496	351	145	248.4
70.8	493	349	144	247.1
70.8	490	347	143	245.7
70.8	489	346	143	244.8
70.8	487	345	142	244.4
70.8	486	344	142	243.5
70.8	483	342	141	242.2
70.8	480	340	140	240.8
70.8	479	339	140	239.9
70.8	476	337	139	238.6
70.8	473	335	138	237.3
70.8	472	334	138	236.3
70.8	469	332	137	235.0
70.8	466	330	136	233.7
70.8	465	329	136	232.8
70.8	463	328	135	232.4
70.8	462	327	135	231.4
70.8	459	325	134	230.1
70.8	456	323	133	228.8
70.8	455	322	133	227.9
70.8	452	320	132	226.5
70.8	449	318	131	225.2
70.8	448	317	131	224.3
70.8	445	315	130	223.0
70.8	442	313	129	221.6
70.8	439	311	128	220.3
70.8	438	310	128	219.4
70.8	435	308	127	218.1
70.8	432	306	126	216.8
70.8	431	305	126	215.8
70.8	428	303	125	214.5
70.8	425	301	124	213.2
70.8	424	300	124	212.3
70.8	421	298	123	210.9
70.8	418	296	122	209.6
70.8	415	294	121	208.3
70.8	414	293	121	207.4
70.8	411	291	120	206.0
70.8	408	289	119	204.7
70.8	407	288	119	203.8
70.8	404	286	118	202.5
70.8	401	284	117	201.1
70.8	400	283	117	200.2
70.8	397	281	116	198.9
70.8	394	279	115	197.6
70.8	391	277	114	196.2
70.8	390	276	114	195.3
70.8	387	274	113	194.0
70.8	384	272	112	192.7
70.8	383	271	112	191.8
70.8	380	269	111	190.4
70.8	377	267	110	189.1
70.8	373	264	109	186.9
70.8	370	262	108	185.5
70.8	367	260	107	184.2
70.8	366	259	107	183.3
70.8	363	257	106	182.0
70.8	360	255	105	180.6
70.8	359	254	105	179.7
70.8	356	252	104	178.4
70.8	353	250	103	177.1
70.8	349	247	102	174.8
70.8	346	245	101	173.5
70.8	343	243	100	172.2
70.8	342	242	100	171.2
70.8	339	240	99	169.9
70.8	336	238	98	168.6
70.8	332	235	97	166.3
70.8	329	233	96	165.0
70.8	325	230	95	162.8
70.8	322	228	94	161.4
70.8	319	226	93	160.1
70.8	318	225	93	159.2
70.8	315	223	92	157.9
70.8	312	221	91	156.5
70.8	308	218	90	154.3
70.8	305	216	89	153.0
70.8	301	213	88	150.7
70.8	298	211	87	149.4
70.8	295	209	86	148.1
70.8	291	206	85	145.8
70.8	288	204	84	144.5
70.8	284	201	83	142.3
70.8	281	199	82	140.9
70.8	277	196	81	138.7
70.8	274	194	80	137.4
70.8	271	192	79	136.0
70.8	267	189	78	133.8
70.8	264	187	77	132.5
70.8	260	184	76	130.2
70.8	257	182	75	128.9
70.8	253	179	74	126.6
70.8	250	177	73	125.3
70.8	243	172	71	121.7
70.8	240	170	70	120.4
70.8	236	167	69	118.2
70.8	233	165	68	116.8
70.8	226	160	66	113.3
70.8	219	155	64	109.7
70.8	216	153	63	108.4
70.8	212	150	62	106.1
70.8	209	148	61	104.8
70.8	202	143	59	101.2
70.8	195	138	57	97.7
70.8	192	136	56	96.3
70.8	185	131	54	92.8
70.8	178	126	52	89.2
70.8	171	121	50	85.6
70.8	168	119	49	84.3
70.8	161	114	47	80.7
70.8	154	109	45	77.1
70.8	144	102	42	72.3
70.8	137	97	40	68.7
70.8	130	92	38	65.1
70.8	120	85	35	60.2
70.8	113	80	33	56.6
70.8	106	75	31	53.1
70.8	96	68	28	48.2
70.8	89	63	26	44.6
70.8	72	51	21	36.1
70.8	65	46	19	32.6
70.8	48	34	14	24.1
70.8	24	17	7	12.0
70.7	499	353	146	249.7
70.7	498	352	146	248.8
70.7	495	350	145	247.5
70.7	492	348	144	246.1
70.7	491	347	144	245.2
70.7	488	345	143	243.9
70.7	485	343	142	242.6
70.7	484	342	142	241.7
70.7	482	341	141	241.2
70.7	481	340	141	240.3
70.7	478	338	140	239.0
70.7	477	337	140	238.1
70.7	475	336	139	237.7
70.7	474	335	139	236.8
70.7	471	333	138	235.4
70.7	468	331	137	234.1
70.7	467	330	137	233.2
70.7	464	328	136	231.9
70.7	461	326	135	230.5
70.7	460	325	135	229.6
70.7	458	324	134	229.2
70.7	457	323	134	228.3
70.7	454	321	133	227.0
70.7	451	319	132	225.6
70.7	450	318	132	224.7
70.7	447	316	131	223.4
70.7	444	314	130	222.1
70.7	443	313	130	221.1
70.7	441	312	129	220.7
70.7	440	311	129	219.8
70.7	437	309	128	218.5
70.7	434	307	127	217.2
70.7	433	306	127	216.2
70.7	430	304	126	214.9
70.7	427	302	125	213.6
70.7	426	301	125	212.7
70.7	423	299	124	211.3
70.7	420	297	123	210.0
70.7	417	295	122	208.7
70.7	416	294	122	207.8
70.7	413	292	121	206.5
70.7	410	290	120	205.1
70.7	409	289	120	204.2
70.7	406	287	119	202.9
70.7	403	285	118	201.6
70.7	399	282	117	199.3
70.7	396	280	116	198.0
70.7	393	278	115	196.7
70.7	392	277	115	195.7
70.7	389	275	114	194.4
70.7	386	273	113	193.1
70.7	382	270	112	190.8
70.7	379	268	111	189.5
70.7	376	266	110	188.2
70.7	375	265	110	187.3
70.7	372	263	109	185.9
70.7	369	261	108	184.6
70.7	368	260	108	183.7
70.7	365	258	107	182.4
70.7	362	256	106	181.0
70.7	358	253	105	178.8
70.7	355	251	104	177.5
70.7	352	249	103	176.1
70.7	351	248	103	175.2
70.7	348	246	102	173.9
70.7	345	244	101	172.6
70.7	341	241	100	170.3
70.7	338	239	99	169.0
70.7	335	237	98	167.7
70.7	334	236	98	166.8
70.7	331	234	97	165.4
70.7	328	232	96	164.1
70.7	324	229	95	161.9
70.7	321	227	94	160.5
70.7	317	224	93	158.3
70.7	314	222	92	157.0
70.7	311	220	91	155.6
70.7	307	217	90	153.4
70.7	304	215	89	152.1
70.7	300	212	88	149.8
70.7	297	210	87	148.5
70.7	294	208	86	147.2
70.7	290	205	85	144.9
70.7	287	203	84	143.6
70.7	283	200	83	141.3
70.7	280	198	82	140.0
70.7	276	195	81	137.8
70.7	273	193	80	136.4
70.7	270	191	79	135.1
70.7	266	188	78	132.9
70.7	263	186	77	131.5
70.7	259	183	76	129.3
70.7	256	181	75	128.0
70.7	249	176	73	124.4
70.7	246	174	72	123.1
70.7	242	171	71	120.8
70.7	239	169	70	119.5
70.7	232	164	68	115.9
70.7	229	162	67	114.6
70.7	225	159	66	112.4
70.7	222	157	65	111.0
70.7	215	152	63	107.5
70.7	208	147	61	103.9
70.7	205	145	60	102.6
70.7	198	140	58	99.0
70.7	191	135	56	95.4
70.7	188	133	55	94.1
70.7	184	130	54	91.8
70.7	181	128	53	90.5
70.7	174	123	51	86.9
70.7	167	118	49	83.4
70.7	164	116	48	82.0
70.7	157	111	46	78.5
70.7	150	106	44	74.9

%E	M1	M2	DM	M*
70.7	147	104	43	73.6
70.7	140	99	41	70.0
70.7	133	94	39	66.4
70.7	123	87	36	61.5
70.7	116	82	34	58.0
70.7	99	70	29	49.5
70.7	92	65	27	45.9
70.7	82	58	24	41.0
70.7	75	53	22	37.5
70.7	58	41	17	29.0
70.7	41	29	12	20.5
70.6	500	353	147	249.2
70.6	497	351	146	247.9
70.6	496	350	146	247.0
70.6	494	349	145	246.6
70.6	493	348	145	245.6
70.6	490	346	144	244.3
70.6	489	345	144	243.4
70.6	487	344	143	243.0
70.6	486	343	143	242.1
70.6	483	341	142	240.7
70.6	480	339	141	239.4
70.6	479	338	141	238.5
70.6	476	336	140	237.2
70.6	473	334	139	235.8
70.6	472	333	139	234.9
70.6	470	332	138	234.5
70.6	469	331	138	233.6
70.6	466	329	137	232.3
70.6	463	327	136	230.9
70.6	462	326	136	230.0
70.6	459	324	135	228.7
70.6	456	322	134	227.4
70.6	453	320	133	226.0
70.6	452	319	133	225.1
70.6	449	317	132	223.8
70.6	446	315	131	222.5
70.6	445	314	131	221.6
70.6	442	312	130	220.2
70.6	439	310	129	218.9
70.6	436	308	128	217.6
70.6	435	307	128	216.7
70.6	432	305	127	215.3
70.6	429	303	126	214.0
70.6	428	302	126	213.1
70.6	425	300	125	211.8
70.6	422	298	124	210.4
70.6	419	296	123	209.1
70.6	418	295	123	208.2
70.6	415	293	122	206.9

%E	M1	M2	DM	M*
70.6	412	291	121	205.5
70.6	411	290	121	204.6
70.6	408	288	120	203.3
70.6	405	286	119	202.0
70.6	402	284	118	200.6
70.6	401	283	118	199.7
70.6	398	281	117	198.4
70.6	395	279	116	197.1
70.6	394	278	116	196.2
70.6	391	276	115	194.8
70.6	388	274	114	193.5
70.6	385	272	113	192.2
70.6	384	271	113	191.3
70.6	381	269	112	189.9
70.6	378	267	111	188.6
70.6	377	266	111	187.7
70.6	374	264	110	186.4
70.6	371	262	109	185.0
70.6	367	259	108	182.8
70.6	364	257	107	181.5
70.6	361	255	106	180.1
70.6	360	254	106	179.2
70.6	357	252	105	177.9
70.6	354	250	104	176.6
70.6	350	247	103	174.3
70.6	347	245	102	173.0
70.6	344	243	101	171.7
70.6	343	242	101	170.7
70.6	340	240	100	169.4
70.6	337	238	99	168.1
70.6	333	235	98	165.8
70.6	330	233	97	164.5
70.6	327	231	96	163.2
70.6	326	230	96	162.3
70.6	323	228	95	160.9
70.6	320	226	94	159.6
70.6	316	223	93	157.4
70.6	313	221	92	156.0
70.6	310	219	91	154.7
70.6	309	218	91	153.8
70.6	306	216	90	152.5
70.6	303	214	89	151.1
70.6	299	211	88	148.9
70.6	296	209	87	147.6
70.6	293	207	86	146.2
70.6	289	204	85	144.0
70.6	286	202	84	142.7
70.6	282	199	83	140.4
70.6	279	197	82	139.1
70.6	272	192	80	135.5

%E	M1	M2	DM	M*
70.6	269	190	79	134.2
70.6	265	187	78	132.0
70.6	262	185	77	130.6
70.6	255	180	75	127.1
70.6	252	178	74	125.7
70.6	248	175	73	123.5
70.6	245	173	72	122.2
70.6	238	168	70	118.6
70.6	235	166	69	117.3
70.6	231	163	68	115.0
70.6	228	161	67	113.7
70.6	221	156	65	110.1
70.6	218	154	64	108.8
70.6	214	151	63	106.5
70.6	211	149	62	105.2
70.6	204	144	60	101.6
70.6	201	142	59	100.3
70.6	197	139	58	98.1
70.6	194	137	57	96.7
70.6	187	132	55	93.2
70.6	180	127	53	89.6
70.6	177	125	52	88.3
70.6	170	120	50	84.7
70.6	163	115	48	81.1
70.6	160	113	47	79.8
70.6	153	108	45	76.2
70.6	143	101	42	71.3
70.6	136	96	40	67.8
70.6	126	89	37	62.9
70.6	119	84	35	59.3
70.6	109	77	32	54.4
70.6	102	72	30	50.8
70.6	85	60	25	42.4
70.6	68	48	20	33.9
70.6	51	36	15	25.4
70.6	34	24	10	16.9
70.6	17	12	5	8.5
70.5	499	352	147	248.3
70.5	498	351	147	247.4
70.5	495	349	146	246.1
70.5	492	347	145	244.7
70.5	491	346	145	243.8
70.5	488	344	144	242.5
70.5	485	342	143	241.2
70.5	484	341	143	240.3
70.5	482	340	142	239.8
70.5	481	339	142	238.9
70.5	478	337	141	237.6
70.5	475	335	140	236.3
70.5	474	334	140	235.4

%E	M1	M2	DM	M*
70.5	471	332	139	234.0
70.5	468	330	138	232.7
70.5	467	329	138	231.8
70.5	465	328	137	231.4
70.5	464	327	137	230.5
70.5	461	325	136	229.1
70.5	458	323	135	227.8
70.5	457	322	135	226.9
70.5	455	321	134	226.5
70.5	454	320	134	225.6
70.5	451	318	133	224.2
70.5	448	316	132	222.9
70.5	447	315	132	222.0
70.5	444	313	131	220.7
70.5	441	311	130	219.3
70.5	440	310	130	218.4
70.5	438	309	129	218.0
70.5	437	308	129	217.1
70.5	434	306	128	215.8
70.5	431	304	127	214.4
70.5	430	303	127	213.5
70.5	427	301	126	212.2
70.5	424	299	125	210.9
70.5	421	297	124	209.5
70.5	420	296	124	208.6
70.5	417	294	123	207.3
70.5	414	292	122	206.0
70.5	413	291	122	205.0
70.5	410	289	121	203.7
70.5	407	287	120	202.4
70.5	404	285	119	201.1
70.5	403	284	119	200.1
70.5	400	282	118	198.8
70.5	397	280	117	197.5
70.5	396	279	117	196.6
70.5	393	277	116	195.2
70.5	390	275	115	193.9
70.5	387	273	114	192.6
70.5	386	272	114	191.7
70.5	383	270	113	190.3
70.5	380	268	112	189.0
70.5	376	265	111	186.8
70.5	373	263	110	185.4
70.5	370	261	109	184.1
70.5	369	260	109	183.2
70.5	366	258	108	181.9
70.5	363	256	107	180.5
70.5	359	253	106	178.3
70.5	356	251	105	177.0
70.5	353	249	104	175.6

%E	M1	M2	DM	M*
70.5	352	248	104	174.7
70.5	349	246	103	173.4
70.5	346	244	102	172.1
70.5	342	241	101	169.8
70.5	339	239	100	168.5
70.5	336	237	99	167.2
70.5	332	234	98	164.9
70.5	329	232	97	163.6
70.5	325	229	96	161.4
70.5	322	227	95	160.0
70.5	319	225	94	158.7
70.5	315	222	93	156.5
70.5	312	220	92	155.1
70.5	308	217	91	152.9
70.5	305	215	90	151.6
70.5	302	213	89	150.2
70.5	298	210	88	148.0
70.5	295	208	87	146.7
70.5	292	206	86	145.3
70.5	288	203	85	143.1
70.5	285	201	84	141.8
70.5	281	198	83	139.5
70.5	278	196	82	138.2
70.5	275	194	81	136.9
70.5	271	191	80	134.6
70.5	268	189	79	133.3
70.5	264	186	78	131.0
70.5	261	184	77	129.7
70.5	258	182	76	128.4
70.5	254	179	75	126.1
70.5	251	177	74	124.8
70.5	244	172	72	121.2
70.5	241	170	71	119.9
70.5	237	167	70	117.7
70.5	234	165	69	116.3
70.5	227	160	67	112.8
70.5	224	158	66	111.4
70.5	220	155	65	109.2
70.5	217	153	64	107.9
70.5	210	148	62	104.3
70.5	207	146	61	103.0
70.5	200	141	59	99.4
70.5	193	136	57	95.8
70.5	190	134	56	94.5
70.5	183	129	54	90.9
70.5	176	124	52	87.4
70.5	173	122	51	86.0
70.5	166	117	49	82.5
70.5	156	110	46	77.6
70.5	149	105	44	74.0

%E	M1	M2	DM	M*
70.5	146	103	43	72.7
70.5	139	98	41	69.1
70.5	132	93	39	65.5
70.5	129	91	38	64.2
70.5	122	86	36	60.6
70.5	112	79	33	55.7
70.5	105	74	31	52.2
70.5	95	67	28	47.3
70.5	88	62	26	43.7
70.5	78	55	23	38.8
70.5	61	43	18	30.3
70.5	44	31	13	21.8
70.4	500	352	148	247.8
70.4	497	350	147	246.5
70.4	496	349	147	245.6
70.4	494	348	146	245.1
70.4	493	347	146	244.2
70.4	490	345	145	242.9
70.4	487	343	144	241.6
70.4	486	342	144	240.7
70.4	483	340	143	239.3
70.4	480	338	142	238.0
70.4	479	337	142	237.1
70.4	477	336	141	236.7
70.4	476	335	141	235.8
70.4	473	333	140	234.4
70.4	470	331	139	233.1
70.4	469	330	139	232.2
70.4	466	328	138	230.9
70.4	463	326	137	229.5
70.4	460	324	136	228.2
70.4	459	323	136	227.3
70.4	456	321	135	226.0
70.4	453	319	134	224.6
70.4	452	318	134	223.7
70.4	450	317	133	223.3
70.4	449	316	133	222.4
70.4	446	314	132	221.1
70.4	443	312	131	219.7
70.4	442	311	131	218.8
70.4	439	309	130	217.5
70.4	436	307	129	216.2
70.4	433	305	128	214.8
70.4	432	304	128	213.9
70.4	429	302	127	212.6
70.4	426	300	126	211.3
70.4	425	299	126	210.4
70.4	423	298	125	209.9
70.4	422	297	125	209.0
70.4	419	295	124	207.7
70.4	416	293	123	206.4
70.4	415	292	123	205.5
70.4	412	290	122	204.1
70.4	409	288	121	202.8
70.4	406	286	120	201.5
70.4	405	285	120	200.6
70.4	402	283	119	199.2
70.4	399	281	118	197.9
70.4	398	280	118	197.0
70.4	395	278	117	195.7
70.4	392	276	116	194.3
70.4	389	274	115	193.0
70.4	388	273	115	192.1
70.4	385	271	114	190.8
70.4	382	269	113	189.4
70.4	379	267	112	188.1
70.4	378	266	112	187.2
70.4	375	264	111	185.9
70.4	372	262	110	184.5
70.4	371	261	110	183.6
70.4	368	259	109	182.3
70.4	365	257	108	181.0
70.4	362	255	107	179.6
70.4	361	254	107	178.7
70.4	358	252	106	177.4
70.4	355	250	105	176.1
70.4	351	247	104	173.8
70.4	348	245	103	172.5
70.4	345	243	102	171.2
70.4	341	240	101	168.9
70.4	338	238	100	167.6
70.4	335	236	99	166.3
70.4	334	235	99	165.3
70.4	331	233	98	164.0
70.4	328	231	97	162.7
70.4	324	228	96	160.4
70.4	321	226	95	159.1
70.4	318	224	94	157.8
70.4	314	221	93	155.5
70.4	311	219	92	154.2
70.4	307	216	91	152.0
70.4	304	214	90	150.6
70.4	301	212	89	149.3
70.4	297	209	88	147.1
70.4	294	207	87	145.7
70.4	291	205	86	144.4
70.4	287	202	85	142.2
70.4	284	200	84	140.8
70.4	280	197	83	138.6
70.4	277	195	82	137.3
70.4	274	193	81	135.9
70.4	270	190	80	133.7
70.4	267	188	79	132.4
70.4	260	183	77	128.8
70.4	257	181	76	127.5
70.4	253	178	75	125.2
70.4	250	176	74	123.9
70.4	247	174	73	122.6
70.4	243	171	72	120.3
70.4	240	169	71	119.0
70.4	233	164	69	115.4
70.4	230	162	68	114.1
70.4	226	159	67	111.9
70.4	223	157	66	110.5
70.4	216	152	64	107.0
70.4	213	150	63	105.6
70.4	206	145	61	102.1
70.4	203	143	60	100.7
70.4	199	140	59	98.5
70.4	196	138	58	97.2
70.4	189	133	56	93.6
70.4	186	131	55	92.3
70.4	179	126	53	88.7
70.4	169	119	50	83.8
70.4	162	114	48	80.2
70.4	159	112	47	78.9
70.4	152	107	45	75.3
70.4	142	100	42	70.4
70.4	135	95	40	66.9
70.4	125	88	37	62.0
70.4	115	81	34	57.1
70.4	108	76	32	53.5
70.4	98	69	29	48.6
70.4	81	57	24	40.1
70.4	71	50	21	35.2
70.4	54	38	16	26.7
70.4	27	19	8	13.4
70.3	499	351	148	246.9
70.3	498	350	148	246.0
70.3	495	348	147	244.7
70.3	492	346	146	243.3
70.3	491	345	146	242.4
70.3	489	344	145	242.0
70.3	488	343	145	241.1
70.3	485	341	144	239.8
70.3	482	339	143	238.4
70.3	481	338	143	237.5
70.3	478	336	142	236.2
70.3	475	334	141	234.9
70.3	474	333	141	233.9
70.3	472	332	140	233.5
70.3	471	331	140	232.6
70.3	468	329	139	231.3
70.3	465	327	138	230.0
70.3	464	326	138	229.0
70.3	462	325	137	228.6
70.3	461	324	137	227.7
70.3	458	322	136	226.4
70.3	455	320	135	225.1
70.3	454	319	135	224.1
70.3	451	317	134	222.8
70.3	448	315	133	221.5
70.3	445	313	132	220.2
70.3	444	312	132	219.2
70.3	441	310	131	217.9
70.3	438	308	130	216.6
70.3	437	307	130	215.7
70.3	435	306	129	215.3
70.3	434	305	129	214.3
70.3	431	303	128	213.0
70.3	428	301	127	211.7
70.3	427	300	127	210.8
70.3	424	298	126	209.4
70.3	421	296	125	208.1
70.3	418	294	124	206.8
70.3	417	293	124	205.9
70.3	414	291	123	204.5
70.3	411	289	122	203.2
70.3	408	287	121	201.9
70.3	407	286	121	201.0
70.3	404	284	120	199.6
70.3	401	282	119	198.3
70.3	400	281	119	197.4
70.3	397	279	118	196.1
70.3	394	277	117	194.7
70.3	391	275	116	193.4
70.3	390	274	116	192.5
70.3	387	272	115	191.2
70.3	384	270	114	189.8
70.3	381	268	113	188.5
70.3	380	267	113	187.6
70.3	377	265	112	186.3
70.3	374	263	111	184.9
70.3	370	260	110	182.7
70.3	367	258	109	181.4
70.3	364	256	108	180.0
70.3	360	253	107	177.8
70.3	357	251	106	176.5
70.3	354	249	105	175.1
70.3	353	248	105	174.2
70.3	350	246	104	172.9
70.3	347	244	103	171.6
70.3	344	242	102	170.2
70.3	343	241	102	169.3
70.3	340	239	101	168.0
70.3	337	237	100	166.7
70.3	333	234	99	164.4
70.3	330	232	98	163.1
70.3	327	230	97	161.8
70.3	323	227	96	159.5
70.3	320	225	95	158.2
70.3	317	223	94	156.9
70.3	316	222	94	156.0
70.3	313	220	93	154.6
70.3	310	218	92	153.3
70.3	306	215	91	151.1
70.3	303	213	90	149.7
70.3	300	211	89	148.4
70.3	296	208	88	146.2
70.3	293	206	87	144.8
70.3	290	204	86	143.5
70.3	286	201	85	141.3
70.3	283	199	84	139.9
70.3	279	196	83	137.7
70.3	276	194	82	136.4
70.3	273	192	81	135.0
70.3	269	189	80	132.8
70.3	266	187	79	131.5
70.3	263	185	78	130.1
70.3	259	182	77	127.9
70.3	256	180	76	126.6
70.3	249	175	74	123.0
70.3	246	173	73	121.7
70.3	239	168	71	118.1
70.3	236	166	70	116.8
70.3	232	163	69	114.5
70.3	229	161	68	113.2
70.3	222	156	66	109.6
70.3	219	154	65	108.3
70.3	212	149	63	104.7
70.3	209	147	62	103.4
70.3	202	142	60	99.8
70.3	195	137	58	96.3
70.3	192	135	57	94.9
70.3	185	130	55	91.4
70.3	182	128	54	90.0
70.3	175	123	52	86.5
70.3	172	121	51	85.1
70.3	165	116	49	81.6
70.3	158	111	47	78.0

%E	M1	M2	DM	M*
70.3	155	109	46	76.7
70.3	148	104	44	73.1
70.3	145	102	43	71.8
70.3	138	97	41	68.2
70.3	128	90	38	63.3
70.3	118	83	35	58.4
70.3	111	78	33	54.8
70.3	101	71	30	49.9
70.3	91	64	27	45.0
70.3	74	52	22	36.5
70.3	64	45	19	31.6
70.3	37	26	11	18.3
70.2	500	351	149	246.4
70.2	497	349	148	245.1
70.2	496	348	148	244.2
70.2	494	347	147	243.7
70.2	493	346	147	242.8
70.2	490	344	146	241.5
70.2	487	342	145	240.2
70.2	486	341	145	239.3
70.2	484	340	144	238.8
70.2	483	339	144	237.9
70.2	480	337	143	236.6
70.2	477	335	142	235.3
70.2	476	334	142	234.4
70.2	473	332	141	233.0
70.2	470	330	140	231.7
70.2	467	328	139	230.4
70.2	466	327	139	229.5
70.2	463	325	138	228.1
70.2	460	323	137	226.8
70.2	459	322	137	225.9
70.2	457	321	136	225.5
70.2	456	320	136	224.6
70.2	453	318	135	223.2
70.2	450	316	134	221.9
70.2	449	315	134	221.0
70.2	447	314	133	220.6
70.2	446	313	133	219.7
70.2	443	311	132	218.3
70.2	440	309	131	217.0
70.2	439	308	131	216.1
70.2	436	306	130	214.8
70.2	433	304	129	213.4
70.2	430	302	128	212.1
70.2	429	301	128	211.2
70.2	426	299	127	209.9
70.2	423	297	126	208.5
70.2	420	295	125	207.2
70.2	419	294	125	206.3
70.2	416	292	124	205.0
70.2	413	290	123	203.6
70.2	410	288	122	202.3
70.2	409	287	122	201.4
70.2	406	285	121	200.1
70.2	403	283	120	198.7
70.2	399	280	119	196.5
70.2	396	278	118	195.2
70.2	393	276	117	193.8
70.2	392	275	117	192.9
70.2	389	273	116	191.6
70.2	386	271	115	190.3
70.2	383	269	114	188.9
70.2	382	268	114	188.0
70.2	379	266	113	186.7
70.2	376	264	112	185.4
70.2	373	262	111	184.0
70.2	372	261	111	183.1
70.2	369	259	110	181.8
70.2	366	257	109	180.5
70.2	363	255	108	179.1
70.2	362	254	108	178.2
70.2	359	252	107	176.9
70.2	356	250	106	175.6
70.2	352	247	105	173.3
70.2	349	245	104	172.0
70.2	346	243	103	170.7
70.2	342	240	102	168.4
70.2	339	238	101	167.1
70.2	336	236	100	165.8
70.2	332	233	99	163.5
70.2	329	231	98	162.2
70.2	326	229	97	160.9
70.2	325	228	97	160.0
70.2	322	226	96	158.6
70.2	319	224	95	157.3
70.2	315	221	94	155.1
70.2	312	219	93	153.7
70.2	309	217	92	152.4
70.2	305	214	91	150.2
70.2	302	212	90	148.8
70.2	299	210	89	147.5
70.2	295	207	88	145.3
70.2	292	205	87	143.9
70.2	289	203	86	142.6
70.2	285	200	85	140.4
70.2	282	198	84	139.0
70.2	275	193	82	135.5
70.2	272	191	81	134.1
70.2	265	186	79	130.6
70.2	262	184	78	129.2
70.2	258	181	77	127.0
70.2	255	179	76	125.7
70.2	252	177	75	124.3
70.2	248	174	74	122.1
70.2	245	172	73	120.8
70.2	242	170	72	119.4
70.2	238	167	71	117.2
70.2	235	165	70	115.9
70.2	228	160	68	112.3
70.2	225	158	67	111.0
70.2	218	153	65	107.4
70.2	215	151	64	106.1
70.2	208	146	62	102.5
70.2	205	144	61	101.2
70.2	198	139	59	97.6
70.2	191	134	57	94.0
70.2	188	132	56	92.7
70.2	181	127	54	89.1
70.2	178	125	53	87.8
70.2	171	120	51	84.2
70.2	168	118	50	82.9
70.2	161	113	48	79.3
70.2	151	106	45	74.4
70.2	141	99	42	69.5
70.2	131	92	39	64.6
70.2	124	87	37	61.0
70.2	121	85	36	59.7
70.2	114	80	34	56.1
70.2	104	73	31	51.2
70.2	94	66	28	46.3
70.2	84	59	25	41.4
70.2	57	40	17	28.1
70.2	47	33	14	23.2
70.1	499	350	149	245.5
70.1	498	349	149	244.6
70.1	495	347	148	243.3
70.1	492	345	147	241.9
70.1	491	344	147	241.0
70.1	489	343	146	240.6
70.1	488	342	146	239.7
70.1	485	340	145	238.4
70.1	482	338	144	237.0
70.1	481	337	144	236.1
70.1	479	336	143	235.7
70.1	478	335	143	234.8
70.1	475	333	142	233.5
70.1	472	331	141	232.1
70.1	471	330	141	231.2
70.1	469	329	140	230.8
70.1	468	328	140	229.9
70.1	465	326	139	228.6
70.1	462	324	138	227.2
70.1	461	323	138	226.3
70.1	458	321	137	225.0
70.1	455	319	136	223.7
70.1	452	317	135	222.3
70.1	451	316	135	221.4
70.1	448	314	134	220.1
70.1	445	312	133	218.8
70.1	442	310	132	217.4
70.1	441	309	132	216.5
70.1	438	307	131	215.2
70.1	435	305	130	213.9
70.1	432	303	129	212.5
70.1	431	302	129	211.6
70.1	428	300	128	210.3
70.1	425	298	127	209.0
70.1	422	296	126	207.6
70.1	421	295	126	206.7
70.1	418	293	125	205.4
70.1	415	291	124	204.1
70.1	412	289	123	202.7
70.1	411	288	123	201.8
70.1	408	286	122	200.5
70.1	405	284	121	199.2
70.1	404	283	121	198.2
70.1	402	282	120	197.8
70.1	401	281	120	196.9
70.1	398	279	119	195.6
70.1	395	277	118	194.3
70.1	394	276	118	193.3
70.1	391	274	117	192.0
70.1	388	272	116	190.7
70.1	385	270	115	189.4
70.1	384	269	115	188.4
70.1	381	267	114	187.1
70.1	378	265	113	185.8
70.1	375	263	112	184.5
70.1	374	262	112	183.5
70.1	371	260	111	182.2
70.1	368	258	110	180.9
70.1	365	256	109	179.6
70.1	364	255	109	178.6
70.1	361	253	108	177.3
70.1	358	251	107	176.0
70.1	355	249	106	174.7
70.1	354	248	106	173.7
70.1	351	246	105	172.4
70.1	348	244	104	171.1
70.1	345	242	103	169.8
70.1	344	241	103	168.8
70.1	341	239	102	167.5
70.1	338	237	101	166.2
70.1	335	235	100	164.9
70.1	334	234	100	163.9
70.1	331	232	99	162.6
70.1	328	230	98	161.3
70.1	324	227	97	159.0
70.1	321	225	96	157.7
70.1	318	223	95	156.4
70.1	314	220	94	154.1
70.1	311	218	93	152.8
70.1	308	216	92	151.5
70.1	304	213	91	149.2
70.1	301	211	90	147.9
70.1	298	209	89	146.6
70.1	294	206	88	144.3
70.1	291	204	87	143.0
70.1	288	202	86	141.7
70.1	284	199	85	139.4
70.1	281	197	84	138.1
70.1	278	195	83	136.8
70.1	274	192	82	134.5
70.1	271	190	81	133.2
70.1	268	188	80	131.9
70.1	264	185	79	129.6
70.1	261	183	78	128.3
70.1	254	178	76	124.7
70.1	251	176	75	123.4
70.1	244	171	73	119.8
70.1	241	169	72	118.5
70.1	234	164	70	114.9
70.1	231	162	69	113.6
70.1	224	157	67	110.0
70.1	221	155	66	108.7
70.1	214	150	64	105.1
70.1	211	148	63	103.8
70.1	204	143	61	100.2
70.1	201	141	60	98.9
70.1	197	138	59	96.7
70.1	194	136	58	95.3
70.1	187	131	56	91.8
70.1	184	129	55	90.4
70.1	177	124	53	86.9
70.1	174	122	52	85.5
70.1	167	117	50	82.0
70.1	164	115	49	80.6
70.1	157	110	47	77.1
70.1	154	108	46	75.7

%E	M1	M2	DM	M*
70.1	147	103	44	72.2
70.1	144	101	43	70.8
70.1	137	96	41	67.3
70.1	134	94	40	65.9
70.1	127	89	38	62.4
70.1	117	82	35	57.5
70.1	107	75	32	52.6
70.1	97	68	29	47.7
70.1	87	61	26	42.8
70.1	77	54	23	37.9
70.1	67	47	20	33.0
70.0	500	350	150	245.0
70.0	497	348	149	243.7
70.0	496	347	149	242.8
70.0	494	346	148	242.3
70.0	493	345	148	241.4
70.0	490	343	147	240.1
70.0	487	341	146	238.8
70.0	486	340	146	237.9
70.0	484	339	145	237.4
70.0	483	338	145	236.5
70.0	480	336	144	235.2
70.0	477	334	143	233.9
70.0	476	333	143	233.0
70.0	474	332	142	232.5
70.0	473	331	142	231.6
70.0	470	329	141	230.3
70.0	467	327	140	229.0
70.0	466	326	140	228.1
70.0	464	325	139	227.6
70.0	463	324	139	226.7
70.0	460	322	138	225.4
70.0	457	320	137	224.1
70.0	456	319	137	223.2
70.0	454	318	136	222.7
70.0	453	317	136	221.8
70.0	450	315	135	220.5
70.0	447	313	134	219.2
70.0	446	312	134	218.3
70.0	444	311	133	217.8
70.0	443	310	133	216.9
70.0	440	308	132	215.6
70.0	437	306	131	214.3
70.0	436	305	131	213.4
70.0	434	304	130	212.9
70.0	433	303	130	212.0
70.0	430	301	129	210.7
70.0	427	299	128	209.4
70.0	426	298	128	208.5
70.0	424	297	127	208.0

%E	M1	M2	DM	M*
70.0	423	296	127	207.1
70.0	420	294	126	205.8
70.0	417	292	125	204.5
70.0	416	291	125	203.6
70.0	414	290	124	203.1
70.0	413	289	124	202.2
70.0	410	287	123	200.9
70.0	407	285	122	199.6
70.0	406	284	122	198.7
70.0	403	282	121	197.3
70.0	400	280	120	196.0
70.0	397	278	119	194.7
70.0	393	275	118	192.4
70.0	390	273	117	191.1
70.0	387	271	116	189.8
70.0	383	268	115	187.5
70.0	380	266	114	186.2
70.0	377	264	113	184.9
70.0	373	261	112	182.6
70.0	370	259	111	181.3
70.0	367	257	110	180.0
70.0	363	254	109	177.7
70.0	360	252	108	176.4
70.0	357	250	107	175.1
70.0	353	247	106	172.8
70.0	350	245	105	171.5
70.0	347	243	104	170.2
70.0	343	240	103	167.9
70.0	340	238	102	166.6
70.0	337	236	101	165.3
70.0	333	233	100	163.0
70.0	330	231	99	161.7
70.0	327	229	98	160.4
70.0	323	226	97	158.1
70.0	320	224	96	156.8
70.0	317	222	95	155.5
70.0	313	219	94	153.2
70.0	310	217	93	151.9
70.0	307	215	92	150.6
70.0	303	212	91	148.3
70.0	300	210	90	147.0
70.0	297	208	89	145.7
70.0	293	205	88	143.4
70.0	290	203	87	142.1
70.0	287	201	86	140.8
70.0	283	198	85	138.5
70.0	280	196	84	137.2
70.0	277	194	83	135.9
70.0	273	191	82	133.6
70.0	270	189	81	132.3

%E	M1	M2	DM	M*
70.0	267	187	80	131.0
70.0	263	184	79	128.7
70.0	260	182	78	127.4
70.0	257	180	77	126.1
70.0	253	177	76	123.8
70.0	250	175	75	122.5
70.0	247	173	74	121.2
70.0	243	170	73	118.9
70.0	240	168	72	117.6
70.0	237	166	71	116.3
70.0	233	163	70	114.0
70.0	230	161	69	112.7
70.0	227	159	68	111.4
70.0	223	156	67	109.1
70.0	220	154	66	107.8
70.0	217	152	65	106.5
70.0	213	149	64	104.2
70.0	210	147	63	102.9
70.0	207	145	62	101.6
70.0	203	142	61	99.3
70.0	200	140	60	98.0
70.0	190	133	57	93.1
70.0	180	126	54	88.2
70.0	170	119	51	83.3
70.0	160	112	48	78.4
70.0	150	105	45	73.5
70.0	140	98	42	68.6
70.0	130	91	39	63.7
70.0	120	84	36	58.8
70.0	110	77	33	53.9
70.0	100	70	30	49.0
70.0	90	63	27	44.1
70.0	80	56	24	39.2
70.0	70	49	21	34.3
70.0	60	42	18	29.4
70.0	50	35	15	24.5
70.0	40	28	12	19.6
70.0	30	21	9	14.7
70.0	20	14	6	9.8
70.0	10	7	3	4.9
69.9	499	349	150	244.1
69.9	498	348	150	243.2
69.9	495	346	149	241.9
69.9	492	344	148	240.5
69.9	491	343	148	239.6
69.9	489	342	147	239.2
69.9	488	341	147	238.3
69.9	485	339	146	237.0
69.9	482	337	145	235.6
69.9	481	336	145	234.7

%E	M1	M2	DM	M*
69.9	479	335	144	234.3
69.9	478	334	144	233.4
69.9	475	332	143	232.1
69.9	472	330	142	230.7
69.9	471	329	142	229.8
69.9	469	328	141	229.4
69.9	468	327	141	228.5
69.9	465	325	140	227.2
69.9	462	323	139	225.8
69.9	459	321	138	224.5
69.9	458	320	138	223.6
69.9	455	318	137	222.3
69.9	452	316	136	220.9
69.9	449	314	135	219.6
69.9	448	313	135	218.7
69.9	445	311	134	217.4
69.9	442	309	133	216.0
69.9	439	307	132	214.7
69.9	438	306	132	213.8
69.9	435	304	131	212.5
69.9	432	302	130	211.1
69.9	429	300	129	209.8
69.9	428	299	129	208.9
69.9	425	297	128	207.6
69.9	422	295	127	206.2
69.9	419	293	126	204.9
69.9	418	292	126	204.0
69.9	415	290	125	202.7
69.9	412	288	124	201.3
69.9	409	286	123	200.0
69.9	408	285	123	199.1
69.9	405	283	122	197.8
69.9	402	281	121	196.4
69.9	399	279	120	195.1
69.9	396	277	119	193.8
69.9	395	276	119	192.9
69.9	392	274	118	191.5
69.9	389	272	117	190.2
69.9	386	270	116	188.9
69.9	385	269	116	188.0
69.9	382	267	115	186.6
69.9	379	265	114	185.3
69.9	376	263	113	184.0
69.9	375	262	113	183.1
69.9	372	260	112	181.7
69.9	369	258	111	180.4
69.9	366	256	110	179.1
69.9	365	255	110	178.2
69.9	362	253	109	176.8
69.9	359	251	108	175.5

%E	M1	M2	DM	M*
69.9	356	249	107	174.2
69.9	355	248	107	173.3
69.9	352	246	106	171.9
69.9	349	244	105	170.6
69.9	346	242	104	169.3
69.9	345	241	104	168.4
69.9	342	239	103	167.0
69.9	339	237	102	165.7
69.9	336	235	101	164.4
69.9	335	234	101	163.5
69.9	332	232	100	162.1
69.9	329	230	99	160.8
69.9	326	228	98	159.5
69.9	322	225	97	157.2
69.9	319	223	96	155.9
69.9	316	221	95	154.6
69.9	312	218	94	152.3
69.9	309	216	93	151.0
69.9	306	214	92	149.7
69.9	302	211	91	147.4
69.9	299	209	90	146.1
69.9	296	207	89	144.8
69.9	292	204	88	142.5
69.9	289	202	87	141.2
69.9	286	200	86	139.9
69.9	282	197	85	137.6
69.9	279	195	84	136.3
69.9	276	193	83	135.0
69.9	272	190	82	132.7
69.9	269	188	81	131.4
69.9	266	186	80	130.1
69.9	259	181	78	126.5
69.9	256	179	77	125.2
69.9	249	174	75	121.6
69.9	246	172	74	120.3
69.9	239	167	72	116.7
69.9	236	165	71	115.4
69.9	229	160	69	111.8
69.9	226	158	68	110.5
69.9	219	153	66	106.9
69.9	216	151	65	105.6
69.9	209	146	63	102.0
69.9	206	144	62	100.7
69.9	196	137	59	95.8
69.9	193	135	58	94.4
69.9	186	130	56	90.9
69.9	183	128	55	89.5
69.9	176	123	53	86.0
69.9	173	121	52	84.6
69.9	166	116	50	81.1

%E	M1	M2	DM	M*
69.9	163	114	49	79.7
69.9	156	109	47	76.2
69.9	153	107	46	74.8
69.9	146	102	44	71.3
69.9	143	100	43	69.9
69.9	136	95	41	66.4
69.9	133	93	40	65.0
69.9	123	86	37	60.1
69.9	113	79	34	55.2
69.9	103	72	31	50.3
69.9	93	65	28	45.4
69.9	83	58	25	40.5
69.9	73	51	22	35.6
69.8	500	349	151	243.6
69.8	497	347	150	242.3
69.8	496	346	150	241.4
69.8	494	345	149	240.9
69.8	493	344	149	240.0
69.8	490	342	148	238.7
69.8	487	340	147	237.4
69.8	486	339	147	236.5
69.8	484	338	146	236.0
69.8	483	337	146	235.1
69.8	480	335	145	233.8
69.8	477	333	144	232.5
69.8	474	331	143	231.1
69.8	473	330	143	230.2
69.8	470	328	142	228.9
69.8	467	326	141	227.6
69.8	464	324	140	226.2
69.8	463	323	140	225.3
69.8	461	322	139	224.9
69.8	460	321	139	224.0
69.8	457	319	138	222.7
69.8	454	317	137	221.3
69.8	453	316	137	220.4
69.8	451	315	136	220.0
69.8	450	314	136	219.1
69.8	447	312	135	217.8
69.8	444	310	134	216.4
69.8	443	309	134	215.5
69.8	441	308	133	215.1
69.8	440	307	133	214.2
69.8	437	305	132	212.9
69.8	434	303	131	211.5
69.8	431	301	130	210.2
69.8	430	300	130	209.3
69.8	427	298	129	208.0
69.8	424	296	128	206.6
69.8	421	294	127	205.3

%E	M1	M2	DM	M*
69.8	420	293	127	204.4
69.8	417	291	126	203.1
69.8	414	289	125	201.7
69.8	411	287	124	200.4
69.8	410	286	124	199.5
69.8	407	284	123	198.2
69.8	404	282	122	196.8
69.8	401	280	121	195.5
69.8	400	279	121	194.6
69.8	398	278	120	194.2
69.8	397	277	120	193.3
69.8	394	275	119	191.9
69.8	391	273	118	190.6
69.8	388	271	117	189.3
69.8	387	270	117	188.4
69.8	384	268	116	187.0
69.8	381	266	115	185.7
69.8	378	264	114	184.4
69.8	377	263	114	183.5
69.8	374	261	113	182.1
69.8	371	259	112	180.8
69.8	368	257	111	179.5
69.8	367	256	111	178.6
69.8	364	254	110	177.2
69.8	361	252	109	175.9
69.8	358	250	108	174.6
69.8	354	247	107	172.3
69.8	351	245	106	171.0
69.8	348	243	105	169.7
69.8	344	240	104	167.4
69.8	341	238	103	166.1
69.8	338	236	102	164.8
69.8	334	233	101	162.5
69.8	331	231	100	161.2
69.8	328	229	99	159.9
69.8	325	227	98	158.6
69.8	324	226	98	157.6
69.8	321	224	97	156.3
69.8	318	222	96	155.0
69.8	315	220	95	153.7
69.8	311	217	94	151.4
69.8	308	215	93	150.1
69.8	305	213	92	148.8
69.8	301	210	91	146.5
69.8	298	208	90	145.2
69.8	295	206	89	143.9
69.8	291	203	88	141.6
69.8	288	201	87	140.3
69.8	285	199	86	139.0
69.8	281	196	85	136.7

%E	M1	M2	DM	M*
69.8	278	194	84	135.4
69.8	275	192	83	134.1
69.8	268	187	81	130.5
69.8	265	185	80	129.2
69.8	262	183	79	127.8
69.8	258	180	78	125.6
69.8	255	178	77	124.3
69.8	252	176	76	122.9
69.8	248	173	75	120.7
69.8	245	171	74	119.4
69.8	242	169	73	118.0
69.8	235	164	71	114.5
69.8	232	162	70	113.1
69.8	225	157	68	109.6
69.8	222	155	67	108.2
69.8	215	150	65	104.7
69.8	212	148	64	103.3
69.8	205	143	62	99.8
69.8	202	141	61	98.4
69.8	199	139	60	97.1
69.8	192	134	58	93.5
69.8	189	132	57	92.2
69.8	182	127	55	88.6
69.8	179	125	54	87.3
69.8	172	120	52	83.7
69.8	169	118	51	82.4
69.8	162	113	49	78.8
69.8	159	111	48	77.5
69.8	149	104	45	72.6
69.8	139	97	42	67.7
69.8	129	90	39	62.8
69.8	126	88	38	61.5
69.8	116	81	35	56.6
69.8	106	74	32	51.7
69.8	96	67	29	46.8
69.8	86	60	26	41.9
69.8	63	44	19	30.7
69.8	53	37	16	25.8
69.8	43	30	13	20.9
69.7	499	348	151	242.7
69.7	498	347	151	241.8
69.7	495	345	150	240.5
69.7	492	343	149	239.1
69.7	491	342	149	238.2
69.7	489	341	148	237.8
69.7	488	340	148	236.9
69.7	485	338	147	235.6
69.7	482	336	146	234.2
69.7	479	334	145	232.9
69.7	478	333	145	232.0

%E	M1	M2	DM	M*
69.7	476	332	144	231.6
69.7	475	331	144	230.7
69.7	472	329	143	229.3
69.7	469	327	142	228.0
69.7	468	326	142	227.1
69.7	466	325	141	226.7
69.7	465	324	141	225.8
69.7	462	322	140	224.4
69.7	459	320	139	223.1
69.7	458	319	139	222.2
69.7	456	318	138	221.8
69.7	455	317	138	220.9
69.7	452	315	137	219.5
69.7	449	313	136	218.2
69.7	446	311	135	216.9
69.7	445	310	135	216.0
69.7	442	308	134	214.6
69.7	439	306	133	213.3
69.7	436	304	132	212.0
69.7	435	303	132	211.1
69.7	433	302	131	210.6
69.7	432	301	131	209.7
69.7	429	299	130	208.4
69.7	426	297	129	207.1
69.7	423	295	128	205.7
69.7	422	294	128	204.8
69.7	419	292	127	203.5
69.7	416	290	126	202.2
69.7	413	288	125	200.8
69.7	412	287	125	199.9
69.7	409	285	124	198.6
69.7	406	283	123	197.3
69.7	403	281	122	195.9
69.7	402	280	122	195.0
69.7	399	278	121	193.7
69.7	396	276	120	192.4
69.7	393	274	119	191.0
69.7	390	272	118	189.7
69.7	389	271	118	188.8
69.7	386	269	117	187.5
69.7	383	267	116	186.1
69.7	380	265	115	184.8
69.7	379	264	115	183.9
69.7	376	262	114	182.6
69.7	373	260	113	181.2
69.7	370	258	112	179.9
69.7	366	255	111	177.7
69.7	363	253	110	176.3
69.7	360	251	109	175.0
69.7	357	249	108	173.7

%E	M1	M2	DM	M*
69.7	356	248	108	172.8
69.7	353	246	107	171.4
69.7	350	244	106	170.1
69.7	347	242	105	168.8
69.7	346	241	105	167.9
69.7	343	239	104	166.5
69.7	340	237	103	165.2
69.7	337	235	102	163.9
69.7	333	232	101	161.6
69.7	330	230	100	160.3
69.7	327	228	99	159.0
69.7	323	225	98	156.7
69.7	320	223	97	155.4
69.7	317	221	96	154.1
69.7	314	219	95	152.7
69.7	310	216	94	150.5
69.7	307	214	93	149.2
69.7	304	212	92	147.8
69.7	300	209	91	145.6
69.7	297	207	90	144.3
69.7	294	205	89	142.9
69.7	290	202	88	140.7
69.7	287	200	87	139.4
69.7	284	198	86	138.0
69.7	277	193	84	134.5
69.7	274	191	83	133.1
69.7	271	189	82	131.8
69.7	267	186	81	129.6
69.7	264	184	80	128.2
69.7	261	182	79	126.9
69.7	257	179	78	124.7
69.7	254	177	77	123.3
69.7	251	175	76	122.0
69.7	244	170	74	118.4
69.7	241	168	73	117.1
69.7	238	166	72	115.8
69.7	234	163	71	113.5
69.7	231	161	70	112.2
69.7	228	159	69	110.9
69.7	221	154	67	107.3
69.7	218	152	66	106.0
69.7	211	147	64	102.4
69.7	208	145	63	101.1
69.7	201	140	61	97.5
69.7	198	138	60	96.2
69.7	195	136	59	94.9
69.7	188	131	57	91.3
69.7	185	129	56	90.0
69.7	178	124	54	86.4
69.7	175	122	53	85.1

%E	M1	M2	DM	M*
69.7	165	115	50	80.2
69.7	155	108	47	75.3
69.7	152	106	46	73.9
69.7	145	101	44	70.4
69.7	142	99	43	69.0
69.7	132	92	40	64.1
69.7	122	85	37	59.2
69.7	119	83	36	57.9
69.7	109	76	33	53.0
69.7	99	69	30	48.1
69.7	89	62	27	43.2
69.7	76	53	23	37.0
69.7	66	46	20	32.1
69.7	33	23	10	16.0
69.6	500	348	152	242.2
69.6	497	346	151	240.9
69.6	496	345	151	240.0
69.6	494	344	150	239.5
69.6	493	343	150	238.6
69.6	490	341	149	237.3
69.6	487	339	148	236.0
69.6	484	337	147	234.6
69.6	483	336	147	233.7
69.6	481	335	146	233.3
69.6	480	334	146	232.4
69.6	477	332	145	231.1
69.6	474	330	144	229.7
69.6	473	329	144	228.8
69.6	471	328	143	228.4
69.6	470	327	143	227.5
69.6	467	325	142	226.2
69.6	464	323	141	224.8
69.6	461	321	140	223.5
69.6	460	320	140	222.6
69.6	457	318	139	221.3
69.6	454	316	138	219.9
69.6	451	314	137	218.6
69.6	450	313	137	217.7
69.6	448	312	136	217.3
69.6	447	311	136	216.4
69.6	444	309	135	215.0
69.6	441	307	134	213.7
69.6	438	305	133	212.4
69.6	437	304	133	211.5
69.6	434	302	132	210.1
69.6	431	300	131	208.8
69.6	428	298	130	207.5
69.6	427	297	130	206.6
69.6	425	296	129	206.2
69.6	424	295	129	205.2

%E	M1	M2	DM	M*
69.6	421	293	128	203.9
69.6	418	291	127	202.6
69.6	415	289	126	201.3
69.6	414	288	126	200.3
69.6	411	286	125	199.0
69.6	408	284	124	197.7
69.6	405	282	123	196.4
69.6	404	281	123	195.4
69.6	401	279	122	194.1
69.6	398	277	121	192.8
69.6	395	275	120	191.5
69.6	392	273	119	190.1
69.6	391	272	119	189.2
69.6	388	270	118	187.9
69.6	385	268	117	186.6
69.6	382	266	116	185.2
69.6	381	265	116	184.3
69.6	378	263	115	183.0
69.6	375	261	114	181.7
69.6	372	259	113	180.3
69.6	369	257	112	179.0
69.6	368	256	112	178.1
69.6	365	254	111	176.8
69.6	362	252	110	175.4
69.6	359	250	109	174.1
69.6	358	249	109	173.2
69.6	355	247	108	171.9
69.6	352	245	107	170.5
69.6	349	243	106	169.2
69.6	345	240	105	167.0
69.6	342	238	104	165.6
69.6	339	236	103	164.3
69.6	336	234	102	163.0
69.6	335	233	102	162.1
69.6	332	231	101	160.7
69.6	329	229	100	159.4
69.6	326	227	99	158.1
69.6	322	224	98	155.8
69.6	319	222	97	154.5
69.6	316	220	96	153.2
69.6	313	218	95	151.8
69.6	312	217	95	150.9
69.6	309	215	94	149.6
69.6	306	213	93	148.3
69.6	303	211	92	146.9
69.6	299	208	91	144.7
69.6	296	206	90	143.4
69.6	293	204	89	142.0
69.6	289	201	88	139.8
69.6	286	199	87	138.5

%E	M1	M2	DM	M*
69.6	283	197	86	137.1
69.6	280	195	85	135.8
69.6	276	192	84	133.6
69.6	273	190	83	132.2
69.6	270	188	82	130.9
69.6	263	183	80	127.3
69.6	260	181	79	126.0
69.6	253	176	77	122.4
69.6	250	174	76	121.1
69.6	247	172	75	119.8
69.6	240	167	73	116.2
69.6	237	165	72	114.9
69.6	230	160	70	111.3
69.6	227	158	69	110.0
69.6	224	156	68	108.6
69.6	217	151	66	105.1
69.6	214	149	65	103.7
69.6	207	144	63	100.2
69.6	204	142	62	98.8
69.6	194	135	59	93.9
69.6	191	133	58	92.6
69.6	184	128	56	89.0
69.6	181	126	55	87.7
69.6	171	119	52	82.8
69.6	168	117	51	81.5
69.6	161	112	49	77.9
69.6	158	110	48	76.6
69.6	148	103	45	71.7
69.6	138	96	42	66.8
69.6	135	94	41	65.5
69.6	125	87	38	60.6
69.6	115	80	35	55.7
69.6	112	78	34	54.3
69.6	102	71	31	49.4
69.6	92	64	28	44.5
69.6	79	55	24	38.3
69.6	69	48	21	33.4
69.6	56	39	17	27.2
69.6	46	32	14	22.3
69.6	23	16	7	11.1
69.5	499	347	152	241.3
69.5	498	346	152	240.4
69.5	495	344	151	239.1
69.5	492	342	150	237.7
69.5	491	341	150	236.8
69.5	489	340	149	236.4
69.5	488	339	149	235.5
69.5	486	338	148	235.1
69.5	485	337	148	234.2
69.5	482	335	147	232.8

%E	M1	M2	DM	M*
69.5	479	333	146	231.5
69.5	478	332	146	230.6
69.5	476	331	145	230.2
69.5	475	330	145	229.3
69.5	472	328	144	227.9
69.5	469	326	143	226.6
69.5	466	324	142	225.3
69.5	465	323	142	224.4
69.5	463	322	141	223.9
69.5	462	321	141	223.0
69.5	459	319	140	221.7
69.5	456	317	139	220.4
69.5	455	316	139	219.5
69.5	453	315	138	219.0
69.5	452	314	138	218.1
69.5	449	312	137	216.8
69.5	446	310	136	215.5
69.5	443	308	135	214.1
69.5	442	307	135	213.2
69.5	440	306	134	212.8
69.5	439	305	134	211.9
69.5	436	303	133	210.6
69.5	433	301	132	209.2
69.5	430	299	131	207.9
69.5	429	298	131	207.0
69.5	426	296	130	205.7
69.5	423	294	129	204.3
69.5	420	292	128	203.0
69.5	419	291	128	202.1
69.5	417	290	127	201.7
69.5	416	289	127	200.8
69.5	413	287	126	199.4
69.5	410	285	125	198.1
69.5	407	283	124	196.8
69.5	406	282	124	195.9
69.5	403	280	123	194.5
69.5	400	278	122	193.2
69.5	397	276	121	191.9
69.5	394	274	120	190.5
69.5	393	273	120	189.6
69.5	390	271	119	188.3
69.5	387	269	118	187.0
69.5	384	267	117	185.6
69.5	383	266	117	184.7
69.5	380	264	116	183.4
69.5	377	262	115	182.1
69.5	374	260	114	180.7
69.5	371	258	113	179.4
69.5	370	257	113	178.5
69.5	367	255	112	177.2

%E	M1	M2	DM	M*
69.5	364	253	111	175.8
69.5	361	251	110	174.5
69.5	357	248	109	172.3
69.5	354	246	108	170.9
69.5	351	244	107	169.6
69.5	348	242	106	168.3
69.5	347	241	106	167.4
69.5	344	239	105	166.0
69.5	341	237	104	164.7
69.5	338	235	103	163.4
69.5	334	232	102	161.1
69.5	331	230	101	159.8
69.5	328	228	100	158.5
69.5	325	226	99	157.2
69.5	321	223	98	154.9
69.5	318	221	97	153.6
69.5	315	219	96	152.3
69.5	311	216	95	150.0
69.5	308	214	94	148.7
69.5	305	212	93	147.4
69.5	302	210	92	146.0
69.5	298	207	91	143.8
69.5	295	205	90	142.5
69.5	292	203	89	141.1
69.5	285	198	87	137.6
69.5	282	196	86	136.2
69.5	279	194	85	134.9
69.5	275	191	84	132.7
69.5	272	189	83	131.3
69.5	269	187	82	130.0
69.5	266	185	81	128.7
69.5	262	182	80	126.4
69.5	259	180	79	125.1
69.5	256	178	78	123.8
69.5	249	173	76	120.2
69.5	246	171	75	118.9
69.5	243	169	74	117.5
69.5	239	166	73	115.3
69.5	236	164	72	114.0
69.5	233	162	71	112.6
69.5	226	157	69	109.1
69.5	223	155	68	107.7
69.5	220	153	67	106.4
69.5	213	148	65	102.8
69.5	210	146	64	101.5
69.5	203	141	62	97.9
69.5	200	139	61	96.6
69.5	197	137	60	95.3
69.5	190	132	58	91.7
69.5	187	130	57	90.4

%E	M1	M2	DM	M*
69.5	177	123	54	85.5
69.5	174	121	53	84.1
69.5	167	116	51	80.6
69.5	164	114	50	79.2
69.5	154	107	47	74.3
69.5	151	105	46	73.0
69.5	141	98	43	68.1
69.5	131	91	40	63.2
69.5	128	89	39	61.9
69.5	118	82	36	57.0
69.5	105	73	32	50.8
69.5	95	66	29	45.9
69.5	82	57	25	39.6
69.5	59	41	18	28.5
69.4	500	347	153	240.8
69.4	497	345	152	239.5
69.4	496	344	152	238.6
69.4	494	343	151	238.2
69.4	493	342	151	237.2
69.4	490	340	150	235.9
69.4	487	338	149	234.6
69.4	484	336	148	233.3
69.4	483	335	148	232.3
69.4	481	334	147	231.9
69.4	480	333	147	231.0
69.4	477	331	146	229.7
69.4	474	329	145	228.4
69.4	471	327	144	227.0
69.4	470	326	144	226.1
69.4	468	325	143	225.7
69.4	467	324	143	224.8
69.4	464	322	142	223.5
69.4	461	320	141	222.1
69.4	458	318	140	220.8
69.4	457	317	140	219.9
69.4	454	315	139	218.6
69.4	451	313	138	217.2
69.4	448	311	137	215.9
69.4	447	310	137	215.0
69.4	445	309	136	214.6
69.4	444	308	136	213.7
69.4	441	306	135	212.3
69.4	438	304	134	211.0
69.4	435	302	133	209.7
69.4	434	301	133	208.8
69.4	432	300	132	208.3
69.4	431	299	132	207.4
69.4	428	297	131	206.1
69.4	425	295	130	204.8
69.4	422	293	129	203.4

%E	M1	M2	DM	M*
69.4	421	292	129	202.5
69.4	418	290	128	201.2
69.4	415	288	127	199.9
69.4	412	286	126	198.5
69.4	409	284	125	197.2
69.4	408	283	125	196.3
69.4	405	281	124	195.0
69.4	402	279	123	193.6
69.4	399	277	122	192.3
69.4	396	275	121	191.0
69.4	395	274	121	190.1
69.4	392	272	120	188.7
69.4	389	270	119	187.4
69.4	386	268	118	186.1
69.4	385	267	118	185.2
69.4	382	265	117	183.8
69.4	379	263	116	182.5
69.4	376	261	115	181.2
69.4	373	259	114	179.8
69.4	372	258	114	178.9
69.4	369	256	113	177.6
69.4	366	254	112	176.3
69.4	363	252	111	174.9
69.4	360	250	110	173.6
69.4	359	249	110	172.7
69.4	356	247	109	171.4
69.4	353	245	108	170.0
69.4	350	243	107	168.7
69.4	346	240	106	166.5
69.4	343	238	105	165.1
69.4	340	236	104	163.8
69.4	337	234	103	162.5
69.4	333	231	102	160.2
69.4	330	229	101	158.9
69.4	327	227	100	157.6
69.4	324	225	99	156.3
69.4	323	224	99	155.3
69.4	320	222	98	154.0
69.4	317	220	97	152.7
69.4	314	218	96	151.4
69.4	310	215	95	149.1
69.4	307	213	94	147.8
69.4	304	211	93	146.5
69.4	301	209	92	145.1
69.4	297	206	91	142.9
69.4	294	204	90	141.6
69.4	291	202	89	140.2
69.4	288	200	88	138.9
69.4	284	197	87	136.7
69.4	281	195	86	135.3

%E	M1	M2	DM	M*
69.4	278	193	85	134.0
69.4	271	188	83	130.4
69.4	268	186	82	129.1
69.4	265	184	81	127.8
69.4	258	179	79	124.2
69.4	255	177	78	122.9
69.4	252	175	77	121.5
69.4	248	172	76	119.3
69.4	245	170	75	118.0
69.4	242	168	74	116.6
69.4	235	163	72	113.1
69.4	232	161	71	111.7
69.4	229	159	70	110.4
69.4	222	154	68	106.8
69.4	219	152	67	105.5
69.4	216	150	66	104.2
69.4	209	145	64	100.6
69.4	206	143	63	99.3
69.4	196	136	60	94.4
69.4	193	134	59	93.0
69.4	186	129	57	89.5
69.4	183	127	56	88.1
69.4	180	125	55	86.8
69.4	173	120	53	83.2
69.4	170	118	52	81.9
69.4	160	111	49	77.0
69.4	157	109	48	75.7
69.4	147	102	45	70.8
69.4	144	100	44	69.4
69.4	134	93	41	64.5
69.4	124	86	38	59.6
69.4	121	84	37	58.3
69.4	111	77	34	53.4
69.4	108	75	33	52.1
69.4	98	68	30	47.2
69.4	85	59	26	41.0
69.4	72	50	22	34.7
69.4	62	43	19	29.8
69.4	49	34	15	23.6
69.4	36	25	11	17.4
69.3	499	346	153	239.9
69.3	498	345	153	239.0
69.3	495	343	152	237.7
69.3	492	341	151	236.3
69.3	489	339	150	235.0
69.3	488	338	150	234.1
69.3	486	337	149	233.7
69.3	485	336	149	232.8
69.3	482	334	148	231.4
69.3	479	332	147	230.1

%E	M1	M2	DM	M*
69.3	476	330	146	228.8
69.3	475	329	146	227.9
69.3	473	328	145	227.5
69.3	472	327	145	226.5
69.3	469	325	144	225.2
69.3	466	323	143	223.9
69.3	463	321	142	222.6
69.3	462	320	142	221.6
69.3	460	319	141	221.2
69.3	459	318	141	220.3
69.3	456	316	140	219.0
69.3	453	314	139	217.7
69.3	450	312	138	216.3
69.3	449	311	138	215.4
69.3	446	309	137	214.1
69.3	443	307	136	212.8
69.3	440	305	135	211.4
69.3	437	303	134	210.1
69.3	436	302	134	209.2
69.3	433	300	133	207.9
69.3	430	298	132	206.5
69.3	427	296	131	205.2
69.3	424	294	130	203.9
69.3	423	293	130	203.0
69.3	420	291	129	201.6
69.3	417	289	128	200.3
69.3	414	287	127	199.0
69.3	411	285	126	197.6
69.3	410	284	126	196.7
69.3	407	282	125	195.4
69.3	404	280	124	194.1
69.3	401	278	123	192.7
69.3	400	277	123	191.8
69.3	398	276	122	191.4
69.3	397	275	122	190.5
69.3	394	273	121	189.2
69.3	391	271	120	187.8
69.3	388	269	119	186.5
69.3	387	268	119	185.6
69.3	384	266	118	184.3
69.3	381	264	117	182.9
69.3	378	262	116	181.6
69.3	375	260	115	180.3
69.3	374	259	115	179.4
69.3	371	257	114	178.0
69.3	368	255	113	176.7
69.3	365	253	112	175.4
69.3	362	251	111	174.0
69.3	361	250	111	173.1
69.3	358	248	110	171.8

%E	M1	M2	DM	M*
69.3	355	246	109	170.5
69.3	352	244	108	169.1
69.3	349	242	107	167.8
69.3	348	241	107	166.9
69.3	345	239	106	165.6
69.3	342	237	105	164.2
69.3	339	235	104	162.9
69.3	336	233	103	161.6
69.3	335	232	103	160.7
69.3	332	230	102	159.3
69.3	329	228	101	158.0
69.3	326	226	100	156.7
69.3	322	223	99	154.4
69.3	319	221	98	153.1
69.3	316	219	97	151.8
69.3	313	217	96	150.4
69.3	309	214	95	148.2
69.3	306	212	94	146.9
69.3	303	210	93	145.5
69.3	300	208	92	144.2
69.3	296	205	91	142.0
69.3	293	203	90	140.6
69.3	290	201	89	139.3
69.3	287	199	88	138.0
69.3	283	196	87	135.7
69.3	280	194	86	134.4
69.3	277	192	85	133.1
69.3	274	190	84	131.8
69.3	270	187	83	129.5
69.3	267	185	82	128.2
69.3	264	183	81	126.9
69.3	261	181	80	125.5
69.3	257	178	79	123.3
69.3	254	176	78	122.0
69.3	251	174	77	120.6
69.3	244	169	75	117.1
69.3	241	167	74	115.7
69.3	238	165	73	114.4
69.3	231	160	71	110.8
69.3	228	158	70	109.5
69.3	225	156	69	108.2
69.3	218	151	67	104.6
69.3	215	149	66	103.3
69.3	212	147	65	101.9
69.3	205	142	63	98.4
69.3	202	140	62	97.0
69.3	199	138	61	95.7
69.3	192	133	59	92.1
69.3	189	131	58	90.8
69.3	179	124	55	85.9

%E	M1	M2	DM	M*
69.3	176	122	54	84.6
69.3	166	115	51	79.7
69.3	163	113	50	78.3
69.3	153	106	47	73.4
69.3	150	104	46	72.1
69.3	140	97	43	67.2
69.3	137	95	42	65.9
69.3	127	88	39	61.0
69.3	114	79	35	54.7
69.3	101	70	31	48.5
69.3	88	61	27	42.3
69.3	75	52	23	36.1
69.2	500	346	154	239.4
69.2	497	344	153	238.1
69.2	496	343	153	237.2
69.2	494	342	152	236.8
69.2	493	341	152	235.9
69.2	491	340	151	235.4
69.2	490	339	151	234.5
69.2	487	337	150	233.2
69.2	484	335	149	231.9
69.2	483	334	149	231.0
69.2	481	333	148	230.5
69.2	480	332	148	229.6
69.2	478	331	147	229.2
69.2	477	330	147	228.3
69.2	474	328	146	227.0
69.2	471	326	145	225.6
69.2	468	324	144	224.3
69.2	467	323	144	223.4
69.2	465	322	143	223.0
69.2	464	321	143	222.1
69.2	461	319	142	220.7
69.2	458	317	141	219.4
69.2	455	315	140	218.1
69.2	454	314	140	217.2
69.2	452	313	139	216.7
69.2	451	312	139	215.8
69.2	448	310	138	214.5
69.2	445	308	137	213.2
69.2	442	306	136	211.8
69.2	441	305	136	210.9
69.2	439	304	135	210.5
69.2	438	303	135	209.6
69.2	435	301	134	208.3
69.2	432	299	133	206.9
69.2	429	297	132	205.6
69.2	428	296	132	204.7
69.2	426	295	131	204.3
69.2	425	294	131	203.4

%E	M1	M2	DM	M*
69.2	422	292	130	202.0
69.2	419	290	129	200.7
69.2	416	288	128	199.4
69.2	415	287	128	198.5
69.2	413	286	127	198.1
69.2	412	285	127	197.1
69.2	409	283	126	195.8
69.2	406	281	125	194.5
69.2	403	279	124	193.2
69.2	402	278	124	192.2
69.2	399	276	123	190.9
69.2	396	274	122	189.6
69.2	393	272	121	188.3
69.2	390	270	120	186.9
69.2	389	269	120	186.0
69.2	386	267	119	184.7
69.2	383	265	118	183.4
69.2	380	263	117	182.0
69.2	377	261	116	180.7
69.2	373	258	115	178.5
69.2	370	256	114	177.1
69.2	367	254	113	175.8
69.2	364	252	112	174.5
69.2	360	249	111	172.2
69.2	357	247	110	170.9
69.2	354	245	109	169.6
69.2	351	243	108	168.2
69.2	347	240	107	166.0
69.2	344	238	106	164.7
69.2	341	236	105	163.3
69.2	338	234	104	162.0
69.2	334	231	103	159.8
69.2	331	229	102	158.4
69.2	328	227	101	157.1
69.2	325	225	100	155.8
69.2	321	222	99	153.5
69.2	318	220	98	152.2
69.2	315	218	97	150.9
69.2	312	216	96	149.5
69.2	308	213	95	147.3
69.2	305	211	94	146.0
69.2	302	209	93	144.6
69.2	299	207	92	143.3
69.2	295	204	91	141.1
69.2	292	202	90	139.7
69.2	289	200	89	138.4
69.2	286	198	88	137.1
69.2	279	193	86	133.5
69.2	276	191	85	132.2
69.2	273	189	84	130.8

%E	M1	M2	DM	M*
69.2	266	184	82	127.3
69.2	263	182	81	125.9
69.2	260	180	80	124.6
69.2	253	175	78	121.0
69.2	250	173	77	119.7
69.2	247	171	76	118.4
69.2	240	166	74	114.8
69.2	237	164	73	113.5
69.2	234	162	72	112.2
69.2	227	157	70	108.6
69.2	224	155	69	107.3
69.2	221	153	68	105.9
69.2	214	148	66	102.4
69.2	211	146	65	101.0
69.2	208	144	64	99.7
69.2	201	139	62	96.1
69.2	198	137	61	94.8
69.2	195	135	60	93.5
69.2	185	128	57	88.6
69.2	182	126	56	87.2
69.2	172	119	53	82.3
69.2	169	117	52	81.0
69.2	159	110	49	76.1
69.2	156	108	48	74.8
69.2	146	101	45	69.9
69.2	143	99	44	68.5
69.2	133	92	41	63.6
69.2	130	90	40	62.3
69.2	120	83	37	57.4
69.2	117	81	36	56.1
69.2	107	74	33	51.2
69.2	104	72	32	49.8
69.2	91	63	28	43.6
69.2	78	54	24	37.4
69.2	65	45	20	31.2
69.2	52	36	16	24.9
69.2	39	27	12	18.7
69.2	26	18	8	12.5
69.2	13	9	4	6.2
69.1	499	345	154	238.5
69.1	498	344	154	237.6
69.1	495	342	153	236.3
69.1	492	340	152	235.0
69.1	489	338	151	233.6
69.1	488	337	151	232.7
69.1	486	336	150	232.3
69.1	485	335	150	231.4
69.1	482	333	149	230.1
69.1	479	331	148	228.7
69.1	476	329	147	227.4

%E	M1	M2	DM	M*
69.1	475	328	147	226.5
69.1	473	327	146	226.1
69.1	472	326	146	225.2
69.1	470	325	145	224.7
69.1	469	324	145	223.8
69.1	466	322	144	222.5
69.1	463	320	143	221.2
69.1	460	318	142	219.8
69.1	459	317	142	218.9
69.1	457	316	141	218.5
69.1	456	315	141	217.6
69.1	453	313	140	216.3
69.1	450	311	139	214.9
69.1	447	309	138	213.6
69.1	446	308	138	212.7
69.1	444	307	137	212.3
69.1	443	306	137	211.4
69.1	440	304	136	210.0
69.1	437	302	135	208.7
69.1	434	300	134	207.4
69.1	433	299	134	206.5
69.1	431	298	133	206.0
69.1	430	297	133	205.1
69.1	427	295	132	203.8
69.1	424	293	131	202.5
69.1	421	291	130	201.1
69.1	418	289	129	199.8
69.1	417	288	129	198.9
69.1	414	286	128	197.6
69.1	411	284	127	196.2
69.1	408	282	126	194.9
69.1	405	280	125	193.6
69.1	404	279	125	192.7
69.1	401	277	124	191.3
69.1	398	275	123	190.0
69.1	395	273	122	188.7
69.1	392	271	121	187.3
69.1	391	270	121	186.4
69.1	388	268	120	185.1
69.1	385	266	119	183.8
69.1	382	264	118	182.5
69.1	379	262	117	181.1
69.1	376	260	116	179.8
69.1	375	259	116	178.9
69.1	372	257	115	177.6
69.1	369	255	114	176.2
69.1	366	253	113	174.9
69.1	363	251	112	173.6
69.1	362	250	112	172.7
69.1	359	248	111	171.3

%E	M1	M2	DM	M*
69.1	356	246	110	170.0
69.1	353	244	109	168.7
69.1	350	242	108	167.3
69.1	349	241	108	166.4
69.1	346	239	107	165.1
69.1	343	237	106	163.8
69.1	340	235	105	162.4
69.1	337	233	104	161.1
69.1	333	230	103	158.9
69.1	330	228	102	157.5
69.1	327	226	101	156.2
69.1	324	224	100	154.9
69.1	320	221	99	152.6
69.1	317	219	98	151.3
69.1	314	217	97	150.0
69.1	311	215	96	148.6
69.1	307	212	95	146.4
69.1	304	210	94	145.1
69.1	301	208	93	143.7
69.1	298	206	92	142.4
69.1	291	201	90	138.8
69.1	288	199	89	137.5
69.1	285	197	88	136.2
69.1	282	195	87	134.8
69.1	278	192	86	132.6
69.1	275	190	85	131.3
69.1	272	188	84	129.9
69.1	269	186	83	128.6
69.1	265	183	82	126.4
69.1	262	181	81	125.0
69.1	259	179	80	123.7
69.1	256	177	79	122.4
69.1	249	172	77	118.8
69.1	246	170	76	117.5
69.1	243	168	75	116.1
69.1	236	163	73	112.6
69.1	233	161	72	111.2
69.1	230	159	71	109.9
69.1	223	154	69	106.3
69.1	220	152	68	105.0
69.1	217	150	67	103.7
69.1	207	143	64	98.8
69.1	204	141	63	97.5
69.1	194	134	60	92.6
69.1	191	132	59	91.2
69.1	188	130	58	89.9
69.1	181	125	56	86.3
69.1	178	123	55	85.0
69.1	175	121	54	83.7
69.1	165	114	51	78.8

%E	M1	M2	DM	M*
69.1	162	112	50	77.4
69.1	152	105	47	72.5
69.1	149	103	46	71.2
69.1	139	96	43	66.3
69.1	136	94	42	65.0
69.1	123	85	38	58.7
69.1	110	76	34	52.5
69.1	97	67	30	46.3
69.1	94	65	29	44.9
69.1	81	56	25	38.7
69.1	68	47	21	32.5
69.1	55	38	17	26.3
69.0	500	345	155	238.0
69.0	497	343	154	236.7
69.0	496	342	154	235.8
69.0	494	341	153	235.4
69.0	493	340	153	234.5
69.0	491	339	152	234.1
69.0	490	338	152	233.2
69.0	487	336	151	231.8
69.0	484	334	150	230.5
69.0	481	332	149	229.2
69.0	480	331	149	228.3
69.0	478	330	148	227.8
69.0	477	329	148	226.9
69.0	474	327	147	225.6
69.0	471	325	146	224.3
69.0	468	323	145	222.9
69.0	467	322	145	222.0
69.0	465	321	144	221.6
69.0	464	320	144	220.7
69.0	462	319	143	220.3
69.0	461	318	143	219.4
69.0	458	316	142	218.0
69.0	455	314	141	216.7
69.0	452	312	140	215.4
69.0	451	311	140	214.5
69.0	449	310	139	214.0
69.0	448	309	139	213.1
69.0	445	307	138	211.8
69.0	442	305	137	210.5
69.0	439	303	136	209.1
69.0	438	302	136	208.2
69.0	436	301	135	207.8
69.0	435	300	135	206.9
69.0	432	298	134	205.6
69.0	429	296	133	204.2
69.0	426	294	132	202.9
69.0	423	292	131	201.6
69.0	422	291	131	200.7

%E	M1	M2	DM	M*
69.0	420	290	130	200.2
69.0	419	289	130	199.3
69.0	416	287	129	198.0
69.0	413	285	128	196.7
69.0	410	283	127	195.3
69.0	407	281	126	194.0
69.0	406	280	126	193.1
69.0	403	278	125	191.8
69.0	400	276	124	190.4
69.0	397	274	123	189.1
69.0	394	272	122	187.8
69.0	393	271	122	186.9
69.0	390	269	121	185.5
69.0	387	267	120	184.2
69.0	384	265	119	182.9
69.0	381	263	118	181.5
69.0	378	261	117	180.2
69.0	377	260	117	179.3
69.0	374	258	116	178.0
69.0	371	256	115	176.6
69.0	368	254	114	175.3
69.0	365	252	113	174.0
69.0	364	251	113	173.1
69.0	361	249	112	171.7
69.0	358	247	111	170.4
69.0	355	245	110	169.1
69.0	352	243	109	167.8
69.0	348	240	108	165.5
69.0	345	238	107	164.2
69.0	342	236	106	162.9
69.0	339	234	105	161.5
69.0	336	232	104	160.2
69.0	335	231	104	159.3
69.0	332	229	103	158.0
69.0	329	227	102	156.6
69.0	326	225	101	155.3
69.0	323	223	100	154.0
69.0	319	220	99	151.7
69.0	316	218	98	150.4
69.0	313	216	97	149.1
69.0	310	214	96	147.7
69.0	306	211	95	145.5
69.0	303	209	94	144.2
69.0	300	207	93	142.8
69.0	297	205	92	141.5
69.0	294	203	91	140.2
69.0	290	200	90	137.9
69.0	287	198	89	136.6
69.0	284	196	88	135.3
69.0	281	194	87	133.9

%E	M1	M2	DM	M*
69.0	277	191	86	131.7
69.0	274	189	85	130.4
69.0	271	187	84	129.0
69.0	268	185	83	127.7
69.0	261	180	81	124.1
69.0	258	178	80	122.8
69.0	255	176	79	121.5
69.0	252	174	78	120.1
69.0	248	171	77	117.9
69.0	245	169	76	116.6
69.0	242	167	75	115.2
69.0	239	165	74	113.9
69.0	232	160	72	110.3
69.0	229	158	71	109.0
69.0	226	156	70	107.7
69.0	219	151	68	104.1
69.0	216	149	67	102.8
69.0	213	147	66	101.5
69.0	210	145	65	100.1
69.0	203	140	63	96.6
69.0	200	138	62	95.2
69.0	197	136	61	93.9
69.0	187	129	58	89.0
69.0	184	127	57	87.7
69.0	174	120	54	82.8
69.0	171	118	53	81.4
69.0	168	116	52	80.1
69.0	158	109	49	75.2
69.0	155	107	48	73.9
69.0	145	100	45	69.0
69.0	142	98	44	67.6
69.0	129	89	40	61.4
69.0	126	87	39	60.1
69.0	116	80	36	55.2
69.0	113	78	35	53.8
69.0	100	69	31	47.6
69.0	87	60	27	41.4
69.0	84	58	26	40.0
69.0	71	49	22	33.8
69.0	58	40	18	27.6
69.0	42	29	13	20.0
69.0	29	20	9	13.8
68.9	499	344	155	237.1
68.9	498	343	155	236.2
68.9	495	341	154	234.9
68.9	492	339	153	233.6
68.9	489	337	152	232.2
68.9	488	336	152	231.3
68.9	486	335	151	230.9
68.9	485	334	151	230.0

%E	M1	M2	DM	M*
68.9	483	333	150	229.6
68.9	482	332	150	228.7
68.9	479	330	149	227.3
68.9	476	328	148	226.0
68.9	473	326	147	224.7
68.9	472	325	147	223.8
68.9	470	324	146	223.4
68.9	469	323	146	222.4
68.9	466	321	145	221.1
68.9	463	319	144	219.8
68.9	460	317	143	218.5
68.9	457	315	142	217.1
68.9	456	314	142	216.2
68.9	454	313	141	215.8
68.9	453	312	141	214.9
68.9	450	310	140	213.6
68.9	447	308	139	212.2
68.9	444	306	138	210.9
68.9	441	304	137	209.6
68.9	440	303	137	208.7
68.9	437	301	136	207.3
68.9	434	299	135	206.0
68.9	431	297	134	204.7
68.9	428	295	133	203.3
68.9	427	294	133	202.4
68.9	425	293	132	202.0
68.9	424	292	132	201.1
68.9	421	290	131	199.8
68.9	418	288	130	198.4
68.9	415	286	129	197.1
68.9	412	284	128	195.8
68.9	411	283	128	194.9
68.9	409	282	127	194.4
68.9	408	281	127	193.5
68.9	405	279	126	192.2
68.9	402	277	125	190.9
68.9	399	275	124	189.5
68.9	396	273	123	188.2
68.9	395	272	123	187.3
68.9	392	270	122	186.0
68.9	389	268	121	184.6
68.9	386	266	120	183.3
68.9	383	264	119	182.0
68.9	380	262	118	180.6
68.9	379	261	118	179.7
68.9	376	259	117	178.4
68.9	373	257	116	177.1
68.9	370	255	115	175.7
68.9	367	253	114	174.4
68.9	366	252	114	173.5

%E	M1	M2	DM	M*
68.9	363	250	113	172.2
68.9	360	248	112	170.8
68.9	357	246	111	169.5
68.9	354	244	110	168.2
68.9	351	242	109	166.8
68.9	350	241	109	165.9
68.9	347	239	108	164.6
68.9	344	237	107	163.3
68.9	341	235	106	162.0
68.9	338	233	105	160.6
68.9	334	230	104	158.4
68.9	331	228	103	157.1
68.9	328	226	102	155.7
68.9	325	224	101	154.4
68.9	322	222	100	153.1
68.9	318	219	99	150.8
68.9	315	217	98	149.5
68.9	312	215	97	148.2
68.9	309	213	96	146.8
68.9	305	210	95	144.6
68.9	302	208	94	143.3
68.9	299	206	93	141.9
68.9	296	204	92	140.6
68.9	293	202	91	139.3
68.9	289	199	90	137.0
68.9	286	197	89	135.7
68.9	283	195	88	134.4
68.9	280	193	87	133.0
68.9	273	188	85	129.5
68.9	270	186	84	128.1
68.9	267	184	83	126.8
68.9	264	182	82	125.5
68.9	257	177	80	121.9
68.9	254	175	79	120.6
68.9	251	173	78	119.2
68.9	244	168	76	115.7
68.9	241	166	75	114.3
68.9	238	164	74	113.0
68.9	235	162	73	111.7
68.9	228	157	71	108.1
68.9	225	155	70	106.8
68.9	222	153	69	105.4
68.9	212	146	66	100.5
68.9	209	144	65	99.2
68.9	206	142	64	97.9
68.9	196	135	61	93.0
68.9	193	133	60	91.7
68.9	190	131	59	90.3
68.9	183	126	57	86.8
68.9	180	124	56	85.4

%E	M1	M2	DM	M*
68.9	177	122	55	84.1
68.9	167	115	52	79.2
68.9	164	113	51	77.9
68.9	161	111	50	76.5
68.9	151	104	47	71.6
68.9	148	102	46	70.3
68.9	135	93	42	64.1
68.9	132	91	41	62.7
68.9	122	84	38	57.8
68.9	119	82	37	56.5
68.9	106	73	33	50.3
68.9	103	71	32	48.9
68.9	90	62	28	42.7
68.9	74	51	23	35.1
68.9	61	42	19	28.9
68.9	45	31	14	21.4
68.8	500	344	156	236.7
68.8	497	342	155	235.3
68.8	496	341	155	234.4
68.8	494	340	154	234.0
68.8	493	339	154	233.1
68.8	491	338	153	232.7
68.8	490	337	153	231.8
68.8	487	335	152	230.4
68.8	484	333	151	229.1
68.8	481	331	150	227.8
68.8	480	330	150	226.9
68.8	478	329	149	226.4
68.8	477	328	149	225.5
68.8	475	327	148	225.1
68.8	474	326	148	224.2
68.8	471	324	147	222.9
68.8	468	322	146	221.5
68.8	465	320	145	220.2
68.8	464	319	145	219.3
68.8	462	318	144	218.9
68.8	461	317	144	218.0
68.8	459	316	143	217.6
68.8	458	315	143	216.6
68.8	455	313	142	215.3
68.8	452	311	141	214.0
68.8	449	309	140	212.7
68.8	448	308	140	211.8
68.8	446	307	139	211.3
68.8	445	306	139	210.4
68.8	443	305	138	210.0
68.8	442	304	138	209.1
68.8	439	302	137	207.8
68.8	436	300	136	206.4
68.8	433	298	135	205.1
68.8	432	297	135	204.2
68.8	430	296	134	203.8
68.8	429	295	134	202.9
68.8	426	293	133	201.5
68.8	423	291	132	200.2
68.8	420	289	131	198.9
68.8	417	287	130	197.5
68.8	416	286	130	196.6
68.8	414	285	129	196.2
68.8	413	284	129	195.3
68.8	410	282	128	194.0
68.8	407	280	127	192.6
68.8	404	278	126	191.3
68.8	401	276	125	190.0
68.8	400	275	125	189.1
68.8	398	274	124	188.6
68.8	397	273	124	187.7
68.8	394	271	123	186.4
68.8	391	269	122	185.1
68.8	388	267	121	183.7
68.8	385	265	120	182.4
68.8	384	264	120	181.5
68.8	382	263	119	181.1
68.8	381	262	119	180.2
68.8	378	260	118	178.8
68.8	375	258	117	177.5
68.8	372	256	116	176.2
68.8	369	254	115	174.8
68.8	368	253	115	173.9
68.8	365	251	114	172.6
68.8	362	249	113	171.3
68.8	359	247	112	169.9
68.8	356	245	111	168.6
68.8	353	243	110	167.3
68.8	352	242	110	166.4
68.8	349	240	109	165.0
68.8	346	238	108	163.7
68.8	343	236	107	162.4
68.8	340	234	106	161.0
68.8	337	232	105	159.7
68.8	336	231	105	158.8
68.8	333	229	104	157.5
68.8	330	227	103	156.1
68.8	327	225	102	154.8
68.8	324	223	101	153.5
68.8	321	221	100	152.2
68.8	320	220	100	151.3
68.8	317	218	99	149.9
68.8	314	216	98	148.6
68.8	311	214	97	147.3
68.8	308	212	96	145.9
68.8	304	209	95	143.7
68.8	301	207	94	142.4
68.8	298	205	93	141.0
68.8	295	203	92	139.7
68.8	292	201	91	138.4
68.8	288	198	90	136.1
68.8	285	196	89	134.8
68.8	282	194	88	133.5
68.8	279	192	87	132.1
68.8	276	190	86	130.8
68.8	272	187	85	128.6
68.8	269	185	84	127.2
68.8	266	183	83	125.9
68.8	263	181	82	124.6
68.8	260	179	81	123.2
68.8	256	176	80	121.0
68.8	253	174	79	119.7
68.8	250	172	78	118.3
68.8	247	170	77	117.0
68.8	240	165	75	113.4
68.8	237	163	74	112.1
68.8	234	161	73	110.8
68.8	231	159	72	109.4
68.8	224	154	70	105.9
68.8	221	152	69	104.5
68.8	218	150	68	103.2
68.8	215	148	67	101.9
68.8	208	143	65	98.3
68.8	205	141	64	97.0
68.8	202	139	63	95.6
68.8	199	137	62	94.3
68.8	192	132	60	90.8
68.8	189	130	59	89.4
68.8	186	128	58	88.1
68.8	176	121	55	83.2
68.8	173	119	54	81.9
68.8	170	117	53	80.5
68.8	160	110	50	75.6
68.8	157	108	49	74.3
68.8	154	106	48	73.0
68.8	144	99	45	68.1
68.8	141	97	44	66.7
68.8	138	95	43	65.4
68.8	128	88	40	60.5
68.8	125	86	39	59.2
68.8	112	77	35	52.9
68.8	109	75	34	51.6
68.8	96	66	30	45.4
68.8	93	64	29	44.0
68.8	80	55	25	37.8
68.8	77	53	24	36.5
68.8	64	44	20	30.3
68.8	48	33	15	22.7
68.8	32	22	10	15.1
68.8	16	11	5	7.6
68.7	499	343	156	235.8
68.7	498	342	156	234.9
68.7	495	340	155	233.5
68.7	492	338	154	232.2
68.7	489	336	153	230.9
68.7	486	334	152	229.5
68.7	485	333	152	228.6
68.7	483	332	151	228.2
68.7	482	331	151	227.3
68.7	479	329	150	226.0
68.7	476	327	149	224.6
68.7	473	325	148	223.3
68.7	470	323	147	222.0
68.7	469	322	147	221.1
68.7	467	321	146	220.6
68.7	466	320	146	219.7
68.7	463	318	145	218.4
68.7	460	316	144	217.1
68.7	457	314	143	215.7
68.7	454	312	142	214.4
68.7	453	311	142	213.5
68.7	451	310	141	213.1
68.7	450	309	141	212.2
68.7	447	307	140	210.8
68.7	444	305	139	209.5
68.7	441	303	138	208.2
68.7	438	301	137	206.9
68.7	437	300	137	205.9
68.7	435	299	136	205.5
68.7	434	298	136	204.6
68.7	431	296	135	203.3
68.7	428	294	134	202.0
68.7	425	292	133	200.6
68.7	422	290	132	199.3
68.7	419	288	131	198.0
68.7	418	287	131	197.1
68.7	415	285	130	195.7
68.7	412	283	129	194.4
68.7	409	281	128	193.1
68.7	406	279	127	191.7
68.7	403	277	126	190.4
68.7	402	276	126	189.5
68.7	399	274	125	188.2
68.7	396	272	124	186.8
68.7	393	270	123	185.5
68.7	390	268	122	184.2
68.7	387	266	121	182.8
68.7	386	265	121	181.9
68.7	383	263	120	180.6
68.7	380	261	119	179.3
68.7	377	259	118	177.9
68.7	374	257	117	176.6
68.7	371	255	116	175.3
68.7	367	252	115	173.0
68.7	364	250	114	171.7
68.7	361	248	113	170.4
68.7	358	246	112	169.0
68.7	355	244	111	167.7
68.7	351	241	110	165.5
68.7	348	239	109	164.1
68.7	345	237	108	162.8
68.7	342	235	107	161.5
68.7	339	233	106	160.1
68.7	335	230	105	157.9
68.7	332	228	104	156.6
68.7	329	226	103	155.2
68.7	326	224	102	153.9
68.7	323	222	101	152.6
68.7	319	219	100	150.3
68.7	316	217	99	149.0
68.7	313	215	98	147.7
68.7	310	213	97	146.4
68.7	307	211	96	145.0
68.7	300	206	94	141.5
68.7	297	204	93	140.1
68.7	294	202	92	138.8
68.7	291	200	91	137.5
68.7	284	195	89	133.9
68.7	281	193	88	132.6
68.7	278	191	87	131.2
68.7	275	189	86	129.9
68.7	268	184	84	126.3
68.7	265	182	83	125.0
68.7	262	180	82	123.7
68.7	259	178	81	122.3
68.7	252	173	79	118.8
68.7	249	171	78	117.4
68.7	246	169	77	116.1
68.7	243	167	76	114.8
68.7	233	160	73	109.9
68.7	230	158	72	108.5
68.7	227	156	71	107.2
68.7	217	149	68	102.3
68.7	214	147	67	101.0

%E	M1	M2	DM	M*
68.7	211	145	66	99.6
68.7	201	138	63	94.7
68.7	198	136	62	93.4
68.7	195	134	61	92.1
68.7	182	125	57	85.9
68.7	179	123	56	84.5
68.7	166	114	52	78.3
68.7	163	112	51	77.0
68.7	150	103	47	70.7
68.7	147	101	46	69.4
68.7	134	92	42	63.2
68.7	131	90	41	61.8
68.7	115	79	36	54.3
68.7	99	68	31	46.7
68.7	83	57	26	39.1
68.7	67	46	21	31.6
68.6	500	343	157	235.3
68.6	497	341	156	234.0
68.6	494	339	155	232.6
68.6	493	338	155	231.7
68.6	491	337	154	231.3
68.6	490	336	154	230.4
68.6	488	335	153	230.0
68.6	487	334	153	229.1
68.6	484	332	152	227.7
68.6	481	330	151	226.4
68.6	478	328	150	225.1
68.6	477	327	150	224.2
68.6	475	326	149	223.7
68.6	474	325	149	222.8
68.6	472	324	148	222.4
68.6	471	323	148	221.5
68.6	468	321	147	220.2
68.6	465	319	146	218.8
68.6	462	317	145	217.5
68.6	459	315	144	216.2
68.6	458	314	144	215.3
68.6	456	313	143	214.8
68.6	455	312	143	213.9
68.6	452	310	142	212.6
68.6	449	308	141	211.3
68.6	446	306	140	209.9
68.6	443	304	139	208.6
68.6	442	303	139	207.7
68.6	440	302	138	207.3
68.6	439	301	138	206.4
68.6	436	299	137	205.0
68.6	433	297	136	203.7
68.6	430	295	135	202.4
68.6	427	293	134	201.1

%E	M1	M2	DM	M*
68.6	424	291	133	199.7
68.6	423	290	133	198.8
68.6	421	289	132	198.4
68.6	420	288	132	197.5
68.6	417	286	131	196.2
68.6	414	284	130	194.8
68.6	411	282	129	193.5
68.6	408	280	128	192.2
68.6	407	279	128	191.3
68.6	405	278	127	190.8
68.6	404	277	127	189.9
68.6	401	275	126	188.6
68.6	398	273	125	187.3
68.6	395	271	124	185.9
68.6	392	269	123	184.6
68.6	389	267	122	183.3
68.6	388	266	122	182.4
68.6	385	264	121	181.0
68.6	382	262	120	179.7
68.6	379	260	119	178.4
68.6	376	258	118	177.0
68.6	373	256	117	175.7
68.6	370	254	116	174.4
68.6	369	253	116	173.5
68.6	366	251	115	172.1
68.6	363	249	114	170.8
68.6	360	247	113	169.5
68.6	357	245	112	168.1
68.6	354	243	111	166.8
68.6	353	242	111	165.9
68.6	350	240	110	164.6
68.6	347	238	109	163.2
68.6	344	236	108	161.9
68.6	341	234	107	160.6
68.6	338	232	106	159.2
68.6	334	229	105	157.0
68.6	331	227	104	155.7
68.6	328	225	103	154.3
68.6	325	223	102	153.0
68.6	322	221	101	151.7
68.6	318	218	100	149.4
68.6	315	216	99	148.1
68.6	312	214	98	146.8
68.6	309	212	97	145.4
68.6	306	210	96	144.1
68.6	303	208	95	142.8
68.6	299	205	94	140.6
68.6	296	203	93	139.2
68.6	293	201	92	137.9
68.6	290	199	91	136.6

%E	M1	M2	DM	M*
68.6	287	197	90	135.2
68.6	283	194	89	133.0
68.6	280	192	88	131.7
68.6	277	190	87	130.3
68.6	274	188	86	129.0
68.6	271	186	85	127.7
68.6	264	181	83	124.1
68.6	261	179	82	122.8
68.6	258	177	81	121.4
68.6	255	175	80	120.1
68.6	245	168	77	115.2
68.6	242	166	76	113.9
68.6	239	164	75	112.5
68.6	236	162	74	111.2
68.6	229	157	72	107.6
68.6	226	155	71	106.3
68.6	223	153	70	105.0
68.6	220	151	69	103.6
68.6	210	144	66	98.7
68.6	207	142	65	97.4
68.6	204	140	64	96.1
68.6	194	133	61	91.2
68.6	191	131	60	89.8
68.6	188	129	59	88.5
68.6	185	127	58	87.2
68.6	175	120	55	82.3
68.6	172	118	54	81.0
68.6	169	116	53	79.6
68.6	159	109	50	74.7
68.6	156	107	49	73.4
68.6	153	105	48	72.1
68.6	140	96	44	65.8
68.6	137	94	43	64.5
68.6	121	83	38	56.9
68.6	118	81	37	55.6
68.6	105	72	33	49.4
68.6	102	70	32	48.0
68.6	86	59	27	40.5
68.6	70	48	22	32.9
68.6	51	35	16	24.0
68.6	35	24	11	16.5
68.5	499	342	157	234.4
68.5	498	341	157	233.5
68.5	496	340	156	233.1
68.5	495	339	156	232.2
68.5	492	337	155	230.8
68.5	489	335	154	229.5
68.5	486	333	153	228.2
68.5	485	332	153	227.3
68.5	483	331	152	226.8

%E	M1	M2	DM	M*
68.5	482	330	152	225.9
68.5	480	329	151	225.5
68.5	479	328	151	224.6
68.5	476	326	150	223.3
68.5	473	324	149	221.9
68.5	470	322	148	220.6
68.5	467	320	147	219.3
68.5	466	319	147	218.4
68.5	464	318	146	217.9
68.5	463	317	146	217.0
68.5	461	316	145	216.6
68.5	460	315	145	215.7
68.5	457	313	144	214.4
68.5	454	311	143	213.0
68.5	451	309	142	211.7
68.5	448	307	141	210.4
68.5	447	306	141	209.5
68.5	445	305	140	209.0
68.5	444	304	140	208.1
68.5	441	302	139	206.8
68.5	438	300	138	205.5
68.5	435	298	137	204.1
68.5	432	296	136	202.8
68.5	429	294	135	201.5
68.5	428	293	135	200.6
68.5	426	292	134	200.2
68.5	425	291	134	199.2
68.5	422	289	133	197.9
68.5	419	287	132	196.6
68.5	416	285	131	195.3
68.5	413	283	130	193.9
68.5	410	281	129	192.6
68.5	409	280	129	191.7
68.5	406	278	128	190.4
68.5	403	276	127	189.0
68.5	400	274	126	187.7
68.5	397	272	125	186.4
68.5	394	270	124	185.0
68.5	391	268	123	183.7
68.5	390	267	123	182.8
68.5	387	265	122	181.5
68.5	384	263	121	180.1
68.5	381	261	120	178.8
68.5	378	259	119	177.5
68.5	375	257	118	176.1
68.5	372	255	117	174.8
68.5	371	254	117	173.9
68.5	368	252	116	172.6
68.5	365	250	115	171.2
68.5	362	248	114	169.9

%E	M1	M2	DM	M*
68.5	359	246	113	168.6
68.5	356	244	112	167.2
68.5	355	243	112	166.3
68.5	352	241	111	165.0
68.5	349	239	110	163.7
68.5	346	237	109	162.3
68.5	343	235	108	161.0
68.5	340	233	107	159.7
68.5	337	231	106	158.3
68.5	336	230	106	157.4
68.5	333	228	105	156.1
68.5	330	226	104	154.8
68.5	327	224	103	153.4
68.5	324	222	102	152.1
68.5	321	220	101	150.8
68.5	317	217	100	148.5
68.5	314	215	99	147.2
68.5	311	213	98	145.9
68.5	308	211	97	144.5
68.5	305	209	96	143.2
68.5	302	207	95	141.9
68.5	298	204	94	139.7
68.5	295	202	93	138.3
68.5	292	200	92	137.0
68.5	289	198	91	135.7
68.5	286	196	90	134.3
68.5	279	191	88	130.8
68.5	276	189	87	129.4
68.5	273	187	86	128.1
68.5	270	185	85	126.8
68.5	267	183	84	125.4
68.5	260	178	82	121.9
68.5	257	176	81	120.5
68.5	254	174	80	119.2
68.5	251	172	79	117.9
68.5	248	170	78	116.5
68.5	241	165	76	113.0
68.5	238	163	75	111.6
68.5	235	161	74	110.3
68.5	232	159	73	109.0
68.5	222	152	70	104.1
68.5	219	150	69	102.7
68.5	216	148	68	101.4
68.5	213	146	67	100.1
68.5	203	139	64	95.2
68.5	200	137	63	93.8
68.5	197	135	62	92.5
68.5	184	126	58	86.3
68.5	181	124	57	85.0
68.5	178	122	56	83.6

%E	M1	M2	DM	M*
68.5	168	115	53	78.7
68.5	165	113	52	77.4
68.5	162	111	51	76.1
68.5	149	102	47	69.8
68.5	146	100	46	68.5
68.5	143	98	45	67.2
68.5	130	89	41	60.9
68.5	127	87	40	59.6
68.5	124	85	39	58.3
68.5	111	76	35	52.0
68.5	108	74	34	50.7
68.5	92	63	29	43.1
68.5	89	61	28	41.8
68.5	73	50	23	34.2
68.5	54	37	17	25.4
68.4	500	342	158	233.9
68.4	497	340	157	232.6
68.4	494	338	156	231.3
68.4	493	337	156	230.4
68.4	491	336	155	229.9
68.4	490	335	155	229.0
68.4	488	334	154	228.6
68.4	487	333	154	227.7
68.4	484	331	153	226.4
68.4	481	329	152	225.0
68.4	478	327	151	223.7
68.4	475	325	150	222.4
68.4	474	324	150	221.5
68.4	472	323	149	221.0
68.4	471	322	149	220.1
68.4	469	321	148	219.7
68.4	468	320	148	218.8
68.4	465	318	147	217.5
68.4	462	316	146	216.1
68.4	459	314	145	214.8
68.4	456	312	144	213.5
68.4	455	311	144	212.6
68.4	453	310	143	212.1
68.4	452	309	143	211.2
68.4	450	308	142	210.8
68.4	449	307	142	209.9
68.4	446	305	141	208.6
68.4	443	303	140	207.2
68.4	440	301	139	205.9
68.4	437	299	138	204.6
68.4	434	297	137	203.2
68.4	433	296	137	202.3
68.4	431	295	136	201.9
68.4	430	294	136	201.0
68.4	427	292	135	199.7
68.4	424	290	134	198.3
68.4	421	288	133	197.0
68.4	418	286	132	195.7
68.4	415	284	131	194.4
68.4	414	283	131	193.5
68.4	412	282	130	193.0
68.4	411	281	130	192.1
68.4	408	279	129	190.8
68.4	405	277	128	189.5
68.4	402	275	127	188.1
68.4	399	273	126	186.8
68.4	396	271	125	185.5
68.4	395	270	125	184.6
68.4	393	269	124	184.1
68.4	392	268	124	183.2
68.4	389	266	123	181.9
68.4	386	264	122	180.6
68.4	383	262	121	179.2
68.4	380	260	120	177.9
68.4	377	258	119	176.6
68.4	376	257	119	175.7
68.4	374	256	118	175.2
68.4	373	255	118	174.3
68.4	370	253	117	173.0
68.4	367	251	116	171.7
68.4	364	249	115	170.3
68.4	361	247	114	169.0
68.4	358	245	113	167.7
68.4	354	242	112	165.4
68.4	351	240	111	164.1
68.4	348	238	110	162.8
68.4	345	236	109	161.4
68.4	342	234	108	160.1
68.4	339	232	107	158.8
68.4	335	229	106	156.5
68.4	332	227	105	155.2
68.4	329	225	104	153.9
68.4	326	223	103	152.5
68.4	323	221	102	151.2
68.4	320	219	101	149.9
68.4	316	216	100	147.6
68.4	313	214	99	146.3
68.4	310	212	98	145.0
68.4	307	210	97	143.6
68.4	304	208	96	142.3
68.4	301	206	95	141.0
68.4	297	203	94	138.8
68.4	294	201	93	137.4
68.4	291	199	92	136.1
68.4	288	197	91	134.8
68.4	285	195	90	133.4
68.4	282	193	89	132.1
68.4	275	188	87	128.5
68.4	272	186	86	127.2
68.4	269	184	85	125.9
68.4	266	182	84	124.5
68.4	263	180	83	123.2
68.4	256	175	81	119.6
68.4	253	173	80	118.3
68.4	250	171	79	117.0
68.4	247	169	78	115.6
68.4	244	167	77	114.3
68.4	237	162	75	110.7
68.4	234	160	74	109.4
68.4	231	158	73	108.1
68.4	228	156	72	106.7
68.4	225	154	71	105.4
68.4	215	147	68	100.5
68.4	212	145	67	99.2
68.4	209	143	66	97.8
68.4	206	141	65	96.5
68.4	196	134	62	91.6
68.4	193	132	61	90.3
68.4	190	130	60	88.9
68.4	187	128	59	87.6
68.4	177	121	56	82.7
68.4	174	119	55	81.4
68.4	171	117	54	80.1
68.4	158	108	50	73.8
68.4	155	106	49	72.5
68.4	152	104	48	71.2
68.4	136	93	43	63.6
68.4	133	91	42	62.3
68.4	117	80	37	54.7
68.4	114	78	36	53.4
68.4	98	67	31	45.8
68.4	95	65	30	44.5
68.4	79	54	25	36.9
68.4	76	52	24	35.6
68.4	57	39	18	26.7
68.4	38	26	12	17.8
68.4	19	13	6	8.9
68.3	499	341	158	233.0
68.3	498	340	158	232.1
68.3	496	339	157	231.7
68.3	495	338	157	230.8
68.3	492	336	156	229.5
68.3	489	334	155	228.1
68.3	486	332	154	226.8
68.3	483	330	153	225.5
68.3	482	329	153	224.6
68.3	480	328	152	224.1
68.3	479	327	152	223.2
68.3	477	326	151	222.8
68.3	476	325	151	221.9
68.3	473	323	150	220.6
68.3	470	321	149	219.2
68.3	467	319	148	217.9
68.3	464	317	147	216.6
68.3	463	316	147	215.7
68.3	461	315	146	215.2
68.3	460	314	146	214.3
68.3	458	313	145	213.9
68.3	457	312	145	213.0
68.3	454	310	144	211.7
68.3	451	308	143	210.3
68.3	448	306	142	209.0
68.3	445	304	141	207.7
68.3	442	302	140	206.3
68.3	441	301	140	205.4
68.3	439	300	139	205.0
68.3	438	299	139	204.1
68.3	436	298	138	203.7
68.3	435	297	138	202.8
68.3	432	295	137	201.4
68.3	429	293	136	200.1
68.3	426	291	135	198.8
68.3	423	289	134	197.4
68.3	420	287	133	196.1
68.3	419	286	133	195.2
68.3	417	285	132	194.8
68.3	416	284	132	193.9
68.3	413	282	131	192.6
68.3	410	280	130	191.2
68.3	407	278	129	189.9
68.3	404	276	128	188.6
68.3	401	274	127	187.2
68.3	400	273	127	186.3
68.3	398	272	126	185.9
68.3	397	271	126	185.0
68.3	394	269	125	183.7
68.3	391	267	124	182.3
68.3	388	265	123	181.0
68.3	385	263	122	179.7
68.3	382	261	121	178.3
68.3	379	259	120	177.0
68.3	378	258	120	176.1
68.3	375	256	119	174.8
68.3	372	254	118	173.4
68.3	369	252	117	172.1
68.3	366	250	116	170.8
68.3	363	248	115	169.4
68.3	360	246	114	168.1
68.3	357	244	113	166.8
68.3	356	243	113	165.9
68.3	353	241	112	164.5
68.3	350	239	111	163.2
68.3	347	237	110	161.9
68.3	344	235	109	160.5
68.3	341	233	108	159.2
68.3	338	231	107	157.9
68.3	334	228	106	155.6
68.3	331	226	105	154.3
68.3	328	224	104	153.0
68.3	325	222	103	151.6
68.3	322	220	102	150.3
68.3	319	218	101	149.0
68.3	315	215	100	146.7
68.3	312	213	99	145.4
68.3	309	211	98	144.1
68.3	306	209	97	142.7
68.3	303	207	96	141.4
68.3	300	205	95	140.1
68.3	293	200	93	136.5
68.3	290	198	92	135.2
68.3	287	196	91	133.9
68.3	284	194	90	132.5
68.3	281	192	89	131.2
68.3	278	190	88	129.9
68.3	271	185	86	126.3
68.3	268	183	85	125.0
68.3	265	181	84	123.6
68.3	262	179	83	122.3
68.3	259	177	82	121.0
68.3	252	172	80	117.4
68.3	249	170	79	116.1
68.3	246	168	78	114.7
68.3	243	166	77	113.4
68.3	240	164	76	112.1
68.3	230	157	73	107.2
68.3	227	155	72	105.8
68.3	224	153	71	104.5
68.3	221	151	70	103.2
68.3	218	149	69	101.8
68.3	208	142	66	96.9
68.3	205	140	65	95.6
68.3	202	138	64	94.3
68.3	199	136	63	92.9
68.3	189	129	60	88.0
68.3	186	127	59	86.7

%E	M1	M2	DM	M*
68.3	183	125	58	85.4
68.3	180	123	57	84.0
68.3	167	114	53	77.8
68.3	164	112	52	76.5
68.3	161	110	51	75.2
68.3	145	99	46	67.6
68.3	142	97	45	66.3
68.3	139	95	44	64.9
68.3	126	86	40	58.7
68.3	123	84	39	57.4
68.3	120	82	38	56.0
68.3	104	71	33	48.5
68.3	101	69	32	47.1
68.3	82	56	26	38.2
68.3	63	43	20	29.3
68.3	60	41	19	28.0
68.3	41	28	13	19.1
68.2	500	341	159	232.6
68.2	497	339	158	231.2
68.2	494	337	157	229.9
68.2	493	336	157	229.0
68.2	491	335	156	228.6
68.2	490	334	156	227.7
68.2	488	333	155	227.2
68.2	487	332	155	226.3
68.2	485	331	154	225.9
68.2	484	330	154	225.0
68.2	481	328	153	223.7
68.2	478	326	152	222.3
68.2	475	324	151	221.0
68.2	472	322	150	219.7
68.2	471	321	150	218.8
68.2	469	320	149	218.3
68.2	468	319	149	217.4
68.2	466	318	148	217.0
68.2	465	317	148	216.1
68.2	462	315	147	214.8
68.2	459	313	146	213.4
68.2	456	311	145	212.1
68.2	453	309	144	210.8
68.2	450	307	143	209.4
68.2	449	306	143	208.5
68.2	447	305	142	208.1
68.2	446	304	142	207.2
68.2	444	303	141	206.8
68.2	443	302	141	205.9
68.2	440	300	140	204.5
68.2	437	298	139	203.2
68.2	434	296	138	201.9
68.2	431	294	137	200.5

%E	M1	M2	DM	M*
68.2	428	292	136	199.2
68.2	427	291	136	198.3
68.2	425	290	135	197.9
68.2	424	289	135	197.0
68.2	422	288	134	196.5
68.2	421	287	134	195.7
68.2	418	285	133	194.3
68.2	415	283	132	193.0
68.2	412	281	131	191.7
68.2	409	279	130	190.3
68.2	406	277	129	189.0
68.2	403	275	128	187.7
68.2	402	274	128	186.8
68.2	399	272	127	185.4
68.2	396	270	126	184.1
68.2	393	268	125	182.8
68.2	390	266	124	181.4
68.2	387	264	123	180.1
68.2	384	262	122	178.8
68.2	381	260	121	177.4
68.2	380	259	121	176.5
68.2	377	257	120	175.2
68.2	374	255	119	173.9
68.2	371	253	118	172.5
68.2	368	251	117	171.2
68.2	365	249	116	169.9
68.2	362	247	115	168.5
68.2	359	245	114	167.2
68.2	358	244	114	166.3
68.2	355	242	113	165.0
68.2	352	240	112	163.6
68.2	349	238	111	162.3
68.2	346	236	110	161.0
68.2	343	234	109	159.6
68.2	340	232	108	158.3
68.2	337	230	107	157.0
68.2	336	229	107	156.1
68.2	333	227	106	154.7
68.2	330	225	105	153.4
68.2	327	223	104	152.1
68.2	324	221	103	150.7
68.2	321	219	102	149.4
68.2	318	217	101	148.1
68.2	314	214	100	145.8
68.2	311	212	99	144.5
68.2	308	210	98	143.2
68.2	305	208	97	141.8
68.2	302	206	96	140.5
68.2	299	204	95	139.2
68.2	296	202	94	137.9

%E	M1	M2	DM	M*
68.2	292	199	93	135.6
68.2	289	197	92	134.3
68.2	286	195	91	133.0
68.2	283	193	90	131.6
68.2	280	191	89	130.3
68.2	277	189	88	129.0
68.2	274	187	87	127.6
68.2	267	182	85	124.1
68.2	264	180	84	122.7
68.2	261	178	83	121.4
68.2	258	176	82	120.1
68.2	255	174	81	118.7
68.2	245	167	78	113.8
68.2	242	165	77	112.5
68.2	239	163	76	111.2
68.2	236	161	75	109.8
68.2	233	159	74	108.5
68.2	223	152	71	103.6
68.2	220	150	70	102.3
68.2	217	148	69	100.9
68.2	214	146	68	99.6
68.2	211	144	67	98.3
68.2	201	137	64	93.4
68.2	198	135	63	92.0
68.2	195	133	62	90.7
68.2	192	131	61	89.4
68.2	179	122	57	83.2
68.2	176	120	56	81.8
68.2	173	118	55	80.5
68.2	170	116	54	79.2
68.2	157	107	50	72.9
68.2	154	105	49	71.6
68.2	151	103	48	70.3
68.2	148	101	47	68.9
68.2	132	90	42	61.4
68.2	129	88	41	60.0
68.2	110	75	35	51.1
68.2	107	73	34	49.8
68.2	88	60	28	40.9
68.2	85	58	27	39.6
68.2	66	45	21	30.7
68.2	44	30	14	20.5
68.2	22	15	7	10.2
68.1	499	340	159	231.7
68.1	498	339	159	230.8
68.1	496	338	158	230.3
68.1	495	337	158	229.4
68.1	492	335	157	228.1
68.1	489	333	156	226.8
68.1	486	331	155	225.4

%E	M1	M2	DM	M*
68.1	483	329	154	224.1
68.1	482	328	154	223.2
68.1	480	327	153	222.8
68.1	479	326	153	221.9
68.1	477	325	152	221.4
68.1	476	324	152	220.5
68.1	474	323	151	220.1
68.1	473	322	151	219.2
68.1	470	320	150	217.9
68.1	467	318	149	216.5
68.1	464	316	148	215.2
68.1	461	314	147	213.9
68.1	458	312	146	212.5
68.1	457	311	146	211.6
68.1	455	310	145	211.2
68.1	454	309	145	210.3
68.1	452	308	144	209.9
68.1	451	307	144	209.0
68.1	448	305	143	207.6
68.1	445	303	142	206.3
68.1	442	301	141	205.0
68.1	439	299	140	203.6
68.1	436	297	139	202.3
68.1	433	295	138	201.0
68.1	432	294	138	200.1
68.1	430	293	137	199.6
68.1	429	292	137	198.8
68.1	426	290	136	197.4
68.1	423	288	135	196.1
68.1	420	286	134	194.8
68.1	417	284	133	193.4
68.1	414	282	132	192.1
68.1	411	280	131	190.8
68.1	408	278	130	189.4
68.1	407	277	130	188.5
68.1	405	276	129	188.1
68.1	404	275	129	187.2
68.1	401	273	128	185.9
68.1	398	271	127	184.5
68.1	395	269	126	183.2
68.1	392	267	125	181.9
68.1	389	265	124	180.5
68.1	386	263	123	179.2
68.1	385	262	123	178.3
68.1	383	261	122	177.9
68.1	382	260	122	177.0
68.1	379	258	121	175.6
68.1	376	256	120	174.3
68.1	373	254	119	173.0
68.1	370	252	118	171.6

%E	M1	M2	DM	M*
68.1	367	250	117	170.3
68.1	364	248	116	169.0
68.1	361	246	115	167.6
68.1	360	245	115	166.7
68.1	357	243	114	165.4
68.1	354	241	113	164.1
68.1	351	239	112	162.7
68.1	348	237	111	161.4
68.1	345	235	110	160.1
68.1	342	233	109	158.7
68.1	339	231	108	157.4
68.1	335	228	107	155.2
68.1	332	226	106	153.8
68.1	329	224	105	152.5
68.1	326	222	104	151.2
68.1	323	220	103	149.8
68.1	320	218	102	148.5
68.1	317	216	101	147.2
68.1	313	213	100	144.9
68.1	310	211	99	143.6
68.1	307	209	98	142.3
68.1	304	207	97	141.0
68.1	301	205	96	139.6
68.1	298	203	95	138.3
68.1	295	201	94	137.0
68.1	288	196	92	133.4
68.1	285	194	91	132.1
68.1	282	192	90	130.7
68.1	279	190	89	129.4
68.1	276	188	88	128.1
68.1	273	186	87	126.7
68.1	270	184	86	125.4
68.1	263	179	84	121.8
68.1	260	177	83	120.5
68.1	257	175	82	119.2
68.1	254	173	81	117.8
68.1	251	171	80	116.5
68.1	248	169	79	115.2
68.1	241	164	77	111.6
68.1	238	162	76	110.3
68.1	235	160	75	108.9
68.1	232	158	74	107.6
68.1	229	156	73	106.3
68.1	226	154	72	104.9
68.1	216	147	69	100.0
68.1	213	145	68	98.7
68.1	210	143	67	97.4
68.1	207	141	66	96.0
68.1	204	139	65	94.7
68.1	191	130	61	88.5

%E	M1	M2	DM	M*
68.1	188	128	60	87.1
68.1	185	126	59	85.8
68.1	182	124	58	84.5
68.1	166	113	53	76.9
68.1	163	111	52	75.6
68.1	160	109	51	74.3
68.1	144	98	46	66.7
68.1	141	96	45	65.4
68.1	138	94	44	64.0
68.1	135	92	43	62.7
68.1	119	81	38	55.1
68.1	116	79	37	53.8
68.1	113	77	36	52.5
68.1	94	64	30	43.6
68.1	91	62	29	42.2
68.1	72	49	23	33.3
68.1	69	47	22	32.0
68.1	47	32	15	21.8
68.0	500	340	160	231.2
68.0	497	338	159	229.9
68.0	494	336	158	228.5
68.0	493	335	158	227.6
68.0	491	334	157	227.2
68.0	490	333	157	226.3
68.0	488	332	156	225.9
68.0	487	331	156	225.0
68.0	485	330	155	224.5
68.0	484	329	155	223.6
68.0	481	327	154	222.3
68.0	478	325	153	221.0
68.0	475	323	152	219.6
68.0	472	321	151	218.3
68.0	469	319	150	217.0
68.0	466	317	149	215.6
68.0	465	316	149	214.7
68.0	463	315	148	214.3
68.0	462	314	148	213.4
68.0	460	313	147	213.0
68.0	459	312	147	212.1
68.0	456	310	146	210.7
68.0	453	308	145	209.4
68.0	450	306	144	208.1
68.0	447	304	143	206.7
68.0	444	302	142	205.4
68.0	441	300	141	204.1
68.0	440	299	141	203.2
68.0	438	298	140	202.7
68.0	437	297	140	201.9
68.0	435	296	139	201.4
68.0	434	295	139	200.5

%E	M1	M2	DM	M*
68.0	431	293	138	199.2
68.0	428	291	137	197.9
68.0	425	289	136	196.5
68.0	422	287	135	195.2
68.0	419	285	134	193.9
68.0	416	283	133	192.5
68.0	415	282	133	191.6
68.0	413	281	132	191.2
68.0	412	280	132	190.3
68.0	410	279	131	189.9
68.0	409	278	131	189.0
68.0	406	276	130	187.6
68.0	403	274	129	186.3
68.0	400	272	128	185.0
68.0	397	270	127	183.6
68.0	394	268	126	182.3
68.0	391	266	125	181.0
68.0	388	264	124	179.6
68.0	387	263	124	178.7
68.0	384	261	123	177.4
68.0	381	259	122	176.1
68.0	378	257	121	174.7
68.0	375	255	120	173.4
68.0	372	253	119	172.1
68.0	369	251	118	170.7
68.0	366	249	117	169.4
68.0	363	247	116	168.1
68.0	362	246	116	167.2
68.0	359	244	115	165.8
68.0	356	242	114	164.5
68.0	353	240	113	163.2
68.0	350	238	112	161.8
68.0	347	236	111	160.5
68.0	344	234	110	159.2
68.0	341	232	109	157.8
68.0	338	230	108	156.5
68.0	337	229	108	155.6
68.0	334	227	107	154.3
68.0	331	225	106	152.9
68.0	328	223	105	151.6
68.0	325	221	104	150.3
68.0	322	219	103	148.9
68.0	319	217	102	147.6
68.0	316	215	101	146.3
68.0	309	210	99	142.7
68.0	306	208	98	141.4
68.0	303	206	97	140.1
68.0	300	204	96	138.7
68.0	297	202	95	137.4
68.0	294	200	94	136.1

%E	M1	M2	DM	M*
68.0	291	198	93	134.7
68.0	284	193	91	131.2
68.0	281	191	90	129.8
68.0	278	189	89	128.5
68.0	275	187	88	127.2
68.0	272	185	87	125.8
68.0	269	183	86	124.5
68.0	266	181	85	123.2
68.0	259	176	83	119.6
68.0	256	174	82	118.3
68.0	253	172	81	116.9
68.0	250	170	80	115.6
68.0	247	168	79	114.3
68.0	244	166	78	112.9
68.0	231	157	74	106.7
68.0	228	155	73	105.4
68.0	225	153	72	104.0
68.0	222	151	71	102.7
68.0	219	149	70	101.4
68.0	206	140	66	95.1
68.0	203	138	65	93.8
68.0	200	136	64	92.5
68.0	197	134	63	91.1
68.0	194	132	62	89.8
68.0	181	123	58	83.6
68.0	178	121	57	82.3
68.0	175	119	56	80.9
68.0	172	117	55	79.6
68.0	169	115	54	78.3
68.0	153	104	49	70.7
68.0	150	102	48	69.4
68.0	147	100	47	68.0
68.0	128	87	41	59.1
68.0	125	85	40	57.8
68.0	122	83	39	56.5
68.0	103	70	33	47.6
68.0	100	68	32	46.2
68.0	97	66	31	44.9
68.0	75	51	24	34.7
68.0	50	34	16	23.1
68.0	25	17	8	11.6
67.9	499	339	160	230.3
67.9	498	338	160	229.4
67.9	496	337	159	229.0
67.9	495	336	159	228.1
67.9	492	334	158	226.7
67.9	489	332	157	225.4
67.9	486	330	156	224.1
67.9	483	328	155	222.7
67.9	480	326	154	221.4

%E	M1	M2	DM	M*
67.9	479	325	154	220.5
67.9	477	324	153	220.1
67.9	476	323	153	219.2
67.9	474	322	152	218.7
67.9	473	321	152	217.8
67.9	471	320	151	217.4
67.9	470	319	151	216.5
67.9	468	318	150	216.1
67.9	467	317	150	215.2
67.9	464	315	149	213.8
67.9	461	313	148	212.5
67.9	458	311	147	211.2
67.9	455	309	146	209.8
67.9	452	307	145	208.5
67.9	449	305	144	207.2
67.9	448	304	144	206.3
67.9	446	303	143	205.8
67.9	445	302	143	205.0
67.9	443	301	142	204.5
67.9	442	300	142	203.6
67.9	439	298	141	202.3
67.9	436	296	140	201.0
67.9	433	294	139	199.6
67.9	430	292	138	198.3
67.9	427	290	137	197.0
67.9	424	288	136	195.6
67.9	421	286	135	194.3
67.9	420	285	135	193.4
67.9	418	284	134	193.0
67.9	417	283	134	192.1
67.9	414	281	133	190.7
67.9	411	279	132	189.4
67.9	408	277	131	188.1
67.9	405	275	130	186.7
67.9	402	273	129	185.4
67.9	399	271	128	184.1
67.9	396	269	127	182.7
67.9	393	267	126	181.4
67.9	392	266	126	180.5
67.9	390	265	125	180.1
67.9	389	264	125	179.2
67.9	386	262	124	177.8
67.9	383	260	123	176.5
67.9	380	258	122	175.2
67.9	377	256	121	173.8
67.9	374	254	120	172.5
67.9	371	252	119	171.2
67.9	368	250	118	169.8
67.9	365	248	117	168.5
67.9	364	247	117	167.6

%E	M1	M2	DM	M*
67.9	361	245	116	166.3
67.9	358	243	115	164.9
67.9	355	241	114	163.6
67.9	352	239	113	162.3
67.9	349	237	112	160.9
67.9	346	235	111	159.6
67.9	343	233	110	158.3
67.9	340	231	109	156.9
67.9	336	228	108	154.7
67.9	333	226	107	153.4
67.9	330	224	106	152.0
67.9	327	222	105	150.7
67.9	324	220	104	149.4
67.9	321	218	103	148.0
67.9	318	216	102	146.7
67.9	315	214	101	145.4
67.9	312	212	100	144.1
67.9	308	209	99	141.8
67.9	305	207	98	140.5
67.9	302	205	97	139.2
67.9	299	203	96	137.8
67.9	296	201	95	136.5
67.9	293	199	94	135.2
67.9	290	197	93	133.8
67.9	287	195	92	132.5
67.9	280	190	90	128.9
67.9	277	188	89	127.6
67.9	274	186	88	126.3
67.9	271	184	87	124.9
67.9	268	182	86	123.6
67.9	265	180	85	122.3
67.9	262	178	84	120.9
67.9	252	171	81	116.0
67.9	249	169	80	114.7
67.9	246	167	79	113.4
67.9	243	165	78	112.0
67.9	240	163	77	110.7
67.9	237	161	76	109.4
67.9	234	159	75	108.0
67.9	224	152	72	103.1
67.9	221	150	71	101.8
67.9	218	148	70	100.5
67.9	215	146	69	99.1
67.9	212	144	68	97.8
67.9	209	142	67	96.5
67.9	196	133	63	90.3
67.9	193	131	62	88.9
67.9	190	129	61	87.6
67.9	187	127	60	86.3
67.9	184	125	59	84.9

%E	M1	M2	DM	M*
67.9	168	114	54	77.4
67.9	165	112	53	76.0
67.9	162	110	52	74.7
67.9	159	108	51	73.4
67.9	156	106	50	72.0
67.9	140	95	45	64.5
67.9	137	93	44	63.1
67.9	134	91	43	61.8
67.9	131	89	42	60.5
67.9	112	76	36	51.6
67.9	109	74	35	50.2
67.9	106	72	34	48.9
67.9	84	57	27	38.7
67.9	81	55	26	37.3
67.9	78	53	25	36.0
67.9	56	38	18	25.8
67.9	53	36	17	24.5
67.9	28	19	9	12.9
67.8	500	339	161	229.8
67.8	497	337	160	228.5
67.8	494	335	159	227.2
67.8	491	333	158	225.8
67.8	490	332	158	224.9
67.8	488	331	157	224.5
67.8	487	330	157	223.6
67.8	485	329	156	223.2
67.8	484	328	156	222.3
67.8	482	327	155	221.8
67.8	481	326	155	220.9
67.8	478	324	154	219.6
67.8	475	322	153	218.3
67.8	472	320	152	216.9
67.8	469	318	151	215.6
67.8	466	316	150	214.3
67.8	463	314	149	213.0
67.8	460	312	148	211.6
67.8	459	311	148	210.7
67.8	457	310	147	210.3
67.8	456	309	147	209.4
67.8	454	308	146	209.0
67.8	453	307	146	208.1
67.8	451	306	145	207.6
67.8	450	305	145	206.7
67.8	447	303	144	205.4
67.8	444	301	143	204.1
67.8	441	299	142	202.7
67.8	438	297	141	201.4
67.8	435	295	140	200.1
67.8	432	293	139	198.7
67.8	429	291	138	197.4

%E	M1	M2	DM	M*
67.8	428	290	138	196.5
67.8	426	289	137	196.1
67.8	425	288	137	195.2
67.8	423	287	136	194.7
67.8	422	286	136	193.8
67.8	419	284	135	192.5
67.8	416	282	134	191.2
67.8	413	280	133	189.8
67.8	410	278	132	188.5
67.8	407	276	131	187.2
67.8	404	274	130	185.8
67.8	401	272	129	184.5
67.8	400	271	129	183.6
67.8	398	270	128	183.2
67.8	397	269	128	182.3
67.8	395	268	127	181.8
67.8	394	267	127	180.9
67.8	391	265	126	179.6
67.8	388	263	125	178.3
67.8	385	261	124	176.9
67.8	382	259	123	175.6
67.8	379	257	122	174.3
67.8	376	255	121	172.9
67.8	373	253	120	171.6
67.8	370	251	119	170.3
67.8	369	250	119	169.4
67.8	367	249	118	168.9
67.8	366	248	118	168.0
67.8	363	246	117	166.7
67.8	360	244	116	165.4
67.8	357	242	115	164.0
67.8	354	240	114	162.7
67.8	351	238	113	161.4
67.8	348	236	112	160.0
67.8	345	234	111	158.7
67.8	342	232	110	157.4
67.8	339	230	109	156.0
67.8	338	229	109	155.2
67.8	335	227	108	153.8
67.8	332	225	107	152.5
67.8	329	223	106	151.2
67.8	326	221	105	149.8
67.8	323	219	104	148.5
67.8	320	217	103	147.2
67.8	317	215	102	145.8
67.8	314	213	101	144.5
67.8	311	211	100	143.2
67.8	307	208	99	140.9
67.8	304	206	98	139.6
67.8	301	204	97	138.3

%E	M1	M2	DM	M*
67.8	298	202	96	136.9
67.8	295	200	95	135.6
67.8	292	198	94	134.3
67.8	289	196	93	132.9
67.8	286	194	92	131.6
67.8	283	192	91	130.3
67.8	276	187	89	126.7
67.8	273	185	88	125.4
67.8	270	183	87	124.0
67.8	267	181	86	122.7
67.8	264	179	85	121.4
67.8	261	177	84	120.0
67.8	258	175	83	118.7
67.8	255	173	82	117.4
67.8	245	166	79	112.5
67.8	242	164	78	111.1
67.8	239	162	77	109.8
67.8	236	160	76	108.5
67.8	233	158	75	107.1
67.8	230	156	74	105.8
67.8	227	154	73	104.5
67.8	214	145	69	98.2
67.8	211	143	68	96.9
67.8	208	141	67	95.6
67.8	205	139	66	94.2
67.8	202	137	65	92.9
67.8	199	135	64	91.6
67.8	183	124	59	84.0
67.8	180	122	58	82.7
67.8	177	120	57	81.4
67.8	174	118	56	80.0
67.8	171	116	55	78.7
67.8	152	103	49	69.8
67.8	149	101	48	68.5
67.8	146	99	47	67.1
67.8	143	97	46	65.8
67.8	121	82	39	55.6
67.8	118	80	38	54.2
67.8	115	78	37	52.9
67.8	90	61	29	41.3
67.8	87	59	28	40.0
67.8	59	40	19	27.1
67.7	499	338	161	228.9
67.7	498	337	161	228.1
67.7	496	336	160	227.6
67.7	495	335	160	226.7
67.7	493	334	159	226.3
67.7	492	333	159	225.4
67.7	489	331	158	224.1
67.7	486	329	157	222.7

%E	M1	M2	DM	M*
67.7	483	327	156	221.4
67.7	480	325	155	220.1
67.7	477	323	154	218.7
67.7	474	321	153	217.4
67.7	473	320	153	216.5
67.7	471	319	152	216.1
67.7	470	318	152	215.2
67.7	468	317	151	214.7
67.7	467	316	151	213.8
67.7	465	315	150	213.4
67.7	464	314	150	212.5
67.7	462	313	149	212.1
67.7	461	312	149	211.2
67.7	458	310	148	209.8
67.7	455	308	147	208.5
67.7	452	306	146	207.2
67.7	449	304	145	205.8
67.7	446	302	144	204.5
67.7	443	300	143	203.2
67.7	440	298	142	201.8
67.7	439	297	142	200.9
67.7	437	296	141	200.5
67.7	436	295	141	199.6
67.7	434	294	140	199.2
67.7	433	293	140	198.3
67.7	431	292	139	197.8
67.7	430	291	139	196.9
67.7	427	289	138	195.6
67.7	424	287	137	194.3
67.7	421	285	136	192.9
67.7	418	283	135	191.6
67.7	415	281	134	190.3
67.7	412	279	133	188.9
67.7	409	277	132	187.6
67.7	406	275	131	186.3
67.7	405	274	131	185.4
67.7	403	273	130	184.9
67.7	402	272	130	184.0
67.7	399	270	129	182.7
67.7	396	268	128	181.4
67.7	393	266	127	180.0
67.7	390	264	126	178.7
67.7	387	262	125	177.4
67.7	384	260	124	176.0
67.7	381	258	123	174.7
67.7	378	256	122	173.4
67.7	375	254	121	172.0
67.7	372	252	120	170.7
67.7	371	251	120	169.8
67.7	368	249	119	168.5

%E	M1	M2	DM	M*
67.7	365	247	118	167.1
67.7	362	245	117	165.8
67.7	359	243	116	164.5
67.7	356	241	115	163.1
67.7	353	239	114	161.8
67.7	350	237	113	160.5
67.7	347	235	112	159.1
67.7	344	233	111	157.8
67.7	341	231	110	156.5
67.7	337	228	109	154.3
67.7	334	226	108	152.9
67.7	331	224	107	151.6
67.7	328	222	106	150.3
67.7	325	220	105	148.9
67.7	322	218	104	147.6
67.7	319	216	103	146.3
67.7	316	214	102	144.9
67.7	313	212	101	143.6
67.7	310	210	100	142.3
67.7	303	205	98	138.7
67.7	300	203	97	137.4
67.7	297	201	96	136.0
67.7	294	199	95	134.7
67.7	291	197	94	133.4
67.7	288	195	93	132.0
67.7	285	193	92	130.7
67.7	282	191	91	129.4
67.7	279	189	90	128.0
67.7	269	182	87	123.1
67.7	266	180	86	121.8
67.7	263	178	85	120.5
67.7	260	176	84	119.1
67.7	257	174	83	117.8
67.7	254	172	82	116.5
67.7	251	170	81	115.1
67.7	248	168	80	113.8
67.7	235	159	76	107.6
67.7	232	157	75	106.2
67.7	229	155	74	104.9
67.7	226	153	73	103.6
67.7	223	151	72	102.2
67.7	220	149	71	100.9
67.7	217	147	70	99.6
67.7	201	136	65	92.0
67.7	198	134	64	90.7
67.7	195	132	63	89.4
67.7	192	130	62	88.0
67.7	189	128	61	86.7
67.7	186	126	60	85.4
67.7	167	113	54	76.5

%E	M1	M2	DM	M*
67.7	164	111	53	75.1
67.7	161	109	52	73.8
67.7	158	107	51	72.5
67.7	155	105	50	71.1
67.7	133	90	43	60.9
67.7	130	88	42	59.6
67.7	127	86	41	58.2
67.7	124	84	40	56.9
67.7	99	67	32	45.3
67.7	96	65	31	44.0
67.7	93	63	30	42.7
67.7	65	44	21	29.8
67.7	62	42	20	28.5
67.7	31	21	10	14.2
67.6	500	338	162	228.5
67.6	497	336	161	227.2
67.6	494	334	160	225.8
67.6	491	332	159	224.5
67.6	490	331	159	223.6
67.6	488	330	158	223.2
67.6	487	329	158	222.3
67.6	485	328	157	221.8
67.6	484	327	157	220.9
67.6	482	326	156	220.5
67.6	481	325	156	219.6
67.6	479	324	155	219.2
67.6	478	323	155	218.3
67.6	476	322	154	217.8
67.6	475	321	154	216.9
67.6	472	319	153	215.6
67.6	469	317	152	214.3
67.6	466	315	151	212.9
67.6	463	313	150	211.6
67.6	460	311	149	210.3
67.6	457	309	148	208.9
67.6	454	307	147	207.6
67.6	453	306	147	206.7
67.6	451	305	146	206.3
67.6	450	304	146	205.4
67.6	448	303	145	204.9
67.6	447	302	145	204.0
67.6	445	301	144	203.6
67.6	444	300	144	202.7
67.6	442	299	143	202.3
67.6	441	298	143	201.4
67.6	438	296	142	200.0
67.6	435	294	141	198.7
67.6	432	292	140	197.4
67.6	429	290	139	196.0
67.6	426	288	138	194.7
67.6	423	286	137	193.4
67.6	420	284	136	192.0
67.6	417	282	135	190.7
67.6	414	280	134	189.4
67.6	413	279	134	188.5
67.6	411	278	133	188.0
67.6	410	277	133	187.1
67.6	408	276	132	186.7
67.6	407	275	132	185.8
67.6	404	273	131	184.5
67.6	401	271	130	183.1
67.6	398	269	129	181.8
67.6	395	267	128	180.5
67.6	392	265	127	179.1
67.6	389	263	126	177.8
67.6	386	261	125	176.5
67.6	383	259	124	175.1
67.6	380	257	123	173.8
67.6	377	255	122	172.5
67.6	376	254	122	171.6
67.6	374	253	121	171.1
67.6	373	252	121	170.3
67.6	370	250	120	168.9
67.6	367	248	119	167.6
67.6	364	246	118	166.3
67.6	361	244	117	164.9
67.6	358	242	116	163.6
67.6	355	240	115	162.3
67.6	352	238	114	160.9
67.6	349	236	113	159.6
67.6	346	234	112	158.3
67.6	343	232	111	156.9
67.6	340	230	110	155.6
67.6	339	229	110	154.7
67.6	336	227	109	153.4
67.6	333	225	108	152.0
67.6	330	223	107	150.7
67.6	327	221	106	149.4
67.6	324	219	105	148.0
67.6	321	217	104	146.7
67.6	318	215	103	145.4
67.6	315	213	102	144.0
67.6	312	211	101	142.7
67.6	309	209	100	141.4
67.6	306	207	99	140.0
67.6	302	204	98	137.8
67.6	299	202	97	136.5
67.6	296	200	96	135.1
67.6	293	198	95	133.8
67.6	290	196	94	132.5
67.6	287	194	93	131.1
67.6	284	192	92	129.8
67.6	281	190	91	128.5
67.6	278	188	90	127.1
67.6	275	186	89	125.8
67.6	272	184	88	124.5
67.6	262	177	85	119.6
67.6	259	175	84	118.2
67.6	256	173	83	116.9
67.6	253	171	82	115.6
67.6	250	169	81	114.2
67.6	247	167	80	112.9
67.6	244	165	79	111.6
67.6	241	163	78	110.2
67.6	238	161	77	108.9
67.6	225	152	73	102.7
67.6	222	150	72	101.4
67.6	219	148	71	100.0
67.6	216	146	70	98.7
67.6	213	144	69	97.4
67.6	210	142	68	96.0
67.6	207	140	67	94.7
67.6	204	138	66	93.4
67.6	188	127	61	85.8
67.6	185	125	60	84.5
67.6	182	123	59	83.1
67.6	179	121	58	81.8
67.6	176	119	57	80.5
67.6	173	117	56	79.1
67.6	170	115	55	77.8
67.6	151	102	49	68.9
67.6	148	100	48	67.6
67.6	145	98	47	66.2
67.6	142	96	46	64.9
67.6	139	94	45	63.6
67.6	136	92	44	62.2
67.6	111	75	36	50.7
67.6	108	73	35	49.3
67.6	105	71	34	48.0
67.6	102	69	33	46.7
67.6	74	50	24	33.8
67.6	71	48	23	32.5
67.6	68	46	22	31.1
67.6	37	25	12	16.9
67.6	34	23	11	15.6
67.5	499	337	162	227.6
67.5	498	336	162	226.7
67.5	496	335	161	226.3
67.5	495	334	161	225.4
67.5	493	333	160	224.9
67.5	492	332	160	224.0
67.5	489	330	159	222.7
67.5	486	328	158	221.4
67.5	483	326	157	220.0
67.5	480	324	156	218.7
67.5	477	322	155	217.4
67.5	474	320	154	216.0
67.5	471	318	153	214.7
67.5	468	316	152	213.4
67.5	467	315	152	212.5
67.5	465	314	151	212.0
67.5	464	313	151	211.1
67.5	462	312	150	210.7
67.5	461	311	150	209.8
67.5	459	310	149	209.4
67.5	458	309	149	208.5
67.5	456	308	148	208.0
67.5	455	307	148	207.1
67.5	452	305	147	205.8
67.5	449	303	146	204.5
67.5	446	301	145	203.1
67.5	443	299	144	201.8
67.5	440	297	143	200.5
67.5	437	295	142	199.1
67.5	434	293	141	197.8
67.5	431	291	140	196.5
67.5	428	289	139	195.1
67.5	425	287	138	193.8
67.5	424	286	138	192.9
67.5	422	285	137	192.5
67.5	421	284	137	191.6
67.5	419	283	136	191.1
67.5	418	282	136	190.2
67.5	416	281	135	189.8
67.5	415	280	135	188.9
67.5	412	278	134	187.6
67.5	409	276	133	186.2
67.5	406	274	132	184.9
67.5	403	272	131	183.6
67.5	400	270	130	182.3
67.5	397	268	129	180.9
67.5	394	266	128	179.6
67.5	391	264	127	178.3
67.5	388	262	126	176.9
67.5	385	260	125	175.6
67.5	382	258	124	174.3
67.5	381	257	124	173.4
67.5	379	256	123	172.9
67.5	378	255	123	172.0
67.5	375	253	122	170.7
67.5	372	251	121	169.4
67.5	369	249	120	168.0
67.5	366	247	119	166.7
67.5	363	245	118	165.4
67.5	360	243	117	164.0
67.5	357	241	116	162.7
67.5	354	239	115	161.4
67.5	351	237	114	160.0
67.5	348	235	113	158.7
67.5	345	233	112	157.4
67.5	342	231	111	156.0
67.5	338	228	110	153.8
67.5	335	226	109	152.5
67.5	332	224	108	151.1
67.5	329	222	107	149.8
67.5	326	220	106	148.5
67.5	323	218	105	147.1
67.5	320	216	104	145.8
67.5	317	214	103	144.5
67.5	314	212	102	143.1
67.5	311	210	101	141.8
67.5	308	208	100	140.5
67.5	305	206	99	139.1
67.5	298	201	97	135.6
67.5	295	199	96	134.2
67.5	292	197	95	132.9
67.5	289	195	94	131.6
67.5	286	193	93	130.2
67.5	283	191	92	128.9
67.5	280	189	91	127.6
67.5	277	187	90	126.2
67.5	274	185	89	124.9
67.5	271	183	88	123.6
67.5	268	181	87	122.2
67.5	265	179	86	120.9
67.5	255	172	83	116.0
67.5	252	170	82	114.7
67.5	249	168	81	113.3
67.5	246	166	80	112.0
67.5	243	164	79	110.7
67.5	240	162	78	109.3
67.5	237	160	77	108.0
67.5	234	158	76	106.7
67.5	231	156	75	105.4
67.5	228	154	74	104.0
67.5	212	143	69	96.5
67.5	209	141	68	95.1
67.5	206	139	67	93.8
67.5	203	137	66	92.5
67.5	200	135	65	91.1

%E	M1	M2	DM	M*
67.5	197	133	64	89.8
67.5	194	131	63	88.5
67.5	191	129	62	87.1
67.5	169	114	55	76.9
67.5	166	112	54	75.6
67.5	163	110	53	74.2
67.5	160	108	52	72.9
67.5	157	106	51	71.6
67.5	154	104	50	70.2
67.5	126	85	41	57.3
67.5	123	83	40	56.0
67.5	120	81	39	54.7
67.5	117	79	38	53.3
67.5	114	77	37	52.0
67.5	83	56	27	37.8
67.5	80	54	26	36.4
67.5	77	52	25	35.1
67.5	40	27	13	18.2
67.4	500	337	163	227.1
67.4	497	335	162	225.8
67.4	494	333	161	224.5
67.4	491	331	160	223.1
67.4	488	329	159	221.8
67.4	487	328	159	220.9
67.4	485	327	158	220.5
67.4	484	326	158	219.6
67.4	482	325	157	219.1
67.4	481	324	157	218.2
67.4	479	323	156	217.8
67.4	478	322	156	216.9
67.4	476	321	155	216.5
67.4	475	320	155	215.6
67.4	473	319	154	215.1
67.4	472	318	154	214.2
67.4	470	317	153	213.8
67.4	469	316	153	212.9
67.4	466	314	152	211.6
67.4	463	312	151	210.2
67.4	460	310	150	208.9
67.4	457	308	149	207.6
67.4	454	306	148	206.2
67.4	451	304	147	204.9
67.4	448	302	146	203.6
67.4	445	300	145	202.2
67.4	442	298	144	200.9
67.4	439	296	143	199.6
67.4	438	295	143	198.7
67.4	436	294	142	198.2
67.4	435	293	142	197.4
67.4	433	292	141	196.9

%E	M1	M2	DM	M*
67.4	432	291	141	196.0
67.4	430	290	140	195.6
67.4	429	289	140	194.7
67.4	427	288	139	194.2
67.4	426	287	139	193.4
67.4	423	285	138	192.0
67.4	420	283	137	190.7
67.4	417	281	136	189.4
67.4	414	279	135	188.0
67.4	411	277	134	186.7
67.4	408	275	133	185.4
67.4	405	273	132	184.0
67.4	402	271	131	182.7
67.4	399	269	130	181.4
67.4	396	267	129	180.0
67.4	393	265	128	178.7
67.4	390	263	127	177.4
67.4	389	262	127	176.5
67.4	387	261	126	176.0
67.4	386	260	126	175.1
67.4	384	259	125	174.7
67.4	383	258	125	173.8
67.4	380	256	124	172.5
67.4	377	254	123	171.1
67.4	374	252	122	169.8
67.4	371	250	121	168.5
67.4	368	248	120	167.1
67.4	365	246	119	165.8
67.4	362	244	118	164.5
67.4	359	242	117	163.1
67.4	356	240	116	161.8
67.4	353	238	115	160.5
67.4	350	236	114	159.1
67.4	347	234	113	157.8
67.4	344	232	112	156.5
67.4	341	230	111	155.1
67.4	340	229	111	154.2
67.4	337	227	110	152.9
67.4	334	225	109	151.6
67.4	331	223	108	150.2
67.4	328	221	107	148.9
67.4	325	219	106	147.6
67.4	322	217	105	146.2
67.4	319	215	104	144.9
67.4	316	213	103	143.6
67.4	313	211	102	142.2
67.4	310	209	101	140.9
67.4	307	207	100	139.6
67.4	304	205	99	138.2
67.4	301	203	98	136.9

%E	M1	M2	DM	M*
67.4	291	196	95	132.0
67.4	288	194	94	130.7
67.4	285	192	93	129.3
67.4	282	190	92	128.0
67.4	279	188	91	126.7
67.4	276	186	90	125.3
67.4	273	184	89	124.0
67.4	270	182	88	122.7
67.4	267	180	87	121.3
67.4	264	178	86	120.0
67.4	261	176	85	118.7
67.4	258	174	84	117.3
67.4	242	163	79	109.8
67.4	239	161	78	108.5
67.4	236	159	77	107.1
67.4	233	157	76	105.8
67.4	230	155	75	104.5
67.4	227	153	74	103.1
67.4	224	151	73	101.8
67.4	221	149	72	100.5
67.4	218	147	71	99.1
67.4	215	145	70	97.8
67.4	193	130	63	87.6
67.4	190	128	62	86.2
67.4	187	126	61	84.9
67.4	184	124	60	83.6
67.4	181	122	59	82.2
67.4	178	120	58	80.9
67.4	175	118	57	79.6
67.4	172	116	56	78.2
67.4	144	97	47	65.3
67.4	141	95	46	64.0
67.4	138	93	45	62.7
67.4	135	91	44	61.3
67.4	132	89	43	60.0
67.4	129	87	42	58.7
67.4	95	64	31	43.1
67.4	92	62	30	41.8
67.4	89	60	29	40.4
67.4	86	58	28	39.1
67.4	46	31	15	20.9
67.4	43	29	14	19.6
67.3	499	336	163	226.2
67.3	498	335	163	225.4
67.3	496	334	162	224.9
67.3	495	333	162	224.0
67.3	493	332	161	223.6
67.3	492	331	161	222.7
67.3	490	330	160	222.2
67.3	489	329	160	221.4

%E	M1	M2	DM	M*
67.3	486	327	159	220.0
67.3	483	325	158	218.7
67.3	480	323	157	217.4
67.3	477	321	156	216.0
67.3	474	319	155	214.7
67.3	471	317	154	213.4
67.3	468	315	153	212.0
67.3	465	313	152	210.7
67.3	462	311	151	209.4
67.3	459	309	150	208.0
67.3	456	307	149	206.7
67.3	455	306	149	205.8
67.3	453	305	148	205.4
67.3	452	304	148	204.5
67.3	450	303	147	204.0
67.3	449	302	147	203.1
67.3	447	301	146	202.7
67.3	446	300	146	201.8
67.3	444	299	145	201.4
67.3	443	298	145	200.5
67.3	441	297	144	200.0
67.3	440	296	144	199.1
67.3	437	294	143	197.8
67.3	434	292	142	196.5
67.3	431	290	141	195.1
67.3	428	288	140	193.8
67.3	425	286	139	192.5
67.3	422	284	138	191.1
67.3	419	282	137	189.8
67.3	416	280	136	188.5
67.3	413	278	135	187.1
67.3	410	276	134	185.8
67.3	407	274	133	184.5
67.3	404	272	132	183.1
67.3	401	270	131	181.8
67.3	400	269	131	180.9
67.3	398	268	130	180.5
67.3	397	267	130	179.6
67.3	395	266	129	179.1
67.3	394	265	129	178.2
67.3	392	264	128	177.8
67.3	391	263	128	176.9
67.3	388	261	127	175.6
67.3	385	259	126	174.2
67.3	382	257	125	172.9
67.3	379	255	124	171.6
67.3	376	253	123	170.2
67.3	373	251	122	168.9
67.3	370	249	121	167.6
67.3	367	247	120	166.2

%E	M1	M2	DM	M*
67.3	364	245	119	164.9
67.3	361	243	118	163.6
67.3	358	241	117	162.2
67.3	355	239	116	160.9
67.3	352	237	115	159.6
67.3	349	235	114	158.2
67.3	346	233	113	156.9
67.3	343	231	112	155.6
67.3	342	230	112	154.7
67.3	339	228	111	153.3
67.3	336	226	110	152.0
67.3	333	224	109	150.7
67.3	330	222	108	149.3
67.3	327	220	107	148.0
67.3	324	218	106	146.7
67.3	321	216	105	145.3
67.3	318	214	104	144.0
67.3	315	212	103	142.7
67.3	312	210	102	141.3
67.3	309	208	101	140.0
67.3	306	206	100	138.7
67.3	303	204	99	137.3
67.3	300	202	98	136.0
67.3	297	200	97	134.7
67.3	294	198	96	133.3
67.3	284	191	93	128.5
67.3	281	189	92	127.1
67.3	278	187	91	125.8
67.3	275	185	90	124.5
67.3	272	183	89	123.1
67.3	269	181	88	121.8
67.3	266	179	87	120.5
67.3	263	177	86	119.1
67.3	260	175	85	117.8
67.3	257	173	84	116.5
67.3	254	171	83	115.1
67.3	251	169	82	113.8
67.3	248	167	81	112.5
67.3	245	165	80	111.1
67.3	226	152	74	102.2
67.3	223	150	73	100.9
67.3	220	148	72	99.6
67.3	217	146	71	98.2
67.3	214	144	70	96.9
67.3	211	142	69	95.6
67.3	208	140	68	94.2
67.3	205	138	67	92.9
67.3	202	136	66	91.6
67.3	199	134	65	90.2
67.3	196	132	64	88.9

%E	M1	M2	DM	M*
67.3	171	115	56	77.3
67.3	168	113	55	76.0
67.3	165	111	54	74.7
67.3	162	109	53	73.3
67.3	159	107	52	72.0
67.3	156	105	51	70.7
67.3	153	103	50	69.3
67.3	150	101	49	68.0
67.3	147	99	48	66.7
67.3	113	76	37	51.1
67.3	110	74	36	49.8
67.3	107	72	35	48.4
67.3	104	70	34	47.1
67.3	101	68	33	45.8
67.3	98	66	32	44.4
67.3	55	37	18	24.9
67.3	52	35	17	23.6
67.3	49	33	16	22.2
67.2	500	336	164	225.8
67.2	497	334	163	224.5
67.2	494	332	162	223.1
67.2	491	330	161	221.8
67.2	488	328	160	220.5
67.2	485	326	159	219.1
67.2	482	324	158	217.8
67.2	481	323	158	216.9
67.2	479	322	157	216.5
67.2	478	321	157	215.6
67.2	476	320	156	215.1
67.2	475	319	156	214.2
67.2	473	318	155	213.8
67.2	472	317	155	212.9
67.2	470	316	154	212.5
67.2	469	315	154	211.6
67.2	467	314	153	211.1
67.2	466	313	153	210.2
67.2	464	312	152	209.8
67.2	463	311	152	208.9
67.2	461	310	151	208.5
67.2	460	309	151	207.6
67.2	458	308	150	207.1
67.2	457	307	150	206.2
67.2	454	305	149	204.9
67.2	451	303	148	203.6
67.2	448	301	147	202.2
67.2	445	299	146	200.9
67.2	442	297	145	199.6
67.2	439	295	144	198.2
67.2	436	293	143	196.9
67.2	433	291	142	195.6

%E	M1	M2	DM	M*
67.2	430	289	141	194.2
67.2	427	287	140	192.9
67.2	424	285	139	191.6
67.2	421	283	138	190.2
67.2	418	281	137	188.9
67.2	415	279	136	187.6
67.2	414	278	136	186.7
67.2	412	277	135	186.2
67.2	411	276	135	185.3
67.2	409	275	134	184.9
67.2	408	274	134	184.0
67.2	406	273	133	183.6
67.2	405	272	133	182.7
67.2	403	271	132	182.2
67.2	402	270	132	181.3
67.2	399	268	131	180.0
67.2	396	266	130	178.7
67.2	393	264	129	177.3
67.2	390	262	128	176.0
67.2	387	260	127	174.7
67.2	384	258	126	173.3
67.2	381	256	125	172.0
67.2	378	254	124	170.7
67.2	375	252	123	169.3
67.2	372	250	122	168.0
67.2	369	248	121	166.7
67.2	366	246	120	165.3
67.2	363	244	119	164.0
67.2	360	242	118	162.7
67.2	357	240	117	161.3
67.2	354	238	116	160.0
67.2	351	236	115	158.7
67.2	348	234	114	157.3
67.2	345	232	113	156.0
67.2	344	231	113	155.1
67.2	341	229	112	153.8
67.2	338	227	111	152.5
67.2	335	225	110	151.1
67.2	332	223	109	149.8
67.2	329	221	108	148.5
67.2	326	219	107	147.1
67.2	323	217	106	145.8
67.2	320	215	105	144.5
67.2	317	213	104	143.1
67.2	314	211	103	141.8
67.2	311	209	102	140.5
67.2	308	207	101	139.1
67.2	305	205	100	137.8
67.2	302	203	99	136.5
67.2	299	201	98	135.1

%E	M1	M2	DM	M*
67.2	296	199	97	133.8
67.2	293	197	96	132.5
67.2	290	195	95	131.1
67.2	287	193	94	129.8
67.2	274	184	90	123.6
67.2	271	182	89	122.2
67.2	268	180	88	120.9
67.2	265	178	87	119.6
67.2	262	176	86	118.2
67.2	259	174	85	116.9
67.2	256	172	84	115.6
67.2	253	170	83	114.2
67.2	250	168	82	112.9
67.2	247	166	81	111.6
67.2	244	164	80	110.2
67.2	241	162	79	108.9
67.2	238	160	78	107.6
67.2	235	158	77	106.2
67.2	232	156	76	104.9
67.2	229	154	75	103.6
67.2	207	139	68	93.3
67.2	204	137	67	92.0
67.2	201	135	66	90.7
67.2	198	133	65	89.3
67.2	195	131	64	88.0
67.2	192	129	63	86.7
67.2	189	127	62	85.3
67.2	186	125	61	84.0
67.2	183	123	60	82.7
67.2	180	121	59	81.3
67.2	177	119	58	80.0
67.2	174	117	57	78.7
67.2	137	92	45	61.8
67.2	134	90	44	60.4
67.2	131	88	43	59.1
67.2	128	86	42	57.8
67.2	125	84	41	56.4
67.2	122	82	40	55.1
67.2	119	80	39	53.8
67.2	116	78	38	52.4
67.2	67	45	22	30.2
67.2	64	43	21	28.9
67.2	61	41	20	27.6
67.2	58	39	19	26.2
67.1	499	335	164	224.9
67.1	498	334	164	224.0
67.1	496	333	163	223.6
67.1	495	332	163	222.7
67.1	493	331	162	222.2
67.1	492	330	162	221.3

%E	M1	M2	DM	M*
67.1	490	329	161	220.9
67.1	489	328	161	220.0
67.1	487	327	160	219.6
67.1	486	326	160	218.7
67.1	484	325	159	218.2
67.1	483	324	159	217.3
67.1	480	322	158	216.0
67.1	477	320	157	214.7
67.1	474	318	156	213.3
67.1	471	316	155	212.0
67.1	468	314	154	210.7
67.1	465	312	153	209.3
67.1	462	310	152	208.0
67.1	459	308	151	206.7
67.1	456	306	150	205.3
67.1	453	304	149	204.0
67.1	450	302	148	202.7
67.1	447	300	147	201.3
67.1	444	298	146	200.0
67.1	441	296	145	198.7
67.1	438	294	144	197.3
67.1	435	292	143	196.0
67.1	434	291	143	195.1
67.1	432	290	142	194.7
67.1	431	289	142	193.8
67.1	429	288	141	193.3
67.1	428	287	141	192.5
67.1	426	286	140	192.0
67.1	425	285	140	191.1
67.1	423	284	139	190.7
67.1	422	283	139	189.8
67.1	420	282	138	189.3
67.1	419	281	138	188.5
67.1	417	280	137	188.0
67.1	416	279	137	187.1
67.1	413	277	136	185.8
67.1	410	275	135	184.5
67.1	407	273	134	183.1
67.1	404	271	133	181.8
67.1	401	269	132	180.5
67.1	398	267	131	179.1
67.1	395	265	130	177.8
67.1	392	263	129	176.5
67.1	389	261	128	175.1
67.1	386	259	127	173.8
67.1	383	257	126	172.5
67.1	380	255	125	171.1
67.1	377	253	124	169.8
67.1	374	251	123	168.5
67.1	371	249	122	167.1

%E	M1	M2	DM	M*
67.1	368	247	121	165.8
67.1	365	245	120	164.5
67.1	362	243	119	163.1
67.1	359	241	118	161.8
67.1	356	239	117	160.5
67.1	353	237	116	159.1
67.1	350	235	115	157.8
67.1	347	233	114	156.5
67.1	346	232	114	155.6
67.1	343	230	113	154.2
67.1	340	228	112	152.9
67.1	337	226	111	151.6
67.1	334	224	110	150.2
67.1	331	222	109	148.9
67.1	328	220	108	147.6
67.1	325	218	107	146.2
67.1	322	216	106	144.9
67.1	319	214	105	143.6
67.1	316	212	104	142.2
67.1	313	210	103	140.9
67.1	310	208	102	139.6
67.1	307	206	101	138.2
67.1	304	204	100	136.9
67.1	301	202	99	135.6
67.1	298	200	98	134.2
67.1	295	198	97	132.9
67.1	292	196	96	131.6
67.1	289	194	95	130.2
67.1	286	192	94	128.9
67.1	283	190	93	127.6
67.1	280	188	92	126.2
67.1	277	186	91	124.9
67.1	261	175	86	117.3
67.1	258	173	85	116.0
67.1	255	171	84	114.7
67.1	252	169	83	113.3
67.1	249	167	82	112.0
67.1	246	165	81	110.7
67.1	243	163	80	109.3
67.1	240	161	79	108.0
67.1	237	159	78	106.7
67.1	234	157	77	105.3
67.1	231	155	76	104.0
67.1	228	153	75	102.7
67.1	225	151	74	101.3
67.1	222	149	73	100.0
67.1	219	147	72	98.7
67.1	216	145	71	97.3
67.1	213	143	70	96.0
67.1	210	141	69	94.7

%E	M1	M2	DM	M*
67.1	173	116	57	77.8
67.1	170	114	56	76.4
67.1	167	112	55	75.1
67.1	164	110	54	73.8
67.1	161	108	53	72.4
67.1	158	106	52	71.1
67.1	155	104	51	69.8
67.1	152	102	50	68.4
67.1	149	100	49	67.1
67.1	146	98	48	65.8
67.1	143	96	47	64.4
67.1	140	94	46	63.1
67.1	85	57	28	38.2
67.1	82	55	27	36.9
67.1	79	53	26	35.6
67.1	76	51	25	34.2
67.1	73	49	24	32.9
67.1	70	47	23	31.6
67.0	500	335	165	224.4
67.0	497	333	164	223.1
67.0	494	331	163	221.8
67.0	491	329	162	220.5
67.0	488	327	161	219.1
67.0	485	325	160	217.8
67.0	482	323	159	216.5
67.0	479	321	158	215.1
67.0	476	319	157	213.8
67.0	473	317	156	212.5
67.0	470	315	155	211.1
67.0	469	314	155	210.2
67.0	467	313	154	209.8
67.0	466	312	154	208.9
67.0	464	311	153	208.5
67.0	463	310	153	207.6
67.0	461	309	152	207.1
67.0	460	308	152	206.2
67.0	458	307	151	205.8
67.0	457	306	151	204.9
67.0	455	305	150	204.5
67.0	454	304	150	203.6
67.0	452	303	149	203.1
67.0	451	302	149	202.2
67.0	449	301	148	201.8
67.0	448	300	148	200.9
67.0	446	299	147	200.5
67.0	445	298	147	199.6
67.0	443	297	146	199.1
67.0	442	296	146	198.2
67.0	440	295	145	197.8
67.0	439	294	145	196.9

%E	M1	M2	DM	M*
67.0	437	293	144	196.5
67.0	436	292	144	195.6
67.0	433	290	143	194.2
67.0	430	288	142	192.9
67.0	427	286	141	191.6
67.0	424	284	140	190.2
67.0	421	282	139	188.9
67.0	418	280	138	187.6
67.0	415	278	137	186.2
67.0	412	276	136	184.9
67.0	409	274	135	183.6
67.0	406	272	134	182.2
67.0	403	270	133	180.9
67.0	400	268	132	179.6
67.0	397	266	131	178.2
67.0	394	264	130	176.9
67.0	391	262	129	175.6
67.0	388	260	128	174.2
67.0	385	258	127	172.9
67.0	382	256	126	171.6
67.0	379	254	125	170.2
67.0	376	252	124	168.9
67.0	373	250	123	167.6
67.0	370	248	122	166.2
67.0	367	246	121	164.9
67.0	364	244	120	163.6
67.0	361	242	119	162.2
67.0	358	240	118	160.9
67.0	355	238	117	159.6
67.0	352	236	116	158.2
67.0	351	235	116	157.3
67.0	349	234	115	156.9
67.0	348	233	115	156.0
67.0	345	231	114	154.7
67.0	342	229	113	153.3
67.0	339	227	112	152.0
67.0	336	225	111	150.7
67.0	333	223	110	149.3
67.0	330	221	109	148.0
67.0	327	219	108	146.7
67.0	324	217	107	145.3
67.0	321	215	106	144.0
67.0	318	213	105	142.7
67.0	315	211	104	141.3
67.0	312	209	103	140.0
67.0	309	207	102	138.7
67.0	306	205	101	137.3
67.0	303	203	100	136.0
67.0	300	201	99	134.7
67.0	297	199	98	133.3

%E	M1	M2	DM	M*
67.0	294	197	97	132.0
67.0	291	195	96	130.7
67.0	288	193	95	129.3
67.0	285	191	94	128.0
67.0	282	189	93	126.7
67.0	279	187	92	125.3
67.0	276	185	91	124.0
67.0	273	183	90	122.7
67.0	270	181	89	121.3
67.0	267	179	88	120.0
67.0	264	177	87	118.7
67.0	233	156	77	104.4
67.0	230	154	76	103.1
67.0	227	152	75	101.8
67.0	224	150	74	100.4
67.0	221	148	73	99.1
67.0	218	146	72	97.8
67.0	215	144	71	96.4
67.0	212	142	70	95.1
67.0	209	140	69	93.8
67.0	206	138	68	92.4
67.0	203	136	67	91.1
67.0	200	134	66	89.8
67.0	197	132	65	88.4
67.0	194	130	64	87.1
67.0	191	128	63	85.8
67.0	188	126	62	84.4
67.0	185	124	61	83.1
67.0	182	122	60	81.8
67.0	179	120	59	80.4
67.0	176	118	58	79.1
67.0	115	77	38	51.6
67.0	112	75	37	50.2
67.0	109	73	36	48.9
67.0	106	71	35	47.6
67.0	103	69	34	46.2
67.0	100	67	33	44.9
67.0	97	65	32	43.6
67.0	94	63	31	42.2
67.0	91	61	30	40.9
67.0	88	59	29	39.6
66.9	499	334	165	223.6
66.9	498	333	165	222.7
66.9	496	332	164	222.2
66.9	495	331	164	221.3
66.9	493	330	163	220.9
66.9	492	329	163	220.0
66.9	490	328	162	219.6
66.9	489	327	162	218.7
66.9	487	326	161	218.2

%E	M1	M2	DM	M*
66.9	486	325	161	217.3
66.9	484	324	160	216.9
66.9	483	323	160	216.0
66.9	481	322	159	215.6
66.9	480	321	159	214.7
66.9	478	320	158	214.2
66.9	477	319	158	213.3
66.9	475	318	157	212.9
66.9	474	317	157	212.0
66.9	472	316	156	211.6
66.9	471	315	156	210.7
66.9	468	313	155	209.3
66.9	465	311	154	208.0
66.9	462	309	153	206.7
66.9	459	307	152	205.3
66.9	456	305	151	204.0
66.9	453	303	150	202.7
66.9	450	301	149	201.3
66.9	447	299	148	200.0
66.9	444	297	147	198.7
66.9	441	295	146	197.3
66.9	438	293	145	196.0
66.9	435	291	144	194.7
66.9	432	289	143	193.3
66.9	429	287	142	192.0
66.9	426	285	141	190.7
66.9	423	283	140	189.3
66.9	420	281	139	188.0
66.9	417	279	138	186.7
66.9	414	277	137	185.3
66.9	411	275	136	184.0
66.9	408	273	135	182.7
66.9	405	271	134	181.3
66.9	402	269	133	180.0
66.9	399	267	132	178.7
66.9	396	265	131	177.3
66.9	393	263	130	176.0
66.9	390	261	129	174.7
66.9	387	259	128	173.3
66.9	384	257	127	172.0
66.9	381	255	126	170.7
66.9	378	253	125	169.3
66.9	375	251	124	168.0
66.9	372	249	123	166.7
66.9	369	247	122	165.3
66.9	366	245	121	164.0
66.9	363	243	120	162.7
66.9	362	242	120	161.8
66.9	360	241	119	161.3
66.9	359	240	119	160.4

%E	M1	M2	DM	M*
66.9	357	239	118	160.0
66.9	356	238	118	159.1
66.9	354	237	117	158.7
66.9	353	236	117	157.8
66.9	350	234	116	156.4
66.9	347	232	115	155.1
66.9	344	230	114	153.8
66.9	341	228	113	152.4
66.9	338	226	112	151.1
66.9	335	224	111	149.8
66.9	332	222	110	148.4
66.9	329	220	109	147.1
66.9	326	218	108	145.8
66.9	323	216	107	144.4
66.9	320	214	106	143.1
66.9	317	212	105	141.8
66.9	314	210	104	140.4
66.9	311	208	103	139.1
66.9	308	206	102	137.8
66.9	305	204	101	136.4
66.9	302	202	100	135.1
66.9	299	200	99	133.8
66.9	296	198	98	132.4
66.9	293	196	97	131.1
66.9	290	194	96	129.8
66.9	287	192	95	128.4
66.9	284	190	94	127.1
66.9	281	188	93	125.8
66.9	278	186	92	124.4
66.9	275	184	91	123.1
66.9	272	182	90	121.8
66.9	269	180	89	120.4
66.9	266	178	88	119.1
66.9	263	176	87	117.8
66.9	260	174	86	116.4
66.9	257	172	85	115.1
66.9	254	170	84	113.8
66.9	251	168	83	112.4
66.9	248	166	82	111.1
66.9	245	164	81	109.8
66.9	242	162	80	108.4
66.9	239	160	79	107.1
66.9	236	158	78	105.8
66.9	181	121	60	80.9
66.9	178	119	59	79.6
66.9	175	117	58	78.2
66.9	172	115	57	76.9
66.9	169	113	56	75.6
66.9	166	111	55	74.2
66.9	163	109	54	72.9

%E	M1	M2	DM	M*
66.9	160	107	53	71.6
66.9	157	105	52	70.2
66.9	154	103	51	68.9
66.9	151	101	50	67.6
66.9	148	99	49	66.2
66.9	145	97	48	64.9
66.9	142	95	47	63.6
66.9	139	93	46	62.2
66.9	136	91	45	60.9
66.9	133	89	44	59.6
66.9	130	87	43	58.2
66.9	127	85	42	56.9
66.9	124	83	41	55.6
66.9	121	81	40	54.2
66.9	118	79	39	52.9
66.8	500	334	166	223.1
66.8	497	332	165	221.8
66.8	494	330	164	220.4
66.8	491	328	163	219.1
66.8	488	326	162	217.8
66.8	485	324	161	216.4
66.8	482	322	160	215.1
66.8	479	320	159	213.8
66.8	476	318	158	212.4
66.8	473	316	157	211.1
66.8	470	314	156	209.8
66.8	467	312	155	208.4
66.8	464	310	154	207.1
66.8	461	308	153	205.8
66.8	458	306	152	204.4
66.8	455	304	151	203.1
66.8	452	302	150	201.8
66.8	449	300	149	200.4
66.8	446	298	148	199.1
66.8	443	296	147	197.8
66.8	440	294	146	196.4
66.8	437	292	145	195.1
66.8	434	290	144	193.8
66.8	431	288	143	192.4
66.8	428	286	142	191.1
66.8	425	284	141	189.8
66.8	422	282	140	188.4
66.8	419	280	139	187.1
66.8	416	278	138	185.8
66.8	413	276	137	184.4
66.8	410	274	136	183.1
66.8	407	272	135	181.8
66.8	404	270	134	180.4
66.8	401	268	133	179.1
66.8	400	267	133	178.2

%E	M1	M2	DM	M*
66.8	398	266	132	177.8
66.8	397	265	132	176.9
66.8	395	264	131	176.4
66.8	394	263	131	175.6
66.8	392	262	130	175.1
66.8	391	261	130	174.2
66.8	389	260	129	173.8
66.8	388	259	129	172.9
66.8	386	258	128	172.4
66.8	385	257	128	171.6
66.8	383	256	127	171.1
66.8	382	255	127	170.2
66.8	380	254	126	169.8
66.8	379	253	126	168.9
66.8	377	252	125	168.4
66.8	376	251	125	167.6
66.8	374	250	124	167.1
66.8	373	249	124	166.2
66.8	371	248	123	165.8
66.8	370	247	123	164.9
66.8	368	246	122	164.4
66.8	367	245	122	163.6
66.8	365	244	121	163.1
66.8	364	243	121	162.2
66.8	361	241	120	160.9
66.8	358	239	119	159.6
66.8	355	237	118	158.2
66.8	352	235	117	156.9
66.8	349	233	116	155.6
66.8	346	231	115	154.2
66.8	343	229	114	152.9
66.8	340	227	113	151.6
66.8	337	225	112	150.2
66.8	334	223	111	148.9
66.8	331	221	110	147.6
66.8	328	219	109	146.2
66.8	325	217	108	144.9
66.8	322	215	107	143.6
66.8	319	213	106	142.2
66.8	316	211	105	140.9
66.8	313	209	104	139.6
66.8	310	207	103	138.2
66.8	307	205	102	136.9
66.8	304	203	101	135.6
66.8	301	201	100	134.2
66.8	298	199	99	132.9
66.8	295	197	98	131.6
66.8	292	195	97	130.2
66.8	289	193	96	128.9
66.8	286	191	95	127.6

%E	M1	M2	DM	M*
66.8	283	189	94	126.2
66.8	280	187	93	124.9
66.8	277	185	92	123.6
66.8	274	183	91	122.2
66.8	271	181	90	120.9
66.8	268	179	89	119.6
66.8	265	177	88	118.2
66.8	262	175	87	116.9
66.8	259	173	86	115.6
66.8	256	171	85	114.2
66.8	253	169	84	112.9
66.8	250	167	83	111.6
66.8	247	165	82	110.2
66.8	244	163	81	108.9
66.8	241	161	80	107.6
66.8	238	159	79	106.2
66.8	235	157	78	104.9
66.8	232	155	77	103.6
66.8	229	153	76	102.2
66.8	226	151	75	100.9
66.8	223	149	74	99.6
66.8	220	147	73	98.2
66.8	217	145	72	96.9
66.8	214	143	71	95.6
66.8	211	141	70	94.2
66.8	208	139	69	92.9
66.8	205	137	68	91.6
66.8	202	135	67	90.2
66.8	199	133	66	88.9
66.8	196	131	65	87.6
66.8	193	129	64	86.2
66.8	190	127	63	84.9
66.8	187	125	62	83.6
66.8	184	123	61	82.2
66.7	499	333	166	222.2
66.7	498	332	166	221.3
66.7	496	331	165	220.9
66.7	495	330	165	220.0
66.7	493	329	164	219.6
66.7	492	328	164	218.7
66.7	490	327	163	218.2
66.7	489	326	163	217.3
66.7	487	325	162	216.9
66.7	486	324	162	216.0
66.7	484	323	161	215.6
66.7	483	322	161	214.7
66.7	481	321	160	214.2
66.7	480	320	160	213.3
66.7	478	319	159	212.9
66.7	477	318	159	212.0

%E	M1	M2	DM	M*
66.7	475	317	158	211.6
66.7	474	316	158	210.7
66.7	472	315	157	210.2
66.7	471	314	157	209.3
66.7	469	313	156	208.9
66.7	468	312	156	208.0
66.7	466	311	155	207.6
66.7	465	310	155	206.7
66.7	463	309	154	206.2
66.7	462	308	154	205.3
66.7	460	307	153	204.9
66.7	459	306	153	204.0
66.7	457	305	152	203.6
66.7	456	304	152	202.7
66.7	454	303	151	202.2
66.7	453	302	151	201.3
66.7	451	301	150	200.9
66.7	450	300	150	200.0
66.7	448	299	149	199.6
66.7	447	298	149	198.7
66.7	445	297	148	198.2
66.7	444	296	148	197.3
66.7	442	295	147	196.9
66.7	441	294	147	196.0
66.7	439	293	146	195.6
66.7	438	292	146	194.7
66.7	436	291	145	194.2
66.7	435	290	145	193.3
66.7	433	289	144	192.9
66.7	432	288	144	192.0
66.7	430	287	143	191.6
66.7	429	286	143	190.7
66.7	427	285	142	190.2
66.7	426	284	142	189.3
66.7	424	283	141	188.9
66.7	423	282	141	188.0
66.7	421	281	140	187.6
66.7	420	280	140	186.7
66.7	418	279	139	186.2
66.7	417	278	139	185.3
66.7	415	277	138	184.9
66.7	414	276	138	184.0
66.7	412	275	137	183.6
66.7	411	274	137	182.7
66.7	409	273	136	182.2
66.7	408	272	136	181.3
66.7	406	271	135	180.9
66.7	405	270	135	180.0
66.7	403	269	134	179.6
66.7	402	268	134	178.7

%E	M1	M2	DM	M*
66.7	399	266	133	177.3
66.7	396	264	132	176.0
66.7	393	262	131	174.7
66.7	390	260	130	173.3
66.7	387	258	129	172.0
66.7	384	256	128	170.7
66.7	381	254	127	169.3
66.7	378	252	126	168.0
66.7	375	250	125	166.7
66.7	372	248	124	165.3
66.7	369	246	123	164.0
66.7	366	244	122	162.7
66.7	363	242	121	161.3
66.7	360	240	120	160.0
66.7	357	238	119	158.7
66.7	354	236	118	157.3
66.7	351	234	117	156.0
66.7	348	232	116	154.7
66.7	345	230	115	153.3
66.7	342	228	114	152.0
66.7	339	226	113	150.7
66.7	336	224	112	149.3
66.7	333	222	111	148.0
66.7	330	220	110	146.7
66.7	327	218	109	145.3
66.7	324	216	108	144.0
66.7	321	214	107	142.7
66.7	318	212	106	141.3
66.7	315	210	105	140.0
66.7	312	208	104	138.7
66.7	309	206	103	137.3
66.7	306	204	102	136.0
66.7	303	202	101	134.7
66.7	300	200	100	133.3
66.7	297	198	99	132.0
66.7	294	196	98	130.7
66.7	291	194	97	129.3
66.7	288	192	96	128.0
66.7	285	190	95	126.7
66.7	282	188	94	125.3
66.7	279	186	93	124.0
66.7	276	184	92	122.7
66.7	273	182	91	121.3
66.7	270	180	90	120.0
66.7	267	178	89	118.7
66.7	264	176	88	117.3
66.7	261	174	87	116.0
66.7	258	172	86	114.7
66.7	255	170	85	113.3
66.7	252	168	84	112.0

%E	M1	M2	DM	M*
66.7	249	166	83	110.7
66.7	246	164	82	109.3
66.7	243	162	81	108.0
66.7	240	160	80	106.7
66.7	237	158	79	105.3
66.7	234	156	78	104.0
66.7	231	154	77	102.7
66.7	228	152	76	101.3
66.7	225	150	75	100.0
66.7	222	148	74	98.7
66.7	219	146	73	97.3
66.7	216	144	72	96.0
66.7	213	142	71	94.7
66.7	210	140	70	93.3
66.7	207	138	69	92.0
66.7	204	136	68	90.7
66.7	201	134	67	89.3
66.7	198	132	66	88.0
66.7	195	130	65	86.7
66.7	192	128	64	85.3
66.7	189	126	63	84.0
66.7	186	124	62	82.7
66.7	183	122	61	81.3
66.7	180	120	60	80.0
66.7	177	118	59	78.7
66.7	174	116	58	77.3
66.7	171	114	57	76.0
66.7	168	112	56	74.7
66.7	165	110	55	73.3
66.7	162	108	54	72.0
66.7	159	106	53	70.7
66.7	156	104	52	69.3
66.7	153	102	51	68.0
66.7	150	100	50	66.7
66.7	147	98	49	65.3
66.7	144	96	48	64.0
66.7	141	94	47	62.7
66.7	138	92	46	61.3
66.7	135	90	45	60.0
66.7	132	88	44	58.7
66.7	129	86	43	57.3
66.7	126	84	42	56.0
66.7	123	82	41	54.7
66.7	120	80	40	53.3
66.7	117	78	39	52.0
66.7	114	76	38	50.7
66.7	111	74	37	49.3
66.7	108	72	36	48.0
66.7	105	70	35	46.7
66.7	102	68	34	45.3

%E	M1	M2	DM	M*
66.7	99	66	33	44.0
66.7	96	64	32	42.7
66.7	93	62	31	41.3
66.7	90	60	30	40.0
66.7	87	58	29	38.7
66.7	84	56	28	37.3
66.7	81	54	27	36.0
66.7	78	52	26	34.7
66.7	75	50	25	33.3
66.7	72	48	24	32.0
66.7	69	46	23	30.7
66.7	66	44	22	29.3
66.7	63	42	21	28.0
66.7	60	40	20	26.7
66.7	57	38	19	25.3
66.7	54	36	18	24.0
66.7	51	34	17	22.7
66.7	48	32	16	21.3
66.7	45	30	15	20.0
66.7	42	28	14	18.7
66.7	39	26	13	17.3
66.7	36	24	12	16.0
66.7	33	22	11	14.7
66.7	30	20	10	13.3
66.7	27	18	9	12.0
66.7	24	16	8	10.7
66.7	21	14	7	9.3
66.7	18	12	6	8.0
66.7	15	10	5	6.7
66.7	12	8	4	5.3
66.7	9	6	3	4.0
66.7	6	4	2	2.7
66.7	3	2	1	1.3
66.6	500	333	167	221.8
66.6	497	331	166	220.4
66.6	494	329	165	219.1
66.6	491	327	164	217.8
66.6	488	325	163	216.4
66.6	485	323	162	215.1
66.6	482	321	161	213.8
66.6	479	319	160	212.4
66.6	476	317	159	211.1
66.6	473	315	158	209.8
66.6	470	313	157	208.4
66.6	467	311	156	207.1
66.6	464	309	155	205.8
66.6	461	307	154	204.4
66.6	458	305	153	203.1
66.6	455	303	152	201.8
66.6	452	301	151	200.4

%E	M1	M2	DM	M*
66.6	449	299	150	199.1
66.6	446	297	149	197.8
66.6	443	295	148	196.4
66.6	440	293	147	195.1
66.6	437	291	146	193.8
66.6	434	289	145	192.4
66.6	431	287	144	191.1
66.6	428	285	143	189.8
66.6	425	283	142	188.4
66.6	422	281	141	187.1
66.6	419	279	140	185.8
66.6	416	277	139	184.4
66.6	413	275	138	183.1
66.6	410	273	137	181.8
66.6	407	271	136	180.4
66.6	404	269	135	179.1
66.6	401	267	134	177.8
66.6	398	265	133	176.4
66.6	395	263	132	175.1
66.6	392	261	131	173.8
66.6	389	259	130	172.4
66.6	386	257	129	171.1
66.6	383	255	128	169.8
66.6	380	253	127	168.4
66.6	377	251	126	167.1
66.6	374	249	125	165.8
66.6	371	247	124	164.4
66.6	368	245	123	163.1
66.6	365	243	122	161.8
66.6	362	241	121	160.4
66.6	359	239	120	159.1
66.6	356	237	119	157.8
66.6	353	235	118	156.4
66.6	350	233	117	155.1
66.6	347	231	116	153.8
66.6	344	229	115	152.4
66.6	341	227	114	151.1
66.6	338	225	113	149.8
66.6	335	223	112	148.4
66.6	332	221	111	147.1
66.6	329	219	110	145.8
66.6	326	217	109	144.4
66.6	323	215	108	143.1
66.6	320	213	107	141.8
66.6	317	211	106	140.4
66.6	314	209	105	139.1
66.6	311	207	104	137.8
66.6	308	205	103	136.4
66.6	305	203	102	135.1
66.6	302	201	101	133.8

%E	M1	M2	DM	M*
66.6	299	199	100	132.4
66.6	296	197	99	131.1
66.6	293	195	98	129.8
66.6	290	193	97	128.4
66.6	287	191	96	127.1
66.5	499	332	167	220.9
66.5	498	331	167	220.0
66.5	496	330	166	219.6
66.5	495	329	166	218.7
66.5	493	328	165	218.2
66.5	492	327	165	217.3
66.5	490	326	164	216.9
66.5	489	325	164	216.0
66.5	487	324	163	215.6
66.5	486	323	163	214.7
66.5	484	322	162	214.2
66.5	483	321	162	213.3
66.5	481	320	161	212.9
66.5	480	319	161	212.0
66.5	478	318	160	211.6
66.5	477	317	160	210.7
66.5	475	316	159	210.2
66.5	474	315	159	209.3
66.5	472	314	158	208.9
66.5	471	313	158	208.0
66.5	469	312	157	207.6
66.5	468	311	157	206.7
66.5	466	310	156	206.2
66.5	465	309	156	205.3
66.5	463	308	155	204.9
66.5	462	307	155	204.0
66.5	460	306	154	203.6
66.5	457	304	153	202.2
66.5	454	302	152	200.9
66.5	451	300	151	199.6
66.5	448	298	150	198.2
66.5	445	296	149	196.9
66.5	442	294	148	195.6
66.5	439	292	147	194.2
66.5	436	290	146	192.9
66.5	433	288	145	191.6
66.5	430	286	144	190.2
66.5	427	284	143	188.9
66.5	424	282	142	187.6
66.5	421	280	141	186.2
66.5	418	278	140	184.9
66.5	415	276	139	183.6
66.5	412	274	138	182.2
66.5	409	272	137	180.9
66.5	406	270	136	179.6

%E	M1	M2	DM	M*
66.5	403	268	135	178.2
66.5	400	266	134	176.9
66.5	397	264	133	175.6
66.5	394	262	132	174.2
66.5	391	260	131	172.9
66.5	388	258	130	171.6
66.5	385	256	129	170.2
66.5	382	254	128	168.9
66.5	379	252	127	167.6
66.5	376	250	126	166.2
66.5	373	248	125	164.9
66.5	370	246	124	163.6
66.5	367	244	123	162.2
66.5	364	242	122	160.9
66.5	361	240	121	159.6
66.5	358	238	120	158.2
66.5	355	236	119	156.9
66.5	352	234	118	155.6
66.5	349	232	117	154.2
66.5	346	230	116	152.9
66.5	343	228	115	151.6
66.5	340	226	114	150.2
66.5	337	224	113	148.9
66.5	334	222	112	147.6
66.5	331	220	111	146.2
66.5	328	218	110	144.9
66.5	325	216	109	143.6
66.5	322	214	108	142.2
66.5	319	212	107	140.9
66.5	316	210	106	139.6
66.5	313	208	105	138.2
66.5	310	206	104	136.9
66.5	307	204	103	135.6
66.5	284	189	95	125.8
66.5	281	187	94	124.4
66.5	278	185	93	123.1
66.5	275	183	92	121.8
66.5	272	181	91	120.4
66.5	269	179	90	119.1
66.5	266	177	89	117.8
66.5	263	175	88	116.4
66.5	260	173	87	115.1
66.5	257	171	86	113.8
66.5	254	169	85	112.4
66.5	251	167	84	111.1
66.5	248	165	83	109.8
66.5	245	163	82	108.4
66.5	242	161	81	107.1
66.5	239	159	80	105.8
66.5	236	157	79	104.4

%E	M1	M2	DM	M*
66.5	233	155	78	103.1
66.5	230	153	77	101.8
66.5	227	151	76	100.4
66.5	224	149	75	99.1
66.5	221	147	74	97.8
66.5	218	145	73	96.4
66.5	215	143	72	95.1
66.5	212	141	71	93.8
66.5	209	139	70	92.4
66.5	206	137	69	91.1
66.5	203	135	68	89.8
66.5	200	133	67	88.4
66.5	197	131	66	87.1
66.5	194	129	65	85.8
66.5	191	127	64	84.4
66.5	188	125	63	83.1
66.5	185	123	62	81.8
66.5	182	121	61	80.4
66.5	179	119	60	79.1
66.5	176	117	59	77.8
66.5	173	115	58	76.4
66.5	170	113	57	75.1
66.5	167	111	56	73.8
66.5	164	109	55	72.4
66.5	161	107	54	71.1
66.5	158	105	53	69.8
66.5	155	103	52	68.4
66.4	500	332	168	220.4
66.4	497	330	167	219.1
66.4	494	328	166	217.8
66.4	491	326	165	216.4
66.4	488	324	164	215.1
66.4	485	322	163	213.8
66.4	482	320	162	212.4
66.4	479	318	161	211.1
66.4	476	316	160	209.8
66.4	473	314	159	208.4
66.4	470	312	158	207.1
66.4	467	310	157	205.8
66.4	464	308	156	204.4
66.4	461	306	155	203.1
66.4	459	305	154	202.7
66.4	458	304	154	201.8
66.4	456	303	153	201.3
66.4	455	302	153	200.4
66.4	453	301	152	200.0
66.4	452	300	152	199.1
66.4	450	299	151	198.7
66.4	449	298	151	197.8
66.4	447	297	150	197.3

%E	M1	M2	DM	M*
66.4	446	296	150	196.4
66.4	444	295	149	196.0
66.4	443	294	149	195.1
66.4	441	293	148	194.7
66.4	440	292	148	193.8
66.4	438	291	147	193.3
66.4	437	290	147	192.4
66.4	435	289	146	192.0
66.4	434	288	146	191.1
66.4	432	287	145	190.7
66.4	431	286	145	189.8
66.4	429	285	144	189.3
66.4	428	284	144	188.4
66.4	426	283	143	188.0
66.4	425	282	143	187.1
66.4	423	281	142	186.7
66.4	422	280	142	185.8
66.4	420	279	141	185.3
66.4	417	277	140	184.0
66.4	414	275	139	182.7
66.4	411	273	138	181.3
66.4	408	271	137	180.0
66.4	405	269	136	178.7
66.4	402	267	135	177.3
66.4	399	265	134	176.0
66.4	396	263	133	174.7
66.4	393	261	132	173.3
66.4	390	259	131	172.0
66.4	387	257	130	170.7
66.4	384	255	129	169.3
66.4	381	253	128	168.0
66.4	378	251	127	166.7
66.4	375	249	126	165.3
66.4	372	247	125	164.0
66.4	369	245	124	162.7
66.4	366	243	123	161.3
66.4	363	241	122	160.0
66.4	360	239	121	158.7
66.4	357	237	120	157.3
66.4	354	235	119	156.0
66.4	351	233	118	154.7
66.4	348	231	117	153.3
66.4	345	229	116	152.0
66.4	342	227	115	150.7
66.4	339	225	114	149.3
66.4	336	223	113	148.0
66.4	333	221	112	146.7
66.4	330	219	111	145.3
66.4	327	217	110	144.0
66.4	324	215	109	142.7

%E	M1	M2	DM	M*
66.4	321	213	108	141.3
66.4	318	211	107	140.0
66.4	304	202	102	134.2
66.4	301	200	101	132.9
66.4	298	198	100	131.6
66.4	295	196	99	130.2
66.4	292	194	98	128.9
66.4	289	192	97	127.6
66.4	286	190	96	126.2
66.4	283	188	95	124.9
66.4	280	186	94	123.6
66.4	277	184	93	122.2
66.4	274	182	92	120.9
66.4	271	180	91	119.6
66.4	268	178	90	118.2
66.4	265	176	89	116.9
66.4	262	174	88	115.6
66.4	259	172	87	114.2
66.4	256	170	86	112.9
66.4	253	168	85	111.6
66.4	250	166	84	110.2
66.4	247	164	83	108.9
66.4	244	162	82	107.6
66.4	241	160	81	106.2
66.4	238	158	80	104.9
66.4	235	156	79	103.6
66.4	232	154	78	102.2
66.4	229	152	77	100.9
66.4	226	150	76	99.6
66.4	223	148	75	98.2
66.4	220	146	74	96.9
66.4	217	144	73	95.6
66.4	214	142	72	94.2
66.4	211	140	71	92.9
66.4	152	101	51	67.1
66.4	149	99	50	65.8
66.4	146	97	49	64.4
66.4	143	95	48	63.1
66.4	140	93	47	61.8
66.4	137	91	46	60.4
66.4	134	89	45	59.1
66.4	131	87	44	57.8
66.4	128	85	43	56.4
66.4	125	83	42	55.1
66.4	122	81	41	53.8
66.4	119	79	40	52.4
66.4	116	77	39	51.1
66.4	113	75	38	49.8
66.4	110	73	37	48.4
66.4	107	71	36	47.1

%E	M1	M2	DM	M*
66.3	499	331	168	219.6
66.3	498	330	168	218.7
66.3	496	329	167	218.2
66.3	495	328	167	217.3
66.3	493	327	166	216.9
66.3	492	326	166	216.0
66.3	490	325	165	215.6
66.3	489	324	165	214.7
66.3	487	323	164	214.2
66.3	486	322	164	213.3
66.3	484	321	163	212.9
66.3	483	320	163	212.0
66.3	481	319	162	211.6
66.3	480	318	162	210.7
66.3	478	317	161	210.2
66.3	475	315	160	208.9
66.3	472	313	159	207.6
66.3	469	311	158	206.2
66.3	466	309	157	204.9
66.3	463	307	156	203.6
66.3	460	305	155	202.2
66.3	457	303	154	200.9
66.3	454	301	153	199.6
66.3	451	299	152	198.2
66.3	448	297	151	196.9
66.3	445	295	150	195.6
66.3	442	293	149	194.2
66.3	439	291	148	192.9
66.3	436	289	147	191.6
66.3	433	287	146	190.2
66.3	430	285	145	188.9
66.3	427	283	144	187.6
66.3	424	281	143	186.2
66.3	421	279	142	184.9
66.3	419	278	141	184.4
66.3	418	277	141	183.6
66.3	416	276	140	183.1
66.3	415	275	140	182.2
66.3	413	274	139	181.8
66.3	412	273	139	180.9
66.3	410	272	138	180.4
66.3	409	271	138	179.6
66.3	407	270	137	179.1
66.3	406	269	137	178.2
66.3	404	268	136	177.8
66.3	403	267	136	176.9
66.3	401	266	135	176.4
66.3	400	265	135	175.6
66.3	398	264	134	175.1
66.3	395	262	133	173.8

%E	M1	M2	DM	M*
66.3	392	260	132	172.4
66.3	389	258	131	171.1
66.3	386	256	130	169.8
66.3	383	254	129	168.4
66.3	380	252	128	167.1
66.3	377	250	127	165.8
66.3	374	248	126	164.4
66.3	371	246	125	163.1
66.3	368	244	124	161.8
66.3	365	242	123	160.4
66.3	362	240	122	159.1
66.3	359	238	121	157.8
66.3	356	236	120	156.4
66.3	353	234	119	155.1
66.3	350	232	118	153.8
66.3	347	230	117	152.4
66.3	344	228	116	151.1
66.3	341	226	115	149.8
66.3	338	224	114	148.4
66.3	335	222	113	147.1
66.3	332	220	112	145.8
66.3	329	218	111	144.4
66.3	326	216	110	143.1
66.3	323	214	109	141.8
66.3	320	212	108	140.4
66.3	315	209	106	138.7
66.3	312	207	105	137.3
66.3	309	205	104	136.0
66.3	306	203	103	134.7
66.3	303	201	102	133.3
66.3	300	199	101	132.0
66.3	297	197	100	130.7
66.3	294	195	99	129.3
66.3	291	193	98	128.0
66.3	288	191	97	126.7
66.3	285	189	96	125.3
66.3	282	187	95	124.0
66.3	279	185	94	122.7
66.3	276	183	93	121.3
66.3	273	181	92	120.0
66.3	270	179	91	118.7
66.3	267	177	90	117.3
66.3	264	175	89	116.0
66.3	261	173	88	114.7
66.3	258	171	87	113.3
66.3	255	169	86	112.0
66.3	252	167	85	110.7
66.3	249	165	84	109.3
66.3	246	163	83	108.0
66.3	243	161	82	106.7

%E	M1	M2	DM	M*
66.3	240	159	81	105.3
66.3	208	138	70	91.6
66.3	205	136	69	90.2
66.3	202	134	68	88.9
66.3	199	132	67	87.6
66.3	196	130	66	86.2
66.3	193	128	65	84.9
66.3	190	126	64	83.6
66.3	187	124	63	82.2
66.3	184	122	62	80.9
66.3	181	120	61	79.6
66.3	178	118	60	78.2
66.3	175	116	59	76.9
66.3	172	114	58	75.6
66.3	169	112	57	74.2
66.3	166	110	56	72.9
66.3	163	108	55	71.6
66.3	160	106	54	70.2
66.3	104	69	35	45.8
66.3	101	67	34	44.4
66.3	98	65	33	43.1
66.3	95	63	32	41.8
66.3	92	61	31	40.4
66.3	89	59	30	39.1
66.3	86	57	29	37.8
66.3	83	55	28	36.4
66.3	80	53	27	35.1
66.2	500	331	169	219.1
66.2	497	329	168	217.8
66.2	494	327	167	216.5
66.2	491	325	166	215.1
66.2	488	323	165	213.8
66.2	485	321	164	212.5
66.2	482	319	163	211.1
66.2	479	317	162	209.8
66.2	477	316	161	209.3
66.2	476	315	161	208.5
66.2	474	314	160	208.0
66.2	473	313	160	207.1
66.2	471	312	159	206.7
66.2	470	311	159	205.8
66.2	468	310	158	205.3
66.2	467	309	158	204.5
66.2	465	308	157	204.0
66.2	464	307	157	203.1
66.2	462	306	156	202.7
66.2	461	305	156	201.8
66.2	459	304	155	201.3
66.2	458	303	155	200.5
66.2	456	302	154	200.0

%E	M1	M2	DM	M*
66.2	455	301	154	199.1
66.2	453	300	153	198.7
66.2	452	299	153	197.8
66.2	450	298	152	197.3
66.2	447	296	151	196.0
66.2	444	294	150	194.7
66.2	441	292	149	193.3
66.2	438	290	148	192.0
66.2	435	288	147	190.7
66.2	432	286	146	189.3
66.2	429	284	145	188.0
66.2	426	282	144	186.7
66.2	423	280	143	185.3
66.2	420	278	142	184.0
66.2	417	276	141	182.7
66.2	414	274	140	181.3
66.2	411	272	139	180.0
66.2	408	270	138	178.7
66.2	405	268	137	177.3
66.2	402	266	136	176.0
66.2	399	264	135	174.7
66.2	397	263	134	174.2
66.2	396	262	134	173.3
66.2	394	261	133	172.9
66.2	393	260	133	172.0
66.2	391	259	132	171.6
66.2	390	258	132	170.7
66.2	388	257	131	170.2
66.2	387	256	131	169.3
66.2	385	255	130	168.9
66.2	382	253	129	167.6
66.2	379	251	128	166.2
66.2	376	249	127	164.9
66.2	373	247	126	163.6
66.2	370	245	125	162.2
66.2	367	243	124	160.9
66.2	364	241	123	159.6
66.2	361	239	122	158.2
66.2	358	237	121	156.9
66.2	355	235	120	155.6
66.2	352	233	119	154.2
66.2	349	231	118	152.9
66.2	346	229	117	151.6
66.2	343	227	116	150.2
66.2	340	225	115	148.9
66.2	337	223	114	147.6
66.2	334	221	113	146.2
66.2	331	219	112	144.9
66.2	328	217	111	143.6
66.2	325	215	110	142.2

%E	M1	M2	DM	M*
66.2	317	210	107	139.1
66.2	314	208	106	137.8
66.2	311	206	105	136.5
66.2	308	204	104	135.1
66.2	305	202	103	133.8
66.2	302	200	102	132.5
66.2	299	198	101	131.1
66.2	296	196	100	129.8
66.2	293	194	99	128.5
66.2	290	192	98	127.1
66.2	287	190	97	125.8
66.2	284	188	96	124.5
66.2	281	186	95	123.1
66.2	278	184	94	121.8
66.2	275	182	93	120.5
66.2	272	180	92	119.1
66.2	269	178	91	117.8
66.2	266	176	90	116.5
66.2	263	174	89	115.1
66.2	260	172	88	113.8
66.2	237	157	80	104.0
66.2	234	155	79	102.7
66.2	231	153	78	101.3
66.2	228	151	77	100.0
66.2	225	149	76	98.7
66.2	222	147	75	97.3
66.2	219	145	74	96.0
66.2	216	143	73	94.7
66.2	213	141	72	93.3
66.2	210	139	71	92.0
66.2	207	137	70	90.7
66.2	204	135	69	89.3
66.2	201	133	68	88.0
66.2	198	131	67	86.7
66.2	195	129	66	85.3
66.2	157	104	53	68.9
66.2	154	102	52	67.6
66.2	151	100	51	66.2
66.2	148	98	50	64.9
66.2	145	96	49	63.6
66.2	142	94	48	62.2
66.2	139	92	47	60.9
66.2	136	90	46	59.6
66.2	133	88	45	58.2
66.2	130	86	44	56.9
66.2	77	51	26	33.8
66.2	74	49	25	32.4
66.2	71	47	24	31.1
66.2	68	45	23	29.8
66.2	65	43	22	28.4

%E	M1	M2	DM	M*
66.1	499	330	169	218.2
66.1	498	329	169	217.4
66.1	496	328	168	216.9
66.1	495	327	168	216.0
66.1	493	326	167	215.6
66.1	492	325	167	214.7
66.1	490	324	166	214.2
66.1	489	323	166	213.4
66.1	487	322	165	212.9
66.1	484	320	164	211.6
66.1	481	318	163	210.2
66.1	478	316	162	208.9
66.1	475	314	161	207.6
66.1	472	312	160	206.2
66.1	469	310	159	204.9
66.1	466	308	158	203.6
66.1	463	306	157	202.2
66.1	460	304	156	200.9
66.1	457	302	155	199.6
66.1	454	300	154	198.2
66.1	451	298	153	196.9
66.1	449	297	152	196.5
66.1	448	296	152	195.6
66.1	446	295	151	195.1
66.1	445	294	151	194.2
66.1	443	293	150	193.8
66.1	442	292	150	192.9
66.1	440	291	149	192.5
66.1	439	290	149	191.6
66.1	437	289	148	191.1
66.1	436	288	148	190.2
66.1	434	287	147	189.8
66.1	433	286	147	188.9
66.1	431	285	146	188.5
66.1	428	283	145	187.1
66.1	425	281	144	185.8
66.1	422	279	143	184.5
66.1	419	277	142	183.1
66.1	416	275	141	181.8
66.1	413	273	140	180.5
66.1	410	271	139	179.1
66.1	407	269	138	177.8
66.1	404	267	137	176.5
66.1	401	265	136	175.1
66.1	398	263	135	173.8
66.1	395	261	134	172.5
66.1	392	259	133	171.1
66.1	389	257	132	169.8
66.1	386	255	131	168.5
66.1	384	254	130	168.0

%E	M1	M2	DM	M*
66.1	383	253	130	167.1
66.1	381	252	129	166.7
66.1	380	251	129	165.8
66.1	378	250	128	165.3
66.1	375	248	127	164.0
66.1	372	246	126	162.7
66.1	369	244	125	161.3
66.1	366	242	124	160.0
66.1	363	240	123	158.7
66.1	360	238	122	157.3
66.1	357	236	121	156.0
66.1	354	234	120	154.7
66.1	351	232	119	153.3
66.1	348	230	118	152.0
66.1	345	228	117	150.7
66.1	342	226	116	149.3
66.1	339	224	115	148.0
66.1	336	222	114	146.7
66.1	333	220	113	145.3
66.1	330	218	112	144.0
66.1	327	216	111	142.7
66.1	322	213	109	140.9
66.1	319	211	108	139.6
66.1	316	209	107	138.2
66.1	313	207	106	136.9
66.1	310	205	105	135.6
66.1	307	203	104	134.2
66.1	304	201	103	132.9
66.1	301	199	102	131.6
66.1	298	197	101	130.2
66.1	295	195	100	128.9
66.1	292	193	99	127.6
66.1	289	191	98	126.2
66.1	286	189	97	124.9
66.1	283	187	96	123.6
66.1	280	185	95	122.2
66.1	277	183	94	120.9
66.1	274	181	93	119.6
66.1	271	179	92	118.2
66.1	257	170	87	112.5
66.1	254	168	86	111.1
66.1	251	166	85	109.8
66.1	248	164	84	108.5
66.1	245	162	83	107.1
66.1	242	160	82	105.8
66.1	239	158	81	104.5
66.1	236	156	80	103.1
66.1	233	154	79	101.8
66.1	230	152	78	100.5
66.1	227	150	77	99.1

%E	M1	M2	DM	M*
66.1	224	148	76	97.8
66.1	221	146	75	96.5
66.1	218	144	74	95.1
66.1	192	127	65	84.0
66.1	189	125	64	82.7
66.1	186	123	63	81.3
66.1	183	121	62	80.0
66.1	180	119	61	78.7
66.1	177	117	60	77.3
66.1	174	115	59	76.0
66.1	171	113	58	74.7
66.1	168	111	57	73.3
66.1	165	109	56	72.0
66.1	127	84	43	55.6
66.1	124	82	42	54.2
66.1	121	80	41	52.9
66.1	118	78	40	51.6
66.1	115	76	39	50.2
66.1	112	74	38	48.9
66.1	109	72	37	47.6
66.1	62	41	21	27.1
66.1	59	39	20	25.8
66.1	56	37	19	24.4
66.0	500	330	170	217.8
66.0	497	328	169	216.5
66.0	494	326	168	215.1
66.0	491	324	167	213.8
66.0	488	322	166	212.5
66.0	486	321	165	212.0
66.0	485	320	165	211.1
66.0	483	319	164	210.7
66.0	482	318	164	209.8
66.0	480	317	163	209.4
66.0	479	316	163	208.5
66.0	477	315	162	208.0
66.0	476	314	162	207.1
66.0	474	313	161	206.7
66.0	473	312	161	205.8
66.0	471	311	160	205.4
66.0	470	310	160	204.5
66.0	468	309	159	204.0
66.0	467	308	159	203.1
66.0	465	307	158	202.7
66.0	462	305	157	201.4
66.0	459	303	156	200.0
66.0	456	301	155	198.7
66.0	453	299	154	197.4
66.0	450	297	153	196.0
66.0	447	295	152	194.7
66.0	444	293	151	193.4
66.0	441	291	150	192.0
66.0	438	289	149	190.7
66.0	435	287	148	189.4
66.0	432	285	147	188.0
66.0	430	284	146	187.6
66.0	429	283	146	186.7
66.0	427	282	145	186.2
66.0	426	281	145	185.4
66.0	424	280	144	184.9
66.0	423	279	144	184.0
66.0	421	278	143	183.6
66.0	420	277	143	182.7
66.0	418	276	142	182.2
66.0	415	274	141	180.9
66.0	412	272	140	179.6
66.0	409	270	139	178.2
66.0	406	268	138	176.9
66.0	403	266	137	175.6
66.0	400	264	136	174.2
66.0	397	262	135	172.9
66.0	394	260	134	171.6
66.0	391	258	133	170.2
66.0	388	256	132	168.9
66.0	385	254	131	167.6
66.0	382	252	130	166.2
66.0	379	250	129	164.9
66.0	377	249	128	164.5
66.0	376	248	128	163.6
66.0	374	247	127	163.1
66.0	373	246	127	162.2
66.0	371	245	126	161.8
66.0	368	243	125	160.5
66.0	365	241	124	159.1
66.0	362	239	123	157.8
66.0	359	237	122	156.5
66.0	356	235	121	155.1
66.0	353	233	120	153.8
66.0	350	231	119	152.5
66.0	347	229	118	151.1
66.0	344	227	117	149.8
66.0	341	225	116	148.5
66.0	338	223	115	147.1
66.0	335	221	114	145.8
66.0	332	219	113	144.5
66.0	329	217	112	143.1
66.0	326	215	111	141.8
66.0	324	214	110	141.3
66.0	321	212	109	140.0
66.0	318	210	108	138.7
66.0	315	208	107	137.3
66.0	312	206	106	136.0
66.0	309	204	105	134.7
66.0	306	202	104	133.3
66.0	303	200	103	132.0
66.0	300	198	102	130.7
66.0	297	196	101	129.3
66.0	294	194	100	128.0
66.0	291	192	99	126.7
66.0	288	190	98	125.3
66.0	285	188	97	124.0
66.0	282	186	96	122.7
66.0	279	184	95	121.3
66.0	268	177	91	116.9
66.0	265	175	90	115.6
66.0	262	173	89	114.2
66.0	259	171	88	112.9
66.0	256	169	87	111.6
66.0	253	167	86	110.2
66.0	250	165	85	108.9
66.0	247	163	84	107.6
66.0	244	161	83	106.2
66.0	241	159	82	104.9
66.0	238	157	81	103.6
66.0	235	155	80	102.2
66.0	215	142	73	93.8
66.0	212	140	72	92.5
66.0	209	138	71	91.1
66.0	206	136	70	89.8
66.0	203	134	69	88.5
66.0	200	132	68	87.1
66.0	197	130	67	85.8
66.0	194	128	66	84.5
66.0	191	126	65	83.1
66.0	188	124	64	81.8
66.0	162	107	55	70.7
66.0	159	105	54	69.3
66.0	156	103	53	68.0
66.0	153	101	52	66.7
66.0	150	99	51	65.3
66.0	147	97	50	64.0
66.0	144	95	49	62.7
66.0	141	93	48	61.3
66.0	106	70	36	46.2
66.0	103	68	35	44.9
66.0	100	66	34	43.6
66.0	97	64	33	42.2
66.0	94	62	32	40.9
66.0	53	35	18	23.1
66.0	50	33	17	21.8
66.0	47	31	16	20.4
65.9	499	329	170	216.9
65.9	498	328	170	216.0
65.9	496	327	169	215.6
65.9	495	326	169	214.7
65.9	493	325	168	214.2
65.9	492	324	168	213.4
65.9	490	323	167	212.9
65.9	487	321	166	211.6
65.9	484	319	165	210.3
65.9	481	317	164	208.9
65.9	478	315	163	207.6
65.9	475	313	162	206.3
65.9	472	311	161	204.9
65.9	469	309	160	203.6
65.9	466	307	159	202.3
65.9	464	306	158	201.8
65.9	463	305	158	200.9
65.9	461	304	157	200.5
65.9	460	303	157	199.6
65.9	458	302	156	199.1
65.9	457	301	156	198.3
65.9	455	300	155	197.8
65.9	454	299	155	196.9
65.9	452	298	154	196.5
65.9	451	297	154	195.6
65.9	449	296	153	195.1
65.9	446	294	152	193.8
65.9	443	292	151	192.5
65.9	440	290	150	191.1
65.9	437	288	149	189.8
65.9	434	286	148	188.5
65.9	431	284	147	187.1
65.9	428	282	146	185.8
65.9	425	280	145	184.5
65.9	422	278	144	183.1
65.9	419	276	143	181.8
65.9	417	275	142	181.4
65.9	416	274	142	180.5
65.9	414	273	141	180.0
65.9	413	272	141	179.1
65.9	411	271	140	178.7
65.9	410	270	140	177.8
65.9	408	269	139	177.4
65.9	405	267	138	176.0
65.9	402	265	137	174.7
65.9	399	263	136	173.4
65.9	396	261	135	172.0
65.9	393	259	134	170.7
65.9	390	257	133	169.4
65.9	387	255	132	168.0
65.9	384	253	131	166.7
65.9	381	251	130	165.4
65.9	378	249	129	164.0
65.9	375	247	128	162.7
65.9	372	245	127	161.4
65.9	370	244	126	160.9
65.9	369	243	126	160.0
65.9	367	242	125	159.6
65.9	364	240	124	158.2
65.9	361	238	123	156.9
65.9	358	236	122	155.6
65.9	355	234	121	154.2
65.9	352	232	120	152.9
65.9	349	230	119	151.6
65.9	346	228	118	150.2
65.9	343	226	117	148.9
65.9	340	224	116	147.6
65.9	337	222	115	146.2
65.9	334	220	114	144.9
65.9	331	218	113	143.6
65.9	328	216	112	142.2
65.9	323	213	110	140.5
65.9	320	211	109	139.1
65.9	317	209	108	137.8
65.9	314	207	107	136.5
65.9	311	205	106	135.1
65.9	308	203	105	133.8
65.9	305	201	104	132.5
65.9	302	199	103	131.1
65.9	299	197	102	129.8
65.9	296	195	101	128.5
65.9	293	193	100	127.1
65.9	290	191	99	125.8
65.9	287	189	98	124.5
65.9	276	182	94	120.0
65.9	273	180	93	118.7
65.9	270	178	92	117.3
65.9	267	176	91	116.0
65.9	264	174	90	114.7
65.9	261	172	89	113.3
65.9	258	170	88	112.0
65.9	255	168	87	110.7
65.9	252	166	86	109.3
65.9	249	164	85	108.0
65.9	246	162	84	106.7
65.9	232	153	79	100.9
65.9	229	151	78	99.6
65.9	226	149	77	98.2
65.9	223	147	76	96.9
65.9	220	145	75	95.6

%E	M1	M2	DM	M*
65.9	217	143	74	94.2
65.9	214	141	73	92.9
65.9	211	139	72	91.6
65.9	208	137	71	90.2
65.9	205	135	70	88.9
65.9	185	122	63	80.5
65.9	182	120	62	79.1
65.9	179	118	61	77.8
65.9	176	116	60	76.5
65.9	173	114	59	75.1
65.9	170	112	58	73.8
65.9	167	110	57	72.5
65.9	164	108	56	71.1
65.9	138	91	47	60.0
65.9	135	89	46	58.7
65.9	132	87	45	57.3
65.9	129	85	44	56.0
65.9	126	83	43	54.7
65.9	123	81	42	53.3
65.9	91	60	31	39.6
65.9	88	58	30	38.2
65.9	85	56	29	36.9
65.9	82	54	28	35.6
65.9	44	29	15	19.1
65.9	41	27	14	17.8
65.8	500	329	171	216.5
65.8	497	327	170	215.1
65.8	494	325	169	213.8
65.8	491	323	168	212.5
65.8	489	322	167	212.0
65.8	488	321	167	211.1
65.8	486	320	166	210.7
65.8	485	319	166	209.8
65.8	483	318	165	209.4
65.8	482	317	165	208.5
65.8	480	316	164	208.0
65.8	479	315	164	207.2
65.8	477	314	163	206.7
65.8	476	313	163	205.8
65.8	474	312	162	205.4
65.8	473	311	162	204.5
65.8	471	310	161	204.0
65.8	468	308	160	202.7
65.8	465	306	159	201.4
65.8	462	304	158	200.0
65.8	459	302	157	198.7
65.8	456	300	156	197.4
65.8	453	298	155	196.0
65.8	450	296	154	194.7
65.8	448	295	153	194.3

%E	M1	M2	DM	M*
65.8	447	294	153	193.4
65.8	445	293	152	192.9
65.8	444	292	152	192.0
65.8	442	291	151	191.6
65.8	441	290	151	190.7
65.8	439	289	150	190.3
65.8	438	288	150	189.4
65.8	436	287	149	188.9
65.8	433	285	148	187.6
65.8	430	283	147	186.3
65.8	427	281	146	184.9
65.8	424	279	145	183.6
65.8	421	277	144	182.3
65.8	418	275	143	180.9
65.8	415	273	142	179.6
65.8	412	271	141	178.3
65.8	409	269	140	176.9
65.8	407	268	139	176.5
65.8	406	267	139	175.6
65.8	404	266	138	175.1
65.8	403	265	138	174.3
65.8	401	264	137	173.8
65.8	400	263	137	172.9
65.8	398	262	136	172.5
65.8	395	260	135	171.1
65.8	392	258	134	169.8
65.8	389	256	133	168.5
65.8	386	254	132	167.1
65.8	383	252	131	165.8
65.8	380	250	130	164.5
65.8	377	248	129	163.1
65.8	374	246	128	161.8
65.8	371	244	127	160.5
65.8	368	242	126	159.1
65.8	366	241	125	158.7
65.8	365	240	125	157.8
65.8	363	239	124	157.4
65.8	360	237	123	156.0
65.8	357	235	122	154.7
65.8	354	233	121	153.4
65.8	351	231	120	152.0
65.8	348	229	119	150.7
65.8	345	227	118	149.4
65.8	342	225	117	148.0
65.8	339	223	116	146.7
65.8	336	221	115	145.4
65.8	333	219	114	144.0
65.8	330	217	113	142.7
65.8	325	214	111	140.9
65.8	322	212	110	139.6

%E	M1	M2	DM	M*
65.8	319	210	109	138.2
65.8	316	208	108	136.9
65.8	313	206	107	135.6
65.8	310	204	106	134.2
65.8	307	202	105	132.9
65.8	304	200	104	131.6
65.8	301	198	103	130.2
65.8	298	196	102	128.9
65.8	295	194	101	127.6
65.8	292	192	100	126.2
65.8	284	187	97	123.1
65.8	281	185	96	121.8
65.8	278	183	95	120.5
65.8	275	181	94	119.1
65.8	272	179	93	117.8
65.8	269	177	92	116.5
65.8	266	175	91	115.1
65.8	263	173	90	113.8
65.8	260	171	89	112.5
65.8	257	169	88	111.1
65.8	243	160	83	105.3
65.8	240	158	82	104.0
65.8	237	156	81	102.7
65.8	234	154	80	101.4
65.8	231	152	79	100.0
65.8	228	150	78	98.7
65.8	225	148	77	97.4
65.8	222	146	76	96.0
65.8	219	144	75	94.7
65.8	202	133	69	87.6
65.8	199	131	68	86.2
65.8	196	129	67	84.9
65.8	193	127	66	83.6
65.8	190	125	65	82.2
65.8	187	123	64	80.9
65.8	184	121	63	79.6
65.8	161	106	55	69.8
65.8	158	104	54	68.5
65.8	155	102	53	67.1
65.8	152	100	52	65.8
65.8	149	98	51	64.5
65.8	146	96	50	63.1
65.8	120	79	41	52.0
65.8	117	77	40	50.7
65.8	114	75	39	49.3
65.8	111	73	38	48.0
65.8	79	52	27	34.2
65.8	76	50	26	32.9
65.8	73	48	25	31.6
65.8	38	25	13	16.4

%E	M1	M2	DM	M*
65.7	499	328	171	215.6
65.7	498	327	171	214.7
65.7	496	326	170	214.3
65.7	495	325	170	213.4
65.7	493	324	169	212.9
65.7	492	323	169	212.1
65.7	490	322	168	211.6
65.7	487	320	167	210.3
65.7	484	318	166	208.9
65.7	481	316	165	207.6
65.7	478	314	164	206.3
65.7	475	312	163	204.9
65.7	472	310	162	203.6
65.7	470	309	161	203.2
65.7	469	308	161	202.3
65.7	467	307	160	201.8
65.7	466	306	160	200.9
65.7	464	305	159	200.5
65.7	463	304	159	199.6
65.7	461	303	158	199.2
65.7	460	302	158	198.3
65.7	458	301	157	197.8
65.7	455	299	156	196.5
65.7	452	297	155	195.2
65.7	449	295	154	193.8
65.7	446	293	153	192.5
65.7	443	291	152	191.2
65.7	440	289	151	189.8
65.7	437	287	150	188.5
65.7	435	286	149	188.0
65.7	434	285	149	187.2
65.7	432	284	148	186.7
65.7	431	283	148	185.8
65.7	429	282	147	185.4
65.7	428	281	147	184.5
65.7	426	280	146	184.0
65.7	423	278	145	182.7
65.7	420	276	144	181.4
65.7	417	274	143	180.0
65.7	414	272	142	178.7
65.7	411	270	141	177.4
65.7	408	268	140	176.0
65.7	405	266	139	174.7
65.7	402	264	138	173.4
65.7	399	262	137	172.0
65.7	397	261	136	171.6
65.7	396	260	136	170.7
65.7	394	259	135	170.3
65.7	391	257	134	168.9
65.7	388	255	133	167.6

%E	M1	M2	DM	M*
65.7	385	253	132	166.3
65.7	382	251	131	164.9
65.7	379	249	130	163.6
65.7	376	247	129	162.3
65.7	373	245	128	160.9
65.7	370	243	127	159.6
65.7	367	241	126	158.3
65.7	364	239	125	156.9
65.7	362	238	124	156.5
65.7	361	237	124	155.6
65.7	359	236	123	155.1
65.7	356	234	122	153.8
65.7	353	232	121	152.5
65.7	350	230	120	151.1
65.7	347	228	119	149.8
65.7	344	226	118	148.5
65.7	341	224	117	147.1
65.7	338	222	116	145.8
65.7	335	220	115	144.5
65.7	332	218	114	143.1
65.7	329	216	113	141.8
65.7	327	215	112	141.4
65.7	324	213	111	140.0
65.7	321	211	110	138.7
65.7	318	209	109	137.4
65.7	315	207	108	136.0
65.7	312	205	107	134.7
65.7	309	203	106	133.4
65.7	306	201	105	132.0
65.7	303	199	104	130.7
65.7	300	197	103	129.4
65.7	297	195	102	128.0
65.7	289	190	99	124.9
65.7	286	188	98	123.6
65.7	283	186	97	122.2
65.7	280	184	96	120.9
65.7	277	182	95	119.6
65.7	274	180	94	118.2
65.7	271	178	93	116.9
65.7	268	176	92	115.6
65.7	265	174	91	114.2
65.7	254	167	87	109.8
65.7	251	165	86	108.5
65.7	248	163	85	107.1
65.7	245	161	84	105.8
65.7	242	159	83	104.5
65.7	239	157	82	103.1
65.7	236	155	81	101.8
65.7	233	153	80	100.5
65.7	230	151	79	99.1

%E	M1	M2	DM	M*
65.7	216	142	74	93.4
65.7	213	140	73	92.0
65.7	210	138	72	90.7
65.7	207	136	71	89.4
65.7	204	134	70	88.0
65.7	201	132	69	86.7
65.7	198	130	68	85.4
65.7	181	119	62	78.2
65.7	178	117	61	76.9
65.7	175	115	60	75.6
65.7	172	113	59	74.2
65.7	169	111	58	72.9
65.7	166	109	57	71.6
65.7	143	94	49	61.8
65.7	140	92	48	60.5
65.7	137	90	47	59.1
65.7	134	88	46	57.8
65.7	108	71	37	46.7
65.7	105	69	36	45.3
65.7	102	67	35	44.0
65.7	99	65	34	42.7
65.7	70	46	24	30.2
65.7	67	44	23	28.9
65.7	35	23	12	15.1
65.6	500	328	172	215.2
65.6	497	326	171	213.8
65.6	494	324	170	212.5
65.6	491	322	169	211.2
65.6	489	321	168	210.7
65.6	488	320	168	209.8
65.6	486	319	167	209.4
65.6	485	318	167	208.5
65.6	483	317	166	208.1
65.6	482	316	166	207.2
65.6	480	315	165	206.7
65.6	479	314	165	205.8
65.6	477	313	164	205.4
65.6	474	311	163	204.1
65.6	471	309	162	202.7
65.6	468	307	161	201.4
65.6	465	305	160	200.1
65.6	462	303	159	198.7
65.6	459	301	158	197.4
65.6	457	300	157	196.9
65.6	456	299	157	196.1
65.6	454	298	156	195.6
65.6	453	297	156	194.7
65.6	451	296	155	194.3
65.6	450	295	155	193.4
65.6	448	294	154	192.9

%E	M1	M2	DM	M*
65.6	445	292	153	191.6
65.6	442	290	152	190.3
65.6	439	288	151	188.9
65.6	436	286	150	187.6
65.6	433	284	149	186.3
65.6	430	282	148	184.9
65.6	427	280	147	183.6
65.6	425	279	146	183.2
65.6	424	278	146	182.3
65.6	422	277	145	181.8
65.6	421	276	145	180.9
65.6	419	275	144	180.5
65.6	418	274	144	179.6
65.6	416	273	143	179.2
65.6	413	271	142	177.8
65.6	410	269	141	176.5
65.6	407	267	140	175.2
65.6	404	265	139	173.8
65.6	401	263	138	172.5
65.6	398	261	137	171.2
65.6	395	259	136	169.8
65.6	393	258	135	169.4
65.6	392	257	135	168.5
65.6	390	256	134	168.0
65.6	389	255	134	167.2
65.6	387	254	133	166.7
65.6	384	252	132	165.4
65.6	381	250	131	164.0
65.6	378	248	130	162.7
65.6	375	246	129	161.4
65.6	372	244	128	160.0
65.6	369	242	127	158.7
65.6	366	240	126	157.4
65.6	363	238	125	156.0
65.6	360	236	124	154.7
65.6	358	235	123	154.3
65.6	355	233	122	152.9
65.6	352	231	121	151.6
65.6	349	229	120	150.3
65.6	346	227	119	148.9
65.6	343	225	118	147.6
65.6	340	223	117	146.3
65.6	337	221	116	144.9
65.6	334	219	115	143.6
65.6	331	217	114	142.3
65.6	326	214	112	140.5
65.6	323	212	111	139.1
65.6	320	210	110	137.8
65.6	317	208	109	136.5
65.6	314	206	108	135.1

%E	M1	M2	DM	M*
65.6	311	204	107	133.8
65.6	308	202	106	132.5
65.6	305	200	105	131.1
65.6	302	198	104	129.8
65.6	299	196	103	128.5
65.6	294	193	101	126.7
65.6	291	191	100	125.4
65.6	288	189	99	124.0
65.6	285	187	98	122.7
65.6	282	185	97	121.4
65.6	279	183	96	120.0
65.6	276	181	95	118.7
65.6	273	179	94	117.4
65.6	270	177	93	116.0
65.6	262	172	90	112.9
65.6	259	170	89	111.6
65.6	256	168	88	110.3
65.6	253	166	87	108.9
65.6	250	164	86	107.6
65.6	247	162	85	106.3
65.6	244	160	84	104.9
65.6	241	158	83	103.6
65.6	227	149	78	97.8
65.6	224	147	77	96.5
65.6	221	145	76	95.1
65.6	218	143	75	93.8
65.6	215	141	74	92.5
65.6	212	139	73	91.1
65.6	209	137	72	89.8
65.6	195	128	67	84.0
65.6	192	126	66	82.7
65.6	189	124	65	81.4
65.6	186	122	64	80.0
65.6	183	120	63	78.7
65.6	180	118	62	77.4
65.6	163	107	56	70.2
65.6	160	105	55	68.9
65.6	157	103	54	67.6
65.6	154	101	53	66.2
65.6	151	99	52	64.9
65.6	131	86	45	56.5
65.6	128	84	44	55.1
65.6	125	82	43	53.8
65.6	122	80	42	52.5
65.6	96	63	33	41.3
65.6	93	61	32	40.0
65.6	90	59	31	38.7
65.6	64	42	22	27.6
65.6	61	40	21	26.2
65.6	32	21	11	13.8

%E	M1	M2	DM	M*
65.5	499	327	172	214.3
65.5	498	326	172	213.4
65.5	496	325	171	213.0
65.5	495	324	171	212.1
65.5	493	323	170	211.6
65.5	490	321	169	210.3
65.5	487	319	168	209.0
65.5	484	317	167	207.6
65.5	481	315	166	206.3
65.5	478	313	165	205.0
65.5	476	312	164	204.5
65.5	475	311	164	203.6
65.5	473	310	163	203.2
65.5	472	309	163	202.3
65.5	470	308	162	201.8
65.5	469	307	162	201.0
65.5	467	306	161	200.5
65.5	466	305	161	199.6
65.5	464	304	160	199.2
65.5	461	302	159	197.8
65.5	458	300	158	196.5
65.5	455	298	157	195.2
65.5	452	296	156	193.8
65.5	449	294	155	192.5
65.5	447	293	154	192.1
65.5	446	292	154	191.2
65.5	444	291	153	190.7
65.5	443	290	153	189.8
65.5	441	289	152	189.4
65.5	440	288	152	188.5
65.5	438	287	151	188.1
65.5	435	285	150	186.7
65.5	432	283	149	185.4
65.5	429	281	148	184.1
65.5	426	279	147	182.7
65.5	423	277	146	181.4
65.5	420	275	145	180.1
65.5	417	273	144	178.7
65.5	415	272	143	178.3
65.5	414	271	143	177.4
65.5	412	270	142	176.9
65.5	411	269	142	176.1
65.5	409	268	141	175.6
65.5	406	266	140	174.3
65.5	403	264	139	172.9
65.5	400	262	138	171.6
65.5	397	260	137	170.3
65.5	394	258	136	168.9
65.5	391	256	135	167.6
65.5	388	254	134	166.3

%E	M1	M2	DM	M*
65.5	386	253	133	165.8
65.5	385	252	133	164.9
65.5	383	251	132	164.5
65.5	380	249	131	163.2
65.5	377	247	130	161.8
65.5	374	245	129	160.5
65.5	371	243	128	159.2
65.5	368	241	127	157.8
65.5	365	239	126	156.5
65.5	362	237	125	155.2
65.5	359	235	124	153.8
65.5	357	234	123	153.4
65.5	354	232	122	152.0
65.5	351	230	121	150.7
65.5	348	228	120	149.4
65.5	345	226	119	148.0
65.5	342	224	118	146.7
65.5	339	222	117	145.4
65.5	336	220	116	144.0
65.5	333	218	115	142.7
65.5	330	216	114	141.4
65.5	328	215	113	140.9
65.5	325	213	112	139.6
65.5	322	211	111	138.3
65.5	319	209	110	136.9
65.5	316	207	109	135.6
65.5	313	205	108	134.3
65.5	310	203	107	132.9
65.5	307	201	106	131.6
65.5	304	199	105	130.3
65.5	296	194	102	127.1
65.5	293	192	101	125.8
65.5	290	190	100	124.5
65.5	287	188	99	123.1
65.5	284	186	98	121.8
65.5	281	184	97	120.5
65.5	278	182	96	119.2
65.5	275	180	95	117.8
65.5	267	175	92	114.7
65.5	264	173	91	113.4
65.5	261	171	90	112.0
65.5	258	169	89	110.7
65.5	255	167	88	109.4
65.5	252	165	87	108.0
65.5	249	163	86	106.7
65.5	238	156	82	102.3
65.5	235	154	81	100.9
65.5	232	152	80	99.6
65.5	229	150	79	98.3
65.5	226	148	78	96.9

%E	M1	M2	DM	M*
65.5	223	146	77	95.6
65.5	220	144	76	94.3
65.5	206	135	71	88.5
65.5	203	133	70	87.1
65.5	200	131	69	85.8
65.5	197	129	68	84.5
65.5	194	127	67	83.1
65.5	177	116	61	76.0
65.5	174	114	60	74.7
65.5	171	112	59	73.4
65.5	168	110	58	72.0
65.5	165	108	57	70.7
65.5	148	97	51	63.6
65.5	145	95	50	62.2
65.5	142	93	49	60.9
65.5	139	91	48	59.6
65.5	119	78	41	51.1
65.5	116	76	40	49.8
65.5	113	74	39	48.5
65.5	110	72	38	47.1
65.5	87	57	30	37.3
65.5	84	55	29	36.0
65.5	58	38	20	24.9
65.5	55	36	19	23.6
65.5	29	19	10	12.4
65.4	500	327	173	213.9
65.4	497	325	172	212.5
65.4	494	323	171	211.2
65.4	492	322	170	210.7
65.4	491	321	170	209.9
65.4	489	320	169	209.4
65.4	488	319	169	208.5
65.4	486	318	168	208.1
65.4	485	317	168	207.2
65.4	483	316	167	206.7
65.4	482	315	167	205.9
65.4	480	314	166	205.4
65.4	477	312	165	204.1
65.4	474	310	164	202.7
65.4	471	308	163	201.4
65.4	468	306	162	200.1
65.4	465	304	161	198.7
65.4	463	303	160	198.3
65.4	462	302	160	197.4
65.4	460	301	159	197.0
65.4	459	300	159	196.1
65.4	457	299	158	195.6
65.4	456	298	158	194.7
65.4	454	297	157	194.3
65.4	451	295	156	193.0

%E	M1	M2	DM	M*
65.4	448	293	155	191.6
65.4	445	291	154	190.3
65.4	442	289	153	189.0
65.4	439	287	152	187.6
65.4	437	286	151	187.2
65.4	436	285	151	186.3
65.4	434	284	150	185.8
65.4	433	283	150	185.0
65.4	431	282	149	184.5
65.4	428	280	148	183.2
65.4	425	278	147	181.8
65.4	422	276	146	180.5
65.4	419	274	145	179.2
65.4	416	272	144	177.8
65.4	413	270	143	176.5
65.4	410	268	142	175.2
65.4	408	267	141	174.7
65.4	407	266	141	173.8
65.4	405	265	140	173.4
65.4	402	263	139	172.1
65.4	399	261	138	170.7
65.4	396	259	137	169.4
65.4	393	257	136	168.1
65.4	390	255	135	166.7
65.4	387	253	134	165.4
65.4	384	251	133	164.1
65.4	382	250	132	163.6
65.4	381	249	132	162.7
65.4	379	248	131	162.3
65.4	376	246	130	160.9
65.4	373	244	129	159.6
65.4	370	242	128	158.3
65.4	367	240	127	156.9
65.4	364	238	126	155.6
65.4	361	236	125	154.3
65.4	358	234	124	152.9
65.4	356	233	123	152.5
65.4	355	232	123	151.6
65.4	353	231	122	151.2
65.4	350	229	121	149.8
65.4	347	227	120	148.5
65.4	344	225	119	147.2
65.4	341	223	118	145.8
65.4	338	221	117	144.5
65.4	335	219	116	143.2
65.4	332	217	115	141.8
65.4	329	215	114	140.5
65.4	327	214	113	140.0
65.4	324	212	112	138.7
65.4	321	210	111	137.4

%E	M1	M2	DM	M*
65.4	318	208	110	136.1
65.4	315	206	109	134.7
65.4	312	204	108	133.4
65.4	309	202	107	132.1
65.4	306	200	106	130.7
65.4	301	197	104	128.9
65.4	298	195	103	127.6
65.4	295	193	102	126.3
65.4	292	191	101	124.9
65.4	289	189	100	123.6
65.4	286	187	99	122.3
65.4	283	185	98	120.9
65.4	280	183	97	119.6
65.4	272	178	94	116.5
65.4	269	176	93	115.2
65.4	266	174	92	113.8
65.4	263	172	91	112.5
65.4	260	170	90	111.2
65.4	257	168	89	109.8
65.4	254	166	88	108.5
65.4	246	161	85	105.4
65.4	243	159	84	104.0
65.4	240	157	83	102.7
65.4	237	155	82	101.4
65.4	234	153	81	100.0
65.4	231	151	80	98.7
65.4	228	149	79	97.4
65.4	217	142	75	92.9
65.4	214	140	74	91.6
65.4	211	138	73	90.3
65.4	208	136	72	88.9
65.4	205	134	71	87.6
65.4	191	125	66	81.8
65.4	188	123	65	80.5
65.4	185	121	64	79.1
65.4	182	119	63	77.8
65.4	179	117	62	76.5
65.4	162	106	56	69.4
65.4	159	104	55	68.0
65.4	156	102	54	66.7
65.4	153	100	53	65.4
65.4	136	89	47	58.2
65.4	133	87	46	56.9
65.4	130	85	45	55.6
65.4	127	83	44	54.2
65.4	107	70	37	45.8
65.4	104	68	36	44.5
65.4	81	53	28	34.7
65.4	78	51	27	33.3
65.4	52	34	18	22.2

%E	M1	M2	DM	M*
65.4	26	17	9	11.1
65.3	499	326	173	213.0
65.3	498	325	173	212.1
65.3	496	324	172	211.6
65.3	495	323	172	210.8
65.3	493	322	171	210.3
65.3	490	320	170	209.0
65.3	487	318	169	207.6
65.3	484	316	168	206.3
65.3	481	314	167	205.0
65.3	479	313	166	204.5
65.3	478	312	166	203.6
65.3	476	311	165	203.2
65.3	475	310	165	202.3
65.3	473	309	164	201.9
65.3	472	308	164	201.0
65.3	470	307	163	200.5
65.3	467	305	162	199.2
65.3	464	303	161	197.9
65.3	461	301	160	196.5
65.3	458	299	159	195.2
65.3	455	297	158	193.9
65.3	453	296	157	193.4
65.3	452	295	157	192.5
65.3	450	294	156	192.1
65.3	449	293	156	191.2
65.3	447	292	155	190.7
65.3	444	290	154	189.4
65.3	441	288	153	188.1
65.3	438	286	152	186.7
65.3	435	284	151	185.4
65.3	432	282	150	184.1
65.3	430	281	149	183.6
65.3	429	280	149	182.8
65.3	427	279	148	182.3
65.3	426	278	148	181.4
65.3	424	277	147	181.0
65.3	421	275	146	179.6
65.3	418	273	145	178.3
65.3	415	271	144	177.0
65.3	412	269	143	175.6
65.3	409	267	142	174.3
65.3	406	265	141	173.0
65.3	404	264	140	172.5
65.3	403	263	140	171.6
65.3	401	262	139	171.2
65.3	400	261	139	170.3
65.3	398	260	138	169.8
65.3	395	258	137	168.5
65.3	392	256	136	167.2

%E	M1	M2	DM	M*
65.3	389	254	135	165.9
65.3	386	252	134	164.5
65.3	383	250	133	163.2
65.3	380	248	132	161.9
65.3	378	247	131	161.4
65.3	377	246	131	160.5
65.3	375	245	130	160.1
65.3	372	243	129	158.7
65.3	369	241	128	157.4
65.3	366	239	127	156.1
65.3	363	237	126	154.7
65.3	360	235	125	153.4
65.3	357	233	124	152.1
65.3	354	231	123	150.7
65.3	352	230	122	150.3
65.3	349	228	121	149.0
65.3	346	226	120	147.6
65.3	343	224	119	146.3
65.3	340	222	118	145.0
65.3	337	220	117	143.6
65.3	334	218	116	142.3
65.3	331	216	115	141.0
65.3	326	213	113	139.2
65.3	323	211	112	137.8
65.3	320	209	111	136.5
65.3	317	207	110	135.2
65.3	314	205	109	133.8
65.3	311	203	108	132.5
65.3	308	201	107	131.2
65.3	303	198	105	129.4
65.3	300	196	104	128.1
65.3	297	194	103	126.7
65.3	294	192	102	125.4
65.3	291	190	101	124.1
65.3	288	188	100	122.7
65.3	285	186	99	121.4
65.3	277	181	96	118.3
65.3	274	179	95	116.9
65.3	271	177	94	115.6
65.3	268	175	93	114.3
65.3	265	173	92	112.9
65.3	262	171	91	111.6
65.3	259	169	90	110.3
65.3	251	164	87	107.2
65.3	248	162	86	105.8
65.3	245	160	85	104.5
65.3	242	158	84	103.2
65.3	239	156	83	101.8
65.3	236	154	82	100.5
65.3	225	147	78	96.0

%E	M1	M2	DM	M*
65.3	222	145	77	94.7
65.3	219	143	76	93.4
65.3	216	141	75	92.0
65.3	213	139	74	90.7
65.3	202	132	70	86.3
65.3	199	130	69	84.9
65.3	196	128	68	83.6
65.3	193	126	67	82.3
65.3	190	124	66	80.9
65.3	176	115	61	75.1
65.3	173	113	60	73.8
65.3	170	111	59	72.5
65.3	167	109	58	71.1
65.3	150	98	52	64.0
65.3	147	96	51	62.7
65.3	144	94	50	61.4
65.3	124	81	43	52.9
65.3	121	79	42	51.6
65.3	118	77	41	50.2
65.3	101	66	35	43.1
65.3	98	64	34	41.8
65.3	95	62	33	40.5
65.3	75	49	26	32.0
65.3	72	47	25	30.7
65.3	49	32	17	20.9
65.2	500	326	174	212.6
65.2	497	324	173	211.2
65.2	494	322	172	209.9
65.2	492	321	171	209.4
65.2	491	320	171	208.6
65.2	489	319	170	208.1
65.2	488	318	170	207.2
65.2	486	317	169	206.8
65.2	485	316	169	205.9
65.2	483	315	168	205.4
65.2	480	313	167	204.1
65.2	477	311	166	202.8
65.2	474	309	165	201.4
65.2	471	307	164	200.1
65.2	469	306	163	199.7
65.2	468	305	163	198.8
65.2	466	304	162	198.3
65.2	465	303	162	197.4
65.2	463	302	161	197.0
65.2	462	301	161	196.1
65.2	460	300	160	195.7
65.2	457	298	159	194.3
65.2	454	296	158	193.0
65.2	451	294	157	191.7
65.2	448	292	156	190.3

%E	M1	M2	DM	M*
65.2	446	291	155	189.9
65.2	445	290	155	189.0
65.2	443	289	154	188.5
65.2	442	288	154	187.7
65.2	440	287	153	187.2
65.2	437	285	152	185.9
65.2	434	283	151	184.5
65.2	431	281	150	183.2
65.2	428	279	149	181.9
65.2	425	277	148	180.5
65.2	423	276	147	180.1
65.2	422	275	147	179.2
65.2	420	274	146	178.8
65.2	419	273	146	177.9
65.2	417	272	145	177.4
65.2	414	270	144	176.1
65.2	411	268	143	174.8
65.2	408	266	142	173.4
65.2	405	264	141	172.1
65.2	402	262	140	170.8
65.2	399	260	139	169.4
65.2	397	259	138	169.0
65.2	396	258	138	168.1
65.2	394	257	137	167.6
65.2	391	255	136	166.3
65.2	388	253	135	165.0
65.2	385	251	134	163.6
65.2	382	249	133	162.3
65.2	379	247	132	161.0
65.2	376	245	131	159.6
65.2	374	244	130	159.2
65.2	371	242	129	157.9
65.2	368	240	128	156.5
65.2	365	238	127	155.2
65.2	362	236	126	153.9
65.2	359	234	125	152.5
65.2	356	232	124	151.2
65.2	353	230	123	149.9
65.2	351	229	122	149.4
65.2	348	227	121	148.1
65.2	345	225	120	146.7
65.2	342	223	119	145.4
65.2	339	221	118	144.1
65.2	336	219	117	142.7
65.2	333	217	116	141.4
65.2	330	215	115	140.1
65.2	328	214	114	139.6
65.2	325	212	113	138.3
65.2	322	210	112	137.0
65.2	319	208	111	135.6

%E	M1	M2	DM	M*
65.2	316	206	110	134.3
65.2	313	204	109	133.0
65.2	310	202	108	131.6
65.2	305	199	106	129.8
65.2	302	197	105	128.5
65.2	299	195	104	127.2
65.2	296	193	103	125.8
65.2	293	191	102	124.5
65.2	290	189	101	123.2
65.2	287	187	100	121.8
65.2	282	184	98	120.1
65.2	279	182	97	118.7
65.2	276	180	96	117.4
65.2	273	178	95	116.1
65.2	270	176	94	114.7
65.2	267	174	93	113.4
65.2	264	172	92	112.1
65.2	256	167	89	108.9
65.2	253	165	88	107.6
65.2	250	163	87	106.3
65.2	247	161	86	104.9
65.2	244	159	85	103.6
65.2	233	152	81	99.2
65.2	230	150	80	97.8
65.2	227	148	79	96.5
65.2	224	146	78	95.2
65.2	221	144	77	93.8
65.2	210	137	73	89.4
65.2	207	135	72	88.0
65.2	204	133	71	86.7
65.2	201	131	70	85.4
65.2	198	129	69	84.0
65.2	187	122	65	79.6
65.2	184	120	64	78.3
65.2	181	118	63	76.9
65.2	178	116	62	75.6
65.2	164	107	57	69.8
65.2	161	105	56	68.5
65.2	158	103	55	67.1
65.2	155	101	54	65.8
65.2	141	92	49	60.0
65.2	138	90	48	58.7
65.2	135	88	47	57.4
65.2	132	86	46	56.0
65.2	115	75	40	48.9
65.2	112	73	39	47.6
65.2	92	60	32	39.1
65.2	89	58	31	37.8
65.2	69	45	24	29.3
65.2	66	43	23	28.0

%E	M1	M2	DM	M*
65.2	46	30	16	19.6
65.2	23	15	8	9.8
65.1	499	325	174	211.7
65.1	498	324	174	210.8
65.1	496	323	173	210.3
65.1	495	322	173	209.5
65.1	493	321	172	209.0
65.1	490	319	171	207.7
65.1	487	317	170	206.3
65.1	484	315	169	205.0
65.1	482	314	168	204.6
65.1	481	313	168	203.7
65.1	479	312	167	203.2
65.1	478	311	167	202.3
65.1	476	310	166	201.9
65.1	475	309	166	201.0
65.1	473	308	165	200.6
65.1	470	306	164	199.2
65.1	467	304	163	197.9
65.1	464	302	162	196.6
65.1	461	300	161	195.2
65.1	459	299	160	194.8
65.1	458	298	160	193.9
65.1	456	297	159	193.4
65.1	455	296	159	192.6
65.1	453	295	158	192.1
65.1	450	293	157	190.8
65.1	447	291	156	189.4
65.1	444	289	155	188.1
65.1	441	287	154	186.8
65.1	439	286	153	186.3
65.1	438	285	153	185.4
65.1	436	284	152	185.0
65.1	435	283	152	184.1
65.1	433	282	151	183.7
65.1	430	280	150	182.3
65.1	427	278	149	181.0
65.1	424	276	148	179.7
65.1	421	274	147	178.3
65.1	418	272	146	177.0
65.1	416	271	145	176.5
65.1	415	270	145	175.7
65.1	413	269	144	175.2
65.1	410	267	143	173.9
65.1	407	265	142	172.5
65.1	404	263	141	171.2
65.1	401	261	140	169.9
65.1	398	259	139	168.5
65.1	395	257	138	167.2
65.1	393	256	137	166.8

%E	M1	M2	DM	M*
65.1	392	255	137	165.9
65.1	390	254	136	165.4
65.1	387	252	135	164.1
65.1	384	250	134	162.8
65.1	381	248	133	161.4
65.1	378	246	132	160.1
65.1	375	244	131	158.8
65.1	373	243	130	158.3
65.1	372	242	130	157.4
65.1	370	241	129	157.0
65.1	367	239	128	155.6
65.1	364	237	127	154.3
65.1	361	235	126	153.0
65.1	358	233	125	151.6
65.1	355	231	124	150.3
65.1	352	229	123	149.0
65.1	350	228	122	148.5
65.1	347	226	121	147.2
65.1	344	224	120	145.9
65.1	341	222	119	144.5
65.1	338	220	118	143.2
65.1	335	218	117	141.9
65.1	332	216	116	140.5
65.1	327	213	114	138.7
65.1	324	211	113	137.4
65.1	321	209	112	136.1
65.1	318	207	111	134.7
65.1	315	205	110	133.4
65.1	312	203	109	132.1
65.1	307	200	107	130.3
65.1	304	198	106	129.0
65.1	301	196	105	127.6
65.1	298	194	104	126.3
65.1	295	192	103	125.0
65.1	292	190	102	123.6
65.1	289	188	101	122.3
65.1	284	185	99	120.5
65.1	281	183	98	119.2
65.1	278	181	97	117.8
65.1	275	179	96	116.5
65.1	272	177	95	115.2
65.1	269	175	94	113.8
65.1	261	170	91	110.7
65.1	258	168	90	109.4
65.1	255	166	89	108.1
65.1	252	164	88	106.7
65.1	249	162	87	105.4
65.1	241	157	84	102.3
65.1	238	155	83	100.9
65.1	235	153	82	99.6

%E	M1	M2	DM	M*
65.1	232	151	81	98.3
65.1	229	149	80	96.9
65.1	218	142	76	92.5
65.1	215	140	75	91.2
65.1	212	138	74	89.8
65.1	209	136	73	88.5
65.1	195	127	68	82.7
65.1	192	125	67	81.4
65.1	189	123	66	80.0
65.1	186	121	65	78.7
65.1	175	114	61	74.3
65.1	172	112	60	72.9
65.1	169	110	59	71.6
65.1	166	108	58	70.3
65.1	152	99	53	64.5
65.1	149	97	52	63.1
65.1	146	95	51	61.8
65.1	129	84	45	54.7
65.1	126	82	44	53.4
65.1	109	71	38	46.2
65.1	106	69	37	44.9
65.1	86	56	30	36.5
65.1	83	54	29	35.1
65.1	63	41	22	26.7
65.1	43	28	15	18.2
65.0	500	325	175	211.3
65.0	497	323	174	209.9
65.0	494	321	173	208.6
65.0	492	320	172	208.1
65.0	491	319	172	207.3
65.0	489	318	171	206.8
65.0	488	317	171	205.9
65.0	486	316	170	205.5
65.0	483	314	169	204.1
65.0	480	312	168	202.8
65.0	477	310	167	201.5
65.0	474	308	166	200.1
65.0	472	307	165	199.7
65.0	471	306	165	198.8
65.0	469	305	164	198.3
65.0	468	304	164	197.5
65.0	466	303	163	197.0
65.0	463	301	162	195.7
65.0	460	299	161	194.3
65.0	457	297	160	193.0
65.0	454	295	159	191.7
65.0	452	294	158	191.2
65.0	451	293	158	190.4
65.0	449	292	157	189.9
65.0	448	291	157	189.0

%E	M1	M2	DM	M*
65.0	446	290	156	188.6
65.0	443	288	155	187.2
65.0	440	286	154	185.9
65.0	437	284	153	184.6
65.0	434	282	152	183.2
65.0	432	281	151	182.8
65.0	431	280	151	181.9
65.0	429	279	150	181.4
65.0	428	278	150	180.6
65.0	426	277	149	180.1
65.0	423	275	148	178.8
65.0	420	273	147	177.4
65.0	417	271	146	176.1
65.0	414	269	145	174.8
65.0	412	268	144	174.3
65.0	411	267	144	173.5
65.0	409	266	143	173.0
65.0	408	265	143	172.1
65.0	406	264	142	171.7
65.0	403	262	141	170.3
65.0	400	260	140	169.0
65.0	397	258	139	167.7
65.0	394	256	138	166.3
65.0	391	254	137	165.0
65.0	389	253	136	164.5
65.0	386	251	135	163.2
65.0	383	249	134	161.9
65.0	380	247	133	160.5
65.0	377	245	132	159.2
65.0	374	243	131	157.9
65.0	371	241	130	156.6
65.0	369	240	129	156.1
65.0	366	238	128	154.8
65.0	363	236	127	153.4
65.0	360	234	126	152.1
65.0	357	232	125	150.8
65.0	354	230	124	149.4
65.0	351	228	123	148.1
65.0	349	227	122	147.6
65.0	346	225	121	146.3
65.0	343	223	120	145.0
65.0	340	221	119	143.6
65.0	337	219	118	142.3
65.0	334	217	117	141.0
65.0	331	215	116	139.7
65.0	329	214	115	139.2
65.0	326	212	114	137.9
65.0	323	210	113	136.5
65.0	320	208	112	135.2
65.0	317	206	111	133.9

%E	M1	M2	DM	M*
65.0	314	204	110	132.5
65.0	311	202	109	131.2
65.0	309	201	108	130.7
65.0	306	199	107	129.4
65.0	303	197	106	128.1
65.0	300	195	105	126.8
65.0	297	193	104	125.4
65.0	294	191	103	124.1
65.0	286	186	100	121.0
65.0	283	184	99	119.6
65.0	280	182	98	118.3
65.0	277	180	97	117.0
65.0	274	178	96	115.6
65.0	266	173	93	112.5
65.0	263	171	92	111.2
65.0	260	169	91	109.8
65.0	257	167	90	108.5
65.0	254	165	89	107.2
65.0	246	160	86	104.1
65.0	243	158	85	102.7
65.0	240	156	84	101.4
65.0	237	154	83	100.1
65.0	234	152	82	98.7
65.0	226	147	79	95.6
65.0	223	145	78	94.3
65.0	220	143	77	92.9
65.0	217	141	76	91.6
65.0	214	139	75	90.3
65.0	206	134	72	87.2
65.0	203	132	71	85.8
65.0	200	130	70	84.5
65.0	197	128	69	83.2
65.0	183	119	64	77.4
65.0	180	117	63	76.0
65.0	177	115	62	74.7
65.0	163	106	57	68.9
65.0	160	104	56	67.6
65.0	157	102	55	66.3
65.0	143	93	50	60.5
65.0	140	91	49	59.1
65.0	137	89	48	57.8
65.0	123	80	43	52.0
65.0	120	78	42	50.7
65.0	117	76	41	49.4
65.0	103	67	36	43.6
65.0	100	65	35	42.3
65.0	80	52	28	33.8
65.0	60	39	21	25.3
65.0	40	26	14	16.9
65.0	20	13	7	8.4

%E	M1	M2	DM	M*
64.9	499	324	175	210.4
64.9	498	323	175	209.5
64.9	496	322	174	209.0
64.9	493	320	173	207.7
64.9	490	318	172	206.4
64.9	487	316	171	205.0
64.9	485	315	170	204.6
64.9	484	314	170	203.7
64.9	482	313	169	203.3
64.9	481	312	169	202.4
64.9	479	311	168	201.9
64.9	478	310	168	201.0
64.9	476	309	167	200.6
64.9	473	307	166	199.3
64.9	470	305	165	197.9
64.9	467	303	164	196.6
64.9	465	302	163	196.1
64.9	464	301	163	195.3
64.9	462	300	162	194.8
64.9	461	299	162	193.9
64.9	459	298	161	193.5
64.9	456	296	160	192.1
64.9	453	294	159	190.8
64.9	450	292	158	189.5
64.9	447	290	157	188.1
64.9	445	289	156	187.7
64.9	444	288	156	186.8
64.9	442	287	155	186.4
64.9	441	286	155	185.5
64.9	439	285	154	185.0
64.9	436	283	153	183.7
64.9	433	281	152	182.4
64.9	430	279	151	181.0
64.9	427	277	150	179.7
64.9	425	276	149	179.2
64.9	424	275	149	178.4
64.9	422	274	148	177.9
64.9	419	272	147	176.6
64.9	416	270	146	175.2
64.9	413	268	145	173.9
64.9	410	266	144	172.6
64.9	407	264	143	171.2
64.9	405	263	142	170.8
64.9	404	262	142	169.9
64.9	402	261	141	169.5
64.9	399	259	140	168.1
64.9	396	257	139	166.8
64.9	393	255	138	165.5
64.9	390	253	137	164.1
64.9	388	252	136	163.7

%E	M1	M2	DM	M*
64.9	387	251	136	162.8
64.9	385	250	135	162.3
64.9	382	248	134	161.0
64.9	379	246	133	159.7
64.9	376	244	132	158.3
64.9	373	242	131	157.0
64.9	370	240	130	155.7
64.9	368	239	129	155.2
64.9	367	238	129	154.3
64.9	365	237	128	153.9
64.9	362	235	127	152.6
64.9	359	233	126	151.2
64.9	356	231	125	149.9
64.9	353	229	124	148.6
64.9	350	227	123	147.2
64.9	348	226	122	146.8
64.9	345	224	121	145.4
64.9	342	222	120	144.1
64.9	339	220	119	142.8
64.9	336	218	118	141.4
64.9	333	216	117	140.1
64.9	328	213	115	138.3
64.9	325	211	114	137.0
64.9	322	209	113	135.7
64.9	319	207	112	134.3
64.9	316	205	111	133.0
64.9	313	203	110	131.7
64.9	308	200	108	129.9
64.9	305	198	107	128.5
64.9	302	196	106	127.2
64.9	299	194	105	125.9
64.9	296	192	104	124.5
64.9	291	189	102	122.8
64.9	288	187	101	121.4
64.9	285	185	100	120.1
64.9	282	183	99	118.8
64.9	279	181	98	117.4
64.9	276	179	97	116.1
64.9	271	176	95	114.3
64.9	268	174	94	113.0
64.9	265	172	93	111.6
64.9	262	170	92	110.3
64.9	259	168	91	109.0
64.9	251	163	88	105.9
64.9	248	161	87	104.5
64.9	245	159	86	103.2
64.9	242	157	85	101.9
64.9	239	155	84	100.5
64.9	231	150	81	97.4
64.9	228	148	80	96.1

%E	M1	M2	DM	M*
64.9	225	146	79	94.7
64.9	222	144	78	93.4
64.9	211	137	74	89.0
64.9	208	135	73	87.6
64.9	205	133	72	86.3
64.9	202	131	71	85.0
64.9	194	126	68	81.8
64.9	191	124	67	80.5
64.9	188	122	66	79.2
64.9	185	120	65	77.8
64.9	174	113	61	73.4
64.9	171	111	60	72.1
64.9	168	109	59	70.7
64.9	154	100	54	64.9
64.9	151	98	53	63.6
64.9	148	96	52	62.3
64.9	134	87	47	56.5
64.9	131	85	46	55.2
64.9	114	74	40	48.0
64.9	111	72	39	46.7
64.9	97	63	34	40.9
64.9	94	61	33	39.6
64.9	77	50	27	32.5
64.9	74	48	26	31.1
64.9	57	37	20	24.0
64.9	37	24	13	15.6
64.8	500	324	176	210.0
64.8	497	322	175	208.6
64.8	495	321	174	208.2
64.8	494	320	174	207.3
64.8	492	319	173	206.8
64.8	491	318	173	206.0
64.8	489	317	172	205.5
64.8	488	316	172	204.6
64.8	486	315	171	204.2
64.8	483	313	170	202.8
64.8	480	311	169	201.5
64.8	477	309	168	200.2
64.8	475	308	167	199.7
64.8	474	307	167	198.8
64.8	472	306	166	198.4
64.8	471	305	166	197.5
64.8	469	304	165	197.0
64.8	466	302	164	195.7
64.8	463	300	163	194.4
64.8	460	298	162	193.1
64.8	458	297	161	192.6
64.8	457	296	161	191.7
64.8	455	295	160	191.3
64.8	454	294	160	190.4

%E	M1	M2	DM	M*
64.8	452	293	159	189.9
64.8	449	291	158	188.6
64.8	446	289	157	187.3
64.8	443	287	156	185.9
64.8	440	285	155	184.6
64.8	438	284	154	184.1
64.8	437	283	154	183.3
64.8	435	282	153	182.8
64.8	432	280	152	181.5
64.8	429	278	151	180.1
64.8	426	276	150	178.8
64.8	423	274	149	177.5
64.8	421	273	148	177.0
64.8	420	272	148	176.2
64.8	418	271	147	175.7
64.8	415	269	146	174.4
64.8	412	267	145	173.0
64.8	409	265	144	171.7
64.8	406	263	143	170.4
64.8	403	261	142	169.0
64.8	401	260	141	168.6
64.8	400	259	141	167.7
64.8	398	258	140	167.2
64.8	395	256	139	165.9
64.8	392	254	138	164.6
64.8	389	252	137	163.2
64.8	386	250	136	161.9
64.8	384	249	135	161.5
64.8	383	248	135	160.6
64.8	381	247	134	160.1
64.8	378	245	133	158.8
64.8	375	243	132	157.5
64.8	372	241	131	156.1
64.8	369	239	130	154.8
64.8	366	237	129	153.5
64.8	364	236	128	153.0
64.8	361	234	127	151.7
64.8	358	232	126	150.3
64.8	355	230	125	149.0
64.8	352	228	124	147.7
64.8	349	226	123	146.3
64.8	347	225	122	145.9
64.8	344	223	121	144.6
64.8	341	221	120	143.2
64.8	338	219	119	141.9
64.8	335	217	118	140.6
64.8	332	215	117	139.2
64.8	330	214	116	138.8
64.8	327	212	115	137.4
64.8	324	210	114	136.1

%E	M1	M2	DM	M*
64.8	321	208	113	134.8
64.8	318	206	112	133.4
64.8	315	204	111	132.1
64.8	310	201	109	130.3
64.8	307	199	108	129.0
64.8	304	197	107	127.7
64.8	301	195	106	126.3
64.8	298	193	105	125.0
64.8	293	190	103	123.2
64.8	290	188	102	121.9
64.8	287	186	101	120.5
64.8	284	184	100	119.2
64.8	281	182	99	117.9
64.8	273	177	96	114.8
64.8	270	175	95	113.4
64.8	267	173	94	112.1
64.8	264	171	93	110.8
64.8	261	169	92	109.4
64.8	256	166	90	107.6
64.8	253	164	89	106.3
64.8	250	162	88	105.0
64.8	247	160	87	103.6
64.8	244	158	86	102.3
64.8	236	153	83	99.2
64.8	233	151	82	97.9
64.8	230	149	81	96.5
64.8	227	147	80	95.2
64.8	219	142	77	92.1
64.8	216	140	76	90.7
64.8	213	138	75	89.4
64.8	210	136	74	88.1
64.8	199	129	70	83.6
64.8	196	127	69	82.3
64.8	193	125	68	81.0
64.8	182	118	64	76.5
64.8	179	116	63	75.2
64.8	176	114	62	73.8
64.8	165	107	58	69.4
64.8	162	105	57	68.1
64.8	159	103	56	66.7
64.8	145	94	51	60.9
64.8	142	92	50	59.6
64.8	128	83	45	53.8
64.8	125	81	44	52.5
64.8	122	79	43	51.2
64.8	108	70	38	45.4
64.8	105	68	37	44.0
64.8	91	59	32	38.3
64.8	88	57	31	36.9
64.8	71	46	25	29.8

%E	M1	M2	DM	M*
64.8	54	35	19	22.7
64.7	499	323	176	209.1
64.7	498	322	176	208.2
64.7	496	321	175	207.7
64.7	493	319	174	206.4
64.7	490	317	173	205.1
64.7	487	315	172	203.7
64.7	485	314	171	203.3
64.7	484	313	171	202.4
64.7	482	312	170	202.0
64.7	481	311	170	201.1
64.7	479	310	169	200.6
64.7	476	308	168	199.3
64.7	473	306	167	198.0
64.7	470	304	166	196.6
64.7	468	303	165	196.2
64.7	467	302	165	195.3
64.7	465	301	164	194.8
64.7	464	300	164	194.0
64.7	462	299	163	193.5
64.7	459	297	162	192.2
64.7	456	295	161	190.8
64.7	453	293	160	189.5
64.7	451	292	159	189.1
64.7	450	291	159	188.2
64.7	448	290	158	187.7
64.7	447	289	158	186.8
64.7	445	288	157	186.4
64.7	442	286	156	185.1
64.7	439	284	155	183.7
64.7	436	282	154	182.4
64.7	434	281	153	181.9
64.7	433	280	153	181.1
64.7	431	279	152	180.6
64.7	430	278	152	179.7
64.7	428	277	151	179.3
64.7	425	275	150	177.9
64.7	422	273	149	176.6
64.7	419	271	148	175.3
64.7	417	270	147	174.8
64.7	416	269	147	173.9
64.7	414	268	146	173.5
64.7	411	266	145	172.2
64.7	408	264	144	170.8
64.7	405	262	143	169.5
64.7	402	260	142	168.2
64.7	399	258	141	166.8
64.7	397	257	140	166.4
64.7	394	255	139	165.0
64.7	391	253	138	163.7

%E	M1	M2	DM	M*
64.7	388	251	137	162.4
64.7	385	249	136	161.0
64.7	382	247	135	159.7
64.7	380	246	134	159.3
64.7	377	244	133	157.9
64.7	374	242	132	156.6
64.7	371	240	131	155.3
64.7	368	238	130	153.9
64.7	365	236	129	152.6
64.7	363	235	128	152.1
64.7	360	233	127	150.8
64.7	357	231	126	149.5
64.7	354	229	125	148.1
64.7	351	227	124	146.8
64.7	348	225	123	145.5
64.7	346	224	122	145.0
64.7	343	222	121	143.7
64.7	340	220	120	142.4
64.7	337	218	119	141.0
64.7	334	216	118	139.7
64.7	331	214	117	138.4
64.7	329	213	116	137.9
64.7	326	211	115	136.6
64.7	323	209	114	135.2
64.7	320	207	113	133.9
64.7	317	205	112	132.6
64.7	314	203	111	131.2
64.7	312	202	110	130.8
64.7	309	200	109	129.4
64.7	306	198	108	128.1
64.7	303	196	107	126.8
64.7	300	194	106	125.5
64.7	295	191	104	123.7
64.7	292	189	103	122.3
64.7	289	187	102	121.0
64.7	286	185	101	119.7
64.7	283	183	100	118.3
64.7	278	180	98	116.5
64.7	275	178	97	115.2
64.7	272	176	96	113.9
64.7	269	174	95	112.6
64.7	266	172	94	111.2
64.7	258	167	91	108.1
64.7	255	165	90	106.8
64.7	252	163	89	105.4
64.7	249	161	88	104.1
64.7	241	156	85	101.0
64.7	238	154	84	99.6
64.7	235	152	83	98.3
64.7	232	150	82	97.0

%E	M1	M2	DM	M*
64.7	224	145	79	93.9
64.7	221	143	78	92.5
64.7	218	141	77	91.2
64.7	215	139	76	89.9
64.7	207	134	73	86.7
64.7	204	132	72	85.4
64.7	201	130	71	84.1
64.7	190	123	67	79.6
64.7	187	121	66	78.3
64.7	184	119	65	77.0
64.7	173	112	61	72.5
64.7	170	110	60	71.2
64.7	167	108	59	69.8
64.7	156	101	55	65.4
64.7	153	99	54	64.1
64.7	150	97	53	62.7
64.7	139	90	49	58.3
64.7	136	88	48	56.9
64.7	133	86	47	55.6
64.7	119	77	42	49.8
64.7	116	75	41	48.5
64.7	102	66	36	42.7
64.7	85	55	30	35.6
64.7	68	44	24	28.5
64.7	51	33	18	21.4
64.7	34	22	12	14.2
64.7	17	11	6	7.1
64.6	500	323	177	208.7
64.6	497	321	176	207.3
64.6	495	320	175	206.9
64.6	494	319	175	206.0
64.6	492	318	174	205.5
64.6	491	317	174	204.7
64.6	489	316	173	204.2
64.6	486	314	172	202.9
64.6	483	312	171	201.5
64.6	480	310	170	200.2
64.6	478	309	169	199.8
64.6	477	308	169	198.9
64.6	475	307	168	198.4
64.6	474	306	168	197.5
64.6	472	305	167	197.1
64.6	469	303	166	195.8
64.6	466	301	165	194.4
64.6	463	299	164	193.1
64.6	461	298	163	192.6
64.6	460	297	163	191.8
64.6	458	296	162	191.3
64.6	457	295	162	190.4
64.6	455	294	161	190.0

%E	M1	M2	DM	M*
64.6	452	292	160	188.6
64.6	449	290	159	187.3
64.6	446	288	158	186.0
64.6	444	287	157	185.5
64.6	443	286	157	184.6
64.6	441	285	156	184.2
64.6	438	283	155	182.9
64.6	435	281	154	181.5
64.6	432	279	153	180.2
64.6	429	277	152	178.9
64.6	427	276	151	178.4
64.6	426	275	151	177.5
64.6	424	274	150	177.1
64.6	421	272	149	175.7
64.6	418	270	148	174.4
64.6	415	268	147	173.1
64.6	413	267	146	172.6
64.6	412	266	146	171.7
64.6	410	265	145	171.3
64.6	407	263	144	169.9
64.6	404	261	143	168.6
64.6	401	259	142	167.3
64.6	398	257	141	166.0
64.6	396	256	140	165.5
64.6	395	255	140	164.6
64.6	393	254	139	164.2
64.6	390	252	138	162.8
64.6	387	250	137	161.5
64.6	384	248	136	160.2
64.6	381	246	135	158.8
64.6	379	245	134	158.4
64.6	378	244	134	157.5
64.6	376	243	133	157.0
64.6	373	241	132	155.7
64.6	370	239	131	154.4
64.6	367	237	130	153.0
64.6	364	235	129	151.7
64.6	362	234	128	151.3
64.6	359	232	127	149.9
64.6	356	230	126	148.6
64.6	353	228	125	147.3
64.6	350	226	124	145.9
64.6	347	224	123	144.6
64.6	345	223	122	144.1
64.6	342	221	121	142.8
64.6	339	219	120	141.5
64.6	336	217	119	140.1
64.6	333	215	118	138.8
64.6	328	212	116	137.0
64.6	325	210	115	135.7

%E	M1	M2	DM	M*
64.6	322	208	114	134.4
64.6	319	206	113	133.0
64.6	316	204	112	131.7
64.6	311	201	110	129.9
64.6	308	199	109	128.6
64.6	305	197	108	127.2
64.6	302	195	107	125.9
64.6	297	192	105	124.1
64.6	294	190	104	122.8
64.6	291	188	103	121.5
64.6	288	186	102	120.1
64.6	285	184	101	118.8
64.6	280	181	99	117.0
64.6	277	179	98	115.7
64.6	274	177	97	114.3
64.6	271	175	96	113.0
64.6	268	173	95	111.7
64.6	263	170	93	109.9
64.6	260	168	92	108.6
64.6	257	166	91	107.2
64.6	254	164	90	105.9
64.6	246	159	87	102.8
64.6	243	157	86	101.4
64.6	240	155	85	100.1
64.6	237	153	84	98.8
64.6	229	148	81	95.7
64.6	226	146	80	94.3
64.6	223	144	79	93.0
64.6	212	137	75	88.5
64.6	209	135	74	87.2
64.6	206	133	73	85.9
64.6	198	128	70	82.7
64.6	195	126	69	81.4
64.6	192	124	68	80.1
64.6	189	122	67	78.8
64.6	181	117	64	75.6
64.6	178	115	63	74.3
64.6	175	113	62	73.0
64.6	164	106	58	68.5
64.6	161	104	57	67.2
64.6	158	102	56	65.8
64.6	147	95	52	61.4
64.6	144	93	51	60.1
64.6	130	84	46	54.3
64.6	127	82	45	52.9
64.6	113	73	40	47.2
64.6	99	64	35	41.4
64.6	96	62	34	40.0
64.6	82	53	29	34.3
64.6	79	51	28	32.9

%E	M1	M2	DM	M*
64.6	65	42	23	27.1
64.6	48	31	17	20.0
64.5	499	322	177	207.8
64.5	498	321	177	206.9
64.5	496	320	176	206.5
64.5	493	318	175	205.1
64.5	490	316	174	203.8
64.5	488	315	173	203.3
64.5	487	314	173	202.5
64.5	485	313	172	202.0
64.5	484	312	172	201.1
64.5	482	311	171	200.7
64.5	479	309	170	199.3
64.5	476	307	169	198.0
64.5	473	305	168	196.7
64.5	471	304	167	196.2
64.5	470	303	167	195.3
64.5	468	302	166	194.9
64.5	467	301	166	194.0
64.5	465	300	165	193.5
64.5	462	298	164	192.2
64.5	459	296	163	190.9
64.5	456	294	162	189.6
64.5	454	293	161	189.1
64.5	453	292	161	188.2
64.5	451	291	160	187.8
64.5	448	289	159	186.4
64.5	445	287	158	185.1
64.5	442	285	157	183.8
64.5	440	284	156	183.3
64.5	439	283	156	182.4
64.5	437	282	155	182.0
64.5	436	281	155	181.1
64.5	434	280	154	180.6
64.5	431	278	153	179.3
64.5	428	276	152	178.0
64.5	425	274	151	176.6
64.5	423	273	150	176.2
64.5	422	272	150	175.3
64.5	420	271	149	174.9
64.5	417	269	148	173.5
64.5	414	267	147	172.2
64.5	411	265	146	170.9
64.5	409	264	145	170.4
64.5	408	263	145	169.5
64.5	406	262	144	169.1
64.5	403	260	143	167.7
64.5	400	258	142	166.4
64.5	397	256	141	165.1
64.5	394	254	140	163.7

%E	M1	M2	DM	M*
64.5	392	253	139	163.3
64.5	391	252	139	162.4
64.5	389	251	138	162.0
64.5	386	249	137	160.6
64.5	383	247	136	159.3
64.5	380	245	135	158.0
64.5	377	243	134	156.6
64.5	375	242	133	156.2
64.5	372	240	132	154.8
64.5	369	238	131	153.5
64.5	366	236	130	152.2
64.5	363	234	129	150.8
64.5	361	233	128	150.4
64.5	358	231	127	149.1
64.5	355	229	126	147.7
64.5	352	227	125	146.4
64.5	349	225	124	145.1
64.5	346	223	123	143.7
64.5	344	222	122	143.3
64.5	341	220	121	141.9
64.5	338	218	120	140.6
64.5	335	216	119	139.3
64.5	332	214	118	137.9
64.5	330	213	117	137.5
64.5	327	211	116	136.1
64.5	324	209	115	134.8
64.5	321	207	114	133.5
64.5	318	205	113	132.2
64.5	313	202	111	130.4
64.5	310	200	110	129.0
64.5	307	198	109	127.7
64.5	304	196	108	126.4
64.5	301	194	107	125.0
64.5	299	193	106	124.6
64.5	296	191	105	123.2
64.5	293	189	104	121.9
64.5	290	187	103	120.6
64.5	287	185	102	119.3
64.5	282	182	100	117.5
64.5	279	180	99	116.1
64.5	276	178	98	114.8
64.5	273	176	97	113.5
64.5	265	171	94	110.3
64.5	262	169	93	109.0
64.5	259	167	92	107.7
64.5	256	165	91	106.3
64.5	251	162	89	104.6
64.5	248	160	88	103.2
64.5	245	158	87	101.9
64.5	242	156	86	100.6

%E	M1	M2	DM	M*
64.5	234	151	83	97.4
64.5	231	149	82	96.1
64.5	228	147	81	94.8
64.5	220	142	78	91.7
64.5	217	140	77	90.3
64.5	214	138	76	89.0
64.5	211	136	75	87.7
64.5	203	131	72	84.5
64.5	200	129	71	83.2
64.5	197	127	70	81.9
64.5	186	120	66	77.4
64.5	183	118	65	76.1
64.5	172	111	61	71.6
64.5	169	109	60	70.3
64.5	166	107	59	69.0
64.5	155	100	55	64.5
64.5	152	98	54	63.2
64.5	141	91	50	58.7
64.5	138	89	49	57.4
64.5	124	80	44	51.6
64.5	121	78	43	50.3
64.5	110	71	39	45.8
64.5	107	69	38	44.5
64.5	93	60	33	38.7
64.5	76	49	27	31.6
64.5	62	40	22	25.8
64.5	31	20	11	12.9
64.4	500	322	178	207.4
64.4	497	320	177	206.0
64.4	495	319	176	205.6
64.4	494	318	176	204.7
64.4	492	317	175	204.2
64.4	491	316	175	203.4
64.4	489	315	174	202.9
64.4	486	313	173	201.6
64.4	483	311	172	200.3
64.4	481	310	171	199.8
64.4	480	309	171	198.9
64.4	478	308	170	198.5
64.4	477	307	170	197.6
64.4	475	306	169	197.1
64.4	472	304	168	195.8
64.4	469	302	167	194.5
64.4	466	300	166	193.1
64.4	464	299	165	192.7
64.4	463	298	165	191.8
64.4	461	297	164	191.3
64.4	458	295	163	190.0
64.4	455	293	162	188.7
64.4	452	291	161	187.3

%E	M1	M2	DM	M*
64.4	450	290	160	186.9
64.4	449	289	160	186.0
64.4	447	288	159	185.6
64.4	446	287	159	184.7
64.4	444	286	158	184.2
64.4	441	284	157	182.9
64.4	438	282	156	181.6
64.4	435	280	155	180.2
64.4	433	279	154	179.8
64.4	432	278	154	178.9
64.4	430	277	153	178.4
64.4	427	275	152	177.1
64.4	424	273	151	175.8
64.4	421	271	150	174.4
64.4	419	270	149	174.0
64.4	418	269	149	173.1
64.4	416	268	148	172.7
64.4	413	266	147	171.3
64.4	410	264	146	170.0
64.4	407	262	145	168.7
64.4	405	261	144	168.2
64.4	404	260	144	167.3
64.4	402	259	143	166.9
64.4	399	257	142	165.5
64.4	396	255	141	164.2
64.4	393	253	140	162.9
64.4	390	251	139	161.5
64.4	388	250	138	161.1
64.4	385	248	137	159.8
64.4	382	246	136	158.4
64.4	379	244	135	157.1
64.4	376	242	134	155.8
64.4	374	241	133	155.3
64.4	371	239	132	154.0
64.4	368	237	131	152.6
64.4	365	235	130	151.3
64.4	362	233	129	150.0
64.4	360	232	128	149.5
64.4	357	230	127	148.2
64.4	354	228	126	146.8
64.4	351	226	125	145.5
64.4	348	224	124	144.2
64.4	343	221	122	142.4
64.4	340	219	121	141.1
64.4	337	217	120	139.7
64.4	334	215	119	138.4
64.4	331	213	118	137.1
64.4	329	212	117	136.6
64.4	326	210	116	135.3
64.4	323	208	115	133.9

%E	M1	M2	DM	M*
64.4	320	206	114	132.6
64.4	317	204	113	131.3
64.4	315	203	112	130.8
64.4	312	201	111	129.5
64.4	309	199	110	128.2
64.4	306	197	109	126.8
64.4	303	195	108	125.5
64.4	298	192	106	123.7
64.4	295	190	105	122.4
64.4	292	188	104	121.0
64.4	289	186	103	119.7
64.4	284	183	101	117.9
64.4	281	181	100	116.6
64.4	278	179	99	115.3
64.4	275	177	98	113.9
64.4	270	174	96	112.1
64.4	267	172	95	110.8
64.4	264	170	94	109.5
64.4	261	168	93	108.1
64.4	253	163	90	105.0
64.4	250	161	89	103.7
64.4	247	159	88	102.4
64.4	239	154	85	99.2
64.4	236	152	84	97.9
64.4	233	150	83	96.6
64.4	225	145	80	93.4
64.4	222	143	79	92.1
64.4	219	141	78	90.8
64.4	216	139	77	89.4
64.4	208	134	74	86.3
64.4	205	132	73	85.0
64.4	202	130	72	83.7
64.4	194	125	69	80.5
64.4	191	123	68	79.2
64.4	188	121	67	77.9
64.4	180	116	64	74.8
64.4	177	114	63	73.4
64.4	174	112	62	72.1
64.4	163	105	58	67.6
64.4	160	103	57	66.3
64.4	149	96	53	61.9
64.4	146	94	52	60.5
64.4	135	87	48	56.1
64.4	132	85	47	54.7
64.4	118	76	42	48.9
64.4	104	67	37	43.2
64.4	101	65	36	41.8
64.4	90	58	32	37.4
64.4	87	56	31	36.0
64.4	73	47	26	30.3

%E	M1	M2	DM	M*
64.4	59	38	21	24.5
64.4	45	29	16	18.7
64.3	499	321	178	206.5
64.3	498	320	178	205.6
64.3	496	319	177	205.2
64.3	493	317	176	203.8
64.3	490	315	175	202.5
64.3	488	314	174	202.0
64.3	487	313	174	201.2
64.3	485	312	173	200.7
64.3	484	311	173	199.8
64.3	482	310	172	199.4
64.3	479	308	171	198.0
64.3	476	306	170	196.7
64.3	474	305	169	196.3
64.3	473	304	169	195.4
64.3	471	303	168	194.9
64.3	470	302	168	194.1
64.3	468	301	167	193.6
64.3	465	299	166	192.3
64.3	462	297	165	190.9
64.3	460	296	164	190.5
64.3	459	295	164	189.6
64.3	457	294	163	189.1
64.3	456	293	163	188.3
64.3	454	292	162	187.8
64.3	451	290	161	186.5
64.3	448	288	160	185.1
64.3	445	286	159	183.8
64.3	443	285	158	183.4
64.3	442	284	158	182.5
64.3	440	283	157	182.0
64.3	437	281	156	180.7
64.3	434	279	155	179.4
64.3	431	277	154	178.0
64.3	429	276	153	177.6
64.3	428	275	153	176.7
64.3	426	274	152	176.2
64.3	423	272	151	174.9
64.3	420	270	150	173.6
64.3	417	268	149	172.2
64.3	415	267	148	171.8
64.3	414	266	148	170.9
64.3	412	265	147	170.4
64.3	409	263	146	169.1
64.3	406	261	145	167.8
64.3	403	259	144	166.5
64.3	401	258	143	166.0
64.3	400	257	143	165.1
64.3	398	256	142	164.7

%E	M1	M2	DM	M*
64.3	395	254	141	163.3
64.3	392	252	140	162.0
64.3	389	250	139	160.7
64.3	387	249	138	160.2
64.3	384	247	137	158.9
64.3	381	245	136	157.5
64.3	378	243	135	156.2
64.3	375	241	134	154.9
64.3	373	240	133	154.4
64.3	370	238	132	153.1
64.3	367	236	131	151.8
64.3	364	234	130	150.4
64.3	361	232	129	149.1
64.3	359	231	128	148.6
64.3	356	229	127	147.3
64.3	353	227	126	146.0
64.3	350	225	125	144.6
64.3	347	223	124	143.3
64.3	345	222	123	142.9
64.3	342	220	122	141.5
64.3	339	218	121	140.2
64.3	336	216	120	138.9
64.3	333	214	119	137.5
64.3	328	211	117	135.7
64.3	325	209	116	134.4
64.3	322	207	115	133.1
64.3	319	205	114	131.7
64.3	314	202	112	129.9
64.3	311	200	111	128.6
64.3	308	198	110	127.3
64.3	305	196	109	126.0
64.3	300	193	107	124.2
64.3	297	191	106	122.8
64.3	294	189	105	121.5
64.3	291	187	104	120.2
64.3	286	184	102	118.4
64.3	283	182	101	117.0
64.3	280	180	100	115.7
64.3	277	178	99	114.4
64.3	272	175	97	112.6
64.3	269	173	96	111.3
64.3	266	171	95	109.9
64.3	263	169	94	108.6
64.3	258	166	92	106.8
64.3	255	164	91	105.5
64.3	252	162	90	104.1
64.3	249	160	89	102.8
64.3	244	157	87	101.0
64.3	241	155	86	99.7
64.3	238	153	85	98.4

%E	M1	M2	DM	M*
64.3	235	151	84	97.0
64.3	230	148	82	95.2
64.3	227	146	81	93.9
64.3	224	144	80	92.6
64.3	221	142	79	91.2
64.3	213	137	76	88.1
64.3	210	135	75	86.8
64.3	207	133	74	85.5
64.3	199	128	71	82.3
64.3	196	126	70	81.0
64.3	185	119	66	76.5
64.3	182	117	65	75.2
64.3	171	110	61	70.8
64.3	168	108	60	69.4
64.3	157	101	56	65.0
64.3	154	99	55	63.6
64.3	143	92	51	59.2
64.3	140	90	50	57.9
64.3	129	83	46	53.4
64.3	126	81	45	52.1
64.3	115	74	41	47.6
64.3	112	72	40	46.3
64.3	98	63	35	40.5
64.3	84	54	30	34.7
64.3	70	45	25	28.9
64.3	56	36	20	23.1
64.3	42	27	15	17.4
64.3	28	18	10	11.6
64.3	14	9	5	5.8
64.2	500	321	179	206.1
64.2	497	319	178	204.8
64.2	495	318	177	204.3
64.2	494	317	177	203.4
64.2	492	316	176	203.0
64.2	491	315	176	202.1
64.2	489	314	175	201.6
64.2	486	312	174	200.3
64.2	483	310	173	199.0
64.2	481	309	172	198.5
64.2	480	308	172	197.6
64.2	478	307	171	197.2
64.2	477	306	171	196.3
64.2	475	305	170	195.8
64.2	472	303	169	194.5
64.2	469	301	168	193.2
64.2	467	300	167	192.7
64.2	466	299	167	191.8
64.2	464	298	166	191.4
64.2	461	296	165	190.1
64.2	458	294	164	188.7

%E	M1	M2	DM	M*
64.2	455	292	163	187.4
64.2	453	291	162	186.9
64.2	452	290	162	186.1
64.2	450	289	161	185.6
64.2	447	287	160	184.3
64.2	444	285	159	182.9
64.2	441	283	158	181.6
64.2	439	282	157	181.1
64.2	438	281	157	180.3
64.2	436	280	156	179.8
64.2	433	278	155	178.5
64.2	430	276	154	177.2
64.2	427	274	153	175.8
64.2	425	273	152	175.4
64.2	424	272	152	174.5
64.2	422	271	151	174.0
64.2	419	269	150	172.7
64.2	416	267	149	171.4
64.2	413	265	148	170.0
64.2	411	264	147	169.6
64.2	408	262	146	168.2
64.2	405	260	145	166.9
64.2	402	258	144	165.6
64.2	399	256	143	164.3
64.2	397	255	142	163.8
64.2	394	253	141	162.5
64.2	391	251	140	161.1
64.2	388	249	139	159.8
64.2	386	248	138	159.3
64.2	385	247	138	158.5
64.2	383	246	137	158.0
64.2	380	244	136	156.7
64.2	377	242	135	155.3
64.2	374	240	134	154.0
64.2	372	239	133	153.6
64.2	371	238	133	152.7
64.2	369	237	132	152.2
64.2	366	235	131	150.9
64.2	363	233	130	149.6
64.2	360	231	129	148.2
64.2	358	230	128	147.8
64.2	355	228	127	146.4
64.2	352	226	126	145.1
64.2	349	224	125	143.8
64.2	346	222	124	142.4
64.2	344	221	123	142.0
64.2	341	219	122	140.6
64.2	338	217	121	139.3
64.2	335	215	120	138.0
64.2	332	213	119	136.7

%E	M1	M2	DM	M*
64.2	330	212	118	136.2
64.2	327	210	117	134.9
64.2	324	208	116	133.5
64.2	321	206	115	132.2
64.2	318	204	114	130.9
64.2	316	203	113	130.4
64.2	313	201	112	129.1
64.2	310	199	111	127.7
64.2	307	197	110	126.4
64.2	302	194	108	124.6
64.2	299	192	107	123.3
64.2	296	190	106	122.0
64.2	293	188	105	120.6
64.2	288	185	103	118.8
64.2	285	183	102	117.5
64.2	282	181	101	116.2
64.2	279	179	100	114.8
64.2	274	176	98	113.1
64.2	271	174	97	111.7
64.2	268	172	96	110.4
64.2	265	170	95	109.1
64.2	260	167	93	107.3
64.2	257	165	92	105.9
64.2	254	163	91	104.6
64.2	246	158	88	101.5
64.2	243	156	87	100.1
64.2	240	154	86	98.8
64.2	232	149	83	95.7
64.2	229	147	82	94.4
64.2	226	145	81	93.0
64.2	218	140	78	89.9
64.2	215	138	77	88.6
64.2	212	136	76	87.2
64.2	204	131	73	84.1
64.2	201	129	72	82.8
64.2	193	124	69	79.7
64.2	190	122	68	78.3
64.2	187	120	67	77.0
64.2	179	115	64	73.9
64.2	176	113	63	72.6
64.2	173	111	62	71.2
64.2	165	106	59	68.1
64.2	162	104	58	66.8
64.2	159	102	57	65.4
64.2	151	97	54	62.3
64.2	148	95	53	61.0
64.2	137	88	49	56.5
64.2	134	86	48	55.2
64.2	123	79	44	50.7
64.2	120	77	43	49.4

%E	M1	M2	DM	M*
64.2	109	70	39	45.0
64.2	106	68	38	43.6
64.2	95	61	34	39.2
64.2	81	52	29	33.4
64.2	67	43	24	27.6
64.2	53	34	19	21.8
64.1	499	320	179	205.2
64.1	498	319	179	204.3
64.1	496	318	178	203.9
64.1	493	316	177	202.5
64.1	490	314	176	201.2
64.1	488	313	175	200.8
64.1	487	312	175	199.9
64.1	485	311	174	199.4
64.1	484	310	174	198.6
64.1	482	309	173	198.1
64.1	479	307	172	196.8
64.1	476	305	171	195.4
64.1	474	304	170	195.0
64.1	473	303	170	194.1
64.1	471	302	169	193.6
64.1	468	300	168	192.3
64.1	465	298	167	191.0
64.1	463	297	166	190.5
64.1	462	296	166	189.6
64.1	460	295	165	189.2
64.1	459	294	165	188.3
64.1	457	293	164	187.9
64.1	454	291	163	186.5
64.1	451	289	162	185.2
64.1	449	288	161	184.7
64.1	448	287	161	183.9
64.1	446	286	160	183.4
64.1	443	284	159	182.1
64.1	440	282	158	180.7
64.1	437	280	157	179.4
64.1	435	279	156	178.9
64.1	434	278	156	178.1
64.1	432	277	155	177.6
64.1	429	275	154	176.3
64.1	426	273	153	175.0
64.1	423	271	152	173.6
64.1	421	270	151	173.2
64.1	418	268	150	171.8
64.1	415	266	149	170.5
64.1	412	264	148	169.2
64.1	410	263	147	168.7
64.1	409	262	147	167.8
64.1	407	261	146	167.4
64.1	404	259	145	166.0

%E	M1	M2	DM	M*
64.1	401	257	144	164.7
64.1	398	255	143	163.4
64.1	396	254	142	162.9
64.1	395	253	142	162.0
64.1	393	252	141	161.6
64.1	390	250	140	160.3
64.1	387	248	139	158.9
64.1	384	246	138	157.6
64.1	382	245	137	157.1
64.1	379	243	136	155.8
64.1	376	241	135	154.5
64.1	373	239	134	153.1
64.1	370	237	133	151.8
64.1	368	236	132	151.3
64.1	365	234	131	150.0
64.1	362	232	130	148.7
64.1	359	230	129	147.4
64.1	357	229	128	146.9
64.1	354	227	127	145.6
64.1	351	225	126	144.2
64.1	348	223	125	142.9
64.1	345	221	124	141.6
64.1	343	220	123	141.1
64.1	340	218	122	139.8
64.1	337	216	121	138.4
64.1	334	214	120	137.1
64.1	329	211	118	135.3
64.1	326	209	117	134.0
64.1	323	207	116	132.7
64.1	320	205	115	131.3
64.1	315	202	113	129.5
64.1	312	200	112	128.2
64.1	309	198	111	126.9
64.1	306	196	110	125.5
64.1	304	195	109	125.1
64.1	301	193	108	123.8
64.1	298	191	107	122.4
64.1	295	189	106	121.1
64.1	290	186	104	119.3
64.1	287	184	103	118.0
64.1	284	182	102	116.6
64.1	281	180	101	115.3
64.1	276	177	99	113.5
64.1	273	175	98	112.2
64.1	270	173	97	110.8
64.1	262	168	94	107.7
64.1	259	166	93	106.4
64.1	256	164	92	105.1
64.1	251	161	90	103.3
64.1	248	159	89	101.9

%E	M1	M2	DM	M*
64.1	245	157	88	100.6
64.1	242	155	87	99.3
64.1	237	152	85	97.5
64.1	234	150	84	96.2
64.1	231	148	83	94.8
64.1	223	143	80	91.7
64.1	220	141	79	90.4
64.1	217	139	78	89.0
64.1	209	134	75	85.9
64.1	206	132	74	84.6
64.1	198	127	71	81.5
64.1	195	125	70	80.1
64.1	192	123	69	78.8
64.1	184	118	66	75.7
64.1	181	116	65	74.3
64.1	170	109	61	69.9
64.1	167	107	60	68.6
64.1	156	100	56	64.1
64.1	153	98	55	62.8
64.1	145	93	52	59.6
64.1	142	91	51	58.3
64.1	131	84	47	53.9
64.1	128	82	46	52.5
64.1	117	75	42	48.1
64.1	103	66	37	42.3
64.1	92	59	33	37.8
64.1	78	50	28	32.1
64.1	64	41	23	26.3
64.1	39	25	14	16.0
64.0	500	320	180	204.8
64.0	497	318	179	203.5
64.0	495	317	178	203.0
64.0	494	316	178	202.1
64.0	492	315	177	201.7
64.0	491	314	177	200.8
64.0	489	313	176	200.3
64.0	486	311	175	199.0
64.0	483	309	174	197.7
64.0	481	308	173	197.2
64.0	480	307	173	196.4
64.0	478	306	172	195.9
64.0	475	304	171	194.6
64.0	472	302	170	193.2
64.0	470	301	169	192.8
64.0	469	300	169	191.9
64.0	467	299	168	191.4
64.0	464	297	167	190.1
64.0	461	295	166	188.8
64.0	458	293	165	187.4
64.0	456	292	164	187.0
64.0	455	291	164	186.1
64.0	453	290	163	185.7
64.0	450	288	162	184.3
64.0	447	286	161	183.0
64.0	445	285	160	182.5
64.0	444	284	160	181.7
64.0	442	283	159	181.2
64.0	439	281	158	179.9
64.0	436	279	157	178.5
64.0	433	277	156	177.2
64.0	431	276	155	176.7
64.0	430	275	155	175.9
64.0	428	274	154	175.4
64.0	425	272	153	174.1
64.0	422	270	152	172.7
64.0	420	269	151	172.3
64.0	419	268	151	171.4
64.0	417	267	150	171.0
64.0	414	265	149	169.6
64.0	411	263	148	168.3
64.0	408	261	147	167.0
64.0	406	260	146	166.5
64.0	405	259	146	165.6
64.0	403	258	145	165.2
64.0	400	256	144	163.8
64.0	397	254	143	162.5
64.0	394	252	142	161.2
64.0	392	251	141	160.7
64.0	389	249	140	159.4
64.0	386	247	139	158.1
64.0	383	245	138	156.7
64.0	381	244	137	156.3
64.0	378	242	136	154.9
64.0	375	240	135	153.6
64.0	372	238	134	152.3
64.0	369	236	133	150.9
64.0	367	235	132	150.5
64.0	364	233	131	149.1
64.0	361	231	130	147.8
64.0	358	229	129	146.5
64.0	356	228	128	146.0
64.0	353	226	127	144.7
64.0	350	224	126	143.4
64.0	347	222	125	142.0
64.0	344	220	124	140.7
64.0	342	219	123	140.2
64.0	339	217	122	138.9
64.0	336	215	121	137.6
64.0	333	213	120	136.2
64.0	331	212	119	135.8
64.0	328	210	118	134.5
64.0	325	208	117	133.1
64.0	322	206	116	131.8
64.0	319	204	115	130.5
64.0	317	203	114	130.0
64.0	314	201	113	128.7
64.0	311	199	112	127.3
64.0	308	197	111	126.0
64.0	303	194	109	124.2
64.0	300	192	108	122.9
64.0	297	190	107	121.5
64.0	292	187	105	119.8
64.0	289	185	104	118.4
64.0	286	183	103	117.1
64.0	283	181	102	115.8
64.0	278	178	100	114.0
64.0	275	176	99	112.6
64.0	272	174	98	111.3
64.0	267	171	96	109.5
64.0	264	169	95	108.2
64.0	261	167	94	106.9
64.0	258	165	93	105.5
64.0	253	162	91	103.7
64.0	250	160	90	102.4
64.0	247	158	89	101.1
64.0	239	153	86	97.9
64.0	236	151	85	96.6
64.0	228	146	82	93.5
64.0	225	144	81	92.2
64.0	222	142	80	90.8
64.0	214	137	77	87.7
64.0	211	135	76	86.4
64.0	203	130	73	83.3
64.0	200	128	72	81.9
64.0	197	126	71	80.6
64.0	189	121	68	77.5
64.0	186	119	67	76.1
64.0	178	114	64	73.0
64.0	175	112	63	71.7
64.0	172	110	62	70.3
64.0	164	105	59	67.2
64.0	161	103	58	65.9
64.0	150	96	54	61.4
64.0	139	89	50	57.0
64.0	136	87	49	55.7
64.0	125	80	45	51.2
64.0	114	73	41	46.7
64.0	111	71	40	45.4
64.0	100	64	36	41.0
64.0	89	57	32	36.5
64.0	86	55	31	35.2
64.0	75	48	27	30.7
64.0	50	32	18	20.5
64.0	25	16	9	10.2
63.9	499	319	180	203.9
63.9	498	318	180	203.1
63.9	496	317	179	202.6
63.9	493	315	178	201.3
63.9	490	313	177	199.9
63.9	488	312	176	199.5
63.9	487	311	176	198.6
63.9	485	310	175	198.1
63.9	482	308	174	196.8
63.9	479	306	173	195.5
63.9	477	305	172	195.0
63.9	476	304	172	194.2
63.9	474	303	171	193.7
63.9	471	301	170	192.4
63.9	468	299	169	191.0
63.9	466	298	168	190.6
63.9	465	297	168	189.7
63.9	463	296	167	189.2
63.9	462	295	167	188.4
63.9	460	294	166	187.9
63.9	457	292	165	186.6
63.9	454	290	164	185.2
63.9	452	289	163	184.8
63.9	451	288	163	183.9
63.9	449	287	162	183.4
63.9	446	285	161	182.1
63.9	443	283	160	180.8
63.9	441	282	159	180.3
63.9	440	281	159	179.5
63.9	438	280	158	179.0
63.9	435	278	157	177.7
63.9	432	276	156	176.3
63.9	429	274	155	175.0
63.9	427	273	154	174.5
63.9	426	272	154	173.7
63.9	424	271	153	173.2
63.9	421	269	152	171.9
63.9	418	267	151	170.5
63.9	416	266	150	170.1
63.9	415	265	150	169.2
63.9	413	264	149	168.8
63.9	410	262	148	167.4
63.9	407	260	147	166.1
63.9	404	258	146	164.8
63.9	402	257	145	164.3
63.9	399	255	144	163.0
63.9	396	253	143	161.6
63.9	393	251	142	160.3
63.9	391	250	141	159.8
63.9	388	248	140	158.5
63.9	385	246	139	157.2
63.9	382	244	138	155.9
63.9	380	243	137	155.4
63.9	379	242	137	154.5
63.9	377	241	136	154.1
63.9	374	239	135	152.7
63.9	371	237	134	151.4
63.9	368	235	133	150.1
63.9	366	234	132	149.6
63.9	363	232	131	148.3
63.9	360	230	130	146.9
63.9	357	228	129	145.6
63.9	355	227	128	145.2
63.9	352	225	127	143.8
63.9	349	223	126	142.5
63.9	346	221	125	141.2
63.9	341	218	123	139.4
63.9	338	216	122	138.0
63.9	335	214	121	136.7
63.9	332	212	120	135.4
63.9	330	211	119	134.9
63.9	327	209	118	133.6
63.9	324	207	117	132.3
63.9	321	205	116	130.9
63.9	316	202	114	129.1
63.9	313	200	113	127.8
63.9	310	198	112	126.5
63.9	305	195	110	124.7
63.9	302	193	109	123.3
63.9	299	191	108	122.0
63.9	296	189	107	120.7
63.9	294	188	106	120.2
63.9	291	186	105	118.9
63.9	288	184	104	117.6
63.9	285	182	103	116.2
63.9	280	179	101	114.4
63.9	277	177	100	113.1
63.9	274	175	99	111.8
63.9	269	172	97	110.0
63.9	266	170	96	108.6
63.9	263	168	95	107.3
63.9	255	163	92	104.2
63.9	252	161	91	102.9
63.9	249	159	90	101.5
63.9	244	156	88	99.7
63.9	241	154	87	98.4

%E	M1	M2	DM	M*
63.9	238	152	86	97.1
63.9	233	149	84	95.3
63.9	230	147	83	94.0
63.9	227	145	82	92.6
63.9	219	140	79	89.5
63.9	216	138	78	88.2
63.9	213	136	77	86.8
63.9	208	133	75	85.0
63.9	205	131	74	83.7
63.9	202	129	73	82.4
63.9	194	124	70	79.3
63.9	191	122	69	77.9
63.9	183	117	66	74.8
63.9	180	115	65	73.5
63.9	169	108	61	69.0
63.9	166	106	60	67.7
63.9	158	101	57	64.6
63.9	155	99	56	63.2
63.9	147	94	53	60.1
63.9	144	92	52	58.8
63.9	133	85	48	54.3
63.9	122	78	44	49.9
63.9	119	76	43	48.5
63.9	108	69	39	44.1
63.9	97	62	35	39.6
63.9	83	53	30	33.8
63.9	72	46	26	29.4
63.9	61	39	22	24.9
63.9	36	23	13	14.7
63.8	500	319	181	203.5
63.8	497	317	180	202.2
63.8	495	316	179	201.7
63.8	494	315	179	200.9
63.8	492	314	178	200.4
63.8	489	312	177	199.1
63.8	486	310	176	197.7
63.8	484	309	175	197.3
63.8	483	308	175	196.4
63.8	481	307	174	195.9
63.8	480	306	174	195.1
63.8	478	305	173	194.6
63.8	475	303	172	193.3
63.8	473	302	171	192.8
63.8	472	301	171	192.0
63.8	470	300	170	191.5
63.8	469	299	170	190.6
63.8	467	298	169	190.2
63.8	464	296	168	188.8
63.8	461	294	167	187.5
63.8	459	293	166	187.0

%E	M1	M2	DM	M*
63.8	458	292	166	186.2
63.8	456	291	165	185.7
63.8	453	289	164	184.4
63.8	450	287	163	183.0
63.8	448	286	162	182.6
63.8	447	285	162	181.7
63.8	445	284	161	181.2
63.8	442	282	160	179.9
63.8	439	280	159	178.6
63.8	437	279	158	178.1
63.8	436	278	158	177.3
63.8	434	277	157	176.8
63.8	431	275	156	175.5
63.8	428	273	155	174.1
63.8	425	271	154	172.8
63.8	423	270	153	172.3
63.8	420	268	152	171.0
63.8	417	266	151	169.7
63.8	414	264	150	168.3
63.8	412	263	149	167.9
63.8	409	261	148	166.6
63.8	406	259	147	165.2
63.8	403	257	146	163.9
63.8	401	256	145	163.4
63.8	400	255	145	162.6
63.8	398	254	144	162.1
63.8	395	252	143	160.8
63.8	392	250	142	159.4
63.8	390	249	141	159.0
63.8	389	248	141	158.1
63.8	387	247	140	157.6
63.8	384	245	139	156.3
63.8	381	243	138	155.0
63.8	378	241	137	153.7
63.8	376	240	136	153.2
63.8	373	238	135	151.9
63.8	370	236	134	150.5
63.8	367	234	133	149.2
63.8	365	233	132	148.7
63.8	362	231	131	147.4
63.8	359	229	130	146.1
63.8	356	227	129	144.7
63.8	354	226	128	144.3
63.8	351	224	127	143.0
63.8	348	222	126	141.6
63.8	345	220	125	140.3
63.8	343	219	124	139.8
63.8	340	217	123	138.5
63.8	337	215	122	137.2
63.8	334	213	121	135.8

%E	M1	M2	DM	M*
63.8	329	210	119	134.0
63.8	326	208	118	132.7
63.8	323	206	117	131.4
63.8	320	204	116	130.0
63.8	318	203	115	129.6
63.8	315	201	114	128.3
63.8	312	199	113	126.9
63.8	309	197	112	125.6
63.8	307	196	111	125.1
63.8	304	194	110	123.8
63.8	301	192	109	122.5
63.8	298	190	108	121.1
63.8	293	187	106	119.3
63.8	290	185	105	118.0
63.8	287	183	104	116.7
63.8	282	180	102	114.9
63.8	279	178	101	113.6
63.8	276	176	100	112.2
63.8	271	173	98	110.4
63.8	268	171	97	109.1
63.8	265	169	96	107.8
63.8	260	166	94	106.0
63.8	257	164	93	104.7
63.8	254	162	92	103.3
63.8	246	157	89	100.2
63.8	243	155	88	98.9
63.8	240	153	87	97.5
63.8	235	150	85	95.7
63.8	232	148	84	94.4
63.8	229	146	83	93.1
63.8	224	143	81	91.3
63.8	221	141	80	90.0
63.8	218	139	79	88.6
63.8	210	134	76	85.5
63.8	207	132	75	84.2
63.8	199	127	72	81.1
63.8	196	125	71	79.7
63.8	188	120	68	76.6
63.8	185	118	67	75.3
63.8	177	113	64	72.1
63.8	174	111	63	70.8
63.8	163	104	59	66.4
63.8	160	102	58	65.0
63.8	152	97	55	61.9
63.8	149	95	54	60.6
63.8	141	90	51	57.4
63.8	138	88	50	56.1
63.8	130	83	47	53.0
63.8	127	81	46	51.7
63.8	116	74	42	47.2

%E	M1	M2	DM	M*
63.8	105	67	38	42.8
63.8	94	60	34	38.3
63.8	80	51	29	32.5
63.8	69	44	25	28.1
63.8	58	37	21	23.6
63.8	47	30	17	19.1
63.7	499	318	181	202.7
63.7	498	317	181	201.8
63.7	496	316	180	201.3
63.7	493	314	179	200.0
63.7	491	313	178	199.5
63.7	490	312	178	198.7
63.7	488	311	177	198.2
63.7	487	310	177	197.3
63.7	485	309	176	196.9
63.7	482	307	175	195.5
63.7	479	305	174	194.2
63.7	477	304	173	193.7
63.7	476	303	173	192.9
63.7	474	302	172	192.4
63.7	471	300	171	191.1
63.7	468	298	170	189.8
63.7	466	297	169	189.3
63.7	465	296	169	188.4
63.7	463	295	168	188.0
63.7	460	293	167	186.6
63.7	457	291	166	185.3
63.7	455	290	165	184.8
63.7	454	289	165	184.0
63.7	452	288	164	183.5
63.7	449	286	163	182.2
63.7	446	284	162	180.8
63.7	444	283	161	180.4
63.7	443	282	161	179.5
63.7	441	281	160	179.0
63.7	438	279	159	177.7
63.7	435	277	158	176.4
63.7	433	276	157	175.9
63.7	432	275	157	175.1
63.7	430	274	156	174.6
63.7	427	272	155	173.3
63.7	424	270	154	171.9
63.7	422	269	153	171.5
63.7	421	268	153	170.6
63.7	419	267	152	170.1
63.7	416	265	151	168.8
63.7	413	263	150	167.5
63.7	411	262	149	167.0
63.7	410	261	149	166.1
63.7	408	260	148	165.7

%E	M1	M2	DM	M*
63.7	405	258	147	164.4
63.7	402	256	146	163.0
63.7	399	254	145	161.7
63.7	397	253	144	161.2
63.7	394	251	143	159.9
63.7	391	249	142	158.6
63.7	388	247	141	157.2
63.7	386	246	140	156.8
63.7	383	244	139	155.4
63.7	380	242	138	154.1
63.7	377	240	137	152.8
63.7	375	239	136	152.3
63.7	372	237	135	151.0
63.7	369	235	134	149.7
63.7	366	233	133	148.3
63.7	364	232	132	147.9
63.7	361	230	131	146.5
63.7	358	228	130	145.2
63.7	355	226	129	143.9
63.7	353	225	128	143.4
63.7	350	223	127	142.1
63.7	347	221	126	140.8
63.7	344	219	125	139.4
63.7	342	218	124	139.0
63.7	339	216	123	137.6
63.7	336	214	122	136.3
63.7	333	212	121	135.0
63.7	331	211	120	134.5
63.7	328	209	119	133.2
63.7	325	207	118	131.8
63.7	322	205	117	130.5
63.7	317	202	115	128.7
63.7	314	200	114	127.4
63.7	311	198	113	126.1
63.7	306	195	111	124.3
63.7	303	193	110	122.9
63.7	300	191	109	121.6
63.7	295	188	107	119.8
63.7	292	186	106	118.5
63.7	289	184	105	117.1
63.7	284	181	103	115.4
63.7	281	179	102	114.0
63.7	278	177	101	112.7
63.7	273	174	99	110.9
63.7	270	172	98	109.6
63.7	267	170	97	108.2
63.7	262	167	95	106.4
63.7	259	165	94	105.1
63.7	256	163	93	103.8
63.7	251	160	91	102.0

%E	M1	M2	DM	M*
63.7	248	158	90	100.7
63.7	245	156	89	99.3
63.7	237	151	86	96.2
63.7	234	149	85	94.9
63.7	226	144	82	91.8
63.7	223	142	81	90.4
63.7	215	137	78	87.3
63.7	212	135	77	86.0
63.7	204	130	74	82.8
63.7	201	128	73	81.5
63.7	193	123	70	78.4
63.7	190	121	69	77.1
63.7	182	116	66	73.9
63.7	179	114	65	72.6
63.7	171	109	62	69.5
63.7	168	107	61	68.1
63.7	157	100	57	63.7
63.7	146	93	53	59.2
63.7	135	86	49	54.8
63.7	124	79	45	50.3
63.7	113	72	41	45.9
63.7	102	65	37	41.4
63.7	91	58	33	37.0
63.6	500	318	182	202.2
63.6	497	316	181	200.9
63.6	495	315	180	200.5
63.6	494	314	180	199.6
63.6	492	313	179	199.1
63.6	489	311	178	197.8
63.6	486	309	177	196.5
63.6	484	308	176	196.0
63.6	483	307	176	195.1
63.6	481	306	175	194.7
63.6	478	304	174	193.3
63.6	475	302	173	192.0
63.6	473	301	172	191.5
63.6	472	300	172	190.7
63.6	470	299	171	190.2
63.6	467	297	170	188.9
63.6	464	295	169	187.6
63.6	462	294	168	187.1
63.6	461	293	168	186.2
63.6	459	292	167	185.8
63.6	456	290	166	184.4
63.6	453	288	165	183.1
63.6	451	287	164	182.6
63.6	450	286	164	181.8
63.6	448	285	163	181.3
63.6	445	283	162	180.0
63.6	442	281	161	178.6

%E	M1	M2	DM	M*
63.6	440	280	160	178.2
63.6	439	279	160	177.3
63.6	437	278	159	176.9
63.6	434	276	158	175.5
63.6	431	274	157	174.2
63.6	429	273	156	173.7
63.6	428	272	156	172.9
63.6	426	271	155	172.4
63.6	423	269	154	171.1
63.6	420	267	153	169.7
63.6	418	266	152	169.3
63.6	415	264	151	167.9
63.6	412	262	150	166.6
63.6	409	260	149	165.3
63.6	407	259	148	164.8
63.6	404	257	147	163.5
63.6	401	255	146	162.2
63.6	398	253	145	160.8
63.6	396	252	144	160.4
63.6	393	250	143	159.0
63.6	390	248	142	157.7
63.6	387	246	141	156.4
63.6	385	245	140	155.9
63.6	382	243	139	154.6
63.6	379	241	138	153.2
63.6	376	239	137	151.9
63.6	374	238	136	151.5
63.6	371	236	135	150.1
63.6	368	234	134	148.8
63.6	365	232	133	147.5
63.6	363	231	132	147.0
63.6	360	229	131	145.7
63.6	357	227	130	144.3
63.6	354	225	129	143.0
63.6	352	224	128	142.5
63.6	349	222	127	141.2
63.6	346	220	126	139.9
63.6	343	218	125	138.6
63.6	341	217	124	138.1
63.6	338	215	123	136.8
63.6	335	213	122	135.4
63.6	332	211	121	134.1
63.6	330	210	120	133.6
63.6	327	208	119	132.3
63.6	324	206	118	131.0
63.6	321	204	117	129.6
63.6	319	203	116	129.2
63.6	316	201	115	127.9
63.6	313	199	114	126.5
63.6	308	196	112	124.7

%E	M1	M2	DM	M*
63.6	305	194	111	123.4
63.6	302	192	110	122.1
63.6	297	189	108	120.3
63.6	294	187	107	118.9
63.6	291	185	106	117.6
63.6	286	182	104	115.8
63.6	283	180	103	114.5
63.6	280	178	102	113.2
63.6	275	175	100	111.4
63.6	272	173	99	110.0
63.6	269	171	98	108.7
63.6	264	168	96	106.9
63.6	261	166	95	105.6
63.6	258	164	94	104.2
63.6	253	161	92	102.5
63.6	250	159	91	101.1
63.6	247	157	90	99.8
63.6	242	154	88	98.0
63.6	239	152	87	96.7
63.6	236	150	86	95.3
63.6	231	147	84	93.5
63.6	228	145	83	92.2
63.6	225	143	82	90.9
63.6	220	140	80	89.1
63.6	217	138	79	87.8
63.6	214	136	78	86.4
63.6	209	133	76	84.6
63.6	206	131	75	83.3
63.6	198	126	72	80.2
63.6	195	124	71	78.9
63.6	187	119	68	75.7
63.6	184	117	67	74.4
63.6	176	112	64	71.3
63.6	173	110	63	69.9
63.6	165	105	60	66.8
63.6	162	103	59	65.5
63.6	154	98	56	62.4
63.6	151	96	55	61.0
63.6	143	91	52	57.9
63.6	140	89	51	56.6
63.6	132	84	48	53.5
63.6	129	82	47	52.1
63.6	121	77	44	49.0
63.6	118	75	43	47.7
63.6	110	70	40	44.5
63.6	107	68	39	43.2
63.6	99	63	36	40.1
63.6	88	56	32	35.6
63.6	77	49	28	31.2
63.6	66	42	24	26.7

%E	M1	M2	DM	M*
63.6	55	35	20	22.3
63.6	44	28	16	17.8
63.6	33	21	12	13.4
63.6	22	14	8	8.9
63.6	11	7	4	4.5
63.5	499	317	182	201.4
63.5	498	316	182	200.5
63.5	496	315	181	200.1
63.5	493	313	180	198.7
63.5	491	312	179	198.3
63.5	490	311	179	197.4
63.5	488	310	178	196.9
63.5	487	309	178	196.1
63.5	485	308	177	195.6
63.5	482	306	176	194.3
63.5	480	305	175	193.8
63.5	479	304	175	192.9
63.5	477	303	174	192.5
63.5	474	301	173	191.1
63.5	471	299	172	189.8
63.5	469	298	171	189.3
63.5	468	297	171	188.5
63.5	466	296	170	188.0
63.5	463	294	169	186.7
63.5	460	292	168	185.4
63.5	458	291	167	184.9
63.5	457	290	167	184.0
63.5	455	289	166	183.6
63.5	452	287	165	182.2
63.5	449	285	164	180.9
63.5	447	284	163	180.4
63.5	446	283	163	179.6
63.5	444	282	162	179.1
63.5	441	280	161	177.8
63.5	438	278	160	176.4
63.5	436	277	159	176.0
63.5	433	275	158	174.7
63.5	430	273	157	173.3
63.5	427	271	156	172.0
63.5	425	270	155	171.5
63.5	422	268	154	170.2
63.5	419	266	153	168.9
63.5	417	265	152	168.4
63.5	416	264	152	167.5
63.5	414	263	151	167.1
63.5	411	261	150	165.7
63.5	408	259	149	164.4
63.5	406	258	148	164.0
63.5	405	257	148	163.1
63.5	403	256	147	162.6

%E	M1	M2	DM	M*
63.5	400	254	146	161.3
63.5	397	252	145	160.0
63.5	395	251	144	159.5
63.5	394	250	144	158.6
63.5	392	249	143	158.2
63.5	389	247	142	156.8
63.5	386	245	141	155.5
63.5	384	244	140	155.0
63.5	381	242	139	153.7
63.5	378	240	138	152.4
63.5	375	238	137	151.1
63.5	373	237	136	150.6
63.5	370	235	135	149.3
63.5	367	233	134	147.9
63.5	364	231	133	146.6
63.5	362	230	132	146.1
63.5	359	228	131	144.8
63.5	356	226	130	143.5
63.5	353	224	129	142.1
63.5	351	223	128	141.7
63.5	348	221	127	140.3
63.5	345	219	126	139.0
63.5	342	217	125	137.7
63.5	340	216	124	137.2
63.5	337	214	123	135.9
63.5	334	212	122	134.6
63.5	329	209	120	132.8
63.5	326	207	119	131.4
63.5	323	205	118	130.1
63.5	318	202	116	128.3
63.5	315	200	115	127.0
63.5	312	198	114	125.7
63.5	310	197	113	125.2
63.5	307	195	112	123.9
63.5	304	193	111	122.5
63.5	301	191	110	121.2
63.5	299	190	109	120.7
63.5	296	188	108	119.4
63.5	293	186	107	118.1
63.5	288	183	105	116.3
63.5	285	181	104	115.0
63.5	282	179	103	113.6
63.5	277	176	101	111.8
63.5	274	174	100	110.5
63.5	271	172	99	109.2
63.5	266	169	97	107.4
63.5	263	167	96	106.0
63.5	260	165	95	104.7
63.5	255	162	93	102.9
63.5	252	160	92	101.6

%E	M1	M2	DM	M*
63.5	249	158	91	100.3
63.5	244	155	89	98.5
63.5	241	153	88	97.1
63.5	233	148	85	94.0
63.5	230	146	84	92.7
63.5	222	141	81	89.6
63.5	219	139	80	88.2
63.5	211	134	77	85.1
63.5	208	132	76	83.8
63.5	203	129	74	82.0
63.5	200	127	73	80.6
63.5	197	125	72	79.3
63.5	192	122	70	77.5
63.5	189	120	69	76.2
63.5	181	115	66	73.1
63.5	178	113	65	71.7
63.5	170	108	62	68.6
63.5	167	106	61	67.3
63.5	159	101	58	64.2
63.5	156	99	57	62.8
63.5	148	94	54	59.7
63.5	137	87	50	55.2
63.5	126	80	46	50.8
63.5	115	73	42	46.3
63.5	104	66	38	41.9
63.5	96	61	35	38.8
63.5	85	54	31	34.3
63.5	74	47	27	29.9
63.5	63	40	23	25.4
63.5	52	33	19	20.9
63.4	500	317	183	201.0
63.4	497	315	182	199.6
63.4	495	314	181	199.2
63.4	494	313	181	198.3
63.4	492	312	180	197.9
63.4	489	310	179	196.5
63.4	486	308	178	195.2
63.4	484	307	177	194.7
63.4	483	306	177	193.9
63.4	481	305	176	193.4
63.4	478	303	175	192.1
63.4	476	302	174	191.6
63.4	475	301	174	190.7
63.4	473	300	173	190.3
63.4	470	298	172	188.9
63.4	467	296	171	187.6
63.4	465	295	170	187.2
63.4	464	294	170	186.3
63.4	462	293	169	185.8
63.4	459	291	168	134.5

%E	M1	M2	DM	M*
63.4	456	289	167	183.2
63.4	454	288	166	182.7
63.4	453	287	166	181.8
63.4	451	286	165	181.4
63.4	448	284	164	180.0
63.4	445	282	163	178.7
63.4	443	281	162	178.2
63.4	440	279	161	176.9
63.4	437	277	160	175.6
63.4	435	276	159	175.1
63.4	434	275	159	174.3
63.4	432	274	158	173.8
63.4	429	272	157	172.5
63.4	426	270	156	171.1
63.4	424	269	155	170.7
63.4	423	268	155	169.8
63.4	421	267	154	169.3
63.4	418	265	153	168.0
63.4	415	263	152	166.7
63.4	413	262	151	166.2
63.4	412	261	151	165.3
63.4	410	260	150	164.9
63.4	407	258	149	163.5
63.4	404	256	148	162.2
63.4	402	255	147	161.8
63.4	399	253	146	160.4
63.4	396	251	145	159.1
63.4	393	249	144	157.8
63.4	391	248	143	157.3
63.4	388	246	142	156.0
63.4	385	244	141	154.6
63.4	383	243	140	154.2
63.4	382	242	140	153.3
63.4	380	241	139	152.8
63.4	377	239	138	151.5
63.4	374	237	137	150.2
63.4	372	236	136	149.7
63.4	369	234	135	148.4
63.4	366	232	134	147.1
63.4	363	230	133	145.7
63.4	361	229	132	145.3
63.4	358	227	131	143.9
63.4	355	225	130	142.6
63.4	352	223	129	141.3
63.4	350	222	128	140.8
63.4	347	220	127	139.5
63.4	344	218	126	138.2
63.4	339	215	124	136.4
63.4	336	213	123	135.0
63.4	333	211	122	133.7

%E	M1	M2	DM	M*
63.4	331	210	121	133.2
63.4	328	208	120	131.9
63.4	325	206	119	130.6
63.4	322	204	118	129.2
63.4	320	203	117	128.8
63.4	317	201	116	127.4
63.4	314	199	115	126.1
63.4	309	196	113	124.3
63.4	306	194	112	123.0
63.4	303	192	111	121.7
63.4	298	189	109	119.9
63.4	295	187	108	118.5
63.4	292	185	107	117.2
63.4	290	184	106	116.7
63.4	287	182	105	115.4
63.4	284	180	104	114.1
63.4	279	177	102	112.3
63.4	276	175	101	111.0
63.4	273	173	100	109.6
63.4	268	170	98	107.8
63.4	265	168	97	106.5
63.4	262	166	96	105.2
63.4	257	163	94	103.4
63.4	254	161	93	102.1
63.4	246	156	90	98.9
63.4	243	154	89	97.6
63.4	238	151	87	95.8
63.4	235	149	86	94.5
63.4	232	147	85	93.1
63.4	227	144	83	91.3
63.4	224	142	82	90.0
63.4	216	137	79	86.9
63.4	213	135	78	85.6
63.4	205	130	75	82.4
63.4	202	128	74	81.1
63.4	194	123	71	78.0
63.4	191	121	70	76.7
63.4	186	118	68	74.9
63.4	183	116	67	73.5
63.4	175	111	64	70.4
63.4	172	109	63	69.1
63.4	164	104	60	66.0
63.4	161	102	59	64.6
63.4	153	97	56	61.5
63.4	145	92	53	58.4
63.4	142	90	52	57.0
63.4	134	85	49	53.9
63.4	131	83	48	52.6
63.4	123	78	45	49.5
63.4	112	71	41	45.0

%E	M1	M2	DM	M*
63.4	101	64	37	40.6
63.4	93	59	34	37.4
63.4	82	52	30	33.0
63.4	71	45	26	28.5
63.4	41	26	15	16.5
63.3	499	316	183	200.1
63.3	498	315	183	199.2
63.3	496	314	182	198.8
63.3	493	312	181	197.5
63.3	491	311	180	197.0
63.3	490	310	180	196.1
63.3	488	309	179	195.7
63.3	485	307	178	194.3
63.3	482	305	177	193.0
63.3	480	304	176	192.5
63.3	479	303	176	191.7
63.3	477	302	175	191.2
63.3	474	300	174	189.9
63.3	472	299	173	189.4
63.3	471	298	173	188.5
63.3	469	297	172	188.1
63.3	466	295	171	186.7
63.3	463	293	170	185.4
63.3	461	292	169	185.0
63.3	460	291	169	184.1
63.3	458	290	168	183.6
63.3	455	288	167	182.3
63.3	452	286	166	181.0
63.3	450	285	165	180.5
63.3	449	284	165	179.6
63.3	447	283	164	179.2
63.3	444	281	163	177.8
63.3	442	280	162	177.4
63.3	441	279	162	176.5
63.3	439	278	161	176.0
63.3	436	276	160	174.7
63.3	433	274	159	173.4
63.3	431	273	158	172.9
63.3	430	272	158	172.1
63.3	428	271	157	171.6
63.3	425	269	156	170.3
63.3	422	267	155	168.9
63.3	420	266	154	168.5
63.3	417	264	153	167.1
63.3	414	262	152	165.8
63.3	411	260	151	164.5
63.3	409	259	150	164.0
63.3	406	257	149	162.7
63.3	403	255	148	161.4
63.3	401	254	147	160.9

%E	M1	M2	DM	M*
63.3	400	253	147	160.0
63.3	398	252	146	159.6
63.3	395	250	145	158.2
63.3	392	248	144	156.9
63.3	390	247	143	156.4
63.3	387	245	142	155.1
63.3	384	243	141	153.8
63.3	381	241	140	152.4
63.3	379	240	139	152.0
63.3	376	238	138	150.6
63.3	373	236	137	149.3
63.3	371	235	136	148.9
63.3	368	233	135	147.5
63.3	365	231	134	146.2
63.3	362	229	133	144.9
63.3	360	228	132	144.4
63.3	357	226	131	143.1
63.3	354	224	130	141.7
63.3	349	221	128	139.9
63.3	346	219	127	138.6
63.3	343	217	126	137.3
63.3	341	216	125	136.8
63.3	338	214	124	135.5
63.3	335	212	123	134.2
63.3	332	210	122	132.8
63.3	330	209	121	132.4
63.3	327	207	120	131.0
63.3	324	205	119	129.7
63.3	319	202	117	127.9
63.3	316	200	116	126.6
63.3	313	198	115	125.3
63.3	311	197	114	124.8
63.3	308	195	113	123.5
63.3	305	193	112	122.1
63.3	300	190	110	120.3
63.3	297	188	109	119.0
63.3	294	186	108	117.7
63.3	289	183	106	115.9
63.3	286	181	105	114.5
63.3	283	179	104	113.2
63.3	281	178	103	112.8
63.3	278	176	102	111.4
63.3	275	174	101	110.1
63.3	270	171	99	108.3
63.3	267	169	98	107.0
63.3	264	167	97	105.6
63.3	259	164	95	103.8
63.3	256	162	94	102.5
63.3	251	159	92	100.7
63.3	248	157	91	99.4

%E	M1	M2	DM	M*
63.3	245	155	90	98.1
63.3	240	152	88	96.3
63.3	237	150	87	94.9
63.3	229	145	84	91.8
63.3	226	143	83	90.5
63.3	221	140	81	88.7
63.3	218	138	80	87.4
63.3	215	136	79	86.0
63.3	210	133	77	84.2
63.3	207	131	76	82.9
63.3	199	126	73	79.8
63.3	196	124	72	78.4
63.3	188	119	69	75.3
63.3	180	114	66	72.2
63.3	177	112	65	70.9
63.3	169	107	62	67.7
63.3	166	105	61	66.4
63.3	158	100	58	63.3
63.3	150	95	55	60.2
63.3	147	93	54	58.8
63.3	139	88	51	55.7
63.3	128	81	47	51.3
63.3	120	76	44	48.1
63.3	109	69	40	43.7
63.3	98	62	36	39.2
63.3	90	57	33	36.1
63.3	79	50	29	31.6
63.3	60	38	22	24.1
63.3	49	31	18	19.6
63.3	30	19	11	12.0
63.2	500	316	184	199.7
63.2	497	314	183	198.4
63.2	495	313	182	197.9
63.2	494	312	182	197.1
63.2	492	311	181	196.6
63.2	489	309	180	195.3
63.2	487	308	179	194.8
63.2	486	307	179	193.9
63.2	484	306	178	193.5
63.2	481	304	177	192.1
63.2	478	302	176	190.8
63.2	476	301	175	190.3
63.2	475	300	175	189.5
63.2	473	299	174	189.0
63.2	470	297	173	187.7
63.2	468	296	172	187.2
63.2	467	295	172	186.3
63.2	465	294	171	185.9
63.2	462	292	170	184.6
63.2	459	290	169	183.2

%E	M1	M2	DM	M*
63.2	457	289	168	182.8
63.2	456	288	168	181.9
63.2	454	287	167	181.4
63.2	451	285	166	180.1
63.2	448	283	165	178.8
63.2	446	282	164	178.3
63.2	443	280	163	177.0
63.2	440	278	162	175.6
63.2	438	277	161	175.2
63.2	437	276	161	174.3
63.2	435	275	160	173.9
63.2	432	273	159	172.5
63.2	429	271	158	171.2
63.2	427	270	157	170.7
63.2	424	268	156	169.4
63.2	421	266	155	168.1
63.2	419	265	154	167.6
63.2	418	264	154	166.7
63.2	416	263	153	166.3
63.2	413	261	152	164.9
63.2	410	259	151	163.6
63.2	408	258	150	163.1
63.2	405	256	149	161.8
63.2	402	254	148	160.5
63.2	399	252	147	159.2
63.2	397	251	146	158.7
63.2	394	249	145	157.4
63.2	391	247	144	156.0
63.2	389	246	143	155.6
63.2	386	244	142	154.2
63.2	383	242	141	152.9
63.2	380	240	140	151.6
63.2	378	239	139	151.1
63.2	375	237	138	149.8
63.2	372	235	137	148.5
63.2	370	234	136	148.0
63.2	367	232	135	146.7
63.2	364	230	134	145.3
63.2	361	228	133	144.0
63.2	359	227	132	143.5
63.2	356	225	131	142.2
63.2	353	223	130	140.9
63.2	351	222	129	140.4
63.2	348	220	128	139.1
63.2	345	218	127	137.8
63.2	342	216	126	136.4
63.2	340	215	125	136.0
63.2	337	213	124	134.6
63.2	334	211	123	133.3
63.2	329	208	121	131.5

%E	M1	M2	DM	M*
63.2	326	206	120	130.2
63.2	323	204	119	128.8
63.2	321	203	118	128.4
63.2	318	201	117	127.0
63.2	315	199	116	125.7
63.2	310	196	114	123.9
63.2	307	194	113	122.6
63.2	304	192	112	121.3
63.2	302	191	111	120.8
63.2	299	189	110	119.5
63.2	296	187	109	118.1
63.2	291	184	107	116.3
63.2	288	182	106	115.0
63.2	285	180	105	113.7
63.2	280	177	103	111.9
63.2	277	175	102	110.6
63.2	272	172	100	108.8
63.2	269	170	99	107.4
63.2	266	168	98	106.1
63.2	261	165	96	104.3
63.2	258	163	95	103.0
63.2	253	160	93	101.2
63.2	250	158	92	99.9
63.2	247	156	91	98.5
63.2	242	153	89	96.7
63.2	239	151	88	95.4
63.2	234	148	86	93.6
63.2	231	146	85	92.3
63.2	228	144	84	90.9
63.2	223	141	82	89.2
63.2	220	139	81	87.8
63.2	212	134	78	84.7
63.2	209	132	77	83.4
63.2	204	129	75	81.6
63.2	201	127	74	80.2
63.2	193	122	71	77.1
63.2	190	120	70	75.8
63.2	185	117	68	74.0
63.2	182	115	67	72.7
63.2	174	110	64	69.5
63.2	171	108	63	68.2
63.2	163	103	60	65.1
63.2	155	98	57	62.0
63.2	152	96	56	60.6
63.2	144	91	53	57.5
63.2	136	86	50	54.4
63.2	133	84	49	53.1
63.2	125	79	46	49.9
63.2	117	74	43	46.8
63.2	114	72	42	45.5

%E	M1	M2	DM	M*
63.2	106	67	39	42.3
63.2	95	60	35	37.9
63.2	87	55	32	34.8
63.2	76	48	28	30.3
63.2	68	43	25	27.2
63.2	57	36	21	22.7
63.2	38	24	14	15.2
63.2	19	12	7	7.6
63.1	499	315	184	198.8
63.1	498	314	184	198.0
63.1	496	313	183	197.5
63.1	493	311	182	196.2
63.1	491	310	181	195.7
63.1	490	309	181	194.9
63.1	488	308	180	194.4
63.1	485	306	179	193.1
63.1	483	305	178	192.6
63.1	482	304	178	191.7
63.1	480	303	177	191.3
63.1	477	301	176	189.9
63.1	474	299	175	188.6
63.1	472	298	174	188.1
63.1	471	297	174	187.3
63.1	469	296	173	186.8
63.1	466	294	172	185.5
63.1	464	293	171	185.0
63.1	463	292	171	184.2
63.1	461	291	170	183.7
63.1	458	289	169	182.4
63.1	455	287	168	181.0
63.1	453	286	167	180.6
63.1	452	285	167	179.7
63.1	450	284	166	179.2
63.1	447	282	165	177.9
63.1	445	281	164	177.4
63.1	444	280	164	176.6
63.1	442	279	163	176.1
63.1	439	277	162	174.8
63.1	436	275	161	173.5
63.1	434	274	160	173.0
63.1	431	272	159	171.7
63.1	428	270	158	170.3
63.1	426	269	157	169.9
63.1	425	268	157	169.0
63.1	423	267	156	168.5
63.1	420	265	155	167.2
63.1	417	263	154	165.9
63.1	415	262	153	165.4
63.1	412	260	152	164.1
63.1	409	258	151	162.7

%E	M1	M2	DM	M*
63.1	407	257	150	162.3
63.1	406	256	150	161.4
63.1	404	255	149	161.0
63.1	401	253	148	159.6
63.1	398	251	147	158.3
63.1	396	250	146	157.8
63.1	393	248	145	156.5
63.1	390	246	144	155.2
63.1	388	245	143	154.7
63.1	385	243	142	153.4
63.1	382	241	141	152.0
63.1	379	239	140	150.7
63.1	377	238	139	150.2
63.1	374	236	138	148.9
63.1	371	234	137	147.6
63.1	369	233	136	147.1
63.1	366	231	135	145.8
63.1	363	229	134	144.5
63.1	360	227	133	143.1
63.1	358	226	132	142.7
63.1	355	224	131	141.3
63.1	352	222	130	140.0
63.1	350	221	129	139.5
63.1	347	219	128	138.2
63.1	344	217	127	136.9
63.1	341	215	126	135.6
63.1	339	214	125	135.1
63.1	336	212	124	133.8
63.1	333	210	123	132.4
63.1	331	209	122	132.0
63.1	328	207	121	130.6
63.1	325	205	120	129.3
63.1	320	202	118	127.5
63.1	317	200	117	126.2
63.1	314	198	116	124.9
63.1	312	197	115	124.4
63.1	309	195	114	123.1
63.1	306	193	113	121.7
63.1	301	190	111	119.9
63.1	298	188	110	118.6
63.1	295	186	109	117.3
63.1	293	185	108	116.8
63.1	290	183	107	115.5
63.1	287	181	106	114.1
63.1	282	178	104	112.4
63.1	279	176	103	111.0
63.1	274	173	101	109.2
63.1	271	171	100	107.9
63.1	268	169	99	106.6
63.1	263	166	97	104.8

%E	M1	M2	DM	M*
63.1	260	164	96	103.4
63.1	255	161	94	101.7
63.1	252	159	93	100.3
63.1	249	157	92	99.0
63.1	244	154	90	97.2
63.1	241	152	89	95.9
63.1	236	149	87	94.1
63.1	233	147	86	92.7
63.1	225	142	83	89.6
63.1	222	140	82	88.3
63.1	217	137	80	86.5
63.1	214	135	79	85.2
63.1	206	130	76	82.0
63.1	203	128	75	80.7
63.1	198	125	73	78.9
63.1	195	123	72	77.6
63.1	187	118	69	74.5
63.1	179	113	66	71.3
63.1	176	111	65	70.0
63.1	168	106	62	66.9
63.1	160	101	59	63.8
63.1	157	99	58	62.4
63.1	149	94	55	59.3
63.1	141	89	52	56.2
63.1	130	82	48	51.7
63.1	122	77	45	48.6
63.1	111	70	41	44.1
63.1	103	65	38	41.0
63.1	84	53	31	33.4
63.1	65	41	24	25.9
63.0	500	315	185	198.4
63.0	497	313	184	197.1
63.0	495	312	183	196.7
63.0	494	311	183	195.8
63.0	492	310	182	195.3
63.0	489	308	181	194.0
63.0	487	307	180	193.5
63.0	486	306	180	192.7
63.0	484	305	179	192.2
63.0	481	303	178	190.9
63.0	479	302	177	190.4
63.0	478	301	177	189.5
63.0	476	300	176	189.1
63.0	473	298	175	187.7
63.0	470	296	174	186.4
63.0	468	295	173	186.0
63.0	467	294	173	185.1
63.0	465	293	172	184.6
63.0	462	291	171	183.3
63.0	460	290	170	182.8

%E	M1	M2	DM	M*
63.0	459	289	170	182.0
63.0	457	288	169	181.5
63.0	454	286	168	180.2
63.0	451	284	167	178.8
63.0	449	283	166	178.4
63.0	446	281	165	177.0
63.0	443	279	164	175.7
63.0	441	278	163	175.2
63.0	440	277	163	174.4
63.0	438	276	162	173.9
63.0	435	274	161	172.6
63.0	433	273	160	172.1
63.0	432	272	160	171.3
63.0	430	271	159	170.8
63.0	427	269	158	169.5
63.0	424	267	157	168.1
63.0	422	266	156	167.7
63.0	419	264	155	166.3
63.0	416	262	154	165.0
63.0	414	261	153	164.5
63.0	413	260	153	163.7
63.0	411	259	152	163.2
63.0	408	257	151	161.9
63.0	405	255	150	160.6
63.0	403	254	149	160.1
63.0	400	252	148	158.8
63.0	397	250	147	157.4
63.0	395	249	146	157.0
63.0	392	247	145	155.6
63.0	389	245	144	154.3
63.0	387	244	143	153.8
63.0	386	243	143	153.0
63.0	384	242	142	152.5
63.0	381	240	141	151.2
63.0	378	238	140	149.9
63.0	376	237	139	149.4
63.0	373	235	138	148.1
63.0	370	233	137	146.7
63.0	368	232	136	146.3
63.0	365	230	135	144.9
63.0	362	228	134	143.6
63.0	359	226	133	142.3
63.0	357	225	132	141.8
63.0	354	223	131	140.5
63.0	351	221	130	139.1
63.0	349	220	129	138.7
63.0	346	218	128	137.4
63.0	343	216	127	136.0
63.0	338	213	125	134.2
63.0	335	211	124	132.9

%E	M1	M2	DM	M*
63.0	332	209	123	131.6
63.0	330	208	122	131.1
63.0	327	206	121	129.8
63.0	324	204	120	128.4
63.0	322	203	119	128.0
63.0	319	201	118	126.6
63.0	316	199	117	125.3
63.0	311	196	115	123.5
63.0	308	194	114	122.2
63.0	305	192	113	120.9
63.0	303	191	112	120.4
63.0	300	189	111	119.1
63.0	297	187	110	117.7
63.0	292	184	108	115.9
63.0	289	182	107	114.6
63.0	284	179	105	112.8
63.0	281	177	104	111.5
63.0	278	175	103	110.2
63.0	276	174	102	109.7
63.0	273	172	101	108.4
63.0	270	170	100	107.0
63.0	265	167	98	105.2
63.0	262	165	97	103.9
63.0	257	162	95	102.1
63.0	254	160	94	100.8
63.0	246	155	91	97.7
63.0	243	153	90	96.3
63.0	238	150	88	94.5
63.0	235	148	87	93.2
63.0	230	145	85	91.4
63.0	227	143	84	90.1
63.0	219	138	81	87.0
63.0	216	136	80	85.6
63.0	211	133	78	83.8
63.0	208	131	77	82.5
63.0	200	126	74	79.4
63.0	192	121	71	76.3
63.0	189	119	70	74.9
63.0	184	116	68	73.1
63.0	181	114	67	71.8
63.0	173	109	64	68.7
63.0	165	104	61	65.6
63.0	162	102	60	64.2
63.0	154	97	57	61.1
63.0	146	92	54	58.0
63.0	138	87	51	54.8
63.0	135	85	50	53.5
63.0	127	80	47	50.4
63.0	119	75	44	47.3
63.0	108	68	40	42.8

%E	M1	M2	DM	M*
63.0	100	63	37	39.7
63.0	92	58	34	36.6
63.0	81	51	30	32.1
63.0	73	46	27	29.0
63.0	54	34	20	21.4
63.0	46	29	17	18.3
63.0	27	17	10	10.7
62.9	499	314	185	197.6
62.9	498	313	185	196.7
62.9	496	312	184	196.3
62.9	493	310	183	194.9
62.9	491	309	182	194.5
62.9	490	308	182	193.6
62.9	488	307	181	193.1
62.9	485	305	180	191.8
62.9	483	304	179	191.3
62.9	482	303	179	190.5
62.9	480	302	178	190.0
62.9	477	300	177	188.7
62.9	475	299	176	188.2
62.9	474	298	176	187.4
62.9	472	297	175	186.9
62.9	469	295	174	185.6
62.9	466	293	173	184.2
62.9	464	292	172	183.8
62.9	463	291	172	182.9
62.9	461	290	171	182.4
62.9	458	288	170	181.1
62.9	456	287	169	180.6
62.9	455	286	169	179.8
62.9	453	285	168	179.3
62.9	450	283	167	178.0
62.9	448	282	166	177.5
62.9	447	281	166	176.6
62.9	445	280	165	176.2
62.9	442	278	164	174.9
62.9	439	276	163	173.5
62.9	437	275	162	173.1
62.9	434	273	161	171.7
62.9	431	271	160	170.4
62.9	429	270	159	169.9
62.9	428	269	159	169.1
62.9	426	268	158	168.6
62.9	423	266	157	167.3
62.9	421	265	156	166.8
62.9	420	264	156	165.9
62.9	418	263	155	165.5
62.9	415	261	154	164.1
62.9	412	259	153	162.8
62.9	410	258	152	162.4

%E	M1	M2	DM	M*
62.9	407	256	151	161.0
62.9	404	254	150	159.7
62.9	402	253	149	159.2
62.9	399	251	148	157.9
62.9	396	249	147	156.6
62.9	394	248	146	156.1
62.9	393	247	146	155.2
62.9	391	246	145	154.8
62.9	388	244	144	153.4
62.9	385	242	143	152.1
62.9	383	241	142	151.6
62.9	380	239	141	150.3
62.9	377	237	140	149.0
62.9	375	236	139	148.5
62.9	372	234	138	147.2
62.9	369	232	137	145.9
62.9	367	231	136	145.4
62.9	364	229	135	144.1
62.9	361	227	134	142.7
62.9	356	224	132	140.9
62.9	353	222	131	139.6
62.9	350	220	130	138.3
62.9	348	219	129	137.8
62.9	345	217	128	136.5
62.9	342	215	127	135.2
62.9	340	214	126	134.7
62.9	337	212	125	133.4
62.9	334	210	124	132.0
62.9	329	207	122	130.2
62.9	326	205	121	128.9
62.9	321	202	119	127.1
62.9	318	200	118	125.8
62.9	315	198	117	124.5
62.9	313	197	116	124.0
62.9	310	195	115	122.7
62.9	307	193	114	121.3
62.9	302	190	112	119.5
62.9	299	188	111	118.2
62.9	294	185	109	116.4
62.9	291	183	108	115.1
62.9	286	180	106	113.3
62.9	283	178	105	112.0
62.9	280	176	104	110.6
62.9	275	173	102	108.8
62.9	272	171	101	107.5
62.9	267	168	99	105.7
62.9	264	166	98	104.4
62.9	259	163	96	102.6
62.9	256	161	95	101.3
62.9	251	158	93	99.5

%E	M1	M2	DM	M*
62.9	248	156	92	98.1
62.9	245	154	91	96.8
62.9	240	151	89	95.0
62.9	237	149	88	93.7
62.9	232	146	86	91.9
62.9	229	144	85	90.6
62.9	224	141	83	88.8
62.9	221	139	82	87.4
62.9	213	134	79	84.3
62.9	210	132	78	83.0
62.9	205	129	76	81.2
62.9	202	127	75	79.8
62.9	197	124	73	78.1
62.9	194	122	72	76.7
62.9	186	117	69	73.6
62.9	178	112	66	70.5
62.9	175	110	65	69.1
62.9	170	107	63	67.3
62.9	167	105	62	66.0
62.9	159	100	59	62.9
62.9	151	95	56	59.8
62.9	143	90	53	56.6
62.9	140	88	52	55.3
62.9	132	83	49	52.2
62.9	124	78	46	49.1
62.9	116	73	43	45.9
62.9	105	66	39	41.5
62.9	97	61	36	38.4
62.9	89	56	33	35.2
62.9	70	44	26	27.7
62.9	62	39	23	24.5
62.9	35	22	13	13.8
62.8	500	314	186	197.2
62.8	497	312	185	195.9
62.8	495	311	184	195.4
62.8	494	310	184	194.5
62.8	492	309	183	194.1
62.8	489	307	182	192.7
62.8	487	306	181	192.3
62.8	486	305	181	191.4
62.8	484	304	180	190.9
62.8	481	302	179	189.6
62.8	479	301	178	189.1
62.8	478	300	178	188.3
62.8	476	299	177	187.8
62.8	473	297	176	186.5
62.8	471	296	175	186.0
62.8	470	295	175	185.2
62.8	468	294	174	184.7
62.8	465	292	173	183.4

%E	M1	M2	DM	M*
62.8	462	290	172	182.0
62.8	460	289	171	181.6
62.8	457	287	170	180.2
62.8	454	285	169	178.9
62.8	452	284	168	178.4
62.8	449	282	167	177.1
62.8	446	280	166	175.8
62.8	444	279	165	175.3
62.8	443	278	165	174.5
62.8	441	277	164	174.0
62.8	438	275	163	172.7
62.8	436	274	162	172.2
62.8	435	273	162	171.3
62.8	433	272	161	170.9
62.8	430	270	160	169.5
62.8	427	268	159	168.2
62.8	425	267	158	167.7
62.8	422	265	157	166.4
62.8	419	263	156	165.1
62.8	417	262	155	164.6
62.8	414	260	154	163.3
62.8	411	258	153	162.0
62.8	409	257	152	161.5
62.8	406	255	151	160.2
62.8	403	253	150	158.8
62.8	401	252	149	158.4
62.8	400	251	149	157.5
62.8	398	250	148	157.0
62.8	395	248	147	155.7
62.8	392	246	146	154.4
62.8	390	245	145	153.9
62.8	387	243	144	152.6
62.8	384	241	143	151.3
62.8	382	240	142	150.8
62.8	379	238	141	149.5
62.8	376	236	140	148.1
62.8	374	235	139	147.7
62.8	371	233	138	146.3
62.8	368	231	137	145.0
62.8	366	230	136	144.5
62.8	363	228	135	143.2
62.8	360	226	134	141.9
62.8	358	225	133	141.4
62.8	355	223	132	140.1
62.8	352	221	131	138.8
62.8	349	219	130	137.4
62.8	347	218	129	137.0
62.8	344	216	128	135.6
62.8	341	214	127	134.3
62.8	339	213	126	133.8

%E	M1	M2	DM	M*
62.8	336	211	125	132.5
62.8	333	209	124	131.2
62.8	331	208	123	130.7
62.8	328	206	122	129.4
62.8	325	204	121	128.0
62.8	323	203	120	127.6
62.8	320	201	119	126.3
62.8	317	199	118	124.9
62.8	312	196	116	123.1
62.8	309	194	115	121.8
62.8	304	191	113	120.0
62.8	301	189	112	118.7
62.8	298	187	111	117.3
62.8	296	186	110	116.9
62.8	293	184	109	115.5
62.8	290	182	108	114.2
62.8	288	181	107	113.8
62.8	285	179	106	112.4
62.8	282	177	105	111.1
62.8	277	174	103	109.3
62.8	274	172	102	108.0
62.8	269	169	100	106.2
62.8	266	167	99	104.8
62.8	261	164	97	103.0
62.8	258	162	96	101.7
62.8	253	159	94	99.9
62.8	250	157	93	98.6
62.8	247	155	92	97.3
62.8	242	152	90	95.5
62.8	239	150	89	94.1
62.8	234	147	87	92.3
62.8	231	145	86	91.0
62.8	226	142	84	89.2
62.8	223	140	83	87.9
62.8	218	137	81	86.1
62.8	215	135	80	84.8
62.8	207	130	77	81.6
62.8	199	125	74	78.5
62.8	196	123	73	77.2
62.8	191	120	71	75.4
62.8	188	118	70	74.1
62.8	183	115	68	72.3
62.8	180	113	67	70.9
62.8	172	108	64	67.8
62.8	164	103	61	64.7
62.8	156	98	58	61.6
62.8	148	93	55	58.4
62.8	145	91	54	57.1
62.8	137	86	51	54.0
62.8	129	81	48	50.9

%E	M1	M2	DM	M*
62.8	121	76	45	47.7
62.8	113	71	42	44.6
62.8	94	59	35	37.0
62.8	86	54	32	33.9
62.8	78	49	29	30.8
62.8	43	27	16	17.0
62.7	499	313	186	196.3
62.7	498	312	186	195.5
62.7	496	311	185	195.0
62.7	493	309	184	193.7
62.7	491	308	183	193.2
62.7	490	307	183	192.3
62.7	488	306	182	191.9
62.7	485	304	181	190.5
62.7	483	303	180	190.1
62.7	482	302	180	189.2
62.7	480	301	179	188.8
62.7	477	299	178	187.4
62.7	475	298	177	187.0
62.7	474	297	177	186.1
62.7	472	296	176	185.6
62.7	469	294	175	184.3
62.7	467	293	174	183.8
62.7	466	292	174	183.0
62.7	464	291	173	182.5
62.7	461	289	172	181.2
62.7	459	288	171	180.7
62.7	458	287	171	179.8
62.7	456	286	170	179.4
62.7	453	284	169	178.0
62.7	451	283	168	177.6
62.7	450	282	168	176.7
62.7	448	281	167	176.3
62.7	445	279	166	174.9
62.7	442	277	165	173.6
62.7	440	276	164	173.1
62.7	437	274	163	171.8
62.7	434	272	162	170.5
62.7	432	271	161	170.0
62.7	429	269	160	168.7
62.7	426	267	159	167.3
62.7	424	266	158	166.9
62.7	421	264	157	165.5
62.7	418	262	156	164.2
62.7	416	261	155	163.8
62.7	415	260	155	162.9
62.7	413	259	154	162.4
62.7	410	257	153	161.1
62.7	408	256	152	160.6
62.7	407	255	152	159.8

%E	M1	M2	DM	M*
62.7	405	254	151	159.3
62.7	402	252	150	158.0
62.7	399	250	149	156.6
62.7	397	249	148	156.2
62.7	394	247	147	154.8
62.7	391	245	146	153.5
62.7	389	244	145	153.0
62.7	386	242	144	151.7
62.7	383	240	143	150.4
62.7	381	239	142	149.9
62.7	378	237	141	148.6
62.7	375	235	140	147.3
62.7	373	234	139	146.8
62.7	370	232	138	145.5
62.7	367	230	137	144.1
62.7	365	229	136	143.7
62.7	362	227	135	142.3
62.7	359	225	134	141.0
62.7	357	224	133	140.5
62.7	354	222	132	139.2
62.7	351	220	131	137.9
62.7	346	217	129	136.1
62.7	343	215	128	134.8
62.7	338	212	126	133.0
62.7	335	210	125	131.6
62.7	332	208	124	130.3
62.7	330	207	123	129.8
62.7	327	205	122	128.5
62.7	324	203	121	127.2
62.7	322	202	120	126.7
62.7	319	200	119	125.4
62.7	316	198	118	124.1
62.7	314	197	117	123.6
62.7	311	195	116	122.3
62.7	308	193	115	120.9
62.7	306	192	114	120.5
62.7	303	190	113	119.1
62.7	300	188	112	117.8
62.7	295	185	110	116.0
62.7	292	183	109	114.7
62.7	287	180	107	112.9
62.7	284	178	106	111.6
62.7	279	175	104	109.8
62.7	276	173	103	108.4
62.7	271	170	101	106.6
62.7	268	168	100	105.3
62.7	263	165	98	103.5
62.7	260	163	97	102.2
62.7	255	160	95	100.4
62.7	252	158	94	99.1

%E	M1	M2	DM	M*
62.7	249	156	93	97.7
62.7	244	153	91	95.9
62.7	241	151	90	94.6
62.7	236	148	88	92.8
62.7	233	146	87	91.5
62.7	228	143	85	89.7
62.7	225	141	84	88.4
62.7	220	138	82	86.6
62.7	217	136	81	85.2
62.7	212	133	79	83.4
62.7	209	131	78	82.1
62.7	204	128	76	80.3
62.7	201	126	75	79.0
62.7	193	121	72	75.9
62.7	185	116	69	72.7
62.7	177	111	66	69.6
62.7	169	106	63	66.5
62.7	166	104	62	65.2
62.7	161	101	60	63.4
62.7	158	99	59	62.0
62.7	153	96	57	60.2
62.7	150	94	56	58.9
62.7	142	89	53	55.8
62.7	134	84	50	52.7
62.7	126	79	47	49.5
62.7	118	74	44	46.4
62.7	110	69	41	43.3
62.7	102	64	38	40.2
62.7	83	52	31	32.6
62.7	75	47	28	29.5
62.7	67	42	25	26.3
62.7	59	37	22	23.2
62.7	51	32	19	20.1
62.6	500	313	187	195.9
62.6	497	311	186	194.6
62.6	495	310	185	194.1
62.6	494	309	185	193.3
62.6	492	308	184	192.8
62.6	489	306	183	191.5
62.6	487	305	182	191.0
62.6	486	304	182	190.2
62.6	484	303	181	189.7
62.6	481	301	180	188.4
62.6	479	300	179	187.9
62.6	478	299	179	187.0
62.6	476	298	178	186.6
62.6	473	296	177	185.2
62.6	471	295	176	184.8
62.6	470	294	176	183.9
62.6	468	293	175	183.4
62.6	465	291	174	182.1
62.6	463	290	173	181.6
62.6	462	289	173	180.8
62.6	460	288	172	180.3
62.6	457	286	171	179.0
62.6	455	285	170	178.5
62.6	454	284	170	177.7
62.6	452	283	169	177.2
62.6	449	281	168	175.9
62.6	447	280	167	175.4
62.6	446	279	167	174.5
62.6	444	278	166	174.1
62.6	441	276	165	172.7
62.6	439	275	164	172.3
62.6	438	274	164	171.4
62.6	436	273	163	170.9
62.6	433	271	162	169.6
62.6	431	270	161	169.1
62.6	430	269	161	168.3
62.6	428	268	160	167.8
62.6	425	266	159	166.5
62.6	423	265	158	166.0
62.6	422	264	158	165.2
62.6	420	263	157	164.7
62.6	417	261	156	163.4
62.6	414	259	155	162.0
62.6	412	258	154	161.6
62.6	409	256	153	160.2
62.6	406	254	152	158.9
62.6	404	253	151	158.4
62.6	401	251	150	157.1
62.6	398	249	149	155.8
62.6	396	248	148	155.3
62.6	393	246	147	154.0
62.6	390	244	146	152.7
62.6	388	243	145	152.2
62.6	385	241	144	150.9
62.6	382	239	143	149.5
62.6	380	238	142	149.1
62.6	377	236	141	147.7
62.6	374	234	140	146.4
62.6	372	233	139	145.9
62.6	369	231	138	144.6
62.6	366	229	137	143.3
62.6	364	228	136	142.8
62.6	361	226	135	141.5
62.6	358	224	134	140.2
62.6	356	223	133	139.7
62.6	353	221	132	138.4
62.6	350	219	131	137.0
62.6	348	218	130	136.6
62.6	345	216	129	135.2
62.6	342	214	128	133.9
62.6	340	213	127	133.4
62.6	337	211	126	132.1
62.6	334	209	125	130.8
62.6	329	206	123	129.0
62.6	326	204	122	127.7
62.6	321	201	120	125.9
62.6	318	199	119	124.5
62.6	313	196	117	122.7
62.6	310	194	116	121.4
62.6	305	191	114	119.6
62.6	302	189	113	118.3
62.6	297	186	111	116.5
62.6	294	184	110	115.2
62.6	289	181	108	113.4
62.6	286	179	107	112.0
62.6	281	176	105	110.2
62.6	278	174	104	108.9
62.6	273	171	102	107.1
62.6	270	169	101	105.8
62.6	265	166	99	104.0
62.6	262	164	98	102.7
62.6	257	161	96	100.9
62.6	254	159	95	99.5
62.6	251	157	94	98.2
62.6	246	154	92	96.4
62.6	243	152	91	95.1
62.6	238	149	89	93.3
62.6	235	147	88	92.0
62.6	230	144	86	90.2
62.6	227	142	85	88.8
62.6	222	139	83	87.0
62.6	219	137	82	85.7
62.6	214	134	80	83.9
62.6	211	132	79	82.6
62.6	206	129	77	80.8
62.6	203	127	76	79.5
62.6	198	124	74	77.7
62.6	195	122	73	76.3
62.6	190	119	71	74.5
62.6	187	117	70	73.2
62.6	182	114	68	71.4
62.6	179	112	67	70.1
62.6	174	109	65	68.3
62.6	171	107	64	67.0
62.6	163	102	61	63.8
62.6	155	97	58	60.7
62.6	147	92	55	57.6
62.6	139	87	52	54.5
62.6	131	82	49	51.3
62.6	123	77	46	48.2
62.6	115	72	43	45.1
62.6	107	67	40	42.0
62.6	99	62	37	38.8
62.6	91	57	34	35.7
62.5	499	312	187	195.1
62.5	498	311	187	194.2
62.5	496	310	186	193.8
62.5	493	308	185	192.4
62.5	491	307	184	192.0
62.5	488	305	183	190.6
62.5	485	303	182	189.3
62.5	483	302	181	188.8
62.5	480	300	180	187.5
62.5	477	298	179	186.2
62.5	475	297	178	185.7
62.5	472	295	177	184.4
62.5	469	293	176	183.0
62.5	467	292	175	182.6
62.5	464	290	174	181.3
62.5	461	288	173	179.9
62.5	459	287	172	179.5
62.5	456	285	171	178.1
62.5	453	283	170	176.8
62.5	451	282	169	176.3
62.5	448	280	168	175.0
62.5	445	278	167	173.7
62.5	443	277	166	173.2
62.5	440	275	165	171.9
62.5	437	273	164	170.5
62.5	435	272	163	170.1
62.5	432	270	162	168.8
62.5	429	268	161	167.4
62.5	427	267	160	167.0
62.5	424	265	159	165.6
62.5	421	263	158	164.3
62.5	419	262	157	163.8
62.5	416	260	156	162.5
62.5	413	258	155	161.2
62.5	411	257	154	160.7
62.5	408	255	153	159.4
62.5	405	253	152	158.0
62.5	403	252	151	157.6
62.5	400	250	150	156.3
62.5	397	248	149	154.9
62.5	395	247	148	154.5
62.5	392	245	147	153.1
62.5	389	243	146	151.8
62.5	387	242	145	151.3
62.5	384	240	144	150.0
62.5	381	238	143	148.7
62.5	379	237	142	148.2
62.5	376	235	141	146.9
62.5	373	233	140	145.5
62.5	371	232	139	145.1
62.5	368	230	138	143.8
62.5	365	228	137	142.4
62.5	363	227	136	142.0
62.5	360	225	135	140.6
62.5	357	223	134	139.3
62.5	355	222	133	138.8
62.5	352	220	132	137.5
62.5	349	218	131	136.2
62.5	347	217	130	135.7
62.5	344	215	129	134.4
62.5	341	213	128	133.0
62.5	339	212	127	132.6
62.5	336	210	126	131.3
62.5	333	208	125	129.9
62.5	331	207	124	129.5
62.5	328	205	123	128.1
62.5	325	203	122	126.8
62.5	323	202	121	126.3
62.5	320	200	120	125.0
62.5	317	198	119	123.7
62.5	315	197	118	123.2
62.5	312	195	117	121.9
62.5	309	193	116	120.5
62.5	307	192	115	120.1
62.5	304	190	114	118.8
62.5	301	188	113	117.4
62.5	299	187	112	117.0
62.5	296	185	111	115.6
62.5	293	183	110	114.3
62.5	291	182	109	113.8
62.5	288	180	108	112.5
62.5	285	178	107	111.2
62.5	283	177	106	110.7
62.5	280	175	105	109.4
62.5	277	173	104	108.0
62.5	275	172	103	107.6
62.5	272	170	102	106.3
62.5	269	168	101	104.9
62.5	267	167	100	104.5
62.5	264	165	99	103.1
62.5	261	163	98	101.8
62.5	259	162	97	101.3
62.5	256	160	96	100.0

%E	M1	M2	DM	M*
62.5	253	158	95	98.7
62.5	248	155	93	96.9
62.5	240	150	90	93.8
62.5	232	145	87	90.6
62.5	224	140	84	87.5
62.5	216	135	81	84.4
62.5	208	130	78	81.3
62.5	200	125	75	78.1
62.5	192	120	72	75.0
62.5	184	115	69	71.9
62.5	176	110	66	68.8
62.5	168	105	63	65.6
62.5	160	100	60	62.5
62.5	152	95	57	59.4
62.5	144	90	54	56.3
62.5	136	85	51	53.1
62.5	128	80	48	50.0
62.5	120	75	45	46.9
62.5	112	70	42	43.8
62.5	104	65	39	40.6
62.5	96	60	36	37.5
62.5	88	55	33	34.4
62.5	80	50	30	31.3
62.5	72	45	27	28.1
62.5	64	40	24	25.0
62.5	56	35	21	21.9
62.5	48	30	18	18.8
62.5	40	25	15	15.6
62.5	32	20	12	12.5
62.5	24	15	9	9.4
62.5	16	10	6	6.3
62.5	8	5	3	3.1
62.4	500	312	188	194.7
62.4	497	310	187	193.4
62.4	495	309	186	192.9
62.4	492	307	185	191.6
62.4	490	306	184	191.1
62.4	489	305	184	190.2
62.4	487	304	183	189.8
62.4	484	302	182	188.4
62.4	482	301	181	188.0
62.4	481	300	181	187.1
62.4	479	299	180	186.6
62.4	476	297	179	185.3
62.4	474	296	178	184.8
62.4	473	295	178	184.0
62.4	471	294	177	183.5
62.4	468	292	176	182.2
62.4	466	291	175	181.7
62.4	465	290	175	180.9

%E	M1	M2	DM	M*
62.4	463	289	174	180.4
62.4	460	287	173	179.1
62.4	458	286	172	178.6
62.4	457	285	172	177.7
62.4	455	284	171	177.3
62.4	452	282	170	175.9
62.4	450	281	169	175.5
62.4	449	280	169	174.6
62.4	447	279	168	174.1
62.4	444	277	167	172.8
62.4	442	276	166	172.3
62.4	441	275	166	171.5
62.4	439	274	165	171.0
62.4	436	272	164	169.7
62.4	434	271	163	169.2
62.4	433	270	163	168.4
62.4	431	269	162	167.9
62.4	428	267	161	166.6
62.4	426	266	160	166.1
62.4	425	265	160	165.2
62.4	423	264	159	164.8
62.4	420	262	158	163.4
62.4	418	261	157	163.0
62.4	417	260	157	162.1
62.4	415	259	156	161.6
62.4	412	257	155	160.3
62.4	410	256	154	159.8
62.4	407	254	153	158.5
62.4	404	252	152	157.2
62.4	402	251	151	156.7
62.4	399	249	150	155.4
62.4	396	247	149	154.1
62.4	394	246	148	153.6
62.4	391	244	147	152.3
62.4	388	242	146	150.9
62.4	386	241	145	150.5
62.4	383	239	144	149.1
62.4	380	237	143	147.8
62.4	378	236	142	147.3
62.4	375	234	141	146.0
62.4	372	232	140	144.7
62.4	370	231	139	144.2
62.4	367	229	138	142.9
62.4	364	227	137	141.6
62.4	362	226	136	141.1
62.4	359	224	135	139.8
62.4	356	222	134	138.4
62.4	354	221	133	138.0
62.4	351	219	132	136.6
62.4	348	217	131	135.3

%E	M1	M2	DM	M*
62.4	346	216	130	134.8
62.4	343	214	129	133.5
62.4	340	212	128	132.2
62.4	338	211	127	131.7
62.4	335	209	126	130.4
62.4	330	206	124	128.6
62.4	327	204	123	127.3
62.4	322	201	121	125.5
62.4	319	199	120	124.1
62.4	314	196	118	122.3
62.4	311	194	117	121.0
62.4	306	191	115	119.2
62.4	303	189	114	117.9
62.4	298	186	112	116.1
62.4	295	184	111	114.8
62.4	290	181	109	113.0
62.4	287	179	108	111.6
62.4	282	176	106	109.8
62.4	279	174	105	108.5
62.4	274	171	103	106.7
62.4	271	169	102	105.4
62.4	266	166	100	103.6
62.4	263	164	99	102.3
62.4	258	161	97	100.5
62.4	255	159	96	99.1
62.4	250	156	94	97.3
62.4	245	153	92	95.5
62.4	242	151	91	94.2
62.4	237	148	89	92.4
62.4	234	146	88	91.1
62.4	229	143	86	89.3
62.4	226	141	85	88.0
62.4	221	138	83	86.2
62.4	218	136	82	84.8
62.4	213	133	80	83.0
62.4	210	131	79	81.7
62.4	205	128	77	79.9
62.4	202	126	76	78.6
62.4	197	123	74	76.8
62.4	194	121	73	75.5
62.4	189	118	71	73.7
62.4	186	116	70	72.3
62.4	181	113	68	70.5
62.4	178	111	67	69.2
62.4	173	108	65	67.4
62.4	170	106	64	66.1
62.4	165	103	62	64.3
62.4	157	98	59	61.2
62.4	149	93	56	58.0
62.4	141	88	53	54.9

%E	M1	M2	DM	M*
62.4	133	83	50	51.8
62.4	125	78	47	48.7
62.4	117	73	44	45.5
62.4	109	68	41	42.4
62.4	101	63	38	39.3
62.4	93	58	35	36.2
62.4	85	53	32	33.0
62.3	499	311	188	193.8
62.3	496	309	187	192.5
62.3	494	308	186	192.0
62.3	493	307	186	191.2
62.3	491	306	185	190.7
62.3	488	304	184	189.4
62.3	486	303	183	188.9
62.3	485	302	183	188.0
62.3	483	301	182	187.6
62.3	480	299	181	186.3
62.3	478	298	180	185.8
62.3	477	297	180	184.9
62.3	475	296	179	184.5
62.3	472	294	178	183.1
62.3	470	293	177	182.7
62.3	469	292	177	181.8
62.3	467	291	176	181.3
62.3	464	289	175	180.0
62.3	462	288	174	179.5
62.3	461	287	174	178.7
62.3	459	286	173	178.2
62.3	456	284	172	176.9
62.3	454	283	171	176.4
62.3	453	282	171	175.5
62.3	451	281	170	175.1
62.3	448	279	169	173.8
62.3	446	278	168	173.3
62.3	443	276	167	172.0
62.3	440	274	166	170.6
62.3	438	273	165	170.2
62.3	435	271	164	168.8
62.3	432	269	163	167.5
62.3	430	268	162	167.0
62.3	427	266	161	165.7
62.3	424	264	160	164.4
62.3	422	263	159	163.9
62.3	419	261	158	162.6
62.3	416	259	157	161.3
62.3	414	258	156	160.8
62.3	411	256	155	159.5
62.3	409	255	154	159.0
62.3	408	254	154	158.1
62.3	406	253	153	157.7

%E	M1	M2	DM	M*
62.3	403	251	152	156.3
62.3	401	250	151	155.9
62.3	400	249	151	155.0
62.3	398	248	150	154.5
62.3	395	246	149	153.2
62.3	393	245	148	152.7
62.3	390	243	147	151.4
62.3	387	241	146	150.1
62.3	385	240	145	149.6
62.3	382	238	144	148.3
62.3	379	236	143	147.0
62.3	377	235	142	146.5
62.3	374	233	141	145.2
62.3	371	231	140	143.8
62.3	369	230	139	143.4
62.3	366	228	138	142.0
62.3	363	226	137	140.7
62.3	361	225	136	140.2
62.3	358	223	135	138.9
62.3	355	221	134	137.6
62.3	353	220	133	137.1
62.3	350	218	132	135.8
62.3	345	215	130	134.0
62.3	342	213	129	132.7
62.3	337	210	127	130.9
62.3	334	208	126	129.5
62.3	332	207	125	129.1
62.3	329	205	124	127.7
62.3	326	203	123	126.4
62.3	324	202	122	125.9
62.3	321	200	121	124.6
62.3	318	198	120	123.3
62.3	316	197	119	122.8
62.3	313	195	118	121.5
62.3	310	193	117	120.2
62.3	308	192	116	119.7
62.3	305	190	115	118.4
62.3	302	188	114	117.0
62.3	300	187	113	116.6
62.3	297	185	112	115.2
62.3	292	182	110	113.4
62.3	289	180	109	112.1
62.3	284	177	107	110.3
62.3	281	175	106	109.0
62.3	276	172	104	107.2
62.3	273	170	103	105.9
62.3	268	167	101	104.1
62.3	265	165	100	102.7
62.3	260	162	98	100.9
62.3	257	160	97	99.6

%E	M1	M2	DM	M*
62.3	252	157	95	97.8
62.3	247	154	93	96.0
62.3	244	152	92	94.7
62.3	239	149	90	92.9
62.3	236	147	89	91.6
62.3	231	144	87	89.8
62.3	228	142	86	88.4
62.3	223	139	84	86.6
62.3	220	137	83	85.3
62.3	215	134	81	83.5
62.3	212	132	80	82.2
62.3	207	129	78	80.4
62.3	204	127	77	79.1
62.3	199	124	75	77.3
62.3	191	119	72	74.1
62.3	183	114	69	71.0
62.3	175	109	66	67.9
62.3	167	104	63	64.8
62.3	162	101	61	63.0
62.3	159	99	60	61.6
62.3	154	96	58	59.8
62.3	151	94	57	58.5
62.3	146	91	55	56.7
62.3	138	86	52	53.6
62.3	130	81	49	50.5
62.3	122	76	46	47.3
62.3	114	71	43	44.2
62.3	106	66	40	41.1
62.3	77	48	29	29.9
62.3	69	43	26	26.8
62.3	61	38	23	23.7
62.3	53	33	20	20.5
62.2	500	311	189	193.4
62.2	498	310	188	193.0
62.2	497	309	188	192.1
62.2	495	308	187	191.6
62.2	492	306	186	190.3
62.2	490	305	185	189.8
62.2	489	304	185	189.0
62.2	487	303	184	188.5
62.2	484	301	183	187.2
62.2	482	300	182	186.7
62.2	481	299	182	185.9
62.2	479	298	181	185.4
62.2	476	296	180	184.1
62.2	474	295	179	183.6
62.2	473	294	179	182.7
62.2	471	293	178	182.3
62.2	468	291	177	180.9
62.2	466	290	176	180.5

%E	M1	M2	DM	M*
62.2	465	289	176	179.6
62.2	463	288	175	179.1
62.2	460	286	174	177.8
62.2	458	285	173	177.3
62.2	455	283	172	176.0
62.2	452	281	171	174.7
62.2	450	280	170	174.2
62.2	447	278	169	172.9
62.2	445	277	168	172.4
62.2	444	276	168	171.6
62.2	442	275	167	171.1
62.2	439	273	166	169.8
62.2	437	272	165	169.3
62.2	436	271	165	168.4
62.2	434	270	164	168.0
62.2	431	268	163	166.6
62.2	429	267	162	166.2
62.2	428	266	162	165.3
62.2	426	265	161	164.8
62.2	423	263	160	163.5
62.2	421	262	159	163.0
62.2	418	260	158	161.7
62.2	415	258	157	160.4
62.2	413	257	156	159.9
62.2	410	255	155	158.6
62.2	407	253	154	157.3
62.2	405	252	153	156.8
62.2	402	250	152	155.5
62.2	399	248	151	154.1
62.2	397	247	150	153.7
62.2	394	245	149	152.3
62.2	392	244	148	151.9
62.2	389	242	147	150.6
62.2	386	240	146	149.2
62.2	384	239	145	148.8
62.2	381	237	144	147.4
62.2	378	235	143	146.1
62.2	376	234	142	145.6
62.2	373	232	141	144.3
62.2	370	230	140	143.0
62.2	368	229	139	142.5
62.2	365	227	138	141.2
62.2	362	225	137	139.8
62.2	360	224	136	139.4
62.2	357	222	135	138.1
62.2	352	219	133	136.3
62.2	349	217	132	134.9
62.2	347	216	131	134.5
62.2	344	214	130	133.1
62.2	341	212	129	131.8

%E	M1	M2	DM	M*
62.2	339	211	128	131.3
62.2	336	209	127	130.0
62.2	333	207	126	128.7
62.2	331	206	125	128.2
62.2	328	204	124	126.9
62.2	325	202	123	125.6
62.2	323	201	122	125.1
62.2	320	199	121	123.8
62.2	315	196	119	122.0
62.2	312	194	118	120.6
62.2	307	191	116	118.8
62.2	304	189	115	117.5
62.2	299	186	113	115.7
62.2	296	184	112	114.4
62.2	294	183	111	113.9
62.2	291	181	110	112.6
62.2	288	179	109	111.3
62.2	286	178	108	110.8
62.2	283	176	107	109.5
62.2	278	173	105	107.7
62.2	275	171	104	106.3
62.2	270	168	102	104.5
62.2	267	166	101	103.2
62.2	262	163	99	101.4
62.2	259	161	98	100.1
62.2	254	158	96	98.3
62.2	251	156	95	97.0
62.2	249	155	94	96.5
62.2	246	153	93	95.2
62.2	241	150	91	93.4
62.2	238	148	90	92.0
62.2	233	145	88	90.2
62.2	230	143	87	88.9
62.2	225	140	85	87.1
62.2	222	138	84	85.8
62.2	217	135	82	84.0
62.2	214	133	81	82.7
62.2	209	130	79	80.9
62.2	201	125	76	77.7
62.2	196	122	74	75.9
62.2	193	120	73	74.6
62.2	188	117	71	72.8
62.2	185	115	70	71.5
62.2	180	112	68	69.7
62.2	172	107	65	66.6
62.2	164	102	62	63.4
62.2	156	97	59	60.3
62.2	148	92	56	57.2
62.2	143	89	54	55.4
62.2	135	84	51	52.3

%E	M1	M2	DM	M*
62.2	127	79	48	49.1
62.2	119	74	45	46.0
62.2	111	69	42	42.9
62.2	98	61	37	38.0
62.2	90	56	34	34.8
62.2	82	51	31	31.7
62.2	74	46	28	28.6
62.2	45	28	17	17.4
62.2	37	23	14	14.3
62.1	499	310	189	192.6
62.1	496	308	188	191.3
62.1	494	307	187	190.8
62.1	493	306	187	189.9
62.1	491	305	186	189.5
62.1	488	303	185	188.1
62.1	486	302	184	187.7
62.1	485	301	184	186.8
62.1	483	300	183	186.3
62.1	480	298	182	185.0
62.1	478	297	181	184.5
62.1	477	296	181	183.7
62.1	475	295	180	183.2
62.1	472	293	179	181.9
62.1	470	292	178	181.4
62.1	467	290	177	180.1
62.1	464	288	176	178.8
62.1	462	287	175	178.3
62.1	459	285	174	177.0
62.1	457	284	173	176.5
62.1	456	283	173	175.6
62.1	454	282	172	175.2
62.1	451	280	171	173.8
62.1	449	279	170	173.4
62.1	448	278	170	172.5
62.1	446	277	169	172.0
62.1	443	275	168	170.7
62.1	441	274	167	170.2
62.1	438	272	166	168.9
62.1	435	270	165	167.6
62.1	433	269	164	167.1
62.1	430	267	163	165.8
62.1	427	265	162	164.5
62.1	425	264	161	164.0
62.1	422	262	160	162.7
62.1	420	261	159	162.2
62.1	419	260	159	161.3
62.1	417	259	158	160.9
62.1	414	257	157	159.5
62.1	412	256	156	159.1
62.1	409	254	155	157.7

%E	M1	M2	DM	M*
62.1	406	252	154	156.4
62.1	404	251	153	155.9
62.1	401	249	152	154.6
62.1	398	247	151	153.3
62.1	396	246	150	152.8
62.1	393	244	149	151.5
62.1	391	243	148	151.0
62.1	390	242	148	150.2
62.1	388	241	147	149.7
62.1	385	239	146	148.4
62.1	383	238	145	147.9
62.1	380	236	144	146.6
62.1	377	234	143	145.2
62.1	375	233	142	144.8
62.1	372	231	141	143.4
62.1	369	229	140	142.1
62.1	367	228	139	141.6
62.1	364	226	138	140.3
62.1	361	224	137	139.0
62.1	359	223	136	138.5
62.1	356	221	135	137.2
62.1	354	220	134	136.7
62.1	351	218	133	135.4
62.1	348	216	132	134.1
62.1	346	215	131	133.6
62.1	343	213	130	132.3
62.1	340	211	129	130.9
62.1	338	210	128	130.5
62.1	335	208	127	129.1
62.1	330	205	125	127.3
62.1	327	203	124	126.0
62.1	322	200	122	124.2
62.1	319	198	121	122.9
62.1	317	197	120	122.4
62.1	314	195	119	121.1
62.1	311	193	118	119.8
62.1	309	192	117	119.3
62.1	306	190	116	118.0
62.1	301	187	114	116.2
62.1	298	185	113	114.8
62.1	293	182	111	113.1
62.1	290	180	110	111.7
62.1	285	177	108	109.9
62.1	282	175	107	108.6
62.1	280	174	106	108.1
62.1	277	172	105	106.8
62.1	272	169	103	105.0
62.1	269	167	102	103.7
62.1	264	164	100	101.9
62.1	261	162	99	100.6

%E	M1	M2	DM	M*
62.1	256	159	97	98.8
62.1	253	157	96	97.4
62.1	248	154	94	95.6
62.1	243	151	92	93.8
62.1	240	149	91	92.5
62.1	235	146	89	90.7
62.1	232	144	88	89.4
62.1	227	141	86	87.6
62.1	224	139	85	86.3
62.1	219	136	83	84.5
62.1	211	131	80	81.3
62.1	206	128	78	79.5
62.1	203	126	77	78.2
62.1	198	123	75	76.4
62.1	195	121	74	75.1
62.1	190	118	72	73.3
62.1	182	113	69	70.2
62.1	177	110	67	68.4
62.1	174	108	66	67.0
62.1	169	105	64	65.2
62.1	161	100	61	62.1
62.1	153	95	58	59.0
62.1	145	90	55	55.9
62.1	140	87	53	54.1
62.1	132	82	50	50.9
62.1	124	77	47	47.8
62.1	116	72	44	44.7
62.1	103	64	39	39.8
62.1	95	59	36	36.6
62.1	87	54	33	33.5
62.1	66	41	25	25.5
62.1	58	36	22	22.3
62.1	29	18	11	11.2
62.0	500	310	190	192.2
62.0	498	309	189	191.7
62.0	497	308	189	190.9
62.0	495	307	188	190.4
62.0	492	305	187	189.1
62.0	490	304	186	188.6
62.0	489	303	186	187.7
62.0	487	302	185	187.3
62.0	484	300	184	186.0
62.0	482	299	183	185.5
62.0	481	298	183	184.6
62.0	479	297	182	184.2
62.0	476	295	181	182.8
62.0	474	294	180	182.4
62.0	471	292	179	181.0
62.0	469	291	178	180.6
62.0	468	290	178	179.7
62.0	466	289	177	179.2
62.0	463	287	176	177.9
62.0	461	286	175	177.4
62.0	460	285	175	176.6
62.0	458	284	174	176.1
62.0	455	282	173	174.8
62.0	453	281	172	174.3
62.0	450	279	171	173.0
62.0	447	277	170	171.7
62.0	445	276	169	171.2
62.0	442	274	168	169.9
62.0	440	273	167	169.4
62.0	439	272	167	168.5
62.0	437	271	166	168.1
62.0	434	269	165	166.7
62.0	432	268	164	166.3
62.0	429	266	163	164.9
62.0	426	264	162	163.6
62.0	424	263	161	163.1
62.0	421	261	160	161.8
62.0	418	259	159	160.5
62.0	416	258	158	160.0
62.0	413	256	157	158.7
62.0	411	255	156	158.2
62.0	410	254	156	157.4
62.0	408	253	155	156.9
62.0	405	251	154	155.6
62.0	403	250	153	155.1
62.0	400	248	152	153.8
62.0	397	246	151	152.4
62.0	395	245	150	152.0
62.0	392	243	149	150.6
62.0	389	241	148	149.3
62.0	387	240	147	148.8
62.0	384	238	146	147.5
62.0	382	237	145	147.0
62.0	379	235	144	145.7
62.0	376	233	143	144.4
62.0	374	232	142	143.9
62.0	371	230	141	142.6
62.0	368	228	140	141.3
62.0	366	227	139	140.8
62.0	363	225	138	139.5
62.0	358	222	136	137.7
62.0	355	220	135	136.3
62.0	353	219	134	135.9
62.0	350	217	133	134.5
62.0	347	215	132	133.2
62.0	345	214	131	132.7
62.0	342	212	130	131.4
62.0	337	209	128	129.6
62.0	334	207	127	128.3
62.0	332	206	126	127.8
62.0	329	204	125	126.5
62.0	326	202	124	125.2
62.0	324	201	123	124.7
62.0	321	199	122	123.4
62.0	318	197	121	122.0
62.0	316	196	120	121.6
62.0	313	194	119	120.2
62.0	308	191	117	118.4
62.0	305	189	116	117.1
62.0	303	188	115	116.6
62.0	300	186	114	115.3
62.0	297	184	113	114.0
62.0	295	183	112	113.5
62.0	292	181	111	112.2
62.0	287	178	109	110.4
62.0	284	176	108	109.1
62.0	279	173	106	107.3
62.0	276	171	105	105.9
62.0	274	170	104	105.5
62.0	271	168	103	104.1
62.0	266	165	101	102.3
62.0	263	163	100	101.0
62.0	258	160	98	99.2
62.0	255	158	97	97.9
62.0	250	155	95	96.1
62.0	245	152	93	94.3
62.0	242	150	92	93.0
62.0	237	147	90	91.2
62.0	234	145	89	89.9
62.0	229	142	87	88.1
62.0	221	137	84	84.9
62.0	216	134	82	83.1
62.0	213	132	81	81.8
62.0	208	129	79	80.0
62.0	205	127	78	78.7
62.0	200	124	76	76.9
62.0	192	119	73	73.8
62.0	187	116	71	72.0
62.0	184	114	70	70.6
62.0	179	111	68	68.8
62.0	171	106	65	65.7
62.0	166	103	63	63.9
62.0	163	101	62	62.6
62.0	158	98	60	60.8
62.0	150	93	57	57.7
62.0	142	88	54	54.5
62.0	137	85	52	52.7
62.0	129	80	49	49.6
62.0	121	75	46	46.5
62.0	108	67	41	41.6
62.0	100	62	38	38.4
62.0	92	57	35	35.3
62.0	79	49	30	30.4
62.0	71	44	27	27.3
62.0	50	31	19	19.2
61.9	499	309	190	191.3
61.9	496	307	189	190.0
61.9	494	306	188	189.5
61.9	493	305	188	188.7
61.9	491	304	187	188.2
61.9	488	302	186	186.9
61.9	486	301	185	186.4
61.9	485	300	185	185.6
61.9	483	299	184	185.1
61.9	480	297	183	183.8
61.9	478	296	182	183.3
61.9	475	294	181	182.0
61.9	473	293	180	181.5
61.9	472	292	180	180.6
61.9	470	291	179	180.2
61.9	467	289	178	178.8
61.9	465	288	177	178.4
61.9	464	287	177	177.5
61.9	462	286	176	177.0
61.9	459	284	175	175.7
61.9	457	283	174	175.2
61.9	454	281	173	173.9
61.9	452	280	172	173.5
61.9	451	279	172	172.6
61.9	449	278	171	172.1
61.9	446	276	170	170.8
61.9	444	275	169	170.3
61.9	443	274	169	169.5
61.9	441	273	168	169.0
61.9	438	271	167	167.7
61.9	436	270	166	167.2
61.9	433	268	165	165.9
61.9	431	267	164	165.4
61.9	430	266	164	164.5
61.9	428	265	163	164.1
61.9	425	263	162	162.8
61.9	423	262	161	162.3
61.9	420	260	160	161.0
61.9	417	258	159	159.6
61.9	415	257	158	159.2
61.9	412	255	157	157.8
61.9	409	253	156	156.5
61.9	407	252	155	156.0
61.9	404	250	154	154.7
61.9	402	249	153	154.2
61.9	399	247	152	152.9
61.9	396	245	151	151.6
61.9	394	244	150	151.1
61.9	391	242	149	149.8
61.9	388	240	148	148.5
61.9	386	239	147	148.0
61.9	383	237	146	146.7
61.9	381	236	145	146.2
61.9	378	234	144	144.9
61.9	375	232	143	143.5
61.9	373	231	142	143.1
61.9	370	229	141	141.7
61.9	367	227	140	140.4
61.9	365	226	139	139.9
61.9	362	224	138	138.6
61.9	360	223	137	138.1
61.9	357	221	136	136.8
61.9	354	219	135	135.5
61.9	352	218	134	135.0
61.9	349	216	133	133.7
61.9	346	214	132	132.4
61.9	344	213	131	131.9
61.9	341	211	130	130.6
61.9	339	210	129	130.1
61.9	336	208	128	128.8
61.9	333	206	127	127.4
61.9	331	205	126	127.0
61.9	328	203	125	125.6
61.9	323	200	123	123.8
61.9	320	198	122	122.5
61.9	315	195	120	120.7
61.9	312	193	119	119.4
61.9	310	192	118	118.9
61.9	307	190	117	117.6
61.9	302	187	115	115.8
61.9	299	185	114	114.5
61.9	294	182	112	112.7
61.9	291	180	111	111.3
61.9	289	179	110	110.9
61.9	286	177	109	109.5
61.9	281	174	107	107.7
61.9	278	172	106	106.4
61.9	273	169	104	104.6
61.9	270	167	103	103.3
61.9	268	166	102	102.8
61.9	265	164	101	101.5
61.9	260	161	99	99.7

%E	M1	M2	DM	M*
61.9	257	159	98	98.4
61.9	252	156	96	96.6
61.9	247	153	94	94.8
61.9	244	151	93	93.4
61.9	239	148	91	91.6
61.9	236	146	90	90.3
61.9	231	143	88	88.5
61.9	226	140	86	86.7
61.9	223	138	85	85.4
61.9	218	135	83	83.6
61.9	215	133	82	82.3
61.9	210	130	80	80.5
61.9	202	125	77	77.4
61.9	197	122	75	75.6
61.9	194	120	74	74.2
61.9	189	117	72	72.4
61.9	181	112	69	69.3
61.9	176	109	67	67.5
61.9	173	107	66	66.2
61.9	168	104	64	64.4
61.9	160	99	61	61.3
61.9	155	96	59	59.5
61.9	147	91	56	56.3
61.9	139	86	53	53.2
61.9	134	83	51	51.4
61.9	126	78	48	48.3
61.9	118	73	45	45.2
61.9	113	70	43	43.4
61.9	105	65	40	40.2
61.9	97	60	37	37.1
61.9	84	52	32	32.2
61.9	63	39	24	24.1
61.9	42	26	16	16.1
61.9	21	13	8	8.0
61.8	500	309	191	191.0
61.8	498	308	190	190.5
61.8	497	307	190	189.6
61.8	495	306	189	189.2
61.8	492	304	188	187.8
61.8	490	303	187	187.4
61.8	489	302	187	186.5
61.8	487	301	186	186.0
61.8	484	299	185	184.7
61.8	482	298	184	184.2
61.8	479	296	183	182.9
61.8	477	295	182	182.4
61.8	476	294	182	181.6
61.8	474	293	181	181.1
61.8	471	291	180	179.8
61.8	469	290	179	179.3
61.8	468	289	179	178.5
61.8	466	288	178	178.0
61.8	463	286	177	176.7
61.8	461	285	176	176.2
61.8	458	283	175	174.9
61.8	456	282	174	174.4
61.8	455	281	174	173.5
61.8	453	280	173	173.1
61.8	450	278	172	171.7
61.8	448	277	171	171.3
61.8	445	275	170	169.9
61.8	442	273	169	168.6
61.8	440	272	168	168.1
61.8	437	270	167	166.8
61.8	435	269	166	166.3
61.8	434	268	166	165.5
61.8	432	267	165	165.0
61.8	429	265	164	163.7
61.8	427	264	163	163.2
61.8	424	262	162	161.9
61.8	422	261	161	161.4
61.8	421	260	161	160.6
61.8	419	259	160	160.1
61.8	416	257	159	158.8
61.8	414	256	158	158.3
61.8	411	254	157	157.0
61.8	408	252	156	155.6
61.8	406	251	155	155.2
61.8	403	249	154	153.8
61.8	401	248	153	153.4
61.8	400	247	153	152.5
61.8	398	246	152	152.1
61.8	395	244	151	150.7
61.8	393	243	150	150.3
61.8	390	241	149	148.9
61.8	387	239	148	147.6
61.8	385	238	147	147.1
61.8	382	236	146	145.8
61.8	380	235	145	145.3
61.8	377	233	144	144.0
61.8	374	231	143	142.7
61.8	372	230	142	142.2
61.8	369	228	141	140.9
61.8	364	225	139	139.1
61.8	361	223	138	137.8
61.8	359	222	137	137.3
61.8	356	220	136	136.0
61.8	353	218	135	134.6
61.8	351	217	134	134.2
61.8	348	215	133	132.8
61.8	343	212	131	131.0
61.8	340	210	130	129.7
61.8	338	209	129	129.2
61.8	335	207	128	127.9
61.8	330	204	126	126.1
61.8	327	202	125	124.8
61.8	325	201	124	124.3
61.8	322	199	123	123.0
61.8	319	197	122	121.7
61.8	317	196	121	121.2
61.8	314	194	120	119.9
61.8	309	191	118	118.1
61.8	306	189	117	116.7
61.8	304	188	116	116.3
61.8	301	186	115	114.9
61.8	296	183	113	113.1
61.8	293	181	112	111.8
61.8	288	178	110	110.0
61.8	285	176	109	108.7
61.8	283	175	108	108.2
61.8	280	173	107	106.9
61.8	275	170	105	105.1
61.8	272	168	104	103.8
61.8	267	165	102	102.0
61.8	262	162	100	100.2
61.8	259	160	99	98.8
61.8	254	157	97	97.0
61.8	251	155	96	95.7
61.8	249	154	95	95.2
61.8	246	152	94	93.9
61.8	241	149	92	92.1
61.8	238	147	91	90.8
61.8	233	144	89	89.0
61.8	228	141	87	87.2
61.8	225	139	86	85.9
61.8	220	136	84	84.1
61.8	217	134	83	82.7
61.8	212	131	81	80.9
61.8	207	128	79	79.1
61.8	204	126	78	77.8
61.8	199	123	76	76.0
61.8	191	118	73	72.9
61.8	186	115	71	71.1
61.8	178	110	68	68.0
61.8	170	105	65	64.9
61.8	165	102	63	63.1
61.8	157	97	60	59.9
61.8	152	94	58	58.1
61.8	144	89	55	55.0
61.8	136	84	52	51.9
61.8	131	81	50	50.1
61.8	123	76	47	47.0
61.8	110	68	42	42.0
61.8	102	63	39	38.9
61.8	89	55	34	34.0
61.8	76	47	29	29.1
61.8	68	42	26	25.9
61.8	55	34	21	21.0
61.8	34	21	13	13.0
61.7	499	308	191	190.1
61.7	496	306	190	188.8
61.7	494	305	189	188.3
61.7	493	304	189	187.5
61.7	491	303	188	187.0
61.7	488	301	187	185.7
61.7	486	300	186	185.2
61.7	483	298	185	183.9
61.7	481	297	184	183.4
61.7	480	296	184	182.5
61.7	478	295	183	182.1
61.7	475	293	182	180.7
61.7	473	292	181	180.3
61.7	472	291	181	179.4
61.7	470	290	180	178.9
61.7	467	288	179	177.6
61.7	465	287	178	177.1
61.7	462	285	177	175.8
61.7	460	284	176	175.3
61.7	459	283	176	174.5
61.7	457	282	175	174.0
61.7	454	280	174	172.7
61.7	452	279	173	172.2
61.7	449	277	172	170.9
61.7	447	276	171	170.4
61.7	446	275	171	169.6
61.7	444	274	170	169.1
61.7	441	272	169	167.8
61.7	439	271	168	167.3
61.7	436	269	167	166.0
61.7	433	267	166	164.6
61.7	431	266	165	164.2
61.7	428	264	164	162.8
61.7	426	263	163	162.4
61.7	423	261	162	161.0
61.7	420	259	161	159.7
61.7	418	258	160	159.2
61.7	415	256	159	157.9
61.7	413	255	158	157.4
61.7	412	254	158	156.6
61.7	410	253	157	156.1
61.7	407	251	156	154.8
61.7	405	250	155	154.3
61.7	402	248	154	153.0
61.7	399	246	153	151.7
61.7	397	245	152	151.2
61.7	394	243	151	149.9
61.7	392	242	150	149.4
61.7	389	240	149	148.1
61.7	386	238	148	146.7
61.7	384	237	147	146.3
61.7	381	235	146	144.9
61.7	379	234	145	144.5
61.7	376	232	144	143.1
61.7	373	230	143	141.8
61.7	371	229	142	141.4
61.7	368	227	141	140.0
61.7	366	226	140	139.6
61.7	363	224	139	138.2
61.7	360	222	138	136.9
61.7	358	221	137	136.4
61.7	355	219	136	135.1
61.7	350	216	134	133.3
61.7	347	214	133	132.0
61.7	345	213	132	131.5
61.7	342	211	131	130.2
61.7	339	209	130	128.9
61.7	337	208	129	128.4
61.7	334	206	128	127.1
61.7	332	205	127	126.6
61.7	329	203	126	125.3
61.7	326	201	125	123.9
61.7	324	200	124	123.5
61.7	321	198	123	122.1
61.7	316	195	121	120.3
61.7	313	193	120	119.0
61.7	311	192	119	118.5
61.7	308	190	118	117.2
61.7	303	187	116	115.4
61.7	300	185	115	114.1
61.7	298	184	114	113.6
61.7	295	182	113	112.3
61.7	290	179	111	110.5
61.7	287	177	110	109.2
61.7	282	174	108	107.4
61.7	277	171	106	105.6
61.7	274	169	105	104.2
61.7	269	166	103	102.4
61.7	266	164	102	101.1
61.7	264	163	101	100.6
61.7	261	161	100	99.3

%E	M1	M2	DM	M*
61.7	256	158	98	97.5
61.7	253	156	97	96.2
61.7	248	153	95	94.4
61.7	243	150	93	92.6
61.7	240	148	92	91.3
61.7	235	145	90	89.5
61.7	230	142	88	87.7
61.7	227	140	87	86.3
61.7	222	137	85	84.5
61.7	214	132	82	81.4
61.7	209	129	80	79.6
61.7	206	127	79	78.3
61.7	201	124	77	76.5
61.7	196	121	75	74.7
61.7	193	119	74	73.4
61.7	188	116	72	71.6
61.7	183	113	70	69.8
61.7	180	111	69	68.4
61.7	175	108	67	66.7
61.7	167	103	64	63.5
61.7	162	100	62	61.7
61.7	154	95	59	58.6
61.7	149	92	57	56.8
61.7	141	87	54	53.7
61.7	133	82	51	50.6
61.7	128	79	49	48.8
61.7	120	74	46	45.6
61.7	115	71	44	43.8
61.7	107	66	41	40.7
61.7	94	58	36	35.8
61.7	81	50	31	30.9
61.7	60	37	23	22.8
61.7	47	29	18	17.9
61.6	500	308	192	189.7
61.6	498	307	191	189.3
61.6	497	306	191	188.4
61.6	495	305	190	187.9
61.6	492	303	189	186.6
61.6	490	302	188	186.1
61.6	489	301	188	185.3
61.6	487	300	187	184.8
61.6	485	299	186	184.3
61.6	484	298	186	183.5
61.6	482	297	185	183.0
61.6	479	295	184	181.7
61.6	477	294	183	181.2
61.6	476	293	183	180.4
61.6	474	292	182	179.9
61.6	471	290	181	178.6
61.6	469	289	180	178.1

%E	M1	M2	DM	M*
61.6	466	287	179	176.8
61.6	464	286	178	176.3
61.6	463	285	178	175.4
61.6	461	284	177	175.0
61.6	458	282	176	173.6
61.6	456	281	175	173.2
61.6	453	279	174	171.8
61.6	451	278	173	171.4
61.6	450	277	173	170.5
61.6	448	276	172	170.0
61.6	445	274	171	168.7
61.6	443	273	170	168.2
61.6	440	271	169	166.9
61.6	438	270	168	166.4
61.6	437	269	168	165.6
61.6	435	268	167	165.1
61.6	432	266	166	163.8
61.6	430	265	165	163.3
61.6	427	263	164	162.0
61.6	425	262	163	161.5
61.6	424	261	163	160.7
61.6	422	260	162	160.2
61.6	419	258	161	158.9
61.6	417	257	160	158.4
61.6	414	255	159	157.1
61.6	411	253	158	155.7
61.6	409	252	157	155.3
61.6	406	250	156	153.9
61.6	404	249	155	153.5
61.6	401	247	154	152.1
61.6	398	245	153	150.8
61.6	396	244	152	150.3
61.6	393	242	151	149.0
61.6	391	241	150	148.5
61.6	388	239	149	147.2
61.6	385	237	148	145.9
61.6	383	236	147	145.4
61.6	380	234	146	144.1
61.6	378	233	145	143.6
61.6	375	231	144	142.3
61.6	372	229	143	141.0
61.6	370	228	142	140.5
61.6	367	226	141	139.2
61.6	365	225	140	138.7
61.6	362	223	139	137.4
61.6	359	221	138	136.0
61.6	357	220	137	135.6
61.6	354	218	136	134.2
61.6	352	217	135	133.8
61.6	349	215	134	132.4

%E	M1	M2	DM	M*
61.6	346	213	133	131.1
61.6	344	212	132	130.7
61.6	341	210	131	129.3
61.6	336	207	129	127.5
61.6	333	205	128	126.2
61.6	331	204	127	125.7
61.6	328	202	126	124.4
61.6	323	199	124	122.6
61.6	320	197	123	121.3
61.6	318	196	122	120.8
61.6	315	194	121	119.5
61.6	310	191	119	117.7
61.6	307	189	118	116.4
61.6	305	188	117	115.9
61.6	302	186	116	114.6
61.6	297	183	114	112.8
61.6	294	181	113	111.4
61.6	292	180	112	111.0
61.6	289	178	111	109.6
61.6	284	175	109	107.8
61.6	281	173	108	106.5
61.6	279	172	107	106.0
61.6	276	170	106	104.7
61.6	271	167	104	102.9
61.6	268	165	103	101.6
61.6	263	162	101	99.8
61.6	258	159	99	98.0
61.6	255	157	98	96.7
61.6	250	154	96	94.9
61.6	245	151	94	93.1
61.6	242	149	93	91.7
61.6	237	146	91	89.9
61.6	232	143	89	88.1
61.6	229	141	88	86.8
61.6	224	138	86	85.0
61.6	219	135	84	83.2
61.6	216	133	83	81.9
61.6	211	130	81	80.1
61.6	203	125	78	77.0
61.6	198	122	76	75.2
61.6	190	117	73	72.0
61.6	185	114	71	70.2
61.6	177	109	68	67.1
61.6	172	106	66	65.3
61.6	164	101	63	62.2
61.6	159	98	61	60.4
61.6	151	93	58	57.3
61.6	146	90	56	55.5
61.6	138	85	53	52.4
61.6	125	77	48	47.4

%E	M1	M2	DM	M*
61.6	112	69	43	42.5
61.6	99	61	38	37.6
61.6	86	53	33	32.7
61.6	73	45	28	27.7
61.5	499	307	192	188.9
61.5	496	305	191	187.6
61.5	494	304	190	187.1
61.5	493	303	190	186.2
61.5	491	302	189	185.8
61.5	488	300	188	184.4
61.5	486	299	187	184.0
61.5	483	297	186	182.6
61.5	481	296	185	182.2
61.5	480	295	185	181.3
61.5	478	294	184	180.8
61.5	475	292	183	179.5
61.5	473	291	182	179.0
61.5	470	289	181	177.7
61.5	468	288	180	177.2
61.5	467	287	180	176.4
61.5	465	286	179	175.9
61.5	462	284	178	174.6
61.5	460	283	177	174.1
61.5	457	281	176	172.8
61.5	455	280	175	172.3
61.5	454	279	175	171.5
61.5	452	278	174	171.0
61.5	449	276	173	169.7
61.5	447	275	172	169.2
61.5	444	273	171	167.9
61.5	442	272	170	167.4
61.5	441	271	170	166.5
61.5	439	270	169	166.1
61.5	436	268	168	164.7
61.5	434	267	167	164.3
61.5	431	265	166	162.9
61.5	429	264	165	162.5
61.5	426	262	164	161.1
61.5	423	260	163	159.8
61.5	421	259	162	159.3
61.5	418	257	161	158.0
61.5	416	256	160	157.5
61.5	413	254	159	156.2
61.5	410	252	158	154.9
61.5	408	251	157	154.4
61.5	405	249	156	153.1
61.5	403	248	155	152.6
61.5	400	246	154	151.3
61.5	397	244	153	150.0
61.5	395	243	152	149.5

%E	M1	M2	DM	M*
61.5	392	241	151	148.2
61.5	390	240	150	147.7
61.5	387	238	149	146.4
61.5	384	236	148	145.0
61.5	382	235	147	144.6
61.5	379	233	146	143.2
61.5	377	232	145	142.8
61.5	374	230	144	141.4
61.5	371	228	143	140.1
61.5	369	227	142	139.6
61.5	366	225	141	138.3
61.5	364	224	140	137.8
61.5	361	222	139	136.5
61.5	358	220	138	135.2
61.5	356	219	137	134.7
61.5	353	217	136	133.4
61.5	351	216	135	132.9
61.5	348	214	134	131.6
61.5	343	211	132	129.8
61.5	340	209	131	128.5
61.5	338	208	130	128.0
61.5	335	206	129	126.7
61.5	330	203	127	124.9
61.5	327	201	126	123.6
61.5	325	200	125	123.1
61.5	322	198	124	121.8
61.5	317	195	122	120.0
61.5	314	193	121	118.6
61.5	312	192	120	118.2
61.5	309	190	119	116.8
61.5	304	187	117	115.0
61.5	301	185	116	113.7
61.5	299	184	115	113.2
61.5	296	182	114	111.9
61.5	291	179	112	110.1
61.5	288	177	111	108.8
61.5	286	176	110	108.3
61.5	283	174	109	107.0
61.5	278	171	107	105.2
61.5	275	169	106	103.9
61.5	273	168	105	103.4
61.5	270	166	104	102.1
61.5	265	163	102	100.3
61.5	262	161	101	98.9
61.5	260	160	100	98.5
61.5	257	158	99	97.1
61.5	252	155	97	95.3
61.5	247	152	95	93.5
61.5	244	150	94	92.2
61.5	239	147	92	90.4

%E	M1	M2	DM	M*
61.5	234	144	90	88.6
61.5	231	142	89	87.3
61.5	226	139	87	85.5
61.5	221	136	85	83.7
61.5	218	134	84	82.4
61.5	213	131	82	80.6
61.5	208	128	80	78.8
61.5	205	126	79	77.4
61.5	200	123	77	75.6
61.5	195	120	75	73.8
61.5	192	118	74	72.5
61.5	187	115	72	70.7
61.5	182	112	70	68.9
61.5	179	110	69	67.6
61.5	174	107	67	65.8
61.5	169	104	65	64.0
61.5	161	99	62	60.9
61.5	156	96	60	59.1
61.5	148	91	57	56.0
61.5	143	88	55	54.2
61.5	135	83	52	51.0
61.5	130	80	50	49.2
61.5	122	75	47	46.1
61.5	117	72	45	44.3
61.5	109	67	42	41.2
61.5	104	64	40	39.4
61.5	96	59	37	36.3
61.5	91	56	35	34.5
61.5	78	48	30	29.5
61.5	65	40	25	24.6
61.5	52	32	20	19.7
61.5	39	24	15	14.8
61.5	26	16	10	9.8
61.5	13	8	5	4.9
61.4	500	307	193	188.5
61.4	498	306	192	188.0
61.4	497	305	192	187.2
61.4	495	304	191	186.7
61.4	492	302	190	185.4
61.4	490	301	189	184.9
61.4	489	300	189	184.0
61.4	487	299	188	183.6
61.4	485	298	187	183.1
61.4	484	297	187	182.3
61.4	482	296	186	181.8
61.4	479	294	185	180.5
61.4	477	293	184	180.0
61.4	474	291	183	178.7
61.4	472	290	182	178.2
61.4	471	289	182	177.3

%E	M1	M2	DM	M*
61.4	469	288	181	176.9
61.4	466	286	180	175.5
61.4	464	285	179	175.1
61.4	461	283	178	173.7
61.4	459	282	177	173.3
61.4	458	281	177	172.4
61.4	456	280	176	171.9
61.4	453	278	175	170.6
61.4	451	277	174	170.1
61.4	448	275	173	168.8
61.4	446	274	172	168.3
61.4	443	272	171	167.0
61.4	440	270	170	165.7
61.4	438	269	169	165.2
61.4	435	267	168	163.9
61.4	433	266	167	163.4
61.4	430	264	166	162.1
61.4	428	263	165	161.6
61.4	427	262	165	160.8
61.4	425	261	164	160.3
61.4	422	259	163	159.0
61.4	420	258	162	158.5
61.4	417	256	161	157.2
61.4	415	255	160	156.7
61.4	414	254	160	155.8
61.4	412	253	159	155.4
61.4	409	251	158	154.0
61.4	407	250	157	153.6
61.4	404	248	156	152.2
61.4	402	247	155	151.8
61.4	399	245	154	150.4
61.4	396	243	153	149.1
61.4	394	242	152	148.6
61.4	391	240	151	147.3
61.4	389	239	150	146.8
61.4	386	237	149	145.5
61.4	383	235	148	144.2
61.4	381	234	147	143.7
61.4	378	232	146	142.4
61.4	376	231	145	141.9
61.4	373	229	144	140.6
61.4	370	227	143	139.3
61.4	368	226	142	138.8
61.4	365	224	141	137.5
61.4	363	223	140	137.0
61.4	360	221	139	135.7
61.4	355	218	137	133.9
61.4	352	216	136	132.5
61.4	350	215	135	132.1
61.4	347	213	134	130.7

%E	M1	M2	DM	M*
61.4	345	212	133	130.3
61.4	342	210	132	128.9
61.4	339	208	131	127.6
61.4	337	207	130	127.1
61.4	334	205	129	125.8
61.4	332	204	128	125.3
61.4	329	202	127	124.0
61.4	326	200	126	122.7
61.4	324	199	125	122.2
61.4	321	197	124	120.9
61.4	319	196	123	120.4
61.4	316	194	122	119.1
61.4	311	191	120	117.3
61.4	308	189	119	116.0
61.4	306	188	118	115.5
61.4	303	186	117	114.2
61.4	298	183	115	112.4
61.4	295	181	114	111.1
61.4	293	180	113	110.6
61.4	290	178	112	109.3
61.4	285	175	110	107.5
61.4	280	172	108	105.7
61.4	277	170	107	104.3
61.4	272	167	105	102.5
61.4	267	164	103	100.7
61.4	264	162	102	99.4
61.4	259	159	100	97.6
61.4	254	156	98	95.8
61.4	251	154	97	94.5
61.4	249	153	96	94.0
61.4	246	151	95	92.7
61.4	241	148	93	90.9
61.4	236	145	91	89.1
61.4	233	143	90	87.8
61.4	228	140	88	86.0
61.4	223	137	86	84.2
61.4	220	135	85	82.8
61.4	215	132	83	81.0
61.4	210	129	81	79.2
61.4	207	127	80	77.9
61.4	202	124	78	76.1
61.4	197	121	76	74.3
61.4	189	116	73	71.2
61.4	184	113	71	69.4
61.4	176	108	68	66.3
61.4	171	105	66	64.5
61.4	166	102	64	62.7
61.4	163	100	63	61.3
61.4	158	97	61	59.6
61.4	153	94	59	57.8

%E	M1	M2	DM	M*
61.4	145	89	56	54.6
61.4	140	86	54	52.8
61.4	132	81	51	49.7
61.4	127	78	49	47.9
61.4	114	70	44	43.0
61.4	101	62	39	38.1
61.4	88	54	34	33.1
61.4	83	51	32	31.3
61.4	70	43	27	26.4
61.4	57	35	22	21.5
61.4	44	27	17	16.6
61.3	499	306	193	187.6
61.3	496	304	192	186.3
61.3	494	303	191	185.8
61.3	493	302	191	185.0
61.3	491	301	190	184.5
61.3	488	299	189	183.2
61.3	486	298	188	182.7
61.3	483	296	187	181.4
61.3	481	295	186	180.9
61.3	480	294	186	180.1
61.3	478	293	185	179.6
61.3	476	292	184	179.1
61.3	475	291	184	178.3
61.3	473	290	183	177.8
61.3	470	288	182	176.5
61.3	468	287	181	176.0
61.3	465	285	180	174.7
61.3	463	284	179	174.2
61.3	462	283	179	173.4
61.3	460	282	178	172.9
61.3	457	280	177	171.6
61.3	455	279	176	171.1
61.3	452	277	175	169.8
61.3	450	276	174	169.3
61.3	447	274	173	168.0
61.3	445	273	172	167.5
61.3	444	272	172	166.6
61.3	442	271	171	166.2
61.3	439	269	170	164.8
61.3	437	268	169	164.4
61.3	434	266	168	163.0
61.3	432	265	167	162.6
61.3	431	264	167	161.7
61.3	429	263	166	161.2
61.3	426	261	165	159.9
61.3	424	260	164	159.4
61.3	421	258	163	158.1
61.3	419	257	162	157.6
61.3	416	255	161	156.3

%E	M1	M2	DM	M*
61.3	413	253	160	155.0
61.3	411	252	159	154.5
61.3	408	250	158	153.2
61.3	406	249	157	152.7
61.3	403	247	156	151.4
61.3	401	246	155	150.9
61.3	400	245	155	150.1
61.3	398	244	154	149.6
61.3	395	242	153	148.3
61.3	393	241	152	147.8
61.3	390	239	151	146.5
61.3	388	238	150	146.0
61.3	385	236	149	144.7
61.3	382	234	148	143.3
61.3	380	233	147	142.9
61.3	377	231	146	141.5
61.3	375	230	145	141.1
61.3	372	228	144	139.7
61.3	367	225	142	137.9
61.3	364	223	141	136.6
61.3	362	222	140	136.1
61.3	359	220	139	134.8
61.3	357	219	138	134.3
61.3	354	217	137	133.0
61.3	351	215	136	131.7
61.3	349	214	135	131.2
61.3	346	212	134	129.9
61.3	344	211	133	129.4
61.3	341	209	132	128.1
61.3	336	206	130	126.3
61.3	333	204	129	125.0
61.3	331	203	128	124.5
61.3	328	201	127	123.2
61.3	323	198	125	121.4
61.3	320	196	124	120.0
61.3	318	195	123	119.6
61.3	315	193	122	118.3
61.3	313	192	121	117.8
61.3	310	190	120	116.5
61.3	305	187	118	114.7
61.3	302	185	117	113.3
61.3	300	184	116	112.9
61.3	297	182	115	111.5
61.3	292	179	113	109.7
61.3	287	176	111	107.9
61.3	284	174	110	106.6
61.3	282	173	109	106.1
61.3	279	171	108	104.8
61.3	274	168	106	103.0
61.3	271	166	105	101.7

%E	M1	M2	DM	M*
61.3	269	165	104	101.2
61.3	266	163	103	99.9
61.3	261	160	101	98.1
61.3	256	157	99	96.3
61.3	253	155	98	95.0
61.3	248	152	96	93.2
61.3	243	149	94	91.4
61.3	240	147	93	90.0
61.3	238	146	92	89.6
61.3	235	144	91	88.2
61.3	230	141	89	86.4
61.3	225	138	87	84.6
61.3	222	136	86	83.3
61.3	217	133	84	81.5
61.3	212	130	82	79.7
61.3	204	125	79	76.6
61.3	199	122	77	74.8
61.3	194	119	75	73.0
61.3	191	117	74	71.7
61.3	186	114	72	69.9
61.3	181	111	70	68.1
61.3	173	106	67	64.9
61.3	168	103	65	63.1
61.3	160	98	62	60.0
61.3	155	95	60	58.2
61.3	150	92	58	56.4
61.3	142	87	55	53.3
61.3	137	84	53	51.5
61.3	124	76	48	46.6
61.3	119	73	46	44.8
61.3	111	68	43	41.7
61.3	106	65	41	39.9
61.3	93	57	36	34.9
61.3	80	49	31	30.0
61.3	75	46	29	28.2
61.3	62	38	24	23.3
61.3	31	19	12	11.6
61.2	500	306	194	187.3
61.2	498	305	193	186.8
61.2	497	304	193	185.9
61.2	495	303	192	185.5
61.2	492	301	191	184.1
61.2	490	300	190	183.7
61.2	487	298	189	182.3
61.2	485	297	188	181.9
61.2	484	296	188	181.0
61.2	482	295	187	180.5
61.2	479	293	186	179.2
61.2	477	292	185	178.8
61.2	474	290	184	177.4

%E	M1	M2	DM	M*
61.2	472	289	183	177.0
61.2	469	287	182	175.6
61.2	467	286	181	175.2
61.2	466	285	181	174.3
61.2	464	284	180	173.8
61.2	461	282	179	172.5
61.2	459	281	178	172.0
61.2	456	279	177	170.7
61.2	454	278	176	170.2
61.2	451	276	175	168.9
61.2	449	275	174	168.4
61.2	448	274	174	167.6
61.2	446	273	173	167.1
61.2	443	271	172	165.8
61.2	441	270	171	165.3
61.2	438	268	170	164.0
61.2	436	267	169	163.5
61.2	433	265	168	162.2
61.2	430	263	167	160.9
61.2	428	262	166	160.4
61.2	425	260	165	159.1
61.2	423	259	164	158.6
61.2	420	257	163	157.3
61.2	418	256	162	156.8
61.2	417	255	162	155.9
61.2	415	254	161	155.5
61.2	412	252	160	154.1
61.2	410	251	159	153.7
61.2	407	249	158	152.3
61.2	405	248	157	151.9
61.2	402	246	156	150.5
61.2	399	244	155	149.2
61.2	397	243	154	148.7
61.2	394	241	153	147.4
61.2	392	240	152	146.9
61.2	389	238	151	145.6
61.2	387	237	150	145.1
61.2	384	235	149	143.8
61.2	381	233	148	142.5
61.2	379	232	147	142.0
61.2	376	230	146	140.7
61.2	374	229	145	140.2
61.2	371	227	144	138.9
61.2	369	226	143	138.4
61.2	366	224	142	137.1
61.2	363	222	141	135.8
61.2	361	221	140	135.3
61.2	358	219	139	134.0
61.2	356	218	138	133.5
61.2	353	216	137	132.2

%E	M1	M2	DM	M*
61.2	348	213	135	130.4
61.2	345	211	134	129.0
61.2	343	210	133	128.6
61.2	340	208	132	127.2
61.2	338	207	131	126.8
61.2	335	205	130	125.4
61.2	330	202	128	123.6
61.2	327	200	127	122.3
61.2	325	199	126	121.8
61.2	322	197	125	120.5
61.2	317	194	123	118.7
61.2	312	191	121	116.9
61.2	309	189	120	115.6
61.2	307	188	119	115.1
61.2	304	186	118	113.8
61.2	299	183	116	112.0
61.2	294	180	114	110.2
61.2	291	178	113	108.9
61.2	289	177	112	108.4
61.2	286	175	111	107.1
61.2	281	172	109	105.3
61.2	278	170	108	104.0
61.2	276	169	107	103.5
61.2	273	167	106	102.2
61.2	268	164	104	100.4
61.2	263	161	102	98.6
61.2	260	159	101	97.2
61.2	258	158	100	96.8
61.2	255	156	99	95.4
61.2	250	153	97	93.6
61.2	245	150	95	91.8
61.2	242	148	94	90.5
61.2	237	145	92	88.7
61.2	232	142	90	86.9
61.2	227	139	88	85.1
61.2	224	137	87	83.8
61.2	219	134	85	82.0
61.2	214	131	83	80.2
61.2	209	128	81	78.4
61.2	206	126	80	77.1
61.2	201	123	78	75.3
61.2	196	120	76	73.5
61.2	188	115	73	70.3
61.2	183	112	71	68.5
61.2	178	109	69	66.7
61.2	170	104	66	63.6
61.2	165	101	64	61.8
61.2	152	93	59	56.9
61.2	147	90	57	55.1
61.2	139	85	54	52.0

%E	M1	M2	DM	M*
61.2	134	82	52	50.2
61.2	129	79	50	48.4
61.2	121	74	47	45.3
61.2	116	71	45	43.5
61.2	103	63	40	38.5
61.2	98	60	38	36.7
61.2	85	52	33	31.8
61.2	67	41	26	25.1
61.2	49	30	19	18.4
61.1	499	305	194	186.4
61.1	496	303	193	185.1
61.1	494	302	192	184.6
61.1	493	301	192	183.8
61.1	491	300	191	183.3
61.1	489	299	190	182.8
61.1	488	298	190	182.0
61.1	486	297	189	181.5
61.1	483	295	188	180.2
61.1	481	294	187	179.7
61.1	478	292	186	178.4
61.1	476	291	185	177.9
61.1	475	290	185	177.1
61.1	473	289	184	176.6
61.1	471	288	183	176.1
61.1	470	287	183	175.3
61.1	468	286	182	174.8
61.1	465	284	181	173.5
61.1	463	283	180	173.0
61.1	460	281	179	171.7
61.1	458	280	178	171.2
61.1	457	279	178	170.3
61.1	455	278	177	169.9
61.1	453	277	176	169.4
61.1	452	276	176	168.5
61.1	450	275	175	168.1
61.1	447	273	174	166.7
61.1	445	272	173	166.3
61.1	442	270	172	164.9
61.1	440	269	171	164.5
61.1	437	267	170	163.1
61.1	435	266	169	162.7
61.1	434	265	169	161.8
61.1	432	264	168	161.3
61.1	429	262	167	160.0
61.1	427	261	166	159.5
61.1	424	259	165	158.2
61.1	422	258	164	157.7
61.1	419	256	163	156.4
61.1	416	254	162	155.1
61.1	414	253	161	154.6

%E	M1	M2	DM	M*
61.1	411	251	160	153.3
61.1	409	250	159	152.8
61.1	406	248	158	151.5
61.1	404	247	157	151.0
61.1	401	245	156	149.7
61.1	398	243	155	148.4
61.1	396	242	154	147.9
61.1	393	240	153	146.6
61.1	391	239	152	146.1
61.1	388	237	151	144.8
61.1	386	236	150	144.3
61.1	383	234	149	143.0
61.1	380	232	148	141.6
61.1	378	231	147	141.2
61.1	375	229	146	139.8
61.1	373	228	145	139.4
61.1	370	226	144	138.0
61.1	368	225	143	137.6
61.1	365	223	142	136.2
61.1	362	221	141	134.9
61.1	360	220	140	134.4
61.1	357	218	139	133.1
61.1	355	217	138	132.6
61.1	352	215	137	131.3
61.1	350	214	136	130.8
61.1	347	212	135	129.5
61.1	342	209	133	127.7
61.1	339	207	132	126.4
61.1	337	206	131	125.9
61.1	334	204	130	124.6
61.1	332	203	129	124.1
61.1	329	201	128	122.8
61.1	324	198	126	121.0
61.1	321	196	125	119.7
61.1	319	195	124	119.2
61.1	316	193	123	117.9
61.1	314	192	122	117.4
61.1	311	190	121	116.1
61.1	306	187	119	114.3
61.1	303	185	118	113.0
61.1	301	184	117	112.5
61.1	298	182	116	111.2
61.1	296	181	115	110.7
61.1	293	179	114	109.4
61.1	288	176	112	107.6
61.1	285	174	111	106.2
61.1	283	173	110	105.8
61.1	280	171	109	104.4
61.1	275	168	107	102.6
61.1	270	165	105	100.8

%E	M1	M2	DM	M*
61.1	265	162	103	99.0
61.1	262	160	102	97.7
61.1	257	157	100	95.9
61.1	252	154	98	94.1
61.1	247	151	96	92.3
61.1	244	149	95	91.0
61.1	239	146	93	89.2
61.1	234	143	91	87.4
61.1	229	140	89	85.6
61.1	226	138	88	84.3
61.1	221	135	86	82.5
61.1	216	132	84	80.7
61.1	211	129	82	78.9
61.1	208	127	81	77.5
61.1	203	124	79	75.7
61.1	198	121	77	73.9
61.1	193	118	75	72.1
61.1	190	116	74	70.8
61.1	185	113	72	69.0
61.1	180	110	70	67.2
61.1	175	107	68	65.4
61.1	167	102	65	62.3
61.1	162	99	63	60.5
61.1	157	96	61	58.7
61.1	149	91	58	55.6
61.1	144	88	56	53.8
61.1	131	80	51	48.9
61.1	126	77	49	47.1
61.1	113	69	44	42.1
61.1	108	66	42	40.3
61.1	95	58	37	35.4
61.1	90	55	35	33.6
61.1	72	44	28	26.9
61.1	54	33	21	20.2
61.1	36	22	14	13.4
61.1	18	11	7	6.7
61.0	500	305	195	186.0
61.0	498	304	194	185.6
61.0	497	303	194	184.7
61.0	495	302	193	184.3
61.0	492	300	192	182.9
61.0	490	299	191	182.5
61.0	487	297	190	181.1
61.0	485	296	189	180.7
61.0	484	295	189	179.8
61.0	482	294	188	179.3
61.0	480	293	187	178.9
61.0	479	292	187	178.0
61.0	477	291	186	177.5
61.0	474	289	185	176.2

%E	M1	M2	DM	M*
61.0	472	288	184	175.7
61.0	469	286	183	174.4
61.0	467	285	182	173.9
61.0	464	283	181	172.6
61.0	462	282	180	172.1
61.0	461	281	180	171.3
61.0	459	280	179	170.8
61.0	456	278	178	169.5
61.0	454	277	177	169.0
61.0	451	275	176	167.7
61.0	449	274	175	167.2
61.0	446	272	174	165.9
61.0	444	271	173	165.4
61.0	441	269	172	164.1
61.0	439	268	171	163.6
61.0	438	267	171	162.8
61.0	436	266	170	162.3
61.0	433	264	169	161.0
61.0	431	263	168	160.5
61.0	428	261	167	159.2
61.0	426	260	166	158.7
61.0	423	258	165	157.4
61.0	421	257	164	156.9
61.0	420	256	164	156.0
61.0	418	255	163	155.6
61.0	415	253	162	154.2
61.0	413	252	161	153.8
61.0	410	250	160	152.4
61.0	408	249	159	152.0
61.0	405	247	158	150.6
61.0	403	246	157	150.2
61.0	400	244	156	148.8
61.0	397	242	155	147.5
61.0	395	241	154	147.0
61.0	392	239	153	145.7
61.0	390	238	152	145.2
61.0	387	236	151	143.9
61.0	385	235	150	143.4
61.0	382	233	149	142.1
61.0	379	231	148	140.8
61.0	377	230	147	140.3
61.0	374	228	146	139.0
61.0	372	227	145	138.5
61.0	369	225	144	137.2
61.0	367	224	143	136.7
61.0	364	222	142	135.4
61.0	359	219	140	133.6
61.0	356	217	139	132.3
61.0	354	216	138	131.8
61.0	351	214	137	130.5

%E	M1	M2	DM	M*
61.0	349	213	136	130.0
61.0	346	211	135	128.7
61.0	344	210	134	128.2
61.0	341	208	133	126.9
61.0	336	205	131	125.1
61.0	333	203	130	123.8
61.0	331	202	129	123.3
61.0	328	200	128	122.0
61.0	326	199	127	121.5
61.0	323	197	126	120.2
61.0	318	194	124	118.4
61.0	315	192	123	117.0
61.0	313	191	122	116.6
61.0	310	189	121	115.2
61.0	308	188	120	114.8
61.0	305	186	119	113.4
61.0	300	183	117	111.6
61.0	295	180	115	109.8
61.0	292	178	114	108.5
61.0	290	177	113	108.0
61.0	287	175	112	106.7
61.0	282	172	110	104.9
61.0	277	169	108	103.1
61.0	272	166	106	101.3
61.0	269	164	105	100.0
61.0	267	163	104	99.5
61.0	264	161	103	98.2
61.0	259	158	101	96.4
61.0	254	155	99	94.6
61.0	251	153	98	93.3
61.0	249	152	97	92.8
61.0	246	150	96	91.5
61.0	241	147	94	89.7
61.0	236	144	92	87.9
61.0	231	141	90	86.1
61.0	228	139	89	84.7
61.0	223	136	87	82.9
61.0	218	133	85	81.1
61.0	213	130	83	79.3
61.0	210	128	82	78.0
61.0	205	125	80	76.2
61.0	200	122	78	74.4
61.0	195	119	76	72.6
61.0	187	114	73	69.5
61.0	182	111	71	67.7
61.0	177	108	69	65.9
61.0	172	105	67	64.1
61.0	164	100	64	61.0
61.0	159	97	62	59.2
61.0	154	94	60	57.4

%E	M1	M2	DM	M*
61.0	146	89	57	54.3
61.0	141	86	55	52.5
61.0	136	83	53	50.7
61.0	123	75	48	45.7
61.0	118	72	46	43.9
61.0	105	64	41	39.0
61.0	100	61	39	37.2
61.0	82	50	32	30.5
61.0	77	47	30	28.7
61.0	59	36	23	22.0
61.0	41	25	16	15.2
60.9	499	304	195	185.2
60.9	496	302	194	183.9
60.9	494	301	193	183.4
60.9	493	300	193	182.6
60.9	491	299	192	182.1
60.9	489	298	191	181.6
60.9	488	297	191	180.8
60.9	486	296	190	180.3
60.9	483	294	189	179.0
60.9	481	293	188	178.5
60.9	478	291	187	177.2
60.9	476	290	186	176.7
60.9	473	288	185	175.4
60.9	471	287	184	174.9
60.9	470	286	184	174.0
60.9	468	285	183	173.6
60.9	466	284	182	173.1
60.9	465	283	182	172.2
60.9	463	282	181	171.8
60.9	460	280	180	170.4
60.9	458	279	179	170.0
60.9	455	277	178	168.6
60.9	453	276	177	168.2
60.9	450	274	176	166.8
60.9	448	273	175	166.4
60.9	447	272	175	165.5
60.9	445	271	174	165.0
60.9	443	270	173	164.6
60.9	442	269	173	163.7
60.9	440	268	172	163.2
60.9	437	266	171	161.9
60.9	435	265	170	161.4
60.9	432	263	169	160.1
60.9	430	262	168	159.6
60.9	427	260	167	158.3
60.9	425	259	166	157.8
60.9	422	257	165	156.5
60.9	419	255	164	155.2
60.9	417	254	163	154.7

%E	M1	M2	DM	M*
60.9	414	252	162	153.4
60.9	412	251	161	152.9
60.9	409	249	160	151.6
60.9	407	248	159	151.1
60.9	404	246	158	149.8
60.9	402	245	157	149.3
60.9	399	243	156	148.0
60.9	396	241	155	146.7
60.9	394	240	154	146.2
60.9	391	238	153	144.9
60.9	389	237	152	144.4
60.9	386	235	151	143.1
60.9	384	234	150	142.6
60.9	381	232	149	141.3
60.9	376	229	147	139.5
60.9	373	227	146	138.1
60.9	371	226	145	137.7
60.9	368	224	144	136.3
60.9	366	223	143	135.9
60.9	363	221	142	134.5
60.9	361	220	141	134.1
60.9	358	218	140	132.7
60.9	353	215	138	130.9
60.9	350	213	137	129.6
60.9	348	212	136	129.1
60.9	345	210	135	127.8
60.9	343	209	134	127.3
60.9	340	207	133	126.0
60.9	338	206	132	125.6
60.9	335	204	131	124.2
60.9	330	201	129	122.4
60.9	327	199	128	121.1
60.9	325	198	127	120.6
60.9	322	196	126	119.3
60.9	320	195	125	118.8
60.9	317	193	124	117.5
60.9	312	190	122	115.7
60.9	307	187	120	113.9
60.9	304	185	119	112.6
60.9	302	184	118	112.1
60.9	299	182	117	110.8
60.9	297	181	116	110.3
60.9	294	179	115	109.0
60.9	289	176	113	107.2
60.9	284	173	111	105.4
60.9	281	171	110	104.1
60.9	279	170	109	103.6
60.9	276	168	108	102.3
60.9	274	167	107	101.8
60.9	271	165	106	100.5

%E	M1	M2	DM	M*
60.9	266	162	104	98.7
60.9	261	159	102	96.9
60.9	258	157	101	95.5
60.9	256	156	100	95.1
60.9	253	154	99	93.7
60.9	248	151	97	91.9
60.9	243	148	95	90.1
60.9	238	145	93	88.3
60.9	235	143	92	87.0
60.9	233	142	91	86.5
60.9	230	140	90	85.2
60.9	225	137	88	83.4
60.9	220	134	86	81.6
60.9	215	131	84	79.8
60.9	207	126	81	76.7
60.9	202	123	79	74.9
60.9	197	120	77	73.1
60.9	192	117	75	71.3
60.9	184	112	72	68.2
60.9	179	109	70	66.4
60.9	174	106	68	64.6
60.9	169	103	66	62.8
60.9	161	98	63	59.7
60.9	156	95	61	57.9
60.9	151	92	59	56.1
60.9	138	84	54	51.1
60.9	133	81	52	49.3
60.9	128	78	50	47.5
60.9	115	70	45	42.6
60.9	110	67	43	40.8
60.9	92	56	36	34.1
60.9	87	53	34	32.3
60.9	69	42	27	25.6
60.9	64	39	25	23.8
60.9	46	28	18	17.0
60.9	23	14	9	8.5
60.8	500	304	196	184.8
60.8	498	303	195	184.4
60.8	497	302	195	183.5
60.8	495	301	194	183.0
60.8	492	299	193	181.7
60.8	490	298	192	181.2
60.8	487	296	191	179.9
60.8	485	295	190	179.4
60.8	482	293	189	178.1
60.8	480	292	188	177.6
60.8	479	291	188	176.8
60.8	477	290	187	176.3
60.8	475	289	186	175.8
60.8	474	288	186	175.0

%E	M1	M2	DM	M*
60.8	472	287	185	174.5
60.8	469	285	184	173.2
60.8	467	284	183	172.7
60.8	464	282	182	171.4
60.8	462	281	181	170.9
60.8	459	279	180	169.6
60.8	457	278	179	169.1
60.8	454	276	178	167.8
60.8	452	275	177	167.3
60.8	451	274	177	166.5
60.8	449	273	176	166.0
60.8	446	271	175	164.7
60.8	444	270	174	164.2
60.8	441	268	173	162.9
60.8	439	267	172	162.4
60.8	436	265	171	161.1
60.8	434	264	170	160.6
60.8	431	262	169	159.3
60.8	429	261	168	158.8
60.8	426	259	167	157.5
60.8	424	258	166	157.0
60.8	423	257	166	156.1
60.8	421	256	165	155.7
60.8	418	254	164	154.3
60.8	416	253	163	153.9
60.8	413	251	162	152.5
60.8	411	250	161	152.1
60.8	408	248	160	150.7
60.8	406	247	159	150.3
60.8	403	245	158	148.9
60.8	401	244	157	148.5
60.8	400	243	157	147.6
60.8	398	242	156	147.1
60.8	395	240	155	145.8
60.8	393	239	154	145.3
60.8	390	237	153	144.0
60.8	388	236	152	143.5
60.8	385	234	151	142.2
60.8	383	233	150	141.7
60.8	380	231	149	140.4
60.8	378	230	148	139.9
60.8	375	228	147	138.6
60.8	372	226	146	137.3
60.8	370	225	145	136.8
60.8	367	223	144	135.5
60.8	365	222	143	135.0
60.8	362	220	142	133.7
60.8	360	219	141	133.2
60.8	357	217	140	131.9
60.8	355	216	139	131.4

%E	M1	M2	DM	M*
60.8	352	214	138	130.1
60.8	347	211	136	128.3
60.8	344	209	135	127.0
60.8	342	208	134	126.5
60.8	339	206	133	125.2
60.8	337	205	132	124.7
60.8	334	203	131	123.4
60.8	332	202	130	122.9
60.8	329	200	129	121.6
60.8	324	197	127	119.8
60.8	319	194	125	118.0
60.8	316	192	124	116.7
60.8	314	191	123	116.2
60.8	311	189	122	114.9
60.8	309	188	121	114.4
60.8	306	186	120	113.1
60.8	301	183	118	111.3
60.8	296	180	116	109.5
60.8	293	178	115	108.1
60.8	291	177	114	107.7
60.8	288	175	113	106.3
60.8	286	174	112	105.9
60.8	283	172	111	104.5
60.8	278	169	109	102.7
60.8	273	166	107	100.9
60.8	268	163	105	99.1
60.8	265	161	104	97.8
60.8	263	160	103	97.3
60.8	260	158	102	96.0
60.8	255	155	100	94.2
60.8	250	152	98	92.4
60.8	245	149	96	90.6
60.8	240	146	94	88.8
60.8	237	144	93	87.5
60.8	232	141	91	85.7
60.8	227	138	89	83.9
60.8	222	135	87	82.1
60.8	217	132	85	80.3
60.8	212	129	83	78.5
60.8	209	127	82	77.2
60.8	204	124	80	75.4
60.8	199	121	78	73.6
60.8	194	118	76	71.8
60.8	189	115	74	70.0
60.8	186	113	73	68.7
60.8	181	110	71	66.9
60.8	176	107	69	65.1
60.8	171	104	67	63.3
60.8	166	101	65	61.5
60.8	158	96	62	58.3

%E	M1	M2	DM	M*
60.8	153	93	60	56.5
60.8	148	90	58	54.7
60.8	143	87	56	52.9
60.8	130	79	51	48.0
60.8	125	76	49	46.2
60.8	120	73	47	44.4
60.8	102	62	40	37.7
60.8	97	59	38	35.9
60.8	79	48	31	29.2
60.8	74	45	29	27.4
60.8	51	31	20	18.8
60.7	499	303	196	184.0
60.7	496	301	195	182.7
60.7	494	300	194	182.2
60.7	491	298	193	180.9
60.7	489	297	192	180.4
60.7	488	296	192	179.5
60.7	486	295	191	179.1
60.7	484	294	190	178.6
60.7	483	293	190	177.7
60.7	481	292	189	177.3
60.7	478	290	188	175.9
60.7	476	289	187	175.5
60.7	473	287	186	174.1
60.7	471	286	185	173.7
60.7	468	284	184	172.3
60.7	466	283	183	171.9
60.7	463	281	182	170.5
60.7	461	280	181	170.1
60.7	460	279	181	169.2
60.7	458	278	180	168.7
60.7	456	277	179	168.3
60.7	455	276	179	167.4
60.7	453	275	178	166.9
60.7	450	273	177	165.6
60.7	448	272	176	165.1
60.7	445	270	175	163.8
60.7	443	269	174	163.3
60.7	440	267	173	162.0
60.7	438	266	172	161.5
60.7	435	264	171	160.2
60.7	433	263	170	159.7
60.7	430	261	169	158.4
60.7	428	260	168	157.9
60.7	427	259	168	157.1
60.7	425	258	167	156.6
60.7	422	256	166	155.3
60.7	420	255	165	154.8
60.7	417	253	164	153.5
60.7	415	252	163	153.0

%E	M1	M2	DM	M*
60.7	412	250	162	151.7
60.7	410	249	161	151.2
60.7	407	247	160	149.9
60.7	405	246	159	149.4
60.7	402	244	158	148.1
60.7	399	242	157	146.8
60.7	397	241	156	146.3
60.7	394	239	155	145.0
60.7	392	238	154	144.5
60.7	389	236	153	143.2
60.7	387	235	152	142.7
60.7	384	233	151	141.4
60.7	382	232	150	140.9
60.7	379	230	149	139.6
60.7	377	229	148	139.1
60.7	374	227	147	137.8
60.7	369	224	145	136.0
60.7	366	222	144	134.7
60.7	364	221	143	134.2
60.7	361	219	142	132.9
60.7	359	218	141	132.4
60.7	356	216	140	131.1
60.7	354	215	139	130.6
60.7	351	213	138	129.3
60.7	349	212	137	128.8
60.7	346	210	136	127.5
60.7	341	207	134	125.7
60.7	338	205	133	124.3
60.7	336	204	132	123.9
60.7	333	202	131	122.5
60.7	331	201	130	122.1
60.7	328	199	129	120.7
60.7	326	198	128	120.3
60.7	323	196	127	118.9
60.7	321	195	126	118.5
60.7	318	193	125	117.1
60.7	313	190	123	115.3
60.7	308	187	121	113.5
60.7	305	185	120	112.2
60.7	303	184	119	111.7
60.7	300	182	118	110.4
60.7	298	181	117	109.9
60.7	295	179	116	108.6
60.7	290	176	114	106.8
60.7	285	173	112	105.0
60.7	280	170	110	103.2
60.7	277	168	109	101.9
60.7	275	167	108	101.4
60.7	272	165	107	100.1
60.7	270	164	106	99.6

%E	M1	M2	DM	M*
60.7	267	162	105	98.3
60.7	262	159	103	96.5
60.7	257	156	101	94.7
60.7	252	153	99	92.9
60.7	247	150	97	91.1
60.7	244	148	96	89.8
60.7	242	147	95	89.3
60.7	239	145	94	88.0
60.7	234	142	92	86.2
60.7	229	139	90	84.4
60.7	224	136	88	82.6
60.7	219	133	86	80.8
60.7	214	130	84	79.0
60.7	211	128	83	77.6
60.7	206	125	81	75.8
60.7	201	122	79	74.0
60.7	196	119	77	72.3
60.7	191	116	75	70.5
60.7	183	111	72	67.3
60.7	178	108	70	65.5
60.7	173	105	68	63.7
60.7	168	102	66	61.9
60.7	163	99	64	60.1
60.7	150	91	59	55.2
60.7	145	88	57	53.4
60.7	140	85	55	51.6
60.7	135	82	53	49.8
60.7	122	74	48	44.9
60.7	117	71	46	43.1
60.7	112	68	44	41.3
60.7	107	65	42	39.5
60.7	89	54	35	32.8
60.7	84	51	33	31.0
60.7	61	37	24	22.4
60.7	56	34	22	20.6
60.7	28	17	11	10.3
60.6	500	303	197	183.6
60.6	498	302	196	183.1
60.6	497	301	196	182.3
60.6	495	300	195	181.8
60.6	493	299	194	181.3
60.6	492	298	194	180.5
60.6	490	297	193	180.0
60.6	487	295	192	178.7
60.6	485	294	191	178.2
60.6	482	292	190	176.9
60.6	480	291	189	176.4
60.6	477	289	188	175.1
60.6	475	288	187	174.6
60.6	472	286	186	173.3

%E	M1	M2	DM	M*
60.6	470	285	185	172.8
60.6	469	284	185	172.0
60.6	467	283	184	171.5
60.6	465	282	183	171.0
60.6	464	281	183	170.2
60.6	462	280	182	169.7
60.6	459	278	181	168.4
60.6	457	277	180	167.9
60.6	454	275	179	166.6
60.6	452	274	178	166.1
60.6	449	272	177	164.8
60.6	447	271	176	164.3
60.6	444	269	175	163.0
60.6	442	268	174	162.5
60.6	439	266	173	161.2
60.6	437	265	172	160.7
60.6	436	264	172	159.9
60.6	434	263	171	159.4
60.6	432	262	170	158.9
60.6	431	261	170	158.1
60.6	429	260	169	157.6
60.6	426	258	168	156.3
60.6	424	257	167	155.8
60.6	421	255	166	154.5
60.6	419	254	165	154.0
60.6	416	252	164	152.7
60.6	414	251	163	152.2
60.6	411	249	162	150.9
60.6	409	248	161	150.4
60.6	406	246	160	149.1
60.6	404	245	159	148.6
60.6	401	243	158	147.3
60.6	398	241	157	145.9
60.6	396	240	156	145.5
60.6	393	238	155	144.1
60.6	391	237	154	143.7
60.6	388	235	153	142.3
60.6	386	234	152	141.9
60.6	383	232	151	140.5
60.6	381	231	150	140.1
60.6	378	229	149	138.7
60.6	376	228	148	138.3
60.6	373	226	147	136.9
60.6	371	225	146	136.5
60.6	368	223	145	135.1
60.6	363	220	143	133.3
60.6	360	218	142	132.0
60.6	358	217	141	131.5
60.6	355	215	140	130.2
60.6	353	214	139	129.7

%E	M1	M2	DM	M*
60.6	350	212	138	128.4
60.6	348	211	137	127.9
60.6	345	209	136	126.6
60.6	343	208	135	126.1
60.6	340	206	134	124.8
60.6	335	203	132	123.0
60.6	330	200	130	121.2
60.6	327	198	129	119.9
60.6	325	197	128	119.4
60.6	322	195	127	118.1
60.6	320	194	126	117.6
60.6	317	192	125	116.3
60.6	315	191	124	115.8
60.6	312	189	123	114.5
60.6	310	188	122	114.0
60.6	307	186	121	112.7
60.6	302	183	119	110.9
60.6	297	180	117	109.1
60.6	292	177	115	107.3
60.6	289	175	114	106.0
60.6	287	174	113	105.5
60.6	284	172	112	104.2
60.6	282	171	111	103.7
60.6	279	169	110	102.4
60.6	274	166	108	100.6
60.6	269	163	106	98.8
60.6	264	160	104	97.0
60.6	259	157	102	95.2
60.6	254	154	100	93.4
60.6	251	152	99	92.0
60.6	249	151	98	91.6
60.6	246	149	97	90.2
60.6	241	146	95	88.4
60.6	236	143	93	86.6
60.6	231	140	91	84.8
60.6	226	137	89	83.0
60.6	221	134	87	81.2
60.6	218	132	86	79.9
60.6	216	131	85	79.4
60.6	213	129	84	78.1
60.6	208	126	82	76.3
60.6	203	123	80	74.5
60.6	198	120	78	72.7
60.6	193	117	76	70.9
60.6	188	114	74	69.1
60.6	180	109	71	66.0
60.6	175	106	69	64.2
60.6	170	103	67	62.4
60.6	165	100	65	60.6
60.6	160	97	63	58.8

%E	M1	M2	DM	M*
60.6	155	94	61	57.0
60.6	142	86	56	52.1
60.6	137	83	54	50.3
60.6	132	80	52	48.5
60.6	127	77	50	46.7
60.6	109	66	43	40.0
60.6	104	63	41	38.2
60.6	99	60	39	36.4
60.6	94	57	37	34.6
60.6	71	43	28	26.0
60.6	66	40	26	24.2
60.6	33	20	13	12.1
60.5	499	302	197	182.8
60.5	496	300	196	181.5
60.5	494	299	195	181.0
60.5	491	297	194	179.7
60.5	489	296	193	179.2
60.5	488	295	193	178.3
60.5	486	294	192	177.9
60.5	484	293	191	177.4
60.5	483	292	191	176.5
60.5	481	291	190	176.1
60.5	479	290	189	175.6
60.5	478	289	189	174.7
60.5	476	288	188	174.3
60.5	474	287	187	173.8
60.5	473	286	187	172.9
60.5	471	285	186	172.5
60.5	468	283	185	171.1
60.5	466	282	184	170.7
60.5	463	280	183	169.3
60.5	461	279	182	168.9
60.5	458	277	181	167.5
60.5	456	276	180	167.1
60.5	453	274	179	165.7
60.5	451	273	178	165.3
60.5	448	271	177	163.9
60.5	446	270	176	163.5
60.5	443	268	175	162.1
60.5	441	267	174	161.7
60.5	440	266	174	160.8
60.5	438	265	173	160.3
60.5	435	263	172	159.0
60.5	433	262	171	158.5
60.5	430	260	170	157.2
60.5	428	259	169	156.7
60.5	425	257	168	155.4
60.5	423	256	167	154.9
60.5	420	254	166	153.6
60.5	418	253	165	153.1

%E	M1	M2	DM	M*
60.5	415	251	164	151.8
60.5	413	250	163	151.3
60.5	410	248	162	150.0
60.5	408	247	161	149.5
60.5	405	245	160	148.2
60.5	403	244	159	147.7
60.5	400	242	158	146.4
60.5	397	240	157	145.1
60.5	395	239	156	144.6
60.5	392	237	155	143.3
60.5	390	236	154	142.8
60.5	387	234	153	141.5
60.5	385	233	152	141.0
60.5	382	231	151	139.7
60.5	380	230	150	139.2
60.5	377	228	149	137.9
60.5	375	227	148	137.4
60.5	372	225	147	136.1
60.5	370	224	146	135.6
60.5	367	222	145	134.3
60.5	365	221	144	133.8
60.5	362	219	143	132.5
60.5	357	216	141	130.7
60.5	354	214	140	129.4
60.5	352	213	139	128.9
60.5	349	211	138	127.6
60.5	347	210	137	127.1
60.5	344	208	136	125.8
60.5	342	207	135	125.3
60.5	339	205	134	124.0
60.5	337	204	133	123.5
60.5	334	202	132	122.2
60.5	332	201	131	121.7
60.5	329	199	130	120.4
60.5	324	196	128	118.6
60.5	319	193	126	116.8
60.5	314	190	124	115.0
60.5	311	188	123	113.6
60.5	309	187	122	113.2
60.5	306	185	121	111.8
60.5	304	184	120	111.4
60.5	301	182	119	110.0
60.5	299	181	118	109.6
60.5	296	179	117	108.2
60.5	294	178	116	107.8
60.5	291	176	115	106.4
60.5	286	173	113	104.6
60.5	281	170	111	102.8
60.5	276	167	109	101.0
60.5	271	164	107	99.2

%E	M1	M2	DM	M*
60.5	266	161	105	97.4
60.5	263	159	104	96.1
60.5	261	158	103	95.6
60.5	258	156	102	94.3
60.5	256	155	101	93.8
60.5	253	153	100	92.5
60.5	248	150	98	90.7
60.5	243	147	96	88.9
60.5	238	144	94	87.1
60.5	233	141	92	85.3
60.5	228	138	90	83.5
60.5	223	135	88	81.7
60.5	220	133	87	80.4
60.5	215	130	85	78.6
60.5	210	127	83	76.8
60.5	205	124	81	75.0
60.5	200	121	79	73.2
60.5	195	118	77	71.4
60.5	190	115	75	69.6
60.5	185	112	73	67.8
60.5	177	107	70	64.7
60.5	172	104	68	62.9
60.5	167	101	66	61.1
60.5	162	98	64	59.3
60.5	157	95	62	57.5
60.5	152	92	60	55.7
60.5	147	89	58	53.9
60.5	129	78	51	47.2
60.5	124	75	49	45.4
60.5	119	72	47	43.6
60.5	114	69	45	41.8
60.5	86	52	34	31.4
60.5	81	49	32	29.6
60.5	76	46	30	27.8
60.5	43	26	17	15.7
60.5	38	23	15	13.9
60.4	500	302	198	182.4
60.4	498	301	197	181.9
60.4	497	300	197	181.1
60.4	495	299	196	180.6
60.4	493	298	195	180.1
60.4	492	297	195	179.3
60.4	490	296	194	178.8
60.4	487	294	193	177.5
60.4	485	293	192	177.0
60.4	482	291	191	175.7
60.4	480	290	190	175.2
60.4	477	288	189	173.9
60.4	475	287	188	173.4
60.4	472	285	187	172.1
60.4	470	284	186	171.6
60.4	467	282	185	170.3
60.4	465	281	184	169.8
60.4	462	279	183	168.5
60.4	460	278	182	168.0
60.4	457	276	181	166.7
60.4	455	275	180	166.2
60.4	454	274	180	165.4
60.4	452	273	179	164.9
60.4	450	272	178	164.4
60.4	449	271	178	163.6
60.4	447	270	177	163.1
60.4	445	269	176	162.6
60.4	444	268	176	161.8
60.4	442	267	175	161.3
60.4	439	265	174	160.0
60.4	437	264	173	159.5
60.4	434	262	172	158.2
60.4	432	261	171	157.7
60.4	429	259	170	156.4
60.4	427	258	169	155.9
60.4	424	256	168	154.6
60.4	422	255	167	154.1
60.4	419	253	166	152.8
60.4	417	252	165	152.3
60.4	414	250	164	151.0
60.4	412	249	163	150.5
60.4	409	247	162	149.2
60.4	407	246	161	148.7
60.4	404	244	160	147.4
60.4	402	243	159	146.9
60.4	399	241	158	145.6
60.4	396	239	157	144.2
60.4	394	238	156	143.8
60.4	391	236	155	142.4
60.4	389	235	154	142.0
60.4	386	233	153	140.6
60.4	384	232	152	140.2
60.4	381	230	151	138.8
60.4	379	229	150	138.4
60.4	376	227	149	137.0
60.4	374	226	148	136.6
60.4	371	224	147	135.2
60.4	369	223	146	134.8
60.4	366	221	145	133.4
60.4	364	220	144	133.0
60.4	361	218	143	131.6
60.4	359	217	142	131.2
60.4	356	215	141	129.8
60.4	351	212	139	128.0
60.4	346	209	137	126.2
60.4	343	207	136	124.9
60.4	341	206	135	124.4
60.4	338	204	134	123.1
60.4	336	203	133	122.6
60.4	333	201	132	121.3
60.4	331	200	131	120.8
60.4	328	198	130	119.5
60.4	326	197	129	119.0
60.4	323	195	128	117.7
60.4	321	194	127	117.2
60.4	318	192	126	115.9
60.4	316	191	125	115.4
60.4	313	189	124	114.1
60.4	308	186	122	112.3
60.4	303	183	120	110.5
60.4	298	180	118	108.7
60.4	293	177	116	106.9
60.4	288	174	114	105.1
60.4	285	172	113	103.8
60.4	283	171	112	103.3
60.4	280	169	111	102.0
60.4	278	168	110	101.5
60.4	275	166	109	100.2
60.4	273	165	108	99.7
60.4	270	163	107	98.4
60.4	268	162	106	97.9
60.4	265	160	105	96.6
60.4	260	157	103	94.8
60.4	255	154	101	93.0
60.4	250	151	99	91.2
60.4	245	148	97	89.4
60.4	240	145	95	87.6
60.4	235	142	93	85.8
60.4	230	139	91	84.0
60.4	227	137	90	82.7
60.4	225	136	89	82.2
60.4	222	134	88	80.9
60.4	217	131	86	79.1
60.4	212	128	84	77.3
60.4	207	125	82	75.5
60.4	202	122	80	73.7
60.4	197	119	78	71.9
60.4	192	116	76	70.1
60.4	187	113	74	68.3
60.4	182	110	72	66.5
60.4	169	102	67	61.6
60.4	164	99	65	59.8
60.4	159	96	63	58.0
60.4	154	93	61	56.2
60.4	149	90	59	54.4
60.4	144	87	57	52.6
60.4	139	84	55	50.8
60.4	134	81	53	49.0
60.4	111	67	44	40.4
60.4	106	64	42	38.6
60.4	101	61	40	36.8
60.4	96	58	38	35.0
60.4	91	55	36	33.2
60.4	53	32	21	19.3
60.4	48	29	19	17.5
60.3	499	301	198	181.6
60.3	496	299	197	180.2
60.3	494	298	196	179.8
60.3	491	296	195	178.4
60.3	489	295	194	178.0
60.3	486	293	193	176.6
60.3	484	292	192	176.2
60.3	481	290	191	174.8
60.3	479	289	190	174.4
60.3	478	288	190	173.5
60.3	476	287	189	173.0
60.3	474	286	188	172.6
60.3	473	285	188	171.7
60.3	471	284	187	171.2
60.3	469	283	186	170.8
60.3	468	282	186	169.9
60.3	466	281	185	169.4
60.3	464	280	184	169.0
60.3	463	279	184	168.1
60.3	461	278	183	167.6
60.3	459	277	182	167.2
60.3	458	276	182	166.3
60.3	456	275	181	165.8
60.3	453	273	180	164.5
60.3	451	272	179	164.0
60.3	448	270	178	162.7
60.3	446	269	177	162.2
60.3	443	267	176	160.9
60.3	441	266	175	160.4
60.3	438	264	174	159.1
60.3	436	263	173	158.6
60.3	433	261	172	157.3
60.3	431	260	171	156.8
60.3	428	258	170	155.5
60.3	426	257	169	155.0
60.3	423	255	168	153.7
60.3	421	254	167	153.2
60.3	418	252	166	151.9
60.3	416	251	165	151.4
60.3	413	249	164	150.1
60.3	411	248	163	149.6
60.3	408	246	162	148.3
60.3	406	245	161	147.8
60.3	403	243	160	146.5
60.3	401	242	159	146.0
60.3	400	241	159	145.2
60.3	398	240	158	144.7
60.3	395	238	157	143.4
60.3	393	237	156	142.9
60.3	390	235	155	141.6
60.3	388	234	154	141.1
60.3	385	232	153	139.8
60.3	383	231	152	139.3
60.3	380	229	151	138.0
60.3	378	228	150	137.5
60.3	375	226	149	136.2
60.3	373	225	148	135.7
60.3	370	223	147	134.4
60.3	368	222	146	133.9
60.3	365	220	145	132.6
60.3	363	219	144	132.1
60.3	360	217	143	130.8
60.3	358	216	142	130.3
60.3	355	214	141	129.0
60.3	353	213	140	128.5
60.3	350	211	139	127.2
60.3	348	210	138	126.7
60.3	345	208	137	125.4
60.3	340	205	135	123.6
60.3	335	202	133	121.8
60.3	330	199	131	120.0
60.3	325	196	129	118.2
60.3	320	193	127	116.4
60.3	317	191	126	115.1
60.3	315	190	125	114.6
60.3	312	188	124	113.3
60.3	310	187	123	112.8
60.3	307	185	122	111.5
60.3	305	184	121	111.0
60.3	302	182	120	109.7
60.3	300	181	119	109.2
60.3	297	179	118	107.9
60.3	295	178	117	107.4
60.3	292	176	116	106.1
60.3	290	175	115	105.6
60.3	287	173	114	104.3
60.3	282	170	112	102.5
60.3	277	167	110	100.7
60.3	272	164	108	98.9

%E	M1	M2	DM	M*
60.3	267	161	106	97.1
60.3	262	158	104	95.3
60.3	257	155	102	93.5
60.3	252	152	100	91.7
60.3	247	149	98	89.9
60.3	242	146	96	88.1
60.3	239	144	95	86.8
60.3	237	143	94	86.3
60.3	234	141	93	85.0
60.3	232	140	92	84.5
60.3	229	138	91	83.2
60.3	224	135	89	81.4
60.3	219	132	87	79.6
60.3	214	129	85	77.8
60.3	209	126	83	76.0
60.3	204	123	81	74.2
60.3	199	120	79	72.4
60.3	194	117	77	70.6
60.3	189	114	75	68.8
60.3	184	111	73	67.0
60.3	179	108	71	65.2
60.3	174	105	69	63.4
60.3	156	94	62	56.6
60.3	151	91	60	54.8
60.3	146	88	58	53.0
60.3	141	85	56	51.2
60.3	136	82	54	49.4
60.3	131	79	52	47.6
60.3	126	76	50	45.8
60.3	121	73	48	44.0
60.3	116	70	46	42.2
60.3	78	47	31	28.3
60.3	73	44	29	26.5
60.3	68	41	27	24.7
60.3	63	38	25	22.9
60.3	58	35	23	21.1
60.2	500	301	199	181.2
60.2	498	300	198	180.7
60.2	497	299	198	179.9
60.2	495	298	197	179.4
60.2	493	297	196	178.9
60.2	492	296	196	178.1
60.2	490	295	195	177.6
60.2	488	294	194	177.1
60.2	487	293	194	176.3
60.2	485	292	193	175.8
60.2	483	291	192	175.3
60.2	482	290	192	174.5
60.2	480	289	191	174.0
60.2	477	287	190	172.7

%E	M1	M2	DM	M*
60.2	475	286	189	172.2
60.2	472	284	188	170.9
60.2	470	283	187	170.4
60.2	467	281	186	169.1
60.2	465	280	185	168.6
60.2	462	278	184	167.3
60.2	460	277	183	166.8
60.2	457	275	182	165.5
60.2	455	274	181	165.0
60.2	452	272	180	163.7
60.2	450	271	179	163.2
60.2	447	269	178	161.9
60.2	445	268	177	161.4
60.2	442	266	176	160.1
60.2	440	265	175	159.6
60.2	437	263	174	158.3
60.2	435	262	173	157.8
60.2	432	260	172	156.5
60.2	430	259	171	156.0
60.2	427	257	170	154.7
60.2	425	256	169	154.2
60.2	422	254	168	152.9
60.2	420	253	167	152.4
60.2	417	251	166	151.1
60.2	415	250	165	150.6
60.2	412	248	164	149.3
60.2	410	247	163	148.8
60.2	407	245	162	147.5
60.2	405	244	161	147.0
60.2	402	242	160	145.7
60.2	399	240	159	144.4
60.2	397	239	158	143.9
60.2	394	237	157	142.6
60.2	392	236	156	142.1
60.2	389	234	155	140.8
60.2	387	233	154	140.3
60.2	384	231	153	139.0
60.2	382	230	152	138.5
60.2	379	228	151	137.2
60.2	377	227	150	136.7
60.2	374	225	149	135.4
60.2	372	224	148	134.9
60.2	369	222	147	133.6
60.2	367	221	146	133.1
60.2	364	219	145	131.8
60.2	362	218	144	131.3
60.2	359	216	143	130.0
60.2	357	215	142	129.5
60.2	354	213	141	128.2
60.2	352	212	140	127.7

%E	M1	M2	DM	M*
60.2	349	210	139	126.4
60.2	347	209	138	125.9
60.2	344	207	137	124.6
60.2	342	206	136	124.1
60.2	339	204	135	122.8
60.2	337	203	134	122.3
60.2	334	201	133	121.0
60.2	332	200	132	120.5
60.2	329	198	131	119.2
60.2	327	197	130	118.7
60.2	324	195	129	117.4
60.2	322	194	128	116.9
60.2	319	192	127	115.6
60.2	314	189	125	113.8
60.2	309	186	123	112.0
60.2	304	183	121	110.2
60.2	299	180	119	108.4
60.2	294	177	117	106.6
60.2	289	174	115	104.8
60.2	284	171	113	103.0
60.2	279	168	111	101.2
60.2	274	165	109	99.4
60.2	269	162	107	97.6
60.2	266	160	106	96.2
60.2	264	159	105	95.8
60.2	261	157	104	94.4
60.2	259	156	103	94.0
60.2	256	154	102	92.6
60.2	254	153	101	92.2
60.2	251	151	100	90.8
60.2	249	150	99	90.4
60.2	246	148	98	89.0
60.2	244	147	97	88.6
60.2	241	145	96	87.2
60.2	236	142	94	85.4
60.2	231	139	92	83.6
60.2	226	136	90	81.8
60.2	221	133	88	80.0
60.2	216	130	86	78.2
60.2	211	127	84	76.4
60.2	206	124	82	74.6
60.2	201	121	80	72.8
60.2	196	118	78	71.0
60.2	191	115	76	69.2
60.2	186	112	74	67.4
60.2	181	109	72	65.6
60.2	176	106	70	63.8
60.2	171	103	68	62.0
60.2	166	100	66	60.2
60.2	161	97	64	58.4

%E	M1	M2	DM	M*
60.2	133	80	53	48.1
60.2	128	77	51	46.3
60.2	123	74	49	44.5
60.2	118	71	47	42.7
60.2	113	68	45	40.9
60.2	108	65	43	39.1
60.2	103	62	41	37.3
60.2	98	59	39	35.5
60.2	93	56	37	33.7
60.2	88	53	35	31.9
60.2	83	50	33	30.1
60.1	499	300	199	180.4
60.1	496	298	198	179.0
60.1	494	297	197	178.6
60.1	491	295	196	177.2
60.1	489	294	195	176.8
60.1	486	292	194	175.4
60.1	484	291	193	175.0
60.1	481	289	192	173.6
60.1	479	288	191	173.2
60.1	476	286	190	171.8
60.1	474	285	189	171.4
60.1	471	283	188	170.0
60.1	469	282	187	169.6
60.1	466	280	186	168.2
60.1	464	279	185	167.8
60.1	461	277	184	166.4
60.1	459	276	183	166.0
60.1	456	274	182	164.6
60.1	454	273	181	164.2
60.1	451	271	180	162.8
60.1	449	270	179	162.4
60.1	446	268	178	161.0
60.1	444	267	177	160.6
60.1	441	265	176	159.2
60.1	439	264	175	158.8
60.1	436	262	174	157.4
60.1	434	261	173	157.0
60.1	431	259	172	155.6
60.1	429	258	171	155.2
60.1	426	256	170	153.8
60.1	424	255	169	153.4
60.1	421	253	168	152.0
60.1	419	252	167	151.6
60.1	416	250	166	150.2
60.1	414	249	165	149.8
60.1	411	247	164	148.4
60.1	409	246	163	148.0
60.1	406	244	162	146.6
60.1	404	243	161	146.2

%E	M1	M2	DM	M*
60.1	403	242	161	145.3
60.1	401	241	160	144.8
60.1	398	239	159	143.5
60.1	396	238	158	143.0
60.1	393	236	157	141.7
60.1	391	235	156	141.2
60.1	388	233	155	139.9
60.1	386	232	154	139.4
60.1	383	230	153	138.1
60.1	381	229	152	137.6
60.1	378	227	151	136.3
60.1	376	226	150	135.8
60.1	373	224	149	134.5
60.1	371	223	148	134.0
60.1	368	221	147	132.7
60.1	366	220	146	132.2
60.1	363	218	145	130.9
60.1	361	217	144	130.4
60.1	358	215	143	129.1
60.1	356	214	142	128.6
60.1	353	212	141	127.3
60.1	351	211	140	126.8
60.1	348	209	139	125.5
60.1	346	208	138	125.0
60.1	343	206	137	123.7
60.1	341	205	136	123.2
60.1	338	203	135	121.9
60.1	336	202	134	121.4
60.1	333	200	133	120.1
60.1	331	199	132	119.6
60.1	328	197	131	118.3
60.1	326	196	130	117.8
60.1	323	194	129	116.5
60.1	321	193	128	116.0
60.1	318	191	127	114.7
60.1	316	190	126	114.2
60.1	313	188	125	112.9
60.1	311	187	124	112.4
60.1	308	185	123	111.1
60.1	306	184	122	110.6
60.1	303	182	121	109.3
60.1	301	181	120	108.8
60.1	298	179	119	107.5
60.1	296	178	118	107.0
60.1	293	176	117	105.7
60.1	291	175	116	105.2
60.1	288	173	115	103.9
60.1	286	172	114	103.4
60.1	283	170	113	102.1
60.1	281	169	112	101.6

%E	M1	M2	DM	M*
60.1	278	167	111	100.3
60.1	276	166	110	99.8
60.1	273	164	109	98.5
60.1	271	163	108	98.0
60.1	268	161	107	96.7
60.1	263	158	105	94.9
60.1	258	155	103	93.1
60.1	253	152	101	91.3
60.1	248	149	99	89.5
60.1	243	146	97	87.7
60.1	238	143	95	85.9
60.1	233	140	93	84.1
60.1	228	137	91	82.3
60.1	223	134	89	80.5
60.1	218	131	87	78.7
60.1	213	128	35	76.9
60.1	208	125	83	75.1
60.1	203	122	81	73.3
60.1	198	119	79	71.5
60.1	193	116	77	69.7
60.1	188	113	75	67.9
60.1	183	110	73	66.1
60.1	178	107	71	64.3
60.1	173	104	69	62.5
60.1	168	101	67	60.7
60.1	163	98	65	58.9
60.1	158	95	63	57.1
60.1	153	92	61	55.3
60.1	148	89	59	53.5
60.1	143	86	57	51.7
60.1	138	83	55	49.9
60.0	500	300	200	180.0
60.0	498	299	199	179.5
60.0	497	298	199	178.7
60.0	495	297	198	178.2
60.0	493	296	197	177.7
60.0	492	295	197	176.9
60.0	490	294	196	176.4
60.0	488	293	195	175.9
60.0	487	292	195	175.1
60.0	485	291	194	174.6
60.0	483	290	193	174.1
60.0	482	289	193	173.3
60.0	480	288	192	172.8
60.0	478	287	191	172.3
60.0	477	286	191	171.5
60.0	475	285	190	171.0
60.0	473	284	189	170.5
60.0	472	283	189	169.7
60.0	470	282	188	169.2

%E	M1	M2	DM	M*
60.0	468	281	187	168.7
60.0	467	280	187	167.9
60.0	465	279	186	167.4
60.0	463	278	185	166.9
60.0	462	277	185	166.1
60.0	460	276	184	165.6
60.0	458	275	183	165.1
60.0	457	274	183	164.3
60.0	455	273	182	163.8
60.0	453	272	181	163.3
60.0	452	271	181	162.5
60.0	450	270	180	162.0
60.0	448	269	179	161.5
60.0	447	268	179	160.7
60.0	445	267	178	160.2
60.0	443	266	177	159.7
60.0	442	265	177	158.9
60.0	440	264	176	158.4
60.0	438	263	175	157.9
60.0	437	262	175	157.1
60.0	435	261	174	156.6
60.0	433	260	173	156.1
60.0	432	259	173	155.3
60.0	430	258	172	154.8
60.0	428	257	171	154.3
60.0	427	256	171	153.5
60.0	425	255	170	153.0
60.0	423	254	169	152.5
60.0	422	253	169	151.7
60.0	420	252	168	151.2
60.0	418	251	167	150.7
60.0	417	250	167	149.9
60.0	415	249	166	149.4
60.0	413	248	165	148.9
60.0	412	247	165	148.1
60.0	410	246	164	147.6
60.0	408	245	163	147.1
60.0	407	244	163	146.3
60.0	405	243	162	145.8
60.0	402	241	161	144.5
60.0	400	240	160	144.0
60.0	397	238	159	142.7
60.0	395	237	158	142.2
60.0	390	234	156	140.4
60.0	385	231	154	138.6
60.0	380	228	152	136.8
60.0	375	225	150	135.0
60.0	370	222	148	133.2
60.0	365	219	146	131.4
60.0	360	216	144	129.6

%E	M1	M2	DM	M*
60.0	355	213	142	127.8
60.0	350	210	140	126.0
60.0	345	207	138	124.2
60.0	340	204	136	122.4
60.0	335	201	134	120.6
60.0	330	198	132	118.8
60.0	325	195	130	117.0
60.0	320	192	128	115.2
60.0	315	189	126	113.4
60.0	310	186	124	111.6
60.0	305	183	122	109.8
60.0	300	180	120	108.0
60.0	295	177	118	106.2
60.0	290	174	116	104.4
60.0	285	171	114	102.6
60.0	280	168	112	100.8
60.0	275	165	110	99.0
60.0	270	162	108	97.2
60.0	265	159	106	95.4
60.0	260	156	104	93.6
60.0	255	153	102	91.8
60.0	250	150	100	90.0
60.0	245	147	98	88.2
60.0	240	144	96	86.4
60.0	235	141	94	84.6
60.0	230	138	92	82.8
60.0	225	135	90	81.0
60.0	220	132	88	79.2
60.0	215	129	86	77.4
60.0	210	126	84	75.6
60.0	205	123	82	73.8
60.0	200	120	80	72.0
60.0	195	117	78	70.2
60.0	190	114	76	68.4
60.0	185	111	74	66.6
60.0	180	108	72	64.8
60.0	175	105	70	63.0
60.0	170	102	68	61.2
60.0	165	99	66	59.4
60.0	160	96	64	57.6
60.0	155	93	62	55.8
60.0	150	90	60	54.0
60.0	145	87	58	52.2
60.0	140	84	56	50.4
60.0	135	81	54	48.6
60.0	130	78	52	46.8
60.0	125	75	50	45.0
60.0	120	72	48	43.2
60.0	115	69	46	41.4
60.0	110	66	44	39.6

%E	M1	M2	DM	M*
60.0	105	63	42	37.8
60.0	100	60	40	36.0
60.0	95	57	38	34.2
60.0	90	54	36	32.4
60.0	85	51	34	30.6
60.0	80	48	32	28.8
60.0	75	45	30	27.0
60.0	70	42	28	25.2
60.0	65	39	26	23.4
60.0	60	36	24	21.6
60.0	55	33	22	19.8
60.0	50	30	20	18.0
60.0	45	27	18	16.2
60.0	40	24	16	14.4
60.0	35	21	14	12.6
60.0	30	18	12	10.8
60.0	25	15	10	9.0
60.0	20	12	8	7.2
60.0	15	9	6	5.4
60.0	10	6	4	3.6
60.0	5	3	2	1.8
59.9	499	299	200	179.2
59.9	496	297	199	177.8
59.9	494	296	198	177.4
59.9	491	294	197	176.0
59.9	489	293	196	175.6
59.9	486	291	195	174.2
59.9	484	290	194	173.8
59.9	481	288	193	172.4
59.9	479	287	192	172.0
59.9	476	285	191	170.6
59.9	474	284	190	170.2
59.9	471	282	189	168.8
59.9	469	281	188	168.4
59.9	466	279	187	167.0
59.9	464	278	186	166.6
59.9	461	276	185	165.2
59.9	459	275	184	164.8
59.9	456	273	183	163.4
59.9	454	272	182	163.0
59.9	451	270	181	161.6
59.9	449	269	180	161.2
59.9	446	267	179	159.8
59.9	444	266	178	159.4
59.9	441	264	177	158.0
59.9	439	263	176	157.6
59.9	436	261	175	156.2
59.9	434	260	174	155.8
59.9	431	258	173	154.4
59.9	429	257	172	154.0

%E	M1	M2	DM	M*
59.9	426	255	171	152.6
59.9	424	254	170	152.2
59.9	421	252	169	150.8
59.9	419	251	168	150.4
59.9	416	249	167	149.0
59.9	414	248	166	148.6
59.9	411	246	165	147.2
59.9	409	245	164	146.8
59.9	406	243	163	145.4
59.9	404	242	162	145.0
59.9	401	240	161	143.6
59.9	399	239	160	143.2
59.9	394	236	158	141.4
59.9	392	235	157	140.9
59.9	389	233	156	139.6
59.9	387	232	155	139.1
59.9	384	230	154	137.8
59.9	382	229	153	137.3
59.9	379	227	152	136.0
59.9	377	226	151	135.5
59.9	374	224	150	134.2
59.9	372	223	149	133.7
59.9	369	221	148	132.4
59.9	367	220	147	131.9
59.9	364	218	146	130.6
59.9	362	217	145	130.1
59.9	359	215	144	128.8
59.9	357	214	143	128.3
59.9	354	212	142	127.0
59.9	352	211	141	126.5
59.9	349	209	140	125.2
59.9	347	208	139	124.7
59.9	344	206	138	123.4
59.9	342	205	137	122.9
59.9	339	203	136	121.6
59.9	337	202	135	121.1
59.9	334	200	134	119.8
59.9	332	199	133	119.3
59.9	329	197	132	118.0
59.9	327	196	131	117.5
59.9	324	194	130	116.2
59.9	322	193	129	115.7
59.9	319	191	128	114.4
59.9	317	190	127	113.9
59.9	314	188	126	112.6
59.9	312	187	125	112.1
59.9	309	185	124	110.8
59.9	307	184	123	110.3
59.9	304	182	122	109.0
59.9	302	181	121	108.5

%E	M1	M2	DM	M*
59.9	299	179	120	107.2
59.9	297	178	119	106.7
59.9	294	176	118	105.4
59.9	292	175	117	104.9
59.9	289	173	116	103.6
59.9	287	172	115	103.1
59.9	284	170	114	101.8
59.9	282	169	113	101.3
59.9	279	167	112	100.0
59.9	277	166	111	99.5
59.9	274	164	110	98.2
59.9	272	163	109	97.7
59.9	269	161	108	96.4
59.9	267	160	107	95.9
59.9	262	157	105	94.1
59.9	257	154	103	92.3
59.9	252	151	101	90.5
59.9	247	148	99	88.7
59.9	242	145	97	86.9
59.9	237	142	95	85.1
59.9	232	139	93	83.3
59.9	227	136	91	81.5
59.9	222	133	89	79.7
59.9	217	130	87	77.9
59.9	212	127	85	76.1
59.9	207	124	83	74.3
59.9	202	121	81	72.5
59.9	197	118	79	70.7
59.9	192	115	77	68.9
59.9	187	112	75	67.1
59.9	182	109	73	65.3
59.9	177	106	71	63.5
59.9	172	103	69	61.7
59.9	167	100	67	59.9
59.9	162	97	65	58.1
59.9	157	94	63	56.3
59.9	152	91	61	54.5
59.9	147	88	59	52.7
59.9	142	85	57	50.9
59.9	137	82	55	49.1
59.8	500	299	201	178.8
59.8	498	298	200	178.3
59.8	497	297	200	177.5
59.8	495	296	199	177.0
59.8	493	295	198	176.5
59.8	492	294	198	175.7
59.8	490	293	197	175.2
59.8	488	292	196	174.7
59.8	487	291	196	173.9
59.8	485	290	195	173.4

%E	M1	M2	DM	M*
59.8	483	289	194	172.9
59.8	482	288	194	172.1
59.8	480	287	193	171.6
59.8	478	286	192	171.1
59.8	475	284	191	169.8
59.8	473	283	190	169.3
59.8	470	281	189	168.0
59.8	468	280	188	167.5
59.8	465	278	187	166.2
59.8	463	277	186	165.7
59.8	460	275	185	164.4
59.8	458	274	184	163.9
59.8	455	272	183	162.6
59.8	453	271	182	162.1
59.8	450	269	181	160.8
59.8	448	268	180	160.3
59.8	445	266	179	159.0
59.8	443	265	178	158.5
59.8	440	263	177	157.2
59.8	438	262	176	156.7
59.8	435	260	175	155.4
59.8	433	259	174	154.9
59.8	430	257	173	153.6
59.8	428	256	172	153.1
59.8	425	254	171	151.8
59.8	423	253	170	151.3
59.8	420	251	169	150.0
59.8	418	250	168	149.5
59.8	415	248	167	148.2
59.8	413	247	166	147.7
59.8	410	245	165	146.4
59.8	408	244	164	145.9
59.8	405	242	163	144.6
59.8	403	241	162	144.1
59.8	400	239	161	142.8
59.8	398	238	160	142.3
59.8	396	237	159	141.8
59.8	393	235	158	140.5
59.8	391	234	157	140.0
59.8	388	232	156	138.7
59.8	386	231	155	138.2
59.8	383	229	154	136.9
59.8	381	228	153	136.4
59.8	378	226	152	135.1
59.8	376	225	151	134.6
59.8	373	223	150	133.3
59.8	371	222	149	132.8
59.8	368	220	148	131.5
59.8	366	219	147	131.0
59.8	363	217	146	129.7

%E	M1	M2	DM	M*
59.8	361	216	145	129.2
59.8	358	214	144	127.9
59.8	356	213	143	127.4
59.8	353	211	142	126.1
59.8	351	210	141	125.6
59.8	348	208	140	124.3
59.8	346	207	139	123.8
59.8	343	205	138	122.5
59.8	341	204	137	122.0
59.8	338	202	136	120.7
59.8	336	201	135	120.2
59.8	333	199	134	118.9
59.8	331	198	133	118.4
59.8	328	196	132	117.1
59.8	326	195	131	116.6
59.8	323	193	130	115.3
59.8	321	192	129	114.8
59.8	316	189	127	113.0
59.8	311	186	125	111.2
59.8	306	183	123	109.4
59.8	301	180	121	107.6
59.8	296	177	119	105.8
59.8	291	174	117	104.0
59.8	286	171	115	102.2
59.8	281	168	113	100.4
59.8	276	165	111	98.6
59.8	271	162	109	96.8
59.8	266	159	107	95.0
59.8	264	158	106	94.6
59.8	261	156	105	93.2
59.8	259	155	104	92.8
59.8	256	153	103	91.4
59.8	254	152	102	91.0
59.8	251	150	101	89.6
59.8	249	149	100	89.2
59.8	246	147	99	87.8
59.8	244	146	98	87.4
59.8	241	144	97	86.0
59.8	239	143	96	85.6
59.8	234	140	94	83.8
59.8	229	137	92	82.0
59.8	224	134	90	80.2
59.8	219	131	88	78.4
59.8	214	128	86	76.6
59.8	209	125	84	74.8
59.8	204	122	82	73.0
59.8	199	119	80	71.2
59.8	194	116	78	69.4
59.8	189	113	76	67.6
59.8	184	110	74	65.8

%E	M1	M2	DM	M*
59.8	179	107	72	64.0
59.8	174	104	70	62.2
59.8	169	101	68	60.4
59.8	164	98	66	58.6
59.8	132	79	53	47.3
59.8	127	76	51	45.5
59.8	122	73	49	43.7
59.8	117	70	47	41.9
59.8	112	67	45	40.1
59.8	107	64	43	38.3
59.8	102	61	41	36.5
59.8	97	58	39	34.7
59.8	92	55	37	32.9
59.8	87	52	35	31.1
59.8	82	49	33	29.3
59.7	499	298	201	178.0
59.7	496	296	200	176.6
59.7	494	295	199	176.2
59.7	491	293	198	174.8
59.7	489	292	197	174.4
59.7	486	290	196	173.0
59.7	484	289	195	172.6
59.7	481	287	194	171.2
59.7	479	286	193	170.8
59.7	477	285	192	170.3
59.7	476	284	192	169.4
59.7	474	283	191	169.0
59.7	472	282	190	168.5
59.7	471	281	190	167.6
59.7	469	280	189	167.2
59.7	467	279	188	166.7
59.7	466	278	188	165.8
59.7	464	277	187	165.4
59.7	462	276	186	164.9
59.7	461	275	186	164.0
59.7	459	274	185	163.6
59.7	457	273	184	163.1
59.7	454	271	183	161.8
59.7	452	270	182	161.3
59.7	449	268	181	160.0
59.7	447	267	180	159.5
59.7	444	265	179	158.2
59.7	442	264	178	157.7
59.7	439	262	177	156.4
59.7	437	261	176	155.9
59.7	434	259	175	154.6
59.7	432	258	174	154.1
59.7	429	256	173	152.8
59.7	427	255	172	152.3
59.7	424	253	171	151.0

%E	M1	M2	DM	M*
59.7	422	252	170	150.5
59.7	419	250	169	149.2
59.7	417	249	168	148.7
59.7	414	247	167	147.4
59.7	412	246	166	146.9
59.7	409	244	165	145.6
59.7	407	243	164	145.1
59.7	404	241	163	143.8
59.7	402	240	162	143.3
59.7	397	237	160	141.5
59.7	395	236	159	141.0
59.7	392	234	158	139.7
59.7	390	233	157	139.2
59.7	387	231	156	137.9
59.7	385	230	155	137.4
59.7	382	228	154	136.1
59.7	380	227	153	135.6
59.7	377	225	152	134.3
59.7	375	224	151	133.8
59.7	372	222	150	132.5
59.7	370	221	149	132.0
59.7	367	219	148	130.7
59.7	365	218	147	130.2
59.7	362	216	146	128.9
59.7	360	215	145	128.4
59.7	357	213	144	127.1
59.7	355	212	143	126.6
59.7	352	210	142	125.3
59.7	350	209	141	124.8
59.7	347	207	140	123.5
59.7	345	206	139	123.0
59.7	340	203	137	121.2
59.7	335	200	135	119.4
59.7	330	197	133	117.6
59.7	325	194	131	115.8
59.7	320	191	129	114.0
59.7	318	190	128	113.5
59.7	315	188	127	112.2
59.7	313	187	126	111.7
59.7	310	185	125	110.4
59.7	308	184	124	109.9
59.7	305	182	123	108.6
59.7	303	181	122	108.1
59.7	300	179	121	106.8
59.7	298	178	120	106.3
59.7	295	176	119	105.0
59.7	293	175	118	104.5
59.7	290	173	117	103.2
59.7	288	172	116	102.7
59.7	283	169	114	100.9

%E	M1	M2	DM	M*
59.7	278	166	112	99.1
59.7	273	163	110	97.3
59.7	268	160	108	95.5
59.7	263	157	106	93.7
59.7	258	154	104	91.9
59.7	253	151	102	90.1
59.7	248	148	100	88.3
59.7	243	145	98	86.5
59.7	238	142	96	84.7
59.7	236	141	95	84.2
59.7	233	139	94	82.9
59.7	231	138	93	82.4
59.7	226	135	91	80.6
59.7	221	132	89	78.8
59.7	216	129	87	77.0
59.7	211	126	85	75.2
59.7	206	123	83	73.4
59.7	201	120	81	71.6
59.7	196	117	79	69.8
59.7	191	114	77	68.0
59.7	186	111	75	66.2
59.7	181	108	73	64.4
59.7	176	105	71	62.6
59.7	159	95	64	56.8
59.7	154	92	62	55.0
59.7	149	89	60	53.2
59.7	144	86	58	51.4
59.7	139	83	56	49.6
59.7	134	80	54	47.8
59.7	129	77	52	46.0
59.7	124	74	50	44.2
59.7	119	71	48	42.4
59.7	77	46	31	27.5
59.7	72	43	29	25.7
59.7	67	40	27	23.9
59.7	62	37	25	22.1
59.6	500	298	202	177.6
59.6	498	297	201	177.1
59.6	497	296	201	176.3
59.6	495	295	200	175.8
59.6	493	294	199	175.3
59.6	492	293	199	174.5
59.6	490	292	198	174.0
59.6	488	291	197	173.5
59.6	485	289	196	172.2
59.6	483	288	195	171.7
59.6	480	286	194	170.4
59.6	478	285	193	169.9
59.6	475	283	192	168.6
59.6	473	282	191	168.1

%E	M1	M2	DM	M*
59.6	470	280	190	166.8
59.6	468	279	189	166.3
59.6	465	277	188	165.0
59.6	463	276	187	164.5
59.6	460	274	186	163.2
59.6	458	273	185	162.7
59.6	456	272	184	162.2
59.6	455	271	184	161.4
59.6	453	270	183	160.9
59.6	451	269	182	160.4
59.6	450	268	182	159.6
59.6	448	267	181	159.1
59.6	446	266	180	158.6
59.6	445	265	180	157.8
59.6	443	264	179	157.3
59.6	441	263	178	156.8
59.6	438	261	177	155.5
59.6	436	260	176	155.0
59.6	433	258	175	153.7
59.6	431	257	174	153.2
59.6	428	255	173	151.9
59.6	426	254	172	151.4
59.6	423	252	171	150.1
59.6	421	251	170	149.6
59.6	418	249	169	148.3
59.6	416	248	168	147.8
59.6	413	246	167	146.5
59.6	411	245	166	146.0
59.6	408	243	165	144.7
59.6	406	242	164	144.2
59.6	403	240	163	142.9
59.6	401	239	162	142.4
59.6	399	238	161	142.0
59.6	396	236	160	140.6
59.6	394	235	159	140.2
59.6	391	233	158	138.8
59.6	389	232	157	138.4
59.6	386	230	156	137.0
59.6	384	229	155	136.6
59.6	381	227	154	135.2
59.6	379	226	153	134.8
59.6	376	224	152	133.4
59.6	374	223	151	133.0
59.6	371	221	150	131.6
59.6	369	220	149	131.2
59.6	366	218	148	129.8
59.6	364	217	147	129.4
59.6	361	215	146	128.0
59.6	359	214	145	127.6
59.6	356	212	144	126.2

%E	M1	M2	DM	M*
59.6	354	211	143	125.8
59.6	349	208	141	124.0
59.6	344	205	139	122.2
59.6	342	204	138	121.7
59.6	339	202	137	120.4
59.6	337	201	136	119.9
59.6	334	199	135	118.6
59.6	332	198	134	118.1
59.6	329	196	133	116.8
59.6	327	195	132	116.3
59.6	324	193	131	115.0
59.6	322	192	130	114.5
59.6	319	190	129	113.2
59.6	317	189	128	112.7
59.6	314	187	127	111.4
59.6	312	186	126	110.9
59.6	307	183	124	109.1
59.6	302	180	122	107.3
59.6	297	177	120	105.5
59.6	292	174	118	103.7
59.6	287	171	116	101.9
59.6	285	170	115	101.4
59.6	282	168	114	100.1
59.6	280	167	113	99.6
59.6	277	165	112	98.3
59.6	275	164	111	97.8
59.6	272	162	110	96.5
59.6	270	161	109	96.0
59.6	267	159	108	94.7
59.6	265	158	107	94.2
59.6	260	155	105	92.4
59.6	255	152	103	90.6
59.6	250	149	101	88.8
59.6	245	146	99	87.0
59.6	240	143	97	85.2
59.6	235	140	95	83.4
59.6	230	137	93	81.6
59.6	228	136	92	81.1
59.6	225	134	91	79.8
59.6	223	133	90	79.3
59.6	218	130	88	77.5
59.6	213	127	86	75.7
59.6	208	124	84	73.9
59.6	203	121	82	72.1
59.6	198	118	80	70.3
59.6	193	115	78	68.5
59.6	188	112	76	66.7
59.6	183	109	74	64.9
59.6	178	106	72	63.1
59.6	171	102	69	60.8

%E	M1	M2	DM	M*
59.6	166	99	67	59.0
59.6	161	96	65	57.2
59.6	156	93	63	55.4
59.6	151	90	61	53.6
59.6	146	87	59	51.8
59.6	141	84	57	50.0
59.6	136	81	55	48.2
59.6	114	68	46	40.6
59.6	109	65	44	38.8
59.6	104	62	42	37.0
59.6	99	59	40	35.2
59.6	94	56	38	33.4
59.6	89	53	36	31.6
59.6	57	34	23	20.3
59.6	52	31	21	18.5
59.6	47	28	19	16.7
59.5	499	297	202	176.8
59.5	496	295	201	175.5
59.5	494	294	200	175.0
59.5	491	292	199	173.7
59.5	489	291	198	173.2
59.5	487	290	197	172.7
59.5	486	289	197	171.9
59.5	484	288	196	171.4
59.5	482	287	195	170.9
59.5	481	286	195	170.1
59.5	479	285	194	169.6
59.5	477	284	193	169.1
59.5	476	283	193	168.3
59.5	474	282	192	167.8
59.5	472	281	191	167.3
59.5	469	279	190	166.0
59.5	467	278	189	165.5
59.5	464	276	188	164.2
59.5	462	275	187	163.7
59.5	459	273	186	162.4
59.5	457	272	185	161.9
59.5	454	270	184	160.6
59.5	452	269	183	160.1
59.5	449	267	182	158.8
59.5	447	266	181	158.3
59.5	444	264	180	157.0
59.5	442	263	179	156.5
59.5	440	262	178	156.0
59.5	439	261	178	155.2
59.5	437	260	177	154.7
59.5	435	259	176	154.2
59.5	432	257	175	152.9
59.5	430	256	174	152.4
59.5	427	254	173	151.1

%E	M1	M2	DM	M*
59.5	425	253	172	150.6
59.5	422	251	171	149.3
59.5	420	250	170	148.8
59.5	417	248	169	147.5
59.5	415	247	168	147.0
59.5	412	245	167	145.7
59.5	410	244	166	145.2
59.5	407	242	165	143.9
59.5	405	241	164	143.4
59.5	402	239	163	142.1
59.5	400	238	162	141.6
59.5	398	237	161	141.1
59.5	395	235	160	139.8
59.5	393	234	159	139.3
59.5	390	232	158	138.0
59.5	388	231	157	137.5
59.5	385	229	156	136.2
59.5	383	228	155	135.7
59.5	380	226	154	134.4
59.5	378	225	153	133.9
59.5	375	223	152	132.6
59.5	373	222	151	132.1
59.5	370	220	150	130.8
59.5	368	219	149	130.3
59.5	365	217	148	129.0
59.5	363	216	147	128.5
59.5	358	213	145	126.7
59.5	353	210	143	124.9
59.5	351	209	142	124.4
59.5	348	207	141	123.1
59.5	346	206	140	122.6
59.5	343	204	139	121.3
59.5	341	203	138	120.8
59.5	338	201	137	119.5
59.5	336	200	136	119.0
59.5	333	198	135	117.7
59.5	331	197	134	117.2
59.5	328	195	133	115.9
59.5	326	194	132	115.4
59.5	321	191	130	113.6
59.5	316	188	128	111.8
59.5	311	185	126	110.0
59.5	309	184	125	109.6
59.5	306	182	124	108.2
59.5	304	181	123	107.8
59.5	301	179	122	106.4
59.5	299	178	121	106.0
59.5	296	176	120	104.6
59.5	294	175	119	104.2
59.5	291	173	118	102.8

%E	M1	M2	DM	M*
59.5	289	172	117	102.4
59.5	284	169	115	100.6
59.5	279	166	113	98.8
59.5	274	163	111	97.0
59.5	269	160	109	95.2
59.5	264	157	107	93.4
59.5	262	156	106	92.9
59.5	259	154	105	91.6
59.5	257	153	104	91.1
59.5	252	150	102	89.3
59.5	247	147	100	87.5
59.5	242	144	98	85.7
59.5	237	141	96	83.9
59.5	232	138	94	82.1
59.5	227	135	92	80.3
59.5	222	132	90	78.5
59.5	220	131	89	78.0
59.5	215	128	87	76.2
59.5	210	125	85	74.4
59.5	205	122	83	72.6
59.5	200	119	81	70.8
59.5	195	116	79	69.0
59.5	190	113	77	67.2
59.5	185	110	75	65.4
59.5	173	103	70	61.3
59.5	168	100	68	59.5
59.5	163	97	66	57.7
59.5	158	94	64	55.9
59.5	153	91	62	54.1
59.5	148	88	60	52.3
59.5	131	78	53	46.4
59.5	126	75	51	44.6
59.5	121	72	49	42.8
59.5	116	69	47	41.0
59.5	111	66	45	39.2
59.5	84	50	34	29.8
59.5	79	47	32	28.0
59.5	74	44	30	26.2
59.5	42	25	17	14.9
59.5	37	22	15	13.1
59.4	500	297	203	176.4
59.4	498	296	202	175.9
59.4	497	295	202	175.1
59.4	495	294	201	174.6
59.4	493	293	200	174.1
59.4	492	292	200	173.3
59.4	490	291	199	172.8
59.4	488	290	198	172.3
59.4	485	288	197	171.0
59.4	483	287	196	170.5

%E	M1	M2	DM	M*
59.4	480	285	195	169.2
59.4	478	284	194	168.7
59.4	475	282	193	167.4
59.4	473	281	192	166.9
59.4	471	280	191	166.5
59.4	470	279	191	165.6
59.4	468	278	190	165.1
59.4	466	277	189	164.7
59.4	465	276	189	163.8
59.4	463	275	188	163.3
59.4	461	274	187	162.9
59.4	458	272	186	161.5
59.4	456	271	185	161.1
59.4	453	269	184	159.7
59.4	451	268	183	159.3
59.4	448	266	182	157.9
59.4	446	265	181	157.5
59.4	443	263	180	156.1
59.4	441	262	179	155.7
59.4	438	260	178	154.3
59.4	436	259	177	153.9
59.4	434	258	176	153.4
59.4	433	257	176	152.5
59.4	431	256	175	152.1
59.4	429	255	174	151.6
59.4	426	253	173	150.3
59.4	424	252	172	149.8
59.4	421	250	171	148.5
59.4	419	249	170	148.0
59.4	416	247	169	146.7
59.4	414	246	168	146.2
59.4	411	244	167	144.9
59.4	409	243	166	144.4
59.4	406	241	165	143.1
59.4	404	240	164	142.6
59.4	401	238	163	141.3
59.4	399	237	162	140.8
59.4	397	236	161	140.3
59.4	394	234	160	139.0
59.4	392	233	159	138.5
59.4	389	231	158	137.2
59.4	387	230	157	136.7
59.4	384	228	156	135.4
59.4	382	227	155	134.9
59.4	379	225	154	133.6
59.4	377	224	153	133.1
59.4	374	222	152	131.8
59.4	372	221	151	131.3
59.4	369	219	150	130.0
59.4	367	218	149	129.5

%E	M1	M2	DM	M*
59.4	362	215	147	127.7
59.4	360	214	146	127.2
59.4	357	212	145	125.9
59.4	355	211	144	125.4
59.4	352	209	143	124.1
59.4	350	208	142	123.6
59.4	347	206	141	122.3
59.4	345	205	140	121.8
59.4	342	203	139	120.5
59.4	340	202	138	120.0
59.4	335	199	136	118.2
59.4	330	196	134	116.4
59.4	325	193	132	114.6
59.4	323	192	131	114.1
59.4	320	190	130	112.8
59.4	318	189	129	112.3
59.4	315	187	128	111.0
59.4	313	186	127	110.5
59.4	310	184	126	109.2
59.4	308	183	125	108.7
59.4	303	180	123	106.9
59.4	298	177	121	105.1
59.4	293	174	119	103.3
59.4	288	171	117	101.5
59.4	286	170	116	101.0
59.4	283	168	115	99.7
59.4	281	167	114	99.2
59.4	278	165	113	97.9
59.4	276	164	112	97.4
59.4	271	161	110	95.6
59.4	266	158	108	93.8
59.4	261	155	106	92.0
59.4	256	152	104	90.3
59.4	254	151	103	89.8
59.4	251	149	102	88.5
59.4	249	148	101	88.0
59.4	246	146	100	86.7
59.4	244	145	99	86.2
59.4	239	142	97	84.4
59.4	234	139	95	82.6
59.4	229	136	93	80.8
59.4	224	133	91	79.0
59.4	219	130	89	77.2
59.4	217	129	88	76.7
59.4	212	126	86	74.9
59.4	207	123	84	73.1
59.4	202	120	82	71.3
59.4	197	117	80	69.5
59.4	192	114	78	67.7
59.4	187	111	76	65.9

%E	M1	M2	DM	M*
59.4	180	107	73	63.6
59.4	175	104	71	61.8
59.4	170	101	69	60.0
59.4	165	98	67	58.2
59.4	160	95	65	56.4
59.4	155	92	63	54.6
59.4	143	85	58	50.5
59.4	138	82	56	48.7
59.4	133	79	54	46.9
59.4	128	76	52	45.1
59.4	123	73	50	43.3
59.4	106	63	43	37.4
59.4	101	60	41	35.6
59.4	96	57	39	33.8
59.4	69	41	28	24.4
59.4	64	38	26	22.6
59.4	32	19	13	11.3
59.3	499	296	203	175.6
59.3	496	294	202	174.3
59.3	494	293	201	173.8
59.3	491	291	200	172.5
59.3	489	290	199	172.0
59.3	487	289	198	171.5
59.3	486	288	198	170.7
59.3	484	287	197	170.2
59.3	482	286	196	169.7
59.3	481	285	196	168.9
59.3	479	284	195	168.4
59.3	477	283	194	167.9
59.3	474	281	193	166.6
59.3	472	280	192	166.1
59.3	469	278	191	164.8
59.3	467	277	190	164.3
59.3	464	275	189	163.0
59.3	462	274	188	162.5
59.3	460	273	187	162.0
59.3	459	272	187	161.2
59.3	457	271	186	160.7
59.3	455	270	185	160.2
59.3	454	269	185	159.4
59.3	452	268	184	158.9
59.3	450	267	183	158.4
59.3	447	265	182	157.1
59.3	445	264	181	156.6
59.3	442	262	180	155.3
59.3	440	261	179	154.8
59.3	437	259	178	153.5
59.3	435	258	177	153.0
59.3	432	256	176	151.7
59.3	430	255	175	151.2

%E	M1	M2	DM	M*
59.3	428	254	174	150.7
59.3	427	253	174	149.9
59.3	425	252	173	149.4
59.3	423	251	172	148.9
59.3	420	249	171	147.6
59.3	418	248	170	147.1
59.3	415	246	169	145.8
59.3	413	245	168	145.3
59.3	410	243	167	144.0
59.3	408	242	166	143.5
59.3	405	240	165	142.2
59.3	403	239	164	141.7
59.3	400	237	163	140.4
59.3	398	236	162	139.9
59.3	396	235	161	139.5
59.3	393	233	160	138.1
59.3	391	232	159	137.7
59.3	388	230	158	136.3
59.3	386	229	157	135.9
59.3	383	227	156	134.5
59.3	381	226	155	134.1
59.3	378	224	154	132.7
59.3	376	223	153	132.3
59.3	371	220	151	130.5
59.3	366	217	149	128.7
59.3	364	216	148	128.2
59.3	361	214	147	126.9
59.3	359	213	146	126.4
59.3	356	211	145	125.1
59.3	354	210	144	124.6
59.3	351	208	143	123.3
59.3	349	207	142	122.8
59.3	344	204	140	121.0
59.3	339	201	138	119.2
59.3	337	200	137	118.7
59.3	334	198	136	117.4
59.3	332	197	135	116.9
59.3	329	195	134	115.6
59.3	327	194	133	115.1
59.3	324	192	132	113.8
59.3	322	191	131	113.3
59.3	317	188	129	111.5
59.3	312	185	127	109.7
59.3	307	182	125	107.9
59.3	305	181	124	107.4
59.3	302	179	123	106.1
59.3	300	178	122	105.6
59.3	297	176	121	104.3
59.3	295	175	120	103.8
59.3	290	172	118	102.0

%E	M1	M2	DM	M*
59.3	285	169	116	100.2
59.3	280	166	114	98.4
59.3	275	163	112	96.6
59.3	273	162	111	96.1
59.3	270	160	110	94.8
59.3	268	159	109	94.3
59.3	263	156	107	92.5
59.3	258	153	105	90.7
59.3	253	150	103	88.9
59.3	248	147	101	87.1
59.3	243	144	99	85.3
59.3	241	143	98	84.9
59.3	236	140	96	83.1
59.3	231	137	94	81.3
59.3	226	134	92	79.5
59.3	221	131	90	77.7
59.3	216	128	88	75.9
59.3	214	127	87	75.4
59.3	209	124	85	73.6
59.3	204	121	83	71.8
59.3	199	118	81	70.0
59.3	194	115	79	68.2
59.3	189	112	77	66.4
59.3	182	108	74	64.1
59.3	177	105	72	62.3
59.3	172	102	70	60.5
59.3	167	99	68	58.7
59.3	162	96	66	56.9
59.3	150	89	61	52.8
59.3	145	86	59	51.0
59.3	140	83	57	49.2
59.3	135	80	55	47.4
59.3	118	70	48	41.5
59.3	113	67	46	39.7
59.3	108	64	44	37.9
59.3	91	54	37	32.0
59.3	86	51	35	30.2
59.3	81	48	33	28.4
59.3	59	35	24	20.8
59.3	54	32	22	19.0
59.3	27	16	11	9.5
59.2	500	296	204	175.2
59.2	498	295	203	174.7
59.2	497	294	203	173.9
59.2	495	293	202	173.4
59.2	493	292	201	172.9
59.2	490	290	200	171.6
59.2	488	289	199	171.1
59.2	485	287	198	169.8
59.2	483	286	197	169.3

%E	M1	M2	DM	M*
59.2	480	284	196	168.0
59.2	478	283	195	167.6
59.2	476	282	194	167.1
59.2	475	281	194	166.2
59.2	473	280	193	165.8
59.2	471	279	192	165.3
59.2	468	277	191	164.0
59.2	466	276	190	163.5
59.2	463	274	189	162.2
59.2	461	273	188	161.7
59.2	458	271	187	160.4
59.2	456	270	186	159.9
59.2	453	268	185	158.6
59.2	451	267	184	158.1
59.2	449	266	183	157.6
59.2	448	265	183	156.8
59.2	446	264	182	156.3
59.2	444	263	181	155.8
59.2	441	261	180	154.5
59.2	439	260	179	154.0
59.2	436	258	178	152.7
59.2	434	257	177	152.2
59.2	431	255	176	150.9
59.2	429	254	175	150.4
59.2	426	252	174	149.1
59.2	424	251	173	148.6
59.2	422	250	172	148.1
59.2	419	248	171	146.8
59.2	417	247	170	146.3
59.2	414	245	169	145.0
59.2	412	244	168	144.5
59.2	409	242	167	143.2
59.2	407	241	166	142.7
59.2	404	239	165	141.4
59.2	402	238	164	140.9
59.2	397	235	162	139.1
59.2	395	234	161	138.6
59.2	392	232	160	137.3
59.2	390	231	159	136.8
59.2	387	229	158	135.5
59.2	385	228	157	135.0
59.2	382	226	156	133.7
59.2	380	225	155	133.2
59.2	377	223	154	131.9
59.2	375	222	153	131.4
59.2	373	221	152	130.9
59.2	370	219	151	129.6
59.2	368	218	150	129.1
59.2	365	216	149	127.8
59.2	363	215	148	127.3

%E	M1	M2	DM	M*
59.2	360	213	147	126.0
59.2	358	212	146	125.5
59.2	355	210	145	124.2
59.2	353	209	144	123.7
59.2	348	206	142	121.9
59.2	346	205	141	121.5
59.2	343	203	140	120.1
59.2	341	202	139	119.7
59.2	338	200	138	118.3
59.2	336	199	137	117.9
59.2	333	197	136	116.5
59.2	331	196	135	116.1
59.2	326	193	133	114.3
59.2	321	190	131	112.5
59.2	319	189	130	112.0
59.2	316	187	129	110.7
59.2	314	186	128	110.2
59.2	311	184	127	108.9
59.2	309	183	126	108.4
59.2	306	181	125	107.1
59.2	304	180	124	106.6
59.2	299	177	122	104.8
59.2	294	174	120	103.0
59.2	292	173	119	102.5
59.2	289	171	118	101.2
59.2	287	170	117	100.7
59.2	284	168	116	99.4
59.2	282	167	115	98.9
59.2	277	164	113	97.1
59.2	272	161	111	95.3
59.2	267	158	109	93.5
59.2	265	157	108	93.0
59.2	262	155	107	91.7
59.2	260	154	106	91.2
59.2	255	151	104	89.4
59.2	250	148	102	87.6
59.2	245	145	100	85.8
59.2	240	142	98	84.0
59.2	238	141	97	83.5
59.2	233	138	95	81.7
59.2	228	135	93	79.9
59.2	223	132	91	78.1
59.2	218	129	89	76.3
59.2	213	126	87	74.5
59.2	211	125	86	74.1
59.2	206	122	84	72.3
59.2	201	119	82	70.5
59.2	196	116	80	68.7
59.2	191	113	78	66.9
59.2	184	109	75	64.6

%E	M1	M2	DM	M*
59.2	179	106	73	62.8
59.2	174	103	71	61.0
59.2	169	100	69	59.2
59.2	157	93	64	55.1
59.2	152	90	62	53.3
59.2	147	87	60	51.5
59.2	142	84	58	49.7
59.2	130	77	53	45.6
59.2	125	74	51	43.8
59.2	120	71	49	42.0
59.2	103	61	42	36.1
59.2	98	58	40	34.3
59.2	76	45	31	26.6
59.2	71	42	29	24.8
59.2	49	29	20	17.2
59.1	499	295	204	174.4
59.1	496	293	203	173.1
59.1	494	292	202	172.6
59.1	492	291	201	172.1
59.1	491	290	201	171.3
59.1	489	289	200	170.8
59.1	487	288	199	170.3
59.1	486	287	199	169.5
59.1	484	286	198	169.0
59.1	482	285	197	168.5
59.1	479	283	196	167.2
59.1	477	282	195	166.7
59.1	474	280	194	165.4
59.1	472	279	193	164.9
59.1	470	278	192	164.4
59.1	469	277	192	163.6
59.1	467	276	191	163.1
59.1	465	275	190	162.6
59.1	464	274	190	161.8
59.1	462	273	189	161.3
59.1	460	272	188	160.8
59.1	457	270	187	159.5
59.1	455	269	186	159.0
59.1	452	267	185	157.7
59.1	450	266	184	157.2
59.1	447	264	183	155.9
59.1	445	263	182	155.4
59.1	443	262	181	155.0
59.1	442	261	181	154.1
59.1	440	260	180	153.6
59.1	438	259	179	153.2
59.1	435	257	178	151.8
59.1	433	256	177	151.4
59.1	430	254	176	150.0
59.1	428	253	175	149.6

%E	M1	M2	DM	M*
59.1	425	251	174	148.2
59.1	423	250	173	147.8
59.1	421	249	172	147.3
59.1	418	247	171	146.0
59.1	416	246	170	145.5
59.1	413	244	169	144.2
59.1	411	243	168	143.7
59.1	408	241	167	142.4
59.1	406	240	166	141.9
59.1	403	238	165	140.6
59.1	401	237	164	140.1
59.1	399	236	163	139.6
59.1	396	234	162	138.3
59.1	394	233	161	137.8
59.1	391	231	160	136.5
59.1	389	230	159	136.0
59.1	386	228	158	134.7
59.1	384	227	157	134.2
59.1	381	225	156	132.9
59.1	379	224	155	132.4
59.1	374	221	153	130.6
59.1	372	220	152	130.1
59.1	369	218	151	128.8
59.1	367	217	150	128.3
59.1	364	215	149	127.0
59.1	362	214	148	126.5
59.1	359	212	147	125.2
59.1	357	211	146	124.7
59.1	352	208	144	122.9
59.1	350	207	143	122.4
59.1	347	205	142	121.1
59.1	345	204	141	120.6
59.1	342	202	140	119.3
59.1	340	201	139	118.8
59.1	337	199	138	117.5
59.1	335	198	137	117.0
59.1	330	195	135	115.2
59.1	328	194	134	114.7
59.1	325	192	133	113.4
59.1	323	191	132	112.9
59.1	320	189	131	111.6
59.1	318	188	130	111.1
59.1	313	185	128	109.3
59.1	308	182	126	107.5
59.1	303	179	124	105.7
59.1	301	178	123	105.3
59.1	298	176	122	103.9
59.1	296	175	121	103.5
59.1	291	172	119	101.7
59.1	286	169	117	99.9

%E	M1	M2	DM	M*	%E	M1	M2	DM	M*	%E	M1	M2	DM	M*	%E	M1	M2	DM	M*	%E	M1	M2	DM	M*
59.1	281	166	115	98.1	59.0	481	284	197	167.7	59.0	363	214	149	126.2	59.0	178	105	73	61.9	58.9	426	251	175	147.9
59.1	279	165	114	97.6	59.0	480	283	197	166.9	59.0	361	213	148	125.7	59.0	173	102	71	60.1	58.9	423	249	174	146.6
59.1	276	163	113	96.3	59.0	478	282	196	166.4	59.0	356	210	146	123.9	59.0	166	98	68	57.9	58.9	421	248	173	146.1
59.1	274	162	112	95.8	59.0	476	281	195	165.9	59.0	354	209	145	123.4	59.0	161	95	66	56.1	58.9	418	246	172	144.8
59.1	269	159	110	94.0	59.0	473	279	194	164.6	59.0	351	207	144	122.1	59.0	156	92	64	54.3	58.9	416	245	171	144.3
59.1	264	156	108	92.2	59.0	471	278	193	164.1	59.0	349	206	143	121.6	59.0	144	85	59	50.2	58.9	414	244	170	143.8
59.1	259	153	106	90.4	59.0	468	276	192	162.8	59.0	346	204	142	120.3	59.0	139	82	57	48.4	58.9	411	242	169	142.5
59.1	257	152	105	89.9	59.0	466	275	191	162.3	59.0	344	203	141	119.8	59.0	134	79	55	46.6	58.9	409	241	168	142.0
59.1	254	150	104	88.6	59.0	463	273	190	161.0	59.0	339	200	139	118.0	59.0	122	72	50	42.5	58.9	406	239	167	140.7
59.1	252	149	103	88.1	59.0	461	272	189	160.5	59.0	334	197	137	116.2	59.0	117	69	48	40.7	58.9	404	238	166	140.2
59.1	247	146	101	86.3	59.0	459	271	188	160.0	59.0	332	196	136	115.7	59.0	105	62	43	36.6	58.9	401	236	165	138.9
59.1	242	143	99	84.5	59.0	458	270	188	159.2	59.0	329	194	135	114.4	59.0	100	59	41	34.8	58.9	399	235	164	138.4
59.1	237	140	97	82.7	59.0	456	269	187	158.7	59.0	327	193	134	113.9	59.0	83	49	34	28.9	58.9	397	234	163	137.9
59.1	235	139	96	82.2	59.0	454	268	186	158.2	59.0	324	191	133	112.6	59.0	78	46	32	27.1	58.9	394	232	162	136.6
59.1	232	137	95	80.9	59.0	451	266	185	156.9	59.0	322	190	132	112.1	59.0	61	36	25	21.2	58.9	392	231	161	136.1
59.1	230	136	94	80.4	59.0	449	265	184	156.4	59.0	317	187	130	110.3	59.0	39	23	16	13.6	58.9	389	229	160	134.8
59.1	225	133	92	78.6	59.0	446	263	183	155.1	59.0	315	186	129	109.8	58.9	499	294	205	173.2	58.9	387	228	159	134.3
59.1	220	130	90	76.8	59.0	444	262	182	154.6	59.0	312	184	128	108.5	58.9	496	292	204	171.9	58.9	384	226	158	133.0
59.1	215	127	88	75.0	59.0	441	260	181	153.3	59.0	310	183	127	108.0	58.9	494	291	203	171.4	58.9	382	225	157	132.5
59.1	208	123	85	72.7	59.0	439	259	180	152.8	59.0	307	181	126	106.7	58.9	492	290	202	170.9	58.9	380	224	156	132.0
59.1	203	120	83	70.9	59.0	437	258	179	152.3	59.0	305	180	125	106.2	58.9	491	289	202	170.1	58.9	377	222	155	130.7
59.1	198	117	81	69.1	59.0	434	256	178	151.0	59.0	300	177	123	104.4	58.9	489	288	201	169.6	58.9	375	221	154	130.2
59.1	193	114	79	67.3	59.0	432	255	177	150.5	59.0	295	174	121	102.6	58.9	487	287	200	169.1	58.9	372	219	153	128.9
59.1	186	110	76	65.1	59.0	429	253	176	149.2	59.0	293	173	120	102.1	58.9	484	285	199	167.8	58.9	370	218	152	128.4
59.1	181	107	74	63.3	59.0	427	252	175	148.7	59.0	290	171	119	100.8	58.9	482	284	198	167.3	58.9	367	216	151	127.1
59.1	176	104	72	61.5	59.0	424	250	174	147.4	59.0	288	170	118	100.3	58.9	479	282	197	166.0	58.9	365	215	150	126.6
59.1	171	101	70	59.7	59.0	422	249	173	146.9	59.0	283	167	116	98.5	58.9	477	281	196	165.5	58.9	360	212	148	124.8
59.1	164	97	67	57.4	59.0	420	248	172	146.4	59.0	278	164	114	96.7	58.9	475	280	195	165.1	58.9	358	211	147	124.4
59.1	159	94	65	55.6	59.0	419	247	172	145.6	59.0	273	161	112	94.9	58.9	474	279	195	164.2	58.9	355	209	146	123.0
59.1	154	91	63	53.8	59.0	417	246	171	145.1	59.0	271	160	111	94.5	58.9	472	278	194	163.7	58.9	353	208	145	122.6
59.1	149	88	61	52.0	59.0	415	245	170	144.6	59.0	268	158	110	93.1	58.9	470	277	193	163.3	58.9	350	206	144	121.2
59.1	137	81	56	47.9	59.0	412	243	169	143.3	59.0	266	157	109	92.7	58.9	467	275	192	161.9	58.9	348	205	143	120.8
59.1	132	78	54	46.1	59.0	410	242	168	142.8	59.0	261	154	107	90.9	58.9	465	274	191	161.5	58.9	343	202	141	119.0
59.1	127	75	52	44.3	59.0	407	240	167	141.5	59.0	256	151	105	89.1	58.9	462	272	190	160.1	58.9	341	201	140	118.5
59.1	115	68	47	40.2	59.0	405	239	166	141.0	59.0	251	148	103	87.3	58.9	460	271	189	159.7	58.9	338	199	139	117.2
59.1	110	65	45	38.4	59.0	402	237	165	139.7	59.0	249	147	102	86.8	58.9	457	269	188	158.3	58.9	336	198	138	116.7
59.1	93	55	38	32.5	59.0	400	236	164	139.2	59.0	244	144	100	85.0	58.9	455	268	187	157.9	58.9	333	196	137	115.4
59.1	88	52	36	30.7	59.0	398	235	163	138.8	59.0	239	141	98	83.2	58.9	453	267	186	157.4	58.9	331	195	136	114.9
59.1	66	39	27	23.0	59.0	395	233	162	137.4	59.0	234	138	96	81.4	58.9	452	266	186	156.5	58.9	326	192	134	113.1
59.1	44	26	18	15.4	59.0	393	232	161	137.0	59.0	229	135	94	79.6	58.9	450	265	185	156.1	58.9	321	189	132	111.3
59.1	22	13	9	7.7	59.0	390	230	160	135.6	59.0	227	134	93	79.1	58.9	448	264	184	155.6	58.9	319	188	131	110.8
59.0	500	295	205	174.0	59.0	388	229	159	135.2	59.0	222	131	91	77.3	58.9	445	262	183	154.3	58.9	316	186	130	109.5
59.0	498	294	204	173.6	59.0	385	227	158	133.8	59.0	217	128	89	75.5	58.9	443	261	182	153.8	58.9	314	185	129	109.0
59.0	497	293	204	172.7	59.0	383	226	157	133.4	59.0	212	125	87	73.7	58.9	440	259	181	152.5	58.9	309	182	127	107.2
59.0	495	292	203	172.3	59.0	378	223	155	131.6	59.0	210	124	86	73.2	58.9	438	258	180	152.0	58.9	304	179	125	105.4
59.0	493	291	202	171.8	59.0	376	222	154	131.1	59.0	205	121	84	71.4	58.9	436	257	179	151.5	58.9	302	178	124	104.9
59.0	490	289	201	170.5	59.0	373	220	153	129.8	59.0	200	118	82	69.6	58.9	435	256	179	150.7	58.9	299	176	123	103.6
59.0	488	288	200	170.0	59.0	371	219	152	129.3	59.0	195	115	80	67.8	58.9	433	255	178	150.2	58.9	297	175	122	103.1
59.0	485	286	199	168.7	59.0	368	217	151	128.0	59.0	188	111	77	65.5	58.9	431	254	177	149.7	58.9	292	172	120	101.3
59.0	483	285	198	168.2	59.0	366	216	150	127.5	59.0	183	108	75	63.7	58.9	428	252	176	148.4	58.9	287	169	118	99.5

%E	M1	M2	DM	M*
58.9	285	168	117	99.0
58.9	282	166	116	97.7
58.9	280	165	115	97.2
58.9	275	162	113	95.4
58.9	270	159	111	93.6
58.9	265	156	109	91.8
58.9	263	155	108	91.3
58.9	258	152	106	89.6
58.9	253	149	104	87.8
58.9	248	146	102	86.0
58.9	246	145	101	85.5
58.9	241	142	99	83.7
58.9	236	139	97	81.9
58.9	231	136	95	80.1
58.9	226	133	93	78.3
58.9	224	132	92	77.8
58.9	219	129	90	76.0
58.9	214	126	88	74.2
58.9	209	123	86	72.4
58.9	207	122	85	71.9
58.9	202	119	83	70.1
58.9	197	116	81	68.3
58.9	192	113	79	66.5
58.9	190	112	78	66.0
58.9	185	109	76	64.2
58.9	180	106	74	62.4
58.9	175	103	72	60.6
58.9	168	99	69	58.3
58.9	163	96	67	56.5
58.9	158	93	65	54.7
58.9	151	89	62	52.5
58.9	146	86	60	50.7
58.9	141	83	58	48.9
58.9	129	76	53	44.8
58.9	124	73	51	43.0
58.9	112	66	46	38.9
58.9	107	63	44	37.1
58.9	95	56	39	33.0
58.9	90	53	37	31.2
58.9	73	43	30	25.3
58.9	56	33	23	19.4
58.8	500	294	206	172.9
58.8	498	293	205	172.4
58.8	497	292	205	171.6
58.8	495	291	204	171.1
58.8	493	290	203	170.6
58.8	490	288	202	169.3
58.8	488	287	201	168.8
58.8	486	286	200	168.3
58.8	485	285	200	167.5

%E	M1	M2	DM	M*
58.8	483	284	199	167.0
58.8	481	283	198	166.5
58.8	480	282	198	165.7
58.8	478	281	197	165.2
58.8	476	280	196	164.7
58.8	473	278	195	163.4
58.8	471	277	194	162.9
58.8	469	276	193	162.4
58.8	468	275	193	161.6
58.8	466	274	192	161.1
58.8	464	273	191	160.6
58.8	461	271	190	159.3
58.8	459	270	189	158.8
58.8	456	268	188	157.5
58.8	454	267	187	157.0
58.8	451	265	186	155.7
58.8	449	264	185	155.2
58.8	447	263	184	154.7
58.8	444	261	183	153.4
58.8	442	260	182	152.9
58.8	439	258	181	151.6
58.8	437	257	180	151.1
58.8	434	255	179	149.8
58.8	432	254	178	149.3
58.8	430	253	177	148.9
58.8	427	251	176	147.5
58.8	425	250	175	147.1
58.8	422	248	174	145.7
58.8	420	247	173	145.3
58.8	417	245	172	143.9
58.8	415	244	171	143.5
58.8	413	243	170	143.0
58.8	410	241	169	141.7
58.8	408	240	168	141.2
58.8	405	238	167	139.9
58.8	403	237	166	139.4
58.8	400	235	165	138.1
58.8	398	234	164	137.6
58.8	396	233	163	137.1
58.8	393	231	162	135.8
58.8	391	230	161	135.3
58.8	388	228	160	134.0
58.8	386	227	159	133.5
58.8	381	224	157	131.7
58.8	379	223	156	131.2
58.8	376	221	155	129.9
58.8	374	220	154	129.4
58.8	371	218	153	128.1
58.8	369	217	152	127.6
58.8	364	214	150	125.8

%E	M1	M2	DM	M*
58.8	362	213	149	125.3
58.8	359	211	148	124.0
58.8	357	210	147	123.5
58.8	354	208	146	122.2
58.8	352	207	145	121.7
58.8	347	204	143	119.9
58.8	345	203	142	119.4
58.8	342	201	141	118.1
58.8	340	200	140	117.6
58.8	337	198	139	116.3
58.8	335	197	138	115.8
58.8	330	194	136	114.0
58.8	328	193	135	113.6
58.8	325	191	134	112.2
58.8	323	190	133	111.8
58.8	320	188	132	110.4
58.8	318	187	131	110.0
58.8	313	184	129	108.2
58.8	311	183	128	107.7
58.8	308	181	127	106.4
58.8	306	180	126	105.9
58.8	301	177	124	104.1
58.8	296	174	122	102.3
58.8	294	173	121	101.8
58.8	291	171	120	100.5
58.8	289	170	119	100.0
58.8	284	167	117	98.2
58.8	279	164	115	96.4
58.8	277	163	114	95.9
58.8	274	161	113	94.6
58.8	272	160	112	94.1
58.8	267	157	110	92.3
58.8	262	154	108	90.5
58.8	260	153	107	90.0
58.8	257	151	106	88.7
58.8	255	150	105	88.2
58.8	250	147	103	86.4
58.8	245	144	101	84.6
58.8	243	143	100	84.2
58.8	240	141	99	82.8
58.8	238	140	98	82.4
58.8	233	137	96	80.6
58.8	228	134	94	78.8
58.8	221	130	91	76.5
58.8	216	127	89	74.7
58.8	211	124	87	72.9
58.8	204	120	84	70.6
58.8	199	117	82	68.8
58.8	194	114	80	67.0
58.8	187	110	77	64.7

%E	M1	M2	DM	M*
58.8	182	107	75	62.9
58.8	177	104	73	61.1
58.8	170	100	70	58.8
58.8	165	97	68	57.0
58.8	160	94	66	55.2
58.8	153	90	63	52.9
58.8	148	87	61	51.1
58.8	136	80	56	47.1
58.8	131	77	54	45.3
58.8	119	70	49	41.2
58.8	114	67	47	39.4
58.8	102	60	42	35.3
58.8	97	57	40	33.5
58.8	85	50	35	29.4
58.8	80	47	33	27.6
58.8	68	40	28	23.5
58.8	51	30	21	17.6
58.8	34	20	14	11.8
58.8	17	10	7	5.9
58.7	499	293	206	172.0
58.7	496	291	205	170.7
58.7	494	290	204	170.2
58.7	492	289	203	169.8
58.7	491	288	203	168.9
58.7	489	287	202	168.4
58.7	487	286	201	168.0
58.7	484	284	200	166.6
58.7	482	283	199	166.2
58.7	479	281	198	164.8
58.7	477	280	197	164.4
58.7	475	279	196	163.9
58.7	474	278	196	163.0
58.7	472	277	195	162.6
58.7	470	276	194	162.1
58.7	467	274	193	160.8
58.7	465	273	192	160.3
58.7	463	272	191	159.8
58.7	462	271	191	159.0
58.7	460	270	190	158.5
58.7	458	269	189	158.0
58.7	455	267	188	156.7
58.7	453	266	187	156.2
58.7	450	264	186	154.9
58.7	448	263	185	154.4
58.7	446	262	184	153.9
58.7	445	261	184	153.1
58.7	443	260	183	152.6
58.7	441	259	182	152.1
58.7	438	257	181	150.8
58.7	436	256	180	150.3

%E	M1	M2	DM	M*
58.7	433	254	179	149.0
58.7	431	253	178	148.5
58.7	429	252	177	148.0
58.7	426	250	176	146.7
58.7	424	249	175	146.2
58.7	421	247	174	144.9
58.7	419	246	173	144.4
58.7	416	244	172	143.1
58.7	414	243	171	142.6
58.7	412	242	170	142.1
58.7	409	240	169	140.8
58.7	407	239	168	140.3
58.7	404	237	167	139.0
58.7	402	236	166	138.5
58.7	397	233	164	136.7
58.7	395	232	163	136.3
58.7	392	230	162	134.9
58.7	390	229	161	134.5
58.7	387	227	160	133.1
58.7	385	226	159	132.7
58.7	383	225	158	132.2
58.7	380	223	157	130.9
58.7	378	222	156	130.4
58.7	375	220	155	129.1
58.7	373	219	154	128.6
58.7	368	216	152	126.8
58.7	366	215	151	126.3
58.7	363	213	150	125.0
58.7	361	212	149	124.5
58.7	358	210	148	123.2
58.7	356	209	147	122.7
58.7	351	206	145	120.9
58.7	349	205	144	120.4
58.7	346	203	143	119.1
58.7	344	202	142	118.6
58.7	341	200	141	117.3
58.7	339	199	140	116.8
58.7	334	196	138	115.0
58.7	332	195	137	114.5
58.7	329	193	136	113.2
58.7	327	192	135	112.7
58.7	322	189	133	110.9
58.7	317	186	131	109.1
58.7	315	185	130	108.7
58.7	312	183	129	107.3
58.7	310	182	128	106.9
58.7	305	179	126	105.1
58.7	303	178	125	104.6
58.7	300	176	124	103.3
58.7	298	175	123	102.8

%E	M1	M2	DM	M*
58.7	293	172	121	101.0
58.7	288	169	119	99.2
58.7	286	168	118	98.7
58.7	283	166	117	97.4
58.7	281	165	116	96.9
58.7	276	162	114	95.1
58.7	271	159	112	93.3
58.7	269	158	111	92.8
58.7	264	155	109	91.0
58.7	259	152	107	89.2
58.7	254	149	105	87.4
58.7	252	148	104	86.9
58.7	247	145	102	85.1
58.7	242	142	100	83.3
58.7	237	139	98	81.5
58.7	235	138	97	81.0
58.7	230	135	95	79.2
58.7	225	132	93	77.4
58.7	223	131	92	77.0
58.7	218	128	90	75.2
58.7	213	125	88	73.4
58.7	208	122	86	71.6
58.7	206	121	85	71.1
58.7	201	118	83	69.3
58.7	196	115	81	67.5
58.7	189	111	78	65.2
58.7	184	108	76	63.4
58.7	179	105	74	61.6
58.7	172	101	71	59.3
58.7	167	98	69	57.5
58.7	155	91	64	53.4
58.7	150	88	62	51.6
58.7	143	84	59	49.3
58.7	138	81	57	47.5
58.7	126	74	52	43.5
58.7	121	71	50	41.7
58.7	109	64	45	37.6
58.7	104	61	43	35.8
58.7	92	54	38	31.7
58.7	75	44	31	25.8
58.7	63	37	26	21.7
58.7	46	27	19	15.8
58.6	500	293	207	171.7
58.6	498	292	206	171.2
58.6	497	291	206	170.4
58.6	495	290	205	169.9
58.6	493	289	204	169.4
58.6	490	287	203	168.1
58.6	488	286	202	167.6
58.6	486	285	201	167.1
58.6	485	284	201	166.3
58.6	483	283	200	165.8
58.6	481	282	199	165.3
58.6	478	280	198	164.0
58.6	476	279	197	163.5
58.6	473	277	196	162.2
58.6	471	276	195	161.7
58.6	469	275	194	161.2
58.6	466	273	193	159.9
58.6	464	272	192	159.4
58.6	461	270	191	158.1
58.6	459	269	190	157.6
58.6	457	268	189	157.2
58.6	456	267	189	156.3
58.6	454	266	188	155.9
58.6	452	265	187	155.4
58.6	449	263	186	154.1
58.6	447	262	185	153.6
58.6	444	260	184	152.3
58.6	442	259	183	151.8
58.6	440	258	182	151.3
58.6	437	256	181	150.0
58.6	435	255	180	149.5
58.6	432	253	179	148.2
58.6	430	252	178	147.7
58.6	428	251	177	147.2
58.6	425	249	176	145.9
58.6	423	248	175	145.4
58.6	420	246	174	144.1
58.6	418	245	173	143.6
58.6	415	243	172	142.3
58.6	413	242	171	141.8
58.6	411	241	170	141.3
58.6	408	239	169	140.0
58.6	406	238	168	139.5
58.6	403	236	167	138.2
58.6	401	235	166	137.7
58.6	399	234	165	137.2
58.6	396	232	164	135.9
58.6	394	231	163	135.4
58.6	391	229	162	134.1
58.6	389	228	161	133.6
58.6	384	225	159	131.8
58.6	382	224	158	131.4
58.6	379	222	157	130.0
58.6	377	221	156	129.6
58.6	374	219	155	128.2
58.6	372	218	154	127.8
58.6	370	217	153	127.3
58.6	367	215	152	126.0
58.6	365	214	151	125.5
58.6	362	212	150	124.2
58.6	360	211	149	123.7
58.6	355	208	147	121.9
58.6	353	207	146	121.4
58.6	350	205	145	120.1
58.6	348	204	144	119.6
58.6	345	202	143	118.3
58.6	343	201	142	117.8
58.6	338	198	140	116.0
58.6	336	197	139	115.5
58.6	333	195	138	114.2
58.6	331	194	137	113.7
58.6	326	191	135	111.9
58.6	324	190	134	111.4
58.6	321	188	133	110.1
58.6	319	187	132	109.6
58.6	314	184	130	107.8
58.6	309	181	128	106.0
58.6	307	180	127	105.5
58.6	304	178	126	104.2
58.6	302	177	125	103.7
58.6	297	174	123	101.9
58.6	295	173	122	101.5
58.6	292	171	121	100.1
58.6	290	170	120	99.7
58.6	285	167	118	97.9
58.6	280	164	116	96.1
58.6	278	163	115	95.6
58.6	273	160	113	93.8
58.6	268	157	111	92.0
58.6	266	156	110	91.5
58.6	263	154	109	90.2
58.6	261	153	108	89.7
58.6	256	150	106	87.9
58.6	251	147	104	86.1
58.6	249	146	103	85.6
58.6	244	143	101	83.8
58.6	239	140	99	82.0
58.6	232	136	96	79.7
58.6	227	133	94	77.9
58.6	222	130	92	76.1
58.6	220	129	91	75.6
58.6	215	126	89	73.8
58.6	210	123	87	72.0
58.6	203	119	84	69.8
58.6	198	116	82	68.0
58.6	191	112	79	65.7
58.6	186	109	77	63.9
58.6	181	106	75	62.1
58.6	174	102	72	59.8
58.6	169	99	70	58.0
58.6	162	95	67	55.7
58.6	157	92	65	53.9
58.6	152	89	63	52.1
58.6	145	85	60	49.8
58.6	140	82	58	48.0
58.6	133	78	55	45.7
58.6	128	75	53	43.9
58.6	116	68	48	39.9
58.6	111	65	46	38.1
58.6	99	58	41	34.0
58.6	87	51	36	29.9
58.6	70	41	29	24.0
58.6	58	34	24	19.9
58.6	29	17	12	10.0
58.5	499	292	207	170.9
58.5	496	290	206	169.6
58.5	494	289	205	169.1
58.5	492	288	204	168.6
58.5	491	287	204	167.8
58.5	489	286	203	167.3
58.5	487	285	202	166.8
58.5	484	283	201	165.5
58.5	482	282	200	165.0
58.5	480	281	199	164.5
58.5	479	280	199	163.7
58.5	477	279	198	163.2
58.5	475	278	197	162.7
58.5	472	276	196	161.4
58.5	470	275	195	160.9
58.5	468	274	194	160.4
58.5	467	273	194	159.6
58.5	465	272	193	159.1
58.5	463	271	192	158.6
58.5	460	269	191	157.3
58.5	458	268	190	156.8
58.5	455	266	189	155.5
58.5	453	265	188	155.0
58.5	451	264	187	154.5
58.5	448	262	186	153.2
58.5	446	261	185	152.7
58.5	443	259	184	151.4
58.5	441	258	183	150.9
58.5	439	257	182	150.5
58.5	436	255	181	149.1
58.5	434	254	180	148.7
58.5	431	252	179	147.3
58.5	429	251	178	146.9
58.5	427	250	177	146.4
58.5	426	249	177	145.5
58.5	424	248	176	145.1
58.5	422	247	175	144.6
58.5	419	245	174	143.3
58.5	417	244	173	142.8
58.5	414	242	172	141.5
58.5	412	241	171	141.0
58.5	410	240	170	140.5
58.5	407	238	169	139.2
58.5	405	237	168	138.7
58.5	402	235	167	137.4
58.5	400	234	166	136.9
58.5	398	233	165	136.4
58.5	395	231	164	135.1
58.5	393	230	163	134.6
58.5	390	228	162	133.3
58.5	388	227	161	132.8
58.5	386	226	160	132.3
58.5	383	224	159	131.0
58.5	381	223	158	130.5
58.5	378	221	157	129.2
58.5	376	220	156	128.7
58.5	371	217	154	126.9
58.5	369	216	153	126.4
58.5	366	214	152	125.1
58.5	364	213	151	124.6
58.5	359	210	149	122.8
58.5	357	209	148	122.4
58.5	354	207	147	121.0
58.5	352	206	146	120.6
58.5	349	204	145	119.2
58.5	347	203	144	118.8
58.5	342	200	142	117.0
58.5	340	199	141	116.5
58.5	337	197	140	115.2
58.5	335	196	139	114.7
58.5	330	193	137	112.9
58.5	328	192	136	112.4
58.5	325	190	135	111.1
58.5	323	189	134	110.6
58.5	318	186	132	108.8
58.5	316	185	131	108.3
58.5	313	183	130	107.0
58.5	311	182	129	106.5
58.5	306	179	127	104.7
58.5	301	176	125	102.9
58.5	299	175	124	102.4
58.5	294	172	122	100.6
58.5	289	169	120	98.8
58.5	287	168	119	98.3

%E	M1	M2	DM	M*
58.5	284	166	118	97.0
58.5	282	165	117	96.5
58.5	277	162	115	94.7
58.5	275	161	114	94.3
58.5	272	159	113	92.9
58.5	270	158	112	92.5
58.5	265	155	110	90.7
58.5	260	152	108	88.9
58.5	258	151	107	88.4
58.5	253	148	105	86.6
58.5	248	145	103	84.8
58.5	246	144	102	84.3
58.5	241	141	100	82.5
58.5	236	138	98	80.7
58.5	234	137	97	80.2
58.5	229	134	95	78.4
58.5	224	131	93	76.6
58.5	217	127	90	74.3
58.5	212	124	88	72.5
58.5	207	121	86	70.7
58.5	205	120	85	70.2
58.5	200	117	83	68.4
58.5	195	114	81	66.6
58.5	193	113	80	66.2
58.5	188	110	78	64.4
58.5	183	107	76	62.6
58.5	176	103	73	60.3
58.5	171	100	71	58.5
58.5	164	96	68	56.2
58.5	159	93	66	54.4
58.5	147	86	61	50.3
58.5	142	83	59	48.5
58.5	135	79	56	46.2
58.5	130	76	54	44.4
58.5	123	72	51	42.1
58.5	118	69	49	40.3
58.5	106	62	44	36.3
58.5	94	55	39	32.2
58.5	82	48	34	28.1
58.5	65	38	27	22.2
58.5	53	31	22	18.1
58.5	41	24	17	14.0
58.4	500	292	208	170.5
58.4	498	291	207	170.0
58.4	497	290	207	169.2
58.4	495	289	206	168.7
58.4	493	288	205	168.2
58.4	490	286	204	166.9
58.4	488	285	203	166.4
58.4	486	284	202	166.0

%E	M1	M2	DM	M*
58.4	485	283	202	165.1
58.4	483	282	201	164.6
58.4	481	281	200	164.2
58.4	478	279	199	162.8
58.4	476	278	198	162.4
58.4	474	277	197	161.9
58.4	473	276	197	161.0
58.4	471	275	196	160.6
58.4	469	274	195	160.1
58.4	466	272	194	158.8
58.4	464	271	193	158.3
58.4	462	270	192	157.8
58.4	461	269	192	157.0
58.4	459	268	191	156.5
58.4	457	267	190	156.0
58.4	454	265	189	154.7
58.4	452	264	188	154.2
58.4	450	263	187	153.7
58.4	449	262	187	152.9
58.4	447	261	186	152.4
58.4	445	260	185	151.9
58.4	442	258	184	150.6
58.4	440	257	183	150.1
58.4	438	256	182	149.6
58.4	437	255	182	148.8
58.4	435	254	181	148.3
58.4	433	253	180	147.8
58.4	430	251	179	146.5
58.4	428	250	178	146.0
58.4	425	248	177	144.7
58.4	423	247	176	144.2
58.4	421	246	175	143.7
58.4	418	244	174	142.4
58.4	416	243	173	141.9
58.4	413	241	172	140.6
58.4	411	240	171	140.1
58.4	409	239	170	139.7
58.4	406	237	169	138.3
58.4	404	236	168	137.9
58.4	401	234	167	136.5
58.4	399	233	166	136.1
58.4	397	232	165	135.6
58.4	394	230	164	134.3
58.4	392	229	163	133.8
58.4	389	227	162	132.5
58.4	387	226	161	132.0
58.4	385	225	160	131.5
58.4	382	223	159	130.2
58.4	380	222	158	129.7
58.4	377	220	157	128.4

%E	M1	M2	DM	M*
58.4	375	219	156	127.9
58.4	373	218	155	127.4
58.4	370	216	154	126.1
58.4	368	215	153	125.6
58.4	365	213	152	124.3
58.4	363	212	151	123.8
58.4	361	211	150	123.3
58.4	358	209	149	122.0
58.4	356	208	148	121.5
58.4	353	206	147	120.2
58.4	351	205	146	119.7
58.4	346	202	144	117.9
58.4	344	201	143	117.4
58.4	341	199	142	116.1
58.4	339	198	141	115.6
58.4	334	195	139	113.8
58.4	332	194	138	113.4
58.4	329	192	137	112.0
58.4	327	191	136	111.6
58.4	322	188	134	109.8
58.4	320	187	133	109.3
58.4	317	185	132	108.0
58.4	315	184	131	107.5
58.4	310	181	129	105.7
58.4	308	180	128	105.2
58.4	305	178	127	103.9
58.4	303	177	126	103.4
58.4	298	174	124	101.6
58.4	296	173	123	101.1
58.4	293	171	122	99.8
58.4	291	170	121	99.3
58.4	286	167	119	97.5
58.4	281	164	117	95.7
58.4	279	163	116	95.2
58.4	274	160	114	93.4
58.4	269	157	112	91.6
58.4	267	156	111	91.1
58.4	262	153	109	89.3
58.4	257	150	107	87.5
58.4	255	149	106	87.1
58.4	250	146	104	85.3
58.4	245	143	102	83.5
58.4	243	142	101	83.0
58.4	238	139	99	81.2
58.4	233	136	97	79.4
58.4	231	135	96	78.9
58.4	226	132	94	77.1
58.4	221	129	92	75.3
58.4	219	128	91	74.8
58.4	214	125	89	73.0

%E	M1	M2	DM	M*
58.4	209	122	87	71.2
58.4	202	118	84	68.9
58.4	197	115	82	67.1
58.4	190	111	79	64.8
58.4	185	108	77	63.0
58.4	178	104	74	60.8
58.4	173	101	72	59.0
58.4	166	97	69	56.7
58.4	161	94	67	54.9
58.4	154	90	64	52.6
58.4	149	87	62	50.8
58.4	137	80	57	46.7
58.4	125	73	52	42.6
58.4	113	66	47	38.5
58.4	101	59	42	34.5
58.4	89	52	37	30.4
58.4	77	45	32	26.3
58.3	499	291	208	169.7
58.3	496	289	207	168.4
58.3	494	288	206	167.9
58.3	492	287	205	167.4
58.3	489	285	204	166.1
58.3	487	284	203	165.6
58.3	484	282	202	164.3
58.3	482	281	201	163.8
58.3	480	280	200	163.3
58.3	477	278	199	162.0
58.3	475	277	198	161.5
58.3	472	275	197	160.2
58.3	470	274	196	159.7
58.3	468	273	195	159.3
58.3	465	271	194	157.9
58.3	463	270	193	157.5
58.3	460	268	192	156.1
58.3	458	267	191	155.7
58.3	456	266	190	155.2
58.3	453	264	189	153.9
58.3	451	263	188	153.4
58.3	448	261	187	152.1
58.3	446	260	186	151.6
58.3	444	259	185	151.1
58.3	441	257	184	149.8
58.3	439	256	183	149.3
58.3	436	254	182	148.0
58.3	434	253	181	147.5
58.3	432	252	180	147.0
58.3	429	250	179	145.7
58.3	427	249	178	145.2
58.3	424	247	177	143.9
58.3	422	246	176	143.4

%E	M1	M2	DM	M*
58.3	420	245	175	142.9
58.3	417	243	174	141.6
58.3	415	242	173	141.1
58.3	412	240	172	139.8
58.3	410	239	171	139.3
58.3	408	238	170	138.8
58.3	405	236	169	137.5
58.3	403	235	168	137.0
58.3	400	233	167	135.7
58.3	398	232	166	135.2
58.3	396	231	165	134.8
58.3	393	229	164	133.4
58.3	391	228	163	133.0
58.3	386	225	161	131.2
58.3	384	224	160	130.7
58.3	381	222	159	129.4
58.3	379	221	158	128.9
58.3	374	218	156	127.1
58.3	372	217	155	126.6
58.3	369	215	154	125.3
58.3	367	214	153	124.8
58.3	362	211	151	123.0
58.3	360	210	150	122.5
58.3	357	208	149	121.2
58.3	355	207	148	120.7
58.3	350	204	146	118.9
58.3	348	203	145	118.4
58.3	345	201	144	117.1
58.3	343	200	143	116.6
58.3	338	197	141	114.8
58.3	336	196	140	114.3
58.3	333	194	139	113.0
58.3	331	193	138	112.5
58.3	326	190	136	110.7
58.3	324	189	135	110.3
58.3	321	187	134	108.9
58.3	319	186	133	108.5
58.3	314	183	131	106.7
58.3	312	182	130	106.2
58.3	309	180	129	104.9
58.3	307	179	128	104.4
58.3	302	176	126	102.6
58.3	300	175	125	102.1
58.3	295	172	123	100.3
58.3	290	169	121	98.5
58.3	288	168	120	98.0
58.3	283	165	118	96.2
58.3	278	162	116	94.4
58.3	276	161	115	93.9
58.3	271	158	113	92.1

%E	M1	M2	DM	M*
58.3	266	155	111	90.3
58.3	264	154	110	89.8
58.3	259	151	108	88.0
58.3	254	148	106	86.2
58.3	252	147	105	85.8
58.3	247	144	103	84.0
58.3	242	141	101	82.2
58.3	240	140	100	81.7
58.3	235	137	98	79.9
58.3	230	134	96	78.1
58.3	228	133	95	77.6
58.3	223	130	93	75.8
58.3	218	127	91	74.0
58.3	216	126	90	73.5
58.3	211	123	88	71.7
58.3	206	120	86	69.9
58.3	204	119	85	69.4
58.3	199	116	83	67.6
58.3	192	112	80	65.3
58.3	187	109	78	63.5
58.3	180	105	75	61.3
58.3	175	102	73	59.5
58.3	168	98	70	57.2
58.3	163	95	68	55.4
58.3	156	91	65	53.1
58.3	151	88	63	51.3
58.3	144	84	60	49.0
58.3	139	81	58	47.2
58.3	132	77	55	44.9
58.3	127	74	53	43.1
58.3	120	70	50	40.8
58.3	115	67	48	39.0
58.3	108	63	45	36.8
58.3	103	60	43	35.0
58.3	96	56	40	32.7
58.3	84	49	35	28.6
58.3	72	42	30	24.5
58.3	60	35	25	20.4
58.3	48	28	20	16.3
58.3	36	21	15	12.3
58.3	24	14	10	8.2
58.3	12	7	5	4.1
58.2	500	291	209	169.4
58.2	498	290	208	168.9
58.2	495	288	207	167.6
58.2	493	287	206	167.1
58.2	491	286	205	166.6
58.2	490	285	205	165.8
58.2	488	284	204	165.3
58.2	486	283	203	164.8
58.2	483	281	202	163.5
58.2	481	280	201	163.0
58.2	479	279	200	162.5
58.2	478	278	200	161.7
58.2	476	277	199	161.2
58.2	474	276	198	160.7
58.2	471	274	197	159.4
58.2	469	273	196	158.9
58.2	467	272	195	158.4
58.2	466	271	195	157.6
58.2	464	270	194	157.1
58.2	462	269	193	156.6
58.2	459	267	192	155.3
58.2	457	266	191	154.8
58.2	455	265	190	154.3
58.2	454	264	190	153.5
58.2	452	263	189	153.0
58.2	450	262	188	152.5
58.2	447	260	187	151.2
58.2	445	259	186	150.7
58.2	443	258	185	150.3
58.2	440	256	184	148.9
58.2	438	255	183	148.5
58.2	435	253	182	147.1
58.2	433	252	181	146.7
58.2	431	251	180	146.2
58.2	428	249	179	144.9
58.2	426	248	178	144.4
58.2	423	246	177	143.1
58.2	421	245	176	142.6
58.2	419	244	175	142.1
58.2	416	242	174	140.8
58.2	414	241	173	140.3
58.2	411	239	172	139.0
58.2	409	238	171	138.5
58.2	407	237	170	138.0
58.2	404	235	169	136.7
58.2	402	234	168	136.2
58.2	397	231	166	134.4
58.2	395	230	165	133.9
58.2	392	228	164	132.6
58.2	390	227	163	132.1
58.2	388	226	162	131.6
58.2	385	224	161	130.3
58.2	383	223	160	129.8
58.2	380	221	159	128.5
58.2	378	220	158	128.0
58.2	376	219	157	127.6
58.2	373	217	156	126.2
58.2	371	216	155	125.8
58.2	368	214	154	124.4
58.2	366	213	153	124.0
58.2	364	212	152	123.5
58.2	361	210	151	122.2
58.2	359	209	150	121.7
58.2	354	206	148	119.9
58.2	352	205	147	119.4
58.2	349	203	146	118.1
58.2	347	202	145	117.6
58.2	342	199	143	115.8
58.2	340	198	142	115.3
58.2	337	196	141	114.0
58.2	335	195	140	113.5
58.2	330	192	138	111.7
58.2	328	191	137	111.2
58.2	325	189	136	109.9
58.2	323	188	135	109.4
58.2	318	185	133	107.6
58.2	316	184	132	107.1
58.2	311	181	130	105.3
58.2	306	178	128	103.5
58.2	304	177	127	103.1
58.2	299	174	125	101.3
58.2	297	173	124	100.8
58.2	294	171	123	99.5
58.2	292	170	122	99.0
58.2	287	167	120	97.2
58.2	285	166	119	96.7
58.2	282	164	118	95.4
58.2	280	163	117	94.9
58.2	275	160	115	93.1
58.2	273	159	114	92.6
58.2	268	156	112	90.8
58.2	263	153	110	89.0
58.2	261	152	109	88.5
58.2	256	149	107	86.7
58.2	251	146	105	84.9
58.2	249	145	104	84.4
58.2	244	142	102	82.6
58.2	239	139	100	80.8
58.2	237	138	99	80.4
58.2	232	135	97	78.6
58.2	227	132	95	76.8
58.2	225	131	94	76.3
58.2	220	128	92	74.5
58.2	213	124	89	72.2
58.2	208	121	87	70.4
58.2	201	117	84	68.1
58.2	196	114	82	66.3
58.2	194	113	81	65.8
58.2	189	110	79	64.0
58.2	184	107	77	62.2
58.2	182	106	76	61.7
58.2	177	103	74	59.9
58.2	170	99	71	57.7
58.2	165	96	69	55.9
58.2	158	92	66	53.6
58.2	153	89	64	51.8
58.2	146	85	61	49.5
58.2	141	82	59	47.7
58.2	134	78	56	45.4
58.2	122	71	51	41.3
58.2	110	64	46	37.2
58.2	98	57	41	33.2
58.2	91	53	38	30.9
58.2	79	46	33	26.8
58.2	67	39	28	22.7
58.2	55	32	23	18.6
58.1	499	290	209	168.5
58.1	497	289	208	168.1
58.1	496	288	208	167.2
58.1	494	287	207	166.7
58.1	492	286	206	166.3
58.1	489	284	205	164.9
58.1	487	283	204	164.5
58.1	485	282	203	164.0
58.1	484	281	203	163.1
58.1	482	280	202	162.7
58.1	480	279	201	162.2
58.1	477	277	200	160.9
58.1	475	276	199	160.4
58.1	473	275	198	159.9
58.1	472	274	198	159.1
58.1	470	273	197	158.6
58.1	468	272	196	158.1
58.1	465	270	195	156.8
58.1	463	269	194	156.3
58.1	461	268	193	155.8
58.1	458	266	192	154.5
58.1	456	265	191	154.0
58.1	453	263	190	152.7
58.1	451	262	189	152.2
58.1	449	261	188	151.7
58.1	446	259	187	150.4
58.1	444	258	186	149.9
58.1	442	257	185	149.4
58.1	441	256	185	148.6
58.1	439	255	184	148.1
58.1	437	254	183	147.6
58.1	434	252	182	146.3
58.1	432	251	181	145.8
58.1	430	250	180	145.3
58.1	427	248	179	144.0
58.1	425	247	178	143.6
58.1	422	245	177	142.2
58.1	420	244	176	141.8
58.1	418	243	175	141.3
58.1	415	241	174	140.0
58.1	413	240	173	139.5
58.1	408	237	171	137.7
58.1	406	236	170	137.2
58.1	403	234	169	135.9
58.1	401	233	168	135.4
58.1	399	232	167	134.9
58.1	396	230	166	133.6
58.1	394	229	165	133.1
58.1	391	227	164	131.8
58.1	389	226	163	131.3
58.1	387	225	162	130.8
58.1	384	223	161	129.5
58.1	382	222	160	129.0
58.1	377	219	158	127.2
58.1	375	218	157	126.7
58.1	372	216	156	125.4
58.1	370	215	155	124.9
58.1	365	212	153	123.1
58.1	363	211	152	122.6
58.1	360	209	151	121.3
58.1	358	208	150	120.8
58.1	356	207	149	120.4
58.1	353	205	148	119.1
58.1	351	204	147	118.6
58.1	346	201	145	116.8
58.1	344	200	144	116.3
58.1	341	198	143	115.0
58.1	339	197	142	114.5
58.1	334	194	140	112.7
58.1	332	193	139	112.2
58.1	329	191	138	110.9
58.1	327	190	137	110.4
58.1	322	187	135	108.6
58.1	320	186	134	108.1
58.1	315	183	132	106.3
58.1	313	182	131	105.8
58.1	310	180	130	104.5
58.1	308	179	129	104.0
58.1	303	176	127	102.2
58.1	301	175	126	101.7
58.1	298	173	125	100.4
58.1	296	172	124	99.9

%E	M1	M2	DM	M*
58.1	291	169	122	98.1
58.1	289	168	121	97.7
58.1	284	165	119	95.9
58.1	279	162	117	94.1
58.1	277	161	116	93.6
58.1	272	158	114	91.8
58.1	270	157	113	91.3
58.1	267	155	112	90.0
58.1	265	154	111	89.5
58.1	260	151	109	87.7
58.1	258	150	108	87.2
58.1	253	147	106	85.4
58.1	248	144	104	83.6
58.1	246	143	103	83.1
58.1	241	140	101	81.3
58.1	236	137	99	79.5
58.1	234	136	98	79.0
58.1	229	133	96	77.2
58.1	222	129	93	75.0
58.1	217	126	91	73.2
58.1	215	125	90	72.7
58.1	210	122	88	70.9
58.1	203	118	85	68.6
58.1	198	115	83	66.8
58.1	191	111	80	64.5
58.1	186	108	78	62.7
58.1	179	104	75	60.4
58.1	172	100	72	58.1
58.1	167	97	70	56.3
58.1	160	93	67	54.1
58.1	155	90	65	52.3
58.1	148	86	62	50.0
58.1	136	79	57	45.9
58.1	129	75	54	43.6
58.1	124	72	52	41.8
58.1	117	68	49	39.5
58.1	105	61	44	35.4
58.1	93	54	39	31.4
58.1	86	50	36	29.1
58.1	74	43	31	25.0
58.1	62	36	26	20.9
58.1	43	25	18	14.5
58.1	31	18	13	10.5
58.0	500	290	210	168.2
58.0	498	289	209	167.7
58.0	495	287	208	166.4
58.0	493	286	207	165.9
58.0	491	285	206	165.4
58.0	490	284	206	164.6
58.0	488	283	205	164.1

%E	M1	M2	DM	M*
58.0	486	282	204	163.6
58.0	483	280	203	162.3
58.0	481	279	202	161.8
58.0	479	278	201	161.3
58.0	478	277	201	160.5
58.0	476	276	200	160.0
58.0	474	275	199	159.5
58.0	471	273	198	158.2
58.0	469	272	197	157.7
58.0	467	271	196	157.3
58.0	464	269	195	156.0
58.0	462	268	194	155.5
58.0	460	267	193	155.0
58.0	459	266	193	154.2
58.0	457	265	192	153.7
58.0	455	264	191	153.2
58.0	452	262	190	151.9
58.0	450	261	189	151.4
58.0	448	260	188	150.9
58.0	445	258	187	149.6
58.0	443	257	186	149.1
58.0	440	255	185	147.8
58.0	438	254	184	147.3
58.0	436	253	183	146.8
58.0	433	251	182	145.5
58.0	431	250	181	145.0
58.0	429	249	180	144.5
58.0	426	247	179	143.2
58.0	424	246	178	142.7
58.0	421	244	177	141.4
58.0	419	243	176	140.9
58.0	417	242	175	140.4
58.0	414	240	174	139.1
58.0	412	239	173	138.6
58.0	410	238	172	138.2
58.0	407	236	171	136.8
58.0	405	235	170	136.4
58.0	402	233	169	135.0
58.0	400	232	168	134.6
58.0	398	231	167	134.1
58.0	395	229	166	132.8
58.0	393	228	165	132.3
58.0	388	225	163	130.5
58.0	386	224	162	130.0
58.0	383	222	161	128.7
58.0	381	221	160	128.2
58.0	379	220	159	127.7
58.0	376	218	158	126.4
58.0	374	217	157	125.9
58.0	371	215	156	124.6

%E	M1	M2	DM	M*
58.0	369	214	155	124.1
58.0	367	213	154	123.6
58.0	364	211	153	122.3
58.0	362	210	152	121.8
58.0	357	207	150	120.0
58.0	355	206	149	119.5
58.0	352	204	148	118.2
58.0	350	203	147	117.7
58.0	348	202	146	117.3
58.0	345	200	145	115.9
58.0	343	199	144	115.5
58.0	338	196	142	113.7
58.0	336	195	141	113.2
58.0	333	193	140	111.9
58.0	331	192	139	111.4
58.0	326	189	137	109.6
58.0	324	188	136	109.1
58.0	319	185	134	107.3
58.0	317	184	133	106.8
58.0	314	182	132	105.5
58.0	312	181	131	105.0
58.0	307	178	129	103.2
58.0	305	177	128	102.7
58.0	300	174	126	100.9
58.0	295	171	124	99.1
58.0	293	170	123	98.6
58.0	288	167	121	96.8
58.0	286	166	120	96.3
58.0	283	164	119	95.0
58.0	281	163	118	94.6
58.0	276	160	116	92.8
58.0	274	159	115	92.3
58.0	269	156	113	90.5
58.0	264	153	111	88.7
58.0	262	152	110	88.2
58.0	257	149	108	86.4
58.0	255	148	107	85.9
58.0	250	145	105	84.1
58.0	245	142	103	82.3
58.0	243	141	102	81.8
58.0	238	138	100	80.0
58.0	231	134	97	77.7
58.0	226	131	95	75.9
58.0	224	130	94	75.4
58.0	219	127	92	73.6
58.0	212	123	89	71.4
58.0	207	120	87	69.6
58.0	205	119	86	69.1
58.0	200	116	84	67.3
58.0	193	112	81	65.0

%E	M1	M2	DM	M*
58.0	188	109	79	63.2
58.0	181	105	76	60.9
58.0	176	102	74	59.1
58.0	174	101	73	58.6
58.0	169	98	71	56.8
58.0	162	94	68	54.5
58.0	157	91	66	52.7
58.0	150	87	63	50.5
58.0	143	83	60	48.2
58.0	138	80	58	46.4
58.0	131	76	55	44.1
58.0	119	69	50	40.0
58.0	112	65	47	37.7
58.0	100	58	42	33.6
58.0	88	51	37	29.6
58.0	81	47	34	27.3
58.0	69	40	29	23.2
58.0	50	29	21	16.8
57.9	499	289	210	167.4
57.9	497	288	209	166.9
57.9	496	287	209	166.1
57.9	494	286	208	165.6
57.9	492	285	207	165.1
57.9	489	283	206	163.8
57.9	487	282	205	163.3
57.9	485	281	204	162.8
57.9	484	280	204	162.0
57.9	482	279	203	161.5
57.9	480	278	202	161.0
57.9	477	276	201	159.7
57.9	475	275	200	159.2
57.9	473	274	199	158.7
57.9	470	272	198	157.4
57.9	468	271	197	156.9
57.9	466	270	196	156.4
57.9	463	268	195	155.1
57.9	461	267	194	154.6
57.9	458	265	193	153.3
57.9	456	264	192	152.8
57.9	454	263	191	152.4
57.9	451	261	190	151.0
57.9	449	260	189	150.6
57.9	447	259	188	150.1
57.9	444	257	187	148.8
57.9	442	256	186	148.3
57.9	439	254	185	147.0
57.9	437	253	184	146.5
57.9	435	252	183	146.0
57.9	432	250	182	144.7
57.9	430	249	181	144.2

%E	M1	M2	DM	M*
57.9	428	248	180	143.7
57.9	425	246	179	142.4
57.9	423	245	178	141.9
57.9	420	243	177	140.6
57.9	418	242	176	140.1
57.9	416	241	175	139.6
57.9	413	239	174	138.3
57.9	411	238	173	137.8
57.9	409	237	172	137.3
57.9	406	235	171	136.0
57.9	404	234	170	135.5
57.9	401	232	169	134.2
57.9	399	231	168	133.7
57.9	397	230	167	133.2
57.9	394	228	166	131.9
57.9	392	227	165	131.5
57.9	390	226	164	131.0
57.9	387	224	163	129.7
57.9	385	223	162	129.2
57.9	382	221	161	127.9
57.9	380	220	160	127.4
57.9	378	219	159	126.9
57.9	375	217	158	125.6
57.9	373	216	157	125.1
57.9	368	213	155	123.3
57.9	366	212	154	122.8
57.9	363	210	153	121.5
57.9	361	209	152	121.0
57.9	359	208	151	120.5
57.9	356	206	150	119.2
57.9	354	205	149	118.7
57.9	349	202	147	116.9
57.9	347	201	146	116.4
57.9	342	198	144	114.6
57.9	340	197	143	114.1
57.9	337	195	142	112.8
57.9	335	194	141	112.3
57.9	330	191	139	110.5
57.9	328	190	138	110.1
57.9	323	187	136	108.3
57.9	321	186	135	107.8
57.9	318	184	134	106.5
57.9	316	183	133	106.0
57.9	311	180	131	104.2
57.9	309	179	130	103.7
57.9	304	176	128	101.9
57.9	302	175	127	101.4
57.9	299	173	126	100.1
57.9	297	172	125	99.6
57.9	292	169	123	97.8

%E	M1	M2	DM	M*
57.9	290	168	122	97.3
57.9	285	165	120	95.5
57.9	280	162	118	93.7
57.9	278	161	117	93.2
57.9	273	158	115	91.4
57.9	271	157	114	91.0
57.9	266	154	112	89.2
57.9	261	151	110	87.4
57.9	259	150	109	86.9
57.9	254	147	107	85.1
57.9	252	146	106	84.6
57.9	247	143	104	82.8
57.9	242	140	102	81.0
57.9	240	139	101	80.5
57.9	235	136	99	78.7
57.9	233	135	98	78.2
57.9	228	132	96	76.4
57.9	221	128	93	74.1
57.9	216	125	91	72.3
57.9	214	124	90	71.9
57.9	209	121	88	70.1
57.9	202	117	85	67.8
57.9	197	114	83	66.0
57.9	195	113	82	65.5
57.9	190	110	80	63.7
57.9	183	106	77	61.4
57.9	178	103	75	59.6
57.9	171	99	72	57.3
57.9	164	95	69	55.0
57.9	159	92	67	53.2
57.9	152	88	64	50.9
57.9	145	84	61	48.7
57.9	140	81	59	46.9
57.9	133	77	56	44.6
57.9	126	73	53	42.3
57.9	121	70	51	40.5
57.9	114	66	48	38.2
57.9	107	62	45	35.9
57.9	95	55	40	31.8
57.9	76	44	32	25.5
57.9	57	33	24	19.1
57.9	38	22	16	12.7
57.9	19	11	8	6.4
57.8	500	289	211	167.0
57.8	498	288	210	166.6
57.8	495	286	209	165.2
57.8	493	285	208	164.8
57.8	491	284	207	164.3
57.8	490	283	207	163.4
57.8	488	282	206	163.0

%E	M1	M2	DM	M*
57.8	486	281	205	162.5
57.8	483	279	204	161.2
57.8	481	278	203	160.7
57.8	479	277	202	160.2
57.8	476	275	201	158.9
57.8	474	274	200	158.4
57.8	472	273	199	157.9
57.8	469	271	198	156.6
57.8	467	270	197	156.1
57.8	465	269	196	155.6
57.8	464	268	196	154.8
57.8	462	267	195	154.3
57.8	460	266	194	153.8
57.8	457	264	193	152.5
57.8	455	263	192	152.0
57.8	453	262	191	151.5
57.8	450	260	190	150.2
57.8	448	259	189	149.7
57.8	446	258	188	149.2
57.8	445	257	188	148.4
57.8	443	256	187	147.9
57.8	441	255	186	147.4
57.8	438	253	185	146.1
57.8	436	252	184	145.7
57.8	434	251	183	145.2
57.8	431	249	182	143.9
57.8	429	248	181	143.4
57.8	427	247	180	142.9
57.8	424	245	179	141.6
57.8	422	244	178	141.1
57.8	419	242	177	139.8
57.8	417	241	176	139.3
57.8	415	240	175	138.8
57.8	412	238	174	137.5
57.8	410	237	173	137.0
57.8	408	236	172	136.5
57.8	405	234	171	135.2
57.8	403	233	170	134.7
57.8	400	231	169	133.4
57.8	398	230	168	132.9
57.8	396	229	167	132.4
57.8	393	227	166	131.1
57.8	391	226	165	130.6
57.8	389	225	164	130.1
57.8	386	223	163	128.8
57.8	384	222	162	128.3
57.8	379	219	160	126.5
57.8	377	218	159	126.1
57.8	374	216	158	124.7
57.8	372	215	157	124.3

%E	M1	M2	DM	M*
57.8	370	214	156	123.8
57.8	367	212	155	122.5
57.8	365	211	154	122.0
57.8	360	208	152	120.2
57.8	358	207	151	119.7
57.8	353	204	149	117.9
57.8	351	203	148	117.4
57.8	348	201	147	116.1
57.8	346	200	146	115.6
57.8	344	199	145	115.1
57.8	341	197	144	113.8
57.8	339	196	143	113.3
57.8	334	193	141	111.5
57.8	332	192	140	111.0
57.8	329	190	139	109.7
57.8	327	189	138	109.2
57.8	325	188	137	108.8
57.8	322	186	136	107.4
57.8	320	185	135	107.0
57.8	315	182	133	105.2
57.8	313	181	132	104.7
57.8	308	178	130	102.9
57.8	306	177	129	102.4
57.8	303	175	128	101.1
57.8	301	174	127	100.6
57.8	296	171	125	98.8
57.8	294	170	124	98.3
57.8	289	167	122	96.5
57.8	287	166	121	96.0
57.8	282	163	119	94.2
57.8	277	160	117	92.4
57.8	275	159	116	91.9
57.8	270	156	114	90.1
57.8	268	155	113	89.6
57.8	263	152	111	87.8
57.8	258	149	109	86.1
57.8	256	148	108	85.6
57.8	251	145	106	83.8
57.8	249	144	105	83.3
57.8	244	141	103	81.5
57.8	237	137	100	79.2
57.8	232	134	98	77.4
57.8	230	133	97	76.9
57.8	225	130	95	75.1
57.8	223	129	94	74.6
57.8	218	126	92	72.8
57.8	211	122	89	70.5
57.8	206	119	87	68.7
57.8	204	118	86	68.3
57.8	199	115	84	66.5

%E	M1	M2	DM	M*
57.8	192	111	81	64.2
57.8	187	108	79	62.4
57.8	185	107	78	61.9
57.8	180	104	76	60.1
57.8	173	100	73	57.8
57.8	166	96	70	55.5
57.8	161	93	68	53.7
57.8	154	89	65	51.4
57.8	147	85	62	49.1
57.8	135	78	57	45.1
57.8	128	74	54	42.8
57.8	116	67	49	38.7
57.8	109	63	46	36.4
57.8	102	59	43	34.1
57.8	90	52	38	30.0
57.8	83	48	35	27.8
57.8	64	37	27	21.4
57.8	45	26	19	15.0
57.7	499	288	211	166.2
57.7	497	287	210	165.7
57.7	496	286	210	164.9
57.7	494	285	209	164.4
57.7	492	284	208	163.9
57.7	489	282	207	162.6
57.7	487	281	206	162.1
57.7	485	280	205	161.6
57.7	482	278	204	160.3
57.7	480	277	203	159.9
57.7	478	276	202	159.4
57.7	477	275	202	158.5
57.7	475	274	201	158.1
57.7	473	273	200	157.6
57.7	471	272	199	157.1
57.7	470	271	199	156.3
57.7	468	270	198	155.8
57.7	466	269	197	155.3
57.7	463	267	196	154.0
57.7	461	266	195	153.5
57.7	459	265	194	153.0
57.7	456	263	193	151.7
57.7	454	262	192	151.2
57.7	452	261	191	150.7
57.7	451	260	191	149.9
57.7	449	259	190	149.4
57.7	447	258	189	148.9
57.7	444	256	188	147.6
57.7	442	255	187	147.1
57.7	440	254	186	146.6
57.7	437	252	185	145.3
57.7	435	251	184	144.8

%E	M1	M2	DM	M*
57.7	433	250	183	144.3
57.7	430	248	182	143.0
57.7	428	247	181	142.5
57.7	426	246	180	142.1
57.7	423	244	179	140.7
57.7	421	243	178	140.3
57.7	418	241	177	138.9
57.7	416	240	176	138.5
57.7	414	239	175	138.0
57.7	411	237	174	136.7
57.7	409	236	173	136.2
57.7	407	235	172	135.7
57.7	404	233	171	134.4
57.7	402	232	170	133.9
57.7	397	229	168	132.1
57.7	395	228	167	131.6
57.7	392	226	166	130.3
57.7	390	225	165	129.8
57.7	388	224	164	129.3
57.7	385	222	163	128.0
57.7	383	221	162	127.5
57.7	381	220	161	127.0
57.7	378	218	160	125.7
57.7	376	217	159	125.2
57.7	371	214	157	123.4
57.7	369	213	156	123.0
57.7	366	211	155	121.6
57.7	364	210	154	121.2
57.7	362	209	153	120.7
57.7	359	207	152	119.4
57.7	357	206	151	118.9
57.7	355	205	150	118.4
57.7	352	203	149	117.1
57.7	350	202	148	116.6
57.7	345	199	146	114.8
57.7	343	198	145	114.3
57.7	338	195	143	112.5
57.7	336	194	142	112.0
57.7	333	192	141	110.7
57.7	331	191	140	110.2
57.7	326	188	138	108.4
57.7	324	187	137	107.9
57.7	319	184	135	106.1
57.7	317	183	134	105.6
57.7	312	180	132	103.8
57.7	310	179	131	103.4
57.7	307	177	130	102.0
57.7	305	176	129	101.6
57.7	300	173	127	99.8
57.7	298	172	126	99.3

%E	M1	M2	DM	M*	%E	M1	M2	DM	M*	%E	M1	M2	DM	M*	%E	M1	M2	DM	M*	%E	M1	M2	DM	M*
57.7	293	169	124	97.5	57.6	488	281	207	161.8	57.6	370	213	157	122.6	57.6	191	110	81	63.4	57.5	433	249	184	143.2
57.7	291	168	123	97.0	57.6	486	280	206	161.3	57.6	368	212	156	122.1	57.6	184	106	78	61.1	57.5	431	248	183	142.7
57.7	286	165	121	95.2	57.6	484	279	205	160.8	57.6	363	209	154	120.3	57.6	177	102	75	58.8	57.5	428	246	182	141.4
57.7	284	164	120	94.7	57.6	483	278	205	160.0	57.6	361	208	153	119.8	57.6	172	99	73	57.0	57.5	426	245	181	140.9
57.7	281	162	119	93.4	57.6	481	277	204	159.5	57.6	356	205	151	118.0	57.6	170	98	72	56.5	57.5	424	244	180	140.4
57.7	279	161	118	92.9	57.6	479	276	203	159.0	57.6	354	204	150	117.6	57.6	165	95	70	54.7	57.5	421	242	179	139.1
57.7	274	158	116	91.1	57.6	476	274	202	157.7	57.6	351	202	149	116.3	57.6	158	91	67	52.4	57.5	419	241	178	138.6
57.7	272	157	115	90.6	57.6	474	273	201	157.2	57.6	349	201	148	115.8	57.6	151	87	64	50.1	57.5	416	239	177	137.3
57.7	267	154	113	88.8	57.6	472	272	200	156.7	57.6	347	200	147	115.3	57.6	144	83	61	47.8	57.5	414	238	176	136.8
57.7	265	153	112	88.3	57.6	469	270	199	155.4	57.6	344	198	146	114.0	57.6	139	80	59	46.0	57.5	412	237	175	136.3
57.7	260	150	110	86.5	57.6	467	269	198	154.9	57.6	342	197	145	113.5	57.6	132	76	56	43.8	57.5	409	235	174	135.0
57.7	253	146	107	84.3	57.6	465	268	197	154.5	57.6	340	196	144	113.0	57.6	125	72	53	41.5	57.5	407	234	173	134.5
57.7	248	143	105	82.5	57.6	462	266	196	153.2	57.6	337	194	143	111.7	57.6	118	68	50	39.2	57.5	405	233	172	134.0
57.7	246	142	104	82.0	57.6	460	265	195	152.7	57.6	335	193	142	111.2	57.6	99	57	42	32.8	57.5	402	231	171	132.7
57.7	241	139	102	80.2	57.6	458	264	194	152.2	57.6	330	190	140	109.4	57.6	92	53	39	30.5	57.5	400	230	170	132.3
57.7	239	138	101	79.7	57.6	455	262	193	150.9	57.6	328	189	139	108.9	57.6	85	49	36	28.2	57.5	398	229	169	131.8
57.7	234	135	99	77.9	57.6	453	261	192	150.4	57.6	323	186	137	107.1	57.6	66	38	28	21.9	57.5	395	227	168	130.5
57.7	227	131	96	75.6	57.6	450	259	191	149.1	57.6	321	185	136	106.6	57.6	59	34	25	19.6	57.5	393	226	167	130.0
57.7	222	128	94	73.8	57.6	448	258	190	148.6	57.6	316	182	134	104.8	57.6	33	19	14	10.9	57.5	391	225	166	129.5
57.7	220	127	93	73.3	57.6	446	257	189	148.1	57.6	314	181	133	104.3	57.5	499	287	212	165.1	57.5	388	223	165	128.2
57.7	215	124	91	71.5	57.6	443	255	188	146.8	57.6	311	179	132	103.0	57.5	497	286	211	164.6	57.5	386	222	164	127.7
57.7	213	123	90	71.0	57.6	441	254	187	146.3	57.6	309	178	131	102.5	57.5	496	285	211	163.8	57.5	381	219	162	125.9
57.7	208	120	88	69.2	57.6	439	253	186	145.8	57.6	304	175	129	100.7	57.5	494	284	210	163.3	57.5	379	218	161	125.4
57.7	201	116	85	66.9	57.6	436	251	185	144.5	57.6	302	174	128	100.3	57.5	492	283	209	162.8	57.5	374	215	159	123.6
57.7	196	113	83	65.1	57.6	434	250	184	144.0	57.6	297	171	126	98.5	57.5	489	281	208	161.5	57.5	372	214	158	123.1
57.7	194	112	82	64.7	57.6	432	249	183	143.5	57.6	295	170	125	98.0	57.5	487	280	207	161.0	57.5	369	212	157	121.8
57.7	189	109	80	62.9	57.6	429	247	182	142.2	57.6	290	167	123	96.2	57.5	485	279	206	160.5	57.5	367	211	156	121.3
57.7	182	105	77	60.6	57.6	427	246	181	141.7	57.6	288	166	122	95.7	57.5	482	277	205	159.2	57.5	365	210	155	120.8
57.7	175	101	74	58.3	57.6	425	245	180	141.2	57.6	283	163	120	93.9	57.5	480	276	204	158.7	57.5	362	208	154	119.5
57.7	168	97	71	56.0	57.6	422	243	179	139.9	57.6	278	160	118	92.1	57.5	478	275	203	158.2	57.5	360	207	153	119.0
57.7	163	94	69	54.2	57.6	420	242	178	139.4	57.6	276	159	117	91.6	57.5	475	273	202	156.9	57.5	358	206	152	118.5
57.7	156	90	66	51.9	57.6	417	240	177	138.1	57.6	271	156	115	89.8	57.5	473	272	201	156.4	57.5	355	204	151	117.2
57.7	149	86	63	49.6	57.6	415	239	176	137.6	57.6	269	155	114	89.3	57.5	471	271	200	155.9	57.5	353	203	150	116.7
57.7	142	82	60	47.4	57.6	413	238	175	137.2	57.6	264	152	112	87.5	57.5	468	269	199	154.6	57.5	348	200	148	114.9
57.7	137	79	58	45.6	57.6	410	236	174	135.8	57.6	262	151	111	87.0	57.5	466	268	198	154.1	57.5	346	199	147	114.5
57.7	130	75	55	43.3	57.6	408	235	173	135.4	57.6	257	148	109	85.2	57.5	464	267	197	153.6	57.5	341	196	145	112.7
57.7	123	71	52	41.0	57.6	406	234	172	134.9	57.6	255	147	108	84.7	57.5	463	266	197	152.8	57.5	339	195	144	112.2
57.7	111	64	47	36.9	57.6	403	232	171	133.6	57.6	250	144	106	82.9	57.5	461	265	196	152.3	57.5	334	192	142	110.4
57.7	104	60	44	34.6	57.6	401	231	170	133.1	57.6	245	141	104	81.1	57.5	459	264	195	151.8	57.5	332	191	141	109.9
57.7	97	56	41	32.3	57.6	399	230	169	132.6	57.6	243	140	103	80.7	57.5	457	263	194	151.4	57.5	327	188	139	108.1
57.7	78	45	33	26.0	57.6	396	228	168	131.3	57.6	238	137	101	78.9	57.5	456	262	194	150.5	57.5	325	187	138	107.6
57.7	71	41	30	23.7	57.6	394	227	167	130.8	57.6	236	136	100	78.4	57.5	454	261	193	150.0	57.5	322	185	137	106.3
57.7	52	30	22	17.3	57.6	389	224	165	129.0	57.6	231	133	98	76.6	57.5	452	260	192	149.6	57.5	320	184	136	105.8
57.7	26	15	11	8.7	57.6	387	223	164	128.5	57.6	229	132	97	76.1	57.5	449	258	191	148.2	57.5	318	183	135	105.3
57.6	500	288	212	165.9	57.6	384	221	163	127.2	57.6	224	129	95	74.3	57.5	447	257	190	147.8	57.5	315	181	134	104.0
57.6	498	287	211	165.4	57.6	382	220	162	126.7	57.6	217	125	92	72.0	57.5	445	256	189	147.3	57.5	313	180	133	103.5
57.6	495	285	210	164.1	57.6	380	219	161	126.2	57.6	210	121	89	69.7	57.5	442	254	188	146.0	57.5	308	177	131	101.7
57.6	493	284	209	163.6	57.6	377	217	160	124.9	57.6	205	118	87	67.9	57.5	440	253	187	145.5	57.5	306	176	130	101.2
57.6	491	283	208	163.1	57.6	375	216	159	124.4	57.6	203	117	86	67.4	57.5	438	252	186	145.0	57.5	301	173	128	99.4
57.6	490	282	208	162.3	57.6	373	215	158	123.9	57.6	198	114	84	65.6	57.5	435	250	185	143.7	57.5	299	172	127	98.9

%E	M1	M2	DM	M*
57.5	294	169	125	97.1
57.5	292	168	124	96.7
57.5	287	165	122	94.9
57.5	285	164	121	94.4
57.5	280	161	119	92.6
57.5	275	158	117	90.8
57.5	273	157	116	90.3
57.5	268	154	114	88.5
57.5	266	153	113	88.0
57.5	261	150	111	86.2
57.5	259	149	110	85.7
57.5	254	146	108	83.9
57.5	252	145	107	83.4
57.5	247	142	105	81.6
57.5	240	138	102	79.3
57.5	233	134	99	77.1
57.5	228	131	97	75.3
57.5	226	130	96	74.8
57.5	221	127	94	73.0
57.5	219	126	93	72.5
57.5	214	123	91	70.7
57.5	212	122	90	70.2
57.5	207	119	88	68.4
57.5	200	115	85	66.1
57.5	193	111	82	63.8
57.5	186	107	79	61.6
57.5	181	104	77	59.8
57.5	179	103	76	59.3
57.5	174	100	74	57.5
57.5	167	96	71	55.2
57.5	160	92	68	52.9
57.5	153	88	65	50.6
57.5	146	84	62	48.3
57.5	134	77	57	44.2
57.5	127	73	54	42.0
57.5	120	69	51	39.7
57.5	113	65	48	37.4
57.5	106	61	45	35.1
57.5	87	50	37	28.7
57.5	80	46	34	26.4
57.5	73	42	31	24.2
57.5	40	23	17	13.2
57.4	500	287	213	164.7
57.4	498	286	212	164.2
57.4	495	284	211	162.9
57.4	493	283	210	162.5
57.4	491	282	209	162.0
57.4	488	280	208	160.7
57.4	486	279	207	160.2
57.4	484	278	206	159.7

%E	M1	M2	DM	M*
57.4	483	277	206	158.9
57.4	481	276	205	158.4
57.4	479	275	204	157.9
57.4	477	274	203	157.4
57.4	476	273	203	156.6
57.4	474	272	202	156.1
57.4	472	271	201	155.6
57.4	470	270	200	155.1
57.4	469	269	200	154.3
57.4	467	268	199	153.8
57.4	465	267	198	153.3
57.4	462	265	197	152.0
57.4	460	264	196	151.5
57.4	458	263	195	151.0
57.4	455	261	194	149.7
57.4	453	260	193	149.2
57.4	451	259	192	148.7
57.4	448	257	191	147.4
57.4	446	256	190	146.9
57.4	444	255	189	146.5
57.4	441	253	188	145.1
57.4	439	252	187	144.7
57.4	437	251	186	144.2
57.4	434	249	185	142.9
57.4	432	248	184	142.4
57.4	430	247	183	141.9
57.4	427	245	182	140.6
57.4	425	244	181	140.1
57.4	423	243	180	139.6
57.4	420	241	179	138.3
57.4	418	240	178	137.8
57.4	413	237	176	136.0
57.4	411	236	175	135.5
57.4	408	234	174	134.2
57.4	406	233	173	133.7
57.4	404	232	172	133.2
57.4	401	230	171	131.9
57.4	399	229	170	131.4
57.4	397	228	169	130.9
57.4	394	226	168	129.6
57.4	392	225	167	129.1
57.4	390	224	166	128.7
57.4	387	222	165	127.3
57.4	385	221	164	126.9
57.4	383	220	163	126.4
57.4	380	218	162	125.1
57.4	378	217	161	124.6
57.4	376	216	160	124.1
57.4	373	214	159	122.8
57.4	371	213	158	122.3

%E	M1	M2	DM	M*
57.4	366	210	156	120.5
57.4	364	209	155	120.0
57.4	359	206	153	118.2
57.4	357	205	152	117.7
57.4	352	202	150	115.9
57.4	350	201	149	115.4
57.4	345	198	147	113.6
57.4	343	197	146	113.1
57.4	340	195	145	111.8
57.4	338	194	144	111.3
57.4	336	193	143	110.9
57.4	333	191	142	109.6
57.4	331	190	141	109.1
57.4	329	189	140	108.6
57.4	326	187	139	107.3
57.4	324	186	138	106.8
57.4	319	183	136	105.0
57.4	317	182	135	104.5
57.4	312	179	133	102.7
57.4	310	178	132	102.2
57.4	305	175	130	100.4
57.4	303	174	129	99.9
57.4	298	171	127	98.1
57.4	296	170	126	97.6
57.4	291	167	124	95.8
57.4	289	166	123	95.3
57.4	284	163	121	93.6
57.4	282	162	120	93.1
57.4	277	159	118	91.3
57.4	272	156	116	89.5
57.4	270	155	115	89.0
57.4	265	152	113	87.2
57.4	263	151	112	86.7
57.4	258	148	110	84.9
57.4	256	147	109	84.4
57.4	251	144	107	82.6
57.4	249	143	106	82.1
57.4	244	140	104	80.3
57.4	242	139	103	79.8
57.4	237	136	101	78.0
57.4	235	135	100	77.6
57.4	230	132	98	75.8
57.4	223	128	95	73.5
57.4	216	124	92	71.2
57.4	209	120	89	68.9
57.4	204	117	87	67.1
57.4	202	116	86	66.6
57.4	197	113	84	64.8
57.4	195	112	83	64.3
57.4	190	109	81	62.5

%E	M1	M2	DM	M*
57.4	188	108	80	62.0
57.4	183	105	78	60.2
57.4	176	101	75	58.0
57.4	169	97	72	55.7
57.4	162	93	69	53.4
57.4	155	89	66	51.1
57.4	148	85	63	48.8
57.4	141	81	60	46.5
57.4	136	78	58	44.7
57.4	129	74	55	42.4
57.4	122	70	52	40.2
57.4	115	66	49	37.9
57.4	108	62	46	35.6
57.4	101	58	43	33.3
57.4	94	54	40	31.0
57.4	68	39	29	22.4
57.4	61	35	26	20.1
57.4	54	31	23	17.8
57.4	47	27	20	15.5
57.3	499	286	213	163.9
57.3	497	285	212	163.4
57.3	496	284	212	162.6
57.3	494	283	211	162.1
57.3	492	282	210	161.6
57.3	490	281	209	161.1
57.3	489	280	209	160.3
57.3	487	279	208	159.8
57.3	485	278	207	159.3
57.3	482	276	206	158.0
57.3	480	275	205	157.6
57.3	478	274	204	157.1
57.3	475	272	203	155.8
57.3	473	271	202	155.3
57.3	471	270	201	154.8
57.3	468	268	200	153.5
57.3	466	267	199	153.0
57.3	464	266	198	152.5
57.3	461	264	197	151.2
57.3	459	263	196	150.7
57.3	457	262	195	150.2
57.3	454	260	194	148.9
57.3	452	259	193	148.4
57.3	450	258	192	147.9
57.3	447	256	191	146.6
57.3	445	255	190	146.1
57.3	443	254	189	145.6
57.3	440	252	188	144.3
57.3	438	251	187	143.8
57.3	436	250	186	143.3
57.3	433	248	185	142.0

%E	M1	M2	DM	M*
57.3	431	247	184	141.6
57.3	429	246	183	141.1
57.3	426	244	182	139.8
57.3	424	243	181	139.3
57.3	422	242	180	138.8
57.3	419	240	179	137.5
57.3	417	239	178	137.0
57.3	415	238	177	136.5
57.3	412	236	176	135.2
57.3	410	235	175	134.7
57.3	405	232	173	132.9
57.3	403	231	172	132.4
57.3	400	229	171	131.1
57.3	398	228	170	130.6
57.3	396	227	169	130.1
57.3	393	225	168	128.8
57.3	391	224	167	128.3
57.3	389	223	166	127.8
57.3	386	221	165	126.5
57.3	384	220	164	126.0
57.3	382	219	163	125.6
57.3	379	217	162	124.2
57.3	377	216	161	123.8
57.3	375	215	160	123.3
57.3	372	213	159	122.0
57.3	370	212	158	121.5
57.3	368	211	157	121.0
57.3	365	209	156	119.7
57.3	363	208	155	119.2
57.3	361	207	154	118.7
57.3	358	205	153	117.4
57.3	356	204	152	116.9
57.3	354	203	151	116.4
57.3	351	201	150	115.1
57.3	349	200	149	114.6
57.3	347	199	148	114.1
57.3	344	197	147	112.8
57.3	342	196	146	112.3
57.3	337	193	144	110.5
57.3	335	192	143	110.0
57.3	330	189	141	108.2
57.3	328	188	140	107.8
57.3	323	185	138	106.0
57.3	321	184	137	105.5
57.3	316	181	135	103.7
57.3	314	180	134	103.2
57.3	309	177	132	101.4
57.3	307	176	131	100.9
57.3	302	173	129	99.1
57.3	300	172	128	98.6

%E	M1	M2	DM	M*	%E	M1	M2	DM	M*	%E	M1	M2	DM	M*	%E	M1	M2	DM	M*	%E	M1	M2	DM	M*
57.3	295	169	126	96.8	57.2	484	277	207	158.5	57.2	348	199	149	113.8	57.1	487	278	209	158.7	57.1	380	217	163	123.9
57.3	293	168	125	96.3	57.2	481	275	206	157.2	57.2	346	198	148	113.3	57.1	485	277	208	158.2	57.1	378	216	162	123.4
57.3	288	165	123	94.5	57.2	479	274	205	156.7	57.2	341	195	146	111.5	57.1	483	276	207	157.7	57.1	375	214	161	122.1
57.3	286	164	122	94.0	57.2	477	273	204	156.2	57.2	339	194	145	111.0	57.1	482	275	207	156.9	57.1	373	213	160	121.6
57.3	281	161	120	92.2	57.2	474	271	203	154.9	57.2	334	191	143	109.2	57.1	480	274	206	156.4	57.1	371	212	159	121.1
57.3	279	160	119	91.8	57.2	472	270	202	154.4	57.2	332	190	142	108.7	57.1	478	273	205	155.9	57.1	368	210	158	119.8
57.3	274	157	117	90.0	57.2	470	269	201	154.0	57.2	327	187	140	106.9	57.1	476	272	204	155.4	57.1	366	209	157	119.3
57.3	267	153	114	87.7	57.2	467	267	200	152.7	57.2	325	186	139	106.4	57.1	475	271	204	154.6	57.1	364	208	156	118.9
57.3	262	150	112	85.9	57.2	465	266	199	152.2	57.2	320	183	137	104.7	57.1	473	270	203	154.1	57.1	361	206	155	117.6
57.3	260	149	111	85.4	57.2	463	265	198	151.7	57.2	318	182	136	104.2	57.1	471	269	202	153.6	57.1	359	205	154	117.1
57.3	255	146	109	83.6	57.2	460	263	197	150.4	57.2	313	179	134	102.4	57.1	469	268	201	153.1	57.1	357	204	153	116.6
57.3	253	145	108	83.1	57.2	458	262	196	149.9	57.2	311	178	133	101.9	57.1	468	267	201	152.3	57.1	354	202	152	115.3
57.3	248	142	106	81.3	57.2	456	261	195	149.4	57.2	306	175	131	100.1	57.1	466	266	200	151.8	57.1	352	201	151	114.8
57.3	246	141	105	80.8	57.2	453	259	194	148.1	57.2	304	174	130	99.6	57.1	464	265	199	151.3	57.1	350	200	150	114.3
57.3	241	138	103	79.0	57.2	451	258	193	147.6	57.2	299	171	128	97.8	57.1	462	264	198	150.9	57.1	347	198	149	113.0
57.3	239	137	102	78.5	57.2	449	257	192	147.1	57.2	297	170	127	97.3	57.1	461	263	198	150.0	57.1	345	197	148	112.5
57.3	234	134	100	76.7	57.2	446	255	191	145.8	57.2	292	167	125	95.5	57.1	459	262	197	149.6	57.1	343	196	147	112.0
57.3	232	133	99	76.2	57.2	444	254	190	145.3	57.2	290	166	124	95.0	57.1	457	261	196	149.1	57.1	340	194	146	110.7
57.3	227	130	97	74.4	57.2	442	253	189	144.8	57.2	285	163	122	93.2	57.1	455	260	195	148.6	57.1	338	193	145	110.2
57.3	225	129	96	74.0	57.2	439	251	188	143.5	57.2	283	162	121	92.7	57.1	452	258	194	147.3	57.1	336	192	144	109.7
57.3	220	126	94	72.2	57.2	437	250	187	143.0	57.2	278	159	119	90.9	57.1	450	257	193	146.8	57.1	333	190	143	108.4
57.3	218	125	93	71.7	57.2	435	249	186	142.5	57.2	276	158	118	90.4	57.1	448	256	192	146.3	57.1	331	189	142	107.9
57.3	213	122	91	69.9	57.2	432	247	185	141.2	57.2	271	155	116	88.7	57.1	445	254	191	145.0	57.1	329	188	141	107.4
57.3	211	121	90	69.4	57.2	430	246	184	140.7	57.2	269	154	115	88.2	57.1	443	253	190	144.5	57.1	326	186	140	106.1
57.3	206	118	88	67.6	57.2	428	245	183	140.2	57.2	264	151	113	86.4	57.1	441	252	189	144.0	57.1	324	185	139	105.6
57.3	199	114	85	65.3	57.2	425	243	182	138.9	57.2	257	147	110	84.1	57.1	438	250	188	142.7	57.1	322	184	138	105.1
57.3	192	110	82	63.0	57.2	423	242	181	138.4	57.2	250	143	107	81.8	57.1	436	249	187	142.2	57.1	319	182	137	103.8
57.3	185	106	79	60.7	57.2	421	241	180	138.0	57.2	243	139	104	79.5	57.1	434	248	186	141.7	57.1	317	181	136	103.3
57.3	178	102	76	58.4	57.2	418	239	179	136.7	57.2	236	135	101	77.2	57.1	431	246	185	140.4	57.1	315	180	135	102.9
57.3	171	98	73	56.2	57.2	416	238	178	136.2	57.2	229	131	98	74.9	57.1	429	245	184	139.9	57.1	312	178	134	101.6
57.3	164	94	70	53.9	57.2	414	237	177	135.7	57.2	222	127	95	72.7	57.1	427	244	183	139.4	57.1	310	177	133	101.1
57.3	157	90	67	51.6	57.2	411	235	176	134.4	57.2	215	123	92	70.4	57.1	424	242	182	138.1	57.1	308	176	132	100.6
57.3	150	86	64	49.3	57.2	409	234	175	133.9	57.2	208	119	89	68.1	57.1	422	241	181	137.6	57.1	303	173	130	98.8
57.3	143	82	61	47.0	57.2	407	233	174	133.4	57.2	201	115	86	65.8	57.1	420	240	180	137.1	57.1	301	172	129	98.3
57.3	131	75	56	42.9	57.2	404	231	173	132.1	57.2	194	111	83	63.5	57.1	417	238	179	135.8	57.1	296	169	127	96.5
57.3	124	71	53	40.7	57.2	402	230	172	131.6	57.2	187	107	80	61.2	57.1	415	237	178	135.3	57.1	294	168	126	96.0
57.3	117	67	50	38.4	57.2	397	227	170	129.8	57.2	180	103	77	58.9	57.1	413	236	177	134.9	57.1	289	165	124	94.2
57.3	110	63	47	36.1	57.2	395	226	169	129.3	57.2	173	99	74	56.7	57.1	410	234	176	133.6	57.1	287	164	123	93.7
57.3	103	59	44	33.8	57.2	390	223	167	127.5	57.2	166	95	71	54.4	57.1	408	233	175	133.1	57.1	282	161	121	91.9
57.3	96	55	41	31.5	57.2	388	222	166	127.0	57.2	159	91	68	52.1	57.1	406	232	174	132.6	57.1	280	160	120	91.4
57.3	89	51	38	29.2	57.2	383	219	164	125.2	57.2	152	87	65	49.8	57.1	403	230	173	131.3	57.1	275	157	118	89.6
57.3	82	47	35	26.9	57.2	381	218	163	124.7	57.2	145	83	62	47.5	57.1	401	229	172	130.8	57.1	273	156	117	89.1
57.3	75	43	32	24.7	57.2	376	215	161	122.9	57.2	138	79	59	45.2	57.1	399	228	171	130.3	57.1	268	153	115	87.3
57.2	500	286	214	163.6	57.2	374	214	160	122.4	57.1	499	285	214	162.8	57.1	396	226	170	129.0	57.1	266	152	114	86.9
57.2	498	285	213	163.1	57.2	369	211	158	120.7	57.1	497	284	213	162.3	57.1	394	225	169	128.5	57.1	261	149	112	85.1
57.2	495	283	212	161.8	57.2	367	210	157	120.2	57.1	496	283	213	161.5	57.1	392	224	168	128.0	57.1	259	148	111	84.6
57.2	493	282	211	161.3	57.2	362	207	155	118.4	57.1	494	282	212	161.0	57.1	389	222	167	126.7	57.1	254	145	109	82.8
57.2	491	281	210	160.8	57.2	360	206	154	117.9	57.1	492	281	211	160.5	57.1	387	221	166	126.2	57.1	252	144	108	82.3
57.2	488	279	209	159.5	57.2	355	203	152	116.1	57.1	490	280	210	160.0	57.1	385	220	165	125.7	57.1	247	141	106	80.5
57.2	486	278	208	159.0	57.2	353	202	151	115.6	57.1	489	279	210	159.2	57.1	382	218	164	124.4	57.1	245	140	105	80.0

%E	M1	M2	DM	M*
57.1	240	137	103	78.2
57.1	238	136	102	77.7
57.1	233	133	100	75.9
57.1	231	132	99	75.4
57.1	226	129	97	73.6
57.1	224	128	96	73.1
57.1	219	125	94	71.3
57.1	217	124	93	70.9
57.1	212	121	91	69.1
57.1	210	120	90	68.6
57.1	205	117	88	66.8
57.1	203	116	87	66.3
57.1	198	113	85	64.5
57.1	196	112	84	64.0
57.1	191	109	82	62.2
57.1	189	108	81	61.7
57.1	184	105	79	59.9
57.1	182	104	78	59.4
57.1	177	101	76	57.6
57.1	175	100	75	57.1
57.1	170	97	73	55.3
57.1	168	96	72	54.9
57.1	163	93	70	53.1
57.1	161	92	69	52.6
57.1	156	89	67	50.8
57.1	154	88	66	50.3
57.1	147	84	63	48.0
57.1	140	80	60	45.7
57.1	133	76	57	43.4
57.1	126	72	54	41.1
57.1	119	68	51	38.9
57.1	112	64	48	36.6
57.1	105	60	45	34.3
57.1	98	56	42	32.0
57.1	91	52	39	29.7
57.1	84	48	36	27.4
57.1	77	44	33	25.1
57.1	70	40	30	22.9
57.1	63	36	27	20.6
57.1	56	32	24	18.3
57.1	49	28	21	16.0
57.1	42	24	18	13.7
57.1	35	20	15	11.4
57.1	28	16	12	9.1
57.1	21	12	9	6.9
57.1	14	8	6	4.6
57.1	7	4	3	2.3
57.0	500	285	215	162.4
57.0	498	284	214	162.0
57.0	495	282	213	160.7

%E	M1	M2	DM	M*
57.0	493	281	212	160.2
57.0	491	280	211	159.7
57.0	488	278	210	158.4
57.0	486	277	209	157.9
57.0	484	276	208	157.4
57.0	481	274	207	156.1
57.0	479	273	206	155.6
57.0	477	272	205	155.1
57.0	474	270	204	153.8
57.0	472	269	203	153.3
57.0	470	268	202	152.8
57.0	467	266	201	151.5
57.0	465	265	200	151.0
57.0	463	264	199	150.5
57.0	460	262	198	149.2
57.0	458	261	197	148.7
57.0	456	260	196	148.2
57.0	454	259	195	147.8
57.0	453	258	195	146.9
57.0	451	257	194	146.5
57.0	449	256	193	146.0
57.0	447	255	192	145.5
57.0	446	254	192	144.7
57.0	444	253	191	144.2
57.0	442	252	190	143.7
57.0	440	251	189	143.2
57.0	437	249	188	141.9
57.0	435	248	187	141.4
57.0	433	247	186	140.9
57.0	430	245	185	139.6
57.0	428	244	184	139.1
57.0	426	243	183	138.6
57.0	423	241	182	137.3
57.0	421	240	181	136.8
57.0	419	239	180	136.3
57.0	416	237	179	135.0
57.0	414	236	178	134.5
57.0	412	235	177	134.0
57.0	409	233	176	132.7
57.0	407	232	175	132.2
57.0	405	231	174	131.8
57.0	402	229	173	130.5
57.0	400	228	172	130.0
57.0	398	227	171	129.5
57.0	395	225	170	128.2
57.0	393	224	169	127.7
57.0	391	223	168	127.2
57.0	388	221	167	125.9
57.0	386	220	166	125.4
57.0	384	219	165	124.9

%E	M1	M2	DM	M*
57.0	381	217	164	123.6
57.0	379	216	163	123.1
57.0	377	215	162	122.6
57.0	374	213	161	121.3
57.0	372	212	160	120.8
57.0	370	211	159	120.3
57.0	365	208	157	118.5
57.0	363	207	156	118.0
57.0	358	204	154	116.2
57.0	356	203	153	115.8
57.0	351	200	151	114.0
57.0	349	199	150	113.5
57.0	344	196	148	111.7
57.0	342	195	147	111.2
57.0	337	192	145	109.4
57.0	335	191	144	108.9
57.0	330	188	142	107.1
57.0	328	187	141	106.6
57.0	323	184	139	104.8
57.0	321	183	138	104.3
57.0	316	180	136	102.5
57.0	314	179	135	102.0
57.0	309	176	133	100.2
57.0	307	175	132	99.8
57.0	305	174	131	99.3
57.0	302	172	130	98.0
57.0	300	171	129	97.5
57.0	298	170	128	97.0
57.0	293	167	126	95.2
57.0	291	166	125	94.7
57.0	286	163	123	92.9
57.0	284	162	122	92.4
57.0	279	159	120	90.6
57.0	277	158	119	90.1
57.0	272	155	117	88.3
57.0	270	154	116	87.8
57.0	265	151	114	86.0
57.0	263	150	113	85.6
57.0	258	147	111	83.8
57.0	256	146	110	83.3
57.0	251	143	108	81.5
57.0	249	142	107	81.0
57.0	244	139	105	79.2
57.0	242	138	104	78.7
57.0	237	135	102	76.9
57.0	235	134	101	76.4
57.0	230	131	99	74.6
57.0	228	130	98	74.1
57.0	223	127	96	72.3
57.0	221	126	95	71.8

%E	M1	M2	DM	M*
57.0	214	122	92	69.6
57.0	207	118	89	67.3
57.0	200	114	86	65.0
57.0	193	110	83	62.7
57.0	186	106	80	60.4
57.0	179	102	77	58.1
57.0	172	98	74	55.8
57.0	165	94	71	53.6
57.0	158	90	68	51.3
57.0	151	86	65	49.0
57.0	149	85	64	48.5
57.0	142	81	61	46.2
57.0	135	77	58	43.9
57.0	128	73	55	41.6
57.0	121	69	52	39.3
57.0	114	65	49	37.1
57.0	107	61	46	34.8
57.0	100	57	43	32.5
57.0	93	53	40	30.2
57.0	86	49	37	27.9
57.0	79	45	34	25.6
56.9	499	284	215	161.6
56.9	497	283	214	161.1
56.9	496	282	214	160.3
56.9	494	281	213	159.8
56.9	492	280	212	159.3
56.9	490	279	211	158.9
56.9	489	278	211	158.0
56.9	487	277	210	157.6
56.9	485	276	209	157.1
56.9	483	275	208	156.6
56.9	480	273	207	155.3
56.9	478	272	206	154.8
56.9	476	271	205	154.3
56.9	473	269	204	153.0
56.9	471	268	203	152.5
56.9	469	267	202	152.0
56.9	466	265	201	150.7
56.9	464	264	200	150.2
56.9	462	263	199	149.7
56.9	459	261	198	148.4
56.9	457	260	197	147.9
56.9	455	259	196	147.4
56.9	452	257	195	146.1
56.9	450	256	194	145.6
56.9	448	255	193	145.1
56.9	445	253	192	143.8
56.9	443	252	191	143.3
56.9	441	251	190	142.9
56.9	439	250	189	142.4

%E	M1	M2	DM	M*
56.9	436	248	188	141.1
56.9	434	247	187	140.6
56.9	432	246	186	140.1
56.9	429	244	185	138.8
56.9	427	243	184	138.3
56.9	425	242	183	137.8
56.9	422	240	182	136.5
56.9	420	239	181	136.0
56.9	418	238	180	135.5
56.9	415	236	179	134.2
56.9	413	235	178	133.7
56.9	411	234	177	133.2
56.9	408	232	176	131.9
56.9	406	231	175	131.4
56.9	404	230	174	130.9
56.9	401	228	173	129.6
56.9	399	227	172	129.1
56.9	397	226	171	128.7
56.9	394	224	170	127.4
56.9	392	223	169	126.9
56.9	390	222	168	126.4
56.9	385	219	166	124.6
56.9	383	218	165	124.1
56.9	378	215	163	122.3
56.9	376	214	162	121.8
56.9	371	211	160	120.0
56.9	369	210	159	119.5
56.9	367	209	158	119.0
56.9	364	207	157	117.7
56.9	362	206	156	117.2
56.9	360	205	155	116.7
56.9	357	203	154	115.4
56.9	355	202	153	114.9
56.9	353	201	152	114.5
56.9	350	199	151	113.1
56.9	348	198	150	112.7
56.9	346	197	149	112.2
56.9	343	195	148	110.9
56.9	341	194	147	110.4
56.9	339	193	146	109.9
56.9	334	190	144	108.1
56.9	332	189	143	107.6
56.9	327	186	141	105.8
56.9	325	185	140	105.3
56.9	320	182	138	103.5
56.9	318	181	137	103.0
56.9	313	178	135	101.2
56.9	311	177	134	100.7
56.9	306	174	132	98.9
56.9	304	173	131	98.5

%E	M1	M2	DM	M*
56.9	299	170	129	96.7
56.9	297	169	128	96.2
56.9	295	168	127	95.7
56.9	290	165	125	93.9
56.9	288	164	124	93.4
56.9	283	161	122	91.6
56.9	281	160	121	91.1
56.9	276	157	119	89.3
56.9	274	156	118	88.8
56.9	269	153	116	87.0
56.9	267	152	115	86.5
56.9	262	149	113	84.7
56.9	260	148	112	84.2
56.9	255	145	110	82.5
56.9	253	144	109	82.0
56.9	248	141	107	80.2
56.9	246	140	106	79.7
56.9	239	136	103	77.4
56.9	232	132	100	75.1
56.9	225	128	97	72.8
56.9	218	124	94	70.5
56.9	216	123	93	70.0
56.9	211	120	91	68.2
56.9	209	119	90	67.8
56.9	204	116	88	66.0
56.9	202	115	87	65.5
56.9	197	112	85	63.7
56.9	195	111	84	63.2
56.9	188	107	81	60.9
56.9	181	103	78	58.6
56.9	174	99	75	56.3
56.9	167	95	72	54.0
56.9	160	91	69	51.8
56.9	153	87	66	49.5
56.9	144	82	62	46.7
56.9	137	78	59	44.4
56.9	130	74	56	42.1
56.9	123	70	53	39.8
56.9	116	66	50	37.6
56.9	109	62	47	35.3
56.9	102	58	44	33.0
56.9	72	41	31	23.3
56.9	65	37	28	21.1
56.9	58	33	25	18.8
56.9	51	29	22	16.5
56.8	500	284	216	161.3
56.8	498	283	215	160.8
56.8	495	281	214	159.5
56.8	493	280	213	159.0
56.8	491	279	212	158.5

%E	M1	M2	DM	M*
56.8	488	277	211	157.2
56.8	486	276	210	156.7
56.8	484	275	209	156.3
56.8	482	274	208	155.8
56.8	481	273	208	154.9
56.8	479	272	207	154.5
56.8	477	271	206	154.0
56.8	475	270	205	153.5
56.8	474	269	205	152.7
56.8	472	268	204	152.2
56.8	470	267	203	151.7
56.8	468	266	202	151.2
56.8	465	264	201	149.9
56.8	463	263	200	149.4
56.8	461	262	199	148.9
56.8	458	260	198	147.6
56.8	456	259	197	147.1
56.8	454	258	196	146.6
56.8	451	256	195	145.3
56.8	449	255	194	144.8
56.8	447	254	193	144.3
56.8	444	252	192	143.0
56.8	442	251	191	142.5
56.8	440	250	190	142.0
56.8	438	249	189	141.6
56.8	437	248	189	140.7
56.8	435	247	188	140.3
56.8	433	246	187	139.8
56.8	431	245	186	139.3
56.8	428	243	185	138.0
56.8	426	242	184	137.5
56.8	424	241	183	137.0
56.8	421	239	182	135.7
56.8	419	238	181	135.2
56.8	417	237	180	134.7
56.8	414	235	179	133.4
56.8	412	234	178	132.9
56.8	410	233	177	132.4
56.8	407	231	176	131.1
56.8	405	230	175	130.6
56.8	403	229	174	130.1
56.8	400	227	173	128.8
56.8	398	226	172	128.3
56.8	396	225	171	127.8
56.8	391	222	169	126.0
56.8	389	221	168	125.6
56.8	387	220	167	125.1
56.8	384	218	166	123.8
56.8	382	217	165	123.3
56.8	380	216	164	122.8

%E	M1	M2	DM	M*
56.8	377	214	163	121.5
56.8	375	213	162	121.0
56.8	373	212	161	120.5
56.8	370	210	160	119.2
56.8	368	209	159	118.7
56.8	366	208	158	118.2
56.8	361	205	156	116.4
56.8	359	204	155	115.9
56.8	354	201	153	114.1
56.8	352	200	152	113.6
56.8	347	197	150	111.8
56.8	345	196	149	111.4
56.8	340	193	147	109.6
56.8	338	192	146	109.1
56.8	336	191	145	108.6
56.8	333	189	144	107.3
56.8	331	188	143	106.8
56.8	329	187	142	106.3
56.8	324	184	140	104.5
56.8	322	183	139	104.0
56.8	317	180	137	102.2
56.8	315	179	136	101.7
56.8	310	176	134	99.9
56.8	308	175	133	99.4
56.8	303	172	131	97.6
56.8	301	171	130	97.1
56.8	296	168	128	95.4
56.8	294	167	127	94.9
56.8	292	166	126	94.4
56.8	287	163	124	92.6
56.8	285	162	123	92.1
56.8	280	159	121	90.3
56.8	278	158	120	89.8
56.8	273	155	118	88.0
56.8	271	154	117	87.5
56.8	266	151	115	85.7
56.8	264	150	114	85.2
56.8	259	147	112	83.4
56.8	257	146	111	82.9
56.8	250	142	108	80.7
56.8	243	138	105	78.4
56.8	241	137	104	77.9
56.8	236	134	102	76.1
56.8	234	133	101	75.6
56.8	229	130	99	73.8
56.8	227	129	98	73.3
56.8	222	126	96	71.5
56.8	220	125	95	71.0
56.8	213	121	92	68.7
56.8	206	117	89	66.5

%E	M1	M2	DM	M*
56.8	199	113	86	64.2
56.8	192	109	83	61.9
56.8	190	108	82	61.4
56.8	185	105	80	59.6
56.8	183	104	79	59.1
56.8	176	100	76	56.8
56.8	169	96	73	54.5
56.8	162	92	70	52.2
56.8	155	88	67	50.0
56.8	148	84	64	47.7
56.8	146	83	63	47.2
56.8	139	79	60	44.9
56.8	132	75	57	42.6
56.8	125	71	54	40.3
56.8	118	67	51	38.0
56.8	111	63	48	35.8
56.8	95	54	41	30.7
56.8	88	50	38	28.4
56.8	81	46	35	26.1
56.8	74	42	32	23.8
56.8	44	25	19	14.2
56.8	37	21	16	11.9
56.7	499	283	216	160.5
56.7	497	282	215	160.0
56.7	496	281	215	159.2
56.7	494	280	214	158.7
56.7	492	279	213	158.2
56.7	490	278	212	157.7
56.7	487	276	211	156.4
56.7	485	275	210	155.9
56.7	483	274	209	155.4
56.7	480	272	208	154.1
56.7	478	271	207	153.6
56.7	476	270	206	153.2
56.7	473	268	205	151.8
56.7	471	267	204	151.4
56.7	469	266	203	150.9
56.7	467	265	202	150.4
56.7	466	264	202	149.6
56.7	464	263	201	149.1
56.7	462	262	200	148.6
56.7	460	261	199	148.1
56.7	457	259	198	146.8
56.7	455	258	197	146.3
56.7	453	257	196	145.8
56.7	450	255	195	144.5
56.7	448	254	194	144.0
56.7	446	253	193	143.5
56.7	443	251	192	142.2
56.7	441	250	191	141.7

%E	M1	M2	DM	M*
56.7	439	249	190	141.2
56.7	436	247	189	139.9
56.7	434	246	188	139.4
56.7	432	245	187	138.9
56.7	430	244	186	138.5
56.7	427	242	185	137.2
56.7	425	241	184	136.7
56.7	423	240	183	136.2
56.7	420	238	182	134.9
56.7	418	237	181	134.4
56.7	416	236	180	133.9
56.7	413	234	179	132.6
56.7	411	233	178	132.1
56.7	409	232	177	131.6
56.7	406	230	176	130.3
56.7	404	229	175	129.8
56.7	402	228	174	129.3
56.7	397	225	172	127.5
56.7	395	224	171	127.0
56.7	393	223	170	126.5
56.7	390	221	169	125.2
56.7	388	220	168	124.7
56.7	386	219	167	124.3
56.7	383	217	166	122.9
56.7	381	216	165	122.5
56.7	379	215	164	122.0
56.7	374	212	162	120.2
56.7	372	211	161	119.7
56.7	367	208	159	117.9
56.7	365	207	158	117.4
56.7	363	206	157	116.9
56.7	360	204	156	115.6
56.7	358	203	155	115.1
56.7	356	202	154	114.6
56.7	353	200	153	113.3
56.7	351	199	152	112.8
56.7	349	198	151	112.3
56.7	344	195	149	110.5
56.7	342	194	148	110.0
56.7	337	191	146	108.3
56.7	335	190	145	107.8
56.7	330	187	143	106.0
56.7	328	186	142	105.5
56.7	326	185	141	105.0
56.7	323	183	140	103.7
56.7	321	182	139	103.2
56.7	319	181	138	102.7
56.7	314	178	136	100.9
56.7	312	177	135	100.4
56.7	307	174	133	98.6

%E	M1	M2	DM	M*
56.7	305	173	132	98.1
56.7	300	170	130	96.3
56.7	298	169	129	95.8
56.7	293	166	127	94.0
56.7	291	165	126	93.6
56.7	289	164	125	93.1
56.7	284	161	123	91.3
56.7	282	160	122	90.8
56.7	277	157	120	89.0
56.7	275	156	119	88.5
56.7	270	153	117	86.7
56.7	268	152	116	86.2
56.7	263	149	114	84.4
56.7	261	148	113	83.9
56.7	254	144	110	81.6
56.7	252	143	109	81.1
56.7	247	140	107	79.4
56.7	245	139	106	78.9
56.7	240	136	104	77.1
56.7	238	135	103	76.6
56.7	233	132	101	74.8
56.7	231	131	100	74.3
56.7	224	127	97	72.0
56.7	217	123	94	69.7
56.7	215	122	93	69.2
56.7	210	119	91	67.4
56.7	208	118	90	66.9
56.7	203	115	88	65.1
56.7	201	114	87	64.7
56.7	194	110	84	62.4
56.7	187	106	81	60.1
56.7	180	102	78	57.8
56.7	178	101	77	57.3
56.7	171	97	74	55.0
56.7	164	93	71	52.7
56.7	157	89	68	50.5
56.7	150	85	65	48.2
56.7	141	80	61	45.4
56.7	134	76	58	43.1
56.7	127	72	55	40.8
56.7	120	68	52	38.5
56.7	104	59	45	33.5
56.7	97	55	42	31.2
56.7	90	51	39	28.9
56.7	67	38	29	21.6
56.7	60	34	26	19.3
56.7	30	17	13	9.6
56.6	500	283	217	160.2
56.6	498	282	216	159.7
56.6	495	280	215	158.4

%E	M1	M2	DM	M*
56.6	493	279	214	157.9
56.6	491	278	213	157.4
56.6	489	277	212	156.9
56.6	488	276	212	156.1
56.6	486	275	211	155.6
56.6	484	274	210	155.1
56.6	482	273	209	154.6
56.6	479	271	208	153.3
56.6	477	270	207	152.8
56.6	475	269	206	152.3
56.6	472	267	205	151.0
56.6	470	266	204	150.5
56.6	468	265	203	150.1
56.6	465	263	202	148.8
56.6	463	262	201	148.3
56.6	461	261	200	147.8
56.6	459	260	199	147.3
56.6	458	259	199	146.5
56.6	456	258	198	146.0
56.6	454	257	197	145.5
56.6	452	256	196	145.0
56.6	449	254	195	143.7
56.6	447	253	194	143.2
56.6	445	252	193	142.7
56.6	442	250	192	141.4
56.6	440	249	191	140.9
56.6	438	248	190	140.4
56.6	435	246	189	139.1
56.6	433	245	188	138.6
56.6	431	244	187	138.1
56.6	429	243	186	137.6
56.6	426	241	185	136.3
56.6	424	240	184	135.8
56.6	422	239	183	135.4
56.6	419	237	182	134.1
56.6	417	236	181	133.6
56.6	415	235	180	133.1
56.6	412	233	179	131.8
56.6	410	232	178	131.3
56.6	408	231	177	130.8
56.6	403	228	175	129.0
56.6	401	227	174	128.5
56.6	399	226	173	128.0
56.6	396	224	172	126.7
56.6	394	223	171	126.2
56.6	392	222	170	125.7
56.6	389	220	169	124.4
56.6	387	219	168	123.9
56.6	385	218	167	123.4
56.6	380	215	165	121.6

%E	M1	M2	DM	M*
56.6	378	214	164	121.2
56.6	376	213	163	120.7
56.6	373	211	162	119.4
56.6	371	210	161	118.9
56.6	369	209	160	118.4
56.6	366	207	159	117.1
56.6	364	206	158	116.6
56.6	362	205	157	116.1
56.6	357	202	155	114.3
56.6	355	201	154	113.8
56.6	350	198	152	112.0
56.6	348	197	151	111.5
56.6	346	196	150	111.0
56.6	343	194	149	109.7
56.6	341	193	148	109.2
56.6	339	192	147	108.7
56.6	334	189	145	106.9
56.6	332	188	144	106.5
56.6	327	185	142	104.7
56.6	325	184	141	104.2
56.6	320	181	139	102.4
56.6	318	180	138	101.9
56.6	316	179	137	101.4
56.6	313	177	136	100.1
56.6	311	176	135	99.6
56.6	309	175	134	99.1
56.6	304	172	132	97.3
56.6	302	171	131	96.8
56.6	297	168	129	95.0
56.6	295	167	128	94.5
56.6	290	164	126	92.7
56.6	288	163	125	92.3
56.6	286	162	124	91.8
56.6	281	159	122	90.0
56.6	279	158	121	89.5
56.6	274	155	119	87.7
56.6	272	154	118	87.2
56.6	267	151	116	85.4
56.6	265	150	115	84.9
56.6	258	146	112	82.6
56.6	256	145	111	82.1
56.6	251	142	109	80.3
56.6	249	141	108	79.8
56.6	244	138	106	78.0
56.6	242	137	105	77.6
56.6	235	133	102	75.3
56.6	228	129	99	73.0
56.6	226	128	98	72.5
56.6	221	125	96	70.7
56.6	219	124	95	70.2

%E	M1	M2	DM	M*
56.6	212	120	92	67.9
56.6	205	116	89	65.6
56.6	198	112	86	63.4
56.6	196	111	85	62.9
56.6	189	107	82	60.6
56.6	182	103	79	58.3
56.6	175	99	76	56.0
56.6	173	98	75	55.5
56.6	166	94	72	53.2
56.6	159	90	69	50.9
56.6	152	86	66	48.7
56.6	145	82	63	46.4
56.6	143	81	62	45.9
56.6	136	77	59	43.6
56.6	129	73	56	41.3
56.6	122	69	53	39.0
56.6	113	64	49	36.2
56.6	106	60	46	34.0
56.6	99	56	43	31.7
56.6	83	47	36	26.6
56.6	76	43	33	24.3
56.6	53	30	23	17.0
56.5	499	282	217	159.4
56.5	497	281	216	158.9
56.5	496	280	216	158.1
56.5	494	279	215	157.6
56.5	492	278	214	157.1
56.5	490	277	213	156.6
56.5	487	275	212	155.3
56.5	485	274	211	154.8
56.5	483	273	210	154.3
56.5	481	272	209	153.8
56.5	480	271	209	153.0
56.5	478	270	208	152.5
56.5	476	269	207	152.0
56.5	474	268	206	151.5
56.5	471	266	205	150.2
56.5	469	265	204	149.7
56.5	467	264	203	149.2
56.5	464	262	202	147.9
56.5	462	261	201	147.4
56.5	460	260	200	147.0
56.5	457	258	199	145.7
56.5	455	257	198	145.2
56.5	453	256	197	144.7
56.5	451	255	196	144.2
56.5	448	253	195	142.9
56.5	446	252	194	142.4
56.5	444	251	193	141.9
56.5	441	249	192	140.6

%E	M1	M2	DM	M*
56.5	439	248	191	140.1
56.5	437	247	190	139.6
56.5	434	245	189	138.3
56.5	432	244	188	137.8
56.5	430	243	187	137.3
56.5	428	242	186	136.8
56.5	425	240	185	135.5
56.5	423	239	184	135.0
56.5	421	238	183	134.5
56.5	418	236	182	133.2
56.5	416	235	181	132.8
56.5	414	234	180	132.3
56.5	409	231	178	130.5
56.5	407	230	177	130.0
56.5	405	229	176	129.5
56.5	402	227	175	128.2
56.5	400	226	174	127.7
56.5	398	225	173	127.2
56.5	395	223	172	125.9
56.5	393	222	171	125.4
56.5	391	221	170	124.9
56.5	386	218	168	123.1
56.5	384	217	167	122.6
56.5	382	216	166	122.1
56.5	379	214	165	120.8
56.5	377	213	164	120.3
56.5	375	212	163	119.9
56.5	372	210	162	118.5
56.5	370	209	161	118.1
56.5	368	208	160	117.6
56.5	363	205	158	115.8
56.5	361	204	157	115.3
56.5	359	203	156	114.8
56.5	356	201	155	113.5
56.5	354	200	154	113.0
56.5	352	199	153	112.5
56.5	347	196	151	110.7
56.5	345	195	150	110.2
56.5	340	192	148	108.4
56.5	338	191	147	107.9
56.5	336	190	146	107.4
56.5	333	188	145	106.1
56.5	331	187	144	105.6
56.5	329	186	143	105.2
56.5	324	183	141	103.4
56.5	322	182	140	102.9
56.5	317	179	138	101.1
56.5	315	178	137	100.6
56.5	310	175	135	98.8
56.5	308	174	134	98.3

%E	M1	M2	DM	M*
56.5	306	173	133	97.8
56.5	301	170	131	96.0
56.5	299	169	130	95.5
56.5	294	166	128	93.7
56.5	292	165	127	93.2
56.5	285	161	124	91.0
56.5	283	160	123	90.5
56.5	278	157	121	88.7
56.5	276	156	120	88.2
56.5	271	153	118	86.4
56.5	269	152	117	85.9
56.5	262	148	114	83.6
56.5	260	147	113	83.1
56.5	255	144	111	81.3
56.5	253	143	110	80.8
56.5	248	140	108	79.0
56.5	246	139	107	78.5
56.5	239	135	104	76.3
56.5	237	134	103	75.8
56.5	232	131	101	74.0
56.5	230	130	100	73.5
56.5	223	126	97	71.2
56.5	216	122	94	68.9
56.5	214	121	93	68.4
56.5	209	118	91	66.6
56.5	207	117	90	66.1
56.5	200	113	87	63.8
56.5	193	109	84	61.6
56.5	191	108	83	61.1
56.5	186	105	81	59.3
56.5	184	104	80	58.8
56.5	177	100	77	56.5
56.5	170	96	74	54.2
56.5	168	95	73	53.7
56.5	161	91	70	51.4
56.5	154	87	67	49.1
56.5	147	83	64	46.9
56.5	138	78	60	44.1
56.5	131	74	57	41.8
56.5	124	70	54	39.5
56.5	115	65	50	36.7
56.5	108	61	47	34.5
56.5	92	52	40	29.4
56.5	85	48	37	27.1
56.5	69	39	30	22.0
56.5	62	35	27	19.8
56.5	46	26	20	14.7
56.5	23	13	10	7.3
56.4	500	282	218	159.0
56.4	498	281	217	158.6

%E	M1	M2	DM	M*
56.4	495	279	216	157.3
56.4	493	278	215	156.8
56.4	491	277	214	156.3
56.4	489	276	213	155.8
56.4	488	275	213	155.0
56.4	486	274	212	154.5
56.4	484	273	211	154.0
56.4	482	272	210	153.5
56.4	479	270	209	152.2
56.4	477	269	208	151.7
56.4	475	268	207	151.2
56.4	473	267	206	150.7
56.4	472	266	206	149.9
56.4	470	265	205	149.4
56.4	468	264	204	148.9
56.4	466	263	203	148.4
56.4	463	261	202	147.1
56.4	461	260	201	146.6
56.4	459	259	200	146.1
56.4	456	257	199	144.8
56.4	454	256	198	144.4
56.4	452	255	197	143.9
56.4	450	254	196	143.4
56.4	447	252	195	142.1
56.4	445	251	194	141.6
56.4	443	250	193	141.1
56.4	440	248	192	139.8
56.4	438	247	191	139.3
56.4	436	246	190	138.8
56.4	433	244	189	137.5
56.4	431	243	188	137.0
56.4	429	242	187	136.5
56.4	427	241	186	136.0
56.4	424	239	185	134.7
56.4	422	238	184	134.2
56.4	420	237	183	133.7
56.4	417	235	182	132.4
56.4	415	234	181	131.9
56.4	413	233	180	131.5
56.4	411	232	179	131.0
56.4	408	230	178	129.7
56.4	406	229	177	129.2
56.4	404	228	176	128.7
56.4	401	226	175	127.4
56.4	399	225	174	126.9
56.4	397	224	173	126.4
56.4	392	221	171	124.6
56.4	390	220	170	124.1
56.4	388	219	169	123.6
56.4	385	217	168	122.3

%E	M1	M2	DM	M*
56.4	383	216	167	121.8
56.4	381	215	166	121.3
56.4	376	212	164	119.5
56.4	374	211	163	119.0
56.4	369	208	161	117.2
56.4	367	207	160	116.8
56.4	365	206	159	116.3
56.4	362	204	158	115.0
56.4	360	203	157	114.5
56.4	358	202	156	114.0
56.4	353	199	154	112.2
56.4	351	198	153	111.7
56.4	349	197	152	111.2
56.4	346	195	151	109.9
56.4	344	194	150	109.4
56.4	342	193	149	108.9
56.4	337	190	147	107.1
56.4	335	189	146	106.6
56.4	330	186	144	104.8
56.4	328	185	143	104.3
56.4	326	184	142	103.9
56.4	321	181	140	102.1
56.4	319	180	139	101.6
56.4	314	177	137	99.8
56.4	312	176	136	99.3
56.4	307	173	134	97.5
56.4	305	172	133	97.0
56.4	303	171	132	96.5
56.4	298	168	130	94.7
56.4	296	167	129	94.2
56.4	291	164	127	92.4
56.4	289	163	126	91.9
56.4	287	162	125	91.4
56.4	282	159	123	89.6
56.4	280	158	122	89.2
56.4	275	155	120	87.4
56.4	273	154	119	86.9
56.4	266	150	116	84.6
56.4	264	149	115	84.1
56.4	259	146	113	82.3
56.4	257	145	112	81.8
56.4	250	141	109	79.5
56.4	243	137	106	77.2
56.4	241	136	105	76.7
56.4	236	133	103	75.0
56.4	234	132	102	74.5
56.4	227	128	99	72.2
56.4	225	127	98	71.7
56.4	220	124	96	69.9
56.4	218	123	95	69.4

%E	M1	M2	DM	M*
56.4	211	119	92	67.1
56.4	204	115	89	64.8
56.4	202	114	88	64.3
56.4	195	110	85	62.1
56.4	188	106	82	59.8
56.4	181	102	79	57.5
56.4	179	101	78	57.0
56.4	172	97	75	54.7
56.4	165	93	72	52.4
56.4	163	92	71	51.9
56.4	156	88	68	49.6
56.4	149	84	65	47.4
56.4	140	79	61	44.6
56.4	133	75	58	42.3
56.4	117	66	51	37.2
56.4	110	62	48	34.9
56.4	101	57	44	32.2
56.4	94	53	41	29.9
56.4	78	44	34	24.8
56.4	55	31	24	17.5
56.4	39	22	17	12.4
56.3	499	281	218	158.2
56.3	497	280	217	157.7
56.3	496	279	217	156.9
56.3	494	278	216	156.4
56.3	492	277	215	156.0
56.3	490	276	214	155.5
56.3	487	274	213	154.2
56.3	485	273	212	153.7
56.3	483	272	211	153.2
56.3	481	271	210	152.7
56.3	480	270	210	151.9
56.3	478	269	209	151.4
56.3	476	268	208	150.9
56.3	474	267	207	150.4
56.3	471	265	206	149.1
56.3	469	264	205	148.6
56.3	467	263	204	148.1
56.3	465	262	203	147.6
56.3	464	261	203	146.8
56.3	462	260	202	146.3
56.3	460	259	201	145.8
56.3	458	258	200	145.3
56.3	455	256	199	144.0
56.3	453	255	198	143.5
56.3	451	254	197	143.1
56.3	449	253	196	142.6
56.3	448	252	196	141.8
56.3	446	251	195	141.3
56.3	444	250	194	140.8

%E	M1	M2	DM	M*
56.3	442	249	193	140.3
56.3	439	247	192	139.0
56.3	437	246	191	138.5
56.3	435	245	190	138.0
56.3	432	243	189	136.7
56.3	430	242	188	136.2
56.3	428	241	187	135.7
56.3	426	240	186	135.2
56.3	423	238	185	133.9
56.3	421	237	184	133.4
56.3	419	236	183	132.9
56.3	416	234	182	131.6
56.3	414	233	181	131.1
56.3	412	232	180	130.6
56.3	410	231	179	130.1
56.3	407	229	178	128.8
56.3	405	228	177	128.4
56.3	403	227	176	127.9
56.3	400	225	175	126.6
56.3	398	224	174	126.1
56.3	396	223	173	125.6
56.3	394	222	172	125.1
56.3	391	220	171	123.8
56.3	389	219	170	123.3
56.3	387	218	169	122.8
56.3	384	216	168	121.5
56.3	382	215	167	121.0
56.3	380	214	166	120.5
56.3	378	213	165	120.0
56.3	375	211	164	118.7
56.3	373	210	163	118.2
56.3	371	209	162	117.7
56.3	368	207	161	116.4
56.3	366	206	160	115.9
56.3	364	205	159	115.5
56.3	359	202	157	113.7
56.3	357	201	156	113.2
56.3	355	200	155	112.7
56.3	352	198	154	111.4
56.3	350	197	153	110.9
56.3	348	196	152	110.4
56.3	343	193	150	108.6
56.3	341	192	149	108.1
56.3	339	191	148	107.6
56.3	336	189	147	106.3
56.3	334	188	146	105.8
56.3	332	187	145	105.3
56.3	327	184	143	103.5
56.3	325	183	142	103.0
56.3	323	182	141	102.6

%E	M1	M2	DM	M*
56.3	320	180	140	101.3
56.3	318	179	139	100.8
56.3	316	178	138	100.3
56.3	311	175	136	98.5
56.3	309	174	135	98.0
56.3	304	171	133	96.2
56.3	302	170	132	95.7
56.3	300	169	131	95.2
56.3	295	166	129	93.4
56.3	293	165	128	92.9
56.3	288	162	126	91.1
56.3	286	161	125	90.6
56.3	284	160	124	90.1
56.3	279	157	122	88.3
56.3	277	156	121	87.9
56.3	272	153	119	86.1
56.3	270	152	118	85.6
56.3	268	151	117	85.1
56.3	263	148	115	83.3
56.3	261	147	114	82.8
56.3	256	144	112	81.0
56.3	254	143	111	80.5
56.3	252	142	110	80.0
56.3	247	139	108	78.2
56.3	245	138	107	77.7
56.3	240	135	105	75.9
56.3	238	134	104	75.4
56.3	231	130	101	73.2
56.3	229	129	100	72.7
56.3	224	126	98	70.9
56.3	222	125	97	70.4
56.3	215	121	94	68.1
56.3	213	120	93	67.6
56.3	208	117	91	65.8
56.3	206	116	90	65.3
56.3	199	112	87	63.0
56.3	197	111	86	62.5
56.3	192	108	84	60.8
56.3	190	107	83	60.3
56.3	183	103	80	58.0
56.3	176	99	77	55.7
56.3	174	98	76	55.2
56.3	167	94	73	52.9
56.3	160	90	70	50.6
56.3	158	89	69	50.1
56.3	151	85	66	47.8
56.3	144	81	63	45.6
56.3	142	80	62	45.1
56.3	135	76	59	42.8
56.3	128	72	56	40.5

%E	M1	M2	DM	M*
56.3	126	71	55	40.0
56.3	119	67	52	37.7
56.3	112	63	49	35.4
56.3	103	58	45	32.7
56.3	96	54	42	30.4
56.3	87	49	38	27.6
56.3	80	45	35	25.3
56.3	71	40	31	22.5
56.3	64	36	28	20.3
56.3	48	27	21	15.2
56.3	32	18	14	10.1
56.3	16	9	7	5.1
56.2	500	281	219	157.9
56.2	498	280	218	157.4
56.2	495	278	217	156.1
56.2	493	277	216	155.6
56.2	491	276	215	155.1
56.2	489	275	214	154.7
56.2	486	273	213	153.4
56.2	484	272	212	152.9
56.2	482	271	211	152.4
56.2	479	269	210	151.1
56.2	477	268	209	150.6
56.2	475	267	208	150.1
56.2	473	266	207	149.6
56.2	470	264	206	148.3
56.2	468	263	205	147.8
56.2	466	262	204	147.3
56.2	463	260	203	146.0
56.2	461	259	202	145.5
56.2	459	258	201	145.0
56.2	457	257	200	144.5
56.2	454	255	199	143.2
56.2	452	254	198	142.7
56.2	450	253	197	142.2
56.2	447	251	196	140.9
56.2	445	250	195	140.4
56.2	443	249	194	140.0
56.2	441	248	193	139.5
56.2	438	246	192	138.2
56.2	436	245	191	137.7
56.2	434	244	190	137.2
56.2	429	241	188	135.4
56.2	427	240	187	134.9
56.2	425	239	186	134.4
56.2	422	237	185	133.1
56.2	420	236	184	132.6
56.2	418	235	183	132.1
56.2	413	232	181	130.3
56.2	411	231	180	129.8

%E	M1	M2	DM	M*
56.2	409	230	179	129.3
56.2	406	228	178	128.0
56.2	404	227	177	127.5
56.2	402	226	176	127.1
56.2	397	223	174	125.3
56.2	395	222	173	124.8
56.2	393	221	172	124.3
56.2	390	219	171	123.0
56.2	388	218	170	122.5
56.2	386	217	169	122.0
56.2	381	214	167	120.2
56.2	379	213	166	119.7
56.2	377	212	165	119.2
56.2	374	210	164	117.9
56.2	372	209	163	117.4
56.2	370	208	162	116.9
56.2	365	205	160	115.1
56.2	363	204	159	114.6
56.2	361	203	158	114.2
56.2	356	200	156	112.4
56.2	354	199	155	111.9
56.2	349	196	153	110.1
56.2	347	195	152	109.6
56.2	345	194	151	109.1
56.2	340	191	149	107.3
56.2	338	190	148	106.8
56.2	333	187	146	105.0
56.2	331	186	145	104.5
56.2	329	185	144	104.0
56.2	324	182	142	102.2
56.2	322	181	141	101.7
56.2	317	178	139	99.9
56.2	315	177	138	99.5
56.2	313	176	137	99.0
56.2	308	173	135	97.2
56.2	306	172	134	96.7
56.2	299	168	131	94.4
56.2	297	167	130	93.9
56.2	292	164	128	92.1
56.2	290	163	127	91.6
56.2	283	159	124	89.3
56.2	281	158	123	88.8
56.2	276	155	121	87.0
56.2	274	154	120	86.6
56.2	267	150	117	84.3
56.2	265	149	116	83.8
56.2	260	146	114	82.0
56.2	258	145	113	81.5
56.2	251	141	110	79.2
56.2	249	140	109	78.7

%E	M1	M2	DM	M*
56.2	242	136	106	76.4
56.2	235	132	103	74.1
56.2	233	131	102	73.7
56.2	226	127	99	71.4
56.2	219	123	96	69.1
56.2	217	122	95	68.6
56.2	210	118	92	66.3
56.2	203	114	89	64.0
56.2	201	113	88	63.5
56.2	194	109	85	61.2
56.2	187	105	82	59.0
56.2	185	104	81	58.5
56.2	178	100	78	56.2
56.2	169	95	74	53.4
56.2	162	91	71	51.1
56.2	153	86	67	48.3
56.2	146	82	64	46.1
56.2	137	77	60	43.3
56.2	130	73	57	41.0
56.2	121	68	53	38.2
56.2	105	59	46	33.2
56.2	89	50	39	28.1
56.2	73	41	32	23.0
56.1	499	280	219	157.1
56.1	497	279	218	156.6
56.1	494	277	217	155.3
56.1	492	276	216	154.8
56.1	490	275	215	154.3
56.1	488	274	214	153.8
56.1	487	273	214	153.0
56.1	485	272	213	152.5
56.1	483	271	212	152.1
56.1	481	270	211	151.6
56.1	478	268	210	150.3
56.1	476	267	209	149.8
56.1	474	266	208	149.3
56.1	472	265	207	148.8
56.1	471	264	207	148.0
56.1	469	263	206	147.5
56.1	467	262	205	147.0
56.1	465	261	204	146.5
56.1	462	259	203	145.2
56.1	460	258	202	144.7
56.1	458	257	201	144.2
56.1	456	256	200	143.7
56.1	453	254	199	142.4
56.1	451	253	198	141.9
56.1	449	252	197	141.4
56.1	446	250	196	140.1
56.1	444	249	195	139.6

%E	M1	M2	DM	M*
56.1	442	248	194	139.1
56.1	440	247	193	138.7
56.1	437	245	192	137.4
56.1	435	244	191	136.9
56.1	433	243	190	136.4
56.1	431	242	189	135.9
56.1	428	240	188	134.6
56.1	426	239	187	134.1
56.1	424	238	186	133.6
56.1	421	236	185	132.3
56.1	419	235	184	131.8
56.1	417	234	183	131.3
56.1	415	233	182	130.8
56.1	412	231	181	129.5
56.1	410	230	180	129.0
56.1	408	229	179	128.5
56.1	403	226	177	126.7
56.1	401	225	176	126.2
56.1	399	224	175	125.8
56.1	396	222	174	124.5
56.1	394	221	173	124.0
56.1	392	220	172	123.5
56.1	387	217	170	121.7
56.1	385	216	169	121.2
56.1	383	215	168	120.7
56.1	380	213	167	119.4
56.1	378	212	166	118.9
56.1	376	211	165	118.4
56.1	371	208	163	116.6
56.1	369	207	162	116.1
56.1	367	206	161	115.6
56.1	362	203	159	113.8
56.1	360	202	158	113.3
56.1	358	201	157	112.9
56.1	355	199	156	111.6
56.1	353	198	155	111.1
56.1	351	197	154	110.6
56.1	346	194	152	108.8
56.1	344	193	151	108.3
56.1	342	192	150	107.8
56.1	337	189	148	106.0
56.1	335	188	147	105.5
56.1	330	185	145	103.7
56.1	328	184	144	103.2
56.1	326	183	143	102.7
56.1	321	180	141	100.9
56.1	319	179	140	100.4
56.1	314	176	138	98.6
56.1	312	175	137	98.2
56.1	310	174	136	97.7

%E	M1	M2	DM	M*
56.1	305	171	134	95.9
56.1	303	170	133	95.4
56.1	301	169	132	94.9
56.1	296	166	130	93.1
56.1	294	165	129	92.6
56.1	289	162	127	90.8
56.1	287	161	126	90.3
56.1	285	160	125	89.8
56.1	280	157	123	88.0
56.1	278	156	122	87.5
56.1	271	152	119	85.3
56.1	269	151	118	84.8
56.1	264	148	116	83.0
56.1	262	147	115	82.5
56.1	255	143	112	80.2
56.1	253	142	111	79.7
56.1	246	138	108	77.4
56.1	244	137	107	76.9
56.1	239	134	105	75.1
56.1	237	133	104	74.6
56.1	230	129	101	72.4
56.1	228	128	100	71.9
56.1	223	125	98	70.1
56.1	221	124	97	69.6
56.1	214	120	94	67.3
56.1	212	119	93	66.8
56.1	205	115	90	64.5
56.1	198	111	87	62.2
56.1	196	110	86	61.7
56.1	189	106	83	59.4
56.1	180	101	79	56.7
56.1	173	97	76	54.4
56.1	171	96	75	53.9
56.1	164	92	72	51.6
56.1	157	88	69	49.3
56.1	155	87	68	48.8
56.1	148	83	65	46.5
56.1	139	78	61	43.8
56.1	132	74	58	41.5
56.1	123	69	54	38.7
56.1	114	64	50	35.9
56.1	107	60	47	33.6
56.1	98	55	43	30.9
56.1	82	46	36	25.8
56.1	66	37	29	20.7
56.1	57	32	25	18.0
56.1	41	23	18	12.9
56.0	500	280	220	156.8
56.0	498	279	219	156.3
56.0	496	278	218	155.8

%E	M1	M2	DM	M*
56.0	495	277	218	155.0
56.0	493	276	217	154.5
56.0	491	275	216	154.0
56.0	489	274	215	153.5
56.0	486	272	214	152.2
56.0	484	271	213	151.7
56.0	482	270	212	151.2
56.0	480	269	211	150.8
56.0	479	268	211	149.9
56.0	477	267	210	149.5
56.0	475	266	209	149.0
56.0	473	265	208	148.5
56.0	470	263	207	147.2
56.0	468	262	206	146.7
56.0	466	261	205	146.2
56.0	464	260	204	145.7
56.0	461	258	203	144.4
56.0	459	257	202	143.9
56.0	457	256	201	143.4
56.0	455	255	200	142.9
56.0	452	253	199	141.6
56.0	450	252	198	141.1
56.0	448	251	197	140.6
56.0	445	249	196	139.3
56.0	443	248	195	138.8
56.0	441	247	194	138.3
56.0	439	246	193	137.8
56.0	436	244	192	136.6
56.0	434	243	191	136.1
56.0	432	242	190	135.6
56.0	430	241	189	135.1
56.0	427	239	188	133.8
56.0	425	238	187	133.3
56.0	423	237	186	132.8
56.0	420	235	185	131.5
56.0	418	234	184	131.0
56.0	416	233	183	130.5
56.0	414	232	182	130.0
56.0	411	230	181	128.7
56.0	409	229	180	128.2
56.0	407	228	179	127.7
56.0	405	227	178	127.2
56.0	402	225	177	125.9
56.0	400	224	176	125.4
56.0	398	223	175	124.9
56.0	393	220	173	123.2
56.0	391	219	172	122.7
56.0	389	218	171	122.2
56.0	386	216	170	120.9
56.0	384	215	169	120.4

%E	M1	M2	DM	M*
56.0	382	214	168	119.9
56.0	377	211	166	118.1
56.0	375	210	165	117.6
56.0	373	209	164	117.1
56.0	368	206	162	115.3
56.0	366	205	161	114.8
56.0	364	204	160	114.3
56.0	361	202	159	113.0
56.0	359	201	158	112.5
56.0	357	200	157	112.0
56.0	352	197	155	110.3
56.0	350	196	154	109.8
56.0	348	195	153	109.3
56.0	343	192	151	107.5
56.0	341	191	150	107.0
56.0	339	190	149	106.5
56.0	336	188	148	105.2
56.0	334	187	147	104.7
56.0	332	186	146	104.2
56.0	327	183	144	102.4
56.0	325	182	143	101.9
56.0	323	181	142	101.4
56.0	318	178	140	99.6
56.0	316	177	139	99.1
56.0	309	173	136	96.9
56.0	307	172	135	96.4
56.0	302	169	133	94.6
56.0	300	168	132	94.1
56.0	298	167	131	93.6
56.0	293	164	129	91.8
56.0	291	163	128	91.3
56.0	284	159	125	89.0
56.0	282	158	124	88.5
56.0	277	155	122	86.7
56.0	275	154	121	86.2
56.0	273	153	120	85.7
56.0	268	150	118	84.0
56.0	266	149	117	83.5
56.0	259	145	114	81.2
56.0	257	144	113	80.7
56.0	252	141	111	78.9
56.0	250	140	110	78.4
56.0	248	139	109	77.9
56.0	243	136	107	76.1
56.0	241	135	106	75.6
56.0	234	131	103	73.3
56.0	232	130	102	72.8
56.0	225	126	99	70.6
56.0	218	122	96	68.3
56.0	216	121	95	67.8

%E	M1	M2	DM	M*
56.0	209	117	92	65.5
56.0	207	116	91	65.0
56.0	200	112	88	62.7
56.0	193	108	85	60.4
56.0	191	107	84	59.9
56.0	184	103	81	57.7
56.0	182	102	80	57.2
56.0	175	98	77	54.9
56.0	168	94	74	52.6
56.0	166	93	73	52.1
56.0	159	89	70	49.8
56.0	150	84	66	47.0
56.0	141	79	62	44.3
56.0	134	75	59	42.0
56.0	125	70	55	39.2
56.0	116	65	51	36.4
56.0	109	61	48	34.1
56.0	100	56	44	31.4
56.0	91	51	40	28.6
56.0	84	47	37	26.3
56.0	75	42	33	23.5
56.0	50	28	22	15.7
56.0	25	14	11	7.8
55.9	499	279	220	156.0
55.9	497	278	219	155.5
55.9	494	276	218	154.2
55.9	492	275	217	153.7
55.9	490	274	216	153.2
55.9	488	273	215	152.7
55.9	487	272	215	151.9
55.9	485	271	214	151.4
55.9	483	270	213	150.9
55.9	481	269	212	150.4
55.9	478	267	211	149.1
55.9	476	266	210	148.6
55.9	474	265	209	148.2
55.9	472	264	208	147.7
55.9	469	262	207	146.4
55.9	467	261	206	145.9
55.9	465	260	205	145.4
55.9	463	259	204	144.9
55.9	460	257	203	143.6
55.9	458	256	202	143.1
55.9	456	255	201	142.6
55.9	454	254	200	142.1
55.9	453	253	200	141.3
55.9	451	252	199	140.8
55.9	449	251	198	140.3
55.9	447	250	197	139.8
55.9	444	248	196	138.5

%E	M1	M2	DM	M*
55.9	442	247	195	138.0
55.9	440	246	194	137.5
55.9	438	245	193	137.0
55.9	435	243	192	135.7
55.9	433	242	191	135.3
55.9	431	241	190	134.8
55.9	429	240	189	134.3
55.9	426	238	188	133.0
55.9	424	237	187	132.5
55.9	422	236	186	132.0
55.9	417	233	184	130.2
55.9	415	232	183	129.7
55.9	413	231	182	129.2
55.9	410	229	181	127.9
55.9	408	228	180	127.4
55.9	406	227	179	126.9
55.9	404	226	178	126.4
55.9	401	224	177	125.1
55.9	399	223	176	124.6
55.9	397	222	175	124.1
55.9	395	221	174	123.6
55.9	392	219	173	122.3
55.9	390	218	172	121.9
55.9	388	217	171	121.4
55.9	383	214	169	119.6
55.9	381	213	168	119.1
55.9	379	212	167	118.6
55.9	376	210	166	117.3
55.9	374	209	165	116.8
55.9	372	208	164	116.3
55.9	370	207	163	115.8
55.9	367	205	162	114.5
55.9	365	204	161	114.0
55.9	363	203	160	113.5
55.9	358	200	158	111.7
55.9	356	199	157	111.2
55.9	354	198	156	110.7
55.9	349	195	154	109.0
55.9	347	194	153	108.5
55.9	345	193	152	108.0
55.9	340	190	150	106.2
55.9	338	189	149	105.7
55.9	333	186	147	103.9
55.9	331	185	146	103.4
55.9	329	184	145	102.9
55.9	324	181	143	101.1
55.9	322	180	142	100.6
55.9	320	179	141	100.1
55.9	315	176	139	98.3
55.9	313	175	138	97.8

%E	M1	M2	DM	M*
55.9	311	174	137	97.4
55.9	306	171	135	95.6
55.9	304	170	134	95.1
55.9	299	167	132	93.3
55.9	297	166	131	92.8
55.9	295	165	130	92.3
55.9	290	162	128	90.5
55.9	288	161	127	90.0
55.9	286	160	126	89.5
55.9	281	157	124	87.7
55.9	279	156	123	87.2
55.9	272	152	120	84.9
55.9	270	151	119	84.4
55.9	263	147	116	82.2
55.9	261	146	115	81.7
55.9	256	143	113	79.9
55.9	254	142	112	79.4
55.9	247	138	109	77.1
55.9	245	137	108	76.6
55.9	238	133	105	74.3
55.9	236	132	104	73.8
55.9	229	128	101	71.5
55.9	227	127	100	71.1
55.9	222	124	98	69.3
55.9	220	123	97	68.8
55.9	213	119	94	66.5
55.9	211	118	93	66.0
55.9	204	114	90	63.7
55.9	202	113	89	63.2
55.9	195	109	86	60.9
55.9	188	105	83	58.6
55.9	186	104	82	58.2
55.9	179	100	79	55.9
55.9	177	99	78	55.4
55.9	170	95	75	53.1
55.9	161	90	71	50.3
55.9	152	85	67	47.5
55.9	145	81	64	45.2
55.9	143	80	63	44.8
55.9	136	76	60	42.5
55.9	127	71	56	39.7
55.9	118	66	52	36.9
55.9	111	62	49	34.6
55.9	102	57	45	31.9
55.9	93	52	41	29.1
55.9	68	38	30	21.2
55.9	59	33	26	18.5
55.9	34	19	15	10.6
55.8	500	279	221	155.7
55.8	498	278	220	155.2

%E	M1	M2	DM	M*
55.8	496	277	219	154.7
55.8	495	276	219	153.9
55.8	493	275	218	153.4
55.8	491	274	217	152.9
55.8	489	273	216	152.4
55.8	486	271	215	151.1
55.8	484	270	214	150.6
55.8	482	269	213	150.1
55.8	480	268	212	149.6
55.8	477	266	211	148.3
55.8	475	265	210	147.8
55.8	473	264	209	147.3
55.8	471	263	208	146.9
55.8	468	261	207	145.6
55.8	466	260	206	145.1
55.8	464	259	205	144.6
55.8	462	258	204	144.1
55.8	459	256	203	142.8
55.8	457	255	202	142.3
55.8	455	254	201	141.8
55.8	452	252	200	140.5
55.8	450	251	199	140.0
55.8	448	250	198	139.5
55.8	446	249	197	139.0
55.8	443	247	196	137.7
55.8	441	246	195	137.2
55.8	439	245	194	136.7
55.8	437	244	193	136.2
55.8	434	242	192	134.9
55.8	432	241	191	134.4
55.8	430	240	190	134.0
55.8	428	239	189	133.5
55.8	425	237	188	132.2
55.8	423	236	187	131.7
55.8	421	235	186	131.2
55.8	419	234	185	130.7
55.8	416	232	184	129.4
55.8	414	231	183	128.9
55.8	412	230	182	128.4
55.8	407	227	180	126.6
55.8	405	226	179	126.1
55.8	403	225	178	125.6
55.8	400	223	177	124.3
55.8	398	222	176	123.8
55.8	396	221	175	123.3
55.8	394	220	174	122.8
55.8	391	218	173	121.5
55.8	389	217	172	121.1
55.8	387	216	171	120.6
55.8	385	215	170	120.1

%E	M1	M2	DM	M*
55.8	382	213	169	118.8
55.8	380	212	168	118.3
55.8	378	211	167	117.8
55.8	373	208	165	116.0
55.8	371	207	164	115.5
55.8	369	206	163	115.0
55.8	364	203	161	113.2
55.8	362	202	160	112.7
55.8	360	201	159	112.2
55.8	355	198	157	110.4
55.8	353	197	156	109.9
55.8	351	196	155	109.4
55.8	346	193	153	107.7
55.8	344	192	152	107.2
55.8	342	191	151	106.7
55.8	339	189	150	105.4
55.8	337	188	149	104.9
55.8	335	187	148	104.4
55.8	330	184	146	102.6
55.8	328	183	145	102.1
55.8	326	182	144	101.6
55.8	321	179	142	99.8
55.8	319	178	141	99.3
55.8	317	177	140	98.8
55.8	312	174	138	97.0
55.8	310	173	137	96.5
55.8	308	172	136	96.1
55.8	303	169	134	94.3
55.8	301	168	133	93.8
55.8	294	164	130	91.5
55.8	292	163	129	91.0
55.8	285	159	126	88.7
55.8	283	158	125	88.2
55.8	278	155	123	86.4
55.8	276	154	122	85.9
55.8	274	153	121	85.4
55.8	269	150	119	83.6
55.8	267	149	118	83.1
55.8	265	148	117	82.7
55.8	260	145	115	80.9
55.8	258	144	114	80.4
55.8	251	140	111	78.1
55.8	249	139	110	77.6
55.8	242	135	107	75.3
55.8	240	134	106	74.8
55.8	233	130	103	72.5
55.8	231	129	102	72.0
55.8	226	126	100	70.2
55.8	224	125	99	69.8
55.8	217	121	96	67.5

%E	M1	M2	DM	M*
55.8	215	120	95	67.0
55.8	208	116	92	64.7
55.8	206	115	91	64.2
55.8	199	111	88	61.9
55.8	197	110	87	61.4
55.8	190	106	84	59.1
55.8	181	101	80	56.4
55.8	172	96	76	53.6
55.8	165	92	73	51.3
55.8	163	91	72	50.8
55.8	156	87	69	48.5
55.8	154	86	68	48.0
55.8	147	82	65	45.7
55.8	138	77	61	43.0
55.8	129	72	57	40.2
55.8	120	67	53	37.4
55.8	113	63	50	35.1
55.8	104	58	46	32.3
55.8	95	53	42	29.6
55.8	86	48	38	26.8
55.8	77	43	34	24.0
55.8	52	29	23	16.2
55.8	43	24	19	13.4
55.7	499	278	221	154.9
55.7	497	277	220	154.4
55.7	494	275	219	153.1
55.7	492	274	218	152.6
55.7	490	273	217	152.1
55.7	488	272	216	151.6
55.7	485	270	215	150.3
55.7	483	269	214	149.8
55.7	481	268	213	149.3
55.7	479	267	212	148.8
55.7	476	265	211	147.5
55.7	474	264	210	147.0
55.7	472	263	209	146.5
55.7	470	262	208	146.1
55.7	469	261	208	145.2
55.7	467	260	207	144.8
55.7	465	259	206	144.3
55.7	463	258	205	143.8
55.7	461	257	204	143.3
55.7	460	256	204	142.5
55.7	458	255	203	142.0
55.7	456	254	202	141.5
55.7	454	253	201	141.0
55.7	451	251	200	139.7
55.7	449	250	199	139.2
55.7	447	249	198	138.7
55.7	445	248	197	138.2

%E	M1	M2	DM	M*
55.7	442	246	196	136.9
55.7	440	245	195	136.4
55.7	438	244	194	135.9
55.7	436	243	193	135.4
55.7	433	241	192	134.1
55.7	431	240	191	133.6
55.7	429	239	190	133.1
55.7	427	238	189	132.7
55.7	424	236	188	131.4
55.7	422	235	187	130.9
55.7	420	234	186	130.4
55.7	418	233	185	129.9
55.7	415	231	184	128.6
55.7	413	230	183	128.1
55.7	411	229	182	127.6
55.7	409	228	181	127.1
55.7	406	226	180	125.8
55.7	404	225	179	125.3
55.7	402	224	178	124.8
55.7	397	221	176	123.0
55.7	395	220	175	122.5
55.7	393	219	174	122.0
55.7	388	216	172	120.2
55.7	386	215	171	119.8
55.7	384	214	170	119.3
55.7	379	211	168	117.5
55.7	377	210	167	117.0
55.7	375	209	166	116.5
55.7	370	206	164	114.7
55.7	368	205	163	114.2
55.7	366	204	162	113.7
55.7	361	201	160	111.9
55.7	359	200	159	111.4
55.7	357	199	158	110.9
55.7	354	197	157	109.6
55.7	352	196	156	109.1
55.7	350	195	155	108.6
55.7	348	194	154	108.1
55.7	345	192	153	106.9
55.7	343	191	152	106.4
55.7	341	190	151	105.9
55.7	336	187	149	104.1
55.7	334	186	148	103.6
55.7	332	185	147	103.1
55.7	327	182	145	101.3
55.7	325	181	144	100.8
55.7	323	180	143	100.3
55.7	318	177	141	98.5
55.7	316	176	140	98.0
55.7	314	175	139	97.5

%E	M_1	M_2	DM	M*
55.7	309	172	137	95.7
55.7	307	171	136	95.2
55.7	305	170	135	94.8
55.7	300	167	133	93.0
55.7	298	166	132	92.5
55.7	296	165	131	92.0
55.7	291	162	129	90.2
55.7	289	161	128	89.7
55.7	287	160	127	89.2
55.7	282	157	125	87.4
55.7	280	156	124	86.9
55.7	273	152	121	84.6
55.7	271	151	120	84.1
55.7	264	147	117	81.9
55.7	262	146	116	81.4
55.7	255	142	113	79.1
55.7	253	141	112	78.6
55.7	246	137	109	76.3
55.7	244	136	108	75.8
55.7	237	132	105	73.5
55.7	235	131	104	73.0
55.7	230	128	102	71.2
55.7	228	127	101	70.7
55.7	221	123	98	68.5
55.7	219	122	97	68.0
55.7	212	118	94	65.7
55.7	210	117	93	65.2
55.7	203	113	90	62.9
55.7	201	112	89	62.4
55.7	194	108	86	60.1
55.7	192	107	85	59.6
55.7	185	103	82	57.3
55.7	183	102	81	56.9
55.7	176	98	78	54.6
55.7	174	97	77	54.1
55.7	167	93	74	51.8
55.7	158	88	70	49.0
55.7	149	83	66	46.2
55.7	140	78	62	43.5
55.7	131	73	58	40.7
55.7	122	68	54	37.9
55.7	115	64	51	35.6
55.7	106	59	47	32.8
55.7	97	54	43	30.1
55.7	88	49	39	27.3
55.7	79	44	35	24.5
55.7	70	39	31	21.7
55.7	61	34	27	19.0
55.6	500	278	222	154.6
55.6	498	277	221	154.1

%E	M_1	M_2	DM	M*
55.6	496	276	220	153.6
55.6	495	275	220	152.8
55.6	493	274	219	152.3
55.6	491	273	218	151.8
55.6	489	272	217	151.3
55.6	487	271	216	150.8
55.6	486	270	216	150.0
55.6	484	269	215	149.5
55.6	482	268	214	149.0
55.6	480	267	213	148.5
55.6	478	266	212	148.0
55.6	477	265	212	147.2
55.6	475	264	211	146.7
55.6	473	263	210	146.2
55.6	471	262	209	145.7
55.6	468	260	208	144.4
55.6	466	259	207	144.0
55.6	464	258	206	143.5
55.6	462	257	205	143.0
55.6	459	255	204	141.7
55.6	457	254	203	141.2
55.6	455	253	202	140.7
55.6	453	252	201	140.2
55.6	450	250	200	138.9
55.6	448	249	199	138.4
55.6	446	248	198	137.9
55.6	444	247	197	137.4
55.6	441	245	196	136.1
55.6	439	244	195	135.6
55.6	437	243	194	135.1
55.6	435	242	193	134.6
55.6	432	240	192	133.3
55.6	430	239	191	132.8
55.6	428	238	190	132.3
55.6	426	237	189	131.9
55.6	423	235	188	130.6
55.6	421	234	187	130.1
55.6	419	233	186	129.6
55.6	417	232	185	129.1
55.6	414	230	184	127.8
55.6	412	229	183	127.3
55.6	410	228	182	126.8
55.6	408	227	181	126.3
55.6	405	225	180	125.0
55.6	403	224	179	124.5
55.6	401	223	178	124.0
55.6	399	222	177	123.5
55.6	396	220	176	122.2
55.6	394	219	175	121.7
55.6	392	218	174	121.2

%E	M_1	M_2	DM	M*
55.6	390	217	173	120.7
55.6	387	215	172	119.4
55.6	385	214	171	119.0
55.6	383	213	170	118.5
55.6	381	212	169	118.0
55.6	378	210	168	116.7
55.6	376	209	167	116.2
55.6	374	208	166	115.7
55.6	372	207	165	115.2
55.6	369	205	164	113.9
55.6	367	204	163	113.4
55.6	365	203	162	112.9
55.6	363	202	161	112.4
55.6	360	200	160	111.1
55.6	358	199	159	110.6
55.6	356	198	158	110.1
55.6	351	195	156	108.3
55.6	349	194	155	107.8
55.6	347	193	154	107.3
55.6	342	190	152	105.6
55.6	340	189	151	105.1
55.6	338	188	150	104.6
55.6	333	185	148	102.8
55.6	331	184	147	102.3
55.6	329	183	146	101.8
55.6	324	180	144	100.0
55.6	322	179	143	99.5
55.6	320	178	142	99.0
55.6	315	175	140	97.2
55.6	313	174	139	96.7
55.6	311	173	138	96.2
55.6	306	170	136	94.4
55.6	304	169	135	94.0
55.6	302	168	134	93.5
55.6	297	165	132	91.7
55.6	295	164	131	91.2
55.6	293	163	130	90.7
55.6	288	160	128	88.9
55.6	286	159	127	88.4
55.6	284	158	126	87.9
55.6	279	155	124	86.1
55.6	277	154	123	85.6
55.6	275	153	122	85.1
55.6	270	150	120	83.3
55.6	268	149	119	82.8
55.6	266	148	118	82.3
55.6	261	145	116	80.6
55.6	259	144	115	80.1
55.6	257	143	114	79.6
55.6	252	140	112	77.8

%E	M_1	M_2	DM	M*
55.6	250	139	111	77.3
55.6	248	138	110	76.8
55.6	243	135	108	75.0
55.6	241	134	107	74.5
55.6	239	133	106	74.0
55.6	234	130	104	72.2
55.6	232	129	103	71.7
55.6	225	125	100	69.4
55.6	223	124	99	69.0
55.6	216	120	96	66.7
55.6	214	119	95	66.2
55.6	207	115	92	63.9
55.6	205	114	91	63.4
55.6	198	110	88	61.1
55.6	196	109	87	60.6
55.6	189	105	84	58.3
55.6	187	104	83	57.8
55.6	180	100	80	55.6
55.6	178	99	79	55.1
55.6	171	95	76	52.8
55.6	169	94	75	52.3
55.6	162	90	72	50.0
55.6	160	89	71	49.5
55.6	153	85	68	47.2
55.6	151	84	67	46.7
55.6	144	80	64	44.4
55.6	142	79	63	44.0
55.6	135	75	60	41.7
55.6	133	74	59	41.2
55.6	126	70	56	38.9
55.6	124	69	55	38.4
55.6	117	65	52	36.1
55.6	108	60	48	33.3
55.6	99	55	44	30.6
55.6	90	50	40	27.8
55.6	81	45	36	25.0
55.6	72	40	32	22.2
55.6	63	35	28	19.4
55.6	54	30	24	16.7
55.6	45	25	20	13.9
55.6	36	20	16	11.1
55.6	27	15	12	8.3
55.6	18	10	8	5.6
55.6	9	5	4	2.8
55.5	499	277	222	153.8
55.5	497	276	221	153.3
55.5	494	274	220	152.0
55.5	492	273	219	151.5
55.5	490	272	218	151.0
55.5	488	271	217	150.5

%E	M_1	M_2	DM	M*
55.5	485	269	216	149.2
55.5	483	268	215	148.7
55.5	481	267	214	148.2
55.5	479	266	213	147.7
55.5	476	264	212	146.4
55.5	474	263	211	145.9
55.5	472	262	210	145.4
55.5	470	261	209	144.9
55.5	467	259	208	143.6
55.5	465	258	207	143.1
55.5	463	257	206	142.7
55.5	461	256	205	142.2
55.5	458	254	204	140.9
55.5	456	253	203	140.4
55.5	454	252	202	139.9
55.5	452	251	201	139.4
55.5	449	249	200	138.1
55.5	447	248	199	137.6
55.5	445	247	198	137.1
55.5	443	246	197	136.6
55.5	440	244	196	135.3
55.5	438	243	195	134.8
55.5	436	242	194	134.3
55.5	434	241	193	133.8
55.5	431	239	192	132.5
55.5	429	238	191	132.0
55.5	427	237	190	131.5
55.5	425	236	189	131.0
55.5	422	234	188	129.8
55.5	420	233	187	129.3
55.5	418	232	186	128.8
55.5	416	231	185	128.3
55.5	411	228	183	126.5
55.5	409	227	182	126.0
55.5	407	226	181	125.5
55.5	402	223	179	123.7
55.5	400	222	178	123.2
55.5	398	221	177	122.7
55.5	393	218	175	120.9
55.5	391	217	174	120.4
55.5	389	216	173	119.9
55.5	384	213	171	118.1
55.5	382	212	170	117.7
55.5	380	211	169	117.2
55.5	375	208	167	115.4
55.5	373	207	166	114.9
55.5	371	206	165	114.4
55.5	366	203	163	112.6
55.5	364	202	162	112.1
55.5	362	201	161	111.6

%E	M1	M2	DM	M*
55.5	357	198	159	109.8
55.5	355	197	158	109.3
55.5	353	196	157	108.8
55.5	348	193	155	107.0
55.5	346	192	154	106.5
55.5	344	191	153	106.0
55.5	339	188	151	104.3
55.5	337	187	150	103.8
55.5	335	186	149	103.3
55.5	330	183	147	101.5
55.5	328	182	146	101.0
55.5	326	181	145	100.5
55.5	321	178	143	98.7
55.5	319	177	142	98.2
55.5	317	176	141	97.7
55.5	310	172	138	95.4
55.5	308	171	137	94.9
55.5	301	167	134	92.7
55.5	299	166	133	92.2
55.5	292	162	130	89.9
55.5	290	161	129	89.4
55.5	283	157	126	87.1
55.5	281	156	125	86.6
55.5	274	152	122	84.3
55.5	272	151	121	83.8
55.5	265	147	118	81.5
55.5	263	146	117	81.0
55.5	256	142	114	78.8
55.5	254	141	113	78.3
55.5	247	137	110	76.0
55.5	245	136	109	75.5
55.5	238	132	106	73.2
55.5	236	131	105	72.7
55.5	229	127	102	70.4
55.5	227	126	101	69.9
55.5	220	122	98	67.7
55.5	218	121	97	67.2
55.5	211	117	94	64.9
55.5	209	116	93	64.4
55.5	200	111	89	61.6
55.5	191	106	85	58.8
55.5	182	101	81	56.0
55.5	173	96	77	53.3
55.5	164	91	73	50.5
55.5	155	86	69	47.7
55.5	146	81	65	44.9
55.5	137	76	61	42.2
55.5	128	71	57	39.4
55.5	119	66	53	36.6
55.5	110	61	49	33.8

%E	M1	M2	DM	M*
55.4	500	277	223	153.5
55.4	498	276	222	153.0
55.4	496	275	221	152.5
55.4	495	274	221	151.7
55.4	493	273	220	151.2
55.4	491	272	219	150.7
55.4	489	271	218	150.2
55.4	487	270	217	149.7
55.4	486	269	217	148.9
55.4	484	268	216	148.4
55.4	482	267	215	147.9
55.4	480	266	214	147.4
55.4	478	265	213	146.9
55.4	475	263	212	145.6
55.4	473	262	211	145.1
55.4	471	261	210	144.6
55.4	469	260	209	144.1
55.4	466	258	208	142.8
55.4	464	257	207	142.3
55.4	462	256	206	141.9
55.4	460	255	205	141.4
55.4	457	253	204	140.1
55.4	455	252	203	139.6
55.4	453	251	202	139.1
55.4	451	250	201	138.6
55.4	448	248	200	137.3
55.4	446	247	199	136.8
55.4	444	246	198	136.3
55.4	442	245	197	135.8
55.4	439	243	196	134.5
55.4	437	242	195	134.0
55.4	435	241	194	133.5
55.4	433	240	193	133.0
55.4	428	237	191	131.2
55.4	426	236	190	130.7
55.4	424	235	189	130.2
55.4	419	232	187	128.5
55.4	417	231	186	128.0
55.4	415	230	185	127.5
55.4	413	229	184	127.0
55.4	410	227	183	125.7
55.4	408	226	182	125.2
55.4	406	225	181	124.7
55.4	404	224	180	124.2
55.4	401	222	179	122.9
55.4	399	221	178	122.4
55.4	397	220	177	121.9
55.4	395	219	176	121.4
55.4	392	217	175	120.1
55.4	390	216	174	119.6

%E	M1	M2	DM	M*
55.4	388	215	173	119.1
55.4	386	214	172	118.6
55.4	383	212	171	117.3
55.4	381	211	170	116.9
55.4	379	210	169	116.4
55.4	377	209	168	115.9
55.4	372	206	166	114.1
55.4	370	205	165	113.6
55.4	368	204	164	113.1
55.4	363	201	162	111.3
55.4	361	200	161	110.8
55.4	359	199	160	110.3
55.4	354	196	158	108.5
55.4	352	195	157	108.0
55.4	350	194	156	107.5
55.4	345	191	154	105.7
55.4	343	190	153	105.2
55.4	341	189	152	104.8
55.4	336	186	150	103.0
55.4	334	185	149	102.5
55.4	332	184	148	102.0
55.4	327	181	146	100.2
55.4	325	180	145	99.7
55.4	323	179	144	99.2
55.4	316	175	141	96.9
55.4	314	174	140	96.4
55.4	312	173	139	95.9
55.4	307	170	137	94.1
55.4	305	169	136	93.6
55.4	303	168	135	93.1
55.4	298	165	133	91.4
55.4	296	164	132	90.9
55.4	294	163	131	90.4
55.4	289	160	129	88.6
55.4	287	159	128	88.1
55.4	285	158	127	87.6
55.4	280	155	125	85.8
55.4	278	154	124	85.3
55.4	276	153	123	84.8
55.4	271	150	121	83.0
55.4	269	149	120	82.5
55.4	267	148	119	82.0
55.4	260	144	116	79.8
55.4	258	143	115	79.3
55.4	251	139	112	77.0
55.4	249	138	111	76.5
55.4	242	134	108	74.2
55.4	240	133	107	73.7
55.4	233	129	104	71.4
55.4	231	128	103	70.9

%E	M1	M2	DM	M*
55.4	224	124	100	68.6
55.4	222	123	99	68.1
55.4	213	118	95	65.4
55.4	204	113	91	62.6
55.4	202	112	90	62.1
55.4	195	108	87	59.8
55.4	193	107	86	59.3
55.4	186	103	83	57.0
55.4	184	102	82	56.5
55.4	177	98	79	54.3
55.4	175	97	78	53.8
55.4	168	93	75	51.5
55.4	166	92	74	51.0
55.4	157	87	70	48.2
55.4	148	82	66	45.4
55.4	139	77	62	42.7
55.4	130	72	58	39.9
55.4	121	67	54	37.1
55.4	112	62	50	34.3
55.4	101	56	45	31.0
55.4	92	51	41	28.3
55.4	83	46	37	25.5
55.4	74	41	33	22.7
55.4	65	36	29	19.9
55.4	56	31	25	17.2
55.3	499	276	223	152.7
55.3	497	275	222	152.2
55.3	494	273	221	150.9
55.3	492	272	220	150.4
55.3	490	271	219	149.9
55.3	488	270	218	149.4
55.3	485	268	217	148.1
55.3	483	267	216	147.6
55.3	481	266	215	147.1
55.3	479	265	214	146.6
55.3	477	264	213	146.1
55.3	476	263	213	145.3
55.3	474	262	212	144.8
55.3	472	261	211	144.3
55.3	470	260	210	143.8
55.3	468	259	209	143.3
55.3	465	257	208	142.0
55.3	463	256	207	141.5
55.3	461	255	206	141.1
55.3	459	254	205	140.6
55.3	456	252	204	139.3
55.3	454	251	203	138.8
55.3	452	250	202	138.3
55.3	450	249	201	137.8
55.3	447	247	200	136.5

%E	M1	M2	DM	M*
55.3	445	246	199	136.0
55.3	443	245	198	135.5
55.3	441	244	197	135.0
55.3	438	242	196	133.7
55.3	436	241	195	133.2
55.3	434	240	194	132.7
55.3	432	239	193	132.2
55.3	430	238	192	131.7
55.3	427	236	191	130.4
55.3	425	235	190	129.9
55.3	423	234	189	129.4
55.3	421	233	188	129.0
55.3	418	231	187	127.7
55.3	416	230	186	127.2
55.3	414	229	185	126.7
55.3	412	228	184	126.2
55.3	409	226	183	124.9
55.3	407	225	182	124.4
55.3	405	224	181	123.9
55.3	403	223	180	123.4
55.3	400	221	179	122.1
55.3	398	220	178	121.6
55.3	396	219	177	121.1
55.3	394	218	176	120.6
55.3	389	215	174	118.8
55.3	387	214	173	118.3
55.3	385	213	172	117.8
55.3	380	210	170	116.1
55.3	378	209	169	115.6
55.3	376	208	168	115.1
55.3	374	207	167	114.6
55.3	371	205	166	113.3
55.3	369	204	165	112.8
55.3	367	203	164	112.3
55.3	365	202	163	111.8
55.3	360	199	161	110.0
55.3	358	198	160	109.5
55.3	356	197	159	109.0
55.3	351	194	157	107.2
55.3	349	193	156	106.7
55.3	347	192	155	106.2
55.3	342	189	153	104.4
55.3	340	188	152	104.0
55.3	338	187	151	103.5
55.3	333	184	149	101.7
55.3	331	183	148	101.2
55.3	329	182	147	100.7
55.3	322	178	144	98.4
55.3	320	177	143	97.9
55.3	318	176	142	97.4

%E	M1	M2	DM	M*
55.3	313	173	140	95.6
55.3	311	172	139	95.1
55.3	309	171	138	94.6
55.3	304	168	136	92.8
55.3	302	167	135	92.3
55.3	300	166	134	91.9
55.3	295	163	132	90.1
55.3	293	162	131	89.6
55.3	291	161	130	89.1
55.3	284	157	127	86.8
55.3	282	156	126	86.3
55.3	275	152	123	84.0
55.3	273	151	122	83.5
55.3	266	147	119	81.2
55.3	264	146	118	80.7
55.3	262	145	117	80.2
55.3	257	142	115	78.5
55.3	255	141	114	78.0
55.3	253	140	113	77.5
55.3	246	136	110	75.2
55.3	244	135	109	74.7
55.3	237	131	106	72.4
55.3	235	130	105	71.9
55.3	228	126	102	69.6
55.3	226	125	101	69.1
55.3	219	121	98	66.9
55.3	217	120	97	66.4
55.3	215	119	96	65.9
55.3	208	115	93	63.6
55.3	206	114	92	63.1
55.3	199	110	89	60.8
55.3	197	109	88	60.3
55.3	190	105	85	58.0
55.3	188	104	84	57.5
55.3	179	99	80	54.8
55.3	170	94	76	52.0
55.3	161	89	72	49.2
55.3	159	88	71	48.7
55.3	152	84	68	46.4
55.3	150	83	67	45.9
55.3	141	78	63	43.1
55.3	132	73	59	40.4
55.3	123	68	55	37.6
55.3	114	63	51	34.8
55.3	103	57	46	31.5
55.3	94	52	42	28.8
55.3	85	47	38	26.0
55.3	76	42	34	23.2
55.3	47	26	21	14.4
55.3	38	21	17	11.6

%E	M1	M2	DM	M*
55.2	500	276	224	152.4
55.2	498	275	223	151.9
55.2	496	274	222	151.4
55.2	495	273	222	150.6
55.2	493	272	221	150.1
55.2	491	271	220	149.6
55.2	489	270	219	149.1
55.2	487	269	218	148.6
55.2	484	267	217	147.3
55.2	482	266	216	146.8
55.2	480	265	215	146.3
55.2	478	264	214	145.8
55.2	475	262	213	144.5
55.2	473	261	212	144.0
55.2	471	260	211	143.5
55.2	469	259	210	143.0
55.2	467	258	209	142.5
55.2	466	257	209	141.7
55.2	464	256	208	141.2
55.2	462	255	207	140.7
55.2	460	254	206	140.3
55.2	458	253	205	139.8
55.2	455	251	204	138.5
55.2	453	250	203	138.0
55.2	451	249	202	137.5
55.2	449	248	201	137.0
55.2	446	246	200	135.7
55.2	444	245	199	135.2
55.2	442	244	198	134.7
55.2	440	243	197	134.2
55.2	435	240	195	132.4
55.2	433	239	194	131.9
55.2	431	238	193	131.4
55.2	429	237	192	130.9
55.2	426	235	191	129.6
55.2	424	234	190	129.1
55.2	422	233	189	128.6
55.2	420	232	188	128.2
55.2	417	230	187	126.9
55.2	415	229	186	126.4
55.2	413	228	185	125.9
55.2	411	227	184	125.4
55.2	406	224	182	123.6
55.2	404	223	181	123.1
55.2	402	222	180	122.6
55.2	397	219	178	120.8
55.2	395	218	177	120.3
55.2	393	217	176	119.8
55.2	391	216	175	119.3
55.2	388	214	174	118.0

%E	M1	M2	DM	M*
55.2	386	213	173	117.5
55.2	384	212	172	117.0
55.2	382	211	171	116.5
55.2	377	208	169	114.8
55.2	375	207	168	114.3
55.2	373	206	167	113.8
55.2	368	203	165	112.0
55.2	366	202	164	111.5
55.2	364	201	163	111.0
55.2	362	200	162	110.5
55.2	359	198	161	109.2
55.2	357	197	160	108.7
55.2	355	196	159	108.2
55.2	353	195	158	107.7
55.2	348	192	156	105.9
55.2	346	191	155	105.4
55.2	344	190	154	104.9
55.2	339	187	152	103.2
55.2	337	186	151	102.7
55.2	335	185	150	102.2
55.2	330	182	148	100.4
55.2	328	181	147	99.9
55.2	326	180	146	99.4
55.2	324	179	145	98.9
55.2	319	176	143	97.1
55.2	317	175	142	96.6
55.2	315	174	141	96.1
55.2	310	171	139	94.3
55.2	308	170	138	93.8
55.2	306	169	137	93.3
55.2	301	166	135	91.5
55.2	299	165	134	91.1
55.2	297	164	133	90.6
55.2	290	160	130	88.3
55.2	288	159	129	87.8
55.2	286	158	128	87.3
55.2	281	155	126	85.5
55.2	279	154	125	85.0
55.2	277	153	124	84.5
55.2	270	149	121	82.2
55.2	268	148	120	81.7
55.2	261	144	117	79.4
55.2	259	143	116	79.0
55.2	252	139	113	76.7
55.2	250	138	112	76.2
55.2	248	137	111	75.7
55.2	241	133	108	73.4
55.2	239	132	107	72.9
55.2	232	128	104	70.6
55.2	230	127	103	70.1

%E	M1	M2	DM	M*
55.2	223	123	100	67.8
55.2	221	122	99	67.3
55.2	212	117	95	64.6
55.2	210	116	94	64.1
55.2	203	112	91	61.8
55.2	201	111	90	61.3
55.2	194	107	87	59.0
55.2	192	106	86	58.5
55.2	183	101	82	55.7
55.2	181	100	81	55.2
55.2	174	96	78	53.0
55.2	172	95	77	52.5
55.2	165	91	74	50.2
55.2	163	90	73	49.7
55.2	154	85	69	46.9
55.2	145	80	65	44.1
55.2	143	79	64	43.6
55.2	134	74	60	40.9
55.2	125	69	56	38.1
55.2	116	64	52	35.3
55.2	105	58	47	32.0
55.2	96	53	43	29.3
55.2	87	48	39	26.5
55.2	67	37	30	20.4
55.2	58	32	26	17.7
55.2	29	16	13	8.8
55.1	499	275	224	151.6
55.1	497	274	223	151.1
55.1	494	272	222	149.8
55.1	492	271	221	149.3
55.1	490	270	220	148.8
55.1	488	269	219	148.3
55.1	486	268	218	147.8
55.1	485	267	218	147.0
55.1	483	266	217	146.5
55.1	481	265	216	146.0
55.1	479	264	215	145.5
55.1	477	263	214	145.0
55.1	474	261	213	143.7
55.1	472	260	212	143.2
55.1	470	259	211	142.7
55.1	468	258	210	142.2
55.1	465	256	209	140.9
55.1	463	255	208	140.4
55.1	461	254	207	139.9
55.1	459	253	206	139.5
55.1	457	252	205	139.0
55.1	454	250	204	137.7
55.1	452	249	203	137.2
55.1	450	248	202	136.7

%E	M1	M2	DM	M*
55.1	448	247	201	136.2
55.1	445	245	200	134.9
55.1	443	244	199	134.4
55.1	441	243	198	133.9
55.1	439	242	197	133.4
55.1	437	241	196	132.9
55.1	434	239	195	131.6
55.1	432	238	194	131.1
55.1	430	237	193	130.6
55.1	428	236	192	130.1
55.1	425	234	191	128.8
55.1	423	233	190	128.3
55.1	421	232	189	127.8
55.1	419	231	188	127.4
55.1	414	228	186	125.6
55.1	412	227	185	125.1
55.1	410	226	184	124.6
55.1	408	225	183	124.1
55.1	405	223	182	122.8
55.1	403	222	181	122.3
55.1	401	221	180	121.8
55.1	399	220	179	121.3
55.1	396	218	178	120.0
55.1	394	217	177	119.5
55.1	392	216	176	119.0
55.1	390	215	175	118.5
55.1	385	212	173	116.7
55.1	383	211	172	116.2
55.1	381	210	171	115.7
55.1	379	209	170	115.3
55.1	376	207	169	114.0
55.1	374	206	168	113.5
55.1	372	205	167	113.0
55.1	370	204	166	112.5
55.1	365	201	164	110.7
55.1	363	200	163	110.2
55.1	361	199	162	109.7
55.1	356	196	160	107.9
55.1	354	195	159	107.4
55.1	352	194	158	106.9
55.1	350	193	157	106.4
55.1	345	190	155	104.6
55.1	343	189	154	104.1
55.1	341	188	153	103.6
55.1	336	185	151	101.9
55.1	334	184	150	101.4
55.1	332	183	149	100.9
55.1	325	179	146	98.6
55.1	323	178	145	98.1
55.1	321	177	144	97.6

%E	M1	M2	DM	M*	%E	M1	M2	DM	M*	%E	M1	M2	DM	M*	%E	M1	M2	DM	M*	%E	M1	M2	DM	M*
55.1	316	174	142	95.8	55.0	500	275	225	151.3	55.0	387	213	174	117.2	55.0	220	121	99	66.5	54.9	448	246	202	135.1
55.1	314	173	141	95.3	55.0	498	274	224	150.8	55.0	382	210	172	115.4	55.0	218	120	98	66.1	54.9	446	245	201	134.6
55.1	312	172	140	94.8	55.0	496	273	223	150.3	55.0	380	209	171	114.9	55.0	211	116	95	63.8	54.9	443	243	200	133.3
55.1	305	168	137	92.5	55.0	493	271	222	149.0	55.0	378	208	170	114.5	55.0	209	115	94	63.3	54.9	441	242	199	132.8
55.1	303	167	136	92.0	55.0	491	270	221	148.5	55.0	373	205	168	112.7	55.0	202	111	91	61.0	54.9	439	241	198	132.3
55.1	296	163	133	89.8	55.0	489	269	220	148.0	55.0	371	204	167	112.2	55.0	200	110	90	60.5	54.9	437	240	197	131.8
55.1	294	162	132	89.3	55.0	487	268	219	147.5	55.0	369	203	166	111.7	55.0	191	105	86	57.7	54.9	435	239	196	131.3
55.1	292	161	131	88.8	55.0	484	266	218	146.2	55.0	367	202	165	111.2	55.0	189	104	85	57.2	54.9	432	237	195	130.0
55.1	287	158	129	87.0	55.0	482	265	217	145.7	55.0	362	199	163	109.4	55.0	180	99	81	54.4	54.9	430	236	194	129.5
55.1	285	157	128	86.5	55.0	480	264	216	145.2	55.0	360	198	162	108.9	55.0	171	94	77	51.7	54.9	428	235	193	129.0
55.1	283	156	127	86.0	55.0	478	263	215	144.7	55.0	358	197	161	108.4	55.0	169	93	76	51.2	54.9	426	234	192	128.5
55.1	276	152	124	83.7	55.0	476	262	214	144.2	55.0	353	194	159	106.6	55.0	160	88	72	48.4	54.9	421	231	190	126.7
55.1	274	151	123	83.2	55.0	473	260	213	142.9	55.0	351	193	158	106.1	55.0	151	83	68	45.6	54.9	419	230	189	126.3
55.1	272	150	122	82.7	55.0	471	259	212	142.4	55.0	349	192	157	105.6	55.0	149	82	67	45.1	54.9	417	229	188	125.8
55.1	267	147	120	80.9	55.0	469	258	211	141.9	55.0	347	191	156	105.1	55.0	140	77	63	42.3	54.9	415	228	187	125.3
55.1	265	146	119	80.4	55.0	467	257	210	141.4	55.0	342	188	154	103.3	55.0	131	72	59	39.6	54.9	412	226	186	124.0
55.1	263	145	118	79.9	55.0	464	255	209	140.1	55.0	340	187	153	102.8	55.0	129	71	58	39.1	54.9	410	225	185	123.5
55.1	256	141	115	77.7	55.0	462	254	208	139.6	55.0	338	186	152	102.4	55.0	120	66	54	36.3	54.9	408	224	184	123.0
55.1	254	140	114	77.2	55.0	460	253	207	139.1	55.0	333	183	150	100.6	55.0	111	61	50	33.5	54.9	406	223	183	122.5
55.1	247	136	111	74.9	55.0	458	252	206	138.7	55.0	331	182	149	100.1	55.0	109	60	49	33.0	54.9	401	220	181	120.7
55.1	245	135	110	74.4	55.0	456	251	205	138.2	55.0	329	181	148	99.6	55.0	100	55	45	30.3	54.9	399	219	180	120.2
55.1	243	134	109	73.9	55.0	453	249	204	136.9	55.0	327	180	147	99.1	55.0	80	44	36	24.2	54.9	397	218	179	119.7
55.1	236	130	106	71.6	55.0	451	248	203	136.4	55.0	322	177	145	97.3	55.0	60	33	27	18.1	54.9	395	217	178	119.2
55.1	234	129	105	71.1	55.0	449	247	202	135.9	55.0	320	176	144	96.8	55.0	40	22	18	12.1	54.9	390	214	176	117.4
55.1	227	125	102	68.8	55.0	447	246	201	135.4	55.0	318	175	143	96.3	55.0	20	11	9	6.0	54.9	388	213	175	116.9
55.1	225	124	101	68.3	55.0	444	244	200	134.1	55.0	313	172	141	94.5	54.9	499	274	225	150.5	54.9	386	212	174	116.4
55.1	216	119	97	65.6	55.0	442	243	199	133.6	55.0	311	171	140	94.0	54.9	497	273	224	150.0	54.9	384	211	173	115.9
55.1	214	118	96	65.1	55.0	440	242	198	133.1	55.0	309	170	139	93.5	54.9	495	272	223	149.5	54.9	381	209	172	114.6
55.1	207	114	93	62.8	55.0	438	241	197	132.6	55.0	307	169	138	93.0	54.9	494	271	223	148.7	54.9	379	208	171	114.2
55.1	205	113	92	62.3	55.0	436	240	196	132.1	55.0	302	166	136	91.2	54.9	492	270	222	148.2	54.9	377	207	170	113.7
55.1	198	109	89	60.0	55.0	433	238	195	130.8	55.0	300	165	135	90.8	54.9	490	269	221	147.7	54.9	375	206	169	113.2
55.1	196	108	88	59.5	55.0	431	237	194	130.3	55.0	298	164	134	90.3	54.9	488	268	220	147.2	54.9	370	203	167	111.4
55.1	187	103	84	56.7	55.0	429	236	193	129.8	55.0	291	160	131	88.0	54.9	486	267	219	146.7	54.9	368	202	166	110.9
55.1	185	102	83	56.2	55.0	427	235	192	129.3	55.0	289	159	130	87.5	54.9	483	265	218	145.4	54.9	366	201	165	110.4
55.1	178	98	80	54.0	55.0	424	233	191	128.0	55.0	282	155	127	85.2	54.9	481	264	217	144.9	54.9	364	200	164	109.9
55.1	176	97	79	53.5	55.0	422	232	190	127.5	55.0	280	154	126	84.7	54.9	479	263	216	144.4	54.9	359	197	162	108.1
55.1	167	92	75	50.7	55.0	420	231	189	127.0	55.0	278	153	125	84.2	54.9	477	262	215	143.9	54.9	357	196	161	107.6
55.1	158	87	71	47.9	55.0	418	230	188	126.6	55.0	271	149	122	81.9	54.9	475	261	214	143.4	54.9	355	195	160	107.1
55.1	156	86	70	47.4	55.0	416	229	187	126.1	55.0	269	148	121	81.4	54.9	474	260	214	142.6	54.9	350	192	158	105.3
55.1	147	81	66	44.6	55.0	413	227	186	124.8	55.0	262	144	118	79.1	54.9	472	259	213	142.1	54.9	348	191	157	104.8
55.1	138	76	62	41.9	55.0	411	226	185	124.3	55.0	260	143	117	78.6	54.9	470	258	212	141.6	54.9	346	190	156	104.3
55.1	136	75	61	41.4	55.0	409	225	184	123.8	55.0	258	142	116	78.2	54.9	468	257	211	141.1	54.9	344	189	155	103.8
55.1	127	70	57	38.6	55.0	407	224	183	123.3	55.0	251	138	113	75.9	54.9	466	256	210	140.6	54.9	339	186	153	102.1
55.1	118	65	53	35.8	55.0	404	222	182	122.0	55.0	249	137	112	75.4	54.9	463	254	209	139.3	54.9	337	185	152	101.6
55.1	107	59	48	32.5	55.0	402	221	181	121.5	55.0	242	133	109	73.1	54.9	461	253	208	138.8	54.9	335	184	151	101.1
55.1	98	54	44	29.8	55.0	400	220	180	121.0	55.0	240	132	108	72.6	54.9	459	252	207	138.4	54.9	328	180	148	98.8
55.1	89	49	40	27.0	55.0	398	219	179	120.5	55.0	238	131	107	72.1	54.9	457	251	206	137.9	54.9	326	179	147	98.3
55.1	78	43	35	23.7	55.0	393	216	177	118.7	55.0	231	127	104	69.8	54.9	455	250	205	137.4	54.9	324	178	146	97.8
55.1	69	38	31	20.9	55.0	391	215	176	118.2	55.0	229	126	103	69.3	54.9	452	248	204	136.1	54.9	319	175	144	96.0
55.1	49	27	22	14.9	55.0	389	214	175	117.7	55.0	222	122	100	67.0	54.9	450	247	203	135.6	54.9	317	174	143	95.5

%E	M1	M2	DM	M*
54.9	315	173	142	95.0
54.9	308	169	139	92.7
54.9	306	168	138	92.2
54.9	304	167	137	91.7
54.9	297	163	134	89.5
54.9	295	162	133	89.0
54.9	293	161	132	88.5
54.9	288	158	130	86.7
54.9	286	157	129	86.2
54.9	284	156	128	85.7
54.9	277	152	125	83.4
54.9	275	151	124	82.9
54.9	273	150	123	82.4
54.9	268	147	121	80.6
54.9	266	146	120	80.1
54.9	264	145	119	79.6
54.9	257	141	116	77.4
54.9	255	140	115	76.9
54.9	253	139	114	76.4
54.9	246	135	111	74.1
54.9	244	134	110	73.6
54.9	237	130	107	71.3
54.9	235	129	106	70.8
54.9	233	128	105	70.3
54.9	226	124	102	68.0
54.9	224	123	101	67.5
54.9	215	118	97	64.8
54.9	213	117	96	64.3
54.9	206	113	93	62.0
54.9	204	112	92	61.5
54.9	195	107	88	58.7
54.9	193	106	87	58.2
54.9	184	101	83	55.4
54.9	182	100	82	54.9
54.9	175	96	79	52.7
54.9	173	95	78	52.2
54.9	164	90	74	49.4
54.9	162	89	73	48.9
54.9	153	84	69	46.1
54.9	144	79	65	43.3
54.9	142	78	64	42.8
54.9	133	73	60	40.1
54.9	122	67	55	36.8
54.9	113	62	51	34.0
54.9	102	56	46	30.7
54.9	91	50	41	27.5
54.9	82	45	37	24.7
54.9	71	39	32	21.4
54.9	51	28	23	15.4
54.8	500	274	226	150.2
54.8	498	273	225	149.7
54.8	496	272	224	149.2
54.8	493	270	223	147.9
54.8	491	269	222	147.4
54.8	489	268	221	146.9
54.8	487	267	220	146.4
54.8	485	266	219	145.9
54.8	484	265	219	145.1
54.8	482	264	218	144.6
54.8	480	263	217	144.1
54.8	478	262	216	143.6
54.8	476	261	215	143.1
54.8	473	259	214	141.8
54.8	471	258	213	141.3
54.8	469	257	212	140.8
54.8	467	256	211	140.3
54.8	465	255	210	139.8
54.8	462	253	209	138.5
54.8	460	252	208	138.1
54.8	458	251	207	137.6
54.8	456	250	206	137.1
54.8	454	249	205	136.6
54.8	451	247	204	135.3
54.8	449	246	203	134.8
54.8	447	245	202	134.3
54.8	445	244	201	133.8
54.8	442	242	200	132.5
54.8	440	241	199	132.0
54.8	438	240	198	131.5
54.8	436	239	197	131.0
54.8	434	238	196	130.5
54.8	431	236	195	129.2
54.8	429	235	194	128.7
54.8	427	234	193	128.2
54.8	425	233	192	127.7
54.8	423	232	191	127.2
54.8	420	230	190	126.0
54.8	418	229	189	125.5
54.8	416	228	188	125.0
54.8	414	227	187	124.5
54.8	409	224	185	122.7
54.8	407	223	184	122.2
54.8	405	222	183	121.7
54.8	403	221	182	121.2
54.8	400	219	181	119.9
54.8	398	218	180	119.4
54.8	396	217	179	118.9
54.8	394	216	178	118.4
54.8	392	215	177	117.9
54.8	389	213	176	116.6
54.8	387	212	175	116.1
54.8	385	211	174	115.6
54.8	383	210	173	115.1
54.8	378	207	171	113.4
54.8	376	206	170	112.9
54.8	374	205	169	112.4
54.8	372	204	168	111.9
54.8	367	201	166	110.1
54.8	365	200	165	109.6
54.8	363	199	164	109.1
54.8	361	198	163	108.6
54.8	356	195	161	106.8
54.8	354	194	160	106.3
54.8	352	193	159	105.8
54.8	347	190	157	104.0
54.8	345	189	156	103.5
54.8	343	188	155	103.0
54.8	341	187	154	102.5
54.8	336	184	152	100.8
54.8	334	183	151	100.3
54.8	332	182	150	99.8
54.8	330	181	149	99.3
54.8	325	178	147	97.5
54.8	323	177	146	97.0
54.8	321	176	145	96.5
54.8	314	172	142	94.2
54.8	312	171	141	93.7
54.8	310	170	140	93.2
54.8	305	167	138	91.4
54.8	303	166	137	90.9
54.8	301	165	136	90.4
54.8	299	164	135	90.0
54.8	294	161	133	88.2
54.8	292	160	132	87.7
54.8	290	159	131	87.2
54.8	283	155	128	84.9
54.8	281	154	127	84.4
54.8	279	153	126	83.9
54.8	272	149	123	81.6
54.8	270	148	122	81.1
54.8	263	144	119	78.8
54.8	261	143	118	78.3
54.8	259	142	117	77.9
54.8	252	138	114	75.6
54.8	250	137	113	75.1
54.8	248	136	112	74.6
54.8	241	132	109	72.3
54.8	239	131	108	71.8
54.8	230	126	104	69.0
54.8	228	125	103	68.5
54.8	221	121	100	66.2
54.8	219	120	99	65.8
54.8	217	119	98	65.3
54.8	210	115	95	63.0
54.8	208	114	94	62.5
54.8	199	109	90	59.7
54.8	197	108	89	59.2
54.8	188	103	85	56.4
54.8	186	102	84	55.9
54.8	177	97	80	53.2
54.8	168	92	76	50.4
54.8	166	91	75	49.9
54.8	157	86	71	47.1
54.8	155	85	70	46.6
54.8	146	80	66	43.8
54.8	135	74	61	40.6
54.8	126	69	57	37.8
54.8	124	68	56	37.3
54.8	115	63	52	34.5
54.8	104	57	47	31.2
54.8	93	51	42	28.0
54.8	84	46	38	25.2
54.8	73	40	33	21.9
54.8	62	34	28	18.6
54.8	42	23	19	12.6
54.8	31	17	14	9.3
54.7	499	273	226	149.4
54.7	497	272	225	148.9
54.7	495	271	224	148.4
54.7	494	270	224	147.6
54.7	492	269	223	147.1
54.7	490	268	222	146.6
54.7	488	267	221	146.1
54.7	486	266	220	145.6
54.7	483	264	219	144.3
54.7	481	263	218	143.8
54.7	479	262	217	143.3
54.7	477	261	216	142.8
54.7	475	260	215	142.3
54.7	472	258	214	141.0
54.7	470	257	213	140.5
54.7	468	256	212	140.0
54.7	466	255	211	139.5
54.7	464	254	210	139.0
54.7	461	252	209	137.8
54.7	459	251	208	137.3
54.7	457	250	207	136.8
54.7	455	249	206	136.3
54.7	453	248	205	135.8
54.7	450	246	204	134.5
54.7	448	245	203	134.0
54.7	446	244	202	133.5
54.7	444	243	201	133.0
54.7	439	240	199	131.2
54.7	437	239	198	130.7
54.7	435	238	197	130.2
54.7	433	237	196	129.7
54.7	430	235	195	128.4
54.7	428	234	194	127.9
54.7	426	233	193	127.4
54.7	424	232	192	126.9
54.7	422	231	191	126.4
54.7	419	229	190	125.2
54.7	417	228	189	124.7
54.7	415	227	188	124.2
54.7	413	226	187	123.7
54.7	411	225	186	123.2
54.7	408	223	185	121.9
54.7	406	222	184	121.4
54.7	404	221	183	120.9
54.7	402	220	182	120.4
54.7	397	217	180	118.6
54.7	395	216	179	118.1
54.7	393	215	178	117.6
54.7	391	214	177	117.1
54.7	386	211	175	115.3
54.7	384	210	174	114.8
54.7	382	209	173	114.3
54.7	380	208	172	113.9
54.7	375	205	170	112.1
54.7	373	204	169	111.6
54.7	371	203	168	111.1
54.7	369	202	167	110.6
54.7	364	199	165	108.8
54.7	362	198	164	108.3
54.7	360	197	163	107.8
54.7	358	196	162	107.3
54.7	353	193	160	105.5
54.7	351	192	159	105.0
54.7	349	191	158	104.5
54.7	344	188	156	102.7
54.7	342	187	155	102.2
54.7	340	186	154	101.8
54.7	338	185	153	101.3
54.7	333	182	151	99.5
54.7	331	181	150	99.0
54.7	329	180	149	98.5
54.7	327	179	148	98.0
54.7	322	176	146	96.2
54.7	320	175	145	95.7

%E	M1	M2	DM	M*
54.7	318	174	144	95.2
54.7	316	173	143	94.7
54.7	311	170	141	92.9
54.7	309	169	140	92.4
54.7	307	168	139	91.9
54.7	300	164	136	89.7
54.7	298	163	135	89.2
54.7	296	162	134	88.7
54.7	289	158	131	86.4
54.7	287	157	130	85.9
54.7	285	156	129	85.4
54.7	278	152	126	83.1
54.7	276	151	125	82.6
54.7	274	150	124	82.1
54.7	267	146	121	79.8
54.7	265	145	120	79.3
54.7	258	141	117	77.1
54.7	256	140	116	76.6
54.7	254	139	115	76.1
54.7	247	135	112	73.8
54.7	245	134	111	73.3
54.7	243	133	110	72.8
54.7	236	129	107	70.5
54.7	234	128	106	70.0
54.7	232	127	105	69.5
54.7	225	123	102	67.2
54.7	223	122	101	66.7
54.7	214	117	97	64.0
54.7	212	116	96	63.5
54.7	203	111	92	60.7
54.7	201	110	91	60.2
54.7	192	105	87	57.4
54.7	190	104	86	56.9
54.7	181	99	82	54.1
54.7	179	98	81	53.7
54.7	172	94	78	51.4
54.7	170	93	77	50.9
54.7	161	88	73	48.1
54.7	159	87	72	47.6
54.7	150	82	68	44.8
54.7	148	81	67	44.3
54.7	139	76	63	41.6
54.7	137	75	62	41.1
54.7	128	70	58	38.3
54.7	117	64	53	35.0
54.7	106	58	48	31.7
54.7	95	52	43	28.5
54.7	86	47	39	25.7
54.7	75	41	34	22.4
54.7	64	35	29	19.1

%E	M1	M2	DM	M*
54.7	53	29	24	15.9
54.6	500	273	227	149.1
54.6	498	272	226	148.6
54.6	496	271	225	148.1
54.6	493	269	224	146.8
54.6	491	268	223	146.3
54.6	489	267	222	145.8
54.6	487	266	221	145.3
54.6	485	265	220	144.8
54.6	482	263	219	143.5
54.6	480	262	218	143.0
54.6	478	261	217	142.5
54.6	476	260	216	142.0
54.6	474	259	215	141.5
54.6	471	257	214	140.2
54.6	469	256	213	139.7
54.6	467	255	212	139.2
54.6	465	254	211	138.7
54.6	463	253	210	138.2
54.6	460	251	209	137.0
54.6	458	250	208	136.5
54.6	456	249	207	136.0
54.6	454	248	206	135.5
54.6	452	247	205	135.0
54.6	449	245	204	133.7
54.6	447	244	203	133.2
54.6	445	243	202	132.7
54.6	443	242	201	132.2
54.6	441	241	200	131.7
54.6	438	239	199	130.4
54.6	436	238	198	129.9
54.6	434	237	197	129.4
54.6	432	236	196	128.9
54.6	427	233	194	127.1
54.6	425	232	193	126.6
54.6	423	231	192	126.1
54.6	421	230	191	125.7
54.6	416	227	189	123.9
54.6	414	226	188	123.4
54.6	412	225	187	122.9
54.6	410	224	186	122.4
54.6	405	221	184	120.6
54.6	403	220	183	120.1
54.6	401	219	182	119.6
54.6	399	218	181	119.1
54.6	394	215	179	117.3
54.6	392	214	178	116.8
54.6	390	213	177	116.3
54.6	388	212	176	115.8
54.6	383	209	174	114.0

%E	M1	M2	DM	M*
54.6	381	208	173	113.6
54.6	379	207	172	113.1
54.6	377	206	171	112.6
54.6	372	203	169	110.8
54.6	370	202	168	110.3
54.6	368	201	167	109.8
54.6	366	200	166	109.3
54.6	361	197	164	107.5
54.6	359	196	163	107.0
54.6	357	195	162	106.5
54.6	355	194	161	106.0
54.6	350	191	159	104.2
54.6	348	190	158	103.7
54.6	346	189	157	103.2
54.6	339	185	154	101.0
54.6	337	184	153	100.5
54.6	335	183	152	100.0
54.6	328	179	149	97.7
54.6	326	178	148	97.2
54.6	324	177	147	96.7
54.6	317	173	144	94.4
54.6	315	172	143	93.9
54.6	313	171	142	93.4
54.6	306	167	139	91.1
54.6	304	166	138	90.6
54.6	302	165	137	90.1
54.6	295	161	134	87.9
54.6	293	160	133	87.4
54.6	291	159	132	86.9
54.6	284	155	129	84.6
54.6	282	154	128	84.1
54.6	280	153	127	83.6
54.6	273	149	124	81.3
54.6	271	148	123	80.8
54.6	269	147	122	80.3
54.6	262	143	119	78.0
54.6	260	142	118	77.6
54.6	251	137	114	74.8
54.6	249	136	113	74.3
54.6	240	131	109	71.5
54.6	238	130	108	71.0
54.6	229	125	104	68.2
54.6	227	124	103	67.7
54.6	218	119	99	65.0
54.6	216	118	98	64.5
54.6	207	113	94	61.7
54.6	205	112	93	61.2
54.6	196	107	89	58.4
54.6	194	106	88	57.9
54.6	185	101	84	55.1

%E	M1	M2	DM	M*
54.6	183	100	83	54.6
54.6	174	95	79	51.9
54.6	163	89	74	48.6
54.6	152	83	69	45.3
54.6	141	77	64	42.0
54.6	130	71	59	38.8
54.6	119	65	54	35.5
54.6	108	59	49	32.2
54.6	97	53	44	29.0
54.5	499	272	227	148.3
54.5	497	271	226	147.8
54.5	495	270	225	147.3
54.5	494	269	225	146.5
54.5	492	268	224	146.0
54.5	490	267	223	145.5
54.5	488	266	222	145.0
54.5	486	265	221	144.5
54.5	484	264	220	144.0
54.5	483	263	220	143.2
54.5	481	262	219	142.7
54.5	479	261	218	142.2
54.5	477	260	217	141.7
54.5	475	259	216	141.2
54.5	473	258	215	140.7
54.5	470	256	214	139.4
54.5	468	255	213	138.9
54.5	466	254	212	138.4
54.5	464	253	211	138.0
54.5	462	252	210	137.5
54.5	459	250	209	136.2
54.5	457	249	208	135.7
54.5	455	248	207	135.2
54.5	453	247	206	134.7
54.5	451	246	205	134.2
54.5	448	244	204	132.9
54.5	446	243	203	132.4
54.5	444	242	202	131.9
54.5	442	241	201	131.4
54.5	440	240	200	130.9
54.5	437	238	199	129.6
54.5	435	237	198	129.1
54.5	433	236	197	128.6
54.5	431	235	196	128.1
54.5	429	234	195	127.6
54.5	426	232	194	126.3
54.5	424	231	193	125.9
54.5	422	230	192	125.4
54.5	420	229	191	124.9
54.5	418	228	190	124.4
54.5	415	226	189	123.1

%E	M1	M2	DM	M*
54.5	413	225	188	122.6
54.5	411	224	187	122.1
54.5	409	223	186	121.6
54.5	407	222	185	121.1
54.5	404	220	184	119.8
54.5	402	219	183	119.3
54.5	400	218	182	118.8
54.5	398	217	181	118.3
54.5	396	216	180	117.8
54.5	393	214	179	116.5
54.5	391	213	178	116.0
54.5	389	212	177	115.5
54.5	387	211	176	115.0
54.5	385	210	175	114.5
54.5	382	208	174	113.3
54.5	380	207	173	112.8
54.5	378	206	172	112.3
54.5	376	205	171	111.8
54.5	374	204	170	111.3
54.5	369	201	168	109.5
54.5	367	200	167	109.0
54.5	365	199	166	108.5
54.5	363	198	165	108.0
54.5	358	195	163	106.2
54.5	356	194	162	105.7
54.5	354	193	161	105.2
54.5	352	192	160	104.7
54.5	347	189	158	102.9
54.5	345	188	157	102.4
54.5	343	187	156	102.0
54.5	341	186	155	101.5
54.5	336	183	153	99.7
54.5	334	182	152	99.2
54.5	332	181	151	98.7
54.5	330	180	150	98.2
54.5	325	177	148	96.4
54.5	323	176	147	95.9
54.5	321	175	146	95.4
54.5	319	174	145	94.9
54.5	314	171	143	93.1
54.5	312	170	142	92.6
54.5	310	169	141	92.1
54.5	308	168	140	91.6
54.5	303	165	138	89.9
54.5	301	164	137	89.4
54.5	299	163	136	88.9
54.5	297	162	135	88.4
54.5	292	159	133	86.6
54.5	290	158	132	86.1
54.5	288	157	131	85.6

%E	M1	M2	DM	M*
54.5	286	156	130	85.1
54.5	279	152	127	82.8
54.5	277	151	126	82.3
54.5	275	150	125	81.8
54.5	268	146	122	79.5
54.5	266	145	121	79.0
54.5	264	144	120	78.5
54.5	257	140	117	76.3
54.5	255	139	116	75.8
54.5	253	138	115	75.3
54.5	246	134	112	73.0
54.5	244	133	111	72.5
54.5	242	132	110	72.0
54.5	235	128	107	69.7
54.5	233	127	106	69.2
54.5	231	126	105	68.7
54.5	224	122	102	66.4
54.5	222	121	101	66.0
54.5	220	120	100	65.5
54.5	213	116	97	63.2
54.5	211	115	96	62.7
54.5	209	114	95	62.2
54.5	202	110	92	59.9
54.5	200	109	91	59.4
54.5	198	108	90	58.9
54.5	191	104	87	56.6
54.5	189	103	86	56.1
54.5	187	102	85	55.6
54.5	178	97	81	52.9
54.5	176	96	80	52.4
54.5	167	91	76	49.6
54.5	165	90	75	49.1
54.5	156	85	71	46.3
54.5	154	84	70	45.8
54.5	145	79	66	43.0
54.5	143	78	65	42.5
54.5	134	73	61	39.8
54.5	132	72	60	39.3
54.5	123	67	56	36.5
54.5	121	66	55	36.0
54.5	112	61	51	33.2
54.5	110	60	50	32.7
54.5	101	55	46	30.0
54.5	99	54	45	29.5
54.5	88	48	40	26.2
54.5	77	42	35	22.9
54.5	66	36	30	19.6
54.5	55	30	25	16.4
54.5	44	24	20	13.1
54.5	33	18	15	9.8

%E	M1	M2	DM	M*
54.5	22	12	10	6.5
54.5	11	6	5	3.3
54.4	500	272	228	148.0
54.4	498	271	227	147.5
54.4	496	270	226	147.0
54.4	493	268	225	145.7
54.4	491	267	224	145.2
54.4	489	266	223	144.7
54.4	487	265	222	144.2
54.4	485	264	221	143.7
54.4	482	262	220	142.4
54.4	480	261	219	141.9
54.4	478	260	218	141.4
54.4	476	259	217	140.9
54.4	474	258	216	140.4
54.4	472	257	215	139.9
54.4	471	256	215	139.1
54.4	469	255	214	138.6
54.4	467	254	213	138.1
54.4	465	253	212	137.7
54.4	463	252	211	137.2
54.4	461	251	210	136.7
54.4	458	249	209	135.4
54.4	456	248	208	134.9
54.4	454	247	207	134.4
54.4	452	246	206	133.9
54.4	450	245	205	133.4
54.4	447	243	204	132.1
54.4	445	242	203	131.6
54.4	443	241	202	131.1
54.4	441	240	201	130.6
54.4	439	239	200	130.1
54.4	436	237	199	128.8
54.4	434	236	198	128.3
54.4	432	235	197	127.8
54.4	430	234	196	127.3
54.4	428	233	195	126.8
54.4	425	231	194	125.6
54.4	423	230	193	125.1
54.4	421	229	192	124.6
54.4	419	228	191	124.1
54.4	417	227	190	123.6
54.4	412	224	188	121.8
54.4	410	223	187	121.3
54.4	408	222	186	120.8
54.4	406	221	185	120.3
54.4	401	218	183	118.5
54.4	399	217	182	118.0
54.4	397	216	181	117.5
54.4	395	215	180	117.0

%E	M1	M2	DM	M*
54.4	390	212	178	115.2
54.4	388	211	177	114.7
54.4	386	210	176	114.2
54.4	384	209	175	113.8
54.4	379	206	173	112.0
54.4	377	205	172	111.5
54.4	375	204	171	111.0
54.4	373	203	170	110.5
54.4	371	202	169	110.0
54.4	366	199	167	108.2
54.4	364	198	166	107.7
54.4	362	197	165	107.2
54.4	360	196	164	106.7
54.4	355	193	162	104.9
54.4	353	192	161	104.4
54.4	351	191	160	103.9
54.4	349	190	159	103.4
54.4	344	187	157	101.7
54.4	342	186	156	101.2
54.4	340	185	155	100.7
54.4	338	184	154	100.2
54.4	333	181	152	98.4
54.4	331	180	151	97.9
54.4	329	179	150	97.4
54.4	327	178	149	96.9
54.4	320	174	146	94.6
54.4	318	173	145	94.1
54.4	316	172	144	93.6
54.4	309	168	141	91.3
54.4	307	167	140	90.8
54.4	305	166	139	90.3
54.4	298	162	136	88.1
54.4	296	161	135	87.6
54.4	294	160	134	87.1
54.4	287	156	131	84.8
54.4	285	155	130	84.3
54.4	283	154	129	83.8
54.4	281	153	128	83.3
54.4	274	149	125	81.0
54.4	272	148	124	80.5
54.4	270	147	123	80.0
54.4	263	143	120	77.8
54.4	261	142	119	77.3
54.4	259	141	118	76.8
54.4	252	137	115	74.5
54.4	250	136	114	74.0
54.4	248	135	113	73.5
54.4	241	131	110	71.2
54.4	239	130	109	70.7
54.4	237	129	108	70.2

%E	M1	M2	DM	M*
54.4	228	124	104	67.4
54.4	226	123	103	66.9
54.4	217	118	99	64.2
54.4	215	117	98	63.7
54.4	206	112	94	60.9
54.4	204	111	93	60.4
54.4	195	106	89	57.6
54.4	193	105	88	57.1
54.4	182	99	83	53.9
54.4	180	98	82	53.4
54.4	171	93	78	50.6
54.4	169	92	77	50.1
54.4	160	87	73	47.3
54.4	158	86	72	46.8
54.4	149	81	68	44.0
54.4	147	80	67	43.5
54.4	136	74	62	40.3
54.4	125	68	57	37.0
54.4	114	62	52	33.7
54.4	103	56	47	30.4
54.4	90	49	41	26.7
54.4	79	43	36	23.4
54.4	68	37	31	20.1
54.4	57	31	26	16.9
54.3	499	271	228	147.2
54.3	497	270	227	146.7
54.3	495	269	226	146.2
54.3	494	268	226	145.4
54.3	492	267	225	144.9
54.3	490	266	224	144.4
54.3	488	265	223	143.9
54.3	486	264	222	143.4
54.3	484	263	221	142.9
54.3	481	261	220	141.6
54.3	479	260	219	141.1
54.3	477	259	218	140.6
54.3	475	258	217	140.1
54.3	473	257	216	139.6
54.3	470	255	215	138.4
54.3	468	254	214	137.9
54.3	466	253	213	137.4
54.3	464	252	212	136.9
54.3	462	251	211	136.4
54.3	460	250	210	135.9
54.3	457	248	209	134.6
54.3	455	247	208	134.1
54.3	453	246	207	133.6
54.3	451	245	206	133.1
54.3	449	244	205	132.6
54.3	446	242	204	131.3

%E	M1	M2	DM	M*
54.3	444	241	203	130.8
54.3	442	240	202	130.3
54.3	440	239	201	129.8
54.3	438	238	200	129.3
54.3	435	236	199	128.0
54.3	433	235	198	127.5
54.3	431	234	197	127.0
54.3	429	233	196	126.5
54.3	427	232	195	126.1
54.3	422	229	193	124.3
54.3	420	228	192	123.8
54.3	418	227	191	123.3
54.3	416	226	190	122.8
54.3	414	225	189	122.3
54.3	411	223	188	121.0
54.3	409	222	187	120.5
54.3	407	221	186	120.0
54.3	405	220	185	119.5
54.3	403	219	184	119.0
54.3	400	217	183	117.7
54.3	398	216	182	117.2
54.3	396	215	181	116.7
54.3	394	214	180	116.2
54.3	392	213	179	115.7
54.3	387	210	177	114.0
54.3	385	209	176	113.5
54.3	383	208	175	113.0
54.3	381	207	174	112.5
54.3	376	204	172	110.7
54.3	374	203	171	110.2
54.3	372	202	170	109.7
54.3	370	201	169	109.2
54.3	368	200	168	108.7
54.3	363	197	166	106.9
54.3	361	196	165	106.4
54.3	359	195	164	105.9
54.3	357	194	163	105.4
54.3	352	191	161	103.6
54.3	350	190	160	103.1
54.3	348	189	159	102.6
54.3	346	188	158	102.2
54.3	341	185	156	100.4
54.3	339	184	155	99.9
54.3	337	183	154	99.4
54.3	335	182	153	98.9
54.3	328	178	150	96.6
54.3	326	177	149	96.1
54.3	324	176	148	95.6
54.3	322	175	147	95.1
54.3	317	172	145	93.3

%E	M1	M2	DM	M*
54.3	315	171	144	92.8
54.3	313	170	143	92.3
54.3	311	169	142	91.8
54.3	304	165	139	89.6
54.3	302	164	138	89.1
54.3	300	163	137	88.6
54.3	293	159	134	86.3
54.3	291	158	133	85.8
54.3	289	157	132	85.3
54.3	282	153	129	83.0
54.3	280	152	128	82.5
54.3	278	151	127	82.0
54.3	276	150	126	81.5
54.3	269	146	123	79.2
54.3	267	145	122	78.7
54.3	265	144	121	78.2
54.3	258	140	118	76.0
54.3	256	139	117	75.5
54.3	254	138	116	75.0
54.3	247	134	113	72.7
54.3	245	133	112	72.2
54.3	243	132	111	71.7
54.3	234	127	107	68.9
54.3	232	126	106	68.4
54.3	230	125	105	67.9
54.3	223	121	102	65.7
54.3	221	120	101	65.2
54.3	219	119	100	64.7
54.3	210	114	96	61.9
54.3	208	113	95	61.4
54.3	199	108	91	58.6
54.3	197	107	90	58.1
54.3	188	102	86	55.3
54.3	186	101	85	54.8
54.3	184	100	84	54.3
54.3	175	95	80	51.6
54.3	173	94	79	51.1
54.3	164	89	75	48.3
54.3	162	88	74	47.8
54.3	151	82	69	44.5
54.3	140	76	64	41.3
54.3	138	75	63	40.8
54.3	129	70	59	38.0
54.3	127	69	58	37.5
54.3	116	63	53	34.2
54.3	105	57	48	30.9
54.3	94	51	43	27.7
54.3	92	50	42	27.2
54.3	81	44	37	23.9
54.3	70	38	32	20.6

%E	M1	M2	DM	M*
54.3	46	25	21	13.6
54.3	35	19	16	10.3
54.2	500	271	229	146.9
54.2	498	270	228	146.4
54.2	496	269	227	145.9
54.2	493	267	226	144.6
54.2	491	266	225	144.1
54.2	489	265	224	143.6
54.2	487	264	223	143.1
54.2	485	263	222	142.6
54.2	483	262	221	142.1
54.2	480	260	220	140.8
54.2	478	259	219	140.3
54.2	476	258	218	139.8
54.2	474	257	217	139.3
54.2	472	256	216	138.8
54.2	469	254	215	137.6
54.2	467	253	214	137.1
54.2	465	252	213	136.6
54.2	463	251	212	136.1
54.2	461	250	211	135.6
54.2	459	249	210	135.1
54.2	456	247	209	133.8
54.2	454	246	208	133.3
54.2	452	245	207	132.8
54.2	450	244	206	132.3
54.2	448	243	205	131.8
54.2	445	241	204	130.5
54.2	443	240	203	130.0
54.2	441	239	202	129.5
54.2	439	238	201	129.0
54.2	437	237	200	128.5
54.2	432	234	198	126.8
54.2	430	233	197	126.3
54.2	428	232	196	125.8
54.2	426	231	195	125.3
54.2	424	230	194	124.8
54.2	421	228	193	123.5
54.2	419	227	192	123.0
54.2	417	226	191	122.5
54.2	415	225	190	122.0
54.2	413	224	189	121.5
54.2	408	221	187	119.7
54.2	406	220	186	119.2
54.2	404	219	185	118.7
54.2	402	218	184	118.2
54.2	397	215	182	116.4
54.2	395	214	181	115.9
54.2	393	213	180	115.4
54.2	391	212	179	114.9

%E	M1	M2	DM	M*
54.2	389	211	178	114.4
54.2	384	208	176	112.7
54.2	382	207	175	112.2
54.2	380	206	174	111.7
54.2	378	205	173	111.2
54.2	373	202	171	109.4
54.2	371	201	170	108.9
54.2	369	200	169	108.4
54.2	367	199	168	107.9
54.2	365	198	167	107.4
54.2	360	195	165	105.6
54.2	358	194	164	105.1
54.2	356	193	163	104.6
54.2	354	192	162	104.1
54.2	349	189	160	102.4
54.2	347	188	159	101.9
54.2	345	187	158	101.4
54.2	343	186	157	100.9
54.2	336	182	154	98.6
54.2	334	181	153	98.1
54.2	332	180	152	97.6
54.2	330	179	151	97.1
54.2	325	176	149	95.3
54.2	323	175	148	94.8
54.2	321	174	147	94.3
54.2	319	173	146	93.8
54.2	312	169	143	91.5
54.2	310	168	142	91.0
54.2	308	167	141	90.5
54.2	306	166	140	90.1
54.2	301	163	138	88.3
54.2	299	162	137	87.8
54.2	297	161	136	87.3
54.2	295	160	135	86.8
54.2	288	156	132	84.5
54.2	286	155	131	84.0
54.2	284	154	130	83.5
54.2	277	150	127	81.2
54.2	275	149	126	80.7
54.2	273	148	125	80.2
54.2	271	147	124	79.7
54.2	264	143	121	77.5
54.2	262	142	120	77.0
54.2	260	141	119	76.5
54.2	253	137	116	74.2
54.2	251	136	115	73.7
54.2	249	135	114	73.2
54.2	240	130	110	70.4
54.2	238	129	109	69.9
54.2	236	128	108	69.4

%E	M1	M2	DM	M*
54.2	227	123	104	66.6
54.2	225	122	103	66.2
54.2	216	117	99	63.4
54.2	214	116	98	62.9
54.2	212	115	97	62.4
54.2	203	110	93	59.6
54.2	201	109	92	59.1
54.2	192	104	88	56.3
54.2	190	103	87	55.8
54.2	179	97	82	52.6
54.2	177	96	81	52.1
54.2	168	91	77	49.3
54.2	166	90	76	48.8
54.2	155	84	71	45.5
54.2	153	83	70	45.0
54.2	144	78	66	42.3
54.2	142	77	65	41.8
54.2	131	71	60	38.5
54.2	120	65	55	35.2
54.2	118	64	54	34.7
54.2	107	58	49	31.4
54.2	96	52	44	28.2
54.2	83	45	38	24.4
54.2	72	39	33	21.1
54.2	59	32	27	17.4
54.2	48	26	22	14.1
54.2	24	13	11	7.0
54.1	499	270	229	146.1
54.1	497	269	228	145.6
54.1	495	268	227	145.1
54.1	492	266	226	143.8
54.1	490	265	225	143.3
54.1	488	264	224	142.8
54.1	486	263	223	142.3
54.1	484	262	222	141.8
54.1	482	261	221	141.3
54.1	481	260	221	140.5
54.1	479	259	220	140.0
54.1	477	258	219	139.5
54.1	475	257	218	139.1
54.1	473	256	217	138.6
54.1	471	255	216	138.1
54.1	468	253	215	136.8
54.1	466	252	214	136.3
54.1	464	251	213	135.8
54.1	462	250	212	135.3
54.1	460	249	211	134.8
54.1	458	248	210	134.3
54.1	455	246	209	133.0
54.1	453	245	208	132.5

%E	M1	M2	DM	M*
54.1	451	244	207	132.0
54.1	449	243	206	131.5
54.1	447	242	205	131.0
54.1	444	240	204	129.7
54.1	442	239	203	129.2
54.1	440	238	202	128.7
54.1	438	237	201	128.2
54.1	436	236	200	127.7
54.1	434	235	199	127.2
54.1	431	233	198	126.0
54.1	429	232	197	125.5
54.1	427	231	196	125.0
54.1	425	230	195	124.5
54.1	423	229	194	124.0
54.1	418	226	192	122.2
54.1	416	225	191	121.7
54.1	414	224	190	121.2
54.1	412	223	189	120.7
54.1	410	222	188	120.2
54.1	407	220	187	118.9
54.1	405	219	186	118.4
54.1	403	218	185	117.9
54.1	401	217	184	117.4
54.1	399	216	183	116.9
54.1	394	213	181	115.1
54.1	392	212	180	114.7
54.1	390	211	179	114.2
54.1	388	210	178	113.7
54.1	386	209	177	113.2
54.1	381	206	175	111.4
54.1	379	205	174	110.9
54.1	377	204	173	110.4
54.1	375	203	172	109.9
54.1	370	200	170	108.1
54.1	368	199	169	107.6
54.1	366	198	168	107.1
54.1	364	197	167	106.6
54.1	362	196	166	106.1
54.1	357	193	164	104.3
54.1	355	192	163	103.8
54.1	353	191	162	103.3
54.1	351	190	161	102.8
54.1	344	186	158	100.6
54.1	342	185	157	100.1
54.1	340	184	156	99.6
54.1	338	183	155	99.1
54.1	333	180	153	97.3
54.1	331	179	152	96.8
54.1	329	178	151	96.3
54.1	327	177	150	95.8

%E	M1	M2	DM	M*
54.1	320	173	147	93.5
54.1	318	172	146	93.0
54.1	316	171	145	92.5
54.1	314	170	144	92.0
54.1	307	166	141	89.8
54.1	305	165	140	89.3
54.1	303	164	139	88.8
54.1	296	160	136	86.5
54.1	294	159	135	86.0
54.1	292	158	134	85.5
54.1	290	157	133	85.0
54.1	283	153	130	82.7
54.1	281	152	129	82.2
54.1	279	151	128	81.7
54.1	270	146	124	78.9
54.1	268	145	123	78.5
54.1	266	144	122	78.0
54.1	259	140	119	75.7
54.1	257	139	118	75.2
54.1	255	138	117	74.7
54.1	246	133	113	71.9
54.1	244	132	112	71.4
54.1	242	131	111	70.9
54.1	233	126	107	68.1
54.1	231	125	106	67.6
54.1	229	124	105	67.1
54.1	222	120	102	64.9
54.1	220	119	101	64.4
54.1	218	118	100	63.9
54.1	209	113	96	61.1
54.1	207	112	95	60.6
54.1	205	111	94	60.1
54.1	196	106	90	57.3
54.1	194	105	89	56.8
54.1	185	100	85	54.1
54.1	183	99	84	53.6
54.1	181	98	83	53.1
54.1	172	93	79	50.3
54.1	170	92	78	49.8
54.1	159	86	73	46.5
54.1	157	85	72	46.0
54.1	148	80	68	43.2
54.1	146	79	67	42.7
54.1	135	73	62	39.5
54.1	133	72	61	39.0
54.1	122	66	56	35.7
54.1	111	60	51	32.4
54.1	109	59	50	31.9
54.1	98	53	45	28.7
54.1	85	46	39	24.9

%E	M1	M2	DM	M*
54.1	74	40	34	21.6
54.1	61	33	28	17.9
54.1	37	20	17	10.8
54.0	500	270	230	145.8
54.0	498	269	229	145.3
54.0	496	268	228	144.8
54.0	494	267	227	144.3
54.0	493	266	227	143.5
54.0	491	265	226	143.0
54.0	489	264	225	142.5
54.0	487	263	224	142.0
54.0	485	262	223	141.5
54.0	483	261	222	141.0
54.0	480	259	221	139.8
54.0	478	258	220	139.3
54.0	476	257	219	138.8
54.0	474	256	218	138.3
54.0	472	255	217	137.8
54.0	470	254	216	137.3
54.0	467	252	215	136.0
54.0	465	251	214	135.5
54.0	463	250	213	135.0
54.0	461	249	212	134.5
54.0	459	248	211	134.0
54.0	457	247	210	133.5
54.0	454	245	209	132.2
54.0	452	244	208	131.7
54.0	450	243	207	131.2
54.0	448	242	206	130.7
54.0	446	241	205	130.2
54.0	443	239	204	128.9
54.0	441	238	203	128.4
54.0	439	237	202	127.9
54.0	437	236	201	127.5
54.0	435	235	200	127.0
54.0	433	234	199	126.5
54.0	430	232	198	125.2
54.0	428	231	197	124.7
54.0	426	230	196	124.2
54.0	424	229	195	123.7
54.0	422	228	194	123.2
54.0	420	227	193	122.7
54.0	417	225	192	121.4
54.0	415	224	191	120.9
54.0	413	223	190	120.4
54.0	411	222	189	119.9
54.0	409	221	188	119.4
54.0	404	218	186	117.6
54.0	402	217	185	117.1
54.0	400	216	184	116.6

%E	M1	M2	DM	M*
54.0	398	215	183	116.1
54.0	396	214	182	115.6
54.0	391	211	180	113.9
54.0	389	210	179	113.4
54.0	387	209	178	112.9
54.0	385	208	177	112.4
54.0	383	207	176	111.9
54.0	378	204	174	110.1
54.0	376	203	173	109.6
54.0	374	202	172	109.1
54.0	372	201	171	108.6
54.0	367	198	169	106.8
54.0	365	197	168	106.3
54.0	363	196	167	105.8
54.0	361	195	166	105.3
54.0	359	194	165	104.8
54.0	354	191	163	103.1
54.0	352	190	162	102.6
54.0	350	189	161	102.1
54.0	348	188	160	101.6
54.0	346	187	159	101.1
54.0	341	184	157	99.3
54.0	339	183	156	98.8
54.0	337	182	155	98.3
54.0	335	181	154	97.8
54.0	328	177	151	95.5
54.0	326	176	150	95.0
54.0	324	175	149	94.5
54.0	322	174	148	94.0
54.0	315	170	145	91.7
54.0	313	169	144	91.2
54.0	311	168	143	90.8
54.0	309	167	142	90.3
54.0	302	163	139	88.0
54.0	300	162	138	87.5
54.0	298	161	137	87.0
54.0	291	157	134	84.7
54.0	289	156	133	84.2
54.0	287	155	132	83.7
54.0	285	154	131	83.2
54.0	278	150	128	80.9
54.0	276	149	127	80.4
54.0	274	148	126	79.9
54.0	272	147	125	79.4
54.0	265	143	122	77.2
54.0	263	142	121	76.7
54.0	261	141	120	76.2
54.0	252	136	116	73.4
54.0	250	135	115	72.9
54.0	248	134	114	72.4

%E	M1	M2	DM	M*
54.0	239	129	110	69.6
54.0	237	128	109	69.1
54.0	235	127	108	68.6
54.0	226	122	104	65.9
54.0	224	121	103	65.4
54.0	215	116	99	62.6
54.0	213	115	98	62.1
54.0	211	114	97	61.6
54.0	202	109	93	58.8
54.0	200	108	92	58.3
54.0	198	107	91	57.8
54.0	189	102	87	55.0
54.0	187	101	86	54.6
54.0	176	95	81	51.3
54.0	174	94	80	50.8
54.0	163	88	75	47.5
54.0	161	87	74	47.0
54.0	150	81	69	43.7
54.0	139	75	64	40.5
54.0	137	74	63	40.0
54.0	126	68	58	36.7
54.0	124	67	57	36.2
54.0	113	61	52	32.9
54.0	100	54	46	29.2
54.0	87	47	40	25.4
54.0	63	34	29	18.3
54.0	50	27	23	14.6
53.9	499	269	230	145.0
53.9	497	268	229	144.5
53.9	495	267	228	144.0
53.9	492	265	227	142.7
53.9	490	264	226	142.2
53.9	488	263	225	141.7
53.9	486	262	224	141.2
53.9	484	261	223	140.7
53.9	482	260	222	140.2
53.9	479	258	221	139.0
53.9	477	257	220	138.5
53.9	475	256	219	138.0
53.9	473	255	218	137.5
53.9	471	254	217	137.0
53.9	469	253	216	136.5
53.9	466	251	215	135.2
53.9	464	250	214	134.7
53.9	462	249	213	134.2
53.9	460	248	212	133.7
53.9	458	247	211	133.2
53.9	456	246	210	132.7
53.9	453	244	209	131.4
53.9	451	243	208	130.9

%E	M1	M2	DM	M*
53.9	449	242	207	130.4
53.9	447	241	206	129.9
53.9	445	240	205	129.4
53.9	440	237	203	127.7
53.9	438	236	202	127.2
53.9	436	235	201	126.7
53.9	434	234	200	126.2
53.9	432	233	199	125.7
53.9	427	230	197	123.9
53.9	425	229	196	123.4
53.9	423	228	195	122.9
53.9	421	227	194	122.4
53.9	419	226	193	121.9
53.9	414	223	191	120.1
53.9	412	222	190	119.6
53.9	410	221	189	119.1
53.9	408	220	188	118.6
53.9	406	219	187	118.1
53.9	401	216	185	116.3
53.9	399	215	184	115.9
53.9	397	214	183	115.4
53.9	395	213	182	114.9
53.9	393	212	181	114.4
53.9	388	209	179	112.6
53.9	386	208	178	112.1
53.9	384	207	177	111.6
53.9	382	206	176	111.1
53.9	380	205	175	110.6
53.9	375	202	173	108.8
53.9	373	201	172	108.3
53.9	371	200	171	107.8
53.9	369	199	170	107.3
53.9	362	195	167	105.0
53.9	360	194	166	104.5
53.9	358	193	165	104.0
53.9	356	192	164	103.6
53.9	349	188	161	101.3
53.9	347	187	160	100.8
53.9	345	186	159	100.3
53.9	343	185	158	99.8
53.9	336	181	155	97.5
53.9	334	180	154	97.0
53.9	332	179	153	96.5
53.9	330	178	152	96.0
53.9	323	174	149	93.7
53.9	321	173	148	93.2
53.9	319	172	147	92.7
53.9	317	171	146	92.2
53.9	310	167	143	90.0
53.9	308	166	142	89.5

%E	M1	M2	DM	M*
53.9	306	165	141	89.0
53.9	304	164	140	88.5
53.9	297	160	137	86.2
53.9	295	159	136	85.7
53.9	293	158	135	85.2
53.9	284	153	131	82.4
53.9	282	152	130	81.9
53.9	280	151	129	81.4
53.9	271	146	125	78.7
53.9	269	145	124	78.2
53.9	267	144	123	77.7
53.9	258	139	119	74.9
53.9	256	138	118	74.4
53.9	254	137	117	73.9
53.9	245	132	113	71.1
53.9	243	131	112	70.6
53.9	241	130	111	70.1
53.9	232	125	107	67.3
53.9	230	124	106	66.9
53.9	228	123	105	66.4
53.9	219	118	101	63.6
53.9	217	117	100	63.1
53.9	206	111	95	59.8
53.9	204	110	94	59.3
53.9	193	104	89	56.0
53.9	191	103	88	55.5
53.9	180	97	83	52.3
53.9	178	96	82	51.8
53.9	167	90	77	48.5
53.9	165	89	76	48.0
53.9	154	83	71	44.7
53.9	152	82	70	44.2
53.9	141	76	65	41.0
53.9	128	69	59	37.2
53.9	115	62	53	33.4
53.9	102	55	47	29.7
53.9	89	48	41	25.9
53.9	76	41	35	22.1
53.8	500	269	231	144.7
53.8	498	268	230	144.2
53.8	496	267	229	143.7
53.8	494	266	228	143.2
53.8	493	265	228	142.4
53.8	491	264	227	141.9
53.8	489	263	226	141.4
53.8	487	262	225	141.0
53.8	485	261	224	140.5
53.8	483	260	223	140.0
53.8	481	259	222	139.5
53.8	480	258	222	138.7

%E	M1	M2	DM	M*
53.8	478	257	221	138.2
53.8	476	256	220	137.7
53.8	474	255	219	137.2
53.8	472	254	218	136.7
53.8	470	253	217	136.2
53.8	468	252	216	135.7
53.8	465	250	215	134.4
53.8	463	249	214	133.9
53.8	461	248	213	133.4
53.8	459	247	212	132.9
53.8	457	246	211	132.4
53.8	455	245	210	131.9
53.8	452	243	209	130.6
53.8	450	242	208	130.1
53.8	448	241	207	129.6
53.8	446	240	206	129.1
53.8	444	239	205	128.7
53.8	442	238	204	128.2
53.8	439	236	203	126.9
53.8	437	235	202	126.4
53.8	435	234	201	125.9
53.8	433	233	200	125.4
53.8	431	232	199	124.9
53.8	429	231	198	124.4
53.8	426	229	197	123.1
53.8	424	228	196	122.6
53.8	422	227	195	122.1
53.8	420	226	194	121.6
53.8	418	225	193	121.1
53.8	416	224	192	120.6
53.8	413	222	191	119.3
53.8	411	221	190	118.8
53.8	409	220	189	118.3
53.8	407	219	188	117.8
53.8	405	218	187	117.3
53.8	403	217	186	116.8
53.8	400	215	185	115.6
53.8	398	214	184	115.1
53.8	396	213	183	114.6
53.8	394	212	182	114.1
53.8	392	211	181	113.6
53.8	390	210	180	113.1
53.8	385	207	178	111.3
53.8	383	206	177	110.8
53.8	381	205	176	110.3
53.8	379	204	175	109.8
53.8	377	203	174	109.3
53.8	372	200	172	107.5
53.8	370	199	171	107.0
53.8	368	198	170	106.5

%E	M1	M2	DM	M*
53.8	366	197	169	106.0
53.8	364	196	168	105.5
53.8	359	193	166	103.8
53.8	357	192	165	103.3
53.8	355	191	164	102.8
53.8	353	190	163	102.3
53.8	351	189	162	101.8
53.8	346	186	160	100.0
53.8	344	185	159	99.5
53.8	342	184	158	99.0
53.8	340	183	157	98.5
53.8	338	182	156	98.0
53.8	333	179	154	96.2
53.8	331	178	153	95.7
53.8	329	177	152	95.2
53.8	327	176	151	94.7
53.8	325	175	150	94.2
53.8	320	172	148	92.4
53.8	318	171	147	92.0
53.8	316	170	146	91.5
53.8	314	169	145	91.0
53.8	312	168	144	90.5
53.8	305	164	141	88.2
53.8	303	163	140	87.7
53.8	301	162	139	87.2
53.8	299	161	138	86.7
53.8	292	157	135	84.4
53.8	290	156	134	83.9
53.8	288	155	133	83.4
53.8	286	154	132	82.9
53.8	279	150	129	80.6
53.8	277	149	128	80.1
53.8	275	148	127	79.7
53.8	273	147	126	79.2
53.8	266	143	123	76.9
53.8	264	142	122	76.4
53.8	262	141	121	75.9
53.8	260	140	120	75.4
53.8	253	136	117	73.1
53.8	251	135	116	72.6
53.8	249	134	115	72.1
53.8	247	133	114	71.6
53.8	240	129	111	69.3
53.8	238	128	110	68.8
53.8	236	127	109	68.3
53.8	234	126	108	67.8
53.8	225	121	104	65.1
53.8	223	120	103	64.6
53.8	221	119	102	64.1
53.8	212	114	98	61.3

%E	M1	M2	DM	M*
53.8	210	113	97	60.8
53.8	208	112	96	60.3
53.8	199	107	92	57.5
53.8	197	106	91	57.0
53.8	195	105	90	56.5
53.8	186	100	86	53.8
53.8	184	99	85	53.3
53.8	182	98	84	52.8
53.8	173	93	80	50.0
53.8	171	92	79	49.5
53.8	169	91	78	49.0
53.8	160	86	74	46.2
53.8	158	85	73	45.7
53.8	156	84	72	45.2
53.8	145	78	67	42.0
53.8	143	77	66	41.5
53.8	132	71	61	38.2
53.8	130	70	60	37.7
53.8	119	64	55	34.4
53.8	117	63	54	33.9
53.8	106	57	49	30.7
53.8	104	56	48	30.2
53.8	93	50	43	26.9
53.8	91	49	42	26.4
53.8	80	43	37	23.1
53.8	78	42	36	22.6
53.8	65	35	30	18.8
53.8	52	28	24	15.1
53.8	39	21	18	11.3
53.8	26	14	12	7.5
53.8	13	7	6	3.8
53.7	499	268	231	143.9
53.7	497	267	230	143.4
53.7	495	266	229	142.9
53.7	492	264	228	141.7
53.7	490	263	227	141.2
53.7	488	262	226	140.7
53.7	486	261	225	140.2
53.7	484	260	224	139.7
53.7	482	259	223	139.2
53.7	479	257	222	137.9
53.7	477	256	221	137.4
53.7	475	255	220	136.9
53.7	473	254	219	136.4
53.7	471	253	218	135.9
53.7	469	252	217	135.4
53.7	467	251	216	134.9
53.7	464	249	215	133.6
53.7	462	248	214	133.1
53.7	460	247	213	132.6

%E	M1	M2	DM	M*
53.7	458	246	212	132.1
53.7	456	245	211	131.6
53.7	454	244	210	131.1
53.7	451	242	209	129.9
53.7	449	241	208	129.4
53.7	447	240	207	128.9
53.7	445	239	206	128.4
53.7	443	238	205	127.9
53.7	441	237	204	127.4
53.7	438	235	203	126.1
53.7	436	234	202	125.6
53.7	434	233	201	125.1
53.7	432	232	200	124.6
53.7	430	231	199	124.1
53.7	428	230	198	123.6
53.7	423	227	196	121.8
53.7	421	226	195	121.3
53.7	419	225	194	120.8
53.7	417	224	193	120.3
53.7	415	223	192	119.8
53.7	410	220	190	118.0
53.7	408	219	189	117.6
53.7	406	218	188	117.1
53.7	404	217	187	116.6
53.7	402	216	186	116.1
53.7	397	213	184	114.3
53.7	395	212	183	113.8
53.7	393	211	182	113.3
53.7	391	210	181	112.8
53.7	389	209	180	112.3
53.7	387	208	179	111.8
53.7	382	205	177	110.0
53.7	380	204	176	109.5
53.7	378	203	175	109.0
53.7	376	202	174	108.5
53.7	374	201	173	108.0
53.7	369	198	171	106.2
53.7	367	197	170	105.7
53.7	365	196	169	105.2
53.7	363	195	168	104.8
53.7	361	194	167	104.3
53.7	356	191	165	102.5
53.7	354	190	164	102.0
53.7	352	189	163	101.5
53.7	350	188	162	101.0
53.7	348	187	161	100.5
53.7	341	183	158	98.2
53.7	339	182	157	97.7
53.7	337	181	156	97.2
53.7	335	180	155	96.7

%E	M1	M2	DM	M*
53.7	328	176	152	94.4
53.7	326	175	151	93.9
53.7	324	174	150	93.4
53.7	322	173	149	92.9
53.7	315	169	146	90.7
53.7	313	168	145	90.2
53.7	311	167	144	89.7
53.7	309	166	143	89.2
53.7	307	165	142	88.7
53.7	300	161	139	86.4
53.7	298	160	138	85.9
53.7	296	159	137	85.4
53.7	294	158	136	84.9
53.7	287	154	133	82.6
53.7	285	153	132	82.1
53.7	283	152	131	81.6
53.7	281	151	130	81.1
53.7	274	147	127	78.9
53.7	272	146	126	78.4
53.7	270	145	125	77.9
53.7	268	144	124	77.4
53.7	259	139	120	74.6
53.7	257	138	119	74.1
53.7	255	137	118	73.6
53.7	246	132	114	70.8
53.7	244	131	113	70.3
53.7	242	130	112	69.8
53.7	231	124	107	66.6
53.7	229	123	106	66.1
53.7	227	122	105	65.6
53.7	218	117	101	62.8
53.7	216	116	100	62.3
53.7	214	115	99	61.8
53.7	205	110	95	59.0
53.7	203	109	94	58.5
53.7	201	108	93	58.0
53.7	190	102	88	54.8
53.7	188	101	87	54.3
53.7	177	95	82	51.0
53.7	175	94	81	50.5
53.7	164	88	76	47.2
53.7	162	87	75	46.7
53.7	149	80	69	43.0
53.7	147	79	68	42.5
53.7	136	73	63	39.2
53.7	134	72	62	38.7
53.7	123	66	57	35.4
53.7	121	65	56	34.9
53.7	108	58	50	31.1
53.7	95	51	44	27.4

%E	M1	M2	DM	M*
53.7	82	44	38	23.6
53.7	67	36	31	19.3
53.7	54	29	25	15.6
53.7	41	22	19	11.8
53.6	500	268	232	143.6
53.6	498	267	231	143.2
53.6	496	266	230	142.7
53.6	494	265	229	142.2
53.6	493	264	229	141.4
53.6	491	263	228	140.9
53.6	489	262	227	140.4
53.6	487	261	226	139.9
53.6	485	260	225	139.4
53.6	483	259	224	138.9
53.6	481	258	223	138.4
53.6	478	256	222	137.1
53.6	476	255	221	136.6
53.6	474	254	220	136.1
53.6	472	253	219	135.6
53.6	470	252	218	135.1
53.6	468	251	217	134.6
53.6	466	250	216	134.1
53.6	463	248	215	132.8
53.6	461	247	214	132.3
53.6	459	246	213	131.8
53.6	457	245	212	131.3
53.6	455	244	211	130.8
53.6	453	243	210	130.4
53.6	450	241	209	129.1
53.6	448	240	208	128.6
53.6	446	239	207	128.1
53.6	444	238	206	127.6
53.6	442	237	205	127.1
53.6	440	236	204	126.6
53.6	435	233	202	124.8
53.6	433	232	201	124.3
53.6	431	231	200	123.8
53.6	429	230	199	123.3
53.6	427	229	198	122.8
53.6	425	228	197	122.3
53.6	422	226	196	121.0
53.6	420	225	195	120.5
53.6	418	224	194	120.0
53.6	416	223	193	119.5
53.6	414	222	192	119.0
53.6	412	221	191	118.5
53.6	407	218	189	116.8
53.6	405	217	188	116.3
53.6	403	216	187	115.8
53.6	401	215	186	115.3

%E	M1	M2	DM	M*
53.6	399	214	185	114.8
53.6	394	211	183	113.0
53.6	392	210	182	112.5
53.6	390	209	181	112.0
53.6	388	208	180	111.5
53.6	386	207	179	111.0
53.6	384	206	178	110.5
53.6	379	203	176	108.7
53.6	377	202	175	108.2
53.6	375	201	174	107.7
53.6	373	200	173	107.2
53.6	371	199	172	106.7
53.6	366	196	170	105.0
53.6	364	195	169	104.5
53.6	362	194	168	104.0
53.6	360	193	167	103.5
53.6	358	192	166	103.0
53.6	351	188	163	100.7
53.6	349	187	162	100.2
53.6	347	186	161	99.7
53.6	345	185	160	99.2
53.6	343	184	159	98.7
53.6	338	181	157	96.9
53.6	336	180	156	96.4
53.6	334	179	155	95.9
53.6	332	178	154	95.4
53.6	330	177	153	94.9
53.6	323	173	150	92.7
53.6	321	172	149	92.2
53.6	319	171	148	91.7
53.6	317	170	147	91.2
53.6	308	165	143	88.4
53.6	306	164	142	87.9
53.6	304	163	141	87.4
53.6	302	162	140	86.9
53.6	295	158	137	84.6
53.6	293	157	136	84.1
53.6	291	156	135	83.6
53.6	289	155	134	83.1
53.6	280	150	130	80.4
53.6	278	149	129	79.9
53.6	276	148	128	79.4
53.6	267	143	124	76.6
53.6	265	142	123	76.1
53.6	263	141	122	75.6
53.6	261	140	121	75.1
53.6	252	135	117	72.3
53.6	250	134	116	71.8
53.6	248	133	115	71.3
53.6	239	128	111	68.6

%E	M1	M2	DM	M*
53.6	237	127	110	68.1
53.6	235	126	109	67.6
53.6	233	125	108	67.1
53.6	224	120	104	64.3
53.6	222	119	103	63.8
53.6	220	118	102	63.3
53.6	211	113	98	60.5
53.6	209	112	97	60.0
53.6	207	111	96	59.5
53.6	196	105	91	56.3
53.6	194	104	90	55.8
53.6	192	103	89	55.3
53.6	183	98	85	52.5
53.6	181	97	84	52.0
53.6	179	96	83	51.5
53.6	168	90	78	48.2
53.6	166	89	77	47.7
53.6	153	82	71	43.9
53.6	151	81	70	43.5
53.6	140	75	65	40.2
53.6	138	74	64	39.7
53.6	125	67	58	35.9
53.6	112	60	52	32.1
53.6	110	59	51	31.6
53.6	97	52	45	27.9
53.6	84	45	39	24.1
53.6	69	37	32	19.8
53.6	56	30	26	16.1
53.6	28	15	13	8.0
53.5	499	267	232	142.9
53.5	497	266	231	142.4
53.5	495	265	230	141.9
53.5	492	263	229	140.6
53.5	490	262	228	140.1
53.5	488	261	227	139.6
53.5	486	260	226	139.1
53.5	484	259	225	138.6
53.5	482	258	224	138.1
53.5	480	257	223	137.6
53.5	477	255	222	136.3
53.5	475	254	221	135.8
53.5	473	253	220	135.3
53.5	471	252	219	134.8
53.5	469	251	218	134.3
53.5	467	250	217	133.8
53.5	465	249	216	133.3
53.5	462	247	215	132.1
53.5	460	246	214	131.6
53.5	458	245	213	131.1
53.5	456	244	212	130.6

%E	M1	M2	DM	M*
53.5	454	243	211	130.1
53.5	452	242	210	129.6
53.5	449	240	209	128.3
53.5	447	239	208	127.8
53.5	445	238	207	127.3
53.5	443	237	206	126.8
53.5	441	236	205	126.3
53.5	439	235	204	125.8
53.5	437	234	203	125.3
53.5	434	232	202	124.0
53.5	432	231	201	123.5
53.5	430	230	200	123.0
53.5	428	229	199	122.5
53.5	426	228	198	122.0
53.5	424	227	197	121.5
53.5	419	224	195	119.8
53.5	417	223	194	119.3
53.5	415	222	193	118.8
53.5	413	221	192	118.3
53.5	411	220	191	117.8
53.5	409	219	190	117.3
53.5	404	216	188	115.5
53.5	402	215	187	115.0
53.5	400	214	186	114.5
53.5	398	213	185	114.0
53.5	396	212	184	113.5
53.5	391	209	182	111.7
53.5	389	208	181	111.2
53.5	387	207	180	110.7
53.5	385	206	179	110.2
53.5	383	205	178	109.7
53.5	381	204	177	109.2
53.5	376	201	175	107.4
53.5	374	200	174	107.0
53.5	372	199	173	106.5
53.5	370	198	172	106.0
53.5	368	197	171	105.5
53.5	361	193	168	103.2
53.5	359	192	167	102.7
53.5	357	191	166	102.2
53.5	355	190	165	101.7
53.5	353	189	164	101.2
53.5	346	185	161	98.9
53.5	344	184	160	98.4
53.5	342	183	159	97.9
53.5	340	182	158	97.4
53.5	333	178	155	95.1
53.5	331	177	154	94.6
53.5	329	176	153	94.2
53.5	327	175	152	93.7

%E	M1	M2	DM	M*
53.5	325	174	151	93.2
53.5	318	170	148	90.9
53.5	316	169	147	90.4
53.5	314	168	146	89.9
53.5	312	167	145	89.4
53.5	310	166	144	88.9
53.5	303	162	141	86.6
53.5	301	161	140	86.1
53.5	299	160	139	85.6
53.5	297	159	138	85.1
53.5	288	154	134	82.3
53.5	286	153	133	81.8
53.5	284	152	132	81.4
53.5	282	151	131	80.9
53.5	275	147	128	78.6
53.5	273	146	127	78.1
53.5	271	145	126	77.6
53.5	269	144	125	77.1
53.5	260	139	121	74.3
53.5	258	138	120	73.8
53.5	256	137	119	73.3
53.5	254	136	118	72.8
53.5	245	131	114	70.0
53.5	243	130	113	69.5
53.5	241	129	112	69.0
53.5	230	123	107	65.8
53.5	228	122	106	65.3
53.5	226	121	105	64.8
53.5	217	116	101	62.0
53.5	215	115	100	61.5
53.5	213	114	99	61.0
53.5	202	108	94	57.7
53.5	200	107	93	57.2
53.5	198	106	92	56.7
53.5	187	100	87	53.5
53.5	185	99	86	53.0
53.5	172	92	80	49.2
53.5	170	91	79	48.7
53.5	159	85	74	45.4
53.5	157	84	73	44.9
53.5	155	83	72	44.4
53.5	144	77	67	41.2
53.5	142	76	66	40.7
53.5	129	69	60	36.9
53.5	127	68	59	36.4
53.5	114	61	53	32.6
53.5	101	54	47	28.9
53.5	99	53	46	28.4
53.5	86	46	40	24.6
53.5	71	38	33	20.3

%E	M1	M2	DM	M*
53.5	43	23	20	12.3
53.4	500	267	233	142.6
53.4	498	266	232	142.1
53.4	496	265	231	141.6
53.4	494	264	230	141.1
53.4	491	262	229	139.8
53.4	489	261	228	139.3
53.4	487	260	227	138.8
53.4	485	259	226	138.3
53.4	483	258	225	137.8
53.4	481	257	224	137.3
53.4	479	256	223	136.8
53.4	476	254	222	135.5
53.4	474	253	221	135.0
53.4	472	252	220	134.5
53.4	470	251	219	134.0
53.4	468	250	218	133.5
53.4	466	249	217	133.0
53.4	464	248	216	132.6
53.4	461	246	215	131.3
53.4	459	245	214	130.8
53.4	457	244	213	130.3
53.4	455	243	212	129.8
53.4	453	242	211	129.3
53.4	451	241	210	128.8
53.4	446	238	208	127.0
53.4	444	237	207	126.5
53.4	442	236	206	126.0
53.4	440	235	205	125.5
53.4	438	234	204	125.0
53.4	436	233	203	124.5
53.4	431	230	201	122.7
53.4	429	229	200	122.2
53.4	427	228	199	121.7
53.4	425	227	198	121.2
53.4	423	226	197	120.7
53.4	421	225	196	120.2
53.4	416	222	194	118.5
53.4	414	221	193	118.0
53.4	412	220	192	117.5
53.4	410	219	191	117.0
53.4	408	218	190	116.5
53.4	406	217	189	116.0
53.4	403	215	188	114.7
53.4	401	214	187	114.2
53.4	399	213	186	113.7
53.4	397	212	185	113.2
53.4	395	211	184	112.7
53.4	393	210	183	112.2
53.4	388	207	181	110.4

%E	M1	M2	DM	M*
53.4	386	206	180	109.9
53.4	384	205	179	109.4
53.4	382	204	178	108.9
53.4	380	203	177	108.4
53.4	378	202	176	107.9
53.4	373	199	174	106.2
53.4	371	198	173	105.7
53.4	369	197	172	105.2
53.4	367	196	171	104.7
53.4	365	195	170	104.2
53.4	363	194	169	103.7
53.4	358	191	167	101.9
53.4	356	190	166	101.4
53.4	354	189	165	100.9
53.4	352	188	164	100.4
53.4	350	187	163	99.9
53.4	348	186	162	99.4
53.4	343	183	160	97.6
53.4	341	182	159	97.1
53.4	339	181	158	96.6
53.4	337	180	157	96.1
53.4	335	179	156	95.6
53.4	328	175	153	93.4
53.4	326	174	152	92.9
53.4	324	173	151	92.4
53.4	322	172	150	91.9
53.4	320	171	149	91.4
53.4	313	167	146	89.1
53.4	311	166	145	88.6
53.4	309	165	144	88.1
53.4	307	164	143	87.6
53.4	305	163	142	87.1
53.4	298	159	139	84.8
53.4	296	158	138	84.3
53.4	294	157	137	83.8
53.4	292	156	136	83.3
53.4	290	155	135	82.8
53.4	283	151	132	80.6
53.4	281	150	131	80.1
53.4	279	149	130	79.6
53.4	277	148	129	79.1
53.4	268	143	125	76.3
53.4	266	142	124	75.8
53.4	264	141	123	75.3
53.4	262	140	122	74.8
53.4	253	135	118	72.0
53.4	251	134	117	71.5
53.4	249	133	116	71.0
53.4	247	132	115	70.5
53.4	238	127	111	67.8

%E	M1	M2	DM	M*
53.4	236	126	110	67.3
53.4	234	125	109	66.8
53.4	232	124	108	66.3
53.4	223	119	104	63.5
53.4	221	118	103	63.0
53.4	219	117	102	62.5
53.4	208	111	97	59.2
53.4	206	110	96	58.7
53.4	204	109	95	58.2
53.4	193	103	90	55.0
53.4	191	102	89	54.5
53.4	189	101	88	54.0
53.4	178	95	83	50.7
53.4	176	94	82	50.2
53.4	174	93	81	49.7
53.4	163	87	76	46.4
53.4	161	86	75	45.9
53.4	148	79	69	42.2
53.4	146	78	68	41.7
53.4	133	71	62	37.9
53.4	131	70	61	37.4
53.4	118	63	55	33.6
53.4	116	62	54	33.1
53.4	103	55	48	29.4
53.4	88	47	41	25.1
53.4	73	39	34	20.8
53.4	58	31	27	16.6
53.3	499	266	233	141.8
53.3	497	265	232	141.3
53.3	495	264	231	140.8
53.3	493	263	230	140.3
53.3	492	262	230	139.5
53.3	490	261	229	139.0
53.3	488	260	228	138.5
53.3	486	259	227	138.0
53.3	484	258	226	137.5
53.3	482	257	225	137.0
53.3	480	256	224	136.5
53.3	478	255	223	136.0
53.3	475	253	222	134.8
53.3	473	252	221	134.3
53.3	471	251	220	133.8
53.3	469	250	219	133.3
53.3	467	249	218	132.8
53.3	465	248	217	132.3
53.3	463	247	216	131.8
53.3	460	245	215	130.5
53.3	458	244	214	130.0
53.3	456	243	213	129.5
53.3	454	242	212	129.0

%E	M1	M2	DM	M*
53.3	452	241	211	128.5
53.3	450	240	210	128.0
53.3	448	239	209	127.5
53.3	445	237	208	126.2
53.3	443	236	207	125.7
53.3	441	235	206	125.2
53.3	439	234	205	124.7
53.3	437	233	204	124.2
53.3	435	232	203	123.7
53.3	433	231	202	123.2
53.3	430	229	201	122.0
53.3	428	228	200	121.5
53.3	426	227	199	121.0
53.3	424	226	198	120.5
53.3	422	225	197	120.0
53.3	420	224	196	119.5
53.3	418	223	195	119.0
53.3	415	221	194	117.7
53.3	413	220	193	117.2
53.3	411	219	192	116.7
53.3	409	218	191	116.2
53.3	407	217	190	115.7
53.3	405	216	189	115.2
53.3	400	213	187	113.4
53.3	398	212	186	112.9
53.3	396	211	185	112.4
53.3	394	210	184	111.9
53.3	392	209	183	111.4
53.3	390	208	182	110.9
53.3	383	204	179	108.7
53.3	381	203	178	108.2
53.3	379	202	177	107.7
53.3	377	201	176	107.2
53.3	375	200	175	106.7
53.3	368	196	172	104.4
53.3	366	195	171	103.9
53.3	364	194	170	103.4
53.3	362	193	169	102.9
53.3	360	192	168	102.4
53.3	353	188	165	100.1
53.3	351	187	164	99.6
53.3	349	186	163	99.1
53.3	347	185	162	98.6
53.3	345	184	161	98.1
53.3	338	180	158	95.9
53.3	336	179	157	95.4
53.3	334	178	156	94.9
53.3	332	177	155	94.4
53.3	330	176	154	93.9
53.3	323	172	151	91.6

%E	M1	M2	DM	M*
53.3	321	171	150	91.1
53.3	319	170	149	90.6
53.3	317	169	148	90.1
53.3	315	168	147	89.6
53.3	306	163	143	86.8
53.3	304	162	142	86.3
53.3	302	161	141	85.8
53.3	300	160	140	85.3
53.3	291	155	136	82.6
53.3	289	154	135	82.1
53.3	287	153	134	81.6
53.3	285	152	133	81.1
53.3	276	147	129	78.3
53.3	274	146	128	77.8
53.3	272	145	127	77.3
53.3	270	144	126	76.8
53.3	261	139	122	74.0
53.3	259	138	121	73.5
53.3	257	137	120	73.0
53.3	255	136	119	72.5
53.3	246	131	115	69.8
53.3	244	130	114	69.3
53.3	242	129	113	68.8
53.3	240	128	112	68.3
53.3	229	122	107	65.0
53.3	227	121	106	64.5
53.3	225	120	105	64.0
53.3	214	114	100	60.7
53.3	212	113	99	60.2
53.3	210	112	98	59.7
53.3	199	106	93	56.5
53.3	197	105	92	56.0
53.3	195	104	91	55.5
53.3	184	98	86	52.2
53.3	182	97	85	51.7
53.3	180	96	84	51.2
53.3	169	90	79	47.9
53.3	167	89	78	47.4
53.3	165	88	77	46.9
53.3	152	81	71	43.2
53.3	150	80	70	42.7
53.3	137	73	64	38.9
53.3	135	72	63	38.4
53.3	122	65	57	34.6
53.3	120	64	56	34.1
53.3	107	57	50	30.4
53.3	105	56	49	29.9
53.3	92	49	43	26.1
53.3	90	48	42	25.6
53.3	75	40	35	21.3
53.3	60	32	28	17.1
53.3	45	24	21	12.8
53.3	30	16	14	8.5
53.3	15	8	7	4.3
53.2	500	266	234	141.5
53.2	498	265	233	141.0
53.2	496	264	232	140.5
53.2	494	263	231	140.0
53.2	491	261	230	138.7
53.2	489	260	229	138.2
53.2	487	259	228	137.7
53.2	485	258	227	137.2
53.2	483	257	226	136.7
53.2	481	256	225	136.2
53.2	479	255	224	135.8
53.2	477	254	223	135.3
53.2	476	253	223	134.5
53.2	474	252	222	134.0
53.2	472	251	221	133.5
53.2	470	250	220	133.0
53.2	468	249	219	132.5
53.2	466	248	218	132.0
53.2	464	247	217	131.5
53.2	462	246	216	131.0
53.2	459	244	215	129.7
53.2	457	243	214	129.2
53.2	455	242	213	128.7
53.2	453	241	212	128.2
53.2	451	240	211	127.7
53.2	449	239	210	127.2
53.2	447	238	209	126.7
53.2	444	236	208	125.4
53.2	442	235	207	124.9
53.2	440	234	206	124.4
53.2	438	233	205	123.9
53.2	436	232	204	123.4
53.2	434	231	203	123.0
53.2	432	230	202	122.5
53.2	427	227	200	120.7
53.2	425	226	199	120.2
53.2	423	225	198	119.7
53.2	421	224	197	119.2
53.2	419	223	196	118.7
53.2	417	222	195	118.2
53.2	412	219	193	116.4
53.2	410	218	192	115.9
53.2	408	217	191	115.4
53.2	406	216	190	114.9
53.2	404	215	189	114.4
53.2	402	214	188	113.9
53.2	395	210	185	111.6
53.2	393	209	184	111.1
53.2	391	208	183	110.6
53.2	389	207	182	110.2
53.2	387	206	181	109.7
53.2	385	205	180	109.2
53.2	380	202	178	107.4
53.2	378	201	177	106.9
53.2	376	200	176	106.4
53.2	374	199	175	105.9
53.2	372	198	174	105.4
53.2	370	197	173	104.9
53.2	365	194	171	103.1
53.2	363	193	170	102.6
53.2	361	192	169	102.1
53.2	359	191	168	101.6
53.2	357	190	167	101.1
53.2	355	189	166	100.6
53.2	348	185	163	98.3
53.2	346	184	162	97.8
53.2	344	183	161	97.4
53.2	342	182	160	96.9
53.2	340	181	159	96.4
53.2	333	177	156	94.1
53.2	331	176	155	93.6
53.2	329	175	154	93.1
53.2	327	174	153	92.6
53.2	325	173	152	92.1
53.2	316	168	148	89.3
53.2	314	167	147	88.8
53.2	312	166	146	88.3
53.2	310	165	145	87.8
53.2	308	164	144	87.3
53.2	301	160	141	85.0
53.2	299	159	140	84.6
53.2	297	158	139	84.1
53.2	295	157	138	83.6
53.2	293	156	137	83.1
53.2	284	151	133	80.3
53.2	282	150	132	79.8
53.2	280	149	131	79.3
53.2	278	148	130	78.8
53.2	269	143	126	76.0
53.2	267	142	125	75.5
53.2	265	141	124	75.0
53.2	263	140	123	74.5
53.2	254	135	119	71.8
53.2	252	134	118	71.3
53.2	250	133	117	70.8
53.2	248	132	116	70.3
53.2	237	126	111	67.0
53.2	235	125	110	66.5
53.2	233	124	109	66.0
53.2	231	123	108	65.5
53.2	222	118	104	62.7
53.2	220	117	103	62.2
53.2	218	116	102	61.7
53.2	216	115	101	61.2
53.2	205	109	96	58.0
53.2	203	108	95	57.5
53.2	201	107	94	57.0
53.2	190	101	89	53.7
53.2	188	100	88	53.2
53.2	186	99	87	52.7
53.2	173	92	81	48.9
53.2	171	91	80	48.4
53.2	158	84	74	44.7
53.2	156	83	73	44.2
53.2	154	82	72	43.7
53.2	141	75	66	39.9
53.2	139	74	65	39.4
53.2	126	67	59	35.6
53.2	124	66	58	35.1
53.2	111	59	52	31.4
53.2	109	58	51	30.9
53.2	94	50	44	26.6
53.2	79	42	37	22.3
53.2	77	41	36	21.8
53.2	62	33	29	17.6
53.2	47	25	22	13.3
53.1	499	265	234	140.7
53.1	497	264	233	140.2
53.1	495	263	232	139.7
53.1	493	262	231	139.2
53.1	490	260	230	138.0
53.1	488	259	229	137.5
53.1	486	258	228	137.0
53.1	484	257	227	136.5
53.1	482	256	226	136.0
53.1	480	255	225	135.5
53.1	478	254	224	135.0
53.1	475	252	223	133.7
53.1	473	251	222	133.2
53.1	471	250	221	132.7
53.1	469	249	220	132.2
53.1	467	248	219	131.7
53.1	465	247	218	131.2
53.1	463	246	217	130.7
53.1	461	245	216	130.2
53.1	458	243	215	128.9
53.1	456	242	214	128.4
53.1	454	241	213	127.9
53.1	452	240	212	127.4
53.1	450	239	211	126.9
53.1	448	238	210	126.4
53.1	446	237	209	125.9
53.1	441	234	207	124.2
53.1	439	233	206	123.7
53.1	437	232	205	123.2
53.1	435	231	204	122.7
53.1	433	230	203	122.2
53.1	431	229	202	121.7
53.1	429	228	201	121.2
53.1	426	226	200	119.9
53.1	424	225	199	119.4
53.1	422	224	198	118.9
53.1	420	223	197	118.4
53.1	418	222	196	117.9
53.1	416	221	195	117.4
53.1	414	220	194	116.9
53.1	409	217	192	115.1
53.1	407	216	191	114.6
53.1	405	215	190	114.1
53.1	403	214	189	113.6
53.1	401	213	188	113.1
53.1	399	212	187	112.6
53.1	397	211	186	112.1
53.1	392	208	184	110.4
53.1	390	207	183	109.9
53.1	388	206	182	109.4
53.1	386	205	181	108.9
53.1	384	204	180	108.4
53.1	382	203	179	107.9
53.1	377	200	177	106.1
53.1	375	199	176	105.6
53.1	373	198	175	105.1
53.1	371	197	174	104.6
53.1	369	196	173	104.1
53.1	367	195	172	103.6
53.1	360	191	169	101.3
53.1	358	190	168	100.8
53.1	356	189	167	100.3
53.1	354	188	166	99.8
53.1	352	187	165	99.3
53.1	350	186	164	98.8
53.1	343	182	161	96.6
53.1	341	181	160	96.1
53.1	339	180	159	95.6
53.1	337	179	158	95.1
53.1	335	178	157	94.6

%E	M1	M2	DM	M*
53.1	326	173	153	91.8
53.1	324	172	152	91.3
53.1	322	171	151	90.8
53.1	320	170	150	90.3
53.1	318	169	149	89.8
53.1	311	165	146	87.5
53.1	309	164	145	87.0
53.1	307	163	144	86.5
53.1	305	162	143	86.0
53.1	303	161	142	85.5
53.1	294	156	138	82.8
53.1	292	155	137	82.3
53.1	290	154	136	81.8
53.1	288	153	135	81.3
53.1	286	152	134	80.8
53.1	277	147	130	78.0
53.1	275	146	129	77.5
53.1	273	145	128	77.0
53.1	271	144	127	76.5
53.1	262	139	123	73.7
53.1	260	138	122	73.2
53.1	258	137	121	72.7
53.1	256	136	120	72.3
53.1	245	130	115	69.0
53.1	243	129	114	68.5
53.1	241	128	113	68.0
53.1	239	127	112	67.5
53.1	228	121	107	64.2
53.1	226	120	106	63.7
53.1	224	119	105	63.2
53.1	213	113	100	59.9
53.1	211	112	99	59.5
53.1	209	111	98	59.0
53.1	207	110	97	58.5
53.1	196	104	92	55.2
53.1	194	103	91	54.7
53.1	192	102	90	54.2
53.1	179	95	84	50.4
53.1	177	94	83	49.9
53.1	175	93	82	49.4
53.1	162	86	76	45.7
53.1	160	85	75	45.2
53.1	147	78	69	41.4
53.1	145	77	68	40.9
53.1	143	76	67	40.4
53.1	130	69	61	36.6
53.1	128	68	60	36.1
53.1	113	60	53	31.9
53.1	98	52	46	27.6
53.1	96	51	45	27.1

%E	M1	M2	DM	M*
53.1	81	43	38	22.8
53.1	64	34	30	18.1
53.1	49	26	23	13.8
53.1	32	17	15	9.0
53.0	500	265	235	140.4
53.0	498	264	234	140.0
53.0	496	263	233	139.5
53.0	494	262	232	139.0
53.0	492	261	231	138.5
53.0	491	260	231	137.7
53.0	489	259	230	137.2
53.0	487	258	229	136.7
53.0	485	257	228	136.2
53.0	483	256	227	135.7
53.0	481	255	226	135.2
53.0	479	254	225	134.7
53.0	477	253	224	134.2
53.0	474	251	223	132.9
53.0	472	250	222	132.4
53.0	470	249	221	131.9
53.0	468	248	220	131.4
53.0	466	247	219	130.9
53.0	464	246	218	130.4
53.0	462	245	217	129.9
53.0	460	244	216	129.4
53.0	457	242	215	128.1
53.0	455	241	214	127.7
53.0	453	240	213	127.2
53.0	451	239	212	126.7
53.0	449	238	211	126.2
53.0	447	237	210	125.7
53.0	445	236	209	125.2
53.0	443	235	208	124.7
53.0	440	233	207	123.4
53.0	438	232	206	122.9
53.0	436	231	205	122.4
53.0	434	230	204	121.9
53.0	432	229	203	121.4
53.0	430	228	202	120.9
53.0	428	227	201	120.4
53.0	423	224	199	118.6
53.0	421	223	198	118.1
53.0	419	222	197	117.6
53.0	417	221	196	117.1
53.0	415	220	195	116.6
53.0	413	219	194	116.1
53.0	411	218	193	115.6
53.0	406	215	191	113.9
53.0	404	214	190	113.4
53.0	402	213	189	112.9

%E	M1	M2	DM	M*
53.0	400	212	188	112.4
53.0	398	211	187	111.9
53.0	396	210	186	111.4
53.0	394	209	185	110.9
53.0	389	206	183	109.1
53.0	387	205	182	108.6
53.0	385	204	181	108.1
53.0	383	203	180	107.6
53.0	381	202	179	107.1
53.0	379	201	178	106.6
53.0	372	197	175	104.3
53.0	370	196	174	103.8
53.0	368	195	173	103.3
53.0	366	194	172	102.8
53.0	364	193	171	102.3
53.0	362	192	170	101.8
53.0	355	188	167	99.6
53.0	353	187	166	99.1
53.0	351	186	165	98.6
53.0	349	185	164	98.1
53.0	347	184	163	97.6
53.0	345	183	162	97.1
53.0	338	179	159	94.8
53.0	336	178	158	94.3
53.0	334	177	157	93.8
53.0	332	176	156	93.3
53.0	330	175	155	92.8
53.0	328	174	154	92.3
53.0	321	170	151	90.0
53.0	319	169	150	89.5
53.0	317	168	149	89.0
53.0	315	167	148	88.5
53.0	313	166	147	88.0
53.0	304	161	143	85.3
53.0	302	160	142	84.8
53.0	300	159	141	84.3
53.0	298	158	140	83.8
53.0	296	157	139	83.3
53.0	287	152	135	80.5
53.0	285	151	134	80.0
53.0	283	150	133	79.5
53.0	281	149	132	79.0
53.0	279	148	131	78.5
53.0	270	143	127	75.7
53.0	268	142	126	75.2
53.0	266	141	125	74.7
53.0	264	140	124	74.2
53.0	253	134	119	71.0
53.0	251	133	118	70.5
53.0	249	132	117	70.0

%E	M1	M2	DM	M*
53.0	247	131	116	69.5
53.0	236	125	111	66.2
53.0	234	124	110	65.7
53.0	232	123	109	65.2
53.0	230	122	108	64.7
53.0	219	116	103	61.4
53.0	217	115	102	60.9
53.0	215	114	101	60.4
53.0	202	107	95	56.7
53.0	200	106	94	56.2
53.0	198	105	93	55.7
53.0	185	98	87	51.9
53.0	183	97	86	51.4
53.0	181	96	85	50.9
53.0	168	89	79	47.1
53.0	166	88	78	46.7
53.0	164	87	77	46.2
53.0	151	80	71	42.4
53.0	149	79	70	41.9
53.0	134	71	63	37.6
53.0	132	70	62	37.1
53.0	117	62	55	32.9
53.0	115	61	54	32.4
53.0	100	53	47	28.1
53.0	83	44	39	23.3
53.0	66	35	31	18.6
52.9	499	264	235	139.7
52.9	497	263	234	139.2
52.9	495	262	233	138.7
52.9	493	261	232	138.2
52.9	490	259	231	136.9
52.9	488	258	230	136.4
52.9	486	257	229	135.9
52.9	484	256	228	135.4
52.9	482	255	227	134.9
52.9	480	254	226	134.4
52.9	478	253	225	133.9
52.9	476	252	224	133.4
52.9	473	250	223	132.1
52.9	471	249	222	131.6
52.9	469	248	221	131.1
52.9	467	247	220	130.6
52.9	465	246	219	130.1
52.9	463	245	218	129.6
52.9	461	244	217	129.1
52.9	459	243	216	128.6
52.9	456	241	215	127.4
52.9	454	240	214	126.9
52.9	452	239	213	126.4
52.9	450	238	212	125.9

%E	M1	M2	DM	M*
52.9	448	237	211	125.4
52.9	446	236	210	124.9
52.9	444	235	209	124.4
52.9	442	234	208	123.9
52.9	437	231	206	122.1
52.9	435	230	205	121.6
52.9	433	229	204	121.1
52.9	431	228	203	120.6
52.9	429	227	202	120.1
52.9	427	226	201	119.6
52.9	425	225	200	119.1
52.9	420	222	198	117.3
52.9	418	221	197	116.8
52.9	416	220	196	116.3
52.9	414	219	195	115.8
52.9	412	218	194	115.3
52.9	410	217	193	114.9
52.9	408	216	192	114.4
52.9	403	213	190	112.6
52.9	401	212	189	112.1
52.9	399	211	188	111.6
52.9	397	210	187	111.1
52.9	395	209	186	110.6
52.9	393	208	185	110.1
52.9	391	207	184	109.6
52.9	386	204	182	107.8
52.9	384	203	181	107.3
52.9	382	202	180	106.8
52.9	380	201	179	106.3
52.9	378	200	178	105.8
52.9	376	199	177	105.3
52.9	374	198	176	104.8
52.9	367	194	173	102.6
52.9	365	193	172	102.1
52.9	363	192	171	101.6
52.9	361	191	170	101.1
52.9	359	190	169	100.6
52.9	357	189	168	100.1
52.9	350	185	165	97.8
52.9	348	184	164	97.3
52.9	346	183	163	96.8
52.9	344	182	162	96.3
52.9	342	181	161	95.8
52.9	340	180	160	95.3
52.9	333	176	157	93.0
52.9	331	175	156	92.5
52.9	329	174	155	92.0
52.9	327	173	154	91.5
52.9	325	172	153	91.0
52.9	323	171	152	90.5

%E	M1	M2	DM	M*
52•9	314	166	148	87•8
52•9	312	165	147	87•3
52•9	310	164	146	86•8
52•9	308	163	145	86•3
52•9	306	162	144	85•8
52•9	297	157	140	83•0
52•9	295	156	139	82•5
52•9	293	155	138	82•0
52•9	291	154	137	81•5
52•9	289	153	136	81•0
52•9	280	148	132	78•2
52•9	278	147	131	77•7
52•9	276	146	130	77•2
52•9	274	145	129	76•7
52•9	272	144	128	76•2
52•9	263	139	124	73•5
52•9	261	138	123	73•0
52•9	259	137	122	72•5
52•9	257	136	121	72•0
52•9	255	135	120	71•5
52•9	244	129	115	68•2
52•9	242	128	114	67•7
52•9	240	127	113	67•2
52•9	238	126	112	66•7
52•9	227	120	107	63•4
52•9	225	119	106	62•9
52•9	223	118	105	62•4
52•9	221	117	104	61•9
52•9	210	111	99	58•7
52•9	208	110	98	58•2
52•9	206	109	97	57•7
52•9	204	108	96	57•2
52•9	193	102	91	53•9
52•9	191	101	90	53•4
52•9	189	100	89	52•9
52•9	187	99	88	52•4
52•9	174	92	82	48•6
52•9	172	91	81	48•1
52•9	170	90	80	47•6
52•9	157	83	74	43•9
52•9	155	82	73	43•4
52•9	153	81	72	42•9
52•9	140	74	66	39•1
52•9	138	73	65	38•6
52•9	136	72	64	38•1
52•9	121	64	57	33•9
52•9	119	63	56	33•4
52•9	104	55	49	29•1
52•9	102	54	48	28•6
52•9	87	46	41	24•3

%E	M1	M2	DM	M*
52•9	85	45	40	23•8
52•9	70	37	33	19•6
52•9	68	36	32	19•1
52•9	51	27	24	14•3
52•9	34	18	16	9•5
52•9	17	9	8	4•8
52•8	500	264	236	139•4
52•8	498	263	235	138•9
52•8	496	262	234	138•4
52•8	494	261	233	137•9
52•8	492	260	232	137•4
52•8	489	258	231	136•1
52•8	487	257	230	135•6
52•8	485	256	229	135•1
52•8	483	255	228	134•6
52•8	481	254	227	134•1
52•8	479	253	226	133•6
52•8	477	252	225	133•1
52•8	475	251	224	132•6
52•8	472	249	223	131•4
52•8	470	248	222	130•9
52•8	468	247	221	130•4
52•8	466	246	220	129•9
52•8	464	245	219	129•4
52•8	462	244	218	128•9
52•8	460	243	217	128•4
52•8	458	242	216	127•9
52•8	453	239	214	126•1
52•8	451	238	213	125•6
52•8	449	237	212	125•1
52•8	447	236	211	124•6
52•8	445	235	210	124•1
52•8	443	234	209	123•6
52•8	441	233	208	123•1
52•8	439	232	207	122•6
52•8	436	230	206	121•3
52•8	434	229	205	120•8
52•8	432	228	204	120•3
52•8	430	227	203	119•8
52•8	428	226	202	119•3
52•8	426	225	201	118•8
52•8	424	224	200	118•3
52•8	422	223	199	117•8
52•8	417	220	197	116•1
52•8	415	219	196	115•6
52•8	413	218	195	115•1
52•8	411	217	194	114•6
52•8	409	216	193	114•1
52•8	407	215	192	113•6
52•8	405	214	191	113•1

%E	M1	M2	DM	M*
52•8	400	211	189	111•3
52•8	398	210	188	110•8
52•8	396	209	187	110•3
52•8	394	208	186	109•8
52•8	392	207	185	109•3
52•8	390	206	184	108•8
52•8	388	205	183	108•3
52•8	381	201	180	106•0
52•8	379	200	179	105•5
52•8	377	199	178	105•0
52•8	375	198	177	104•5
52•8	373	197	176	104•0
52•8	371	196	175	103•5
52•8	369	195	174	103•0
52•8	362	191	171	100•8
52•8	360	190	170	100•3
52•8	358	189	169	99•8
52•8	356	188	168	99•3
52•8	354	187	167	98•8
52•8	352	186	166	98•3
52•8	345	182	163	96•0
52•8	343	181	162	95•5
52•8	341	180	161	95•0
52•8	339	179	160	94•5
52•8	337	178	159	94•0
52•8	335	177	158	93•5
52•8	326	172	154	90•7
52•8	324	171	153	90•3
52•8	322	170	152	89•8
52•8	320	169	151	89•3
52•8	318	168	150	88•8
52•8	316	167	149	88•3
52•8	309	163	146	86•0
52•8	307	162	145	85•5
52•8	305	161	144	85•0
52•8	303	160	143	84•5
52•8	301	159	142	84•0
52•8	299	158	141	83•5
52•8	290	153	137	80•7
52•8	288	152	136	80•2
52•8	286	151	135	79•7
52•8	284	150	134	79•2
52•8	282	149	133	78•7
52•8	271	143	128	75•5
52•8	269	142	127	75•0
52•8	267	141	126	74•5
52•8	265	140	125	74•0
52•8	254	134	120	70•7
52•8	252	133	119	70•2
52•8	250	132	118	69•7

%E	M1	M2	DM	M*
52•8	248	131	117	69•2
52•8	246	130	116	68•7
52•8	235	124	111	65•4
52•8	233	123	110	64•9
52•8	231	122	109	64•4
52•8	229	121	108	63•9
52•8	218	115	103	60•7
52•8	216	114	102	60•2
52•8	214	113	101	59•7
52•8	212	112	100	59•2
52•8	199	105	94	55•4
52•8	197	104	93	54•9
52•8	195	103	92	54•4
52•8	180	95	85	50•1
52•8	178	94	84	49•6
52•8	176	93	83	49•1
52•8	163	86	77	45•4
52•8	161	85	76	44•9
52•8	159	84	75	44•4
52•8	144	76	68	40•1
52•8	142	75	67	39•6
52•8	127	67	60	35•3
52•8	125	66	59	34•8
52•8	123	65	58	34•3
52•8	108	57	51	30•1
52•8	106	56	50	29•6
52•8	89	47	42	24•8
52•8	72	38	34	20•1
52•8	53	28	25	14•8
52•8	36	19	17	10•0
52•7	499	263	236	138•6
52•7	497	262	235	138•1
52•7	495	261	234	137•6
52•7	493	260	233	137•1
52•7	491	259	232	136•6
52•7	490	258	232	135•8
52•7	488	257	231	135•3
52•7	486	256	230	134•8
52•7	484	255	229	134•3
52•7	482	254	228	133•9
52•7	480	253	227	133•4
52•7	478	252	226	132•9
52•7	476	251	225	132•4
52•7	474	250	224	131•9
52•7	471	248	223	130•6
52•7	469	247	222	130•1
52•7	467	246	221	129•6
52•7	465	245	220	129•1
52•7	463	244	219	128•6
52•7	461	243	218	128•1

%E	M1	M2	DM	M*
52•7	459	242	217	127•6
52•7	457	241	216	127•1
52•7	455	240	215	126•6
52•7	452	238	214	125•3
52•7	450	237	213	124•8
52•7	448	236	212	124•3
52•7	446	235	211	123•8
52•7	444	234	210	123•3
52•7	442	233	209	122•8
52•7	440	232	208	122•3
52•7	438	231	207	121•8
52•7	433	228	205	120•1
52•7	431	227	204	119•6
52•7	429	226	203	119•1
52•7	427	225	202	118•6
52•7	425	224	201	118•1
52•7	423	223	200	117•6
52•7	421	222	199	117•1
52•7	419	221	198	116•6
52•7	414	218	196	114•8
52•7	412	217	195	114•3
52•7	410	216	194	113•8
52•7	408	215	193	113•3
52•7	406	214	192	112•8
52•7	404	213	191	112•3
52•7	402	212	190	111•8
52•7	395	208	187	109•5
52•7	393	207	186	109•0
52•7	391	206	185	108•5
52•7	389	205	184	108•0
52•7	387	204	183	107•5
52•7	385	203	182	107•0
52•7	383	202	181	106•5
52•7	376	198	178	104•3
52•7	374	197	177	103•8
52•7	372	196	176	103•3
52•7	370	195	175	102•8
52•7	368	194	174	102•3
52•7	366	193	173	101•8
52•7	364	192	172	101•3
52•7	357	188	169	99•0
52•7	355	187	168	98•5
52•7	353	186	167	98•0
52•7	351	185	166	97•5
52•7	349	184	165	97•0
52•7	347	183	164	96•5
52•7	338	178	160	93•7
52•7	336	177	159	93•2
52•7	334	176	158	92•7
52•7	332	175	157	92•2

%E	M1	M2	DM	M*
52.7	330	174	156	91.7
52.7	328	173	155	91.2
52.7	319	168	151	88.5
52.7	317	167	150	88.0
52.7	315	166	149	87.5
52.7	313	165	148	87.0
52.7	311	164	147	86.5
52.7	300	158	142	83.2
52.7	298	157	141	82.7
52.7	296	156	140	82.2
52.7	294	155	139	81.7
52.7	292	154	138	81.2
52.7	283	149	134	78.4
52.7	281	148	133	78.0
52.7	279	147	132	77.5
52.7	277	146	131	77.0
52.7	275	145	130	76.5
52.7	273	144	129	76.0
52.7	264	139	125	73.2
52.7	262	138	124	72.7
52.7	260	137	123	72.2
52.7	258	136	122	71.7
52.7	256	135	121	71.2
52.7	245	129	116	67.9
52.7	243	128	115	67.4
52.7	241	127	114	66.9
52.7	239	126	113	66.4
52.7	237	125	112	65.9
52.7	226	119	107	62.7
52.7	224	118	106	62.2
52.7	222	117	105	61.7
52.7	220	116	104	61.2
52.7	207	109	98	57.4
52.7	205	108	97	56.9
52.7	203	107	96	56.4
52.7	201	106	95	55.9
52.7	188	99	89	52.1
52.7	186	98	88	51.6
52.7	184	97	87	51.1
52.7	182	96	86	50.6
52.7	169	89	80	46.9
52.7	167	88	79	46.4
52.7	165	87	78	45.9
52.7	150	79	71	41.6
52.7	148	78	70	41.1
52.7	146	77	69	40.6
52.7	131	69	62	36.3
52.7	129	68	61	35.8
52.7	112	59	53	31.1
52.7	110	58	52	30.6

%E	M1	M2	DM	M*
52.7	93	49	44	25.8
52.7	91	48	43	25.3
52.7	74	39	35	20.6
52.7	55	29	26	15.3
52.6	500	263	237	138.3
52.6	498	262	236	137.8
52.6	496	261	235	137.3
52.6	494	260	234	136.8
52.6	492	259	233	136.3
52.6	489	257	232	135.1
52.6	487	256	231	134.6
52.6	485	255	230	134.1
52.6	483	254	229	133.6
52.6	481	253	228	133.1
52.6	479	252	227	132.6
52.6	477	251	226	132.1
52.6	475	250	225	131.6
52.6	473	249	224	131.1
52.6	470	247	223	129.8
52.6	468	246	222	129.3
52.6	466	245	221	128.8
52.6	464	244	220	128.3
52.6	462	243	219	127.8
52.6	460	242	218	127.3
52.6	458	241	217	126.8
52.6	456	240	216	126.3
52.6	454	239	215	125.8
52.6	451	237	214	124.5
52.6	449	236	213	124.0
52.6	447	235	212	123.5
52.6	445	234	211	123.0
52.6	443	233	210	122.5
52.6	441	232	209	122.0
52.6	439	231	208	121.6
52.6	437	230	207	121.1
52.6	435	229	206	120.6
52.6	430	226	204	118.8
52.6	428	225	203	118.3
52.6	426	224	202	117.8
52.6	424	223	201	117.3
52.6	422	222	200	116.8
52.6	420	221	199	116.3
52.6	418	220	198	115.8
52.6	416	219	197	115.3
52.6	411	216	195	113.5
52.6	409	215	194	113.0
52.6	407	214	193	112.5
52.6	405	213	192	112.0
52.6	403	212	191	111.5
52.6	401	211	190	111.0

%E	M1	M2	DM	M*
52.6	399	210	189	110.5
52.6	397	209	188	110.0
52.6	392	206	186	108.3
52.6	390	205	185	107.8
52.6	388	204	184	107.3
52.6	386	203	183	106.8
52.6	384	202	182	106.3
52.6	382	201	181	105.8
52.6	380	200	180	105.3
52.6	378	199	179	104.8
52.6	371	195	176	102.5
52.6	369	194	175	102.0
52.6	367	193	174	101.5
52.6	365	192	173	101.0
52.6	363	191	172	100.5
52.6	361	190	171	100.0
52.6	359	189	170	99.5
52.6	352	185	167	97.2
52.6	350	184	166	96.7
52.6	348	183	165	96.2
52.6	346	182	164	95.7
52.6	344	181	163	95.2
52.6	342	180	162	94.7
52.6	340	179	161	94.2
52.6	333	175	158	92.0
52.6	331	174	157	91.5
52.6	329	173	156	91.0
52.6	327	172	155	90.5
52.6	325	171	154	90.0
52.6	323	170	153	89.5
52.6	321	169	152	89.0
52.6	312	164	148	86.2
52.6	310	163	147	85.7
52.6	308	162	146	85.2
52.6	306	161	145	84.7
52.6	304	160	144	84.2
52.6	302	159	143	83.7
52.6	293	154	139	80.9
52.6	291	153	138	80.4
52.6	289	152	137	79.9
52.6	287	151	136	79.4
52.6	285	150	135	78.9
52.6	274	144	130	75.7
52.6	272	143	129	75.2
52.6	270	142	128	74.7
52.6	268	141	127	74.2
52.6	266	140	126	73.7
52.6	253	133	120	69.9
52.6	251	132	119	69.4
52.6	249	131	118	68.9

%E	M1	M2	DM	M*
52.6	247	130	117	68.4
52.6	234	123	111	64.7
52.6	232	122	110	64.2
52.6	230	121	109	63.7
52.6	228	120	108	63.2
52.6	215	113	102	59.4
52.6	213	112	101	58.9
52.6	211	111	100	58.4
52.6	209	110	99	57.9
52.6	196	103	93	54.1
52.6	194	102	92	53.6
52.6	192	101	91	53.1
52.6	190	100	90	52.6
52.6	175	92	83	48.4
52.6	173	91	82	47.9
52.6	171	90	81	47.4
52.6	156	82	74	43.1
52.6	154	81	73	42.6
52.6	152	80	72	42.1
52.6	137	72	65	37.8
52.6	135	71	64	37.3
52.6	133	70	63	36.8
52.6	116	61	55	32.1
52.6	114	60	54	31.6
52.6	97	51	46	26.8
52.6	95	50	45	26.3
52.6	78	41	37	21.6
52.6	76	40	36	21.1
52.6	57	30	27	15.8
52.6	38	20	18	10.5
52.6	19	10	9	5.3
52.5	499	262	237	137.6
52.5	497	261	236	137.1
52.5	495	260	235	136.6
52.5	493	259	234	136.1
52.5	491	258	233	135.6
52.5	488	256	232	134.3
52.5	486	255	231	133.8
52.5	484	254	230	133.3
52.5	482	253	229	132.8
52.5	480	252	228	132.3
52.5	478	251	227	131.8
52.5	476	250	226	131.3
52.5	474	249	225	130.8
52.5	472	248	224	130.3
52.5	469	246	223	129.0
52.5	467	245	222	128.5
52.5	465	244	221	128.0
52.5	463	243	220	127.5
52.5	461	242	219	127.0

%E	M1	M2	DM	M*
52.5	459	241	218	126.5
52.5	457	240	217	126.0
52.5	455	239	216	125.5
52.5	453	238	215	125.0
52.5	448	235	213	123.3
52.5	446	234	212	122.8
52.5	444	233	211	122.3
52.5	442	232	210	121.8
52.5	440	231	209	121.3
52.5	438	230	208	120.8
52.5	436	229	207	120.3
52.5	434	228	206	119.8
52.5	432	227	205	119.3
52.5	427	224	203	117.5
52.5	425	223	202	117.0
52.5	423	222	201	116.5
52.5	421	221	200	116.0
52.5	419	220	199	115.5
52.5	417	219	198	115.0
52.5	415	218	197	114.5
52.5	413	217	196	114.0
52.5	408	214	194	112.2
52.5	406	213	193	111.7
52.5	404	212	192	111.2
52.5	402	211	191	110.7
52.5	400	210	190	110.3
52.5	398	209	189	109.8
52.5	396	208	188	109.3
52.5	394	207	187	108.8
52.5	387	203	184	106.5
52.5	385	202	183	106.0
52.5	383	201	182	105.5
52.5	381	200	181	105.0
52.5	379	199	180	104.5
52.5	377	198	179	104.0
52.5	375	197	178	103.5
52.5	373	196	177	103.0
52.5	366	192	174	100.7
52.5	364	191	173	100.2
52.5	362	190	172	99.7
52.5	360	189	171	99.2
52.5	358	188	170	98.7
52.5	356	187	169	98.2
52.5	354	186	168	97.7
52.5	347	182	165	95.5
52.5	345	181	164	95.0
52.5	343	180	163	94.5
52.5	341	179	162	94.0
52.5	339	178	161	93.5
52.5	337	177	160	93.0

%E	M1	M2	DM	M*
52.5	335	176	159	92.5
52.5	326	171	155	89.7
52.5	324	170	154	89.2
52.5	322	169	153	88.7
52.5	320	168	152	88.2
52.5	318	167	151	87.7
52.5	316	166	150	87.2
52.5	314	165	149	86.7
52.5	305	160	145	83.9
52.5	303	159	144	83.4
52.5	301	158	143	82.9
52.5	299	157	142	82.4
52.5	297	156	141	81.9
52.5	295	155	140	81.4
52.5	284	149	135	78.2
52.5	282	148	134	77.7
52.5	280	147	133	77.2
52.5	278	146	132	76.7
52.5	276	145	131	76.2
52.5	265	139	126	72.9
52.5	263	138	125	72.4
52.5	261	137	124	71.9
52.5	259	136	123	71.4
52.5	257	135	122	70.9
52.5	255	134	121	70.4
52.5	244	128	116	67.1
52.5	242	127	115	66.6
52.5	240	126	114	66.1
52.5	238	125	113	65.7
52.5	236	124	112	65.2
52.5	223	117	106	61.4
52.5	221	116	105	60.9
52.5	219	115	104	60.4
52.5	217	114	103	59.9
52.5	204	107	97	56.1
52.5	202	106	96	55.6
52.5	200	105	95	55.1
52.5	198	104	94	54.6
52.5	183	96	87	50.4
52.5	181	95	86	49.9
52.5	179	94	85	49.4
52.5	177	93	84	48.9
52.5	162	85	77	44.6
52.5	160	84	76	44.1
52.5	158	83	75	43.6
52.5	141	74	67	38.8
52.5	139	73	66	38.3
52.5	122	64	58	33.6
52.5	120	63	57	33.1
52.5	118	62	56	32.6

%E	M1	M2	DM	M*
52.5	101	53	48	27.8
52.5	99	52	47	27.3
52.5	80	42	38	22.0
52.5	61	32	29	16.8
52.5	59	31	28	16.3
52.5	40	21	19	11.0
52.4	500	262	238	137.3
52.4	498	261	237	136.8
52.4	496	260	236	136.3
52.4	494	259	235	135.8
52.4	492	258	234	135.3
52.4	490	257	233	134.8
52.4	489	256	233	134.0
52.4	487	255	232	133.5
52.4	485	254	231	133.0
52.4	483	253	230	132.5
52.4	481	252	229	132.0
52.4	479	251	228	131.5
52.4	477	250	227	131.0
52.4	475	249	226	130.5
52.4	473	248	225	130.0
52.4	471	247	224	129.5
52.4	468	245	223	128.3
52.4	466	244	222	127.8
52.4	464	243	221	127.3
52.4	462	242	220	126.8
52.4	460	241	219	126.3
52.4	458	240	218	125.8
52.4	456	239	217	125.3
52.4	454	238	216	124.8
52.4	452	237	215	124.3
52.4	450	236	214	123.8
52.4	445	233	212	122.0
52.4	443	232	211	121.5
52.4	441	231	210	121.0
52.4	439	230	209	120.5
52.4	437	229	208	120.0
52.4	435	228	207	119.5
52.4	433	227	206	119.0
52.4	431	226	205	118.5
52.4	429	225	204	118.0
52.4	424	222	202	116.2
52.4	422	221	201	115.7
52.4	420	220	200	115.2
52.4	418	219	199	114.7
52.4	416	218	198	114.2
52.4	414	217	197	113.7
52.4	412	216	196	113.2
52.4	410	215	195	112.7
52.4	403	211	192	110.5

%E	M1	M2	DM	M*
52.4	401	210	191	110.0
52.4	399	209	190	109.5
52.4	397	208	189	109.0
52.4	395	207	188	108.5
52.4	393	206	187	108.0
52.4	391	205	186	107.5
52.4	389	204	185	107.0
52.4	382	200	182	104.7
52.4	380	199	181	104.2
52.4	378	198	180	103.7
52.4	376	197	179	103.2
52.4	374	196	178	102.7
52.4	372	195	177	102.2
52.4	370	194	176	101.7
52.4	368	193	175	101.2
52.4	361	189	172	99.0
52.4	359	188	171	98.5
52.4	357	187	170	98.0
52.4	355	186	169	97.5
52.4	353	185	168	97.0
52.4	351	184	167	96.5
52.4	349	183	166	96.0
52.4	340	178	162	93.2
52.4	338	177	161	92.7
52.4	336	176	160	92.2
52.4	334	175	159	91.7
52.4	332	174	158	91.2
52.4	330	173	157	90.7
52.4	328	172	156	90.2
52.4	319	167	152	87.4
52.4	317	166	151	86.9
52.4	315	165	150	86.4
52.4	313	164	149	85.9
52.4	311	163	148	85.4
52.4	309	162	147	84.9
52.4	307	161	146	84.4
52.4	296	155	141	81.2
52.4	294	154	140	80.7
52.4	292	153	139	80.2
52.4	290	152	138	79.7
52.4	288	151	137	79.2
52.4	286	150	136	78.7
52.4	275	144	131	75.4
52.4	273	143	130	74.9
52.4	271	142	129	74.4
52.4	269	141	128	73.9
52.4	267	140	127	73.4
52.4	254	133	121	69.6
52.4	252	132	120	69.1
52.4	250	131	119	68.6

%E	M1	M2	DM	M*
52.4	248	130	118	68.1
52.4	246	129	117	67.6
52.4	233	122	111	63.9
52.4	231	121	110	63.4
52.4	229	120	109	62.9
52.4	227	119	108	62.4
52.4	225	118	107	61.9
52.4	212	111	101	58.1
52.4	210	110	100	57.6
52.4	208	109	99	57.1
52.4	206	108	98	56.6
52.4	191	100	91	52.4
52.4	189	99	90	51.9
52.4	187	98	89	51.4
52.4	185	97	88	50.9
52.4	170	89	81	46.6
52.4	168	88	80	46.1
52.4	166	87	79	45.6
52.4	164	86	78	45.1
52.4	147	77	70	40.3
52.4	145	76	69	39.8
52.4	143	75	68	39.3
52.4	126	66	60	34.6
52.4	124	65	59	34.1
52.4	105	55	50	28.8
52.4	103	54	49	28.3
52.4	84	44	40	23.0
52.4	82	43	39	22.5
52.4	63	33	30	17.3
52.4	42	22	20	11.5
52.4	21	11	10	5.8
52.3	499	261	238	136.5
52.3	497	260	237	136.0
52.3	495	259	236	135.5
52.3	493	258	235	135.0
52.3	491	257	234	134.5
52.3	488	255	233	133.2
52.3	486	254	232	132.7
52.3	484	253	231	132.3
52.3	482	252	230	131.8
52.3	480	251	229	131.3
52.3	478	250	228	130.8
52.3	476	249	227	130.3
52.3	474	248	226	129.8
52.3	472	247	225	129.3
52.3	470	246	224	128.8
52.3	465	243	222	127.0
52.3	463	242	221	126.5
52.3	461	241	220	126.0
52.3	459	240	219	125.5

%E	M1	M2	DM	M*
52.3	457	239	218	125.0
52.3	455	238	217	124.5
52.3	453	237	216	124.0
52.3	451	236	215	123.5
52.3	449	235	214	123.0
52.3	447	234	213	122.5
52.3	444	232	212	121.2
52.3	442	231	211	120.7
52.3	440	230	210	120.2
52.3	438	229	209	119.7
52.3	436	228	208	119.2
52.3	434	227	207	118.7
52.3	432	226	206	118.2
52.3	430	225	205	117.7
52.3	428	224	204	117.2
52.3	426	223	203	116.7
52.3	421	220	201	115.0
52.3	419	219	200	114.5
52.3	417	218	199	114.0
52.3	415	217	198	113.5
52.3	413	216	197	113.0
52.3	411	215	196	112.5
52.3	409	214	195	112.0
52.3	407	213	194	111.5
52.3	405	212	193	111.0
52.3	400	209	191	109.2
52.3	398	208	190	108.7
52.3	396	207	189	108.2
52.3	394	206	188	107.7
52.3	392	205	187	107.2
52.3	390	204	186	106.7
52.3	388	203	185	106.2
52.3	386	202	184	105.7
52.3	384	201	183	105.2
52.3	377	197	180	102.9
52.3	375	196	179	102.4
52.3	373	195	178	101.9
52.3	371	194	177	101.4
52.3	369	193	176	100.9
52.3	367	192	175	100.4
52.3	365	191	174	99.9
52.3	363	190	173	99.4
52.3	354	185	169	96.7
52.3	352	184	168	96.2
52.3	350	183	167	95.7
52.3	348	182	166	95.2
52.3	346	181	165	94.7
52.3	344	180	164	94.2
52.3	342	179	163	93.7
52.3	333	174	159	90.9

%E	M1	M2	DM	M*
52.3	331	173	158	90.4
52.3	329	172	157	89.9
52.3	327	171	156	89.4
52.3	325	170	155	88.9
52.3	323	169	154	88.4
52.3	321	168	153	87.9
52.3	310	162	148	84.7
52.3	308	161	147	84.2
52.3	306	160	146	83.7
52.3	304	159	145	83.2
52.3	302	158	144	82.7
52.3	300	157	143	82.2
52.3	298	156	142	81.7
52.3	287	150	137	78.4
52.3	285	149	136	77.9
52.3	283	148	135	77.4
52.3	281	147	134	76.9
52.3	279	146	133	76.4
52.3	277	145	132	75.9
52.3	266	139	127	72.6
52.3	264	138	126	72.1
52.3	262	137	125	71.6
52.3	260	136	124	71.1
52.3	258	135	123	70.6
52.3	256	134	122	70.1
52.3	243	127	116	66.4
52.3	241	126	115	65.9
52.3	239	125	114	65.4
52.3	237	124	113	64.9
52.3	235	123	112	64.4
52.3	222	116	106	60.6
52.3	220	115	105	60.1
52.3	218	114	104	59.6
52.3	216	113	103	59.1
52.3	214	112	102	58.6
52.3	199	104	95	54.4
52.3	197	103	94	53.9
52.3	195	102	93	53.4
52.3	193	101	92	52.9
52.3	176	92	84	48.1
52.3	174	91	83	47.6
52.3	172	90	82	47.1
52.3	155	81	74	42.3
52.3	153	80	73	41.8
52.3	151	79	72	41.3
52.3	149	78	71	40.8
52.3	132	69	63	36.1
52.3	130	68	62	35.6
52.3	128	67	61	35.1
52.3	111	58	53	30.3

%E	M1	M2	DM	M*
52.3	109	57	52	29.8
52.3	107	56	51	29.3
52.3	88	46	42	24.0
52.3	86	45	41	23.5
52.3	65	34	31	17.8
52.3	44	23	21	12.0
52.2	500	261	239	136.2
52.2	498	260	238	135.7
52.2	496	259	237	135.2
52.2	494	258	236	134.7
52.2	492	257	235	134.2
52.2	490	256	234	133.7
52.2	487	254	233	132.5
52.2	485	253	232	132.0
52.2	483	252	231	131.5
52.2	481	251	230	131.0
52.2	479	250	229	130.5
52.2	477	249	228	130.0
52.2	475	248	227	129.5
52.2	473	247	226	129.0
52.2	471	246	225	128.5
52.2	469	245	224	128.0
52.2	467	244	223	127.5
52.2	464	242	222	126.2
52.2	462	241	221	125.7
52.2	460	240	220	125.2
52.2	458	239	219	124.7
52.2	456	238	218	124.2
52.2	454	237	217	123.7
52.2	452	236	216	123.2
52.2	450	235	215	122.7
52.2	448	234	214	122.2
52.2	446	233	213	121.7
52.2	441	230	211	120.0
52.2	439	229	210	119.5
52.2	437	228	209	119.0
52.2	435	227	208	118.5
52.2	433	226	207	118.0
52.2	431	225	206	117.5
52.2	429	224	205	117.0
52.2	427	223	204	116.5
52.2	425	222	203	116.0
52.2	423	221	202	115.5
52.2	418	218	200	113.7
52.2	416	217	199	113.2
52.2	414	216	198	112.7
52.2	412	215	197	112.2
52.2	410	214	196	111.7
52.2	408	213	195	111.2
52.2	406	212	194	110.7

%E	M1	M2	DM	M*
52.2	404	211	193	110.2
52.2	402	210	192	109.7
52.2	395	206	189	107.4
52.2	393	205	188	106.9
52.2	391	204	187	106.4
52.2	389	203	186	105.9
52.2	387	202	185	105.4
52.2	385	201	184	104.9
52.2	383	200	183	104.4
52.2	381	199	182	103.9
52.2	379	198	181	103.4
52.2	372	194	178	101.2
52.2	370	193	177	100.7
52.2	368	192	176	100.2
52.2	366	191	175	99.7
52.2	364	190	174	99.2
52.2	362	189	173	98.7
52.2	360	188	172	98.2
52.2	358	187	171	97.7
52.2	356	186	170	97.2
52.2	347	181	166	94.4
52.2	345	180	165	93.9
52.2	343	179	164	93.4
52.2	341	178	163	92.9
52.2	339	177	162	92.4
52.2	337	176	161	91.9
52.2	335	175	160	91.4
52.2	324	169	155	88.2
52.2	322	168	154	87.7
52.2	320	167	153	87.2
52.2	318	166	152	86.7
52.2	316	165	151	86.2
52.2	314	164	150	85.7
52.2	312	163	149	85.2
52.2	301	157	144	81.9
52.2	299	156	143	81.4
52.2	297	155	142	80.9
52.2	295	154	141	80.4
52.2	293	153	140	79.9
52.2	291	152	139	79.4
52.2	289	151	138	78.9
52.2	278	145	133	75.6
52.2	276	144	132	75.1
52.2	274	143	131	74.6
52.2	272	142	130	74.1
52.2	270	141	129	73.6
52.2	268	140	128	73.1
52.2	255	133	122	69.4
52.2	253	132	121	68.9
52.2	251	131	120	68.4

%E	M1	M2	DM	M*
52.2	249	130	119	67.9
52.2	247	129	118	67.4
52.2	245	128	117	66.9
52.2	232	121	111	63.1
52.2	230	120	110	62.6
52.2	228	119	109	62.1
52.2	226	118	108	61.6
52.2	224	117	107	61.1
52.2	209	109	100	56.8
52.2	207	108	99	56.3
52.2	205	107	98	55.8
52.2	203	106	97	55.3
52.2	201	105	96	54.9
52.2	186	97	89	50.6
52.2	184	96	88	50.1
52.2	182	95	87	49.6
52.2	180	94	86	49.1
52.2	178	93	85	48.6
52.2	161	84	77	43.8
52.2	159	83	76	43.3
52.2	157	82	75	42.8
52.2	138	72	66	37.6
52.2	136	71	65	37.1
52.2	134	70	64	36.6
52.2	115	60	55	31.3
52.2	113	59	54	30.8
52.2	92	48	44	25.0
52.2	90	47	43	24.5
52.2	69	36	33	18.8
52.2	67	35	32	18.3
52.2	46	24	22	12.5
52.2	23	12	11	6.3
52.1	499	260	239	135.5
52.1	497	259	238	135.0
52.1	495	258	237	134.5
52.1	493	257	236	134.0
52.1	491	256	235	133.5
52.1	489	255	234	133.0
52.1	486	253	233	131.7
52.1	484	252	232	131.2
52.1	482	251	231	130.7
52.1	480	250	230	130.2
52.1	478	249	229	129.7
52.1	476	248	228	129.2
52.1	474	247	227	128.7
52.1	472	246	226	128.2
52.1	470	245	225	127.7
52.1	468	244	224	127.2
52.1	466	243	223	126.7
52.1	463	241	222	125.4

%E	M1	M2	DM	M*
52.1	461	240	221	124.9
52.1	459	239	220	124.4
52.1	457	238	219	123.9
52.1	455	237	218	123.4
52.1	453	236	217	122.9
52.1	451	235	216	122.5
52.1	449	234	215	122.0
52.1	447	233	214	121.5
52.1	445	232	213	121.0
52.1	443	231	212	120.5
52.1	438	228	210	118.7
52.1	436	227	209	118.2
52.1	434	226	208	117.7
52.1	432	225	207	117.2
52.1	430	224	206	116.7
52.1	428	223	205	116.2
52.1	426	222	204	115.7
52.1	424	221	203	115.2
52.1	422	220	202	114.7
52.1	420	219	201	114.2
52.1	413	215	198	111.9
52.1	411	214	197	111.4
52.1	409	213	196	110.9
52.1	407	212	195	110.4
52.1	405	211	194	109.9
52.1	403	210	193	109.4
52.1	401	209	192	108.9
52.1	399	208	191	108.4
52.1	397	207	190	107.9
52.1	390	203	187	105.7
52.1	388	202	186	105.2
52.1	386	201	185	104.7
52.1	384	200	184	104.2
52.1	382	199	183	103.7
52.1	380	198	182	103.2
52.1	378	197	181	102.7
52.1	376	196	180	102.2
52.1	374	195	179	101.7
52.1	365	190	175	98.9
52.1	363	189	174	98.4
52.1	361	188	173	97.9
52.1	359	187	172	97.4
52.1	357	186	171	96.9
52.1	355	185	170	96.4
52.1	353	184	169	95.9
52.1	351	183	168	95.4
52.1	349	182	167	94.9
52.1	340	177	163	92.1
52.1	338	176	162	91.6
52.1	336	175	161	91.1

%E	M1	M2	DM	M*
52.1	334	174	160	90.6
52.1	332	173	159	90.1
52.1	330	172	158	89.6
52.1	328	171	157	89.1
52.1	326	170	156	88.7
52.1	317	165	152	85.9
52.1	315	164	151	85.4
52.1	313	163	150	84.9
52.1	311	162	149	84.4
52.1	309	161	148	83.9
52.1	307	160	147	83.4
52.1	305	159	146	82.9
52.1	303	158	145	82.4
52.1	292	152	140	79.1
52.1	290	151	139	78.6
52.1	288	150	138	78.1
52.1	286	149	137	77.6
52.1	284	148	136	77.1
52.1	282	147	135	76.6
52.1	280	146	134	76.1
52.1	267	139	128	72.4
52.1	265	138	127	71.9
52.1	263	137	126	71.4
52.1	261	136	125	70.9
52.1	259	135	124	70.4
52.1	257	134	123	69.9
52.1	242	126	116	65.6
52.1	240	125	115	65.1
52.1	238	124	114	64.6
52.1	236	123	113	64.1
52.1	234	122	112	63.6
52.1	219	114	105	59.3
52.1	217	113	104	58.8
52.1	215	112	103	58.3
52.1	213	111	102	57.8
52.1	211	110	101	57.3
52.1	194	101	93	52.6
52.1	192	100	92	52.1
52.1	190	99	91	51.6
52.1	188	98	90	51.1
52.1	169	88	81	45.8
52.1	167	87	80	45.3
52.1	165	86	79	44.8
52.1	163	85	78	44.3
52.1	146	76	70	39.6
52.1	144	75	69	39.1
52.1	142	74	68	38.6
52.1	140	73	67	38.1
52.1	121	63	58	32.8
52.1	119	62	57	32.3

%E	M1	M2	DM	M*
52.1	117	61	56	31.8
52.1	96	50	46	26.0
52.1	94	49	45	25.5
52.1	73	38	35	19.8
52.1	71	37	34	19.3
52.1	48	25	23	13.0
52.0	500	260	240	135.2
52.0	498	259	239	134.7
52.0	496	258	238	134.2
52.0	494	257	237	133.7
52.0	492	256	236	133.2
52.0	490	255	235	132.7
52.0	488	254	234	132.2
52.0	487	253	234	131.4
52.0	485	252	233	130.9
52.0	483	251	232	130.4
52.0	481	250	231	129.9
52.0	479	249	230	129.4
52.0	477	248	229	128.9
52.0	475	247	228	128.4
52.0	473	246	227	127.9
52.0	471	245	226	127.4
52.0	469	244	225	126.9
52.0	467	243	224	126.4
52.0	465	242	223	125.9
52.0	460	239	221	124.2
52.0	458	238	220	123.7
52.0	456	237	219	123.2
52.0	454	236	218	122.7
52.0	452	235	217	122.2
52.0	450	234	216	121.7
52.0	448	233	215	121.2
52.0	446	232	214	120.7
52.0	444	231	213	120.2
52.0	442	230	212	119.7
52.0	440	229	211	119.2
52.0	435	226	209	117.4
52.0	433	225	208	116.9
52.0	431	224	207	116.4
52.0	429	223	206	115.9
52.0	427	222	205	115.4
52.0	425	221	204	114.9
52.0	423	220	203	114.4
52.0	421	219	202	113.9
52.0	419	218	201	113.4
52.0	417	217	200	112.9
52.0	415	216	199	112.4
52.0	410	213	197	110.7
52.0	408	212	196	110.2
52.0	406	211	195	109.7

%E	M1	M2	DM	M*
52.0	404	210	194	109.2
52.0	402	209	193	108.7
52.0	400	208	192	108.2
52.0	398	207	191	107.7
52.0	396	206	190	107.2
52.0	394	205	189	106.7
52.0	392	204	188	106.2
52.0	383	199	184	103.4
52.0	381	198	183	102.9
52.0	379	197	182	102.4
52.0	377	196	181	101.9
52.0	375	195	180	101.4
52.0	373	194	179	100.9
52.0	371	193	178	100.4
52.0	369	192	177	99.9
52.0	367	191	176	99.4
52.0	358	186	172	96.6
52.0	356	185	171	96.1
52.0	354	184	170	95.6
52.0	352	183	169	95.1
52.0	350	182	168	94.6
52.0	348	181	167	94.1
52.0	346	180	166	93.6
52.0	344	179	165	93.1
52.0	342	178	164	92.6
52.0	333	173	160	89.9
52.0	331	172	159	89.4
52.0	329	171	158	88.9
52.0	327	170	157	88.4
52.0	325	169	156	87.9
52.0	323	168	155	87.4
52.0	321	167	154	86.9
52.0	319	166	153	86.4
52.0	306	159	147	82.6
52.0	304	158	146	82.1
52.0	302	157	145	81.6
52.0	300	156	144	81.1
52.0	298	155	143	80.6
52.0	296	154	142	80.1
52.0	294	153	141	79.6
52.0	281	146	135	75.9
52.0	279	145	134	75.4
52.0	277	144	133	74.9
52.0	275	143	132	74.4
52.0	273	142	131	73.9
52.0	271	141	130	73.4
52.0	269	140	129	72.9
52.0	256	133	123	69.1
52.0	254	132	122	68.6
52.0	252	131	121	68.1

%E	M1	M2	DM	M*
52.0	250	130	120	67.6
52.0	248	129	119	67.1
52.0	246	128	118	66.6
52.0	244	127	117	66.1
52.0	229	119	110	61.8
52.0	227	118	109	61.3
52.0	225	117	108	60.8
52.0	223	116	107	60.3
52.0	221	115	106	59.8
52.0	204	106	98	55.1
52.0	202	105	97	54.6
52.0	200	104	96	54.1
52.0	198	103	95	53.6
52.0	196	102	94	53.1
52.0	179	93	86	48.3
52.0	177	92	85	47.8
52.0	175	91	84	47.3
52.0	173	90	83	46.8
52.0	171	89	82	46.3
52.0	152	79	73	41.1
52.0	150	78	72	40.6
52.0	148	77	71	40.1
52.0	127	66	61	34.3
52.0	125	65	60	33.8
52.0	123	64	59	33.3
52.0	102	53	49	27.5
52.0	100	52	48	27.0
52.0	98	51	47	26.5
52.0	75	39	36	20.3
52.0	50	26	24	13.5
52.0	25	13	12	6.8
51.9	499	259	240	134.4
51.9	497	258	239	133.9
51.9	495	257	238	133.4
51.9	493	256	237	132.9
51.9	491	255	236	132.4
51.9	489	254	235	131.9
51.9	486	252	234	130.7
51.9	484	251	233	130.2
51.9	482	250	232	129.7
51.9	480	249	231	129.2
51.9	478	248	230	128.7
51.9	476	247	229	128.2
51.9	474	246	228	127.7
51.9	472	245	227	127.2
51.9	470	244	226	126.7
51.9	468	243	225	126.2
51.9	466	242	224	125.7
51.9	464	241	223	125.2
51.9	462	240	222	124.7

%E	M1	M2	DM	M*
51.9	459	238	221	123.4
51.9	457	237	220	122.9
51.9	455	236	219	122.4
51.9	453	235	218	121.9
51.9	451	234	217	121.4
51.9	449	233	216	120.9
51.9	447	232	215	120.4
51.9	445	231	214	119.9
51.9	443	230	213	119.4
51.9	441	229	212	118.9
51.9	439	228	211	118.4
51.9	437	227	210	117.9
51.9	432	224	208	116.1
51.9	430	223	207	115.6
51.9	428	222	206	115.1
51.9	426	221	205	114.7
51.9	424	220	204	114.2
51.9	422	219	203	113.7
51.9	420	218	202	113.2
51.9	418	217	201	112.7
51.9	416	216	200	112.2
51.9	414	215	199	111.7
51.9	412	214	198	111.2
51.9	405	210	195	108.9
51.9	403	209	194	108.4
51.9	401	208	193	107.9
51.9	399	207	192	107.4
51.9	397	206	191	106.9
51.9	395	205	190	106.4
51.9	393	204	189	105.9
51.9	391	203	188	105.4
51.9	389	202	187	104.9
51.9	387	201	186	104.4
51.9	385	200	185	103.9
51.9	378	196	182	101.6
51.9	376	195	181	101.1
51.9	374	194	180	100.6
51.9	372	193	179	100.1
51.9	370	192	178	99.6
51.9	368	191	177	99.1
51.9	366	190	176	98.6
51.9	364	189	175	98.1
51.9	362	188	174	97.6
51.9	360	187	173	97.1
51.9	351	182	169	94.4
51.9	349	181	168	93.9
51.9	347	180	167	93.4
51.9	345	179	166	92.9
51.9	343	178	165	92.4
51.9	341	177	164	91.9

%E	M1	M2	DM	M*
51.9	339	176	163	91.4
51.9	337	175	162	90.9
51.9	335	174	161	90.4
51.9	324	168	156	87.1
51.9	322	167	155	86.6
51.9	320	166	154	86.1
51.9	318	165	153	85.6
51.9	316	164	152	85.1
51.9	314	163	151	84.6
51.9	312	162	150	84.1
51.9	310	161	149	83.6
51.9	308	160	148	83.1
51.9	297	154	143	79.9
51.9	295	153	142	79.4
51.9	293	152	141	78.9
51.9	291	151	140	78.4
51.9	289	150	139	77.9
51.9	287	149	138	77.4
51.9	285	148	137	76.9
51.9	283	147	136	76.4
51.9	270	140	130	72.6
51.9	268	139	129	72.1
51.9	266	138	128	71.6
51.9	264	137	127	71.1
51.9	262	136	126	70.6
51.9	260	135	125	70.1
51.9	258	134	124	69.6
51.9	243	126	117	65.3
51.9	241	125	116	64.8
51.9	239	124	115	64.3
51.9	237	123	114	63.8
51.9	235	122	113	63.3
51.9	233	121	112	62.8
51.9	231	120	111	62.3
51.9	216	112	104	58.1
51.9	214	111	103	57.6
51.9	212	110	102	57.1
51.9	210	109	101	56.6
51.9	208	108	100	56.1
51.9	206	107	99	55.6
51.9	189	98	91	50.8
51.9	187	97	90	50.3
51.9	185	96	89	49.8
51.9	183	95	88	49.3
51.9	181	94	87	48.8
51.9	162	84	78	43.6
51.9	160	83	77	43.1
51.9	158	82	76	42.6
51.9	156	81	75	42.1
51.9	154	80	74	41.6

%E	M1	M2	DM	M*
51.9	135	70	65	36.3
51.9	133	69	64	35.8
51.9	131	68	63	35.3
51.9	129	67	62	34.8
51.9	108	56	52	29.0
51.9	106	55	51	28.5
51.9	104	54	50	28.0
51.9	81	42	39	21.8
51.9	79	41	38	21.3
51.9	77	40	37	20.8
51.9	54	28	26	14.5
51.9	52	27	25	14.0
51.9	27	14	13	7.3
51.8	500	259	241	134.2
51.8	498	258	240	133.7
51.8	496	257	239	133.2
51.8	494	256	238	132.7
51.8	492	255	237	132.2
51.8	490	254	236	131.7
51.8	488	253	235	131.2
51.8	485	251	234	129.9
51.8	483	250	233	129.4
51.8	481	249	232	128.9
51.8	479	248	231	128.4
51.8	477	247	230	127.9
51.8	475	246	229	127.4
51.8	473	245	228	126.9
51.8	471	244	227	126.4
51.8	469	243	226	125.9
51.8	467	242	225	125.4
51.8	465	241	224	124.9
51.8	463	240	223	124.4
51.8	461	239	222	123.9
51.8	456	236	220	122.1
51.8	454	235	219	121.6
51.8	452	234	218	121.1
51.8	450	233	217	120.6
51.8	448	232	216	120.1
51.8	446	231	215	119.6
51.8	444	230	214	119.1
51.8	442	229	213	118.6
51.8	440	228	212	118.1
51.8	438	227	211	117.6
51.8	436	226	210	117.1
51.8	434	225	209	116.6
51.8	427	221	206	114.4
51.8	425	220	205	113.9
51.8	423	219	204	113.4
51.8	421	218	203	112.9
51.8	419	217	202	112.4

%E	M1	M2	DM	M*
51.8	417	216	201	111.9
51.8	415	215	200	111.4
51.8	413	214	199	110.9
51.8	411	213	198	110.4
51.8	409	212	197	109.9
51.8	407	211	196	109.4
51.8	400	207	193	107.1
51.8	398	206	192	106.6
51.8	396	205	191	106.1
51.8	394	204	190	105.6
51.8	392	203	189	105.1
51.8	390	202	188	104.6
51.8	388	201	187	104.1
51.8	386	200	186	103.6
51.8	384	199	185	103.1
51.8	382	198	184	102.6
51.8	380	197	183	102.1
51.8	371	192	179	99.4
51.8	369	191	178	98.9
51.8	367	190	177	98.4
51.8	365	189	176	97.9
51.8	363	188	175	97.4
51.8	361	187	174	96.9
51.8	359	186	173	96.4
51.8	357	185	172	95.9
51.8	355	184	171	95.4
51.8	353	183	170	94.9
51.8	342	177	165	91.6
51.8	340	176	164	91.1
51.8	338	175	163	90.6
51.8	336	174	162	90.1
51.8	334	173	161	89.6
51.8	332	172	160	89.1
51.8	330	171	159	88.6
51.8	328	170	158	88.1
51.8	326	169	157	87.6
51.8	313	162	151	83.8
51.8	311	161	150	83.3
51.8	309	160	149	82.8
51.8	307	159	148	82.3
51.8	305	158	147	81.8
51.8	303	157	146	81.3
51.8	301	156	145	80.9
51.8	299	155	144	80.4
51.8	284	147	137	76.1
51.8	282	146	136	75.6
51.8	280	145	135	75.1
51.8	278	144	134	74.6
51.8	276	143	133	74.1
51.8	274	142	132	73.6

%E	M1	M2	DM	M*
51.8	272	141	131	73.1
51.8	257	133	124	68.8
51.8	255	132	123	68.3
51.8	253	131	122	67.8
51.8	251	130	121	67.3
51.8	249	129	120	66.8
51.8	247	128	119	66.3
51.8	245	127	118	65.8
51.8	228	118	110	61.1
51.8	226	117	109	60.6
51.8	224	116	108	60.1
51.8	222	115	107	59.6
51.8	220	114	106	59.1
51.8	218	113	105	58.6
51.8	199	103	96	53.3
51.8	197	102	95	52.8
51.8	195	101	94	52.3
51.8	193	100	93	51.8
51.8	191	99	92	51.3
51.8	170	88	82	45.6
51.8	168	87	81	45.1
51.8	166	86	80	44.6
51.8	164	85	79	44.1
51.8	141	73	68	37.8
51.8	139	72	67	37.3
51.8	137	71	66	36.8
51.8	114	59	55	30.5
51.8	112	58	54	30.0
51.8	110	57	53	29.5
51.8	85	44	41	22.8
51.8	83	43	40	22.3
51.8	56	29	27	15.0
51.7	499	258	241	133.4
51.7	497	257	240	132.9
51.7	495	256	239	132.4
51.7	493	255	238	131.9
51.7	491	254	237	131.4
51.7	489	253	236	130.9
51.7	487	252	235	130.4
51.7	484	250	234	129.1
51.7	482	249	233	128.6
51.7	480	248	232	128.1
51.7	478	247	231	127.6
51.7	476	246	230	127.1
51.7	474	245	229	126.6
51.7	472	244	228	126.1
51.7	470	243	227	125.6
51.7	468	242	226	125.1
51.7	466	241	225	124.6
51.7	464	240	224	124.1

%E	M1	M2	DM	M*
51.7	462	239	223	123.6
51.7	460	238	222	123.1
51.7	458	237	221	122.6
51.7	453	234	219	120.9
51.7	451	233	218	120.4
51.7	449	232	217	119.9
51.7	447	231	216	119.4
51.7	445	230	215	118.9
51.7	443	229	214	118.4
51.7	441	228	213	117.9
51.7	439	227	212	117.4
51.7	437	226	211	116.9
51.7	435	225	210	116.4
51.7	433	224	209	115.9
51.7	431	223	208	115.4
51.7	429	222	207	114.9
51.7	424	219	205	113.1
51.7	422	218	204	112.6
51.7	420	217	203	112.1
51.7	418	216	202	111.6
51.7	416	215	201	111.1
51.7	414	214	200	110.6
51.7	412	213	199	110.1
51.7	410	212	198	109.6
51.7	408	211	197	109.1
51.7	406	210	196	108.6
51.7	404	209	195	108.1
51.7	402	208	194	107.6
51.7	393	203	190	104.9
51.7	391	202	189	104.4
51.7	389	201	188	103.9
51.7	387	200	187	103.4
51.7	385	199	186	102.9
51.7	383	198	185	102.4
51.7	381	197	184	101.9
51.7	379	196	183	101.4
51.7	377	195	182	100.9
51.7	375	194	181	100.4
51.7	373	193	180	99.9
51.7	362	187	175	96.6
51.7	360	186	174	96.1
51.7	358	185	173	95.6
51.7	356	184	172	95.1
51.7	354	183	171	94.6
51.7	352	182	170	94.1
51.7	350	181	169	93.6
51.7	348	180	168	93.1
51.7	346	179	167	92.6
51.7	344	178	166	92.1
51.7	333	172	161	88.8

%E	M1	M2	DM	M*
51.7	331	171	160	88.3
51.7	329	170	159	87.8
51.7	327	169	158	87.3
51.7	325	168	157	86.8
51.7	323	167	156	86.3
51.7	321	166	155	85.8
51.7	319	165	154	85.3
51.7	317	164	153	84.8
51.7	315	163	152	84.3
51.7	302	156	146	80.6
51.7	300	155	145	80.1
51.7	298	154	144	79.6
51.7	296	153	143	79.1
51.7	294	152	142	78.6
51.7	292	151	141	78.1
51.7	290	150	140	77.6
51.7	288	149	139	77.1
51.7	286	148	138	76.6
51.7	271	140	131	72.3
51.7	269	139	130	71.8
51.7	267	138	129	71.3
51.7	265	137	128	70.8
51.7	263	136	127	70.3
51.7	261	135	126	69.8
51.7	259	134	125	69.3
51.7	242	125	117	64.6
51.7	240	124	116	64.1
51.7	238	123	115	63.6
51.7	236	122	114	63.1
51.7	234	121	113	62.6
51.7	232	120	112	62.1
51.7	230	119	111	61.6
51.7	211	109	102	56.3
51.7	209	108	101	55.8
51.7	207	107	100	55.3
51.7	205	106	99	54.8
51.7	203	105	98	54.3
51.7	201	104	97	53.8
51.7	180	93	87	48.0
51.7	178	92	86	47.6
51.7	176	91	85	47.1
51.7	174	90	84	46.6
51.7	172	89	83	46.1
51.7	151	78	73	40.3
51.7	149	77	72	39.8
51.7	147	76	71	39.3
51.7	145	75	70	38.8
51.7	143	74	69	38.3
51.7	120	62	58	32.0
51.7	118	61	57	31.5
51.7	116	60	56	31.0
51.7	89	46	43	23.8
51.7	87	45	42	23.3
51.7	60	31	29	16.0
51.7	58	30	28	15.5
51.7	29	15	14	7.8
51.6	500	258	242	133.1
51.6	498	257	241	132.6
51.6	496	256	240	132.1
51.6	494	255	239	131.6
51.6	492	254	238	131.1
51.6	490	253	237	130.6
51.6	488	252	236	130.1
51.6	486	251	235	129.6
51.6	483	249	234	128.4
51.6	481	248	233	127.9
51.6	479	247	232	127.4
51.6	477	246	231	126.9
51.6	475	245	230	126.4
51.6	473	244	229	125.9
51.6	471	243	228	125.4
51.6	469	242	227	124.9
51.6	467	241	226	124.4
51.6	465	240	225	123.9
51.6	463	239	224	123.4
51.6	461	238	223	122.9
51.6	459	237	222	122.4
51.6	457	236	221	121.9
51.6	455	235	220	121.4
51.6	450	232	218	119.6
51.6	448	231	217	119.1
51.6	446	230	216	118.6
51.6	444	229	215	118.1
51.6	442	228	214	117.6
51.6	440	227	213	117.1
51.6	438	226	212	116.6
51.6	436	225	211	116.1
51.6	434	224	210	115.6
51.6	432	223	209	115.1
51.6	430	222	208	114.6
51.6	428	221	207	114.1
51.6	426	220	206	113.6
51.6	419	216	203	111.4
51.6	417	215	202	110.9
51.6	415	214	201	110.4
51.6	413	213	200	109.9
51.6	411	212	199	109.4
51.6	409	211	198	108.9
51.6	407	210	197	108.4
51.6	405	209	196	107.9
51.6	403	208	195	107.4
51.6	401	207	194	106.9
51.6	399	206	193	106.4
51.6	397	205	192	105.9
51.6	395	204	191	105.4
51.6	386	199	187	102.6
51.6	384	198	186	102.1
51.6	382	197	185	101.6
51.6	380	196	184	101.1
51.6	378	195	183	100.6
51.6	376	194	182	100.1
51.6	374	193	181	99.6
51.6	372	192	180	99.1
51.6	370	191	179	98.6
51.6	368	190	178	98.1
51.6	366	189	177	97.6
51.6	364	188	176	97.1
51.6	353	182	171	93.8
51.6	351	181	170	93.3
51.6	349	180	169	92.8
51.6	347	179	168	92.3
51.6	345	178	167	91.8
51.6	343	177	166	91.3
51.6	341	176	165	90.8
51.6	339	175	164	90.3
51.6	337	174	163	89.8
51.6	335	173	162	89.3
51.6	322	166	156	85.6
51.6	320	165	155	85.1
51.6	318	164	154	84.6
51.6	316	163	153	84.1
51.6	314	162	152	83.6
51.6	312	161	151	83.1
51.6	310	160	150	82.6
51.6	308	159	149	82.1
51.6	306	158	148	81.6
51.6	304	157	147	81.1
51.6	289	149	140	76.8
51.6	287	148	139	76.3
51.6	285	147	138	75.8
51.6	283	146	137	75.3
51.6	281	145	136	74.8
51.6	279	144	135	74.3
51.6	277	143	134	73.8
51.6	275	142	133	73.3
51.6	273	141	132	72.8
51.6	258	133	125	68.6
51.6	256	132	124	68.1
51.6	254	131	123	67.6
51.6	252	130	122	67.1
51.6	250	129	121	66.6
51.6	248	128	120	66.1
51.6	246	127	119	65.6
51.6	244	126	118	65.1
51.6	225	116	109	59.8
51.6	223	115	108	59.3
51.6	221	114	107	58.8
51.6	219	113	106	58.3
51.6	217	112	105	57.8
51.6	215	111	104	57.3
51.6	213	110	103	56.8
51.6	192	99	93	51.0
51.6	190	98	92	50.5
51.6	188	97	91	50.0
51.6	186	96	90	49.5
51.6	184	95	89	49.0
51.6	182	94	88	48.5
51.6	161	83	78	42.8
51.6	159	82	77	42.3
51.6	157	81	76	41.8
51.6	155	80	75	41.3
51.6	153	79	74	40.8
51.6	128	66	62	34.0
51.6	126	65	61	33.5
51.6	124	64	60	33.0
51.6	122	63	59	32.5
51.6	95	49	46	25.3
51.6	93	48	45	24.8
51.6	91	47	44	24.3
51.6	64	33	31	17.0
51.6	62	32	30	16.5
51.6	31	16	15	8.3
51.5	499	257	242	132.4
51.5	497	256	241	131.9
51.5	495	255	240	131.4
51.5	493	254	239	130.9
51.5	491	253	238	130.4
51.5	489	252	237	129.9
51.5	487	251	236	129.4
51.5	485	250	235	128.9
51.5	482	248	234	127.6
51.5	480	247	233	127.1
51.5	478	246	232	126.6
51.5	476	245	231	126.1
51.5	474	244	230	125.6
51.5	472	243	229	125.1
51.5	470	242	228	124.6
51.5	468	241	227	124.1
51.5	466	240	226	123.6
51.5	464	239	225	123.1
51.5	462	238	224	122.6
51.5	460	237	223	122.1
51.5	458	236	222	121.6
51.5	456	235	221	121.1
51.5	454	234	220	120.6
51.5	452	233	219	120.1
51.5	447	230	217	118.3
51.5	445	229	216	117.8
51.5	443	228	215	117.3
51.5	441	227	214	116.8
51.5	439	226	213	116.3
51.5	437	225	212	115.8
51.5	435	224	211	115.3
51.5	433	223	210	114.8
51.5	431	222	209	114.3
51.5	429	221	208	113.8
51.5	427	220	207	113.3
51.5	425	219	206	112.8
51.5	423	218	205	112.3
51.5	421	217	204	111.9
51.5	412	212	200	109.1
51.5	410	211	199	108.6
51.5	408	210	198	108.1
51.5	406	209	197	107.6
51.5	404	208	196	107.1
51.5	402	207	195	106.6
51.5	400	206	194	106.1
51.5	398	205	193	105.6
51.5	396	204	192	105.1
51.5	394	203	191	104.6
51.5	392	202	190	104.1
51.5	390	201	189	103.6
51.5	388	200	188	103.1
51.5	379	195	184	100.3
51.5	377	194	183	99.8
51.5	375	193	182	99.3
51.5	373	192	181	98.8
51.5	371	191	180	98.3
51.5	369	190	179	97.8
51.5	367	189	178	97.3
51.5	365	188	177	96.8
51.5	363	187	176	96.3
51.5	361	186	175	95.8
51.5	359	185	174	95.3
51.5	357	184	173	94.8
51.5	355	183	172	94.3
51.5	344	177	167	91.1
51.5	342	176	166	90.6
51.5	340	175	165	90.1
51.5	338	174	164	89.6

%E	M1	M2	DM	M*
51.5	336	173	163	89.1
51.5	334	172	162	88.6
51.5	332	171	161	88.1
51.5	330	170	160	87.6
51.5	328	169	159	87.1
51.5	326	168	158	86.6
51.5	324	167	157	86.1
51.5	309	159	150	81.8
51.5	307	158	149	81.3
51.5	305	157	148	80.8
51.5	303	156	147	80.3
51.5	301	155	146	79.8
51.5	299	154	145	79.3
51.5	297	153	144	78.8
51.5	295	152	143	78.3
51.5	293	151	142	77.8
51.5	291	150	141	77.3
51.5	274	141	133	72.6
51.5	272	140	132	72.1
51.5	270	139	131	71.6
51.5	268	138	130	71.1
51.5	266	137	129	70.6
51.5	264	136	128	70.1
51.5	262	135	127	69.6
51.5	260	134	126	69.1
51.5	241	124	117	63.8
51.5	239	123	116	63.3
51.5	237	122	115	62.8
51.5	235	121	114	62.3
51.5	233	120	113	61.8
51.5	231	119	112	61.3
51.5	229	118	111	60.8
51.5	227	117	110	60.3
51.5	206	106	100	54.5
51.5	204	105	99	54.0
51.5	202	104	98	53.5
51.5	200	103	97	53.0
51.5	198	102	96	52.5
51.5	196	101	95	52.0
51.5	194	100	94	51.5
51.5	171	88	83	45.3
51.5	169	87	82	44.8
51.5	167	86	81	44.3
51.5	165	85	80	43.8
51.5	163	84	79	43.3
51.5	136	70	66	36.0
51.5	134	69	65	35.5
51.5	132	68	64	35.0
51.5	130	67	63	34.5
51.5	103	53	50	27.3
51.5	101	52	49	26.8
51.5	99	51	48	26.3
51.5	97	50	47	25.8
51.5	68	35	33	18.0
51.5	66	34	32	17.5
51.5	33	17	16	8.8
51.4	500	257	243	132.1
51.4	498	256	242	131.6
51.4	496	255	241	131.1
51.4	494	254	240	130.6
51.4	492	253	239	130.1
51.4	490	252	238	129.6
51.4	488	251	237	129.1
51.4	486	250	236	128.6
51.4	484	249	235	128.1
51.4	481	247	234	126.8
51.4	479	246	233	126.3
51.4	477	245	232	125.8
51.4	475	244	231	125.3
51.4	473	243	230	124.8
51.4	471	242	229	124.3
51.4	469	241	228	123.8
51.4	467	240	227	123.3
51.4	465	239	226	122.8
51.4	463	238	225	122.3
51.4	461	237	224	121.8
51.4	459	236	223	121.3
51.4	457	235	222	120.8
51.4	455	234	221	120.3
51.4	453	233	220	119.8
51.4	451	232	219	119.3
51.4	449	231	218	118.8
51.4	444	228	216	117.1
51.4	442	227	215	116.6
51.4	440	226	214	116.1
51.4	438	225	213	115.6
51.4	436	224	212	115.1
51.4	434	223	211	114.6
51.4	432	222	210	114.1
51.4	430	221	209	113.6
51.4	428	220	208	113.1
51.4	426	219	207	112.6
51.4	424	218	206	112.1
51.4	422	217	205	111.6
51.4	420	216	204	111.1
51.4	418	215	203	110.6
51.4	416	214	202	110.1
51.4	414	213	201	109.6
51.4	407	209	198	107.3
51.4	405	208	197	106.8
51.4	403	207	196	106.3
51.4	401	206	195	105.8
51.4	399	205	194	105.3
51.4	397	204	193	104.8
51.4	395	203	192	104.3
51.4	393	202	191	103.8
51.4	391	201	190	103.3
51.4	389	200	189	102.8
51.4	387	199	188	102.3
51.4	385	198	187	101.8
51.4	383	197	186	101.3
51.4	381	196	185	100.8
51.4	370	190	180	97.6
51.4	368	189	179	97.1
51.4	366	188	178	96.6
51.4	364	187	177	96.1
51.4	362	186	176	95.6
51.4	360	185	175	95.1
51.4	358	184	174	94.6
51.4	356	183	173	94.1
51.4	354	182	172	93.6
51.4	352	181	171	93.1
51.4	350	180	170	92.6
51.4	348	179	169	92.1
51.4	346	178	168	91.6
51.4	333	171	162	87.8
51.4	331	170	161	87.3
51.4	329	169	160	86.8
51.4	327	168	159	86.3
51.4	325	167	158	85.8
51.4	323	166	157	85.3
51.4	321	165	156	84.8
51.4	319	164	155	84.3
51.4	317	163	154	83.8
51.4	315	162	153	83.3
51.4	313	161	152	82.8
51.4	311	160	151	82.3
51.4	296	152	144	78.1
51.4	294	151	143	77.6
51.4	292	150	142	77.1
51.4	290	149	141	76.6
51.4	288	148	140	76.1
51.4	286	147	139	75.6
51.4	284	146	138	75.1
51.4	282	145	137	74.6
51.4	280	144	136	74.1
51.4	278	143	135	73.6
51.4	276	142	134	73.1
51.4	259	133	126	68.3
51.4	257	132	125	67.8
51.4	255	131	124	67.3
51.4	253	130	123	66.8
51.4	251	129	122	66.3
51.4	249	128	121	65.8
51.4	247	127	120	65.3
51.4	245	126	119	64.8
51.4	243	125	118	64.3
51.4	222	114	108	58.5
51.4	220	113	107	58.0
51.4	218	112	106	57.5
51.4	216	111	105	57.0
51.4	214	110	104	56.5
51.4	212	109	103	56.0
51.4	210	108	102	55.5
51.4	208	107	101	55.0
51.4	185	95	90	48.8
51.4	183	94	89	48.3
51.4	181	93	88	47.8
51.4	179	92	87	47.3
51.4	177	91	86	46.8
51.4	175	90	85	46.3
51.4	173	89	84	45.8
51.4	148	76	72	39.0
51.4	146	75	71	38.5
51.4	144	74	70	38.0
51.4	142	73	69	37.5
51.4	140	72	68	37.0
51.4	138	71	67	36.5
51.4	111	57	54	29.3
51.4	109	56	53	28.8
51.4	107	55	52	28.3
51.4	105	54	51	27.8
51.4	74	38	36	19.5
51.4	72	37	35	19.0
51.4	70	36	34	18.5
51.4	37	19	18	9.8
51.4	35	18	17	9.3
51.3	499	256	243	131.3
51.3	497	255	242	130.8
51.3	495	254	241	130.3
51.3	493	253	240	129.8
51.3	491	252	239	129.3
51.3	489	251	238	128.8
51.3	487	250	237	128.3
51.3	485	249	236	127.8
51.3	483	248	235	127.3
51.3	480	246	234	126.1
51.3	478	245	233	125.6
51.3	476	244	232	125.1
51.3	474	243	231	124.6
51.3	472	242	230	124.1
51.3	470	241	229	123.6
51.3	468	240	228	123.1
51.3	466	239	227	122.6
51.3	464	238	226	122.1
51.3	462	237	225	121.6
51.3	460	236	224	121.1
51.3	458	235	223	120.6
51.3	456	234	222	120.1
51.3	454	233	221	119.6
51.3	452	232	220	119.1
51.3	450	231	219	118.6
51.3	448	230	218	118.1
51.3	446	229	217	117.6
51.3	439	225	214	115.3
51.3	437	224	213	114.8
51.3	435	223	212	114.3
51.3	433	222	211	113.8
51.3	431	221	210	113.3
51.3	429	220	209	112.8
51.3	427	219	208	112.3
51.3	425	218	207	111.8
51.3	423	217	206	111.3
51.3	421	216	205	110.8
51.3	419	215	204	110.3
51.3	417	214	203	109.8
51.3	415	213	202	109.3
51.3	413	212	201	108.8
51.3	411	211	200	108.3
51.3	409	210	199	107.8
51.3	400	205	195	105.1
51.3	398	204	194	104.6
51.3	396	203	193	104.1
51.3	394	202	192	103.6
51.3	392	201	191	103.1
51.3	390	200	190	102.6
51.3	388	199	189	102.1
51.3	386	198	188	101.6
51.3	384	197	187	101.1
51.3	382	196	186	100.6
51.3	380	195	185	100.1
51.3	378	194	184	99.6
51.3	376	193	183	99.1
51.3	374	192	182	98.6
51.3	372	191	181	98.1
51.3	359	184	175	94.3
51.3	357	183	174	93.8
51.3	355	182	173	93.3
51.3	353	181	172	92.8
51.3	351	180	171	92.3

%E	M1	M2	DM	M*
51.3	349	179	170	91.8
51.3	347	178	169	91.3
51.3	345	177	168	90.8
51.3	343	176	167	90.3
51.3	341	175	166	89.8
51.3	339	174	165	89.3
51.3	337	173	164	88.8
51.3	335	172	163	88.3
51.3	320	164	156	84.0
51.3	318	163	155	83.6
51.3	316	162	154	83.1
51.3	314	161	153	82.6
51.3	312	160	152	82.1
51.3	310	159	151	81.6
51.3	308	158	150	81.1
51.3	306	157	149	80.6
51.3	304	156	148	80.1
51.3	302	155	147	79.6
51.3	300	154	146	79.1
51.3	298	153	145	78.6
51.3	279	143	136	73.3
51.3	277	142	135	72.8
51.3	275	141	134	72.3
51.3	273	140	133	71.8
51.3	271	139	132	71.3
51.3	269	138	131	70.8
51.3	267	137	130	70.3
51.3	265	136	129	69.8
51.3	263	135	128	69.3
51.3	261	134	127	68.8
51.3	240	123	117	63.0
51.3	238	122	116	62.5
51.3	236	121	115	62.0
51.3	234	120	114	61.5
51.3	232	119	113	61.0
51.3	230	118	112	60.5
51.3	228	117	111	60.0
51.3	226	116	110	59.5
51.3	224	115	109	59.0
51.3	199	102	97	52.3
51.3	197	101	96	51.8
51.3	195	100	95	51.3
51.3	193	99	94	50.8
51.3	191	98	93	50.3
51.3	189	97	92	49.8
51.3	187	96	91	49.3
51.3	160	82	78	42.0
51.3	158	81	77	41.5
51.3	156	80	76	41.0
51.3	154	79	75	40.5

%E	M1	M2	DM	M*
51.3	152	78	74	40.0
51.3	150	77	73	39.5
51.3	119	61	58	31.3
51.3	117	60	57	30.8
51.3	115	59	56	30.3
51.3	113	58	55	29.8
51.3	80	41	39	21.0
51.3	78	40	38	20.5
51.3	76	39	37	20.0
51.3	39	20	19	10.3
51.2	500	256	244	131.1
51.2	498	255	243	130.6
51.2	496	254	242	130.1
51.2	494	253	241	129.6
51.2	492	252	240	129.1
51.2	490	251	239	128.6
51.2	488	250	238	128.1
51.2	486	249	237	127.6
51.2	484	248	236	127.1
51.2	482	247	235	126.6
51.2	477	244	233	124.8
51.2	475	243	232	124.3
51.2	473	242	231	123.8
51.2	471	241	230	123.3
51.2	469	240	229	122.8
51.2	467	239	228	122.3
51.2	465	238	227	121.8
51.2	463	237	226	121.3
51.2	461	236	225	120.8
51.2	459	235	224	120.3
51.2	457	234	223	119.8
51.2	455	233	222	119.3
51.2	453	232	221	118.8
51.2	451	231	220	118.3
51.2	449	230	219	117.8
51.2	447	229	218	117.3
51.2	445	228	217	116.8
51.2	443	227	216	116.3
51.2	441	226	215	115.8
51.2	434	222	212	113.6
51.2	432	221	211	113.1
51.2	430	220	210	112.6
51.2	428	219	209	112.1
51.2	426	218	208	111.6
51.2	424	217	207	111.1
51.2	422	216	206	110.6
51.2	420	215	205	110.1
51.2	418	214	204	109.6
51.2	416	213	203	109.1
51.2	414	212	202	108.6

%E	M1	M2	DM	M*
51.2	412	211	201	108.1
51.2	410	210	200	107.6
51.2	408	209	199	107.1
51.2	406	208	198	106.6
51.2	404	207	197	106.1
51.2	402	206	196	105.6
51.2	391	200	191	102.3
51.2	389	199	190	101.8
51.2	387	198	189	101.3
51.2	385	197	188	100.8
51.2	383	196	187	100.3
51.2	381	195	186	99.8
51.2	379	194	185	99.3
51.2	377	193	184	98.8
51.2	375	192	183	98.3
51.2	373	191	182	97.8
51.2	371	190	181	97.3
51.2	369	189	180	96.8
51.2	367	188	179	96.3
51.2	365	187	178	95.8
51.2	363	186	177	95.3
51.2	361	185	176	94.8
51.2	346	177	169	90.5
51.2	344	176	168	90.0
51.2	342	175	167	89.5
51.2	340	174	166	89.0
51.2	338	173	165	88.5
51.2	336	172	164	88.0
51.2	334	171	163	87.5
51.2	332	170	162	87.0
51.2	330	169	161	86.5
51.2	328	168	160	86.0
51.2	326	167	159	85.5
51.2	324	166	158	85.0
51.2	322	165	157	84.5
51.2	303	155	148	79.3
51.2	301	154	147	78.8
51.2	299	153	146	78.3
51.2	297	152	145	77.8
51.2	295	151	144	77.3
51.2	293	150	143	76.8
51.2	291	149	142	76.3
51.2	289	148	141	75.8
51.2	287	147	140	75.3
51.2	285	146	139	74.8
51.2	283	145	138	74.3
51.2	281	144	137	73.8
51.2	260	133	127	68.0
51.2	258	132	126	67.5
51.2	256	131	125	67.0

%E	M1	M2	DM	M*
51.2	254	130	124	66.5
51.2	252	129	123	66.0
51.2	250	128	122	65.5
51.2	248	127	121	65.0
51.2	246	126	120	64.5
51.2	244	125	119	64.0
51.2	242	124	118	63.5
51.2	217	111	106	56.8
51.2	215	110	105	56.3
51.2	213	109	104	55.8
51.2	211	108	103	55.3
51.2	209	107	102	54.8
51.2	207	106	101	54.3
51.2	205	105	100	53.8
51.2	203	104	99	53.3
51.2	201	103	98	52.8
51.2	172	88	84	45.0
51.2	170	87	83	44.5
51.2	168	86	82	44.0
51.2	166	85	81	43.5
51.2	164	84	80	43.0
51.2	162	83	79	42.5
51.2	129	66	63	33.8
51.2	127	65	62	33.3
51.2	125	64	61	32.8
51.2	123	63	60	32.3
51.2	121	62	59	31.8
51.2	86	44	42	22.5
51.2	84	43	41	22.0
51.2	82	42	40	21.5
51.2	43	22	21	11.3
51.2	41	21	20	10.8
51.1	499	255	244	130.3
51.1	497	254	243	129.8
51.1	495	253	242	129.3
51.1	493	252	241	128.8
51.1	491	251	240	128.3
51.1	489	250	239	127.8
51.1	487	249	238	127.3
51.1	485	248	237	126.8
51.1	483	247	236	126.3
51.1	481	246	235	125.8
51.1	479	245	234	125.3
51.1	476	243	233	124.1
51.1	474	242	232	123.6
51.1	472	241	231	123.1
51.1	470	240	230	122.6
51.1	468	239	229	122.1
51.1	466	238	228	121.6
51.1	464	237	227	121.1

%E	M1	M2	DM	M*
51.1	462	236	226	120.6
51.1	460	235	225	120.1
51.1	458	234	224	119.6
51.1	456	233	223	119.1
51.1	454	232	222	118.6
51.1	452	231	221	118.1
51.1	450	230	220	117.6
51.1	448	229	219	117.1
51.1	446	228	218	116.6
51.1	444	227	217	116.1
51.1	442	226	216	115.6
51.1	440	225	215	115.1
51.1	438	224	214	114.6
51.1	436	223	213	114.1
51.1	427	218	209	111.3
51.1	425	217	208	110.8
51.1	423	216	207	110.3
51.1	421	215	206	109.8
51.1	419	214	205	109.3
51.1	417	213	204	108.8
51.1	415	212	203	108.3
51.1	413	211	202	107.8
51.1	411	210	201	107.3
51.1	409	209	200	106.8
51.1	407	208	199	106.3
51.1	405	207	198	105.8
51.1	403	206	197	105.3
51.1	401	205	196	104.8
51.1	399	204	195	104.3
51.1	397	203	194	103.8
51.1	395	202	193	103.3
51.1	393	201	192	102.8
51.1	380	194	186	99.0
51.1	378	193	185	98.5
51.1	376	192	184	98.0
51.1	374	191	183	97.5
51.1	372	190	182	97.0
51.1	370	189	181	96.5
51.1	368	188	180	96.0
51.1	366	187	179	95.5
51.1	364	186	178	95.0
51.1	362	185	177	94.5
51.1	360	184	176	94.0
51.1	358	183	175	93.5
51.1	356	182	174	93.0
51.1	354	181	173	92.5
51.1	352	180	172	92.0
51.1	350	179	171	91.5
51.1	348	178	170	91.0
51.1	333	170	163	86.8

%E	M1	M2	DM	M*
51.1	331	169	162	86.3
51.1	329	168	161	85.8
51.1	327	167	160	85.3
51.1	325	166	159	84.8
51.1	323	165	158	84.3
51.1	321	164	157	83.8
51.1	319	163	156	83.3
51.1	317	162	155	82.8
51.1	315	161	154	82.3
51.1	313	160	153	81.8
51.1	311	159	152	81.3
51.1	309	158	151	80.8
51.1	307	157	150	80.3
51.1	305	156	149	79.8
51.1	284	145	139	74.0
51.1	282	144	138	73.5
51.1	280	143	137	73.0
51.1	278	142	136	72.5
51.1	276	141	135	72.0
51.1	274	140	134	71.5
51.1	272	139	133	71.0
51.1	270	138	132	70.5
51.1	268	137	131	70.0
51.1	266	136	130	69.5
51.1	264	135	129	69.0
51.1	262	134	128	68.5
51.1	237	121	116	61.8
51.1	235	120	115	61.3
51.1	233	119	114	60.8
51.1	231	118	113	60.3
51.1	229	117	112	59.8
51.1	227	116	111	59.3
51.1	225	115	110	58.8
51.1	223	114	109	58.3
51.1	221	113	108	57.8
51.1	219	112	107	57.3
51.1	190	97	93	49.5
51.1	188	96	92	49.0
51.1	186	95	91	48.5
51.1	184	94	90	48.0
51.1	182	93	89	47.5
51.1	180	92	88	47.0
51.1	178	91	87	46.5
51.1	176	90	86	46.0
51.1	174	89	85	45.5
51.1	141	72	69	36.8
51.1	139	71	68	36.3
51.1	137	70	67	35.8
51.1	135	69	66	35.3
51.1	133	68	65	34.8

%E	M1	M2	DM	M*
51.1	131	67	64	34.3
51.1	94	48	46	24.5
51.1	92	47	45	24.0
51.1	90	46	44	23.5
51.1	88	45	43	23.0
51.1	47	24	23	12.3
51.1	45	23	22	11.8
51.0	500	255	245	130.0
51.0	498	254	244	129.6
51.0	496	253	243	129.1
51.0	494	252	242	128.6
51.0	492	251	241	128.1
51.0	490	250	240	127.6
51.0	488	249	239	127.1
51.0	486	248	238	126.6
51.0	484	247	237	126.1
51.0	482	246	236	125.6
51.0	480	245	235	125.1
51.0	478	244	234	124.6
51.0	473	241	232	122.8
51.0	471	240	231	122.3
51.0	469	239	230	121.8
51.0	467	238	229	121.3
51.0	465	237	228	120.8
51.0	463	236	227	120.3
51.0	461	235	226	119.8
51.0	459	234	225	119.3
51.0	457	233	224	118.8
51.0	455	232	223	118.3
51.0	453	231	222	117.8
51.0	451	230	221	117.3
51.0	449	229	220	116.8
51.0	447	228	219	116.3
51.0	445	227	218	115.8
51.0	443	226	217	115.3
51.0	441	225	216	114.8
51.0	439	224	215	114.3
51.0	437	223	214	113.8
51.0	435	222	213	113.3
51.0	433	221	212	112.8
51.0	431	220	211	112.3
51.0	429	219	210	111.8
51.0	420	214	206	109.0
51.0	418	213	205	108.5
51.0	416	212	204	108.0
51.0	414	211	203	107.5
51.0	412	210	202	107.0
51.0	410	209	201	106.5
51.0	408	208	200	106.0
51.0	406	207	199	105.5

%E	M1	M2	DM	M*
51.0	404	206	198	105.0
51.0	402	205	197	104.5
51.0	400	204	196	104.0
51.0	398	203	195	103.5
51.0	396	202	194	103.0
51.0	394	201	193	102.5
51.0	392	200	192	102.0
51.0	390	199	191	101.5
51.0	388	198	190	101.0
51.0	386	197	189	100.5
51.0	384	196	188	100.0
51.0	382	195	187	99.5
51.0	367	187	180	95.3
51.0	365	186	179	94.8
51.0	363	185	178	94.3
51.0	361	184	177	93.8
51.0	359	183	176	93.3
51.0	357	182	175	92.8
51.0	355	181	174	92.3
51.0	353	180	173	91.8
51.0	351	179	172	91.3
51.0	349	178	171	90.8
51.0	347	177	170	90.3
51.0	345	176	169	89.8
51.0	343	175	168	89.3
51.0	341	174	167	88.8
51.0	339	173	166	88.3
51.0	337	172	165	87.8
51.0	335	171	164	87.3
51.0	314	160	154	81.5
51.0	312	159	153	81.0
51.0	310	158	152	80.5
51.0	308	157	151	80.0
51.0	306	156	150	79.5
51.0	304	155	149	79.0
51.0	302	154	148	78.5
51.0	300	153	147	78.0
51.0	298	152	146	77.5
51.0	296	151	145	77.0
51.0	294	150	144	76.5
51.0	292	149	143	76.0
51.0	290	148	142	75.5
51.0	288	147	141	75.0
51.0	286	146	140	74.5
51.0	263	134	129	68.3
51.0	261	133	128	67.8
51.0	259	132	127	67.3
51.0	257	131	126	66.8
51.0	255	130	125	66.3
51.0	253	129	124	65.8

%E	M1	M2	DM	M*
51.0	251	128	123	65.3
51.0	249	127	122	64.8
51.0	247	126	121	64.3
51.0	245	125	120	63.8
51.0	243	124	119	63.3
51.0	241	123	118	62.8
51.0	239	122	117	62.3
51.0	210	107	103	54.5
51.0	208	106	102	54.0
51.0	206	105	101	53.5
51.0	204	104	100	53.0
51.0	202	103	99	52.5
51.0	200	102	98	52.0
51.0	198	101	97	51.5
51.0	196	100	96	51.0
51.0	194	99	95	50.5
51.0	192	98	94	50.0
51.0	157	80	77	40.8
51.0	155	79	76	40.3
51.0	153	78	75	39.8
51.0	151	77	74	39.3
51.0	149	76	73	38.8
51.0	147	75	72	38.3
51.0	145	74	71	37.8
51.0	143	73	70	37.3
51.0	104	53	51	27.0
51.0	102	52	50	26.5
51.0	100	51	49	26.0
51.0	98	50	48	25.5
51.0	96	49	47	25.0
51.0	51	26	25	13.3
51.0	49	25	24	12.8
50.9	499	254	245	129.3
50.9	497	253	244	128.8
50.9	495	252	243	128.3
50.9	493	251	242	127.8
50.9	491	250	241	127.3
50.9	489	249	240	126.8
50.9	487	248	239	126.3
50.9	485	247	238	125.8
50.9	483	246	237	125.3
50.9	481	245	236	124.8
50.9	479	244	235	124.3
50.9	477	243	234	123.8
50.9	475	242	233	123.3
50.9	470	239	231	121.5
50.9	468	238	230	121.0
50.9	466	237	229	120.5
50.9	464	236	228	120.0
50.9	462	235	227	119.5

%E	M1	M2	DM	M*
50.9	460	234	226	119.0
50.9	458	233	225	118.5
50.9	456	232	224	118.0
50.9	454	231	223	117.5
50.9	452	230	222	117.0
50.9	450	229	221	116.5
50.9	448	228	220	116.0
50.9	446	227	219	115.5
50.9	444	226	218	115.0
50.9	442	225	217	114.5
50.9	440	224	216	114.0
50.9	438	223	215	113.5
50.9	436	222	214	113.0
50.9	434	221	213	112.5
50.9	432	220	212	112.0
50.9	430	219	211	111.5
50.9	428	218	210	111.0
50.9	426	217	209	110.5
50.9	424	216	208	110.0
50.9	422	215	207	109.5
50.9	411	209	202	106.3
50.9	409	208	201	105.8
50.9	407	207	200	105.3
50.9	405	206	199	104.8
50.9	403	205	198	104.3
50.9	401	204	197	103.8
50.9	399	203	196	103.3
50.9	397	202	195	102.8
50.9	395	201	194	102.3
50.9	393	200	193	101.8
50.9	391	199	192	101.3
50.9	389	198	191	100.8
50.9	387	197	190	100.3
50.9	385	196	189	99.8
50.9	383	195	188	99.3
50.9	381	194	187	98.8
50.9	379	193	186	98.3
50.9	377	192	185	97.8
50.9	375	191	184	97.3
50.9	373	190	183	96.8
50.9	371	189	182	96.3
50.9	369	188	181	95.8
50.9	352	179	173	91.0
50.9	350	178	172	90.5
50.9	348	177	171	90.0
50.9	346	176	170	89.5
50.9	344	175	169	89.0
50.9	342	174	168	88.5
50.9	340	173	167	88.0
50.9	338	172	166	87.5

%E	M1	M2	DM	M*
50.9	336	171	165	87.0
50.9	334	170	164	86.5
50.9	332	169	163	86.0
50.9	330	168	162	85.5
50.9	328	167	161	85.0
50.9	326	166	160	84.5
50.9	324	165	159	84.0
50.9	322	164	158	83.5
50.9	320	163	157	83.0
50.9	318	162	156	82.5
50.9	316	161	155	82.0
50.9	293	149	144	75.8
50.9	291	148	143	75.3
50.9	289	147	142	74.8
50.9	287	146	141	74.3
50.9	285	145	140	73.8
50.9	283	144	139	73.3
50.9	281	143	138	72.8
50.9	279	142	137	72.3
50.9	277	141	136	71.8
50.9	275	140	135	71.3
50.9	273	139	134	70.8
50.9	271	138	133	70.3
50.9	269	137	132	69.8
50.9	267	136	131	69.3
50.9	265	135	130	68.8
50.9	234	119	115	60.5
50.9	232	118	114	60.0
50.9	230	117	113	59.5
50.9	228	116	112	59.0
50.9	226	115	111	58.5
50.9	224	114	110	58.0
50.9	222	113	109	57.5
50.9	220	112	108	57.0
50.9	218	111	107	56.5
50.9	216	110	106	56.0
50.9	214	109	105	55.5
50.9	212	108	104	55.0
50.9	175	89	86	45.3
50.9	173	88	85	44.8
50.9	171	87	84	44.3
50.9	169	86	83	43.8
50.9	167	85	82	43.3
50.9	165	84	81	42.8
50.9	163	83	80	42.3
50.9	161	82	79	41.8
50.9	159	81	78	41.3
50.9	116	59	57	30.0
50.9	114	58	56	29.5
50.9	112	57	55	29.0

%E	M1	M2	DM	M*
50.9	110	56	54	28.5
50.9	108	55	53	28.0
50.9	106	54	52	27.5
50.9	57	29	28	14.8
50.9	55	28	27	14.3
50.9	53	27	26	13.8
50.8	500	254	246	129.0
50.8	498	253	245	128.5
50.8	496	252	244	128.0
50.8	494	251	243	127.5
50.8	492	250	242	127.0
50.8	490	249	241	126.5
50.8	488	248	240	126.0
50.8	486	247	239	125.5
50.8	484	246	238	125.0
50.8	482	245	237	124.5
50.8	480	244	236	124.0
50.8	478	243	235	123.5
50.8	476	242	234	123.0
50.8	474	241	233	122.5
50.8	472	240	232	122.0
50.8	465	236	229	119.8
50.8	463	235	228	119.3
50.8	461	234	227	118.8
50.8	459	233	226	118.3
50.8	457	232	225	117.8
50.8	455	231	224	117.3
50.8	453	230	223	116.8
50.8	451	229	222	116.3
50.8	449	228	221	115.8
50.8	447	227	220	115.3
50.8	445	226	219	114.8
50.8	443	225	218	114.3
50.8	441	224	217	113.8
50.8	439	223	216	113.3
50.8	437	222	215	112.8
50.8	435	221	214	112.3
50.8	433	220	213	111.8
50.8	431	219	212	111.3
50.8	429	218	211	110.8
50.8	427	217	210	110.3
50.8	425	216	209	109.8
50.8	423	215	208	109.3
50.8	421	214	207	108.8
50.8	419	213	206	108.3
50.8	417	212	205	107.8
50.8	415	211	204	107.3
50.8	413	210	203	106.8
50.8	400	203	197	103.0
50.8	398	202	196	102.5

%E	M1	M2	DM	M*
50.8	396	201	195	102.0
50.8	394	200	194	101.5
50.8	392	199	193	101.0
50.8	390	198	192	100.5
50.8	388	197	191	100.0
50.8	386	196	190	99.5
50.8	384	195	189	99.0
50.8	382	194	188	98.5
50.8	380	193	187	98.0
50.8	378	192	186	97.5
50.8	376	191	185	97.0
50.8	374	190	184	96.5
50.8	372	189	183	96.0
50.8	370	188	182	95.5
50.8	368	187	181	95.0
50.8	366	186	180	94.5
50.8	364	185	179	94.0
50.8	362	184	178	93.5
50.8	360	183	177	93.0
50.8	358	182	176	92.5
50.8	356	181	175	92.0
50.8	354	180	174	91.5
50.8	333	169	164	85.8
50.8	331	168	163	85.3
50.8	329	167	162	84.8
50.8	327	166	161	84.3
50.8	325	165	160	83.8
50.8	323	164	159	83.3
50.8	321	163	158	82.8
50.8	319	162	157	82.3
50.8	317	161	156	81.8
50.8	315	160	155	81.3
50.8	313	159	154	80.8
50.8	311	158	153	80.3
50.8	309	157	152	79.8
50.8	307	156	151	79.3
50.8	305	155	150	78.8
50.8	303	154	149	78.3
50.8	301	153	148	77.8
50.8	299	152	147	77.3
50.8	297	151	146	76.8
50.8	295	150	145	76.3
50.8	266	135	131	68.5
50.8	264	134	130	68.0
50.8	262	133	129	67.5
50.8	260	132	128	67.0
50.8	258	131	127	66.5
50.8	256	130	126	66.0
50.8	254	129	125	65.5
50.8	252	128	124	65.0

%E	M1	M2	DM	M*
50.8	250	127	123	64.5
50.8	248	126	122	64.0
50.8	246	125	121	63.5
50.8	244	124	120	63.0
50.8	242	123	119	62.5
50.8	240	122	118	62.0
50.8	238	121	117	61.5
50.8	236	120	116	61.0
50.8	199	101	98	51.3
50.8	197	100	97	50.8
50.8	195	99	96	50.3
50.8	193	98	95	49.8
50.8	191	97	94	49.3
50.8	189	96	93	48.8
50.8	187	95	92	48.3
50.8	185	94	91	47.8
50.8	183	93	90	47.3
50.8	181	92	89	46.8
50.8	179	91	88	46.3
50.8	177	90	87	45.8
50.8	132	67	65	34.0
50.8	130	66	64	33.5
50.8	128	65	63	33.0
50.8	126	64	62	32.5
50.8	124	63	61	32.0
50.8	122	62	60	31.5
50.8	120	61	59	31.0
50.8	118	60	58	30.5
50.8	65	33	32	16.8
50.8	63	32	31	16.3
50.8	61	31	30	15.8
50.8	59	30	29	15.3
50.7	499	253	246	128.3
50.7	497	252	245	127.8
50.7	495	251	244	127.3
50.7	493	250	243	126.8
50.7	491	249	242	126.3
50.7	489	248	241	125.8
50.7	487	247	240	125.3
50.7	485	246	239	124.8
50.7	483	245	238	124.3
50.7	481	244	237	123.8
50.7	479	243	236	123.3
50.7	477	242	235	122.8
50.7	475	241	234	122.3
50.7	473	240	233	121.8
50.7	471	239	232	121.3
50.7	469	238	231	120.8
50.7	467	237	230	120.3
50.7	460	233	227	118.0

%E	M1	M2	DM	M*
50.7	458	232	226	117.5
50.7	456	231	225	117.0
50.7	454	230	224	116.5
50.7	452	229	223	116.0
50.7	450	228	222	115.5
50.7	448	227	221	115.0
50.7	446	226	220	114.5
50.7	444	225	219	114.0
50.7	442	224	218	113.5
50.7	440	223	217	113.0
50.7	438	222	216	112.5
50.7	436	221	215	112.0
50.7	434	220	214	111.5
50.7	432	219	213	111.0
50.7	430	218	212	110.5
50.7	428	217	211	110.0
50.7	426	216	210	109.5
50.7	424	215	209	109.0
50.7	422	214	208	108.5
50.7	420	213	207	108.0
50.7	418	212	206	107.5
50.7	416	211	205	107.0
50.7	414	210	204	106.5
50.7	412	209	203	106.0
50.7	410	208	202	105.5
50.7	408	207	201	105.0
50.7	406	206	200	104.5
50.7	404	205	199	104.0
50.7	402	204	198	103.5
50.7	383	194	189	98.3
50.7	381	193	188	97.8
50.7	379	192	187	97.3
50.7	377	191	186	96.8
50.7	375	190	185	96.3
50.7	373	189	184	95.8
50.7	371	188	183	95.3
50.7	369	187	182	94.8
50.7	367	186	181	94.3
50.7	365	185	180	93.8
50.7	363	184	179	93.3
50.7	361	183	178	92.8
50.7	359	182	177	92.3
50.7	357	181	176	91.8
50.7	355	180	175	91.3
50.7	353	179	174	90.8
50.7	351	178	173	90.3
50.7	349	177	172	89.8
50.7	347	176	171	89.3
50.7	345	175	170	88.8
50.7	343	174	169	88.3

%E	M1	M2	DM	M*
50.7	341	173	168	87.8
50.7	339	172	167	87.3
50.7	337	171	166	86.8
50.7	335	170	165	86.3
50.7	306	155	151	78.5
50.7	304	154	150	78.0
50.7	302	153	149	77.5
50.7	300	152	148	77.0
50.7	298	151	147	76.5
50.7	296	150	146	76.0
50.7	294	149	145	75.5
50.7	292	148	144	75.0
50.7	290	147	143	74.5
50.7	288	146	142	74.0
50.7	286	145	141	73.5
50.7	284	144	140	73.0
50.7	282	143	139	72.5
50.7	280	142	138	72.0
50.7	278	141	137	71.5
50.7	276	140	136	71.0
50.7	274	139	135	70.5
50.7	272	138	134	70.0
50.7	270	137	133	69.5
50.7	268	136	132	69.0
50.7	229	116	113	58.8
50.7	227	115	112	58.3
50.7	225	114	111	57.8
50.7	223	113	110	57.3
50.7	221	112	109	56.8
50.7	219	111	108	56.3
50.7	217	110	107	55.8
50.7	215	109	106	55.3
50.7	213	108	105	54.8
50.7	211	107	104	54.3
50.7	209	106	103	53.8
50.7	207	105	102	53.3
50.7	205	104	101	52.8
50.7	203	103	100	52.3
50.7	201	102	99	51.8
50.7	152	77	75	39.0
50.7	150	76	74	38.5
50.7	148	75	73	38.0
50.7	146	74	72	37.5
50.7	144	73	71	37.0
50.7	142	72	70	36.5
50.7	140	71	69	36.0
50.7	138	70	68	35.5
50.7	136	69	67	35.0
50.7	134	68	66	34.5
50.7	75	38	37	19.3

%E	M1	M2	DM	M*
50.7	73	37	36	18.8
50.7	71	36	35	18.3
50.7	69	35	34	17.8
50.7	67	34	33	17.3
50.6	500	253	247	128.0
50.6	498	252	246	127.5
50.6	496	251	245	127.0
50.6	494	250	244	126.5
50.6	492	249	243	126.0
50.6	490	248	242	125.5
50.6	488	247	241	125.0
50.6	486	246	240	124.5
50.6	484	245	239	124.0
50.6	482	244	238	123.5
50.6	480	243	237	123.0
50.6	478	242	236	122.5
50.6	476	241	235	122.0
50.6	474	240	234	121.5
50.6	472	239	233	121.0
50.6	470	238	232	120.5
50.6	468	237	231	120.0
50.6	466	236	230	119.5
50.6	464	235	229	119.0
50.6	462	234	228	118.5
50.6	453	229	224	115.8
50.6	451	228	223	115.3
50.6	449	227	222	114.8
50.6	447	226	221	114.3
50.6	445	225	220	113.8
50.6	443	224	219	113.3
50.6	441	223	218	112.8
50.6	439	222	217	112.3
50.6	437	221	216	111.8
50.6	435	220	215	111.3
50.6	433	219	214	110.8
50.6	431	218	213	110.3
50.6	429	217	212	109.8
50.6	427	216	211	109.3
50.6	425	215	210	108.8
50.6	423	214	209	108.3
50.6	421	213	208	107.8
50.6	419	212	207	107.3
50.6	417	211	206	106.8
50.6	415	210	205	106.3
50.6	413	209	204	105.8
50.6	411	208	203	105.3
50.6	409	207	202	104.8
50.6	407	206	201	104.3
50.6	405	205	200	103.8
50.6	403	204	199	103.3

%E	M1	M2	DM	M*
50.6	401	203	198	102.8
50.6	399	202	197	102.3
50.6	397	201	196	101.8
50.6	395	200	195	101.3
50.6	393	199	194	100.8
50.6	391	198	193	100.3
50.6	389	197	192	99.8
50.6	387	196	191	99.3
50.6	385	195	190	98.8
50.6	362	183	179	92.5
50.6	360	182	178	92.0
50.6	358	181	177	91.5
50.6	356	180	176	91.0
50.6	354	179	175	90.5
50.6	352	178	174	90.0
50.6	350	177	173	89.5
50.6	348	176	172	89.0
50.6	346	175	171	88.5
50.6	344	174	170	88.0
50.6	342	173	169	87.5
50.6	340	172	168	87.0
50.6	338	171	167	86.5
50.6	336	170	166	86.0
50.6	334	169	165	85.5
50.6	332	168	164	85.0
50.6	330	167	163	84.5
50.6	328	166	162	84.0
50.6	326	165	161	83.5
50.6	324	164	160	83.0
50.6	322	163	159	82.5
50.6	320	162	158	82.0
50.6	318	161	157	81.5
50.6	316	160	156	81.0
50.6	314	159	155	80.5
50.6	312	158	154	80.0
50.6	310	157	153	79.5
50.6	308	156	152	79.0
50.6	271	137	134	69.3
50.6	269	136	133	68.8
50.6	267	135	132	68.3
50.6	265	134	131	67.8
50.6	263	133	130	67.3
50.6	261	132	129	66.8
50.6	259	131	128	66.3
50.6	257	130	127	65.8
50.6	255	129	126	65.3
50.6	253	128	125	64.8
50.6	251	127	124	64.3
50.6	249	126	123	63.8
50.6	247	125	122	63.3

%E	M1	M2	DM	M*
50.6	245	124	121	62.8
50.6	243	123	120	62.3
50.6	241	122	119	61.8
50.6	239	121	118	61.3
50.6	237	120	117	60.8
50.6	235	119	116	60.3
50.6	233	118	115	59.8
50.6	231	117	114	59.3
50.6	180	91	89	46.0
50.6	178	90	88	45.5
50.6	176	89	87	45.0
50.6	174	88	86	44.5
50.6	172	87	85	44.0
50.6	170	86	84	43.5
50.6	168	85	83	43.0
50.6	166	84	82	42.5
50.6	164	83	81	42.0
50.6	162	82	80	41.5
50.6	160	81	79	41.0
50.6	158	80	78	40.5
50.6	156	79	77	40.0
50.6	154	78	76	39.5
50.6	89	45	44	22.8
50.6	87	44	43	22.3
50.6	85	43	42	21.8
50.6	83	42	41	21.3
50.6	81	41	40	20.8
50.6	79	40	39	20.3
50.6	77	39	38	19.8
50.5	499	252	247	127.3
50.5	497	251	246	126.8
50.5	495	250	245	126.3
50.5	493	249	244	125.8
50.5	491	248	243	125.3
50.5	489	247	242	124.8
50.5	487	246	241	124.3
50.5	485	245	240	123.8
50.5	483	244	239	123.3
50.5	481	243	238	122.8
50.5	479	242	237	122.3
50.5	477	241	236	121.8
50.5	475	240	235	121.3
50.5	473	239	234	120.8
50.5	471	238	233	120.3
50.5	469	237	232	119.8
50.5	467	236	231	119.3
50.5	465	235	230	118.8
50.5	463	234	229	118.3
50.5	461	233	228	117.8
50.5	459	232	227	117.3

%E	M1	M2	DM	M*
50.5	457	231	226	116.8
50.5	455	230	225	116.3
50.5	444	224	220	113.0
50.5	442	223	219	112.5
50.5	440	222	218	112.0
50.5	438	221	217	111.5
50.5	436	220	216	111.0
50.5	434	219	215	110.5
50.5	432	218	214	110.0
50.5	430	217	213	109.5
50.5	428	216	212	109.0
50.5	426	215	211	108.5
50.5	424	214	210	108.0
50.5	422	213	209	107.5
50.5	420	212	208	107.0
50.5	418	211	207	106.5
50.5	416	210	206	106.0
50.5	414	209	205	105.5
50.5	412	208	204	105.0
50.5	410	207	203	104.5
50.5	408	206	202	104.0
50.5	406	205	201	103.5
50.5	404	204	200	103.0
50.5	402	203	199	102.5
50.5	400	202	198	102.0
50.5	398	201	197	101.5
50.5	396	200	196	101.0
50.5	394	199	195	100.5
50.5	392	198	194	100.0
50.5	390	197	193	99.5
50.5	388	196	192	99.0
50.5	386	195	191	98.5
50.5	384	194	190	98.0
50.5	382	193	189	97.5
50.5	380	192	188	97.0
50.5	378	191	187	96.5
50.5	376	190	186	96.0
50.5	374	189	185	95.5
50.5	372	188	184	95.0
50.5	370	187	183	94.5
50.5	368	186	182	94.0
50.5	366	185	181	93.5
50.5	364	184	180	93.0
50.5	333	168	165	84.8
50.5	331	167	164	84.3
50.5	329	166	163	83.8
50.5	327	165	162	83.3
50.5	325	164	161	82.8
50.5	323	163	160	82.3
50.5	321	162	159	81.8

%E	M1	M2	DM	M*
50.5	319	161	158	81.3
50.5	317	160	157	80.8
50.5	315	159	156	80.3
50.5	313	158	155	79.8
50.5	311	157	154	79.3
50.5	309	156	153	78.8
50.5	307	155	152	78.3
50.5	305	154	151	77.8
50.5	303	153	150	77.3
50.5	301	152	149	76.8
50.5	299	151	148	76.3
50.5	297	150	147	75.8
50.5	295	149	146	75.3
50.5	293	148	145	74.8
50.5	291	147	144	74.3
50.5	289	146	143	73.8
50.5	287	145	142	73.3
50.5	285	144	141	72.8
50.5	283	143	140	72.3
50.5	281	142	139	71.8
50.5	279	141	138	71.3
50.5	277	140	137	70.8
50.5	275	139	136	70.3
50.5	273	138	135	69.8
50.5	222	112	110	56.5
50.5	220	111	109	56.0
50.5	218	110	108	55.5
50.5	216	109	107	55.0
50.5	214	108	106	54.5
50.5	212	107	105	54.0
50.5	210	106	104	53.5
50.5	208	105	103	53.0
50.5	206	104	102	52.5
50.5	204	103	101	52.0
50.5	202	102	100	51.5
50.5	200	101	99	51.0
50.5	198	100	98	50.5
50.5	196	99	97	50.0
50.5	194	98	96	49.5
50.5	192	97	95	49.0
50.5	190	96	94	48.5
50.5	188	95	93	48.0
50.5	186	94	92	47.5
50.5	184	93	91	47.0
50.5	182	92	90	46.5
50.5	111	56	55	28.3
50.5	109	55	54	27.8
50.5	107	54	53	27.3
50.5	105	53	52	26.8
50.5	103	52	51	26.3

%E	M1	M2	DM	M*
50.5	101	51	50	25.8
50.5	99	50	49	25.3
50.5	97	49	48	24.8
50.5	95	48	47	24.3
50.5	93	47	46	23.8
50.5	91	46	45	23.3
50.4	500	252	248	127.0
50.4	498	251	247	126.5
50.4	496	250	246	126.0
50.4	494	249	245	125.5
50.4	492	248	244	125.0
50.4	490	247	243	124.5
50.4	488	246	242	124.0
50.4	486	245	241	123.5
50.4	484	244	240	123.0
50.4	482	243	239	122.5
50.4	480	242	238	122.0
50.4	478	241	237	121.5
50.4	476	240	236	121.0
50.4	474	239	235	120.5
50.4	472	238	234	120.0
50.4	470	237	233	119.5
50.4	468	236	232	119.0
50.4	466	235	231	118.5
50.4	464	234	230	118.0
50.4	462	233	229	117.5
50.4	460	232	228	117.0
50.4	458	231	227	116.5
50.4	456	230	226	116.0
50.4	454	229	225	115.5
50.4	452	228	224	115.0
50.4	450	227	223	114.5
50.4	448	226	222	114.0
50.4	446	225	221	113.5
50.4	429	216	213	108.8
50.4	427	215	212	108.3
50.4	425	214	211	107.8
50.4	423	213	210	107.3
50.4	421	212	209	106.8
50.4	419	211	208	106.3
50.4	417	210	207	105.8
50.4	415	209	206	105.3
50.4	413	208	205	104.8
50.4	411	207	204	104.3
50.4	409	206	203	103.8
50.4	407	205	202	103.3
50.4	405	204	201	102.8
50.4	403	203	200	102.3
50.4	401	202	199	101.8
50.4	399	201	198	101.3

%E	M1	M2	DM	M*
50.4	397	200	197	100.8
50.4	395	199	196	100.3
50.4	393	198	195	99.8
50.4	391	197	194	99.3
50.4	389	196	193	98.8
50.4	387	195	192	98.3
50.4	385	194	191	97.8
50.4	383	193	190	97.3
50.4	381	192	189	96.8
50.4	379	191	188	96.3
50.4	377	190	187	95.8
50.4	375	189	186	95.3
50.4	373	188	185	94.8
50.4	371	187	184	94.3
50.4	369	186	183	93.8
50.4	367	185	182	93.3
50.4	365	184	181	92.8
50.4	363	183	180	92.3
50.4	361	182	179	91.8
50.4	359	181	178	91.3
50.4	357	180	177	90.8
50.4	355	179	176	90.3
50.4	353	178	175	89.8
50.4	351	177	174	89.3
50.4	349	176	173	88.8
50.4	347	175	172	88.3
50.4	345	174	171	87.8
50.4	343	173	170	87.3
50.4	341	172	169	86.8
50.4	339	171	168	86.3
50.4	337	170	167	85.8
50.4	335	169	166	85.3
50.4	286	144	142	72.5
50.4	284	143	141	72.0
50.4	282	142	140	71.5
50.4	280	141	139	71.0
50.4	278	140	138	70.5
50.4	276	139	137	70.0
50.4	274	138	136	69.5
50.4	272	137	135	69.0
50.4	270	136	134	68.5
50.4	268	135	133	68.0
50.4	266	134	132	67.5
50.4	264	133	131	67.0
50.4	262	132	130	66.5
50.4	260	131	129	66.0
50.4	258	130	128	65.5
50.4	256	129	127	65.0
50.4	254	128	126	64.5
50.4	252	127	125	64.0

%E	M1	M2	DM	M*
50.4	250	126	124	63.5
50.4	248	125	123	63.0
50.4	246	124	122	62.5
50.4	244	123	121	62.0
50.4	242	122	120	61.5
50.4	240	121	119	61.0
50.4	238	120	118	60.5
50.4	236	119	117	60.0
50.4	234	118	116	59.5
50.4	232	117	115	59.0
50.4	230	116	114	58.5
50.4	228	115	113	58.0
50.4	226	114	112	57.5
50.4	224	113	111	57.0
50.4	143	72	71	36.3
50.4	141	71	70	35.8
50.4	139	70	69	35.3
50.4	137	69	68	34.8
50.4	135	68	67	34.3
50.4	133	67	66	33.8
50.4	131	66	65	33.3
50.4	129	65	64	32.8
50.4	127	64	63	32.3
50.4	125	63	62	31.8
50.4	123	62	61	31.3
50.4	121	61	60	30.8
50.4	119	60	59	30.3
50.4	117	59	58	29.8
50.4	115	58	57	29.3
50.4	113	57	56	28.8
50.3	499	251	248	126.3
50.3	497	250	247	125.8
50.3	495	249	246	125.3
50.3	493	248	245	124.8
50.3	491	247	244	124.3
50.3	489	246	243	123.8
50.3	487	245	242	123.3
50.3	485	244	241	122.8
50.3	483	243	240	122.3
50.3	481	242	239	121.8
50.3	479	241	238	121.3
50.3	477	240	237	120.8
50.3	475	239	236	120.3
50.3	473	238	235	119.8
50.3	471	237	234	119.3
50.3	469	236	233	118.8
50.3	467	235	232	118.3
50.3	465	234	231	117.8
50.3	463	233	230	117.3
50.3	461	232	229	116.8

%E	M1	M2	DM	M*
50.3	459	231	228	116.3
50.3	457	230	227	115.8
50.3	455	229	226	115.3
50.3	453	228	225	114.8
50.3	451	227	224	114.3
50.3	449	226	223	113.8
50.3	447	225	222	113.3
50.3	445	224	221	112.8
50.3	443	223	220	112.3
50.3	441	222	219	111.8
50.3	439	221	218	111.3
50.3	437	220	217	110.8
50.3	435	219	216	110.3
50.3	433	218	215	109.8
50.3	431	217	214	109.3
50.3	400	201	199	101.0
50.3	398	200	198	100.5
50.3	396	199	197	100.0
50.3	394	198	196	99.5
50.3	392	197	195	99.0
50.3	390	196	194	98.5
50.3	388	195	193	98.0
50.3	386	194	192	97.5
50.3	384	193	191	97.0
50.3	382	192	190	96.5
50.3	380	191	189	96.0
50.3	378	190	188	95.5
50.3	376	189	187	95.0
50.3	374	188	186	94.5
50.3	372	187	185	94.0
50.3	370	186	184	93.5
50.3	368	185	183	93.0
50.3	366	184	182	92.5
50.3	364	183	181	92.0
50.3	362	182	180	91.5
50.3	360	181	179	91.0
50.3	358	180	178	90.5
50.3	356	179	177	90.0
50.3	354	178	176	89.5
50.3	352	177	175	89.0
50.3	350	176	174	88.5
50.3	348	175	173	88.0
50.3	346	174	172	87.5
50.3	344	173	171	87.0
50.3	342	172	170	86.5
50.3	340	171	169	86.0
50.3	338	170	168	85.5
50.3	336	169	167	85.0
50.3	334	168	166	84.5
50.3	332	167	165	84.0

%E	M1	M2	DM	M*
50.3	330	166	164	83.5
50.3	328	165	163	83.0
50.3	326	164	162	82.5
50.3	324	163	161	82.0
50.3	322	162	160	81.5
50.3	320	161	159	81.0
50.3	318	160	158	80.5
50.3	316	159	157	80.0
50.3	314	158	156	79.5
50.3	312	157	155	79.0
50.3	310	156	154	78.5
50.3	308	155	153	78.0
50.3	306	154	152	77.5
50.3	304	153	151	77.0
50.3	302	152	150	76.5
50.3	300	151	149	76.0
50.3	298	150	148	75.5
50.3	296	149	147	75.0
50.3	294	148	146	74.5
50.3	292	147	145	74.0
50.3	290	146	144	73.5
50.3	288	145	143	73.0
50.3	199	100	99	50.3
50.3	197	99	98	49.8
50.3	195	98	97	49.3
50.3	193	97	96	48.8
50.3	191	96	95	48.3
50.3	189	95	94	47.8
50.3	187	94	93	47.3
50.3	185	93	92	46.8
50.3	183	92	91	46.3
50.3	181	91	90	45.8
50.3	179	90	89	45.3
50.3	177	89	88	44.8
50.3	175	88	87	44.3
50.3	173	87	86	43.8
50.3	171	86	85	43.3
50.3	169	85	84	42.8
50.3	167	84	83	42.3
50.3	165	83	82	41.8
50.3	163	82	81	41.3
50.3	161	81	80	40.8
50.3	159	80	79	40.3
50.3	157	79	78	39.8
50.3	155	78	77	39.3
50.3	153	77	76	38.8
50.3	151	76	75	38.3
50.3	149	75	74	37.8
50.3	147	74	73	37.3
50.3	145	73	72	36.8

%E	M1	M2	DM	M*
50.2	500	251	249	126.0
50.2	498	250	248	125.5
50.2	496	249	247	125.0
50.2	494	248	246	124.5
50.2	492	247	245	124.0
50.2	490	246	244	123.5
50.2	488	245	243	123.0
50.2	486	244	242	122.5
50.2	484	243	241	122.0
50.2	482	242	240	121.5
50.2	480	241	239	121.0
50.2	478	240	238	120.5
50.2	476	239	237	120.0
50.2	474	238	236	119.5
50.2	472	237	235	119.0
50.2	470	236	234	118.5
50.2	468	235	233	118.0
50.2	466	234	232	117.5
50.2	464	233	231	117.0
50.2	462	232	230	116.5
50.2	460	231	229	116.0
50.2	458	230	228	115.5
50.2	456	229	227	115.0
50.2	454	228	226	114.5
50.2	452	227	225	114.0
50.2	450	226	224	113.5
50.2	448	225	223	113.0
50.2	446	224	222	112.5
50.2	444	223	221	112.0
50.2	442	222	220	111.5
50.2	440	221	219	111.0
50.2	438	220	218	110.5
50.2	436	219	217	110.0
50.2	434	218	216	109.5
50.2	432	217	215	109.0
50.2	430	216	214	108.5
50.2	428	215	213	108.0
50.2	426	214	212	107.5
50.2	424	213	211	107.0
50.2	422	212	210	106.5
50.2	420	211	209	106.0
50.2	418	210	208	105.5
50.2	416	209	207	105.0
50.2	414	208	206	104.5
50.2	412	207	205	104.0
50.2	410	206	204	103.5
50.2	408	205	203	103.0
50.2	406	204	202	102.5
50.2	404	203	201	102.0
50.2	402	202	200	101.5

%E	M1	M2	DM	M*
50.2	333	167	166	83.8
50.2	331	166	165	83.3
50.2	329	165	164	82.8
50.2	327	164	163	82.3
50.2	325	163	162	81.8
50.2	323	162	161	81.3
50.2	321	161	160	80.8
50.2	319	160	159	80.3
50.2	317	159	158	79.8
50.2	315	158	157	79.3
50.2	313	157	156	78.8
50.2	311	156	155	78.3
50.2	309	155	154	77.8
50.2	307	154	153	77.3
50.2	305	153	152	76.8
50.2	303	152	151	76.3
50.2	301	151	150	75.8
50.2	299	150	149	75.3
50.2	297	149	148	74.8
50.2	295	148	147	74.3
50.2	293	147	146	73.8
50.2	291	146	145	73.3
50.2	289	145	144	72.8
50.2	287	144	143	72.3
50.2	285	143	142	71.8
50.2	283	142	141	71.3
50.2	281	141	140	70.8
50.2	279	140	139	70.3
50.2	277	139	138	69.8
50.2	275	138	137	69.3
50.2	273	137	136	68.8
50.2	271	136	135	68.3
50.2	269	135	134	67.8
50.2	267	134	133	67.3
50.2	265	133	132	66.8
50.2	263	132	131	66.3
50.2	261	131	130	65.8
50.2	259	130	129	65.3
50.2	257	129	128	64.8
50.2	255	128	127	64.3
50.2	253	127	126	63.8
50.2	251	126	125	63.3
50.2	249	125	124	62.8
50.2	247	124	123	62.3
50.2	245	123	122	61.8
50.2	243	122	121	61.3
50.2	241	121	120	60.8
50.2	239	120	119	60.3
50.2	237	119	118	59.8
50.2	235	118	117	59.3

%E	M1	M2	DM	M*
50.2	233	117	116	58.8
50.2	231	116	115	58.3
50.2	229	115	114	57.8
50.2	227	114	113	57.3
50.2	225	113	112	56.8
50.2	223	112	111	56.3
50.2	221	111	110	55.8
50.2	219	110	109	55.3
50.2	217	109	108	54.8
50.2	215	108	107	54.3
50.2	213	107	106	53.8
50.2	211	106	105	53.3
50.2	209	105	104	52.8
50.2	207	104	103	52.3
50.2	205	103	102	51.8
50.2	203	102	101	51.3
50.2	201	101	100	50.8
50.1	499	250	249	125.3
50.1	497	249	248	124.8
50.1	495	248	247	124.3
50.1	493	247	246	123.8
50.1	491	246	245	123.3
50.1	489	245	244	122.8
50.1	487	244	243	122.3
50.1	485	243	242	121.8
50.1	483	242	241	121.3
50.1	481	241	240	120.8
50.1	479	240	239	120.3
50.1	477	239	238	119.8
50.1	475	238	237	119.3
50.1	473	237	236	118.8
50.1	471	236	235	118.3
50.1	469	235	234	117.8
50.1	467	234	233	117.3
50.1	465	233	232	116.8
50.1	463	232	231	116.3
50.1	461	231	230	115.8
50.1	459	230	229	115.3
50.1	457	229	228	114.8
50.1	455	228	227	114.3
50.1	453	227	226	113.8
50.1	451	226	225	113.3
50.1	449	225	224	112.8
50.1	447	224	223	112.3
50.1	445	223	222	111.8
50.1	443	222	221	111.3
50.1	441	221	220	110.8
50.1	439	220	219	110.3
50.1	437	219	218	109.8
50.1	435	218	217	109.3

%E	M1	M2	DM	M*
50.1	433	217	216	108.8
50.1	431	216	215	108.3
50.1	429	215	214	107.8
50.1	427	214	213	107.3
50.1	425	213	212	106.8
50.1	423	212	211	106.3
50.1	421	211	210	105.8
50.1	419	210	209	105.3
50.1	417	209	208	104.8
50.1	415	208	207	104.3
50.1	413	207	206	103.8
50.1	411	206	205	103.3
50.1	409	205	204	102.8
50.1	407	204	203	102.3
50.1	405	203	202	101.8
50.1	403	202	201	101.3
50.1	401	201	200	100.8
50.1	399	200	199	100.3
50.1	397	199	198	99.8
50.1	395	198	197	99.3
50.1	393	197	196	98.8
50.1	391	196	195	98.3
50.1	389	195	194	97.8
50.1	387	194	193	97.3
50.1	385	193	192	96.8
50.1	383	192	191	96.3
50.1	381	191	190	95.8
50.1	379	190	189	95.3
50.1	377	189	188	94.8
50.1	375	188	187	94.3
50.1	373	187	186	93.8
50.1	371	186	185	93.3
50.1	369	185	184	92.8
50.1	367	184	183	92.3
50.1	365	183	182	91.8
50.1	363	182	181	91.3
50.1	361	181	180	90.8
50.1	359	180	179	90.3
50.1	357	179	178	89.8
50.1	355	178	177	89.3
50.1	353	177	176	88.8
50.1	351	176	175	88.3
50.1	349	175	174	87.8
50.1	347	174	173	87.3
50.1	345	173	172	86.8
50.1	343	172	171	86.3
50.1	341	171	170	85.8
50.1	339	170	169	85.3
50.1	337	169	168	84.8
50.1	335	168	167	84.3

%E	M1	M2	DM	M*
50.0	500	250	250	125.0
50.0	498	249	249	124.5
50.0	496	248	248	124.0
50.0	494	247	247	123.5
50.0	492	246	246	123.0
50.0	490	245	245	122.5
50.0	488	244	244	122.0
50.0	486	243	243	121.5
50.0	484	242	242	121.0
50.0	482	241	241	120.5
50.0	480	240	240	120.0
50.0	478	239	239	119.5
50.0	476	238	238	119.0
50.0	474	237	237	118.5
50.0	472	236	236	118.0
50.0	470	235	235	117.5
50.0	468	234	234	117.0
50.0	466	233	233	116.5
50.0	464	232	232	116.0
50.0	462	231	231	115.5
50.0	460	230	230	115.0
50.0	458	229	229	114.5
50.0	456	228	228	114.0
50.0	454	227	227	113.5
50.0	452	226	226	113.0
50.0	450	225	225	112.5
50.0	448	224	224	112.0
50.0	446	223	223	111.5
50.0	444	222	222	111.0
50.0	442	221	221	110.5
50.0	440	220	220	110.0
50.0	438	219	219	109.5
50.0	436	218	218	109.0
50.0	434	217	217	108.5
50.0	432	216	216	108.0
50.0	430	215	215	107.5
50.0	428	214	214	107.0
50.0	426	213	213	106.5
50.0	424	212	212	106.0
50.0	422	211	211	105.5
50.0	420	210	210	105.0
50.0	418	209	209	104.5
50.0	416	208	208	104.0
50.0	414	207	207	103.5
50.0	412	206	206	103.0
50.0	410	205	205	102.5
50.0	408	204	204	102.0
50.0	406	203	203	101.5
50.0	404	202	202	101.0
50.0	402	201	201	100.5

%E	M1	M2	DM	M*
50.0	400	200	200	100.0
50.0	398	199	199	99.5
50.0	396	198	198	99.0
50.0	394	197	197	98.5
50.0	392	196	196	98.0
50.0	390	195	195	97.5
50.0	388	194	194	97.0
50.0	386	193	193	96.5
50.0	384	192	192	96.0
50.0	382	191	191	95.5
50.0	380	190	190	95.0
50.0	378	189	189	94.5
50.0	376	188	188	94.0
50.0	374	187	187	93.5
50.0	372	186	186	93.0
50.0	370	185	185	92.5
50.0	368	184	184	92.0
50.0	366	183	183	91.5
50.0	364	182	182	91.0
50.0	362	181	181	90.5
50.0	360	180	180	90.0
50.0	358	179	179	89.5
50.0	356	178	178	89.0
50.0	354	177	177	88.5
50.0	352	176	176	88.0
50.0	350	175	175	87.5
50.0	348	174	174	87.0
50.0	346	173	173	86.5
50.0	344	172	172	86.0
50.0	342	171	171	85.5
50.0	340	170	170	85.0
50.0	338	169	169	84.5
50.0	336	168	168	84.0
50.0	334	167	167	83.5
50.0	332	166	166	83.0
50.0	330	165	165	82.5
50.0	328	164	164	82.0
50.0	326	163	163	81.5
50.0	324	162	162	81.0
50.0	322	161	161	80.5
50.0	320	160	160	80.0
50.0	318	159	159	79.5
50.0	316	158	158	79.0
50.0	314	157	157	78.5
50.0	312	156	156	78.0
50.0	310	155	155	77.5
50.0	308	154	154	77.0
50.0	306	153	153	76.5
50.0	304	152	152	76.0
50.0	302	151	151	75.5

%E	M1	M2	DM	M*
50.0	300	150	150	75.0
50.0	298	149	149	74.5
50.0	296	148	148	74.0
50.0	294	147	147	73.5
50.0	292	146	146	73.0
50.0	290	145	145	72.5
50.0	288	144	144	72.0
50.0	286	143	143	71.5
50.0	284	142	142	71.0
50.0	282	141	141	70.5
50.0	280	140	140	70.0
50.0	278	139	139	69.5
50.0	276	138	138	69.0
50.0	274	137	137	68.5
50.0	272	136	136	68.0
50.0	270	135	135	67.5
50.0	268	134	134	67.0
50.0	266	133	133	66.5
50.0	264	132	132	66.0
50.0	262	131	131	65.5
50.0	260	130	130	65.0
50.0	258	129	129	64.5
50.0	256	128	128	64.0
50.0	254	127	127	63.5
50.0	252	126	126	63.0
50.0	250	125	125	62.5
50.0	248	124	124	62.0
50.0	246	123	123	61.5
50.0	244	122	122	61.0
50.0	242	121	121	60.5
50.0	240	120	120	60.0
50.0	238	119	119	59.5
50.0	236	118	118	59.0
50.0	234	117	117	58.5
50.0	232	116	116	58.0
50.0	230	115	115	57.5
50.0	228	114	114	57.0
50.0	226	113	113	56.5
50.0	224	112	112	56.0
50.0	222	111	111	55.5
50.0	220	110	110	55.0
50.0	218	109	109	54.5
50.0	216	108	108	54.0
50.0	214	107	107	53.5
50.0	212	106	106	53.0
50.0	210	105	105	52.5
50.0	208	104	104	52.0
50.0	206	103	103	51.5
50.0	204	102	102	51.0
50.0	202	101	101	50.5

%E	M1	M2	DM	M*
50.0	200	100	100	50.0
50.0	198	99	99	49.5
50.0	196	98	98	49.0
50.0	194	97	97	48.5
50.0	192	96	96	48.0
50.0	190	95	95	47.5
50.0	188	94	94	47.0
50.0	186	93	93	46.5
50.0	184	92	92	46.0
50.0	182	91	91	45.5
50.0	180	90	90	45.0
50.0	178	89	89	44.5
50.0	176	88	88	44.0
50.0	174	87	87	43.5
50.0	172	86	86	43.0
50.0	170	85	85	42.5
50.0	168	84	84	42.0
50.0	166	83	83	41.5
50.0	164	82	82	41.0
50.0	162	81	81	40.5
50.0	160	80	80	40.0
50.0	158	79	79	39.5
50.0	156	78	78	39.0
50.0	154	77	77	38.5
50.0	152	76	76	38.0
50.0	150	75	75	37.5
50.0	148	74	74	37.0
50.0	146	73	73	36.5
50.0	144	72	72	36.0
50.0	142	71	71	35.5
50.0	140	70	70	35.0
50.0	138	69	69	34.5
50.0	136	68	68	34.0
50.0	134	67	67	33.5
50.0	132	66	66	33.0
50.0	130	65	65	32.5
50.0	128	64	64	32.0
50.0	126	63	63	31.5
50.0	124	62	62	31.0
50.0	122	61	61	30.5
50.0	120	60	60	30.0
50.0	118	59	59	29.5
50.0	116	58	58	29.0
50.0	114	57	57	28.5
50.0	112	56	56	28.0
50.0	110	55	55	27.5
50.0	108	54	54	27.0
50.0	106	53	53	26.5
50.0	104	52	52	26.0
50.0	102	51	51	25.5

%E	M1	M2	DM	M*
50.0	100	50	50	25.0
50.0	98	49	49	24.5
50.0	96	48	48	24.0
50.0	94	47	47	23.5
50.0	92	46	46	23.0
50.0	90	45	45	22.5
50.0	88	44	44	22.0
50.0	86	43	43	21.5
50.0	84	42	42	21.0
50.0	82	41	41	20.5
50.0	80	40	40	20.0
50.0	78	39	39	19.5
50.0	76	38	38	19.0
50.0	74	37	37	18.5
50.0	72	36	36	18.0
50.0	70	35	35	17.5
50.0	68	34	34	17.0
50.0	66	33	33	16.5
50.0	64	32	32	16.0
50.0	62	31	31	15.5
50.0	60	30	30	15.0
50.0	58	29	29	14.5
50.0	56	28	28	14.0
50.0	54	27	27	13.5
50.0	52	26	26	13.0
50.0	50	25	25	12.5
50.0	48	24	24	12.0
50.0	46	23	23	11.5
50.0	44	22	22	11.0
50.0	42	21	21	10.5
50.0	40	20	20	10.0
50.0	38	19	19	9.5
50.0	36	18	18	9.0
50.0	34	17	17	8.5
50.0	32	16	16	8.0
50.0	30	15	15	7.5
50.0	28	14	14	7.0
50.0	26	13	13	6.5
50.0	24	12	12	6.0
50.0	22	11	11	5.5
50.0	20	10	10	5.0
50.0	18	9	9	4.5
50.0	16	8	8	4.0
50.0	14	7	7	3.5
50.0	12	6	6	3.0
50.0	10	5	5	2.5
50.0	8	4	4	2.0
50.0	6	3	3	1.5
50.0	4	2	2	1.0
50.0	2	1	1	0.5

%E	M1	M2	DM	M*
49.9	499	249	250	124.3
49.9	497	248	249	123.8
49.9	495	247	248	123.3
49.9	493	246	247	122.8
49.9	491	245	246	122.3
49.9	489	244	245	121.8
49.9	487	243	244	121.3
49.9	485	242	243	120.8
49.9	483	241	242	120.3
49.9	481	240	241	119.8
49.9	479	239	240	119.3
49.9	477	238	239	118.8
49.9	475	237	238	118.3
49.9	473	236	237	117.8
49.9	471	235	236	117.3
49.9	469	234	235	116.8
49.9	467	233	234	116.3
49.9	465	232	233	115.8
49.9	463	231	232	115.3
49.9	461	230	231	114.8
49.9	459	229	230	114.3
49.9	457	228	229	113.8
49.9	455	227	228	113.3
49.9	453	226	227	112.8
49.9	451	225	226	112.3
49.9	449	224	225	111.8
49.9	447	223	224	111.3
49.9	445	222	223	110.8
49.9	443	221	222	110.3
49.9	441	220	221	109.8
49.9	439	219	220	109.3
49.9	437	218	219	108.8
49.9	435	217	218	108.3
49.9	433	216	217	107.8
49.9	431	215	216	107.3
49.9	429	214	215	106.8
49.9	427	213	214	106.3
49.9	425	212	213	105.8
49.9	423	211	212	105.3
49.9	421	210	211	104.8
49.9	419	209	210	104.3
49.9	417	208	209	103.8
49.9	415	207	208	103.3
49.9	413	206	207	102.8
49.9	411	205	206	102.3
49.9	409	204	205	101.8
49.9	407	203	204	101.3
49.9	405	202	203	100.8
49.9	403	201	202	100.3
49.9	401	200	201	99.8

%E	M1	M2	DM	M*
49.9	399	199	200	99.3
49.9	397	198	199	98.8
49.9	395	197	198	98.3
49.9	393	196	197	97.8
49.9	391	195	196	97.3
49.9	389	194	195	96.8
49.9	387	193	194	96.3
49.9	385	192	193	95.8
49.9	383	191	192	95.3
49.9	381	190	191	94.8
49.9	379	189	190	94.3
49.9	377	188	189	93.8
49.9	375	187	188	93.3
49.9	373	186	187	92.8
49.9	371	185	186	92.3
49.9	369	184	185	91.8
49.9	367	183	184	91.3
49.9	365	182	183	90.8
49.9	363	181	182	90.3
49.9	361	180	181	89.8
49.9	359	179	180	89.3
49.9	357	178	179	88.8
49.9	355	177	178	88.3
49.9	353	176	177	87.8
49.9	351	175	176	87.3
49.9	349	174	175	86.8
49.9	347	173	174	86.3
49.9	345	172	173	85.8
49.9	343	171	172	85.3
49.9	341	170	171	84.8
49.9	339	169	170	84.3
49.9	337	168	169	83.8
49.9	335	167	168	83.3
49.9	333	166	167	82.8
49.8	500	249	251	124.0
49.8	498	248	250	123.5
49.8	496	247	249	123.0
49.8	494	246	248	122.5
49.8	492	245	247	122.0
49.8	490	244	246	121.5
49.8	488	243	245	121.0
49.8	486	242	244	120.5
49.8	484	241	243	120.0
49.8	482	240	242	119.5
49.8	480	239	241	119.0
49.8	478	238	240	118.5
49.8	476	237	239	118.0
49.8	474	236	238	117.5
49.8	472	235	237	117.0
49.8	470	234	236	116.5

%E	M1	M2	DM	M*
49.8	468	233	235	116.0
49.8	466	232	234	115.5
49.8	464	231	233	115.0
49.8	462	230	232	114.5
49.8	460	229	231	114.0
49.8	458	228	230	113.5
49.8	456	227	229	113.0
49.8	454	226	228	112.5
49.8	452	225	227	112.0
49.8	450	224	226	111.5
49.8	448	223	225	111.0
49.8	446	222	224	110.5
49.8	444	221	223	110.0
49.8	442	220	222	109.5
49.8	440	219	221	109.0
49.8	438	218	220	108.5
49.8	436	217	219	108.0
49.8	434	216	218	107.5
49.8	432	215	217	107.0
49.8	430	214	216	106.5
49.8	428	213	215	106.0
49.8	426	212	214	105.5
49.8	424	211	213	105.0
49.8	422	210	212	104.5
49.8	420	209	211	104.0
49.8	418	208	210	103.5
49.8	416	207	209	103.0
49.8	414	206	208	102.5
49.8	412	205	207	102.0
49.8	410	204	206	101.5
49.8	408	203	205	101.0
49.8	406	202	204	100.5
49.8	404	201	203	100.0
49.8	402	200	202	99.5
49.8	400	199	201	99.0
49.8	331	165	166	82.3
49.8	329	164	165	81.8
49.8	327	163	164	81.3
49.8	325	162	163	80.8
49.8	323	161	162	80.3
49.8	321	160	161	79.8
49.8	319	159	160	79.3
49.8	317	158	159	78.8
49.8	315	157	158	78.3
49.8	313	156	157	77.8
49.8	311	155	156	77.3
49.8	309	154	155	76.8
49.8	307	153	154	76.3
49.8	305	152	153	75.8
49.8	303	151	152	75.3

%E	M1	M2	DM	M*
49.8	301	150	151	74.8
49.8	299	149	150	74.3
49.8	297	148	149	73.8
49.8	295	147	148	73.3
49.8	293	146	147	72.8
49.8	291	145	146	72.3
49.8	289	144	145	71.8
49.8	287	143	144	71.3
49.8	285	142	143	70.8
49.8	283	141	142	70.3
49.8	281	140	141	69.8
49.8	279	139	140	69.3
49.8	277	138	139	68.8
49.8	275	137	138	68.3
49.8	273	136	137	67.8
49.8	271	135	136	67.3
49.8	269	134	135	66.8
49.8	267	133	134	66.3
49.8	265	132	133	65.8
49.8	263	131	132	65.3
49.8	261	130	131	64.8
49.8	259	129	130	64.3
49.8	257	128	129	63.8
49.8	255	127	128	63.3
49.8	253	126	127	62.8
49.8	251	125	126	62.3
49.8	249	124	125	61.8
49.8	247	123	124	61.3
49.8	245	122	123	60.8
49.8	243	121	122	60.3
49.8	241	120	121	59.8
49.8	239	119	120	59.3
49.8	237	118	119	58.8
49.8	235	117	118	58.3
49.8	233	116	117	57.8
49.8	231	115	116	57.3
49.8	229	114	115	56.8
49.8	227	113	114	56.3
49.8	225	112	113	55.8
49.8	223	111	112	55.3
49.8	221	110	111	54.8
49.8	219	109	110	54.3
49.8	217	108	109	53.8
49.8	215	107	108	53.3
49.8	213	106	107	52.8
49.8	211	105	106	52.3
49.8	209	104	105	51.8
49.8	207	103	104	51.3
49.8	205	102	103	50.8
49.8	203	101	102	50.3

%E	M1	M2	DM	M*
49.8	201	100	101	49.8
49.7	499	248	251	123.3
49.7	497	247	250	122.8
49.7	495	246	249	122.3
49.7	493	245	248	121.8
49.7	491	244	247	121.3
49.7	489	243	246	120.8
49.7	487	242	245	120.3
49.7	485	241	244	119.8
49.7	483	240	243	119.3
49.7	481	239	242	118.8
49.7	479	238	241	118.3
49.7	477	237	240	117.8
49.7	475	236	239	117.3
49.7	473	235	238	116.8
49.7	471	234	237	116.3
49.7	469	233	236	115.8
49.7	467	232	235	115.3
49.7	465	231	234	114.8
49.7	463	230	233	114.3
49.7	461	229	232	113.8
49.7	459	228	231	113.3
49.7	457	227	230	112.8
49.7	455	226	229	112.3
49.7	453	225	228	111.8
49.7	451	224	227	111.3
49.7	449	223	226	110.8
49.7	447	222	225	110.3
49.7	445	221	224	109.8
49.7	443	220	223	109.3
49.7	441	219	222	108.8
49.7	439	218	221	108.3
49.7	437	217	220	107.8
49.7	435	216	219	107.3
49.7	433	215	218	106.8
49.7	431	214	217	106.3
49.7	429	213	216	105.8
49.7	398	198	200	98.5
49.7	396	197	199	98.0
49.7	394	196	198	97.5
49.7	392	195	197	97.0
49.7	390	194	196	96.5
49.7	388	193	195	96.0
49.7	386	192	194	95.5
49.7	384	191	193	95.0
49.7	382	190	192	94.5
49.7	380	189	191	94.0
49.7	378	188	190	93.5
49.7	376	187	189	93.0
49.7	374	186	188	92.5

%E	M1	M2	DM	M*
49.7	372	185	187	92.0
49.7	370	184	186	91.5
49.7	368	183	185	91.0
49.7	366	182	184	90.5
49.7	364	181	183	90.0
49.7	362	180	182	89.5
49.7	360	179	181	89.0
49.7	358	178	180	88.5
49.7	356	177	179	88.0
49.7	354	176	178	87.5
49.7	352	175	177	87.0
49.7	350	174	176	86.5
49.7	348	173	175	86.0
49.7	346	172	174	85.5
49.7	344	171	173	85.0
49.7	342	170	172	84.5
49.7	340	169	171	84.0
49.7	338	168	170	83.5
49.7	336	167	169	83.0
49.7	334	166	168	82.5
49.7	332	165	167	82.0
49.7	330	164	166	81.5
49.7	328	163	165	81.0
49.7	326	162	164	80.5
49.7	324	161	163	80.0
49.7	322	160	162	79.5
49.7	320	159	161	79.0
49.7	318	158	160	78.5
49.7	316	157	159	78.0
49.7	314	156	158	77.5
49.7	312	155	157	77.0
49.7	310	154	156	76.5
49.7	308	153	155	76.0
49.7	306	152	154	75.5
49.7	304	151	153	75.0
49.7	302	150	152	74.5
49.7	300	149	151	74.0
49.7	298	148	150	73.5
49.7	296	147	149	73.0
49.7	294	146	148	72.5
49.7	292	145	147	72.0
49.7	290	144	146	71.5
49.7	288	143	145	71.0
49.7	286	142	144	70.5
49.7	199	99	100	49.3
49.7	197	98	99	48.8
49.7	195	97	98	48.3
49.7	193	96	97	47.8
49.7	191	95	96	47.3
49.7	189	94	95	46.8
49.7	187	93	94	46.3
49.7	185	92	93	45.8
49.7	183	91	92	45.3
49.7	181	90	91	44.8
49.7	179	89	90	44.3
49.7	177	88	89	43.8
49.7	175	87	88	43.3
49.7	173	86	87	42.8
49.7	171	85	86	42.3
49.7	169	84	85	41.8
49.7	167	83	84	41.3
49.7	165	82	83	40.8
49.7	163	81	82	40.3
49.7	161	80	81	39.8
49.7	159	79	80	39.3
49.7	157	78	79	38.8
49.7	155	77	78	38.3
49.7	153	76	77	37.8
49.7	151	75	76	37.3
49.7	149	74	75	36.8
49.7	147	73	74	36.3
49.7	145	72	73	35.8
49.7	143	71	72	35.3
49.6	500	248	252	123.0
49.6	498	247	251	122.5
49.6	496	246	250	122.0
49.6	494	245	249	121.5
49.6	492	244	248	121.0
49.6	490	243	247	120.5
49.6	488	242	246	120.0
49.6	486	241	245	119.5
49.6	484	240	244	119.0
49.6	482	239	243	118.5
49.6	480	238	242	118.0
49.6	478	237	241	117.5
49.6	476	236	240	117.0
49.6	474	235	239	116.5
49.6	472	234	238	116.0
49.6	470	233	237	115.5
49.6	468	232	236	115.0
49.6	466	231	235	114.5
49.6	464	230	234	114.0
49.6	462	229	233	113.5
49.6	460	228	232	113.0
49.6	458	227	231	112.5
49.6	456	226	230	112.0
49.6	454	225	229	111.5
49.6	452	224	228	111.0
49.6	450	223	227	110.5
49.6	448	222	226	110.0
49.6	446	221	225	109.5
49.6	444	220	224	109.0
49.6	427	212	215	105.3
49.6	425	211	214	104.8
49.6	423	210	213	104.3
49.6	421	209	212	103.8
49.6	419	208	211	103.3
49.6	417	207	210	102.8
49.6	415	206	209	102.3
49.6	413	205	208	101.8
49.6	411	204	207	101.3
49.6	409	203	206	100.8
49.6	407	202	205	100.3
49.6	405	201	204	99.8
49.6	403	200	203	99.3
49.6	401	199	202	98.8
49.6	399	198	201	98.3
49.6	397	197	200	97.8
49.6	395	196	199	97.3
49.6	393	195	198	96.8
49.6	391	194	197	96.3
49.6	389	193	196	95.8
49.6	387	192	195	95.3
49.6	385	191	194	94.8
49.6	383	190	193	94.3
49.6	381	189	192	93.8
49.6	379	188	191	93.3
49.6	377	187	190	92.8
49.6	375	186	189	92.3
49.6	373	185	188	91.8
49.6	371	184	187	91.3
49.6	369	183	186	90.8
49.6	367	182	185	90.3
49.6	365	181	184	89.8
49.6	363	180	183	89.3
49.6	361	179	182	88.8
49.6	359	178	181	88.3
49.6	357	177	180	87.8
49.6	355	176	179	87.3
49.6	353	175	178	86.8
49.6	351	174	177	86.3
49.6	349	173	176	85.8
49.6	347	172	175	85.3
49.6	345	171	174	84.8
49.6	343	170	173	84.3
49.6	341	169	172	83.8
49.6	339	168	171	83.3
49.6	337	167	170	82.8
49.6	335	166	169	82.3
49.6	333	165	168	81.8
49.6	284	141	143	70.0
49.6	282	140	142	69.5
49.6	280	139	141	69.0
49.6	278	138	140	68.5
49.6	276	137	139	68.0
49.6	274	136	138	67.5
49.6	272	135	137	67.0
49.6	270	134	136	66.5
49.6	268	133	135	66.0
49.6	266	132	134	65.5
49.6	264	131	133	65.0
49.6	262	130	132	64.5
49.6	260	129	131	64.0
49.6	258	128	130	63.5
49.6	256	127	129	63.0
49.6	254	126	128	62.5
49.6	252	125	127	62.0
49.6	250	124	126	61.5
49.6	248	123	125	61.0
49.6	246	122	124	60.5
49.6	244	121	123	60.0
49.6	242	120	122	59.5
49.6	240	119	121	59.0
49.6	238	118	120	58.5
49.6	236	117	119	58.0
49.6	234	116	118	57.5
49.6	232	115	117	57.0
49.6	230	114	116	56.5
49.6	228	113	115	56.0
49.6	226	112	114	55.5
49.6	224	111	113	55.0
49.6	222	110	112	54.5
49.6	141	70	71	34.8
49.6	139	69	70	34.3
49.6	137	68	69	33.8
49.6	135	67	68	33.3
49.6	133	66	67	32.8
49.6	131	65	66	32.3
49.6	129	64	65	31.8
49.6	127	63	64	31.3
49.6	125	62	63	30.8
49.6	123	61	62	30.3
49.6	121	60	61	29.8
49.6	119	59	60	29.3
49.6	117	58	59	28.8
49.6	115	57	58	28.3
49.6	113	56	57	27.8
49.6	111	55	56	27.3
49.5	499	247	252	122.3
49.5	497	246	251	121.8
49.5	495	245	250	121.3
49.5	493	244	249	120.8
49.5	491	243	248	120.3
49.5	489	242	247	119.8
49.5	487	241	246	119.3
49.5	485	240	245	118.8
49.5	483	239	244	118.3
49.5	481	238	243	117.8
49.5	479	237	242	117.3
49.5	477	236	241	116.8
49.5	475	235	240	116.3
49.5	473	234	239	115.8
49.5	471	233	238	115.3
49.5	469	232	237	114.8
49.5	467	231	236	114.3
49.5	465	230	235	113.8
49.5	463	229	234	113.3
49.5	461	228	233	112.8
49.5	459	227	232	112.3
49.5	457	226	231	111.8
49.5	455	225	230	111.3
49.5	442	219	223	108.5
49.5	440	218	222	108.0
49.5	438	217	221	107.5
49.5	436	216	220	107.0
49.5	434	215	219	106.5
49.5	432	214	218	106.0
49.5	430	213	217	105.5
49.5	428	212	216	105.0
49.5	426	211	215	104.5
49.5	424	210	214	104.0
49.5	422	209	213	103.5
49.5	420	208	212	103.0
49.5	418	207	211	102.5
49.5	416	206	210	102.0
49.5	414	205	209	101.5
49.5	412	204	208	101.0
49.5	410	203	207	100.5
49.5	408	202	206	100.0
49.5	406	201	205	99.5
49.5	404	200	204	99.0
49.5	402	199	203	98.5
49.5	400	198	202	98.0
49.5	398	197	201	97.5
49.5	396	196	200	97.0
49.5	394	195	199	96.5
49.5	392	194	198	96.0
49.5	390	193	197	95.5
49.5	388	192	196	95.0
49.5	386	191	195	94.5

%E	M1	M2	DM	M*
49.5	384	190	194	94.0
49.5	382	189	193	93.5
49.5	380	188	192	93.0
49.5	378	187	191	92.5
49.5	376	186	190	92.0
49.5	374	185	189	91.5
49.5	372	184	188	91.0
49.5	370	183	187	90.5
49.5	368	182	186	90.0
49.5	366	181	185	89.5
49.5	364	180	184	89.0
49.5	331	164	167	81.3
49.5	329	163	166	80.8
49.5	327	162	165	80.3
49.5	325	161	164	79.8
49.5	323	160	163	79.3
49.5	321	159	162	78.8
49.5	319	158	161	78.3
49.5	317	157	160	77.8
49.5	315	156	159	77.3
49.5	313	155	158	76.8
49.5	311	154	157	76.3
49.5	309	153	156	75.8
49.5	307	152	155	75.3
49.5	305	151	154	74.8
49.5	303	150	153	74.3
49.5	301	149	152	73.8
49.5	299	148	151	73.3
49.5	297	147	150	72.8
49.5	295	146	149	72.3
49.5	293	145	148	71.8
49.5	291	144	147	71.3
49.5	289	143	146	70.8
49.5	287	142	145	70.3
49.5	285	141	144	69.8
49.5	283	140	143	69.3
49.5	281	139	142	68.8
49.5	279	138	141	68.3
49.5	277	137	140	67.8
49.5	275	136	139	67.3
49.5	273	135	138	66.8
49.5	220	109	111	54.0
49.5	218	108	110	53.5
49.5	216	107	109	53.0
49.5	214	106	108	52.5
49.5	212	105	107	52.0
49.5	210	104	106	51.5
49.5	208	103	105	51.0
49.5	206	102	104	50.5
49.5	204	101	103	50.0

%E	M1	M2	DM	M*
49.5	202	100	102	49.5
49.5	200	99	101	49.0
49.5	198	98	100	48.5
49.5	196	97	99	48.0
49.5	194	96	98	47.5
49.5	192	95	97	47.0
49.5	190	94	96	46.5
49.5	188	93	95	46.0
49.5	186	92	94	45.5
49.5	184	91	93	45.0
49.5	182	90	92	44.5
49.5	109	54	55	26.8
49.5	107	53	54	26.3
49.5	105	52	53	25.8
49.5	103	51	52	25.3
49.5	101	50	51	24.8
49.5	99	49	50	24.3
49.5	97	48	49	23.8
49.5	95	47	48	23.3
49.5	93	46	47	22.8
49.5	91	45	46	22.3
49.4	500	247	253	122.0
49.4	498	246	252	121.5
49.4	496	245	251	121.0
49.4	494	244	250	120.5
49.4	492	243	249	120.0
49.4	490	242	248	119.5
49.4	488	241	247	119.0
49.4	486	240	246	118.5
49.4	484	239	245	118.0
49.4	482	238	244	117.5
49.4	480	237	243	117.0
49.4	478	236	242	116.5
49.4	476	235	241	116.0
49.4	474	234	240	115.5
49.4	472	233	239	115.0
49.4	470	232	238	114.5
49.4	468	231	237	114.0
49.4	466	230	236	113.5
49.4	464	229	235	113.0
49.4	462	228	234	112.5
49.4	453	224	229	110.8
49.4	451	223	228	110.3
49.4	449	222	227	109.8
49.4	447	221	226	109.3
49.4	445	220	225	108.8
49.4	443	219	224	108.3
49.4	441	218	223	107.8
49.4	439	217	222	107.3
49.4	437	216	221	106.8

%E	M1	M2	DM	M*
49.4	435	215	220	106.3
49.4	433	214	219	105.8
49.4	431	213	218	105.3
49.4	429	212	217	104.8
49.4	427	211	216	104.3
49.4	425	210	215	103.8
49.4	423	209	214	103.3
49.4	421	208	213	102.8
49.4	419	207	212	102.3
49.4	417	206	211	101.8
49.4	415	205	210	101.3
49.4	413	204	209	100.8
49.4	411	203	208	100.3
49.4	409	202	207	99.8
49.4	407	201	206	99.3
49.4	405	200	205	98.8
49.4	403	199	204	98.3
49.4	401	198	203	97.8
49.4	399	197	202	97.3
49.4	397	196	201	96.8
49.4	395	195	200	96.3
49.4	393	194	199	95.8
49.4	391	193	198	95.3
49.4	389	192	197	94.8
49.4	387	191	196	94.3
49.4	385	190	195	93.8
49.4	362	179	183	88.5
49.4	360	178	182	88.0
49.4	358	177	181	87.5
49.4	356	176	180	87.0
49.4	354	175	179	86.5
49.4	352	174	178	86.0
49.4	350	173	177	85.5
49.4	348	172	176	85.0
49.4	346	171	175	84.5
49.4	344	170	174	84.0
49.4	342	169	173	83.5
49.4	340	168	172	83.0
49.4	338	167	171	82.5
49.4	336	166	170	82.0
49.4	334	165	169	81.5
49.4	332	164	168	81.0
49.4	330	163	167	80.5
49.4	328	162	166	80.0
49.4	326	161	165	79.5
49.4	324	160	164	79.0
49.4	322	159	163	78.5
49.4	320	158	162	78.0
49.4	318	157	161	77.5
49.4	316	156	160	77.0

%E	M1	M2	DM	M*
49.4	314	155	159	76.5
49.4	312	154	158	76.0
49.4	310	153	157	75.5
49.4	308	152	156	75.0
49.4	271	134	137	66.3
49.4	269	133	136	65.8
49.4	267	132	135	65.3
49.4	265	131	134	64.8
49.4	263	130	133	64.3
49.4	261	129	132	63.8
49.4	259	128	131	63.3
49.4	257	127	130	62.8
49.4	255	126	129	62.3
49.4	253	125	128	61.8
49.4	251	124	127	61.3
49.4	249	123	126	60.8
49.4	247	122	125	60.3
49.4	245	121	124	59.8
49.4	243	120	123	59.3
49.4	241	119	122	58.8
49.4	239	118	121	58.3
49.4	237	117	120	57.8
49.4	235	116	119	57.3
49.4	233	115	118	56.8
49.4	231	114	117	56.3
49.4	180	89	91	44.0
49.4	178	88	90	43.5
49.4	176	87	89	43.0
49.4	174	86	88	42.5
49.4	172	85	87	42.0
49.4	170	84	86	41.5
49.4	168	83	85	41.0
49.4	166	82	84	40.5
49.4	164	81	83	40.0
49.4	162	80	82	39.5
49.4	160	79	81	39.0
49.4	158	78	80	38.5
49.4	156	77	79	38.0
49.4	154	76	78	37.5
49.4	89	44	45	21.8
49.4	87	43	44	21.3
49.4	85	42	43	20.8
49.4	83	41	42	20.3
49.4	81	40	41	19.8
49.4	79	39	40	19.3
49.4	77	38	39	18.8
49.3	499	246	253	121.3
49.3	497	245	252	120.8
49.3	495	244	251	120.3
49.3	493	243	250	119.8

%E	M1	M2	DM	M*
49.3	491	242	249	119.3
49.3	489	241	248	118.8
49.3	487	240	247	118.3
49.3	485	239	246	117.8
49.3	483	238	245	117.3
49.3	481	237	244	116.8
49.3	479	236	243	116.3
49.3	477	235	242	115.8
49.3	475	234	241	115.3
49.3	473	233	240	114.8
49.3	471	232	239	114.3
49.3	469	231	238	113.8
49.3	467	230	237	113.3
49.3	460	227	233	112.0
49.3	458	226	232	111.5
49.3	456	225	231	111.0
49.3	454	224	230	110.5
49.3	452	223	229	110.0
49.3	450	222	228	109.5
49.3	448	221	227	109.0
49.3	446	220	226	108.5
49.3	444	219	225	108.0
49.3	442	218	224	107.5
49.3	440	217	223	107.0
49.3	438	216	222	106.5
49.3	436	215	221	106.0
49.3	434	214	220	105.5
49.3	432	213	219	105.0
49.3	430	212	218	104.5
49.3	428	211	217	104.0
49.3	426	210	216	103.5
49.3	424	209	215	103.0
49.3	422	208	214	102.5
49.3	420	207	213	102.0
49.3	418	206	212	101.5
49.3	416	205	211	101.0
49.3	414	204	210	100.5
49.3	412	203	209	100.0
49.3	410	202	208	99.5
49.3	408	201	207	99.0
49.3	406	200	206	98.5
49.3	404	199	205	98.0
49.3	402	198	204	97.5
49.3	400	197	203	97.0
49.3	383	189	194	93.3
49.3	381	188	193	92.8
49.3	379	187	192	92.3
49.3	377	186	191	91.8
49.3	375	185	190	91.3
49.3	373	184	189	90.8

%E	M1	M2	DM	M*
49.3	371	183	188	90.3
49.3	369	182	187	89.8
49.3	367	181	186	89.3
49.3	365	180	185	88.8
49.3	363	179	184	88.3
49.3	361	178	183	87.8
49.3	359	177	182	87.3
49.3	357	176	181	86.8
49.3	355	175	180	86.3
49.3	353	174	179	85.8
49.3	351	173	178	85.3
49.3	349	172	177	84.8
49.3	347	171	176	84.3
49.3	345	170	175	83.8
49.3	343	169	174	83.3
49.3	341	168	173	82.8
49.3	339	167	172	82.3
49.3	337	166	171	81.8
49.3	335	165	170	81.3
49.3	306	151	155	74.5
49.3	304	150	154	74.0
49.3	302	149	153	73.5
49.3	300	148	152	73.0
49.3	298	147	151	72.5
49.3	296	146	150	72.0
49.3	294	145	149	71.5
49.3	292	144	148	71.0
49.3	290	143	147	70.5
49.3	288	142	146	70.0
49.3	286	141	145	69.5
49.3	284	140	144	69.0
49.3	282	139	143	68.5
49.3	280	138	142	68.0
49.3	278	137	141	67.5
49.3	276	136	140	67.0
49.3	274	135	139	66.5
49.3	272	134	138	66.0
49.3	270	133	137	65.5
49.3	268	132	136	65.0
49.3	229	113	116	55.8
49.3	227	112	115	55.3
49.3	225	111	114	54.8
49.3	223	110	113	54.3
49.3	221	109	112	53.8
49.3	219	108	111	53.3
49.3	217	107	110	52.8
49.3	215	106	109	52.3
49.3	213	105	108	51.8
49.3	211	104	107	51.3
49.3	209	103	106	50.8

%E	M1	M2	DM	M*
49.3	207	102	105	50.3
49.3	205	101	104	49.8
49.3	203	100	103	49.3
49.3	201	99	102	48.8
49.3	152	75	77	37.0
49.3	150	74	76	36.5
49.3	148	73	75	36.0
49.3	146	72	74	35.5
49.3	144	71	73	35.0
49.3	142	70	72	34.5
49.3	140	69	71	34.0
49.3	138	68	70	33.5
49.3	136	67	69	33.0
49.3	134	66	68	32.5
49.3	75	37	38	18.3
49.3	73	36	37	17.8
49.3	71	35	36	17.3
49.3	69	34	35	16.8
49.3	67	33	34	16.3
49.2	500	246	254	121.0
49.2	498	245	253	120.5
49.2	496	244	252	120.0
49.2	494	243	251	119.5
49.2	492	242	250	119.0
49.2	490	241	249	118.5
49.2	488	240	248	118.0
49.2	486	239	247	117.5
49.2	484	238	246	117.0
49.2	482	237	245	116.5
49.2	480	236	244	116.0
49.2	478	235	243	115.5
49.2	476	234	242	115.0
49.2	474	233	241	114.5
49.2	472	232	240	114.0
49.2	465	229	236	112.8
49.2	463	228	235	112.3
49.2	461	227	234	111.8
49.2	459	226	233	111.3
49.2	457	225	232	110.8
49.2	455	224	231	110.3
49.2	453	223	230	109.8
49.2	451	222	229	109.3
49.2	449	221	228	108.8
49.2	447	220	227	108.3
49.2	445	219	226	107.8
49.2	443	218	225	107.3
49.2	441	217	224	106.8
49.2	439	216	223	106.3
49.2	437	215	222	105.8
49.2	435	214	221	105.3

%E	M1	M2	DM	M*
49.2	433	213	220	104.8
49.2	431	212	219	104.3
49.2	429	211	218	103.8
49.2	427	210	217	103.3
49.2	425	209	216	102.8
49.2	423	208	215	102.3
49.2	421	207	214	101.8
49.2	419	206	213	101.3
49.2	417	205	212	100.8
49.2	415	204	211	100.3
49.2	413	203	210	99.8
49.2	398	196	202	96.5
49.2	396	195	201	96.0
49.2	394	194	200	95.5
49.2	392	193	199	95.0
49.2	390	192	198	94.5
49.2	388	191	197	94.0
49.2	386	190	196	93.5
49.2	384	189	195	93.0
49.2	382	188	194	92.5
49.2	380	187	193	92.0
49.2	378	186	192	91.5
49.2	376	185	191	91.0
49.2	374	184	190	90.5
49.2	372	183	189	90.0
49.2	370	182	188	89.5
49.2	368	181	187	89.0
49.2	366	180	186	88.5
49.2	364	179	185	88.0
49.2	362	178	184	87.5
49.2	360	177	183	87.0
49.2	358	176	182	86.5
49.2	356	175	181	86.0
49.2	354	174	180	85.5
49.2	333	164	169	80.8
49.2	331	163	168	80.3
49.2	329	162	167	79.8
49.2	327	161	166	79.3
49.2	325	160	165	78.8
49.2	323	159	164	78.3
49.2	321	158	163	77.8
49.2	319	157	162	77.3
49.2	317	156	161	76.8
49.2	315	155	160	76.3
49.2	313	154	159	75.8
49.2	311	153	158	75.3
49.2	309	152	157	74.8
49.2	307	151	156	74.3
49.2	305	150	155	73.8
49.2	303	149	154	73.3

%E	M1	M2	DM	M*
49.2	301	148	153	72.8
49.2	299	147	152	72.3
49.2	297	146	151	71.8
49.2	295	145	150	71.3
49.2	266	131	135	64.5
49.2	264	130	134	64.0
49.2	262	129	133	63.5
49.2	260	128	132	63.0
49.2	258	127	131	62.5
49.2	256	126	130	62.0
49.2	254	125	129	61.5
49.2	252	124	128	61.0
49.2	250	123	127	60.5
49.2	248	122	126	60.0
49.2	246	121	125	59.5
49.2	244	120	124	59.0
49.2	242	119	123	58.5
49.2	240	118	122	58.0
49.2	238	117	121	57.5
49.2	236	116	120	57.0
49.2	199	98	101	48.3
49.2	197	97	100	47.8
49.2	195	96	99	47.3
49.2	193	95	98	46.8
49.2	191	94	97	46.3
49.2	189	93	96	45.8
49.2	187	92	95	45.3
49.2	185	91	94	44.8
49.2	183	90	93	44.3
49.2	181	89	92	43.8
49.2	179	88	91	43.3
49.2	177	87	90	42.8
49.2	132	65	67	32.0
49.2	130	64	66	31.5
49.2	128	63	65	31.0
49.2	126	62	64	30.5
49.2	124	61	63	30.0
49.2	122	60	62	29.5
49.2	120	59	61	29.0
49.2	118	58	60	28.5
49.2	65	32	33	15.8
49.2	63	31	32	15.3
49.2	61	30	31	14.8
49.2	59	29	30	14.3
49.1	499	245	254	120.3
49.1	497	244	253	119.8
49.1	495	243	252	119.3
49.1	493	242	251	118.8
49.1	491	241	250	118.3
49.1	489	240	249	117.8

%E	M1	M2	DM	M*
49.1	487	239	248	117.3
49.1	485	238	247	116.8
49.1	483	237	246	116.3
49.1	481	236	245	115.8
49.1	479	235	244	115.3
49.1	477	234	243	114.8
49.1	475	233	242	114.3
49.1	470	231	239	113.5
49.1	468	230	238	113.0
49.1	466	229	237	112.5
49.1	464	228	236	112.0
49.1	462	227	235	111.5
49.1	460	226	234	111.0
49.1	458	225	233	110.5
49.1	456	224	232	110.0
49.1	454	223	231	109.5
49.1	452	222	230	109.0
49.1	450	221	229	108.5
49.1	448	220	228	108.0
49.1	446	219	227	107.5
49.1	444	218	226	107.0
49.1	442	217	225	106.5
49.1	440	216	224	106.0
49.1	438	215	223	105.5
49.1	436	214	222	105.0
49.1	434	213	221	104.5
49.1	432	212	220	104.0
49.1	430	211	219	103.5
49.1	428	210	218	103.0
49.1	426	209	217	102.5
49.1	424	208	216	102.0
49.1	422	207	215	101.5
49.1	411	202	209	99.3
49.1	409	201	208	98.8
49.1	407	200	207	98.3
49.1	405	199	206	97.8
49.1	403	198	205	97.3
49.1	401	197	204	96.8
49.1	399	196	203	96.3
49.1	397	195	202	95.8
49.1	395	194	201	95.3
49.1	393	193	200	94.8
49.1	391	192	199	94.3
49.1	389	191	198	93.8
49.1	387	190	197	93.3
49.1	385	189	196	92.8
49.1	383	188	195	92.3
49.1	381	187	194	91.8
49.1	379	186	193	91.3
49.1	377	185	192	90.8

%E	M1	M2	DM	M*
49.1	375	184	191	90.3
49.1	373	183	190	89.8
49.1	371	182	189	89.3
49.1	369	181	188	88.8
49.1	352	173	179	85.0
49.1	350	172	178	84.5
49.1	348	171	177	84.0
49.1	346	170	176	83.5
49.1	344	169	175	83.0
49.1	342	168	174	82.5
49.1	340	167	173	82.0
49.1	338	166	172	81.5
49.1	336	165	171	81.0
49.1	334	164	170	80.5
49.1	332	163	169	80.0
49.1	330	162	168	79.5
49.1	328	161	167	79.0
49.1	326	160	166	78.5
49.1	324	159	165	78.0
49.1	322	158	164	77.5
49.1	320	157	163	77.0
49.1	318	156	162	76.5
49.1	316	155	161	76.0
49.1	293	144	149	70.8
49.1	291	143	148	70.3
49.1	289	142	147	69.8
49.1	287	141	146	69.3
49.1	285	140	145	68.8
49.1	283	139	144	68.3
49.1	281	138	143	67.8
49.1	279	137	142	67.3
49.1	277	136	141	66.8
49.1	275	135	140	66.3
49.1	273	134	139	65.8
49.1	271	133	138	65.3
49.1	269	132	137	64.8
49.1	267	131	136	64.3
49.1	265	130	135	63.8
49.1	234	115	119	56.5
49.1	232	114	118	56.0
49.1	230	113	117	55.5
49.1	228	112	116	55.0
49.1	226	111	115	54.5
49.1	224	110	114	54.0
49.1	222	109	113	53.5
49.1	220	108	112	53.0
49.1	218	107	111	52.5
49.1	216	106	110	52.0
49.1	214	105	109	51.5
49.1	212	104	108	51.0

%E	M1	M2	DM	M*
49.1	175	86	89	42.3
49.1	173	85	88	41.8
49.1	171	84	87	41.3
49.1	169	83	86	40.8
49.1	167	82	85	40.3
49.1	165	81	84	39.8
49.1	163	80	83	39.3
49.1	161	79	82	38.8
49.1	159	78	81	38.3
49.1	116	57	59	28.0
49.1	114	56	58	27.5
49.1	112	55	57	27.0
49.1	110	54	56	26.5
49.1	108	53	55	26.0
49.1	106	52	54	25.5
49.1	57	28	29	13.8
49.1	55	27	28	13.3
49.1	53	26	27	12.8
49.0	500	245	255	120.0
49.0	498	244	254	119.6
49.0	496	243	253	119.1
49.0	494	242	252	118.6
49.0	492	241	251	118.1
49.0	490	240	250	117.6
49.0	488	239	249	117.1
49.0	486	238	248	116.6
49.0	484	237	247	116.1
49.0	482	236	246	115.6
49.0	480	235	245	115.1
49.0	478	234	244	114.6
49.0	476	233	243	114.1
49.0	473	232	241	113.8
49.0	471	231	240	113.3
49.0	469	230	239	112.8
49.0	467	229	238	112.3
49.0	465	228	237	111.8
49.0	463	227	236	111.3
49.0	461	226	235	110.8
49.0	459	225	234	110.3
49.0	457	224	233	109.8
49.0	455	223	232	109.3
49.0	453	222	231	108.8
49.0	451	221	230	108.3
49.0	449	220	229	107.8
49.0	447	219	228	107.3
49.0	445	218	227	106.8
49.0	443	217	226	106.3
49.0	441	216	225	105.8
49.0	439	215	224	105.3
49.0	437	214	223	104.8

%E	M1	M2	DM	M*
49.0	435	213	222	104.3
49.0	433	212	221	103.8
49.0	431	211	220	103.3
49.0	429	210	219	102.8
49.0	420	206	214	101.0
49.0	418	205	213	100.5
49.0	416	204	212	100.0
49.0	414	203	211	99.5
49.0	412	202	210	99.0
49.0	410	201	209	98.5
49.0	408	200	208	98.0
49.0	406	199	207	97.5
49.0	404	198	206	97.0
49.0	402	197	205	96.5
49.0	400	196	204	96.0
49.0	398	195	203	95.5
49.0	396	194	202	95.0
49.0	394	193	201	94.5
49.0	392	192	200	94.0
49.0	390	191	199	93.5
49.0	388	190	198	93.0
49.0	386	189	197	92.5
49.0	384	188	196	92.0
49.0	382	187	195	91.5
49.0	367	180	187	88.3
49.0	365	179	186	87.8
49.0	363	178	185	87.3
49.0	361	177	184	86.8
49.0	359	176	183	86.3
49.0	357	175	182	85.8
49.0	355	174	181	85.3
49.0	353	173	180	84.8
49.0	351	172	179	84.3
49.0	349	171	178	83.8
49.0	347	170	177	83.3
49.0	345	169	176	82.8
49.0	343	168	175	82.3
49.0	341	167	174	81.8
49.0	339	166	173	81.3
49.0	337	165	172	80.8
49.0	335	164	171	80.3
49.0	314	154	160	75.5
49.0	312	153	159	75.0
49.0	310	152	158	74.5
49.0	308	151	157	74.0
49.0	306	150	156	73.5
49.0	304	149	155	73.0
49.0	302	148	154	72.5
49.0	300	147	153	72.0
49.0	298	146	152	71.5

%E	M1	M2	DM	M*
49.0	296	145	151	71.0
49.0	294	144	150	70.5
49.0	292	143	149	70.0
49.0	290	142	148	69.5
49.0	288	141	147	69.0
49.0	286	140	146	68.5
49.0	263	129	134	63.3
49.0	261	128	133	62.8
49.0	259	127	132	62.3
49.0	257	126	131	61.8
49.0	255	125	130	61.3
49.0	253	124	129	60.8
49.0	251	123	128	60.3
49.0	249	122	127	59.8
49.0	247	121	126	59.3
49.0	245	120	125	58.8
49.0	243	119	124	58.3
49.0	241	118	123	57.8
49.0	239	117	122	57.3
49.0	210	103	107	50.5
49.0	208	102	106	50.0
49.0	206	101	105	49.5
49.0	204	100	104	49.0
49.0	202	99	103	48.5
49.0	200	98	102	48.0
49.0	198	97	101	47.5
49.0	196	96	100	47.0
49.0	194	95	99	46.5
49.0	192	94	98	46.0
49.0	157	77	80	37.8
49.0	155	76	79	37.3
49.0	153	75	78	36.8
49.0	151	74	77	36.3
49.0	149	73	76	35.8
49.0	147	72	75	35.3
49.0	145	71	74	34.8
49.0	143	70	73	34.3
49.0	104	51	53	25.0
49.0	102	50	52	24.5
49.0	100	49	51	24.0
49.0	98	48	50	23.5
49.0	96	47	49	23.0
49.0	51	25	26	12.3
49.0	49	24	25	11.8
48.9	499	244	255	119.3
48.9	497	243	254	118.8
48.9	495	242	253	118.3
48.9	493	241	252	117.8
48.9	491	240	251	117.3
48.9	489	239	250	116.8

%E	M1	M2	DM	M*
48.9	487	238	249	116.3
48.9	485	237	248	115.8
48.9	483	236	247	115.3
48.9	481	235	246	114.8
48.9	479	234	245	114.3
48.9	474	232	242	113.6
48.9	472	231	241	113.1
48.9	470	230	240	112.6
48.9	468	229	239	112.1
48.9	466	228	238	111.6
48.9	464	227	237	111.1
48.9	462	226	236	110.6
48.9	460	225	235	110.1
48.9	458	224	234	109.6
48.9	456	223	233	109.1
48.9	454	222	232	108.6
48.9	452	221	231	108.1
48.9	450	220	230	107.6
48.9	448	219	229	107.1
48.9	446	218	228	106.6
48.9	444	217	227	106.1
48.9	442	216	226	105.6
48.9	440	215	225	105.1
48.9	438	214	224	104.6
48.9	436	213	223	104.1
48.9	427	209	218	102.3
48.9	425	208	217	101.8
48.9	423	207	216	101.3
48.9	421	206	215	100.8
48.9	419	205	214	100.3
48.9	417	204	213	99.8
48.9	415	203	212	99.3
48.9	413	202	211	98.8
48.9	411	201	210	98.3
48.9	409	200	209	97.8
48.9	407	199	208	97.3
48.9	405	198	207	96.8
48.9	403	197	206	96.3
48.9	401	196	205	95.8
48.9	399	195	204	95.3
48.9	397	194	203	94.8
48.9	395	193	202	94.3
48.9	393	192	201	93.8
48.9	380	186	194	91.0
48.9	378	185	193	90.5
48.9	376	184	192	90.0
48.9	374	183	191	89.5
48.9	372	182	190	89.0
48.9	370	181	189	88.5
48.9	368	180	188	88.0

%E	M1	M2	DM	M*
48.9	366	179	187	87.5
48.9	364	178	186	87.0
48.9	362	177	185	86.5
48.9	360	176	184	86.0
48.9	358	175	183	85.5
48.9	356	174	182	85.0
48.9	354	173	181	84.5
48.9	352	172	180	84.0
48.9	350	171	179	83.5
48.9	348	170	178	83.0
48.9	333	163	170	79.8
48.9	331	162	169	79.3
48.9	329	161	168	78.8
48.9	327	160	167	78.3
48.9	325	159	166	77.8
48.9	323	158	165	77.3
48.9	321	157	164	76.8
48.9	319	156	163	76.3
48.9	317	155	162	75.8
48.9	315	154	161	75.3
48.9	313	153	160	74.8
48.9	311	152	159	74.3
48.9	309	151	158	73.8
48.9	307	150	157	73.3
48.9	305	149	156	72.8
48.9	284	139	145	68.0
48.9	282	138	144	67.5
48.9	280	137	143	67.0
48.9	278	136	142	66.5
48.9	276	135	141	66.0
48.9	274	134	140	65.5
48.9	272	133	139	65.0
48.9	270	132	138	64.5
48.9	268	131	137	64.0
48.9	266	130	136	63.5
48.9	264	129	135	63.0
48.9	262	128	134	62.5
48.9	237	116	121	56.8
48.9	235	115	120	56.3
48.9	233	114	119	55.8
48.9	231	113	118	55.3
48.9	229	112	117	54.8
48.9	227	111	116	54.3
48.9	225	110	115	53.8
48.9	223	109	114	53.3
48.9	221	108	113	52.8
48.9	219	107	112	52.3
48.9	190	93	97	45.5
48.9	188	92	96	45.0
48.9	186	91	95	44.5

%E	M1	M2	DM	M*
48.9	184	90	94	44.0
48.9	182	89	93	43.5
48.9	180	88	92	43.0
48.9	178	87	91	42.5
48.9	176	86	90	42.0
48.9	174	85	89	41.5
48.9	141	69	72	33.8
48.9	139	68	71	33.3
48.9	137	67	70	32.8
48.9	135	66	69	32.3
48.9	133	65	68	31.8
48.9	131	64	67	31.3
48.9	94	46	48	22.5
48.9	92	45	47	22.0
48.9	90	44	46	21.5
48.9	88	43	45	21.0
48.9	47	23	24	11.3
48.9	45	22	23	10.8
48.8	500	244	256	119.1
48.8	498	243	255	118.6
48.8	496	242	254	118.1
48.8	494	241	253	117.6
48.8	492	240	252	117.1
48.8	490	239	251	116.6
48.8	488	238	250	116.1
48.8	486	237	249	115.6
48.8	484	236	248	115.1
48.8	482	235	247	114.6
48.8	480	234	246	114.1
48.8	477	233	244	113.8
48.8	475	232	243	113.3
48.8	473	231	242	112.8
48.8	471	230	241	112.3
48.8	469	229	240	111.8
48.8	467	228	239	111.3
48.8	465	227	238	110.8
48.8	463	226	237	110.3
48.8	461	225	236	109.8
48.8	459	224	235	109.3
48.8	457	223	234	108.8
48.8	455	222	233	108.3
48.8	453	221	232	107.8
48.8	451	220	231	107.3
48.8	449	219	230	106.8
48.8	447	218	229	106.3
48.8	445	217	228	105.8
48.8	443	216	227	105.3
48.8	441	215	226	104.8
48.8	434	212	222	103.6
48.8	432	211	221	103.1

%E	M1	M2	DM	M*
48.8	430	210	220	102.6
48.8	428	209	219	102.1
48.8	426	208	218	101.6
48.8	424	207	217	101.1
48.8	422	206	216	100.6
48.8	420	205	215	100.1
48.8	418	204	214	99.6
48.8	416	203	213	99.1
48.8	414	202	212	98.6
48.8	412	201	211	98.1
48.8	410	200	210	97.6
48.8	408	199	209	97.1
48.8	406	198	208	96.6
48.8	404	197	207	96.1
48.8	402	196	206	95.6
48.8	400	195	205	95.1
48.8	391	191	200	93.3
48.8	389	190	199	92.8
48.8	387	189	198	92.3
48.8	385	188	197	91.8
48.8	383	187	196	91.3
48.8	381	186	195	90.8
48.8	379	185	194	90.3
48.8	377	184	193	89.8
48.8	375	183	192	89.3
48.8	373	182	191	88.8
48.8	371	181	190	88.3
48.8	369	180	189	87.8
48.8	367	179	188	87.3
48.8	365	178	187	86.8
48.8	363	177	186	86.3
48.8	361	176	185	85.8
48.8	346	169	177	82.5
48.8	344	168	176	82.0
48.8	342	167	175	81.5
48.8	340	166	174	81.0
48.8	338	165	173	80.5
48.8	336	164	172	80.0
48.8	334	163	171	79.5
48.8	332	162	170	79.0
48.8	330	161	169	78.5
48.8	328	160	168	78.0
48.8	326	159	167	77.5
48.8	324	158	166	77.0
48.8	322	157	165	76.5
48.8	320	156	164	76.0
48.8	303	148	155	72.3
48.8	301	147	154	71.8
48.8	299	146	153	71.3
48.8	297	145	152	70.8

%E	M1	M2	DM	M*
48.8	295	144	151	70.3
48.8	293	143	150	69.8
48.8	291	142	149	69.3
48.8	289	141	148	68.8
48.8	287	140	147	68.3
48.8	285	139	146	67.8
48.8	283	138	145	67.3
48.8	281	137	144	66.8
48.8	260	127	133	62.0
48.8	258	126	132	61.5
48.8	256	125	131	61.0
48.8	254	124	130	60.5
48.8	252	123	129	60.0
48.8	250	122	128	59.5
48.8	248	121	127	59.0
48.8	246	120	126	58.5
48.8	244	119	125	58.0
48.8	242	118	124	57.5
48.8	240	117	123	57.0
48.8	217	106	111	51.8
48.8	215	105	110	51.3
48.8	213	104	109	50.8
48.8	211	103	108	50.3
48.8	209	102	107	49.8
48.8	207	101	106	49.3
48.8	205	100	105	48.8
48.8	203	99	104	48.3
48.8	201	98	103	47.8
48.8	172	84	88	41.0
48.8	170	83	87	40.5
48.8	168	82	86	40.0
48.8	166	81	85	39.5
48.8	164	80	84	39.0
48.8	162	79	83	38.5
48.8	160	78	82	38.0
48.8	129	63	66	30.8
48.8	127	62	65	30.3
48.8	125	61	64	29.8
48.8	123	60	63	29.3
48.8	121	59	62	28.8
48.8	86	42	44	20.5
48.8	84	41	43	20.0
48.8	82	40	42	19.5
48.8	80	39	41	19.0
48.8	43	21	22	10.3
48.8	41	20	21	9.8
48.7	499	243	256	118.3
48.7	497	242	255	117.8
48.7	495	241	254	117.3
48.7	493	240	253	116.8

%E	M1	M2	DM	M*
48.7	491	239	252	116.3
48.7	489	238	251	115.8
48.7	487	237	250	115.3
48.7	485	236	249	114.8
48.7	483	235	248	114.3
48.7	478	233	245	113.6
48.7	476	232	244	113.1
48.7	474	231	243	112.6
48.7	472	230	242	112.1
48.7	470	229	241	111.6
48.7	468	228	240	111.1
48.7	466	227	239	110.6
48.7	464	226	238	110.1
48.7	462	225	237	109.6
48.7	460	224	236	109.1
48.7	458	223	235	108.6
48.7	456	222	234	108.1
48.7	454	221	233	107.6
48.7	452	220	232	107.1
48.7	450	219	231	106.6
48.7	448	218	230	106.1
48.7	446	217	229	105.6
48.7	439	214	225	104.3
48.7	437	213	224	103.8
48.7	435	212	223	103.3
48.7	433	211	222	102.8
48.7	431	210	221	102.3
48.7	429	209	220	101.8
48.7	427	208	219	101.3
48.7	425	207	218	100.8
48.7	423	206	217	100.3
48.7	421	205	216	99.8
48.7	419	204	215	99.3
48.7	417	203	214	98.8
48.7	415	202	213	98.3
48.7	413	201	212	97.8
48.7	411	200	211	97.3
48.7	409	199	210	96.8
48.7	398	194	204	94.6
48.7	396	193	203	94.1
48.7	394	192	202	93.6
48.7	392	191	201	93.1
48.7	390	190	200	92.6
48.7	388	189	199	92.1
48.7	386	188	198	91.6
48.7	384	187	197	91.1
48.7	382	186	196	90.6
48.7	380	185	195	90.1
48.7	378	184	194	89.6
48.7	376	183	193	89.1

%E	M1	M2	DM	M*
48.7	374	182	192	88.6
48.7	372	181	191	88.1
48.7	359	175	184	85.3
48.7	357	174	183	84.8
48.7	355	173	182	84.3
48.7	353	172	181	83.8
48.7	351	171	180	83.3
48.7	349	170	179	82.8
48.7	347	169	178	82.3
48.7	345	168	177	81.8
48.7	343	167	176	81.3
48.7	341	166	175	80.8
48.7	339	165	174	80.3
48.7	337	164	173	79.8
48.7	335	163	172	79.3
48.7	318	155	163	75.6
48.7	316	154	162	75.1
48.7	314	153	161	74.6
48.7	312	152	160	74.1
48.7	310	151	159	73.6
48.7	308	150	158	73.1
48.7	306	149	157	72.6
48.7	304	148	156	72.1
48.7	302	147	155	71.6
48.7	300	146	154	71.1
48.7	298	145	153	70.6
48.7	279	136	143	66.3
48.7	277	135	142	65.8
48.7	275	134	141	65.3
48.7	273	133	140	64.8
48.7	271	132	139	64.3
48.7	269	131	138	63.8
48.7	267	130	137	63.3
48.7	265	129	136	62.8
48.7	263	128	135	62.3
48.7	261	127	134	61.8
48.7	238	116	122	56.5
48.7	236	115	121	56.0
48.7	234	114	120	55.5
48.7	232	113	119	55.0
48.7	230	112	118	54.5
48.7	228	111	117	54.0
48.7	226	110	116	53.5
48.7	224	109	115	53.0
48.7	199	97	102	47.3
48.7	197	96	101	46.8
48.7	195	95	100	46.3
48.7	193	94	99	45.8
48.7	191	93	98	45.3
48.7	189	92	97	44.8

%E	M1	M2	DM	M*
48.7	187	91	96	44.3
48.7	158	77	81	37.5
48.7	156	76	80	37.0
48.7	154	75	79	36.5
48.7	152	74	78	36.0
48.7	150	73	77	35.5
48.7	119	58	61	28.3
48.7	117	57	60	27.8
48.7	115	56	59	27.3
48.7	113	55	58	26.8
48.7	78	38	40	18.5
48.7	76	37	39	18.0
48.7	39	19	20	9.3
48.6	500	243	257	118.1
48.6	498	242	256	117.6
48.6	496	241	255	117.1
48.6	494	240	254	116.6
48.6	492	239	253	116.1
48.6	490	238	252	115.6
48.6	488	237	251	115.1
48.6	486	236	250	114.6
48.6	484	235	249	114.1
48.6	481	234	247	113.8
48.6	479	233	246	113.3
48.6	477	232	245	112.8
48.6	475	231	244	112.3
48.6	473	230	243	111.8
48.6	471	229	242	111.3
48.6	469	228	241	110.8
48.6	467	227	240	110.3
48.6	465	226	239	109.8
48.6	463	225	238	109.3
48.6	461	224	237	108.8
48.6	459	223	236	108.3
48.6	457	222	235	107.8
48.6	455	221	234	107.3
48.6	453	220	233	106.8
48.6	451	219	232	106.3
48.6	449	218	231	105.8
48.6	444	216	228	105.1
48.6	442	215	227	104.6
48.6	440	214	226	104.1
48.6	438	213	225	103.6
48.6	436	212	224	103.1
48.6	434	211	223	102.6
48.6	432	210	222	102.1
48.6	430	209	221	101.6
48.6	428	208	220	101.1
48.6	426	207	219	100.6
48.6	424	206	218	100.1

%E	M1	M2	DM	M*
48.6	422	205	217	99.6
48.6	420	204	216	99.1
48.6	418	203	215	98.6
48.6	416	202	214	98.1
48.6	414	201	213	97.6
48.6	407	198	209	96.3
48.6	405	197	208	95.8
48.6	403	196	207	95.3
48.6	401	195	206	94.8
48.6	399	194	205	94.3
48.6	397	193	204	93.8
48.6	395	192	203	93.3
48.6	393	191	202	92.8
48.6	391	190	201	92.3
48.6	389	189	200	91.8
48.6	387	188	199	91.3
48.6	385	187	198	90.8
48.6	383	186	197	90.3
48.6	381	185	196	89.8
48.6	370	180	190	87.6
48.6	368	179	189	87.1
48.6	366	178	188	86.6
48.6	364	177	187	86.1
48.6	362	176	186	85.6
48.6	360	175	185	85.1
48.6	358	174	184	84.6
48.6	356	173	183	84.1
48.6	354	172	182	83.6
48.6	352	171	181	83.1
48.6	350	170	180	82.6
48.6	348	169	179	82.1
48.6	346	168	178	81.6
48.6	333	162	171	78.8
48.6	331	161	170	78.3
48.6	329	160	169	77.8
48.6	327	159	168	77.3
48.6	325	158	167	76.8
48.6	323	157	166	76.3
48.6	321	156	165	75.8
48.6	319	155	164	75.3
48.6	317	154	163	74.8
48.6	315	153	162	74.3
48.6	313	152	161	73.8
48.6	311	151	160	73.3
48.6	296	144	152	70.1
48.6	294	143	151	69.6
48.6	292	142	150	69.1
48.6	290	141	149	68.6
48.6	288	140	148	68.1
48.6	286	139	147	67.6

%E	M1	M2	DM	M*
48.6	284	138	146	67.1
48.6	282	137	145	66.6
48.6	280	136	144	66.1
48.6	278	135	143	65.6
48.6	276	134	142	65.1
48.6	259	126	133	61.3
48.6	257	125	132	60.8
48.6	255	124	131	60.3
48.6	253	123	130	59.8
48.6	251	122	129	59.3
48.6	249	121	128	58.8
48.6	247	120	127	58.3
48.6	245	119	126	57.8
48.6	243	118	125	57.3
48.6	222	108	114	52.5
48.6	220	107	113	52.0
48.6	218	106	112	51.5
48.6	216	105	111	51.0
48.6	214	104	110	50.5
48.6	212	103	109	50.0
48.6	210	102	108	49.5
48.6	208	101	107	49.0
48.6	185	90	95	43.8
48.6	183	89	94	43.3
48.6	181	88	93	42.8
48.6	179	87	92	42.3
48.6	177	86	91	41.8
48.6	175	85	90	41.3
48.6	173	84	89	40.8
48.6	148	72	76	35.0
48.6	146	71	75	34.5
48.6	144	70	74	34.0
48.6	142	69	73	33.5
48.6	140	68	72	33.0
48.6	138	67	71	32.5
48.6	111	54	57	26.3
48.6	109	53	56	25.8
48.6	107	52	55	25.3
48.6	105	51	54	24.8
48.6	74	36	38	17.5
48.6	72	35	37	17.0
48.6	70	34	36	16.5
48.6	37	18	19	8.8
48.6	35	17	18	8.3
48.5	499	242	257	117.4
48.5	497	241	256	116.9
48.5	495	240	255	116.4
48.5	493	239	254	115.9
48.5	491	238	253	115.4
48.5	489	237	252	114.9

%E	M1	M2	DM	M*
48.5	487	236	251	114.4
48.5	485	235	250	113.9
48.5	482	234	248	113.6
48.5	480	233	247	113.1
48.5	478	232	246	112.6
48.5	476	231	245	112.1
48.5	474	230	244	111.6
48.5	472	229	243	111.1
48.5	470	228	242	110.6
48.5	468	227	241	110.1
48.5	466	226	240	109.6
48.5	464	225	239	109.1
48.5	462	224	238	108.6
48.5	460	223	237	108.1
48.5	458	222	236	107.6
48.5	456	221	235	107.1
48.5	454	220	234	106.6
48.5	452	219	233	106.1
48.5	447	217	230	105.3
48.5	445	216	229	104.8
48.5	443	215	228	104.3
48.5	441	214	227	103.8
48.5	439	213	226	103.3
48.5	437	212	225	102.8
48.5	435	211	224	102.3
48.5	433	210	223	101.8
48.5	431	209	222	101.3
48.5	429	208	221	100.8
48.5	427	207	220	100.3
48.5	425	206	219	99.8
48.5	423	205	218	99.3
48.5	421	204	217	98.9
48.5	412	200	212	97.1
48.5	410	199	211	96.6
48.5	408	198	210	96.1
48.5	406	197	209	95.6
48.5	404	196	208	95.1
48.5	402	195	207	94.6
48.5	400	194	206	94.1
48.5	398	193	205	93.6
48.5	396	192	204	93.1
48.5	394	191	203	92.6
48.5	392	190	202	92.1
48.5	390	189	201	91.6
48.5	388	188	200	91.1
48.5	379	184	195	89.3
48.5	377	183	194	88.8
48.5	375	182	193	88.3
48.5	373	181	192	87.8
48.5	371	180	191	87.3

%E	M1	M2	DM	M*
48.5	369	179	190	86.8
48.5	367	178	189	86.3
48.5	365	177	188	85.8
48.5	363	176	187	85.3
48.5	361	175	186	84.8
48.5	359	174	185	84.3
48.5	357	173	184	83.8
48.5	355	172	183	83.3
48.5	344	167	177	81.1
48.5	342	166	176	80.6
48.5	340	165	175	80.1
48.5	338	164	174	79.6
48.5	336	163	173	79.1
48.5	334	162	172	78.6
48.5	332	161	171	78.1
48.5	330	160	170	77.6
48.5	328	159	169	77.1
48.5	326	158	168	76.6
48.5	324	157	167	76.1
48.5	309	150	159	72.8
48.5	307	149	158	72.3
48.5	305	148	157	71.8
48.5	303	147	156	71.3
48.5	301	146	155	70.8
48.5	299	145	154	70.3
48.5	297	144	153	69.8
48.5	295	143	152	69.3
48.5	293	142	151	68.8
48.5	291	141	150	68.3
48.5	274	133	141	64.6
48.5	272	132	140	64.1
48.5	270	131	139	63.6
48.5	268	130	138	63.1
48.5	266	129	137	62.6
48.5	264	128	136	62.1
48.5	262	127	135	61.6
48.5	260	126	134	61.1
48.5	258	125	133	60.6
48.5	241	117	124	56.8
48.5	239	116	123	56.3
48.5	237	115	122	55.8
48.5	235	114	121	55.3
48.5	233	113	120	54.8
48.5	231	112	119	54.3
48.5	229	111	118	53.8
48.5	227	110	117	53.3
48.5	206	100	106	48.5
48.5	204	99	105	48.0
48.5	202	98	104	47.5
48.5	200	97	103	47.0
48.5	198	96	102	46.5
48.5	196	95	101	46.0
48.5	194	94	100	45.5
48.5	171	83	88	40.3
48.5	169	82	87	39.8
48.5	167	81	86	39.3
48.5	165	80	85	38.8
48.5	163	79	84	38.3
48.5	136	66	70	32.0
48.5	134	65	69	31.5
48.5	132	64	68	31.0
48.5	130	63	67	30.5
48.5	103	50	53	24.3
48.5	101	49	52	23.8
48.5	99	48	51	23.3
48.5	97	47	50	22.8
48.5	68	33	35	16.0
48.5	66	32	34	15.5
48.5	33	16	17	7.8
48.4	500	242	258	117.1
48.4	498	241	257	116.6
48.4	496	240	256	116.1
48.4	494	239	255	115.6
48.4	492	238	254	115.1
48.4	490	237	253	114.6
48.4	488	236	252	114.1
48.4	486	235	251	113.6
48.4	483	234	249	113.4
48.4	481	233	248	112.9
48.4	479	232	247	112.4
48.4	477	231	246	111.9
48.4	475	230	245	111.4
48.4	473	229	244	110.9
48.4	471	228	243	110.4
48.4	469	227	242	109.9
48.4	467	226	241	109.4
48.4	465	225	240	108.9
48.4	463	224	239	108.4
48.4	461	223	238	107.9
48.4	459	222	237	107.4
48.4	457	221	236	106.9
48.4	455	220	235	106.4
48.4	450	218	232	105.6
48.4	448	217	231	105.1
48.4	446	216	230	104.6
48.4	444	215	229	104.1
48.4	442	214	228	103.6
48.4	440	213	227	103.1
48.4	438	212	226	102.6
48.4	436	211	225	102.1
48.4	434	210	224	101.6
48.4	432	209	223	101.1
48.4	430	208	222	100.6
48.4	428	207	221	100.1
48.4	426	206	220	99.6
48.4	419	203	216	98.4
48.4	417	202	215	97.9
48.4	415	201	214	97.4
48.4	413	200	213	96.9
48.4	411	199	212	96.4
48.4	409	198	211	95.9
48.4	407	197	210	95.4
48.4	405	196	209	94.9
48.4	403	195	208	94.4
48.4	401	194	207	93.9
48.4	399	193	206	93.4
48.4	397	192	205	92.9
48.4	395	191	204	92.4
48.4	386	187	199	90.6
48.4	384	186	198	90.1
48.4	382	185	197	89.6
48.4	380	184	196	89.1
48.4	378	183	195	88.6
48.4	376	182	194	88.1
48.4	374	181	193	87.6
48.4	372	180	192	87.1
48.4	370	179	191	86.6
48.4	368	178	190	86.1
48.4	366	177	189	85.6
48.4	364	176	188	85.1
48.4	353	171	182	82.8
48.4	351	170	181	82.3
48.4	349	169	180	81.8
48.4	347	168	179	81.3
48.4	345	167	178	80.8
48.4	343	166	177	80.3
48.4	341	165	176	79.8
48.4	339	164	175	79.3
48.4	337	163	174	78.8
48.4	335	162	173	78.3
48.4	322	156	166	75.6
48.4	320	155	165	75.1
48.4	318	154	164	74.6
48.4	316	153	163	74.1
48.4	314	152	162	73.6
48.4	312	151	161	73.1
48.4	310	150	160	72.6
48.4	308	149	159	72.1
48.4	306	148	158	71.6
48.4	304	147	157	71.1
48.4	289	140	149	67.8
48.4	287	139	148	67.3
48.4	285	138	147	66.8
48.4	283	137	146	66.3
48.4	281	136	145	65.8
48.4	279	135	144	65.3
48.4	277	134	143	64.8
48.4	275	133	142	64.3
48.4	273	132	141	63.8
48.4	256	124	132	60.1
48.4	254	123	131	59.6
48.4	252	122	130	59.1
48.4	250	121	129	58.6
48.4	248	120	128	58.1
48.4	246	119	127	57.6
48.4	244	118	126	57.1
48.4	225	109	116	52.8
48.4	223	108	115	52.3
48.4	221	107	114	51.8
48.4	219	106	113	51.3
48.4	217	105	112	50.8
48.4	215	104	111	50.3
48.4	213	103	110	49.8
48.4	192	93	99	45.0
48.4	190	92	98	44.5
48.4	188	91	97	44.0
48.4	186	90	96	43.5
48.4	184	89	95	43.0
48.4	182	88	94	42.5
48.4	161	78	83	37.8
48.4	159	77	82	37.3
48.4	157	76	81	36.8
48.4	155	75	80	36.3
48.4	153	74	79	35.8
48.4	128	62	66	30.0
48.4	126	61	65	29.5
48.4	124	60	64	29.0
48.4	122	59	63	28.5
48.4	95	46	49	22.3
48.4	93	45	48	21.8
48.4	91	44	47	21.3
48.4	64	31	33	15.0
48.4	62	30	32	14.5
48.4	31	15	16	7.3
48.3	499	241	258	116.4
48.3	497	240	257	115.9
48.3	495	239	256	115.4
48.3	493	238	255	114.9
48.3	491	237	254	114.4
48.3	489	236	253	113.9
48.3	487	235	252	113.4
48.3	484	234	250	113.1
48.3	482	233	249	112.6
48.3	480	232	248	112.1
48.3	478	231	247	111.6
48.3	476	230	246	111.1
48.3	474	229	245	110.6
48.3	472	228	244	110.1
48.3	470	227	243	109.6
48.3	468	226	242	109.1
48.3	466	225	241	108.6
48.3	464	224	240	108.1
48.3	462	223	239	107.6
48.3	460	222	238	107.1
48.3	458	221	237	106.6
48.3	453	219	234	105.9
48.3	451	218	233	105.4
48.3	449	217	232	104.9
48.3	447	216	231	104.4
48.3	445	215	230	103.9
48.3	443	214	229	103.4
48.3	441	213	228	102.9
48.3	439	212	227	102.4
48.3	437	211	226	101.9
48.3	435	210	225	101.4
48.3	433	209	224	100.9
48.3	431	208	223	100.4
48.3	429	207	222	99.9
48.3	424	205	219	99.1
48.3	422	204	218	98.6
48.3	420	203	217	98.1
48.3	418	202	216	97.6
48.3	416	201	215	97.1
48.3	414	200	214	96.6
48.3	412	199	213	96.1
48.3	410	198	212	95.6
48.3	408	197	211	95.1
48.3	406	196	210	94.6
48.3	404	195	209	94.1
48.3	402	194	208	93.6
48.3	400	193	207	93.1
48.3	393	190	203	91.9
48.3	391	189	202	91.4
48.3	389	188	201	90.9
48.3	387	187	200	90.4
48.3	385	186	199	89.9
48.3	383	185	198	89.4
48.3	381	184	197	88.9
48.3	379	183	196	88.4
48.3	377	182	195	87.9

%E	M1	M2	DM	M*
48.3	375	181	194	87.4
48.3	373	180	193	86.9
48.3	362	175	187	84.6
48.3	360	174	186	84.1
48.3	358	173	185	83.6
48.3	356	172	184	83.1
48.3	354	171	183	82.6
48.3	352	170	182	82.1
48.3	350	169	181	81.6
48.3	348	168	180	81.1
48.3	346	167	179	80.6
48.3	344	166	178	80.1
48.3	333	161	172	77.8
48.3	331	160	171	77.3
48.3	329	159	170	76.8
48.3	327	158	169	76.3
48.3	325	157	168	75.8
48.3	323	156	167	75.3
48.3	321	155	166	74.8
48.3	319	154	165	74.3
48.3	317	153	164	73.8
48.3	315	152	163	73.3
48.3	302	146	156	70.6
48.3	300	145	155	70.1
48.3	298	144	154	69.6
48.3	296	143	153	69.1
48.3	294	142	152	68.6
48.3	292	141	151	68.1
48.3	290	140	150	67.6
48.3	288	139	149	67.1
48.3	286	138	148	66.6
48.3	271	131	140	63.3
48.3	269	130	139	62.8
48.3	267	129	138	62.3
48.3	265	128	137	61.8
48.3	263	127	136	61.3
48.3	261	126	135	60.8
48.3	259	125	134	60.3
48.3	242	117	125	56.6
48.3	240	116	124	56.1
48.3	238	115	123	55.6
48.3	236	114	122	55.1
48.3	234	113	121	54.6
48.3	232	112	120	54.1
48.3	230	111	119	53.6
48.3	211	102	109	49.3
48.3	209	101	108	48.8
48.3	207	100	107	48.3
48.3	205	99	106	47.8
48.3	203	98	105	47.3

%E	M1	M2	DM	M*
48.3	201	97	104	46.8
48.3	180	87	93	42.0
48.3	178	86	92	41.6
48.3	176	85	91	41.1
48.3	174	84	90	40.6
48.3	172	83	89	40.1
48.3	151	73	78	35.3
48.3	149	72	77	34.8
48.3	147	71	76	34.3
48.3	145	70	75	33.8
48.3	143	69	74	33.3
48.3	120	58	62	28.0
48.3	118	57	61	27.5
48.3	116	56	60	27.0
48.3	89	43	46	20.8
48.3	87	42	45	20.3
48.3	60	29	31	14.0
48.3	58	28	30	13.5
48.3	29	14	15	6.8
48.2	500	241	259	116.2
48.2	498	240	258	115.7
48.2	496	239	257	115.2
48.2	494	238	256	114.7
48.2	492	237	255	114.2
48.2	490	236	254	113.7
48.2	488	235	253	113.2
48.2	485	234	251	112.9
48.2	483	233	250	112.4
48.2	481	232	249	111.9
48.2	479	231	248	111.4
48.2	477	230	247	110.9
48.2	475	229	246	110.4
48.2	473	228	245	109.9
48.2	471	227	244	109.4
48.2	469	226	243	108.9
48.2	467	225	242	108.4
48.2	465	224	241	107.9
48.2	463	223	240	107.4
48.2	461	222	239	106.9
48.2	456	220	236	106.1
48.2	454	219	235	105.6
48.2	452	218	234	105.1
48.2	450	217	233	104.6
48.2	448	216	232	104.1
48.2	446	215	231	103.6
48.2	444	214	230	103.1
48.2	442	213	229	102.6
48.2	440	212	228	102.1
48.2	438	211	227	101.6
48.2	436	210	226	101.1

%E	M1	M2	DM	M*
48.2	434	209	225	100.6
48.2	427	206	221	99.4
48.2	425	205	220	98.9
48.2	423	204	219	98.4
48.2	421	203	218	97.9
48.2	419	202	217	97.4
48.2	417	201	216	96.9
48.2	415	200	215	96.4
48.2	413	199	214	95.9
48.2	411	198	213	95.4
48.2	409	197	212	94.9
48.2	407	196	211	94.4
48.2	398	192	206	92.6
48.2	396	191	205	92.1
48.2	394	190	204	91.6
48.2	392	189	203	91.1
48.2	390	188	202	90.6
48.2	388	187	201	90.1
48.2	386	186	200	89.6
48.2	384	185	199	89.1
48.2	382	184	198	88.6
48.2	380	183	197	88.1
48.2	371	179	192	86.4
48.2	369	178	191	85.9
48.2	367	177	190	85.4
48.2	365	176	189	84.9
48.2	363	175	188	84.4
48.2	361	174	187	83.9
48.2	359	173	186	83.4
48.2	357	172	185	82.9
48.2	355	171	184	82.4
48.2	353	170	183	81.9
48.2	342	165	177	79.6
48.2	340	164	176	79.1
48.2	338	163	175	78.6
48.2	336	162	174	78.1
48.2	334	161	173	77.6
48.2	332	160	172	77.1
48.2	330	159	171	76.6
48.2	328	158	170	76.1
48.2	326	157	169	75.6
48.2	313	151	162	72.8
48.2	311	150	161	72.3
48.2	309	149	160	71.8
48.2	307	148	159	71.3
48.2	305	147	158	70.8
48.2	303	146	157	70.3
48.2	301	145	156	69.9
48.2	299	144	155	69.4
48.2	284	137	147	66.1

%E	M1	M2	DM	M*
48.2	282	136	146	65.6
48.2	280	135	145	65.1
48.2	278	134	144	64.6
48.2	276	133	143	64.1
48.2	274	132	142	63.6
48.2	272	131	141	63.1
48.2	257	124	133	59.8
48.2	255	123	132	59.3
48.2	253	122	131	58.8
48.2	251	121	130	58.3
48.2	249	120	129	57.8
48.2	247	119	128	57.3
48.2	245	118	127	56.8
48.2	228	110	118	53.1
48.2	226	109	117	52.6
48.2	224	108	116	52.1
48.2	222	107	115	51.6
48.2	220	106	114	51.1
48.2	218	105	113	50.6
48.2	199	96	103	46.3
48.2	197	95	102	45.8
48.2	195	94	101	45.3
48.2	193	93	100	44.8
48.2	191	92	99	44.3
48.2	170	82	88	39.6
48.2	168	81	87	39.1
48.2	166	80	86	38.6
48.2	164	79	85	38.1
48.2	141	68	73	32.8
48.2	139	67	72	32.3
48.2	137	66	71	31.8
48.2	114	55	59	26.5
48.2	112	54	58	26.0
48.2	110	53	57	25.5
48.2	85	41	44	19.8
48.2	83	40	43	19.3
48.2	56	27	29	13.0
48.1	499	240	259	115.4
48.1	497	239	258	114.9
48.1	495	238	257	114.4
48.1	493	237	256	113.9
48.1	491	236	255	113.4
48.1	489	235	254	112.9
48.1	486	234	252	112.7
48.1	484	233	251	112.2
48.1	482	232	250	111.7
48.1	480	231	249	111.2
48.1	478	230	248	110.7
48.1	476	229	247	110.2
48.1	474	228	246	109.7

%E	M1	M2	DM	M*
48.1	472	227	245	109.2
48.1	470	226	244	108.7
48.1	468	225	243	108.2
48.1	466	224	242	107.7
48.1	464	223	241	107.2
48.1	462	222	240	106.7
48.1	459	221	238	106.4
48.1	457	220	237	105.9
48.1	455	219	236	105.4
48.1	453	218	235	104.9
48.1	451	217	234	104.4
48.1	449	216	233	103.9
48.1	447	215	232	103.4
48.1	445	214	231	102.9
48.1	443	213	230	102.4
48.1	441	212	229	101.9
48.1	439	211	228	101.4
48.1	437	210	227	100.9
48.1	432	208	224	100.1
48.1	430	207	223	99.6
48.1	428	206	222	99.1
48.1	426	205	221	98.7
48.1	424	204	220	98.2
48.1	422	203	219	97.7
48.1	420	202	218	97.2
48.1	418	201	217	96.7
48.1	416	200	216	96.2
48.1	414	199	215	95.7
48.1	412	198	214	95.2
48.1	405	195	210	93.9
48.1	403	194	209	93.4
48.1	401	193	208	92.9
48.1	399	192	207	92.4
48.1	397	191	206	91.9
48.1	395	190	205	91.4
48.1	393	189	204	90.9
48.1	391	188	203	90.4
48.1	389	187	202	89.9
48.1	387	186	201	89.4
48.1	385	185	200	88.9
48.1	378	182	196	87.6
48.1	376	181	195	87.1
48.1	374	180	194	86.6
48.1	372	179	193	86.1
48.1	370	178	192	85.6
48.1	368	177	191	85.1
48.1	366	176	190	84.6
48.1	364	175	189	84.1
48.1	362	174	188	83.6
48.1	360	173	187	83.1

%E	M1	M2	DM	M*
48.1	351	169	182	81.4
48.1	349	168	181	80.9
48.1	347	167	180	80.4
48.1	345	166	179	79.9
48.1	343	165	178	79.4
48.1	341	164	177	78.9
48.1	339	163	176	78.4
48.1	337	162	175	77.9
48.1	335	161	174	77.4
48.1	324	156	168	75.1
48.1	322	155	167	74.6
48.1	320	154	166	74.1
48.1	318	153	165	73.6
48.1	316	152	164	73.1
48.1	314	151	163	72.6
48.1	312	150	162	72.1
48.1	310	149	161	71.6
48.1	308	148	160	71.1
48.1	297	143	154	68.9
48.1	295	142	153	68.4
48.1	293	141	152	67.9
48.1	291	140	151	67.4
48.1	289	139	150	66.9
48.1	287	138	149	66.4
48.1	285	137	148	65.9
48.1	283	136	147	65.4
48.1	270	130	140	62.6
48.1	268	129	139	62.1
48.1	266	128	138	61.6
48.1	264	127	137	61.1
48.1	262	126	136	60.6
48.1	260	125	135	60.1
48.1	258	124	134	59.6
48.1	243	117	126	56.3
48.1	241	116	125	55.8
48.1	239	115	124	55.3
48.1	237	114	123	54.8
48.1	235	113	122	54.3
48.1	233	112	121	53.8
48.1	231	111	120	53.3
48.1	216	104	112	50.1
48.1	214	103	111	49.6
48.1	212	102	110	49.1
48.1	210	101	109	48.6
48.1	208	100	108	48.1
48.1	206	99	107	47.6
48.1	189	91	98	43.8
48.1	187	90	97	43.3
48.1	185	89	96	42.8
48.1	183	88	95	42.3

%E	M1	M2	DM	M*
48.1	181	87	94	41.8
48.1	162	78	84	37.6
48.1	160	77	83	37.1
48.1	158	76	82	36.6
48.1	156	75	81	36.1
48.1	154	74	80	35.6
48.1	135	65	70	31.3
48.1	133	64	69	30.8
48.1	131	63	68	30.3
48.1	129	62	67	29.8
48.1	108	52	56	25.0
48.1	106	51	55	24.5
48.1	104	50	54	24.0
48.1	81	39	42	18.8
48.1	79	38	41	18.3
48.1	77	37	40	17.8
48.1	54	26	28	12.5
48.1	52	25	27	12.0
48.1	27	13	14	6.3
48.0	500	240	260	115.2
48.0	498	239	259	114.7
48.0	496	238	258	114.2
48.0	494	237	257	113.7
48.0	492	236	256	113.2
48.0	490	235	255	112.7
48.0	488	234	254	112.2
48.0	487	234	253	112.4
48.0	485	233	252	111.9
48.0	483	232	251	111.4
48.0	481	231	250	110.9
48.0	479	230	249	110.4
48.0	477	229	248	109.9
48.0	475	228	247	109.4
48.0	473	227	246	108.9
48.0	471	226	245	108.4
48.0	469	225	244	107.9
48.0	467	224	243	107.4
48.0	465	223	242	106.9
48.0	460	221	239	106.2
48.0	458	220	238	105.7
48.0	456	219	237	105.2
48.0	454	218	236	104.7
48.0	452	217	235	104.2
48.0	450	216	234	103.7
48.0	448	215	233	103.2
48.0	446	214	232	102.7
48.0	444	213	231	102.2
48.0	442	212	230	101.7
48.0	440	211	229	101.2
48.0	435	209	226	100.4

%E	M1	M2	DM	M*
48.0	433	208	225	99.9
48.0	431	207	224	99.4
48.0	429	206	223	98.9
48.0	427	205	222	98.4
48.0	425	204	221	97.9
48.0	423	203	220	97.4
48.0	421	202	219	96.9
48.0	419	201	218	96.4
48.0	417	200	217	95.9
48.0	415	199	216	95.4
48.0	410	197	213	94.7
48.0	408	196	212	94.2
48.0	406	195	211	93.7
48.0	404	194	210	93.2
48.0	402	193	209	92.7
48.0	400	192	208	92.2
48.0	398	191	207	91.7
48.0	396	190	206	91.2
48.0	394	189	205	90.7
48.0	392	188	204	90.2
48.0	383	184	199	88.4
48.0	381	183	198	87.9
48.0	379	182	197	87.4
48.0	377	181	196	86.9
48.0	375	180	195	86.4
48.0	373	179	194	85.9
48.0	371	178	193	85.4
48.0	369	177	192	84.9
48.0	367	176	191	84.4
48.0	358	172	186	82.6
48.0	356	171	185	82.1
48.0	354	170	184	81.6
48.0	352	169	183	81.1
48.0	350	168	182	80.6
48.0	348	167	181	80.1
48.0	346	166	180	79.6
48.0	344	165	179	79.1
48.0	342	164	178	78.6
48.0	333	160	173	76.9
48.0	331	159	172	76.4
48.0	329	158	171	75.9
48.0	327	157	170	75.4
48.0	325	156	169	74.9
48.0	323	155	168	74.4
48.0	321	154	167	73.9
48.0	319	153	166	73.4
48.0	317	152	165	72.9
48.0	306	147	159	70.6
48.0	304	146	158	70.1
48.0	302	145	157	69.6

%E	M1	M2	DM	M*
48.0	300	144	156	69.1
48.0	298	143	155	68.6
48.0	296	142	154	68.1
48.0	294	141	153	67.6
48.0	281	135	146	64.9
48.0	279	134	145	64.4
48.0	277	133	144	63.9
48.0	275	132	143	63.4
48.0	273	131	142	62.9
48.0	271	130	141	62.4
48.0	269	129	140	61.9
48.0	256	123	133	59.1
48.0	254	122	132	58.6
48.0	252	121	131	58.1
48.0	250	120	130	57.6
48.0	248	119	129	57.1
48.0	246	118	128	56.6
48.0	244	117	127	56.1
48.0	229	110	119	52.8
48.0	227	109	118	52.3
48.0	225	108	117	51.8
48.0	223	107	116	51.3
48.0	221	106	115	50.8
48.0	204	98	106	47.1
48.0	202	97	105	46.6
48.0	200	96	104	46.1
48.0	198	95	103	45.6
48.0	196	94	102	45.1
48.0	179	86	93	41.3
48.0	177	85	92	40.8
48.0	175	84	91	40.3
48.0	173	83	90	39.8
48.0	171	82	89	39.3
48.0	152	73	79	35.1
48.0	150	72	78	34.6
48.0	148	71	77	34.1
48.0	127	61	66	29.3
48.0	125	60	65	28.8
48.0	123	59	64	28.3
48.0	102	49	53	23.5
48.0	100	48	52	23.0
48.0	98	47	51	22.5
48.0	75	36	39	17.3
48.0	50	24	26	11.5
48.0	25	12	13	5.8
47.9	499	239	260	114.5
47.9	497	238	259	114.0
47.9	495	237	258	113.5
47.9	493	236	257	113.0
47.9	491	235	256	112.5

%E	M1	M2	DM	M*
47.9	489	234	255	112.0
47.9	486	233	253	111.7
47.9	484	232	252	111.2
47.9	482	231	251	110.7
47.9	480	230	250	110.2
47.9	478	229	249	109.7
47.9	476	228	248	109.2
47.9	474	227	247	108.7
47.9	472	226	246	108.2
47.9	470	225	245	107.7
47.9	468	224	244	107.2
47.9	466	223	243	106.7
47.9	463	222	241	106.4
47.9	461	221	240	105.9
47.9	459	220	239	105.4
47.9	457	219	238	104.9
47.9	455	218	237	104.4
47.9	453	217	236	103.9
47.9	451	216	235	103.5
47.9	449	215	234	103.0
47.9	447	214	233	102.5
47.9	445	213	232	102.0
47.9	443	212	231	101.5
47.9	438	210	228	100.7
47.9	436	209	227	100.2
47.9	434	208	226	99.7
47.9	432	207	225	99.2
47.9	430	206	224	98.7
47.9	428	205	223	98.2
47.9	426	204	222	97.7
47.9	424	203	221	97.2
47.9	422	202	220	96.7
47.9	420	201	219	96.2
47.9	413	198	215	94.9
47.9	411	197	214	94.4
47.9	409	196	213	93.9
47.9	407	195	212	93.4
47.9	405	194	211	92.9
47.9	403	193	210	92.4
47.9	401	192	209	91.9
47.9	399	191	208	91.4
47.9	397	190	207	90.9
47.9	390	187	203	89.7
47.9	388	186	202	89.2
47.9	386	185	201	88.7
47.9	384	184	200	88.2
47.9	382	183	199	87.7
47.9	380	182	198	87.2
47.9	378	181	197	86.7
47.9	376	180	196	86.2

%E	M1	M2	DM	M*
47.9	374	179	195	85.7
47.9	365	175	190	83.9
47.9	363	174	189	83.4
47.9	361	173	188	82.9
47.9	359	172	187	82.4
47.9	357	171	186	81.9
47.9	355	170	185	81.4
47.9	353	169	184	80.9
47.9	351	168	183	80.4
47.9	349	167	182	79.9
47.9	340	163	177	78.1
47.9	338	162	176	77.6
47.9	336	161	175	77.1
47.9	334	160	174	76.6
47.9	332	159	173	76.1
47.9	330	158	172	75.6
47.9	328	157	171	75.1
47.9	326	156	170	74.7
47.9	315	151	164	72.4
47.9	313	150	163	71.9
47.9	311	149	162	71.4
47.9	309	148	161	70.9
47.9	307	147	160	70.4
47.9	305	146	159	69.9
47.9	303	145	158	69.4
47.9	292	140	152	67.1
47.9	290	139	151	66.6
47.9	288	138	150	66.1
47.9	286	137	149	65.6
47.9	284	136	148	65.1
47.9	282	135	147	64.6
47.9	280	134	146	64.1
47.9	267	128	139	61.4
47.9	265	127	138	60.9
47.9	263	126	137	60.4
47.9	261	125	136	59.9
47.9	259	124	135	59.4
47.9	257	123	134	58.9
47.9	242	116	126	55.6
47.9	240	115	125	55.1
47.9	238	114	124	54.6
47.9	236	113	123	54.1
47.9	234	112	122	53.6
47.9	219	105	114	50.3
47.9	217	104	113	49.8
47.9	215	103	112	49.3
47.9	213	102	111	48.8
47.9	211	101	110	48.3
47.9	194	93	101	44.6
47.9	192	92	100	44.1

%E	M1	M2	DM	M*
47.9	190	91	99	43.6
47.9	188	90	98	43.1
47.9	169	81	88	38.8
47.9	167	80	87	38.3
47.9	165	79	86	37.8
47.9	163	78	85	37.3
47.9	146	70	76	33.6
47.9	144	69	75	33.1
47.9	142	68	74	32.6
47.9	140	67	73	32.1
47.9	121	58	63	27.8
47.9	119	57	62	27.3
47.9	117	56	61	26.8
47.9	96	46	50	22.0
47.9	94	45	49	21.5
47.9	73	35	38	16.8
47.9	71	34	37	16.3
47.9	48	23	25	11.0
47.8	500	239	261	114.2
47.8	498	238	260	113.7
47.8	496	237	259	113.2
47.8	494	236	258	112.7
47.8	492	235	257	112.2
47.8	490	234	256	111.7
47.8	487	233	254	111.5
47.8	485	232	253	111.0
47.8	483	231	252	110.5
47.8	481	230	251	110.0
47.8	479	229	250	109.5
47.8	477	228	249	109.0
47.8	475	227	248	108.5
47.8	473	226	247	108.0
47.8	471	225	246	107.5
47.8	469	224	245	107.0
47.8	467	223	244	106.5
47.8	464	222	242	106.2
47.8	462	221	241	105.7
47.8	460	220	240	105.2
47.8	458	219	239	104.7
47.8	456	218	238	104.2
47.8	454	217	237	103.7
47.8	452	216	236	103.2
47.8	450	215	235	102.7
47.8	448	214	234	102.2
47.8	446	213	233	101.7
47.8	441	211	230	101.0
47.8	439	210	229	100.5
47.8	437	209	228	100.0
47.8	435	208	227	99.5
47.8	433	207	226	99.0

%E	M1	M2	DM	M*
47.8	431	206	225	98.5
47.8	429	205	224	98.0
47.8	427	204	223	97.5
47.8	425	203	222	97.0
47.8	423	202	221	96.5
47.8	418	200	218	95.7
47.8	416	199	217	95.2
47.8	414	198	216	94.7
47.8	412	197	215	94.2
47.8	410	196	214	93.7
47.8	408	195	213	93.2
47.8	406	194	212	92.7
47.8	404	193	211	92.2
47.8	402	192	210	91.7
47.8	400	191	209	91.2
47.8	395	189	206	90.4
47.8	393	188	205	89.9
47.8	391	187	204	89.4
47.8	389	186	203	88.9
47.8	387	185	202	88.4
47.8	385	184	201	87.9
47.8	383	183	200	87.4
47.8	381	182	199	86.9
47.8	379	181	198	86.4
47.8	372	178	194	85.2
47.8	370	177	193	84.7
47.8	368	176	192	84.2
47.8	366	175	191	83.7
47.8	364	174	190	83.2
47.8	362	173	189	82.7
47.8	360	172	188	82.2
47.8	358	171	187	81.7
47.8	356	170	186	81.2
47.8	347	166	181	79.4
47.8	345	165	180	78.9
47.8	343	164	179	78.4
47.8	341	163	178	77.9
47.8	339	162	177	77.4
47.8	337	161	176	76.9
47.8	335	160	175	76.4
47.8	324	155	169	74.2
47.8	322	154	168	73.7
47.8	320	153	167	73.2
47.8	318	152	166	72.7
47.8	316	151	165	72.2
47.8	314	150	164	71.7
47.8	312	149	163	71.2
47.8	301	144	157	68.9
47.8	299	143	156	68.4
47.8	297	142	155	67.9

%E	M1	M2	DM	M*
47.8	295	141	154	67.4
47.8	293	140	153	66.9
47.8	291	139	152	66.4
47.8	289	138	151	65.9
47.8	278	133	145	63.6
47.8	276	132	144	63.1
47.8	274	131	143	62.6
47.8	272	130	142	62.1
47.8	270	129	141	61.6
47.8	268	128	140	61.1
47.8	255	122	133	58.4
47.8	253	121	132	57.9
47.8	251	120	131	57.4
47.8	249	119	130	56.9
47.8	247	118	129	56.4
47.8	245	117	128	55.9
47.8	232	111	121	53.1
47.8	230	110	120	52.6
47.8	228	109	119	52.1
47.8	226	108	118	51.6
47.8	224	107	117	51.1
47.8	209	100	109	47.8
47.8	207	99	108	47.3
47.8	205	98	107	46.8
47.8	203	97	106	46.3
47.8	201	96	105	45.9
47.8	186	89	97	42.6
47.8	184	88	96	42.1
47.8	182	87	95	41.6
47.8	180	86	94	41.1
47.8	178	85	93	40.6
47.8	161	77	84	36.8
47.8	159	76	83	36.3
47.8	157	75	82	35.8
47.8	138	66	72	31.6
47.8	136	65	71	31.1
47.8	134	64	70	30.6
47.8	115	55	60	26.3
47.8	113	54	59	25.8
47.8	92	44	48	21.0
47.8	90	43	47	20.5
47.8	69	33	36	15.8
47.8	67	32	35	15.3
47.8	46	22	24	10.5
47.8	23	11	12	5.3
47.7	499	238	261	113.5
47.7	497	237	260	113.0
47.7	495	236	259	112.5
47.7	493	235	258	112.0
47.7	491	234	257	111.5

%E	M1	M2	DM	M*
47.7	488	233	255	111.2
47.7	486	232	254	110.7
47.7	484	231	253	110.3
47.7	482	230	252	109.8
47.7	480	229	251	109.3
47.7	478	228	250	108.8
47.7	476	227	249	108.3
47.7	474	226	248	107.8
47.7	472	225	247	107.3
47.7	470	224	246	106.8
47.7	468	223	245	106.3
47.7	465	222	243	106.0
47.7	463	221	242	105.5
47.7	461	220	241	105.0
47.7	459	219	240	104.5
47.7	457	218	239	104.0
47.7	455	217	238	103.5
47.7	453	216	237	103.0
47.7	451	215	236	102.5
47.7	449	214	235	102.0
47.7	447	213	234	101.5
47.7	444	212	232	101.2
47.7	442	211	231	100.7
47.7	440	210	230	100.2
47.7	438	209	229	99.7
47.7	436	208	228	99.2
47.7	434	207	227	98.7
47.7	432	206	226	98.2
47.7	430	205	225	97.7
47.7	428	204	224	97.2
47.7	426	203	223	96.7
47.7	421	201	220	96.0
47.7	419	200	219	95.5
47.7	417	199	218	95.0
47.7	415	198	217	94.5
47.7	413	197	216	94.0
47.7	411	196	215	93.5
47.7	409	195	214	93.0
47.7	407	194	213	92.5
47.7	405	193	212	92.0
47.7	398	190	208	90.7
47.7	396	189	207	90.2
47.7	394	188	206	89.7
47.7	392	187	205	89.2
47.7	390	186	204	88.7
47.7	388	185	203	88.2
47.7	386	184	202	87.7
47.7	384	183	201	87.2
47.7	377	180	197	85.9
47.7	375	179	196	85.4

%E	M1	M2	DM	M*
47.7	373	178	195	84.9
47.7	371	177	194	84.4
47.7	369	176	193	83.9
47.7	367	175	192	83.4
47.7	365	174	191	82.9
47.7	363	173	190	82.4
47.7	354	169	185	80.7
47.7	352	168	184	80.2
47.7	350	167	183	79.7
47.7	348	166	182	79.2
47.7	346	165	181	78.7
47.7	344	164	180	78.2
47.7	342	163	179	77.7
47.7	333	159	174	75.9
47.7	331	158	173	75.4
47.7	329	157	172	74.9
47.7	327	156	171	74.4
47.7	325	155	170	73.9
47.7	323	154	169	73.4
47.7	321	153	168	72.9
47.7	310	148	162	70.7
47.7	308	147	161	70.2
47.7	306	146	160	69.7
47.7	304	145	159	69.2
47.7	302	144	158	68.7
47.7	300	143	157	68.2
47.7	298	142	156	67.7
47.7	287	137	150	65.4
47.7	285	136	149	64.9
47.7	283	135	148	64.4
47.7	281	134	147	63.9
47.7	279	133	146	63.4
47.7	277	132	145	62.9
47.7	266	127	139	60.6
47.7	264	126	138	60.1
47.7	262	125	137	59.6
47.7	260	124	136	59.1
47.7	258	123	135	58.6
47.7	256	122	134	58.1
47.7	243	116	127	55.4
47.7	241	115	126	54.9
47.7	239	114	125	54.4
47.7	237	113	124	53.9
47.7	235	112	123	53.4
47.7	222	106	116	50.6
47.7	220	105	115	50.1
47.7	218	104	114	49.6
47.7	216	103	113	49.1
47.7	214	102	112	48.6
47.7	199	95	104	45.4

%E	M1	M2	DM	M*
47.7	197	94	103	44.9
47.7	195	93	102	44.4
47.7	193	92	101	43.9
47.7	176	84	92	40.1
47.7	174	83	91	39.6
47.7	172	82	90	39.1
47.7	155	74	81	35.3
47.7	153	73	80	34.8
47.7	151	72	79	34.3
47.7	149	71	78	33.8
47.7	132	63	69	30.1
47.7	130	62	68	29.6
47.7	128	61	67	29.1
47.7	111	53	58	25.3
47.7	109	52	57	24.8
47.7	107	51	56	24.3
47.7	88	42	46	20.0
47.7	86	41	45	19.5
47.7	65	31	34	14.8
47.7	44	21	23	10.0
47.6	500	238	262	113.3
47.6	498	237	261	112.8
47.6	496	236	260	112.3
47.6	494	235	259	111.8
47.6	492	234	258	111.3
47.6	490	233	257	110.8
47.6	489	233	256	111.0
47.6	487	232	255	110.5
47.6	485	231	254	110.0
47.6	483	230	253	109.5
47.6	481	229	252	109.0
47.6	479	228	251	108.5
47.6	477	227	250	108.0
47.6	475	226	249	107.5
47.6	473	225	248	107.0
47.6	471	224	247	106.5
47.6	466	222	244	105.8
47.6	464	221	243	105.3
47.6	462	220	242	104.8
47.6	460	219	241	104.3
47.6	458	218	240	103.8
47.6	456	217	239	103.3
47.6	454	216	238	102.8
47.6	452	215	237	102.3
47.6	450	214	236	101.8
47.6	445	212	233	101.0
47.6	443	211	232	100.5
47.6	441	210	231	100.0
47.6	439	209	230	99.5
47.6	437	208	229	99.0

%E	M1	M2	DM	M*
47.6	435	207	228	98.5
47.6	433	206	227	98.0
47.6	431	205	226	97.5
47.6	429	204	225	97.0
47.6	424	202	222	96.2
47.6	422	201	221	95.7
47.6	420	200	220	95.2
47.6	418	199	219	94.7
47.6	416	198	218	94.2
47.6	414	197	217	93.7
47.6	412	196	216	93.2
47.6	410	195	215	92.7
47.6	403	192	211	91.5
47.6	401	191	210	91.0
47.6	399	190	209	90.5
47.6	397	189	208	90.0
47.6	395	188	207	89.5
47.6	393	187	206	89.0
47.6	391	186	205	88.5
47.6	389	185	204	88.0
47.6	382	182	200	86.7
47.6	380	181	199	86.2
47.6	378	180	198	85.7
47.6	376	179	197	85.2
47.6	374	178	196	84.7
47.6	372	177	195	84.2
47.6	370	176	194	83.7
47.6	368	175	193	83.2
47.6	361	172	189	82.0
47.6	359	171	188	81.5
47.6	357	170	187	81.0
47.6	355	169	186	80.5
47.6	353	168	185	80.0
47.6	351	167	184	79.5
47.6	349	166	183	79.0
47.6	347	165	182	78.5
47.6	340	162	178	77.2
47.6	338	161	177	76.7
47.6	336	160	176	76.2
47.6	334	159	175	75.7
47.6	332	158	174	75.2
47.6	330	157	173	74.7
47.6	328	156	172	74.2
47.6	319	152	167	72.4
47.6	317	151	166	71.9
47.6	315	150	165	71.4
47.6	313	149	164	70.9
47.6	311	148	163	70.4
47.6	309	147	162	69.9
47.6	307	146	161	69.4

%E	M1	M2	DM	M*
47.6	296	141	155	67.2
47.6	294	140	154	66.7
47.6	292	139	153	66.2
47.6	290	138	152	65.7
47.6	288	137	151	65.2
47.6	286	136	150	64.7
47.6	275	131	144	62.4
47.6	273	130	143	61.9
47.6	271	129	142	61.4
47.6	269	128	141	60.9
47.6	267	127	140	60.4
47.6	254	121	133	57.6
47.6	252	120	132	57.1
47.6	250	119	131	56.6
47.6	248	118	130	56.1
47.6	246	117	129	55.6
47.6	233	111	122	52.9
47.6	231	110	121	52.4
47.6	229	109	120	51.9
47.6	227	108	119	51.4
47.6	225	107	118	50.9
47.6	212	101	111	48.1
47.6	210	100	110	47.6
47.6	208	99	109	47.1
47.6	206	98	108	46.6
47.6	191	91	100	43.4
47.6	189	90	99	42.9
47.6	187	89	98	42.4
47.6	185	88	97	41.9
47.6	170	81	89	38.6
47.6	168	80	88	38.1
47.6	166	79	87	37.6
47.6	164	78	86	37.1
47.6	147	70	77	33.3
47.6	145	69	76	32.8
47.6	143	68	75	32.3
47.6	126	60	66	28.6
47.6	124	59	65	28.1
47.6	105	50	55	23.8
47.6	103	49	54	23.3
47.6	84	40	44	19.0
47.6	82	39	43	18.5
47.6	63	30	33	14.3
47.6	42	20	22	9.5
47.6	21	10	11	4.8
47.5	499	237	262	112.6
47.5	497	236	261	112.1
47.5	495	235	260	111.6
47.5	493	234	259	111.1
47.5	491	233	258	110.6

%E	M1	M2	DM	M*
47.5	488	232	256	110.3
47.5	486	231	255	109.8
47.5	484	230	254	109.3
47.5	482	229	253	108.8
47.5	480	228	252	108.3
47.5	478	227	251	107.8
47.5	476	226	250	107.3
47.5	474	225	249	106.8
47.5	472	224	248	106.3
47.5	469	223	246	106.0
47.5	467	222	245	105.5
47.5	465	221	244	105.0
47.5	463	220	243	104.5
47.5	461	219	242	104.0
47.5	459	218	241	103.5
47.5	457	217	240	103.0
47.5	455	216	239	102.5
47.5	453	215	238	102.0
47.5	451	214	237	101.5
47.5	448	213	235	101.3
47.5	446	212	234	100.8
47.5	444	211	233	100.3
47.5	442	210	232	99.8
47.5	440	209	231	99.3
47.5	438	208	230	98.8
47.5	436	207	229	98.3
47.5	434	206	228	97.8
47.5	432	205	227	97.3
47.5	427	203	224	96.5
47.5	425	202	223	96.0
47.5	423	201	222	95.5
47.5	421	200	221	95.0
47.5	419	199	220	94.5
47.5	417	198	219	94.0
47.5	415	197	218	93.5
47.5	413	196	217	93.0
47.5	408	194	214	92.2
47.5	406	193	213	91.7
47.5	404	192	212	91.2
47.5	402	191	211	90.7
47.5	400	190	210	90.3
47.5	398	189	209	89.8
47.5	396	188	208	89.3
47.5	394	187	207	88.8
47.5	387	184	203	87.5
47.5	385	183	202	87.0
47.5	383	182	201	86.5
47.5	381	181	200	86.0
47.5	379	180	199	85.5
47.5	377	179	198	85.0

%E	M1	M2	DM	M*
47.5	375	178	197	84.5
47.5	373	177	196	84.0
47.5	366	174	192	82.7
47.5	364	173	191	82.2
47.5	362	172	190	81.7
47.5	360	171	189	81.2
47.5	358	170	188	80.7
47.5	356	169	187	80.2
47.5	354	168	186	79.7
47.5	345	164	181	78.0
47.5	343	163	180	77.5
47.5	341	162	179	77.0
47.5	339	161	178	76.5
47.5	337	160	177	76.0
47.5	335	159	176	75.5
47.5	326	155	171	73.7
47.5	324	154	170	73.2
47.5	322	153	169	72.7
47.5	320	152	168	72.2
47.5	318	151	167	71.7
47.5	316	150	166	71.2
47.5	314	149	165	70.7
47.5	305	145	160	68.9
47.5	303	144	159	68.4
47.5	301	143	158	67.9
47.5	299	142	157	67.4
47.5	297	141	156	66.9
47.5	295	140	155	66.4
47.5	284	135	149	64.2
47.5	282	134	148	63.7
47.5	280	133	147	63.2
47.5	278	132	146	62.7
47.5	276	131	145	62.2
47.5	265	126	139	59.9
47.5	263	125	138	59.4
47.5	261	124	137	58.9
47.5	259	123	136	58.4
47.5	257	122	135	57.9
47.5	255	121	134	57.4
47.5	244	116	128	55.1
47.5	242	115	127	54.6
47.5	240	114	126	54.1
47.5	238	113	125	53.7
47.5	236	112	124	53.2
47.5	223	106	117	50.4
47.5	221	105	116	49.9
47.5	219	104	115	49.4
47.5	217	103	114	48.9
47.5	204	97	107	46.1
47.5	202	96	106	45.6

%E	M1	M2	DM	M*
47.5	200	95	105	45.1
47.5	198	94	104	44.6
47.5	183	87	96	41.4
47.5	181	86	95	40.9
47.5	179	85	94	40.4
47.5	177	84	93	39.9
47.5	162	77	85	36.6
47.5	160	76	84	36.1
47.5	158	75	83	35.6
47.5	141	67	74	31.8
47.5	139	66	73	31.3
47.5	122	58	64	27.6
47.5	120	57	63	27.1
47.5	118	56	62	26.6
47.5	101	48	53	22.8
47.5	99	47	52	22.3
47.5	80	38	42	18.0
47.5	61	29	32	13.8
47.5	59	28	31	13.3
47.5	40	19	21	9.0
47.4	500	237	263	112.3
47.4	498	236	262	111.8
47.4	496	235	261	111.3
47.4	494	234	260	110.8
47.4	492	233	259	110.3
47.4	489	232	257	110.1
47.4	487	231	256	109.6
47.4	485	230	255	109.1
47.4	483	229	254	108.6
47.4	481	228	253	108.1
47.4	479	227	252	107.6
47.4	477	226	251	107.1
47.4	475	225	250	106.6
47.4	473	224	249	106.1
47.4	470	223	247	105.8
47.4	468	222	246	105.3
47.4	466	221	245	104.8
47.4	464	220	244	104.3
47.4	462	219	243	103.8
47.4	460	218	242	103.3
47.4	458	217	241	102.8
47.4	456	216	240	102.3
47.4	454	215	239	101.8
47.4	449	213	236	101.0
47.4	447	212	235	100.5
47.4	445	211	234	100.0
47.4	443	210	233	99.5
47.4	441	209	232	99.0
47.4	439	208	231	98.6
47.4	437	207	230	98.1

%E	M1	M2	DM	M*
47.4	435	206	229	97.6
47.4	430	204	226	96.8
47.4	428	203	225	96.3
47.4	426	202	224	95.8
47.4	424	201	223	95.3
47.4	422	200	222	94.8
47.4	420	199	221	94.3
47.4	418	198	220	93.8
47.4	416	197	219	93.3
47.4	411	195	216	92.5
47.4	409	194	215	92.0
47.4	407	193	214	91.5
47.4	405	192	213	91.0
47.4	403	191	212	90.5
47.4	401	190	211	90.0
47.4	399	189	210	89.5
47.4	397	188	209	89.0
47.4	392	186	206	88.3
47.4	390	185	205	87.8
47.4	388	184	204	87.3
47.4	386	183	203	86.8
47.4	384	182	202	86.3
47.4	382	181	201	85.8
47.4	380	180	200	85.3
47.4	378	179	199	84.8
47.4	371	176	195	83.5
47.4	369	175	194	83.0
47.4	367	174	193	82.5
47.4	365	173	192	82.0
47.4	363	172	191	81.5
47.4	361	171	190	81.0
47.4	359	170	189	80.5
47.4	352	167	185	79.2
47.4	350	166	184	78.7
47.4	348	165	183	78.2
47.4	346	164	182	77.7
47.4	344	163	181	77.2
47.4	342	162	180	76.7
47.4	340	161	179	76.2
47.4	333	158	175	75.0
47.4	331	157	174	74.5
47.4	329	156	173	74.0
47.4	327	155	172	73.5
47.4	325	154	171	73.0
47.4	323	153	170	72.5
47.4	321	152	169	72.0
47.4	312	148	164	70.2
47.4	310	147	163	69.7
47.4	308	146	162	69.2
47.4	306	145	161	68.7

%E	M1	M2	DM	M*
47.4	304	144	160	68.2
47.4	302	143	159	67.7
47.4	293	139	154	65.9
47.4	291	138	153	65.4
47.4	289	137	152	64.9
47.4	287	136	151	64.4
47.4	285	135	150	63.9
47.4	283	134	149	63.4
47.4	274	130	144	61.7
47.4	272	129	143	61.2
47.4	270	128	142	60.7
47.4	268	127	141	60.2
47.4	266	126	140	59.7
47.4	253	120	133	56.9
47.4	251	119	132	56.4
47.4	249	118	131	55.9
47.4	247	117	130	55.4
47.4	234	111	123	52.7
47.4	232	110	122	52.2
47.4	230	109	121	51.7
47.4	228	108	120	51.2
47.4	215	102	113	48.4
47.4	213	101	112	47.9
47.4	211	100	111	47.4
47.4	209	99	110	46.9
47.4	196	93	103	44.1
47.4	194	92	102	43.6
47.4	192	91	101	43.1
47.4	190	90	100	42.6
47.4	175	83	92	39.4
47.4	173	82	91	38.9
47.4	171	81	90	38.4
47.4	156	74	82	35.1
47.4	154	73	81	34.6
47.4	152	72	80	34.1
47.4	137	65	72	30.8
47.4	135	64	71	30.3
47.4	133	63	70	29.8
47.4	116	55	61	26.1
47.4	114	54	60	25.6
47.4	97	46	51	21.8
47.4	95	45	50	21.3
47.4	78	37	41	17.6
47.4	76	36	40	17.1
47.4	57	27	30	12.8
47.4	38	18	20	8.5
47.4	19	9	10	4.3
47.3	499	236	263	111.6
47.3	497	235	262	111.1
47.3	495	234	261	110.6

%E	M1	M2	DM	M*
47.3	493	233	260	110.1
47.3	491	232	259	109.6
47.3	490	232	258	109.8
47.3	488	231	257	109.3
47.3	486	230	256	108.8
47.3	484	229	255	108.3
47.3	482	228	254	107.9
47.3	480	227	253	107.4
47.3	478	226	252	106.9
47.3	476	225	251	106.4
47.3	474	224	250	105.9
47.3	471	223	248	105.6
47.3	469	222	247	105.1
47.3	467	221	246	104.6
47.3	465	220	245	104.1
47.3	463	219	244	103.6
47.3	461	218	243	103.1
47.3	459	217	242	102.6
47.3	457	216	241	102.1
47.3	455	215	240	101.6
47.3	452	214	238	101.3
47.3	450	213	237	100.8
47.3	448	212	236	100.3
47.3	446	211	235	99.8
47.3	444	210	234	99.3
47.3	442	209	233	98.8
47.3	440	208	232	98.3
47.3	438	207	231	97.8
47.3	433	205	228	97.1
47.3	431	204	227	96.6
47.3	429	203	226	96.1
47.3	427	202	225	95.6
47.3	425	201	224	95.1
47.3	423	200	223	94.6
47.3	421	199	222	94.1
47.3	419	198	221	93.6
47.3	414	196	218	92.8
47.3	412	195	217	92.3
47.3	410	194	216	91.8
47.3	408	193	215	91.3
47.3	406	192	214	90.8
47.3	404	191	213	90.3
47.3	402	190	212	89.8
47.3	400	189	211	89.3
47.3	395	187	208	88.5
47.3	393	186	207	88.0
47.3	391	185	206	87.5
47.3	389	184	205	87.0
47.3	387	183	204	86.5
47.3	385	182	203	86.0

%E	M1	M2	DM	M*
47.3	383	181	202	85.5
47.3	376	178	198	84.3
47.3	374	177	197	83.8
47.3	372	176	196	83.3
47.3	370	175	195	82.8
47.3	368	174	194	82.3
47.3	366	173	193	81.8
47.3	364	172	192	81.3
47.3	357	169	188	80.0
47.3	355	168	187	79.5
47.3	353	167	186	79.0
47.3	351	166	185	78.5
47.3	349	165	184	78.0
47.3	347	164	183	77.5
47.3	338	160	178	75.7
47.3	336	159	177	75.2
47.3	334	158	176	74.7
47.3	332	157	175	74.2
47.3	330	156	174	73.7
47.3	328	155	173	73.2
47.3	319	151	168	71.5
47.3	317	150	167	71.0
47.3	315	149	166	70.5
47.3	313	148	165	70.0
47.3	311	147	164	69.5
47.3	300	142	158	67.2
47.3	298	141	157	66.7
47.3	296	140	156	66.2
47.3	294	139	155	65.7
47.3	292	138	154	65.2
47.3	281	133	148	63.0
47.3	279	132	147	62.5
47.3	277	131	146	62.0
47.3	275	130	145	61.5
47.3	273	129	144	61.0
47.3	264	125	139	59.2
47.3	262	124	138	58.7
47.3	260	123	137	58.2
47.3	258	122	136	57.7
47.3	256	121	135	57.2
47.3	245	116	129	54.9
47.3	243	115	128	54.4
47.3	241	114	127	53.9
47.3	239	113	126	53.4
47.3	237	112	125	52.9
47.3	226	107	119	50.7
47.3	224	106	118	50.2
47.3	222	105	117	49.7
47.3	220	104	116	49.2
47.3	207	98	109	46.4

%E	M1	M2	DM	M*
47.3	205	97	108	45.9
47.3	203	96	107	45.4
47.3	201	95	106	44.9
47.3	188	89	99	42.1
47.3	186	88	98	41.6
47.3	184	87	97	41.1
47.3	182	86	96	40.6
47.3	169	80	89	37.9
47.3	167	79	88	37.4
47.3	165	78	87	36.9
47.3	150	71	79	33.6
47.3	148	70	78	33.1
47.3	146	69	77	32.6
47.3	131	62	69	29.3
47.3	129	61	68	28.8
47.3	112	53	59	25.1
47.3	110	52	58	24.6
47.3	93	44	49	20.8
47.3	91	43	48	20.3
47.3	74	35	39	16.6
47.3	55	26	29	12.3
47.2	500	236	264	111.4
47.2	498	235	263	110.9
47.2	496	234	262	110.4
47.2	494	233	261	109.9
47.2	492	232	260	109.4
47.2	489	231	258	109.1
47.2	487	230	257	108.6
47.2	485	229	256	108.1
47.2	483	228	255	107.6
47.2	481	227	254	107.1
47.2	479	226	253	106.6
47.2	477	225	252	106.1
47.2	475	224	251	105.6
47.2	472	223	249	105.4
47.2	470	222	248	104.9
47.2	468	221	247	104.4
47.2	466	220	246	103.9
47.2	464	219	245	103.4
47.2	462	218	244	102.9
47.2	460	217	243	102.4
47.2	458	216	242	101.9
47.2	453	214	239	101.1
47.2	451	213	238	100.6
47.2	449	212	237	100.1
47.2	447	211	236	99.6
47.2	445	210	235	99.1
47.2	443	209	234	98.6
47.2	441	208	233	98.1
47.2	439	207	232	97.6

%E	M1	M2	DM	M*
47.2	436	206	230	97.3
47.2	434	205	229	96.8
47.2	432	204	228	96.3
47.2	430	203	227	95.8
47.2	428	202	226	95.3
47.2	426	201	225	94.8
47.2	424	200	224	94.3
47.2	422	199	223	93.8
47.2	417	197	220	93.1
47.2	415	196	219	92.6
47.2	413	195	218	92.1
47.2	411	194	217	91.6
47.2	409	193	216	91.1
47.2	407	192	215	90.6
47.2	405	191	214	90.1
47.2	398	188	210	88.8
47.2	396	187	209	88.3
47.2	394	186	208	87.8
47.2	392	185	207	87.3
47.2	390	184	206	86.8
47.2	388	183	205	86.3
47.2	386	182	204	85.8
47.2	381	180	201	85.0
47.2	379	179	200	84.5
47.2	377	178	199	84.0
47.2	375	177	198	83.5
47.2	373	176	197	83.0
47.2	371	175	196	82.5
47.2	369	174	195	82.0
47.2	362	171	191	80.8
47.2	360	170	190	80.3
47.2	358	169	189	79.8
47.2	356	168	188	79.3
47.2	354	167	187	78.8
47.2	352	166	186	78.3
47.2	345	163	182	77.0
47.2	343	162	181	76.5
47.2	341	161	180	76.0
47.2	339	160	179	75.5
47.2	337	159	178	75.0
47.2	335	158	177	74.5
47.2	326	154	172	72.7
47.2	324	153	171	72.3
47.2	322	152	170	71.8
47.2	320	151	169	71.3
47.2	318	150	168	70.8
47.2	316	149	167	70.3
47.2	309	146	163	69.0
47.2	307	145	162	68.5
47.2	305	144	161	68.0

%E	M1	M2	DM	M*
47.2	303	143	160	67.5
47.2	301	142	159	67.0
47.2	299	141	158	66.5
47.2	290	137	153	64.7
47.2	288	136	152	64.2
47.2	286	135	151	63.7
47.2	284	134	150	63.2
47.2	282	133	149	62.7
47.2	271	128	143	60.5
47.2	269	127	142	60.0
47.2	267	126	141	59.5
47.2	265	125	140	59.0
47.2	254	120	134	56.7
47.2	252	119	133	56.2
47.2	250	118	132	55.7
47.2	248	117	131	55.2
47.2	246	116	130	54.7
47.2	235	111	124	52.4
47.2	233	110	123	51.9
47.2	231	109	122	51.4
47.2	229	108	121	50.9
47.2	218	103	115	48.7
47.2	216	102	114	48.2
47.2	214	101	113	47.7
47.2	212	100	112	47.2
47.2	199	94	105	44.4
47.2	197	93	104	43.9
47.2	195	92	103	43.4
47.2	193	91	102	42.9
47.2	180	85	95	40.1
47.2	178	84	94	39.6
47.2	176	83	93	39.1
47.2	163	77	86	36.4
47.2	161	76	85	35.9
47.2	159	75	84	35.4
47.2	144	68	76	32.1
47.2	142	67	75	31.6
47.2	127	60	67	28.3
47.2	125	59	66	27.8
47.2	123	58	65	27.3
47.2	108	51	57	24.1
47.2	106	50	56	23.6
47.2	89	42	47	19.8
47.2	72	34	38	16.1
47.2	53	25	28	11.8
47.2	36	17	19	8.0
47.1	499	235	264	110.7
47.1	497	234	263	110.2
47.1	495	233	262	109.7
47.1	493	232	261	109.2

%E	M1	M2	DM	M*
47.1	490	231	259	108.9
47.1	488	230	258	108.4
47.1	486	229	257	107.9
47.1	484	228	256	107.4
47.1	482	227	255	106.9
47.1	480	226	254	106.4
47.1	478	225	253	105.9
47.1	476	224	252	105.4
47.1	473	223	250	105.1
47.1	471	222	249	104.6
47.1	469	221	248	104.1
47.1	467	220	247	103.6
47.1	465	219	246	103.1
47.1	463	218	245	102.6
47.1	461	217	244	102.1
47.1	459	216	243	101.6
47.1	456	215	241	101.4
47.1	454	214	240	100.9
47.1	452	213	239	100.4
47.1	450	212	238	99.9
47.1	448	211	237	99.4
47.1	446	210	236	98.9
47.1	444	209	235	98.4
47.1	442	208	234	97.9
47.1	437	206	231	97.1
47.1	435	205	230	96.6
47.1	433	204	229	96.1
47.1	431	203	228	95.6
47.1	429	202	227	95.1
47.1	427	201	226	94.6
47.1	425	200	225	94.1
47.1	420	198	222	93.3
47.1	418	197	221	92.8
47.1	416	196	220	92.3
47.1	414	195	219	91.8
47.1	412	194	218	91.3
47.1	410	193	217	90.9
47.1	408	192	216	90.4
47.1	403	190	213	89.6
47.1	401	189	212	89.1
47.1	399	188	211	88.6
47.1	397	187	210	88.1
47.1	395	186	209	87.6
47.1	393	185	208	87.1
47.1	391	184	207	86.6
47.1	384	181	203	85.3
47.1	382	180	202	84.8
47.1	380	179	201	84.3
47.1	378	178	200	83.8
47.1	376	177	199	83.3

%E	M1	M2	DM	M*
47.1	374	176	198	82.8
47.1	367	173	194	81.6
47.1	365	172	193	81.1
47.1	363	171	192	80.6
47.1	361	170	191	80.1
47.1	359	169	190	79.6
47.1	357	168	189	79.1
47.1	350	165	185	77.8
47.1	348	164	184	77.3
47.1	346	163	183	76.8
47.1	344	162	182	76.3
47.1	342	161	181	75.8
47.1	340	160	180	75.3
47.1	333	157	176	74.0
47.1	331	156	175	73.5
47.1	329	155	174	73.0
47.1	327	154	173	72.5
47.1	325	153	172	72.0
47.1	323	152	171	71.5
47.1	314	148	166	69.8
47.1	312	147	165	69.3
47.1	310	146	164	68.8
47.1	308	145	163	68.3
47.1	306	144	162	67.8
47.1	297	140	157	66.0
47.1	295	139	156	65.5
47.1	293	138	155	65.0
47.1	291	137	154	64.5
47.1	289	136	153	64.0
47.1	280	132	148	62.2
47.1	278	131	147	61.7
47.1	276	130	146	61.2
47.1	274	129	145	60.7
47.1	272	128	144	60.2
47.1	263	124	139	58.5
47.1	261	123	138	58.0
47.1	259	122	137	57.5
47.1	257	121	136	57.0
47.1	255	120	135	56.5
47.1	244	115	129	54.2
47.1	242	114	128	53.7
47.1	240	113	127	53.2
47.1	238	112	126	52.7
47.1	227	107	120	50.4
47.1	225	106	119	49.9
47.1	223	105	118	49.4
47.1	221	104	117	48.9
47.1	210	99	111	46.7
47.1	208	98	110	46.2
47.1	206	97	109	45.7

%E	M1	M2	DM	M*
47.1	204	96	108	45.2
47.1	191	90	101	42.4
47.1	189	89	100	41.9
47.1	187	88	99	41.4
47.1	174	82	92	38.6
47.1	172	81	91	38.1
47.1	170	80	90	37.6
47.1	157	74	83	34.9
47.1	155	73	82	34.4
47.1	153	72	81	33.9
47.1	140	66	74	31.1
47.1	138	65	73	30.6
47.1	136	64	72	30.1
47.1	121	57	64	26.9
47.1	119	56	63	26.4
47.1	104	49	55	23.1
47.1	102	48	54	22.6
47.1	87	41	46	19.3
47.1	85	40	45	18.8
47.1	70	33	37	15.6
47.1	68	32	36	15.1
47.1	51	24	27	11.3
47.1	34	16	18	7.5
47.1	17	8	9	3.8
47.0	500	235	265	110.4
47.0	498	234	264	110.0
47.0	496	233	263	109.5
47.0	494	232	262	109.0
47.0	492	231	261	108.5
47.0	491	231	260	108.7
47.0	489	230	259	108.2
47.0	487	229	258	107.7
47.0	485	228	257	107.2
47.0	483	227	256	106.7
47.0	481	226	255	106.2
47.0	479	225	254	105.7
47.0	477	224	253	105.2
47.0	474	223	251	104.9
47.0	472	222	250	104.4
47.0	470	221	249	103.9
47.0	468	220	248	103.4
47.0	466	219	247	102.9
47.0	464	218	246	102.4
47.0	462	217	245	101.9
47.0	460	216	244	101.4
47.0	457	215	242	101.1
47.0	455	214	241	100.7
47.0	453	213	240	100.2
47.0	451	212	239	99.7
47.0	449	211	238	99.2

%E	M1	M2	DM	M*
47.0	447	210	237	98.7
47.0	445	209	236	98.2
47.0	443	208	235	97.7
47.0	440	207	233	97.4
47.0	438	206	232	96.9
47.0	436	205	231	96.4
47.0	434	204	230	95.9
47.0	432	203	229	95.4
47.0	430	202	228	94.9
47.0	428	201	227	94.4
47.0	423	199	224	93.6
47.0	421	198	223	93.1
47.0	419	197	222	92.6
47.0	417	196	221	92.1
47.0	415	195	220	91.6
47.0	413	194	219	91.1
47.0	411	193	218	90.6
47.0	406	191	215	89.9
47.0	404	190	214	89.4
47.0	402	189	213	88.9
47.0	400	188	212	88.4
47.0	398	187	211	87.9
47.0	396	186	210	87.4
47.0	394	185	209	86.9
47.0	389	183	206	86.1
47.0	387	182	205	85.6
47.0	385	181	204	85.1
47.0	383	180	203	84.6
47.0	381	179	202	84.1
47.0	379	178	201	83.6
47.0	377	177	200	83.1
47.0	372	175	197	82.3
47.0	370	174	196	81.8
47.0	368	173	195	81.3
47.0	366	172	194	80.8
47.0	364	171	193	80.3
47.0	362	170	192	79.8
47.0	355	167	188	78.6
47.0	353	166	187	78.1
47.0	351	165	186	77.6
47.0	349	164	185	77.1
47.0	347	163	184	76.6
47.0	345	162	183	76.1
47.0	338	159	179	74.8
47.0	336	158	178	74.3
47.0	334	157	177	73.8
47.0	332	156	176	73.3
47.0	330	155	175	72.8
47.0	328	154	174	72.3
47.0	321	151	170	71.0

%E	M1	M2	DM	M*
47.0	319	150	169	70.5
47.0	317	149	168	70.0
47.0	315	148	167	69.5
47.0	313	147	166	69.0
47.0	304	143	161	67.3
47.0	302	142	160	66.8
47.0	300	141	159	66.3
47.0	298	140	158	65.8
47.0	296	139	157	65.3
47.0	287	135	152	63.5
47.0	285	134	151	63.0
47.0	283	133	150	62.5
47.0	281	132	149	62.0
47.0	279	131	148	61.5
47.0	270	127	143	59.7
47.0	268	126	142	59.2
47.0	266	125	141	58.7
47.0	264	124	140	58.2
47.0	253	119	134	56.0
47.0	251	118	133	55.5
47.0	249	117	132	55.0
47.0	247	116	131	54.5
47.0	236	111	125	52.2
47.0	234	110	124	51.7
47.0	232	109	123	51.2
47.0	230	108	122	50.7
47.0	219	103	116	48.4
47.0	217	102	115	47.9
47.0	215	101	114	47.4
47.0	202	95	107	44.7
47.0	200	94	106	44.2
47.0	198	93	105	43.7
47.0	185	87	98	40.9
47.0	183	86	97	40.4
47.0	181	85	96	39.9
47.0	168	79	89	37.1
47.0	166	78	88	36.7
47.0	164	77	87	36.2
47.0	151	71	80	33.4
47.0	149	70	79	32.9
47.0	134	63	71	29.6
47.0	132	62	70	29.1
47.0	117	55	62	25.9
47.0	115	54	61	25.4
47.0	100	47	53	22.1
47.0	83	39	44	18.3
47.0	66	31	35	14.6
46.9	499	234	265	109.7
46.9	497	233	264	109.2
46.9	495	232	263	108.7

%E	M1	M2	DM	M*
46.9	493	231	262	108.2
46.9	490	230	260	108.0
46.9	488	229	259	107.5
46.9	486	228	258	107.0
46.9	484	227	257	106.5
46.9	482	226	256	106.0
46.9	480	225	255	105.5
46.9	478	224	254	105.0
46.9	475	223	252	104.7
46.9	473	222	251	104.2
46.9	471	221	250	103.7
46.9	469	220	249	103.2
46.9	467	219	248	102.7
46.9	465	218	247	102.2
46.9	463	217	246	101.7
46.9	461	216	245	101.2
46.9	458	215	243	100.9
46.9	456	214	242	100.4
46.9	454	213	241	99.9
46.9	452	212	240	99.4
46.9	450	211	239	98.9
46.9	448	210	238	98.4
46.9	446	209	237	97.9
46.9	441	207	234	97.2
46.9	439	206	233	96.7
46.9	437	205	232	96.2
46.9	435	204	231	95.7
46.9	433	203	230	95.2
46.9	431	202	229	94.7
46.9	429	201	228	94.2
46.9	426	200	226	93.9
46.9	424	199	225	93.4
46.9	422	198	224	92.9
46.9	420	197	223	92.4
46.9	418	196	222	91.9
46.9	416	195	221	91.4
46.9	414	194	220	90.9
46.9	409	192	217	90.1
46.9	407	191	216	89.6
46.9	405	190	215	89.1
46.9	403	189	214	88.6
46.9	401	188	213	88.1
46.9	399	187	212	87.6
46.9	397	186	211	87.1
46.9	392	184	208	86.4
46.9	390	183	207	85.9
46.9	388	182	206	85.4
46.9	386	181	205	84.9
46.9	384	180	204	84.4
46.9	382	179	203	83.9

%E	M1	M2	DM	M*
46.9	375	176	199	82.6
46.9	373	175	198	82.1
46.9	371	174	197	81.6
46.9	369	173	196	81.1
46.9	367	172	195	80.6
46.9	360	169	191	79.3
46.9	358	168	190	78.8
46.9	356	167	189	78.3
46.9	354	166	188	77.8
46.9	352	165	187	77.3
46.9	350	164	186	76.8
46.9	343	161	182	75.6
46.9	341	160	181	75.1
46.9	339	159	180	74.6
46.9	337	158	179	74.1
46.9	335	157	178	73.6
46.9	326	153	173	71.8
46.9	324	152	172	71.3
46.9	322	151	171	70.8
46.9	320	150	170	70.3
46.9	318	149	169	69.8
46.9	311	146	165	68.5
46.9	309	145	164	68.0
46.9	307	144	163	67.5
46.9	305	143	162	67.0
46.9	303	142	161	66.5
46.9	294	138	156	64.8
46.9	292	137	155	64.3
46.9	290	136	154	63.8
46.9	288	135	153	63.3
46.9	286	134	152	62.8
46.9	277	130	147	61.0
46.9	275	129	146	60.5
46.9	273	128	145	60.0
46.9	271	127	144	59.5
46.9	262	123	139	57.7
46.9	260	122	138	57.2
46.9	258	121	137	56.7
46.9	256	120	136	56.3
46.9	254	119	135	55.8
46.9	245	115	130	54.0
46.9	243	114	129	53.5
46.9	241	113	128	53.0
46.9	239	112	127	52.5
46.9	228	107	121	50.2
46.9	226	106	120	49.7
46.9	224	105	119	49.2
46.9	213	100	113	46.9
46.9	211	99	112	46.5
46.9	209	98	111	46.0

%E	M1	M2	DM	M*
46.9	207	97	110	45.5
46.9	196	92	104	43.2
46.9	194	91	103	42.7
46.9	192	90	102	42.2
46.9	179	84	95	39.4
46.9	177	83	94	38.9
46.9	175	82	93	38.4
46.9	162	76	86	35.7
46.9	160	75	85	35.2
46.9	147	69	78	32.4
46.9	145	68	77	31.9
46.9	143	67	76	31.4
46.9	130	61	69	28.6
46.9	128	60	68	28.1
46.9	113	53	60	24.9
46.9	98	46	52	21.6
46.9	96	45	51	21.1
46.9	81	38	43	17.8
46.9	64	30	34	14.1
46.9	49	23	26	10.8
46.9	32	15	17	7.0
46.8	500	234	266	109.5
46.8	498	233	265	109.0
46.8	496	232	264	108.5
46.8	494	231	263	108.0
46.8	491	230	261	107.7
46.8	489	229	260	107.2
46.8	487	228	259	106.7
46.8	485	227	258	106.2
46.8	483	226	257	105.7
46.8	481	225	256	105.2
46.8	479	224	255	104.8
46.8	477	223	254	104.3
46.8	476	223	253	104.5
46.8	474	222	252	104.0
46.8	472	221	251	103.5
46.8	470	220	250	103.0
46.8	468	219	249	102.5
46.8	466	218	248	102.0
46.8	464	217	247	101.5
46.8	462	216	246	101.0
46.8	459	215	244	100.7
46.8	457	214	243	100.2
46.8	455	213	242	99.7
46.8	453	212	241	99.2
46.8	451	211	240	98.7
46.8	449	210	239	98.2
46.8	447	209	238	97.7
46.8	444	208	236	97.4
46.8	442	207	235	96.9

%E	M1	M2	DM	M*
46.8	440	206	234	96.4
46.8	438	205	233	95.9
46.8	436	204	232	95.4
46.8	434	203	231	95.0
46.8	432	202	230	94.5
46.8	427	200	227	93.7
46.8	425	199	226	93.2
46.8	423	198	225	92.7
46.8	421	197	224	92.2
46.8	419	196	223	91.7
46.8	417	195	222	91.2
46.8	412	193	219	90.4
46.8	410	192	218	89.9
46.8	408	191	217	89.4
46.8	406	190	216	88.9
46.8	404	189	215	88.4
46.8	402	188	214	87.9
46.8	400	187	213	87.4
46.8	395	185	210	86.6
46.8	393	184	209	86.1
46.8	391	183	208	85.6
46.8	389	182	207	85.2
46.8	387	181	206	84.7
46.8	385	180	205	84.2
46.8	380	178	202	83.4
46.8	378	177	201	82.9
46.8	376	176	200	82.4
46.8	374	175	199	81.9
46.8	372	174	198	81.4
46.8	370	173	197	80.9
46.8	365	171	194	80.1
46.8	363	170	193	79.6
46.8	361	169	192	79.1
46.8	359	168	191	78.6
46.8	357	167	190	78.1
46.8	355	166	189	77.6
46.8	348	163	185	76.3
46.8	346	162	184	75.8
46.8	344	161	183	75.4
46.8	342	160	182	74.9
46.8	340	159	181	74.4
46.8	333	156	177	73.1
46.8	331	155	176	72.6
46.8	329	154	175	72.1
46.8	327	153	174	71.6
46.8	325	152	173	71.1
46.8	316	148	168	69.3
46.8	314	147	167	68.8
46.8	312	146	166	68.3
46.8	310	145	165	67.8

%E	M1	M2	DM	M*
46.8	308	144	164	67.3
46.8	301	141	160	66.0
46.8	299	140	159	65.6
46.8	297	139	158	65.1
46.8	295	138	157	64.6
46.8	293	137	156	64.1
46.8	284	133	151	62.3
46.8	282	132	150	61.8
46.8	280	131	149	61.3
46.8	278	130	148	60.8
46.8	269	126	143	59.0
46.8	267	125	142	58.5
46.8	265	124	141	58.0
46.8	263	123	140	57.5
46.8	252	118	134	55.3
46.8	250	117	133	54.8
46.8	248	116	132	54.3
46.8	237	111	126	52.0
46.8	235	110	125	51.5
46.8	233	109	124	51.0
46.8	231	108	123	50.5
46.8	222	104	118	48.7
46.8	220	103	117	48.2
46.8	218	102	116	47.7
46.8	216	101	115	47.2
46.8	205	96	109	45.0
46.8	203	95	108	44.5
46.8	201	94	107	44.0
46.8	190	89	101	41.7
46.8	188	88	100	41.2
46.8	186	87	99	40.7
46.8	173	81	92	37.9
46.8	171	80	91	37.4
46.8	158	74	84	34.7
46.8	156	73	83	34.2
46.8	154	72	82	33.7
46.8	141	66	75	30.9
46.8	139	65	74	30.4
46.8	126	59	67	27.6
46.8	124	58	66	27.1
46.8	111	52	59	24.4
46.8	109	51	58	23.9
46.8	94	44	50	20.6
46.8	79	37	42	17.3
46.8	77	36	41	16.8
46.8	62	29	33	13.6
46.8	47	22	25	10.3
46.7	499	233	266	108.8
46.7	497	232	265	108.3
46.7	495	231	264	107.8

%E	M1	M2	DM	M*
46.7	493	230	263	107.3
46.7	492	230	262	107.5
46.7	490	229	261	107.0
46.7	488	228	260	106.5
46.7	486	227	259	106.0
46.7	484	226	258	105.5
46.7	482	225	257	105.0
46.7	480	224	256	104.5
46.7	478	223	255	104.0
46.7	475	222	253	103.8
46.7	473	221	252	103.3
46.7	471	220	251	102.8
46.7	469	219	250	102.3
46.7	467	218	249	101.8
46.7	465	217	248	101.3
46.7	463	216	247	100.8
46.7	460	215	245	100.5
46.7	458	214	244	100.0
46.7	456	213	243	99.5
46.7	454	212	242	99.0
46.7	452	211	241	98.5
46.7	450	210	240	98.0
46.7	448	209	239	97.5
46.7	445	208	237	97.2
46.7	443	207	236	96.7
46.7	441	206	235	96.2
46.7	439	205	234	95.7
46.7	437	204	233	95.2
46.7	435	203	232	94.7
46.7	433	202	231	94.2
46.7	430	201	229	94.0
46.7	428	200	228	93.5
46.7	426	199	227	93.0
46.7	424	198	226	92.5
46.7	422	197	225	92.0
46.7	420	196	224	91.5
46.7	418	195	223	91.0
46.7	415	194	221	90.7
46.7	413	193	220	90.2
46.7	411	192	219	89.7
46.7	409	191	218	89.2
46.7	407	190	217	88.7
46.7	405	189	216	88.2
46.7	403	188	215	87.7
46.7	398	186	212	86.9
46.7	396	185	211	86.4
46.7	394	184	210	85.9
46.7	392	183	209	85.4
46.7	390	182	208	84.9
46.7	383	179	204	83.7

%E	M1	M2	DM	M*
46.7	381	178	203	83.2
46.7	379	177	202	82.7
46.7	377	176	201	82.2
46.7	375	175	200	81.7
46.7	368	172	196	80.4
46.7	366	171	195	79.9
46.7	364	170	194	79.4
46.7	362	169	193	78.9
46.7	360	168	192	78.4
46.7	353	165	188	77.1
46.7	351	164	187	76.6
46.7	349	163	186	76.1
46.7	347	162	185	75.6
46.7	345	161	184	75.1
46.7	338	158	180	73.9
46.7	336	157	179	73.4
46.7	334	156	178	72.9
46.7	332	155	177	72.4
46.7	330	154	176	71.9
46.7	323	151	172	70.6
46.7	321	150	171	70.1
46.7	319	149	170	69.6
46.7	317	148	169	69.1
46.7	315	147	168	68.6
46.7	306	143	163	66.8
46.7	304	142	162	66.3
46.7	302	141	161	65.8
46.7	300	140	160	65.3
46.7	291	136	155	63.6
46.7	289	135	154	63.1
46.7	287	134	153	62.6
46.7	285	133	152	62.1
46.7	276	129	147	60.3
46.7	274	128	146	59.8
46.7	272	127	145	59.3
46.7	270	126	144	58.8
46.7	261	122	139	57.0
46.7	259	121	138	56.5
46.7	257	120	137	56.0
46.7	255	119	136	55.5
46.7	246	115	131	53.8
46.7	244	114	130	53.3
46.7	242	113	129	52.8
46.7	240	112	128	52.3
46.7	229	107	122	50.0
46.7	227	106	121	49.5
46.7	225	105	120	49.0
46.7	214	100	114	46.7
46.7	212	99	113	46.2
46.7	210	98	112	45.7

%E	M1	M2	DM	M*
46.7	199	93	106	43.5
46.7	197	92	105	43.0
46.7	195	91	104	42.5
46.7	184	86	98	40.2
46.7	182	85	97	39.7
46.7	180	84	96	39.2
46.7	169	79	90	36.9
46.7	167	78	89	36.4
46.7	165	77	88	35.9
46.7	152	71	81	33.2
46.7	150	70	80	32.7
46.7	137	64	73	29.9
46.7	135	63	72	29.4
46.7	122	57	65	26.6
46.7	120	56	64	26.1
46.7	107	50	57	23.4
46.7	105	49	56	22.9
46.7	92	43	49	20.1
46.7	90	42	48	19.6
46.7	75	35	40	16.3
46.7	60	28	32	13.1
46.7	45	21	24	9.8
46.7	30	14	16	6.5
46.7	15	7	8	3.3
46.6	500	233	267	108.6
46.6	498	232	266	108.1
46.6	496	231	265	107.6
46.6	494	230	264	107.1
46.6	491	229	262	106.8
46.6	489	228	261	106.3
46.6	487	227	260	105.8
46.6	485	226	259	105.3
46.6	483	225	258	104.8
46.6	481	224	257	104.3
46.6	479	223	256	103.8
46.6	476	222	254	103.5
46.6	474	221	253	103.0
46.6	472	220	252	102.5
46.6	470	219	251	102.0
46.6	468	218	250	101.5
46.6	466	217	249	101.0
46.6	464	216	248	100.6
46.6	461	215	246	100.3
46.6	459	214	245	99.8
46.6	457	213	244	99.3
46.6	455	212	243	98.8
46.6	453	211	242	98.3
46.6	451	210	241	97.8
46.6	446	208	238	97.0
46.6	444	207	237	96.5

%E	M1	M2	DM	M*
46.6	442	206	236	96.0
46.6	440	205	235	95.5
46.6	438	204	234	95.0
46.6	436	203	233	94.5
46.6	431	201	230	93.7
46.6	429	200	229	93.2
46.6	427	199	228	92.7
46.6	425	198	227	92.2
46.6	423	197	226	91.7
46.6	421	196	225	91.2
46.6	416	194	222	90.5
46.6	414	193	221	90.0
46.6	412	192	220	89.5
46.6	410	191	219	89.0
46.6	408	190	218	88.5
46.6	406	189	217	88.0
46.6	401	187	214	87.2
46.6	399	186	213	86.7
46.6	397	185	212	86.2
46.6	395	184	211	85.7
46.6	393	183	210	85.2
46.6	388	181	207	84.4
46.6	386	180	206	83.9
46.6	384	179	205	83.4
46.6	382	178	204	82.9
46.6	380	177	203	82.4
46.6	378	176	202	81.9
46.6	373	174	199	81.2
46.6	371	173	198	80.7
46.6	369	172	197	80.2
46.6	367	171	196	79.7
46.6	365	170	195	79.2
46.6	363	169	194	78.7
46.6	358	167	191	77.9
46.6	356	166	190	77.4
46.6	354	165	189	76.9
46.6	352	164	188	76.4
46.6	350	163	187	75.9
46.6	348	162	186	75.4
46.6	343	160	183	74.6
46.6	341	159	182	74.1
46.6	339	158	181	73.6
46.6	337	157	180	73.1
46.6	335	156	179	72.6
46.6	328	153	175	71.4
46.6	326	152	174	70.9
46.6	324	151	173	70.4
46.6	322	150	172	69.9
46.6	320	149	171	69.4
46.6	313	146	167	68.1

%E	M1	M2	DM	M*
46.6	311	145	166	67.6
46.6	309	144	165	67.1
46.6	307	143	164	66.6
46.6	305	142	163	66.1
46.6	298	139	159	64.8
46.6	296	138	158	64.3
46.6	294	137	157	63.8
46.6	292	136	156	63.3
46.6	290	135	155	62.8
46.6	283	132	151	61.6
46.6	281	131	150	61.1
46.6	279	130	149	60.6
46.6	277	129	148	60.1
46.6	268	125	143	58.3
46.6	266	124	142	57.8
46.6	264	123	141	57.3
46.6	262	122	140	56.8
46.6	253	118	135	55.0
46.6	251	117	134	54.5
46.6	249	116	133	54.0
46.6	247	115	132	53.5
46.6	238	111	127	51.8
46.6	236	110	126	51.3
46.6	234	109	125	50.8
46.6	232	108	124	50.3
46.6	223	104	119	48.5
46.6	221	103	118	48.0
46.6	219	102	117	47.5
46.6	208	97	111	45.2
46.6	206	96	110	44.7
46.6	204	95	109	44.2
46.6	193	90	103	42.0
46.6	191	89	102	41.5
46.6	189	88	101	41.0
46.6	178	83	95	38.7
46.6	176	82	94	38.2
46.6	174	81	93	37.7
46.6	163	76	87	35.4
46.6	161	75	86	34.9
46.6	148	69	79	32.2
46.6	146	68	78	31.7
46.6	133	62	71	28.9
46.6	131	61	70	28.4
46.6	118	55	63	25.6
46.6	116	54	62	25.1
46.6	103	48	55	22.4
46.6	88	41	47	19.1
46.6	73	34	39	15.8
46.6	58	27	31	12.6
46.5	499	232	267	107.9

%E	M1	M2	DM	M*
46.5	497	231	266	107.4
46.5	495	230	265	106.9
46.5	493	229	264	106.4
46.5	492	229	263	106.6
46.5	490	228	262	106.1
46.5	488	227	261	105.6
46.5	486	226	260	105.1
46.5	484	225	259	104.6
46.5	482	224	258	104.1
46.5	480	223	257	103.6
46.5	477	222	255	103.3
46.5	475	221	254	102.8
46.5	473	220	253	102.3
46.5	471	219	252	101.8
46.5	469	218	251	101.3
46.5	467	217	250	100.8
46.5	465	216	249	100.3
46.5	462	215	247	100.1
46.5	460	214	246	99.6
46.5	458	213	245	99.1
46.5	456	212	244	98.6
46.5	454	211	243	98.1
46.5	452	210	242	97.6
46.5	449	209	240	97.3
46.5	447	208	239	96.8
46.5	445	207	238	96.3
46.5	443	206	237	95.8
46.5	441	205	236	95.3
46.5	439	204	235	94.8
46.5	437	203	234	94.3
46.5	434	202	232	94.0
46.5	432	201	231	93.5
46.5	430	200	230	93.0
46.5	428	199	229	92.5
46.5	426	198	228	92.0
46.5	424	197	227	91.5
46.5	419	195	224	90.8
46.5	417	194	223	90.3
46.5	415	193	222	89.8
46.5	413	192	221	89.3
46.5	411	191	220	88.8
46.5	409	190	219	88.3
46.5	404	188	216	87.5
46.5	402	187	215	87.0
46.5	400	186	214	86.5
46.5	398	185	213	86.0
46.5	396	184	212	85.5
46.5	391	182	209	84.7
46.5	389	181	208	84.2
46.5	387	180	207	83.7

%E	M1	M2	DM	M*
46.5	385	179	206	83.2
46.5	383	178	205	82.7
46.5	381	177	204	82.2
46.5	376	175	201	81.4
46.5	374	174	200	81.0
46.5	372	173	199	80.5
46.5	370	172	198	80.0
46.5	368	171	197	79.5
46.5	361	168	193	78.2
46.5	359	167	192	77.7
46.5	357	166	191	77.2
46.5	355	165	190	76.7
46.5	353	164	189	76.2
46.5	346	161	185	74.9
46.5	344	160	184	74.4
46.5	342	159	183	73.9
46.5	340	158	182	73.4
46.5	338	157	181	72.9
46.5	333	155	178	72.1
46.5	331	154	177	71.6
46.5	329	153	176	71.2
46.5	327	152	175	70.7
46.5	325	151	174	70.2
46.5	318	148	170	68.9
46.5	316	147	169	68.4
46.5	314	146	168	67.9
46.5	312	145	167	67.4
46.5	310	144	166	66.9
46.5	303	141	162	65.6
46.5	301	140	161	65.1
46.5	299	139	160	64.6
46.5	297	138	159	64.1
46.5	288	134	154	62.3
46.5	286	133	153	61.8
46.5	284	132	152	61.4
46.5	282	131	151	60.9
46.5	275	128	147	59.6
46.5	273	127	146	59.1
46.5	271	126	145	58.6
46.5	269	125	144	58.1
46.5	260	121	139	56.3
46.5	258	120	138	55.8
46.5	256	119	137	55.3
46.5	254	118	136	54.8
46.5	245	114	131	53.0
46.5	243	113	130	52.5
46.5	241	112	129	52.0
46.5	230	107	123	49.8
46.5	228	106	122	49.3
46.5	226	105	121	48.8
46.5	217	101	116	47.0
46.5	215	100	115	46.5
46.5	213	99	114	46.0
46.5	202	94	108	43.7
46.5	200	93	107	43.2
46.5	198	92	106	42.7
46.5	187	87	100	40.5
46.5	185	86	99	40.0
46.5	172	80	92	37.2
46.5	170	79	91	36.7
46.5	159	74	85	34.4
46.5	157	73	84	33.9
46.5	155	72	83	33.4
46.5	144	67	77	31.2
46.5	142	66	76	30.7
46.5	129	60	69	27.9
46.5	127	59	68	27.4
46.5	114	53	61	24.6
46.5	101	47	54	21.9
46.5	99	46	53	21.4
46.5	86	40	46	18.6
46.5	71	33	38	15.3
46.5	43	20	23	9.3
46.4	500	232	268	107.6
46.4	498	231	267	107.2
46.4	496	230	266	106.7
46.4	494	229	265	106.2
46.4	491	228	263	105.9
46.4	489	227	262	105.4
46.4	487	226	261	104.9
46.4	485	225	260	104.4
46.4	483	224	259	103.9
46.4	481	223	258	103.4
46.4	478	222	256	103.1
46.4	476	221	255	102.6
46.4	474	220	254	102.1
46.4	472	219	253	101.6
46.4	470	218	252	101.1
46.4	468	217	251	100.6
46.4	466	216	250	100.1
46.4	463	215	248	99.8
46.4	461	214	247	99.3
46.4	459	213	246	98.8
46.4	457	212	245	98.3
46.4	455	211	244	97.8
46.4	453	210	243	97.4
46.4	450	209	241	97.1
46.4	448	208	240	96.6
46.4	446	207	239	96.1
46.4	444	206	238	95.6
46.4	442	205	237	95.1
46.4	440	204	236	94.6
46.4	435	202	233	93.8
46.4	433	201	232	93.3
46.4	431	200	231	92.8
46.4	429	199	230	92.3
46.4	427	198	229	91.8
46.4	425	197	228	91.3
46.4	422	196	226	91.0
46.4	420	195	225	90.5
46.4	418	194	224	90.0
46.4	416	193	223	89.5
46.4	414	192	222	89.0
46.4	412	191	221	88.5
46.4	407	189	218	87.8
46.4	405	188	217	87.3
46.4	403	187	216	86.8
46.4	401	186	215	86.3
46.4	399	185	214	85.8
46.4	394	183	211	85.0
46.4	392	182	210	84.5
46.4	390	181	209	84.0
46.4	388	180	208	83.5
46.4	386	179	207	83.0
46.4	384	178	206	82.5
46.4	379	176	203	81.7
46.4	377	175	202	81.2
46.4	375	174	201	80.7
46.4	373	173	200	80.2
46.4	371	172	199	79.7
46.4	366	170	196	79.0
46.4	364	169	195	78.5
46.4	362	168	194	78.0
46.4	360	167	193	77.5
46.4	358	166	192	77.0
46.4	351	163	188	75.7
46.4	349	162	187	75.2
46.4	347	161	186	74.7
46.4	345	160	185	74.2
46.4	343	159	184	73.7
46.4	336	156	180	72.4
46.4	334	155	179	71.9
46.4	332	154	178	71.4
46.4	330	153	177	70.9
46.4	323	150	173	69.7
46.4	321	149	172	69.2
46.4	319	148	171	68.7
46.4	317	147	170	68.2
46.4	308	143	165	66.4
46.4	306	142	164	65.9
46.4	304	141	163	65.4
46.4	302	140	162	64.9
46.4	295	137	158	63.6
46.4	293	136	157	63.1
46.4	291	135	156	62.6
46.4	289	134	155	62.1
46.4	280	130	150	60.4
46.4	278	129	149	59.9
46.4	276	128	148	59.4
46.4	274	127	147	58.9
46.4	267	124	143	57.6
46.4	265	123	142	57.1
46.4	263	122	141	56.6
46.4	261	121	140	56.1
46.4	252	117	135	54.3
46.4	250	116	134	53.8
46.4	248	115	133	53.3
46.4	239	111	128	51.6
46.4	237	110	127	51.1
46.4	235	109	126	50.6
46.4	233	108	125	50.1
46.4	224	104	120	48.3
46.4	222	103	119	47.8
46.4	220	102	118	47.3
46.4	211	98	113	45.5
46.4	209	97	112	45.0
46.4	207	96	111	44.5
46.4	196	91	105	42.3
46.4	194	90	104	41.8
46.4	192	89	103	41.3
46.4	183	85	98	39.5
46.4	181	84	97	39.0
46.4	179	83	96	38.5
46.4	168	78	90	36.2
46.4	166	77	89	35.7
46.4	153	71	82	32.9
46.4	151	70	81	32.5
46.4	140	65	75	30.2
46.4	138	64	74	29.7
46.4	125	58	67	26.9
46.4	112	52	60	24.1
46.4	110	51	59	23.6
46.4	97	45	52	20.9
46.4	84	39	45	18.1
46.4	69	32	37	14.8
46.4	56	26	30	12.1
46.4	28	13	15	6.0
46.3	499	231	268	106.9
46.3	497	230	267	106.4
46.3	495	229	266	105.9
46.3	492	228	264	105.7
46.3	490	227	263	105.2
46.3	488	226	262	104.7
46.3	486	225	261	104.2
46.3	484	224	260	103.7
46.3	482	223	259	103.2
46.3	480	222	258	102.7
46.3	479	222	257	102.9
46.3	477	221	256	102.4
46.3	475	220	255	101.9
46.3	473	219	254	101.4
46.3	471	218	253	100.9
46.3	469	217	252	100.4
46.3	467	216	251	99.9
46.3	464	215	249	99.6
46.3	462	214	248	99.1
46.3	460	213	247	98.6
46.3	458	212	246	98.1
46.3	456	211	245	97.6
46.3	454	210	244	97.1
46.3	451	209	242	96.9
46.3	449	208	241	96.4
46.3	447	207	240	95.9
46.3	445	206	239	95.4
46.3	443	205	238	94.9
46.3	441	204	237	94.4
46.3	438	203	235	94.1
46.3	436	202	234	93.6
46.3	434	201	233	93.1
46.3	432	200	232	92.6
46.3	430	199	231	92.1
46.3	428	198	230	91.6
46.3	423	196	227	90.8
46.3	421	195	226	90.3
46.3	419	194	225	89.8
46.3	417	193	224	89.3
46.3	415	192	223	88.8
46.3	410	190	220	88.0
46.3	408	189	219	87.6
46.3	406	188	218	87.1
46.3	404	187	217	86.6
46.3	402	186	216	86.1
46.3	400	185	215	85.6
46.3	397	184	213	85.3
46.3	395	183	212	84.8
46.3	393	182	211	84.3
46.3	391	181	210	83.8
46.3	389	180	209	83.3
46.3	387	179	208	82.8
46.3	382	177	205	82.0

%E	M1	M2	DM	M*
46.3	380	176	204	81.5
46.3	378	175	203	81.0
46.3	376	174	202	80.5
46.3	374	173	201	80.0
46.3	369	171	198	79.2
46.3	367	170	197	78.7
46.3	365	169	196	78.2
46.3	363	168	195	77.8
46.3	361	167	194	77.3
46.3	356	165	191	76.5
46.3	354	164	190	76.0
46.3	352	163	189	75.5
46.3	350	162	188	75.0
46.3	348	161	187	74.5
46.3	341	158	183	73.2
46.3	339	157	182	72.7
46.3	337	156	181	72.2
46.3	335	155	180	71.7
46.3	328	152	176	70.4
46.3	326	151	175	69.9
46.3	324	150	174	69.4
46.3	322	149	173	68.9
46.3	320	148	172	68.4
46.3	315	146	169	67.7
46.3	313	145	168	67.2
46.3	311	144	167	66.7
46.3	309	143	166	66.2
46.3	307	142	165	65.7
46.3	300	139	161	64.4
46.3	298	138	160	63.9
46.3	296	137	159	63.4
46.3	294	136	158	62.9
46.3	287	133	154	61.6
46.3	285	132	153	61.1
46.3	283	131	152	60.6
46.3	281	130	151	60.1
46.3	272	126	146	58.4
46.3	270	125	145	57.9
46.3	268	124	144	57.4
46.3	259	120	139	55.6
46.3	257	119	138	55.1
46.3	255	118	137	54.6
46.3	246	114	132	52.8
46.3	244	113	131	52.3
46.3	242	112	130	51.8
46.3	240	111	129	51.3
46.3	231	107	124	49.6
46.3	229	106	123	49.1
46.3	227	105	122	48.6
46.3	218	101	117	46.8
46.3	216	100	116	46.3
46.3	214	99	115	45.8
46.3	205	95	110	44.0
46.3	203	94	109	43.5
46.3	201	93	108	43.0
46.3	190	88	102	40.8
46.3	188	87	101	40.3
46.3	177	82	95	38.0
46.3	175	81	94	37.5
46.3	164	76	88	35.2
46.3	162	75	87	34.7
46.3	160	74	86	34.2
46.3	149	69	80	32.0
46.3	147	68	79	31.5
46.3	136	63	73	29.2
46.3	134	62	72	28.7
46.3	123	57	66	26.4
46.3	121	56	65	25.9
46.3	108	50	58	23.1
46.3	95	44	51	20.4
46.3	82	38	44	17.6
46.3	80	37	43	17.1
46.3	67	31	36	14.3
46.3	54	25	29	11.6
46.3	41	19	22	8.8
46.2	500	231	269	106.7
46.2	498	230	268	106.2
46.2	496	229	267	105.7
46.2	494	228	266	105.2
46.2	493	228	265	105.4
46.2	491	227	264	104.9
46.2	489	226	263	104.4
46.2	487	225	262	104.0
46.2	485	224	261	103.5
46.2	483	223	260	103.0
46.2	481	222	259	102.5
46.2	478	221	257	102.2
46.2	476	220	256	101.7
46.2	474	219	255	101.2
46.2	472	218	254	100.7
46.2	470	217	253	100.2
46.2	468	216	252	99.7
46.2	465	215	250	99.4
46.2	463	214	249	98.9
46.2	461	213	248	98.4
46.2	459	212	247	97.9
46.2	457	211	246	97.4
46.2	455	210	245	96.9
46.2	452	209	243	96.6
46.2	450	208	242	96.1
46.2	448	207	241	95.6
46.2	446	206	240	95.1
46.2	444	205	239	94.7
46.2	442	204	238	94.2
46.2	439	203	236	93.9
46.2	437	202	235	93.4
46.2	435	201	234	92.9
46.2	433	200	233	92.4
46.2	431	199	232	91.9
46.2	429	198	231	91.4
46.2	426	197	229	91.1
46.2	424	196	228	90.6
46.2	422	195	227	90.1
46.2	420	194	226	89.6
46.2	418	193	225	89.1
46.2	416	192	224	88.6
46.2	413	191	222	88.3
46.2	411	190	221	87.8
46.2	409	189	220	87.3
46.2	407	188	219	86.8
46.2	405	187	218	86.3
46.2	403	186	217	85.8
46.2	398	184	214	85.1
46.2	396	183	213	84.6
46.2	394	182	212	84.1
46.2	392	181	211	83.6
46.2	390	180	210	83.1
46.2	385	178	207	82.3
46.2	383	177	206	81.8
46.2	381	176	205	81.3
46.2	379	175	204	80.8
46.2	377	174	203	80.3
46.2	372	172	200	79.5
46.2	370	171	199	79.0
46.2	368	170	198	78.5
46.2	366	169	197	78.0
46.2	364	168	196	77.5
46.2	359	166	193	76.8
46.2	357	165	192	76.3
46.2	355	164	191	75.8
46.2	353	163	190	75.3
46.2	351	162	189	74.8
46.2	346	160	186	74.0
46.2	344	159	185	73.5
46.2	342	158	184	73.0
46.2	340	157	183	72.5
46.2	338	156	182	72.0
46.2	333	154	179	71.2
46.2	331	153	178	70.7
46.2	329	152	177	70.2
46.2	327	151	176	69.7
46.2	325	150	175	69.2
46.2	318	147	171	68.0
46.2	316	146	170	67.5
46.2	314	145	169	67.0
46.2	312	144	168	66.5
46.2	305	141	164	65.2
46.2	303	140	163	64.7
46.2	301	139	162	64.2
46.2	299	138	161	63.7
46.2	292	135	157	62.4
46.2	290	134	156	61.9
46.2	288	133	155	61.4
46.2	286	132	154	60.9
46.2	279	129	150	59.6
46.2	277	128	149	59.1
46.2	275	127	148	58.7
46.2	273	126	147	58.2
46.2	266	123	143	56.9
46.2	264	122	142	56.4
46.2	262	121	141	55.9
46.2	260	120	140	55.4
46.2	253	117	136	54.1
46.2	251	116	135	53.6
46.2	249	115	134	53.1
46.2	247	114	133	52.6
46.2	238	110	128	50.8
46.2	236	109	127	50.3
46.2	234	108	126	49.8
46.2	225	104	121	48.1
46.2	223	103	120	47.6
46.2	221	102	119	47.1
46.2	212	98	114	45.3
46.2	210	97	113	44.8
46.2	208	96	112	44.3
46.2	199	92	107	42.5
46.2	197	91	106	42.0
46.2	195	90	105	41.5
46.2	186	86	100	39.8
46.2	184	85	99	39.3
46.2	182	84	98	38.8
46.2	173	80	93	37.0
46.2	171	79	92	36.5
46.2	169	78	91	36.0
46.2	158	73	85	33.7
46.2	156	72	84	33.2
46.2	145	67	78	31.0
46.2	143	66	77	30.5
46.2	132	61	71	28.2
46.2	130	60	70	27.7
46.2	119	55	64	25.4
46.2	117	54	63	24.9
46.2	106	49	57	22.7
46.2	104	48	56	22.2
46.2	93	43	50	19.9
46.2	91	42	49	19.4
46.2	78	36	42	16.6
46.2	65	30	35	13.8
46.2	52	24	28	11.1
46.2	39	18	21	8.3
46.2	26	12	14	5.5
46.2	13	6	7	2.8
46.1	499	230	269	106.0
46.1	497	229	268	105.5
46.1	495	228	267	105.0
46.1	492	227	265	104.7
46.1	490	226	264	104.2
46.1	488	225	263	103.7
46.1	486	224	262	103.2
46.1	484	223	261	102.7
46.1	482	222	260	102.2
46.1	479	221	258	102.0
46.1	477	220	257	101.5
46.1	475	219	256	101.0
46.1	473	218	255	100.5
46.1	471	217	254	100.0
46.1	469	216	253	99.5
46.1	466	215	251	99.2
46.1	464	214	250	98.7
46.1	462	213	249	98.2
46.1	460	212	248	97.7
46.1	458	211	247	97.2
46.1	456	210	246	96.7
46.1	453	209	244	96.4
46.1	451	208	243	95.9
46.1	449	207	242	95.4
46.1	447	206	241	94.9
46.1	445	205	240	94.4
46.1	443	204	239	93.9
46.1	440	203	237	93.7
46.1	438	202	236	93.2
46.1	436	201	235	92.7
46.1	434	200	234	92.2
46.1	432	199	233	91.7
46.1	427	197	230	90.9
46.1	425	196	229	90.4
46.1	423	195	228	89.9
46.1	421	194	227	89.4
46.1	419	193	226	88.9
46.1	414	191	223	88.1

%E	M1	M2	DM	M*
46.1	412	190	222	87.6
46.1	410	189	221	87.1
46.1	408	188	220	86.6
46.1	406	187	219	86.1
46.1	401	185	216	85.3
46.1	399	184	215	84.9
46.1	397	183	214	84.4
46.1	395	182	213	83.9
46.1	393	181	212	83.4
46.1	388	179	209	82.6
46.1	386	178	208	82.1
46.1	384	177	207	81.6
46.1	382	176	206	81.1
46.1	380	175	205	80.6
46.1	375	173	202	79.8
46.1	373	172	201	79.3
46.1	371	171	200	78.8
46.1	369	170	199	78.3
46.1	362	167	195	77.0
46.1	360	166	194	76.5
46.1	358	165	193	76.0
46.1	356	164	192	75.6
46.1	349	161	188	74.3
46.1	347	160	187	73.8
46.1	345	159	186	73.3
46.1	343	158	185	72.8
46.1	336	155	181	71.5
46.1	334	154	180	71.0
46.1	332	153	179	70.5
46.1	330	152	178	70.0
46.1	323	149	174	68.7
46.1	321	148	173	68.2
46.1	319	147	172	67.7
46.1	317	146	171	67.2
46.1	310	143	167	66.0
46.1	308	142	166	65.5
46.1	306	141	165	65.0
46.1	304	140	164	64.5
46.1	297	137	160	63.2
46.1	295	136	159	62.7
46.1	293	135	158	62.2
46.1	284	131	153	60.4
46.1	282	130	152	59.9
46.1	280	129	151	59.4
46.1	271	125	146	57.7
46.1	269	124	145	57.2
46.1	267	123	144	56.7
46.1	258	119	139	54.9
46.1	256	118	138	54.4
46.1	254	117	137	53.9

%E	M1	M2	DM	M*
46.1	245	113	132	52.1
46.1	243	112	131	51.6
46.1	241	111	130	51.1
46.1	232	107	125	49.3
46.1	230	106	124	48.9
46.1	228	105	123	48.4
46.1	219	101	118	46.6
46.1	217	100	117	46.1
46.1	206	95	111	43.8
46.1	204	94	110	43.3
46.1	193	89	104	41.0
46.1	191	88	103	40.5
46.1	180	83	97	38.3
46.1	178	82	96	37.8
46.1	167	77	90	35.5
46.1	165	76	89	35.0
46.1	154	71	83	32.7
46.1	152	70	82	32.2
46.1	141	65	76	30.0
46.1	128	59	69	27.2
46.1	115	53	62	24.4
46.1	102	47	55	21.7
46.1	89	41	48	18.9
46.1	76	35	41	16.1
46.0	500	230	270	105.8
46.0	498	229	269	105.3
46.0	496	228	268	104.8
46.0	494	227	267	104.3
46.0	493	227	266	104.5
46.0	491	226	265	104.0
46.0	489	225	264	103.5
46.0	487	224	263	103.0
46.0	485	223	262	102.5
46.0	483	222	261	102.0
46.0	480	221	259	101.8
46.0	478	220	258	101.3
46.0	476	219	257	100.8
46.0	474	218	256	100.3
46.0	472	217	255	99.8
46.0	470	216	254	99.3
46.0	467	215	252	99.0
46.0	465	214	251	98.5
46.0	463	213	250	98.0
46.0	461	212	249	97.5
46.0	459	211	248	97.0
46.0	457	210	247	96.5
46.0	454	209	245	96.2
46.0	452	208	244	95.7
46.0	450	207	243	95.2
46.0	448	206	242	94.7

%E	M1	M2	DM	M*
46.0	446	205	241	94.2
46.0	441	203	238	93.4
46.0	439	202	237	92.9
46.0	437	201	236	92.5
46.0	435	200	235	92.0
46.0	433	199	234	91.5
46.0	430	198	232	91.2
46.0	428	197	231	90.7
46.0	426	196	230	90.2
46.0	424	195	229	89.7
46.0	422	194	228	89.2
46.0	420	193	227	88.7
46.0	417	192	225	88.4
46.0	415	191	224	87.9
46.0	413	190	223	87.4
46.0	411	189	222	86.9
46.0	409	188	221	86.4
46.0	404	186	218	85.6
46.0	402	185	217	85.1
46.0	400	184	216	84.6
46.0	398	183	215	84.1
46.0	396	182	214	83.6
46.0	391	180	211	82.9
46.0	389	179	210	82.4
46.0	387	178	209	81.9
46.0	385	177	208	81.4
46.0	383	176	207	80.9
46.0	378	174	204	80.1
46.0	376	173	203	79.6
46.0	374	172	202	79.1
46.0	372	171	201	78.6
46.0	367	169	198	77.8
46.0	365	168	197	77.3
46.0	363	167	196	76.8
46.0	361	166	195	76.3
46.0	359	165	194	75.8
46.0	354	163	191	75.1
46.0	352	162	190	74.6
46.0	350	161	189	74.1
46.0	348	160	188	73.6
46.0	346	159	187	73.1
46.0	341	157	184	72.3
46.0	339	156	183	71.8
46.0	337	155	182	71.3
46.0	335	154	181	70.8
46.0	328	151	177	69.5
46.0	326	150	176	69.0
46.0	324	149	175	68.5
46.0	322	148	174	68.0
46.0	315	145	170	66.7

%E	M1	M2	DM	M*
46.0	313	144	169	66.2
46.0	311	143	168	65.8
46.0	309	142	167	65.3
46.0	302	139	163	64.0
46.0	300	138	162	63.5
46.0	298	137	161	63.0
46.0	291	134	157	61.7
46.0	289	133	156	61.2
46.0	287	132	155	60.7
46.0	285	131	154	60.2
46.0	278	128	150	58.9
46.0	276	127	149	58.4
46.0	274	126	148	57.9
46.0	272	125	147	57.4
46.0	265	122	143	56.2
46.0	263	121	142	55.7
46.0	261	120	141	55.2
46.0	252	116	136	53.4
46.0	250	115	135	52.9
46.0	248	114	134	52.4
46.0	239	110	129	50.6
46.0	237	109	128	50.1
46.0	235	108	127	49.6
46.0	226	104	122	47.9
46.0	224	103	121	47.4
46.0	215	99	116	45.6
46.0	213	98	115	45.1
46.0	211	97	114	44.6
46.0	202	93	109	42.8
46.0	200	92	108	42.3
46.0	198	91	107	41.8
46.0	189	87	102	40.0
46.0	187	86	101	39.6
46.0	176	81	95	37.3
46.0	174	80	94	36.8
46.0	163	75	88	34.5
46.0	161	74	87	34.0
46.0	150	69	81	31.7
46.0	139	64	75	29.5
46.0	137	63	74	29.0
46.0	126	58	68	26.7
46.0	124	57	67	26.2
46.0	113	52	61	23.9
46.0	100	46	54	21.2
46.0	87	40	47	18.4
46.0	63	29	34	13.3
46.0	50	23	27	10.6
45.9	499	229	270	105.1
45.9	497	228	269	104.6
45.9	495	227	268	104.1

%E	M1	M2	DM	M*
45.9	492	226	266	103.8
45.9	490	225	265	103.3
45.9	488	224	264	102.8
45.9	486	223	263	102.3
45.9	484	222	262	101.8
45.9	482	221	261	101.3
45.9	481	221	260	101.5
45.9	479	220	259	101.0
45.9	477	219	258	100.5
45.9	475	218	257	100.1
45.9	473	217	256	99.6
45.9	471	216	255	99.1
45.9	468	215	253	98.8
45.9	466	214	252	98.3
45.9	464	213	251	97.8
45.9	462	212	250	97.3
45.9	460	211	249	96.8
45.9	458	210	248	96.3
45.9	455	209	246	96.0
45.9	453	208	245	95.5
45.9	451	207	244	95.0
45.9	449	206	243	94.5
45.9	447	205	242	94.0
45.9	444	204	240	93.7
45.9	442	203	239	93.2
45.9	440	202	238	92.7
45.9	438	201	237	92.2
45.9	436	200	236	91.7
45.9	434	199	235	91.2
45.9	431	198	233	91.0
45.9	429	197	232	90.5
45.9	427	196	231	90.0
45.9	425	195	230	89.5
45.9	423	194	229	89.0
45.9	418	192	226	88.2
45.9	416	191	225	87.7
45.9	414	190	224	87.2
45.9	412	189	223	86.7
45.9	410	188	222	86.2
45.9	407	187	220	85.9
45.9	405	186	219	85.4
45.9	403	185	218	84.9
45.9	401	184	217	84.4
45.9	399	183	216	83.9
45.9	394	181	213	83.1
45.9	392	180	212	82.7
45.9	390	179	211	82.2
45.9	388	178	210	81.7
45.9	386	177	209	81.2
45.9	381	175	206	80.4

%E	M1	M2	DM	M*
45.9	379	174	205	79.9
45.9	377	173	204	79.4
45.9	375	172	203	78.9
45.9	370	170	200	78.1
45.9	368	169	199	77.6
45.9	366	168	198	77.1
45.9	364	167	197	76.6
45.9	362	166	196	76.1
45.9	357	164	193	75.3
45.9	355	163	192	74.8
45.9	353	162	191	74.3
45.9	351	161	190	73.8
45.9	344	158	186	72.6
45.9	342	157	185	72.1
45.9	340	156	184	71.6
45.9	338	155	183	71.1
45.9	333	153	180	70.3
45.9	331	152	179	69.8
45.9	329	151	178	69.3
45.9	327	150	177	68.8
45.9	320	147	173	67.5
45.9	318	146	172	67.0
45.9	316	145	171	66.5
45.9	314	144	170	66.0
45.9	307	141	166	64.8
45.9	305	140	165	64.3
45.9	303	139	164	63.8
45.9	296	136	160	62.5
45.9	294	135	159	62.0
45.9	292	134	158	61.5
45.9	290	133	157	61.0
45.9	283	130	153	59.7
45.9	281	129	152	59.2
45.9	279	128	151	58.7
45.9	270	124	146	56.9
45.9	268	123	145	56.5
45.9	266	122	144	56.0
45.9	259	119	140	54.7
45.9	257	118	139	54.2
45.9	255	117	138	53.7
45.9	253	116	137	53.2
45.9	246	113	133	51.9
45.9	244	112	132	51.4
45.9	242	111	131	50.9
45.9	233	107	126	49.1
45.9	231	106	125	48.6
45.9	229	105	124	48.1
45.9	222	102	120	46.9
45.9	220	101	119	46.4
45.9	218	100	118	45.9

%E	M1	M2	DM	M*
45.9	209	96	113	44.1
45.9	207	95	112	43.6
45.9	205	94	111	43.1
45.9	196	90	106	41.3
45.9	194	89	105	40.8
45.9	185	85	100	39.1
45.9	183	84	99	38.6
45.9	181	83	98	38.1
45.9	172	79	93	36.3
45.9	170	78	92	35.8
45.9	159	73	86	33.5
45.9	157	72	85	33.0
45.9	148	68	80	31.2
45.9	146	67	79	30.7
45.9	135	62	73	28.5
45.9	133	61	72	28.0
45.9	122	56	66	25.7
45.9	111	51	60	23.4
45.9	109	50	59	22.9
45.9	98	45	53	20.7
45.9	85	39	46	17.9
45.9	74	34	40	15.6
45.9	61	28	33	12.9
45.9	37	17	20	7.8
45.8	500	229	271	104.9
45.8	498	228	270	104.4
45.8	496	227	269	103.9
45.8	493	226	267	103.6
45.8	491	225	266	103.1
45.8	489	224	265	102.6
45.8	487	223	264	102.1
45.8	485	222	263	101.6
45.8	483	221	262	101.1
45.8	480	220	260	100.8
45.8	478	219	259	100.3
45.8	476	218	258	99.8
45.8	474	217	257	99.3
45.8	472	216	256	98.8
45.8	469	215	254	98.6
45.8	467	214	253	98.1
45.8	465	213	252	97.6
45.8	463	212	251	97.1
45.8	461	211	250	96.6
45.8	459	210	249	96.1
45.8	456	209	247	95.8
45.8	454	208	246	95.3
45.8	452	207	245	94.8
45.8	450	206	244	94.3
45.8	448	205	243	93.8
45.8	445	204	241	93.5

%E	M1	M2	DM	M*
45.8	443	203	240	93.0
45.8	441	202	239	92.5
45.8	439	201	238	92.0
45.8	437	200	237	91.5
45.8	432	198	234	90.8
45.8	430	197	233	90.3
45.8	428	196	232	89.8
45.8	426	195	231	89.3
45.8	424	194	230	88.8
45.8	421	193	228	88.5
45.8	419	192	227	88.0
45.8	417	191	226	87.5
45.8	415	190	225	87.0
45.8	413	189	224	86.5
45.8	408	187	221	85.7
45.8	406	186	220	85.2
45.8	404	185	219	84.7
45.8	402	184	218	84.2
45.8	400	183	217	83.7
45.8	397	182	215	83.4
45.8	395	181	214	82.9
45.8	393	180	213	82.4
45.8	391	179	212	81.9
45.8	389	178	211	81.4
45.8	384	176	208	80.7
45.8	382	175	207	80.2
45.8	380	174	206	79.7
45.8	378	173	205	79.2
45.8	373	171	202	78.4
45.8	371	170	201	77.9
45.8	369	169	200	77.4
45.8	367	168	199	76.9
45.8	365	167	198	76.4
45.8	360	165	195	75.6
45.8	358	164	194	75.1
45.8	356	163	193	74.6
45.8	354	162	192	74.1
45.8	349	160	189	73.4
45.8	347	159	188	72.9
45.8	345	158	187	72.4
45.8	343	157	186	71.9
45.8	336	154	182	70.6
45.8	334	153	181	70.1
45.8	332	152	180	69.6
45.8	330	151	179	69.1
45.8	325	149	176	68.3
45.8	323	148	175	67.8
45.8	321	147	174	67.3
45.8	319	146	173	66.8
45.8	312	143	169	65.5

%E	M1	M2	DM	M*
45.8	310	142	168	65.0
45.8	308	141	167	64.5
45.8	306	140	166	64.1
45.8	301	138	163	63.3
45.8	299	137	162	62.8
45.8	297	136	161	62.3
45.8	295	135	160	61.8
45.8	288	132	156	60.5
45.8	286	131	155	60.0
45.8	284	130	154	59.5
45.8	277	127	150	58.2
45.8	275	126	149	57.7
45.8	273	125	148	57.2
45.8	271	124	147	56.7
45.8	264	121	143	55.5
45.8	262	120	142	55.0
45.8	260	119	141	54.5
45.8	251	115	136	52.7
45.8	249	114	135	52.2
45.8	240	110	130	50.4
45.8	238	109	129	49.9
45.8	236	108	128	49.4
45.8	227	104	123	47.6
45.8	225	103	122	47.2
45.8	216	99	117	45.4
45.8	214	98	116	44.9
45.8	212	97	115	44.4
45.8	203	93	110	42.6
45.8	201	92	109	42.1
45.8	192	88	104	40.3
45.8	190	87	103	39.8
45.8	179	82	97	37.6
45.8	177	81	96	37.1
45.8	168	77	91	35.3
45.8	166	76	90	34.8
45.8	155	71	84	32.5
45.8	153	70	83	32.0
45.8	144	66	78	30.3
45.8	142	65	77	29.8
45.8	131	60	71	27.5
45.8	120	55	65	25.2
45.8	118	54	64	24.7
45.8	107	49	58	22.4
45.8	96	44	52	20.2
45.8	83	38	45	17.4
45.8	72	33	39	15.1
45.8	59	27	32	12.4
45.8	48	22	26	10.1
45.8	24	11	13	5.0
45.7	499	228	271	104.2

%E	M1	M2	DM	M*
45.7	497	227	270	103.7
45.7	495	226	269	103.2
45.7	494	226	268	103.4
45.7	492	225	267	102.9
45.7	490	224	266	102.4
45.7	488	223	265	101.9
45.7	486	222	264	101.4
45.7	484	221	263	100.9
45.7	481	220	261	100.6
45.7	479	219	260	100.1
45.7	477	218	259	99.6
45.7	475	217	258	99.1
45.7	473	216	257	98.6
45.7	470	215	255	98.4
45.7	468	214	254	97.9
45.7	466	213	253	97.4
45.7	464	212	252	96.9
45.7	462	211	251	96.4
45.7	460	210	250	95.9
45.7	457	209	248	95.6
45.7	455	208	247	95.1
45.7	453	207	246	94.6
45.7	451	206	245	94.1
45.7	449	205	244	93.6
45.7	446	204	242	93.3
45.7	444	203	241	92.8
45.7	442	202	240	92.3
45.7	440	201	239	91.8
45.7	438	200	238	91.3
45.7	435	199	236	91.0
45.7	433	198	235	90.5
45.7	431	197	234	90.0
45.7	429	196	233	89.5
45.7	427	195	232	89.1
45.7	422	193	229	88.3
45.7	420	192	228	87.8
45.7	418	191	227	87.3
45.7	416	190	226	86.8
45.7	414	189	225	86.3
45.7	411	188	223	86.0
45.7	409	187	222	85.5
45.7	407	186	221	85.0
45.7	405	185	220	84.5
45.7	403	184	219	84.0
45.7	398	182	216	83.2
45.7	396	181	215	82.7
45.7	394	180	214	82.2
45.7	392	179	213	81.7
45.7	387	177	210	81.0
45.7	385	176	209	80.5

%E	M1	M2	DM	M*
45.7	383	175	208	80.0
45.7	381	174	207	79.5
45.7	376	172	204	78.7
45.7	374	171	203	78.2
45.7	372	170	202	77.7
45.7	370	169	201	77.2
45.7	368	168	200	76.7
45.7	363	166	197	75.9
45.7	361	165	196	75.4
45.7	359	164	195	74.9
45.7	357	163	194	74.4
45.7	352	161	191	73.6
45.7	350	160	190	73.1
45.7	348	159	189	72.6
45.7	346	158	188	72.2
45.7	341	156	185	71.4
45.7	339	155	184	70.9
45.7	337	154	183	70.4
45.7	335	153	182	69.9
45.7	328	150	178	68.6
45.7	326	149	177	68.1
45.7	324	148	176	67.6
45.7	322	147	175	67.1
45.7	317	145	172	66.3
45.7	315	144	171	65.8
45.7	313	143	170	65.3
45.7	311	142	169	64.8
45.7	304	139	165	63.6
45.7	302	138	164	63.1
45.7	300	137	163	62.6
45.7	293	134	159	61.3
45.7	291	133	158	60.8
45.7	289	132	157	60.3
45.7	282	129	153	59.0
45.7	280	128	152	58.5
45.7	278	127	151	58.0
45.7	276	126	150	57.5
45.7	269	123	146	56.2
45.7	267	122	145	55.7
45.7	265	121	144	55.2
45.7	258	118	140	54.0
45.7	256	117	139	53.5
45.7	254	116	138	53.0
45.7	247	113	134	51.7
45.7	245	112	133	51.2
45.7	243	111	132	50.7
45.7	234	107	127	48.9
45.7	232	106	126	48.4
45.7	230	105	125	47.9
45.7	223	102	121	46.7

%E	M1	M2	DM	M*
45.7	221	101	120	46.2
45.7	219	100	119	45.7
45.7	210	96	114	43.9
45.7	208	95	113	43.4
45.7	199	91	108	41.6
45.7	197	90	107	41.1
45.7	188	86	102	39.3
45.7	186	85	101	38.8
45.7	184	84	100	38.3
45.7	175	80	95	36.6
45.7	173	79	94	36.1
45.7	164	75	89	34.3
45.7	162	74	88	33.8
45.7	151	69	82	31.5
45.7	140	64	76	29.3
45.7	138	63	75	28.8
45.7	129	59	70	27.0
45.7	127	58	69	26.5
45.7	116	53	63	24.2
45.7	105	48	57	21.9
45.7	94	43	51	19.7
45.7	92	42	50	19.2
45.7	81	37	44	16.9
45.7	70	32	38	14.6
45.7	46	21	25	9.6
45.7	35	16	19	7.3
45.6	500	228	272	104.0
45.6	498	227	271	103.5
45.6	496	226	270	103.0
45.6	493	225	268	102.7
45.6	491	224	267	102.2
45.6	489	223	266	101.7
45.6	487	222	265	101.2
45.6	485	221	264	100.7
45.6	482	220	262	100.4
45.6	480	219	261	99.9
45.6	478	218	260	99.4
45.6	476	217	259	98.9
45.6	474	216	258	98.4
45.6	472	215	257	97.9
45.6	471	215	256	98.1
45.6	469	214	255	97.6
45.6	467	213	254	97.1
45.6	465	212	253	96.7
45.6	463	211	252	96.2
45.6	461	210	251	95.7
45.6	458	209	249	95.4
45.6	456	208	248	94.9
45.6	454	207	247	94.4
45.6	452	206	246	93.9

%E	M1	M2	DM	M*
45.6	450	205	245	93.4
45.6	447	204	243	93.1
45.6	445	203	242	92.6
45.6	443	202	241	92.1
45.6	441	201	240	91.6
45.6	439	200	239	91.1
45.6	436	199	237	90.8
45.6	434	198	236	90.3
45.6	432	197	235	89.8
45.6	430	196	234	89.3
45.6	428	195	233	88.8
45.6	425	194	231	88.6
45.6	423	193	230	88.1
45.6	421	192	229	87.6
45.6	419	191	228	87.1
45.6	417	190	227	86.6
45.6	412	188	224	85.8
45.6	410	187	223	85.3
45.6	408	186	222	84.8
45.6	406	185	221	84.3
45.6	401	183	218	83.5
45.6	399	182	217	83.0
45.6	397	181	216	82.5
45.6	395	180	215	82.0
45.6	390	178	212	81.2
45.6	388	177	211	80.7
45.6	386	176	210	80.2
45.6	384	175	209	79.8
45.6	382	174	208	79.3
45.6	379	173	206	79.0
45.6	377	172	205	78.5
45.6	375	171	204	78.0
45.6	373	170	203	77.5
45.6	371	169	202	77.0
45.6	366	167	199	76.2
45.6	364	166	198	75.7
45.6	362	165	197	75.2
45.6	360	164	196	74.7
45.6	355	162	193	73.9
45.6	353	161	192	73.4
45.6	351	160	191	72.9
45.6	349	159	190	72.4
45.6	344	157	187	71.7
45.6	342	156	186	71.2
45.6	340	155	185	70.7
45.6	338	154	184	70.2
45.6	333	152	181	69.4
45.6	331	151	180	68.9
45.6	329	150	179	68.4
45.6	327	149	178	67.9

%E	M1	M2	DM	M*
45.6	320	146	174	66.6
45.6	318	145	173	66.1
45.6	316	144	172	65.6
45.6	309	141	168	64.3
45.6	307	140	167	63.8
45.6	305	139	166	63.3
45.6	298	136	162	62.1
45.6	296	135	161	61.6
45.6	294	134	160	61.1
45.6	287	131	156	59.8
45.6	285	130	155	59.3
45.6	283	129	154	58.8
45.6	281	128	153	58.3
45.6	274	125	149	57.0
45.6	272	124	148	56.5
45.6	270	123	147	56.0
45.6	263	120	143	54.8
45.6	261	119	142	54.3
45.6	259	118	141	53.8
45.6	252	115	137	52.5
45.6	250	114	136	52.0
45.6	248	113	135	51.5
45.6	241	110	131	50.2
45.6	239	109	130	49.7
45.6	237	108	129	49.2
45.6	228	104	124	47.4
45.6	226	103	123	46.9
45.6	217	99	118	45.2
45.6	215	98	117	44.7
45.6	206	94	112	42.9
45.6	204	93	111	42.4
45.6	195	89	106	40.6
45.6	193	88	105	40.1
45.6	191	87	104	39.6
45.6	182	83	99	37.9
45.6	180	82	98	37.4
45.6	171	78	93	35.6
45.6	169	77	92	35.1
45.6	160	73	87	33.3
45.6	158	72	86	32.8
45.6	149	68	81	31.0
45.6	147	67	80	30.5
45.6	136	62	74	28.3
45.6	125	57	68	26.0
45.6	114	52	62	23.7
45.6	103	47	56	21.4
45.6	90	41	49	18.7
45.6	79	36	43	16.4
45.6	68	31	37	14.1
45.6	57	26	31	11.9

%E	M1	M2	DM	M*
45.5	499	227	272	103.3
45.5	497	226	271	102.8
45.5	495	225	270	102.3
45.5	494	225	269	102.5
45.5	492	224	268	102.0
45.5	490	223	267	101.5
45.5	488	222	266	101.0
45.5	486	221	265	100.5
45.5	484	220	264	100.0
45.5	483	220	263	100.2
45.5	481	219	262	99.7
45.5	479	218	261	99.2
45.5	477	217	260	98.7
45.5	475	216	259	98.2
45.5	473	215	258	97.7
45.5	470	214	256	97.4
45.5	468	213	255	96.9
45.5	466	212	254	96.4
45.5	464	211	253	96.0
45.5	462	210	252	95.5
45.5	459	209	250	95.2
45.5	457	208	249	94.7
45.5	455	207	248	94.2
45.5	453	206	247	93.7
45.5	451	205	246	93.2
45.5	448	204	244	92.9
45.5	446	203	243	92.4
45.5	444	202	242	91.9
45.5	442	201	241	91.4
45.5	440	200	240	90.9
45.5	437	199	238	90.6
45.5	435	198	237	90.1
45.5	433	197	236	89.6
45.5	431	196	235	89.1
45.5	429	195	234	88.6
45.5	426	194	232	88.3
45.5	424	193	231	87.9
45.5	422	192	230	87.4
45.5	420	191	229	86.9
45.5	418	190	228	86.4
45.5	415	189	226	86.1
45.5	413	188	225	85.6
45.5	411	187	224	85.1
45.5	409	186	223	84.6
45.5	407	185	222	84.1
45.5	404	184	220	83.8
45.5	402	183	219	83.3
45.5	400	182	218	82.8
45.5	398	181	217	82.3
45.5	396	180	216	81.8

%E	M1	M2	DM	M*
45.5	393	179	214	81.5
45.5	391	178	213	81.0
45.5	389	177	212	80.5
45.5	387	176	211	80.0
45.5	385	175	210	79.5
45.5	380	173	207	78.8
45.5	378	172	206	78.3
45.5	376	171	205	77.8
45.5	374	170	204	77.3
45.5	369	168	201	76.5
45.5	367	167	200	76.0
45.5	365	166	199	75.5
45.5	363	165	198	75.0
45.5	358	163	195	74.2
45.5	356	162	194	73.7
45.5	354	161	193	73.2
45.5	352	160	192	72.7
45.5	347	158	189	71.9
45.5	345	157	188	71.4
45.5	343	156	187	71.0
45.5	341	155	186	70.5
45.5	336	153	183	69.7
45.5	334	152	182	69.2
45.5	332	151	181	68.7
45.5	330	150	180	68.2
45.5	325	148	177	67.4
45.5	323	147	176	66.9
45.5	321	146	175	66.4
45.5	319	145	174	65.9
45.5	314	143	171	65.1
45.5	312	142	170	64.6
45.5	310	141	169	64.1
45.5	308	140	168	63.6
45.5	303	138	165	62.9
45.5	301	137	164	62.4
45.5	299	136	163	61.9
45.5	297	135	162	61.4
45.5	292	133	159	60.6
45.5	290	132	158	60.1
45.5	288	131	157	59.6
45.5	286	130	156	59.1
45.5	279	127	152	57.8
45.5	277	126	151	57.3
45.5	275	125	150	56.8
45.5	268	122	146	55.5
45.5	266	121	145	55.0
45.5	264	120	144	54.5
45.5	257	117	140	53.3
45.5	255	116	139	52.8
45.5	253	115	138	52.3

%E	M1	M2	DM	M*
45.5	246	112	134	51.0
45.5	244	111	133	50.5
45.5	242	110	132	50.0
45.5	235	107	128	48.7
45.5	233	106	127	48.2
45.5	231	105	126	47.7
45.5	224	102	122	46.4
45.5	222	101	121	46.0
45.5	220	100	120	45.5
45.5	213	97	116	44.2
45.5	211	96	115	43.7
45.5	209	95	114	43.2
45.5	202	92	110	41.9
45.5	200	91	109	41.4
45.5	198	90	108	40.9
45.5	189	86	103	39.1
45.5	187	85	102	38.6
45.5	178	81	97	36.9
45.5	176	80	96	36.4
45.5	167	76	91	34.6
45.5	165	75	90	34.1
45.5	156	71	85	32.3
45.5	154	70	84	31.8
45.5	145	66	79	30.0
45.5	143	65	78	29.5
45.5	134	61	73	27.8
45.5	132	60	72	27.3
45.5	123	56	67	25.5
45.5	121	55	66	25.0
45.5	112	51	61	23.2
45.5	110	50	60	22.7
45.5	101	46	55	21.0
45.5	99	45	54	20.5
45.5	88	40	48	18.2
45.5	77	35	42	15.9
45.5	66	30	36	13.6
45.5	55	25	30	11.4
45.5	44	20	24	9.1
45.5	33	15	18	6.8
45.5	22	10	12	4.5
45.5	11	5	6	2.3
45.4	500	227	273	103.1
45.4	498	226	272	102.6
45.4	496	225	271	102.1
45.4	493	224	269	101.8
45.4	491	223	268	101.3
45.4	489	222	267	100.8
45.4	487	221	266	100.3
45.4	485	220	265	99.8
45.4	482	219	263	99.5

%E	M1	M2	DM	M*
45.4	480	218	262	99.0
45.4	478	217	261	98.5
45.4	476	216	260	98.0
45.4	474	215	259	97.5
45.4	471	214	257	97.2
45.4	469	213	256	96.7
45.4	467	212	255	96.2
45.4	465	211	254	95.7
45.4	463	210	253	95.2
45.4	460	209	251	95.0
45.4	458	208	250	94.5
45.4	456	207	249	94.0
45.4	454	206	248	93.5
45.4	452	205	247	93.0
45.4	449	204	245	92.7
45.4	447	203	244	92.2
45.4	445	202	243	91.7
45.4	443	201	242	91.2
45.4	441	200	241	90.7
45.4	438	199	239	90.4
45.4	436	198	238	89.9
45.4	434	197	237	89.4
45.4	432	196	236	88.9
45.4	427	194	233	88.1
45.4	425	193	232	87.6
45.4	423	192	231	87.1
45.4	421	191	230	86.7
45.4	416	189	227	85.9
45.4	414	188	226	85.4
45.4	412	187	225	84.9
45.4	410	186	224	84.4
45.4	405	184	221	83.6
45.4	403	183	220	83.1
45.4	401	182	219	82.6
45.4	399	181	218	82.1
45.4	394	179	215	81.3
45.4	392	178	214	80.8
45.4	390	177	213	80.3
45.4	388	176	212	79.8
45.4	383	174	209	79.0
45.4	381	173	208	78.6
45.4	379	172	207	78.1
45.4	377	171	206	77.6
45.4	372	169	203	76.8
45.4	370	168	202	76.3
45.4	368	167	201	75.8
45.4	366	166	200	75.3
45.4	361	164	197	74.5
45.4	359	163	196	74.0
45.4	357	162	195	73.5

%E	M1	M2	DM	M*
45.4	355	161	194	73.0
45.4	350	159	191	72.2
45.4	348	158	190	71.7
45.4	346	157	189	71.2
45.4	339	154	185	70.0
45.4	337	153	184	69.5
45.4	335	152	183	69.0
45.4	328	149	179	67.7
45.4	326	148	178	67.2
45.4	324	147	177	66.7
45.4	317	144	173	65.4
45.4	315	143	172	64.9
45.4	313	142	171	64.4
45.4	306	139	167	63.1
45.4	304	138	166	62.6
45.4	302	137	165	62.1
45.4	295	134	161	60.9
45.4	293	133	160	60.4
45.4	291	132	159	59.9
45.4	284	129	155	58.6
45.4	282	128	154	58.1
45.4	280	127	153	57.6
45.4	273	124	149	56.3
45.4	271	123	148	55.8
45.4	269	122	147	55.3
45.4	262	119	143	54.0
45.4	260	118	142	53.6
45.4	251	114	137	51.8
45.4	249	113	136	51.3
45.4	240	109	131	49.5
45.4	238	108	130	49.0
45.4	229	104	125	47.2
45.4	227	103	124	46.7
45.4	218	99	119	45.0
45.4	216	98	118	44.5
45.4	207	94	113	42.7
45.4	205	93	112	42.2
45.4	196	89	107	40.4
45.4	194	88	106	39.9
45.4	185	84	101	38.1
45.4	183	83	100	37.6
45.4	174	79	95	35.9
45.4	163	74	89	33.6
45.4	152	69	83	31.3
45.4	141	64	77	29.0
45.4	130	59	71	26.8
45.4	119	54	65	24.5
45.4	108	49	59	22.2
45.4	97	44	53	20.0
45.3	499	226	273	102.4

%E	M1	M2	DM	M*
45.3	497	225	272	101.9
45.3	495	224	271	101.4
45.3	494	224	270	101.6
45.3	492	223	269	101.1
45.3	490	222	268	100.6
45.3	488	221	267	100.1
45.3	486	220	266	99.6
45.3	483	219	264	99.3
45.3	481	218	263	98.8
45.3	479	217	262	98.3
45.3	477	216	261	97.8
45.3	475	215	260	97.3
45.3	472	214	258	97.0
45.3	470	213	257	96.5
45.3	468	212	256	96.0
45.3	466	211	255	95.5
45.3	464	210	254	95.0
45.3	461	209	252	94.8
45.3	459	208	251	94.3
45.3	457	207	250	93.8
45.3	455	206	249	93.3
45.3	453	205	248	92.8
45.3	450	204	246	92.5
45.3	448	203	245	92.0
45.3	446	202	244	91.5
45.3	444	201	243	91.0
45.3	439	199	240	90.2
45.3	437	198	239	89.7
45.3	435	197	238	89.2
45.3	433	196	237	88.7
45.3	430	195	235	88.4
45.3	428	194	234	87.9
45.3	426	193	233	87.4
45.3	424	192	232	86.9
45.3	422	191	231	86.4
45.3	419	190	229	86.2
45.3	417	189	228	85.7
45.3	415	188	227	85.2
45.3	413	187	226	84.7
45.3	411	186	225	84.2
45.3	408	185	223	83.9
45.3	406	184	222	83.4
45.3	404	183	221	82.9
45.3	402	182	220	82.4
45.3	400	181	219	81.9
45.3	397	180	217	81.6
45.3	395	179	216	81.1
45.3	393	178	215	80.6
45.3	391	177	214	80.1
45.3	386	175	211	79.3

%E	M1	M2	DM	M*
45.3	384	174	210	78.8
45.3	382	173	209	78.3
45.3	380	172	208	77.9
45.3	375	170	205	77.1
45.3	373	169	204	76.6
45.3	371	168	203	76.1
45.3	369	167	202	75.6
45.3	364	165	199	74.8
45.3	362	164	198	74.3
45.3	360	163	197	73.8
45.3	358	162	196	73.3
45.3	353	160	193	72.5
45.3	351	159	192	72.0
45.3	349	158	191	71.5
45.3	344	156	188	70.7
45.3	342	155	187	70.2
45.3	340	154	186	69.8
45.3	338	153	185	69.3
45.3	333	151	182	68.5
45.3	331	150	181	68.0
45.3	329	149	180	67.5
45.3	327	148	179	67.0
45.3	322	146	176	66.2
45.3	320	145	175	65.7
45.3	318	144	174	65.2
45.3	316	143	173	64.7
45.3	311	141	170	63.9
45.3	309	140	169	63.4
45.3	307	139	168	62.9
45.3	300	136	164	61.7
45.3	298	135	163	61.2
45.3	296	134	162	60.7
45.3	289	131	158	59.4
45.3	287	130	157	58.9
45.3	285	129	156	58.4
45.3	278	126	152	57.1
45.3	276	125	151	56.6
45.3	274	124	150	56.1
45.3	267	121	146	54.8
45.3	265	120	145	54.3
45.3	258	117	141	53.1
45.3	256	116	140	52.6
45.3	254	115	139	52.1
45.3	247	112	135	50.8
45.3	245	111	134	50.3
45.3	243	110	133	49.8
45.3	236	107	129	48.5
45.3	234	106	128	48.0
45.3	232	105	127	47.5
45.3	225	102	123	46.2

%E	M1	M2	DM	M*
45.3	223	101	122	45.7
45.3	214	97	117	44.0
45.3	212	96	116	43.5
45.3	203	92	111	41.7
45.3	201	91	110	41.2
45.3	192	87	105	39.4
45.3	190	86	104	38.9
45.3	181	82	99	37.1
45.3	179	81	98	36.7
45.3	172	78	94	35.4
45.3	170	77	93	34.9
45.3	161	73	88	33.1
45.3	159	72	87	32.6
45.3	150	68	82	30.8
45.3	148	67	81	30.3
45.3	139	63	76	28.6
45.3	137	62	75	28.1
45.3	128	58	70	26.3
45.3	117	53	64	24.0
45.3	106	48	58	21.7
45.3	95	43	52	19.5
45.3	86	39	47	17.7
45.3	75	34	41	15.4
45.3	64	29	35	13.1
45.3	53	24	29	10.9
45.2	500	226	274	102.2
45.2	498	225	273	101.7
45.2	496	224	272	101.2
45.2	493	223	270	100.9
45.2	491	222	269	100.4
45.2	489	221	268	99.9
45.2	487	220	267	99.4
45.2	485	219	266	98.9
45.2	484	219	265	99.1
45.2	482	218	264	98.6
45.2	480	217	263	98.1
45.2	478	216	262	97.6
45.2	476	215	261	97.1
45.2	473	214	259	96.8
45.2	471	213	258	96.3
45.2	469	212	257	95.8
45.2	467	211	256	95.3
45.2	465	210	255	94.8
45.2	462	209	253	94.5
45.2	460	208	252	94.1
45.2	458	207	251	93.6
45.2	456	206	250	93.1
45.2	454	205	249	92.6
45.2	451	204	247	92.3
45.2	449	203	246	91.8

%E	M1	M2	DM	M*
45.2	447	202	245	91.3
45.2	445	201	244	90.8
45.2	442	200	242	90.5
45.2	440	199	241	90.0
45.2	438	198	240	89.5
45.2	436	197	239	89.0
45.2	434	196	238	88.5
45.2	431	195	236	88.2
45.2	429	194	235	87.7
45.2	427	193	234	87.2
45.2	425	192	233	86.7
45.2	423	191	232	86.2
45.2	420	190	230	86.0
45.2	418	189	229	85.5
45.2	416	188	228	85.0
45.2	414	187	227	84.5
45.2	409	185	224	83.7
45.2	407	184	223	83.2
45.2	405	183	222	82.7
45.2	403	182	221	82.2
45.2	398	180	218	81.4
45.2	396	179	217	80.9
45.2	394	178	216	80.4
45.2	392	177	215	79.9
45.2	389	176	213	79.6
45.2	387	175	212	79.1
45.2	385	174	211	78.6
45.2	383	173	210	78.1
45.2	378	171	207	77.4
45.2	376	170	206	76.9
45.2	374	169	205	76.4
45.2	372	168	204	75.9
45.2	367	166	201	75.1
45.2	365	165	200	74.6
45.2	363	164	199	74.1
45.2	361	163	198	73.6
45.2	356	161	195	72.8
45.2	354	160	194	72.3
45.2	352	159	193	71.8
45.2	347	157	190	71.0
45.2	345	156	189	70.5
45.2	343	155	188	70.0
45.2	341	154	187	69.5
45.2	336	152	184	68.8
45.2	334	151	183	68.3
45.2	332	150	182	67.8
45.2	330	149	181	67.3
45.2	325	147	178	66.5
45.2	323	146	177	66.0
45.2	321	145	176	65.5

%E	M1	M2	DM	M*
45.2	314	142	172	64.2
45.2	312	141	171	63.7
45.2	310	140	170	63.2
45.2	305	138	167	62.4
45.2	303	137	166	61.9
45.2	301	136	165	61.4
45.2	299	135	164	61.0
45.2	294	133	161	60.2
45.2	292	132	160	59.7
45.2	290	131	159	59.2
45.2	283	128	155	57.9
45.2	281	127	154	57.4
45.2	279	126	153	56.9
45.2	272	123	149	55.6
45.2	270	122	148	55.1
45.2	263	119	144	53.8
45.2	261	118	143	53.3
45.2	259	117	142	52.9
45.2	252	114	138	51.6
45.2	250	113	137	51.1
45.2	248	112	136	50.6
45.2	241	109	132	49.3
45.2	239	108	131	48.8
45.2	230	104	126	47.0
45.2	228	103	125	46.5
45.2	221	100	121	45.2
45.2	219	99	120	44.8
45.2	217	98	119	44.3
45.2	210	95	115	43.0
45.2	208	94	114	42.5
45.2	199	90	109	40.7
45.2	197	89	108	40.2
45.2	188	85	103	38.4
45.2	186	84	102	37.9
45.2	177	80	97	36.2
45.2	168	76	92	34.4
45.2	166	75	91	33.9
45.2	157	71	86	32.1
45.2	155	70	85	31.6
45.2	146	66	80	29.8
45.2	135	61	74	27.6
45.2	126	57	69	25.8
45.2	124	56	68	25.3
45.2	115	52	63	23.5
45.2	104	47	57	21.2
45.2	93	42	51	19.0
45.2	84	38	46	17.2
45.2	73	33	40	14.9
45.2	62	28	34	12.6
45.2	42	19	23	8.6

%E	M1	M2	DM	M*
45.2	31	14	17	6.3
45.1	499	225	274	101.5
45.1	497	224	273	101.0
45.1	495	223	272	100.5
45.1	494	223	271	100.7
45.1	492	222	270	100.2
45.1	490	221	269	99.7
45.1	488	220	268	99.2
45.1	486	219	267	98.7
45.1	483	218	265	98.4
45.1	481	217	264	97.9
45.1	479	216	263	97.4
45.1	477	215	262	96.9
45.1	475	214	261	96.4
45.1	474	214	260	96.6
45.1	472	213	259	96.1
45.1	470	212	258	95.6
45.1	468	211	257	95.1
45.1	466	210	256	94.6
45.1	463	209	254	94.3
45.1	461	208	253	93.8
45.1	459	207	252	93.4
45.1	457	206	251	92.9
45.1	455	205	250	92.4
45.1	452	204	248	92.1
45.1	450	203	247	91.6
45.1	448	202	246	91.1
45.1	446	201	245	90.6
45.1	443	200	243	90.3
45.1	441	199	242	89.8
45.1	439	198	241	89.3
45.1	437	197	240	88.8
45.1	435	196	239	88.3
45.1	432	195	237	88.0
45.1	430	194	236	87.5
45.1	428	193	235	87.0
45.1	426	192	234	86.5
45.1	421	190	231	85.7
45.1	419	189	230	85.3
45.1	417	188	229	84.8
45.1	415	187	228	84.3
45.1	412	186	226	84.0
45.1	410	185	225	83.5
45.1	408	184	224	83.0
45.1	406	183	223	82.5
45.1	404	182	222	82.0
45.1	401	181	220	81.7
45.1	399	180	219	81.2
45.1	397	179	218	80.7
45.1	395	178	217	80.2

%E	M1	M2	DM	M*
45.1	390	176	214	79.4
45.1	388	175	213	78.9
45.1	386	174	212	78.4
45.1	384	173	211	77.9
45.1	381	172	209	77.6
45.1	379	171	208	77.2
45.1	377	170	207	76.7
45.1	375	169	206	76.2
45.1	370	167	203	75.4
45.1	368	166	202	74.9
45.1	366	165	201	74.4
45.1	364	164	200	73.9
45.1	359	162	197	73.1
45.1	357	161	196	72.6
45.1	355	160	195	72.1
45.1	350	158	192	71.3
45.1	348	157	191	70.8
45.1	346	156	190	70.3
45.1	344	155	189	69.8
45.1	339	153	186	69.1
45.1	337	152	185	68.6
45.1	335	151	184	68.1
45.1	328	148	180	66.8
45.1	326	147	179	66.3
45.1	324	146	178	65.8
45.1	319	144	175	65.0
45.1	317	143	174	64.5
45.1	315	142	173	64.0
45.1	308	139	169	62.7
45.1	306	138	168	62.2
45.1	304	137	167	61.7
45.1	297	134	163	60.5
45.1	295	133	162	60.0
45.1	293	132	161	59.5
45.1	288	130	158	58.7
45.1	286	129	157	58.2
45.1	284	128	156	57.7
45.1	277	125	152	56.4
45.1	275	124	151	55.9
45.1	273	123	150	55.4
45.1	268	121	147	54.6
45.1	266	120	146	54.1
45.1	264	119	145	53.6
45.1	257	116	141	52.4
45.1	255	115	140	51.9
45.1	253	114	139	51.4
45.1	246	111	135	50.1
45.1	244	110	134	49.6
45.1	237	107	130	48.3
45.1	235	106	129	47.8

%E	M1	M2	DM	M*
45.1	233	105	128	47.3
45.1	226	102	124	46.0
45.1	224	101	123	45.5
45.1	215	97	118	43.8
45.1	213	96	117	43.3
45.1	206	93	113	42.0
45.1	204	92	112	41.5
45.1	202	91	111	41.0
45.1	195	88	107	39.7
45.1	193	87	106	39.2
45.1	184	83	101	37.4
45.1	182	82	100	36.9
45.1	175	79	96	35.7
45.1	173	78	95	35.2
45.1	164	74	90	33.4
45.1	162	73	89	32.9
45.1	153	69	84	31.1
45.1	144	65	79	29.3
45.1	142	64	78	28.8
45.1	133	60	73	27.1
45.1	122	55	67	24.8
45.1	113	51	62	23.0
45.1	102	46	56	20.7
45.1	91	41	50	18.5
45.1	82	37	45	16.7
45.1	71	32	39	14.4
45.1	51	23	28	10.4
45.0	500	225	275	101.3
45.0	498	224	274	100.8
45.0	496	223	273	100.3
45.0	493	222	271	100.0
45.0	491	221	270	99.5
45.0	489	220	269	99.0
45.0	487	219	268	98.5
45.0	484	218	266	98.2
45.0	482	217	265	97.7
45.0	480	216	264	97.2
45.0	478	215	263	96.7
45.0	476	214	262	96.2
45.0	473	213	260	95.9
45.0	471	212	259	95.4
45.0	469	211	258	94.9
45.0	467	210	257	94.4
45.0	464	209	255	94.1
45.0	462	208	254	93.6
45.0	460	207	253	93.1
45.0	458	206	252	92.7
45.0	456	205	251	92.2
45.0	453	204	249	91.9
45.0	451	203	248	91.4

%E	M1	M2	DM	M*
45.0	449	202	247	90.9
45.0	447	201	246	90.4
45.0	444	200	244	90.1
45.0	442	199	243	89.6
45.0	440	198	242	89.1
45.0	438	197	241	88.6
45.0	436	196	240	88.1
45.0	433	195	238	87.8
45.0	431	194	237	87.3
45.0	429	193	236	86.8
45.0	427	192	235	86.3
45.0	424	191	233	86.0
45.0	422	190	232	85.5
45.0	420	189	231	85.0
45.0	418	188	230	84.6
45.0	416	187	229	84.1
45.0	413	186	227	83.8
45.0	411	185	226	83.3
45.0	409	184	225	82.8
45.0	407	183	224	82.3
45.0	402	181	221	81.5
45.0	400	180	220	81.0
45.0	398	179	219	80.5
45.0	393	177	216	79.7
45.0	391	176	215	79.2
45.0	389	175	214	78.7
45.0	387	174	213	78.2
45.0	382	172	210	77.4
45.0	380	171	209	76.9
45.0	378	170	208	76.5
45.0	373	168	205	75.7
45.0	371	167	204	75.2
45.0	369	166	203	74.7
45.0	367	165	202	74.2
45.0	362	163	199	73.4
45.0	360	162	198	72.9
45.0	358	161	197	72.4
45.0	353	159	194	71.6
45.0	351	158	193	71.1
45.0	349	157	192	70.6
45.0	347	156	191	70.1
45.0	342	154	188	69.3
45.0	340	153	187	68.8
45.0	338	152	186	68.4
45.0	333	150	183	67.6
45.0	331	149	182	67.1
45.0	329	148	181	66.6
45.0	327	147	180	66.1
45.0	322	145	177	65.3
45.0	320	144	176	64.8

%E	M1	M2	DM	M*
45.0	318	143	175	64.3
45.0	313	141	172	63.5
45.0	311	140	171	63.0
45.0	309	139	170	62.5
45.0	307	138	169	62.0
45.0	302	136	166	61.2
45.0	300	135	165	60.8
45.0	298	134	164	60.3
45.0	291	131	160	59.0
45.0	289	130	159	58.5
45.0	282	127	155	57.2
45.0	280	126	154	56.7
45.0	278	125	153	56.2
45.0	271	122	149	54.9
45.0	269	121	148	54.4
45.0	262	118	144	53.1
45.0	260	117	143	52.6
45.0	258	116	142	52.2
45.0	251	113	138	50.9
45.0	249	112	137	50.4
45.0	242	109	133	49.1
45.0	240	108	132	48.6
45.0	238	107	131	48.1
45.0	231	104	127	46.8
45.0	229	103	126	46.3
45.0	222	100	122	45.0
45.0	220	99	121	44.5
45.0	218	98	120	44.1
45.0	211	95	116	42.8
45.0	209	94	115	42.3
45.0	200	90	110	40.5
45.0	191	86	105	38.7
45.0	189	85	104	38.2
45.0	180	81	99	36.4
45.0	171	77	94	34.7
45.0	169	76	93	34.2
45.0	160	72	88	32.4
45.0	151	68	83	30.6
45.0	149	67	82	30.1
45.0	140	63	77	28.3
45.0	131	59	72	26.6
45.0	129	58	71	26.1
45.0	120	54	66	24.3
45.0	111	50	61	22.5
45.0	109	49	60	22.0
45.0	100	45	55	20.3
45.0	80	36	44	16.2
45.0	60	27	33	12.1
45.0	40	18	22	8.1
45.0	20	9	11	4.0

%E	M1	M2	DM	M*
44.9	499	224	275	100.6
44.9	497	223	274	100.1
44.9	494	222	272	99.8
44.9	492	221	271	99.3
44.9	490	220	270	98.8
44.9	488	219	269	98.3
44.9	486	218	268	97.8
44.9	485	218	267	98.0
44.9	483	217	266	97.5
44.9	481	216	265	97.0
44.9	479	215	264	96.5
44.9	477	214	263	96.0
44.9	474	213	261	95.7
44.9	472	212	260	95.2
44.9	470	211	259	94.7
44.9	468	210	258	94.2
44.9	466	209	257	93.7
44.9	465	209	256	93.9
44.9	463	208	255	93.4
44.9	461	207	254	92.9
44.9	459	206	253	92.5
44.9	457	205	252	92.0
44.9	454	204	250	91.7
44.9	452	203	249	91.2
44.9	450	202	248	90.7
44.9	448	201	247	90.2
44.9	445	200	245	89.9
44.9	443	199	244	89.4
44.9	441	198	243	88.9
44.9	439	197	242	88.4
44.9	437	196	241	87.9
44.9	434	195	239	87.6
44.9	432	194	238	87.1
44.9	430	193	237	86.6
44.9	428	192	236	86.1
44.9	425	191	234	85.8
44.9	423	190	233	85.3
44.9	421	189	232	84.8
44.9	419	188	231	84.4
44.9	414	186	228	83.6
44.9	412	185	227	83.1
44.9	410	184	226	82.6
44.9	408	183	225	82.1
44.9	405	182	223	81.8
44.9	403	181	222	81.3
44.9	401	180	221	80.8
44.9	399	179	220	80.3
44.9	396	178	218	80.0
44.9	394	177	217	79.5
44.9	392	176	216	79.0

%E	M1	M2	DM	M*
44.9	390	175	215	78.5
44.9	385	173	212	77.7
44.9	383	172	211	77.2
44.9	381	171	210	76.7
44.9	379	170	209	76.3
44.9	376	169	207	76.0
44.9	374	168	206	75.5
44.9	372	167	205	75.0
44.9	370	166	204	74.5
44.9	365	164	201	73.7
44.9	363	163	200	73.2
44.9	361	162	199	72.7
44.9	356	160	196	71.9
44.9	354	159	195	71.4
44.9	352	158	194	70.9
44.9	350	157	193	70.4
44.9	345	155	190	69.6
44.9	343	154	189	69.1
44.9	341	153	188	68.6
44.9	336	151	185	67.9
44.9	334	150	184	67.4
44.9	332	149	183	66.9
44.9	325	146	179	65.6
44.9	323	145	178	65.1
44.9	321	144	177	64.6
44.9	316	142	174	63.8
44.9	314	141	173	63.3
44.9	312	140	172	62.8
44.9	305	137	168	61.5
44.9	303	136	167	61.0
44.9	301	135	166	60.5
44.9	296	133	163	59.8
44.9	294	132	162	59.3
44.9	292	131	161	58.8
44.9	287	129	158	58.0
44.9	285	128	157	57.5
44.9	283	127	156	57.0
44.9	276	124	152	55.7
44.9	274	123	151	55.2
44.9	272	122	150	54.7
44.9	267	120	147	53.9
44.9	265	119	146	53.4
44.9	263	118	145	52.9
44.9	256	115	141	51.7
44.9	254	114	140	51.2
44.9	247	111	136	49.9
44.9	245	110	135	49.4
44.9	243	109	134	48.9
44.9	236	106	130	47.6
44.9	234	105	129	47.1

%E	M1	M2	DM	M*
44.9	227	102	125	45.8
44.9	225	101	124	45.3
44.9	216	97	119	43.6
44.9	214	96	118	43.1
44.9	207	93	114	41.8
44.9	205	92	113	41.3
44.9	198	89	109	40.0
44.9	196	88	108	39.5
44.9	187	84	103	37.7
44.9	185	83	102	37.2
44.9	178	80	98	36.0
44.9	176	79	97	35.5
44.9	167	75	92	33.7
44.9	158	71	87	31.9
44.9	156	70	86	31.4
44.9	147	66	81	29.6
44.9	138	62	76	27.9
44.9	136	61	75	27.4
44.9	127	57	70	25.6
44.9	118	53	65	23.8
44.9	107	48	59	21.5
44.9	98	44	54	19.8
44.9	89	40	49	18.0
44.9	78	35	43	15.7
44.9	69	31	38	13.9
44.9	49	22	27	9.9
44.8	500	224	276	100.4
44.8	498	223	275	99.9
44.8	496	222	274	99.4
44.8	495	222	273	99.6
44.8	493	221	272	99.1
44.8	491	220	271	98.6
44.8	489	219	270	98.1
44.8	487	218	269	97.6
44.8	484	217	267	97.3
44.8	482	216	266	96.8
44.8	480	215	265	96.3
44.8	478	214	264	95.8
44.8	475	213	262	95.5
44.8	473	212	261	95.0
44.8	471	211	260	94.5
44.8	469	210	259	94.0
44.8	467	209	258	93.5
44.8	464	208	256	93.2
44.8	462	207	255	92.7
44.8	460	206	254	92.3
44.8	458	205	253	91.8
44.8	455	204	251	91.5
44.8	453	203	250	91.0
44.8	451	202	249	90.5

%E	M1	M2	DM	M*
44.8	449	201	248	90.0
44.8	446	200	246	89.7
44.8	444	199	245	89.2
44.8	442	198	244	88.7
44.8	440	197	243	88.2
44.8	435	195	240	87.4
44.8	433	194	239	86.9
44.8	431	193	238	86.4
44.8	429	192	237	85.9
44.8	426	191	235	85.6
44.8	424	190	234	85.1
44.8	422	189	233	84.6
44.8	420	188	232	84.2
44.8	417	187	230	83.9
44.8	415	186	229	83.4
44.8	413	185	228	82.9
44.8	411	184	227	82.4
44.8	406	182	224	81.6
44.8	404	181	223	81.1
44.8	402	180	222	80.6
44.8	400	179	221	80.1
44.8	397	178	219	79.8
44.8	395	177	218	79.3
44.8	393	176	217	78.8
44.8	391	175	216	78.3
44.8	388	174	214	78.0
44.8	386	173	213	77.5
44.8	384	172	212	77.0
44.8	382	171	211	76.5
44.8	377	169	208	75.8
44.8	375	168	207	75.3
44.8	373	167	206	74.8
44.8	368	165	203	74.0
44.8	366	164	202	73.5
44.8	364	163	201	73.0
44.8	362	162	200	72.5
44.8	359	161	198	72.2
44.8	357	160	197	71.7
44.8	355	159	196	71.2
44.8	353	158	195	70.7
44.8	348	156	192	69.9
44.8	346	155	191	69.4
44.8	344	154	190	68.9
44.8	339	152	187	68.2
44.8	337	151	186	67.7
44.8	335	150	185	67.2
44.8	330	148	182	66.4
44.8	328	147	181	65.9
44.8	326	146	180	65.4
44.8	324	145	179	64.9

%E	M1	M2	DM	M*
44.8	319	143	176	64.1
44.8	317	142	175	63.6
44.8	315	141	174	63.1
44.8	310	139	171	62.3
44.8	308	138	170	61.8
44.8	306	137	169	61.3
44.8	299	134	165	60.1
44.8	297	133	164	59.6
44.8	290	130	160	58.3
44.8	288	129	159	57.8
44.8	286	128	158	57.3
44.8	281	126	155	56.5
44.8	279	125	154	56.0
44.8	277	124	153	55.5
44.8	270	121	149	54.2
44.8	268	120	148	53.7
44.8	261	117	144	52.4
44.8	259	116	143	52.0
44.8	252	113	139	50.7
44.8	250	112	138	50.2
44.8	248	111	137	49.7
44.8	241	108	133	48.4
44.8	239	107	132	47.9
44.8	232	104	128	46.6
44.8	230	103	127	46.1
44.8	223	100	123	44.8
44.8	221	99	122	44.3
44.8	212	95	117	42.6
44.8	210	94	116	42.1
44.8	203	91	112	40.8
44.8	201	90	111	40.3
44.8	194	87	107	39.0
44.8	192	86	106	38.5
44.8	183	82	101	36.7
44.8	181	81	100	36.2
44.8	174	78	96	35.0
44.8	172	77	95	34.5
44.8	165	74	91	33.2
44.8	163	73	90	32.7
44.8	154	69	85	30.9
44.8	145	65	80	29.1
44.8	143	64	79	28.6
44.8	134	60	74	26.9
44.8	125	56	69	25.1
44.8	116	52	64	23.3
44.8	105	47	58	21.0
44.8	96	43	53	19.3
44.8	87	39	48	17.5
44.8	67	30	37	13.4
44.8	58	26	32	11.7

%E	M1	M2	DM	M*
44.8	29	13	16	5.8
44.7	499	223	276	99.7
44.7	497	222	275	99.2
44.7	494	221	273	98.9
44.7	492	220	272	98.4
44.7	490	219	271	97.9
44.7	488	218	270	97.4
44.7	486	217	269	96.9
44.7	485	217	268	97.1
44.7	483	216	267	96.6
44.7	481	215	266	96.1
44.7	479	214	265	95.6
44.7	477	213	264	95.1
44.7	476	213	263	95.3
44.7	474	212	262	94.8
44.7	472	211	261	94.3
44.7	470	210	260	93.8
44.7	468	209	259	93.3
44.7	465	208	257	93.0
44.7	463	207	256	92.5
44.7	461	206	255	92.1
44.7	459	205	254	91.6
44.7	456	204	252	91.3
44.7	454	203	251	90.8
44.7	452	202	250	90.3
44.7	450	201	249	89.8
44.7	447	200	247	89.5
44.7	445	199	246	89.0
44.7	443	198	245	88.5
44.7	441	197	244	88.0
44.7	438	196	242	87.7
44.7	436	195	241	87.2
44.7	434	194	240	86.7
44.7	432	193	239	86.2
44.7	430	192	238	85.7
44.7	427	191	236	85.4
44.7	425	190	235	84.9
44.7	423	189	234	84.4
44.7	421	188	233	84.0
44.7	418	187	231	83.7
44.7	416	186	230	83.2
44.7	414	185	229	82.7
44.7	412	184	228	82.2
44.7	409	183	226	81.9
44.7	407	182	225	81.4
44.7	405	181	224	80.9
44.7	403	180	223	80.4
44.7	398	178	220	79.6
44.7	396	177	219	79.1
44.7	394	176	218	78.6

%E	M1	M2	DM	M*	%E	M1	M2	DM	M*	%E	M1	M2	DM	M*	%E	M1	M2	DM	M*	%E	M1	M2	DM	M*
44.7	389	174	215	77.8	44.7	226	101	125	45.1	44.6	448	200	248	89.3	44.6	316	141	175	62.9	44.6	65	29	36	12.9
44.7	387	173	214	77.3	44.7	219	98	121	43.9	44.6	446	199	247	88.8	44.6	314	140	174	62.4	44.6	56	25	31	11.2
44.7	385	172	213	76.8	44.7	217	97	120	43.4	44.6	444	198	246	88.3	44.6	312	139	173	61.9	44.5	499	222	277	98.8
44.7	380	170	210	76.1	44.7	215	96	119	42.9	44.6	442	197	245	87.8	44.6	307	137	170	61.1	44.5	497	221	276	98.3
44.7	378	169	209	75.6	44.7	208	93	115	41.6	44.6	439	196	243	87.5	44.6	305	136	169	60.6	44.5	494	220	274	98.0
44.7	376	168	208	75.1	44.7	206	92	114	41.1	44.6	437	195	242	87.0	44.6	303	135	168	60.1	44.5	492	219	273	97.5
44.7	374	167	207	74.6	44.7	199	89	110	39.8	44.6	435	194	241	86.5	44.6	298	133	165	59.4	44.5	490	218	272	97.0
44.7	371	166	205	74.3	44.7	197	88	109	39.3	44.6	433	193	240	86.0	44.6	296	132	164	58.9	44.5	488	217	271	96.5
44.7	369	165	204	73.8	44.7	190	85	105	38.0	44.6	428	191	237	85.2	44.6	294	131	163	58.4	44.5	485	216	269	96.2
44.7	367	164	203	73.3	44.7	188	84	104	37.5	44.6	426	190	236	84.7	44.6	289	129	160	57.6	44.5	483	215	268	95.7
44.7	365	163	202	72.8	44.7	179	80	99	35.8	44.6	424	189	235	84.2	44.6	287	128	159	57.1	44.5	481	214	267	95.2
44.7	360	161	199	72.0	44.7	170	76	94	34.0	44.6	422	188	234	83.8	44.6	285	127	158	56.6	44.5	479	213	266	94.7
44.7	358	160	198	71.5	44.7	161	72	89	32.2	44.6	419	187	232	83.5	44.6	280	125	155	55.8	44.5	476	212	264	94.4
44.7	356	159	197	71.0	44.7	159	71	88	31.7	44.6	417	186	231	83.0	44.6	278	124	154	55.3	44.5	474	211	263	93.9
44.7	351	157	194	70.2	44.7	152	68	84	30.4	44.6	415	185	230	82.5	44.6	276	123	153	54.8	44.5	472	210	262	93.4
44.7	349	156	193	69.7	44.7	150	67	83	29.9	44.6	413	184	229	82.0	44.6	271	121	150	54.0	44.5	470	209	261	92.9
44.7	347	155	192	69.2	44.7	141	63	78	28.1	44.6	410	183	227	81.7	44.6	269	120	149	53.5	44.5	467	208	259	92.6
44.7	342	153	189	68.4	44.7	132	59	73	26.4	44.6	408	182	226	81.2	44.6	267	119	148	53.0	44.5	465	207	258	92.1
44.7	340	152	188	68.0	44.7	123	55	68	24.6	44.6	406	181	225	80.7	44.6	260	116	144	51.8	44.5	463	206	257	91.7
44.7	338	151	187	67.5	44.7	114	51	63	22.8	44.6	404	180	224	80.2	44.6	258	115	143	51.3	44.5	461	205	256	91.2
44.7	333	149	184	66.7	44.7	103	46	57	20.5	44.6	401	179	222	79.9	44.6	251	112	139	50.0	44.5	458	204	254	90.9
44.7	331	148	183	66.2	44.7	94	42	52	18.8	44.6	399	178	221	79.4	44.6	249	111	138	49.5	44.5	456	203	253	90.4
44.7	329	147	182	65.7	44.7	85	38	47	17.0	44.6	397	177	220	78.9	44.6	242	108	134	48.2	44.5	454	202	252	89.9
44.7	322	144	178	64.4	44.7	76	34	42	15.2	44.6	395	176	219	78.4	44.6	240	107	133	47.7	44.5	452	201	251	89.4
44.7	320	143	177	63.9	44.7	47	21	26	9.4	44.6	392	175	217	78.1	44.6	233	104	129	46.4	44.5	449	200	249	89.1
44.7	318	142	176	63.4	44.7	38	17	21	7.6	44.6	390	174	216	77.6	44.6	231	103	128	45.9	44.5	447	199	248	88.6
44.7	313	140	173	62.6	44.6	500	223	277	99.5	44.6	388	173	215	77.1	44.6	224	100	124	44.6	44.5	445	198	247	88.1
44.7	311	139	172	62.1	44.6	498	222	276	99.0	44.6	386	172	214	76.6	44.6	222	99	123	44.1	44.5	443	197	246	87.6
44.7	309	138	171	61.6	44.6	496	221	275	98.5	44.6	383	171	212	76.3	44.6	213	95	118	42.4	44.5	440	196	244	87.3
44.7	304	136	168	60.8	44.6	495	221	274	98.7	44.6	381	170	211	75.9	44.6	211	94	117	41.9	44.5	438	195	243	86.8
44.7	302	135	167	60.3	44.6	493	220	273	98.2	44.6	379	169	210	75.4	44.6	204	91	113	40.6	44.5	436	194	242	86.3
44.7	300	134	166	59.9	44.6	491	219	272	97.7	44.6	377	168	209	74.9	44.6	202	90	112	40.1	44.5	434	193	241	85.8
44.7	295	132	163	59.1	44.6	489	218	271	97.2	44.6	372	166	206	74.1	44.6	195	87	108	38.8	44.5	431	192	239	85.5
44.7	293	131	162	58.6	44.6	487	217	270	96.7	44.6	370	165	205	73.6	44.6	193	86	107	38.3	44.5	429	191	238	85.0
44.7	291	130	161	58.1	44.6	484	216	268	96.4	44.6	368	164	204	73.1	44.6	186	83	103	37.0	44.5	427	190	237	84.5
44.7	284	127	157	56.8	44.6	482	215	267	95.9	44.6	363	162	201	72.3	44.6	184	82	102	36.5	44.5	425	189	236	84.0
44.7	282	126	156	56.3	44.6	480	214	266	95.4	44.6	361	161	200	71.8	44.6	177	79	98	35.3	44.5	420	187	233	83.3
44.7	275	123	152	55.0	44.6	478	213	265	94.9	44.6	359	160	199	71.3	44.6	175	78	97	34.8	44.5	418	186	232	82.8
44.7	273	122	151	54.5	44.6	475	212	263	94.6	44.6	354	158	196	70.5	44.6	168	75	93	33.5	44.5	416	185	231	82.3
44.7	266	119	147	53.2	44.6	473	211	262	94.1	44.6	352	157	195	70.0	44.6	166	74	92	33.0	44.5	411	183	228	81.5
44.7	264	118	146	52.7	44.6	471	210	261	93.6	44.6	350	156	194	69.5	44.6	157	70	87	31.2	44.5	409	182	227	81.0
44.7	262	117	145	52.2	44.6	469	209	260	93.1	44.6	345	154	191	68.7	44.6	148	66	82	29.4	44.5	407	181	226	80.5
44.7	257	115	142	51.5	44.6	466	208	258	92.8	44.6	343	153	190	68.2	44.6	139	62	77	27.7	44.5	402	179	223	79.7
44.7	255	114	141	51.0	44.6	464	207	257	92.3	44.6	341	152	189	67.8	44.6	130	58	72	25.9	44.5	400	178	222	79.2
44.7	253	113	140	50.5	44.6	462	206	256	91.9	44.6	336	150	186	67.0	44.6	121	54	67	24.1	44.5	398	177	221	78.7
44.7	246	110	136	49.2	44.6	460	205	255	91.4	44.6	334	149	185	66.5	44.6	112	50	62	22.3	44.5	393	175	218	77.9
44.7	244	109	135	48.7	44.6	457	204	253	91.1	44.6	332	148	184	66.0	44.6	101	45	56	20.0	44.5	391	174	217	77.4
44.7	237	106	131	47.4	44.6	455	203	252	90.6	44.6	327	146	181	65.2	44.6	92	41	51	18.3	44.5	389	173	216	76.9
44.7	235	105	130	46.9	44.6	453	202	251	90.1	44.6	325	145	180	64.7	44.6	83	37	46	16.5	44.5	384	171	213	76.1
44.7	228	102	126	45.6	44.6	451	201	250	89.6	44.6	323	144	179	64.2	44.6	74	33	41	14.7	44.5	382	170	212	75.7

%E	M1	M2	DM	M*
44.5	380	169	211	75.2
44.5	375	167	208	74.4
44.5	373	166	207	73.9
44.5	371	165	206	73.4
44.5	366	163	203	72.6
44.5	364	162	202	72.1
44.5	362	161	201	71.6
44.5	357	159	198	70.8
44.5	355	158	197	70.3
44.5	353	157	196	69.8
44.5	348	155	193	69.0
44.5	346	154	192	68.5
44.5	344	153	191	68.0
44.5	339	151	188	67.3
44.5	337	150	187	66.8
44.5	335	149	186	66.3
44.5	330	147	183	65.5
44.5	328	146	182	65.0
44.5	326	145	181	64.5
44.5	321	143	178	63.7
44.5	319	142	177	63.2
44.5	317	141	176	62.7
44.5	310	138	172	61.4
44.5	308	137	171	60.9
44.5	301	134	167	59.7
44.5	299	133	166	59.2
44.5	292	130	162	57.9
44.5	290	129	161	57.4
44.5	283	126	157	56.1
44.5	281	125	156	55.6
44.5	274	122	152	54.3
44.5	272	121	151	53.8
44.5	265	118	147	52.5
44.5	263	117	146	52.0
44.5	256	114	142	50.8
44.5	254	113	141	50.3
44.5	247	110	137	49.0
44.5	245	109	136	48.5
44.5	238	106	132	47.2
44.5	236	105	131	46.7
44.5	229	102	127	45.4
44.5	227	101	126	44.9
44.5	220	98	122	43.7
44.5	218	97	121	43.2
44.5	209	93	116	41.4
44.5	200	89	111	39.6
44.5	191	85	106	37.8
44.5	182	81	101	36.0
44.5	173	77	96	34.3
44.5	164	73	91	32.5
44.5	155	69	86	30.7
44.5	146	65	81	28.9
44.5	137	61	76	27.2
44.5	128	57	71	25.4
44.5	119	53	66	23.6
44.5	110	49	61	21.8
44.4	500	222	278	98.6
44.4	498	221	277	98.1
44.4	496	220	276	97.6
44.4	495	220	275	97.8
44.4	493	219	274	97.3
44.4	491	218	273	96.8
44.4	489	217	272	96.3
44.4	487	216	271	95.8
44.4	486	216	270	96.0
44.4	484	215	269	95.5
44.4	482	214	268	95.0
44.4	480	213	267	94.5
44.4	478	212	266	94.0
44.4	477	212	265	94.2
44.4	475	211	264	93.7
44.4	473	210	263	93.2
44.4	471	209	262	92.7
44.4	469	208	261	92.2
44.4	468	208	260	92.4
44.4	466	207	259	92.0
44.4	464	206	258	91.5
44.4	462	205	257	91.0
44.4	459	204	255	90.7
44.4	457	203	254	90.2
44.4	455	202	253	89.7
44.4	453	201	252	89.2
44.4	450	200	250	88.9
44.4	448	199	249	88.4
44.4	446	198	248	87.9
44.4	444	197	247	87.4
44.4	441	196	245	87.1
44.4	439	195	244	86.6
44.4	437	194	243	86.1
44.4	435	193	242	85.6
44.4	432	192	240	85.3
44.4	430	191	239	84.8
44.4	428	190	238	84.3
44.4	426	189	237	83.9
44.4	423	188	235	83.6
44.4	421	187	234	83.1
44.4	419	186	233	82.6
44.4	417	185	232	82.1
44.4	414	184	230	81.8
44.4	412	183	229	81.3
44.4	410	182	228	80.8
44.4	408	181	227	80.3
44.4	405	180	225	80.0
44.4	403	179	224	79.5
44.4	401	178	223	79.0
44.4	399	177	222	78.5
44.4	396	176	220	78.2
44.4	394	175	219	77.7
44.4	392	174	218	77.2
44.4	390	173	217	76.7
44.4	387	172	215	76.4
44.4	385	171	214	76.0
44.4	383	170	213	75.5
44.4	381	169	212	75.0
44.4	378	168	210	74.7
44.4	376	167	209	74.2
44.4	374	166	208	73.7
44.4	372	165	207	73.2
44.4	369	164	205	72.9
44.4	367	163	204	72.4
44.4	365	162	203	71.9
44.4	363	161	202	71.4
44.4	360	160	200	71.1
44.4	358	159	199	70.6
44.4	356	158	198	70.1
44.4	354	157	197	69.6
44.4	351	156	195	69.3
44.4	349	155	194	68.8
44.4	347	154	193	68.3
44.4	342	152	190	67.6
44.4	340	151	189	67.1
44.4	338	150	188	66.6
44.4	333	148	185	65.8
44.4	331	147	184	65.3
44.4	329	146	183	64.8
44.4	324	144	180	64.0
44.4	322	143	179	63.5
44.4	320	142	178	63.0
44.4	315	140	175	62.2
44.4	313	139	174	61.7
44.4	311	138	173	61.2
44.4	306	136	170	60.4
44.4	304	135	169	60.0
44.4	302	134	168	59.5
44.4	297	132	165	58.7
44.4	295	131	164	58.2
44.4	293	130	163	57.7
44.4	288	128	160	56.9
44.4	286	127	159	56.4
44.4	284	126	158	55.9
44.4	279	124	155	55.1
44.4	277	123	154	54.6
44.4	275	122	153	54.1
44.4	270	120	150	53.3
44.4	268	119	149	52.8
44.4	266	118	148	52.3
44.4	261	116	145	51.6
44.4	259	115	144	51.1
44.4	257	114	143	50.6
44.4	252	112	140	49.8
44.4	250	111	139	49.3
44.4	248	110	138	48.8
44.4	243	108	135	48.0
44.4	241	107	134	47.5
44.4	239	106	133	47.0
44.4	234	104	130	46.2
44.4	232	103	129	45.7
44.4	225	100	125	44.4
44.4	223	99	124	44.0
44.4	216	96	120	42.7
44.4	214	95	119	42.2
44.4	207	92	115	40.9
44.4	205	91	114	40.4
44.4	198	88	110	39.1
44.4	196	87	109	38.6
44.4	189	84	105	37.3
44.4	187	83	104	36.8
44.4	180	80	100	35.6
44.4	178	79	99	35.1
44.4	171	76	95	33.8
44.4	169	75	94	33.3
44.4	162	72	90	32.0
44.4	160	71	89	31.5
44.4	153	68	85	30.2
44.4	151	67	84	29.7
44.4	144	64	80	28.4
44.4	142	63	79	28.0
44.4	135	60	75	26.7
44.4	133	59	74	26.2
44.4	126	56	70	24.9
44.4	124	55	69	24.4
44.4	117	52	65	23.1
44.4	108	48	60	21.3
44.4	99	44	55	19.6
44.4	90	40	50	17.8
44.4	81	36	45	16.0
44.4	72	32	40	14.2
44.4	63	28	35	12.4
44.4	54	24	30	10.7
44.4	45	20	25	8.9
44.4	36	16	20	7.1
44.4	27	12	15	5.3
44.4	18	8	10	3.6
44.4	9	4	5	1.8
44.3	499	221	278	97.9
44.3	497	220	277	97.4
44.3	494	219	275	97.1
44.3	492	218	274	96.6
44.3	490	217	273	96.1
44.3	488	216	272	95.6
44.3	485	215	270	95.3
44.3	483	214	269	94.8
44.3	481	213	268	94.3
44.3	479	212	267	93.8
44.3	476	211	265	93.5
44.3	474	210	264	93.0
44.3	472	209	263	92.5
44.3	470	208	262	92.1
44.3	467	207	260	91.8
44.3	465	206	259	91.3
44.3	463	205	258	90.8
44.3	461	204	257	90.3
44.3	460	204	256	90.5
44.3	458	203	255	90.0
44.3	456	202	254	89.5
44.3	454	201	253	89.0
44.3	451	200	251	88.7
44.3	449	199	250	88.2
44.3	447	198	249	87.7
44.3	445	197	248	87.2
44.3	442	196	246	86.9
44.3	440	195	245	86.4
44.3	438	194	244	85.9
44.3	436	193	243	85.4
44.3	433	192	241	85.1
44.3	431	191	240	84.6
44.3	429	190	239	84.1
44.3	427	189	238	83.7
44.3	424	188	236	83.4
44.3	422	187	235	82.9
44.3	420	186	234	82.4
44.3	418	185	233	81.9
44.3	415	184	231	81.6
44.3	413	183	230	81.1
44.3	411	182	229	80.6
44.3	409	181	228	80.1
44.3	406	180	226	79.8
44.3	404	179	225	79.3
44.3	402	178	224	78.8
44.3	400	177	223	78.3

%E	M1	M2	DM	M*
44.3	397	176	221	78.0
44.3	395	175	220	77.5
44.3	393	174	219	77.0
44.3	388	172	216	76.2
44.3	386	171	215	75.8
44.3	384	170	214	75.3
44.3	379	168	211	74.5
44.3	377	167	210	74.0
44.3	375	166	209	73.5
44.3	370	164	206	72.7
44.3	368	163	205	72.2
44.3	366	162	204	71.7
44.3	361	160	201	70.9
44.3	359	159	200	70.4
44.3	357	158	199	69.9
44.3	352	156	196	69.1
44.3	350	155	195	68.6
44.3	348	154	194	68.1
44.3	345	153	192	67.9
44.3	343	152	191	67.4
44.3	341	151	190	66.9
44.3	336	149	187	66.1
44.3	334	148	186	65.6
44.3	332	147	185	65.1
44.3	327	145	182	64.3
44.3	325	144	181	63.8
44.3	323	143	180	63.3
44.3	318	141	177	62.5
44.3	316	140	176	62.0
44.3	314	139	175	61.5
44.3	309	137	172	60.7
44.3	307	136	171	60.2
44.3	305	135	170	59.8
44.3	300	133	167	59.0
44.3	298	132	166	58.5
44.3	296	131	165	58.0
44.3	291	129	162	57.2
44.3	289	128	161	56.7
44.3	287	127	160	56.2
44.3	282	125	157	55.4
44.3	280	124	156	54.9
44.3	273	121	152	53.6
44.3	271	120	151	53.1
44.3	264	117	147	51.9
44.3	262	116	146	51.4
44.3	255	113	142	50.1
44.3	253	112	141	49.6
44.3	246	109	137	48.3
44.3	244	108	136	47.8
44.3	237	105	132	46.5

%E	M1	M2	DM	M*
44.3	235	104	131	46.0
44.3	230	102	128	45.2
44.3	228	101	127	44.7
44.3	221	98	123	43.5
44.3	219	97	122	43.0
44.3	212	94	118	41.7
44.3	210	93	117	41.2
44.3	203	90	113	39.9
44.3	201	89	112	39.4
44.3	194	86	108	38.1
44.3	192	85	107	37.6
44.3	185	82	103	36.3
44.3	183	81	102	35.9
44.3	176	78	98	34.6
44.3	174	77	97	34.1
44.3	167	74	93	32.8
44.3	158	70	88	31.0
44.3	149	66	83	29.2
44.3	140	62	78	27.5
44.3	131	58	73	25.7
44.3	122	54	68	23.9
44.3	115	51	64	22.6
44.3	106	47	59	20.8
44.3	97	43	54	19.1
44.3	88	39	49	17.3
44.3	79	35	44	15.5
44.3	70	31	39	13.7
44.3	61	27	34	12.0
44.2	500	221	279	97.7
44.2	498	220	278	97.2
44.2	496	219	277	96.7
44.2	495	219	276	96.9
44.2	493	218	275	96.4
44.2	491	217	274	95.9
44.2	489	216	273	95.4
44.2	486	215	271	95.1
44.2	484	214	270	94.6
44.2	482	213	269	94.1
44.2	480	212	268	93.6
44.2	477	211	266	93.3
44.2	475	210	265	92.8
44.2	473	209	264	92.3
44.2	471	208	263	91.9
44.2	468	207	261	91.6
44.2	466	206	260	91.1
44.2	464	205	259	90.6
44.2	462	204	258	90.1
44.2	459	203	256	89.8
44.2	457	202	255	89.3
44.2	455	201	254	88.8

%E	M1	M2	DM	M*
44.2	453	200	253	88.3
44.2	452	200	252	88.5
44.2	450	199	251	88.0
44.2	448	198	250	87.5
44.2	446	197	249	87.0
44.2	443	196	247	86.7
44.2	441	195	246	86.2
44.2	439	194	245	85.7
44.2	437	193	244	85.2
44.2	434	192	242	84.9
44.2	432	191	241	84.4
44.2	430	190	240	84.0
44.2	428	189	239	83.5
44.2	425	188	237	83.2
44.2	423	187	236	82.7
44.2	421	186	235	82.2
44.2	419	185	234	81.7
44.2	416	184	232	81.4
44.2	414	183	231	80.9
44.2	412	182	230	80.4
44.2	407	180	227	79.6
44.2	405	179	226	79.1
44.2	403	178	225	78.6
44.2	398	176	222	77.8
44.2	396	175	221	77.3
44.2	394	174	220	76.8
44.2	391	173	218	76.5
44.2	389	172	217	76.1
44.2	387	171	216	75.6
44.2	385	170	215	75.1
44.2	382	169	213	74.8
44.2	380	168	212	74.3
44.2	378	167	211	73.8
44.2	373	165	208	73.0
44.2	371	164	207	72.5
44.2	369	163	206	72.0
44.2	364	161	203	71.2
44.2	362	160	202	70.7
44.2	360	159	201	70.2
44.2	355	157	198	69.4
44.2	353	156	197	68.9
44.2	351	155	196	68.4
44.2	346	153	193	67.7
44.2	344	152	192	67.2
44.2	342	151	191	66.7
44.2	339	150	189	66.4
44.2	337	149	188	65.9
44.2	335	148	187	65.4
44.2	330	146	184	64.6
44.2	328	145	183	64.1

%E	M1	M2	DM	M*
44.2	326	144	182	63.6
44.2	321	142	179	62.8
44.2	319	141	178	62.3
44.2	317	140	177	61.8
44.2	312	138	174	61.0
44.2	310	137	173	60.5
44.2	308	136	172	60.1
44.2	303	134	169	59.3
44.2	301	133	168	58.8
44.2	294	130	164	57.5
44.2	292	129	163	57.0
44.2	285	126	159	55.7
44.2	283	125	158	55.2
44.2	278	123	155	54.4
44.2	276	122	154	53.9
44.2	274	121	153	53.4
44.2	269	119	150	52.6
44.2	267	118	149	52.1
44.2	265	117	148	51.7
44.2	260	115	145	50.9
44.2	258	114	144	50.4
44.2	251	111	140	49.1
44.2	249	110	139	48.6
44.2	242	107	135	47.3
44.2	240	106	134	46.8
44.2	233	103	130	45.5
44.2	231	102	129	45.0
44.2	226	100	126	44.2
44.2	224	99	125	43.8
44.2	217	96	121	42.5
44.2	215	95	120	42.0
44.2	208	92	116	40.7
44.2	206	91	115	40.2
44.2	199	88	111	38.9
44.2	197	87	110	38.4
44.2	190	84	106	37.1
44.2	181	80	101	35.4
44.2	172	76	96	33.6
44.2	165	73	92	32.3
44.2	163	72	91	31.8
44.2	156	69	87	30.5
44.2	154	68	86	30.0
44.2	147	65	82	28.7
44.2	138	61	77	27.0
44.2	129	57	72	25.2
44.2	120	53	67	23.4
44.2	113	50	63	22.1
44.2	104	46	58	20.3
44.2	95	42	53	18.6
44.2	86	38	48	16.8

%E	M1	M2	DM	M*
44.2	77	34	43	15.0
44.2	52	23	29	10.2
44.2	43	19	24	8.4
44.1	499	220	279	97.0
44.1	497	219	278	96.5
44.1	494	218	276	96.2
44.1	492	217	275	95.7
44.1	490	216	274	95.2
44.1	488	215	273	94.7
44.1	487	215	272	94.9
44.1	485	214	271	94.4
44.1	483	213	270	93.9
44.1	481	212	269	93.4
44.1	479	211	268	92.9
44.1	478	211	267	93.1
44.1	476	210	266	92.6
44.1	474	209	265	92.2
44.1	472	208	264	91.7
44.1	469	207	262	91.4
44.1	467	206	261	90.9
44.1	465	205	260	90.4
44.1	463	204	259	89.9
44.1	460	203	257	89.6
44.1	458	202	256	89.1
44.1	456	201	255	88.6
44.1	454	200	254	88.1
44.1	451	199	252	87.8
44.1	449	198	251	87.3
44.1	447	197	250	86.8
44.1	444	196	248	86.5
44.1	442	195	247	86.0
44.1	440	194	246	85.5
44.1	438	193	245	85.0
44.1	435	192	243	84.7
44.1	433	191	242	84.3
44.1	431	190	241	83.8
44.1	429	189	240	83.3
44.1	426	188	238	83.0
44.1	424	187	237	82.5
44.1	422	186	236	82.0
44.1	417	184	233	81.2
44.1	415	183	232	80.7
44.1	413	182	231	80.2
44.1	410	181	229	79.9
44.1	408	180	228	79.4
44.1	406	179	227	78.9
44.1	404	178	226	78.4
44.1	401	177	224	78.1
44.1	399	176	223	77.6
44.1	397	175	222	77.1

%E	M1	M2	DM	M*
44.1	395	174	221	76.6
44.1	392	173	219	76.3
44.1	390	172	218	75.9
44.1	388	171	217	75.4
44.1	383	169	214	74.6
44.1	381	168	213	74.1
44.1	379	167	212	73.6
44.1	376	166	210	73.3
44.1	374	165	209	72.8
44.1	372	164	208	72.3
44.1	370	163	207	71.8
44.1	367	162	205	71.5
44.1	365	161	204	71.0
44.1	363	160	203	70.5
44.1	358	158	200	69.7
44.1	356	157	199	69.2
44.1	354	156	198	68.7
44.1	349	154	195	68.0
44.1	347	153	194	67.5
44.1	345	152	193	67.0
44.1	340	150	190	66.2
44.1	338	149	189	65.7
44.1	333	147	186	64.9
44.1	331	146	185	64.4
44.1	329	145	184	63.9
44.1	324	143	181	63.1
44.1	322	142	180	62.6
44.1	320	141	179	62.1
44.1	315	139	176	61.3
44.1	313	138	175	60.8
44.1	311	137	174	60.4
44.1	306	135	171	59.6
44.1	304	134	170	59.1
44.1	299	132	167	58.3
44.1	297	131	166	57.8
44.1	295	130	165	57.3
44.1	290	128	162	56.5
44.1	288	127	161	56.0
44.1	286	126	160	55.5
44.1	281	124	157	54.7
44.1	279	123	156	54.2
44.1	272	120	152	52.9
44.1	270	119	151	52.4
44.1	263	116	147	51.2
44.1	261	115	146	50.7
44.1	256	113	143	49.9
44.1	254	112	142	49.4
44.1	247	109	138	48.1
44.1	245	108	137	47.6
44.1	238	105	133	46.3

%E	M1	M2	DM	M*
44.1	236	104	132	45.8
44.1	229	101	128	44.5
44.1	227	100	127	44.1
44.1	222	98	124	43.3
44.1	220	97	123	42.8
44.1	213	94	119	41.5
44.1	211	93	118	41.0
44.1	204	90	114	39.7
44.1	202	89	113	39.2
44.1	195	86	109	37.9
44.1	188	83	105	36.6
44.1	186	82	104	36.2
44.1	179	79	100	34.9
44.1	177	78	99	34.4
44.1	170	75	95	33.1
44.1	161	71	90	31.3
44.1	152	67	85	29.5
44.1	145	64	81	28.2
44.1	143	63	80	27.8
44.1	136	60	76	26.5
44.1	127	56	71	24.7
44.1	118	52	66	22.9
44.1	111	49	62	21.6
44.1	102	45	57	19.9
44.1	93	41	52	18.1
44.1	68	30	38	13.2
44.1	59	26	33	11.5
44.1	34	15	19	6.6
44.0	500	220	280	96.8
44.0	498	219	279	96.3
44.0	496	218	278	95.8
44.0	495	218	277	96.0
44.0	493	217	276	95.5
44.0	491	216	275	95.0
44.0	489	215	274	94.5
44.0	486	214	272	94.2
44.0	484	213	271	93.7
44.0	482	212	270	93.2
44.0	480	211	269	92.8
44.0	477	210	267	92.5
44.0	475	209	266	92.0
44.0	473	208	265	91.5
44.0	470	207	263	91.2
44.0	468	206	262	90.7
44.0	466	205	261	90.2
44.0	464	204	260	89.7
44.0	461	203	258	89.4
44.0	459	202	257	88.9
44.0	457	201	256	88.4
44.0	455	200	255	87.9

%E	M1	M2	DM	M*
44.0	452	199	253	87.6
44.0	450	198	252	87.1
44.0	448	197	251	86.6
44.0	445	196	249	86.3
44.0	443	195	248	85.8
44.0	441	194	247	85.3
44.0	439	193	246	84.8
44.0	436	192	244	84.6
44.0	434	191	243	84.1
44.0	432	190	242	83.6
44.0	430	189	241	83.1
44.0	427	188	239	82.8
44.0	425	187	238	82.3
44.0	423	186	237	81.8
44.0	420	185	235	81.5
44.0	418	184	234	81.0
44.0	416	183	233	80.5
44.0	414	182	232	80.0
44.0	411	181	230	79.7
44.0	409	180	229	79.2
44.0	407	179	228	78.7
44.0	405	178	227	78.2
44.0	402	177	225	77.9
44.0	400	176	224	77.4
44.0	398	175	223	76.9
44.0	393	173	220	76.2
44.0	391	172	219	75.7
44.0	389	171	218	75.2
44.0	386	170	216	74.9
44.0	384	169	215	74.4
44.0	382	168	214	73.9
44.0	377	166	211	73.1
44.0	375	165	210	72.6
44.0	373	164	209	72.1
44.0	368	162	206	71.3
44.0	366	161	205	70.8
44.0	364	160	204	70.3
44.0	361	159	202	70.0
44.0	359	158	201	69.5
44.0	357	157	200	69.0
44.0	352	155	197	68.3
44.0	350	154	196	67.8
44.0	348	153	195	67.3
44.0	343	151	192	66.5
44.0	341	150	191	66.0
44.0	339	149	190	65.5
44.0	336	148	188	65.2
44.0	334	147	187	64.7
44.0	332	146	186	64.2
44.0	327	144	183	63.4

%E	M1	M2	DM	M*
44.0	325	143	182	62.9
44.0	323	142	181	62.4
44.0	318	140	178	61.6
44.0	316	139	177	61.1
44.0	309	136	173	59.9
44.0	307	135	172	59.4
44.0	302	133	169	58.6
44.0	300	132	168	58.1
44.0	298	131	167	57.6
44.0	293	129	164	56.8
44.0	291	128	163	56.3
44.0	284	125	159	55.0
44.0	282	124	158	54.5
44.0	277	122	155	53.7
44.0	275	121	154	53.2
44.0	273	120	153	52.7
44.0	268	118	150	52.0
44.0	266	117	149	51.5
44.0	259	114	145	50.2
44.0	257	113	144	49.7
44.0	252	111	141	48.9
44.0	250	110	140	48.4
44.0	248	109	139	47.9
44.0	243	107	136	47.1
44.0	241	106	135	46.6
44.0	234	103	131	45.3
44.0	232	102	130	44.8
44.0	225	99	126	43.6
44.0	218	96	122	42.3
44.0	216	95	121	41.8
44.0	209	92	117	40.5
44.0	207	91	116	40.0
44.0	200	88	112	38.7
44.0	193	85	108	37.4
44.0	191	84	107	36.9
44.0	184	81	103	35.7
44.0	182	80	102	35.2
44.0	175	77	98	33.9
44.0	168	74	94	32.6
44.0	166	73	93	32.1
44.0	159	70	89	30.8
44.0	150	66	84	29.0
44.0	141	62	79	27.3
44.0	134	59	75	26.0
44.0	125	55	70	24.2
44.0	116	51	65	22.4
44.0	109	48	61	21.1
44.0	100	44	56	19.4
44.0	91	40	51	17.6
44.0	84	37	47	16.3

%E	M1	M2	DM	M*
44.0	75	33	42	14.5
44.0	50	22	28	9.7
44.0	25	11	14	4.8
43.9	499	219	280	96.1
43.9	497	218	279	95.6
43.9	494	217	277	95.3
43.9	492	216	276	94.8
43.9	490	215	275	94.3
43.9	488	214	274	93.8
43.9	487	214	273	94.0
43.9	485	213	272	93.5
43.9	483	212	271	93.1
43.9	481	211	270	92.6
43.9	478	210	268	92.3
43.9	476	209	267	91.8
43.9	474	208	266	91.3
43.9	472	207	265	90.8
43.9	471	207	264	91.0
43.9	469	206	263	90.5
43.9	467	205	262	90.0
43.9	465	204	261	89.5
43.9	462	203	259	89.2
43.9	460	202	258	88.7
43.9	458	201	257	88.2
43.9	456	200	256	87.7
43.9	453	199	254	87.4
43.9	451	198	253	86.9
43.9	449	197	252	86.4
43.9	446	196	250	86.1
43.9	444	195	249	85.6
43.9	442	194	248	85.1
43.9	440	193	247	84.7
43.9	437	192	245	84.4
43.9	435	191	244	83.9
43.9	433	190	243	83.4
43.9	431	189	242	82.9
43.9	428	188	240	82.6
43.9	426	187	239	82.1
43.9	424	186	238	81.6
43.9	421	185	236	81.3
43.9	419	184	235	80.8
43.9	417	183	234	80.3
43.9	415	182	233	79.8
43.9	412	181	231	79.5
43.9	410	180	230	79.0
43.9	408	179	229	78.5
43.9	403	177	226	77.7
43.9	401	176	225	77.2
43.9	399	175	224	76.8
43.9	396	174	222	76.5

%E	M1	M2	DM	M*
43.9	394	173	221	76.0
43.9	392	172	220	75.5
43.9	387	170	217	74.7
43.9	385	169	216	74.2
43.9	383	168	215	73.7
43.9	380	167	213	73.4
43.9	378	166	212	72.9
43.9	376	165	211	72.4
43.9	374	164	210	71.9
43.9	371	163	208	71.6
43.9	369	162	207	71.1
43.9	367	161	206	70.6
43.9	362	159	203	69.8
43.9	360	158	202	69.3
43.9	358	157	201	68.9
43.9	355	156	199	68.6
43.9	353	155	198	68.1
43.9	351	154	197	67.6
43.9	346	152	194	66.8
43.9	344	151	193	66.3
43.9	342	150	192	65.8
43.9	337	148	189	65.0
43.9	335	147	188	64.5
43.9	330	145	185	63.7
43.9	328	144	184	63.2
43.9	326	143	183	62.7
43.9	321	141	180	61.9
43.9	319	140	179	61.4
43.9	314	138	176	60.6
43.9	312	137	175	60.2
43.9	310	136	174	59.7
43.9	305	134	171	58.9
43.9	303	133	170	58.4
43.9	301	132	169	57.9
43.9	296	130	166	57.1
43.9	294	129	165	56.6
43.9	289	127	162	55.8
43.9	287	126	161	55.3
43.9	285	125	160	54.8
43.9	280	123	157	54.0
43.9	278	122	156	53.5
43.9	271	119	152	52.3
43.9	269	118	151	51.8
43.9	264	116	148	51.0
43.9	262	115	147	50.5
43.9	255	112	143	49.2
43.9	253	111	142	48.7
43.9	246	108	138	47.4
43.9	244	107	137	46.9
43.9	239	105	134	46.1

%E	M1	M2	DM	M*
43.9	237	104	133	45.6
43.9	230	101	129	44.4
43.9	228	100	128	43.9
43.9	223	98	125	43.1
43.9	221	97	124	42.6
43.9	214	94	120	41.3
43.9	212	93	119	40.8
43.9	205	90	115	39.5
43.9	198	87	111	38.2
43.9	196	86	110	37.7
43.9	189	83	106	36.4
43.9	187	82	105	36.0
43.9	180	79	101	34.7
43.9	173	76	97	33.4
43.9	171	75	96	32.9
43.9	164	72	92	31.6
43.9	157	69	88	30.3
43.9	155	68	87	29.8
43.9	148	65	83	28.5
43.9	139	61	78	26.8
43.9	132	58	74	25.5
43.9	123	54	69	23.7
43.9	114	50	64	21.9
43.9	107	47	60	20.6
43.9	98	43	55	18.9
43.9	82	36	46	15.8
43.9	66	29	37	12.7
43.9	57	25	32	11.0
43.9	41	18	23	7.9
43.8	500	219	281	95.9
43.8	498	218	280	95.4
43.8	496	217	279	94.9
43.8	495	217	278	95.1
43.8	493	216	277	94.6
43.8	491	215	276	94.1
43.8	489	214	275	93.7
43.8	486	213	273	93.4
43.8	484	212	272	92.9
43.8	482	211	271	92.4
43.8	480	210	270	91.9
43.8	479	210	269	92.1
43.8	477	209	268	91.6
43.8	475	208	267	91.1
43.8	473	207	266	90.6
43.8	470	206	264	90.3
43.8	468	205	263	89.8
43.8	466	204	262	89.3
43.8	464	203	261	88.8
43.8	463	203	260	89.0
43.8	461	202	259	88.5

%E	M1	M2	DM	M*
43.8	459	201	258	88.0
43.8	457	200	257	87.5
43.8	454	199	255	87.2
43.8	452	198	254	86.7
43.8	450	197	253	86.2
43.8	448	196	252	85.8
43.8	447	196	251	85.9
43.8	445	195	250	85.4
43.8	443	194	249	85.0
43.8	441	193	248	84.5
43.8	438	192	246	84.2
43.8	436	191	245	83.7
43.8	434	190	244	83.2
43.8	432	189	243	82.7
43.8	429	188	241	82.4
43.8	427	187	240	81.9
43.8	425	186	239	81.4
43.8	422	185	237	81.1
43.8	420	184	236	80.6
43.8	418	183	235	80.1
43.8	416	182	234	79.6
43.8	413	181	232	79.3
43.8	411	180	231	78.8
43.8	409	179	230	78.3
43.8	406	178	228	78.0
43.8	404	177	227	77.5
43.8	402	176	226	77.1
43.8	400	175	225	76.6
43.8	397	174	223	76.3
43.8	395	173	222	75.8
43.8	393	172	221	75.3
43.8	390	171	219	75.0
43.8	388	170	218	74.5
43.8	386	169	217	74.0
43.8	384	168	216	73.5
43.8	381	167	214	73.2
43.8	379	166	213	72.7
43.8	377	165	212	72.2
43.8	372	163	209	71.4
43.8	370	162	208	70.9
43.8	368	161	207	70.4
43.8	365	160	205	70.1
43.8	363	159	204	69.6
43.8	361	158	203	69.2
43.8	356	156	200	68.4
43.8	354	155	199	67.9
43.8	352	154	198	67.4
43.8	349	153	196	67.1
43.8	347	152	195	66.6
43.8	345	151	194	66.1

%E	M1	M2	DM	M*
43.8	340	149	191	65.3
43.8	338	148	190	64.8
43.8	336	147	189	64.3
43.8	333	146	187	64.0
43.8	331	145	186	63.5
43.8	329	144	185	63.0
43.8	324	142	182	62.2
43.8	322	141	181	61.7
43.8	320	140	180	61.3
43.8	317	139	178	60.9
43.8	315	138	177	60.5
43.8	313	137	176	60.0
43.8	308	135	173	59.2
43.8	306	134	172	58.7
43.8	304	133	171	58.2
43.8	299	131	168	57.4
43.8	297	130	167	56.9
43.8	292	128	164	56.1
43.8	290	127	163	55.6
43.8	288	126	162	55.1
43.8	283	124	159	54.3
43.8	281	123	158	53.8
43.8	276	121	155	53.0
43.8	274	120	154	52.6
43.8	272	119	153	52.1
43.8	267	117	150	51.3
43.8	265	116	149	50.8
43.8	260	114	146	50.0
43.8	258	113	145	49.5
43.8	256	112	144	49.0
43.8	251	110	141	48.2
43.8	249	109	140	47.7
43.8	242	106	136	46.4
43.8	240	105	135	45.9
43.8	235	103	132	45.1
43.8	233	102	131	44.7
43.8	226	99	127	43.4
43.8	224	98	126	42.9
43.8	219	96	123	42.1
43.8	217	95	122	41.6
43.8	210	92	118	40.3
43.8	208	91	117	39.8
43.8	203	89	114	39.0
43.8	201	88	113	38.5
43.8	194	85	109	37.2
43.8	192	84	108	36.8
43.8	185	81	104	35.5
43.8	178	78	100	34.2
43.8	176	77	99	33.7
43.8	169	74	95	32.4

%E	M1	M2	DM	M*
43.8	162	71	91	31.1
43.8	160	70	90	30.6
43.8	153	67	86	29.3
43.8	146	64	82	28.1
43.8	144	63	81	27.6
43.8	137	60	77	26.3
43.8	130	57	73	25.0
43.8	128	56	72	24.5
43.8	121	53	68	23.2
43.8	112	49	63	21.4
43.8	105	46	59	20.2
43.8	96	42	54	18.4
43.8	89	39	50	17.1
43.8	80	35	45	15.3
43.8	73	32	41	14.0
43.8	64	28	36	12.3
43.8	48	21	27	9.2
43.8	32	14	18	6.1
43.8	16	7	9	3.1
43.7	499	218	281	95.2
43.7	497	217	280	94.7
43.7	494	216	278	94.4
43.7	492	215	277	94.0
43.7	490	214	276	93.5
43.7	487	213	274	93.2
43.7	485	212	273	92.7
43.7	483	211	272	92.2
43.7	481	210	271	91.7
43.7	478	209	269	91.4
43.7	476	208	268	90.9
43.7	474	207	267	90.4
43.7	471	206	265	90.1
43.7	469	205	264	89.6
43.7	467	204	263	89.1
43.7	465	203	262	88.6
43.7	462	202	260	88.3
43.7	460	201	259	87.8
43.7	458	200	258	87.3
43.7	455	199	256	87.0
43.7	453	198	255	86.5
43.7	451	197	254	86.1
43.7	449	196	253	85.6
43.7	446	195	251	85.3
43.7	444	194	250	84.8
43.7	442	193	249	84.3
43.7	439	192	247	84.0
43.7	437	191	246	83.5
43.7	435	190	245	83.0
43.7	430	188	242	82.2
43.7	428	187	241	81.7

%E	M1	M2	DM	M*
43.7	426	186	240	81.2
43.7	423	185	238	80.9
43.7	421	184	237	80.4
43.7	419	183	236	79.9
43.7	414	181	233	79.1
43.7	412	180	232	78.6
43.7	410	179	231	78.1
43.7	407	178	229	77.8
43.7	405	177	228	77.4
43.7	403	176	227	76.9
43.7	398	174	224	76.1
43.7	396	173	223	75.6
43.7	394	172	222	75.1
43.7	391	171	220	74.8
43.7	389	170	219	74.3
43.7	387	169	218	73.8
43.7	382	167	215	73.0
43.7	380	166	214	72.5
43.7	378	165	213	72.0
43.7	375	164	211	71.7
43.7	373	163	210	71.2
43.7	371	162	209	70.7
43.7	366	160	206	69.9
43.7	364	159	205	69.5
43.7	359	157	202	68.7
43.7	357	156	201	68.2
43.7	355	155	200	67.7
43.7	350	153	197	66.9
43.7	348	152	196	66.4
43.7	343	150	193	65.6
43.7	341	149	192	65.1
43.7	339	148	191	64.6
43.7	334	146	188	63.8
43.7	332	145	187	63.3
43.7	327	143	184	62.5
43.7	325	142	183	62.0
43.7	323	141	182	61.6
43.7	318	139	179	60.8
43.7	316	138	178	60.3
43.7	311	136	175	59.5
43.7	309	135	174	59.0
43.7	302	132	170	57.7
43.7	300	131	169	57.2
43.7	295	129	166	56.4
43.7	293	128	165	55.9
43.7	286	125	161	54.6
43.7	284	124	160	54.1
43.7	279	122	157	53.3
43.7	277	121	156	52.9
43.7	270	118	152	51.6

%E	M1	M2	DM	M*
43.7	268	117	151	51.1
43.7	263	115	148	50.3
43.7	261	114	147	49.8
43.7	254	111	143	48.5
43.7	252	110	142	48.0
43.7	247	108	139	47.2
43.7	245	107	138	46.7
43.7	238	104	134	45.4
43.7	231	101	130	44.2
43.7	229	100	129	43.7
43.7	222	97	125	42.4
43.7	215	94	121	41.1
43.7	213	93	120	40.6
43.7	206	90	116	39.3
43.7	199	87	112	38.0
43.7	197	86	111	37.5
43.7	190	83	107	36.3
43.7	183	80	103	35.0
43.7	174	76	98	33.2
43.7	167	73	94	31.9
43.7	158	69	89	30.1
43.7	151	66	85	28.8
43.7	142	62	80	27.1
43.7	135	59	76	25.8
43.7	126	55	71	24.0
43.7	119	52	67	22.7
43.7	103	45	58	19.7
43.7	87	38	49	16.6
43.7	71	31	40	13.5
43.6	500	218	282	95.0
43.6	498	217	281	94.6
43.6	495	216	279	94.3
43.6	493	215	278	93.8
43.6	491	214	277	93.3
43.6	489	213	276	92.8
43.6	488	213	275	93.0
43.6	486	212	274	92.5
43.6	484	211	273	92.0
43.6	482	210	272	91.5
43.6	479	209	270	91.2
43.6	477	208	269	90.7
43.6	475	207	268	90.2
43.6	473	206	267	89.7
43.6	472	206	266	89.9
43.6	470	205	265	89.4
43.6	468	204	264	88.9
43.6	466	203	263	88.4
43.6	463	202	261	88.1
43.6	461	201	260	87.6
43.6	459	200	259	87.1

%E	M1	M2	DM	M*
43.6	456	199	257	86.8
43.6	454	198	256	86.4
43.6	452	197	255	85.9
43.6	450	196	254	85.4
43.6	447	195	252	85.1
43.6	445	194	251	84.6
43.6	443	193	250	84.1
43.6	440	192	248	83.8
43.6	438	191	247	83.3
43.6	436	190	246	82.8
43.6	433	189	244	82.5
43.6	431	188	243	82.0
43.6	429	187	242	81.5
43.6	427	186	241	81.0
43.6	424	185	239	80.7
43.6	422	184	238	80.2
43.6	420	183	237	79.7
43.6	417	182	235	79.4
43.6	415	181	234	78.9
43.6	413	180	233	78.5
43.6	411	179	232	78.0
43.6	408	178	230	77.7
43.6	406	177	229	77.2
43.6	404	176	228	76.7
43.6	401	175	226	76.4
43.6	399	174	225	75.9
43.6	397	173	224	75.4
43.6	392	171	221	74.6
43.6	390	170	220	74.1
43.6	388	169	219	73.6
43.6	385	168	217	73.3
43.6	383	167	216	72.8
43.6	381	166	215	72.3
43.6	376	164	212	71.5
43.6	374	163	211	71.0
43.6	369	161	208	70.2
43.6	367	160	207	69.8
43.6	365	159	206	69.3
43.6	362	158	204	69.0
43.6	360	157	203	68.5
43.6	358	156	202	68.0
43.6	353	154	199	67.2
43.6	351	153	198	66.7
43.6	349	152	197	66.2
43.6	346	151	195	65.9
43.6	344	150	194	65.4
43.6	342	149	193	64.9
43.6	337	147	190	64.1
43.6	335	146	189	63.6
43.6	330	144	186	62.8

%E	M1	M2	DM	M*
43.6	328	143	185	62.3
43.6	326	142	184	61.9
43.6	321	140	181	61.1
43.6	319	139	180	60.6
43.6	314	137	177	59.8
43.6	312	136	176	59.3
43.6	307	134	173	58.5
43.6	305	133	172	58.0
43.6	303	132	171	57.5
43.6	298	130	168	56.7
43.6	296	129	167	56.2
43.6	291	127	164	55.4
43.6	289	126	163	54.9
43.6	287	125	162	54.4
43.6	282	123	159	53.6
43.6	280	122	158	53.2
43.6	275	120	155	52.4
43.6	273	119	154	51.9
43.6	266	116	150	50.6
43.6	264	115	149	50.1
43.6	259	113	146	49.3
43.6	257	112	145	48.8
43.6	250	109	141	47.5
43.6	243	106	137	46.2
43.6	241	105	136	45.7
43.6	236	103	133	45.0
43.6	234	102	132	44.5
43.6	227	99	128	43.2
43.6	225	98	127	42.7
43.6	220	96	124	41.9
43.6	218	95	123	41.4
43.6	211	92	119	40.1
43.6	204	89	115	38.8
43.6	202	88	114	38.3
43.6	195	85	110	37.1
43.6	188	82	106	35.8
43.6	181	79	102	34.5
43.6	179	78	101	34.0
43.6	172	75	97	32.7
43.6	165	72	93	31.4
43.6	163	71	92	30.9
43.6	156	68	88	29.6
43.6	149	65	84	28.4
43.6	140	61	79	26.6
43.6	133	58	75	25.3
43.6	117	51	66	22.2
43.6	110	48	62	20.9
43.6	101	44	57	19.2
43.6	94	41	53	17.9
43.6	78	34	44	14.8

%E	M1	M2	DM	M*
43.6	55	24	31	10.5
43.6	39	17	22	7.4
43.5	499	217	282	94.4
43.5	497	216	281	93.9
43.5	496	216	280	94.1
43.5	494	215	279	93.6
43.5	492	214	278	93.1
43.5	490	213	277	92.6
43.5	487	212	275	92.3
43.5	485	211	274	91.8
43.5	483	210	273	91.3
43.5	481	209	272	90.8
43.5	480	209	271	91.0
43.5	478	208	270	90.5
43.5	476	207	269	90.0
43.5	474	206	268	89.5
43.5	471	205	266	89.2
43.5	469	204	265	88.7
43.5	467	203	264	88.2
43.5	464	202	262	87.9
43.5	462	201	261	87.4
43.5	460	200	260	87.0
43.5	458	199	259	86.5
43.5	457	199	258	86.7
43.5	455	198	257	86.2
43.5	453	197	256	85.7
43.5	451	196	255	85.2
43.5	448	195	253	84.9
43.5	446	194	252	84.4
43.5	444	193	251	83.9
43.5	441	192	249	83.6
43.5	439	191	248	83.1
43.5	437	190	247	82.6
43.5	434	189	245	82.3
43.5	432	188	244	81.8
43.5	430	187	243	81.3
43.5	428	186	242	80.8
43.5	425	185	240	80.5
43.5	423	184	239	80.0
43.5	421	183	238	79.5
43.5	418	182	236	79.2
43.5	416	181	235	78.8
43.5	414	180	234	78.3
43.5	409	178	231	77.5
43.5	407	177	230	77.0
43.5	405	176	229	76.5
43.5	402	175	227	76.2
43.5	400	174	226	75.7
43.5	398	173	225	75.2
43.5	395	172	223	74.9

%E	M1	M2	DM	M*
43.5	393	171	222	74.4
43.5	391	170	221	73.9
43.5	386	168	218	73.1
43.5	384	167	217	72.6
43.5	382	166	216	72.1
43.5	379	165	214	71.8
43.5	377	164	213	71.3
43.5	375	163	212	70.9
43.5	372	162	210	70.5
43.5	370	161	209	70.1
43.5	368	160	208	69.6
43.5	363	158	205	68.8
43.5	361	157	204	68.3
43.5	359	156	203	67.8
43.5	356	155	201	67.5
43.5	354	154	200	67.0
43.5	352	153	199	66.5
43.5	347	151	196	65.7
43.5	345	150	195	65.2
43.5	340	148	192	64.4
43.5	338	147	191	63.9
43.5	336	146	190	63.4
43.5	333	145	188	63.1
43.5	331	144	187	62.6
43.5	329	143	186	62.2
43.5	324	141	183	61.4
43.5	322	140	182	60.9
43.5	317	138	179	60.1
43.5	315	137	178	59.6
43.5	313	136	177	59.1
43.5	310	135	175	58.8
43.5	308	134	174	58.3
43.5	306	133	173	57.8
43.5	301	131	170	57.0
43.5	299	130	169	56.5
43.5	294	128	166	55.7
43.5	292	127	165	55.2
43.5	285	124	161	54.0
43.5	283	123	160	53.5
43.5	278	121	157	52.7
43.5	276	120	156	52.2
43.5	271	118	153	51.4
43.5	269	117	152	50.9
43.5	262	114	148	49.6
43.5	260	113	147	49.1
43.5	255	111	144	48.3
43.5	253	110	143	47.8
43.5	248	108	140	47.0
43.5	246	107	139	46.5
43.5	239	104	135	45.3

%E	M1	M2	DM	M*
43.5	237	103	134	44.8
43.5	232	101	131	44.0
43.5	230	100	130	43.5
43.5	223	97	126	42.2
43.5	216	94	122	40.9
43.5	214	93	121	40.4
43.5	209	91	118	39.6
43.5	207	90	117	39.1
43.5	200	87	113	37.8
43.5	193	84	109	36.6
43.5	191	83	108	36.1
43.5	186	81	105	35.3
43.5	184	80	104	34.8
43.5	177	77	100	33.5
43.5	170	74	96	32.2
43.5	168	73	95	31.7
43.5	161	70	91	30.4
43.5	154	67	87	29.1
43.5	147	64	83	27.9
43.5	138	60	78	26.1
43.5	131	57	74	24.8
43.5	124	54	70	23.5
43.5	115	50	65	21.7
43.5	108	47	61	20.5
43.5	92	40	52	17.4
43.5	85	37	48	16.1
43.5	69	30	39	13.0
43.5	62	27	35	11.8
43.5	46	20	26	8.7
43.5	23	10	13	4.3
43.4	500	217	283	94.2
43.4	498	216	282	93.7
43.4	495	215	280	93.4
43.4	493	214	279	92.9
43.4	491	213	278	92.4
43.4	489	212	277	91.9
43.4	488	212	276	92.1
43.4	486	211	275	91.6
43.4	484	210	274	91.1
43.4	482	209	273	90.6
43.4	479	208	271	90.3
43.4	477	207	270	89.8
43.4	475	206	269	89.3
43.4	472	205	267	89.0
43.4	470	204	266	88.5
43.4	468	203	265	88.1
43.4	465	202	263	87.8
43.4	463	201	262	87.3
43.4	461	200	261	86.8
43.4	459	199	260	86.3

%E	M1	M2	DM	M*
43.4	456	198	258	86.0
43.4	454	197	257	85.5
43.4	452	196	256	85.0
43.4	449	195	254	84.7
43.4	447	194	253	84.2
43.4	445	193	252	83.7
43.4	442	192	250	83.4
43.4	440	191	249	82.9
43.4	438	190	248	82.4
43.4	435	189	246	82.1
43.4	433	188	245	81.6
43.4	431	187	244	81.1
43.4	429	186	243	80.6
43.4	426	185	241	80.3
43.4	424	184	240	79.8
43.4	422	183	239	79.4
43.4	419	182	237	79.1
43.4	417	181	236	78.6
43.4	415	180	235	78.1
43.4	412	179	233	77.8
43.4	410	178	232	77.3
43.4	408	177	231	76.8
43.4	406	176	230	76.3
43.4	403	175	228	76.0
43.4	401	174	227	75.5
43.4	399	173	226	75.0
43.4	396	172	224	74.7
43.4	394	171	223	74.2
43.4	392	170	222	73.7
43.4	389	169	220	73.4
43.4	387	168	219	72.9
43.4	385	167	218	72.4
43.4	380	165	215	71.6
43.4	378	164	214	71.2
43.4	376	163	213	70.7
43.4	373	162	211	70.4
43.4	371	161	210	69.9
43.4	369	160	209	69.4
43.4	366	159	207	69.1
43.4	364	158	206	68.6
43.4	362	157	205	68.1
43.4	357	155	202	67.3
43.4	355	154	201	66.8
43.4	350	152	198	66.0
43.4	348	151	197	65.5
43.4	346	150	196	65.0
43.4	343	149	194	64.7
43.4	341	148	193	64.2
43.4	339	147	192	63.7
43.4	334	145	189	62.9

%E	M1	M2	DM	M*
43.4	332	144	188	62.5
43.4	327	142	185	61.7
43.4	325	141	184	61.2
43.4	320	139	181	60.4
43.4	318	138	180	59.9
43.4	316	137	179	59.4
43.4	311	135	176	58.6
43.4	309	134	175	58.1
43.4	304	132	172	57.3
43.4	302	131	171	56.8
43.4	297	129	168	56.0
43.4	295	128	167	55.5
43.4	290	126	164	54.7
43.4	288	125	163	54.3
43.4	286	124	162	53.8
43.4	281	122	159	53.0
43.4	279	121	158	52.5
43.4	274	119	155	51.7
43.4	272	118	154	51.2
43.4	267	116	151	50.4
43.4	265	115	150	49.9
43.4	258	112	146	48.6
43.4	256	111	145	48.1
43.4	251	109	142	47.3
43.4	249	108	141	46.8
43.4	244	106	138	46.0
43.4	242	105	137	45.6
43.4	235	102	133	44.3
43.4	228	99	129	43.0
43.4	226	98	128	42.5
43.4	221	96	125	41.7
43.4	219	95	124	41.2
43.4	212	92	120	39.9
43.4	205	89	116	38.6
43.4	203	88	115	38.1
43.4	198	86	112	37.4
43.4	196	85	111	36.9
43.4	189	82	107	35.6
43.4	182	79	103	34.3
43.4	175	76	99	33.0
43.4	173	75	98	32.5
43.4	166	72	94	31.2
43.4	159	69	90	29.9
43.4	152	66	86	28.7
43.4	145	63	82	27.4
43.4	143	62	81	26.9
43.4	136	59	77	25.6
43.4	129	56	73	24.3
43.4	122	53	69	23.0
43.4	113	49	64	21.2

%E	M1	M2	DM	M*
43.4	106	46	60	20.0
43.4	99	43	56	18.7
43.4	83	36	47	15.6
43.4	76	33	43	14.3
43.4	53	23	30	10.0
43.3	499	216	283	93.5
43.3	497	215	282	93.0
43.3	496	215	281	93.2
43.3	494	214	280	92.7
43.3	492	213	279	92.2
43.3	490	212	278	91.7
43.3	487	211	276	91.4
43.3	485	210	275	90.9
43.3	483	209	274	90.4
43.3	480	208	272	90.1
43.3	478	207	271	89.6
43.3	476	206	270	89.2
43.3	473	205	268	88.8
43.3	471	204	267	88.4
43.3	469	203	266	87.9
43.3	467	202	265	87.4
43.3	466	202	264	87.6
43.3	464	201	263	87.1
43.3	462	200	262	86.6
43.3	460	199	261	86.1
43.3	457	198	259	85.8
43.3	455	197	258	85.3
43.3	453	196	257	84.8
43.3	450	195	255	84.5
43.3	448	194	254	84.0
43.3	446	193	253	83.5
43.3	443	192	251	83.2
43.3	441	191	250	82.7
43.3	439	190	249	82.2
43.3	436	189	247	81.9
43.3	434	188	246	81.4
43.3	432	187	245	80.9
43.3	430	186	244	80.5
43.3	427	185	242	80.2
43.3	425	184	241	79.7
43.3	423	183	240	79.2
43.3	420	182	238	78.9
43.3	418	181	237	78.4
43.3	416	180	236	77.9
43.3	413	179	234	77.6
43.3	411	178	233	77.1
43.3	409	177	232	76.6
43.3	404	175	229	75.8
43.3	402	174	228	75.3
43.3	400	173	227	74.8

%E	M1	M2	DM	M*
43.3	397	172	225	74.5
43.3	395	171	224	74.0
43.3	393	170	223	73.5
43.3	390	169	221	73.2
43.3	388	168	220	72.7
43.3	386	167	219	72.3
43.3	383	166	217	71.9
43.3	381	165	216	71.5
43.3	379	164	215	71.0
43.3	374	162	212	70.2
43.3	372	161	211	69.7
43.3	367	159	208	68.9
43.3	365	158	207	68.4
43.3	363	157	206	67.9
43.3	360	156	204	67.6
43.3	358	155	203	67.1
43.3	356	154	202	66.6
43.3	353	153	200	66.3
43.3	351	152	199	65.8
43.3	349	151	198	65.3
43.3	344	149	195	64.5
43.3	342	148	194	64.0
43.3	337	146	191	63.3
43.3	335	145	190	62.8
43.3	330	143	187	62.0
43.3	328	142	186	61.5
43.3	326	141	185	61.0
43.3	323	140	183	60.7
43.3	321	139	182	60.2
43.3	319	138	181	59.7
43.3	314	136	178	58.9
43.3	312	135	177	58.4
43.3	307	133	174	57.6
43.3	305	132	173	57.1
43.3	300	130	170	56.3
43.3	298	129	169	55.8
43.3	293	127	166	55.0
43.3	291	126	165	54.6
43.3	289	125	164	54.1
43.3	284	123	161	53.3
43.3	282	122	160	52.8
43.3	277	120	157	52.0
43.3	275	119	156	51.5
43.3	270	117	153	50.7
43.3	268	116	152	50.2
43.3	263	114	149	49.4
43.3	261	113	148	48.9
43.3	254	110	144	47.6
43.3	252	109	143	47.1
43.3	247	107	140	46.4
43.3	245	106	139	45.9
43.3	240	104	136	45.1
43.3	238	103	135	44.6
43.3	233	101	132	43.8
43.3	231	100	131	43.3
43.3	224	97	127	42.0
43.3	217	94	123	40.7
43.3	215	93	122	40.2
43.3	210	91	119	39.4
43.3	208	90	118	38.9
43.3	201	87	114	37.7
43.3	194	84	110	36.4
43.3	187	81	106	35.1
43.3	180	78	102	33.8
43.3	178	77	101	33.3
43.3	171	74	97	32.0
43.3	164	71	93	30.7
43.3	157	68	89	29.5
43.3	150	65	85	28.2
43.3	141	61	80	26.4
43.3	134	58	76	25.1
43.3	127	55	72	23.8
43.3	120	52	68	22.5
43.3	104	45	59	19.5
43.3	97	42	55	18.2
43.3	90	39	51	16.9
43.3	67	29	38	12.6
43.3	60	26	34	11.3
43.3	30	13	17	5.6
43.2	500	216	284	93.3
43.2	498	215	283	92.8
43.2	495	214	281	92.5
43.2	493	213	280	92.0
43.2	491	212	279	91.5
43.2	488	211	277	91.2
43.2	486	210	276	90.7
43.2	484	209	275	90.3
43.2	482	208	274	89.8
43.2	481	208	273	89.9
43.2	479	207	272	89.5
43.2	477	206	271	89.0
43.2	475	205	270	88.5
43.2	474	205	269	88.7
43.2	472	204	268	88.2
43.2	470	203	267	87.7
43.2	468	202	266	87.2
43.2	465	201	264	86.9
43.2	463	200	263	86.4
43.2	461	199	262	85.9
43.2	458	198	260	85.6
43.2	456	197	259	85.1
43.2	454	196	258	84.6
43.2	451	195	256	84.3
43.2	449	194	255	83.8
43.2	447	193	254	83.3
43.2	444	192	252	83.0
43.2	442	191	251	82.5
43.2	440	190	250	82.0
43.2	438	189	249	81.6
43.2	437	189	248	81.7
43.2	435	188	247	81.3
43.2	433	187	246	80.8
43.2	431	186	245	80.3
43.2	428	185	243	80.0
43.2	426	184	242	79.5
43.2	424	183	241	79.0
43.2	421	182	239	78.7
43.2	419	181	238	78.2
43.2	417	180	237	77.7
43.2	414	179	235	77.4
43.2	412	178	234	76.9
43.2	410	177	233	76.4
43.2	407	176	231	76.1
43.2	405	175	230	75.6
43.2	403	174	229	75.1
43.2	398	172	226	74.3
43.2	396	171	225	73.8
43.2	391	169	222	73.0
43.2	389	168	221	72.6
43.2	387	167	220	72.1
43.2	384	166	218	71.8
43.2	382	165	217	71.3
43.2	380	164	216	70.8
43.2	377	163	214	70.5
43.2	375	162	213	70.0
43.2	373	161	212	69.5
43.2	370	160	210	69.2
43.2	368	159	209	68.7
43.2	366	158	208	68.2
43.2	361	156	205	67.4
43.2	359	155	204	66.9
43.2	354	153	201	66.1
43.2	352	152	200	65.6
43.2	347	150	197	64.8
43.2	345	149	196	64.4
43.2	340	147	193	63.6
43.2	338	146	192	63.1
43.2	336	145	191	62.6
43.2	333	144	189	62.3
43.2	331	143	188	61.8
43.2	329	142	187	61.3
43.2	324	140	184	60.5
43.2	322	139	183	60.0
43.2	317	137	180	59.2
43.2	315	136	179	58.7
43.2	310	134	176	57.9
43.2	308	133	175	57.4
43.2	303	131	172	56.6
43.2	301	130	171	56.1
43.2	296	128	168	55.4
43.2	294	127	167	54.9
43.2	292	126	166	54.4
43.2	287	124	163	53.6
43.2	285	123	162	53.1
43.2	280	121	159	52.3
43.2	278	120	158	51.8
43.2	273	118	155	51.0
43.2	271	117	154	50.5
43.2	266	115	151	49.7
43.2	264	114	150	49.2
43.2	259	112	147	48.4
43.2	257	111	146	47.9
43.2	250	108	142	46.7
43.2	243	105	138	45.4
43.2	241	104	137	44.9
43.2	236	102	134	44.1
43.2	234	101	133	43.6
43.2	229	99	130	42.8
43.2	227	98	129	42.3
43.2	222	96	126	41.5
43.2	220	95	125	41.0
43.2	213	92	121	39.7
43.2	206	89	117	38.5
43.2	199	86	113	37.2
43.2	192	83	109	35.9
43.2	190	82	108	35.4
43.2	185	80	105	34.6
43.2	183	79	104	34.1
43.2	176	76	100	32.8
43.2	169	73	96	31.5
43.2	162	70	92	30.2
43.2	155	67	88	29.0
43.2	148	64	84	27.7
43.2	146	63	83	27.2
43.2	139	60	79	25.9
43.2	132	57	75	24.6
43.2	125	54	71	23.3
43.2	118	51	67	22.0
43.2	111	48	63	20.8
43.2	95	41	54	17.7
43.2	88	38	50	16.4
43.2	81	35	46	15.1
43.2	74	32	42	13.8
43.2	44	19	25	8.2
43.2	37	16	21	6.9
43.1	499	215	284	92.6
43.1	497	214	283	92.1
43.1	496	214	282	92.3
43.1	494	213	281	91.8
43.1	492	212	280	91.3
43.1	490	211	279	90.9
43.1	489	211	278	91.0
43.1	487	210	277	90.6
43.1	485	209	276	90.1
43.1	483	208	275	89.6
43.1	480	207	273	89.3
43.1	478	206	272	88.8
43.1	476	205	271	88.3
43.1	473	204	269	88.0
43.1	471	203	268	87.5
43.1	469	202	267	87.0
43.1	466	201	265	86.7
43.1	464	200	264	86.2
43.1	462	199	263	85.7
43.1	459	198	261	85.4
43.1	457	197	260	84.9
43.1	455	196	259	84.4
43.1	452	195	257	84.1
43.1	450	194	256	83.6
43.1	448	193	255	83.1
43.1	445	192	253	82.8
43.1	443	191	252	82.3
43.1	441	190	251	81.9
43.1	439	189	250	81.4
43.1	436	188	248	81.1
43.1	434	187	247	80.6
43.1	432	186	246	80.1
43.1	429	185	244	79.8
43.1	427	184	243	79.3
43.1	425	183	242	78.8
43.1	422	182	240	78.5
43.1	420	181	239	78.0
43.1	418	180	238	77.5
43.1	415	179	236	77.2
43.1	413	178	235	76.7
43.1	411	177	234	76.2
43.1	408	176	232	75.9
43.1	406	175	231	75.4
43.1	404	174	230	74.9
43.1	401	173	228	74.6

%E	M1	M2	DM	M*	%E	M1	M2	DM	M*	%E	M1	M2	DM	M*	%E	M1	M2	DM	M*	%E	M1	M2	DM	M*
43.1	399	172	227	74.1	43.1	246	106	140	45.7	43.0	454	195	259	83.8	43.0	328	141	187	60.6	43.0	100	43	57	18.5
43.1	397	171	226	73.7	43.1	239	103	136	44.4	43.0	453	195	258	83.9	43.0	323	139	184	59.8	43.0	93	40	53	17.2
43.1	394	170	224	73.4	43.1	232	100	132	43.1	43.0	451	194	257	83.5	43.0	321	138	183	59.3	43.0	86	37	49	15.9
43.1	392	169	223	72.9	43.1	225	97	128	41.8	43.0	449	193	256	83.0	43.0	316	136	180	58.5	43.0	79	34	45	14.6
43.1	390	168	222	72.4	43.1	218	94	124	40.5	43.0	447	192	255	82.5	43.0	314	135	179	58.0	42.9	499	214	285	91.8
43.1	385	166	219	71.6	43.1	216	93	123	40.0	43.0	446	192	254	82.7	43.0	309	133	176	57.2	42.9	497	213	284	91.3
43.1	383	165	218	71.1	43.1	211	91	120	39.2	43.0	444	191	253	82.2	43.0	307	132	175	56.8	42.9	496	213	283	91.5
43.1	378	163	215	70.3	43.1	209	90	119	38.8	43.0	442	190	252	81.7	43.0	305	131	174	56.3	42.9	494	212	282	91.0
43.1	376	162	214	69.8	43.1	204	88	116	38.0	43.0	440	189	251	81.2	43.0	302	130	172	56.0	42.9	492	211	281	90.5
43.1	371	160	211	69.0	43.1	202	87	115	37.5	43.0	437	188	249	80.9	43.0	300	129	171	55.5	42.9	490	210	280	90.0
43.1	369	159	210	68.5	43.1	197	85	112	36.7	43.0	435	187	248	80.4	43.0	298	128	170	55.0	42.9	489	210	279	90.2
43.1	367	158	209	68.0	43.1	195	84	111	36.2	43.0	433	186	247	79.9	43.0	293	126	167	54.2	42.9	487	209	278	89.7
43.1	364	157	207	67.7	43.1	188	81	107	34.9	43.0	430	185	245	79.6	43.0	291	125	166	53.7	42.9	485	208	277	89.2
43.1	362	156	206	67.2	43.1	181	78	103	33.6	43.0	428	184	244	79.1	43.0	286	123	163	52.9	42.9	483	207	276	88.7
43.1	360	155	205	66.7	43.1	174	75	99	32.3	43.0	426	183	243	78.6	43.0	284	122	162	52.4	42.9	482	207	275	88.9
43.1	357	154	203	66.4	43.1	167	72	95	31.0	43.0	423	182	241	78.3	43.0	279	120	159	51.6	42.9	480	206	274	88.4
43.1	355	153	202	65.9	43.1	160	69	91	29.8	43.0	421	181	240	77.8	43.0	277	119	158	51.1	42.9	478	205	273	87.9
43.1	353	152	201	65.5	43.1	153	66	87	28.5	43.0	419	180	239	77.3	43.0	272	117	155	50.3	42.9	476	204	272	87.4
43.1	350	151	199	65.1	43.1	144	62	82	26.7	43.0	416	179	237	77.0	43.0	270	116	154	49.8	42.9	475	204	271	87.6
43.1	348	150	198	64.7	43.1	137	59	78	25.4	43.0	414	178	236	76.5	43.0	265	114	151	49.0	42.9	473	203	270	87.1
43.1	346	149	197	64.2	43.1	130	56	74	24.1	43.0	412	177	235	76.0	43.0	263	113	150	48.6	42.9	471	202	269	86.6
43.1	343	148	195	63.9	43.1	123	53	70	22.8	43.0	409	176	233	75.7	43.0	258	111	147	47.8	42.9	469	201	268	86.1
43.1	341	147	194	63.4	43.1	116	50	66	21.6	43.0	407	175	232	75.2	43.0	256	110	146	47.3	42.9	468	201	267	86.3
43.1	339	146	193	62.9	43.1	109	47	62	20.3	43.0	405	174	231	74.8	43.0	251	108	143	46.5	42.9	466	200	266	85.8
43.1	334	144	190	62.1	43.1	102	44	58	19.0	43.0	402	173	229	74.5	43.0	249	107	142	46.0	42.9	464	199	265	85.3
43.1	332	143	189	61.6	43.1	72	31	41	13.3	43.0	400	172	228	74.0	43.0	244	105	139	45.2	42.9	462	198	264	84.9
43.1	327	141	186	60.8	43.1	65	28	37	12.1	43.0	398	171	227	73.5	43.0	242	104	138	44.7	42.9	459	197	262	84.6
43.1	325	140	185	60.3	43.1	58	25	33	10.8	43.0	395	170	225	73.2	43.0	237	102	135	43.9	42.9	457	196	261	84.1
43.1	320	138	182	59.5	43.1	51	22	29	9.5	43.0	393	169	224	72.7	43.0	235	101	134	43.4	42.9	455	195	260	83.6
43.1	318	137	181	59.0	43.0	500	215	285	92.4	43.0	391	168	223	72.2	43.0	230	99	131	42.6	42.9	452	194	258	83.3
43.1	313	135	178	58.2	43.0	498	214	284	92.0	43.0	388	167	221	71.9	43.0	228	98	130	42.1	42.9	450	193	257	82.8
43.1	311	134	177	57.7	43.0	495	213	282	91.7	43.0	386	166	220	71.4	43.0	223	96	127	41.3	42.9	448	192	256	82.3
43.1	306	132	174	56.9	43.0	493	212	281	91.2	43.0	384	165	219	70.9	43.0	221	95	126	40.8	42.9	445	191	254	82.0
43.1	304	131	173	56.5	43.0	491	211	280	90.7	43.0	381	164	217	70.6	43.0	214	92	122	39.6	42.9	443	190	253	81.5
43.1	299	129	170	55.7	43.0	488	210	278	90.4	43.0	379	163	216	70.1	43.0	207	89	118	38.3	42.9	441	189	252	81.0
43.1	297	128	169	55.2	43.0	486	209	277	89.9	43.0	377	162	215	69.6	43.0	200	86	114	37.0	42.9	438	188	250	80.7
43.1	295	127	168	54.7	43.0	484	208	276	89.4	43.0	374	161	213	69.3	43.0	193	83	110	35.7	42.9	436	187	249	80.2
43.1	290	125	165	53.9	43.0	481	207	274	89.1	43.0	372	160	212	68.8	43.0	186	80	106	34.4	42.9	434	186	248	79.7
43.1	288	124	164	53.4	43.0	479	206	273	88.6	43.0	370	159	211	68.3	43.0	179	77	102	33.1	42.9	431	185	246	79.4
43.1	283	122	161	52.6	43.0	477	205	272	88.1	43.0	365	157	208	67.5	43.0	172	74	98	31.8	42.9	429	184	245	78.9
43.1	281	121	160	52.1	43.0	474	204	270	87.8	43.0	363	156	207	67.0	43.0	165	71	94	30.6	42.9	427	183	244	78.4
43.1	276	119	157	51.3	43.0	472	203	269	87.3	43.0	358	154	204	66.2	43.0	158	68	90	29.3	42.9	424	182	242	78.1
43.1	274	118	156	50.8	43.0	470	202	268	86.8	43.0	356	153	203	65.8	43.0	151	65	86	28.0	42.9	422	181	241	77.6
43.1	269	116	153	50.0	43.0	467	201	266	86.5	43.0	351	151	200	65.0	43.0	149	64	85	27.5	42.9	420	180	240	77.1
43.1	267	115	152	49.5	43.0	465	200	265	86.0	43.0	349	150	199	64.5	43.0	142	61	81	26.2	42.9	417	179	238	76.8
43.1	262	113	149	48.7	43.0	463	199	264	85.5	43.0	344	148	196	63.7	43.0	135	58	77	24.9	42.9	415	178	237	76.3
43.1	260	112	148	48.2	43.0	461	198	263	85.0	43.0	342	147	195	63.2	43.0	128	55	73	23.6	42.9	413	177	236	75.9
43.1	255	110	145	47.5	43.0	460	198	262	85.2	43.0	337	145	192	62.4	43.0	121	52	69	22.3	42.9	410	176	234	75.6
43.1	253	109	144	47.0	43.0	458	197	261	84.7	43.0	335	144	191	61.9	43.0	114	49	65	21.1	42.9	408	175	233	75.1
43.1	248	107	141	46.2	43.0	456	196	260	84.2	43.0	330	142	188	61.1	43.0	107	46	61	19.8	42.9	406	174	232	74.6

%E	M1	M2	DM	M*	%E	M1	M2	DM	M*	%E	M1	M2	DM	M*	%E	M1	M2	DM	M*	%E	M1	M2	DM	M*
42.9	403	173	230	74.3	42.9	275	118	157	50.6	42.9	49	21	28	9.0	42.8	400	171	229	73.1	42.8	187	80	107	34.2
42.9	401	172	229	73.8	42.9	273	117	156	50.1	42.9	42	18	24	7.7	42.8	397	170	227	72.8	42.8	180	77	103	32.9
42.9	399	171	228	73.3	42.9	268	115	153	49.3	42.9	35	15	20	6.4	42.8	395	169	226	72.3	42.8	173	74	99	31.7
42.9	396	170	226	73.0	42.9	266	114	152	48.9	42.9	28	12	16	5.1	42.8	390	167	223	71.5	42.8	166	71	95	30.4
42.9	394	169	225	72.5	42.9	261	112	149	48.1	42.9	21	9	12	3.9	42.8	388	166	222	71.0	42.8	159	68	91	29.1
42.9	392	168	224	72.0	42.9	259	111	148	47.6	42.9	14	6	8	2.6	42.8	383	164	219	70.2	42.8	152	65	87	27.8
42.9	389	167	222	71.7	42.9	254	109	145	46.8	42.9	7	3	4	1.3	42.8	381	163	218	69.7	42.8	145	62	83	26.5
42.9	387	166	221	71.2	42.9	252	108	144	46.3	42.8	500	214	286	91.6	42.8	376	161	215	68.9	42.8	138	59	79	25.2
42.9	385	165	220	70.7	42.9	247	106	141	45.5	42.8	498	213	285	91.1	42.8	374	160	214	68.4	42.7	499	213	286	90.9
42.9	382	164	218	70.4	42.9	245	105	140	45.0	42.8	495	212	283	90.8	42.8	369	158	211	67.7	42.7	497	212	285	90.4
42.9	380	163	217	69.9	42.9	240	103	137	44.2	42.8	493	211	282	90.3	42.8	367	157	210	67.2	42.7	496	212	284	90.6
42.9	378	162	216	69.4	42.9	238	102	136	43.7	42.8	491	210	281	89.8	42.8	362	155	207	66.4	42.7	494	211	283	90.1
42.9	375	161	214	69.1	42.9	233	100	133	42.9	42.8	488	209	279	89.5	42.8	360	154	206	65.9	42.7	492	210	282	89.6
42.9	373	160	213	68.6	42.9	231	99	132	42.4	42.8	486	208	278	89.0	42.8	355	152	203	65.1	42.7	490	209	281	89.1
42.9	371	159	212	68.1	42.9	226	97	129	41.6	42.8	484	207	277	88.5	42.8	353	151	202	64.6	42.7	489	209	280	89.3
42.9	368	158	210	67.8	42.9	224	96	128	41.1	42.8	481	206	275	88.2	42.8	348	149	199	63.8	42.7	487	208	279	88.8
42.9	366	157	209	67.3	42.9	219	94	125	40.3	42.8	479	205	274	87.7	42.8	346	148	198	63.3	42.7	485	207	278	88.3
42.9	364	156	208	66.9	42.9	217	93	124	39.9	42.8	477	204	273	87.2	42.8	341	146	195	62.5	42.7	483	206	277	87.9
42.9	361	155	206	66.6	42.9	212	91	121	39.1	42.8	474	203	271	86.9	42.8	339	145	194	62.0	42.7	482	206	276	88.0
42.9	359	154	205	66.1	42.9	210	90	120	38.6	42.8	472	202	270	86.4	42.8	334	143	191	61.2	42.7	480	205	275	87.6
42.9	357	153	204	65.6	42.9	205	88	117	37.8	42.8	470	201	269	86.0	42.8	332	142	190	60.7	42.7	478	204	274	87.1
42.9	354	152	202	65.3	42.9	203	87	116	37.3	42.8	467	200	267	85.7	42.8	327	140	187	59.9	42.7	475	203	272	86.8
42.9	352	151	201	64.8	42.9	198	85	113	36.5	42.8	465	199	266	85.2	42.8	325	139	186	59.4	42.7	473	202	271	86.3
42.9	350	150	200	64.3	42.9	196	84	112	36.0	42.8	463	198	265	84.7	42.8	320	137	183	58.7	42.7	471	201	270	85.8
42.9	347	149	198	64.0	42.9	191	82	109	35.2	42.8	460	197	263	84.4	42.8	318	136	182	58.2	42.7	468	200	268	85.5
42.9	345	148	197	63.5	42.9	189	81	108	34.7	42.8	458	196	262	83.9	42.8	313	134	179	57.4	42.7	466	199	267	85.0
42.9	343	147	196	63.0	42.9	184	79	105	33.9	42.8	456	195	261	83.4	42.8	311	133	178	56.9	42.7	464	198	266	84.5
42.9	340	146	194	62.7	42.9	182	78	104	33.4	42.8	453	194	259	83.1	42.8	306	131	175	56.1	42.7	461	197	264	84.2
42.9	338	145	193	62.2	42.9	177	76	101	32.6	42.8	451	193	258	82.6	42.8	304	130	174	55.6	42.7	459	196	263	83.7
42.9	336	144	192	61.7	42.9	175	75	100	32.1	42.8	449	192	257	82.1	42.8	299	128	171	54.8	42.7	457	195	262	83.2
42.9	333	143	190	61.4	42.9	170	73	97	31.3	42.8	446	191	255	81.8	42.8	297	127	170	54.3	42.7	454	194	260	82.9
42.9	331	142	189	60.9	42.9	168	72	96	30.9	42.8	444	190	254	81.3	42.8	292	125	167	53.5	42.7	452	193	259	82.4
42.9	329	141	188	60.4	42.9	163	70	93	30.1	42.8	442	189	253	80.8	42.8	290	124	166	53.0	42.7	450	192	258	81.9
42.9	326	140	186	60.1	42.9	161	69	92	29.6	42.8	439	188	251	80.5	42.8	285	122	163	52.2	42.7	447	191	256	81.6
42.9	324	139	185	59.6	42.9	156	67	89	28.8	42.8	437	187	250	80.0	42.8	283	121	162	51.7	42.7	445	190	255	81.1
42.9	322	138	184	59.1	42.9	154	66	88	28.3	42.8	435	186	249	79.5	42.8	278	119	159	50.9	42.7	443	189	254	80.6
42.9	319	137	182	58.8	42.9	147	63	84	27.0	42.8	432	185	247	79.2	42.8	276	118	158	50.4	42.7	440	188	252	80.3
42.9	317	136	181	58.3	42.9	140	60	80	25.7	42.8	430	184	246	78.7	42.8	271	116	155	49.7	42.7	438	187	251	79.8
42.9	315	135	180	57.9	42.9	133	57	76	24.4	42.8	428	183	245	78.2	42.8	269	115	154	49.2	42.7	436	186	250	79.3
42.9	312	134	178	57.6	42.9	126	54	72	23.1	42.8	425	182	243	77.9	42.8	264	113	151	48.4	42.7	433	185	248	79.0
42.9	310	133	177	57.1	42.9	119	51	68	21.9	42.8	423	181	242	77.4	42.8	257	110	147	47.1	42.7	431	184	247	78.6
42.9	308	132	176	56.6	42.9	112	48	64	20.6	42.8	421	180	241	77.0	42.8	250	107	143	45.8	42.7	429	183	246	78.1
42.9	303	130	173	55.8	42.9	105	45	60	19.3	42.8	418	179	239	76.7	42.8	243	104	139	44.5	42.7	426	182	244	77.8
42.9	301	129	172	55.3	42.9	98	42	56	18.0	42.8	416	178	238	76.2	42.8	236	101	135	43.2	42.7	424	181	243	77.3
42.9	296	127	169	54.5	42.9	91	39	52	16.7	42.8	414	177	237	75.7	42.8	229	98	131	41.9	42.7	422	180	242	76.8
42.9	294	126	168	54.0	42.9	84	36	48	15.4	42.8	411	176	235	75.4	42.8	222	95	127	40.7	42.7	419	179	240	76.5
42.9	289	124	165	53.2	42.9	77	33	44	14.1	42.8	409	175	234	74.9	42.8	215	92	123	39.4	42.7	417	178	239	76.0
42.9	287	123	164	52.7	42.9	70	30	40	12.9	42.8	407	174	233	74.4	42.8	208	89	119	38.1	42.7	415	177	238	75.5
42.9	282	121	161	51.9	42.9	63	27	36	11.6	42.8	404	173	231	74.1	42.8	201	86	115	36.8	42.7	412	176	236	75.2
42.9	280	120	160	51.4	42.9	56	24	32	10.3	42.8	402	172	230	73.6	42.8	194	83	111	35.5	42.7	410	175	235	74.7

%E	M1	M2	DM	M*
42.7	405	173	232	73.9
42.7	403	172	231	73.4
42.7	398	170	228	72.6
42.7	396	169	227	72.1
42.7	393	168	225	71.8
42.7	391	167	224	71.3
42.7	389	166	223	70.8
42.7	386	165	221	70.5
42.7	384	164	220	70.0
42.7	382	163	219	69.6
42.7	379	162	217	69.2
42.7	377	161	216	68.8
42.7	375	160	215	68.3
42.7	372	159	213	68.0
42.7	370	158	212	67.5
42.7	368	157	211	67.0
42.7	365	156	209	66.7
42.7	363	155	208	66.2
42.7	361	154	207	65.7
42.7	358	153	205	65.4
42.7	356	152	204	64.9
42.7	354	151	203	64.4
42.7	351	150	201	64.1
42.7	349	149	200	63.6
42.7	347	148	199	63.1
42.7	344	147	197	62.8
42.7	342	146	196	62.3
42.7	337	144	193	61.5
42.7	335	143	192	61.0
42.7	330	141	189	60.2
42.7	328	140	188	59.8
42.7	323	138	185	59.0
42.7	321	137	184	58.5
42.7	316	135	181	57.7
42.7	314	134	180	57.2
42.7	309	132	177	56.4
42.7	307	131	176	55.9
42.7	302	129	173	55.1
42.7	300	128	172	54.6
42.7	295	126	169	53.8
42.7	293	125	168	53.3
42.7	288	123	165	52.5
42.7	286	122	164	52.0
42.7	281	120	161	51.2
42.7	279	119	160	50.8
42.7	274	117	157	50.0
42.7	267	114	153	48.7
42.7	262	112	150	47.9
42.7	260	111	149	47.4
42.7	255	109	146	46.6

%E	M1	M2	DM	M*
42.7	253	108	145	46.1
42.7	248	106	142	45.3
42.7	246	105	141	44.8
42.7	241	103	138	44.0
42.7	239	102	137	43.5
42.7	234	100	134	42.7
42.7	232	99	133	42.2
42.7	227	97	130	41.4
42.7	225	96	129	41.0
42.7	220	94	126	40.2
42.7	218	93	125	39.7
42.7	213	91	122	38.9
42.7	211	90	121	38.4
42.7	206	88	118	37.6
42.7	199	85	114	36.3
42.7	192	82	110	35.0
42.7	185	79	106	33.7
42.7	178	76	102	32.4
42.7	171	73	98	31.2
42.7	164	70	94	29.9
42.7	157	67	90	28.6
42.7	150	64	86	27.3
42.7	143	61	82	26.0
42.7	131	56	75	23.9
42.7	124	53	71	22.7
42.7	117	50	67	21.4
42.7	110	47	63	20.1
42.7	103	44	59	18.8
42.7	96	41	55	17.5
42.7	89	38	51	16.2
42.7	82	35	47	14.9
42.7	75	32	43	13.7
42.6	500	213	287	90.7
42.6	498	212	286	90.2
42.6	495	211	284	89.9
42.6	493	210	283	89.5
42.6	491	209	282	89.0
42.6	488	208	280	88.7
42.6	486	207	279	88.2
42.6	484	206	278	87.7
42.6	481	205	276	87.4
42.6	479	204	275	86.9
42.6	477	203	274	86.4
42.6	476	203	273	86.6
42.6	474	202	272	86.1
42.6	472	201	271	85.6
42.6	470	200	270	85.1
42.6	469	200	269	85.3
42.6	467	199	268	84.8
42.6	465	198	267	84.3

%E	M1	M2	DM	M*
42.6	462	197	265	84.0
42.6	460	196	264	83.5
42.6	458	195	263	83.0
42.6	455	194	261	82.7
42.6	453	193	260	82.2
42.6	451	192	259	81.7
42.6	448	191	257	81.4
42.6	446	190	256	80.9
42.6	444	189	255	80.5
42.6	441	188	253	80.1
42.6	439	187	252	79.7
42.6	437	186	251	79.2
42.6	434	185	249	78.9
42.6	432	184	248	78.4
42.6	430	183	247	77.9
42.6	427	182	245	77.6
42.6	425	181	244	77.1
42.6	423	180	243	76.6
42.6	420	179	241	76.3
42.6	418	178	240	75.8
42.6	413	176	237	75.0
42.6	411	175	236	74.5
42.6	408	174	234	74.2
42.6	406	173	233	73.7
42.6	404	172	232	73.2
42.6	401	171	230	72.9
42.6	399	170	229	72.4
42.6	397	169	228	71.9
42.6	394	168	226	71.6
42.6	392	167	225	71.1
42.6	390	166	224	70.7
42.6	387	165	222	70.3
42.6	385	164	221	69.9
42.6	383	163	220	69.4
42.6	380	162	218	69.1
42.6	378	161	217	68.6
42.6	376	160	216	68.1
42.6	373	159	214	67.8
42.6	371	158	213	67.3
42.6	366	156	210	66.5
42.6	364	155	209	66.0
42.6	359	153	206	65.2
42.6	357	152	205	64.7
42.6	352	150	202	63.9
42.6	350	149	201	63.4
42.6	345	147	198	62.6
42.6	343	146	197	62.1
42.6	340	145	195	61.8
42.6	338	144	194	61.3
42.6	336	143	193	60.9

%E	M1	M2	DM	M*
42.6	333	142	191	60.6
42.6	331	141	190	60.1
42.6	329	140	189	59.6
42.6	326	139	187	59.3
42.6	324	138	186	58.8
42.6	319	136	183	58.0
42.6	317	135	182	57.5
42.6	312	133	179	56.7
42.6	310	132	178	56.2
42.6	305	130	175	55.4
42.6	303	129	174	54.9
42.6	298	127	171	54.1
42.6	296	126	170	53.6
42.6	291	124	167	52.8
42.6	289	123	166	52.3
42.6	284	121	163	51.6
42.6	282	120	162	51.1
42.6	277	118	159	50.3
42.6	272	116	156	49.5
42.6	270	115	155	49.0
42.6	265	113	152	48.2
42.6	263	112	151	47.7
42.6	258	110	148	46.9
42.6	256	109	147	46.4
42.6	251	107	144	45.6
42.6	249	106	143	45.1
42.6	244	104	140	44.3
42.6	242	103	139	43.8
42.6	237	101	136	43.0
42.6	235	100	135	42.6
42.6	230	98	132	41.8
42.6	223	95	128	40.5
42.6	216	92	124	39.2
42.6	209	89	120	37.9
42.6	204	87	117	37.1
42.6	202	86	116	36.6
42.6	197	84	113	35.8
42.6	195	83	112	35.3
42.6	190	81	109	34.5
42.6	188	80	108	34.0
42.6	183	78	105	33.2
42.6	176	75	101	32.0
42.6	169	72	97	30.7
42.6	162	69	93	29.4
42.6	155	66	89	28.1
42.6	148	63	85	26.8
42.6	141	60	81	25.5
42.6	136	58	78	24.7
42.6	129	55	74	23.4
42.6	122	52	70	22.2

%E	M1	M2	DM	M*
42.6	115	49	66	20.9
42.6	108	46	62	19.6
42.6	101	43	58	18.3
42.6	94	40	54	17.0
42.6	68	29	39	12.4
42.6	61	26	35	11.1
42.6	54	23	31	9.8
42.6	47	20	27	8.5
42.5	499	212	287	90.1
42.5	497	211	286	89.6
42.5	496	211	285	89.8
42.5	494	210	284	89.3
42.5	492	209	283	88.8
42.5	489	208	281	88.5
42.5	487	207	280	88.0
42.5	485	206	279	87.5
42.5	482	205	277	87.2
42.5	480	204	276	86.7
42.5	478	203	275	86.2
42.5	475	202	273	85.9
42.5	473	201	272	85.4
42.5	471	200	271	84.9
42.5	468	199	269	84.6
42.5	466	198	268	84.1
42.5	464	197	267	83.6
42.5	463	197	266	83.8
42.5	461	196	265	83.3
42.5	459	195	264	82.8
42.5	457	194	263	82.4
42.5	456	194	262	82.5
42.5	454	193	261	82.0
42.5	452	192	260	81.6
42.5	449	191	258	81.2
42.5	447	190	257	80.8
42.5	445	189	256	80.3
42.5	442	188	254	80.0
42.5	440	187	253	79.5
42.5	438	186	252	79.0
42.5	435	185	250	78.7
42.5	433	184	249	78.2
42.5	431	183	248	77.7
42.5	428	182	246	77.4
42.5	426	181	245	76.9
42.5	424	180	244	76.4
42.5	421	179	242	76.1
42.5	419	178	241	75.6
42.5	416	177	239	75.3
42.5	414	176	238	74.8
42.5	412	175	237	74.3
42.5	409	174	235	74.0

%E	M1	M2	DM	M*
42.5	407	173	234	73.5
42.5	405	172	233	73.0
42.5	402	171	231	72.7
42.5	400	170	230	72.3
42.5	398	169	229	71.8
42.5	395	168	227	71.5
42.5	393	167	226	71.0
42.5	391	166	225	70.5
42.5	388	165	223	70.2
42.5	386	164	222	69.7
42.5	381	162	219	68.9
42.5	379	161	218	68.4
42.5	374	159	215	67.6
42.5	372	158	214	67.1
42.5	369	157	212	66.8
42.5	367	156	211	66.3
42.5	365	155	210	65.8
42.5	362	154	208	65.5
42.5	360	153	207	65.0
42.5	358	152	206	64.5
42.5	355	151	204	64.2
42.5	353	150	203	63.7
42.5	351	149	202	63.3
42.5	348	148	200	62.9
42.5	346	147	199	62.5
42.5	341	145	196	61.7
42.5	339	144	195	61.2
42.5	334	142	192	60.4
42.5	332	141	191	59.9
42.5	327	139	188	59.1
42.5	325	138	187	58.6
42.5	322	137	185	58.3
42.5	320	136	184	57.8
42.5	318	135	183	57.3
42.5	315	134	181	57.0
42.5	313	133	180	56.5
42.5	308	131	177	55.7
42.5	306	130	176	55.2
42.5	301	128	173	54.4
42.5	299	127	172	53.9
42.5	294	125	169	53.1
42.5	292	124	168	52.7
42.5	287	122	165	51.9
42.5	285	121	164	51.4
42.5	280	119	161	50.6
42.5	275	117	158	49.8
42.5	273	116	157	49.3
42.5	268	114	154	48.5
42.5	266	113	153	48.0
42.5	261	111	150	47.2

%E	M1	M2	DM	M*
42.5	259	110	149	46.7
42.5	254	108	146	45.9
42.5	252	107	145	45.4
42.5	247	105	142	44.6
42.5	240	102	138	43.3
42.5	233	99	134	42.1
42.5	228	97	131	41.3
42.5	226	96	130	40.8
42.5	221	94	127	40.0
42.5	219	93	126	39.5
42.5	214	91	123	38.7
42.5	212	90	122	38.2
42.5	207	88	119	37.4
42.5	200	85	115	36.1
42.5	193	82	111	34.8
42.5	186	79	107	33.6
42.5	181	77	104	32.8
42.5	179	76	103	32.3
42.5	174	74	100	31.5
42.5	167	71	96	30.2
42.5	160	68	92	28.9
42.5	153	65	88	27.6
42.5	146	62	84	26.3
42.5	134	57	77	24.2
42.5	127	54	73	23.0
42.5	120	51	69	21.7
42.5	113	48	65	20.4
42.5	106	45	61	19.1
42.5	87	37	50	15.7
42.5	80	34	46	14.4
42.5	73	31	42	13.2
42.5	40	17	23	7.2
42.4	500	212	288	89.9
42.4	498	211	287	89.4
42.4	495	210	285	89.1
42.4	493	209	284	88.6
42.4	491	208	283	88.1
42.4	490	208	282	88.3
42.4	488	207	281	87.8
42.4	486	206	280	87.3
42.4	484	205	279	86.8
42.4	483	205	278	87.0
42.4	481	204	277	86.5
42.4	479	203	276	86.0
42.4	476	202	274	85.7
42.4	474	201	273	85.2
42.4	472	200	272	84.7
42.4	469	199	270	84.4
42.4	467	198	269	83.9
42.4	465	197	268	83.5

%E	M1	M2	DM	M*
42.4	462	196	266	83.2
42.4	460	195	265	82.7
42.4	458	194	264	82.2
42.4	455	193	262	81.9
42.4	453	192	261	81.4
42.4	451	191	260	80.9
42.4	450	191	259	81.1
42.4	448	190	258	80.6
42.4	446	189	257	80.1
42.4	443	188	255	79.8
42.4	441	187	254	79.3
42.4	439	186	253	78.8
42.4	436	185	251	78.5
42.4	434	184	250	78.0
42.4	432	183	249	77.5
42.4	429	182	247	77.2
42.4	427	181	246	76.7
42.4	425	180	245	76.2
42.4	422	179	243	75.9
42.4	420	178	242	75.4
42.4	417	177	240	75.1
42.4	415	176	239	74.6
42.4	413	175	238	74.2
42.4	410	174	236	73.8
42.4	408	173	235	73.4
42.4	406	172	234	72.9
42.4	403	171	232	72.6
42.4	401	170	231	72.1
42.4	399	169	230	71.6
42.4	396	168	228	71.3
42.4	394	167	227	70.8
42.4	389	165	224	70.0
42.4	387	164	223	69.5
42.4	384	163	221	69.2
42.4	382	162	220	68.7
42.4	380	161	219	68.2
42.4	377	160	217	67.9
42.4	375	159	216	67.4
42.4	373	158	215	66.9
42.4	370	157	213	66.6
42.4	368	156	212	66.1
42.4	366	155	211	65.6
42.4	363	154	209	65.3
42.4	361	153	208	64.8
42.4	356	151	205	64.0
42.4	354	150	204	63.6
42.4	349	148	201	62.8
42.4	347	147	200	62.3
42.4	344	146	198	62.0
42.4	342	145	197	61.5

%E	M1	M2	DM	M*
42.4	340	144	196	61.0
42.4	337	143	194	60.7
42.4	335	142	193	60.2
42.4	330	140	190	59.4
42.4	328	139	189	58.9
42.4	323	137	186	58.1
42.4	321	136	185	57.6
42.4	316	134	182	56.8
42.4	314	133	181	56.3
42.4	311	132	179	56.0
42.4	309	131	178	55.5
42.4	304	129	175	54.7
42.4	302	128	174	54.3
42.4	297	126	171	53.5
42.4	295	125	170	53.0
42.4	290	123	167	52.2
42.4	288	122	166	51.7
42.4	283	120	163	50.9
42.4	278	118	160	50.1
42.4	276	117	159	49.6
42.4	271	115	156	48.8
42.4	269	114	155	48.3
42.4	264	112	152	47.5
42.4	262	111	151	47.0
42.4	257	109	148	46.2
42.4	255	108	147	45.7
42.4	250	106	144	44.9
42.4	245	104	141	44.1
42.4	243	103	140	43.7
42.4	238	101	137	42.9
42.4	236	100	136	42.4
42.4	231	98	133	41.6
42.4	229	97	132	41.1
42.4	224	95	129	40.3
42.4	217	92	125	39.0
42.4	210	89	121	37.7
42.4	205	87	118	36.9
42.4	203	86	117	36.4
42.4	198	84	114	35.6
42.4	191	81	110	34.4
42.4	184	78	106	33.1
42.4	177	75	102	31.8
42.4	172	73	99	31.0
42.4	170	72	98	30.5
42.4	165	70	95	29.7
42.4	158	67	91	28.4
42.4	151	64	87	27.1
42.4	144	61	83	25.8
42.4	139	59	80	25.0
42.4	132	56	76	23.8

%E	M1	M2	DM	M*
42.4	125	53	72	22.5
42.4	118	50	68	21.2
42.4	99	42	57	17.8
42.4	92	39	53	16.5
42.4	85	36	49	15.2
42.4	66	28	38	11.9
42.4	59	25	34	10.6
42.4	33	14	19	5.9
42.3	499	211	288	89.2
42.3	497	210	287	88.7
42.3	496	210	286	88.9
42.3	494	209	285	88.4
42.3	492	208	284	87.9
42.3	489	207	282	87.6
42.3	487	206	281	87.1
42.3	485	205	280	86.6
42.3	482	204	278	86.3
42.3	480	203	277	85.9
42.3	478	202	276	85.4
42.3	477	202	275	85.5
42.3	475	201	274	85.1
42.3	473	200	273	84.6
42.3	471	199	272	84.1
42.3	470	199	271	84.3
42.3	468	198	270	83.8
42.3	466	197	269	83.3
42.3	463	196	267	83.0
42.3	461	195	266	82.5
42.3	459	194	265	82.0
42.3	456	193	263	81.7
42.3	454	192	262	81.2
42.3	452	191	261	80.7
42.3	449	190	259	80.4
42.3	447	189	258	79.9
42.3	444	188	256	79.6
42.3	442	187	255	79.1
42.3	440	186	254	78.6
42.3	437	185	252	78.3
42.3	435	184	251	77.8
42.3	433	183	250	77.3
42.3	430	182	248	77.0
42.3	428	181	247	76.5
42.3	426	180	246	76.1
42.3	423	179	244	75.7
42.3	421	178	243	75.3
42.3	418	177	241	74.9
42.3	416	176	240	74.5
42.3	414	175	239	74.0
42.3	411	174	237	73.7
42.3	409	173	236	73.2

%E	M1	M2	DM	M*
42.3	407	172	235	72.7
42.3	404	171	233	72.4
42.3	402	170	232	71.9
42.3	400	169	231	71.4
42.3	397	168	229	71.1
42.3	395	167	228	70.6
42.3	392	166	226	70.3
42.3	390	165	225	69.8
42.3	388	164	224	69.3
42.3	385	163	222	69.0
42.3	383	162	221	68.5
42.3	381	161	220	68.0
42.3	378	160	218	67.7
42.3	376	159	217	67.2
42.3	371	157	214	66.4
42.3	369	156	213	66.0
42.3	364	154	210	65.2
42.3	362	153	209	64.7
42.3	359	152	207	64.4
42.3	357	151	206	63.9
42.3	355	150	205	63.4
42.3	352	149	203	63.1
42.3	350	148	202	62.6
42.3	345	146	199	61.8
42.3	343	145	198	61.3
42.3	338	143	195	60.5
42.3	336	142	194	60.0
42.3	333	141	192	59.7
42.3	331	140	191	59.2
42.3	326	138	188	58.4
42.3	324	137	187	57.9
42.3	319	135	184	57.1
42.3	317	134	183	56.6
42.3	312	132	180	55.8
42.3	310	131	179	55.4
42.3	307	130	177	55.0
42.3	305	129	176	54.6
42.3	300	127	173	53.8
42.3	298	126	172	53.3
42.3	293	124	169	52.5
42.3	291	123	168	52.0
42.3	286	121	165	51.2
42.3	284	120	164	50.7
42.3	281	119	162	50.4
42.3	279	118	161	49.9
42.3	274	116	158	49.1
42.3	272	115	157	48.6
42.3	267	113	154	47.8
42.3	265	112	153	47.3
42.3	260	110	150	46.5

%E	M1	M2	DM	M*
42.3	253	107	146	45.3
42.3	248	105	143	44.5
42.3	246	104	142	44.0
42.3	241	102	139	43.2
42.3	239	101	138	42.7
42.3	234	99	135	41.9
42.3	227	96	131	40.6
42.3	222	94	128	39.8
42.3	220	93	127	39.3
42.3	215	91	124	38.5
42.3	213	90	123	38.0
42.3	208	88	120	37.2
42.3	201	85	116	35.9
42.3	196	83	113	35.1
42.3	194	82	112	34.7
42.3	189	80	109	33.9
42.3	182	77	105	32.6
42.3	175	74	101	31.3
42.3	168	71	97	30.0
42.3	163	69	94	29.2
42.3	156	66	90	27.9
42.3	149	63	86	26.6
42.3	142	60	82	25.4
42.3	137	58	79	24.6
42.3	130	55	75	23.3
42.3	123	52	71	22.0
42.3	111	47	64	19.9
42.3	104	44	60	18.6
42.3	97	41	56	17.3
42.3	78	33	45	14.0
42.3	71	30	41	12.7
42.3	52	22	30	9.3
42.3	26	11	15	4.7
42.2	500	211	289	89.0
42.2	498	210	288	88.6
42.2	495	209	286	88.2
42.2	493	208	285	87.8
42.2	491	207	284	87.3
42.2	490	207	283	87.4
42.2	488	206	282	87.0
42.2	486	205	281	86.5
42.2	483	204	279	86.2
42.2	481	203	278	85.7
42.2	479	202	277	85.2
42.2	476	201	275	84.9
42.2	474	200	274	84.4
42.2	472	199	273	83.9
42.2	469	198	271	83.6
42.2	467	197	270	83.1
42.2	465	196	269	82.6

%E	M1	M2	DM	M*
42.2	464	196	268	82.8
42.2	462	195	267	82.3
42.2	460	194	266	81.8
42.2	457	193	264	81.5
42.2	455	192	263	81.0
42.2	453	191	262	80.5
42.2	450	190	260	80.2
42.2	448	189	259	79.7
42.2	446	188	258	79.2
42.2	445	188	257	79.4
42.2	443	187	256	78.9
42.2	441	186	255	78.4
42.2	438	185	253	78.1
42.2	436	184	252	77.7
42.2	434	183	251	77.2
42.2	431	182	249	76.9
42.2	429	181	248	76.4
42.2	427	180	247	75.9
42.2	424	179	245	75.6
42.2	422	178	244	75.1
42.2	419	177	242	74.8
42.2	417	176	241	74.3
42.2	415	175	240	73.8
42.2	412	174	238	73.5
42.2	410	173	237	73.0
42.2	408	172	236	72.5
42.2	405	171	234	72.2
42.2	403	170	233	71.7
42.2	398	168	230	70.9
42.2	396	167	229	70.4
42.2	393	166	227	70.1
42.2	391	165	226	69.6
42.2	389	164	225	69.1
42.2	386	163	223	68.8
42.2	384	162	222	68.3
42.2	379	160	219	67.5
42.2	377	159	218	67.1
42.2	374	158	216	66.7
42.2	372	157	215	66.3
42.2	370	156	214	65.8
42.2	367	155	212	65.5
42.2	365	154	211	65.0
42.2	360	152	208	64.2
42.2	358	151	207	63.7
42.2	353	149	204	62.9
42.2	351	148	203	62.4
42.2	348	147	201	62.1
42.2	346	146	200	61.6
42.2	344	145	199	61.1
42.2	341	144	197	60.8

%E	M1	M2	DM	M*
42.2	339	143	196	60.3
42.2	334	141	193	59.5
42.2	332	140	192	59.0
42.2	329	139	190	58.7
42.2	327	138	189	58.2
42.2	325	137	188	57.8
42.2	322	136	186	57.4
42.2	320	135	185	57.0
42.2	315	133	182	56.2
42.2	313	132	181	55.7
42.2	308	130	178	54.9
42.2	306	129	177	54.4
42.2	303	128	175	54.1
42.2	301	127	174	53.6
42.2	296	125	171	52.8
42.2	294	124	170	52.3
42.2	289	122	167	51.5
42.2	287	121	166	51.0
42.2	282	119	163	50.2
42.2	277	117	160	49.4
42.2	275	116	159	48.9
42.2	270	114	156	48.1
42.2	268	113	155	47.6
42.2	263	111	152	46.8
42.2	258	109	149	46.1
42.2	256	108	148	45.6
42.2	251	106	145	44.8
42.2	249	105	144	44.3
42.2	244	103	141	43.5
42.2	237	100	137	42.2
42.2	232	98	134	41.4
42.2	230	97	133	40.9
42.2	225	95	130	40.1
42.2	223	94	129	39.6
42.2	218	92	126	38.8
42.2	211	89	122	37.5
42.2	206	87	119	36.7
42.2	204	86	118	36.3
42.2	199	84	115	35.5
42.2	192	81	111	34.2
42.2	187	79	108	33.4
42.2	185	78	107	32.9
42.2	180	76	104	32.1
42.2	173	73	100	30.8
42.2	166	70	96	29.5
42.2	161	68	93	28.7
42.2	154	65	89	27.4
42.2	147	62	85	26.1
42.2	135	57	78	24.1
42.2	128	54	74	22.8

%E	M1	M2	DM	M*
42.2	116	49	67	20.7
42.2	109	46	63	19.4
42.2	102	43	59	18.1
42.2	90	38	52	16.0
42.2	83	35	48	14.8
42.2	64	27	37	11.4
42.2	45	19	26	8.0
42.1	499	210	289	88.4
42.1	497	209	288	87.9
42.1	496	209	287	88.1
42.1	494	208	286	87.6
42.1	492	207	285	87.1
42.1	489	206	283	86.8
42.1	487	205	282	86.3
42.1	485	204	281	85.8
42.1	484	204	280	86.0
42.1	482	203	279	85.5
42.1	480	202	278	85.0
42.1	478	201	277	84.5
42.1	477	201	276	84.7
42.1	475	200	275	84.2
42.1	473	199	274	83.7
42.1	470	198	272	83.4
42.1	468	197	271	82.9
42.1	466	196	270	82.4
42.1	463	195	268	82.1
42.1	461	194	267	81.6
42.1	458	193	265	81.3
42.1	456	192	264	80.8
42.1	454	191	263	80.4
42.1	451	190	261	80.0
42.1	449	189	260	79.6
42.1	447	188	259	79.1
42.1	444	187	257	78.8
42.1	442	186	256	78.3
42.1	439	185	254	78.0
42.1	437	184	253	77.5
42.1	435	183	252	77.0
42.1	432	182	250	76.7
42.1	430	181	249	76.2
42.1	428	180	248	75.7
42.1	425	179	246	75.4
42.1	423	178	245	74.9
42.1	420	177	243	74.6
42.1	418	176	242	74.1
42.1	416	175	241	73.6
42.1	413	174	239	73.3
42.1	411	173	238	72.8
42.1	409	172	237	72.3
42.1	406	171	235	72.0

%E	M1	M2	DM	M*	%E	M1	M2	DM	M*	%E	M1	M2	DM	M*	%E	M1	M2	DM	M*	%E	M1	M2	DM	M*
42.1	404	170	234	71.5	42.1	252	106	146	44.6	42.0	462	194	268	81.5	42.0	343	144	199	60.5	42.0	131	55	76	23.1
42.1	401	169	232	71.2	42.1	247	104	143	43.8	42.0	460	193	267	81.0	42.0	338	142	196	59.7	42.0	119	50	69	21.0
42.1	399	168	231	70.7	42.1	242	102	140	43.0	42.0	459	193	266	81.2	42.0	336	141	195	59.2	42.0	112	47	65	19.7
42.1	397	167	230	70.2	42.1	240	101	139	42.5	42.0	457	192	265	80.7	42.0	333	140	193	58.9	42.0	100	42	58	17.6
42.1	394	166	228	69.9	42.1	235	99	136	41.7	42.0	455	191	264	80.2	42.0	331	139	192	58.4	42.0	88	37	51	15.6
42.1	392	165	227	69.5	42.1	233	98	135	41.2	42.0	452	190	262	79.9	42.0	326	137	189	57.6	42.0	81	34	47	14.3
42.1	390	164	226	69.0	42.1	228	96	132	40.4	42.0	450	189	261	79.4	42.0	324	136	188	57.1	42.0	69	29	40	12.2
42.1	387	163	224	68.7	42.1	221	93	128	39.1	42.0	448	188	260	78.9	42.0	319	134	185	56.3	42.0	50	21	29	8.8
42.1	385	162	223	68.2	42.1	216	91	125	38.3	42.0	445	187	258	78.6	42.0	317	133	184	55.8	41.9	499	209	290	87.5
42.1	382	161	221	67.9	42.1	214	90	124	37.9	42.0	443	186	257	78.1	42.0	314	132	182	55.5	41.9	497	208	289	87.1
42.1	380	160	220	67.4	42.1	209	88	121	37.1	42.0	441	185	256	77.6	42.0	312	131	181	55.0	41.9	496	208	288	87.2
42.1	378	159	219	66.9	42.1	202	85	117	35.8	42.0	440	185	255	77.8	42.0	307	129	178	54.2	41.9	494	207	287	86.7
42.1	375	158	217	66.6	42.1	197	83	114	35.0	42.0	438	184	254	77.3	42.0	305	128	177	53.7	41.9	492	206	286	86.3
42.1	373	157	216	66.1	42.1	195	82	113	34.5	42.0	436	183	253	76.8	42.0	300	126	174	52.9	41.9	489	205	284	85.9
42.1	368	155	213	65.3	42.1	190	80	110	33.7	42.0	433	182	251	76.5	42.0	295	124	171	52.1	41.9	487	204	283	85.5
42.1	366	154	212	64.8	42.1	183	77	106	32.4	42.0	431	181	250	76.0	42.0	293	123	170	51.6	41.9	485	203	282	85.0
42.1	363	153	210	64.5	42.1	178	75	103	31.6	42.0	429	180	249	75.5	42.0	288	121	167	50.8	41.9	484	203	281	85.1
42.1	361	152	209	64.0	42.1	171	72	99	30.3	42.0	426	179	247	75.2	42.0	286	120	166	50.3	41.9	482	202	280	84.7
42.1	359	151	208	63.5	42.1	164	69	95	29.0	42.0	424	178	246	74.7	42.0	283	119	164	50.0	41.9	480	201	279	84.2
42.1	356	150	206	63.2	42.1	159	67	92	28.2	42.0	421	177	244	74.4	42.0	281	118	163	49.6	41.9	477	200	277	83.9
42.1	354	149	205	62.7	42.1	152	64	88	26.9	42.0	419	176	243	73.9	42.0	276	116	160	48.8	41.9	475	199	276	83.4
42.1	349	147	202	61.9	42.1	145	61	84	25.7	42.0	417	175	242	73.4	42.0	274	115	159	48.3	41.9	473	198	275	82.9
42.1	347	146	201	61.4	42.1	140	59	81	24.9	42.0	414	174	240	73.1	42.0	269	113	156	47.5	41.9	472	198	274	83.1
42.1	342	144	198	60.6	42.1	133	56	77	23.6	42.0	412	173	239	72.6	42.0	264	111	153	46.7	41.9	470	197	273	82.6
42.1	340	143	197	60.1	42.1	126	53	73	22.3	42.0	410	172	238	72.2	42.0	262	110	152	46.2	41.9	468	196	272	82.1
42.1	337	142	195	59.8	42.1	121	51	70	21.5	42.0	407	171	236	71.8	42.0	257	108	149	45.4	41.9	465	195	270	81.8
42.1	335	141	194	59.3	42.1	114	48	66	20.2	42.0	405	170	235	71.4	42.0	255	107	148	44.9	41.9	463	194	269	81.3
42.1	330	139	191	58.5	42.1	107	45	62	18.9	42.0	402	169	233	71.0	42.0	250	105	145	44.1	41.9	461	193	268	80.8
42.1	328	138	190	58.1	42.1	95	40	55	16.8	42.0	400	168	232	70.6	42.0	245	103	142	43.3	41.9	458	192	266	80.5
42.1	323	136	187	57.3	42.1	76	32	44	13.5	42.0	398	167	231	70.1	42.0	243	102	141	42.8	41.9	456	191	265	80.0
42.1	321	135	186	56.8	42.1	57	24	33	10.1	42.0	395	166	229	69.8	42.0	238	100	138	42.0	41.9	454	190	264	79.5
42.1	318	134	184	56.5	42.1	38	16	22	6.7	42.0	393	165	228	69.3	42.0	231	97	134	40.7	41.9	453	190	263	79.7
42.1	316	133	183	56.0	42.1	19	8	11	3.4	42.0	388	163	225	68.5	42.0	226	95	131	39.9	41.9	451	189	262	79.2
42.1	311	131	180	55.2	42.0	500	210	290	88.2	42.0	386	162	224	68.0	42.0	224	94	130	39.4	41.9	449	188	261	78.7
42.1	309	130	179	54.7	42.0	498	209	289	87.7	42.0	383	161	222	67.7	42.0	219	92	127	38.6	41.9	446	187	259	78.4
42.1	304	128	176	53.9	42.0	495	208	287	87.4	42.0	381	160	221	67.2	42.0	212	89	123	37.4	41.9	444	186	258	77.9
42.1	302	127	175	53.4	42.0	493	207	286	86.9	42.0	379	159	220	66.7	42.0	207	87	120	36.6	41.9	442	185	257	77.4
42.1	299	126	173	53.1	42.0	491	206	285	86.4	42.0	376	158	218	66.4	42.0	205	86	119	36.1	41.9	439	184	255	77.1
42.1	297	125	172	52.6	42.0	490	206	284	86.6	42.0	374	157	217	65.9	42.0	200	84	116	35.3	41.9	437	183	254	76.6
42.1	292	123	169	51.8	42.0	488	205	283	86.1	42.0	371	156	215	65.6	42.0	193	81	112	34.0	41.9	434	182	252	76.3
42.1	290	122	168	51.3	42.0	486	204	282	85.6	42.0	369	155	214	65.1	42.0	188	79	109	33.2	41.9	432	181	251	75.8
42.1	285	120	165	50.5	42.0	483	203	280	85.3	42.0	367	154	213	64.6	42.0	181	76	105	31.9	41.9	430	180	250	75.3
42.1	280	118	162	49.7	42.0	481	202	279	84.8	42.0	364	153	211	64.3	42.0	176	74	102	31.1	41.9	427	179	248	75.0
42.1	278	117	161	49.2	42.0	479	201	278	84.3	42.0	362	152	210	63.8	42.0	174	73	101	30.6	41.9	425	178	247	74.6
42.1	273	115	158	48.4	42.0	476	200	276	84.0	42.0	357	150	207	63.0	42.0	169	71	98	29.8	41.9	422	177	245	74.2
42.1	271	114	157	48.0	42.0	474	199	275	83.5	42.0	355	149	206	62.5	42.0	162	68	94	28.5	41.9	420	176	244	73.8
42.1	266	112	154	47.2	42.0	471	198	273	83.2	42.0	352	148	204	62.2	42.0	157	66	91	27.7	41.9	418	175	243	73.3
42.1	261	110	151	46.4	42.0	469	197	272	82.7	42.0	350	147	203	61.7	42.0	150	63	87	26.5	41.9	415	174	241	73.0
42.1	259	109	150	45.9	42.0	467	196	271	82.3	42.0	348	146	202	61.3	42.0	143	60	83	25.2	41.9	413	173	240	72.5
42.1	254	107	147	45.1	42.0	464	195	269	82.0	42.0	345	145	200	60.9	42.0	138	58	80	24.4	41.9	408	171	237	71.7

%E	M1	M2	DM	M*
41.9	406	170	236	71.2
41.9	403	169	234	70.9
41.9	401	168	233	70.4
41.9	399	167	232	69.9
41.9	396	166	230	69.6
41.9	394	165	229	69.1
41.9	391	164	227	68.8
41.9	389	163	226	68.3
41.9	387	162	225	67.8
41.9	384	161	223	67.5
41.9	382	160	222	67.0
41.9	377	158	219	66.2
41.9	375	157	218	65.7
41.9	372	156	216	65.4
41.9	370	155	215	64.9
41.9	365	153	212	64.1
41.9	363	152	211	63.6
41.9	360	151	209	63.3
41.9	358	150	208	62.8
41.9	356	149	207	62.4
41.9	353	148	205	62.1
41.9	351	147	204	61.6
41.9	346	145	201	60.8
41.9	344	144	200	60.3
41.9	341	143	198	60.0
41.9	339	142	197	59.5
41.9	334	140	194	58.7
41.9	332	139	193	58.2
41.9	329	138	191	57.9
41.9	327	137	190	57.4
41.9	322	135	187	56.6
41.9	320	134	186	56.1
41.9	315	132	183	55.3
41.9	313	131	182	54.8
41.9	310	130	180	54.5
41.9	308	129	179	54.0
41.9	303	127	176	53.2
41.9	301	126	175	52.7
41.9	298	125	173	52.4
41.9	296	124	172	51.9
41.9	291	122	169	51.1
41.9	289	121	168	50.7
41.9	284	119	165	49.9
41.9	279	117	162	49.1
41.9	277	116	161	48.6
41.9	272	114	158	47.8
41.9	270	113	157	47.3
41.9	267	112	155	47.0
41.9	265	111	154	46.5
41.9	260	109	151	45.7
41.9	258	108	150	45.2
41.9	253	106	147	44.4
41.9	248	104	144	43.6
41.9	246	103	143	43.1
41.9	241	101	140	42.3
41.9	236	99	137	41.5
41.9	234	98	136	41.0
41.9	229	96	133	40.2
41.9	227	95	132	39.8
41.9	222	93	129	39.0
41.9	217	91	126	38.2
41.9	215	90	125	37.7
41.9	210	88	122	36.9
41.9	203	85	118	35.6
41.9	198	83	115	34.8
41.9	191	80	111	33.5
41.9	186	78	108	32.7
41.9	179	75	104	31.4
41.9	172	72	100	30.1
41.9	167	70	97	29.3
41.9	160	67	93	28.1
41.9	155	65	90	27.3
41.9	148	62	86	26.0
41.9	136	57	79	23.9
41.9	129	54	75	22.6
41.9	124	52	72	21.8
41.9	117	49	68	20.5
41.9	105	44	61	18.4
41.9	93	39	54	16.4
41.9	86	36	50	15.1
41.9	74	31	43	13.0
41.9	62	26	36	10.9
41.9	43	18	25	7.5
41.9	31	13	18	5.5
41.8	500	209	291	87.4
41.8	498	208	290	86.9
41.8	495	207	288	86.6
41.8	493	206	287	86.1
41.8	491	205	286	85.6
41.8	490	205	285	85.8
41.8	488	204	284	85.3
41.8	486	203	283	84.8
41.8	483	202	281	84.5
41.8	481	201	280	84.0
41.8	479	200	279	83.5
41.8	478	200	278	83.7
41.8	476	199	277	83.2
41.8	474	198	276	82.7
41.8	471	197	274	82.4
41.8	469	196	273	81.9
41.8	467	195	272	81.4
41.8	466	195	271	81.6
41.8	464	194	270	81.1
41.8	462	193	269	80.6
41.8	459	192	267	80.3
41.8	457	191	266	79.8
41.8	455	190	265	79.3
41.8	452	189	263	79.0
41.8	450	188	262	78.5
41.8	447	187	260	78.2
41.8	445	186	259	77.7
41.8	443	185	258	77.3
41.8	440	184	256	76.9
41.8	438	183	255	76.5
41.8	435	182	253	76.1
41.8	433	181	252	75.7
41.8	431	180	251	75.2
41.8	428	179	249	74.9
41.8	426	178	248	74.4
41.8	423	177	246	74.1
41.8	421	176	245	73.6
41.8	419	175	244	73.1
41.8	416	174	242	72.8
41.8	414	173	241	72.3
41.8	411	172	239	72.0
41.8	409	171	238	71.5
41.8	407	170	237	71.0
41.8	404	169	235	70.7
41.8	402	168	234	70.2
41.8	400	167	233	69.7
41.8	397	166	231	69.4
41.8	395	165	230	68.9
41.8	392	164	228	68.6
41.8	390	163	227	68.1
41.8	388	162	226	67.6
41.8	385	161	224	67.3
41.8	383	160	223	66.8
41.8	380	159	221	66.5
41.8	378	158	220	66.0
41.8	376	157	219	65.6
41.8	373	156	217	65.2
41.8	371	155	216	64.8
41.8	368	154	214	64.4
41.8	366	153	213	64.0
41.8	364	152	212	63.5
41.8	361	151	210	63.2
41.8	359	150	209	62.7
41.8	354	148	206	61.9
41.8	352	147	205	61.4
41.8	349	146	203	61.1
41.8	347	145	202	60.6
41.8	342	143	199	59.8
41.8	340	142	198	59.3
41.8	337	141	196	59.0
41.8	335	140	195	58.5
41.8	330	138	192	57.7
41.8	328	137	191	57.2
41.8	325	136	189	56.9
41.8	323	135	188	56.4
41.8	318	133	185	55.6
41.8	316	132	184	55.1
41.8	311	130	181	54.3
41.8	306	128	178	53.5
41.8	304	127	177	53.1
41.8	299	125	174	52.3
41.8	297	124	173	51.8
41.8	294	123	171	51.5
41.8	292	122	170	51.0
41.8	287	120	167	50.2
41.8	285	119	166	49.7
41.8	282	118	164	49.4
41.8	280	117	163	48.9
41.8	275	115	160	48.1
41.8	273	114	159	47.6
41.8	268	112	156	46.8
41.8	263	110	153	46.0
41.8	261	109	152	45.5
41.8	256	107	149	44.7
41.8	251	105	146	43.9
41.8	249	104	145	43.4
41.8	244	102	142	42.6
41.8	239	100	139	41.8
41.8	237	99	138	41.4
41.8	232	97	135	40.6
41.8	225	94	131	39.3
41.8	220	92	128	38.5
41.8	213	89	124	37.2
41.8	208	87	121	36.4
41.8	201	84	117	35.1
41.8	196	82	114	34.3
41.8	194	81	113	33.8
41.8	189	79	110	33.0
41.8	184	77	107	32.2
41.8	182	76	106	31.7
41.8	177	74	103	30.9
41.8	170	71	99	29.7
41.8	165	69	96	28.9
41.8	158	66	92	27.6
41.8	153	64	89	26.8
41.8	146	61	85	25.5
41.8	141	59	82	24.7
41.8	134	56	78	23.4
41.8	122	51	71	21.3
41.8	110	46	64	19.2
41.8	98	41	57	17.2
41.8	91	38	53	15.9
41.8	79	33	46	13.8
41.8	67	28	39	11.7
41.8	55	23	32	9.6
41.7	499	208	291	86.7
41.7	497	207	290	86.2
41.7	496	207	289	86.4
41.7	494	206	288	85.9
41.7	492	205	287	85.4
41.7	489	204	285	85.1
41.7	487	203	284	84.6
41.7	484	202	282	84.3
41.7	482	201	281	83.8
41.7	480	200	280	83.3
41.7	477	199	278	83.0
41.7	475	198	277	82.5
41.7	472	197	275	82.2
41.7	470	196	274	81.7
41.7	468	195	273	81.3
41.7	465	194	271	80.9
41.7	463	193	270	80.5
41.7	460	192	268	80.1
41.7	458	191	267	79.7
41.7	456	190	266	79.2
41.7	453	189	264	78.9
41.7	451	188	263	78.4
41.7	448	187	261	78.1
41.7	446	186	260	77.6
41.7	444	185	259	77.1
41.7	441	184	257	76.8
41.7	439	183	256	76.3
41.7	436	182	254	76.0
41.7	434	181	253	75.5
41.7	432	180	252	75.0
41.7	429	179	250	74.7
41.7	427	178	249	74.2
41.7	424	177	247	73.9
41.7	422	176	246	73.4
41.7	420	175	245	72.9
41.7	417	174	243	72.6
41.7	415	173	242	72.1
41.7	412	172	240	71.8
41.7	410	171	239	71.3
41.7	408	170	238	70.8
41.7	405	169	236	70.5

%E	M1	M2	DM	M*
41.7	403	168	235	70.0
41.7	398	166	232	69.2
41.7	396	165	231	68.8
41.7	393	164	229	68.4
41.7	391	163	228	68.0
41.7	386	161	225	67.2
41.7	384	160	224	66.7
41.7	381	159	222	66.4
41.7	379	158	221	65.9
41.7	374	156	218	65.1
41.7	372	155	217	64.6
41.7	369	154	215	64.3
41.7	367	153	214	63.8
41.7	362	151	211	63.0
41.7	360	150	210	62.5
41.7	357	149	208	62.2
41.7	355	148	207	61.7
41.7	350	146	204	60.9
41.7	348	145	203	60.4
41.7	345	144	201	60.1
41.7	343	143	200	59.6
41.7	338	141	197	58.8
41.7	336	140	196	58.3
41.7	333	139	194	58.0
41.7	331	138	193	57.5
41.7	326	136	190	56.7
41.7	324	135	189	56.3
41.7	321	134	187	55.9
41.7	319	133	186	55.5
41.7	314	131	183	54.7
41.7	312	130	182	54.2
41.7	309	129	180	53.9
41.7	307	128	179	53.4
41.7	302	126	176	52.6
41.7	300	125	175	52.1
41.7	295	123	172	51.3
41.7	290	121	169	50.5
41.7	288	120	168	50.0
41.7	283	118	165	49.2
41.7	278	116	162	48.4
41.7	276	115	161	47.9
41.7	271	113	158	47.1
41.7	266	111	155	46.3
41.7	264	110	154	45.8
41.7	259	108	151	45.0
41.7	254	106	148	44.2
41.7	252	105	147	43.8
41.7	247	103	144	43.0
41.7	242	101	141	42.2
41.7	240	100	140	41.7

%E	M1	M2	DM	M*
41.7	235	98	137	40.9
41.7	230	96	134	40.1
41.7	228	95	133	39.6
41.7	223	93	130	38.8
41.7	218	91	127	38.0
41.7	216	90	126	37.5
41.7	211	88	123	36.7
41.7	206	86	120	35.9
41.7	204	85	119	35.4
41.7	199	83	116	34.6
41.7	192	80	112	33.3
41.7	187	78	109	32.5
41.7	180	75	105	31.3
41.7	175	73	102	30.5
41.7	168	70	98	29.2
41.7	163	68	95	28.4
41.7	156	65	91	27.1
41.7	151	63	88	26.3
41.7	144	60	84	25.0
41.7	139	58	81	24.2
41.7	132	55	77	22.9
41.7	127	53	74	22.1
41.7	120	50	70	20.8
41.7	115	48	67	20.0
41.7	108	45	63	18.8
41.7	103	43	60	18.0
41.7	96	40	56	16.7
41.7	84	35	49	14.6
41.7	72	30	42	12.5
41.7	60	25	35	10.4
41.7	48	20	28	8.3
41.7	36	15	21	6.3
41.7	24	10	14	4.2
41.7	12	5	7	2.1
41.6	500	208	292	86.5
41.6	498	207	291	86.0
41.6	495	206	289	85.7
41.6	493	205	288	85.2
41.6	490	204	286	84.9
41.6	488	203	285	84.4
41.6	486	202	284	84.0
41.6	485	202	283	84.1
41.6	483	201	282	83.6
41.6	481	200	281	83.2
41.6	478	199	279	82.8
41.6	476	198	278	82.4
41.6	474	197	277	81.9
41.6	473	197	276	82.0
41.6	471	196	275	81.6
41.6	469	195	274	81.1

%E	M1	M2	DM	M*
41.6	466	194	272	80.8
41.6	464	193	271	80.3
41.6	462	192	270	79.8
41.6	461	192	269	80.0
41.6	459	191	268	79.5
41.6	457	190	267	79.0
41.6	454	189	265	78.7
41.6	452	188	264	78.2
41.6	450	187	263	77.7
41.6	449	187	262	77.9
41.6	447	186	261	77.4
41.6	445	185	260	76.9
41.6	442	184	258	76.6
41.6	440	183	257	76.1
41.6	438	182	256	75.6
41.6	437	182	255	75.8
41.6	435	181	254	75.3
41.6	433	180	253	74.8
41.6	430	179	251	74.5
41.6	428	178	250	74.0
41.6	425	177	248	73.7
41.6	423	176	247	73.2
41.6	421	175	246	72.7
41.6	418	174	244	72.4
41.6	416	173	243	71.9
41.6	413	172	241	71.6
41.6	411	171	240	71.1
41.6	409	170	239	70.7
41.6	406	169	237	70.3
41.6	404	168	236	69.9
41.6	401	167	234	69.5
41.6	399	166	233	69.1
41.6	397	165	232	68.6
41.6	394	164	230	68.3
41.6	392	163	229	67.8
41.6	389	162	227	67.5
41.6	387	161	226	67.0
41.6	385	160	225	66.5
41.6	382	159	223	66.2
41.6	380	158	222	65.7
41.6	377	157	220	65.4
41.6	375	156	219	64.9
41.6	373	155	218	64.4
41.6	370	154	216	64.1
41.6	368	153	215	63.6
41.6	365	152	213	63.3
41.6	363	151	212	62.8
41.6	361	150	211	62.3
41.6	358	149	209	62.0
41.6	356	148	208	61.5

%E	M1	M2	DM	M*
41.6	353	147	206	61.2
41.6	351	146	205	60.7
41.6	346	144	202	59.9
41.6	344	143	201	59.4
41.6	341	142	199	59.1
41.6	339	141	198	58.6
41.6	334	139	195	57.8
41.6	332	138	194	57.4
41.6	329	137	192	57.0
41.6	327	136	191	56.6
41.6	322	134	188	55.8
41.6	320	133	187	55.3
41.6	317	132	185	55.0
41.6	315	131	184	54.5
41.6	310	129	181	53.7
41.6	308	128	180	53.2
41.6	305	127	178	52.9
41.6	303	126	177	52.4
41.6	298	124	174	51.6
41.6	296	123	173	51.1
41.6	293	122	171	50.8
41.6	291	121	170	50.3
41.6	286	119	167	49.5
41.6	281	117	164	48.7
41.6	279	116	163	48.2
41.6	274	114	160	47.4
41.6	269	112	157	46.6
41.6	267	111	156	46.1
41.6	262	109	153	45.3
41.6	257	107	150	44.5
41.6	255	106	149	44.1
41.6	250	104	146	43.3
41.6	245	102	143	42.5
41.6	243	101	142	42.0
41.6	238	99	139	41.2
41.6	233	97	136	40.4
41.6	231	96	135	39.9
41.6	226	94	132	39.1
41.6	221	92	129	38.3
41.6	219	91	128	37.8
41.6	214	89	125	37.0
41.6	209	87	122	36.2
41.6	202	84	118	34.9
41.6	197	82	115	34.1
41.6	190	79	111	32.8
41.6	185	77	108	32.0
41.6	178	74	104	30.8
41.6	173	72	101	30.0
41.6	166	69	97	28.7
41.6	161	67	94	27.9

%E	M1	M2	DM	M*
41.6	154	64	90	26.6
41.6	149	62	87	25.8
41.6	137	57	80	23.7
41.6	125	52	73	21.6
41.6	113	47	66	19.5
41.6	101	42	59	17.5
41.6	89	37	52	15.4
41.6	77	32	45	13.3
41.5	499	207	292	85.9
41.5	496	206	290	85.6
41.5	494	205	289	85.1
41.5	492	204	288	84.6
41.5	491	204	287	84.8
41.5	489	203	286	84.3
41.5	487	202	285	83.8
41.5	484	201	283	83.5
41.5	482	200	282	83.0
41.5	480	199	281	82.5
41.5	479	199	280	82.7
41.5	477	198	279	82.2
41.5	475	197	278	81.7
41.5	472	196	276	81.4
41.5	470	195	275	80.9
41.5	468	194	274	80.4
41.5	467	194	273	80.6
41.5	465	193	272	80.1
41.5	463	192	271	79.6
41.5	460	191	269	79.3
41.5	458	190	268	78.8
41.5	455	189	266	78.5
41.5	453	188	265	78.0
41.5	451	187	264	77.5
41.5	448	186	262	77.2
41.5	446	185	261	76.7
41.5	443	184	259	76.4
41.5	441	183	258	75.9
41.5	439	182	257	75.5
41.5	436	181	255	75.1
41.5	434	180	254	74.7
41.5	431	179	252	74.3
41.5	429	178	251	73.9
41.5	427	177	250	73.4
41.5	426	177	249	73.5
41.5	424	176	248	73.1
41.5	422	175	247	72.6
41.5	419	174	245	72.3
41.5	417	173	244	71.8
41.5	414	172	242	71.5
41.5	412	171	241	71.0
41.5	410	170	240	70.5

%E	M1	M2	DM	M*
41.5	407	169	238	70.2
41.5	405	168	237	69.7
41.5	402	167	235	69.4
41.5	400	166	234	68.9
41.5	398	165	233	68.4
41.5	395	164	231	68.1
41.5	393	163	230	67.6
41.5	390	162	228	67.3
41.5	388	161	227	66.8
41.5	386	160	226	66.3
41.5	383	159	224	66.0
41.5	381	158	223	65.5
41.5	378	157	221	65.2
41.5	376	156	220	64.7
41.5	371	154	217	63.9
41.5	369	153	216	63.4
41.5	366	152	214	63.1
41.5	364	151	213	62.6
41.5	359	149	210	61.8
41.5	357	148	209	61.4
41.5	354	147	207	61.0
41.5	352	146	206	60.6
41.5	349	145	204	60.2
41.5	347	144	203	59.8
41.5	342	142	200	59.0
41.5	340	141	199	58.5
41.5	337	140	197	58.2
41.5	335	139	196	57.7
41.5	330	137	193	56.9
41.5	328	136	192	56.4
41.5	325	135	190	56.1
41.5	323	134	189	55.6
41.5	318	132	186	54.8
41.5	316	131	185	54.3
41.5	313	130	183	54.0
41.5	311	129	182	53.5
41.5	306	127	179	52.7
41.5	301	125	176	51.9
41.5	299	124	175	51.4
41.5	294	122	172	50.6
41.5	289	120	169	49.8
41.5	287	119	168	49.3
41.5	284	118	166	49.0
41.5	282	117	165	48.5
41.5	277	115	162	47.7
41.5	275	114	161	47.3
41.5	272	113	159	46.9
41.5	270	112	158	46.5
41.5	265	110	155	45.7
41.5	260	108	152	44.9

%E	M1	M2	DM	M*
41.5	258	107	151	44.4
41.5	253	105	148	43.6
41.5	248	103	145	42.8
41.5	246	102	144	42.3
41.5	241	100	141	41.5
41.5	236	98	138	40.7
41.5	234	97	137	40.2
41.5	229	95	134	39.4
41.5	224	93	131	38.6
41.5	217	90	127	37.3
41.5	212	88	124	36.5
41.5	207	86	121	35.7
41.5	205	85	120	35.2
41.5	200	83	117	34.4
41.5	195	81	114	33.6
41.5	193	80	113	33.2
41.5	188	78	110	32.4
41.5	183	76	107	31.6
41.5	176	73	103	30.3
41.5	171	71	100	29.5
41.5	164	68	96	28.2
41.5	159	66	93	27.4
41.5	147	61	86	25.3
41.5	142	59	83	24.5
41.5	135	56	79	23.2
41.5	130	54	76	22.4
41.5	123	51	72	21.1
41.5	118	49	69	20.3
41.5	106	44	62	18.3
41.5	94	39	55	16.2
41.5	82	34	48	14.1
41.5	65	27	38	11.2
41.5	53	22	31	9.1
41.5	41	17	24	7.0
41.4	500	207	293	85.7
41.4	498	206	292	85.2
41.4	497	206	291	85.4
41.4	495	205	290	84.9
41.4	493	204	289	84.4
41.4	490	203	287	84.1
41.4	488	202	286	83.6
41.4	486	201	285	83.1
41.4	485	201	284	83.3
41.4	483	200	283	82.8
41.4	481	199	282	82.3
41.4	478	198	280	82.0
41.4	476	197	279	81.5
41.4	474	196	278	81.0
41.4	473	196	277	81.2
41.4	471	195	276	80.7

%E	M1	M2	DM	M*
41.4	469	194	275	80.2
41.4	466	193	273	79.9
41.4	464	192	272	79.4
41.4	461	191	270	79.1
41.4	459	190	269	78.6
41.4	457	189	268	78.2
41.4	456	189	267	78.3
41.4	454	188	266	77.9
41.4	452	187	265	77.4
41.4	449	186	263	77.1
41.4	447	185	262	76.6
41.4	444	184	260	76.3
41.4	442	183	259	75.8
41.4	440	182	258	75.3
41.4	437	181	256	75.0
41.4	435	180	255	74.5
41.4	432	179	253	74.2
41.4	430	178	252	73.7
41.4	428	177	251	73.2
41.4	425	176	249	72.9
41.4	423	175	248	72.4
41.4	420	174	246	72.1
41.4	418	173	245	71.6
41.4	415	172	243	71.3
41.4	413	171	242	70.8
41.4	411	170	241	70.3
41.4	408	169	239	70.0
41.4	406	168	238	69.5
41.4	403	167	236	69.2
41.4	401	166	235	68.7
41.4	399	165	234	68.2
41.4	396	164	232	67.9
41.4	394	163	231	67.4
41.4	391	162	229	67.1
41.4	389	161	228	66.6
41.4	384	159	225	65.8
41.4	382	158	224	65.4
41.4	379	157	222	65.0
41.4	377	156	221	64.6
41.4	374	155	219	64.2
41.4	372	154	218	63.8
41.4	370	153	217	63.3
41.4	367	152	215	63.0
41.4	365	151	214	62.5
41.4	362	150	212	62.2
41.4	360	149	211	61.7
41.4	355	147	208	60.9
41.4	353	146	207	60.4
41.4	350	145	205	60.1
41.4	348	144	204	59.6

%E	M1	M2	DM	M*
41.4	345	143	202	59.3
41.4	343	142	201	58.8
41.4	338	140	198	58.0
41.4	336	139	197	57.5
41.4	333	138	195	57.2
41.4	331	137	194	56.7
41.4	326	135	191	55.9
41.4	324	134	190	55.4
41.4	321	133	188	55.1
41.4	319	132	187	54.6
41.4	314	130	184	53.8
41.4	309	128	181	53.0
41.4	307	127	180	52.5
41.4	304	126	178	52.2
41.4	302	125	177	51.7
41.4	297	123	174	50.9
41.4	295	122	173	50.5
41.4	292	121	171	50.1
41.4	290	120	170	49.7
41.4	285	118	167	48.9
41.4	280	116	164	48.1
41.4	278	115	163	47.6
41.4	273	113	160	46.8
41.4	268	111	157	46.0
41.4	266	110	156	45.5
41.4	263	109	154	45.2
41.4	261	108	153	44.7
41.4	256	106	150	43.9
41.4	251	104	147	43.1
41.4	249	103	146	42.6
41.4	244	101	143	41.8
41.4	239	99	140	41.0
41.4	237	98	139	40.5
41.4	232	96	136	39.7
41.4	227	94	133	38.9
41.4	222	92	130	38.1
41.4	220	91	129	37.6
41.4	215	89	126	36.8
41.4	210	87	123	36.0
41.4	203	84	119	34.8
41.4	198	82	116	34.0
41.4	191	79	112	32.7
41.4	186	77	109	31.9
41.4	181	75	106	31.1
41.4	174	72	102	29.8
41.4	169	70	99	29.0
41.4	162	67	95	27.7
41.4	157	65	92	26.9
41.4	152	63	89	26.1
41.4	145	60	85	24.8

%E	M1	M2	DM	M*
41.4	140	58	82	24.0
41.4	133	55	78	22.7
41.4	128	53	75	21.9
41.4	116	48	68	19.9
41.4	111	46	65	19.1
41.4	99	41	58	17.0
41.4	87	36	51	14.9
41.4	70	29	41	12.0
41.4	58	24	34	9.9
41.4	29	12	17	5.0
41.3	499	206	293	85.0
41.3	496	205	291	84.7
41.3	494	204	290	84.2
41.3	492	203	289	83.8
41.3	491	203	288	83.9
41.3	489	202	287	83.4
41.3	487	201	286	83.0
41.3	484	200	284	82.6
41.3	482	199	283	82.2
41.3	480	198	282	81.7
41.3	479	198	281	81.8
41.3	477	197	280	81.4
41.3	475	196	279	80.9
41.3	472	195	277	80.6
41.3	470	194	276	80.1
41.3	467	193	274	79.8
41.3	465	192	273	79.3
41.3	463	191	272	78.8
41.3	462	191	271	79.0
41.3	460	190	270	78.5
41.3	458	189	269	78.0
41.3	455	188	267	77.7
41.3	453	187	266	77.2
41.3	450	186	264	76.9
41.3	448	185	263	76.4
41.3	446	184	262	75.9
41.3	445	184	261	76.1
41.3	443	183	260	75.6
41.3	441	182	259	75.1
41.3	438	181	257	74.8
41.3	436	180	256	74.3
41.3	433	179	254	74.0
41.3	431	178	253	73.5
41.3	429	177	252	73.0
41.3	426	176	250	72.7
41.3	424	175	249	72.2
41.3	421	174	247	71.9
41.3	419	173	246	71.4
41.3	416	172	244	71.1
41.3	414	171	243	70.6

%E	M1	M2	DM	M*
41.3	412	170	242	70.1
41.3	409	169	240	69.8
41.3	407	168	239	69.3
41.3	404	167	237	69.0
41.3	402	166	236	68.5
41.3	400	165	235	68.1
41.3	397	164	233	67.7
41.3	395	163	232	67.3
41.3	392	162	230	66.9
41.3	390	161	229	66.5
41.3	387	160	227	66.1
41.3	385	159	226	65.7
41.3	383	158	225	65.2
41.3	380	157	223	64.9
41.3	378	156	222	64.4
41.3	375	155	220	64.1
41.3	373	154	219	63.6
41.3	368	152	216	62.8
41.3	366	151	215	62.3
41.3	363	150	213	62.0
41.3	361	149	212	61.5
41.3	358	148	210	61.2
41.3	356	147	209	60.7
41.3	351	145	206	59.9
41.3	349	144	205	59.4
41.3	346	143	203	59.1
41.3	344	142	202	58.6
41.3	341	141	200	58.3
41.3	339	140	199	57.8
41.3	334	138	196	57.0
41.3	332	137	195	56.5
41.3	329	136	193	56.2
41.3	327	135	192	55.7
41.3	322	133	189	54.9
41.3	320	132	188	54.4
41.3	317	131	186	54.1
41.3	315	130	185	53.7
41.3	312	129	183	53.3
41.3	310	128	182	52.9
41.3	305	126	179	52.1
41.3	303	125	178	51.6
41.3	300	124	176	51.3
41.3	298	123	175	50.8
41.3	293	121	172	50.0
41.3	288	119	169	49.2
41.3	286	118	168	48.7
41.3	283	117	166	48.4
41.3	281	116	165	47.9
41.3	276	114	162	47.1
41.3	271	112	159	46.3

%E	M1	M2	DM	M*
41.3	269	111	158	45.8
41.3	264	109	155	45.0
41.3	259	107	152	44.2
41.3	254	105	149	43.4
41.3	252	104	148	42.9
41.3	247	102	145	42.1
41.3	242	100	142	41.3
41.3	240	99	141	40.8
41.3	235	97	138	40.0
41.3	230	95	135	39.2
41.3	225	93	132	38.4
41.3	223	92	131	38.0
41.3	218	90	128	37.2
41.3	213	88	125	36.4
41.3	208	86	122	35.6
41.3	206	85	121	35.1
41.3	201	83	118	34.3
41.3	196	81	115	33.5
41.3	189	78	111	32.2
41.3	184	76	108	31.4
41.3	179	74	105	30.6
41.3	172	71	101	29.3
41.3	167	69	98	28.5
41.3	160	66	94	27.2
41.3	155	64	91	26.4
41.3	150	62	88	25.6
41.3	143	59	84	24.3
41.3	138	57	81	23.5
41.3	126	52	74	21.5
41.3	121	50	71	20.7
41.3	109	45	64	18.6
41.3	104	43	61	17.8
41.3	92	38	54	15.7
41.3	80	33	47	13.6
41.3	75	31	44	12.8
41.3	63	26	37	10.7
41.3	46	19	27	7.8
41.2	500	206	294	84.9
41.2	498	205	293	84.4
41.2	497	205	292	84.6
41.2	495	204	291	84.1
41.2	493	203	290	83.6
41.2	490	202	288	83.3
41.2	488	201	287	82.8
41.2	486	200	286	82.3
41.2	485	200	285	82.5
41.2	483	199	284	82.0
41.2	481	198	283	81.5
41.2	478	197	281	81.2
41.2	476	196	280	80.7

%E	M1	M2	DM	M*
41.2	473	195	278	80.4
41.2	471	194	277	79.9
41.2	469	193	276	79.4
41.2	468	193	275	79.6
41.2	466	192	274	79.1
41.2	464	191	273	78.6
41.2	461	190	271	78.3
41.2	459	189	270	77.8
41.2	456	188	268	77.5
41.2	454	187	267	77.0
41.2	452	186	266	76.5
41.2	451	186	265	76.7
41.2	449	185	264	76.2
41.2	447	184	263	75.7
41.2	444	183	261	75.4
41.2	442	182	260	74.9
41.2	439	181	258	74.6
41.2	437	180	257	74.1
41.2	434	179	255	73.8
41.2	432	178	254	73.3
41.2	430	177	253	72.9
41.2	427	176	251	72.5
41.2	425	175	250	72.1
41.2	422	174	248	71.7
41.2	420	173	247	71.3
41.2	417	172	245	70.9
41.2	415	171	244	70.5
41.2	413	170	243	70.0
41.2	410	169	241	69.7
41.2	408	168	240	69.2
41.2	405	167	238	68.9
41.2	403	166	237	68.4
41.2	398	164	234	67.6
41.2	396	163	233	67.1
41.2	393	162	231	66.8
41.2	391	161	230	66.3
41.2	388	160	228	66.0
41.2	386	159	227	65.5
41.2	381	157	224	64.7
41.2	379	156	223	64.2
41.2	376	155	221	63.9
41.2	374	154	220	63.4
41.2	371	153	218	63.1
41.2	369	152	217	62.6
41.2	364	150	214	61.8
41.2	362	149	213	61.3
41.2	359	148	211	61.0
41.2	357	147	210	60.5
41.2	354	146	208	60.2
41.2	352	145	207	59.7

%E	M1	M2	DM	M*
41.2	347	143	204	58.9
41.2	345	142	203	58.4
41.2	342	141	201	58.1
41.2	340	140	200	57.6
41.2	337	139	198	57.3
41.2	335	138	197	56.8
41.2	330	136	194	56.0
41.2	328	135	193	55.6
41.2	325	134	191	55.2
41.2	323	133	190	54.8
41.2	318	131	187	54.0
41.2	313	129	184	53.2
41.2	311	128	183	52.7
41.2	308	127	181	52.4
41.2	306	126	180	51.9
41.2	301	124	177	51.1
41.2	296	122	174	50.3
41.2	294	121	173	49.8
41.2	291	120	171	49.5
41.2	289	119	170	49.0
41.2	284	117	167	48.2
41.2	279	115	164	47.4
41.2	277	114	163	46.9
41.2	274	113	161	46.6
41.2	272	112	160	46.1
41.2	267	110	157	45.3
41.2	262	108	154	44.5
41.2	260	107	153	44.0
41.2	257	106	151	43.7
41.2	255	105	150	43.2
41.2	250	103	147	42.4
41.2	245	101	144	41.6
41.2	243	100	143	41.2
41.2	238	98	140	40.4
41.2	233	96	137	39.6
41.2	228	94	134	38.8
41.2	226	93	133	38.3
41.2	221	91	130	37.5
41.2	216	89	127	36.7
41.2	211	87	124	35.9
41.2	204	84	120	34.6
41.2	199	82	117	33.8
41.2	194	80	114	33.0
41.2	187	77	110	31.7
41.2	182	75	107	30.9
41.2	177	73	104	30.1
41.2	170	70	100	28.8
41.2	165	68	97	28.0
41.2	153	63	90	25.9
41.2	148	61	87	25.1

%E	M1	M2	DM	M*
41.2	136	56	80	23.1
41.2	131	54	77	22.3
41.2	119	49	70	20.2
41.2	114	47	67	19.4
41.2	102	42	60	17.3
41.2	97	40	57	16.5
41.2	85	35	50	14.4
41.2	68	28	40	11.5
41.2	51	21	30	8.6
41.2	34	14	20	5.8
41.2	17	7	10	2.9
41.1	499	205	294	84.2
41.1	496	204	292	83.9
41.1	494	203	291	83.4
41.1	492	202	290	82.9
41.1	491	202	289	83.1
41.1	489	201	288	82.6
41.1	487	200	287	82.1
41.1	484	199	285	81.8
41.1	482	198	284	81.3
41.1	479	197	282	81.0
41.1	477	196	281	80.5
41.1	475	195	280	80.1
41.1	474	195	279	80.2
41.1	472	194	278	79.7
41.1	470	193	277	79.3
41.1	467	192	275	78.9
41.1	465	191	274	78.5
41.1	462	190	272	78.1
41.1	460	189	271	77.7
41.1	457	188	269	77.3
41.1	455	187	268	76.9
41.1	453	186	267	76.4
41.1	450	185	265	76.1
41.1	448	184	264	75.6
41.1	445	183	262	75.3
41.1	443	182	261	74.8
41.1	440	181	259	74.5
41.1	438	180	258	74.0
41.1	436	179	257	73.5
41.1	435	179	256	73.7
41.1	433	178	255	73.2
41.1	431	177	254	72.7
41.1	428	176	252	72.4
41.1	426	175	251	71.9
41.1	423	174	249	71.6
41.1	421	173	248	71.1
41.1	419	172	247	70.6
41.1	418	172	246	70.8
41.1	416	171	245	70.3

%E	M1	M2	DM	M*
41.1	414	170	244	69.8
41.1	411	169	242	69.5
41.1	409	168	241	69.0
41.1	406	167	239	68.7
41.1	404	166	238	68.2
41.1	401	165	236	67.9
41.1	399	164	235	67.4
41.1	397	163	234	66.9
41.1	394	162	232	66.6
41.1	392	161	231	66.1
41.1	389	160	229	65.8
41.1	387	159	228	65.3
41.1	384	158	226	65.0
41.1	382	157	225	64.5
41.1	380	156	224	64.0
41.1	377	155	222	63.7
41.1	375	154	221	63.2
41.1	372	153	219	62.9
41.1	370	152	218	62.4
41.1	367	151	216	62.1
41.1	365	150	215	61.6
41.1	360	148	212	60.8
41.1	358	147	211	60.4
41.1	355	146	209	60.0
41.1	353	145	208	59.6
41.1	350	144	206	59.2
41.1	348	143	205	58.8
41.1	343	141	202	58.0
41.1	341	140	201	57.5
41.1	338	139	199	57.2
41.1	336	138	198	56.7
41.1	333	137	196	56.4
41.1	331	136	195	55.9
41.1	326	134	192	55.1
41.1	321	132	189	54.3
41.1	319	131	188	53.8
41.1	316	130	186	53.5
41.1	314	129	185	53.0
41.1	309	127	182	52.2
41.1	304	125	179	51.4
41.1	302	124	178	50.9
41.1	299	123	176	50.6
41.1	297	122	175	50.1
41.1	292	120	172	49.3
41.1	287	118	169	48.5
41.1	285	117	168	48.0
41.1	282	116	166	47.7
41.1	280	115	165	47.2
41.1	275	113	162	46.4
41.1	270	111	159	45.6

%E	M1	M2	DM	M*
41.1	265	109	156	44.8
41.1	263	108	155	44.3
41.1	258	106	152	43.6
41.1	253	104	149	42.8
41.1	248	102	146	42.0
41.1	246	101	145	41.5
41.1	241	99	142	40.7
41.1	236	97	139	39.9
41.1	231	95	136	39.1
41.1	224	92	132	37.8
41.1	219	90	129	37.0
41.1	214	88	126	36.2
41.1	209	86	123	35.4
41.1	207	85	122	34.9
41.1	202	83	119	34.1
41.1	197	81	116	33.3
41.1	192	79	113	32.5
41.1	190	78	112	32.0
41.1	185	76	109	31.2
41.1	180	74	106	30.4
41.1	175	72	103	29.6
41.1	168	69	99	28.3
41.1	163	67	96	27.5
41.1	158	65	93	26.7
41.1	151	62	89	25.5
41.1	146	60	86	24.7
41.1	141	58	83	23.9
41.1	129	53	76	21.8
41.1	124	51	73	21.0
41.1	112	46	66	18.9
41.1	107	44	63	18.1
41.1	95	39	56	16.0
41.1	90	37	53	15.2
41.1	73	30	43	12.3
41.1	56	23	33	9.4
41.0	500	205	295	84.0
41.0	498	204	294	83.6
41.0	497	204	293	83.7
41.0	495	203	292	83.3
41.0	493	202	291	82.8
41.0	490	201	289	82.5
41.0	488	200	288	82.0
41.0	485	199	286	81.7
41.0	483	198	285	81.2
41.0	481	197	284	80.7
41.0	480	197	283	80.9
41.0	478	196	282	80.4
41.0	476	195	281	79.9
41.0	473	194	279	79.6
41.0	471	193	278	79.1

%E	M1	M2	DM	M*
41.0	468	192	276	78.8
41.0	466	191	275	78.3
41.0	463	190	273	78.0
41.0	461	189	272	77.5
41.0	459	188	271	77.0
41.0	458	188	270	77.2
41.0	456	187	269	76.7
41.0	454	186	268	76.2
41.0	451	185	266	75.9
41.0	449	184	265	75.4
41.0	446	183	263	75.1
41.0	444	182	262	74.6
41.0	442	181	261	74.1
41.0	441	181	260	74.3
41.0	439	180	259	73.8
41.0	437	179	258	73.3
41.0	434	178	256	73.0
41.0	432	177	255	72.5
41.0	429	176	253	72.2
41.0	427	175	252	71.7
41.0	424	174	250	71.4
41.0	422	173	249	70.9
41.0	420	172	248	70.4
41.0	417	171	246	70.1
41.0	415	170	245	69.6
41.0	412	169	243	69.3
41.0	410	168	242	68.8
41.0	407	167	240	68.5
41.0	405	166	239	68.0
41.0	402	165	237	67.7
41.0	400	164	236	67.2
41.0	398	163	235	66.8
41.0	395	162	233	66.4
41.0	393	161	232	66.0
41.0	390	160	230	65.6
41.0	388	159	229	65.2
41.0	385	158	227	64.8
41.0	383	157	226	64.4
41.0	378	155	223	63.6
41.0	376	154	222	63.1
41.0	373	153	220	62.8
41.0	371	152	219	62.3
41.0	368	151	217	62.0
41.0	366	150	216	61.5
41.0	363	149	214	61.2
41.0	361	148	213	60.7
41.0	356	146	210	59.9
41.0	354	145	209	59.4
41.0	351	144	207	59.1
41.0	349	143	206	58.6

%E	M1	M2	DM	M*
41.0	346	142	204	58.3
41.0	344	141	203	57.8
41.0	339	139	200	57.0
41.0	337	138	199	56.5
41.0	334	137	197	56.2
41.0	332	136	196	55.7
41.0	329	135	194	55.4
41.0	327	134	193	54.9
41.0	324	133	191	54.6
41.0	322	132	190	54.1
41.0	317	130	187	53.3
41.0	315	129	186	52.8
41.0	312	128	184	52.5
41.0	310	127	183	52.0
41.0	307	126	181	51.7
41.0	305	125	180	51.2
41.0	300	123	177	50.4
41.0	295	121	174	49.6
41.0	293	120	173	49.1
41.0	290	119	171	48.8
41.0	288	118	170	48.3
41.0	283	116	167	47.5
41.0	278	114	164	46.7
41.0	273	112	161	45.9
41.0	271	111	160	45.5
41.0	268	110	158	45.1
41.0	266	109	157	44.7
41.0	261	107	154	43.9
41.0	256	105	151	43.1
41.0	251	103	148	42.3
41.0	249	102	147	41.8
41.0	244	100	144	41.0
41.0	239	98	141	40.2
41.0	234	96	138	39.4
41.0	229	94	135	38.6
41.0	227	93	134	38.1
41.0	222	91	131	37.3
41.0	217	89	128	36.5
41.0	212	87	125	35.7
41.0	210	86	124	35.2
41.0	205	84	121	34.4
41.0	200	82	118	33.6
41.0	195	80	115	32.8
41.0	188	77	111	31.5
41.0	183	75	108	30.7
41.0	178	73	105	29.9
41.0	173	71	102	29.1
41.0	166	68	98	27.9
41.0	161	66	95	27.1
41.0	156	64	92	26.3

%E	M1	M2	DM	M*
41.0	144	59	85	24.2
41.0	139	57	82	23.4
41.0	134	55	79	22.6
41.0	122	50	72	20.5
41.0	117	48	69	19.7
41.0	105	43	62	17.6
41.0	100	41	59	16.8
41.0	83	34	49	13.9
41.0	78	32	46	13.1
41.0	61	25	36	10.2
41.0	39	16	23	6.6
40.9	499	204	295	83.4
40.9	496	203	293	83.1
40.9	494	202	292	82.6
40.9	492	201	291	82.1
40.9	491	201	290	82.3
40.9	489	200	289	81.8
40.9	487	199	288	81.3
40.9	486	199	287	81.5
40.9	484	198	286	81.0
40.9	482	197	285	80.5
40.9	479	196	283	80.2
40.9	477	195	282	79.7
40.9	474	194	280	79.4
40.9	472	193	279	78.9
40.9	470	192	278	78.4
40.9	469	192	277	78.6
40.9	467	191	276	78.1
40.9	465	190	275	77.6
40.9	464	190	274	77.8
40.9	462	189	273	77.3
40.9	460	188	272	76.8
40.9	457	187	270	76.5
40.9	455	186	269	76.0
40.9	452	185	267	75.7
40.9	450	184	266	75.2
40.9	447	183	264	74.9
40.9	445	182	263	74.4
40.9	443	181	262	74.0
40.9	440	180	260	73.6
40.9	438	179	259	73.2
40.9	435	178	257	72.8
40.9	433	177	256	72.4
40.9	430	176	254	72.0
40.9	428	175	253	71.6
40.9	425	174	251	71.2
40.9	423	173	250	70.8
40.9	421	172	249	70.3
40.9	418	171	247	70.0
40.9	416	170	246	69.5

%E	M1	M2	DM	M*
40.9	413	169	244	69.2
40.9	411	168	243	68.7
40.9	408	167	241	68.4
40.9	406	166	240	67.9
40.9	403	165	238	67.6
40.9	401	164	237	67.1
40.9	399	163	236	66.6
40.9	396	162	234	66.3
40.9	394	161	233	65.8
40.9	391	160	231	65.5
40.9	389	159	230	65.0
40.9	386	158	228	64.7
40.9	384	157	227	64.2
40.9	381	156	225	63.9
40.9	379	155	224	63.4
40.9	374	153	221	62.6
40.9	372	152	220	62.1
40.9	369	151	218	61.8
40.9	367	150	217	61.3
40.9	364	149	215	61.0
40.9	362	148	214	60.5
40.9	359	147	212	60.2
40.9	357	146	211	59.7
40.9	352	144	208	58.9
40.9	350	143	207	58.4
40.9	347	142	205	58.1
40.9	345	141	204	57.6
40.9	342	140	202	57.3
40.9	340	139	201	56.8
40.9	335	137	198	56.0
40.9	330	135	195	55.2
40.9	328	134	194	54.7
40.9	325	133	192	54.4
40.9	323	132	191	53.9
40.9	320	131	189	53.6
40.9	318	130	188	53.1
40.9	313	128	185	52.3
40.9	308	126	182	51.5
40.9	306	125	181	51.1
40.9	303	124	179	50.7
40.9	301	123	178	50.3
40.9	298	122	176	49.9
40.9	296	121	175	49.5
40.9	291	119	172	48.7
40.9	286	117	169	47.9
40.9	281	115	166	47.1
40.9	279	114	165	46.6
40.9	276	113	163	46.3
40.9	274	112	162	45.8
40.9	269	110	159	45.0

%E	M1	M2	DM	M*
40.9	264	108	156	44.2
40.9	259	106	153	43.4
40.9	257	105	152	42.9
40.9	254	104	150	42.6
40.9	252	103	149	42.1
40.9	247	101	146	41.3
40.9	242	99	143	40.5
40.9	237	97	140	39.7
40.9	235	96	139	39.2
40.9	232	95	137	38.9
40.9	230	94	136	38.4
40.9	225	92	133	37.6
40.9	220	90	130	36.8
40.9	215	88	127	36.0
40.9	208	85	123	34.7
40.9	203	83	120	33.9
40.9	198	81	117	33.1
40.9	193	79	114	32.3
40.9	186	76	110	31.1
40.9	181	74	107	30.3
40.9	176	72	104	29.5
40.9	171	70	101	28.7
40.9	164	67	97	27.4
40.9	159	65	94	26.6
40.9	154	63	91	25.8
40.9	149	61	88	25.0
40.9	137	56	81	22.9
40.9	132	54	78	22.1
40.9	127	52	75	21.3
40.9	115	47	68	19.2
40.9	110	45	65	18.4
40.9	93	38	55	15.5
40.9	88	36	52	14.7
40.9	66	27	39	11.0
40.9	44	18	26	7.4
40.9	22	9	13	3.7
40.8	500	204	296	83.2
40.8	498	203	295	82.7
40.8	497	203	294	82.9
40.8	495	202	293	82.4
40.8	493	201	292	81.9
40.8	490	200	290	81.6
40.8	488	199	289	81.1
40.8	485	198	287	80.8
40.8	483	197	286	80.3
40.8	480	196	284	80.0
40.8	478	195	283	79.6
40.8	476	194	282	79.1
40.8	475	194	281	79.2
40.8	473	193	280	78.8

%E	M1	M2	DM	M*
40.8	471	192	279	78.3
40.8	468	191	277	78.0
40.8	466	190	276	77.5
40.8	463	189	274	77.2
40.8	461	188	273	76.7
40.8	458	187	271	76.4
40.8	456	186	270	75.9
40.8	453	185	268	75.6
40.8	451	184	267	75.1
40.8	449	183	266	74.6
40.8	448	183	265	74.8
40.8	446	182	264	74.3
40.8	444	181	263	73.8
40.8	441	180	261	73.5
40.8	439	179	260	73.0
40.8	436	178	258	72.7
40.8	434	177	257	72.2
40.8	431	176	255	71.9
40.8	429	175	254	71.4
40.8	426	174	252	71.1
40.8	424	173	251	70.6
40.8	422	172	250	70.1
40.8	419	171	248	69.8
40.8	417	170	247	69.3
40.8	414	169	245	69.0
40.8	412	168	244	68.5
40.8	409	167	242	68.2
40.8	407	166	241	67.7
40.8	404	165	239	67.4
40.8	402	164	238	66.9
40.8	400	163	237	66.4
40.8	397	162	235	66.1
40.8	395	161	234	65.6
40.8	392	160	232	65.3
40.8	390	159	231	64.8
40.8	387	158	229	64.5
40.8	385	157	228	64.0
40.8	382	156	226	63.7
40.8	380	155	225	63.2
40.8	377	154	223	62.9
40.8	375	153	222	62.4
40.8	373	152	221	61.9
40.8	370	151	219	61.6
40.8	368	150	218	61.1
40.8	365	149	216	60.8
40.8	363	148	215	60.3
40.8	360	147	213	60.0
40.8	358	146	212	59.5
40.8	355	145	210	59.2
40.8	353	144	209	58.7

%E	M1	M2	DM	M*
40.8	348	142	206	57.9
40.8	346	141	205	57.5
40.8	343	140	203	57.1
40.8	341	139	202	56.7
40.8	338	138	200	56.3
40.8	336	137	199	55.9
40.8	333	136	197	55.5
40.8	331	135	196	55.1
40.8	326	133	193	54.3
40.8	321	131	190	53.5
40.8	319	130	189	53.0
40.8	316	129	187	52.7
40.8	314	128	186	52.2
40.8	311	127	184	51.9
40.8	309	126	183	51.4
40.8	304	124	180	50.6
40.8	299	122	177	49.8
40.8	294	120	174	49.0
40.8	292	119	173	48.5
40.8	289	118	171	48.2
40.8	287	117	170	47.7
40.8	284	116	168	47.4
40.8	282	115	167	46.9
40.8	277	113	164	46.1
40.8	272	111	161	45.3
40.8	267	109	158	44.5
40.8	265	108	157	44.0
40.8	262	107	155	43.7
40.8	260	106	154	43.2
40.8	255	104	151	42.4
40.8	250	102	148	41.6
40.8	245	100	145	40.8
40.8	240	98	142	40.0
40.8	238	97	141	39.5
40.8	233	95	138	38.7
40.8	228	93	135	37.9
40.8	223	91	132	37.1
40.8	218	89	129	36.3
40.8	213	87	126	35.5
40.8	211	86	125	35.1
40.8	206	84	122	34.3
40.8	201	82	119	33.5
40.8	196	80	116	32.7
40.8	191	78	113	31.9
40.8	184	75	109	30.6
40.8	179	73	106	29.8
40.8	174	71	103	29.0
40.8	169	69	100	28.2
40.8	157	64	93	26.1
40.8	152	62	90	25.3

%E	M1	M2	DM	M*
40.8	147	60	87	24.5
40.8	142	58	84	23.7
40.8	130	53	77	21.6
40.8	125	51	74	20.8
40.8	120	49	71	20.0
40.8	103	42	61	17.1
40.8	98	40	58	16.3
40.8	76	31	45	12.6
40.8	71	29	42	11.8
40.8	49	20	29	8.2
40.7	499	203	296	82.6
40.7	496	202	294	82.3
40.7	494	201	293	81.8
40.7	492	200	292	81.3
40.7	491	200	291	81.5
40.7	489	199	290	81.0
40.7	487	198	289	80.5
40.7	486	198	288	80.7
40.7	484	197	287	80.2
40.7	482	196	286	79.7
40.7	481	196	285	79.9
40.7	479	195	284	79.4
40.7	477	194	283	78.9
40.7	474	193	281	78.6
40.7	472	192	280	78.1
40.7	469	191	278	77.8
40.7	467	190	277	77.3
40.7	464	189	275	77.0
40.7	462	188	274	76.5
40.7	460	187	273	76.0
40.7	459	187	272	76.2
40.7	457	186	271	75.7
40.7	455	185	270	75.2
40.7	454	185	269	75.4
40.7	452	184	268	74.9
40.7	450	183	267	74.4
40.7	447	182	265	74.1
40.7	445	181	264	73.6
40.7	442	180	262	73.3
40.7	440	179	261	72.8
40.7	437	178	259	72.5
40.7	435	177	258	72.0
40.7	432	176	256	71.7
40.7	430	175	255	71.2
40.7	428	174	254	70.7
40.7	427	174	253	70.9
40.7	425	173	252	70.4
40.7	423	172	251	69.9
40.7	420	171	249	69.6
40.7	418	170	248	69.1

%E	M1	M2	DM	M*
40.7	415	169	246	68.8
40.7	413	168	245	68.3
40.7	410	167	243	68.0
40.7	408	166	242	67.5
40.7	405	165	240	67.2
40.7	403	164	239	66.7
40.7	398	162	236	65.9
40.7	396	161	235	65.5
40.7	393	160	233	65.1
40.7	391	159	232	64.7
40.7	388	158	230	64.3
40.7	386	157	229	63.9
40.7	383	156	227	63.5
40.7	381	155	226	63.1
40.7	378	154	224	62.7
40.7	376	153	223	62.3
40.7	371	151	220	61.5
40.7	369	150	219	61.0
40.7	366	149	217	60.7
40.7	364	148	216	60.2
40.7	361	147	214	59.9
40.7	359	146	213	59.4
40.7	356	145	211	59.1
40.7	354	144	210	58.6
40.7	351	143	208	58.3
40.7	349	142	207	57.8
40.7	344	140	204	57.0
40.7	339	138	201	56.2
40.7	337	137	200	55.7
40.7	334	136	198	55.4
40.7	332	135	197	54.9
40.7	329	134	195	54.6
40.7	327	133	194	54.1
40.7	324	132	192	53.8
40.7	322	131	191	53.3
40.7	317	129	188	52.5
40.7	312	127	185	51.7
40.7	307	125	182	50.9
40.7	305	124	181	50.4
40.7	302	123	179	50.1
40.7	300	122	178	49.6
40.7	297	121	176	49.3
40.7	295	120	175	48.8
40.7	290	118	172	48.0
40.7	285	116	169	47.2
40.7	280	114	166	46.4
40.7	275	112	163	45.6
40.7	273	111	162	45.1
40.7	270	110	160	44.8
40.7	268	109	159	44.3
40.7	263	107	156	43.5
40.7	258	105	153	42.7
40.7	253	103	150	41.9
40.7	248	101	147	41.1
40.7	246	100	146	40.7
40.7	243	99	144	40.3
40.7	241	98	143	39.9
40.7	236	96	140	39.1
40.7	231	94	137	38.3
40.7	226	92	134	37.5
40.7	221	90	131	36.7
40.7	216	88	128	35.9
40.7	214	87	127	35.4
40.7	209	85	124	34.6
40.7	204	83	121	33.8
40.7	199	81	118	33.0
40.7	194	79	115	32.2
40.7	189	77	112	31.4
40.7	182	74	108	30.1
40.7	177	72	105	29.3
40.7	172	70	102	28.5
40.7	167	68	99	27.7
40.7	162	66	96	26.9
40.7	150	61	89	24.8
40.7	145	59	86	24.0
40.7	140	57	83	23.2
40.7	135	55	80	22.4
40.7	123	50	73	20.3
40.7	118	48	70	19.5
40.7	113	46	67	18.7
40.7	108	44	64	17.9
40.7	91	37	54	15.0
40.7	86	35	51	14.2
40.7	81	33	48	13.4
40.7	59	24	35	9.8
40.7	54	22	32	9.0
40.7	27	11	16	4.5
40.6	500	203	297	82.4
40.6	498	202	296	81.9
40.6	497	202	295	82.1
40.6	495	201	294	81.6
40.6	493	200	293	81.1
40.6	490	199	291	80.8
40.6	488	198	290	80.3
40.6	485	197	288	80.0
40.6	483	196	287	79.5
40.6	480	195	285	79.2
40.6	478	194	284	78.7
40.6	475	193	282	78.4
40.6	473	192	281	77.9
40.6	471	191	280	77.5
40.6	470	191	279	77.6
40.6	468	190	278	77.1
40.6	466	189	277	76.7
40.6	465	189	276	76.8
40.6	463	188	275	76.3
40.6	461	187	274	75.9
40.6	458	186	272	75.5
40.6	456	185	271	75.1
40.6	453	184	269	74.7
40.6	451	183	268	74.3
40.6	448	182	266	73.9
40.6	446	181	265	73.5
40.6	443	180	263	73.1
40.6	441	179	262	72.7
40.6	438	178	260	72.3
40.6	436	177	259	71.9
40.6	434	176	258	71.4
40.6	433	176	257	71.5
40.6	431	175	256	71.1
40.6	429	174	255	70.6
40.6	426	173	253	70.3
40.6	424	172	252	69.8
40.6	421	171	250	69.5
40.6	419	170	249	69.0
40.6	416	169	247	68.7
40.6	414	168	246	68.2
40.6	411	167	244	67.9
40.6	409	166	243	67.4
40.6	406	165	241	67.1
40.6	404	164	240	66.6
40.6	401	163	238	66.3
40.6	399	162	237	65.8
40.6	397	161	236	65.3
40.6	394	160	234	65.0
40.6	392	159	233	64.5
40.6	389	158	231	64.2
40.6	387	157	230	63.7
40.6	384	156	228	63.4
40.6	382	155	227	62.9
40.6	379	154	225	62.6
40.6	377	153	224	62.1
40.6	374	152	222	61.8
40.6	372	151	221	61.3
40.6	367	149	218	60.5
40.6	362	147	215	59.7
40.6	360	146	214	59.2
40.6	357	145	212	58.9
40.6	355	144	211	58.4
40.6	352	143	209	58.1
40.6	350	142	208	57.6
40.6	347	141	206	57.3
40.6	345	140	205	56.8
40.6	342	139	203	56.5
40.6	340	138	202	56.0
40.6	335	136	199	55.2
40.6	330	134	196	54.4
40.6	325	132	193	53.6
40.6	323	131	192	53.1
40.6	320	130	190	52.8
40.6	318	129	189	52.3
40.6	315	128	187	52.0
40.6	313	127	186	51.5
40.6	310	126	184	51.2
40.6	308	125	183	50.7
40.6	303	123	180	49.9
40.6	298	121	177	49.1
40.6	293	119	174	48.3
40.6	291	118	173	47.8
40.6	288	117	171	47.5
40.6	286	116	170	47.0
40.6	283	115	168	46.7
40.6	281	114	167	46.2
40.6	278	113	165	45.9
40.6	276	112	164	45.4
40.6	271	110	161	44.6
40.6	266	108	158	43.8
40.6	261	106	155	43.0
40.6	256	104	152	42.3
40.6	254	103	151	41.8
40.6	251	102	149	41.5
40.6	249	101	148	41.0
40.6	244	99	145	40.2
40.6	239	97	142	39.4
40.6	234	95	139	38.6
40.6	229	93	136	37.8
40.6	224	91	133	37.0
40.6	219	89	130	36.2
40.6	217	88	129	35.7
40.6	212	86	126	34.9
40.6	207	84	123	34.1
40.6	202	82	120	33.3
40.6	197	80	117	32.5
40.6	192	78	114	31.7
40.6	187	76	111	30.9
40.6	180	73	107	29.6
40.6	175	71	104	28.8
40.6	170	69	101	28.0
40.6	165	67	98	27.2
40.6	160	65	95	26.4
40.6	155	63	92	25.6
40.6	143	58	85	23.5
40.6	138	56	82	22.7
40.6	133	54	79	21.9
40.6	128	52	76	21.1
40.6	106	43	63	17.4
40.6	101	41	60	16.6
40.6	96	39	57	15.8
40.6	69	28	41	11.4
40.6	64	26	38	10.6
40.6	32	13	19	5.3
40.5	499	202	297	81.8
40.5	496	201	295	81.5
40.5	494	200	294	81.0
40.5	491	199	292	80.7
40.5	489	198	291	80.2
40.5	487	197	290	79.7
40.5	486	197	289	79.9
40.5	484	196	288	79.4
40.5	482	195	287	78.9
40.5	481	195	286	79.1
40.5	479	194	285	78.6
40.5	477	193	284	78.1
40.5	476	193	283	78.3
40.5	474	192	282	77.8
40.5	472	191	281	77.3
40.5	469	190	279	77.0
40.5	467	189	278	76.5
40.5	464	188	276	76.2
40.5	462	187	275	75.7
40.5	459	186	273	75.4
40.5	457	185	272	74.9
40.5	454	184	270	74.6
40.5	452	183	269	74.1
40.5	449	182	267	73.8
40.5	447	181	266	73.3
40.5	444	180	264	73.0
40.5	442	179	263	72.5
40.5	440	178	262	72.0
40.5	439	178	261	72.2
40.5	437	177	260	71.7
40.5	435	176	259	71.2
40.5	432	175	257	70.9
40.5	430	174	256	70.4
40.5	427	173	254	70.1
40.5	425	172	253	69.6
40.5	422	171	251	69.3
40.5	420	170	250	68.8
40.5	417	169	248	68.5
40.5	415	168	247	68.0

%E	M1	M2	DM	M*
40.5	412	167	245	67.7
40.5	410	166	244	67.2
40.5	407	165	242	66.9
40.5	405	164	241	66.4
40.5	402	163	239	66.1
40.5	400	162	238	65.6
40.5	398	161	237	65.1
40.5	395	160	235	64.8
40.5	393	159	234	64.3
40.5	390	158	232	64.0
40.5	388	157	231	63.5
40.5	385	156	229	63.2
40.5	383	155	228	62.7
40.5	380	154	226	62.4
40.5	378	153	225	61.9
40.5	375	152	223	61.6
40.5	373	151	222	61.1
40.5	370	150	220	60.8
40.5	368	149	219	60.3
40.5	365	148	217	60.0
40.5	363	147	216	59.5
40.5	358	145	213	58.7
40.5	353	143	210	57.9
40.5	351	142	209	57.4
40.5	348	141	207	57.1
40.5	346	140	206	56.6
40.5	343	139	204	56.3
40.5	341	138	203	55.8
40.5	338	137	201	55.5
40.5	336	136	200	55.0
40.5	333	135	198	54.7
40.5	331	134	197	54.2
40.5	328	133	195	53.9
40.5	326	132	194	53.4
40.5	321	130	191	52.6
40.5	316	128	188	51.8
40.5	311	126	185	51.0
40.5	309	125	184	50.6
40.5	306	124	182	50.2
40.5	304	123	181	49.8
40.5	301	122	179	49.4
40.5	299	121	178	49.0
40.5	296	120	176	48.6
40.5	294	119	175	48.2
40.5	289	117	172	47.4
40.5	284	115	169	46.6
40.5	279	113	166	45.8
40.5	274	111	163	45.0
40.5	269	109	160	44.2
40.5	264	107	157	43.4

%E	M1	M2	DM	M*
40.5	262	106	156	42.9
40.5	259	105	154	42.6
40.5	257	104	153	42.1
40.5	252	102	150	41.3
40.5	247	100	147	40.5
40.5	242	98	144	39.7
40.5	237	96	141	38.9
40.5	232	94	138	38.1
40.5	227	92	135	37.3
40.5	222	90	132	36.5
40.5	220	89	131	36.0
40.5	215	87	128	35.2
40.5	210	85	125	34.4
40.5	205	83	122	33.6
40.5	200	81	119	32.8
40.5	195	79	116	32.0
40.5	190	77	113	31.2
40.5	185	75	110	30.4
40.5	173	70	103	28.3
40.5	168	68	100	27.5
40.5	163	66	97	26.7
40.5	158	64	94	25.9
40.5	153	62	91	25.1
40.5	148	60	88	24.3
40.5	131	53	78	21.4
40.5	126	51	75	20.6
40.5	121	49	72	19.8
40.5	116	47	69	19.0
40.5	111	45	66	18.2
40.5	84	34	50	13.8
40.5	79	32	47	13.0
40.5	74	30	44	12.2
40.5	42	17	25	6.9
40.5	37	15	22	6.1
40.4	500	202	298	81.6
40.4	498	201	297	81.1
40.4	497	201	296	81.3
40.4	495	200	295	80.8
40.4	493	199	294	80.3
40.4	492	199	293	80.5
40.4	490	198	292	80.0
40.4	488	197	291	79.5
40.4	485	196	289	79.2
40.4	483	195	288	78.7
40.4	480	194	286	78.4
40.4	478	193	285	77.9
40.4	475	192	283	77.6
40.4	473	191	282	77.1
40.4	470	190	280	76.8
40.4	468	189	279	76.3

%E	M1	M2	DM	M*
40.4	465	188	277	76.0
40.4	463	187	276	75.5
40.4	460	186	274	75.2
40.4	458	185	273	74.7
40.4	456	184	272	74.2
40.4	455	184	271	74.4
40.4	453	183	270	73.9
40.4	451	182	269	73.4
40.4	450	182	268	73.6
40.4	448	181	267	73.1
40.4	446	180	266	72.6
40.4	445	180	265	72.8
40.4	443	179	264	72.3
40.4	441	178	263	71.8
40.4	438	177	261	71.5
40.4	436	176	260	71.0
40.4	433	175	258	70.7
40.4	431	174	257	70.2
40.4	428	173	255	69.9
40.4	426	172	254	69.4
40.4	423	171	252	69.1
40.4	421	170	251	68.6
40.4	418	169	249	68.3
40.4	416	168	248	67.8
40.4	413	167	246	67.5
40.4	411	166	245	67.0
40.4	408	165	243	66.7
40.4	406	164	242	66.2
40.4	403	163	240	65.9
40.4	401	162	239	65.4
40.4	399	161	238	65.0
40.4	396	160	236	64.6
40.4	394	159	235	64.2
40.4	391	158	233	63.8
40.4	389	157	232	63.4
40.4	386	156	230	63.0
40.4	384	155	229	62.6
40.4	381	154	227	62.2
40.4	379	153	226	61.8
40.4	376	152	224	61.4
40.4	374	151	223	61.0
40.4	371	150	221	60.6
40.4	369	149	220	60.2
40.4	366	148	218	59.8
40.4	364	147	217	59.4
40.4	361	146	215	59.0
40.4	359	145	214	58.6
40.4	356	144	212	58.2
40.4	354	143	211	57.8
40.4	349	141	208	57.0

%E	M1	M2	DM	M*
40.4	344	139	205	56.2
40.4	342	138	204	55.7
40.4	339	137	202	55.4
40.4	337	136	201	54.9
40.4	334	135	199	54.6
40.4	332	134	198	54.1
40.4	329	133	196	53.8
40.4	327	132	195	53.3
40.4	324	131	193	53.0
40.4	322	130	192	52.5
40.4	319	129	190	52.2
40.4	317	128	189	51.7
40.4	314	127	187	51.4
40.4	312	126	186	50.9
40.4	307	124	183	50.1
40.4	302	122	180	49.3
40.4	297	120	177	48.5
40.4	292	118	174	47.7
40.4	287	116	171	46.9
40.4	285	115	170	46.4
40.4	282	114	168	46.1
40.4	280	113	167	45.6
40.4	277	112	165	45.3
40.4	275	111	164	44.8
40.4	272	110	162	44.5
40.4	270	109	161	44.0
40.4	267	108	159	43.7
40.4	265	107	158	43.2
40.4	260	105	155	42.4
40.4	255	103	152	41.6
40.4	250	101	149	40.8
40.4	245	99	146	40.0
40.4	240	97	143	39.2
40.4	235	95	140	38.4
40.4	230	93	137	37.6
40.4	228	92	136	37.1
40.4	225	91	134	36.8
40.4	223	90	133	36.3
40.4	218	88	130	35.5
40.4	213	86	127	34.7
40.4	208	84	124	33.9
40.4	203	82	121	33.1
40.4	198	80	118	32.3
40.4	193	78	115	31.5
40.4	188	76	112	30.7
40.4	183	74	109	29.9
40.4	178	72	106	29.1
40.4	171	69	102	27.8
40.4	166	67	99	27.0
40.4	161	65	96	26.2

%E	M1	M2	DM	M*
40.4	156	63	93	25.4
40.4	151	61	90	24.6
40.4	146	59	87	23.8
40.4	141	57	84	23.0
40.4	136	55	81	22.2
40.4	114	46	68	18.6
40.4	109	44	65	17.8
40.4	104	42	62	17.0
40.4	99	40	59	16.2
40.4	94	38	56	15.4
40.4	89	36	53	14.6
40.4	57	23	34	9.3
40.4	52	21	31	8.5
40.4	47	19	28	7.7
40.3	499	201	298	81.0
40.3	496	200	296	80.6
40.3	494	199	295	80.2
40.3	491	198	293	79.8
40.3	489	197	292	79.4
40.3	486	196	290	79.0
40.3	484	195	289	78.6
40.3	481	194	287	78.2
40.3	479	193	286	77.8
40.3	477	192	285	77.3
40.3	476	192	284	77.4
40.3	474	191	283	77.0
40.3	472	190	282	76.5
40.3	471	190	281	76.6
40.3	469	189	280	76.2
40.3	467	188	279	75.7
40.3	466	188	278	75.8
40.3	464	187	277	75.4
40.3	462	186	276	74.9
40.3	461	186	275	75.0
40.3	459	185	274	74.6
40.3	457	184	273	74.1
40.3	454	183	271	73.8
40.3	452	182	270	73.3
40.3	449	181	268	73.0
40.3	447	180	267	72.5
40.3	444	179	265	72.2
40.3	442	178	264	71.7
40.3	439	177	262	71.4
40.3	437	176	261	70.9
40.3	434	175	259	70.6
40.3	432	174	258	70.1
40.3	429	173	256	69.8
40.3	427	172	255	69.3
40.3	424	171	253	69.0
40.3	422	170	252	68.5

%E	M1	M2	DM	M*
40.3	419	169	250	68.2
40.3	417	168	249	67.7
40.3	414	167	247	67.4
40.3	412	166	246	66.9
40.3	409	165	244	66.6
40.3	407	164	243	66.1
40.3	404	163	241	65.8
40.3	402	162	240	65.3
40.3	400	161	239	64.8
40.3	397	160	237	64.5
40.3	395	159	236	64.0
40.3	392	158	234	63.7
40.3	390	157	233	63.2
40.3	387	156	231	62.9
40.3	385	155	230	62.4
40.3	382	154	228	62.1
40.3	380	153	227	61.6
40.3	377	152	225	61.3
40.3	375	151	224	60.8
40.3	372	150	222	60.5
40.3	370	149	221	60.0
40.3	367	148	219	59.7
40.3	365	147	218	59.2
40.3	362	146	216	58.9
40.3	360	145	215	58.4
40.3	357	144	213	58.1
40.3	355	143	212	57.6
40.3	352	142	210	57.3
40.3	350	141	209	56.8
40.3	347	140	207	56.5
40.3	345	139	206	56.0
40.3	340	137	203	55.2
40.3	335	135	200	54.4
40.3	330	133	197	53.6
40.3	325	131	194	52.8
40.3	320	129	191	52.0
40.3	318	128	190	51.5
40.3	315	127	188	51.2
40.3	313	126	187	50.7
40.3	310	125	185	50.4
40.3	308	124	184	49.9
40.3	305	123	182	49.6
40.3	303	122	181	49.1
40.3	300	121	179	48.8
40.3	298	120	178	48.3
40.3	295	119	176	48.0
40.3	293	118	175	47.5
40.3	290	117	173	47.2
40.3	288	116	172	46.7
40.3	283	114	169	45.9

%E	M1	M2	DM	M*
40.3	278	112	166	45.1
40.3	273	110	163	44.3
40.3	268	108	160	43.5
40.3	263	106	157	42.7
40.3	258	104	154	41.9
40.3	253	102	151	41.1
40.3	248	100	148	40.3
40.3	243	98	145	39.5
40.3	238	96	142	38.7
40.3	236	95	141	38.2
40.3	233	94	139	37.9
40.3	231	93	138	37.4
40.3	226	91	135	36.6
40.3	221	89	132	35.8
40.3	216	87	129	35.0
40.3	211	85	126	34.2
40.3	206	83	123	33.4
40.3	201	81	120	32.6
40.3	196	79	117	31.8
40.3	191	77	114	31.0
40.3	186	75	111	30.2
40.3	181	73	108	29.4
40.3	176	71	105	28.6
40.3	159	64	95	25.8
40.3	154	62	92	25.0
40.3	149	60	89	24.2
40.3	144	58	86	23.4
40.3	139	56	83	22.6
40.3	134	54	80	21.8
40.3	129	52	77	21.0
40.3	124	50	74	20.2
40.3	119	48	71	19.4
40.3	77	31	46	12.5
40.3	72	29	43	11.7
40.3	67	27	40	10.9
40.3	62	25	37	10.1
40.2	500	201	299	80.8
40.2	498	200	298	80.3
40.2	497	200	297	80.5
40.2	495	199	296	80.0
40.2	493	198	295	79.5
40.2	492	198	294	79.7
40.2	490	197	293	79.2
40.2	488	196	292	78.7
40.2	487	196	291	78.9
40.2	485	195	290	78.4
40.2	483	194	289	77.9
40.2	482	194	288	78.1
40.2	480	193	287	77.6
40.2	478	192	286	77.1

%E	M1	M2	DM	M*
40.2	475	191	284	76.8
40.2	473	190	283	76.3
40.2	470	189	281	76.0
40.2	468	188	280	75.5
40.2	465	187	278	75.2
40.2	463	186	277	74.7
40.2	460	185	275	74.4
40.2	458	184	274	73.9
40.2	455	183	272	73.6
40.2	453	182	271	73.1
40.2	450	181	269	72.8
40.2	448	180	268	72.3
40.2	445	179	266	72.0
40.2	443	178	265	71.5
40.2	440	177	263	71.2
40.2	438	176	262	70.7
40.2	435	175	260	70.4
40.2	433	174	259	69.9
40.2	430	173	257	69.6
40.2	428	172	256	69.1
40.2	425	171	254	68.8
40.2	423	170	253	68.3
40.2	420	169	251	68.0
40.2	418	168	250	67.5
40.2	415	167	248	67.2
40.2	413	166	247	66.7
40.2	410	165	245	66.4
40.2	408	164	244	65.9
40.2	405	163	242	65.6
40.2	403	162	241	65.1
40.2	401	161	240	64.6
40.2	398	160	238	64.3
40.2	396	159	237	63.8
40.2	393	158	235	63.5
40.2	391	157	234	63.0
40.2	388	156	232	62.7
40.2	386	155	231	62.2
40.2	383	154	229	61.9
40.2	381	153	228	61.4
40.2	378	152	226	61.1
40.2	376	151	225	60.6
40.2	373	150	223	60.3
40.2	371	149	222	59.8
40.2	368	148	220	59.5
40.2	366	147	219	59.0
40.2	363	146	217	58.7
40.2	361	145	216	58.2
40.2	358	144	214	57.9
40.2	356	143	213	57.4
40.2	353	142	211	57.1

%E	M1	M2	DM	M*
40.2	351	141	210	56.6
40.2	348	140	208	56.3
40.2	346	139	207	55.8
40.2	343	138	205	55.5
40.2	341	137	204	55.0
40.2	338	136	202	54.7
40.2	336	135	201	54.2
40.2	333	134	199	53.9
40.2	331	133	198	53.4
40.2	328	132	196	53.1
40.2	326	131	195	52.6
40.2	323	130	193	52.3
40.2	321	129	192	51.8
40.2	316	127	189	51.0
40.2	311	125	186	50.2
40.2	306	123	183	49.4
40.2	301	121	180	48.6
40.2	296	119	177	47.8
40.2	291	117	174	47.0
40.2	286	115	171	46.2
40.2	281	113	168	45.4
40.2	276	111	165	44.6
40.2	271	109	162	43.8
40.2	266	107	159	43.0
40.2	264	106	158	42.6
40.2	261	105	156	42.2
40.2	259	104	155	41.8
40.2	256	103	153	41.4
40.2	254	102	152	41.0
40.2	251	101	150	40.6
40.2	249	100	149	40.2
40.2	246	99	147	39.8
40.2	244	98	146	39.4
40.2	241	97	144	39.0
40.2	239	96	143	38.6
40.2	234	94	140	37.8
40.2	229	92	137	37.0
40.2	224	90	134	36.2
40.2	219	88	131	35.4
40.2	214	86	128	34.6
40.2	209	84	125	33.8
40.2	204	82	122	33.0
40.2	199	80	119	32.2
40.2	194	78	116	31.4
40.2	189	76	113	30.6
40.2	184	74	110	29.8
40.2	179	72	107	29.0
40.2	174	70	104	28.2
40.2	169	68	101	27.4
40.2	164	66	98	26.6

%E	M1	M2	DM	M*
40.2	132	53	79	21.3
40.2	127	51	76	20.5
40.2	122	49	73	19.7
40.2	117	47	70	18.9
40.2	112	45	67	18.1
40.2	107	43	64	17.3
40.2	102	41	61	16.5
40.2	97	39	58	15.7
40.2	92	37	55	14.9
40.2	87	35	52	14.1
40.2	82	33	49	13.3
40.1	499	200	299	80.2
40.1	496	199	297	79.8
40.1	494	198	296	79.4
40.1	491	197	294	79.0
40.1	489	196	293	78.6
40.1	486	195	291	78.2
40.1	484	194	290	77.8
40.1	481	193	288	77.4
40.1	479	192	287	77.0
40.1	476	191	285	76.6
40.1	474	190	284	76.2
40.1	471	189	282	75.8
40.1	469	188	281	75.4
40.1	466	187	279	75.0
40.1	464	186	278	74.6
40.1	461	185	276	74.2
40.1	459	184	275	73.8
40.1	456	183	273	73.4
40.1	454	182	272	73.0
40.1	451	181	270	72.6
40.1	449	180	269	72.2
40.1	446	179	267	71.8
40.1	444	178	266	71.4
40.1	441	177	264	71.0
40.1	439	176	263	70.6
40.1	436	175	261	70.2
40.1	434	174	260	69.8
40.1	431	173	258	69.4
40.1	429	172	257	69.0
40.1	426	171	255	68.6
40.1	424	170	254	68.2
40.1	421	169	252	67.8
40.1	419	168	251	67.4
40.1	416	167	249	67.0
40.1	414	166	248	66.6
40.1	411	165	246	66.2
40.1	409	164	245	65.8
40.1	406	163	243	65.4
40.1	404	162	242	65.0

%E	M1	M2	DM	M*
40•1	402	161	241	64•5
40•1	399	160	239	64•2
40•1	397	159	238	63•7
40•1	394	158	236	63•4
40•1	392	157	235	62•9
40•1	389	156	233	62•6
40•1	387	155	232	62•1
40•1	384	154	230	61•8
40•1	382	153	229	61•3
40•1	379	152	227	61•0
40•1	377	151	226	60•5
40•1	374	150	224	60•2
40•1	372	149	223	59•7
40•1	369	148	221	59•4
40•1	367	147	220	58•9
40•1	364	146	218	58•6
40•1	362	145	217	58•1
40•1	359	144	215	57•8
40•1	357	143	214	57•3
40•1	354	142	212	57•0
40•1	352	141	211	56•5
40•1	349	140	209	56•2
40•1	347	139	208	55•7
40•1	344	138	206	55•4
40•1	342	137	205	54•9
40•1	339	136	203	54•6
40•1	337	135	202	54•1
40•1	334	134	200	53•8
40•1	332	133	199	53•3
40•1	329	132	197	53•0
40•1	327	131	196	52•5
40•1	324	130	194	52•2
40•1	322	129	193	51•7
40•1	319	128	191	51•4
40•1	317	127	190	50•9
40•1	314	126	188	50•6
40•1	312	125	187	50•1
40•1	309	124	185	49•8
40•1	307	123	184	49•3
40•1	304	122	182	49•0
40•1	302	121	181	48•5
40•1	299	120	179	48•2
40•1	297	119	178	47•7
40•1	294	118	176	47•4
40•1	292	117	175	46•9
40•1	289	116	173	46•6
40•1	287	115	172	46•1
40•1	284	114	170	45•8
40•1	282	113	169	45•3
40•1	279	112	167	45•0

%E	M1	M2	DM	M*
40•1	277	111	166	44•5
40•1	274	110	164	44•2
40•1	272	109	163	43•7
40•1	269	108	161	43•4
40•1	267	107	160	42•9
40•1	262	105	157	42•1
40•1	257	103	154	41•3
40•1	252	101	151	40•5
40•1	247	99	148	39•7
40•1	242	97	145	38•9
40•1	237	95	142	38•1
40•1	232	93	139	37•3
40•1	227	91	136	36•5
40•1	222	89	133	35•7
40•1	217	87	130	34•9
40•1	212	85	127	34•1
40•1	207	83	124	33•3
40•1	202	81	121	32•5
40•1	197	79	118	31•7
40•1	192	77	115	30•9
40•1	187	75	112	30•1
40•1	182	73	109	29•3
40•1	177	71	106	28•5
40•1	172	69	103	27•7
40•1	167	67	100	26•9
40•1	162	65	97	26•1
40•1	157	63	94	25•3
40•1	152	61	91	24•5
40•1	147	59	88	23•7
40•1	142	57	85	22•9
40•1	137	55	82	22•1
40•0	500	200	300	80•0
40•0	498	199	299	79•5
40•0	497	199	298	79•7
40•0	495	198	297	79•2
40•0	493	197	296	78•7
40•0	492	197	295	78•9
40•0	490	196	294	78•4
40•0	488	195	293	77•9
40•0	487	195	292	78•1
40•0	485	194	291	77•6
40•0	483	193	290	77•1
40•0	482	193	289	77•3
40•0	480	192	288	76•8
40•0	478	191	287	76•3
40•0	477	191	286	76•5
40•0	475	190	285	76•0
40•0	473	189	284	75•5
40•0	472	189	283	75•7
40•0	470	188	282	75•2

%E	M1	M2	DM	M*
40•0	468	187	281	74•7
40•0	467	187	280	74•9
40•0	465	186	279	74•4
40•0	463	185	278	73•9
40•0	462	185	277	74•1
40•0	460	184	276	73•6
40•0	458	183	275	73•1
40•0	457	183	274	73•3
40•0	455	182	273	72•8
40•0	453	181	272	72•3
40•0	452	181	271	72•5
40•0	450	180	270	72•0
40•0	448	179	269	71•5
40•0	447	179	268	71•7
40•0	445	178	267	71•2
40•0	443	177	266	70•7
40•0	442	177	265	70•9
40•0	440	176	264	70•4
40•0	438	175	263	69•9
40•0	437	175	262	70•1
40•0	435	174	261	69•6
40•0	433	173	260	69•1
40•0	432	173	259	69•3
40•0	430	172	258	68•8
40•0	428	171	257	68•3
40•0	427	171	256	68•5
40•0	425	170	255	68•0
40•0	423	169	254	67•5
40•0	422	169	253	67•7
40•0	420	168	252	67•2
40•0	418	167	251	66•7
40•0	417	167	250	66•9
40•0	415	166	249	66•4
40•0	413	165	248	65•9
40•0	412	165	247	66•1
40•0	410	164	246	65•6
40•0	408	163	245	65•1
40•0	407	163	244	65•3
40•0	405	162	243	64•8
40•0	403	161	242	64•3
40•0	400	160	240	64•0
40•0	398	159	239	63•5
40•0	395	158	237	63•2
40•0	390	156	234	62•4
40•0	385	154	231	61•6
40•0	380	152	228	60•8
40•0	375	150	225	60•0
40•0	370	148	222	59•2
40•0	365	146	219	58•4
40•0	360	144	216	57•6

%E	M1	M2	DM	M*
40•0	355	142	213	56•8
40•0	350	140	210	56•0
40•0	345	138	207	55•2
40•0	340	136	204	54•4
40•0	335	134	201	53•6
40•0	330	132	198	52•8
40•0	325	130	195	52•0
40•0	320	128	192	51•2
40•0	315	126	189	50•4
40•0	310	124	186	49•6
40•0	305	122	183	48•8
40•0	300	120	180	48•0
40•0	295	118	177	47•2
40•0	290	116	174	46•4
40•0	285	114	171	45•6
40•0	280	112	168	44•8
40•0	275	110	165	44•0
40•0	270	108	162	43•2
40•0	265	106	159	42•4
40•0	260	104	156	41•6
40•0	255	102	153	40•8
40•0	250	100	150	40•0
40•0	245	98	147	39•2
40•0	240	96	144	38•4
40•0	235	94	141	37•6
40•0	230	92	138	36•8
40•0	225	90	135	36•0
40•0	220	88	132	35•2
40•0	215	86	129	34•4
40•0	210	84	126	33•6
40•0	205	82	123	32•8
40•0	200	80	120	32•0
40•0	195	78	117	31•2
40•0	190	76	114	30•4
40•0	185	74	111	29•6
40•0	180	72	108	28•8
40•0	175	70	105	28•0
40•0	170	68	102	27•2
40•0	165	66	99	26•4
40•0	160	64	96	25•6
40•0	155	62	93	24•8
40•0	150	60	90	24•0
40•0	145	58	87	23•2
40•0	140	56	84	22•4
40•0	135	54	81	21•6
40•0	130	52	78	20•8
40•0	125	50	75	20•0
40•0	120	48	72	19•2
40•0	115	46	69	18•4
40•0	110	44	66	17•6

%E	M1	M2	DM	M*
40•0	105	42	63	16•8
40•0	100	40	60	16•0
40•0	95	38	57	15•2
40•0	90	36	54	14•4
40•0	85	34	51	13•6
40•0	80	32	48	12•8
40•0	75	30	45	12•0
40•0	70	28	42	11•2
40•0	65	26	39	10•4
40•0	60	24	36	9•6
40•0	55	22	33	8•8
40•0	50	20	30	8•0
40•0	45	18	27	7•2
40•0	40	16	24	6•4
40•0	35	14	21	5•6
40•0	30	12	18	4•8
40•0	25	10	15	4•0
40•0	20	8	12	3•2
40•0	15	6	9	2•4
40•0	10	4	6	1•6
40•0	5	2	3	0•8
39•9	499	199	300	79•4
39•9	496	198	298	79•0
39•9	494	197	297	78•6
39•9	491	196	295	78•2
39•9	489	195	294	77•8
39•9	486	194	292	77•4
39•9	484	193	291	77•0
39•9	481	192	289	76•6
39•9	479	191	288	76•2
39•9	476	190	286	75•8
39•9	474	189	285	75•4
39•9	471	188	283	75•0
39•9	469	187	282	74•6
39•9	466	186	280	74•2
39•9	464	185	279	73•8
39•9	461	184	277	73•4
39•9	459	183	276	73•0
39•9	456	182	274	72•6
39•9	454	181	273	72•2
39•9	451	180	271	71•8
39•9	449	179	270	71•4
39•9	446	178	268	71•0
39•9	444	177	267	70•6
39•9	441	176	265	70•2
39•9	439	175	264	69•8
39•9	436	174	262	69•4
39•9	434	173	261	69•0
39•9	431	172	259	68•6
39•9	429	171	258	68•2

%E	M1	M2	DM	M*	%E	M1	M2	DM	M*	%E	M1	M2	DM	M*	%E	M1	M2	DM	M*	%E	M1	M2	DM	M*
39.9	426	170	256	67.8	39.9	301	120	181	47.8	39.8	487	194	293	77.3	39.8	364	145	219	57.8	39.8	181	72	109	28.6
39.9	424	169	255	67.4	39.9	298	119	179	47.5	39.8	485	193	292	76.8	39.8	362	144	218	57.3	39.8	176	70	106	27.8
39.9	421	168	253	67.0	39.9	296	118	178	47.0	39.8	483	192	291	76.3	39.8	359	143	216	57.0	39.8	171	68	103	27.0
39.9	419	167	252	66.6	39.9	293	117	176	46.7	39.8	482	192	290	76.5	39.8	357	142	215	56.5	39.8	166	66	100	26.2
39.9	416	166	250	66.2	39.9	291	116	175	46.2	39.8	480	191	289	76.0	39.8	354	141	213	56.2	39.8	161	64	97	25.4
39.9	414	165	249	65.8	39.9	288	115	173	45.9	39.8	477	190	287	75.7	39.8	352	140	212	55.7	39.8	128	51	77	20.3
39.9	411	164	247	65.4	39.9	286	114	172	45.4	39.8	475	189	286	75.2	39.8	349	139	210	55.4	39.8	123	49	74	19.5
39.9	409	163	246	65.0	39.9	283	113	170	45.1	39.8	472	188	284	74.9	39.8	347	138	209	54.9	39.8	118	47	71	18.7
39.9	406	162	244	64.6	39.9	281	112	169	44.6	39.8	470	187	283	74.4	39.8	344	137	207	54.6	39.8	113	45	68	17.9
39.9	404	161	243	64.2	39.9	278	111	167	44.3	39.8	467	186	281	74.1	39.8	342	136	206	54.1	39.8	108	43	65	17.1
39.9	401	160	241	63.8	39.9	276	110	166	43.8	39.8	465	185	280	73.6	39.8	339	135	204	53.8	39.8	103	41	62	16.3
39.9	399	159	240	63.4	39.9	273	109	164	43.5	39.8	462	184	278	73.3	39.8	337	134	203	53.3	39.8	98	39	59	15.5
39.9	396	158	238	63.0	39.9	271	108	163	43.0	39.8	460	183	277	72.8	39.8	334	133	201	53.0	39.8	93	37	56	14.7
39.9	393	157	236	62.7	39.9	268	107	161	42.7	39.8	457	182	275	72.5	39.8	332	132	200	52.5	39.8	88	35	53	13.9
39.9	391	156	235	62.2	39.9	266	106	160	42.2	39.8	455	181	274	72.0	39.8	329	131	198	52.2	39.8	83	33	50	13.1
39.9	388	155	233	61.9	39.9	263	105	158	41.9	39.8	452	180	272	71.7	39.8	327	130	197	51.7	39.7	499	198	301	78.6
39.9	386	154	232	61.4	39.9	258	103	155	41.1	39.8	450	179	271	71.2	39.8	324	129	195	51.4	39.7	496	197	299	78.2
39.9	383	153	230	61.1	39.9	253	101	152	40.3	39.8	447	178	269	70.9	39.8	322	128	194	50.9	39.7	494	196	298	77.8
39.9	381	152	229	60.6	39.9	248	99	149	39.5	39.8	445	177	268	70.4	39.8	319	127	192	50.6	39.7	491	195	296	77.4
39.9	378	151	227	60.3	39.9	243	97	146	38.7	39.8	442	176	266	70.1	39.8	314	125	189	49.8	39.7	489	194	295	77.0
39.9	376	150	226	59.8	39.9	238	95	143	37.9	39.8	440	175	265	69.6	39.8	309	123	186	49.0	39.7	486	193	293	76.6
39.9	373	149	224	59.5	39.9	233	93	140	37.1	39.8	437	174	263	69.3	39.8	304	121	183	48.2	39.7	484	192	292	76.2
39.9	371	148	223	59.0	39.9	228	91	137	36.3	39.8	435	173	262	68.8	39.8	299	119	180	47.4	39.7	481	191	290	75.8
39.9	368	147	221	58.7	39.9	223	89	134	35.5	39.8	432	172	260	68.5	39.8	294	117	177	46.6	39.7	479	190	289	75.4
39.9	366	146	220	58.2	39.9	218	87	131	34.7	39.8	430	171	259	68.0	39.8	289	115	174	45.8	39.7	478	190	288	75.5
39.9	363	145	218	57.9	39.9	213	85	128	33.9	39.8	427	170	257	67.7	39.8	284	113	171	45.0	39.7	476	189	287	75.0
39.9	361	144	217	57.4	39.9	208	83	125	33.1	39.8	425	169	256	67.2	39.8	279	111	168	44.2	39.7	474	188	286	74.6
39.9	358	143	215	57.1	39.9	203	81	122	32.3	39.8	422	168	254	66.9	39.8	274	109	165	43.4	39.7	473	188	285	74.7
39.9	356	142	214	56.6	39.9	198	79	119	31.5	39.8	420	167	253	66.4	39.8	269	107	162	42.6	39.7	471	187	284	74.2
39.9	353	141	212	56.3	39.9	193	77	116	30.7	39.8	417	166	251	66.1	39.8	264	105	159	41.8	39.7	469	186	283	73.8
39.9	351	140	211	55.8	39.9	188	75	113	29.9	39.8	415	165	250	65.6	39.8	261	104	157	41.4	39.7	468	186	282	73.9
39.9	348	139	209	55.5	39.9	183	73	110	29.1	39.8	412	164	248	65.3	39.8	259	103	156	41.0	39.7	466	185	281	73.4
39.9	346	138	208	55.0	39.9	178	71	107	28.3	39.8	410	163	247	64.8	39.8	256	102	154	40.6	39.7	464	184	280	73.0
39.9	343	137	206	54.7	39.9	173	69	104	27.5	39.8	407	162	245	64.5	39.8	254	101	153	40.2	39.7	463	184	279	73.1
39.9	341	136	205	54.2	39.9	168	67	101	26.7	39.8	405	161	244	64.0	39.8	251	100	151	39.8	39.7	461	183	278	72.6
39.9	338	135	203	53.9	39.9	163	65	98	25.9	39.8	402	160	242	63.7	39.8	249	99	150	39.4	39.7	459	182	277	72.2
39.9	336	134	202	53.4	39.9	158	63	95	25.1	39.8	400	159	241	63.2	39.8	246	98	148	39.0	39.7	458	182	276	72.3
39.9	333	133	200	53.1	39.9	153	61	92	24.3	39.8	397	158	239	62.9	39.8	244	97	147	38.6	39.7	456	181	275	71.8
39.9	331	132	19[illegible]	52.6	39.9	148	59	89	23.5	39.8	394	157	237	62.6	39.8	241	96	145	38.2	39.7	453	180	273	71.5
39.9	328	131	197	52.3	39.9	143	57	86	22.7	39.8	392	156	236	62.1	39.8	236	94	142	37.4	39.7	451	179	272	71.0
39.9	326	130	196	51.8	39.9	138	55	83	21.9	39.8	389	155	234	61.8	39.8	231	92	139	36.6	39.7	448	178	270	70.7
39.9	323	129	194	51.5	39.9	133	53	80	21.1	39.8	387	154	233	61.3	39.8	226	90	136	35.8	39.7	446	177	269	70.2
39.9	321	128	193	51.0	39.8	500	199	301	79.2	39.8	384	153	231	61.0	39.8	221	88	133	35.0	39.7	443	176	267	69.9
39.9	318	127	191	50.7	39.8	498	198	300	78.7	39.8	382	152	230	60.5	39.8	216	86	130	34.2	39.7	441	175	266	69.4
39.9	316	126	190	50.2	39.8	497	198	299	78.9	39.8	379	151	228	60.2	39.8	211	84	127	33.4	39.7	438	174	264	69.1
39.9	313	125	188	49.9	39.8	495	197	298	78.4	39.8	377	150	227	59.7	39.8	206	82	124	32.6	39.7	436	173	263	68.6
39.9	311	124	187	49.4	39.8	493	196	297	77.9	39.8	374	149	225	59.4	39.8	201	80	121	31.8	39.7	433	172	261	68.3
39.9	308	123	185	49.1	39.8	492	196	296	78.1	39.8	372	148	224	58.9	39.8	196	78	118	31.0	39.7	431	171	260	67.8
39.9	306	122	184	48.6	39.8	490	195	295	77.6	39.8	369	147	222	58.6	39.8	191	76	115	30.2	39.7	428	170	258	67.5
39.9	303	121	182	48.3	39.8	488	194	294	77.1	39.8	367	146	221	58.1	39.8	186	74	112	29.4	39.7	426	169	257	67.0

%E	M1	M2	DM	M*
39.7	423	168	255	66.7
39.7	421	167	254	66.2
39.7	418	166	252	65.9
39.7	416	165	251	65.4
39.7	413	164	249	65.1
39.7	411	163	248	64.6
39.7	408	162	246	64.3
39.7	406	161	245	63.8
39.7	403	160	243	63.5
39.7	401	159	242	63.0
39.7	398	158	240	62.7
39.7	395	157	238	62.4
39.7	393	156	237	61.9
39.7	390	155	235	61.6
39.7	388	154	234	61.1
39.7	385	153	232	60.8
39.7	383	152	231	60.3
39.7	380	151	229	60.0
39.7	378	150	228	59.5
39.7	375	149	226	59.2
39.7	373	148	225	58.7
39.7	370	147	223	58.4
39.7	368	146	222	57.9
39.7	365	145	220	57.6
39.7	363	144	219	57.1
39.7	360	143	217	56.8
39.7	358	142	216	56.3
39.7	355	141	214	56.0
39.7	353	140	213	55.5
39.7	350	139	211	55.2
39.7	348	138	210	54.7
39.7	345	137	208	54.4
39.7	343	136	207	53.9
39.7	340	135	205	53.6
39.7	335	133	202	52.8
39.7	330	131	199	52.0
39.7	325	129	196	51.2
39.7	320	127	193	50.4
39.7	317	126	191	50.1
39.7	315	125	190	49.6
39.7	312	124	188	49.3
39.7	310	123	187	48.8
39.7	307	122	185	48.5
39.7	305	121	184	48.0
39.7	302	120	182	47.7
39.7	300	119	181	47.2
39.7	297	118	179	46.9
39.7	295	117	178	46.4
39.7	292	116	176	46.1
39.7	290	115	175	45.6
39.7	287	114	173	45.3
39.7	282	112	170	44.5
39.7	277	110	167	43.7
39.7	272	108	164	42.9
39.7	267	106	161	42.1
39.7	262	104	158	41.3
39.7	257	102	155	40.5
39.7	252	100	152	39.7
39.7	247	98	149	38.9
39.7	242	96	146	38.1
39.7	239	95	144	37.8
39.7	237	94	143	37.3
39.7	234	93	141	37.0
39.7	232	92	140	36.5
39.7	229	91	138	36.2
39.7	224	89	135	35.4
39.7	219	87	132	34.6
39.7	214	85	129	33.8
39.7	209	83	126	33.0
39.7	204	81	123	32.2
39.7	199	79	120	31.4
39.7	194	77	117	30.6
39.7	189	75	114	29.8
39.7	184	73	111	29.0
39.7	179	71	108	28.2
39.7	174	69	105	27.4
39.7	156	62	94	24.6
39.7	151	60	91	23.8
39.7	146	58	88	23.0
39.7	141	56	85	22.2
39.7	136	54	82	21.4
39.7	131	52	79	20.6
39.7	126	50	76	19.8
39.7	121	48	73	19.0
39.7	116	46	70	18.2
39.7	78	31	47	12.3
39.7	73	29	44	11.5
39.7	68	27	41	10.7
39.7	63	25	38	9.9
39.7	58	23	35	9.1
39.6	500	198	302	78.4
39.6	498	197	301	77.9
39.6	497	197	300	78.1
39.6	495	196	299	77.6
39.6	493	195	298	77.1
39.6	492	195	297	77.3
39.6	490	194	296	76.8
39.6	487	193	294	76.5
39.6	485	192	293	76.0
39.6	482	191	291	75.7
39.6	480	190	290	75.2
39.6	477	189	288	74.9
39.6	475	188	287	74.4
39.6	472	187	285	74.1
39.6	470	186	284	73.6
39.6	467	185	282	73.3
39.6	465	184	281	72.8
39.6	462	183	279	72.5
39.6	460	182	278	72.0
39.6	457	181	276	71.7
39.6	455	180	275	71.2
39.6	454	180	274	71.4
39.6	452	179	273	70.9
39.6	450	178	272	70.4
39.6	449	178	271	70.6
39.6	447	177	270	70.1
39.6	445	176	269	69.6
39.6	444	176	268	69.8
39.6	442	175	267	69.3
39.6	439	174	265	69.0
39.6	437	173	264	68.5
39.6	434	172	262	68.2
39.6	432	171	261	67.7
39.6	429	170	259	67.4
39.6	427	169	258	66.9
39.6	424	168	256	66.6
39.6	422	167	255	66.1
39.6	419	166	253	65.8
39.6	417	165	252	65.3
39.6	414	164	250	65.0
39.6	412	163	249	64.5
39.6	409	162	247	64.2
39.6	407	161	246	63.7
39.6	404	160	244	63.4
39.6	402	159	243	62.9
39.6	399	158	241	62.6
39.6	396	157	239	62.2
39.6	394	156	238	61.8
39.6	391	155	236	61.4
39.6	389	154	235	61.0
39.6	386	153	233	60.6
39.6	384	152	232	60.2
39.6	381	151	230	59.8
39.6	379	150	229	59.4
39.6	376	149	227	59.0
39.6	374	148	226	58.6
39.6	371	147	224	58.2
39.6	369	146	223	57.8
39.6	366	145	221	57.4
39.6	364	144	220	57.0
39.6	361	143	218	56.6
39.6	359	142	217	56.2
39.6	356	141	215	55.8
39.6	351	139	212	55.0
39.6	346	137	209	54.2
39.6	341	135	206	53.4
39.6	338	134	204	53.1
39.6	336	133	203	52.6
39.6	333	132	201	52.3
39.6	331	131	200	51.8
39.6	328	130	198	51.5
39.6	326	129	197	51.0
39.6	323	128	195	50.7
39.6	321	127	194	50.2
39.6	318	126	192	49.9
39.6	316	125	191	49.4
39.6	313	124	189	49.1
39.6	311	123	188	48.6
39.6	308	122	186	48.3
39.6	303	120	183	47.5
39.6	298	118	180	46.7
39.6	293	116	177	45.9
39.6	288	114	174	45.1
39.6	285	113	172	44.8
39.6	283	112	171	44.3
39.6	280	111	169	44.0
39.6	278	110	168	43.5
39.6	275	109	166	43.2
39.6	273	108	165	42.7
39.6	270	107	163	42.4
39.6	268	106	162	41.9
39.6	265	105	160	41.6
39.6	260	103	157	40.8
39.6	255	101	154	40.0
39.6	250	99	151	39.2
39.6	245	97	148	38.4
39.6	240	95	145	37.6
39.6	235	93	142	36.8
39.6	230	91	139	36.0
39.6	227	90	137	35.7
39.6	225	89	136	35.2
39.6	222	88	134	34.9
39.6	217	86	131	34.1
39.6	212	84	128	33.3
39.6	207	82	125	32.5
39.6	202	80	122	31.7
39.6	197	78	119	30.9
39.6	192	76	116	30.1
39.6	187	74	113	29.3
39.6	182	72	110	28.5
39.6	169	67	102	26.6
39.6	164	65	99	25.8
39.6	159	63	96	25.0
39.6	154	61	93	24.2
39.6	149	59	90	23.4
39.6	144	57	87	22.6
39.6	139	55	84	21.8
39.6	134	53	81	21.0
39.6	111	44	67	17.4
39.6	106	42	64	16.6
39.6	101	40	61	15.8
39.6	96	38	58	15.0
39.6	91	36	55	14.2
39.6	53	21	32	8.3
39.6	48	19	29	7.5
39.5	499	197	302	77.8
39.5	496	196	300	77.5
39.5	494	195	299	77.0
39.5	491	194	297	76.7
39.5	489	193	296	76.2
39.5	488	193	295	76.3
39.5	486	192	294	75.9
39.5	484	191	293	75.4
39.5	483	191	292	75.5
39.5	481	190	291	75.1
39.5	479	189	290	74.6
39.5	478	189	289	74.7
39.5	476	188	288	74.3
39.5	474	187	287	73.8
39.5	473	187	286	73.9
39.5	471	186	285	73.5
39.5	468	185	283	73.1
39.5	466	184	282	72.7
39.5	463	183	280	72.3
39.5	461	182	279	71.9
39.5	458	181	277	71.5
39.5	456	180	276	71.1
39.5	453	179	274	70.7
39.5	451	178	273	70.3
39.5	448	177	271	69.9
39.5	446	176	270	69.5
39.5	443	175	268	69.1
39.5	441	174	267	68.7
39.5	440	174	266	68.8
39.5	438	173	265	68.3
39.5	436	172	264	67.9
39.5	435	172	263	68.0
39.5	433	171	262	67.5
39.5	430	170	260	67.2
39.5	428	169	259	66.7

%E	M1	M2	DM	M*
39.5	425	168	257	66.4
39.5	423	167	256	65.9
39.5	420	166	254	65.6
39.5	418	165	253	65.1
39.5	415	164	251	64.8
39.5	413	163	250	64.3
39.5	410	162	248	64.0
39.5	408	161	247	63.5
39.5	405	160	245	63.2
39.5	403	159	244	62.7
39.5	400	158	242	62.4
39.5	397	157	240	62.1
39.5	395	156	239	61.6
39.5	392	155	237	61.3
39.5	390	154	236	60.8
39.5	387	153	234	60.5
39.5	385	152	233	60.0
39.5	382	151	231	59.7
39.5	380	150	230	59.2
39.5	377	149	228	58.9
39.5	375	148	227	58.4
39.5	372	147	225	58.1
39.5	370	146	224	57.6
39.5	367	145	222	57.3
39.5	365	144	221	56.8
39.5	362	143	219	56.5
39.5	357	141	216	55.7
39.5	354	140	214	55.4
39.5	352	139	213	54.9
39.5	349	138	211	54.6
39.5	347	137	210	54.1
39.5	344	136	208	53.8
39.5	342	135	207	53.3
39.5	339	134	205	53.0
39.5	337	133	204	52.5
39.5	334	132	202	52.2
39.5	332	131	201	51.7
39.5	329	130	199	51.4
39.5	327	129	198	50.9
39.5	324	128	196	50.6
39.5	319	126	193	49.8
39.5	314	124	190	49.0
39.5	309	122	187	48.2
39.5	306	121	185	47.8
39.5	304	120	184	47.4
39.5	301	119	182	47.0
39.5	299	118	181	46.6
39.5	296	117	179	46.2
39.5	294	116	178	45.8
39.5	291	115	176	45.4
39.5	286	113	173	44.6
39.5	281	111	170	43.8
39.5	276	109	167	43.0
39.5	271	107	164	42.2
39.5	266	105	161	41.4
39.5	263	104	159	41.1
39.5	261	103	158	40.6
39.5	258	102	156	40.3
39.5	256	101	155	39.8
39.5	253	100	153	39.5
39.5	248	98	150	38.7
39.5	243	96	147	37.9
39.5	238	94	144	37.1
39.5	233	92	141	36.3
39.5	228	90	138	35.5
39.5	223	88	135	34.7
39.5	220	87	133	34.4
39.5	218	86	132	33.9
39.5	215	85	130	33.6
39.5	210	83	127	32.8
39.5	205	81	124	32.0
39.5	200	79	121	31.2
39.5	195	77	118	30.4
39.5	190	75	115	29.6
39.5	185	73	112	28.8
39.5	177	70	107	27.7
39.5	172	68	104	26.9
39.5	167	66	101	26.1
39.5	162	64	98	25.3
39.5	157	62	95	24.5
39.5	152	60	92	23.7
39.5	147	58	89	22.9
39.5	129	51	78	20.2
39.5	124	49	75	19.4
39.5	119	47	72	18.6
39.5	114	45	69	17.8
39.5	109	43	66	17.0
39.5	86	34	52	13.4
39.5	81	32	49	12.6
39.5	76	30	46	11.8
39.5	43	17	26	6.7
39.5	38	15	23	5.9
39.4	500	197	303	77.6
39.4	498	196	302	77.1
39.4	497	196	301	77.3
39.4	495	195	300	76.8
39.4	493	194	299	76.3
39.4	492	194	298	76.5
39.4	490	193	297	76.0
39.4	487	192	295	75.7
39.4	485	191	294	75.2
39.4	482	190	292	74.9
39.4	480	189	291	74.4
39.4	477	188	289	74.1
39.4	475	187	288	73.6
39.4	472	186	286	73.3
39.4	470	185	285	72.8
39.4	469	185	284	73.0
39.4	467	184	283	72.5
39.4	465	183	282	72.0
39.4	464	183	281	72.2
39.4	462	182	280	71.7
39.4	459	181	278	71.4
39.4	457	180	277	70.9
39.4	454	179	275	70.6
39.4	452	178	274	70.1
39.4	449	177	272	69.8
39.4	447	176	271	69.3
39.4	444	175	269	69.0
39.4	442	174	268	68.5
39.4	439	173	266	68.2
39.4	437	172	265	67.7
39.4	434	171	263	67.4
39.4	432	170	262	66.9
39.4	431	170	261	67.1
39.4	429	169	260	66.6
39.4	426	168	258	66.3
39.4	424	167	257	65.8
39.4	421	166	255	65.5
39.4	419	165	254	65.0
39.4	416	164	252	64.7
39.4	414	163	251	64.2
39.4	411	162	249	63.9
39.4	409	161	248	63.4
39.4	406	160	246	63.1
39.4	404	159	245	62.6
39.4	401	158	243	62.3
39.4	398	157	241	61.9
39.4	396	156	240	61.5
39.4	393	155	238	61.1
39.4	391	154	237	60.7
39.4	388	153	235	60.3
39.4	386	152	234	59.9
39.4	383	151	232	59.5
39.4	381	150	231	59.1
39.4	378	149	229	58.7
39.4	376	148	228	58.3
39.4	373	147	226	57.9
39.4	371	146	225	57.5
39.4	368	145	223	57.1
39.4	363	143	220	56.3
39.4	360	142	218	56.0
39.4	358	141	217	55.5
39.4	355	140	215	55.2
39.4	353	139	214	54.7
39.4	350	138	212	54.4
39.4	348	137	211	53.9
39.4	345	136	209	53.6
39.4	343	135	208	53.1
39.4	340	134	206	52.8
39.4	335	132	203	52.0
39.4	330	130	200	51.2
39.4	325	128	197	50.4
39.4	322	127	195	50.1
39.4	320	126	194	49.6
39.4	317	125	192	49.3
39.4	315	124	191	48.8
39.4	312	123	189	48.5
39.4	310	122	188	48.0
39.4	307	121	186	47.7
39.4	302	119	183	46.9
39.4	297	117	180	46.1
39.4	292	115	177	45.3
39.4	289	114	175	45.0
39.4	287	113	174	44.5
39.4	284	112	172	44.2
39.4	282	111	171	43.7
39.4	279	110	169	43.4
39.4	277	109	168	42.9
39.4	274	108	166	42.6
39.4	269	106	163	41.8
39.4	264	104	160	41.0
39.4	259	102	157	40.2
39.4	254	100	154	39.4
39.4	251	99	152	39.0
39.4	249	98	151	38.6
39.4	246	97	149	38.2
39.4	241	95	146	37.4
39.4	236	93	143	36.6
39.4	231	91	140	35.8
39.4	226	89	137	35.0
39.4	221	87	134	34.2
39.4	216	85	131	33.4
39.4	213	84	129	33.1
39.4	208	82	126	32.3
39.4	203	80	123	31.5
39.4	198	78	120	30.7
39.4	193	76	117	29.9
39.4	188	74	114	29.1
39.4	180	71	109	28.0
39.4	175	69	106	27.2
39.4	170	67	103	26.4
39.4	165	65	100	25.6
39.4	160	63	97	24.8
39.4	155	61	94	24.0
39.4	142	56	86	22.1
39.4	137	54	83	21.3
39.4	132	52	80	20.5
39.4	127	50	77	19.7
39.4	104	41	63	16.2
39.4	99	39	60	15.4
39.4	94	37	57	14.6
39.4	71	28	43	11.0
39.4	66	26	40	10.2
39.4	33	13	20	5.1
39.3	499	196	303	77.0
39.3	496	195	301	76.7
39.3	494	194	300	76.2
39.3	491	193	298	75.9
39.3	489	192	297	75.4
39.3	488	192	296	75.5
39.3	486	191	295	75.1
39.3	484	190	294	74.6
39.3	483	190	293	74.7
39.3	481	189	292	74.3
39.3	478	188	290	73.9
39.3	476	187	289	73.5
39.3	473	186	287	73.1
39.3	471	185	286	72.7
39.3	468	184	284	72.3
39.3	466	183	283	71.9
39.3	463	182	281	71.5
39.3	461	181	280	71.1
39.3	460	181	279	71.2
39.3	458	180	278	70.7
39.3	456	179	277	70.3
39.3	455	179	276	70.4
39.3	453	178	275	69.9
39.3	450	177	273	69.6
39.3	448	176	272	69.1
39.3	445	175	270	68.8
39.3	443	174	269	68.3
39.3	440	173	267	68.0
39.3	438	172	266	67.5
39.3	435	171	264	67.2
39.3	433	170	263	66.7
39.3	430	169	261	66.4
39.3	428	168	260	65.9
39.3	427	168	259	66.1
39.3	425	167	258	65.6

%E	M1	M2	DM	M*
39•3	422	166	256	65•3
39•3	420	165	255	64•8
39•3	417	164	253	64•5
39•3	415	163	252	64•0
39•3	412	162	250	63•7
39•3	410	161	249	63•2
39•3	407	160	247	62•9
39•3	405	159	246	62•4
39•3	402	158	244	62•1
39•3	400	157	243	61•6
39•3	399	157	242	61•8
39•3	397	156	241	61•3
39•3	394	155	239	61•0
39•3	392	154	238	60•5
39•3	389	153	236	60•2
39•3	387	152	235	59•7
39•3	384	151	233	59•4
39•3	382	150	232	58•9
39•3	379	149	230	58•6
39•3	377	148	229	58•1
39•3	374	147	227	57•8
39•3	369	145	224	57•0
39•3	366	144	222	56•7
39•3	364	143	221	56•2
39•3	361	142	219	55•9
39•3	359	141	218	55•4
39•3	356	140	216	55•1
39•3	354	139	215	54•6
39•3	351	138	213	54•3
39•3	349	137	212	53•8
39•3	346	136	210	53•5
39•3	341	134	207	52•7
39•3	338	133	205	52•3
39•3	336	132	204	51•9
39•3	333	131	202	51•5
39•3	331	130	201	51•1
39•3	328	129	199	50•7
39•3	326	128	198	50•3
39•3	323	127	196	49•9
39•3	321	126	195	49•5
39•3	318	125	193	49•1
39•3	313	123	190	48•3
39•3	308	121	187	47•5
39•3	305	120	185	47•2
39•3	303	119	184	46•7
39•3	300	118	182	46•4
39•3	298	117	181	45•9
39•3	295	116	179	45•6
39•3	290	114	176	44•8
39•3	285	112	173	44•0

%E	M1	M2	DM	M*
39•3	280	110	170	43•2
39•3	275	108	167	42•4
39•3	272	107	165	42•1
39•3	270	106	164	41•6
39•3	267	105	162	41•3
39•3	262	103	159	40•5
39•3	257	101	156	39•7
39•3	252	99	153	38•9
39•3	247	97	150	38•1
39•3	244	96	148	37•8
39•3	242	95	147	37•3
39•3	239	94	145	37•0
39•3	234	92	142	36•2
39•3	229	90	139	35•4
39•3	224	88	136	34•6
39•3	219	86	133	33•8
39•3	214	84	130	33•0
39•3	211	83	128	32•6
39•3	206	81	125	31•8
39•3	201	79	122	31•0
39•3	196	77	119	30•3
39•3	191	75	116	29•5
39•3	183	72	111	28•3
39•3	178	70	108	27•5
39•3	173	68	105	26•7
39•3	168	66	102	25•9
39•3	163	64	99	25•1
39•3	150	59	91	23•2
39•3	145	57	88	22•4
39•3	140	55	85	21•6
39•3	135	53	82	20•8
39•3	122	48	74	18•9
39•3	117	46	71	18•1
39•3	112	44	68	17•3
39•3	107	42	65	16•5
39•3	89	35	54	13•8
39•3	84	33	51	13•0
39•3	61	24	37	9•4
39•3	56	22	34	8•6
39•3	28	11	17	4•3
39•2	500	196	304	76•8
39•2	498	195	303	76•4
39•2	497	195	302	76•5
39•2	495	194	301	76•0
39•2	492	193	299	75•7
39•2	490	192	298	75•2
39•2	487	191	296	74•9
39•2	485	190	295	74•4
39•2	482	189	293	74•1
39•2	480	188	292	73•6

%E	M1	M2	DM	M*
39•2	479	188	291	73•8
39•2	477	187	290	73•3
39•2	475	186	289	72•8
39•2	474	186	288	73•0
39•2	472	185	287	72•5
39•2	469	184	285	72•2
39•2	467	183	284	71•7
39•2	464	182	282	71•4
39•2	462	181	281	70•9
39•2	459	180	279	70•6
39•2	457	179	278	70•1
39•2	454	178	276	69•8
39•2	452	177	275	69•3
39•2	451	177	274	69•5
39•2	449	176	273	69•0
39•2	447	175	272	68•5
39•2	446	175	271	68•7
39•2	444	174	270	68•2
39•2	441	173	268	67•9
39•2	439	172	267	67•4
39•2	436	171	265	67•1
39•2	434	170	264	66•6
39•2	431	169	262	66•3
39•2	429	168	261	65•8
39•2	426	167	259	65•5
39•2	424	166	258	65•0
39•2	423	166	257	65•1
39•2	421	165	256	64•7
39•2	418	164	254	64•3
39•2	416	163	253	63•9
39•2	413	162	251	63•5
39•2	411	161	250	63•1
39•2	408	160	248	62•7
39•2	406	159	247	62•3
39•2	403	158	245	61•9
39•2	401	157	244	61•5
39•2	398	156	242	61•1
39•2	395	155	240	60•8
39•2	393	154	239	60•3
39•2	390	153	237	60•0
39•2	388	152	236	59•5
39•2	385	151	234	59•2
39•2	383	150	233	58•7
39•2	380	149	231	58•4
39•2	378	148	230	57•9
39•2	375	147	228	57•6
39•2	372	146	226	57•3
39•2	370	145	225	56•8
39•2	367	144	223	56•5
39•2	365	143	222	56•0

%E	M1	M2	DM	M*
39•2	362	142	220	55•7
39•2	360	141	219	55•2
39•2	357	140	217	54•9
39•2	355	139	216	54•4
39•2	352	138	214	54•1
39•2	347	136	211	53•3
39•2	344	135	209	53•0
39•2	342	134	208	52•5
39•2	339	133	206	52•2
39•2	337	132	205	51•7
39•2	334	131	203	51•4
39•2	332	130	202	50•9
39•2	329	129	200	50•6
39•2	324	127	197	49•8
39•2	319	125	194	49•0
39•2	316	124	192	48•7
39•2	314	123	191	48•2
39•2	311	122	189	47•9
39•2	309	121	188	47•4
39•2	306	120	186	47•1
39•2	301	118	183	46•3
39•2	296	116	180	45•5
39•2	293	115	178	45•1
39•2	291	114	177	44•7
39•2	288	113	175	44•3
39•2	286	112	174	43•9
39•2	283	111	172	43•5
39•2	278	109	169	42•7
39•2	273	107	166	41•9
39•2	268	105	163	41•1
39•2	265	104	161	40•8
39•2	263	103	160	40•3
39•2	260	102	158	40•0
39•2	255	100	155	39•2
39•2	250	98	152	38•4
39•2	245	96	149	37•6
39•2	240	94	146	36•8
39•2	237	93	144	36•5
39•2	232	91	141	35•7
39•2	227	89	138	34•9
39•2	222	87	135	34•1
39•2	217	85	132	33•3
39•2	212	83	129	32•5
39•2	209	82	127	32•2
39•2	204	80	124	31•4
39•2	199	78	121	30•6
39•2	194	76	118	29•8
39•2	189	74	115	29•0
39•2	186	73	113	28•7
39•2	181	71	110	27•9

%E	M1	M2	DM	M*
39•2	176	69	107	27•1
39•2	171	67	104	26•3
39•2	166	65	101	25•5
39•2	158	62	96	24•3
39•2	153	60	93	23•5
39•2	148	58	90	22•7
39•2	143	56	87	21•9
39•2	130	51	79	20•0
39•2	125	49	76	19•2
39•2	120	47	73	18•4
39•2	102	40	62	15•7
39•2	97	38	59	14•9
39•2	79	31	48	12•2
39•2	74	29	45	11•4
39•2	51	20	31	7•8
39•1	499	195	304	76•2
39•1	496	194	302	75•9
39•1	494	193	301	75•4
39•1	493	193	300	75•6
39•1	491	192	299	75•1
39•1	489	191	298	74•6
39•1	488	191	297	74•8
39•1	486	190	296	74•3
39•1	484	189	295	73•8
39•1	483	189	294	74•0
39•1	481	188	293	73•5
39•1	478	187	291	73•2
39•1	476	186	290	72•7
39•1	473	185	288	72•4
39•1	471	184	287	71•9
39•1	470	184	286	72•0
39•1	468	183	285	71•6
39•1	466	182	284	71•1
39•1	465	182	283	71•2
39•1	463	181	282	70•8
39•1	460	180	280	70•4
39•1	458	179	279	70•0
39•1	455	178	277	69•6
39•1	453	177	276	69•2
39•1	450	176	274	68•8
39•1	448	175	273	68•4
39•1	445	174	271	68•0
39•1	443	173	270	67•6
39•1	442	173	269	67•7
39•1	440	172	268	67•2
39•1	437	171	266	66•9
39•1	435	170	265	66•4
39•1	432	169	263	66•1
39•1	430	168	262	65•6
39•1	427	167	260	65•3

%E	M1	M2	DM	M*
39·1	425	166	259	64·8
39·1	422	165	257	64·5
39·1	419	164	255	64·2
39·1	417	163	254	63·7
39·1	414	162	252	63·4
39·1	412	161	251	62·9
39·1	409	160	249	62·6
39·1	407	159	248	62·1
39·1	404	158	246	61·8
39·1	402	157	245	61·3
39·1	399	156	243	61·0
39·1	396	155	241	60·7
39·1	394	154	240	60·2
39·1	391	153	238	59·9
39·1	389	152	237	59·4
39·1	386	151	235	59·1
39·1	384	150	234	58·6
39·1	381	149	232	58·3
39·1	379	148	231	57·8
39·1	376	147	229	57·5
39·1	373	146	227	57·1
39·1	371	145	226	56·7
39·1	368	144	224	56·3
39·1	366	143	223	55·9
39·1	363	142	221	55·5
39·1	361	141	220	55·1
39·1	358	140	218	54·7
39·1	353	138	215	53·9
39·1	350	137	213	53·6
39·1	348	136	212	53·1
39·1	345	135	210	52·8
39·1	343	134	209	52·3
39·1	340	133	207	52·0
39·1	338	132	206	51·6
39·1	335	131	204	51·2
39·1	330	129	201	50·4
39·1	327	128	199	50·1
39·1	325	127	198	49·6
39·1	322	126	196	49·3
39·1	320	125	195	48·8
39·1	317	124	193	48·5
39·1	312	122	190	47·7
39·1	307	120	187	46·9
39·1	304	119	185	46·6
39·1	302	118	184	46·1
39·1	299	117	182	45·8
39·1	297	116	181	45·3
39·1	294	115	179	45·0
39·1	289	113	176	44·2
39·1	284	111	173	43·4

%E	M1	M2	DM	M*
39·1	281	110	171	43·1
39·1	279	109	170	42·6
39·1	276	108	168	42·3
39·1	274	107	167	41·8
39·1	271	106	165	41·5
39·1	266	104	162	40·7
39·1	261	102	159	39·9
39·1	258	101	157	39·5
39·1	256	100	156	39·1
39·1	253	99	154	38·7
39·1	248	97	151	37·9
39·1	243	95	148	37·1
39·1	238	93	145	36·3
39·1	235	92	143	36·0
39·1	233	91	142	35·5
39·1	230	90	140	35·2
39·1	225	88	137	34·4
39·1	220	86	134	33·6
39·1	215	84	131	32·8
39·1	207	81	126	31·7
39·1	202	79	123	30·9
39·1	197	77	120	30·1
39·1	192	75	117	29·3
39·1	184	72	112	28·2
39·1	179	70	109	27·4
39·1	174	68	106	26·6
39·1	169	66	103	25·8
39·1	161	63	98	24·7
39·1	156	61	95	23·9
39·1	151	59	92	23·1
39·1	138	54	84	21·1
39·1	133	52	81	20·3
39·1	128	50	78	19·5
39·1	115	45	70	17·6
39·1	110	43	67	16·8
39·1	92	36	56	14·1
39·1	87	34	53	13·3
39·1	69	27	42	10·6
39·1	64	25	39	9·8
39·1	46	18	28	7·0
39·1	23	9	14	3·5
39·0	500	195	305	76·0
39·0	498	194	304	75·6
39·0	497	194	303	75·7
39·0	495	193	302	75·3
39·0	492	192	300	74·9
39·0	490	191	299	74·5
39·0	487	190	297	74·1
39·0	485	189	296	73·7
39·0	482	188	294	73·3

%E	M1	M2	DM	M*
39·0	480	187	293	72·9
39·0	479	187	292	73·0
39·0	477	186	291	72·5
39·0	474	185	289	72·2
39·0	472	184	288	71·7
39·0	469	183	286	71·4
39·0	467	182	285	70·9
39·0	464	181	283	70·6
39·0	462	180	282	70·1
39·0	461	180	281	70·3
39·0	459	179	280	69·8
39·0	457	178	279	69·3
39·0	456	178	278	69·5
39·0	454	177	277	69·0
39·0	451	176	275	68·7
39·0	449	175	274	68·2
39·0	446	174	272	67·9
39·0	444	173	271	67·4
39·0	441	172	269	67·1
39·0	439	171	268	66·6
39·0	438	171	267	66·8
39·0	436	170	266	66·3
39·0	433	169	264	66·0
39·0	431	168	263	65·5
39·0	428	167	261	65·2
39·0	426	166	260	64·7
39·0	423	165	258	64·4
39·0	421	164	257	63·9
39·0	420	164	256	64·0
39·0	418	163	255	63·6
39·0	415	162	253	63·2
39·0	413	161	252	62·8
39·0	410	160	250	62·4
39·0	408	159	249	62·0
39·0	405	158	247	61·6
39·0	403	157	246	61·2
39·0	400	156	244	60·8
39·0	397	155	242	60·5
39·0	395	154	241	60·0
39·0	392	153	239	59·7
39·0	390	152	238	59·2
39·0	387	151	236	58·9
39·0	385	150	235	58·4
39·0	382	149	233	58·1
39·0	377	147	230	57·3
39·0	374	146	228	57·0
39·0	372	145	227	56·5
39·0	369	144	225	56·2
39·0	367	143	224	55·7
39·0	364	142	222	55·4

%E	M1	M2	DM	M*
39·0	362	141	221	54·9
39·0	359	140	219	54·6
39·0	356	139	217	54·3
39·0	354	138	216	53·8
39·0	351	137	214	53·5
39·0	349	136	213	53·0
39·0	346	135	211	52·7
39·0	344	134	210	52·2
39·0	341	133	208	51·9
39·0	336	131	205	51·1
39·0	333	130	203	50·8
39·0	331	129	202	50·3
39·0	328	128	200	50·0
39·0	326	127	199	49·5
39·0	323	126	197	49·2
39·0	318	124	194	48·4
39·0	315	123	192	48·0
39·0	313	122	191	47·6
39·0	310	121	189	47·2
39·0	308	120	188	46·8
39·0	305	119	186	46·4
39·0	300	117	183	45·6
39·0	295	115	180	44·8
39·0	292	114	178	44·5
39·0	290	113	177	44·0
39·0	287	112	175	43·7
39·0	282	110	172	42·9
39·0	277	108	169	42·1
39·0	272	106	166	41·3
39·0	269	105	164	41·0
39·0	267	104	163	40·5
39·0	264	103	161	40·2
39·0	259	101	158	39·4
39·0	254	99	155	38·6
39·0	251	98	153	38·3
39·0	249	97	152	37·8
39·0	246	96	150	37·5
39·0	241	94	147	36·7
39·0	236	92	144	35·9
39·0	231	90	141	35·1
39·0	228	89	139	34·7
39·0	223	87	136	33·9
39·0	218	85	133	33·1
39·0	213	83	130	32·3
39·0	210	82	128	32·0
39·0	205	80	125	31·2
39·0	200	78	122	30·4
39·0	195	76	119	29·6
39·0	187	73	114	28·5
39·0	182	71	111	27·7

%E	M1	M2	DM	M*
39·0	177	69	108	26·9
39·0	172	67	105	26·1
39·0	164	64	100	25·0
39·0	159	62	97	24·2
39·0	154	60	94	23·4
39·0	146	57	89	22·3
39·0	141	55	86	21·5
39·0	136	53	83	20·7
39·0	123	48	75	18·7
39·0	118	46	72	17·9
39·0	105	41	64	16·0
39·0	100	39	61	15·2
39·0	82	32	50	12·5
39·0	77	30	47	11·7
39·0	59	23	36	9·0
39·0	41	16	25	6·2
38·9	499	194	305	75·4
38·9	496	193	303	75·1
38·9	494	192	302	74·6
38·9	493	192	301	74·8
38·9	491	191	300	74·3
38·9	489	190	299	73·8
38·9	488	190	298	74·0
38·9	486	189	297	73·5
38·9	483	188	295	73·2
38·9	481	187	294	72·7
38·9	478	186	292	72·4
38·9	476	185	291	71·9
38·9	475	185	290	72·1
38·9	473	184	289	71·6
38·9	471	183	288	71·1
38·9	470	183	287	71·3
38·9	468	182	286	70·8
38·9	465	181	284	70·5
38·9	463	180	283	70·0
38·9	460	179	281	69·7
38·9	458	178	280	69·2
38·9	455	177	278	68·9
38·9	453	176	277	68·4
38·9	452	176	276	68·5
38·9	450	175	275	68·1
38·9	447	174	273	67·7
38·9	445	173	272	67·3
38·9	442	172	270	66·9
38·9	440	171	269	66·5
38·9	437	170	267	66·1
38·9	435	169	266	65·7
38·9	434	169	265	65·8
38·9	432	168	264	65·3
38·9	429	167	262	65·0

%E	M1	M2	DM	M*
38.9	427	166	261	64.5
38.9	424	165	259	64.2
38.9	422	164	258	63.7
38.9	419	163	256	63.4
38.9	416	162	254	63.1
38.9	414	161	253	62.6
38.9	411	160	251	62.3
38.9	409	159	250	61.8
38.9	406	158	248	61.5
38.9	404	157	247	61.0
38.9	401	156	245	60.7
38.9	398	155	243	60.4
38.9	396	154	242	59.9
38.9	393	153	240	59.6
38.9	391	152	239	59.1
38.9	388	151	237	58.8
38.9	386	150	236	58.3
38.9	383	149	234	58.0
38.9	380	148	232	57.6
38.9	378	147	231	57.2
38.9	375	146	229	56.8
38.9	373	145	228	56.4
38.9	370	144	226	56.0
38.9	368	143	225	55.6
38.9	365	142	223	55.2
38.9	360	140	220	54.4
38.9	357	139	218	54.1
38.9	355	138	217	53.6
38.9	352	137	215	53.3
38.9	350	136	214	52.8
38.9	347	135	212	52.5
38.9	342	133	209	51.7
38.9	339	132	207	51.4
38.9	337	131	206	50.9
38.9	334	130	204	50.6
38.9	332	129	203	50.1
38.9	329	128	201	49.8
38.9	324	126	198	49.0
38.9	321	125	196	48.7
38.9	319	124	195	48.2
38.9	316	123	193	47.9
38.9	314	122	192	47.4
38.9	311	121	190	47.1
38.9	306	119	187	46.3
38.9	303	118	185	46.0
38.9	301	117	184	45.5
38.9	298	116	182	45.2
38.9	296	115	181	44.7
38.9	293	114	179	44.4
38.9	288	112	176	43.6

%E	M1	M2	DM	M*
38.9	285	111	174	43.2
38.9	283	110	173	42.8
38.9	280	109	171	42.4
38.9	275	107	168	41.6
38.9	270	105	165	40.8
38.9	265	103	162	40.0
38.9	262	102	160	39.7
38.9	257	100	157	38.9
38.9	252	98	154	38.1
38.9	247	96	151	37.3
38.9	244	95	149	37.0
38.9	239	93	146	36.2
38.9	234	91	143	35.4
38.9	229	89	140	34.6
38.9	226	88	138	34.3
38.9	221	86	135	33.5
38.9	216	84	132	32.7
38.9	211	82	129	31.9
38.9	208	81	127	31.5
38.9	203	79	124	30.7
38.9	198	77	121	29.9
38.9	193	75	118	29.1
38.9	190	74	116	28.8
38.9	185	72	113	28.0
38.9	180	70	110	27.2
38.9	175	68	107	26.4
38.9	167	65	102	25.3
38.9	162	63	99	24.5
38.9	157	61	96	23.7
38.9	149	58	91	22.6
38.9	144	56	88	21.8
38.9	131	51	80	19.9
38.9	126	49	77	19.1
38.9	113	44	69	17.1
38.9	108	42	66	16.3
38.9	95	37	58	14.4
38.9	90	35	55	13.6
38.9	72	28	44	10.9
38.9	54	21	33	8.2
38.9	36	14	22	5.4
38.9	18	7	11	2.7
38.8	500	194	306	75.3
38.8	498	193	305	74.8
38.8	497	193	304	74.9
38.8	495	192	303	74.5
38.8	492	191	301	74.1
38.8	490	190	300	73.7
38.8	487	189	298	73.3
38.8	485	188	297	72.9
38.8	484	188	296	73.0

%E	M1	M2	DM	M*
38.8	482	187	295	72.5
38.8	480	186	294	72.1
38.8	479	186	293	72.2
38.8	477	185	292	71.8
38.8	474	184	290	71.4
38.8	472	183	289	71.0
38.8	469	182	287	70.6
38.8	467	181	286	70.2
38.8	466	181	285	70.3
38.8	464	180	284	69.8
38.8	461	179	282	69.5
38.8	459	178	281	69.0
38.8	456	177	279	68.7
38.8	454	176	278	68.2
38.8	451	175	276	67.9
38.8	449	174	275	67.4
38.8	448	174	274	67.6
38.8	446	173	273	67.1
38.8	443	172	271	66.8
38.8	441	171	270	66.3
38.8	438	170	268	66.0
38.8	436	169	267	65.5
38.8	433	168	265	65.2
38.8	430	167	263	64.9
38.8	428	166	262	64.4
38.8	425	165	260	64.1
38.8	423	164	259	63.6
38.8	420	163	257	63.3
38.8	418	162	256	62.8
38.8	417	162	255	62.9
38.8	415	161	254	62.5
38.8	412	160	252	62.1
38.8	410	159	251	61.7
38.8	407	158	249	61.3
38.8	405	157	248	60.9
38.8	402	156	246	60.5
38.8	400	155	245	60.1
38.8	399	155	244	60.2
38.8	397	154	243	59.7
38.8	394	153	241	59.4
38.8	392	152	240	58.9
38.8	389	151	238	58.6
38.8	387	150	237	58.1
38.8	384	149	235	57.8
38.8	381	148	233	57.5
38.8	379	147	232	57.0
38.8	376	146	230	56.7
38.8	374	145	229	56.2
38.8	371	144	227	55.9
38.8	369	143	226	55.4

%E	M1	M2	DM	M*
38.8	366	142	224	55.1
38.8	363	141	222	54.8
38.8	361	140	221	54.3
38.8	358	139	219	54.0
38.8	356	138	218	53.5
38.8	353	137	216	53.2
38.8	348	135	213	52.4
38.8	345	134	211	52.0
38.8	343	133	210	51.6
38.8	340	132	208	51.2
38.8	338	131	207	50.8
38.8	335	130	205	50.4
38.8	330	128	202	49.6
38.8	327	127	200	49.3
38.8	325	126	199	48.8
38.8	322	125	197	48.5
38.8	320	124	196	48.0
38.8	317	123	194	47.7
38.8	312	121	191	46.9
38.8	309	120	189	46.6
38.8	307	119	188	46.1
38.8	304	118	186	45.8
38.8	299	116	183	45.0
38.8	294	114	180	44.2
38.8	291	113	178	43.9
38.8	289	112	177	43.4
38.8	286	111	175	43.1
38.8	281	109	172	42.3
38.8	278	108	170	42.0
38.8	276	107	169	41.5
38.8	273	106	167	41.2
38.8	268	104	164	40.4
38.8	263	102	161	39.6
38.8	260	101	159	39.2
38.8	258	100	158	38.8
38.8	255	99	156	38.4
38.8	250	97	153	37.6
38.8	245	95	150	36.8
38.8	242	94	148	36.5
38.8	240	93	147	36.0
38.8	237	92	145	35.7
38.8	232	90	142	34.9
38.8	227	88	139	34.1
38.8	224	87	137	33.8
38.8	219	85	134	33.0
38.8	214	83	131	32.2
38.8	209	81	128	31.4
38.8	206	80	126	31.1
38.8	201	78	123	30.3
38.8	196	76	120	29.5

%E	M1	M2	DM	M*
38.8	188	73	115	28.3
38.8	183	71	112	27.5
38.8	178	69	109	26.7
38.8	170	66	104	25.6
38.8	165	64	101	24.8
38.8	160	62	98	24.0
38.8	152	59	93	22.9
38.8	147	57	90	22.1
38.8	139	54	85	21.0
38.8	134	52	82	20.2
38.8	129	50	79	19.4
38.8	121	47	74	18.3
38.8	116	45	71	17.5
38.8	103	40	63	15.5
38.8	98	38	60	14.7
38.8	85	33	52	12.8
38.8	80	31	49	12.0
38.8	67	26	41	10.1
38.8	49	19	30	7.4
38.7	499	193	306	74.6
38.7	496	192	304	74.3
38.7	494	191	303	73.8
38.7	493	191	302	74.0
38.7	491	190	301	73.5
38.7	489	189	300	73.0
38.7	488	189	299	73.2
38.7	486	188	298	72.7
38.7	483	187	296	72.4
38.7	481	186	295	71.9
38.7	478	185	293	71.6
38.7	476	184	292	71.1
38.7	475	184	291	71.3
38.7	473	183	290	70.8
38.7	470	182	288	70.5
38.7	468	181	287	70.0
38.7	465	180	285	69.7
38.7	463	179	284	69.2
38.7	462	179	283	69.4
38.7	460	178	282	68.9
38.7	457	177	280	68.6
38.7	455	176	279	68.1
38.7	452	175	277	67.8
38.7	450	174	276	67.3
38.7	447	173	274	67.0
38.7	445	172	273	66.5
38.7	444	172	272	66.6
38.7	442	171	271	66.2
38.7	439	170	269	65.8
38.7	437	169	268	65.4
38.7	434	168	266	65.0

%E	M1	M2	DM	M*
38.7	432	167	265	64.6
38.7	431	167	264	64.7
38.7	429	166	263	64.2
38.7	426	165	261	63.9
38.7	424	164	260	63.4
38.7	421	163	258	63.1
38.7	419	162	257	62.6
38.7	416	161	255	62.3
38.7	413	160	253	62.0
38.7	411	159	252	61.5
38.7	408	158	250	61.2
38.7	406	157	249	60.7
38.7	403	156	247	60.4
38.7	401	155	246	59.9
38.7	398	154	244	59.6
38.7	395	153	242	59.3
38.7	393	152	241	58.8
38.7	390	151	239	58.5
38.7	388	150	238	58.0
38.7	385	149	236	57.7
38.7	382	148	234	57.3
38.7	380	147	233	56.9
38.7	377	146	231	56.5
38.7	375	145	230	56.1
38.7	372	144	228	55.7
38.7	367	142	225	54.9
38.7	364	141	223	54.6
38.7	362	140	222	54.1
38.7	359	139	220	53.8
38.7	357	138	219	53.3
38.7	354	137	217	53.0
38.7	351	136	215	52.7
38.7	349	135	214	52.2
38.7	346	134	212	51.9
38.7	344	133	211	51.4
38.7	341	132	209	51.1
38.7	336	130	206	50.3
38.7	333	129	204	50.0
38.7	331	128	203	49.5
38.7	328	127	201	49.2
38.7	326	126	200	48.7
38.7	323	125	198	48.4
38.7	318	123	195	47.6
38.7	315	122	193	47.3
38.7	313	121	192	46.8
38.7	310	120	190	46.5
38.7	305	118	187	45.7
38.7	302	117	185	45.3
38.7	300	116	184	44.9
38.7	297	115	182	44.5

%E	M1	M2	DM	M*
38.7	292	113	179	43.7
38.7	287	111	176	42.9
38.7	284	110	174	42.6
38.7	282	109	173	42.1
38.7	279	108	171	41.8
38.7	274	106	168	41.0
38.7	271	105	166	40.7
38.7	269	104	165	40.2
38.7	266	103	163	39.9
38.7	261	101	160	39.1
38.7	256	99	157	38.3
38.7	253	98	155	38.0
38.7	248	96	152	37.2
38.7	243	94	149	36.4
38.7	238	92	146	35.6
38.7	235	91	144	35.2
38.7	230	89	141	34.4
38.7	225	87	138	33.6
38.7	222	86	136	33.3
38.7	217	84	133	32.5
38.7	212	82	130	31.7
38.7	204	79	125	30.6
38.7	199	77	122	29.8
38.7	194	75	119	29.0
38.7	191	74	117	28.7
38.7	186	72	114	27.9
38.7	181	70	111	27.1
38.7	173	67	106	25.9
38.7	168	65	103	25.1
38.7	163	63	100	24.3
38.7	155	60	95	23.2
38.7	150	58	92	22.4
38.7	142	55	87	21.3
38.7	137	53	84	20.5
38.7	124	48	76	18.6
38.7	119	46	73	17.8
38.7	111	43	68	16.7
38.7	106	41	65	15.9
38.7	93	36	57	13.9
38.7	75	29	46	11.2
38.7	62	24	38	9.3
38.7	31	12	19	4.6
38.6	500	193	307	74.5
38.6	498	192	306	74.0
38.6	497	192	305	74.2
38.6	495	191	304	73.7
38.6	492	190	302	73.4
38.6	490	189	301	72.9
38.6	487	188	299	72.6
38.6	485	187	298	72.1

%E	M1	M2	DM	M*
38.6	484	187	297	72.3
38.6	482	186	296	71.8
38.6	479	185	294	71.5
38.6	477	184	293	71.0
38.6	474	183	291	70.7
38.6	472	182	290	70.2
38.6	471	182	289	70.3
38.6	469	181	288	69.9
38.6	466	180	286	69.5
38.6	464	179	285	69.1
38.6	461	178	283	68.7
38.6	459	177	282	68.3
38.6	458	177	281	68.4
38.6	456	176	280	67.9
38.6	453	175	278	67.6
38.6	451	174	277	67.1
38.6	448	173	275	66.8
38.6	446	172	274	66.3
38.6	443	171	272	66.0
38.6	440	170	270	65.7
38.6	438	169	269	65.2
38.6	435	168	267	64.9
38.6	433	167	266	64.4
38.6	430	166	264	64.1
38.6	428	165	263	63.6
38.6	427	165	262	63.8
38.6	425	164	261	63.3
38.6	422	163	259	63.0
38.6	420	162	258	62.5
38.6	417	161	256	62.2
38.6	415	160	255	61.7
38.6	414	160	254	61.8
38.6	412	159	253	61.4
38.6	409	158	251	61.0
38.6	407	157	250	60.6
38.6	404	156	248	60.2
38.6	402	155	247	59.8
38.6	399	154	245	59.4
38.6	396	153	243	59.1
38.6	394	152	242	58.6
38.6	391	151	240	58.3
38.6	389	150	239	57.8
38.6	386	149	237	57.5
38.6	383	148	235	57.2
38.6	381	147	234	56.7
38.6	378	146	232	56.4
38.6	376	145	231	55.9
38.6	373	144	229	55.6
38.6	370	143	227	55.3
38.6	368	142	226	54.8

%E	M1	M2	DM	M*
38.6	365	141	224	54.5
38.6	363	140	223	54.0
38.6	360	139	221	53.7
38.6	355	137	218	52.9
38.6	352	136	216	52.5
38.6	350	135	215	52.1
38.6	347	134	213	51.7
38.6	345	133	212	51.3
38.6	342	132	210	50.9
38.6	339	131	208	50.6
38.6	337	130	207	50.1
38.6	334	129	205	49.8
38.6	332	128	204	49.3
38.6	329	127	202	49.0
38.6	324	125	199	48.2
38.6	321	124	197	47.9
38.6	319	123	196	47.4
38.6	316	122	194	47.1
38.6	311	120	191	46.3
38.6	308	119	189	46.0
38.6	306	118	188	45.5
38.6	303	117	186	45.2
38.6	298	115	183	44.4
38.6	295	114	181	44.1
38.6	293	113	180	43.6
38.6	290	112	178	43.3
38.6	285	110	175	42.5
38.6	280	108	172	41.7
38.6	277	107	170	41.3
38.6	272	105	167	40.5
38.6	267	103	164	39.7
38.6	264	102	162	39.4
38.6	262	101	161	38.9
38.6	259	100	159	38.6
38.6	254	98	156	37.8
38.6	251	97	154	37.5
38.6	249	96	153	37.0
38.6	246	95	151	36.7
38.6	241	93	148	35.9
38.6	236	91	145	35.1
38.6	233	90	143	34.8
38.6	228	88	140	34.0
38.6	223	86	137	33.2
38.6	220	85	135	32.8
38.6	215	83	132	32.0
38.6	210	81	129	31.2
38.6	207	80	127	30.9
38.6	202	78	124	30.1
38.6	197	76	121	29.3
38.6	189	73	116	28.2

%E	M1	M2	DM	M*
38.6	184	71	113	27.4
38.6	176	68	108	26.3
38.6	171	66	105	25.5
38.6	166	64	102	24.7
38.6	158	61	97	23.6
38.6	153	59	94	22.8
38.6	145	56	89	21.6
38.6	140	54	86	20.8
38.6	132	51	81	19.7
38.6	127	49	78	18.9
38.6	114	44	70	17.0
38.6	101	39	62	15.1
38.6	88	34	54	13.1
38.6	83	32	51	12.3
38.6	70	27	43	10.4
38.6	57	22	35	8.5
38.6	44	17	27	6.6
38.5	499	192	307	73.9
38.5	496	191	305	73.6
38.5	494	190	304	73.1
38.5	493	190	303	73.2
38.5	491	189	302	72.8
38.5	488	188	300	72.4
38.5	486	187	299	72.0
38.5	483	186	297	71.6
38.5	481	185	296	71.2
38.5	480	185	295	71.3
38.5	478	184	294	70.8
38.5	475	183	292	70.5
38.5	473	182	291	70.0
38.5	470	181	289	69.7
38.5	468	180	288	69.2
38.5	467	180	287	69.4
38.5	465	179	286	68.9
38.5	462	178	284	68.6
38.5	460	177	283	68.1
38.5	457	176	281	67.8
38.5	455	175	280	67.3
38.5	454	175	279	67.5
38.5	452	174	278	67.0
38.5	449	173	276	66.7
38.5	447	172	275	66.2
38.5	444	171	273	65.9
38.5	442	170	272	65.4
38.5	441	170	271	65.5
38.5	439	169	270	65.1
38.5	436	168	268	64.7
38.5	434	167	267	64.3
38.5	431	166	265	63.9
38.5	429	165	264	63.5

%E	M1	M2	DM	M*	%E	M1	M2	DM	M*	%E	M1	M2	DM	M*	%E	M1	M2	DM	M*	%E	M1	M2	DM	M*
38.5	426	164	262	63.1	38.5	283	109	174	42.0	38.4	492	189	303	72.6	38.4	383	147	236	56.4	38.4	216	83	133	31.9
38.5	423	163	260	62.8	38.5	278	107	171	41.2	38.4	490	188	302	72.1	38.4	380	146	234	56.1	38.4	211	81	130	31.1
38.5	421	162	259	62.3	38.5	275	106	169	40.9	38.4	489	188	301	72.3	38.4	378	145	233	55.6	38.4	206	79	127	30.3
38.5	418	161	257	62.0	38.5	273	105	168	40.4	38.4	487	187	300	71.8	38.4	375	144	231	55.3	38.4	203	78	125	30.0
38.5	416	160	256	61.5	38.5	270	104	166	40.1	38.4	485	186	299	71.3	38.4	372	143	229	55.0	38.4	198	76	122	29.2
38.5	413	159	254	61.2	38.5	265	102	163	39.3	38.4	484	186	298	71.5	38.4	370	142	228	54.5	38.4	190	73	117	28.0
38.5	410	158	252	60.9	38.5	260	100	160	38.5	38.4	482	185	297	71.0	38.4	367	141	226	54.2	38.4	185	71	114	27.2
38.5	408	157	251	60.4	38.5	257	99	158	38.1	38.4	479	184	295	70.7	38.4	365	140	225	53.7	38.4	177	68	109	26.1
38.5	405	156	249	60.1	38.5	252	97	155	37.3	38.4	477	183	294	70.2	38.4	362	139	223	53.4	38.4	172	66	106	25.3
38.5	403	155	248	59.6	38.5	247	95	152	36.5	38.4	476	183	293	70.4	38.4	359	138	221	53.0	38.4	164	63	101	24.2
38.5	400	154	246	59.3	38.5	244	94	150	36.2	38.4	474	182	292	69.9	38.4	357	137	220	52.6	38.4	159	61	98	23.4
38.5	397	153	244	59.0	38.5	239	92	147	35.4	38.4	471	181	290	69.6	38.4	354	136	218	52.2	38.4	151	58	93	22.3
38.5	395	152	243	58.5	38.5	234	90	144	34.6	38.4	469	180	289	69.1	38.4	352	135	217	51.8	38.4	146	56	90	21.5
38.5	392	151	241	58.2	38.5	231	89	142	34.3	38.4	466	179	287	68.8	38.4	349	134	215	51.4	38.4	138	53	85	20.4
38.5	390	150	240	57.7	38.5	226	87	139	33.5	38.4	464	178	286	68.3	38.4	346	133	213	51.1	38.4	125	48	77	18.4
38.5	387	149	238	57.4	38.5	221	85	136	32.7	38.4	463	178	285	68.4	38.4	344	132	212	50.7	38.4	112	43	69	16.5
38.5	384	148	236	57.0	38.5	218	84	134	32.4	38.4	461	177	284	68.0	38.4	341	131	210	50.3	38.4	99	38	61	14.6
38.5	382	147	235	56.6	38.5	213	82	131	31.6	38.4	458	176	282	67.6	38.4	336	129	207	49.5	38.4	86	33	53	12.7
38.5	379	146	233	56.2	38.5	208	80	128	30.8	38.4	456	175	281	67.2	38.4	333	128	205	49.2	38.4	73	28	45	10.7
38.5	377	145	232	55.8	38.5	205	79	126	30.4	38.4	453	174	279	66.8	38.4	331	127	204	48.7	38.3	499	191	308	73.1
38.5	374	144	230	55.4	38.5	200	77	123	29.6	38.4	451	173	278	66.4	38.4	328	126	202	48.4	38.3	496	190	306	72.8
38.5	371	143	228	55.1	38.5	195	75	120	28.8	38.4	450	173	277	66.5	38.4	323	124	199	47.6	38.3	494	189	305	72.3
38.5	369	142	227	54.6	38.5	192	74	118	28.5	38.4	448	172	276	66.0	38.4	320	123	197	47.3	38.3	493	189	304	72.5
38.5	366	141	225	54.3	38.5	187	72	115	27.7	38.4	445	171	274	65.7	38.4	318	122	196	46.8	38.3	491	188	303	72.0
38.5	364	140	224	53.8	38.5	182	70	112	26.9	38.4	443	170	273	65.2	38.4	315	121	194	46.5	38.3	488	187	301	71.7
38.5	361	139	222	53.5	38.5	179	69	110	26.6	38.4	440	169	271	64.9	38.4	310	119	191	45.7	38.3	486	186	300	71.2
38.5	358	138	220	53.2	38.5	174	67	107	25.8	38.4	438	168	270	64.4	38.4	307	118	189	45.4	38.3	483	185	298	70.9
38.5	356	137	219	52.7	38.5	169	65	104	25.0	38.4	437	168	269	64.6	38.4	305	117	188	44.9	38.3	481	184	297	70.4
38.5	353	136	217	52.4	38.5	161	62	99	23.9	38.4	435	167	268	64.1	38.4	302	116	186	44.6	38.3	480	184	296	70.5
38.5	351	135	216	51.9	38.5	156	60	96	23.1	38.4	432	166	266	63.8	38.4	297	114	183	43.8	38.3	478	183	295	70.1
38.5	348	134	214	51.6	38.5	148	57	91	22.0	38.4	430	165	265	63.3	38.4	294	113	181	43.4	38.3	475	182	293	69.7
38.5	343	132	211	50.8	38.5	143	55	88	21.2	38.4	427	164	263	63.0	38.4	292	112	180	43.0	38.3	473	181	292	69.3
38.5	340	131	209	50.5	38.5	135	52	83	20.0	38.4	425	163	262	62.5	38.4	289	111	178	42.6	38.3	472	181	291	69.4
38.5	338	130	208	50.0	38.5	130	50	80	19.2	38.4	424	163	261	62.7	38.4	284	109	175	41.8	38.3	470	180	290	68.9
38.5	335	129	206	49.7	38.5	122	47	75	18.1	38.4	422	162	260	62.2	38.4	281	108	173	41.5	38.3	467	179	288	68.6
38.5	330	127	203	48.9	38.5	117	45	72	17.3	38.4	419	161	258	61.9	38.4	279	107	172	41.0	38.3	465	178	287	68.1
38.5	327	126	201	48.6	38.5	109	42	67	16.2	38.4	417	160	257	61.4	38.4	276	106	170	40.7	38.3	462	177	285	67.8
38.5	325	125	200	48.1	38.5	104	40	64	15.4	38.4	414	159	255	61.1	38.4	271	104	167	39.9	38.3	460	176	284	67.3
38.5	322	124	198	47.8	38.5	96	37	59	14.3	38.4	412	158	254	60.6	38.4	268	103	165	39.6	38.3	459	176	283	67.5
38.5	317	122	195	47.0	38.5	91	35	56	13.5	38.4	411	158	253	60.7	38.4	263	101	162	38.8	38.3	457	175	282	67.0
38.5	314	121	193	46.6	38.5	78	30	48	11.5	38.4	409	157	252	60.3	38.4	258	99	159	38.0	38.3	454	174	280	66.7
38.5	312	120	192	46.2	38.5	65	25	40	9.6	38.4	406	156	250	59.9	38.4	255	98	157	37.7	38.3	452	173	279	66.2
38.5	309	119	190	45.8	38.5	52	20	32	7.7	38.4	404	155	249	59.5	38.4	250	96	154	36.9	38.3	449	172	277	65.9
38.5	304	117	187	45.0	38.5	39	15	24	5.8	38.4	401	154	247	59.1	38.4	245	94	151	36.1	38.3	447	171	276	65.4
38.5	301	116	185	44.7	38.5	26	10	16	3.8	38.4	398	153	245	58.8	38.4	242	93	149	35.7	38.3	446	171	275	65.6
38.5	299	115	184	44.2	38.5	13	5	8	1.9	38.4	396	152	244	58.3	38.4	237	91	146	34.9	38.3	444	170	274	65.1
38.5	296	114	182	43.9	38.4	500	192	308	73.7	38.4	393	151	242	58.0	38.4	232	89	143	34.1	38.3	441	169	272	64.8
38.5	291	112	179	43.1	38.4	498	191	307	73.3	38.4	391	150	241	57.5	38.4	229	88	141	33.8	38.3	439	168	271	64.3
38.5	288	111	177	42.8	38.4	497	191	306	73.4	38.4	388	149	239	57.2	38.4	224	86	138	33.0	38.3	436	167	269	64.0
38.5	286	110	176	42.3	38.4	495	190	305	72.9	38.4	385	148	237	56.9	38.4	219	84	135	32.2	38.3	433	166	267	63.6

%E	M1	M2	DM	M*	%E	M1	M2	DM	M*	%E	M1	M2	DM	M*	%E	M1	M2	DM	M*	%E	M1	M2	DM	M*
38.3	431	165	266	63.2	38.3	295	113	182	43.3	38.2	484	185	299	70.7	38.2	364	139	225	53.1	38.2	186	71	115	27.1
38.3	428	164	264	62.8	38.3	290	111	179	42.5	38.2	482	184	298	70.2	38.2	361	138	223	52.8	38.2	178	68	110	26.0
38.3	426	163	263	62.4	38.3	287	110	177	42.2	38.2	479	183	296	69.9	38.2	359	137	222	52.3	38.2	173	66	107	25.2
38.3	423	162	261	62.0	38.3	282	108	174	41.4	38.2	477	182	295	69.4	38.2	356	136	220	52.0	38.2	170	65	105	24.9
38.3	420	161	259	61.7	38.3	277	106	171	40.6	38.2	476	182	294	69.6	38.2	353	135	218	51.6	38.2	165	63	102	24.1
38.3	418	160	258	61.2	38.3	274	105	169	40.2	38.2	474	181	293	69.1	38.2	351	134	217	51.2	38.2	157	60	97	22.9
38.3	415	159	256	60.9	38.3	269	103	166	39.4	38.2	471	180	291	68.8	38.2	348	133	215	50.8	38.2	152	58	94	22.1
38.3	413	158	255	60.4	38.3	266	102	164	39.1	38.2	469	179	290	68.3	38.2	346	132	214	50.4	38.2	144	55	89	21.0
38.3	410	157	253	60.1	38.3	264	101	163	38.6	38.2	468	179	289	68.5	38.2	343	131	212	50.0	38.2	136	52	84	19.9
38.3	407	156	251	59.8	38.3	261	100	161	38.3	38.2	466	178	288	68.0	38.2	340	130	210	49.7	38.2	131	50	81	19.1
38.3	405	155	250	59.3	38.3	256	98	158	37.5	38.2	463	177	286	67.7	38.2	338	129	209	49.2	38.2	123	47	76	18.0
38.3	402	154	248	59.0	38.3	253	97	156	37.2	38.2	461	176	285	67.2	38.2	335	128	207	48.9	38.2	110	42	68	16.0
38.3	400	153	247	58.5	38.3	248	95	153	36.4	38.2	458	175	283	66.9	38.2	330	126	204	48.1	38.2	102	39	63	14.9
38.3	399	153	246	58.7	38.3	243	93	150	35.6	38.2	456	174	282	66.4	38.2	327	125	202	47.8	38.2	89	34	55	13.0
38.3	397	152	245	58.2	38.3	240	92	148	35.3	38.2	455	174	281	66.5	38.2	325	124	201	47.3	38.2	76	29	47	11.1
38.3	394	151	243	57.9	38.3	235	90	145	34.5	38.2	453	173	280	66.1	38.2	322	123	199	47.0	38.2	68	26	42	9.9
38.3	392	150	242	57.4	38.3	230	88	142	33.7	38.2	450	172	278	65.7	38.2	319	122	197	46.7	38.2	55	21	34	8.0
38.3	389	149	240	57.1	38.3	227	87	140	33.3	38.2	448	171	277	65.3	38.2	317	121	196	46.2	38.2	34	13	21	5.0
38.3	386	148	238	56.7	38.3	222	85	137	32.5	38.2	445	170	275	64.9	38.2	314	120	194	45.9	38.1	499	190	309	72.3
38.3	384	147	237	56.3	38.3	214	82	132	31.4	38.2	442	169	273	64.6	38.2	309	118	191	45.1	38.1	496	189	307	72.0
38.3	381	146	235	55.9	38.3	209	80	129	30.6	38.2	440	168	272	64.1	38.2	306	117	189	44.7	38.1	494	188	306	71.5
38.3	379	145	234	55.5	38.3	201	77	124	29.5	38.2	437	167	270	63.8	38.2	304	116	188	44.3	38.1	493	188	305	71.7
38.3	376	144	232	55.1	38.3	196	75	121	28.7	38.2	435	166	269	63.3	38.2	301	115	186	43.9	38.1	491	187	304	71.2
38.3	373	143	230	54.8	38.3	193	74	119	28.4	38.2	434	166	268	63.5	38.2	296	113	183	43.1	38.1	488	186	302	70.9
38.3	371	142	229	54.4	38.3	188	72	116	27.6	38.2	432	165	267	63.0	38.2	293	112	181	42.8	38.1	486	185	301	70.4
38.3	368	141	227	54.0	38.3	183	70	113	26.8	38.2	429	164	265	62.7	38.2	288	110	178	42.0	38.1	485	185	300	70.6
38.3	366	140	226	53.6	38.3	180	69	111	26.4	38.2	427	163	264	62.2	38.2	285	109	176	41.7	38.1	483	184	299	70.1
38.3	363	139	224	53.2	38.3	175	67	108	25.7	38.2	424	162	262	61.9	38.2	283	108	175	41.2	38.1	480	183	297	69.8
38.3	360	138	222	52.9	38.3	167	64	103	24.5	38.2	422	161	261	61.4	38.2	280	107	173	40.9	38.1	478	182	296	69.3
38.3	358	137	221	52.4	38.3	162	62	100	23.7	38.2	421	161	260	61.6	38.2	275	105	170	40.1	38.1	475	181	294	69.0
38.3	355	136	219	52.1	38.3	154	59	95	22.6	38.2	419	160	259	61.1	38.2	272	104	168	39.8	38.1	473	180	293	68.5
38.3	350	134	216	51.3	38.3	149	57	92	21.8	38.2	416	159	257	60.8	38.2	267	102	165	39.0	38.1	472	180	292	68.6
38.3	347	133	214	51.0	38.3	141	54	87	20.7	38.2	414	158	256	60.3	38.2	262	100	162	38.2	38.1	470	179	291	68.2
38.3	345	132	213	50.5	38.3	133	51	82	19.6	38.2	411	157	254	60.0	38.2	259	99	160	37.8	38.1	467	178	289	67.8
38.3	342	131	211	50.2	38.3	128	49	79	18.8	38.2	408	156	252	59.6	38.2	254	97	157	37.0	38.1	465	177	288	67.4
38.3	339	130	209	49.9	38.3	120	46	74	17.6	38.2	406	155	251	59.2	38.2	251	96	155	36.7	38.1	464	177	287	67.5
38.3	337	129	208	49.4	38.3	115	44	71	16.8	38.2	403	154	249	58.8	38.2	249	95	154	36.2	38.1	462	176	286	67.0
38.3	334	128	206	49.1	38.3	107	41	66	15.7	38.2	401	153	248	58.4	38.2	246	94	152	35.9	38.1	459	175	284	66.7
38.3	332	127	205	48.6	38.3	94	36	58	13.8	38.2	398	152	246	58.1	38.2	241	92	149	35.1	38.1	457	174	283	66.2
38.3	329	126	203	48.3	38.3	81	31	50	11.9	38.2	395	151	244	57.7	38.2	238	91	147	34.8	38.1	454	173	281	65.9
38.3	326	125	201	47.9	38.3	60	23	37	8.8	38.2	393	150	243	57.3	38.2	233	89	144	34.0	38.1	452	172	280	65.5
38.3	324	124	200	47.5	38.3	47	18	29	6.9	38.2	390	149	241	56.9	38.2	228	87	141	33.2	38.1	451	172	279	65.6
38.3	321	123	198	47.1	38.2	500	191	309	73.0	38.2	387	148	239	56.6	38.2	225	86	139	32.9	38.1	449	171	278	65.1
38.3	316	121	195	46.3	38.2	498	190	308	72.5	38.2	385	147	238	56.1	38.2	220	84	136	32.1	38.1	446	170	276	64.8
38.3	313	120	193	46.0	38.2	497	190	307	72.6	38.2	382	146	236	55.8	38.2	217	83	134	31.7	38.1	444	169	275	64.3
38.3	311	119	192	45.5	38.2	495	189	306	72.2	38.2	380	145	235	55.3	38.2	212	81	131	30.9	38.1	443	169	274	64.5
38.3	308	118	190	45.2	38.2	492	188	304	71.8	38.2	377	144	233	55.0	38.2	207	79	128	30.1	38.1	441	168	273	64.0
38.3	303	116	187	44.4	38.2	490	187	303	71.4	38.2	374	143	231	54.7	38.2	204	78	126	29.8	38.1	438	167	271	63.7
38.3	300	115	185	44.1	38.2	489	187	302	71.5	38.2	372	142	230	54.2	38.2	199	76	123	29.0	38.1	436	166	270	63.2
38.3	298	114	184	43.6	38.2	487	186	301	71.0	38.2	369	141	228	53.9	38.2	191	73	118	27.9	38.1	433	165	268	62.9

%E	M1	M2	DM	M*
38.1	431	164	267	62.4
38.1	430	164	266	62.5
38.1	428	163	265	62.1
38.1	425	162	263	61.8
38.1	423	161	262	61.3
38.1	420	160	260	61.0
38.1	417	159	258	60.6
38.1	415	158	257	60.2
38.1	412	157	255	59.8
38.1	409	156	253	59.5
38.1	407	155	252	59.0
38.1	404	154	250	58.7
38.1	402	153	249	58.2
38.1	399	152	247	57.9
38.1	396	151	245	57.6
38.1	394	150	244	57.1
38.1	391	149	242	56.8
38.1	388	148	240	56.5
38.1	386	147	239	56.0
38.1	383	146	237	55.7
38.1	381	145	236	55.2
38.1	378	144	234	54.9
38.1	375	143	232	54.5
38.1	373	142	231	54.1
38.1	370	141	229	53.7
38.1	367	140	227	53.4
38.1	365	139	226	52.9
38.1	362	138	224	52.6
38.1	360	137	223	52.1
38.1	357	136	221	51.8
38.1	354	135	219	51.5
38.1	352	134	218	51.0
38.1	349	133	216	50.7
38.1	344	131	213	49.9
38.1	341	130	211	49.6
38.1	339	129	210	49.1
38.1	336	128	208	48.8
38.1	333	127	206	48.4
38.1	331	126	205	48.0
38.1	328	125	203	47.6
38.1	323	123	200	46.8
38.1	320	122	198	46.5
38.1	318	121	197	46.0
38.1	315	120	195	45.7
38.1	312	119	193	45.4
38.1	310	118	192	44.9
38.1	307	117	190	44.6
38.1	302	115	187	43.8
38.1	299	114	185	43.5
38.1	294	112	182	42.7

%E	M1	M2	DM	M*
38.1	291	111	180	42.3
38.1	289	110	179	41.9
38.1	286	109	177	41.5
38.1	281	107	174	40.7
38.1	278	106	172	40.4
38.1	273	104	169	39.6
38.1	270	103	167	39.3
38.1	268	102	166	38.8
38.1	265	101	164	38.5
38.1	260	99	161	37.7
38.1	257	98	159	37.4
38.1	252	96	156	36.6
38.1	247	94	153	35.8
38.1	244	93	151	35.4
38.1	239	91	148	34.6
38.1	236	90	146	34.3
38.1	231	88	143	33.5
38.1	226	86	140	32.7
38.1	223	85	138	32.4
38.1	218	83	135	31.6
38.1	215	82	133	31.3
38.1	210	80	130	30.5
38.1	202	77	125	29.4
38.1	197	75	122	28.6
38.1	194	74	120	28.2
38.1	189	72	117	27.4
38.1	181	69	112	26.3
38.1	176	67	109	25.5
38.1	168	64	104	24.4
38.1	160	61	99	23.3
38.1	155	59	96	22.5
38.1	147	56	91	21.3
38.1	139	53	86	20.2
38.1	134	51	83	19.4
38.1	126	48	78	18.3
38.1	118	45	73	17.2
38.1	113	43	70	16.4
38.1	105	40	65	15.2
38.1	97	37	60	14.1
38.1	84	32	52	12.2
38.1	63	24	39	9.1
38.1	42	16	26	6.1
38.1	21	8	13	3.0
38.0	500	190	310	72.2
38.0	498	189	309	71.7
38.0	497	189	308	71.9
38.0	495	188	307	71.4
38.0	492	187	305	71.1
38.0	490	186	304	70.6
38.0	489	186	303	70.7

%E	M1	M2	DM	M*
38.0	487	185	302	70.3
38.0	484	184	300	70.0
38.0	482	183	299	69.5
38.0	481	183	298	69.6
38.0	479	182	297	69.2
38.0	476	181	295	68.8
38.0	474	180	294	68.4
38.0	471	179	292	68.0
38.0	469	178	291	67.6
38.0	468	178	290	67.7
38.0	466	177	289	67.2
38.0	463	176	287	66.9
38.0	461	175	286	66.4
38.0	460	175	285	66.6
38.0	458	174	284	66.1
38.0	455	173	282	65.8
38.0	453	172	281	65.3
38.0	450	171	279	65.0
38.0	447	170	277	64.7
38.0	445	169	276	64.2
38.0	442	168	274	63.9
38.0	440	167	273	63.4
38.0	439	167	272	63.5
38.0	437	166	271	63.1
38.0	434	165	269	62.7
38.0	432	164	268	62.3
38.0	429	163	266	61.9
38.0	426	162	264	61.6
38.0	424	161	263	61.1
38.0	421	160	261	60.8
38.0	418	159	259	60.5
38.0	416	158	258	60.0
38.0	413	157	256	59.7
38.0	411	156	255	59.2
38.0	410	156	254	59.4
38.0	408	155	253	58.9
38.0	405	154	251	58.6
38.0	403	153	250	58.1
38.0	400	152	248	57.8
38.0	397	151	246	57.4
38.0	395	150	245	57.0
38.0	392	149	243	56.6
38.0	389	148	241	56.3
38.0	387	147	240	55.8
38.0	384	146	238	55.5
38.0	382	145	237	55.0
38.0	379	144	235	54.7
38.0	376	143	233	54.4
38.0	374	142	232	53.9
38.0	371	141	230	53.6

%E	M1	M2	DM	M*
38.0	368	140	228	53.3
38.0	366	139	227	52.8
38.0	363	138	225	52.5
38.0	361	137	224	52.0
38.0	358	136	222	51.7
38.0	355	135	220	51.3
38.0	353	134	219	50.9
38.0	350	133	217	50.5
38.0	347	132	215	50.2
38.0	345	131	214	49.7
38.0	342	130	212	49.4
38.0	337	128	209	48.6
38.0	334	127	207	48.3
38.0	332	126	206	47.8
38.0	329	125	204	47.5
38.0	326	124	202	47.2
38.0	324	123	201	46.7
38.0	321	122	199	46.4
38.0	316	120	196	45.6
38.0	313	119	194	45.2
38.0	308	117	191	44.4
38.0	305	116	189	44.1
38.0	303	115	188	43.6
38.0	300	114	186	43.3
38.0	297	113	184	43.0
38.0	295	112	183	42.5
38.0	292	111	181	42.2
38.0	287	109	178	41.4
38.0	284	108	176	41.1
38.0	279	106	173	40.3
38.0	276	105	171	39.9
38.0	274	104	170	39.5
38.0	271	103	168	39.1
38.0	266	101	165	38.3
38.0	263	100	163	38.0
38.0	258	98	160	37.2
38.0	255	97	158	36.9
38.0	250	95	155	36.1
38.0	245	93	152	35.3
38.0	242	92	150	35.0
38.0	237	90	147	34.2
38.0	234	89	145	33.9
38.0	229	87	142	33.1
38.0	221	84	137	31.9
38.0	216	82	134	31.1
38.0	213	81	132	30.8
38.0	208	79	129	30.0
38.0	205	78	127	29.7
38.0	200	76	124	28.9
38.0	192	73	119	27.8

%E	M1	M2	DM	M*
38.0	187	71	116	27.0
38.0	184	70	114	26.6
38.0	179	68	111	25.8
38.0	171	65	106	24.7
38.0	166	63	103	23.9
38.0	163	62	101	23.6
38.0	158	60	98	22.8
38.0	150	57	93	21.7
38.0	142	54	88	20.5
38.0	137	52	85	19.7
38.0	129	49	80	18.6
38.0	121	46	75	17.5
38.0	108	41	67	15.6
38.0	100	38	62	14.4
38.0	92	35	57	13.3
38.0	79	30	49	11.4
38.0	71	27	44	10.3
38.0	50	19	31	7.2
37.9	499	189	310	71.6
37.9	496	188	308	71.3
37.9	494	187	307	70.8
37.9	493	187	306	70.9
37.9	491	186	305	70.5
37.9	488	185	303	70.1
37.9	486	184	302	69.7
37.9	485	184	301	69.8
37.9	483	183	300	69.3
37.9	480	182	298	69.0
37.9	478	181	297	68.5
37.9	477	181	296	68.7
37.9	475	180	295	68.2
37.9	472	179	293	67.9
37.9	470	178	292	67.4
37.9	467	177	290	67.1
37.9	464	176	288	66.8
37.9	462	175	287	66.3
37.9	459	174	285	66.0
37.9	457	173	284	65.5
37.9	456	173	283	65.6
37.9	454	172	282	65.2
37.9	451	171	280	64.8
37.9	449	170	279	64.4
37.9	448	170	278	64.5
37.9	446	169	277	64.0
37.9	443	168	275	63.7
37.9	441	167	274	63.2
37.9	438	166	272	62.9
37.9	435	165	270	62.6
37.9	433	164	269	62.1
37.9	430	163	267	61.8

%E	M1	M2	DM	M*
37.9	428	162	266	61.3
37.9	427	162	265	61.5
37.9	425	161	264	61.0
37.9	422	160	262	60.7
37.9	420	159	261	60.2
37.9	419	159	260	60.3
37.9	417	158	259	59.9
37.9	414	157	257	59.5
37.9	412	156	256	59.1
37.9	409	155	254	58.7
37.9	406	154	252	58.4
37.9	404	153	251	57.9
37.9	401	152	249	57.6
37.9	398	151	247	57.3
37.9	396	150	246	56.8
37.9	393	149	244	56.5
37.9	391	148	243	56.0
37.9	390	148	242	56.2
37.9	388	147	241	55.7
37.9	385	146	239	55.4
37.9	383	145	238	54.9
37.9	380	144	236	54.6
37.9	377	143	234	54.2
37.9	375	142	233	53.8
37.9	372	141	231	53.4
37.9	369	140	229	53.1
37.9	367	139	228	52.6
37.9	364	138	226	52.3
37.9	359	136	223	51.5
37.9	356	135	221	51.2
37.9	354	134	220	50.7
37.9	351	133	218	50.4
37.9	348	132	216	50.1
37.9	346	131	215	49.6
37.9	343	130	213	49.3
37.9	340	129	211	48.9
37.9	338	128	210	48.5
37.9	335	127	208	48.1
37.9	330	125	205	47.3
37.9	327	124	203	47.0
37.9	322	122	200	46.2
37.9	319	121	198	45.9
37.9	317	120	197	45.4
37.9	314	119	195	45.1
37.9	311	118	193	44.8
37.9	309	117	192	44.3
37.9	306	116	190	44.0
37.9	301	114	187	43.2
37.9	298	113	185	42.8
37.9	293	111	182	42.1

%E	M1	M2	DM	M*
37.9	290	110	180	41.7
37.9	285	108	177	40.9
37.9	282	107	175	40.6
37.9	280	106	174	40.1
37.9	277	105	172	39.8
37.9	272	103	169	39.0
37.9	269	102	167	38.7
37.9	264	100	164	37.9
37.9	261	99	162	37.6
37.9	256	97	159	36.8
37.9	253	96	157	36.4
37.9	248	94	154	35.6
37.9	243	92	151	34.8
37.9	240	91	149	34.5
37.9	235	89	146	33.7
37.9	232	88	144	33.4
37.9	227	86	141	32.6
37.9	224	85	139	32.3
37.9	219	83	136	31.5
37.9	214	81	133	30.7
37.9	211	80	131	30.3
37.9	206	78	128	29.5
37.9	203	77	126	29.2
37.9	198	75	123	28.4
37.9	195	74	121	28.1
37.9	190	72	118	27.3
37.9	182	69	113	26.2
37.9	177	67	110	25.4
37.9	174	66	108	25.0
37.9	169	64	105	24.2
37.9	161	61	100	23.1
37.9	153	58	95	22.0
37.9	145	55	90	20.9
37.9	140	53	87	20.1
37.9	132	50	82	18.9
37.9	124	47	77	17.8
37.9	116	44	72	16.7
37.9	103	39	64	14.8
37.9	95	36	59	13.6
37.9	87	33	54	12.5
37.9	66	25	41	9.5
37.9	58	22	36	8.3
37.9	29	11	18	4.2
37.8	500	189	311	71.4
37.8	498	188	310	71.0
37.8	497	188	309	71.1
37.8	495	187	308	70.6
37.8	492	186	306	70.3
37.8	490	185	305	69.8
37.8	489	185	304	70.0

%E	M1	M2	DM	M*
37.8	487	184	303	69.5
37.8	484	183	301	69.2
37.8	482	182	300	68.7
37.8	481	182	299	68.9
37.8	479	181	298	68.4
37.8	476	180	296	68.1
37.8	474	179	295	67.6
37.8	473	179	294	67.7
37.8	471	178	293	67.3
37.8	468	177	291	66.9
37.8	466	176	290	66.5
37.8	465	176	289	66.6
37.8	463	175	288	66.1
37.8	460	174	286	65.8
37.8	458	173	285	65.3
37.8	455	172	283	65.0
37.8	452	171	281	64.7
37.8	450	170	280	64.2
37.8	447	169	278	63.9
37.8	445	168	277	63.4
37.8	444	168	276	63.6
37.8	442	167	275	63.1
37.8	439	166	273	62.8
37.8	437	165	272	62.3
37.8	436	165	271	62.4
37.8	434	164	270	62.0
37.8	431	163	268	61.6
37.8	429	162	267	61.2
37.8	426	161	265	60.8
37.8	423	160	263	60.5
37.8	421	159	262	60.0
37.8	418	158	260	59.7
37.8	415	157	258	59.4
37.8	413	156	257	58.9
37.8	410	155	255	58.6
37.8	407	154	253	58.3
37.8	405	153	252	57.8
37.8	402	152	250	57.5
37.8	400	151	249	57.0
37.8	399	151	248	57.1
37.8	397	150	247	56.7
37.8	394	149	245	56.3
37.8	392	148	244	55.9
37.8	389	147	242	55.6
37.8	386	146	240	55.2
37.8	384	145	239	54.8
37.8	381	144	237	54.4
37.8	378	143	235	54.1
37.8	376	142	234	53.6
37.8	373	141	232	53.3

%E	M1	M2	DM	M*
37.8	370	140	230	53.0
37.8	368	139	229	52.5
37.8	365	138	227	52.2
37.8	362	137	225	51.8
37.8	360	136	224	51.4
37.8	357	135	222	51.1
37.8	352	133	219	50.3
37.8	349	132	217	49.9
37.8	347	131	216	49.5
37.8	344	130	214	49.1
37.8	341	129	212	48.8
37.8	339	128	211	48.3
37.8	336	127	209	48.0
37.8	333	126	207	47.7
37.8	331	125	206	47.2
37.8	328	124	204	46.9
37.8	325	123	202	46.6
37.8	323	122	201	46.1
37.8	320	121	199	45.8
37.8	315	119	196	45.0
37.8	312	118	194	44.6
37.8	307	116	191	43.8
37.8	304	115	189	43.5
37.8	299	113	186	42.7
37.8	296	112	184	42.4
37.8	294	111	183	41.9
37.8	291	110	181	41.6
37.8	288	109	179	41.3
37.8	286	108	178	40.8
37.8	283	107	176	40.5
37.8	278	105	173	39.7
37.8	275	104	171	39.3
37.8	270	102	168	38.5
37.8	267	101	166	38.2
37.8	262	99	163	37.4
37.8	259	98	161	37.1
37.8	254	96	158	36.3
37.8	251	95	156	36.0
37.8	249	94	155	35.5
37.8	246	93	153	35.2
37.8	241	91	150	34.4
37.8	238	90	148	34.0
37.8	233	88	145	33.2
37.8	230	87	143	32.9
37.8	225	85	140	32.1
37.8	222	84	138	31.8
37.8	217	82	135	31.0
37.8	209	79	130	29.9
37.8	201	76	125	28.7
37.8	196	74	122	27.9

%E	M1	M2	DM	M*
37.8	193	73	120	27.6
37.8	188	71	117	26.8
37.8	185	70	115	26.5
37.8	180	68	112	25.7
37.8	172	65	107	24.6
37.8	164	62	102	23.4
37.8	156	59	97	22.3
37.8	148	56	92	21.2
37.8	143	54	89	20.4
37.8	135	51	84	19.3
37.8	127	48	79	18.1
37.8	119	45	74	17.0
37.8	111	42	69	15.9
37.8	98	37	61	14.0
37.8	90	34	56	12.8
37.8	82	31	51	11.7
37.8	74	28	46	10.6
37.8	45	17	28	6.4
37.8	37	14	23	5.3
37.7	499	188	311	70.8
37.7	496	187	309	70.5
37.7	494	186	308	70.0
37.7	493	186	307	70.2
37.7	491	185	306	69.7
37.7	488	184	304	69.4
37.7	486	183	303	68.9
37.7	485	183	302	69.0
37.7	483	182	301	68.6
37.7	480	181	299	68.3
37.7	478	180	298	67.8
37.7	477	180	297	67.9
37.7	475	179	296	67.5
37.7	472	178	294	67.1
37.7	470	177	293	66.7
37.7	469	177	292	66.8
37.7	467	176	291	66.3
37.7	464	175	289	66.0
37.7	462	174	288	65.5
37.7	461	174	287	65.7
37.7	459	173	286	65.2
37.7	456	172	284	64.9
37.7	454	171	283	64.4
37.7	453	171	282	64.5
37.7	451	170	281	64.1
37.7	448	169	279	63.8
37.7	446	168	278	63.3
37.7	443	167	276	63.0
37.7	440	166	274	62.6
37.7	438	165	273	62.2
37.7	435	164	271	61.8

%E	M1	M2	DM	M*
37.7	432	163	269	61.5
37.7	430	162	268	61.0
37.7	427	161	266	60.7
37.7	424	160	264	60.4
37.7	422	159	263	59.9
37.7	419	158	261	59.6
37.7	417	157	260	59.1
37.7	416	157	259	59.3
37.7	414	156	258	58.8
37.7	411	155	256	58.5
37.7	409	154	255	58.0
37.7	408	154	254	58.1
37.7	406	153	253	57.7
37.7	403	152	251	57.3
37.7	401	151	250	56.9
37.7	398	150	248	56.5
37.7	395	149	246	56.2
37.7	393	148	245	55.7
37.7	390	147	243	55.4
37.7	387	146	241	55.1
37.7	385	145	240	54.6
37.7	382	144	238	54.3
37.7	379	143	236	54.0
37.7	377	142	235	53.5
37.7	374	141	233	53.2
37.7	371	140	231	52.8
37.7	369	139	230	52.4
37.7	366	138	228	52.0
37.7	363	137	226	51.7
37.7	361	136	225	51.2
37.7	358	135	223	50.9
37.7	355	134	221	50.6
37.7	353	133	220	50.1
37.7	350	132	218	49.8
37.7	345	130	215	49.0
37.7	342	129	213	48.7
37.7	337	127	210	47.9
37.7	334	126	208	47.5
37.7	332	125	207	47.1
37.7	329	124	205	46.7
37.7	326	123	203	46.4
37.7	324	122	202	45.9
37.7	321	121	200	45.6
37.7	318	120	198	45.3
37.7	316	119	197	44.8
37.7	313	118	195	44.5
37.7	310	117	193	44.2
37.7	308	116	192	43.7
37.7	305	115	190	43.4
37.7	302	114	188	43.0

%E	M1	M2	DM	M*
37.7	300	113	187	42.6
37.7	297	112	185	42.2
37.7	292	110	182	41.4
37.7	289	109	180	41.1
37.7	284	107	177	40.3
37.7	281	106	175	40.0
37.7	276	104	172	39.2
37.7	273	103	170	38.9
37.7	268	101	167	38.1
37.7	265	100	165	37.7
37.7	260	98	162	36.9
37.7	257	97	160	36.6
37.7	252	95	157	35.8
37.7	247	93	154	35.0
37.7	244	92	152	34.7
37.7	239	90	149	33.9
37.7	236	89	147	33.6
37.7	231	87	144	32.8
37.7	228	86	142	32.4
37.7	223	84	139	31.6
37.7	220	83	137	31.3
37.7	215	81	134	30.5
37.7	212	80	132	30.2
37.7	207	78	129	29.4
37.7	204	77	127	29.1
37.7	199	75	124	28.3
37.7	191	72	119	27.1
37.7	183	69	114	26.0
37.7	175	66	109	24.9
37.7	167	63	104	23.8
37.7	162	61	101	23.0
37.7	159	60	99	22.6
37.7	154	58	96	21.8
37.7	151	57	94	21.5
37.7	146	55	91	20.7
37.7	138	52	86	19.6
37.7	130	49	81	18.5
37.7	122	46	76	17.3
37.7	114	43	71	16.2
37.7	106	40	66	15.1
37.7	77	29	48	10.9
37.7	69	26	43	9.8
37.7	61	23	38	8.7
37.7	53	20	33	7.5
37.6	500	188	312	70.7
37.6	498	187	311	70.2
37.6	497	187	310	70.4
37.6	495	186	309	69.9
37.6	492	185	307	69.6
37.6	490	184	306	69.1

%E	M1	M2	DM	M*
37.6	489	184	305	69.2
37.6	487	183	304	68.8
37.6	484	182	302	68.4
37.6	482	181	301	68.0
37.6	481	181	300	68.1
37.6	479	180	299	67.6
37.6	476	179	297	67.3
37.6	474	178	296	66.8
37.6	473	178	295	67.0
37.6	471	177	294	66.5
37.6	468	176	292	66.2
37.6	466	175	291	65.7
37.6	465	175	290	65.9
37.6	463	174	289	65.4
37.6	460	173	287	65.1
37.6	458	172	286	64.6
37.6	457	172	285	64.7
37.6	455	171	284	64.3
37.6	452	170	282	63.9
37.6	450	169	281	63.5
37.6	449	169	280	63.6
37.6	447	168	279	63.1
37.6	444	167	277	62.8
37.6	442	166	276	62.3
37.6	441	166	275	62.5
37.6	439	165	274	62.0
37.6	436	164	272	61.7
37.6	434	163	271	61.2
37.6	433	163	270	61.4
37.6	431	162	269	60.9
37.6	428	161	267	60.6
37.6	426	160	266	60.1
37.6	425	160	265	60.2
37.6	423	159	264	59.8
37.6	420	158	262	59.4
37.6	418	157	261	59.0
37.6	415	156	259	58.6
37.6	412	155	257	58.3
37.6	410	154	256	57.8
37.6	407	153	254	57.5
37.6	404	152	252	57.2
37.6	402	151	251	56.7
37.6	399	150	249	56.4
37.6	396	149	247	56.1
37.6	394	148	246	55.6
37.6	391	147	244	55.3
37.6	388	146	242	54.9
37.6	386	145	241	54.5
37.6	383	144	239	54.1
37.6	380	143	237	53.8

%E	M1	M2	DM	M*
37.6	378	142	236	53.3
37.6	375	141	234	53.0
37.6	372	140	232	52.7
37.6	370	139	231	52.2
37.6	367	138	229	51.9
37.6	364	137	227	51.6
37.6	362	136	226	51.1
37.6	359	135	224	50.8
37.6	356	134	222	50.4
37.6	354	133	221	50.0
37.6	351	132	219	49.6
37.6	348	131	217	49.3
37.6	346	130	216	48.8
37.6	343	129	214	48.5
37.6	340	128	212	48.2
37.6	338	127	211	47.7
37.6	335	126	209	47.4
37.6	330	124	206	46.6
37.6	327	123	204	46.3
37.6	322	121	201	45.5
37.6	319	120	199	45.1
37.6	314	118	196	44.3
37.6	311	117	194	44.0
37.6	306	115	191	43.2
37.6	303	114	189	42.9
37.6	298	112	186	42.1
37.6	295	111	184	41.8
37.6	290	109	181	41.0
37.6	287	108	179	40.6
37.6	282	106	176	39.8
37.6	279	105	174	39.5
37.6	274	103	171	38.7
37.6	271	102	169	38.4
37.6	266	100	166	37.6
37.6	263	99	164	37.3
37.6	258	97	161	36.5
37.6	255	96	159	36.1
37.6	250	94	156	35.3
37.6	245	92	153	34.5
37.6	242	91	151	34.2
37.6	237	89	148	33.4
37.6	234	88	146	33.1
37.6	229	86	143	32.3
37.6	226	85	141	32.0
37.6	221	83	138	31.2
37.6	218	82	136	30.8
37.6	213	80	133	30.0
37.6	210	79	131	29.7
37.6	205	77	128	28.9
37.6	202	76	126	28.6

%E	M1	M2	DM	M*
37.6	197	74	123	27.8
37.6	194	73	121	27.5
37.6	189	71	118	26.7
37.6	186	70	116	26.3
37.6	181	68	113	25.5
37.6	178	67	111	25.2
37.6	173	65	108	24.4
37.6	170	64	106	24.1
37.6	165	62	103	23.3
37.6	157	59	98	22.2
37.6	149	56	93	21.0
37.6	141	53	88	19.9
37.6	133	50	83	18.8
37.6	125	47	78	17.7
37.6	117	44	73	16.5
37.6	109	41	68	15.4
37.6	101	38	63	14.3
37.6	93	35	58	13.2
37.6	85	32	53	12.0
37.5	499	187	312	70.1
37.5	496	186	310	69.8
37.5	493	185	308	69.4
37.5	491	184	307	69.0
37.5	488	183	305	68.6
37.5	485	182	303	68.3
37.5	483	181	302	67.8
37.5	480	180	300	67.5
37.5	477	179	298	67.2
37.5	475	178	297	66.7
37.5	472	177	295	66.4
37.5	469	176	293	66.0
37.5	467	175	292	65.6
37.5	464	174	290	65.3
37.5	461	173	288	64.9
37.5	459	172	287	64.5
37.5	456	171	285	64.1
37.5	453	170	283	63.8
37.5	451	169	282	63.3
37.5	448	168	280	63.0
37.5	445	167	278	62.7
37.5	443	166	277	62.2
37.5	440	165	275	61.9
37.5	437	164	273	61.5
37.5	435	163	272	61.1
37.5	432	162	270	60.8
37.5	429	161	268	60.4
37.5	427	160	267	60.0
37.5	424	159	265	59.6
37.5	421	158	263	59.3
37.5	419	157	262	58.8

%E	M1	M2	DM	M*
37.5	416	156	260	58.5
37.5	413	155	258	58.2
37.5	411	154	257	57.7
37.5	408	153	255	57.4
37.5	405	152	253	57.0
37.5	403	151	252	56.6
37.5	400	150	250	56.3
37.5	397	149	248	55.9
37.5	395	148	247	55.5
37.5	392	147	245	55.1
37.5	389	146	243	54.8
37.5	387	145	242	54.3
37.5	384	144	240	54.0
37.5	381	143	238	53.7
37.5	379	142	237	53.2
37.5	376	141	235	52.9
37.5	373	140	233	52.5
37.5	371	139	232	52.1
37.5	368	138	230	51.8
37.5	365	137	228	51.4
37.5	363	136	227	51.0
37.5	360	135	225	50.6
37.5	357	134	223	50.3
37.5	355	133	222	49.8
37.5	352	132	220	49.5
37.5	349	131	218	49.2
37.5	347	130	217	48.7
37.5	344	129	215	48.4
37.5	341	128	213	48.0
37.5	339	127	212	47.6
37.5	336	126	210	47.3
37.5	333	125	208	46.9
37.5	331	124	207	46.5
37.5	328	123	205	46.1
37.5	325	122	203	45.8
37.5	323	121	202	45.3
37.5	320	120	200	45.0
37.5	317	119	198	44.7
37.5	315	118	197	44.2
37.5	312	117	195	43.9
37.5	309	116	193	43.5
37.5	307	115	192	43.1
37.5	304	114	190	42.8
37.5	301	113	188	42.4
37.5	299	112	187	42.0
37.5	296	111	185	41.6
37.5	293	110	183	41.3
37.5	291	109	182	40.8
37.5	288	108	180	40.5
37.5	285	107	178	40.2

%E	M1	M2	DM	M*
37.5	283	106	177	39.7
37.5	280	105	175	39.4
37.5	277	104	173	39.0
37.5	275	103	172	38.6
37.5	272	102	170	38.3
37.5	269	101	168	37.9
37.5	267	100	167	37.5
37.5	264	99	165	37.1
37.5	261	98	163	36.8
37.5	259	97	162	36.3
37.5	256	96	160	36.0
37.5	253	95	158	35.7
37.5	251	94	157	35.2
37.5	248	93	155	34.9
37.5	240	90	150	33.8
37.5	232	87	145	32.6
37.5	224	84	140	31.5
37.5	216	81	135	30.4
37.5	208	78	130	29.3
37.5	200	75	125	28.1
37.5	192	72	120	27.0
37.5	184	69	115	25.9
37.5	176	66	110	24.8
37.5	168	63	105	23.6
37.5	160	60	100	22.5
37.5	152	57	95	21.4
37.5	144	54	90	20.3
37.5	136	51	85	19.1
37.5	128	48	80	18.0
37.5	120	45	75	16.9
37.5	112	42	70	15.8
37.5	104	39	65	14.6
37.5	96	36	60	13.5
37.5	88	33	55	12.4
37.5	80	30	50	11.3
37.5	72	27	45	10.1
37.5	64	24	40	9.0
37.5	56	21	35	7.9
37.5	48	18	30	6.8
37.5	40	15	25	5.6
37.5	32	12	20	4.5
37.5	24	9	15	3.4
37.5	16	6	10	2.3
37.5	8	3	5	1.1
37.4	500	187	313	69.9
37.4	497	186	311	69.6
37.4	495	185	310	69.1
37.4	494	185	309	69.3
37.4	492	184	308	68.8
37.4	489	183	306	68.5

%E	M1	M2	DM	M*
37.4	487	182	305	68.0
37.4	486	182	304	68.2
37.4	484	181	303	67.7
37.4	481	180	301	67.4
37.4	479	179	300	66.9
37.4	478	179	299	67.0
37.4	476	178	298	66.6
37.4	473	177	296	66.2
37.4	471	176	295	65.8
37.4	470	176	294	65.9
37.4	468	175	293	65.4
37.4	465	174	291	65.1
37.4	463	173	290	64.6
37.4	462	173	289	64.8
37.4	460	172	288	64.3
37.4	457	171	286	64.0
37.4	455	170	285	63.5
37.4	454	170	284	63.7
37.4	452	169	283	63.2
37.4	449	168	281	62.9
37.4	447	167	280	62.4
37.4	446	167	279	62.5
37.4	444	166	278	62.1
37.4	441	165	276	61.7
37.4	439	164	275	61.3
37.4	438	164	274	61.4
37.4	436	163	273	60.9
37.4	433	162	271	60.6
37.4	431	161	270	60.1
37.4	430	161	269	60.3
37.4	428	160	268	59.8
37.4	425	159	266	59.5
37.4	423	158	265	59.0
37.4	422	158	264	59.2
37.4	420	157	263	58.7
37.4	417	156	261	58.4
37.4	414	155	259	58.0
37.4	412	154	258	57.6
37.4	409	153	256	57.2
37.4	406	152	254	56.9
37.4	404	151	253	56.4
37.4	401	150	251	56.1
37.4	398	149	249	55.8
37.4	396	148	248	55.3
37.4	393	147	246	55.0
37.4	390	146	244	54.7
37.4	388	145	243	54.2
37.4	385	144	241	53.9
37.4	382	143	239	53.5
37.4	380	142	238	53.1

%E	M1	M2	DM	M*
37.4	377	141	236	52.7
37.4	374	140	234	52.4
37.4	372	139	233	51.9
37.4	369	138	231	51.6
37.4	366	137	229	51.3
37.4	364	136	228	50.8
37.4	361	135	226	50.5
37.4	358	134	224	50.2
37.4	356	133	223	49.7
37.4	353	132	221	49.4
37.4	350	131	219	49.0
37.4	348	130	218	48.6
37.4	345	129	216	48.2
37.4	342	128	214	47.9
37.4	340	127	213	47.4
37.4	337	126	211	47.1
37.4	334	125	209	46.8
37.4	329	123	206	46.0
37.4	326	122	204	45.7
37.4	321	120	201	44.9
37.4	318	119	199	44.5
37.4	313	117	196	43.7
37.4	310	116	194	43.4
37.4	305	114	191	42.6
37.4	302	113	189	42.3
37.4	297	111	186	41.5
37.4	294	110	184	41.2
37.4	289	108	181	40.4
37.4	286	107	179	40.0
37.4	281	105	176	39.2
37.4	278	104	174	38.9
37.4	273	102	171	38.1
37.4	270	101	169	37.8
37.4	265	99	166	37.0
37.4	262	98	164	36.7
37.4	257	96	161	35.9
37.4	254	95	159	35.5
37.4	246	92	154	34.4
37.4	243	91	152	34.1
37.4	238	89	149	33.3
37.4	235	88	147	33.0
37.4	230	86	144	32.2
37.4	227	85	142	31.8
37.4	222	83	139	31.0
37.4	219	82	137	30.7
37.4	214	80	134	29.9
37.4	211	79	132	29.6
37.4	206	77	129	28.8
37.4	203	76	127	28.5
37.4	198	74	124	27.7

%E	M1	M2	DM	M*
37.4	195	73	122	27.3
37.4	190	71	119	26.5
37.4	187	70	117	26.2
37.4	182	68	114	25.4
37.4	179	67	112	25.1
37.4	174	65	109	24.3
37.4	171	64	107	24.0
37.4	163	61	102	22.8
37.4	155	58	97	21.7
37.4	147	55	92	20.6
37.4	139	52	87	19.5
37.4	131	49	82	18.3
37.4	123	46	77	17.2
37.4	115	43	72	16.1
37.4	107	40	67	15.0
37.4	99	37	62	13.8
37.4	91	34	57	12.7
37.3	499	186	313	69.3
37.3	498	186	312	69.5
37.3	496	185	311	69.0
37.3	493	184	309	68.7
37.3	491	183	308	68.2
37.3	490	183	307	68.3
37.3	488	182	306	67.9
37.3	485	181	304	67.5
37.3	483	180	303	67.1
37.3	482	180	302	67.2
37.3	480	179	301	66.8
37.3	477	178	299	66.4
37.3	475	177	298	66.0
37.3	474	177	297	66.1
37.3	472	176	296	65.6
37.3	469	175	294	65.3
37.3	467	174	293	64.8
37.3	466	174	292	65.0
37.3	464	173	291	64.5
37.3	461	172	289	64.2
37.3	459	171	288	63.7
37.3	458	171	287	63.8
37.3	456	170	286	63.4
37.3	453	169	284	63.0
37.3	451	168	283	62.6
37.3	450	168	282	62.7
37.3	448	167	281	62.3
37.3	445	166	279	61.9
37.3	442	165	277	61.6
37.3	440	164	276	61.1
37.3	437	163	274	60.8
37.3	434	162	272	60.5
37.3	432	161	271	60.0

%E	M1	M2	DM	M*
37.3	429	160	269	59.7
37.3	426	159	267	59.3
37.3	424	158	266	58.9
37.3	421	157	264	58.5
37.3	418	156	262	58.2
37.3	416	155	261	57.8
37.3	415	155	260	57.9
37.3	413	154	259	57.4
37.3	410	153	257	57.1
37.3	408	152	256	56.6
37.3	407	152	255	56.8
37.3	405	151	254	56.3
37.3	402	150	252	56.0
37.3	400	149	251	55.5
37.3	399	149	250	55.6
37.3	397	148	249	55.2
37.3	394	147	247	54.8
37.3	391	146	245	54.5
37.3	389	145	244	54.0
37.3	386	144	242	53.7
37.3	383	143	240	53.4
37.3	381	142	239	52.9
37.3	378	141	237	52.6
37.3	375	140	235	52.3
37.3	373	139	234	51.8
37.3	370	138	232	51.5
37.3	367	137	230	51.1
37.3	365	136	229	50.7
37.3	362	135	227	50.3
37.3	359	134	225	50.0
37.3	357	133	224	49.5
37.3	354	132	222	49.2
37.3	351	131	220	48.9
37.3	346	129	217	48.1
37.3	343	128	215	47.8
37.3	338	126	212	47.0
37.3	335	125	210	46.6
37.3	332	124	208	46.3
37.3	330	123	207	45.8
37.3	327	122	205	45.5
37.3	324	121	203	45.2
37.3	322	120	202	44.7
37.3	319	119	200	44.4
37.3	316	118	198	44.1
37.3	314	117	197	43.6
37.3	311	116	195	43.3
37.3	308	115	193	42.9
37.3	306	114	192	42.5
37.3	303	113	190	42.1
37.3	300	112	188	41.8

%E	M1	M2	DM	M*
37.3	295	110	185	41.0
37.3	292	109	183	40.7
37.3	287	107	180	39.9
37.3	284	106	178	39.6
37.3	279	104	175	38.8
37.3	276	103	173	38.4
37.3	271	101	170	37.6
37.3	268	100	168	37.3
37.3	263	98	165	36.5
37.3	260	97	163	36.2
37.3	255	95	160	35.4
37.3	252	94	158	35.1
37.3	249	93	156	34.7
37.3	244	91	153	33.9
37.3	241	90	151	33.6
37.3	236	88	148	32.8
37.3	233	87	146	32.5
37.3	228	85	143	31.7
37.3	225	84	141	31.4
37.3	220	82	138	30.6
37.3	217	81	136	30.2
37.3	212	79	133	29.4
37.3	209	78	131	29.1
37.3	204	76	128	28.3
37.3	201	75	126	28.0
37.3	193	72	121	26.9
37.3	185	69	116	25.7
37.3	177	66	111	24.6
37.3	169	63	106	23.5
37.3	166	62	104	23.2
37.3	161	60	101	22.4
37.3	158	59	99	22.0
37.3	153	57	96	21.2
37.3	150	56	94	20.9
37.3	142	53	89	19.8
37.3	134	50	84	18.7
37.3	126	47	79	17.5
37.3	118	44	74	16.4
37.3	110	41	69	15.3
37.3	102	38	64	14.2
37.3	83	31	52	11.6
37.3	75	28	47	10.5
37.3	67	25	42	9.3
37.3	59	22	37	8.2
37.3	51	19	32	7.1
37.2	500	186	314	69.2
37.2	497	185	312	68.9
37.2	495	184	311	68.4
37.2	494	184	310	68.5
37.2	492	183	309	68.1

%E	M1	M2	DM	M*
37.2	489	182	307	67.7
37.2	487	181	306	67.3
37.2	486	181	305	67.4
37.2	484	180	304	66.9
37.2	481	179	302	66.6
37.2	479	178	301	66.1
37.2	478	178	300	66.3
37.2	476	177	299	65.8
37.2	473	176	297	65.5
37.2	471	175	296	65.0
37.2	470	175	295	65.2
37.2	468	174	294	64.7
37.2	465	173	292	64.4
37.2	462	172	290	64.0
37.2	460	171	289	63.6
37.2	457	170	287	63.2
37.2	454	169	285	62.9
37.2	452	168	284	62.4
37.2	449	167	282	62.1
37.2	446	166	280	61.8
37.2	444	165	279	61.3
37.2	443	165	278	61.5
37.2	441	164	277	61.0
37.2	438	163	275	60.7
37.2	436	162	274	60.2
37.2	435	162	273	60.3
37.2	433	161	272	59.9
37.2	430	160	270	59.5
37.2	428	159	269	59.1
37.2	427	159	268	59.2
37.2	425	158	267	58.7
37.2	422	157	265	58.4
37.2	419	156	263	58.1
37.2	417	155	262	57.6
37.2	414	154	260	57.3
37.2	411	153	258	57.0
37.2	409	152	257	56.5
37.2	406	151	255	56.2
37.2	403	150	253	55.8
37.2	401	149	252	55.4
37.2	398	148	250	55.0
37.2	395	147	248	54.7
37.2	393	146	247	54.2
37.2	392	146	246	54.4
37.2	390	145	245	53.9
37.2	387	144	243	53.6
37.2	384	143	241	53.3
37.2	382	142	240	52.8
37.2	379	141	238	52.5
37.2	376	140	236	52.1

%E	M1	M2	DM	M*
37.2	374	139	235	51.7
37.2	371	138	233	51.3
37.2	368	137	231	51.0
37.2	366	136	230	50.5
37.2	363	135	228	50.2
37.2	360	134	226	49.9
37.2	358	133	225	49.4
37.2	355	132	223	49.1
37.2	352	131	221	48.8
37.2	349	130	219	48.4
37.2	347	129	218	48.0
37.2	344	128	216	47.6
37.2	341	127	214	47.3
37.2	339	126	213	46.8
37.2	336	125	211	46.5
37.2	333	124	209	46.2
37.2	331	123	208	45.7
37.2	328	122	206	45.4
37.2	325	121	204	45.0
37.2	323	120	203	44.6
37.2	320	119	201	44.3
37.2	317	118	199	43.9
37.2	312	116	196	43.1
37.2	309	115	194	42.8
37.2	304	113	191	42.0
37.2	301	112	189	41.7
37.2	298	111	187	41.3
37.2	296	110	186	40.9
37.2	293	109	184	40.5
37.2	290	108	182	40.2
37.2	288	107	181	39.8
37.2	285	106	179	39.4
37.2	282	105	177	39.1
37.2	277	103	174	38.3
37.2	274	102	172	38.0
37.2	269	100	169	37.2
37.2	266	99	167	36.8
37.2	261	97	164	36.0
37.2	258	96	162	35.7
37.2	253	94	159	34.9
37.2	250	93	157	34.6
37.2	247	92	155	34.3
37.2	242	90	152	33.5
37.2	239	89	150	33.1
37.2	234	87	147	32.3
37.2	231	86	145	32.0
37.2	226	84	142	31.2
37.2	223	83	140	30.9
37.2	218	81	137	30.1
37.2	215	80	135	29.8

%E	M1	M2	DM	M*
37.2	207	77	130	28.6
37.2	199	74	125	27.5
37.2	196	73	123	27.2
37.2	191	71	120	26.4
37.2	188	70	118	26.1
37.2	183	68	115	25.3
37.2	180	67	113	24.9
37.2	172	64	108	23.8
37.2	164	61	103	22.7
37.2	156	58	98	21.6
37.2	148	55	93	20.4
37.2	145	54	91	20.1
37.2	137	51	86	19.0
37.2	129	48	81	17.9
37.2	121	45	76	16.7
37.2	113	42	71	15.6
37.2	94	35	59	13.0
37.2	86	32	54	11.9
37.2	78	29	49	10.8
37.2	43	16	27	6.0
37.1	499	185	314	68.6
37.1	498	185	313	68.7
37.1	496	184	312	68.3
37.1	493	183	310	67.9
37.1	491	182	309	67.5
37.1	490	182	308	67.6
37.1	488	181	307	67.1
37.1	485	180	305	66.8
37.1	483	179	304	66.3
37.1	482	179	303	66.5
37.1	480	178	302	66.0
37.1	477	177	300	65.7
37.1	475	176	299	65.2
37.1	474	176	298	65.4
37.1	472	175	297	64.9
37.1	469	174	295	64.6
37.1	466	173	293	64.2
37.1	464	172	292	63.8
37.1	463	172	291	63.9
37.1	461	171	290	63.4
37.1	458	170	288	63.1
37.1	456	169	287	62.6
37.1	455	169	286	62.8
37.1	453	168	285	62.3
37.1	450	167	283	62.0
37.1	448	166	282	61.5
37.1	447	166	281	61.6
37.1	445	165	280	61.2
37.1	442	164	278	60.9
37.1	439	163	276	60.5

%E	M1	M2	DM	M*
37.1	437	162	275	60.1
37.1	434	161	273	59.7
37.1	431	160	271	59.4
37.1	429	159	270	58.9
37.1	426	158	268	58.6
37.1	423	157	266	58.3
37.1	421	156	265	57.8
37.1	420	156	264	57.9
37.1	418	155	263	57.5
37.1	415	154	261	57.1
37.1	412	153	259	56.8
37.1	410	152	258	56.4
37.1	407	151	256	56.0
37.1	404	150	254	55.7
37.1	402	149	253	55.2
37.1	399	148	251	54.9
37.1	396	147	249	54.6
37.1	394	146	248	54.1
37.1	391	145	246	53.8
37.1	388	144	244	53.4
37.1	385	143	242	53.1
37.1	383	142	241	52.6
37.1	380	141	239	52.3
37.1	377	140	237	52.0
37.1	375	139	236	51.5
37.1	372	138	234	51.2
37.1	369	137	232	50.9
37.1	367	136	231	50.4
37.1	364	135	229	50.1
37.1	361	134	227	49.7
37.1	356	132	224	48.9
37.1	353	131	222	48.6
37.1	350	130	220	48.3
37.1	348	129	219	47.8
37.1	345	128	217	47.5
37.1	342	127	215	47.2
37.1	340	126	214	46.7
37.1	337	125	212	46.4
37.1	334	124	210	46.0
37.1	329	122	207	45.2
37.1	326	121	205	44.9
37.1	321	119	202	44.1
37.1	318	118	200	43.8
37.1	315	117	198	43.5
37.1	313	116	197	43.0
37.1	310	115	195	42.7
37.1	307	114	193	42.3
37.1	302	112	190	41.5
37.1	299	111	188	41.2
37.1	294	109	185	40.4
37.1	291	108	183	40.1
37.1	286	106	180	39.3
37.1	283	105	178	39.0
37.1	280	104	176	38.6
37.1	278	103	175	38.2
37.1	275	102	173	37.8
37.1	272	101	171	37.5
37.1	267	99	168	36.7
37.1	264	98	166	36.4
37.1	259	96	163	35.6
37.1	256	95	161	35.3
37.1	251	93	158	34.5
37.1	248	92	156	34.1
37.1	245	91	154	33.8
37.1	240	89	151	33.0
37.1	237	88	149	32.7
37.1	232	86	146	31.9
37.1	229	85	144	31.6
37.1	224	83	141	30.8
37.1	221	82	139	30.4
37.1	213	79	134	29.3
37.1	210	78	132	29.0
37.1	205	76	129	28.2
37.1	202	75	127	27.8
37.1	197	73	124	27.1
37.1	194	72	122	26.7
37.1	186	69	117	25.6
37.1	178	66	112	24.5
37.1	175	65	110	24.1
37.1	170	63	107	23.3
37.1	167	62	105	23.0
37.1	159	59	100	21.9
37.1	151	56	95	20.8
37.1	143	53	90	19.6
37.1	140	52	88	19.3
37.1	132	49	83	18.2
37.1	124	46	78	17.1
37.1	116	43	73	15.9
37.1	105	39	66	14.5
37.1	97	36	61	13.4
37.1	89	33	56	12.2
37.1	70	26	44	9.7
37.1	62	23	39	8.5
37.1	35	13	22	4.8
37.0	500	185	315	68.4
37.0	497	184	313	68.1
37.0	495	183	312	67.7
37.0	494	183	311	67.8
37.0	492	182	310	67.3
37.0	489	181	308	67.0
37.0	487	180	307	66.5
37.0	486	180	306	66.7
37.0	484	179	305	66.2
37.0	481	178	303	65.9
37.0	479	177	302	65.4
37.0	478	177	301	65.5
37.0	476	176	300	65.1
37.0	473	175	298	64.7
37.0	470	174	296	64.4
37.0	468	173	295	64.0
37.0	467	173	294	64.1
37.0	465	172	293	63.6
37.0	462	171	291	63.3
37.0	460	170	290	62.8
37.0	459	170	289	63.0
37.0	457	169	288	62.5
37.0	454	168	286	62.2
37.0	451	167	284	61.8
37.0	449	166	283	61.4
37.0	446	165	281	61.0
37.0	443	164	279	60.7
37.0	441	163	278	60.2
37.0	440	163	277	60.4
37.0	438	162	276	59.9
37.0	435	161	274	59.6
37.0	433	160	273	59.1
37.0	432	160	272	59.3
37.0	430	159	271	58.8
37.0	427	158	269	58.5
37.0	424	157	267	58.1
37.0	422	156	266	57.7
37.0	419	155	264	57.3
37.0	416	154	262	57.0
37.0	414	153	261	56.5
37.0	413	153	260	56.7
37.0	411	152	259	56.2
37.0	408	151	257	55.9
37.0	405	150	255	55.6
37.0	403	149	254	55.1
37.0	400	148	252	54.8
37.0	397	147	250	54.4
37.0	395	146	249	54.0
37.0	392	145	247	53.6
37.0	389	144	245	53.3
37.0	387	143	244	52.8
37.0	386	143	243	53.0
37.0	384	142	242	52.5
37.0	381	141	240	52.2
37.0	378	140	238	51.9
37.0	376	139	237	51.4
37.0	373	138	235	51.1
37.0	370	137	233	50.7
37.0	368	136	232	50.3
37.0	365	135	230	49.9
37.0	362	134	228	49.6
37.0	359	133	226	49.3
37.0	357	132	225	48.8
37.0	354	131	223	48.5
37.0	351	130	221	48.1
37.0	349	129	220	47.7
37.0	346	128	218	47.4
37.0	343	127	216	47.0
37.0	341	126	215	46.6
37.0	338	125	213	46.2
37.0	335	124	211	45.9
37.0	332	123	209	45.6
37.0	330	122	208	45.1
37.0	327	121	206	44.8
37.0	324	120	204	44.4
37.0	322	119	203	44.0
37.0	319	118	201	43.6
37.0	316	117	199	43.3
37.0	311	115	196	42.5
37.0	308	114	194	42.2
37.0	305	113	192	41.9
37.0	303	112	191	41.4
37.0	300	111	189	41.1
37.0	297	110	187	40.7
37.0	292	108	184	39.9
37.0	289	107	182	39.6
37.0	284	105	179	38.8
37.0	281	104	177	38.5
37.0	276	102	174	37.7
37.0	273	101	172	37.4
37.0	270	100	170	37.0
37.0	265	98	167	36.2
37.0	262	97	165	35.9
37.0	257	95	162	35.1
37.0	254	94	160	34.8
37.0	246	91	155	33.7
37.0	243	90	153	33.3
37.0	238	88	150	32.5
37.0	235	87	148	32.2
37.0	230	85	145	31.4
37.0	227	84	143	31.1
37.0	219	81	138	30.0
37.0	216	80	136	29.6
37.0	211	78	133	28.8
37.0	208	77	131	28.5
37.0	200	74	126	27.4
37.0	192	71	121	26.3
37.0	189	70	119	25.9
37.0	184	68	116	25.1
37.0	181	67	114	24.8
37.0	173	64	109	23.7
37.0	165	61	104	22.6
37.0	162	60	102	22.2
37.0	154	57	97	21.1
37.0	146	54	92	20.0
37.0	138	51	87	18.8
37.0	135	50	85	18.5
37.0	127	47	80	17.4
37.0	119	44	75	16.3
37.0	108	40	68	14.8
37.0	100	37	63	13.7
37.0	92	34	58	12.6
37.0	81	30	51	11.1
37.0	73	27	46	10.0
37.0	54	20	34	7.4
37.0	46	17	29	6.3
37.0	27	10	17	3.7
36.9	499	184	315	67.8
36.9	498	184	314	68.0
36.9	496	183	313	67.5
36.9	493	182	311	67.2
36.9	491	181	310	66.7
36.9	490	181	309	66.9
36.9	488	180	308	66.4
36.9	485	179	306	66.1
36.9	483	178	305	65.6
36.9	482	178	304	65.7
36.9	480	177	303	65.3
36.9	477	176	301	64.9
36.9	474	175	299	64.6
36.9	472	174	298	64.1
36.9	471	174	297	64.3
36.9	469	173	296	63.8
36.9	466	172	294	63.5
36.9	464	171	293	63.0
36.9	463	171	292	63.2
36.9	461	170	291	62.7
36.9	458	169	289	62.4
36.9	455	168	287	62.0
36.9	453	167	286	61.6
36.9	452	167	285	61.7
36.9	450	166	284	61.2
36.9	447	165	282	60.9
36.9	445	164	281	60.4
36.9	444	164	280	60.6
36.9	442	163	279	60.1

%E	M1	M2	DM	M*
36.9	439	162	277	59.8
36.9	436	161	275	59.5
36.9	434	160	274	59.0
36.9	431	159	272	58.7
36.9	428	158	270	58.3
36.9	426	157	269	57.9
36.9	425	157	268	58.0
36.9	423	156	267	57.5
36.9	420	155	265	57.2
36.9	417	154	263	56.9
36.9	415	153	262	56.4
36.9	412	152	260	56.1
36.9	409	151	258	55.7
36.9	407	150	257	55.3
36.9	406	150	256	55.4
36.9	404	149	255	55.0
36.9	401	148	253	54.6
36.9	398	147	251	54.3
36.9	396	146	250	53.8
36.9	393	145	248	53.5
36.9	390	144	246	53.2
36.9	388	143	245	52.7
36.9	385	142	243	52.4
36.9	382	141	241	52.0
36.9	379	140	239	51.7
36.9	377	139	238	51.2
36.9	374	138	236	50.9
36.9	371	137	234	50.6
36.9	369	136	233	50.1
36.9	366	135	231	49.8
36.9	363	134	229	49.5
36.9	360	133	227	49.1
36.9	358	132	226	48.7
36.9	355	131	224	48.3
36.9	352	130	222	48.0
36.9	350	129	221	47.5
36.9	347	128	219	47.2
36.9	344	127	217	46.9
36.9	339	125	214	46.1
36.9	336	124	212	45.8
36.9	333	123	210	45.4
36.9	331	122	209	45.0
36.9	328	121	207	44.6
36.9	325	120	205	44.3
36.9	320	118	202	43.5
36.9	317	117	200	43.2
36.9	314	116	198	42.9
36.9	312	115	197	42.4
36.9	309	114	195	42.1
36.9	306	113	193	41.7

%E	M1	M2	DM	M*
36.9	301	111	190	40.9
36.9	298	110	188	40.6
36.9	295	109	186	40.3
36.9	293	108	185	39.8
36.9	290	107	183	39.5
36.9	287	106	181	39.1
36.9	282	104	178	38.4
36.9	279	103	176	38.0
36.9	274	101	173	37.2
36.9	271	100	171	36.9
36.9	268	99	169	36.6
36.9	263	97	166	35.8
36.9	260	96	164	35.4
36.9	255	94	161	34.7
36.9	252	93	159	34.3
36.9	249	92	157	34.0
36.9	244	90	154	33.2
36.9	241	89	152	32.9
36.9	236	87	149	32.1
36.9	233	86	147	31.7
36.9	225	83	142	30.6
36.9	222	82	140	30.3
36.9	217	80	137	29.5
36.9	214	79	135	29.2
36.9	206	76	130	28.0
36.9	203	75	128	27.7
36.9	198	73	125	26.9
36.9	195	72	123	26.6
36.9	187	69	118	25.5
36.9	179	66	113	24.3
36.9	176	65	111	24.0
36.9	168	62	106	22.9
36.9	160	59	101	21.8
36.9	157	58	99	21.4
36.9	149	55	94	20.3
36.9	141	52	89	19.2
36.9	130	48	82	17.7
36.9	122	45	77	16.6
36.9	111	41	70	15.1
36.9	103	38	65	14.0
36.9	84	31	53	11.4
36.9	65	24	41	8.9
36.8	500	184	316	67.7
36.8	497	183	314	67.4
36.8	495	182	313	66.9
36.8	494	182	312	67.1
36.8	492	181	311	66.6
36.8	489	180	309	66.3
36.8	487	179	308	65.8
36.8	486	179	307	65.9

%E	M1	M2	DM	M*
36.8	484	178	306	65.5
36.8	481	177	304	65.1
36.8	478	176	302	64.8
36.8	476	175	301	64.3
36.8	475	175	300	64.5
36.8	473	174	299	64.0
36.8	470	173	297	63.7
36.8	468	172	296	63.2
36.8	467	172	295	63.3
36.8	465	171	294	62.9
36.8	462	170	292	62.6
36.8	459	169	290	62.2
36.8	457	168	289	61.8
36.8	456	168	288	61.9
36.8	454	167	287	61.4
36.8	451	166	285	61.1
36.8	448	165	283	60.8
36.8	446	164	282	60.3
36.8	443	163	280	60.0
36.8	440	162	278	59.6
36.8	438	161	277	59.2
36.8	437	161	276	59.3
36.8	435	160	275	58.9
36.8	432	159	273	58.5
36.8	429	158	271	58.2
36.8	427	157	270	57.7
36.8	424	156	268	57.4
36.8	421	155	266	57.1
36.8	419	154	265	56.6
36.8	418	154	264	56.7
36.8	416	153	263	56.3
36.8	413	152	261	55.9
36.8	410	151	259	55.6
36.8	408	150	258	55.1
36.8	405	149	256	54.8
36.8	402	148	254	54.5
36.8	400	147	253	54.0
36.8	399	147	252	54.2
36.8	397	146	251	53.7
36.8	394	145	249	53.4
36.8	391	144	247	53.0
36.8	389	143	246	52.6
36.8	386	142	244	52.2
36.8	383	141	242	51.9
36.8	380	140	240	51.6
36.8	378	139	239	51.1
36.8	375	138	237	50.8
36.8	372	137	235	50.5
36.8	370	136	234	50.0
36.8	367	135	232	49.7

%E	M1	M2	DM	M*
36.8	364	134	230	49.3
36.8	361	133	228	49.0
36.8	359	132	227	48.5
36.8	356	131	225	48.2
36.8	353	130	223	47.9
36.8	351	129	222	47.4
36.8	348	128	220	47.1
36.8	345	127	218	46.8
36.8	342	126	216	46.4
36.8	340	125	215	46.0
36.8	337	124	213	45.6
36.8	334	123	211	45.3
36.8	329	121	208	44.5
36.8	326	120	206	44.2
36.8	323	119	204	43.8
36.8	321	118	203	43.4
36.8	318	117	201	43.0
36.8	315	116	199	42.7
36.8	310	114	196	41.9
36.8	307	113	194	41.6
36.8	304	112	192	41.3
36.8	302	111	191	40.8
36.8	299	110	189	40.5
36.8	296	109	187	40.1
36.8	291	107	184	39.3
36.8	288	106	182	39.0
36.8	285	105	180	38.7
36.8	280	103	177	37.9
36.8	277	102	175	37.6
36.8	272	100	172	36.8
36.8	269	99	170	36.4
36.8	266	98	168	36.1
36.8	261	96	165	35.3
36.8	258	95	163	35.0
36.8	253	93	160	34.2
36.8	250	92	158	33.9
36.8	247	91	156	33.5
36.8	242	89	153	32.7
36.8	239	88	151	32.4
36.8	234	86	148	31.6
36.8	231	85	146	31.3
36.8	228	84	144	30.9
36.8	223	82	141	30.2
36.8	220	81	139	29.8
36.8	212	78	134	28.7
36.8	209	77	132	28.4
36.8	204	75	129	27.6
36.8	201	74	127	27.2
36.8	193	71	122	26.1
36.8	190	70	120	25.8

%E	M1	M2	DM	M*
36.8	185	68	117	25.0
36.8	182	67	115	24.7
36.8	174	64	110	23.5
36.8	171	63	108	23.2
36.8	163	60	103	22.1
36.8	155	57	98	21.0
36.8	152	56	96	20.6
36.8	144	53	91	19.5
36.8	136	50	86	18.4
36.8	133	49	84	18.1
36.8	125	46	79	16.9
36.8	117	43	74	15.8
36.8	114	42	72	15.5
36.8	106	39	67	14.3
36.8	95	35	60	12.9
36.8	87	32	55	11.8
36.8	76	28	48	10.3
36.8	68	25	43	9.2
36.8	57	21	36	7.7
36.8	38	14	24	5.2
36.8	19	7	12	2.6
36.7	499	183	316	67.1
36.7	498	183	315	67.2
36.7	496	182	314	66.8
36.7	493	181	312	66.5
36.7	491	180	311	66.0
36.7	490	180	310	66.1
36.7	488	179	309	65.7
36.7	485	178	307	65.3
36.7	482	177	305	65.0
36.7	480	176	304	64.5
36.7	479	176	303	64.7
36.7	477	175	302	64.2
36.7	474	174	300	63.9
36.7	472	173	299	63.4
36.7	471	173	298	63.5
36.7	469	172	297	63.1
36.7	466	171	295	62.7
36.7	463	170	293	62.4
36.7	461	169	292	62.0
36.7	460	169	291	62.1
36.7	458	168	290	61.6
36.7	455	167	288	61.3
36.7	452	166	286	61.0
36.7	450	165	285	60.5
36.7	449	165	284	60.6
36.7	447	164	283	60.2
36.7	444	163	281	59.8
36.7	442	162	280	59.4
36.7	441	162	279	59.5

%E	M1	M2	DM	M*
36.7	439	161	278	59.0
36.7	436	160	276	58.7
36.7	433	159	274	58.4
36.7	431	158	273	57.9
36.7	430	158	272	58.1
36.7	428	157	271	57.6
36.7	425	156	269	57.3
36.7	422	155	267	56.9
36.7	420	154	266	56.5
36.7	417	153	264	56.1
36.7	414	152	262	55.8
36.7	412	151	261	55.3
36.7	411	151	260	55.5
36.7	409	150	259	55.0
36.7	406	149	257	54.7
36.7	403	148	255	54.4
36.7	401	147	254	53.9
36.7	398	146	252	53.6
36.7	395	145	250	53.2
36.7	392	144	248	52.9
36.7	390	143	247	52.4
36.7	387	142	245	52.1
36.7	384	141	243	51.8
36.7	381	140	241	51.4
36.7	379	139	240	51.0
36.7	376	138	238	50.6
36.7	373	137	236	50.3
36.7	371	136	235	49.9
36.7	368	135	233	49.5
36.7	365	134	231	49.2
36.7	362	133	229	48.9
36.7	360	132	228	48.4
36.7	357	131	226	48.1
36.7	354	130	224	47.7
36.7	349	128	221	46.9
36.7	346	127	219	46.6
36.7	343	126	217	46.3
36.7	341	125	216	45.8
36.7	338	124	214	45.5
36.7	335	123	212	45.2
36.7	332	122	210	44.8
36.7	330	121	209	44.4
36.7	327	120	207	44.0
36.7	324	119	205	43.7
36.7	319	117	202	42.9
36.7	316	116	200	42.6
36.7	313	115	198	42.3
36.7	311	114	197	41.8
36.7	308	113	195	41.5
36.7	305	112	193	41.1

%E	M1	M2	DM	M*
36.7	300	110	190	40.3
36.7	297	109	188	40.0
36.7	294	108	186	39.7
36.7	289	106	183	38.9
36.7	286	105	181	38.5
36.7	283	104	179	38.2
36.7	281	103	178	37.8
36.7	278	102	176	37.4
36.7	275	101	174	37.1
36.7	270	99	171	36.3
36.7	267	98	169	36.0
36.7	264	97	167	35.6
36.7	259	95	164	34.8
36.7	256	94	162	34.5
36.7	251	92	159	33.7
36.7	248	91	157	33.4
36.7	245	90	155	33.1
36.7	240	88	152	32.3
36.7	237	87	150	31.9
36.7	229	84	145	30.8
36.7	226	83	143	30.5
36.7	221	81	140	29.7
36.7	218	80	138	29.4
36.7	215	79	136	29.0
36.7	210	77	133	28.2
36.7	207	76	131	27.9
36.7	199	73	126	26.8
36.7	196	72	124	26.4
36.7	188	69	119	25.3
36.7	180	66	114	24.2
36.7	177	65	112	23.9
36.7	169	62	107	22.7
36.7	166	61	105	22.4
36.7	158	58	100	21.3
36.7	150	55	95	20.2
36.7	147	54	93	19.8
36.7	139	51	88	18.7
36.7	128	47	81	17.3
36.7	120	44	76	16.1
36.7	109	40	69	14.7
36.7	98	36	62	13.2
36.7	90	33	57	12.1
36.7	79	29	50	10.6
36.7	60	22	38	8.1
36.7	49	18	31	6.6
36.7	30	11	19	4.0
36.6	500	183	317	67.0
36.6	497	182	315	66.6
36.6	495	181	314	66.2
36.6	494	181	313	66.3

%E	M1	M2	DM	M*
36.6	492	180	312	65.9
36.6	489	179	310	65.5
36.6	487	178	309	65.1
36.6	486	178	308	65.2
36.6	484	177	307	64.7
36.6	483	177	306	64.9
36.6	481	176	305	64.4
36.6	478	175	303	64.1
36.6	476	174	302	63.6
36.6	475	174	301	63.7
36.6	473	173	300	63.3
36.6	470	172	298	62.9
36.6	467	171	296	62.6
36.6	465	170	295	62.2
36.6	464	170	294	62.3
36.6	462	169	293	61.8
36.6	459	168	291	61.5
36.6	456	167	289	61.2
36.6	454	166	288	60.7
36.6	453	166	287	60.8
36.6	451	165	286	60.4
36.6	448	164	284	60.0
36.6	445	163	282	59.7
36.6	443	162	281	59.2
36.6	440	161	279	58.9
36.6	437	160	277	58.6
36.6	435	159	276	58.1
36.6	434	159	275	58.3
36.6	432	158	274	57.8
36.6	429	157	272	57.5
36.6	426	156	270	57.1
36.6	424	155	269	56.7
36.6	423	155	268	56.8
36.6	421	154	267	56.3
36.6	418	153	265	56.0
36.6	415	152	263	55.7
36.6	413	151	262	55.2
36.6	410	150	260	54.9
36.6	407	149	258	54.5
36.6	404	148	256	54.2
36.6	402	147	255	53.8
36.6	399	146	253	53.4
36.6	396	145	251	53.1
36.6	393	144	249	52.8
36.6	391	143	248	52.3
36.6	388	142	246	52.0
36.6	385	141	244	51.6
36.6	383	140	243	51.2
36.6	382	140	242	51.3
36.6	380	139	241	50.8

%E	M1	M2	DM	M*
36.6	377	138	239	50.5
36.6	374	137	237	50.2
36.6	372	136	236	49.7
36.6	369	135	234	49.4
36.6	366	134	232	49.1
36.6	363	133	230	48.7
36.6	361	132	229	48.3
36.6	358	131	227	47.9
36.6	355	130	225	47.6
36.6	352	129	223	47.3
36.6	350	128	222	46.8
36.6	347	127	220	46.5
36.6	344	126	218	46.2
36.6	342	125	217	45.7
36.6	339	124	215	45.4
36.6	336	123	213	45.0
36.6	333	122	211	44.7
36.6	331	121	210	44.2
36.6	328	120	208	43.9
36.6	325	119	206	43.6
36.6	322	118	204	43.2
36.6	320	117	203	42.8
36.6	317	116	201	42.4
36.6	314	115	199	42.1
36.6	309	113	196	41.3
36.6	306	112	194	41.0
36.6	303	111	192	40.7
36.6	298	109	189	39.9
36.6	295	108	187	39.5
36.6	292	107	185	39.2
36.6	290	106	184	38.7
36.6	287	105	182	38.4
36.6	284	104	180	38.1
36.6	279	102	177	37.3
36.6	276	101	175	37.0
36.6	273	100	173	36.6
36.6	268	98	170	35.8
36.6	265	97	168	35.5
36.6	262	96	166	35.2
36.6	257	94	163	34.4
36.6	254	93	161	34.1
36.6	246	90	156	32.9
36.6	243	89	154	32.6
36.6	238	87	151	31.8
36.6	235	86	149	31.5
36.6	232	85	147	31.1
36.6	227	83	144	30.3
36.6	224	82	142	30.0
36.6	216	79	137	28.9
36.6	213	78	135	28.6

%E	M1	M2	DM	M*
36.6	205	75	130	27.4
36.6	202	74	128	27.1
36.6	194	71	123	26.0
36.6	191	70	121	25.7
36.6	186	68	118	24.9
36.6	183	67	116	24.5
36.6	175	64	111	23.4
36.6	172	63	109	23.1
36.6	164	60	104	22.0
36.6	161	59	102	21.6
36.6	153	56	97	20.5
36.6	145	53	92	19.4
36.6	142	52	90	19.0
36.6	134	49	85	17.9
36.6	131	48	83	17.6
36.6	123	45	78	16.5
36.6	112	41	71	15.0
36.6	101	37	64	13.6
36.6	93	34	59	12.4
36.6	82	30	52	11.0
36.6	71	26	45	9.5
36.6	41	15	26	5.5
36.5	499	182	317	66.4
36.5	498	182	316	66.5
36.5	496	181	315	66.1
36.5	493	180	313	65.7
36.5	491	179	312	65.3
36.5	490	179	311	65.4
36.5	488	178	310	64.9
36.5	485	177	308	64.6
36.5	482	176	306	64.3
36.5	480	175	305	63.8
36.5	479	175	304	63.9
36.5	477	174	303	63.5
36.5	474	173	301	63.1
36.5	471	172	299	62.8
36.5	469	171	298	62.3
36.5	468	171	297	62.5
36.5	466	170	296	62.0
36.5	463	169	294	61.7
36.5	460	168	292	61.4
36.5	458	167	291	60.9
36.5	457	167	290	61.0
36.5	455	166	289	60.6
36.5	452	165	287	60.2
36.5	449	164	285	59.9
36.5	447	163	284	59.4
36.5	446	163	283	59.6
36.5	444	162	282	59.1
36.5	441	161	280	58.8

%E	M1	M2	DM	M*
36.5	438	160	278	58.4
36.5	436	159	277	58.0
36.5	433	158	275	57.7
36.5	430	157	273	57.3
36.5	427	156	271	57.0
36.5	425	155	270	56.5
36.5	422	154	268	56.2
36.5	419	153	266	55.9
36.5	417	152	265	55.4
36.5	416	152	264	55.5
36.5	414	151	263	55.1
36.5	411	150	261	54.7
36.5	408	149	259	54.4
36.5	406	148	258	54.0
36.5	405	148	257	54.1
36.5	403	147	256	53.6
36.5	400	146	254	53.3
36.5	397	145	252	53.0
36.5	395	144	251	52.5
36.5	394	144	250	52.6
36.5	392	143	249	52.2
36.5	389	142	247	51.8
36.5	386	141	245	51.5
36.5	384	140	244	51.0
36.5	381	139	242	50.7
36.5	378	138	240	50.4
36.5	375	137	238	50.1
36.5	373	136	237	49.6
36.5	370	135	235	49.3
36.5	367	134	233	48.9
36.5	364	133	231	48.6
36.5	362	132	230	48.1
36.5	359	131	228	47.8
36.5	356	130	226	47.5
36.5	353	129	224	47.1
36.5	351	128	223	46.7
36.5	348	127	221	46.3
36.5	345	126	219	46.0
36.5	340	124	216	45.2
36.5	337	123	214	44.9
36.5	334	122	212	44.6
36.5	329	120	209	43.8
36.5	326	119	207	43.4
36.5	323	118	205	43.1
36.5	318	116	202	42.3
36.5	315	115	200	42.0
36.5	312	114	198	41.7
36.5	310	113	197	41.2
36.5	307	112	195	40.9
36.5	304	111	193	40.5

%E	M1	M2	DM	M*
36.5	301	110	191	40.2
36.5	299	109	190	39.7
36.5	296	108	188	39.4
36.5	293	107	186	39.1
36.5	288	105	183	38.3
36.5	285	104	181	38.0
36.5	282	103	179	37.6
36.5	277	101	176	36.8
36.5	274	100	174	36.5
36.5	271	99	172	36.2
36.5	266	97	169	35.4
36.5	263	96	167	35.0
36.5	260	95	165	34.7
36.5	255	93	162	33.9
36.5	252	92	160	33.6
36.5	249	91	158	33.3
36.5	244	89	155	32.5
36.5	241	88	153	32.1
36.5	233	85	148	31.0
36.5	230	84	146	30.7
36.5	222	81	141	29.6
36.5	219	80	139	29.2
36.5	211	77	134	28.1
36.5	208	76	132	27.8
36.5	203	74	129	27.0
36.5	200	73	127	26.6
36.5	197	72	125	26.3
36.5	192	70	122	25.5
36.5	189	69	120	25.2
36.5	181	66	115	24.1
36.5	178	65	113	23.7
36.5	170	62	108	22.6
36.5	167	61	106	22.3
36.5	159	58	101	21.2
36.5	156	57	99	20.8
36.5	148	54	94	19.7
36.5	137	50	87	18.2
36.5	126	46	80	16.8
36.5	115	42	73	15.3
36.5	104	38	66	13.9
36.5	96	35	61	12.8
36.5	85	31	54	11.3
36.5	74	27	47	9.9
36.5	63	23	40	8.4
36.5	52	19	33	6.9
36.4	500	182	318	66.2
36.4	497	181	316	65.9
36.4	495	180	315	65.5
36.4	494	180	314	65.6
36.4	492	179	313	65.1

%E	M1	M2	DM	M*
36.4	489	178	311	64.8
36.4	486	177	309	64.5
36.4	484	176	308	64.0
36.4	483	176	307	64.1
36.4	481	175	306	63.7
36.4	478	174	304	63.3
36.4	475	173	302	63.0
36.4	473	172	301	62.5
36.4	472	172	300	62.7
36.4	470	171	299	62.2
36.4	467	170	297	61.9
36.4	464	169	295	61.6
36.4	462	168	294	61.1
36.4	461	168	293	61.2
36.4	459	167	292	60.8
36.4	456	166	290	60.4
36.4	453	165	288	60.1
36.4	451	164	287	59.6
36.4	450	164	286	59.8
36.4	448	163	285	59.3
36.4	445	162	283	59.0
36.4	442	161	281	58.6
36.4	440	160	280	58.2
36.4	439	160	279	58.3
36.4	437	159	278	57.9
36.4	434	158	276	57.5
36.4	431	157	274	57.2
36.4	429	156	273	56.7
36.4	428	156	272	56.9
36.4	426	155	271	56.4
36.4	423	154	269	56.1
36.4	420	153	267	55.7
36.4	418	152	266	55.3
36.4	415	151	264	54.9
36.4	412	150	262	54.6
36.4	409	149	260	54.3
36.4	407	148	259	53.8
36.4	404	147	257	53.5
36.4	401	146	255	53.2
36.4	398	145	253	52.8
36.4	396	144	252	52.4
36.4	393	143	250	52.0
36.4	390	142	248	51.7
36.4	387	141	246	51.4
36.4	385	140	245	50.9
36.4	382	139	243	50.6
36.4	379	138	241	50.2
36.4	376	137	239	49.9
36.4	374	136	238	49.5
36.4	371	135	236	49.1

%E	M1	M2	DM	M*
36.4	368	134	234	48.8
36.4	365	133	232	48.5
36.4	363	132	231	48.0
36.4	360	131	229	47.7
36.4	357	130	227	47.3
36.4	354	129	225	47.0
36.4	352	128	224	46.5
36.4	349	127	222	46.2
36.4	346	126	220	45.9
36.4	343	125	218	45.6
36.4	341	124	217	45.1
36.4	338	123	215	44.8
36.4	335	122	213	44.4
36.4	332	121	211	44.1
36.4	330	120	210	43.6
36.4	327	119	208	43.3
36.4	324	118	206	43.0
36.4	321	117	204	42.6
36.4	319	116	203	42.2
36.4	316	115	201	41.9
36.4	313	114	199	41.5
36.4	308	112	196	40.7
36.4	305	111	194	40.4
36.4	302	110	192	40.1
36.4	297	108	189	39.3
36.4	294	107	187	38.9
36.4	291	106	185	38.6
36.4	286	104	182	37.8
36.4	283	103	180	37.5
36.4	280	102	178	37.2
36.4	275	100	175	36.4
36.4	272	99	173	36.0
36.4	269	98	171	35.7
36.4	264	96	168	34.9
36.4	261	95	166	34.6
36.4	258	94	164	34.2
36.4	253	92	161	33.5
36.4	250	91	159	33.1
36.4	247	90	157	32.8
36.4	242	88	154	32.0
36.4	239	87	152	31.7
36.4	236	86	150	31.3
36.4	231	84	147	30.5
36.4	228	83	145	30.2
36.4	225	82	143	29.9
36.4	220	80	140	29.1
36.4	217	79	138	28.8
36.4	214	78	136	28.4
36.4	209	76	133	27.6
36.4	206	75	131	27.3

%E	M1	M2	DM	M*
36.4	198	72	126	26.2
36.4	195	71	124	25.9
36.4	187	68	119	24.7
36.4	184	67	117	24.4
36.4	176	64	112	23.3
36.4	173	63	110	22.9
36.4	165	60	105	21.8
36.4	162	59	103	21.5
36.4	154	56	98	20.4
36.4	151	55	96	20.0
36.4	143	52	91	18.9
36.4	140	51	89	18.6
36.4	132	48	84	17.5
36.4	129	47	82	17.1
36.4	121	44	77	16.0
36.4	118	43	75	15.7
36.4	110	40	70	14.5
36.4	107	39	68	14.2
36.4	99	36	63	13.1
36.4	88	32	56	11.6
36.4	77	28	49	10.2
36.4	66	24	42	8.7
36.4	55	20	35	7.3
36.4	44	16	28	5.8
36.4	33	12	21	4.4
36.4	22	8	14	2.9
36.4	11	4	7	1.5
36.3	499	181	318	65.7
36.3	498	181	317	65.8
36.3	496	180	316	65.3
36.3	493	179	314	65.0
36.3	491	178	313	64.5
36.3	490	178	312	64.7
36.3	488	177	311	64.2
36.3	487	177	310	64.3
36.3	485	176	309	63.9
36.3	482	175	307	63.5
36.3	480	174	306	63.1
36.3	479	174	305	63.2
36.3	477	173	304	62.7
36.3	476	173	303	62.9
36.3	474	172	302	62.4
36.3	471	171	300	62.1
36.3	468	170	298	61.8
36.3	466	169	297	61.3
36.3	465	169	296	61.4
36.3	463	168	295	61.0
36.3	460	167	293	60.6
36.3	457	166	291	60.3
36.3	455	165	290	59.8

%E	M1	M2	DM	M*
36.3	454	165	289	60.0
36.3	452	164	288	59.5
36.3	449	163	286	59.2
36.3	446	162	284	58.8
36.3	444	161	283	58.4
36.3	443	161	282	58.5
36.3	441	160	281	58.0
36.3	438	159	279	57.7
36.3	435	158	277	57.4
36.3	433	157	276	56.9
36.3	432	157	275	57.1
36.3	430	156	274	56.6
36.3	427	155	272	56.3
36.3	424	154	270	55.9
36.3	422	153	269	55.5
36.3	421	153	268	55.6
36.3	419	152	267	55.1
36.3	416	151	265	54.8
36.3	413	150	263	54.5
36.3	411	149	262	54.0
36.3	410	149	261	54.1
36.3	408	148	260	53.7
36.3	405	147	258	53.4
36.3	402	146	256	53.0
36.3	400	145	255	52.6
36.3	399	145	254	52.7
36.3	397	144	253	52.2
36.3	394	143	251	51.9
36.3	391	142	249	51.6
36.3	388	141	247	51.2
36.3	386	140	246	50.8
36.3	383	139	244	50.4
36.3	380	138	242	50.1
36.3	377	137	240	49.8
36.3	375	136	239	49.3
36.3	372	135	237	49.0
36.3	369	134	235	48.7
36.3	366	133	233	48.3
36.3	364	132	232	47.9
36.3	361	131	230	47.5
36.3	358	130	228	47.2
36.3	355	129	226	46.9
36.3	353	128	225	46.4
36.3	350	127	223	46.1
36.3	347	126	221	45.8
36.3	344	125	219	45.4
36.3	342	124	218	45.0
36.3	339	123	216	44.6
36.3	336	122	214	44.3
36.3	333	121	212	44.0

%E	M1	M2	DM	M*
36.3	331	120	211	43.5
36.3	328	119	209	43.2
36.3	325	118	207	42.8
36.3	322	117	205	42.5
36.3	320	116	204	42.0
36.3	317	115	202	41.7
36.3	314	114	200	41.4
36.3	311	113	198	41.1
36.3	306	111	195	40.3
36.3	303	110	193	39.9
36.3	300	109	191	39.6
36.3	295	107	188	38.8
36.3	292	106	186	38.5
36.3	289	105	184	38.1
36.3	284	103	181	37.4
36.3	281	102	179	37.0
36.3	278	101	177	36.7
36.3	273	99	174	35.9
36.3	270	98	172	35.6
36.3	267	97	170	35.2
36.3	262	95	167	34.4
36.3	259	94	165	34.1
36.3	256	93	163	33.8
36.3	251	91	160	33.0
36.3	248	90	158	32.7
36.3	245	89	156	32.3
36.3	240	87	153	31.5
36.3	237	86	151	31.2
36.3	234	85	149	30.9
36.3	226	82	144	29.8
36.3	223	81	142	29.4
36.3	215	78	137	28.3
36.3	212	77	135	28.0
36.3	204	74	130	26.8
36.3	201	73	128	26.5
36.3	193	70	123	25.4
36.3	190	69	121	25.1
36.3	182	66	116	23.9
36.3	179	65	114	23.6
36.3	171	62	109	22.5
36.3	168	61	107	22.1
36.3	160	58	102	21.0
36.3	157	57	100	20.7
36.3	146	53	93	19.2
36.3	135	49	86	17.8
36.3	124	45	79	16.3
36.3	113	41	72	14.9
36.3	102	37	65	13.4
36.3	91	33	58	12.0
36.3	80	29	51	10.5

%E	M1	M2	DM	M*
36.2	500	181	319	65.5
36.2	497	180	317	65.2
36.2	495	179	316	64.7
36.2	494	179	315	64.9
36.2	492	178	314	64.4
36.2	489	177	312	64.1
36.2	486	176	310	63.7
36.2	484	175	309	63.3
36.2	483	175	308	63.4
36.2	481	174	307	62.9
36.2	478	173	305	62.6
36.2	475	172	303	62.3
36.2	473	171	302	61.8
36.2	472	171	301	62.0
36.2	470	170	300	61.5
36.2	469	170	299	61.6
36.2	467	169	298	61.2
36.2	464	168	296	60.8
36.2	461	167	294	60.5
36.2	459	166	293	60.0
36.2	458	166	292	60.2
36.2	456	165	291	59.7
36.2	453	164	289	59.4
36.2	450	163	287	59.0
36.2	448	162	286	58.6
36.2	447	162	285	58.7
36.2	445	161	284	58.2
36.2	442	160	282	57.9
36.2	439	159	280	57.6
36.2	437	158	279	57.1
36.2	436	158	278	57.3
36.2	434	157	277	56.8
36.2	431	156	275	56.5
36.2	428	155	273	56.1
36.2	426	154	272	55.7
36.2	425	154	271	55.8
36.2	423	153	270	55.3
36.2	420	152	268	55.0
36.2	417	151	266	54.7
36.2	414	150	264	54.3
36.2	412	149	263	53.9
36.2	409	148	261	53.6
36.2	406	147	259	53.2
36.2	403	146	257	52.9
36.2	401	145	256	52.4
36.2	398	144	254	52.1
36.2	395	143	252	51.8
36.2	392	142	250	51.4
36.2	390	141	249	51.0
36.2	389	141	248	51.1

%E	M1	M2	DM	M*
36.2	387	140	247	50.6
36.2	384	139	245	50.3
36.2	381	138	243	50.0
36.2	378	137	241	49.7
36.2	376	136	240	49.2
36.2	373	135	238	48.9
36.2	370	134	236	48.5
36.2	367	133	234	48.2
36.2	365	132	233	47.7
36.2	362	131	231	47.4
36.2	359	130	229	47.1
36.2	356	129	227	46.7
36.2	354	128	226	46.3
36.2	351	127	224	46.0
36.2	348	126	222	45.6
36.2	345	125	220	45.3
36.2	343	124	219	44.8
36.2	340	123	217	44.5
36.2	337	122	215	44.2
36.2	334	121	213	43.8
36.2	329	119	210	43.0
36.2	326	118	208	42.7
36.2	323	117	206	42.4
36.2	318	115	203	41.6
36.2	315	114	201	41.3
36.2	312	113	199	40.9
36.2	309	112	197	40.6
36.2	307	111	196	40.1
36.2	304	110	194	39.8
36.2	301	109	192	39.5
36.2	298	108	190	39.1
36.2	293	106	187	38.3
36.2	290	105	185	38.0
36.2	287	104	183	37.7
36.2	282	102	180	36.9
36.2	279	101	178	36.6
36.2	276	100	176	36.2
36.2	271	98	173	35.4
36.2	268	97	171	35.1
36.2	265	96	169	34.8
36.2	260	94	166	34.0
36.2	257	93	164	33.7
36.2	254	92	162	33.3
36.2	246	89	157	32.2
36.2	243	88	155	31.9
36.2	235	85	150	30.7
36.2	232	84	148	30.4
36.2	229	83	146	30.1
36.2	224	81	143	29.3
36.2	221	80	141	29.0

%E	M1	M2	DM	M*
36.2	218	79	139	28.6
36.2	213	77	136	27.8
36.2	210	76	134	27.5
36.2	207	75	132	27.2
36.2	199	72	127	26.1
36.2	196	71	125	25.7
36.2	188	68	120	24.6
36.2	185	67	118	24.3
36.2	177	64	113	23.1
36.2	174	63	111	22.8
36.2	163	59	104	21.4
36.2	152	55	97	19.9
36.2	149	54	95	19.6
36.2	141	51	90	18.4
36.2	138	50	88	18.1
36.2	130	47	83	17.0
36.2	127	46	81	16.7
36.2	116	42	74	15.2
36.2	105	38	67	13.8
36.2	94	34	60	12.3
36.2	69	25	44	9.1
36.2	58	21	37	7.6
36.2	47	17	30	6.1
36.1	499	180	319	64.9
36.1	498	180	318	65.1
36.1	496	179	317	64.6
36.1	493	178	315	64.3
36.1	490	177	313	63.9
36.1	488	176	312	63.5
36.1	487	176	311	63.6
36.1	485	175	310	63.1
36.1	482	174	308	62.8
36.1	479	173	306	62.5
36.1	477	172	305	62.0
36.1	476	172	304	62.2
36.1	474	171	303	61.7
36.1	471	170	301	61.4
36.1	468	169	299	61.0
36.1	466	168	298	60.6
36.1	465	168	297	60.7
36.1	463	167	296	60.2
36.1	462	167	295	60.4
36.1	460	166	294	59.9
36.1	457	165	292	59.6
36.1	454	164	290	59.2
36.1	452	163	289	58.8
36.1	451	163	288	58.9
36.1	449	162	287	58.4
36.1	446	161	285	58.1
36.1	443	160	283	57.8

%E	M1	M2	DM	M*
36.1	441	159	282	57.3
36.1	440	159	281	57.5
36.1	438	158	280	57.0
36.1	435	157	278	56.7
36.1	432	156	276	56.3
36.1	429	155	274	56.0
36.1	427	154	273	55.5
36.1	424	153	271	55.2
36.1	421	152	269	54.9
36.1	418	151	267	54.5
36.1	416	150	266	54.1
36.1	415	150	265	54.2
36.1	413	149	264	53.8
36.1	410	148	262	53.4
36.1	407	147	260	53.1
36.1	404	146	258	52.8
36.1	402	145	257	52.3
36.1	399	144	255	52.0
36.1	396	143	253	51.6
36.1	393	142	251	51.3
36.1	391	141	250	50.8
36.1	388	140	248	50.5
36.1	385	139	246	50.2
36.1	382	138	244	49.9
36.1	380	137	243	49.4
36.1	379	137	242	49.5
36.1	377	136	241	49.1
36.1	374	135	239	48.7
36.1	371	134	237	48.4
36.1	368	133	235	48.1
36.1	366	132	234	47.6
36.1	363	131	232	47.3
36.1	360	130	230	46.9
36.1	357	129	228	46.6
36.1	355	128	227	46.2
36.1	352	127	225	45.8
36.1	349	126	223	45.5
36.1	346	125	221	45.2
36.1	341	123	218	44.4
36.1	338	122	216	44.0
36.1	335	121	214	43.7
36.1	332	120	212	43.4
36.1	330	119	211	42.9
36.1	327	118	209	42.6
36.1	324	117	207	42.3
36.1	321	116	205	41.9
36.1	319	115	204	41.5
36.1	316	114	202	41.1
36.1	313	113	200	40.8
36.1	310	112	198	40.5
36.1	305	110	195	39.7
36.1	302	109	193	39.3
36.1	299	108	191	39.0
36.1	296	107	189	38.7
36.1	294	106	188	38.2
36.1	291	105	186	37.9
36.1	288	104	184	37.6
36.1	285	103	182	37.2
36.1	280	101	179	36.4
36.1	277	100	177	36.1
36.1	274	99	175	35.8
36.1	269	97	172	35.0
36.1	266	96	170	34.6
36.1	263	95	168	34.3
36.1	255	92	163	33.2
36.1	252	91	161	32.9
36.1	249	90	159	32.5
36.1	244	88	156	31.7
36.1	241	87	154	31.4
36.1	238	86	152	31.1
36.1	233	84	149	30.3
36.1	230	83	147	30.0
36.1	227	82	145	29.6
36.1	219	79	140	28.5
36.1	216	78	138	28.2
36.1	208	75	133	27.0
36.1	205	74	131	26.7
36.1	202	73	129	26.4
36.1	194	70	124	25.3
36.1	191	69	122	24.9
36.1	183	66	117	23.8
36.1	180	65	115	23.5
36.1	169	61	108	22.0
36.1	166	60	106	21.7
36.1	158	57	101	20.6
36.1	155	56	99	20.2
36.1	147	53	94	19.1
36.1	144	52	92	18.8
36.1	133	48	85	17.3
36.1	122	44	78	15.9
36.1	119	43	76	15.5
36.1	108	39	69	14.1
36.1	97	35	62	12.6
36.1	83	30	53	10.8
36.1	72	26	46	9.4
36.1	61	22	39	7.9
36.1	36	13	23	4.7
36.0	500	180	320	64.8
36.0	497	179	318	64.5
36.0	495	178	317	64.0
36.0	494	178	316	64.1
36.0	492	177	315	63.7
36.0	491	177	314	63.8
36.0	489	176	313	63.3
36.0	486	175	311	63.0
36.0	484	174	310	62.6
36.0	483	174	309	62.7
36.0	481	173	308	62.2
36.0	480	173	307	62.4
36.0	478	172	306	61.9
36.0	475	171	304	61.6
36.0	472	170	302	61.2
36.0	470	169	301	60.8
36.0	469	169	300	60.9
36.0	467	168	299	60.4
36.0	464	167	297	60.1
36.0	461	166	295	59.8
36.0	458	165	293	59.4
36.0	456	164	292	59.0
36.0	455	164	291	59.1
36.0	453	163	290	58.7
36.0	450	162	288	58.3
36.0	447	161	286	58.0
36.0	445	160	285	57.5
36.0	444	160	284	57.7
36.0	442	159	283	57.2
36.0	439	158	281	56.9
36.0	436	157	279	56.5
36.0	433	156	277	56.2
36.0	431	155	276	55.7
36.0	430	155	275	55.9
36.0	428	154	274	55.4
36.0	425	153	272	55.1
36.0	422	152	270	54.7
36.0	420	151	269	54.3
36.0	419	151	268	54.4
36.0	417	150	267	54.0
36.0	414	149	265	53.6
36.0	411	148	263	53.3
36.0	408	147	261	53.0
36.0	406	146	260	52.5
36.0	405	146	259	52.6
36.0	403	145	258	52.2
36.0	400	144	256	51.8
36.0	397	143	254	51.5
36.0	394	142	252	51.2
36.0	392	141	251	50.7
36.0	389	140	249	50.4
36.0	386	139	247	50.1
36.0	383	138	245	49.7
36.0	381	137	244	49.3
36.0	378	136	242	48.9
36.0	375	135	240	48.6
36.0	372	134	238	48.3
36.0	369	133	236	47.9
36.0	367	132	235	47.5
36.0	364	131	233	47.1
36.0	361	130	231	46.8
36.0	358	129	229	46.5
36.0	356	128	228	46.0
36.0	353	127	226	45.7
36.0	350	126	224	45.4
36.0	347	125	222	45.0
36.0	344	124	220	44.7
36.0	342	123	219	44.2
36.0	339	122	217	43.9
36.0	336	121	215	43.6
36.0	333	120	213	43.2
36.0	331	119	212	42.8
36.0	328	118	210	42.5
36.0	325	117	208	42.1
36.0	322	116	206	41.8
36.0	317	114	203	41.0
36.0	314	113	201	40.7
36.0	311	112	199	40.3
36.0	308	111	197	40.0
36.0	303	109	194	39.2
36.0	300	108	192	38.9
36.0	297	107	190	38.5
36.0	292	105	187	37.8
36.0	289	104	185	37.4
36.0	286	103	183	37.1
36.0	283	102	181	36.8
36.0	278	100	178	36.0
36.0	275	99	176	35.6
36.0	272	98	174	35.3
36.0	267	96	171	34.5
36.0	264	95	169	34.2
36.0	261	94	167	33.9
36.0	258	93	165	33.5
36.0	253	91	162	32.7
36.0	250	90	160	32.4
36.0	247	89	158	32.1
36.0	242	87	155	31.3
36.0	239	86	153	30.9
36.0	236	85	151	30.6
36.0	228	82	146	29.5
36.0	225	81	144	29.2
36.0	222	80	142	28.8
36.0	214	77	137	27.7
36.0	211	76	135	27.4
36.0	203	73	130	26.3
36.0	200	72	128	25.9
36.0	197	71	126	25.6
36.0	189	68	121	24.5
36.0	186	67	119	24.1
36.0	178	64	114	23.0
36.0	175	63	112	22.7
36.0	172	62	110	22.3
36.0	164	59	105	21.2
36.0	161	58	103	20.9
36.0	150	54	96	19.4
36.0	139	50	89	18.0
36.0	136	49	87	17.7
36.0	125	45	80	16.2
36.0	114	41	73	14.7
36.0	111	40	71	14.4
36.0	100	36	64	13.0
36.0	89	32	57	11.5
36.0	86	31	55	11.2
36.0	75	27	48	9.7
36.0	50	18	32	6.5
36.0	25	9	16	3.2
35.9	499	179	320	64.2
35.9	498	179	319	64.3
35.9	496	178	318	63.9
35.9	493	177	316	63.5
35.9	490	176	314	63.2
35.9	488	175	313	62.8
35.9	487	175	312	62.9
35.9	485	174	311	62.4
35.9	482	173	309	62.1
35.9	479	172	307	61.8
35.9	476	171	305	61.4
35.9	474	170	304	61.0
35.9	473	170	303	61.1
35.9	471	169	302	60.6
35.9	468	168	300	60.3
35.9	465	167	298	60.0
35.9	463	166	297	59.5
35.9	462	166	296	59.6
35.9	460	165	295	59.2
35.9	459	165	294	59.3
35.9	457	164	293	58.9
35.9	454	163	291	58.5
35.9	451	162	289	58.2
35.9	449	161	288	57.7
35.9	448	161	287	57.9
35.9	446	160	286	57.4
35.9	443	159	284	57.1

%E	M1	M2	DM	M*
35.9	440	158	282	56.7
35.9	437	157	280	56.4
35.9	435	156	279	55.9
35.9	434	156	278	56.1
35.9	432	155	277	55.6
35.9	429	154	275	55.3
35.9	426	153	273	55.0
35.9	423	152	271	54.6
35.9	421	151	270	54.2
35.9	418	150	268	53.8
35.9	415	149	266	53.5
35.9	412	148	264	53.2
35.9	410	147	263	52.7
35.9	409	147	262	52.8
35.9	407	146	261	52.4
35.9	404	145	259	52.0
35.9	401	144	257	51.7
35.9	398	143	255	51.4
35.9	396	142	254	50.9
35.9	395	142	253	51.0
35.9	393	141	252	50.6
35.9	390	140	250	50.3
35.9	387	139	248	49.9
35.9	384	138	246	49.6
35.9	382	137	245	49.1
35.9	379	136	243	48.8
35.9	376	135	241	48.5
35.9	373	134	239	48.1
35.9	370	133	237	47.8
35.9	368	132	236	47.3
35.9	365	131	234	47.0
35.9	362	130	232	46.7
35.9	359	129	230	46.4
35.9	357	128	229	45.9
35.9	354	127	227	45.6
35.9	351	126	225	45.2
35.9	348	125	223	44.9
35.9	345	124	221	44.6
35.9	343	123	220	44.1
35.9	340	122	218	43.8
35.9	337	121	216	43.4
35.9	334	120	214	43.1
35.9	329	118	211	42.3
35.9	326	117	209	42.0
35.9	323	116	207	41.7
35.9	320	115	205	41.3
35.9	315	113	202	40.5
35.9	312	112	200	40.2
35.9	309	111	198	39.9
35.9	306	110	196	39.5
35.9	304	109	195	39.1
35.9	301	108	193	38.8
35.9	298	107	191	38.4
35.9	295	106	189	38.1
35.9	290	104	186	37.3
35.9	287	103	184	37.0
35.9	284	102	182	36.6
35.9	281	101	180	36.3
35.9	276	99	177	35.5
35.9	273	98	175	35.2
35.9	270	97	173	34.8
35.9	262	94	168	33.7
35.9	259	93	166	33.4
35.9	256	92	164	33.1
35.9	251	90	161	32.3
35.9	248	89	159	31.9
35.9	245	88	157	31.6
35.9	237	85	152	30.5
35.9	234	84	150	30.2
35.9	231	83	148	29.8
35.9	223	80	143	28.7
35.9	220	79	141	28.4
35.9	217	78	139	28.0
35.9	209	75	134	26.9
35.9	206	74	132	26.6
35.9	198	71	127	25.5
35.9	195	70	125	25.1
35.9	192	69	123	24.8
35.9	184	66	118	23.7
35.9	181	65	116	23.3
35.9	170	61	109	21.9
35.9	167	60	107	21.6
35.9	156	56	100	20.1
35.9	153	55	98	19.8
35.9	145	52	93	18.6
35.9	142	51	91	18.3
35.9	131	47	84	16.9
35.9	128	46	82	16.5
35.9	117	42	75	15.1
35.9	103	37	66	13.3
35.9	92	33	59	11.8
35.9	78	28	50	10.1
35.9	64	23	41	8.3
35.9	39	14	25	5.0
35.8	500	179	321	64.1
35.8	497	178	319	63.8
35.8	495	177	318	63.3
35.8	494	177	317	63.4
35.8	492	176	316	63.0
35.8	491	176	315	63.1
35.8	489	175	314	62.6
35.8	486	174	312	62.3
35.8	483	173	310	62.0
35.8	481	172	309	61.5
35.8	480	172	308	61.6
35.8	478	171	307	61.2
35.8	477	171	306	61.3
35.8	475	170	305	60.8
35.8	472	169	303	60.5
35.8	469	168	301	60.2
35.8	467	167	300	59.7
35.8	466	167	299	59.8
35.8	464	166	298	59.4
35.8	461	165	296	59.1
35.8	458	164	294	58.7
35.8	455	163	292	58.4
35.8	453	162	291	57.9
35.8	452	162	290	58.1
35.8	450	161	289	57.6
35.8	447	160	287	57.3
35.8	444	159	285	56.9
35.8	441	158	283	56.6
35.8	439	157	282	56.1
35.8	438	157	281	56.3
35.8	436	156	280	55.8
35.8	433	155	278	55.5
35.8	430	154	276	55.2
35.8	427	153	274	54.8
35.8	425	152	273	54.4
35.8	424	152	272	54.5
35.8	422	151	271	54.0
35.8	419	150	269	53.7
35.8	416	149	267	53.4
35.8	413	148	265	53.0
35.8	411	147	264	52.6
35.8	408	146	262	52.2
35.8	405	145	260	51.9
35.8	402	144	258	51.6
35.8	400	143	257	51.1
35.8	399	143	256	51.3
35.8	397	142	255	50.8
35.8	394	141	253	50.5
35.8	391	140	251	50.1
35.8	388	139	249	49.8
35.8	386	138	248	49.3
35.8	385	138	247	49.5
35.8	383	137	246	49.0
35.8	380	136	244	48.7
35.8	377	135	242	48.3
35.8	374	134	240	48.0
35.8	372	133	239	47.6
35.8	371	133	238	47.7
35.8	369	132	237	47.2
35.8	366	131	235	46.9
35.8	363	130	233	46.6
35.8	360	129	231	46.2
35.8	358	128	230	45.8
35.8	355	127	228	45.4
35.8	352	126	226	45.1
35.8	349	125	224	44.8
35.8	346	124	222	44.4
35.8	344	123	221	44.0
35.8	341	122	219	43.6
35.8	338	121	217	43.3
35.8	335	120	215	43.0
35.8	332	119	213	42.7
35.8	330	118	212	42.2
35.8	327	117	210	41.9
35.8	324	116	208	41.5
35.8	321	115	206	41.2
35.8	318	114	204	40.9
35.8	316	113	203	40.4
35.8	313	112	201	40.1
35.8	310	111	199	39.7
35.8	307	110	197	39.4
35.8	302	108	194	38.6
35.8	299	107	192	38.3
35.8	296	106	190	38.0
35.8	293	105	188	37.6
35.8	288	103	185	36.8
35.8	285	102	183	36.5
35.8	282	101	181	36.2
35.8	279	100	179	35.8
35.8	274	98	176	35.1
35.8	271	97	174	34.7
35.8	268	96	172	34.4
35.8	265	95	170	34.1
35.8	260	93	167	33.3
35.8	257	92	165	32.9
35.8	254	91	163	32.6
35.8	246	88	158	31.5
35.8	243	87	156	31.1
35.8	240	86	154	30.8
35.8	232	83	149	29.7
35.8	229	82	147	29.4
35.8	226	81	145	29.0
35.8	218	78	140	27.9
35.8	215	77	138	27.6
35.8	212	76	136	27.2
35.8	204	73	131	26.1
35.8	201	72	129	25.8
35.8	193	69	124	24.7
35.8	190	68	122	24.3
35.8	187	67	120	24.0
35.8	179	64	115	22.9
35.8	176	63	113	22.6
35.8	173	62	111	22.2
35.8	165	59	106	21.1
35.8	162	58	104	20.8
35.8	159	57	102	20.4
35.8	151	54	97	19.3
35.8	148	53	95	19.0
35.8	137	49	88	17.5
35.8	134	48	86	17.2
35.8	123	44	79	15.7
35.8	120	43	77	15.4
35.8	109	39	70	14.0
35.8	106	38	68	13.6
35.8	95	34	61	12.2
35.8	81	29	52	10.4
35.8	67	24	43	8.6
35.8	53	19	34	6.8
35.7	499	178	321	63.5
35.7	498	178	320	63.6
35.7	496	177	319	63.2
35.7	493	176	317	62.8
35.7	490	175	315	62.5
35.7	488	174	314	62.0
35.7	487	174	313	62.2
35.7	485	173	312	61.7
35.7	484	173	311	61.8
35.7	482	172	310	61.4
35.7	479	171	308	61.0
35.7	476	170	306	60.7
35.7	474	169	305	60.3
35.7	473	169	304	60.4
35.7	471	168	303	59.9
35.7	470	168	302	60.1
35.7	468	167	301	59.6
35.7	465	166	299	59.3
35.7	462	165	297	58.9
35.7	460	164	296	58.5
35.7	459	164	295	58.6
35.7	457	163	294	58.1
35.7	456	163	293	58.3
35.7	454	162	292	57.8
35.7	451	161	290	57.5
35.7	448	160	288	57.1
35.7	446	159	287	56.7
35.7	445	159	286	56.8

%E	M1	M2	DM	M*
35.7	443	158	285	56.4
35.7	442	158	284	56.5
35.7	440	157	283	56.0
35.7	437	156	281	55.7
35.7	434	155	279	55.4
35.7	431	154	277	55.0
35.7	429	153	276	54.6
35.7	428	153	275	54.7
35.7	426	152	274	54.2
35.7	423	151	272	53.9
35.7	420	150	270	53.6
35.7	417	149	268	53.2
35.7	415	148	267	52.8
35.7	414	148	266	52.9
35.7	412	147	265	52.4
35.7	409	146	263	52.1
35.7	406	145	261	51.8
35.7	403	144	259	51.5
35.7	401	143	258	51.0
35.7	398	142	256	50.7
35.7	395	141	254	50.3
35.7	392	140	252	50.0
35.7	389	139	250	49.7
35.7	387	138	249	49.2
35.7	384	137	247	48.9
35.7	381	136	245	48.5
35.7	378	135	243	48.2
35.7	375	134	241	47.9
35.7	373	133	240	47.4
35.7	370	132	238	47.1
35.7	367	131	236	46.8
35.7	364	130	234	46.4
35.7	361	129	232	46.1
35.7	359	128	231	45.6
35.7	356	127	229	45.3
35.7	353	126	227	45.0
35.7	350	125	225	44.6
35.7	347	124	223	44.3
35.7	345	123	222	43.9
35.7	342	122	220	43.5
35.7	339	121	218	43.2
35.7	336	120	216	42.9
35.7	333	119	214	42.5
35.7	331	118	213	42.1
35.7	328	117	211	41.7
35.7	325	116	209	41.4
35.7	322	115	207	41.1
35.7	319	114	205	40.7
35.7	314	112	202	39.9
35.7	311	111	200	39.6

%E	M1	M2	DM	M*
35.7	308	110	198	39.3
35.7	305	109	196	39.0
35.7	300	107	193	38.2
35.7	297	106	191	37.8
35.7	294	105	189	37.5
35.7	291	104	187	37.2
35.7	286	102	184	36.4
35.7	283	101	182	36.0
35.7	280	100	180	35.7
35.7	277	99	178	35.4
35.7	272	97	175	34.6
35.7	269	96	173	34.3
35.7	266	95	171	33.9
35.7	263	94	169	33.6
35.7	258	92	166	32.8
35.7	255	91	164	32.5
35.7	252	90	162	32.1
35.7	249	89	160	31.8
35.7	244	87	157	31.0
35.7	241	86	155	30.7
35.7	238	85	153	30.4
35.7	235	84	151	30.0
35.7	230	82	148	29.2
35.7	227	81	146	28.9
35.7	224	80	144	28.6
35.7	221	79	142	28.2
35.7	213	76	137	27.1
35.7	210	75	135	26.8
35.7	207	74	133	26.5
35.7	199	71	128	25.3
35.7	196	70	126	25.0
35.7	185	66	119	23.5
35.7	182	65	117	23.2
35.7	171	61	110	21.8
35.7	168	60	108	21.4
35.7	157	56	101	20.0
35.7	154	55	99	19.6
35.7	143	51	92	18.2
35.7	140	50	90	17.9
35.7	129	46	83	16.4
35.7	126	45	81	16.1
35.7	115	41	74	14.6
35.7	112	40	72	14.3
35.7	98	35	63	12.5
35.7	84	30	54	10.7
35.7	70	25	45	8.9
35.7	56	20	36	7.1
35.7	42	15	27	5.4
35.7	28	10	18	3.6
35.7	14	5	9	1.8

%E	M1	M2	DM	M*
35.6	500	178	322	63.4
35.6	497	177	320	63.0
35.6	495	176	319	62.6
35.6	494	176	318	62.7
35.6	492	175	317	62.2
35.6	491	175	316	62.4
35.6	489	174	315	61.9
35.6	486	173	313	61.6
35.6	483	172	311	61.3
35.6	481	171	310	60.8
35.6	480	171	309	60.9
35.6	478	170	308	60.5
35.6	477	170	307	60.6
35.6	475	169	306	60.1
35.6	472	168	304	59.8
35.6	469	167	302	59.5
35.6	466	166	300	59.1
35.6	464	165	299	58.7
35.6	463	165	298	58.8
35.6	461	164	297	58.3
35.6	458	163	295	58.0
35.6	455	162	293	57.7
35.6	452	161	291	57.3
35.6	450	160	290	56.9
35.6	449	160	289	57.0
35.6	447	159	288	56.6
35.6	444	158	286	56.2
35.6	441	157	284	55.9
35.6	438	156	282	55.6
35.6	436	155	281	55.1
35.6	435	155	280	55.2
35.6	433	154	279	54.8
35.6	432	154	278	54.9
35.6	430	153	277	54.4
35.6	427	152	275	54.1
35.6	424	151	273	53.8
35.6	421	150	271	53.4
35.6	419	149	270	53.0
35.6	418	149	269	53.1
35.6	416	148	268	52.7
35.6	413	147	266	52.3
35.6	410	146	264	52.0
35.6	407	145	262	51.7
35.6	405	144	261	51.2
35.6	404	144	260	51.3
35.6	402	143	259	50.9
35.6	399	142	257	50.5
35.6	396	141	255	50.2
35.6	393	140	253	49.9
35.6	391	139	252	49.4

%E	M1	M2	DM	M*
35.6	390	139	251	49.5
35.6	388	138	250	49.1
35.6	385	137	248	48.8
35.6	382	136	246	48.4
35.6	379	135	244	48.1
35.6	376	134	242	47.8
35.6	374	133	241	47.3
35.6	371	132	239	47.0
35.6	368	131	237	46.6
35.6	365	130	235	46.3
35.6	362	129	233	46.0
35.6	360	128	232	45.5
35.6	357	127	230	45.2
35.6	354	126	228	44.8
35.6	351	125	226	44.5
35.6	348	124	224	44.2
35.6	343	122	221	43.4
35.6	340	121	219	43.1
35.6	337	120	217	42.7
35.6	334	119	215	42.4
35.6	329	117	212	41.6
35.6	326	116	210	41.3
35.6	323	115	208	40.9
35.6	320	114	206	40.6
35.6	317	113	204	40.3
35.6	315	112	203	39.8
35.6	312	111	201	39.5
35.6	309	110	199	39.2
35.6	306	109	197	38.8
35.6	303	108	195	38.5
35.6	298	106	192	37.7
35.6	295	105	190	37.4
35.6	292	104	188	37.0
35.6	289	103	186	36.7
35.6	284	101	183	35.9
35.6	281	100	181	35.6
35.6	278	99	179	35.3
35.6	275	98	177	34.9
35.6	270	96	174	34.1
35.6	267	95	172	33.8
35.6	264	94	170	33.5
35.6	261	93	168	33.1
35.6	253	90	163	32.0
35.6	250	89	161	31.7
35.6	247	88	159	31.4
35.6	239	85	154	30.2
35.6	236	84	152	29.9
35.6	233	83	150	29.6
35.6	225	80	145	28.4
35.6	222	79	143	28.1

%E	M1	M2	DM	M*
35.6	219	78	141	27.8
35.6	216	77	139	27.4
35.6	208	74	134	26.3
35.6	205	73	132	26.0
35.6	202	72	130	25.7
35.6	194	69	125	24.5
35.6	191	68	123	24.2
35.6	188	67	121	23.9
35.6	180	64	116	22.8
35.6	177	63	114	22.4
35.6	174	62	112	22.1
35.6	163	58	105	20.6
35.6	160	57	103	20.3
35.6	149	53	96	18.9
35.6	146	52	94	18.5
35.6	135	48	87	17.1
35.6	132	47	85	16.7
35.6	118	42	76	14.9
35.6	104	37	67	13.2
35.6	101	36	65	12.8
35.6	90	32	58	11.4
35.6	87	31	56	11.0
35.6	73	26	47	9.3
35.6	59	21	38	7.5
35.6	45	16	29	5.7
35.5	499	177	322	62.8
35.5	498	177	321	62.9
35.5	496	176	320	62.5
35.5	493	175	318	62.1
35.5	490	174	316	61.8
35.5	488	173	315	61.3
35.5	487	173	314	61.5
35.5	485	172	313	61.0
35.5	484	172	312	61.1
35.5	482	171	311	60.7
35.5	479	170	309	60.3
35.5	476	169	307	60.0
35.5	473	168	305	59.7
35.5	471	167	304	59.2
35.5	470	167	303	59.3
35.5	468	166	302	58.9
35.5	467	166	301	59.0
35.5	465	165	300	58.5
35.5	462	164	298	58.2
35.5	459	163	296	57.9
35.5	456	162	294	57.6
35.5	454	161	293	57.1
35.5	453	161	292	57.2
35.5	451	160	291	56.8
35.5	448	159	289	56.4

%E	M1	M2	DM	M*
35.5	445	158	287	56.1
35.5	442	157	285	55.8
35.5	440	156	284	55.3
35.5	439	156	283	55.4
35.5	437	155	282	55.0
35.5	434	154	280	54.6
35.5	431	153	278	54.3
35.5	428	152	276	54.0
35.5	425	151	274	53.6
35.5	423	150	273	53.2
35.5	422	150	272	53.3
35.5	420	149	271	52.9
35.5	417	148	269	52.5
35.5	414	147	267	52.2
35.5	411	146	265	51.9
35.5	409	145	264	51.4
35.5	408	145	263	51.5
35.5	406	144	262	51.1
35.5	403	143	260	50.7
35.5	400	142	258	50.4
35.5	397	141	256	50.1
35.5	394	140	254	49.7
35.5	392	139	253	49.3
35.5	389	138	251	49.0
35.5	386	137	249	48.6
35.5	383	136	247	48.3
35.5	380	135	245	48.0
35.5	378	134	244	47.5
35.5	377	134	243	47.6
35.5	375	133	242	47.2
35.5	372	132	240	46.8
35.5	369	131	238	46.5
35.5	366	130	236	46.2
35.5	363	129	234	45.8
35.5	361	128	233	45.4
35.5	358	127	231	45.1
35.5	355	126	229	44.7
35.5	352	125	227	44.4
35.5	349	124	225	44.1
35.5	346	123	223	43.7
35.5	344	122	222	43.3
35.5	341	121	220	42.9
35.5	338	120	218	42.6
35.5	335	119	216	42.3
35.5	332	118	214	41.9
35.5	330	117	213	41.5
35.5	327	116	211	41.1
35.5	324	115	209	40.8
35.5	321	114	207	40.5
35.5	318	113	205	40.2

%E	M1	M2	DM	M*
35.5	313	111	202	39.4
35.5	310	110	200	39.0
35.5	307	109	198	38.7
35.5	304	108	196	38.4
35.5	301	107	194	38.0
35.5	299	106	193	37.6
35.5	296	105	191	37.2
35.5	293	104	189	36.9
35.5	290	103	187	36.6
35.5	287	102	185	36.3
35.5	282	100	182	35.5
35.5	279	99	180	35.1
35.5	276	98	178	34.8
35.5	273	97	176	34.5
35.5	265	94	171	33.3
35.5	262	93	169	33.0
35.5	259	92	167	32.7
35.5	256	91	165	32.3
35.5	251	89	162	31.6
35.5	248	88	160	31.2
35.5	245	87	158	30.9
35.5	242	86	156	30.6
35.5	234	83	151	29.4
35.5	231	82	149	29.1
35.5	228	81	147	28.8
35.5	220	78	142	27.7
35.5	217	77	140	27.3
35.5	214	76	138	27.0
35.5	211	75	136	26.7
35.5	203	72	131	25.5
35.5	200	71	129	25.2
35.5	197	70	127	24.9
35.5	189	67	122	23.8
35.5	186	66	120	23.4
35.5	183	65	118	23.1
35.5	172	61	111	21.6
35.5	169	60	109	21.3
35.5	166	59	107	21.0
35.5	155	55	100	19.5
35.5	152	54	98	19.2
35.5	141	50	91	17.7
35.5	138	49	89	17.4
35.5	124	44	80	15.6
35.5	121	43	78	15.3
35.5	110	39	71	13.8
35.5	107	38	69	13.5
35.5	93	33	60	11.7
35.5	76	27	49	9.6
35.5	62	22	40	7.8
35.5	31	11	20	3.9

%E	M1	M2	DM	M*
35.4	500	177	323	62.7
35.4	497	176	321	62.3
35.4	495	175	320	61.9
35.4	494	175	319	62.0
35.4	492	174	318	61.5
35.4	491	174	317	61.7
35.4	489	173	316	61.2
35.4	486	172	314	60.9
35.4	483	171	312	60.5
35.4	480	170	310	60.2
35.4	478	169	309	59.8
35.4	477	169	308	59.9
35.4	475	168	307	59.4
35.4	474	168	306	59.5
35.4	472	167	305	59.1
35.4	469	166	303	58.8
35.4	466	165	301	58.4
35.4	463	164	299	58.1
35.4	461	163	298	57.6
35.4	460	163	297	57.8
35.4	458	162	296	57.3
35.4	457	162	295	57.4
35.4	455	161	294	57.0
35.4	452	160	292	56.6
35.4	449	159	290	56.3
35.4	446	158	288	56.0
35.4	444	157	287	55.5
35.4	443	157	286	55.6
35.4	441	156	285	55.2
35.4	438	155	283	54.9
35.4	435	154	281	54.5
35.4	432	153	279	54.2
35.4	429	152	277	53.9
35.4	427	151	276	53.4
35.4	426	151	275	53.5
35.4	424	150	274	53.1
35.4	421	149	272	52.7
35.4	418	148	270	52.4
35.4	415	147	268	52.1
35.4	413	146	267	51.6
35.4	412	146	266	51.7
35.4	410	145	265	51.3
35.4	407	144	263	50.9
35.4	404	143	261	50.6
35.4	401	142	259	50.3
35.4	398	141	257	50.0
35.4	396	140	256	49.5
35.4	395	140	255	49.6
35.4	393	139	254	49.2
35.4	390	138	252	48.8

%E	M1	M2	DM	M*
35.4	387	137	250	48.5
35.4	384	136	248	48.2
35.4	381	135	246	47.8
35.4	379	134	245	47.4
35.4	376	133	243	47.0
35.4	373	132	241	46.7
35.4	370	131	239	46.4
35.4	367	130	237	46.0
35.4	364	129	235	45.7
35.4	362	128	234	45.3
35.4	359	127	232	44.9
35.4	356	126	230	44.6
35.4	353	125	228	44.3
35.4	350	124	226	43.9
35.4	347	123	224	43.6
35.4	345	122	223	43.1
35.4	342	121	221	42.8
35.4	339	120	219	42.5
35.4	336	119	217	42.1
35.4	333	118	215	41.8
35.4	328	116	212	41.0
35.4	325	115	210	40.7
35.4	322	114	208	40.4
35.4	319	113	206	40.0
35.4	316	112	204	39.7
35.4	314	111	203	39.2
35.4	311	110	201	38.9
35.4	308	109	199	38.6
35.4	305	108	197	38.2
35.4	302	107	195	37.9
35.4	297	105	192	37.1
35.4	294	104	190	36.8
35.4	291	103	188	36.5
35.4	288	102	186	36.1
35.4	285	101	184	35.8
35.4	280	99	181	35.0
35.4	277	98	179	34.7
35.4	274	97	177	34.3
35.4	271	96	175	34.0
35.4	268	95	173	33.7
35.4	263	93	170	32.9
35.4	260	92	168	32.6
35.4	257	91	166	32.2
35.4	254	90	164	31.9
35.4	246	87	159	30.8
35.4	243	86	157	30.4
35.4	240	85	155	30.1
35.4	237	84	153	29.8
35.4	229	81	148	28.7
35.4	226	80	146	28.3

%E	M1	M2	DM	M*
35.4	223	79	144	28.0
35.4	212	75	137	26.5
35.4	209	74	135	26.2
35.4	206	73	133	25.9
35.4	198	70	128	24.7
35.4	195	69	126	24.4
35.4	192	68	124	24.1
35.4	181	64	117	22.6
35.4	178	63	115	22.3
35.4	175	62	113	22.0
35.4	164	58	106	20.5
35.4	161	57	104	20.2
35.4	158	56	102	19.8
35.4	147	52	95	18.4
35.4	144	51	93	18.1
35.4	130	46	84	16.3
35.4	127	45	82	15.9
35.4	113	40	73	14.2
35.4	99	35	64	12.4
35.4	96	34	62	12.0
35.4	82	29	53	10.3
35.4	79	28	51	9.9
35.4	65	23	42	8.1
35.4	48	17	31	6.0
35.3	499	176	323	62.1
35.3	498	176	322	62.2
35.3	496	175	321	61.7
35.3	493	174	319	61.4
35.3	490	173	317	61.1
35.3	487	172	315	60.7
35.3	485	171	314	60.3
35.3	484	171	313	60.4
35.3	482	170	312	60.0
35.3	481	170	311	60.1
35.3	479	169	310	59.6
35.3	476	168	308	59.3
35.3	473	167	306	59.0
35.3	470	166	304	58.6
35.3	468	165	303	58.2
35.3	467	165	302	58.3
35.3	465	164	301	57.8
35.3	464	164	300	58.0
35.3	462	163	299	57.5
35.3	459	162	297	57.2
35.3	456	161	295	56.8
35.3	453	160	293	56.5
35.3	451	159	292	56.1
35.3	450	159	291	56.2
35.3	448	158	290	55.7
35.3	447	158	289	55.8

%E	M1	M2	DM	M*
35.3	445	157	288	55.4
35.3	442	156	286	55.1
35.3	439	155	284	54.7
35.3	436	154	282	54.4
35.3	434	153	281	53.9
35.3	433	153	280	54.1
35.3	431	152	279	53.6
35.3	430	152	278	53.7
35.3	428	151	277	53.3
35.3	425	150	275	52.9
35.3	422	149	273	52.6
35.3	419	148	271	52.3
35.3	417	147	270	51.8
35.3	416	147	269	51.9
35.3	414	146	268	51.5
35.3	411	145	266	51.2
35.3	408	144	264	50.8
35.3	405	143	262	50.5
35.3	402	142	260	50.2
35.3	400	141	259	49.7
35.3	399	141	258	49.8
35.3	397	140	257	49.4
35.3	394	139	255	49.0
35.3	391	138	253	48.7
35.3	388	137	251	48.4
35.3	385	136	249	48.0
35.3	382	135	247	47.7
35.3	380	134	246	47.3
35.3	377	133	244	46.9
35.3	374	132	242	46.6
35.3	371	131	240	46.3
35.3	368	130	238	45.9
35.3	365	129	236	45.6
35.3	363	128	235	45.1
35.3	360	127	233	44.8
35.3	357	126	231	44.5
35.3	354	125	229	44.1
35.3	351	124	227	43.8
35.3	348	123	225	43.5
35.3	346	122	224	43.0
35.3	343	121	222	42.7
35.3	340	120	220	42.4
35.3	337	119	218	42.0
35.3	334	118	216	41.7
35.3	331	117	214	41.4
35.3	329	116	213	40.9
35.3	326	115	211	40.6
35.3	323	114	209	40.2
35.3	320	113	207	39.9
35.3	317	112	205	39.6

%E	M1	M2	DM	M*
35.3	312	110	202	38.8
35.3	309	109	200	38.4
35.3	306	108	198	38.1
35.3	303	107	196	37.8
35.3	300	106	194	37.5
35.3	295	104	191	36.7
35.3	292	103	189	36.3
35.3	289	102	187	36.0
35.3	286	101	185	35.7
35.3	283	100	183	35.3
35.3	278	98	180	34.5
35.3	275	97	178	34.2
35.3	272	96	176	33.9
35.3	269	95	174	33.6
35.3	266	94	172	33.2
35.3	258	91	167	32.1
35.3	255	90	165	31.8
35.3	252	89	163	31.4
35.3	249	88	161	31.1
35.3	241	85	156	30.0
35.3	238	84	154	29.6
35.3	235	83	152	29.3
35.3	232	82	150	29.0
35.3	224	79	145	27.9
35.3	221	78	143	27.5
35.3	218	77	141	27.2
35.3	215	76	139	26.9
35.3	207	73	134	25.7
35.3	204	72	132	25.4
35.3	201	71	130	25.1
35.3	190	67	123	23.6
35.3	187	66	121	23.3
35.3	184	65	119	23.0
35.3	173	61	112	21.5
35.3	170	60	110	21.2
35.3	167	59	108	20.8
35.3	156	55	101	19.4
35.3	153	54	99	19.1
35.3	150	53	97	18.7
35.3	139	49	90	17.3
35.3	136	48	88	16.9
35.3	133	47	86	16.6
35.3	119	42	77	14.8
35.3	116	41	75	14.5
35.3	102	36	66	12.7
35.3	85	30	55	10.6
35.3	68	24	44	8.5
35.3	51	18	33	6.4
35.3	34	12	22	4.2
35.3	17	6	11	2.1

%E	M1	M2	DM	M*
35.2	500	176	324	62.0
35.2	497	175	322	61.6
35.2	495	174	321	61.2
35.2	494	174	320	61.3
35.2	492	173	319	60.8
35.2	491	173	318	61.0
35.2	489	172	317	60.5
35.2	488	172	316	60.6
35.2	486	171	315	60.2
35.2	483	170	313	59.8
35.2	480	169	311	59.5
35.2	477	168	309	59.2
35.2	475	167	308	58.7
35.2	474	167	307	58.8
35.2	472	166	306	58.4
35.2	471	166	305	58.5
35.2	469	165	304	58.0
35.2	466	164	302	57.7
35.2	463	163	300	57.4
35.2	460	162	298	57.1
35.2	458	161	297	56.6
35.2	457	161	296	56.7
35.2	455	160	295	56.3
35.2	454	160	294	56.4
35.2	452	159	293	55.9
35.2	449	158	291	55.6
35.2	446	157	289	55.3
35.2	443	156	287	54.9
35.2	440	155	285	54.6
35.2	438	154	284	54.1
35.2	437	154	283	54.3
35.2	435	153	282	53.8
35.2	432	152	280	53.5
35.2	429	151	278	53.1
35.2	426	150	276	52.8
35.2	423	149	274	52.5
35.2	421	148	273	52.0
35.2	420	148	272	52.2
35.2	418	147	271	51.7
35.2	415	146	269	51.4
35.2	412	145	267	51.0
35.2	409	144	265	50.7
35.2	406	143	263	50.4
35.2	403	142	261	50.0
35.2	401	141	260	49.6
35.2	398	140	258	49.2
35.2	395	139	256	48.9
35.2	392	138	254	48.6
35.2	389	137	252	48.2
35.2	386	136	250	47.9

%E	M1	M2	DM	M*
35.2	384	135	249	47.5
35.2	383	135	248	47.6
35.2	381	134	247	47.1
35.2	378	133	245	46.8
35.2	375	132	243	46.5
35.2	372	131	241	46.1
35.2	369	130	239	45.8
35.2	367	129	238	45.3
35.2	366	129	237	45.5
35.2	364	128	236	45.0
35.2	361	127	234	44.7
35.2	358	126	232	44.3
35.2	355	125	230	44.0
35.2	352	124	228	43.7
35.2	349	123	226	43.3
35.2	347	122	225	42.9
35.2	344	121	223	42.6
35.2	341	120	221	42.2
35.2	338	119	219	41.9
35.2	335	118	217	41.6
35.2	332	117	215	41.2
35.2	330	116	214	40.8
35.2	327	115	212	40.4
35.2	324	114	210	40.1
35.2	321	113	208	39.8
35.2	318	112	206	39.4
35.2	315	111	204	39.1
35.2	310	109	201	38.3
35.2	307	108	199	38.0
35.2	304	107	197	37.7
35.2	301	106	195	37.3
35.2	298	105	193	37.0
35.2	293	103	190	36.2
35.2	290	102	188	35.9
35.2	287	101	186	35.5
35.2	284	100	184	35.2
35.2	281	99	182	34.9
35.2	273	96	177	33.8
35.2	270	95	175	33.4
35.2	267	94	173	33.1
35.2	264	93	171	32.8
35.2	261	92	169	32.4
35.2	256	90	166	31.6
35.2	253	89	164	31.3
35.2	250	88	162	31.0
35.2	247	87	160	30.6
35.2	244	86	158	30.3
35.2	236	83	153	29.2
35.2	233	82	151	28.9
35.2	230	81	149	28.5

%E	M1	M2	DM	M*
35.2	227	80	147	28.2
35.2	219	77	142	27.1
35.2	216	76	140	26.7
35.2	213	75	138	26.4
35.2	210	74	136	26.1
35.2	199	70	129	24.6
35.2	196	69	127	24.3
35.2	193	68	125	24.0
35.2	182	64	118	22.5
35.2	179	63	116	22.2
35.2	176	62	114	21.8
35.2	165	58	107	20.4
35.2	162	57	105	20.1
35.2	159	56	103	19.7
35.2	145	51	94	17.9
35.2	142	50	92	17.6
35.2	128	45	83	15.8
35.2	125	44	81	15.5
35.2	122	43	79	15.2
35.2	108	38	70	13.4
35.2	105	37	68	13.0
35.2	91	32	59	11.3
35.2	88	31	57	10.9
35.2	71	25	46	8.8
35.2	54	19	35	6.7
35.1	499	175	324	61.4
35.1	498	175	323	61.5
35.1	496	174	322	61.0
35.1	493	173	320	60.7
35.1	490	172	318	60.4
35.1	487	171	316	60.0
35.1	485	170	315	59.6
35.1	484	170	314	59.7
35.1	482	169	313	59.3
35.1	481	169	312	59.4
35.1	479	168	311	58.9
35.1	478	168	310	59.0
35.1	476	167	309	58.6
35.1	473	166	307	58.3
35.1	470	165	305	57.9
35.1	467	164	303	57.6
35.1	465	163	302	57.1
35.1	464	163	301	57.3
35.1	462	162	300	56.8
35.1	461	162	299	56.9
35.1	459	161	298	56.5
35.1	456	160	296	56.1
35.1	453	159	294	55.8
35.1	450	158	292	55.5
35.1	447	157	290	55.1

%E	M1	M2	DM	M*
35.1	445	156	289	54.7
35.1	444	156	288	54.8
35.1	442	155	287	54.4
35.1	441	155	286	54.5
35.1	439	154	285	54.0
35.1	436	153	283	53.7
35.1	433	152	281	53.4
35.1	430	151	279	53.0
35.1	427	150	277	52.7
35.1	425	149	276	52.2
35.1	424	149	275	52.4
35.1	422	148	274	51.9
35.1	419	147	272	51.6
35.1	416	146	270	51.2
35.1	413	145	268	50.9
35.1	410	144	266	50.6
35.1	407	143	264	50.2
35.1	405	142	263	49.8
35.1	404	142	262	49.9
35.1	402	141	261	49.5
35.1	399	140	259	49.1
35.1	396	139	257	48.8
35.1	393	138	255	48.5
35.1	390	137	253	48.1
35.1	388	136	252	47.7
35.1	387	136	251	47.8
35.1	385	135	250	47.3
35.1	382	134	248	47.0
35.1	379	133	246	46.7
35.1	376	132	244	46.3
35.1	373	131	242	46.0
35.1	370	130	240	45.7
35.1	368	129	239	45.2
35.1	365	128	237	44.9
35.1	362	127	235	44.6
35.1	359	126	233	44.2
35.1	356	125	231	43.9
35.1	353	124	229	43.6
35.1	350	123	227	43.2
35.1	348	122	226	42.8
35.1	345	121	224	42.4
35.1	342	120	222	42.1
35.1	339	119	220	41.8
35.1	336	118	218	41.4
35.1	333	117	216	41.1
35.1	328	115	213	40.3
35.1	325	114	211	40.0
35.1	322	113	209	39.7
35.1	319	112	207	39.3
35.1	316	111	205	39.0

%E	M1	M2	DM	M*
35.1	313	110	203	38.7
35.1	308	108	200	37.9
35.1	305	107	198	37.5
35.1	302	106	196	37.2
35.1	299	105	194	36.9
35.1	296	104	192	36.5
35.1	291	102	189	35.8
35.1	288	101	187	35.4
35.1	285	100	185	35.1
35.1	282	99	183	34.8
35.1	279	98	181	34.4
35.1	276	97	179	34.1
35.1	271	95	176	33.3
35.1	268	94	174	33.0
35.1	265	93	172	32.6
35.1	262	92	170	32.3
35.1	259	91	168	32.0
35.1	251	88	163	30.9
35.1	248	87	161	30.5
35.1	245	86	159	30.2
35.1	242	85	157	29.9
35.1	239	84	155	29.5
35.1	231	81	150	28.4
35.1	228	80	148	28.1
35.1	225	79	146	27.7
35.1	222	78	144	27.4
35.1	211	74	137	26.0
35.1	208	73	135	25.6
35.1	205	72	133	25.3
35.1	202	71	131	25.0
35.1	194	68	126	23.8
35.1	191	67	124	23.5
35.1	188	66	122	23.2
35.1	185	65	120	22.8
35.1	174	61	113	21.4
35.1	171	60	111	21.1
35.1	168	59	109	20.7
35.1	154	54	100	18.9
35.1	151	53	98	18.6
35.1	148	52	96	18.3
35.1	134	47	87	16.5
35.1	131	46	85	16.2
35.1	114	40	74	14.0
35.1	111	39	72	13.7
35.1	97	34	63	11.9
35.1	94	33	61	11.6
35.1	77	27	50	9.5
35.1	74	26	48	9.1
35.1	57	20	37	7.0
35.1	37	13	24	4.6

%E	M1	M2	DM	M*
35.0	500	175	325	61.3
35.0	497	174	323	60.9
35.0	494	173	321	60.6
35.0	492	172	320	60.1
35.0	491	172	319	60.3
35.0	489	171	318	59.8
35.0	488	171	317	59.9
35.0	486	170	316	59.5
35.0	483	169	314	59.1
35.0	480	168	312	58.8
35.0	477	167	310	58.5
35.0	474	166	308	58.1
35.0	472	165	307	57.7
35.0	471	165	306	57.8
35.0	469	164	305	57.3
35.0	468	164	304	57.5
35.0	466	163	303	57.0
35.0	463	162	301	56.7
35.0	460	161	299	56.3
35.0	457	160	297	56.0
35.0	454	159	295	55.7
35.0	452	158	294	55.2
35.0	451	158	293	55.4
35.0	449	157	292	54.9
35.0	448	157	291	55.0
35.0	446	156	290	54.6
35.0	443	155	288	54.2
35.0	440	154	286	53.9
35.0	437	153	284	53.6
35.0	434	152	282	53.2
35.0	432	151	281	52.8
35.0	431	151	280	52.9
35.0	429	150	279	52.4
35.0	428	150	278	52.6
35.0	426	149	277	52.1
35.0	423	148	275	51.8
35.0	420	147	273	51.4
35.0	417	146	271	51.1
35.0	414	145	269	50.8
35.0	412	144	268	50.3
35.0	411	144	267	50.5
35.0	409	143	266	50.0
35.0	408	143	265	50.1
35.0	406	142	264	49.7
35.0	403	141	262	49.3
35.0	400	140	260	49.0
35.0	397	139	258	48.7
35.0	394	138	256	48.3
35.0	391	137	254	48.0
35.0	389	136	253	47.5

%E	M1	M2	DM	M*
35.0	386	135	251	47.2
35.0	383	134	249	46.9
35.0	380	133	247	46.5
35.0	377	132	245	46.2
35.0	374	131	243	45.9
35.0	371	130	241	45.6
35.0	369	129	240	45.1
35.0	366	128	238	44.8
35.0	363	127	236	44.4
35.0	360	126	234	44.1
35.0	357	125	232	43.8
35.0	354	124	230	43.4
35.0	351	123	228	43.1
35.0	349	122	227	42.6
35.0	346	121	225	42.3
35.0	343	120	223	42.0
35.0	340	119	221	41.6
35.0	337	118	219	41.3
35.0	334	117	217	41.0
35.0	331	116	215	40.7
35.0	329	115	214	40.2
35.0	326	114	212	39.9
35.0	323	113	210	39.5
35.0	320	112	208	39.2
35.0	317	111	206	38.9
35.0	314	110	204	38.5
35.0	311	109	202	38.2
35.0	309	108	201	37.7
35.0	306	107	199	37.4
35.0	303	106	197	37.1
35.0	300	105	195	36.8
35.0	297	104	193	36.4
35.0	294	103	191	36.1
35.0	286	100	186	35.0
35.0	283	99	184	34.6
35.0	280	98	182	34.3
35.0	277	97	180	34.0
35.0	274	96	178	33.6
35.0	266	93	173	32.5
35.0	263	92	171	32.2
35.0	260	91	169	31.8
35.0	257	90	167	31.5
35.0	254	89	165	31.2
35.0	246	86	160	30.1
35.0	243	85	158	29.7
35.0	240	84	156	29.4
35.0	237	83	154	29.1
35.0	234	82	152	28.7
35.0	226	79	147	27.6
35.0	223	78	145	27.3

%E	M1	M2	DM	M*
35.0	220	77	143	26.9
35.0	217	76	141	26.6
35.0	214	75	139	26.3
35.0	206	72	134	25.2
35.0	203	71	132	24.8
35.0	200	70	130	24.5
35.0	197	69	128	24.2
35.0	183	64	119	22.4
35.0	180	63	117	22.0
35.0	177	62	115	21.7
35.0	163	57	106	19.9
35.0	160	56	104	19.6
35.0	157	55	102	19.3
35.0	143	50	93	17.5
35.0	140	49	91	17.1
35.0	137	48	89	16.8
35.0	123	43	80	15.0
35.0	120	42	78	14.7
35.0	117	41	76	14.4
35.0	103	36	67	12.6
35.0	100	35	65	12.3
35.0	80	28	52	9.8
35.0	60	21	39	7.3
35.0	40	14	26	4.9
35.0	20	7	13	2.4
34.9	499	174	325	60.7
34.9	498	174	324	60.8
34.9	496	173	323	60.3
34.9	495	173	322	60.5
34.9	493	172	321	60.0
34.9	490	171	319	59.7
34.9	487	170	317	59.3
34.9	484	169	315	59.0
34.9	482	168	314	58.6
34.9	481	168	313	58.7
34.9	479	167	312	58.2
34.9	478	167	311	58.3
34.9	476	166	310	57.9
34.9	475	166	309	58.0
34.9	473	165	308	57.6
34.9	470	164	306	57.2
34.9	467	163	304	56.9
34.9	464	162	302	56.6
34.9	461	161	300	56.2
34.9	459	160	299	55.8
34.9	458	160	298	55.9
34.9	456	159	297	55.4
34.9	455	159	296	55.6
34.9	453	158	295	55.1
34.9	450	157	293	54.8

%E	M1	M2	DM	M*
34.9	447	156	291	54.4
34.9	444	155	289	54.1
34.9	441	154	287	53.8
34.9	439	153	286	53.3
34.9	438	153	285	53.4
34.9	436	152	284	53.0
34.9	435	152	283	53.1
34.9	433	151	282	52.7
34.9	430	150	280	52.3
34.9	427	149	278	52.0
34.9	424	148	276	51.7
34.9	421	147	274	51.3
34.9	418	146	272	51.0
34.9	416	145	271	50.5
34.9	415	145	270	50.7
34.9	413	144	269	50.2
34.9	410	143	267	49.9
34.9	407	142	265	49.5
34.9	404	141	263	49.2
34.9	401	140	261	48.9
34.9	398	139	259	48.5
34.9	395	138	257	48.2
34.9	393	137	256	47.8
34.9	392	137	255	47.9
34.9	390	136	254	47.4
34.9	387	135	252	47.1
34.9	384	134	250	46.8
34.9	381	133	248	46.4
34.9	378	132	246	46.1
34.9	375	131	244	45.8
34.9	373	130	243	45.3
34.9	372	130	242	45.4
34.9	370	129	241	45.0
34.9	367	128	239	44.6
34.9	364	127	237	44.3
34.9	361	126	235	44.0
34.9	358	125	233	43.6
34.9	355	124	231	43.3
34.9	352	123	229	43.0
34.9	350	122	228	42.5
34.9	347	121	226	42.2
34.9	344	120	224	41.9
34.9	341	119	222	41.5
34.9	338	118	220	41.2
34.9	335	117	218	40.9
34.9	332	116	216	40.5
34.9	327	114	213	39.7
34.9	324	113	211	39.4
34.9	321	112	209	39.1
34.9	318	111	207	38.7

%E	M1	M2	DM	M*
34.9	315	110	205	38.4
34.9	312	109	203	38.1
34.9	307	107	200	37.3
34.9	304	106	198	37.0
34.9	301	105	196	36.6
34.9	298	104	194	36.3
34.9	295	103	192	36.0
34.9	292	102	190	35.6
34.9	289	101	188	35.3
34.9	284	99	185	34.5
34.9	281	98	183	34.2
34.9	278	97	181	33.8
34.9	275	96	179	33.5
34.9	272	95	177	33.2
34.9	269	94	175	32.8
34.9	261	91	170	31.7
34.9	258	90	168	31.4
34.9	255	89	166	31.1
34.9	252	88	164	30.7
34.9	249	87	162	30.4
34.9	241	84	157	29.3
34.9	238	83	155	28.9
34.9	235	82	153	28.6
34.9	232	81	151	28.3
34.9	229	80	149	27.9
34.9	218	76	142	26.5
34.9	215	75	140	26.2
34.9	212	74	138	25.8
34.9	209	73	136	25.5
34.9	195	68	127	23.7
34.9	192	67	125	23.4
34.9	189	66	123	23.0
34.9	186	65	121	22.7
34.9	175	61	114	21.3
34.9	172	60	112	20.9
34.9	169	59	110	20.6
34.9	166	58	108	20.3
34.9	152	53	99	18.5
34.9	149	52	97	18.1
34.9	146	51	95	17.8
34.9	129	45	84	15.7
34.9	126	44	82	15.4
34.9	109	38	71	13.2
34.9	106	37	69	12.9
34.9	86	30	56	10.5
34.9	83	29	54	10.1
34.9	63	22	41	7.7
34.9	43	15	28	5.2
34.8	500	174	326	60.6
34.8	497	173	324	60.2

%E	M1	M2	DM	M*
34.8	494	172	322	59.9
34.8	492	171	321	59.4
34.8	491	171	320	59.6
34.8	489	170	319	59.1
34.8	488	170	318	59.2
34.8	486	169	317	58.8
34.8	485	169	316	58.9
34.8	483	168	315	58.4
34.8	480	167	313	58.1
34.8	477	166	311	57.8
34.8	474	165	309	57.4
34.8	471	164	307	57.1
34.8	469	163	306	56.7
34.8	468	163	305	56.8
34.8	466	162	304	56.3
34.8	465	162	303	56.4
34.8	463	161	302	56.0
34.8	462	161	301	56.1
34.8	460	160	300	55.7
34.8	457	159	298	55.3
34.8	454	158	296	55.0
34.8	451	157	294	54.7
34.8	448	156	292	54.3
34.8	446	155	291	53.9
34.8	445	155	290	54.0
34.8	443	154	289	53.5
34.8	442	154	288	53.7
34.8	440	153	287	53.2
34.8	437	152	285	52.9
34.8	434	151	283	52.5
34.8	431	150	281	52.2
34.8	428	149	279	51.9
34.8	425	148	277	51.5
34.8	423	147	276	51.1
34.8	422	147	275	51.2
34.8	420	146	274	50.8
34.8	419	146	273	50.9
34.8	417	145	272	50.4
34.8	414	144	270	50.1
34.8	411	143	268	49.8
34.8	408	142	266	49.4
34.8	405	141	264	49.1
34.8	402	140	262	48.8
34.8	400	139	261	48.3
34.8	399	139	260	48.4
34.8	397	138	259	48.0
34.8	396	138	258	48.1
34.8	394	137	257	47.6
34.8	391	136	255	47.3
34.8	388	135	253	47.0

%E	M1	M2	DM	M*
34.8	385	134	251	46.6
34.8	382	133	249	46.3
34.8	379	132	247	46.0
34.8	376	131	245	45.6
34.8	374	130	244	45.2
34.8	371	129	242	44.9
34.8	368	128	240	44.5
34.8	365	127	238	44.2
34.8	362	126	236	43.9
34.8	359	125	234	43.5
34.8	356	124	232	43.2
34.8	353	123	230	42.9
34.8	351	122	229	42.4
34.8	348	121	227	42.1
34.8	345	120	225	41.7
34.8	342	119	223	41.4
34.8	339	118	221	41.1
34.8	336	117	219	40.7
34.8	333	116	217	40.4
34.8	330	115	215	40.1
34.8	328	114	214	39.6
34.8	325	113	212	39.3
34.8	322	112	210	39.0
34.8	319	111	208	38.6
34.8	316	110	206	38.3
34.8	313	109	204	38.0
34.8	310	108	202	37.6
34.8	305	106	199	36.8
34.8	302	105	197	36.5
34.8	299	104	195	36.2
34.8	296	103	193	35.8
34.8	293	102	191	35.5
34.8	290	101	189	35.2
34.8	287	100	187	34.8
34.8	282	98	184	34.1
34.8	279	97	182	33.7
34.8	276	96	180	33.4
34.8	273	95	178	33.1
34.8	270	94	176	32.7
34.8	267	93	174	32.4
34.8	264	92	172	32.1
34.8	256	89	167	30.9
34.8	253	88	165	30.6
34.8	250	87	163	30.3
34.8	247	86	161	29.9
34.8	244	85	159	29.6
34.8	233	81	152	28.2
34.8	230	80	150	27.8
34.8	227	79	148	27.5
34.8	224	78	146	27.2

%E	M1	M2	DM	M*
34.8	221	77	144	26.8
34.8	210	73	137	25.4
34.8	207	72	135	25.0
34.8	204	71	133	24.7
34.8	201	70	131	24.4
34.8	198	69	129	24.0
34.8	187	65	122	22.6
34.8	184	64	120	22.3
34.8	181	63	118	21.9
34.8	178	62	116	21.6
34.8	164	57	107	19.8
34.8	161	56	105	19.5
34.8	158	55	103	19.1
34.8	155	54	101	18.8
34.8	141	49	92	17.0
34.8	138	48	90	16.7
34.8	135	47	88	16.4
34.8	132	46	86	16.0
34.8	115	40	75	13.9
34.8	112	39	73	13.6
34.8	92	32	60	11.1
34.8	89	31	58	10.8
34.8	69	24	45	8.3
34.8	66	23	43	8.0
34.8	46	16	30	5.6
34.8	23	8	15	2.8
34.7	499	173	326	60.0
34.7	498	173	325	60.1
34.7	496	172	324	59.6
34.7	495	172	323	59.8
34.7	493	171	322	59.3
34.7	490	170	320	59.0
34.7	487	169	318	58.6
34.7	484	168	316	58.3
34.7	481	167	314	58.0
34.7	479	166	313	57.5
34.7	478	166	312	57.6
34.7	476	165	311	57.2
34.7	475	165	310	57.3
34.7	473	164	309	56.9
34.7	472	164	308	57.0
34.7	470	163	307	56.5
34.7	467	162	305	56.2
34.7	464	161	303	55.9
34.7	461	160	301	55.5
34.7	458	159	299	55.2
34.7	455	158	297	54.9
34.7	453	157	296	54.4
34.7	452	157	295	54.5
34.7	450	156	294	54.1

%E	M1	M2	DM	M*
34.7	449	156	293	54.2
34.7	447	155	292	53.7
34.7	444	154	290	53.4
34.7	441	153	288	53.1
34.7	438	152	286	52.7
34.7	435	151	284	52.4
34.7	432	150	282	52.1
34.7	430	149	281	51.6
34.7	429	149	280	51.8
34.7	427	148	279	51.3
34.7	426	148	278	51.4
34.7	424	147	277	51.0
34.7	421	146	275	50.6
34.7	418	145	273	50.3
34.7	415	144	271	50.0
34.7	412	143	269	49.6
34.7	409	142	267	49.3
34.7	406	141	265	49.0
34.7	404	140	264	48.5
34.7	403	140	263	48.6
34.7	401	139	262	48.2
34.7	398	138	260	47.8
34.7	395	137	258	47.5
34.7	392	136	256	47.2
34.7	389	135	254	46.9
34.7	386	134	252	46.5
34.7	383	133	250	46.2
34.7	380	132	248	45.9
34.7	378	131	247	45.4
34.7	377	131	246	45.5
34.7	375	130	245	45.1
34.7	372	129	243	44.7
34.7	369	128	241	44.4
34.7	366	127	239	44.1
34.7	363	126	237	43.7
34.7	360	125	235	43.4
34.7	357	124	233	43.1
34.7	354	123	231	42.7
34.7	352	122	230	42.3
34.7	349	121	228	42.0
34.7	346	120	226	41.6
34.7	343	119	224	41.3
34.7	340	118	222	41.0
34.7	337	117	220	40.6
34.7	334	116	218	40.3
34.7	331	115	216	40.0
34.7	329	114	215	39.5
34.7	326	113	213	39.2
34.7	323	112	211	38.8
34.7	320	111	209	38.5

%E	M1	M2	DM	M*
34.7	317	110	207	38.2
34.7	314	109	205	37.8
34.7	311	108	203	37.5
34.7	308	107	201	37.2
34.7	303	105	198	36.4
34.7	300	104	196	36.1
34.7	297	103	194	35.7
34.7	294	102	192	35.4
34.7	291	101	190	35.1
34.7	288	100	188	34.7
34.7	285	99	186	34.4
34.7	277	96	181	33.3
34.7	274	95	179	32.9
34.7	271	94	177	32.6
34.7	268	93	175	32.3
34.7	265	92	173	31.9
34.7	262	91	171	31.6
34.7	259	90	169	31.3
34.7	251	87	164	30.2
34.7	248	86	162	29.8
34.7	245	85	160	29.5
34.7	242	84	158	29.2
34.7	239	83	156	28.8
34.7	236	82	154	28.5
34.7	225	78	147	27.0
34.7	222	77	145	26.7
34.7	219	76	143	26.4
34.7	216	75	141	26.0
34.7	213	74	139	25.7
34.7	202	70	132	24.3
34.7	199	69	130	23.9
34.7	196	68	128	23.6
34.7	193	67	126	23.3
34.7	190	66	124	22.9
34.7	176	61	115	21.1
34.7	173	60	113	20.8
34.7	170	59	111	20.5
34.7	167	58	109	20.1
34.7	150	52	98	18.0
34.7	147	51	96	17.7
34.7	144	50	94	17.4
34.7	124	43	81	14.9
34.7	121	42	79	14.6
34.7	118	41	77	14.2
34.7	101	35	66	12.1
34.7	98	34	64	11.8
34.7	95	33	62	11.5
34.7	75	26	49	9.0
34.7	72	25	47	8.7
34.7	49	17	32	5.9

%E	M1	M2	DM	M*
34.6	500	173	327	59.9
34.6	497	172	325	59.5
34.6	494	171	323	59.2
34.6	492	170	322	58.7
34.6	491	170	321	58.9
34.6	489	169	320	58.4
34.6	488	169	319	58.5
34.6	486	168	318	58.1
34.6	485	168	317	58.2
34.6	483	167	316	57.7
34.6	482	167	315	57.9
34.6	480	166	314	57.4
34.6	477	165	312	57.1
34.6	474	164	310	56.7
34.6	471	163	308	56.4
34.6	468	162	306	56.1
34.6	465	161	304	55.7
34.6	463	160	303	55.3
34.6	462	160	302	55.4
34.6	460	159	301	55.0
34.6	459	159	300	55.1
34.6	457	158	299	54.6
34.6	456	158	298	54.7
34.6	454	157	297	54.3
34.6	451	156	295	54.0
34.6	448	155	293	53.6
34.6	445	154	291	53.3
34.6	442	153	289	53.0
34.6	439	152	287	52.6
34.6	437	151	286	52.2
34.6	436	151	285	52.3
34.6	434	150	284	51.8
34.6	433	150	283	52.0
34.6	431	149	282	51.5
34.6	428	148	280	51.2
34.6	425	147	278	50.8
34.6	422	146	276	50.5
34.6	419	145	274	50.2
34.6	416	144	272	49.8
34.6	413	143	270	49.5
34.6	411	142	269	49.1
34.6	410	142	268	49.2
34.6	408	141	267	48.7
34.6	407	141	266	48.8
34.6	405	140	265	48.4
34.6	402	139	263	48.1
34.6	399	138	261	47.7
34.6	396	137	259	47.4
34.6	393	136	257	47.1
34.6	390	135	255	46.7

%E	M1	M2	DM	M*
34.6	387	134	253	46.4
34.6	384	133	251	46.1
34.6	382	132	250	45.6
34.6	381	132	249	45.7
34.6	379	131	248	45.3
34.6	376	130	246	44.9
34.6	373	129	244	44.6
34.6	370	128	242	44.3
34.6	367	127	240	43.9
34.6	364	126	238	43.6
34.6	361	125	236	43.3
34.6	358	124	234	42.9
34.6	356	123	233	42.5
34.6	355	123	232	42.6
34.6	353	122	231	42.2
34.6	350	121	229	41.8
34.6	347	120	227	41.5
34.6	344	119	225	41.2
34.6	341	118	223	40.8
34.6	338	117	221	40.5
34.6	335	116	219	40.2
34.6	332	115	217	39.8
34.6	327	113	214	39.0
34.6	324	112	212	38.7
34.6	321	111	210	38.4
34.6	318	110	208	38.1
34.6	315	109	206	37.7
34.6	312	108	204	37.4
34.6	309	107	202	37.1
34.6	306	106	200	36.7
34.6	301	104	197	35.9
34.6	298	103	195	35.6
34.6	295	102	193	35.3
34.6	292	101	191	34.9
34.6	289	100	189	34.6
34.6	286	99	187	34.3
34.6	283	98	185	33.9
34.6	280	97	183	33.6
34.6	272	94	178	32.5
34.6	269	93	176	32.2
34.6	266	92	174	31.8
34.6	263	91	172	31.5
34.6	260	90	170	31.2
34.6	257	89	168	30.8
34.6	254	88	166	30.5
34.6	246	85	161	29.4
34.6	243	84	159	29.0
34.6	240	83	157	28.7
34.6	237	82	155	28.4
34.6	234	81	153	28.0

%E	M1	M2	DM	M*
34.6	231	80	151	27.7
34.6	228	79	149	27.4
34.6	217	75	142	25.9
34.6	214	74	140	25.6
34.6	211	73	138	25.3
34.6	208	72	136	24.9
34.6	205	71	134	24.6
34.6	191	66	125	22.8
34.6	188	65	123	22.5
34.6	185	64	121	22.1
34.6	182	63	119	21.8
34.6	179	62	117	21.5
34.6	162	56	106	19.4
34.6	159	55	104	19.0
34.6	156	54	102	18.7
34.6	153	53	100	18.4
34.6	136	47	89	16.2
34.6	133	46	87	15.9
34.6	130	45	85	15.6
34.6	127	44	83	15.2
34.6	107	37	70	12.8
34.6	104	36	68	12.5
34.6	81	28	53	9.7
34.6	78	27	51	9.3
34.6	52	18	34	6.2
34.6	26	9	17	3.1
34.5	499	172	327	59.3
34.5	498	172	326	59.4
34.5	496	171	325	59.0
34.5	495	171	324	59.1
34.5	493	170	323	58.6
34.5	490	169	321	58.3
34.5	487	168	319	58.0
34.5	484	167	317	57.6
34.5	481	166	315	57.3
34.5	478	165	313	57.0
34.5	476	164	312	56.5
34.5	475	164	311	56.6
34.5	473	163	310	56.2
34.5	472	163	309	56.3
34.5	470	162	308	55.8
34.5	469	162	307	56.0
34.5	467	161	306	55.5
34.5	466	161	305	55.6
34.5	464	160	304	55.2
34.5	461	159	302	54.8
34.5	458	158	300	54.5
34.5	455	157	298	54.2
34.5	452	156	296	53.8
34.5	449	155	294	53.5

%E	M1	M2	DM	M*
34.5	447	154	293	53.1
34.5	446	154	292	53.2
34.5	444	153	291	52.7
34.5	443	153	290	52.8
34.5	441	152	289	52.4
34.5	440	152	288	52.5
34.5	438	151	287	52.1
34.5	435	150	285	51.7
34.5	432	149	283	51.4
34.5	429	148	281	51.1
34.5	426	147	279	50.7
34.5	423	146	277	50.4
34.5	420	145	275	50.1
34.5	418	144	274	49.6
34.5	417	144	273	49.7
34.5	415	143	272	49.3
34.5	414	143	271	49.4
34.5	412	142	270	48.9
34.5	409	141	268	48.6
34.5	406	140	266	48.3
34.5	403	139	264	47.9
34.5	400	138	262	47.6
34.5	397	137	260	47.3
34.5	394	136	258	46.9
34.5	391	135	256	46.6
34.5	388	134	254	46.3
34.5	386	133	253	45.8
34.5	385	133	252	45.9
34.5	383	132	251	45.5
34.5	380	131	249	45.2
34.5	377	130	247	44.8
34.5	374	129	245	44.5
34.5	371	128	243	44.2
34.5	368	127	241	43.8
34.5	365	126	239	43.5
34.5	362	125	237	43.2
34.5	359	124	235	42.8
34.5	357	123	234	42.4
34.5	354	122	232	42.0
34.5	351	121	230	41.7
34.5	348	120	228	41.4
34.5	345	119	226	41.0
34.5	342	118	224	40.7
34.5	339	117	222	40.4
34.5	336	116	220	40.0
34.5	333	115	218	39.7
34.5	330	114	216	39.4
34.5	328	113	215	38.9
34.5	325	112	213	38.6
34.5	322	111	211	38.3

%E	M1	M2	DM	M*
34.5	319	110	209	37.9
34.5	316	109	207	37.6
34.5	313	108	205	37.3
34.5	310	107	203	36.9
34.5	307	106	201	36.6
34.5	304	105	199	36.3
34.5	296	102	194	35.1
34.5	293	101	192	34.8
34.5	290	100	190	34.5
34.5	287	99	188	34.1
34.5	284	98	186	33.8
34.5	281	97	184	33.5
34.5	278	96	182	33.2
34.5	275	95	180	32.8
34.5	267	92	175	31.7
34.5	264	91	173	31.4
34.5	261	90	171	31.0
34.5	258	89	169	30.7
34.5	255	88	167	30.4
34.5	252	87	165	30.0
34.5	249	86	163	29.7
34.5	238	82	156	28.3
34.5	235	81	154	27.9
34.5	232	80	152	27.6
34.5	229	79	150	27.3
34.5	226	78	148	26.9
34.5	223	77	146	26.6
34.5	220	76	144	26.3
34.5	209	72	137	24.8
34.5	206	71	135	24.5
34.5	203	70	133	24.1
34.5	200	69	131	23.8
34.5	197	68	129	23.5
34.5	194	67	127	23.1
34.5	177	61	116	21.0
34.5	174	60	114	20.7
34.5	171	59	112	20.4
34.5	168	58	110	20.0
34.5	165	57	108	19.7
34.5	148	51	97	17.6
34.5	145	50	95	17.2
34.5	142	49	93	16.9
34.5	139	48	91	16.6
34.5	119	41	78	14.1
34.5	116	40	76	13.8
34.5	113	39	74	13.5
34.5	110	38	72	13.1
34.5	87	30	57	10.3
34.5	84	29	55	10.0
34.5	58	20	38	6.9

%E	M1	M2	DM	M*
34.5	55	19	36	6.6
34.5	29	10	19	3.4
34.4	500	172	328	59.2
34.4	497	171	326	58.8
34.4	494	170	324	58.5
34.4	492	169	323	58.1
34.4	491	169	322	58.2
34.4	489	168	321	57.7
34.4	488	168	320	57.8
34.4	486	167	319	57.4
34.4	485	167	318	57.5
34.4	483	166	317	57.1
34.4	482	166	316	57.2
34.4	480	165	315	56.7
34.4	479	165	314	56.8
34.4	477	164	313	56.4
34.4	474	163	311	56.1
34.4	471	162	309	55.7
34.4	468	161	307	55.4
34.4	465	160	305	55.1
34.4	462	159	303	54.7
34.4	459	158	301	54.4
34.4	457	157	300	53.9
34.4	456	157	299	54.1
34.4	454	156	298	53.6
34.4	453	156	297	53.7
34.4	451	155	296	53.3
34.4	450	155	295	53.4
34.4	448	154	294	52.9
34.4	445	153	292	52.6
34.4	442	152	290	52.3
34.4	439	151	288	51.9
34.4	436	150	286	51.6
34.4	433	149	284	51.3
34.4	430	148	282	50.9
34.4	427	147	280	50.6
34.4	425	146	279	50.2
34.4	424	146	278	50.3
34.4	422	145	277	49.8
34.4	421	145	276	49.9
34.4	419	144	275	49.5
34.4	416	143	273	49.2
34.4	413	142	271	48.8
34.4	410	141	269	48.5
34.4	407	140	267	48.2
34.4	404	139	265	47.8
34.4	401	138	263	47.5
34.4	398	137	261	47.2
34.4	395	136	259	46.8
34.4	393	135	258	46.4

%E	M1	M2	DM	M*
34.4	392	135	257	46.5
34.4	390	134	256	46.0
34.4	389	134	255	46.2
34.4	387	133	254	45.7
34.4	384	132	252	45.4
34.4	381	131	250	45.0
34.4	378	130	248	44.7
34.4	375	129	246	44.4
34.4	372	128	244	44.0
34.4	369	127	242	43.7
34.4	366	126	240	43.4
34.4	363	125	238	43.0
34.4	360	124	236	42.7
34.4	358	123	235	42.3
34.4	355	122	233	41.9
34.4	352	121	231	41.6
34.4	349	120	229	41.3
34.4	346	119	227	40.9
34.4	343	118	225	40.6
34.4	340	117	223	40.3
34.4	337	116	221	39.9
34.4	334	115	219	39.6
34.4	331	114	217	39.3
34.4	326	112	214	38.5
34.4	323	111	212	38.1
34.4	320	110	210	37.8
34.4	317	109	208	37.5
34.4	314	108	206	37.1
34.4	311	107	204	36.8
34.4	308	106	202	36.5
34.4	305	105	200	36.1
34.4	302	104	198	35.8
34.4	299	103	196	35.5
34.4	294	101	193	34.7
34.4	291	100	191	34.4
34.4	288	99	189	34.0
34.4	285	98	187	33.7
34.4	282	97	185	33.4
34.4	279	96	183	33.0
34.4	276	95	181	32.7
34.4	273	94	179	32.4
34.4	270	93	177	32.0
34.4	262	90	172	30.9
34.4	259	89	170	30.6
34.4	256	88	168	30.3
34.4	253	87	166	29.9
34.4	250	86	164	29.6
34.4	247	85	162	29.3
34.4	244	84	160	28.9
34.4	241	83	158	28.6

%E	M1	M2	DM	M*
34.4	227	78	149	26.8
34.4	224	77	147	26.5
34.4	221	76	145	26.1
34.4	218	75	143	25.8
34.4	215	74	141	25.5
34.4	212	73	139	25.1
34.4	195	67	128	23.0
34.4	192	66	126	22.7
34.4	189	65	124	22.4
34.4	186	64	122	22.0
34.4	183	63	120	21.7
34.4	180	62	118	21.4
34.4	163	56	107	19.2
34.4	160	55	105	18.9
34.4	157	54	103	18.6
34.4	154	53	101	18.2
34.4	151	52	99	17.9
34.4	131	45	86	15.5
34.4	128	44	84	15.1
34.4	125	43	82	14.8
34.4	122	42	80	14.5
34.4	96	33	63	11.3
34.4	93	32	61	11.0
34.4	90	31	59	10.7
34.4	64	22	42	7.6
34.4	61	21	40	7.2
34.4	32	11	21	3.8
34.3	499	171	328	58.6
34.3	498	171	327	58.7
34.3	496	170	326	58.3
34.3	495	170	325	58.4
34.3	493	169	324	57.9
34.3	490	168	322	57.6
34.3	487	167	320	57.3
34.3	484	166	318	56.9
34.3	481	165	316	56.6
34.3	478	164	314	56.3
34.3	475	163	312	55.9
34.3	472	162	310	55.6
34.3	470	161	309	55.2
34.3	469	161	308	55.3
34.3	467	160	307	54.8
34.3	466	160	306	54.9
34.3	464	159	305	54.5
34.3	463	159	304	54.6
34.3	461	158	303	54.2
34.3	460	158	302	54.3
34.3	458	157	301	53.8
34.3	455	156	299	53.5
34.3	452	155	297	53.2

%E	M1	M2	DM	M*
34.3	449	154	295	52.8
34.3	446	153	293	52.5
34.3	443	152	291	52.2
34.3	440	151	289	51.8
34.3	437	150	287	51.5
34.3	435	149	286	51.0
34.3	434	149	285	51.2
34.3	432	148	284	50.7
34.3	431	148	283	50.8
34.3	429	147	282	50.4
34.3	428	147	281	50.5
34.3	426	146	280	50.0
34.3	423	145	278	49.7
34.3	420	144	276	49.4
34.3	417	143	274	49.0
34.3	414	142	272	48.7
34.3	411	141	270	48.4
34.3	408	140	268	48.0
34.3	405	139	266	47.7
34.3	402	138	264	47.4
34.3	400	137	263	46.9
34.3	399	137	262	47.0
34.3	397	136	261	46.6
34.3	396	136	260	46.7
34.3	394	135	259	46.3
34.3	391	134	257	45.9
34.3	388	133	255	45.6
34.3	385	132	253	45.3
34.3	382	131	251	44.9
34.3	379	130	249	44.6
34.3	376	129	247	44.3
34.3	373	128	245	43.9
34.3	370	127	243	43.6
34.3	367	126	241	43.3
34.3	364	125	239	42.9
34.3	362	124	238	42.5
34.3	361	124	237	42.6
34.3	359	123	236	42.1
34.3	356	122	234	41.8
34.3	353	121	232	41.5
34.3	350	120	230	41.1
34.3	347	119	228	40.8
34.3	344	118	226	40.5
34.3	341	117	224	40.1
34.3	338	116	222	39.8
34.3	335	115	220	39.5
34.3	332	114	218	39.1
34.3	329	113	216	38.8
34.3	327	112	215	38.4
34.3	324	111	213	38.0

%E	M1	M2	DM	M*
34.3	321	110	211	37.7
34.3	318	109	209	37.4
34.3	315	108	207	37.0
34.3	312	107	205	36.7
34.3	309	106	203	36.4
34.3	306	105	201	36.0
34.3	303	104	199	35.7
34.3	300	103	197	35.4
34.3	297	102	195	35.0
34.3	289	99	190	33.9
34.3	286	98	188	33.6
34.3	283	97	186	33.2
34.3	280	96	184	32.9
34.3	277	95	182	32.6
34.3	274	94	180	32.2
34.3	271	93	178	31.9
34.3	268	92	176	31.6
34.3	265	91	174	31.2
34.3	254	87	167	29.8
34.3	251	86	165	29.5
34.3	248	85	163	29.1
34.3	245	84	161	28.8
34.3	242	83	159	28.5
34.3	239	82	157	28.1
34.3	236	81	155	27.8
34.3	233	80	153	27.5
34.3	230	79	151	27.1
34.3	216	74	142	25.4
34.3	213	73	140	25.0
34.3	210	72	138	24.7
34.3	207	71	136	24.4
34.3	204	70	134	24.0
34.3	201	69	132	23.7
34.3	198	68	130	23.4
34.3	181	62	119	21.2
34.3	178	61	117	20.9
34.3	175	60	115	20.6
34.3	172	59	113	20.2
34.3	169	58	111	19.9
34.3	166	57	109	19.6
34.3	143	49	94	16.8
34.3	140	48	92	16.5
34.3	137	47	90	16.1
34.3	134	46	88	15.8
34.3	108	37	71	12.7
34.3	105	36	69	12.3
34.3	102	35	67	12.0
34.3	99	34	65	11.7
34.3	70	24	46	8.2
34.3	67	23	44	7.9

%E	M1	M2	DM	M*
34.3	35	12	23	4.1
34.2	500	171	329	58.5
34.2	497	170	327	58.1
34.2	494	169	325	57.8
34.2	491	168	323	57.5
34.2	489	167	322	57.0
34.2	488	167	321	57.1
34.2	486	166	320	56.7
34.2	485	166	319	56.8
34.2	483	165	318	56.4
34.2	482	165	317	56.5
34.2	480	164	316	56.0
34.2	479	164	315	56.2
34.2	477	163	314	55.7
34.2	476	163	313	55.8
34.2	474	162	312	55.4
34.2	473	162	311	55.5
34.2	471	161	310	55.0
34.2	468	160	308	54.7
34.2	465	159	306	54.4
34.2	462	158	304	54.0
34.2	459	157	302	53.7
34.2	456	156	300	53.4
34.2	453	155	298	53.0
34.2	450	154	296	52.7
34.2	448	153	295	52.3
34.2	447	153	294	52.4
34.2	445	152	293	51.9
34.2	444	152	292	52.0
34.2	442	151	291	51.6
34.2	441	151	290	51.7
34.2	439	150	289	51.3
34.2	438	150	288	51.4
34.2	436	149	287	50.9
34.2	433	148	285	50.6
34.2	430	147	283	50.3
34.2	427	146	281	49.9
34.2	424	145	279	49.6
34.2	421	144	277	49.3
34.2	418	143	275	48.9
34.2	415	142	273	48.6
34.2	412	141	271	48.3
34.2	409	140	269	47.9
34.2	407	139	268	47.5
34.2	406	139	267	47.6
34.2	404	138	266	47.1
34.2	403	138	265	47.3
34.2	401	137	264	46.8
34.2	398	136	262	46.5
34.2	395	135	260	46.1

%E	M1	M2	DM	M*
34.2	392	134	258	45.8
34.2	389	133	256	45.5
34.2	386	132	254	45.1
34.2	383	131	252	44.8
34.2	380	130	250	44.5
34.2	377	129	248	44.1
34.2	374	128	246	43.8
34.2	371	127	244	43.5
34.2	368	126	242	43.1
34.2	366	125	241	42.7
34.2	365	125	240	42.8
34.2	363	124	239	42.4
34.2	360	123	237	42.0
34.2	357	122	235	41.7
34.2	354	121	233	41.4
34.2	351	120	231	41.0
34.2	348	119	229	40.7
34.2	345	118	227	40.4
34.2	342	117	225	40.0
34.2	339	116	223	39.7
34.2	336	115	221	39.4
34.2	333	114	219	39.0
34.2	330	113	217	38.7
34.2	325	111	214	37.9
34.2	322	110	212	37.6
34.2	319	109	210	37.2
34.2	316	108	208	36.9
34.2	313	107	206	36.6
34.2	310	106	204	36.2
34.2	307	105	202	35.9
34.2	304	104	200	35.6
34.2	301	103	198	35.2
34.2	298	102	196	34.9
34.2	295	101	194	34.6
34.2	292	100	192	34.2
34.2	284	97	187	33.1
34.2	281	96	185	32.8
34.2	278	95	183	32.5
34.2	275	94	181	32.1
34.2	272	93	179	31.8
34.2	269	92	177	31.5
34.2	266	91	175	31.1
34.2	263	90	173	30.8
34.2	260	89	171	30.5
34.2	257	88	169	30.1
34.2	243	83	160	28.3
34.2	240	82	158	28.0
34.2	237	81	156	27.7
34.2	234	80	154	27.4
34.2	231	79	152	27.0

%E	M1	M2	DM	M*
34.2	228	78	150	26.7
34.2	225	77	148	26.4
34.2	222	76	146	26.0
34.2	219	75	144	25.7
34.2	202	69	133	23.6
34.2	199	68	131	23.2
34.2	196	67	129	22.9
34.2	193	66	127	22.6
34.2	190	65	125	22.2
34.2	187	64	123	21.9
34.2	184	63	121	21.6
34.2	161	55	106	18.8
34.2	158	54	104	18.5
34.2	155	53	102	18.1
34.2	152	52	100	17.8
34.2	149	51	98	17.5
34.2	146	50	96	17.1
34.2	120	41	79	14.0
34.2	117	40	77	13.7
34.2	114	39	75	13.3
34.2	111	38	73	13.0
34.2	79	27	52	9.2
34.2	76	26	50	8.9
34.2	73	25	48	8.6
34.2	38	13	25	4.4
34.1	499	170	329	57.9
34.1	498	170	328	58.0
34.1	496	169	327	57.6
34.1	495	169	326	57.7
34.1	493	168	325	57.2
34.1	492	168	324	57.4
34.1	490	167	323	56.9
34.1	487	166	321	56.6
34.1	484	165	319	56.3
34.1	481	164	317	55.9
34.1	478	163	315	55.6
34.1	475	162	313	55.3
34.1	472	161	311	54.9
34.1	469	160	309	54.6
34.1	466	159	307	54.3
34.1	464	158	306	53.8
34.1	463	158	305	53.9
34.1	461	157	304	53.5
34.1	460	157	303	53.6
34.1	458	156	302	53.1
34.1	457	156	301	53.3
34.1	455	155	300	52.8
34.1	454	155	299	52.9
34.1	452	154	298	52.5
34.1	451	154	297	52.6

%E	M1	M2	DM	M*
34.1	449	153	296	52.1
34.1	446	152	294	51.8
34.1	443	151	292	51.5
34.1	440	150	290	51.1
34.1	437	149	288	50.8
34.1	434	148	286	50.5
34.1	431	147	284	50.1
34.1	428	146	282	49.8
34.1	425	145	280	49.5
34.1	422	144	278	49.1
34.1	419	143	276	48.8
34.1	417	142	275	48.4
34.1	416	142	274	48.5
34.1	414	141	273	48.0
34.1	413	141	272	48.1
34.1	411	140	271	47.7
34.1	410	140	270	47.8
34.1	408	139	269	47.4
34.1	405	138	267	47.0
34.1	402	137	265	46.7
34.1	399	136	263	46.4
34.1	396	135	261	46.0
34.1	393	134	259	45.7
34.1	390	133	257	45.4
34.1	387	132	255	45.0
34.1	384	131	253	44.7
34.1	381	130	251	44.4
34.1	378	129	249	44.0
34.1	375	128	247	43.7
34.1	372	127	245	43.4
34.1	370	126	244	42.9
34.1	369	126	243	43.0
34.1	367	125	242	42.6
34.1	364	124	240	42.2
34.1	361	123	238	41.9
34.1	358	122	236	41.6
34.1	355	121	234	41.2
34.1	352	120	232	40.9
34.1	349	119	230	40.6
34.1	346	118	228	40.2
34.1	343	117	226	39.9
34.1	340	116	224	39.6
34.1	337	115	222	39.2
34.1	334	114	220	38.9
34.1	331	113	218	38.6
34.1	328	112	216	38.2
34.1	323	110	213	37.5
34.1	320	109	211	37.1
34.1	317	108	209	36.8
34.1	314	107	207	36.5

%E	M1	M2	DM	M*
34.1	311	106	205	36.1
34.1	308	105	203	35.8
34.1	305	104	201	35.5
34.1	302	103	199	35.1
34.1	299	102	197	34.8
34.1	296	101	195	34.5
34.1	293	100	193	34.1
34.1	290	99	191	33.8
34.1	287	98	189	33.5
34.1	279	95	184	32.3
34.1	276	94	182	32.0
34.1	273	93	180	31.7
34.1	270	92	178	31.3
34.1	267	91	176	31.0
34.1	264	90	174	30.7
34.1	261	89	172	30.3
34.1	258	88	170	30.0
34.1	255	87	168	29.7
34.1	252	86	166	29.3
34.1	249	85	164	29.0
34.1	246	84	162	28.7
34.1	232	79	153	26.9
34.1	229	78	151	26.6
34.1	226	77	149	26.2
34.1	223	76	147	25.9
34.1	220	75	145	25.6
34.1	217	74	143	25.2
34.1	214	73	141	24.9
34.1	211	72	139	24.6
34.1	208	71	137	24.2
34.1	205	70	135	23.9
34.1	185	63	122	21.5
34.1	182	62	120	21.1
34.1	179	61	118	20.8
34.1	176	60	116	20.5
34.1	173	59	114	20.1
34.1	170	58	112	19.8
34.1	167	57	110	19.5
34.1	164	56	108	19.1
34.1	138	47	91	16.0
34.1	135	46	89	15.7
34.1	132	45	87	15.3
34.1	129	44	85	15.0
34.1	126	43	83	14.7
34.1	123	42	81	14.3
34.1	91	31	60	10.6
34.1	88	30	58	10.2
34.1	85	29	56	9.9
34.1	82	28	54	9.6
34.1	44	15	29	5.1

%E	M1	M2	DM	M*
34.1	41	14	27	4.8
34.0	500	170	330	57.8
34.0	497	169	328	57.5
34.0	494	168	326	57.1
34.0	491	167	324	56.8
34.0	488	166	322	56.5
34.0	486	165	321	56.0
34.0	485	165	320	56.1
34.0	483	164	319	55.7
34.0	482	164	318	55.8
34.0	480	163	317	55.4
34.0	479	163	316	55.5
34.0	477	162	315	55.0
34.0	476	162	314	55.1
34.0	474	161	313	54.7
34.0	473	161	312	54.8
34.0	471	160	311	54.4
34.0	470	160	310	54.5
34.0	468	159	309	54.0
34.0	467	159	308	54.1
34.0	465	158	307	53.7
34.0	462	157	305	53.4
34.0	459	156	303	53.0
34.0	456	155	301	52.7
34.0	453	154	299	52.4
34.0	450	153	297	52.0
34.0	447	152	295	51.7
34.0	444	151	293	51.4
34.0	441	150	291	51.0
34.0	438	149	289	50.7
34.0	435	148	287	50.4
34.0	432	147	285	50.0
34.0	430	146	284	49.6
34.0	429	146	283	49.7
34.0	427	145	282	49.2
34.0	426	145	281	49.4
34.0	424	144	280	48.9
34.0	423	144	279	49.0
34.0	421	143	278	48.6
34.0	420	143	277	48.7
34.0	418	142	276	48.2
34.0	415	141	274	47.9
34.0	412	140	272	47.6
34.0	409	139	270	47.2
34.0	406	138	268	46.9
34.0	403	137	266	46.6
34.0	400	136	264	46.2
34.0	397	135	262	45.9
34.0	394	134	260	45.6
34.0	391	133	258	45.2

%E	M1	M2	DM	M*
34.0	388	132	256	44.9
34.0	385	131	254	44.6
34.0	382	130	252	44.2
34.0	379	129	250	43.9
34.0	377	128	249	43.5
34.0	376	128	248	43.6
34.0	374	127	247	43.1
34.0	373	127	246	43.2
34.0	371	126	245	42.8
34.0	368	125	243	42.5
34.0	365	124	241	42.1
34.0	362	123	239	41.8
34.0	359	122	237	41.5
34.0	356	121	235	41.1
34.0	353	120	233	40.8
34.0	350	119	231	40.5
34.0	347	118	229	40.1
34.0	344	117	227	39.8
34.0	341	116	225	39.5
34.0	338	115	223	39.1
34.0	335	114	221	38.8
34.0	332	113	219	38.5
34.0	329	112	217	38.1
34.0	326	111	215	37.8
34.0	324	110	214	37.3
34.0	321	109	212	37.0
34.0	318	108	210	36.7
34.0	315	107	208	36.3
34.0	312	106	206	36.0
34.0	309	105	204	35.7
34.0	306	104	202	35.3
34.0	303	103	200	35.0
34.0	300	102	198	34.7
34.0	297	101	196	34.3
34.0	294	100	194	34.0
34.0	291	99	192	33.7
34.0	288	98	190	33.3
34.0	285	97	188	33.0
34.0	282	96	186	32.7
34.0	268	91	177	30.9
34.0	265	90	175	30.6
34.0	262	89	173	30.2
34.0	259	88	171	29.9
34.0	256	87	169	29.6
34.0	253	86	167	29.2
34.0	250	85	165	28.9
34.0	247	84	163	28.6
34.0	244	83	161	28.2
34.0	241	82	159	27.9
34.0	238	81	157	27.6

%E	M1	M2	DM	M*
34.0	235	80	155	27.2
34.0	215	73	142	24.8
34.0	212	72	140	24.5
34.0	209	71	138	24.1
34.0	206	70	136	23.8
34.0	203	69	134	23.5
34.0	200	68	132	23.1
34.0	197	67	130	22.8
34.0	194	66	128	22.5
34.0	191	65	126	22.1
34.0	188	64	124	21.8
34.0	162	55	107	18.7
34.0	159	54	105	18.3
34.0	156	53	103	18.0
34.0	153	52	101	17.7
34.0	150	51	99	17.3
34.0	147	50	97	17.0
34.0	144	49	95	16.7
34.0	141	48	93	16.3
34.0	106	36	70	12.2
34.0	103	35	68	11.9
34.0	100	34	66	11.6
34.0	97	33	64	11.2
34.0	94	32	62	10.9
34.0	53	18	35	6.1
34.0	50	17	33	5.8
34.0	47	16	31	5.4
33.9	499	169	330	57.2
33.9	498	169	329	57.4
33.9	496	168	328	56.9
33.9	495	168	327	57.0
33.9	493	167	326	56.6
33.9	492	167	325	56.7
33.9	490	166	324	56.2
33.9	489	166	323	56.4
33.9	487	165	322	55.9
33.9	484	164	320	55.6
33.9	481	163	318	55.2
33.9	478	162	316	54.9
33.9	475	161	314	54.6
33.9	472	160	312	54.2
33.9	469	159	310	53.9
33.9	466	158	308	53.6
33.9	463	157	306	53.2
33.9	460	156	304	52.9
33.9	457	155	302	52.6
33.9	454	154	300	52.2
33.9	452	153	299	51.8
33.9	451	153	298	51.9
33.9	449	152	297	51.5

%E	M1	M2	DM	M*
33.9	448	152	296	51.6
33.9	446	151	295	51.1
33.9	445	151	294	51.2
33.9	443	150	293	50.8
33.9	442	150	292	50.9
33.9	440	149	291	50.5
33.9	439	149	290	50.6
33.9	437	148	289	50.1
33.9	436	148	288	50.2
33.9	434	147	287	49.8
33.9	433	147	286	49.9
33.9	431	146	285	49.5
33.9	428	145	283	49.1
33.9	425	144	281	48.8
33.9	422	143	279	48.5
33.9	419	142	277	48.1
33.9	416	141	275	47.8
33.9	413	140	273	47.5
33.9	410	139	271	47.1
33.9	407	138	269	46.8
33.9	404	137	267	46.5
33.9	401	136	265	46.1
33.9	398	135	263	45.8
33.9	395	134	261	45.5
33.9	392	133	259	45.1
33.9	389	132	257	44.8
33.9	387	131	256	44.3
33.9	386	131	255	44.5
33.9	384	130	254	44.0
33.9	383	130	253	44.1
33.9	381	129	252	43.7
33.9	380	129	251	43.8
33.9	378	128	250	43.3
33.9	375	127	248	43.0
33.9	372	126	246	42.7
33.9	369	125	244	42.3
33.9	366	124	242	42.0
33.9	363	123	240	41.7
33.9	360	122	238	41.3
33.9	357	121	236	41.0
33.9	354	120	234	40.7
33.9	351	119	232	40.3
33.9	348	118	230	40.0
33.9	345	117	228	39.7
33.9	342	116	226	39.3
33.9	339	115	224	39.0
33.9	336	114	222	38.7
33.9	333	113	220	38.3
33.9	330	112	218	38.0
33.9	327	111	216	37.7

%E	M1	M2	DM	M*
33.9	322	109	213	36.9
33.9	319	108	211	36.6
33.9	316	107	209	36.2
33.9	313	106	207	35.9
33.9	310	105	205	35.6
33.9	307	104	203	35.2
33.9	304	103	201	34.9
33.9	301	102	199	34.6
33.9	298	101	197	34.2
33.9	295	100	195	33.9
33.9	292	99	193	33.6
33.9	289	98	191	33.2
33.9	286	97	189	32.9
33.9	283	96	187	32.6
33.9	280	95	185	32.2
33.9	277	94	183	31.9
33.9	274	93	181	31.6
33.9	271	92	179	31.2
33.9	257	87	170	29.5
33.9	254	86	168	29.1
33.9	251	85	166	28.8
33.9	248	84	164	28.5
33.9	245	83	162	28.1
33.9	242	82	160	27.8
33.9	239	81	158	27.5
33.9	236	80	156	27.1
33.9	233	79	154	26.8
33.9	230	78	152	26.5
33.9	227	77	150	26.1
33.9	224	76	148	25.8
33.9	221	75	146	25.5
33.9	218	74	144	25.1
33.9	192	65	127	22.0
33.9	189	64	125	21.7
33.9	186	63	123	21.3
33.9	183	62	121	21.0
33.9	180	61	119	20.7
33.9	177	60	117	20.3
33.9	174	59	115	20.0
33.9	171	58	113	19.7
33.9	168	57	111	19.3
33.9	165	56	109	19.0
33.9	127	43	84	14.6
33.9	124	42	82	14.2
33.9	121	41	80	13.9
33.9	118	40	78	13.6
33.9	115	39	76	13.2
33.9	112	38	74	12.9
33.9	109	37	72	12.6
33.9	62	21	41	7.1

%E	M1	M2	DM	M*
33.9	59	20	39	6.8
33.9	56	19	37	6.4
33.8	500	169	331	57.1
33.8	497	168	329	56.8
33.8	494	167	327	56.5
33.8	491	166	325	56.1
33.8	488	165	323	55.8
33.8	485	164	321	55.5
33.8	482	163	319	55.1
33.8	480	162	318	54.7
33.8	479	162	317	54.8
33.8	477	161	316	54.3
33.8	476	161	315	54.5
33.8	474	160	314	54.0
33.8	473	160	313	54.1
33.8	471	159	312	53.7
33.8	470	159	311	53.8
33.8	468	158	310	53.3
33.8	467	158	309	53.5
33.8	465	157	308	53.0
33.8	464	157	307	53.1
33.8	462	156	306	52.7
33.8	461	156	305	52.8
33.8	459	155	304	52.3
33.8	458	155	303	52.5
33.8	456	154	302	52.0
33.8	455	154	301	52.1
33.8	453	153	300	51.7
33.8	450	152	298	51.3
33.8	447	151	296	51.0
33.8	444	150	294	50.7
33.8	441	149	292	50.3
33.8	438	148	290	50.0
33.8	435	147	288	49.7
33.8	432	146	286	49.3
33.8	429	145	284	49.0
33.8	426	144	282	48.7
33.8	423	143	280	48.3
33.8	420	142	278	48.0
33.8	417	141	276	47.7
33.8	414	140	274	47.3
33.8	411	139	272	47.0
33.8	408	138	270	46.7
33.8	405	137	268	46.3
33.8	402	136	266	46.0
33.8	400	135	265	45.6
33.8	399	135	264	45.7
33.8	397	134	263	45.2
33.8	396	134	262	45.3
33.8	394	133	261	44.9

%E	M1	M2	DM	M*
33.8	393	133	260	45.0
33.8	391	132	259	44.6
33.8	390	132	258	44.7
33.8	388	131	257	44.2
33.8	385	130	255	43.9
33.8	382	129	253	43.6
33.8	379	128	251	43.2
33.8	376	127	249	42.9
33.8	373	126	247	42.6
33.8	370	125	245	42.2
33.8	367	124	243	41.9
33.8	364	123	241	41.6
33.8	361	122	239	41.2
33.8	358	121	237	40.9
33.8	355	120	235	40.6
33.8	352	119	233	40.2
33.8	349	118	231	39.9
33.8	346	117	229	39.6
33.8	343	116	227	39.2
33.8	340	115	225	38.9
33.8	337	114	223	38.6
33.8	334	113	221	38.2
33.8	331	112	219	37.9
33.8	328	111	217	37.6
33.8	325	110	215	37.2
33.8	320	108	212	36.4
33.8	317	107	210	36.1
33.8	314	106	208	35.8
33.8	311	105	206	35.5
33.8	308	104	204	35.1
33.8	305	103	202	34.8
33.8	302	102	200	34.5
33.8	299	101	198	34.1
33.8	296	100	196	33.8
33.8	293	99	194	33.5
33.8	290	98	192	33.1
33.8	287	97	190	32.8
33.8	284	96	188	32.5
33.8	281	95	186	32.1
33.8	278	94	184	31.8
33.8	275	93	182	31.5
33.8	272	92	180	31.1
33.8	269	91	178	30.8
33.8	266	90	176	30.5
33.8	263	89	174	30.1
33.8	260	88	172	29.8
33.8	240	81	159	27.3
33.8	237	80	157	27.0
33.8	234	79	155	26.7
33.8	231	78	153	26.3

%E	M1	M2	DM	M*
33.8	228	77	151	26.0
33.8	225	76	149	25.7
33.8	222	75	147	25.3
33.8	219	74	145	25.0
33.8	216	73	143	24.7
33.8	213	72	141	24.3
33.8	210	71	139	24.0
33.8	207	70	137	23.7
33.8	204	69	135	23.3
33.8	201	68	133	23.0
33.8	198	67	131	22.7
33.8	195	66	129	22.3
33.8	160	54	106	18.2
33.8	157	53	104	17.9
33.8	154	52	102	17.6
33.8	151	51	100	17.2
33.8	148	50	98	16.9
33.8	145	49	96	16.6
33.8	142	48	94	16.2
33.8	139	47	92	15.9
33.8	136	46	90	15.6
33.8	133	45	88	15.2
33.8	130	44	86	14.9
33.8	80	27	53	9.1
33.8	77	26	51	8.8
33.8	74	25	49	8.4
33.8	71	24	47	8.1
33.8	68	23	45	7.8
33.8	65	22	43	7.4
33.7	499	168	331	56.6
33.7	498	168	330	56.7
33.7	496	167	329	56.2
33.7	495	167	328	56.3
33.7	493	166	327	55.9
33.7	492	166	326	56.0
33.7	490	165	325	55.6
33.7	489	165	324	55.7
33.7	487	164	323	55.2
33.7	486	164	322	55.3
33.7	484	163	321	54.9
33.7	483	163	320	55.0
33.7	481	162	319	54.6
33.7	478	161	317	54.2
33.7	475	160	315	53.9
33.7	472	159	313	53.6
33.7	469	158	311	53.2
33.7	466	157	309	52.9
33.7	463	156	307	52.6
33.7	460	155	305	52.2
33.7	457	154	303	51.9

%E	M1	M2	DM	M*	%E	M1	M2	DM	M*	%E	M1	M2	DM	M*	%E	M1	M2	DM	M*	%E	M1	M2	DM	M*
33.7	454	153	301	51.6	33.7	323	109	214	36.8	33.6	500	168	332	56.4	33.6	393	132	261	44.3	33.6	235	79	156	26.6
33.7	451	152	299	51.2	33.7	315	106	209	35.7	33.6	497	167	330	56.1	33.6	390	131	259	44.0	33.6	232	78	154	26.2
33.7	448	151	297	50.9	33.7	312	105	207	35.3	33.6	494	166	328	55.8	33.6	387	130	257	43.7	33.6	229	77	152	25.9
33.7	445	150	295	50.6	33.7	309	104	205	35.0	33.6	491	165	326	55.4	33.6	384	129	255	43.3	33.6	226	76	150	25.6
33.7	442	149	293	50.2	33.7	306	103	203	34.7	33.6	488	164	324	55.1	33.6	381	128	253	43.0	33.6	223	75	148	25.2
33.7	439	148	291	49.9	33.7	303	102	201	34.3	33.6	485	163	322	54.8	33.6	378	127	251	42.7	33.6	220	74	146	24.9
33.7	436	147	289	49.6	33.7	300	101	199	34.0	33.6	482	162	320	54.4	33.6	375	126	249	42.3	33.6	217	73	144	24.6
33.7	433	146	287	49.2	33.7	297	100	197	33.7	33.6	479	161	318	54.1	33.6	372	125	247	42.0	33.6	214	72	142	24.2
33.7	430	145	285	48.9	33.7	294	99	195	33.3	33.6	476	160	316	53.8	33.6	369	124	245	41.7	33.6	211	71	140	23.9
33.7	427	144	283	48.6	33.7	291	98	193	33.0	33.6	473	159	314	53.4	33.6	366	123	243	41.3	33.6	152	51	101	17.1
33.7	424	143	281	48.2	33.7	288	97	191	32.7	33.6	470	158	312	53.1	33.6	363	122	241	41.0	33.6	149	50	99	16.8
33.7	421	142	279	47.9	33.7	285	96	189	32.3	33.6	467	157	310	52.8	33.6	360	121	239	40.7	33.6	146	49	97	16.4
33.7	419	141	278	47.4	33.7	282	95	187	32.0	33.6	464	156	308	52.4	33.6	357	120	237	40.3	33.6	143	48	95	16.1
33.7	418	141	277	47.6	33.7	279	94	185	31.7	33.6	462	155	307	52.0	33.6	354	119	235	40.0	33.6	140	47	93	15.8
33.7	416	140	276	47.1	33.7	276	93	183	31.3	33.6	461	155	306	52.1	33.6	351	118	233	39.7	33.6	137	46	91	15.4
33.7	415	140	275	47.2	33.7	273	92	181	31.0	33.6	459	154	305	51.7	33.6	348	117	231	39.3	33.6	134	45	89	15.1
33.7	413	139	274	46.8	33.7	270	91	179	30.7	33.6	458	154	304	51.8	33.6	345	116	229	39.0	33.6	131	44	87	14.8
33.7	412	139	273	46.9	33.7	267	90	177	30.3	33.6	456	153	303	51.3	33.6	342	115	227	38.7	33.6	128	43	85	14.4
33.7	410	138	272	46.4	33.7	264	89	175	30.0	33.6	455	153	302	51.4	33.6	339	114	225	38.3	33.6	125	42	83	14.1
33.7	409	138	271	46.6	33.7	261	88	173	29.7	33.6	453	152	301	51.0	33.6	336	113	223	38.0	33.6	122	41	81	13.8
33.7	407	137	270	46.1	33.7	258	87	171	29.3	33.6	452	152	300	51.1	33.6	333	112	221	37.7	33.6	119	40	79	13.4
33.7	406	137	269	46.2	33.7	255	86	169	29.0	33.6	450	151	299	50.7	33.6	330	111	219	37.3	33.6	116	39	77	13.1
33.7	404	136	268	45.8	33.7	252	85	167	28.7	33.6	449	151	298	50.8	33.6	327	110	217	37.0	33.6	113	38	75	12.8
33.7	403	136	267	45.9	33.7	249	84	165	28.3	33.6	447	150	297	50.3	33.6	324	109	215	36.7	33.6	110	37	73	12.4
33.7	401	135	266	45.4	33.7	246	83	163	28.0	33.6	446	150	296	50.4	33.6	321	108	213	36.3	33.6	107	36	71	12.1
33.7	398	134	264	45.1	33.7	243	82	161	27.7	33.6	444	149	295	50.0	33.6	318	107	211	36.0	33.5	499	167	332	55.9
33.7	395	133	262	44.8	33.7	208	70	138	23.6	33.6	443	149	294	50.1	33.6	307	103	204	34.6	33.5	498	167	331	56.0
33.7	392	132	260	44.4	33.7	205	69	136	23.2	33.6	441	148	293	49.7	33.6	304	102	202	34.2	33.5	496	166	330	55.6
33.7	389	131	258	44.1	33.7	202	68	134	22.9	33.6	440	148	292	49.8	33.6	301	101	200	33.9	33.5	495	166	329	55.7
33.7	386	130	256	43.8	33.7	199	67	132	22.6	33.6	438	147	291	49.3	33.6	298	100	198	33.6	33.5	493	165	328	55.2
33.7	383	129	254	43.4	33.7	196	66	130	22.2	33.6	437	147	290	49.4	33.6	295	99	196	33.2	33.5	492	165	327	55.3
33.7	380	128	252	43.1	33.7	193	65	128	21.9	33.6	435	146	289	49.0	33.6	292	98	194	32.9	33.5	490	164	326	54.9
33.7	377	127	250	42.8	33.7	190	64	126	21.6	33.6	434	146	288	49.1	33.6	289	97	192	32.6	33.5	489	164	325	55.0
33.7	374	126	248	42.4	33.7	187	63	124	21.2	33.6	432	145	287	48.7	33.6	286	96	190	32.2	33.5	487	163	324	54.6
33.7	371	125	246	42.1	33.7	184	62	122	20.9	33.6	431	145	286	48.8	33.6	283	95	188	31.9	33.5	486	163	323	54.7
33.7	368	124	244	41.8	33.7	181	61	120	20.6	33.6	429	144	285	48.3	33.6	280	94	186	31.6	33.5	484	162	322	54.2
33.7	365	123	242	41.4	33.7	178	60	118	20.2	33.6	428	144	284	48.4	33.6	277	93	184	31.2	33.5	483	162	321	54.3
33.7	362	122	240	41.1	33.7	175	59	116	19.9	33.6	426	143	283	48.0	33.6	274	92	182	30.9	33.5	481	161	320	53.9
33.7	359	121	238	40.8	33.7	172	58	114	19.6	33.6	425	143	282	48.1	33.6	271	91	180	30.6	33.5	480	161	319	54.0
33.7	356	120	236	40.4	33.7	169	57	112	19.2	33.6	423	142	281	47.7	33.6	268	90	178	30.2	33.5	478	160	318	53.6
33.7	353	119	234	40.1	33.7	166	56	110	18.9	33.6	422	142	280	47.8	33.6	265	89	176	29.9	33.5	477	160	317	53.7
33.7	350	118	232	39.8	33.7	163	55	108	18.6	33.6	420	141	279	47.3	33.6	262	88	174	29.6	33.5	475	159	316	53.2
33.7	347	117	230	39.4	33.7	104	35	69	11.8	33.6	417	140	277	47.0	33.6	259	87	172	29.2	33.5	474	159	315	53.3
33.7	344	116	228	39.1	33.7	101	34	67	11.4	33.6	414	139	275	46.7	33.6	256	86	170	28.9	33.5	472	158	314	52.9
33.7	341	115	226	38.8	33.7	98	33	65	11.1	33.6	411	138	273	46.3	33.6	253	85	168	28.6	33.5	471	158	313	53.0
33.7	338	114	224	38.4	33.7	95	32	63	10.8	33.6	408	137	271	46.0	33.6	250	84	166	28.2	33.5	469	157	312	52.6
33.7	335	113	222	38.1	33.7	92	31	61	10.4	33.6	405	136	269	45.7	33.6	247	83	164	27.9	33.5	468	157	311	52.7
33.7	332	112	220	37.8	33.7	89	30	59	10.1	33.6	402	135	267	45.3	33.6	244	82	162	27.6	33.5	466	156	310	52.2
33.7	329	111	218	37.4	33.7	86	29	57	9.8	33.6	399	134	265	45.0	33.6	241	81	160	27.2	33.5	465	156	309	52.3
33.7	326	110	216	37.1	33.7	83	28	55	9.4	33.6	396	133	263	44.7	33.6	238	80	158	26.9	33.5	463	155	308	51.9

%E	M1	M2	DM	M*
33.5	460	154	306	51.6
33.5	457	153	304	51.2
33.5	454	152	302	50.9
33.5	451	151	300	50.6
33.5	448	150	298	50.2
33.5	445	149	296	49.9
33.5	442	148	294	49.6
33.5	439	147	292	49.2
33.5	436	146	290	48.9
33.5	433	145	288	48.6
33.5	430	144	286	48.2
33.5	427	143	284	47.9
33.5	424	142	282	47.6
33.5	421	141	280	47.2
33.5	418	140	278	46.9
33.5	415	139	276	46.6
33.5	412	138	274	46.2
33.5	409	137	272	45.9
33.5	406	136	270	45.6
33.5	403	135	268	45.2
33.5	400	134	266	44.9
33.5	397	133	264	44.6
33.5	394	132	262	44.2
33.5	391	131	260	43.9
33.5	388	130	258	43.6
33.5	385	129	256	43.2
33.5	382	128	254	42.9
33.5	379	127	252	42.6
33.5	376	126	250	42.2
33.5	373	125	248	41.9
33.5	370	124	246	41.6
33.5	367	123	244	41.2
33.5	364	122	242	40.9
33.5	361	121	240	40.6
33.5	358	120	238	40.2
33.5	355	119	236	39.9
33.5	352	118	234	39.6
33.5	349	117	232	39.2
33.5	346	116	230	38.9
33.5	343	115	228	38.6
33.5	340	114	226	38.2
33.5	337	113	224	37.9
33.5	334	112	222	37.6
33.5	331	111	220	37.2
33.5	328	110	218	36.9
33.5	325	109	216	36.6
33.5	322	108	214	36.2
33.5	319	107	212	35.9
33.5	316	106	210	35.6
33.5	313	105	208	35.2

%E	M1	M2	DM	M*
33.5	310	104	206	34.9
33.5	284	95	189	31.8
33.5	281	94	187	31.4
33.5	278	93	185	31.1
33.5	275	92	183	30.8
33.5	272	91	181	30.4
33.5	269	90	179	30.1
33.5	266	89	177	29.8
33.5	263	88	175	29.4
33.5	260	87	173	29.1
33.5	257	86	171	28.8
33.5	254	85	169	28.4
33.5	251	84	167	28.1
33.5	248	83	165	27.8
33.5	245	82	163	27.4
33.5	242	81	161	27.1
33.5	239	80	159	26.8
33.5	236	79	157	26.4
33.5	233	78	155	26.1
33.5	230	77	153	25.8
33.5	227	76	151	25.4
33.5	224	75	149	25.1
33.5	221	74	147	24.8
33.5	218	73	145	24.4
33.5	215	72	143	24.1
33.5	212	71	141	23.8
33.5	209	70	139	23.4
33.5	206	69	137	23.1
33.5	203	68	135	22.8
33.5	200	67	133	22.4
33.5	197	66	131	22.1
33.5	194	65	129	21.8
33.5	191	64	127	21.4
33.5	188	63	125	21.1
33.5	185	62	123	20.8
33.5	182	61	121	20.4
33.5	179	60	119	20.1
33.5	176	59	117	19.8
33.5	173	58	115	19.4
33.5	170	57	113	19.1
33.5	167	56	111	18.8
33.5	164	55	109	18.4
33.5	161	54	107	18.1
33.5	158	53	105	17.8
33.5	155	52	103	17.4
33.4	500	167	333	55.8
33.4	497	166	331	55.4
33.4	494	165	329	55.1
33.4	491	164	327	54.8
33.4	488	163	325	54.4

%E	M1	M2	DM	M*
33.4	485	162	323	54.1
33.4	482	161	321	53.8
33.4	479	160	319	53.4
33.4	476	159	317	53.1
33.4	473	158	315	52.8
33.4	470	157	313	52.4
33.4	467	156	311	52.1
33.4	464	155	309	51.8
33.4	461	154	307	51.4
33.4	458	153	305	51.1
33.4	455	152	303	50.8
33.4	452	151	301	50.4
33.4	449	150	299	50.1
33.4	446	149	297	49.8
33.4	443	148	295	49.4
33.4	440	147	293	49.1
33.4	437	146	291	48.8
33.4	434	145	289	48.4
33.4	431	144	287	48.1
33.4	428	143	285	47.8
33.4	425	142	283	47.4
33.4	422	141	281	47.1
33.4	419	140	279	46.8
33.4	416	139	277	46.4
33.4	413	138	275	46.1
33.4	410	137	273	45.8
33.4	407	136	271	45.4
33.4	404	135	269	45.1
33.4	401	134	267	44.8
33.4	398	133	265	44.4
33.4	395	132	263	44.1
33.4	392	131	261	43.8
33.4	389	130	259	43.4
33.4	386	129	257	43.1
33.4	383	128	255	42.8
33.4	380	127	253	42.4
33.4	377	126	251	42.1
33.4	374	125	249	41.8
33.4	371	124	247	41.4
33.4	368	123	245	41.1
33.4	365	122	243	40.8
33.4	362	121	241	40.4
33.4	359	120	239	40.1
33.4	356	119	237	39.8
33.4	353	118	235	39.4
33.4	350	117	233	39.1
33.4	347	116	231	38.8
33.4	344	115	229	38.4
33.4	341	114	227	38.1
33.4	338	113	225	37.8

%E	M1	M2	DM	M*
33.4	335	112	223	37.4
33.4	332	111	221	37.1
33.4	329	110	219	36.8
33.4	326	109	217	36.4
33.4	323	108	215	36.1
33.4	320	107	213	35.8
33.4	317	106	211	35.4
33.4	314	105	209	35.1
33.4	311	104	207	34.8
33.4	308	103	205	34.4
33.4	305	102	203	34.1
33.4	302	101	201	33.8
33.4	299	100	199	33.4
33.4	296	99	197	33.1
33.4	293	98	195	32.8
33.4	290	97	193	32.4
33.4	287	96	191	32.1
33.3	499	166	333	55.2
33.3	498	166	332	55.3
33.3	496	165	331	54.9
33.3	495	165	330	55.0
33.3	493	164	329	54.6
33.3	492	164	328	54.7
33.3	490	163	327	54.2
33.3	489	163	326	54.3
33.3	487	162	325	53.9
33.3	486	162	324	54.0
33.3	484	161	323	53.6
33.3	483	161	322	53.7
33.3	481	160	321	53.2
33.3	480	160	320	53.3
33.3	478	159	319	52.9
33.3	477	159	318	53.0
33.3	475	158	317	52.6
33.3	474	158	316	52.7
33.3	472	157	315	52.2
33.3	471	157	314	52.3
33.3	469	156	313	51.9
33.3	468	156	312	52.0
33.3	466	155	311	51.6
33.3	465	155	310	51.7
33.3	463	154	309	51.2
33.3	462	154	308	51.3
33.3	460	153	307	50.9
33.3	459	153	306	51.0
33.3	457	152	305	50.6
33.3	456	152	304	50.7
33.3	454	151	303	50.2
33.3	453	151	302	50.3
33.3	451	150	301	49.9

%E	M1	M2	DM	M*
33.3	450	150	300	50.0
33.3	448	149	299	49.6
33.3	447	149	298	49.7
33.3	445	148	297	49.2
33.3	444	148	296	49.3
33.3	442	147	295	48.9
33.3	441	147	294	49.0
33.3	439	146	293	48.6
33.3	438	146	292	48.7
33.3	436	145	291	48.2
33.3	435	145	290	48.3
33.3	433	144	289	47.9
33.3	432	144	288	48.0
33.3	430	143	287	47.6
33.3	429	143	286	47.7
33.3	427	142	285	47.2
33.3	426	142	284	47.3
33.3	424	141	283	46.9
33.3	423	141	282	47.0
33.3	421	140	281	46.6
33.3	420	140	280	46.7
33.3	418	139	279	46.2
33.3	417	139	278	46.3
33.3	415	138	277	45.9
33.3	414	138	276	46.0
33.3	412	137	275	45.6
33.3	411	137	274	45.7
33.3	409	136	273	45.2
33.3	408	136	272	45.3
33.3	406	135	271	44.9
33.3	405	135	270	45.0
33.3	403	134	269	44.6
33.3	402	134	268	44.7
33.3	400	133	267	44.2
33.3	399	133	266	44.3
33.3	396	132	264	44.0
33.3	393	131	262	43.7
33.3	390	130	260	43.3
33.3	387	129	258	43.0
33.3	384	128	256	42.7
33.3	381	127	254	42.3
33.3	378	126	252	42.0
33.3	375	125	250	41.7
33.3	372	124	248	41.3
33.3	369	123	246	41.0
33.3	366	122	244	40.7
33.3	363	121	242	40.3
33.3	360	120	240	40.0
33.3	357	119	238	39.7
33.3	354	118	236	39.3

%E	M1	M2	DM	M*
33.3	351	117	234	39.0
33.3	348	116	232	38.7
33.3	345	115	230	38.3
33.3	342	114	228	38.0
33.3	339	113	226	37.7
33.3	336	112	224	37.3
33.3	333	111	222	37.0
33.3	330	110	220	36.7
33.3	327	109	218	36.3
33.3	324	108	216	36.0
33.3	321	107	214	35.7
33.3	318	106	212	35.3
33.3	315	105	210	35.0
33.3	312	104	208	34.7
33.3	309	103	206	34.3
33.3	306	102	204	34.0
33.3	303	101	202	33.7
33.3	300	100	200	33.3
33.3	297	99	198	33.0
33.3	294	98	196	32.7
33.3	291	97	194	32.3
33.3	288	96	192	32.0
33.3	285	95	190	31.7
33.3	282	94	188	31.3
33.3	279	93	186	31.0
33.3	276	92	184	30.7
33.3	273	91	182	30.3
33.3	270	90	180	30.0
33.3	267	89	178	29.7
33.3	264	88	176	29.3
33.3	261	87	174	29.0
33.3	258	86	172	28.7
33.3	255	85	170	28.3
33.3	252	84	168	28.0
33.3	249	83	166	27.7
33.3	246	82	164	27.3
33.3	243	81	162	27.0
33.3	240	80	160	26.7
33.3	237	79	158	26.3
33.3	234	78	156	26.0
33.3	231	77	154	25.7
33.3	228	76	152	25.3
33.3	225	75	150	25.0
33.3	222	74	148	24.7
33.3	219	73	146	24.3
33.3	216	72	144	24.0
33.3	213	71	142	23.7
33.3	210	70	140	23.3
33.3	207	69	138	23.0
33.3	204	68	136	22.7
33.3	201	67	134	22.3
33.3	198	66	132	22.0
33.3	195	65	130	21.7
33.3	192	64	128	21.3
33.3	189	63	126	21.0
33.3	186	62	124	20.7
33.3	183	61	122	20.3
33.3	180	60	120	20.0
33.3	177	59	118	19.7
33.3	174	58	116	19.3
33.3	171	57	114	19.0
33.3	168	56	112	18.7
33.3	165	55	110	18.3
33.3	162	54	108	18.0
33.3	159	53	106	17.7
33.3	156	52	104	17.3
33.3	153	51	102	17.0
33.3	150	50	100	16.7
33.3	147	49	98	16.3
33.3	144	48	96	16.0
33.3	141	47	94	15.7
33.3	138	46	92	15.3
33.3	135	45	90	15.0
33.3	132	44	88	14.7
33.3	129	43	86	14.3
33.3	126	42	84	14.0
33.3	123	41	82	13.7
33.3	120	40	80	13.3
33.3	117	39	78	13.0
33.3	114	38	76	12.7
33.3	111	37	74	12.3
33.3	108	36	72	12.0
33.3	105	35	70	11.7
33.3	102	34	68	11.3
33.3	99	33	66	11.0
33.3	96	32	64	10.7
33.3	93	31	62	10.3
33.3	90	30	60	10.0
33.3	87	29	58	9.7
33.3	84	28	56	9.3
33.3	81	27	54	9.0
33.3	78	26	52	8.7
33.3	75	25	50	8.3
33.3	72	24	48	8.0
33.3	69	23	46	7.7
33.3	66	22	44	7.3
33.3	63	21	42	7.0
33.3	60	20	40	6.7
33.3	57	19	38	6.3
33.3	54	18	36	6.0
33.3	51	17	34	5.7
33.3	48	16	32	5.3
33.3	45	15	30	5.0
33.3	42	14	28	4.7
33.3	39	13	26	4.3
33.3	36	12	24	4.0
33.3	33	11	22	3.7
33.3	30	10	20	3.3
33.3	27	9	18	3.0
33.3	24	8	16	2.7
33.3	21	7	14	2.3
33.3	18	6	12	2.0
33.3	15	5	10	1.7
33.3	12	4	8	1.3
33.3	9	3	6	1.0
33.3	6	2	4	0.7
33.3	3	1	2	0.3
33.2	500	166	334	55.1
33.2	497	165	332	54.8
33.2	494	164	330	54.4
33.2	491	163	328	54.1
33.2	488	162	326	53.8
33.2	485	161	324	53.4
33.2	482	160	322	53.1
33.2	479	159	320	52.8
33.2	476	158	318	52.4
33.2	473	157	316	52.1
33.2	470	156	314	51.8
33.2	467	155	312	51.4
33.2	464	154	310	51.1
33.2	461	153	308	50.8
33.2	458	152	306	50.4
33.2	455	151	304	50.1
33.2	452	150	302	49.8
33.2	449	149	300	49.4
33.2	446	148	298	49.1
33.2	443	147	296	48.8
33.2	440	146	294	48.4
33.2	437	145	292	48.1
33.2	434	144	290	47.8
33.2	431	143	288	47.4
33.2	428	142	286	47.1
33.2	425	141	284	46.8
33.2	422	140	282	46.4
33.2	419	139	280	46.1
33.2	416	138	278	45.8
33.2	413	137	276	45.4
33.2	410	136	274	45.1
33.2	407	135	272	44.8
33.2	404	134	270	44.4
33.2	401	133	268	44.1
33.2	398	132	266	43.8
33.2	397	132	265	43.9
33.2	395	131	264	43.4
33.2	394	131	263	43.6
33.2	392	130	262	43.1
33.2	391	130	261	43.2
33.2	389	129	260	42.8
33.2	388	129	259	42.9
33.2	386	128	258	42.4
33.2	385	128	257	42.6
33.2	383	127	256	42.1
33.2	382	127	255	42.2
33.2	380	126	254	41.8
33.2	379	126	253	41.9
33.2	377	125	252	41.4
33.2	376	125	251	41.6
33.2	374	124	250	41.1
33.2	373	124	249	41.2
33.2	371	123	248	40.8
33.2	370	123	247	40.9
33.2	368	122	246	40.4
33.2	367	122	245	40.6
33.2	365	121	244	40.1
33.2	364	121	243	40.2
33.2	361	120	241	39.9
33.2	358	119	239	39.6
33.2	355	118	237	39.2
33.2	352	117	235	38.9
33.2	349	116	233	38.6
33.2	346	115	231	38.2
33.2	343	114	229	37.9
33.2	340	113	227	37.6
33.2	337	112	225	37.2
33.2	334	111	223	36.9
33.2	331	110	221	36.6
33.2	328	109	219	36.2
33.2	325	108	217	35.9
33.2	322	107	215	35.6
33.2	319	106	213	35.2
33.2	316	105	211	34.9
33.2	313	104	209	34.6
33.2	310	103	207	34.2
33.2	307	102	205	33.9
33.2	304	101	203	33.6
33.2	301	100	201	33.2
33.2	298	99	199	32.9
33.2	295	98	197	32.6
33.2	292	97	195	32.2
33.2	289	96	193	31.9
33.2	286	95	191	31.6
33.2	283	94	189	31.2
33.2	280	93	187	30.9
33.2	277	92	185	30.6
33.2	274	91	183	30.2
33.2	271	90	181	29.9
33.2	268	89	179	29.6
33.2	265	88	177	29.2
33.2	262	87	175	28.9
33.2	259	86	173	28.6
33.2	256	85	171	28.2
33.2	253	84	169	27.9
33.2	250	83	167	27.6
33.2	247	82	165	27.2
33.2	244	81	163	26.9
33.2	241	80	161	26.6
33.2	238	79	159	26.2
33.2	235	78	157	25.9
33.2	232	77	155	25.6
33.2	229	76	153	25.2
33.2	226	75	151	24.9
33.2	223	74	149	24.6
33.2	220	73	147	24.2
33.2	217	72	145	23.9
33.2	214	71	143	23.6
33.2	211	70	141	23.2
33.2	208	69	139	22.9
33.2	205	68	137	22.6
33.2	202	67	135	22.2
33.2	199	66	133	21.9
33.2	196	65	131	21.6
33.2	193	64	129	21.2
33.2	190	63	127	20.9
33.2	187	62	125	20.6
33.2	184	61	123	20.2
33.1	499	165	334	54.6
33.1	498	165	333	54.7
33.1	496	164	332	54.2
33.1	495	164	331	54.3
33.1	493	163	330	53.9
33.1	492	163	329	54.0
33.1	490	162	328	53.6
33.1	489	162	327	53.7
33.1	487	161	326	53.2
33.1	486	161	325	53.3
33.1	484	160	324	52.9
33.1	483	160	323	53.0
33.1	481	159	322	52.6
33.1	480	159	321	52.7
33.1	478	158	320	52.2

%E	M1	M2	DM	M*
33.1	477	158	319	52.3
33.1	475	157	318	51.9
33.1	474	157	317	52.0
33.1	472	156	316	51.6
33.1	471	156	315	51.7
33.1	468	155	313	51.3
33.1	465	154	311	51.0
33.1	462	153	309	50.7
33.1	459	152	307	50.3
33.1	456	151	305	50.0
33.1	453	150	303	49.7
33.1	450	149	301	49.3
33.1	447	148	299	49.0
33.1	444	147	297	48.7
33.1	441	146	295	48.3
33.1	438	145	293	48.0
33.1	435	144	291	47.7
33.1	432	143	289	47.3
33.1	429	142	287	47.0
33.1	426	141	285	46.7
33.1	423	140	283	46.3
33.1	420	139	281	46.0
33.1	417	138	279	45.7
33.1	414	137	277	45.3
33.1	411	136	275	45.0
33.1	408	135	273	44.7
33.1	405	134	271	44.3
33.1	402	133	269	44.0
33.1	399	132	267	43.7
33.1	396	131	265	43.3
33.1	393	130	263	43.0
33.1	390	129	261	42.7
33.1	387	128	259	42.3
33.1	384	127	257	42.0
33.1	381	126	255	41.7
33.1	378	125	253	41.3
33.1	375	124	251	41.0
33.1	372	123	249	40.7
33.1	369	122	247	40.3
33.1	366	121	245	40.0
33.1	363	120	243	39.7
33.1	362	120	242	39.8
33.1	360	119	241	39.3
33.1	359	119	240	39.4
33.1	357	118	239	39.0
33.1	356	118	238	39.1
33.1	354	117	237	38.7
33.1	353	117	236	38.8
33.1	350	116	234	38.4
33.1	347	115	232	38.1

%E	M1	M2	DM	M*
33.1	344	114	230	37.8
33.1	341	113	228	37.4
33.1	338	112	226	37.1
33.1	335	111	224	36.8
33.1	332	110	222	36.4
33.1	329	109	220	36.1
33.1	326	108	218	35.8
33.1	323	107	216	35.4
33.1	320	106	214	35.1
33.1	317	105	212	34.8
33.1	314	104	210	34.4
33.1	311	103	208	34.1
33.1	308	102	206	33.8
33.1	305	101	204	33.4
33.1	302	100	202	33.1
33.1	299	99	200	32.8
33.1	296	98	198	32.4
33.1	293	97	196	32.1
33.1	290	96	194	31.8
33.1	287	95	192	31.4
33.1	284	94	190	31.1
33.1	281	93	188	30.8
33.1	278	92	186	30.4
33.1	275	91	184	30.1
33.1	272	90	182	29.8
33.1	269	89	180	29.4
33.1	266	88	178	29.1
33.1	263	87	176	28.8
33.1	260	86	174	28.4
33.1	257	85	172	28.1
33.1	254	84	170	27.8
33.1	251	83	168	27.4
33.1	248	82	166	27.1
33.1	245	81	164	26.8
33.1	242	80	162	26.4
33.1	239	79	160	26.1
33.1	236	78	158	25.8
33.1	181	60	121	19.9
33.1	178	59	119	19.6
33.1	175	58	117	19.2
33.1	172	57	115	18.9
33.1	169	56	113	18.6
33.1	166	55	111	18.2
33.1	163	54	109	17.9
33.1	160	53	107	17.6
33.1	157	52	105	17.2
33.1	154	51	103	16.9
33.1	151	50	101	16.6
33.1	148	49	99	16.2
33.1	145	48	97	15.9

%E	M1	M2	DM	M*
33.1	142	47	95	15.6
33.1	139	46	93	15.2
33.1	136	45	91	14.9
33.1	133	44	89	14.6
33.1	130	43	87	14.2
33.1	127	42	85	13.9
33.1	124	41	83	13.6
33.1	121	40	81	13.2
33.1	118	39	79	12.9
33.0	500	165	335	54.4
33.0	497	164	333	54.1
33.0	494	163	331	53.8
33.0	491	162	329	53.5
33.0	488	161	327	53.1
33.0	485	160	325	52.8
33.0	482	159	323	52.5
33.0	479	158	321	52.1
33.0	476	157	319	51.8
33.0	473	156	317	51.5
33.0	470	155	315	51.1
33.0	469	155	314	51.2
33.0	467	154	313	50.8
33.0	466	154	312	50.9
33.0	464	153	311	50.5
33.0	463	153	310	50.6
33.0	461	152	309	50.1
33.0	460	152	308	50.2
33.0	458	151	307	49.8
33.0	457	151	306	49.9
33.0	455	150	305	49.5
33.0	454	150	304	49.6
33.0	452	149	303	49.1
33.0	451	149	302	49.2
33.0	449	148	301	48.8
33.0	448	148	300	48.9
33.0	446	147	299	48.5
33.0	445	147	298	48.6
33.0	443	146	297	48.1
33.0	442	146	296	48.2
33.0	440	145	295	47.8
33.0	439	145	294	47.9
33.0	437	144	293	47.5
33.0	436	144	292	47.6
33.0	433	143	290	47.2
33.0	430	142	288	46.9
33.0	427	141	286	46.6
33.0	424	140	284	46.2
33.0	421	139	282	45.9
33.0	418	138	280	45.6
33.0	415	137	278	45.2

%E	M1	M2	DM	M*
33.0	412	136	276	44.9
33.0	409	135	274	44.6
33.0	406	134	272	44.2
33.0	403	133	270	43.9
33.0	400	132	268	43.6
33.0	397	131	266	43.2
33.0	394	130	264	42.9
33.0	391	129	262	42.6
33.0	388	128	260	42.2
33.0	385	127	258	41.9
33.0	382	126	256	41.6
33.0	379	125	254	41.2
33.0	376	124	252	40.9
33.0	373	123	250	40.6
33.0	370	122	248	40.2
33.0	367	121	246	39.9
33.0	364	120	244	39.6
33.0	361	119	242	39.2
33.0	358	118	240	38.9
33.0	355	117	238	38.6
33.0	352	116	236	38.2
33.0	351	116	235	38.3
33.0	349	115	234	37.9
33.0	348	115	233	38.0
33.0	345	114	231	37.7
33.0	342	113	229	37.3
33.0	339	112	227	37.0
33.0	336	111	225	36.7
33.0	333	110	223	36.3
33.0	330	109	221	36.0
33.0	327	108	219	35.7
33.0	324	107	217	35.3
33.0	321	106	215	35.0
33.0	318	105	213	34.7
33.0	315	104	211	34.3
33.0	312	103	209	34.0
33.0	309	102	207	33.7
33.0	306	101	205	33.3
33.0	303	100	203	33.0
33.0	300	99	201	32.7
33.0	297	98	199	32.3
33.0	294	97	197	32.0
33.0	291	96	195	31.7
33.0	288	95	193	31.3
33.0	285	94	191	31.0
33.0	282	93	189	30.7
33.0	279	92	187	30.3
33.0	276	91	185	30.0
33.0	273	90	183	29.7
33.0	270	89	181	29.3

%E	M1	M2	DM	M*
33.0	267	88	179	29.0
33.0	264	87	177	28.7
33.0	261	86	175	28.3
33.0	233	77	156	25.4
33.0	230	76	154	25.1
33.0	227	75	152	24.8
33.0	224	74	150	24.4
33.0	221	73	148	24.1
33.0	218	72	146	23.8
33.0	215	71	144	23.4
33.0	212	70	142	23.1
33.0	209	69	140	22.8
33.0	206	68	138	22.4
33.0	203	67	136	22.1
33.0	200	66	134	21.8
33.0	197	65	132	21.4
33.0	194	64	130	21.1
33.0	191	63	128	20.8
33.0	188	62	126	20.4
33.0	185	61	124	20.1
33.0	182	60	122	19.8
33.0	179	59	120	19.4
33.0	176	58	118	19.1
33.0	115	38	77	12.6
33.0	112	37	75	12.2
33.0	109	36	73	11.9
33.0	106	35	71	11.6
33.0	103	34	69	11.2
33.0	100	33	67	10.9
33.0	97	32	65	10.6
33.0	94	31	63	10.2
33.0	91	30	61	9.9
33.0	88	29	59	9.6
32.9	499	164	335	53.9
32.9	498	164	334	54.0
32.9	496	163	333	53.6
32.9	495	163	332	53.7
32.9	493	162	331	53.2
32.9	492	162	330	53.3
32.9	490	161	329	52.9
32.9	489	161	328	53.0
32.9	487	160	327	52.6
32.9	486	160	326	52.7
32.9	484	159	325	52.2
32.9	483	159	324	52.3
32.9	480	158	322	52.0
32.9	477	157	320	51.7
32.9	474	156	318	51.3
32.9	471	155	316	51.0
32.9	468	154	314	50.7

%E	M1	M2	DM	M*
32.9	465	153	312	50.3
32.9	462	152	310	50.0
32.9	459	151	308	49.7
32.9	456	150	306	49.3
32.9	453	149	304	49.0
32.9	450	148	302	48.7
32.9	447	147	300	48.3
32.9	444	146	298	48.0
32.9	441	145	296	47.7
32.9	438	144	294	47.3
32.9	435	143	292	47.0
32.9	434	143	291	47.1
32.9	432	142	290	46.7
32.9	431	142	289	46.8
32.9	429	141	288	46.3
32.9	428	141	287	46.5
32.9	426	140	286	46.0
32.9	425	140	285	46.1
32.9	423	139	284	45.7
32.9	422	139	283	45.8
32.9	420	138	282	45.3
32.9	419	138	281	45.5
32.9	417	137	280	45.0
32.9	416	137	279	45.1
32.9	414	136	278	44.7
32.9	413	136	277	44.8
32.9	410	135	275	44.5
32.9	407	134	273	44.1
32.9	404	133	271	43.8
32.9	401	132	269	43.5
32.9	398	131	267	43.1
32.9	395	130	265	42.8
32.9	392	129	263	42.5
32.9	389	128	261	42.1
32.9	386	127	259	41.8
32.9	383	126	257	41.5
32.9	380	125	255	41.1
32.9	377	124	253	40.8
32.9	374	123	251	40.5
32.9	371	122	249	40.1
32.9	368	121	247	39.8
32.9	365	120	245	39.5
32.9	362	119	243	39.1
32.9	359	118	241	38.8
32.9	356	117	239	38.5
32.9	353	116	237	38.1
32.9	350	115	235	37.8
32.9	347	114	233	37.5
32.9	346	114	232	37.6
32.9	343	113	230	37.2

%E	M1	M2	DM	M*
32.9	340	112	228	36.9
32.9	337	111	226	36.6
32.9	334	110	224	36.2
32.9	331	109	222	35.9
32.9	328	108	220	35.6
32.9	325	107	218	35.2
32.9	322	106	216	34.9
32.9	319	105	214	34.6
32.9	316	104	212	34.2
32.9	313	103	210	33.9
32.9	310	102	208	33.6
32.9	307	101	206	33.2
32.9	304	100	204	32.9
32.9	301	99	202	32.6
32.9	298	98	200	32.2
32.9	295	97	198	31.9
32.9	292	96	196	31.6
32.9	289	95	194	31.2
32.9	286	94	192	30.9
32.9	283	93	190	30.6
32.9	280	92	188	30.2
32.9	277	91	186	29.9
32.9	258	85	173	28.0
32.9	255	84	171	27.7
32.9	252	83	169	27.3
32.9	249	82	167	27.0
32.9	246	81	165	26.7
32.9	243	80	163	26.3
32.9	240	79	161	26.0
32.9	237	78	159	25.7
32.9	234	77	157	25.3
32.9	231	76	155	25.0
32.9	228	75	153	24.7
32.9	225	74	151	24.3
32.9	222	73	149	24.0
32.9	219	72	147	23.7
32.9	216	71	145	23.3
32.9	213	70	143	23.0
32.9	210	69	141	22.7
32.9	207	68	139	22.3
32.9	173	57	116	18.8
32.9	170	56	114	18.4
32.9	167	55	112	18.1
32.9	164	54	110	17.8
32.9	161	53	108	17.4
32.9	158	52	106	17.1
32.9	155	51	104	16.8
32.9	152	50	102	16.4
32.9	149	49	100	16.1
32.9	146	48	98	15.8

%E	M1	M2	DM	M*
32.9	143	47	96	15.4
32.9	140	46	94	15.1
32.9	85	28	57	9.2
32.9	82	27	55	8.9
32.9	79	26	53	8.6
32.9	76	25	51	8.2
32.9	73	24	49	7.9
32.9	70	23	47	7.6
32.8	500	164	336	53.8
32.8	497	163	334	53.5
32.8	494	162	332	53.1
32.8	491	161	330	52.8
32.8	488	160	328	52.5
32.8	485	159	326	52.1
32.8	482	158	324	51.8
32.8	481	158	323	51.9
32.8	479	157	322	51.5
32.8	478	157	321	51.6
32.8	476	156	320	51.1
32.8	475	156	319	51.2
32.8	473	155	318	50.8
32.8	472	155	317	50.9
32.8	470	154	316	50.5
32.8	469	154	315	50.6
32.8	467	153	314	50.1
32.8	466	153	313	50.2
32.8	464	152	312	49.8
32.8	463	152	311	49.9
32.8	461	151	310	49.5
32.8	460	151	309	49.6
32.8	458	150	308	49.1
32.8	457	150	307	49.2
32.8	454	149	305	48.9
32.8	451	148	303	48.6
32.8	448	147	301	48.2
32.8	445	146	299	47.9
32.8	442	145	297	47.6
32.8	439	144	295	47.2
32.8	436	143	293	46.9
32.8	433	142	291	46.6
32.8	430	141	289	46.2
32.8	427	140	287	45.9
32.8	424	139	285	45.6
32.8	421	138	283	45.2
32.8	418	137	281	44.9
32.8	415	136	279	44.6
32.8	412	135	277	44.2
32.8	411	135	276	44.3
32.8	409	134	275	43.9
32.8	408	134	274	44.0

%E	M1	M2	DM	M*
32.8	406	133	273	43.6
32.8	405	133	272	43.7
32.8	403	132	271	43.2
32.8	402	132	270	43.3
32.8	400	131	269	42.9
32.8	399	131	268	43.0
32.8	396	130	266	42.7
32.8	393	129	264	42.3
32.8	390	128	262	42.0
32.8	387	127	260	41.7
32.8	384	126	258	41.3
32.8	381	125	256	41.0
32.8	378	124	254	40.7
32.8	375	123	252	40.3
32.8	372	122	250	40.0
32.8	369	121	248	39.7
32.8	366	120	246	39.3
32.8	363	119	244	39.0
32.8	360	118	242	38.7
32.8	357	117	240	38.3
32.8	354	116	238	38.0
32.8	351	115	236	37.7
32.8	348	114	234	37.3
32.8	345	113	232	37.0
32.8	344	113	231	37.1
32.8	341	112	229	36.8
32.8	338	111	227	36.5
32.8	335	110	225	36.1
32.8	332	109	223	35.8
32.8	329	108	221	35.5
32.8	326	107	219	35.1
32.8	323	106	217	34.8
32.8	320	105	215	34.5
32.8	317	104	213	34.1
32.8	314	103	211	33.8
32.8	311	102	209	33.5
32.8	308	101	207	33.1
32.8	305	100	205	32.8
32.8	302	99	203	32.5
32.8	299	98	201	32.1
32.8	296	97	199	31.8
32.8	293	96	197	31.5
32.8	290	95	195	31.1
32.8	287	94	193	30.8
32.8	274	90	184	29.6
32.8	271	89	182	29.2
32.8	268	88	180	28.9
32.8	265	87	178	28.6
32.8	262	86	176	28.2
32.8	259	85	174	27.9

%E	M1	M2	DM	M*
32.8	256	84	172	27.6
32.8	253	83	170	27.2
32.8	250	82	168	26.9
32.8	247	81	166	26.6
32.8	244	80	164	26.2
32.8	241	79	162	25.9
32.8	238	78	160	25.6
32.8	235	77	158	25.2
32.8	232	76	156	24.9
32.8	229	75	154	24.6
32.8	204	67	137	22.0
32.8	201	66	135	21.7
32.8	198	65	133	21.3
32.8	195	64	131	21.0
32.8	192	63	129	20.7
32.8	189	62	127	20.3
32.8	186	61	125	20.0
32.8	183	60	123	19.7
32.8	180	59	121	19.3
32.8	177	58	119	19.0
32.8	174	57	117	18.7
32.8	137	45	92	14.8
32.8	134	44	90	14.4
32.8	131	43	88	14.1
32.8	128	42	86	13.8
32.8	125	41	84	13.4
32.8	122	40	82	13.1
32.8	119	39	80	12.8
32.8	116	38	78	12.4
32.8	67	22	45	7.2
32.8	64	21	43	6.9
32.8	61	20	41	6.6
32.8	58	19	39	6.2
32.7	499	163	336	53.2
32.7	498	163	335	53.4
32.7	496	162	334	52.9
32.7	495	162	333	53.0
32.7	493	161	332	52.6
32.7	492	161	331	52.7
32.7	490	160	330	52.2
32.7	489	160	329	52.4
32.7	486	159	327	52.0
32.7	483	158	325	51.7
32.7	480	157	323	51.4
32.7	477	156	321	51.0
32.7	474	155	319	50.7
32.7	471	154	317	50.4
32.7	468	153	315	50.0
32.7	465	152	313	49.7
32.7	462	151	311	49.4

%E	M1	M2	DM	M*
32.7	459	150	309	49.0
32.7	456	149	307	48.7
32.7	455	149	306	48.8
32.7	453	148	305	48.4
32.7	452	148	304	48.5
32.7	450	147	303	48.0
32.7	449	147	302	48.1
32.7	447	146	301	47.7
32.7	446	146	300	47.8
32.7	444	145	299	47.4
32.7	443	145	298	47.5
32.7	441	144	297	47.0
32.7	440	144	296	47.1
32.7	437	143	294	46.8
32.7	434	142	292	46.5
32.7	431	141	290	46.1
32.7	428	140	288	45.8
32.7	425	139	286	45.5
32.7	422	138	284	45.1
32.7	419	137	282	44.8
32.7	416	136	280	44.5
32.7	413	135	278	44.1
32.7	410	134	276	43.8
32.7	407	133	274	43.5
32.7	404	132	272	43.1
32.7	401	131	270	42.8
32.7	398	130	268	42.5
32.7	397	130	267	42.6
32.7	395	129	266	42.1
32.7	394	129	265	42.2
32.7	392	128	264	41.8
32.7	391	128	263	41.9
32.7	388	127	261	41.6
32.7	385	126	259	41.2
32.7	382	125	257	40.9
32.7	379	124	255	40.6
32.7	376	123	253	40.2
32.7	373	122	251	39.9
32.7	370	121	249	39.6
32.7	367	120	247	39.2
32.7	364	119	245	38.9
32.7	361	118	243	38.6
32.7	358	117	241	38.2
32.7	355	116	239	37.9
32.7	352	115	237	37.6
32.7	349	114	235	37.2
32.7	346	113	233	36.9
32.7	343	112	231	36.6
32.7	342	112	230	36.7
32.7	339	111	228	36.3

%E	M1	M2	DM	M*
32.7	336	110	226	36.0
32.7	333	109	224	35.7
32.7	330	108	222	35.3
32.7	327	107	220	35.0
32.7	324	106	218	34.7
32.7	321	105	216	34.3
32.7	318	104	214	34.0
32.7	315	103	212	33.7
32.7	312	102	210	33.3
32.7	309	101	208	33.0
32.7	306	100	206	32.7
32.7	303	99	204	32.3
32.7	300	98	202	32.0
32.7	297	97	200	31.7
32.7	294	96	198	31.3
32.7	284	93	191	30.5
32.7	281	92	189	30.1
32.7	278	91	187	29.8
32.7	275	90	185	29.5
32.7	272	89	183	29.1
32.7	269	88	181	28.8
32.7	266	87	179	28.5
32.7	263	86	177	28.1
32.7	260	85	175	27.8
32.7	257	84	173	27.5
32.7	254	83	171	27.1
32.7	251	82	169	26.8
32.7	248	81	167	26.5
32.7	245	80	165	26.1
32.7	226	74	152	24.2
32.7	223	73	150	23.9
32.7	220	72	148	23.6
32.7	217	71	146	23.2
32.7	214	70	144	22.9
32.7	211	69	142	22.6
32.7	208	68	140	22.2
32.7	205	67	138	21.9
32.7	202	66	136	21.6
32.7	199	65	134	21.2
32.7	196	64	132	20.9
32.7	171	56	115	18.3
32.7	168	55	113	18.0
32.7	165	54	111	17.7
32.7	162	53	109	17.3
32.7	159	52	107	17.0
32.7	156	51	105	16.7
32.7	153	50	103	16.3
32.7	150	49	101	16.0
32.7	147	48	99	15.7
32.7	113	37	76	12.1

%E	M1	M2	DM	M*
32.7	110	36	74	11.8
32.7	107	35	72	11.4
32.7	104	34	70	11.1
32.7	101	33	68	10.8
32.7	98	32	66	10.4
32.7	55	18	37	5.9
32.7	52	17	35	5.6
32.7	49	16	33	5.2
32.6	500	163	337	53.1
32.6	497	162	335	52.8
32.6	494	161	333	52.5
32.6	491	160	331	52.1
32.6	488	159	329	51.8
32.6	487	159	328	51.9
32.6	485	158	327	51.5
32.6	484	158	326	51.6
32.6	482	157	325	51.1
32.6	481	157	324	51.2
32.6	479	156	323	50.8
32.6	478	156	322	50.9
32.6	476	155	321	50.5
32.6	475	155	320	50.6
32.6	473	154	319	50.1
32.6	472	154	318	50.2
32.6	470	153	317	49.8
32.6	469	153	316	49.9
32.6	466	152	314	49.6
32.6	463	151	312	49.2
32.6	460	150	310	48.9
32.6	457	149	308	48.6
32.6	454	148	306	48.2
32.6	451	147	304	47.9
32.6	448	146	302	47.6
32.6	445	145	300	47.2
32.6	442	144	298	46.9
32.6	439	143	296	46.6
32.6	438	143	295	46.7
32.6	436	142	294	46.2
32.6	435	142	293	46.4
32.6	433	141	292	45.9
32.6	432	141	291	46.0
32.6	430	140	290	45.6
32.6	429	140	289	45.7
32.6	427	139	288	45.2
32.6	426	139	287	45.4
32.6	423	138	285	45.0
32.6	420	137	283	44.7
32.6	417	136	281	44.4
32.6	414	135	279	44.0
32.6	411	134	277	43.7

%E	M1	M2	DM	M*
32.6	408	133	275	43.4
32.6	405	132	273	43.0
32.6	402	131	271	42.7
32.6	399	130	269	42.4
32.6	396	129	267	42.0
32.6	393	128	265	41.7
32.6	390	127	263	41.4
32.6	389	127	262	41.5
32.6	387	126	261	41.0
32.6	386	126	260	41.1
32.6	384	125	259	40.7
32.6	383	125	258	40.8
32.6	380	124	256	40.5
32.6	377	123	254	40.1
32.6	374	122	252	39.8
32.6	371	121	250	39.5
32.6	368	120	248	39.1
32.6	365	119	246	38.8
32.6	362	118	244	38.5
32.6	359	117	242	38.1
32.6	356	116	240	37.8
32.6	353	115	238	37.5
32.6	350	114	236	37.1
32.6	347	113	234	36.8
32.6	344	112	232	36.5
32.6	341	111	230	36.1
32.6	340	111	229	36.2
32.6	337	110	227	35.9
32.6	334	109	225	35.6
32.6	331	108	223	35.2
32.6	328	107	221	34.9
32.6	325	106	219	34.6
32.6	322	105	217	34.2
32.6	319	104	215	33.9
32.6	316	103	213	33.6
32.6	313	102	211	33.2
32.6	310	101	209	32.9
32.6	307	100	207	32.6
32.6	304	99	205	32.2
32.6	301	98	203	31.9
32.6	298	97	201	31.6
32.6	291	95	196	31.0
32.6	288	94	194	30.7
32.6	285	93	192	30.3
32.6	282	92	190	30.0
32.6	279	91	188	29.7
32.6	276	90	186	29.3
32.6	273	89	184	29.0
32.6	270	88	182	28.7
32.6	267	87	180	28.3

%E	M1	M2	DM	M*
32.6	264	86	178	28.0
32.6	261	85	176	27.7
32.6	258	84	174	27.3
32.6	242	79	163	25.8
32.6	239	78	161	25.5
32.6	236	77	159	25.1
32.6	233	76	157	24.8
32.6	230	75	155	24.5
32.6	227	74	153	24.1
32.6	224	73	151	23.8
32.6	221	72	149	23.5
32.6	218	71	147	23.1
32.6	215	70	145	22.8
32.6	193	63	130	20.6
32.6	190	62	128	20.2
32.6	187	61	126	19.9
32.6	184	60	124	19.6
32.6	181	59	122	19.2
32.6	178	58	120	18.9
32.6	175	57	118	18.6
32.6	172	56	116	18.2
32.6	144	47	97	15.3
32.6	141	46	95	15.0
32.6	138	45	93	14.7
32.6	135	44	91	14.3
32.6	132	43	89	14.0
32.6	129	42	87	13.7
32.6	95	31	64	10.1
32.6	92	30	62	9.8
32.6	89	29	60	9.4
32.6	86	28	58	9.1
32.6	46	15	31	4.9
32.6	43	14	29	4.6
32.5	499	162	337	52.6
32.5	498	162	336	52.7
32.5	496	161	335	52.3
32.5	495	161	334	52.4
32.5	493	160	333	51.9
32.5	492	160	332	52.0
32.5	489	159	330	51.7
32.5	486	158	328	51.4
32.5	483	157	326	51.0
32.5	480	156	324	50.7
32.5	477	155	322	50.4
32.5	474	154	320	50.0
32.5	471	153	318	49.7
32.5	468	152	316	49.4
32.5	467	152	315	49.5
32.5	465	151	314	49.0
32.5	464	151	313	49.1

%E	M1	M2	DM	M*
32.5	462	150	312	48.7
32.5	461	150	311	48.8
32.5	459	149	310	48.4
32.5	458	149	309	48.5
32.5	456	148	308	48.0
32.5	455	148	307	48.1
32.5	453	147	306	47.7
32.5	452	147	305	47.8
32.5	449	146	303	47.5
32.5	446	145	301	47.1
32.5	443	144	299	46.8
32.5	440	143	297	46.5
32.5	437	142	295	46.1
32.5	434	141	293	45.8
32.5	431	140	291	45.5
32.5	428	139	289	45.1
32.5	425	138	287	44.8
32.5	424	138	286	44.9
32.5	422	137	285	44.5
32.5	421	137	284	44.6
32.5	419	136	283	44.1
32.5	418	136	282	44.2
32.5	416	135	281	43.8
32.5	415	135	280	43.9
32.5	412	134	278	43.6
32.5	409	133	276	43.2
32.5	406	132	274	42.9
32.5	403	131	272	42.6
32.5	400	130	270	42.3
32.5	397	129	268	41.9
32.5	394	128	266	41.6
32.5	391	127	264	41.3
32.5	388	126	262	40.9
32.5	385	125	260	40.6
32.5	382	124	258	40.3
32.5	381	124	257	40.4
32.5	379	123	256	39.9
32.5	378	123	255	40.0
32.5	375	122	253	39.7
32.5	372	121	251	39.4
32.5	369	120	249	39.0
32.5	366	119	247	38.7
32.5	363	118	245	38.4
32.5	360	117	243	38.0
32.5	357	116	241	37.7
32.5	354	115	239	37.4
32.5	351	114	237	37.0
32.5	348	113	235	36.7
32.5	345	112	233	36.4
32.5	342	111	231	36.0

%E	M1	M2	DM	M*
32.5	338	110	228	35.8
32.5	335	109	226	35.5
32.5	332	108	224	35.1
32.5	329	107	222	34.8
32.5	326	106	220	34.5
32.5	323	105	218	34.1
32.5	320	104	216	33.8
32.5	317	103	214	33.5
32.5	314	102	212	33.1
32.5	311	101	210	32.8
32.5	308	100	208	32.5
32.5	305	99	206	32.1
32.5	302	98	204	31.8
32.5	295	96	199	31.2
32.5	292	95	197	30.9
32.5	289	94	195	30.6
32.5	286	93	193	30.2
32.5	283	92	191	29.9
32.5	280	91	189	29.6
32.5	277	90	187	29.2
32.5	274	89	185	28.9
32.5	271	88	183	28.6
32.5	268	87	181	28.2
32.5	265	86	179	27.9
32.5	255	83	172	27.0
32.5	252	82	170	26.7
32.5	249	81	168	26.3
32.5	246	80	166	26.0
32.5	243	79	164	25.7
32.5	240	78	162	25.3
32.5	237	77	160	25.0
32.5	234	76	158	24.7
32.5	231	75	156	24.4
32.5	228	74	154	24.0
32.5	212	69	143	22.5
32.5	209	68	141	22.1
32.5	206	67	139	21.8
32.5	203	66	137	21.5
32.5	200	65	135	21.1
32.5	197	64	133	20.8
32.5	194	63	131	20.5
32.5	191	62	129	20.1
32.5	169	55	114	17.9
32.5	166	54	112	17.6
32.5	163	53	110	17.2
32.5	160	52	108	16.9
32.5	157	51	106	16.6
32.5	154	50	104	16.2
32.5	151	49	102	15.9
32.5	126	41	85	13.3

%E	M1	M2	DM	M*
32.5	123	40	83	13.0
32.5	120	39	81	12.7
32.5	117	38	79	12.3
32.5	114	37	77	12.0
32.5	83	27	56	8.8
32.5	80	26	54	8.4
32.5	77	25	52	8.1
32.5	40	13	27	4.2
32.4	500	162	338	52.5
32.4	497	161	336	52.2
32.4	494	160	334	51.8
32.4	491	159	332	51.5
32.4	490	159	331	51.6
32.4	488	158	330	51.2
32.4	487	158	329	51.3
32.4	485	157	328	50.8
32.4	484	157	327	50.9
32.4	482	156	326	50.5
32.4	481	156	325	50.6
32.4	479	155	324	50.2
32.4	478	155	323	50.3
32.4	476	154	322	49.8
32.4	475	154	321	49.9
32.4	472	153	319	49.6
32.4	469	152	317	49.3
32.4	466	151	315	48.9
32.4	463	150	313	48.6
32.4	460	149	311	48.3
32.4	457	148	309	47.9
32.4	454	147	307	47.6
32.4	451	146	305	47.3
32.4	450	146	304	47.4
32.4	448	145	303	46.9
32.4	447	145	302	47.0
32.4	445	144	301	46.6
32.4	444	144	300	46.7
32.4	442	143	299	46.3
32.4	441	143	298	46.4
32.4	438	142	296	46.0
32.4	435	141	294	45.7
32.4	432	140	292	45.4
32.4	429	139	290	45.0
32.4	426	138	288	44.7
32.4	423	137	286	44.4
32.4	420	136	284	44.0
32.4	417	135	282	43.7
32.4	414	134	280	43.4
32.4	413	134	279	43.5
32.4	411	133	278	43.0
32.4	410	133	277	43.1

%E	M1	M2	DM	M*
32.4	408	132	276	42.7
32.4	407	132	275	42.8
32.4	404	131	273	42.5
32.4	401	130	271	42.1
32.4	398	129	269	41.8
32.4	395	128	267	41.5
32.4	392	127	265	41.1
32.4	389	126	263	40.8
32.4	386	125	261	40.5
32.4	383	124	259	40.1
32.4	380	123	257	39.8
32.4	377	122	255	39.5
32.4	376	122	254	39.6
32.4	374	121	253	39.1
32.4	373	121	252	39.3
32.4	370	120	250	38.9
32.4	367	119	248	38.6
32.4	364	118	246	38.3
32.4	361	117	244	37.9
32.4	358	116	242	37.6
32.4	355	115	240	37.3
32.4	352	114	238	36.9
32.4	349	113	236	36.6
32.4	346	112	234	36.3
32.4	343	111	232	35.9
32.4	340	110	230	35.6
32.4	339	110	229	35.7
32.4	336	109	227	35.4
32.4	333	108	225	35.0
32.4	330	107	223	34.7
32.4	327	106	221	34.4
32.4	324	105	219	34.0
32.4	321	104	217	33.7
32.4	318	103	215	33.4
32.4	315	102	213	33.0
32.4	312	101	211	32.7
32.4	309	100	209	32.4
32.4	306	99	207	32.0
32.4	299	97	202	31.5
32.4	296	96	200	31.1
32.4	293	95	198	30.8
32.4	290	94	196	30.5
32.4	287	93	194	30.1
32.4	284	92	192	29.8
32.4	281	91	190	29.5
32.4	278	90	188	29.1
32.4	275	89	186	28.8
32.4	272	88	184	28.5
32.4	262	85	177	27.6
32.4	259	84	175	27.2

%E	M1	M2	DM	M*
32.4	256	83	173	26.9
32.4	253	82	171	26.6
32.4	250	81	169	26.2
32.4	247	80	167	25.9
32.4	244	79	165	25.6
32.4	241	78	163	25.2
32.4	238	77	161	24.9
32.4	225	73	152	23.7
32.4	222	72	150	23.4
32.4	219	71	148	23.0
32.4	216	70	146	22.7
32.4	213	69	144	22.4
32.4	210	68	142	22.0
32.4	207	67	140	21.7
32.4	204	66	138	21.4
32.4	188	61	127	19.8
32.4	185	60	125	19.5
32.4	182	59	123	19.1
32.4	179	58	121	18.8
32.4	176	57	119	18.5
32.4	173	56	117	18.1
32.4	170	55	115	17.8
32.4	148	48	100	15.6
32.4	145	47	98	15.2
32.4	142	46	96	14.9
32.4	139	45	94	14.6
32.4	136	44	92	14.2
32.4	111	36	75	11.7
32.4	108	35	73	11.3
32.4	105	34	71	11.0
32.4	102	33	69	10.7
32.4	74	24	50	7.8
32.4	71	23	48	7.5
32.4	68	22	46	7.1
32.4	37	12	25	3.9
32.4	34	11	23	3.6
32.3	499	161	338	51.9
32.3	498	161	337	52.1
32.3	496	160	336	51.6
32.3	495	160	335	51.7
32.3	493	159	334	51.3
32.3	492	159	333	51.4
32.3	489	158	331	51.1
32.3	486	157	329	50.7
32.3	483	156	327	50.4
32.3	480	155	325	50.1
32.3	477	154	323	49.7
32.3	474	153	321	49.4
32.3	473	153	320	49.5
32.3	471	152	319	49.1

%E	M1	M2	DM	M*
32.3	470	152	318	49.2
32.3	468	151	317	48.7
32.3	467	151	316	48.8
32.3	465	150	315	48.4
32.3	464	150	314	48.5
32.3	462	149	313	48.1
32.3	461	149	312	48.2
32.3	458	148	310	47.8
32.3	455	147	308	47.5
32.3	452	146	306	47.2
32.3	449	145	304	46.8
32.3	446	144	302	46.5
32.3	443	143	300	46.2
32.3	440	142	298	45.8
32.3	439	142	297	45.9
32.3	437	141	296	45.5
32.3	436	141	295	45.6
32.3	434	140	294	45.2
32.3	433	140	293	45.3
32.3	431	139	292	44.8
32.3	430	139	291	44.9
32.3	427	138	289	44.6
32.3	424	137	287	44.3
32.3	421	136	285	43.9
32.3	418	135	283	43.6
32.3	415	134	281	43.3
32.3	412	133	279	42.9
32.3	409	132	277	42.6
32.3	406	131	275	42.3
32.3	405	131	274	42.4
32.3	403	130	273	41.9
32.3	402	130	272	42.0
32.3	400	129	271	41.6
32.3	399	129	270	41.7
32.3	396	128	268	41.4
32.3	393	127	266	41.0
32.3	390	126	264	40.7
32.3	387	125	262	40.4
32.3	384	124	260	40.0
32.3	381	123	258	39.7
32.3	378	122	256	39.4
32.3	375	121	254	39.0
32.3	372	120	252	38.7
32.3	371	120	251	38.8
32.3	368	119	249	38.5
32.3	365	118	247	38.1
32.3	362	117	245	37.8
32.3	359	116	243	37.5
32.3	356	115	241	37.1
32.3	353	114	239	36.8

%E	M1	M2	DM	M*
32.3	350	113	237	36.5
32.3	347	112	235	36.1
32.3	344	111	233	35.8
32.3	341	110	231	35.5
32.3	337	109	228	35.3
32.3	334	108	226	34.9
32.3	331	107	224	34.6
32.3	328	106	222	34.3
32.3	325	105	220	33.9
32.3	322	104	218	33.6
32.3	319	103	216	33.3
32.3	316	102	214	32.9
32.3	313	101	212	32.6
32.3	310	100	210	32.3
32.3	303	98	205	31.7
32.3	300	97	203	31.4
32.3	297	96	201	31.0
32.3	294	95	199	30.7
32.3	291	94	197	30.4
32.3	288	93	195	30.0
32.3	285	92	193	29.7
32.3	282	91	191	29.4
32.3	279	90	189	29.0
32.3	269	87	182	28.1
32.3	266	86	180	27.8
32.3	263	85	178	27.5
32.3	260	84	176	27.1
32.3	257	83	174	26.8
32.3	254	82	172	26.5
32.3	251	81	170	26.1
32.3	248	80	168	25.8
32.3	235	76	159	24.6
32.3	232	75	157	24.2
32.3	229	74	155	23.9
32.3	226	73	153	23.6
32.3	223	72	151	23.2
32.3	220	71	149	22.9
32.3	217	70	147	22.6
32.3	201	65	136	21.0
32.3	198	64	134	20.7
32.3	195	63	132	20.4
32.3	192	62	130	20.0
32.3	189	61	128	19.7
32.3	186	60	126	19.4
32.3	167	54	113	17.5
32.3	164	53	111	17.1
32.3	161	52	109	16.8
32.3	158	51	107	16.5
32.3	155	50	105	16.1
32.3	133	43	90	13.9

%E	M1	M2	DM	M*
32.3	130	42	88	13.6
32.3	127	41	86	13.2
32.3	124	40	84	12.9
32.3	99	32	67	10.3
32.3	96	31	65	10.0
32.3	93	30	63	9.7
32.3	65	21	44	6.8
32.3	62	20	42	6.5
32.3	31	10	21	3.2
32.2	500	161	339	51.8
32.2	497	160	337	51.5
32.2	494	159	335	51.2
32.2	491	158	333	50.8
32.2	490	158	332	50.9
32.2	488	157	331	50.5
32.2	487	157	330	50.6
32.2	485	156	329	50.2
32.2	484	156	328	50.3
32.2	482	155	327	49.8
32.2	481	155	326	49.9
32.2	479	154	325	49.5
32.2	478	154	324	49.6
32.2	475	153	322	49.3
32.2	472	152	320	48.9
32.2	469	151	318	48.6
32.2	466	150	316	48.3
32.2	463	149	314	48.0
32.2	460	148	312	47.6
32.2	459	148	311	47.7
32.2	457	147	310	47.3
32.2	456	147	309	47.4
32.2	454	146	308	47.0
32.2	453	146	307	47.1
32.2	451	145	306	46.6
32.2	450	145	305	46.7
32.2	447	144	303	46.4
32.2	444	143	301	46.1
32.2	441	142	299	45.7
32.2	438	141	297	45.4
32.2	435	140	295	45.1
32.2	432	139	293	44.7
32.2	429	138	291	44.4
32.2	428	138	290	44.5
32.2	426	137	289	44.1
32.2	425	137	288	44.2
32.2	423	136	287	43.7
32.2	422	136	286	43.8
32.2	419	135	284	43.5
32.2	416	134	282	43.2
32.2	413	133	280	42.8

%E	M1	M2	DM	M*
32.2	410	132	278	42.5
32.2	407	131	276	42.2
32.2	404	130	274	41.8
32.2	401	129	272	41.5
32.2	398	128	270	41.2
32.2	397	128	269	41.3
32.2	395	127	268	40.8
32.2	394	127	267	40.9
32.2	391	126	265	40.6
32.2	388	125	263	40.3
32.2	385	124	261	39.9
32.2	382	123	259	39.6
32.2	379	122	257	39.3
32.2	376	121	255	38.9
32.2	373	120	253	38.6
32.2	370	119	251	38.3
32.2	369	119	250	38.4
32.2	367	118	249	37.9
32.2	366	118	248	38.0
32.2	363	117	246	37.7
32.2	360	116	244	37.4
32.2	357	115	242	37.0
32.2	354	114	240	36.7
32.2	351	113	238	36.4
32.2	348	112	236	36.0
32.2	345	111	234	35.7
32.2	342	110	232	35.4
32.2	339	109	230	35.0
32.2	338	109	229	35.2
32.2	335	108	227	34.8
32.2	332	107	225	34.5
32.2	329	106	223	34.2
32.2	326	105	221	33.8
32.2	323	104	219	33.5
32.2	320	103	217	33.2
32.2	317	102	215	32.8
32.2	314	101	213	32.5
32.2	311	100	211	32.2
32.2	307	99	208	31.9
32.2	304	98	206	31.6
32.2	301	97	204	31.3
32.2	298	96	202	30.9
32.2	295	95	200	30.6
32.2	292	94	198	30.3
32.2	289	93	196	29.9
32.2	286	92	194	29.6
32.2	283	91	192	29.3
32.2	276	89	187	28.7
32.2	273	88	185	28.4
32.2	270	87	183	28.0

%E	M1	M2	DM	M*
32.2	267	86	181	27.7
32.2	264	85	179	27.4
32.2	261	84	177	27.0
32.2	258	83	175	26.7
32.2	255	82	173	26.4
32.2	245	79	166	25.5
32.2	242	78	164	25.1
32.2	239	77	162	24.8
32.2	236	76	160	24.5
32.2	233	75	158	24.1
32.2	230	74	156	23.8
32.2	227	73	154	23.5
32.2	214	69	145	22.2
32.2	211	68	143	21.9
32.2	208	67	141	21.6
32.2	205	66	139	21.2
32.2	202	65	137	20.9
32.2	199	64	135	20.6
32.2	183	59	124	19.0
32.2	180	58	122	18.7
32.2	177	57	120	18.4
32.2	174	56	118	18.0
32.2	171	55	116	17.7
32.2	152	49	103	15.8
32.2	149	48	101	15.5
32.2	146	47	99	15.1
32.2	143	46	97	14.8
32.2	121	39	82	12.6
32.2	118	38	80	12.2
32.2	115	37	78	11.9
32.2	90	29	61	9.3
32.2	87	28	59	9.0
32.2	59	19	40	6.1
32.1	499	160	339	51.3
32.1	498	160	338	51.4
32.1	496	159	337	51.0
32.1	495	159	336	51.1
32.1	492	158	334	50.7
32.1	489	157	332	50.4
32.1	486	156	330	50.1
32.1	483	155	328	49.7
32.1	480	154	326	49.4
32.1	477	153	324	49.1
32.1	476	153	323	49.2
32.1	474	152	322	48.7
32.1	473	152	321	48.8
32.1	471	151	320	48.4
32.1	470	151	319	48.5
32.1	468	150	318	48.1
32.1	467	150	317	48.2

%E	M1	M2	DM	M*
32•1	464	149	315	47•8
32•1	461	148	313	47•5
32•1	458	147	311	47•2
32•1	455	146	309	46•8
32•1	452	145	307	46•5
32•1	449	144	305	46•2
32•1	448	144	304	46•3
32•1	446	143	303	45•8
32•1	445	143	302	46•0
32•1	443	142	301	45•5
32•1	442	142	300	45•6
32•1	439	141	298	45•3
32•1	436	140	296	45•0
32•1	433	139	294	44•6
32•1	430	138	292	44•3
32•1	427	137	290	44•0
32•1	424	136	288	43•6
32•1	421	135	286	43•3
32•1	420	135	285	43•4
32•1	418	134	284	43•0
32•1	417	134	283	43•1
32•1	414	133	281	42•7
32•1	411	132	279	42•4
32•1	408	131	277	42•1
32•1	405	130	275	41•7
32•1	402	129	273	41•4
32•1	399	128	271	41•1
32•1	396	127	269	40•7
32•1	393	126	267	40•4
32•1	392	126	266	40•5
32•1	390	125	265	40•1
32•1	389	125	264	40•2
32•1	386	124	262	39•8
32•1	383	123	260	39•5
32•1	380	122	258	39•2
32•1	377	121	256	38•8
32•1	374	120	254	38•5
32•1	371	119	252	38•2
32•1	368	118	250	37•8
32•1	365	117	248	37•5
32•1	364	117	247	37•6
32•1	361	116	245	37•3
32•1	358	115	243	36•9
32•1	355	114	241	36•6
32•1	352	113	239	36•3
32•1	349	112	237	35•9
32•1	346	111	235	35•6
32•1	343	110	233	35•3
32•1	340	109	231	34•9
32•1	336	108	228	34•7

%E	M1	M2	DM	M*
32•1	333	107	226	34•4
32•1	330	106	224	34•0
32•1	327	105	222	33•7
32•1	324	104	220	33•4
32•1	321	103	218	33•0
32•1	318	102	216	32•7
32•1	315	101	214	32•4
32•1	312	100	212	32•1
32•1	308	99	209	31•8
32•1	305	98	207	31•5
32•1	302	97	205	31•2
32•1	299	96	203	30•8
32•1	296	95	201	30•5
32•1	293	94	199	30•2
32•1	290	93	197	29•8
32•1	287	92	195	29•5
32•1	280	90	190	28•9
32•1	277	89	188	28•6
32•1	274	88	186	28•3
32•1	271	87	184	27•9
32•1	268	86	182	27•6
32•1	265	85	180	27•3
32•1	262	84	178	26•9
32•1	252	81	171	26•0
32•1	249	80	169	25•7
32•1	246	79	167	25•4
32•1	243	78	165	25•0
32•1	240	77	163	24•7
32•1	237	76	161	24•4
32•1	234	75	159	24•0
32•1	224	72	152	23•1
32•1	221	71	150	22•8
32•1	218	70	148	22•5
32•1	215	69	146	22•1
32•1	212	68	144	21•8
32•1	209	67	142	21•5
32•1	196	63	133	20•3
32•1	193	62	131	19•9
32•1	190	61	129	19•6
32•1	187	60	127	19•3
32•1	184	59	125	18•9
32•1	168	54	114	17•4
32•1	165	53	112	17•0
32•1	162	52	110	16•7
32•1	159	51	108	16•4
32•1	156	50	106	16•0
32•1	140	45	95	14•5
32•1	137	44	93	14•1
32•1	134	43	91	13•8
32•1	131	42	89	13•5

%E	M1	M2	DM	M*
32•1	112	36	76	11•6
32•1	109	35	74	11•2
32•1	106	34	72	10•9
32•1	84	27	57	8•7
32•1	81	26	55	8•3
32•1	78	25	53	8•0
32•1	56	18	38	5•8
32•1	53	17	36	5•5
32•1	28	9	19	2•9
32•0	500	160	340	51•2
32•0	497	159	338	50•9
32•0	494	158	336	50•5
32•0	493	158	335	50•6
32•0	491	157	334	50•2
32•0	490	157	333	50•3
32•0	488	156	332	49•9
32•0	487	156	331	50•0
32•0	485	155	330	49•5
32•0	484	155	329	49•6
32•0	482	154	328	49•2
32•0	481	154	327	49•3
32•0	478	153	325	49•0
32•0	475	152	323	48•6
32•0	472	151	321	48•3
32•0	469	150	319	48•0
32•0	466	149	317	47•6
32•0	465	149	316	47•7
32•0	463	148	315	47•3
32•0	462	148	314	47•4
32•0	460	147	313	47•0
32•0	459	147	312	47•1
32•0	456	146	310	46•7
32•0	453	145	308	46•4
32•0	450	144	306	46•1
32•0	447	143	304	45•7
32•0	444	142	302	45•4
32•0	441	141	300	45•1
32•0	440	141	299	45•2
32•0	438	140	298	44•7
32•0	437	140	297	44•9
32•0	435	139	296	44•4
32•0	434	139	295	44•5
32•0	431	138	293	44•2
32•0	428	137	291	43•9
32•0	425	136	289	43•5
32•0	422	135	287	43•2
32•0	419	134	285	42•9
32•0	416	133	283	42•5
32•0	415	133	282	42•6
32•0	413	132	281	42•2

%E	M1	M2	DM	M*
32•0	412	132	280	42•3
32•0	410	131	279	41•9
32•0	409	131	278	42•0
32•0	406	130	276	41•6
32•0	403	129	274	41•3
32•0	400	128	272	41•0
32•0	397	127	270	40•6
32•0	394	126	268	40•3
32•0	391	125	266	40•0
32•0	388	124	264	39•6
32•0	387	124	263	39•7
32•0	384	123	261	39•4
32•0	381	122	259	39•1
32•0	378	121	257	38•7
32•0	375	120	255	38•4
32•0	372	119	253	38•1
32•0	369	118	251	37•7
32•0	366	117	249	37•4
32•0	363	116	247	37•1
32•0	362	116	246	37•2
32•0	359	115	244	36•8
32•0	356	114	242	36•5
32•0	353	113	240	36•2
32•0	350	112	238	35•8
32•0	347	111	236	35•5
32•0	344	110	234	35•2
32•0	341	109	232	34•8
32•0	338	108	230	34•5
32•0	337	108	229	34•6
32•0	334	107	227	34•3
32•0	331	106	225	33•9
32•0	328	105	223	33•6
32•0	325	104	221	33•3
32•0	322	103	219	32•9
32•0	319	102	217	32•6
32•0	316	101	215	32•3
32•0	309	99	210	31•7
32•0	306	98	208	31•4
32•0	303	97	206	31•1
32•0	300	96	204	30•7
32•0	297	95	202	30•4
32•0	294	94	200	30•1
32•0	291	93	198	29•7
32•0	284	91	193	29•2
32•0	281	90	191	28•8
32•0	278	89	189	28•5
32•0	275	88	187	28•2
32•0	272	87	185	27•8
32•0	269	86	183	27•5
32•0	266	85	181	27•2

%E	M1	M2	DM	M*
32•0	259	83	176	26•6
32•0	256	82	174	26•3
32•0	253	81	172	25•9
32•0	250	80	170	25•6
32•0	247	79	168	25•3
32•0	244	78	166	24•9
32•0	241	77	164	24•6
32•0	231	74	157	23•7
32•0	228	73	155	23•4
32•0	225	72	153	23•0
32•0	222	71	151	22•7
32•0	219	70	149	22•4
32•0	206	66	140	21•1
32•0	203	65	138	20•8
32•0	200	64	136	20•5
32•0	197	63	134	20•1
32•0	194	62	132	19•8
32•0	181	58	123	18•6
32•0	178	57	121	18•3
32•0	175	56	119	17•9
32•0	172	55	117	17•6
32•0	169	54	115	17•3
32•0	153	49	104	15•7
32•0	150	48	102	15•4
32•0	147	47	100	15•0
32•0	128	41	87	13•1
32•0	125	40	85	12•8
32•0	122	39	83	12•5
32•0	103	33	70	10•6
32•0	100	32	68	10•2
32•0	97	31	66	9•9
32•0	75	24	51	7•7
32•0	50	16	34	5•1
32•0	25	8	17	2•6
31•9	499	159	340	50•7
31•9	498	159	339	50•8
31•9	496	158	338	50•3
31•9	495	158	337	50•4
31•9	492	157	335	50•1
31•9	489	156	333	49•8
31•9	486	155	331	49•4
31•9	483	154	329	49•1
31•9	480	153	327	48•8
31•9	479	153	326	48•9
31•9	477	152	325	48•4
31•9	476	152	324	48•5
31•9	474	151	323	48•1
31•9	473	151	322	48•2
31•9	470	150	320	47•9
31•9	467	149	318	47•5

%E	M1	M2	DM	M*
31.9	464	148	316	47.2
31.9	461	147	314	46.9
31.9	458	146	312	46.5
31.9	457	146	311	46.6
31.9	455	145	310	46.2
31.9	454	145	309	46.3
31.9	452	144	308	45.9
31.9	451	144	307	46.0
31.9	448	143	305	45.6
31.9	445	142	303	45.3
31.9	442	141	301	45.0
31.9	439	140	299	44.6
31.9	436	139	297	44.3
31.9	433	138	295	44.0
31.9	432	138	294	44.1
31.9	430	137	293	43.6
31.9	429	137	292	43.8
31.9	427	136	291	43.3
31.9	426	136	290	43.4
31.9	423	135	288	43.1
31.9	420	134	286	42.8
31.9	417	133	284	42.4
31.9	414	132	282	42.1
31.9	411	131	280	41.8
31.9	408	130	278	41.4
31.9	407	130	277	41.5
31.9	405	129	276	41.1
31.9	404	129	275	41.2
31.9	401	128	273	40.9
31.9	398	127	271	40.5
31.9	395	126	269	40.2
31.9	392	125	267	39.9
31.9	389	124	265	39.5
31.9	386	123	263	39.2
31.9	385	123	262	39.3
31.9	383	122	261	38.9
31.9	382	122	260	39.0
31.9	379	121	258	38.6
31.9	376	120	256	38.3
31.9	373	119	254	38.0
31.9	370	118	252	37.6
31.9	367	117	250	37.3
31.9	364	116	248	37.0
31.9	361	115	246	36.6
31.9	360	115	245	36.7
31.9	357	114	243	36.4
31.9	354	113	241	36.1
31.9	351	112	239	35.7
31.9	348	111	237	35.4
31.9	345	110	235	35.1
31.9	342	109	233	34.7
31.9	339	108	231	34.4
31.9	335	107	228	34.2
31.9	332	106	226	33.8
31.9	329	105	224	33.5
31.9	326	104	222	33.2
31.9	323	103	220	32.8
31.9	320	102	218	32.5
31.9	317	101	216	32.2
31.9	313	100	213	31.9
31.9	310	99	211	31.6
31.9	307	98	209	31.3
31.9	304	97	207	31.0
31.9	301	96	205	30.6
31.9	298	95	203	30.3
31.9	295	94	201	30.0
31.9	288	92	196	29.4
31.9	285	91	194	29.1
31.9	282	90	192	28.7
31.9	279	89	190	28.4
31.9	276	88	188	28.1
31.9	273	87	186	27.7
31.9	270	86	184	27.4
31.9	263	84	179	26.8
31.9	260	83	177	26.5
31.9	257	82	175	26.2
31.9	254	81	173	25.8
31.9	251	80	171	25.5
31.9	248	79	169	25.2
31.9	238	76	162	24.3
31.9	235	75	160	23.9
31.9	232	74	158	23.6
31.9	229	73	156	23.3
31.9	226	72	154	22.9
31.9	216	69	147	22.0
31.9	213	68	145	21.7
31.9	210	67	143	21.4
31.9	207	66	141	21.0
31.9	204	65	139	20.7
31.9	191	61	130	19.5
31.9	188	60	128	19.1
31.9	185	59	126	18.8
31.9	182	58	124	18.5
31.9	166	53	113	16.9
31.9	163	52	111	16.6
31.9	160	51	109	16.3
31.9	144	46	98	14.7
31.9	141	45	96	14.4
31.9	138	44	94	14.0
31.9	135	43	92	13.7
31.9	119	38	81	12.1
31.9	116	37	79	11.8
31.9	113	36	77	11.5
31.9	94	30	64	9.6
31.9	91	29	62	9.2
31.9	72	23	49	7.3
31.9	69	22	47	7.0
31.9	47	15	32	4.8
31.8	500	159	341	50.6
31.8	497	158	339	50.2
31.8	494	157	337	49.9
31.8	493	157	336	50.0
31.8	491	156	335	49.6
31.8	490	156	334	49.7
31.8	488	155	333	49.2
31.8	487	155	332	49.3
31.8	485	154	331	48.9
31.8	484	154	330	49.0
31.8	481	153	328	48.7
31.8	478	152	326	48.3
31.8	475	151	324	48.0
31.8	472	150	322	47.7
31.8	471	150	321	47.8
31.8	469	149	320	47.3
31.8	468	149	319	47.4
31.8	466	148	318	47.0
31.8	465	148	317	47.1
31.8	462	147	315	46.8
31.8	459	146	313	46.4
31.8	456	145	311	46.1
31.8	453	144	309	45.8
31.8	450	143	307	45.4
31.8	449	143	306	45.5
31.8	447	142	305	45.1
31.8	446	142	304	45.2
31.8	444	141	303	44.8
31.8	443	141	302	44.9
31.8	440	140	300	44.5
31.8	437	139	298	44.2
31.8	434	138	296	43.9
31.8	431	137	294	43.5
31.8	428	136	292	43.2
31.8	425	135	290	42.9
31.8	424	135	289	43.0
31.8	422	134	288	42.5
31.8	421	134	287	42.7
31.8	418	133	285	42.3
31.8	415	132	283	42.0
31.8	412	131	281	41.7
31.8	409	130	279	41.3
31.8	406	129	277	41.0
31.8	403	128	275	40.7
31.8	402	128	274	40.8
31.8	400	127	273	40.3
31.8	399	127	272	40.4
31.8	396	126	270	40.1
31.8	393	125	268	39.8
31.8	390	124	266	39.4
31.8	387	123	264	39.1
31.8	384	122	262	38.8
31.8	381	121	260	38.4
31.8	380	121	259	38.5
31.8	377	120	257	38.2
31.8	374	119	255	37.9
31.8	371	118	253	37.5
31.8	368	117	251	37.2
31.8	365	116	249	36.9
31.8	362	115	247	36.5
31.8	359	114	245	36.2
31.8	358	114	244	36.3
31.8	355	113	242	36.0
31.8	352	112	240	35.6
31.8	349	111	238	35.3
31.8	346	110	236	35.0
31.8	343	109	234	34.6
31.8	340	108	232	34.3
31.8	337	107	230	34.0
31.8	336	107	229	34.1
31.8	333	106	227	33.7
31.8	330	105	225	33.4
31.8	327	104	223	33.1
31.8	324	103	221	32.7
31.8	321	102	219	32.4
31.8	318	101	217	32.1
31.8	314	100	214	31.8
31.8	311	99	212	31.5
31.8	308	98	210	31.2
31.8	305	97	208	30.8
31.8	302	96	206	30.5
31.8	299	95	204	30.2
31.8	296	94	202	29.9
31.8	292	93	199	29.6
31.8	289	92	197	29.3
31.8	286	91	195	29.0
31.8	283	90	193	28.6
31.8	280	89	191	28.3
31.8	277	88	189	28.0
31.8	274	87	187	27.6
31.8	267	85	182	27.1
31.8	264	84	180	26.7
31.8	261	83	178	26.4
31.8	258	82	176	26.1
31.8	255	81	174	25.7
31.8	245	78	167	24.8
31.8	242	77	165	24.5
31.8	239	76	163	24.2
31.8	236	75	161	23.8
31.8	233	74	159	23.5
31.8	223	71	152	22.6
31.8	220	70	150	22.3
31.8	217	69	148	21.9
31.8	214	68	146	21.6
31.8	211	67	144	21.3
31.8	201	64	137	20.4
31.8	198	63	135	20.0
31.8	195	62	133	19.7
31.8	192	61	131	19.4
31.8	179	57	122	18.2
31.8	176	56	120	17.8
31.8	173	55	118	17.5
31.8	170	54	116	17.2
31.8	157	50	107	15.9
31.8	154	49	105	15.6
31.8	151	48	103	15.3
31.8	148	47	101	14.9
31.8	132	42	90	13.4
31.8	129	41	88	13.0
31.8	110	35	75	11.1
31.8	107	34	73	10.8
31.8	88	28	60	8.9
31.8	85	27	58	8.6
31.8	66	21	45	6.7
31.8	44	14	30	4.5
31.8	22	7	15	2.2
31.7	499	158	341	50.0
31.7	498	158	340	50.1
31.7	496	157	339	49.7
31.7	495	157	338	49.8
31.7	492	156	336	49.5
31.7	489	155	334	49.1
31.7	486	154	332	48.8
31.7	483	153	330	48.5
31.7	482	153	329	48.6
31.7	480	152	328	48.1
31.7	479	152	327	48.2
31.7	477	151	326	47.8
31.7	476	151	325	47.9
31.7	473	150	323	47.6
31.7	470	149	321	47.2
31.7	467	148	319	46.9

%E	M1	M2	DM	M*
31.7	464	147	317	46.6
31.7	463	147	316	46.7
31.7	461	146	315	46.2
31.7	460	146	314	46.3
31.7	458	145	313	45.9
31.7	457	145	312	46.0
31.7	454	144	310	45.7
31.7	451	143	308	45.3
31.7	448	142	306	45.0
31.7	445	141	304	44.7
31.7	442	140	302	44.3
31.7	441	140	301	44.4
31.7	439	139	300	44.0
31.7	438	139	299	44.1
31.7	436	138	298	43.7
31.7	435	138	297	43.8
31.7	432	137	295	43.4
31.7	429	136	293	43.1
31.7	426	135	291	42.8
31.7	423	134	289	42.4
31.7	420	133	287	42.1
31.7	419	133	286	42.2
31.7	417	132	285	41.8
31.7	416	132	284	41.9
31.7	413	131	282	41.6
31.7	410	130	280	41.2
31.7	407	129	278	40.9
31.7	404	128	276	40.6
31.7	401	127	274	40.2
31.7	398	126	272	39.9
31.7	397	126	271	40.0
31.7	394	125	269	39.7
31.7	391	124	267	39.3
31.7	388	123	265	39.0
31.7	385	122	263	38.7
31.7	382	121	261	38.3
31.7	379	120	259	38.0
31.7	378	120	258	38.1
31.7	375	119	256	37.8
31.7	372	118	254	37.4
31.7	369	117	252	37.1
31.7	366	116	250	36.8
31.7	363	115	248	36.4
31.7	360	114	246	36.1
31.7	357	113	244	35.8
31.7	356	113	243	35.9
31.7	353	112	241	35.5
31.7	350	111	239	35.2
31.7	347	110	237	34.9
31.7	344	109	235	34.5

%E	M1	M2	DM	M*
31.7	341	108	233	34.2
31.7	338	107	231	33.9
31.7	334	106	228	33.6
31.7	331	105	226	33.3
31.7	328	104	224	33.0
31.7	325	103	222	32.6
31.7	322	102	220	32.3
31.7	319	101	218	32.0
31.7	315	100	215	31.7
31.7	312	99	213	31.4
31.7	309	98	211	31.1
31.7	306	97	209	30.7
31.7	303	96	207	30.4
31.7	300	95	205	30.1
31.7	297	94	203	29.8
31.7	293	93	200	29.5
31.7	290	92	198	29.2
31.7	287	91	196	28.9
31.7	284	90	194	28.5
31.7	281	89	192	28.2
31.7	278	88	190	27.9
31.7	271	86	185	27.3
31.7	268	85	183	27.0
31.7	265	84	181	26.6
31.7	262	83	179	26.3
31.7	259	82	177	26.0
31.7	252	80	172	25.4
31.7	249	79	170	25.1
31.7	246	78	168	24.7
31.7	243	77	166	24.4
31.7	240	76	164	24.1
31.7	230	73	157	23.2
31.7	227	72	155	22.8
31.7	224	71	153	22.5
31.7	221	70	151	22.2
31.7	218	69	149	21.8
31.7	208	66	142	20.9
31.7	205	65	140	20.6
31.7	202	64	138	20.3
31.7	199	63	136	19.9
31.7	189	60	129	19.0
31.7	186	59	127	18.7
31.7	183	58	125	18.4
31.7	180	57	123	18.0
31.7	167	53	114	16.8
31.7	164	52	112	16.5
31.7	161	51	110	16.2
31.7	145	46	99	14.6
31.7	142	45	97	14.3
31.7	139	44	95	13.9

%E	M1	M2	DM	M*
31.7	126	40	86	12.7
31.7	123	39	84	12.4
31.7	120	38	82	12.0
31.7	104	33	71	10.5
31.7	101	32	69	10.1
31.7	82	26	56	8.2
31.7	63	20	43	6.3
31.7	60	19	41	6.0
31.7	41	13	28	4.1
31.6	500	158	342	49.9
31.6	497	157	340	49.6
31.6	494	156	338	49.3
31.6	493	156	337	49.4
31.6	491	155	336	48.9
31.6	490	155	335	49.0
31.6	488	154	334	48.6
31.6	487	154	333	48.7
31.6	484	153	331	48.4
31.6	481	152	329	48.0
31.6	478	151	327	47.7
31.6	475	150	325	47.4
31.6	474	150	324	47.5
31.6	472	149	323	47.0
31.6	471	149	322	47.1
31.6	469	148	321	46.7
31.6	468	148	320	46.8
31.6	465	147	318	46.5
31.6	462	146	316	46.1
31.6	459	145	314	45.8
31.6	456	144	312	45.5
31.6	455	144	311	45.6
31.6	453	143	310	45.1
31.6	452	143	309	45.2
31.6	450	142	308	44.8
31.6	449	142	307	44.9
31.6	446	141	305	44.6
31.6	443	140	303	44.2
31.6	440	139	301	43.9
31.6	437	138	299	43.6
31.6	434	137	297	43.2
31.6	433	137	296	43.3
31.6	431	136	295	42.9
31.6	430	136	294	43.0
31.6	427	135	292	42.7
31.6	424	134	290	42.3
31.6	421	133	288	42.0
31.6	418	132	286	41.7
31.6	415	131	284	41.4
31.6	414	131	283	41.5
31.6	412	130	282	41.0

%E	M1	M2	DM	M*
31.6	411	130	281	41.1
31.6	408	129	279	40.8
31.6	405	128	277	40.5
31.6	402	127	275	40.1
31.6	399	126	273	39.8
31.6	396	125	271	39.5
31.6	395	125	270	39.6
31.6	393	124	269	39.1
31.6	392	124	268	39.2
31.6	389	123	266	38.9
31.6	386	122	264	38.6
31.6	383	121	262	38.2
31.6	380	120	260	37.9
31.6	377	119	258	37.6
31.6	376	119	257	37.7
31.6	374	118	256	37.2
31.6	373	118	255	37.3
31.6	370	117	253	37.0
31.6	367	116	251	36.7
31.6	364	115	249	36.3
31.6	361	114	247	36.0
31.6	358	113	245	35.7
31.6	354	112	242	35.4
31.6	351	111	240	35.1
31.6	348	110	238	34.8
31.6	345	109	236	34.4
31.6	342	108	234	34.1
31.6	339	107	232	33.8
31.6	335	106	229	33.5
31.6	332	105	227	33.2
31.6	329	104	225	32.9
31.6	326	103	223	32.5
31.6	323	102	221	32.2
31.6	320	101	219	31.9
31.6	316	100	216	31.6
31.6	313	99	214	31.3
31.6	310	98	212	31.0
31.6	307	97	210	30.6
31.6	304	96	208	30.3
31.6	301	95	206	30.0
31.6	294	93	201	29.4
31.6	291	92	199	29.1
31.6	288	91	197	28.8
31.6	285	90	195	28.4
31.6	282	89	193	28.1
31.6	275	87	188	27.5
31.6	272	86	186	27.2
31.6	269	85	184	26.9
31.6	266	84	182	26.5
31.6	263	83	180	26.2

%E	M1	M2	DM	M*
31.6	256	81	175	25.6
31.6	253	80	173	25.3
31.6	250	79	171	25.0
31.6	247	78	169	24.6
31.6	244	77	167	24.3
31.6	237	75	162	23.7
31.6	234	74	160	23.4
31.6	231	73	158	23.1
31.6	228	72	156	22.7
31.6	225	71	154	22.4
31.6	215	68	147	21.5
31.6	212	67	145	21.2
31.6	209	66	143	20.8
31.6	206	65	141	20.5
31.6	196	62	134	19.6
31.6	193	61	132	19.3
31.6	190	60	130	18.9
31.6	187	59	128	18.6
31.6	177	56	121	17.7
31.6	174	55	119	17.4
31.6	171	54	117	17.1
31.6	158	50	108	15.8
31.6	155	49	106	15.5
31.6	152	48	104	15.2
31.6	136	43	93	13.6
31.6	133	42	91	13.3
31.6	117	37	80	11.7
31.6	114	36	78	11.4
31.6	98	31	67	9.8
31.6	95	30	65	9.5
31.6	79	25	54	7.9
31.6	76	24	52	7.6
31.6	57	18	39	5.7
31.6	38	12	26	3.8
31.6	19	6	13	1.9
31.5	499	157	342	49.4
31.5	498	157	341	49.5
31.5	496	156	340	49.1
31.5	495	156	339	49.2
31.5	492	155	337	48.8
31.5	489	154	335	48.5
31.5	486	153	333	48.2
31.5	485	153	332	48.3
31.5	483	152	331	47.8
31.5	482	152	330	47.9
31.5	480	151	329	47.5
31.5	479	151	328	47.6
31.5	476	150	326	47.3
31.5	473	149	324	46.9
31.5	470	148	322	46.6

%E	M1	M2	DM	M*
31.5	467	147	320	46.3
31.5	466	147	319	46.4
31.5	464	146	318	45.9
31.5	463	146	317	46.0
31.5	461	145	316	45.6
31.5	460	145	315	45.7
31.5	457	144	313	45.4
31.5	454	143	311	45.0
31.5	451	142	309	44.7
31.5	448	141	307	44.4
31.5	447	141	306	44.5
31.5	445	140	305	44.0
31.5	444	140	304	44.1
31.5	441	139	302	43.8
31.5	438	138	300	43.5
31.5	435	137	298	43.1
31.5	432	136	296	42.8
31.5	429	135	294	42.5
31.5	428	135	293	42.6
31.5	426	134	292	42.2
31.5	425	134	291	42.2
31.5	422	133	289	41.9
31.5	419	132	287	41.6
31.5	416	131	285	41.3
31.5	413	130	283	40.9
31.5	410	129	281	40.6
31.5	409	129	280	40.7
31.5	407	128	279	40.3
31.5	406	128	278	40.4
31.5	403	127	276	40.0
31.5	400	126	274	39.7
31.5	397	125	272	39.4
31.5	394	124	270	39.0
31.5	391	123	268	38.7
31.5	390	123	267	38.8
31.5	387	122	265	38.5
31.5	384	121	263	38.1
31.5	381	120	261	37.8
31.5	378	119	259	37.5
31.5	375	118	257	37.1
31.5	372	117	255	36.8
31.5	371	117	254	36.9
31.5	368	116	252	36.6
31.5	365	115	250	36.2
31.5	362	114	248	35.9
31.5	359	113	246	35.6
31.5	356	112	244	35.2
31.5	355	112	243	35.3
31.5	352	111	241	35.0
31.5	349	110	239	34.7

%E	M1	M2	DM	M*
31.5	346	109	237	34.3
31.5	343	108	235	34.0
31.5	340	107	233	33.7
31.5	337	106	231	33.3
31.5	336	106	230	33.4
31.5	333	105	228	33.1
31.5	330	104	226	32.8
31.5	327	103	224	32.4
31.5	324	102	222	32.1
31.5	321	101	220	31.8
31.5	317	100	217	31.5
31.5	314	99	215	31.2
31.5	311	98	213	30.9
31.5	308	97	211	30.5
31.5	305	96	209	30.2
31.5	302	95	207	29.9
31.5	298	94	204	29.7
31.5	295	93	202	29.3
31.5	292	92	200	29.0
31.5	289	91	198	28.7
31.5	286	90	196	28.3
31.5	279	88	191	27.8
31.5	276	87	189	27.4
31.5	273	86	187	27.1
31.5	270	85	185	26.8
31.5	267	84	183	26.4
31.5	260	82	178	25.9
31.5	257	81	176	25.5
31.5	254	80	174	25.2
31.5	251	79	172	24.9
31.5	248	78	170	24.5
31.5	241	76	165	24.0
31.5	238	75	163	23.6
31.5	235	74	161	23.3
31.5	232	73	159	23.0
31.5	222	70	152	22.1
31.5	219	69	150	21.7
31.5	216	68	148	21.4
31.5	213	67	146	21.1
31.5	203	64	139	20.2
31.5	200	63	137	19.8
31.5	197	62	135	19.5
31.5	184	58	126	18.3
31.5	181	57	124	18.0
31.5	178	56	122	17.6
31.5	168	53	115	16.7
31.5	165	52	113	16.4
31.5	162	51	111	16.1
31.5	149	47	102	14.8
31.5	146	46	100	14.5

%E	M1	M2	DM	M*
31.5	143	45	98	14.2
31.5	130	41	89	12.9
31.5	127	40	87	12.6
31.5	124	39	85	12.3
31.5	111	35	76	11.0
31.5	108	34	74	10.7
31.5	92	29	63	9.1
31.5	89	28	61	8.8
31.5	73	23	50	7.2
31.5	54	17	37	5.4
31.4	500	157	343	49.3
31.4	497	156	341	49.0
31.4	494	155	339	48.6
31.4	493	155	338	48.7
31.4	491	154	337	48.3
31.4	490	154	336	48.4
31.4	488	153	335	48.0
31.4	487	153	334	48.1
31.4	484	152	332	47.7
31.4	481	151	330	47.4
31.4	478	150	328	47.1
31.4	477	150	327	47.2
31.4	475	149	326	46.7
31.4	474	149	325	46.8
31.4	472	148	324	46.4
31.4	471	148	323	46.5
31.4	468	147	321	46.2
31.4	465	146	319	45.8
31.4	462	145	317	45.5
31.4	459	144	315	45.2
31.4	458	144	314	45.3
31.4	456	143	313	44.8
31.4	455	143	312	44.9
31.4	452	142	310	44.6
31.4	449	141	308	44.3
31.4	446	140	306	43.9
31.4	443	139	304	43.6
31.4	442	139	303	43.7
31.4	440	138	302	43.3
31.4	439	138	301	43.4
31.4	437	137	300	42.9
31.4	436	137	299	43.0
31.4	433	136	297	42.7
31.4	430	135	295	42.4
31.4	427	134	293	42.1
31.4	424	133	291	41.7
31.4	423	133	290	41.8
31.4	421	132	289	41.4
31.4	420	132	288	41.5
31.4	417	131	286	41.2

%E	M1	M2	DM	M*
31.4	414	130	284	40.8
31.4	411	129	282	40.5
31.4	408	128	280	40.2
31.4	405	127	278	39.8
31.4	404	127	277	39.9
31.4	401	126	275	39.6
31.4	398	125	273	39.3
31.4	395	124	271	38.9
31.4	392	123	269	38.6
31.4	389	122	267	38.3
31.4	388	122	266	38.4
31.4	385	121	264	38.0
31.4	382	120	262	37.7
31.4	379	119	260	37.4
31.4	376	118	258	37.0
31.4	373	117	256	36.7
31.4	370	116	254	36.4
31.4	369	116	253	36.5
31.4	366	115	251	36.1
31.4	363	114	249	35.8
31.4	360	113	247	35.5
31.4	357	112	245	35.1
31.4	354	111	243	34.8
31.4	353	111	242	34.9
31.4	350	110	240	34.6
31.4	347	109	238	34.2
31.4	344	108	236	33.9
31.4	341	107	234	33.6
31.4	338	106	232	33.2
31.4	334	105	229	33.0
31.4	331	104	227	32.7
31.4	328	103	225	32.3
31.4	325	102	223	32.0
31.4	322	101	221	31.7
31.4	318	100	218	31.4
31.4	315	99	216	31.1
31.4	312	98	214	30.8
31.4	309	97	212	30.4
31.4	306	96	210	30.1
31.4	303	95	208	29.8
31.4	299	94	205	29.6
31.4	296	93	203	29.2
31.4	293	92	201	28.9
31.4	290	91	199	28.6
31.4	287	90	197	28.2
31.4	283	89	194	28.0
31.4	280	88	192	27.7
31.4	277	87	190	27.3
31.4	274	86	188	27.0
31.4	271	85	186	26.7

%E	M1	M2	DM	M*
31.4	264	83	181	26.1
31.4	261	82	179	25.8
31.4	258	81	177	25.4
31.4	255	80	175	25.1
31.4	245	77	168	24.2
31.4	242	76	166	23.9
31.4	239	75	164	23.5
31.4	236	74	162	23.2
31.4	229	72	157	22.6
31.4	226	71	155	22.3
31.4	223	70	153	22.0
31.4	220	69	151	21.6
31.4	210	66	144	20.7
31.4	207	65	142	20.4
31.4	204	64	140	20.1
31.4	194	61	133	19.2
31.4	191	60	131	18.8
31.4	188	59	129	18.5
31.4	185	58	127	18.2
31.4	175	55	120	17.3
31.4	172	54	118	17.0
31.4	169	53	116	16.6
31.4	159	50	109	15.7
31.4	156	49	107	15.4
31.4	153	48	105	15.1
31.4	140	44	96	13.8
31.4	137	43	94	13.5
31.4	121	38	83	11.9
31.4	118	37	81	11.6
31.4	105	33	72	10.4
31.4	102	32	70	10.0
31.4	86	27	59	8.5
31.4	70	22	48	6.9
31.4	51	16	35	5.0
31.4	35	11	24	3.5
31.3	499	156	343	48.8
31.3	498	156	342	48.9
31.3	496	155	341	48.4
31.3	495	155	340	48.5
31.3	492	154	338	48.2
31.3	489	153	336	47.9
31.3	486	152	334	47.5
31.3	485	152	333	47.6
31.3	483	151	332	47.2
31.3	482	151	331	47.3
31.3	480	150	330	46.9
31.3	479	150	329	47.0
31.3	476	149	327	46.6
31.3	473	148	325	46.3
31.3	470	147	323	46.0

%E	M1	M2	DM	M*
31.3	469	147	322	46.1
31.3	467	146	321	45.6
31.3	466	146	320	45.7
31.3	464	145	319	45.3
31.3	463	145	318	45.4
31.3	460	144	316	45.1
31.3	457	143	314	44.7
31.3	454	142	312	44.4
31.3	453	142	311	44.5
31.3	451	141	310	44.1
31.3	450	141	309	44.2
31.3	448	140	308	43.8
31.3	447	140	307	43.8
31.3	444	139	305	43.5
31.3	441	138	303	43.2
31.3	438	137	301	42.9
31.3	435	136	299	42.5
31.3	434	136	298	42.6
31.3	432	135	297	42.2
31.3	431	135	296	42.3
31.3	428	134	294	42.0
31.3	425	133	292	41.6
31.3	422	132	290	41.3
31.3	419	131	288	41.0
31.3	418	131	287	41.1
31.3	416	130	286	40.6
31.3	415	130	285	40.7
31.3	412	129	283	40.4
31.3	409	128	281	40.1
31.3	406	127	279	39.7
31.3	403	126	277	39.4
31.3	402	126	276	39.5
31.3	400	125	275	39.1
31.3	399	125	274	39.2
31.3	396	124	272	38.8
31.3	393	123	270	38.5
31.3	390	122	268	38.2
31.3	387	121	266	37.8
31.3	386	121	265	37.9
31.3	384	120	264	37.5
31.3	383	120	263	37.6
31.3	380	119	261	37.3
31.3	377	118	259	36.9
31.3	374	117	257	36.6
31.3	371	116	255	36.3
31.3	368	115	253	35.9
31.3	367	115	252	36.0
31.3	364	114	250	35.7
31.3	361	113	248	35.4
31.3	358	112	246	35.0

%E	M1	M2	DM	M*
31.3	355	111	244	34.7
31.3	352	110	242	34.4
31.3	351	110	241	34.5
31.3	348	109	239	34.1
31.3	345	108	237	33.8
31.3	342	107	235	33.5
31.3	339	106	233	33.1
31.3	336	105	231	32.8
31.3	335	105	230	32.9
31.3	332	104	228	32.6
31.3	329	103	226	32.2
31.3	326	102	224	31.9
31.3	323	101	222	31.6
31.3	320	100	220	31.3
31.3	319	100	219	31.3
31.3	316	99	217	31.0
31.3	313	98	215	30.7
31.3	310	97	213	30.4
31.3	307	96	211	30.0
31.3	304	95	209	29.7
31.3	300	94	206	29.5
31.3	297	93	204	29.1
31.3	294	92	202	28.8
31.3	291	91	200	28.5
31.3	288	90	198	28.1
31.3	284	89	195	27.9
31.3	281	88	193	27.6
31.3	278	87	191	27.2
31.3	275	86	189	26.9
31.3	272	85	187	26.6
31.3	268	84	184	26.3
31.3	265	83	182	26.0
31.3	262	82	180	25.7
31.3	259	81	178	25.3
31.3	256	80	176	25.0
31.3	252	79	173	24.8
31.3	249	78	171	24.4
31.3	246	77	169	24.1
31.3	243	76	167	23.8
31.3	240	75	165	23.4
31.3	233	73	160	22.9
31.3	230	72	158	22.5
31.3	227	71	156	22.2
31.3	224	70	154	21.9
31.3	217	68	149	21.3
31.3	214	67	147	21.0
31.3	211	66	145	20.6
31.3	208	65	143	20.3
31.3	201	63	138	19.7
31.3	198	62	136	19.4

%E	M1	M2	DM	M*
31.3	195	61	134	19.1
31.3	192	60	132	18.8
31.3	182	57	125	17.9
31.3	179	56	123	17.5
31.3	176	55	121	17.2
31.3	166	52	114	16.3
31.3	163	51	112	16.0
31.3	160	50	110	15.6
31.3	150	47	103	14.7
31.3	147	46	101	14.4
31.3	144	45	99	14.1
31.3	134	42	92	13.2
31.3	131	41	90	12.8
31.3	128	40	88	12.5
31.3	115	36	79	11.3
31.3	112	35	77	10.9
31.3	99	31	68	9.7
31.3	96	30	66	9.4
31.3	83	26	57	8.1
31.3	80	25	55	7.8
31.3	67	21	46	6.6
31.3	64	20	44	6.3
31.3	48	15	33	4.7
31.3	32	10	22	3.1
31.3	16	5	11	1.6
31.2	500	156	344	48.7
31.2	497	155	342	48.3
31.2	494	154	340	48.0
31.2	493	154	339	48.1
31.2	491	153	338	47.7
31.2	490	153	337	47.8
31.2	487	152	335	47.4
31.2	484	151	333	47.1
31.2	481	150	331	46.8
31.2	478	149	329	46.4
31.2	477	149	328	46.5
31.2	475	148	327	46.1
31.2	474	148	326	46.2
31.2	471	147	324	45.9
31.2	468	146	322	45.5
31.2	465	145	320	45.2
31.2	462	144	318	44.9
31.2	461	144	317	45.0
31.2	459	143	316	44.6
31.2	458	143	315	44.6
31.2	455	142	313	44.3
31.2	452	141	311	44.0
31.2	449	140	309	43.7
31.2	446	139	307	43.3
31.2	445	139	306	43.4

%E	M1	M2	DM	M*
31.2	443	138	305	43.0
31.2	442	138	304	43.1
31.2	439	137	302	42.8
31.2	436	136	300	42.4
31.2	433	135	298	42.1
31.2	430	134	296	41.8
31.2	429	134	295	41.9
31.2	426	133	293	41.5
31.2	423	132	291	41.2
31.2	420	131	289	40.9
31.2	417	130	287	40.5
31.2	414	129	285	40.2
31.2	413	129	284	40.3
31.2	410	128	282	40.0
31.2	407	127	280	39.6
31.2	404	126	278	39.3
31.2	401	125	276	39.0
31.2	398	124	274	38.6
31.2	397	124	273	38.7
31.2	394	123	271	38.4
31.2	391	122	269	38.1
31.2	388	121	267	37.7
31.2	385	120	265	37.4
31.2	382	119	263	37.1
31.2	381	119	262	37.2
31.2	378	118	260	36.8
31.2	375	117	258	36.5
31.2	372	116	256	36.2
31.2	369	115	254	35.8
31.2	365	114	251	35.6
31.2	362	113	249	35.3
31.2	359	112	247	34.9
31.2	356	111	245	34.6
31.2	353	110	243	34.3
31.2	349	109	240	34.0
31.2	346	108	238	33.7
31.2	343	107	236	33.4
31.2	340	106	234	33.0
31.2	337	105	232	32.7
31.2	333	104	229	32.5
31.2	330	103	227	32.1
31.2	327	102	225	31.8
31.2	324	101	223	31.5
31.2	321	100	221	31.2
31.2	317	99	218	30.9
31.2	314	98	216	30.6
31.2	311	97	214	30.3
31.2	308	96	212	29.9
31.2	301	94	207	29.4
31.2	298	93	205	29.0

%E	M1	M2	DM	M*
31.2	295	92	203	28.7
31.2	292	91	201	28.4
31.2	285	89	196	27.8
31.2	282	88	194	27.5
31.2	279	87	192	27.1
31.2	276	86	190	26.8
31.2	269	84	185	26.2
31.2	266	83	183	25.9
31.2	263	82	181	25.6
31.2	260	81	179	25.2
31.2	253	79	174	24.7
31.2	250	78	172	24.3
31.2	247	77	170	24.0
31.2	237	74	163	23.1
31.2	234	73	161	22.8
31.2	231	72	159	22.4
31.2	221	69	152	21.5
31.2	218	68	150	21.2
31.2	215	67	148	20.9
31.2	205	64	141	20.0
31.2	202	63	139	19.6
31.2	199	62	137	19.3
31.2	189	59	130	18.4
31.2	186	58	128	18.1
31.2	173	54	119	16.9
31.2	170	53	117	16.5
31.2	157	49	108	15.3
31.2	154	48	106	15.0
31.2	141	44	97	13.7
31.2	138	43	95	13.4
31.2	125	39	86	12.2
31.2	109	34	75	10.6
31.2	93	29	64	9.0
31.2	77	24	53	7.5
31.1	499	155	344	48.1
31.1	498	155	343	48.2
31.1	495	154	341	47.9
31.1	492	153	339	47.6
31.1	489	152	337	47.2
31.1	488	152	336	47.3
31.1	486	151	335	46.9
31.1	485	151	334	47.0
31.1	483	150	333	46.6
31.1	482	150	332	46.7
31.1	479	149	330	46.3
31.1	476	148	328	46.0
31.1	473	147	326	45.7
31.1	472	147	325	45.8
31.1	470	146	324	45.4
31.1	469	146	323	45.4

%E	M1	M2	DM	M*
31.1	466	145	321	45.1
31.1	463	144	319	44.8
31.1	460	143	317	44.5
31.1	457	142	315	44.1
31.1	456	142	314	44.2
31.1	454	141	313	43.8
31.1	453	141	312	43.9
31.1	450	140	310	43.6
31.1	447	139	308	43.2
31.1	444	138	306	42.9
31.1	441	137	304	42.6
31.1	440	137	303	42.7
31.1	438	136	302	42.2
31.1	437	136	301	42.3
31.1	434	135	299	42.0
31.1	431	134	297	41.7
31.1	428	133	295	41.3
31.1	427	133	294	41.4
31.1	425	132	293	41.0
31.1	424	132	292	41.1
31.1	421	131	290	40.8
31.1	418	130	288	40.4
31.1	415	129	286	40.1
31.1	412	128	284	39.8
31.1	411	128	283	39.9
31.1	409	127	282	39.4
31.1	408	127	281	39.5
31.1	405	126	279	39.2
31.1	402	125	277	38.9
31.1	399	124	275	38.5
31.1	396	123	273	38.2
31.1	395	123	272	38.3
31.1	392	122	270	38.0
31.1	389	121	268	37.6
31.1	386	120	266	37.3
31.1	383	119	264	37.0
31.1	380	118	262	36.6
31.1	379	118	261	36.7
31.1	376	117	259	36.4
31.1	373	116	257	36.1
31.1	370	115	255	35.7
31.1	367	114	253	35.4
31.1	366	114	252	35.5
31.1	363	113	250	35.2
31.1	360	112	248	34.8
31.1	357	111	246	34.5
31.1	354	110	244	34.2
31.1	351	109	242	33.8
31.1	350	109	241	33.9
31.1	347	108	239	33.6

%E	M1	M2	DM	M*
31.1	344	107	237	33.3
31.1	341	106	235	33.0
31.1	338	105	233	32.6
31.1	334	104	230	32.4
31.1	331	103	228	32.1
31.1	328	102	226	31.7
31.1	325	101	224	31.4
31.1	322	100	222	31.1
31.1	318	99	219	30.8
31.1	315	98	217	30.5
31.1	312	97	215	30.2
31.1	309	96	213	29.8
31.1	305	95	210	29.6
31.1	302	94	208	29.3
31.1	299	93	206	28.9
31.1	296	92	204	28.6
31.1	293	91	202	28.3
31.1	289	90	199	28.0
31.1	286	89	197	27.7
31.1	283	88	195	27.4
31.1	280	87	193	27.0
31.1	273	85	188	26.5
31.1	270	84	186	26.1
31.1	267	83	184	25.8
31.1	264	82	182	25.5
31.1	257	80	177	24.9
31.1	254	79	175	24.6
31.1	251	78	173	24.2
31.1	244	76	168	23.7
31.1	241	75	166	23.3
31.1	238	74	164	23.0
31.1	235	73	162	22.7
31.1	228	71	157	22.1
31.1	225	70	155	21.8
31.1	222	69	153	21.4
31.1	219	68	151	21.1
31.1	212	66	146	20.5
31.1	209	65	144	20.2
31.1	206	64	142	19.9
31.1	196	61	135	19.0
31.1	193	60	133	18.7
31.1	190	59	131	18.3
31.1	183	57	126	17.8
31.1	180	56	124	17.4
31.1	177	55	122	17.1
31.1	167	52	115	16.2
31.1	164	51	113	15.9
31.1	161	50	111	15.5
31.1	151	47	104	14.6
31.1	148	46	102	14.3

%E	M1	M2	DM	M*
31.1	135	42	93	13.1
31.1	132	41	91	12.7
31.1	122	38	84	11.8
31.1	119	37	82	11.5
31.1	106	33	73	10.3
31.1	103	32	71	9.9
31.1	90	28	62	8.7
31.1	74	23	51	7.1
31.1	61	19	42	5.9
31.1	45	14	31	4.4
31.0	500	155	345	48.0
31.0	497	154	343	47.7
31.0	496	154	342	47.8
31.0	494	153	341	47.4
31.0	493	153	340	47.5
31.0	491	152	339	47.1
31.0	490	152	338	47.2
31.0	487	151	336	46.8
31.0	484	150	334	46.5
31.0	481	149	332	46.2
31.0	480	149	331	46.3
31.0	478	148	330	45.8
31.0	477	148	329	45.9
31.0	474	147	327	45.6
31.0	471	146	325	45.3
31.0	468	145	323	44.9
31.0	467	145	322	45.0
31.0	465	144	321	44.6
31.0	464	144	320	44.7
31.0	462	143	319	44.3
31.0	461	143	318	44.4
31.0	458	142	316	44.0
31.0	455	141	314	43.7
31.0	452	140	312	43.4
31.0	451	140	311	43.5
31.0	449	139	310	43.0
31.0	448	139	309	43.1
31.0	445	138	307	42.8
31.0	442	137	305	42.5
31.0	439	136	303	42.1
31.0	436	135	301	41.8
31.0	435	135	300	41.9
31.0	432	134	298	41.6
31.0	429	133	296	41.2
31.0	426	132	294	40.9
31.0	423	131	292	40.6
31.0	422	131	291	40.7
31.0	420	130	290	40.2
31.0	419	130	289	40.3
31.0	416	129	287	40.0

%E	M1	M2	DM	M*
31.0	413	128	285	39.7
31.0	410	127	283	39.3
31.0	407	126	281	39.0
31.0	406	126	280	39.1
31.0	403	125	278	38.8
31.0	400	124	276	38.4
31.0	397	123	274	38.1
31.0	394	122	272	37.8
31.0	393	122	271	37.9
31.0	390	121	269	37.5
31.0	387	120	267	37.2
31.0	384	119	265	36.9
31.0	381	118	263	36.5
31.0	378	117	261	36.2
31.0	377	117	260	36.3
31.0	374	116	258	36.0
31.0	371	115	256	35.6
31.0	368	114	254	35.3
31.0	365	113	252	35.0
31.0	364	113	251	35.1
31.0	361	112	249	34.7
31.0	358	111	247	34.4
31.0	355	110	245	34.1
31.0	352	109	243	33.8
31.0	348	108	240	33.5
31.0	345	107	238	33.2
31.0	342	106	236	32.9
31.0	339	105	234	32.5
31.0	336	104	232	32.2
31.0	335	104	231	32.3
31.0	332	103	229	32.0
31.0	329	102	227	31.6
31.0	326	101	225	31.3
31.0	323	100	223	31.0
31.0	319	99	220	30.7
31.0	316	98	218	30.4
31.0	313	97	216	30.1
31.0	310	96	214	29.7
31.0	306	95	211	29.5
31.0	303	94	209	29.2
31.0	300	93	207	28.8
31.0	297	92	205	28.5
31.0	294	91	203	28.2
31.0	290	90	200	27.9
31.0	287	89	198	27.6
31.0	284	88	196	27.3
31.0	281	87	194	26.9
31.0	277	86	191	26.7
31.0	274	85	189	26.4
31.0	271	84	187	26.0

%E	M1	M2	DM	M*
31.0	268	83	185	25.7
31.0	261	81	180	25.1
31.0	258	80	178	24.8
31.0	255	79	176	24.5
31.0	252	78	174	24.1
31.0	248	77	171	23.9
31.0	245	76	169	23.6
31.0	242	75	167	23.2
31.0	239	74	165	22.9
31.0	232	72	160	22.3
31.0	229	71	158	22.0
31.0	226	70	156	21.7
31.0	216	67	149	20.8
31.0	213	66	147	20.5
31.0	210	65	145	20.1
31.0	203	63	140	19.6
31.0	200	62	138	19.2
31.0	197	61	136	18.9
31.0	187	58	129	18.0
31.0	184	57	127	17.7
31.0	174	54	120	16.8
31.0	171	53	118	16.4
31.0	168	52	116	16.1
31.0	158	49	109	15.2
31.0	155	48	107	14.9
31.0	145	45	100	14.0
31.0	142	44	98	13.6
31.0	129	40	89	12.4
31.0	126	39	87	12.1
31.0	116	36	80	11.2
31.0	113	35	78	10.8
31.0	100	31	69	9.6
31.0	87	27	60	8.4
31.0	84	26	58	8.0
31.0	71	22	49	6.8
31.0	58	18	40	5.6
31.0	42	13	29	4.0
31.0	29	9	20	2.8
30.9	499	154	345	47.5
30.9	498	154	344	47.6
30.9	495	153	342	47.3
30.9	492	152	340	47.0
30.9	489	151	338	46.6
30.9	488	151	337	46.7
30.9	486	150	336	46.3
30.9	485	150	335	46.4
30.9	482	149	333	46.1
30.9	479	148	331	45.7
30.9	476	147	329	45.4
30.9	475	147	328	45.5

%E	M1	M2	DM	M*
30.9	473	146	327	45.1
30.9	472	146	326	45.2
30.9	470	145	325	44.7
30.9	469	145	324	44.8
30.9	466	144	322	44.5
30.9	463	143	320	44.2
30.9	460	142	318	43.8
30.9	459	142	317	43.9
30.9	457	141	316	43.5
30.9	456	141	315	43.6
30.9	453	140	313	43.3
30.9	450	139	311	42.9
30.9	447	138	309	42.6
30.9	446	138	308	42.7
30.9	444	137	307	42.3
30.9	443	137	306	42.4
30.9	440	136	304	42.0
30.9	437	135	302	41.7
30.9	434	134	300	41.4
30.9	433	134	299	41.5
30.9	431	133	298	41.0
30.9	430	133	297	41.1
30.9	427	132	295	40.8
30.9	424	131	293	40.5
30.9	421	130	291	40.1
30.9	418	129	289	39.8
30.9	417	129	288	39.9
30.9	414	128	286	39.6
30.9	411	127	284	39.2
30.9	408	126	282	38.9
30.9	405	125	280	38.6
30.9	404	125	279	38.7
30.9	401	124	277	38.3
30.9	398	123	275	38.0
30.9	395	122	273	37.7
30.9	392	121	271	37.3
30.9	391	121	270	37.4
30.9	388	120	268	37.1
30.9	385	119	266	36.8
30.9	382	118	264	36.5
30.9	379	117	262	36.1
30.9	376	116	260	35.8
30.9	375	116	259	35.9
30.9	372	115	257	35.6
30.9	369	114	255	35.2
30.9	366	113	253	34.9
30.9	363	112	251	34.6
30.9	362	112	250	34.7
30.9	359	111	248	34.3
30.9	356	110	246	34.0

%E	M1	M2	DM	M*
30.9	353	109	244	33.7
30.9	350	108	242	33.3
30.9	349	108	241	33.4
30.9	346	107	239	33.1
30.9	343	106	237	32.8
30.9	340	105	235	32.4
30.9	337	104	233	32.1
30.9	333	103	230	31.9
30.9	330	102	228	31.5
30.9	327	101	226	31.2
30.9	324	100	224	30.9
30.9	320	99	221	30.6
30.9	317	98	219	30.3
30.9	314	97	217	30.0
30.9	311	96	215	29.6
30.9	307	95	212	29.4
30.9	304	94	210	29.1
30.9	301	93	208	28.7
30.9	298	92	206	28.4
30.9	291	90	201	27.8
30.9	288	89	199	27.5
30.9	285	88	197	27.2
30.9	282	87	195	26.8
30.9	278	86	192	26.6
30.9	275	85	190	26.3
30.9	272	84	188	25.9
30.9	269	83	186	25.6
30.9	265	82	183	25.4
30.9	262	81	181	25.0
30.9	259	80	179	24.7
30.9	256	79	177	24.4
30.9	249	77	172	23.8
30.9	246	76	170	23.5
30.9	243	75	168	23.1
30.9	236	73	163	22.6
30.9	233	72	161	22.2
30.9	230	71	159	21.9
30.9	223	69	154	21.3
30.9	220	68	152	21.0
30.9	217	67	150	20.7
30.9	207	64	143	19.8
30.9	204	63	141	19.5
30.9	194	60	134	18.6
30.9	191	59	132	18.2
30.9	188	58	130	17.9
30.9	181	56	125	17.3
30.9	178	55	123	17.0
30.9	175	54	121	16.7
30.9	165	51	114	15.8
30.9	162	50	112	15.4

%E	M1	M2	DM	M*
30.9	152	47	105	14.5
30.9	149	46	103	14.2
30.9	139	43	96	13.3
30.9	136	42	94	13.0
30.9	123	38	85	11.7
30.9	110	34	76	10.5
30.9	97	30	67	9.3
30.9	94	29	65	8.9
30.9	81	25	56	7.7
30.9	68	21	47	6.5
30.9	55	17	38	5.3
30.8	500	154	346	47.4
30.8	497	153	344	47.1
30.8	496	153	343	47.2
30.8	494	152	342	46.8
30.8	493	152	341	46.9
30.8	491	151	340	46.4
30.8	490	151	339	46.5
30.8	487	150	337	46.2
30.8	484	149	335	45.9
30.8	483	149	334	46.0
30.8	481	148	333	45.5
30.8	480	148	332	45.6
30.8	478	147	331	45.2
30.8	477	147	330	45.3
30.8	474	146	328	45.0
30.8	471	145	326	44.6
30.8	468	144	324	44.3
30.8	467	144	323	44.4
30.8	465	143	322	44.0
30.8	464	143	321	44.1
30.8	461	142	319	43.7
30.8	458	141	317	43.4
30.8	455	140	315	43.1
30.8	454	140	314	43.2
30.8	452	139	313	42.7
30.8	451	139	312	42.8
30.8	448	138	310	42.5
30.8	445	137	308	42.2
30.8	442	136	306	41.8
30.8	441	136	305	41.9
30.8	439	135	304	41.5
30.8	438	135	303	41.6
30.8	435	134	301	41.3
30.8	432	133	299	40.9
30.8	429	132	297	40.6
30.8	428	132	296	40.7
30.8	426	131	295	40.3
30.8	425	131	294	40.4
30.8	422	130	292	40.0

%E	M1	M2	DM	M*
30.8	419	129	290	39.7
30.8	416	128	288	39.4
30.8	415	128	287	39.5
30.8	413	127	286	39.1
30.8	412	127	285	39.1
30.8	409	126	283	38.8
30.8	406	125	281	38.5
30.8	403	124	279	38.2
30.8	402	124	278	38.2
30.8	400	123	277	37.8
30.8	399	123	276	37.9
30.8	396	122	274	37.6
30.8	393	121	272	37.3
30.8	390	120	270	36.9
30.8	389	120	269	37.0
30.8	386	119	267	36.7
30.8	383	118	265	36.4
30.8	380	117	263	36.0
30.8	377	116	261	35.7
30.8	373	115	258	35.5
30.8	370	114	256	35.1
30.8	367	113	254	34.8
30.8	364	112	252	34.5
30.8	360	111	249	34.2
30.8	357	110	247	33.9
30.8	354	109	245	33.6
30.8	351	108	243	33.2
30.8	347	107	240	33.0
30.8	344	106	238	32.7
30.8	341	105	236	32.3
30.8	338	104	234	32.0
30.8	334	103	231	31.8
30.8	331	102	229	31.4
30.8	328	101	227	31.1
30.8	325	100	225	30.8
30.8	321	99	222	30.5
30.8	318	98	220	30.2
30.8	315	97	218	29.9
30.8	312	96	216	29.5
30.8	308	95	213	29.3
30.8	305	94	211	29.0
30.8	302	93	209	28.6
30.8	299	92	207	28.3
30.8	295	91	204	28.1
30.8	292	90	202	27.7
30.8	289	89	200	27.4
30.8	286	88	198	27.1
30.8	279	86	193	26.5
30.8	276	85	191	26.2
30.8	273	84	189	25.8

%E	M1	M2	DM	M*
30.8	266	82	184	25.3
30.8	263	81	182	24.9
30.8	260	80	180	24.6
30.8	253	78	175	24.0
30.8	250	77	173	23.7
30.8	247	76	171	23.4
30.8	240	74	166	22.8
30.8	237	73	164	22.5
30.8	234	72	162	22.2
30.8	227	70	157	21.6
30.8	224	69	155	21.3
30.8	221	68	153	20.9
30.8	214	66	148	20.4
30.8	211	65	146	20.0
30.8	208	64	144	19.7
30.8	201	62	139	19.1
30.8	198	61	137	18.8
30.8	195	60	135	18.5
30.8	185	57	128	17.6
30.8	182	56	126	17.2
30.8	172	53	119	16.3
30.8	169	52	117	16.0
30.8	159	49	110	15.1
30.8	156	48	108	14.8
30.8	146	45	101	13.9
30.8	143	44	99	13.5
30.8	133	41	92	12.6
30.8	130	40	90	12.3
30.8	120	37	83	11.4
30.8	117	36	81	11.1
30.8	107	33	74	10.2
30.8	104	32	72	9.8
30.8	91	28	63	8.6
30.8	78	24	54	7.4
30.8	65	20	45	6.2
30.8	52	16	36	4.9
30.8	39	12	27	3.7
30.8	26	8	18	2.5
30.8	13	4	9	1.2
30.7	499	153	346	46.9
30.7	498	153	345	47.0
30.7	495	152	343	46.7
30.7	492	151	341	46.3
30.7	489	150	339	46.0
30.7	488	150	338	46.1
30.7	486	149	337	45.7
30.7	485	149	336	45.8
30.7	482	148	334	45.4
30.7	479	147	332	45.1
30.7	476	146	330	44.8

%E	M1	M2	DM	M*	%E	M1	M2	DM	M*	%E	M1	M2	DM	M*	%E	M1	M2	DM	M*	%E	M1	M2	DM	M*
30.7	475	146	329	44.9	30.7	349	107	242	32.8	30.7	166	51	115	15.7	30.6	415	127	288	38.9	30.6	265	81	184	24.8
30.7	473	145	328	44.5	30.7	348	107	241	32.9	30.7	163	50	113	15.3	30.6	412	126	286	38.5	30.6	258	79	179	24.2
30.7	472	145	327	44.5	30.7	345	106	239	32.6	30.7	153	47	106	14.4	30.6	409	125	284	38.2	30.6	255	78	177	23.9
30.7	469	144	325	44.2	30.7	342	105	237	32.2	30.7	150	46	104	14.1	30.6	408	125	283	38.3	30.6	252	77	175	23.5
30.7	466	143	323	43.9	30.7	339	104	235	31.9	30.7	140	43	97	13.2	30.6	405	124	281	38.0	30.6	248	76	172	23.3
30.7	463	142	321	43.6	30.7	336	103	233	31.6	30.7	137	42	95	12.9	30.6	402	123	279	37.6	30.6	245	75	170	23.0
30.7	462	142	320	43.6	30.7	335	103	232	31.7	30.7	127	39	88	12.0	30.6	399	122	277	37.3	30.6	242	74	168	22.6
30.7	460	141	319	43.2	30.7	332	102	230	31.3	30.7	114	35	79	10.7	30.6	396	121	275	37.0	30.6	235	72	163	22.1
30.7	459	141	318	43.3	30.7	329	101	228	31.0	30.7	101	31	70	9.5	30.6	395	121	274	37.1	30.6	232	71	161	21.7
30.7	456	140	316	43.0	30.7	326	100	226	30.7	30.7	88	27	61	8.3	30.6	392	120	272	36.7	30.6	229	70	159	21.4
30.7	453	139	314	42.7	30.7	323	99	224	30.3	30.7	75	23	52	7.1	30.6	389	119	270	36.4	30.6	222	68	154	20.8
30.7	450	138	312	42.3	30.7	322	99	223	30.4	30.6	500	153	347	46.8	30.6	386	118	268	36.1	30.6	219	67	152	20.5
30.7	449	138	311	42.4	30.7	319	98	221	30.1	30.6	497	152	345	46.5	30.6	385	118	267	36.2	30.6	216	66	150	20.2
30.7	446	137	309	42.1	30.7	316	97	219	29.8	30.6	496	152	344	46.6	30.6	382	117	265	35.8	30.6	209	64	145	19.6
30.7	443	136	307	41.8	30.7	313	96	217	29.4	30.6	494	151	343	46.2	30.6	379	116	263	35.5	30.6	206	63	143	19.3
30.7	440	135	305	41.4	30.7	309	95	214	29.2	30.6	493	151	342	46.2	30.6	376	115	261	35.2	30.6	196	60	136	18.4
30.7	437	134	303	41.1	30.7	306	94	212	28.9	30.6	491	150	341	45.8	30.6	373	114	259	34.8	30.6	193	59	134	18.0
30.7	436	134	302	41.2	30.7	303	93	210	28.5	30.6	490	150	340	45.9	30.6	372	114	258	34.9	30.6	186	57	129	17.5
30.7	433	133	300	40.9	30.7	300	92	208	28.2	30.6	487	149	338	45.6	30.6	369	113	256	34.6	30.6	183	56	127	17.1
30.7	430	132	298	40.5	30.7	296	91	205	28.0	30.6	484	148	336	45.3	30.6	366	112	254	34.3	30.6	180	55	125	16.8
30.7	427	131	296	40.2	30.7	293	90	203	27.6	30.6	483	148	335	45.3	30.6	363	111	252	33.9	30.6	173	53	120	16.2
30.7	424	130	294	39.9	30.7	290	89	201	27.3	30.6	481	147	334	44.9	30.6	360	110	250	33.6	30.6	170	52	118	15.9
30.7	423	130	293	40.0	30.7	287	88	199	27.0	30.6	480	147	333	45.0	30.6	359	110	249	33.7	30.6	160	49	111	15.0
30.7	420	129	291	39.6	30.7	283	87	196	26.7	30.6	477	146	331	44.7	30.6	356	109	247	33.4	30.6	157	48	109	14.7
30.7	417	128	289	39.3	30.7	280	86	194	26.4	30.6	474	145	329	44.4	30.6	353	108	245	33.0	30.6	147	45	102	13.8
30.7	414	127	287	39.0	30.7	277	85	192	26.1	30.6	471	144	327	44.0	30.6	350	107	243	32.7	30.6	144	44	100	13.4
30.7	411	126	285	38.6	30.7	274	84	190	25.8	30.6	470	144	326	44.1	30.6	346	106	240	32.5	30.6	134	41	93	12.5
30.7	410	126	284	38.7	30.7	270	83	187	25.5	30.6	468	143	325	43.7	30.6	343	105	238	32.1	30.6	124	38	86	11.6
30.7	407	125	282	38.4	30.7	267	82	185	25.2	30.6	467	143	324	43.8	30.6	340	104	236	31.8	30.6	121	37	84	11.3
30.7	404	124	280	38.1	30.7	264	81	183	24.9	30.6	464	142	322	43.5	30.6	337	103	234	31.5	30.6	111	34	77	10.4
30.7	401	123	278	37.7	30.7	261	80	181	24.5	30.6	461	141	320	43.1	30.6	333	102	231	31.2	30.6	108	33	75	10.1
30.7	398	122	276	37.4	30.7	257	79	178	24.3	30.6	458	140	318	42.8	30.6	330	101	229	30.9	30.6	98	30	68	9.2
30.7	397	122	275	37.5	30.7	254	78	176	24.0	30.6	457	140	317	42.9	30.6	327	100	227	30.6	30.6	85	26	59	8.0
30.7	394	121	273	37.2	30.7	251	77	174	23.6	30.6	454	139	315	42.6	30.6	324	99	225	30.3	30.6	72	22	50	6.7
30.7	391	120	271	36.8	30.7	244	75	169	23.1	30.6	451	138	313	42.2	30.6	320	98	222	30.0	30.6	62	19	43	5.8
30.7	388	119	269	36.5	30.7	241	74	167	22.7	30.6	448	137	311	41.9	30.6	317	97	220	29.7	30.6	49	15	34	4.6
30.7	387	119	268	36.6	30.7	238	73	165	22.4	30.6	447	137	310	42.0	30.6	314	96	218	29.4	30.6	36	11	25	3.4
30.7	384	118	266	36.3	30.7	231	71	160	21.8	30.6	445	136	309	41.6	30.6	310	95	215	29.1	30.5	499	152	347	46.3
30.7	381	117	264	35.9	30.7	228	70	158	21.5	30.6	444	136	308	41.7	30.6	307	94	213	28.8	30.5	498	152	346	46.4
30.7	378	116	262	35.6	30.7	225	69	156	21.2	30.6	441	135	306	41.3	30.6	304	93	211	28.5	30.5	495	151	344	46.1
30.7	375	115	260	35.3	30.7	218	67	151	20.6	30.6	438	134	304	41.0	30.6	301	92	209	28.1	30.5	492	150	342	45.7
30.7	374	115	259	35.4	30.7	215	66	149	20.3	30.6	435	133	302	40.7	30.6	297	91	206	27.9	30.5	489	149	340	45.4
30.7	371	114	257	35.0	30.7	212	65	147	19.9	30.6	434	133	301	40.8	30.6	294	90	204	27.6	30.5	488	149	339	45.5
30.7	368	113	255	34.7	30.7	205	63	142	19.4	30.6	432	132	300	40.3	30.6	291	89	202	27.2	30.5	486	148	338	45.1
30.7	365	112	253	34.4	30.7	202	62	140	19.0	30.6	431	132	299	40.4	30.6	288	88	200	26.9	30.5	485	148	337	45.2
30.7	362	111	251	34.0	30.7	199	61	138	18.7	30.6	428	131	297	40.1	30.6	284	87	197	26.7	30.5	482	147	335	44.8
30.7	361	111	250	34.1	30.7	192	59	133	18.1	30.6	425	130	295	39.8	30.6	281	86	195	26.3	30.5	479	146	333	44.5
30.7	358	110	248	33.8	30.7	189	58	131	17.8	30.6	422	129	293	39.4	30.6	278	85	193	26.0	30.5	478	146	332	44.6
30.7	355	109	246	33.5	30.7	179	55	124	16.9	30.6	421	129	292	39.5	30.6	271	83	188	25.4	30.5	476	145	331	44.2
30.7	352	108	244	33.1	30.7	176	54	122	16.6	30.6	418	128	290	39.2	30.6	268	82	186	25.1	30.5	475	145	330	44.3

%E	M1	M2	DM	M*
30.5	472	144	328	43.9
30.5	469	143	326	43.6
30.5	466	142	324	43.3
30.5	465	142	323	43.4
30.5	463	141	322	42.9
30.5	462	141	321	43.0
30.5	459	140	319	42.7
30.5	456	139	317	42.4
30.5	455	139	316	42.5
30.5	453	138	315	42.0
30.5	452	138	314	42.1
30.5	449	137	312	41.8
30.5	446	136	310	41.5
30.5	443	135	308	41.1
30.5	442	135	307	41.2
30.5	440	134	306	40.8
30.5	439	134	305	40.9
30.5	436	133	303	40.6
30.5	433	132	301	40.2
30.5	430	131	299	39.9
30.5	429	131	298	40.0
30.5	426	130	296	39.7
30.5	423	129	294	39.3
30.5	420	128	292	39.0
30.5	419	128	291	39.1
30.5	417	127	290	38.7
30.5	416	127	289	38.8
30.5	413	126	287	38.4
30.5	410	125	285	38.1
30.5	407	124	283	37.8
30.5	406	124	282	37.9
30.5	403	123	280	37.5
30.5	400	122	278	37.2
30.5	397	121	276	36.9
30.5	394	120	274	36.5
30.5	393	120	273	36.6
30.5	390	119	271	36.3
30.5	387	118	269	36.0
30.5	384	117	267	35.6
30.5	383	117	266	35.7
30.5	380	116	264	35.4
30.5	377	115	262	35.1
30.5	374	114	260	34.7
30.5	371	113	258	34.4
30.5	370	113	257	34.5
30.5	367	112	255	34.2
30.5	364	111	253	33.8
30.5	361	110	251	33.5
30.5	357	109	248	33.3
30.5	354	108	246	32.9
30.5	351	107	244	32.6
30.5	348	106	242	32.3
30.5	347	106	241	32.4
30.5	344	105	239	32.0
30.5	341	104	237	31.7
30.5	338	103	235	31.4
30.5	334	102	232	31.1
30.5	331	101	230	30.8
30.5	328	100	228	30.5
30.5	325	99	226	30.2
30.5	321	98	223	29.9
30.5	318	97	221	29.6
30.5	315	96	219	29.3
30.5	311	95	216	29.0
30.5	308	94	214	28.7
30.5	305	93	212	28.4
30.5	302	92	210	28.0
30.5	298	91	207	27.8
30.5	295	90	205	27.5
30.5	292	89	203	27.1
30.5	289	88	201	26.8
30.5	285	87	198	26.6
30.5	282	86	196	26.2
30.5	279	85	194	25.9
30.5	275	84	191	25.7
30.5	272	83	189	25.3
30.5	269	82	187	25.0
30.5	266	81	185	24.7
30.5	262	80	182	24.4
30.5	259	79	180	24.1
30.5	256	78	178	23.8
30.5	249	76	173	23.2
30.5	246	75	171	22.9
30.5	243	74	169	22.5
30.5	239	73	166	22.3
30.5	236	72	164	22.0
30.5	233	71	162	21.6
30.5	226	69	157	21.1
30.5	223	68	155	20.7
30.5	220	67	153	20.4
30.5	213	65	148	19.8
30.5	210	64	146	19.5
30.5	203	62	141	18.9
30.5	200	61	139	18.6
30.5	197	60	137	18.3
30.5	190	58	132	17.7
30.5	187	57	130	17.4
30.5	177	54	123	16.5
30.5	174	53	121	16.1
30.5	167	51	116	15.6
30.5	164	50	114	15.2
30.5	154	47	107	14.3
30.5	151	46	105	14.0
30.5	141	43	98	13.1
30.5	131	40	91	12.2
30.5	128	39	89	11.9
30.5	118	36	82	11.0
30.5	105	32	73	9.8
30.5	95	29	66	8.9
30.5	82	25	57	7.6
30.5	59	18	41	5.5
30.4	500	152	348	46.2
30.4	497	151	346	45.9
30.4	496	151	345	46.0
30.4	494	150	344	45.5
30.4	493	150	343	45.6
30.4	490	149	341	45.3
30.4	487	148	339	45.0
30.4	484	147	337	44.6
30.4	483	147	336	44.7
30.4	481	146	335	44.3
30.4	480	146	334	44.4
30.4	477	145	332	44.1
30.4	474	144	330	43.7
30.4	473	144	329	43.8
30.4	471	143	328	43.4
30.4	470	143	327	43.5
30.4	467	142	325	43.2
30.4	464	141	323	42.8
30.4	461	140	321	42.5
30.4	460	140	320	42.6
30.4	457	139	318	42.3
30.4	454	138	316	41.9
30.4	451	137	314	41.6
30.4	450	137	313	41.7
30.4	448	136	312	41.3
30.4	447	136	311	41.4
30.4	444	135	309	41.0
30.4	441	134	307	40.7
30.4	438	133	305	40.4
30.4	437	133	304	40.5
30.4	434	132	302	40.1
30.4	431	131	300	39.8
30.4	428	130	298	39.5
30.4	427	130	297	39.6
30.4	425	129	296	39.2
30.4	424	129	295	39.2
30.4	421	128	293	38.9
30.4	418	127	291	38.6
30.4	415	126	289	38.3
30.4	414	126	288	38.3
30.4	411	125	286	38.0
30.4	408	124	284	37.7
30.4	405	123	282	37.4
30.4	404	123	281	37.4
30.4	401	122	279	37.1
30.4	398	121	277	36.8
30.4	395	120	275	36.5
30.4	392	119	273	36.1
30.4	391	119	272	36.2
30.4	388	118	270	35.9
30.4	385	117	268	35.6
30.4	382	116	266	35.2
30.4	381	116	265	35.3
30.4	378	115	263	35.0
30.4	375	114	261	34.7
30.4	372	113	259	34.3
30.4	369	112	257	34.0
30.4	368	112	256	34.1
30.4	365	111	254	33.8
30.4	362	110	252	33.4
30.4	359	109	250	33.1
30.4	358	109	249	33.2
30.4	355	108	247	32.9
30.4	352	107	245	32.5
30.4	349	106	243	32.2
30.4	345	105	240	32.0
30.4	342	104	238	31.6
30.4	339	103	236	31.3
30.4	336	102	234	31.0
30.4	335	102	233	31.1
30.4	332	101	231	30.7
30.4	329	100	229	30.4
30.4	326	99	227	30.1
30.4	322	98	224	29.8
30.4	319	97	222	29.5
30.4	316	96	220	29.2
30.4	313	95	218	28.8
30.4	312	95	217	28.9
30.4	309	94	215	28.6
30.4	306	93	213	28.3
30.4	303	92	211	27.9
30.4	299	91	208	27.7
30.4	296	90	206	27.4
30.4	293	89	204	27.0
30.4	286	87	199	26.5
30.4	283	86	197	26.1
30.4	280	85	195	25.8
30.4	276	84	192	25.6
30.4	273	83	190	25.2
30.4	270	82	188	24.9
30.4	263	80	183	24.3
30.4	260	79	181	24.0
30.4	257	78	179	23.7
30.4	253	77	176	23.4
30.4	250	76	174	23.1
30.4	247	75	172	22.8
30.4	240	73	167	22.2
30.4	237	72	165	21.9
30.4	230	70	160	21.3
30.4	227	69	158	21.0
30.4	224	68	156	20.6
30.4	217	66	151	20.1
30.4	214	65	149	19.7
30.4	207	63	144	19.2
30.4	204	62	142	18.8
30.4	194	59	135	17.9
30.4	191	58	133	17.6
30.4	184	56	128	17.0
30.4	181	55	126	16.7
30.4	171	52	119	15.8
30.4	168	51	117	15.5
30.4	161	49	112	14.9
30.4	158	48	110	14.6
30.4	148	45	103	13.7
30.4	138	42	96	12.8
30.4	135	41	94	12.5
30.4	125	38	87	11.6
30.4	115	35	80	10.7
30.4	112	34	78	10.3
30.4	102	31	71	9.4
30.4	92	28	64	8.5
30.4	79	24	55	7.3
30.4	69	21	48	6.4
30.4	56	17	39	5.2
30.4	46	14	32	4.3
30.4	23	7	16	2.1
30.3	499	151	348	45.7
30.3	498	151	347	45.8
30.3	495	150	345	45.5
30.3	492	149	343	45.1
30.3	491	149	342	45.2
30.3	489	148	341	44.8
30.3	488	148	340	44.9
30.3	485	147	338	44.6
30.3	482	146	336	44.2
30.3	479	145	334	43.9
30.3	478	145	333	44.0
30.3	476	144	332	43.6
30.3	475	144	331	43.7

%E	M1	M2	DM	M*
30.3	472	143	329	43.3
30.3	469	142	327	43.0
30.3	468	142	326	43.1
30.3	466	141	325	42.7
30.3	465	141	324	42.8
30.3	462	140	322	42.4
30.3	459	139	320	42.1
30.3	458	139	319	42.2
30.3	456	138	318	41.8
30.3	455	138	317	41.9
30.3	452	137	315	41.5
30.3	449	136	313	41.2
30.3	446	135	311	40.9
30.3	445	135	310	41.0
30.3	442	134	308	40.6
30.3	439	133	306	40.3
30.3	436	132	304	40.0
30.3	435	132	303	40.1
30.3	433	131	302	39.6
30.3	432	131	301	39.7
30.3	429	130	299	39.4
30.3	426	129	297	39.1
30.3	423	128	295	38.7
30.3	422	128	294	38.8
30.3	419	127	292	38.5
30.3	416	126	290	38.2
30.3	413	125	288	37.8
30.3	412	125	287	37.9
30.3	409	124	285	37.6
30.3	406	123	283	37.3
30.3	403	122	281	36.9
30.3	402	122	280	37.0
30.3	400	121	279	36.6
30.3	399	121	278	36.7
30.3	396	120	276	36.4
30.3	393	119	274	36.0
30.3	390	118	272	35.7
30.3	389	118	271	35.8
30.3	386	117	269	35.5
30.3	383	116	267	35.1
30.3	380	115	265	34.8
30.3	379	115	264	34.9
30.3	376	114	262	34.6
30.3	373	113	260	34.2
30.3	370	112	258	33.9
30.3	366	111	255	33.7
30.3	363	110	253	33.3
30.3	360	109	251	33.0
30.3	357	108	249	32.7
30.3	356	108	248	32.8

%E	M1	M2	DM	M*
30.3	353	107	246	32.4
30.3	350	106	244	32.1
30.3	347	105	242	31.8
30.3	346	105	241	31.9
30.3	343	104	239	31.5
30.3	340	103	237	31.2
30.3	337	102	235	30.9
30.3	333	101	232	30.6
30.3	330	100	230	30.3
30.3	327	99	228	30.0
30.3	323	98	225	29.7
30.3	320	97	223	29.4
30.3	317	96	221	29.1
30.3	314	95	219	28.7
30.3	310	94	216	28.5
30.3	307	93	214	28.2
30.3	304	92	212	27.8
30.3	300	91	209	27.6
30.3	297	90	207	27.3
30.3	294	89	205	26.9
30.3	290	88	202	26.7
30.3	287	87	200	26.4
30.3	284	86	198	26.0
30.3	277	84	193	25.5
30.3	274	83	191	25.1
30.3	271	82	189	24.8
30.3	267	81	186	24.6
30.3	264	80	184	24.2
30.3	261	79	182	23.9
30.3	254	77	177	23.3
30.3	251	76	175	23.0
30.3	244	74	170	22.4
30.3	241	73	168	22.1
30.3	238	72	166	21.8
30.3	234	71	163	21.5
30.3	231	70	161	21.2
30.3	228	69	159	20.9
30.3	221	67	154	20.3
30.3	218	66	152	20.0
30.3	211	64	147	19.4
30.3	208	63	145	19.1
30.3	201	61	140	18.5
30.3	198	60	138	18.2
30.3	195	59	136	17.9
30.3	188	57	131	17.3
30.3	185	56	129	17.0
30.3	178	54	124	16.4
30.3	175	53	122	16.1
30.3	165	50	115	15.2
30.3	155	47	108	14.3

%E	M1	M2	DM	M*
30.3	152	46	106	13.9
30.3	145	44	101	13.4
30.3	142	43	99	13.0
30.3	132	40	92	12.1
30.3	122	37	85	11.2
30.3	119	36	83	10.9
30.3	109	33	76	10.0
30.3	99	30	69	9.1
30.3	89	27	62	8.2
30.3	76	23	53	7.0
30.3	66	20	46	6.1
30.3	33	10	23	3.0
30.2	500	151	349	45.6
30.2	497	150	347	45.3
30.2	496	150	346	45.4
30.2	494	149	345	44.9
30.2	493	149	344	45.0
30.2	490	148	342	44.7
30.2	487	147	340	44.4
30.2	486	147	339	44.5
30.2	484	146	338	44.0
30.2	483	146	337	44.1
30.2	480	145	335	43.8
30.2	477	144	333	43.5
30.2	474	143	331	43.1
30.2	473	143	330	43.2
30.2	470	142	328	42.9
30.2	467	141	326	42.6
30.2	464	140	324	42.2
30.2	463	140	323	42.3
30.2	461	139	322	41.9
30.2	460	139	321	42.0
30.2	457	138	319	41.7
30.2	454	137	317	41.3
30.2	453	137	316	41.4
30.2	451	136	315	41.0
30.2	450	136	314	41.1
30.2	447	135	312	40.8
30.2	444	134	310	40.4
30.2	443	134	309	40.5
30.2	441	133	308	40.1
30.2	440	133	307	40.2
30.2	437	132	305	39.9
30.2	434	131	303	39.5
30.2	431	130	301	39.2
30.2	430	130	300	39.3
30.2	427	129	298	39.0
30.2	424	128	296	38.6
30.2	421	127	294	38.3
30.2	420	127	293	38.4

%E	M1	M2	DM	M*
30.2	417	126	291	38.1
30.2	414	125	289	37.7
30.2	411	124	287	37.4
30.2	410	124	286	37.5
30.2	407	123	284	37.2
30.2	404	122	282	36.8
30.2	401	121	280	36.5
30.2	398	120	278	36.2
30.2	397	120	277	36.3
30.2	394	119	275	35.9
30.2	391	118	273	35.6
30.2	388	117	271	35.3
30.2	387	117	270	35.4
30.2	384	116	268	35.0
30.2	381	115	266	34.7
30.2	378	114	264	34.4
30.2	377	114	263	34.5
30.2	374	113	261	34.1
30.2	371	112	259	33.8
30.2	368	111	257	33.5
30.2	367	111	256	33.6
30.2	364	110	254	33.2
30.2	361	109	252	32.9
30.2	358	108	250	32.6
30.2	354	107	247	32.3
30.2	351	106	245	32.0
30.2	348	105	243	31.7
30.2	344	104	240	31.4
30.2	341	103	238	31.1
30.2	338	102	236	30.8
30.2	334	101	233	30.5
30.2	331	100	231	30.2
30.2	328	99	229	29.9
30.2	325	98	227	29.6
30.2	324	98	226	29.6
30.2	321	97	224	29.3
30.2	318	96	222	29.0
30.2	315	95	220	28.7
30.2	311	94	217	28.4
30.2	308	93	215	28.1
30.2	305	92	213	27.8
30.2	301	91	210	27.5
30.2	298	90	208	27.2
30.2	295	89	206	26.9
30.2	291	88	203	26.6
30.2	288	87	201	26.3
30.2	285	86	199	26.0
30.2	281	85	196	25.7
30.2	278	84	194	25.4
30.2	275	83	192	25.1

%E	M1	M2	DM	M*
30.2	268	81	187	24.5
30.2	265	80	185	24.2
30.2	262	79	183	23.8
30.2	258	78	180	23.6
30.2	255	77	178	23.3
30.2	252	76	176	22.9
30.2	248	75	173	22.7
30.2	245	74	171	22.4
30.2	242	73	169	22.0
30.2	235	71	164	21.5
30.2	232	70	162	21.1
30.2	225	68	157	20.6
30.2	222	67	155	20.2
30.2	215	65	150	19.7
30.2	212	64	148	19.3
30.2	205	62	143	18.8
30.2	202	61	141	18.4
30.2	199	60	139	18.1
30.2	192	58	134	17.5
30.2	189	57	132	17.2
30.2	182	55	127	16.6
30.2	179	54	125	16.3
30.2	172	52	120	15.7
30.2	169	51	118	15.4
30.2	162	49	113	14.8
30.2	159	48	111	14.5
30.2	149	45	104	13.6
30.2	139	42	97	12.7
30.2	129	39	90	11.8
30.2	126	38	88	11.5
30.2	116	35	81	10.6
30.2	106	32	74	9.7
30.2	96	29	67	8.8
30.2	86	26	60	7.9
30.2	63	19	44	5.7
30.2	53	16	37	4.8
30.2	43	13	30	3.9
30.1	499	150	349	45.1
30.1	498	150	348	45.2
30.1	495	149	346	44.9
30.1	492	148	344	44.5
30.1	491	148	343	44.6
30.1	489	147	342	44.2
30.1	488	147	341	44.3
30.1	485	146	339	44.0
30.1	482	145	337	43.6
30.1	481	145	336	43.7
30.1	479	144	335	43.3
30.1	478	144	334	43.4
30.1	475	143	332	43.1

%E	M1	M2	DM	M*
30.1	472	142	330	42.7
30.1	471	142	329	42.8
30.1	469	141	328	42.4
30.1	468	141	327	42.5
30.1	465	140	325	42.2
30.1	462	139	323	41.8
30.1	459	138	321	41.5
30.1	458	138	320	41.6
30.1	455	137	318	41.3
30.1	452	136	316	40.9
30.1	449	135	314	40.6
30.1	448	135	313	40.7
30.1	445	134	311	40.4
30.1	442	133	309	40.0
30.1	439	132	307	39.7
30.1	438	132	306	39.8
30.1	435	131	304	39.5
30.1	432	130	302	39.1
30.1	429	129	300	38.8
30.1	428	129	299	38.9
30.1	425	128	297	38.6
30.1	422	127	295	38.2
30.1	419	126	293	37.9
30.1	418	126	292	38.0
30.1	415	125	290	37.7
30.1	412	124	288	37.3
30.1	409	123	286	37.0
30.1	408	123	285	37.1
30.1	405	122	283	36.8
30.1	402	121	281	36.4
30.1	399	120	279	36.1
30.1	396	119	277	35.8
30.1	395	119	276	35.9
30.1	392	118	274	35.5
30.1	389	117	272	35.2
30.1	386	116	270	34.9
30.1	385	116	269	35.0
30.1	382	115	267	34.6
30.1	379	114	265	34.3
30.1	376	113	263	34.0
30.1	375	113	262	34.1
30.1	372	112	260	33.7
30.1	369	111	258	33.4
30.1	366	110	256	33.1
30.1	365	110	255	33.2
30.1	362	109	253	32.8
30.1	359	108	251	32.5
30.1	356	107	249	32.2
30.1	355	107	248	32.3
30.1	352	106	246	31.9

%E	M1	M2	DM	M*
30.1	349	105	244	31.6
30.1	346	104	242	31.3
30.1	345	104	241	31.4
30.1	342	103	239	31.0
30.1	339	102	237	30.7
30.1	336	101	235	30.4
30.1	335	101	234	30.5
30.1	332	100	232	30.1
30.1	329	99	230	29.8
30.1	326	98	228	29.5
30.1	322	97	225	29.2
30.1	319	96	223	28.9
30.1	316	95	221	28.6
30.1	312	94	218	28.3
30.1	309	93	216	28.0
30.1	306	92	214	27.7
30.1	302	91	211	27.4
30.1	299	90	209	27.1
30.1	296	89	207	26.8
30.1	292	88	204	26.5
30.1	289	87	202	26.2
30.1	286	86	200	25.9
30.1	282	85	197	25.6
30.1	279	84	195	25.3
30.1	276	83	193	25.0
30.1	272	82	190	24.7
30.1	269	81	188	24.4
30.1	266	80	186	24.1
30.1	259	78	181	23.5
30.1	256	77	179	23.2
30.1	249	75	174	22.6
30.1	246	74	172	22.3
30.1	239	72	167	21.7
30.1	236	71	165	21.4
30.1	229	69	160	20.8
30.1	226	68	158	20.5
30.1	219	66	153	19.9
30.1	216	65	151	19.6
30.1	209	63	146	19.0
30.1	206	62	144	18.7
30.1	196	59	137	17.8
30.1	193	58	135	17.4
30.1	186	56	130	16.9
30.1	183	55	128	16.5
30.1	176	53	123	16.0
30.1	173	52	121	15.6
30.1	166	50	116	15.1
30.1	163	49	114	14.7
30.1	156	47	109	14.2
30.1	153	46	107	13.8

%E	M1	M2	DM	M*
30.1	146	44	102	13.3
30.1	143	43	100	12.9
30.1	136	41	95	12.4
30.1	133	40	93	12.0
30.1	123	37	86	11.1
30.1	113	34	79	10.2
30.1	103	31	72	9.3
30.1	93	28	65	8.4
30.1	83	25	58	7.5
30.1	73	22	51	6.6
30.0	500	150	350	45.0
30.0	497	149	348	44.7
30.0	496	149	347	44.8
30.0	494	148	346	44.3
30.0	493	148	345	44.4
30.0	490	147	343	44.1
30.0	487	146	341	43.8
30.0	486	146	340	43.9
30.0	484	145	339	43.4
30.0	483	145	338	43.5
30.0	480	144	336	43.2
30.0	477	143	334	42.9
30.0	476	143	333	43.0
30.0	474	142	332	42.5
30.0	473	142	331	42.6
30.0	470	141	329	42.3
30.0	467	140	327	42.0
30.0	466	140	326	42.1
30.0	464	139	325	41.6
30.0	463	139	324	41.7
30.0	460	138	322	41.4
30.0	457	137	320	41.1
30.0	456	137	319	41.2
30.0	454	136	318	40.7
30.0	453	136	317	40.8
30.0	450	135	315	40.5
30.0	447	134	313	40.2
30.0	446	134	312	40.3
30.0	444	133	311	39.8
30.0	443	133	310	39.9
30.0	440	132	308	39.6
30.0	437	131	306	39.3
30.0	436	131	305	39.4
30.0	434	130	304	38.9
30.0	433	130	303	39.0
30.0	430	129	301	38.7
30.0	427	128	299	38.4
30.0	426	128	298	38.5
30.0	424	127	297	38.0
30.0	423	127	296	38.1

%E	M1	M2	DM	M*
30.0	420	126	294	37.8
30.0	417	125	292	37.5
30.0	416	125	291	37.6
30.0	414	124	290	37.1
30.0	413	124	289	37.2
30.0	410	123	287	36.9
30.0	407	122	285	36.6
30.0	406	122	284	36.7
30.0	404	121	283	36.2
30.0	403	121	282	36.3
30.0	400	120	280	36.0
30.0	397	119	278	35.7
30.0	393	118	275	35.4
30.0	390	117	273	35.1
30.0	387	116	271	34.8
30.0	383	115	268	34.5
30.0	380	114	266	34.2
30.0	377	113	264	33.9
30.0	373	112	261	33.6
30.0	370	111	259	33.3
30.0	367	110	257	33.0
30.0	363	109	254	32.7
30.0	360	108	252	32.4
30.0	357	107	250	32.1
30.0	353	106	247	31.8
30.0	350	105	245	31.5
30.0	347	104	243	31.2
30.0	343	103	240	30.9
30.0	340	102	238	30.6
30.0	337	101	236	30.3
30.0	333	100	233	30.0
30.0	330	99	231	29.7
30.0	327	98	229	29.4
30.0	323	97	226	29.1
30.0	320	96	224	28.8
30.0	317	95	222	28.5
30.0	313	94	219	28.2
30.0	310	93	217	27.9
30.0	307	92	215	27.6
30.0	303	91	212	27.3
30.0	300	90	210	27.0
30.0	297	89	208	26.7
30.0	293	88	205	26.4
30.0	290	87	203	26.1
30.0	287	86	201	25.8
30.0	283	85	198	25.5
30.0	280	84	196	25.2
30.0	277	83	194	24.9
30.0	273	82	191	24.6
30.0	270	81	189	24.3

%E	M1	M2	DM	M*
30.0	267	80	187	24.0
30.0	263	79	184	23.7
30.0	260	78	182	23.4
30.0	257	77	180	23.1
30.0	253	76	177	22.8
30.0	250	75	175	22.5
30.0	247	74	173	22.2
30.0	243	73	170	21.9
30.0	240	72	168	21.6
30.0	237	71	166	21.3
30.0	233	70	163	21.0
30.0	230	69	161	20.7
30.0	227	68	159	20.4
30.0	223	67	156	20.1
30.0	220	66	154	19.8
30.0	217	65	152	19.5
30.0	213	64	149	19.2
30.0	210	63	147	18.9
30.0	207	62	145	18.6
30.0	203	61	142	18.3
30.0	200	60	140	18.0
30.0	190	57	133	17.1
30.0	180	54	126	16.2
30.0	170	51	119	15.3
30.0	160	48	112	14.4
30.0	150	45	105	13.5
30.0	140	42	98	12.6
30.0	130	39	91	11.7
30.0	120	36	84	10.8
30.0	110	33	77	9.9
30.0	100	30	70	9.0
30.0	90	27	63	8.1
30.0	80	24	56	7.2
30.0	70	21	49	6.3
30.0	60	18	42	5.4
30.0	50	15	35	4.5
30.0	40	12	28	3.6
30.0	30	9	21	2.7
30.0	20	6	14	1.8
30.0	10	3	7	0.9
29.9	499	149	350	44.5
29.9	498	149	349	44.6
29.9	495	148	347	44.3
29.9	492	147	345	43.9
29.9	491	147	344	44.0
29.9	489	146	343	43.6
29.9	488	146	342	43.7
29.9	485	145	340	43.4
29.9	482	144	338	43.0
29.9	481	144	337	43.1

%E	M1	M2	DM	M*
29.9	479	143	336	42.7
29.9	478	143	335	42.8
29.9	475	142	333	42.5
29.9	472	141	331	42.1
29.9	471	141	330	42.2
29.9	469	140	329	41.8
29.9	468	140	328	41.9
29.9	465	139	326	41.6
29.9	462	138	324	41.2
29.9	461	138	323	41.3
29.9	458	137	321	41.0
29.9	455	136	319	40.7
29.9	452	135	317	40.3
29.9	451	135	316	40.4
29.9	448	134	314	40.1
29.9	445	133	312	39.8
29.9	442	132	310	39.4
29.9	441	132	309	39.5
29.9	438	131	307	39.2
29.9	435	130	305	38.9
29.9	432	129	303	38.5
29.9	431	129	302	38.6
29.9	428	128	300	38.3
29.9	425	127	298	38.0
29.9	422	126	296	37.6
29.9	421	126	295	37.7
29.9	418	125	293	37.4
29.9	415	124	291	37.1
29.9	412	123	289	36.7
29.9	411	123	288	36.8
29.9	408	122	286	36.5
29.9	405	121	284	36.2
29.9	402	120	282	35.8
29.9	401	120	281	35.9
29.9	398	119	279	35.6
29.9	395	118	277	35.3
29.9	394	118	276	35.3
29.9	391	117	274	35.0
29.9	388	116	272	34.7
29.9	385	115	270	34.4
29.9	384	115	269	34.4
29.9	381	114	267	34.1
29.9	378	113	265	33.8
29.9	375	112	263	33.5
29.9	374	112	262	33.5
29.9	371	111	260	33.2
29.9	368	110	258	32.9
29.9	365	109	256	32.6
29.9	364	109	255	32.6
29.9	361	108	253	32.3
29.9	358	107	251	32.0
29.9	355	106	249	31.7
29.9	354	106	248	31.7
29.9	351	105	246	31.4
29.9	348	104	244	31.1
29.9	345	103	242	30.8
29.9	344	103	241	30.8
29.9	341	102	239	30.5
29.9	338	101	237	30.2
29.9	335	100	235	29.9
29.9	334	100	234	29.9
29.9	331	99	232	29.6
29.9	328	98	230	29.3
29.9	324	97	227	29.0
29.9	321	96	225	28.7
29.9	318	95	223	28.4
29.9	314	94	220	28.1
29.9	311	93	218	27.8
29.9	308	92	216	27.5
29.9	304	91	213	27.2
29.9	301	90	211	26.9
29.9	298	89	209	26.6
29.9	294	88	206	26.3
29.9	291	87	204	26.0
29.9	288	86	202	25.7
29.9	284	85	199	25.4
29.9	281	84	197	25.1
29.9	278	83	195	24.8
29.9	274	82	192	24.5
29.9	271	81	190	24.2
29.9	268	80	188	23.9
29.9	264	79	185	23.6
29.9	261	78	183	23.3
29.9	254	76	178	22.7
29.9	251	75	176	22.4
29.9	244	73	171	21.8
29.9	241	72	169	21.5
29.9	234	70	164	20.9
29.9	231	69	162	20.6
29.9	224	67	157	20.0
29.9	221	66	155	19.7
29.9	214	64	150	19.1
29.9	211	63	148	18.8
29.9	204	61	143	18.2
29.9	201	60	141	17.9
29.9	197	59	138	17.7
29.9	194	58	136	17.3
29.9	187	56	131	16.8
29.9	184	55	129	16.4
29.9	177	53	124	15.9
29.9	174	52	122	15.5
29.9	167	50	117	15.0
29.9	164	49	115	14.6
29.9	157	47	110	14.1
29.9	154	46	108	13.7
29.9	147	44	103	13.2
29.9	144	43	101	12.8
29.9	137	41	96	12.3
29.9	134	40	94	11.9
29.9	127	38	89	11.4
29.9	117	35	82	10.5
29.9	107	32	75	9.6
29.9	97	29	68	8.7
29.9	87	26	61	7.8
29.9	77	23	54	6.9
29.9	67	20	47	6.0
29.8	500	149	351	44.4
29.8	497	148	349	44.1
29.8	496	148	348	44.2
29.8	494	147	347	43.7
29.8	493	147	346	43.8
29.8	490	146	344	43.5
29.8	487	145	342	43.2
29.8	486	145	341	43.3
29.8	484	144	340	42.8
29.8	483	144	339	42.9
29.8	480	143	337	42.6
29.8	477	142	335	42.3
29.8	476	142	334	42.4
29.8	473	141	332	42.0
29.8	470	140	330	41.7
29.8	467	139	328	41.4
29.8	466	139	327	41.5
29.8	463	138	325	41.1
29.8	460	137	323	40.8
29.8	459	137	322	40.9
29.8	457	136	321	40.5
29.8	456	136	320	40.6
29.8	453	135	318	40.2
29.8	450	134	316	39.9
29.8	449	134	315	40.0
29.8	447	133	314	39.6
29.8	446	133	313	39.7
29.8	443	132	311	39.3
29.8	440	131	309	39.0
29.8	439	131	308	39.1
29.8	436	130	306	38.8
29.8	433	129	304	38.4
29.8	430	128	302	38.1
29.8	429	128	301	38.2
29.8	426	127	299	37.9
29.8	423	126	297	37.5
29.8	420	125	295	37.2
29.8	419	125	294	37.3
29.8	416	124	292	37.0
29.8	413	123	290	36.6
29.8	410	122	288	36.3
29.8	409	122	287	36.4
29.8	406	121	285	36.1
29.8	403	120	283	35.7
29.8	400	119	281	35.4
29.8	399	119	280	35.5
29.8	396	118	278	35.2
29.8	393	117	276	34.8
29.8	392	117	275	34.9
29.8	389	116	273	34.6
29.8	386	115	271	34.3
29.8	383	114	269	33.9
29.8	382	114	268	34.0
29.8	379	113	266	33.7
29.8	376	112	264	33.4
29.8	373	111	262	33.0
29.8	372	111	261	33.1
29.8	369	110	259	32.8
29.8	366	109	257	32.5
29.8	363	108	255	32.1
29.8	362	108	254	32.2
29.8	359	107	252	31.9
29.8	356	106	250	31.6
29.8	352	105	247	31.3
29.8	349	104	245	31.0
29.8	346	103	243	30.7
29.8	342	102	240	30.4
29.8	339	101	238	30.1
29.8	336	100	236	29.8
29.8	332	99	233	29.5
29.8	329	98	231	29.2
29.8	326	97	229	28.9
29.8	325	97	228	29.0
29.8	322	96	226	28.6
29.8	319	95	224	28.3
29.8	315	94	221	28.1
29.8	312	93	219	27.7
29.8	309	92	217	27.4
29.8	305	91	214	27.2
29.8	302	90	212	26.8
29.8	299	89	210	26.5
29.8	295	88	207	26.3
29.8	292	87	205	25.9
29.8	289	86	203	25.6
29.8	285	85	200	25.4
29.8	282	84	198	25.0
29.8	275	82	193	24.5
29.8	272	81	191	24.1
29.8	265	79	186	23.6
29.8	262	78	184	23.2
29.8	258	77	181	23.0
29.8	255	76	179	22.7
29.8	252	75	177	22.3
29.8	248	74	174	22.1
29.8	245	73	172	21.8
29.8	242	72	170	21.4
29.8	238	71	167	21.2
29.8	235	70	165	20.9
29.8	228	68	160	20.3
29.8	225	67	158	20.0
29.8	218	65	153	19.4
29.8	215	64	151	19.1
29.8	208	62	146	18.5
29.8	205	61	144	18.2
29.8	198	59	139	17.6
29.8	191	57	134	17.0
29.8	188	56	132	16.7
29.8	181	54	127	16.1
29.8	178	53	125	15.8
29.8	171	51	120	15.2
29.8	168	50	118	14.9
29.8	161	48	113	14.3
29.8	151	45	106	13.4
29.8	141	42	99	12.5
29.8	131	39	92	11.6
29.8	124	37	87	11.0
29.8	121	36	85	10.7
29.8	114	34	80	10.1
29.8	104	31	73	9.2
29.8	94	28	66	8.3
29.8	84	25	59	7.4
29.8	57	17	40	5.1
29.8	47	14	33	4.2
29.7	499	148	351	43.9
29.7	498	148	350	44.0
29.7	495	147	348	43.7
29.7	492	146	346	43.3
29.7	491	146	345	43.4
29.7	489	145	344	43.0
29.7	488	145	343	43.1
29.7	485	144	341	42.8
29.7	482	143	339	42.4
29.7	481	143	338	42.5
29.7	478	142	336	42.2

%E	M1	M2	DM	M*
29.7	475	141	334	41.9
29.7	474	141	333	41.9
29.7	472	140	332	41.5
29.7	471	140	331	41.6
29.7	468	139	329	41.3
29.7	465	138	327	41.0
29.7	464	138	326	41.0
29.7	462	137	325	40.6
29.7	461	137	324	40.7
29.7	458	136	322	40.4
29.7	455	135	320	40.1
29.7	454	135	319	40.1
29.7	451	134	317	39.8
29.7	448	133	315	39.5
29.7	445	132	313	39.2
29.7	444	132	312	39.2
29.7	441	131	310	38.9
29.7	438	130	308	38.6
29.7	437	130	307	38.7
29.7	435	129	306	38.3
29.7	434	129	305	38.3
29.7	431	128	303	38.0
29.7	428	127	301	37.7
29.7	427	127	300	37.8
29.7	424	126	298	37.4
29.7	421	125	296	37.1
29.7	418	124	294	36.8
29.7	417	124	293	36.9
29.7	414	123	291	36.5
29.7	411	122	289	36.2
29.7	408	121	287	35.9
29.7	407	121	286	36.0
29.7	404	120	284	35.6
29.7	401	119	282	35.3
29.7	397	118	279	35.1
29.7	394	117	277	34.7
29.7	391	116	275	34.4
29.7	390	116	274	34.5
29.7	387	115	272	34.2
29.7	384	114	270	33.8
29.7	381	113	268	33.5
29.7	380	113	267	33.6
29.7	377	112	265	33.3
29.7	374	111	263	32.9
29.7	371	110	261	32.6
29.7	370	110	260	32.7
29.7	367	109	258	32.4
29.7	364	108	256	32.0
29.7	360	107	253	31.8
29.7	357	106	251	31.5
29.7	354	105	249	31.1
29.7	353	105	248	31.2
29.7	350	104	246	30.9
29.7	347	103	244	30.6
29.7	344	102	242	30.2
29.7	343	102	241	30.3
29.7	340	101	239	30.0
29.7	337	100	237	29.7
29.7	333	99	234	29.4
29.7	330	98	232	29.1
29.7	327	97	230	28.8
29.7	323	96	227	28.5
29.7	320	95	225	28.2
29.7	317	94	223	27.9
29.7	316	94	222	28.0
29.7	313	93	220	27.6
29.7	310	92	218	27.3
29.7	306	91	215	27.1
29.7	303	90	213	26.7
29.7	300	89	211	26.4
29.7	296	88	208	26.2
29.7	293	87	206	25.8
29.7	290	86	204	25.5
29.7	286	85	201	25.3
29.7	283	84	199	24.9
29.7	279	83	196	24.7
29.7	276	82	194	24.4
29.7	273	81	192	24.0
29.7	269	80	189	23.8
29.7	266	79	187	23.5
29.7	263	78	185	23.1
29.7	259	77	182	22.9
29.7	256	76	180	22.6
29.7	249	74	175	22.0
29.7	246	73	173	21.7
29.7	239	71	168	21.1
29.7	236	70	166	20.8
29.7	232	69	163	20.5
29.7	229	68	161	20.2
29.7	222	66	156	19.6
29.7	219	65	154	19.3
29.7	212	63	149	18.7
29.7	209	62	147	18.4
29.7	202	60	142	17.8
29.7	195	58	137	17.3
29.7	192	57	135	16.9
29.7	185	55	130	16.4
29.7	182	54	128	16.0
29.7	175	52	123	15.5
29.7	172	51	121	15.1
29.7	165	49	116	14.6
29.7	158	47	111	14.0
29.7	155	46	109	13.7
29.7	148	44	104	13.1
29.7	145	43	102	12.8
29.7	138	41	97	12.2
29.7	128	38	90	11.3
29.7	118	35	83	10.4
29.7	111	33	78	9.8
29.7	101	30	71	8.9
29.7	91	27	64	8.0
29.7	74	22	52	6.5
29.7	64	19	45	5.6
29.7	37	11	26	3.3
29.6	500	148	352	43.8
29.6	497	147	350	43.5
29.6	496	147	349	43.6
29.6	494	146	348	43.1
29.6	493	146	347	43.2
29.6	490	145	345	42.9
29.6	487	144	343	42.6
29.6	486	144	342	42.7
29.6	483	143	340	42.3
29.6	480	142	338	42.0
29.6	479	142	337	42.1
29.6	477	141	336	41.7
29.6	476	141	335	41.8
29.6	473	140	333	41.4
29.6	470	139	331	41.1
29.6	469	139	330	41.2
29.6	467	138	329	40.8
29.6	466	138	328	40.9
29.6	463	137	326	40.5
29.6	460	136	324	40.2
29.6	459	136	323	40.3
29.6	456	135	321	40.0
29.6	453	134	319	39.6
29.6	452	134	318	39.7
29.6	450	133	317	39.3
29.6	449	133	316	39.4
29.6	446	132	314	39.1
29.6	443	131	312	38.7
29.6	442	131	311	38.8
29.6	439	130	309	38.5
29.6	436	129	307	38.2
29.6	433	128	305	37.8
29.6	432	128	304	37.9
29.6	429	127	302	37.6
29.6	426	126	300	37.3
29.6	425	126	299	37.4
29.6	423	125	298	36.9
29.6	422	125	297	37.0
29.6	419	124	295	36.7
29.6	416	123	293	36.4
29.6	415	123	292	36.5
29.6	412	122	290	36.1
29.6	409	121	288	35.8
29.6	406	120	286	35.5
29.6	405	120	285	35.6
29.6	402	119	283	35.2
29.6	399	118	281	34.9
29.6	398	118	280	35.0
29.6	395	117	278	34.7
29.6	392	116	276	34.3
29.6	389	115	274	34.0
29.6	388	115	273	34.1
29.6	385	114	271	33.8
29.6	382	113	269	33.4
29.6	379	112	267	33.1
29.6	378	112	266	33.2
29.6	375	111	264	32.9
29.6	372	110	262	32.5
29.6	368	109	259	32.3
29.6	365	108	257	32.0
29.6	362	107	255	31.6
29.6	361	107	254	31.7
29.6	358	106	252	31.4
29.6	355	105	250	31.1
29.6	351	104	247	30.8
29.6	348	103	245	30.5
29.6	345	102	243	30.2
29.6	341	101	240	29.9
29.6	338	100	238	29.6
29.6	335	99	236	29.3
29.6	334	99	235	29.3
29.6	331	98	233	29.0
29.6	328	97	231	28.7
29.6	324	96	228	28.4
29.6	321	95	226	28.1
29.6	318	94	224	27.8
29.6	314	93	221	27.5
29.6	311	92	219	27.2
29.6	307	91	216	27.0
29.6	304	90	214	26.6
29.6	301	89	212	26.3
29.6	297	88	209	26.1
29.6	294	87	207	25.7
29.6	291	86	205	25.4
29.6	287	85	202	25.2
29.6	284	84	200	24.8
29.6	280	83	197	24.6
29.6	277	82	195	24.3
29.6	274	81	193	23.9
29.6	270	80	190	23.7
29.6	267	79	188	23.4
29.6	260	77	183	22.8
29.6	257	76	181	22.5
29.6	253	75	178	22.2
29.6	250	74	176	21.9
29.6	247	73	174	21.6
29.6	243	72	171	21.3
29.6	240	71	169	21.0
29.6	233	69	164	20.4
29.6	230	68	162	20.1
29.6	226	67	159	19.9
29.6	223	66	157	19.5
29.6	216	64	152	19.0
29.6	213	63	150	18.6
29.6	206	61	145	18.1
29.6	203	60	143	17.7
29.6	199	59	140	17.5
29.6	196	58	138	17.2
29.6	189	56	133	16.6
29.6	186	55	131	16.3
29.6	179	53	126	15.7
29.6	169	50	119	14.8
29.6	162	48	114	14.2
29.6	159	47	112	13.9
29.6	152	45	107	13.3
29.6	142	42	100	12.4
29.6	135	40	95	11.9
29.6	125	37	88	11.0
29.6	115	34	81	10.1
29.6	108	32	76	9.5
29.6	98	29	69	8.6
29.6	81	24	57	7.1
29.6	71	21	50	6.2
29.6	54	16	38	4.7
29.6	27	8	19	2.4
29.5	499	147	352	43.3
29.5	498	147	351	43.4
29.5	495	146	349	43.1
29.5	492	145	347	42.7
29.5	491	145	346	42.8
29.5	488	144	344	42.5
29.5	485	143	342	42.2
29.5	484	143	341	42.3
29.5	482	142	340	41.8
29.5	481	142	339	41.9
29.5	478	141	337	41.6

%E	M1	M2	DM	M*
29.5	475	140	335	41.3
29.5	474	140	334	41.4
29.5	471	139	332	41.0
29.5	468	138	330	40.7
29.5	465	137	328	40.4
29.5	464	137	327	40.5
29.5	461	136	325	40.1
29.5	458	135	323	39.8
29.5	457	135	322	39.9
29.5	455	134	321	39.5
29.5	454	134	320	39.6
29.5	451	133	318	39.2
29.5	448	132	316	38.9
29.5	447	132	315	39.0
29.5	444	131	313	38.7
29.5	441	130	311	38.3
29.5	440	130	310	38.4
29.5	438	129	309	38.0
29.5	437	129	308	38.1
29.5	434	128	306	37.8
29.5	431	127	304	37.4
29.5	430	127	303	37.5
29.5	427	126	301	37.2
29.5	424	125	299	36.9
29.5	421	124	297	36.5
29.5	420	124	296	36.6
29.5	417	123	294	36.3
29.5	414	122	292	36.0
29.5	413	122	291	36.0
29.5	410	121	289	35.7
29.5	407	120	287	35.4
29.5	404	119	285	35.1
29.5	403	119	284	35.1
29.5	400	118	282	34.8
29.5	397	117	280	34.5
29.5	396	117	279	34.6
29.5	393	116	277	34.2
29.5	390	115	275	33.9
29.5	387	114	273	33.6
29.5	386	114	272	33.7
29.5	383	113	270	33.3
29.5	380	112	268	33.0
29.5	376	111	265	32.8
29.5	373	110	263	32.4
29.5	370	109	261	32.1
29.5	369	109	260	32.2
29.5	366	108	258	31.9
29.5	363	107	256	31.5
29.5	359	106	253	31.3
29.5	356	105	251	31.0

%E	M1	M2	DM	M*
29.5	353	104	249	30.6
29.5	352	104	248	30.7
29.5	349	103	246	30.4
29.5	346	102	244	30.1
29.5	342	101	241	29.8
29.5	339	100	239	29.5
29.5	336	99	237	29.2
29.5	332	98	234	28.9
29.5	329	97	232	28.6
29.5	325	96	229	28.4
29.5	322	95	227	28.0
29.5	319	94	225	27.7
29.5	315	93	222	27.5
29.5	312	92	220	27.1
29.5	309	91	218	26.8
29.5	308	91	217	26.9
29.5	305	90	215	26.6
29.5	302	89	213	26.2
29.5	298	88	210	26.0
29.5	295	87	208	25.7
29.5	292	86	206	25.3
29.5	288	85	203	25.1
29.5	285	84	201	24.8
29.5	281	83	198	24.5
29.5	278	82	196	24.2
29.5	275	81	194	23.9
29.5	271	80	191	23.6
29.5	268	79	189	23.3
29.5	264	78	186	23.0
29.5	261	77	184	22.7
29.5	258	76	182	22.4
29.5	254	75	179	22.1
29.5	251	74	177	21.8
29.5	244	72	172	21.2
29.5	241	71	170	20.9
29.5	237	70	167	20.7
29.5	234	69	165	20.3
29.5	227	67	160	19.8
29.5	224	66	158	19.4
29.5	220	65	155	19.2
29.5	217	64	153	18.9
29.5	210	62	148	18.3
29.5	207	61	146	18.0
29.5	200	59	141	17.4
29.5	193	57	136	16.8
29.5	190	56	134	16.5
29.5	183	54	129	15.9
29.5	176	52	124	15.4
29.5	173	51	122	15.0
29.5	166	49	117	14.5

%E	M1	M2	DM	M*
29.5	156	46	110	13.6
29.5	149	44	105	13.0
29.5	146	43	103	12.7
29.5	139	41	98	12.1
29.5	132	39	93	11.5
29.5	129	38	91	11.2
29.5	122	36	86	10.6
29.5	112	33	79	9.7
29.5	105	31	74	9.2
29.5	95	28	67	8.3
29.5	88	26	62	7.7
29.5	78	23	55	6.8
29.5	61	18	43	5.3
29.5	44	13	31	3.8
29.4	500	147	353	43.2
29.4	497	146	351	42.9
29.4	496	146	350	43.0
29.4	494	145	349	42.6
29.4	493	145	348	42.6
29.4	490	144	346	42.3
29.4	489	144	345	42.4
29.4	487	143	344	42.0
29.4	486	143	343	42.1
29.4	483	142	341	41.7
29.4	480	141	339	41.4
29.4	479	141	338	41.5
29.4	477	140	337	41.1
29.4	476	140	336	41.2
29.4	473	139	334	40.8
29.4	472	139	333	40.9
29.4	470	138	332	40.5
29.4	469	138	331	40.6
29.4	466	137	329	40.3
29.4	463	136	327	39.9
29.4	462	136	326	40.0
29.4	459	135	324	39.7
29.4	456	134	322	39.4
29.4	453	133	320	39.0
29.4	452	133	319	39.1
29.4	449	132	317	38.8
29.4	446	131	315	38.5
29.4	445	131	314	38.6
29.4	442	130	312	38.2
29.4	439	129	310	37.9
29.4	436	128	308	37.6
29.4	435	128	307	37.7
29.4	432	127	305	37.3
29.4	429	126	303	37.0
29.4	428	126	302	37.1
29.4	425	125	300	36.8

%E	M1	M2	DM	M*
29.4	422	124	298	36.4
29.4	419	123	296	36.1
29.4	418	123	295	36.2
29.4	415	122	293	35.9
29.4	412	121	291	35.5
29.4	411	121	290	35.6
29.4	408	120	288	35.3
29.4	405	119	286	35.0
29.4	402	118	284	34.6
29.4	401	118	283	34.7
29.4	398	117	281	34.4
29.4	395	116	279	34.1
29.4	394	116	278	34.2
29.4	391	115	276	33.8
29.4	388	114	274	33.5
29.4	385	113	272	33.2
29.4	384	113	271	33.3
29.4	381	112	269	32.9
29.4	378	111	267	32.6
29.4	377	111	266	32.7
29.4	374	110	264	32.4
29.4	371	109	262	32.0
29.4	367	108	259	31.8
29.4	364	107	257	31.5
29.4	361	106	255	31.1
29.4	360	106	254	31.2
29.4	357	105	252	30.9
29.4	354	104	250	30.6
29.4	350	103	247	30.3
29.4	347	102	245	30.0
29.4	344	101	243	29.7
29.4	343	101	242	29.7
29.4	340	100	240	29.4
29.4	337	99	238	29.1
29.4	333	98	235	28.8
29.4	330	97	233	28.5
29.4	327	96	231	28.2
29.4	326	96	230	28.3
29.4	323	95	228	27.9
29.4	320	94	226	27.6
29.4	316	93	223	27.4
29.4	313	92	221	27.0
29.4	310	91	219	26.7
29.4	306	90	216	26.5
29.4	303	89	214	26.1
29.4	299	88	211	25.9
29.4	296	87	209	25.6
29.4	293	86	207	25.2
29.4	289	85	204	25.0
29.4	286	84	202	24.7

%E	M1	M2	DM	M*
29.4	282	83	199	24.4
29.4	279	82	197	24.1
29.4	272	80	192	23.5
29.4	269	79	190	23.2
29.4	265	78	187	23.0
29.4	262	77	185	22.6
29.4	255	75	180	22.1
29.4	252	74	178	21.7
29.4	248	73	175	21.5
29.4	245	72	173	21.2
29.4	238	70	168	20.6
29.4	235	69	166	20.3
29.4	231	68	163	20.0
29.4	228	67	161	19.7
29.4	221	65	156	19.1
29.4	218	64	154	18.8
29.4	214	63	151	18.5
29.4	211	62	149	18.2
29.4	204	60	144	17.6
29.4	201	59	142	17.3
29.4	197	58	139	17.1
29.4	194	57	137	16.7
29.4	187	55	132	16.2
29.4	180	53	127	15.6
29.4	177	52	125	15.3
29.4	170	50	120	14.7
29.4	163	48	115	14.1
29.4	160	47	113	13.8
29.4	153	45	108	13.2
29.4	143	42	101	12.3
29.4	136	40	96	11.8
29.4	126	37	89	10.9
29.4	119	35	84	10.3
29.4	109	32	77	9.4
29.4	102	30	72	8.8
29.4	85	25	60	7.4
29.4	68	20	48	5.9
29.4	51	15	36	4.4
29.4	34	10	24	2.9
29.4	17	5	12	1.5
29.3	499	146	353	42.7
29.3	498	146	352	42.8
29.3	495	145	350	42.5
29.3	492	144	348	42.1
29.3	491	144	347	42.2
29.3	488	143	345	41.9
29.3	485	142	343	41.6
29.3	484	142	342	41.7
29.3	482	141	341	41.2
29.3	481	141	340	41.3

%E	M1	M2	DM	M*
29.3	478	140	338	41.0
29.3	475	139	336	40.7
29.3	474	139	335	40.8
29.3	471	138	333	40.4
29.3	468	137	331	40.1
29.3	467	137	330	40.2
29.3	464	136	328	39.9
29.3	461	135	326	39.5
29.3	460	135	325	39.6
29.3	458	134	324	39.2
29.3	457	134	323	39.3
29.3	454	133	321	39.0
29.3	451	132	319	38.6
29.3	450	132	318	38.7
29.3	447	131	316	38.4
29.3	444	130	314	38.1
29.3	443	130	313	38.1
29.3	441	129	312	37.7
29.3	440	129	311	37.8
29.3	437	128	309	37.5
29.3	434	127	307	37.2
29.3	433	127	306	37.2
29.3	430	126	304	36.9
29.3	427	125	302	36.6
29.3	426	125	301	36.7
29.3	423	124	299	36.3
29.3	420	123	297	36.0
29.3	417	122	295	35.7
29.3	416	122	294	35.8
29.3	413	121	292	35.5
29.3	410	120	290	35.1
29.3	409	120	289	35.2
29.3	406	119	287	34.9
29.3	403	118	285	34.6
29.3	400	117	283	34.2
29.3	399	117	282	34.3
29.3	396	116	280	34.0
29.3	393	115	278	33.7
29.3	392	115	277	33.7
29.3	389	114	275	33.4
29.3	386	113	273	33.1
29.3	382	112	270	32.8
29.3	379	111	268	32.5
29.3	376	110	266	32.2
29.3	375	110	265	32.3
29.3	372	109	263	31.9
29.3	369	108	261	31.6
29.3	368	108	260	31.7
29.3	365	107	258	31.4
29.3	362	106	256	31.0
29.3	358	105	253	30.8
29.3	355	104	251	30.5
29.3	352	103	249	30.1
29.3	351	103	248	30.2
29.3	348	102	246	29.9
29.3	345	101	244	29.6
29.3	341	100	241	29.3
29.3	338	99	239	29.0
29.3	335	98	237	28.7
29.3	334	98	236	28.8
29.3	331	97	234	28.4
29.3	328	96	232	28.1
29.3	324	95	229	27.9
29.3	321	94	227	27.5
29.3	317	93	224	27.3
29.3	314	92	222	27.0
29.3	311	91	220	26.6
29.3	307	90	217	26.4
29.3	304	89	215	26.1
29.3	300	88	212	25.8
29.3	297	87	210	25.5
29.3	294	86	208	25.2
29.3	290	85	205	24.9
29.3	287	84	203	24.6
29.3	283	83	200	24.3
29.3	280	82	198	24.0
29.3	276	81	195	23.8
29.3	273	80	193	23.4
29.3	270	79	191	23.1
29.3	266	78	188	22.9
29.3	263	77	186	22.5
29.3	259	76	183	22.3
29.3	256	75	181	22.0
29.3	249	73	176	21.4
29.3	246	72	174	21.1
29.3	242	71	171	20.8
29.3	239	70	169	20.5
29.3	232	68	164	19.9
29.3	229	67	162	19.6
29.3	225	66	159	19.4
29.3	222	65	157	19.0
29.3	215	63	152	18.5
29.3	208	61	147	17.9
29.3	205	60	145	17.6
29.3	198	58	140	17.0
29.3	191	56	135	16.4
29.3	188	55	133	16.1
29.3	184	54	130	15.8
29.3	181	53	128	15.5
29.3	174	51	123	14.9
29.3	167	49	118	14.4
29.3	164	48	116	14.0
29.3	157	46	111	13.5
29.3	150	44	106	12.9
29.3	147	43	104	12.6
29.3	140	41	99	12.0
29.3	133	39	94	11.4
29.3	123	36	87	10.5
29.3	116	34	82	10.0
29.3	99	29	70	8.5
29.3	92	27	65	7.9
29.3	82	24	58	7.0
29.3	75	22	53	6.5
29.3	58	17	41	5.0
29.3	41	12	29	3.5
29.2	500	146	354	42.6
29.2	497	145	352	42.3
29.2	496	145	351	42.4
29.2	494	144	350	42.0
29.2	493	144	349	42.1
29.2	490	143	347	41.7
29.2	489	143	346	41.8
29.2	487	142	345	41.4
29.2	486	142	344	41.5
29.2	483	141	342	41.2
29.2	480	140	340	40.8
29.2	479	140	339	40.9
29.2	476	139	337	40.6
29.2	473	138	335	40.3
29.2	472	138	334	40.3
29.2	469	137	332	40.0
29.2	466	136	330	39.7
29.2	465	136	329	39.8
29.2	463	135	328	39.4
29.2	462	135	327	39.4
29.2	459	134	325	39.1
29.2	456	133	323	38.8
29.2	455	133	322	38.9
29.2	452	132	320	38.5
29.2	449	131	318	38.2
29.2	448	131	317	38.3
29.2	445	130	315	38.0
29.2	442	129	313	37.6
29.2	439	128	311	37.3
29.2	438	128	310	37.4
29.2	435	127	308	37.1
29.2	432	126	306	36.8
29.2	431	126	305	36.8
29.2	428	125	303	36.5
29.2	425	124	301	36.2
29.2	424	124	300	36.3
29.2	421	123	298	35.9
29.2	418	122	296	35.6
29.2	415	121	294	35.3
29.2	414	121	293	35.4
29.2	411	120	291	35.0
29.2	408	119	289	34.7
29.2	407	119	288	34.8
29.2	404	118	286	34.5
29.2	401	117	284	34.1
29.2	397	116	281	33.9
29.2	394	115	279	33.6
29.2	391	114	277	33.2
29.2	390	114	276	33.3
29.2	387	113	274	33.0
29.2	384	112	272	32.7
29.2	383	112	271	32.8
29.2	380	111	269	32.4
29.2	377	110	267	32.1
29.2	373	109	264	31.9
29.2	370	108	262	31.5
29.2	367	107	260	31.2
29.2	366	107	259	31.3
29.2	363	106	257	31.0
29.2	360	105	255	30.6
29.2	359	105	254	30.7
29.2	356	104	252	30.4
29.2	353	103	250	30.1
29.2	349	102	247	29.8
29.2	346	101	245	29.5
29.2	343	100	243	29.2
29.2	342	100	242	29.2
29.2	339	99	240	28.9
29.2	336	98	238	28.6
29.2	332	97	235	28.3
29.2	329	96	233	28.0
29.2	325	95	230	27.8
29.2	322	94	228	27.4
29.2	319	93	226	27.1
29.2	318	93	225	27.2
29.2	315	92	223	26.9
29.2	312	91	221	26.5
29.2	308	90	218	26.3
29.2	305	89	216	26.0
29.2	301	88	213	25.7
29.2	298	87	211	25.4
29.2	295	86	209	25.1
29.2	291	85	206	24.8
29.2	288	84	204	24.5
29.2	284	83	201	24.3
29.2	281	82	199	23.9
29.2	277	81	196	23.7
29.2	274	80	194	23.4
29.2	271	79	192	23.0
29.2	267	78	189	22.8
29.2	264	77	187	22.5
29.2	260	76	184	22.2
29.2	257	75	182	21.9
29.2	253	74	179	21.6
29.2	250	73	177	21.3
29.2	247	72	175	21.0
29.2	243	71	172	20.7
29.2	240	70	170	20.4
29.2	236	69	167	20.2
29.2	233	68	165	19.8
29.2	226	66	160	19.3
29.2	219	64	155	18.7
29.2	216	63	153	18.4
29.2	212	62	150	18.1
29.2	209	61	148	17.8
29.2	202	59	143	17.2
29.2	195	57	138	16.7
29.2	192	56	136	16.3
29.2	185	54	131	15.8
29.2	178	52	126	15.2
29.2	171	50	121	14.6
29.2	168	49	119	14.3
29.2	161	47	114	13.7
29.2	154	45	109	13.1
29.2	144	42	102	12.3
29.2	137	40	97	11.7
29.2	130	38	92	11.1
29.2	120	35	85	10.2
29.2	113	33	80	9.6
29.2	106	31	75	9.1
29.2	96	28	68	8.2
29.2	89	26	63	7.6
29.2	72	21	51	6.1
29.2	65	19	46	5.6
29.2	48	14	34	4.1
29.2	24	7	17	2.0
29.1	499	145	354	42.1
29.1	498	145	353	42.2
29.1	495	144	351	41.9
29.1	492	143	349	41.6
29.1	491	143	348	41.6
29.1	488	142	346	41.3
29.1	485	141	344	41.0
29.1	484	141	343	41.1
29.1	481	140	341	40.7

%E	M1	M2	DM	M*
29.1	478	139	339	40.4
29.1	477	139	338	40.5
29.1	475	138	337	40.1
29.1	474	138	336	40.2
29.1	471	137	334	39.8
29.1	470	137	333	39.9
29.1	468	136	332	39.5
29.1	467	136	331	39.6
29.1	464	135	329	39.3
29.1	461	134	327	39.0
29.1	460	134	326	39.0
29.1	457	133	324	38.7
29.1	454	132	322	38.4
29.1	453	132	321	38.5
29.1	450	131	319	38.1
29.1	447	130	317	37.8
29.1	446	130	316	37.9
29.1	444	129	315	37.5
29.1	443	129	314	37.6
29.1	440	128	312	37.2
29.1	437	127	310	36.9
29.1	436	127	309	37.0
29.1	433	126	307	36.7
29.1	430	125	305	36.3
29.1	429	125	304	36.4
29.1	426	124	302	36.1
29.1	423	123	300	35.8
29.1	422	123	299	35.9
29.1	419	122	297	35.5
29.1	416	121	295	35.2
29.1	413	120	293	34.9
29.1	412	120	292	35.0
29.1	409	119	290	34.6
29.1	406	118	288	34.3
29.1	405	118	287	34.4
29.1	402	117	285	34.1
29.1	399	116	283	33.7
29.1	398	116	282	33.8
29.1	395	115	280	33.5
29.1	392	114	278	33.2
29.1	388	113	275	32.9
29.1	385	112	273	32.6
29.1	382	111	271	32.3
29.1	381	111	270	32.3
29.1	378	110	268	32.0
29.1	375	109	266	31.7
29.1	374	109	265	31.8
29.1	371	108	263	31.4
29.1	368	107	261	31.1
29.1	364	106	258	30.9

%E	M1	M2	DM	M*
29.1	361	105	256	30.5
29.1	358	104	254	30.2
29.1	357	104	253	30.3
29.1	354	103	251	30.0
29.1	351	102	249	29.6
29.1	350	102	248	29.7
29.1	347	101	246	29.4
29.1	344	100	244	29.1
29.1	340	99	241	28.8
29.1	337	98	239	28.5
29.1	333	97	236	28.3
29.1	330	96	234	27.9
29.1	327	95	232	27.6
29.1	326	95	231	27.7
29.1	323	94	229	27.4
29.1	320	93	227	27.0
29.1	316	92	224	26.8
29.1	313	91	222	26.5
29.1	309	90	219	26.2
29.1	306	89	217	25.9
29.1	302	88	214	25.6
29.1	299	87	212	25.3
29.1	296	86	210	25.0
29.1	292	85	207	24.7
29.1	289	84	205	24.4
29.1	285	83	202	24.2
29.1	282	82	200	23.8
29.1	278	81	197	23.6
29.1	275	80	195	23.3
29.1	268	78	190	22.7
29.1	265	77	188	22.4
29.1	261	76	185	22.1
29.1	258	75	183	21.8
29.1	254	74	180	21.6
29.1	251	73	178	21.2
29.1	244	71	173	20.7
29.1	237	69	168	20.1
29.1	234	68	166	19.8
29.1	230	67	163	19.5
29.1	227	66	161	19.2
29.1	223	65	158	18.9
29.1	220	64	156	18.6
29.1	213	62	151	18.0
29.1	206	60	146	17.5
29.1	203	59	144	17.1
29.1	199	58	141	16.9
29.1	196	57	139	16.6
29.1	189	55	134	16.0
29.1	182	53	129	15.4
29.1	179	52	127	15.1

%E	M1	M2	DM	M*
29.1	175	51	124	14.9
29.1	172	50	122	14.5
29.1	165	48	117	14.0
29.1	158	46	112	13.4
29.1	151	44	107	12.8
29.1	148	43	105	12.5
29.1	141	41	100	11.9
29.1	134	39	95	11.4
29.1	127	37	90	10.8
29.1	117	34	83	9.9
29.1	110	32	78	9.3
29.1	103	30	73	8.7
29.1	86	25	61	7.3
29.1	79	23	56	6.7
29.1	55	16	39	4.7
29.0	500	145	355	42.0
29.0	497	144	353	41.7
29.0	496	144	352	41.8
29.0	493	143	350	41.5
29.0	490	142	348	41.2
29.0	489	142	347	41.2
29.0	487	141	346	40.8
29.0	486	141	345	40.9
29.0	483	140	343	40.6
29.0	482	140	342	40.7
29.0	480	139	341	40.3
29.0	479	139	340	40.3
29.0	476	138	338	40.0
29.0	473	137	336	39.7
29.0	472	137	335	39.8
29.0	469	136	333	39.4
29.0	466	135	331	39.1
29.0	465	135	330	39.2
29.0	462	134	328	38.9
29.0	459	133	326	38.5
29.0	458	133	325	38.6
29.0	455	132	323	38.3
29.0	452	131	321	38.0
29.0	451	131	320	38.1
29.0	449	130	319	37.6
29.0	448	130	318	37.7
29.0	445	129	316	37.4
29.0	442	128	314	37.1
29.0	441	128	313	37.2
29.0	438	127	311	36.8
29.0	435	126	309	36.5
29.0	434	126	308	36.6
29.0	431	125	306	36.3
29.0	428	124	304	35.9
29.0	427	124	303	36.0

%E	M1	M2	DM	M*
29.0	424	123	301	35.7
29.0	421	122	299	35.4
29.0	420	122	298	35.4
29.0	417	121	296	35.1
29.0	414	120	294	34.8
29.0	411	119	292	34.5
29.0	410	119	291	34.5
29.0	407	118	289	34.2
29.0	404	117	287	33.9
29.0	403	117	286	34.0
29.0	400	116	284	33.6
29.0	397	115	282	33.3
29.0	396	115	281	33.4
29.0	393	114	279	33.1
29.0	390	113	277	32.7
29.0	389	113	276	32.8
29.0	386	112	274	32.5
29.0	383	111	272	32.2
29.0	379	110	269	31.9
29.0	376	109	267	31.6
29.0	373	108	265	31.3
29.0	372	108	264	31.4
29.0	369	107	262	31.0
29.0	366	106	260	30.7
29.0	365	106	259	30.8
29.0	362	105	257	30.5
29.0	359	104	255	30.1
29.0	355	103	252	29.9
29.0	352	102	250	29.6
29.0	348	101	247	29.3
29.0	345	100	245	29.0
29.0	341	99	242	28.7
29.0	338	98	240	28.4
29.0	335	97	238	28.1
29.0	334	97	237	28.2
29.0	331	96	235	27.8
29.0	328	95	233	27.5
29.0	324	94	230	27.3
29.0	321	93	228	26.9
29.0	317	92	225	26.7
29.0	314	91	223	26.4
29.0	310	90	220	26.1
29.0	307	89	218	25.8
29.0	303	88	215	25.6
29.0	300	87	213	25.2
29.0	297	86	211	24.9
29.0	293	85	208	24.7
29.0	290	84	206	24.3
29.0	286	83	203	24.1
29.0	283	82	201	23.8

%E	M1	M2	DM	M*
29.0	279	81	198	23.5
29.0	276	80	196	23.2
29.0	272	79	193	22.9
29.0	269	78	191	22.6
29.0	262	76	186	22.0
29.0	259	75	184	21.7
29.0	255	74	181	21.5
29.0	252	73	179	21.1
29.0	248	72	176	20.9
29.0	245	71	174	20.6
29.0	241	70	171	20.3
29.0	238	69	169	20.0
29.0	231	67	164	19.4
29.0	224	65	159	18.9
29.0	221	64	157	18.5
29.0	217	63	154	18.3
29.0	214	62	152	18.0
29.0	210	61	149	17.7
29.0	207	60	147	17.4
29.0	200	58	142	16.8
29.0	193	56	137	16.2
29.0	186	54	132	15.7
29.0	183	53	130	15.3
29.0	176	51	125	14.8
29.0	169	49	120	14.2
29.0	162	47	115	13.6
29.0	155	45	110	13.1
29.0	145	42	103	12.2
29.0	138	40	98	11.6
29.0	131	38	93	11.0
29.0	124	36	88	10.5
29.0	107	31	76	9.0
29.0	100	29	71	8.4
29.0	93	27	66	7.8
29.0	69	20	49	5.8
29.0	62	18	44	5.2
29.0	31	9	22	2.6
28.9	499	144	355	41.6
28.9	498	144	354	41.6
28.9	495	143	352	41.3
28.9	494	143	351	41.4
28.9	492	142	350	41.0
28.9	491	142	349	41.1
28.9	488	141	347	40.7
28.9	485	140	345	40.4
28.9	484	140	344	40.5
28.9	481	139	342	40.2
28.9	478	138	340	39.8
28.9	477	138	339	39.9
28.9	474	137	337	39.6

%E	M1	M2	DM	M*
28.9	471	136	335	39.3
28.9	470	136	334	39.4
28.9	467	135	332	39.0
28.9	464	134	330	38.7
28.9	463	134	329	38.8
28.9	461	133	328	38.4
28.9	460	133	327	38.5
28.9	457	132	325	38.1
28.9	456	132	324	38.2
28.9	454	131	323	37.8
28.9	453	131	322	37.9
28.9	450	130	320	37.6
28.9	447	129	318	37.2
28.9	446	129	317	37.3
28.9	443	128	315	37.0
28.9	440	127	313	36.7
28.9	439	127	312	36.7
28.9	436	126	310	36.4
28.9	433	125	308	36.1
28.9	432	125	307	36.2
28.9	429	124	305	35.8
28.9	426	123	303	35.5
28.9	425	123	302	35.6
28.9	422	122	300	35.3
28.9	419	121	298	34.9
28.9	418	121	297	35.0
28.9	415	120	295	34.7
28.9	412	119	293	34.4
28.9	409	118	291	34.0
28.9	408	118	290	34.1
28.9	405	117	288	33.8
28.9	402	116	286	33.5
28.9	401	116	285	33.6
28.9	398	115	283	33.2
28.9	395	114	281	32.9
28.9	394	114	280	33.0
28.9	391	113	278	32.7
28.9	388	112	276	32.3
28.9	387	112	275	32.4
28.9	384	111	273	32.1
28.9	381	110	271	31.8
28.9	380	110	270	31.8
28.9	377	109	268	31.5
28.9	374	108	266	31.2
28.9	370	107	263	30.9
28.9	367	106	261	30.6
28.9	363	105	258	30.4
28.9	360	104	256	30.0
28.9	357	103	254	29.7
28.9	356	103	253	29.8
28.9	353	102	251	29.5
28.9	350	101	249	29.1
28.9	349	101	248	29.2
28.9	346	100	246	28.9
28.9	343	99	244	28.6
28.9	342	99	243	28.7
28.9	339	98	241	28.3
28.9	336	97	239	28.0
28.9	332	96	236	27.8
28.9	329	95	234	27.4
28.9	325	94	231	27.2
28.9	322	93	229	26.9
28.9	318	92	226	26.6
28.9	315	91	224	26.3
28.9	311	90	221	26.0
28.9	308	89	219	25.7
28.9	305	88	217	25.4
28.9	304	88	216	25.5
28.9	301	87	214	25.1
28.9	298	86	212	24.8
28.9	294	85	209	24.6
28.9	291	84	207	24.2
28.9	287	83	204	24.0
28.9	284	82	202	23.7
28.9	280	81	199	23.4
28.9	277	80	197	23.1
28.9	273	79	194	22.9
28.9	270	78	192	22.5
28.9	266	77	189	22.3
28.9	263	76	187	22.0
28.9	256	74	182	21.4
28.9	253	73	180	21.1
28.9	249	72	177	20.8
28.9	246	71	175	20.5
28.9	242	70	172	20.2
28.9	239	69	170	19.9
28.9	235	68	167	19.7
28.9	232	67	165	19.3
28.9	228	66	162	19.1
28.9	225	65	160	18.8
28.9	218	63	155	18.2
28.9	211	61	150	17.6
28.9	204	59	145	17.1
28.9	201	58	143	16.7
28.9	197	57	140	16.5
28.9	194	56	138	16.2
28.9	190	55	135	15.9
28.9	187	54	133	15.6
28.9	180	52	128	15.0
28.9	173	50	123	14.5
28.9	166	48	118	13.9
28.9	159	46	113	13.3
28.9	152	44	108	12.7
28.9	149	43	106	12.4
28.9	142	41	101	11.8
28.9	135	39	96	11.3
28.9	128	37	91	10.7
28.9	121	35	86	10.1
28.9	114	33	81	9.6
28.9	97	28	69	8.1
28.9	90	26	64	7.5
28.9	83	24	59	6.9
28.9	76	22	54	6.4
28.9	45	13	32	3.8
28.9	38	11	27	3.2
28.8	500	144	356	41.5
28.8	497	143	354	41.1
28.8	496	143	353	41.2
28.8	493	142	351	40.9
28.8	490	141	349	40.6
28.8	489	141	348	40.7
28.8	486	140	346	40.3
28.8	483	139	344	40.0
28.8	482	139	343	40.1
28.8	480	138	342	39.7
28.8	479	138	341	39.8
28.8	476	137	339	39.4
28.8	475	137	338	39.5
28.8	473	136	337	39.1
28.8	472	136	336	39.2
28.8	469	135	334	38.9
28.8	468	135	333	38.9
28.8	466	134	332	38.5
28.8	465	134	331	38.6
28.8	462	133	329	38.3
28.8	459	132	327	38.0
28.8	458	132	326	38.0
28.8	455	131	324	37.7
28.8	452	130	322	37.4
28.8	451	130	321	37.5
28.8	448	129	319	37.1
28.8	445	128	317	36.8
28.8	444	128	316	36.9
28.8	441	127	314	36.6
28.8	438	126	312	36.2
28.8	437	126	311	36.3
28.8	434	125	309	36.0
28.8	431	124	307	35.7
28.8	430	124	306	35.8
28.8	427	123	304	35.4
28.8	424	122	302	35.1
28.8	423	122	301	35.2
28.8	420	121	299	34.9
28.8	417	120	297	34.5
28.8	416	120	296	34.6
28.8	413	119	294	34.3
28.8	410	118	292	34.0
28.8	406	117	289	33.7
28.8	403	116	287	33.4
28.8	400	115	285	33.1
28.8	399	115	284	33.1
28.8	396	114	282	32.8
28.8	393	113	280	32.5
28.8	392	113	279	32.6
28.8	389	112	277	32.2
28.8	386	111	275	31.9
28.8	385	111	274	32.0
28.8	382	110	272	31.7
28.8	379	109	270	31.3
28.8	378	109	269	31.4
28.8	375	108	267	31.1
28.8	372	107	265	30.8
28.8	371	107	264	30.9
28.8	368	106	262	30.5
28.8	365	105	260	30.2
28.8	364	105	259	30.3
28.8	361	104	257	30.0
28.8	358	103	255	29.6
28.8	354	102	252	29.4
28.8	351	101	250	29.1
28.8	347	100	247	28.8
28.8	344	99	245	28.5
28.8	340	98	242	28.2
28.8	337	97	240	27.9
28.8	333	96	237	27.7
28.8	330	95	235	27.3
28.8	326	94	232	27.1
28.8	323	93	230	26.8
28.8	320	92	228	26.4
28.8	319	92	227	26.5
28.8	316	91	225	26.2
28.8	313	90	223	25.9
28.8	312	90	222	26.0
28.8	309	89	220	25.6
28.8	306	88	218	25.3
28.8	302	87	215	25.1
28.8	299	86	213	24.7
28.8	295	85	210	24.5
28.8	292	84	208	24.2
28.8	288	83	205	23.9
28.8	285	82	203	23.6
28.8	281	81	200	23.3
28.8	278	80	198	23.0
28.8	274	79	195	22.8
28.8	271	78	193	22.5
28.8	267	77	190	22.2
28.8	264	76	188	21.9
28.8	260	75	185	21.6
28.8	257	74	183	21.3
28.8	250	72	178	20.7
28.8	243	70	173	20.2
28.8	240	69	171	19.8
28.8	236	68	168	19.6
28.8	233	67	166	19.3
28.8	229	66	163	19.0
28.8	226	65	161	18.7
28.8	222	64	158	18.5
28.8	219	63	156	18.1
28.8	215	62	153	17.9
28.8	212	61	151	17.6
28.8	208	60	148	17.3
28.8	205	59	146	17.0
28.8	198	57	141	16.4
28.8	191	55	136	15.8
28.8	184	53	131	15.3
28.8	177	51	126	14.7
28.8	170	49	121	14.1
28.8	163	47	116	13.6
28.8	160	46	114	13.2
28.8	156	45	111	13.0
28.8	153	44	109	12.7
28.8	146	42	104	12.1
28.8	139	40	99	11.5
28.8	132	38	94	10.9
28.8	125	36	89	10.4
28.8	118	34	84	9.8
28.8	111	32	79	9.2
28.8	104	30	74	8.7
28.8	80	23	57	6.6
28.8	73	21	52	6.0
28.8	66	19	47	5.5
28.8	59	17	42	4.9
28.8	52	15	37	4.3
28.7	499	143	356	41.0
28.7	498	143	355	41.1
28.7	495	142	353	40.7
28.7	494	142	352	40.8
28.7	492	141	351	40.4
28.7	491	141	350	40.5
28.7	488	140	348	40.2

%E	M1	M2	DM	M*	%E	M1	M2	DM	M*	%E	M1	M2	DM	M*	%E	M1	M2	DM	M*	%E	M1	M2	DM	M*
28.7	487	140	347	40.2	28.7	359	103	256	29.6	28.7	171	49	122	14.0	28.6	416	119	297	34.0	28.6	266	76	190	21.7
28.7	485	139	346	39.8	28.7	356	102	254	29.2	28.7	167	48	119	13.8	28.6	413	118	295	33.7	28.6	262	75	187	21.5
28.7	484	139	345	39.9	28.7	355	102	253	29.3	28.7	164	47	117	13.5	28.6	412	118	294	33.8	28.6	259	74	185	21.1
28.7	481	138	343	39.6	28.7	352	101	251	29.0	28.7	157	45	112	12.9	28.6	409	117	292	33.5	28.6	255	73	182	20.9
28.7	478	137	341	39.3	28.7	349	100	249	28.7	28.7	150	43	107	12.3	28.6	406	116	290	33.1	28.6	252	72	180	20.6
28.7	477	137	340	39.3	28.7	348	100	248	28.7	28.7	143	41	102	11.8	28.6	405	116	289	33.2	28.6	248	71	177	20.3
28.7	474	136	338	39.0	28.7	345	99	246	28.4	28.7	136	39	97	11.2	28.6	402	115	287	32.9	28.6	245	70	175	20.0
28.7	471	135	336	38.7	28.7	342	98	244	28.1	28.7	129	37	92	10.6	28.6	399	114	285	32.6	28.6	241	69	172	19.8
28.7	470	135	335	38.8	28.7	341	98	243	28.2	28.7	122	35	87	10.0	28.6	398	114	284	32.7	28.6	238	68	170	19.4
28.7	467	134	333	38.4	28.7	338	97	241	27.8	28.7	115	33	82	9.5	28.6	395	113	282	32.3	28.6	234	67	167	19.2
28.7	464	133	331	38.1	28.7	335	96	239	27.5	28.7	108	31	77	8.9	28.6	392	112	280	32.0	28.6	231	66	165	18.9
28.7	463	133	330	38.2	28.7	334	96	238	27.6	28.7	101	29	72	8.3	28.6	391	112	279	32.1	28.6	227	65	162	18.6
28.7	460	132	328	37.9	28.7	331	95	236	27.3	28.7	94	27	67	7.8	28.6	388	111	277	31.8	28.6	224	64	160	18.3
28.7	457	131	326	37.6	28.7	328	94	234	26.9	28.7	87	25	62	7.2	28.6	385	110	275	31.4	28.6	220	63	157	18.0
28.7	456	131	325	37.6	28.7	327	94	233	27.0	28.6	500	143	357	40.9	28.6	384	110	274	31.5	28.6	217	62	155	17.7
28.7	453	130	323	37.3	28.7	324	93	231	26.7	28.6	497	142	355	40.6	28.6	381	109	272	31.2	28.6	213	61	152	17.5
28.7	450	129	321	37.0	28.7	321	92	229	26.4	28.6	496	142	354	40.7	28.6	378	108	270	30.9	28.6	210	60	150	17.1
28.7	449	129	320	37.1	28.7	317	91	226	26.1	28.6	493	141	352	40.3	28.6	377	108	269	30.9	28.6	206	59	147	16.9
28.7	446	128	318	36.7	28.7	314	90	224	25.8	28.6	490	140	350	40.0	28.6	374	107	267	30.6	28.6	203	58	145	16.6
28.7	443	127	316	36.4	28.7	310	89	221	25.6	28.6	489	140	349	40.1	28.6	371	106	265	30.3	28.6	199	57	142	16.3
28.7	442	127	315	36.5	28.7	307	88	219	25.2	28.6	486	139	347	39.8	28.6	370	106	264	30.4	28.6	196	56	140	16.0
28.7	439	126	313	36.2	28.7	303	87	216	25.0	28.6	483	138	345	39.4	28.6	367	105	262	30.0	28.6	192	55	137	15.8
28.7	436	125	311	35.8	28.7	300	86	214	24.7	28.6	482	138	344	39.5	28.6	364	104	260	29.7	28.6	189	54	135	15.4
28.7	435	125	310	35.9	28.7	296	85	211	24.4	28.6	479	137	342	39.2	28.6	360	103	257	29.5	28.6	185	53	132	15.2
28.7	432	124	308	35.6	28.7	293	84	209	24.1	28.6	476	136	340	38.9	28.6	357	102	255	29.1	28.6	182	52	130	14.9
28.7	429	123	306	35.3	28.7	289	83	206	23.8	28.6	475	136	339	38.9	28.6	353	101	252	28.9	28.6	175	50	125	14.3
28.7	428	123	305	35.3	28.7	286	82	204	23.5	28.6	472	135	337	38.6	28.6	350	100	250	28.6	28.6	168	48	120	13.7
28.7	425	122	303	35.0	28.7	282	81	201	23.3	28.6	469	134	335	38.3	28.6	346	99	247	28.3	28.6	161	46	115	13.1
28.7	422	121	301	34.7	28.7	279	80	199	22.9	28.6	468	134	334	38.4	28.6	343	98	245	28.0	28.6	154	44	110	12.6
28.7	421	121	300	34.8	28.7	275	79	196	22.7	28.6	465	133	332	38.0	28.6	339	97	242	27.8	28.6	147	42	105	12.0
28.7	418	120	298	34.4	28.7	272	78	194	22.4	28.6	462	132	330	37.7	28.6	336	96	240	27.4	28.6	140	40	100	11.4
28.7	415	119	296	34.1	28.7	268	77	191	22.1	28.6	461	132	329	37.8	28.6	332	95	237	27.2	28.6	133	38	95	10.9
28.7	414	119	295	34.2	28.7	265	76	189	21.8	28.6	458	131	327	37.5	28.6	329	94	235	26.9	28.6	126	36	90	10.3
28.7	411	118	293	33.9	28.7	261	75	186	21.6	28.6	455	130	325	37.1	28.6	325	93	232	26.6	28.6	119	34	85	9.7
28.7	408	117	291	33.6	28.7	258	74	184	21.2	28.6	454	130	324	37.2	28.6	322	92	230	26.3	28.6	112	32	80	9.1
28.7	407	117	290	33.6	28.7	254	73	181	21.0	28.6	451	129	322	36.9	28.6	318	91	227	26.0	28.6	105	30	75	8.6
28.7	404	116	288	33.3	28.7	251	72	179	20.7	28.6	448	128	320	36.6	28.6	315	90	225	25.7	28.6	98	28	70	8.0
28.7	401	115	286	33.0	28.7	247	71	176	20.4	28.6	447	128	319	36.7	28.6	311	89	222	25.5	28.6	91	26	65	7.4
28.7	397	114	283	32.7	28.7	244	70	174	20.1	28.6	444	127	317	36.3	28.6	308	88	220	25.1	28.6	84	24	60	6.9
28.7	394	113	281	32.4	28.7	237	68	169	19.5	28.6	441	126	315	36.0	28.6	304	87	217	24.9	28.6	77	22	55	6.3
28.7	390	112	278	32.2	28.7	230	66	164	18.9	28.6	440	126	314	36.1	28.6	301	86	215	24.6	28.6	70	20	50	5.7
28.7	387	111	276	31.8	28.7	223	64	159	18.4	28.6	437	125	312	35.8	28.6	297	85	212	24.3	28.6	63	18	45	5.1
28.7	383	110	273	31.6	28.7	216	62	154	17.8	28.6	434	124	310	35.4	28.6	294	84	210	24.0	28.6	56	16	40	4.6
28.7	380	109	271	31.3	28.7	209	60	149	17.2	28.6	433	124	309	35.5	28.6	290	83	207	23.8	28.6	49	14	35	4.0
28.7	376	108	268	31.0	28.7	202	58	144	16.7	28.6	430	123	307	35.2	28.6	287	82	205	23.4	28.6	42	12	30	3.4
28.7	373	107	266	30.7	28.7	195	56	139	16.1	28.6	427	122	305	34.9	28.6	283	81	202	23.2	28.6	35	10	25	2.9
28.7	369	106	263	30.4	28.7	188	54	134	15.5	28.6	426	122	304	34.9	28.6	280	80	200	22.9	28.6	28	8	20	2.3
28.7	366	105	261	30.1	28.7	181	52	129	14.9	28.6	423	121	302	34.6	28.6	276	79	197	22.6	28.6	21	6	15	1.7
28.7	363	104	259	29.8	28.7	178	51	127	14.6	28.6	420	120	300	34.3	28.6	273	78	195	22.3	28.6	14	4	10	1.1
28.7	362	104	258	29.9	28.7	174	50	124	14.4	28.6	419	120	299	34.4	28.6	269	77	192	22.0	28.6	7	2	5	0.6

%E	M1	M2	DM	M*	%E	M1	M2	DM	M*	%E	M1	M2	DM	M*	%E	M1	M2	DM	M*	%E	M1	M2	DM	M*
28.5	499	142	357	40.4	28.5	393	112	281	31.9	28.5	239	68	171	19.3	28.4	430	122	308	34.6	28.4	292	83	209	23.6
28.5	498	142	356	40.5	28.5	390	111	279	31.6	28.5	235	67	168	19.1	28.4	429	122	307	34.7	28.4	289	82	207	23.3
28.5	495	141	354	40.2	28.5	389	111	278	31.7	28.5	228	65	163	18.5	28.4	426	121	305	34.4	28.4	285	81	204	23.0
28.5	494	141	353	40.2	28.5	386	110	276	31.3	28.5	221	63	158	18.0	28.4	423	120	303	34.0	28.4	282	80	202	22.7
28.5	492	140	352	39.8	28.5	383	109	274	31.0	28.5	214	61	153	17.4	28.4	422	120	302	34.1	28.4	278	79	199	22.4
28.5	491	140	351	39.9	28.5	382	109	273	31.1	28.5	207	59	148	16.8	28.4	419	119	300	33.8	28.4	275	78	197	22.1
28.5	488	139	349	39.6	28.5	379	108	271	30.8	28.5	200	57	143	16.2	28.4	416	118	298	33.5	28.4	271	77	194	21.9
28.5	487	139	348	39.7	28.5	376	107	269	30.4	28.5	193	55	138	15.7	28.4	415	118	297	33.6	28.4	268	76	192	21.6
28.5	485	138	347	39.3	28.5	375	107	268	30.5	28.5	186	53	133	15.1	28.4	412	117	295	33.2	28.4	264	75	189	21.3
28.5	484	138	346	39.3	28.5	372	106	266	30.2	28.5	179	51	128	14.5	28.4	409	116	293	32.9	28.4	261	74	187	21.0
28.5	481	137	344	39.0	28.5	369	105	264	29.9	28.5	172	49	123	14.0	28.4	408	116	292	33.0	28.4	257	73	184	20.7
28.5	480	137	343	39.1	28.5	368	105	263	30.0	28.5	165	47	118	13.4	28.4	405	115	290	32.7	28.4	250	71	179	20.2
28.5	478	136	342	38.7	28.5	365	104	261	29.6	28.5	158	45	113	12.8	28.4	402	114	288	32.3	28.4	243	69	174	19.6
28.5	477	136	341	38.8	28.5	362	103	259	29.3	28.5	151	43	108	12.2	28.4	401	114	287	32.4	28.4	236	67	169	19.0
28.5	474	135	339	38.4	28.5	361	103	258	29.4	28.5	144	41	103	11.7	28.4	398	113	285	32.1	28.4	232	66	166	18.8
28.5	473	135	338	38.5	28.5	358	102	256	29.1	28.5	137	39	98	11.1	28.4	395	112	283	31.8	28.4	229	65	164	18.4
28.5	471	134	337	38.1	28.5	355	101	254	28.7	28.5	130	37	93	10.5	28.4	394	112	282	31.8	28.4	225	64	161	18.2
28.5	470	134	336	38.2	28.5	354	101	253	28.8	28.5	123	35	88	10.0	28.4	391	111	280	31.5	28.4	222	63	159	17.9
28.5	467	133	334	37.9	28.5	351	100	251	28.5	28.4	500	142	358	40.3	28.4	388	110	278	31.2	28.4	218	62	156	17.6
28.5	466	133	333	38.0	28.5	347	99	248	28.2	28.4	497	141	356	40.0	28.4	387	110	277	31.3	28.4	215	61	154	17.3
28.5	463	132	331	37.6	28.5	344	98	246	27.9	28.4	496	141	355	40.1	28.4	384	109	275	30.9	28.4	211	60	151	17.1
28.5	460	131	329	37.3	28.5	340	97	243	27.7	28.4	493	140	353	39.8	28.4	380	108	272	30.7	28.4	208	59	149	16.7
28.5	459	131	328	37.4	28.5	337	96	241	27.3	28.4	490	139	351	39.4	28.4	377	107	270	30.4	28.4	204	58	146	16.5
28.5	456	130	326	37.1	28.5	333	95	238	27.1	28.4	489	139	350	39.5	28.4	373	106	267	30.1	28.4	201	57	144	16.2
28.5	453	129	324	36.7	28.5	330	94	236	26.8	28.4	486	138	348	39.2	28.4	370	105	265	29.8	28.4	197	56	141	15.9
28.5	452	129	323	36.8	28.5	326	93	233	26.5	28.4	483	137	346	38.9	28.4	366	104	262	29.6	28.4	194	55	139	15.6
28.5	449	128	321	36.5	28.5	323	92	231	26.2	28.4	482	137	345	38.9	28.4	363	103	260	29.2	28.4	190	54	136	15.3
28.5	446	127	319	36.2	28.5	319	91	228	26.0	28.4	479	136	343	38.6	28.4	359	102	257	29.0	28.4	183	52	131	14.8
28.5	445	127	318	36.2	28.5	316	90	226	25.6	28.4	476	135	341	38.3	28.4	356	101	255	28.7	28.4	176	50	126	14.2
28.5	442	126	316	35.9	28.5	312	89	223	25.4	28.4	475	135	340	38.4	28.4	352	100	252	28.4	28.4	169	48	121	13.6
28.5	439	125	314	35.6	28.5	309	88	221	25.1	28.4	472	134	338	38.0	28.4	349	99	250	28.1	28.4	162	46	116	13.1
28.5	438	125	313	35.7	28.5	305	87	218	24.8	28.4	469	133	336	37.7	28.4	348	99	249	28.2	28.4	155	44	111	12.5
28.5	435	124	311	35.3	28.5	302	86	216	24.5	28.4	468	133	335	37.8	28.4	345	98	247	27.8	28.4	148	42	106	11.9
28.5	432	123	309	35.0	28.5	298	85	213	24.2	28.4	465	132	333	37.5	28.4	342	97	245	27.5	28.4	141	40	101	11.3
28.5	431	123	308	35.1	28.5	295	84	211	23.9	28.4	464	132	332	37.6	28.4	341	97	244	27.6	28.4	134	38	96	10.8
28.5	428	122	306	34.8	28.5	291	83	208	23.7	28.4	462	131	331	37.1	28.4	338	96	242	27.3	28.4	116	33	83	9.4
28.5	425	121	304	34.4	28.5	288	82	206	23.3	28.4	461	131	330	37.2	28.4	335	95	240	26.9	28.4	109	31	78	8.8
28.5	424	121	303	34.5	28.5	284	81	203	23.1	28.4	458	130	328	36.9	28.4	334	95	239	27.0	28.4	102	29	73	8.2
28.5	421	120	301	34.2	28.5	281	80	201	22.8	28.4	457	130	327	37.0	28.4	331	94	237	26.7	28.4	95	27	68	7.7
28.5	418	119	299	33.9	28.5	277	79	198	22.5	28.4	455	129	326	36.6	28.4	328	93	235	26.4	28.4	88	25	63	7.1
28.5	417	119	298	34.0	28.5	274	78	196	22.2	28.4	454	129	325	36.7	28.4	327	93	234	26.4	28.4	81	23	58	6.5
28.5	414	118	296	33.6	28.5	270	77	193	22.0	28.4	451	128	323	36.3	28.4	324	92	232	26.1	28.4	74	21	53	6.0
28.5	411	117	294	33.3	28.5	267	76	191	21.6	28.4	450	128	322	36.4	28.4	320	91	229	25.9	28.4	67	19	48	5.4
28.5	410	117	293	33.4	28.5	263	75	188	21.4	28.4	447	127	320	36.1	28.4	317	90	227	25.6	28.3	499	141	358	39.8
28.5	407	116	291	33.1	28.5	260	74	186	21.1	28.4	444	126	318	35.8	28.4	313	89	224	25.3	28.3	498	141	357	39.9
28.5	404	115	289	32.7	28.5	256	73	183	20.8	28.4	443	126	317	35.8	28.4	310	88	222	25.0	28.3	495	140	355	39.6
28.5	403	115	288	32.8	28.5	253	72	181	20.5	28.4	440	125	315	35.5	28.4	306	87	219	24.7	28.3	494	140	354	39.7
28.5	400	114	286	32.5	28.5	249	71	178	20.2	28.4	437	124	313	35.2	28.4	303	86	217	24.4	28.3	492	139	353	39.3
28.5	397	113	284	32.2	28.5	246	70	176	19.9	28.4	436	124	312	35.3	28.4	299	85	214	24.2	28.3	491	139	352	39.4
28.5	396	113	283	32.2	28.5	242	69	173	19.7	28.4	433	123	310	34.9	28.4	296	84	212	23.8	28.3	488	138	350	39.0

%E	M1	M2	DM	M*
28.3	487	138	349	39.1
28.3	484	137	347	38.8
28.3	481	136	345	38.5
28.3	480	136	344	38.5
28.3	477	135	342	38.2
28.3	474	134	340	37.9
28.3	473	134	339	38.0
28.3	470	133	337	37.6
28.3	467	132	335	37.3
28.3	466	132	334	37.4
28.3	463	131	332	37.1
28.3	460	130	330	36.7
28.3	459	130	329	36.8
28.3	456	129	327	36.5
28.3	453	128	325	36.2
28.3	452	128	324	36.2
28.3	449	127	322	35.9
28.3	448	127	321	36.0
28.3	446	126	320	35.6
28.3	445	126	319	35.7
28.3	442	125	317	35.4
28.3	441	125	316	35.4
28.3	438	124	314	35.1
28.3	435	123	312	34.8
28.3	434	123	311	34.9
28.3	431	122	309	34.5
28.3	428	121	307	34.2
28.3	427	121	306	34.3
28.3	424	120	304	34.0
28.3	421	119	302	33.6
28.3	420	119	301	33.7
28.3	417	118	299	33.4
28.3	414	117	297	33.1
28.3	413	117	296	33.1
28.3	410	116	294	32.8
28.3	407	115	292	32.5
28.3	406	115	291	32.6
28.3	403	114	289	32.2
28.3	400	113	287	31.9
28.3	399	113	286	32.0
28.3	396	112	284	31.7
28.3	392	111	281	31.4
28.3	389	110	279	31.1
28.3	385	109	276	30.9
28.3	382	108	274	30.5
28.3	381	108	273	30.6
28.3	378	107	271	30.3
28.3	375	106	269	30.0
28.3	374	106	268	30.0
28.3	371	105	266	29.7

%E	M1	M2	DM	M*
28.3	368	104	264	29.4
28.3	367	104	263	29.5
28.3	364	103	261	29.1
28.3	361	102	259	28.8
28.3	360	102	258	28.9
28.3	357	101	256	28.6
28.3	353	100	253	28.3
28.3	350	99	251	28.0
28.3	346	98	248	27.8
28.3	343	97	246	27.4
28.3	339	96	243	27.2
28.3	336	95	241	26.9
28.3	332	94	238	26.6
28.3	329	93	236	26.3
28.3	325	92	233	26.0
28.3	322	91	231	25.7
28.3	321	91	230	25.8
28.3	318	90	228	25.5
28.3	315	89	226	25.1
28.3	314	89	225	25.2
28.3	311	88	223	24.9
28.3	307	87	220	24.7
28.3	304	86	218	24.3
28.3	300	85	215	24.1
28.3	297	84	213	23.8
28.3	293	83	210	23.5
28.3	290	82	208	23.2
28.3	286	81	205	22.9
28.3	283	80	203	22.6
28.3	279	79	200	22.4
28.3	276	78	198	22.0
28.3	272	77	195	21.8
28.3	269	76	193	21.5
28.3	265	75	190	21.2
28.3	258	73	185	20.7
28.3	254	72	182	20.4
28.3	251	71	180	20.1
28.3	247	70	177	19.8
28.3	244	69	175	19.5
28.3	240	68	172	19.3
28.3	237	67	170	18.9
28.3	233	66	167	18.7
28.3	230	65	165	18.4
28.3	226	64	162	18.1
28.3	223	63	160	17.8
28.3	219	62	157	17.6
28.3	212	60	152	17.0
28.3	205	58	147	16.4
28.3	198	56	142	15.8
28.3	191	54	137	15.3

%E	M1	M2	DM	M*
28.3	187	53	134	15.0
28.3	184	52	132	14.7
28.3	180	51	129	14.4
28.3	173	49	124	13.9
28.3	166	47	119	13.3
28.3	159	45	114	12.7
28.3	152	43	109	12.2
28.3	145	41	104	11.6
28.3	138	39	99	11.0
28.3	127	36	91	10.2
28.3	120	34	86	9.6
28.3	113	32	81	9.1
28.3	106	30	76	8.5
28.3	99	28	71	7.9
28.3	92	26	66	7.3
28.3	60	17	43	4.8
28.3	53	15	38	4.2
28.3	46	13	33	3.7
28.2	500	141	359	39.8
28.2	497	140	357	39.4
28.2	496	140	356	39.5
28.2	493	139	354	39.2
28.2	490	138	352	38.9
28.2	489	138	351	38.9
28.2	486	137	349	38.6
28.2	485	137	348	38.7
28.2	483	136	347	38.3
28.2	482	136	346	38.4
28.2	479	135	344	38.0
28.2	478	135	343	38.1
28.2	476	134	342	37.7
28.2	475	134	341	37.8
28.2	472	133	339	37.5
28.2	471	133	338	37.6
28.2	468	132	336	37.2
28.2	465	131	334	36.9
28.2	464	131	333	37.0
28.2	461	130	331	36.7
28.2	458	129	329	36.3
28.2	457	129	328	36.4
28.2	454	128	326	36.1
28.2	451	127	324	35.8
28.2	450	127	323	35.8
28.2	447	126	321	35.5
28.2	444	125	319	35.2
28.2	443	125	318	35.3
28.2	440	124	316	34.9
28.2	439	124	315	35.0
28.2	436	123	313	34.7
28.2	433	122	311	34.4

%E	M1	M2	DM	M*
28.2	432	122	310	34.5
28.2	429	121	308	34.1
28.2	426	120	306	33.8
28.2	425	120	305	33.9
28.2	422	119	303	33.6
28.2	419	118	301	33.2
28.2	418	118	300	33.3
28.2	415	117	298	33.0
28.2	412	116	296	32.7
28.2	411	116	295	32.7
28.2	408	115	293	32.4
28.2	404	114	290	32.2
28.2	401	113	288	31.8
28.2	397	112	285	31.6
28.2	394	111	283	31.3
28.2	393	111	282	31.4
28.2	390	110	280	31.0
28.2	387	109	278	30.7
28.2	386	109	277	30.8
28.2	383	108	275	30.5
28.2	380	107	273	30.1
28.2	379	107	272	30.2
28.2	376	106	270	29.9
28.2	373	105	268	29.6
28.2	372	105	267	29.6
28.2	369	104	265	29.3
28.2	365	103	262	29.1
28.2	362	102	260	28.7
28.2	358	101	257	28.5
28.2	355	100	255	28.2
28.2	354	100	254	28.2
28.2	351	99	252	27.9
28.2	348	98	250	27.6
28.2	347	98	249	27.7
28.2	344	97	247	27.4
28.2	341	96	245	27.0
28.2	340	96	244	27.1
28.2	337	95	242	26.8
28.2	333	94	239	26.5
28.2	330	93	237	26.2
28.2	326	92	234	26.0
28.2	323	91	232	25.6
28.2	319	90	229	25.4
28.2	316	89	227	25.1
28.2	312	88	224	24.8
28.2	309	87	222	24.5
28.2	308	87	221	24.6
28.2	305	86	219	24.2
28.2	301	85	216	24.0
28.2	298	84	214	23.7

%E	M1	M2	DM	M*
28.2	294	83	211	23.4
28.2	291	82	209	23.1
28.2	287	81	206	22.9
28.2	284	80	204	22.5
28.2	280	79	201	22.3
28.2	277	78	199	22.0
28.2	273	77	196	21.7
28.2	266	75	191	21.1
28.2	262	74	188	20.9
28.2	259	73	186	20.6
28.2	255	72	183	20.3
28.2	252	71	181	20.0
28.2	248	70	178	19.8
28.2	245	69	176	19.4
28.2	241	68	173	19.2
28.2	238	67	171	18.9
28.2	234	66	168	18.6
28.2	227	64	163	18.0
28.2	220	62	158	17.5
28.2	216	61	155	17.2
28.2	213	60	153	16.9
28.2	209	59	150	16.7
28.2	206	58	148	16.3
28.2	202	57	145	16.1
28.2	195	55	140	15.5
28.2	188	53	135	14.9
28.2	181	51	130	14.4
28.2	177	50	127	14.1
28.2	174	49	125	13.8
28.2	170	48	122	13.6
28.2	163	46	117	13.0
28.2	156	44	112	12.4
28.2	149	42	107	11.8
28.2	142	40	102	11.3
28.2	131	37	94	10.5
28.2	124	35	89	9.9
28.2	117	33	84	9.3
28.2	110	31	79	8.7
28.2	103	29	74	8.2
28.2	85	24	61	6.8
28.2	78	22	56	6.2
28.2	71	20	51	5.6
28.2	39	11	28	3.1
28.1	499	140	359	39.3
28.1	498	140	358	39.4
28.1	495	139	356	39.0
28.1	494	139	355	39.1
28.1	491	138	353	38.8
28.1	488	137	351	38.5
28.1	487	137	350	38.5

%E	M1	M2	DM	M*
28.1	484	136	348	38.2
28.1	481	135	346	37.9
28.1	480	135	345	38.0
28.1	477	134	343	37.6
28.1	474	133	341	37.3
28.1	473	133	340	37.4
28.1	470	132	338	37.1
28.1	469	132	337	37.2
28.1	467	131	336	36.7
28.1	466	131	335	36.8
28.1	463	130	333	36.5
28.1	462	130	332	36.6
28.1	459	129	330	36.3
28.1	456	128	328	35.9
28.1	455	128	327	36.0
28.1	452	127	325	35.7
28.1	449	126	323	35.4
28.1	448	126	322	35.4
28.1	445	125	320	35.1
28.1	442	124	318	34.8
28.1	441	124	317	34.9
28.1	438	123	315	34.5
28.1	437	123	314	34.6
28.1	434	122	312	34.3
28.1	431	121	310	34.0
28.1	430	121	309	34.0
28.1	427	120	307	33.7
28.1	424	119	305	33.4
28.1	423	119	304	33.5
28.1	420	118	302	33.2
28.1	417	117	300	32.8
28.1	416	117	299	32.9
28.1	413	116	297	32.6
28.1	409	115	294	32.3
28.1	406	114	292	32.0
28.1	405	114	291	32.1
28.1	402	113	289	31.8
28.1	399	112	287	31.4
28.1	398	112	286	31.5
28.1	395	111	284	31.2
28.1	392	110	282	30.9
28.1	391	110	281	30.9
28.1	388	109	279	30.6
28.1	385	108	277	30.3
28.1	384	108	276	30.4
28.1	381	107	274	30.0
28.1	377	106	271	29.8
28.1	374	105	269	29.5
28.1	370	104	266	29.2
28.1	367	103	264	28.9

%E	M1	M2	DM	M*
28.1	366	103	263	29.0
28.1	363	102	261	28.7
28.1	360	101	259	28.3
28.1	359	101	258	28.4
28.1	356	100	256	28.1
28.1	352	99	253	27.8
28.1	349	98	251	27.5
28.1	345	97	248	27.3
28.1	342	96	246	26.9
28.1	338	95	243	26.7
28.1	335	94	241	26.4
28.1	334	94	240	26.5
28.1	331	93	238	26.1
28.1	327	92	235	25.9
28.1	324	91	233	25.6
28.1	320	90	230	25.3
28.1	317	89	228	25.0
28.1	313	88	225	24.7
28.1	310	87	223	24.4
28.1	306	86	220	24.2
28.1	303	85	218	23.8
28.1	302	85	217	23.9
28.1	299	84	215	23.6
28.1	295	83	212	23.4
28.1	292	82	210	23.0
28.1	288	81	207	22.8
28.1	285	80	205	22.5
28.1	281	79	202	22.2
28.1	278	78	200	21.9
28.1	274	77	197	21.6
28.1	270	76	194	21.4
28.1	267	75	192	21.1
28.1	263	74	189	20.8
28.1	260	73	187	20.5
28.1	256	72	184	20.3
28.1	253	71	182	19.9
28.1	249	70	179	19.7
28.1	242	68	174	19.1
28.1	235	66	169	18.5
28.1	231	65	166	18.3
28.1	228	64	164	18.0
28.1	224	63	161	17.7
28.1	221	62	159	17.4
28.1	217	61	156	17.1
28.1	210	59	151	16.6
28.1	203	57	146	16.0
28.1	199	56	143	15.8
28.1	196	55	141	15.4
28.1	192	54	138	15.2
28.1	185	52	133	14.6

%E	M1	M2	DM	M*
28.1	178	50	128	14.0
28.1	171	48	123	13.5
28.1	167	47	120	13.2
28.1	160	45	115	12.7
28.1	153	43	110	12.1
28.1	146	41	105	11.5
28.1	139	39	100	10.9
28.1	135	38	97	10.7
28.1	128	36	92	10.1
28.1	121	34	87	9.6
28.1	114	32	82	9.0
28.1	96	27	69	7.6
28.1	89	25	64	7.0
28.1	64	18	46	5.1
28.1	57	16	41	4.5
28.1	32	9	23	2.5
28.0	500	140	360	39.2
28.0	497	139	358	38.9
28.0	496	139	357	39.0
28.0	493	138	355	38.6
28.0	492	138	354	38.7
28.0	490	137	353	38.3
28.0	489	137	352	38.4
28.0	486	136	350	38.1
28.0	485	136	349	38.1
28.0	483	135	348	37.7
28.0	482	135	347	37.8
28.0	479	134	345	37.5
28.0	478	134	344	37.6
28.0	475	133	342	37.2
28.0	472	132	340	36.9
28.0	471	132	339	37.0
28.0	468	131	337	36.7
28.0	465	130	335	36.3
28.0	464	130	334	36.4
28.0	461	129	332	36.1
28.0	460	129	331	36.2
28.0	457	128	329	35.9
28.0	454	127	327	35.5
28.0	453	127	326	35.6
28.0	450	126	324	35.3
28.0	447	125	322	35.0
28.0	446	125	321	35.0
28.0	443	124	319	34.7
28.0	440	123	317	34.4
28.0	439	123	316	34.5
28.0	436	122	314	34.1
28.0	435	122	313	34.2
28.0	432	121	311	33.9
28.0	429	120	309	33.6

%E	M1	M2	DM	M*
28.0	428	120	308	33.6
28.0	425	119	306	33.3
28.0	422	118	304	33.0
28.0	421	118	303	33.1
28.0	418	117	301	32.7
28.0	415	116	299	32.4
28.0	414	116	298	32.5
28.0	411	115	296	32.2
28.0	410	115	295	32.3
28.0	407	114	293	31.9
28.0	404	113	291	31.6
28.0	403	113	290	31.7
28.0	400	112	288	31.4
28.0	397	111	286	31.0
28.0	396	111	285	31.1
28.0	393	110	283	30.8
28.0	389	109	280	30.5
28.0	386	108	278	30.2
28.0	382	107	275	30.0
28.0	379	106	273	29.6
28.0	378	106	272	29.7
28.0	375	105	270	29.4
28.0	372	104	268	29.1
28.0	371	104	267	29.2
28.0	368	103	265	28.8
28.0	364	102	262	28.6
28.0	361	101	260	28.3
28.0	357	100	257	28.0
28.0	354	99	255	27.7
28.0	353	99	254	27.8
28.0	350	98	252	27.4
28.0	347	97	250	27.1
28.0	346	97	249	27.2
28.0	343	96	247	26.9
28.0	339	95	244	26.6
28.0	336	94	242	26.3
28.0	332	93	239	26.1
28.0	329	92	237	25.7
28.0	328	92	236	25.8
28.0	325	91	234	25.5
28.0	322	90	232	25.2
28.0	321	90	231	25.2
28.0	318	89	229	24.9
28.0	314	88	226	24.7
28.0	311	87	224	24.3
28.0	307	86	221	24.1
28.0	304	85	219	23.8
28.0	300	84	216	23.5
28.0	296	83	213	23.3
28.0	293	82	211	22.9

%E	M1	M2	DM	M*
28.0	289	81	208	22.7
28.0	286	80	206	22.4
28.0	282	79	203	22.1
28.0	279	78	201	21.8
28.0	275	77	198	21.6
28.0	271	76	195	21.3
28.0	268	75	193	21.0
28.0	264	74	190	20.7
28.0	261	73	188	20.4
28.0	257	72	185	20.2
28.0	254	71	183	19.8
28.0	250	70	180	19.6
28.0	246	69	177	19.4
28.0	243	68	175	19.0
28.0	239	67	172	18.8
28.0	236	66	170	18.5
28.0	232	65	167	18.2
28.0	225	63	162	17.6
28.0	218	61	157	17.1
28.0	214	60	154	16.8
28.0	211	59	152	16.5
28.0	207	58	149	16.3
28.0	200	56	144	15.7
28.0	193	54	139	15.1
28.0	189	53	136	14.9
28.0	186	52	134	14.5
28.0	182	51	131	14.3
28.0	175	49	126	13.7
28.0	168	47	121	13.1
28.0	164	46	118	12.9
28.0	161	45	116	12.6
28.0	157	44	113	12.3
28.0	150	42	108	11.8
28.0	143	40	103	11.2
28.0	132	37	95	10.4
28.0	125	35	90	9.8
28.0	118	33	85	9.2
28.0	107	30	77	8.4
28.0	100	28	72	7.8
28.0	93	26	67	7.3
28.0	82	23	59	6.5
28.0	75	21	54	5.9
28.0	50	14	36	3.9
28.0	25	7	18	2.0
27.9	499	139	360	38.7
27.9	498	139	359	38.8
27.9	495	138	357	38.5
27.9	494	138	356	38.6
27.9	491	137	354	38.2
27.9	488	136	352	37.9

%E	M1	M2	DM	M*
27.9	487	136	351	38.0
27.9	484	135	349	37.7
27.9	481	134	347	37.3
27.9	480	134	346	37.4
27.9	477	133	344	37.1
27.9	476	133	343	37.2
27.9	473	132	341	36.8
27.9	470	131	339	36.5
27.9	469	131	338	36.6
27.9	466	130	336	36.3
27.9	463	129	334	35.9
27.9	462	129	333	36.0
27.9	459	128	331	35.7
27.9	458	128	330	35.8
27.9	456	127	329	35.4
27.9	455	127	328	35.4
27.9	452	126	326	35.1
27.9	451	126	325	35.2
27.9	448	125	323	34.9
27.9	445	124	321	34.6
27.9	444	124	320	34.6
27.9	441	123	318	34.3
27.9	438	122	316	34.0
27.9	437	122	315	34.1
27.9	434	121	313	33.7
27.9	433	121	312	33.8
27.9	430	120	310	33.5
27.9	427	119	308	33.2
27.9	426	119	307	33.2
27.9	423	118	305	32.9
27.9	420	117	303	32.6
27.9	419	117	302	32.7
27.9	416	116	300	32.3
27.9	412	115	297	32.1
27.9	409	114	295	31.8
27.9	408	114	294	31.9
27.9	405	113	292	31.5
27.9	402	112	290	31.2
27.9	401	112	289	31.3
27.9	398	111	287	31.0
27.9	394	110	284	30.7
27.9	391	109	282	30.4
27.9	390	109	281	30.5
27.9	387	108	279	30.1
27.9	384	107	277	29.8
27.9	383	107	276	29.9
27.9	380	106	274	29.6
27.9	377	105	272	29.2
27.9	376	105	271	29.3
27.9	373	104	269	29.0

%E	M1	M2	DM	M*
27.9	369	103	266	28.8
27.9	366	102	264	28.4
27.9	365	102	263	28.5
27.9	362	101	261	28.2
27.9	359	100	259	27.9
27.9	358	100	258	27.9
27.9	355	99	256	27.6
27.9	351	98	253	27.4
27.9	348	97	251	27.0
27.9	344	96	248	26.8
27.9	341	95	246	26.5
27.9	340	95	245	26.5
27.9	337	94	243	26.2
27.9	333	93	240	26.0
27.9	330	92	238	25.6
27.9	326	91	235	25.4
27.9	323	90	233	25.1
27.9	319	89	230	24.8
27.9	315	88	227	24.6
27.9	312	87	225	24.3
27.9	308	86	222	24.0
27.9	305	85	220	23.7
27.9	301	84	217	23.4
27.9	298	83	215	23.1
27.9	297	83	214	23.2
27.9	294	82	212	22.9
27.9	290	81	209	22.6
27.9	287	80	207	22.3
27.9	283	79	204	22.1
27.9	280	78	202	21.7
27.9	276	77	199	21.5
27.9	272	76	196	21.2
27.9	269	75	194	20.9
27.9	265	74	191	20.7
27.9	262	73	189	20.3
27.9	258	72	186	20.1
27.9	251	70	181	19.5
27.9	247	69	178	19.3
27.9	244	68	176	19.0
27.9	240	67	173	18.7
27.9	233	65	168	18.1
27.9	229	64	165	17.9
27.9	226	63	163	17.6
27.9	222	62	160	17.3
27.9	219	61	158	17.0
27.9	215	60	155	16.7
27.9	208	58	150	16.2
27.9	204	57	147	15.9
27.9	201	56	145	15.6
27.9	197	55	142	15.4

%E	M1	M2	DM	M*
27.9	190	53	137	14.8
27.9	183	51	132	14.2
27.9	179	50	129	14.0
27.9	172	48	124	13.4
27.9	165	46	119	12.8
27.9	154	43	111	12.0
27.9	147	41	106	11.4
27.9	140	39	101	10.9
27.9	136	38	98	10.6
27.9	129	36	93	10.0
27.9	122	34	88	9.5
27.9	111	31	80	8.7
27.9	104	29	75	8.1
27.9	86	24	62	6.7
27.9	68	19	49	5.3
27.9	61	17	44	4.7
27.9	43	12	31	3.3
27.8	500	139	361	38.6
27.8	497	138	359	38.3
27.8	496	138	358	38.4
27.8	493	137	356	38.1
27.8	492	137	355	38.1
27.8	490	136	354	37.7
27.8	489	136	353	37.8
27.8	486	135	351	37.5
27.8	485	135	350	37.6
27.8	482	134	348	37.3
27.8	479	133	346	36.9
27.8	478	133	345	37.0
27.8	475	132	343	36.7
27.8	474	132	342	36.8
27.8	472	131	341	36.4
27.8	471	131	340	36.4
27.8	468	130	338	36.1
27.8	467	130	337	36.2
27.8	464	129	335	35.9
27.8	461	128	333	35.5
27.8	460	128	332	35.6
27.8	457	127	330	35.3
27.8	454	126	328	35.0
27.8	453	126	327	35.0
27.8	450	125	325	34.7
27.8	449	125	324	34.8
27.8	446	124	322	34.5
27.8	443	123	320	34.2
27.8	442	123	319	34.2
27.8	439	122	317	33.9
27.8	436	121	315	33.6
27.8	435	121	314	33.7
27.8	432	120	312	33.3

%E	M1	M2	DM	M*
27.8	431	120	311	33.4
27.8	428	119	309	33.1
27.8	425	118	307	32.8
27.8	424	118	306	32.8
27.8	421	117	304	32.5
27.8	418	116	302	32.2
27.8	417	116	301	32.3
27.8	414	115	299	31.9
27.8	413	115	298	32.0
27.8	410	114	296	31.7
27.8	407	113	294	31.4
27.8	406	113	293	31.5
27.8	403	112	291	31.1
27.8	400	111	289	30.8
27.8	399	111	288	30.9
27.8	396	110	286	30.6
27.8	395	110	285	30.6
27.8	392	109	283	30.3
27.8	389	108	281	30.0
27.8	388	108	280	30.1
27.8	385	107	278	29.7
27.8	381	106	275	29.5
27.8	378	105	273	29.2
27.8	374	104	270	28.9
27.8	371	103	268	28.6
27.8	370	103	267	28.7
27.8	367	102	265	28.3
27.8	363	101	262	28.1
27.8	360	100	260	27.8
27.8	356	99	257	27.5
27.8	353	98	255	27.2
27.8	352	98	254	27.3
27.8	349	97	252	27.0
27.8	345	96	249	26.7
27.8	342	95	247	26.4
27.8	338	94	244	26.1
27.8	335	93	242	25.8
27.8	334	93	241	25.9
27.8	331	92	239	25.6
27.8	327	91	236	25.3
27.8	324	90	234	25.0
27.8	320	89	231	24.8
27.8	317	88	229	24.4
27.8	316	88	228	24.5
27.8	313	87	226	24.2
27.8	309	86	223	23.9
27.8	306	85	221	23.6
27.8	302	84	218	23.4
27.8	299	83	216	23.0
27.8	295	82	213	22.8

%E	M1	M2	DM	M*
27.8	291	81	210	22.5
27.8	288	80	208	22.2
27.8	284	79	205	22.0
27.8	281	78	203	21.7
27.8	277	77	200	21.4
27.8	273	76	197	21.2
27.8	270	75	195	20.8
27.8	266	74	192	20.6
27.8	263	73	190	20.3
27.8	259	72	187	20.0
27.8	255	71	184	19.8
27.8	252	70	182	19.4
27.8	248	69	179	19.2
27.8	245	68	177	18.9
27.8	241	67	174	18.6
27.8	237	66	171	18.4
27.8	234	65	169	18.1
27.8	230	64	166	17.8
27.8	227	63	164	17.5
27.8	223	62	161	17.2
27.8	216	60	156	16.7
27.8	212	59	153	16.4
27.8	209	58	151	16.1
27.8	205	57	148	15.8
27.8	198	55	143	15.3
27.8	194	54	140	15.0
27.8	187	52	135	14.5
27.8	180	50	130	13.9
27.8	176	49	127	13.6
27.8	169	47	122	13.1
27.8	162	45	117	12.5
27.8	158	44	114	12.3
27.8	151	42	109	11.7
27.8	144	40	104	11.1
27.8	133	37	96	10.3
27.8	126	35	91	9.7
27.8	115	32	83	8.9
27.8	108	30	78	8.3
27.8	97	27	70	7.5
27.8	90	25	65	6.9
27.8	79	22	57	6.1
27.8	72	20	52	5.6
27.8	54	15	39	4.2
27.8	36	10	26	2.8
27.8	18	5	13	1.4
27.7	499	138	361	38.2
27.7	498	138	360	38.2
27.7	495	137	358	37.9
27.7	494	137	357	38.0
27.7	491	136	355	37.7

%E	M1	M2	DM	M*	%E	M1	M2	DM	M*	%E	M1	M2	DM	M*	%E	M1	M2	DM	M*	%E	M1	M2	DM	M*
27.7	488	135	353	37.3	27.7	372	103	269	28.5	27.7	195	54	141	15.0	27.6	435	120	315	33.1	27.6	297	82	215	22.6
27.7	487	135	352	37.4	27.7	368	102	266	28.3	27.7	191	53	138	14.7	27.6	431	119	312	32.9	27.6	294	81	213	22.3
27.7	484	134	350	37.1	27.7	365	101	264	27.9	27.7	188	52	136	14.4	27.6	428	118	310	32.5	27.6	293	81	212	22.4
27.7	483	134	349	37.2	27.7	364	101	263	28.0	27.7	184	51	133	14.1	27.6	427	118	309	32.6	27.6	290	80	210	22.1
27.7	481	133	348	36.8	27.7	361	100	261	27.7	27.7	177	49	128	13.6	27.6	424	117	307	32.3	27.6	286	79	207	21.8
27.7	480	133	347	36.9	27.7	358	99	259	27.4	27.7	173	48	125	13.3	27.6	421	116	305	32.0	27.6	283	78	205	21.5
27.7	477	132	345	36.5	27.7	357	99	258	27.5	27.7	166	46	120	12.7	27.6	420	116	304	32.0	27.6	279	77	202	21.3
27.7	476	132	344	36.6	27.7	354	98	256	27.1	27.7	159	44	115	12.2	27.6	417	115	302	31.7	27.6	275	76	199	21.0
27.7	473	131	342	36.3	27.7	350	97	253	26.9	27.7	155	43	112	11.9	27.6	416	115	301	31.8	27.6	272	75	197	20.7
27.7	470	130	340	36.0	27.7	347	96	251	26.6	27.7	148	41	107	11.4	27.6	413	114	299	31.5	27.6	268	74	194	20.4
27.7	469	130	339	36.0	27.7	346	96	250	26.6	27.7	141	39	102	10.8	27.6	410	113	297	31.1	27.6	261	72	189	19.9
27.7	466	129	337	35.7	27.7	343	95	248	26.3	27.7	137	38	99	10.5	27.6	409	113	296	31.2	27.6	257	71	186	19.6
27.7	465	129	336	35.8	27.7	339	94	245	26.1	27.7	130	36	94	10.0	27.6	406	112	294	30.9	27.6	254	70	184	19.3
27.7	462	128	334	35.5	27.7	336	93	243	25.7	27.7	119	33	86	9.2	27.6	402	111	291	30.6	27.6	250	69	181	19.0
27.7	459	127	332	35.1	27.7	332	92	240	25.5	27.7	112	31	81	8.6	27.6	399	110	289	30.3	27.6	246	68	178	18.8
27.7	458	127	331	35.2	27.7	329	91	238	25.2	27.7	101	28	73	7.8	27.6	398	110	288	30.4	27.6	243	67	176	18.5
27.7	455	126	329	34.9	27.7	328	91	237	25.2	27.7	94	26	68	7.2	27.6	395	109	286	30.1	27.6	239	66	173	18.2
27.7	452	125	327	34.6	27.7	325	90	235	24.9	27.7	83	23	60	6.4	27.6	392	108	284	29.8	27.6	232	64	168	17.7
27.7	451	125	326	34.6	27.7	321	89	232	24.7	27.7	65	18	47	5.0	27.6	391	108	283	29.8	27.6	228	63	165	17.4
27.7	448	124	324	34.3	27.7	318	88	230	24.4	27.7	47	13	34	3.6	27.6	388	107	281	29.5	27.6	225	62	163	17.1
27.7	447	124	323	34.4	27.7	314	87	227	24.1	27.6	500	138	362	38.1	27.6	387	107	280	29.6	27.6	221	61	160	16.8
27.7	444	123	321	34.1	27.7	311	86	225	23.8	27.6	497	137	360	37.8	27.6	384	106	278	29.3	27.6	214	59	155	16.3
27.7	441	122	319	33.8	27.7	310	86	224	23.9	27.6	496	137	359	37.8	27.6	381	105	276	28.9	27.6	210	58	152	16.0
27.7	440	122	318	33.8	27.7	307	85	222	23.5	27.6	493	136	357	37.5	27.6	380	105	275	29.0	27.6	203	56	147	15.4
27.7	437	121	316	33.5	27.7	303	84	219	23.3	27.6	492	136	356	37.6	27.6	377	104	273	28.7	27.6	199	55	144	15.2
27.7	434	120	314	33.2	27.7	300	83	217	23.0	27.6	490	135	355	37.2	27.6	373	103	270	28.4	27.6	196	54	142	14.9
27.7	433	120	313	33.3	27.7	296	82	214	22.7	27.6	489	135	354	37.3	27.6	370	102	268	28.1	27.6	192	53	139	14.6
27.7	430	119	311	32.9	27.7	292	81	211	22.5	27.6	486	134	352	36.9	27.6	369	102	267	28.2	27.6	185	51	134	14.1
27.7	429	119	310	33.0	27.7	289	80	209	22.1	27.6	485	134	351	37.0	27.6	366	101	265	27.9	27.6	181	50	131	13.8
27.7	426	118	308	32.7	27.7	285	79	206	21.9	27.6	482	133	349	36.7	27.6	362	100	262	27.6	27.6	174	48	126	13.2
27.7	423	117	306	32.4	27.7	282	78	204	21.6	27.6	479	132	347	36.4	27.6	359	99	260	27.3	27.6	170	47	123	13.0
27.7	422	117	305	32.4	27.7	278	77	201	21.3	27.6	478	132	346	36.5	27.6	355	98	257	27.1	27.6	163	45	118	12.4
27.7	419	116	303	32.1	27.7	274	76	198	21.1	27.6	475	131	344	36.1	27.6	352	97	255	26.7	27.6	156	43	113	11.9
27.7	415	115	300	31.9	27.7	271	75	196	20.8	27.6	474	131	343	36.2	27.6	351	97	254	26.8	27.6	152	42	110	11.6
27.7	412	114	298	31.5	27.7	267	74	193	20.5	27.6	471	130	341	35.9	27.6	348	96	252	26.5	27.6	145	40	105	11.0
27.7	411	114	297	31.6	27.7	264	73	191	20.2	27.6	468	129	339	35.6	27.6	344	95	249	26.2	27.6	134	37	97	10.2
27.7	408	113	295	31.3	27.7	260	72	188	19.9	27.6	467	129	338	35.6	27.6	341	94	247	25.9	27.6	127	35	92	9.6
27.7	405	112	293	31.0	27.7	256	71	185	19.7	27.6	464	128	336	35.3	27.6	340	94	246	26.0	27.6	123	34	89	9.4
27.7	404	112	292	31.0	27.7	253	70	183	19.4	27.6	463	128	335	35.4	27.6	337	93	244	25.7	27.6	116	32	84	8.8
27.7	401	111	290	30.7	27.7	249	69	180	19.1	27.6	460	127	333	35.1	27.6	333	92	241	25.4	27.6	105	29	76	8.0
27.7	397	110	287	30.5	27.7	242	67	175	18.5	27.6	457	126	331	34.7	27.6	330	91	239	25.1	27.6	98	27	71	7.4
27.7	394	109	285	30.2	27.7	238	66	172	18.3	27.6	456	126	330	34.8	27.6	326	90	236	24.8	27.6	87	24	63	6.6
27.7	393	109	284	30.2	27.7	235	65	170	18.0	27.6	453	125	328	34.5	27.6	323	89	234	24.5	27.6	76	21	55	5.8
27.7	390	108	282	29.9	27.7	231	64	167	17.7	27.6	450	124	326	34.2	27.6	322	89	233	24.6	27.6	58	16	42	4.4
27.7	386	107	279	29.7	27.7	224	62	162	17.2	27.6	449	124	325	34.2	27.6	319	88	231	24.3	27.6	29	8	21	2.2
27.7	383	106	277	29.3	27.7	220	61	159	16.9	27.6	446	123	323	33.9	27.6	315	87	228	24.0	27.5	499	137	362	37.6
27.7	382	106	276	29.4	27.7	217	60	157	16.6	27.6	445	123	322	34.0	27.6	312	86	226	23.7	27.5	498	137	361	37.7
27.7	379	105	274	29.1	27.7	213	59	154	16.3	27.6	442	122	320	33.7	27.6	308	85	223	23.5	27.5	495	136	359	37.4
27.7	376	104	272	28.8	27.7	206	57	149	15.8	27.6	439	121	318	33.4	27.6	304	84	220	23.2	27.5	494	136	358	37.4
27.7	375	104	271	28.8	27.7	202	56	146	15.5	27.6	438	121	317	33.4	27.6	301	83	218	22.9	27.5	491	135	356	37.1

%E	M1	M2	DM	M*	%E	M1	M2	DM	M*	%E	M1	M2	DM	M*	%E	M1	M2	DM	M*	%E	M1	M2	DM	M*
27.5	488	134	354	36.8	27.5	374	103	271	28.4	27.5	204	56	148	15.4	27.4	441	121	320	33.2	27.4	307	84	223	23.0
27.5	487	134	353	36.9	27.5	371	102	269	28.0	27.5	200	55	145	15.1	27.4	438	120	318	32.9	27.4	303	83	220	22.7
27.5	484	133	351	36.5	27.5	367	101	266	27.8	27.5	193	53	140	14.6	27.4	435	119	316	32.6	27.4	299	82	217	22.5
27.5	483	133	350	36.6	27.5	364	100	264	27.5	27.5	189	52	137	14.3	27.4	434	119	315	32.6	27.4	296	81	215	22.2
27.5	480	132	348	36.3	27.5	363	100	263	27.5	27.5	182	50	132	13.7	27.4	431	118	313	32.3	27.4	292	80	212	21.9
27.5	477	131	346	36.0	27.5	360	99	261	27.2	27.5	178	49	129	13.5	27.4	430	118	312	32.4	27.4	288	79	209	21.7
27.5	476	131	345	36.1	27.5	357	98	259	26.9	27.5	171	47	124	12.9	27.4	427	117	310	32.1	27.4	285	78	207	21.3
27.5	473	130	343	35.7	27.5	356	98	258	27.0	27.5	167	46	121	12.7	27.4	424	116	308	31.7	27.4	281	77	204	21.1
27.5	472	130	342	35.8	27.5	353	97	256	26.7	27.5	160	44	116	12.1	27.4	423	116	307	31.8	27.4	277	76	201	20.9
27.5	469	129	340	35.5	27.5	349	96	253	26.4	27.5	153	42	111	11.5	27.4	420	115	305	31.5	27.4	274	75	199	20.5
27.5	466	128	338	35.2	27.5	346	95	251	26.1	27.5	149	41	108	11.3	27.4	419	115	304	31.6	27.4	270	74	196	20.3
27.5	465	128	337	35.2	27.5	345	95	250	26.2	27.5	142	39	103	10.7	27.4	416	114	302	31.2	27.4	266	73	193	20.0
27.5	462	127	335	34.9	27.5	342	94	248	25.8	27.5	138	38	100	10.5	27.4	413	113	300	30.9	27.4	263	72	191	19.7
27.5	461	127	334	35.0	27.5	338	93	245	25.6	27.5	131	36	95	9.9	27.4	412	113	299	31.0	27.4	259	71	188	19.5
27.5	459	126	333	34.6	27.5	335	92	243	25.3	27.5	120	33	87	9.1	27.4	409	112	297	30.7	27.4	252	69	183	18.9
27.5	458	126	332	34.7	27.5	334	92	242	25.3	27.5	109	30	79	8.3	27.4	405	111	294	30.4	27.4	248	68	180	18.6
27.5	455	125	330	34.3	27.5	331	91	240	25.0	27.5	102	28	74	7.7	27.4	402	110	292	30.1	27.4	241	66	175	18.1
27.5	454	125	329	34.4	27.5	327	90	237	24.8	27.5	91	25	66	6.9	27.4	401	110	291	30.2	27.4	237	65	172	17.8
27.5	451	124	327	34.1	27.5	324	89	235	24.4	27.5	80	22	58	6.0	27.4	398	109	289	29.9	27.4	234	64	170	17.5
27.5	448	123	325	33.8	27.5	320	88	232	24.2	27.5	69	19	50	5.2	27.4	394	108	286	29.6	27.4	230	63	167	17.3
27.5	447	123	324	33.8	27.5	316	87	229	24.0	27.5	51	14	37	3.8	27.4	391	107	284	29.3	27.4	226	62	164	17.0
27.5	444	122	322	33.5	27.5	313	86	227	23.6	27.5	40	11	29	3.0	27.4	390	107	283	29.4	27.4	223	61	162	16.7
27.5	443	122	321	33.6	27.5	309	85	224	23.4	27.4	500	137	363	37.5	27.4	387	106	281	29.0	27.4	219	60	159	16.4
27.5	440	121	319	33.3	27.5	306	84	222	23.1	27.4	497	136	361	37.2	27.4	383	105	278	28.8	27.4	215	59	156	16.2
27.5	437	120	317	33.0	27.5	305	84	221	23.1	27.4	496	136	360	37.3	27.4	380	104	276	28.5	27.4	212	58	154	15.9
27.5	436	120	316	33.0	27.5	302	83	219	22.8	27.4	493	135	358	37.0	27.4	379	104	275	28.5	27.4	208	57	151	15.6
27.5	433	119	314	32.7	27.5	298	82	216	22.6	27.4	492	135	357	37.0	27.4	376	103	273	28.2	27.4	201	55	146	15.0
27.5	432	119	313	32.8	27.5	295	81	214	22.2	27.4	489	134	355	36.7	27.4	372	102	270	28.0	27.4	197	54	143	14.8
27.5	429	118	311	32.5	27.5	291	80	211	22.0	27.4	486	133	353	36.4	27.4	369	101	268	27.6	27.4	190	52	138	14.2
27.5	426	117	309	32.1	27.5	287	79	208	21.7	27.4	485	133	352	36.5	27.4	368	101	267	27.7	27.4	186	51	135	14.0
27.5	425	117	308	32.2	27.5	284	78	206	21.4	27.4	482	132	350	36.1	27.4	365	100	265	27.4	27.4	179	49	130	13.4
27.5	422	116	306	31.9	27.5	280	77	203	21.2	27.4	481	132	349	36.2	27.4	361	99	262	27.1	27.4	175	48	127	13.2
27.5	418	115	303	31.6	27.5	276	76	200	20.9	27.4	478	131	347	35.9	27.4	358	98	260	26.8	27.4	168	46	122	12.6
27.5	415	114	301	31.3	27.5	273	75	198	20.6	27.4	475	130	345	35.6	27.4	354	97	257	26.6	27.4	164	45	119	12.3
27.5	414	114	300	31.4	27.5	269	74	195	20.4	27.4	474	130	344	35.7	27.4	351	96	255	26.3	27.4	157	43	114	11.8
27.5	411	113	298	31.1	27.5	265	73	192	20.1	27.4	471	129	342	35.3	27.4	350	96	254	26.3	27.4	146	40	106	11.0
27.5	408	112	296	30.7	27.5	262	72	190	19.8	27.4	470	129	341	35.4	27.4	347	95	252	26.0	27.4	135	37	98	10.1
27.5	407	112	295	30.8	27.5	258	71	187	19.5	27.4	468	128	340	35.0	27.4	343	94	249	25.8	27.4	124	34	90	9.3
27.5	404	111	293	30.5	27.5	255	70	185	19.2	27.4	467	128	339	35.1	27.4	340	93	247	25.4	27.4	117	32	85	8.8
27.5	403	111	292	30.6	27.5	251	69	182	19.0	27.4	464	127	337	34.8	27.4	339	93	246	25.5	27.4	113	31	82	8.5
27.5	400	110	290	30.3	27.5	247	68	179	18.7	27.4	463	127	336	34.8	27.4	336	92	244	25.2	27.4	106	29	77	7.9
27.5	397	109	288	29.9	27.5	244	67	177	18.4	27.4	460	126	334	34.5	27.4	332	91	241	24.9	27.4	95	26	69	7.1
27.5	396	109	287	30.0	27.5	240	66	174	18.1	27.4	457	125	332	34.2	27.4	329	90	239	24.6	27.4	84	23	61	6.3
27.5	393	108	285	29.7	27.5	236	65	171	17.9	27.4	456	125	331	34.3	27.4	328	90	238	24.7	27.4	73	20	53	5.5
27.5	389	107	282	29.4	27.5	233	64	169	17.6	27.4	453	124	329	33.9	27.4	325	89	236	24.4	27.4	62	17	45	4.7
27.5	386	106	280	29.1	27.5	229	63	166	17.3	27.4	452	124	328	34.0	27.4	321	88	233	24.1	27.3	499	136	363	37.1
27.5	385	106	279	29.2	27.5	222	61	161	16.8	27.4	449	123	326	33.7	27.4	318	87	231	23.8	27.3	498	136	362	37.1
27.5	382	105	277	28.9	27.5	218	60	158	16.5	27.4	446	122	324	33.4	27.4	317	87	230	23.9	27.3	495	135	360	36.8
27.5	378	104	274	28.6	27.5	211	58	153	15.9	27.4	445	122	323	33.4	27.4	314	86	228	23.6	27.3	494	135	359	36.9
27.5	375	103	272	28.3	27.5	207	57	150	15.7	27.4	442	121	321	33.1	27.4	310	85	225	23.3	27.3	491	134	357	36.6

%E	M1	M2	DM	M*
27.3	490	134	356	36.6
27.3	488	133	355	36.2
27.3	487	133	354	36.3
27.3	484	132	352	36.0
27.3	483	132	351	36.1
27.3	480	131	349	35.8
27.3	479	131	348	35.8
27.3	477	130	347	35.4
27.3	476	130	346	35.5
27.3	473	129	344	35.2
27.3	472	129	343	35.3
27.3	469	128	341	34.9
27.3	466	127	339	34.6
27.3	465	127	338	34.7
27.3	462	126	336	34.4
27.3	461	126	335	34.4
27.3	458	125	333	34.1
27.3	455	124	331	33.8
27.3	454	124	330	33.9
27.3	451	123	328	33.5
27.3	450	123	327	33.6
27.3	447	122	325	33.3
27.3	444	121	323	33.0
27.3	443	121	322	33.0
27.3	440	120	320	32.7
27.3	439	120	319	32.8
27.3	436	119	317	32.5
27.3	433	118	315	32.2
27.3	432	118	314	32.2
27.3	429	117	312	31.9
27.3	428	117	311	32.0
27.3	425	116	309	31.7
27.3	422	115	307	31.3
27.3	421	115	306	31.4
27.3	418	114	304	31.1
27.3	417	114	303	31.2
27.3	414	113	301	30.8
27.3	411	112	299	30.5
27.3	410	112	298	30.6
27.3	407	111	296	30.3
27.3	406	111	295	30.3
27.3	403	110	293	30.0
27.3	400	109	291	29.7
27.3	399	109	290	29.8
27.3	396	108	288	29.5
27.3	395	108	287	29.5
27.3	392	107	285	29.2
27.3	388	106	282	29.0
27.3	385	105	280	28.6
27.3	384	105	279	28.7

%E	M1	M2	DM	M*
27.3	381	104	277	28.4
27.3	377	103	274	28.1
27.3	374	102	272	27.8
27.3	373	102	271	27.9
27.3	370	101	269	27.6
27.3	366	100	266	27.3
27.3	363	99	264	27.0
27.3	362	99	263	27.1
27.3	359	98	261	26.8
27.3	355	97	258	26.5
27.3	352	96	256	26.2
27.3	348	95	253	25.9
27.3	344	94	250	25.7
27.3	341	93	248	25.4
27.3	337	92	245	25.1
27.3	333	91	242	24.9
27.3	330	90	240	24.5
27.3	326	89	237	24.3
27.3	322	88	234	24.0
27.3	319	87	232	23.7
27.3	315	86	229	23.5
27.3	311	85	226	23.2
27.3	308	84	224	22.9
27.3	304	83	221	22.7
27.3	300	82	218	22.4
27.3	297	81	216	22.1
27.3	293	80	213	21.8
27.3	289	79	210	21.6
27.3	286	78	208	21.3
27.3	282	77	205	21.0
27.3	278	76	202	20.8
27.3	275	75	200	20.5
27.3	271	74	197	20.2
27.3	267	73	194	20.0
27.3	264	72	192	19.6
27.3	260	71	189	19.4
27.3	256	70	186	19.1
27.3	253	69	184	18.8
27.3	249	68	181	18.6
27.3	245	67	178	18.3
27.3	242	66	176	18.0
27.3	238	65	173	17.8
27.3	231	63	168	17.2
27.3	227	62	165	16.9
27.3	220	60	160	16.4
27.3	216	59	157	16.1
27.3	209	57	152	15.5
27.3	205	56	149	15.3
27.3	198	54	144	14.7
27.3	194	53	141	14.5

%E	M1	M2	DM	M*
27.3	187	51	136	13.9
27.3	183	50	133	13.7
27.3	176	48	128	13.1
27.3	172	47	125	12.8
27.3	165	45	120	12.3
27.3	161	44	117	12.0
27.3	154	42	112	11.5
27.3	150	41	109	11.2
27.3	143	39	104	10.6
27.3	139	38	101	10.4
27.3	132	36	96	9.8
27.3	128	35	93	9.6
27.3	121	33	88	9.0
27.3	110	30	80	8.2
27.3	99	27	72	7.4
27.3	88	24	64	6.5
27.3	77	21	56	5.7
27.3	66	18	48	4.9
27.3	55	15	40	4.1
27.3	44	12	32	3.3
27.3	33	9	24	2.5
27.3	22	6	16	1.6
27.3	11	3	8	0.8
27.2	500	136	364	37.0
27.2	497	135	362	36.7
27.2	496	135	361	36.7
27.2	493	134	359	36.4
27.2	492	134	358	36.5
27.2	489	133	356	36.2
27.2	486	132	354	35.9
27.2	485	132	353	35.9
27.2	482	131	351	35.6
27.2	481	131	350	35.7
27.2	478	130	348	35.4
27.2	475	129	346	35.0
27.2	474	129	345	35.1
27.2	471	128	343	34.8
27.2	470	128	342	34.9
27.2	467	127	340	34.5
27.2	464	126	338	34.2
27.2	463	126	337	34.3
27.2	460	125	335	34.0
27.2	459	125	334	34.0
27.2	456	124	332	33.7
27.2	453	123	330	33.4
27.2	452	123	329	33.5
27.2	449	122	327	33.1
27.2	448	122	326	33.2
27.2	445	121	324	32.9
27.2	441	120	321	32.7

%E	M1	M2	DM	M*
27.2	438	119	319	32.3
27.2	437	119	318	32.4
27.2	434	118	316	32.1
27.2	430	117	313	31.8
27.2	427	116	311	31.5
27.2	426	116	310	31.6
27.2	423	115	308	31.3
27.2	419	114	305	31.0
27.2	416	113	303	30.7
27.2	415	113	302	30.8
27.2	412	112	300	30.4
27.2	408	111	297	30.2
27.2	405	110	295	29.9
27.2	404	110	294	30.0
27.2	401	109	292	29.6
27.2	397	108	289	29.4
27.2	394	107	287	29.1
27.2	393	107	286	29.1
27.2	390	106	284	28.8
27.2	389	106	283	28.9
27.2	386	105	281	28.6
27.2	383	104	279	28.2
27.2	382	104	278	28.3
27.2	379	103	276	28.0
27.2	378	103	275	28.1
27.2	375	102	273	27.7
27.2	372	101	271	27.4
27.2	371	101	270	27.5
27.2	368	100	268	27.2
27.2	367	100	267	27.2
27.2	364	99	265	26.9
27.2	360	98	262	26.7
27.2	357	97	260	26.4
27.2	356	97	259	26.4
27.2	353	96	257	26.1
27.2	349	95	254	25.9
27.2	346	94	252	25.5
27.2	345	94	251	25.6
27.2	342	93	249	25.3
27.2	338	92	246	25.0
27.2	335	91	244	24.7
27.2	334	91	243	24.8
27.2	331	90	241	24.5
27.2	327	89	238	24.2
27.2	324	88	236	23.9
27.2	323	88	235	24.0
27.2	320	87	233	23.7
27.2	316	86	230	23.4
27.2	313	85	228	23.1
27.2	312	85	227	23.2

%E	M1	M2	DM	M*
27.2	309	84	225	22.8
27.2	305	83	222	22.6
27.2	302	82	220	22.3
27.2	301	82	219	22.3
27.2	298	81	217	22.0
27.2	294	80	214	21.8
27.2	290	79	211	21.5
27.2	287	78	209	21.2
27.2	283	77	206	21.0
27.2	279	76	203	20.7
27.2	276	75	201	20.4
27.2	272	74	198	20.1
27.2	268	73	195	19.9
27.2	265	72	193	19.6
27.2	261	71	190	19.3
27.2	257	70	187	19.1
27.2	254	69	185	18.7
27.2	250	68	182	18.5
27.2	246	67	179	18.2
27.2	243	66	177	17.9
27.2	239	65	174	17.7
27.2	235	64	171	17.4
27.2	232	63	169	17.1
27.2	228	62	166	16.9
27.2	224	61	163	16.6
27.2	217	59	158	16.0
27.2	213	58	155	15.8
27.2	206	56	150	15.2
27.2	202	55	147	15.0
27.2	195	53	142	14.4
27.2	191	52	139	14.2
27.2	184	50	134	13.6
27.2	180	49	131	13.3
27.2	173	47	126	12.8
27.2	169	46	123	12.5
27.2	162	44	118	12.0
27.2	158	43	115	11.7
27.2	151	41	110	11.1
27.2	147	40	107	10.9
27.2	136	37	99	10.1
27.2	125	34	91	9.2
27.2	114	31	83	8.4
27.2	103	28	75	7.6
27.2	92	25	67	6.8
27.2	81	22	59	6.0
27.1	499	135	364	36.5
27.1	498	135	363	36.6
27.1	495	134	361	36.3
27.1	494	134	360	36.3
27.1	491	133	358	36.0

%E	M1	M2	DM	M*
27.1	490	133	357	36.1
27.1	487	132	355	35.8
27.1	484	131	353	35.5
27.1	483	131	352	35.5
27.1	480	130	350	35.2
27.1	479	130	349	35.3
27.1	476	129	347	35.0
27.1	473	128	345	34.6
27.1	472	128	344	34.7
27.1	469	127	342	34.4
27.1	468	127	341	34.5
27.1	465	126	339	34.1
27.1	462	125	337	33.8
27.1	461	125	336	33.9
27.1	458	124	334	33.6
27.1	457	124	333	33.6
27.1	454	123	331	33.3
27.1	451	122	329	33.0
27.1	450	122	328	33.1
27.1	447	121	326	32.8
27.1	446	121	325	32.8
27.1	443	120	323	32.5
27.1	442	120	322	32.6
27.1	439	119	320	32.3
27.1	436	118	318	31.9
27.1	435	118	317	32.0
27.1	432	117	315	31.7
27.1	431	117	314	31.8
27.1	428	116	312	31.4
27.1	425	115	310	31.1
27.1	424	115	309	31.2
27.1	421	114	307	30.9
27.1	420	114	306	30.9
27.1	417	113	304	30.6
27.1	414	112	302	30.3
27.1	413	112	301	30.4
27.1	410	111	299	30.1
27.1	409	111	298	30.1
27.1	406	110	296	29.8
27.1	402	109	293	29.6
27.1	399	108	291	29.2
27.1	398	108	290	29.3
27.1	395	107	288	29.0
27.1	391	106	285	28.7
27.1	388	105	283	28.4
27.1	387	105	282	28.5
27.1	384	104	280	28.2
27.1	380	103	277	27.9
27.1	377	102	275	27.6
27.1	376	102	274	27.7
27.1	373	101	272	27.3
27.1	369	100	269	27.1
27.1	365	99	266	26.9
27.1	362	98	264	26.5
27.1	361	98	263	26.6
27.1	358	97	261	26.3
27.1	354	96	258	26.0
27.1	351	95	256	25.7
27.1	350	95	255	25.8
27.1	347	94	253	25.5
27.1	343	93	250	25.2
27.1	340	92	248	24.9
27.1	339	92	247	25.0
27.1	336	91	245	24.6
27.1	332	90	242	24.4
27.1	329	89	240	24.1
27.1	328	89	239	24.1
27.1	325	88	237	23.8
27.1	321	87	234	23.6
27.1	317	86	231	23.3
27.1	314	85	229	23.0
27.1	310	84	226	22.8
27.1	306	83	223	22.5
27.1	303	82	221	22.2
27.1	299	81	218	21.9
27.1	295	80	215	21.7
27.1	292	79	213	21.4
27.1	291	79	212	21.4
27.1	288	78	210	21.1
27.1	284	77	207	20.9
27.1	280	76	204	20.6
27.1	277	75	202	20.3
27.1	273	74	199	20.1
27.1	269	73	196	19.8
27.1	266	72	194	19.5
27.1	262	71	191	19.2
27.1	258	70	188	19.0
27.1	255	69	186	18.7
27.1	251	68	183	18.4
27.1	247	67	180	18.2
27.1	240	65	175	17.6
27.1	236	64	172	17.4
27.1	229	62	167	16.8
27.1	225	61	164	16.5
27.1	221	60	161	16.3
27.1	218	59	159	16.0
27.1	214	58	156	15.7
27.1	210	57	153	15.5
27.1	207	56	151	15.1
27.1	203	55	148	14.9
27.1	199	54	145	14.7
27.1	192	52	140	14.1
27.1	188	51	137	13.8
27.1	181	49	132	13.3
27.1	177	48	129	13.0
27.1	170	46	124	12.4
27.1	166	45	121	12.2
27.1	155	42	113	11.4
27.1	144	39	105	10.6
27.1	140	38	102	10.3
27.1	133	36	97	9.7
27.1	129	35	94	9.5
27.1	118	32	86	8.7
27.1	107	29	78	7.9
27.1	96	26	70	7.0
27.1	85	23	62	6.2
27.1	70	19	51	5.2
27.1	59	16	43	4.3
27.1	48	13	35	3.5
27.0	500	135	365	36.4
27.0	497	134	363	36.1
27.0	496	134	362	36.2
27.0	493	133	360	35.9
27.0	492	133	359	36.0
27.0	489	132	357	35.6
27.0	488	132	356	35.7
27.0	486	131	355	35.3
27.0	485	131	354	35.4
27.0	482	130	352	35.1
27.0	481	130	351	35.1
27.0	478	129	349	34.8
27.0	477	129	348	34.9
27.0	474	128	346	34.6
27.0	471	127	344	34.2
27.0	470	127	343	34.3
27.0	467	126	341	34.0
27.0	466	126	340	34.1
27.0	463	125	338	33.7
27.0	460	124	336	33.4
27.0	459	124	335	33.5
27.0	456	123	333	33.2
27.0	455	123	332	33.3
27.0	452	122	330	32.9
27.0	448	121	327	32.7
27.0	445	120	325	32.4
27.0	444	120	324	32.4
27.0	441	119	322	32.1
27.0	440	119	321	32.2
27.0	437	118	319	31.9
27.0	434	117	317	31.5
27.0	433	117	316	31.6
27.0	430	116	314	31.3
27.0	429	116	313	31.4
27.0	426	115	311	31.0
27.0	423	114	309	30.7
27.0	422	114	308	30.8
27.0	419	113	306	30.5
27.0	418	113	305	30.5
27.0	415	112	303	30.2
27.0	411	111	300	30.0
27.0	408	110	298	29.7
27.0	407	110	297	29.7
27.0	404	109	295	29.4
27.0	403	109	294	29.5
27.0	400	108	292	29.2
27.0	397	107	290	28.8
27.0	396	107	289	28.9
27.0	393	106	287	28.6
27.0	392	106	286	28.7
27.0	389	105	284	28.3
27.0	385	104	281	28.1
27.0	382	103	279	27.8
27.0	381	103	278	27.8
27.0	378	102	276	27.5
27.0	374	101	273	27.3
27.0	371	100	271	27.0
27.0	370	100	270	27.0
27.0	367	99	268	26.7
27.0	366	99	267	26.8
27.0	363	98	265	26.5
27.0	359	97	262	26.2
27.0	356	96	260	25.9
27.0	355	96	259	26.0
27.0	352	95	257	25.6
27.0	348	94	254	25.4
27.0	345	93	252	25.1
27.0	344	93	251	25.1
27.0	341	92	249	24.8
27.0	337	91	246	24.6
27.0	333	90	243	24.3
27.0	330	89	241	24.0
27.0	326	88	238	23.8
27.0	322	87	235	23.5
27.0	319	86	233	23.2
27.0	318	86	232	23.3
27.0	315	85	230	22.9
27.0	311	84	227	22.7
27.0	307	83	224	22.4
27.0	304	82	222	22.1
27.0	300	81	219	21.9
27.0	296	80	216	21.6
27.0	293	79	214	21.3
27.0	289	78	211	21.1
27.0	285	77	208	20.8
27.0	282	76	206	20.5
27.0	281	76	205	20.6
27.0	278	75	203	20.2
27.0	274	74	200	20.0
27.0	270	73	197	19.7
27.0	267	72	195	19.4
27.0	263	71	192	19.2
27.0	259	70	189	18.9
27.0	256	69	187	18.6
27.0	252	68	184	18.3
27.0	248	67	181	18.1
27.0	244	66	178	17.9
27.0	241	65	176	17.5
27.0	237	64	173	17.3
27.0	233	63	170	17.0
27.0	230	62	168	16.7
27.0	226	61	165	16.5
27.0	222	60	162	16.2
27.0	215	58	157	15.6
27.0	211	57	154	15.4
27.0	204	55	149	14.8
27.0	200	54	146	14.6
27.0	196	53	143	14.3
27.0	189	51	138	13.8
27.0	185	50	135	13.5
27.0	178	48	130	12.9
27.0	174	47	127	12.7
27.0	163	44	119	11.9
27.0	159	43	116	11.6
27.0	152	41	111	11.1
27.0	148	40	108	10.8
27.0	141	38	103	10.2
27.0	137	37	100	10.0
27.0	126	34	92	9.2
27.0	122	33	89	8.9
27.0	115	31	84	8.4
27.0	111	30	81	8.1
27.0	100	27	73	7.3
27.0	89	24	65	6.5
27.0	74	20	54	5.4
27.0	63	17	46	4.6
27.0	37	10	27	2.7
26.9	499	134	365	36.0
26.9	498	134	364	36.1
26.9	495	133	362	35.7
26.9	494	133	361	35.8

%E	M1	M2	DM	M*
26.9	491	132	359	35.5
26.9	490	132	358	35.6
26.9	487	131	356	35.2
26.9	484	130	354	34.9
26.9	483	130	353	35.0
26.9	480	129	351	34.7
26.9	479	129	350	34.7
26.9	476	128	348	34.4
26.9	475	128	347	34.5
26.9	473	127	346	34.1
26.9	472	127	345	34.2
26.9	469	126	343	33.9
26.9	468	126	342	33.9
26.9	465	125	340	33.6
26.9	464	125	339	33.7
26.9	461	124	337	33.4
26.9	458	123	335	33.0
26.9	457	123	334	33.1
26.9	454	122	332	32.8
26.9	453	122	331	32.9
26.9	450	121	329	32.5
26.9	449	121	328	32.6
26.9	446	120	326	32.3
26.9	443	119	324	32.0
26.9	442	119	323	32.0
26.9	439	118	321	31.7
26.9	438	118	320	31.8
26.9	435	117	318	31.5
26.9	432	116	316	31.1
26.9	431	116	315	31.2
26.9	428	115	313	30.9
26.9	427	115	312	31.0
26.9	424	114	310	30.7
26.9	420	113	307	30.4
26.9	417	112	305	30.1
26.9	416	112	304	30.2
26.9	413	111	302	29.8
26.9	412	111	301	29.9
26.9	409	110	299	29.6
26.9	405	109	296	29.3
26.9	402	108	294	29.0
26.9	401	108	293	29.1
26.9	398	107	291	28.8
26.9	394	106	288	28.5
26.9	391	105	286	28.2
26.9	390	105	285	28.3
26.9	387	104	283	27.9
26.9	386	104	282	28.0
26.9	383	103	280	27.7
26.9	379	102	277	27.5

%E	M1	M2	DM	M*
26.9	376	101	275	27.1
26.9	375	101	274	27.2
26.9	372	100	272	26.9
26.9	368	99	269	26.6
26.9	364	98	266	26.4
26.9	361	97	264	26.1
26.9	360	97	263	26.1
26.9	357	96	261	25.8
26.9	353	95	258	25.6
26.9	350	94	256	25.2
26.9	349	94	255	25.3
26.9	346	93	253	25.0
26.9	342	92	250	24.7
26.9	338	91	247	24.5
26.9	335	90	245	24.2
26.9	334	90	244	24.3
26.9	331	89	242	23.9
26.9	327	88	239	23.7
26.9	324	87	237	23.4
26.9	323	87	236	23.4
26.9	320	86	234	23.1
26.9	316	85	231	22.9
26.9	312	84	228	22.6
26.9	309	83	226	22.3
26.9	308	83	225	22.4
26.9	305	82	223	22.0
26.9	301	81	220	21.8
26.9	297	80	217	21.5
26.9	294	79	215	21.2
26.9	290	78	212	21.0
26.9	286	77	209	20.7
26.9	283	76	207	20.4
26.9	279	75	204	20.2
26.9	275	74	201	19.9
26.9	271	73	198	19.7
26.9	268	72	196	19.3
26.9	264	71	193	19.1
26.9	260	70	190	18.8
26.9	253	68	185	18.3
26.9	249	67	182	18.0
26.9	245	66	179	17.8
26.9	242	65	177	17.5
26.9	238	64	174	17.2
26.9	234	63	171	17.0
26.9	227	61	166	16.4
26.9	223	60	163	16.1
26.9	219	59	160	15.9
26.9	216	58	158	15.6
26.9	212	57	155	15.3
26.9	208	56	152	15.1

%E	M1	M2	DM	M*
26.9	201	54	147	14.5
26.9	197	53	144	14.3
26.9	193	52	141	14.0
26.9	186	50	136	13.4
26.9	182	49	133	13.2
26.9	175	47	128	12.6
26.9	171	46	125	12.4
26.9	167	45	122	12.1
26.9	160	43	117	11.6
26.9	156	42	114	11.3
26.9	145	39	106	10.5
26.9	134	36	98	9.7
26.9	130	35	95	9.4
26.9	119	32	87	8.6
26.9	108	29	79	7.8
26.9	104	28	76	7.5
26.9	93	25	68	6.7
26.9	78	21	57	5.7
26.9	67	18	49	4.8
26.9	52	14	38	3.8
26.9	26	7	19	1.9
26.8	500	134	366	35.9
26.8	497	133	364	35.6
26.8	496	133	363	35.7
26.8	493	132	361	35.3
26.8	492	132	360	35.4
26.8	489	131	358	35.1
26.8	488	131	357	35.2
26.8	485	130	355	34.8
26.8	482	129	353	34.5
26.8	481	129	352	34.6
26.8	478	128	350	34.3
26.8	477	128	349	34.3
26.8	474	127	347	34.0
26.8	471	126	345	33.7
26.8	470	126	344	33.8
26.8	467	125	342	33.5
26.8	466	125	341	33.5
26.8	463	124	339	33.2
26.8	462	124	338	33.3
26.8	459	123	336	33.0
26.8	456	122	334	32.6
26.8	455	122	333	32.7
26.8	452	121	331	32.4
26.8	451	121	330	32.5
26.8	448	120	328	32.1
26.8	447	120	327	32.2
26.8	444	119	325	31.9
26.8	441	118	323	31.6
26.8	440	118	322	31.6

%E	M1	M2	DM	M*
26.8	437	117	320	31.3
26.8	436	117	319	31.4
26.8	433	116	317	31.1
26.8	429	115	314	30.8
26.8	426	114	312	30.5
26.8	425	114	311	30.6
26.8	422	113	309	30.3
26.8	421	113	308	30.3
26.8	418	112	306	30.0
26.8	414	111	303	29.8
26.8	411	110	301	29.4
26.8	410	110	300	29.5
26.8	407	109	298	29.2
26.8	406	109	297	29.3
26.8	403	108	295	28.9
26.8	400	107	293	28.6
26.8	399	107	292	28.7
26.8	396	106	290	28.4
26.8	395	106	289	28.4
26.8	392	105	287	28.1
26.8	388	104	284	27.9
26.8	385	103	282	27.6
26.8	384	103	281	27.6
26.8	381	102	279	27.3
26.8	380	102	278	27.4
26.8	377	101	276	27.1
26.8	373	100	273	26.8
26.8	370	99	271	26.5
26.8	369	99	270	26.6
26.8	366	98	268	26.2
26.8	365	98	267	26.3
26.8	362	97	265	26.0
26.8	358	96	262	25.7
26.8	355	95	260	25.4
26.8	354	95	259	25.5
26.8	351	94	257	25.2
26.8	347	93	254	24.9
26.8	343	92	251	24.7
26.8	340	91	249	24.4
26.8	339	91	248	24.4
26.8	336	90	246	24.1
26.8	332	89	243	23.9
26.8	328	88	240	23.6
26.8	325	87	238	23.3
26.8	321	86	235	23.0
26.8	317	85	232	22.8
26.8	314	84	230	22.5
26.8	313	84	229	22.5
26.8	310	83	227	22.2
26.8	306	82	224	22.0

%E	M1	M2	DM	M*
26.8	302	81	221	21.7
26.8	299	80	219	21.4
26.8	298	80	218	21.5
26.8	295	79	216	21.2
26.8	291	78	213	20.9
26.8	287	77	210	20.7
26.8	284	76	208	20.3
26.8	280	75	205	20.1
26.8	276	74	202	19.8
26.8	272	73	199	19.6
26.8	269	72	197	19.3
26.8	265	71	194	19.0
26.8	261	70	191	18.8
26.8	257	69	188	18.5
26.8	254	68	186	18.2
26.8	250	67	183	18.0
26.8	246	66	180	17.7
26.8	239	64	175	17.1
26.8	235	63	172	16.9
26.8	231	62	169	16.6
26.8	228	61	167	16.3
26.8	224	60	164	16.1
26.8	220	59	161	15.8
26.8	213	57	156	15.3
26.8	209	56	153	15.0
26.8	205	55	150	14.8
26.8	198	53	145	14.2
26.8	194	52	142	13.9
26.8	190	51	139	13.7
26.8	183	49	134	13.1
26.8	179	48	131	12.9
26.8	168	45	123	12.1
26.8	164	44	120	11.8
26.8	157	42	115	11.2
26.8	153	41	112	11.0
26.8	149	40	109	10.7
26.8	142	38	104	10.2
26.8	138	37	101	9.9
26.8	127	34	93	9.1
26.8	123	33	90	8.9
26.8	112	30	82	8.0
26.8	97	26	71	7.0
26.8	82	22	60	5.9
26.8	71	19	52	5.1
26.8	56	15	41	4.0
26.8	41	11	30	3.0
26.7	499	133	366	35.4
26.7	498	133	365	35.5
26.7	495	132	363	35.2
26.7	494	132	362	35.3

%E	M1	M2	DM	M*
26.7	491	131	360	35.0
26.7	490	131	359	35.0
26.7	487	130	357	34.7
26.7	486	130	356	34.8
26.7	484	129	355	34.4
26.7	483	129	354	34.5
26.7	480	128	352	34.1
26.7	479	128	351	34.2
26.7	476	127	349	33.9
26.7	475	127	348	34.0
26.7	472	126	346	33.6
26.7	469	125	344	33.3
26.7	468	125	343	33.4
26.7	465	124	341	33.1
26.7	464	124	340	33.1
26.7	461	123	338	32.8
26.7	460	123	337	32.9
26.7	457	122	335	32.6
26.7	454	121	333	32.2
26.7	453	121	332	32.3
26.7	450	120	330	32.0
26.7	449	120	329	32.1
26.7	446	119	327	31.8
26.7	445	119	326	31.8
26.7	442	118	324	31.5
26.7	439	117	322	31.2
26.7	438	117	321	31.3
26.7	435	116	319	30.9
26.7	434	116	318	31.0
26.7	431	115	316	30.7
26.7	430	115	315	30.8
26.7	427	114	313	30.4
26.7	424	113	311	30.1
26.7	423	113	310	30.2
26.7	420	112	308	29.9
26.7	419	112	307	29.9
26.7	416	111	305	29.6
26.7	415	111	304	29.7
26.7	412	110	302	29.4
26.7	409	109	300	29.0
26.7	408	109	299	29.1
26.7	405	108	297	28.8
26.7	404	108	296	28.9
26.7	401	107	294	28.6
26.7	397	106	291	28.3
26.7	394	105	289	28.0
26.7	393	105	288	28.1
26.7	390	104	286	27.7
26.7	389	104	285	27.8
26.7	386	103	283	27.5

%E	M1	M2	DM	M*
26.7	382	102	280	27.2
26.7	378	101	277	27.0
26.7	375	100	275	26.7
26.7	374	100	274	26.7
26.7	371	99	272	26.4
26.7	367	98	269	26.2
26.7	363	97	266	25.9
26.7	360	96	264	25.6
26.7	359	96	263	25.7
26.7	356	95	261	25.4
26.7	352	94	258	25.1
26.7	348	93	255	24.9
26.7	345	92	253	24.5
26.7	344	92	252	24.6
26.7	341	91	250	24.3
26.7	337	90	247	24.0
26.7	333	89	244	23.8
26.7	330	88	242	23.5
26.7	329	88	241	23.5
26.7	326	87	239	23.2
26.7	322	86	236	23.0
26.7	318	85	233	22.7
26.7	315	84	231	22.4
26.7	311	83	228	22.2
26.7	307	82	225	21.9
26.7	303	81	222	21.7
26.7	300	80	220	21.3
26.7	296	79	217	21.1
26.7	292	78	214	20.8
26.7	288	77	211	20.6
26.7	285	76	209	20.3
26.7	281	75	206	20.0
26.7	277	74	203	19.8
26.7	273	73	200	19.5
26.7	270	72	198	19.2
26.7	266	71	195	19.0
26.7	262	70	192	18.7
26.7	258	69	189	18.5
26.7	255	68	187	18.1
26.7	251	67	184	17.9
26.7	247	66	181	17.6
26.7	243	65	178	17.4
26.7	240	64	176	17.1
26.7	236	63	173	16.8
26.7	232	62	170	16.6
26.7	225	60	165	16.0
26.7	221	59	162	15.8
26.7	217	58	159	15.5
26.7	210	56	154	14.9
26.7	206	55	151	14.7

%E	M1	M2	DM	M*
26.7	202	54	148	14.4
26.7	195	52	143	13.9
26.7	191	51	140	13.6
26.7	187	50	137	13.4
26.7	180	48	132	12.8
26.7	176	47	129	12.6
26.7	172	46	126	12.3
26.7	165	44	121	11.7
26.7	161	43	118	11.5
26.7	150	40	110	10.7
26.7	146	39	107	10.4
26.7	135	36	99	9.6
26.7	131	35	96	9.4
26.7	120	32	88	8.5
26.7	116	31	85	8.3
26.7	105	28	77	7.5
26.7	101	27	74	7.2
26.7	90	24	66	6.4
26.7	86	23	63	6.2
26.7	75	20	55	5.3
26.7	60	16	44	4.3
26.7	45	12	33	3.2
26.7	30	8	22	2.1
26.7	15	4	11	1.1
26.6	500	133	367	35.4
26.6	497	132	365	35.1
26.6	496	132	364	35.1
26.6	493	131	362	34.8
26.6	492	131	361	34.9
26.6	489	130	359	34.6
26.6	488	130	358	34.6
26.6	485	129	356	34.3
26.6	482	128	354	34.0
26.6	481	128	353	34.1
26.6	478	127	351	33.7
26.6	477	127	350	33.8
26.6	474	126	348	33.5
26.6	473	126	347	33.6
26.6	470	125	345	33.2
26.6	467	124	343	32.9
26.6	466	124	342	33.0
26.6	463	123	340	32.7
26.6	462	123	339	32.7
26.6	459	122	337	32.4
26.6	458	122	336	32.5
26.6	455	121	334	32.2
26.6	451	120	331	31.9
26.6	448	119	329	31.6
26.6	447	119	328	31.7
26.6	444	118	326	31.4

%E	M1	M2	DM	M*
26.6	443	118	325	31.4
26.6	440	117	323	31.1
26.6	436	116	320	30.9
26.6	433	115	318	30.5
26.6	432	115	317	30.6
26.6	429	114	315	30.3
26.6	428	114	314	30.4
26.6	425	113	312	30.0
26.6	421	112	309	29.8
26.6	418	111	307	29.5
26.6	417	111	306	29.5
26.6	414	110	304	29.2
26.6	413	110	303	29.3
26.6	410	109	301	29.0
26.6	406	108	298	28.7
26.6	403	107	296	28.4
26.6	402	107	295	28.5
26.6	399	106	293	28.2
26.6	398	106	292	28.2
26.6	395	105	290	27.9
26.6	391	104	287	27.7
26.6	387	103	284	27.4
26.6	384	102	282	27.1
26.6	383	102	281	27.2
26.6	380	101	279	26.8
26.6	379	101	278	26.9
26.6	376	100	276	26.6
26.6	372	99	273	26.3
26.6	369	98	271	26.0
26.6	368	98	270	26.1
26.6	365	97	268	25.8
26.6	364	97	267	25.8
26.6	361	96	265	25.5
26.6	357	95	262	25.3
26.6	354	94	260	25.0
26.6	353	94	259	25.0
26.6	350	93	257	24.7
26.6	349	93	256	24.8
26.6	346	92	254	24.5
26.6	342	91	251	24.2
26.6	338	90	248	24.0
26.6	335	89	246	23.6
26.6	334	89	245	23.7
26.6	331	88	243	23.4
26.6	327	87	240	23.1
26.6	323	86	237	22.9
26.6	320	85	235	22.6
26.6	319	85	234	22.6
26.6	316	84	232	22.3
26.6	312	83	229	22.1

%E	M1	M2	DM	M*
26.6	308	82	226	21.8
26.6	305	81	224	21.5
26.6	304	81	223	21.6
26.6	301	80	221	21.3
26.6	297	79	218	21.0
26.6	293	78	215	20.8
26.6	290	77	213	20.4
26.6	289	77	212	20.5
26.6	286	76	210	20.2
26.6	282	75	207	19.9
26.6	278	74	204	19.7
26.6	274	73	201	19.4
26.6	271	72	199	19.1
26.6	267	71	196	18.9
26.6	263	70	193	18.6
26.6	259	69	190	18.4
26.6	256	68	188	18.1
26.6	252	67	185	17.8
26.6	248	66	182	17.6
26.6	244	65	179	17.3
26.6	241	64	177	17.0
26.6	237	63	174	16.7
26.6	233	62	171	16.5
26.6	229	61	168	16.2
26.6	222	59	163	15.7
26.6	218	58	160	15.4
26.6	214	57	157	15.2
26.6	207	55	152	14.6
26.6	203	54	149	14.4
26.6	199	53	146	14.1
26.6	192	51	141	13.5
26.6	188	50	138	13.3
26.6	184	49	135	13.0
26.6	177	47	130	12.5
26.6	173	46	127	12.2
26.6	169	45	124	12.0
26.6	158	42	116	11.2
26.6	154	41	113	10.9
26.6	143	38	105	10.1
26.6	139	37	102	9.8
26.6	128	34	94	9.0
26.6	124	33	91	8.8
26.6	109	29	80	7.7
26.6	94	25	69	6.6
26.6	79	21	58	5.6
26.6	64	17	47	4.5
26.5	499	132	367	34.9
26.5	498	132	366	35.0
26.5	495	131	364	34.7
26.5	494	131	363	34.7

%E	M1	M2	DM	M*
26.5	491	130	361	34.4
26.5	490	130	360	34.5
26.5	487	129	358	34.2
26.5	486	129	357	34.2
26.5	483	128	355	33.9
26.5	480	127	353	33.6
26.5	479	127	352	33.7
26.5	476	126	350	33.4
26.5	475	126	349	33.4
26.5	472	125	347	33.1
26.5	471	125	346	33.2
26.5	468	124	344	32.9
26.5	465	123	342	32.5
26.5	464	123	341	32.6
26.5	461	122	339	32.3
26.5	460	122	338	32.4
26.5	457	121	336	32.0
26.5	456	121	335	32.1
26.5	453	120	333	31.8
26.5	452	120	332	31.9
26.5	449	119	330	31.5
26.5	446	118	328	31.2
26.5	445	118	327	31.3
26.5	442	117	325	31.0
26.5	441	117	324	31.0
26.5	438	116	322	30.7
26.5	437	116	321	30.8
26.5	434	115	319	30.5
26.5	431	114	317	30.2
26.5	430	114	316	30.2
26.5	427	113	314	29.9
26.5	426	113	313	30.0
26.5	423	112	311	29.7
26.5	422	112	310	29.7
26.5	419	111	308	29.4
26.5	415	110	305	29.2
26.5	412	109	303	28.8
26.5	411	109	302	28.9
26.5	408	108	300	28.6
26.5	407	108	299	28.7
26.5	404	107	297	28.3
26.5	400	106	294	28.1
26.5	396	105	291	27.8
26.5	393	104	289	27.5
26.5	392	104	288	27.6
26.5	389	103	286	27.3
26.5	388	103	285	27.3
26.5	385	102	283	27.0
26.5	381	101	280	26.8
26.5	378	100	278	26.5

%E	M1	M2	DM	M*
26.5	377	100	277	26.5
26.5	374	99	275	26.2
26.5	373	99	274	26.3
26.5	370	98	272	26.0
26.5	366	97	269	25.7
26.5	362	96	266	25.5
26.5	359	95	264	25.1
26.5	358	95	263	25.2
26.5	355	94	261	24.9
26.5	351	93	258	24.6
26.5	347	92	255	24.4
26.5	344	91	253	24.1
26.5	343	91	252	24.1
26.5	340	90	250	23.8
26.5	339	90	249	23.9
26.5	336	89	247	23.6
26.5	332	88	244	23.3
26.5	328	87	241	23.1
26.5	325	86	239	22.8
26.5	324	86	238	22.8
26.5	321	85	236	22.5
26.5	317	84	233	22.3
26.5	313	83	230	22.0
26.5	310	82	228	21.7
26.5	309	82	227	21.8
26.5	306	81	225	21.4
26.5	302	80	222	21.2
26.5	298	79	219	20.9
26.5	294	78	216	20.7
26.5	291	77	214	20.4
26.5	287	76	211	20.1
26.5	283	75	208	19.9
26.5	279	74	205	19.6
26.5	275	73	202	19.4
26.5	272	72	200	19.1
26.5	268	71	197	18.8
26.5	264	70	194	18.6
26.5	260	69	191	18.3
26.5	257	68	189	18.0
26.5	253	67	186	17.7
26.5	249	66	183	17.5
26.5	245	65	180	17.2
26.5	238	63	175	16.7
26.5	234	62	172	16.4
26.5	230	61	169	16.2
26.5	226	60	166	15.9
26.5	223	59	164	15.6
26.5	219	58	161	15.4
26.5	215	57	158	15.1
26.5	211	56	155	14.9

%E	M1	M2	DM	M*
26.5	204	54	150	14.3
26.5	200	53	147	14.0
26.5	196	52	144	13.8
26.5	189	50	139	13.2
26.5	185	49	136	13.0
26.5	181	48	133	12.7
26.5	170	45	125	11.9
26.5	166	44	122	11.7
26.5	162	43	119	11.4
26.5	155	41	114	10.8
26.5	151	40	111	10.6
26.5	147	39	108	10.3
26.5	136	36	100	9.5
26.5	132	35	97	9.3
26.5	117	31	86	8.2
26.5	113	30	83	8.0
26.5	102	27	75	7.1
26.5	98	26	72	6.9
26.5	83	22	61	5.8
26.5	68	18	50	4.8
26.5	49	13	36	3.4
26.5	34	9	25	2.4
26.4	500	132	368	34.8
26.4	497	131	366	34.5
26.4	496	131	365	34.6
26.4	493	130	363	34.3
26.4	492	130	362	34.3
26.4	489	129	360	34.0
26.4	488	129	359	34.1
26.4	485	128	357	33.8
26.4	484	128	356	33.9
26.4	481	127	354	33.5
26.4	478	126	352	33.2
26.4	477	126	351	33.3
26.4	474	125	349	33.0
26.4	473	125	348	33.0
26.4	470	124	346	32.7
26.4	469	124	345	32.8
26.4	466	123	343	32.5
26.4	463	122	341	32.1
26.4	462	122	340	32.2
26.4	459	121	338	31.9
26.4	458	121	337	32.0
26.4	455	120	335	31.6
26.4	454	120	334	31.7
26.4	451	119	332	31.4
26.4	450	119	331	31.5
26.4	447	118	329	31.1
26.4	444	117	327	30.8
26.4	443	117	326	30.9

%E	M1	M2	DM	M*
26.4	440	116	324	30.6
26.4	439	116	323	30.7
26.4	436	115	321	30.3
26.4	435	115	320	30.4
26.4	432	114	318	30.1
26.4	428	113	315	29.8
26.4	425	112	313	29.5
26.4	424	112	312	29.6
26.4	421	111	310	29.3
26.4	420	111	309	29.3
26.4	417	110	307	29.0
26.4	416	110	306	29.1
26.4	413	109	304	28.8
26.4	409	108	301	28.5
26.4	406	107	299	28.2
26.4	405	107	298	28.3
26.4	402	106	296	28.0
26.4	401	106	295	28.0
26.4	398	105	293	27.7
26.4	397	105	292	27.8
26.4	394	104	290	27.5
26.4	390	103	287	27.2
26.4	387	102	285	26.9
26.4	386	102	284	27.0
26.4	383	101	282	26.6
26.4	382	101	281	26.7
26.4	379	100	279	26.4
26.4	375	99	276	26.1
26.4	371	98	273	25.9
26.4	368	97	271	25.6
26.4	367	97	270	25.6
26.4	364	96	268	25.3
26.4	363	96	267	25.4
26.4	360	95	265	25.1
26.4	356	94	262	24.8
26.4	352	93	259	24.6
26.4	349	92	257	24.3
26.4	348	92	256	24.3
26.4	345	91	254	24.0
26.4	341	90	251	23.8
26.4	337	89	248	23.5
26.4	333	88	245	23.3
26.4	330	87	243	22.9
26.4	329	87	242	23.0
26.4	326	86	240	22.7
26.4	322	85	237	22.4
26.4	318	84	234	22.2
26.4	314	83	231	21.9
26.4	311	82	229	21.6
26.4	307	81	226	21.4

%E	M1	M2	DM	M*
26.4	303	80	223	21.1
26.4	299	79	220	20.9
26.4	296	78	218	20.6
26.4	295	78	217	20.6
26.4	292	77	215	20.3
26.4	288	76	212	20.1
26.4	284	75	209	19.8
26.4	280	74	206	19.6
26.4	277	73	204	19.2
26.4	276	73	203	19.3
26.4	273	72	201	19.0
26.4	269	71	198	18.7
26.4	265	70	195	18.5
26.4	261	69	192	18.2
26.4	258	68	190	17.9
26.4	254	67	187	17.7
26.4	250	66	184	17.4
26.4	246	65	181	17.2
26.4	242	64	178	16.9
26.4	239	63	176	16.6
26.4	235	62	173	16.4
26.4	231	61	170	16.1
26.4	227	60	167	15.9
26.4	220	58	162	15.3
26.4	216	57	159	15.0
26.4	212	56	156	14.8
26.4	208	55	153	14.5
26.4	201	53	148	14.0
26.4	197	52	145	13.7
26.4	193	51	142	13.5
26.4	182	48	134	12.7
26.4	178	47	131	12.4
26.4	174	46	128	12.2
26.4	163	43	120	11.3
26.4	159	42	117	11.1
26.4	148	39	109	10.3
26.4	144	38	106	10.0
26.4	140	37	103	9.8
26.4	129	34	95	9.0
26.4	125	33	92	8.7
26.4	121	32	89	8.5
26.4	110	29	81	7.6
26.4	106	28	78	7.4
26.4	91	24	67	6.3
26.4	87	23	64	6.1
26.4	72	19	53	5.0
26.4	53	14	39	3.7
26.3	499	131	368	34.4
26.3	498	131	367	34.5
26.3	495	130	365	34.1

%E	M1	M2	DM	M*
26.3	494	130	364	34.2
26.3	491	129	362	33.9
26.3	490	129	361	34.0
26.3	487	128	359	33.6
26.3	486	128	358	33.7
26.3	483	127	356	33.4
26.3	482	127	355	33.5
26.3	480	126	354	33.1
26.3	479	126	353	33.1
26.3	476	125	351	32.8
26.3	475	125	350	32.9
26.3	472	124	348	32.6
26.3	471	124	347	32.6
26.3	468	123	345	32.3
26.3	467	123	344	32.4
26.3	464	122	342	32.1
26.3	460	121	339	31.8
26.3	457	120	337	31.5
26.3	456	120	336	31.6
26.3	453	119	334	31.3
26.3	452	119	333	31.3
26.3	449	118	331	31.0
26.3	448	118	330	31.1
26.3	445	117	328	30.8
26.3	441	116	325	30.5
26.3	438	115	323	30.2
26.3	437	115	322	30.3
26.3	434	114	320	29.9
26.3	433	114	319	30.0
26.3	430	113	317	29.7
26.3	429	113	316	29.8
26.3	426	112	314	29.4
26.3	422	111	311	29.2
26.3	419	110	309	28.9
26.3	418	110	308	28.9
26.3	415	109	306	28.6
26.3	414	109	305	28.7
26.3	411	108	303	28.4
26.3	410	108	302	28.4
26.3	407	107	300	28.1
26.3	403	106	297	27.9
26.3	400	105	295	27.6
26.3	399	105	294	27.6
26.3	396	104	292	27.3
26.3	395	104	291	27.4
26.3	392	103	289	27.1
26.3	391	103	288	27.1
26.3	388	102	286	26.8
26.3	384	101	283	26.6
26.3	380	100	280	26.3

%E	M1	M2	DM	M*
26.3	377	99	278	26.0
26.3	376	99	277	26.1
26.3	373	98	275	25.7
26.3	372	98	274	25.8
26.3	369	97	272	25.5
26.3	365	96	269	25.2
26.3	361	95	266	25.0
26.3	358	94	264	24.7
26.3	357	94	263	24.8
26.3	354	93	261	24.4
26.3	353	93	260	24.5
26.3	350	92	258	24.2
26.3	346	91	255	23.9
26.3	342	90	252	23.7
26.3	339	89	250	23.4
26.3	338	89	249	23.4
26.3	335	88	247	23.1
26.3	334	88	246	23.2
26.3	331	87	244	22.9
26.3	327	86	241	22.6
26.3	323	85	238	22.4
26.3	320	84	236	22.0
26.3	319	84	235	22.1
26.3	316	83	233	21.8
26.3	315	83	232	21.9
26.3	312	82	230	21.6
26.3	308	81	227	21.3
26.3	304	80	224	21.1
26.3	300	79	221	20.8
26.3	297	78	219	20.5
26.3	293	77	216	20.2
26.3	289	76	213	20.0
26.3	285	75	210	19.7
26.3	281	74	207	19.5
26.3	278	73	205	19.2
26.3	274	72	202	18.9
26.3	270	71	199	18.7
26.3	266	70	196	18.4
26.3	262	69	193	18.2
26.3	259	68	191	17.9
26.3	255	67	188	17.6
26.3	251	66	185	17.4
26.3	247	65	182	17.1
26.3	243	64	179	16.9
26.3	240	63	177	16.5
26.3	236	62	174	16.3
26.3	232	61	171	16.0
26.3	228	60	168	15.8
26.3	224	59	165	15.5
26.3	217	57	160	15.0

%E	M1	M2	DM	M*
26.3	213	56	157	14.7
26.3	209	55	154	14.5
26.3	205	54	151	14.2
26.3	198	52	146	13.7
26.3	194	51	143	13.4
26.3	190	50	140	13.2
26.3	186	49	137	12.9
26.3	179	47	132	12.3
26.3	175	46	129	12.1
26.3	171	45	126	11.8
26.3	167	44	123	11.6
26.3	160	42	118	11.0
26.3	156	41	115	10.8
26.3	152	40	112	10.5
26.3	137	36	101	9.5
26.3	133	35	98	9.2
26.3	118	31	87	8.1
26.3	114	30	84	7.9
26.3	99	26	73	6.8
26.3	95	25	70	6.6
26.3	80	21	59	5.5
26.3	76	20	56	5.3
26.3	57	15	42	3.9
26.3	38	10	28	2.6
26.3	19	5	14	1.3
26.2	500	131	369	34.3
26.2	497	130	367	34.0
26.2	496	130	366	34.1
26.2	493	129	364	33.8
26.2	492	129	363	33.8
26.2	489	128	361	33.5
26.2	488	128	360	33.6
26.2	485	127	358	33.3
26.2	484	127	357	33.3
26.2	481	126	355	33.0
26.2	478	125	353	32.7
26.2	477	125	352	32.8
26.2	474	124	350	32.4
26.2	473	124	349	32.5
26.2	470	123	347	32.2
26.2	469	123	346	32.3
26.2	466	122	344	31.9
26.2	465	122	343	32.0
26.2	462	121	341	31.7
26.2	461	121	340	31.8
26.2	458	120	338	31.4
26.2	455	119	336	31.1
26.2	454	119	335	31.2
26.2	451	118	333	30.9
26.2	450	118	332	30.9

%E	M1	M2	DM	M*
26.2	447	117	330	30.6
26.2	446	117	329	30.7
26.2	443	116	327	30.4
26.2	442	116	326	30.4
26.2	439	115	324	30.1
26.2	435	114	321	29.9
26.2	432	113	319	29.6
26.2	431	113	318	29.6
26.2	428	112	316	29.3
26.2	427	112	315	29.4
26.2	424	111	313	29.1
26.2	423	111	312	29.1
26.2	420	110	310	28.8
26.2	416	109	307	28.6
26.2	413	108	305	28.2
26.2	412	108	304	28.3
26.2	409	107	302	28.0
26.2	408	107	301	28.1
26.2	405	106	299	27.7
26.2	404	106	298	27.8
26.2	401	105	296	27.5
26.2	397	104	293	27.2
26.2	393	103	290	27.0
26.2	390	102	288	26.7
26.2	389	102	287	26.7
26.2	386	101	285	26.4
26.2	385	101	284	26.5
26.2	382	100	282	26.2
26.2	381	100	281	26.2
26.2	378	99	279	25.9
26.2	374	98	276	25.7
26.2	370	97	273	25.4
26.2	367	96	271	25.1
26.2	366	96	270	25.2
26.2	363	95	268	24.9
26.2	362	95	267	24.9
26.2	359	94	265	24.6
26.2	355	93	262	24.4
26.2	351	92	259	24.1
26.2	347	91	256	23.9
26.2	344	90	254	23.5
26.2	343	90	253	23.6
26.2	340	89	251	23.3
26.2	336	88	248	23.0
26.2	332	87	245	22.8
26.2	328	86	242	22.5
26.2	325	85	240	22.2
26.2	324	85	239	22.3
26.2	321	84	237	22.0
26.2	317	83	234	21.7

%E	M1	M2	DM	M*
26.2	313	82	231	21.5
26.2	309	81	228	21.2
26.2	305	80	225	21.0
26.2	302	79	223	20.7
26.2	301	79	222	20.7
26.2	298	78	220	20.4
26.2	294	77	217	20.2
26.2	290	76	214	19.9
26.2	286	75	211	19.7
26.2	282	74	208	19.4
26.2	279	73	206	19.1
26.2	275	72	203	18.9
26.2	271	71	200	18.6
26.2	267	70	197	18.4
26.2	263	69	194	18.1
26.2	260	68	192	17.8
26.2	256	67	189	17.5
26.2	252	66	186	17.3
26.2	248	65	183	17.0
26.2	244	64	180	16.8
26.2	237	62	175	16.2
26.2	233	61	172	16.0
26.2	229	60	169	15.7
26.2	225	59	166	15.5
26.2	221	58	163	15.2
26.2	214	56	158	14.7
26.2	210	55	155	14.4
26.2	206	54	152	14.2
26.2	202	53	149	13.9
26.2	195	51	144	13.3
26.2	191	50	141	13.1
26.2	187	49	138	12.8
26.2	183	48	135	12.6
26.2	172	45	127	11.8
26.2	168	44	124	11.5
26.2	164	43	121	11.3
26.2	149	39	110	10.2
26.2	145	38	107	10.0
26.2	141	37	104	9.7
26.2	130	34	96	8.9
26.2	126	33	93	8.6
26.2	122	32	90	8.4
26.2	107	28	79	7.3
26.2	103	27	76	7.1
26.2	84	22	62	5.8
26.2	65	17	48	4.4
26.2	61	16	45	4.2
26.2	42	11	31	2.9
26.1	499	130	369	33.9
26.1	498	130	368	33.9

%E	M1	M2	DM	M*
26.1	495	129	366	33.6
26.1	494	129	365	33.7
26.1	491	128	363	33.4
26.1	490	128	362	33.4
26.1	487	127	360	33.1
26.1	486	127	359	33.2
26.1	483	126	357	32.9
26.1	482	126	356	32.9
26.1	479	125	354	32.6
26.1	476	124	352	32.3
26.1	475	124	351	32.4
26.1	472	123	349	32.1
26.1	471	123	348	32.1
26.1	468	122	346	31.8
26.1	467	122	345	31.9
26.1	464	121	343	31.6
26.1	463	121	342	31.6
26.1	460	120	340	31.3
26.1	459	120	339	31.4
26.1	456	119	337	31.1
26.1	452	118	334	30.8
26.1	449	117	332	30.5
26.1	448	117	331	30.6
26.1	445	116	329	30.2
26.1	444	116	328	30.3
26.1	441	115	326	30.0
26.1	440	115	325	30.1
26.1	437	114	323	29.7
26.1	436	114	322	29.8
26.1	433	113	320	29.5
26.1	429	112	317	29.2
26.1	426	111	315	28.9
26.1	425	111	314	29.0
26.1	422	110	312	28.7
26.1	421	110	311	28.7
26.1	418	109	309	28.4
26.1	417	109	308	28.5
26.1	414	108	306	28.2
26.1	410	107	303	27.9
26.1	406	106	300	27.7
26.1	403	105	298	27.4
26.1	402	105	297	27.4
26.1	399	104	295	27.1
26.1	398	104	294	27.2
26.1	395	103	292	26.9
26.1	394	103	291	26.9
26.1	391	102	289	26.6
26.1	387	101	286	26.4
26.1	383	100	283	26.1
26.1	380	99	281	25.8

%E	M1	M2	DM	M*
26.1	379	99	280	25.9
26.1	376	98	278	25.5
26.1	375	98	277	25.6
26.1	372	97	275	25.3
26.1	371	97	274	25.4
26.1	368	96	272	25.0
26.1	364	95	269	24.8
26.1	360	94	266	24.5
26.1	357	93	264	24.2
26.1	356	93	263	24.3
26.1	353	92	261	24.0
26.1	352	92	260	24.0
26.1	349	91	258	23.7
26.1	348	91	257	23.8
26.1	345	90	255	23.5
26.1	341	89	252	23.2
26.1	337	88	249	23.0
26.1	333	87	246	22.7
26.1	330	86	244	22.4
26.1	329	86	243	22.5
26.1	326	85	241	22.2
26.1	322	84	238	21.9
26.1	318	83	235	21.7
26.1	314	82	232	21.4
26.1	310	81	229	21.2
26.1	307	80	227	20.8
26.1	306	80	226	20.9
26.1	303	79	224	20.6
26.1	299	78	221	20.3
26.1	295	77	218	20.1
26.1	291	76	215	19.8
26.1	287	75	212	19.6
26.1	284	74	210	19.3
26.1	283	74	209	19.3
26.1	280	73	207	19.0
26.1	276	72	204	18.8
26.1	272	71	201	18.5
26.1	268	70	198	18.3
26.1	264	69	195	18.0
26.1	261	68	193	17.7
26.1	257	67	190	17.5
26.1	253	66	187	17.2
26.1	249	65	184	17.0
26.1	245	64	181	16.7
26.1	241	63	178	16.5
26.1	238	62	176	16.2
26.1	234	61	173	15.9
26.1	230	60	170	15.7
26.1	226	59	167	15.4
26.1	222	58	164	15.2

%E	M1	M2	DM	M*
26.1	218	57	161	14.9
26.1	211	55	156	14.3
26.1	207	54	153	14.1
26.1	203	53	150	13.8
26.1	199	52	147	13.6
26.1	188	49	139	12.8
26.1	184	48	136	12.5
26.1	180	47	133	12.3
26.1	176	46	130	12.0
26.1	165	43	122	11.2
26.1	161	42	119	11.0
26.1	157	41	116	10.7
26.1	153	40	113	10.5
26.1	142	37	105	9.6
26.1	138	36	102	9.4
26.1	134	35	99	9.1
26.1	119	31	88	8.1
26.1	115	30	85	7.8
26.1	111	29	82	7.6
26.1	92	24	68	6.3
26.1	88	23	65	6.0
26.1	69	18	51	4.7
26.1	46	12	34	3.1
26.1	23	6	17	1.6
26.0	500	130	370	33.8
26.0	497	129	368	33.5
26.0	496	129	367	33.6
26.0	493	128	365	33.2
26.0	492	128	364	33.3
26.0	489	127	362	33.0
26.0	488	127	361	33.1
26.0	485	126	359	32.7
26.0	484	126	358	32.8
26.0	481	125	356	32.5
26.0	480	125	355	32.6
26.0	477	124	353	32.2
26.0	473	123	350	32.0
26.0	470	122	348	31.7
26.0	469	122	347	31.7
26.0	466	121	345	31.4
26.0	465	121	344	31.5
26.0	462	120	342	31.2
26.0	461	120	341	31.2
26.0	458	119	339	30.9
26.0	457	119	338	31.0
26.0	454	118	336	30.7
26.0	453	118	335	30.7
26.0	450	117	333	30.4
26.0	447	116	331	30.1
26.0	446	116	330	30.2

%E	M1	M2	DM	M*
26.0	443	115	328	29.9
26.0	442	115	327	29.9
26.0	439	114	325	29.6
26.0	438	114	324	29.7
26.0	435	113	322	29.4
26.0	434	113	321	29.4
26.0	431	112	319	29.1
26.0	430	112	318	29.2
26.0	427	111	316	28.9
26.0	423	110	313	28.6
26.0	420	109	311	28.3
26.0	419	109	310	28.4
26.0	416	108	308	28.0
26.0	415	108	307	28.1
26.0	412	107	305	27.8
26.0	411	107	304	27.9
26.0	408	106	302	27.5
26.0	407	106	301	27.6
26.0	404	105	299	27.3
26.0	400	104	296	27.0
26.0	396	103	293	26.8
26.0	393	102	291	26.5
26.0	392	102	290	26.5
26.0	389	101	288	26.2
26.0	388	101	287	26.3
26.0	385	100	285	26.0
26.0	384	100	284	26.0
26.0	381	99	282	25.7
26.0	377	98	279	25.5
26.0	373	97	276	25.2
26.0	369	96	273	25.0
26.0	366	95	271	24.7
26.0	365	95	270	24.7
26.0	362	94	268	24.4
26.0	361	94	267	24.5
26.0	358	93	265	24.2
26.0	354	92	262	23.9
26.0	350	91	259	23.7
26.0	346	90	256	23.4
26.0	342	89	253	23.2
26.0	339	88	251	22.8
26.0	338	88	250	22.9
26.0	335	87	248	22.6
26.0	334	87	247	22.7
26.0	331	86	245	22.3
26.0	327	85	242	22.1
26.0	323	84	239	21.8
26.0	319	83	236	21.6
26.0	315	82	233	21.3
26.0	312	81	231	21.0

%E	M1	M2	DM	M*
26.0	311	81	230	21.1
26.0	308	80	228	20.8
26.0	304	79	225	20.5
26.0	300	78	222	20.3
26.0	296	77	219	20.0
26.0	292	76	216	19.8
26.0	289	75	214	19.5
26.0	288	75	213	19.5
26.0	285	74	211	19.2
26.0	281	73	208	19.0
26.0	277	72	205	18.7
26.0	273	71	202	18.5
26.0	269	70	199	18.2
26.0	265	69	196	18.0
26.0	262	68	194	17.6
26.0	258	67	191	17.4
26.0	254	66	188	17.1
26.0	250	65	185	16.9
26.0	246	64	182	16.7
26.0	242	63	179	16.4
26.0	235	61	174	15.8
26.0	231	60	171	15.6
26.0	227	59	168	15.3
26.0	223	58	165	15.1
26.0	219	57	162	14.8
26.0	215	56	159	14.6
26.0	208	54	154	14.0
26.0	204	53	151	13.8
26.0	200	52	148	13.5
26.0	196	51	145	13.3
26.0	192	50	142	13.0
26.0	181	47	134	12.2
26.0	177	46	131	12.0
26.0	173	45	128	11.7
26.0	169	44	125	11.5
26.0	154	40	114	10.4
26.0	150	39	111	10.1
26.0	146	38	108	9.9
26.0	131	34	97	8.8
26.0	127	33	94	8.6
26.0	123	32	91	8.3
26.0	104	27	77	7.0
26.0	100	26	74	6.8
26.0	96	25	71	6.5
26.0	77	20	57	5.2
26.0	73	19	54	4.9
26.0	50	13	37	3.4
25.9	499	129	370	33.3
25.9	498	129	369	33.4
25.9	495	128	367	33.1

%E	M1	M2	DM	M*
25.9	494	128	366	33.2
25.9	491	127	364	32.8
25.9	490	127	363	32.9
25.9	487	126	361	32.6
25.9	486	126	360	32.7
25.9	483	125	358	32.3
25.9	482	125	357	32.4
25.9	479	124	355	32.1
25.9	478	124	354	32.2
25.9	475	123	352	31.9
25.9	474	123	351	31.9
25.9	471	122	349	31.6
25.9	468	121	347	31.3
25.9	467	121	346	31.4
25.9	464	120	344	31.0
25.9	463	120	343	31.1
25.9	460	119	341	30.8
25.9	459	119	340	30.9
25.9	456	118	338	30.5
25.9	455	118	337	30.6
25.9	452	117	335	30.3
25.9	451	117	334	30.4
25.9	448	116	332	30.0
25.9	444	115	329	29.8
25.9	441	114	327	29.5
25.9	440	114	326	29.5
25.9	437	113	324	29.2
25.9	436	113	323	29.3
25.9	433	112	321	29.0
25.9	432	112	320	29.0
25.9	429	111	318	28.7
25.9	428	111	317	28.8
25.9	425	110	315	28.5
25.9	424	110	314	28.5
25.9	421	109	312	28.2
25.9	417	108	309	28.0
25.9	413	107	306	27.7
25.9	410	106	304	27.4
25.9	409	106	303	27.5
25.9	406	105	301	27.2
25.9	405	105	300	27.2
25.9	402	104	298	26.9
25.9	401	104	297	27.0
25.9	398	103	295	26.7
25.9	397	103	294	26.7
25.9	394	102	292	26.4
25.9	390	101	289	26.2
25.9	386	100	286	25.9
25.9	382	99	283	25.7
25.9	379	98	281	25.3

%E	M1	M2	DM	M*
25.9	378	98	280	25.4
25.9	375	97	278	25.1
25.9	374	97	277	25.2
25.9	371	96	275	24.8
25.9	370	96	274	24.9
25.9	367	95	272	24.6
25.9	363	94	269	24.3
25.9	359	93	266	24.1
25.9	355	92	263	23.8
25.9	352	91	261	23.5
25.9	351	91	260	23.6
25.9	348	90	258	23.3
25.9	347	90	257	23.3
25.9	344	89	255	23.0
25.9	343	89	254	23.1
25.9	340	88	252	22.8
25.9	336	87	249	22.5
25.9	332	86	246	22.3
25.9	328	85	243	22.0
25.9	324	84	240	21.8
25.9	321	83	238	21.5
25.9	320	83	237	21.5
25.9	317	82	235	21.2
25.9	316	82	234	21.3
25.9	313	81	232	21.0
25.9	309	80	229	20.7
25.9	305	79	226	20.5
25.9	301	78	223	20.2
25.9	297	77	220	20.0
25.9	294	76	218	19.6
25.9	293	76	217	19.7
25.9	290	75	215	19.4
25.9	286	74	212	19.1
25.9	282	73	209	18.9
25.9	278	72	206	18.6
25.9	274	71	203	18.4
25.9	270	70	200	18.1
25.9	266	69	197	17.9
25.9	263	68	195	17.6
25.9	259	67	192	17.3
25.9	255	66	189	17.1
25.9	251	65	186	16.8
25.9	247	64	183	16.6
25.9	243	63	180	16.3
25.9	239	62	177	16.1
25.9	232	60	172	15.5
25.9	228	59	169	15.3
25.9	224	58	166	15.0
25.9	220	57	163	14.8
25.9	216	56	160	14.5

%E	M1	M2	DM	M*
25.9	212	55	157	14.3
25.9	205	53	152	13.7
25.9	201	52	149	13.5
25.9	197	51	146	13.2
25.9	193	50	143	13.0
25.9	189	49	140	12.7
25.9	185	48	137	12.5
25.9	174	45	129	11.6
25.9	170	44	126	11.4
25.9	166	43	123	11.1
25.9	162	42	120	10.9
25.9	158	41	117	10.6
25.9	147	38	109	9.8
25.9	143	37	106	9.6
25.9	139	36	103	9.3
25.9	135	35	100	9.1
25.9	116	30	86	7.8
25.9	112	29	83	7.5
25.9	108	28	80	7.3
25.9	85	22	63	5.7
25.9	81	21	60	5.4
25.9	58	15	43	3.9
25.9	54	14	40	3.6
25.9	27	7	20	1.8
25.8	500	129	371	33.3
25.8	497	128	369	33.0
25.8	496	128	368	33.0
25.8	493	127	366	32.7
25.8	492	127	365	32.8
25.8	489	126	363	32.5
25.8	488	126	362	32.5
25.8	485	125	360	32.2
25.8	484	125	359	32.3
25.8	481	124	357	32.0
25.8	480	124	356	32.0
25.8	477	123	354	31.7
25.8	476	123	353	31.8
25.8	473	122	351	31.5
25.8	472	122	350	31.5
25.8	469	121	348	31.2
25.8	466	120	346	30.9
25.8	465	120	345	31.0
25.8	462	119	343	30.7
25.8	461	119	342	30.7
25.8	458	118	340	30.4
25.8	457	118	339	30.5
25.8	454	117	337	30.2
25.8	453	117	336	30.2
25.8	450	116	334	29.9
25.8	449	116	333	30.0

%E	M1	M2	DM	M*
25.8	446	115	331	29.7
25.8	445	115	330	29.7
25.8	442	114	328	29.4
25.8	438	113	325	29.2
25.8	434	112	322	28.9
25.8	431	111	320	28.6
25.8	430	111	319	28.7
25.8	427	110	317	28.3
25.8	426	110	316	28.4
25.8	423	109	314	28.1
25.8	422	109	313	28.2
25.8	419	108	311	27.8
25.8	418	108	310	27.9
25.8	415	107	308	27.6
25.8	414	107	307	27.7
25.8	411	106	305	27.3
25.8	407	105	302	27.1
25.8	403	104	299	26.8
25.8	400	103	297	26.5
25.8	399	103	296	26.6
25.8	396	102	294	26.3
25.8	395	102	293	26.3
25.8	392	101	291	26.0
25.8	391	101	290	26.1
25.8	388	100	288	25.8
25.8	387	100	287	25.8
25.8	384	99	285	25.5
25.8	383	99	284	25.6
25.8	380	98	282	25.3
25.8	376	97	279	25.0
25.8	372	96	276	24.8
25.8	368	95	273	24.5
25.8	365	94	271	24.2
25.8	364	94	270	24.3
25.8	361	93	268	24.0
25.8	360	93	267	24.0
25.8	357	92	265	23.7
25.8	356	92	264	23.8
25.8	353	91	262	23.5
25.8	349	90	259	23.2
25.8	345	89	256	23.0
25.8	341	88	253	22.7
25.8	337	87	250	22.5
25.8	333	86	247	22.2
25.8	330	85	245	21.9
25.8	329	85	244	22.0
25.8	326	84	242	21.6
25.8	325	84	241	21.7
25.8	322	83	239	21.4
25.8	318	82	236	21.1

%E	M1	M2	DM	M*
25.8	314	81	233	20.9
25.8	310	80	230	20.6
25.8	306	79	227	20.4
25.8	302	78	224	20.1
25.8	299	77	222	19.8
25.8	298	77	221	19.9
25.8	295	76	219	19.6
25.8	291	75	216	19.3
25.8	287	74	213	19.1
25.8	283	73	210	18.8
25.8	279	72	207	18.6
25.8	275	71	204	18.3
25.8	271	70	201	18.1
25.8	267	69	198	17.8
25.8	264	68	196	17.5
25.8	260	67	193	17.3
25.8	256	66	190	17.0
25.8	252	65	187	16.8
25.8	248	64	184	16.5
25.8	244	63	181	16.3
25.8	240	62	178	16.0
25.8	236	61	175	15.8
25.8	233	60	173	15.5
25.8	229	59	170	15.2
25.8	225	58	167	15.0
25.8	221	57	164	14.7
25.8	217	56	161	14.5
25.8	213	55	158	14.2
25.8	209	54	155	14.0
25.8	198	51	147	13.1
25.8	194	50	144	12.9
25.8	190	49	141	12.6
25.8	186	48	138	12.4
25.8	182	47	135	12.1
25.8	178	46	132	11.9
25.8	163	42	121	10.8
25.8	159	41	118	10.6
25.8	155	40	115	10.3
25.8	151	39	112	10.1
25.8	132	34	98	8.8
25.8	128	33	95	8.5
25.8	124	32	92	8.3
25.8	120	31	89	8.0
25.8	97	25	72	6.4
25.8	93	24	69	6.2
25.8	89	23	66	5.9
25.8	66	17	49	4.4
25.8	62	16	46	4.1
25.8	31	8	23	2.1
25.7	499	128	371	32.8

%E	M1	M2	DM	M*
25.7	498	128	370	32.9
25.7	495	127	368	32.6
25.7	494	127	367	32.6
25.7	491	126	365	32.3
25.7	490	126	364	32.4
25.7	487	125	362	32.1
25.7	486	125	361	32.2
25.7	483	124	359	31.8
25.7	482	124	358	31.9
25.7	479	123	356	31.6
25.7	478	123	355	31.7
25.7	475	122	353	31.3
25.7	474	122	352	31.4
25.7	471	121	350	31.1
25.7	470	121	349	31.2
25.7	467	120	347	30.8
25.7	463	119	344	30.6
25.7	460	118	342	30.3
25.7	459	118	341	30.3
25.7	456	117	339	30.0
25.7	455	117	338	30.1
25.7	452	116	336	29.8
25.7	451	116	335	29.8
25.7	448	115	333	29.5
25.7	447	115	332	29.6
25.7	444	114	330	29.3
25.7	443	114	329	29.3
25.7	440	113	327	29.0
25.7	439	113	326	29.1
25.7	436	112	324	28.8
25.7	435	112	323	28.8
25.7	432	111	321	28.5
25.7	428	110	318	28.3
25.7	424	109	315	28.0
25.7	421	108	313	27.7
25.7	420	108	312	27.8
25.7	417	107	310	27.5
25.7	416	107	309	27.5
25.7	413	106	307	27.2
25.7	412	106	306	27.3
25.7	409	105	304	27.0
25.7	408	105	303	27.0
25.7	405	104	301	26.7
25.7	404	104	300	26.8
25.7	401	103	298	26.5
25.7	397	102	295	26.2
25.7	393	101	292	26.0
25.7	389	100	289	25.7
25.7	385	99	286	25.5
25.7	382	98	284	25.1
25.7	381	98	283	25.2
25.7	378	97	281	24.9
25.7	377	97	280	25.0
25.7	374	96	278	24.6
25.7	373	96	277	24.7
25.7	370	95	275	24.4
25.7	369	95	274	24.5
25.7	366	94	272	24.1
25.7	362	93	269	23.9
25.7	358	92	266	23.6
25.7	354	91	263	23.4
25.7	350	90	260	23.1
25.7	346	89	257	22.9
25.7	343	88	255	22.6
25.7	342	88	254	22.6
25.7	339	87	252	22.3
25.7	338	87	251	22.4
25.7	335	86	249	22.1
25.7	334	86	248	22.1
25.7	331	85	246	21.8
25.7	327	84	243	21.6
25.7	323	83	240	21.3
25.7	319	82	237	21.1
25.7	315	81	234	20.8
25.7	311	80	231	20.6
25.7	307	79	228	20.3
25.7	304	78	226	20.0
25.7	303	78	225	20.1
25.7	300	77	223	19.8
25.7	296	76	220	19.5
25.7	292	75	217	19.3
25.7	288	74	214	19.0
25.7	284	73	211	18.8
25.7	280	72	208	18.5
25.7	276	71	205	18.3
25.7	272	70	202	18.0
25.7	269	69	200	17.7
25.7	268	69	199	17.8
25.7	265	68	197	17.4
25.7	261	67	194	17.2
25.7	257	66	191	16.9
25.7	253	65	188	16.7
25.7	249	64	185	16.4
25.7	245	63	182	16.2
25.7	241	62	179	16.0
25.7	237	61	176	15.7
25.7	230	59	171	15.1
25.7	226	58	168	14.9
25.7	222	57	165	14.6
25.7	218	56	162	14.4
25.7	214	55	159	14.1
25.7	210	54	156	13.9
25.7	206	53	153	13.6
25.7	202	52	150	13.4
25.7	191	49	142	12.6
25.7	187	48	139	12.3
25.7	183	47	136	12.1
25.7	179	46	133	11.8
25.7	175	45	130	11.6
25.7	171	44	127	11.3
25.7	167	43	124	11.1
25.7	152	39	113	10.0
25.7	148	38	110	9.8
25.7	144	37	107	9.5
25.7	140	36	104	9.3
25.7	136	35	101	9.0
25.7	113	29	84	7.4
25.7	109	28	81	7.2
25.7	105	27	78	6.9
25.7	101	26	75	6.7
25.7	74	19	55	4.9
25.7	70	18	52	4.6
25.7	35	9	26	2.3
25.6	500	128	372	32.8
25.6	497	127	370	32.5
25.6	496	127	369	32.5
25.6	493	126	367	32.2
25.6	492	126	366	32.3
25.6	489	125	364	32.0
25.6	488	125	363	32.0
25.6	485	124	361	31.7
25.6	484	124	360	31.8
25.6	481	123	358	31.5
25.6	480	123	357	31.5
25.6	477	122	355	31.2
25.6	476	122	354	31.3
25.6	473	121	352	31.0
25.6	472	121	351	31.0
25.6	469	120	349	30.7
25.6	468	120	348	30.8
25.6	465	119	346	30.5
25.6	464	119	345	30.5
25.6	461	118	343	30.2
25.6	457	117	340	30.0
25.6	454	116	338	29.6
25.6	453	116	337	29.7
25.6	450	115	335	29.4
25.6	449	115	334	29.5
25.6	446	114	332	29.1
25.6	445	114	331	29.2
25.6	442	113	329	28.9
25.6	441	113	328	29.0
25.6	438	112	326	28.6
25.6	437	112	325	28.7
25.6	434	111	323	28.4
25.6	433	111	322	28.5
25.6	430	110	320	28.1
25.6	429	110	319	28.2
25.6	426	109	317	27.9
25.6	425	109	316	28.0
25.6	422	108	314	27.6
25.6	418	107	311	27.4
25.6	414	106	308	27.1
25.6	410	105	305	26.9
25.6	407	104	303	26.6
25.6	406	104	302	26.6
25.6	403	103	300	26.3
25.6	402	103	299	26.4
25.6	399	102	297	26.1
25.6	398	102	296	26.1
25.6	395	101	294	25.8
25.6	394	101	293	25.9
25.6	391	100	291	25.6
25.6	390	100	290	25.6
25.6	387	99	288	25.3
25.6	386	99	287	25.4
25.6	383	98	285	25.1
25.6	379	97	282	24.8
25.6	375	96	279	24.6
25.6	371	95	276	24.3
25.6	367	94	273	24.1
25.6	363	93	270	23.8
25.6	360	92	268	23.5
25.6	359	92	267	23.6
25.6	356	91	265	23.3
25.6	355	91	264	23.3
25.6	352	90	262	23.0
25.6	351	90	261	23.1
25.6	348	89	259	22.8
25.6	347	89	258	22.8
25.6	344	88	256	22.5
25.6	340	87	253	22.3
25.6	336	86	250	22.0
25.6	332	85	247	21.8
25.6	328	84	244	21.5
25.6	324	83	241	21.3
25.6	320	82	238	21.0
25.6	317	81	236	20.7
25.6	316	81	235	20.8
25.6	313	80	233	20.4
25.6	312	80	232	20.5
25.6	309	79	230	20.2
25.6	308	79	229	20.3
25.6	305	78	227	19.9
25.6	301	77	224	19.7
25.6	297	76	221	19.4
25.6	293	75	218	19.2
25.6	289	74	215	18.9
25.6	285	73	212	18.7
25.6	281	72	209	18.4
25.6	277	71	206	18.2
25.6	273	70	203	17.9
25.6	270	69	201	17.6
25.6	266	68	198	17.4
25.6	262	67	195	17.1
25.6	258	66	192	16.9
25.6	254	65	189	16.6
25.6	250	64	186	16.4
25.6	246	63	183	16.1
25.6	242	62	180	15.9
25.6	238	61	177	15.6
25.6	234	60	174	15.4
25.6	227	58	169	14.8
25.6	223	57	166	14.6
25.6	219	56	163	14.3
25.6	215	55	160	14.1
25.6	211	54	157	13.8
25.6	207	53	154	13.6
25.6	203	52	151	13.3
25.6	199	51	148	13.1
25.6	195	50	145	12.8
25.6	180	46	134	11.8
25.6	176	45	131	11.5
25.6	172	44	128	11.3
25.6	168	43	125	11.0
25.6	164	42	122	10.8
25.6	160	41	119	10.5
25.6	156	40	116	10.3
25.6	133	34	99	8.7
25.6	129	33	96	8.4
25.6	125	32	93	8.2
25.6	121	31	90	7.9
25.6	117	30	87	7.7
25.6	90	23	67	5.9
25.6	86	22	64	5.6
25.6	82	21	61	5.4
25.6	78	20	58	5.1
25.6	43	11	32	2.8
25.6	39	10	29	2.6
25.5	499	127	372	32.3

%E	M1	M2	DM	M*
25.5	498	127	371	32.4
25.5	495	126	369	32.1
25.5	494	126	368	32.1
25.5	491	125	366	31.8
25.5	490	125	365	31.9
25.5	487	124	363	31.6
25.5	486	124	362	31.6
25.5	483	123	360	31.3
25.5	482	123	359	31.4
25.5	479	122	357	31.1
25.5	478	122	356	31.1
25.5	475	121	354	30.8
25.5	474	121	353	30.9
25.5	471	120	351	30.6
25.5	470	120	350	30.6
25.5	467	119	348	30.3
25.5	466	119	347	30.4
25.5	463	118	345	30.1
25.5	462	118	344	30.1
25.5	459	117	342	29.8
25.5	458	117	341	29.9
25.5	455	116	339	29.6
25.5	451	115	336	29.3
25.5	447	114	333	29.1
25.5	444	113	331	28.8
25.5	443	113	330	28.8
25.5	440	112	328	28.5
25.5	439	112	327	28.6
25.5	436	111	325	28.3
25.5	435	111	324	28.3
25.5	432	110	322	28.0
25.5	431	110	321	28.1
25.5	428	109	319	27.8
25.5	427	109	318	27.8
25.5	424	108	316	27.5
25.5	423	108	315	27.6
25.5	420	107	313	27.3
25.5	419	107	312	27.3
25.5	416	106	310	27.0
25.5	415	106	309	27.1
25.5	412	105	307	26.8
25.5	411	105	306	26.8
25.5	408	104	304	26.5
25.5	404	103	301	26.3
25.5	400	102	298	26.0
25.5	396	101	295	25.8
25.5	392	100	292	25.5
25.5	389	99	290	25.2
25.5	388	99	289	25.3
25.5	385	98	287	24.9

%E	M1	M2	DM	M*
25.5	384	98	286	25.0
25.5	381	97	284	24.7
25.5	380	97	283	24.8
25.5	377	96	281	24.4
25.5	376	96	280	24.5
25.5	373	95	278	24.2
25.5	372	95	277	24.3
25.5	369	94	275	23.9
25.5	368	94	274	24.0
25.5	365	93	272	23.7
25.5	364	93	271	23.8
25.5	361	92	269	23.4
25.5	357	91	266	23.2
25.5	353	90	263	22.9
25.5	349	89	260	22.7
25.5	345	88	257	22.4
25.5	341	87	254	22.2
25.5	337	86	251	21.9
25.5	333	85	248	21.7
25.5	330	84	246	21.4
25.5	329	84	245	21.4
25.5	326	83	243	21.1
25.5	325	83	242	21.2
25.5	322	82	240	20.9
25.5	321	82	239	20.9
25.5	318	81	237	20.6
25.5	314	80	234	20.4
25.5	310	79	231	20.1
25.5	306	78	228	19.9
25.5	302	77	225	19.6
25.5	298	76	222	19.4
25.5	294	75	219	19.1
25.5	290	74	216	18.9
25.5	286	73	213	18.6
25.5	282	72	210	18.4
25.5	278	71	207	18.1
25.5	275	70	205	17.8
25.5	274	70	204	17.9
25.5	271	69	202	17.6
25.5	267	68	199	17.3
25.5	263	67	196	17.1
25.5	259	66	193	16.8
25.5	255	65	190	16.6
25.5	251	64	187	16.3
25.5	247	63	184	16.1
25.5	243	62	181	15.8
25.5	239	61	178	15.6
25.5	235	60	175	15.3
25.5	231	59	172	15.1
25.5	220	56	164	14.3

%E	M1	M2	DM	M*
25.5	216	55	161	14.0
25.5	212	54	158	13.8
25.5	208	53	155	13.5
25.5	204	52	152	13.3
25.5	200	51	149	13.0
25.5	196	50	146	12.8
25.5	192	49	143	12.5
25.5	188	48	140	12.3
25.5	184	47	137	12.0
25.5	165	42	123	10.7
25.5	161	41	120	10.4
25.5	157	40	117	10.2
25.5	153	39	114	9.9
25.5	149	38	111	9.7
25.5	145	37	108	9.4
25.5	141	36	105	9.2
25.5	137	35	102	8.9
25.5	110	28	82	7.1
25.5	106	27	79	6.9
25.5	102	26	76	6.6
25.5	98	25	73	6.4
25.5	94	24	70	6.1
25.5	55	14	41	3.6
25.5	51	13	38	3.3
25.5	47	12	35	3.1
25.4	500	127	373	32.3
25.4	497	126	371	31.9
25.4	496	126	370	32.0
25.4	493	125	368	31.7
25.4	492	125	367	31.8
25.4	489	124	365	31.4
25.4	488	124	364	31.5
25.4	485	123	362	31.2
25.4	484	123	361	31.3
25.4	481	122	359	30.9
25.4	480	122	358	31.0
25.4	477	121	356	30.7
25.4	476	121	355	30.8
25.4	473	120	353	30.4
25.4	472	120	352	30.5
25.4	469	119	350	30.2
25.4	468	119	349	30.3
25.4	465	118	347	29.9
25.4	464	118	346	30.0
25.4	461	117	344	29.7
25.4	460	117	343	29.8
25.4	457	116	341	29.4
25.4	456	116	340	29.5
25.4	453	115	338	29.2
25.4	452	115	337	29.3

%E	M1	M2	DM	M*
25.4	449	114	335	28.9
25.4	448	114	334	29.0
25.4	445	113	332	28.7
25.4	441	112	329	28.4
25.4	437	111	326	28.2
25.4	433	110	323	27.9
25.4	429	109	320	27.7
25.4	426	108	318	27.4
25.4	425	108	317	27.4
25.4	422	107	315	27.1
25.4	421	107	314	27.2
25.4	418	106	312	26.9
25.4	417	106	311	26.9
25.4	414	105	309	26.6
25.4	413	105	308	26.7
25.4	410	104	306	26.4
25.4	409	104	305	26.4
25.4	406	103	303	26.1
25.4	405	103	302	26.2
25.4	402	102	300	25.9
25.4	401	102	299	25.9
25.4	398	101	297	25.6
25.4	397	101	296	25.7
25.4	394	100	294	25.4
25.4	393	100	293	25.4
25.4	390	99	291	25.1
25.4	386	98	288	24.9
25.4	382	97	285	24.6
25.4	378	96	282	24.4
25.4	374	95	279	24.1
25.4	370	94	276	23.9
25.4	366	93	273	23.6
25.4	362	92	270	23.4
25.4	358	91	267	23.1
25.4	355	90	265	22.8
25.4	354	90	264	22.9
25.4	351	89	262	22.6
25.4	350	89	261	22.6
25.4	347	88	259	22.3
25.4	346	88	258	22.4
25.4	343	87	256	22.1
25.4	342	87	255	22.1
25.4	339	86	253	21.8
25.4	338	86	252	21.9
25.4	335	85	250	21.6
25.4	334	85	249	21.6
25.4	331	84	247	21.3
25.4	327	83	244	21.1
25.4	323	82	241	20.8
25.4	319	81	238	20.6

%E	M1	M2	DM	M*
25.4	315	80	235	20.3
25.4	311	79	232	20.1
25.4	307	78	229	19.8
25.4	303	77	226	19.6
25.4	299	76	223	19.3
25.4	295	75	220	19.1
25.4	291	74	217	18.8
25.4	287	73	214	18.6
25.4	284	72	212	18.3
25.4	283	72	211	18.3
25.4	280	71	209	18.0
25.4	279	71	208	18.1
25.4	276	70	206	17.8
25.4	272	69	203	17.5
25.4	268	68	200	17.3
25.4	264	67	197	17.0
25.4	260	66	194	16.8
25.4	256	65	191	16.5
25.4	252	64	188	16.3
25.4	248	63	185	16.0
25.4	244	62	182	15.8
25.4	240	61	179	15.5
25.4	236	60	176	15.3
25.4	232	59	173	15.0
25.4	228	58	170	14.8
25.4	224	57	167	14.5
25.4	213	54	159	13.7
25.4	209	53	156	13.4
25.4	205	52	153	13.2
25.4	201	51	150	12.9
25.4	197	50	147	12.7
25.4	193	49	144	12.4
25.4	189	48	141	12.2
25.4	185	47	138	11.9
25.4	181	46	135	11.7
25.4	177	45	132	11.4
25.4	173	44	129	11.2
25.4	169	43	126	10.9
25.4	142	36	106	9.1
25.4	138	35	103	8.9
25.4	134	34	100	8.6
25.4	130	33	97	8.4
25.4	126	32	94	8.1
25.4	122	31	91	7.9
25.4	118	30	88	7.6
25.4	114	29	85	7.4
25.4	71	18	53	4.6
25.4	67	17	50	4.3
25.4	63	16	47	4.1
25.4	59	15	44	3.8

%E	M1	M2	DM	M*
25.3	499	126	373	31.8
25.3	498	126	372	31.9
25.3	495	125	370	31.6
25.3	494	125	369	31.6
25.3	491	124	367	31.3
25.3	490	124	366	31.4
25.3	487	123	364	31.1
25.3	486	123	363	31.1
25.3	483	122	361	30.8
25.3	482	122	360	30.9
25.3	479	121	358	30.6
25.3	478	121	357	30.6
25.3	475	120	355	30.3
25.3	474	120	354	30.4
25.3	471	119	352	30.1
25.3	470	119	351	30.1
25.3	467	118	349	29.8
25.3	466	118	348	29.9
25.3	463	117	346	29.6
25.3	462	117	345	29.6
25.3	459	116	343	29.3
25.3	458	116	342	29.4
25.3	455	115	340	29.1
25.3	454	115	339	29.1
25.3	451	114	337	28.8
25.3	450	114	336	28.9
25.3	447	113	334	28.6
25.3	446	113	333	28.6
25.3	443	112	331	28.3
25.3	442	112	330	28.4
25.3	439	111	328	28.1
25.3	438	111	327	28.1
25.3	435	110	325	27.8
25.3	434	110	324	27.9
25.3	431	109	322	27.6
25.3	430	109	321	27.6
25.3	427	108	319	27.3
25.3	423	107	316	27.1
25.3	419	106	313	26.8
25.3	415	105	310	26.6
25.3	411	104	307	26.3
25.3	407	103	304	26.1
25.3	403	102	301	25.8
25.3	400	101	299	25.5
25.3	399	101	298	25.6
25.3	396	100	296	25.3
25.3	395	100	295	25.3
25.3	392	99	293	25.0
25.3	391	99	292	25.1
25.3	388	98	290	24.8

%E	M1	M2	DM	M*
25.3	387	98	289	24.8
25.3	384	97	287	24.5
25.3	383	97	286	24.6
25.3	380	96	284	24.3
25.3	379	96	283	24.3
25.3	376	95	281	24.0
25.3	375	95	280	24.1
25.3	372	94	278	23.8
25.3	371	94	277	23.8
25.3	368	93	275	23.5
25.3	367	93	274	23.6
25.3	364	92	272	23.3
25.3	363	92	271	23.3
25.3	360	91	269	23.0
25.3	359	91	268	23.1
25.3	356	90	266	22.8
25.3	352	89	263	22.5
25.3	348	88	260	22.3
25.3	344	87	257	22.0
25.3	340	86	254	21.8
25.3	336	85	251	21.5
25.3	332	84	248	21.3
25.3	328	83	245	21.0
25.3	324	82	242	20.8
25.3	320	81	239	20.5
25.3	316	80	236	20.3
25.3	312	79	233	20.0
25.3	308	78	230	19.8
25.3	304	77	227	19.5
25.3	300	76	224	19.3
25.3	297	75	222	18.9
25.3	296	75	221	19.0
25.3	293	74	219	18.7
25.3	292	74	218	18.8
25.3	289	73	216	18.4
25.3	288	73	215	18.5
25.3	285	72	213	18.2
25.3	281	71	210	17.9
25.3	277	70	207	17.7
25.3	273	69	204	17.4
25.3	269	68	201	17.2
25.3	265	67	198	16.9
25.3	261	66	195	16.7
25.3	257	65	192	16.4
25.3	253	64	189	16.2
25.3	249	63	186	15.9
25.3	245	62	183	15.7
25.3	241	61	180	15.4
25.3	237	60	177	15.2
25.3	233	59	174	14.9

%E	M1	M2	DM	M*
25.3	229	58	171	14.7
25.3	225	57	168	14.4
25.3	221	56	165	14.2
25.3	217	55	162	13.9
25.3	198	50	148	12.6
25.3	194	49	145	12.4
25.3	190	48	142	12.1
25.3	186	47	139	11.9
25.3	182	46	136	11.6
25.3	178	45	133	11.4
25.3	174	44	130	11.1
25.3	170	43	127	10.9
25.3	166	42	124	10.6
25.3	162	41	121	10.4
25.3	158	40	118	10.1
25.3	154	39	115	9.9
25.3	150	38	112	9.6
25.3	146	37	109	9.4
25.3	99	25	74	6.3
25.3	95	24	71	6.1
25.3	91	23	68	5.8
25.3	87	22	65	5.6
25.3	83	21	62	5.3
25.3	79	20	59	5.1
25.3	75	19	56	4.8
25.2	500	126	374	31.8
25.2	497	125	372	31.4
25.2	496	125	371	31.5
25.2	493	124	369	31.2
25.2	492	124	368	31.3
25.2	489	123	366	30.9
25.2	488	123	365	31.0
25.2	485	122	363	30.7
25.2	484	122	362	30.8
25.2	481	121	360	30.4
25.2	480	121	359	30.5
25.2	477	120	357	30.2
25.2	476	120	356	30.3
25.2	473	119	354	29.9
25.2	472	119	353	30.0
25.2	469	118	351	29.7
25.2	468	118	350	29.8
25.2	465	117	348	29.4
25.2	464	117	347	29.5
25.2	461	116	345	29.2
25.2	460	116	344	29.3
25.2	457	115	342	28.9
25.2	456	115	341	29.0
25.2	453	114	339	28.7
25.2	452	114	338	28.8

%E	M1	M2	DM	M*
25.2	449	113	336	28.4
25.2	448	113	335	28.5
25.2	445	112	333	28.2
25.2	444	112	332	28.3
25.2	441	111	330	27.9
25.2	440	111	329	28.0
25.2	437	110	327	27.7
25.2	436	110	326	27.8
25.2	433	109	324	27.4
25.2	432	109	323	27.5
25.2	429	108	321	27.2
25.2	428	108	320	27.3
25.2	425	107	318	26.9
25.2	424	107	317	27.0
25.2	421	106	315	26.7
25.2	420	106	314	26.8
25.2	417	105	312	26.4
25.2	416	105	311	26.5
25.2	413	104	309	26.2
25.2	412	104	308	26.3
25.2	409	103	306	25.9
25.2	408	103	305	26.0
25.2	405	102	303	25.7
25.2	404	102	302	25.8
25.2	401	101	300	25.4
25.2	397	100	297	25.2
25.2	393	99	294	24.9
25.2	389	98	291	24.7
25.2	385	97	288	24.4
25.2	381	96	285	24.2
25.2	377	95	282	23.9
25.2	373	94	279	23.7
25.2	369	93	276	23.4
25.2	365	92	273	23.2
25.2	361	91	270	22.9
25.2	357	90	267	22.7
25.2	353	89	264	22.4
25.2	349	88	261	22.2
25.2	345	87	258	21.9
25.2	341	86	255	21.7
25.2	337	85	252	21.4
25.2	334	84	250	21.1
25.2	333	84	249	21.2
25.2	330	83	247	20.9
25.2	329	83	246	20.9
25.2	326	82	244	20.6
25.2	325	82	243	20.7
25.2	322	81	241	20.4
25.2	321	81	240	20.4
25.2	318	80	238	20.1

%E	M1	M2	DM	M*
25.2	317	80	237	20.2
25.2	314	79	235	19.9
25.2	313	79	234	19.9
25.2	310	78	232	19.6
25.2	309	78	231	19.7
25.2	306	77	229	19.4
25.2	305	77	228	19.4
25.2	302	76	226	19.1
25.2	301	76	225	19.2
25.2	298	75	223	18.9
25.2	294	74	220	18.6
25.2	290	73	217	18.4
25.2	286	72	214	18.1
25.2	282	71	211	17.9
25.2	278	70	208	17.6
25.2	274	69	205	17.4
25.2	270	68	202	17.1
25.2	266	67	199	16.9
25.2	262	66	196	16.6
25.2	258	65	193	16.4
25.2	254	64	190	16.1
25.2	250	63	187	15.9
25.2	246	62	184	15.6
25.2	242	61	181	15.4
25.2	238	60	178	15.1
25.2	234	59	175	14.9
25.2	230	58	172	14.6
25.2	226	57	169	14.4
25.2	222	56	166	14.1
25.2	218	55	163	13.9
25.2	214	54	160	13.6
25.2	210	53	157	13.4
25.2	206	52	154	13.1
25.2	202	51	151	12.9
25.2	167	42	125	10.6
25.2	163	41	122	10.3
25.2	159	40	119	10.1
25.2	155	39	116	9.8
25.2	151	38	113	9.6
25.2	147	37	110	9.3
25.2	143	36	107	9.1
25.2	139	35	104	8.8
25.2	135	34	101	8.6
25.2	131	33	98	8.3
25.2	127	32	95	8.1
25.2	123	31	92	7.8
25.2	119	30	89	7.6
25.2	115	29	86	7.3
25.2	111	28	83	7.1
25.2	107	27	80	6.8

%E	M1	M2	DM	M*
25.2	103	26	77	6.6
25.1	499	125	374	31.3
25.1	498	125	373	31.4
25.1	495	124	371	31.1
25.1	494	124	370	31.1
25.1	491	123	368	30.8
25.1	490	123	367	30.9
25.1	487	122	365	30.6
25.1	486	122	364	30.6
25.1	483	121	362	30.3
25.1	482	121	361	30.4
25.1	479	120	359	30.1
25.1	478	120	358	30.1
25.1	475	119	356	29.8
25.1	474	119	355	29.9
25.1	471	118	353	29.6
25.1	470	118	352	29.6
25.1	467	117	350	29.3
25.1	466	117	349	29.4
25.1	463	116	347	29.1
25.1	462	116	346	29.1
25.1	459	115	344	28.8
25.1	458	115	343	28.9
25.1	455	114	341	28.6
25.1	454	114	340	28.6
25.1	451	113	338	28.3
25.1	450	113	337	28.4
25.1	447	112	335	28.1
25.1	446	112	334	28.1
25.1	443	111	332	27.8
25.1	442	111	331	27.9
25.1	439	110	329	27.6
25.1	438	110	328	27.6
25.1	435	109	326	27.3
25.1	434	109	325	27.4
25.1	431	108	323	27.1
25.1	430	108	322	27.1
25.1	427	107	320	26.8
25.1	426	107	319	26.9
25.1	423	106	317	26.6
25.1	422	106	316	26.6
25.1	419	105	314	26.3
25.1	418	105	313	26.4
25.1	415	104	311	26.1
25.1	414	104	310	26.1
25.1	411	103	308	25.8
25.1	410	103	307	25.9
25.1	407	102	305	25.6
25.1	406	102	304	25.6
25.1	403	101	302	25.3

%E	M1	M2	DM	M*
25.1	402	101	301	25.4
25.1	399	100	299	25.1
25.1	398	100	298	25.1
25.1	395	99	296	24.8
25.1	394	99	295	24.9
25.1	391	98	293	24.6
25.1	390	98	292	24.6
25.1	387	97	290	24.3
25.1	386	97	289	24.4
25.1	383	96	287	24.1
25.1	382	96	286	24.1
25.1	379	95	284	23.8
25.1	378	95	283	23.9
25.1	375	94	281	23.6
25.1	374	94	280	23.6
25.1	371	93	278	23.3
25.1	370	93	277	23.4
25.1	367	92	275	23.1
25.1	366	92	274	23.1
25.1	363	91	272	22.8
25.1	362	91	271	22.9
25.1	359	90	269	22.6
25.1	358	90	268	22.6
25.1	355	89	266	22.3
25.1	354	89	265	22.4
25.1	351	88	263	22.1
25.1	350	88	262	22.1
25.1	347	87	260	21.8
25.1	346	87	259	21.9
25.1	343	86	257	21.6
25.1	342	86	256	21.6
25.1	339	85	254	21.3
25.1	338	85	253	21.4
25.1	335	84	251	21.1
25.1	331	83	248	20.8
25.1	327	82	245	20.6
25.1	323	81	242	20.3
25.1	319	80	239	20.1
25.1	315	79	236	19.8
25.1	311	78	233	19.6
25.1	307	77	230	19.3
25.1	303	76	227	19.1
25.1	299	75	224	18.8
25.1	295	74	221	18.6
25.1	291	73	218	18.3
25.1	287	72	215	18.1
25.1	283	71	212	17.8
25.1	279	70	209	17.6
25.1	275	69	206	17.3
25.1	271	68	203	17.1

%E	M1	M2	DM	M*
25.1	267	67	200	16.8
25.1	263	66	197	16.6
25.1	259	65	194	16.3
25.1	255	64	191	16.1
25.1	251	63	188	15.8
25.1	247	62	185	15.6
25.1	243	61	182	15.3
25.1	239	60	179	15.1
25.1	235	59	176	14.8
25.1	231	58	173	14.6
25.1	227	57	170	14.3
25.1	223	56	167	14.1
25.1	219	55	164	13.8
25.1	215	54	161	13.6
25.1	211	53	158	13.3
25.1	207	52	155	13.1
25.1	203	51	152	12.8
25.1	199	50	149	12.6
25.1	195	49	146	12.3
25.1	191	48	143	12.1
25.1	187	47	140	11.8
25.1	183	46	137	11.6
25.1	179	45	134	11.3
25.1	175	44	131	11.1
25.1	171	43	128	10.8
25.0	500	125	375	31.3
25.0	497	124	373	30.9
25.0	496	124	372	31.0
25.0	492	123	369	30.8
25.0	488	122	366	30.5
25.0	484	121	363	30.3
25.0	480	120	360	30.0
25.0	476	119	357	29.8
25.0	472	118	354	29.5
25.0	468	117	351	29.3
25.0	464	116	348	29.0
25.0	460	115	345	28.8
25.0	456	114	342	28.5
25.0	452	113	339	28.3
25.0	448	112	336	28.0
25.0	444	111	333	27.8
25.0	440	110	330	27.5
25.0	436	109	327	27.3
25.0	432	108	324	27.0
25.0	428	107	321	26.8
25.0	424	106	318	26.5
25.0	420	105	315	26.3
25.0	416	104	312	26.0
25.0	412	103	309	25.8
25.0	408	102	306	25.5

%E	M1	M2	DM	M*
25.0	404	101	303	25.3
25.0	400	100	300	25.0
25.0	396	99	297	24.8
25.0	392	98	294	24.5
25.0	388	97	291	24.3
25.0	384	96	288	24.0
25.0	380	95	285	23.8
25.0	376	94	282	23.5
25.0	372	93	279	23.3
25.0	368	92	276	23.0
25.0	364	91	273	22.8
25.0	360	90	270	22.5
25.0	356	89	267	22.3
25.0	352	88	264	22.0
25.0	348	87	261	21.8
25.0	344	86	258	21.5
25.0	340	85	255	21.3
25.0	336	84	252	21.0
25.0	332	83	249	20.8
25.0	328	82	246	20.5
25.0	324	81	243	20.3
25.0	320	80	240	20.0
25.0	316	79	237	19.8
25.0	312	78	234	19.5
25.0	308	77	231	19.3
25.0	304	76	228	19.0
25.0	300	75	225	18.8
25.0	296	74	222	18.5
25.0	292	73	219	18.3
25.0	288	72	216	18.0
25.0	284	71	213	17.8
25.0	280	70	210	17.5
25.0	276	69	207	17.3
25.0	272	68	204	17.0
25.0	268	67	201	16.8
25.0	264	66	198	16.5
25.0	260	65	195	16.3
25.0	256	64	192	16.0
25.0	252	63	189	15.8
25.0	248	62	186	15.5
25.0	244	61	183	15.3
25.0	240	60	180	15.0
25.0	236	59	177	14.8
25.0	232	58	174	14.5
25.0	228	57	171	14.3
25.0	224	56	168	14.0
25.0	220	55	165	13.8
25.0	216	54	162	13.5
25.0	212	53	159	13.3
25.0	208	52	156	13.0

%E	M1	M2	DM	M*
25.0	204	51	153	12.8
25.0	200	50	150	12.5
25.0	196	49	147	12.3
25.0	192	48	144	12.0
25.0	188	47	141	11.8
25.0	184	46	138	11.5
25.0	180	45	135	11.3
25.0	176	44	132	11.0
25.0	172	43	129	10.8
25.0	168	42	126	10.5
25.0	164	41	123	10.3
25.0	160	40	120	10.0
25.0	156	39	117	9.8
25.0	152	38	114	9.5
25.0	148	37	111	9.3
25.0	144	36	108	9.0
25.0	140	35	105	8.8
25.0	136	34	102	8.5
25.0	132	33	99	8.3
25.0	128	32	96	8.0
25.0	124	31	93	7.8
25.0	120	30	90	7.5
25.0	116	29	87	7.3
25.0	112	28	84	7.0
25.0	108	27	81	6.8
25.0	104	26	78	6.5
25.0	100	25	75	6.3
25.0	96	24	72	6.0
25.0	92	23	69	5.8
25.0	88	22	66	5.5
25.0	84	21	63	5.3
25.0	80	20	60	5.0
25.0	76	19	57	4.8
25.0	72	18	54	4.5
25.0	68	17	51	4.3
25.0	64	16	48	4.0
25.0	60	15	45	3.8
25.0	56	14	42	3.5
25.0	52	13	39	3.3
25.0	48	12	36	3.0
25.0	44	11	33	2.8
25.0	40	10	30	2.5
25.0	36	9	27	2.3
25.0	32	8	24	2.0
25.0	28	7	21	1.8
25.0	24	6	18	1.5
25.0	20	5	15	1.3
25.0	16	4	12	1.0
25.0	12	3	9	0.8
25.0	8	2	6	0.5

%E	M1	M2	DM	M*
25.0	4	1	3	0.3
24.9	499	124	375	30.8
24.9	498	124	374	30.9
24.9	494	123	371	30.6
24.9	493	123	370	30.7
24.9	490	122	368	30.4
24.9	489	122	367	30.4
24.9	486	121	365	30.1
24.9	485	121	364	30.2
24.9	482	120	362	29.9
24.9	481	120	361	29.9
24.9	478	119	359	29.6
24.9	477	119	358	29.7
24.9	474	118	356	29.4
24.9	473	118	355	29.4
24.9	470	117	353	29.1
24.9	469	117	352	29.2
24.9	466	116	350	28.9
24.9	465	116	349	28.9
24.9	462	115	347	28.6
24.9	461	115	346	28.7
24.9	458	114	344	28.4
24.9	457	114	343	28.4
24.9	454	113	341	28.1
24.9	453	113	340	28.2
24.9	450	112	338	27.9
24.9	449	112	337	27.9
24.9	446	111	335	27.6
24.9	445	111	334	27.7
24.9	442	110	332	27.4
24.9	441	110	331	27.4
24.9	438	109	329	27.1
24.9	437	109	328	27.2
24.9	434	108	326	26.9
24.9	433	108	325	26.9
24.9	430	107	323	26.6
24.9	429	107	322	26.7
24.9	426	106	320	26.4
24.9	425	106	319	26.4
24.9	422	105	317	26.1
24.9	421	105	316	26.2
24.9	418	104	314	25.9
24.9	417	104	313	25.9
24.9	414	103	311	25.6
24.9	413	103	310	25.7
24.9	410	102	308	25.4
24.9	409	102	307	25.4
24.9	406	101	305	25.1
24.9	405	101	304	25.2
24.9	402	100	302	24.9

%E	M1	M2	DM	M*
24.9	401	100	301	24.9
24.9	398	99	299	24.6
24.9	397	99	298	24.7
24.9	394	98	296	24.4
24.9	393	98	295	24.4
24.9	390	97	293	24.1
24.9	389	97	292	24.2
24.9	386	96	290	23.9
24.9	385	96	289	23.9
24.9	382	95	287	23.6
24.9	381	95	286	23.7
24.9	378	94	284	23.4
24.9	377	94	283	23.4
24.9	374	93	281	23.1
24.9	373	93	280	23.2
24.9	370	92	278	22.9
24.9	369	92	277	22.9
24.9	366	91	275	22.6
24.9	365	91	274	22.7
24.9	362	90	272	22.4
24.9	361	90	271	22.4
24.9	358	89	269	22.1
24.9	357	89	268	22.2
24.9	354	88	266	21.9
24.9	353	88	265	21.9
24.9	350	87	263	21.6
24.9	349	87	262	21.7
24.9	346	86	260	21.4
24.9	345	86	259	21.4
24.9	342	85	257	21.1
24.9	341	85	256	21.2
24.9	338	84	254	20.9
24.9	337	84	253	20.9
24.9	334	83	251	20.6
24.9	333	83	250	20.7
24.9	329	82	247	20.4
24.9	325	81	244	20.2
24.9	321	80	241	19.9
24.9	317	79	238	19.7
24.9	313	78	235	19.4
24.9	309	77	232	19.2
24.9	305	76	229	18.9
24.9	301	75	226	18.7
24.9	297	74	223	18.4
24.9	293	73	220	18.2
24.9	289	72	217	17.9
24.9	285	71	214	17.7
24.9	281	70	211	17.4
24.9	277	69	208	17.2
24.9	273	68	205	16.9

%E	M1	M2	DM	M*
24.9	269	67	202	16.7
24.9	265	66	199	16.4
24.9	261	65	196	16.2
24.9	257	64	193	15.9
24.9	253	63	190	15.7
24.9	249	62	187	15.4
24.9	245	61	184	15.2
24.9	241	60	181	14.9
24.9	237	59	178	14.7
24.9	233	58	175	14.4
24.9	229	57	172	14.2
24.9	225	56	169	13.9
24.9	221	55	166	13.7
24.9	217	54	163	13.4
24.9	213	53	160	13.2
24.9	209	52	157	12.9
24.9	205	51	154	12.7
24.9	201	50	151	12.4
24.9	197	49	148	12.2
24.9	193	48	145	11.9
24.9	189	47	142	11.7
24.9	185	46	139	11.4
24.9	181	45	136	11.2
24.9	177	44	133	10.9
24.9	173	43	130	10.7
24.9	169	42	127	10.4
24.8	500	124	376	30.8
24.8	496	123	373	30.5
24.8	495	123	372	30.6
24.8	492	122	370	30.3
24.8	491	122	369	30.3
24.8	488	121	367	30.0
24.8	487	121	366	30.1
24.8	484	120	364	29.8
24.8	483	120	363	29.8
24.8	480	119	361	29.5
24.8	479	119	360	29.6
24.8	476	118	358	29.3
24.8	475	118	357	29.3
24.8	472	117	355	29.0
24.8	471	117	354	29.1
24.8	468	116	352	28.8
24.8	467	116	351	28.8
24.8	464	115	349	28.5
24.8	463	115	348	28.6
24.8	460	114	346	28.3
24.8	459	114	345	28.3
24.8	456	113	343	28.0
24.8	455	113	342	28.1
24.8	452	112	340	27.8

%E	M1	M2	DM	M*
24.8	451	112	339	27.8
24.8	448	111	337	27.5
24.8	447	111	336	27.6
24.8	444	110	334	27.3
24.8	443	110	333	27.3
24.8	440	109	331	27.0
24.8	439	109	330	27.1
24.8	436	108	328	26.8
24.8	435	108	327	26.8
24.8	432	107	325	26.5
24.8	431	107	324	26.6
24.8	428	106	322	26.3
24.8	427	106	321	26.3
24.8	424	105	319	26.0
24.8	423	105	318	26.1
24.8	420	104	316	25.8
24.8	419	104	315	25.8
24.8	416	103	313	25.5
24.8	415	103	312	25.6
24.8	412	102	310	25.3
24.8	411	102	309	25.3
24.8	408	101	307	25.0
24.8	407	101	306	25.1
24.8	404	100	304	24.8
24.8	403	100	303	24.8
24.8	400	99	301	24.5
24.8	399	99	300	24.6
24.8	395	98	297	24.3
24.8	391	97	294	24.1
24.8	387	96	291	23.8
24.8	383	95	288	23.6
24.8	379	94	285	23.3
24.8	375	93	282	23.1
24.8	371	92	279	22.8
24.8	367	91	276	22.6
24.8	363	90	273	22.3
24.8	359	89	270	22.1
24.8	355	88	267	21.8
24.8	351	87	264	21.6
24.8	347	86	261	21.3
24.8	343	85	258	21.1
24.8	339	84	255	20.8
24.8	335	83	252	20.6
24.8	331	82	249	20.3
24.8	330	82	248	20.4
24.8	327	81	246	20.1
24.8	326	81	245	20.1
24.8	323	80	243	19.8
24.8	322	80	242	19.9
24.8	319	79	240	19.6

%E	M1	M2	DM	M*
24.8	318	79	239	19.6
24.8	315	78	237	19.3
24.8	314	78	236	19.4
24.8	311	77	234	19.1
24.8	310	77	233	19.1
24.8	307	76	231	18.8
24.8	306	76	230	18.9
24.8	303	75	228	18.6
24.8	302	75	227	18.6
24.8	298	74	224	18.4
24.8	294	73	221	18.1
24.8	290	72	218	17.9
24.8	286	71	215	17.6
24.8	282	70	212	17.4
24.8	278	69	209	17.1
24.8	274	68	206	16.9
24.8	270	67	203	16.6
24.8	266	66	200	16.4
24.8	262	65	197	16.1
24.8	258	64	194	15.9
24.8	254	63	191	15.6
24.8	250	62	188	15.4
24.8	246	61	185	15.1
24.8	242	60	182	14.9
24.8	238	59	179	14.6
24.8	234	58	176	14.4
24.8	230	57	173	14.1
24.8	226	56	170	13.9
24.8	222	55	167	13.6
24.8	218	54	164	13.4
24.8	214	53	161	13.1
24.8	210	52	158	12.9
24.8	206	51	155	12.6
24.8	202	50	152	12.4
24.8	165	41	124	10.2
24.8	161	40	121	9.9
24.8	157	39	118	9.7
24.8	153	38	115	9.4
24.8	149	37	112	9.2
24.8	145	36	109	8.9
24.8	141	35	106	8.7
24.8	137	34	103	8.4
24.8	133	33	100	8.2
24.8	129	32	97	7.9
24.8	125	31	94	7.7
24.8	121	30	91	7.4
24.8	117	29	88	7.2
24.8	113	28	85	6.9
24.8	109	27	82	6.7
24.8	105	26	79	6.4

%E	M1	M2	DM	M*
24.8	101	25	76	6.2
24.7	498	123	375	30.4
24.7	497	123	374	30.4
24.7	494	122	372	30.1
24.7	493	122	371	30.2
24.7	490	121	369	29.9
24.7	489	121	368	29.9
24.7	486	120	366	29.6
24.7	485	120	365	29.7
24.7	482	119	363	29.4
24.7	481	119	362	29.4
24.7	478	118	360	29.1
24.7	477	118	359	29.2
24.7	474	117	357	28.9
24.7	473	117	356	28.9
24.7	470	116	354	28.6
24.7	469	116	353	28.7
24.7	466	115	351	28.4
24.7	465	115	350	28.4
24.7	462	114	348	28.1
24.7	461	114	347	28.2
24.7	458	113	345	27.9
24.7	457	113	344	27.9
24.7	454	112	342	27.6
24.7	453	112	341	27.7
24.7	450	111	339	27.4
24.7	449	111	338	27.4
24.7	446	110	336	27.1
24.7	445	110	335	27.2
24.7	442	109	333	26.9
24.7	441	109	332	26.9
24.7	438	108	330	26.6
24.7	437	108	329	26.7
24.7	434	107	327	26.4
24.7	433	107	326	26.4
24.7	430	106	324	26.1
24.7	429	106	323	26.2
24.7	425	105	320	25.9
24.7	421	104	317	25.7
24.7	417	103	314	25.4
24.7	413	102	311	25.2
24.7	409	101	308	24.9
24.7	405	100	305	24.7
24.7	401	99	302	24.4
24.7	397	98	299	24.2
24.7	396	98	298	24.3
24.7	393	97	296	23.9
24.7	392	97	295	24.0
24.7	389	96	293	23.7
24.7	388	96	292	23.8

%E	M1	M2	DM	M*
24.7	385	95	290	23.4
24.7	384	95	289	23.5
24.7	381	94	287	23.2
24.7	380	94	286	23.3
24.7	377	93	284	22.9
24.7	376	93	283	23.0
24.7	373	92	281	22.7
24.7	372	92	280	22.8
24.7	369	91	278	22.4
24.7	368	91	277	22.5
24.7	365	90	275	22.2
24.7	364	90	274	22.3
24.7	361	89	272	21.9
24.7	360	89	271	22.0
24.7	357	88	269	21.7
24.7	356	88	268	21.8
24.7	352	87	265	21.5
24.7	348	86	262	21.3
24.7	344	85	259	21.0
24.7	340	84	256	20.8
24.7	336	83	253	20.5
24.7	332	82	250	20.3
24.7	328	81	247	20.0
24.7	324	80	244	19.8
24.7	320	79	241	19.5
24.7	316	78	238	19.3
24.7	312	77	235	19.0
24.7	308	76	232	18.8
24.7	304	75	229	18.5
24.7	300	74	226	18.3
24.7	299	74	225	18.3
24.7	296	73	223	18.0
24.7	295	73	222	18.1
24.7	292	72	220	17.8
24.7	291	72	219	17.8
24.7	288	71	217	17.5
24.7	287	71	216	17.6
24.7	283	70	213	17.3
24.7	279	69	210	17.1
24.7	275	68	207	16.8
24.7	271	67	204	16.6
24.7	267	66	201	16.3
24.7	263	65	198	16.1
24.7	259	64	195	15.8
24.7	255	63	192	15.6
24.7	251	62	189	15.3
24.7	247	61	186	15.1
24.7	243	60	183	14.8
24.7	239	59	180	14.6
24.7	235	58	177	14.3

%E	M1	M2	DM	M*
24.7	231	57	174	14.1
24.7	227	56	171	13.8
24.7	223	55	168	13.6
24.7	219	54	165	13.3
24.7	215	53	162	13.1
24.7	198	49	149	12.1
24.7	194	48	146	11.9
24.7	190	47	143	11.6
24.7	186	46	140	11.4
24.7	182	45	137	11.1
24.7	178	44	134	10.9
24.7	174	43	131	10.6
24.7	170	42	128	10.4
24.7	166	41	125	10.1
24.7	162	40	122	9.9
24.7	158	39	119	9.6
24.7	154	38	116	9.4
24.7	150	37	113	9.1
24.7	146	36	110	8.9
24.7	97	24	73	5.9
24.7	93	23	70	5.7
24.7	89	22	67	5.4
24.7	85	21	64	5.2
24.7	81	20	61	4.9
24.7	77	19	58	4.7
24.7	73	18	55	4.4
24.6	500	123	377	30.3
24.6	499	123	376	30.3
24.6	496	122	374	30.0
24.6	495	122	373	30.1
24.6	492	121	371	29.8
24.6	491	121	370	29.8
24.6	488	120	368	29.5
24.6	487	120	367	29.6
24.6	484	119	365	29.3
24.6	483	119	364	29.3
24.6	480	118	362	29.0
24.6	479	118	361	29.1
24.6	476	117	359	28.8
24.6	475	117	358	28.8
24.6	472	116	356	28.5
24.6	471	116	355	28.6
24.6	468	115	353	28.3
24.6	467	115	352	28.3
24.6	464	114	350	28.0
24.6	463	114	349	28.1
24.6	460	113	347	27.8
24.6	459	113	346	27.8
24.6	456	112	344	27.5
24.6	455	112	343	27.6

%E	M1	M2	DM	M*
24.6	452	111	341	27.3
24.6	451	111	340	27.3
24.6	448	110	338	27.0
24.6	447	110	337	27.1
24.6	444	109	335	26.8
24.6	443	109	334	26.8
24.6	439	108	331	26.6
24.6	435	107	328	26.3
24.6	431	106	325	26.1
24.6	427	105	322	25.8
24.6	426	105	321	25.9
24.6	423	104	319	25.6
24.6	422	104	318	25.6
24.6	419	103	316	25.3
24.6	418	103	315	25.4
24.6	415	102	313	25.1
24.6	414	102	312	25.1
24.6	411	101	310	24.8
24.6	410	101	309	24.9
24.6	407	100	307	24.6
24.6	406	100	306	24.6
24.6	403	99	304	24.3
24.6	402	99	303	24.4
24.6	399	98	301	24.1
24.6	398	98	300	24.1
24.6	395	97	298	23.8
24.6	394	97	297	23.9
24.6	391	96	295	23.6
24.6	390	96	294	23.6
24.6	386	95	291	23.4
24.6	382	94	288	23.1
24.6	378	93	285	22.9
24.6	374	92	282	22.6
24.6	370	91	279	22.4
24.6	366	90	276	22.1
24.6	362	89	273	21.9
24.6	358	88	270	21.6
24.6	354	87	267	21.4
24.6	353	87	266	21.4
24.6	350	86	264	21.1
24.6	349	86	263	21.2
24.6	346	85	261	20.9
24.6	345	85	260	20.9
24.6	342	84	258	20.6
24.6	341	84	257	20.7
24.6	338	83	255	20.4
24.6	337	83	254	20.4
24.6	334	82	252	20.1
24.6	333	82	251	20.2
24.6	329	81	248	19.9

%E	M1	M2	DM	M*
24.6	325	80	245	19.7
24.6	321	79	242	19.4
24.6	317	78	239	19.2
24.6	313	77	236	18.9
24.6	309	76	233	18.7
24.6	305	75	230	18.4
24.6	301	74	227	18.2
24.6	297	73	224	17.9
24.6	293	72	221	17.7
24.6	289	71	218	17.4
24.6	285	70	215	17.2
24.6	284	70	214	17.3
24.6	281	69	212	16.9
24.6	280	69	211	17.0
24.6	276	68	208	16.8
24.6	272	67	205	16.5
24.6	268	66	202	16.3
24.6	264	65	199	16.0
24.6	260	64	196	15.8
24.6	256	63	193	15.5
24.6	252	62	190	15.3
24.6	248	61	187	15.0
24.6	244	60	184	14.8
24.6	240	59	181	14.5
24.6	236	58	178	14.3
24.6	232	57	175	14.0
24.6	228	56	172	13.8
24.6	224	55	169	13.5
24.6	211	52	159	12.8
24.6	207	51	156	12.6
24.6	203	50	153	12.3
24.6	199	49	150	12.1
24.6	195	48	147	11.8
24.6	191	47	144	11.6
24.6	187	46	141	11.3
24.6	183	45	138	11.1
24.6	179	44	135	10.8
24.6	175	43	132	10.6
24.6	171	42	129	10.3
24.6	167	41	126	10.1
24.6	142	35	107	8.6
24.6	138	34	104	8.4
24.6	134	33	101	8.1
24.6	130	32	98	7.9
24.6	126	31	95	7.6
24.6	122	30	92	7.4
24.6	118	29	89	7.1
24.6	114	28	86	6.9
24.6	69	17	52	4.2
24.6	65	16	49	3.9

%E	M1	M2	DM	M*
24.6	61	15	46	3.7
24.6	57	14	43	3.4
24.5	498	122	376	29.9
24.5	497	122	375	29.9
24.5	494	121	373	29.6
24.5	493	121	372	29.7
24.5	490	120	370	29.4
24.5	489	120	369	29.4
24.5	486	119	367	29.1
24.5	485	119	366	29.2
24.5	482	118	364	28.9
24.5	481	118	363	28.9
24.5	478	117	361	28.6
24.5	477	117	360	28.7
24.5	474	116	358	28.4
24.5	473	116	357	28.4
24.5	470	115	355	28.1
24.5	469	115	354	28.2
24.5	466	114	352	27.9
24.5	465	114	351	27.9
24.5	462	113	349	27.6
24.5	461	113	348	27.7
24.5	458	112	346	27.4
24.5	457	112	345	27.4
24.5	453	111	342	27.2
24.5	449	110	339	26.9
24.5	445	109	336	26.7
24.5	441	108	333	26.4
24.5	440	108	332	26.5
24.5	437	107	330	26.2
24.5	436	107	329	26.3
24.5	433	106	327	25.9
24.5	432	106	326	26.0
24.5	429	105	324	25.7
24.5	428	105	323	25.8
24.5	425	104	321	25.4
24.5	424	104	320	25.5
24.5	421	103	318	25.2
24.5	420	103	317	25.3
24.5	417	102	315	24.9
24.5	416	102	314	25.0
24.6	413	101	312	24.7
24.5	412	101	311	24.8
24.5	409	100	309	24.4
24.5	408	100	308	24.5
24.5	404	99	305	24.3
24.5	400	98	302	24.0
24.5	396	97	299	23.8
24.5	392	96	296	23.5
24.5	388	95	293	23.3

%E	M1	M2	DM	M*
24.5	387	95	292	23.3
24.5	384	94	290	23.0
24.5	383	94	289	23.1
24.5	380	93	287	22.8
24.5	379	93	286	22.8
24.5	376	92	284	22.5
24.5	375	92	283	22.6
24.5	372	91	281	22.3
24.5	371	91	280	22.3
24.5	368	90	278	22.0
24.5	367	90	277	22.1
24.5	364	89	275	21.8
24.5	363	89	274	21.8
24.5	359	88	271	21.6
24.5	355	87	268	21.3
24.5	351	86	265	21.1
24.5	347	85	262	20.8
24.5	343	84	259	20.6
24.5	339	83	256	20.3
24.5	335	82	253	20.1
24.5	331	81	250	19.8
24.5	330	81	249	19.9
24.5	327	80	247	19.6
24.5	326	80	246	19.6
24.5	323	79	244	19.3
24.5	322	79	243	19.4
24.5	319	78	241	19.1
24.5	318	78	240	19.1
24.5	314	77	237	18.9
24.5	310	76	234	18.6
24.5	306	75	231	18.4
24.5	302	74	228	18.1
24.5	298	73	225	17.9
24.5	294	72	222	17.6
24.5	290	71	219	17.4
24.5	286	70	216	17.1
24.5	282	69	213	16.9
24.5	278	68	210	16.6
24.5	277	68	209	16.7
24.5	274	67	207	16.4
24.5	273	67	206	16.4
24.5	269	66	203	16.2
24.5	265	65	200	15.9
24.5	261	64	197	15.7
24.5	257	63	194	15.4
24.5	253	62	191	15.2
24.5	249	61	188	14.9
24.5	245	60	185	14.7
24.5	241	59	182	14.4
24.5	237	58	179	14.2

%E	M1	M2	DM	M*
24.5	233	57	176	13.9
24.5	229	56	173	13.7
24.5	220	54	166	13.3
24.5	216	53	163	13.0
24.5	212	52	160	12.8
24.5	208	51	157	12.5
24.5	204	50	154	12.3
24.5	200	49	151	12.0
24.5	196	48	148	11.8
24.5	192	47	145	11.5
24.5	188	46	142	11.3
24.5	184	45	139	11.0
24.5	163	40	123	9.8
24.5	159	39	120	9.6
24.5	155	38	117	9.3
24.5	151	37	114	9.1
24.5	147	36	111	8.8
24.5	143	35	108	8.6
24.5	139	34	105	8.3
24.5	110	27	83	6.6
24.5	106	26	80	6.4
24.5	102	25	77	6.1
24.5	98	24	74	5.9
24.5	94	23	71	5.6
24.5	53	13	40	3.2
24.5	49	12	37	2.9
24.4	500	122	378	29.8
24.4	499	122	377	29.8
24.4	496	121	375	29.5
24.4	495	121	374	29.6
24.4	492	120	372	29.3
24.4	491	120	371	29.3
24.4	488	119	369	29.0
24.4	487	119	368	29.1
24.4	484	118	366	28.8
24.4	483	118	365	28.8
24.4	480	117	363	28.5
24.4	479	117	362	28.6
24.4	476	116	360	28.3
24.4	475	116	359	28.3
24.4	472	115	357	28.0
24.4	471	115	356	28.1
24.4	468	114	354	27.8
24.4	467	114	353	27.8
24.4	464	113	351	27.5
24.4	463	113	350	27.6
24.4	459	112	347	27.3
24.4	455	111	344	27.1
24.4	454	111	343	27.1
24.4	451	110	341	26.8

%E	M1	M2	DM	M*
24.4	450	110	340	26.9
24.4	447	109	338	26.6
24.4	446	109	337	26.6
24.4	443	108	335	26.3
24.4	442	108	334	26.4
24.4	439	107	332	26.1
24.4	438	107	331	26.1
24.4	435	106	329	25.8
24.4	434	106	328	25.9
24.4	431	105	326	25.6
24.4	430	105	325	25.6
24.4	427	104	323	25.3
24.4	426	104	322	25.4
24.4	423	103	320	25.1
24.4	422	103	319	25.1
24.4	418	102	316	24.9
24.4	414	101	313	24.6
24.4	410	100	310	24.4
24.4	406	99	307	24.1
24.4	405	99	306	24.2
24.4	402	98	304	23.9
24.4	401	98	303	24.0
24.4	398	97	301	23.6
24.4	397	97	300	23.7
24.4	394	96	298	23.4
24.4	393	96	297	23.5
24.4	390	95	295	23.1
24.4	389	95	294	23.2
24.4	386	94	292	22.9
24.4	385	94	291	23.0
24.4	381	93	288	22.7
24.4	377	92	285	22.5
24.4	373	91	282	22.2
24.4	369	90	279	22.0
24.4	365	89	276	21.7
24.4	361	88	273	21.5
24.4	360	88	272	21.5
24.4	357	87	270	21.2
24.4	356	87	269	21.3
24.4	353	86	267	21.0
24.4	352	86	266	21.0
24.4	349	85	264	20.7
24.4	348	85	263	20.8
24.4	344	84	260	20.5
24.4	340	83	257	20.3
24.4	336	82	254	20.0
24.4	332	81	251	19.8
24.4	328	80	248	19.5
24.4	324	79	245	19.3
24.4	320	78	242	19.0

%E	M1	M2	DM	M*
24.4	316	77	239	18.8
24.4	315	77	238	18.8
24.4	312	76	236	18.5
24.4	311	76	235	18.6
24.4	308	75	233	18.3
24.4	307	75	232	18.3
24.4	303	74	229	18.1
24.4	299	73	226	17.8
24.4	295	72	223	17.6
24.4	291	71	220	17.3
24.4	287	70	217	17.1
24.4	283	69	214	16.8
24.4	279	68	211	16.6
24.4	275	67	208	16.3
24.4	271	66	205	16.1
24.4	270	66	204	16.1
24.4	266	65	201	15.9
24.4	262	64	198	15.6
24.4	258	63	195	15.4
24.4	254	62	192	15.1
24.4	250	61	189	14.9
24.4	246	60	186	14.6
24.4	242	59	183	14.4
24.4	238	58	180	14.1
24.4	234	57	177	13.9
24.4	225	55	170	13.4
24.4	221	54	167	13.2
24.4	217	53	164	12.9
24.4	213	52	161	12.7
24.4	209	51	158	12.4
24.4	205	50	155	12.2
24.4	201	49	152	11.9
24.4	197	48	149	11.7
24.4	193	47	146	11.4
24.4	180	44	136	10.8
24.4	176	43	133	10.5
24.4	172	42	130	10.3
24.4	168	41	127	10.0
24.4	164	40	124	9.8
24.4	160	39	121	9.5
24.4	156	38	118	9.3
24.4	135	33	102	8.1
24.4	131	32	99	7.8
24.4	127	31	96	7.6
24.4	123	30	93	7.3
24.4	119	29	90	7.1
24.4	90	22	68	5.4
24.4	86	21	65	5.1
24.4	82	20	62	4.9
24.4	78	19	59	4.6

%E	M1	M2	DM	M*
24.4	45	11	34	2.7
24.4	41	10	31	2.4
24.3	498	121	377	29.4
24.3	497	121	376	29.5
24.3	494	120	374	29.1
24.3	493	120	373	29.2
24.3	490	119	371	28.9
24.3	489	119	370	29.0
24.3	486	118	368	28.7
24.3	485	118	367	28.7
24.3	482	117	365	28.4
24.3	481	117	364	28.5
24.3	478	116	362	28.2
24.3	477	116	361	28.2
24.3	474	115	359	27.9
24.3	473	115	358	28.0
24.3	470	114	356	27.7
24.3	469	114	355	27.7
24.3	465	113	352	27.5
24.3	461	112	349	27.2
24.3	460	112	348	27.3
24.3	457	111	346	27.0
24.3	456	111	345	27.0
24.3	453	110	343	26.7
24.3	452	110	342	26.8
24.3	449	109	340	26.5
24.3	448	109	339	26.5
24.3	445	108	337	26.2
24.3	444	108	336	26.3
24.3	441	107	334	26.0
24.3	440	107	333	26.0
24.3	437	106	331	25.7
24.3	436	106	330	25.8
24.3	432	105	327	25.5
24.3	428	104	324	25.3
24.3	424	103	321	25.0
24.3	420	102	318	24.8
24.3	419	102	317	24.8
24.3	416	101	315	24.5
24.3	415	101	314	24.6
24.3	412	100	312	24.3
24.3	411	100	311	24.3
24.3	408	99	309	24.0
24.3	407	99	308	24.1
24.3	404	98	306	23.8
24.3	403	98	305	23.8
24.3	400	97	303	23.5
24.3	399	97	302	23.6
24.3	395	96	299	23.3
24.3	391	95	296	23.1

%E	M1	M2	DM	M*
24.3	387	94	293	22.8
24.3	383	93	290	22.6
24.3	382	93	289	22.6
24.3	379	92	287	22.3
24.3	378	92	286	22.4
24.3	375	91	284	22.1
24.3	374	91	283	22.1
24.3	371	90	281	21.8
24.3	370	90	280	21.9
24.3	367	89	278	21.6
24.3	366	89	277	21.6
24.3	362	88	274	21.4
24.3	358	87	271	21.1
24.3	354	86	268	20.9
24.3	350	85	265	20.6
24.3	346	84	262	20.4
24.3	345	84	261	20.5
24.3	342	83	259	20.1
24.3	341	83	258	20.2
24.3	338	82	256	19.9
24.3	337	82	255	20.0
24.3	334	81	253	19.6
24.3	333	81	252	19.7
24.3	329	80	249	19.5
24.3	325	79	246	19.2
24.3	321	78	243	19.0
24.3	317	77	240	18.7
24.3	313	76	237	18.5
24.3	309	75	234	18.2
24.3	305	74	231	18.0
24.3	304	74	230	18.0
24.3	301	73	228	17.7
24.3	300	73	227	17.8
24.3	296	72	224	17.5
24.3	292	71	221	17.3
24.3	288	70	218	17.0
24.3	284	69	215	16.8
24.3	280	68	212	16.5
24.3	276	67	209	16.3
24.3	272	66	206	16.0
24.3	268	65	203	15.8
24.3	267	65	202	15.8
24.3	263	64	199	15.6
24.3	259	63	196	15.3
24.3	255	62	193	15.1
24.3	251	61	190	14.8
24.3	247	60	187	14.6
24.3	243	59	184	14.3
24.3	239	58	181	14.1
24.3	235	57	178	13.8

%E	M1	M2	DM	M*
24.3	230	56	174	13.6
24.3	226	55	171	13.4
24.3	222	54	168	13.1
24.3	218	53	165	12.9
24.3	214	52	162	12.6
24.3	210	51	159	12.4
24.3	206	50	156	12.1
24.3	202	49	153	11.9
24.3	189	46	143	11.2
24.3	185	45	140	10.9
24.3	181	44	137	10.7
24.3	177	43	134	10.4
24.3	173	42	131	10.2
24.3	169	41	128	9.9
24.3	152	37	115	9.0
24.3	148	36	112	8.8
24.3	144	35	109	8.5
24.3	140	34	106	8.3
24.3	136	33	103	8.0
24.3	115	28	87	6.8
24.3	111	27	84	6.6
24.3	107	26	81	6.3
24.3	103	25	78	6.1
24.3	74	18	56	4.4
24.3	70	17	53	4.1
24.3	37	9	28	2.2
24.2	500	121	379	29.3
24.2	499	121	378	29.3
24.2	496	120	376	29.0
24.2	495	120	375	29.1
24.2	492	119	373	28.8
24.2	491	119	372	28.8
24.2	488	118	370	28.5
24.2	487	118	369	28.6
24.2	484	117	367	28.3
24.2	483	117	366	28.3
24.2	480	116	364	28.0
24.2	479	116	363	28.1
24.2	476	115	361	27.8
24.2	475	115	360	27.8
24.2	472	114	358	27.5
24.2	471	114	357	27.6
24.2	467	113	354	27.3
24.2	466	113	353	27.4
24.2	463	112	351	27.1
24.2	462	112	350	27.2
24.2	459	111	348	26.8
24.2	458	111	347	26.9
24.2	455	110	345	26.6
24.2	454	110	344	26.7

%E	M1	M2	DM	M*
24.2	451	109	342	26.3
24.2	450	109	341	26.4
24.2	447	108	339	26.1
24.2	446	108	338	26.2
24.2	443	107	336	25.8
24.2	442	107	335	25.9
24.2	438	106	332	25.7
24.2	434	105	329	25.4
24.2	433	105	328	25.5
24.2	430	104	326	25.2
24.2	429	104	325	25.2
24.2	426	103	323	24.9
24.2	425	103	322	25.0
24.2	422	102	320	24.7
24.2	421	102	319	24.7
24.2	418	101	317	24.4
24.2	417	101	316	24.5
24.2	414	100	314	24.2
24.2	413	100	313	24.2
24.2	409	99	310	24.0
24.2	405	98	307	23.7
24.2	401	97	304	23.5
24.2	397	96	301	23.2
24.2	396	96	300	23.3
24.2	393	95	298	23.0
24.2	392	95	297	23.0
24.2	389	94	295	22.7
24.2	388	94	294	22.8
24.2	385	93	292	22.5
24.2	384	93	291	22.5
24.2	380	92	288	22.3
24.2	376	91	285	22.0
24.2	372	90	282	21.8
24.2	368	89	279	21.5
24.2	364	88	276	21.3
24.2	363	88	275	21.3
24.2	360	87	273	21.0
24.2	359	87	272	21.1
24.2	356	86	270	20.8
24.2	355	86	269	20.8
24.2	351	85	266	20.6
24.2	347	84	263	20.3
24.2	343	83	260	20.1
24.2	339	82	257	19.8
24.2	335	81	254	19.6
24.2	331	80	251	19.3
24.2	330	80	250	19.4
24.2	327	79	248	19.1
24.2	326	79	247	19.1
24.2	322	78	244	18.9

%E	M1	M2	DM	M*
24.2	318	77	241	18.6
24.2	314	76	238	18.4
24.2	310	75	235	18.1
24.2	306	74	232	17.9
24.2	302	73	229	17.6
24.2	298	72	226	17.4
24.2	297	72	225	17.5
24.2	294	71	223	17.1
24.2	293	71	222	17.2
24.2	289	70	219	17.0
24.2	285	69	216	16.7
24.2	281	68	213	16.5
24.2	277	67	210	16.2
24.2	273	66	207	16.0
24.2	269	65	204	15.7
24.2	265	64	201	15.5
24.2	264	64	200	15.5
24.2	260	63	197	15.3
24.2	256	62	194	15.0
24.2	252	61	191	14.8
24.2	248	60	188	14.5
24.2	244	59	185	14.3
24.2	240	58	182	14.0
24.2	236	57	179	13.8
24.2	231	56	175	13.6
24.2	227	55	172	13.3
24.2	223	54	169	13.1
24.2	219	53	166	12.8
24.2	215	52	163	12.6
24.2	211	51	160	12.3
24.2	207	50	157	12.1
24.2	198	48	150	11.6
24.2	194	47	147	11.4
24.2	190	46	144	11.1
24.2	186	45	141	10.9
24.2	182	44	138	10.6
24.2	178	43	135	10.4
24.2	165	40	125	9.7
24.2	161	39	122	9.4
24.2	157	38	119	9.2
24.2	153	37	116	8.9
24.2	149	36	113	8.7
24.2	132	32	100	7.8
24.2	128	31	97	7.5
24.2	124	30	94	7.3
24.2	120	29	91	7.0
24.2	99	24	75	5.8
24.2	95	23	72	5.6
24.2	91	22	69	5.3
24.2	66	16	50	3.9

%E	M1	M2	DM	M*
24.2	62	15	47	3.6
24.2	33	8	25	1.9
24.1	498	120	378	28.9
24.1	497	120	377	29.0
24.1	494	119	375	28.7
24.1	493	119	374	28.7
24.1	490	118	372	28.4
24.1	489	118	371	28.5
24.1	486	117	369	28.2
24.1	485	117	368	28.2
24.1	482	116	366	27.9
24.1	481	116	365	28.0
24.1	478	115	363	27.7
24.1	477	115	362	27.7
24.1	474	114	360	27.4
24.1	473	114	359	27.5
24.1	469	113	356	27.2
24.1	468	113	355	27.3
24.1	465	112	353	27.0
24.1	464	112	352	27.0
24.1	461	111	350	26.7
24.1	460	111	349	26.8
24.1	457	110	347	26.5
24.1	456	110	346	26.5
24.1	453	109	344	26.2
24.1	452	109	343	26.3
24.1	449	108	341	26.0
24.1	448	108	340	26.0
24.1	444	107	337	25.8
24.1	440	106	334	25.5
24.1	439	106	333	25.6
24.1	436	105	331	25.3
24.1	435	105	330	25.3
24.1	432	104	328	25.0
24.1	431	104	327	25.1
24.1	428	103	325	24.8
24.1	427	103	324	24.8
24.1	424	102	322	24.5
24.1	423	102	321	24.6
24.1	419	101	318	24.3
24.1	415	100	315	24.1
24.1	411	99	312	23.8
24.1	410	99	311	23.9
24.1	407	98	309	23.6
24.1	406	98	308	23.7
24.1	403	97	306	23.3
24.1	402	97	305	23.4
24.1	399	96	303	23.1
24.1	398	96	302	23.2
24.1	395	95	300	22.8

%E	M1	M2	DM	M*
24.1	394	95	299	22.9
24.1	390	94	296	22.7
24.1	386	93	293	22.4
24.1	382	92	290	22.2
24.1	381	92	289	22.2
24.1	378	91	287	21.9
24.1	377	91	286	22.0
24.1	374	90	284	21.7
24.1	373	90	283	21.7
24.1	370	89	281	21.4
24.1	369	89	280	21.5
24.1	365	88	277	21.2
24.1	361	87	274	21.0
24.1	357	86	271	20.7
24.1	353	85	268	20.5
24.1	352	85	267	20.5
24.1	349	84	265	20.2
24.1	348	84	264	20.3
24.1	345	83	262	20.0
24.1	344	83	261	20.0
24.1	340	82	258	19.8
24.1	336	81	255	19.5
24.1	332	80	252	19.3
24.1	328	79	249	19.0
24.1	324	78	246	18.8
24.1	323	78	245	18.8
24.1	320	77	243	18.5
24.1	319	77	242	18.6
24.1	316	76	240	18.3
24.1	315	76	239	18.3
24.1	311	75	236	18.1
24.1	307	74	233	17.8
24.1	303	73	230	17.6
24.1	299	72	227	17.3
24.1	295	71	224	17.1
24.1	291	70	221	16.8
24.1	290	70	220	16.9
24.1	286	69	217	16.6
24.1	282	68	214	16.4
24.1	278	67	211	16.1
24.1	274	66	208	15.9
24.1	270	65	205	15.6
24.1	266	64	202	15.4
24.1	261	63	198	15.2
24.1	257	62	195	15.0
24.1	253	61	192	14.7
24.1	249	60	189	14.5
24.1	245	59	186	14.2
24.1	241	58	183	14.0
24.1	237	57	180	13.7

%E	M1	M2	DM	M*
24.1	232	56	176	13.5
24.1	228	55	173	13.3
24.1	224	54	170	13.0
24.1	220	53	167	12.8
24.1	216	52	164	12.5
24.1	212	51	161	12.3
24.1	203	49	154	11.8
24.1	199	48	151	11.6
24.1	195	47	148	11.3
24.1	191	46	145	11.1
24.1	187	45	142	10.8
24.1	174	42	132	10.1
24.1	170	41	129	9.9
24.1	166	40	126	9.6
24.1	162	39	123	9.4
24.1	158	38	120	9.1
24.1	145	35	110	8.4
24.1	141	34	107	8.2
24.1	137	33	104	7.9
24.1	133	32	101	7.7
24.1	116	28	88	6.8
24.1	112	27	85	6.5
24.1	108	26	82	6.3
24.1	87	21	66	5.1
24.1	83	20	63	4.8
24.1	79	19	60	4.6
24.1	58	14	44	3.4
24.1	54	13	41	3.1
24.1	29	7	22	1.7
24.0	500	120	380	28.8
24.0	499	120	379	28.9
24.0	496	119	377	28.6
24.0	495	119	376	28.6
24.0	492	118	374	28.3
24.0	491	118	373	28.4
24.0	488	117	371	28.1
24.0	487	117	370	28.1
24.0	484	116	368	27.8
24.0	483	116	367	27.9
24.0	480	115	365	27.6
24.0	479	115	364	27.6
24.0	476	114	362	27.3
24.0	475	114	361	27.4
24.0	471	113	358	27.1
24.0	470	113	357	27.2
24.0	467	112	355	26.9
24.0	466	112	354	26.9
24.0	463	111	352	26.6
24.0	462	111	351	26.7
24.0	459	110	349	26.4

%E	M1	M2	DM	M*
24.0	458	110	348	26.4
24.0	455	109	346	26.1
24.0	454	109	345	26.2
24.0	450	108	342	25.9
24.0	446	107	339	25.7
24.0	445	107	338	25.7
24.0	442	106	336	25.4
24.0	441	106	335	25.5
24.0	438	105	333	25.2
24.0	437	105	332	25.2
24.0	434	104	330	24.9
24.0	433	104	329	25.0
24.0	430	103	327	24.7
24.0	429	103	326	24.7
24.0	425	102	323	24.5
24.0	421	101	320	24.2
24.0	420	101	319	24.3
24.0	417	100	317	24.0
24.0	416	100	316	24.0
24.0	413	99	314	23.7
24.0	412	99	313	23.8
24.0	409	98	311	23.5
24.0	408	98	310	23.5
24.0	405	97	308	23.2
24.0	404	97	307	23.3
24.0	400	96	304	23.0
24.0	396	95	301	22.8
24.0	392	94	298	22.5
24.0	391	94	297	22.6
24.0	388	93	295	22.3
24.0	387	93	294	22.3
24.0	384	92	292	22.0
24.0	383	92	291	22.1
24.0	379	91	288	21.8
24.0	375	90	285	21.6
24.0	371	89	282	21.4
24.0	367	88	279	21.1
24.0	366	88	278	21.2
24.0	363	87	276	20.9
24.0	362	87	275	20.9
24.0	359	86	273	20.6
24.0	358	86	272	20.7
24.0	354	85	269	20.4
24.0	350	84	266	20.2
24.0	346	83	263	19.9
24.0	342	82	260	19.7
24.0	341	82	259	19.7
24.0	338	81	257	19.4
24.0	337	81	256	19.5
24.0	334	80	254	19.2

%E	M1	M2	DM	M*
24.0	333	80	253	19.2
24.0	329	79	250	19.0
24.0	325	78	247	18.7
24.0	321	77	244	18.5
24.0	317	76	241	18.2
24.0	313	75	238	18.0
24.0	312	75	237	18.0
24.0	308	74	234	17.8
24.0	304	73	231	17.5
24.0	300	72	228	17.3
24.0	296	71	225	17.0
24.0	292	70	222	16.8
24.0	288	69	219	16.5
24.0	287	69	218	16.6
24.0	283	68	215	16.3
24.0	279	67	212	16.1
24.0	275	66	209	15.8
24.0	271	65	206	15.6
24.0	267	64	203	15.3
24.0	263	63	200	15.1
24.0	262	63	199	15.1
24.0	258	62	196	14.9
24.0	254	61	193	14.6
24.0	250	60	190	14.4
24.0	246	59	187	14.2
24.0	242	58	184	13.9
24.0	238	57	181	13.7
24.0	233	56	177	13.5
24.0	229	55	174	13.2
24.0	225	54	171	13.0
24.0	221	53	168	12.7
24.0	217	52	165	12.5
24.0	208	50	158	12.0
24.0	204	49	155	11.8
24.0	200	48	152	11.5
24.0	196	47	149	11.3
24.0	192	46	146	11.0
24.0	183	44	139	10.6
24.0	179	43	136	10.3
24.0	175	42	133	10.1
24.0	171	41	130	9.8
24.0	167	40	127	9.6
24.0	154	37	117	8.9
24.0	150	36	114	8.6
24.0	146	35	111	8.4
24.0	129	31	98	7.4
24.0	125	30	95	7.2
24.0	121	29	92	7.0
24.0	104	25	79	6.0
24.0	100	24	76	5.8

%E	M1	M2	DM	M*
24.0	96	23	73	5.5
24.0	75	18	57	4.3
24.0	50	12	38	2.9
24.0	25	6	19	1.4
23.9	498	119	379	28.4
23.9	497	119	378	28.5
23.9	494	118	376	28.2
23.9	493	118	375	28.2
23.9	490	117	373	27.9
23.9	489	117	372	28.0
23.9	486	116	370	27.7
23.9	485	116	369	27.7
23.9	482	115	367	27.4
23.9	481	115	366	27.5
23.9	477	114	363	27.2
23.9	473	113	360	27.0
23.9	472	113	359	27.1
23.9	469	112	357	26.7
23.9	468	112	356	26.8
23.9	465	111	354	26.5
23.9	464	111	353	26.6
23.9	461	110	351	26.2
23.9	460	110	350	26.3
23.9	457	109	348	26.0
23.9	456	109	347	26.1
23.9	452	108	344	25.8
23.9	451	108	343	25.9
23.9	448	107	341	25.6
23.9	447	107	340	25.6
23.9	444	106	338	25.3
23.9	443	106	337	25.4
23.9	440	105	335	25.1
23.9	439	105	334	25.1
23.9	436	104	332	24.6
23.9	435	104	331	24.9
23.9	431	103	328	24.6
23.9	427	102	325	24.4
23.9	426	102	324	24.4
23.9	423	101	322	24.1
23.9	422	101	321	24.2
23.9	419	100	319	23.9
23.9	418	100	318	23.9
23.9	415	99	316	23.6
23.9	414	99	315	23.7
23.9	410	98	312	23.4
23.9	406	97	309	23.2
23.9	402	96	306	22.9
23.9	401	96	305	23.0
23.9	398	95	303	22.7
23.9	397	95	302	22.7

%E	M1	M2	DM	M*
23.9	394	94	300	22.4
23.9	393	94	299	22.5
23.9	389	93	296	22.2
23.9	385	92	293	22.0
23.9	381	91	290	21.7
23.9	380	91	289	21.8
23.9	377	90	287	21.5
23.9	376	90	286	21.5
23.9	373	89	284	21.2
23.9	372	89	283	21.3
23.9	368	88	280	21.0
23.9	364	87	277	20.8
23.9	360	86	274	20.5
23.9	356	85	271	20.3
23.9	355	85	270	20.4
23.9	352	84	268	20.0
23.9	351	84	267	20.1
23.9	348	83	265	19.8
23.9	347	83	264	19.9
23.9	343	82	261	19.6
23.9	339	81	258	19.4
23.9	335	80	255	19.1
23.9	331	79	252	18.9
23.9	330	79	251	18.9
23.9	327	78	249	18.6
23.9	326	78	248	18.7
23.9	322	77	245	18.4
23.9	318	76	242	18.2
23.9	314	75	239	17.9
23.9	310	74	236	17.7
23.9	309	74	235	17.7
23.9	306	73	233	17.4
23.9	305	73	232	17.5
23.9	301	72	229	17.2
23.9	297	71	226	17.0
23.9	293	70	223	16.7
23.9	289	69	220	16.5
23.9	285	68	217	16.2
23.9	284	68	216	16.3
23.9	280	67	213	16.0
23.9	276	66	210	15.8
23.9	272	65	207	15.5
23.9	268	64	204	15.3
23.9	264	63	201	15.0
23.9	259	62	197	14.8
23.9	255	61	194	14.6
23.9	251	60	191	14.3
23.9	247	59	188	14.1
23.9	243	58	185	13.8
23.9	234	56	178	13.4

%E	M1	M2	DM	M*
23.9	230	55	175	13.2
23.9	226	54	172	12.9
23.9	222	53	169	12.7
23.9	218	52	166	12.4
23.9	213	51	162	12.2
23.9	209	50	159	12.0
23.9	205	49	156	11.7
23.9	201	48	153	11.5
23.9	197	47	150	11.2
23.9	188	45	143	10.8
23.9	184	44	140	10.5
23.9	180	43	137	10.3
23.9	176	42	134	10.0
23.9	163	39	124	9.3
23.9	159	38	121	9.1
23.9	155	37	118	8.8
23.9	142	34	108	8.1
23.9	138	33	105	7.9
23.9	134	32	102	7.6
23.9	117	28	89	6.7
23.9	113	27	86	6.5
23.9	109	26	83	6.2
23.9	92	22	70	5.3
23.9	88	21	67	5.0
23.9	71	17	54	4.1
23.9	67	16	51	3.8
23.9	46	11	35	2.6
23.8	500	119	381	28.3
23.8	499	119	380	28.4
23.8	496	118	378	28.1
23.8	495	118	377	28.1
23.8	492	117	375	27.8
23.8	491	117	374	27.9
23.8	488	116	372	27.6
23.8	487	116	371	27.6
23.8	484	115	369	27.3
23.8	483	115	368	27.4
23.8	480	114	366	27.1
23.8	479	114	365	27.1
23.8	478	114	364	27.2
23.8	475	113	362	26.9
23.8	474	113	361	26.9
23.8	471	112	359	26.6
23.8	470	112	358	26.7
23.8	467	111	356	26.4
23.8	466	111	355	26.4
23.8	463	110	353	26.1
23.8	462	110	352	26.2
23.8	458	109	349	25.9
23.8	454	108	346	25.7

%E	M1	M2	DM	M*
23.8	453	108	345	25.7
23.8	450	107	343	25.4
23.8	449	107	342	25.5
23.8	446	106	340	25.2
23.8	445	106	339	25.2
23.8	442	105	337	24.9
23.8	441	105	336	25.0
23.8	437	104	333	24.8
23.8	433	103	330	24.5
23.8	432	103	329	24.6
23.8	429	102	327	24.3
23.8	428	102	326	24.3
23.8	425	101	324	24.0
23.8	424	101	323	24.1
23.8	421	100	321	23.8
23.8	420	100	320	23.8
23.8	416	99	317	23.6
23.8	412	98	314	23.3
23.8	411	98	313	23.4
23.8	408	97	311	23.1
23.8	407	97	310	23.1
23.8	404	96	308	22.8
23.8	403	96	307	22.9
23.8	400	95	305	22.6
23.8	399	95	304	22.6
23.8	395	94	301	22.4
23.8	391	93	298	22.1
23.8	390	93	297	22.2
23.8	387	92	295	21.9
23.8	386	92	294	21.9
23.8	383	91	292	21.6
23.8	382	91	291	21.7
23.8	378	90	288	21.4
23.8	374	89	285	21.2
23.8	370	88	282	20.9
23.8	369	88	281	21.0
23.8	366	87	279	20.7
23.8	365	87	278	20.7
23.8	362	86	276	20.4
23.8	361	86	275	20.5
23.8	357	85	272	20.2
23.8	353	84	269	20.0
23.8	349	83	266	19.7
23.8	345	82	263	19.5
23.8	344	82	262	19.5
23.8	341	81	260	19.2
23.8	340	81	259	19.3
23.8	336	80	256	19.0
23.8	332	79	253	18.8
23.8	328	78	250	18.5

%E	M1	M2	DM	M*
23.8	324	77	247	18.3
23.8	323	77	246	18.4
23.8	320	76	244	18.0
23.8	319	76	243	18.1
23.8	315	75	240	17.9
23.8	311	74	237	17.6
23.8	307	73	234	17.4
23.8	303	72	231	17.1
23.8	302	72	230	17.2
23.8	298	71	227	16.9
23.8	294	70	224	16.7
23.8	290	69	221	16.4
23.8	286	68	218	16.2
23.8	282	67	215	15.9
23.8	281	67	214	16.0
23.8	277	66	211	15.7
23.8	273	65	208	15.5
23.8	269	64	205	15.2
23.8	265	63	202	15.0
23.8	261	62	199	14.7
23.8	260	62	198	14.8
23.8	256	61	195	14.5
23.8	252	60	192	14.3
23.8	248	59	189	14.0
23.8	244	58	186	13.8
23.8	240	57	183	13.5
23.8	239	57	182	13.6
23.8	235	56	179	13.3
23.8	231	55	176	13.1
23.8	227	54	173	12.8
23.8	223	53	170	12.6
23.8	214	51	163	12.2
23.8	210	50	160	11.9
23.8	206	49	157	11.7
23.8	202	48	154	11.4
23.8	193	46	147	11.0
23.8	189	45	144	10.7
23.8	185	44	141	10.5
23.8	181	43	138	10.2
23.8	172	41	131	9.8
23.8	168	40	128	9.5
23.8	164	39	125	9.3
23.8	160	38	122	9.0
23.8	151	36	115	8.6
23.8	147	35	112	8.3
23.8	143	34	109	8.1
23.8	130	31	99	7.4
23.8	126	30	96	7.1
23.8	122	29	93	6.9
23.8	105	25	80	6.0

%E	M1	M2	DM	M*
23.8	101	24	77	5.7
23.8	84	20	64	4.8
23.8	80	19	61	4.5
23.8	63	15	48	3.6
23.8	42	10	32	2.4
23.8	21	5	16	1.2
23.7	498	118	380	28.0
23.7	497	118	379	28.0
23.7	494	117	377	27.7
23.7	493	117	376	27.8
23.7	490	116	374	27.5
23.7	489	116	373	27.5
23.7	486	115	371	27.2
23.7	485	115	370	27.3
23.7	482	114	368	27.0
23.7	481	114	367	27.0
23.7	477	113	364	26.8
23.7	476	113	363	26.8
23.7	473	112	361	26.5
23.7	472	112	360	26.6
23.7	469	111	358	26.3
23.7	468	111	357	26.3
23.7	465	110	355	26.0
23.7	464	110	354	26.1
23.7	460	109	351	25.8
23.7	459	109	350	25.9
23.7	456	108	348	25.6
23.7	455	108	347	25.6
23.7	452	107	345	25.3
23.7	451	107	344	25.4
23.7	448	106	342	25.1
23.7	447	106	341	25.1
23.7	443	105	338	24.9
23.7	439	104	335	24.6
23.7	438	104	334	24.7
23.7	435	103	332	24.4
23.7	434	103	331	24.4
23.7	431	102	329	24.1
23.7	430	102	328	24.2
23.7	427	101	326	23.9
23.7	426	101	325	23.9
23.7	422	100	322	23.7
23.7	418	99	319	23.4
23.7	417	99	318	23.5
23.7	414	98	316	23.2
23.7	413	98	315	23.3
23.7	410	97	313	22.9
23.7	409	97	312	23.0
23.7	405	96	309	22.8
23.7	401	95	306	22.5

%E	M1	M2	DM	M*
23.7	397	94	303	22.3
23.7	396	94	302	22.3
23.7	393	93	300	22.0
23.7	392	93	299	22.1
23.7	389	92	297	21.8
23.7	388	92	296	21.8
23.7	384	91	293	21.6
23.7	380	90	290	21.3
23.7	379	90	289	21.4
23.7	376	89	287	21.1
23.7	375	89	286	21.1
23.7	372	88	284	20.8
23.7	371	88	283	20.9
23.7	367	87	280	20.6
23.7	363	86	277	20.4
23.7	359	85	274	20.1
23.7	358	85	273	20.2
23.7	355	84	271	19.9
23.7	354	84	270	19.9
23.7	350	83	267	19.7
23.7	346	82	264	19.4
23.7	342	81	261	19.2
23.7	338	80	258	18.9
23.7	337	80	257	19.0
23.7	334	79	255	18.7
23.7	333	79	254	18.7
23.7	329	78	251	18.5
23.7	325	77	248	18.2
23.7	321	76	245	18.0
23.7	317	75	242	17.7
23.7	316	75	241	17.8
23.7	312	74	238	17.6
23.7	308	73	235	17.3
23.7	304	72	232	17.1
23.7	300	71	229	16.8
23.7	299	71	228	16.9
23.7	295	70	225	16.6
23.7	291	69	222	16.4
23.7	287	68	219	16.1
23.7	283	67	216	15.9
23.7	279	66	213	15.6
23.7	278	66	212	15.7
23.7	274	65	209	15.4
23.7	270	64	206	15.2
23.7	266	63	203	14.9
23.7	262	62	200	14.7
23.7	257	61	196	14.5
23.7	253	60	193	14.2
23.7	249	59	190	14.0
23.7	245	58	187	13.7

%E	M1	M2	DM	M*
23.7	241	57	184	13.5
23.7	236	56	180	13.3
23.7	232	55	177	13.0
23.7	228	54	174	12.8
23.7	224	53	171	12.5
23.7	219	52	167	12.3
23.7	215	51	164	12.1
23.7	211	50	161	11.8
23.7	207	49	158	11.6
23.7	198	47	151	11.2
23.7	194	46	148	10.9
23.7	190	45	145	10.7
23.7	186	44	142	10.4
23.7	177	42	135	10.0
23.7	173	41	132	9.7
23.7	169	40	129	9.5
23.7	156	37	119	8.8
23.7	152	36	116	8.5
23.7	139	33	106	7.8
23.7	135	32	103	7.6
23.7	131	31	100	7.3
23.7	118	28	90	6.6
23.7	114	27	87	6.4
23.7	97	23	74	5.5
23.7	93	22	71	5.2
23.7	76	18	58	4.3
23.7	59	14	45	3.3
23.7	38	9	29	2.1
23.6	500	118	382	27.8
23.6	499	118	381	27.9
23.6	496	117	379	27.6
23.6	495	117	378	27.7
23.6	492	116	376	27.3
23.6	491	116	375	27.4
23.6	488	115	373	27.1
23.6	487	115	372	27.2
23.6	484	114	370	26.9
23.6	483	114	369	26.9
23.6	479	113	366	26.7
23.6	478	113	365	26.7
23.6	475	112	363	26.4
23.6	474	112	362	26.5
23.6	471	111	360	26.2
23.6	470	111	359	26.2
23.6	467	110	357	25.9
23.6	466	110	356	26.0
23.6	462	109	353	25.7
23.6	461	109	352	25.8
23.6	458	108	350	25.5
23.6	457	108	349	25.5

%E	M1	M2	DM	M*
23.6	454	107	347	25.2
23.6	453	107	346	25.3
23.6	450	106	344	25.0
23.6	449	106	343	25.0
23.6	445	105	340	24.8
23.6	444	105	339	24.8
23.6	441	104	337	24.5
23.6	440	104	336	24.6
23.6	437	103	334	24.3
23.6	436	103	333	24.3
23.6	433	102	331	24.0
23.6	432	102	330	24.1
23.6	428	101	327	23.8
23.6	424	100	324	23.6
23.6	423	100	323	23.6
23.6	420	99	321	23.3
23.6	419	99	320	23.4
23.6	416	98	318	23.1
23.6	415	98	317	23.1
23.6	411	97	314	22.9
23.6	407	96	311	22.6
23.6	406	96	310	22.7
23.6	403	95	308	22.4
23.6	402	95	307	22.5
23.6	399	94	305	22.1
23.6	398	94	304	22.2
23.6	394	93	301	22.0
23.6	390	92	298	21.7
23.6	386	91	295	21.5
23.6	385	91	294	21.5
23.6	382	90	292	21.2
23.6	381	90	291	21.3
23.6	377	89	288	21.0
23.6	373	88	285	20.8
23.6	369	87	282	20.5
23.6	368	87	281	20.6
23.6	365	86	279	20.3
23.6	364	86	278	20.3
23.6	360	85	275	20.1
23.6	356	84	272	19.8
23.6	352	83	269	19.6
23.6	351	83	268	19.6
23.6	348	82	266	19.3
23.6	347	82	265	19.4
23.6	343	81	262	19.1
23.6	339	80	259	18.9
23.6	335	79	256	18.6
23.6	331	78	253	18.4
23.6	330	78	252	18.4
23.6	326	77	249	18.2

%E	M1	M2	DM	M*
23.6	322	76	246	17.9
23.6	318	75	243	17.7
23.6	314	74	240	17.4
23.6	313	74	239	17.5
23.6	309	73	236	17.2
23.6	305	72	233	17.0
23.6	301	71	230	16.7
23.6	297	70	227	16.5
23.6	296	70	226	16.6
23.6	292	69	223	16.3
23.6	288	68	220	16.1
23.6	284	67	217	15.8
23.6	280	66	214	15.6
23.6	276	65	211	15.3
23.6	275	65	210	15.4
23.6	271	64	207	15.1
23.6	267	63	204	14.9
23.6	263	62	201	14.6
23.6	259	61	198	14.4
23.6	258	61	197	14.4
23.6	254	60	194	14.2
23.6	250	59	191	13.9
23.6	246	58	188	13.7
23.6	242	57	185	13.4
23.6	237	56	181	13.2
23.6	233	55	178	13.0
23.6	229	54	175	12.7
23.6	225	53	172	12.5
23.6	220	52	168	12.3
23.6	216	51	165	12.0
23.6	212	50	162	11.8
23.6	208	49	159	11.5
23.6	203	48	155	11.3
23.6	199	47	152	11.1
23.6	195	46	149	10.9
23.6	191	45	146	10.6
23.6	182	43	139	10.2
23.6	178	42	136	9.9
23.6	174	41	133	9.7
23.6	165	39	126	9.2
23.6	161	38	123	9.0
23.6	157	37	120	8.7
23.6	148	35	113	8.3
23.6	144	34	110	8.0
23.6	140	33	107	7.8
23.6	127	30	97	7.1
23.6	123	29	94	6.8
23.6	110	26	84	6.1
23.6	106	25	81	5.9
23.6	89	21	68	5.0

%E	M1	M2	DM	M*
23.6	72	17	55	4.0
23.6	55	13	42	3.1
23.5	498	117	381	27.5
23.5	497	117	380	27.5
23.5	494	116	378	27.2
23.5	493	116	377	27.3
23.5	490	115	375	27.0
23.5	489	115	374	27.0
23.5	486	114	372	26.7
23.5	485	114	371	26.8
23.5	481	113	368	26.5
23.5	480	113	367	26.6
23.5	477	112	365	26.3
23.5	476	112	364	26.4
23.5	473	111	362	26.0
23.5	472	111	361	26.1
23.5	469	110	359	25.8
23.5	468	110	358	25.9
23.5	464	109	355	25.6
23.5	463	109	354	25.7
23.5	460	108	352	25.4
23.5	459	108	351	25.4
23.5	456	107	349	25.1
23.5	455	107	348	25.2
23.5	452	106	346	24.9
23.5	451	106	345	24.9
23.5	447	105	342	24.7
23.5	446	105	341	24.7
23.5	443	104	339	24.4
23.5	442	104	338	24.5
23.5	439	103	336	24.2
23.5	438	103	335	24.2
23.5	434	102	332	24.0
23.5	430	101	329	23.7
23.5	429	101	328	23.8
23.5	426	100	326	23.5
23.5	425	100	325	23.5
23.5	422	99	323	23.2
23.5	421	99	322	23.3
23.5	417	98	319	23.0
23.5	413	97	316	22.8
23.5	412	97	315	22.8
23.5	409	96	313	22.5
23.5	408	96	312	22.6
23.5	405	95	310	22.3
23.5	404	95	309	22.3
23.5	400	94	306	22.1
23.5	396	93	303	21.8
23.5	395	93	302	21.9
23.5	392	92	300	21.6

%E	M1	M2	DM	M*
23.5	391	92	299	21.6
23.5	388	91	297	21.3
23.5	387	91	296	21.4
23.5	383	90	293	21.1
23.5	379	89	290	20.9
23.5	378	89	289	21.0
23.5	375	88	287	20.7
23.5	374	88	286	20.7
23.5	371	87	284	20.4
23.5	370	87	283	20.5
23.5	366	86	280	20.2
23.5	362	85	277	20.0
23.5	361	85	276	20.0
23.5	358	84	274	19.7
23.5	357	84	273	19.8
23.5	353	83	270	19.5
23.5	349	82	267	19.3
23.5	345	81	264	19.0
23.5	344	81	263	19.1
23.5	341	80	261	18.8
23.5	340	80	260	18.8
23.5	336	79	257	18.6
23.5	332	78	254	18.3
23.5	328	77	251	18.1
23.5	327	77	250	18.1
23.5	324	76	248	17.8
23.5	323	76	247	17.9
23.5	319	75	244	17.6
23.5	315	74	241	17.4
23.5	311	73	238	17.1
23.5	310	73	237	17.2
23.5	307	72	235	16.9
23.5	306	72	234	16.9
23.5	302	71	231	16.7
23.5	298	70	228	16.4
23.5	294	69	225	16.2
23.5	293	69	224	16.2
23.5	289	68	221	16.0
23.5	285	67	218	15.8
23.5	281	66	215	15.5
23.5	277	65	212	15.3
23.5	272	64	208	15.1
23.5	268	63	205	14.8
23.5	264	62	202	14.6
23.5	260	61	199	14.3
23.5	255	60	195	14.1
23.5	251	59	192	13.9
23.5	247	58	189	13.6
23.5	243	57	186	13.4
23.5	238	56	182	13.2

%E	M1	M2	DM	M*
23.5	234	55	179	12.9
23.5	230	54	176	12.7
23.5	226	53	173	12.4
23.5	221	52	169	12.2
23.5	217	51	166	12.0
23.5	213	50	163	11.7
23.5	204	48	156	11.3
23.5	200	47	153	11.0
23.5	196	46	150	10.8
23.5	187	44	143	10.4
23.5	183	43	140	10.1
23.5	179	42	137	9.9
23.5	170	40	130	9.4
23.5	166	39	127	9.2
23.5	162	38	124	8.9
23.5	153	36	117	8.5
23.5	149	35	114	8.2
23.5	136	32	104	7.5
23.5	132	31	101	7.3
23.5	119	28	91	6.6
23.5	115	27	88	6.3
23.5	102	24	78	5.6
23.5	98	23	75	5.4
23.5	85	20	65	4.7
23.5	81	19	62	4.5
23.5	68	16	52	3.8
23.5	51	12	39	2.8
23.5	34	8	26	1.9
23.5	17	4	13	0.9
23.4	500	117	383	27.4
23.4	499	117	382	27.4
23.4	496	116	380	27.1
23.4	495	116	379	27.2
23.4	492	115	377	26.9
23.4	491	115	376	26.9
23.4	488	114	374	26.6
23.4	487	114	373	26.7
23.4	483	113	370	26.4
23.4	482	113	369	26.5
23.4	479	112	367	26.2
23.4	478	112	366	26.2
23.4	475	111	364	25.9
23.4	474	111	363	26.0
23.4	471	110	361	25.7
23.4	470	110	360	25.7
23.4	466	109	357	25.5
23.4	465	109	356	25.6
23.4	462	108	354	25.2
23.4	461	108	353	25.3
23.4	458	107	351	25.0

%E	M1	M2	DM	M*
23.4	457	107	350	25.1
23.4	453	106	347	24.8
23.4	449	105	344	24.6
23.4	448	105	343	24.6
23.4	445	104	341	24.3
23.4	444	104	340	24.4
23.4	441	103	338	24.1
23.4	440	103	337	24.1
23.4	436	102	334	23.9
23.4	435	102	333	23.9
23.4	432	101	331	23.6
23.4	431	101	330	23.7
23.4	428	100	328	23.4
23.4	427	100	327	23.4
23.4	423	99	324	23.2
23.4	419	98	321	22.9
23.4	418	98	320	23.0
23.4	415	97	318	22.7
23.4	414	97	317	22.7
23.4	411	96	315	22.4
23.4	410	96	314	22.5
23.4	406	95	311	22.2
23.4	402	94	308	22.0
23.4	401	94	307	22.0
23.4	398	93	305	21.7
23.4	397	93	304	21.8
23.4	394	92	302	21.5
23.4	393	92	301	21.5
23.4	389	91	298	21.3
23.4	385	90	295	21.0
23.4	384	90	294	21.1
23.4	381	89	292	20.8
23.4	380	89	291	20.8
23.4	376	88	288	20.6
23.4	372	87	285	20.3
23.4	368	86	282	20.1
23.4	367	86	281	20.2
23.4	364	85	279	19.8
23.4	363	85	278	19.9
23.4	359	84	275	19.7
23.4	355	83	272	19.4
23.4	354	83	271	19.5
23.4	351	82	269	19.2
23.4	350	82	268	19.2
23.4	346	81	265	19.0
23.4	342	80	262	18.7
23.4	338	79	259	18.5
23.4	337	79	258	18.5
23.4	334	78	256	18.2
23.4	333	78	255	18.3

%E	M1	M2	DM	M*
23.4	329	77	252	18.0
23.4	325	76	249	17.8
23.4	321	75	246	17.5
23.4	320	75	245	17.6
23.4	316	74	242	17.3
23.4	312	73	239	17.1
23.4	308	72	236	16.8
23.4	304	71	233	16.6
23.4	303	71	232	16.6
23.4	299	70	229	16.4
23.4	295	69	226	16.1
23.4	291	68	223	15.9
23.4	290	68	222	15.9
23.4	286	67	219	15.7
23.4	282	66	216	15.4
23.4	278	65	213	15.2
23.4	274	64	210	14.9
23.4	273	64	209	15.0
23.4	269	63	206	14.8
23.4	265	62	203	14.5
23.4	261	61	200	14.3
23.4	256	60	196	14.1
23.4	252	59	193	13.8
23.4	248	58	190	13.6
23.4	244	57	187	13.3
23.4	239	56	183	13.1
23.4	235	55	180	12.9
23.4	231	54	177	12.6
23.4	222	52	170	12.2
23.4	218	51	167	11.9
23.4	214	50	164	11.7
23.4	209	49	160	11.5
23.4	205	48	157	11.2
23.4	201	47	154	11.0
23.4	197	46	151	10.7
23.4	192	45	147	10.5
23.4	188	44	144	10.3
23.4	184	43	141	10.0
23.4	175	41	134	9.6
23.4	171	40	131	9.4
23.4	167	39	128	9.1
23.4	158	37	121	8.7
23.4	154	36	118	8.4
23.4	145	34	111	8.0
23.4	141	33	108	7.7
23.4	137	32	105	7.5
23.4	128	30	98	7.0
23.4	124	29	95	6.8
23.4	111	26	85	6.1
23.4	107	25	82	5.8

%E	M1	M2	DM	M*
23.4	94	22	72	5.1
23.4	77	18	59	4.2
23.4	64	15	49	3.5
23.4	47	11	36	2.6
23.3	498	116	382	27.0
23.3	497	116	381	27.1
23.3	494	115	379	26.8
23.3	493	115	378	26.8
23.3	490	114	376	26.5
23.3	489	114	375	26.6
23.3	486	113	373	26.3
23.3	485	113	372	26.3
23.3	484	113	371	26.4
23.3	481	112	369	26.1
23.3	480	112	368	26.1
23.3	477	111	366	25.8
23.3	476	111	365	25.9
23.3	473	110	363	25.6
23.3	472	110	362	25.6
23.3	468	109	359	25.4
23.3	467	109	358	25.4
23.3	464	108	356	25.1
23.3	463	108	355	25.2
23.3	460	107	353	24.9
23.3	459	107	352	24.9
23.3	455	106	349	24.7
23.3	454	106	348	24.7
23.3	451	105	346	24.4
23.3	450	105	345	24.5
23.3	447	104	343	24.2
23.3	446	104	342	24.3
23.3	443	103	340	23.9
23.3	442	103	339	24.0
23.3	438	102	336	23.8
23.3	437	102	335	23.8
23.3	434	101	333	23.5
23.3	433	101	332	23.6
23.3	430	100	330	23.3
23.3	429	100	329	23.3
23.3	425	99	326	23.1
23.3	424	99	325	23.1
23.3	421	98	323	22.8
23.3	420	98	322	22.9
23.3	417	97	320	22.6
23.3	416	97	319	22.6
23.3	412	96	316	22.4
23.3	408	95	313	22.1
23.3	407	95	312	22.2
23.3	404	94	310	21.9
23.3	403	94	309	21.9
23.3	400	93	307	21.6
23.3	399	93	306	21.7
23.3	395	92	303	21.4
23.3	391	91	300	21.2
23.3	390	91	299	21.2
23.3	387	90	297	20.9
23.3	386	90	296	21.0
23.3	382	89	293	20.7
23.3	378	88	290	20.5
23.3	377	88	289	20.5
23.3	374	87	287	20.2
23.3	373	87	286	20.3
23.3	369	86	283	20.0
23.3	365	85	280	19.8
23.3	361	84	277	19.5
23.3	360	84	276	19.6
23.3	356	83	273	19.4
23.3	352	82	270	19.1
23.3	348	81	267	18.9
23.3	347	81	266	18.9
23.3	344	80	264	18.6
23.3	343	80	263	18.7
23.3	339	79	260	18.4
23.3	335	78	257	18.2
23.3	331	77	254	17.9
23.3	330	77	253	18.0
23.3	326	76	250	17.7
23.3	322	75	247	17.5
23.3	318	74	244	17.2
23.3	317	74	243	17.3
23.3	313	73	240	17.0
23.3	309	72	237	16.8
23.3	305	71	234	16.5
23.3	301	70	231	16.3
23.3	300	70	230	16.3
23.3	296	69	227	16.1
23.3	292	68	224	15.8
23.3	288	67	221	15.6
23.3	287	67	220	15.6
23.3	283	66	217	15.4
23.3	279	65	214	15.1
23.3	275	64	211	14.9
23.3	270	63	207	14.7
23.3	266	62	204	14.5
23.3	262	61	201	14.2
23.3	258	60	198	14.0
23.3	257	60	197	14.0
23.3	253	59	194	13.8
23.3	249	58	191	13.5
23.3	245	57	188	13.3
23.3	240	56	184	13.1
23.3	236	55	181	12.8
23.3	232	54	178	12.6
23.3	227	53	174	12.4
23.3	223	52	171	12.1
23.3	219	51	168	11.9
23.3	215	50	165	11.6
23.3	210	49	161	11.4
23.3	206	48	158	11.2
23.3	202	47	155	10.9
23.3	193	45	148	10.5
23.3	189	44	145	10.2
23.3	180	42	138	9.8
23.3	176	41	135	9.6
23.3	172	40	132	9.3
23.3	163	38	125	8.9
23.3	159	37	122	8.6
23.3	150	35	115	8.2
23.3	146	34	112	7.9
23.3	133	31	102	7.2
23.3	129	30	99	7.0
23.3	120	28	92	6.5
23.3	116	27	89	6.3
23.3	103	24	79	5.6
23.3	90	21	69	4.9
23.3	86	20	66	4.7
23.3	73	17	56	4.0
23.3	60	14	46	3.3
23.3	43	10	33	2.3
23.3	30	7	23	1.6
23.2	500	116	384	26.9
23.2	499	116	383	27.0
23.2	496	115	381	26.7
23.2	495	115	380	26.7
23.2	492	114	378	26.4
23.2	491	114	377	26.5
23.2	488	113	375	26.2
23.2	487	113	374	26.2
23.2	483	112	371	26.0
23.2	482	112	370	26.0
23.2	479	111	368	25.7
23.2	478	111	367	25.8
23.2	475	110	365	25.5
23.2	474	110	364	25.5
23.2	470	109	361	25.3
23.2	469	109	360	25.3
23.2	466	108	358	25.0
23.2	465	108	357	25.1
23.2	462	107	355	24.8
23.2	461	107	354	24.8
23.2	457	106	351	24.6
23.2	456	106	350	24.6
23.2	453	105	348	24.3
23.2	452	105	347	24.4
23.2	449	104	345	24.1
23.2	448	104	344	24.1
23.2	444	103	341	23.9
23.2	440	102	338	23.6
23.2	439	102	337	23.7
23.2	436	101	335	23.4
23.2	435	101	334	23.5
23.2	431	100	331	23.2
23.2	427	99	328	23.0
23.2	426	99	327	23.0
23.2	423	98	325	22.7
23.2	422	98	324	22.8
23.2	419	97	322	22.5
23.2	418	97	321	22.5
23.2	414	96	318	22.3
23.2	413	96	317	22.3
23.2	410	95	315	22.0
23.2	409	95	314	22.1
23.2	406	94	312	21.8
23.2	405	94	311	21.8
23.2	401	93	308	21.6
23.2	397	92	305	21.3
23.2	396	92	304	21.4
23.2	393	91	302	21.1
23.2	392	91	301	21.1
23.2	388	90	298	20.9
23.2	384	89	295	20.6
23.2	383	89	294	20.7
23.2	380	88	292	20.4
23.2	379	88	291	20.4
23.2	375	87	288	20.2
23.2	371	86	285	19.9
23.2	370	86	284	20.0
23.2	367	85	282	19.7
23.2	366	85	281	19.7
23.2	362	84	278	19.5
23.2	358	83	275	19.2
23.2	357	83	274	19.3
23.2	354	82	272	19.0
23.2	353	82	271	19.0
23.2	349	81	268	18.8
23.2	345	80	265	18.6
23.2	341	79	262	18.3
23.2	340	79	261	18.4
23.2	336	78	258	18.1
23.2	332	77	255	17.9
23.2	328	76	252	17.6
23.2	327	76	251	17.7
23.2	323	75	248	17.4
23.2	319	74	245	17.2
23.2	315	73	242	16.9
23.2	314	73	241	17.0
23.2	311	72	239	16.7
23.2	310	72	238	16.7
23.2	306	71	235	16.5
23.2	302	70	232	16.2
23.2	298	69	229	16.0
23.2	297	69	228	16.0
23.2	293	68	225	15.8
23.2	289	67	222	15.5
23.2	285	66	219	15.3
23.2	284	66	218	15.3
23.2	280	65	215	15.1
23.2	276	64	212	14.8
23.2	272	63	209	14.6
23.2	271	63	208	14.6
23.2	267	62	205	14.4
23.2	263	61	202	14.1
23.2	259	60	199	13.9
23.2	254	59	195	13.7
23.2	250	58	192	13.5
23.2	246	57	189	13.2
23.2	241	56	185	13.0
23.2	237	55	182	12.8
23.2	233	54	179	12.5
23.2	228	53	175	12.3
23.2	224	52	172	12.1
23.2	220	51	169	11.8
23.2	211	49	162	11.4
23.2	207	48	159	11.1
23.2	203	47	156	10.9
23.2	198	46	152	10.7
23.2	194	45	149	10.4
23.2	190	44	146	10.2
23.2	185	43	142	10.0
23.2	181	42	139	9.7
23.2	177	41	136	9.5
23.2	168	39	129	9.1
23.2	164	38	126	8.8
23.2	155	36	119	8.4
23.2	151	35	116	8.1
23.2	142	33	109	7.7
23.2	138	32	106	7.4
23.2	125	29	96	6.7
23.2	112	26	86	6.0
23.2	99	23	76	5.3

%E	M1	M2	DM	M*
23.2	95	22	73	5.1
23.2	82	19	63	4.4
23.2	69	16	53	3.7
23.2	56	13	43	3.0
23.1	498	115	383	26.6
23.1	497	115	382	26.6
23.1	494	114	380	26.3
23.1	493	114	379	26.4
23.1	490	113	377	26.1
23.1	489	113	376	26.1
23.1	485	112	373	25.9
23.1	484	112	372	25.9
23.1	481	111	370	25.6
23.1	480	111	369	25.7
23.1	477	110	367	25.4
23.1	476	110	366	25.4
23.1	472	109	363	25.2
23.1	471	109	362	25.2
23.1	468	108	360	24.9
23.1	467	108	359	25.0
23.1	464	107	357	24.7
23.1	463	107	356	24.7
23.1	459	106	353	24.5
23.1	458	106	352	24.5
23.1	455	105	350	24.2
23.1	454	105	349	24.3
23.1	451	104	347	24.0
23.1	450	104	346	24.0
23.1	446	103	343	23.8
23.1	445	103	342	23.8
23.1	442	102	340	23.5
23.1	441	102	339	23.6
23.1	438	101	337	23.3
23.1	437	101	336	23.3
23.1	433	100	333	23.1
23.1	432	100	332	23.1
23.1	429	99	330	22.8
23.1	428	99	329	22.9
23.1	425	98	327	22.6
23.1	424	98	326	22.7
23.1	420	97	323	22.4
23.1	416	96	320	22.2
23.1	415	96	319	22.2
23.1	412	95	317	21.9
23.1	411	95	316	22.0
23.1	407	94	313	21.7
23.1	403	93	310	21.5
23.1	402	93	309	21.5
23.1	399	92	307	21.2
23.1	398	92	306	21.3

%E	M1	M2	DM	M*
23.1	394	91	303	21.0
23.1	390	90	300	20.8
23.1	389	90	299	20.8
23.1	386	89	297	20.5
23.1	385	89	296	20.6
23.1	381	88	293	20.3
23.1	377	87	290	20.1
23.1	376	87	289	20.1
23.1	373	86	287	19.8
23.1	372	86	286	19.9
23.1	368	85	283	19.6
23.1	364	84	280	19.4
23.1	363	84	279	19.4
23.1	360	83	277	19.1
23.1	359	83	276	19.2
23.1	355	82	273	18.9
23.1	351	81	270	18.7
23.1	350	81	269	18.7
23.1	347	80	267	18.4
23.1	346	80	266	18.5
23.1	342	79	263	18.2
23.1	338	78	260	18.0
23.1	337	78	259	18.1
23.1	334	77	257	17.8
23.1	333	77	256	17.8
23.1	329	76	253	17.6
23.1	325	75	250	17.3
23.1	324	75	249	17.4
23.1	321	74	247	17.1
23.1	320	74	246	17.1
23.1	316	73	243	16.9
23.1	312	72	240	16.6
23.1	308	71	237	16.4
23.1	307	71	236	16.4
23.1	303	70	233	16.2
23.1	299	69	230	15.9
23.1	295	68	227	15.7
23.1	294	68	226	15.7
23.1	290	67	223	15.5
23.1	286	66	220	15.2
23.1	282	65	217	15.0
23.1	281	65	216	15.0
23.1	277	64	213	14.8
23.1	273	63	210	14.5
23.1	268	62	206	14.3
23.1	264	61	203	14.1
23.1	260	60	200	13.8
23.1	255	59	196	13.7
23.1	251	58	193	13.4
23.1	247	57	190	13.2

%E	M1	M2	DM	M*
23.1	242	56	186	13.0
23.1	238	55	183	12.7
23.1	234	54	180	12.5
23.1	229	53	176	12.3
23.1	225	52	173	12.0
23.1	221	51	170	11.8
23.1	216	50	166	11.6
23.1	212	49	163	11.3
23.1	208	48	160	11.1
23.1	199	46	153	10.6
23.1	195	45	150	10.4
23.1	186	43	143	9.9
23.1	182	42	140	9.7
23.1	173	40	133	9.2
23.1	169	39	130	9.0
23.1	160	37	123	8.6
23.1	156	36	120	8.3
23.1	147	34	113	7.9
23.1	143	33	110	7.6
23.1	134	31	103	7.2
23.1	130	30	100	6.9
23.1	121	28	93	6.5
23.1	117	27	90	6.2
23.1	108	25	83	5.8
23.1	104	24	80	5.5
23.1	91	21	70	4.8
23.1	78	18	60	4.2
23.1	65	15	50	3.5
23.1	52	12	40	2.8
23.1	39	9	30	2.1
23.1	26	6	20	1.4
23.1	13	3	10	0.7
23.0	500	115	385	26.4
23.0	499	115	384	26.5
23.0	496	114	382	26.2
23.0	495	114	381	26.3
23.0	492	113	379	26.0
23.0	491	113	378	26.0
23.0	488	112	376	25.7
23.0	487	112	375	25.8
23.0	486	112	374	25.8
23.0	483	111	372	25.5
23.0	482	111	371	25.6
23.0	479	110	369	25.3
23.0	478	110	368	25.3
23.0	474	109	365	25.1
23.0	473	109	364	25.1
23.0	470	108	362	24.8
23.0	469	108	361	24.9
23.0	466	107	359	24.6

%E	M1	M2	DM	M*
23.0	465	107	358	24.6
23.0	461	106	355	24.4
23.0	460	106	354	24.4
23.0	457	105	352	24.1
23.0	456	105	351	24.2
23.0	453	104	349	23.9
23.0	452	104	348	23.9
23.0	448	103	345	23.7
23.0	447	103	344	23.7
23.0	444	102	342	23.4
23.0	443	102	341	23.5
23.0	440	101	339	23.2
23.0	439	101	338	23.2
23.0	435	100	335	23.0
23.0	434	100	334	23.0
23.0	431	99	332	22.7
23.0	430	99	331	22.8
23.0	427	98	329	22.5
23.0	426	98	328	22.5
23.0	422	97	325	22.3
23.0	421	97	324	22.3
23.0	418	96	322	22.0
23.0	417	96	321	22.1
23.0	413	95	318	21.9
23.0	409	94	315	21.6
23.0	408	94	314	21.7
23.0	405	93	312	21.4
23.0	404	93	311	21.4
23.0	400	92	308	21.2
23.0	396	91	305	20.9
23.0	395	91	304	21.0
23.0	392	90	302	20.7
23.0	391	90	301	20.7
23.0	387	89	298	20.5
23.0	383	88	295	20.2
23.0	382	88	294	20.3
23.0	379	87	292	20.0
23.0	378	87	291	20.0
23.0	374	86	288	19.8
23.0	370	85	285	19.5
23.0	369	85	284	19.6
23.0	366	84	282	19.3
23.0	365	84	281	19.3
23.0	361	83	278	19.1
23.0	357	82	275	18.8
23.0	356	82	274	18.9
23.0	352	81	271	18.6
23.0	348	80	268	18.4
23.0	344	79	265	18.1
23.0	343	79	264	18.2

%E	M1	M2	DM	M*
23.0	339	78	261	17.9
23.0	335	77	258	17.7
23.0	331	76	255	17.5
23.0	330	76	254	17.5
23.0	326	75	251	17.3
23.0	322	74	248	17.0
23.0	318	73	245	16.8
23.0	317	73	244	16.8
23.0	313	72	241	16.6
23.0	309	71	238	16.3
23.0	305	70	235	16.1
23.0	304	70	234	16.1
23.0	300	69	231	15.9
23.0	296	68	228	15.6
23.0	291	67	224	15.4
23.0	287	66	221	15.2
23.0	283	65	218	14.9
23.0	278	64	214	14.7
23.0	274	63	211	14.5
23.0	270	62	208	14.2
23.0	269	62	207	14.3
23.0	265	61	204	14.0
23.0	261	60	201	13.8
23.0	257	59	198	13.5
23.0	256	59	197	13.6
23.0	252	58	194	13.3
23.0	248	57	191	13.1
23.0	244	56	188	12.9
23.0	243	56	187	12.9
23.0	239	55	184	12.7
23.0	235	54	181	12.4
23.0	230	53	177	12.2
23.0	226	52	174	12.0
23.0	222	51	171	11.7
23.0	217	50	167	11.5
23.0	213	49	164	11.3
23.0	209	48	161	11.0
23.0	204	47	157	10.8
23.0	200	46	154	10.6
23.0	196	45	151	10.3
23.0	191	44	147	10.1
23.0	187	43	144	9.9
23.0	183	42	141	9.6
23.0	178	41	137	9.4
23.0	174	40	134	9.2
23.0	165	38	127	8.8
23.0	161	37	124	8.5
23.0	152	35	117	8.1
23.0	148	34	114	7.8
23.0	139	32	107	7.4

%E	M1	M2	DM	M*
23.0	135	31	104	7.1
23.0	126	29	97	6.7
23.0	122	28	94	6.4
23.0	113	26	87	6.0
23.0	100	23	77	5.3
23.0	87	20	67	4.6
23.0	74	17	57	3.9
23.0	61	14	47	3.2
22.9	498	114	384	26.1
22.9	497	114	383	26.1
22.9	494	113	381	25.8
22.9	493	113	380	25.9
22.9	490	112	378	25.6
22.9	489	112	377	25.7
22.9	485	111	374	25.4
22.9	484	111	373	25.5
22.9	481	110	371	25.2
22.9	480	110	370	25.2
22.9	477	109	368	24.9
22.9	476	109	367	25.0
22.9	475	109	366	25.0
22.9	472	108	364	24.7
22.9	471	108	363	24.8
22.9	468	107	361	24.5
22.9	467	107	360	24.5
22.9	463	106	357	24.3
22.9	462	106	356	24.3
22.9	459	105	354	24.0
22.9	458	105	353	24.1
22.9	455	104	351	23.8
22.9	454	104	350	23.8
22.9	450	103	347	23.6
22.9	449	103	346	23.6
22.9	446	102	344	23.3
22.9	445	102	343	23.4
22.9	442	101	341	23.1
22.9	441	101	340	23.1
22.9	437	100	337	22.9
22.9	436	100	336	22.9
22.9	433	99	334	22.6
22.9	432	99	333	22.7
22.9	428	98	330	22.4
22.9	424	97	327	22.2
22.9	423	97	326	22.2
22.9	420	96	324	21.9
22.9	419	96	323	22.0
22.9	415	95	320	21.7
22.9	414	95	319	21.8
22.9	411	94	317	21.5
22.9	410	94	316	21.6
22.9	407	93	314	21.3
22.9	406	93	313	21.3
22.9	402	92	310	21.1
22.9	401	92	309	21.1
22.9	398	91	307	20.8
22.9	397	91	306	20.9
22.9	393	90	303	20.6
22.9	389	89	300	20.4
22.9	388	89	299	20.4
22.9	385	88	297	20.1
22.9	384	88	296	20.2
22.9	380	87	293	19.9
22.9	376	86	290	19.7
22.9	375	86	289	19.7
22.9	371	85	286	19.5
22.9	367	84	283	19.2
22.9	363	83	280	19.0
22.9	362	83	279	19.0
22.9	358	82	276	18.8
22.9	354	81	273	18.5
22.9	353	81	272	18.6
22.9	350	80	270	18.3
22.9	349	80	269	18.3
22.9	345	79	266	18.1
22.9	341	78	263	17.8
22.9	340	78	262	17.9
22.9	336	77	259	17.6
22.9	332	76	256	17.4
22.9	328	75	253	17.1
22.9	327	75	252	17.2
22.9	323	74	249	17.0
22.9	319	73	246	16.7
22.9	315	72	243	16.5
22.9	314	72	242	16.5
22.9	310	71	239	16.3
22.9	306	70	236	16.0
22.9	301	69	232	15.8
22.9	297	68	229	15.6
22.9	293	67	226	15.3
22.9	292	67	225	15.4
22.9	288	66	222	15.1
22.9	284	65	219	14.9
22.9	280	64	216	14.6
22.9	279	64	215	14.7
22.9	275	63	212	14.4
22.9	271	62	209	14.2
22.9	266	61	205	14.0
22.9	262	60	202	13.7
22.9	258	59	199	13.5
22.9	253	58	195	13.3
22.9	249	57	192	13.0
22.9	245	56	189	12.8
22.9	240	55	185	12.6
22.9	236	54	182	12.4
22.9	231	53	178	12.2
22.9	227	52	175	11.9
22.9	223	51	172	11.7
22.9	218	50	168	11.5
22.9	214	49	165	11.2
22.9	210	48	162	11.0
22.9	205	47	158	10.8
22.9	201	46	155	10.5
22.9	192	44	148	10.1
22.9	188	43	145	9.8
22.9	179	41	138	9.4
22.9	175	40	135	9.1
22.9	170	39	131	8.9
22.9	166	38	128	8.7
22.9	157	36	121	8.3
22.9	153	35	118	8.0
22.9	144	33	111	7.6
22.9	140	32	108	7.3
22.9	131	30	101	6.9
22.9	118	27	91	6.2
22.9	109	25	84	5.7
22.9	105	24	81	5.5
22.9	96	22	74	5.0
22.9	83	19	64	4.3
22.9	70	16	54	3.7
22.9	48	11	37	2.5
22.9	35	8	27	1.8
22.8	500	114	386	26.0
22.8	499	114	385	26.0
22.8	496	113	383	25.7
22.8	495	113	382	25.8
22.8	492	112	380	25.5
22.8	491	112	379	25.5
22.8	487	111	376	25.3
22.8	486	111	375	25.4
22.8	483	110	373	25.1
22.8	482	110	372	25.1
22.8	479	109	370	24.8
22.8	478	109	369	24.9
22.8	474	108	366	24.6
22.8	473	108	365	24.7
22.8	470	107	363	24.4
22.8	469	107	362	24.4
22.8	465	106	359	24.2
22.8	464	106	358	24.2
22.8	461	105	356	23.9
22.8	460	105	355	24.0
22.8	457	104	353	23.7
22.8	456	104	352	23.7
22.8	452	103	349	23.5
22.8	451	103	348	23.5
22.8	448	102	346	23.2
22.8	447	102	345	23.3
22.8	443	101	342	23.0
22.8	439	100	339	22.8
22.8	438	100	338	22.8
22.8	435	99	336	22.5
22.8	434	99	335	22.6
22.8	430	98	332	22.3
22.8	429	98	331	22.4
22.8	426	97	329	22.1
22.8	425	97	328	22.1
22.8	421	96	325	21.9
22.8	417	95	322	21.6
22.8	416	95	321	21.7
22.8	413	94	319	21.4
22.8	412	94	318	21.4
22.8	408	93	315	21.2
22.8	404	92	312	21.0
22.8	403	92	311	21.0
22.8	400	91	309	20.7
22.8	399	91	308	20.8
22.8	395	90	305	20.5
22.8	394	90	304	20.6
22.8	391	89	302	20.3
22.8	390	89	301	20.3
22.8	386	88	298	20.1
22.8	382	87	295	19.8
22.8	381	87	294	19.9
22.8	378	86	292	19.6
22.8	377	86	291	19.6
22.8	373	85	288	19.4
22.8	372	85	287	19.4
22.8	369	84	285	19.1
22.8	368	84	284	19.2
22.8	364	83	281	18.9
22.8	360	82	278	18.7
22.8	359	82	277	18.7
22.8	356	81	275	18.4
22.8	355	81	274	18.5
22.8	351	80	271	18.2
22.8	347	79	268	18.0
22.8	346	79	267	18.0
22.8	342	78	264	17.8
22.8	338	77	261	17.5
22.8	337	77	260	17.6
22.8	334	76	258	17.3
22.8	333	76	257	17.3
22.8	329	75	254	17.1
22.8	325	74	251	16.8
22.8	324	74	250	16.9
22.8	320	73	247	16.7
22.8	316	72	244	16.4
22.8	312	71	241	16.2
22.8	311	71	240	16.2
22.8	307	70	237	16.0
22.8	303	69	234	15.7
22.8	302	69	233	15.8
22.8	298	68	230	15.5
22.8	294	67	227	15.3
22.8	290	66	224	15.0
22.8	289	66	223	15.1
22.8	285	65	220	14.8
22.8	281	64	217	14.6
22.8	276	63	213	14.4
22.8	272	62	210	14.1
22.8	268	61	207	13.9
22.8	267	61	206	13.9
22.8	263	60	203	13.7
22.8	259	59	200	13.4
22.8	254	58	196	13.2
22.8	250	57	193	13.0
22.8	246	56	190	12.7
22.8	241	55	186	12.6
22.8	237	54	183	12.3
22.8	232	53	179	12.1
22.8	228	52	176	11.9
22.8	224	51	173	11.6
22.8	219	50	169	11.4
22.8	215	49	166	11.2
22.8	206	47	159	10.7
22.8	202	46	156	10.5
22.8	197	45	152	10.3
22.8	193	44	149	10.0
22.8	189	43	146	9.8
22.8	184	42	142	9.6
22.8	180	41	139	9.3
22.8	171	39	132	8.9
22.8	167	38	129	8.6
22.8	162	37	125	8.5
22.8	158	36	122	8.2
22.8	149	34	115	7.8
22.8	145	33	112	7.5
22.8	136	31	105	7.1
22.8	127	29	98	6.6
22.8	123	28	95	6.4

%E	M1	M2	DM	M*
22.8	114	26	88	5.9
22.8	101	23	78	5.2
22.8	92	21	71	4.8
22.8	79	18	61	4.1
22.8	57	13	44	3.0
22.7	498	113	385	25.6
22.7	497	113	384	25.7
22.7	494	112	382	25.4
22.7	493	112	381	25.4
22.7	490	111	379	25.1
22.7	489	111	378	25.2
22.7	488	111	377	25.2
22.7	485	110	375	24.9
22.7	484	110	374	25.0
22.7	481	109	372	24.7
22.7	480	109	371	24.8
22.7	476	108	368	24.5
22.7	475	108	367	24.6
22.7	472	107	365	24.3
22.7	471	107	364	24.3
22.7	468	106	362	24.0
22.7	467	106	361	24.1
22.7	466	106	360	24.1
22.7	463	105	358	23.8
22.7	462	105	357	23.9
22.7	459	104	355	23.6
22.7	458	104	354	23.6
22.7	454	103	351	23.4
22.7	453	103	350	23.4
22.7	450	102	348	23.1
22.7	449	102	347	23.2
22.7	445	101	344	22.9
22.7	444	101	343	23.0
22.7	441	100	341	22.7
22.7	440	100	340	22.7
22.7	437	99	338	22.4
22.7	436	99	337	22.5
22.7	432	98	334	22.2
22.7	431	98	333	22.3
22.7	428	97	331	22.0
22.7	427	97	330	22.0
22.7	423	96	327	21.8
22.7	422	96	326	21.8
22.7	419	95	324	21.5
22.7	418	95	323	21.6
22.7	415	94	321	21.3
22.7	414	94	320	21.3
22.7	410	93	317	21.1
22.7	409	93	316	21.1
22.7	406	92	314	20.8

%E	M1	M2	DM	M*
22.7	405	92	313	20.9
22.7	401	91	310	20.7
22.7	397	90	307	20.4
22.7	396	90	306	20.5
22.7	392	89	303	20.2
22.7	388	88	300	20.0
22.7	387	88	299	20.0
22.7	384	87	297	19.7
22.7	383	87	296	19.8
22.7	379	86	293	19.5
22.7	375	85	290	19.3
22.7	374	85	289	19.3
22.7	370	84	286	19.1
22.7	366	83	283	18.8
22.7	365	83	282	18.9
22.7	362	82	280	18.6
22.7	361	82	279	18.6
22.7	357	81	276	18.4
22.7	353	80	273	18.1
22.7	352	80	272	18.2
22.7	348	79	269	17.9
22.7	344	78	266	17.7
22.7	343	78	265	17.7
22.7	339	77	262	17.5
22.7	335	76	259	17.2
22.7	331	75	256	17.0
22.7	330	75	255	17.0
22.7	326	74	252	16.8
22.7	322	73	249	16.5
22.7	321	73	248	16.6
22.7	317	72	245	16.4
22.7	313	71	242	16.1
22.7	309	70	239	15.9
22.7	308	70	238	15.9
22.7	304	69	235	15.7
22.7	300	68	232	15.4
22.7	299	68	231	15.5
22.7	295	67	228	15.2
22.7	291	66	225	15.0
22.7	286	65	221	14.8
22.7	282	64	218	14.5
22.7	278	63	215	14.3
22.7	277	63	214	14.3
22.7	273	62	211	14.1
22.7	269	61	208	13.8
22.7	264	60	204	13.6
22.7	260	59	201	13.4
22.7	256	58	198	13.1
22.7	255	58	197	13.2
22.7	251	57	194	12.9

%E	M1	M2	DM	M*
22.7	247	56	191	12.7
22.7	242	55	187	12.5
22.7	238	54	184	12.3
22.7	234	53	181	12.0
22.7	233	53	180	12.1
22.7	229	52	177	11.8
22.7	225	51	174	11.6
22.7	220	50	170	11.4
22.7	216	49	167	11.1
22.7	211	48	163	10.9
22.7	207	47	160	10.7
22.7	203	46	157	10.4
22.7	198	45	153	10.2
22.7	194	44	150	10.0
22.7	185	42	143	9.5
22.7	181	41	140	9.3
22.7	176	40	136	9.1
22.7	172	39	133	8.8
22.7	163	37	126	8.4
22.7	154	35	119	8.0
22.7	150	34	116	7.7
22.7	141	32	109	7.3
22.7	132	30	102	6.8
22.7	128	29	99	6.6
22.7	119	27	92	6.1
22.7	110	25	85	5.7
22.7	97	22	75	5.0
22.7	88	20	68	4.5
22.7	75	17	58	3.9
22.7	66	15	51	3.4
22.7	44	10	34	2.3
22.7	22	5	17	1.1
22.6	500	113	387	25.5
22.6	499	113	386	25.6
22.6	496	112	384	25.3
22.6	495	112	383	25.3
22.6	492	111	381	25.0
22.6	491	111	380	25.1
22.6	487	110	377	24.8
22.6	486	110	376	24.9
22.6	483	109	374	24.6
22.6	482	109	373	24.6
22.6	478	108	370	24.4
22.6	477	108	369	24.5
22.6	474	107	367	24.2
22.6	473	107	366	24.2
22.6	470	106	364	23.9
22.6	469	106	363	24.0
22.6	465	105	360	23.7
22.6	464	105	359	23.8

%E	M1	M2	DM	M*
22.6	461	104	357	23.5
22.6	460	104	356	23.5
22.6	456	103	353	23.3
22.6	455	103	352	23.3
22.6	452	102	350	23.0
22.6	451	102	349	23.1
22.6	447	101	346	22.8
22.6	446	101	345	22.9
22.6	443	100	343	22.6
22.6	442	100	342	22.6
22.6	439	99	340	22.3
22.6	438	99	339	22.4
22.6	434	98	336	22.1
22.6	433	98	335	22.2
22.6	430	97	333	21.9
22.6	429	97	332	21.9
22.6	425	96	329	21.7
22.6	424	96	328	21.7
22.6	421	95	326	21.4
22.6	420	95	325	21.5
22.6	416	94	322	21.2
22.6	412	93	319	21.0
22.6	411	93	318	21.0
22.6	407	92	315	20.8
22.6	403	91	312	20.5
22.6	402	91	311	20.6
22.6	399	90	309	20.3
22.6	398	90	308	20.4
22.6	394	89	305	20.1
22.6	393	89	304	20.2
22.6	390	88	302	19.9
22.6	389	88	301	19.9
22.6	385	87	298	19.7
22.6	381	86	295	19.4
22.6	380	86	294	19.5
22.6	376	85	291	19.2
22.6	372	84	288	19.0
22.6	371	84	287	19.0
22.6	368	83	285	18.7
22.6	367	83	284	18.8
22.6	363	82	281	18.5
22.6	359	81	278	18.3
22.6	358	81	277	18.3
22.6	354	80	274	18.1
22.6	350	79	271	17.8
22.6	349	79	270	17.9
22.6	345	78	267	17.6
22.6	341	77	264	17.4
22.6	340	77	263	17.4
22.6	337	76	261	17.1

%E	M1	M2	DM	M*
22.6	336	76	260	17.2
22.6	332	75	257	16.9
22.6	328	74	254	16.7
22.6	327	74	253	16.7
22.6	323	73	250	16.5
22.6	319	72	247	16.3
22.6	318	72	246	16.3
22.6	314	71	243	16.1
22.6	310	70	240	15.8
22.6	305	69	236	15.6
22.6	301	68	233	15.4
22.6	297	67	230	15.1
22.6	296	67	229	15.2
22.6	292	66	226	14.9
22.6	288	65	223	14.7
22.6	287	65	222	14.7
22.6	283	64	219	14.5
22.6	279	63	216	14.2
22.6	274	62	212	14.0
22.6	270	61	209	13.8
22.6	266	60	206	13.5
22.6	265	60	205	13.6
22.6	261	59	202	13.3
22.6	257	58	199	13.1
22.6	252	57	195	12.9
22.6	248	56	192	12.6
22.6	243	55	188	12.4
22.6	239	54	185	12.2
22.6	235	53	182	12.0
22.6	230	52	178	11.8
22.6	226	51	175	11.5
22.6	221	50	171	11.3
22.6	217	49	168	11.1
22.6	212	48	164	10.9
22.6	208	47	161	10.6
22.6	199	45	154	10.2
22.6	195	44	151	9.9
22.6	190	43	147	9.7
22.6	186	42	144	9.5
22.6	177	40	137	9.0
22.6	168	38	130	8.6
22.6	164	37	127	8.3
22.6	159	36	123	8.2
22.6	155	35	120	7.9
22.6	146	33	113	7.5
22.6	137	31	106	7.0
22.6	133	30	103	6.8
22.6	124	28	96	6.3
22.6	115	26	89	5.9
22.6	106	24	82	5.4

%E	M1	M2	DM	M*
22.6	93	21	72	4.7
22.6	84	19	65	4.3
22.6	62	14	48	3.2
22.6	53	12	41	2.7
22.6	31	7	24	1.6
22.5	498	112	386	25.2
22.5	497	112	385	25.2
22.5	494	111	383	24.9
22.5	493	111	382	25.0
22.5	489	110	379	24.7
22.5	488	110	378	24.8
22.5	485	109	376	24.5
22.5	484	109	375	24.5
22.5	481	108	373	24.2
22.5	480	108	372	24.3
22.5	479	108	371	24.4
22.5	476	107	369	24.1
22.5	475	107	368	24.1
22.5	472	106	366	23.8
22.5	471	106	365	23.9
22.5	467	105	362	23.6
22.5	466	105	361	23.7
22.5	463	104	359	23.4
22.5	462	104	358	23.4
22.5	458	103	355	23.2
22.5	457	103	354	23.2
22.5	454	102	352	22.9
22.5	453	102	351	23.0
22.5	449	101	348	22.7
22.5	448	101	347	22.8
22.5	445	100	345	22.5
22.5	444	100	344	22.5
22.5	440	99	341	22.3
22.5	436	98	338	22.0
22.5	435	98	337	22.1
22.5	432	97	335	21.8
22.5	431	97	334	21.8
22.5	427	96	331	21.6
22.5	426	96	330	21.6
22.5	423	95	328	21.3
22.5	422	95	327	21.4
22.5	418	94	324	21.1
22.5	417	94	323	21.2
22.5	414	93	321	20.9
22.5	413	93	320	20.9
22.5	409	92	317	20.7
22.5	408	92	316	20.7
22.5	405	91	314	20.4
22.5	404	91	313	20.5
22.5	400	90	310	20.3

%E	M1	M2	DM	M*
22.5	396	89	307	20.0
22.5	395	89	306	20.1
22.5	391	88	303	19.8
22.5	387	87	300	19.6
22.5	386	87	299	19.6
22.5	383	86	297	19.3
22.5	382	86	296	19.4
22.5	378	85	293	19.1
22.5	377	85	292	19.2
22.5	374	84	290	18.9
22.5	373	84	289	18.9
22.5	369	83	286	18.7
22.5	365	82	283	18.4
22.5	364	82	282	18.5
22.5	360	81	279	18.2
22.5	356	80	276	18.0
22.5	355	80	275	18.0
22.5	351	79	272	17.8
22.5	347	78	269	17.5
22.5	346	78	268	17.6
22.5	342	77	265	17.3
22.5	338	76	262	17.1
22.5	334	75	259	16.8
22.5	333	75	258	16.9
22.5	329	74	255	16.6
22.5	325	73	252	16.4
22.5	324	73	251	16.4
22.5	320	72	248	16.2
22.5	316	71	245	16.0
22.5	315	71	244	16.0
22.5	311	70	241	15.8
22.5	307	69	238	15.5
22.5	306	69	237	15.6
22.5	302	68	234	15.3
22.5	298	67	231	15.1
22.5	293	66	227	14.9
22.5	289	65	224	14.6
22.5	285	64	221	14.4
22.5	284	64	220	14.4
22.5	280	63	217	14.2
22.5	276	62	214	13.9
22.5	275	62	213	14.0
22.5	271	61	210	13.7
22.5	267	60	207	13.5
22.5	262	59	203	13.3
22.5	258	58	200	13.0
22.5	253	57	196	12.8
22.5	249	56	193	12.6
22.5	244	55	189	12.4
22.5	240	54	186	12.1

%E	M1	M2	DM	M*
22.5	236	53	183	11.9
22.5	231	52	179	11.7
22.5	227	51	176	11.5
22.5	222	50	172	11.3
22.5	218	49	169	11.0
22.5	213	48	165	10.8
22.5	209	47	162	10.6
22.5	204	46	158	10.4
22.5	200	45	155	10.1
22.5	191	43	148	9.7
22.5	187	42	145	9.4
22.5	182	41	141	9.2
22.5	178	40	138	9.0
22.5	173	39	134	8.8
22.5	169	38	131	8.5
22.5	160	36	124	8.1
22.5	151	34	117	7.7
22.5	142	32	110	7.2
22.5	138	31	107	7.0
22.5	129	29	100	6.5
22.5	120	27	93	6.1
22.5	111	25	86	5.6
22.5	102	23	79	5.2
22.5	89	20	69	4.5
22.5	80	18	62	4.0
22.5	71	16	55	3.6
22.5	40	9	31	2.0
22.4	500	112	388	25.1
22.4	499	112	387	25.1
22.4	496	111	385	24.8
22.4	495	111	384	24.9
22.4	492	110	382	24.6
22.4	491	110	381	24.6
22.4	490	110	380	24.7
22.4	487	109	378	24.4
22.4	486	109	377	24.4
22.4	483	108	375	24.1
22.4	482	108	374	24.2
22.4	478	107	371	24.0
22.4	477	107	370	24.0
22.4	474	106	368	23.7
22.4	473	106	367	23.8
22.4	469	105	364	23.5
22.4	468	105	363	23.6
22.4	465	104	361	23.3
22.4	464	104	360	23.3
22.4	460	103	357	23.1
22.4	459	103	356	23.1
22.4	456	102	354	22.8
22.4	455	102	353	22.9

%E	M1	M2	DM	M*
22.4	451	101	350	22.6
22.4	450	101	349	22.7
22.4	447	100	347	22.4
22.4	446	100	346	22.4
22.4	442	99	343	22.2
22.4	441	99	342	22.2
22.4	438	98	340	21.9
22.4	437	98	339	22.0
22.4	434	97	337	21.7
22.4	433	97	336	21.7
22.4	429	96	333	21.5
22.4	428	96	332	21.5
22.4	425	95	330	21.2
22.4	424	95	329	21.3
22.4	420	94	326	21.0
22.4	419	94	325	21.1
22.4	416	93	323	20.8
22.4	415	93	322	20.8
22.4	411	92	319	20.6
22.4	410	92	318	20.6
22.4	407	91	316	20.3
22.4	406	91	315	20.4
22.4	402	90	312	20.1
22.4	401	90	311	20.2
22.4	398	89	309	19.9
22.4	397	89	308	20.0
22.4	393	88	305	19.7
22.4	392	88	304	19.8
22.4	389	87	302	19.5
22.4	388	87	301	19.5
22.4	384	86	298	19.3
22.4	380	85	295	19.0
22.4	379	85	294	19.1
22.4	375	84	291	18.8
22.4	371	83	288	18.6
22.4	370	83	287	18.6
22.4	366	82	284	18.4
22.4	362	81	281	18.1
22.4	361	81	280	18.2
22.4	357	80	277	17.9
22.4	353	79	274	17.7
22.4	352	79	273	17.7
22.4	349	78	271	17.4
22.4	348	78	270	17.5
22.4	344	77	267	17.2
22.4	343	77	266	17.3
22.4	340	76	264	17.0
22.4	339	76	263	17.0
22.4	335	75	260	16.8
22.4	331	74	257	16.5

%E	M1	M2	DM	M*
22.4	330	74	256	16.6
22.4	326	73	253	16.3
22.4	322	72	250	16.1
22.4	321	72	249	16.1
22.4	317	71	246	15.9
22.4	313	70	243	15.7
22.4	312	70	242	15.7
22.4	308	69	239	15.5
22.4	304	68	236	15.2
22.4	303	68	235	15.3
22.4	299	67	232	15.0
22.4	295	66	229	14.8
22.4	294	66	228	14.8
22.4	290	65	225	14.6
22.4	286	64	222	14.3
22.4	281	63	218	14.1
22.4	277	62	215	13.9
22.4	272	61	211	13.7
22.4	268	60	208	13.4
22.4	263	59	204	13.2
22.4	259	58	201	13.0
22.4	255	57	198	12.7
22.4	254	57	197	12.8
22.4	250	56	194	12.5
22.4	246	55	191	12.3
22.4	245	55	190	12.3
22.4	241	54	187	12.1
22.4	237	53	184	11.9
22.4	232	52	180	11.7
22.4	228	51	177	11.4
22.4	223	50	173	11.2
22.4	219	49	170	11.0
22.4	214	48	166	10.8
22.4	210	47	163	10.5
22.4	205	46	159	10.3
22.4	201	45	156	10.1
22.4	196	44	152	9.9
22.4	192	43	149	9.6
22.4	183	41	142	9.2
22.4	174	39	135	8.7
22.4	170	38	132	8.5
22.4	165	37	128	8.3
22.4	161	36	125	8.0
22.4	156	35	121	7.9
22.4	152	34	118	7.6
22.4	147	33	114	7.4
22.4	143	32	111	7.2
22.4	134	30	104	6.7
22.4	125	28	97	6.3
22.4	116	26	90	5.8

%E	M1	M2	DM	M*
22.4	107	24	83	5.4
22.4	98	22	76	4.9
22.4	85	19	66	4.2
22.4	76	17	59	3.8
22.4	67	15	52	3.4
22.4	58	13	45	2.9
22.4	49	11	38	2.5
22.3	498	111	387	24.7
22.3	497	111	386	24.8
22.3	494	110	384	24.5
22.3	493	110	383	24.5
22.3	489	109	380	24.3
22.3	488	109	379	24.3
22.3	485	108	377	24.0
22.3	484	108	376	24.1
22.3	480	107	373	23.9
22.3	479	107	372	23.9
22.3	476	106	370	23.6
22.3	475	106	369	23.7
22.3	471	105	366	23.4
22.3	470	105	365	23.5
22.3	467	104	363	23.2
22.3	466	104	362	23.2
22.3	462	103	359	23.0
22.3	461	103	358	23.0
22.3	458	102	356	22.7
22.3	457	102	355	22.8
22.3	453	101	352	22.5
22.3	452	101	351	22.6
22.3	449	100	349	22.3
22.3	448	100	348	22.3
22.3	444	99	345	22.1
22.3	443	99	344	22.1
22.3	440	98	342	21.8
22.3	439	98	341	21.9
22.3	435	97	338	21.6
22.3	431	96	335	21.4
22.3	430	96	334	21.4
22.3	426	95	331	21.2
22.3	422	94	328	20.9
22.3	421	94	327	21.0
22.3	417	93	324	20.7
22.3	413	92	321	20.5
22.3	412	92	320	20.5
22.3	408	91	317	20.3
22.3	404	90	314	20.0
22.3	403	90	313	20.1
22.3	400	89	311	19.8
22.3	399	89	310	19.9
22.3	395	88	307	19.6

%E	M1	M2	DM	M*
22.3	394	88	306	19.7
22.3	391	87	304	19.4
22.3	390	87	303	19.4
22.3	386	86	300	19.2
22.3	385	86	299	19.2
22.3	382	85	297	18.9
22.3	381	85	296	19.0
22.3	377	84	293	18.7
22.3	376	84	292	18.8
22.3	373	83	290	18.5
22.3	372	83	289	18.5
22.3	368	82	286	18.3
22.3	367	82	285	18.3
22.3	364	81	283	18.0
22.3	363	81	282	18.1
22.3	359	80	279	17.8
22.3	358	80	278	17.9
22.3	355	79	276	17.6
22.3	354	79	275	17.6
22.3	350	78	272	17.4
22.3	346	77	269	17.1
22.3	345	77	268	17.2
22.3	341	76	265	16.9
22.3	337	75	262	16.7
22.3	336	75	261	16.7
22.3	332	74	258	16.5
22.3	328	73	255	16.2
22.3	327	73	254	16.3
22.3	323	72	251	16.0
22.3	319	71	248	15.8
22.3	318	71	247	15.9
22.3	314	70	244	15.6
22.3	310	69	241	15.4
22.3	309	69	240	15.4
22.3	305	68	237	15.2
22.3	301	67	234	14.9
22.3	300	67	233	15.0
22.3	296	66	230	14.7
22.3	292	65	227	14.5
22.3	291	65	226	14.5
22.3	287	64	223	14.3
22.3	283	63	220	14.0
22.3	282	63	219	14.1
22.3	278	62	216	13.8
22.3	274	61	213	13.6
22.3	273	61	212	13.6
22.3	269	60	209	13.4
22.3	265	59	206	13.1
22.3	264	59	205	13.2
22.3	260	58	202	12.9

%E	M1	M2	DM	M*
22.3	256	57	199	12.7
22.3	251	56	195	12.5
22.3	247	55	192	12.2
22.3	242	54	188	12.0
22.3	238	53	185	11.8
22.3	233	52	181	11.6
22.3	229	51	178	11.4
22.3	224	50	174	11.2
22.3	220	49	171	10.9
22.3	215	48	167	10.7
22.3	211	47	164	10.5
22.3	206	46	160	10.3
22.3	202	45	157	10.0
22.3	197	44	153	9.8
22.3	193	43	150	9.6
22.3	188	42	146	9.4
22.3	184	41	143	9.1
22.3	179	40	139	8.9
22.3	175	39	136	8.7
22.3	166	37	129	8.2
22.3	157	35	122	7.8
22.3	148	33	115	7.4
22.3	139	31	108	6.9
22.3	130	29	101	6.5
22.3	121	27	94	6.0
22.3	112	25	87	5.6
22.3	103	23	80	5.1
22.3	94	21	73	4.7
22.2	500	111	389	24.6
22.2	499	111	388	24.7
22.2	496	110	386	24.4
22.2	495	110	385	24.4
22.2	492	109	383	24.1
22.2	491	109	382	24.2
22.2	490	109	381	24.2
22.2	487	108	379	24.0
22.2	486	108	378	24.0
22.2	483	107	376	23.7
22.2	482	107	375	23.8
22.2	481	107	374	23.8
22.2	478	106	372	23.5
22.2	477	106	371	23.6
22.2	474	105	369	23.3
22.2	473	105	368	23.3
22.2	472	105	367	23.4
22.2	469	104	365	23.1
22.2	468	104	364	23.1
22.2	465	103	362	22.8
22.2	464	103	361	22.9
22.2	463	103	360	22.9

%E	M1	M2	DM	M*
22.2	460	102	358	22.6
22.2	459	102	357	22.7
22.2	455	101	354	22.4
22.2	454	101	353	22.5
22.2	451	100	351	22.2
22.2	450	100	350	22.2
22.2	446	99	347	22.0
22.2	445	99	346	22.0
22.2	442	98	344	21.7
22.2	441	98	343	21.8
22.2	437	97	340	21.5
22.2	436	97	339	21.6
22.2	433	96	337	21.3
22.2	432	96	336	21.3
22.2	428	95	333	21.1
22.2	427	95	332	21.1
22.2	424	94	330	20.8
22.2	423	94	329	20.9
22.2	419	93	326	20.6
22.2	418	93	325	20.7
22.2	415	92	323	20.4
22.2	414	92	322	20.4
22.2	410	91	319	20.2
22.2	409	91	318	20.2
22.2	406	90	316	20.0
22.2	405	90	315	20.0
22.2	401	89	312	19.8
22.2	397	88	309	19.5
22.2	396	88	308	19.6
22.2	392	87	305	19.3
22.2	388	86	302	19.1
22.2	387	86	301	19.1
22.2	383	85	298	18.9
22.2	379	84	295	18.6
22.2	378	84	294	18.7
22.2	374	83	291	18.4
22.2	370	82	288	18.2
22.2	369	82	287	18.2
22.2	365	81	284	18.0
22.2	361	80	281	17.7
22.2	360	80	280	17.8
22.2	356	79	277	17.5
22.2	352	78	274	17.3
22.2	351	78	273	17.3
22.2	347	77	270	17.1
22.2	343	76	267	16.8
22.2	342	76	266	16.9
22.2	338	75	263	16.6
22.2	334	74	260	16.4
22.2	333	74	259	16.4

%E	M1	M2	DM	M*
22.2	329	73	256	16.2
22.2	325	72	253	16.0
22.2	324	72	252	16.0
22.2	320	71	249	15.8
22.2	316	70	246	15.5
22.2	315	70	245	15.6
22.2	311	69	242	15.3
22.2	307	68	239	15.1
22.2	306	68	238	15.1
22.2	302	67	235	14.9
22.2	297	66	231	14.7
22.2	293	65	228	14.4
22.2	288	64	224	14.2
22.2	284	63	221	14.0
22.2	279	62	217	13.8
22.2	275	61	214	13.5
22.2	270	60	210	13.3
22.2	266	59	207	13.1
22.2	261	58	203	12.9
22.2	257	57	200	12.6
22.2	252	56	196	12.4
22.2	248	55	193	12.2
22.2	243	54	189	12.0
22.2	239	53	186	11.8
22.2	234	52	182	11.6
22.2	230	51	179	11.3
22.2	225	50	175	11.1
22.2	221	49	172	10.9
22.2	216	48	168	10.7
22.2	212	47	165	10.4
22.2	207	46	161	10.2
22.2	203	45	158	10.0
22.2	198	44	154	9.8
22.2	194	43	151	9.5
22.2	189	42	147	9.3
22.2	185	41	144	9.1
22.2	180	40	140	8.9
22.2	176	39	137	8.6
22.2	171	38	133	8.4
22.2	167	37	130	8.2
22.2	162	36	126	8.0
22.2	158	35	123	7.8
22.2	153	34	119	7.6
22.2	144	32	112	7.1
22.2	135	30	105	6.7
22.2	126	28	98	6.2
22.2	117	26	91	5.8
22.2	108	24	84	5.3
22.2	99	22	77	4.9
22.2	90	20	70	4.4

%E	M1	M2	DM	M*
22.2	81	18	63	4.0
22.2	72	16	56	3.6
22.2	63	14	49	3.1
22.2	54	12	42	2.7
22.2	45	10	35	2.2
22.2	36	8	28	1.8
22.2	27	6	21	1.3
22.2	18	4	14	0.9
22.2	9	2	7	0.4
22.1	498	110	388	24.3
22.1	497	110	387	24.3
22.1	494	109	385	24.1
22.1	493	109	384	24.1
22.1	489	108	381	23.9
22.1	488	108	380	23.9
22.1	485	107	378	23.6
22.1	484	107	377	23.7
22.1	480	106	374	23.4
22.1	479	106	373	23.5
22.1	476	105	371	23.2
22.1	475	105	370	23.2
22.1	471	104	367	23.0
22.1	470	104	366	23.0
22.1	467	103	364	22.7
22.1	466	103	363	22.8
22.1	462	102	360	22.5
22.1	461	102	359	22.6
22.1	458	101	357	22.3
22.1	457	101	356	22.3
22.1	456	101	355	22.4
22.1	453	100	353	22.1
22.1	452	100	352	22.1
22.1	448	99	349	21.9
22.1	447	99	348	21.9
22.1	444	98	346	21.6
22.1	443	98	345	21.7
22.1	439	97	342	21.4
22.1	438	97	341	21.5
22.1	435	96	339	21.2
22.1	434	96	338	21.2
22.1	430	95	335	21.0
22.1	429	95	334	21.0
22.1	426	94	332	20.7
22.1	425	94	331	20.8
22.1	421	93	328	20.5
22.1	420	93	327	20.6
22.1	417	92	325	20.3
22.1	416	92	324	20.3
22.1	412	91	321	20.1
22.1	411	91	320	20.1

%E	M1	M2	DM	M*
22.1	408	90	318	19.9
22.1	407	90	317	19.9
22.1	403	89	314	19.7
22.1	402	89	313	19.7
22.1	399	88	311	19.4
22.1	398	88	310	19.5
22.1	394	87	307	19.2
22.1	393	87	306	19.3
22.1	390	86	304	19.0
22.1	389	86	303	19.0
22.1	385	85	300	18.8
22.1	384	85	299	18.8
22.1	380	84	296	18.6
22.1	376	83	293	18.3
22.1	375	83	292	18.4
22.1	371	82	289	18.1
22.1	367	81	286	17.9
22.1	366	81	285	17.9
22.1	362	80	282	17.7
22.1	358	79	279	17.4
22.1	357	79	278	17.5
22.1	353	78	275	17.2
22.1	349	77	272	17.0
22.1	348	77	271	17.0
22.1	344	76	268	16.8
22.1	340	75	265	16.5
22.1	339	75	264	16.6
22.1	335	74	261	16.3
22.1	331	73	258	16.1
22.1	330	73	257	16.1
22.1	326	72	254	15.9
22.1	322	71	251	15.7
22.1	321	71	250	15.7
22.1	317	70	247	15.5
22.1	312	69	243	15.3
22.1	308	68	240	15.0
22.1	303	67	236	14.8
22.1	299	66	233	14.6
22.1	298	66	232	14.6
22.1	294	65	229	14.4
22.1	290	64	226	14.1
22.1	289	64	225	14.2
22.1	285	63	222	13.9
22.1	281	62	219	13.7
22.1	280	62	218	13.7
22.1	276	61	215	13.5
22.1	272	60	212	13.2
22.1	271	60	211	13.3
22.1	267	59	208	13.0
22.1	263	58	205	12.8

%E	M1	M2	DM	M*
22.1	262	58	204	12.8
22.1	258	57	201	12.6
22.1	253	56	197	12.4
22.1	249	55	194	12.1
22.1	244	54	190	12.0
22.1	240	53	187	11.7
22.1	235	52	183	11.5
22.1	231	51	180	11.3
22.1	226	50	176	11.1
22.1	222	49	173	10.8
22.1	217	48	169	10.6
22.1	213	47	166	10.4
22.1	208	46	162	10.2
22.1	204	45	159	9.9
22.1	199	44	155	9.7
22.1	195	43	152	9.5
22.1	190	42	148	9.3
22.1	181	40	141	8.8
22.1	172	38	134	8.4
22.1	163	36	127	8.0
22.1	154	34	120	7.5
22.1	149	33	116	7.3
22.1	145	32	113	7.1
22.1	140	31	109	6.9
22.1	136	30	106	6.6
22.1	131	29	102	6.4
22.1	122	27	95	6.0
22.1	113	25	88	5.5
22.1	104	23	81	5.1
22.1	95	21	74	4.6
22.1	86	19	67	4.2
22.1	77	17	60	3.8
22.1	68	15	53	3.3
22.0	500	110	390	24.2
22.0	499	110	389	24.2
22.0	496	109	387	24.0
22.0	495	109	386	24.0
22.0	492	108	384	23.7
22.0	491	108	383	23.8
22.0	490	108	382	23.8
22.0	487	107	380	23.5
22.0	486	107	379	23.6
22.0	482	106	376	23.3
22.0	481	106	375	23.4
22.0	478	105	373	23.1
22.0	477	105	372	23.1
22.0	473	104	369	22.9
22.0	472	104	368	22.9
22.0	469	103	366	22.6
22.0	468	103	365	22.7

%E	M1	M2	DM	M*
22.0	464	102	362	22.4
22.0	463	102	361	22.5
22.0	460	101	359	22.2
22.0	459	101	358	22.2
22.0	455	100	355	22.0
22.0	454	100	354	22.0
22.0	451	99	352	21.7
22.0	450	99	351	21.8
22.0	449	99	350	21.8
22.0	446	98	348	21.5
22.0	445	98	347	21.6
22.0	441	97	344	21.3
22.0	440	97	343	21.4
22.0	437	96	341	21.1
22.0	436	96	340	21.1
22.0	432	95	337	20.9
22.0	431	95	336	20.9
22.0	428	94	334	20.6
22.0	427	94	333	20.7
22.0	423	93	330	20.4
22.0	422	93	329	20.5
22.0	419	92	327	20.2
22.0	418	92	326	20.2
22.0	414	91	323	20.0
22.0	413	91	322	20.1
22.0	410	90	320	19.8
22.0	409	90	319	19.8
22.0	405	89	316	19.6
22.0	404	89	315	19.6
22.0	400	88	312	19.4
22.0	396	87	309	19.1
22.0	395	87	308	19.2
22.0	391	86	305	18.9
22.0	387	85	302	18.7
22.0	386	85	301	18.7
22.0	382	84	298	18.5
22.0	381	84	297	18.5
22.0	378	83	295	18.2
22.0	377	83	294	18.3
22.0	373	82	291	18.0
22.0	372	82	290	18.1
22.0	369	81	288	17.8
22.0	368	81	287	17.8
22.0	364	80	284	17.6
22.0	363	80	283	17.6
22.0	359	79	280	17.4
22.0	355	78	277	17.1
22.0	354	78	276	17.2
22.0	350	77	273	16.9
22.0	346	76	270	16.7

%E	M1	M2	DM	M*
22.0	345	76	269	16.7
22.0	341	75	266	16.5
22.0	337	74	263	16.2
22.0	336	74	262	16.3
22.0	332	73	259	16.1
22.0	328	72	256	15.8
22.0	327	72	255	15.9
22.0	323	71	252	15.6
22.0	318	70	248	15.4
22.0	314	69	245	15.2
22.0	313	69	244	15.2
22.0	309	68	241	15.0
22.0	305	67	238	14.7
22.0	304	67	237	14.8
22.0	300	66	234	14.5
22.0	296	65	231	14.3
22.0	295	65	230	14.3
22.0	291	64	227	14.1
22.0	287	63	224	13.8
22.0	286	63	223	13.9
22.0	282	62	220	13.6
22.0	277	61	216	13.4
22.0	273	60	213	13.2
22.0	268	59	209	13.0
22.0	264	58	206	12.7
22.0	259	57	202	12.5
22.0	255	56	199	12.3
22.0	254	56	198	12.3
22.0	250	55	195	12.1
22.0	246	54	192	11.9
22.0	245	54	191	11.9
22.0	241	53	188	11.7
22.0	236	52	184	11.5
22.0	232	51	181	11.2
22.0	227	50	177	11.0
22.0	223	49	174	10.8
22.0	218	48	170	10.6
22.0	214	47	167	10.3
22.0	209	46	163	10.1
22.0	205	45	160	9.9
22.0	200	44	156	9.7
22.0	191	42	149	9.2
22.0	186	41	145	9.0
22.0	182	40	142	8.8
22.0	177	39	138	8.6
22.0	173	38	135	8.3
22.0	168	37	131	8.1
22.0	164	36	128	7.9
22.0	159	35	124	7.7
22.0	150	33	117	7.3

%E	M1	M2	DM	M*
22.0	141	31	110	6.8
22.0	132	29	103	6.4
22.0	127	28	99	6.2
22.0	123	27	96	5.9
22.0	118	26	92	5.7
22.0	109	24	85	5.3
22.0	100	22	78	4.8
22.0	91	20	71	4.4
22.0	82	18	64	4.0
22.0	59	13	46	2.9
22.0	50	11	39	2.4
22.0	41	9	32	2.0
21.9	498	109	389	23.9
21.9	497	109	388	23.9
21.9	494	108	386	23.6
21.9	493	108	385	23.7
21.9	489	107	382	23.4
21.9	488	107	381	23.5
21.9	485	106	379	23.2
21.9	484	106	378	23.2
21.9	483	106	377	23.3
21.9	480	105	375	23.0
21.9	479	105	374	23.0
21.9	475	104	371	22.8
21.9	474	104	370	22.8
21.9	471	103	368	22.5
21.9	470	103	367	22.6
21.9	466	102	364	22.3
21.9	465	102	363	22.4
21.9	462	101	361	22.1
21.9	461	101	360	22.1
21.9	457	100	357	21.9
21.9	456	100	356	21.9
21.9	453	99	354	21.6
21.9	452	99	353	21.7
21.9	448	98	350	21.4
21.9	447	98	349	21.5
21.9	443	97	346	21.2
21.9	442	97	345	21.3
21.9	439	96	343	21.0
21.9	438	96	342	21.0
21.9	434	95	339	20.8
21.9	433	95	338	20.8
21.9	430	94	336	20.5
21.9	429	94	335	20.6
21.9	425	93	332	20.4
21.9	424	93	331	20.4
21.9	421	92	329	20.1
21.9	420	92	328	20.2
21.9	416	91	325	19.9

%E	M1	M2	DM	M*
21.9	415	91	324	20.0
21.9	411	90	321	19.7
21.9	407	89	318	19.5
21.9	406	89	317	19.5
21.9	402	88	314	19.3
21.9	401	88	313	19.3
21.9	398	87	311	19.0
21.9	397	87	310	19.1
21.9	393	86	307	18.8
21.9	392	86	306	18.9
21.9	389	85	304	18.6
21.9	388	85	303	18.6
21.9	384	84	300	18.4
21.9	383	84	299	18.4
21.9	379	83	296	18.2
21.9	375	82	293	17.9
21.9	374	82	292	18.0
21.9	370	81	289	17.7
21.9	366	80	286	17.5
21.9	365	80	285	17.5
21.9	361	79	282	17.3
21.9	360	79	281	17.3
21.9	356	78	278	17.1
21.9	352	77	275	16.8
21.9	351	77	274	16.9
21.9	347	76	271	16.6
21.9	343	75	268	16.4
21.9	342	75	267	16.4
21.9	338	74	264	16.2
21.9	334	73	261	16.0
21.9	333	73	260	16.0
21.9	329	72	257	15.8
21.9	324	71	253	15.6
21.9	320	70	250	15.3
21.9	319	70	249	15.4
21.9	315	69	246	15.1
21.9	311	68	243	14.9
21.9	310	68	242	14.9
21.9	306	67	239	14.7
21.9	302	66	236	14.4
21.9	301	66	235	14.5
21.9	297	65	232	14.2
21.9	292	64	228	14.0
21.9	288	63	225	13.8
21.9	283	62	221	13.6
21.9	279	61	218	13.3
21.9	278	61	217	13.4
21.9	274	60	214	13.1
21.9	270	59	211	12.9
21.9	269	59	210	12.9

%E	M1	M2	DM	M*
21.9	265	58	207	12.7
21.9	260	57	203	12.5
21.9	256	56	200	12.3
21.9	251	55	196	12.1
21.9	247	54	193	11.8
21.9	242	53	189	11.6
21.9	237	52	185	11.4
21.9	233	51	182	11.2
21.9	228	50	178	11.0
21.9	224	49	175	10.7
21.9	219	48	171	10.5
21.9	215	47	168	10.3
21.9	210	46	164	10.1
21.9	201	44	157	9.6
21.9	196	43	153	9.4
21.9	192	42	150	9.2
21.9	187	41	146	9.0
21.9	183	40	143	8.7
21.9	178	39	139	8.5
21.9	169	37	132	8.1
21.9	160	35	125	7.7
21.9	155	34	121	7.5
21.9	151	33	118	7.2
21.9	146	32	114	7.0
21.9	137	30	107	6.6
21.9	128	28	100	6.1
21.9	114	25	89	5.5
21.9	105	23	82	5.0
21.9	96	21	75	4.6
21.9	73	16	57	3.5
21.9	64	14	50	3.1
21.9	32	7	25	1.5
21.8	500	109	391	23.8
21.8	499	109	390	23.8
21.8	496	108	388	23.5
21.8	495	108	387	23.6
21.8	491	107	384	23.3
21.8	490	107	383	23.4
21.8	487	106	381	23.1
21.8	486	106	380	23.1
21.8	482	105	377	22.9
21.8	481	105	376	22.9
21.8	478	104	374	22.6
21.8	477	104	373	22.7
21.8	476	104	372	22.7
21.8	473	103	370	22.4
21.8	472	103	369	22.5
21.8	468	102	366	22.2
21.8	467	102	365	22.3
21.8	464	101	363	22.0

%E	M1	M2	DM	M*
21.8	463	101	362	22.0
21.8	459	100	359	21.8
21.8	458	100	358	21.8
21.8	455	99	356	21.5
21.8	454	99	355	21.6
21.8	450	98	352	21.3
21.8	449	98	351	21.4
21.8	445	97	348	21.1
21.8	444	97	347	21.2
21.8	441	96	345	20.9
21.8	440	96	344	20.9
21.8	436	95	341	20.7
21.8	435	95	340	20.7
21.8	432	94	338	20.5
21.8	431	94	337	20.5
21.8	427	93	334	20.3
21.8	426	93	333	20.3
21.8	422	92	330	20.1
21.8	418	91	327	19.8
21.8	417	91	326	19.9
21.8	413	90	323	19.6
21.8	412	90	322	19.7
21.8	409	89	320	19.4
21.8	408	89	319	19.4
21.8	404	88	316	19.2
21.8	403	88	315	19.2
21.8	400	87	313	18.9
21.8	399	87	312	19.0
21.8	395	86	309	18.7
21.8	394	86	308	18.8
21.8	390	85	305	18.5
21.8	386	84	302	18.3
21.8	385	84	301	18.3
21.8	381	83	298	18.1
21.8	380	83	297	18.1
21.8	377	82	295	17.8
21.8	376	82	294	17.9
21.8	372	81	291	17.6
21.8	371	81	290	17.7
21.8	367	80	287	17.4
21.8	363	79	284	17.2
21.8	362	79	283	17.2
21.8	358	78	280	17.0
21.8	357	78	279	17.0
21.8	354	77	277	16.7
21.8	353	77	276	16.8
21.8	349	76	273	16.6
21.8	348	76	272	16.6
21.8	344	75	269	16.4
21.8	340	74	266	16.1

%E	M1	M2	DM	M*
21.8	339	74	265	16.2
21.8	335	73	262	15.9
21.8	331	72	259	15.7
21.8	330	72	258	15.7
21.8	326	71	255	15.5
21.8	325	71	254	15.5
21.8	321	70	251	15.3
21.8	317	69	248	15.0
21.8	316	69	247	15.1
21.8	312	68	244	14.8
21.8	308	67	241	14.6
21.8	307	67	240	14.6
21.8	303	66	237	14.4
21.8	298	65	233	14.2
21.8	294	64	230	13.9
21.8	293	64	229	14.0
21.8	289	63	226	13.7
21.8	285	62	223	13.5
21.8	284	62	222	13.5
21.8	280	61	219	13.3
21.8	275	60	215	13.1
21.8	271	59	212	12.8
21.8	266	58	208	12.6
21.8	262	57	205	12.4
21.8	261	57	204	12.4
21.8	257	56	201	12.2
21.8	252	55	197	12.0
21.8	248	54	194	11.8
21.8	243	53	190	11.6
21.8	239	52	187	11.3
21.8	238	52	186	11.4
21.8	234	51	183	11.1
21.8	229	50	179	10.9
21.8	225	49	176	10.7
21.8	220	48	172	10.5
21.8	216	47	169	10.2
21.8	211	46	165	10.0
21.8	206	45	161	9.8
21.8	202	44	158	9.6
21.8	197	43	154	9.4
21.8	193	42	151	9.1
21.8	188	41	147	8.9
21.8	179	39	140	8.5
21.8	174	38	136	8.3
21.8	170	37	133	8.1
21.8	165	36	129	7.9
21.8	156	34	122	7.4
21.8	147	32	115	7.0
21.8	142	31	111	6.8
21.8	133	29	104	6.3

%E	M1	M2	DM	M*
21.8	124	27	97	5.9
21.8	119	26	93	5.7
21.8	110	24	86	5.2
21.8	101	22	79	4.8
21.8	87	19	68	4.1
21.8	78	17	61	3.7
21.8	55	12	43	2.6
21.7	498	108	390	23.4
21.7	497	108	389	23.5
21.7	494	107	387	23.2
21.7	493	107	386	23.2
21.7	492	107	385	23.3
21.7	489	106	383	23.0
21.7	488	106	382	23.0
21.7	484	105	379	22.8
21.7	483	105	378	22.8
21.7	480	104	376	22.5
21.7	479	104	375	22.6
21.7	475	103	372	22.3
21.7	474	103	371	22.4
21.7	471	102	369	22.1
21.7	470	102	368	22.1
21.7	469	102	367	22.2
21.7	466	101	365	21.9
21.7	465	101	364	21.9
21.7	461	100	361	21.7
21.7	460	100	360	21.7
21.7	457	99	358	21.4
21.7	456	99	357	21.5
21.7	452	98	354	21.2
21.7	451	98	353	21.3
21.7	448	97	351	21.0
21.7	447	97	350	21.0
21.7	446	97	349	21.1
21.7	443	96	347	20.8
21.7	442	96	346	20.9
21.7	438	95	343	20.6
21.7	437	95	342	20.7
21.7	434	94	340	20.4
21.7	433	94	339	20.4
21.7	429	93	336	20.2
21.7	428	93	335	20.2
21.7	424	92	332	20.0
21.7	423	92	331	20.0
21.7	420	91	329	19.7
21.7	419	91	328	19.8
21.7	415	90	325	19.5
21.7	414	90	324	19.6
21.7	411	89	322	19.3
21.7	410	89	321	19.3

%E	M1	M2	DM	M*
21.7	406	88	318	19.1
21.7	405	88	317	19.1
21.7	401	87	314	18.9
21.7	397	86	311	18.6
21.7	396	86	310	18.7
21.7	392	85	307	18.4
21.7	391	85	306	18.5
21.7	387	84	303	18.2
21.7	383	83	300	18.0
21.7	382	83	299	18.0
21.7	378	82	296	17.8
21.7	374	81	293	17.5
21.7	373	81	292	17.6
21.7	369	80	289	17.3
21.7	368	80	288	17.4
21.7	364	79	285	17.1
21.7	360	78	282	16.9
21.7	359	78	281	16.9
21.7	355	77	278	16.7
21.7	351	76	275	16.5
21.7	350	76	274	16.5
21.7	346	75	271	16.3
21.7	345	75	270	16.3
21.7	341	74	267	16.1
21.7	337	73	264	15.8
21.7	336	73	263	15.9
21.7	332	72	260	15.6
21.7	327	71	256	15.4
21.7	323	70	253	15.2
21.7	322	70	252	15.2
21.7	318	69	249	15.0
21.7	314	68	246	14.7
21.7	313	68	245	14.8
21.7	309	67	242	14.5
21.7	304	66	238	14.3
21.7	300	65	235	14.1
21.7	299	65	234	14.1
21.7	295	64	231	13.9
21.7	290	63	227	13.7
21.7	286	62	224	13.4
21.7	281	61	220	13.2
21.7	277	60	217	13.0
21.7	276	60	216	13.0
21.7	272	59	213	12.8
21.7	267	58	209	12.6
21.7	263	57	206	12.4
21.7	258	56	202	12.2
21.7	254	55	199	11.9
21.7	253	55	198	12.0
21.7	249	54	195	11.7

%E	M1	M2	DM	M*
21.7	244	53	191	11.5
21.7	240	52	188	11.3
21.7	235	51	184	11.1
21.7	230	50	180	10.9
21.7	226	49	177	10.6
21.7	221	48	173	10.4
21.7	217	47	170	10.2
21.7	212	46	166	10.0
21.7	207	45	162	9.8
21.7	203	44	159	9.5
21.7	198	43	155	9.3
21.7	189	41	148	8.9
21.7	184	40	144	8.7
21.7	180	39	141	8.4
21.7	175	38	137	8.3
21.7	166	36	130	7.8
21.7	161	35	126	7.6
21.7	157	34	123	7.4
21.7	152	33	119	7.2
21.7	143	31	112	6.7
21.7	138	30	108	6.5
21.7	129	28	101	6.1
21.7	120	26	94	5.6
21.7	115	25	90	5.4
21.7	106	23	83	5.0
21.7	92	20	72	4.3
21.7	83	18	65	3.9
21.7	69	15	54	3.3
21.7	60	13	47	2.8
21.7	46	10	36	2.2
21.7	23	5	18	1.1
21.6	500	108	392	23.3
21.6	499	108	391	23.4
21.6	496	107	389	23.1
21.6	495	107	388	23.1
21.6	491	106	385	22.9
21.6	490	106	384	22.9
21.6	487	105	382	22.6
21.6	486	105	381	22.7
21.6	485	105	380	22.7
21.6	482	104	378	22.4
21.6	481	104	377	22.5
21.6	477	103	374	22.2
21.6	476	103	373	22.3
21.6	473	102	371	22.0
21.6	472	102	370	22.0
21.6	468	101	367	21.8
21.6	467	101	366	21.8
21.6	464	100	364	21.6
21.6	463	100	363	21.6

%E	M1	M2	DM	M*
21.6	462	100	362	21.6
21.6	459	99	360	21.4
21.6	458	99	359	21.4
21.6	454	98	356	21.2
21.6	453	98	355	21.2
21.6	450	97	353	20.9
21.6	449	97	352	21.0
21.6	445	96	349	20.7
21.6	444	96	348	20.8
21.6	440	95	345	20.5
21.6	439	95	344	20.6
21.6	436	94	342	20.3
21.6	435	94	341	20.3
21.6	431	93	338	20.1
21.6	430	93	337	20.1
21.6	426	92	334	19.9
21.6	425	92	333	19.9
21.6	422	91	331	19.6
21.6	421	91	330	19.7
21.6	417	90	327	19.4
21.6	416	90	326	19.5
21.6	413	89	324	19.2
21.6	412	89	323	19.2
21.6	408	88	320	19.0
21.6	407	88	319	19.0
21.6	403	87	316	18.8
21.6	402	87	315	18.8
21.6	399	86	313	18.5
21.6	398	86	312	18.6
21.6	394	85	309	18.3
21.6	393	85	308	18.4
21.6	389	84	305	18.1
21.6	388	84	304	18.2
21.6	385	83	302	17.9
21.6	384	83	301	17.9
21.6	380	82	298	17.7
21.6	379	82	297	17.7
21.6	375	81	294	17.5
21.6	371	80	291	17.3
21.6	370	80	290	17.3
21.6	366	79	287	17.1
21.6	365	79	286	17.1
21.6	361	78	283	16.9
21.6	357	77	280	16.6
21.6	356	77	279	16.7
21.6	352	76	276	16.4
21.6	348	75	273	16.2
21.6	347	75	272	16.2
21.6	343	74	269	16.0
21.6	342	74	268	16.0

%E	M1	M2	DM	M*
21.6	338	73	265	15.8
21.6	334	72	262	15.5
21.6	333	72	261	15.6
21.6	329	71	258	15.3
21.6	328	71	257	15.4
21.6	324	70	254	15.1
21.6	320	69	251	14.9
21.6	319	69	250	14.9
21.6	315	68	247	14.7
21.6	310	67	243	14.5
21.6	306	66	240	14.2
21.6	305	66	239	14.3
21.6	301	65	236	14.0
21.6	296	64	232	13.8
21.6	292	63	229	13.6
21.6	291	63	228	13.6
21.6	287	62	225	13.4
21.6	283	61	222	13.1
21.6	282	61	221	13.2
21.6	278	60	218	12.9
21.6	273	59	214	12.8
21.6	269	58	211	12.5
21.6	268	58	210	12.6
21.6	264	57	207	12.3
21.6	259	56	203	12.1
21.6	255	55	200	11.9
21.6	250	54	196	11.7
21.6	245	53	192	11.5
21.6	241	52	189	11.2
21.6	236	51	185	11.0
21.6	232	50	182	10.8
21.6	231	50	181	10.8
21.6	227	49	178	10.6
21.6	222	48	174	10.4
21.6	218	47	171	10.1
21.6	213	46	167	9.9
21.6	208	45	163	9.7
21.6	204	44	160	9.5
21.6	199	43	156	9.3
21.6	194	42	152	9.1
21.6	190	41	149	8.8
21.6	185	40	145	8.6
21.6	176	38	138	8.2
21.6	171	37	134	8.0
21.6	167	36	131	7.8
21.6	162	35	127	7.6
21.6	153	33	120	7.1
21.6	148	32	116	6.9
21.6	139	30	109	6.5
21.6	134	29	105	6.3

%E	M1	M2	DM	M*
21.6	125	27	98	5.8
21.6	116	25	91	5.4
21.6	111	24	87	5.2
21.6	102	22	80	4.7
21.6	97	21	76	4.5
21.6	88	19	69	4.1
21.6	74	16	58	3.5
21.6	51	11	40	2.4
21.6	37	8	29	1.7
21.5	498	107	391	23.0
21.5	497	107	390	23.0
21.5	494	106	388	22.7
21.5	493	106	387	22.8
21.5	492	106	386	22.8
21.5	489	105	384	22.5
21.5	488	105	383	22.6
21.5	484	104	380	22.3
21.5	483	104	379	22.4
21.5	480	103	377	22.1
21.5	479	103	376	22.1
21.5	478	103	375	22.2
21.5	475	102	373	21.9
21.5	474	102	372	21.9
21.5	470	101	369	21.7
21.5	469	101	368	21.8
21.5	466	100	366	21.5
21.5	465	100	365	21.5
21.5	461	99	362	21.3
21.5	460	99	361	21.3
21.5	456	98	358	21.1
21.5	455	98	357	21.1
21.5	452	97	355	20.8
21.5	451	97	354	20.9
21.5	447	96	351	20.6
21.5	446	96	350	20.7
21.5	442	95	347	20.4
21.5	441	95	346	20.5
21.5	438	94	344	20.2
21.5	437	94	343	20.2
21.5	433	93	340	20.0
21.5	432	93	339	20.0
21.5	428	92	336	19.8
21.5	427	92	335	19.8
21.5	424	91	333	19.5
21.5	423	91	332	19.6
21.5	419	90	329	19.3
21.5	418	90	328	19.4
21.5	414	89	325	19.1
21.5	410	88	322	18.9
21.5	409	88	321	18.9

%E	M1	M2	DM	M*
21.5	405	87	318	18.7
21.5	404	87	317	18.7
21.5	400	86	314	18.5
21.5	396	85	311	18.2
21.5	395	85	310	18.3
21.5	391	84	307	18.0
21.5	390	84	306	18.1
21.5	386	83	303	17.8
21.5	382	82	300	17.6
21.5	381	82	299	17.6
21.5	377	81	296	17.4
21.5	376	81	295	17.4
21.5	372	80	292	17.2
21.5	368	79	289	17.0
21.5	367	79	288	17.0
21.5	363	78	285	16.8
21.5	362	78	284	16.8
21.5	358	77	281	16.6
21.5	354	76	278	16.3
21.5	353	76	277	16.4
21.5	349	75	274	16.1
21.5	344	74	270	15.9
21.5	340	73	267	15.7
21.5	339	73	266	15.7
21.5	335	72	263	15.5
21.5	331	71	260	15.2
21.5	330	71	259	15.3
21.5	326	70	256	15.0
21.5	325	70	255	15.1
21.5	321	69	252	14.8
21.5	317	68	249	14.6
21.5	316	68	248	14.6
21.5	312	67	245	14.4
21.5	311	67	244	14.4
21.5	307	66	241	14.2
21.5	303	65	238	13.9
21.5	302	65	237	14.0
21.5	298	64	234	13.7
21.5	297	64	233	13.8
21.5	293	63	230	13.5
21.5	289	62	227	13.3
21.5	288	62	226	13.3
21.5	284	61	223	13.1
21.5	279	60	219	12.9
21.5	275	59	216	12.7
21.5	274	59	215	12.7
21.5	270	58	212	12.5
21.5	265	57	208	12.3
21.5	261	56	205	12.0
21.5	260	56	204	12.1

%E	M1	M2	DM	M*
21.5	256	55	201	11.8
21.5	251	54	197	11.6
21.5	247	53	194	11.4
21.5	246	53	193	11.4
21.5	242	52	190	11.2
21.5	237	51	186	11.0
21.5	233	50	183	10.7
21.5	228	49	179	10.5
21.5	223	48	175	10.3
21.5	219	47	172	10.1
21.5	214	46	168	9.9
21.5	209	45	164	9.7
21.5	205	44	161	9.4
21.5	200	43	157	9.2
21.5	195	42	153	9.0
21.5	191	41	150	8.8
21.5	186	40	146	8.6
21.5	181	39	142	8.4
21.5	177	38	139	8.2
21.5	172	37	135	8.0
21.5	163	35	128	7.5
21.5	158	34	124	7.3
21.5	149	32	117	6.9
21.5	144	31	113	6.7
21.5	135	29	106	6.2
21.5	130	28	102	6.0
21.5	121	26	95	5.6
21.5	107	23	84	4.9
21.5	93	20	73	4.3
21.5	79	17	62	3.7
21.5	65	14	51	3.0
21.4	500	107	393	22.9
21.4	499	107	392	22.9
21.4	496	106	390	22.7
21.4	495	106	389	22.7
21.4	491	105	386	22.5
21.4	490	105	385	22.5
21.4	487	104	383	22.2
21.4	486	104	382	22.3
21.4	485	104	381	22.3
21.4	482	103	379	22.0
21.4	481	103	378	22.1
21.4	477	102	375	21.8
21.4	476	102	374	21.9
21.4	473	101	372	21.6
21.4	472	101	371	21.6
21.4	471	101	370	21.7
21.4	468	100	368	21.4
21.4	467	100	367	21.4
21.4	463	99	364	21.2

%E	M1	M2	DM	M*
21.4	462	99	363	21.2
21.4	459	98	361	20.9
21.4	458	98	360	21.0
21.4	457	98	359	21.0
21.4	454	97	357	20.7
21.4	453	97	356	20.8
21.4	449	96	353	20.5
21.4	448	96	352	20.6
21.4	444	95	349	20.3
21.4	443	95	348	20.4
21.4	440	94	346	20.1
21.4	439	94	345	20.1
21.4	435	93	342	19.9
21.4	434	93	341	19.9
21.4	430	92	338	19.7
21.4	429	92	337	19.7
21.4	426	91	335	19.4
21.4	425	91	334	19.5
21.4	421	90	331	19.2
21.4	420	90	330	19.3
21.4	416	89	327	19.0
21.4	415	89	326	19.1
21.4	412	88	324	18.8
21.4	411	88	323	18.8
21.4	407	87	320	18.6
21.4	406	87	319	18.6
21.4	402	86	316	18.4
21.4	401	86	315	18.4
21.4	398	85	313	18.2
21.4	397	85	312	18.2
21.4	393	84	309	18.0
21.4	392	84	308	18.0
21.4	388	83	305	17.8
21.4	387	83	304	17.8
21.4	384	82	302	17.5
21.4	383	82	301	17.6
21.4	379	81	298	17.3
21.4	378	81	297	17.4
21.4	374	80	294	17.1
21.4	373	80	293	17.2
21.4	370	79	291	16.9
21.4	369	79	290	16.9
21.4	365	78	287	16.7
21.4	364	78	286	16.7
21.4	360	77	283	16.5
21.4	359	77	282	16.5
21.4	355	76	279	16.3
21.4	351	75	276	16.0
21.4	350	75	275	16.1
21.4	346	74	272	15.8

%E	M1	M2	DM	M*
21.4	345	74	271	15.9
21.4	341	73	268	15.6
21.4	337	72	265	15.4
21.4	336	72	264	15.4
21.4	332	71	261	15.2
21.4	327	70	257	15.0
21.4	323	69	254	14.7
21.4	322	69	253	14.8
21.4	318	68	250	14.5
21.4	313	67	246	14.3
21.4	309	66	243	14.1
21.4	308	66	242	14.1
21.4	304	65	239	13.9
21.4	299	64	235	13.7
21.4	295	63	232	13.5
21.4	294	63	231	13.5
21.4	290	62	228	13.3
21.4	285	61	224	13.1
21.4	281	60	221	12.8
21.4	280	60	220	12.9
21.4	276	59	217	12.6
21.4	271	58	213	12.4
21.4	266	57	209	12.2
21.4	262	56	206	12.0
21.4	257	55	202	11.8
21.4	252	54	198	11.6
21.4	248	53	195	11.3
21.4	243	52	191	11.1
21.4	238	51	187	10.9
21.4	234	50	184	10.7
21.4	229	49	180	10.5
21.4	224	48	176	10.3
21.4	220	47	173	10.0
21.4	215	46	169	9.8
21.4	210	45	165	9.6
21.4	206	44	162	9.4
21.4	201	43	158	9.2
21.4	196	42	154	9.0
21.4	192	41	151	8.8
21.4	187	40	147	8.6
21.4	182	39	143	8.4
21.4	173	37	136	7.9
21.4	168	36	132	7.7
21.4	159	34	125	7.3
21.4	154	33	121	7.1
21.4	145	31	114	6.6
21.4	140	30	110	6.4
21.4	131	28	103	6.0
21.4	126	27	99	5.8
21.4	117	25	92	5.3

%E	M1	M2	DM	M*
21.4	112	24	88	5.1
21.4	103	22	81	4.7
21.4	98	21	77	4.5
21.4	84	18	66	3.9
21.4	70	15	55	3.2
21.4	56	12	44	2.6
21.4	42	9	33	1.9
21.4	28	6	22	1.3
21.4	14	3	11	0.6
21.3	498	106	392	22.6
21.3	497	106	391	22.6
21.3	494	105	389	22.3
21.3	493	105	388	22.4
21.3	492	105	387	22.4
21.3	489	104	385	22.1
21.3	488	104	384	22.2
21.3	484	103	381	21.9
21.3	483	103	380	22.0
21.3	480	102	378	21.7
21.3	479	102	377	21.7
21.3	478	102	376	21.8
21.3	475	101	374	21.5
21.3	474	101	373	21.5
21.3	470	100	370	21.3
21.3	469	100	369	21.3
21.3	465	99	366	21.1
21.3	464	99	365	21.1
21.3	461	98	363	20.8
21.3	460	98	362	20.9
21.3	456	97	359	20.6
21.3	455	97	358	20.7
21.3	451	96	355	20.4
21.3	450	96	354	20.5
21.3	447	95	352	20.2
21.3	446	95	351	20.2
21.3	445	95	350	20.3
21.3	442	94	348	20.0
21.3	441	94	347	20.0
21.3	437	93	344	19.8
21.3	436	93	343	19.8
21.3	432	92	340	19.6
21.3	431	92	339	19.6
21.3	428	91	337	19.3
21.3	427	91	336	19.4
21.3	423	90	333	19.1
21.3	422	90	332	19.2
21.3	418	89	329	18.9
21.3	417	89	328	19.0
21.3	414	88	326	18.7
21.3	413	88	325	18.8

%E	M1	M2	DM	M*
21.3	409	87	322	18.5
21.3	408	87	321	18.6
21.3	404	86	318	18.3
21.3	403	86	317	18.4
21.3	400	85	315	18.1
21.3	399	85	314	18.1
21.3	395	84	311	17.9
21.3	394	84	310	17.9
21.3	390	83	307	17.7
21.3	389	83	306	17.7
21.3	385	82	303	17.5
21.3	381	81	300	17.2
21.3	380	81	299	17.3
21.3	376	80	296	17.0
21.3	375	80	295	17.1
21.3	371	79	292	16.8
21.3	367	78	289	16.6
21.3	366	78	288	16.6
21.3	362	77	285	16.4
21.3	361	77	284	16.4
21.3	357	76	281	16.2
21.3	356	76	280	16.2
21.3	352	75	277	16.0
21.3	348	74	274	15.7
21.3	347	74	273	15.8
21.3	343	73	270	15.5
21.3	342	73	269	15.6
21.3	338	72	266	15.3
21.3	334	71	263	15.1
21.3	333	71	262	15.1
21.3	329	70	259	14.9
21.3	328	70	258	14.9
21.3	324	69	255	14.7
21.3	320	68	252	14.4
21.3	319	68	251	14.5
21.3	315	67	248	14.3
21.3	314	67	247	14.3
21.3	310	66	244	14.1
21.3	305	65	240	13.9
21.3	301	64	237	13.6
21.3	300	64	236	13.7
21.3	296	63	233	13.4
21.3	291	62	229	13.2
21.3	287	61	226	13.0
21.3	286	61	225	13.0
21.3	282	60	222	12.8
21.3	277	59	218	12.6
21.3	272	58	214	12.4
21.3	268	57	211	12.1
21.3	267	57	210	12.2

%E	M1	M2	DM	M*
21.3	263	56	207	11.9
21.3	258	55	203	11.7
21.3	254	54	200	11.5
21.3	253	54	199	11.5
21.3	249	53	196	11.3
21.3	244	52	192	11.1
21.3	240	51	189	10.8
21.3	239	51	188	10.9
21.3	235	50	185	10.6
21.3	230	49	181	10.4
21.3	225	48	177	10.2
21.3	221	47	174	10.0
21.3	216	46	170	9.8
21.3	211	45	166	9.6
21.3	207	44	163	9.4
21.3	202	43	159	9.2
21.3	197	42	155	9.0
21.3	188	40	148	8.5
21.3	183	39	144	8.3
21.3	178	38	140	8.1
21.3	174	37	137	7.9
21.3	169	36	133	7.7
21.3	164	35	129	7.5
21.3	160	34	126	7.2
21.3	155	33	122	7.0
21.3	150	32	118	6.8
21.3	141	30	111	6.4
21.3	136	29	107	6.2
21.3	127	27	100	5.7
21.3	122	26	96	5.5
21.3	108	23	85	4.9
21.3	94	20	74	4.3
21.3	89	19	70	4.1
21.3	80	17	63	3.6
21.3	75	16	59	3.4
21.3	61	13	48	2.8
21.3	47	10	37	2.1
21.2	500	106	394	22.5
21.2	499	106	393	22.5
21.2	496	105	391	22.2
21.2	495	105	390	22.3
21.2	491	104	387	22.0
21.2	490	104	386	22.1
21.2	487	103	384	21.8
21.2	486	103	383	21.8
21.2	485	103	382	21.9
21.2	482	102	380	21.6
21.2	481	102	379	21.6
21.2	477	101	376	21.4
21.2	476	101	375	21.4

%E	M1	M2	DM	M*
21.2	472	100	372	21.2
21.2	471	100	371	21.2
21.2	468	99	369	20.9
21.2	467	99	368	21.0
21.2	466	99	367	21.0
21.2	463	98	365	20.7
21.2	462	98	364	20.8
21.2	458	97	361	20.5
21.2	457	97	360	20.6
21.2	453	96	357	20.3
21.2	452	96	356	20.4
21.2	449	95	354	20.1
21.2	448	95	353	20.1
21.2	444	94	350	19.9
21.2	443	94	349	19.9
21.2	439	93	346	19.7
21.2	438	93	345	19.7
21.2	434	92	342	19.5
21.2	433	92	341	19.5
21.2	430	91	339	19.3
21.2	429	91	338	19.3
21.2	425	90	335	19.1
21.2	424	90	334	19.1
21.2	420	89	331	18.9
21.2	419	89	330	18.9
21.2	416	88	328	18.6
21.2	415	88	327	18.7
21.2	411	87	324	18.4
21.2	410	87	323	18.5
21.2	406	86	320	18.2
21.2	405	86	319	18.3
21.2	401	85	316	18.0
21.2	397	84	313	17.8
21.2	396	84	312	17.8
21.2	392	83	309	17.6
21.2	391	83	308	17.6
21.2	387	82	305	17.4
21.2	386	82	304	17.4
21.2	382	81	301	17.2
21.2	378	80	298	16.9
21.2	377	80	297	17.0
21.2	373	79	294	16.7
21.2	372	79	293	16.8
21.2	368	78	290	16.5
21.2	364	77	287	16.3
21.2	363	77	286	16.3
21.2	359	76	283	16.1
21.2	358	76	282	16.1
21.2	354	75	279	15.9
21.2	353	75	278	15.9

%E	M1	M2	DM	M*
21.2	349	74	275	15.7
21.2	345	73	272	15.4
21.2	344	73	271	15.5
21.2	340	72	268	15.2
21.2	339	72	267	15.3
21.2	335	71	264	15.0
21.2	330	70	260	14.8
21.2	326	69	257	14.6
21.2	325	69	256	14.6
21.2	321	68	253	14.4
21.2	316	67	249	14.2
21.2	312	66	246	14.0
21.2	311	66	245	14.0
21.2	307	65	242	13.8
21.2	306	65	241	13.8
21.2	302	64	238	13.6
21.2	297	63	234	13.4
21.2	293	62	231	13.1
21.2	292	62	230	13.2
21.2	288	61	227	12.9
21.2	283	60	223	12.7
21.2	278	59	219	12.5
21.2	274	58	216	12.3
21.2	273	58	215	12.3
21.2	269	57	212	12.1
21.2	264	56	208	11.9
21.2	260	55	205	11.6
21.2	259	55	204	11.7
21.2	255	54	201	11.4
21.2	250	53	197	11.2
21.2	245	52	193	11.0
21.2	241	51	190	10.8
21.2	236	50	186	10.6
21.2	231	49	182	10.4
21.2	226	48	178	10.2
21.2	222	47	175	10.0
21.2	217	46	171	9.8
21.2	212	45	167	9.6
21.2	208	44	164	9.3
21.2	203	43	160	9.1
21.2	198	42	156	8.9
21.2	193	41	152	8.7
21.2	189	40	149	8.5
21.2	184	39	145	8.3
21.2	179	38	141	8.1
21.2	170	36	134	7.6
21.2	165	35	130	7.4
21.2	156	33	123	7.0
21.2	151	32	119	6.8
21.2	146	31	115	6.6

%E	M1	M2	DM	M*
21.2	137	29	108	6.1
21.2	132	28	104	5.9
21.2	118	25	93	5.3
21.2	113	24	89	5.1
21.2	104	22	82	4.7
21.2	99	21	78	4.5
21.2	85	18	67	3.8
21.2	66	14	52	3.0
21.2	52	11	41	2.3
21.2	33	7	26	1.5
21.1	498	105	393	22.1
21.1	497	105	392	22.2
21.1	494	104	390	21.9
21.1	493	104	389	21.9
21.1	492	104	388	22.0
21.1	489	103	386	21.7
21.1	488	103	385	21.7
21.1	484	102	382	21.5
21.1	483	102	381	21.5
21.1	479	101	378	21.3
21.1	478	101	377	21.3
21.1	475	100	375	21.1
21.1	474	100	374	21.1
21.1	473	100	373	21.1
21.1	470	99	371	20.9
21.1	469	99	370	20.9
21.1	465	98	367	20.7
21.1	464	98	366	20.7
21.1	460	97	363	20.5
21.1	459	97	362	20.5
21.1	456	96	360	20.2
21.1	455	96	359	20.3
21.1	454	96	358	20.3
21.1	451	95	356	20.0
21.1	450	95	355	20.1
21.1	446	94	352	19.8
21.1	445	94	351	19.9
21.1	441	93	348	19.6
21.1	440	93	347	19.7
21.1	437	92	345	19.4
21.1	436	92	344	19.4
21.1	435	92	343	19.5
21.1	432	91	341	19.2
21.1	431	91	340	19.2
21.1	427	90	337	19.0
21.1	426	90	336	19.0
21.1	422	89	333	18.8
21.1	421	89	332	18.8
21.1	418	88	330	18.5
21.1	417	88	329	18.6

%E	M1	M2	DM	M*
21.1	413	87	326	18.3
21.1	412	87	325	18.4
21.1	408	86	322	18.1
21.1	407	86	321	18.2
21.1	403	85	318	17.9
21.1	402	85	317	18.0
21.1	399	84	315	17.7
21.1	398	84	314	17.7
21.1	394	83	311	17.5
21.1	393	83	310	17.5
21.1	389	82	307	17.3
21.1	388	82	306	17.3
21.1	384	81	303	17.1
21.1	383	81	302	17.1
21.1	380	80	300	16.8
21.1	379	80	299	16.9
21.1	375	79	296	16.6
21.1	374	79	295	16.7
21.1	370	78	292	16.4
21.1	369	78	291	16.5
21.1	365	77	288	16.2
21.1	361	76	285	16.0
21.1	360	76	284	16.0
21.1	356	75	281	15.8
21.1	355	75	280	15.8
21.1	351	74	277	15.6
21.1	350	74	276	15.6
21.1	346	73	273	15.4
21.1	342	72	270	15.2
21.1	341	72	269	15.2
21.1	337	71	266	15.0
21.1	336	71	265	15.0
21.1	332	70	262	14.8
21.1	331	70	261	14.8
21.1	327	69	258	14.6
21.1	323	68	255	14.3
21.1	322	68	254	14.4
21.1	318	67	251	14.1
21.1	317	67	250	14.2
21.1	313	66	247	13.9
21.1	308	65	243	13.7
21.1	304	64	240	13.5
21.1	303	64	239	13.5
21.1	299	63	236	13.3
21.1	298	63	235	13.3
21.1	294	62	232	13.1
21.1	289	61	228	12.9
21.1	285	60	225	12.6
21.1	284	60	224	12.7
21.1	280	59	221	12.4

%E	M1	M2	DM	M*
21.1	279	59	220	12.5
21.1	275	58	217	12.2
21.1	270	57	213	12.0
21.1	266	56	210	11.8
21.1	265	56	209	11.8
21.1	261	55	206	11.6
21.1	256	54	202	11.4
21.1	251	53	198	11.2
21.1	247	52	195	10.9
21.1	246	52	194	11.0
21.1	242	51	191	10.7
21.1	237	50	187	10.5
21.1	232	49	183	10.3
21.1	228	48	180	10.1
21.1	227	48	179	10.1
21.1	223	47	176	9.9
21.1	218	46	172	9.7
21.1	213	45	168	9.5
21.1	209	44	165	9.3
21.1	204	43	161	9.1
21.1	199	42	157	8.9
21.1	194	41	153	8.7
21.1	190	40	150	8.4
21.1	185	39	146	8.2
21.1	180	38	142	8.0
21.1	175	37	138	7.8
21.1	171	36	135	7.6
21.1	166	35	131	7.4
21.1	161	34	127	7.2
21.1	152	32	120	6.7
21.1	147	31	116	6.5
21.1	142	30	112	6.3
21.1	133	28	105	5.9
21.1	128	27	101	5.7
21.1	123	26	97	5.5
21.1	114	24	90	5.1
21.1	109	23	86	4.9
21.1	95	20	75	4.2
21.1	90	19	71	4.0
21.1	76	16	60	3.4
21.1	71	15	56	3.2
21.1	57	12	45	2.5
21.1	38	8	30	1.7
21.1	19	4	15	0.8
21.0	500	105	395	22.0
21.0	499	105	394	22.1
21.0	496	104	392	21.8
21.0	495	104	391	21.9
21.0	491	103	388	21.6
21.0	490	103	387	21.7

%E	M1	M2	DM	M*
21.0	486	102	384	21.4
21.0	485	102	383	21.5
21.0	482	101	381	21.2
21.0	481	101	380	21.2
21.0	480	101	379	21.3
21.0	477	100	377	21.0
21.0	476	100	376	21.0
21.0	472	99	373	20.8
21.0	471	99	372	20.8
21.0	467	98	369	20.6
21.0	466	98	368	20.6
21.0	463	97	366	20.3
21.0	462	97	365	20.4
21.0	461	97	364	20.4
21.0	458	96	362	20.1
21.0	457	96	361	20.2
21.0	453	95	358	19.9
21.0	452	95	357	20.0
21.0	448	94	354	19.7
21.0	447	94	353	19.8
21.0	443	93	350	19.5
21.0	442	93	349	19.6
21.0	439	92	347	19.3
21.0	438	92	346	19.3
21.0	434	91	343	19.1
21.0	433	91	342	19.1
21.0	429	90	339	18.9
21.0	428	90	338	18.9
21.0	424	89	335	18.7
21.0	423	89	334	18.7
21.0	420	88	332	18.4
21.0	419	88	331	18.5
21.0	415	87	328	18.2
21.0	414	87	327	18.3
21.0	410	86	324	18.0
21.0	409	86	323	18.1
21.0	405	85	320	17.8
21.0	404	85	319	17.9
21.0	400	84	316	17.6
21.0	396	83	313	17.4
21.0	395	83	312	17.4
21.0	391	82	309	17.2
21.0	390	82	308	17.2
21.0	386	81	305	17.0
21.0	385	81	304	17.0
21.0	381	80	301	16.8
21.0	377	79	298	16.6
21.0	376	79	297	16.6
21.0	372	78	294	16.4
21.0	371	78	293	16.4

%E	M1	M2	DM	M*
21.0	367	77	290	16.2
21.0	366	77	289	16.2
21.0	362	76	286	16.0
21.0	358	75	283	15.7
21.0	357	75	282	15.8
21.0	353	74	279	15.5
21.0	352	74	278	15.6
21.0	348	73	275	15.3
21.0	347	73	274	15.4
21.0	343	72	271	15.1
21.0	338	71	267	14.9
21.0	334	70	264	14.7
21.0	333	70	263	14.7
21.0	329	69	260	14.5
21.0	328	69	259	14.5
21.0	324	68	256	14.3
21.0	319	67	252	14.1
21.0	315	66	249	13.8
21.0	314	66	248	13.9
21.0	310	65	245	13.6
21.0	309	65	244	13.7
21.0	305	64	241	13.4
21.0	300	63	237	13.2
21.0	295	62	233	13.0
21.0	291	61	230	12.8
21.0	290	61	229	12.8
21.0	286	60	226	12.6
21.0	281	59	222	12.4
21.0	276	58	218	12.2
21.0	272	57	215	11.9
21.0	271	57	214	12.0
21.0	267	56	211	11.7
21.0	262	55	207	11.5
21.0	257	54	203	11.3
21.0	252	53	199	11.1
21.0	248	52	196	10.9
21.0	243	51	192	10.7
21.0	238	50	188	10.5
21.0	233	49	184	10.3
21.0	229	48	181	10.1
21.0	224	47	177	9.9
21.0	219	46	173	9.7
21.0	214	45	169	9.5
21.0	210	44	166	9.2
21.0	205	43	162	9.0
21.0	200	42	158	8.8
21.0	195	41	154	8.6
21.0	186	39	147	8.2
21.0	181	38	143	8.0
21.0	176	37	139	7.8

%E	M1	M2	DM	M*
21.0	167	35	132	7.3
21.0	162	34	128	7.1
21.0	157	33	124	6.9
21.0	143	30	113	6.3
21.0	138	29	109	6.1
21.0	124	26	98	5.5
21.0	119	25	94	5.3
21.0	105	22	83	4.6
21.0	100	21	79	4.4
21.0	81	17	64	3.6
21.0	62	13	49	2.7
20.9	498	104	394	21.7
20.9	497	104	393	21.8
20.9	494	103	391	21.5
20.9	493	103	390	21.5
20.9	492	103	389	21.6
20.9	489	102	387	21.3
20.9	488	102	386	21.3
20.9	487	102	385	21.4
20.9	484	101	383	21.1
20.9	483	101	382	21.1
20.9	479	100	379	20.9
20.9	478	100	378	20.9
20.9	474	99	375	20.7
20.9	473	99	374	20.7
20.9	470	98	372	20.4
20.9	469	98	371	20.5
20.9	468	98	370	20.5
20.9	465	97	368	20.2
20.9	464	97	367	20.3
20.9	460	96	364	20.0
20.9	459	96	363	20.1
20.9	455	95	360	19.8
20.9	454	95	359	19.9
20.9	450	94	356	19.6
20.9	449	94	355	19.7
20.9	446	93	353	19.4
20.9	445	93	352	19.4
20.9	444	93	351	19.5
20.9	441	92	349	19.2
20.9	440	92	348	19.2
20.9	436	91	345	19.0
20.9	435	91	344	19.0
20.9	431	90	341	18.8
20.9	430	90	340	18.8
20.9	426	89	337	18.6
20.9	425	89	336	18.6
20.9	422	88	334	18.4
20.9	421	88	333	18.4
20.9	417	87	330	18.2
20.9	416	87	329	18.2
20.9	412	86	326	18.0
20.9	411	86	325	18.0
20.9	407	85	322	17.8
20.9	406	85	321	17.8
20.9	402	84	318	17.6
20.9	401	84	317	17.6
20.9	398	83	315	17.3
20.9	397	83	314	17.4
20.9	393	82	311	17.1
20.9	392	82	310	17.2
20.9	388	81	307	16.9
20.9	387	81	306	17.0
20.9	383	80	303	16.7
20.9	382	80	302	16.8
20.9	378	79	299	16.5
20.9	374	78	296	16.3
20.9	373	78	295	16.3
20.9	369	77	292	16.1
20.9	368	77	291	16.1
20.9	364	76	288	15.9
20.9	363	76	287	15.9
20.9	359	75	284	15.7
20.9	354	74	280	15.5
20.9	350	73	277	15.2
20.9	349	73	276	15.3
20.9	345	72	273	15.0
20.9	344	72	272	15.1
20.9	340	71	269	14.8
20.9	339	71	268	14.9
20.9	335	70	265	14.6
20.9	330	69	261	14.4
20.9	326	68	258	14.2
20.9	325	68	257	14.2
20.9	321	67	254	14.0
20.9	320	67	253	14.0
20.9	316	66	250	13.8
20.9	311	65	246	13.6
20.9	306	64	242	13.4
20.9	302	63	239	13.1
20.9	301	63	238	13.2
20.9	297	62	235	12.9
20.9	296	62	234	13.0
20.9	292	61	231	12.7
20.9	287	60	227	12.5
20.9	282	59	223	12.3
20.9	278	58	220	12.1
20.9	277	58	219	12.1
20.9	273	57	216	11.9
20.9	268	56	212	11.7
20.9	263	55	208	11.5
20.9	258	54	204	11.3
20.9	254	53	201	11.1
20.9	253	53	200	11.1
20.9	249	52	197	10.9
20.9	244	51	193	10.7
20.9	239	50	189	10.5
20.9	235	49	186	10.2
20.9	234	49	185	10.3
20.9	230	48	182	10.0
20.9	225	47	178	9.8
20.9	220	46	174	9.6
20.9	215	45	170	9.4
20.9	211	44	167	9.2
20.9	206	43	163	9.0
20.9	201	42	159	8.8
20.9	196	41	155	8.6
20.9	191	40	151	8.4
20.9	187	39	148	8.1
20.9	182	38	144	7.9
20.9	177	37	140	7.7
20.9	172	36	136	7.5
20.9	163	34	129	7.1
20.9	158	33	125	6.9
20.9	153	32	121	6.7
20.9	148	31	117	6.5
20.9	139	29	110	6.1
20.9	134	28	106	5.9
20.9	129	27	102	5.7
20.9	115	24	91	5.0
20.9	110	23	87	4.8
20.9	91	19	72	4.0
20.9	86	18	68	3.8
20.9	67	14	53	2.9
20.9	43	9	34	1.9
20.8	500	104	396	21.6
20.8	499	104	395	21.7
20.8	496	103	393	21.4
20.8	495	103	392	21.4
20.8	491	102	389	21.2
20.8	490	102	388	21.2
20.8	486	101	385	21.0
20.8	485	101	384	21.0
20.8	481	100	381	20.8
20.8	480	100	380	20.8
20.8	477	99	378	20.5
20.8	476	99	377	20.6
20.8	475	99	376	20.6
20.8	472	98	374	20.3
20.8	471	98	373	20.4
20.8	467	97	370	20.1
20.8	466	97	369	20.2
20.8	462	96	366	19.9
20.8	461	96	365	20.0
20.8	457	95	362	19.7
20.8	456	95	361	19.8
20.8	453	94	359	19.5
20.8	452	94	358	19.5
20.8	451	94	357	19.6
20.8	448	93	355	19.3
20.8	447	93	354	19.3
20.8	443	92	351	19.1
20.8	442	92	350	19.1
20.8	438	91	347	18.9
20.8	437	91	346	18.9
20.8	433	90	343	18.7
20.8	432	90	342	18.8
20.8	428	89	339	18.5
20.8	427	89	338	18.6
20.8	424	88	336	18.3
20.8	423	88	335	18.3
20.8	419	87	332	18.1
20.8	418	87	331	18.1
20.8	414	86	328	17.9
20.8	413	86	327	17.9
20.8	409	85	324	17.7
20.8	408	85	323	17.7
20.8	404	84	320	17.5
20.8	403	84	319	17.5
20.8	400	83	317	17.2
20.8	399	83	316	17.3
20.8	395	82	313	17.0
20.8	394	82	312	17.1
20.8	390	81	309	16.8
20.8	389	81	308	16.9
20.8	385	80	305	16.6
20.8	384	80	304	16.7
20.8	380	79	301	16.4
20.8	379	79	300	16.5
20.8	375	78	297	16.2
20.8	371	77	294	16.0
20.8	370	77	293	16.0
20.8	366	76	290	15.8
20.8	365	76	289	15.8
20.8	361	75	286	15.6
20.8	360	75	285	15.6
20.8	356	74	282	15.4
20.8	355	74	281	15.4
20.8	351	73	278	15.2
20.8	346	72	274	15.0
20.8	342	71	271	14.7
20.8	341	71	270	14.8
20.8	337	70	267	14.5
20.8	336	70	266	14.6
20.8	332	69	263	14.3
20.8	331	69	262	14.4
20.8	327	68	259	14.1
20.8	322	67	255	13.9
20.8	318	66	252	13.7
20.8	317	66	251	13.7
20.8	313	65	248	13.5
20.8	312	65	247	13.5
20.8	308	64	244	13.3
20.8	307	64	243	13.3
20.8	303	63	240	13.1
20.8	298	62	236	12.9
20.8	293	61	232	12.7
20.8	289	60	229	12.5
20.8	288	60	228	12.5
20.8	284	59	225	12.3
20.8	283	59	224	12.3
20.8	279	58	221	12.1
20.8	274	57	217	11.9
20.8	269	56	213	11.7
20.8	265	55	210	11.4
20.8	264	55	209	11.5
20.8	260	54	206	11.2
20.8	259	54	205	11.3
20.8	255	53	202	11.0
20.8	250	52	198	10.8
20.8	245	51	194	10.6
20.8	240	50	190	10.4
20.8	236	49	187	10.2
20.8	231	48	183	10.0
20.8	226	47	179	9.8
20.8	221	46	175	9.6
20.8	216	45	171	9.4
20.8	212	44	168	9.1
20.8	207	43	164	8.9
20.8	202	42	160	8.7
20.8	197	41	156	8.5
20.8	192	40	152	8.3
20.8	183	38	145	7.9
20.8	178	37	141	7.7
20.8	173	36	137	7.5
20.8	168	35	133	7.3
20.8	159	33	126	6.8
20.8	154	32	122	6.6
20.8	149	31	118	6.4
20.8	144	30	114	6.3

%E	M1	M2	DM	M*	%E	M1	M2	DM	M*	%E	M1	M2	DM	M*	%E	M1	M2	DM	M*	%E	M1	M2	DM	M*
20.8	130	27	103	5.6	20.7	415	86	329	17.8	20.7	270	56	214	11.6	20.6	471	97	374	20.0	20.6	349	72	277	14.9
20.8	125	26	99	5.4	20.7	411	85	326	17.6	20.7	266	55	211	11.4	20.6	470	97	373	20.0	20.6	345	71	274	14.6
20.8	120	25	95	5.2	20.7	410	85	325	17.6	20.7	261	54	207	11.2	20.6	467	96	371	19.7	20.6	344	71	273	14.7
20.8	106	22	84	4.6	20.7	406	84	322	17.4	20.7	256	53	203	11.0	20.6	466	96	370	19.8	20.6	340	70	270	14.4
20.8	101	21	80	4.4	20.7	405	84	321	17.4	20.7	251	52	199	10.8	20.6	465	96	369	19.8	20.6	339	70	269	14.5
20.8	96	20	76	4.2	20.7	401	83	318	17.2	20.7	246	51	195	10.6	20.6	462	95	367	19.5	20.6	335	69	266	14.2
20.8	77	16	61	3.3	20.7	397	82	315	16.9	20.7	242	50	192	10.3	20.6	461	95	366	19.6	20.6	330	68	262	14.0
20.8	72	15	57	3.1	20.7	396	82	314	17.0	20.7	241	50	191	10.4	20.6	457	94	363	19.3	20.6	326	67	259	13.8
20.8	53	11	42	2.3	20.7	392	81	311	16.7	20.7	237	49	188	10.1	20.6	456	94	362	19.4	20.6	325	67	258	13.8
20.8	48	10	38	2.1	20.7	391	81	310	16.8	20.7	232	48	184	9.9	20.6	452	93	359	19.1	20.6	321	66	255	13.6
20.8	24	5	19	1.0	20.7	387	80	307	16.5	20.7	227	47	180	9.7	20.6	451	93	358	19.2	20.6	320	66	254	13.6
20.7	498	103	395	21.3	20.7	386	80	306	16.6	20.7	222	46	176	9.5	20.6	447	92	355	18.9	20.6	316	65	251	13.4
20.7	497	103	394	21.3	20.7	382	79	303	16.3	20.7	217	45	172	9.3	20.6	446	92	354	19.0	20.6	315	65	250	13.4
20.7	493	102	391	21.1	20.7	381	79	302	16.4	20.7	213	44	169	9.1	20.6	442	91	351	18.7	20.6	311	64	247	13.2
20.7	492	102	390	21.1	20.7	377	78	299	16.1	20.7	208	43	165	8.9	20.6	441	91	350	18.8	20.6	310	64	246	13.2
20.7	489	101	388	20.9	20.7	376	78	298	16.2	20.7	203	42	161	8.7	20.6	437	90	347	18.5	20.6	306	63	243	13.0
20.7	488	101	387	20.9	20.7	372	77	295	15.9	20.7	198	41	157	8.5	20.6	436	90	346	18.6	20.6	301	62	239	12.8
20.7	487	101	386	20.9	20.7	368	76	292	15.7	20.7	193	40	153	8.3	20.6	433	89	344	18.3	20.6	296	61	235	12.6
20.7	484	100	384	20.7	20.7	367	76	291	15.7	20.7	188	39	149	8.1	20.6	432	89	343	18.3	20.6	291	60	231	12.4
20.7	483	100	383	20.7	20.7	363	75	288	15.5	20.7	184	38	146	7.8	20.6	428	88	340	18.1	20.6	287	59	228	12.1
20.7	482	100	382	20.7	20.7	362	75	287	15.5	20.7	179	37	142	7.6	20.6	427	88	339	18.1	20.6	286	59	227	12.2
20.7	479	99	380	20.5	20.7	358	74	284	15.3	20.7	174	36	138	7.4	20.6	423	87	336	17.9	20.6	282	58	224	11.9
20.7	478	99	379	20.5	20.7	357	74	283	15.3	20.7	169	35	134	7.2	20.6	422	87	335	17.9	20.6	281	58	223	12.0
20.7	474	98	376	20.3	20.7	353	73	280	15.1	20.7	164	34	130	7.0	20.6	418	86	332	17.7	20.6	277	57	220	11.7
20.7	473	98	375	20.3	20.7	352	73	279	15.1	20.7	150	31	119	6.4	20.6	417	86	331	17.7	20.6	272	56	216	11.5
20.7	469	97	372	20.1	20.7	348	72	276	14.9	20.7	145	30	115	6.2	20.6	413	85	328	17.5	20.6	267	55	212	11.3
20.7	468	97	371	20.1	20.7	347	72	275	14.9	20.7	140	29	111	6.0	20.6	412	85	327	17.5	20.6	262	54	208	11.1
20.7	464	96	368	19.9	20.7	343	71	272	14.7	20.7	135	28	107	5.8	20.6	408	84	324	17.3	20.6	257	53	204	10.9
20.7	463	96	367	19.9	20.7	338	70	268	14.5	20.7	121	25	96	5.2	20.6	407	84	323	17.3	20.6	253	52	201	10.7
20.7	460	95	365	19.6	20.7	334	69	265	14.3	20.7	116	24	92	5.0	20.6	403	83	320	17.1	20.6	252	52	200	10.7
20.7	459	95	364	19.7	20.7	333	69	264	14.3	20.7	111	23	88	4.8	20.6	402	83	319	17.1	20.6	248	51	197	10.5
20.7	458	95	363	19.7	20.7	329	68	261	14.1	20.7	92	19	73	3.9	20.6	399	82	317	16.9	20.6	247	51	196	10.5
20.7	455	94	361	19.4	20.7	328	68	260	14.1	20.7	87	18	69	3.7	20.6	398	82	316	16.9	20.6	243	50	193	10.3
20.7	454	94	360	19.5	20.7	324	67	257	13.9	20.7	82	17	65	3.5	20.6	394	81	313	16.7	20.6	238	49	189	10.1
20.7	450	93	357	19.2	20.7	323	67	256	13.9	20.7	58	12	46	2.5	20.6	393	81	312	16.7	20.6	233	48	185	9.9
20.7	449	93	356	19.3	20.7	319	66	253	13.7	20.7	29	6	23	1.2	20.6	389	80	309	16.5	20.6	228	47	181	9.7
20.7	445	92	353	19.0	20.7	314	65	249	13.5	20.6	500	103	397	21.2	20.6	388	80	308	16.5	20.6	223	46	177	9.5
20.7	444	92	352	19.1	20.7	309	64	245	13.3	20.6	499	103	396	21.3	20.6	384	79	305	16.3	20.6	218	45	173	9.3
20.7	440	91	349	18.8	20.7	305	63	242	13.0	20.6	496	102	394	21.0	20.6	383	79	304	16.3	20.6	214	44	170	9.0
20.7	439	91	348	18.9	20.7	304	63	241	13.1	20.6	495	102	393	21.0	20.6	379	78	301	16.1	20.6	209	43	166	8.8
20.7	435	90	345	18.6	20.7	300	62	238	12.8	20.6	494	102	392	21.1	20.6	378	78	300	16.1	20.6	204	42	162	8.6
20.7	434	90	344	18.7	20.7	299	62	237	12.9	20.6	491	101	390	20.8	20.6	374	77	297	15.9	20.6	199	41	158	8.4
20.7	431	89	342	18.4	20.7	295	61	234	12.6	20.6	490	101	389	20.0	20.6	373	77	296	15.9	20.6	194	40	154	8.2
20.7	430	89	341	18.4	20.7	294	61	233	12.7	20.6	486	100	386	20.6	20.6	369	76	293	15.7	20.6	189	39	150	8.0
20.7	429	89	340	18.5	20.7	290	60	230	12.4	20.6	485	100	385	20.6	20.6	364	75	289	15.5	20.6	180	37	143	7.6
20.7	426	88	338	18.2	20.7	285	59	226	12.2	20.6	481	99	382	20.4	20.6	360	74	286	15.2	20.6	175	36	139	7.4
20.7	425	88	337	18.2	20.7	280	58	222	12.0	20.6	480	99	381	20.4	20.6	359	74	285	15.3	20.6	170	35	135	7.2
20.7	421	87	334	18.0	20.7	276	57	219	11.8	20.6	476	98	378	20.2	20.6	355	73	282	15.0	20.6	165	34	131	7.0
20.7	420	87	333	18.0	20.7	275	57	218	11.8	20.6	475	98	377	20.2	20.6	354	73	281	15.1	20.6	160	33	127	6.8
20.7	416	86	330	17.8	20.7	271	56	215	11.6	20.6	472	97	375	19.9	20.6	350	72	278	14.8	20.6	155	32	123	6.6

%E	M1	M2	DM	M*
20.6	141	29	112	6.0
20.6	136	28	108	5.8
20.6	131	27	104	5.6
20.6	126	26	100	5.4
20.6	107	22	85	4.5
20.6	102	21	81	4.3
20.6	97	20	77	4.1
20.6	68	14	54	2.9
20.6	63	13	50	2.7
20.6	34	7	27	1.4
20.5	498	102	396	20.9
20.5	497	102	395	20.9
20.5	493	101	392	20.7
20.5	492	101	391	20.7
20.5	489	100	389	20.4
20.5	488	100	388	20.5
20.5	487	100	387	20.5
20.5	484	99	385	20.3
20.5	483	99	384	20.3
20.5	482	99	383	20.3
20.5	479	98	381	20.1
20.5	478	98	380	20.1
20.5	477	98	379	20.1
20.5	474	97	377	19.9
20.5	473	97	376	19.9
20.5	469	96	373	19.7
20.5	468	96	372	19.7
20.5	464	95	369	19.5
20.5	463	95	368	19.5
20.5	459	94	365	19.3
20.5	458	94	364	19.3
20.5	454	93	361	19.1
20.5	453	93	360	19.1
20.5	449	92	357	18.9
20.5	448	92	356	18.9
20.5	444	91	353	18.7
20.5	443	91	352	18.7
20.5	440	90	350	18.4
20.5	439	90	349	18.5
20.5	438	90	348	18.5
20.5	435	89	346	18.2
20.5	434	89	345	18.3
20.5	430	88	342	18.0
20.5	429	88	341	18.1
20.5	425	87	338	17.8
20.5	424	87	337	17.9
20.5	420	86	334	17.6
20.5	419	86	333	17.7
20.5	415	85	330	17.4
20.5	414	85	329	17.5

%E	M1	M2	DM	M*
20.5	410	84	326	17.2
20.5	409	84	325	17.3
20.5	405	83	322	17.0
20.5	404	83	321	17.1
20.5	400	82	318	16.8
20.5	396	81	315	16.6
20.5	395	81	314	16.6
20.5	391	80	311	16.4
20.5	390	80	310	16.4
20.5	386	79	307	16.2
20.5	385	79	306	16.2
20.5	381	78	303	16.0
20.5	380	78	302	16.0
20.5	376	77	299	15.8
20.5	375	77	298	15.8
20.5	371	76	295	15.6
20.5	370	76	294	15.6
20.5	366	75	291	15.4
20.5	365	75	290	15.4
20.5	361	74	287	15.2
20.5	356	73	283	15.0
20.5	352	72	280	14.7
20.5	351	72	279	14.8
20.5	347	71	276	14.5
20.5	346	71	275	14.6
20.5	342	70	272	14.3
20.5	341	70	271	14.4
20.5	337	69	268	14.1
20.5	336	69	267	14.2
20.5	332	68	264	13.9
20.5	331	68	263	14.0
20.5	327	67	260	13.7
20.5	322	66	256	13.5
20.5	317	65	252	13.3
20.5	312	64	248	13.1
20.5	308	63	245	12.9
20.5	307	63	244	12.9
20.5	303	62	241	12.7
20.5	302	62	240	12.7
20.5	298	61	237	12.5
20.5	297	61	236	12.5
20.5	293	60	233	12.3
20.5	292	60	232	12.3
20.5	288	59	229	12.1
20.5	283	58	225	11.9
20.5	278	57	221	11.7
20.5	273	56	217	11.5
20.5	268	55	213	11.3
20.5	264	54	210	11.0
20.5	263	54	209	11.1

%E	M1	M2	DM	M*
20.5	259	53	206	10.8
20.5	258	53	205	10.9
20.5	254	52	202	10.6
20.5	249	51	198	10.4
20.5	244	50	194	10.2
20.5	239	49	190	10.0
20.5	234	48	186	9.8
20.5	229	47	182	9.6
20.5	224	46	178	9.4
20.5	220	45	175	9.2
20.5	219	45	174	9.2
20.5	215	44	171	9.0
20.5	210	43	167	8.8
20.5	205	42	163	8.6
20.5	200	41	159	8.4
20.5	195	40	155	8.2
20.5	190	39	151	8.0
20.5	185	38	147	7.8
20.5	176	36	140	7.4
20.5	171	35	136	7.2
20.5	166	34	132	7.0
20.5	161	33	128	6.8
20.5	156	32	124	6.6
20.5	151	31	120	6.4
20.5	146	30	116	6.2
20.5	132	27	105	5.5
20.5	127	26	101	5.3
20.5	122	25	97	5.1
20.5	117	24	93	4.9
20.5	112	23	89	4.7
20.5	88	18	70	3.7
20.5	83	17	66	3.5
20.5	78	16	62	3.3
20.5	73	15	58	3.1
20.5	44	9	35	1.8
20.5	39	8	31	1.6
20.4	500	102	398	20.8
20.4	499	102	397	20.8
20.4	496	101	395	20.6
20.4	495	101	394	20.6
20.4	494	101	393	20.6
20.4	491	100	391	20.4
20.4	490	100	390	20.4
20.4	486	99	387	20.2
20.4	485	99	386	20.2
20.4	481	98	383	20.0
20.4	480	98	382	20.0
20.4	476	97	379	19.8
20.4	475	97	378	19.8
20.4	471	96	375	19.6

%E	M1	M2	DM	M*
20.4	470	96	374	19.6
20.4	466	95	371	19.4
20.4	465	95	370	19.4
20.4	461	94	367	19.2
20.4	460	94	366	19.2
20.4	457	93	364	18.9
20.4	456	93	363	19.0
20.4	455	93	362	19.0
20.4	452	92	360	18.7
20.4	451	92	359	18.8
20.4	450	92	358	18.8
20.4	447	91	356	18.5
20.4	446	91	355	18.6
20.4	445	91	354	18.6
20.4	442	90	352	18.3
20.4	441	90	351	18.4
20.4	437	89	348	18.1
20.4	436	89	347	18.2
20.4	432	88	344	17.9
20.4	431	88	343	18.0
20.4	427	87	340	17.7
20.4	426	87	339	17.8
20.4	422	86	336	17.5
20.4	421	86	335	17.6
20.4	417	85	332	17.3
20.4	416	85	331	17.4
20.4	412	84	328	17.1
20.4	411	84	327	17.2
20.4	407	83	324	16.9
20.4	406	83	323	17.0
20.4	402	82	320	16.7
20.4	401	82	319	16.8
20.4	398	81	317	16.5
20.4	397	81	316	16.5
20.4	393	80	313	16.3
20.4	392	80	312	16.3
20.4	388	79	309	16.1
20.4	387	79	308	16.1
20.4	383	78	305	15.9
20.4	382	78	304	15.9
20.4	378	77	301	15.7
20.4	377	77	300	15.7
20.4	373	76	297	15.5
20.4	372	76	296	15.5
20.4	368	75	293	15.3
20.4	367	75	292	15.3
20.4	363	74	289	15.1
20.4	362	74	288	15.1
20.4	358	73	285	14.9
20.4	357	73	284	14.9

%E	M1	M2	DM	M*
20.4	353	72	281	14.7
20.4	348	71	277	14.5
20.4	343	70	273	14.3
20.4	339	69	270	14.0
20.4	338	69	269	14.1
20.4	334	68	266	13.8
20.4	333	68	265	13.9
20.4	329	67	262	13.6
20.4	328	67	261	13.7
20.4	324	66	258	13.4
20.4	323	66	257	13.5
20.4	319	65	254	13.2
20.4	318	65	253	13.3
20.4	314	64	250	13.0
20.4	313	64	249	13.1
20.4	309	63	246	12.8
20.4	304	62	242	12.6
20.4	299	61	238	12.4
20.4	294	60	234	12.2
20.4	289	59	230	12.0
20.4	285	58	227	11.8
20.4	284	58	226	11.8
20.4	280	57	223	11.6
20.4	279	57	222	11.6
20.4	275	56	219	11.4
20.4	274	56	218	11.4
20.4	270	55	215	11.2
20.4	269	55	214	11.2
20.4	265	54	211	11.0
20.4	260	53	207	10.8
20.4	255	52	203	10.6
20.4	250	51	199	10.4
20.4	245	50	195	10.2
20.4	240	49	191	10.0
20.4	235	48	187	9.8
20.4	230	47	183	9.6
20.4	226	46	180	9.4
20.4	225	46	179	9.4
20.4	221	45	176	9.2
20.4	216	44	172	9.0
20.4	211	43	168	8.8
20.4	206	42	164	8.6
20.4	201	41	160	8.4
20.4	196	40	156	8.2
20.4	191	39	152	8.0
20.4	186	38	148	7.8
20.4	181	37	144	7.6
20.4	167	34	133	6.9
20.4	162	33	129	6.7
20.4	157	32	125	6.5

%E	M1	M2	DM	M*
20.4	152	31	121	6.3
20.4	147	30	117	6.1
20.4	142	29	113	5.9
20.4	137	28	109	5.7
20.4	113	23	90	4.7
20.4	108	22	86	4.5
20.4	103	21	82	4.3
20.4	98	20	78	4.1
20.4	93	19	74	3.9
20.4	54	11	43	2.2
20.4	49	10	39	2.0
20.3	498	101	397	20.5
20.3	497	101	396	20.5
20.3	493	100	393	20.3
20.3	492	100	392	20.3
20.3	488	99	389	20.1
20.3	487	99	388	20.1
20.3	483	98	385	19.9
20.3	482	98	384	19.9
20.3	479	97	382	19.6
20.3	478	97	381	19.7
20.3	477	97	380	19.7
20.3	474	96	378	19.4
20.3	473	96	377	19.5
20.3	472	96	376	19.5
20.3	469	95	374	19.2
20.3	468	95	373	19.3
20.3	467	95	372	19.3
20.3	464	94	370	19.0
20.3	463	94	369	19.1
20.3	462	94	368	19.1
20.3	459	93	366	18.8
20.3	458	93	365	18.9
20.3	454	92	362	18.6
20.3	453	92	361	18.7
20.3	449	91	358	18.4
20.3	448	91	357	18.5
20.3	444	90	354	18.2
20.3	443	90	353	18.3
20.3	439	89	350	18.0
20.3	438	89	349	18.1
20.3	434	88	346	17.8
20.3	433	88	345	17.9
20.3	429	87	342	17.6
20.3	428	87	341	17.7
20.3	424	86	338	17.4
20.3	423	86	337	17.5
20.3	419	85	334	17.2
20.3	418	85	333	17.3
20.3	414	84	330	17.0

%E	M1	M2	DM	M*
20.3	413	84	329	17.1
20.3	409	83	326	16.8
20.3	408	83	325	16.9
20.3	404	82	322	16.6
20.3	403	82	321	16.7
20.3	400	81	319	16.4
20.3	399	81	318	16.4
20.3	395	80	315	16.2
20.3	394	80	314	16.2
20.3	390	79	311	16.0
20.3	389	79	310	16.0
20.3	385	78	307	15.8
20.3	384	78	306	15.8
20.3	380	77	303	15.6
20.3	379	77	302	15.6
20.3	375	76	299	15.4
20.3	374	76	298	15.4
20.3	370	75	295	15.2
20.3	369	75	294	15.2
20.3	365	74	291	15.0
20.3	364	74	290	15.0
20.3	360	73	287	14.8
20.3	359	73	286	14.8
20.3	355	72	283	14.6
20.3	354	72	282	14.6
20.3	350	71	279	14.4
20.3	349	71	278	14.4
20.3	345	70	275	14.2
20.3	344	70	274	14.2
20.3	340	69	271	14.0
20.3	335	68	267	13.8
20.3	330	67	263	13.6
20.3	325	66	259	13.4
20.3	320	65	255	13.2
20.3	316	64	252	13.0
20.3	315	64	251	13.0
20.3	311	63	248	12.8
20.3	310	63	247	12.8
20.3	306	62	244	12.6
20.3	305	62	243	12.6
20.3	301	61	240	12.4
20.3	300	61	239	12.4
20.3	296	60	236	12.2
20.3	295	60	235	12.2
20.3	291	59	232	12.0
20.3	290	59	231	12.0
20.3	286	58	228	11.8
20.3	281	57	224	11.6
20.3	276	56	220	11.4
20.3	271	55	216	11.2

%E	M1	M2	DM	M*
20.3	266	54	212	11.0
20.3	261	53	208	10.8
20.3	256	52	204	10.6
20.3	251	51	200	10.4
20.3	246	50	196	10.2
20.3	241	49	192	10.0
20.3	237	48	189	9.7
20.3	236	48	188	9.8
20.3	232	47	185	9.5
20.3	231	47	184	9.6
20.3	227	46	181	9.3
20.3	222	45	177	9.1
20.3	217	44	173	8.9
20.3	212	43	169	8.7
20.3	207	42	165	8.5
20.3	202	41	161	8.3
20.3	197	40	157	8.1
20.3	192	39	153	7.9
20.3	187	38	149	7.7
20.3	182	37	145	7.5
20.3	177	36	141	7.3
20.3	172	35	137	7.1
20.3	158	32	126	6.5
20.3	153	31	122	6.3
20.3	148	30	118	6.1
20.3	143	29	114	5.9
20.3	138	28	110	5.7
20.3	133	27	106	5.5
20.3	128	26	102	5.3
20.3	123	25	98	5.1
20.3	118	24	94	4.9
20.3	79	16	63	3.2
20.3	74	15	59	3.0
20.3	69	14	55	2.8
20.3	64	13	51	2.6
20.3	59	12	47	2.4
20.2	500	101	399	20.4
20.2	499	101	398	20.4
20.2	496	100	396	20.2
20.2	495	100	395	20.2
20.2	494	100	394	20.2
20.2	491	99	392	20.0
20.2	490	99	391	20.0
20.2	489	99	390	20.0
20.2	486	98	388	19.8
20.2	485	98	387	19.8
20.2	484	98	386	19.8
20.2	481	97	384	19.6
20.2	480	97	383	19.6
20.2	476	96	380	19.4

%E	M1	M2	DM	M*
20.2	475	96	379	19.4
20.2	471	95	376	19.2
20.2	470	95	375	19.2
20.2	466	94	372	19.0
20.2	465	94	371	19.0
20.2	461	93	368	18.8
20.2	460	93	367	18.8
20.2	456	92	364	18.6
20.2	455	92	363	18.6
20.2	451	91	360	18.4
20.2	450	91	359	18.4
20.2	446	90	356	18.2
20.2	445	90	355	18.2
20.2	441	89	352	18.0
20.2	440	89	351	18.0
20.2	436	88	348	17.8
20.2	435	88	347	17.8
20.2	431	87	344	17.6
20.2	430	87	343	17.6
20.2	426	86	340	17.4
20.2	425	86	339	17.4
20.2	421	85	336	17.2
20.2	420	85	335	17.2
20.2	416	84	332	17.0
20.2	415	84	331	17.0
20.2	411	83	328	16.8
20.2	410	83	327	16.8
20.2	406	82	324	16.6
20.2	405	82	323	16.6
20.2	401	81	320	16.4
20.2	397	80	317	16.1
20.2	396	80	316	16.2
20.2	392	79	313	15.9
20.2	391	79	312	16.0
20.2	387	78	309	15.7
20.2	386	78	308	15.8
20.2	382	77	305	15.5
20.2	381	77	304	15.6
20.2	377	76	301	15.3
20.2	376	76	300	15.4
20.2	372	75	297	15.1
20.2	371	75	296	15.2
20.2	367	74	293	14.9
20.2	366	74	292	15.0
20.2	362	73	289	14.7
20.2	361	73	288	14.8
20.2	357	72	285	14.5
20.2	356	72	284	14.6
20.2	352	71	281	14.3
20.2	351	71	280	14.4

%E	M1	M2	DM	M*
20.2	347	70	277	14.1
20.2	346	70	276	14.2
20.2	342	69	273	13.9
20.2	341	69	272	14.0
20.2	337	68	269	13.7
20.2	336	68	268	13.8
20.2	332	67	265	13.5
20.2	331	67	264	13.6
20.2	327	66	261	13.3
20.2	326	66	260	13.4
20.2	322	65	257	13.1
20.2	321	65	256	13.2
20.2	317	64	253	12.9
20.2	312	63	249	12.7
20.2	307	62	245	12.5
20.2	302	61	241	12.3
20.2	297	60	237	12.1
20.2	292	59	233	11.9
20.2	287	58	229	11.7
20.2	282	57	225	11.5
20.2	277	56	221	11.3
20.2	272	55	217	11.1
20.2	267	54	213	10.9
20.2	263	53	210	10.7
20.2	262	53	209	10.7
20.2	258	52	206	10.5
20.2	257	52	205	10.5
20.2	253	51	202	10.3
20.2	252	51	201	10.3
20.2	248	50	198	10.1
20.2	247	50	197	10.1
20.2	243	49	194	9.9
20.2	242	49	193	9.9
20.2	238	48	190	9.7
20.2	233	47	186	9.5
20.2	228	46	182	9.3
20.2	223	45	178	9.1
20.2	218	44	174	8.9
20.2	213	43	170	8.7
20.2	208	42	166	8.5
20.2	203	41	162	8.3
20.2	198	40	158	8.1
20.2	193	39	154	7.9
20.2	188	38	150	7.7
20.2	183	37	146	7.5
20.2	178	36	142	7.3
20.2	173	35	138	7.1
20.2	168	34	134	6.9
20.2	163	33	130	6.7
20.2	129	26	103	5.2

%E	M1	M2	DM	M*
20.2	124	25	99	5.0
20.2	119	24	95	4.8
20.2	114	23	91	4.6
20.2	109	22	87	4.4
20.2	104	21	83	4.2
20.2	99	20	79	4.0
20.2	94	19	75	3.8
20.2	89	18	71	3.6
20.2	84	17	67	3.4
20.1	498	100	398	20.1
20.1	497	100	397	20.1
20.1	493	99	394	19.9
20.1	492	99	393	19.9
20.1	488	98	390	19.7
20.1	487	98	389	19.7
20.1	483	97	386	19.5
20.1	482	97	385	19.5
20.1	478	96	382	19.3
20.1	477	96	381	19.3
20.1	473	95	378	19.1
20.1	472	95	377	19.1
20.1	468	94	374	18.9
20.1	467	94	373	18.9
20.1	463	93	370	18.7
20.1	462	93	369	18.7
20.1	458	92	366	18.5
20.1	457	92	365	18.5
20.1	453	91	362	18.3
20.1	452	91	361	18.3
20.1	448	90	358	18.1
20.1	447	90	357	18.1
20.1	443	89	354	17.9
20.1	442	89	353	17.9
20.1	438	88	350	17.7
20.1	437	88	349	17.7
20.1	433	87	346	17.5
20.1	432	87	345	17.5
20.1	428	86	342	17.3
20.1	427	86	341	17.3
20.1	423	85	338	17.1
20.1	422	85	337	17.1
20.1	418	84	334	16.9
20.1	417	84	333	16.9
20.1	413	83	330	16.7
20.1	412	83	329	16.7
20.1	408	82	326	16.5
20.1	407	82	325	16.5
20.1	404	81	323	16.2
20.1	403	81	322	16.3
20.1	402	81	321	16.3

%E	M1	M2	DM	M*
20.1	399	80	319	16.0
20.1	398	80	318	16.1
20.1	394	79	315	15.8
20.1	393	79	314	15.9
20.1	389	78	311	15.6
20.1	388	78	310	15.7
20.1	384	77	307	15.4
20.1	383	77	306	15.5
20.1	379	76	303	15.2
20.1	378	76	302	15.3
20.1	374	75	299	15.0
20.1	373	75	298	15.1
20.1	369	74	295	14.8
20.1	368	74	294	14.9
20.1	364	73	291	14.6
20.1	363	73	290	14.7
20.1	359	72	287	14.4
20.1	358	72	286	14.5
20.1	354	71	283	14.2
20.1	353	71	282	14.3
20.1	349	70	279	14.0
20.1	348	70	278	14.1
20.1	344	69	275	13.8
20.1	343	69	274	13.9
20.1	339	68	271	13.6
20.1	338	68	270	13.7
20.1	334	67	267	13.4
20.1	333	67	266	13.5
20.1	329	66	263	13.2
20.1	328	66	262	13.3
20.1	324	65	259	13.0
20.1	323	65	258	13.1
20.1	319	64	255	12.8
20.1	318	64	254	12.9
20.1	314	63	251	12.6
20.1	313	63	250	12.7
20.1	309	62	247	12.4
20.1	308	62	246	12.5
20.1	304	61	243	12.2
20.1	303	61	242	12.3
20.1	299	60	239	12.0
20.1	298	60	238	12.1
20.1	294	59	235	11.8
20.1	293	59	234	11.9
20.1	289	58	231	11.6
20.1	288	58	230	11.7
20.1	284	57	227	11.4
20.1	283	57	226	11.5
20.1	279	56	223	11.2
20.1	278	56	222	11.3

%E	M1	M2	DM	M*
20.1	274	55	219	11.0
20.1	273	55	218	11.1
20.1	269	54	215	10.8
20.1	268	54	214	10.9
20.1	264	53	211	10.6
20.1	259	52	207	10.4
20.1	254	51	203	10.2
20.1	249	50	199	10.0
20.1	244	49	195	9.8
20.1	239	48	191	9.6
20.1	234	47	187	9.4
20.1	229	46	183	9.2
20.1	224	45	179	9.0
20.1	219	44	175	8.8
20.1	214	43	171	8.6
20.1	209	42	167	8.4
20.1	204	41	163	8.2
20.1	199	40	159	8.0
20.1	194	39	155	7.8
20.1	189	38	151	7.6
20.1	184	37	147	7.4
20.1	179	36	143	7.2
20.1	174	35	139	7.0
20.1	169	34	135	6.8
20.1	164	33	131	6.6
20.1	159	32	127	6.4
20.1	154	31	123	6.2
20.1	149	30	119	6.0
20.1	144	29	115	5.8
20.1	139	28	111	5.6
20.1	134	27	107	5.4
20.0	500	100	400	20.0
20.0	499	100	399	20.0
20.0	496	99	397	19.8
20.0	495	99	396	19.8
20.0	494	99	395	19.8
20.0	491	98	393	19.6
20.0	490	98	392	19.6
20.0	489	98	391	19.6
20.0	486	97	389	19.4
20.0	485	97	388	19.4
20.0	484	97	387	19.4
20.0	481	96	385	19.2
20.0	480	96	384	19.2
20.0	479	96	383	19.2
20.0	476	95	381	19.0
20.0	475	95	380	19.0
20.0	474	95	379	19.0
20.0	471	94	377	18.8
20.0	470	94	376	18.8

%E	M1	M2	DM	M*
20.0	469	94	375	18.8
20.0	466	93	373	18.6
20.0	465	93	372	18.6
20.0	464	93	371	18.6
20.0	461	92	369	18.4
20.0	460	92	368	18.4
20.0	459	92	367	18.4
20.0	456	91	365	18.2
20.0	455	91	364	18.2
20.0	454	91	363	18.2
20.0	451	90	361	18.0
20.0	450	90	360	18.0
20.0	449	90	359	18.0
20.0	446	89	357	17.8
20.0	445	89	356	17.8
20.0	444	89	355	17.8
20.0	441	88	353	17.6
20.0	440	88	352	17.6
20.0	439	88	351	17.6
20.0	436	87	349	17.4
20.0	435	87	348	17.4
20.0	434	87	347	17.4
20.0	431	86	345	17.2
20.0	430	86	344	17.2
20.0	429	86	343	17.2
20.0	426	85	341	17.0
20.0	425	85	340	17.0
20.0	424	85	339	17.0
20.0	421	84	337	16.8
20.0	420	84	336	16.8
20.0	419	84	335	16.8
20.0	416	83	333	16.6
20.0	415	83	332	16.6
20.0	414	83	331	16.6
20.0	411	82	329	16.4
20.0	410	82	328	16.4
20.0	409	82	327	16.4
20.0	406	81	325	16.2
20.0	405	81	324	16.2
20.0	401	80	321	16.0
20.0	400	80	320	16.0
20.0	395	79	316	15.8
20.0	390	78	312	15.6
20.0	385	77	308	15.4
20.0	380	76	304	15.2
20.0	375	75	300	15.0
20.0	370	74	296	14.8
20.0	365	73	292	14.6
20.0	360	72	288	14.4
20.0	355	71	284	14.2

%E	M1	M2	DM	M*
20.0	350	70	280	14.0
20.0	345	69	276	13.8
20.0	340	68	272	13.6
20.0	335	67	268	13.4
20.0	330	66	264	13.2
20.0	325	65	260	13.0
20.0	320	64	256	12.8
20.0	315	63	252	12.6
20.0	310	62	248	12.4
20.0	305	61	244	12.2
20.0	300	60	240	12.0
20.0	295	59	236	11.8
20.0	290	58	232	11.6
20.0	285	57	228	11.4
20.0	280	56	224	11.2
20.0	275	55	220	11.0
20.0	270	54	216	10.8
20.0	265	53	212	10.6
20.0	260	52	208	10.4
20.0	255	51	204	10.2
20.0	250	50	200	10.0
20.0	245	49	196	9.8
20.0	240	48	192	9.6
20.0	235	47	188	9.4
20.0	230	46	184	9.2
20.0	225	45	180	9.0
20.0	220	44	176	8.8
20.0	215	43	172	8.6
20.0	210	42	168	8.4
20.0	205	41	164	8.2
20.0	200	40	160	8.0
20.0	195	39	156	7.8
20.0	190	38	152	7.6
20.0	185	37	148	7.4
20.0	180	36	144	7.2
20.0	175	35	140	7.0
20.0	170	34	136	6.8
20.0	165	33	132	6.6
20.0	160	32	128	6.4
20.0	155	31	124	6.2
20.0	150	30	120	6.0
20.0	145	29	116	5.8
20.0	140	28	112	5.6
20.0	135	27	108	5.4
20.0	130	26	104	5.2
20.0	125	25	100	5.0
20.0	120	24	96	4.8
20.0	115	23	92	4.6
20.0	110	22	88	4.4
20.0	105	21	84	4.2

%E	M1	M2	DM	M*
20.0	100	20	80	4.0
20.0	95	19	76	3.8
20.0	90	18	72	3.6
20.0	85	17	68	3.4
20.0	80	16	64	3.2
20.0	75	15	60	3.0
20.0	70	14	56	2.8
20.0	65	13	52	2.6
20.0	60	12	48	2.4
20.0	55	11	44	2.2
20.0	50	10	40	2.0
20.0	45	9	36	1.8
20.0	40	8	32	1.6
20.0	35	7	28	1.4
20.0	30	6	24	1.2
20.0	25	5	20	1.0
20.0	20	4	16	0.8
20.0	15	3	12	0.6
20.0	10	2	8	0.4
20.0	5	1	4	0.2
19.9	498	99	399	19.7
19.9	497	99	398	19.7
19.9	493	98	395	19.5
19.9	492	98	394	19.5
19.9	488	97	391	19.3
19.9	487	97	390	19.3
19.9	483	96	387	19.1
19.9	482	96	386	19.1
19.9	478	95	383	18.9
19.9	477	95	382	18.9
19.9	473	94	379	18.7
19.9	472	94	378	18.7
19.9	468	93	375	18.5
19.9	467	93	374	18.5
19.9	463	92	371	18.3
19.9	462	92	370	18.3
19.9	458	91	367	18.1
19.9	457	91	366	18.1
19.9	453	90	363	17.9
19.9	452	90	362	17.9
19.9	448	89	359	17.7
19.9	447	89	358	17.7
19.9	443	88	355	17.5
19.9	442	88	354	17.5
19.9	438	87	351	17.3
19.9	437	87	350	17.3
19.9	433	86	347	17.1
19.9	432	86	346	17.1
19.9	428	85	343	16.9
19.9	427	85	342	16.9

%E	M1	M2	DM	M*
19.9	423	84	339	16.7
19.9	422	84	338	16.7
19.9	418	83	335	16.5
19.9	417	83	334	16.5
19.9	413	82	331	16.3
19.9	412	82	330	16.3
19.9	408	81	327	16.1
19.9	407	81	326	16.1
19.9	403	80	323	15.9
19.9	402	80	322	15.9
19.9	397	79	318	15.7
19.9	396	79	317	15.8
19.9	392	78	314	15.5
19.9	391	78	313	15.6
19.9	387	77	310	15.3
19.9	386	77	309	15.4
19.9	382	76	306	15.1
19.9	381	76	305	15.2
19.9	377	75	302	14.9
19.9	376	75	301	15.0
19.9	372	74	298	14.7
19.9	371	74	297	14.8
19.9	367	73	294	14.5
19.9	366	73	293	14.6
19.9	362	72	290	14.3
19.9	361	72	289	14.4
19.9	357	71	286	14.1
19.9	356	71	285	14.2
19.9	352	70	282	13.9
19.9	351	70	281	14.0
19.9	347	69	278	13.7
19.9	346	69	277	13.8
19.9	342	68	274	13.5
19.9	341	68	273	13.6
19.9	337	67	270	13.3
19.9	336	67	269	13.4
19.9	332	66	266	13.1
19.9	331	66	265	13.2
19.9	327	65	262	12.9
19.9	326	65	261	13.0
19.9	322	64	258	12.7
19.9	321	64	257	12.8
19.9	317	63	254	12.5
19.9	316	63	253	12.6
19.9	312	62	250	12.3
19.9	311	62	249	12.4
19.9	307	61	246	12.1
19.9	306	61	245	12.2
19.9	302	60	242	11.9
19.9	301	60	241	12.0

%E	M1	M2	DM	M*
19.9	297	59	238	11.7
19.9	296	59	237	11.8
19.9	292	58	234	11.5
19.9	291	58	233	11.6
19.9	287	57	230	11.3
19.9	286	57	229	11.4
19.9	282	56	226	11.1
19.9	281	56	225	11.2
19.9	277	55	222	10.9
19.9	276	55	221	11.0
19.9	272	54	218	10.7
19.9	271	54	217	10.8
19.9	267	53	214	10.5
19.9	266	53	213	10.6
19.9	261	52	209	10.4
19.9	256	51	205	10.2
19.9	251	50	201	10.0
19.9	246	49	197	9.8
19.9	241	48	193	9.6
19.9	236	47	189	9.4
19.9	231	46	185	9.2
19.9	226	45	181	9.0
19.9	221	44	177	8.8
19.9	216	43	173	8.6
19.9	211	42	169	8.4
19.9	206	41	165	8.2
19.9	201	40	161	8.0
19.9	196	39	157	7.8
19.9	191	38	153	7.6
19.9	186	37	149	7.4
19.9	181	36	145	7.2
19.9	176	35	141	7.0
19.9	171	34	137	6.8
19.9	166	33	133	6.6
19.9	161	32	129	6.4
19.9	156	31	125	6.2
19.9	151	30	121	6.0
19.9	146	29	117	5.8
19.9	141	28	113	5.6
19.9	136	27	109	5.4
19.8	500	99	401	19.6
19.8	499	99	400	19.6
19.8	496	98	398	19.4
19.8	495	98	397	19.4
19.8	494	98	396	19.4
19.8	491	97	394	19.2
19.8	490	97	393	19.2
19.8	489	97	392	19.2
19.8	486	96	390	19.0
19.8	485	96	389	19.0

%E	M1	M2	DM	M*
19.8	484	96	388	19.0
19.8	481	95	386	18.8
19.8	480	95	385	18.8
19.8	479	95	384	18.8
19.8	475	94	381	18.6
19.8	474	94	380	18.6
19.8	470	93	377	18.4
19.8	469	93	376	18.4
19.8	465	92	373	18.2
19.8	464	92	372	18.2
19.8	460	91	369	18.0
19.8	459	91	368	18.0
19.8	455	90	365	17.8
19.8	454	90	364	17.8
19.8	450	89	361	17.6
19.8	449	89	360	17.6
19.8	445	88	357	17.4
19.8	444	88	356	17.4
19.8	440	87	353	17.2
19.8	439	87	352	17.2
19.8	435	86	349	17.0
19.8	434	86	348	17.0
19.8	430	85	345	16.8
19.8	429	85	344	16.8
19.8	425	84	341	16.6
19.8	424	84	340	16.6
19.8	420	83	337	16.4
19.8	419	83	336	16.4
19.8	415	82	333	16.2
19.8	414	82	332	16.2
19.8	410	81	329	16.0
19.8	409	81	328	16.0
19.8	405	80	325	15.8
19.8	404	80	324	15.8
19.8	400	79	321	15.6
19.8	399	79	320	15.6
19.8	398	79	319	15.7
19.8	394	78	316	15.4
19.8	393	78	315	15.5
19.8	389	77	312	15.2
19.8	388	77	311	15.3
19.8	384	76	308	15.0
19.8	383	76	307	15.1
19.8	379	75	304	14.8
19.8	378	75	303	14.9
19.8	374	74	300	14.6
19.8	373	74	299	14.7
19.8	369	73	296	14.4
19.8	368	73	295	14.5
19.8	364	72	292	14.2

%E	M1	M2	DM	M*
19.8	363	72	291	14.3
19.8	359	71	288	14.0
19.8	358	71	287	14.1
19.8	354	70	284	13.8
19.8	353	70	283	13.9
19.8	349	69	280	13.6
19.8	348	69	279	13.7
19.8	344	68	276	13.4
19.8	343	68	275	13.5
19.8	339	67	272	13.2
19.8	338	67	271	13.3
19.8	334	66	268	13.0
19.8	333	66	267	13.1
19.8	329	65	264	12.8
19.8	328	65	263	12.9
19.8	324	64	260	12.6
19.8	323	64	259	12.7
19.8	318	63	255	12.5
19.8	313	62	251	12.3
19.8	308	61	247	12.1
19.8	303	60	243	11.9
19.8	298	59	239	11.7
19.8	293	58	235	11.5
19.8	288	57	231	11.3
19.8	283	56	227	11.1
19.8	278	55	223	10.9
19.8	273	54	219	10.7
19.8	268	53	215	10.5
19.8	263	52	211	10.3
19.8	262	52	210	10.3
19.8	258	51	207	10.1
19.8	257	51	206	10.1
19.8	253	50	203	9.9
19.8	252	50	202	9.9
19.8	248	49	199	9.7
19.8	247	49	198	9.7
19.8	243	48	195	9.5
19.8	242	48	194	9.5
19.8	237	47	190	9.3
19.8	232	46	186	9.1
19.8	227	45	182	8.9
19.8	222	44	178	8.7
19.8	217	43	174	8.5
19.8	212	42	170	8.3
19.8	207	41	166	8.1
19.8	202	40	162	7.9
19.8	197	39	158	7.7
19.8	192	38	154	7.5
19.8	187	37	150	7.3
19.8	182	36	146	7.1

%E	M1	M2	DM	M*
19.8	177	35	142	6.9
19.8	172	34	138	6.7
19.8	167	33	134	6.5
19.8	162	32	130	6.3
19.8	131	26	105	5.2
19.8	126	25	101	5.0
19.8	121	24	97	4.8
19.8	116	23	93	4.6
19.8	111	22	89	4.4
19.8	106	21	85	4.2
19.8	101	20	81	4.0
19.8	96	19	77	3.8
19.8	91	18	73	3.6
19.8	86	17	69	3.4
19.8	81	16	65	3.2
19.7	498	98	400	19.3
19.7	497	98	399	19.3
19.7	493	97	396	19.1
19.7	492	97	395	19.1
19.7	488	96	392	18.9
19.7	487	96	391	18.9
19.7	483	95	388	18.7
19.7	482	95	387	18.7
19.7	478	94	384	18.5
19.7	477	94	383	18.5
19.7	476	94	382	18.6
19.7	473	93	380	18.3
19.7	472	93	379	18.3
19.7	471	93	378	18.4
19.7	468	92	376	18.1
19.7	467	92	375	18.1
19.7	466	92	374	18.2
19.7	463	91	372	17.9
19.7	462	91	371	17.9
19.7	461	91	370	18.0
19.7	458	90	368	17.7
19.7	457	90	367	17.7
19.7	456	90	366	17.8
19.7	452	89	363	17.5
19.7	451	89	362	17.6
19.7	447	88	359	17.3
19.7	446	88	358	17.4
19.7	442	87	355	17.1
19.7	441	87	354	17.2
19.7	437	86	351	16.9
19.7	436	86	350	17.0
19.7	432	85	347	16.7
19.7	431	85	346	16.8
19.7	427	84	343	16.5
19.7	426	84	342	16.6

%E	M1	M2	DM	M*
19.7	422	83	339	16.3
19.7	421	83	338	16.4
19.7	417	82	335	16.1
19.7	416	82	334	16.2
19.7	412	81	331	15.9
19.7	411	81	330	16.0
19.7	407	80	327	15.7
19.7	406	80	326	15.8
19.7	402	79	323	15.5
19.7	401	79	322	15.6
19.7	396	78	318	15.4
19.7	395	78	317	15.4
19.7	391	77	314	15.2
19.7	390	77	313	15.2
19.7	386	76	310	15.0
19.7	385	76	309	15.0
19.7	381	75	306	14.8
19.7	380	75	305	14.8
19.7	376	74	302	14.6
19.7	375	74	301	14.6
19.7	371	73	298	14.4
19.7	370	73	297	14.4
19.7	366	72	294	14.2
19.7	365	72	293	14.2
19.7	361	71	290	14.0
19.7	360	71	289	14.0
19.7	356	70	286	13.8
19.7	355	70	285	13.8
19.7	351	69	282	13.6
19.7	350	69	281	13.6
19.7	346	68	278	13.4
19.7	345	68	277	13.4
19.7	340	67	273	13.2
19.7	335	66	269	13.0
19.7	330	65	265	12.8
19.7	325	64	261	12.6
19.7	320	63	257	12.4
19.7	319	63	256	12.4
19.7	315	62	253	12.2
19.7	314	62	252	12.2
19.7	310	61	249	12.0
19.7	309	61	248	12.0
19.7	305	60	245	11.8
19.7	304	60	244	11.8
19.7	300	59	241	11.6
19.7	299	59	240	11.6
19.7	295	58	237	11.4
19.7	294	58	236	11.4
19.7	290	57	233	11.2
19.7	289	57	232	11.2

%E	M1	M2	DM	M*
19.7	284	56	228	11.0
19.7	279	55	224	10.8
19.7	274	54	220	10.6
19.7	269	53	216	10.4
19.7	264	52	212	10.2
19.7	259	51	208	10.0
19.7	254	50	204	9.8
19.7	249	49	200	9.6
19.7	244	48	196	9.4
19.7	239	47	192	9.2
19.7	238	47	191	9.3
19.7	234	46	188	9.0
19.7	233	46	187	9.1
19.7	229	45	184	8.8
19.7	228	45	183	8.9
19.7	223	44	179	8.7
19.7	218	43	175	8.5
19.7	213	42	171	8.3
19.7	208	41	167	8.1
19.7	203	40	163	7.9
19.7	198	39	159	7.7
19.7	193	38	155	7.5
19.7	188	37	151	7.3
19.7	183	36	147	7.1
19.7	178	35	143	6.9
19.7	173	34	139	6.7
19.7	157	31	126	6.1
19.7	152	30	122	5.9
19.7	147	29	118	5.7
19.7	142	28	114	5.5
19.7	137	27	110	5.3
19.7	132	26	106	5.1
19.7	127	25	102	4.9
19.7	122	24	98	4.7
19.7	117	23	94	4.5
19.7	76	15	61	3.0
19.7	71	14	57	2.8
19.7	66	13	53	2.6
19.7	61	12	49	2.4
19.6	500	98	402	19.2
19.6	499	98	401	19.2
19.6	496	97	399	19.0
19.6	495	97	398	19.0
19.6	494	97	397	19.0
19.6	491	96	395	18.8
19.6	490	96	394	18.8
19.6	489	96	393	18.8
19.6	485	95	390	18.6
19.6	484	95	389	18.6
19.6	480	94	386	18.4

%E	M1	M2	DM	M*
19.6	479	94	385	18.4
19.6	475	93	382	18.2
19.6	474	93	381	18.2
19.6	470	92	378	18.0
19.6	469	92	377	18.0
19.6	465	91	374	17.8
19.6	464	91	373	17.8
19.6	460	90	370	17.6
19.6	459	90	369	17.6
19.6	455	89	366	17.4
19.6	454	89	365	17.4
19.6	453	89	364	17.5
19.6	450	88	362	17.2
19.6	449	88	361	17.2
19.6	448	88	360	17.3
19.6	445	87	358	17.0
19.6	444	87	357	17.0
19.6	443	87	356	17.1
19.6	439	86	353	16.8
19.6	438	86	352	16.9
19.6	434	85	349	16.6
19.6	433	85	348	16.7
19.6	429	84	345	16.4
19.6	428	84	344	16.5
19.6	424	83	341	16.2
19.6	423	83	340	16.3
19.6	419	82	337	16.0
19.6	418	82	336	16.1
19.6	414	81	333	15.8
19.6	413	81	332	15.9
19.6	409	80	329	15.6
19.6	408	80	328	15.7
19.6	404	79	325	15.4
19.6	403	79	324	15.5
19.6	398	78	320	15.3
19.6	397	78	319	15.3
19.6	393	77	316	15.1
19.6	392	77	315	15.1
19.6	388	76	312	14.9
19.6	387	76	311	14.9
19.6	383	75	308	14.7
19.6	382	75	307	14.7
19.6	378	74	304	14.5
19.6	377	74	303	14.5
19.6	373	73	300	14.3
19.6	372	73	299	14.3
19.6	368	72	296	14.1
19.6	367	72	295	14.1
19.6	363	71	292	13.9
19.6	362	71	291	13.9

%E	M1	M2	DM	M*
19.6	358	70	288	13.7
19.6	357	70	287	13.7
19.6	352	69	283	13.5
19.6	347	68	279	13.3
19.6	342	67	275	13.1
19.6	341	67	274	13.2
19.6	337	66	271	12.9
19.6	336	66	270	13.0
19.6	332	65	267	12.7
19.6	331	65	266	12.8
19.6	327	64	263	12.5
19.6	326	64	262	12.6
19.6	322	63	259	12.3
19.6	321	63	258	12.4
19.6	317	62	255	12.1
19.6	316	62	254	12.2
19.6	312	61	251	11.9
19.6	311	61	250	12.0
19.6	306	60	246	11.8
19.6	301	59	242	11.6
19.6	296	58	238	11.4
19.6	291	57	234	11.2
19.6	286	56	230	11.0
19.6	285	56	229	11.0
19.6	281	55	226	10.8
19.6	280	55	225	10.8
19.6	276	54	222	10.6
19.6	275	54	221	10.6
19.6	271	53	218	10.4
19.6	270	53	217	10.4
19.6	265	52	213	10.2
19.6	260	51	209	10.0
19.6	255	50	205	9.8
19.6	250	49	201	9.6
19.6	245	48	197	9.4
19.6	240	47	193	9.2
19.6	235	46	189	9.0
19.6	230	45	185	8.8
19.6	225	44	181	8.6
19.6	224	44	180	8.6
19.6	219	43	176	8.4
19.6	214	42	172	8.2
19.6	209	41	168	8.0
19.6	204	40	164	7.8
19.6	199	39	160	7.6
19.6	194	38	156	7.4
19.6	189	37	152	7.2
19.6	184	36	148	7.0
19.6	179	35	144	6.8
19.6	168	33	135	6.5

%E	M1	M2	DM	M*
19.6	163	32	131	6.3
19.6	158	31	127	6.1
19.6	153	30	123	5.9
19.6	148	29	119	5.7
19.6	143	28	115	5.5
19.6	138	27	111	5.3
19.6	112	22	90	4.3
19.6	107	21	86	4.1
19.6	102	20	82	3.9
19.6	97	19	78	3.7
19.6	92	18	74	3.5
19.6	56	11	45	2.2
19.6	51	10	41	2.0
19.6	46	9	37	1.8
19.5	498	97	401	18.9
19.5	497	97	400	18.9
19.5	493	96	397	18.7
19.5	492	96	396	18.7
19.5	488	95	393	18.5
19.5	487	95	392	18.5
19.5	486	95	391	18.6
19.5	483	94	389	18.3
19.5	482	94	388	18.3
19.5	481	94	387	18.4
19.5	478	93	385	18.1
19.5	477	93	384	18.1
19.5	476	93	383	18.2
19.5	473	92	381	17.9
19.5	472	92	380	17.9
19.5	471	92	379	18.0
19.5	467	91	376	17.7
19.5	466	91	375	17.8
19.5	462	90	372	17.5
19.5	461	90	371	17.6
19.5	457	89	368	17.3
19.5	456	89	367	17.4
19.5	452	88	364	17.1
19.5	451	88	363	17.2
19.5	447	87	360	16.9
19.5	446	87	359	17.0
19.5	442	86	356	16.7
19.5	441	86	355	16.8
19.5	440	86	354	16.8
19.5	437	85	352	16.5
19.5	436	85	351	16.6
19.5	435	85	350	16.6
19.5	431	84	347	16.4
19.5	430	84	346	16.4
19.5	426	83	343	16.2
19.5	425	83	342	16.2

%E	M1	M2	DM	M*
19.5	421	82	339	16.0
19.5	420	82	338	16.0
19.5	416	81	335	15.8
19.5	415	81	334	15.8
19.5	411	80	331	15.6
19.5	410	80	330	15.6
19.5	406	79	327	15.4
19.5	405	79	326	15.4
19.5	401	78	323	15.2
19.5	400	78	322	15.2
19.5	399	78	321	15.2
19.5	395	77	318	15.0
19.5	394	77	317	15.0
19.5	390	76	314	14.8
19.5	389	76	313	14.8
19.5	385	75	310	14.6
19.5	384	75	309	14.6
19.5	380	74	306	14.4
19.5	379	74	305	14.4
19.5	375	73	302	14.2
19.5	374	73	301	14.2
19.5	370	72	298	14.0
19.5	369	72	297	14.0
19.5	365	71	294	13.8
19.5	364	71	293	13.8
19.5	359	70	289	13.6
19.5	354	69	285	13.4
19.5	353	69	284	13.5
19.5	349	68	281	13.2
19.5	348	68	280	13.3
19.5	344	67	277	13.0
19.5	343	67	276	13.1
19.5	339	66	273	12.8
19.5	338	66	272	12.9
19.5	334	65	269	12.6
19.5	333	65	268	12.7
19.5	329	64	265	12.4
19.5	328	64	264	12.5
19.5	323	63	260	12.3
19.5	318	62	256	12.1
19.5	313	61	252	11.9
19.5	308	60	248	11.7
19.5	307	60	247	11.7
19.5	303	59	244	11.5
19.5	302	59	243	11.5
19.5	298	58	240	11.3
19.5	297	58	239	11.3
19.5	293	57	236	11.1
19.5	292	57	235	11.1
19.5	287	56	231	10.9

%E	M1	M2	DM	M*
19.5	282	55	227	10.7
19.5	277	54	223	10.5
19.5	272	53	219	10.3
19.5	267	52	215	10.1
19.5	266	52	214	10.2
19.5	262	51	211	9.9
19.5	261	51	210	10.0
19.5	257	50	207	9.7
19.5	256	50	206	9.8
19.5	251	49	202	9.6
19.5	246	48	198	9.4
19.5	241	47	194	9.2
19.5	236	46	190	9.0
19.5	231	45	186	8.8
19.5	226	44	182	8.6
19.5	221	43	178	8.4
19.5	220	43	177	8.4
19.5	215	42	173	8.2
19.5	210	41	169	8.0
19.5	205	40	165	7.8
19.5	200	39	161	7.6
19.5	195	38	157	7.4
19.5	190	37	153	7.2
19.5	185	36	149	7.0
19.5	174	34	140	6.6
19.5	169	33	136	6.4
19.5	164	32	132	6.2
19.5	159	31	128	6.0
19.5	154	30	124	5.8
19.5	149	29	120	5.6
19.5	133	26	107	5.1
19.5	128	25	103	4.9
19.5	123	24	99	4.7
19.5	118	23	95	4.5
19.5	113	22	91	4.3
19.5	87	17	70	3.3
19.5	82	16	66	3.1
19.5	77	15	62	2.9
19.5	41	8	33	1.6
19.4	500	97	403	18.8
19.4	499	97	402	18.9
19.4	496	96	400	18.6
19.4	495	96	399	18.6
19.4	494	96	398	18.7
19.4	490	95	395	18.4
19.4	489	95	394	18.5
19.4	485	94	391	18.2
19.4	484	94	390	18.3
19.4	480	93	387	18.0
19.4	479	93	386	18.1

%E	M1	M2	DM	M*
19.4	475	92	383	17.8
19.4	474	92	382	17.9
19.4	470	91	379	17.6
19.4	469	91	378	17.7
19.4	468	91	377	17.7
19.4	465	90	375	17.4
19.4	464	90	374	17.5
19.4	463	90	373	17.5
19.4	459	89	370	17.3
19.4	458	89	369	17.3
19.4	454	88	366	17.1
19.4	453	88	365	17.1
19.4	449	87	362	16.9
19.4	448	87	361	16.9
19.4	444	86	358	16.7
19.4	443	86	357	16.7
19.4	439	85	354	16.5
19.4	438	85	353	16.5
19.4	434	84	350	16.3
19.4	433	84	349	16.3
19.4	432	84	348	16.3
19.4	428	83	345	16.1
19.4	427	83	344	16.1
19.4	423	82	341	15.9
19.4	422	82	340	15.9
19.4	418	81	337	15.7
19.4	417	81	336	15.7
19.4	413	80	333	15.5
19.4	412	80	332	15.5
19.4	408	79	329	15.3
19.4	407	79	328	15.3
19.4	403	78	325	15.1
19.4	402	78	324	15.1
19.4	397	77	320	14.9
19.4	396	77	319	15.0
19.4	392	76	316	14.7
19.4	391	76	315	14.8
19.4	387	75	312	14.5
19.4	386	75	311	14.6
19.4	382	74	308	14.3
19.4	381	74	307	14.4
19.4	377	73	304	14.1
19.4	376	73	303	14.2
19.4	372	72	300	13.9
19.4	371	72	299	14.0
19.4	366	71	295	13.8
19.4	361	70	291	13.6
19.4	360	70	290	13.6
19.4	356	69	287	13.4
19.4	355	69	286	13.4

%E	M1	M2	DM	M*
19.4	351	68	283	13.2
19.4	350	68	282	13.2
19.4	346	67	279	13.0
19.4	345	67	278	13.0
19.4	341	66	275	12.8
19.4	340	66	274	12.8
19.4	335	65	270	12.6
19.4	330	64	266	12.4
19.4	325	63	262	12.2
19.4	324	63	261	12.3
19.4	320	62	258	12.0
19.4	319	62	257	12.1
19.4	315	61	254	11.8
19.4	314	61	253	11.9
19.4	310	60	250	11.6
19.4	309	60	249	11.7
19.4	304	59	245	11.5
19.4	299	58	241	11.3
19.4	294	57	237	11.1
19.4	289	56	233	10.9
19.4	288	56	232	10.9
19.4	284	55	229	10.7
19.4	283	55	228	10.7
19.4	279	54	225	10.5
19.4	278	54	224	10.5
19.4	273	53	220	10.3
19.4	268	52	216	10.1
19.4	263	51	212	9.9
19.4	258	50	208	9.7
19.4	253	49	204	9.5
19.4	252	49	203	9.5
19.4	248	48	200	9.3
19.4	247	48	199	9.3
19.4	242	47	195	9.1
19.4	237	46	191	8.9
19.4	232	45	187	8.7
19.4	227	44	183	8.5
19.4	222	43	179	8.3
19.4	217	42	175	8.1
19.4	216	42	174	8.2
19.4	211	41	170	8.0
19.4	206	40	166	7.8
19.4	201	39	162	7.6
19.4	196	38	158	7.4
19.4	191	37	154	7.2
19.4	186	36	150	7.0
19.4	180	35	145	6.8
19.4	175	34	141	6.6
19.4	170	33	137	6.4
19.4	165	32	133	6.2

%E	M1	M2	DM	M*
19.4	160	31	129	6.0
19.4	155	30	125	5.8
19.4	144	28	116	5.4
19.4	139	27	112	5.2
19.4	134	26	108	5.0
19.4	129	25	104	4.8
19.4	124	24	100	4.6
19.4	108	21	87	4.1
19.4	103	20	83	3.9
19.4	98	19	79	3.7
19.4	93	18	75	3.5
19.4	72	14	58	2.7
19.4	67	13	54	2.5
19.4	62	12	50	2.3
19.4	36	7	29	1.4
19.4	31	6	25	1.2
19.3	498	96	402	18.5
19.3	497	96	401	18.5
19.3	493	95	398	18.3
19.3	492	95	397	18.3
19.3	491	95	396	18.4
19.3	488	94	394	18.1
19.3	487	94	393	18.1
19.3	486	94	392	18.2
19.3	483	93	390	17.9
19.3	482	93	389	17.9
19.3	481	93	388	18.0
19.3	477	92	385	17.7
19.3	476	92	384	17.8
19.3	472	91	381	17.5
19.3	471	91	380	17.6
19.3	467	90	377	17.3
19.3	466	90	376	17.4
19.3	462	89	373	17.1
19.3	461	89	372	17.2
19.3	460	89	371	17.2
19.3	457	88	369	16.9
19.3	456	88	368	17.0
19.3	455	88	367	17.0
19.3	451	87	364	16.8
19.3	450	87	363	16.8
19.3	446	86	360	16.6
19.3	445	86	359	16.6
19.3	441	85	356	16.4
19.3	440	85	355	16.4
19.3	436	84	352	16.2
19.3	435	84	351	16.2
19.3	431	83	348	16.0
19.3	430	83	347	16.0
19.3	429	83	346	16.1

%E	M1	M2	DM	M*
19.3	425	82	343	15.8
19.3	424	82	342	15.9
19.3	420	81	339	15.6
19.3	419	81	338	15.7
19.3	415	80	335	15.4
19.3	414	80	334	15.5
19.3	410	79	331	15.2
19.3	409	79	330	15.3
19.3	405	78	327	15.0
19.3	404	78	326	15.1
19.3	400	77	323	14.8
19.3	399	77	322	14.9
19.3	398	77	321	14.9
19.3	394	76	318	14.7
19.3	393	76	317	14.7
19.3	389	75	314	14.5
19.3	388	75	313	14.5
19.3	384	74	310	14.3
19.3	383	74	309	14.3
19.3	379	73	306	14.1
19.3	378	73	305	14.1
19.3	374	72	302	13.9
19.3	373	72	301	13.9
19.3	368	71	297	13.7
19.3	367	71	296	13.7
19.3	363	70	293	13.5
19.3	362	70	292	13.5
19.3	358	69	289	13.3
19.3	357	69	288	13.3
19.3	353	68	285	13.1
19.3	352	68	284	13.1
19.3	348	67	281	12.9
19.3	347	67	280	12.9
19.3	342	66	276	12.7
19.3	337	65	272	12.5
19.3	336	65	271	12.6
19.3	332	64	268	12.3
19.3	331	64	267	12.4
19.3	327	63	264	12.1
19.3	326	63	263	12.2
19.3	322	62	260	11.9
19.3	321	62	259	12.0
19.3	316	61	255	11.8
19.3	311	60	251	11.6
19.3	306	59	247	11.4
19.3	305	59	246	11.4
19.3	301	58	243	11.2
19.3	300	58	242	11.2
19.3	296	57	239	11.0
19.3	295	57	238	11.0

%E	M1	M2	DM	M*
19.3	290	56	234	10.8
19.3	285	55	230	10.6
19.3	280	54	226	10.4
19.3	275	53	222	10.2
19.3	274	53	221	10.3
19.3	270	52	218	10.0
19.3	269	52	217	10.1
19.3	264	51	213	9.9
19.3	259	50	209	9.7
19.3	254	49	205	9.5
19.3	249	48	201	9.3
19.3	244	47	197	9.1
19.3	243	47	196	9.1
19.3	238	46	192	8.9
19.3	233	45	188	8.7
19.3	228	44	184	8.5
19.3	223	43	180	8.3
19.3	218	42	176	8.1
19.3	212	41	171	7.9
19.3	207	40	167	7.7
19.3	202	39	163	7.5
19.3	197	38	159	7.3
19.3	192	37	155	7.1
19.3	187	36	151	6.9
19.3	181	35	146	6.8
19.3	176	34	142	6.6
19.3	171	33	138	6.4
19.3	166	32	134	6.2
19.3	161	31	130	6.0
19.3	150	29	121	5.6
19.3	145	28	117	5.4
19.3	140	27	113	5.2
19.3	135	26	109	5.0
19.3	119	23	96	4.4
19.3	114	22	92	4.2
19.3	109	21	88	4.0
19.3	88	17	71	3.3
19.3	83	16	67	3.1
19.3	57	11	46	2.1
19.2	500	96	404	18.4
19.2	499	96	403	18.5
19.2	496	95	401	18.2
19.2	495	95	400	18.2
19.2	494	95	399	18.3
19.2	490	94	396	18.0
19.2	489	94	395	18.1
19.2	485	93	392	17.8
19.2	484	93	391	17.9
19.2	480	92	388	17.6
19.2	479	92	387	17.7

%E	M1	M2	DM	M*
19.2	478	92	386	17.7
19.2	475	91	384	17.4
19.2	474	91	383	17.5
19.2	473	91	382	17.5
19.2	469	90	379	17.3
19.2	468	90	378	17.3
19.2	464	89	375	17.1
19.2	463	89	374	17.1
19.2	459	88	371	16.9
19.2	458	88	370	16.9
19.2	454	87	367	16.7
19.2	453	87	366	16.7
19.2	452	87	365	16.7
19.2	449	86	363	16.5
19.2	448	86	362	16.5
19.2	447	86	361	16.5
19.2	443	85	358	16.3
19.2	442	85	357	16.3
19.2	438	84	354	16.1
19.2	437	84	353	16.1
19.2	433	83	350	15.9
19.2	432	83	349	15.9
19.2	428	82	346	15.7
19.2	427	82	345	15.7
19.2	426	82	344	15.8
19.2	422	81	341	15.5
19.2	421	81	340	15.6
19.2	417	80	337	15.3
19.2	416	80	336	15.4
19.2	412	79	333	15.1
19.2	411	79	332	15.2
19.2	407	78	329	14.9
19.2	406	78	328	15.0
19.2	402	77	325	14.7
19.2	401	77	324	14.8
19.2	396	76	320	14.6
19.2	395	76	319	14.6
19.2	391	75	316	14.4
19.2	390	75	315	14.4
19.2	386	74	312	14.2
19.2	385	74	311	14.2
19.2	381	73	308	14.0
19.2	380	73	307	14.0
19.2	375	72	303	13.8
19.2	370	71	299	13.6
19.2	369	71	298	13.7
19.2	365	70	295	13.4
19.2	364	70	294	13.5
19.2	360	69	291	13.2
19.2	359	69	290	13.3

%E	M1	M2	DM	M*
19.2	355	68	287	13.0
19.2	354	68	286	13.1
19.2	349	67	282	12.9
19.2	344	66	278	12.7
19.2	343	66	277	12.7
19.2	339	65	274	12.5
19.2	338	65	273	12.5
19.2	334	64	270	12.3
19.2	333	64	269	12.3
19.2	328	63	265	12.1
19.2	323	62	261	11.9
19.2	318	61	257	11.7
19.2	317	61	256	11.7
19.2	313	60	253	11.5
19.2	312	60	252	11.5
19.2	308	59	249	11.3
19.2	307	59	248	11.3
19.2	302	58	244	11.1
19.2	297	57	240	10.9
19.2	292	56	236	10.7
19.2	291	56	235	10.8
19.2	287	55	232	10.5
19.2	286	55	231	10.6
19.2	281	54	227	10.4
19.2	276	53	223	10.2
19.2	271	52	219	10.0
19.2	266	51	215	9.8
19.2	265	51	214	9.8
19.2	261	50	211	9.6
19.2	260	50	210	9.6
19.2	255	49	206	9.4
19.2	250	48	202	9.2
19.2	245	47	198	9.0
19.2	240	46	194	8.8
19.2	239	46	193	8.9
19.2	234	45	189	8.7
19.2	229	44	185	8.5
19.2	224	43	181	8.3
19.2	219	42	177	8.1
19.2	214	41	173	7.9
19.2	213	41	172	7.9
19.2	208	40	168	7.7
19.2	203	39	164	7.5
19.2	198	38	160	7.3
19.2	193	37	156	7.1
19.2	182	35	147	6.7
19.2	177	34	143	6.5
19.2	172	33	139	6.3
19.2	167	32	135	6.1
19.2	156	30	126	5.8

%E	M1	M2	DM	M*
19.2	151	29	122	5.6
19.2	146	28	118	5.4
19.2	130	25	105	4.8
19.2	125	24	101	4.6
19.2	120	23	97	4.4
19.2	104	20	84	3.8
19.2	99	19	80	3.6
19.2	78	15	63	2.9
19.2	73	14	59	2.7
19.2	52	10	42	1.9
19.2	26	5	21	1.0
19.1	498	95	403	18.1
19.1	497	95	402	18.2
19.1	493	94	399	17.9
19.1	492	94	398	18.0
19.1	491	94	397	18.0
19.1	488	93	395	17.7
19.1	487	93	394	17.8
19.1	486	93	393	17.8
19.1	482	92	390	17.6
19.1	481	92	389	17.6
19.1	477	91	386	17.4
19.1	476	91	385	17.4
19.1	472	90	382	17.2
19.1	471	90	381	17.2
19.1	470	90	380	17.2
19.1	467	89	378	17.0
19.1	466	89	377	17.0
19.1	465	89	376	17.0
19.1	461	88	373	16.8
19.1	460	88	372	16.8
19.1	456	87	369	16.6
19.1	455	87	368	16.6
19.1	451	86	365	16.4
19.1	450	86	364	16.4
19.1	446	85	361	16.2
19.1	445	85	360	16.2
19.1	444	85	359	16.3
19.1	440	84	356	16.0
19.1	439	84	355	16.1
19.1	435	83	352	15.8
19.1	434	83	351	15.9
19.1	430	82	348	15.6
19.1	429	82	347	15.7
19.1	425	81	344	15.4
19.1	424	81	343	15.5
19.1	423	81	342	15.5
19.1	419	80	339	15.3
19.1	418	80	338	15.3
19.1	414	79	335	15.1

%E	M1	M2	DM	M*
19.1	413	79	334	15.1
19.1	409	78	331	14.9
19.1	408	78	330	14.9
19.1	404	77	327	14.7
19.1	403	77	326	14.7
19.1	398	76	322	14.5
19.1	397	76	321	14.5
19.1	393	75	318	14.3
19.1	392	75	317	14.3
19.1	388	74	314	14.1
19.1	387	74	313	14.1
19.1	383	73	310	13.9
19.1	382	73	309	14.0
19.1	377	72	305	13.8
19.1	376	72	304	13.8
19.1	372	71	301	13.6
19.1	371	71	300	13.6
19.1	367	70	297	13.4
19.1	366	70	296	13.4
19.1	362	69	293	13.2
19.1	361	69	292	13.2
19.1	356	68	288	13.0
19.1	351	67	284	12.8
19.1	350	67	283	12.8
19.1	346	66	280	12.6
19.1	345	66	279	12.6
19.1	341	65	276	12.4
19.1	340	65	275	12.4
19.1	335	64	271	12.2
19.1	330	63	267	12.0
19.1	329	63	266	12.1
19.1	325	62	263	11.8
19.1	324	62	262	11.9
19.1	320	61	259	11.6
19.1	319	61	258	11.7
19.1	314	60	254	11.5
19.1	309	59	250	11.3
19.1	304	58	246	11.1
19.1	303	58	245	11.1
19.1	299	57	242	10.9
19.1	298	57	241	10.9
19.1	293	56	237	10.7
19.1	288	55	233	10.5
19.1	283	54	229	10.3
19.1	282	54	228	10.3
19.1	278	53	225	10.1
19.1	277	53	224	10.1
19.1	272	52	220	9.9
19.1	267	51	216	9.7
19.1	262	50	212	9.5

%E	M1	M2	DM	M*
19.1	257	49	208	9.3
19.1	256	49	207	9.4
19.1	251	48	203	9.2
19.1	246	47	199	9.0
19.1	241	46	195	8.8
19.1	236	45	191	8.6
19.1	235	45	190	8.6
19.1	230	44	186	8.4
19.1	225	43	182	8.2
19.1	220	42	178	8.0
19.1	215	41	174	7.8
19.1	209	40	169	7.7
19.1	204	39	165	7.5
19.1	199	38	161	7.3
19.1	194	37	157	7.1
19.1	188	36	152	6.9
19.1	183	35	148	6.7
19.1	178	34	144	6.5
19.1	173	33	140	6.3
19.1	162	31	131	5.9
19.1	157	30	127	5.7
19.1	152	29	123	5.5
19.1	141	27	114	5.2
19.1	136	26	110	5.0
19.1	131	25	106	4.8
19.1	115	22	93	4.2
19.1	110	21	89	4.0
19.1	94	18	76	3.4
19.1	89	17	72	3.2
19.1	68	13	55	2.5
19.1	47	9	38	1.7
19.0	500	95	405	18.0
19.0	499	95	404	18.1
19.0	496	94	402	17.8
19.0	495	94	401	17.9
19.0	494	94	400	17.9
19.0	490	93	397	17.7
19.0	489	93	396	17.7
19.0	485	92	393	17.5
19.0	484	92	392	17.5
19.0	483	92	391	17.5
19.0	480	91	389	17.3
19.0	479	91	388	17.3
19.0	478	91	387	17.3
19.0	474	90	384	17.1
19.0	473	90	383	17.1
19.0	469	89	380	16.9
19.0	468	89	379	16.9
19.0	464	88	376	16.7
19.0	463	88	375	16.7

%E	M1	M2	DM	M*
19.0	462	88	374	16.8
19.0	459	87	372	16.5
19.0	458	87	371	16.5
19.0	457	87	370	16.6
19.0	453	86	367	16.3
19.0	452	86	366	16.4
19.0	448	85	363	16.1
19.0	447	85	362	16.2
19.0	443	84	359	15.9
19.0	442	84	358	16.0
19.0	441	84	357	16.0
19.0	438	83	355	15.7
19.0	437	83	354	15.8
19.0	436	83	353	15.8
19.0	432	82	350	15.6
19.0	431	82	349	15.6
19.0	427	81	346	15.4
19.0	426	81	345	15.4
19.0	422	80	342	15.2
19.0	421	80	341	15.2
19.0	420	80	340	15.2
19.0	416	79	337	15.0
19.0	415	79	336	15.0
19.0	411	78	333	14.8
19.0	410	78	332	14.8
19.0	406	77	329	14.6
19.0	405	77	328	14.6
19.0	401	76	325	14.4
19.0	400	76	324	14.4
19.0	399	76	323	14.5
19.0	395	75	320	14.2
19.0	394	75	319	14.3
19.0	390	74	316	14.0
19.0	389	74	315	14.1
19.0	385	73	312	13.8
19.0	384	73	311	13.9
19.0	379	72	307	13.7
19.0	378	72	306	13.7
19.0	374	71	303	13.5
19.0	373	71	302	13.5
19.0	369	70	299	13.3
19.0	368	70	298	13.3
19.0	364	69	295	13.1
19.0	363	69	294	13.1
19.0	358	68	290	12.9
19.0	357	68	289	13.0
19.0	353	67	286	12.7
19.0	352	67	285	12.8
19.0	348	66	282	12.5
19.0	347	66	281	12.6

%E	M1	M2	DM	M*
19.0	343	65	278	12.3
19.0	342	65	277	12.4
19.0	337	64	273	12.2
19.0	336	64	272	12.2
19.0	332	63	269	12.0
19.0	331	63	268	12.0
19.0	327	62	265	11.8
19.0	326	62	264	11.8
19.0	321	61	260	11.6
19.0	316	60	256	11.4
19.0	315	60	255	11.4
19.0	311	59	252	11.2
19.0	310	59	251	11.2
19.0	306	58	248	11.0
19.0	305	58	247	11.0
19.0	300	57	243	10.8
19.0	295	56	239	10.6
19.0	294	56	238	10.7
19.0	290	55	235	10.4
19.0	289	55	234	10.5
19.0	284	54	230	10.3
19.0	279	53	226	10.1
19.0	274	52	222	9.9
19.0	273	52	221	9.9
19.0	269	51	218	9.7
19.0	268	51	217	9.7
19.0	263	50	213	9.5
19.0	258	49	209	9.3
19.0	253	48	205	9.1
19.0	252	48	204	9.1
19.0	248	47	201	8.9
19.0	247	47	200	8.9
19.0	242	46	196	8.7
19.0	237	45	192	8.5
19.0	232	44	188	8.3
19.0	231	44	187	8.4
19.0	226	43	183	8.2
19.0	221	42	179	8.0
19.0	216	41	175	7.8
19.0	211	40	171	7.6
19.0	210	40	170	7.6
19.0	205	39	166	7.4
19.0	200	38	162	7.2
19.0	195	37	158	7.0
19.0	189	36	153	6.9
19.0	184	35	149	6.7
19.0	179	34	145	6.5
19.0	174	33	141	6.3
19.0	168	32	136	6.1
19.0	163	31	132	5.9

%E	M1	M2	DM	M*
19.0	158	30	128	5.7
19.0	153	29	124	5.5
19.0	147	28	119	5.3
19.0	142	27	115	5.1
19.0	137	26	111	4.9
19.0	126	24	102	4.6
19.0	121	23	98	4.4
19.0	116	22	94	4.2
19.0	105	20	85	3.8
19.0	100	19	81	3.6
19.0	84	16	68	3.0
19.0	79	15	64	2.8
19.0	63	12	51	2.3
19.0	58	11	47	2.1
19.0	42	8	34	1.5
19.0	21	4	17	0.8
18.9	498	94	404	17.7
18.9	497	94	403	17.8
18.9	493	93	400	17.5
18.9	492	93	399	17.6
18.9	491	93	398	17.6
18.9	488	92	396	17.3
18.9	487	92	395	17.4
18.9	486	92	394	17.4
18.9	482	91	391	17.2
18.9	481	91	390	17.2
18.9	477	90	387	17.0
18.9	476	90	386	17.0
18.9	475	90	385	17.1
18.9	472	89	383	16.8
18.9	471	89	382	16.8
18.9	470	89	381	16.9
18.9	466	88	378	16.6
18.9	465	88	377	16.7
18.9	461	87	374	16.4
18.9	460	87	373	16.5
18.9	456	86	370	16.2
18.9	455	86	369	16.3
18.9	454	86	368	16.3
18.9	450	85	365	16.1
18.9	449	85	364	16.1
18.9	445	84	361	15.9
18.9	444	84	360	15.9
18.9	440	83	357	15.7
18.9	439	83	356	15.7
18.9	435	82	353	15.5
18.9	434	82	352	15.5
18.9	433	82	351	15.5
18.9	429	81	348	15.3
18.9	428	81	347	15.3

%E	M1	M2	DM	M*
18.9	424	80	344	15.1
18.9	423	80	343	15.1
18.9	419	79	340	14.9
18.9	418	79	339	14.9
18.9	417	79	338	15.0
18.9	413	78	335	14.7
18.9	412	78	334	14.8
18.9	408	77	331	14.5
18.9	407	77	330	14.6
18.9	403	76	327	14.3
18.9	402	76	326	14.4
18.9	397	75	322	14.2
18.9	396	75	321	14.2
18.9	392	74	318	14.0
18.9	391	74	317	14.0
18.9	387	73	314	13.8
18.9	386	73	313	13.8
18.9	381	72	309	13.6
18.9	380	72	308	13.6
18.9	376	71	305	13.4
18.9	375	71	304	13.4
18.9	371	70	301	13.2
18.9	370	70	300	13.2
18.9	366	69	297	13.0
18.9	365	69	296	13.0
18.9	360	68	292	12.8
18.9	359	68	291	12.9
18.9	355	67	288	12.6
18.9	354	67	287	12.7
18.9	350	66	284	12.4
18.9	349	66	283	12.5
18.9	344	65	279	12.3
18.9	339	64	275	12.1
18.9	338	64	274	12.1
18.9	334	63	271	11.9
18.9	333	63	270	11.9
18.9	328	62	266	11.7
18.9	323	61	262	11.5
18.9	322	61	261	11.6
18.9	318	60	258	11.3
18.9	317	60	257	11.4
18.9	313	59	254	11.1
18.9	312	59	253	11.2
18.9	307	58	249	11.0
18.9	302	57	245	10.8
18.9	301	57	244	10.8
18.9	297	56	241	10.6
18.9	296	56	240	10.6
18.9	291	55	236	10.4
18.9	286	54	232	10.2

%E	M1	M2	DM	M*
18.9	285	54	231	10.2
18.9	281	53	228	10.0
18.9	280	53	227	10.0
18.9	275	52	223	9.8
18.9	270	51	219	9.6
18.9	265	50	215	9.4
18.9	264	50	214	9.5
18.9	259	49	210	9.3
18.9	254	48	206	9.1
18.9	249	47	202	8.9
18.9	244	46	198	8.7
18.9	243	46	197	8.7
18.9	238	45	193	8.5
18.9	233	44	189	8.3
18.9	228	43	185	8.1
18.9	227	43	184	8.1
18.9	222	42	180	7.9
18.9	217	41	176	7.7
18.9	212	40	172	7.5
18.9	206	39	167	7.4
18.9	201	38	163	7.2
18.9	196	37	159	7.0
18.9	190	36	154	6.8
18.9	185	35	150	6.6
18.9	180	34	146	6.4
18.9	175	33	142	6.2
18.9	169	32	137	6.1
18.9	164	31	133	5.9
18.9	159	30	129	5.7
18.9	148	28	120	5.3
18.9	143	27	116	5.1
18.9	132	25	107	4.7
18.9	127	24	103	4.5
18.9	122	23	99	4.3
18.9	111	21	90	4.0
18.9	106	20	86	3.8
18.9	95	18	77	3.4
18.9	90	17	73	3.2
18.9	74	14	60	2.6
18.9	53	10	43	1.9
18.9	37	7	30	1.3
18.8	500	94	406	17.7
18.8	499	94	405	17.7
18.8	496	93	403	17.4
18.8	495	93	402	17.5
18.8	494	93	401	17.5
18.8	490	92	398	17.3
18.8	489	92	397	17.3
18.8	485	91	394	17.1
18.8	484	91	393	17.1

%E	M1	M2	DM	M*
18.8	483	91	392	17.1
18.8	480	90	390	16.9
18.8	479	90	389	16.9
18.8	478	90	388	16.9
18.8	474	89	385	16.7
18.8	473	89	384	16.7
18.8	469	88	381	16.5
18.8	468	88	380	16.5
18.8	467	88	379	16.6
18.8	464	87	377	16.3
18.8	463	87	376	16.3
18.8	462	87	375	16.4
18.8	458	86	372	16.1
18.8	457	86	371	16.2
18.8	453	85	368	15.9
18.8	452	85	367	16.0
18.8	451	85	366	16.0
18.8	448	84	364	15.8
18.8	447	84	363	15.8
18.8	446	84	362	15.8
18.8	442	83	359	15.6
18.8	441	83	358	15.6
18.8	437	82	355	15.4
18.8	436	82	354	15.4
18.8	432	81	351	15.2
18.8	431	81	350	15.2
18.8	430	81	349	15.3
18.8	426	80	346	15.0
18.8	425	80	345	15.1
18.8	421	79	342	14.8
18.8	420	79	341	14.9
18.8	416	78	338	14.6
18.8	415	78	337	14.7
18.8	414	78	336	14.7
18.8	410	77	333	14.5
18.8	409	77	332	14.5
18.8	405	76	329	14.3
18.8	404	76	328	14.3
18.8	400	75	325	14.1
18.8	399	75	324	14.1
18.8	398	75	323	14.1
18.8	394	74	320	13.9
18.8	393	74	319	13.9
18.8	389	73	316	13.7
18.8	388	73	315	13.7
18.8	384	72	312	13.5
18.8	383	72	311	13.5
18.8	382	72	310	13.6
18.8	378	71	307	13.3
18.8	377	71	306	13.4

%E	M1	M2	DM	M*
18.8	373	70	303	13.1
18.8	372	70	302	13.2
18.8	368	69	299	12.9
18.8	367	69	298	13.0
18.8	362	68	294	12.8
18.8	361	68	293	12.8
18.8	357	67	290	12.6
18.8	356	67	289	12.6
18.8	352	66	286	12.4
18.8	351	66	285	12.4
18.8	346	65	281	12.2
18.8	345	65	280	12.2
18.8	341	64	277	12.0
18.8	340	64	276	12.0
18.8	336	63	273	11.8
18.8	335	63	272	11.8
18.8	330	62	268	11.6
18.8	329	62	267	11.7
18.8	325	61	264	11.4
18.8	324	61	263	11.5
18.8	320	60	260	11.3
18.8	319	60	259	11.3
18.8	314	59	255	11.1
18.8	309	58	251	10.9
18.8	308	58	250	10.9
18.8	304	57	247	10.7
18.8	303	57	246	10.7
18.8	298	56	242	10.5
18.8	293	55	238	10.3
18.8	292	55	237	10.4
18.8	288	54	234	10.1
18.8	287	54	233	10.2
18.8	282	53	229	10.0
18.8	277	52	225	9.8
18.8	276	52	224	9.8
18.8	272	51	221	9.6
18.8	271	51	220	9.6
18.8	266	50	216	9.4
18.8	261	49	212	9.2
18.8	260	49	211	9.2
18.8	256	48	208	9.0
18.8	255	48	207	9.0
18.8	250	47	203	8.8
18.8	245	46	199	8.6
18.8	240	45	195	8.4
18.8	239	45	194	8.5
18.8	234	44	190	8.3
18.8	229	43	186	8.1
18.8	224	42	182	7.9
18.8	223	42	181	7.9

%E	M1	M2	DM	M*
18.8	218	41	177	7.7
18.8	213	40	173	7.5
18.8	208	39	169	7.3
18.8	207	39	168	7.3
18.8	202	38	164	7.1
18.8	197	37	160	6.9
18.8	192	36	156	6.8
18.8	191	36	155	6.8
18.8	186	35	151	6.6
18.8	181	34	147	6.4
18.8	176	33	143	6.2
18.8	170	32	138	6.0
18.8	165	31	134	5.8
18.8	160	30	130	5.6
18.8	154	29	125	5.5
18.8	149	28	121	5.3
18.8	144	27	117	5.1
18.8	138	26	112	4.9
18.8	133	25	108	4.7
18.8	128	24	104	4.5
18.8	117	22	95	4.1
18.8	112	21	91	3.9
18.8	101	19	82	3.6
18.8	96	18	78	3.4
18.8	85	16	69	3.0
18.8	80	15	65	2.8
18.8	69	13	56	2.4
18.8	64	12	52	2.3
18.8	48	9	39	1.7
18.8	32	6	26	1.1
18.8	16	3	13	0.6
18.7	498	93	405	17.4
18.7	497	93	404	17.4
18.7	493	92	401	17.2
18.7	492	92	400	17.2
18.7	491	92	399	17.2
18.7	487	91	396	17.0
18.7	486	91	395	17.0
18.7	482	90	392	16.8
18.7	481	90	391	16.8
18.7	477	89	388	16.6
18.7	476	89	387	16.6
18.7	475	89	386	16.7
18.7	471	88	383	16.4
18.7	470	88	382	16.5
18.7	466	87	379	16.2
18.7	465	87	378	16.3
18.7	461	86	375	16.0
18.7	460	86	374	16.1
18.7	459	86	373	16.1

%E	M1	M2	DM	M*
18.7	455	85	370	15.9
18.7	454	85	369	15.9
18.7	450	84	366	15.7
18.7	449	84	365	15.7
18.7	445	83	362	15.5
18.7	444	83	361	15.5
18.7	443	83	360	15.6
18.7	439	82	357	15.3
18.7	438	82	356	15.4
18.7	434	81	353	15.1
18.7	433	81	352	15.2
18.7	428	80	348	15.0
18.7	427	80	347	15.0
18.7	423	79	344	14.8
18.7	422	79	343	14.8
18.7	418	78	340	14.6
18.7	417	78	339	14.6
18.7	412	77	335	14.4
18.7	411	77	334	14.4
18.7	407	76	331	14.2
18.7	406	76	330	14.2
18.7	402	75	327	14.0
18.7	401	75	326	14.0
18.7	396	74	322	13.8
18.7	395	74	321	13.9
18.7	391	73	318	13.6
18.7	390	73	317	13.7
18.7	386	72	314	13.4
18.7	385	72	313	13.5
18.7	380	71	309	13.3
18.7	379	71	308	13.3
18.7	375	70	305	13.1
18.7	374	70	304	13.1
18.7	369	69	300	12.9
18.7	364	68	296	12.7
18.7	363	68	295	12.7
18.7	359	67	292	12.5
18.7	358	67	291	12.5
18.7	353	66	287	12.3
18.7	348	65	283	12.1
18.7	347	65	282	12.2
18.7	343	64	279	11.9
18.7	342	64	278	12.0
18.7	337	63	274	11.8
18.7	332	62	270	11.6
18.7	331	62	269	11.6
18.7	327	61	266	11.4
18.7	326	61	265	11.4
18.7	321	60	261	11.2
18.7	316	59	257	11.0

%E	M1	M2	DM	M*
18.7	315	59	256	11.1
18.7	311	58	253	10.8
18.7	310	58	252	10.9
18.7	305	57	248	10.7
18.7	300	56	244	10.5
18.7	299	56	243	10.5
18.7	294	55	239	10.3
18.7	289	54	235	10.1
18.7	284	53	231	9.9
18.7	283	53	230	9.9
18.7	278	52	226	9.7
18.7	273	51	222	9.5
18.7	268	50	218	9.3
18.7	267	50	217	9.4
18.7	262	49	213	9.2
18.7	257	48	209	9.0
18.7	252	47	205	8.8
18.7	251	47	204	8.8
18.7	246	46	200	8.6
18.7	241	45	196	8.4
18.7	235	44	191	8.2
18.7	230	43	187	8.0
18.7	225	42	183	7.8
18.7	219	41	178	7.7
18.7	214	40	174	7.5
18.7	209	39	170	7.3
18.7	203	38	165	7.1
18.7	198	37	161	6.9
18.7	193	36	157	6.7
18.7	187	35	152	6.6
18.7	182	34	148	6.4
18.7	171	32	139	6.0
18.7	166	31	135	5.8
18.7	155	29	126	5.4
18.7	150	28	122	5.2
18.7	139	26	113	4.9
18.7	134	25	109	4.7
18.7	123	23	100	4.3
18.7	107	20	87	3.7
18.7	91	17	74	3.2
18.7	75	14	61	2.6
18.6	500	93	407	17.3
18.6	499	93	406	17.3
18.6	495	92	403	17.1
18.6	494	92	402	17.1
18.6	490	91	399	16.9
18.6	489	91	398	16.9
18.6	488	91	397	17.0
18.6	485	90	395	16.7
18.6	484	90	394	16.7

%E	M1	M2	DM	M*
18.6	483	90	393	16.8
18.6	479	89	390	16.5
18.6	478	89	389	16.6
18.6	474	88	386	16.3
18.6	473	88	385	16.4
18.6	472	88	384	16.4
18.6	469	87	382	16.1
18.6	468	87	381	16.2
18.6	467	87	380	16.2
18.6	463	86	377	16.0
18.6	462	86	376	16.0
18.6	458	85	373	15.8
18.6	457	85	372	15.8
18.6	456	85	371	15.8
18.6	452	84	368	15.6
18.6	451	84	367	15.6
18.6	447	83	364	15.4
18.6	446	83	363	15.4
18.6	442	82	360	15.2
18.6	441	82	359	15.2
18.6	440	82	358	15.3
18.6	436	81	355	15.0
18.6	435	81	354	15.1
18.6	431	80	351	14.8
18.6	430	80	350	14.9
18.6	429	80	349	14.9
18.6	425	79	346	14.7
18.6	424	79	345	14.7
18.6	420	78	342	14.5
18.6	419	78	341	14.5
18.6	415	77	338	14.3
18.6	414	77	337	14.3
18.6	413	77	336	14.4
18.6	409	76	333	14.1
18.6	408	76	332	14.2
18.6	404	75	329	13.9
18.6	403	75	328	14.0
18.6	398	74	324	13.8
18.6	397	74	323	13.8
18.6	393	73	320	13.6
18.6	392	73	319	13.6
18.6	388	72	316	13.4
18.6	387	72	315	13.4
18.6	382	71	311	13.2
18.6	381	71	310	13.2
18.6	377	70	307	13.0
18.6	376	70	306	13.0
18.6	371	69	302	12.8
18.6	370	69	301	12.9
18.6	366	68	298	12.6

%E	M1	M2	DM	M*
18.6	365	68	297	12.7
18.6	361	67	294	12.4
18.6	360	67	293	12.5
18.6	355	66	289	12.3
18.6	354	66	288	12.3
18.6	350	65	285	12.1
18.6	349	65	284	12.1
18.6	345	64	281	11.9
18.6	344	64	280	11.9
18.6	339	63	276	11.7
18.6	338	63	275	11.7
18.6	334	62	272	11.5
18.6	333	62	271	11.5
18.6	328	61	267	11.3
18.6	323	60	263	11.1
18.6	322	60	262	11.2
18.6	318	59	259	10.9
18.6	317	59	258	11.0
18.6	312	58	254	10.8
18.6	307	57	250	10.6
18.6	306	57	249	10.6
18.6	301	56	245	10.4
18.6	296	55	241	10.2
18.6	295	55	240	10.3
18.6	291	54	237	10.0
18.6	290	54	236	10.1
18.6	285	53	232	9.9
18.6	280	52	228	9.7
18.6	279	52	227	9.7
18.6	274	51	223	9.5
18.6	269	50	219	9.3
18.6	264	49	215	9.1
18.6	263	49	214	9.1
18.6	258	48	210	8.9
18.6	253	47	206	8.7
18.6	247	46	201	8.6
18.6	242	45	197	8.4
18.6	237	44	193	8.2
18.6	236	44	192	8.2
18.6	231	43	188	8.0
18.6	226	42	184	7.8
18.6	221	41	180	7.6
18.6	220	41	179	7.6
18.6	215	40	175	7.4
18.6	210	39	171	7.2
18.6	204	38	166	7.1
18.6	199	37	162	6.9
18.6	194	36	158	6.7
18.6	188	35	153	6.5
18.6	183	34	149	6.3

%E	M1	M2	DM	M*
18.6	177	33	144	6.2
18.6	172	32	140	6.0
18.6	167	31	136	5.8
18.6	161	30	131	5.6
18.6	156	29	127	5.4
18.6	145	27	118	5.0
18.6	140	26	114	4.8
18.6	129	24	105	4.5
18.6	118	22	96	4.1
18.6	113	21	92	3.9
18.6	102	19	83	3.5
18.6	97	18	79	3.3
18.6	86	16	70	3.0
18.6	70	13	57	2.4
18.6	59	11	48	2.1
18.6	43	8	35	1.5
18.5	498	92	406	17.0
18.5	497	92	405	17.0
18.5	496	92	404	17.1
18.5	493	91	402	16.8
18.5	492	91	401	16.8
18.5	491	91	400	16.9
18.5	487	90	397	16.6
18.5	486	90	396	16.7
18.5	482	89	393	16.4
18.5	481	89	392	16.5
18.5	480	89	391	16.5
18.5	476	88	388	16.3
18.5	475	88	387	16.3
18.5	471	87	384	16.1
18.5	470	87	383	16.1
18.5	466	86	380	15.9
18.5	465	86	379	15.9
18.5	464	86	378	15.9
18.5	460	85	375	15.7
18.5	459	85	374	15.7
18.5	455	84	371	15.5
18.5	454	84	370	15.5
18.5	453	84	369	15.6
18.5	449	83	366	15.3
18.5	448	83	365	15.4
18.5	444	82	362	15.1
18.5	443	82	361	15.2
18.5	439	81	358	14.9
18.5	438	81	357	15.0
18.5	437	81	356	15.0
18.5	433	80	353	14.8
18.5	432	80	352	14.8
18.5	428	79	349	14.6
18.5	427	79	348	14.6

%E	M1	M2	DM	M*
18.5	426	79	347	14.7
18.5	422	78	344	14.4
18.5	421	78	343	14.5
18.5	417	77	340	14.2
18.5	416	77	339	14.3
18.5	411	76	335	14.1
18.5	410	76	334	14.1
18.5	406	75	331	13.9
18.5	405	75	330	13.9
18.5	401	74	327	13.7
18.5	400	74	326	13.7
18.5	399	74	325	13.7
18.5	395	73	322	13.5
18.5	394	73	321	13.5
18.5	390	72	318	13.3
18.5	389	72	317	13.3
18.5	384	71	313	13.1
18.5	383	71	312	13.2
18.5	379	70	309	12.9
18.5	378	70	308	13.0
18.5	373	69	304	12.8
18.5	372	69	303	12.8
18.5	368	68	300	12.6
18.5	367	68	299	12.6
18.5	363	67	296	12.4
18.5	362	67	295	12.4
18.5	357	66	291	12.2
18.5	356	66	290	12.2
18.5	352	65	287	12.0
18.5	351	65	286	12.0
18.5	346	64	282	11.8
18.5	341	63	278	11.6
18.5	340	63	277	11.7
18.5	336	62	274	11.4
18.5	335	62	273	11.5
18.5	330	61	269	11.3
18.5	329	61	268	11.3
18.5	325	60	265	11.1
18.5	324	60	264	11.1
18.5	319	59	260	10.9
18.5	314	58	256	10.7
18.5	313	58	255	10.7
18.5	308	57	251	10.5
18.5	303	56	247	10.3
18.5	302	56	246	10.4
18.5	298	55	243	10.2
18.5	297	55	242	10.2
18.5	292	54	238	10.0
18.5	287	53	234	9.8
18.5	286	53	233	9.8

%E	M1	M2	DM	M*
18.5	281	52	229	9.6
18.5	276	51	225	9.4
18.5	275	51	224	9.5
18.5	271	50	221	9.2
18.5	270	50	220	9.3
18.5	265	49	216	9.1
18.5	260	48	212	8.9
18.5	259	48	211	8.9
18.5	254	47	207	8.7
18.5	249	46	203	8.5
18.5	248	46	202	8.5
18.5	243	45	198	8.3
18.5	238	44	194	8.1
18.5	233	43	190	7.9
18.5	232	43	189	8.0
18.5	227	42	185	7.8
18.5	222	41	181	7.6
18.5	216	40	176	7.4
18.5	211	39	172	7.2
18.5	205	38	167	7.0
18.5	200	37	163	6.8
18.5	195	36	159	6.6
18.5	189	35	154	6.5
18.5	184	34	150	6.3
18.5	178	33	145	6.1
18.5	173	32	141	5.9
18.5	168	31	137	5.7
18.5	162	30	132	5.6
18.5	157	29	128	5.4
18.5	151	28	123	5.2
18.5	146	27	119	5.0
18.5	135	25	110	4.6
18.5	130	24	106	4.4
18.5	124	23	101	4.3
18.5	119	22	97	4.1
18.5	108	20	88	3.7
18.5	92	17	75	3.1
18.5	81	15	66	2.8
18.5	65	12	53	2.2
18.5	54	10	44	1.9
18.5	27	5	22	0.9
18.4	500	92	408	16.9
18.4	499	92	407	17.0
18.4	495	91	404	16.7
18.4	494	91	403	16.8
18.4	490	90	400	16.5
18.4	489	90	399	16.6
18.4	488	90	398	16.6
18.4	485	89	396	16.3
18.4	484	89	395	16.4

%E	M1	M2	DM	M*
18.4	483	89	394	16.4
18.4	479	88	391	16.2
18.4	478	88	390	16.2
18.4	477	88	389	16.2
18.4	474	87	387	16.0
18.4	473	87	386	16.0
18.4	472	87	385	16.0
18.4	468	86	382	15.8
18.4	467	86	381	15.8
18.4	463	85	378	15.6
18.4	462	85	377	15.6
18.4	461	85	376	15.7
18.4	457	84	373	15.4
18.4	456	84	372	15.5
18.4	452	83	369	15.2
18.4	451	83	368	15.3
18.4	450	83	367	15.3
18.4	446	82	364	15.1
18.4	445	82	363	15.1
18.4	441	81	360	14.9
18.4	440	81	359	14.9
18.4	435	80	355	14.7
18.4	434	80	354	14.7
18.4	430	79	351	14.5
18.4	429	79	350	14.5
18.4	425	78	347	14.3
18.4	424	78	346	14.3
18.4	423	78	345	14.4
18.4	419	77	342	14.2
18.4	418	77	341	14.2
18.4	414	76	338	14.0
18.4	413	76	337	14.0
18.4	412	76	336	14.0
18.4	408	75	333	13.8
18.4	407	75	332	13.8
18.4	403	74	329	13.6
18.4	402	74	328	13.6
18.4	397	73	324	13.4
18.4	396	73	323	13.5
18.4	392	72	320	13.2
18.4	391	72	319	13.3
18.4	386	71	315	13.1
18.4	385	71	314	13.1
18.4	381	70	311	12.9
18.4	380	70	310	12.9
18.4	376	69	307	12.7
18.4	375	69	306	12.7
18.4	374	69	305	12.7
18.4	370	68	302	12.5
18.4	369	68	301	12.5

%E	M1	M2	DM	M*
18.4	365	67	298	12.3
18.4	364	67	297	12.3
18.4	359	66	293	12.1
18.4	358	66	292	12.2
18.4	354	65	289	11.9
18.4	353	65	288	12.0
18.4	348	64	284	11.8
18.4	347	64	283	11.8
18.4	343	63	280	11.6
18.4	342	63	279	11.6
18.4	337	62	275	11.4
18.4	332	61	271	11.2
18.4	331	61	270	11.2
18.4	326	60	266	11.0
18.4	321	59	262	10.8
18.4	320	59	261	10.9
18.4	316	58	258	10.6
18.4	315	58	257	10.7
18.4	310	57	253	10.5
18.4	309	57	252	10.5
18.4	305	56	249	10.3
18.4	304	56	248	10.3
18.4	299	55	244	10.1
18.4	294	54	240	9.9
18.4	293	54	239	10.0
18.4	288	53	235	9.8
18.4	283	52	231	9.6
18.4	282	52	230	9.6
18.4	277	51	226	9.4
18.4	272	50	222	9.2
18.4	267	49	218	9.0
18.4	266	49	217	9.0
18.4	261	48	213	8.8
18.4	256	47	209	8.6
18.4	255	47	208	8.7
18.4	250	46	204	8.5
18.4	245	45	200	8.3
18.4	244	45	199	8.3
18.4	239	44	195	8.1
18.4	234	43	191	7.9
18.4	228	42	186	7.7
18.4	223	41	182	7.5
18.4	217	40	177	7.4
18.4	212	39	173	7.2
18.4	207	38	169	7.0
18.4	206	38	168	7.0
18.4	201	37	164	6.8
18.4	196	36	160	6.6
18.4	190	35	155	6.4
18.4	185	34	151	6.2

%E	M1	M2	DM	M*
18.4	179	33	146	6.1
18.4	174	32	142	5.9
18.4	163	30	133	5.5
18.4	158	29	129	5.3
18.4	152	28	124	5.2
18.4	147	27	120	5.0
18.4	141	26	115	4.8
18.4	136	25	111	4.6
18.4	125	23	102	4.2
18.4	114	21	93	3.9
18.4	103	19	84	3.5
18.4	98	18	80	3.3
18.4	87	16	71	2.9
18.4	76	14	62	2.6
18.4	49	9	40	1.7
18.4	38	7	31	1.3
18.3	498	91	407	16.6
18.3	497	91	406	16.7
18.3	496	91	405	16.7
18.3	493	90	403	16.4
18.3	492	90	402	16.5
18.3	491	90	401	16.5
18.3	487	89	398	16.3
18.3	486	89	397	16.3
18.3	482	88	394	16.1
18.3	481	88	393	16.1
18.3	480	88	392	16.1
18.3	476	87	389	15.9
18.3	475	87	388	15.9
18.3	471	86	385	15.7
18.3	470	86	384	15.7
18.3	469	86	383	15.8
18.3	465	85	380	15.5
18.3	464	85	379	15.6
18.3	460	84	376	15.3
18.3	459	84	375	15.4
18.3	458	84	374	15.4
18.3	454	83	371	15.2
18.3	453	83	370	15.2
18.3	449	82	367	15.0
18.3	448	82	366	15.0
18.3	447	82	365	15.0
18.3	443	81	362	14.8
18.3	442	81	361	14.8
18.3	438	80	358	14.6
18.3	437	80	357	14.6
18.3	436	80	356	14.7
18.3	432	79	353	14.4
18.3	431	79	352	14.5
18.3	427	78	349	14.2
18.3	426	78	348	14.3
18.3	421	77	344	14.1
18.3	420	77	343	14.1
18.3	416	76	340	13.9
18.3	415	76	339	13.9
18.3	410	75	335	13.7
18.3	409	75	334	13.8
18.3	405	74	331	13.5
18.3	404	74	330	13.6
18.3	400	73	327	13.3
18.3	399	73	326	13.4
18.3	398	73	325	13.4
18.3	394	72	322	13.2
18.3	393	72	321	13.2
18.3	389	71	318	13.0
18.3	388	71	317	13.0
18.3	387	71	316	13.0
18.3	383	70	313	12.8
18.3	382	70	312	12.8
18.3	378	69	309	12.6
18.3	377	69	308	12.6
18.3	372	68	304	12.4
18.3	371	68	303	12.5
18.3	367	67	300	12.2
18.3	366	67	299	12.3
18.3	361	66	295	12.1
18.3	360	66	294	12.1
18.3	356	65	291	11.9
18.3	355	65	290	11.9
18.3	350	64	286	11.7
18.3	349	64	285	11.7
18.3	345	63	282	11.5
18.3	344	63	281	11.5
18.3	339	62	277	11.3
18.3	338	62	276	11.4
18.3	334	61	273	11.1
18.3	333	61	272	11.2
18.3	328	60	268	11.0
18.3	327	60	267	11.0
18.3	323	59	264	10.8
18.3	322	59	263	10.8
18.3	317	58	259	10.6
18.3	312	57	255	10.4
18.3	311	57	254	10.4
18.3	306	56	250	10.2
18.3	301	55	246	10.0
18.3	300	55	245	10.1
18.3	295	54	241	9.9
18.3	290	53	237	9.7
18.3	289	53	236	9.7
18.3	284	52	232	9.5
18.3	279	51	228	9.3
18.3	278	51	227	9.4
18.3	273	50	223	9.2
18.3	268	49	219	9.0
18.3	263	48	215	8.8
18.3	262	48	214	8.8
18.3	257	47	210	8.6
18.3	252	46	206	8.4
18.3	251	46	205	8.4
18.3	246	45	201	8.2
18.3	241	44	197	8.0
18.3	240	44	196	8.1
18.3	235	43	192	7.9
18.3	230	42	188	7.7
18.3	229	42	187	7.7
18.3	224	41	183	7.5
18.3	219	40	179	7.3
18.3	218	40	178	7.3
18.3	213	39	174	7.1
18.3	208	38	170	6.9
18.3	202	37	165	6.8
18.3	197	36	161	6.6
18.3	191	35	156	6.4
18.3	186	34	152	6.2
18.3	180	33	147	6.0
18.3	175	32	143	5.9
18.3	169	31	138	5.7
18.3	164	30	134	5.5
18.3	153	28	125	5.1
18.3	142	26	116	4.8
18.3	131	24	107	4.4
18.3	126	23	103	4.2
18.3	120	22	98	4.0
18.3	115	21	94	3.8
18.3	109	20	89	3.7
18.3	104	19	85	3.5
18.3	93	17	76	3.1
18.3	82	15	67	2.7
18.3	71	13	58	2.4
18.3	60	11	49	2.0
18.2	500	91	409	16.6
18.2	499	91	408	16.6
18.2	495	90	405	16.4
18.2	494	90	404	16.4
18.2	490	89	401	16.2
18.2	489	89	400	16.2
18.2	488	89	399	16.2
18.2	484	88	396	16.0
18.2	483	88	395	16.0
18.2	479	87	392	15.8
18.2	478	87	391	15.8
18.2	477	87	390	15.9
18.2	473	86	387	15.6
18.2	472	86	386	15.7
18.2	468	85	383	15.4
18.2	467	85	382	15.5
18.2	466	85	381	15.5
18.2	462	84	378	15.3
18.2	461	84	377	15.3
18.2	457	83	374	15.1
18.2	456	83	373	15.1
18.2	455	83	372	15.1
18.2	451	82	369	14.9
18.2	450	82	368	14.9
18.2	446	81	365	14.7
18.2	445	81	364	14.7
18.2	444	81	363	14.8
18.2	440	80	360	14.5
18.2	439	80	359	14.6
18.2	435	79	356	14.3
18.2	434	79	355	14.4
18.2	433	79	354	14.4
18.2	429	78	351	14.2
18.2	428	78	350	14.2
18.2	424	77	347	14.0
18.2	423	77	346	14.0
18.2	422	77	345	14.0
18.2	418	76	342	13.8
18.2	417	76	341	13.9
18.2	413	75	338	13.6
18.2	412	75	337	13.7
18.2	411	75	336	13.7
18.2	407	74	333	13.5
18.2	406	74	332	13.5
18.2	402	73	329	13.3
18.2	401	73	328	13.3
18.2	396	72	324	13.1
18.2	395	72	323	13.1
18.2	391	71	320	12.9
18.2	390	71	319	12.9
18.2	385	70	315	12.7
18.2	384	70	314	12.8
18.2	380	69	311	12.5
18.2	379	69	310	12.6
18.2	374	68	306	12.4
18.2	373	68	305	12.4
18.2	369	67	302	12.2
18.2	368	67	301	12.2
18.2	363	66	297	12.0
18.2	362	66	296	12.0
18.2	358	65	293	11.8
18.2	357	65	292	11.8
18.2	352	64	288	11.6
18.2	351	64	287	11.7
18.2	347	63	284	11.4
18.2	346	63	283	11.5
18.2	341	62	279	11.3
18.2	340	62	278	11.3
18.2	336	61	275	11.1
18.2	335	61	274	11.1
18.2	330	60	270	10.9
18.2	329	60	269	10.9
18.2	325	59	266	10.7
18.2	324	59	265	10.7
18.2	319	58	261	10.5
18.2	318	58	260	10.6
18.2	314	57	257	10.3
18.2	313	57	256	10.4
18.2	308	56	252	10.2
18.2	307	56	251	10.2
18.2	303	55	248	10.0
18.2	302	55	247	10.0
18.2	297	54	243	9.8
18.2	296	54	242	9.9
18.2	292	53	239	9.6
18.2	291	53	238	9.7
18.2	286	52	234	9.5
18.2	285	52	233	9.5
18.2	280	51	229	9.3
18.2	275	50	225	9.1
18.2	274	50	224	9.1
18.2	269	49	220	8.9
18.2	264	48	216	8.7
18.2	258	47	211	8.6
18.2	253	46	207	8.4
18.2	247	45	202	8.2
18.2	242	44	198	8.0
18.2	236	43	193	7.8
18.2	231	42	189	7.6
18.2	225	41	184	7.5
18.2	220	40	180	7.3
18.2	214	39	175	7.1
18.2	209	38	171	6.9
18.2	203	37	166	6.7
18.2	198	36	162	6.5
18.2	192	35	157	6.4
18.2	187	34	153	6.2
18.2	181	33	148	6.0
18.2	176	32	144	5.8

%E	M1	M2	DM	M*
18.2	170	31	139	5.7
18.2	165	30	135	5.5
18.2	159	29	130	5.3
18.2	154	28	126	5.1
18.2	148	27	121	4.9
18.2	143	26	117	4.7
18.2	137	25	112	4.6
18.2	132	24	108	4.4
18.2	121	22	99	4.0
18.2	110	20	90	3.6
18.2	99	18	81	3.3
18.2	88	16	72	2.9
18.2	77	14	63	2.5
18.2	66	12	54	2.2
18.2	55	10	45	1.8
18.2	44	8	36	1.5
18.2	33	6	27	1.1
18.2	22	4	18	0.7
18.2	11	2	9	0.4
18.1	498	90	408	16.3
18.1	497	90	407	16.3
18.1	496	90	406	16.3
18.1	493	89	404	16.1
18.1	492	89	403	16.1
18.1	491	89	402	16.1
18.1	487	88	399	15.9
18.1	486	88	398	15.9
18.1	485	88	397	16.0
18.1	482	87	395	15.7
18.1	481	87	394	15.7
18.1	480	87	393	15.8
18.1	476	86	390	15.5
18.1	475	86	389	15.6
18.1	474	86	388	15.6
18.1	470	85	385	15.4
18.1	469	85	384	15.4
18.1	465	84	381	15.2
18.1	464	84	380	15.2
18.1	463	84	379	15.2
18.1	459	83	376	15.0
18.1	458	83	375	15.0
18.1	454	82	372	14.8
18.1	453	82	371	14.8
18.1	452	82	370	14.9
18.1	448	81	367	14.6
18.1	447	81	366	14.7
18.1	443	80	363	14.4
18.1	442	80	362	14.5
18.1	441	80	361	14.5
18.1	437	79	358	14.3
18.1	436	79	357	14.3
18.1	432	78	354	14.1
18.1	431	78	353	14.1
18.1	430	78	352	14.1
18.1	426	77	349	13.9
18.1	425	77	348	14.0
18.1	421	76	345	13.7
18.1	420	76	344	13.8
18.1	419	76	343	13.8
18.1	415	75	340	13.6
18.1	414	75	339	13.6
18.1	409	74	335	13.4
18.1	408	74	334	13.4
18.1	404	73	331	13.2
18.1	403	73	330	13.2
18.1	398	72	326	13.0
18.1	397	72	325	13.1
18.1	393	71	322	12.8
18.1	392	71	321	12.9
18.1	387	70	317	12.7
18.1	386	70	316	12.7
18.1	382	69	313	12.5
18.1	381	69	312	12.5
18.1	376	68	308	12.3
18.1	375	68	307	12.3
18.1	371	67	304	12.1
18.1	370	67	303	12.1
18.1	365	66	299	11.9
18.1	364	66	298	12.0
18.1	360	65	295	11.7
18.1	359	65	294	11.8
18.1	354	64	290	11.6
18.1	353	64	289	11.6
18.1	349	63	286	11.4
18.1	348	63	285	11.4
18.1	343	62	281	11.2
18.1	342	62	280	11.2
18.1	337	61	276	11.0
18.1	332	60	272	10.8
18.1	331	60	271	10.9
18.1	326	59	267	10.7
18.1	321	58	263	10.5
18.1	320	58	262	10.5
18.1	315	57	258	10.3
18.1	310	56	254	10.1
18.1	309	56	253	10.1
18.1	304	55	249	10.0
18.1	299	54	245	9.8
18.1	298	54	244	9.8
18.1	293	53	240	9.6
18.1	288	52	236	9.4
18.1	287	52	235	9.4
18.1	282	51	231	9.2
18.1	281	51	230	9.3
18.1	277	50	227	9.0
18.1	276	50	226	9.1
18.1	271	49	222	8.9
18.1	270	49	221	8.9
18.1	265	48	217	8.7
18.1	260	47	213	8.5
18.1	259	47	212	8.5
18.1	254	46	208	8.3
18.1	249	45	204	8.1
18.1	248	45	203	8.2
18.1	243	44	199	8.0
18.1	238	43	195	7.8
18.1	237	43	194	7.8
18.1	232	42	190	7.6
18.1	227	41	186	7.4
18.1	226	41	185	7.4
18.1	221	40	181	7.2
18.1	216	39	177	7.0
18.1	215	39	176	7.1
18.1	210	38	172	6.9
18.1	204	37	167	6.7
18.1	199	36	163	6.5
18.1	193	35	158	6.3
18.1	188	34	154	6.1
18.1	182	33	149	6.0
18.1	177	32	145	5.8
18.1	171	31	140	5.6
18.1	166	30	136	5.4
18.1	160	29	131	5.3
18.1	155	28	127	5.1
18.1	149	27	122	4.9
18.1	144	26	118	4.7
18.1	138	25	113	4.5
18.1	127	23	104	4.2
18.1	116	21	95	3.8
18.1	105	19	86	3.4
18.1	94	17	77	3.1
18.1	83	15	68	2.7
18.1	72	13	59	2.3
18.0	500	90	410	16.2
18.0	499	90	409	16.2
18.0	495	89	406	16.0
18.0	494	89	405	16.0
18.0	490	88	402	15.8
18.0	489	88	401	15.8
18.0	488	88	400	15.9
18.0	484	87	397	15.6
18.0	483	87	396	15.7
18.0	479	86	393	15.4
18.0	478	86	392	15.5
18.0	477	86	391	15.5
18.0	473	85	388	15.3
18.0	472	85	387	15.3
18.0	471	85	386	15.3
18.0	467	84	383	15.1
18.0	466	84	382	15.1
18.0	462	83	379	14.9
18.0	461	83	378	14.9
18.0	460	83	377	15.0
18.0	456	82	374	14.7
18.0	455	82	373	14.8
18.0	451	81	370	14.5
18.0	450	81	369	14.6
18.0	449	81	368	14.6
18.0	445	80	365	14.4
18.0	444	80	364	14.4
18.0	440	79	361	14.2
18.0	439	79	360	14.2
18.0	438	79	359	14.2
18.0	434	78	356	14.0
18.0	433	78	355	14.1
18.0	428	77	351	13.9
18.0	427	77	350	13.9
18.0	423	76	347	13.7
18.0	422	76	346	13.7
18.0	417	75	342	13.5
18.0	416	75	341	13.5
18.0	412	74	338	13.3
18.0	411	74	337	13.3
18.0	410	74	336	13.4
18.0	406	73	333	13.1
18.0	405	73	332	13.2
18.0	401	72	329	12.9
18.0	400	72	328	13.0
18.0	399	72	327	13.0
18.0	395	71	324	12.8
18.0	394	71	323	12.8
18.0	389	70	319	12.6
18.0	388	70	318	12.6
18.0	384	69	315	12.4
18.0	383	69	314	12.4
18.0	378	68	310	12.2
18.0	377	68	309	12.3
18.0	373	67	306	12.0
18.0	372	67	305	12.1
18.0	367	66	301	11.9
18.0	366	66	300	11.9
18.0	362	65	297	11.7
18.0	361	65	296	11.7
18.0	356	64	292	11.5
18.0	355	64	291	11.5
18.0	350	63	287	11.3
18.0	345	62	283	11.1
18.0	344	62	282	11.2
18.0	339	61	278	11.0
18.0	338	61	277	11.0
18.0	334	60	274	10.8
18.0	333	60	273	10.8
18.0	328	59	269	10.6
18.0	327	59	268	10.6
18.0	323	58	265	10.4
18.0	322	58	264	10.4
18.0	317	57	260	10.2
18.0	316	57	259	10.3
18.0	311	56	255	10.1
18.0	306	55	251	9.9
18.0	305	55	250	9.9
18.0	300	54	246	9.7
18.0	295	53	242	9.5
18.0	294	53	241	9.6
18.0	289	52	237	9.4
18.0	284	51	233	9.2
18.0	283	51	232	9.2
18.0	278	50	228	9.0
18.0	272	49	223	8.8
18.0	267	48	219	8.6
18.0	266	48	218	8.7
18.0	261	47	214	8.5
18.0	256	46	210	8.3
18.0	255	46	209	8.3
18.0	250	45	205	8.1
18.0	245	44	201	7.9
18.0	244	44	200	7.9
18.0	239	43	196	7.7
18.0	233	42	191	7.6
18.0	228	41	187	7.4
18.0	222	40	182	7.2
18.0	217	39	178	7.0
18.0	211	38	173	6.8
18.0	206	37	169	6.6
18.0	205	37	168	6.7
18.0	200	36	164	6.5
18.0	194	35	159	6.3
18.0	189	34	155	6.1
18.0	183	33	150	6.0
18.0	178	32	146	5.8

%E	M1	M2	DM	M*
18.0	172	31	141	5.6
18.0	167	30	137	5.4
18.0	161	29	132	5.2
18.0	150	27	123	4.9
18.0	139	25	114	4.5
18.0	133	24	109	4.3
18.0	128	23	105	4.1
18.0	122	22	100	4.0
18.0	111	20	91	3.6
18.0	100	18	82	3.2
18.0	89	16	73	2.9
18.0	61	11	50	2.0
18.0	50	9	41	1.6
17.9	498	89	409	15.9
17.9	497	89	408	15.9
17.9	496	89	407	16.0
17.9	493	88	405	15.7
17.9	492	88	404	15.7
17.9	491	88	403	15.8
17.9	487	87	400	15.5
17.9	486	87	399	15.6
17.9	485	87	398	15.6
17.9	481	86	395	15.4
17.9	480	86	394	15.4
17.9	476	85	391	15.2
17.9	475	85	390	15.2
17.9	474	85	389	15.2
17.9	470	84	386	15.0
17.9	469	84	385	15.0
17.9	468	84	384	15.1
17.9	464	83	381	14.8
17.9	463	83	380	14.9
17.9	459	82	377	14.6
17.9	458	82	376	14.7
17.9	457	82	375	14.7
17.9	453	81	372	14.5
17.9	452	81	371	14.5
17.9	448	80	368	14.3
17.9	447	80	367	14.3
17.9	446	80	366	14.3
17.9	442	79	363	14.1
17.9	441	79	362	14.2
17.9	436	78	358	14.0
17.9	435	78	357	14.0
17.9	431	77	354	13.8
17.9	430	77	353	13.8
17.9	429	77	352	13.8
17.9	425	76	349	13.6
17.9	424	76	348	13.6
17.9	420	75	345	13.4

%E	M1	M2	DM	M*
17.9	419	75	344	13.4
17.9	418	75	343	13.5
17.9	414	74	340	13.2
17.9	413	74	339	13.3
17.9	408	73	335	13.1
17.9	407	73	334	13.1
17.9	403	72	331	12.9
17.9	402	72	330	12.9
17.9	397	71	326	12.7
17.9	396	71	325	12.7
17.9	392	70	322	12.5
17.9	391	70	321	12.5
17.9	390	70	320	12.6
17.9	386	69	317	12.3
17.9	385	69	316	12.4
17.9	380	68	312	12.2
17.9	379	68	311	12.2
17.9	375	67	308	12.0
17.9	374	67	307	12.0
17.9	369	66	303	11.8
17.9	368	66	302	11.8
17.9	364	65	299	11.6
17.9	363	65	298	11.6
17.9	358	64	294	11.4
17.9	357	64	293	11.5
17.9	352	63	289	11.3
17.9	351	63	288	11.3
17.9	347	62	285	11.1
17.9	346	62	284	11.1
17.9	341	61	280	10.9
17.9	340	61	279	10.9
17.9	336	60	276	10.7
17.9	335	60	275	10.7
17.9	330	59	271	10.5
17.9	329	59	270	10.6
17.9	324	58	266	10.4
17.9	319	57	262	10.2
17.9	318	57	261	10.2
17.9	313	56	257	10.0
17.9	312	56	256	10.1
17.9	308	55	253	9.8
17.9	307	55	252	9.9
17.9	302	54	248	9.7
17.9	301	54	247	9.7
17.9	296	53	243	9.5
17.9	291	52	239	9.3
17.9	290	52	238	9.3
17.9	285	51	234	9.1
17.9	280	50	230	8.9
17.9	279	50	229	9.0

%E	M1	M2	DM	M*
17.9	274	49	225	8.8
17.9	273	49	224	8.8
17.9	268	48	220	8.6
17.9	263	47	216	8.4
17.9	262	47	215	8.4
17.9	257	46	211	8.2
17.9	252	45	207	8.0
17.9	251	45	206	8.1
17.9	246	44	202	7.9
17.9	240	43	197	7.7
17.9	235	42	193	7.5
17.9	234	42	192	7.5
17.9	229	41	188	7.3
17.9	224	40	184	7.1
17.9	223	40	183	7.2
17.9	218	39	179	7.0
17.9	212	38	174	6.8
17.9	207	37	170	6.6
17.9	201	36	165	6.4
17.9	196	35	161	6.3
17.9	195	35	160	6.3
17.9	190	34	156	6.1
17.9	184	33	151	5.9
17.9	179	32	147	5.7
17.9	173	31	142	5.6
17.9	168	30	138	5.4
17.9	162	29	133	5.2
17.9	156	28	128	5.0
17.9	151	27	124	4.8
17.9	145	26	119	4.7
17.9	140	25	115	4.5
17.9	134	24	110	4.3
17.9	123	22	101	3.9
17.9	117	21	96	3.8
17.9	112	20	92	3.6
17.9	106	19	87	3.4
17.9	95	17	78	3.0
17.9	84	15	69	2.7
17.9	78	14	64	2.5
17.9	67	12	55	2.1
17.9	56	10	46	1.8
17.9	39	7	32	1.3
17.9	28	5	23	0.9
17.8	500	89	411	15.8
17.8	499	89	410	15.9
17.8	495	88	407	15.6
17.8	494	88	406	15.7
17.8	490	87	403	15.4
17.8	489	87	402	15.5
17.8	488	87	401	15.5

%E	M1	M2	DM	M*
17.8	484	86	398	15.3
17.8	483	86	397	15.3
17.8	482	86	396	15.3
17.8	478	85	393	15.1
17.8	477	85	392	15.1
17.8	473	84	389	14.9
17.8	472	84	388	14.9
17.8	471	84	387	15.0
17.8	467	83	384	14.8
17.8	466	83	383	14.8
17.8	465	83	382	14.8
17.8	461	82	379	14.6
17.8	460	82	378	14.6
17.8	456	81	375	14.4
17.8	455	81	374	14.4
17.8	454	81	373	14.5
17.8	450	80	370	14.2
17.8	449	80	369	14.3
17.8	445	79	366	14.0
17.8	444	79	365	14.1
17.8	443	79	364	14.1
17.8	439	78	361	13.9
17.8	438	78	360	13.9
17.8	437	78	359	13.9
17.8	433	77	356	13.7
17.8	432	77	355	13.7
17.8	428	76	352	13.5
17.8	427	76	351	13.5
17.8	426	76	350	13.6
17.8	422	75	347	13.3
17.8	421	75	346	13.4
17.8	416	74	342	13.2
17.8	415	74	341	13.2
17.8	411	73	338	13.0
17.8	410	73	337	13.0
17.8	409	73	336	13.0
17.8	405	72	333	12.8
17.8	404	72	332	12.8
17.8	400	71	329	12.6
17.8	399	71	328	12.6
17.8	398	71	327	12.7
17.8	394	70	324	12.4
17.8	393	70	323	12.5
17.8	388	69	319	12.3
17.8	387	69	318	12.3
17.8	383	68	315	12.1
17.8	382	68	314	12.1
17.8	381	68	313	12.1
17.8	377	67	310	11.9
17.8	376	67	309	11.9

%E	M1	M2	DM	M*
17.8	371	66	305	11.7
17.8	370	66	304	11.8
17.8	366	65	301	11.5
17.8	365	65	300	11.6
17.8	360	64	296	11.4
17.8	359	64	295	11.4
17.8	354	63	291	11.2
17.8	353	63	290	11.2
17.8	349	62	287	11.0
17.8	348	62	286	11.0
17.8	343	61	282	10.8
17.8	342	61	281	10.9
17.8	338	60	278	10.7
17.8	337	60	277	10.7
17.8	332	59	273	10.5
17.8	331	59	272	10.5
17.8	326	58	268	10.3
17.8	325	58	267	10.4
17.8	321	57	264	10.1
17.8	320	57	263	10.2
17.8	315	56	259	10.0
17.8	314	56	258	10.0
17.8	309	55	254	9.8
17.8	304	54	250	9.6
17.8	303	54	249	9.6
17.8	298	53	245	9.4
17.8	297	53	244	9.5
17.8	292	52	240	9.3
17.8	287	51	236	9.1
17.8	286	51	235	9.1
17.8	281	50	231	8.9
17.8	276	49	227	8.7
17.8	275	49	226	8.7
17.8	270	48	222	8.5
17.8	269	48	221	8.6
17.8	264	47	217	8.4
17.8	259	46	213	8.2
17.8	258	46	212	8.2
17.8	253	45	208	8.0
17.8	247	44	203	7.8
17.8	242	43	199	7.6
17.8	241	43	198	7.7
17.8	236	42	194	7.5
17.8	230	41	189	7.3
17.8	225	40	185	7.1
17.8	219	39	180	6.9
17.8	214	38	176	6.7
17.8	213	38	175	6.8
17.8	208	37	171	6.6
17.8	202	36	166	6.4

%E	M1	M2	DM	M*
17.8	197	35	162	6.2
17.8	191	34	157	6.1
17.8	185	33	152	5.9
17.8	180	32	148	5.7
17.8	174	31	143	5.5
17.8	169	30	139	5.3
17.8	163	29	134	5.2
17.8	157	28	129	5.0
17.8	152	27	125	4.8
17.8	146	26	120	4.6
17.8	135	24	111	4.3
17.8	129	23	106	4.1
17.8	118	21	97	3.7
17.8	107	19	88	3.4
17.8	101	18	83	3.2
17.8	90	16	74	2.8
17.8	73	13	60	2.3
17.8	45	8	37	1.4
17.7	498	88	410	15.6
17.7	497	88	409	15.6
17.7	496	88	408	15.6
17.7	492	87	405	15.4
17.7	491	87	404	15.4
17.7	487	86	401	15.2
17.7	486	86	400	15.2
17.7	485	86	399	15.2
17.7	481	85	396	15.0
17.7	480	85	395	15.1
17.7	479	85	394	15.1
17.7	475	84	391	14.9
17.7	474	84	390	14.9
17.7	470	83	387	14.7
17.7	469	83	386	14.7
17.7	468	83	385	14.7
17.7	464	82	382	14.5
17.7	463	82	381	14.5
17.7	462	82	380	14.6
17.7	458	81	377	14.3
17.7	457	81	376	14.4
17.7	453	80	373	14.1
17.7	452	80	372	14.2
17.7	451	80	371	14.2
17.7	447	79	368	14.0
17.7	446	79	367	14.0
17.7	441	78	363	13.8
17.7	440	78	362	13.8
17.7	436	77	359	13.6
17.7	435	77	358	13.6
17.7	434	77	357	13.7
17.7	430	76	354	13.4

%E	M1	M2	DM	M*
17.7	429	76	353	13.5
17.7	424	75	349	13.3
17.7	423	75	348	13.3
17.7	419	74	345	13.1
17.7	418	74	344	13.1
17.7	417	74	343	13.1
17.7	413	73	340	12.9
17.7	412	73	339	12.9
17.7	407	72	335	12.7
17.7	406	72	334	12.8
17.7	402	71	331	12.5
17.7	401	71	330	12.6
17.7	396	70	326	12.4
17.7	395	70	325	12.4
17.7	390	69	321	12.2
17.7	389	69	320	12.2
17.7	385	68	317	12.0
17.7	384	68	316	12.0
17.7	379	67	312	11.8
17.7	378	67	311	11.9
17.7	373	66	307	11.7
17.7	372	66	306	11.7
17.7	368	65	303	11.5
17.7	367	65	302	11.5
17.7	362	64	298	11.3
17.7	361	64	297	11.3
17.7	356	63	293	11.1
17.7	355	63	292	11.2
17.7	351	62	289	11.0
17.7	350	62	288	11.0
17.7	345	61	284	10.8
17.7	344	61	283	10.8
17.7	339	60	279	10.6
17.7	334	59	275	10.4
17.7	333	59	274	10.5
17.7	328	58	270	10.3
17.7	327	58	269	10.3
17.7	322	57	265	10.1
17.7	317	56	261	9.9
17.7	316	56	260	9.9
17.7	311	55	256	9.7
17.7	310	55	255	9.8
17.7	305	54	251	9.6
17.7	300	53	247	9.4
17.7	299	53	246	9.4
17.7	294	52	242	9.2
17.7	293	52	241	9.2
17.7	288	51	237	9.0
17.7	283	50	233	8.8
17.7	282	50	232	8.9

%E	M1	M2	DM	M*
17.7	277	49	228	8.7
17.7	271	48	223	8.5
17.7	266	47	219	8.3
17.7	265	47	218	8.3
17.7	260	46	214	8.1
17.7	254	45	209	8.0
17.7	249	44	205	7.8
17.7	248	44	204	7.8
17.7	243	43	200	7.6
17.7	237	42	195	7.4
17.7	232	41	191	7.2
17.7	231	41	190	7.3
17.7	226	40	186	7.1
17.7	220	39	181	6.9
17.7	215	38	177	6.7
17.7	209	37	172	6.6
17.7	203	36	167	6.4
17.7	198	35	163	6.2
17.7	192	34	158	6.0
17.7	186	33	153	5.9
17.7	181	32	149	5.7
17.7	175	31	144	5.5
17.7	164	29	135	5.1
17.7	158	28	130	5.0
17.7	147	26	121	4.6
17.7	141	25	116	4.4
17.7	130	23	107	4.1
17.7	124	22	102	3.9
17.7	113	20	93	3.5
17.7	96	17	79	3.0
17.7	79	14	65	2.5
17.7	62	11	51	2.0
17.6	500	88	412	15.5
17.6	499	88	411	15.5
17.6	495	87	408	15.3
17.6	494	87	407	15.3
17.6	493	87	406	15.4
17.6	490	86	404	15.1
17.6	489	86	403	15.1
17.6	488	86	402	15.2
17.6	484	85	399	14.9
17.6	483	85	398	15.0
17.6	482	85	397	15.0
17.6	478	84	394	14.8
17.6	477	84	393	14.8
17.6	476	84	392	14.8
17.6	472	83	389	14.6
17.6	471	83	388	14.6
17.6	467	82	385	14.4
17.6	466	82	384	14.4

%E	M1	M2	DM	M*
17.6	465	82	383	14.5
17.6	461	81	380	14.2
17.6	460	81	379	14.3
17.6	459	81	378	14.3
17.6	455	80	375	14.1
17.6	454	80	374	14.1
17.6	450	79	371	13.9
17.6	449	79	370	13.9
17.6	448	79	369	13.9
17.6	444	78	366	13.7
17.6	443	78	365	13.7
17.6	442	78	364	13.8
17.6	438	77	361	13.5
17.6	437	77	360	13.6
17.6	433	76	357	13.3
17.6	432	76	356	13.4
17.6	431	76	355	13.4
17.6	427	75	352	13.2
17.6	426	75	351	13.2
17.6	425	75	350	13.2
17.6	421	74	347	13.0
17.6	420	74	346	13.0
17.6	415	73	342	12.8
17.6	414	73	341	12.9
17.6	410	72	338	12.6
17.6	409	72	337	12.7
17.6	408	72	336	12.7
17.6	404	71	333	12.5
17.6	403	71	332	12.5
17.6	398	70	328	12.3
17.6	397	70	327	12.3
17.6	393	69	324	12.1
17.6	392	69	323	12.1
17.6	391	69	322	12.2
17.6	387	68	319	11.9
17.6	386	68	318	12.0
17.6	381	67	314	11.8
17.6	380	67	313	11.8
17.6	376	66	310	11.6
17.6	375	66	309	11.6
17.6	374	66	308	11.6
17.6	370	65	305	11.4
17.6	369	65	304	11.4
17.6	364	64	300	11.3
17.6	363	64	299	11.3
17.6	358	63	295	11.1
17.6	357	63	294	11.1
17.6	353	62	291	10.9
17.6	352	62	290	10.9
17.6	347	61	286	10.7

%E	M1	M2	DM	M*
17.6	346	61	285	10.8
17.6	341	60	281	10.6
17.6	340	60	280	10.6
17.6	336	59	277	10.4
17.6	335	59	276	10.4
17.6	330	58	272	10.2
17.6	329	58	271	10.2
17.6	324	57	267	10.0
17.6	323	57	266	10.1
17.6	319	56	263	9.8
17.6	318	56	262	9.9
17.6	313	55	258	9.7
17.6	312	55	257	9.7
17.6	307	54	253	9.5
17.6	306	54	252	9.5
17.6	302	53	249	9.3
17.6	301	53	248	9.3
17.6	296	52	244	9.1
17.6	295	52	243	9.2
17.6	290	51	239	9.0
17.6	289	51	238	9.0
17.6	284	50	234	8.8
17.6	279	49	230	8.6
17.6	278	49	229	8.6
17.6	273	48	225	8.4
17.6	272	48	224	8.5
17.6	267	47	220	8.3
17.6	262	46	216	8.1
17.6	261	46	215	8.1
17.6	256	45	211	7.9
17.6	255	45	210	7.9
17.6	250	44	206	7.7
17.6	245	43	202	7.5
17.6	244	43	201	7.6
17.6	239	42	197	7.4
17.6	238	42	196	7.4
17.6	233	41	192	7.2
17.6	227	40	187	7.0
17.6	222	39	183	6.9
17.6	221	39	182	6.9
17.6	216	38	178	6.7
17.6	210	37	173	6.5
17.6	205	36	169	6.3
17.6	204	36	168	6.4
17.6	199	35	164	6.2
17.6	193	34	159	6.0
17.6	188	33	155	5.8
17.6	187	33	154	5.8
17.6	182	32	150	5.6
17.6	176	31	145	5.5

%E	M1	M2	DM	M*
17.6	170	30	140	5.3
17.6	165	29	136	5.1
17.6	159	28	131	4.9
17.6	153	27	126	4.8
17.6	148	26	122	4.6
17.6	142	25	117	4.4
17.6	136	24	112	4.2
17.6	131	23	108	4.0
17.6	125	22	103	3.9
17.6	119	21	98	3.7
17.6	108	19	89	3.3
17.6	102	18	84	3.2
17.6	91	16	75	2.8
17.6	85	15	70	2.6
17.6	74	13	61	2.3
17.6	68	12	56	2.1
17.6	51	9	42	1.6
17.6	34	6	28	1.1
17.6	17	3	14	0.5
17.5	498	87	411	15.2
17.5	497	87	410	15.2
17.5	496	87	409	15.3
17.5	492	86	406	15.0
17.5	491	86	405	15.1
17.5	487	85	402	14.8
17.5	486	85	401	14.9
17.5	485	85	400	14.9
17.5	481	84	397	14.7
17.5	480	84	396	14.7
17.5	479	84	395	14.7
17.5	475	83	392	14.5
17.5	474	83	391	14.5
17.5	473	83	390	14.6
17.5	469	82	387	14.3
17.5	468	82	386	14.4
17.5	464	81	383	14.1
17.5	463	81	382	14.2
17.5	462	81	381	14.2
17.5	458	80	378	14.0
17.5	457	80	377	14.0
17.5	456	80	376	14.0
17.5	452	79	373	13.8
17.5	451	79	372	13.8
17.5	447	78	369	13.6
17.5	446	78	368	13.6
17.5	445	78	367	13.7
17.5	441	77	364	13.4
17.5	440	77	363	13.5
17.5	439	77	362	13.5
17.5	435	76	359	13.3
17.5	434	76	358	13.3
17.5	429	75	354	13.1
17.5	428	75	353	13.1
17.5	424	74	350	12.9
17.5	423	74	349	12.9
17.5	422	74	348	13.0
17.5	418	73	345	12.7
17.5	417	73	344	12.8
17.5	416	73	343	12.8
17.5	412	72	340	12.6
17.5	411	72	339	12.6
17.5	406	71	335	12.4
17.5	405	71	334	12.4
17.5	401	70	331	12.2
17.5	400	70	330	12.3
17.5	399	70	329	12.3
17.5	395	69	326	12.1
17.5	394	69	325	12.1
17.5	389	68	321	11.9
17.5	388	68	320	11.9
17.5	383	67	316	11.7
17.5	382	67	315	11.8
17.5	378	66	312	11.5
17.5	377	66	311	11.6
17.5	372	65	307	11.4
17.5	371	65	306	11.4
17.5	366	64	302	11.2
17.5	365	64	301	11.2
17.5	361	63	298	11.0
17.5	360	63	297	11.0
17.5	359	63	296	11.1
17.5	355	62	293	10.8
17.5	354	62	292	10.9
17.5	349	61	288	10.7
17.5	348	61	287	10.7
17.5	343	60	283	10.5
17.5	342	60	282	10.5
17.5	338	59	279	10.3
17.5	337	59	278	10.3
17.5	332	58	274	10.1
17.5	331	58	273	10.2
17.5	326	57	269	10.0
17.5	325	57	268	10.0
17.5	320	56	264	9.8
17.5	315	55	260	9.6
17.5	314	55	259	9.6
17.5	309	54	255	9.4
17.5	308	54	254	9.5
17.5	303	53	250	9.3
17.5	298	52	246	9.1
17.5	297	52	245	9.1
17.5	292	51	241	8.9
17.5	291	51	240	8.9
17.5	286	50	236	8.7
17.5	285	50	235	8.8
17.5	280	49	231	8.6
17.5	275	48	227	8.4
17.5	274	48	226	8.4
17.5	269	47	222	8.2
17.5	268	47	221	8.2
17.5	263	46	217	8.0
17.5	257	45	212	7.9
17.5	252	44	208	7.7
17.5	251	44	207	7.7
17.5	246	43	203	7.5
17.5	240	42	198	7.3
17.5	234	41	193	7.2
17.5	229	40	189	7.0
17.5	228	40	188	7.0
17.5	223	39	184	6.8
17.5	217	38	179	6.7
17.5	212	37	175	6.5
17.5	211	37	174	6.5
17.5	206	36	170	6.3
17.5	200	35	165	6.1
17.5	194	34	160	6.0
17.5	189	33	156	5.8
17.5	183	32	151	5.6
17.5	177	31	146	5.4
17.5	171	30	141	5.3
17.5	166	29	137	5.1
17.5	160	28	132	4.9
17.5	154	27	127	4.7
17.5	149	26	123	4.5
17.5	143	25	118	4.4
17.5	137	24	113	4.2
17.5	126	22	104	3.8
17.5	120	21	99	3.7
17.5	114	20	94	3.5
17.5	103	18	85	3.1
17.5	97	17	80	3.0
17.5	80	14	66	2.4
17.5	63	11	52	1.9
17.5	57	10	47	1.8
17.5	40	7	33	1.2
17.4	500	87	413	15.1
17.4	499	87	412	15.2
17.4	495	86	409	14.9
17.4	494	86	408	15.0
17.4	493	86	407	15.0
17.4	489	85	404	14.8
17.4	488	85	403	14.8
17.4	484	84	400	14.6
17.4	483	84	399	14.6
17.4	482	84	398	14.6
17.4	478	83	395	14.4
17.4	477	83	394	14.4
17.4	476	83	393	14.5
17.4	472	82	390	14.2
17.4	471	82	389	14.3
17.4	470	82	388	14.3
17.4	466	81	385	14.1
17.4	465	81	384	14.1
17.4	461	80	381	13.9
17.4	460	80	380	13.9
17.4	459	80	379	13.9
17.4	455	79	376	13.7
17.4	454	79	375	13.7
17.4	453	79	374	13.8
17.4	449	78	371	13.6
17.4	448	78	370	13.6
17.4	443	77	366	13.4
17.4	442	77	365	13.4
17.4	438	76	362	13.2
17.4	437	76	361	13.2
17.4	436	76	360	13.2
17.4	432	75	357	13.0
17.4	431	75	356	13.1
17.4	430	75	355	13.1
17.4	426	74	352	12.9
17.4	425	74	351	12.9
17.4	420	73	347	12.7
17.4	419	73	346	12.7
17.4	414	72	342	12.5
17.4	413	72	341	12.6
17.4	409	71	338	12.3
17.4	408	71	337	12.4
17.4	407	71	336	12.4
17.4	403	70	333	12.2
17.4	402	70	332	12.2
17.4	397	69	328	12.0
17.4	396	69	327	12.0
17.4	391	68	323	11.8
17.4	390	68	322	11.9
17.4	386	67	319	11.6
17.4	385	67	318	11.7
17.4	384	67	317	11.7
17.4	380	66	314	11.5
17.4	379	66	313	11.5
17.4	374	65	309	11.3
17.4	373	65	308	11.3
17.4	368	64	304	11.1
17.4	367	64	303	11.2
17.4	363	63	300	10.9
17.4	362	63	299	11.0
17.4	357	62	295	10.8
17.4	356	62	294	10.8
17.4	351	61	290	10.6
17.4	350	61	289	10.6
17.4	345	60	285	10.4
17.4	344	60	284	10.5
17.4	340	59	281	10.2
17.4	339	59	280	10.3
17.4	334	58	276	10.1
17.4	333	58	275	10.1
17.4	328	57	271	9.9
17.4	327	57	270	9.9
17.4	322	56	266	9.7
17.4	321	56	265	9.8
17.4	317	55	262	9.5
17.4	316	55	261	9.6
17.4	311	54	257	9.4
17.4	310	54	256	9.4
17.4	305	53	252	9.2
17.4	304	53	251	9.2
17.4	299	52	247	9.0
17.4	293	51	242	8.9
17.4	288	50	238	8.7
17.4	287	50	237	8.7
17.4	282	49	233	8.5
17.4	281	49	232	8.5
17.4	276	48	228	8.3
17.4	270	47	223	8.2
17.4	265	46	219	8.0
17.4	264	46	218	8.0
17.4	259	45	214	7.8
17.4	258	45	213	7.8
17.4	253	44	209	7.7
17.4	247	43	204	7.5
17.4	242	42	200	7.3
17.4	241	42	199	7.3
17.4	236	41	195	7.1
17.4	235	41	194	7.2
17.4	230	40	190	7.0
17.4	224	39	185	6.8
17.4	219	38	181	6.6
17.4	218	38	180	6.6
17.4	213	37	176	6.4
17.4	207	36	171	6.3
17.4	201	35	166	6.1

%E	M1	M2	DM	M*
17.4	195	34	161	5.9
17.4	190	33	157	5.7
17.4	184	32	152	5.6
17.4	178	31	147	5.4
17.4	172	30	142	5.2
17.4	167	29	138	5.0
17.4	161	28	133	4.9
17.4	155	27	128	4.7
17.4	144	25	119	4.3
17.4	138	24	114	4.2
17.4	132	23	109	4.0
17.4	121	21	100	3.6
17.4	115	20	95	3.5
17.4	109	19	90	3.3
17.4	92	16	76	2.8
17.4	86	15	71	2.6
17.4	69	12	57	2.1
17.4	46	8	38	1.4
17.4	23	4	19	0.7
17.3	498	86	412	14.9
17.3	497	86	411	14.9
17.3	496	86	410	14.9
17.3	492	85	407	14.7
17.3	491	85	406	14.7
17.3	490	85	405	14.7
17.3	486	84	402	14.5
17.3	485	84	401	14.5
17.3	481	83	398	14.3
17.3	480	83	397	14.4
17.3	479	83	396	14.4
17.3	475	82	393	14.2
17.3	474	82	392	14.2
17.3	473	82	391	14.2
17.3	469	81	388	14.0
17.3	468	81	387	14.0
17.3	467	81	386	14.0
17.3	463	80	383	13.8
17.3	462	80	382	13.9
17.3	457	79	378	13.7
17.3	456	79	377	13.7
17.3	452	78	374	13.5
17.3	451	78	373	13.5
17.3	450	78	372	13.5
17.3	446	77	369	13.3
17.3	445	77	368	13.3
17.3	444	77	367	13.4
17.3	440	76	364	13.1
17.3	439	76	363	13.2
17.3	434	75	359	13.0
17.3	433	75	358	13.0

%E	M1	M2	DM	M*
17.3	428	74	354	12.8
17.3	427	74	353	12.8
17.3	423	73	350	12.6
17.3	422	73	349	12.6
17.3	421	73	348	12.7
17.3	417	72	345	12.4
17.3	416	72	344	12.5
17.3	415	72	343	12.5
17.3	411	71	340	12.3
17.3	410	71	339	12.3
17.3	405	70	335	12.1
17.3	404	70	334	12.1
17.3	400	69	331	11.9
17.3	399	69	330	11.9
17.3	398	69	329	12.0
17.3	394	68	326	11.7
17.3	393	68	325	11.8
17.3	392	68	324	11.8
17.3	388	67	321	11.6
17.3	387	67	320	11.6
17.3	382	66	316	11.4
17.3	381	66	315	11.4
17.3	376	65	311	11.2
17.3	375	65	310	11.3
17.3	371	64	307	11.0
17.3	370	64	306	11.1
17.3	369	64	305	11.1
17.3	365	63	302	10.9
17.3	364	63	301	10.9
17.3	359	62	297	10.7
17.3	358	62	296	10.7
17.3	353	61	292	10.5
17.3	352	61	291	10.6
17.3	347	60	287	10.4
17.3	346	60	286	10.4
17.3	342	59	283	10.2
17.3	341	59	282	10.2
17.3	336	58	278	10.0
17.3	335	58	277	10.0
17.3	330	57	273	9.8
17.3	329	57	272	9.9
17.3	324	56	268	9.7
17.3	323	56	267	9.7
17.3	318	55	263	9.5
17.3	313	54	259	9.3
17.3	312	54	258	9.3
17.3	307	53	254	9.1
17.3	306	53	253	9.2
17.3	301	52	249	9.0
17.3	300	52	248	9.0

%E	M1	M2	DM	M*
17.3	295	51	244	8.8
17.3	294	51	243	8.8
17.3	289	50	239	8.7
17.3	284	49	235	8.5
17.3	283	49	234	8.5
17.3	278	48	230	8.3
17.3	277	48	229	8.3
17.3	272	47	225	8.1
17.3	271	47	224	8.2
17.3	266	46	220	8.0
17.3	260	45	215	7.8
17.3	255	44	211	7.6
17.3	254	44	210	7.6
17.3	249	43	206	7.4
17.3	248	43	205	7.5
17.3	243	42	201	7.3
17.3	237	41	196	7.1
17.3	231	40	191	6.9
17.3	226	39	187	6.7
17.3	225	39	186	6.8
17.3	220	38	182	6.6
17.3	214	37	177	6.4
17.3	208	36	172	6.2
17.3	202	35	167	6.1
17.3	197	34	163	5.9
17.3	196	34	162	5.9
17.3	191	33	158	5.7
17.3	185	32	153	5.5
17.3	179	31	148	5.4
17.3	173	30	143	5.2
17.3	168	29	139	5.0
17.3	162	28	134	4.8
17.3	156	27	129	4.7
17.3	150	26	124	4.5
17.3	139	24	115	4.1
17.3	133	23	110	4.0
17.3	127	22	105	3.8
17.3	110	19	91	3.3
17.3	104	18	86	3.1
17.3	98	17	81	2.9
17.3	81	14	67	2.4
17.3	75	13	62	2.3
17.3	52	9	43	1.6
17.2	500	86	414	14.8
17.2	499	86	413	14.8
17.2	495	85	410	14.6
17.2	494	85	409	14.6
17.2	493	85	408	14.7
17.2	489	84	405	14.4
17.2	488	84	404	14.5

%E	M1	M2	DM	M*
17.2	487	84	403	14.5
17.2	483	83	400	14.3
17.2	482	83	399	14.3
17.2	478	82	396	14.1
17.2	477	82	395	14.1
17.2	476	82	394	14.1
17.2	472	81	391	13.9
17.2	471	81	390	13.9
17.2	470	81	389	14.0
17.2	466	80	386	13.7
17.2	465	80	385	13.8
17.2	464	80	384	13.8
17.2	460	79	381	13.6
17.2	459	79	380	13.6
17.2	458	79	379	13.6
17.2	454	78	376	13.4
17.2	453	78	375	13.4
17.2	448	77	371	13.2
17.2	447	77	370	13.3
17.2	443	76	367	13.0
17.2	442	76	366	13.1
17.2	441	76	365	13.1
17.2	437	75	362	12.9
17.2	436	75	361	12.9
17.2	435	75	360	12.9
17.2	431	74	357	12.7
17.2	430	74	356	12.7
17.2	429	74	355	12.8
17.2	425	73	352	12.5
17.2	424	73	351	12.6
17.2	419	72	347	12.4
17.2	418	72	346	12.4
17.2	414	71	343	12.2
17.2	413	71	342	12.2
17.2	412	71	341	12.2
17.2	408	70	338	12.0
17.2	407	70	337	12.0
17.2	406	70	336	12.1
17.2	402	69	333	11.8
17.2	401	69	332	11.9
17.2	396	68	328	11.7
17.2	395	68	327	11.7
17.2	390	67	323	11.5
17.2	389	67	322	11.5
17.2	384	66	318	11.3
17.2	383	66	317	11.4
17.2	379	65	314	11.1
17.2	378	65	313	11.2
17.2	377	65	312	11.2
17.2	373	64	309	11.0

%E	M1	M2	DM	M*
17.2	372	64	308	11.0
17.2	367	63	304	10.8
17.2	366	63	303	10.8
17.2	361	62	299	10.6
17.2	360	62	298	10.7
17.2	355	61	294	10.5
17.2	354	61	293	10.5
17.2	349	60	289	10.3
17.2	348	60	288	10.3
17.2	344	59	285	10.1
17.2	343	59	284	10.1
17.2	338	58	280	10.0
17.2	337	58	279	10.0
17.2	332	57	275	9.8
17.2	331	57	274	9.8
17.2	326	56	270	9.6
17.2	325	56	269	9.6
17.2	320	55	265	9.5
17.2	319	55	264	9.5
17.2	314	54	260	9.3
17.2	309	53	256	9.1
17.2	308	53	255	9.1
17.2	303	52	251	8.9
17.2	302	52	250	9.0
17.2	297	51	246	8.8
17.2	296	51	245	8.8
17.2	291	50	241	8.6
17.2	290	50	240	8.6
17.2	285	49	236	8.4
17.2	279	48	231	8.3
17.2	274	47	227	8.1
17.2	273	47	226	8.1
17.2	268	46	222	7.9
17.2	267	46	221	7.9
17.2	262	45	217	7.7
17.2	261	45	216	7.8
17.2	256	44	212	7.6
17.2	250	43	207	7.4
17.2	244	42	202	7.2
17.2	239	41	198	7.0
17.2	238	41	197	7.1
17.2	233	40	193	6.9
17.2	232	40	192	6.9
17.2	227	39	188	6.7
17.2	221	38	183	6.5
17.2	215	37	178	6.4
17.2	209	36	173	6.2
17.2	204	35	169	6.0
17.2	203	35	168	6.0
17.2	198	34	164	5.8

%E	M1	M2	DM	M*
17.2	192	33	159	5.7
17.2	186	32	154	5.5
17.2	180	31	149	5.3
17.2	174	30	144	5.2
17.2	169	29	140	5.0
17.2	163	28	135	4.8
17.2	157	27	130	4.6
17.2	151	26	125	4.5
17.2	145	25	120	4.3
17.2	134	23	111	3.9
17.2	128	22	106	3.8
17.2	122	21	101	3.6
17.2	116	20	96	3.4
17.2	99	17	82	2.9
17.2	93	16	77	2.8
17.2	87	15	72	2.6
17.2	64	11	53	1.9
17.2	58	10	48	1.7
17.2	29	5	24	0.9
17.1	498	85	413	14.5
17.1	497	85	412	14.5
17.1	496	85	411	14.6
17.1	492	84	408	14.3
17.1	491	84	407	14.4
17.1	490	84	406	14.4
17.1	486	83	403	14.2
17.1	485	83	402	14.2
17.1	484	83	401	14.2
17.1	480	82	398	14.0
17.1	479	82	397	14.0
17.1	475	81	394	13.8
17.1	474	81	393	13.8
17.1	473	81	392	13.9
17.1	469	80	389	13.6
17.1	468	80	388	13.7
17.1	467	80	387	13.7
17.1	463	79	384	13.5
17.1	462	79	383	13.5
17.1	461	79	382	13.5
17.1	457	78	379	13.3
17.1	456	78	378	13.3
17.1	455	78	377	13.4
17.1	451	77	374	13.1
17.1	450	77	373	13.2
17.1	449	77	372	13.2
17.1	445	76	369	13.0
17.1	444	76	368	13.0
17.1	439	75	364	12.8
17.1	438	75	363	12.8
17.1	434	74	360	12.6
17.1	433	74	359	12.6
17.1	432	74	358	12.7
17.1	428	73	355	12.5
17.1	427	73	354	12.5
17.1	426	73	353	12.5
17.1	422	72	350	12.3
17.1	421	72	349	12.3
17.1	420	72	348	12.3
17.1	416	71	345	12.1
17.1	415	71	344	12.1
17.1	410	70	340	12.0
17.1	409	70	339	12.0
17.1	404	69	335	11.8
17.1	403	69	334	11.8
17.1	398	68	330	11.6
17.1	397	68	329	11.6
17.1	392	67	325	11.5
17.1	391	67	324	11.5
17.1	387	66	321	11.3
17.1	386	66	320	11.3
17.1	385	66	319	11.3
17.1	381	65	316	11.1
17.1	380	65	315	11.1
17.1	375	64	311	10.9
17.1	374	64	310	11.0
17.1	369	63	306	10.8
17.1	368	63	305	10.8
17.1	363	62	301	10.6
17.1	362	62	300	10.6
17.1	357	61	296	10.4
17.1	356	61	295	10.5
17.1	351	60	291	10.3
17.1	350	60	290	10.3
17.1	346	59	287	10.1
17.1	345	59	286	10.1
17.1	340	58	282	9.9
17.1	339	58	281	9.9
17.1	334	57	277	9.7
17.1	333	57	276	9.8
17.1	328	56	272	9.6
17.1	327	56	271	9.6
17.1	322	55	267	9.4
17.1	321	55	266	9.4
17.1	316	54	262	9.2
17.1	315	54	261	9.3
17.1	310	53	257	9.1
17.1	304	52	252	8.9
17.1	299	51	248	8.7
17.1	298	51	247	8.7
17.1	293	50	243	8.5
17.1	292	50	242	8.6
17.1	287	49	238	8.4
17.1	286	49	237	8.4
17.1	281	48	233	8.2
17.1	280	48	232	8.2
17.1	275	47	228	8.0
17.1	269	46	223	7.9
17.1	263	45	218	7.7
17.1	258	44	214	7.5
17.1	257	44	213	7.5
17.1	252	43	209	7.3
17.1	251	43	208	7.4
17.1	246	42	204	7.2
17.1	245	42	203	7.2
17.1	240	41	199	7.0
17.1	234	40	194	6.8
17.1	228	39	189	6.7
17.1	222	38	184	6.5
17.1	217	37	180	6.3
17.1	216	37	179	6.3
17.1	211	36	175	6.1
17.1	210	36	174	6.2
17.1	205	35	170	6.0
17.1	199	34	165	5.8
17.1	193	33	160	5.6
17.1	187	32	155	5.5
17.1	181	31	150	5.3
17.1	175	30	145	5.1
17.1	170	29	141	4.9
17.1	164	28	136	4.8
17.1	158	27	131	4.6
17.1	152	26	126	4.4
17.1	146	25	121	4.3
17.1	140	24	116	4.1
17.1	129	22	107	3.8
17.1	123	21	102	3.6
17.1	117	20	97	3.4
17.1	111	19	92	3.3
17.1	105	18	87	3.1
17.1	82	14	68	2.4
17.1	76	13	63	2.2
17.1	70	12	58	2.1
17.1	41	7	34	1.2
17.1	35	6	29	1.0
17.0	500	85	415	14.4
17.0	499	85	414	14.5
17.0	495	84	411	14.3
17.0	494	84	410	14.3
17.0	493	84	409	14.3
17.0	489	83	406	14.1
17.0	488	83	405	14.1
17.0	487	83	404	14.1
17.0	483	82	401	13.9
17.0	482	82	400	14.0
17.0	481	82	399	14.0
17.0	477	81	396	13.8
17.0	476	81	395	13.8
17.0	471	80	391	13.6
17.0	470	80	390	13.6
17.0	466	79	387	13.4
17.0	465	79	386	13.4
17.0	464	79	385	13.5
17.0	460	78	382	13.2
17.0	459	78	381	13.3
17.0	458	78	380	13.3
17.0	454	77	377	13.1
17.0	453	77	376	13.1
17.0	452	77	375	13.1
17.0	448	76	372	12.9
17.0	447	76	371	12.9
17.0	446	76	370	13.0
17.0	442	75	367	12.7
17.0	441	75	366	12.8
17.0	440	75	365	12.8
17.0	436	74	362	12.6
17.0	435	74	361	12.6
17.0	430	73	357	12.4
17.0	429	73	356	12.4
17.0	424	72	352	12.2
17.0	423	72	351	12.3
17.0	418	71	347	12.1
17.0	417	71	346	12.1
17.0	412	70	342	11.9
17.0	411	70	341	11.9
17.0	407	69	338	11.7
17.0	406	69	337	11.7
17.0	405	69	336	11.8
17.0	401	68	333	11.5
17.0	400	68	332	11.6
17.0	399	68	331	11.6
17.0	395	67	328	11.4
17.0	394	67	327	11.4
17.0	393	67	326	11.4
17.0	389	66	323	11.2
17.0	388	66	322	11.2
17.0	383	65	318	11.0
17.0	382	65	317	11.1
17.0	377	64	313	10.9
17.0	376	64	312	10.9
17.0	371	63	308	10.7
17.0	370	63	307	10.7
17.0	365	62	303	10.5
17.0	364	62	302	10.6
17.0	359	61	298	10.4
17.0	358	61	297	10.4
17.0	353	60	293	10.2
17.0	352	60	292	10.2
17.0	348	59	289	10.0
17.0	347	59	288	10.0
17.0	342	58	284	9.8
17.0	341	58	283	9.9
17.0	336	57	279	9.7
17.0	335	57	278	9.7
17.0	330	56	274	9.5
17.0	329	56	273	9.5
17.0	324	55	269	9.3
17.0	323	55	268	9.4
17.0	318	54	264	9.2
17.0	317	54	263	9.2
17.0	312	53	259	9.0
17.0	311	53	258	9.0
17.0	306	52	254	8.8
17.0	305	52	253	8.9
17.0	300	51	249	8.7
17.0	294	50	244	8.5
17.0	289	49	240	8.3
17.0	288	49	239	8.3
17.0	283	48	235	8.1
17.0	282	48	234	8.2
17.0	277	47	230	8.0
17.0	276	47	229	8.0
17.0	271	46	225	7.8
17.0	270	46	224	7.8
17.0	265	45	220	7.6
17.0	264	45	219	7.7
17.0	259	44	215	7.5
17.0	253	43	210	7.3
17.0	247	42	205	7.1
17.0	241	41	200	7.0
17.0	235	40	195	6.8
17.0	230	39	191	6.6
17.0	229	39	190	6.6
17.0	224	38	186	6.4
17.0	223	38	185	6.5
17.0	218	37	181	6.3
17.0	212	36	176	6.1
17.0	206	35	171	5.9
17.0	200	34	166	5.8
17.0	194	33	161	5.6
17.0	188	32	156	5.4

%E	M1	M2	DM	M*
17.0	182	31	151	5.3
17.0	176	30	146	5.1
17.0	171	29	142	4.9
17.0	165	28	137	4.8
17.0	159	27	132	4.6
17.0	153	26	127	4.4
17.0	147	25	122	4.3
17.0	141	24	117	4.1
17.0	135	23	112	3.9
17.0	112	19	93	3.2
17.0	106	18	88	3.1
17.0	100	17	83	2.9
17.0	94	16	78	2.7
17.0	88	15	73	2.6
17.0	53	9	44	1.5
17.0	47	8	39	1.4
16.9	498	84	414	14.2
16.9	497	84	413	14.2
16.9	496	84	412	14.2
16.9	492	83	409	14.0
16.9	491	83	408	14.0
16.9	490	83	407	14.1
16.9	486	82	404	13.8
16.9	485	82	403	13.9
16.9	484	82	402	13.9
16.9	480	81	399	13.7
16.9	479	81	398	13.7
16.9	478	81	397	13.7
16.9	474	80	394	13.5
16.9	473	80	393	13.5
16.9	472	80	392	13.6
16.9	468	79	389	13.3
16.9	467	79	388	13.4
16.9	462	78	384	13.2
16.9	461	78	383	13.2
16.9	456	77	379	13.0
16.9	455	77	378	13.0
16.9	451	76	375	12.8
16.9	450	76	374	12.8
16.9	449	76	373	12.9
16.9	445	75	370	12.6
16.9	444	75	369	12.7
16.9	443	75	368	12.7
16.9	439	74	365	12.5
16.9	438	74	364	12.5
16.9	437	74	363	12.5
16.9	433	73	360	12.3
16.9	432	73	359	12.3
16.9	431	73	358	12.4
16.9	427	72	355	12.1

%E	M1	M2	DM	M*
16.9	426	72	354	12.2
16.9	425	72	353	12.2
16.9	421	71	350	12.0
16.9	420	71	349	12.0
16.9	419	71	348	12.0
16.9	415	70	345	11.8
16.9	414	70	344	11.8
16.9	413	70	343	11.9
16.9	409	69	340	11.6
16.9	408	69	339	11.7
16.9	403	68	335	11.5
16.9	402	68	334	11.5
16.9	397	67	330	11.3
16.9	396	67	329	11.3
16.9	391	66	325	11.1
16.9	390	66	324	11.2
16.9	385	65	320	11.0
16.9	384	65	319	11.0
16.9	379	64	315	10.8
16.9	378	64	314	10.8
16.9	373	63	310	10.6
16.9	372	63	309	10.7
16.9	367	62	305	10.5
16.9	366	62	304	10.5
16.9	362	61	301	10.3
16.9	361	61	300	10.3
16.9	360	61	299	10.3
16.9	356	60	296	10.1
16.9	355	60	295	10.1
16.9	354	60	294	10.2
16.9	350	59	291	9.9
16.9	349	59	290	10.0
16.9	344	58	286	9.8
16.9	343	58	285	9.8
16.9	338	57	281	9.6
16.9	337	57	280	9.6
16.9	332	56	276	9.4
16.9	331	56	275	9.5
16.9	326	55	271	9.3
16.9	325	55	270	9.3
16.9	320	54	266	9.1
16.9	319	54	265	9.1
16.9	314	53	261	8.9
16.9	313	53	260	9.0
16.9	308	52	256	8.8
16.9	307	52	255	8.8
16.9	302	51	251	8.6
16.9	301	51	250	8.6
16.9	296	50	246	8.4
16.9	295	50	245	8.5

%E	M1	M2	DM	M*
16.9	290	49	241	8.3
16.9	284	48	236	8.1
16.9	278	47	231	7.9
16.9	273	46	227	7.8
16.9	272	46	226	7.8
16.9	267	45	222	7.6
16.9	266	45	221	7.6
16.9	261	44	217	7.4
16.9	260	44	216	7.4
16.9	255	43	212	7.3
16.9	254	43	211	7.3
16.9	249	42	207	7.1
16.9	248	42	206	7.1
16.9	243	41	202	6.9
16.9	242	41	201	6.9
16.9	237	40	197	6.8
16.9	236	40	196	6.8
16.9	231	39	192	6.6
16.9	225	38	187	6.4
16.9	219	37	182	6.3
16.9	213	36	177	6.1
16.9	207	35	172	5.9
16.9	201	34	167	5.8
16.9	195	33	162	5.6
16.9	189	32	157	5.4
16.9	183	31	152	5.3
16.9	178	30	148	5.1
16.9	177	30	147	5.1
16.9	172	29	143	4.9
16.9	166	28	138	4.7
16.9	160	27	133	4.6
16.9	154	26	128	4.4
16.9	148	25	123	4.2
16.9	142	24	118	4.1
16.9	136	23	113	3.9
16.9	130	22	108	3.7
16.9	124	21	103	3.6
16.9	118	20	98	3.4
16.9	89	15	74	2.5
16.9	83	14	69	2.4
16.9	77	13	64	2.2
16.9	71	12	59	2.0
16.9	65	11	54	1.9
16.9	59	10	49	1.7
16.8	500	84	416	14.1
16.8	499	84	415	14.1
16.8	495	83	412	13.9
16.8	494	83	411	13.9
16.8	493	83	410	14.0
16.8	489	82	407	13.8

%E	M1	M2	DM	M*
16.8	488	82	406	13.8
16.8	487	82	405	13.8
16.8	483	81	402	13.6
16.8	482	81	401	13.6
16.8	481	81	400	13.6
16.8	477	80	397	13.4
16.8	476	80	396	13.4
16.8	475	80	395	13.5
16.8	471	79	392	13.3
16.8	470	79	391	13.3
16.8	469	79	390	13.3
16.8	465	78	387	13.1
16.8	464	78	386	13.1
16.8	463	78	385	13.1
16.8	459	77	382	12.9
16.8	458	77	381	12.9
16.8	457	77	380	13.0
16.8	453	76	377	12.8
16.8	452	76	376	12.8
16.8	447	75	372	12.6
16.8	446	75	371	12.6
16.8	441	74	367	12.4
16.8	440	74	366	12.4
16.8	435	73	362	12.3
16.8	434	73	361	12.3
16.8	429	72	357	12.1
16.8	428	72	356	12.1
16.8	423	71	352	11.9
16.8	422	71	351	11.9
16.8	417	70	347	11.8
16.8	416	70	346	11.8
16.8	411	69	342	11.6
16.8	410	69	341	11.6
16.8	405	68	337	11.4
16.8	404	68	336	11.4
16.8	400	67	333	11.2
16.8	399	67	332	11.3
16.8	398	67	331	11.3
16.8	394	66	328	11.1
16.8	393	66	327	11.1
16.8	392	66	326	11.1
16.8	388	65	323	10.9
16.8	387	65	322	10.9
16.8	386	65	321	10.9
16.8	382	64	318	10.7
16.8	381	64	317	10.8
16.8	380	64	316	10.8
16.8	376	63	313	10.6
16.8	375	63	312	10.6
16.8	374	63	311	10.6

%E	M1	M2	DM	M*
16.8	370	62	308	10.4
16.8	369	62	307	10.4
16.8	368	62	306	10.4
16.8	364	61	303	10.2
16.8	363	61	302	10.3
16.8	358	60	298	10.1
16.8	357	60	297	10.1
16.8	352	59	293	9.9
16.8	351	59	292	9.9
16.8	346	58	288	9.7
16.8	345	58	287	9.8
16.8	340	57	283	9.6
16.8	339	57	282	9.6
16.8	334	56	278	9.4
16.8	333	56	277	9.4
16.8	328	55	273	9.2
16.8	327	55	272	9.3
16.8	322	54	268	9.1
16.8	321	54	267	9.1
16.8	316	53	263	8.9
16.8	315	53	262	8.9
16.8	310	52	258	8.7
16.8	309	52	257	8.8
16.8	304	51	253	8.6
16.8	303	51	252	8.6
16.8	298	50	248	8.4
16.8	297	50	247	8.4
16.8	292	49	243	8.2
16.8	291	49	242	8.3
16.8	286	48	238	8.1
16.8	285	48	237	8.1
16.8	280	47	233	7.9
16.8	279	47	232	7.9
16.8	274	46	228	7.7
16.8	268	45	223	7.6
16.8	262	44	218	7.4
16.8	256	43	213	7.2
16.8	250	42	208	7.1
16.8	244	41	203	6.9
16.8	238	40	198	6.7
16.8	232	39	193	6.6
16.8	226	38	188	6.4
16.8	220	37	183	6.2
16.8	214	36	178	6.1
16.8	208	35	173	5.9
16.8	202	34	168	5.7
16.8	197	33	164	5.5
16.8	196	33	163	5.6
16.8	191	32	159	5.4
16.8	190	32	158	5.4

%E	M1	M2	DM	M*
16.8	185	31	154	5.2
16.8	184	31	153	5.2
16.8	179	30	149	5.0
16.8	173	29	144	4.9
16.8	167	28	139	4.7
16.8	161	27	134	4.5
16.8	155	26	129	4.4
16.8	149	25	124	4.2
16.8	143	24	119	4.0
16.8	137	23	114	3.9
16.8	131	22	109	3.7
16.8	125	21	104	3.5
16.8	119	20	99	3.4
16.8	113	19	94	3.2
16.8	107	18	89	3.0
16.8	101	17	84	2.9
16.8	95	16	79	2.7
16.7	498	83	415	13.8
16.7	497	83	414	13.9
16.7	496	83	413	13.9
16.7	492	82	410	13.7
16.7	491	82	409	13.7
16.7	490	82	408	13.7
16.7	486	81	405	13.5
16.7	485	81	404	13.5
16.7	484	81	403	13.6
16.7	480	80	400	13.3
16.7	479	80	399	13.4
16.7	478	80	398	13.4
16.7	474	79	395	13.2
16.7	473	79	394	13.2
16.7	472	79	393	13.2
16.7	468	78	390	13.0
16.7	467	78	389	13.0
16.7	466	78	388	13.1
16.7	462	77	385	12.8
16.7	461	77	384	12.9
16.7	460	77	383	12.9
16.7	456	76	380	12.7
16.7	455	76	379	12.7
16.7	454	76	378	12.7
16.7	450	75	375	12.5
16.7	449	75	374	12.5
16.7	448	75	373	12.6
16.7	444	74	370	12.3
16.7	443	74	369	12.4
16.7	442	74	368	12.4
16.7	438	73	365	12.2
16.7	437	73	364	12.2
16.7	436	73	363	12.2
16.7	432	72	360	12.0
16.7	431	72	359	12.0
16.7	430	72	358	12.1
16.7	426	71	355	11.8
16.7	425	71	354	11.9
16.7	424	71	353	11.9
16.7	420	70	350	11.7
16.7	419	70	349	11.7
16.7	418	70	348	11.7
16.7	414	69	345	11.5
16.7	413	69	344	11.5
16.7	412	69	343	11.6
16.7	408	68	340	11.3
16.7	407	68	339	11.4
16.7	406	68	338	11.4
16.7	402	67	335	11.2
16.7	401	67	334	11.2
16.7	396	66	330	11.0
16.7	395	66	329	11.0
16.7	390	65	325	10.8
16.7	389	65	324	10.9
16.7	384	64	320	10.7
16.7	383	64	319	10.7
16.7	378	63	315	10.5
16.7	377	63	314	10.5
16.7	372	62	310	10.3
16.7	371	62	309	10.4
16.7	366	61	305	10.2
16.7	365	61	304	10.2
16.7	360	60	300	10.0
16.7	359	60	299	10.0
16.7	354	59	295	9.8
16.7	353	59	294	9.9
16.7	348	58	290	9.7
16.7	347	58	289	9.7
16.7	342	57	285	9.5
16.7	341	57	284	9.5
16.7	336	56	280	9.3
16.7	335	56	279	9.4
16.7	330	55	275	9.2
16.7	329	55	274	9.2
16.7	324	54	270	9.0
16.7	323	54	269	9.0
16.7	318	53	265	8.8
16.7	317	53	264	8.9
16.7	312	52	260	8.7
16.7	311	52	259	8.7
16.7	306	51	255	8.5
16.7	305	51	254	8.5
16.7	300	50	250	8.3
16.7	299	50	249	8.4
16.7	294	49	245	8.2
16.7	293	49	244	8.2
16.7	288	48	240	8.0
16.7	287	48	239	8.0
16.7	282	47	235	7.8
16.7	281	47	234	7.9
16.7	276	46	230	7.7
16.7	275	46	229	7.7
16.7	270	45	225	7.5
16.7	269	45	224	7.5
16.7	264	44	220	7.3
16.7	263	44	219	7.4
16.7	258	43	215	7.2
16.7	257	43	214	7.2
16.7	252	42	210	7.0
16.7	251	42	209	7.0
16.7	246	41	205	6.8
16.7	245	41	204	6.9
16.7	240	40	200	6.7
16.7	239	40	199	6.7
16.7	234	39	195	6.5
16.7	233	39	194	6.5
16.7	228	38	190	6.3
16.7	227	38	189	6.4
16.7	222	37	185	6.2
16.7	221	37	184	6.2
16.7	216	36	180	6.0
16.7	215	36	179	6.0
16.7	210	35	175	5.8
16.7	209	35	174	5.9
16.7	204	34	170	5.7
16.7	203	34	169	5.7
16.7	198	33	165	5.5
16.7	192	32	160	5.3
16.7	186	31	155	5.2
16.7	180	30	150	5.0
16.7	174	29	145	4.8
16.7	168	28	140	4.7
16.7	162	27	135	4.5
16.7	156	26	130	4.3
16.7	150	25	125	4.2
16.7	144	24	120	4.0
16.7	138	23	115	3.8
16.7	132	22	110	3.7
16.7	126	21	105	3.5
16.7	120	20	100	3.3
16.7	114	19	95	3.2
16.7	108	18	90	3.0
16.7	102	17	85	2.8
16.7	96	16	80	2.7
16.7	90	15	75	2.5
16.7	84	14	70	2.3
16.7	78	13	65	2.2
16.7	72	12	60	2.0
16.7	66	11	55	1.8
16.7	60	10	50	1.7
16.7	54	9	45	1.5
16.7	48	8	40	1.3
16.7	42	7	35	1.2
16.7	36	6	30	1.0
16.7	30	5	25	0.8
16.7	24	4	20	0.7
16.7	18	3	15	0.5
16.7	12	2	10	0.3
16.7	6	1	5	0.2
16.6	500	83	417	13.8
16.6	499	83	416	13.8
16.6	495	82	413	13.6
16.6	494	82	412	13.6
16.6	493	82	411	13.6
16.6	489	81	408	13.4
16.6	488	81	407	13.4
16.6	487	81	406	13.5
16.6	483	80	403	13.3
16.6	482	80	402	13.3
16.6	481	80	401	13.3
16.6	477	79	398	13.1
16.6	476	79	397	13.1
16.6	475	79	396	13.1
16.6	471	78	393	12.9
16.6	470	78	392	12.9
16.6	469	78	391	13.0
16.6	465	77	388	12.8
16.6	464	77	387	12.8
16.6	463	77	386	12.8
16.6	459	76	383	12.6
16.6	458	76	382	12.6
16.6	457	76	381	12.6
16.6	453	75	378	12.4
16.6	452	75	377	12.4
16.6	451	75	376	12.5
16.6	447	74	373	12.3
16.6	446	74	372	12.3
16.6	445	74	371	12.3
16.6	441	73	368	12.1
16.6	440	73	367	12.1
16.6	439	73	366	12.1
16.6	435	72	363	11.9
16.6	434	72	362	11.9
16.6	433	72	361	12.0
16.6	429	71	358	11.8
16.6	428	71	357	11.8
16.6	427	71	356	11.8
16.6	422	70	352	11.6
16.6	421	70	351	11.6
16.6	416	69	347	11.4
16.6	415	69	346	11.5
16.6	410	68	342	11.3
16.6	409	68	341	11.3
16.6	404	67	337	11.1
16.6	403	67	336	11.1
16.6	398	66	332	10.9
16.6	397	66	331	11.0
16.6	392	65	327	10.8
16.6	391	65	326	10.8
16.6	386	64	322	10.6
16.6	385	64	321	10.6
16.6	380	63	317	10.4
16.6	379	63	316	10.5
16.6	374	62	312	10.3
16.6	373	62	311	10.3
16.6	368	61	307	10.1
16.6	367	61	306	10.1
16.6	362	60	302	9.9
16.6	361	60	301	10.0
16.6	356	59	297	9.8
16.6	355	59	296	9.8
16.6	350	58	292	9.6
16.6	349	58	291	9.6
16.6	344	57	287	9.4
16.6	343	57	286	9.5
16.6	338	56	282	9.3
16.6	337	56	281	9.3
16.6	332	55	277	9.1
16.6	331	55	276	9.1
16.6	326	54	272	8.9
16.6	325	54	271	9.0
16.6	320	53	267	8.8
16.6	319	53	266	8.8
16.6	314	52	262	8.6
16.6	313	52	261	8.6
16.6	308	51	257	8.4
16.6	307	51	256	8.5
16.6	302	50	252	8.3
16.6	301	50	251	8.3
16.6	296	49	247	8.1
16.6	295	49	246	8.1
16.6	290	48	242	7.9
16.6	289	48	241	8.0

%E	M1	M2	DM	M*
16.6	283	47	236	7.8
16.6	277	46	231	7.6
16.6	271	45	226	7.5
16.6	265	44	221	7.3
16.6	259	43	216	7.1
16.6	253	42	211	7.0
16.6	247	41	206	6.8
16.6	241	40	201	6.6
16.6	235	39	196	6.5
16.6	229	38	191	6.3
16.6	223	37	186	6.1
16.6	217	36	181	6.0
16.6	211	35	176	5.8
16.6	205	34	171	5.6
16.6	199	33	166	5.5
16.6	193	32	161	5.3
16.6	187	31	156	5.1
16.6	181	30	151	5.0
16.6	175	29	146	4.8
16.6	169	28	141	4.6
16.6	163	27	136	4.5
16.6	157	26	131	4.3
16.6	151	25	126	4.1
16.6	145	24	121	4.0
16.5	498	82	416	13.5
16.5	497	82	415	13.5
16.5	496	82	414	13.6
16.5	492	81	411	13.3
16.5	491	81	410	13.4
16.5	490	81	409	13.4
16.5	486	80	406	13.2
16.5	485	80	405	13.2
16.5	484	80	404	13.2
16.5	480	79	401	13.0
16.5	479	79	400	13.0
16.5	478	79	399	13.1
16.5	474	78	396	12.8
16.5	473	78	395	12.9
16.5	472	78	394	12.9
16.5	468	77	391	12.7
16.5	467	77	390	12.7
16.5	466	77	389	12.7
16.5	462	76	386	12.5
16.5	461	76	385	12.5
16.5	460	76	384	12.6
16.5	455	75	380	12.4
16.5	454	75	379	12.4
16.5	449	74	375	12.2
16.5	448	74	374	12.2
16.5	443	73	370	12.0

%E	M1	M2	DM	M*
16.5	442	73	369	12.1
16.5	437	72	365	11.9
16.5	436	72	364	11.9
16.5	431	71	360	11.7
16.5	430	71	359	11.7
16.5	425	70	355	11.5
16.5	424	70	354	11.6
16.5	423	70	353	11.6
16.5	419	69	350	11.4
16.5	418	69	349	11.4
16.5	417	69	348	11.4
16.5	413	68	345	11.2
16.5	412	68	344	11.2
16.5	411	68	343	11.3
16.5	407	67	340	11.0
16.5	406	67	339	11.1
16.5	405	67	338	11.1
16.5	401	66	335	10.9
16.5	400	66	334	10.9
16.5	399	66	333	10.9
16.5	395	65	330	10.7
16.5	394	65	329	10.7
16.5	393	65	328	10.8
16.5	389	64	325	10.5
16.5	388	64	324	10.6
16.5	387	64	323	10.6
16.5	382	63	319	10.4
16.5	381	63	318	10.4
16.5	376	62	314	10.2
16.5	375	62	313	10.3
16.5	370	61	309	10.1
16.5	369	61	308	10.1
16.5	364	60	304	9.9
16.5	363	60	303	9.9
16.5	358	59	299	9.7
16.5	357	59	298	9.8
16.5	352	58	294	9.6
16.5	351	58	293	9.6
16.5	346	57	289	9.4
16.5	345	57	288	9.4
16.5	340	56	284	9.2
16.5	339	56	283	9.3
16.5	334	55	279	9.1
16.5	333	55	278	9.1
16.5	328	54	274	8.9
16.5	327	54	273	8.9
16.5	322	53	269	8.7
16.5	321	53	268	8.8
16.5	316	52	264	8.6
16.5	315	52	263	8.6

%E	M1	M2	DM	M*
16.5	310	51	259	8.4
16.5	309	51	258	8.4
16.5	303	50	253	8.3
16.5	297	49	248	8.1
16.5	291	48	243	7.9
16.5	285	47	238	7.8
16.5	284	47	237	7.8
16.5	279	46	233	7.6
16.5	278	46	232	7.6
16.5	273	45	228	7.4
16.5	272	45	227	7.4
16.5	267	44	223	7.3
16.5	266	44	222	7.3
16.5	261	43	218	7.1
16.5	260	43	217	7.1
16.5	255	42	213	6.9
16.5	254	42	212	6.9
16.5	249	41	208	6.8
16.5	248	41	207	6.8
16.5	243	40	203	6.6
16.5	242	40	202	6.6
16.5	237	39	198	6.4
16.5	236	39	197	6.4
16.5	231	38	193	6.3
16.5	230	38	192	6.3
16.5	224	37	187	6.1
16.5	218	36	182	5.9
16.5	212	35	177	5.8
16.5	206	34	172	5.6
16.5	200	33	167	5.4
16.5	194	32	162	5.3
16.5	188	31	157	5.1
16.5	182	30	152	4.9
16.5	176	29	147	4.8
16.5	170	28	142	4.6
16.5	164	27	137	4.4
16.5	158	26	132	4.3
16.5	139	23	116	3.8
16.5	133	22	111	3.6
16.5	127	21	106	3.5
16.5	121	20	101	3.3
16.5	115	19	96	3.1
16.5	109	18	91	3.0
16.5	103	17	86	2.8
16.5	97	16	81	2.6
16.5	91	15	76	2.5
16.5	85	14	71	2.3
16.5	79	13	66	2.1
16.4	500	82	418	13.4
16.4	499	82	417	13.5

%E	M1	M2	DM	M*
16.4	495	81	414	13.3
16.4	494	81	413	13.3
16.4	493	81	412	13.3
16.4	489	80	409	13.1
16.4	488	80	408	13.1
16.4	487	80	407	13.1
16.4	483	79	404	12.9
16.4	482	79	403	12.9
16.4	481	79	402	13.0
16.4	477	78	399	12.8
16.4	476	78	398	12.8
16.4	475	78	397	12.8
16.4	470	77	393	12.6
16.4	469	77	392	12.6
16.4	464	76	388	12.4
16.4	463	76	387	12.5
16.4	458	75	383	12.3
16.4	457	75	382	12.3
16.4	456	75	381	12.3
16.4	452	74	378	12.1
16.4	451	74	377	12.1
16.4	450	74	376	12.2
16.4	446	73	373	11.9
16.4	445	73	372	12.0
16.4	444	73	371	12.0
16.4	440	72	368	11.8
16.4	439	72	367	11.8
16.4	438	72	366	11.8
16.4	434	71	363	11.6
16.4	433	71	362	11.6
16.4	432	71	361	11.7
16.4	428	70	358	11.4
16.4	427	70	357	11.5
16.4	426	70	356	11.5
16.4	422	69	353	11.3
16.4	421	69	352	11.3
16.4	420	69	351	11.3
16.4	415	68	347	11.1
16.4	414	68	346	11.2
16.4	409	67	342	11.0
16.4	408	67	341	11.0
16.4	403	66	337	10.8
16.4	402	66	336	10.8
16.4	397	65	332	10.6
16.4	396	65	331	10.7
16.4	391	64	327	10.5
16.4	390	64	326	10.5
16.4	385	63	322	10.3
16.4	384	63	321	10.3
16.4	383	63	320	10.4

%E	M1	M2	DM	M*
16.4	379	62	317	10.1
16.4	378	62	316	10.2
16.4	377	62	315	10.2
16.4	373	61	312	10.0
16.4	372	61	311	10.0
16.4	371	61	310	10.0
16.4	366	60	306	9.8
16.4	365	60	305	9.9
16.4	360	59	301	9.7
16.4	359	59	300	9.7
16.4	354	58	296	9.5
16.4	353	58	295	9.5
16.4	348	57	291	9.3
16.4	347	57	290	9.4
16.4	342	56	286	9.2
16.4	341	56	285	9.2
16.4	336	55	281	9.0
16.4	335	55	280	9.0
16.4	330	54	276	8.8
16.4	329	54	275	8.9
16.4	324	53	271	8.7
16.4	323	53	270	8.7
16.4	318	52	266	8.5
16.4	317	52	265	8.5
16.4	311	51	260	8.4
16.4	305	50	255	8.2
16.4	304	50	254	8.2
16.4	299	49	250	8.0
16.4	298	49	249	8.1
16.4	293	48	245	7.9
16.4	292	48	244	7.9
16.4	287	47	240	7.7
16.4	286	47	239	7.7
16.4	281	46	235	7.5
16.4	280	46	234	7.6
16.4	275	45	230	7.4
16.4	274	45	229	7.4
16.4	269	44	225	7.2
16.4	268	44	224	7.2
16.4	263	43	220	7.0
16.4	262	43	219	7.1
16.4	256	42	214	6.9
16.4	250	41	209	6.7
16.4	244	40	204	6.6
16.4	238	39	199	6.4
16.4	232	38	194	6.2
16.4	226	37	189	6.1
16.4	225	37	188	6.1
16.4	220	36	184	5.9
16.4	219	36	183	5.9

%E	M1	M2	DM	M*
16.4	214	35	179	5.7
16.4	213	35	178	5.8
16.4	207	34	173	5.6
16.4	201	33	168	5.4
16.4	195	32	163	5.3
16.4	189	31	158	5.1
16.4	183	30	153	4.9
16.4	177	29	148	4.8
16.4	171	28	143	4.6
16.4	165	27	138	4.4
16.4	159	26	133	4.3
16.4	152	25	127	4.1
16.4	146	24	122	3.9
16.4	140	23	117	3.8
16.4	134	22	112	3.6
16.4	128	21	107	3.4
16.4	122	20	102	3.3
16.4	116	19	97	3.1
16.4	110	18	92	2.9
16.4	73	12	61	2.0
16.4	67	11	56	1.8
16.4	61	10	51	1.6
16.4	55	9	46	1.5
16.3	498	81	417	13.2
16.3	497	81	416	13.2
16.3	496	81	415	13.2
16.3	492	80	412	13.0
16.3	491	80	411	13.0
16.3	490	80	410	13.1
16.3	486	79	407	12.8
16.3	485	79	406	12.9
16.3	484	79	405	12.9
16.3	480	78	402	12.7
16.3	479	78	401	12.7
16.3	478	78	400	12.7
16.3	473	77	396	12.5
16.3	472	77	395	12.6
16.3	471	77	394	12.6
16.3	467	76	391	12.4
16.3	466	76	390	12.4
16.3	465	76	389	12.4
16.3	461	75	386	12.2
16.3	460	75	385	12.2
16.3	459	75	384	12.3
16.3	455	74	381	12.0
16.3	454	74	380	12.1
16.3	453	74	379	12.1
16.3	449	73	376	11.9
16.3	448	73	375	11.9
16.3	447	73	374	11.9

%E	M1	M2	DM	M*
16.3	443	72	371	11.7
16.3	442	72	370	11.7
16.3	441	72	369	11.8
16.3	436	71	365	11.6
16.3	435	71	364	11.6
16.3	430	70	360	11.4
16.3	429	70	359	11.4
16.3	424	69	355	11.2
16.3	423	69	354	11.3
16.3	418	68	350	11.1
16.3	417	68	349	11.1
16.3	416	68	348	11.1
16.3	412	67	345	10.9
16.3	411	67	344	10.9
16.3	410	67	343	10.9
16.3	406	66	340	10.7
16.3	405	66	339	10.8
16.3	404	66	338	10.8
16.3	400	65	335	10.6
16.3	399	65	334	10.6
16.3	398	65	333	10.6
16.3	393	64	329	10.4
16.3	392	64	328	10.4
16.3	387	63	324	10.3
16.3	386	63	323	10.3
16.3	381	62	319	10.1
16.3	380	62	318	10.1
16.3	375	61	314	9.9
16.3	374	61	313	9.9
16.3	369	60	309	9.8
16.3	368	60	308	9.8
16.3	367	60	307	9.8
16.3	363	59	304	9.6
16.3	362	59	303	9.6
16.3	361	59	302	9.6
16.3	356	58	298	9.4
16.3	355	58	297	9.5
16.3	350	57	293	9.3
16.3	349	57	292	9.3
16.3	344	56	288	9.1
16.3	343	56	287	9.1
16.3	338	55	283	8.9
16.3	337	55	282	9.0
16.3	332	54	278	8.8
16.3	331	54	277	8.8
16.3	326	53	273	8.6
16.3	325	53	272	8.6
16.3	320	52	268	8.4
16.3	319	52	267	8.5
16.3	313	51	262	8.3

%E	M1	M2	DM	M*
16.3	312	51	261	8.3
16.3	307	50	257	8.1
16.3	306	50	256	8.2
16.3	301	49	252	8.0
16.3	300	49	251	8.0
16.3	295	48	247	7.8
16.3	294	48	246	7.8
16.3	289	47	242	7.6
16.3	288	47	241	7.7
16.3	283	46	237	7.5
16.3	282	46	236	7.5
16.3	276	45	231	7.3
16.3	270	44	226	7.2
16.3	264	43	221	7.0
16.3	258	42	216	6.8
16.3	257	42	215	6.9
16.3	252	41	211	6.7
16.3	251	41	210	6.7
16.3	246	40	206	6.5
16.3	245	40	205	6.5
16.3	240	39	201	6.3
16.3	239	39	200	6.4
16.3	233	38	195	6.2
16.3	227	37	190	6.0
16.3	221	36	185	5.9
16.3	215	35	180	5.7
16.3	209	34	175	5.5
16.3	208	34	174	5.6
16.3	203	33	170	5.4
16.3	202	33	169	5.4
16.3	196	32	164	5.2
16.3	190	31	159	5.1
16.3	184	30	154	4.9
16.3	178	29	149	4.7
16.3	172	28	144	4.6
16.3	166	27	139	4.4
16.3	160	26	134	4.2
16.3	153	25	128	4.1
16.3	147	24	123	3.9
16.3	141	23	118	3.8
16.3	135	22	113	3.6
16.3	129	21	108	3.4
16.3	123	20	103	3.3
16.3	104	17	87	2.8
16.3	98	16	82	2.6
16.3	92	15	77	2.4
16.3	86	14	72	2.3
16.3	80	13	67	2.1
16.3	49	8	41	1.3
16.3	43	7	36	1.1

%E	M1	M2	DM	M*
16.2	500	81	419	13.1
16.2	499	81	418	13.1
16.2	495	80	415	12.9
16.2	494	80	414	13.0
16.2	493	80	413	13.0
16.2	489	79	410	12.8
16.2	488	79	409	12.8
16.2	487	79	408	12.8
16.2	482	78	404	12.6
16.2	481	78	403	12.6
16.2	476	77	399	12.5
16.2	475	77	398	12.5
16.2	474	77	397	12.5
16.2	470	76	394	12.3
16.2	469	76	393	12.3
16.2	468	76	392	12.3
16.2	464	75	389	12.1
16.2	463	75	388	12.1
16.2	462	75	387	12.2
16.2	458	74	384	12.0
16.2	457	74	383	12.0
16.2	456	74	382	12.0
16.2	452	73	379	11.8
16.2	451	73	378	11.8
16.2	450	73	377	11.8
16.2	445	72	373	11.6
16.2	444	72	372	11.7
16.2	439	71	368	11.5
16.2	438	71	367	11.5
16.2	437	71	366	11.5
16.2	433	70	363	11.3
16.2	432	70	362	11.3
16.2	431	70	361	11.4
16.2	427	69	358	11.1
16.2	426	69	357	11.2
16.2	425	69	356	11.2
16.2	421	68	353	11.0
16.2	420	68	352	11.0
16.2	419	68	351	11.0
16.2	414	67	347	10.8
16.2	413	67	346	10.9
16.2	408	66	342	10.7
16.2	407	66	341	10.7
16.2	402	65	337	10.5
16.2	401	65	336	10.5
16.2	396	64	332	10.3
16.2	395	64	331	10.4
16.2	394	64	330	10.4
16.2	390	63	327	10.2
16.2	389	63	326	10.2

%E	M1	M2	DM	M*
16.2	388	63	325	10.2
16.2	383	62	321	10.0
16.2	382	62	320	10.1
16.2	377	61	316	9.9
16.2	376	61	315	9.9
16.2	371	60	311	9.7
16.2	370	60	310	9.7
16.2	365	59	306	9.5
16.2	364	59	305	9.6
16.2	359	58	301	9.4
16.2	358	58	300	9.4
16.2	357	58	299	9.4
16.2	352	57	295	9.2
16.2	351	57	294	9.3
16.2	346	56	290	9.1
16.2	345	56	289	9.1
16.2	340	55	285	8.9
16.2	339	55	284	8.9
16.2	334	54	280	8.7
16.2	333	54	279	8.8
16.2	328	53	275	8.6
16.2	327	53	274	8.6
16.2	321	52	269	8.4
16.2	315	51	264	8.3
16.2	314	51	263	8.3
16.2	309	50	259	8.1
16.2	308	50	258	8.1
16.2	303	49	254	7.9
16.2	302	49	253	8.0
16.2	297	48	249	7.8
16.2	296	48	248	7.8
16.2	291	47	244	7.6
16.2	290	47	243	7.6
16.2	284	46	238	7.5
16.2	278	45	233	7.3
16.2	277	45	232	7.3
16.2	272	44	228	7.1
16.2	271	44	227	7.1
16.2	266	43	223	7.0
16.2	265	43	222	7.0
16.2	260	42	218	6.8
16.2	259	42	217	6.8
16.2	253	41	212	6.6
16.2	247	40	207	6.5
16.2	241	39	202	6.3
16.2	235	38	197	6.1
16.2	234	38	196	6.2
16.2	229	37	192	6.0
16.2	228	37	191	6.0
16.2	222	36	186	5.8

%E	M1	M2	DM	M*
16.2	216	35	181	5.7
16.2	210	34	176	5.5
16.2	204	33	171	5.3
16.2	198	32	166	5.2
16.2	197	32	165	5.2
16.2	191	31	160	5.0
16.2	185	30	155	4.9
16.2	179	29	150	4.7
16.2	173	28	145	4.5
16.2	167	27	140	4.4
16.2	154	25	129	4.1
16.2	148	24	124	3.9
16.2	142	23	119	3.7
16.2	136	22	114	3.6
16.2	130	21	109	3.4
16.2	117	19	98	3.1
16.2	111	18	93	2.9
16.2	105	17	88	2.8
16.2	99	16	83	2.6
16.2	74	12	62	1.9
16.2	68	11	57	1.8
16.2	37	6	31	1.0
16.1	498	80	418	12.9
16.1	497	80	417	12.9
16.1	496	80	416	12.9
16.1	492	79	413	12.7
16.1	491	79	412	12.7
16.1	490	79	411	12.7
16.1	485	78	407	12.5
16.1	484	78	406	12.6
16.1	483	78	405	12.6
16.1	479	77	402	12.4
16.1	478	77	401	12.4
16.1	477	77	400	12.4
16.1	473	76	397	12.2
16.1	472	76	396	12.2
16.1	471	76	395	12.3
16.1	467	75	392	12.0
16.1	466	75	391	12.1
16.1	465	75	390	12.1
16.1	461	74	387	11.9
16.1	460	74	386	11.9
16.1	459	74	385	11.9
16.1	454	73	381	11.7
16.1	453	73	380	11.8
16.1	448	72	376	11.6
16.1	447	72	375	11.6
16.1	446	72	374	11.6
16.1	442	71	371	11.4
16.1	441	71	370	11.4
16.1	440	71	369	11.5
16.1	436	70	366	11.2
16.1	435	70	365	11.3
16.1	434	70	364	11.3
16.1	429	69	360	11.1
16.1	428	69	359	11.1
16.1	423	68	355	10.9
16.1	422	68	354	11.0
16.1	417	67	350	10.8
16.1	416	67	349	10.8
16.1	415	67	348	10.8
16.1	411	66	345	10.6
16.1	410	66	344	10.6
16.1	409	66	343	10.7
16.1	404	65	339	10.5
16.1	403	65	338	10.5
16.1	398	64	334	10.3
16.1	397	64	333	10.3
16.1	392	63	329	10.1
16.1	391	63	328	10.2
16.1	386	62	324	10.0
16.1	385	62	323	10.0
16.1	384	62	322	10.0
16.1	380	61	319	9.8
16.1	379	61	318	9.8
16.1	378	61	317	9.8
16.1	373	60	313	9.7
16.1	372	60	312	9.7
16.1	367	59	308	9.5
16.1	366	59	307	9.5
16.1	361	58	303	9.3
16.1	360	58	302	9.3
16.1	355	57	298	9.2
16.1	354	57	297	9.2
16.1	353	57	296	9.2
16.1	348	56	292	9.0
16.1	347	56	291	9.0
16.1	342	55	287	8.8
16.1	341	55	286	8.9
16.1	336	54	282	8.7
16.1	335	54	281	8.7
16.1	330	53	277	8.5
16.1	329	53	276	8.5
16.1	323	52	271	8.4
16.1	322	52	270	8.4
16.1	317	51	266	8.2
16.1	316	51	265	8.2
16.1	311	50	261	8.0
16.1	310	50	260	8.1
16.1	305	49	256	7.9
16.1	304	49	255	7.9
16.1	299	48	251	7.7
16.1	298	48	250	7.7
16.1	292	47	245	7.6
16.1	286	46	240	7.4
16.1	285	46	239	7.4
16.1	280	45	235	7.2
16.1	279	45	234	7.3
16.1	274	44	230	7.1
16.1	273	44	229	7.1
16.1	267	43	224	6.9
16.1	261	42	219	6.8
16.1	255	41	214	6.6
16.1	254	41	213	6.6
16.1	249	40	209	6.4
16.1	248	40	208	6.5
16.1	242	39	203	6.3
16.1	236	38	198	6.1
16.1	230	37	193	6.0
16.1	224	36	188	5.8
16.1	223	36	187	5.8
16.1	218	35	183	5.6
16.1	217	35	182	5.6
16.1	211	34	177	5.5
16.1	205	33	172	5.3
16.1	199	32	167	5.1
16.1	193	31	162	5.0
16.1	192	31	161	5.0
16.1	186	30	156	4.8
16.1	180	29	151	4.7
16.1	174	28	146	4.5
16.1	168	27	141	4.3
16.1	161	26	135	4.2
16.1	155	25	130	4.0
16.1	149	24	125	3.9
16.1	143	23	120	3.7
16.1	137	22	115	3.5
16.1	124	20	104	3.2
16.1	118	19	99	3.1
16.1	112	18	94	2.9
16.1	93	15	78	2.4
16.1	87	14	73	2.3
16.1	62	10	52	1.6
16.1	56	9	47	1.4
16.1	31	5	26	0.8
16.0	500	80	420	12.8
16.0	499	80	419	12.8
16.0	495	79	416	12.6
16.0	494	79	415	12.6
16.0	493	79	414	12.7
16.0	489	78	411	12.4
16.0	488	78	410	12.5
16.0	487	78	409	12.5
16.0	486	78	408	12.5
16.0	482	77	405	12.3
16.0	481	77	404	12.3
16.0	480	77	403	12.4
16.0	476	76	400	12.1
16.0	475	76	399	12.2
16.0	474	76	398	12.2
16.0	470	75	395	12.0
16.0	469	75	394	12.0
16.0	468	75	393	12.0
16.0	463	74	389	11.8
16.0	462	74	388	11.9
16.0	457	73	384	11.7
16.0	456	73	383	11.7
16.0	455	73	382	11.7
16.0	451	72	379	11.5
16.0	450	72	378	11.5
16.0	449	72	377	11.5
16.0	445	71	374	11.3
16.0	444	71	373	11.4
16.0	443	71	372	11.4
16.0	438	70	368	11.2
16.0	437	70	367	11.2
16.0	432	69	363	11.0
16.0	431	69	362	11.0
16.0	430	69	361	11.1
16.0	426	68	358	10.9
16.0	425	68	357	10.9
16.0	424	68	356	10.9
16.0	420	67	353	10.7
16.0	419	67	352	10.7
16.0	418	67	351	10.7
16.0	413	66	347	10.5
16.0	412	66	346	10.6
16.0	407	65	342	10.4
16.0	406	65	341	10.4
16.0	405	65	340	10.4
16.0	401	64	337	10.2
16.0	400	64	336	10.2
16.0	399	64	335	10.3
16.0	394	63	331	10.1
16.0	393	63	330	10.1
16.0	388	62	326	9.9
16.0	387	62	325	9.9
16.0	382	61	321	9.7
16.0	381	61	320	9.8
16.0	376	60	316	9.6
16.0	375	60	315	9.6
16.0	374	60	314	9.6
16.0	369	59	310	9.4
16.0	368	59	309	9.5
16.0	363	58	305	9.3
16.0	362	58	304	9.3
16.0	357	57	300	9.1
16.0	356	57	299	9.1
16.0	351	56	295	8.9
16.0	350	56	294	9.0
16.0	349	56	293	9.0
16.0	344	55	289	8.8
16.0	343	55	288	8.8
16.0	338	54	284	8.6
16.0	337	54	283	8.7
16.0	332	53	279	8.5
16.0	331	53	278	8.5
16.0	326	52	274	8.3
16.0	325	52	273	8.3
16.0	324	52	272	8.3
16.0	319	51	268	8.2
16.0	318	51	267	8.2
16.0	313	50	263	8.0
16.0	312	50	262	8.0
16.0	307	49	258	7.8
16.0	306	49	257	7.8
16.0	300	48	252	7.7
16.0	294	47	247	7.5
16.0	293	47	246	7.5
16.0	288	46	242	7.3
16.0	287	46	241	7.4
16.0	282	45	237	7.2
16.0	281	45	236	7.2
16.0	275	44	231	7.0
16.0	269	43	226	6.9
16.0	268	43	225	6.9
16.0	263	42	221	6.7
16.0	262	42	220	6.7
16.0	257	41	216	6.5
16.0	256	41	215	6.6
16.0	250	40	210	6.4
16.0	244	39	205	6.2
16.0	243	39	204	6.3
16.0	238	38	200	6.1
16.0	237	38	199	6.1
16.0	231	37	194	5.9
16.0	225	36	189	5.8
16.0	219	35	184	5.6
16.0	213	34	179	5.4
16.0	212	34	178	5.5

%E	M1	M2	DM	M*
16.0	206	33	173	5.3
16.0	200	32	168	5.1
16.0	194	31	163	5.0
16.0	188	30	158	4.8
16.0	187	30	157	4.8
16.0	181	29	152	4.6
16.0	175	28	147	4.5
16.0	169	27	142	4.3
16.0	163	26	137	4.1
16.0	162	26	136	4.2
16.0	156	25	131	4.0
16.0	150	24	126	3.8
16.0	144	23	121	3.7
16.0	131	21	110	3.4
16.0	125	20	105	3.2
16.0	119	19	100	3.0
16.0	106	17	89	2.7
16.0	100	16	84	2.6
16.0	94	15	79	2.4
16.0	81	13	68	2.1
16.0	75	12	63	1.9
16.0	50	8	42	1.3
16.0	25	4	21	0.6
15.9	498	79	419	12.5
15.9	497	79	418	12.6
15.9	496	79	417	12.6
15.9	492	78	414	12.4
15.9	491	78	413	12.4
15.9	490	78	412	12.4
15.9	485	77	408	12.2
15.9	484	77	407	12.3
15.9	483	77	406	12.3
15.9	479	76	403	12.1
15.9	478	76	402	12.1
15.9	477	76	401	12.1
15.9	473	75	398	11.9
15.9	472	75	397	11.9
15.9	471	75	396	11.9
15.9	466	74	392	11.8
15.9	465	74	391	11.8
15.9	464	74	390	11.8
15.9	460	73	387	11.6
15.9	459	73	386	11.6
15.9	458	73	385	11.6
15.9	454	72	382	11.4
15.9	453	72	381	11.4
15.9	452	72	380	11.5
15.9	447	71	376	11.3
15.9	446	71	375	11.3
15.9	441	70	371	11.1
15.9	440	70	370	11.1
15.9	439	70	369	11.2
15.9	435	69	366	10.9
15.9	434	69	365	11.0
15.9	433	69	364	11.0
15.9	429	68	361	10.8
15.9	428	68	360	10.8
15.9	427	68	359	10.8
15.9	422	67	355	10.6
15.9	421	67	354	10.7
15.9	416	66	350	10.5
15.9	415	66	349	10.5
15.9	414	66	348	10.5
15.9	410	65	345	10.3
15.9	409	65	344	10.3
15.9	408	65	343	10.4
15.9	403	64	339	10.2
15.9	402	64	338	10.2
15.9	397	63	334	10.0
15.9	396	63	333	10.0
15.9	395	63	332	10.0
15.9	391	62	329	9.8
15.9	390	62	328	9.9
15.9	389	62	327	9.9
15.9	384	61	323	9.7
15.9	383	61	322	9.7
15.9	378	60	318	9.5
15.9	377	60	317	9.5
15.9	372	59	313	9.4
15.9	371	59	312	9.4
15.9	370	59	311	9.4
15.9	365	58	307	9.2
15.9	364	58	306	9.2
15.9	359	57	302	9.1
15.9	358	57	301	9.1
15.9	353	56	297	8.9
15.9	352	56	296	8.9
15.9	347	55	292	8.7
15.9	346	55	291	8.7
15.9	345	55	290	8.8
15.9	340	54	286	8.6
15.9	339	54	285	8.6
15.9	334	53	281	8.4
15.9	333	53	280	8.4
15.9	328	52	276	8.2
15.9	327	52	275	8.3
15.9	321	51	270	8.1
15.9	320	51	269	8.1
15.9	315	50	265	7.9
15.9	314	50	264	8.0
15.9	309	49	260	7.8
15.9	308	49	259	7.8
15.9	302	48	254	7.6
15.9	301	48	253	7.7
15.9	296	47	249	7.5
15.9	295	47	248	7.5
15.9	290	46	244	7.3
15.9	289	46	243	7.3
15.9	283	45	238	7.2
15.9	277	44	233	7.0
15.9	276	44	232	7.0
15.9	271	43	228	6.8
15.9	270	43	227	6.8
15.9	264	42	222	6.7
15.9	258	41	217	6.5
15.9	252	40	212	6.3
15.9	251	40	211	6.4
15.9	246	39	207	6.2
15.9	245	39	206	6.2
15.9	239	38	201	6.0
15.9	233	37	196	5.9
15.9	232	37	195	5.9
15.9	227	36	191	5.7
15.9	226	36	190	5.7
15.9	220	35	185	5.6
15.9	214	34	180	5.4
15.9	208	33	175	5.2
15.9	207	33	174	5.3
15.9	201	32	169	5.1
15.9	195	31	164	4.9
15.9	189	30	159	4.8
15.9	182	29	153	4.6
15.9	176	28	148	4.5
15.9	170	27	143	4.3
15.9	164	26	138	4.1
15.9	157	25	132	4.0
15.9	151	24	127	3.8
15.9	145	23	122	3.6
15.9	138	22	116	3.5
15.9	132	21	111	3.3
15.9	126	20	106	3.2
15.9	113	18	95	2.9
15.9	107	17	90	2.7
15.9	88	14	74	2.2
15.9	82	13	69	2.1
15.9	69	11	58	1.8
15.9	63	10	53	1.6
15.9	44	7	37	1.1
15.8	500	79	421	12.5
15.8	499	79	420	12.5
15.8	495	78	417	12.3
15.8	494	78	416	12.3
15.8	493	78	415	12.3
15.8	488	77	411	12.1
15.8	487	77	410	12.2
15.8	486	77	409	12.2
15.8	482	76	406	12.0
15.8	481	76	405	12.0
15.8	480	76	404	12.0
15.8	476	75	401	11.8
15.8	475	75	400	11.8
15.8	474	75	399	11.9
15.8	469	74	395	11.7
15.8	468	74	394	11.7
15.8	467	74	393	11.7
15.8	463	73	390	11.5
15.8	462	73	389	11.5
15.8	461	73	388	11.6
15.8	457	72	385	11.3
15.8	456	72	384	11.4
15.8	455	72	383	11.4
15.8	450	71	379	11.2
15.8	449	71	378	11.2
15.8	448	71	377	11.3
15.8	444	70	374	11.0
15.8	443	70	373	11.1
15.8	442	70	372	11.1
15.8	438	69	369	10.9
15.8	437	69	368	10.9
15.8	436	69	367	10.9
15.8	431	68	363	10.7
15.8	430	68	362	10.8
15.8	425	67	358	10.6
15.8	424	67	357	10.6
15.8	423	67	356	10.6
15.8	419	66	353	10.4
15.8	418	66	352	10.4
15.8	417	66	351	10.4
15.8	412	65	347	10.3
15.8	411	65	346	10.3
15.8	406	64	342	10.1
15.8	405	64	341	10.1
15.8	404	64	340	10.1
15.8	400	63	337	9.9
15.8	399	63	336	9.9
15.8	398	63	335	10.0
15.8	393	62	331	9.8
15.8	392	62	330	9.8
15.8	387	61	326	9.6
15.8	386	61	325	9.6
15.8	385	61	324	9.7
15.8	380	60	320	9.5
15.8	379	60	319	9.5
15.8	374	59	315	9.3
15.8	373	59	314	9.3
15.8	368	58	310	9.1
15.8	367	58	309	9.2
15.8	366	58	308	9.2
15.8	361	57	304	9.0
15.8	360	57	303	9.0
15.8	355	56	299	8.8
15.8	354	56	298	8.9
15.8	349	55	294	8.7
15.8	348	55	293	8.7
15.8	342	54	288	8.5
15.8	341	54	287	8.6
15.8	336	53	283	8.4
15.8	335	53	282	8.4
15.8	330	52	278	8.2
15.8	329	52	277	8.2
15.8	323	51	272	8.1
15.8	322	51	271	8.1
15.8	317	50	267	7.9
15.8	316	50	266	7.9
15.8	311	49	262	7.7
15.8	310	49	261	7.7
15.8	304	48	256	7.6
15.8	303	48	255	7.6
15.8	298	47	251	7.4
15.8	297	47	250	7.4
15.8	292	46	246	7.2
15.8	291	46	245	7.3
15.8	285	45	240	7.1
15.8	284	45	239	7.1
15.8	279	44	235	6.9
15.8	278	44	234	7.0
15.8	273	43	230	6.8
15.8	272	43	229	6.8
15.8	266	42	224	6.6
15.8	265	42	223	6.7
15.8	260	41	219	6.5
15.8	259	41	218	6.5
15.8	253	40	213	6.3
15.8	247	39	208	6.2
15.8	241	38	203	6.0
15.8	240	38	202	6.0
15.8	234	37	197	5.9
15.8	228	36	192	5.7
15.8	222	35	187	5.5
15.8	221	35	186	5.5

%E	M1	M2	DM	M*
15.8	215	34	181	5.4
15.8	209	33	176	5.2
15.8	203	32	171	5.0
15.8	202	32	170	5.1
15.8	196	31	165	4.9
15.8	190	30	160	4.7
15.8	184	29	155	4.6
15.8	183	29	154	4.6
15.8	177	28	149	4.4
15.8	171	27	144	4.3
15.8	165	26	139	4.1
15.8	158	25	133	4.0
15.8	152	24	128	3.8
15.8	146	23	123	3.6
15.8	139	22	117	3.5
15.8	133	21	112	3.3
15.8	120	19	101	3.0
15.8	114	18	96	2.8
15.8	101	16	85	2.5
15.8	95	15	80	2.4
15.8	76	12	64	1.9
15.8	57	9	48	1.4
15.8	38	6	32	0.9
15.8	19	3	16	0.5
15.7	498	78	420	12.2
15.7	497	78	419	12.2
15.7	496	78	418	12.3
15.7	492	77	415	12.1
15.7	491	77	414	12.1
15.7	490	77	413	12.1
15.7	489	77	412	12.1
15.7	485	76	409	11.9
15.7	484	76	408	11.9
15.7	483	76	407	12.0
15.7	479	75	404	11.7
15.7	478	75	403	11.8
15.7	477	75	402	11.8
15.7	472	74	398	11.6
15.7	471	74	397	11.6
15.7	470	74	396	11.7
15.7	466	73	393	11.4
15.7	465	73	392	11.5
15.7	464	73	391	11.5
15.7	460	72	388	11.3
15.7	459	72	387	11.3
15.7	458	72	386	11.3
15.7	453	71	382	11.1
15.7	452	71	381	11.2
15.7	451	71	380	11.2
15.7	447	70	377	11.0

%E	M1	M2	DM	M*
15.7	446	70	376	11.0
15.7	445	70	375	11.0
15.7	440	69	371	10.8
15.7	439	69	370	10.8
15.7	434	68	366	10.7
15.7	433	68	365	10.7
15.7	432	68	364	10.7
15.7	428	67	361	10.5
15.7	427	67	360	10.5
15.7	426	67	359	10.5
15.7	421	66	355	10.3
15.7	420	66	354	10.4
15.7	415	65	350	10.2
15.7	414	65	349	10.2
15.7	413	65	348	10.2
15.7	408	64	344	10.0
15.7	407	64	343	10.1
15.7	402	63	339	9.9
15.7	401	63	338	9.9
15.7	396	62	334	9.7
15.7	395	62	333	9.7
15.7	394	62	332	9.8
15.7	389	61	328	9.6
15.7	388	61	327	9.6
15.7	383	60	323	9.4
15.7	382	60	322	9.4
15.7	381	60	321	9.4
15.7	377	59	318	9.2
15.7	376	59	317	9.3
15.7	375	59	316	9.3
15.7	370	58	312	9.1
15.7	369	58	311	9.1
15.7	364	57	307	8.9
15.7	363	57	306	9.0
15.7	362	57	305	9.0
15.7	357	56	301	8.8
15.7	356	56	300	8.8
15.7	351	55	296	8.6
15.7	350	55	295	8.6
15.7	345	54	291	8.5
15.7	344	54	290	8.5
15.7	343	54	289	8.5
15.7	338	53	285	8.3
15.7	337	53	284	8.3
15.7	332	52	280	8.1
15.7	331	52	279	8.2
15.7	325	51	274	8.0
15.7	324	51	273	8.0
15.7	319	50	269	7.8
15.7	318	50	268	7.9

%E	M1	M2	DM	M*
15.7	313	49	264	7.7
15.7	312	49	263	7.7
15.7	306	48	258	7.5
15.7	305	48	257	7.6
15.7	300	47	253	7.4
15.7	299	47	252	7.4
15.7	293	46	247	7.2
15.7	287	45	242	7.1
15.7	286	45	241	7.1
15.7	281	44	237	6.9
15.7	280	44	236	6.9
15.7	274	43	231	6.7
15.7	268	42	226	6.6
15.7	267	42	225	6.6
15.7	261	41	220	6.4
15.7	255	40	215	6.3
15.7	254	40	214	6.3
15.7	249	39	210	6.1
15.7	248	39	209	6.1
15.7	242	38	204	6.0
15.7	236	37	199	5.8
15.7	235	37	198	5.8
15.7	230	36	194	5.6
15.7	229	36	193	5.7
15.7	223	35	188	5.5
15.7	217	34	183	5.3
15.7	216	34	182	5.4
15.7	210	33	177	5.2
15.7	204	32	172	5.0
15.7	198	31	167	4.9
15.7	197	31	166	4.9
15.7	191	30	161	4.7
15.7	185	29	156	4.5
15.7	178	28	150	4.4
15.7	172	27	145	4.2
15.7	166	26	140	4.1
15.7	159	25	134	3.9
15.7	153	24	129	3.8
15.7	140	22	118	3.5
15.7	134	21	113	3.3
15.7	127	20	107	3.1
15.7	121	19	102	3.0
15.7	115	18	97	2.8
15.7	108	17	91	2.7
15.7	102	16	86	2.5
15.7	89	14	75	2.2
15.7	83	13	70	2.0
15.7	70	11	59	1.7
15.7	51	8	43	1.3
15.6	500	78	422	12.2

%E	M1	M2	DM	M*
15.6	499	78	421	12.2
15.6	495	77	418	12.0
15.6	494	77	417	12.0
15.6	493	77	416	12.0
15.6	488	76	412	11.8
15.6	487	76	411	11.9
15.6	486	76	410	11.9
15.6	482	75	407	11.7
15.6	481	75	406	11.7
15.6	480	75	405	11.7
15.6	475	74	401	11.5
15.6	474	74	400	11.6
15.6	473	74	399	11.6
15.6	469	73	396	11.4
15.6	468	73	395	11.4
15.6	467	73	394	11.4
15.6	463	72	391	11.2
15.6	462	72	390	11.2
15.6	461	72	389	11.2
15.6	456	71	385	11.1
15.6	455	71	384	11.1
15.6	454	71	383	11.1
15.6	450	70	380	10.9
15.6	449	70	379	10.9
15.6	448	70	378	10.9
15.6	443	69	374	10.7
15.6	442	69	373	10.8
15.6	441	69	372	10.8
15.6	437	68	369	10.6
15.6	436	68	368	10.6
15.6	435	68	367	10.6
15.6	430	67	363	10.4
15.6	429	67	362	10.5
15.6	424	66	358	10.3
15.6	423	66	357	10.3
15.6	422	66	356	10.3
15.6	418	65	353	10.1
15.6	417	65	352	10.1
15.6	416	65	351	10.2
15.6	411	64	347	10.0
15.6	410	64	346	10.0
15.6	409	64	345	10.0
15.6	405	63	342	9.8
15.6	404	63	341	9.8
15.6	403	63	340	9.8
15.6	398	62	336	9.7
15.6	397	62	335	9.7
15.6	392	61	331	9.5
15.6	391	61	330	9.5
15.6	390	61	329	9.5

%E	M1	M2	DM	M*
15.6	385	60	325	9.4
15.6	384	60	324	9.4
15.6	379	59	320	9.2
15.6	378	59	319	9.2
15.6	373	58	315	9.0
15.6	372	58	314	9.0
15.6	371	58	313	9.1
15.6	366	57	309	8.9
15.6	365	57	308	8.9
15.6	360	56	304	8.7
15.6	359	56	303	8.7
15.6	358	56	302	8.8
15.6	353	55	298	8.6
15.6	352	55	297	8.6
15.6	347	54	293	8.4
15.6	346	54	292	8.4
15.6	340	53	287	8.3
15.6	339	53	286	8.3
15.6	334	52	282	8.1
15.6	333	52	281	8.1
15.6	327	51	276	8.0
15.6	326	51	275	8.0
15.6	321	50	271	7.8
15.6	320	50	270	7.8
15.6	315	49	266	7.6
15.6	314	49	265	7.6
15.6	308	48	260	7.5
15.6	307	48	259	7.5
15.6	302	47	255	7.3
15.6	301	47	254	7.3
15.6	295	46	249	7.2
15.6	294	46	248	7.2
15.6	289	45	244	7.0
15.6	288	45	243	7.0
15.6	282	44	238	6.9
15.6	276	43	233	6.7
15.6	275	43	232	6.7
15.6	270	42	228	6.5
15.6	269	42	227	6.6
15.6	263	41	222	6.4
15.6	262	41	221	6.4
15.6	257	40	217	6.2
15.6	256	40	216	6.3
15.6	250	39	211	6.1
15.6	244	38	206	5.9
15.6	243	38	205	5.9
15.6	237	37	200	5.8
15.6	231	36	195	5.6
15.6	225	35	190	5.4
15.6	224	35	189	5.5

%E	M1	M2	DM	M*
15.6	218	34	184	5.3
15.6	212	33	179	5.1
15.6	211	33	178	5.2
15.6	205	32	173	5.0
15.6	199	31	168	4.8
15.6	192	30	162	4.7
15.6	186	29	157	4.5
15.6	180	28	152	4.4
15.6	179	28	151	4.4
15.6	173	27	146	4.2
15.6	167	26	141	4.0
15.6	160	25	135	3.9
15.6	154	24	130	3.7
15.6	147	23	124	3.6
15.6	141	22	119	3.4
15.6	135	21	114	3.3
15.6	128	20	108	3.1
15.6	122	19	103	3.0
15.6	109	17	92	2.7
15.6	96	15	81	2.3
15.6	90	14	76	2.2
15.6	77	12	65	1.9
15.6	64	10	54	1.6
15.6	45	7	38	1.1
15.6	32	5	27	0.8
15.5	498	77	421	11.9
15.5	497	77	420	11.9
15.5	496	77	419	12.0
15.5	491	76	415	11.8
15.5	490	76	414	11.8
15.5	489	76	413	11.8
15.5	485	75	410	11.6
15.5	484	75	409	11.6
15.5	483	75	408	11.6
15.5	478	74	404	11.5
15.5	477	74	403	11.5
15.5	476	74	402	11.5
15.5	472	73	399	11.3
15.5	471	73	398	11.3
15.5	470	73	397	11.3
15.5	466	72	394	11.1
15.5	465	72	393	11.1
15.5	464	72	392	11.2
15.5	459	71	388	11.0
15.5	458	71	387	11.0
15.5	457	71	386	11.0
15.5	453	70	383	10.8
15.5	452	70	382	10.8
15.5	451	70	381	10.9
15.5	446	69	377	10.7

%E	M1	M2	DM	M*
15.5	445	69	376	10.7
15.5	444	69	375	10.7
15.5	440	68	372	10.5
15.5	439	68	371	10.5
15.5	438	68	370	10.6
15.5	433	67	366	10.4
15.5	432	67	365	10.4
15.5	431	67	364	10.4
15.5	427	66	361	10.2
15.5	426	66	360	10.2
15.5	425	66	359	10.2
15.5	420	65	355	10.1
15.5	419	65	354	10.1
15.5	414	64	350	9.9
15.5	413	64	349	9.9
15.5	412	64	348	9.9
15.5	407	63	344	9.8
15.5	406	63	343	9.8
15.5	401	62	339	9.6
15.5	400	62	338	9.6
15.5	399	62	337	9.6
15.5	394	61	333	9.4
15.5	393	61	332	9.5
15.5	388	60	328	9.3
15.5	387	60	327	9.3
15.5	386	60	326	9.3
15.5	381	59	322	9.1
15.5	380	59	321	9.2
15.5	375	58	317	9.0
15.5	374	58	316	9.0
15.5	368	57	311	8.8
15.5	367	57	310	8.9
15.5	362	56	306	8.7
15.5	361	56	305	8.7
15.5	355	55	300	8.5
15.5	354	55	299	8.5
15.5	349	54	295	8.4
15.5	348	54	294	8.4
15.5	343	53	290	8.2
15.5	342	53	289	8.2
15.5	341	53	288	8.2
15.5	336	52	284	8.0
15.5	335	52	283	8.1
15.5	330	51	279	7.9
15.5	329	51	278	7.9
15.5	328	51	277	7.9
15.5	323	50	273	7.7
15.5	322	50	272	7.8
15.5	317	49	268	7.6
15.5	316	49	267	7.6

%E	M1	M2	DM	M*
15.5	310	48	262	7.4
15.5	309	48	261	7.5
15.5	304	47	257	7.3
15.5	303	47	256	7.3
15.5	297	46	251	7.1
15.5	296	46	250	7.1
15.5	291	45	246	7.0
15.5	290	45	245	7.0
15.5	284	44	240	6.8
15.5	283	44	239	6.8
15.5	278	43	235	6.7
15.5	277	43	234	6.7
15.5	271	42	229	6.5
15.5	265	41	224	6.3
15.5	264	41	223	6.4
15.5	258	40	218	6.2
15.5	252	39	213	6.0
15.5	251	39	212	6.1
15.5	245	38	207	5.9
15.5	239	37	202	5.7
15.5	238	37	201	5.8
15.5	233	36	197	5.6
15.5	232	36	196	5.6
15.5	226	35	191	5.4
15.5	220	34	186	5.3
15.5	219	34	185	5.3
15.5	213	33	180	5.1
15.5	207	32	175	4.9
15.5	206	32	174	5.0
15.5	200	31	169	4.8
15.5	194	30	164	4.6
15.5	193	30	163	4.7
15.5	187	29	158	4.5
15.5	181	28	153	4.3
15.5	174	27	147	4.2
15.5	168	26	142	4.0
15.5	161	25	136	3.9
15.5	155	24	131	3.7
15.5	148	23	125	3.6
15.5	142	22	120	3.4
15.5	129	20	109	3.1
15.5	116	18	98	2.8
15.5	110	17	93	2.6
15.5	103	16	87	2.5
15.5	97	15	82	2.3
15.5	84	13	71	2.0
15.5	71	11	60	1.7
15.5	58	9	49	1.4
15.4	500	77	423	11.9
15.4	499	77	422	11.9

%E	M1	M2	DM	M*
15.4	495	76	419	11.7
15.4	494	76	418	11.7
15.4	493	76	417	11.7
15.4	492	76	416	11.7
15.4	488	75	413	11.5
15.4	487	75	412	11.6
15.4	486	75	411	11.6
15.4	482	74	408	11.4
15.4	481	74	407	11.4
15.4	480	74	406	11.4
15.4	479	74	405	11.4
15.4	475	73	402	11.2
15.4	474	73	401	11.2
15.4	473	73	400	11.3
15.4	469	72	397	11.1
15.4	468	72	396	11.1
15.4	467	72	395	11.1
15.4	462	71	391	10.9
15.4	461	71	390	10.9
15.4	460	71	389	11.0
15.4	456	70	386	10.7
15.4	455	70	385	10.8
15.4	454	70	384	10.8
15.4	449	69	380	10.6
15.4	448	69	379	10.6
15.4	447	69	378	10.7
15.4	443	68	375	10.4
15.4	442	68	374	10.5
15.4	441	68	373	10.5
15.4	436	67	369	10.3
15.4	435	67	368	10.3
15.4	434	67	367	10.3
15.4	429	66	363	10.2
15.4	428	66	362	10.2
15.4	423	65	358	10.0
15.4	422	65	357	10.0
15.4	421	65	356	10.0
15.4	416	64	352	9.8
15.4	415	64	351	9.9
15.4	410	63	347	9.7
15.4	409	63	346	9.7
15.4	408	63	345	9.7
15.4	403	62	341	9.5
15.4	402	62	340	9.6
15.4	397	61	336	9.4
15.4	396	61	335	9.4
15.4	395	61	334	9.4
15.4	390	60	330	9.2
15.4	389	60	329	9.3
15.4	384	59	325	9.1

%E	M1	M2	DM	M*
15.4	383	59	324	9.1
15.4	382	59	323	9.1
15.4	377	58	319	8.9
15.4	376	58	318	8.9
15.4	371	57	314	8.8
15.4	370	57	313	8.8
15.4	369	57	312	8.8
15.4	364	56	308	8.6
15.4	363	56	307	8.6
15.4	358	55	303	8.4
15.4	357	55	302	8.5
15.4	356	55	301	8.5
15.4	351	54	297	8.3
15.4	350	54	296	8.3
15.4	345	53	292	8.1
15.4	344	53	291	8.2
15.4	338	52	286	8.0
15.4	337	52	285	8.0
15.4	332	51	281	7.8
15.4	331	51	280	7.9
15.4	325	50	275	7.7
15.4	324	50	274	7.7
15.4	319	49	270	7.5
15.4	318	49	269	7.6
15.4	312	48	264	7.4
15.4	311	48	263	7.4
15.4	306	47	259	7.2
15.4	305	47	258	7.2
15.4	299	46	253	7.1
15.4	298	46	252	7.1
15.4	293	45	248	6.9
15.4	292	45	247	6.9
15.4	286	44	242	6.8
15.4	285	44	241	6.8
15.4	280	43	237	6.6
15.4	279	43	236	6.6
15.4	273	42	231	6.5
15.4	272	42	230	6.5
15.4	267	41	226	6.3
15.4	266	41	225	6.3
15.4	260	40	220	6.2
15.4	259	40	219	6.2
15.4	254	39	215	6.0
15.4	253	39	214	6.0
15.4	247	38	209	5.8
15.4	246	38	208	5.9
15.4	241	37	204	5.7
15.4	240	37	203	5.7
15.4	234	36	198	5.5
15.4	228	35	193	5.4

%E	M1	M2	DM	M*
15.4	227	35	192	5.4
15.4	221	34	187	5.2
15.4	214	33	181	5.1
15.4	208	32	176	4.9
15.4	201	31	170	4.8
15.4	195	30	165	4.6
15.4	188	29	159	4.5
15.4	182	28	154	4.3
15.4	175	27	148	4.2
15.4	169	26	143	4.0
15.4	162	25	137	3.9
15.4	156	24	132	3.7
15.4	149	23	126	3.6
15.4	143	22	121	3.4
15.4	136	21	115	3.2
15.4	130	20	110	3.1
15.4	123	19	104	2.9
15.4	117	18	99	2.8
15.4	104	16	88	2.5
15.4	91	14	77	2.2
15.4	78	12	66	1.8
15.4	65	10	55	1.5
15.4	52	8	44	1.2
15.4	39	6	33	0.9
15.4	26	4	22	0.6
15.4	13	2	11	0.3
15.3	498	76	422	11.6
15.3	497	76	421	11.6
15.3	496	76	420	11.6
15.3	491	75	416	11.5
15.3	490	75	415	11.5
15.3	489	75	414	11.5
15.3	485	74	411	11.3
15.3	484	74	410	11.3
15.3	483	74	409	11.3
15.3	478	73	405	11.1
15.3	477	73	404	11.2
15.3	476	73	403	11.2
15.3	472	72	400	11.0
15.3	471	72	399	11.0
15.3	470	72	398	11.0
15.3	465	71	394	10.8
15.3	464	71	393	10.9
15.3	463	71	392	10.9
15.3	459	70	389	10.7
15.3	458	70	388	10.7
15.3	457	70	387	10.7
15.3	452	69	383	10.5
15.3	451	69	382	10.6
15.3	450	69	381	10.6

%E	M1	M2	DM	M*
15.3	445	68	377	10.4
15.3	444	68	376	10.4
15.3	439	67	372	10.2
15.3	438	67	371	10.2
15.3	437	67	370	10.3
15.3	432	66	366	10.1
15.3	431	66	365	10.1
15.3	430	66	364	10.1
15.3	426	65	361	9.9
15.3	425	65	360	9.9
15.3	424	65	359	10.0
15.3	419	64	355	9.8
15.3	418	64	354	9.8
15.3	417	64	353	9.8
15.3	413	63	350	9.6
15.3	412	63	349	9.6
15.3	411	63	348	9.7
15.3	406	62	344	9.5
15.3	405	62	343	9.5
15.3	404	62	342	9.5
15.3	400	61	339	9.3
15.3	399	61	338	9.3
15.3	398	61	337	9.3
15.3	393	60	333	9.2
15.3	392	60	332	9.2
15.3	391	60	331	9.2
15.3	386	59	327	9.0
15.3	385	59	326	9.0
15.3	380	58	322	8.9
15.3	379	58	321	8.9
15.3	378	58	320	8.9
15.3	373	57	316	8.7
15.3	372	57	315	8.7
15.3	367	56	311	8.5
15.3	366	56	310	8.6
15.3	365	56	309	8.6
15.3	360	55	305	8.4
15.3	359	55	304	8.4
15.3	354	54	300	8.2
15.3	353	54	299	8.3
15.3	352	54	298	8.3
15.3	347	53	294	8.1
15.3	346	53	293	8.1
15.3	340	52	288	8.0
15.3	339	52	287	8.0
15.3	334	51	283	7.8
15.3	333	51	282	7.8
15.3	327	50	277	7.6
15.3	326	50	276	7.7
15.3	321	49	272	7.5

%E	M1	M2	DM	M*
15.3	320	49	271	7.5
15.3	314	48	266	7.3
15.3	313	48	265	7.4
15.3	308	47	261	7.2
15.3	307	47	260	7.2
15.3	301	46	255	7.0
15.3	300	46	254	7.1
15.3	295	45	250	6.9
15.3	294	45	249	6.9
15.3	288	44	244	6.7
15.3	287	44	243	6.7
15.3	281	43	238	6.6
15.3	275	42	233	6.4
15.3	274	42	232	6.4
15.3	268	41	227	6.3
15.3	262	40	222	6.1
15.3	261	40	221	6.1
15.3	255	39	216	6.0
15.3	249	38	211	5.8
15.3	248	38	210	5.8
15.3	242	37	205	5.7
15.3	236	36	200	5.5
15.3	235	36	199	5.5
15.3	229	35	194	5.3
15.3	222	34	188	5.2
15.3	216	33	183	5.0
15.3	215	33	182	5.1
15.3	209	32	177	4.9
15.3	203	31	172	4.7
15.3	202	31	171	4.8
15.3	196	30	166	4.6
15.3	190	29	161	4.4
15.3	189	29	160	4.4
15.3	183	28	155	4.3
15.3	177	27	150	4.1
15.3	176	27	149	4.1
15.3	170	26	144	4.0
15.3	163	25	138	3.8
15.3	157	24	133	3.7
15.3	150	23	127	3.5
15.3	144	22	122	3.4
15.3	137	21	116	3.2
15.3	131	20	111	3.1
15.3	124	19	105	2.9
15.3	118	18	100	2.7
15.3	111	17	94	2.6
15.3	98	15	83	2.3
15.3	85	13	72	2.0
15.3	72	11	61	1.7
15.3	59	9	50	1.4

%E	M1	M2	DM	M*
15.2	500	76	424	11.6
15.2	499	76	423	11.6
15.2	495	75	420	11.4
15.2	494	75	419	11.4
15.2	493	75	418	11.4
15.2	492	75	417	11.4
15.2	488	74	414	11.2
15.2	487	74	413	11.2
15.2	486	74	412	11.3
15.2	481	73	408	11.1
15.2	480	73	407	11.1
15.2	479	73	406	11.1
15.2	475	72	403	10.9
15.2	474	72	402	10.9
15.2	473	72	401	11.0
15.2	468	71	397	10.8
15.2	467	71	396	10.8
15.2	466	71	395	10.8
15.2	462	70	392	10.6
15.2	461	70	391	10.6
15.2	460	70	390	10.7
15.2	455	69	386	10.5
15.2	454	69	385	10.5
15.2	453	69	384	10.5
15.2	448	68	380	10.3
15.2	447	68	379	10.3
15.2	446	68	378	10.4
15.2	442	67	375	10.2
15.2	441	67	374	10.2
15.2	440	67	373	10.2
15.2	435	66	369	10.0
15.2	434	66	368	10.0
15.2	433	66	367	10.1
15.2	429	65	364	9.8
15.2	428	65	363	9.9
15.2	427	65	362	9.9
15.2	422	64	358	9.7
15.2	421	64	357	9.7
15.2	420	64	356	9.8
15.2	415	63	352	9.6
15.2	414	63	351	9.6
15.2	409	62	347	9.4
15.2	408	62	346	9.4
15.2	407	62	345	9.4
15.2	402	61	341	9.3
15.2	401	61	340	9.3
15.2	396	60	336	9.1
15.2	395	60	335	9.1
15.2	394	60	334	9.1
15.2	389	59	330	8.9

%E	M1	M2	DM	M*
15.2	388	59	329	9.0
15.2	387	59	328	9.0
15.2	382	58	324	8.8
15.2	381	58	323	8.8
15.2	376	57	319	8.6
15.2	375	57	318	8.7
15.2	374	57	317	8.7
15.2	369	56	313	8.5
15.2	368	56	312	8.5
15.2	363	55	308	8.3
15.2	362	55	307	8.4
15.2	361	55	306	8.4
15.2	356	54	302	8.2
15.2	355	54	301	8.2
15.2	349	53	296	8.0
15.2	348	53	295	8.1
15.2	343	52	291	7.9
15.2	342	52	290	7.9
15.2	341	52	289	7.9
15.2	336	51	285	7.7
15.2	335	51	284	7.8
15.2	330	50	280	7.6
15.2	329	50	279	7.6
15.2	328	50	278	7.6
15.2	323	49	274	7.4
15.2	322	49	273	7.5
15.2	316	48	268	7.3
15.2	315	48	267	7.3
15.2	310	47	263	7.1
15.2	309	47	262	7.1
15.2	303	46	257	7.0
15.2	302	46	256	7.0
15.2	297	45	252	6.8
15.2	296	45	251	6.8
15.2	290	44	246	6.7
15.2	289	44	245	6.7
15.2	283	43	240	6.5
15.2	282	43	239	6.6
15.2	277	42	235	6.4
15.2	276	42	234	6.4
15.2	270	41	229	6.2
15.2	269	41	228	6.2
15.2	264	40	224	6.1
15.2	263	40	223	6.1
15.2	257	39	218	5.9
15.2	256	39	217	5.9
15.2	250	38	212	5.8
15.2	244	37	207	5.6
15.2	243	37	206	5.6
15.2	237	36	201	5.5

%E	M1	M2	DM	M*
15.2	231	35	196	5.3
15.2	230	35	195	5.3
15.2	224	34	190	5.2
15.2	223	34	189	5.2
15.2	217	33	184	5.0
15.2	211	32	179	4.9
15.2	210	32	178	4.9
15.2	204	31	173	4.7
15.2	198	30	168	4.5
15.2	197	30	167	4.6
15.2	191	29	162	4.4
15.2	184	28	156	4.3
15.2	178	27	151	4.1
15.2	171	26	145	4.0
15.2	165	25	140	3.8
15.2	164	25	139	3.8
15.2	158	24	134	3.6
15.2	151	23	128	3.5
15.2	145	22	123	3.3
15.2	138	21	117	3.2
15.2	132	20	112	3.0
15.2	125	19	106	2.9
15.2	112	17	95	2.6
15.2	105	16	89	2.4
15.2	99	15	84	2.3
15.2	92	14	78	2.1
15.2	79	12	67	1.8
15.2	66	10	56	1.5
15.2	46	7	39	1.1
15.2	33	5	28	0.8
15.1	498	75	423	11.3
15.1	497	75	422	11.3
15.1	496	75	421	11.3
15.1	491	74	417	11.2
15.1	490	74	416	11.2
15.1	489	74	415	11.2
15.1	485	73	412	11.0
15.1	484	73	411	11.0
15.1	483	73	410	11.0
15.1	482	73	409	11.1
15.1	478	72	406	10.8
15.1	477	72	405	10.9
15.1	476	72	404	10.9
15.1	471	71	400	10.7
15.1	470	71	399	10.7
15.1	469	71	398	10.7
15.1	465	70	395	10.5
15.1	464	70	394	10.6
15.1	463	70	393	10.6
15.1	458	69	389	10.4
15.1	457	69	388	10.4
15.1	456	69	387	10.4
15.1	451	68	383	10.3
15.1	450	68	382	10.3
15.1	449	68	381	10.3
15.1	445	67	378	10.1
15.1	444	67	377	10.1
15.1	443	67	376	10.1
15.1	438	66	372	9.9
15.1	437	66	371	10.0
15.1	436	66	370	10.0
15.1	431	65	366	9.8
15.1	430	65	365	9.8
15.1	425	64	361	9.6
15.1	424	64	360	9.7
15.1	423	64	359	9.7
15.1	418	63	355	9.5
15.1	417	63	354	9.5
15.1	416	63	353	9.5
15.1	411	62	349	9.4
15.1	410	62	348	9.4
15.1	405	61	344	9.2
15.1	404	61	343	9.2
15.1	403	61	342	9.2
15.1	398	60	338	9.0
15.1	397	60	337	9.1
15.1	392	59	333	8.9
15.1	391	59	332	8.9
15.1	390	59	331	8.9
15.1	385	58	327	8.7
15.1	384	58	326	8.8
15.1	383	58	325	8.8
15.1	378	57	321	8.6
15.1	377	57	320	8.6
15.1	372	56	316	8.4
15.1	371	56	315	8.5
15.1	370	56	314	8.5
15.1	365	55	310	8.3
15.1	364	55	309	8.3
15.1	358	54	304	8.1
15.1	357	54	303	8.2
15.1	352	53	299	8.0
15.1	351	53	298	8.0
15.1	350	53	297	8.0
15.1	345	52	293	7.8
15.1	344	52	292	7.9
15.1	338	51	287	7.7
15.1	337	51	286	7.7
15.1	332	50	282	7.5
15.1	331	50	281	7.6
15.1	325	49	276	7.4
15.1	324	49	275	7.4
15.1	318	48	270	7.2
15.1	317	48	269	7.3
15.1	312	47	265	7.1
15.1	311	47	264	7.1
15.1	305	46	259	6.9
15.1	304	46	258	7.0
15.1	299	45	254	6.8
15.1	298	45	253	6.8
15.1	292	44	248	6.6
15.1	291	44	247	6.7
15.1	285	43	242	6.5
15.1	284	43	241	6.5
15.1	279	42	237	6.3
15.1	278	42	236	6.3
15.1	272	41	231	6.2
15.1	271	41	230	6.2
15.1	265	40	225	6.0
15.1	259	39	220	5.9
15.1	258	39	219	5.9
15.1	252	38	214	5.7
15.1	251	38	213	5.8
15.1	245	37	208	5.6
15.1	239	36	203	5.4
15.1	238	36	202	5.4
15.1	232	35	197	5.3
15.1	225	34	191	5.1
15.1	219	33	186	5.0
15.1	218	33	185	5.0
15.1	212	32	180	4.8
15.1	205	31	174	4.7
15.1	199	30	169	4.5
15.1	192	29	163	4.4
15.1	186	28	158	4.2
15.1	185	28	157	4.2
15.1	179	27	152	4.1
15.1	172	26	146	3.9
15.1	166	25	141	3.8
15.1	159	24	135	3.6
15.1	152	23	129	3.5
15.1	146	22	124	3.3
15.1	139	21	118	3.2
15.1	126	19	107	2.9
15.1	119	18	101	2.7
15.1	106	16	90	2.4
15.1	93	14	79	2.1
15.1	86	13	73	2.0
15.1	73	11	62	1.7
15.1	53	8	45	1.2
15.0	500	75	425	11.3
15.0	499	75	424	11.3
15.0	494	74	420	11.1
15.0	493	74	419	11.1
15.0	492	74	418	11.1
15.0	488	73	415	10.9
15.0	487	73	414	10.9
15.0	486	73	413	11.0
15.0	481	72	409	10.8
15.0	480	72	408	10.8
15.0	479	72	407	10.8
15.0	474	71	403	10.6
15.0	473	71	402	10.7
15.0	472	71	401	10.7
15.0	468	70	398	10.5
15.0	467	70	397	10.5
15.0	466	70	396	10.5
15.0	461	69	392	10.3
15.0	460	69	391	10.3
15.0	459	69	390	10.4
15.0	454	68	386	10.2
15.0	453	68	385	10.2
15.0	452	68	384	10.2
15.0	448	67	381	10.0
15.0	447	67	380	10.0
15.0	446	67	379	10.1
15.0	441	66	375	9.9
15.0	440	66	374	9.9
15.0	439	66	373	9.9
15.0	434	65	369	9.7
15.0	433	65	368	9.8
15.0	432	65	367	9.8
15.0	428	64	364	9.6
15.0	427	64	363	9.6
15.0	426	64	362	9.6
15.0	421	63	358	9.4
15.0	420	63	357	9.4
15.0	419	63	356	9.5
15.0	414	62	352	9.3
15.0	413	62	351	9.3
15.0	412	62	350	9.3
15.0	408	61	347	9.1
15.0	407	61	346	9.1
15.0	406	61	345	9.2
15.0	401	60	341	9.0
15.0	400	60	340	9.0
15.0	399	60	339	9.0
15.0	394	59	335	8.8
15.0	393	59	334	8.9
15.0	387	58	329	8.7
15.0	386	58	328	8.7
15.0	381	57	324	8.5
15.0	380	57	323	8.5
15.0	379	57	322	8.6
15.0	374	56	318	8.4
15.0	373	56	317	8.4
15.0	367	55	312	8.2
15.0	366	55	311	8.3
15.0	361	54	307	8.1
15.0	360	54	306	8.1
15.0	359	54	305	8.1
15.0	354	53	301	7.9
15.0	353	53	300	8.0
15.0	347	52	295	7.8
15.0	346	52	294	7.8
15.0	341	51	290	7.6
15.0	340	51	289	7.6
15.0	339	51	288	7.7
15.0	334	50	284	7.5
15.0	333	50	283	7.5
15.0	327	49	278	7.3
15.0	326	49	277	7.4
15.0	321	48	273	7.2
15.0	320	48	272	7.2
15.0	319	48	271	7.2
15.0	314	47	267	7.0
15.0	313	47	266	7.1
15.0	307	46	261	6.9
15.0	306	46	260	6.9
15.0	301	45	256	6.7
15.0	300	45	255	6.8
15.0	294	44	250	6.6
15.0	293	44	249	6.6
15.0	287	43	244	6.4
15.0	286	43	243	6.5
15.0	280	42	238	6.3
15.0	274	41	233	6.1
15.0	273	41	232	6.2
15.0	267	40	227	6.0
15.0	266	40	226	6.0
15.0	260	39	221	5.8
15.0	254	38	216	5.7
15.0	253	38	215	5.7
15.0	247	37	210	5.5
15.0	246	37	209	5.6
15.0	240	36	204	5.4
15.0	234	35	199	5.2
15.0	233	35	198	5.3
15.0	227	34	193	5.1
15.0	226	34	192	5.1

%E	M1	M2	DM	M*
15.0	220	33	187	4.9
15.0	214	32	182	4.8
15.0	213	32	181	4.8
15.0	207	31	176	4.6
15.0	206	31	175	4.7
15.0	200	30	170	4.5
15.0	193	29	164	4.4
15.0	187	28	159	4.2
15.0	180	27	153	4.0
15.0	173	26	147	3.9
15.0	167	25	142	3.7
15.0	160	24	136	3.6
15.0	153	23	130	3.5
15.0	147	22	125	3.3
15.0	140	21	119	3.1
15.0	133	20	113	3.0
15.0	127	19	108	2.8
15.0	120	18	102	2.7
15.0	113	17	96	2.6
15.0	107	16	91	2.4
15.0	100	15	85	2.3
15.0	80	12	68	1.8
15.0	60	9	51	1.3
15.0	40	6	34	0.9
15.0	20	3	17	0.4
14.9	498	74	424	11.0
14.9	497	74	423	11.0
14.9	496	74	422	11.0
14.9	495	74	421	11.1
14.9	491	73	418	10.9
14.9	490	73	417	10.9
14.9	489	73	416	10.9
14.9	484	72	412	10.7
14.9	483	72	411	10.7
14.9	482	72	410	10.8
14.9	478	71	407	10.5
14.9	477	71	406	10.6
14.9	476	71	405	10.6
14.9	475	71	404	10.6
14.9	471	70	401	10.4
14.9	470	70	400	10.4
14.9	469	70	399	10.4
14.9	464	69	395	10.3
14.9	463	69	394	10.3
14.9	462	69	393	10.3
14.9	457	68	389	10.1
14.9	456	68	388	10.1
14.9	455	68	387	10.2
14.9	451	67	384	10.0
14.9	450	67	383	10.0

%E	M1	M2	DM	M*
14.9	449	67	382	10.0
14.9	444	66	378	9.8
14.9	443	66	377	9.8
14.9	442	66	376	9.9
14.9	437	65	372	9.7
14.9	436	65	371	9.7
14.9	435	65	370	9.7
14.9	430	64	366	9.5
14.9	429	64	365	9.5
14.9	424	63	361	9.4
14.9	423	63	360	9.4
14.9	422	63	359	9.4
14.9	417	62	355	9.2
14.9	416	62	354	9.2
14.9	415	62	353	9.3
14.9	410	61	349	9.1
14.9	409	61	348	9.1
14.9	404	60	344	8.9
14.9	403	60	343	8.9
14.9	402	60	342	9.0
14.9	397	59	338	8.8
14.9	396	59	337	8.8
14.9	395	59	336	8.8
14.9	390	58	332	8.6
14.9	389	58	331	8.6
14.9	388	58	330	8.7
14.9	383	57	326	8.5
14.9	382	57	325	8.5
14.9	377	56	321	8.3
14.9	376	56	320	8.3
14.9	375	56	319	8.4
14.9	370	55	315	8.2
14.9	369	55	314	8.2
14.9	368	55	313	8.2
14.9	363	54	309	8.0
14.9	362	54	308	8.1
14.9	356	53	303	7.9
14.9	355	53	302	7.9
14.9	350	52	298	7.7
14.9	349	52	297	7.7
14.9	348	52	296	7.8
14.9	343	51	292	7.6
14.9	342	51	291	7.6
14.9	336	50	286	7.4
14.9	335	50	285	7.5
14.9	329	49	280	7.3
14.9	328	49	279	7.3
14.9	323	48	275	7.1
14.9	322	48	274	7.2
14.9	316	47	269	7.0

%E	M1	M2	DM	M*
14.9	315	47	268	7.0
14.9	309	46	263	6.8
14.9	308	46	262	6.9
14.9	303	45	258	6.7
14.9	302	45	257	6.7
14.9	296	44	252	6.5
14.9	295	44	251	6.6
14.9	289	43	246	6.4
14.9	288	43	245	6.4
14.9	282	42	240	6.3
14.9	281	42	239	6.3
14.9	276	41	235	6.1
14.9	275	41	234	6.1
14.9	269	40	229	5.9
14.9	268	40	228	6.0
14.9	262	39	223	5.8
14.9	261	39	222	5.8
14.9	255	38	217	5.7
14.9	249	37	212	5.5
14.9	248	37	211	5.5
14.9	242	36	206	5.4
14.9	241	36	205	5.4
14.9	235	35	200	5.2
14.9	228	34	194	5.1
14.9	222	33	189	4.9
14.9	221	33	188	4.9
14.9	215	32	183	4.8
14.9	208	31	177	4.6
14.9	202	30	172	4.5
14.9	201	30	171	4.5
14.9	195	29	166	4.3
14.9	194	29	165	4.3
14.9	188	28	160	4.2
14.9	181	27	154	4.0
14.9	175	26	149	3.9
14.9	174	26	148	3.9
14.9	168	25	143	3.7
14.9	161	24	137	3.6
14.9	154	23	131	3.4
14.9	148	22	126	3.3
14.9	141	21	120	3.1
14.9	134	20	114	3.0
14.9	121	18	103	2.7
14.9	114	17	97	2.5
14.9	101	15	86	2.2
14.9	94	14	80	2.1
14.9	87	13	74	1.9
14.9	74	11	63	1.6
14.9	67	10	57	1.5
14.9	47	7	40	1.0

%E	M1	M2	DM	M*
14.8	500	74	426	11.0
14.8	499	74	425	11.0
14.8	494	73	421	10.8
14.8	493	73	420	10.8
14.8	492	73	419	10.8
14.8	488	72	416	10.6
14.8	487	72	415	10.6
14.8	486	72	414	10.7
14.8	485	72	413	10.7
14.8	481	71	410	10.5
14.8	480	71	409	10.5
14.8	479	71	408	10.5
14.8	474	70	404	10.3
14.8	473	70	403	10.4
14.8	472	70	402	10.4
14.8	467	69	398	10.2
14.8	466	69	397	10.2
14.8	465	69	396	10.2
14.8	461	68	393	10.0
14.8	460	68	392	10.1
14.8	459	68	391	10.1
14.8	458	68	390	10.1
14.8	454	67	387	9.9
14.8	453	67	386	9.9
14.8	452	67	385	9.9
14.8	447	66	381	9.7
14.8	446	66	380	9.8
14.8	445	66	379	9.8
14.8	440	65	375	9.6
14.8	439	65	374	9.6
14.8	438	65	373	9.6
14.8	433	64	369	9.5
14.8	432	64	368	9.5
14.8	431	64	367	9.5
14.8	427	63	364	9.3
14.8	426	63	363	9.3
14.8	425	63	362	9.3
14.8	420	62	358	9.2
14.8	419	62	357	9.2
14.8	418	62	356	9.2
14.8	413	61	352	9.0
14.8	412	61	351	9.0
14.8	411	61	350	9.1
14.8	406	60	346	8.9
14.8	405	60	345	8.9
14.8	400	59	341	8.7
14.8	399	59	340	8.7
14.8	398	59	339	8.7
14.8	393	58	335	8.6
14.8	392	58	334	8.6

%E	M1	M2	DM	M*
14.8	391	58	333	8.6
14.8	386	57	329	8.4
14.8	385	57	328	8.4
14.8	384	57	327	8.5
14.8	379	56	323	8.3
14.8	378	56	322	8.3
14.8	372	55	317	8.1
14.8	371	55	316	8.2
14.8	366	54	312	8.0
14.8	365	54	311	8.0
14.8	364	54	310	8.0
14.8	359	53	306	7.8
14.8	358	53	305	7.8
14.8	357	53	304	7.9
14.8	352	52	300	7.7
14.8	351	52	299	7.7
14.8	345	51	294	7.5
14.8	344	51	293	7.6
14.8	338	50	288	7.4
14.8	337	50	287	7.4
14.8	332	49	283	7.2
14.8	331	49	282	7.3
14.8	330	49	281	7.3
14.8	325	48	277	7.1
14.8	324	48	276	7.1
14.8	318	47	271	6.9
14.8	317	47	270	7.0
14.8	311	46	265	6.8
14.8	310	46	264	6.8
14.8	305	45	260	6.6
14.8	304	45	259	6.7
14.8	298	44	254	6.5
14.8	297	44	253	6.5
14.8	291	43	248	6.4
14.8	290	43	247	6.4
14.8	284	42	242	6.2
14.8	283	42	241	6.2
14.8	277	41	236	6.1
14.8	271	40	231	5.9
14.8	270	40	230	5.9
14.8	264	39	225	5.8
14.8	263	39	224	5.8
14.8	257	38	219	5.6
14.8	256	38	218	5.6
14.8	250	37	213	5.5
14.8	244	36	208	5.3
14.8	243	36	207	5.3
14.8	237	35	202	5.2
14.8	236	35	201	5.2
14.8	230	34	196	5.0

%E	M1	M2	DM	M*
14.8	229	34	195	5.0
14.8	223	33	190	4.9
14.8	216	32	184	4.7
14.8	210	31	179	4.6
14.8	209	31	178	4.6
14.8	203	30	173	4.4
14.8	196	29	167	4.3
14.8	189	28	161	4.1
14.8	183	27	156	4.0
14.8	182	27	155	4.0
14.8	176	26	150	3.8
14.8	169	25	144	3.7
14.8	162	24	138	3.6
14.8	155	23	132	3.4
14.8	149	22	127	3.2
14.8	142	21	121	3.1
14.8	135	20	115	3.0
14.8	128	19	109	2.8
14.8	122	18	104	2.7
14.8	115	17	98	2.5
14.8	108	16	92	2.4
14.8	88	13	75	1.9
14.8	81	12	69	1.8
14.8	61	9	52	1.3
14.8	54	8	46	1.2
14.8	27	4	23	0.6
14.7	498	73	425	10.7
14.7	497	73	424	10.7
14.7	496	73	423	10.7
14.7	495	73	422	10.8
14.7	491	72	419	10.6
14.7	490	72	418	10.6
14.7	489	72	417	10.6
14.7	484	71	413	10.4
14.7	483	71	412	10.4
14.7	482	71	411	10.5
14.7	477	70	407	10.3
14.7	476	70	406	10.3
14.7	475	70	405	10.3
14.7	471	69	402	10.1
14.7	470	69	401	10.1
14.7	469	69	400	10.2
14.7	468	69	399	10.2
14.7	464	68	396	10.0
14.7	463	68	395	10.0
14.7	462	68	394	10.0
14.7	457	67	390	9.8
14.7	456	67	389	9.8
14.7	455	67	388	9.9
14.7	450	66	384	9.7

%E	M1	M2	DM	M*
14.7	449	66	383	9.7
14.7	448	66	382	9.7
14.7	443	65	378	9.5
14.7	442	65	377	9.6
14.7	441	65	376	9.6
14.7	436	64	372	9.4
14.7	435	64	371	9.4
14.7	434	64	370	9.4
14.7	430	63	367	9.2
14.7	429	63	366	9.3
14.7	428	63	365	9.3
14.7	423	62	361	9.1
14.7	422	62	360	9.1
14.7	421	62	359	9.1
14.7	416	61	355	8.9
14.7	415	61	354	9.0
14.7	414	61	353	9.0
14.7	409	60	349	8.8
14.7	408	60	348	8.8
14.7	407	60	347	8.8
14.7	402	59	343	8.7
14.7	401	59	342	8.7
14.7	395	58	337	8.5
14.7	394	58	336	8.5
14.7	389	57	332	8.4
14.7	388	57	331	8.4
14.7	387	57	330	8.4
14.7	382	56	326	8.2
14.7	381	56	325	8.2
14.7	380	56	324	8.3
14.7	375	55	320	8.1
14.7	374	55	319	8.1
14.7	373	55	318	8.1
14.7	368	54	314	7.9
14.7	367	54	313	7.9
14.7	361	53	308	7.8
14.7	360	53	307	7.8
14.7	354	52	302	7.6
14.7	353	52	301	7.7
14.7	348	51	297	7.5
14.7	347	51	296	7.5
14.7	346	51	295	7.5
14.7	341	50	291	7.3
14.7	340	50	290	7.4
14.7	339	50	289	7.4
14.7	334	49	285	7.2
14.7	333	49	284	7.2
14.7	327	48	279	7.0
14.7	326	48	278	7.1
14.7	320	47	273	6.9

%E	M1	M2	DM	M*
14.7	319	47	272	6.9
14.7	314	46	268	6.7
14.7	313	46	267	6.8
14.7	312	46	266	6.8
14.7	307	45	262	6.6
14.7	306	45	261	6.6
14.7	300	44	256	6.5
14.7	299	44	255	6.5
14.7	293	43	250	6.3
14.7	292	43	249	6.3
14.7	286	42	244	6.2
14.7	285	42	243	6.2
14.7	279	41	238	6.0
14.7	278	41	237	6.0
14.7	273	40	233	5.9
14.7	272	40	232	5.9
14.7	266	39	227	5.7
14.7	265	39	226	5.7
14.7	259	38	221	5.6
14.7	258	38	220	5.6
14.7	252	37	215	5.4
14.7	251	37	214	5.5
14.7	245	36	209	5.3
14.7	238	35	203	5.1
14.7	232	34	198	5.0
14.7	231	34	197	5.0
14.7	225	33	192	4.8
14.7	224	33	191	4.9
14.7	218	32	186	4.7
14.7	217	32	185	4.7
14.7	211	31	180	4.6
14.7	204	30	174	4.4
14.7	197	29	168	4.3
14.7	191	28	163	4.1
14.7	190	28	162	4.1
14.7	184	27	157	4.0
14.7	177	26	151	3.8
14.7	170	25	145	3.7
14.7	163	24	139	3.5
14.7	157	23	134	3.4
14.7	156	23	133	3.4
14.7	150	22	128	3.2
14.7	143	21	122	3.1
14.7	136	20	116	2.9
14.7	129	19	110	2.8
14.7	116	17	99	2.5
14.7	109	16	93	2.3
14.7	102	15	87	2.2
14.7	95	14	81	2.1
14.7	75	11	64	1.6

%E	M1	M2	DM	M*
14.7	68	10	58	1.5
14.7	34	5	29	0.7
14.6	500	73	427	10.7
14.6	499	73	426	10.7
14.6	494	72	422	10.5
14.6	493	72	421	10.5
14.6	492	72	420	10.5
14.6	487	71	416	10.4
14.6	486	71	415	10.4
14.6	485	71	414	10.4
14.6	481	70	411	10.2
14.6	480	70	410	10.2
14.6	479	70	409	10.2
14.6	478	70	408	10.3
14.6	474	69	405	10.0
14.6	473	69	404	10.1
14.6	472	69	403	10.1
14.6	467	68	399	9.9
14.6	466	68	398	9.9
14.6	465	68	397	9.9
14.6	460	67	393	9.8
14.6	459	67	392	9.8
14.6	458	67	391	9.8
14.6	453	66	387	9.6
14.6	452	66	386	9.6
14.6	451	66	385	9.7
14.6	446	65	381	9.5
14.6	445	65	380	9.5
14.6	444	65	379	9.5
14.6	439	64	375	9.3
14.6	438	64	374	9.4
14.6	437	64	373	9.4
14.6	433	63	370	9.2
14.6	432	63	369	9.2
14.6	431	63	368	9.2
14.6	426	62	364	9.0
14.6	425	62	363	9.0
14.6	424	62	362	9.1
14.6	419	61	358	8.9
14.6	418	61	357	8.9
14.6	417	61	356	8.9
14.6	412	60	352	8.7
14.6	411	60	351	8.8
14.6	410	60	350	8.8
14.6	405	59	346	8:6
14.6	404	59	345	8.6
14.6	403	59	344	8.6
14.6	398	58	340	8.5
14.6	397	58	339	8.5
14.6	396	58	338	8.5

%E	M1	M2	DM	M*
14.6	391	57	334	8.3
14.6	390	57	333	8.3
14.6	384	56	328	8.2
14.6	383	56	327	8.2
14.6	378	55	323	8.0
14.6	377	55	322	8.0
14.6	376	55	321	8.0
14.6	371	54	317	7.9
14.6	370	54	316	7.9
14.6	369	54	315	7.9
14.6	364	53	311	7.7
14.6	363	53	310	7.7
14.6	362	53	309	7.8
14.6	357	52	305	7.6
14.6	356	52	304	7.6
14.6	355	52	303	7.6
14.6	350	51	299	7.4
14.6	349	51	298	7.5
14.6	343	50	293	7.3
14.6	342	50	292	7.3
14.6	336	49	287	7.1
14.6	335	49	286	7.2
14.6	329	48	281	7.0
14.6	328	48	280	7.0
14.6	323	47	276	6.8
14.6	322	47	275	6.9
14.6	321	47	274	6.9
14.6	316	46	270	6.7
14.6	315	46	269	6.7
14.6	309	45	264	6.6
14.6	308	45	263	6.6
14.6	302	44	258	6.4
14.6	301	44	257	6.4
14.6	295	43	252	6.3
14.6	294	43	251	6.3
14.6	288	42	246	6.1
14.6	287	42	245	6.1
14.6	281	41	240	6.0
14.6	280	41	239	6.0
14.6	274	40	234	5.8
14.6	268	39	229	5.7
14.6	267	39	228	5.7
14.6	261	38	223	5.5
14.6	260	38	222	5.6
14.6	254	37	217	5.4
14.6	253	37	216	5.4
14.6	247	36	211	5.2
14.6	246	36	210	5.3
14.6	240	35	205	5.1
14.6	239	35	204	5.1

%E	M1	M2	DM	M*
14.6	233	34	199	5.0
14.6	226	33	193	4.8
14.6	219	32	187	4.7
14.6	213	31	182	4.5
14.6	212	31	181	4.5
14.6	206	30	176	4.4
14.6	205	30	175	4.4
14.6	199	29	170	4.2
14.6	198	29	169	4.2
14.6	192	28	164	4.1
14.6	185	27	158	3.9
14.6	178	26	152	3.8
14.6	171	25	146	3.7
14.6	164	24	140	3.5
14.6	158	23	135	3.3
14.6	151	22	129	3.2
14.6	144	21	123	3.1
14.6	137	20	117	2.9
14.6	130	19	111	2.8
14.6	123	18	105	2.6
14.6	103	15	88	2.2
14.6	96	14	82	2.0
14.6	89	13	76	1.9
14.6	82	12	70	1.8
14.6	48	7	41	1.0
14.6	41	6	35	0.9
14.5	498	72	426	10.4
14.5	497	72	425	10.4
14.5	496	72	424	10.5
14.5	495	72	423	10.5
14.5	491	71	420	10.3
14.5	490	71	419	10.3
14.5	489	71	418	10.3
14.5	488	71	417	10.3
14.5	484	70	414	10.1
14.5	483	70	413	10.1
14.5	482	70	412	10.2
14.5	477	69	408	10.0
14.5	476	69	407	10.0
14.5	475	69	406	10.0
14.5	470	68	402	9.8
14.5	469	68	401	9.9
14.5	468	68	400	9.9
14.5	463	67	396	9.7
14.5	462	67	395	9.7
14.5	461	67	394	9.7
14.5	456	66	390	9.6
14.5	455	66	389	9.6
14.5	454	66	388	9.6
14.5	449	65	384	9.4

%E	M1	M2	DM	M*
14.5	448	65	383	9.4
14.5	447	65	382	9.5
14.5	442	64	378	9.3
14.5	441	64	377	9.3
14.5	440	64	376	9.3
14.5	436	63	373	9.1
14.5	435	63	372	9.1
14.5	434	63	371	9.1
14.5	429	62	367	9.0
14.5	428	62	366	9.0
14.5	427	62	365	9.0
14.5	422	61	361	8.8
14.5	421	61	360	8.8
14.5	420	61	359	8.9
14.5	415	60	355	8.7
14.5	414	60	354	8.7
14.5	413	60	353	8.7
14.5	408	59	349	8.5
14.5	407	59	348	8.6
14.5	406	59	347	8.6
14.5	401	58	343	8.4
14.5	400	58	342	8.4
14.5	399	58	341	8.4
14.5	394	57	337	8.2
14.5	393	57	336	8.3
14.5	392	57	335	8.3
14.5	387	56	331	8.1
14.5	386	56	330	8.1
14.5	385	56	329	8.1
14.5	380	55	325	8.0
14.5	379	55	324	8.0
14.5	373	54	319	7.8
14.5	372	54	318	7.8
14.5	366	53	313	7.7
14.5	365	53	312	7.7
14.5	359	52	307	7.5
14.5	358	52	306	7.6
14.5	352	51	301	7.4
14.5	351	51	300	7.4
14.5	346	50	296	7.2
14.5	345	50	295	7.2
14.5	344	50	294	7.3
14.5	339	49	290	7.1
14.5	338	49	289	7.1
14.5	337	49	288	7.1
14.5	332	48	284	6.9
14.5	331	48	283	7.0
14.5	330	48	282	7.0
14.5	325	47	278	6.8
14.5	324	47	277	6.8

%E	M1	M2	DM	M*
14.5	318	46	272	6.7
14.5	317	46	271	6.7
14.5	311	45	266	6.5
14.5	310	45	265	6.5
14.5	304	44	260	6.4
14.5	303	44	259	6.4
14.5	297	43	254	6.2
14.5	296	43	253	6.2
14.5	290	42	248	6.1
14.5	289	42	247	6.1
14.5	283	41	242	5.9
14.5	282	41	241	6.0
14.5	276	40	236	5.8
14.5	275	40	235	5.8
14.5	269	39	230	5.7
14.5	262	38	224	5.5
14.5	256	37	219	5.3
14.5	255	37	218	5.4
14.5	249	36	213	5.2
14.5	248	36	212	5.2
14.5	242	35	207	5.1
14.5	241	35	206	5.1
14.5	235	34	201	4.9
14.5	234	34	200	4.9
14.5	228	33	195	4.8
14.5	227	33	194	4.8
14.5	221	32	189	4.6
14.5	220	32	188	4.7
14.5	214	31	183	4.5
14.5	207	30	177	4.3
14.5	200	29	171	4.2
14.5	193	28	165	4.1
14.5	186	27	159	3.9
14.5	179	26	153	3.8
14.5	173	25	148	3.6
14.5	172	25	147	3.6
14.5	166	24	142	3.5
14.5	165	24	141	3.5
14.5	159	23	136	3.3
14.5	152	22	130	3.2
14.5	145	21	124	3.0
14.5	138	20	118	2.9
14.5	131	19	112	2.8
14.5	124	18	106	2.6
14.5	117	17	100	2.5
14.5	110	16	94	2.3
14.5	83	12	71	1.7
14.5	76	11	65	1.6
14.5	69	10	59	1.4
14.5	62	9	53	1.3

%E	M1	M2	DM	M*
14.5	55	8	47	1.2
14.4	500	72	428	10.4
14.4	499	72	427	10.4
14.4	494	71	423	10.2
14.4	493	71	422	10.2
14.4	492	71	421	10.2
14.4	487	70	417	10.1
14.4	486	70	416	10.1
14.4	485	70	415	10.1
14.4	480	69	411	9.9
14.4	479	69	410	9.9
14.4	478	69	409	10.0
14.4	473	68	405	9.8
14.4	472	68	404	9.8
14.4	471	68	403	9.8
14.4	466	67	399	9.6
14.4	465	67	398	9.7
14.4	464	67	397	9.7
14.4	459	66	393	9.5
14.4	458	66	392	9.5
14.4	457	66	391	9.5
14.4	452	65	387	9.3
14.4	451	65	386	9.4
14.4	450	65	385	9.4
14.4	446	64	382	9.2
14.4	445	64	381	9.2
14.4	444	64	380	9.2
14.4	443	64	379	9.2
14.4	439	63	376	9.0
14.4	438	63	375	9.1
14.4	437	63	374	9.1
14.4	432	62	370	8.9
14.4	431	62	369	8.9
14.4	430	62	368	8.9
14.4	425	61	364	8.8
14.4	424	61	363	8.8
14.4	423	61	362	8.8
14.4	418	60	358	8.6
14.4	417	60	357	8.6
14.4	416	60	356	8.7
14.4	411	59	352	8.5
14.4	410	59	351	8.5
14.4	409	59	350	8.5
14.4	404	58	346	8.3
14.4	403	58	345	8.3
14.4	402	58	344	8.4
14.4	397	57	340	8.2
14.4	396	57	339	8.2
14.4	395	57	338	8.2
14.4	390	56	334	8.0

%E	M1	M2	DM	M*
14.4	389	56	333	8.1
14.4	388	56	332	8.1
14.4	383	55	328	7.9
14.4	382	55	327	7.9
14.4	381	55	326	7.9
14.4	376	54	322	7.8
14.4	375	54	321	7.8
14.4	374	54	320	7.8
14.4	369	53	316	7.6
14.4	368	53	315	7.6
14.4	367	53	314	7.7
14.4	362	52	310	7.5
14.4	361	52	309	7.5
14.4	360	52	308	7.5
14.4	355	51	304	7.3
14.4	354	51	303	7.3
14.4	353	51	302	7.4
14.4	348	50	298	7.2
14.4	347	50	297	7.2
14.4	341	49	292	7.0
14.4	340	49	291	7.1
14.4	334	48	286	6.9
14.4	333	48	285	6.9
14.4	327	47	280	6.8
14.4	326	47	279	6.8
14.4	320	46	274	6.6
14.4	319	46	273	6.6
14.4	313	45	268	6.5
14.4	312	45	267	6.5
14.4	306	44	262	6.3
14.4	305	44	261	6.3
14.4	299	43	256	6.2
14.4	298	43	255	6.2
14.4	292	42	250	6.0
14.4	291	42	249	6.1
14.4	285	41	244	5.9
14.4	284	41	243	5.9
14.4	278	40	238	5.8
14.4	277	40	237	5.8
14.4	271	39	232	5.6
14.4	270	39	231	5.6
14.4	264	38	226	5.5
14.4	263	38	225	5.5
14.4	257	37	220	5.3
14.4	250	36	214	5.2
14.4	243	35	208	5.0
14.4	236	34	202	4.9
14.4	229	33	196	4.8
14.4	223	32	191	4.6
14.4	222	32	190	4.6

%E	M1	M2	DM	M*
14.4	216	31	185	4.4
14.4	215	31	184	4.5
14.4	209	30	179	4.3
14.4	208	30	178	4.3
14.4	202	29	173	4.2
14.4	201	29	172	4.2
14.4	195	28	167	4.0
14.4	194	28	166	4.0
14.4	188	27	161	3.9
14.4	187	27	160	3.9
14.4	181	26	155	3.7
14.4	180	26	154	3.8
14.4	174	25	149	3.6
14.4	167	24	143	3.4
14.4	160	23	137	3.3
14.4	153	22	131	3.2
14.4	146	21	125	3.0
14.4	139	20	119	2.9
14.4	132	19	113	2.7
14.4	125	18	107	2.6
14.4	118	17	101	2.4
14.4	111	16	95	2.3
14.4	104	15	89	2.2
14.4	97	14	83	2.0
14.4	90	13	77	1.9
14.3	498	71	427	10.1
14.3	497	71	426	10.1
14.3	496	71	425	10.2
14.3	495	71	424	10.2
14.3	491	70	421	10.0
14.3	490	70	420	10.0
14.3	489	70	419	10.0
14.3	488	70	418	10.0
14.3	484	69	415	9.8
14.3	483	69	414	9.9
14.3	482	69	413	9.9
14.3	481	69	412	9.9
14.3	477	68	409	9.7
14.3	476	68	408	9.7
14.3	475	68	407	9.7
14.3	474	68	406	9.8
14.3	470	67	403	9.6
14.3	469	67	402	9.6
14.3	468	67	401	9.6
14.3	467	67	400	9.6
14.3	463	66	397	9.4
14.3	462	66	396	9.4
14.3	461	66	395	9.4
14.3	460	66	394	9.5
14.3	456	65	391	9.3

%E	M1	M2	DM	M*
14.3	455	65	390	9.3
14.3	454	65	389	9.3
14.3	453	65	388	9.3
14.3	449	64	385	9.1
14.3	448	64	384	9.1
14.3	447	64	383	9.2
14.3	442	63	379	9.0
14.3	441	63	378	9.0
14.3	440	63	377	9.0
14.3	435	62	373	8.8
14.3	434	62	372	8.9
14.3	433	62	371	8.9
14.3	428	61	367	8.7
14.3	427	61	366	8.7
14.3	426	61	365	8.7
14.3	421	60	361	8.6
14.3	420	60	360	8.6
14.3	419	60	359	8.6
14.3	414	59	355	8.4
14.3	413	59	354	8.4
14.3	412	59	353	8.4
14.3	407	58	349	8.3
14.3	406	58	348	8.3
14.3	405	58	347	8.3
14.3	400	57	343	8.1
14.3	399	57	342	8.1
14.3	398	57	341	8.2
14.3	392	56	336	8.0
14.3	391	56	335	8.0
14.3	385	55	330	7.9
14.3	384	55	329	7.9
14.3	378	54	324	7.7
14.3	377	54	323	7.7
14.3	371	53	318	7.6
14.3	370	53	317	7.6
14.3	364	52	312	7.4
14.3	363	52	311	7.4
14.3	357	51	306	7.3
14.3	356	51	305	7.3
14.3	350	50	300	7.1
14.3	349	50	299	7.2
14.3	343	49	294	7.0
14.3	342	49	293	7.0
14.3	336	48	288	6.9
14.3	335	48	287	6.9
14.3	329	47	282	6.7
14.3	328	47	281	6.7
14.3	322	46	276	6.6
14.3	321	46	275	6.6
14.3	315	45	270	6.4

%E	M1	M2	DM	M*
14.3	314	45	269	6.4
14.3	308	44	264	6.3
14.3	307	44	263	6.3
14.3	301	43	258	6.1
14.3	300	43	257	6.2
14.3	294	42	252	6.0
14.3	293	42	251	6.0
14.3	287	41	246	5.9
14.3	286	41	245	5.9
14.3	280	40	240	5.7
14.3	279	40	239	5.7
14.3	273	39	234	5.6
14.3	272	39	233	5.6
14.3	266	38	228	5.4
14.3	265	38	227	5.4
14.3	259	37	222	5.3
14.3	258	37	221	5.3
14.3	252	36	216	5.1
14.3	251	36	215	5.2
14.3	245	35	210	5.0
14.3	244	35	209	5.0
14.3	238	34	204	4.9
14.3	237	34	203	4.9
14.3	231	33	198	4.7
14.3	230	33	197	4.7
14.3	224	32	192	4.6
14.3	217	31	186	4.4
14.3	210	30	180	4.3
14.3	203	29	174	4.1
14.3	196	28	168	4.0
14.3	189	27	162	3.9
14.3	182	26	156	3.7
14.3	175	25	150	3.6
14.3	168	24	144	3.4
14.3	161	23	138	3.3
14.3	154	22	132	3.1
14.3	147	21	126	3.0
14.3	140	20	120	2.9
14.3	133	19	114	2.7
14.3	126	18	108	2.6
14.3	119	17	102	2.4
14.3	112	16	96	2.3
14.3	105	15	90	2.1
14.3	98	14	84	2.0
14.3	91	13	78	1.9
14.3	84	12	72	1.7
14.3	77	11	66	1.6
14.3	70	10	60	1.4
14.3	63	9	54	1.3
14.3	56	8	48	1.1

%E	M1	M2	DM	M*
14.3	49	7	42	1.0
14.3	42	6	36	0.9
14.3	35	5	30	0.7
14.3	28	4	24	0.6
14.3	21	3	18	0.4
14.3	14	2	12	0.3
14.3	7	1	6	0.1
14.2	500	71	429	10.1
14.2	499	71	428	10.1
14.2	494	70	424	9.9
14.2	493	70	423	9.9
14.2	492	70	422	10.0
14.2	487	69	418	9.8
14.2	486	69	417	9.8
14.2	485	69	416	9.8
14.2	480	68	412	9.6
14.2	479	68	411	9.7
14.2	478	68	410	9.7
14.2	473	67	406	9.5
14.2	472	67	405	9.5
14.2	471	67	404	9.5
14.2	466	66	400	9.3
14.2	465	66	399	9.4
14.2	464	66	398	9.4
14.2	459	65	394	9.2
14.2	458	65	393	9.2
14.2	457	65	392	9.2
14.2	452	64	388	9.1
14.2	451	64	387	9.1
14.2	450	64	386	9.1
14.2	445	63	382	8.9
14.2	444	63	381	8.9
14.2	443	63	380	9.0
14.2	438	62	376	8.8
14.2	437	62	375	8.8
14.2	436	62	374	8.8
14.2	431	61	370	8.6
14.2	430	61	369	8.7
14.2	429	61	368	8.7
14.2	424	60	364	8.5
14.2	423	60	363	8.5
14.2	422	60	362	8.5
14.2	416	59	357	8.4
14.2	415	59	356	8.4
14.2	409	58	351	8.2
14.2	408	58	350	8.2
14.2	402	57	345	8.1
14.2	401	57	344	8.1
14.2	395	56	339	7.9
14.2	394	56	338	8.0

%E	M1	M2	DM	M*
14.2	393	56	337	8.0
14.2	388	55	333	7.8
14.2	387	55	332	7.8
14.2	386	55	331	7.8
14.2	381	54	327	7.7
14.2	380	54	326	7.7
14.2	379	54	325	7.7
14.2	374	53	321	7.5
14.2	373	53	320	7.5
14.2	372	53	319	7.6
14.2	367	52	315	7.4
14.2	366	52	314	7.4
14.2	365	52	313	7.4
14.2	360	51	309	7.2
14.2	359	51	308	7.2
14.2	358	51	307	7.3
14.2	353	50	303	7.1
14.2	352	50	302	7.1
14.2	351	50	301	7.1
14.2	346	49	297	6.9
14.2	345	49	296	7.0
14.2	344	49	295	7.0
14.2	339	48	291	6.8
14.2	338	48	290	6.8
14.2	337	48	289	6.8
14.2	332	47	285	6.7
14.2	331	47	284	6.7
14.2	330	47	283	6.7
14.2	325	46	279	6.5
14.2	324	46	278	6.5
14.2	323	46	277	6.6
14.2	318	45	273	6.4
14.2	317	45	272	6.4
14.2	316	45	271	6.4
14.2	310	44	266	6.2
14.2	309	44	265	6.3
14.2	303	43	260	6.1
14.2	302	43	259	6.1
14.2	296	42	254	6.0
14.2	295	42	253	6.0
14.2	289	41	248	5.8
14.2	288	41	247	5.8
14.2	282	40	242	5.7
14.2	281	40	241	5.7
14.2	275	39	236	5.5
14.2	274	39	235	5.6
14.2	268	38	230	5.4
14.2	267	38	229	5.4
14.2	261	37	224	5.2
14.2	260	37	223	5.3

%E	M1	M2	DM	M*	%E	M1	M2	DM	M*	%E	M1	M2	DM	M*	%E	M1	M2	DM	M*	%E	M1	M2	DM	M*
14.2	254	36	218	5.1	14.1	455	64	391	9.0	14.1	327	46	281	6.5	14.1	92	13	79	1.8	14.0	399	56	343	7.9
14.2	253	36	217	5.1	14.1	454	64	390	9.0	14.1	326	46	280	6.5	14.1	85	12	73	1.7	14.0	394	55	339	7.7
14.2	247	35	212	5.0	14.1	453	64	389	9.0	14.1	320	45	275	6.3	14.1	78	11	67	1.6	14.0	393	55	338	7.7
14.2	246	35	211	5.0	14.1	448	63	385	8.9	14.1	319	45	274	6.3	14.1	71	10	61	1.4	14.0	392	55	337	7.7
14.2	240	34	206	4.8	14.1	447	63	384	8.9	14.1	313	44	269	6.2	14.1	64	9	55	1.3	14.0	387	54	333	7.5
14.2	239	34	205	4.8	14.1	446	63	383	8.9	14.1	312	44	268	6.2	14.0	500	70	430	9.8	14.0	386	54	332	7.6
14.2	233	33	200	4.7	14.1	441	62	379	8.7	14.1	311	44	267	6.2	14.0	499	70	429	9.8	14.0	385	54	331	7.6
14.2	232	33	199	4.7	14.1	440	62	378	8.7	14.1	306	43	263	6.0	14.0	494	69	425	9.6	14.0	379	53	326	7.4
14.2	226	32	194	4.5	14.1	439	62	377	8.8	14.1	305	43	262	6.1	14.0	493	69	424	9.7	14.0	378	53	325	7.4
14.2	225	32	193	4.6	14.1	434	61	373	8.6	14.1	304	43	261	6.1	14.0	492	69	423	9.7	14.0	372	52	320	7.3
14.2	219	31	188	4.4	14.1	433	61	372	8.6	14.1	298	42	256	5.9	14.0	487	68	419	9.5	14.0	371	52	319	7.3
14.2	218	31	187	4.4	14.1	432	61	371	8.6	14.1	297	42	255	5.9	14.0	486	68	418	9.5	14.0	365	51	314	7.1
14.2	212	30	182	4.2	14.1	427	60	367	8.4	14.1	291	41	250	5.8	14.0	485	68	417	9.5	14.0	364	51	313	7.1
14.2	211	30	181	4.3	14.1	426	60	366	8.5	14.1	290	41	249	5.8	14.0	480	67	413	9.4	14.0	358	50	308	7.0
14.2	204	29	175	4.1	14.1	425	60	365	8.5	14.1	284	40	244	5.6	14.0	479	67	412	9.4	14.0	357	50	307	7.0
14.2	197	28	169	4.0	14.1	419	59	360	8.3	14.1	283	40	243	5.7	14.0	478	67	411	9.4	14.0	356	50	306	7.0
14.2	190	27	163	3.8	14.1	418	59	359	8.3	14.1	277	39	238	5.5	14.0	477	67	410	9.4	14.0	351	49	302	6.8
14.2	183	26	157	3.7	14.1	417	59	358	8.3	14.1	276	39	237	5.5	14.0	473	66	407	9.2	14.0	350	49	301	6.9
14.2	176	25	151	3.6	14.1	412	58	354	8.2	14.1	270	38	232	5.3	14.0	472	66	406	9.2	14.0	349	49	300	6.9
14.2	169	24	145	3.4	14.1	411	58	353	8.2	14.1	269	38	231	5.4	14.0	471	66	405	9.2	14.0	344	48	296	6.7
14.2	162	23	139	3.3	14.1	410	58	352	8.2	14.1	263	37	226	5.2	14.0	470	66	404	9.3	14.0	343	48	295	6.7
14.2	155	22	133	3.1	14.1	405	57	348	8.0	14.1	262	37	225	5.2	14.0	465	65	400	9.1	14.0	342	48	294	6.7
14.2	148	21	127	3.0	14.1	404	57	347	8.0	14.1	256	36	220	5.1	14.0	464	65	399	9.1	14.0	336	47	289	6.6
14.2	141	20	121	2.8	14.1	403	57	346	8.1	14.1	255	36	219	5.1	14.0	463	65	398	9.1	14.0	335	47	288	6.6
14.2	134	19	115	2.7	14.1	398	56	342	7.9	14.1	249	35	214	4.9	14.0	458	64	394	8.9	14.0	329	46	283	6.4
14.2	127	18	109	2.6	14.1	397	56	341	7.9	14.1	248	35	213	4.9	14.0	457	64	393	9.0	14.0	328	46	282	6.5
14.2	120	17	103	2.4	14.1	396	56	340	7.9	14.1	242	34	208	4.8	14.0	456	64	392	9.0	14.0	322	45	277	6.3
14.2	113	16	97	2.3	14.1	391	55	336	7.7	14.1	241	34	207	4.8	14.0	451	63	388	8.8	14.0	321	45	276	6.3
14.2	106	15	91	2.1	14.1	390	55	335	7.8	14.1	234	33	201	4.7	14.0	450	63	387	8.8	14.0	315	44	271	6.1
14.1	498	70	428	9.8	14.1	389	55	334	7.8	14.1	227	32	195	4.5	14.0	449	63	386	8.8	14.0	314	44	270	6.2
14.1	497	70	427	9.9	14.1	384	54	330	7.6	14.1	220	31	189	4.4	14.0	444	62	382	8.7	14.0	308	43	265	6.0
14.1	496	70	426	9.9	14.1	383	54	329	7.6	14.1	213	30	183	4.2	14.0	443	62	381	8.7	14.0	307	43	264	6.0
14.1	495	70	425	9.9	14.1	382	54	328	7.6	14.1	206	29	177	4.1	14.0	442	62	380	8.7	14.0	301	42	259	5.9
14.1	491	69	422	9.7	14.1	377	53	324	7.5	14.1	205	29	176	4.1	14.0	437	61	376	8.5	14.0	300	42	258	5.9
14.1	490	69	421	9.7	14.1	376	53	323	7.5	14.1	199	28	171	3.9	14.0	436	61	375	8.5	14.0	299	42	257	5.9
14.1	489	69	420	9.7	14.1	375	53	322	7.5	14.1	198	28	170	4.0	14.0	435	61	374	8.6	14.0	293	41	252	5.7
14.1	488	69	419	9.8	14.1	370	52	318	7.3	14.1	192	27	165	3.8	14.0	430	60	370	8.4	14.0	292	41	251	5.8
14.1	484	68	416	9.6	14.1	369	52	317	7.3	14.1	191	27	164	3.8	14.0	429	60	369	8.4	14.0	286	40	246	5.6
14.1	483	68	415	9.6	14.1	368	52	316	7.3	14.1	185	26	159	3.7	14.0	428	60	368	8.4	14.0	285	40	245	5.6
14.1	482	68	414	9.6	14.1	363	51	312	7.2	14.1	184	26	158	3.7	14.0	422	59	363	8.2	14.0	279	39	240	5.5
14.1	481	68	413	9.6	14.1	362	51	311	7.2	14.1	177	25	152	3.5	14.0	421	59	362	8.3	14.0	278	39	239	5.5
14.1	476	67	409	9.4	14.1	361	51	310	7.2	14.1	170	24	146	3.4	14.0	420	59	361	8.3	14.0	272	38	234	5.3
14.1	475	67	408	9.5	14.1	355	50	305	7.0	14.1	163	23	140	3.2	14.0	415	58	357	8.1	14.0	271	38	233	5.3
14.1	474	67	407	9.5	14.1	354	50	304	7.1	14.1	156	22	134	3.1	14.0	414	58	356	8.1	14.0	265	37	228	5.2
14.1	469	66	403	9.3	14.1	348	49	299	6.9	14.1	149	21	128	3.0	14.0	413	58	355	8.1	14.0	264	37	227	5.2
14.1	468	66	402	9.3	14.1	347	49	298	6.9	14.1	142	20	122	2.8	14.0	408	57	351	8.0	14.0	258	36	222	5.0
14.1	467	66	401	9.3	14.1	341	48	293	6.8	14.1	135	19	116	2.7	14.0	407	57	350	8.0	14.0	257	36	221	5.0
14.1	462	65	397	9.1	14.1	340	48	292	6.8	14.1	128	18	110	2.5	14.0	406	57	349	8.0	14.0	250	35	215	4.9
14.1	461	65	396	9.2	14.1	334	47	287	6.6	14.1	121	17	104	2.4	14.0	401	56	345	7.8	14.0	243	34	209	4.8
14.1	460	65	395	9.2	14.1	333	47	286	6.6	14.1	99	14	85	2.0	14.0	400	56	344	7.8	14.0	236	33	203	4.6

%E	M1	M2	DM	M*
14.0	235	33	202	4.6
14.0	229	32	197	4.5
14.0	228	32	196	4.5
14.0	222	31	191	4.3
14.0	221	31	190	4.3
14.0	215	30	185	4.2
14.0	214	30	184	4.2
14.0	207	29	178	4.1
14.0	200	28	172	3.9
14.0	193	27	166	3.8
14.0	186	26	160	3.6
14.0	179	25	154	3.5
14.0	178	25	153	3.5
14.0	172	24	148	3.3
14.0	171	24	147	3.4
14.0	164	23	141	3.2
14.0	157	22	135	3.1
14.0	150	21	129	2.9
14.0	143	20	123	2.8
14.0	136	19	117	2.7
14.0	129	18	111	2.5
14.0	114	16	98	2.2
14.0	107	15	92	2.1
14.0	100	14	86	2.0
14.0	93	13	80	1.8
14.0	86	12	74	1.7
14.0	57	8	49	1.1
14.0	50	7	43	1.0
14.0	43	6	37	0.8
13.9	498	69	429	9.6
13.9	497	69	428	9.6
13.9	496	69	427	9.6
13.9	495	69	426	9.6
13.9	490	68	422	9.4
13.9	489	68	421	9.5
13.9	488	68	420	9.5
13.9	483	67	416	9.3
13.9	482	67	415	9.3
13.9	481	67	414	9.3
13.9	476	66	410	9.2
13.9	475	66	409	9.2
13.9	474	66	408	9.2
13.9	469	65	404	9.0
13.9	468	65	403	9.0
13.9	467	65	402	9.0
13.9	466	65	401	9.1
13.9	462	64	398	8.9
13.9	461	64	397	8.9
13.9	460	64	396	8.9
13.9	459	64	395	8.9

%E	M1	M2	DM	M*
13.9	454	63	391	8.7
13.9	453	63	390	8.8
13.9	452	63	389	8.8
13.9	447	62	385	8.6
13.9	446	62	384	8.6
13.9	445	62	383	8.6
13.9	440	61	379	8.5
13.9	439	61	378	8.5
13.9	438	61	377	8.5
13.9	433	60	373	8.3
13.9	432	60	372	8.3
13.9	431	60	371	8.4
13.9	426	59	367	8.2
13.9	425	59	366	8.2
13.9	424	59	365	8.2
13.9	423	59	364	8.2
13.9	418	58	360	8.0
13.9	417	58	359	8.1
13.9	416	58	358	8.1
13.9	411	57	354	7.9
13.9	410	57	353	7.9
13.9	409	57	352	7.9
13.9	404	56	348	7.8
13.9	403	56	347	7.8
13.9	402	56	346	7.8
13.9	397	55	342	7.6
13.9	396	55	341	7.6
13.9	395	55	340	7.7
13.9	389	54	335	7.5
13.9	388	54	334	7.5
13.9	382	53	329	7.4
13.9	381	53	328	7.4
13.9	380	53	327	7.4
13.9	375	52	323	7.2
13.9	374	52	322	7.2
13.9	373	52	321	7.2
13.9	368	51	317	7.1
13.9	367	51	316	7.1
13.9	366	51	315	7.1
13.9	361	50	311	6.9
13.9	360	50	310	6.9
13.9	359	50	309	7.0
13.9	353	49	304	6.8
13.9	352	49	303	6.8
13.9	346	48	298	6.7
13.9	345	48	297	6.7
13.9	339	47	292	6.5
13.9	338	47	291	6.5
13.9	337	47	290	6.6
13.9	332	46	286	6.4

%E	M1	M2	DM	M*
13.9	331	46	285	6.4
13.9	330	46	284	6.4
13.9	324	45	279	6.3
13.9	323	45	278	6.3
13.9	317	44	273	6.1
13.9	316	44	272	6.1
13.9	310	43	267	6.0
13.9	309	43	266	6.0
13.9	303	42	261	5.8
13.9	302	42	260	5.8
13.9	296	41	255	5.7
13.9	295	41	254	5.7
13.9	294	41	253	5.7
13.9	288	40	248	5.6
13.9	287	40	247	5.6
13.9	281	39	242	5.4
13.9	280	39	241	5.4
13.9	274	38	236	5.3
13.9	273	38	235	5.3
13.9	267	37	230	5.1
13.9	266	37	229	5.1
13.9	259	36	223	5.0
13.9	252	35	217	4.9
13.9	251	35	216	4.9
13.9	245	34	211	4.7
13.9	244	34	210	4.7
13.9	238	33	205	4.6
13.9	237	33	204	4.6
13.9	231	32	199	4.4
13.9	230	32	198	4.5
13.9	223	31	192	4.3
13.9	216	30	186	4.2
13.9	209	29	180	4.0
13.9	208	29	179	4.0
13.9	202	28	174	3.9
13.9	201	28	173	3.9
13.9	194	27	167	3.8
13.9	187	26	161	3.6
13.9	180	25	155	3.5
13.9	173	24	149	3.3
13.9	166	23	143	3.2
13.9	165	23	142	3.2
13.9	158	22	136	3.1
13.9	151	21	130	2.9
13.9	144	20	124	2.8
13.9	137	19	118	2.6
13.9	122	17	105	2.4
13.9	115	16	99	2.2
13.9	108	15	93	2.1
13.9	101	14	87	1.9

%E	M1	M2	DM	M*
13.9	79	11	68	1.5
13.9	72	10	62	1.4
13.9	36	5	31	0.7
13.8	500	69	431	9.5
13.8	499	69	430	9.5
13.8	494	68	426	9.4
13.8	493	68	425	9.4
13.8	492	68	424	9.4
13.8	491	68	423	9.4
13.8	487	67	420	9.2
13.8	486	67	419	9.2
13.8	485	67	418	9.3
13.8	484	67	417	9.3
13.8	480	66	414	9.1
13.8	479	66	413	9.1
13.8	478	66	412	9.1
13.8	477	66	411	9.1
13.8	472	65	407	9.0
13.8	471	65	406	9.0
13.8	470	65	405	9.0
13.8	465	64	401	8.8
13.8	464	64	400	8.8
13.8	463	64	399	8.8
13.8	458	63	395	8.7
13.8	457	63	394	8.7
13.8	456	63	393	8.7
13.8	455	63	392	8.7
13.8	450	62	388	8.5
13.8	449	62	387	8.6
13.8	448	62	386	8.6
13.8	443	61	382	8.4
13.8	442	61	381	8.4
13.8	441	61	380	8.4
13.8	436	60	376	8.3
13.8	435	60	375	8.3
13.8	434	60	374	8.3
13.8	429	59	370	8.1
13.8	428	59	369	8.1
13.8	427	59	368	8.2
13.8	421	58	363	8.0
13.8	420	58	362	8.0
13.8	419	58	361	8.0
13.8	414	57	357	7.8
13.8	413	57	356	7.9
13.8	412	57	355	7.9
13.8	407	56	351	7.7
13.8	406	56	350	7.7
13.8	405	56	349	7.7
13.8	400	55	345	7.6
13.8	399	55	344	7.6

%E	M1	M2	DM	M*
13.8	398	55	343	7.6
13.8	392	54	338	7.4
13.8	391	54	337	7.5
13.8	390	54	336	7.5
13.8	385	53	332	7.3
13.8	384	53	331	7.3
13.8	383	53	330	7.3
13.8	378	52	326	7.2
13.8	377	52	325	7.2
13.8	376	52	324	7.2
13.8	370	51	319	7.0
13.8	369	51	318	7.0
13.8	363	50	313	6.9
13.8	362	50	312	6.9
13.8	356	49	307	6.7
13.8	355	49	306	6.8
13.8	354	49	305	6.8
13.8	349	48	301	6.6
13.8	348	48	300	6.6
13.8	347	48	299	6.6
13.8	341	47	294	6.5
13.8	340	47	293	6.5
13.8	334	46	288	6.3
13.8	333	46	287	6.4
13.8	327	45	282	6.2
13.8	326	45	281	6.2
13.8	325	45	280	6.2
13.8	320	44	276	6.0
13.8	319	44	275	6.1
13.8	318	44	274	6.1
13.8	312	43	269	5.9
13.8	311	43	268	5.9
13.8	305	42	263	5.8
13.8	304	42	262	5.8
13.8	298	41	257	5.6
13.8	297	41	256	5.7
13.8	290	40	250	5.5
13.8	289	40	249	5.5
13.8	283	39	244	5.4
13.8	282	39	243	5.4
13.8	276	38	238	5.2
13.8	275	38	237	5.3
13.8	269	37	232	5.1
13.8	268	37	231	5.1
13.8	261	36	225	5.0
13.8	260	36	224	5.0
13.8	254	35	219	4.8
13.8	253	35	218	4.8
13.8	247	34	213	4.7
13.8	246	34	212	4.7

%E	M1	M2	DM	M*
13.8	240	33	207	4.5
13.8	239	33	206	4.6
13.8	232	32	200	4.4
13.8	225	31	194	4.3
13.8	224	31	193	4.3
13.8	218	30	188	4.1
13.8	217	30	187	4.1
13.8	210	29	181	4.0
13.8	203	28	175	3.9
13.8	196	27	169	3.7
13.8	195	27	168	3.7
13.8	189	26	163	3.6
13.8	188	26	162	3.6
13.8	181	25	156	3.5
13.8	174	24	150	3.3
13.8	167	23	144	3.2
13.8	160	22	138	3.0
13.8	159	22	137	3.0
13.8	152	21	131	2.9
13.8	145	20	125	2.8
13.8	138	19	119	2.6
13.8	130	18	112	2.5
13.8	123	17	106	2.3
13.8	116	16	100	2.2
13.8	109	15	94	2.1
13.8	94	13	81	1.8
13.8	87	12	75	1.7
13.8	80	11	69	1.5
13.8	65	9	56	1.2
13.8	58	8	50	1.1
13.8	29	4	25	0.6
13.7	498	68	430	9.3
13.7	497	68	429	9.3
13.7	496	68	428	9.3
13.7	495	68	427	9.3
13.7	490	67	423	9.2
13.7	489	67	422	9.2
13.7	488	67	421	9.2
13.7	483	66	417	9.0
13.7	482	66	416	9.0
13.7	481	66	415	9.1
13.7	476	65	411	8.9
13.7	475	65	410	8.9
13.7	474	65	409	8.9
13.7	473	65	408	8.9
13.7	468	64	404	8.8
13.7	467	64	403	8.8
13.7	466	64	402	8.8
13.7	461	63	398	8.6
13.7	460	63	397	8.6

%E	M1	M2	DM	M*
13.7	459	63	396	8.6
13.7	454	62	392	8.5
13.7	453	62	391	8.5
13.7	452	62	390	8.5
13.7	451	62	389	8.5
13.7	446	61	385	8.3
13.7	445	61	384	8.4
13.7	444	61	383	8.4
13.7	439	60	379	8.2
13.7	438	60	378	8.2
13.7	437	60	377	8.2
13.7	432	59	373	8.1
13.7	431	59	372	8.1
13.7	430	59	371	8.1
13.7	424	58	366	7.9
13.7	423	58	365	8.0
13.7	422	58	364	8.0
13.7	417	57	360	7.8
13.7	416	57	359	7.8
13.7	415	57	358	7.8
13.7	410	56	354	7.6
13.7	409	56	353	7.7
13.7	408	56	352	7.7
13.7	402	55	347	7.5
13.7	401	55	346	7.5
13.7	395	54	341	7.4
13.7	394	54	340	7.4
13.7	393	54	339	7.4
13.7	388	53	335	7.2
13.7	387	53	334	7.3
13.7	386	53	333	7.3
13.7	380	52	328	7.1
13.7	379	52	327	7.1
13.7	373	51	322	7.0
13.7	372	51	321	7.0
13.7	371	51	320	7.0
13.7	366	50	316	6.8
13.7	365	50	315	6.8
13.7	364	50	314	6.9
13.7	358	49	309	6.7
13.7	357	49	308	6.7
13.7	351	48	303	6.6
13.7	350	48	302	6.6
13.7	344	47	297	6.4
13.7	343	47	296	6.4
13.7	342	47	295	6.5
13.7	337	46	291	6.3
13.7	336	46	290	6.3
13.7	335	46	289	6.3
13.7	329	45	284	6.2

%E	M1	M2	DM	M*
13.7	328	45	283	6.2
13.7	322	44	278	6.0
13.7	321	44	277	6.0
13.7	315	43	272	5.9
13.7	314	43	271	5.9
13.7	313	43	270	5.9
13.7	307	42	265	5.7
13.7	306	42	264	5.8
13.7	300	41	259	5.6
13.7	299	41	258	5.6
13.7	293	40	253	5.5
13.7	292	40	252	5.5
13.7	291	40	251	5.5
13.7	285	39	246	5.3
13.7	284	39	245	5.4
13.7	278	38	240	5.2
13.7	277	38	239	5.2
13.7	271	37	234	5.1
13.7	270	37	233	5.1
13.7	263	36	227	4.9
13.7	262	36	226	4.9
13.7	256	35	221	4.8
13.7	255	35	220	4.8
13.7	249	34	215	4.6
13.7	248	34	214	4.7
13.7	241	33	208	4.5
13.7	234	32	202	4.4
13.7	233	32	201	4.4
13.7	227	31	196	4.2
13.7	226	31	195	4.3
13.7	219	30	189	4.1
13.7	212	29	183	4.0
13.7	211	29	182	4.0
13.7	205	28	177	3.8
13.7	204	28	176	3.8
13.7	197	27	170	3.7
13.7	190	26	164	3.6
13.7	183	25	158	3.4
13.7	182	25	157	3.4
13.7	175	24	151	3.3
13.7	168	23	145	3.1
13.7	161	22	139	3.0
13.7	153	21	132	2.9
13.7	146	20	126	2.7
13.7	139	19	120	2.6
13.7	131	18	113	2.5
13.7	124	17	107	2.3
13.7	117	16	101	2.2
13.7	102	14	88	1.9
13.7	95	13	82	1.8

%E	M1	M2	DM	M*
13.7	73	10	63	1.4
13.7	51	7	44	1.0
13.6	500	68	432	9.2
13.6	499	68	431	9.3
13.6	494	67	427	9.1
13.6	493	67	426	9.1
13.6	492	67	425	9.1
13.6	491	67	424	9.1
13.6	487	66	421	8.9
13.6	486	66	420	9.0
13.6	485	66	419	9.0
13.6	484	66	418	9.0
13.6	479	65	414	8.8
13.6	478	65	413	8.8
13.6	477	65	412	8.9
13.6	472	64	408	8.7
13.6	471	64	407	8.7
13.6	470	64	406	8.7
13.6	469	64	405	8.7
13.6	464	63	401	8.6
13.6	463	63	400	8.6
13.6	462	63	399	8.6
13.6	457	62	395	8.4
13.6	456	62	394	8.4
13.6	455	62	393	8.4
13.6	450	61	389	8.3
13.6	449	61	388	8.3
13.6	448	61	387	8.3
13.6	447	61	386	8.3
13.6	442	60	382	8.1
13.6	441	60	381	8.2
13.6	440	60	380	8.2
13.6	435	59	376	8.0
13.6	434	59	375	8.0
13.6	433	59	374	8.0
13.6	428	58	370	7.9
13.6	427	58	369	7.9
13.6	426	58	368	7.9
13.6	425	58	367	7.9
13.6	420	57	363	7.7
13.6	419	57	362	7.8
13.6	418	57	361	7.8
13.6	413	56	357	7.6
13.6	412	56	356	7.6
13.6	411	56	355	7.6
13.6	405	55	350	7.5
13.6	404	55	349	7.5
13.6	403	55	348	7.5
13.6	398	54	344	7.3
13.6	397	54	343	7.3

%E	M1	M2	DM	M*
13.6	396	54	342	7.4
13.6	391	53	338	7.2
13.6	390	53	337	7.2
13.6	389	53	336	7.2
13.6	383	52	331	7.1
13.6	382	52	330	7.1
13.6	381	52	329	7.1
13.6	376	51	325	6.9
13.6	375	51	324	6.9
13.6	374	51	323	7.0
13.6	369	50	319	6.8
13.6	368	50	318	6.8
13.6	367	50	317	6.8
13.6	361	49	312	6.7
13.6	360	49	311	6.7
13.6	359	49	310	6.7
13.6	354	48	306	6.5
13.6	353	48	305	6.5
13.6	352	48	304	6.5
13.6	346	47	299	6.4
13.6	345	47	298	6.4
13.6	339	46	293	6.2
13.6	338	46	292	6.3
13.6	332	45	287	6.1
13.6	331	45	286	6.1
13.6	330	45	285	6.1
13.6	324	44	280	6.0
13.6	323	44	279	6.0
13.6	317	43	274	5.8
13.6	316	43	273	5.9
13.6	309	42	267	5.7
13.6	308	42	266	5.7
13.6	302	41	261	5.6
13.6	301	41	260	5.6
13.6	295	40	255	5.4
13.6	294	40	254	5.4
13.6	287	39	248	5.3
13.6	286	39	247	5.3
13.6	280	38	242	5.2
13.6	279	38	241	5.2
13.6	273	37	236	5.0
13.6	272	37	235	5.0
13.6	265	36	229	4.9
13.6	264	36	228	4.9
13.6	258	35	223	4.7
13.6	257	35	222	4.8
13.6	250	34	216	4.6
13.6	243	33	210	4.5
13.6	242	33	209	4.5
13.6	236	32	204	4.3

%E	M1	M2	DM	M*
13.6	235	32	203	4.4
13.6	228	31	197	4.2
13.6	221	30	191	4.1
13.6	220	30	190	4.1
13.6	214	29	185	3.9
13.6	213	29	184	3.9
13.6	206	28	178	3.8
13.6	199	27	172	3.7
13.6	198	27	171	3.7
13.6	191	26	165	3.5
13.6	184	25	159	3.4
13.6	177	24	153	3.3
13.6	176	24	152	3.3
13.6	169	23	146	3.1
13.6	162	22	140	3.0
13.6	154	21	133	2.9
13.6	147	20	127	2.7
13.6	140	19	121	2.6
13.6	132	18	114	2.5
13.6	125	17	108	2.3
13.6	118	16	102	2.2
13.6	110	15	95	2.0
13.6	103	14	89	1.9
13.6	88	12	76	1.6
13.6	81	11	70	1.5
13.6	66	9	57	1.2
13.6	59	8	51	1.1
13.6	44	6	38	0.8
13.6	22	3	19	0.4
13.5	498	67	431	9.0
13.5	497	67	430	9.0
13.5	496	67	429	9.1
13.5	495	67	428	9.1
13.5	490	66	424	8.9
13.5	489	66	423	8.9
13.5	488	66	422	8.9
13.5	483	65	418	8.7
13.5	482	65	417	8.8
13.5	481	65	416	8.8
13.5	480	65	415	8.8
13.5	475	64	411	8.6
13.5	474	64	410	8.6
13.5	473	64	409	8.7
13.5	468	63	405	8.5
13.5	467	63	404	8.5
13.5	466	63	403	8.5
13.5	465	63	402	8.5
13.5	460	62	398	8.4
13.5	459	62	397	8.4
13.5	458	62	396	8.4

%E	M1	M2	DM	M*
13.5	453	61	392	8.2
13.5	452	61	391	8.2
13.5	451	61	390	8.3
13.5	446	60	386	8.1
13.5	445	60	385	8.1
13.5	444	60	384	8.1
13.5	443	60	383	8.1
13.5	438	59	379	7.9
13.5	437	59	378	8.0
13.5	436	59	377	8.0
13.5	431	58	373	7.8
13.5	430	58	372	7.8
13.5	429	58	371	7.8
13.5	423	57	366	7.7
13.5	422	57	365	7.7
13.5	421	57	364	7.7
13.5	416	56	360	7.5
13.5	415	56	359	7.6
13.5	414	56	358	7.6
13.5	408	55	353	7.4
13.5	407	55	352	7.4
13.5	406	55	351	7.5
13.5	401	54	347	7.3
13.5	400	54	346	7.3
13.5	399	54	345	7.3
13.5	394	53	341	7.1
13.5	393	53	340	7.1
13.5	392	53	339	7.2
13.5	386	52	334	7.0
13.5	385	52	333	7.0
13.5	384	52	332	7.0
13.5	379	51	328	6.9
13.5	378	51	327	6.9
13.5	377	51	326	6.9
13.5	371	50	321	6.7
13.5	370	50	320	6.8
13.5	364	49	315	6.6
13.5	363	49	314	6.6
13.5	362	49	313	6.6
13.5	356	48	308	6.5
13.5	355	48	307	6.5
13.5	349	47	302	6.3
13.5	348	47	301	6.3
13.5	347	47	300	6.4
13.5	342	46	296	6.2
13.5	341	46	295	6.2
13.5	340	46	294	6.2
13.5	334	45	289	6.1
13.5	333	45	288	6.1
13.5	327	44	283	5.9

%E	M1	M2	DM	M*
13.5	326	44	282	5.9
13.5	325	44	281	6.0
13.5	319	43	276	5.8
13.5	318	43	275	5.8
13.5	312	42	270	5.7
13.5	311	42	269	5.7
13.5	310	42	268	5.7
13.5	304	41	263	5.5
13.5	303	41	262	5.5
13.5	297	40	257	5.4
13.5	296	40	256	5.4
13.5	289	39	250	5.3
13.5	288	39	249	5.3
13.5	282	38	244	5.1
13.5	281	38	243	5.1
13.5	275	37	238	5.0
13.5	274	37	237	5.0
13.5	267	36	231	4.9
13.5	266	36	230	4.9
13.5	260	35	225	4.7
13.5	259	35	224	4.7
13.5	252	34	218	4.6
13.5	251	34	217	4.6
13.5	245	33	212	4.4
13.5	244	33	211	4.5
13.5	237	32	205	4.3
13.5	230	31	199	4.2
13.5	229	31	198	4.2
13.5	223	30	193	4.0
13.5	222	30	192	4.1
13.5	215	29	186	3.9
13.5	208	28	180	3.8
13.5	207	28	179	3.8
13.5	200	27	173	3.6
13.5	193	26	167	3.5
13.5	192	26	166	3.5
13.5	185	25	160	3.4
13.5	178	24	154	3.2
13.5	171	23	148	3.1
13.5	170	23	147	3.1
13.5	163	22	141	3.0
13.5	156	21	135	2.8
13.5	155	21	134	2.8
13.5	148	20	128	2.7
13.5	141	19	122	2.6
13.5	133	18	115	2.4
13.5	126	17	109	2.3
13.5	111	15	96	2.0
13.5	104	14	90	1.9
13.5	96	13	83	1.8

%E	M1	M2	DM	M*
13.5	89	12	77	1.6
13.5	74	10	64	1.4
13.5	52	7	45	0.9
13.5	37	5	32	0.7
13.4	500	67	433	9.0
13.4	499	67	432	9.0
13.4	494	66	428	8.8
13.4	493	66	427	8.8
13.4	492	66	426	8.9
13.4	491	66	425	8.9
13.4	486	65	421	8.7
13.4	485	65	420	8.7
13.4	484	65	419	8.7
13.4	479	64	415	8.6
13.4	478	64	414	8.6
13.4	477	64	413	8.6
13.4	476	64	412	8.6
13.4	471	63	408	8.4
13.4	470	63	407	8.4
13.4	469	63	406	8.5
13.4	464	62	402	8.3
13.4	463	62	401	8.3
13.4	462	62	400	8.3
13.4	461	62	399	8.3
13.4	456	61	395	8.2
13.4	455	61	394	8.2
13.4	454	61	393	8.2
13.4	449	60	389	8.0
13.4	448	60	388	8.0
13.4	447	60	387	8.1
13.4	441	59	382	7.9
13.4	440	59	381	7.9
13.4	439	59	380	7.9
13.4	434	58	376	7.8
13.4	433	58	375	7.8
13.4	432	58	374	7.8
13.4	426	57	369	7.6
13.4	425	57	368	7.6
13.4	424	57	367	7.7
13.4	419	56	363	7.5
13.4	418	56	362	7.5
13.4	417	56	361	7.5
13.4	412	55	357	7.3
13.4	411	55	356	7.4
13.4	410	55	355	7.4
13.4	409	55	354	7.4
13.4	404	54	350	7.2
13.4	403	54	349	7.2
13.4	402	54	348	7.3
13.4	397	53	344	7.1

%E	M1	M2	DM	M*
13.4	396	53	343	7.1
13.4	395	53	342	7.1
13.4	389	52	337	7.0
13.4	388	52	336	7.0
13.4	387	52	335	7.0
13.4	382	51	331	6.8
13.4	381	51	330	6.8
13.4	380	51	329	6.8
13.4	374	50	324	6.7
13.4	373	50	323	6.7
13.4	372	50	322	6.7
13.4	367	49	318	6.5
13.4	366	49	317	6.6
13.4	365	49	316	6.6
13.4	359	48	311	6.4
13.4	358	48	310	6.4
13.4	357	48	309	6.5
13.4	352	47	305	6.3
13.4	351	47	304	6.3
13.4	350	47	303	6.3
13.4	344	46	298	6.2
13.4	343	46	297	6.2
13.4	337	45	292	6.0
13.4	336	45	291	6.0
13.4	335	45	290	6.0
13.4	329	44	285	5.9
13.4	328	44	284	5.9
13.4	322	43	279	5.7
13.4	321	43	278	5.8
13.4	320	43	277	5.8
13.4	314	42	272	5.6
13.4	313	42	271	5.6
13.4	307	41	266	5.5
13.4	306	41	265	5.5
13.4	305	41	264	5.5
13.4	299	40	259	5.4
13.4	298	40	258	5.4
13.4	292	39	253	5.2
13.4	291	39	252	5.2
13.4	290	39	251	5.2
13.4	284	38	246	5.1
13.4	283	38	245	5.1
13.4	277	37	240	4.9
13.4	276	37	239	5.0
13.4	269	36	233	4.8
13.4	268	36	232	4.8
13.4	262	35	227	4.7
13.4	261	35	226	4.7
13.4	254	34	220	4.6
13.4	253	34	219	4.6

%E	M1	M2	DM	M*
13.4	247	33	214	4.4
13.4	246	33	213	4.4
13.4	239	32	207	4.3
13.4	238	32	206	4.3
13.4	232	31	201	4.1
13.4	231	31	200	4.2
13.4	224	30	194	4.0
13.4	217	29	188	3.9
13.4	216	29	187	3.9
13.4	209	28	181	3.8
13.4	202	27	175	3.6
13.4	201	27	174	3.6
13.4	194	26	168	3.5
13.4	187	25	162	3.3
13.4	186	25	161	3.4
13.4	179	24	155	3.2
13.4	172	23	149	3.1
13.4	164	22	142	3.0
13.4	157	21	136	2.8
13.4	149	20	129	2.7
13.4	142	19	123	2.5
13.4	134	18	116	2.4
13.4	127	17	110	2.3
13.4	119	16	103	2.2
13.4	112	15	97	2.0
13.4	97	13	84	1.7
13.4	82	11	71	1.5
13.4	67	9	58	1.2
13.3	498	66	432	8.7
13.3	497	66	431	8.8
13.3	496	66	430	8.8
13.3	495	66	429	8.8
13.3	490	65	425	8.6
13.3	489	65	424	8.6
13.3	488	65	423	8.7
13.3	487	65	422	8.7
13.3	483	64	419	8.5
13.3	482	64	418	8.5
13.3	481	64	417	8.5
13.3	480	64	416	8.5
13.3	475	63	412	8.4
13.3	474	63	411	8.4
13.3	473	63	410	8.4
13.3	472	63	409	8.4
13.3	467	62	405	8.2
13.3	466	62	404	8.2
13.3	465	62	403	8.3
13.3	460	61	399	8.1
13.3	459	61	398	8.1
13.3	458	61	397	8.1

%E	M1	M2	DM	M*
13.3	457	61	396	8.1
13.3	452	60	392	8.0
13.3	451	60	391	8.0
13.3	450	60	390	8.0
13.3	445	59	386	7.8
13.3	444	59	385	7.8
13.3	443	59	384	7.9
13.3	442	59	383	7.9
13.3	437	58	379	7.7
13.3	436	58	378	7.7
13.3	435	58	377	7.7
13.3	430	57	373	7.6
13.3	429	57	372	7.6
13.3	428	57	371	7.6
13.3	427	57	370	7.6
13.3	422	56	366	7.4
13.3	421	56	365	7.4
13.3	420	56	364	7.5
13.3	415	55	360	7.3
13.3	414	55	359	7.3
13.3	413	55	358	7.3
13.3	407	54	353	7.2
13.3	406	54	352	7.2
13.3	405	54	351	7.2
13.3	400	53	347	7.0
13.3	399	53	346	7.0
13.3	398	53	345	7.1
13.3	392	52	340	6.9
13.3	391	52	339	6.9
13.3	390	52	338	6.9
13.3	384	51	333	6.8
13.3	383	51	332	6.8
13.3	377	50	327	6.6
13.3	376	50	326	6.6
13.3	375	50	325	6.7
13.3	369	49	320	6.5
13.3	368	49	319	6.5
13.3	362	48	314	6.4
13.3	361	48	313	6.4
13.3	360	48	312	6.4
13.3	354	47	307	6.2
13.3	353	47	306	6.3
13.3	347	46	301	6.1
13.3	346	46	300	6.1
13.3	345	46	299	6.1
13.3	339	45	294	6.0
13.3	338	45	293	6.0
13.3	332	44	288	5.8
13.3	331	44	287	5.8
13.3	330	44	286	5.9

%E	M1	M2	DM	M*
13.3	324	43	281	5.7
13.3	323	43	280	5.7
13.3	316	42	274	5.6
13.3	315	42	273	5.6
13.3	309	41	268	5.4
13.3	308	41	267	5.5
13.3	301	40	261	5.3
13.3	300	40	260	5.3
13.3	294	39	255	5.2
13.3	293	39	254	5.2
13.3	286	38	248	5.0
13.3	285	38	247	5.1
13.3	279	37	242	4.9
13.3	278	37	241	4.9
13.3	271	36	235	4.8
13.3	270	36	234	4.8
13.3	264	35	229	4.6
13.3	263	35	228	4.7
13.3	256	34	222	4.5
13.3	255	34	221	4.5
13.3	249	33	216	4.4
13.3	248	33	215	4.4
13.3	241	32	209	4.2
13.3	240	32	208	4.3
13.3	233	31	202	4.1
13.3	226	30	196	4.0
13.3	225	30	195	4.0
13.3	218	29	189	3.9
13.3	211	28	183	3.7
13.3	210	28	182	3.7
13.3	203	27	176	3.6
13.3	196	26	170	3.4
13.3	195	26	169	3.5
13.3	188	25	163	3.3
13.3	181	24	157	3.2
13.3	180	24	156	3.2
13.3	173	23	150	3.1
13.3	166	22	144	2.9
13.3	165	22	143	2.9
13.3	158	21	137	2.8
13.3	150	20	130	2.7
13.3	143	19	124	2.5
13.3	135	18	117	2.4
13.3	128	17	111	2.3
13.3	120	16	104	2.1
13.3	113	15	98	2.0
13.3	105	14	91	1.9
13.3	98	13	85	1.7
13.3	90	12	78	1.6
13.3	83	11	72	1.5

%E	M1	M2	DM	M*
13.3	75	10	65	1.3
13.3	60	8	52	1.1
13.3	45	6	39	0.8
13.3	30	4	26	0.5
13.3	15	2	13	0.3
13.2	500	66	434	8.7
13.2	499	66	433	8.7
13.2	494	65	429	8.6
13.2	493	65	428	8.6
13.2	492	65	427	8.6
13.2	491	65	426	8.6
13.2	486	64	422	8.4
13.2	485	64	421	8.4
13.2	484	64	420	8.5
13.2	479	63	416	8.3
13.2	478	63	415	8.3
13.2	477	63	414	8.3
13.2	476	63	413	8.3
13.2	471	62	409	8.2
13.2	470	62	408	8.2
13.2	469	62	407	8.2
13.2	468	62	406	8.2
13.2	463	61	402	8.0
13.2	462	61	401	8.1
13.2	461	61	400	8.1
13.2	456	60	396	7.9
13.2	455	60	395	7.9
13.2	454	60	394	7.9
13.2	453	60	393	7.9
13.2	448	59	389	7.8
13.2	447	59	388	7.8
13.2	446	59	387	7.8
13.2	441	58	383	7.6
13.2	440	58	382	7.6
13.2	439	58	381	7.7
13.2	438	58	380	7.7
13.2	433	57	376	7.5
13.2	432	57	375	7.5
13.2	431	57	374	7.5
13.2	425	56	369	7.4
13.2	424	56	368	7.4
13.2	423	56	367	7.4
13.2	418	55	363	7.2
13.2	417	55	362	7.3
13.2	416	55	361	7.3
13.2	410	54	356	7.1
13.2	409	54	355	7.1
13.2	408	54	354	7.1
13.2	403	53	350	7.0
13.2	402	53	349	7.0

%E	M1	M2	DM	M*
13.2	401	53	348	7.0
13.2	395	52	343	6.8
13.2	394	52	342	6.9
13.2	393	52	341	6.9
13.2	387	51	336	6.7
13.2	386	51	335	6.7
13.2	385	51	334	6.8
13.2	380	50	330	6.6
13.2	379	50	329	6.6
13.2	378	50	328	6.6
13.2	372	49	323	6.5
13.2	371	49	322	6.5
13.2	370	49	321	6.5
13.2	365	48	317	6.3
13.2	364	48	316	6.3
13.2	363	48	315	6.3
13.2	357	47	310	6.2
13.2	356	47	309	6.2
13.2	355	47	308	6.2
13.2	349	46	303	6.1
13.2	348	46	302	6.1
13.2	342	45	297	5.9
13.2	341	45	296	5.9
13.2	340	45	295	6.0
13.2	334	44	290	5.8
13.2	333	44	289	5.8
13.2	327	43	284	5.7
13.2	326	43	283	5.7
13.2	325	43	282	5.7
13.2	319	42	277	5.5
13.2	318	42	276	5.5
13.2	317	42	275	5.6
13.2	311	41	270	5.4
13.2	310	41	269	5.4
13.2	304	40	264	5.3
13.2	303	40	263	5.3
13.2	302	40	262	5.3
13.2	296	39	257	5.1
13.2	295	39	256	5.2
13.2	288	38	250	5.0
13.2	287	38	249	5.0
13.2	281	37	244	4.9
13.2	280	37	243	4.9
13.2	273	36	237	4.7
13.2	272	36	236	4.8
13.2	266	35	231	4.6
13.2	265	35	230	4.6
13.2	258	34	224	4.5
13.2	257	34	223	4.5
13.2	250	33	217	4.4

%E	M1	M2	DM	M*
13.2	243	32	211	4.2
13.2	242	32	210	4.2
13.2	235	31	204	4.1
13.2	234	31	203	4.1
13.2	228	30	198	3.9
13.2	227	30	197	4.0
13.2	220	29	191	3.8
13.2	219	29	190	3.8
13.2	212	28	184	3.7
13.2	205	27	178	3.6
13.2	204	27	177	3.6
13.2	197	26	171	3.4
13.2	190	25	165	3.3
13.2	189	25	164	3.3
13.2	182	24	158	3.2
13.2	174	23	151	3.0
13.2	167	22	145	2.9
13.2	159	21	138	2.8
13.2	152	20	132	2.6
13.2	151	20	131	2.6
13.2	144	19	125	2.5
13.2	136	18	118	2.4
13.2	129	17	112	2.2
13.2	121	16	105	2.1
13.2	114	15	99	2.0
13.2	106	14	92	1.8
13.2	91	12	79	1.6
13.2	76	10	66	1.3
13.2	68	9	59	1.2
13.2	53	7	46	0.9
13.2	38	5	33	0.7
13.1	498	65	433	8.5
13.1	497	65	432	8.5
13.1	496	65	431	8.5
13.1	495	65	430	8.5
13.1	490	64	426	8.4
13.1	489	64	425	8.4
13.1	488	64	424	8.4
13.1	487	64	423	8.4
13.1	482	63	419	8.2
13.1	481	63	418	8.3
13.1	480	63	417	8.3
13.1	475	62	413	8.1
13.1	474	62	412	8.1
13.1	473	62	411	8.1
13.1	472	62	410	8.1
13.1	467	61	406	8.0
13.1	466	61	405	8.0
13.1	465	61	404	8.0
13.1	464	61	403	8.0

%E	M1	M2	DM	M*
13.1	459	60	399	7.8
13.1	458	60	398	7.9
13.1	457	60	397	7.9
13.1	452	59	393	7.7
13.1	451	59	392	7.7
13.1	450	59	391	7.7
13.1	449	59	390	7.8
13.1	444	58	386	7.6
13.1	443	58	385	7.6
13.1	442	58	384	7.6
13.1	436	57	379	7.5
13.1	435	57	378	7.5
13.1	434	57	377	7.5
13.1	429	56	373	7.3
13.1	428	56	372	7.3
13.1	427	56	371	7.3
13.1	426	56	370	7.4
13.1	421	55	366	7.2
13.1	420	55	365	7.2
13.1	419	55	364	7.2
13.1	413	54	359	7.1
13.1	412	54	358	7.1
13.1	411	54	357	7.1
13.1	406	53	353	6.9
13.1	405	53	352	6.9
13.1	404	53	351	7.0
13.1	398	52	346	6.8
13.1	397	52	345	6.8
13.1	396	52	344	6.8
13.1	390	51	339	6.7
13.1	389	51	338	6.7
13.1	388	51	337	6.7
13.1	383	50	333	6.5
13.1	382	50	332	6.5
13.1	381	50	331	6.6
13.1	375	49	326	6.4
13.1	374	49	325	6.4
13.1	373	49	324	6.4
13.1	367	48	319	6.3
13.1	366	48	318	6.3
13.1	360	47	313	6.1
13.1	359	47	312	6.2
13.1	358	47	311	6.2
13.1	352	46	306	6.0
13.1	351	46	305	6.0
13.1	350	46	304	6.0
13.1	344	45	299	5.9
13.1	343	45	298	5.9
13.1	337	44	293	5.7
13.1	336	44	292	5.8

%E	M1	M2	DM	M*
13.1	335	44	291	5.8
13.1	329	43	286	5.6
13.1	328	43	285	5.6
13.1	321	42	279	5.5
13.1	320	42	278	5.5
13.1	314	41	273	5.4
13.1	313	41	272	5.4
13.1	312	41	271	5.4
13.1	306	40	266	5.2
13.1	305	40	265	5.2
13.1	298	39	259	5.1
13.1	297	39	258	5.1
13.1	291	38	253	5.0
13.1	290	38	252	5.0
13.1	289	38	251	5.0
13.1	283	37	246	4.8
13.1	282	37	245	4.9
13.1	275	36	239	4.7
13.1	274	36	238	4.7
13.1	268	35	233	4.6
13.1	267	35	232	4.6
13.1	260	34	226	4.4
13.1	259	34	225	4.5
13.1	252	33	219	4.3
13.1	251	33	218	4.3
13.1	245	32	213	4.2
13.1	244	32	212	4.2
13.1	237	31	206	4.1
13.1	236	31	205	4.1
13.1	229	30	199	3.9
13.1	222	29	193	3.8
13.1	221	29	192	3.8
13.1	214	28	186	3.7
13.1	213	28	185	3.7
13.1	206	27	179	3.5
13.1	199	26	173	3.4
13.1	198	26	172	3.4
13.1	191	25	166	3.3
13.1	183	24	159	3.1
13.1	176	23	153	3.0
13.1	175	23	152	3.0
13.1	168	22	146	2.9
13.1	160	21	139	2.8
13.1	153	20	133	2.6
13.1	145	19	126	2.5
13.1	137	18	119	2.4
13.1	130	17	113	2.2
13.1	122	16	106	2.1
13.1	107	14	93	1.8
13.1	99	13	86	1.7

%E	M1	M2	DM	M*
13.1	84	11	73	1.4
13.1	61	8	53	1.0
13.0	500	65	435	8.4
13.0	499	65	434	8.5
13.0	494	64	430	8.3
13.0	493	64	429	8.3
13.0	492	64	428	8.3
13.0	491	64	427	8.3
13.0	486	63	423	8.2
13.0	485	63	422	8.2
13.0	484	63	421	8.2
13.0	483	63	420	8.2
13.0	478	62	416	8.0
13.0	477	62	415	8.1
13.0	476	62	414	8.1
13.0	471	61	410	7.9
13.0	470	61	409	7.9
13.0	469	61	408	7.9
13.0	468	61	407	8.0
13.0	463	60	403	7.8
13.0	462	60	402	7.8
13.0	461	60	401	7.8
13.0	460	60	400	7.8
13.0	455	59	396	7.7
13.0	454	59	395	7.7
13.0	453	59	394	7.7
13.0	447	58	389	7.5
13.0	446	58	388	7.5
13.0	445	58	387	7.6
13.0	440	57	383	7.4
13.0	439	57	382	7.4
13.0	438	57	381	7.4
13.0	437	57	380	7.4
13.0	432	56	376	7.3
13.0	431	56	375	7.3
13.0	430	56	374	7.3
13.0	424	55	369	7.1
13.0	423	55	368	7.2
13.0	422	55	367	7.2
13.0	417	54	363	7.0
13.0	416	54	362	7.0
13.0	415	54	361	7.0
13.0	414	54	360	7.0
13.0	409	53	356	6.9
13.0	408	53	355	6.9
13.0	407	53	354	6.9
13.0	401	52	349	6.7
13.0	400	52	348	6.8
13.0	399	52	347	6.8
13.0	393	51	342	6.6

%E	M1	M2	DM	M*
13.0	392	51	341	6.6
13.0	391	51	340	6.7
13.0	386	50	336	6.5
13.0	385	50	335	6.5
13.0	384	50	334	6.5
13.0	378	49	329	6.4
13.0	377	49	328	6.4
13.0	376	49	327	6.4
13.0	370	48	322	6.2
13.0	369	48	321	6.2
13.0	368	48	320	6.3
13.0	362	47	315	6.1
13.0	361	47	314	6.1
13.0	355	46	309	6.0
13.0	354	46	308	6.0
13.0	353	46	307	6.0
13.0	347	45	302	5.8
13.0	346	45	301	5.9
13.0	345	45	300	5.9
13.0	339	44	295	5.7
13.0	338	44	294	5.7
13.0	332	43	289	5.6
13.0	331	43	288	5.6
13.0	330	43	287	5.6
13.0	324	42	282	5.4
13.0	323	42	281	5.5
13.0	322	42	280	5.5
13.0	316	41	275	5.3
13.0	315	41	274	5.3
13.0	308	40	268	5.2
13.0	307	40	267	5.2
13.0	301	39	262	5.1
13.0	300	39	261	5.1
13.0	299	39	260	5.1
13.0	293	38	255	4.9
13.0	292	38	254	4.9
13.0	285	37	248	4.8
13.0	284	37	247	4.8
13.0	278	36	242	4.7
13.0	277	36	241	4.7
13.0	276	36	240	4.7
13.0	270	35	235	4.5
13.0	269	35	234	4.6
13.0	262	34	228	4.4
13.0	261	34	227	4.4
13.0	254	33	221	4.3
13.0	253	33	220	4.3
13.0	247	32	215	4.1
13.0	246	32	214	4.2
13.0	239	31	208	4.0

%E	M1	M2	DM	M*	%E	M1	M2	DM	M*	%E	M1	M2	DM	M*	%E	M1	M2	DM	M*	%E	M1	M2	DM	M*
13.0	238	31	207	4.0	12.9	459	59	400	7.6	12.9	334	43	291	5.5	12.9	93	12	81	1.5	12.8	399	51	348	6.5
13.0	231	30	201	3.9	12.9	458	59	399	7.6	12.9	333	43	290	5.6	12.9	85	11	74	1.4	12.8	398	51	347	6.5
13.0	230	30	200	3.9	12.9	457	59	398	7.6	12.9	326	42	284	5.4	12.9	70	9	61	1.2	12.8	397	51	346	6.6
13.0	223	29	194	3.8	12.9	456	59	397	7.6	12.9	325	42	283	5.4	12.9	62	8	54	1.0	12.8	392	50	342	6.4
13.0	216	28	188	3.6	12.9	451	58	393	7.5	12.9	319	41	278	5.3	12.9	31	4	27	0.5	12.8	391	50	341	6.4
13.0	215	28	187	3.6	12.9	450	58	392	7.5	12.9	318	41	277	5.3	12.8	500	64	436	8.2	12.8	390	50	340	6.4
13.0	208	27	181	3.5	12.9	449	58	391	7.5	12.9	317	41	276	5.3	12.8	499	64	435	8.2	12.8	384	49	335	6.3
13.0	207	27	180	3.5	12.9	448	58	390	7.5	12.9	311	40	271	5.1	12.8	494	63	431	8.0	12.8	383	49	334	6.3
13.0	200	26	174	3.4	12.9	443	57	386	7.3	12.9	310	40	270	5.2	12.8	493	63	430	8.1	12.8	382	49	333	6.3
13.0	193	25	168	3.2	12.9	442	57	385	7.4	12.9	309	40	269	5.2	12.8	492	63	429	8.1	12.8	376	48	328	6.1
13.0	192	25	167	3.3	12.9	441	57	384	7.4	12.9	303	39	264	5.0	12.8	491	63	428	8.1	12.8	375	48	327	6.1
13.0	185	24	161	3.1	12.9	435	56	379	7.2	12.9	302	39	263	5.0	12.8	486	62	424	7.9	12.8	374	48	326	6.2
13.0	184	24	160	3.1	12.9	434	56	378	7.2	12.9	295	38	257	4.9	12.8	485	62	423	7.9	12.8	368	47	321	6.0
13.0	177	23	154	3.0	12.9	433	56	377	7.2	12.9	294	38	256	4.9	12.8	484	62	422	7.9	12.8	367	47	320	6.0
13.0	169	22	147	2.9	12.9	428	55	373	7.1	12.9	287	37	250	4.8	12.8	483	62	421	8.0	12.8	366	47	319	6.0
13.0	162	21	141	2.7	12.9	427	55	372	7.1	12.9	286	37	249	4.8	12.8	478	61	417	7.8	12.8	360	46	314	5.9
13.0	161	21	140	2.7	12.9	426	55	371	7.1	12.9	280	36	244	4.6	12.8	477	61	416	7.8	12.8	359	46	313	5.9
13.0	154	20	134	2.6	12.9	425	55	370	7.1	12.9	279	36	243	4.6	12.8	476	61	415	7.8	12.8	358	46	312	5.9
13.0	146	19	127	2.5	12.9	420	54	366	6.9	12.9	272	35	237	4.5	12.8	475	61	414	7.8	12.8	352	45	307	5.8
13.0	139	18	121	2.3	12.9	419	54	365	7.0	12.9	271	35	236	4.5	12.8	470	60	410	7.7	12.8	351	45	306	5.8
13.0	138	18	120	2.3	12.9	418	54	364	7.0	12.9	264	34	230	4.4	12.8	469	60	409	7.7	12.8	345	44	301	5.6
13.0	131	17	114	2.2	12.9	412	53	359	6.8	12.9	263	34	229	4.4	12.8	468	60	408	7.7	12.8	344	44	300	5.6
13.0	123	16	107	2.1	12.9	411	53	358	6.8	12.9	256	33	223	4.3	12.8	467	60	407	7.7	12.8	343	44	299	5.6
13.0	115	15	100	2.0	12.9	410	53	357	6.9	12.9	255	33	222	4.3	12.8	462	59	403	7.5	12.8	337	43	294	5.5
13.0	108	14	94	1.8	12.9	404	52	352	6.7	12.9	249	32	217	4.1	12.8	461	59	402	7.6	12.8	336	43	293	5.5
13.0	100	13	87	1.7	12.9	403	52	351	6.7	12.9	248	32	216	4.1	12.8	460	59	401	7.6	12.8	335	43	292	5.5
13.0	92	12	80	1.6	12.9	402	52	350	6.7	12.9	241	31	210	4.0	12.8	454	58	396	7.4	12.8	329	42	287	5.4
13.0	77	10	67	1.3	12.9	396	51	345	6.6	12.9	240	31	209	4.0	12.8	453	58	395	7.4	12.8	328	42	286	5.4
13.0	69	9	60	1.2	12.9	395	51	344	6.6	12.9	233	30	203	3.9	12.8	452	58	394	7.4	12.8	327	42	285	5.4
13.0	54	7	47	0.9	12.9	394	51	343	6.6	12.9	232	30	202	3.9	12.8	447	57	390	7.3	12.8	321	41	280	5.2
13.0	46	6	40	0.8	12.9	389	50	339	6.4	12.9	225	29	196	3.7	12.8	446	57	389	7.3	12.8	320	41	279	5.3
13.0	23	3	20	0.4	12.9	388	50	338	6.4	12.9	224	29	195	3.8	12.8	445	57	388	7.3	12.8	313	40	273	5.1
12.9	498	64	434	8.2	12.9	387	50	337	6.5	12.9	217	28	189	3.6	12.8	444	57	387	7.3	12.8	312	40	272	5.1
12.9	497	64	433	8.2	12.9	381	49	332	6.3	12.9	210	27	183	3.5	12.8	439	56	383	7.1	12.8	305	39	266	5.0
12.9	496	64	432	8.3	12.9	380	49	331	6.3	12.9	209	27	182	3.5	12.8	438	56	382	7.2	12.8	304	39	265	5.0
12.9	495	64	431	8.3	12.9	379	49	330	6.3	12.9	202	26	176	3.3	12.8	437	56	381	7.2	12.8	298	38	260	4.8
12.9	490	63	427	8.1	12.9	373	48	325	6.2	12.9	201	26	175	3.4	12.8	436	56	380	7.2	12.8	297	38	259	4.9
12.9	489	63	426	8.1	12.9	372	48	324	6.2	12.9	194	25	169	3.2	12.8	431	55	376	7.0	12.8	296	38	258	4.9
12.9	488	63	425	8.1	12.9	371	48	323	6.2	12.9	186	24	162	3.1	12.8	430	55	375	7.0	12.8	290	37	253	4.7
12.9	487	63	424	8.1	12.9	365	47	318	6.1	12.9	178	23	155	3.0	12.8	429	55	374	7.1	12.8	289	37	252	4.7
12.9	482	62	420	8.0	12.9	364	47	317	6.1	12.9	171	22	149	2.8	12.8	423	54	369	6.9	12.8	288	37	251	4.8
12.9	481	62	419	8.0	12.9	363	47	316	6.1	12.9	170	22	148	2.8	12.8	422	54	368	6.9	12.8	282	36	246	4.6
12.9	480	62	418	8.0	12.9	357	46	311	5.9	12.9	163	21	142	2.7	12.8	421	54	367	6.9	12.8	281	36	245	4.6
12.9	479	62	417	8.0	12.9	356	46	310	5.9	12.9	155	20	135	2.6	12.8	415	53	362	6.8	12.8	274	35	239	4.5
12.9	474	61	413	7.9	12.9	350	45	305	5.8	12.9	147	19	128	2.5	12.8	414	53	361	6.8	12.8	273	35	238	4.5
12.9	473	61	412	7.9	12.9	349	45	304	5.8	12.9	140	18	122	2.3	12.8	413	53	360	6.8	12.8	266	34	232	4.3
12.9	472	61	411	7.9	12.9	348	45	303	5.8	12.9	132	17	115	2.2	12.8	407	52	355	6.6	12.8	265	34	231	4.4
12.9	466	60	406	7.7	12.9	342	44	298	5.7	12.9	124	16	108	2.1	12.8	406	52	354	6.7	12.8	258	33	225	4.2
12.9	465	60	405	7.7	12.9	341	44	297	5.7	12.9	116	15	101	1.9	12.8	405	52	353	6.7	12.8	257	33	224	4.2
12.9	464	60	404	7.8	12.9	340	44	296	5.7	12.9	101	13	88	1.7	12.8	400	51	349	6.5	12.8	250	32	218	4.1

%E	M1	M2	DM	M*
12.8	243	31	212	4.0
12.8	242	31	211	4.0
12.8	235	30	205	3.8
12.8	234	30	204	3.8
12.8	227	29	198	3.7
12.8	226	29	197	3.7
12.8	219	28	191	3.6
12.8	218	28	190	3.6
12.8	211	27	184	3.5
12.8	203	26	177	3.3
12.8	196	25	171	3.2
12.8	195	25	170	3.2
12.8	188	24	164	3.1
12.8	187	24	163	3.1
12.8	180	23	157	2.9
12.8	179	23	156	3.0
12.8	172	22	150	2.8
12.8	164	21	143	2.7
12.8	156	20	136	2.6
12.8	149	19	130	2.4
12.8	148	19	129	2.4
12.8	141	18	123	2.3
12.8	133	17	116	2.2
12.8	125	16	109	2.0
12.8	117	15	102	1.9
12.8	109	14	95	1.8
12.8	94	12	82	1.5
12.8	86	11	75	1.4
12.8	78	10	68	1.3
12.8	47	6	41	0.8
12.8	39	5	34	0.6
12.7	498	63	435	8.0
12.7	497	63	434	8.0
12.7	496	63	433	8.0
12.7	495	63	432	8.0
12.7	490	62	428	7.8
12.7	489	62	427	7.9
12.7	488	62	426	7.9
12.7	487	62	425	7.9
12.7	482	61	421	7.7
12.7	481	61	420	7.7
12.7	480	61	419	7.8
12.7	479	61	418	7.8
12.7	474	60	414	7.6
12.7	473	60	413	7.6
12.7	472	60	412	7.6
12.7	471	60	411	7.6
12.7	466	59	407	7.5
12.7	465	59	406	7.5
12.7	464	59	405	7.5

%E	M1	M2	DM	M*
12.7	463	59	404	7.5
12.7	458	58	400	7.3
12.7	457	58	399	7.4
12.7	456	58	398	7.4
12.7	455	58	397	7.4
12.7	450	57	393	7.2
12.7	449	57	392	7.2
12.7	448	57	391	7.3
12.7	442	56	386	7.1
12.7	441	56	385	7.1
12.7	440	56	384	7.1
12.7	434	55	379	7.0
12.7	433	55	378	7.0
12.7	432	55	377	7.0
12.7	426	54	372	6.8
12.7	425	54	371	6.9
12.7	424	54	370	6.9
12.7	418	53	365	6.7
12.7	417	53	364	6.7
12.7	416	53	363	6.8
12.7	411	52	359	6.6
12.7	410	52	358	6.6
12.7	409	52	357	6.6
12.7	408	52	356	6.6
12.7	403	51	352	6.5
12.7	402	51	351	6.5
12.7	401	51	350	6.5
12.7	395	50	345	6.3
12.7	394	50	344	6.3
12.7	393	50	343	6.4
12.7	387	49	338	6.2
12.7	386	49	337	6.2
12.7	385	49	336	6.2
12.7	379	48	331	6.1
12.7	378	48	330	6.1
12.7	377	48	329	6.1
12.7	371	47	324	6.0
12.7	370	47	323	6.0
12.7	369	47	322	6.0
12.7	363	46	317	5.8
12.7	362	46	316	5.8
12.7	361	46	315	5.9
12.7	355	45	310	5.7
12.7	354	45	309	5.7
12.7	353	45	308	5.7
12.7	347	44	303	5.6
12.7	346	44	302	5.6
12.7	339	43	296	5.5
12.7	338	43	295	5.5
12.7	332	42	290	5.3

%E	M1	M2	DM	M*
12.7	331	42	289	5.3
12.7	330	42	288	5.3
12.7	324	41	283	5.2
12.7	323	41	282	5.2
12.7	322	41	281	5.2
12.7	316	40	276	5.1
12.7	315	40	275	5.1
12.7	314	40	274	5.1
12.7	308	39	269	4.9
12.7	307	39	268	5.0
12.7	306	39	267	5.0
12.7	300	38	262	4.8
12.7	299	38	261	4.8
12.7	292	37	255	4.7
12.7	291	37	254	4.7
12.7	284	36	248	4.6
12.7	283	36	247	4.6
12.7	276	35	241	4.4
12.7	275	35	240	4.5
12.7	268	34	234	4.3
12.7	267	34	233	4.3
12.7	260	33	227	4.2
12.7	259	33	226	4.2
12.7	252	32	220	4.1
12.7	251	32	219	4.1
12.7	245	31	214	3.9
12.7	244	31	213	3.9
12.7	237	30	207	3.8
12.7	236	30	206	3.8
12.7	229	29	200	3.7
12.7	228	29	199	3.7
12.7	221	28	193	3.5
12.7	220	28	192	3.6
12.7	213	27	186	3.4
12.7	212	27	185	3.4
12.7	205	26	179	3.3
12.7	204	26	178	3.3
12.7	197	25	172	3.2
12.7	189	24	165	3.0
12.7	181	23	158	2.9
12.7	173	22	151	2.8
12.7	166	21	145	2.7
12.7	165	21	144	2.7
12.7	158	20	138	2.5
12.7	157	20	137	2.5
12.7	150	19	131	2.4
12.7	142	18	124	2.3
12.7	134	17	117	2.2
12.7	126	16	110	2.0
12.7	118	15	103	1.9

%E	M1	M2	DM	M*
12.7	110	14	96	1.8
12.7	102	13	89	1.7
12.7	79	10	69	1.3
12.7	71	9	62	1.1
12.7	63	8	55	1.0
12.7	55	7	48	0.9
12.6	500	63	437	7.9
12.6	499	63	436	8.0
12.6	494	62	432	7.8
12.6	493	62	431	7.8
12.6	492	62	430	7.8
12.6	491	62	429	7.8
12.6	486	61	425	7.7
12.6	485	61	424	7.7
12.6	484	61	423	7.7
12.6	483	61	422	7.7
12.6	478	60	418	7.5
12.6	477	60	417	7.5
12.6	476	60	416	7.6
12.6	475	60	415	7.6
12.6	470	59	411	7.4
12.6	469	59	410	7.4
12.6	468	59	409	7.4
12.6	467	59	408	7.5
12.6	462	58	404	7.3
12.6	461	58	403	7.3
12.6	460	58	402	7.3
12.6	459	58	401	7.3
12.6	454	57	397	7.2
12.6	453	57	396	7.2
12.6	452	57	395	7.2
12.6	451	57	394	7.2
12.6	446	56	390	7.0
12.6	445	56	389	7.0
12.6	444	56	388	7.1
12.6	443	56	387	7.1
12.6	438	55	383	6.9
12.6	437	55	382	6.9
12.6	436	55	381	6.9
12.6	435	55	380	7.0
12.6	430	54	376	6.8
12.6	429	54	375	6.8
12.6	428	54	374	6.8
12.6	427	54	373	6.8
12.6	422	53	369	6.7
12.6	421	53	368	6.7
12.6	420	53	367	6.7
12.6	419	53	366	6.7
12.6	414	52	362	6.5
12.6	413	52	361	6.5

%E	M1	M2	DM	M*
12.6	412	52	360	6.6
12.6	406	51	355	6.4
12.6	405	51	354	6.4
12.6	404	51	353	6.4
12.6	398	50	348	6.3
12.6	397	50	347	6.3
12.6	396	50	346	6.3
12.6	390	49	341	6.2
12.6	389	49	340	6.2
12.6	388	49	339	6.2
12.6	382	48	334	6.0
12.6	381	48	333	6.0
12.6	380	48	332	6.1
12.6	374	47	327	5.9
12.6	373	47	326	5.9
12.6	372	47	325	5.9
12.6	366	46	320	5.8
12.6	365	46	319	5.8
12.6	364	46	318	5.8
12.6	358	45	313	5.7
12.6	357	45	312	5.7
12.6	356	45	311	5.7
12.6	350	44	306	5.5
12.6	349	44	305	5.5
12.6	348	44	304	5.6
12.6	342	43	299	5.4
12.6	341	43	298	5.4
12.6	340	43	297	5.4
12.6	334	42	292	5.3
12.6	333	42	291	5.3
12.6	326	41	285	5.2
12.6	325	41	284	5.2
12.6	318	40	278	5.0
12.6	317	40	277	5.0
12.6	310	39	271	4.9
12.6	309	39	270	4.9
12.6	302	38	264	4.8
12.6	301	38	263	4.8
12.6	294	37	257	4.7
12.6	293	37	256	4.7
12.6	286	36	250	4.5
12.6	285	36	249	4.5
12.6	278	35	243	4.4
12.6	277	35	242	4.4
12.6	270	34	236	4.3
12.6	269	34	235	4.3
12.6	262	33	229	4.2
12.6	261	33	228	4.2
12.6	254	32	222	4.0
12.6	253	32	221	4.0

%E	M1	M2	DM	M*
12.6	247	31	216	3.9
12.6	246	31	215	3.9
12.6	239	30	209	3.8
12.6	238	30	208	3.8
12.6	231	29	202	3.6
12.6	230	29	201	3.7
12.6	223	28	195	3.5
12.6	222	28	194	3.5
12.6	215	27	188	3.4
12.6	214	27	187	3.4
12.6	207	26	181	3.3
12.6	206	26	180	3.3
12.6	199	25	174	3.1
12.6	198	25	173	3.2
12.6	191	24	167	3.0
12.6	190	24	166	3.0
12.6	183	23	160	2.9
12.6	182	23	159	2.9
12.6	175	22	153	2.8
12.6	174	22	152	2.8
12.6	167	21	146	2.6
12.6	159	20	139	2.5
12.6	151	19	132	2.4
12.6	143	18	125	2.3
12.6	135	17	118	2.1
12.6	127	16	111	2.0
12.6	119	15	104	1.9
12.6	111	14	97	1.8
12.6	103	13	90	1.6
12.6	95	12	83	1.5
12.6	87	11	76	1.4
12.5	498	62	436	7.7
12.5	497	62	435	7.7
12.5	496	62	434	7.8
12.5	495	62	433	7.8
12.5	489	61	428	7.6
12.5	488	61	427	7.6
12.5	487	61	426	7.6
12.5	481	60	421	7.5
12.5	480	60	420	7.5
12.5	479	60	419	7.5
12.5	473	59	414	7.4
12.5	472	59	413	7.4
12.5	471	59	412	7.4
12.5	465	58	407	7.2
12.5	464	58	406	7.3
12.5	463	58	405	7.3
12.5	457	57	400	7.1
12.5	456	57	399	7.1
12.5	455	57	398	7.1

%E	M1	M2	DM	M*
12.5	449	56	393	7.0
12.5	448	56	392	7.0
12.5	447	56	391	7.0
12.5	441	55	386	6.9
12.5	440	55	385	6.9
12.5	439	55	384	6.9
12.5	433	54	379	6.7
12.5	432	54	378	6.8
12.5	431	54	377	6.8
12.5	425	53	372	6.6
12.5	424	53	371	6.6
12.5	423	53	370	6.6
12.5	417	52	365	6.5
12.5	416	52	364	6.5
12.5	415	52	363	6.5
12.5	409	51	358	6.4
12.5	408	51	357	6.4
12.5	407	51	356	6.4
12.5	401	50	351	6.2
12.5	400	50	350	6.3
12.5	399	50	349	6.3
12.5	393	49	344	6.1
12.5	392	49	343	6.1
12.5	391	49	342	6.1
12.5	385	48	337	6.0
12.5	384	48	336	6.0
12.5	383	48	335	6.0
12.5	377	47	330	5.9
12.5	376	47	329	5.9
12.5	375	47	328	5.9
12.5	369	46	323	5.7
12.5	368	46	322	5.8
12.5	367	46	321	5.8
12.5	361	45	316	5.6
12.5	360	45	315	5.6
12.5	359	45	314	5.6
12.5	353	44	309	5.5
12.5	352	44	308	5.5
12.5	351	44	307	5.5
12.5	345	43	302	5.4
12.5	344	43	301	5.4
12.5	343	43	300	5.4
12.5	337	42	295	5.2
12.5	336	42	294	5.3
12.5	335	42	293	5.3
12.5	329	41	288	5.1
12.5	328	41	287	5.1
12.5	327	41	286	5.1
12.5	321	40	281	5.0
12.5	320	40	280	5.0

%E	M1	M2	DM	M*
12.5	319	40	279	5.0
12.5	313	39	274	4.9
12.5	312	39	273	4.9
12.5	311	39	272	4.9
12.5	305	38	267	4.7
12.5	304	38	266	4.8
12.5	303	38	265	4.8
12.5	297	37	260	4.6
12.5	296	37	259	4.6
12.5	295	37	258	4.6
12.5	289	36	253	4.5
12.5	288	36	252	4.5
12.5	287	36	251	4.5
12.5	281	35	246	4.4
12.5	280	35	245	4.4
12.5	279	35	244	4.4
12.5	273	34	239	4.2
12.5	272	34	238	4.3
12.5	271	34	237	4.3
12.5	265	33	232	4.1
12.5	264	33	231	4.1
12.5	263	33	230	4.1
12.5	257	32	225	4.0
12.5	256	32	224	4.0
12.5	255	32	223	4.0
12.5	249	31	218	3.9
12.5	248	31	217	3.9
12.5	240	30	210	3.8
12.5	232	29	203	3.6
12.5	224	28	196	3.5
12.5	216	27	189	3.4
12.5	208	26	182	3.3
12.5	200	25	175	3.1
12.5	192	24	168	3.0
12.5	184	23	161	2.9
12.5	176	22	154	2.8
12.5	168	21	147	2.6
12.5	160	20	140	2.5
12.5	152	19	133	2.4
12.5	144	18	126	2.3
12.5	136	17	119	2.1
12.5	128	16	112	2.0
12.5	120	15	105	1.9
12.5	112	14	98	1.8
12.5	104	13	91	1.6
12.5	96	12	84	1.5
12.5	88	11	77	1.4
12.5	80	10	70	1.3
12.5	72	9	63	1.1
12.5	64	8	56	1.0

%E	M1	M2	DM	M*
12.5	56	7	49	0.9
12.5	48	6	42	0.8
12.5	40	5	35	0.6
12.5	32	4	28	0.5
12.5	24	3	21	0.4
12.5	16	2	14	0.3
12.5	8	1	7	0.1
12.4	500	62	438	7.7
12.4	499	62	437	7.7
12.4	493	61	432	7.5
12.4	492	61	431	7.6
12.4	491	61	430	7.6
12.4	490	61	429	7.6
12.4	485	60	425	7.4
12.4	484	60	424	7.4
12.4	483	60	423	7.5
12.4	482	60	422	7.5
12.4	477	59	418	7.3
12.4	476	59	417	7.3
12.4	475	59	416	7.3
12.4	474	59	415	7.3
12.4	469	58	411	7.2
12.4	468	58	410	7.2
12.4	467	58	409	7.2
12.4	466	58	408	7.2
12.4	461	57	404	7.0
12.4	460	57	403	7.1
12.4	459	57	402	7.1
12.4	458	57	401	7.1
12.4	453	56	397	6.9
12.4	452	56	396	6.9
12.4	451	56	395	7.0
12.4	450	56	394	7.0
12.4	445	55	390	6.8
12.4	444	55	389	6.8
12.4	443	55	388	6.8
12.4	442	55	387	6.8
12.4	437	54	383	6.7
12.4	436	54	382	6.7
12.4	435	54	381	6.7
12.4	434	54	380	6.7
12.4	429	53	376	6.5
12.4	428	53	375	6.6
12.4	427	53	374	6.6
12.4	426	53	373	6.6
12.4	421	52	369	6.4
12.4	420	52	368	6.4
12.4	419	52	367	6.5
12.4	418	52	366	6.5
12.4	412	51	361	6.3

%E	M1	M2	DM	M*
12.4	411	51	360	6.3
12.4	410	51	359	6.3
12.4	404	50	354	6.2
12.4	403	50	353	6.2
12.4	402	50	352	6.2
12.4	396	49	347	6.1
12.4	395	49	346	6.1
12.4	394	49	345	6.1
12.4	388	48	340	5.9
12.4	387	48	339	6.0
12.4	386	48	338	6.0
12.4	380	47	333	5.8
12.4	379	47	332	5.8
12.4	378	47	331	5.8
12.4	372	46	326	5.7
12.4	371	46	325	5.7
12.4	370	46	324	5.7
12.4	364	45	319	5.6
12.4	363	45	318	5.6
12.4	362	45	317	5.6
12.4	356	44	312	5.4
12.4	355	44	311	5.5
12.4	354	44	310	5.5
12.4	348	43	305	5.3
12.4	347	43	304	5.3
12.4	346	43	303	5.3
12.4	340	42	298	5.2
12.4	339	42	297	5.2
12.4	338	42	296	5.2
12.4	331	41	290	5.1
12.4	330	41	289	5.1
12.4	323	40	283	5.0
12.4	322	40	282	5.0
12.4	315	39	276	4.8
12.4	314	39	275	4.8
12.4	307	38	269	4.7
12.4	306	38	268	4.7
12.4	299	37	262	4.6
12.4	298	37	261	4.6
12.4	291	36	255	4.5
12.4	290	36	254	4.5
12.4	283	35	248	4.3
12.4	282	35	247	4.3
12.4	275	34	241	4.2
12.4	274	34	240	4.2
12.4	267	33	234	4.1
12.4	266	33	233	4.1
12.4	259	32	227	4.0
12.4	258	32	226	4.0
12.4	251	31	220	3.8

%E	M1	M2	DM	M*
12.4	250	31	219	3.8
12.4	242	30	212	3.7
12.4	241	30	211	3.7
12.4	234	29	205	3.6
12.4	233	29	204	3.6
12.4	226	28	198	3.5
12.4	225	28	197	3.5
12.4	218	27	191	3.3
12.4	217	27	190	3.4
12.4	210	26	184	3.2
12.4	209	26	183	3.2
12.4	202	25	177	3.1
12.4	201	25	176	3.1
12.4	194	24	170	3.0
12.4	193	24	169	3.0
12.4	186	23	163	2.8
12.4	185	23	162	2.9
12.4	178	22	156	2.7
12.4	177	22	155	2.7
12.4	170	21	149	2.6
12.4	169	21	148	2.6
12.4	161	20	141	2.5
12.4	153	19	134	2.4
12.4	145	18	127	2.2
12.4	137	17	120	2.1
12.4	129	16	113	2.0
12.4	121	15	106	1.9
12.4	113	14	99	1.7
12.4	105	13	92	1.6
12.4	97	12	85	1.5
12.4	89	11	78	1.4
12.3	497	61	436	7.5
12.3	496	61	435	7.5
12.3	495	61	434	7.5
12.3	494	61	433	7.5
12.3	489	60	429	7.4
12.3	488	60	428	7.4
12.3	487	60	427	7.4
12.3	486	60	426	7.4
12.3	481	59	422	7.2
12.3	480	59	421	7.3
12.3	479	59	420	7.3
12.3	478	59	419	7.3
12.3	473	58	415	7.1
12.3	472	58	414	7.1
12.3	471	58	413	7.1
12.3	470	58	412	7.2
12.3	465	57	408	7.0
12.3	464	57	407	7.0
12.3	463	57	406	7.0

%E	M1	M2	DM	M*
12.3	462	57	405	7.0
12.3	457	56	401	6.9
12.3	456	56	400	6.9
12.3	455	56	399	6.9
12.3	454	56	398	6.9
12.3	448	55	393	6.8
12.3	447	55	392	6.8
12.3	446	55	391	6.8
12.3	440	54	386	6.6
12.3	439	54	385	6.6
12.3	438	54	384	6.7
12.3	432	53	379	6.5
12.3	431	53	378	6.5
12.3	430	53	377	6.5
12.3	424	52	372	6.4
12.3	423	52	371	6.4
12.3	422	52	370	6.4
12.3	416	51	365	6.3
12.3	415	51	364	6.3
12.3	414	51	363	6.3
12.3	413	51	362	6.3
12.3	408	50	358	6.1
12.3	407	50	357	6.1
12.3	406	50	356	6.2
12.3	405	50	355	6.2
12.3	400	49	351	6.0
12.3	399	49	350	6.0
12.3	398	49	349	6.0
12.3	397	49	348	6.0
12.3	391	48	343	5.9
12.3	390	48	342	5.9
12.3	389	48	341	5.9
12.3	383	47	336	5.8
12.3	382	47	335	5.8
12.3	381	47	334	5.8
12.3	375	46	329	5.6
12.3	374	46	328	5.7
12.3	373	46	327	5.7
12.3	367	45	322	5.5
12.3	366	45	321	5.5
12.3	365	45	320	5.5
12.3	359	44	315	5.4
12.3	358	44	314	5.4
12.3	357	44	313	5.4
12.3	351	43	308	5.3
12.3	350	43	307	5.3
12.3	349	43	306	5.3
12.3	342	42	300	5.2
12.3	341	42	299	5.2
12.3	334	41	293	5.0

%E	M1	M2	DM	M*
12.3	333	41	292	5.0
12.3	332	41	291	5.1
12.3	326	40	286	4.9
12.3	325	40	285	4.9
12.3	324	40	284	4.9
12.3	318	39	279	4.8
12.3	317	39	278	4.8
12.3	316	39	277	4.8
12.3	310	38	272	4.7
12.3	309	38	271	4.7
12.3	308	38	270	4.7
12.3	302	37	265	4.5
12.3	301	37	264	4.5
12.3	300	37	263	4.6
12.3	293	36	257	4.4
12.3	292	36	256	4.4
12.3	285	35	250	4.3
12.3	284	35	249	4.3
12.3	277	34	243	4.2
12.3	276	34	242	4.2
12.3	269	33	236	4.0
12.3	268	33	235	4.1
12.3	261	32	229	3.9
12.3	260	32	228	3.9
12.3	253	31	222	3.8
12.3	252	31	221	3.8
12.3	244	30	214	3.7
12.3	243	30	213	3.7
12.3	236	29	207	3.6
12.3	235	29	206	3.6
12.3	228	28	200	3.4
12.3	227	28	199	3.5
12.3	220	27	193	3.3
12.3	219	27	192	3.3
12.3	212	26	186	3.2
12.3	211	26	185	3.2
12.3	204	25	179	3.1
12.3	203	25	178	3.1
12.3	195	24	171	3.0
12.3	187	23	164	2.8
12.3	179	22	157	2.7
12.3	171	21	150	2.6
12.3	163	20	143	2.5
12.3	162	20	142	2.5
12.3	155	19	136	2.3
12.3	154	19	135	2.3
12.3	146	18	128	2.2
12.3	138	17	121	2.1
12.3	130	16	114	2.0
12.3	122	15	107	1.8

%E	M1	M2	DM	M*
12.3	114	14	100	1.7
12.3	106	13	93	1.6
12.3	81	10	71	1.2
12.3	73	9	64	1.1
12.3	65	8	57	1.0
12.3	57	7	50	0.9
12.2	500	61	439	7.4
12.2	499	61	438	7.5
12.2	498	61	437	7.5
12.2	493	60	433	7.3
12.2	492	60	432	7.3
12.2	491	60	431	7.3
12.2	490	60	430	7.3
12.2	485	59	426	7.2
12.2	484	59	425	7.2
12.2	483	59	424	7.2
12.2	482	59	423	7.2
12.2	477	58	419	7.1
12.2	476	58	418	7.1
12.2	475	58	417	7.1
12.2	474	58	416	7.1
12.2	469	57	412	6.9
12.2	468	57	411	6.9
12.2	467	57	410	7.0
12.2	466	57	409	7.0
12.2	460	56	404	6.8
12.2	459	56	403	6.8
12.2	458	56	402	6.8
12.2	452	55	397	6.7
12.2	451	55	396	6.7
12.2	450	55	395	6.7
12.2	449	55	394	6.7
12.2	444	54	390	6.6
12.2	443	54	389	6.6
12.2	442	54	388	6.6
12.2	441	54	387	6.6
12.2	436	53	383	6.4
12.2	435	53	382	6.5
12.2	434	53	381	6.5
12.2	433	53	380	6.5
12.2	428	52	376	6.3
12.2	427	52	375	6.3
12.2	426	52	374	6.3
12.2	425	52	373	6.4
12.2	419	51	368	6.2
12.2	418	51	367	6.2
12.2	417	51	366	6.2
12.2	411	50	361	6.1
12.2	410	50	360	6.1
12.2	409	50	359	6.1

%E	M1	M2	DM	M*
12.2	403	49	354	6.0
12.2	402	49	353	6.0
12.2	401	49	352	6.0
12.2	395	48	347	5.8
12.2	394	48	346	5.8
12.2	393	48	345	5.9
12.2	392	48	344	5.9
12.2	386	47	339	5.7
12.2	385	47	338	5.7
12.2	384	47	337	5.8
12.2	378	46	332	5.6
12.2	377	46	331	5.6
12.2	376	46	330	5.6
12.2	370	45	325	5.5
12.2	369	45	324	5.5
12.2	368	45	323	5.5
12.2	362	44	318	5.3
12.2	361	44	317	5.4
12.2	360	44	316	5.4
12.2	353	43	310	5.2
12.2	352	43	309	5.3
12.2	345	42	303	5.1
12.2	344	42	302	5.1
12.2	343	42	301	5.1
12.2	337	41	296	5.0
12.2	336	41	295	5.0
12.2	335	41	294	5.0
12.2	329	40	289	4.9
12.2	328	40	288	4.9
12.2	327	40	287	4.9
12.2	321	39	282	4.7
12.2	320	39	281	4.8
12.2	319	39	280	4.8
12.2	312	38	274	4.6
12.2	311	38	273	4.6
12.2	304	37	267	4.5
12.2	303	37	266	4.5
12.2	296	36	260	4.4
12.2	295	36	259	4.4
12.2	294	36	258	4.4
12.2	288	35	253	4.3
12.2	287	35	252	4.3
12.2	286	35	251	4.3
12.2	279	34	245	4.1
12.2	278	34	244	4.2
12.2	271	33	238	4.0
12.2	270	33	237	4.0
12.2	263	32	231	3.9
12.2	262	32	230	3.9
12.2	255	31	224	3.8

%E	M1	M2	DM	M*
12.2	254	31	223	3.8
12.2	246	30	216	3.7
12.2	245	30	215	3.7
12.2	238	29	209	3.5
12.2	237	29	208	3.5
12.2	230	28	202	3.4
12.2	229	28	201	3.4
12.2	222	27	195	3.3
12.2	221	27	194	3.3
12.2	214	26	188	3.2
12.2	213	26	187	3.2
12.2	205	25	180	3.0
12.2	197	24	173	2.9
12.2	196	24	172	2.9
12.2	189	23	166	2.8
12.2	188	23	165	2.8
12.2	181	22	159	2.7
12.2	180	22	158	2.7
12.2	172	21	151	2.6
12.2	164	20	144	2.4
12.2	156	19	137	2.3
12.2	148	18	130	2.2
12.2	147	18	129	2.2
12.2	139	17	122	2.1
12.2	131	16	115	2.0
12.2	123	15	108	1.8
12.2	115	14	101	1.7
12.2	107	13	94	1.6
12.2	98	12	86	1.5
12.2	90	11	79	1.3
12.2	82	10	72	1.2
12.2	74	9	65	1.1
12.2	49	6	43	0.7
12.2	41	5	36	0.6
12.1	497	60	437	7.2
12.1	496	60	436	7.3
12.1	495	60	435	7.3
12.1	494	60	434	7.3
12.1	489	59	430	7.1
12.1	488	59	429	7.1
12.1	487	59	428	7.1
12.1	486	59	427	7.2
12.1	481	58	423	7.0
12.1	480	58	422	7.0
12.1	479	58	421	7.0
12.1	478	58	420	7.0
12.1	473	57	416	6.9
12.1	472	57	415	6.9
12.1	471	57	414	6.9
12.1	470	57	413	6.9
12.1	464	56	408	6.8
12.1	463	56	407	6.8
12.1	462	56	406	6.8
12.1	461	56	405	6.8
12.1	456	55	401	6.6
12.1	455	55	400	6.6
12.1	454	55	399	6.7
12.1	453	55	398	6.7
12.1	448	54	394	6.5
12.1	447	54	393	6.5
12.1	446	54	392	6.5
12.1	445	54	391	6.6
12.1	439	53	386	6.4
12.1	438	53	385	6.4
12.1	437	53	384	6.4
12.1	431	52	379	6.3
12.1	430	52	378	6.3
12.1	429	52	377	6.3
12.1	423	51	372	6.1
12.1	422	51	371	6.2
12.1	421	51	370	6.2
12.1	420	51	369	6.2
12.1	414	50	364	6.0
12.1	413	50	363	6.1
12.1	412	50	362	6.1
12.1	406	49	357	5.9
12.1	405	49	356	5.9
12.1	404	49	355	5.9
12.1	398	48	350	5.8
12.1	397	48	349	5.8
12.1	396	48	348	5.8
12.1	390	47	343	5.7
12.1	389	47	342	5.7
12.1	388	47	341	5.7
12.1	387	47	340	5.7
12.1	381	46	335	5.6
12.1	380	46	334	5.6
12.1	379	46	333	5.6
12.1	373	45	328	5.4
12.1	372	45	327	5.4
12.1	371	45	326	5.5
12.1	365	44	321	5.3
12.1	364	44	320	5.3
12.1	363	44	319	5.3
12.1	356	43	313	5.2
12.1	355	43	312	5.2
12.1	354	43	311	5.2
12.1	348	42	306	5.1
12.1	347	42	305	5.1
12.1	346	42	304	5.1
12.1	340	41	299	4.9
12.1	339	41	298	5.0
12.1	338	41	297	5.0
12.1	331	40	291	4.8
12.1	330	40	290	4.8
12.1	323	39	284	4.7
12.1	322	39	283	4.7
12.1	315	38	277	4.6
12.1	314	38	276	4.6
12.1	313	38	275	4.6
12.1	307	37	270	4.5
12.1	306	37	269	4.5
12.1	305	37	268	4.5
12.1	298	36	262	4.3
12.1	297	36	261	4.4
12.1	290	35	255	4.2
12.1	289	35	254	4.2
12.1	282	34	248	4.1
12.1	281	34	247	4.1
12.1	280	34	246	4.1
12.1	273	33	240	4.0
12.1	272	33	239	4.0
12.1	265	32	233	3.9
12.1	264	32	232	3.9
12.1	257	31	226	3.7
12.1	256	31	225	3.8
12.1	248	30	218	3.6
12.1	247	30	217	3.6
12.1	240	29	211	3.5
12.1	239	29	210	3.5
12.1	232	28	204	3.4
12.1	231	28	203	3.4
12.1	224	27	197	3.3
12.1	223	27	196	3.3
12.1	215	26	189	3.1
12.1	207	25	182	3.0
12.1	206	25	181	3.0
12.1	199	24	175	2.9
12.1	198	24	174	2.9
12.1	190	23	167	2.8
12.1	182	22	160	2.7
12.1	174	21	153	2.5
12.1	173	21	152	2.5
12.1	165	20	145	2.4
12.1	157	19	138	2.3
12.1	149	18	131	2.2
12.1	141	17	124	2.0
12.1	140	17	123	2.1
12.1	132	16	116	1.9
12.1	124	15	109	1.8
12.1	116	14	102	1.7
12.1	99	12	87	1.5
12.1	91	11	80	1.3
12.1	66	8	58	1.0
12.1	58	7	51	0.8
12.1	33	4	29	0.5
12.0	500	60	440	7.2
12.0	499	60	439	7.2
12.0	498	60	438	7.2
12.0	493	59	434	7.1
12.0	492	59	433	7.1
12.0	491	59	432	7.1
12.0	490	59	431	7.1
12.0	485	58	427	6.9
12.0	484	58	426	7.0
12.0	483	58	425	7.0
12.0	482	58	424	7.0
12.0	477	57	420	6.8
12.0	476	57	419	6.8
12.0	475	57	418	6.8
12.0	474	57	417	6.9
12.0	468	56	412	6.7
12.0	467	56	411	6.7
12.0	466	56	410	6.7
12.0	465	56	409	6.7
12.0	460	55	405	6.6
12.0	459	55	404	6.6
12.0	458	55	403	6.6
12.0	457	55	402	6.6
12.0	451	54	397	6.5
12.0	450	54	396	6.5
12.0	449	54	395	6.5
12.0	443	53	390	6.3
12.0	442	53	389	6.4
12.0	441	53	388	6.4
12.0	440	53	387	6.4
12.0	435	52	383	6.2
12.0	434	52	382	6.2
12.0	433	52	381	6.2
12.0	432	52	380	6.3
12.0	426	51	375	6.1
12.0	425	51	374	6.1
12.0	424	51	373	6.1
12.0	418	50	368	6.0
12.0	417	50	367	6.0
12.0	416	50	366	6.0
12.0	415	50	365	6.0
12.0	410	49	361	5.9
12.0	409	49	360	5.9
12.0	408	49	359	5.9
12.0	407	49	358	5.9
12.0	401	48	353	5.7
12.0	400	48	352	5.8
12.0	399	48	351	5.8
12.0	393	47	346	5.6
12.0	392	47	345	5.6
12.0	391	47	344	5.6
12.0	384	46	338	5.5
12.0	383	46	337	5.5
12.0	382	46	336	5.5
12.0	376	45	331	5.4
12.0	375	45	330	5.4
12.0	374	45	329	5.4
12.0	368	44	324	5.3
12.0	367	44	323	5.3
12.0	366	44	322	5.3
12.0	359	43	316	5.2
12.0	358	43	315	5.2
12.0	357	43	314	5.2
12.0	351	42	309	5.0
12.0	350	42	308	5.0
12.0	349	42	307	5.1
12.0	343	41	302	4.9
12.0	342	41	301	4.9
12.0	341	41	300	4.9
12.0	334	40	294	4.8
12.0	333	40	293	4.8
12.0	332	40	292	4.8
12.0	326	39	287	4.7
12.0	325	39	286	4.7
12.0	324	39	285	4.7
12.0	318	38	280	4.5
12.0	317	38	279	4.6
12.0	316	38	278	4.6
12.0	309	37	272	4.4
12.0	308	37	271	4.4
12.0	301	36	265	4.3
12.0	300	36	264	4.3
12.0	299	36	263	4.3
12.0	292	35	257	4.2
12.0	291	35	256	4.2
12.0	284	34	250	4.1
12.0	283	34	249	4.1
12.0	276	33	243	3.9
12.0	275	33	242	4.0
12.0	274	33	241	4.0
12.0	267	32	235	3.8
12.0	266	32	234	3.8
12.0	259	31	228	3.7
12.0	258	31	227	3.7

%E	M1	M2	DM	M*
12.0	251	30	221	3.6
12.0	250	30	220	3.6
12.0	249	30	219	3.6
12.0	242	29	213	3.5
12.0	241	29	212	3.5
12.0	234	28	206	3.4
12.0	233	28	205	3.4
12.0	225	27	198	3.2
12.0	217	26	191	3.1
12.0	216	26	190	3.1
12.0	209	25	184	3.0
12.0	208	25	183	3.0
12.0	200	24	176	2.9
12.0	192	23	169	2.8
12.0	191	23	168	2.8
12.0	184	22	162	2.6
12.0	183	22	161	2.6
12.0	175	21	154	2.5
12.0	167	20	147	2.4
12.0	166	20	146	2.4
12.0	159	19	140	2.3
12.0	158	19	139	2.3
12.0	150	18	132	2.2
12.0	142	17	125	2.0
12.0	133	16	117	1.9
12.0	125	15	110	1.8
12.0	117	14	103	1.7
12.0	108	13	95	1.6
12.0	100	12	88	1.4
12.0	92	11	81	1.3
12.0	83	10	73	1.2
12.0	75	9	66	1.1
12.0	50	6	44	0.7
12.0	25	3	22	0.4
11.9	497	59	438	7.0
11.9	496	59	437	7.0
11.9	495	59	436	7.0
11.9	494	59	435	7.0
11.9	489	58	431	6.9
11.9	488	58	430	6.9
11.9	487	58	429	6.9
11.9	486	58	428	6.9
11.9	481	57	424	6.8
11.9	480	57	423	6.8
11.9	479	57	422	6.8
11.9	478	57	421	6.8
11.9	472	56	416	6.6
11.9	471	56	415	6.7
11.9	470	56	414	6.7
11.9	469	56	413	6.7

%E	M1	M2	DM	M*
11.9	464	55	409	6.5
11.9	463	55	408	6.5
11.9	462	55	407	6.5
11.9	461	55	406	6.6
11.9	455	54	401	6.4
11.9	454	54	400	6.4
11.9	453	54	399	6.4
11.9	452	54	398	6.5
11.9	447	53	394	6.3
11.9	446	53	393	6.3
11.9	445	53	392	6.3
11.9	444	53	391	6.3
11.9	438	52	386	6.2
11.9	437	52	385	6.2
11.9	436	52	384	6.2
11.9	430	51	379	6.0
11.9	429	51	378	6.1
11.9	428	51	377	6.1
11.9	427	51	376	6.1
11.9	421	50	371	5.9
11.9	420	50	370	6.0
11.9	419	50	369	6.0
11.9	413	49	364	5.8
11.9	412	49	363	5.8
11.9	411	49	362	5.8
11.9	405	48	357	5.7
11.9	404	48	356	5.7
11.9	403	48	355	5.7
11.9	402	48	354	5.7
11.9	396	47	349	5.6
11.9	395	47	348	5.6
11.9	394	47	347	5.6
11.9	388	46	342	5.5
11.9	387	46	341	5.5
11.9	386	46	340	5.5
11.9	385	46	339	5.5
11.9	379	45	334	5.3
11.9	378	45	333	5.4
11.9	377	45	332	5.4
11.9	371	44	327	5.2
11.9	370	44	326	5.2
11.9	369	44	325	5.2
11.9	362	43	319	5.1
11.9	361	43	318	5.1
11.9	360	43	317	5.1
11.9	354	42	312	5.0
11.9	353	42	311	5.0
11.9	352	42	310	5.0
11.9	346	41	305	4.9
11.9	345	41	304	4.9

%E	M1	M2	DM	M*
11.9	344	41	303	4.9
11.9	337	40	297	4.7
11.9	336	40	296	4.8
11.9	335	40	295	4.8
11.9	329	39	290	4.6
11.9	328	39	289	4.6
11.9	327	39	288	4.7
11.9	320	38	282	4.5
11.9	319	38	281	4.5
11.9	312	37	275	4.4
11.9	311	37	274	4.4
11.9	310	37	273	4.4
11.9	303	36	267	4.3
11.9	302	36	266	4.3
11.9	295	35	260	4.2
11.9	294	35	259	4.2
11.9	293	35	258	4.2
11.9	286	34	252	4.0
11.9	285	34	251	4.1
11.9	278	33	245	3.9
11.9	277	33	244	3.9
11.9	270	32	238	3.8
11.9	269	32	237	3.8
11.9	268	32	236	3.8
11.9	261	31	230	3.7
11.9	260	31	229	3.7
11.9	253	30	223	3.6
11.9	252	30	222	3.6
11.9	244	29	215	3.4
11.9	243	29	214	3.5
11.9	236	28	208	3.3
11.9	235	28	207	3.3
11.9	227	27	200	3.2
11.9	226	27	199	3.2
11.9	219	26	193	3.1
11.9	218	26	192	3.1
11.9	210	25	185	3.0
11.9	202	24	178	2.9
11.9	201	24	177	2.9
11.9	194	23	171	2.7
11.9	193	23	170	2.7
11.9	185	22	163	2.6
11.9	177	21	156	2.5
11.9	176	21	155	2.5
11.9	168	20	148	2.4
11.9	160	19	141	2.3
11.9	151	18	133	2.1
11.9	143	17	126	2.0
11.9	135	16	119	1.9
11.9	134	16	118	1.9

%E	M1	M2	DM	M*
11.9	126	15	111	1.8
11.9	118	14	104	1.7
11.9	109	13	96	1.6
11.9	101	12	89	1.4
11.9	84	10	74	1.2
11.9	67	8	59	1.0
11.9	59	7	52	0.8
11.9	42	5	37	0.6
11.8	500	59	441	7.0
11.8	499	59	440	7.0
11.8	498	59	439	7.0
11.8	493	58	435	6.8
11.8	492	58	434	6.8
11.8	491	58	433	6.9
11.8	490	58	432	6.9
11.8	485	57	428	6.7
11.8	484	57	427	6.7
11.8	483	57	426	6.7
11.8	482	57	425	6.7
11.8	476	56	420	6.6
11.8	475	56	419	6.6
11.8	474	56	418	6.6
11.8	473	56	417	6.6
11.8	468	55	413	6.5
11.8	467	55	412	6.5
11.8	466	55	411	6.5
11.8	465	55	410	6.5
11.8	459	54	405	6.4
11.8	458	54	404	6.4
11.8	457	54	403	6.4
11.8	456	54	402	6.4
11.8	451	53	398	6.2
11.8	450	53	397	6.2
11.8	449	53	396	6.3
11.8	448	53	395	6.3
11.8	442	52	390	6.1
11.8	441	52	389	6.1
11.8	440	52	388	6.1
11.8	439	52	387	6.2
11.8	434	51	383	6.0
11.8	433	51	382	6.0
11.8	432	51	381	6.0
11.8	431	51	380	6.0
11.8	425	50	375	5.9
11.8	424	50	374	5.9
11.8	423	50	373	5.9
11.8	422	50	372	5.9
11.8	417	49	368	5.8
11.8	416	49	367	5.8
11.8	415	49	366	5.8

%E	M1	M2	DM	M*
11.8	414	49	365	5.8
11.8	408	48	360	5.6
11.8	407	48	359	5.7
11.8	406	48	358	5.7
11.8	400	47	353	5.5
11.8	399	47	352	5.5
11.8	398	47	351	5.6
11.8	397	47	350	5.6
11.8	391	46	345	5.4
11.8	390	46	344	5.4
11.8	389	46	343	5.4
11.8	382	45	337	5.3
11.8	381	45	336	5.3
11.8	380	45	335	5.3
11.8	374	44	330	5.2
11.8	373	44	329	5.2
11.8	372	44	328	5.2
11.8	365	43	322	5.1
11.8	364	43	321	5.1
11.8	363	43	320	5.1
11.8	357	42	315	4.9
11.8	356	42	314	5.0
11.8	355	42	313	5.0
11.8	348	41	307	4.8
11.8	347	41	306	4.8
11.8	340	40	300	4.7
11.8	339	40	299	4.7
11.8	338	40	298	4.7
11.8	331	39	292	4.6
11.8	330	39	291	4.6
11.8	323	38	285	4.5
11.8	322	38	284	4.5
11.8	321	38	283	4.5
11.8	314	37	277	4.4
11.8	313	37	276	4.4
11.8	306	36	270	4.2
11.8	305	36	269	4.2
11.8	304	36	268	4.3
11.8	297	35	262	4.1
11.8	296	35	261	4.1
11.8	289	34	255	4.0
11.8	288	34	254	4.0
11.8	287	34	253	4.0
11.8	280	33	247	3.9
11.8	279	33	246	3.9
11.8	272	32	240	3.8
11.8	271	32	239	3.8
11.8	263	31	232	3.7
11.8	262	31	231	3.7
11.8	255	30	225	3.5

%E	M1	M2	DM	M*
11.8	254	30	224	3.5
11.8	246	29	217	3.4
11.8	245	29	216	3.4
11.8	238	28	210	3.3
11.8	237	28	209	3.3
11.8	229	27	202	3.2
11.8	228	27	201	3.2
11.8	221	26	195	3.1
11.8	220	26	194	3.1
11.8	212	25	187	2.9
11.8	211	25	186	3.0
11.8	204	24	180	2.8
11.8	203	24	179	2.8
11.8	195	23	172	2.7
11.8	187	22	165	2.6
11.8	186	22	164	2.6
11.8	178	21	157	2.5
11.8	170	20	150	2.4
11.8	169	20	149	2.4
11.8	161	19	142	2.2
11.8	153	18	135	2.1
11.8	152	18	134	2.1
11.8	144	17	127	2.0
11.8	136	16	120	1.9
11.8	127	15	112	1.8
11.8	119	14	105	1.6
11.8	110	13	97	1.5
11.8	102	12	90	1.4
11.8	93	11	82	1.3
11.8	85	10	75	1.2
11.8	76	9	67	1.1
11.8	68	8	60	0.9
11.8	51	6	45	0.7
11.8	34	4	30	0.5
11.8	17	2	15	0.2
11.7	497	58	439	6.8
11.7	496	58	438	6.8
11.7	495	58	437	6.8
11.7	494	58	436	6.8
11.7	489	57	432	6.6
11.7	488	57	431	6.7
11.7	487	57	430	6.7
11.7	486	57	429	6.7
11.7	480	56	424	6.5
11.7	479	56	423	6.5
11.7	478	56	422	6.6
11.7	477	56	421	6.6
11.7	472	55	417	6.4
11.7	471	55	416	6.4
11.7	470	55	415	6.4

%E	M1	M2	DM	M*
11.7	469	55	414	6.4
11.7	463	54	409	6.3
11.7	462	54	408	6.3
11.7	461	54	407	6.3
11.7	460	54	406	6.3
11.7	454	53	401	6.2
11.7	453	53	400	6.2
11.7	452	53	399	6.2
11.7	446	52	394	6.1
11.7	445	52	393	6.1
11.7	444	52	392	6.1
11.7	443	52	391	6.1
11.7	437	51	386	6.0
11.7	436	51	385	6.0
11.7	435	51	384	6.0
11.7	429	50	379	5.8
11.7	428	50	378	5.8
11.7	427	50	377	5.9
11.7	426	50	376	5.9
11.7	420	49	371	5.7
11.7	419	49	370	5.7
11.7	418	49	369	5.7
11.7	412	48	364	5.6
11.7	411	48	363	5.6
11.7	410	48	362	5.6
11.7	409	48	361	5.6
11.7	403	47	356	5.5
11.7	402	47	355	5.5
11.7	401	47	354	5.5
11.7	394	46	348	5.4
11.7	393	46	347	5.4
11.7	392	46	346	5.4
11.7	386	45	341	5.2
11.7	385	45	340	5.3
11.7	384	45	339	5.3
11.7	383	45	338	5.3
11.7	377	44	333	5.1
11.7	376	44	332	5.1
11.7	375	44	331	5.2
11.7	369	43	326	5.0
11.7	368	43	325	5.0
11.7	367	43	324	5.0
11.7	366	43	323	5.1
11.7	360	42	318	4.9
11.7	359	42	317	4.9
11.7	358	42	316	4.9
11.7	351	41	310	4.8
11.7	350	41	309	4.8
11.7	349	41	308	4.8
11.7	343	40	303	4.7

%E	M1	M2	DM	M*
11.7	342	40	302	4.7
11.7	341	40	301	4.7
11.7	334	39	295	4.6
11.7	333	39	294	4.6
11.7	332	39	293	4.6
11.7	326	38	288	4.4
11.7	325	38	287	4.4
11.7	324	38	286	4.5
11.7	317	37	280	4.3
11.7	316	37	279	4.3
11.7	315	37	278	4.3
11.7	309	36	273	4.2
11.7	308	36	272	4.2
11.7	307	36	271	4.2
11.7	300	35	265	4.1
11.7	299	35	264	4.1
11.7	298	35	263	4.1
11.7	291	34	257	4.0
11.7	290	34	256	4.0
11.7	283	33	250	3.8
11.7	282	33	249	3.9
11.7	281	33	248	3.9
11.7	274	32	242	3.7
11.7	273	32	241	3.8
11.7	266	31	235	3.6
11.7	265	31	234	3.6
11.7	264	31	233	3.6
11.7	257	30	227	3.5
11.7	256	30	226	3.5
11.7	248	29	219	3.4
11.7	247	29	218	3.4
11.7	240	28	212	3.3
11.7	239	28	211	3.3
11.7	231	27	204	3.2
11.7	230	27	203	3.2
11.7	223	26	197	3.0
11.7	222	26	196	3.0
11.7	214	25	189	2.9
11.7	213	25	188	2.9
11.7	206	24	182	2.8
11.7	205	24	181	2.8
11.7	197	23	174	2.7
11.7	196	23	173	2.7
11.7	188	22	166	2.6
11.7	180	21	159	2.4
11.7	179	21	158	2.5
11.7	171	20	151	2.3
11.7	163	19	144	2.2
11.7	162	19	143	2.2
11.7	154	18	136	2.1

%E	M1	M2	DM	M*
11.7	145	17	128	2.0
11.7	137	16	121	1.9
11.7	128	15	113	1.8
11.7	120	14	106	1.6
11.7	111	13	98	1.5
11.7	103	12	91	1.4
11.7	94	11	83	1.3
11.7	77	9	68	1.1
11.7	60	7	53	0.8
11.6	500	58	442	6.7
11.6	499	58	441	6.7
11.6	498	58	440	6.8
11.6	493	57	436	6.6
11.6	492	57	435	6.6
11.6	491	57	434	6.6
11.6	490	57	433	6.6
11.6	484	56	428	6.5
11.6	483	56	427	6.5
11.6	482	56	426	6.5
11.6	481	56	425	6.5
11.6	476	55	421	6.4
11.6	475	55	420	6.4
11.6	474	55	419	6.4
11.6	473	55	418	6.4
11.6	467	54	413	6.2
11.6	466	54	412	6.3
11.6	465	54	411	6.3
11.6	464	54	410	6.3
11.6	458	53	405	6.1
11.6	457	53	404	6.1
11.6	456	53	403	6.2
11.6	455	53	402	6.2
11.6	450	52	398	6.0
11.6	449	52	397	6.0
11.6	448	52	396	6.0
11.6	447	52	395	6.0
11.6	441	51	390	5.9
11.6	440	51	389	5.9
11.6	439	51	388	5.9
11.6	438	51	387	5.9
11.6	432	50	382	5.8
11.6	431	50	381	5.8
11.6	430	50	380	5.8
11.6	424	49	375	5.7
11.6	423	49	374	5.7
11.6	422	49	373	5.7
11.6	421	49	372	5.7
11.6	415	48	367	5.6
11.6	414	48	366	5.6
11.6	413	48	365	5.6

%E	M1	M2	DM	M*
11.6	406	47	359	5.4
11.6	405	47	358	5.5
11.6	404	47	357	5.5
11.6	398	46	352	5.3
11.6	397	46	351	5.3
11.6	396	46	350	5.3
11.6	395	46	349	5.4
11.6	389	45	344	5.2
11.6	388	45	343	5.2
11.6	387	45	342	5.2
11.6	380	44	336	5.1
11.6	379	44	335	5.1
11.6	378	44	334	5.1
11.6	372	43	329	5.0
11.6	371	43	328	5.0
11.6	370	43	327	5.0
11.6	363	42	321	4.9
11.6	362	42	320	4.9
11.6	361	42	319	4.9
11.6	354	41	313	4.7
11.6	353	41	312	4.8
11.6	352	41	311	4.8
11.6	346	40	306	4.6
11.6	345	40	305	4.6
11.6	344	40	304	4.7
11.6	337	39	298	4.5
11.6	336	39	297	4.5
11.6	335	39	296	4.5
11.6	329	38	291	4.4
11.6	328	38	290	4.4
11.6	327	38	289	4.4
11.6	320	37	283	4.3
11.6	319	37	282	4.3
11.6	318	37	281	4.3
11.6	311	36	275	4.2
11.6	310	36	274	4.2
11.6	303	35	268	4.0
11.6	302	35	267	4.1
11.6	301	35	266	4.1
11.6	294	34	260	3.9
11.6	293	34	259	3.9
11.6	292	34	258	4.0
11.6	285	33	252	3.8
11.6	284	33	251	3.8
11.6	277	32	245	3.7
11.6	276	32	244	3.7
11.6	275	32	243	3.7
11.6	268	31	237	3.6
11.6	267	31	236	3.6
11.6	259	30	229	3.5

%E	M1	M2	DM	M*
11.6	258	30	228	3.5
11.6	251	29	222	3.4
11.6	250	29	221	3.4
11.6	249	29	220	3.4
11.6	242	28	214	3.2
11.6	241	28	213	3.3
11.6	233	27	206	3.1
11.6	232	27	205	3.1
11.6	225	26	199	3.0
11.6	224	26	198	3.0
11.6	216	25	191	2.9
11.6	215	25	190	2.9
11.6	207	24	183	2.8
11.6	199	23	176	2.7
11.6	198	23	175	2.7
11.6	190	22	168	2.5
11.6	189	22	167	2.6
11.6	181	21	160	2.4
11.6	173	20	153	2.3
11.6	172	20	152	2.3
11.6	164	19	145	2.2
11.6	155	18	137	2.1
11.6	147	17	130	2.0
11.6	146	17	129	2.0
11.6	138	16	122	1.9
11.6	129	15	114	1.7
11.6	121	14	107	1.6
11.6	112	13	99	1.5
11.6	95	11	84	1.3
11.6	86	10	76	1.2
11.6	69	8	61	0.9
11.6	43	5	38	0.6
11.5	497	57	440	6.5
11.5	496	57	439	6.6
11.5	495	57	438	6.6
11.5	494	57	437	6.6
11.5	489	56	433	6.4
11.5	488	56	432	6.4
11.5	487	56	431	6.4
11.5	486	56	430	6.5
11.5	485	56	429	6.5
11.5	480	55	425	6.3
11.5	479	55	424	6.3
11.5	478	55	423	6.3
11.5	477	55	422	6.3
11.5	471	54	417	6.2
11.5	470	54	416	6.2
11.5	469	54	415	6.2
11.5	468	54	414	6.2
11.5	462	53	409	6.1
11.5	461	53	408	6.1
11.5	460	53	407	6.1
11.5	459	53	406	6.1
11.5	454	52	402	6.0
11.5	453	52	401	6.0
11.5	452	52	400	6.0
11.5	451	52	399	6.0
11.5	445	51	394	5.8
11.5	444	51	393	5.9
11.5	443	51	392	5.9
11.5	442	51	391	5.9
11.5	436	50	386	5.7
11.5	435	50	385	5.7
11.5	434	50	384	5.8
11.5	433	50	383	5.8
11.5	427	49	378	5.6
11.5	426	49	377	5.6
11.5	425	49	376	5.6
11.5	419	48	371	5.5
11.5	418	48	370	5.5
11.5	417	48	369	5.5
11.5	416	48	368	5.5
11.5	410	47	363	5.4
11.5	409	47	362	5.4
11.5	408	47	361	5.4
11.5	407	47	360	5.4
11.5	401	46	355	5.3
11.5	400	46	354	5.3
11.5	399	46	353	5.3
11.5	393	45	348	5.2
11.5	392	45	347	5.2
11.5	391	45	346	5.2
11.5	390	45	345	5.2
11.5	384	44	340	5.0
11.5	383	44	339	5.1
11.5	382	44	338	5.1
11.5	381	44	337	5.1
11.5	375	43	332	4.9
11.5	374	43	331	4.9
11.5	373	43	330	5.0
11.5	366	42	324	4.8
11.5	365	42	323	4.8
11.5	364	42	322	4.8
11.5	358	41	317	4.7
11.5	357	41	316	4.7
11.5	356	41	315	4.7
11.5	355	41	314	4.7
11.5	349	40	309	4.6
11.5	348	40	308	4.6
11.5	347	40	307	4.6
11.5	340	39	301	4.5
11.5	339	39	300	4.5
11.5	338	39	299	4.5
11.5	331	38	293	4.4
11.5	330	38	292	4.4
11.5	323	37	286	4.2
11.5	322	37	285	4.3
11.5	321	37	284	4.3
11.5	314	36	278	4.1
11.5	313	36	277	4.1
11.5	312	36	276	4.2
11.5	305	35	270	4.0
11.5	304	35	269	4.0
11.5	296	34	262	3.9
11.5	295	34	261	3.9
11.5	288	33	255	3.8
11.5	287	33	254	3.8
11.5	286	33	253	3.8
11.5	279	32	247	3.7
11.5	278	32	246	3.7
11.5	270	31	239	3.6
11.5	269	31	238	3.6
11.5	262	30	232	3.4
11.5	261	30	231	3.4
11.5	260	30	230	3.5
11.5	253	29	224	3.3
11.5	252	29	223	3.3
11.5	244	28	216	3.2
11.5	243	28	215	3.2
11.5	235	27	208	3.1
11.5	234	27	207	3.1
11.5	227	26	201	3.0
11.5	226	26	200	3.0
11.5	218	25	193	2.9
11.5	217	25	192	2.9
11.5	209	24	185	2.8
11.5	208	24	184	2.8
11.5	200	23	177	2.6
11.5	192	22	170	2.5
11.5	191	22	169	2.5
11.5	183	21	162	2.4
11.5	182	21	161	2.4
11.5	174	20	154	2.3
11.5	165	19	146	2.2
11.5	157	18	139	2.1
11.5	156	18	138	2.1
11.5	148	17	131	2.0
11.5	139	16	123	1.8
11.5	131	15	116	1.7
11.5	130	15	115	1.7
11.5	122	14	108	1.6
11.5	113	13	100	1.5
11.5	104	12	92	1.4
11.5	96	11	85	1.3
11.5	87	10	77	1.1
11.5	78	9	69	1.0
11.5	61	7	54	0.8
11.5	52	6	46	0.7
11.5	26	3	23	0.3
11.4	500	57	443	6.5
11.4	499	57	442	6.5
11.4	498	57	441	6.5
11.4	493	56	437	6.4
11.4	492	56	436	6.4
11.4	491	56	435	6.4
11.4	490	56	434	6.4
11.4	484	55	429	6.3
11.4	483	55	428	6.3
11.4	482	55	427	6.3
11.4	481	55	426	6.3
11.4	475	54	421	6.1
11.4	474	54	420	6.2
11.4	473	54	419	6.2
11.4	472	54	418	6.2
11.4	466	53	413	6.0
11.4	465	53	412	6.0
11.4	464	53	411	6.1
11.4	463	53	410	6.1
11.4	458	52	406	5.9
11.4	457	52	405	5.9
11.4	456	52	404	5.9
11.4	455	52	403	5.9
11.4	449	51	398	5.8
11.4	448	51	397	5.8
11.4	447	51	396	5.8
11.4	446	51	395	5.8
11.4	440	50	390	5.7
11.4	439	50	389	5.7
11.4	438	50	388	5.7
11.4	437	50	387	5.7
11.4	431	49	382	5.6
11.4	430	49	381	5.6
11.4	429	49	380	5.6
11.4	428	49	379	5.6
11.4	422	48	374	5.5
11.4	421	48	373	5.5
11.4	420	48	372	5.5
11.4	414	47	367	5.3
11.4	413	47	366	5.3
11.4	412	47	365	5.4
11.4	411	47	364	5.4
11.4	405	46	359	5.2
11.4	404	46	358	5.2
11.4	403	46	357	5.3
11.4	402	46	356	5.3
11.4	396	45	351	5.1
11.4	395	45	350	5.1
11.4	394	45	349	5.1
11.4	387	44	343	5.0
11.4	386	44	342	5.0
11.4	385	44	341	5.0
11.4	378	43	335	4.9
11.4	377	43	334	4.9
11.4	376	43	333	4.9
11.4	370	42	328	4.8
11.4	369	42	327	4.8
11.4	368	42	326	4.8
11.4	367	42	325	4.8
11.4	361	41	320	4.7
11.4	360	41	319	4.7
11.4	359	41	318	4.7
11.4	352	40	312	4.5
11.4	351	40	311	4.6
11.4	350	40	310	4.6
11.4	343	39	304	4.4
11.4	342	39	303	4.4
11.4	341	39	302	4.5
11.4	334	38	296	4.3
11.4	333	38	295	4.3
11.4	332	38	294	4.3
11.4	326	37	289	4.2
11.4	325	37	288	4.2
11.4	324	37	287	4.2
11.4	317	36	281	4.1
11.4	316	36	280	4.1
11.4	315	36	279	4.1
11.4	308	35	273	4.0
11.4	307	35	272	4.0
11.4	306	35	271	4.0
11.4	299	34	265	3.9
11.4	298	34	264	3.9
11.4	297	34	263	3.9
11.4	290	33	257	3.8
11.4	289	33	256	3.8
11.4	281	32	249	3.6
11.4	280	32	248	3.7
11.4	273	31	242	3.5
11.4	272	31	241	3.5
11.4	271	31	240	3.5
11.4	264	30	234	3.4

%E	M1	M2	DM	M*
11.4	263	30	233	3.4
11.4	255	29	226	3.3
11.4	254	29	225	3.3
11.4	246	28	218	3.2
11.4	245	28	217	3.2
11.4	237	27	210	3.1
11.4	236	27	209	3.1
11.4	229	26	203	3.0
11.4	228	26	202	3.0
11.4	220	25	195	2.8
11.4	219	25	194	2.9
11.4	211	24	187	2.7
11.4	210	24	186	2.7
11.4	202	23	179	2.6
11.4	201	23	178	2.6
11.4	193	22	171	2.5
11.4	185	21	164	2.4
11.4	184	21	163	2.4
11.4	176	20	156	2.3
11.4	175	20	155	2.3
11.4	167	19	148	2.2
11.4	166	19	147	2.2
11.4	158	18	140	2.1
11.4	149	17	132	1.9
11.4	140	16	124	1.8
11.4	132	15	117	1.7
11.4	123	14	109	1.6
11.4	114	13	101	1.5
11.4	105	12	93	1.4
11.4	88	10	78	1.1
11.4	79	9	70	1.0
11.4	70	8	62	0.9
11.4	44	5	39	0.6
11.4	35	4	31	0.5
11.3	497	56	441	6.3
11.3	496	56	440	6.3
11.3	495	56	439	6.3
11.3	494	56	438	6.3
11.3	488	55	433	6.2
11.3	487	55	432	6.2
11.3	486	55	431	6.2
11.3	485	55	430	6.2
11.3	480	54	426	6.1
11.3	479	54	425	6.1
11.3	478	54	424	6.1
11.3	477	54	423	6.1
11.3	476	54	422	6.1
11.3	471	53	418	6.0
11.3	470	53	417	6.0
11.3	469	53	416	6.0
11.3	468	53	415	6.0
11.3	467	53	414	6.0
11.3	462	52	410	5.9
11.3	461	52	409	5.9
11.3	460	52	408	5.9
11.3	459	52	407	5.9
11.3	453	51	402	5.7
11.3	452	51	401	5.8
11.3	451	51	400	5.8
11.3	450	51	399	5.8
11.3	444	50	394	5.6
11.3	443	50	393	5.6
11.3	442	50	392	5.7
11.3	441	50	391	5.7
11.3	435	49	386	5.5
11.3	434	49	385	5.5
11.3	433	49	384	5.5
11.3	432	49	383	5.6
11.3	426	48	378	5.4
11.3	425	48	377	5.4
11.3	424	48	376	5.4
11.3	423	48	375	5.4
11.3	417	47	370	5.3
11.3	416	47	369	5.3
11.3	415	47	368	5.3
11.3	408	46	362	5.2
11.3	407	46	361	5.2
11.3	406	46	360	5.2
11.3	400	45	355	5.1
11.3	399	45	354	5.1
11.3	398	45	353	5.1
11.3	397	45	352	5.1
11.3	391	44	347	5.0
11.3	390	44	346	5.0
11.3	389	44	345	5.0
11.3	388	44	344	5.0
11.3	382	43	339	4.8
11.3	381	43	338	4.9
11.3	380	43	337	4.9
11.3	379	43	336	4.9
11.3	373	42	331	4.7
11.3	372	42	330	4.7
11.3	371	42	329	4.8
11.3	364	41	323	4.6
11.3	363	41	322	4.6
11.3	362	41	321	4.6
11.3	355	40	315	4.5
11.3	354	40	314	4.5
11.3	353	40	313	4.5
11.3	346	39	307	4.4
11.3	345	39	306	4.4
11.3	344	39	305	4.4
11.3	337	38	299	4.3
11.3	336	38	298	4.3
11.3	335	38	297	4.3
11.3	328	37	291	4.2
11.3	327	37	290	4.2
11.3	320	36	284	4.0
11.3	319	36	283	4.1
11.3	318	36	282	4.1
11.3	311	35	276	3.9
11.3	310	35	275	4.0
11.3	309	35	274	4.0
11.3	302	34	268	3.8
11.3	301	34	267	3.8
11.3	300	34	266	3.9
11.3	293	33	260	3.7
11.3	292	33	259	3.7
11.3	291	33	258	3.7
11.3	284	32	252	3.6
11.3	283	32	251	3.6
11.3	282	32	250	3.6
11.3	275	31	244	3.5
11.3	274	31	243	3.5
11.3	266	30	236	3.4
11.3	265	30	235	3.4
11.3	257	29	228	3.3
11.3	256	29	227	3.3
11.3	248	28	220	3.2
11.3	247	28	219	3.2
11.3	240	27	213	3.0
11.3	239	27	212	3.1
11.3	238	27	211	3.1
11.3	231	26	205	2.9
11.3	230	26	204	2.9
11.3	222	25	197	2.8
11.3	221	25	196	2.8
11.3	213	24	189	2.7
11.3	212	24	188	2.7
11.3	204	23	181	2.6
11.3	203	23	180	2.6
11.3	195	22	173	2.5
11.3	194	22	172	2.5
11.3	186	21	165	2.4
11.3	177	20	157	2.3
11.3	168	19	149	2.1
11.3	160	18	142	2.0
11.3	159	18	141	2.0
11.3	151	17	134	1.9
11.3	150	17	133	1.9
11.3	142	16	126	1.8
11.3	141	16	125	1.8
11.3	133	15	118	1.7
11.3	124	14	110	1.6
11.3	115	13	102	1.5
11.3	106	12	94	1.4
11.3	97	11	86	1.2
11.3	80	9	71	1.0
11.3	71	8	63	0.9
11.3	62	7	55	0.8
11.3	53	6	47	0.7
11.2	500	56	444	6.3
11.2	499	56	443	6.3
11.2	498	56	442	6.3
11.2	493	55	438	6.1
11.2	492	55	437	6.1
11.2	491	55	436	6.2
11.2	490	55	435	6.2
11.2	489	55	434	6.2
11.2	484	54	430	6.0
11.2	483	54	429	6.0
11.2	482	54	428	6.0
11.2	481	54	427	6.1
11.2	475	53	422	5.9
11.2	474	53	421	5.9
11.2	473	53	420	5.9
11.2	472	53	419	6.0
11.2	466	52	414	5.8
11.2	465	52	413	5.8
11.2	464	52	412	5.8
11.2	463	52	411	5.8
11.2	457	51	406	5.7
11.2	456	51	405	5.7
11.2	455	51	404	5.7
11.2	454	51	403	5.7
11.2	448	50	398	5.6
11.2	447	50	397	5.6
11.2	446	50	396	5.6
11.2	445	50	395	5.6
11.2	439	49	390	5.5
11.2	438	49	389	5.5
11.2	437	49	388	5.5
11.2	436	49	387	5.5
11.2	430	48	382	5.4
11.2	429	48	381	5.4
11.2	428	48	380	5.4
11.2	427	48	379	5.4
11.2	421	47	374	5.2
11.2	420	47	373	5.3
11.2	419	47	372	5.3
11.2	418	47	371	5.3
11.2	412	46	366	5.1
11.2	411	46	365	5.1
11.2	410	46	364	5.2
11.2	409	46	363	5.2
11.2	403	45	358	5.0
11.2	402	45	357	5.0
11.2	401	45	356	5.0
11.2	394	44	350	4.9
11.2	393	44	349	4.9
11.2	392	44	348	4.9
11.2	385	43	342	4.8
11.2	384	43	341	4.8
11.2	383	43	340	4.8
11.2	376	42	334	4.7
11.2	375	42	333	4.7
11.2	374	42	332	4.7
11.2	367	41	326	4.6
11.2	366	41	325	4.6
11.2	365	41	324	4.6
11.2	358	40	318	4.5
11.2	357	40	317	4.5
11.2	356	40	316	4.5
11.2	349	39	310	4.4
11.2	348	39	309	4.4
11.2	347	39	308	4.4
11.2	340	38	302	4.2
11.2	339	38	301	4.3
11.2	338	38	300	4.3
11.2	331	37	294	4.1
11.2	330	37	293	4.1
11.2	329	37	292	4.2
11.2	322	36	286	4.0
11.2	321	36	285	4.0
11.2	313	35	278	3.9
11.2	312	35	277	3.9
11.2	304	34	270	3.8
11.2	303	34	269	3.8
11.2	295	33	262	3.7
11.2	294	33	261	3.7
11.2	287	32	255	3.6
11.2	286	32	254	3.6
11.2	285	32	253	3.6
11.2	278	31	247	3.5
11.2	277	31	246	3.5
11.2	276	31	245	3.5
11.2	269	30	239	3.3
11.2	268	30	238	3.4
11.2	267	30	237	3.4
11.2	260	29	231	3.2

%E	M1	M2	DM	M*
11.2	259	29	230	3.2
11.2	258	29	229	3.3
11.2	251	28	223	3.1
11.2	250	28	222	3.1
11.2	249	28	221	3.1
11.2	242	27	215	3.0
11.2	241	27	214	3.0
11.2	233	26	207	2.9
11.2	232	26	206	2.9
11.2	224	25	199	2.8
11.2	223	25	198	2.8
11.2	215	24	191	2.7
11.2	214	24	190	2.7
11.2	206	23	183	2.6
11.2	205	23	182	2.6
11.2	197	22	175	2.5
11.2	196	22	174	2.5
11.2	188	21	167	2.3
11.2	187	21	166	2.4
11.2	179	20	159	2.2
11.2	178	20	158	2.2
11.2	170	19	151	2.1
11.2	169	19	150	2.1
11.2	161	18	143	2.0
11.2	152	17	135	1.9
11.2	143	16	127	1.8
11.2	134	15	119	1.7
11.2	125	14	111	1.6
11.2	116	13	103	1.5
11.2	107	12	95	1.3
11.2	98	11	87	1.2
11.2	89	10	79	1.1
11.1	497	55	442	6.1
11.1	496	55	441	6.1
11.1	495	55	440	6.1
11.1	494	55	439	6.1
11.1	488	54	434	6.0
11.1	487	54	433	6.0
11.1	486	54	432	6.0
11.1	485	54	431	6.0
11.1	479	53	426	5.9
11.1	478	53	425	5.9
11.1	477	53	424	5.9
11.1	476	53	423	5.9
11.1	470	52	418	5.8
11.1	469	52	417	5.8
11.1	468	52	416	5.8
11.1	467	52	415	5.8
11.1	461	51	410	5.6
11.1	460	51	409	5.7

%E	M1	M2	DM	M*
11.1	459	51	408	5.7
11.1	458	51	407	5.7
11.1	452	50	402	5.5
11.1	451	50	401	5.5
11.1	450	50	400	5.6
11.1	449	50	399	5.6
11.1	443	49	394	5.4
11.1	442	49	393	5.4
11.1	441	49	392	5.4
11.1	440	49	391	5.5
11.1	434	48	386	5.3
11.1	433	48	385	5.3
11.1	432	48	384	5.3
11.1	431	48	383	5.3
11.1	425	47	378	5.2
11.1	424	47	377	5.2
11.1	423	47	376	5.2
11.1	422	47	375	5.2
11.1	416	46	370	5.1
11.1	415	46	369	5.1
11.1	414	46	368	5.1
11.1	413	46	367	5.1
11.1	407	45	362	5.0
11.1	406	45	361	5.0
11.1	405	45	360	5.0
11.1	404	45	359	5.0
11.1	398	44	354	4.9
11.1	397	44	353	4.9
11.1	396	44	352	4.9
11.1	395	44	351	4.9
11.1	389	43	346	4.8
11.1	388	43	345	4.8
11.1	387	43	344	4.8
11.1	386	43	343	4.8
11.1	380	42	338	4.6
11.1	379	42	337	4.7
11.1	378	42	336	4.7
11.1	377	42	335	4.7
11.1	371	41	330	4.5
11.1	370	41	329	4.5
11.1	369	41	328	4.6
11.1	368	41	327	4.6
11.1	362	40	322	4.4
11.1	361	40	321	4.4
11.1	360	40	320	4.4
11.1	359	40	319	4.5
11.1	352	39	313	4.3
11.1	351	39	312	4.3
11.1	350	39	311	4.3
11.1	343	38	305	4.2

%E	M1	M2	DM	M*
11.1	342	38	304	4.2
11.1	341	38	303	4.2
11.1	334	37	297	4.1
11.1	333	37	296	4.1
11.1	332	37	295	4.1
11.1	325	36	289	4.0
11.1	324	36	288	4.0
11.1	323	36	287	4.0
11.1	316	35	281	3.9
11.1	315	35	280	3.9
11.1	314	35	279	3.9
11.1	307	34	273	3.8
11.1	306	34	272	3.8
11.1	305	34	271	3.8
11.1	298	33	265	3.7
11.1	297	33	264	3.7
11.1	296	33	263	3.7
11.1	289	32	257	3.5
11.1	288	32	256	3.6
11.1	280	31	249	3.4
11.1	279	31	248	3.4
11.1	271	30	241	3.3
11.1	270	30	240	3.3
11.1	262	29	233	3.2
11.1	261	29	232	3.2
11.1	253	28	225	3.1
11.1	252	28	224	3.1
11.1	244	27	217	3.0
11.1	243	27	216	3.0
11.1	235	26	209	2.9
11.1	234	26	208	2.9
11.1	226	25	201	2.8
11.1	225	25	200	2.8
11.1	217	24	193	2.7
11.1	216	24	192	2.7
11.1	208	23	185	2.5
11.1	207	23	184	2.6
11.1	199	22	177	2.4
11.1	198	22	176	2.4
11.1	190	21	169	2.3
11.1	189	21	168	2.3
11.1	181	20	161	2.2
11.1	180	20	160	2.2
11.1	171	19	152	2.1
11.1	162	18	144	2.0
11.1	153	17	136	1.9
11.1	144	16	128	1.8
11.1	135	15	120	1.7
11.1	126	14	112	1.6
11.1	117	13	104	1.4

%E	M1	M2	DM	M*
11.1	108	12	96	1.3
11.1	99	11	88	1.2
11.1	90	10	80	1.1
11.1	81	9	72	1.0
11.1	72	8	64	0.9
11.1	63	7	56	0.8
11.1	54	6	48	0.7
11.1	45	5	40	0.6
11.1	36	4	32	0.4
11.1	27	3	24	0.3
11.1	18	2	16	0.2
11.1	9	1	8	0.1
11.0	500	55	445	6.0
11.0	499	55	444	6.1
11.0	498	55	443	6.1
11.0	493	54	439	5.9
11.0	492	54	438	5.9
11.0	491	54	437	5.9
11.0	490	54	436	6.0
11.0	489	54	435	6.0
11.0	484	53	431	5.8
11.0	483	53	430	5.8
11.0	482	53	429	5.8
11.0	481	53	428	5.8
11.0	480	53	427	5.9
11.0	474	52	422	5.7
11.0	473	52	421	5.7
11.0	472	52	420	5.7
11.0	471	52	419	5.7
11.0	465	51	414	5.6
11.0	464	51	413	5.6
11.0	463	51	412	5.6
11.0	462	51	411	5.6
11.0	456	50	406	5.5
11.0	455	50	405	5.5
11.0	454	50	404	5.5
11.0	453	50	403	5.5
11.0	447	49	398	5.4
11.0	446	49	397	5.4
11.0	445	49	396	5.4
11.0	444	49	395	5.4
11.0	438	48	390	5.3
11.0	437	48	389	5.3
11.0	436	48	388	5.3
11.0	435	48	387	5.3
11.0	429	47	382	5.1
11.0	428	47	381	5.2
11.0	427	47	380	5.2
11.0	426	47	379	5.2
11.0	420	46	374	5.0

%E	M1	M2	DM	M*
11.0	419	46	373	5.1
11.0	418	46	372	5.1
11.0	417	46	371	5.1
11.0	410	45	365	4.9
11.0	409	45	364	5.0
11.0	408	45	363	5.0
11.0	401	44	357	4.8
11.0	400	44	356	4.8
11.0	399	44	355	4.9
11.0	392	43	349	4.7
11.0	391	43	348	4.7
11.0	390	43	347	4.7
11.0	383	42	341	4.6
11.0	382	42	340	4.6
11.0	381	42	339	4.6
11.0	374	41	333	4.5
11.0	373	41	332	4.5
11.0	372	41	331	4.5
11.0	365	40	325	4.4
11.0	364	40	324	4.4
11.0	363	40	323	4.4
11.0	356	39	317	4.3
11.0	355	39	316	4.3
11.0	354	39	315	4.3
11.0	353	39	314	4.3
11.0	347	38	309	4.2
11.0	346	38	308	4.2
11.0	345	38	307	4.2
11.0	344	38	306	4.2
11.0	337	37	300	4.1
11.0	336	37	299	4.1
11.0	335	37	298	4.1
11.0	328	36	292	4.0
11.0	327	36	291	4.0
11.0	326	36	290	4.0
11.0	319	35	284	3.8
11.0	318	35	283	3.9
11.0	317	35	282	3.9
11.0	310	34	276	3.7
11.0	309	34	275	3.7
11.0	308	34	274	3.8
11.0	301	33	268	3.6
11.0	300	33	267	3.6
11.0	299	33	266	3.6
11.0	292	32	260	3.5
11.0	291	32	259	3.5
11.0	290	32	258	3.5
11.0	283	31	252	3.4
11.0	282	31	251	3.4
11.0	281	31	250	3.4

%E	M1	M2	DM	M*
11.0	273	30	243	3.3
11.0	272	30	242	3.3
11.0	264	29	235	3.2
11.0	263	29	234	3.2
11.0	255	28	227	3.1
11.0	254	28	226	3.1
11.0	246	27	219	3.0
11.0	245	27	218	3.0
11.0	237	26	211	2.9
11.0	236	26	210	2.9
11.0	228	25	203	2.7
11.0	227	25	202	2.8
11.0	219	24	195	2.6
11.0	218	24	194	2.6
11.0	210	23	187	2.5
11.0	209	23	186	2.5
11.0	200	22	178	2.4
11.0	191	21	170	2.3
11.0	182	20	162	2.2
11.0	173	19	154	2.1
11.0	172	19	153	2.1
11.0	164	18	146	2.0
11.0	163	18	145	2.0
11.0	155	17	138	1.9
11.0	154	17	137	1.9
11.0	146	16	130	1.8
11.0	145	16	129	1.8
11.0	136	15	121	1.7
11.0	127	14	113	1.5
11.0	118	13	105	1.4
11.0	109	12	97	1.3
11.0	100	11	89	1.2
11.0	91	10	81	1.1
11.0	82	9	73	1.0
11.0	73	8	65	0.9
10.9	497	54	443	5.9
10.9	496	54	442	5.9
10.9	495	54	441	5.9
10.9	494	54	440	5.9
10.9	488	53	435	5.8
10.9	487	53	434	5.8
10.9	486	53	433	5.8
10.9	485	53	432	5.8
10.9	479	52	427	5.6
10.9	478	52	426	5.7
10.9	477	52	425	5.7
10.9	476	52	424	5.7
10.9	475	52	423	5.7
10.9	470	51	419	5.5
10.9	469	51	418	5.5

%E	M1	M2	DM	M*
10.9	468	51	417	5.6
10.9	467	51	416	5.6
10.9	466	51	415	5.6
10.9	460	50	410	5.4
10.9	459	50	409	5.4
10.9	458	50	408	5.5
10.9	457	50	407	5.5
10.9	451	49	402	5.3
10.9	450	49	401	5.3
10.9	449	49	400	5.3
10.9	448	49	399	5.4
10.9	442	48	394	5.2
10.9	441	48	393	5.2
10.9	440	48	392	5.2
10.9	439	48	391	5.2
10.9	433	47	386	5.1
10.9	432	47	385	5.1
10.9	431	47	384	5.1
10.9	430	47	383	5.1
10.9	423	46	377	5.0
10.9	422	46	376	5.0
10.9	421	46	375	5.0
10.9	414	45	369	4.9
10.9	413	45	368	4.9
10.9	412	45	367	4.9
10.9	411	45	366	4.9
10.9	405	44	361	4.8
10.9	404	44	360	4.8
10.9	403	44	359	4.8
10.9	402	44	358	4.8
10.9	396	43	353	4.7
10.9	395	43	352	4.7
10.9	394	43	351	4.7
10.9	393	43	350	4.7
10.9	387	42	345	4.6
10.9	386	42	344	4.6
10.9	385	42	343	4.6
10.9	384	42	342	4.6
10.9	377	41	336	4.5
10.9	376	41	335	4.5
10.9	375	41	334	4.5
10.9	368	40	328	4.3
10.9	367	40	327	4.4
10.9	366	40	326	4.4
10.9	359	39	320	4.2
10.9	358	39	319	4.2
10.9	357	39	318	4.3
10.9	350	38	312	4.1
10.9	349	38	311	4.1
10.9	348	38	310	4.1

%E	M1	M2	DM	M*
10.9	341	37	304	4.0
10.9	340	37	303	4.0
10.9	339	37	302	4.0
10.9	338	37	301	4.1
10.9	331	36	295	3.9
10.9	330	36	294	3.9
10.9	329	36	293	3.9
10.9	322	35	287	3.8
10.9	321	35	286	3.8
10.9	320	35	285	3.8
10.9	313	34	279	3.7
10.9	312	34	278	3.7
10.9	311	34	277	3.7
10.9	304	33	271	3.6
10.9	303	33	270	3.6
10.9	302	33	269	3.6
10.9	294	32	262	3.5
10.9	293	32	261	3.5
10.9	285	31	254	3.4
10.9	284	31	253	3.4
10.9	276	30	246	3.3
10.9	275	30	245	3.3
10.9	274	30	244	3.3
10.9	267	29	238	3.1
10.9	266	29	237	3.2
10.9	265	29	236	3.2
10.9	258	28	230	3.0
10.9	257	28	229	3.1
10.9	256	28	228	3.1
10.9	248	27	221	2.9
10.9	247	27	220	3.0
10.9	239	26	213	2.8
10.9	238	26	212	2.8
10.9	230	25	205	2.7
10.9	229	25	204	2.7
10.9	221	24	197	2.6
10.9	220	24	196	2.6
10.9	211	23	188	2.5
10.9	202	22	180	2.4
10.9	201	22	179	2.4
10.9	193	21	172	2.3
10.9	192	21	171	2.3
10.9	184	20	164	2.2
10.9	183	20	163	2.2
10.9	175	19	156	2.1
10.9	174	19	155	2.1
10.9	165	18	147	2.0
10.9	156	17	139	1.9
10.9	147	16	131	1.7
10.9	138	15	123	1.6

%E	M1	M2	DM	M*
10.9	137	15	122	1.6
10.9	129	14	115	1.5
10.9	128	14	114	1.5
10.9	119	13	106	1.4
10.9	110	12	98	1.3
10.9	101	11	90	1.2
10.9	92	10	82	1.1
10.9	64	7	57	0.8
10.9	55	6	49	0.7
10.9	46	5	41	0.5
10.8	500	54	446	5.8
10.8	499	54	445	5.8
10.8	498	54	444	5.9
10.8	493	53	440	5.7
10.8	492	53	439	5.7
10.8	491	53	438	5.7
10.8	490	53	437	5.7
10.8	489	53	436	5.7
10.8	483	52	431	5.6
10.8	482	52	430	5.6
10.8	481	52	429	5.6
10.8	480	52	428	5.6
10.8	474	51	423	5.5
10.8	473	51	422	5.5
10.8	472	51	421	5.5
10.8	471	51	420	5.5
10.8	465	50	415	5.4
10.8	464	50	414	5.4
10.8	463	50	413	5.4
10.8	462	50	412	5.4
10.8	461	50	411	5.4
10.8	455	49	406	5.3
10.8	454	49	405	5.3
10.8	453	49	404	5.3
10.8	452	49	403	5.3
10.8	446	48	398	5.2
10.8	445	48	397	5.2
10.8	444	48	396	5.2
10.8	443	48	395	5.2
10.8	437	47	390	5.1
10.8	436	47	389	5.1
10.8	435	47	388	5.1
10.8	434	47	387	5.1
10.8	427	46	381	5.0
10.8	426	46	380	5.0
10.8	425	46	379	5.0
10.8	424	46	378	5.0
10.8	418	45	373	4.8
10.8	417	45	372	4.9
10.8	416	45	371	4.9

%E	M1	M2	DM	M*
10.8	415	45	370	4.9
10.8	409	44	365	4.7
10.8	408	44	364	4.7
10.8	407	44	363	4.8
10.8	406	44	362	4.8
10.8	400	43	357	4.6
10.8	399	43	356	4.6
10.8	398	43	355	4.6
10.8	397	43	354	4.7
10.8	390	42	348	4.5
10.8	389	42	347	4.5
10.8	388	42	346	4.5
10.8	381	41	340	4.4
10.8	380	41	339	4.4
10.8	379	41	338	4.4
10.8	378	41	337	4.4
10.8	372	40	332	4.3
10.8	371	40	331	4.3
10.8	370	40	330	4.3
10.8	369	40	329	4.3
10.8	362	39	323	4.2
10.8	361	39	322	4.2
10.8	360	39	321	4.2
10.8	353	38	315	4.1
10.8	352	38	314	4.1
10.8	351	38	313	4.1
10.8	344	37	307	4.0
10.8	343	37	306	4.0
10.8	342	37	305	4.0
10.8	334	36	298	3.9
10.8	333	36	297	3.9
10.8	332	36	296	3.9
10.8	325	35	290	3.8
10.8	324	35	289	3.8
10.8	323	35	288	3.8
10.8	316	34	282	3.7
10.8	315	34	281	3.7
10.8	314	34	280	3.7
10.8	306	33	273	3.6
10.8	305	33	272	3.6
10.8	297	32	265	3.4
10.8	296	32	264	3.5
10.8	295	32	263	3.5
10.8	288	31	257	3.3
10.8	287	31	256	3.3
10.8	286	31	255	3.4
10.8	279	30	249	3.2
10.8	278	30	248	3.2
10.8	277	30	247	3.2
10.8	269	29	240	3.1

%E	M1	M2	DM	M*
10.8	268	29	239	3.1
10.8	260	28	232	3.0
10.8	259	28	231	3.0
10.8	251	27	224	2.9
10.8	250	27	223	2.9
10.8	249	27	222	2.9
10.8	241	26	215	2.8
10.8	240	26	214	2.8
10.8	232	25	207	2.7
10.8	231	25	206	2.7
10.8	223	24	199	2.6
10.8	222	24	198	2.6
10.8	213	23	190	2.5
10.8	212	23	189	2.5
10.8	204	22	182	2.4
10.8	203	22	181	2.4
10.8	195	21	174	2.3
10.8	194	21	173	2.3
10.8	186	20	166	2.2
10.8	185	20	165	2.2
10.8	176	19	157	2.1
10.8	167	18	149	1.9
10.8	166	18	148	2.0
10.8	158	17	141	1.8
10.8	157	17	140	1.8
10.8	148	16	132	1.7
10.8	139	15	124	1.6
10.8	130	14	116	1.5
10.8	120	13	107	1.4
10.8	111	12	99	1.3
10.8	102	11	91	1.2
10.8	93	10	83	1.1
10.8	83	9	74	1.0
10.8	74	8	66	0.9
10.8	65	7	58	0.8
10.8	37	4	33	0.4
10.7	497	53	444	5.7
10.7	496	53	443	5.7
10.7	495	53	442	5.7
10.7	494	53	441	5.7
10.7	488	52	436	5.5
10.7	487	52	435	5.6
10.7	486	52	434	5.6
10.7	485	52	433	5.6
10.7	484	52	432	5.6
10.7	478	51	427	5.4
10.7	477	51	426	5.5
10.7	476	51	425	5.5
10.7	475	51	424	5.5
10.7	469	50	419	5.3
10.7	468	50	418	5.3
10.7	467	50	417	5.4
10.7	466	50	416	5.4
10.7	460	49	411	5.2
10.7	459	49	410	5.2
10.7	458	49	409	5.2
10.7	457	49	408	5.3
10.7	456	49	407	5.3
10.7	450	48	402	5.1
10.7	449	48	401	5.1
10.7	448	48	400	5.1
10.7	447	48	399	5.2
10.7	441	47	394	5.0
10.7	440	47	393	5.0
10.7	439	47	392	5.0
10.7	438	47	391	5.0
10.7	431	46	385	4.9
10.7	430	46	384	4.9
10.7	429	46	383	4.9
10.7	428	46	382	4.9
10.7	422	45	377	4.8
10.7	421	45	376	4.8
10.7	420	45	375	4.8
10.7	419	45	374	4.8
10.7	413	44	369	4.7
10.7	412	44	368	4.7
10.7	411	44	367	4.7
10.7	410	44	366	4.7
10.7	403	43	360	4.6
10.7	402	43	359	4.6
10.7	401	43	358	4.6
10.7	394	42	352	4.5
10.7	393	42	351	4.5
10.7	392	42	350	4.5
10.7	391	42	349	4.5
10.7	384	41	343	4.4
10.7	383	41	342	4.4
10.7	382	41	341	4.4
10.7	375	40	335	4.3
10.7	374	40	334	4.3
10.7	373	40	333	4.3
10.7	366	39	327	4.2
10.7	365	39	326	4.2
10.7	364	39	325	4.2
10.7	363	39	324	4.2
10.7	356	38	318	4.1
10.7	355	38	317	4.1
10.7	354	38	316	4.1
10.7	347	37	310	3.9
10.7	346	37	309	4.0
10.7	345	37	308	4.0
10.7	338	36	302	3.8
10.7	337	36	301	3.8
10.7	336	36	300	3.9
10.7	335	36	299	3.9
10.7	328	35	293	3.7
10.7	327	35	292	3.7
10.7	326	35	291	3.8
10.7	319	34	285	3.6
10.7	318	34	284	3.6
10.7	317	34	283	3.6
10.7	309	33	276	3.5
10.7	308	33	275	3.5
10.7	307	33	274	3.5
10.7	300	32	268	3.4
10.7	299	32	267	3.4
10.7	298	32	266	3.4
10.7	291	31	260	3.3
10.7	290	31	259	3.3
10.7	289	31	258	3.3
10.7	281	30	251	3.2
10.7	280	30	250	3.2
10.7	272	29	243	3.1
10.7	271	29	242	3.1
10.7	270	29	241	3.1
10.7	262	28	234	3.0
10.7	261	28	233	3.0
10.7	253	27	226	2.9
10.7	252	27	225	2.9
10.7	244	26	218	2.8
10.7	243	26	217	2.8
10.7	242	26	216	2.8
10.7	234	25	209	2.7
10.7	233	25	208	2.7
10.7	225	24	201	2.6
10.7	224	24	200	2.6
10.7	215	23	192	2.5
10.7	214	23	191	2.5
10.7	206	22	184	2.3
10.7	205	22	183	2.4
10.7	197	21	176	2.2
10.7	196	21	175	2.3
10.7	187	20	167	2.1
10.7	178	19	159	2.0
10.7	177	19	158	2.0
10.7	169	18	151	1.9
10.7	168	18	150	1.9
10.7	159	17	142	1.8
10.7	150	16	134	1.7
10.7	149	16	133	1.7
10.7	140	15	125	1.6
10.7	131	14	117	1.5
10.7	122	13	109	1.4
10.7	121	13	108	1.4
10.7	112	12	100	1.3
10.7	103	11	92	1.2
10.7	84	9	75	1.0
10.7	75	8	67	0.9
10.7	56	6	50	0.6
10.7	28	3	25	0.3
10.6	500	53	447	5.6
10.6	499	53	446	5.6
10.6	498	53	445	5.6
10.6	492	52	440	5.5
10.6	491	52	439	5.5
10.6	490	52	438	5.5
10.6	489	52	437	5.5
10.6	483	51	432	5.4
10.6	482	51	431	5.4
10.6	481	51	430	5.4
10.6	480	51	429	5.4
10.6	479	51	428	5.4
10.6	473	50	423	5.3
10.6	472	50	422	5.3
10.6	471	50	421	5.3
10.6	470	50	420	5.3
10.6	464	49	415	5.2
10.6	463	49	414	5.2
10.6	462	49	413	5.2
10.6	461	49	412	5.2
10.6	454	48	406	5.1
10.6	453	48	405	5.1
10.6	452	48	404	5.1
10.6	451	48	403	5.1
10.6	445	47	398	5.0
10.6	444	47	397	5.0
10.6	443	47	396	5.0
10.6	442	47	395	5.0
10.6	436	46	390	4.9
10.6	435	46	389	4.9
10.6	434	46	388	4.9
10.6	433	46	387	4.9
10.6	432	46	386	4.9
10.6	426	45	381	4.8
10.6	425	45	380	4.8
10.6	424	45	379	4.8
10.6	423	45	378	4.8
10.6	417	44	373	4.6
10.6	416	44	372	4.7
10.6	415	44	371	4.7
10.6	414	44	370	4.7
10.6	407	43	364	4.5
10.6	406	43	363	4.6
10.6	405	43	362	4.6
10.6	404	43	361	4.6
10.6	398	42	356	4.4
10.6	397	42	355	4.4
10.6	396	42	354	4.5
10.6	395	42	353	4.5
10.6	388	41	347	4.3
10.6	387	41	346	4.3
10.6	386	41	345	4.4
10.6	385	41	344	4.4
10.6	379	40	339	4.2
10.6	378	40	338	4.2
10.6	377	40	337	4.2
10.6	376	40	336	4.3
10.6	369	39	330	4.1
10.6	368	39	329	4.1
10.6	367	39	328	4.1
10.6	360	38	322	4.0
10.6	359	38	321	4.0
10.6	358	38	320	4.0
10.6	357	38	319	4.0
10.6	350	37	313	3.9
10.6	349	37	312	3.9
10.6	348	37	311	3.9
10.6	341	36	305	3.8
10.6	340	36	304	3.8
10.6	339	36	303	3.8
10.6	331	35	296	3.7
10.6	330	35	295	3.7
10.6	329	35	294	3.7
10.6	322	34	288	3.6
10.6	321	34	287	3.6
10.6	320	34	286	3.6
10.6	312	33	279	3.5
10.6	311	33	278	3.5
10.6	310	33	277	3.5
10.6	303	32	271	3.4
10.6	302	32	270	3.4
10.6	301	32	269	3.4
10.6	293	31	262	3.3
10.6	292	31	261	3.3
10.6	284	30	254	3.2
10.6	283	30	253	3.2
10.6	282	30	252	3.2
10.6	274	29	245	3.1
10.6	273	29	244	3.1
10.6	265	28	237	3.0

%E	M1	M2	DM	M*
10.6	264	28	236	3.0
10.6	263	28	235	3.0
10.6	255	27	228	2.9
10.6	254	27	227	2.9
10.6	246	26	220	2.7
10.6	245	26	219	2.8
10.6	236	25	211	2.6
10.6	235	25	210	2.7
10.6	227	24	203	2.5
10.6	226	24	202	2.5
10.6	218	23	195	2.4
10.6	217	23	194	2.4
10.6	216	23	193	2.4
10.6	208	22	186	2.3
10.6	207	22	185	2.3
10.6	199	21	178	2.2
10.6	198	21	177	2.2
10.6	189	20	169	2.1
10.6	188	20	168	2.1
10.6	180	19	161	2.0
10.6	179	19	160	2.0
10.6	170	18	152	1.9
10.6	161	17	144	1.8
10.6	160	17	143	1.8
10.6	151	16	135	1.7
10.6	142	15	127	1.6
10.6	141	15	126	1.6
10.6	132	14	118	1.5
10.6	123	13	110	1.4
10.6	113	12	101	1.3
10.6	104	11	93	1.2
10.6	94	10	84	1.1
10.6	85	9	76	1.0
10.6	66	7	59	0.7
10.6	47	5	42	0.5
10.5	497	52	445	5.4
10.5	496	52	444	5.5
10.5	495	52	443	5.5
10.5	494	52	442	5.5
10.5	493	52	441	5.5
10.5	488	51	437	5.3
10.5	487	51	436	5.3
10.5	486	51	435	5.4
10.5	485	51	434	5.4
10.5	484	51	433	5.4
10.5	478	50	428	5.2
10.5	477	50	427	5.2
10.5	476	50	426	5.3
10.5	475	50	425	5.3
10.5	474	50	424	5.3

%E	M1	M2	DM	M*
10.5	468	49	419	5.1
10.5	467	49	418	5.1
10.5	466	49	417	5.2
10.5	465	49	416	5.2
10.5	459	48	411	5.0
10.5	458	48	410	5.0
10.5	457	48	409	5.0
10.5	456	48	408	5.1
10.5	455	48	407	5.1
10.5	449	47	402	4.9
10.5	448	47	401	4.9
10.5	447	47	400	4.9
10.5	446	47	399	5.0
10.5	440	46	394	4.8
10.5	439	46	393	4.8
10.5	438	46	392	4.8
10.5	437	46	391	4.8
10.5	430	45	385	4.7
10.5	429	45	384	4.7
10.5	428	45	383	4.7
10.5	427	45	382	4.7
10.5	421	44	377	4.6
10.5	420	44	376	4.6
10.5	419	44	375	4.6
10.5	418	44	374	4.6
10.5	411	43	368	4.5
10.5	410	43	367	4.5
10.5	409	43	366	4.5
10.5	408	43	365	4.5
10.5	401	42	359	4.4
10.5	400	42	358	4.4
10.5	399	42	357	4.4
10.5	392	41	351	4.3
10.5	391	41	350	4.3
10.5	390	41	349	4.3
10.5	389	41	348	4.3
10.5	382	40	342	4.2
10.5	381	40	341	4.2
10.5	380	40	340	4.2
10.5	373	39	334	4.1
10.5	372	39	333	4.1
10.5	371	39	332	4.1
10.5	370	39	331	4.1
10.5	363	38	325	4.0
10.5	362	38	324	4.0
10.5	361	38	323	4.0
10.5	354	37	317	3.9
10.5	353	37	316	3.9
10.5	352	37	315	3.9
10.5	351	37	314	3.9

%E	M1	M2	DM	M*
10.5	344	36	308	3.8
10.5	343	36	307	3.8
10.5	342	36	306	3.8
10.5	334	35	299	3.7
10.5	333	35	298	3.7
10.5	332	35	297	3.7
10.5	325	34	291	3.6
10.5	324	34	290	3.6
10.5	323	34	289	3.6
10.5	315	33	282	3.5
10.5	314	33	281	3.5
10.5	313	33	280	3.5
10.5	306	32	274	3.3
10.5	305	32	273	3.4
10.5	304	32	272	3.4
10.5	296	31	265	3.2
10.5	295	31	264	3.3
10.5	294	31	263	3.3
10.5	287	30	257	3.1
10.5	286	30	256	3.1
10.5	285	30	255	3.2
10.5	277	29	248	3.0
10.5	276	29	247	3.0
10.5	275	29	246	3.1
10.5	267	28	239	2.9
10.5	266	28	238	2.9
10.5	258	27	231	2.8
10.5	257	27	230	2.8
10.5	256	27	229	2.8
10.5	248	26	222	2.7
10.5	247	26	221	2.7
10.5	239	25	214	2.6
10.5	238	25	213	2.6
10.5	237	25	212	2.6
10.5	229	24	205	2.5
10.5	228	24	204	2.5
10.5	220	23	197	2.4
10.5	219	23	196	2.4
10.5	210	22	188	2.3
10.5	209	22	187	2.3
10.5	200	21	179	2.2
10.5	191	20	171	2.1
10.5	190	20	170	2.1
10.5	181	19	162	2.0
10.5	172	18	154	1.9
10.5	171	18	153	1.9
10.5	162	17	145	1.8
10.5	153	16	137	1.7
10.5	152	16	136	1.7
10.5	143	15	128	1.6

%E	M1	M2	DM	M*
10.5	133	14	119	1.5
10.5	124	13	111	1.4
10.5	114	12	102	1.3
10.5	105	11	94	1.2
10.5	95	10	85	1.1
10.5	86	9	77	0.9
10.5	76	8	68	0.8
10.5	57	6	51	0.6
10.5	38	4	34	0.4
10.5	19	2	17	0.2
10.4	500	52	448	5.4
10.4	499	52	447	5.4
10.4	498	52	446	5.4
10.4	492	51	441	5.3
10.4	491	51	440	5.3
10.4	490	51	439	5.3
10.4	489	51	438	5.3
10.4	483	50	433	5.2
10.4	482	50	432	5.2
10.4	481	50	431	5.2
10.4	480	50	430	5.2
10.4	479	50	429	5.2
10.4	473	49	424	5.1
10.4	472	49	423	5.1
10.4	471	49	422	5.1
10.4	470	49	421	5.1
10.4	469	49	420	5.1
10.4	463	48	415	5.0
10.4	462	48	414	5.0
10.4	461	48	413	5.0
10.4	460	48	412	5.0
10.4	454	47	407	4.9
10.4	453	47	406	4.9
10.4	452	47	405	4.9
10.4	451	47	404	4.9
10.4	450	47	403	4.9
10.4	444	46	398	4.8
10.4	443	46	397	4.8
10.4	442	46	396	4.8
10.4	441	46	395	4.8
10.4	434	45	389	4.7
10.4	433	45	388	4.7
10.4	432	45	387	4.7
10.4	431	45	386	4.7
10.4	425	44	381	4.6
10.4	424	44	380	4.6
10.4	423	44	379	4.6
10.4	422	44	378	4.6
10.4	415	43	372	4.5
10.4	414	43	371	4.5

%E	M1	M2	DM	M*
10.4	413	43	370	4.5
10.4	412	43	369	4.5
10.4	405	42	363	4.4
10.4	404	42	362	4.4
10.4	403	42	361	4.4
10.4	402	42	360	4.4
10.4	396	41	355	4.2
10.4	395	41	354	4.3
10.4	394	41	353	4.3
10.4	393	41	352	4.3
10.4	386	40	346	4.1
10.4	385	40	345	4.2
10.4	384	40	344	4.2
10.4	383	40	343	4.2
10.4	376	39	337	4.0
10.4	375	39	336	4.1
10.4	374	39	335	4.1
10.4	367	38	329	3.9
10.4	366	38	328	3.9
10.4	365	38	327	4.0
10.4	364	38	326	4.0
10.4	357	37	320	3.8
10.4	356	37	319	3.8
10.4	355	37	318	3.9
10.4	347	36	311	3.7
10.4	346	36	310	3.7
10.4	345	36	309	3.8
10.4	338	35	303	3.6
10.4	337	35	302	3.6
10.4	336	35	301	3.6
10.4	335	35	300	3.7
10.4	328	34	294	3.5
10.4	327	34	293	3.5
10.4	326	34	292	3.5
10.4	318	33	285	3.4
10.4	317	33	284	3.4
10.4	316	33	283	3.4
10.4	309	32	277	3.3
10.4	308	32	276	3.3
10.4	307	32	275	3.3
10.4	299	31	268	3.2
10.4	298	31	267	3.2
10.4	297	31	266	3.2
10.4	289	30	259	3.1
10.4	288	30	258	3.1
10.4	280	29	251	3.0
10.4	279	29	250	3.0
10.4	278	29	249	3.0
10.4	270	28	242	2.9
10.4	269	28	241	2.9

%E	M1	M2	DM	M*
10.4	268	28	240	2.9
10.4	260	27	233	2.8
10.4	259	27	232	2.8
10.4	251	26	225	2.7
10.4	250	26	224	2.7
10.4	249	26	223	2.7
10.4	241	25	216	2.6
10.4	240	25	215	2.6
10.4	231	24	207	2.5
10.4	230	24	206	2.5
10.4	222	23	199	2.4
10.4	221	23	198	2.4
10.4	212	22	190	2.3
10.4	211	22	189	2.3
10.4	202	21	181	2.2
10.4	201	21	180	2.2
10.4	193	20	173	2.1
10.4	192	20	172	2.1
10.4	183	19	164	2.0
10.4	182	19	163	2.0
10.4	173	18	155	1.9
10.4	164	17	147	1.8
10.4	163	17	146	1.8
10.4	154	16	138	1.7
10.4	144	15	129	1.6
10.4	135	14	121	1.5
10.4	134	14	120	1.5
10.4	125	13	112	1.4
10.4	115	12	103	1.3
10.4	106	11	95	1.1
10.4	96	10	86	1.0
10.4	77	8	69	0.8
10.4	67	7	60	0.7
10.4	48	5	43	0.5
10.3	497	51	446	5.2
10.3	496	51	445	5.2
10.3	495	51	444	5.3
10.3	494	51	443	5.3
10.3	493	51	442	5.3
10.3	487	50	437	5.1
10.3	486	50	436	5.1
10.3	485	50	435	5.2
10.3	484	50	434	5.2
10.3	478	49	429	5.0
10.3	477	49	428	5.0
10.3	476	49	427	5.0
10.3	475	49	426	5.1
10.3	474	49	425	5.1
10.3	468	48	420	4.9
10.3	467	48	419	4.9

%E	M1	M2	DM	M*
10.3	466	48	418	4.9
10.3	465	48	417	5.0
10.3	464	48	416	5.0
10.3	458	47	411	4.8
10.3	457	47	410	4.8
10.3	456	47	409	4.8
10.3	455	47	408	4.9
10.3	448	46	402	4.7
10.3	447	46	401	4.7
10.3	446	46	400	4.7
10.3	445	46	399	4.8
10.3	439	45	394	4.6
10.3	438	45	393	4.6
10.3	437	45	392	4.6
10.3	436	45	391	4.6
10.3	435	45	390	4.7
10.3	429	44	385	4.5
10.3	428	44	384	4.5
10.3	427	44	383	4.5
10.3	426	44	382	4.5
10.3	419	43	376	4.4
10.3	418	43	375	4.4
10.3	417	43	374	4.4
10.3	416	43	373	4.4
10.3	409	42	367	4.3
10.3	408	42	366	4.3
10.3	407	42	365	4.3
10.3	406	42	364	4.3
10.3	400	41	359	4.2
10.3	399	41	358	4.2
10.3	398	41	357	4.2
10.3	397	41	356	4.2
10.3	390	40	350	4.1
10.3	389	40	349	4.1
10.3	388	40	348	4.1
10.3	387	40	347	4.1
10.3	380	39	341	4.0
10.3	379	39	340	4.0
10.3	378	39	339	4.0
10.3	377	39	338	4.0
10.3	370	38	332	3.9
10.3	369	38	331	3.9
10.3	368	38	330	3.9
10.3	360	37	323	3.8
10.3	359	37	322	3.8
10.3	358	37	321	3.8
10.3	351	36	315	3.7
10.3	350	36	314	3.7
10.3	349	36	313	3.7
10.3	348	36	312	3.7

%E	M1	M2	DM	M*
10.3	341	35	306	3.6
10.3	340	35	305	3.6
10.3	339	35	304	3.6
10.3	331	34	297	3.5
10.3	330	34	296	3.5
10.3	329	34	295	3.5
10.3	321	33	288	3.4
10.3	320	33	287	3.4
10.3	319	33	286	3.4
10.3	312	32	280	3.3
10.3	311	32	279	3.3
10.3	310	32	278	3.3
10.3	302	31	271	3.2
10.3	301	31	270	3.2
10.3	300	31	269	3.2
10.3	292	30	262	3.1
10.3	291	30	261	3.1
10.3	290	30	260	3.1
10.3	282	29	253	3.0
10.3	281	29	252	3.0
10.3	273	28	245	2.9
10.3	272	28	244	2.9
10.3	271	28	243	2.9
10.3	263	27	236	2.8
10.3	262	27	235	2.8
10.3	261	27	234	2.8
10.3	253	26	227	2.7
10.3	252	26	226	2.7
10.3	243	25	218	2.6
10.3	242	25	217	2.6
10.3	234	24	210	2.5
10.3	233	24	209	2.5
10.3	232	24	208	2.5
10.3	224	23	201	2.4
10.3	223	23	200	2.4
10.3	214	22	192	2.3
10.3	213	22	191	2.3
10.3	204	21	183	2.2
10.3	203	21	182	2.2
10.3	195	20	175	2.1
10.3	194	20	174	2.1
10.3	185	19	166	2.0
10.3	184	19	165	2.0
10.3	175	18	157	1.9
10.3	174	18	156	1.9
10.3	165	17	148	1.8
10.3	156	16	140	1.6
10.3	155	16	139	1.7
10.3	146	15	131	1.5
10.3	145	15	130	1.6

%E	M1	M2	DM	M*
10.3	136	14	122	1.4
10.3	126	13	113	1.3
10.3	117	12	105	1.2
10.3	116	12	104	1.2
10.3	107	11	96	1.1
10.3	97	10	87	1.0
10.3	87	9	78	0.9
10.3	78	8	70	0.8
10.3	68	7	61	0.7
10.3	58	6	52	0.6
10.3	39	4	35	0.4
10.3	29	3	26	0.3
10.2	500	51	449	5.2
10.2	499	51	448	5.2
10.2	498	51	447	5.2
10.2	492	50	442	5.1
10.2	491	50	441	5.1
10.2	490	50	440	5.1
10.2	489	50	439	5.1
10.2	488	50	438	5.1
10.2	482	49	433	5.0
10.2	481	49	432	5.0
10.2	480	49	431	5.0
10.2	479	49	430	5.0
10.2	472	48	424	4.9
10.2	471	48	423	4.9
10.2	470	48	422	4.9
10.2	469	48	421	4.9
10.2	463	47	416	4.8
10.2	462	47	415	4.8
10.2	461	47	414	4.8
10.2	460	47	413	4.8
10.2	459	47	412	4.8
10.2	453	46	407	4.7
10.2	452	46	406	4.7
10.2	451	46	405	4.7
10.2	450	46	404	4.7
10.2	449	46	403	4.7
10.2	443	45	398	4.6
10.2	442	45	397	4.6
10.2	441	45	396	4.6
10.2	440	45	395	4.6
10.2	433	44	389	4.5
10.2	432	44	388	4.5
10.2	431	44	387	4.5
10.2	430	44	386	4.5
10.2	423	43	380	4.4
10.2	422	43	379	4.4
10.2	421	43	378	4.4
10.2	420	43	377	4.4

%E	M1	M2	DM	M*
10.2	413	42	371	4.3
10.2	412	42	370	4.3
10.2	411	42	369	4.3
10.2	410	42	368	4.3
10.2	403	41	362	4.2
10.2	402	41	361	4.2
10.2	401	41	360	4.2
10.2	394	40	354	4.1
10.2	393	40	353	4.1
10.2	392	40	352	4.1
10.2	391	40	351	4.1
10.2	384	39	345	4.0
10.2	383	39	344	4.0
10.2	382	39	343	4.0
10.2	381	39	342	4.0
10.2	374	38	336	3.9
10.2	373	38	335	3.9
10.2	372	38	334	3.9
10.2	371	38	333	3.9
10.2	364	37	327	3.8
10.2	363	37	326	3.8
10.2	362	37	325	3.8
10.2	361	37	324	3.8
10.2	354	36	318	3.7
10.2	353	36	317	3.7
10.2	352	36	316	3.7
10.2	344	35	309	3.6
10.2	343	35	308	3.6
10.2	342	35	307	3.6
10.2	334	34	300	3.5
10.2	333	34	299	3.5
10.2	332	34	298	3.5
10.2	325	33	292	3.4
10.2	324	33	291	3.4
10.2	323	33	290	3.4
10.2	322	33	289	3.4
10.2	315	32	283	3.3
10.2	314	32	282	3.3
10.2	313	32	281	3.3
10.2	305	31	274	3.2
10.2	304	31	273	3.2
10.2	303	31	272	3.2
10.2	295	30	265	3.1
10.2	294	30	264	3.1
10.2	293	30	263	3.1
10.2	285	29	256	3.0
10.2	284	29	255	3.0
10.2	283	29	254	3.0
10.2	275	28	247	2.9
10.2	274	28	246	2.9

%E	M1	M2	DM	M*
10.2	266	27	239	2.7
10.2	265	27	238	2.8
10.2	264	27	237	2.8
10.2	256	26	230	2.6
10.2	255	26	229	2.7
10.2	254	26	228	2.7
10.2	246	25	221	2.5
10.2	245	25	220	2.6
10.2	244	25	219	2.6
10.2	236	24	212	2.4
10.2	235	24	211	2.5
10.2	226	23	203	2.3
10.2	225	23	202	2.4
10.2	216	22	194	2.2
10.2	215	22	193	2.3
10.2	206	21	185	2.1
10.2	205	21	184	2.2
10.2	197	20	177	2.0
10.2	196	20	176	2.0
10.2	187	19	168	1.9
10.2	186	19	167	1.9
10.2	177	18	159	1.8
10.2	176	18	158	1.8
10.2	167	17	150	1.7
10.2	166	17	149	1.7
10.2	157	16	141	1.6
10.2	147	15	132	1.5
10.2	137	14	123	1.4
10.2	128	13	115	1.3
10.2	127	13	114	1.3
10.2	118	12	106	1.2
10.2	108	11	97	1.1
10.2	98	10	88	1.0
10.2	88	9	79	0.9
10.2	59	6	53	0.6
10.2	49	5	44	0.5
10.1	497	50	447	5.0
10.1	496	50	446	5.0
10.1	495	50	445	5.1
10.1	494	50	444	5.1
10.1	493	50	443	5.1
10.1	487	49	438	4.9
10.1	486	49	437	4.9
10.1	485	49	436	5.0
10.1	484	49	435	5.0
10.1	483	49	434	5.0
10.1	477	48	429	4.8
10.1	476	48	428	4.8
10.1	475	48	427	4.9
10.1	474	48	426	4.9

%E	M1	M2	DM	M*
10.1	473	48	425	4.9
10.1	467	47	420	4.7
10.1	466	47	419	4.7
10.1	465	47	418	4.8
10.1	464	47	417	4.8
10.1	457	46	411	4.6
10.1	456	46	410	4.6
10.1	455	46	409	4.7
10.1	454	46	408	4.7
10.1	447	45	402	4.5
10.1	446	45	401	4.5
10.1	445	45	400	4.6
10.1	444	45	399	4.6
10.1	437	44	393	4.4
10.1	436	44	392	4.4
10.1	435	44	391	4.5
10.1	434	44	390	4.5
10.1	427	43	384	4.3
10.1	426	43	383	4.3
10.1	425	43	382	4.4
10.1	424	43	381	4.4
10.1	417	42	375	4.2
10.1	416	42	374	4.2
10.1	415	42	373	4.3
10.1	414	42	372	4.3
10.1	407	41	366	4.1
10.1	406	41	365	4.1
10.1	405	41	364	4.2
10.1	404	41	363	4.2
10.1	398	40	358	4.0
10.1	397	40	357	4.0
10.1	396	40	356	4.0
10.1	395	40	355	4.1
10.1	388	39	349	3.9
10.1	387	39	348	3.9
10.1	386	39	347	3.9
10.1	385	39	346	4.0
10.1	378	38	340	3.8
10.1	377	38	339	3.8
10.1	376	38	338	3.8
10.1	375	38	337	3.9
10.1	368	37	331	3.7
10.1	367	37	330	3.7
10.1	366	37	329	3.7
10.1	365	37	328	3.8
10.1	358	36	322	3.6
10.1	357	36	321	3.6
10.1	356	36	320	3.6
10.1	355	36	319	3.7
10.1	348	35	313	3.5

%E	M1	M2	DM	M*
10.1	347	35	312	3.5
10.1	346	35	311	3.5
10.1	345	35	310	3.6
10.1	338	34	304	3.4
10.1	337	34	303	3.4
10.1	336	34	302	3.4
10.1	335	34	301	3.5
10.1	328	33	295	3.3
10.1	327	33	294	3.3
10.1	326	33	293	3.3
10.1	318	32	286	3.2
10.1	317	32	285	3.2
10.1	316	32	284	3.2
10.1	308	31	277	3.1
10.1	307	31	276	3.1
10.1	306	31	275	3.1
10.1	298	30	268	3.0
10.1	297	30	267	3.0
10.1	296	30	266	3.0
10.1	288	29	259	2.9
10.1	287	29	258	2.9
10.1	286	29	257	2.9
10.1	278	28	250	2.8
10.1	277	28	249	2.8
10.1	276	28	248	2.8
10.1	268	27	241	2.7
10.1	267	27	240	2.7
10.1	258	26	232	2.6
10.1	257	26	231	2.6
10.1	248	25	223	2.5
10.1	247	25	222	2.5
10.1	238	24	214	2.4
10.1	237	24	213	2.4
10.1	228	23	205	2.3
10.1	227	23	204	2.3
10.1	218	22	196	2.2
10.1	217	22	195	2.2
10.1	208	21	187	2.1
10.1	207	21	186	2.1
10.1	199	20	179	2.0
10.1	198	20	178	2.0
10.1	189	19	170	1.9
10.1	188	19	169	1.9
10.1	179	18	161	1.8
10.1	178	18	160	1.8
10.1	169	17	152	1.7
10.1	168	17	151	1.7
10.1	159	16	143	1.6
10.1	158	16	142	1.6
10.1	149	15	134	1.5

%E	M1	M2	DM	M*
10.1	148	15	133	1.5
10.1	139	14	125	1.4
10.1	138	14	124	1.4
10.1	129	13	116	1.3
10.1	119	12	107	1.2
10.1	109	11	98	1.1
10.1	99	10	89	1.0
10.1	89	9	80	0.9
10.1	79	8	71	0.8
10.1	69	7	62	0.7
10.0	500	50	450	5.0
10.0	499	50	449	5.0
10.0	498	50	448	5.0
10.0	492	49	443	4.9
10.0	491	49	442	4.9
10.0	490	49	441	4.9
10.0	489	49	440	4.9
10.0	488	49	439	4.9
10.0	482	48	434	4.8
10.0	481	48	433	4.8
10.0	480	48	432	4.8
10.0	479	48	431	4.8
10.0	478	48	430	4.8
10.0	472	47	425	4.7
10.0	471	47	424	4.7
10.0	470	47	423	4.7
10.0	469	47	422	4.7
10.0	468	47	421	4.7
10.0	462	46	416	4.6
10.0	461	46	415	4.6
10.0	460	46	414	4.6
10.0	459	46	413	4.6
10.0	458	46	412	4.6
10.0	452	45	407	4.5
10.0	451	45	406	4.5
10.0	450	45	405	4.5
10.0	449	45	404	4.5
10.0	448	45	403	4.5
10.0	442	44	398	4.4
10.0	441	44	397	4.4
10.0	440	44	396	4.4
10.0	439	44	395	4.4
10.0	438	44	394	4.4
10.0	432	43	389	4.3
10.0	431	43	388	4.3
10.0	430	43	387	4.3
10.0	429	43	386	4.3
10.0	428	43	385	4.3
10.0	422	42	380	4.2
10.0	421	42	379	4.2

%E	M1	M2	DM	M*
10.0	420	42	378	4.2
10.0	419	42	377	4.2
10.0	418	42	376	4.2
10.0	412	41	371	4.1
10.0	411	41	370	4.1
10.0	410	41	369	4.1
10.0	409	41	368	4.1
10.0	408	41	367	4.1
10.0	402	40	362	4.0
10.0	401	40	361	4.0
10.0	400	40	360	4.0
10.0	399	40	359	4.0
10.0	391	39	352	3.9
10.0	390	39	351	3.9
10.0	389	39	350	3.9
10.0	381	38	343	3.8
10.0	380	38	342	3.8
10.0	379	38	341	3.8
10.0	371	37	334	3.7
10.0	370	37	333	3.7
10.0	369	37	332	3.7
10.0	361	36	325	3.6
10.0	360	36	324	3.6
10.0	359	36	323	3.6
10.0	351	35	316	3.5
10.0	350	35	315	3.5
10.0	349	35	314	3.5
10.0	341	34	307	3.4
10.0	340	34	306	3.4
10.0	339	34	305	3.4
10.0	331	33	298	3.3
10.0	330	33	297	3.3
10.0	329	33	296	3.3
10.0	321	32	289	3.2
10.0	320	32	288	3.2
10.0	319	32	287	3.2
10.0	311	31	280	3.1
10.0	310	31	279	3.1
10.0	309	31	278	3.1
10.0	301	30	271	3.0
10.0	300	30	270	3.0
10.0	299	30	269	3.0
10.0	291	29	262	2.9
10.0	290	29	261	2.9
10.0	289	29	260	2.9
10.0	281	28	253	2.8
10.0	280	28	252	2.8
10.0	279	28	251	2.8
10.0	271	27	244	2.7
10.0	270	27	243	2.7

%E	M1	M2	DM	M*
10.0	269	27	242	2.7
10.0	261	26	235	2.6
10.0	260	26	234	2.6
10.0	259	26	233	2.6
10.0	251	25	226	2.5
10.0	250	25	225	2.5
10.0	249	25	224	2.5
10.0	241	24	217	2.4
10.0	240	24	216	2.4
10.0	239	24	215	2.4
10.0	231	23	208	2.3
10.0	230	23	207	2.3
10.0	229	23	206	2.3
10.0	221	22	199	2.2
10.0	220	22	198	2.2
10.0	219	22	197	2.2
10.0	211	21	190	2.1
10.0	210	21	189	2.1
10.0	209	21	188	2.1
10.0	201	20	181	2.0
10.0	200	20	180	2.0
10.0	190	19	171	1.9
10.0	180	18	162	1.8
10.0	170	17	153	1.7
10.0	160	16	144	1.6
10.0	150	15	135	1.5
10.0	140	14	126	1.4
10.0	130	13	117	1.3
10.0	120	12	108	1.2
10.0	110	11	99	1.1
10.0	100	10	90	1.0
10.0	90	9	81	0.9
10.0	80	8	72	0.8
10.0	70	7	63	0.7
10.0	60	6	54	0.6
10.0	50	5	45	0.5
10.0	40	4	36	0.4
10.0	30	3	27	0.3
10.0	20	2	18	0.2
10.0	10	1	9	0.1
9.9	497	49	448	4.8
9.9	496	49	447	4.8
9.9	495	49	446	4.9
9.9	494	49	445	4.9
9.9	493	49	444	4.9
9.9	487	48	439	4.7
9.9	486	48	438	4.7
9.9	485	48	437	4.8
9.9	484	48	436	4.8
9.9	483	48	435	4.8

%E	M1	M2	DM	M*
9.9	477	47	430	4.6
9.9	476	47	429	4.6
9.9	475	47	428	4.7
9.9	474	47	427	4.7
9.9	473	47	426	4.7
9.9	467	46	421	4.5
9.9	466	46	420	4.5
9.9	465	46	419	4.6
9.9	464	46	418	4.6
9.9	463	46	417	4.6
9.9	456	45	411	4.4
9.9	455	45	410	4.5
9.9	454	45	409	4.5
9.9	453	45	408	4.5
9.9	446	44	402	4.3
9.9	445	44	401	4.4
9.9	444	44	400	4.4
9.9	443	44	399	4.4
9.9	436	43	393	4.2
9.9	435	43	392	4.3
9.9	434	43	391	4.3
9.9	433	43	390	4.3
9.9	426	42	384	4.1
9.9	425	42	383	4.2
9.9	424	42	382	4.2
9.9	423	42	381	4.2
9.9	416	41	375	4.0
9.9	415	41	374	4.1
9.9	414	41	373	4.1
9.9	413	41	372	4.1
9.9	406	40	366	3.9
9.9	405	40	365	4.0
9.9	404	40	364	4.0
9.9	403	40	363	4.0
9.9	395	39	356	3.9
9.9	394	39	355	3.9
9.9	393	39	354	3.9
9.9	392	39	353	3.9
9.9	385	38	347	3.8
9.9	384	38	346	3.8
9.9	383	38	345	3.8
9.9	382	38	344	3.8
9.9	375	37	338	3.7
9.9	374	37	337	3.7
9.9	373	37	336	3.7
9.9	372	37	335	3.7
9.9	365	36	329	3.6
9.9	364	36	328	3.6
9.9	363	36	327	3.6
9.9	362	36	326	3.6

%E	M1	M2	DM	M*
9.9	355	35	320	3.5
9.9	354	35	319	3.5
9.9	353	35	318	3.5
9.9	352	35	317	3.5
9.9	345	34	311	3.4
9.9	344	34	310	3.4
9.9	343	34	309	3.4
9.9	342	34	308	3.4
9.9	335	33	302	3.3
9.9	334	33	301	3.3
9.9	333	33	300	3.3
9.9	332	33	299	3.3
9.9	324	32	292	3.2
9.9	323	32	291	3.2
9.9	322	32	290	3.2
9.9	314	31	283	3.1
9.9	313	31	282	3.1
9.9	312	31	281	3.1
9.9	304	30	274	3.0
9.9	303	30	273	3.0
9.9	302	30	272	3.0
9.9	294	29	265	2.9
9.9	293	29	264	2.9
9.9	292	29	263	2.9
9.9	284	28	256	2.8
9.9	283	28	255	2.8
9.9	282	28	254	2.8
9.9	274	27	247	2.7
9.9	273	27	246	2.7
9.9	272	27	245	2.7
9.9	263	26	237	2.6
9.9	262	26	236	2.6
9.9	253	25	228	2.5
9.9	252	25	227	2.5
9.9	243	24	219	2.4
9.9	242	24	218	2.4
9.9	233	23	210	2.3
9.9	232	23	209	2.3
9.9	223	22	201	2.2
9.9	222	22	200	2.2
9.9	213	21	192	2.1
9.9	212	21	191	2.1
9.9	203	20	183	2.0
9.9	202	20	182	2.0
9.9	192	19	173	1.9
9.9	191	19	172	1.9
9.9	182	18	164	1.8
9.9	181	18	163	1.8
9.9	172	17	155	1.7
9.9	171	17	154	1.7

%E	M1	M2	DM	M*
9.9	162	16	146	1.6
9.9	161	16	145	1.6
9.9	152	15	137	1.5
9.9	151	15	136	1.5
9.9	142	14	128	1.4
9.9	141	14	127	1.4
9.9	131	13	118	1.3
9.9	121	12	109	1.2
9.9	111	11	100	1.1
9.9	101	10	91	1.0
9.9	91	9	82	0.9
9.9	81	8	73	0.8
9.9	71	7	64	0.7
9.8	500	49	451	4.8
9.8	499	49	450	4.8
9.8	498	49	449	4.8
9.8	492	48	444	4.7
9.8	491	48	443	4.7
9.8	490	48	442	4.7
9.8	489	48	441	4.7
9.8	488	48	440	4.7
9.8	482	47	435	4.6
9.8	481	47	434	4.6
9.8	480	47	433	4.6
9.8	479	47	432	4.6
9.8	478	47	431	4.6
9.8	471	46	425	4.5
9.8	470	46	424	4.5
9.8	469	46	423	4.5
9.8	468	46	422	4.5
9.8	461	45	416	4.4
9.8	460	45	415	4.4
9.8	459	45	414	4.4
9.8	458	45	413	4.4
9.8	457	45	412	4.4
9.8	451	44	407	4.3
9.8	450	44	406	4.3
9.8	449	44	405	4.3
9.8	448	44	404	4.3
9.8	447	44	403	4.3
9.8	441	43	398	4.2
9.8	440	43	397	4.2
9.8	439	43	396	4.2
9.8	438	43	395	4.2
9.8	437	43	394	4.2
9.8	430	42	388	4.1
9.8	429	42	387	4.1
9.8	428	42	386	4.1
9.8	427	42	385	4.1
9.8	420	41	379	4.0

%E	M1	M2	DM	M*
9.8	419	41	378	4.0
9.8	418	41	377	4.0
9.8	417	41	376	4.0
9.8	410	40	370	3.9
9.8	409	40	369	3.9
9.8	408	40	368	3.9
9.8	407	40	367	3.9
9.8	400	39	361	3.8
9.8	399	39	360	3.8
9.8	398	39	359	3.8
9.8	397	39	358	3.8
9.8	396	39	357	3.8
9.8	389	38	351	3.7
9.8	388	38	350	3.7
9.8	387	38	349	3.7
9.8	386	38	348	3.7
9.8	379	37	342	3.6
9.8	378	37	341	3.6
9.8	377	37	340	3.6
9.8	376	37	339	3.6
9.8	369	36	333	3.5
9.8	368	36	332	3.5
9.8	367	36	331	3.5
9.8	366	36	330	3.5
9.8	358	35	323	3.4
9.8	357	35	322	3.4
9.8	356	35	321	3.4
9.8	348	34	314	3.3
9.8	347	34	313	3.3
9.8	346	34	312	3.3
9.8	338	33	305	3.2
9.8	337	33	304	3.2
9.8	336	33	303	3.2
9.8	328	32	296	3.1
9.8	327	32	295	3.1
9.8	326	32	294	3.1
9.8	325	32	293	3.2
9.8	317	31	286	3.0
9.8	316	31	285	3.0
9.8	315	31	284	3.1
9.8	307	30	277	2.9
9.8	306	30	276	2.9
9.8	305	30	275	3.0
9.8	297	29	268	2.8
9.8	296	29	267	2.8
9.8	295	29	266	2.9
9.8	287	28	259	2.7
9.8	286	28	258	2.7
9.8	285	28	257	2.8
9.8	276	27	249	2.6

%E	M1	M2	DM	M*
9.8	275	27	248	2.7
9.8	266	26	240	2.5
9.8	265	26	239	2.6
9.8	264	26	238	2.6
9.8	256	25	231	2.4
9.8	255	25	230	2.5
9.8	254	25	229	2.5
9.8	246	24	222	2.3
9.8	245	24	221	2.4
9.8	244	24	220	2.4
9.8	235	23	212	2.3
9.8	234	23	211	2.3
9.8	225	22	203	2.2
9.8	224	22	202	2.2
9.8	215	21	194	2.1
9.8	214	21	193	2.1
9.8	205	20	185	2.0
9.8	204	20	184	2.0
9.8	194	19	175	1.9
9.8	193	19	174	1.9
9.8	184	18	166	1.8
9.8	183	18	165	1.8
9.8	174	17	157	1.7
9.8	173	17	156	1.7
9.8	164	16	148	1.6
9.8	163	16	147	1.6
9.8	153	15	138	1.5
9.8	143	14	129	1.4
9.8	133	13	120	1.3
9.8	132	13	119	1.3
9.8	123	12	111	1.2
9.8	122	12	110	1.2
9.8	112	11	101	1.1
9.8	102	10	92	1.0
9.8	92	9	83	0.9
9.8	82	8	74	0.8
9.8	61	6	55	0.6
9.8	51	5	46	0.5
9.8	41	4	37	0.4
9.7	497	48	449	4.6
9.7	496	48	448	4.6
9.7	495	48	447	4.7
9.7	494	48	446	4.7
9.7	493	48	445	4.7
9.7	487	47	440	4.5
9.7	486	47	439	4.5
9.7	485	47	438	4.6
9.7	484	47	437	4.6
9.7	483	47	436	4.6
9.7	476	46	430	4.4
9.7	475	46	429	4.5
9.7	474	46	428	4.5
9.7	473	46	427	4.5
9.7	472	46	426	4.5
9.7	466	45	421	4.3
9.7	465	45	420	4.4
9.7	464	45	419	4.4
9.7	463	45	418	4.4
9.7	462	45	417	4.4
9.7	455	44	411	4.3
9.7	454	44	410	4.3
9.7	453	44	409	4.3
9.7	452	44	408	4.3
9.7	445	43	402	4.2
9.7	444	43	401	4.2
9.7	443	43	400	4.2
9.7	442	43	399	4.2
9.7	435	42	393	4.1
9.7	434	42	392	4.1
9.7	433	42	391	4.1
9.7	432	42	390	4.1
9.7	431	42	389	4.1
9.7	424	41	383	4.0
9.7	423	41	382	4.0
9.7	422	41	381	4.0
9.7	421	41	380	4.0
9.7	414	40	374	3.9
9.7	413	40	373	3.9
9.7	412	40	372	3.9
9.7	411	40	371	3.9
9.7	404	39	365	3.8
9.7	403	39	364	3.8
9.7	402	39	363	3.8
9.7	401	39	362	3.8
9.7	393	38	355	3.7
9.7	392	38	354	3.7
9.7	391	38	353	3.7
9.7	390	38	352	3.7
9.7	383	37	346	3.6
9.7	382	37	345	3.6
9.7	381	37	344	3.6
9.7	380	37	343	3.6
9.7	373	36	337	3.5
9.7	372	36	336	3.5
9.7	371	36	335	3.5
9.7	370	36	334	3.5
9.7	362	35	327	3.4
9.7	361	35	326	3.4
9.7	360	35	325	3.4
9.7	359	35	324	3.4
9.7	352	34	318	3.3
9.7	351	34	317	3.3
9.7	350	34	316	3.3
9.7	349	34	315	3.3
9.7	341	33	308	3.2
9.7	340	33	307	3.2
9.7	339	33	306	3.2
9.7	331	32	299	3.1
9.7	330	32	298	3.1
9.7	329	32	297	3.1
9.7	321	31	290	3.0
9.7	320	31	289	3.0
9.7	319	31	288	3.0
9.7	318	31	287	3.0
9.7	310	30	280	2.9
9.7	309	30	279	2.9
9.7	308	30	278	2.9
9.7	300	29	271	2.8
9.7	299	29	270	2.8
9.7	298	29	269	2.8
9.7	290	28	262	2.7
9.7	289	28	261	2.7
9.7	288	28	260	2.7
9.7	279	27	252	2.6
9.7	278	27	251	2.6
9.7	277	27	250	2.6
9.7	269	26	243	2.5
9.7	268	26	242	2.5
9.7	267	26	241	2.5
9.7	259	25	234	2.4
9.7	258	25	233	2.4
9.7	257	25	232	2.4
9.7	248	24	224	2.3
9.7	247	24	223	2.3
9.7	238	23	215	2.2
9.7	237	23	214	2.2
9.7	236	23	213	2.2
9.7	227	22	205	2.1
9.7	226	22	204	2.1
9.7	217	21	196	2.0
9.7	216	21	195	2.0
9.7	207	20	187	1.9
9.7	206	20	186	1.9
9.7	196	19	177	1.8
9.7	195	19	176	1.9
9.7	186	18	168	1.7
9.7	185	18	167	1.8
9.7	176	17	159	1.6
9.7	175	17	158	1.7
9.7	165	16	149	1.6
9.7	155	15	140	1.5
9.7	154	15	139	1.5
9.7	145	14	131	1.4
9.7	144	14	130	1.4
9.7	134	13	121	1.3
9.7	124	12	112	1.2
9.7	113	11	102	1.1
9.7	103	10	93	1.0
9.7	93	9	84	0.9
9.7	72	7	65	0.7
9.7	62	6	56	0.6
9.7	31	3	28	0.3
9.6	500	48	452	4.6
9.6	499	48	451	4.6
9.6	498	48	450	4.6
9.6	492	47	445	4.5
9.6	491	47	444	4.5
9.6	490	47	443	4.5
9.6	489	47	442	4.5
9.6	488	47	441	4.5
9.6	481	46	435	4.4
9.6	480	46	434	4.4
9.6	479	46	433	4.4
9.6	478	46	432	4.4
9.6	477	46	431	4.4
9.6	471	45	426	4.3
9.6	470	45	425	4.3
9.6	469	45	424	4.3
9.6	468	45	423	4.3
9.6	467	45	422	4.3
9.6	460	44	416	4.2
9.6	459	44	415	4.2
9.6	458	44	414	4.2
9.6	457	44	413	4.2
9.6	456	44	412	4.2
9.6	450	43	407	4.1
9.6	449	43	406	4.1
9.6	448	43	405	4.1
9.6	447	43	404	4.1
9.6	446	43	403	4.1
9.6	439	42	397	4.0
9.6	438	42	396	4.0
9.6	437	42	395	4.0
9.6	436	42	394	4.0
9.6	429	41	388	3.9
9.6	428	41	387	3.9
9.6	427	41	386	3.9
9.6	426	41	385	3.9
9.6	425	41	384	4.0
9.6	418	40	378	3.8
9.6	417	40	377	3.8
9.6	416	40	376	3.8
9.6	415	40	375	3.9
9.6	408	39	369	3.7
9.6	407	39	368	3.7
9.6	406	39	367	3.7
9.6	405	39	366	3.8
9.6	397	38	359	3.6
9.6	396	38	358	3.6
9.6	395	38	357	3.7
9.6	394	38	356	3.7
9.6	387	37	350	3.5
9.6	386	37	349	3.5
9.6	385	37	348	3.6
9.6	384	37	347	3.6
9.6	376	36	340	3.4
9.6	375	36	339	3.5
9.6	374	36	338	3.5
9.6	366	35	331	3.3
9.6	365	35	330	3.4
9.6	364	35	329	3.4
9.6	363	35	328	3.4
9.6	356	34	322	3.2
9.6	355	34	321	3.3
9.6	354	34	320	3.3
9.6	353	34	319	3.3
9.6	345	33	312	3.2
9.6	344	33	311	3.2
9.6	343	33	310	3.2
9.6	342	33	309	3.2
9.6	335	32	303	3.1
9.6	334	32	302	3.1
9.6	333	32	301	3.1
9.6	332	32	300	3.1
9.6	324	31	293	3.0
9.6	323	31	292	3.0
9.6	322	31	291	3.0
9.6	314	30	284	2.9
9.6	313	30	283	2.9
9.6	312	30	282	2.9
9.6	311	30	281	2.9
9.6	303	29	274	2.8
9.6	302	29	273	2.8
9.6	301	29	272	2.8
9.6	293	28	265	2.7
9.6	292	28	264	2.7
9.6	291	28	263	2.7
9.6	282	27	255	2.6
9.6	281	27	254	2.6
9.6	280	27	253	2.6

%E	M1	M2	DM	M*
9.6	272	26	246	2.5
9.6	271	26	245	2.5
9.6	270	26	244	2.5
9.6	261	25	236	2.4
9.6	260	25	235	2.4
9.6	251	24	227	2.3
9.6	250	24	226	2.3
9.6	249	24	225	2.3
9.6	240	23	217	2.2
9.6	239	23	216	2.2
9.6	230	22	208	2.1
9.6	229	22	207	2.1
9.6	228	22	206	2.1
9.6	219	21	198	2.0
9.6	218	21	197	2.0
9.6	209	20	189	1.9
9.6	208	20	188	1.9
9.6	198	19	179	1.8
9.6	197	19	178	1.8
9.6	188	18	170	1.7
9.6	187	18	169	1.7
9.6	178	17	161	1.6
9.6	177	17	160	1.6
9.6	167	16	151	1.5
9.6	166	16	150	1.5
9.6	157	15	142	1.4
9.6	156	15	141	1.4
9.6	146	14	132	1.3
9.6	136	13	123	1.2
9.6	135	13	122	1.3
9.6	125	12	113	1.2
9.6	115	11	104	1.1
9.6	114	11	103	1.1
9.6	104	10	94	1.0
9.6	94	9	85	0.9
9.6	83	8	75	0.8
9.6	73	7	66	0.7
9.6	52	5	47	0.5
9.5	497	47	450	4.4
9.5	496	47	449	4.5
9.5	495	47	448	4.5
9.5	494	47	447	4.5
9.5	493	47	446	4.5
9.5	486	46	440	4.4
9.5	485	46	439	4.4
9.5	484	46	438	4.4
9.5	483	46	437	4.4
9.5	482	46	436	4.4
9.5	476	45	431	4.3
9.5	475	45	430	4.3

%E	M1	M2	DM	M*
9.5	474	45	429	4.3
9.5	473	45	428	4.3
9.5	472	45	427	4.3
9.5	465	44	421	4.2
9.5	464	44	420	4.2
9.5	463	44	419	4.2
9.5	462	44	418	4.2
9.5	461	44	417	4.2
9.5	455	43	412	4.1
9.5	454	43	411	4.1
9.5	453	43	410	4.1
9.5	452	43	409	4.1
9.5	451	43	408	4.1
9.5	444	42	402	4.0
9.5	443	42	401	4.0
9.5	442	42	400	4.0
9.5	441	42	399	4.0
9.5	440	42	398	4.0
9.5	433	41	392	3.9
9.5	432	41	391	3.9
9.5	431	41	390	3.9
9.5	430	41	389	3.9
9.5	423	40	383	3.8
9.5	422	40	382	3.8
9.5	421	40	381	3.8
9.5	420	40	380	3.8
9.5	419	40	379	3.8
9.5	412	39	373	3.7
9.5	411	39	372	3.7
9.5	410	39	371	3.7
9.5	409	39	370	3.7
9.5	402	38	364	3.6
9.5	401	38	363	3.6
9.5	400	38	362	3.6
9.5	399	38	361	3.6
9.5	398	38	360	3.6
9.5	391	37	354	3.5
9.5	390	37	353	3.5
9.5	389	37	352	3.5
9.5	388	37	351	3.5
9.5	380	36	344	3.4
9.5	379	36	343	3.4
9.5	378	36	342	3.4
9.5	377	36	341	3.4
9.5	370	35	335	3.3
9.5	369	35	334	3.3
9.5	368	35	333	3.3
9.5	367	35	332	3.3
9.5	359	34	325	3.2
9.5	358	34	324	3.2

%E	M1	M2	DM	M*
9.5	357	34	323	3.2
9.5	349	33	316	3.1
9.5	348	33	315	3.1
9.5	347	33	314	3.1
9.5	346	33	313	3.1
9.5	338	32	306	3.0
9.5	337	32	305	3.0
9.5	336	32	304	3.0
9.5	328	31	297	2.9
9.5	327	31	296	2.9
9.5	326	31	295	2.9
9.5	325	31	294	3.0
9.5	317	30	287	2.8
9.5	316	30	286	2.8
9.5	315	30	285	2.9
9.5	306	29	277	2.7
9.5	305	29	276	2.8
9.5	304	29	275	2.8
9.5	296	28	268	2.6
9.5	295	28	267	2.7
9.5	294	28	266	2.7
9.5	285	27	258	2.6
9.5	284	27	257	2.6
9.5	283	27	256	2.6
9.5	275	26	249	2.5
9.5	274	26	248	2.5
9.5	273	26	247	2.5
9.5	264	25	239	2.4
9.5	263	25	238	2.4
9.5	262	25	237	2.4
9.5	253	24	229	2.3
9.5	252	24	228	2.3
9.5	243	23	220	2.2
9.5	242	23	219	2.2
9.5	241	23	218	2.2
9.5	232	22	210	2.1
9.5	231	22	209	2.1
9.5	222	21	201	2.0
9.5	221	21	200	2.0
9.5	220	21	199	2.0
9.5	211	20	191	1.9
9.5	210	20	190	1.9
9.5	201	19	182	1.8
9.5	200	19	181	1.8
9.5	199	19	180	1.8
9.5	190	18	172	1.7
9.5	189	18	171	1.7
9.5	179	17	162	1.6
9.5	169	16	153	1.5
9.5	168	16	152	1.5

%E	M1	M2	DM	M*
9.5	158	15	143	1.4
9.5	148	14	134	1.3
9.5	147	14	133	1.3
9.5	137	13	124	1.2
9.5	126	12	114	1.1
9.5	116	11	105	1.0
9.5	105	10	95	1.0
9.5	95	9	86	0.9
9.5	84	8	76	0.8
9.5	74	7	67	0.7
9.5	63	6	57	0.6
9.5	42	4	38	0.4
9.5	21	2	19	0.2
9.4	500	47	453	4.4
9.4	499	47	452	4.4
9.4	498	47	451	4.4
9.4	492	46	446	4.3
9.4	491	46	445	4.3
9.4	490	46	444	4.3
9.4	489	46	443	4.3
9.4	488	46	442	4.3
9.4	487	46	441	4.3
9.4	481	45	436	4.2
9.4	480	45	435	4.2
9.4	479	45	434	4.2
9.4	478	45	433	4.2
9.4	477	45	432	4.2
9.4	470	44	426	4.1
9.4	469	44	425	4.1
9.4	468	44	424	4.1
9.4	467	44	423	4.1
9.4	466	44	422	4.2
9.4	459	43	416	4.0
9.4	458	43	415	4.0
9.4	457	43	414	4.0
9.4	456	43	413	4.1
9.4	449	42	407	3.9
9.4	448	42	406	3.9
9.4	447	42	405	3.9
9.4	446	42	404	4.0
9.4	445	42	403	4.0
9.4	438	41	397	3.8
9.4	437	41	396	3.8
9.4	436	41	395	3.9
9.4	435	41	394	3.9
9.4	434	41	393	3.9
9.4	427	40	387	3.7
9.4	426	40	386	3.8
9.4	425	40	385	3.8
9.4	424	40	384	3.8

%E	M1	M2	DM	M*
9.4	417	39	378	3.6
9.4	416	39	377	3.7
9.4	415	39	376	3.7
9.4	414	39	375	3.7
9.4	413	39	374	3.7
9.4	406	38	368	3.6
9.4	405	38	367	3.6
9.4	404	38	366	3.6
9.4	403	38	365	3.6
9.4	395	37	358	3.5
9.4	394	37	357	3.5
9.4	393	37	356	3.5
9.4	392	37	355	3.5
9.4	385	36	349	3.4
9.4	384	36	348	3.4
9.4	383	36	347	3.4
9.4	382	36	346	3.4
9.4	381	36	345	3.4
9.4	374	35	339	3.3
9.4	373	35	338	3.3
9.4	372	35	337	3.3
9.4	371	35	336	3.3
9.4	363	34	329	3.2
9.4	362	34	328	3.2
9.4	361	34	327	3.2
9.4	360	34	326	3.2
9.4	352	33	319	3.1
9.4	351	33	318	3.1
9.4	350	33	317	3.1
9.4	342	32	310	3.0
9.4	341	32	309	3.0
9.4	340	32	308	3.0
9.4	339	32	307	3.0
9.4	331	31	300	2.9
9.4	330	31	299	2.9
9.4	329	31	298	2.9
9.4	320	30	290	2.8
9.4	319	30	289	2.8
9.4	318	30	288	2.8
9.4	310	29	281	2.7
9.4	309	29	280	2.7
9.4	308	29	279	2.7
9.4	307	29	278	2.7
9.4	299	28	271	2.6
9.4	298	28	270	2.6
9.4	297	28	269	2.6
9.4	288	27	261	2.5
9.4	287	27	260	2.5
9.4	286	27	259	2.5
9.4	278	26	252	2.4

%E	M1	M2	DM	M*
9.4	277	26	251	2.4
9.4	276	26	250	2.4
9.4	267	25	242	2.3
9.4	266	25	241	2.3
9.4	265	25	240	2.4
9.4	256	24	232	2.3
9.4	255	24	231	2.3
9.4	254	24	230	2.3
9.4	246	23	223	2.2
9.4	245	23	222	2.2
9.4	244	23	221	2.2
9.4	235	22	213	2.1
9.4	234	22	212	2.1
9.4	233	22	211	2.1
9.4	224	21	203	2.0
9.4	223	21	202	2.0
9.4	213	20	193	1.9
9.4	212	20	192	1.9
9.4	203	19	184	1.8
9.4	202	19	183	1.8
9.4	192	18	174	1.7
9.4	191	18	173	1.7
9.4	181	17	164	1.6
9.4	180	17	163	1.6
9.4	171	16	155	1.5
9.4	170	16	154	1.5
9.4	160	15	145	1.4
9.4	159	15	144	1.4
9.4	149	14	135	1.3
9.4	139	13	126	1.2
9.4	138	13	125	1.2
9.4	128	12	116	1.1
9.4	127	12	115	1.1
9.4	117	11	106	1.0
9.4	106	10	96	0.9
9.4	96	9	87	0.8
9.4	85	8	77	0.8
9.4	64	6	58	0.6
9.4	53	5	48	0.5
9.4	32	3	29	0.3
9.3	497	46	451	4.3
9.3	496	46	450	4.3
9.3	495	46	449	4.3
9.3	494	46	448	4.3
9.3	493	46	447	4.3
9.3	486	45	441	4.2
9.3	485	45	440	4.2
9.3	484	45	439	4.2
9.3	483	45	438	4.2
9.3	482	45	437	4.2

%E	M1	M2	DM	M*
9.3	475	44	431	4.1
9.3	474	44	430	4.1
9.3	473	44	429	4.1
9.3	472	44	428	4.1
9.3	471	44	427	4.1
9.3	464	43	421	4.0
9.3	463	43	420	4.0
9.3	462	43	419	4.0
9.3	461	43	418	4.0
9.3	460	43	417	4.0
9.3	454	42	412	3.9
9.3	453	42	411	3.9
9.3	452	42	410	3.9
9.3	451	42	409	3.9
9.3	450	42	408	3.9
9.3	443	41	402	3.8
9.3	442	41	401	3.8
9.3	441	41	400	3.8
9.3	440	41	399	3.8
9.3	439	41	398	3.8
9.3	432	40	392	3.7
9.3	431	40	391	3.7
9.3	430	40	390	3.7
9.3	429	40	389	3.7
9.3	428	40	388	3.7
9.3	421	39	382	3.6
9.3	420	39	381	3.6
9.3	419	39	380	3.6
9.3	418	39	379	3.6
9.3	410	38	372	3.5
9.3	409	38	371	3.5
9.3	408	38	370	3.5
9.3	407	38	369	3.5
9.3	400	37	363	3.4
9.3	399	37	362	3.4
9.3	398	37	361	3.4
9.3	397	37	360	3.4
9.3	396	37	359	3.5
9.3	389	36	353	3.3
9.3	388	36	352	3.3
9.3	387	36	351	3.3
9.3	386	36	350	3.4
9.3	378	35	343	3.2
9.3	377	35	342	3.2
9.3	376	35	341	3.3
9.3	375	35	340	3.3
9.3	367	34	333	3.1
9.3	366	34	332	3.2
9.3	365	34	331	3.2
9.3	364	34	330	3.2

%E	M1	M2	DM	M*
9.3	356	33	323	3.1
9.3	355	33	322	3.1
9.3	354	33	321	3.1
9.3	353	33	320	3.1
9.3	345	32	313	3.0
9.3	344	32	312	3.0
9.3	343	32	311	3.0
9.3	335	31	304	2.9
9.3	334	31	303	2.9
9.3	333	31	302	2.9
9.3	332	31	301	2.9
9.3	324	30	294	2.8
9.3	323	30	293	2.8
9.3	322	30	292	2.8
9.3	321	30	291	2.8
9.3	313	29	284	2.7
9.3	312	29	283	2.7
9.3	311	29	282	2.7
9.3	302	28	274	2.6
9.3	301	28	273	2.6
9.3	300	28	272	2.6
9.3	291	27	264	2.5
9.3	290	27	263	2.5
9.3	289	27	262	2.5
9.3	281	26	255	2.4
9.3	280	26	254	2.4
9.3	279	26	253	2.4
9.3	270	25	245	2.3
9.3	269	25	244	2.3
9.3	268	25	243	2.3
9.3	259	24	235	2.2
9.3	258	24	234	2.2
9.3	257	24	233	2.2
9.3	248	23	225	2.1
9.3	247	23	224	2.1
9.3	237	22	215	2.0
9.3	236	22	214	2.1
9.3	227	21	206	1.9
9.3	226	21	205	2.0
9.3	225	21	204	2.0
9.3	216	20	196	1.9
9.3	215	20	195	1.9
9.3	214	20	194	1.9
9.3	205	19	186	1.8
9.3	204	19	185	1.8
9.3	194	18	176	1.7
9.3	193	18	175	1.7
9.3	183	17	166	1.6
9.3	182	17	165	1.6
9.3	172	16	156	1.5

%E	M1	M2	DM	M*
9.3	162	15	147	1.4
9.3	161	15	146	1.4
9.3	151	14	137	1.3
9.3	150	14	136	1.3
9.3	140	13	127	1.2
9.3	129	12	117	1.1
9.3	118	11	107	1.0
9.3	108	10	98	0.9
9.3	107	10	97	0.9
9.3	97	9	88	0.8
9.3	86	8	78	0.7
9.3	75	7	68	0.7
9.3	54	5	49	0.5
9.3	43	4	39	0.4
9.2	500	46	454	4.2
9.2	499	46	453	4.2
9.2	498	46	452	4.2
9.2	491	45	446	4.1
9.2	490	45	445	4.1
9.2	489	45	444	4.1
9.2	488	45	443	4.1
9.2	487	45	442	4.2
9.2	480	44	436	4.0
9.2	479	44	435	4.0
9.2	478	44	434	4.1
9.2	477	44	433	4.1
9.2	476	44	432	4.1
9.2	469	43	426	3.9
9.2	468	43	425	4.0
9.2	467	43	424	4.0
9.2	466	43	423	4.0
9.2	465	43	422	4.0
9.2	459	42	417	3.8
9.2	458	42	416	3.9
9.2	457	42	415	3.9
9.2	456	42	414	3.9
9.2	455	42	413	3.9
9.2	448	41	407	3.8
9.2	447	41	406	3.8
9.2	446	41	405	3.8
9.2	445	41	404	3.8
9.2	444	41	403	3.8
9.2	437	40	397	3.7
9.2	436	40	396	3.7
9.2	435	40	395	3.7
9.2	434	40	394	3.7
9.2	433	40	393	3.7
9.2	426	39	387	3.6
9.2	425	39	386	3.6
9.2	424	39	385	3.6

%E	M1	M2	DM	M*
9.2	423	39	384	3.6
9.2	422	39	383	3.6
9.2	415	38	377	3.5
9.2	414	38	376	3.5
9.2	413	38	375	3.5
9.2	412	38	374	3.5
9.2	411	38	373	3.5
9.2	404	37	367	3.4
9.2	403	37	366	3.4
9.2	402	37	365	3.4
9.2	401	37	364	3.4
9.2	393	36	357	3.3
9.2	392	36	356	3.3
9.2	391	36	355	3.3
9.2	390	36	354	3.3
9.2	382	35	347	3.2
9.2	381	35	346	3.2
9.2	380	35	345	3.2
9.2	379	35	344	3.2
9.2	371	34	337	3.1
9.2	370	34	336	3.1
9.2	369	34	335	3.1
9.2	368	34	334	3.1
9.2	360	33	327	3.0
9.2	359	33	326	3.0
9.2	358	33	325	3.0
9.2	357	33	324	3.1
9.2	349	32	317	2.9
9.2	348	32	316	2.9
9.2	347	32	315	3.0
9.2	346	32	314	3.0
9.2	338	31	307	2.8
9.2	337	31	306	2.9
9.2	336	31	305	2.9
9.2	327	30	297	2.8
9.2	326	30	296	2.8
9.2	325	30	295	2.8
9.2	316	29	287	2.7
9.2	315	29	286	2.7
9.2	314	29	285	2.7
9.2	306	28	278	2.6
9.2	305	28	277	2.6
9.2	304	28	276	2.6
9.2	303	28	275	2.6
9.2	295	27	268	2.5
9.2	294	27	267	2.5
9.2	293	27	266	2.5
9.2	292	27	265	2.5
9.2	284	26	258	2.4
9.2	283	26	257	2.4

%E	M1	M2	DM	M*
9.2	282	26	256	2.4
9.2	273	25	248	2.3
9.2	272	25	247	2.3
9.2	271	25	246	2.3
9.2	262	24	238	2.2
9.2	261	24	237	2.2
9.2	260	24	236	2.2
9.2	251	23	228	2.1
9.2	250	23	227	2.1
9.2	249	23	226	2.1
9.2	240	22	218	2.0
9.2	239	22	217	2.0
9.2	238	22	216	2.0
9.2	229	21	208	1.9
9.2	228	21	207	1.9
9.2	218	20	198	1.8
9.2	217	20	197	1.8
9.2	207	19	188	1.7
9.2	206	19	187	1.8
9.2	196	18	178	1.7
9.2	195	18	177	1.7
9.2	185	17	168	1.6
9.2	184	17	167	1.6
9.2	174	16	158	1.5
9.2	173	16	157	1.5
9.2	163	15	148	1.4
9.2	153	14	139	1.3
9.2	152	14	138	1.3
9.2	142	13	129	1.2
9.2	141	13	128	1.2
9.2	131	12	119	1.1
9.2	130	12	118	1.1
9.2	120	11	109	1.0
9.2	119	11	108	1.0
9.2	109	10	99	0.9
9.2	98	9	89	0.8
9.2	87	8	79	0.7
9.2	76	7	69	0.6
9.2	65	6	59	0.6
9.1	497	45	452	4.1
9.1	496	45	451	4.1
9.1	495	45	450	4.1
9.1	494	45	449	4.1
9.1	493	45	448	4.1
9.1	492	45	447	4.1
9.1	486	44	442	4.0
9.1	485	44	441	4.0
9.1	484	44	440	4.0
9.1	483	44	439	4.0
9.1	482	44	438	4.0
9.1	481	44	437	4.0
9.1	475	43	432	3.9
9.1	474	43	431	3.9
9.1	473	43	430	3.9
9.1	472	43	429	3.9
9.1	471	43	428	3.9
9.1	470	43	427	3.9
9.1	464	42	422	3.8
9.1	463	42	421	3.8
9.1	462	42	420	3.8
9.1	461	42	419	3.8
9.1	460	42	418	3.8
9.1	453	41	412	3.7
9.1	452	41	411	3.7
9.1	451	41	410	3.7
9.1	450	41	409	3.7
9.1	449	41	408	3.7
9.1	442	40	402	3.6
9.1	441	40	401	3.6
9.1	440	40	400	3.6
9.1	439	40	399	3.6
9.1	438	40	398	3.7
9.1	430	39	391	3.5
9.1	429	39	390	3.5
9.1	428	39	389	3.6
9.1	427	39	388	3.6
9.1	419	38	381	3.4
9.1	418	38	380	3.5
9.1	417	38	379	3.5
9.1	416	38	378	3.5
9.1	408	37	371	3.4
9.1	407	37	370	3.4
9.1	406	37	369	3.4
9.1	405	37	368	3.4
9.1	397	36	361	3.3
9.1	396	36	360	3.3
9.1	395	36	359	3.3
9.1	394	36	358	3.3
9.1	386	35	351	3.2
9.1	385	35	350	3.2
9.1	384	35	349	3.2
9.1	383	35	348	3.2
9.1	375	34	341	3.1
9.1	374	34	340	3.1
9.1	373	34	339	3.1
9.1	372	34	338	3.1
9.1	364	33	331	3.0
9.1	363	33	330	3.0
9.1	362	33	329	3.0
9.1	361	33	328	3.0
9.1	353	32	321	2.9
9.1	352	32	320	2.9
9.1	351	32	319	2.9
9.1	350	32	318	2.9
9.1	342	31	311	2.8
9.1	341	31	310	2.8
9.1	340	31	309	2.8
9.1	339	31	308	2.8
9.1	331	30	301	2.7
9.1	330	30	300	2.7
9.1	329	30	299	2.7
9.1	328	30	298	2.7
9.1	320	29	291	2.6
9.1	319	29	290	2.6
9.1	318	29	289	2.6
9.1	317	29	288	2.7
9.1	309	28	281	2.5
9.1	308	28	280	2.5
9.1	307	28	279	2.6
9.1	298	27	271	2.4
9.1	297	27	270	2.5
9.1	296	27	269	2.5
9.1	287	26	261	2.4
9.1	286	26	260	2.4
9.1	285	26	259	2.4
9.1	276	25	251	2.3
9.1	275	25	250	2.3
9.1	274	25	249	2.3
9.1	265	24	241	2.2
9.1	264	24	240	2.2
9.1	263	24	239	2.2
9.1	254	23	231	2.1
9.1	253	23	230	2.1
9.1	252	23	229	2.1
9.1	243	22	221	2.0
9.1	242	22	220	2.0
9.1	241	22	219	2.0
9.1	232	21	211	1.9
9.1	231	21	210	1.9
9.1	230	21	209	1.9
9.1	221	20	201	1.8
9.1	220	20	200	1.8
9.1	219	20	199	1.8
9.1	209	19	190	1.7
9.1	208	19	189	1.7
9.1	198	18	180	1.6
9.1	197	18	179	1.6
9.1	187	17	170	1.5
9.1	186	17	169	1.6
9.1	176	16	160	1.5
9.1	175	16	159	1.5
9.1	165	15	150	1.4
9.1	164	15	149	1.4
9.1	154	14	140	1.3
9.1	143	13	130	1.2
9.1	132	12	120	1.1
9.1	121	11	110	1.0
9.1	110	10	100	0.9
9.1	99	9	90	0.8
9.1	88	8	80	0.7
9.1	77	7	70	0.6
9.1	66	6	60	0.5
9.1	55	5	50	0.5
9.1	44	4	40	0.4
9.1	33	3	30	0.3
9.1	22	2	20	0.2
9.1	11	1	10	0.1
9.0	500	45	455	4.0
9.0	499	45	454	4.1
9.0	498	45	453	4.1
9.0	491	44	447	3.9
9.0	490	44	446	4.0
9.0	489	44	445	4.0
9.0	488	44	444	4.0
9.0	487	44	443	4.0
9.0	480	43	437	3.9
9.0	479	43	436	3.9
9.0	478	43	435	3.9
9.0	477	43	434	3.9
9.0	476	43	433	3.9
9.0	469	42	427	3.8
9.0	468	42	426	3.8
9.0	467	42	425	3.8
9.0	466	42	424	3.8
9.0	465	42	423	3.8
9.0	458	41	417	3.7
9.0	457	41	416	3.7
9.0	456	41	415	3.7
9.0	455	41	414	3.7
9.0	454	41	413	3.7
9.0	446	40	406	3.6
9.0	445	40	405	3.6
9.0	444	40	404	3.6
9.0	443	40	403	3.6
9.0	435	39	396	3.5
9.0	434	39	395	3.5
9.0	433	39	394	3.5
9.0	432	39	393	3.5
9.0	431	39	392	3.5
9.0	424	38	386	3.4
9.0	423	38	385	3.4
9.0	422	38	384	3.4
9.0	421	38	383	3.4
9.0	420	38	382	3.4
9.0	413	37	376	3.3
9.0	412	37	375	3.3
9.0	411	37	374	3.3
9.0	410	37	373	3.3
9.0	409	37	372	3.3
9.0	402	36	366	3.2
9.0	401	36	365	3.2
9.0	400	36	364	3.2
9.0	399	36	363	3.2
9.0	398	36	362	3.3
9.0	391	35	356	3.1
9.0	390	35	355	3.1
9.0	389	35	354	3.1
9.0	388	35	353	3.2
9.0	387	35	352	3.2
9.0	379	34	345	3.1
9.0	378	34	344	3.1
9.0	377	34	343	3.1
9.0	376	34	342	3.1
9.0	368	33	335	3.0
9.0	367	33	334	3.0
9.0	366	33	333	3.0
9.0	365	33	332	3.0
9.0	357	32	325	2.9
9.0	356	32	324	2.9
9.0	355	32	323	2.9
9.0	354	32	322	2.9
9.0	346	31	315	2.8
9.0	345	31	314	2.8
9.0	344	31	313	2.8
9.0	343	31	312	2.8
9.0	335	30	305	2.7
9.0	334	30	304	2.7
9.0	333	30	303	2.7
9.0	332	30	302	2.7
9.0	324	29	295	2.6
9.0	323	29	294	2.6
9.0	322	29	293	2.6
9.0	321	29	292	2.6
9.0	312	28	284	2.5
9.0	311	28	283	2.5
9.0	310	28	282	2.5
9.0	301	27	274	2.4
9.0	300	27	273	2.4
9.0	299	27	272	2.4
9.0	290	26	264	2.3

%E	M1	M2	DM	M*
9.0	289	26	263	2.3
9.0	288	26	262	2.3
9.0	279	25	254	2.2
9.0	278	25	253	2.2
9.0	277	25	252	2.3
9.0	268	24	244	2.1
9.0	267	24	243	2.2
9.0	266	24	242	2.2
9.0	256	23	233	2.1
9.0	255	23	232	2.1
9.0	245	22	223	2.0
9.0	244	22	222	2.0
9.0	234	21	213	1.9
9.0	233	21	212	1.9
9.0	223	20	203	1.8
9.0	222	20	202	1.8
9.0	212	19	193	1.7
9.0	211	19	192	1.7
9.0	210	19	191	1.7
9.0	201	18	183	1.6
9.0	200	18	182	1.6
9.0	199	18	181	1.6
9.0	189	17	172	1.5
9.0	188	17	171	1.5
9.0	178	16	162	1.4
9.0	177	16	161	1.4
9.0	167	15	152	1.3
9.0	166	15	151	1.4
9.0	156	14	142	1.3
9.0	155	14	141	1.3
9.0	145	13	132	1.2
9.0	144	13	131	1.2
9.0	134	12	122	1.1
9.0	133	12	121	1.1
9.0	122	11	111	1.0
9.0	111	10	101	0.9
9.0	100	9	91	0.8
9.0	89	8	81	0.7
9.0	78	7	71	0.6
9.0	67	6	61	0.5
8.9	497	44	453	3.9
8.9	496	44	452	3.9
8.9	495	44	451	3.9
8.9	494	44	450	3.9
8.9	493	44	449	3.9
8.9	492	44	448	3.9
8.9	485	43	442	3.8
8.9	484	43	441	3.8
8.9	483	43	440	3.8
8.9	482	43	439	3.8
8.9	481	43	438	3.8
8.9	474	42	432	3.7
8.9	473	42	431	3.7
8.9	472	42	430	3.7
8.9	471	42	429	3.7
8.9	470	42	428	3.8
8.9	463	41	422	3.6
8.9	462	41	421	3.6
8.9	461	41	420	3.6
8.9	460	41	419	3.7
8.9	459	41	418	3.7
8.9	452	40	412	3.5
8.9	451	40	411	3.5
8.9	450	40	410	3.6
8.9	449	40	409	3.6
8.9	448	40	408	3.6
8.9	447	40	407	3.6
8.9	440	39	401	3.5
8.9	439	39	400	3.5
8.9	438	39	399	3.5
8.9	437	39	398	3.5
8.9	436	39	397	3.5
8.9	429	38	391	3.4
8.9	428	38	390	3.4
8.9	427	38	389	3.4
8.9	426	38	388	3.4
8.9	425	38	387	3.4
8.9	418	37	381	3.3
8.9	417	37	380	3.3
8.9	416	37	379	3.3
8.9	415	37	378	3.3
8.9	414	37	377	3.3
8.9	406	36	370	3.2
8.9	405	36	369	3.2
8.9	404	36	368	3.2
8.9	403	36	367	3.2
8.9	395	35	360	3.1
8.9	394	35	359	3.1
8.9	393	35	358	3.1
8.9	392	35	357	3.1
8.9	384	34	350	3.0
8.9	383	34	349	3.0
8.9	382	34	348	3.0
8.9	381	34	347	3.0
8.9	380	34	346	3.0
8.9	372	33	339	2.9
8.9	371	33	338	2.9
8.9	370	33	337	2.9
8.9	369	33	336	3.0
8.9	361	32	329	2.8
8.9	360	32	328	2.8
8.9	359	32	327	2.9
8.9	358	32	326	2.9
8.9	350	31	319	2.7
8.9	349	31	318	2.8
8.9	348	31	317	2.8
8.9	347	31	316	2.8
8.9	339	30	309	2.7
8.9	338	30	308	2.7
8.9	337	30	307	2.7
8.9	336	30	306	2.7
8.9	327	29	298	2.6
8.9	326	29	297	2.6
8.9	325	29	296	2.6
8.9	316	28	288	2.5
8.9	315	28	287	2.5
8.9	314	28	286	2.5
8.9	313	28	285	2.5
8.9	305	27	278	2.4
8.9	304	27	277	2.4
8.9	303	27	276	2.4
8.9	302	27	275	2.4
8.9	293	26	267	2.3
8.9	292	26	266	2.3
8.9	291	26	265	2.3
8.9	282	25	257	2.2
8.9	281	25	256	2.2
8.9	280	25	255	2.2
8.9	271	24	247	2.1
8.9	270	24	246	2.1
8.9	269	24	245	2.1
8.9	259	23	236	2.0
8.9	258	23	235	2.1
8.9	257	23	234	2.1
8.9	248	22	226	2.0
8.9	247	22	225	2.0
8.9	246	22	224	2.0
8.9	237	21	216	1.9
8.9	236	21	215	1.9
8.9	235	21	214	1.9
8.9	226	20	206	1.8
8.9	225	20	205	1.8
8.9	224	20	204	1.8
8.9	214	19	195	1.7
8.9	213	19	194	1.7
8.9	203	18	185	1.6
8.9	202	18	184	1.6
8.9	192	17	175	1.5
8.9	191	17	174	1.5
8.9	190	17	173	1.5
8.9	180	16	164	1.4
8.9	179	16	163	1.4
8.9	169	15	154	1.3
8.9	168	15	153	1.3
8.9	158	14	144	1.2
8.9	157	14	143	1.2
8.9	146	13	133	1.2
8.9	135	12	123	1.1
8.9	124	11	113	1.0
8.9	123	11	112	1.0
8.9	113	10	103	0.9
8.9	112	10	102	0.9
8.9	101	9	92	0.8
8.9	90	8	82	0.7
8.9	79	7	72	0.6
8.9	56	5	51	0.4
8.9	45	4	41	0.4
8.8	500	44	456	3.9
8.8	499	44	455	3.9
8.8	498	44	454	3.9
8.8	491	43	448	3.8
8.8	490	43	447	3.8
8.8	489	43	446	3.8
8.8	488	43	445	3.8
8.8	487	43	444	3.8
8.8	486	43	443	3.8
8.8	480	42	438	3.7
8.8	479	42	437	3.7
8.8	478	42	436	3.7
8.8	477	42	435	3.7
8.8	476	42	434	3.7
8.8	475	42	433	3.7
8.8	468	41	427	3.6
8.8	467	41	426	3.6
8.8	466	41	425	3.6
8.8	465	41	424	3.6
8.8	464	41	423	3.6
8.8	457	40	417	3.5
8.8	456	40	416	3.5
8.8	455	40	415	3.5
8.8	454	40	414	3.5
8.8	453	40	413	3.5
8.8	445	39	406	3.4
8.8	444	39	405	3.4
8.8	443	39	404	3.4
8.8	442	39	403	3.4
8.8	441	39	402	3.4
8.8	434	38	396	3.3
8.8	433	38	395	3.3
8.8	432	38	394	3.3
8.8	431	38	393	3.4
8.8	430	38	392	3.4
8.8	422	37	385	3.2
8.8	421	37	384	3.3
8.8	420	37	383	3.3
8.8	419	37	382	3.3
8.8	411	36	375	3.2
8.8	410	36	374	3.2
8.8	409	36	373	3.2
8.8	408	36	372	3.2
8.8	407	36	371	3.2
8.8	400	35	365	3.1
8.8	399	35	364	3.1
8.8	398	35	363	3.1
8.8	397	35	362	3.1
8.8	396	35	361	3.1
8.8	388	34	354	3.0
8.8	387	34	353	3.0
8.8	386	34	352	3.0
8.8	385	34	351	3.0
8.8	377	33	344	2.9
8.8	376	33	343	2.9
8.8	375	33	342	2.9
8.8	374	33	341	2.9
8.8	373	33	340	2.9
8.8	365	32	333	2.8
8.8	364	32	332	2.8
8.8	363	32	331	2.8
8.8	362	32	330	2.8
8.8	354	31	323	2.7
8.8	353	31	322	2.7
8.8	352	31	321	2.7
8.8	351	31	320	2.7
8.8	342	30	312	2.6
8.8	341	30	311	2.6
8.8	340	30	310	2.6
8.8	331	29	302	2.5
8.8	330	29	301	2.5
8.8	329	29	300	2.6
8.8	328	29	299	2.6
8.8	320	28	292	2.4
8.8	319	28	291	2.5
8.8	318	28	290	2.5
8.8	317	28	289	2.5
8.8	308	27	281	2.4
8.8	307	27	280	2.4
8.8	306	27	279	2.4
8.8	297	26	271	2.3
8.8	296	26	270	2.3
8.8	295	26	269	2.3

%E	M1	M2	DM	M*
8.8	294	26	268	2.3
8.8	285	25	260	2.2
8.8	284	25	259	2.2
8.8	283	25	258	2.2
8.8	274	24	250	2.1
8.8	273	24	249	2.1
8.8	272	24	248	2.1
8.8	262	23	239	2.0
8.8	261	23	238	2.0
8.8	260	23	237	2.0
8.8	251	22	229	1.9
8.8	250	22	228	1.9
8.8	249	22	227	1.9
8.8	240	21	219	1.8
8.8	239	21	218	1.8
8.8	238	21	217	1.9
8.8	228	20	208	1.8
8.8	227	20	207	1.8
8.8	217	19	198	1.7
8.8	216	19	197	1.7
8.8	215	19	196	1.7
8.8	205	18	187	1.6
8.8	204	18	186	1.6
8.8	194	17	177	1.5
8.8	193	17	176	1.5
8.8	182	16	166	1.4
8.8	181	16	165	1.4
8.8	171	15	156	1.3
8.8	170	15	155	1.3
8.8	160	14	146	1.2
8.8	159	14	145	1.2
8.8	148	13	135	1.1
8.8	147	13	134	1.1
8.8	137	12	125	1.1
8.8	136	12	124	1.1
8.8	125	11	114	1.0
8.8	114	10	104	0.9
8.8	102	9	93	0.8
8.8	91	8	83	0.7
8.8	80	7	73	0.6
8.8	68	6	62	0.5
8.8	57	5	52	0.4
8.8	34	3	31	0.3
8.7	497	43	454	3.7
8.7	496	43	453	3.7
8.7	495	43	452	3.7
8.7	494	43	451	3.7
8.7	493	43	450	3.8
8.7	492	43	449	3.8
8.7	485	42	443	3.6

%E	M1	M2	DM	M*
8.7	484	42	442	3.6
8.7	483	42	441	3.7
8.7	482	42	440	3.7
8.7	481	42	439	3.7
8.7	474	41	433	3.5
8.7	473	41	432	3.6
8.7	472	41	431	3.6
8.7	471	41	430	3.6
8.7	470	41	429	3.6
8.7	469	41	428	3.6
8.7	462	40	422	3.5
8.7	461	40	421	3.5
8.7	460	40	420	3.5
8.7	459	40	419	3.5
8.7	458	40	418	3.5
8.7	450	39	411	3.4
8.7	449	39	410	3.4
8.7	448	39	409	3.4
8.7	447	39	408	3.4
8.7	446	39	407	3.4
8.7	439	38	401	3.3
8.7	438	38	400	3.3
8.7	437	38	399	3.3
8.7	436	38	398	3.3
8.7	435	38	397	3.3
8.7	427	37	390	3.2
8.7	426	37	389	3.2
8.7	425	37	388	3.2
8.7	424	37	387	3.2
8.7	423	37	386	3.2
8.7	416	36	380	3.1
8.7	415	36	379	3.1
8.7	414	36	378	3.1
8.7	413	36	377	3.1
8.7	412	36	376	3.1
8.7	404	35	369	3.0
8.7	403	35	368	3.0
8.7	402	35	367	3.0
8.7	401	35	366	3.1
8.7	393	34	359	2.9
8.7	392	34	358	2.9
8.7	391	34	357	3.0
8.7	390	34	356	3.0
8.7	389	34	355	3.0
8.7	381	33	348	2.9
8.7	380	33	347	2.9
8.7	379	33	346	2.9
8.7	378	33	345	2.9
8.7	369	32	337	2.8
8.7	368	32	336	2.8

%E	M1	M2	DM	M*
8.7	367	32	335	2.8
8.7	366	32	334	2.8
8.7	358	31	327	2.7
8.7	357	31	326	2.7
8.7	356	31	325	2.7
8.7	355	31	324	2.7
8.7	346	30	316	2.6
8.7	345	30	315	2.6
8.7	344	30	314	2.6
8.7	343	30	313	2.6
8.7	335	29	306	2.5
8.7	334	29	305	2.5
8.7	333	29	304	2.5
8.7	332	29	303	2.5
8.7	323	28	295	2.4
8.7	322	28	294	2.4
8.7	321	28	293	2.4
8.7	312	27	285	2.3
8.7	311	27	284	2.3
8.7	310	27	283	2.4
8.7	309	27	282	2.4
8.7	300	26	274	2.3
8.7	299	26	273	2.3
8.7	298	26	272	2.3
8.7	289	25	264	2.2
8.7	288	25	263	2.2
8.7	287	25	262	2.2
8.7	286	25	261	2.2
8.7	277	24	253	2.1
8.7	276	24	252	2.1
8.7	275	24	251	2.1
8.7	265	23	242	2.0
8.7	264	23	241	2.0
8.7	263	23	240	2.0
8.7	254	22	232	1.9
8.7	253	22	231	1.9
8.7	252	22	230	1.9
8.7	242	21	221	1.8
8.7	241	21	220	1.8
8.7	231	20	211	1.7
8.7	230	20	210	1.7
8.7	229	20	209	1.7
8.7	219	19	200	1.6
8.7	218	19	199	1.7
8.7	208	18	190	1.6
8.7	207	18	189	1.6
8.7	206	18	188	1.6
8.7	196	17	179	1.5
8.7	195	17	178	1.5
8.7	184	16	168	1.4

%E	M1	M2	DM	M*
8.7	183	16	167	1.4
8.7	173	15	158	1.3
8.7	172	15	157	1.3
8.7	161	14	147	1.2
8.7	150	13	137	1.1
8.7	149	13	136	1.1
8.7	138	12	126	1.0
8.7	127	11	116	1.0
8.7	126	11	115	1.0
8.7	115	10	105	0.9
8.7	104	9	95	0.8
8.7	103	9	94	0.8
8.7	92	8	84	0.7
8.7	69	6	63	0.5
8.7	46	4	42	0.3
8.7	23	2	21	0.2
8.6	500	43	457	3.7
8.6	499	43	456	3.7
8.6	498	43	455	3.7
8.6	491	42	449	3.6
8.6	490	42	448	3.6
8.6	489	42	447	3.6
8.6	488	42	446	3.6
8.6	487	42	445	3.6
8.6	486	42	444	3.6
8.6	479	41	438	3.5
8.6	478	41	437	3.5
8.6	477	41	436	3.5
8.6	476	41	435	3.5
8.6	475	41	434	3.5
8.6	467	40	427	3.4
8.6	466	40	426	3.4
8.6	465	40	425	3.4
8.6	464	40	424	3.4
8.6	463	40	423	3.5
8.6	456	39	417	3.3
8.6	455	39	416	3.3
8.6	454	39	415	3.4
8.6	453	39	414	3.4
8.6	452	39	413	3.4
8.6	451	39	412	3.4
8.6	444	38	406	3.3
8.6	443	38	405	3.3
8.6	442	38	404	3.3
8.6	441	38	403	3.3
8.6	440	38	402	3.3
8.6	432	37	395	3.2
8.6	431	37	394	3.2
8.6	430	37	393	3.2
8.6	429	37	392	3.2

%E	M1	M2	DM	M*
8.6	428	37	391	3.2
8.6	421	36	385	3.1
8.6	420	36	384	3.1
8.6	419	36	383	3.1
8.6	418	36	382	3.1
8.6	417	36	381	3.1
8.6	409	35	374	3.0
8.6	408	35	373	3.0
8.6	407	35	372	3.0
8.6	406	35	371	3.0
8.6	405	35	370	3.0
8.6	397	34	363	2.9
8.6	396	34	362	2.9
8.6	395	34	361	2.9
8.6	394	34	360	2.9
8.6	385	33	352	2.8
8.6	384	33	351	2.8
8.6	383	33	350	2.8
8.6	382	33	349	2.9
8.6	374	32	342	2.7
8.6	373	32	341	2.7
8.6	372	32	340	2.8
8.6	371	32	339	2.8
8.6	370	32	338	2.8
8.6	362	31	331	2.7
8.6	361	31	330	2.7
8.6	360	31	329	2.7
8.6	359	31	328	2.7
8.6	350	30	320	2.6
8.6	349	30	319	2.6
8.6	348	30	318	2.6
8.6	347	30	317	2.6
8.6	339	29	310	2.5
8.6	338	29	309	2.5
8.6	337	29	308	2.5
8.6	336	29	307	2.5
8.6	327	28	299	2.4
8.6	326	28	298	2.4
8.6	325	28	297	2.4
8.6	324	28	296	2.4
8.6	315	27	288	2.3
8.6	314	27	287	2.3
8.6	313	27	286	2.3
8.6	304	26	278	2.2
8.6	303	26	277	2.2
8.6	302	26	276	2.2
8.6	301	26	275	2.2
8.6	292	25	267	2.1
8.6	291	25	266	2.1
8.6	290	25	265	2.2

%E	M1	M2	DM	M*
8.6	280	24	256	2.1
8.6	279	24	255	2.1
8.6	278	24	254	2.1
8.6	269	23	246	2.0
8.6	268	23	245	2.0
8.6	267	23	244	2.0
8.6	266	23	243	2.0
8.6	257	22	235	1.9
8.6	256	22	234	1.9
8.6	255	22	233	1.9
8.6	245	21	224	1.8
8.6	244	21	223	1.8
8.6	243	21	222	1.8
8.6	233	20	213	1.7
8.6	232	20	212	1.7
8.6	222	19	203	1.6
8.6	221	19	202	1.6
8.6	220	19	201	1.6
8.6	210	18	192	1.5
8.6	209	18	191	1.6
8.6	198	17	181	1.5
8.6	197	17	180	1.5
8.6	187	16	171	1.4
8.6	186	16	170	1.4
8.6	185	16	169	1.4
8.6	175	15	160	1.3
8.6	174	15	159	1.3
8.6	163	14	149	1.2
8.6	162	14	148	1.2
8.6	152	13	139	1.1
8.6	151	13	138	1.1
8.6	140	12	128	1.0
8.6	139	12	127	1.0
8.6	128	11	117	0.9
8.6	116	10	106	0.9
8.6	105	9	96	0.8
8.6	93	8	85	0.7
8.6	81	7	74	0.6
8.6	70	6	64	0.5
8.6	58	5	53	0.4
8.6	35	3	32	0.3
8.5	497	42	455	3.5
8.5	496	42	454	3.6
8.5	495	42	453	3.6
8.5	494	42	452	3.6
8.5	493	42	451	3.6
8.5	492	42	450	3.6
8.5	485	41	444	3.5
8.5	484	41	443	3.5
8.5	483	41	442	3.5
8.5	482	41	441	3.5
8.5	481	41	440	3.5
8.5	480	41	439	3.5
8.5	473	40	433	3.4
8.5	472	40	432	3.4
8.5	471	40	431	3.4
8.5	470	40	430	3.4
8.5	469	40	429	3.4
8.5	468	40	428	3.4
8.5	461	39	422	3.3
8.5	460	39	421	3.3
8.5	459	39	420	3.3
8.5	458	39	419	3.3
8.5	457	39	418	3.3
8.5	449	38	411	3.2
8.5	448	38	410	3.2
8.5	447	38	409	3.2
8.5	446	38	408	3.2
8.5	445	38	407	3.2
8.5	437	37	400	3.1
8.5	436	37	399	3.1
8.5	435	37	398	3.1
8.5	434	37	397	3.2
8.5	433	37	396	3.2
8.5	426	36	390	3.0
8.5	425	36	389	3.0
8.5	424	36	388	3.1
8.5	423	36	387	3.1
8.5	422	36	386	3.1
8.5	414	35	379	3.0
8.5	413	35	378	3.0
8.5	412	35	377	3.0
8.5	411	35	376	3.0
8.5	410	35	375	3.0
8.5	402	34	368	2.9
8.5	401	34	367	2.9
8.5	400	34	366	2.9
8.5	399	34	365	2.9
8.5	398	34	364	2.9
8.5	390	33	357	2.8
8.5	389	33	356	2.8
8.5	388	33	355	2.8
8.5	387	33	354	2.8
8.5	386	33	353	2.8
8.5	378	32	346	2.7
8.5	377	32	345	2.7
8.5	376	32	344	2.7
8.5	375	32	343	2.7
8.5	366	31	335	2.6
8.5	365	31	334	2.6
8.5	364	31	333	2.6
8.5	363	31	332	2.6
8.5	355	30	325	2.5
8.5	354	30	324	2.5
8.5	353	30	323	2.5
8.5	352	30	322	2.6
8.5	351	30	321	2.6
8.5	343	29	314	2.5
8.5	342	29	313	2.5
8.5	341	29	312	2.5
8.5	340	29	311	2.5
8.5	331	28	303	2.4
8.5	330	28	302	2.4
8.5	329	28	301	2.4
8.5	328	28	300	2.4
8.5	319	27	292	2.3
8.5	318	27	291	2.3
8.5	317	27	290	2.3
8.5	316	27	289	2.3
8.5	307	26	281	2.2
8.5	306	26	280	2.2
8.5	305	26	279	2.2
8.5	295	25	270	2.1
8.5	294	25	269	2.1
8.5	293	25	268	2.1
8.5	284	24	260	2.0
8.5	283	24	259	2.0
8.5	282	24	258	2.0
8.5	281	24	257	2.0
8.5	272	23	249	1.9
8.5	271	23	248	2.0
8.5	270	23	247	2.0
8.5	260	22	238	1.9
8.5	259	22	237	1.9
8.5	258	22	236	1.9
8.5	248	21	227	1.8
8.5	247	21	226	1.8
8.5	246	21	225	1.8
8.5	236	20	216	1.7
8.5	235	20	215	1.7
8.5	234	20	214	1.7
8.5	224	19	205	1.6
8.5	223	19	204	1.6
8.5	213	18	195	1.5
8.5	212	18	194	1.5
8.5	211	18	193	1.5
8.5	201	17	184	1.4
8.5	200	17	183	1.4
8.5	199	17	182	1.5
8.5	189	16	173	1.4
8.5	188	16	172	1.4
8.5	177	15	162	1.3
8.5	176	15	161	1.3
8.5	165	14	151	1.2
8.5	164	14	150	1.2
8.5	153	13	140	1.1
8.5	142	12	130	1.0
8.5	141	12	129	1.0
8.5	130	11	119	0.9
8.5	129	11	118	0.9
8.5	118	10	108	0.8
8.5	117	10	107	0.9
8.5	106	9	97	0.8
8.5	94	8	86	0.7
8.5	82	7	75	0.6
8.5	71	6	65	0.5
8.5	59	5	54	0.4
8.5	47	4	43	0.3
8.4	500	42	458	3.5
8.4	499	42	457	3.5
8.4	498	42	456	3.5
8.4	491	41	450	3.4
8.4	490	41	449	3.4
8.4	489	41	448	3.4
8.4	488	41	447	3.4
8.4	487	41	446	3.5
8.4	486	41	445	3.5
8.4	479	40	439	3.3
8.4	478	40	438	3.3
8.4	477	40	437	3.4
8.4	476	40	436	3.4
8.4	475	40	435	3.4
8.4	474	40	434	3.4
8.4	467	39	428	3.3
8.4	466	39	427	3.3
8.4	465	39	426	3.3
8.4	464	39	425	3.3
8.4	463	39	424	3.3
8.4	462	39	423	3.3
8.4	455	38	417	3.2
8.4	454	38	416	3.2
8.4	453	38	415	3.2
8.4	452	38	414	3.2
8.4	451	38	413	3.2
8.4	450	38	412	3.2
8.4	443	37	406	3.1
8.4	442	37	405	3.1
8.4	441	37	404	3.1
8.4	440	37	403	3.1
8.4	439	37	402	3.1
8.4	438	37	401	3.1
8.4	431	36	395	3.0
8.4	430	36	394	3.0
8.4	429	36	393	3.0
8.4	428	36	392	3.0
8.4	427	36	391	3.0
8.4	419	35	384	2.9
8.4	418	35	383	2.9
8.4	417	35	382	2.9
8.4	416	35	381	2.9
8.4	415	35	380	3.0
8.4	407	34	373	2.8
8.4	406	34	372	2.8
8.4	405	34	371	2.9
8.4	404	34	370	2.9
8.4	403	34	369	2.9
8.4	395	33	362	2.8
8.4	394	33	361	2.8
8.4	393	33	360	2.8
8.4	392	33	359	2.8
8.4	391	33	358	2.8
8.4	383	32	351	2.7
8.4	382	32	350	2.7
8.4	381	32	349	2.7
8.4	380	32	348	2.7
8.4	379	32	347	2.7
8.4	371	31	340	2.6
8.4	370	31	339	2.6
8.4	369	31	338	2.6
8.4	368	31	337	2.6
8.4	367	31	336	2.6
8.4	359	30	329	2.5
8.4	358	30	328	2.5
8.4	357	30	327	2.5
8.4	356	30	326	2.5
8.4	347	29	318	2.4
8.4	346	29	317	2.4
8.4	345	29	316	2.4
8.4	344	29	315	2.4
8.4	335	28	307	2.3
8.4	334	28	306	2.3
8.4	333	28	305	2.4
8.4	332	28	304	2.4
8.4	323	27	296	2.3
8.4	322	27	295	2.3
8.4	321	27	294	2.3
8.4	320	27	293	2.3
8.4	311	26	285	2.2
8.4	310	26	284	2.2
8.4	309	26	283	2.2

%E	M1	M2	DM	M*
8.4	308	26	282	2.2
8.4	299	25	274	2.1
8.4	298	25	273	2.1
8.4	297	25	272	2.1
8.4	296	25	271	2.1
8.4	287	24	263	2.0
8.4	286	24	262	2.0
8.4	285	24	261	2.0
8.4	275	23	252	1.9
8.4	274	23	251	1.9
8.4	273	23	250	1.9
8.4	263	22	241	1.8
8.4	262	22	240	1.8
8.4	261	22	239	1.9
8.4	251	21	230	1.8
8.4	250	21	229	1.8
8.4	249	21	228	1.8
8.4	239	20	219	1.7
8.4	238	20	218	1.7
8.4	237	20	217	1.7
8.4	227	19	208	1.6
8.4	226	19	207	1.6
8.4	225	19	206	1.6
8.4	215	18	197	1.5
8.4	214	18	196	1.5
8.4	203	17	186	1.4
8.4	202	17	185	1.4
8.4	191	16	175	1.3
8.4	190	16	174	1.3
8.4	179	15	164	1.3
8.4	178	15	163	1.3
8.4	167	14	153	1.2
8.4	166	14	152	1.2
8.4	155	13	142	1.1
8.4	154	13	141	1.1
8.4	143	12	131	1.0
8.4	131	11	120	0.9
8.4	119	10	109	0.8
8.4	107	9	98	0.8
8.4	95	8	87	0.7
8.4	83	7	76	0.6
8.3	496	41	455	3.4
8.3	495	41	454	3.4
8.3	494	41	453	3.4
8.3	493	41	452	3.4
8.3	492	41	451	3.4
8.3	484	40	444	3.3
8.3	483	40	443	3.3
8.3	482	40	442	3.3
8.3	481	40	441	3.3

%E	M1	M2	DM	M*
8.3	480	40	440	3.3
8.3	472	39	433	3.2
8.3	471	39	432	3.2
8.3	470	39	431	3.2
8.3	469	39	430	3.2
8.3	468	39	429	3.3
8.3	460	38	422	3.1
8.3	459	38	421	3.1
8.3	458	38	420	3.2
8.3	457	38	419	3.2
8.3	456	38	418	3.2
8.3	448	37	411	3.1
8.3	447	37	410	3.1
8.3	446	37	409	3.1
8.3	445	37	408	3.1
8.3	444	37	407	3.1
8.3	436	36	400	3.0
8.3	435	36	399	3.0
8.3	434	36	398	3.0
8.3	433	36	397	3.0
8.3	432	36	396	3.0
8.3	424	35	389	2.9
8.3	423	35	388	2.9
8.3	422	35	387	2.9
8.3	421	35	386	2.9
8.3	420	35	385	2.9
8.3	412	34	378	2.8
8.3	411	34	377	2.8
8.3	410	34	376	2.8
8.3	409	34	375	2.8
8.3	408	34	374	2.8
8.3	400	33	367	2.7
8.3	399	33	366	2.7
8.3	398	33	365	2.7
8.3	397	33	364	2.7
8.3	396	33	363	2.8
8.3	387	32	355	2.6
8.3	386	32	354	2.7
8.3	385	32	353	2.7
8.3	384	32	352	2.7
8.3	375	31	344	2.6
8.3	374	31	343	2.6
8.3	373	31	342	2.6
8.3	372	31	341	2.6
8.3	363	30	333	2.5
8.3	362	30	332	2.5
8.3	361	30	331	2.5
8.3	360	30	330	2.5
8.3	351	29	322	2.4
8.3	350	29	321	2.4

%E	M1	M2	DM	M*
8.3	349	29	320	2.4
8.3	348	29	319	2.4
8.3	339	28	311	2.3
8.3	338	28	310	2.3
8.3	337	28	309	2.3
8.3	336	28	308	2.3
8.3	327	27	300	2.2
8.3	326	27	299	2.2
8.3	325	27	298	2.2
8.3	324	27	297	2.3
8.3	315	26	289	2.1
8.3	314	26	288	2.2
8.3	313	26	287	2.2
8.3	312	26	286	2.2
8.3	303	25	278	2.1
8.3	302	25	277	2.1
8.3	301	25	276	2.1
8.3	300	25	275	2.1
8.3	290	24	266	2.0
8.3	289	24	265	2.0
8.3	288	24	264	2.0
8.3	278	23	255	1.9
8.3	277	23	254	1.9
8.3	276	23	253	1.9
8.3	266	22	244	1.8
8.3	265	22	243	1.8
8.3	264	22	242	1.8
8.3	254	21	233	1.7
8.3	253	21	232	1.7
8.3	252	21	231	1.8
8.3	242	20	222	1.7
8.3	241	20	221	1.7
8.3	240	20	220	1.7
8.3	230	19	211	1.6
8.3	229	19	210	1.6
8.3	228	19	209	1.6
8.3	218	18	200	1.5
8.3	217	18	199	1.5
8.3	216	18	198	1.5
8.3	206	17	189	1.4
8.3	205	17	188	1.4
8.3	204	17	187	1.4
8.3	193	16	177	1.3
8.3	192	16	176	1.3
8.3	181	15	166	1.2
8.3	180	15	165	1.3
8.3	169	14	155	1.2
8.3	168	14	154	1.2
8.3	157	13	144	1.1
8.3	156	13	143	1.1

%E	M1	M2	DM	M*
8.3	145	12	133	1.0
8.3	144	12	132	1.0
8.3	133	11	122	0.9
8.3	132	11	121	0.9
8.3	121	10	111	0.8
8.3	120	10	110	0.8
8.3	109	9	100	0.7
8.3	108	9	99	0.8
8.3	96	8	88	0.7
8.3	84	7	77	0.6
8.3	72	6	66	0.5
8.3	60	5	55	0.4
8.3	48	4	44	0.3
8.3	36	3	33	0.3
8.3	24	2	22	0.2
8.3	12	1	11	0.1
8.2	500	41	459	3.4
8.2	499	41	458	3.4
8.2	498	41	457	3.4
8.2	497	41	456	3.4
8.2	490	40	450	3.3
8.2	489	40	449	3.3
8.2	488	40	448	3.3
8.2	487	40	447	3.3
8.2	486	40	446	3.3
8.2	485	40	445	3.3
8.2	478	39	439	3.2
8.2	477	39	438	3.2
8.2	476	39	437	3.2
8.2	475	39	436	3.2
8.2	474	39	435	3.2
8.2	473	39	434	3.2
8.2	466	38	428	3.1
8.2	465	38	427	3.1
8.2	464	38	426	3.1
8.2	463	38	425	3.1
8.2	462	38	424	3.1
8.2	461	38	423	3.1
8.2	454	37	417	3.0
8.2	453	37	416	3.0
8.2	452	37	415	3.0
8.2	451	37	414	3.0
8.2	450	37	413	3.0
8.2	449	37	412	3.0
8.2	441	36	405	2.9
8.2	440	36	404	2.9
8.2	439	36	403	3.0
8.2	438	36	402	3.0
8.2	437	36	401	3.0
8.2	429	35	394	2.9

%E	M1	M2	DM	M*
8.2	428	35	393	2.9
8.2	427	35	392	2.9
8.2	426	35	391	2.9
8.2	425	35	390	2.9
8.2	417	34	383	2.8
8.2	416	34	382	2.8
8.2	415	34	381	2.8
8.2	414	34	380	2.8
8.2	413	34	379	2.8
8.2	404	33	371	2.7
8.2	403	33	370	2.7
8.2	402	33	369	2.7
8.2	401	33	368	2.7
8.2	392	32	360	2.6
8.2	391	32	359	2.6
8.2	390	32	358	2.6
8.2	389	32	357	2.6
8.2	388	32	356	2.6
8.2	380	31	349	2.5
8.2	379	31	348	2.5
8.2	378	31	347	2.5
8.2	377	31	346	2.5
8.2	376	31	345	2.6
8.2	368	30	338	2.4
8.2	367	30	337	2.5
8.2	366	30	336	2.5
8.2	365	30	335	2.5
8.2	364	30	334	2.5
8.2	355	29	326	2.4
8.2	354	29	325	2.4
8.2	353	29	324	2.4
8.2	352	29	323	2.4
8.2	343	28	315	2.3
8.2	342	28	314	2.3
8.2	341	28	313	2.3
8.2	340	28	312	2.3
8.2	331	27	304	2.2
8.2	330	27	303	2.2
8.2	329	27	302	2.2
8.2	328	27	301	2.2
8.2	319	26	293	2.1
8.2	318	26	292	2.1
8.2	317	26	291	2.1
8.2	316	26	290	2.1
8.2	306	25	281	2.0
8.2	305	25	280	2.0
8.2	304	25	279	2.1
8.2	294	24	270	2.0
8.2	293	24	269	2.0
8.2	292	24	268	2.0

%E	M1	M2	DM	M*
8.2	291	24	267	2.0
8.2	282	23	259	1.9
8.2	281	23	258	1.9
8.2	280	23	257	1.9
8.2	279	23	256	1.9
8.2	269	22	247	1.8
8.2	268	22	246	1.8
8.2	267	22	245	1.8
8.2	257	21	236	1.7
8.2	256	21	235	1.7
8.2	255	21	234	1.7
8.2	245	20	225	1.6
8.2	244	20	224	1.6
8.2	243	20	223	1.6
8.2	233	19	214	1.5
8.2	232	19	213	1.6
8.2	231	19	212	1.6
8.2	220	18	202	1.5
8.2	219	18	201	1.5
8.2	208	17	191	1.4
8.2	207	17	190	1.4
8.2	196	16	180	1.3
8.2	195	16	179	1.3
8.2	194	16	178	1.3
8.2	184	15	169	1.2
8.2	183	15	168	1.2
8.2	182	15	167	1.2
8.2	171	14	157	1.1
8.2	170	14	156	1.2
8.2	159	13	146	1.1
8.2	158	13	145	1.1
8.2	147	12	135	1.0
8.2	146	12	134	1.0
8.2	134	11	123	0.9
8.2	122	10	112	0.8
8.2	110	9	101	0.7
8.2	98	8	90	0.7
8.2	97	8	89	0.7
8.2	85	7	78	0.6
8.2	73	6	67	0.5
8.2	61	5	56	0.4
8.2	49	4	45	0.3
8.1	496	40	456	3.2
8.1	495	40	455	3.2
8.1	494	40	454	3.2
8.1	493	40	453	3.2
8.1	492	40	452	3.3
8.1	491	40	451	3.3
8.1	484	39	445	3.1
8.1	483	39	444	3.1
8.1	482	39	443	3.2
8.1	481	39	442	3.2
8.1	480	39	441	3.2
8.1	479	39	440	3.2
8.1	472	38	434	3.1
8.1	471	38	433	3.1
8.1	470	38	432	3.1
8.1	469	38	431	3.1
8.1	468	38	430	3.1
8.1	467	38	429	3.1
8.1	459	37	422	3.0
8.1	458	37	421	3.0
8.1	457	37	420	3.0
8.1	456	37	419	3.0
8.1	455	37	418	3.0
8.1	447	36	411	2.9
8.1	446	36	410	2.9
8.1	445	36	409	2.9
8.1	444	36	408	2.9
8.1	443	36	407	2.9
8.1	442	36	406	2.9
8.1	434	35	399	2.8
8.1	433	35	398	2.8
8.1	432	35	397	2.8
8.1	431	35	396	2.8
8.1	430	35	395	2.8
8.1	422	34	388	2.7
8.1	421	34	387	2.7
8.1	420	34	386	2.8
8.1	419	34	385	2.8
8.1	418	34	384	2.8
8.1	409	33	376	2.7
8.1	408	33	375	2.7
8.1	407	33	374	2.7
8.1	406	33	373	2.7
8.1	405	33	372	2.7
8.1	397	32	365	2.6
8.1	396	32	364	2.6
8.1	395	32	363	2.6
8.1	394	32	362	2.6
8.1	393	32	361	2.6
8.1	385	31	354	2.5
8.1	384	31	353	2.5
8.1	383	31	352	2.5
8.1	382	31	351	2.5
8.1	381	31	350	2.5
8.1	372	30	342	2.4
8.1	371	30	341	2.4
8.1	370	30	340	2.4
8.1	369	30	339	2.4
8.1	360	29	331	2.3
8.1	359	29	330	2.3
8.1	358	29	329	2.3
8.1	357	29	328	2.4
8.1	356	29	327	2.4
8.1	347	28	319	2.3
8.1	346	28	318	2.3
8.1	345	28	317	2.3
8.1	344	28	316	2.3
8.1	335	27	308	2.2
8.1	334	27	307	2.2
8.1	333	27	306	2.2
8.1	332	27	305	2.2
8.1	323	26	297	2.1
8.1	322	26	296	2.1
8.1	321	26	295	2.1
8.1	320	26	294	2.1
8.1	310	25	285	2.0
8.1	309	25	284	2.0
8.1	308	25	283	2.0
8.1	307	25	282	2.0
8.1	298	24	274	1.9
8.1	297	24	273	1.9
8.1	296	24	272	1.9
8.1	295	24	271	2.0
8.1	285	23	262	1.9
8.1	284	23	261	1.9
8.1	283	23	260	1.9
8.1	273	22	251	1.8
8.1	272	22	250	1.8
8.1	271	22	249	1.8
8.1	270	22	248	1.8
8.1	260	21	239	1.7
8.1	259	21	238	1.7
8.1	258	21	237	1.7
8.1	248	20	228	1.6
8.1	247	20	227	1.6
8.1	246	20	226	1.6
8.1	236	19	217	1.5
8.1	235	19	216	1.5
8.1	234	19	215	1.5
8.1	223	18	205	1.5
8.1	222	18	204	1.5
8.1	221	18	203	1.5
8.1	211	17	194	1.4
8.1	210	17	193	1.4
8.1	209	17	192	1.4
8.1	198	16	182	1.3
8.1	197	16	181	1.3
8.1	186	15	171	1.2
8.1	185	15	170	1.2
8.1	173	14	159	1.1
8.1	172	14	158	1.1
8.1	161	13	148	1.0
8.1	160	13	147	1.1
8.1	149	12	137	1.0
8.1	148	12	136	1.0
8.1	136	11	125	0.9
8.1	135	11	124	0.9
8.1	124	10	114	0.8
8.1	123	10	113	0.8
8.1	111	9	102	0.7
8.1	99	8	91	0.6
8.1	86	7	79	0.6
8.1	74	6	68	0.5
8.1	62	5	57	0.4
8.1	37	3	34	0.2
8.0	500	40	460	3.2
8.0	499	40	459	3.2
8.0	498	40	458	3.2
8.0	497	40	457	3.2
8.0	490	39	451	3.1
8.0	489	39	450	3.1
8.0	488	39	449	3.1
8.0	487	39	448	3.1
8.0	486	39	447	3.1
8.0	485	39	446	3.1
8.0	478	38	440	3.0
8.0	477	38	439	3.0
8.0	476	38	438	3.0
8.0	475	38	437	3.0
8.0	474	38	436	3.0
8.0	473	38	435	3.1
8.0	465	37	428	2.9
8.0	464	37	427	3.0
8.0	463	37	426	3.0
8.0	462	37	425	3.0
8.0	461	37	424	3.0
8.0	460	37	423	3.0
8.0	452	36	416	2.9
8.0	451	36	415	2.9
8.0	450	36	414	2.9
8.0	449	36	413	2.9
8.0	448	36	412	2.9
8.0	440	35	405	2.8
8.0	439	35	404	2.8
8.0	438	35	403	2.8
8.0	437	35	402	2.8
8.0	436	35	401	2.8
8.0	435	35	400	2.8
8.0	427	34	393	2.7
8.0	426	34	392	2.7
8.0	425	34	391	2.7
8.0	424	34	390	2.7
8.0	423	34	389	2.7
8.0	415	33	382	2.6
8.0	414	33	381	2.6
8.0	413	33	380	2.6
8.0	412	33	379	2.6
8.0	411	33	378	2.6
8.0	410	33	377	2.7
8.0	402	32	370	2.5
8.0	401	32	369	2.6
8.0	400	32	368	2.6
8.0	399	32	367	2.6
8.0	398	32	366	2.6
8.0	389	31	358	2.5
8.0	388	31	357	2.5
8.0	387	31	356	2.5
8.0	386	31	355	2.5
8.0	377	30	347	2.4
8.0	376	30	346	2.4
8.0	375	30	345	2.4
8.0	374	30	344	2.4
8.0	373	30	343	2.4
8.0	364	29	335	2.3
8.0	363	29	334	2.3
8.0	362	29	333	2.3
8.0	361	29	332	2.3
8.0	352	28	324	2.2
8.0	351	28	323	2.2
8.0	350	28	322	2.2
8.0	349	28	321	2.2
8.0	348	28	320	2.3
8.0	339	27	312	2.2
8.0	338	27	311	2.2
8.0	337	27	310	2.2
8.0	336	27	309	2.2
8.0	327	26	301	2.1
8.0	326	26	300	2.1
8.0	325	26	299	2.1
8.0	324	26	298	2.1
8.0	314	25	289	2.0
8.0	313	25	288	2.0
8.0	312	25	287	2.0
8.0	311	25	286	2.0
8.0	301	24	277	1.9
8.0	300	24	276	1.9
8.0	299	24	275	1.9
8.0	289	23	266	1.8

%E	M1	M2	DM	M*
8.0	288	23	265	1.8
8.0	287	23	264	1.8
8.0	286	23	263	1.8
8.0	276	22	254	1.8
8.0	275	22	253	1.8
8.0	274	22	252	1.8
8.0	264	21	243	1.7
8.0	263	21	242	1.7
8.0	262	21	241	1.7
8.0	261	21	240	1.7
8.0	251	20	231	1.6
8.0	250	20	230	1.6
8.0	249	20	229	1.6
8.0	239	19	220	1.5
8.0	238	19	219	1.5
8.0	237	19	218	1.5
8.0	226	18	208	1.4
8.0	225	18	207	1.4
8.0	224	18	206	1.4
8.0	213	17	196	1.4
8.0	212	17	195	1.4
8.0	201	16	185	1.3
8.0	200	16	184	1.3
8.0	199	16	183	1.3
8.0	188	15	173	1.2
8.0	187	15	172	1.2
8.0	176	14	162	1.1
8.0	175	14	161	1.1
8.0	174	14	160	1.1
8.0	163	13	150	1.0
8.0	162	13	149	1.0
8.0	150	12	138	1.0
8.0	138	11	127	0.9
8.0	137	11	126	0.9
8.0	125	10	115	0.8
8.0	113	9	104	0.7
8.0	112	9	103	0.7
8.0	100	8	92	0.6
8.0	88	7	81	0.6
8.0	87	7	80	0.6
8.0	75	6	69	0.5
8.0	50	4	46	0.3
8.0	25	2	23	0.2
7.9	496	39	457	3.1
7.9	495	39	456	3.1
7.9	494	39	455	3.1
7.9	493	39	454	3.1
7.9	492	39	453	3.1
7.9	491	39	452	3.1
7.9	484	38	446	3.0

%E	M1	M2	DM	M*
7.9	483	38	445	3.0
7.9	482	38	444	3.0
7.9	481	38	443	3.0
7.9	480	38	442	3.0
7.9	479	38	441	3.0
7.9	471	37	434	2.9
7.9	470	37	433	2.9
7.9	469	37	432	2.9
7.9	468	37	431	2.9
7.9	467	37	430	2.9
7.9	466	37	429	2.9
7.9	458	36	422	2.8
7.9	457	36	421	2.8
7.9	456	36	420	2.8
7.9	455	36	419	2.8
7.9	454	36	418	2.9
7.9	453	36	417	2.9
7.9	445	35	410	2.8
7.9	444	35	409	2.8
7.9	443	35	408	2.8
7.9	442	35	407	2.8
7.9	441	35	406	2.8
7.9	433	34	399	2.7
7.9	432	34	398	2.7
7.9	431	34	397	2.7
7.9	430	34	396	2.7
7.9	429	34	395	2.7
7.9	428	34	394	2.7
7.9	420	33	387	2.6
7.9	419	33	386	2.6
7.9	418	33	385	2.6
7.9	417	33	384	2.6
7.9	416	33	383	2.6
7.9	407	32	375	2.5
7.9	406	32	374	2.5
7.9	405	32	373	2.5
7.9	404	32	372	2.5
7.9	403	32	371	2.5
7.9	394	31	363	2.4
7.9	393	31	362	2.4
7.9	392	31	361	2.5
7.9	391	31	360	2.5
7.9	390	31	359	2.5
7.9	382	30	352	2.4
7.9	381	30	351	2.4
7.9	380	30	350	2.4
7.9	379	30	349	2.4
7.9	378	30	348	2.4
7.9	369	29	340	2.3
7.9	368	29	339	2.3

%E	M1	M2	DM	M*
7.9	367	29	338	2.3
7.9	366	29	337	2.3
7.9	365	29	336	2.3
7.9	356	28	328	2.2
7.9	355	28	327	2.2
7.9	354	28	326	2.2
7.9	353	28	325	2.2
7.9	343	27	316	2.1
7.9	342	27	315	2.1
7.9	341	27	314	2.1
7.9	340	27	313	2.1
7.9	331	26	305	2.0
7.9	330	26	304	2.0
7.9	329	26	303	2.1
7.9	328	26	302	2.1
7.9	318	25	293	2.0
7.9	317	25	292	2.0
7.9	316	25	291	2.0
7.9	315	25	290	2.0
7.9	305	24	281	1.9
7.9	304	24	280	1.9
7.9	303	24	279	1.9
7.9	302	24	278	1.9
7.9	293	23	270	1.8
7.9	292	23	269	1.8
7.9	291	23	268	1.8
7.9	290	23	267	1.8
7.9	280	22	258	1.7
7.9	279	22	257	1.7
7.9	278	22	256	1.7
7.9	277	22	255	1.7
7.9	267	21	246	1.7
7.9	266	21	245	1.7
7.9	265	21	244	1.7
7.9	254	20	234	1.6
7.9	253	20	233	1.6
7.9	252	20	232	1.6
7.9	242	19	223	1.5
7.9	241	19	222	1.5
7.9	240	19	221	1.5
7.9	229	18	211	1.4
7.9	228	18	210	1.4
7.9	227	18	209	1.4
7.9	216	17	199	1.3
7.9	215	17	198	1.3
7.9	214	17	197	1.4
7.9	203	16	187	1.3
7.9	202	16	186	1.3
7.9	191	15	176	1.2
7.9	190	15	175	1.2

%E	M1	M2	DM	M*
7.9	189	15	174	1.2
7.9	178	14	164	1.1
7.9	177	14	163	1.1
7.9	165	13	152	1.0
7.9	164	13	151	1.0
7.9	152	12	140	0.9
7.9	151	12	139	1.0
7.9	140	11	129	0.9
7.9	139	11	128	0.9
7.9	127	10	117	0.8
7.9	126	10	116	0.8
7.9	114	9	105	0.7
7.9	101	8	93	0.6
7.9	89	7	82	0.6
7.9	76	6	70	0.5
7.9	63	5	58	0.4
7.9	38	3	35	0.2
7.8	500	39	461	3.0
7.8	499	39	460	3.0
7.8	498	39	459	3.1
7.8	497	39	458	3.1
7.8	490	38	452	2.9
7.8	489	38	451	3.0
7.8	488	38	450	3.0
7.8	487	38	449	3.0
7.8	486	38	448	3.0
7.8	485	38	447	3.0
7.8	477	37	440	2.9
7.8	476	37	439	2.9
7.8	475	37	438	2.9
7.8	474	37	437	2.9
7.8	473	37	436	2.9
7.8	472	37	435	2.9
7.8	464	36	428	2.8
7.8	463	36	427	2.8
7.8	462	36	426	2.8
7.8	461	36	425	2.8
7.8	460	36	424	2.8
7.8	459	36	423	2.8
7.8	451	35	416	2.7
7.8	450	35	415	2.7
7.8	449	35	414	2.7
7.8	448	35	413	2.7
7.8	447	35	412	2.7
7.8	446	35	411	2.7
7.8	438	34	404	2.6
7.8	437	34	403	2.6
7.8	436	34	402	2.7
7.8	435	34	401	2.7
7.8	434	34	400	2.7

%E	M1	M2	DM	M*
7.8	425	33	392	2.6
7.8	424	33	391	2.6
7.8	423	33	390	2.6
7.8	422	33	389	2.6
7.8	421	33	388	2.6
7.8	412	32	380	2.5
7.8	411	32	379	2.5
7.8	410	32	378	2.5
7.8	409	32	377	2.5
7.8	408	32	376	2.5
7.8	400	31	369	2.4
7.8	399	31	368	2.4
7.8	398	31	367	2.4
7.8	397	31	366	2.4
7.8	396	31	365	2.4
7.8	395	31	364	2.4
7.8	387	30	357	2.3
7.8	386	30	356	2.3
7.8	385	30	355	2.3
7.8	384	30	354	2.3
7.8	383	30	353	2.3
7.8	374	29	345	2.2
7.8	373	29	344	2.3
7.8	372	29	343	2.3
7.8	371	29	342	2.3
7.8	370	29	341	2.3
7.8	361	28	333	2.2
7.8	360	28	332	2.2
7.8	359	28	331	2.2
7.8	358	28	330	2.2
7.8	357	28	329	2.2
7.8	348	27	321	2.1
7.8	347	27	320	2.1
7.8	346	27	319	2.1
7.8	345	27	318	2.1
7.8	344	27	317	2.1
7.8	335	26	309	2.0
7.8	334	26	308	2.0
7.8	333	26	307	2.0
7.8	332	26	306	2.0
7.8	322	25	297	1.9
7.8	321	25	296	1.9
7.8	320	25	295	2.0
7.8	319	25	294	2.0
7.8	309	24	285	1.9
7.8	308	24	284	1.9
7.8	307	24	283	1.9
7.8	306	24	282	1.9
7.8	296	23	273	1.8
7.8	295	23	272	1.8

%E	M1	M2	DM	M*
7.8	294	23	271	1.8
7.8	283	22	261	1.7
7.8	282	22	260	1.7
7.8	281	22	259	1.7
7.8	270	21	249	1.6
7.8	269	21	248	1.6
7.8	268	21	247	1.6
7.8	258	20	238	1.6
7.8	257	20	237	1.6
7.8	256	20	236	1.6
7.8	255	20	235	1.6
7.8	245	19	226	1.5
7.8	244	19	225	1.5
7.8	243	19	224	1.5
7.8	232	18	214	1.4
7.8	231	18	213	1.4
7.8	230	18	212	1.4
7.8	219	17	202	1.3
7.8	218	17	201	1.3
7.8	217	17	200	1.3
7.8	206	16	190	1.2
7.8	205	16	189	1.2
7.8	204	16	188	1.3
7.8	193	15	178	1.2
7.8	192	15	177	1.2
7.8	180	14	166	1.1
7.8	179	14	165	1.1
7.8	167	13	154	1.0
7.8	166	13	153	1.0
7.8	154	12	142	0.9
7.8	153	12	141	0.9
7.8	141	11	130	0.9
7.8	129	10	119	0.8
7.8	128	10	118	0.8
7.8	116	9	107	0.7
7.8	115	9	106	0.7
7.8	103	8	95	0.6
7.8	102	8	94	0.6
7.8	90	7	83	0.5
7.8	77	6	71	0.5
7.8	64	5	59	0.4
7.8	51	4	47	0.3
7.7	496	38	458	2.9
7.7	495	38	457	2.9
7.7	494	38	456	2.9
7.7	493	38	455	2.9
7.7	492	38	454	2.9
7.7	491	38	453	2.9
7.7	483	37	446	2.8
7.7	482	37	445	2.8

%E	M1	M2	DM	M*
7.7	481	37	444	2.8
7.7	480	37	443	2.9
7.7	479	37	442	2.9
7.7	478	37	441	2.9
7.7	470	36	434	2.8
7.7	469	36	433	2.8
7.7	468	36	432	2.8
7.7	467	36	431	2.8
7.7	466	36	430	2.8
7.7	465	36	429	2.8
7.7	457	35	422	2.7
7.7	456	35	421	2.7
7.7	455	35	420	2.7
7.7	454	35	419	2.7
7.7	453	35	418	2.7
7.7	452	35	417	2.7
7.7	444	34	410	2.6
7.7	443	34	409	2.6
7.7	442	34	408	2.6
7.7	441	34	407	2.6
7.7	440	34	406	2.6
7.7	439	34	405	2.6
7.7	431	33	398	2.5
7.7	430	33	397	2.5
7.7	429	33	396	2.5
7.7	428	33	395	2.5
7.7	427	33	394	2.6
7.7	426	33	393	2.6
7.7	418	32	386	2.4
7.7	417	32	385	2.5
7.7	416	32	384	2.5
7.7	415	32	383	2.5
7.7	414	32	382	2.5
7.7	413	32	381	2.5
7.7	405	31	374	2.4
7.7	404	31	373	2.4
7.7	403	31	372	2.4
7.7	402	31	371	2.4
7.7	401	31	370	2.4
7.7	392	30	362	2.3
7.7	391	30	361	2.3
7.7	390	30	360	2.3
7.7	389	30	359	2.3
7.7	388	30	358	2.3
7.7	379	29	350	2.2
7.7	378	29	349	2.2
7.7	377	29	348	2.2
7.7	376	29	347	2.2
7.7	375	29	346	2.2
7.7	366	28	338	2.1

%E	M1	M2	DM	M*
7.7	365	28	337	2.1
7.7	364	28	336	2.2
7.7	363	28	335	2.2
7.7	362	28	334	2.2
7.7	352	27	325	2.1
7.7	351	27	324	2.1
7.7	350	27	323	2.1
7.7	349	27	322	2.1
7.7	339	26	313	2.0
7.7	338	26	312	2.0
7.7	337	26	311	2.0
7.7	336	26	310	2.0
7.7	326	25	301	1.9
7.7	325	25	300	1.9
7.7	324	25	299	1.9
7.7	323	25	298	1.9
7.7	313	24	289	1.8
7.7	312	24	288	1.8
7.7	311	24	287	1.9
7.7	310	24	286	1.9
7.7	300	23	277	1.8
7.7	299	23	276	1.8
7.7	298	23	275	1.8
7.7	297	23	274	1.8
7.7	287	22	265	1.7
7.7	286	22	264	1.7
7.7	285	22	263	1.7
7.7	284	22	262	1.7
7.7	274	21	253	1.6
7.7	273	21	252	1.6
7.7	272	21	251	1.6
7.7	271	21	250	1.6
7.7	261	20	241	1.5
7.7	260	20	240	1.5
7.7	259	20	239	1.5
7.7	248	19	229	1.5
7.7	247	19	228	1.5
7.7	246	19	227	1.5
7.7	235	18	217	1.4
7.7	234	18	216	1.4
7.7	233	18	215	1.4
7.7	222	17	205	1.3
7.7	221	17	204	1.3
7.7	220	17	203	1.3
7.7	209	16	193	1.2
7.7	208	16	192	1.2
7.7	207	16	191	1.2
7.7	196	15	181	1.1
7.7	195	15	180	1.2
7.7	194	15	179	1.2

%E	M1	M2	DM	M*
7.7	183	14	169	1.1
7.7	182	14	168	1.1
7.7	181	14	167	1.1
7.7	169	13	156	1.0
7.7	168	13	155	1.0
7.7	156	12	144	0.9
7.7	155	12	143	0.9
7.7	143	11	132	0.8
7.7	142	11	131	0.9
7.7	130	10	120	0.8
7.7	117	9	108	0.7
7.7	104	8	96	0.6
7.7	91	7	84	0.5
7.7	78	6	72	0.5
7.7	65	5	60	0.4
7.7	52	4	48	0.3
7.7	39	3	36	0.2
7.7	26	2	24	0.2
7.7	13	1	12	0.1
7.6	500	38	462	2.9
7.6	499	38	461	2.9
7.6	498	38	460	2.9
7.6	497	38	459	2.9
7.6	490	37	453	2.8
7.6	489	37	452	2.8
7.6	488	37	451	2.8
7.6	487	37	450	2.8
7.6	486	37	449	2.8
7.6	485	37	448	2.8
7.6	484	37	447	2.8
7.6	476	36	440	2.7
7.6	475	36	439	2.7
7.6	474	36	438	2.7
7.6	473	36	437	2.7
7.6	472	36	436	2.7
7.6	471	36	435	2.8
7.6	463	35	428	2.6
7.6	462	35	427	2.7
7.6	461	35	426	2.7
7.6	460	35	425	2.7
7.6	459	35	424	2.7
7.6	458	35	423	2.7
7.6	450	34	416	2.6
7.6	449	34	415	2.6
7.6	448	34	414	2.6
7.6	447	34	413	2.6
7.6	446	34	412	2.6
7.6	445	34	411	2.6
7.6	437	33	404	2.5
7.6	436	33	403	2.5

%E	M1	M2	DM	M*
7.6	435	33	402	2.5
7.6	434	33	401	2.5
7.6	433	33	400	2.5
7.6	432	33	399	2.5
7.6	423	32	391	2.4
7.6	422	32	390	2.4
7.6	421	32	389	2.4
7.6	420	32	388	2.4
7.6	419	32	387	2.4
7.6	410	31	379	2.3
7.6	409	31	378	2.3
7.6	408	31	377	2.4
7.6	407	31	376	2.4
7.6	406	31	375	2.4
7.6	397	30	367	2.3
7.6	396	30	366	2.3
7.6	395	30	365	2.3
7.6	394	30	364	2.3
7.6	393	30	363	2.3
7.6	384	29	355	2.2
7.6	383	29	354	2.2
7.6	382	29	353	2.2
7.6	381	29	352	2.2
7.6	380	29	351	2.2
7.6	370	28	342	2.1
7.6	369	28	341	2.1
7.6	368	28	340	2.1
7.6	367	28	339	2.1
7.6	357	27	330	2.0
7.6	356	27	329	2.0
7.6	355	27	328	2.1
7.6	354	27	327	2.1
7.6	353	27	326	2.1
7.6	344	26	318	2.0
7.6	343	26	317	2.0
7.6	342	26	316	2.0
7.6	341	26	315	2.0
7.6	340	26	314	2.0
7.6	331	25	306	1.9
7.6	330	25	305	1.9
7.6	329	25	304	1.9
7.6	328	25	303	1.9
7.6	327	25	302	1.9
7.6	317	24	293	1.8
7.6	316	24	292	1.8
7.6	315	24	291	1.8
7.6	314	24	290	1.8
7.6	304	23	281	1.7
7.6	303	23	280	1.7
7.6	302	23	279	1.8

%E	M1	M2	DM	M*
7.6	301	23	278	1.8
7.6	291	22	269	1.7
7.6	290	22	268	1.7
7.6	289	22	267	1.7
7.6	288	22	266	1.7
7.6	278	21	257	1.6
7.6	277	21	256	1.6
7.6	276	21	255	1.6
7.6	275	21	254	1.6
7.6	264	20	244	1.5
7.6	263	20	243	1.5
7.6	262	20	242	1.5
7.6	251	19	232	1.4
7.6	250	19	231	1.4
7.6	249	19	230	1.4
7.6	238	18	220	1.4
7.6	237	18	219	1.4
7.6	236	18	218	1.4
7.6	225	17	208	1.3
7.6	224	17	207	1.3
7.6	223	17	206	1.3
7.6	211	16	195	1.2
7.6	210	16	194	1.2
7.6	198	15	183	1.1
7.6	197	15	182	1.1
7.6	185	14	171	1.1
7.6	184	14	170	1.1
7.6	172	13	159	1.0
7.6	171	13	158	1.0
7.6	170	13	157	1.0
7.6	158	12	146	0.9
7.6	157	12	145	0.9
7.6	145	11	134	0.8
7.6	144	11	133	0.8
7.6	132	10	122	0.8
7.6	131	10	121	0.8
7.6	119	9	110	0.7
7.6	118	9	109	0.7
7.6	105	8	97	0.6
7.6	92	7	85	0.5
7.6	79	6	73	0.5
7.6	66	5	61	0.4
7.5	496	37	459	2.8
7.5	495	37	458	2.8
7.5	494	37	457	2.8
7.5	493	37	456	2.8
7.5	492	37	455	2.8
7.5	491	37	454	2.8
7.5	483	36	447	2.7
7.5	482	36	446	2.7

%E	M1	M2	DM	M*
7.5	481	36	445	2.7
7.5	480	36	444	2.7
7.5	479	36	443	2.7
7.5	478	36	442	2.7
7.5	477	36	441	2.7
7.5	469	35	434	2.6
7.5	468	35	433	2.6
7.5	467	35	432	2.6
7.5	466	35	431	2.6
7.5	465	35	430	2.6
7.5	464	35	429	2.6
7.5	456	34	422	2.5
7.5	455	34	421	2.5
7.5	454	34	420	2.5
7.5	453	34	419	2.6
7.5	452	34	418	2.6
7.5	451	34	417	2.6
7.5	442	33	409	2.5
7.5	441	33	408	2.5
7.5	440	33	407	2.5
7.5	439	33	406	2.5
7.5	438	33	405	2.5
7.5	429	32	397	2.4
7.5	428	32	396	2.4
7.5	427	32	395	2.4
7.5	426	32	394	2.4
7.5	425	32	393	2.4
7.5	424	32	392	2.4
7.5	416	31	385	2.3
7.5	415	31	384	2.3
7.5	414	31	383	2.3
7.5	413	31	382	2.3
7.5	412	31	381	2.3
7.5	411	31	380	2.3
7.5	402	30	372	2.2
7.5	401	30	371	2.2
7.5	400	30	370	2.3
7.5	399	30	369	2.3
7.5	398	30	368	2.3
7.5	389	29	360	2.2
7.5	388	29	359	2.2
7.5	387	29	358	2.2
7.5	386	29	357	2.2
7.5	385	29	356	2.2
7.5	375	28	347	2.1
7.5	374	28	346	2.1
7.5	373	28	345	2.1
7.5	372	28	344	2.1
7.5	371	28	343	2.1
7.5	362	27	335	2.0

%E	M1	M2	DM	M*
7.5	361	27	334	2.0
7.5	360	27	333	2.0
7.5	359	27	332	2.0
7.5	358	27	331	2.0
7.5	349	26	323	1.9
7.5	348	26	322	1.9
7.5	347	26	321	1.9
7.5	346	26	320	2.0
7.5	345	26	319	2.0
7.5	335	25	310	1.9
7.5	334	25	309	1.9
7.5	333	25	308	1.9
7.5	332	25	307	1.9
7.5	322	24	298	1.8
7.5	321	24	297	1.8
7.5	320	24	296	1.8
7.5	319	24	295	1.8
7.5	318	24	294	1.8
7.5	308	23	285	1.7
7.5	307	23	284	1.7
7.5	306	23	283	1.7
7.5	305	23	282	1.7
7.5	295	22	273	1.6
7.5	294	22	272	1.6
7.5	293	22	271	1.7
7.5	292	22	270	1.7
7.5	281	21	260	1.6
7.5	280	21	259	1.6
7.5	279	21	258	1.6
7.5	268	20	248	1.5
7.5	267	20	247	1.5
7.5	266	20	246	1.5
7.5	265	20	245	1.5
7.5	255	19	236	1.4
7.5	254	19	235	1.4
7.5	253	19	234	1.4
7.5	252	19	233	1.4
7.5	241	18	223	1.3
7.5	240	18	222	1.3
7.5	239	18	221	1.4
7.5	228	17	211	1.3
7.5	227	17	210	1.3
7.5	226	17	209	1.3
7.5	214	16	198	1.2
7.5	213	16	197	1.2
7.5	212	16	196	1.2
7.5	201	15	186	1.1
7.5	200	15	185	1.1
7.5	199	15	184	1.1
7.5	187	14	173	1.0

%E	M1	M2	DM	M*
7.5	186	14	172	1.1
7.5	174	13	161	1.0
7.5	173	13	160	1.0
7.5	161	12	149	0.9
7.5	160	12	148	0.9
7.5	159	12	147	0.9
7.5	147	11	136	0.8
7.5	146	11	135	0.8
7.5	134	10	124	0.7
7.5	133	10	123	0.8
7.5	120	9	111	0.7
7.5	107	8	99	0.6
7.5	106	8	98	0.6
7.5	93	7	86	0.5
7.5	80	6	74	0.4
7.5	67	5	62	0.4
7.5	53	4	49	0.3
7.5	40	3	37	0.2
7.4	500	37	463	2.7
7.4	499	37	462	2.7
7.4	498	37	461	2.7
7.4	497	37	460	2.8
7.4	489	36	453	2.7
7.4	488	36	452	2.7
7.4	487	36	451	2.7
7.4	486	36	450	2.7
7.4	485	36	449	2.7
7.4	484	36	448	2.7
7.4	476	35	441	2.6
7.4	475	35	440	2.6
7.4	474	35	439	2.6
7.4	473	35	438	2.6
7.4	472	35	437	2.6
7.4	471	35	436	2.6
7.4	470	35	435	2.6
7.4	462	34	428	2.5
7.4	461	34	427	2.5
7.4	460	34	426	2.5
7.4	459	34	425	2.5
7.4	458	34	424	2.5
7.4	457	34	423	2.5
7.4	449	33	416	2.4
7.4	448	33	415	2.4
7.4	447	33	414	2.4
7.4	446	33	413	2.4
7.4	445	33	412	2.4
7.4	444	33	411	2.5
7.4	443	33	410	2.5
7.4	435	32	403	2.4
7.4	434	32	402	2.4

%E	M1	M2	DM	M*
7.4	433	32	401	2.4
7.4	432	32	400	2.4
7.4	431	32	399	2.4
7.4	430	32	398	2.4
7.4	421	31	390	2.3
7.4	420	31	389	2.3
7.4	419	31	388	2.3
7.4	418	31	387	2.3
7.4	417	31	386	2.3
7.4	408	30	378	2.2
7.4	407	30	377	2.2
7.4	406	30	376	2.2
7.4	405	30	375	2.2
7.4	404	30	374	2.2
7.4	403	30	373	2.2
7.4	394	29	365	2.1
7.4	393	29	364	2.1
7.4	392	29	363	2.1
7.4	391	29	362	2.2
7.4	390	29	361	2.2
7.4	380	28	352	2.1
7.4	379	28	351	2.1
7.4	378	28	350	2.1
7.4	377	28	349	2.1
7.4	376	28	348	2.1
7.4	367	27	340	2.0
7.4	366	27	339	2.0
7.4	365	27	338	2.0
7.4	364	27	337	2.0
7.4	363	27	336	2.0
7.4	353	26	327	1.9
7.4	352	26	326	1.9
7.4	351	26	325	1.9
7.4	350	26	324	1.9
7.4	340	25	315	1.8
7.4	339	25	314	1.8
7.4	338	25	313	1.8
7.4	337	25	312	1.9
7.4	336	25	311	1.9
7.4	326	24	302	1.8
7.4	325	24	301	1.8
7.4	324	24	300	1.8
7.4	323	24	299	1.8
7.4	312	23	289	1.7
7.4	311	23	288	1.7
7.4	310	23	287	1.7
7.4	309	23	286	1.7
7.4	299	22	277	1.6
7.4	298	22	276	1.6
7.4	297	22	275	1.6

%E	M1	M2	DM	M*
7.4	296	22	274	1.6
7.4	285	21	264	1.5
7.4	284	21	263	1.6
7.4	283	21	262	1.6
7.4	282	21	261	1.6
7.4	272	20	252	1.5
7.4	271	20	251	1.5
7.4	270	20	250	1.5
7.4	269	20	249	1.5
7.4	258	19	239	1.4
7.4	257	19	238	1.4
7.4	256	19	237	1.4
7.4	244	18	226	1.3
7.4	243	18	225	1.3
7.4	242	18	224	1.3
7.4	231	17	214	1.3
7.4	230	17	213	1.3
7.4	229	17	212	1.3
7.4	217	16	201	1.2
7.4	216	16	200	1.2
7.4	215	16	199	1.2
7.4	204	15	189	1.1
7.4	203	15	188	1.1
7.4	202	15	187	1.1
7.4	190	14	176	1.0
7.4	189	14	175	1.0
7.4	188	14	174	1.0
7.4	176	13	163	1.0
7.4	175	13	162	1.0
7.4	163	12	151	0.9
7.4	162	12	150	0.9
7.4	149	11	138	0.8
7.4	148	11	137	0.8
7.4	136	10	126	0.7
7.4	135	10	125	0.7
7.4	122	9	113	0.7
7.4	121	9	112	0.7
7.4	108	8	100	0.6
7.4	95	7	88	0.5
7.4	94	7	87	0.5
7.4	81	6	75	0.4
7.4	68	5	63	0.4
7.4	54	4	50	0.3
7.4	27	2	25	0.1
7.3	496	36	460	2.6
7.3	495	36	459	2.6
7.3	494	36	458	2.6
7.3	493	36	457	2.6
7.3	492	36	456	2.6
7.3	491	36	455	2.6

%E	M1	M2	DM	M*
7.3	490	36	454	2.6
7.3	482	35	447	2.5
7.3	481	35	446	2.5
7.3	480	35	445	2.6
7.3	479	35	444	2.6
7.3	478	35	443	2.6
7.3	477	35	442	2.6
7.3	468	34	434	2.5
7.3	467	34	433	2.5
7.3	466	34	432	2.5
7.3	465	34	431	2.5
7.3	464	34	430	2.5
7.3	463	34	429	2.5
7.3	455	33	422	2.4
7.3	454	33	421	2.4
7.3	453	33	420	2.4
7.3	452	33	419	2.4
7.3	451	33	418	2.4
7.3	450	33	417	2.4
7.3	441	32	409	2.3
7.3	440	32	408	2.3
7.3	439	32	407	2.3
7.3	438	32	406	2.3
7.3	437	32	405	2.3
7.3	436	32	404	2.3
7.3	427	31	396	2.3
7.3	426	31	395	2.3
7.3	425	31	394	2.3
7.3	424	31	393	2.3
7.3	423	31	392	2.3
7.3	422	31	391	2.3
7.3	413	30	383	2.2
7.3	412	30	382	2.2
7.3	411	30	381	2.2
7.3	410	30	380	2.2
7.3	409	30	379	2.2
7.3	400	29	371	2.1
7.3	399	29	370	2.1
7.3	398	29	369	2.1
7.3	397	29	368	2.1
7.3	396	29	367	2.1
7.3	395	29	366	2.1
7.3	386	28	358	2.0
7.3	385	28	357	2.0
7.3	384	28	356	2.0
7.3	383	28	355	2.0
7.3	382	28	354	2.1
7.3	381	28	353	2.1
7.3	372	27	345	2.0
7.3	371	27	344	2.0

%E	M1	M2	DM	M*
7.3	370	27	343	2.0
7.3	369	27	342	2.0
7.3	368	27	341	2.0
7.3	358	26	332	1.9
7.3	357	26	331	1.9
7.3	356	26	330	1.9
7.3	355	26	329	1.9
7.3	354	26	328	1.9
7.3	344	25	319	1.8
7.3	343	25	318	1.8
7.3	342	25	317	1.8
7.3	341	25	316	1.8
7.3	331	24	307	1.7
7.3	330	24	306	1.7
7.3	329	24	305	1.8
7.3	328	24	304	1.8
7.3	327	24	303	1.8
7.3	317	23	294	1.7
7.3	316	23	293	1.7
7.3	315	23	292	1.7
7.3	314	23	291	1.7
7.3	313	23	290	1.7
7.3	303	22	281	1.6
7.3	302	22	280	1.6
7.3	301	22	279	1.6
7.3	300	22	278	1.6
7.3	289	21	268	1.5
7.3	288	21	267	1.5
7.3	287	21	266	1.5
7.3	286	21	265	1.5
7.3	275	20	255	1.5
7.3	274	20	254	1.5
7.3	273	20	253	1.5
7.3	262	19	243	1.4
7.3	261	19	242	1.4
7.3	260	19	241	1.4
7.3	259	19	240	1.4
7.3	248	18	230	1.3
7.3	247	18	229	1.3
7.3	246	18	228	1.3
7.3	245	18	227	1.3
7.3	234	17	217	1.2
7.3	233	17	216	1.2
7.3	232	17	215	1.2
7.3	220	16	204	1.2
7.3	219	16	203	1.2
7.3	218	16	202	1.2
7.3	206	15	191	1.1
7.3	205	15	190	1.1
7.3	193	14	179	1.0

%E	M1	M2	DM	M*
7.3	192	14	178	1.0
7.3	191	14	177	1.0
7.3	179	13	166	0.9
7.3	178	13	165	0.9
7.3	177	13	164	1.0
7.3	165	12	153	0.9
7.3	164	12	152	0.9
7.3	151	11	140	0.8
7.3	150	11	139	0.8
7.3	137	10	127	0.7
7.3	124	9	115	0.7
7.3	123	9	114	0.7
7.3	110	8	102	0.6
7.3	109	8	101	0.6
7.3	96	7	89	0.5
7.3	82	6	76	0.4
7.3	55	4	51	0.3
7.3	41	3	38	0.2
7.2	500	36	464	2.6
7.2	499	36	463	2.6
7.2	498	36	462	2.6
7.2	497	36	461	2.6
7.2	489	35	454	2.5
7.2	488	35	453	2.5
7.2	487	35	452	2.5
7.2	486	35	451	2.5
7.2	485	35	450	2.5
7.2	484	35	449	2.5
7.2	483	35	448	2.5
7.2	475	34	441	2.4
7.2	474	34	440	2.4
7.2	473	34	439	2.4
7.2	472	34	438	2.4
7.2	471	34	437	2.5
7.2	470	34	436	2.5
7.2	469	34	435	2.5
7.2	461	33	428	2.4
7.2	460	33	427	2.4
7.2	459	33	426	2.4
7.2	458	33	425	2.4
7.2	457	33	424	2.4
7.2	456	33	423	2.4
7.2	447	32	415	2.3
7.2	446	32	414	2.3
7.2	445	32	413	2.3
7.2	444	32	412	2.3
7.2	443	32	411	2.3
7.2	442	32	410	2.3
7.2	433	31	402	2.2
7.2	432	31	401	2.2

%E	M1	M2	DM	M*
7.2	431	31	400	2.2
7.2	430	31	399	2.2
7.2	429	31	398	2.2
7.2	428	31	397	2.2
7.2	419	30	389	2.1
7.2	418	30	388	2.2
7.2	417	30	387	2.2
7.2	416	30	386	2.2
7.2	415	30	385	2.2
7.2	414	30	384	2.2
7.2	405	29	376	2.1
7.2	404	29	375	2.1
7.2	403	29	374	2.1
7.2	402	29	373	2.1
7.2	401	29	372	2.1
7.2	391	28	363	2.0
7.2	390	28	362	2.0
7.2	389	28	361	2.0
7.2	388	28	360	2.0
7.2	387	28	359	2.0
7.2	377	27	350	1.9
7.2	376	27	349	1.9
7.2	375	27	348	1.9
7.2	374	27	347	1.9
7.2	373	27	346	2.0
7.2	363	26	337	1.9
7.2	362	26	336	1.9
7.2	361	26	335	1.9
7.2	360	26	334	1.9
7.2	359	26	333	1.9
7.2	349	25	324	1.8
7.2	348	25	323	1.8
7.2	347	25	322	1.8
7.2	346	25	321	1.8
7.2	345	25	320	1.8
7.2	335	24	311	1.7
7.2	334	24	310	1.7
7.2	333	24	309	1.7
7.2	332	24	308	1.7
7.2	321	23	298	1.6
7.2	320	23	297	1.7
7.2	319	23	296	1.7
7.2	318	23	295	1.7
7.2	307	22	285	1.6
7.2	306	22	284	1.6
7.2	305	22	283	1.6
7.2	304	22	282	1.6
7.2	293	21	272	1.5
7.2	292	21	271	1.5
7.2	291	21	270	1.5

%E	M1	M2	DM	M*
7.2	290	21	269	1.5
7.2	279	20	259	1.4
7.2	278	20	258	1.4
7.2	277	20	257	1.4
7.2	276	20	256	1.4
7.2	265	19	246	1.4
7.2	264	19	245	1.4
7.2	263	19	244	1.4
7.2	251	18	233	1.3
7.2	250	18	232	1.3
7.2	249	18	231	1.3
7.2	237	17	220	1.2
7.2	236	17	219	1.2
7.2	235	17	218	1.2
7.2	223	16	207	1.1
7.2	222	16	206	1.2
7.2	221	16	205	1.2
7.2	209	15	194	1.1
7.2	208	15	193	1.1
7.2	207	15	192	1.1
7.2	195	14	181	1.0
7.2	194	14	180	1.0
7.2	181	13	168	0.9
7.2	180	13	167	0.9
7.2	167	12	155	0.9
7.2	166	12	154	0.9
7.2	153	11	142	0.8
7.2	152	11	141	0.8
7.2	139	10	129	0.7
7.2	138	10	128	0.7
7.2	125	9	116	0.6
7.2	111	8	103	0.6
7.2	97	7	90	0.5
7.2	83	6	77	0.4
7.2	69	5	64	0.4
7.1	496	35	461	2.5
7.1	495	35	460	2.5
7.1	494	35	459	2.5
7.1	493	35	458	2.5
7.1	492	35	457	2.5
7.1	491	35	456	2.5
7.1	490	35	455	2.5
7.1	482	34	448	2.4
7.1	481	34	447	2.4
7.1	480	34	446	2.4
7.1	479	34	445	2.4
7.1	478	34	444	2.4
7.1	477	34	443	2.4
7.1	476	34	442	2.4
7.1	468	33	435	2.3

%E	M1	M2	DM	M*
7.1	467	33	434	2.3
7.1	466	33	433	2.3
7.1	465	33	432	2.3
7.1	464	33	431	2.3
7.1	463	33	430	2.4
7.1	462	33	429	2.4
7.1	453	32	421	2.3
7.1	452	32	420	2.3
7.1	451	32	419	2.3
7.1	450	32	418	2.3
7.1	449	32	417	2.3
7.1	448	32	416	2.3
7.1	439	31	408	2.2
7.1	438	31	407	2.2
7.1	437	31	406	2.2
7.1	436	31	405	2.2
7.1	435	31	404	2.2
7.1	434	31	403	2.2
7.1	425	30	395	2.1
7.1	424	30	394	2.1
7.1	423	30	393	2.1
7.1	422	30	392	2.1
7.1	421	30	391	2.1
7.1	420	30	390	2.1
7.1	411	29	382	2.0
7.1	410	29	381	2.1
7.1	409	29	380	2.1
7.1	408	29	379	2.1
7.1	407	29	378	2.1
7.1	406	29	377	2.1
7.1	397	28	369	2.0
7.1	396	28	368	2.0
7.1	395	28	367	2.0
7.1	394	28	366	2.0
7.1	393	28	365	2.0
7.1	392	28	364	2.0
7.1	383	27	356	1.9
7.1	382	27	355	1.9
7.1	381	27	354	1.9
7.1	380	27	353	1.9
7.1	379	27	352	1.9
7.1	378	27	351	1.9
7.1	368	26	342	1.8
7.1	367	26	341	1.8
7.1	366	26	340	1.8
7.1	365	26	339	1.9
7.1	364	26	338	1.9
7.1	354	25	329	1.8
7.1	353	25	328	1.8
7.1	352	25	327	1.8

%E	M1	M2	DM	M*
7.1	351	25	326	1.8
7.1	350	25	325	1.8
7.1	340	24	316	1.7
7.1	339	24	315	1.7
7.1	338	24	314	1.7
7.1	337	24	313	1.7
7.1	336	24	312	1.7
7.1	326	23	303	1.6
7.1	325	23	302	1.6
7.1	324	23	301	1.6
7.1	323	23	300	1.6
7.1	322	23	299	1.6
7.1	312	22	290	1.6
7.1	311	22	289	1.6
7.1	310	22	288	1.6
7.1	309	22	287	1.6
7.1	308	22	286	1.6
7.1	297	21	276	1.5
7.1	296	21	275	1.5
7.1	295	21	274	1.5
7.1	294	21	273	1.5
7.1	283	20	263	1.4
7.1	282	20	262	1.4
7.1	281	20	261	1.4
7.1	280	20	260	1.4
7.1	269	19	250	1.3
7.1	268	19	249	1.3
7.1	267	19	248	1.4
7.1	266	19	247	1.4
7.1	255	18	237	1.3
7.1	254	18	236	1.3
7.1	253	18	235	1.3
7.1	252	18	234	1.3
7.1	241	17	224	1.2
7.1	240	17	223	1.2
7.1	239	17	222	1.2
7.1	238	17	221	1.2
7.1	226	16	210	1.1
7.1	225	16	209	1.1
7.1	224	16	208	1.1
7.1	212	15	197	1.1
7.1	211	15	196	1.1
7.1	210	15	195	1.1
7.1	198	14	184	1.0
7.1	197	14	183	1.0
7.1	196	14	182	1.0
7.1	184	13	171	0.9
7.1	183	13	170	0.9
7.1	182	13	169	0.9
7.1	170	12	158	0.8

%E	M1	M2	DM	M*
7.1	169	12	157	0.9
7.1	168	12	156	0.9
7.1	156	11	145	0.8
7.1	155	11	144	0.8
7.1	154	11	143	0.8
7.1	141	10	131	0.7
7.1	140	10	130	0.7
7.1	127	9	118	0.6
7.1	126	9	117	0.6
7.1	113	8	105	0.6
7.1	112	8	104	0.6
7.1	99	7	92	0.5
7.1	98	7	91	0.5
7.1	85	6	79	0.4
7.1	84	6	78	0.4
7.1	70	5	65	0.4
7.1	56	4	52	0.3
7.1	42	3	39	0.2
7.1	28	2	26	0.1
7.1	14	1	13	0.1
7.0	500	35	465	2.4
7.0	499	35	464	2.5
7.0	498	35	463	2.5
7.0	497	35	462	2.5
7.0	489	34	455	2.4
7.0	488	34	454	2.4
7.0	487	34	453	2.4
7.0	486	34	452	2.4
7.0	485	34	451	2.4
7.0	484	34	450	2.4
7.0	483	34	449	2.4
7.0	474	33	441	2.3
7.0	473	33	440	2.3
7.0	472	33	439	2.3
7.0	471	33	438	2.3
7.0	470	33	437	2.3
7.0	469	33	436	2.3
7.0	460	32	428	2.2
7.0	459	32	427	2.2
7.0	458	32	426	2.2
7.0	457	32	425	2.2
7.0	456	32	424	2.2
7.0	455	32	423	2.3
7.0	454	32	422	2.3
7.0	446	31	415	2.2
7.0	445	31	414	2.2
7.0	444	31	413	2.2
7.0	443	31	412	2.2
7.0	442	31	411	2.2
7.0	441	31	410	2.2

%E	M1	M2	DM	M*
7.0	440	31	409	2.2
7.0	431	30	401	2.1
7.0	430	30	400	2.1
7.0	429	30	399	2.1
7.0	428	30	398	2.1
7.0	427	30	397	2.1
7.0	426	30	396	2.1
7.0	417	29	388	2.0
7.0	416	29	387	2.0
7.0	415	29	386	2.0
7.0	414	29	385	2.0
7.0	413	29	384	2.0
7.0	412	29	383	2.0
7.0	402	28	374	2.0
7.0	401	28	373	2.0
7.0	400	28	372	2.0
7.0	399	28	371	2.0
7.0	398	28	370	2.0
7.0	388	27	361	1.9
7.0	387	27	360	1.9
7.0	386	27	359	1.9
7.0	385	27	358	1.9
7.0	384	27	357	1.9
7.0	374	26	348	1.8
7.0	373	26	347	1.8
7.0	372	26	346	1.8
7.0	371	26	345	1.8
7.0	370	26	344	1.8
7.0	369	26	343	1.8
7.0	359	25	334	1.7
7.0	358	25	333	1.7
7.0	357	25	332	1.8
7.0	356	25	331	1.8
7.0	355	25	330	1.8
7.0	345	24	321	1.7
7.0	344	24	320	1.7
7.0	343	24	319	1.7
7.0	342	24	318	1.7
7.0	341	24	317	1.7
7.0	330	23	307	1.6
7.0	329	23	306	1.6
7.0	328	23	305	1.6
7.0	327	23	304	1.6
7.0	316	22	294	1.5
7.0	315	22	293	1.5
7.0	314	22	292	1.5
7.0	313	22	291	1.5
7.0	302	21	281	1.5
7.0	301	21	280	1.5
7.0	300	21	279	1.5

%E	M1	M2	DM	M*
7.0	299	21	278	1.5
7.0	298	21	277	1.5
7.0	287	20	267	1.4
7.0	286	20	266	1.4
7.0	285	20	265	1.4
7.0	284	20	264	1.4
7.0	273	19	254	1.3
7.0	272	19	253	1.3
7.0	271	19	252	1.3
7.0	270	19	251	1.3
7.0	259	18	241	1.3
7.0	258	18	240	1.3
7.0	257	18	239	1.3
7.0	256	18	238	1.3
7.0	244	17	227	1.2
7.0	243	17	226	1.2
7.0	242	17	225	1.2
7.0	230	16	214	1.1
7.0	229	16	213	1.1
7.0	228	16	212	1.1
7.0	227	16	211	1.1
7.0	215	15	200	1.0
7.0	214	15	199	1.1
7.0	213	15	198	1.1
7.0	201	14	187	1.0
7.0	200	14	186	1.0
7.0	199	14	185	1.0
7.0	187	13	174	0.9
7.0	186	13	173	0.9
7.0	185	13	172	0.9
7.0	172	12	160	0.8
7.0	171	12	159	0.8
7.0	158	11	147	0.8
7.0	157	11	146	0.8
7.0	143	10	133	0.7
7.0	142	10	132	0.7
7.0	129	9	120	0.6
7.0	128	9	119	0.6
7.0	115	8	107	0.6
7.0	114	8	106	0.6
7.0	100	7	93	0.5
7.0	86	6	80	0.4
7.0	71	5	66	0.4
7.0	57	4	53	0.3
7.0	43	3	40	0.2
6.9	496	34	462	2.3
6.9	495	34	461	2.3
6.9	494	34	460	2.3
6.9	493	34	459	2.3
6.9	492	34	458	2.3
6.9	491	34	457	2.4
6.9	490	34	456	2.4
6.9	481	33	448	2.3
6.9	480	33	447	2.3
6.9	479	33	446	2.3
6.9	478	33	445	2.3
6.9	477	33	444	2.3
6.9	476	33	443	2.3
6.9	475	33	442	2.3
6.9	467	32	435	2.2
6.9	466	32	434	2.2
6.9	465	32	433	2.2
6.9	464	32	432	2.2
6.9	463	32	431	2.2
6.9	462	32	430	2.2
6.9	461	32	429	2.2
6.9	452	31	421	2.1
6.9	451	31	420	2.1
6.9	450	31	419	2.1
6.9	449	31	418	2.1
6.9	448	31	417	2.1
6.9	447	31	416	2.1
6.9	437	30	407	2.1
6.9	436	30	406	2.1
6.9	435	30	405	2.1
6.9	434	30	404	2.1
6.9	433	30	403	2.1
6.9	432	30	402	2.1
6.9	423	29	394	2.0
6.9	422	29	393	2.0
6.9	421	29	392	2.0
6.9	420	29	391	2.0
6.9	419	29	390	2.0
6.9	418	29	389	2.0
6.9	408	28	380	1.9
6.9	407	28	379	1.9
6.9	406	28	378	1.9
6.9	405	28	377	1.9
6.9	404	28	376	1.9
6.9	403	28	375	1.9
6.9	394	27	367	1.9
6.9	393	27	366	1.9
6.9	392	27	365	1.9
6.9	391	27	364	1.9
6.9	390	27	363	1.9
6.9	389	27	362	1.9
6.9	379	26	353	1.8
6.9	378	26	352	1.8
6.9	377	26	351	1.8
6.9	376	26	350	1.8
6.9	375	26	349	1.8
6.9	364	25	339	1.7
6.9	363	25	338	1.7
6.9	362	25	337	1.7
6.9	361	25	336	1.7
6.9	360	25	335	1.7
6.9	350	24	326	1.6
6.9	349	24	325	1.7
6.9	348	24	324	1.7
6.9	347	24	323	1.7
6.9	346	24	322	1.7
6.9	335	23	312	1.6
6.9	334	23	311	1.6
6.9	333	23	310	1.6
6.9	332	23	309	1.6
6.9	331	23	308	1.6
6.9	321	22	299	1.5
6.9	320	22	298	1.5
6.9	319	22	297	1.5
6.9	318	22	296	1.5
6.9	317	22	295	1.5
6.9	306	21	285	1.4
6.9	305	21	284	1.4
6.9	304	21	283	1.5
6.9	303	21	282	1.5
6.9	291	20	271	1.4
6.9	290	20	270	1.4
6.9	289	20	269	1.4
6.9	288	20	268	1.4
6.9	277	19	258	1.3
6.9	276	19	257	1.3
6.9	275	19	256	1.3
6.9	274	19	255	1.3
6.9	262	18	244	1.2
6.9	261	18	243	1.2
6.9	260	18	242	1.2
6.9	248	17	231	1.2
6.9	247	17	230	1.2
6.9	246	17	229	1.2
6.9	245	17	228	1.2
6.9	233	16	217	1.1
6.9	232	16	216	1.1
6.9	231	16	215	1.1
6.9	218	15	203	1.0
6.9	217	15	202	1.0
6.9	216	15	201	1.0
6.9	204	14	190	1.0
6.9	203	14	189	1.0
6.9	202	14	188	1.0
6.9	189	13	176	0.9
6.9	188	13	175	0.9
6.9	175	12	163	0.8
6.9	174	12	162	0.8
6.9	173	12	161	0.8
6.9	160	11	149	0.8
6.9	159	11	148	0.8
6.9	145	10	135	0.7
6.9	144	10	134	0.7
6.9	131	9	122	0.6
6.9	130	9	121	0.6
6.9	116	8	108	0.6
6.9	102	7	95	0.5
6.9	101	7	94	0.5
6.9	87	6	81	0.4
6.9	72	5	67	0.3
6.9	58	4	54	0.3
6.9	29	2	27	0.1
6.8	500	34	466	2.3
6.8	499	34	465	2.3
6.8	498	34	464	2.3
6.8	497	34	463	2.3
6.8	488	33	455	2.2
6.8	487	33	454	2.2
6.8	486	33	453	2.2
6.8	485	33	452	2.2
6.8	484	33	451	2.3
6.8	483	33	450	2.3
6.8	482	33	449	2.3
6.8	474	32	442	2.2
6.8	473	32	441	2.2
6.8	472	32	440	2.2
6.8	471	32	439	2.2
6.8	470	32	438	2.2
6.8	469	32	437	2.2
6.8	468	32	436	2.2
6.8	459	31	428	2.1
6.8	458	31	427	2.1
6.8	457	31	426	2.1
6.8	456	31	425	2.1
6.8	455	31	424	2.1
6.8	454	31	423	2.1
6.8	453	31	422	2.1
6.8	444	30	414	2.0
6.8	443	30	413	2.0
6.8	442	30	412	2.0
6.8	441	30	411	2.0
6.8	440	30	410	2.0
6.8	439	30	409	2.1
6.8	438	30	408	2.1
6.8	429	29	400	2.0
6.8	428	29	399	2.0
6.8	427	29	398	2.0
6.8	426	29	397	2.0
6.8	425	29	396	2.0
6.8	424	29	395	2.0
6.8	414	28	386	1.9
6.8	413	28	385	1.9
6.8	412	28	384	1.9
6.8	411	28	383	1.9
6.8	410	28	382	1.9
6.8	409	28	381	1.9
6.8	400	27	373	1.8
6.8	399	27	372	1.8
6.8	398	27	371	1.8
6.8	397	27	370	1.8
6.8	396	27	369	1.8
6.8	395	27	368	1.8
6.8	385	26	359	1.8
6.8	384	26	358	1.8
6.8	383	26	357	1.8
6.8	382	26	356	1.8
6.8	381	26	355	1.8
6.8	380	26	354	1.8
6.8	370	25	345	1.7
6.8	369	25	344	1.7
6.8	368	25	343	1.7
6.8	367	25	342	1.7
6.8	366	25	341	1.7
6.8	365	25	340	1.7
6.8	355	24	331	1.6
6.8	354	24	330	1.6
6.8	353	24	329	1.6
6.8	352	24	328	1.6
6.8	351	24	327	1.6
6.8	340	23	317	1.6
6.8	339	23	316	1.6
6.8	338	23	315	1.6
6.8	337	23	314	1.6
6.8	336	23	313	1.6
6.8	325	22	303	1.5
6.8	324	22	302	1.5
6.8	323	22	301	1.5
6.8	322	22	300	1.5
6.8	311	21	290	1.4
6.8	310	21	289	1.4
6.8	309	21	288	1.4
6.8	308	21	287	1.4
6.8	307	21	286	1.4
6.8	296	20	276	1.4
6.8	295	20	275	1.4

%E	M1	M2	DM	M*
6.8	294	20	274	1.4
6.8	293	20	273	1.4
6.8	292	20	272	1.4
6.8	281	19	262	1.3
6.8	280	19	261	1.3
6.8	279	19	260	1.3
6.8	278	19	259	1.3
6.8	266	18	248	1.2
6.8	265	18	247	1.2
6.8	264	18	246	1.2
6.8	263	18	245	1.2
6.8	251	17	234	1.2
6.8	250	17	233	1.2
6.8	249	17	232	1.2
6.8	237	16	221	1.1
6.8	236	16	220	1.1
6.8	235	16	219	1.1
6.8	234	16	218	1.1
6.8	222	15	207	1.0
6.8	221	15	206	1.0
6.8	220	15	205	1.0
6.8	219	15	204	1.0
6.8	207	14	193	0.9
6.8	206	14	192	1.0
6.8	205	14	191	1.0
6.8	192	13	179	0.9
6.8	191	13	178	0.9
6.8	190	13	177	0.9
6.8	177	12	165	0.8
6.8	176	12	164	0.8
6.8	162	11	151	0.7
6.8	161	11	150	0.8
6.8	148	10	138	0.7
6.8	147	10	137	0.7
6.8	146	10	136	0.7
6.8	133	9	124	0.6
6.8	132	9	123	0.6
6.8	118	8	110	0.5
6.8	117	8	109	0.5
6.8	103	7	96	0.5
6.8	88	6	82	0.4
6.8	74	5	69	0.3
6.8	73	5	68	0.3
6.8	59	4	55	0.3
6.8	44	3	41	0.2
6.7	496	33	463	2.2
6.7	495	33	462	2.2
6.7	494	33	461	2.2
6.7	493	33	460	2.2
6.7	492	33	459	2.2

%E	M1	M2	DM	M*
6.7	491	33	458	2.2
6.7	490	33	457	2.2
6.7	489	33	456	2.2
6.7	481	32	449	2.1
6.7	480	32	448	2.1
6.7	479	32	447	2.1
6.7	478	32	446	2.1
6.7	477	32	445	2.1
6.7	476	32	444	2.2
6.7	475	32	443	2.2
6.7	466	31	435	2.1
6.7	465	31	434	2.1
6.7	464	31	433	2.1
6.7	463	31	432	2.1
6.7	462	31	431	2.1
6.7	461	31	430	2.1
6.7	460	31	429	2.1
6.7	451	30	421	2.0
6.7	450	30	420	2.0
6.7	449	30	419	2.0
6.7	448	30	418	2.0
6.7	447	30	417	2.0
6.7	446	30	416	2.0
6.7	445	30	415	2.0
6.7	436	29	407	1.9
6.7	435	29	406	1.9
6.7	434	29	405	1.9
6.7	433	29	404	1.9
6.7	432	29	403	1.9
6.7	431	29	402	2.0
6.7	430	29	401	2.0
6.7	421	28	393	1.9
6.7	420	28	392	1.9
6.7	419	28	391	1.9
6.7	418	28	390	1.9
6.7	417	28	389	1.9
6.7	416	28	388	1.9
6.7	415	28	387	1.9
6.7	406	27	379	1.8
6.7	405	27	378	1.8
6.7	404	27	377	1.8
6.7	403	27	376	1.8
6.7	402	27	375	1.8
6.7	401	27	374	1.8
6.7	391	26	365	1.7
6.7	390	26	364	1.7
6.7	389	26	363	1.7
6.7	388	26	362	1.7
6.7	387	26	361	1.7
6.7	386	26	360	1.8

%E	M1	M2	DM	M*
6.7	375	25	350	1.7
6.7	374	25	349	1.7
6.7	373	25	348	1.7
6.7	372	25	347	1.7
6.7	371	25	346	1.7
6.7	360	24	336	1.6
6.7	359	24	335	1.6
6.7	358	24	334	1.6
6.7	357	24	333	1.6
6.7	356	24	332	1.6
6.7	345	23	322	1.5
6.7	344	23	321	1.5
6.7	343	23	320	1.5
6.7	342	23	319	1.5
6.7	341	23	318	1.6
6.7	330	22	308	1.5
6.7	329	22	307	1.5
6.7	328	22	306	1.5
6.7	327	22	305	1.5
6.7	326	22	304	1.5
6.7	315	21	294	1.4
6.7	314	21	293	1.4
6.7	313	21	292	1.4
6.7	312	21	291	1.4
6.7	300	20	280	1.3
6.7	299	20	279	1.3
6.7	298	20	278	1.3
6.7	297	20	277	1.3
6.7	285	19	266	1.3
6.7	284	19	265	1.3
6.7	283	19	264	1.3
6.7	282	19	263	1.3
6.7	270	18	252	1.2
6.7	269	18	251	1.2
6.7	268	18	250	1.2
6.7	267	18	249	1.2
6.7	255	17	238	1.1
6.7	254	17	237	1.1
6.7	253	17	236	1.1
6.7	252	17	235	1.1
6.7	240	16	224	1.1
6.7	239	16	223	1.1
6.7	238	16	222	1.1
6.7	225	15	210	1.0
6.7	224	15	209	1.0
6.7	223	15	208	1.0
6.7	210	14	196	0.9
6.7	209	14	195	0.9
6.7	208	14	194	0.9
6.7	195	13	182	0.9

%E	M1	M2	DM	M*
6.7	194	13	181	0.9
6.7	193	13	180	0.9
6.7	180	12	168	0.8
6.7	179	12	167	0.8
6.7	178	12	166	0.8
6.7	165	11	154	0.7
6.7	164	11	153	0.7
6.7	163	11	152	0.7
6.7	150	10	140	0.7
6.7	149	10	139	0.7
6.7	135	9	126	0.6
6.7	134	9	125	0.6
6.7	120	8	112	0.5
6.7	119	8	111	0.5
6.7	105	7	98	0.5
6.7	104	7	97	0.5
6.7	90	6	84	0.4
6.7	89	6	83	0.4
6.7	75	5	70	0.3
6.7	60	4	56	0.3
6.7	45	3	42	0.2
6.7	30	2	28	0.1
6.7	15	1	14	0.1
6.6	500	33	467	2.2
6.6	499	33	466	2.2
6.6	498	33	465	2.2
6.6	497	33	464	2.2
6.6	488	32	456	2.1
6.6	487	32	455	2.1
6.6	486	32	454	2.1
6.6	485	32	453	2.1
6.6	484	32	452	2.1
6.6	483	32	451	2.1
6.6	482	32	450	2.1
6.6	473	31	442	2.0
6.6	472	31	441	2.0
6.6	471	31	440	2.0
6.6	470	31	439	2.0
6.6	469	31	438	2.0
6.6	468	31	437	2.1
6.6	467	31	436	2.1
6.6	458	30	428	2.0
6.6	457	30	427	2.0
6.6	456	30	426	2.0
6.6	455	30	425	2.0
6.6	454	30	424	2.0
6.6	453	30	423	2.0
6.6	452	30	422	2.0
6.6	442	29	413	1.9
6.6	441	29	412	1.9

%E	M1	M2	DM	M*
6.6	440	29	411	1.9
6.6	439	29	410	1.9
6.6	438	29	409	1.9
6.6	437	29	408	1.9
6.6	427	28	399	1.8
6.6	426	28	398	1.8
6.6	425	28	397	1.8
6.6	424	28	396	1.8
6.6	423	28	395	1.9
6.6	422	28	394	1.9
6.6	412	27	385	1.8
6.6	411	27	384	1.8
6.6	410	27	383	1.8
6.6	409	27	382	1.8
6.6	408	27	381	1.8
6.6	407	27	380	1.8
6.6	396	26	370	1.7
6.6	395	26	369	1.7
6.6	394	26	368	1.7
6.6	393	26	367	1.7
6.6	392	26	366	1.7
6.6	381	25	356	1.6
6.6	380	25	355	1.6
6.6	379	25	354	1.6
6.6	378	25	353	1.7
6.6	377	25	352	1.7
6.6	376	25	351	1.7
6.6	366	24	342	1.6
6.6	365	24	341	1.6
6.6	364	24	340	1.6
6.6	363	24	339	1.6
6.6	362	24	338	1.6
6.6	361	24	337	1.6
6.6	351	23	328	1.5
6.6	350	23	327	1.5
6.6	349	23	326	1.5
6.6	348	23	325	1.5
6.6	347	23	324	1.5
6.6	346	23	323	1.5
6.6	335	22	313	1.4
6.6	334	22	312	1.4
6.6	333	22	311	1.5
6.6	332	22	310	1.5
6.6	331	22	309	1.5
6.6	320	21	299	1.4
6.6	319	21	298	1.4
6.6	318	21	297	1.4
6.6	317	21	296	1.4
6.6	316	21	295	1.4
6.6	305	20	285	1.3

%E	M1	M2	DM	M*
6.6	304	20	284	1.3
6.6	303	20	283	1.3
6.6	302	20	282	1.3
6.6	301	20	281	1.3
6.6	290	19	271	1.2
6.6	289	19	270	1.2
6.6	288	19	269	1.3
6.6	287	19	268	1.3
6.6	286	19	267	1.3
6.6	274	18	256	1.2
6.6	273	18	255	1.2
6.6	272	18	254	1.2
6.6	271	18	253	1.2
6.6	259	17	242	1.1
6.6	258	17	241	1.1
6.6	257	17	240	1.1
6.6	256	17	239	1.1
6.6	244	16	228	1.0
6.6	243	16	227	1.1
6.6	242	16	226	1.1
6.6	241	16	225	1.1
6.6	229	15	214	1.0
6.6	228	15	213	1.0
6.6	227	15	212	1.0
6.6	226	15	211	1.0
6.6	213	14	199	0.9
6.6	212	14	198	0.9
6.6	211	14	197	0.9
6.6	198	13	185	0.9
6.6	197	13	184	0.9
6.6	196	13	183	0.9
6.6	183	12	171	0.8
6.6	182	12	170	0.8
6.6	181	12	169	0.8
6.6	167	11	156	0.7
6.6	166	11	155	0.7
6.6	152	10	142	0.7
6.6	151	10	141	0.7
6.6	137	9	128	0.6
6.6	136	9	127	0.6
6.6	122	8	114	0.5
6.6	121	8	113	0.5
6.6	106	7	99	0.5
6.6	91	6	85	0.4
6.6	76	5	71	0.3
6.6	61	4	57	0.3
6.5	496	32	464	2.1
6.5	495	32	463	2.1
6.5	494	32	462	2.1
6.5	493	32	461	2.1

%E	M1	M2	DM	M*
6.5	492	32	460	2.1
6.5	491	32	459	2.1
6.5	490	32	458	2.1
6.5	489	32	457	2.1
6.5	480	31	449	2.0
6.5	479	31	448	2.0
6.5	478	31	447	2.0
6.5	477	31	446	2.0
6.5	476	31	445	2.0
6.5	475	31	444	2.0
6.5	474	31	443	2.0
6.5	465	30	435	1.9
6.5	464	30	434	1.9
6.5	463	30	433	1.9
6.5	462	30	432	1.9
6.5	461	30	431	2.0
6.5	460	30	430	2.0
6.5	459	30	429	2.0
6.5	449	29	420	1.9
6.5	448	29	419	1.9
6.5	447	29	418	1.9
6.5	446	29	417	1.9
6.5	445	29	416	1.9
6.5	444	29	415	1.9
6.5	443	29	414	1.9
6.5	434	28	406	1.8
6.5	433	28	405	1.8
6.5	432	28	404	1.8
6.5	431	28	403	1.8
6.5	430	28	402	1.8
6.5	429	28	401	1.8
6.5	428	28	400	1.8
6.5	418	27	391	1.7
6.5	417	27	390	1.7
6.5	416	27	389	1.8
6.5	415	27	388	1.8
6.5	414	27	387	1.8
6.5	413	27	386	1.8
6.5	403	26	377	1.7
6.5	402	26	376	1.7
6.5	401	26	375	1.7
6.5	400	26	374	1.7
6.5	399	26	373	1.7
6.5	398	26	372	1.7
6.5	397	26	371	1.7
6.5	387	25	362	1.6
6.5	386	25	361	1.6
6.5	385	25	360	1.6
6.5	384	25	359	1.6
6.5	383	25	358	1.6

%E	M1	M2	DM	M*
6.5	382	25	357	1.6
6.5	372	24	348	1.5
6.5	371	24	347	1.6
6.5	370	24	346	1.6
6.5	369	24	345	1.6
6.5	368	24	344	1.6
6.5	367	24	343	1.6
6.5	356	23	333	1.5
6.5	355	23	332	1.5
6.5	354	23	331	1.5
6.5	353	23	330	1.5
6.5	352	23	329	1.5
6.5	341	22	319	1.4
6.5	340	22	318	1.4
6.5	339	22	317	1.4
6.5	338	22	316	1.4
6.5	337	22	315	1.4
6.5	336	22	314	1.4
6.5	325	21	304	1.4
6.5	324	21	303	1.4
6.5	323	21	302	1.4
6.5	322	21	301	1.4
6.5	321	21	300	1.4
6.5	310	20	290	1.3
6.5	309	20	289	1.3
6.5	308	20	288	1.3
6.5	307	20	287	1.3
6.5	306	20	286	1.3
6.5	294	19	275	1.2
6.5	293	19	274	1.2
6.5	292	19	273	1.2
6.5	291	19	272	1.2
6.5	279	18	261	1.2
6.5	278	18	260	1.2
6.5	277	18	259	1.2
6.5	276	18	258	1.2
6.5	275	18	257	1.2
6.5	263	17	246	1.1
6.5	262	17	245	1.1
6.5	261	17	244	1.1
6.5	260	17	243	1.1
6.5	248	16	232	1.0
6.5	247	16	231	1.0
6.5	246	16	230	1.0
6.5	245	16	229	1.0
6.5	232	15	217	1.0
6.5	231	15	216	1.0
6.5	230	15	215	1.0
6.5	217	14	203	0.9
6.5	216	14	202	0.9

%E	M1	M2	DM	M*
6.5	215	14	201	0.9
6.5	214	14	200	0.9
6.5	201	13	188	0.8
6.5	200	13	187	0.8
6.5	199	13	186	0.8
6.5	186	12	174	0.8
6.5	185	12	173	0.8
6.5	184	12	172	0.8
6.5	170	11	159	0.7
6.5	169	11	158	0.7
6.5	168	11	157	0.7
6.5	155	10	145	0.6
6.5	154	10	144	0.6
6.5	153	10	143	0.7
6.5	139	9	130	0.6
6.5	138	9	129	0.6
6.5	124	8	116	0.5
6.5	123	8	115	0.5
6.5	108	7	101	0.5
6.5	107	7	100	0.5
6.5	93	6	87	0.4
6.5	92	6	86	0.4
6.5	77	5	72	0.3
6.5	62	4	58	0.3
6.5	46	3	43	0.2
6.5	31	2	29	0.1
6.4	500	32	468	2.0
6.4	499	32	467	2.1
6.4	498	32	466	2.1
6.4	497	32	465	2.1
6.4	488	31	457	2.0
6.4	487	31	456	2.0
6.4	486	31	455	2.0
6.4	485	31	454	2.0
6.4	484	31	453	2.0
6.4	483	31	452	2.0
6.4	482	31	451	2.0
6.4	481	31	450	2.0
6.4	472	30	442	1.9
6.4	471	30	441	1.9
6.4	470	30	440	1.9
6.4	469	30	439	1.9
6.4	468	30	438	1.9
6.4	467	30	437	1.9
6.4	466	30	436	1.9
6.4	456	29	427	1.8
6.4	455	29	426	1.8
6.4	454	29	425	1.9
6.4	453	29	424	1.9
6.4	452	29	423	1.9

%E	M1	M2	DM	M*
6.4	451	29	422	1.9
6.4	450	29	421	1.9
6.4	440	28	412	1.8
6.4	439	28	411	1.8
6.4	438	28	410	1.8
6.4	437	28	409	1.8
6.4	436	28	408	1.8
6.4	435	28	407	1.8
6.4	425	27	398	1.7
6.4	424	27	397	1.7
6.4	423	27	396	1.7
6.4	422	27	395	1.7
6.4	421	27	394	1.7
6.4	420	27	393	1.7
6.4	419	27	392	1.7
6.4	409	26	383	1.7
6.4	408	26	382	1.7
6.4	407	26	381	1.7
6.4	406	26	380	1.7
6.4	405	26	379	1.7
6.4	404	26	378	1.7
6.4	393	25	368	1.6
6.4	392	25	367	1.6
6.4	391	25	366	1.6
6.4	390	25	365	1.6
6.4	389	25	364	1.6
6.4	388	25	363	1.6
6.4	377	24	353	1.5
6.4	376	24	352	1.5
6.4	375	24	351	1.5
6.4	374	24	350	1.5
6.4	373	24	349	1.5
6.4	362	23	339	1.5
6.4	361	23	338	1.5
6.4	360	23	337	1.5
6.4	359	23	336	1.5
6.4	358	23	335	1.5
6.4	357	23	334	1.5
6.4	346	22	324	1.4
6.4	345	22	323	1.4
6.4	344	22	322	1.4
6.4	343	22	321	1.4
6.4	342	22	320	1.4
6.4	330	21	309	1.3
6.4	329	21	308	1.3
6.4	328	21	307	1.3
6.4	327	21	306	1.3
6.4	326	21	305	1.4
6.4	314	20	294	1.3
6.4	313	20	293	1.3

%E	M1	M2	DM	M*
6.4	312	20	292	1.3
6.4	311	20	291	1.3
6.4	299	19	280	1.2
6.4	298	19	279	1.2
6.4	297	19	278	1.2
6.4	296	19	277	1.2
6.4	295	19	276	1.2
6.4	283	18	265	1.1
6.4	282	18	264	1.1
6.4	281	18	263	1.2
6.4	280	18	262	1.2
6.4	267	17	250	1.1
6.4	266	17	249	1.1
6.4	265	17	248	1.1
6.4	264	17	247	1.1
6.4	251	16	235	1.0
6.4	250	16	234	1.0
6.4	249	16	233	1.0
6.4	236	15	221	1.0
6.4	235	15	220	1.0
6.4	234	15	219	1.0
6.4	233	15	218	1.0
6.4	220	14	206	0.9
6.4	219	14	205	0.9
6.4	218	14	204	0.9
6.4	204	13	191	0.8
6.4	203	13	190	0.8
6.4	202	13	189	0.8
6.4	188	12	176	0.8
6.4	187	12	175	0.8
6.4	173	11	162	0.7
6.4	172	11	161	0.7
6.4	171	11	160	0.7
6.4	157	10	147	0.6
6.4	156	10	146	0.6
6.4	141	9	132	0.6
6.4	140	9	131	0.6
6.4	125	8	117	0.5
6.4	110	7	103	0.4
6.4	109	7	102	0.4
6.4	94	6	88	0.4
6.4	78	5	73	0.3
6.4	47	3	44	0.2
6.3	496	31	465	1.9
6.3	495	31	464	1.9
6.3	494	31	463	1.9
6.3	493	31	462	1.9
6.3	492	31	461	2.C
6.3	491	31	460	2.0
6.3	490	31	459	2.0

%E	M1	M2	DM	M*
6.3	489	31	458	2.0
6.3	480	30	450	1.9
6.3	479	30	449	1.9
6.3	478	30	448	1.9
6.3	477	30	447	1.9
6.3	476	30	446	1.9
6.3	475	30	445	1.9
6.3	474	30	444	1.9
6.3	473	30	443	1.9
6.3	464	29	435	1.8
6.3	463	29	434	1.8
6.3	462	29	433	1.8
6.3	461	29	432	1.8
6.3	460	29	431	1.8
6.3	459	29	430	1.8
6.3	458	29	429	1.8
6.3	457	29	428	1.8
6.3	448	28	420	1.8
6.3	447	28	419	1.8
6.3	446	28	418	1.8
6.3	445	28	417	1.8
6.3	444	28	416	1.8
6.3	443	28	415	1.8
6.3	442	28	414	1.8
6.3	441	28	413	1.8
6.3	432	27	405	1.7
6.3	431	27	404	1.7
6.3	430	27	403	1.7
6.3	429	27	402	1.7
6.3	428	27	401	1.7
6.3	427	27	400	1.7
6.3	426	27	399	1.7
6.3	416	26	390	1.6
6.3	415	26	389	1.6
6.3	414	26	388	1.6
6.3	413	26	387	1.6
6.3	412	26	386	1.6
6.3	411	26	385	1.6
6.3	410	26	384	1.6
6.3	400	25	375	1.6
6.3	399	25	374	1.6
6.3	398	25	373	1.6
6.3	397	25	372	1.6
6.3	396	25	371	1.6
6.3	395	25	370	1.6
6.3	394	25	369	1.6
6.3	384	24	360	1.5
6.3	383	24	359	1.5
6.3	382	24	358	1.5
6.3	381	24	357	1.5

%E	M1	M2	DM	M*
6.3	380	24	356	1.5
6.3	379	24	355	1.5
6.3	378	24	354	1.5
6.3	368	23	345	1.4
6.3	367	23	344	1.4
6.3	366	23	343	1.4
6.3	365	23	342	1.4
6.3	364	23	341	1.5
6.3	363	23	340	1.5
6.3	352	22	330	1.4
6.3	351	22	329	1.4
6.3	350	22	328	1.4
6.3	349	22	327	1.4
6.3	348	22	326	1.4
6.3	347	22	325	1.4
6.3	336	21	315	1.3
6.3	335	21	314	1.3
6.3	334	21	313	1.3
6.3	333	21	312	1.3
6.3	332	21	311	1.3
6.3	331	21	310	1.3
6.3	320	20	300	1.3
6.3	319	20	299	1.3
6.3	318	20	298	1.3
6.3	317	20	297	1.3
6.3	316	20	296	1.3
6.3	315	20	295	1.3
6.3	304	19	285	1.2
6.3	303	19	284	1.2
6.3	302	19	283	1.2
6.3	301	19	282	1.2
6.3	300	19	281	1.2
6.3	288	18	270	1.1
6.3	287	18	269	1.1
6.3	286	18	268	1.1
6.3	285	18	267	1.1
6.3	284	18	266	1.1
6.3	272	17	255	1.1
6.3	271	17	254	1.1
6.3	270	17	253	1.1
6.3	269	17	252	1.1
6.3	268	17	251	1.1
6.3	256	16	240	1.0
6.3	255	16	239	1.0
6.3	254	16	238	1.0
6.3	253	16	237	1.0
6.3	252	16	236	1.0
6.3	240	15	225	0.9
6.3	239	15	224	0.9
6.3	238	15	223	0.9

%E	M1	M2	DM	M*
6.3	237	15	222	0.9
6.3	224	14	210	0.9
6.3	223	14	209	0.9
6.3	222	14	208	0.9
6.3	221	14	207	0.9
6.3	208	13	195	0.8
6.3	207	13	194	0.8
6.3	206	13	193	0.8
6.3	205	13	192	0.8
6.3	192	12	180	0.8
6.3	191	12	179	0.8
6.3	190	12	178	0.8
6.3	189	12	177	0.8
6.3	176	11	165	0.7
6.3	175	11	164	0.7
6.3	174	11	163	0.7
6.3	160	10	150	0.6
6.3	159	10	149	0.6
6.3	158	10	148	0.6
6.3	144	9	135	0.6
6.3	143	9	134	0.6
6.3	142	9	133	0.6
6.3	128	8	120	0.5
6.3	127	8	119	0.5
6.3	126	8	118	0.5
6.3	112	7	105	0.4
6.3	111	7	104	0.4
6.3	96	6	90	0.4
6.3	95	6	89	0.4
6.3	80	5	75	0.3
6.3	79	5	74	0.3
6.3	64	4	60	0.3
6.3	63	4	59	0.3
6.3	48	3	45	0.2
6.3	32	2	30	0.1
6.3	16	1	15	0.1
6.2	500	31	469	1.9
6.2	499	31	468	1.9
6.2	498	31	467	1.9
6.2	497	31	466	1.9
6.2	487	30	457	1.8
6.2	486	30	456	1.9
6.2	485	30	455	1.9
6.2	484	30	454	1.9
6.2	483	30	453	1.9
6.2	482	30	452	1.9
6.2	481	30	451	1.9
6.2	471	29	442	1.8
6.2	470	29	441	1.8
6.2	469	29	440	1.8

%E	M1	M2	DM	M*
6.2	468	29	439	1.8
6.2	467	29	438	1.8
6.2	466	29	437	1.8
6.2	465	29	436	1.8
6.2	455	28	427	1.7
6.2	454	28	426	1.7
6.2	453	28	425	1.7
6.2	452	28	424	1.7
6.2	451	28	423	1.7
6.2	450	28	422	1.7
6.2	449	28	421	1.7
6.2	439	27	412	1.7
6.2	438	27	411	1.7
6.2	437	27	410	1.7
6.2	436	27	409	1.7
6.2	435	27	408	1.7
6.2	434	27	407	1.7
6.2	433	27	406	1.7
6.2	422	26	396	1.6
6.2	421	26	395	1.6
6.2	420	26	394	1.6
6.2	419	26	393	1.6
6.2	418	26	392	1.6
6.2	417	26	391	1.6
6.2	406	25	381	1.5
6.2	405	25	380	1.5
6.2	404	25	379	1.5
6.2	403	25	378	1.6
6.2	402	25	377	1.6
6.2	401	25	376	1.6
6.2	390	24	366	1.5
6.2	389	24	365	1.5
6.2	388	24	364	1.5
6.2	387	24	363	1.5
6.2	386	24	362	1.5
6.2	385	24	361	1.5
6.2	374	23	351	1.4
6.2	373	23	350	1.4
6.2	372	23	349	1.4
6.2	371	23	348	1.4
6.2	370	23	347	1.4
6.2	369	23	346	1.4
6.2	357	22	335	1.4
6.2	356	22	334	1.4
6.2	355	22	333	1.4
6.2	354	22	332	1.4
6.2	353	22	331	1.4
6.2	341	21	320	1.3
6.2	340	21	319	1.3
6.2	339	21	318	1.3

%E	M1	M2	DM	M*
6.2	338	21	317	1.3
6.2	337	21	316	1.3
6.2	325	20	305	1.2
6.2	324	20	304	1.2
6.2	323	20	303	1.2
6.2	322	20	302	1.2
6.2	321	20	301	1.2
6.2	308	19	289	1.2
6.2	307	19	288	1.2
6.2	306	19	287	1.2
6.2	305	19	286	1.2
6.2	292	18	274	1.1
6.2	291	18	273	1.1
6.2	290	18	272	1.1
6.2	289	18	271	1.1
6.2	276	17	259	1.0
6.2	275	17	258	1.1
6.2	274	17	257	1.1
6.2	273	17	256	1.1
6.2	260	16	244	1.0
6.2	259	16	243	1.0
6.2	258	16	242	1.0
6.2	257	16	241	1.0
6.2	243	15	228	0.9
6.2	242	15	227	0.9
6.2	241	15	226	0.9
6.2	227	14	213	0.9
6.2	226	14	212	0.9
6.2	225	14	211	0.9
6.2	211	13	198	0.8
6.2	210	13	197	0.8
6.2	209	13	196	0.8
6.2	195	12	183	0.7
6.2	194	12	182	0.7
6.2	193	12	181	0.7
6.2	178	11	167	0.7
6.2	177	11	166	0.7
6.2	162	10	152	0.6
6.2	161	10	151	0.6
6.2	146	9	137	0.6
6.2	145	9	136	0.6
6.2	130	8	122	0.5
6.2	129	8	121	0.5
6.2	113	7	106	0.4
6.2	97	6	91	0.4
6.2	81	5	76	0.3
6.2	65	4	61	0.2
6.1	495	30	465	1.8
6.1	494	30	464	1.8
6.1	493	30	463	1.8

%E	M1	M2	DM	M*
6.1	492	30	462	1.8
6.1	491	30	461	1.8
6.1	490	30	460	1.8
6.1	489	30	459	1.8
6.1	488	30	458	1.8
6.1	479	29	450	1.8
6.1	478	29	449	1.8
6.1	477	29	448	1.8
6.1	476	29	447	1.8
6.1	475	29	446	1.8
6.1	474	29	445	1.8
6.1	473	29	444	1.8
6.1	472	29	443	1.8
6.1	462	28	434	1.7
6.1	461	28	433	1.7
6.1	460	28	432	1.7
6.1	459	28	431	1.7
6.1	458	28	430	1.7
6.1	457	28	429	1.7
6.1	456	28	428	1.7
6.1	446	27	419	1.6
6.1	445	27	418	1.6
6.1	444	27	417	1.6
6.1	443	27	416	1.6
6.1	442	27	415	1.6
6.1	441	27	414	1.7
6.1	440	27	413	1.7
6.1	429	26	403	1.6
6.1	428	26	402	1.6
6.1	427	26	401	1.6
6.1	426	26	400	1.6
6.1	425	26	399	1.6
6.1	424	26	398	1.6
6.1	423	26	397	1.6
6.1	413	25	388	1.5
6.1	412	25	387	1.5
6.1	411	25	386	1.5
6.1	410	25	385	1.5
6.1	409	25	384	1.5
6.1	408	25	383	1.5
6.1	407	25	382	1.5
6.1	396	24	372	1.5
6.1	395	24	371	1.5
6.1	394	24	370	1.5
6.1	393	24	369	1.5
6.1	392	24	368	1.5
6.1	391	24	367	1.5
6.1	380	23	357	1.4
6.1	379	23	356	1.4
6.1	378	23	355	1.4

%E	M1	M2	DM	M*
6.1	377	23	354	1.4
6.1	376	23	353	1.4
6.1	375	23	352	1.4
6.1	363	22	341	1.3
6.1	362	22	340	1.3
6.1	361	22	339	1.3
6.1	360	22	338	1.3
6.1	359	22	337	1.3
6.1	358	22	336	1.4
6.1	347	21	326	1.3
6.1	346	21	325	1.3
6.1	345	21	324	1.3
6.1	344	21	323	1.3
6.1	343	21	322	1.3
6.1	342	21	321	1.3
6.1	330	20	310	1.2
6.1	329	20	309	1.2
6.1	328	20	308	1.2
6.1	327	20	307	1.2
6.1	326	20	306	1.2
6.1	314	19	295	1.1
6.1	313	19	294	1.2
6.1	312	19	293	1.2
6.1	311	19	292	1.2
6.1	310	19	291	1.2
6.1	309	19	290	1.2
6.1	297	18	279	1.1
6.1	296	18	278	1.1
6.1	295	18	277	1.1
6.1	294	18	276	1.1
6.1	293	18	275	1.1
6.1	281	17	264	1.0
6.1	280	17	263	1.0
6.1	279	17	262	1.0
6.1	278	17	261	1.0
6.1	277	17	260	1.0
6.1	264	16	248	1.0
6.1	263	16	247	1.0
6.1	262	16	246	1.0
6.1	261	16	245	1.0
6.1	247	15	232	0.9
6.1	246	15	231	0.9
6.1	245	15	230	0.9
6.1	244	15	229	0.9
6.1	231	14	217	0.8
6.1	230	14	216	0.9
6.1	229	14	215	0.9
6.1	228	14	214	0.9
6.1	214	13	201	0.8
6.1	213	13	200	0.8

%E	M1	M2	DM	M*
6.1	212	13	199	0.8
6.1	198	12	186	0.7
6.1	197	12	185	0.7
6.1	196	12	184	0.7
6.1	181	11	170	0.7
6.1	180	11	169	0.7
6.1	179	11	168	0.7
6.1	165	10	155	0.6
6.1	164	10	154	0.6
6.1	163	10	153	0.6
6.1	148	9	139	0.5
6.1	147	9	138	0.6
6.1	132	8	124	0.5
6.1	131	8	123	0.5
6.1	115	7	108	0.4
6.1	114	7	107	0.4
6.1	99	6	93	0.4
6.1	98	6	92	0.4
6.1	82	5	77	0.3
6.1	66	4	62	0.2
6.1	49	3	46	0.2
6.1	33	2	31	0.1
6.0	500	30	470	1.8
6.0	499	30	469	1.8
6.0	498	30	468	1.8
6.0	497	30	467	1.8
6.0	496	30	466	1.8
6.0	487	29	458	1.7
6.0	486	29	457	1.7
6.0	485	29	456	1.7
6.0	484	29	455	1.7
6.0	483	29	454	1.7
6.0	482	29	453	1.7
6.0	481	29	452	1.7
6.0	480	29	451	1.8
6.0	470	28	442	1.7
6.0	469	28	441	1.7
6.0	468	28	440	1.7
6.0	467	28	439	1.7
6.0	466	28	438	1.7
6.0	465	28	437	1.7
6.0	464	28	436	1.7
6.0	463	28	435	1.7
6.0	453	27	426	1.6
6.0	452	27	425	1.6
6.0	451	27	424	1.6
6.0	450	27	423	1.6
6.0	449	27	422	1.6
6.0	448	27	421	1.6
6.0	447	27	420	1.6

%E	M1	M2	DM	M*
6.0	437	26	411	1.5
6.0	436	26	410	1.6
6.0	435	26	409	1.6
6.0	434	26	408	1.6
6.0	433	26	407	1.6
6.0	432	26	406	1.6
6.0	431	26	405	1.6
6.0	430	26	404	1.6
6.0	420	25	395	1.5
6.0	419	25	394	1.5
6.0	418	25	393	1.5
6.0	417	25	392	1.5
6.0	416	25	391	1.5
6.0	415	25	390	1.5
6.0	414	25	389	1.5
6.0	403	24	379	1.4
6.0	402	24	378	1.4
6.0	401	24	377	1.4
6.0	400	24	376	1.4
6.0	399	24	375	1.4
6.0	398	24	374	1.4
6.0	397	24	373	1.5
6.0	386	23	363	1.4
6.0	385	23	362	1.4
6.0	384	23	361	1.4
6.0	383	23	360	1.4
6.0	382	23	359	1.4
6.0	381	23	358	1.4
6.0	369	22	347	1.3
6.0	368	22	346	1.3
6.0	367	22	345	1.3
6.0	366	22	344	1.3
6.0	365	22	343	1.3
6.0	364	22	342	1.3
6.0	352	21	331	1.3
6.0	351	21	330	1.3
6.0	350	21	329	1.3
6.0	349	21	328	1.3
6.0	348	21	327	1.3
6.0	336	20	316	1.2
6.0	335	20	315	1.2
6.0	334	20	314	1.2
6.0	333	20	313	1.2
6.0	332	20	312	1.2
6.0	331	20	311	1.2
6.0	319	19	300	1.1
6.0	318	19	299	1.1
6.0	317	19	298	1.1
6.0	316	19	297	1.1
6.0	315	19	296	1.1

%E	M1	M2	DM	M*
6.0	302	18	284	1.1
6.0	301	18	283	1.1
6.0	300	18	282	1.1
6.0	299	18	281	1.1
6.0	298	18	280	1.1
6.0	285	17	268	1.0
6.0	284	17	267	1.0
6.0	283	17	266	1.0
6.0	282	17	265	1.0
6.0	268	16	252	1.0
6.0	267	16	251	1.0
6.0	266	16	250	1.0
6.0	265	16	249	1.0
6.0	252	15	237	0.9
6.0	251	15	236	0.9
6.0	250	15	235	0.9
6.0	249	15	234	0.9
6.0	248	15	233	0.9
6.0	235	14	221	0.8
6.0	234	14	220	0.8
6.0	233	14	219	0.8
6.0	232	14	218	0.8
6.0	218	13	205	0.8
6.0	217	13	204	0.8
6.0	216	13	203	0.8
6.0	215	13	202	0.8
6.0	201	12	189	0.7
6.0	200	12	188	0.7
6.0	199	12	187	0.7
6.0	184	11	173	0.7
6.0	183	11	172	0.7
6.0	182	11	171	0.7
6.0	168	10	158	0.6
6.0	167	10	157	0.6
6.0	166	10	156	0.6
6.0	151	9	142	0.5
6.0	150	9	141	0.5
6.0	149	9	140	0.5
6.0	134	8	126	0.5
6.0	133	8	125	0.5
6.0	117	7	110	0.4
6.0	116	7	109	0.4
6.0	100	6	94	0.4
6.0	84	5	79	0.3
6.0	83	5	78	0.3
6.0	67	4	63	0.2
6.0	50	3	47	0.2
5.9	495	29	466	1.7
5.9	494	29	465	1.7
5.9	493	29	464	1.7

%E	M1	M2	DM	M*
5.9	492	29	463	1.7
5.9	491	29	462	1.7
5.9	490	29	461	1.7
5.9	489	29	460	1.7
5.9	488	29	459	1.7
5.9	478	28	450	1.6
5.9	477	28	449	1.6
5.9	476	28	448	1.6
5.9	475	28	447	1.7
5.9	474	28	446	1.7
5.9	473	28	445	1.7
5.9	472	28	444	1.7
5.9	471	28	443	1.7
5.9	461	27	434	1.6
5.9	460	27	433	1.6
5.9	459	27	432	1.6
5.9	458	27	431	1.6
5.9	457	27	430	1.6
5.9	456	27	429	1.6
5.9	455	27	428	1.6
5.9	454	27	427	1.6
5.9	444	26	418	1.5
5.9	443	26	417	1.5
5.9	442	26	416	1.5
5.9	441	26	415	1.5
5.9	440	26	414	1.5
5.9	439	26	413	1.5
5.9	438	26	412	1.5
5.9	427	25	402	1.5
5.9	426	25	401	1.5
5.9	425	25	400	1.5
5.9	424	25	399	1.5
5.9	423	25	398	1.5
5.9	422	25	397	1.5
5.9	421	25	396	1.5
5.9	410	24	386	1.4
5.9	409	24	385	1.4
5.9	408	24	384	1.4
5.9	407	24	383	1.4
5.9	406	24	382	1.4
5.9	405	24	381	1.4
5.9	404	24	380	1.4
5.9	393	23	370	1.3
5.9	392	23	369	1.3
5.9	391	23	368	1.4
5.9	390	23	367	1.4
5.9	389	23	366	1.4
5.9	388	23	365	1.4
5.9	387	23	364	1.4
5.9	376	22	354	1.3

%E	M1	M2	DM	M*
5.9	375	22	353	1.3
5.9	374	22	352	1.3
5.9	373	22	351	1.3
5.9	372	22	350	1.3
5.9	371	22	349	1.3
5.9	370	22	348	1.3
5.9	359	21	338	1.2
5.9	358	21	337	1.2
5.9	357	21	336	1.2
5.9	356	21	335	1.2
5.9	355	21	334	1.2
5.9	354	21	333	1.2
5.9	353	21	332	1.2
5.9	341	20	321	1.2
5.9	340	20	320	1.2
5.9	339	20	319	1.2
5.9	338	20	318	1.2
5.9	337	20	317	1.2
5.9	324	19	305	1.1
5.9	323	19	304	1.1
5.9	322	19	303	1.1
5.9	321	19	302	1.1
5.9	320	19	301	1.1
5.9	307	18	289	1.1
5.9	306	18	288	1.1
5.9	305	18	287	1.1
5.9	304	18	286	1.1
5.9	303	18	285	1.1
5.9	290	17	273	1.0
5.9	289	17	272	1.0
5.9	288	17	271	1.0
5.9	287	17	270	1.0
5.9	286	17	269	1.0
5.9	273	16	257	0.9
5.9	272	16	256	0.9
5.9	271	16	255	0.9
5.9	270	16	254	0.9
5.9	269	16	253	1.0
5.9	256	15	241	0.9
5.9	255	15	240	0.9
5.9	254	15	239	0.9
5.9	253	15	238	0.9
5.9	239	14	225	0.8
5.9	238	14	224	0.8
5.9	237	14	223	0.8
5.9	236	14	222	0.8
5.9	222	13	209	0.8
5.9	221	13	208	0.8
5.9	220	13	207	0.8
5.9	219	13	206	0.8

%E	M1	M2	DM	M*
5.9	205	12	193	0.7
5.9	204	12	192	0.7
5.9	203	12	191	0.7
5.9	202	12	190	0.7
5.9	188	11	177	0.6
5.9	187	11	176	0.6
5.9	186	11	175	0.7
5.9	185	11	174	0.7
5.9	170	10	160	0.6
5.9	169	10	159	0.6
5.9	153	9	144	0.5
5.9	152	9	143	0.5
5.9	136	8	128	0.5
5.9	135	8	127	0.5
5.9	119	7	112	0.4
5.9	118	7	111	0.4
5.9	102	6	96	0.4
5.9	101	6	95	0.4
5.9	85	5	80	0.3
5.9	68	4	64	0.2
5.9	51	3	48	0.2
5.9	34	2	32	0.1
5.9	17	1	16	0.1
5.8	500	29	471	1.7
5.8	499	29	470	1.7
5.8	498	29	469	1.7
5.8	497	29	468	1.7
5.8	496	29	467	1.7
5.8	486	28	458	1.6
5.8	485	28	457	1.6
5.8	484	28	456	1.6
5.8	483	28	455	1.6
5.8	482	28	454	1.6
5.8	481	28	453	1.6
5.8	480	28	452	1.6
5.8	479	28	451	1.6
5.8	469	27	442	1.6
5.8	468	27	441	1.6
5.8	467	27	440	1.6
5.8	466	27	439	1.6
5.8	465	27	438	1.6
5.8	464	27	437	1.6
5.8	463	27	436	1.6
5.8	462	27	435	1.6
5.8	452	26	426	1.5
5.8	451	26	425	1.5
5.8	450	26	424	1.5
5.8	449	26	423	1.5
5.8	448	26	422	1.5
5.8	447	26	421	1.5

%E	M1	M2	DM	M*
5.8	446	26	420	1.5
5.8	445	26	419	1.5
5.8	434	25	409	1.4
5.8	433	25	408	1.4
5.8	432	25	407	1.4
5.8	431	25	406	1.5
5.8	430	25	405	1.5
5.8	429	25	404	1.5
5.8	428	25	403	1.5
5.8	417	24	393	1.4
5.8	416	24	392	1.4
5.8	415	24	391	1.4
5.8	414	24	390	1.4
5.8	413	24	389	1.4
5.8	412	24	388	1.4
5.8	411	24	387	1.4
5.8	400	23	377	1.3
5.8	399	23	376	1.3
5.8	398	23	375	1.3
5.8	397	23	374	1.3
5.8	396	23	373	1.3
5.8	395	23	372	1.3
5.8	394	23	371	1.3
5.8	382	22	360	1.3
5.8	381	22	359	1.3
5.8	380	22	358	1.3
5.8	379	22	357	1.3
5.8	378	22	356	1.3
5.8	377	22	355	1.3
5.8	365	21	344	1.2
5.8	364	21	343	1.2
5.8	363	21	342	1.2
5.8	362	21	341	1.2
5.8	361	21	340	1.2
5.8	360	21	339	1.2
5.8	347	20	327	1.2
5.8	346	20	326	1.2
5.8	345	20	325	1.2
5.8	344	20	324	1.2
5.8	343	20	323	1.2
5.8	342	20	322	1.2
5.8	330	19	311	1.1
5.8	329	19	310	1.1
5.8	328	19	309	1.1
5.8	327	19	308	1.1
5.8	326	19	307	1.1
5.8	325	19	306	1.1
5.8	313	18	295	1.0
5.8	312	18	294	1.0
5.8	311	18	293	1.0

%E	M1	M2	DM	M*
5.8	310	18	292	1.0
5.8	309	18	291	1.0
5.8	308	18	290	1.1
5.8	295	17	278	1.0
5.8	294	17	277	1.0
5.8	293	17	276	1.0
5.8	292	17	275	1.0
5.8	291	17	274	1.0
5.8	278	16	262	0.9
5.8	277	16	261	0.9
5.8	276	16	260	0.9
5.8	275	16	259	0.9
5.8	274	16	258	0.9
5.8	260	15	245	0.9
5.8	259	15	244	0.9
5.8	258	15	243	0.9
5.8	257	15	242	0.9
5.8	243	14	229	0.8
5.8	242	14	228	0.8
5.8	241	14	227	0.8
5.8	240	14	226	0.8
5.8	226	13	213	0.7
5.8	225	13	212	0.8
5.8	224	13	211	0.8
5.8	223	13	210	0.8
5.8	208	12	196	0.7
5.8	207	12	195	0.7
5.8	206	12	194	0.7
5.8	191	11	180	0.6
5.8	190	11	179	0.6
5.8	189	11	178	0.6
5.8	173	10	163	0.6
5.8	172	10	162	0.6
5.8	171	10	161	0.6
5.8	156	9	147	0.5
5.8	155	9	146	0.5
5.8	154	9	145	0.5
5.8	139	8	131	0.5
5.8	138	8	130	0.5
5.8	137	8	129	0.5
5.8	121	7	114	0.4
5.8	120	7	113	0.4
5.8	104	6	98	0.3
5.8	103	6	97	0.3
5.8	86	5	81	0.3
5.8	69	4	65	0.2
5.8	52	3	49	0.2
5.7	495	28	467	1.6
5.7	494	28	466	1.6
5.7	493	28	465	1.6

%E	M1	M2	DM	M*
5.7	492	28	464	1.6
5.7	491	28	463	1.6
5.7	490	28	462	1.6
5.7	489	28	461	1.6
5.7	488	28	460	1.6
5.7	487	28	459	1.6
5.7	477	27	450	1.5
5.7	476	27	449	1.5
5.7	475	27	448	1.5
5.7	474	27	447	1.5
5.7	473	27	446	1.5
5.7	472	27	445	1.5
5.7	471	27	444	1.5
5.7	470	27	443	1.6
5.7	460	26	434	1.5
5.7	459	26	433	1.5
5.7	458	26	432	1.5
5.7	457	26	431	1.5
5.7	456	26	430	1.5
5.7	455	26	429	1.5
5.7	454	26	428	1.5
5.7	453	26	427	1.5
5.7	442	25	417	1.4
5.7	441	25	416	1.4
5.7	440	25	415	1.4
5.7	439	25	414	1.4
5.7	438	25	413	1.4
5.7	437	25	412	1.4
5.7	436	25	411	1.4
5.7	435	25	410	1.4
5.7	424	24	400	1.4
5.7	423	24	399	1.4
5.7	422	24	398	1.4
5.7	421	24	397	1.4
5.7	420	24	396	1.4
5.7	419	24	395	1.4
5.7	418	24	394	1.4
5.7	407	23	384	1.3
5.7	406	23	383	1.3
5.7	405	23	382	1.3
5.7	404	23	381	1.3
5.7	403	23	380	1.3
5.7	402	23	379	1.3
5.7	401	23	378	1.3
5.7	389	22	367	1.2
5.7	388	22	366	1.2
5.7	387	22	365	1.3
5.7	386	22	364	1.3
5.7	385	22	363	1.3
5.7	384	22	362	1.3

%E	M1	M2	DM	M*
5.7	383	22	361	1.3
5.7	371	21	350	1.2
5.7	370	21	349	1.2
5.7	369	21	348	1.2
5.7	368	21	347	1.2
5.7	367	21	346	1.2
5.7	366	21	345	1.2
5.7	354	20	334	1.1
5.7	353	20	333	1.1
5.7	352	20	332	1.1
5.7	351	20	331	1.1
5.7	350	20	330	1.1
5.7	349	20	329	1.1
5.7	348	20	328	1.1
5.7	336	19	317	1.1
5.7	335	19	316	1.1
5.7	334	19	315	1.1
5.7	333	19	314	1.1
5.7	332	19	313	1.1
5.7	331	19	312	1.1
5.7	318	18	300	1.0
5.7	317	18	299	1.0
5.7	316	18	298	1.0
5.7	315	18	297	1.0
5.7	314	18	296	1.0
5.7	300	17	283	1.0
5.7	299	17	282	1.0
5.7	298	17	281	1.0
5.7	297	17	280	1.0
5.7	296	17	279	1.0
5.7	283	16	267	0.9
5.7	282	16	266	0.9
5.7	281	16	265	0.9
5.7	280	16	264	0.9
5.7	279	16	263	0.9
5.7	265	15	250	0.8
5.7	264	15	249	0.9
5.7	263	15	248	0.9
5.7	262	15	247	0.9
5.7	261	15	246	0.9
5.7	247	14	233	0.8
5.7	246	14	232	0.8
5.7	245	14	231	0.8
5.7	244	14	230	0.8
5.7	230	13	217	0.7
5.7	229	13	216	0.7
5.7	228	13	215	0.7
5.7	227	13	214	0.7
5.7	212	12	200	0.7
5.7	211	12	199	0.7

%E	M1	M2	DM	M*
5.7	210	12	198	0.7
5.7	209	12	197	0.7
5.7	194	11	183	0.6
5.7	193	11	182	0.6
5.7	192	11	181	0.6
5.7	177	10	167	0.6
5.7	176	10	166	0.6
5.7	175	10	165	0.6
5.7	174	10	164	0.6
5.7	159	9	150	0.5
5.7	158	9	149	0.5
5.7	157	9	148	0.5
5.7	141	8	133	0.5
5.7	140	8	132	0.5
5.7	123	7	116	0.4
5.7	122	7	115	0.4
5.7	106	6	100	0.3
5.7	105	6	99	0.3
5.7	88	5	83	0.3
5.7	87	5	82	0.3
5.7	70	4	66	0.2
5.7	53	3	50	0.2
5.7	35	2	33	0.1
5.6	500	28	472	1.6
5.6	499	28	471	1.6
5.6	498	28	470	1.6
5.6	497	28	469	1.6
5.6	496	28	468	1.6
5.6	486	27	459	1.5
5.6	485	27	458	1.5
5.6	484	27	457	1.5
5.6	483	27	456	1.5
5.6	482	27	455	1.5
5.6	481	27	454	1.5
5.6	480	27	453	1.5
5.6	479	27	452	1.5
5.6	478	27	451	1.5
5.6	468	26	442	1.4
5.6	467	26	441	1.4
5.6	466	26	440	1.5
5.6	465	26	439	1.5
5.6	464	26	438	1.5
5.6	463	26	437	1.5
5.6	462	26	436	1.5
5.6	461	26	435	1.5
5.6	450	25	425	1.4
5.6	449	25	424	1.4
5.6	448	25	423	1.4
5.6	447	25	422	1.4
5.6	446	25	421	1.4

%E	M1	M2	DM	M*
5.6	445	25	420	1.4
5.6	444	25	419	1.4
5.6	443	25	418	1.4
5.6	432	24	408	1.3
5.6	431	24	407	1.3
5.6	430	24	406	1.3
5.6	429	24	405	1.3
5.6	428	24	404	1.3
5.6	427	24	403	1.3
5.6	426	24	402	1.4
5.6	425	24	401	1.4
5.6	414	23	391	1.3
5.6	413	23	390	1.3
5.6	412	23	389	1.3
5.6	411	23	388	1.3
5.6	410	23	387	1.3
5.6	409	23	386	1.3
5.6	408	23	385	1.3
5.6	396	22	374	1.2
5.6	395	22	373	1.2
5.6	394	22	372	1.2
5.6	393	22	371	1.2
5.6	392	22	370	1.2
5.6	391	22	369	1.2
5.6	390	22	368	1.2
5.6	378	21	357	1.2
5.6	377	21	356	1.2
5.6	376	21	355	1.2
5.6	375	21	354	1.2
5.6	374	21	353	1.2
5.6	373	21	352	1.2
5.6	372	21	351	1.2
5.6	360	20	340	1.1
5.6	359	20	339	1.1
5.6	358	20	338	1.1
5.6	357	20	337	1.1
5.6	356	20	336	1.1
5.6	355	20	335	1.1
5.6	342	19	323	1.1
5.6	341	19	322	1.1
5.6	340	19	321	1.1
5.6	339	19	320	1.1
5.6	338	19	319	1.1
5.6	337	19	318	1.1
5.6	324	18	306	1.0
5.6	323	18	305	1.0
5.6	322	18	304	1.0
5.6	321	18	303	1.0
5.6	320	18	302	1.0
5.6	319	18	301	1.0

%E	M1	M2	DM	M*
5.6	306	17	289	0.9
5.6	305	17	288	0.9
5.6	304	17	287	1.0
5.6	303	17	286	1.0
5.6	302	17	285	1.0
5.6	301	17	284	1.0
5.6	288	16	272	0.9
5.6	287	16	271	0.9
5.6	286	16	270	0.9
5.6	285	16	269	0.9
5.6	284	16	268	0.9
5.6	270	15	255	0.8
5.6	269	15	254	0.8
5.6	268	15	253	0.8
5.6	267	15	252	0.8
5.6	266	15	251	0.8
5.6	252	14	238	0.8
5.6	251	14	237	0.8
5.6	250	14	236	0.8
5.6	249	14	235	0.8
5.6	248	14	234	0.8
5.6	234	13	221	0.7
5.6	233	13	220	0.7
5.6	232	13	219	0.7
5.6	231	13	218	0.7
5.6	216	12	204	0.7
5.6	215	12	203	0.7
5.6	214	12	202	0.7
5.6	213	12	201	0.7
5.6	198	11	187	0.6
5.6	197	11	186	0.6
5.6	196	11	185	0.6
5.6	195	11	184	0.6
5.6	180	10	170	0.6
5.6	179	10	169	0.6
5.6	178	10	168	0.6
5.6	162	9	153	0.5
5.6	161	9	152	0.5
5.6	160	9	151	0.5
5.6	144	8	136	0.4
5.6	143	8	135	0.4
5.6	142	8	134	0.5
5.6	126	7	119	0.4
5.6	125	7	118	0.4
5.6	124	7	117	0.4
5.6	108	6	102	0.3
5.6	107	6	101	0.3
5.6	90	5	85	0.3
5.6	89	5	84	0.3
5.6	72	4	68	0.2

%E	M1	M2	DM	M*
5.6	71	4	67	0.2
5.6	54	3	51	0.2
5.6	36	2	34	0.1
5.6	18	1	17	0.1
5.5	495	27	468	1.5
5.5	494	27	467	1.5
5.5	493	27	466	1.5
5.5	492	27	465	1.5
5.5	491	27	464	1.5
5.5	490	27	463	1.5
5.5	489	27	462	1.5
5.5	488	27	461	1.5
5.5	487	27	460	1.5
5.5	477	26	451	1.4
5.5	476	26	450	1.4
5.5	475	26	449	1.4
5.5	474	26	448	1.4
5.5	473	26	447	1.4
5.5	472	26	446	1.4
5.5	471	26	445	1.4
5.5	470	26	444	1.4
5.5	469	26	443	1.4
5.5	458	25	433	1.4
5.5	457	25	432	1.4
5.5	456	25	431	1.4
5.5	455	25	430	1.4
5.5	454	25	429	1.4
5.5	453	25	428	1.4
5.5	452	25	427	1.4
5.5	451	25	426	1.4
5.5	440	24	416	1.3
5.5	439	24	415	1.3
5.5	438	24	414	1.3
5.5	437	24	413	1.3
5.5	436	24	412	1.3
5.5	435	24	411	1.3
5.5	434	24	410	1.3
5.5	433	24	409	1.3
5.5	422	23	399	1.3
5.5	421	23	398	1.3
5.5	420	23	397	1.3
5.5	419	23	396	1.3
5.5	418	23	395	1.3
5.5	417	23	394	1.3
5.5	416	23	393	1.3
5.5	415	23	392	1.3
5.5	403	22	381	1.2
5.5	402	22	380	1.2
5.5	401	22	379	1.2
5.5	400	22	378	1.2

%E	M1	M2	DM	M*
5.5	399	22	377	1.2
5.5	398	22	376	1.2
5.5	397	22	375	1.2
5.5	385	21	364	1.1
5.5	384	21	363	1.1
5.5	383	21	362	1.2
5.5	382	21	361	1.2
5.5	381	21	360	1.2
5.5	380	21	359	1.2
5.5	379	21	358	1.2
5.5	367	20	347	1.1
5.5	366	20	346	1.1
5.5	365	20	345	1.1
5.5	364	20	344	1.1
5.5	363	20	343	1.1
5.5	362	20	342	1.1
5.5	361	20	341	1.1
5.5	348	19	329	1.0
5.5	347	19	328	1.0
5.5	346	19	327	1.0
5.5	345	19	326	1.0
5.5	344	19	325	1.0
5.5	343	19	324	1.1
5.5	330	18	312	1.0
5.5	329	18	311	1.0
5.5	328	18	310	1.0
5.5	327	18	309	1.0
5.5	326	18	308	1.0
5.5	325	18	307	1.0
5.5	311	17	294	0.9
5.5	310	17	293	0.9
5.5	309	17	292	0.9
5.5	308	17	291	0.9
5.5	307	17	290	0.9
5.5	293	16	277	0.9
5.5	292	16	276	0.9
5.5	291	16	275	0.9
5.5	290	16	274	0.9
5.5	289	16	273	0.9
5.5	275	15	260	0.8
5.5	274	15	259	0.8
5.5	273	15	258	0.8
5.5	272	15	257	0.8
5.5	271	15	256	0.8
5.5	256	14	242	0.8
5.5	255	14	241	0.8
5.5	254	14	240	0.8
5.5	253	14	239	0.8
5.5	238	13	225	0.7
5.5	237	13	224	0.7

%E	M1	M2	DM	M*
5.5	236	13	223	0.7
5.5	235	13	222	0.7
5.5	220	12	208	0.7
5.5	219	12	207	0.7
5.5	218	12	206	0.7
5.5	217	12	205	0.7
5.5	201	11	190	0.6
5.5	200	11	189	0.6
5.5	199	11	188	0.6
5.5	183	10	173	0.5
5.5	182	10	172	0.5
5.5	181	10	171	0.6
5.5	165	9	156	0.5
5.5	164	9	155	0.5
5.5	163	9	154	0.5
5.5	146	8	138	0.4
5.5	145	8	137	0.4
5.5	128	7	121	0.4
5.5	127	7	120	0.4
5.5	110	6	104	0.3
5.5	109	6	103	0.3
5.5	91	5	86	0.3
5.5	73	4	69	0.2
5.5	55	3	52	0.2
5.4	500	27	473	1.5
5.4	499	27	472	1.5
5.4	498	27	471	1.5
5.4	497	27	470	1.5
5.4	496	27	469	1.5
5.4	486	26	460	1.4
5.4	485	26	459	1.4
5.4	484	26	458	1.4
5.4	483	26	457	1.4
5.4	482	26	456	1.4
5.4	481	26	455	1.4
5.4	480	26	454	1.4
5.4	479	26	453	1.4
5.4	478	26	452	1.4
5.4	467	25	442	1.3
5.4	466	25	441	1.3
5.4	465	25	440	1.3
5.4	464	25	439	1.3
5.4	463	25	438	1.3
5.4	462	25	437	1.4
5.4	461	25	436	1.4
5.4	460	25	435	1.4
5.4	459	25	434	1.4
5.4	448	24	424	1.3
5.4	447	24	423	1.3
5.4	446	24	422	1.3

%E	M1	M2	DM	M*
5.4	445	24	421	1.3
5.4	444	24	420	1.3
5.4	443	24	419	1.3
5.4	442	24	418	1.3
5.4	441	24	417	1.3
5.4	429	23	406	1.2
5.4	428	23	405	1.2
5.4	427	23	404	1.2
5.4	426	23	403	1.2
5.4	425	23	402	1.2
5.4	424	23	401	1.2
5.4	423	23	400	1.3
5.4	411	22	389	1.2
5.4	410	22	388	1.2
5.4	409	22	387	1.2
5.4	408	22	386	1.2
5.4	407	22	385	1.2
5.4	406	22	384	1.2
5.4	405	22	383	1.2
5.4	404	22	382	1.2
5.4	392	21	371	1.1
5.4	391	21	370	1.1
5.4	390	21	369	1.1
5.4	389	21	368	1.1
5.4	388	21	367	1.1
5.4	387	21	366	1.1
5.4	386	21	365	1.1
5.4	373	20	353	1.1
5.4	372	20	352	1.1
5.4	371	20	351	1.1
5.4	370	20	350	1.1
5.4	369	20	349	1.1
5.4	368	20	348	1.1
5.4	355	19	336	1.0
5.4	354	19	335	1.0
5.4	353	19	334	1.0
5.4	352	19	333	1.0
5.4	351	19	332	1.0
5.4	350	19	331	1.0
5.4	349	19	330	1.0
5.4	336	18	318	1.0
5.4	335	18	317	1.0
5.4	334	18	316	1.0
5.4	333	18	315	1.0
5.4	332	18	314	1.0
5.4	331	18	313	1.0
5.4	317	17	300	0.9
5.4	316	17	299	0.9
5.4	315	17	298	0.9
5.4	314	17	297	0.9

%E	M1	M2	DM	M*
5.4	313	17	296	0.9
5.4	312	17	295	0.9
5.4	299	16	283	0.9
5.4	298	16	282	0.9
5.4	297	16	281	0.9
5.4	296	16	280	0.9
5.4	295	16	279	0.9
5.4	294	16	278	0.9
5.4	280	15	265	0.8
5.4	279	15	264	0.8
5.4	278	15	263	0.8
5.4	277	15	262	0.8
5.4	276	15	261	0.8
5.4	261	14	247	0.8
5.4	260	14	246	0.8
5.4	259	14	245	0.8
5.4	258	14	244	0.8
5.4	257	14	243	0.8
5.4	243	13	230	0.7
5.4	242	13	229	0.7
5.4	241	13	228	0.7
5.4	240	13	227	0.7
5.4	239	13	226	0.7
5.4	224	12	212	0.6
5.4	223	12	211	0.6
5.4	222	12	210	0.6
5.4	221	12	209	0.7
5.4	205	11	194	0.6
5.4	204	11	193	0.6
5.4	203	11	192	0.6
5.4	202	11	191	0.6
5.4	186	10	176	0.5
5.4	185	10	175	0.5
5.4	184	10	174	0.5
5.4	168	9	159	0.5
5.4	167	9	158	0.5
5.4	166	9	157	0.5
5.4	149	8	141	0.4
5.4	148	8	140	0.4
5.4	147	8	139	0.4
5.4	130	7	123	0.4
5.4	129	7	122	0.4
5.4	112	6	106	0.3
5.4	111	6	105	0.3
5.4	93	5	88	0.3
5.4	92	5	87	0.3
5.4	74	4	70	0.2
5.4	56	3	53	0.2
5.4	37	2	35	0.1
5.3	495	26	469	1.4

%E	M1	M2	DM	M*
5.3	494	26	468	1.4
5.3	493	26	467	1.4
5.3	492	26	466	1.4
5.3	491	26	465	1.4
5.3	490	26	464	1.4
5.3	489	26	463	1.4
5.3	488	26	462	1.4
5.3	487	26	461	1.4
5.3	476	25	451	1.3
5.3	475	25	450	1.3
5.3	474	25	449	1.3
5.3	473	25	448	1.3
5.3	472	25	447	1.3
5.3	471	25	446	1.3
5.3	470	25	445	1.3
5.3	469	25	444	1.3
5.3	468	25	443	1.3
5.3	457	24	433	1.3
5.3	456	24	432	1.3
5.3	455	24	431	1.3
5.3	454	24	430	1.3
5.3	453	24	429	1.3
5.3	452	24	428	1.3
5.3	451	24	427	1.3
5.3	450	24	426	1.3
5.3	449	24	425	1.3
5.3	438	23	415	1.2
5.3	437	23	414	1.2
5.3	436	23	413	1.2
5.3	435	23	412	1.2
5.3	434	23	411	1.2
5.3	433	23	410	1.2
5.3	432	23	409	1.2
5.3	431	23	408	1.2
5.3	430	23	407	1.2
5.3	419	22	397	1.2
5.3	418	22	396	1.2
5.3	417	22	395	1.2
5.3	416	22	394	1.2
5.3	415	22	393	1.2
5.3	414	22	392	1.2
5.3	413	22	391	1.2
5.3	412	22	390	1.2
5.3	400	21	379	1.1
5.3	399	21	378	1.1
5.3	398	21	377	1.1
5.3	397	21	376	1.1
5.3	396	21	375	1.1
5.3	395	21	374	1.1
5.3	394	21	373	1.1

%E	M1	M2	DM	M*
5.3	393	21	372	1.1
5.3	380	20	360	1.1
5.3	379	20	359	1.1
5.3	378	20	358	1.1
5.3	377	20	357	1.1
5.3	376	20	356	1.1
5.3	375	20	355	1.1
5.3	374	20	354	1.1
5.3	361	19	342	1.0
5.3	360	19	341	1.0
5.3	359	19	340	1.0
5.3	358	19	339	1.0
5.3	357	19	338	1.0
5.3	356	19	337	1.0
5.3	342	18	324	0.9
5.3	341	18	323	1.0
5.3	340	18	322	1.0
5.3	339	18	321	1.0
5.3	338	18	320	1.0
5.3	337	18	319	1.0
5.3	323	17	306	0.9
5.3	322	17	305	0.9
5.3	321	17	304	0.9
5.3	320	17	303	0.9
5.3	319	17	302	0.9
5.3	318	17	301	0.9
5.3	304	16	288	0.8
5.3	303	16	287	0.8
5.3	302	16	286	0.8
5.3	301	16	285	0.9
5.3	300	16	284	0.9
5.3	285	15	270	0.8
5.3	284	15	269	0.8
5.3	283	15	268	0.8
5.3	282	15	267	0.8
5.3	281	15	266	0.8
5.3	266	14	252	0.7
5.3	265	14	251	0.7
5.3	264	14	250	0.7
5.3	263	14	249	0.7
5.3	262	14	248	0.7
5.3	247	13	234	0.7
5.3	246	13	233	0.7
5.3	245	13	232	0.7
5.3	244	13	231	0.7
5.3	228	12	216	0.6
5.3	227	12	215	0.6
5.3	226	12	214	0.6
5.3	225	12	213	0.6
5.3	209	11	198	0.6

%E	M1	M2	DM	M*
5.3	208	11	197	0.6
5.3	207	11	196	0.6
5.3	206	11	195	0.6
5.3	190	10	180	0.5
5.3	189	10	179	0.5
5.3	188	10	178	0.5
5.3	187	10	177	0.5
5.3	171	9	162	0.5
5.3	170	9	161	0.5
5.3	169	9	160	0.5
5.3	152	8	144	0.4
5.3	151	8	143	0.4
5.3	150	8	142	0.4
5.3	133	7	126	0.4
5.3	132	7	125	0.4
5.3	131	7	124	0.4
5.3	114	6	108	0.3
5.3	113	6	107	0.3
5.3	95	5	90	0.3
5.3	94	5	89	0.3
5.3	76	4	72	0.2
5.3	75	4	71	0.2
5.3	57	3	54	0.2
5.3	38	2	36	0.1
5.3	19	1	18	0.1
5.2	500	26	474	1.4
5.2	499	26	473	1.4
5.2	498	26	472	1.4
5.2	497	26	471	1.4
5.2	496	26	470	1.4
5.2	485	25	460	1.3
5.2	484	25	459	1.3
5.2	483	25	458	1.3
5.2	482	25	457	1.3
5.2	481	25	456	1.3
5.2	480	25	455	1.3
5.2	479	25	454	1.3
5.2	478	25	453	1.3
5.2	477	25	452	1.3
5.2	466	24	442	1.2
5.2	465	24	441	1.2
5.2	464	24	440	1.2
5.2	463	24	439	1.2
5.2	462	24	438	1.2
5.2	461	24	437	1.2
5.2	460	24	436	1.3
5.2	459	24	435	1.3
5.2	458	24	434	1.3
5.2	446	23	423	1.2
5.2	445	23	422	1.2

%E	M1	M2	DM	M*
5.2	444	23	421	1.2
5.2	443	23	420	1.2
5.2	442	23	419	1.2
5.2	441	23	418	1.2
5.2	440	23	417	1.2
5.2	439	23	416	1.2
5.2	427	22	405	1.1
5.2	426	22	404	1.1
5.2	425	22	403	1.1
5.2	424	22	402	1.1
5.2	423	22	401	1.1
5.2	422	22	400	1.1
5.2	421	22	399	1.1
5.2	420	22	398	1.2
5.2	407	21	386	1.1
5.2	406	21	385	1.1
5.2	405	21	384	1.1
5.2	404	21	383	1.1
5.2	403	21	382	1.1
5.2	402	21	381	1.1
5.2	401	21	380	1.1
5.2	388	20	368	1.0
5.2	387	20	367	1.0
5.2	386	20	366	1.0
5.2	385	20	365	1.0
5.2	384	20	364	1.0
5.2	383	20	363	1.0
5.2	382	20	362	1.0
5.2	381	20	361	1.0
5.2	368	19	349	1.0
5.2	367	19	348	1.0
5.2	366	19	347	1.0
5.2	365	19	346	1.0
5.2	364	19	345	1.0
5.2	363	19	344	1.0
5.2	362	19	343	1.0
5.2	349	18	331	0.9
5.2	348	18	330	0.9
5.2	347	18	329	0.9
5.2	346	18	328	0.9
5.2	345	18	327	0.9
5.2	344	18	326	0.9
5.2	343	18	325	0.9
5.2	330	17	313	0.9
5.2	329	17	312	0.9
5.2	328	17	311	0.9
5.2	327	17	310	0.9
5.2	326	17	309	0.9
5.2	325	17	308	0.9
5.2	324	17	307	0.9

%E	M1	M2	DM	M*
5·2	310	16	294	0·8
5·2	309	16	293	0·8
5·2	308	16	292	0·8
5·2	307	16	291	0·8
5·2	306	16	290	0·8
5·2	305	16	289	0·8
5·2	291	15	276	0·8
5·2	290	15	275	0·8
5·2	289	15	274	0·8
5·2	288	15	273	0·8
5·2	287	15	272	0·8
5·2	286	15	271	0·8
5·2	271	14	257	0·7
5·2	270	14	256	0·7
5·2	269	14	255	0·7
5·2	268	14	254	0·7
5·2	267	14	253	0·7
5·2	252	13	239	0·7
5·2	251	13	238	0·7
5·2	250	13	237	0·7
5·2	249	13	236	0·7
5·2	248	13	235	0·7
5·2	233	12	221	0·6
5·2	232	12	220	0·6
5·2	231	12	219	0·6
5·2	230	12	218	0·6
5·2	229	12	217	0·6
5·2	213	11	202	0·6
5·2	212	11	201	0·6
5·2	211	11	200	0·6
5·2	210	11	199	0·6
5·2	194	10	184	0·5
5·2	193	10	183	0·5
5·2	192	10	182	0·5
5·2	191	10	181	0·5
5·2	174	9	165	0·5
5·2	173	9	164	0·5
5·2	172	9	163	0·5
5·2	155	8	147	0·4
5·2	154	8	146	0·4
5·2	153	3	145	0·4
5·2	135	7	128	0·4
5·2	134	7	127	0·4
5·2	116	6	110	0·3
5·2	115	6	109	0·3
5·2	97	5	92	0·3
5·2	96	5	91	0·3
5·2	77	4	73	0·2
5·2	58	3	55	0·2
5·1	495	25	470	1·3

%E	M1	M2	DM	M*
5·1	494	25	469	1·3
5·1	493	25	468	1·3
5·1	492	25	467	1·3
5·1	491	25	466	1·3
5·1	490	25	465	1·3
5·1	489	25	464	1·3
5·1	488	25	463	1·3
5·1	487	25	462	1·3
5·1	486	25	461	1·3
5·1	475	24	451	1·2
5·1	474	24	450	1·2
5·1	473	24	449	1·2
5·1	472	24	448	1·2
5·1	471	24	447	1·2
5·1	470	24	446	1·2
5·1	469	24	445	1·2
5·1	468	24	444	1·2
5·1	467	24	443	1·2
5·1	455	23	432	1·2
5·1	454	23	431	1·2
5·1	453	23	430	1·2
5·1	452	23	429	1·2
5·1	451	23	428	1·2
5·1	450	23	427	1·2
5·1	449	23	426	1·2
5·1	448	23	425	1·2
5·1	447	23	424	1·2
5·1	435	22	413	1·1
5·1	434	22	412	1·1
5·1	433	22	411	1·1
5·1	432	22	410	1·1
5·1	431	22	409	1·1
5·1	430	22	408	1·1
5·1	429	22	407	1·1
5·1	428	22	406	1·1
5·1	415	21	394	1·1
5·1	414	21	393	1·1
5·1	413	21	392	1·1
5·1	412	21	391	1·1
5·1	411	21	390	1·1
5·1	410	21	389	1·1
5·1	409	21	388	1·1
5·1	408	21	387	1·1
5·1	396	20	376	1·0
5·1	395	20	375	1·0
5·1	394	20	374	1·0
5·1	393	20	373	1·0
5·1	392	20	372	1·0
5·1	391	20	371	1·0
5·1	390	20	370	1·0

%E	M1	M2	DM	M*
5·1	389	20	369	1·0
5·1	376	19	357	1·0
5·1	375	19	356	1·0
5·1	374	19	355	1·0
5·1	373	19	354	1·0
5·1	372	19	353	1·0
5·1	371	19	352	1·0
5·1	370	19	351	1·0
5·1	369	19	350	1·0
5·1	356	18	338	0·9
5·1	355	18	337	0·9
5·1	354	18	336	0·9
5·1	353	18	335	0·9
5·1	352	18	334	0·9
5·1	351	18	333	0·9
5·1	350	18	332	0·9
5·1	336	17	319	0·9
5·1	335	17	318	0·9
5·1	334	17	317	0·9
5·1	333	17	316	0·9
5·1	332	17	315	0·9
5·1	331	17	314	0·9
5·1	316	16	300	0·8
5·1	315	16	299	0·8
5·1	314	16	298	0·8
5·1	313	16	297	0·8
5·1	312	16	296	0·8
5·1	311	16	295	0·8
5·1	297	15	282	0·8
5·1	296	15	281	0·8
5·1	295	15	280	0·8
5·1	294	15	279	0·8
5·1	293	15	278	0·8
5·1	292	15	277	0·8
5·1	277	14	263	0·7
5·1	276	14	262	0·7
5·1	275	14	261	0·7
5·1	274	14	260	0·7
5·1	273	14	259	0·7
5·1	272	14	258	0·7
5·1	257	13	244	0·7
5·1	256	13	243	0·7
5·1	255	13	242	0·7
5·1	254	13	241	0·7
5·1	253	13	240	0·7
5·1	237	12	225	0·6
5·1	236	12	224	0·6
5·1	235	12	223	0·6
5·1	234	12	222	0·6
5·1	217	11	206	0·6

%E	M1	M2	DM	M*
5·1	216	11	205	0·6
5·1	215	11	204	0·6
5·1	214	11	203	0·6
5·1	198	10	188	0·5
5·1	197	10	187	0·5
5·1	196	10	186	0·5
5·1	195	10	185	0·5
5·1	178	9	169	0·5
5·1	177	9	168	0·5
5·1	176	9	167	0·5
5·1	175	9	166	0·5
5·1	158	8	150	0·4
5·1	157	8	149	0·4
5·1	156	8	148	0·4
5·1	138	7	131	0·4
5·1	137	7	130	0·4
5·1	136	7	129	0·4
5·1	118	6	112	0·3
5·1	117	6	111	0·3
5·1	99	5	94	0·3
5·1	98	5	93	0·3
5·1	79	4	75	0·2
5·1	78	4	74	0·2
5·1	59	3	56	0·2
5·1	39	2	37	0·1
5·0	500	25	475	1·3
5·0	499	25	474	1·3
5·0	498	25	473	1·3
5·0	497	25	472	1·3
5·0	496	25	471	1·3
5·0	484	24	460	1·2
5·0	483	24	459	1·2
5·0	482	24	458	1·2
5·0	481	24	457	1·2
5·0	480	24	456	1·2
5·0	479	24	455	1·2
5·0	478	24	454	1·2
5·0	477	24	453	1·2
5·0	476	24	452	1·2
5·0	464	23	441	1·1
5·0	463	23	440	1·1
5·0	462	23	439	1·1
5·0	461	23	438	1·1
5·0	460	23	437	1·1
5·0	459	23	436	1·2
5·0	458	23	435	1·2
5·0	457	23	434	1·2
5·0	456	23	433	1·2
5·0	444	22	422	1·1
5·0	443	22	421	1·1

%E	M1	M2	DM	M*
5·0	442	22	420	1·1
5·0	441	22	419	1·1
5·0	440	22	418	1·1
5·0	439	22	417	1·1
5·0	438	22	416	1·1
5·0	437	22	415	1·1
5·0	436	22	414	1·1
5·0	424	21	403	1·0
5·0	423	21	402	1·0
5·0	422	21	401	1·0
5·0	421	21	400	1·0
5·0	420	21	399	1·0
5·0	419	21	398	1·1
5·0	418	21	397	1·1
5·0	417	21	396	1·1
5·0	416	21	395	1·1
5·0	404	20	384	1·0
5·0	403	20	383	1·0
5·0	402	20	382	1·0
5·0	401	20	381	1·0
5·0	400	20	380	1·0
5·0	399	20	379	1·0
5·0	398	20	378	1·0
5·0	397	20	377	1·0
5·0	383	19	364	0·9
5·0	382	19	363	0·9
5·0	381	19	362	0·9
5·0	380	19	361	0·9
5·0	379	19	360	1·0
5·0	378	19	359	1·0
5·0	377	19	358	1·0
5·0	363	18	345	0·9
5·0	362	18	344	0·9
5·0	361	18	343	0·9
5·0	360	18	342	0·9
5·0	359	18	341	0·9
5·0	358	18	340	0·9
5·0	357	18	339	0·9
5·0	343	17	326	0·8
5·0	342	17	325	0·8
5·0	341	17	324	0·8
5·0	340	17	323	0·8
5·0	339	17	322	0·9
5·0	338	17	321	0·9
5·0	337	17	320	0·9
5·0	323	16	307	0·8
5·0	322	16	306	0·8
5·0	321	16	305	0·8
5·0	320	16	304	0·8
5·0	319	16	303	0·8

%E	M1	M2	DM	M*
5.0	318	16	302	0.8
5.0	317	16	301	0.8
5.0	303	15	288	0.7
5.0	302	15	287	0.7
5.0	301	15	286	0.7
5.0	300	15	285	0.8
5.0	299	15	284	0.8
5.0	298	15	283	0.8
5.0	282	14	268	0.7
5.0	281	14	267	0.7
5.0	280	14	266	0.7
5.0	279	14	265	0.7
5.0	278	14	264	0.7
5.0	262	13	249	0.6
5.0	261	13	248	0.6
5.0	260	13	247	0.6
5.0	259	13	246	0.7
5.0	258	13	245	0.7
5.0	242	12	230	0.6
5.0	241	12	229	0.6
5.0	240	12	228	0.6
5.0	239	12	227	0.6
5.0	238	12	226	0.6
5.0	222	11	211	0.5
5.0	221	11	210	0.5
5.0	220	11	209	0.5
5.0	219	11	208	0.6
5.0	218	11	207	0.6
5.0	202	10	192	0.5
5.0	201	10	191	0.5
5.0	200	10	190	0.5
5.0	199	10	189	0.5
5.0	181	9	172	0.4
5.0	180	9	171	0.4
5.0	179	9	170	0.5
5.0	161	8	153	0.4
5.0	160	8	152	0.4
5.0	159	8	151	0.4
5.0	141	7	134	0.3
5.0	140	7	133	0.3
5.0	139	7	132	0.4
5.0	121	6	115	0.3
5.0	120	6	114	0.3
5.0	119	6	113	0.3
5.0	101	5	96	0.2
5.0	100	5	95	0.3
5.0	80	4	76	0.2
5.0	60	3	57	0.1
5.0	40	2	38	0.1
5.0	20	1	19	0.0

%E	M1	M2	DM	M*
4.9	494	24	470	1.2
4.9	493	24	469	1.2
4.9	492	24	468	1.2
4.9	491	24	467	1.2
4.9	490	24	466	1.2
4.9	489	24	465	1.2
4.9	488	24	464	1.2
4.9	487	24	463	1.2
4.9	486	24	462	1.2
4.9	485	24	461	1.2
4.9	474	23	451	1.1
4.9	473	23	450	1.1
4.9	472	23	449	1.1
4.9	471	23	448	1.1
4.9	470	23	447	1.1
4.9	469	23	446	1.1
4.9	468	23	445	1.1
4.9	467	23	444	1.1
4.9	466	23	443	1.1
4.9	465	23	442	1.1
4.9	453	22	431	1.1
4.9	452	22	430	1.1
4.9	451	22	429	1.1
4.9	450	22	428	1.1
4.9	449	22	427	1.1
4.9	448	22	426	1.1
4.9	447	22	425	1.1
4.9	446	22	424	1.1
4.9	445	22	423	1.1
4.9	433	21	412	1.0
4.9	432	21	411	1.0
4.9	431	21	410	1.0
4.9	430	21	409	1.0
4.9	429	21	408	1.0
4.9	428	21	407	1.0
4.9	427	21	406	1.0
4.9	426	21	405	1.0
4.9	425	21	404	1.0
4.9	412	20	392	1.0
4.9	411	20	391	1.0
4.9	410	20	390	1.0
4.9	409	20	389	1.0
4.9	408	20	388	1.0
4.9	407	20	387	1.0
4.9	406	20	386	1.0
4.9	405	20	385	1.0
4.9	391	19	372	0.9
4.9	390	19	371	0.9
4.9	389	19	370	0.9
4.9	388	19	369	0.9

%E	M1	M2	DM	M*
4.9	387	19	368	0.9
4.9	386	19	367	0.9
4.9	385	19	366	0.9
4.9	384	19	365	0.9
4.9	371	18	353	0.9
4.9	370	18	352	0.9
4.9	369	18	351	0.9
4.9	368	18	350	0.9
4.9	367	18	349	0.9
4.9	366	18	348	0.9
4.9	365	18	347	0.9
4.9	364	18	346	0.9
4.9	350	17	333	0.8
4.9	349	17	332	0.8
4.9	348	17	331	0.8
4.9	347	17	330	0.8
4.9	346	17	329	0.8
4.9	345	17	328	0.8
4.9	344	17	327	0.8
4.9	329	16	313	0.8
4.9	328	16	312	0.8
4.9	327	16	311	0.8
4.9	326	16	310	0.8
4.9	325	16	309	0.8
4.9	324	16	308	0.8
4.9	309	15	294	0.7
4.9	308	15	293	0.7
4.9	307	15	292	0.7
4.9	306	15	291	0.7
4.9	305	15	290	0.7
4.9	304	15	289	0.7
4.9	288	14	274	0.7
4.9	287	14	273	0.7
4.9	286	14	272	0.7
4.9	285	14	271	0.7
4.9	284	14	270	0.7
4.9	283	14	269	0.7
4.9	268	13	255	0.6
4.9	267	13	254	0.6
4.9	266	13	253	0.6
4.9	265	13	252	0.6
4.9	264	13	251	0.6
4.9	263	13	250	0.6
4.9	247	12	235	0.6
4.9	246	12	234	0.6
4.9	245	12	233	0.6
4.9	244	12	232	0.6
4.9	243	12	231	0.6
4.9	226	11	215	0.5
4.9	225	11	214	0.5

%E	M1	M2	DM	M*
4.9	224	11	213	0.5
4.9	223	11	212	0.5
4.9	206	10	196	0.5
4.9	205	10	195	0.5
4.9	204	10	194	0.5
4.9	203	10	193	0.5
4.9	185	9	176	0.4
4.9	184	9	175	0.4
4.9	183	9	174	0.4
4.9	182	9	173	0.4
4.9	164	8	156	0.4
4.9	163	8	155	0.4
4.9	162	8	154	0.4
4.9	144	7	137	0.3
4.9	143	7	136	0.3
4.9	142	7	135	0.3
4.9	123	6	117	0.3
4.9	122	6	116	0.3
4.9	103	5	98	0.2
4.9	102	5	97	0.2
4.9	82	4	78	0.2
4.9	81	4	77	0.2
4.9	61	3	58	0.1
4.9	41	2	39	0.1
4.8	500	24	476	1.2
4.8	499	24	475	1.2
4.8	498	24	474	1.2
4.8	497	24	473	1.2
4.8	496	24	472	1.2
4.8	495	24	471	1.2
4.8	484	23	461	1.1
4.8	483	23	460	1.1
4.8	482	23	459	1.1
4.8	481	23	458	1.1
4.8	480	23	457	1.1
4.8	479	23	456	1.1
4.8	478	23	455	1.1
4.8	477	23	454	1.1
4.8	476	23	453	1.1
4.8	475	23	452	1.1
4.8	463	22	441	1.0
4.8	462	22	440	1.0
4.8	461	22	439	1.0
4.8	460	22	438	1.1
4.8	459	22	437	1.1
4.8	458	22	436	1.1
4.8	457	22	435	1.1
4.8	456	22	434	1.1
4.8	455	22	433	1.1
4.8	454	22	432	1.1

%E	M1	M2	DM	M*
4.8	442	21	421	1.0
4.8	441	21	420	1.0
4.8	440	21	419	1.0
4.8	439	21	418	1.0
4.8	438	21	417	1.0
4.8	437	21	416	1.0
4.8	436	21	415	1.0
4.8	435	21	414	1.0
4.8	434	21	413	1.0
4.8	421	20	401	1.0
4.8	420	20	400	1.0
4.8	419	20	399	1.0
4.8	418	20	398	1.0
4.8	417	20	397	1.0
4.8	416	20	396	1.0
4.8	415	20	395	1.0
4.8	414	20	394	1.0
4.8	413	20	393	1.0
4.8	400	19	381	0.9
4.8	399	19	380	0.9
4.8	398	19	379	0.9
4.8	397	19	378	0.9
4.8	396	19	377	0.9
4.8	395	19	376	0.9
4.8	394	19	375	0.9
4.8	393	19	374	0.9
4.8	392	19	373	0.9
4.8	378	18	360	0.9
4.8	377	18	359	0.9
4.8	376	18	358	0.9
4.8	375	18	357	0.9
4.8	374	18	356	0.9
4.8	373	18	355	0.9
4.8	372	18	354	0.9
4.8	357	17	340	0.8
4.8	356	17	339	0.8
4.8	355	17	338	0.8
4.8	354	17	337	0.8
4.8	353	17	336	0.8
4.8	352	17	335	0.8
4.8	351	17	334	0.8
4.8	336	16	320	0.8
4.8	335	16	319	0.8
4.8	334	16	318	0.8
4.8	333	16	317	0.8
4.8	332	16	316	0.8
4.8	331	16	315	0.8
4.8	330	16	314	0.8
4.8	315	15	300	0.7
4.8	314	15	299	0.7

%E	M1	M2	DM	M*
4.8	313	15	298	0.7
4.8	312	15	297	0.7
4.8	311	15	296	0.7
4.8	310	15	295	0.7
4.8	294	14	280	0.7
4.8	293	14	279	0.7
4.8	292	14	278	0.7
4.8	291	14	277	0.7
4.8	290	14	276	0.7
4.8	289	14	275	0.7
4.8	273	13	260	0.6
4.8	272	13	259	0.6
4.8	271	13	258	0.6
4.8	270	13	257	0.6
4.8	269	13	256	0.6
4.8	252	12	240	0.6
4.8	251	12	239	0.6
4.8	250	12	238	0.6
4.8	249	12	237	0.6
4.8	248	12	236	0.6
4.8	231	11	220	0.5
4.8	230	11	219	0.5
4.8	229	11	218	0.5
4.8	228	11	217	0.5
4.8	227	11	216	0.5
4.8	210	10	200	0.5
4.8	209	10	199	0.5
4.8	208	10	198	0.5
4.8	207	10	197	0.5
4.8	189	9	180	0.4
4.8	188	9	179	0.4
4.8	187	9	178	0.4
4.8	186	9	177	0.4
4.8	168	8	160	0.4
4.8	167	8	159	0.4
4.8	166	8	158	0.4
4.8	165	8	157	0.4
4.8	147	7	140	0.3
4.8	146	7	139	0.3
4.8	145	7	138	0.3
4.8	126	6	120	0.3
4.8	125	6	119	0.3
4.8	124	6	118	0.3
4.8	105	5	100	0.2
4.8	104	5	99	0.2
4.8	84	4	80	0.2
4.8	83	4	79	0.2
4.8	63	3	60	0.1
4.8	62	3	59	0.1
4.8	42	2	40	0.1

%E	M1	M2	DM	M*
4.8	21	1	20	0.0
4.7	494	23	471	1.1
4.7	493	23	470	1.1
4.7	492	23	469	1.1
4.7	491	23	468	1.1
4.7	490	23	467	1.1
4.7	489	23	466	1.1
4.7	488	23	465	1.1
4.7	487	23	464	1.1
4.7	486	23	463	1.1
4.7	485	23	462	1.1
4.7	473	22	451	1.0
4.7	472	22	450	1.0
4.7	471	22	449	1.0
4.7	470	22	448	1.0
4.7	469	22	447	1.0
4.7	468	22	446	1.0
4.7	467	22	445	1.0
4.7	466	22	444	1.0
4.7	465	22	443	1.0
4.7	464	22	442	1.0
4.7	451	21	430	1.0
4.7	450	21	429	1.0
4.7	449	21	428	1.0
4.7	448	21	427	1.0
4.7	447	21	426	1.0
4.7	446	21	425	1.0
4.7	445	21	424	1.0
4.7	444	21	423	1.0
4.7	443	21	422	1.0
4.7	430	20	410	0.9
4.7	429	20	409	0.9
4.7	428	20	408	0.9
4.7	427	20	407	0.9
4.7	426	20	406	0.9
4.7	425	20	405	0.9
4.7	424	20	404	0.9
4.7	423	20	403	0.9
4.7	422	20	402	0.9
4.7	408	19	389	0.9
4.7	407	19	388	0.9
4.7	406	19	387	0.9
4.7	405	19	386	0.9
4.7	404	19	385	0.9
4.7	403	19	384	0.9
4.7	402	19	383	0.9
4.7	401	19	382	0.9
4.7	387	18	369	0.8
4.7	386	18	368	0.8
4.7	385	18	367	0.8

%E	M1	M2	DM	M*
4.7	384	18	366	0.8
4.7	383	18	365	0.8
4.7	382	18	364	0.8
4.7	381	18	363	0.9
4.7	380	18	362	0.9
4.7	379	18	361	0.9
4.7	365	17	348	0.8
4.7	364	17	347	0.8
4.7	363	17	346	0.8
4.7	362	17	345	0.8
4.7	361	17	344	0.8
4.7	360	17	343	0.8
4.7	359	17	342	0.8
4.7	358	17	341	0.8
4.7	344	16	328	0.7
4.7	343	16	327	0.7
4.7	342	16	326	0.7
4.7	341	16	325	0.8
4.7	340	16	324	0.8
4.7	339	16	323	0.8
4.7	338	16	322	0.8
4.7	337	16	321	0.8
4.7	322	15	307	0.7
4.7	321	15	306	0.7
4.7	320	15	305	0.7
4.7	319	15	304	0.7
4.7	318	15	303	0.7
4.7	317	15	302	0.7
4.7	316	15	301	0.7
4.7	301	14	287	0.7
4.7	300	14	286	0.7
4.7	299	14	285	0.7
4.7	298	14	284	0.7
4.7	297	14	283	0.7
4.7	296	14	282	0.7
4.7	295	14	281	0.7
4.7	279	13	266	0.6
4.7	278	13	265	0.6
4.7	277	13	264	0.6
4.7	276	13	263	0.6
4.7	275	13	262	0.6
4.7	274	13	261	0.6
4.7	258	12	246	0.6
4.7	257	12	245	0.6
4.7	256	12	244	0.6
4.7	255	12	243	0.6
4.7	254	12	242	0.6
4.7	253	12	241	0.6
4.7	236	11	225	0.5
4.7	235	11	224	0.5

%E	M1	M2	DM	M*
4.7	234	11	223	0.5
4.7	233	11	222	0.5
4.7	232	11	221	0.5
4.7	215	10	205	0.5
4.7	214	10	204	0.5
4.7	213	10	203	0.5
4.7	212	10	202	0.5
4.7	211	10	201	0.5
4.7	193	9	184	0.4
4.7	192	9	183	0.4
4.7	191	9	182	0.4
4.7	190	9	181	0.4
4.7	172	8	164	0.4
4.7	171	8	163	0.4
4.7	170	8	162	0.4
4.7	169	8	161	0.4
4.7	150	7	143	0.3
4.7	149	7	142	0.3
4.7	148	7	141	0.3
4.7	129	6	123	0.3
4.7	128	6	122	0.3
4.7	127	6	121	0.3
4.7	107	5	102	0.2
4.7	106	5	101	0.2
4.7	86	4	82	0.2
4.7	85	4	81	0.2
4.7	64	3	61	0.1
4.7	43	2	41	0.1
4.6	500	23	477	1.1
4.6	499	23	476	1.1
4.6	498	23	475	1.1
4.6	497	23	474	1.1
4.6	496	23	473	1.1
4.6	495	23	472	1.1
4.6	483	22	461	1.0
4.6	482	22	460	1.0
4.6	481	22	459	1.0
4.6	480	22	458	1.0
4.6	479	22	457	1.0
4.6	478	22	456	1.0
4.6	477	22	455	1.0
4.6	476	22	454	1.0
4.6	475	22	453	1.0
4.6	474	22	452	1.0
4.6	461	21	440	1.0
4.6	460	21	439	1.0
4.6	459	21	438	1.0
4.6	458	21	437	1.0
4.6	457	21	436	1.0
4.6	456	21	435	1.0

%E	M1	M2	DM	M*
4.6	455	21	434	1.0
4.6	454	21	433	1.0
4.6	453	21	432	1.0
4.6	452	21	431	1.0
4.6	439	20	419	0.9
4.6	438	20	418	0.9
4.6	437	20	417	0.9
4.6	436	20	416	0.9
4.6	435	20	415	0.9
4.6	434	20	414	0.9
4.6	433	20	413	0.9
4.6	432	20	412	0.9
4.6	431	20	411	0.9
4.6	417	19	398	0.9
4.6	416	19	397	0.9
4.6	415	19	396	0.9
4.6	414	19	395	0.9
4.6	413	19	394	0.9
4.6	412	19	393	0.9
4.6	411	19	392	0.9
4.6	410	19	391	0.9
4.6	409	19	390	0.9
4.6	395	18	377	0.8
4.6	394	18	376	0.8
4.6	393	18	375	0.8
4.6	392	18	374	0.8
4.6	391	18	373	0.8
4.6	390	18	372	0.8
4.6	389	18	371	0.8
4.6	388	18	370	0.8
4.6	373	17	356	0.8
4.6	372	17	355	0.8
4.6	371	17	354	0.8
4.6	370	17	353	0.8
4.6	369	17	352	0.8
4.6	368	17	351	0.8
4.6	367	17	350	0.8
4.6	366	17	349	0.8
4.6	351	16	335	0.7
4.6	350	16	334	0.7
4.6	349	16	333	0.7
4.6	348	16	332	0.7
4.6	347	16	331	0.7
4.6	346	16	330	0.7
4.6	345	16	329	0.7
4.6	329	15	314	0.7
4.6	328	15	313	0.7
4.6	327	15	312	0.7
4.6	326	15	311	0.7
4.6	325	15	310	0.7

%E	M1	M2	DM	M*
4.6	324	15	309	0.7
4.6	323	15	308	0.7
4.6	307	14	293	0.6
4.6	306	14	292	0.6
4.6	305	14	291	0.6
4.6	304	14	290	0.6
4.6	303	14	289	0.6
4.6	302	14	288	0.6
4.6	285	13	272	0.6
4.6	284	13	271	0.6
4.6	283	13	270	0.6
4.6	282	13	269	0.6
4.6	281	13	268	0.6
4.6	280	13	267	0.6
4.6	263	12	251	0.5
4.6	262	12	250	0.5
4.6	261	12	249	0.6
4.6	260	12	248	0.6
4.6	259	12	247	0.6
4.6	241	11	230	0.5
4.6	240	11	229	0.5
4.6	239	11	228	0.5
4.6	238	11	227	0.5
4.6	237	11	226	0.5
4.6	219	10	209	0.5
4.6	218	10	208	0.5
4.6	217	10	207	0.5
4.6	216	10	206	0.5
4.6	197	9	188	0.4
4.6	196	9	187	0.4
4.6	195	9	186	0.4
4.6	194	9	185	0.4
4.6	175	8	167	0.4
4.6	174	8	166	0.4
4.6	173	8	165	0.4
4.6	153	7	146	0.3
4.6	152	7	145	0.3
4.6	151	7	144	0.3
4.6	131	6	125	0.3
4.6	130	6	124	0.3
4.6	109	5	104	0.2
4.6	108	5	103	0.2
4.6	87	4	83	0.2
4.6	65	3	62	0.1
4.5	494	22	472	1.0
4.5	493	22	471	1.0
4.5	492	22	470	1.0
4.5	491	22	469	1.0
4.5	490	22	468	1.0
4.5	489	22	467	1.0
4.5	488	22	466	1.0
4.5	487	22	465	1.0
4.5	486	22	464	1.0
4.5	485	22	463	1.0
4.5	484	22	462	1.0
4.5	471	21	450	0.9
4.5	470	21	449	0.9
4.5	469	21	448	0.9
4.5	468	21	447	0.9
4.5	467	21	446	0.9
4.5	466	21	445	0.9
4.5	465	21	444	0.9
4.5	464	21	443	1.0
4.5	463	21	442	1.0
4.5	462	21	441	1.0
4.5	449	20	429	0.9
4.5	448	20	428	0.9
4.5	447	20	427	0.9
4.5	446	20	426	0.9
4.5	445	20	425	0.9
4.5	444	20	424	0.9
4.5	443	20	423	0.9
4.5	442	20	422	0.9
4.5	441	20	421	0.9
4.5	440	20	420	0.9
4.5	427	19	408	0.8
4.5	426	19	407	0.8
4.5	425	19	406	0.8
4.5	424	19	405	0.9
4.5	423	19	404	0.9
4.5	422	19	403	0.9
4.5	421	19	402	0.9
4.5	420	19	401	0.9
4.5	419	19	400	0.9
4.5	418	19	399	0.9
4.5	404	18	386	0.8
4.5	403	18	385	0.8
4.5	402	18	384	0.8
4.5	401	18	383	0.8
4.5	400	18	382	0.8
4.5	399	18	381	0.8
4.5	398	18	380	0.8
4.5	397	18	379	0.8
4.5	396	18	378	0.8
4.5	382	17	365	0.8
4.5	381	17	364	0.8
4.5	380	17	363	0.8
4.5	379	17	362	0.8
4.5	378	17	361	0.8
4.5	377	17	360	0.8
4.5	376	17	359	0.8
4.5	375	17	358	0.8
4.5	374	17	357	0.8
4.5	359	16	343	0.7
4.5	358	16	342	0.7
4.5	357	16	341	0.7
4.5	356	16	340	0.7
4.5	355	16	339	0.7
4.5	354	16	338	0.7
4.5	353	16	337	0.7
4.5	352	16	336	0.7
4.5	337	15	322	0.7
4.5	336	15	321	0.7
4.5	335	15	320	0.7
4.5	334	15	319	0.7
4.5	333	15	318	0.7
4.5	332	15	317	0.7
4.5	331	15	316	0.7
4.5	330	15	315	0.7
4.5	314	14	300	0.6
4.5	313	14	299	0.6
4.5	312	14	298	0.6
4.5	311	14	297	0.6
4.5	310	14	296	0.6
4.5	309	14	295	0.6
4.5	308	14	294	0.6
4.5	292	13	279	0.6
4.5	291	13	278	0.6
4.5	290	13	277	0.6
4.5	289	13	276	0.6
4.5	288	13	275	0.6
4.5	287	13	274	0.6
4.5	286	13	273	0.6
4.5	269	12	257	0.5
4.5	268	12	256	0.5
4.5	267	12	255	0.5
4.5	266	12	254	0.5
4.5	265	12	253	0.5
4.5	264	12	252	0.5
4.5	247	11	236	0.5
4.5	246	11	235	0.5
4.5	245	11	234	0.5
4.5	244	11	233	0.5
4.5	243	11	232	0.5
4.5	242	11	231	0.5
4.5	224	10	214	0.4
4.5	223	10	213	0.4
4.5	222	10	212	0.5
4.5	221	10	211	0.5
4.5	220	10	210	0.5
4.5	202	9	193	0.4
4.5	201	9	192	0.4
4.5	200	9	191	0.4
4.5	199	9	190	0.4
4.5	198	9	189	0.4
4.5	179	8	171	0.4
4.5	178	8	170	0.4
4.5	177	8	169	0.4
4.5	176	8	168	0.4
4.5	157	7	150	0.3
4.5	156	7	149	0.3
4.5	155	7	148	0.3
4.5	154	7	147	0.3
4.5	134	6	128	0.3
4.5	133	6	127	0.3
4.5	132	6	126	0.3
4.5	112	5	107	0.2
4.5	111	5	106	0.2
4.5	110	5	105	0.2
4.5	89	4	85	0.2
4.5	88	4	84	0.2
4.5	67	3	64	0.1
4.5	66	3	63	0.1
4.5	44	2	42	0.1
4.5	22	1	21	0.0
4.4	500	22	478	1.0
4.4	499	22	477	1.0
4.4	498	22	476	1.0
4.4	497	22	475	1.0
4.4	496	22	474	1.0
4.4	495	22	473	1.0
4.4	482	21	461	0.9
4.4	481	21	460	0.9
4.4	480	21	459	0.9
4.4	479	21	458	0.9
4.4	478	21	457	0.9
4.4	477	21	456	0.9
4.4	476	21	455	0.9
4.4	475	21	454	0.9
4.4	474	21	453	0.9
4.4	473	21	452	0.9
4.4	472	21	451	0.9
4.4	459	20	439	0.9
4.4	458	20	438	0.9
4.4	457	20	437	0.9
4.4	456	20	436	0.9
4.4	455	20	435	0.9
4.4	454	20	434	0.9
4.4	453	20	433	0.9
4.4	452	20	432	0.9
4.4	451	20	431	0.9
4.4	450	20	430	0.9
4.4	436	19	417	0.8
4.4	435	19	416	0.8
4.4	434	19	415	0.8
4.4	433	19	414	0.8
4.4	432	19	413	0.8
4.4	431	19	412	0.8
4.4	430	19	411	0.8
4.4	429	19	410	0.8
4.4	428	19	409	0.8
4.4	413	18	395	0.8
4.4	412	18	394	0.8
4.4	411	18	393	0.8
4.4	410	18	392	0.8
4.4	409	18	391	0.8
4.4	408	18	390	0.8
4.4	407	18	389	0.8
4.4	406	18	388	0.8
4.4	405	18	387	0.8
4.4	390	17	373	0.7
4.4	389	17	372	0.7
4.4	388	17	371	0.7
4.4	387	17	370	0.7
4.4	386	17	369	0.7
4.4	385	17	368	0.8
4.4	384	17	367	0.8
4.4	383	17	366	0.8
4.4	367	16	351	0.7
4.4	366	16	350	0.7
4.4	365	16	349	0.7
4.4	364	16	348	0.7
4.4	363	16	347	0.7
4.4	362	16	346	0.7
4.4	361	16	345	0.7
4.4	360	16	344	0.7
4.4	344	15	329	0.7
4.4	343	15	328	0.7
4.4	342	15	327	0.7
4.4	341	15	326	0.7
4.4	340	15	325	0.7
4.4	339	15	324	0.7
4.4	338	15	323	0.7
4.4	321	14	307	0.6
4.4	320	14	306	0.6
4.4	319	14	305	0.6
4.4	318	14	304	0.6
4.4	317	14	303	0.6
4.4	316	14	302	0.6
4.4	315	14	301	0.6

%E	M1	M2	DM	M*
4.4	298	13	285	0.6
4.4	297	13	284	0.6
4.4	296	13	283	0.6
4.4	295	13	282	0.6
4.4	294	13	281	0.6
4.4	293	13	280	0.6
4.4	275	12	263	0.5
4.4	274	12	262	0.5
4.4	273	12	261	0.5
4.4	272	12	260	0.5
4.4	271	12	259	0.5
4.4	270	12	258	0.5
4.4	252	11	241	0.5
4.4	251	11	240	0.5
4.4	250	11	239	0.5
4.4	249	11	238	0.5
4.4	248	11	237	0.5
4.4	229	10	219	0.4
4.4	228	10	218	0.4
4.4	227	10	217	0.4
4.4	226	10	216	0.4
4.4	225	10	215	0.4
4.4	206	9	197	0.4
4.4	205	9	196	0.4
4.4	204	9	195	0.4
4.4	203	9	194	0.4
4.4	183	8	175	0.3
4.4	182	8	174	0.4
4.4	181	8	173	0.4
4.4	180	8	172	0.4
4.4	160	7	153	0.3
4.4	159	7	152	0.3
4.4	158	7	151	0.3
4.4	137	6	131	0.3
4.4	136	6	130	0.3
4.4	135	6	129	0.3
4.4	114	5	109	0.2
4.4	113	5	108	0.2
4.4	91	4	87	0.2
4.4	90	4	86	0.2
4.4	68	3	65	0.1
4.4	45	2	43	0.1
4.3	494	21	473	0.9
4.3	493	21	472	0.9
4.3	492	21	471	0.9
4.3	491	21	470	0.9
4.3	490	21	469	0.9
4.3	489	21	468	0.9
4.3	488	21	467	0.9
4.3	487	21	466	0.9

%E	M1	M2	DM	M*
4.3	486	21	465	0.9
4.3	485	21	464	0.9
4.3	484	21	463	0.9
4.3	483	21	462	0.9
4.3	470	20	450	0.9
4.3	469	20	449	0.9
4.3	468	20	448	0.9
4.3	467	20	447	0.9
4.3	466	20	446	0.9
4.3	465	20	445	0.9
4.3	464	20	444	0.9
4.3	463	20	443	0.9
4.3	462	20	442	0.9
4.3	461	20	441	0.9
4.3	460	20	440	0.9
4.3	447	19	428	0.8
4.3	446	19	427	0.8
4.3	445	19	426	0.8
4.3	444	19	425	0.8
4.3	443	19	424	0.8
4.3	442	19	423	0.8
4.3	441	19	422	0.8
4.3	440	19	421	0.8
4.3	439	19	420	0.8
4.3	438	19	419	0.8
4.3	437	19	418	0.8
4.3	423	18	405	0.8
4.3	422	18	404	0.8
4.3	421	18	403	0.8
4.3	420	18	402	0.8
4.3	419	18	401	0.8
4.3	418	18	400	0.8
4.3	417	18	399	0.8
4.3	416	18	398	0.8
4.3	415	18	397	0.8
4.3	414	18	396	0.8
4.3	400	17	383	0.7
4.3	399	17	382	0.7
4.3	398	17	381	0.7
4.3	397	17	380	0.7
4.3	396	17	379	0.7
4.3	395	17	378	0.7
4.3	394	17	377	0.7
4.3	393	17	376	0.7
4.3	392	17	375	0.7
4.3	391	17	374	0.7
4.3	376	16	360	0.7
4.3	375	16	359	0.7
4.3	374	16	358	0.7
4.3	373	16	357	0.7

%E	M1	M2	DM	M*
4.3	372	16	356	0.7
4.3	371	16	355	0.7
4.3	370	16	354	0.7
4.3	369	16	353	0.7
4.3	368	16	352	0.7
4.3	352	15	337	0.6
4.3	351	15	336	0.6
4.3	350	15	335	0.6
4.3	349	15	334	0.6
4.3	348	15	333	0.6
4.3	347	15	332	0.6
4.3	346	15	331	0.7
4.3	345	15	330	0.7
4.3	329	14	315	0.6
4.3	328	14	314	0.6
4.3	327	14	313	0.6
4.3	326	14	312	0.6
4.3	325	14	311	0.6
4.3	324	14	310	0.6
4.3	323	14	309	0.6
4.3	322	14	308	0.6
4.3	305	13	292	0.6
4.3	304	13	291	0.6
4.3	303	13	290	0.6
4.3	302	13	289	0.6
4.3	301	13	288	0.6
4.3	300	13	287	0.6
4.3	299	13	286	0.6
4.3	282	12	270	0.5
4.3	281	12	269	0.5
4.3	280	12	268	0.5
4.3	279	12	267	0.5
4.3	278	12	266	0.5
4.3	277	12	265	0.5
4.3	276	12	264	0.5
4.3	258	11	247	0.5
4.3	257	11	246	0.5
4.3	256	11	245	0.5
4.3	255	11	244	0.5
4.3	254	11	243	0.5
4.3	253	11	242	0.5
4.3	235	10	225	0.4
4.3	234	10	224	0.4
4.3	233	10	223	0.4
4.3	232	10	222	0.4
4.3	231	10	221	0.4
4.3	230	10	220	0.4
4.3	211	9	202	0.4
4.3	210	9	201	0.4
4.3	209	9	200	0.4

%E	M1	M2	DM	M*
4.3	208	9	199	0.4
4.3	207	9	198	0.4
4.3	188	8	180	0.3
4.3	187	8	179	0.3
4.3	186	8	178	0.3
4.3	185	8	177	0.3
4.3	184	8	176	0.3
4.3	164	7	157	0.3
4.3	163	7	156	0.3
4.3	162	7	155	0.3
4.3	161	7	154	0.3
4.3	141	6	135	0.3
4.3	140	6	134	0.3
4.3	139	6	133	0.3
4.3	138	6	132	0.3
4.3	117	5	112	0.2
4.3	116	5	111	0.2
4.3	115	5	110	0.2
4.3	94	4	90	0.2
4.3	93	4	89	0.2
4.3	92	4	88	0.2
4.3	70	3	67	0.1
4.3	69	3	66	0.1
4.3	47	2	45	0.1
4.3	46	2	44	0.1
4.3	23	1	22	0.0
4.2	500	21	479	0.9
4.2	499	21	478	0.9
4.2	498	21	477	0.9
4.2	497	21	476	0.9
4.2	496	21	475	0.9
4.2	495	21	474	0.9
4.2	481	20	461	0.8
4.2	480	20	460	0.8
4.2	479	20	459	0.8
4.2	478	20	458	0.8
4.2	477	20	457	0.8
4.2	476	20	456	0.8
4.2	475	20	455	0.8
4.2	474	20	454	0.8
4.2	473	20	453	0.8
4.2	472	20	452	0.8
4.2	471	20	451	0.8
4.2	457	19	438	0.8
4.2	456	19	437	0.8
4.2	455	19	436	0.8
4.2	454	19	435	0.8
4.2	453	19	434	0.8
4.2	452	19	433	0.8
4.2	451	19	432	0.8

%E	M1	M2	DM	M*
4.2	450	19	431	0.8
4.2	449	19	430	0.8
4.2	448	19	429	0.8
4.2	433	18	415	0.7
4.2	432	18	414	0.8
4.2	431	18	413	0.8
4.2	430	18	412	0.8
4.2	429	18	411	0.8
4.2	428	18	410	0.8
4.2	427	18	409	0.8
4.2	426	18	408	0.8
4.2	425	18	407	0.8
4.2	424	18	406	0.8
4.2	409	17	392	0.7
4.2	408	17	391	0.7
4.2	407	17	390	0.7
4.2	406	17	389	0.7
4.2	405	17	388	0.7
4.2	404	17	387	0.7
4.2	403	17	386	0.7
4.2	402	17	385	0.7
4.2	401	17	384	0.7
4.2	385	16	369	0.7
4.2	384	16	368	0.7
4.2	383	16	367	0.7
4.2	382	16	366	0.7
4.2	381	16	365	0.7
4.2	380	16	364	0.7
4.2	379	16	363	0.7
4.2	378	16	362	0.7
4.2	377	16	361	0.7
4.2	361	15	346	0.6
4.2	360	15	345	0.6
4.2	359	15	344	0.6
4.2	358	15	343	0.6
4.2	357	15	342	0.6
4.2	356	15	341	0.6
4.2	355	15	340	0.6
4.2	354	15	339	0.6
4.2	353	15	338	0.6
4.2	337	14	323	0.6
4.2	336	14	322	0.6
4.2	335	14	321	0.6
4.2	334	14	320	0.6
4.2	333	14	319	0.6
4.2	332	14	318	0.6
4.2	331	14	317	0.6
4.2	330	14	316	0.6
4.2	313	13	300	0.5
4.2	312	13	299	0.5

%E	M1	M2	DM	M*	%E	M1	M2	DM	M*	%E	M1	M2	DM	M*	%E	M1	M2	DM	M*	%E	M1	M2	DM	M*
4.2	311	13	298	0.5	4.1	493	20	473	0.8	4.1	390	16	374	0.7	4.1	221	9	212	0.4	4.0	450	18	432	0.7
4.2	310	13	297	0.5	4.1	492	20	472	0.8	4.1	389	16	373	0.7	4.1	220	9	211	0.4	4.0	449	18	431	0.7
4.2	309	13	296	0.5	4.1	491	20	471	0.8	4.1	388	16	372	0.7	4.1	219	9	210	0.4	4.0	448	18	430	0.7
4.2	308	13	295	0.5	4.1	490	20	470	0.8	4.1	387	16	371	0.7	4.1	218	9	209	0.4	4.0	447	18	429	0.7
4.2	307	13	294	0.6	4.1	489	20	469	0.8	4.1	386	16	370	0.7	4.1	217	9	208	0.4	4.0	446	18	428	0.7
4.2	306	13	293	0.6	4.1	488	20	468	0.8	4.1	370	15	355	0.6	4.1	197	8	189	0.3	4.0	445	18	427	0.7
4.2	289	12	277	0.5	4.1	487	20	467	0.8	4.1	369	15	354	0.6	4.1	196	8	188	0.3	4.0	430	17	413	0.7
4.2	288	12	276	0.5	4.1	486	20	466	0.8	4.1	368	15	353	0.6	4.1	195	8	187	0.3	4.0	429	17	412	0.7
4.2	287	12	275	0.5	4.1	485	20	465	0.8	4.1	367	15	352	0.6	4.1	194	8	186	0.3	4.0	428	17	411	0.7
4.2	286	12	274	0.5	4.1	484	20	464	0.8	4.1	366	15	351	0.6	4.1	193	8	185	0.3	4.0	427	17	410	0.7
4.2	285	12	273	0.5	4.1	483	20	463	0.8	4.1	365	15	350	0.6	4.1	172	7	165	0.3	4.0	426	17	409	0.7
4.2	284	12	272	0.5	4.1	482	20	462	0.8	4.1	364	15	349	0.6	4.1	171	7	164	0.3	4.0	425	17	408	0.7
4.2	283	12	271	0.5	4.1	469	19	450	0.8	4.1	363	15	348	0.6	4.1	170	7	163	0.3	4.0	424	17	407	0.7
4.2	265	11	254	0.5	4.1	468	19	449	0.8	4.1	362	15	347	0.6	4.1	169	7	162	0.3	4.0	423	17	406	0.7
4.2	264	11	253	0.5	4.1	467	19	448	0.8	4.1	345	14	331	0.6	4.1	148	6	142	0.2	4.0	422	17	405	0.7
4.2	263	11	252	0.5	4.1	466	19	447	0.8	4.1	344	14	330	0.6	4.1	147	6	141	0.2	4.0	421	17	404	0.7
4.2	262	11	251	0.5	4.1	465	19	446	0.8	4.1	343	14	329	0.6	4.1	146	6	140	0.2	4.0	420	17	403	0.7
4.2	261	11	250	0.5	4.1	464	19	445	0.8	4.1	342	14	328	0.6	4.1	145	6	139	0.2	4.0	405	16	389	0.6
4.2	260	11	249	0.5	4.1	463	19	444	0.8	4.1	341	14	327	0.6	4.1	123	5	118	0.2	4.0	404	16	388	0.6
4.2	259	11	248	0.5	4.1	462	19	443	0.8	4.1	340	14	326	0.6	4.1	122	5	117	0.2	4.0	403	16	387	0.6
4.2	240	10	230	0.4	4.1	461	19	442	0.8	4.1	339	14	325	0.6	4.1	121	5	116	0.2	4.0	402	16	386	0.6
4.2	239	10	229	0.4	4.1	460	19	441	0.8	4.1	338	14	324	0.6	4.1	98	4	94	0.2	4.0	401	16	385	0.6
4.2	238	10	228	0.4	4.1	459	19	440	0.8	4.1	321	13	308	0.5	4.1	97	4	93	0.2	4.0	400	16	384	0.6
4.2	237	10	227	0.4	4.1	458	19	439	0.8	4.1	320	13	307	0.5	4.1	74	3	71	0.1	4.0	399	16	383	0.6
4.2	236	10	226	0.4	4.1	444	18	426	0.7	4.1	319	13	306	0.5	4.1	73	3	70	0.1	4.0	398	16	382	0.6
4.2	216	9	207	0.4	4.1	443	18	425	0.7	4.1	318	13	305	0.5	4.1	49	2	47	0.1	4.0	397	16	381	0.6
4.2	215	9	206	0.4	4.1	442	18	424	0.7	4.1	317	13	304	0.5	4.0	500	20	480	0.8	4.0	396	16	380	0.6
4.2	214	9	205	0.4	4.1	441	18	423	0.7	4.1	316	13	303	0.5	4.0	499	20	479	0.8	4.0	379	15	364	0.6
4.2	213	9	204	0.4	4.1	440	18	422	0.7	4.1	315	13	302	0.5	4.0	498	20	478	0.8	4.0	378	15	363	0.6
4.2	212	9	203	0.4	4.1	439	18	421	0.7	4.1	314	13	301	0.5	4.0	497	20	477	0.8	4.0	377	15	362	0.6
4.2	192	8	184	0.3	4.1	438	18	420	0.7	4.1	296	12	284	0.5	4.0	496	20	476	0.8	4.0	376	15	361	0.6
4.2	191	8	183	0.3	4.1	437	18	419	0.7	4.1	295	12	283	0.5	4.0	495	20	475	0.8	4.0	375	15	360	0.6
4.2	190	8	182	0.3	4.1	436	18	418	0.7	4.1	294	12	282	0.5	4.0	494	20	474	0.8	4.0	374	15	359	0.6
4.2	189	8	181	0.3	4.1	435	18	417	0.7	4.1	293	12	281	0.5	4.0	481	19	462	0.8	4.0	373	15	358	0.6
4.2	168	7	161	0.3	4.1	434	18	416	0.7	4.1	292	12	280	0.5	4.0	480	19	461	0.8	4.0	372	15	357	0.6
4.2	167	7	160	0.3	4.1	419	17	402	0.7	4.1	291	12	279	0.5	4.0	479	19	460	0.8	4.0	371	15	356	0.6
4.2	166	7	159	0.3	4.1	418	17	401	0.7	4.1	290	12	278	0.5	4.0	478	19	459	0.8	4.0	354	14	340	0.6
4.2	165	7	158	0.3	4.1	417	17	400	0.7	4.1	271	11	260	0.4	4.0	477	19	458	0.8	4.0	353	14	339	0.6
4.2	144	6	138	0.3	4.1	416	17	399	0.7	4.1	270	11	259	0.4	4.0	476	19	457	0.8	4.0	352	14	338	0.6
4.2	143	6	137	0.3	4.1	415	17	398	0.7	4.1	269	11	258	0.4	4.0	475	19	456	0.8	4.0	351	14	337	0.6
4.2	142	6	136	0.3	4.1	414	17	397	0.7	4.1	268	11	257	0.5	4.0	474	19	455	0.8	4.0	350	14	336	0.6
4.2	120	5	115	0.2	4.1	413	17	396	0.7	4.1	267	11	256	0.5	4.0	473	19	454	0.8	4.0	349	14	335	0.6
4.2	119	5	114	0.2	4.1	412	17	395	0.7	4.1	266	11	255	0.5	4.0	472	19	453	0.8	4.0	348	14	334	0.6
4.2	118	5	113	0.2	4.1	411	17	394	0.7	4.1	246	10	236	0.4	4.0	471	19	452	0.8	4.0	347	14	333	0.6
4.2	96	4	92	0.2	4.1	410	17	393	0.7	4.1	245	10	235	0.4	4.0	470	19	451	0.8	4.0	346	14	332	0.6
4.2	95	4	91	0.2	4.1	395	16	379	0.6	4.1	244	10	234	0.4	4.0	455	18	437	0.7	4.0	329	13	316	0.5
4.2	72	3	69	0.1	4.1	394	16	378	0.6	4.1	243	10	233	0.4	4.0	454	18	436	0.7	4.0	328	13	315	0.5
4.2	71	3	68	0.1	4.1	393	16	377	0.7	4.1	242	10	232	0.4	4.0	453	18	435	0.7	4.0	327	13	314	0.5
4.2	48	2	46	0.1	4.1	392	16	376	0.7	4.1	241	10	231	0.4	4.0	452	18	434	0.7	4.0	326	13	313	0.5
4.2	24	1	23	0.0	4.1	391	16	375	0.7	4.1	222	9	213	0.4	4.0	451	18	433	0.7	4.0	325	13	312	0.5

%E	M1	M2	DM	M*
4.0	324	13	311	0.5
4.0	323	13	310	0.5
4.0	322	13	309	0.5
4.0	303	12	291	0.5
4.0	302	12	290	0.5
4.0	301	12	289	0.5
4.0	300	12	288	0.5
4.0	299	12	287	0.5
4.0	298	12	286	0.5
4.0	297	12	285	0.5
4.0	278	11	267	0.4
4.0	277	11	266	0.4
4.0	276	11	265	0.4
4.0	275	11	264	0.4
4.0	274	11	263	0.4
4.0	273	11	262	0.4
4.0	272	11	261	0.4
4.0	253	10	243	0.4
4.0	252	10	242	0.4
4.0	251	10	241	0.4
4.0	250	10	240	0.4
4.0	249	10	239	0.4
4.0	248	10	238	0.4
4.0	247	10	237	0.4
4.0	227	9	218	0.4
4.0	226	9	217	0.4
4.0	225	9	216	0.4
4.0	224	9	215	0.4
4.0	223	9	214	0.4
4.0	202	8	194	0.3
4.0	201	8	193	0.3
4.0	200	8	192	0.3
4.0	199	8	191	0.3
4.0	198	8	190	0.3
4.0	177	7	170	0.3
4.0	176	7	169	0.3
4.0	175	7	168	0.3
4.0	174	7	167	0.3
4.0	173	7	166	0.3
4.0	151	6	145	0.2
4.0	150	6	144	0.2
4.0	149	6	143	0.2
4.0	126	5	121	0.2
4.0	125	5	120	0.2
4.0	124	5	119	0.2
4.0	101	4	97	0.2
4.0	100	4	96	0.2
4.0	99	4	95	0.2
4.0	75	3	72	0.1
4.0	50	2	48	0.1

%E	M1	M2	DM	M*
4.0	25	1	24	0.0
3.9	493	19	474	0.7
3.9	492	19	473	0.7
3.9	491	19	472	0.7
3.9	490	19	471	0.7
3.9	489	19	470	0.7
3.9	488	19	469	0.7
3.9	487	19	468	0.7
3.9	486	19	467	0.7
3.9	485	19	466	0.7
3.9	484	19	465	0.7
3.9	483	19	464	0.7
3.9	482	19	463	0.7
3.9	467	18	449	0.7
3.9	466	18	448	0.7
3.9	465	18	447	0.7
3.9	464	18	446	0.7
3.9	463	18	445	0.7
3.9	462	18	444	0.7
3.9	461	18	443	0.7
3.9	460	18	442	0.7
3.9	459	18	441	0.7
3.9	458	18	440	0.7
3.9	457	18	439	0.7
3.9	456	18	438	0.7
3.9	441	17	424	0.7
3.9	440	17	423	0.7
3.9	439	17	422	0.7
3.9	438	17	421	0.7
3.9	437	17	420	0.7
3.9	436	17	419	0.7
3.9	435	17	418	0.7
3.9	434	17	417	0.7
3.9	433	17	416	0.7
3.9	432	17	415	0.7
3.9	431	17	414	0.7
3.9	415	16	399	0.6
3.9	414	16	398	0.6
3.9	413	16	397	0.6
3.9	412	16	396	0.6
3.9	411	16	395	0.6
3.9	410	16	394	0.6
3.9	409	16	393	0.6
3.9	408	16	392	0.6
3.9	407	16	391	0.6
3.9	406	16	390	0.6
3.9	389	15	374	0.6
3.9	388	15	373	0.6
3.9	387	15	372	0.6
3.9	386	15	371	0.6

%E	M1	M2	DM	M*
3.9	385	15	370	0.6
3.9	384	15	369	0.6
3.9	383	15	368	0.6
3.9	382	15	367	0.6
3.9	381	15	366	0.6
3.9	380	15	365	0.6
3.9	363	14	349	0.5
3.9	362	14	348	0.5
3.9	361	14	347	0.5
3.9	360	14	346	0.5
3.9	359	14	345	0.5
3.9	358	14	344	0.5
3.9	357	14	343	0.5
3.9	356	14	342	0.6
3.9	355	14	341	0.6
3.9	337	13	324	0.5
3.9	336	13	323	0.5
3.9	335	13	322	0.5
3.9	334	13	321	0.5
3.9	333	13	320	0.5
3.9	332	13	319	0.5
3.9	331	13	318	0.5
3.9	330	13	317	0.5
3.9	311	12	299	0.5
3.9	310	12	298	0.5
3.9	309	12	297	0.5
3.9	308	12	296	0.5
3.9	307	12	295	0.5
3.9	306	12	294	0.5
3.9	305	12	293	0.5
3.9	304	12	292	0.5
3.9	285	11	274	0.4
3.9	284	11	273	0.4
3.9	283	11	272	0.4
3.9	282	11	271	0.4
3.9	281	11	270	0.4
3.9	280	11	269	0.4
3.9	279	11	268	0.4
3.9	259	10	249	0.4
3.9	258	10	248	0.4
3.9	257	10	247	0.4
3.9	256	10	246	0.4
3.9	255	10	245	0.4
3.9	254	10	244	0.4
3.9	233	9	224	0.3
3.9	232	9	223	0.3
3.9	231	9	222	0.4
3.9	230	9	221	0.4
3.9	229	9	220	0.4
3.9	228	9	219	0.4

%E	M1	M2	DM	M*
3.9	207	8	199	0.3
3.9	206	8	198	0.3
3.9	205	8	197	0.3
3.9	204	8	196	0.3
3.9	203	8	195	0.3
3.9	181	7	174	0.3
3.9	180	7	173	0.3
3.9	179	7	172	0.3
3.9	178	7	171	0.3
3.9	155	6	149	0.2
3.9	154	6	148	0.2
3.9	153	6	147	0.2
3.9	152	6	146	0.2
3.9	129	5	124	0.2
3.9	128	5	123	0.2
3.9	127	5	122	0.2
3.9	103	4	99	0.2
3.9	102	4	98	0.2
3.9	77	3	74	0.1
3.9	76	3	73	0.1
3.9	51	2	49	0.1
3.8	500	19	481	0.7
3.8	499	19	480	0.7
3.8	498	19	479	0.7
3.8	497	19	478	0.7
3.8	496	19	477	0.7
3.8	495	19	476	0.7
3.8	494	19	475	0.7
3.8	480	18	462	0.7
3.8	479	18	461	0.7
3.8	478	18	460	0.7
3.8	477	18	459	0.7
3.8	476	18	458	0.7
3.8	475	18	457	0.7
3.8	474	18	456	0.7
3.8	473	18	455	0.7
3.8	472	18	454	0.7
3.8	471	18	453	0.7
3.8	470	18	452	0.7
3.8	469	18	451	0.7
3.8	468	18	450	0.7
3.8	453	17	436	0.6
3.8	452	17	435	0.6
3.8	451	17	434	0.6
3.8	450	17	433	0.6
3.8	449	17	432	0.6
3.8	448	17	431	0.6
3.8	447	17	430	0.6
3.8	446	17	429	0.6
3.8	445	17	428	0.6

%E	M1	M2	DM	M*
3.8	444	17	427	0.7
3.8	443	17	426	0.7
3.8	442	17	425	0.7
3.8	426	16	410	0.6
3.8	425	16	409	0.6
3.8	424	16	408	0.6
3.8	423	16	407	0.6
3.8	422	16	406	0.6
3.8	421	16	405	0.6
3.8	420	16	404	0.6
3.8	419	16	403	0.6
3.8	418	16	402	0.6
3.8	417	16	401	0.6
3.8	416	16	400	0.6
3.8	400	15	385	0.6
3.8	399	15	384	0.6
3.8	398	15	383	0.6
3.8	397	15	382	0.6
3.8	396	15	381	0.6
3.8	395	15	380	0.6
3.8	394	15	379	0.6
3.8	393	15	378	0.6
3.8	392	15	377	0.6
3.8	391	15	376	0.6
3.8	390	15	375	0.6
3.8	373	14	359	0.5
3.8	372	14	358	0.5
3.8	371	14	357	0.5
3.8	370	14	356	0.5
3.8	369	14	355	0.5
3.8	368	14	354	0.5
3.8	367	14	353	0.5
3.8	366	14	352	0.5
3.8	365	14	351	0.5
3.8	364	14	350	0.5
3.8	346	13	333	0.5
3.8	345	13	332	0.5
3.8	344	13	331	0.5
3.8	343	13	330	0.5
3.8	342	13	329	0.5
3.8	341	13	328	0.5
3.8	340	13	327	0.5
3.8	339	13	326	0.5
3.8	338	13	325	0.5
3.8	320	12	308	0.4
3.8	319	12	307	0.5
3.8	318	12	306	0.5
3.8	317	12	305	0.5
3.8	316	12	304	0.5
3.8	315	12	303	0.5

%E	M1	M2	DM	M*
3.8	314	12	302	0.5
3.8	313	12	301	0.5
3.8	312	12	300	0.5
3.8	293	11	282	0.4
3.8	292	11	281	0.4
3.8	291	11	280	0.4
3.8	290	11	279	0.4
3.8	289	11	278	0.4
3.8	288	11	277	0.4
3.8	287	11	276	0.4
3.8	286	11	275	0.4
3.8	266	10	256	0.4
3.8	265	10	255	0.4
3.8	264	10	254	0.4
3.8	263	10	253	0.4
3.8	262	10	252	0.4
3.8	261	10	251	0.4
3.8	260	10	250	0.4
3.8	240	9	231	0.3
3.8	239	9	230	0.3
3.8	238	9	229	0.3
3.8	237	9	228	0.3
3.8	236	9	227	0.3
3.8	235	9	226	0.3
3.8	234	9	225	0.3
3.8	213	8	205	0.3
3.8	212	8	204	0.3
3.8	211	8	203	0.3
3.8	210	8	202	0.3
3.8	209	8	201	0.3
3.8	208	8	200	0.3
3.8	186	7	179	0.3
3.8	185	7	178	0.3
3.8	184	7	177	0.3
3.8	183	7	176	0.3
3.8	182	7	175	0.3
3.8	160	6	154	0.2
3.8	159	6	153	0.2
3.8	158	6	152	0.2
3.8	157	6	151	0.2
3.8	156	6	150	0.2
3.8	133	5	128	0.2
3.8	132	5	127	0.2
3.8	131	5	126	0.2
3.8	130	5	125	0.2
3.8	106	4	102	0.2
3.8	105	4	101	0.2
3.8	104	4	100	0.2
3.8	80	3	77	0.1
3.8	79	3	76	0.1

%E	M1	M2	DM	M*
3.8	78	3	75	0.1
3.8	53	2	51	0.1
3.8	52	2	50	0.1
3.8	26	1	25	0.0
3.7	493	18	475	0.7
3.7	492	18	474	0.7
3.7	491	18	473	0.7
3.7	490	18	472	0.7
3.7	489	18	471	0.7
3.7	488	18	470	0.7
3.7	487	18	469	0.7
3.7	486	18	468	0.7
3.7	485	18	467	0.7
3.7	484	18	466	0.7
3.7	483	18	465	0.7
3.7	482	18	464	0.7
3.7	481	18	463	0.7
3.7	465	17	448	0.6
3.7	464	17	447	0.6
3.7	463	17	446	0.6
3.7	462	17	445	0.6
3.7	461	17	444	0.6
3.7	460	17	443	0.6
3.7	459	17	442	0.6
3.7	458	17	441	0.6
3.7	457	17	440	0.6
3.7	456	17	439	0.6
3.7	455	17	438	0.6
3.7	454	17	437	0.6
3.7	438	16	422	0.6
3.7	437	16	421	0.6
3.7	436	16	420	0.6
3.7	435	16	419	0.6
3.7	434	16	418	0.6
3.7	433	16	417	0.6
3.7	432	16	416	0.6
3.7	431	16	415	0.6
3.7	430	16	414	0.6
3.7	429	16	413	0.6
3.7	428	16	412	0.6
3.7	427	16	411	0.6
3.7	411	15	396	0.5
3.7	410	15	395	0.5
3.7	409	15	394	0.6
3.7	408	15	393	0.6
3.7	407	15	392	0.6
3.7	406	15	391	0.6
3.7	405	15	390	0.6
3.7	404	15	389	0.6
3.7	403	15	388	0.6

%E	M1	M2	DM	M*
3.7	402	15	387	0.6
3.7	401	15	386	0.6
3.7	383	14	369	0.5
3.7	382	14	368	0.5
3.7	381	14	367	0.5
3.7	380	14	366	0.5
3.7	379	14	365	0.5
3.7	378	14	364	0.5
3.7	377	14	363	0.5
3.7	376	14	362	0.5
3.7	375	14	361	0.5
3.7	374	14	360	0.5
3.7	356	13	343	0.5
3.7	355	13	342	0.5
3.7	354	13	341	0.5
3.7	353	13	340	0.5
3.7	352	13	339	0.5
3.7	351	13	338	0.5
3.7	350	13	337	0.5
3.7	349	13	336	0.5
3.7	348	13	335	0.5
3.7	347	13	334	0.5
3.7	328	12	316	0.4
3.7	327	12	315	0.4
3.7	326	12	314	0.4
3.7	325	12	313	0.4
3.7	324	12	312	0.4
3.7	323	12	311	0.4
3.7	322	12	310	0.4
3.7	321	12	309	0.4
3.7	301	11	290	0.4
3.7	300	11	289	0.4
3.7	299	11	288	0.4
3.7	298	11	287	0.4
3.7	297	11	286	0.4
3.7	296	11	285	0.4
3.7	295	11	284	0.4
3.7	294	11	283	0.4
3.7	274	10	264	0.4
3.7	273	10	263	0.4
3.7	272	10	262	0.4
3.7	271	10	261	0.4
3.7	270	10	260	0.4
3.7	269	10	259	0.4
3.7	268	10	258	0.4
3.7	267	10	257	0.4
3.7	246	9	237	0.3
3.7	245	9	236	0.3
3.7	244	9	235	0.3
3.7	243	9	234	0.3

%E	M1	M2	DM	M*
3.7	242	9	233	0.3
3.7	241	9	232	0.3
3.7	219	8	211	0.3
3.7	218	8	210	0.3
3.7	217	8	209	0.3
3.7	216	8	208	0.3
3.7	215	8	207	0.3
3.7	214	8	206	0.3
3.7	191	7	184	0.3
3.7	190	7	183	0.3
3.7	189	7	182	0.3
3.7	188	7	181	0.3
3.7	187	7	180	0.3
3.7	164	6	158	0.2
3.7	163	6	157	0.2
3.7	162	6	156	0.2
3.7	161	6	155	0.2
3.7	137	5	132	0.2
3.7	136	5	131	0.2
3.7	135	5	130	0.2
3.7	134	5	129	0.2
3.7	109	4	105	0.1
3.7	108	4	104	0.1
3.7	107	4	103	0.1
3.7	82	3	79	0.1
3.7	81	3	78	0.1
3.7	54	2	52	0.1
3.7	27	1	26	0.0
3.6	500	18	482	0.6
3.6	499	18	481	0.6
3.6	498	18	480	0.7
3.6	497	18	479	0.7
3.6	496	18	478	0.7
3.6	495	18	477	0.7
3.6	494	18	476	0.7
3.6	478	17	461	0.6
3.6	477	17	460	0.6
3.6	476	17	459	0.6
3.6	475	17	458	0.6
3.6	474	17	457	0.6
3.6	473	17	456	0.6
3.6	472	17	455	0.6
3.6	471	17	454	0.6
3.6	470	17	453	0.6
3.6	469	17	452	0.6
3.6	468	17	451	0.6
3.6	467	17	450	0.6
3.6	466	17	449	0.6
3.6	450	16	434	0.6
3.6	449	16	433	0.6

%E	M1	M2	DM	M*
3.6	448	16	432	0.6
3.6	447	16	431	0.6
3.6	446	16	430	0.6
3.6	445	16	429	0.6
3.6	444	16	428	0.6
3.6	443	16	427	0.6
3.6	442	16	426	0.6
3.6	441	16	425	0.6
3.6	440	16	424	0.6
3.6	439	16	423	0.6
3.6	422	15	407	0.5
3.6	421	15	406	0.5
3.6	420	15	405	0.5
3.6	419	15	404	0.5
3.6	418	15	403	0.5
3.6	417	15	402	0.5
3.6	416	15	401	0.5
3.6	415	15	400	0.5
3.6	414	15	399	0.5
3.6	413	15	398	0.5
3.6	412	15	397	0.5
3.6	394	14	380	0.5
3.6	393	14	379	0.5
3.6	392	14	378	0.5
3.6	391	14	377	0.5
3.6	390	14	376	0.5
3.6	389	14	375	0.5
3.6	388	14	374	0.5
3.6	387	14	373	0.5
3.6	386	14	372	0.5
3.6	385	14	371	0.5
3.6	384	14	370	0.5
3.6	366	13	353	0.5
3.6	365	13	352	0.5
3.6	364	13	351	0.5
3.6	363	13	350	0.5
3.6	362	13	349	0.5
3.6	361	13	348	0.5
3.6	360	13	347	0.5
3.6	359	13	346	0.5
3.6	358	13	345	0.5
3.6	357	13	344	0.5
3.6	338	12	326	0.4
3.6	337	12	325	0.4
3.6	336	12	324	0.4
3.6	335	12	323	0.4
3.6	334	12	322	0.4
3.6	333	12	321	0.4
3.6	332	12	320	0.4
3.6	331	12	319	0.4

%E	M1	M2	DM	M*
3.6	330	12	318	0.4
3.6	329	12	317	0.4
3.6	309	11	298	0.4
3.6	308	11	297	0.4
3.6	307	11	296	0.4
3.6	306	11	295	0.4
3.6	305	11	294	0.4
3.6	304	11	293	0.4
3.6	303	11	292	0.4
3.6	302	11	291	0.4
3.6	281	10	271	0.4
3.6	280	10	270	0.4
3.6	279	10	269	0.4
3.6	278	10	268	0.4
3.6	277	10	267	0.4
3.6	276	10	266	0.4
3.6	275	10	265	0.4
3.6	253	9	244	0.3
3.6	252	9	243	0.3
3.6	251	9	242	0.3
3.6	250	9	241	0.3
3.6	249	9	240	0.3
3.6	248	9	239	0.3
3.6	247	9	238	0.3
3.6	225	8	217	0.3
3.6	224	8	216	0.3
3.6	223	8	215	0.3
3.6	222	8	214	0.3
3.6	221	8	213	0.3
3.6	220	8	212	0.3
3.6	197	7	190	0.2
3.6	196	7	189	0.3
3.6	195	7	188	0.3
3.6	194	7	187	0.3
3.6	193	7	186	0.3
3.6	192	7	185	0.3
3.6	169	6	163	0.2
3.6	168	6	162	0.2
3.6	167	6	161	0.2
3.6	166	6	160	0.2
3.6	165	6	159	0.2
3.6	140	5	135	0.2
3.6	139	5	134	0.2
3.6	138	5	133	0.2
3.6	112	4	108	0.1
3.6	111	4	107	0.1
3.6	110	4	106	0.1
3.6	84	3	81	0.1
3.6	83	3	80	0.1
3.6	56	2	54	0.1

%E	M1	M2	DM	M*
3.6	55	2	53	0.1
3.6	28	1	27	0.0
3.5	492	17	475	0.6
3.5	491	17	474	0.6
3.5	490	17	473	0.6
3.5	489	17	472	0.6
3.5	488	17	471	0.6
3.5	487	17	470	0.6
3.5	486	17	469	0.6
3.5	485	17	468	0.6
3.5	484	17	467	0.6
3.5	483	17	466	0.6
3.5	482	17	465	0.6
3.5	481	17	464	0.6
3.5	480	17	463	0.6
3.5	479	17	462	0.6
3.5	463	16	447	0.6
3.5	462	16	446	0.6
3.5	461	16	445	0.6
3.5	460	16	444	0.6
3.5	459	16	443	0.6
3.5	458	16	442	0.6
3.5	457	16	441	0.6
3.5	456	16	440	0.6
3.5	455	16	439	0.6
3.5	454	16	438	0.6
3.5	453	16	437	0.6
3.5	452	16	436	0.6
3.5	451	16	435	0.6
3.5	434	15	419	0.5
3.5	433	15	418	0.5
3.5	432	15	417	0.5
3.5	431	15	416	0.5
3.5	430	15	415	0.5
3.5	429	15	414	0.5
3.5	428	15	413	0.5
3.5	427	15	412	0.5
3.5	426	15	411	0.5
3.5	425	15	410	0.5
3.5	424	15	409	0.5
3.5	423	15	408	0.5
3.5	405	14	391	0.5
3.5	404	14	390	0.5
3.5	403	14	389	0.5
3.5	402	14	388	0.5
3.5	401	14	387	0.5
3.5	400	14	386	0.5
3.5	399	14	385	0.5
3.5	398	14	384	0.5
3.5	397	14	383	0.5

%E	M1	M2	DM	M*
3.5	396	14	382	0.5
3.5	395	14	381	0.5
3.5	376	13	363	0.4
3.5	375	13	362	0.5
3.5	374	13	361	0.5
3.5	373	13	360	0.5
3.5	372	13	359	0.5
3.5	371	13	358	0.5
3.5	370	13	357	0.5
3.5	369	13	356	0.5
3.5	368	13	355	0.5
3.5	367	13	354	0.5
3.5	347	12	335	0.4
3.5	346	12	334	0.4
3.5	345	12	333	0.4
3.5	344	12	332	0.4
3.5	343	12	331	0.4
3.5	342	12	330	0.4
3.5	341	12	329	0.4
3.5	340	12	328	0.4
3.5	339	12	327	0.4
3.5	318	11	307	0.4
3.5	317	11	306	0.4
3.5	316	11	305	0.4
3.5	315	11	304	0.4
3.5	314	11	303	0.4
3.5	313	11	302	0.4
3.5	312	11	301	0.4
3.5	311	11	300	0.4
3.5	310	11	299	0.4
3.5	289	10	279	0.3
3.5	288	10	278	0.3
3.5	287	10	277	0.3
3.5	286	10	276	0.3
3.5	285	10	275	0.4
3.5	284	10	274	0.4
3.5	283	10	273	0.4
3.5	282	10	272	0.4
3.5	260	9	251	0.3
3.5	259	9	250	0.3
3.5	258	9	249	0.3
3.5	257	9	248	0.3
3.5	256	9	247	0.3
3.5	255	9	246	0.3
3.5	254	9	245	0.3
3.5	231	8	223	0.3
3.5	230	8	222	0.3
3.5	229	8	221	0.3
3.5	228	8	220	0.3
3.5	227	8	219	0.3

%E	M1	M2	DM	M*
3.5	226	8	218	0.3
3.5	202	7	195	0.2
3.5	201	7	194	0.2
3.5	200	7	193	0.2
3.5	199	7	192	0.2
3.5	198	7	191	0.2
3.5	173	6	167	0.2
3.5	172	6	166	0.2
3.5	171	6	165	0.2
3.5	170	6	164	0.2
3.5	144	5	139	0.2
3.5	143	5	138	0.2
3.5	142	5	137	0.2
3.5	141	5	136	0.2
3.5	115	4	111	0.1
3.5	114	4	110	0.1
3.5	113	4	109	0.1
3.5	86	3	83	0.1
3.5	85	3	82	0.1
3.5	57	2	55	0.1
3.4	500	17	483	0.6
3.4	499	17	482	0.6
3.4	498	17	481	0.6
3.4	497	17	480	0.6
3.4	496	17	479	0.6
3.4	495	17	478	0.6
3.4	494	17	477	0.6
3.4	493	17	476	0.6
3.4	477	16	461	0.5
3.4	476	16	460	0.5
3.4	475	16	459	0.5
3.4	474	16	458	0.5
3.4	473	16	457	0.5
3.4	472	16	456	0.5
3.4	471	16	455	0.5
3.4	470	16	454	0.5
3.4	469	16	453	0.5
3.4	468	16	452	0.5
3.4	467	16	451	0.5
3.4	466	16	450	0.5
3.4	465	16	449	0.6
3.4	464	16	448	0.6
3.4	447	15	432	0.5
3.4	446	15	431	0.5
3.4	445	15	430	0.5
3.4	444	15	429	0.5
3.4	443	15	428	0.5
3.4	442	15	427	0.5
3.4	441	15	426	0.5
3.4	440	15	425	0.5

%E	M1	M2	DM	M*
3.4	439	15	424	0.5
3.4	438	15	423	0.5
3.4	437	15	422	0.5
3.4	436	15	421	0.5
3.4	435	15	420	0.5
3.4	417	14	403	0.5
3.4	416	14	402	0.5
3.4	415	14	401	0.5
3.4	414	14	400	0.5
3.4	413	14	399	0.5
3.4	412	14	398	0.5
3.4	411	14	397	0.5
3.4	410	14	396	0.5
3.4	409	14	395	0.5
3.4	408	14	394	0.5
3.4	407	14	393	0.5
3.4	406	14	392	0.5
3.4	388	13	375	0.4
3.4	387	13	374	0.4
3.4	386	13	373	0.4
3.4	385	13	372	0.4
3.4	384	13	371	0.4
3.4	383	13	370	0.4
3.4	382	13	369	0.4
3.4	381	13	368	0.4
3.4	380	13	367	0.4
3.4	379	13	366	0.4
3.4	378	13	365	0.4
3.4	377	13	364	0.4
3.4	358	12	346	0.4
3.4	357	12	345	0.4
3.4	356	12	344	0.4
3.4	355	12	343	0.4
3.4	354	12	342	0.4
3.4	353	12	341	0.4
3.4	352	12	340	0.4
3.4	351	12	339	0.4
3.4	350	12	338	0.4
3.4	349	12	337	0.4
3.4	348	12	336	0.4
3.4	328	11	317	0.4
3.4	327	11	316	0.4
3.4	326	11	315	0.4
3.4	325	11	314	0.4
3.4	324	11	313	0.4
3.4	323	11	312	0.4
3.4	322	11	311	0.4
3.4	321	11	310	0.4
3.4	320	11	309	0.4
3.4	319	11	308	0.4

%E	M1	M2	DM	M*
3.4	298	10	288	0.3
3.4	297	10	287	0.3
3.4	296	10	286	0.3
3.4	295	10	285	0.3
3.4	294	10	284	0.3
3.4	293	10	283	0.3
3.4	292	10	282	0.3
3.4	291	10	281	0.3
3.4	290	10	280	0.3
3.4	268	9	259	0.3
3.4	267	9	258	0.3
3.4	266	9	257	0.3
3.4	265	9	256	0.3
3.4	264	9	255	0.3
3.4	263	9	254	0.3
3.4	262	9	253	0.3
3.4	261	9	252	0.3
3.4	238	8	230	0.3
3.4	237	8	229	0.3
3.4	236	8	228	0.3
3.4	235	8	227	0.3
3.4	234	8	226	0.3
3.4	233	8	225	0.3
3.4	232	8	224	0.3
3.4	208	7	201	0.2
3.4	207	7	200	0.2
3.4	206	7	199	0.2
3.4	205	7	198	0.2
3.4	204	7	197	0.2
3.4	203	7	196	0.2
3.4	179	6	173	0.2
3.4	178	6	172	0.2
3.4	177	6	171	0.2
3.4	176	6	170	0.2
3.4	175	6	169	0.2
3.4	174	6	168	0.2
3.4	149	5	144	0.2
3.4	148	5	143	0.2
3.4	147	5	142	0.2
3.4	146	5	141	0.2
3.4	145	5	140	0.2
3.4	119	4	115	0.1
3.4	118	4	114	0.1
3.4	117	4	113	0.1
3.4	116	4	112	0.1
3.4	89	3	86	0.1
3.4	88	3	85	0.1
3.4	87	3	84	0.1
3.4	59	2	57	0.1
3.4	58	2	56	0.1

%E	M1	M2	DM	M*
3.4	29	1	28	0.0
3.3	492	16	476	0.5
3.3	491	16	475	0.5
3.3	490	16	474	0.5
3.3	489	16	473	0.5
3.3	488	16	472	0.5
3.3	487	16	471	0.5
3.3	486	16	470	0.5
3.3	485	16	469	0.5
3.3	484	16	468	0.5
3.3	483	16	467	0.5
3.3	482	16	466	0.5
3.3	481	16	465	0.5
3.3	480	16	464	0.5
3.3	479	16	463	0.5
3.3	478	16	462	0.5
3.3	461	15	446	0.5
3.3	460	15	445	0.5
3.3	459	15	444	0.5
3.3	458	15	443	0.5
3.3	457	15	442	0.5
3.3	456	15	441	0.5
3.3	455	15	440	0.5
3.3	454	15	439	0.5
3.3	453	15	438	0.5
3.3	452	15	437	0.5
3.3	451	15	436	0.5
3.3	450	15	435	0.5
3.3	449	15	434	0.5
3.3	448	15	433	0.5
3.3	430	14	416	0.5
3.3	429	14	415	0.5
3.3	428	14	414	0.5
3.3	427	14	413	0.5
3.3	426	14	412	0.5
3.3	425	14	411	0.5
3.3	424	14	410	0.5
3.3	423	14	409	0.5
3.3	422	14	408	0.5
3.3	421	14	407	0.5
3.3	420	14	406	0.5
3.3	419	14	405	0.5
3.3	418	14	404	0.5
3.3	400	13	387	0.4
3.3	399	13	386	0.4
3.3	398	13	385	0.4
3.3	397	13	384	0.4
3.3	396	13	383	0.4
3.3	395	13	382	0.4
3.3	394	13	381	0.4

%E	M1	M2	DM	M*
3.3	393	13	380	0.4
3.3	392	13	379	0.4
3.3	391	13	378	0.4
3.3	390	13	377	0.4
3.3	389	13	376	0.4
3.3	369	12	357	0.4
3.3	368	12	356	0.4
3.3	367	12	355	0.4
3.3	366	12	354	0.4
3.3	365	12	353	0.4
3.3	364	12	352	0.4
3.3	363	12	351	0.4
3.3	362	12	350	0.4
3.3	361	12	349	0.4
3.3	360	12	348	0.4
3.3	359	12	347	0.4
3.3	338	11	327	0.4
3.3	337	11	326	0.4
3.3	336	11	325	0.4
3.3	335	11	324	0.4
3.3	334	11	323	0.4
3.3	333	11	322	0.4
3.3	332	11	321	0.4
3.3	331	11	320	0.4
3.3	330	11	319	0.4
3.3	329	11	318	0.4
3.3	307	10	297	0.3
3.3	306	10	296	0.3
3.3	305	10	295	0.3
3.3	304	10	294	0.3
3.3	303	10	293	0.3
3.3	302	10	292	0.3
3.3	301	10	291	0.3
3.3	300	10	290	0.3
3.3	299	10	289	0.3
3.3	276	9	267	0.3
3.3	275	9	266	0.3
3.3	274	9	265	0.3
3.3	273	9	264	0.3
3.3	272	9	263	0.3
3.3	271	9	262	0.3
3.3	270	9	261	0.3
3.3	269	9	260	0.3
3.3	246	8	238	0.3
3.3	245	8	237	0.3
3.3	244	8	236	0.3
3.3	243	8	235	0.3
3.3	242	8	234	0.3
3.3	241	8	233	0.3
3.3	240	8	232	0.3

%E	M1	M2	DM	M*
3.3	239	8	231	0.3
3.3	215	7	208	0.2
3.3	214	7	207	0.2
3.3	213	7	206	0.2
3.3	212	7	205	0.2
3.3	211	7	204	0.2
3.3	210	7	203	0.2
3.3	209	7	202	0.2
3.3	184	6	178	0.2
3.3	183	6	177	0.2
3.3	182	6	176	0.2
3.3	181	6	175	0.2
3.3	180	6	174	0.2
3.3	153	5	148	0.2
3.3	152	5	147	0.2
3.3	151	5	146	0.2
3.3	150	5	145	0.2
3.3	123	4	119	0.1
3.3	122	4	118	0.1
3.3	121	4	117	0.1
3.3	120	4	116	0.1
3.3	92	3	89	0.1
3.3	91	3	88	0.1
3.3	90	3	87	0.1
3.3	61	2	59	0.1
3.3	60	2	58	0.1
3.3	30	1	29	0.0
3.2	500	16	484	0.5
3.2	499	16	483	0.5
3.2	498	16	482	0.5
3.2	497	16	481	0.5
3.2	496	16	480	0.5
3.2	495	16	479	0.5
3.2	494	16	478	0.5
3.2	493	16	477	0.5
3.2	476	15	461	0.5
3.2	475	15	460	0.5
3.2	474	15	459	0.5
3.2	473	15	458	0.5
3.2	472	15	457	0.5
3.2	471	15	456	0.5
3.2	470	15	455	0.5
3.2	469	15	454	0.5
3.2	468	15	453	0.5
3.2	467	15	452	0.5
3.2	466	15	451	0.5
3.2	465	15	450	0.5
3.2	464	15	449	0.5
3.2	463	15	448	0.5
3.2	462	15	447	0.5

%E	M1	M2	DM	M*
3.2	444	14	430	0.4
3.2	443	14	429	0.4
3.2	442	14	428	0.4
3.2	441	14	427	0.4
3.2	440	14	426	0.4
3.2	439	14	425	0.4
3.2	438	14	424	0.4
3.2	437	14	423	0.4
3.2	436	14	422	0.4
3.2	435	14	421	0.5
3.2	434	14	420	0.5
3.2	433	14	419	0.5
3.2	432	14	418	0.5
3.2	431	14	417	0.5
3.2	412	13	399	0.4
3.2	411	13	398	0.4
3.2	410	13	397	0.4
3.2	409	13	396	0.4
3.2	408	13	395	0.4
3.2	407	13	394	0.4
3.2	406	13	393	0.4
3.2	405	13	392	0.4
3.2	404	13	391	0.4
3.2	403	13	390	0.4
3.2	402	13	389	0.4
3.2	401	13	388	0.4
3.2	381	12	369	0.4
3.2	380	12	368	0.4
3.2	379	12	367	0.4
3.2	378	12	366	0.4
3.2	377	12	365	0.4
3.2	376	12	364	0.4
3.2	375	12	363	0.4
3.2	374	12	362	0.4
3.2	373	12	361	0.4
3.2	372	12	360	0.4
3.2	371	12	359	0.4
3.2	370	12	358	0.4
3.2	349	11	338	0.3
3.2	348	11	337	0.3
3.2	347	11	336	0.3
3.2	346	11	335	0.3
3.2	345	11	334	0.4
3.2	344	11	333	0.4
3.2	343	11	332	0.4
3.2	342	11	331	0.4
3.2	341	11	330	0.4
3.2	340	11	329	0.4
3.2	339	11	328	0.4
3.2	317	10	307	0.3

%E	M1	M2	DM	M*
3•2	316	10	306	0•3
3•2	315	10	305	0•3
3•2	314	10	304	0•3
3•2	313	10	303	0•3
3•2	312	10	302	0•3
3•2	311	10	301	0•3
3•2	310	10	300	0•3
3•2	309	10	299	0•3
3•2	308	10	298	0•3
3•2	285	9	276	0•3
3•2	284	9	275	0•3
3•2	283	9	274	0•3
3•2	282	9	273	0•3
3•2	281	9	272	0•3
3•2	280	9	271	0•3
3•2	279	9	270	0•3
3•2	278	9	269	0•3
3•2	277	9	268	0•3
3•2	254	8	246	0•3
3•2	253	8	245	0•3
3•2	252	8	244	0•3
3•2	251	8	243	0•3
3•2	250	8	242	0•3
3•2	249	8	241	0•3
3•2	248	8	240	0•3
3•2	247	8	239	0•3
3•2	222	7	215	0•2
3•2	221	7	214	0•2
3•2	220	7	213	0•2
3•2	219	7	212	0•2
3•2	218	7	211	0•2
3•2	217	7	210	0•2
3•2	216	7	209	0•2
3•2	190	6	184	0•2
3•2	189	6	183	0•2
3•2	188	6	182	0•2
3•2	187	6	181	0•2
3•2	186	6	180	0•2
3•2	185	6	179	0•2
3•2	158	5	153	0•2
3•2	157	5	152	0•2
3•2	156	5	151	0•2
3•2	155	5	150	0•2
3•2	154	5	149	0•2
3•2	127	4	123	0•1
3•2	126	4	122	0•1
3•2	125	4	121	0•1
3•2	124	4	120	0•1
3•2	95	3	92	0•1
3•2	94	3	91	0•1

%E	M1	M2	DM	M*
3•2	93	3	90	0•1
3•2	63	2	61	0•1
3•2	62	2	60	0•1
3•2	31	1	30	0•0
3•1	491	15	476	0•5
3•1	490	15	475	0•5
3•1	489	15	474	0•5
3•1	488	15	473	0•5
3•1	487	15	472	0•5
3•1	486	15	471	0•5
3•1	485	15	470	0•5
3•1	484	15	469	0•5
3•1	483	15	468	0•5
3•1	482	15	467	0•5
3•1	481	15	466	0•5
3•1	480	15	465	0•5
3•1	479	15	464	0•5
3•1	478	15	463	0•5
3•1	477	15	462	0•5
3•1	459	14	445	0•4
3•1	458	14	444	0•4
3•1	457	14	443	0•4
3•1	456	14	442	0•4
3•1	455	14	441	0•4
3•1	454	14	440	0•4
3•1	453	14	439	0•4
3•1	452	14	438	0•4
3•1	451	14	437	0•4
3•1	450	14	436	0•4
3•1	449	14	435	0•4
3•1	448	14	434	0•4
3•1	447	14	433	0•4
3•1	446	14	432	0•4
3•1	445	14	431	0•4
3•1	426	13	413	0•4
3•1	425	13	412	0•4
3•1	424	13	411	0•4
3•1	423	13	410	0•4
3•1	422	13	409	0•4
3•1	421	13	408	0•4
3•1	420	13	407	0•4
3•1	419	13	406	0•4
3•1	418	13	405	0•4
3•1	417	13	404	0•4
3•1	416	13	403	0•4
3•1	415	13	402	0•4
3•1	414	13	401	0•4
3•1	413	13	400	0•4
3•1	393	12	381	0•4
3•1	392	12	380	0•4

%E	M1	M2	DM	M*
3•1	391	12	379	0•4
3•1	390	12	378	0•4
3•1	389	12	377	0•4
3•1	388	12	376	0•4
3•1	387	12	375	0•4
3•1	386	12	374	0•4
3•1	385	12	373	0•4
3•1	384	12	372	0•4
3•1	383	12	371	0•4
3•1	382	12	370	0•4
3•1	360	11	349	0•3
3•1	359	11	348	0•3
3•1	358	11	347	0•3
3•1	357	11	346	0•3
3•1	356	11	345	0•3
3•1	355	11	344	0•3
3•1	354	11	343	0•3
3•1	353	11	342	0•3
3•1	352	11	341	0•3
3•1	351	11	340	0•3
3•1	350	11	339	0•3
3•1	327	10	317	0•3
3•1	326	10	316	0•3
3•1	325	10	315	0•3
3•1	324	10	314	0•3
3•1	323	10	313	0•3
3•1	322	10	312	0•3
3•1	321	10	311	0•3
3•1	320	10	310	0•3
3•1	319	10	309	0•3
3•1	318	10	308	0•3
3•1	295	9	286	0•3
3•1	294	9	285	0•3
3•1	293	9	284	0•3
3•1	292	9	283	0•3
3•1	291	9	282	0•3
3•1	290	9	281	0•3
3•1	289	9	280	0•3
3•1	288	9	279	0•3
3•1	287	9	278	0•3
3•1	286	9	277	0•3
3•1	262	8	254	0•2
3•1	261	8	253	0•2
3•1	260	8	252	0•2
3•1	259	8	251	0•2
3•1	258	8	250	0•2
3•1	257	8	249	0•2
3•1	256	8	248	0•3
3•1	255	8	247	0•3
3•1	229	7	222	0•2

%E	M1	M2	DM	M*
3•1	228	7	221	0•2
3•1	227	7	220	0•2
3•1	226	7	219	0•2
3•1	225	7	218	0•2
3•1	224	7	217	0•2
3•1	223	7	216	0•2
3•1	196	6	190	0•2
3•1	195	6	189	0•2
3•1	194	6	188	0•2
3•1	193	6	187	0•2
3•1	192	6	186	0•2
3•1	191	6	185	0•2
3•1	163	5	158	0•2
3•1	162	5	157	0•2
3•1	161	5	156	0•2
3•1	160	5	155	0•2
3•1	159	5	154	0•2
3•1	131	4	127	0•1
3•1	130	4	126	0•1
3•1	129	4	125	0•1
3•1	128	4	124	0•1
3•1	98	3	95	0•1
3•1	97	3	94	0•1
3•1	96	3	93	0•1
3•1	65	2	63	0•1
3•1	64	2	62	0•1
3•1	32	1	31	0•0
3•0	500	15	485	0•4
3•0	499	15	484	0•5
3•0	498	15	483	0•5
3•0	497	15	482	0•5
3•0	496	15	481	0•5
3•0	495	15	480	0•5
3•0	494	15	479	0•5
3•0	493	15	478	0•5
3•0	492	15	477	0•5
3•0	474	14	460	0•4
3•0	473	14	459	0•4
3•0	472	14	458	0•4
3•0	471	14	457	0•4
3•0	470	14	456	0•4
3•0	469	14	455	0•4
3•0	468	14	454	0•4
3•0	467	14	453	0•4
3•0	466	14	452	0•4
3•0	465	14	451	0•4
3•0	464	14	450	0•4
3•0	463	14	449	0•4
3•0	462	14	448	0•4
3•0	461	14	447	0•4

%E	M1	M2	DM	M*
3•0	460	14	446	0•4
3•0	440	13	427	0•4
3•0	439	13	426	0•4
3•0	438	13	425	0•4
3•0	437	13	424	0•4
3•0	436	13	423	0•4
3•0	435	13	422	0•4
3•0	434	13	421	0•4
3•0	433	13	420	0•4
3•0	432	13	419	0•4
3•0	431	13	418	0•4
3•0	430	13	417	0•4
3•0	429	13	416	0•4
3•0	428	13	415	0•4
3•0	427	13	414	0•4
3•0	406	12	394	0•4
3•0	405	12	393	0•4
3•0	404	12	392	0•4
3•0	403	12	391	0•4
3•0	402	12	390	0•4
3•0	401	12	389	0•4
3•0	400	12	388	0•4
3•0	399	12	387	0•4
3•0	398	12	386	0•4
3•0	397	12	385	0•4
3•0	396	12	384	0•4
3•0	395	12	383	0•4
3•0	394	12	382	0•4
3•0	372	11	361	0•3
3•0	371	11	360	0•3
3•0	370	11	359	0•3
3•0	369	11	358	0•3
3•0	368	11	357	0•3
3•0	367	11	356	0•3
3•0	366	11	355	0•3
3•0	365	11	354	0•3
3•0	364	11	353	0•3
3•0	363	11	352	0•3
3•0	362	11	351	0•3
3•0	361	11	350	0•3
3•0	339	10	329	0•3
3•0	338	10	328	0•3
3•0	337	10	327	0•3
3•0	336	10	326	0•3
3•0	335	10	325	0•3
3•0	334	10	324	0•3
3•0	333	10	323	0•3
3•0	332	10	322	0•3
3•0	331	10	321	0•3
3•0	330	10	320	0•3

%E	M1	M2	DM	M*
3.0	329	10	319	0.3
3.0	328	10	318	0.3
3.0	305	9	296	0.3
3.0	304	9	295	0.3
3.0	303	9	294	0.3
3.0	302	9	293	0.3
3.0	301	9	292	0.3
3.0	300	9	291	0.3
3.0	299	9	290	0.3
3.0	298	9	289	0.3
3.0	297	9	288	0.3
3.0	296	9	287	0.3
3.0	271	8	263	0.2
3.0	270	8	262	0.2
3.0	269	8	261	0.2
3.0	268	8	260	0.2
3.0	267	8	259	0.2
3.0	266	8	258	0.2
3.0	265	8	257	0.2
3.0	264	8	256	0.2
3.0	263	8	255	0.2
3.0	237	7	230	0.2
3.0	236	7	229	0.2
3.0	235	7	228	0.2
3.0	234	7	227	0.2
3.0	233	7	226	0.2
3.0	232	7	225	0.2
3.0	231	7	224	0.2
3.0	230	7	223	0.2
3.0	203	6	197	0.2
3.0	202	6	196	0.2
3.0	201	6	195	0.2
3.0	200	6	194	0.2
3.0	199	6	193	0.2
3.0	198	6	192	0.2
3.0	197	6	191	0.2
3.0	169	5	164	0.1
3.0	168	5	163	0.1
3.0	167	5	162	0.1
3.0	166	5	161	0.2
3.0	165	5	160	0.2
3.0	164	5	159	0.2
3.0	135	4	131	0.1
3.0	134	4	130	0.1
3.0	133	4	129	0.1
3.0	132	4	128	0.1
3.0	101	3	98	0.1
3.0	100	3	97	0.1
3.0	99	3	96	0.1
3.0	67	2	65	0.1

%E	M1	M2	DM	M*
3.0	66	2	64	0.1
3.0	33	1	32	0.0
2.9	491	14	477	0.4
2.9	490	14	476	0.4
2.9	489	14	475	0.4
2.9	488	14	474	0.4
2.9	487	14	473	0.4
2.9	486	14	472	0.4
2.9	485	14	471	0.4
2.9	484	14	470	0.4
2.9	483	14	469	0.4
2.9	482	14	468	0.4
2.9	481	14	467	0.4
2.9	480	14	466	0.4
2.9	479	14	465	0.4
2.9	478	14	464	0.4
2.9	477	14	463	0.4
2.9	476	14	462	0.4
2.9	475	14	461	0.4
2.9	456	13	443	0.4
2.9	455	13	442	0.4
2.9	454	13	441	0.4
2.9	453	13	440	0.4
2.9	452	13	439	0.4
2.9	451	13	438	0.4
2.9	450	13	437	0.4
2.9	449	13	436	0.4
2.9	448	13	435	0.4
2.9	447	13	434	0.4
2.9	446	13	433	0.4
2.9	445	13	432	0.4
2.9	444	13	431	0.4
2.9	443	13	430	0.4
2.9	442	13	429	0.4
2.9	441	13	428	0.4
2.9	421	12	409	0.3
2.9	420	12	408	0.3
2.9	419	12	407	0.3
2.9	418	12	406	0.3
2.9	417	12	405	0.3
2.9	416	12	404	0.3
2.9	415	12	403	0.3
2.9	414	12	402	0.3
2.9	413	12	401	0.3
2.9	412	12	400	0.3
2.9	411	12	399	0.4
2.9	410	12	398	0.4
2.9	409	12	397	0.4
2.9	408	12	396	0.4
2.9	407	12	395	0.4

%E	M1	M2	DM	M*
2.9	386	11	375	0.3
2.9	385	11	374	0.3
2.9	384	11	373	0.3
2.9	383	11	372	0.3
2.9	382	11	371	0.3
2.9	381	11	370	0.3
2.9	380	11	369	0.3
2.9	379	11	368	0.3
2.9	378	11	367	0.3
2.9	377	11	366	0.3
2.9	376	11	365	0.3
2.9	375	11	364	0.3
2.9	374	11	363	0.3
2.9	373	11	362	0.3
2.9	350	10	340	0.3
2.9	349	10	339	0.3
2.9	348	10	338	0.3
2.9	347	10	337	0.3
2.9	346	10	336	0.3
2.9	345	10	335	0.3
2.9	344	10	334	0.3
2.9	343	10	333	0.3
2.9	342	10	332	0.3
2.9	341	10	331	0.3
2.9	340	10	330	0.3
2.9	315	9	306	0.3
2.9	314	9	305	0.3
2.9	313	9	304	0.3
2.9	312	9	303	0.3
2.9	311	9	302	0.3
2.9	310	9	301	0.3
2.9	309	9	300	0.3
2.9	308	9	299	0.3
2.9	307	9	298	0.3
2.9	306	9	297	0.3
2.9	280	8	272	0.2
2.9	279	8	271	0.2
2.9	278	8	270	0.2
2.9	277	8	269	0.2
2.9	276	8	268	0.2
2.9	275	8	267	0.2
2.9	274	8	266	0.2
2.9	273	8	265	0.2
2.9	272	8	264	0.2
2.9	245	7	238	0.2
2.9	244	7	237	0.2
2.9	243	7	236	0.2
2.9	242	7	235	0.2
2.9	241	7	234	0.2
2.9	240	7	233	0.2

%E	M1	M2	DM	M*
2.9	239	7	232	0.2
2.9	238	7	231	0.2
2.9	210	6	204	0.2
2.9	209	6	203	0.2
2.9	208	6	202	0.2
2.9	207	6	201	0.2
2.9	206	6	200	0.2
2.9	205	6	199	0.2
2.9	204	6	198	0.2
2.9	175	5	170	0.1
2.9	174	5	169	0.1
2.9	173	5	168	0.1
2.9	172	5	167	0.1
2.9	171	5	166	0.1
2.9	170	5	165	0.1
2.9	140	4	136	0.1
2.9	139	4	135	0.1
2.9	138	4	134	0.1
2.9	137	4	133	0.1
2.9	136	4	132	0.1
2.9	105	3	102	0.1
2.9	104	3	101	0.1
2.9	103	3	100	0.1
2.9	102	3	99	0.1
2.9	70	2	68	0.1
2.9	69	2	67	0.1
2.9	68	2	66	0.1
2.9	35	1	34	0.0
2.9	34	1	33	0.0
2.8	500	14	486	0.4
2.8	499	14	485	0.4
2.8	498	14	484	0.4
2.8	497	14	483	0.4
2.8	496	14	482	0.4
2.8	495	14	481	0.4
2.8	494	14	480	0.4
2.8	493	14	479	0.4
2.8	492	14	478	0.4
2.8	472	13	459	0.4
2.8	471	13	458	0.4
2.8	470	13	457	0.4
2.8	469	13	456	0.4
2.8	468	13	455	0.4
2.8	467	13	454	0.4
2.8	466	13	453	0.4
2.8	465	13	452	0.4
2.8	464	13	451	0.4
2.8	463	13	450	0.4
2.8	462	13	449	0.4
2.8	461	13	448	0.4

%E	M1	M2	DM	M*
2.8	460	13	447	0.4
2.8	459	13	446	0.4
2.8	458	13	445	0.4
2.8	457	13	444	0.4
2.8	436	12	424	0.3
2.8	435	12	423	0.3
2.8	434	12	422	0.3
2.8	433	12	421	0.3
2.8	432	12	420	0.3
2.8	431	12	419	0.3
2.8	430	12	418	0.3
2.8	429	12	417	0.3
2.8	428	12	416	0.3
2.8	427	12	415	0.3
2.8	426	12	414	0.3
2.8	425	12	413	0.3
2.8	424	12	412	0.3
2.8	423	12	411	0.3
2.8	422	12	410	0.3
2.8	400	11	389	0.3
2.8	399	11	388	0.3
2.8	398	11	387	0.3
2.8	397	11	386	0.3
2.8	396	11	385	0.3
2.8	395	11	384	0.3
2.8	394	11	383	0.3
2.8	393	11	382	0.3
2.8	392	11	381	0.3
2.8	391	11	380	0.3
2.8	390	11	379	0.3
2.8	389	11	378	0.3
2.8	388	11	377	0.3
2.8	387	11	376	0.3
2.8	363	10	353	0.3
2.8	362	10	352	0.3
2.8	361	10	351	0.3
2.8	360	10	350	0.3
2.8	359	10	349	0.3
2.8	358	10	348	0.3
2.8	357	10	347	0.3
2.8	356	10	346	0.3
2.8	355	10	345	0.3
2.8	354	10	344	0.3
2.8	353	10	343	0.3
2.8	352	10	342	0.3
2.8	351	10	341	0.3
2.8	327	9	318	0.2
2.8	326	9	317	0.2
2.8	325	9	316	0.2
2.8	324	9	315	0.3

%E	M1	M2	DM	M*
2.8	323	9	314	0.3
2.8	322	9	313	0.3
2.8	321	9	312	0.3
2.8	320	9	311	0.3
2.8	319	9	310	0.3
2.8	318	9	309	0.3
2.8	317	9	308	0.3
2.8	316	9	307	0.3
2.8	290	8	282	0.2
2.8	289	8	281	0.2
2.8	288	8	280	0.2
2.8	287	8	279	0.2
2.8	286	8	278	0.2
2.8	285	8	277	0.2
2.8	284	8	276	0.2
2.8	283	8	275	0.2
2.8	282	8	274	0.2
2.8	281	8	273	0.2
2.8	254	7	247	0.2
2.8	253	7	246	0.2
2.8	252	7	245	0.2
2.8	251	7	244	0.2
2.8	250	7	243	0.2
2.8	249	7	242	0.2
2.8	248	7	241	0.2
2.8	247	7	240	0.2
2.8	246	7	239	0.2
2.8	218	6	212	0.2
2.8	217	6	211	0.2
2.8	216	6	210	0.2
2.8	215	6	209	0.2
2.8	214	6	208	0.2
2.8	213	6	207	0.2
2.8	212	6	206	0.2
2.8	211	6	205	0.2
2.8	181	5	176	0.1
2.8	180	5	175	0.1
2.8	179	5	174	0.1
2.8	178	5	173	0.1
2.8	177	5	172	0.1
2.8	176	5	171	0.1
2.8	145	4	141	0.1
2.8	144	4	140	0.1
2.8	143	4	139	0.1
2.8	142	4	138	0.1
2.8	141	4	137	0.1
2.8	109	3	106	0.1
2.8	108	3	105	0.1
2.8	107	3	104	0.1
2.8	106	3	103	0.1

%E	M1	M2	DM	M*
2.8	72	2	70	0.1
2.8	71	2	69	0.1
2.8	36	1	35	0.0
2.7	490	13	477	0.3
2.7	489	13	476	0.3
2.7	488	13	475	0.3
2.7	487	13	474	0.3
2.7	486	13	473	0.3
2.7	485	13	472	0.3
2.7	484	13	471	0.3
2.7	483	13	470	0.3
2.7	482	13	469	0.4
2.7	481	13	468	0.4
2.7	480	13	467	0.4
2.7	479	13	466	0.4
2.7	478	13	465	0.4
2.7	477	13	464	0.4
2.7	476	13	463	0.4
2.7	475	13	462	0.4
2.7	474	13	461	0.4
2.7	473	13	460	0.4
2.7	452	12	440	0.3
2.7	451	12	439	0.3
2.7	450	12	438	0.3
2.7	449	12	437	0.3
2.7	448	12	436	0.3
2.7	447	12	435	0.3
2.7	446	12	434	0.3
2.7	445	12	433	0.3
2.7	444	12	432	0.3
2.7	443	12	431	0.3
2.7	442	12	430	0.3
2.7	441	12	429	0.3
2.7	440	12	428	0.3
2.7	439	12	427	0.3
2.7	438	12	426	0.3
2.7	437	12	425	0.3
2.7	415	11	404	0.3
2.7	414	11	403	0.3
2.7	413	11	402	0.3
2.7	412	11	401	0.3
2.7	411	11	400	0.3
2.7	410	11	399	0.3
2.7	409	11	398	0.3
2.7	408	11	397	0.3
2.7	407	11	396	0.3
2.7	406	11	395	0.3
2.7	405	11	394	0.3
2.7	404	11	393	0.3
2.7	403	11	392	0.3

%E	M1	M2	DM	M*
2.7	402	11	391	0.3
2.7	401	11	390	0.3
2.7	377	10	367	0.3
2.7	376	10	366	0.3
2.7	375	10	365	0.3
2.7	374	10	364	0.3
2.7	373	10	363	0.3
2.7	372	10	362	0.3
2.7	371	10	361	0.3
2.7	370	10	360	0.3
2.7	369	10	359	0.3
2.7	368	10	358	0.3
2.7	367	10	357	0.3
2.7	366	10	356	0.3
2.7	365	10	355	0.3
2.7	364	10	354	0.3
2.7	339	9	330	0.2
2.7	338	9	329	0.2
2.7	337	9	328	0.2
2.7	336	9	327	0.2
2.7	335	9	326	0.2
2.7	334	9	325	0.2
2.7	333	9	324	0.2
2.7	332	9	323	0.2
2.7	331	9	322	0.2
2.7	330	9	321	0.2
2.7	329	9	320	0.2
2.7	328	9	319	0.2
2.7	301	8	293	0.2
2.7	300	8	292	0.2
2.7	299	8	291	0.2
2.7	298	8	290	0.2
2.7	297	8	289	0.2
2.7	296	8	288	0.2
2.7	295	8	287	0.2
2.7	294	8	286	0.2
2.7	293	8	285	0.2
2.7	292	8	284	0.2
2.7	291	8	283	0.2
2.7	264	7	257	0.2
2.7	263	7	256	0.2
2.7	262	7	255	0.2
2.7	261	7	254	0.2
2.7	260	7	253	0.2
2.7	259	7	252	0.2
2.7	258	7	251	0.2
2.7	257	7	250	0.2
2.7	256	7	249	0.2
2.7	255	7	248	0.2
2.7	226	6	220	0.2

%E	M1	M2	DM	M*
2.7	225	6	219	0.2
2.7	224	6	218	0.2
2.7	223	6	217	0.2
2.7	222	6	216	0.2
2.7	221	6	215	0.2
2.7	220	6	214	0.2
2.7	219	6	213	0.2
2.7	188	5	183	0.1
2.7	187	5	182	0.1
2.7	186	5	181	0.1
2.7	185	5	180	0.1
2.7	184	5	179	0.1
2.7	183	5	178	0.1
2.7	182	5	177	0.1
2.7	150	4	146	0.1
2.7	149	4	145	0.1
2.7	148	4	144	0.1
2.7	147	4	143	0.1
2.7	146	4	142	0.1
2.7	113	3	110	0.1
2.7	112	3	109	0.1
2.7	111	3	108	0.1
2.7	110	3	107	0.1
2.7	75	2	73	0.1
2.7	74	2	72	0.1
2.7	73	2	71	0.1
2.7	37	1	36	0.0
2.6	500	13	487	0.3
2.6	499	13	486	0.3
2.6	498	13	485	0.3
2.6	497	13	484	0.3
2.6	496	13	483	0.3
2.6	495	13	482	0.3
2.6	494	13	481	0.3
2.6	493	13	480	0.3
2.6	492	13	479	0.3
2.6	491	13	478	0.3
2.6	470	12	458	0.3
2.6	469	12	457	0.3
2.6	468	12	456	0.3
2.6	467	12	455	0.3
2.6	466	12	454	0.3
2.6	465	12	453	0.3
2.6	464	12	452	0.3
2.6	463	12	451	0.3
2.6	462	12	450	0.3
2.6	461	12	449	0.3
2.6	460	12	448	0.3
2.6	459	12	447	0.3
2.6	458	12	446	0.3

%E	M1	M2	DM	M*
2.6	457	12	445	0.3
2.6	456	12	444	0.3
2.6	455	12	443	0.3
2.6	454	12	442	0.3
2.6	453	12	441	0.3
2.6	431	11	420	0.3
2.6	430	11	419	0.3
2.6	429	11	418	0.3
2.6	428	11	417	0.3
2.6	427	11	416	0.3
2.6	426	11	415	0.3
2.6	425	11	414	0.3
2.6	424	11	413	0.3
2.6	423	11	412	0.3
2.6	422	11	411	0.3
2.6	421	11	410	0.3
2.6	420	11	409	0.3
2.6	419	11	408	0.3
2.6	418	11	407	0.3
2.6	417	11	406	0.3
2.6	416	11	405	0.3
2.6	392	10	382	0.3
2.6	391	10	381	0.3
2.6	390	10	380	0.3
2.6	389	10	379	0.3
2.6	388	10	378	0.3
2.6	387	10	377	0.3
2.6	386	10	376	0.3
2.6	385	10	375	0.3
2.6	384	10	374	0.3
2.6	383	10	373	0.3
2.6	382	10	372	0.3
2.6	381	10	371	0.3
2.6	380	10	370	0.3
2.6	379	10	369	0.3
2.6	378	10	368	0.3
2.6	353	9	344	0.2
2.6	352	9	343	0.2
2.6	351	9	342	0.2
2.6	350	9	341	0.2
2.6	349	9	340	0.2
2.6	348	9	339	0.2
2.6	347	9	338	0.2
2.6	346	9	337	0.2
2.6	345	9	336	0.2
2.6	344	9	335	0.2
2.6	343	9	334	0.2
2.6	342	9	333	0.2
2.6	341	9	332	0.2
2.6	340	9	331	0.2

%E	M1	M2	DM	M*
2.6	313	8	305	0.2
2.6	312	8	304	0.2
2.6	311	8	303	0.2
2.6	310	8	302	0.2
2.6	309	8	301	0.2
2.6	308	8	300	0.2
2.6	307	8	299	0.2
2.6	306	8	298	0.2
2.6	305	8	297	0.2
2.6	304	8	296	0.2
2.6	303	8	295	0.2
2.6	302	8	294	0.2
2.6	274	7	267	0.2
2.6	273	7	266	0.2
2.6	272	7	265	0.2
2.6	271	7	264	0.2
2.6	270	7	263	0.2
2.6	269	7	262	0.2
2.6	268	7	261	0.2
2.6	267	7	260	0.2
2.6	266	7	259	0.2
2.6	265	7	258	0.2
2.6	235	6	229	0.2
2.6	234	6	228	0.2
2.6	233	6	227	0.2
2.6	232	6	226	0.2
2.6	231	6	225	0.2
2.6	230	6	224	0.2
2.6	229	6	223	0.2
2.6	228	6	222	0.2
2.6	227	6	221	0.2
2.6	196	5	191	0.1
2.6	195	5	190	0.1
2.6	194	5	189	0.1
2.6	193	5	188	0.1
2.6	192	5	187	0.1
2.6	191	5	186	0.1
2.6	190	5	185	0.1
2.6	189	5	184	0.1
2.6	156	4	152	0.1
2.6	155	4	151	0.1
2.6	154	4	150	0.1
2.6	153	4	149	0.1
2.6	152	4	148	0.1
2.6	151	4	147	0.1
2.6	117	3	114	0.1
2.6	116	3	113	0.1
2.6	115	3	112	0.1
2.6	114	3	111	0.1
2.6	78	2	76	0.1

%E	M1	M2	DM	M*
2.6	77	2	75	0.1
2.6	76	2	74	0.1
2.6	39	1	38	0.0
2.6	38	1	37	0.0
2.5	489	12	477	0.3
2.5	488	12	476	0.3
2.5	487	12	475	0.3
2.5	486	12	474	0.3
2.5	485	12	473	0.3
2.5	484	12	472	0.3
2.5	483	12	471	0.3
2.5	482	12	470	0.3
2.5	481	12	469	0.3
2.5	480	12	468	0.3
2.5	479	12	467	0.3
2.5	478	12	466	0.3
2.5	477	12	465	0.3
2.5	476	12	464	0.3
2.5	475	12	463	0.3
2.5	474	12	462	0.3
2.5	473	12	461	0.3
2.5	472	12	460	0.3
2.5	471	12	459	0.3
2.5	449	11	438	0.3
2.5	448	11	437	0.3
2.5	447	11	436	0.3
2.5	446	11	435	0.3
2.5	445	11	434	0.3
2.5	444	11	433	0.3
2.5	443	11	432	0.3
2.5	442	11	431	0.3
2.5	441	11	430	0.3
2.5	440	11	429	0.3
2.5	439	11	428	0.3
2.5	438	11	427	0.3
2.5	437	11	426	0.3
2.5	436	11	425	0.3
2.5	435	11	424	0.3
2.5	434	11	423	0.3
2.5	433	11	422	0.3
2.5	432	11	421	0.3
2.5	408	10	398	0.2
2.5	407	10	397	0.2
2.5	406	10	396	0.2
2.5	405	10	395	0.2
2.5	404	10	394	0.2
2.5	403	10	393	0.2
2.5	402	10	392	0.2
2.5	401	10	391	0.2
2.5	400	10	390	0.3

%E	M1	M2	DM	M*
2.5	399	10	389	0.3
2.5	398	10	388	0.3
2.5	397	10	387	0.3
2.5	396	10	386	0.3
2.5	395	10	385	0.3
2.5	394	10	384	0.3
2.5	393	10	383	0.3
2.5	367	9	358	0.2
2.5	366	9	357	0.2
2.5	365	9	356	0.2
2.5	364	9	355	0.2
2.5	363	9	354	0.2
2.5	362	9	353	0.2
2.5	361	9	352	0.2
2.5	360	9	351	0.2
2.5	359	9	350	0.2
2.5	358	9	349	0.2
2.5	357	9	348	0.2
2.5	356	9	347	0.2
2.5	355	9	346	0.2
2.5	354	9	345	0.2
2.5	326	8	318	0.2
2.5	325	8	317	0.2
2.5	324	8	316	0.2
2.5	323	8	315	0.2
2.5	322	8	314	0.2
2.5	321	8	313	0.2
2.5	320	8	312	0.2
2.5	319	8	311	0.2
2.5	318	8	310	0.2
2.5	317	8	309	0.2
2.5	316	8	308	0.2
2.5	315	8	307	0.2
2.5	314	8	306	0.2
2.5	285	7	278	0.2
2.5	284	7	277	0.2
2.5	283	7	276	0.2
2.5	282	7	275	0.2
2.5	281	7	274	0.2
2.5	280	7	273	0.2
2.5	279	7	272	0.2
2.5	278	7	271	0.2
2.5	277	7	270	0.2
2.5	276	7	269	0.2
2.5	275	7	268	0.2
2.5	244	6	238	0.1
2.5	243	6	237	0.1
2.5	242	6	236	0.1
2.5	241	6	235	0.1
2.5	240	6	234	0.1

%E	M1	M2	DM	M*
2.5	239	6	233	0.2
2.5	238	6	232	0.2
2.5	237	6	231	0.2
2.5	236	6	230	0.2
2.5	204	5	199	0.1
2.5	203	5	198	0.1
2.5	202	5	197	0.1
2.5	201	5	196	0.1
2.5	200	5	195	0.1
2.5	199	5	194	0.1
2.5	198	5	193	0.1
2.5	197	5	192	0.1
2.5	163	4	159	0.1
2.5	162	4	158	0.1
2.5	161	4	157	0.1
2.5	160	4	156	0.1
2.5	159	4	155	0.1
2.5	158	4	154	0.1
2.5	157	4	153	0.1
2.5	122	3	119	0.1
2.5	121	3	118	0.1
2.5	120	3	117	0.1
2.5	119	3	116	0.1
2.5	118	3	115	0.1
2.5	81	2	79	0.0
2.5	80	2	78	0.0
2.5	79	2	77	0.1
2.5	40	1	39	0.0
2.4	500	12	488	0.3
2.4	499	12	487	0.3
2.4	498	12	486	0.3
2.4	497	12	485	0.3
2.4	496	12	484	0.3
2.4	495	12	483	0.3
2.4	494	12	482	0.3
2.4	493	12	481	0.3
2.4	492	12	480	0.3
2.4	491	12	479	0.3
2.4	490	12	478	0.3
2.4	468	11	457	0.3
2.4	467	11	456	0.3
2.4	466	11	455	0.3
2.4	465	11	454	0.3
2.4	464	11	453	0.3
2.4	463	11	452	0.3
2.4	462	11	451	0.3
2.4	461	11	450	0.3
2.4	460	11	449	0.3
2.4	459	11	448	0.3
2.4	458	11	447	0.3

%E	M1	M2	DM	M*
2.4	457	11	446	0.3
2.4	456	11	445	0.3
2.4	455	11	444	0.3
2.4	454	11	443	0.3
2.4	453	11	442	0.3
2.4	452	11	441	0.3
2.4	451	11	440	0.3
2.4	450	11	439	0.3
2.4	425	10	415	0.2
2.4	424	10	414	0.2
2.4	423	10	413	0.2
2.4	422	10	412	0.2
2.4	421	10	411	0.2
2.4	420	10	410	0.2
2.4	419	10	409	0.2
2.4	418	10	408	0.2
2.4	417	10	407	0.2
2.4	416	10	406	0.2
2.4	415	10	405	0.2
2.4	414	10	404	0.2
2.4	413	10	403	0.2
2.4	412	10	402	0.2
2.4	411	10	401	0.2
2.4	410	10	400	0.2
2.4	409	10	399	0.2
2.4	383	9	374	0.2
2.4	382	9	373	0.2
2.4	381	9	372	0.2
2.4	380	9	371	0.2
2.4	379	9	370	0.2
2.4	378	9	369	0.2
2.4	377	9	368	0.2
2.4	376	9	367	0.2
2.4	375	9	366	0.2
2.4	374	9	365	0.2
2.4	373	9	364	0.2
2.4	372	9	363	0.2
2.4	371	9	362	0.2
2.4	370	9	361	0.2
2.4	369	9	360	0.2
2.4	368	9	359	0.2
2.4	340	8	332	0.2
2.4	339	8	331	0.2
2.4	338	8	330	0.2
2.4	337	8	329	0.2
2.4	336	8	328	0.2
2.4	335	8	327	0.2
2.4	334	8	326	0.2
2.4	333	8	325	0.2
2.4	332	8	324	0.2

%E	M1	M2	DM	M*
2.4	331	8	323	0.2
2.4	330	8	322	0.2
2.4	329	8	321	0.2
2.4	328	8	320	0.2
2.4	327	8	319	0.2
2.4	297	7	290	0.2
2.4	296	7	289	0.2
2.4	295	7	288	0.2
2.4	294	7	287	0.2
2.4	293	7	286	0.2
2.4	292	7	285	0.2
2.4	291	7	284	0.2
2.4	290	7	283	0.2
2.4	289	7	282	0.2
2.4	288	7	281	0.2
2.4	287	7	280	0.2
2.4	286	7	279	0.2
2.4	255	6	249	0.1
2.4	254	6	248	0.1
2.4	253	6	247	0.1
2.4	252	6	246	0.1
2.4	251	6	245	0.1
2.4	250	6	244	0.1
2.4	249	6	243	0.1
2.4	248	6	242	0.1
2.4	247	6	241	0.1
2.4	246	6	240	0.1
2.4	245	6	239	0.1
2.4	212	5	207	0.1
2.4	211	5	206	0.1
2.4	210	5	205	0.1
2.4	209	5	204	0.1
2.4	208	5	203	0.1
2.4	207	5	202	0.1
2.4	206	5	201	0.1
2.4	205	5	200	0.1
2.4	170	4	166	0.1
2.4	169	4	165	0.1
2.4	168	4	164	0.1
2.4	167	4	163	0.1
2.4	166	4	162	0.1
2.4	165	4	161	0.1
2.4	164	4	160	0.1
2.4	127	3	124	0.1
2.4	126	3	123	0.1
2.4	125	3	122	0.1
2.4	124	3	121	0.1
2.4	123	3	120	0.1
2.4	85	2	83	0.0
2.4	84	2	82	0.0

%E	M1	M2	DM	M*
2.4	83	2	81	0.0
2.4	82	2	80	0.0
2.4	42	1	41	0.0
2.4	41	1	40	0.0
2.3	488	11	477	0.2
2.3	487	11	476	0.2
2.3	486	11	475	0.2
2.3	485	11	474	0.2
2.3	484	11	473	0.3
2.3	483	11	472	0.3
2.3	482	11	471	0.3
2.3	481	11	470	0.3
2.3	480	11	469	0.3
2.3	479	11	468	0.3
2.3	478	11	467	0.3
2.3	477	11	466	0.3
2.3	476	11	465	0.3
2.3	475	11	464	0.3
2.3	474	11	463	0.3
2.3	473	11	462	0.3
2.3	472	11	461	0.3
2.3	471	11	460	0.3
2.3	470	11	459	0.3
2.3	469	11	458	0.3
2.3	444	10	434	0.2
2.3	443	10	433	0.2
2.3	442	10	432	0.2
2.3	441	10	431	0.2
2.3	440	10	430	0.2
2.3	439	10	429	0.2
2.3	438	10	428	0.2
2.3	437	10	427	0.2
2.3	436	10	426	0.2
2.3	435	10	425	0.2
2.3	434	10	424	0.2
2.3	433	10	423	0.2
2.3	432	10	422	0.2
2.3	431	10	421	0.2
2.3	430	10	420	0.2
2.3	429	10	419	0.2
2.3	428	10	418	0.2
2.3	427	10	417	0.2
2.3	426	10	416	0.2
2.3	400	9	391	0.2
2.3	399	9	390	0.2
2.3	398	9	389	0.2
2.3	397	9	388	0.2
2.3	396	9	387	0.2
2.3	395	9	386	0.2
2.3	394	9	385	0.2

%E	M1	M2	DM	M*
2.3	393	9	384	0.2
2.3	392	9	383	0.2
2.3	391	9	382	0.2
2.3	390	9	381	0.2
2.3	389	9	380	0.2
2.3	388	9	379	0.2
2.3	387	9	378	0.2
2.3	386	9	377	0.2
2.3	385	9	376	0.2
2.3	384	9	375	0.2
2.3	355	8	347	0.2
2.3	354	8	346	0.2
2.3	353	8	345	0.2
2.3	352	8	344	0.2
2.3	351	8	343	0.2
2.3	350	8	342	0.2
2.3	349	8	341	0.2
2.3	348	8	340	0.2
2.3	347	8	339	0.2
2.3	346	8	338	0.2
2.3	345	8	337	0.2
2.3	344	8	336	0.2
2.3	343	8	335	0.2
2.3	342	8	334	0.2
2.3	341	8	333	0.2
2.3	311	7	304	0.2
2.3	310	7	303	0.2
2.3	309	7	302	0.2
2.3	308	7	301	0.2
2.3	307	7	300	0.2
2.3	306	7	299	0.2
2.3	305	7	298	0.2
2.3	304	7	297	0.2
2.3	303	7	296	0.2
2.3	302	7	295	0.2
2.3	301	7	294	0.2
2.3	300	7	293	0.2
2.3	299	7	292	0.2
2.3	298	7	291	0.2
2.3	266	6	260	0.1
2.3	265	6	259	0.1
2.3	264	6	258	0.1
2.3	263	6	257	0.1
2.3	262	6	256	0.1
2.3	261	6	255	0.1
2.3	260	6	254	0.1
2.3	259	6	253	0.1
2.3	258	6	252	0.1
2.3	257	6	251	0.1
2.3	256	6	250	0.1

%E	M1	M2	DM	M*
2.3	222	5	217	0.1
2.3	221	5	216	0.1
2.3	220	5	215	0.1
2.3	219	5	214	0.1
2.3	218	5	213	0.1
2.3	217	5	212	0.1
2.3	216	5	211	0.1
2.3	215	5	210	0.1
2.3	214	5	209	0.1
2.3	213	5	208	0.1
2.3	177	4	173	0.1
2.3	176	4	172	0.1
2.3	175	4	171	0.1
2.3	174	4	170	0.1
2.3	173	4	169	0.1
2.3	172	4	168	0.1
2.3	171	4	167	0.1
2.3	133	3	130	0.1
2.3	132	3	129	0.1
2.3	131	3	128	0.1
2.3	130	3	127	0.1
2.3	129	3	126	0.1
2.3	128	3	125	0.1
2.3	88	2	86	0.0
2.3	87	2	85	0.0
2.3	86	2	84	0.0
2.3	44	1	43	0.0
2.3	43	1	42	0.0
2.2	500	11	489	0.2
2.2	499	11	488	0.2
2.2	498	11	487	0.2
2.2	497	11	486	0.2
2.2	496	11	485	0.2
2.2	495	11	484	0.2
2.2	494	11	483	0.2
2.2	493	11	482	0.2
2.2	492	11	481	0.2
2.2	491	11	480	0.2
2.2	490	11	479	0.2
2.2	489	11	478	0.2
2.2	465	10	455	0.2
2.2	464	10	454	0.2
2.2	463	10	453	0.2
2.2	462	10	452	0.2
2.2	461	10	451	0.2
2.2	460	10	450	0.2
2.2	459	10	449	0.2
2.2	458	10	448	0.2
2.2	457	10	447	0.2
2.2	456	10	446	0.2

%E	M1	M2	DM	M*
2.2	455	10	445	0.2
2.2	454	10	444	0.2
2.2	453	10	443	0.2
2.2	452	10	442	0.2
2.2	451	10	441	0.2
2.2	450	10	440	0.2
2.2	449	10	439	0.2
2.2	448	10	438	0.2
2.2	447	10	437	0.2
2.2	446	10	436	0.2
2.2	445	10	435	0.2
2.2	418	9	409	0.2
2.2	417	9	408	0.2
2.2	416	9	407	0.2
2.2	415	9	406	0.2
2.2	414	9	405	0.2
2.2	413	9	404	0.2
2.2	412	9	403	0.2
2.2	411	9	402	0.2
2.2	410	9	401	0.2
2.2	409	9	400	0.2
2.2	408	9	399	0.2
2.2	407	9	398	0.2
2.2	406	9	397	0.2
2.2	405	9	396	0.2
2.2	404	9	395	0.2
2.2	403	9	394	0.2
2.2	402	9	393	0.2
2.2	401	9	392	0.2
2.2	372	8	364	0.2
2.2	371	8	363	0.2
2.2	370	8	362	0.2
2.2	369	8	361	0.2
2.2	368	8	360	0.2
2.2	367	8	359	0.2
2.2	366	8	358	0.2
2.2	365	8	357	0.2
2.2	364	8	356	0.2
2.2	363	8	355	0.2
2.2	362	8	354	0.2
2.2	361	8	353	0.2
2.2	360	8	352	0.2
2.2	359	8	351	0.2
2.2	358	8	350	0.2
2.2	357	8	349	0.2
2.2	356	8	348	0.2
2.2	325	7	318	0.2
2.2	324	7	317	0.2
2.2	323	7	316	0.2
2.2	322	7	315	0.2

%E	M1	M2	DM	M*
2•2	321	7	314	0•2
2•2	320	7	313	0•2
2•2	319	7	312	0•2
2•2	318	7	311	0•2
2•2	317	7	310	0•2
2•2	316	7	309	0•2
2•2	315	7	308	0•2
2•2	314	7	307	0•2
2•2	313	7	306	0•2
2•2	312	7	305	0•2
2•2	279	6	273	0•1
2•2	278	6	272	0•1
2•2	277	6	271	0•1
2•2	276	6	270	0•1
2•2	275	6	269	0•1
2•2	274	6	268	0•1
2•2	273	6	267	0•1
2•2	272	6	266	0•1
2•2	271	6	265	0•1
2•2	270	6	264	0•1
2•2	269	6	263	0•1
2•2	268	6	262	0•1
2•2	267	6	261	0•1
2•2	232	5	227	0•1
2•2	231	5	226	0•1
2•2	230	5	225	0•1
2•2	229	5	224	0•1
2•2	228	5	223	0•1
2•2	227	5	222	0•1
2•2	226	5	221	0•1
2•2	225	5	220	0•1
2•2	224	5	219	0•1
2•2	223	5	218	0•1
2•2	186	4	182	0•1
2•2	185	4	181	0•1
2•2	184	4	180	0•1
2•2	183	4	179	0•1
2•2	182	4	178	0•1
2•2	181	4	177	0•1
2•2	180	4	176	0•1
2•2	179	4	175	0•1
2•2	178	4	174	0•1
2•2	139	3	136	0•1
2•2	138	3	135	0•1
2•2	137	3	134	0•1
2•2	136	3	133	0•1
2•2	135	3	132	0•1
2•2	134	3	131	0•1
2•2	93	2	91	0•0
2•2	92	2	90	0•0

%E	M1	M2	DM	M*
2•2	91	2	89	0•0
2•2	90	2	88	0•0
2•2	89	2	87	0•0
2•2	46	1	45	0•0
2•2	45	1	44	0•0
2•1	487	10	477	0•2
2•1	486	10	476	0•2
2•1	485	10	475	0•2
2•1	484	10	474	0•2
2•1	483	10	473	0•2
2•1	482	10	472	0•2
2•1	481	10	471	0•2
2•1	480	10	470	0•2
2•1	479	10	469	0•2
2•1	478	10	468	0•2
2•1	477	10	467	0•2
2•1	476	10	466	0•2
2•1	475	10	465	0•2
2•1	474	10	464	0•2
2•1	473	10	463	0•2
2•1	472	10	462	0•2
2•1	471	10	461	0•2
2•1	470	10	460	0•2
2•1	469	10	459	0•2
2•1	468	10	458	0•2
2•1	467	10	457	0•2
2•1	466	10	456	0•2
2•1	439	9	430	0•2
2•1	438	9	429	0•2
2•1	437	9	428	0•2
2•1	436	9	427	0•2
2•1	435	9	426	0•2
2•1	434	9	425	0•2
2•1	433	9	424	0•2
2•1	432	9	423	0•2
2•1	431	9	422	0•2
2•1	430	9	421	0•2
2•1	429	9	420	0•2
2•1	428	9	419	0•2
2•1	427	9	418	0•2
2•1	426	9	417	0•2
2•1	425	9	416	0•2
2•1	424	9	415	0•2
2•1	423	9	414	0•2
2•1	422	9	413	0•2
2•1	421	9	412	0•2
2•1	420	9	411	0•2
2•1	419	9	410	0•2
2•1	390	8	382	0•2
2•1	389	8	381	0•2

%E	M1	M2	DM	M*
2•1	388	8	380	0•2
2•1	387	8	379	0•2
2•1	386	8	378	0•2
2•1	385	8	377	0•2
2•1	384	8	376	0•2
2•1	383	8	375	0•2
2•1	382	8	374	0•2
2•1	381	8	373	0•2
2•1	380	8	372	0•2
2•1	379	8	371	0•2
2•1	378	8	370	0•2
2•1	377	8	369	0•2
2•1	376	8	368	0•2
2•1	375	8	367	0•2
2•1	374	8	366	0•2
2•1	373	8	365	0•2
2•1	341	7	334	0•1
2•1	340	7	333	0•1
2•1	339	7	332	0•1
2•1	338	7	331	0•1
2•1	337	7	330	0•1
2•1	336	7	329	0•1
2•1	335	7	328	0•1
2•1	334	7	327	0•1
2•1	333	7	326	0•1
2•1	332	7	325	0•1
2•1	331	7	324	0•1
2•1	330	7	323	0•1
2•1	329	7	322	0•1
2•1	328	7	321	0•1
2•1	327	7	320	0•1
2•1	326	7	319	0•2
2•1	292	6	286	0•1
2•1	291	6	285	0•1
2•1	290	6	284	0•1
2•1	289	6	283	0•1
2•1	288	6	282	0•1
2•1	287	6	281	0•1
2•1	286	6	280	0•1
2•1	285	6	279	0•1
2•1	284	6	278	0•1
2•1	283	6	277	0•1
2•1	282	6	276	0•1
2•1	281	6	275	0•1
2•1	280	6	274	0•1
2•1	243	5	238	0•1
2•1	242	5	237	0•1
2•1	241	5	236	0•1
2•1	240	5	235	0•1
2•1	239	5	234	0•1

%E	M1	M2	DM	M*
2•1	238	5	233	0•1
2•1	237	5	232	0•1
2•1	236	5	231	0•1
2•1	235	5	230	0•1
2•1	234	5	229	0•1
2•1	233	5	228	0•1
2•1	195	4	191	0•1
2•1	194	4	190	0•1
2•1	193	4	189	0•1
2•1	192	4	188	0•1
2•1	191	4	187	0•1
2•1	190	4	186	0•1
2•1	189	4	185	0•1
2•1	188	4	184	0•1
2•1	187	4	183	0•1
2•1	146	3	143	0•1
2•1	145	3	142	0•1
2•1	144	3	141	0•1
2•1	143	3	140	0•1
2•1	142	3	139	0•1
2•1	141	3	138	0•1
2•1	140	3	137	0•1
2•1	97	2	95	0•0
2•1	96	2	94	0•0
2•1	95	2	93	0•0
2•1	94	2	92	0•0
2•1	48	1	47	0•0
2•1	47	1	46	0•0
2•0	500	10	490	0•2
2•0	499	10	489	0•2
2•0	498	10	488	0•2
2•0	497	10	487	0•2
2•0	496	10	486	0•2
2•0	495	10	485	0•2
2•0	494	10	484	0•2
2•0	493	10	483	0•2
2•0	492	10	482	0•2
2•0	491	10	481	0•2
2•0	490	10	480	0•2
2•0	489	10	479	0•2
2•0	488	10	478	0•2
2•0	461	9	452	0•2
2•0	460	9	451	0•2
2•0	459	9	450	0•2
2•0	458	9	449	0•2
2•0	457	9	448	0•2
2•0	456	9	447	0•2
2•0	455	9	446	0•2
2•0	454	9	445	0•2
2•0	453	9	444	0•2

%E	M1	M2	DM	M*
2•0	452	9	443	0•2
2•0	451	9	442	0•2
2•0	450	9	441	0•2
2•0	449	9	440	0•2
2•0	448	9	439	0•2
2•0	447	9	438	0•2
2•0	446	9	437	0•2
2•0	445	9	436	0•2
2•0	444	9	435	0•2
2•0	443	9	434	0•2
2•0	442	9	433	0•2
2•0	441	9	432	0•2
2•0	440	9	431	0•2
2•0	410	8	402	0•2
2•0	409	8	401	0•2
2•0	408	8	400	0•2
2•0	407	8	399	0•2
2•0	406	8	398	0•2
2•0	405	8	397	0•2
2•0	404	8	396	0•2
2•0	403	8	395	0•2
2•0	402	8	394	0•2
2•0	401	8	393	0•2
2•0	400	8	392	0•2
2•0	399	8	391	0•2
2•0	398	8	390	0•2
2•0	397	8	389	0•2
2•0	396	8	388	0•2
2•0	395	8	387	0•2
2•0	394	8	386	0•2
2•0	393	8	385	0•2
2•0	392	8	384	0•2
2•0	391	8	383	0•2
2•0	359	7	352	0•1
2•0	358	7	351	0•1
2•0	357	7	350	0•1
2•0	356	7	349	0•1
2•0	355	7	348	0•1
2•0	354	7	347	0•1
2•0	353	7	346	0•1
2•0	352	7	345	0•1
2•0	351	7	344	0•1
2•0	350	7	343	0•1
2•0	349	7	342	0•1
2•0	348	7	341	0•1
2•0	347	7	340	0•1
2•0	346	7	339	0•1
2•0	345	7	338	0•1
2•0	344	7	337	0•1
2•0	343	7	336	0•1

%E	M1	M2	DM	M*
2.0	342	7	335	0.1
2.0	307	6	301	0.1
2.0	306	6	300	0.1
2.0	305	6	299	0.1
2.0	304	6	298	0.1
2.0	303	6	297	0.1
2.0	302	6	296	0.1
2.0	301	6	295	0.1
2.0	300	6	294	0.1
2.0	299	6	293	0.1
2.0	298	6	292	0.1
2.0	297	6	291	0.1
2.0	296	6	290	0.1
2.0	295	6	289	0.1
2.0	294	6	288	0.1
2.0	293	6	287	0.1
2.0	256	5	251	0.1
2.0	255	5	250	0.1
2.0	254	5	249	0.1
2.0	253	5	248	0.1
2.0	252	5	247	0.1
2.0	251	5	246	0.1
2.0	250	5	245	0.1
2.0	249	5	244	0.1
2.0	248	5	243	0.1
2.0	247	5	242	0.1
2.0	246	5	241	0.1
2.0	245	5	240	0.1
2.0	244	5	239	0.1
2.0	205	4	201	0.1
2.0	204	4	200	0.1
2.0	203	4	199	0.1
2.0	202	4	198	0.1
2.0	201	4	197	0.1
2.0	200	4	196	0.1
2.0	199	4	195	0.1
2.0	198	4	194	0.1
2.0	197	4	193	0.1
2.0	196	4	192	0.1
2.0	153	3	150	0.1
2.0	152	3	149	0.1
2.0	151	3	148	0.1
2.0	150	3	147	0.1
2.0	149	3	146	0.1
2.0	148	3	145	0.1
2.0	147	3	144	0.1
2.0	102	2	100	0.0
2.0	101	2	99	0.0
2.0	100	2	98	0.0
2.0	99	2	97	0.0

%E	M1	M2	DM	M*
2.0	98	2	96	0.0
2.0	51	1	50	0.0
2.0	50	1	49	0.0
2.0	49	1	48	0.0
1.9	486	9	477	0.2
1.9	485	9	476	0.2
1.9	484	9	475	0.2
1.9	483	9	474	0.2
1.9	482	9	473	0.2
1.9	481	9	472	0.2
1.9	480	9	471	0.2
1.9	479	9	470	0.2
1.9	478	9	469	0.2
1.9	477	9	468	0.2
1.9	476	9	467	0.2
1.9	475	9	466	0.2
1.9	474	9	465	0.2
1.9	473	9	464	0.2
1.9	472	9	463	0.2
1.9	471	9	462	0.2
1.9	470	9	461	0.2
1.9	469	9	460	0.2
1.9	468	9	459	0.2
1.9	467	9	458	0.2
1.9	466	9	457	0.2
1.9	465	9	456	0.2
1.9	464	9	455	0.2
1.9	463	9	454	0.2
1.9	462	9	453	0.2
1.9	432	8	424	0.1
1.9	431	8	423	0.1
1.9	430	8	422	0.1
1.9	429	8	421	0.1
1.9	428	8	420	0.1
1.9	427	8	419	0.1
1.9	426	8	418	0.2
1.9	425	8	417	0.2
1.9	424	8	416	0.2
1.9	423	8	415	0.2
1.9	422	8	414	0.2
1.9	421	8	413	0.2
1.9	420	8	412	0.2
1.9	419	8	411	0.2
1.9	418	8	410	0.2
1.9	417	8	409	0.2
1.9	416	8	408	0.2
1.9	415	8	407	0.2
1.9	414	8	406	0.2
1.9	413	8	405	0.2
1.9	412	8	404	0.2

%E	M1	M2	DM	M*
1.9	411	8	403	0.2
1.9	378	7	371	0.1
1.9	377	7	370	0.1
1.9	376	7	369	0.1
1.9	375	7	368	0.1
1.9	374	7	367	0.1
1.9	373	7	366	0.1
1.9	372	7	365	0.1
1.9	371	7	364	0.1
1.9	370	7	363	0.1
1.9	369	7	362	0.1
1.9	368	7	361	0.1
1.9	367	7	360	0.1
1.9	366	7	359	0.1
1.9	365	7	358	0.1
1.9	364	7	357	0.1
1.9	363	7	356	0.1
1.9	362	7	355	0.1
1.9	361	7	354	0.1
1.9	360	7	353	0.1
1.9	324	6	318	0.1
1.9	323	6	317	0.1
1.9	322	6	316	0.1
1.9	321	6	315	0.1
1.9	320	6	314	0.1
1.9	319	6	313	0.1
1.9	318	6	312	0.1
1.9	317	6	311	0.1
1.9	316	6	310	0.1
1.9	315	6	309	0.1
1.9	314	6	308	0.1
1.9	313	6	307	0.1
1.9	312	6	306	0.1
1.9	311	6	305	0.1
1.9	310	6	304	0.1
1.9	309	6	303	0.1
1.9	308	6	302	0.1
1.9	270	5	265	0.1
1.9	269	5	264	0.1
1.9	268	5	263	0.1
1.9	267	5	262	0.1
1.9	266	5	261	0.1
1.9	265	5	260	0.1
1.9	264	5	259	0.1
1.9	263	5	258	0.1
1.9	262	5	257	0.1
1.9	261	5	256	0.1
1.9	260	5	255	0.1
1.9	259	5	254	0.1
1.9	258	5	253	0.1

%E	M1	M2	DM	M*
1.9	257	5	252	0.1
1.9	216	4	212	0.1
1.9	215	4	211	0.1
1.9	214	4	210	0.1
1.9	213	4	209	0.1
1.9	212	4	208	0.1
1.9	211	4	207	0.1
1.9	210	4	206	0.1
1.9	209	4	205	0.1
1.9	208	4	204	0.1
1.9	207	4	203	0.1
1.9	206	4	202	0.1
1.9	162	3	159	0.1
1.9	161	3	158	0.1
1.9	160	3	157	0.1
1.9	159	3	156	0.1
1.9	158	3	155	0.1
1.9	157	3	154	0.1
1.9	156	3	153	0.1
1.9	155	3	152	0.1
1.9	154	3	151	0.1
1.9	108	2	106	0.0
1.9	107	2	105	0.0
1.9	106	2	104	0.0
1.9	105	2	103	0.0
1.9	104	2	102	0.0
1.9	103	2	101	0.0
1.9	54	1	53	0.0
1.9	53	1	52	0.0
1.9	52	1	51	0.0
1.8	500	9	491	0.2
1.8	499	9	490	0.2
1.8	498	9	489	0.2
1.8	497	9	488	0.2
1.8	496	9	487	0.2
1.8	495	9	486	0.2
1.8	494	9	485	0.2
1.8	493	9	484	0.2
1.8	492	9	483	0.2
1.8	491	9	482	0.2
1.8	490	9	481	0.2
1.8	489	9	480	0.2
1.8	488	9	479	0.2
1.8	487	9	478	0.2
1.8	457	8	449	0.1
1.8	456	8	448	0.1
1.8	455	8	447	0.1
1.8	454	8	446	0.1
1.8	453	8	445	0.1
1.8	452	8	444	0.1

%E	M1	M2	DM	M*
1.8	451	8	443	0.1
1.8	450	8	442	0.1
1.8	449	8	441	0.1
1.8	448	8	440	0.1
1.8	447	8	439	0.1
1.8	446	8	438	0.1
1.8	445	8	437	0.1
1.8	444	8	436	0.1
1.8	443	8	435	0.1
1.8	442	8	434	0.1
1.8	441	8	433	0.1
1.8	440	8	432	0.1
1.8	439	8	431	0.1
1.8	438	8	430	0.1
1.8	437	8	429	0.1
1.8	436	8	428	0.1
1.8	435	8	427	0.1
1.8	434	8	426	0.1
1.8	433	8	425	0.1
1.8	400	7	393	0.1
1.8	399	7	392	0.1
1.8	398	7	391	0.1
1.8	397	7	390	0.1
1.8	396	7	389	0.1
1.8	395	7	388	0.1
1.8	394	7	387	0.1
1.8	393	7	386	0.1
1.8	392	7	385	0.1
1.8	391	7	384	0.1
1.8	390	7	383	0.1
1.8	389	7	382	0.1
1.8	388	7	381	0.1
1.8	387	7	380	0.1
1.8	386	7	379	0.1
1.8	385	7	378	0.1
1.8	384	7	377	0.1
1.8	383	7	376	0.1
1.8	382	7	375	0.1
1.8	381	7	374	0.1
1.8	380	7	373	0.1
1.8	379	7	372	0.1
1.8	342	6	336	0.1
1.8	341	6	335	0.1
1.8	340	6	334	0.1
1.8	339	6	333	0.1
1.8	338	6	332	0.1
1.8	337	6	331	0.1
1.8	336	6	330	0.1
1.8	335	6	329	0.1
1.8	334	6	328	0.1

%E	M1	M2	DM	M*
1·8	333	6	327	0·1
1·8	332	6	326	0·1
1·8	331	6	325	0·1
1·8	330	6	324	0·1
1·8	329	6	323	0·1
1·8	328	6	322	0·1
1·8	327	6	321	0·1
1·8	326	6	320	0·1
1·8	325	6	319	0·1
1·8	285	5	280	0·1
1·8	284	5	279	0·1
1·8	283	5	278	0·1
1·8	282	5	277	0·1
1·8	281	5	276	0·1
1·8	280	5	275	0·1
1·8	279	5	274	0·1
1·8	278	5	273	0·1
1·8	277	5	272	0·1
1·8	276	5	271	0·1
1·8	275	5	270	0·1
1·8	274	5	269	0·1
1·8	273	5	268	0·1
1·8	272	5	267	0·1
1·8	271	5	266	0·1
1·8	228	4	224	0·1
1·8	227	4	223	0·1
1·8	226	4	222	0·1
1·8	225	4	221	0·1
1·8	224	4	220	0·1
1·8	223	4	219	0·1
1·8	222	4	218	0·1
1·8	221	4	217	0·1
1·8	220	4	216	0·1
1·8	219	4	215	0·1
1·8	218	4	214	0·1
1·8	217	4	213	0·1
1·8	171	3	168	0·1
1·8	170	3	167	0·1
1·8	169	3	166	0·1
1·8	168	3	165	0·1
1·8	167	3	164	0·1
1·8	166	3	163	0·1
1·8	165	3	162	0·1
1·8	164	3	161	0·1
1·8	163	3	160	0·1
1·8	114	2	112	0·0
1·8	113	2	111	0·0
1·8	112	2	110	0·0
1·8	111	2	109	0·0
1·8	110	2	108	0·0

%E	M1	M2	DM	M*
1·8	109	2	107	0·0
1·8	57	1	56	0·0
1·8	56	1	55	0·0
1·8	55	1	54	0·0
1·7	484	8	476	0·1
1·7	483	8	475	0·1
1·7	482	8	474	0·1
1·7	481	8	473	0·1
1·7	480	8	472	0·1
1·7	479	8	471	0·1
1·7	478	8	470	0·1
1·7	477	8	469	0·1
1·7	476	8	468	0·1
1·7	475	8	467	0·1
1·7	474	8	466	0·1
1·7	473	8	465	0·1
1·7	472	8	464	0·1
1·7	471	8	463	0·1
1·7	470	8	462	0·1
1·7	469	8	461	0·1
1·7	468	8	460	0·1
1·7	467	8	459	0·1
1·7	466	8	458	0·1
1·7	465	8	457	0·1
1·7	464	8	456	0·1
1·7	463	8	455	0·1
1·7	462	8	454	0·1
1·7	461	8	453	0·1
1·7	460	8	452	0·1
1·7	459	8	451	0·1
1·7	458	8	450	0·1
1·7	424	7	417	0·1
1·7	423	7	416	0·1
1·7	422	7	415	0·1
1·7	421	7	414	0·1
1·7	420	7	413	0·1
1·7	419	7	412	0·1
1·7	418	7	411	0·1
1·7	417	7	410	0·1
1·7	416	7	409	0·1
1·7	415	7	408	0·1
1·7	414	7	407	0·1
1·7	413	7	406	0·1
1·7	412	7	405	0·1
1·7	411	7	404	0·1
1·7	410	7	403	0·1
1·7	409	7	402	0·1
1·7	408	7	401	0·1
1·7	407	7	400	0·1
1·7	406	7	399	0·1

%E	M1	M2	DM	M*
1·7	405	7	398	0·1
1·7	404	7	397	0·1
1·7	403	7	396	0·1
1·7	402	7	395	0·1
1·7	401	7	394	0·1
1·7	363	6	357	0·1
1·7	362	6	356	0·1
1·7	361	6	355	0·1
1·7	360	6	354	0·1
1·7	359	6	353	0·1
1·7	358	6	352	0·1
1·7	357	6	351	0·1
1·7	356	6	350	0·1
1·7	355	6	349	0·1
1·7	354	6	348	0·1
1·7	353	6	347	0·1
1·7	352	6	346	0·1
1·7	351	6	345	0·1
1·7	350	6	344	0·1
1·7	349	6	343	0·1
1·7	348	6	342	0·1
1·7	347	6	341	0·1
1·7	346	6	340	0·1
1·7	345	6	339	0·1
1·7	344	6	338	0·1
1·7	343	6	337	0·1
1·7	303	5	298	0·1
1·7	302	5	297	0·1
1·7	301	5	296	0·1
1·7	300	5	295	0·1
1·7	299	5	294	0·1
1·7	298	5	293	0·1
1·7	297	5	292	0·1
1·7	296	5	291	0·1
1·7	295	5	290	0·1
1·7	294	5	289	0·1
1·7	293	5	288	0·1
1·7	292	5	287	0·1
1·7	291	5	286	0·1
1·7	290	5	285	0·1
1·7	289	5	284	0·1
1·7	288	5	283	0·1
1·7	287	5	282	0·1
1·7	286	5	281	0·1
1·7	242	4	238	0·1
1·7	241	4	237	0·1
1·7	240	4	236	0·1
1·7	239	4	235	0·1
1·7	238	4	234	0·1
1·7	237	4	233	0·1

%E	M1	M2	DM	M*
1·7	236	4	232	0·1
1·7	235	4	231	0·1
1·7	234	4	230	0·1
1·7	233	4	229	0·1
1·7	232	4	228	0·1
1·7	231	4	227	0·1
1·7	230	4	226	0·1
1·7	229	4	225	0·1
1·7	181	3	178	0·0
1·7	180	3	177	0·0
1·7	179	3	176	0·1
1·7	178	3	175	0·1
1·7	177	3	174	0·1
1·7	176	3	173	0·1
1·7	175	3	172	0·1
1·7	174	3	171	0·1
1·7	173	3	170	0·1
1·7	172	3	169	0·1
1·7	121	2	119	0·0
1·7	120	2	118	0·0
1·7	119	2	117	0·0
1·7	118	2	116	0·0
1·7	117	2	115	0·0
1·7	116	2	114	0·0
1·7	115	2	113	0·0
1·7	60	1	59	0·0
1·7	59	1	58	0·0
1·7	58	1	57	0·0
1·6	500	8	492	0·1
1·6	499	8	491	0·1
1·6	498	8	490	0·1
1·6	497	8	489	0·1
1·6	496	8	488	0·1
1·6	495	8	487	0·1
1·6	494	8	486	0·1
1·6	493	8	485	0·1
1·6	492	8	484	0·1
1·6	491	8	483	0·1
1·6	490	8	482	0·1
1·6	489	8	481	0·1
1·6	488	8	480	0·1
1·6	487	8	479	0·1
1·6	486	8	478	0·1
1·6	485	8	477	0·1
1·6	451	7	444	0·1
1·6	450	7	443	0·1
1·6	449	7	442	0·1
1·6	448	7	441	0·1
1·6	447	7	440	0·1
1·6	446	7	439	0·1

%E	M1	M2	DM	M*
1·6	445	7	438	0·1
1·6	444	7	437	0·1
1·6	443	7	436	0·1
1·6	442	7	435	0·1
1·6	441	7	434	0·1
1·6	440	7	433	0·1
1·6	439	7	432	0·1
1·6	438	7	431	0·1
1·6	437	7	430	0·1
1·6	436	7	429	0·1
1·6	435	7	428	0·1
1·6	434	7	427	0·1
1·6	433	7	426	0·1
1·6	432	7	425	0·1
1·6	431	7	424	0·1
1·6	430	7	423	0·1
1·6	429	7	422	0·1
1·6	428	7	421	0·1
1·6	427	7	420	0·1
1·6	426	7	419	0·1
1·6	425	7	418	0·1
1·6	387	6	381	0·1
1·6	386	6	380	0·1
1·6	385	6	379	0·1
1·6	384	6	378	0·1
1·6	383	6	377	0·1
1·6	382	6	376	0·1
1·6	381	6	375	0·1
1·6	380	6	374	0·1
1·6	379	6	373	0·1
1·6	378	6	372	0·1
1·6	377	6	371	0·1
1·6	376	6	370	0·1
1·6	375	6	369	0·1
1·6	374	6	368	0·1
1·6	373	6	367	0·1
1·6	372	6	366	0·1
1·6	371	6	365	0·1
1·6	370	6	364	0·1
1·6	369	6	363	0·1
1·6	368	6	362	0·1
1·6	367	6	361	0·1
1·6	366	6	360	0·1
1·6	365	6	359	0·1
1·6	364	6	358	0·1
1·6	322	5	317	0·1
1·6	321	5	316	0·1
1·6	320	5	315	0·1
1·6	319	5	314	0·1
1·6	318	5	313	0·1

%E	M1	M2	DM	M*
1.6	317	5	312	0.1
1.6	316	5	311	0.1
1.6	315	5	310	0.1
1.6	314	5	309	0.1
1.6	313	5	308	0.1
1.6	312	5	307	0.1
1.6	311	5	306	0.1
1.6	310	5	305	0.1
1.6	309	5	304	0.1
1.6	308	5	303	0.1
1.6	307	5	302	0.1
1.6	306	5	301	0.1
1.6	305	5	300	0.1
1.6	304	5	299	0.1
1.6	258	4	254	0.1
1.6	257	4	253	0.1
1.6	256	4	252	0.1
1.6	255	4	251	0.1
1.6	254	4	250	0.1
1.6	253	4	249	0.1
1.6	252	4	248	0.1
1.6	251	4	247	0.1
1.6	250	4	246	0.1
1.6	249	4	245	0.1
1.6	248	4	244	0.1
1.6	247	4	243	0.1
1.6	246	4	242	0.1
1.6	245	4	241	0.1
1.6	244	4	240	0.1
1.6	243	4	239	0.1
1.6	193	3	190	0.0
1.6	192	3	189	0.0
1.6	191	3	188	0.0
1.6	190	3	187	0.0
1.6	189	3	186	0.0
1.6	188	3	185	0.0
1.6	187	3	184	0.0
1.6	186	3	183	0.0
1.6	185	3	182	0.0
1.6	184	3	181	0.0
1.6	183	3	180	0.0
1.6	182	3	179	0.0
1.6	129	2	127	0.0
1.6	128	2	126	0.0
1.6	127	2	125	0.0
1.6	126	2	124	0.0
1.6	125	2	123	0.0
1.6	124	2	122	0.0
1.6	123	2	121	0.0
1.6	122	2	120	0.0

%E	M1	M2	DM	M*
1.6	64	1	63	0.0
1.6	63	1	62	0.0
1.6	62	1	61	0.0
1.6	61	1	60	0.0
1.5	482	7	475	0.1
1.5	481	7	474	0.1
1.5	480	7	473	0.1
1.5	479	7	472	0.1
1.5	478	7	471	0.1
1.5	477	7	470	0.1
1.5	476	7	469	0.1
1.5	475	7	468	0.1
1.5	474	7	467	0.1
1.5	473	7	466	0.1
1.5	472	7	465	0.1
1.5	471	7	464	0.1
1.5	470	7	463	0.1
1.5	469	7	462	0.1
1.5	468	7	461	0.1
1.5	467	7	460	0.1
1.5	466	7	459	0.1
1.5	465	7	458	0.1
1.5	464	7	457	0.1
1.5	463	7	456	0.1
1.5	462	7	455	0.1
1.5	461	7	454	0.1
1.5	460	7	453	0.1
1.5	459	7	452	0.1
1.5	458	7	451	0.1
1.5	457	7	450	0.1
1.5	456	7	449	0.1
1.5	455	7	448	0.1
1.5	454	7	447	0.1
1.5	453	7	446	0.1
1.5	452	7	445	0.1
1.5	413	6	407	0.1
1.5	412	6	406	0.1
1.5	411	6	405	0.1
1.5	410	6	404	0.1
1.5	409	6	403	0.1
1.5	408	6	402	0.1
1.5	407	6	401	0.1
1.5	406	6	400	0.1
1.5	405	6	399	0.1
1.5	404	6	398	0.1
1.5	403	6	397	0.1
1.5	402	6	396	0.1
1.5	401	6	395	0.1
1.5	400	6	394	0.1
1.5	399	6	393	0.1

%E	M1	M2	DM	M*
1.5	398	6	392	0.1
1.5	397	6	391	0.1
1.5	396	6	390	0.1
1.5	395	6	389	0.1
1.5	394	6	388	0.1
1.5	393	6	387	0.1
1.5	392	6	386	0.1
1.5	391	6	385	0.1
1.5	390	6	384	0.1
1.5	389	6	383	0.1
1.5	388	6	382	0.1
1.5	344	5	339	0.1
1.5	343	5	338	0.1
1.5	342	5	337	0.1
1.5	341	5	336	0.1
1.5	340	5	335	0.1
1.5	339	5	334	0.1
1.5	338	5	333	0.1
1.5	337	5	332	0.1
1.5	336	5	331	0.1
1.5	335	5	330	0.1
1.5	334	5	329	0.1
1.5	333	5	328	0.1
1.5	332	5	327	0.1
1.5	331	5	326	0.1
1.5	330	5	325	0.1
1.5	329	5	324	0.1
1.5	328	5	323	0.1
1.5	327	5	322	0.1
1.5	326	5	321	0.1
1.5	325	5	320	0.1
1.5	324	5	319	0.1
1.5	323	5	318	0.1
1.5	275	4	271	0.1
1.5	274	4	270	0.1
1.5	273	4	269	0.1
1.5	272	4	268	0.1
1.5	271	4	267	0.1
1.5	270	4	266	0.1
1.5	269	4	265	0.1
1.5	268	4	264	0.1
1.5	267	4	263	0.1
1.5	266	4	262	0.1
1.5	265	4	261	0.1
1.5	264	4	260	0.1
1.5	263	4	259	0.1
1.5	262	4	258	0.1
1.5	261	4	257	0.1
1.5	260	4	256	0.1
1.5	259	4	255	0.1

%E	M1	M2	DM	M*
1.5	206	3	203	0.0
1.5	205	3	202	0.0
1.5	204	3	201	0.0
1.5	203	3	200	0.0
1.5	202	3	199	0.0
1.5	201	3	198	0.0
1.5	200	3	197	0.0
1.5	199	3	196	0.0
1.5	198	3	195	0.0
1.5	197	3	194	0.0
1.5	196	3	193	0.0
1.5	195	3	192	0.0
1.5	194	3	191	0.0
1.5	137	2	135	0.0
1.5	136	2	134	0.0
1.5	135	2	133	0.0
1.5	134	2	132	0.0
1.5	133	2	131	0.0
1.5	132	2	130	0.0
1.5	131	2	129	0.0
1.5	130	2	128	0.0
1.5	68	1	67	0.0
1.5	67	1	66	0.0
1.5	66	1	65	0.0
1.5	65	1	64	0.0
1.4	500	7	493	0.1
1.4	499	7	492	0.1
1.4	498	7	491	0.1
1.4	497	7	490	0.1
1.4	496	7	489	0.1
1.4	495	7	488	0.1
1.4	494	7	487	0.1
1.4	493	7	486	0.1
1.4	492	7	485	0.1
1.4	491	7	484	0.1
1.4	490	7	483	0.1
1.4	489	7	482	0.1
1.4	488	7	481	0.1
1.4	487	7	480	0.1
1.4	486	7	479	0.1
1.4	485	7	478	0.1
1.4	484	7	477	0.1
1.4	483	7	476	0.1
1.4	444	6	438	0.1
1.4	443	6	437	0.1
1.4	442	6	436	0.1
1.4	441	6	435	0.1
1.4	440	6	434	0.1
1.4	439	6	433	0.1
1.4	438	6	432	0.1

%E	M1	M2	DM	M*
1.4	437	6	431	0.1
1.4	436	6	430	0.1
1.4	435	6	429	0.1
1.4	434	6	428	0.1
1.4	433	6	427	0.1
1.4	432	6	426	0.1
1.4	431	6	425	0.1
1.4	430	6	424	0.1
1.4	429	6	423	0.1
1.4	428	6	422	0.1
1.4	427	6	421	0.1
1.4	426	6	420	0.1
1.4	425	6	419	0.1
1.4	424	6	418	0.1
1.4	423	6	417	0.1
1.4	422	6	416	0.1
1.4	421	6	415	0.1
1.4	420	6	414	0.1
1.4	419	6	413	0.1
1.4	418	6	412	0.1
1.4	417	6	411	0.1
1.4	416	6	410	0.1
1.4	415	6	409	0.1
1.4	414	6	408	0.1
1.4	370	5	365	0.1
1.4	369	5	364	0.1
1.4	368	5	363	0.1
1.4	367	5	362	0.1
1.4	366	5	361	0.1
1.4	365	5	360	0.1
1.4	364	5	359	0.1
1.4	363	5	358	0.1
1.4	362	5	357	0.1
1.4	361	5	356	0.1
1.4	360	5	355	0.1
1.4	359	5	354	0.1
1.4	358	5	353	0.1
1.4	357	5	352	0.1
1.4	356	5	351	0.1
1.4	355	5	350	0.1
1.4	354	5	349	0.1
1.4	353	5	348	0.1
1.4	352	5	347	0.1
1.4	351	5	346	0.1
1.4	350	5	345	0.1
1.4	349	5	344	0.1
1.4	348	5	343	0.1
1.4	347	5	342	0.1
1.4	346	5	341	0.1
1.4	345	5	340	0.1

%E	M1	M2	DM	M*
1•4	296	4	292	0•1
1•4	295	4	291	0•1
1•4	294	4	290	0•1
1•4	293	4	289	0•1
1•4	292	4	288	0•1
1•4	291	4	287	0•1
1•4	290	4	286	0•1
1•4	289	4	285	0•1
1•4	288	4	284	0•1
1•4	287	4	283	0•1
1•4	286	4	282	0•1
1•4	285	4	281	0•1
1•4	284	4	280	0•1
1•4	283	4	279	0•1
1•4	282	4	278	0•1
1•4	281	4	277	0•1
1•4	280	4	276	0•1
1•4	279	4	275	0•1
1•4	278	4	274	0•1
1•4	277	4	273	0•1
1•4	276	4	272	0•1
1•4	222	3	219	0•0
1•4	221	3	218	0•0
1•4	220	3	217	0•0
1•4	219	3	216	0•0
1•4	218	3	215	0•0
1•4	217	3	214	0•0
1•4	216	3	213	0•0
1•4	215	3	212	0•0
1•4	214	3	211	0•0
1•4	213	3	210	0•0
1•4	212	3	209	0•0
1•4	211	3	208	0•0
1•4	210	3	207	0•0
1•4	209	3	206	0•0
1•4	208	3	205	0•0
1•4	207	3	204	0•0
1•4	148	2	146	0•0
1•4	147	2	145	0•0
1•4	146	2	144	0•0
1•4	145	2	143	0•0
1•4	144	2	142	0•0
1•4	143	2	141	0•0
1•4	142	2	140	0•0
1•4	141	2	139	0•0
1•4	140	2	138	0•0
1•4	139	2	137	0•0
1•4	138	2	136	0•0
1•4	74	1	73	0•0
1•4	73	1	72	0•0

%E	M1	M2	DM	M*
1•4	72	1	71	0•0
1•4	71	1	70	0•0
1•4	70	1	69	0•0
1•4	69	1	68	0•0
1•3	480	6	474	0•1
1•3	479	6	473	0•1
1•3	478	6	472	0•1
1•3	477	6	471	0•1
1•3	476	6	470	0•1
1•3	475	6	469	0•1
1•3	474	6	468	0•1
1•3	473	6	467	0•1
1•3	472	6	466	0•1
1•3	471	6	465	0•1
1•3	470	6	464	0•1
1•3	469	6	463	0•1
1•3	468	6	462	0•1
1•3	467	6	461	0•1
1•3	466	6	460	0•1
1•3	465	6	459	0•1
1•3	464	6	458	0•1
1•3	463	6	457	0•1
1•3	462	6	456	0•1
1•3	461	6	455	0•1
1•3	460	6	454	0•1
1•3	459	6	453	0•1
1•3	458	6	452	0•1
1•3	457	6	451	0•1
1•3	456	6	450	0•1
1•3	455	6	449	0•1
1•3	454	6	448	0•1
1•3	453	6	447	0•1
1•3	452	6	446	0•1
1•3	451	6	445	0•1
1•3	450	6	444	0•1
1•3	449	6	443	0•1
1•3	448	6	442	0•1
1•3	447	6	441	0•1
1•3	446	6	440	0•1
1•3	445	6	439	0•1
1•3	400	5	395	0•1
1•3	399	5	394	0•1
1•3	398	5	393	0•1
1•3	397	5	392	0•1
1•3	396	5	391	0•1
1•3	395	5	390	0•1
1•3	394	5	389	0•1
1•3	393	5	388	0•1
1•3	392	5	387	0•1
1•3	391	5	386	0•1

%E	M1	M2	DM	M*
1•3	390	5	385	0•1
1•3	389	5	384	0•1
1•3	388	5	383	0•1
1•3	387	5	382	0•1
1•3	386	5	381	0•1
1•3	385	5	380	0•1
1•3	384	5	379	0•1
1•3	383	5	378	0•1
1•3	382	5	377	0•1
1•3	381	5	376	0•1
1•3	380	5	375	0•1
1•3	379	5	374	0•1
1•3	378	5	373	0•1
1•3	377	5	372	0•1
1•3	376	5	371	0•1
1•3	375	5	370	0•1
1•3	374	5	369	0•1
1•3	373	5	368	0•1
1•3	372	5	367	0•1
1•3	371	5	366	0•1
1•3	320	4	316	0•0
1•3	319	4	315	0•1
1•3	318	4	314	0•1
1•3	317	4	313	0•1
1•3	316	4	312	0•1
1•3	315	4	311	0•1
1•3	314	4	310	0•1
1•3	313	4	309	0•1
1•3	312	4	308	0•1
1•3	311	4	307	0•1
1•3	310	4	306	0•1
1•3	309	4	305	0•1
1•3	308	4	304	0•1
1•3	307	4	303	0•1
1•3	306	4	302	0•1
1•3	305	4	301	0•1
1•3	304	4	300	0•1
1•3	303	4	299	0•1
1•3	302	4	298	0•1
1•3	301	4	297	0•1
1•3	300	4	296	0•1
1•3	299	4	295	0•1
1•3	298	4	294	0•1
1•3	297	4	293	0•1
1•3	240	3	237	0•0
1•3	239	3	236	0•0
1•3	238	3	235	0•0
1•3	237	3	234	0•0
1•3	236	3	233	0•0
1•3	235	3	232	0•0

%E	M1	M2	DM	M*
1•3	234	3	231	0•0
1•3	233	3	230	0•0
1•3	232	3	229	0•0
1•3	231	3	228	0•0
1•3	230	3	227	0•0
1•3	229	3	226	0•0
1•3	228	3	225	0•0
1•3	227	3	224	0•0
1•3	226	3	223	0•0
1•3	225	3	222	0•0
1•3	224	3	221	0•0
1•3	223	3	220	0•0
1•3	160	2	158	0•0
1•3	159	2	157	0•0
1•3	158	2	156	0•0
1•3	157	2	155	0•0
1•3	156	2	154	0•0
1•3	155	2	153	0•0
1•3	154	2	152	0•0
1•3	153	2	151	0•0
1•3	152	2	150	0•0
1•3	151	2	149	0•0
1•3	150	2	148	0•0
1•3	149	2	147	0•0
1•3	80	1	79	0•0
1•3	79	1	78	0•0
1•3	78	1	77	0•0
1•3	77	1	76	0•0
1•3	76	1	75	0•0
1•3	75	1	74	0•0
1•2	500	6	494	0•1
1•2	499	6	493	0•1
1•2	498	6	492	0•1
1•2	497	6	491	0•1
1•2	496	6	490	0•1
1•2	495	6	489	0•1
1•2	494	6	488	0•1
1•2	493	6	487	0•1
1•2	492	6	486	0•1
1•2	491	6	485	0•1
1•2	490	6	484	0•1
1•2	489	6	483	0•1
1•2	488	6	482	0•1
1•2	487	6	481	0•1
1•2	486	6	480	0•1
1•2	485	6	479	0•1
1•2	484	6	478	0•1
1•2	483	6	477	0•1
1•2	482	6	476	0•1
1•2	481	6	475	0•1

%E	M1	M2	DM	M*
1•2	434	5	429	0•1
1•2	433	5	428	0•1
1•2	432	5	427	0•1
1•2	431	5	426	0•1
1•2	430	5	425	0•1
1•2	429	5	424	0•1
1•2	428	5	423	0•1
1•2	427	5	422	0•1
1•2	426	5	421	0•1
1•2	425	5	420	0•1
1•2	424	5	419	0•1
1•2	423	5	418	0•1
1•2	422	5	417	0•1
1•2	421	5	416	0•1
1•2	420	5	415	0•1
1•2	419	5	414	0•1
1•2	418	5	413	0•1
1•2	417	5	412	0•1
1•2	416	5	411	0•1
1•2	415	5	410	0•1
1•2	414	5	409	0•1
1•2	413	5	408	0•1
1•2	412	5	407	0•1
1•2	411	5	406	0•1
1•2	410	5	405	0•1
1•2	409	5	404	0•1
1•2	408	5	403	0•1
1•2	407	5	402	0•1
1•2	406	5	401	0•1
1•2	405	5	400	0•1
1•2	404	5	399	0•1
1•2	403	5	398	0•1
1•2	402	5	397	0•1
1•2	401	5	396	0•1
1•2	347	4	343	0•0
1•2	346	4	342	0•0
1•2	345	4	341	0•0
1•2	344	4	340	0•0
1•2	343	4	339	0•0
1•2	342	4	338	0•0
1•2	341	4	337	0•0
1•2	340	4	336	0•0
1•2	339	4	335	0•0
1•2	338	4	334	0•0
1•2	337	4	333	0•0
1•2	336	4	332	0•0
1•2	335	4	331	0•0
1•2	334	4	330	0•0
1•2	333	4	329	0•0
1•2	332	4	328	0•0

%E	M1	M2	DM	M*
1.2	331	4	327	0.0
1.2	330	4	326	0.0
1.2	329	4	325	0.0
1.2	328	4	324	0.0
1.2	327	4	323	0.0
1.2	326	4	322	0.0
1.2	325	4	321	0.0
1.2	324	4	320	0.0
1.2	323	4	319	0.0
1.2	322	4	318	0.0
1.2	321	4	317	0.0
1.2	260	3	257	0.0
1.2	259	3	256	0.0
1.2	258	3	255	0.0
1.2	257	3	254	0.0
1.2	256	3	253	0.0
1.2	255	3	252	0.0
1.2	254	3	251	0.0
1.2	253	3	250	0.0
1.2	252	3	249	0.0
1.2	251	3	248	0.0
1.2	250	3	247	0.0
1.2	249	3	246	0.0
1.2	248	3	245	0.0
1.2	247	3	244	0.0
1.2	246	3	243	0.0
1.2	245	3	242	0.0
1.2	244	3	241	0.0
1.2	243	3	240	0.0
1.2	242	3	239	0.0
1.2	241	3	238	0.0
1.2	173	2	171	0.0
1.2	172	2	170	0.0
1.2	171	2	169	0.0
1.2	170	2	168	0.0
1.2	169	2	167	0.0
1.2	168	2	166	0.0
1.2	167	2	165	0.0
1.2	166	2	164	0.0
1.2	165	2	163	0.0
1.2	164	2	162	0.0
1.2	163	2	161	0.0
1.2	162	2	160	0.0
1.2	161	2	159	0.0
1.2	86	1	85	0.0
1.2	85	1	84	0.0
1.2	84	1	83	0.0
1.2	83	1	82	0.0
1.2	82	1	81	0.0
1.2	81	1	80	0.0

%E	M1	M2	DM	M*
1.1	476	5	471	0.1
1.1	475	5	470	0.1
1.1	474	5	469	0.1
1.1	473	5	468	0.1
1.1	472	5	467	0.1
1.1	471	5	466	0.1
1.1	470	5	465	0.1
1.1	469	5	464	0.1
1.1	468	5	463	0.1
1.1	467	5	462	0.1
1.1	466	5	461	0.1
1.1	465	5	460	0.1
1.1	464	5	459	0.1
1.1	463	5	458	0.1
1.1	462	5	457	0.1
1.1	461	5	456	0.1
1.1	460	5	455	0.1
1.1	459	5	454	0.1
1.1	458	5	453	0.1
1.1	457	5	452	0.1
1.1	456	5	451	0.1
1.1	455	5	450	0.1
1.1	454	5	449	0.1
1.1	453	5	448	0.1
1.1	452	5	447	0.1
1.1	451	5	446	0.1
1.1	450	5	445	0.1
1.1	449	5	444	0.1
1.1	448	5	443	0.1
1.1	447	5	442	0.1
1.1	446	5	441	0.1
1.1	445	5	440	0.1
1.1	444	5	439	0.1
1.1	443	5	438	0.1
1.1	442	5	437	0.1
1.1	441	5	436	0.1
1.1	440	5	435	0.1
1.1	439	5	434	0.1
1.1	438	5	433	0.1
1.1	437	5	432	0.1
1.1	436	5	431	0.1
1.1	435	5	430	0.1
1.1	381	4	377	0.0
1.1	380	4	376	0.0
1.1	379	4	375	0.0
1.1	378	4	374	0.0
1.1	377	4	373	0.0
1.1	376	4	372	0.0
1.1	375	4	371	0.0
1.1	374	4	370	0.0

%E	M1	M2	DM	M*
1.1	373	4	369	0.0
1.1	372	4	368	0.0
1.1	371	4	367	0.0
1.1	370	4	366	0.0
1.1	369	4	365	0.0
1.1	368	4	364	0.0
1.1	367	4	363	0.0
1.1	366	4	362	0.0
1.1	365	4	361	0.0
1.1	364	4	360	0.0
1.1	363	4	359	0.0
1.1	362	4	358	0.0
1.1	361	4	357	0.0
1.1	360	4	356	0.0
1.1	359	4	355	0.0
1.1	358	4	354	0.0
1.1	357	4	353	0.0
1.1	356	4	352	0.0
1.1	355	4	351	0.0
1.1	354	4	350	0.0
1.1	353	4	349	0.0
1.1	352	4	348	0.0
1.1	351	4	347	0.0
1.1	350	4	346	0.0
1.1	349	4	345	0.0
1.1	348	4	344	0.0
1.1	285	3	282	0.0
1.1	284	3	281	0.0
1.1	283	3	280	0.0
1.1	282	3	279	0.0
1.1	281	3	278	0.0
1.1	280	3	277	0.0
1.1	279	3	276	0.0
1.1	278	3	275	0.0
1.1	277	3	274	0.0
1.1	276	3	273	0.0
1.1	275	3	272	0.0
1.1	274	3	271	0.0
1.1	273	3	270	0.0
1.1	272	3	269	0.0
1.1	271	3	268	0.0
1.1	270	3	267	0.0
1.1	269	3	266	0.0
1.1	268	3	265	0.0
1.1	267	3	264	0.0
1.1	266	3	263	0.0
1.1	265	3	262	0.0
1.1	264	3	261	0.0
1.1	263	3	260	0.0
1.1	262	3	259	0.0

%E	M1	M2	DM	M*
1.1	261	3	258	0.0
1.1	190	2	188	0.0
1.1	189	2	187	0.0
1.1	188	2	186	0.0
1.1	187	2	185	0.0
1.1	186	2	184	0.0
1.1	185	2	183	0.0
1.1	184	2	182	0.0
1.1	183	2	181	0.0
1.1	182	2	180	0.0
1.1	181	2	179	0.0
1.1	180	2	178	0.0
1.1	179	2	177	0.0
1.1	178	2	176	0.0
1.1	177	2	175	0.0
1.1	176	2	174	0.0
1.1	175	2	173	0.0
1.1	174	2	172	0.0
1.1	95	1	94	0.0
1.1	94	1	93	0.0
1.1	93	1	92	0.0
1.1	92	1	91	0.0
1.1	91	1	90	0.0
1.1	90	1	89	0.0
1.1	89	1	88	0.0
1.1	88	1	87	0.0
1.1	87	1	86	0.0
1.0	500	5	495	0.0
1.0	499	5	494	0.1
1.0	498	5	493	0.1
1.0	497	5	492	0.1
1.0	496	5	491	0.1
1.0	495	5	490	0.1
1.0	494	5	489	0.1
1.0	493	5	488	0.1
1.0	492	5	487	0.1
1.0	491	5	486	0.1
1.0	490	5	485	0.1
1.0	489	5	484	0.1
1.0	488	5	483	0.1
1.0	487	5	482	0.1
1.0	486	5	481	0.1
1.0	485	5	480	0.1
1.0	484	5	479	0.1
1.0	483	5	478	0.1
1.0	482	5	477	0.1
1.0	481	5	476	0.1
1.0	480	5	475	0.1
1.0	479	5	474	0.1
1.0	478	5	473	0.1

%E	M1	M2	DM	M*
1.0	477	5	472	0.1
1.0	421	4	417	0.0
1.0	420	4	416	0.0
1.0	419	4	415	0.0
1.0	418	4	414	0.0
1.0	417	4	413	0.0
1.0	416	4	412	0.0
1.0	415	4	411	0.0
1.0	414	4	410	0.0
1.0	413	4	409	0.0
1.0	412	4	408	0.0
1.0	411	4	407	0.0
1.0	410	4	406	0.0
1.0	409	4	405	0.0
1.0	408	4	404	0.0
1.0	407	4	403	0.0
1.0	406	4	402	0.0
1.0	405	4	401	0.0
1.0	404	4	400	0.0
1.0	403	4	399	0.0
1.0	402	4	398	0.0
1.0	401	4	397	0.0
1.0	400	4	396	0.0
1.0	399	4	395	0.0
1.0	398	4	394	0.0
1.0	397	4	393	0.0
1.0	396	4	392	0.0
1.0	395	4	391	0.0
1.0	394	4	390	0.0
1.0	393	4	389	0.0
1.0	392	4	388	0.0
1.0	391	4	387	0.0
1.0	390	4	386	0.0
1.0	389	4	385	0.0
1.0	388	4	384	0.0
1.0	387	4	383	0.0
1.0	386	4	382	0.0
1.0	385	4	381	0.0
1.0	384	4	380	0.0
1.0	383	4	379	0.0
1.0	382	4	378	0.0
1.0	315	3	312	0.0
1.0	314	3	311	0.0
1.0	313	3	310	0.0
1.0	312	3	309	0.0
1.0	311	3	308	0.0
1.0	310	3	307	0.0
1.0	309	3	306	0.0
1.0	308	3	305	0.0
1.0	307	3	304	0.0

%E	M1	M2	DM	M*
1.0	306	3	303	0.0
1.0	305	3	302	0.0
1.0	304	3	301	0.0
1.0	303	3	300	0.0
1.0	302	3	299	0.0
1.0	301	3	298	0.0
1.0	300	3	297	0.0
1.0	299	3	296	0.0
1.0	298	3	295	0.0
1.0	297	3	294	0.0
1.0	296	3	293	0.0
1.0	295	3	292	0.0
1.0	294	3	291	0.0
1.0	293	3	290	0.0
1.0	292	3	289	0.0
1.0	291	3	288	0.0
1.0	290	3	287	0.0
1.0	289	3	286	0.0
1.0	288	3	285	0.0
1.0	287	3	284	0.0
1.0	286	3	283	0.0
1.0	210	2	208	0.0
1.0	209	2	207	0.0
1.0	208	2	206	0.0
1.0	207	2	205	0.0
1.0	206	2	204	0.0
1.0	205	2	203	0.0
1.0	204	2	202	0.0
1.0	203	2	201	0.0
1.0	202	2	200	0.0
1.0	201	2	199	0.0
1.0	200	2	198	0.0
1.0	199	2	197	0.0
1.0	198	2	196	0.0
1.0	197	2	195	0.0
1.0	196	2	194	0.0
1.0	195	2	193	0.0
1.0	194	2	192	0.0
1.0	193	2	191	0.0
1.0	192	2	190	0.0
1.0	191	2	189	0.0
1.0	105	1	104	0.0
1.0	104	1	103	0.0
1.0	103	1	102	0.0
1.0	102	1	101	0.0
1.0	101	1	100	0.0
1.0	100	1	99	0.0
1.0	99	1	98	0.0
1.0	98	1	97	0.0
1.0	97	1	96	0.0

%E	M1	M2	DM	M*
1.0	96	1	95	0.0
0.9	470	4	466	0.0
0.9	469	4	465	0.0
0.9	468	4	464	0.0
0.9	467	4	463	0.0
0.9	466	4	462	0.0
0.9	465	4	461	0.0
0.9	464	4	460	0.0
0.9	463	4	459	0.0
0.9	462	4	458	0.0
0.9	461	4	457	0.0
0.9	460	4	456	0.0
0.9	459	4	455	0.0
0.9	458	4	454	0.0
0.9	457	4	453	0.0
0.9	456	4	452	0.0
0.9	455	4	451	0.0
0.9	454	4	450	0.0
0.9	453	4	449	0.0
0.9	452	4	448	0.0
0.9	451	4	447	0.0
0.9	450	4	446	0.0
0.9	449	4	445	0.0
0.9	448	4	444	0.0
0.9	447	4	443	0.0
0.9	446	4	442	0.0
0.9	445	4	441	0.0
0.9	444	4	440	0.0
0.9	443	4	439	0.0
0.9	442	4	438	0.0
0.9	441	4	437	0.0
0.9	440	4	436	0.0
0.9	439	4	435	0.0
0.9	438	4	434	0.0
0.9	437	4	433	0.0
0.9	436	4	432	0.0
0.9	435	4	431	0.0
0.9	434	4	430	0.0
0.9	433	4	429	0.0
0.9	432	4	428	0.0
0.9	431	4	427	0.0
0.9	430	4	426	0.0
0.9	429	4	425	0.0
0.9	428	4	424	0.0
0.9	427	4	423	0.0
0.9	426	4	422	0.0
0.9	425	4	421	0.0
0.9	424	4	420	0.0
0.9	423	4	419	0.0
0.9	422	4	418	0.0

%E	M1	M2	DM	M*
0.9	353	3	350	0.0
0.9	352	3	349	0.0
0.9	351	3	348	0.0
0.9	350	3	347	0.0
0.9	349	3	346	0.0
0.9	348	3	345	0.0
0.9	347	3	344	0.0
0.9	346	3	343	0.0
0.9	345	3	342	0.0
0.9	344	3	341	0.0
0.9	343	3	340	0.0
0.9	342	3	339	0.0
0.9	341	3	338	0.0
0.9	340	3	337	0.0
0.9	339	3	336	0.0
0.9	338	3	335	0.0
0.9	337	3	334	0.0
0.9	336	3	333	0.0
0.9	335	3	332	0.0
0.9	334	3	331	0.0
0.9	333	3	330	0.0
0.9	332	3	329	0.0
0.9	331	3	328	0.0
0.9	330	3	327	0.0
0.9	329	3	326	0.0
0.9	328	3	325	0.0
0.9	327	3	324	0.0
0.9	326	3	323	0.0
0.9	325	3	322	0.0
0.9	324	3	321	0.0
0.9	323	3	320	0.0
0.9	322	3	319	0.0
0.9	321	3	318	0.0
0.9	320	3	317	0.0
0.9	319	3	316	0.0
0.9	318	3	315	0.0
0.9	317	3	314	0.0
0.9	316	3	313	0.0
0.9	235	2	233	0.0
0.9	234	2	232	0.0
0.9	233	2	231	0.0
0.9	232	2	230	0.0
0.9	231	2	229	0.0
0.9	230	2	228	0.0
0.9	229	2	227	0.0
0.9	228	2	226	0.0
0.9	227	2	225	0.0
0.9	226	2	224	0.0
0.9	225	2	223	0.0
0.9	224	2	222	0.0

%E	M1	M2	DM	M*
0.9	223	2	221	0.0
0.9	222	2	220	0.0
0.9	221	2	219	0.0
0.9	220	2	218	0.0
0.9	219	2	217	0.0
0.9	218	2	216	0.0
0.9	217	2	215	0.0
0.9	216	2	214	0.0
0.9	215	2	213	0.0
0.9	214	2	212	0.0
0.9	213	2	211	0.0
0.9	212	2	210	0.0
0.9	211	2	209	0.0
0.9	117	1	116	0.0
0.9	116	1	115	0.0
0.9	115	1	114	0.0
0.9	114	1	113	0.0
0.9	113	1	112	0.0
0.9	112	1	111	0.0
0.9	111	1	110	0.0
0.9	110	1	109	0.0
0.9	109	1	108	0.0
0.9	108	1	107	0.0
0.9	107	1	106	0.0
0.9	106	1	105	0.0
0.8	500	4	496	0.0
0.8	499	4	495	0.0
0.8	498	4	494	0.0
0.8	497	4	493	0.0
0.8	496	4	492	0.0
0.8	495	4	491	0.0
0.8	494	4	490	0.0
0.8	493	4	489	0.0
0.8	492	4	488	0.0
0.8	491	4	487	0.0
0.8	490	4	486	0.0
0.8	489	4	485	0.0
0.8	488	4	484	0.0
0.8	487	4	483	0.0
0.8	486	4	482	0.0
0.8	485	4	481	0.0
0.8	484	4	480	0.0
0.8	483	4	479	0.0
0.8	482	4	478	0.0
0.8	481	4	477	0.0
0.8	480	4	476	0.0
0.8	479	4	475	0.0
0.8	478	4	474	0.0
0.8	477	4	473	0.0
0.8	476	4	472	0.0

%E	M1	M2	DM	M*
0.8	475	4	471	0.0
0.8	474	4	470	0.0
0.8	473	4	469	0.0
0.8	472	4	468	0.0
0.8	471	4	467	0.0
0.8	400	3	397	0.0
0.8	399	3	396	0.0
0.8	398	3	395	0.0
0.8	397	3	394	0.0
0.8	396	3	393	0.0
0.8	395	3	392	0.0
0.8	394	3	391	0.0
0.8	393	3	390	0.0
0.8	392	3	389	0.0
0.8	391	3	388	0.0
0.8	390	3	387	0.0
0.8	389	3	386	0.0
0.8	388	3	385	0.0
0.8	387	3	384	0.0
0.8	386	3	383	0.0
0.8	385	3	382	0.0
0.8	384	3	381	0.0
0.8	383	3	380	0.0
0.8	382	3	379	0.0
0.8	381	3	378	0.0
0.8	380	3	377	0.0
0.8	379	3	376	0.0
0.8	378	3	375	0.0
0.8	377	3	374	0.0
0.8	376	3	373	0.0
0.8	375	3	372	0.0
0.8	374	3	371	0.0
0.8	373	3	370	0.0
0.8	372	3	369	0.0
0.8	371	3	368	0.0
0.8	370	3	367	0.0
0.8	369	3	366	0.0
0.8	368	3	365	0.0
0.8	367	3	364	0.0
0.8	366	3	363	0.0
0.8	365	3	362	0.0
0.8	364	3	361	0.0
0.8	363	3	360	0.0
0.8	362	3	359	0.0
0.8	361	3	358	0.0
0.8	360	3	357	0.0
0.8	359	3	356	0.0
0.8	358	3	355	0.0
0.8	357	3	354	0.0
0.8	356	3	353	0.0

%E	M1	M2	DM	M*
0.8	355	3	352	0.0
0.8	354	3	351	0.0
0.8	266	2	264	0.0
0.8	265	2	263	0.0
0.8	264	2	262	0.0
0.8	263	2	261	0.0
0.8	262	2	260	0.0
0.8	261	2	259	0.0
0.8	260	2	258	0.0
0.8	259	2	257	0.0
0.8	258	2	256	0.0
0.8	257	2	255	0.0
0.8	256	2	254	0.0
0.8	255	2	253	0.0
0.8	254	2	252	0.0
0.8	253	2	251	0.0
0.8	252	2	250	0.0
0.8	251	2	249	0.0
0.8	250	2	248	0.0
0.8	249	2	247	0.0
0.8	248	2	246	0.0
0.8	247	2	245	0.0
0.8	246	2	244	0.0
0.8	245	2	243	0.0
0.8	244	2	242	0.0
0.8	243	2	241	0.0
0.8	242	2	240	0.0
0.8	241	2	239	0.0
0.8	240	2	238	0.0
0.8	239	2	237	0.0
0.8	238	2	236	0.0
0.8	237	2	235	0.0
0.8	236	2	234	0.0
0.8	133	1	132	0.0
0.8	132	1	131	0.0
0.8	131	1	130	0.0
0.8	130	1	129	0.0
0.8	129	1	128	0.0
0.8	128	1	127	0.0
0.8	127	1	126	0.0
0.8	126	1	125	0.0
0.8	125	1	124	0.0
0.8	124	1	123	0.0
0.8	123	1	122	0.0
0.8	122	1	121	0.0
0.8	121	1	120	0.0
0.8	120	1	119	0.0
0.8	119	1	118	0.0
0.8	118	1	117	0.0
0.7	461	3	458	0.0
0.7	460	3	457	0.0
0.7	459	3	456	0.0
0.7	458	3	455	0.0
0.7	457	3	454	0.0
0.7	456	3	453	0.0
0.7	455	3	452	0.0
0.7	454	3	451	0.0
0.7	453	3	450	0.0
0.7	452	3	449	0.0
0.7	451	3	448	0.0
0.7	450	3	447	0.0
0.7	449	3	446	0.0
0.7	448	3	445	0.0
0.7	447	3	444	0.0
0.7	446	3	443	0.0
0.7	445	3	442	0.0
0.7	444	3	441	0.0
0.7	443	3	440	0.0
0.7	442	3	439	0.0
0.7	441	3	438	0.0
0.7	440	3	437	0.0
0.7	439	3	436	0.0
0.7	438	3	435	0.0
0.7	437	3	434	0.0
0.7	436	3	433	0.0
0.7	435	3	432	0.0
0.7	434	3	431	0.0
0.7	433	3	430	0.0
0.7	432	3	429	0.0
0.7	431	3	428	0.0
0.7	430	3	427	0.0
0.7	429	3	426	0.0
0.7	428	3	425	0.0
0.7	427	3	424	0.0
0.7	426	3	423	0.0
0.7	425	3	422	0.0
0.7	424	3	421	0.0
0.7	423	3	420	0.0
0.7	422	3	419	0.0
0.7	421	3	418	0.0
0.7	420	3	417	0.0
0.7	419	3	416	0.0
0.7	418	3	415	0.0
0.7	417	3	414	0.0
0.7	416	3	413	0.0
0.7	415	3	412	0.0
0.7	414	3	411	0.0
0.7	413	3	410	0.0
0.7	412	3	409	0.0
0.7	411	3	408	0.0
0.7	410	3	407	0.0
0.7	409	3	406	0.0
0.7	408	3	405	0.0
0.7	407	3	404	0.0
0.7	406	3	403	0.0
0.7	405	3	402	0.0
0.7	404	3	401	0.0
0.7	403	3	400	0.0
0.7	402	3	399	0.0
0.7	401	3	398	0.0
0.7	307	2	305	0.0
0.7	306	2	304	0.0
0.7	305	2	303	0.0
0.7	304	2	302	0.0
0.7	303	2	301	0.0
0.7	302	2	300	0.0
0.7	301	2	299	0.0
0.7	300	2	298	0.0
0.7	299	2	297	0.0
0.7	298	2	296	0.0
0.7	297	2	295	0.0
0.7	296	2	294	0.0
0.7	295	2	293	0.0
0.7	294	2	292	0.0
0.7	293	2	291	0.0
0.7	292	2	290	0.0
0.7	291	2	289	0.0
0.7	290	2	288	0.0
0.7	289	2	287	0.0
0.7	288	2	286	0.0
0.7	287	2	285	0.0
0.7	286	2	284	0.0
0.7	285	2	283	0.0
0.7	284	2	282	0.0
0.7	283	2	281	0.0
0.7	282	2	280	0.0
0.7	281	2	279	0.0
0.7	280	2	278	0.0
0.7	279	2	277	0.0
0.7	278	2	276	0.0
0.7	277	2	275	0.0
0.7	276	2	274	0.0
0.7	275	2	273	0.0
0.7	274	2	272	0.0
0.7	273	2	271	0.0
0.7	272	2	270	0.0
0.7	271	2	269	0.0
0.7	270	2	268	0.0
0.7	269	2	267	0.0
0.7	268	2	266	0.0
0.7	267	2	265	0.0
0.7	153	1	152	0.0
0.7	152	1	151	0.0
0.7	151	1	150	0.0
0.7	150	1	149	0.0
0.7	149	1	148	0.0
0.7	148	1	147	0.0
0.7	147	1	146	0.0
0.7	146	1	145	0.0
0.7	145	1	144	0.0
0.7	144	1	143	0.0
0.7	143	1	142	0.0
0.7	142	1	141	0.0
0.7	141	1	140	0.0
0.7	140	1	139	0.0
0.7	139	1	138	0.0
0.7	138	1	137	0.0
0.7	137	1	136	0.0
0.7	136	1	135	0.0
0.7	135	1	134	0.0
0.7	134	1	133	0.0
0.6	500	3	497	0.0
0.6	499	3	496	0.0
0.6	498	3	495	0.0
0.6	497	3	494	0.0
0.6	496	3	493	0.0
0.6	495	3	492	0.0
0.6	494	3	491	0.0
0.6	493	3	490	0.0
0.6	492	3	489	0.0
0.6	491	3	488	0.0
0.6	490	3	487	0.0
0.6	489	3	486	0.0
0.6	488	3	485	0.0
0.6	487	3	484	0.0
0.6	486	3	483	0.0
0.6	485	3	482	0.0
0.6	484	3	481	0.0
0.6	483	3	480	0.0
0.6	482	3	479	0.0
0.6	481	3	478	0.0
0.6	480	3	477	0.0
0.6	479	3	476	0.0
0.6	478	3	475	0.0
0.6	477	3	474	0.0
0.6	476	3	473	0.0
0.6	475	3	472	0.0
0.6	474	3	471	0.0
0.6	473	3	470	0.0
0.6	472	3	469	0.0
0.6	471	3	468	0.0
0.6	470	3	467	0.0
0.6	469	3	466	0.0
0.6	468	3	465	0.0
0.6	467	3	464	0.0
0.6	466	3	463	0.0
0.6	465	3	462	0.0
0.6	464	3	461	0.0
0.6	463	3	460	0.0
0.6	462	3	459	0.0
0.6	363	2	361	0.0
0.6	362	2	360	0.0
0.6	361	2	359	0.0
0.6	360	2	358	0.0
0.6	359	2	357	0.0
0.6	358	2	356	0.0
0.6	357	2	355	0.0
0.6	356	2	354	0.0
0.6	355	2	353	0.0
0.6	354	2	352	0.0
0.6	353	2	351	0.0
0.6	352	2	350	0.0
0.6	351	2	349	0.0
0.6	350	2	348	0.0
0.6	349	2	347	0.0
0.6	348	2	346	0.0
0.6	347	2	345	0.0
0.6	346	2	344	0.0
0.6	345	2	343	0.0
0.6	344	2	342	0.0
0.6	343	2	341	0.0
0.6	342	2	340	0.0
0.6	341	2	339	0.0
0.6	340	2	338	0.0
0.6	339	2	337	0.0
0.6	338	2	336	0.0
0.6	337	2	335	0.0
0.6	336	2	334	0.0
0.6	335	2	333	0.0
0.6	334	2	332	0.0
0.6	333	2	331	0.0
0.6	332	2	330	0.0
0.6	331	2	329	0.0
0.6	330	2	328	0.0
0.6	329	2	327	0.0
0.6	328	2	326	0.0
0.6	327	2	325	0.0
0.6	326	2	324	0.0
0.6	325	2	323	0.0
0.6	324	2	322	0.0

%E	M1	M2	DM	M*
0•6	323	2	321	0•0
0•6	322	2	320	0•0
0•6	321	2	319	0•0
0•6	320	2	318	0•0
0•6	319	2	317	0•0
0•6	318	2	316	0•0
0•6	317	2	315	0•0
0•6	316	2	314	0•0
0•6	315	2	313	0•0
0•6	314	2	312	0•0
0•6	313	2	311	0•0
0•6	312	2	310	0•0
0•6	311	2	309	0•0
0•6	310	2	308	0•0
0•6	309	2	307	0•0
0•6	308	2	306	0•0
0•6	181	1	180	0•0
0•6	180	1	179	0•0
0•6	179	1	178	0•0
0•6	178	1	177	0•0
0•6	177	1	176	0•0
0•6	176	1	175	0•0
0•6	175	1	174	0•0
0•6	174	1	173	0•0
0•6	173	1	172	0•0
0•6	172	1	171	0•0
0•6	171	1	170	0•0
0•6	170	1	169	0•0
0•6	169	1	168	0•0
0•6	168	1	167	0•0
0•6	167	1	166	0•0
0•6	166	1	165	0•0
0•6	165	1	164	0•0
0•6	164	1	163	0•0
0•6	163	1	162	0•0
0•6	162	1	161	0•0
0•6	161	1	160	0•0
0•6	160	1	159	0•0
0•6	159	1	158	0•0
0•6	158	1	157	0•0
0•6	157	1	156	0•0
0•6	156	1	155	0•0
0•6	155	1	154	0•0
0•6	154	1	153	0•0
0•5	444	2	442	0•0
0•5	443	2	441	0•0
0•5	442	2	440	0•0
0•5	441	2	439	0•0
0•5	440	2	438	0•0
0•5	439	2	437	0•0

%E	M1	M2	DM	M*
0•5	438	2	436	0•0
0•5	437	2	435	0•0
0•5	436	2	434	0•0
0•5	435	2	433	0•0
0•5	434	2	432	0•0
0•5	433	2	431	0•0
0•5	432	2	430	0•0
0•5	431	2	429	0•0
0•5	430	2	428	0•0
0•5	429	2	427	0•0
0•5	428	2	426	0•0
0•5	427	2	425	0•0
0•5	426	2	424	0•0
0•5	425	2	423	0•0
0•5	424	2	422	0•0
0•5	423	2	421	0•0
0•5	422	2	420	0•0
0•5	421	2	419	0•0
0•5	420	2	418	0•0
0•5	419	2	417	0•0
0•5	418	2	416	0•0
0•5	417	2	415	0•0
0•5	416	2	414	0•0
0•5	415	2	413	0•0
0•5	414	2	412	0•0
0•5	413	2	411	0•0
0•5	412	2	410	0•0
0•5	411	2	409	0•0
0•5	410	2	408	0•0
0•5	409	2	407	0•0
0•5	408	2	406	0•0
0•5	407	2	405	0•0
0•5	406	2	404	0•0
0•5	405	2	403	0•0
0•5	404	2	402	0•0
0•5	403	2	401	0•0
0•5	402	2	400	0•0
0•5	401	2	399	0•0
0•5	400	2	398	0•0
0•5	399	2	397	0•0
0•5	398	2	396	0•0
0•5	397	2	395	0•0
0•5	396	2	394	0•0
0•5	395	2	393	0•0
0•5	394	2	392	0•0
0•5	393	2	391	0•0
0•5	392	2	390	0•0
0•5	391	2	389	0•0
0•5	390	2	388	0•0
0•5	389	2	387	0•0

%E	M1	M2	DM	M*
0•5	388	2	386	0•0
0•5	387	2	385	0•0
0•5	386	2	384	0•0
0•5	385	2	383	0•0
0•5	384	2	382	0•0
0•5	383	2	381	0•0
0•5	382	2	380	0•0
0•5	381	2	379	0•0
0•5	380	2	378	0•0
0•5	379	2	377	0•0
0•5	378	2	376	0•0
0•5	377	2	375	0•0
0•5	376	2	374	0•0
0•5	375	2	373	0•0
0•5	374	2	372	0•0
0•5	373	2	371	0•0
0•5	372	2	370	0•0
0•5	371	2	369	0•0
0•5	370	2	368	0•0
0•5	369	2	367	0•0
0•5	368	2	366	0•0
0•5	367	2	365	0•0
0•5	366	2	364	0•0
0•5	365	2	363	0•0
0•5	364	2	362	0•0
0•5	222	1	221	0•0
0•5	221	1	220	0•0
0•5	220	1	219	0•0
0•5	219	1	218	0•0
0•5	218	1	217	0•0
0•5	217	1	216	0•0
0•5	216	1	215	0•0
0•5	215	1	214	0•0
0•5	214	1	213	0•0
0•5	213	1	212	0•0
0•5	212	1	211	0•0
0•5	211	1	210	0•0
0•5	210	1	209	0•0
0•5	209	1	208	0•0
0•5	208	1	207	0•0
0•5	207	1	206	0•0
0•5	206	1	205	0•0
0•5	205	1	204	0•0
0•5	204	1	203	0•0
0•5	203	1	202	0•0
0•5	202	1	201	0•0
0•5	201	1	200	0•0
0•5	200	1	199	0•0
0•5	199	1	198	0•0
0•5	198	1	197	0•0

%E	M1	M2	DM	M*
0•5	197	1	196	0•0
0•5	196	1	195	0•0
0•5	195	1	194	0•0
0•5	194	1	193	0•0
0•5	193	1	192	0•0
0•5	192	1	191	0•0
0•5	191	1	190	0•0
0•5	190	1	189	0•0
0•5	189	1	188	0•0
0•5	188	1	187	0•0
0•5	187	1	186	0•0
0•5	186	1	185	0•0
0•5	185	1	184	0•0
0•5	184	1	183	0•0
0•5	183	1	182	0•0
0•5	182	1	181	0•0
0•4	500	2	498	0•0
0•4	499	2	497	0•0
0•4	498	2	496	0•0
0•4	497	2	495	0•0
0•4	496	2	494	0•0
0•4	495	2	493	0•0
0•4	494	2	492	0•0
0•4	493	2	491	0•0
0•4	492	2	490	0•0
0•4	491	2	489	0•0
0•4	490	2	488	0•0
0•4	489	2	487	0•0
0•4	488	2	486	0•0
0•4	487	2	485	0•0
0•4	486	2	484	0•0
0•4	485	2	483	0•0
0•4	484	2	482	0•0
0•4	483	2	481	0•0
0•4	482	2	480	0•0
0•4	481	2	479	0•0
0•4	480	2	478	0•0
0•4	479	2	477	0•0
0•4	478	2	476	0•0
0•4	477	2	475	0•0
0•4	476	2	474	0•0
0•4	475	2	473	0•0
0•4	474	2	472	0•0
0•4	473	2	471	0•0
0•4	472	2	470	0•0
0•4	471	2	469	0•0
0•4	470	2	468	0•0
0•4	469	2	467	0•0
0•4	468	2	466	0•0
0•4	467	2	465	0•0

%E	M1	M2	DM	M*
0•4	466	2	464	0•0
0•4	465	2	463	0•0
0•4	464	2	462	0•0
0•4	463	2	461	0•0
0•4	462	2	460	0•0
0•4	461	2	459	0•0
0•4	460	2	458	0•0
0•4	459	2	457	0•0
0•4	458	2	456	0•0
0•4	457	2	455	0•0
0•4	456	2	454	0•0
0•4	455	2	453	0•0
0•4	454	2	452	0•0
0•4	453	2	451	0•0
0•4	452	2	450	0•0
0•4	451	2	449	0•0
0•4	450	2	448	0•0
0•4	449	2	447	0•0
0•4	448	2	446	0•0
0•4	447	2	445	0•0
0•4	446	2	444	0•0
0•4	445	2	443	0•0
0•4	286	1	285	0•0
0•4	285	1	284	0•0
0•4	284	1	283	0•0
0•4	283	1	282	0•0
0•4	282	1	281	0•0
0•4	281	1	280	0•0
0•4	280	1	279	0•0
0•4	279	1	278	0•0
0•4	278	1	277	0•0
0•4	277	1	276	0•0
0•4	276	1	275	0•0
0•4	275	1	274	0•0
0•4	274	1	273	0•0
0•4	273	1	272	0•0
0•4	272	1	271	0•0
0•4	271	1	270	0•0
0•4	270	1	269	0•0
0•4	269	1	268	0•0
0•4	268	1	267	0•0
0•4	267	1	266	0•0
0•4	266	1	265	0•0
0•4	265	1	264	0•0
0•4	264	1	263	0•0
0•4	263	1	262	0•0
0•4	262	1	261	0•0
0•4	261	1	260	0•0
0•4	260	1	259	0•0
0•4	259	1	258	0•0

%E	M1	M2	DM	M*
0.4	258	1	257	0.0
0.4	257	1	256	0.0
0.4	256	1	255	0.0
0.4	255	1	254	0.0
0.4	254	1	253	0.0
0.4	253	1	252	0.0
0.4	252	1	251	0.0
0.4	251	1	250	0.0
0.4	250	1	249	0.0
0.4	249	1	248	0.0
0.4	248	1	247	0.0
0.4	247	1	246	0.0
0.4	246	1	245	0.0
0.4	245	1	244	0.0
0.4	244	1	243	0.0
0.4	243	1	242	0.0
0.4	242	1	241	0.0
0.4	241	1	240	0.0
0.4	240	1	239	0.0
0.4	239	1	238	0.0
0.4	238	1	237	0.0
0.4	237	1	236	0.0
0.4	236	1	235	0.0
0.4	235	1	234	0.0
0.4	234	1	233	0.0
0.4	233	1	232	0.0
0.4	232	1	231	0.0
0.4	231	1	230	0.0
0.4	230	1	229	0.0
0.4	229	1	228	0.0
0.4	228	1	227	0.0
0.4	227	1	226	0.0
0.4	226	1	225	0.0
0.4	225	1	224	0.0
0.4	224	1	223	0.0
0.4	223	1	222	0.0
0.3	400	1	399	0.0
0.3	399	1	398	0.0
0.3	398	1	397	0.0
0.3	397	1	396	0.0
0.3	396	1	395	0.0
0.3	395	1	394	0.0
0.3	394	1	393	0.0
0.3	393	1	392	0.0
0.3	392	1	391	0.0
0.3	391	1	390	0.0
0.3	390	1	389	0.0
0.3	389	1	388	0.0
0.3	388	1	387	0.0
0.3	387	1	386	0.0

%E	M1	M2	DM	M*
0.3	386	1	385	0.0
0.3	385	1	384	0.0
0.3	384	1	383	0.0
0.3	383	1	382	0.0
0.3	382	1	381	0.0
0.3	381	1	380	0.0
0.3	380	1	379	0.0
0.3	379	1	378	0.0
0.3	378	1	377	0.0
0.3	377	1	376	0.0
0.3	376	1	375	0.0
0.3	375	1	374	0.0
0.3	374	1	373	0.0
0.3	373	1	372	0.0
0.3	372	1	371	0.0
0.3	371	1	370	0.0
0.3	370	1	369	0.0
0.3	369	1	368	0.0
0.3	368	1	367	0.0
0.3	367	1	366	0.0
0.3	366	1	365	0.0
0.3	365	1	364	0.0
0.3	364	1	363	0.0
0.3	363	1	362	0.0
0.3	362	1	361	0.0
0.3	361	1	360	0.0
0.3	360	1	359	0.0
0.3	359	1	358	0.0
0.3	358	1	357	0.0
0.3	357	1	356	0.0
0.3	356	1	355	0.0
0.3	355	1	354	0.0
0.3	354	1	353	0.0
0.3	353	1	352	0.0
0.3	352	1	351	0.0
0.3	351	1	350	0.0
0.3	350	1	349	0.0
0.3	349	1	348	0.0
0.3	348	1	347	0.0
0.3	347	1	346	0.0
0.3	346	1	345	0.0
0.3	345	1	344	0.0
0.3	344	1	343	0.0
0.3	343	1	342	0.0
0.3	342	1	341	0.0
0.3	341	1	340	0.0
0.3	340	1	339	0.0
0.3	339	1	338	0.0
0.3	338	1	337	0.0
0.3	337	1	336	0.0

%E	M1	M2	DM	M*
0.3	336	1	335	0.0
0.3	335	1	334	0.0
0.3	334	1	333	0.0
0.3	333	1	332	0.0
0.3	332	1	331	0.0
0.3	331	1	330	0.0
0.3	330	1	329	0.0
0.3	329	1	328	0.0
0.3	328	1	327	0.0
0.3	327	1	326	0.0
0.3	326	1	325	0.0
0.3	325	1	324	0.0
0.3	324	1	323	0.0
0.3	323	1	322	0.0
0.3	322	1	321	0.0
0.3	321	1	320	0.0
0.3	320	1	319	0.0
0.3	319	1	318	0.0
0.3	318	1	317	0.0
0.3	317	1	316	0.0
0.3	316	1	315	0.0
0.3	315	1	314	0.0
0.3	314	1	313	0.0
0.3	313	1	312	0.0
0.3	312	1	311	0.0
0.3	311	1	310	0.0
0.3	310	1	309	0.0
0.3	309	1	308	0.0
0.3	308	1	307	0.0
0.3	307	1	306	0.0
0.3	306	1	305	0.0
0.3	305	1	304	0.0
0.3	304	1	303	0.0
0.3	303	1	302	0.0
0.3	302	1	301	0.0
0.3	301	1	300	0.0
0.3	300	1	299	0.0
0.3	299	1	298	0.0
0.3	298	1	297	0.0
0.3	297	1	296	0.0
0.3	296	1	295	0.0
0.3	295	1	294	0.0
0.3	294	1	293	0.0
0.3	293	1	292	0.0
0.3	292	1	291	0.0
0.3	291	1	290	0.0
0.3	290	1	289	0.0
0.3	289	1	288	0.0
0.3	288	1	287	0.0
0.3	287	1	286	0.0

%E	M1	M2	DM	M*
0.2	500	1	499	0.0
0.2	499	1	498	0.0
0.2	498	1	497	0.0
0.2	497	1	496	0.0
0.2	496	1	495	0.0
0.2	495	1	494	0.0
0.2	494	1	493	0.0
0.2	493	1	492	0.0
0.2	492	1	491	0.0
0.2	491	1	490	0.0
0.2	490	1	489	0.0
0.2	489	1	488	0.0
0.2	488	1	487	0.0
0.2	487	1	486	0.0
0.2	486	1	485	0.0
0.2	485	1	484	0.0
0.2	484	1	483	0.0
0.2	483	1	482	0.0
0.2	482	1	481	0.0
0.2	481	1	480	0.0
0.2	480	1	479	0.0
0.2	479	1	478	0.0
0.2	478	1	477	0.0
0.2	477	1	476	0.0
0.2	476	1	475	0.0
0.2	475	1	474	0.0
0.2	474	1	473	0.0
0.2	473	1	472	0.0
0.2	472	1	471	0.0
0.2	471	1	470	0.0
0.2	470	1	469	0.0
0.2	469	1	468	0.0
0.2	468	1	467	0.0
0.2	467	1	466	0.0
0.2	466	1	465	0.0
0.2	465	1	464	0.0
0.2	464	1	463	0.0
0.2	463	1	462	0.0
0.2	462	1	461	0.0
0.2	461	1	460	0.0
0.2	460	1	459	0.0
0.2	459	1	458	0.0
0.2	458	1	457	0.0
0.2	457	1	456	0.0
0.2	456	1	455	0.0
0.2	455	1	454	0.0
0.2	454	1	453	0.0
0.2	453	1	452	0.0
0.2	452	1	451	0.0
0.2	451	1	450	0.0

%E	M1	M2	DM	M*
0.2	450	1	449	0.0
0.2	449	1	448	0.0
0.2	448	1	447	0.0
0.2	447	1	446	0.0
0.2	446	1	445	0.0
0.2	445	1	444	0.0
0.2	444	1	443	0.0
0.2	443	1	442	0.0
0.2	442	1	441	0.0
0.2	441	1	440	0.0
0.2	440	1	439	0.0
0.2	439	1	438	0.0
0.2	438	1	437	0.0
0.2	437	1	436	0.0
0.2	436	1	435	0.0
0.2	435	1	434	0.0
0.2	434	1	433	0.0
0.2	433	1	432	0.0
0.2	432	1	431	0.0
0.2	431	1	430	0.0
0.2	430	1	429	0.0
0.2	429	1	428	0.0
0.2	428	1	427	0.0
0.2	427	1	426	0.0
0.2	426	1	425	0.0
0.2	425	1	424	0.0
0.2	424	1	423	0.0
0.2	423	1	422	0.0
0.2	422	1	421	0.0
0.2	421	1	420	0.0
0.2	420	1	419	0.0
0.2	419	1	418	0.0
0.2	418	1	417	0.0
0.2	417	1	416	0.0
0.2	416	1	415	0.0
0.2	415	1	414	0.0
0.2	414	1	413	0.0
0.2	413	1	412	0.0
0.2	412	1	411	0.0
0.2	411	1	410	0.0
0.2	410	1	409	0.0
0.2	409	1	408	0.0
0.2	408	1	407	0.0
0.2	407	1	406	0.0
0.2	406	1	405	0.0
0.2	405	1	404	0.0
0.2	404	1	403	0.0
0.2	403	1	402	0.0
0.2	402	1	401	0.0
0.2	401	1	400	0.0